American College of Veterinary Internal Medicine (ACVIM) Forum 2011

Denver, Colorado, USA
15-18 June 2011

Volume 1 of 2

Editors:

Dianne McFarlane **Jeff Toll**

ISBN: 978-1-61839-030-1

Printed from e-media with permission by:

Curran Associates, Inc.
57 Morehouse Lane
Red Hook, NY 12571

proceedings.com

Some format issues inherent in the e-media version may also appear in this print version.

Printed by Curran Associates, Inc. (2011)

For permission requests, please contact the American College of Veterinary Internal Medicine at the address below.

American College of Veterinary Internal Medicine
1997 Wadsworth Boulevard Suite A
Lakewood CO 80214-5293

Phone: (800) 245-9081
Fax: (303) 231-0880

acvim@acvim.org

Additional copies of this publication are available from:

Curran Associates, Inc.
57 Morehouse Lane
Red Hook, NY 12571 USA
Phone: 845-758-0400
Fax: 845-758-2634
Email: curran@proceedings.com
Web: www.proceedings.com

ACVIM Specialty Symposium (Pre-Forum) Proceedings

Cardiology

Large Animal Internal Medicine - Equine

Large Animal Internal Medicine - Food Animal

Neurology

Oncology

Small Animal Internal Medicine

Veterinary Scientific Program Proceedings

AAEP

ABVP

ACVCP/AAVPT

ACVIM Business Management

ACVIM Cardiology

ACVIM Foundation Grant Presentations

ACVIM Large Animal Internal Medicine - Equine

ACVIM Large Animal Internal Medicine - Food Animal

ACVIM Neurology

ACVIM Oncology

ACVIM One Medicine

ACVIM Small Animal Internal Medicine

ASVCP

AVHTM

CGS

Food Animal Internal Medicine - AAVN/ACVN

ISCAID

Liver Study Group

SCE

SVN/U

VCRS (LA)

VCRS (SA)

Session# **Page#**

VECCS

VUS

Technician Program Proceedings

Technician

Neurology

Title / Author

ACVIM FORUM

As far as you want to go.

Research Abstract Program of the 2011 ACVIM Forum

Denver, Colorado,
June 15–18, 2011
Index of Abstracts

ORAL PRESENTATIONS – Thursday, June 16

Time	#	Presenting Author	Abstract Title
SMALL ANIMAL – CARDIOLOGY**			
9:00 am	C-1	Sabine Riesen	Comparison of Short-Term Effects of Ivabradine and Atenolol on Heart Rate and Echocardiographic Variables of Left Heart Function in Healthy Cats
9:15 am	C-2	Gareth Buckley	Randomized, Blinded Comparison of Epinephrine and Vasopressin for Treatment of Naturally Occuring Cardiopulmonary Arrest (CPA) in Dogs
9:30 am	**C-5**	**Eryn Shipley**	**Effect of Pimobendan on Platelet Aggregation in Dogs**
9:45 am	C-4	Rachael Wood	Adiponectin: A Protective Role in Dogs with Congestive Heart Failure?
BREAK			
10:30 am	**C-3**	**Stacey Leach**	**Single-dose and Apparent Steady State Pharmacokinetics and Pharmacodynamics of an Extended Release Diltiazem Formulation in the Dog-A Pilot Study**
10:45 am	**C-6**	**Justin Thomason**	**The Influence of Chronic Combined Enalapril and Spironolactone Administration on Serum Electrolyte Concentrations in Doberman Pinschers with Occult Dilated Cardiomyopathy**
11:00 am	**C-7**	**Rita Singh**	**Occlusion Devices and Approach in Patent Ductus Arteriosus: Comparison of Outcomes**
11:15 am	**C-8**	**Brian Maran**	**Identification of Two Deletion Polymorphisms within the Canine Beta-1 Adrenoceptor Gene**
11:30 am	**C-9**	**Joshua Stern**	**Genome Wide Association Analysis Identifies Location of Interest on Chromosome 21 in Golden Retrievers and Rottweilers with Familial Subvalvular Aortic Stenosis**
11:45 am	**C-10**	**Holly MacLea**	**Canine and Human Degenerative Mitral Valve Disease Mimics Chondrogenesis**
12:00 pm	C-11	Kathryn Meurs	Characteristics of ARVC Boxers with Sudden Death
12:15 pm	**C-12**	**Rita Singh**	**Bronchomalacia in Dogs with Myxomatous Mitral Valve Degeneration**
***Also see Cardiology abstracts C-13 – C-26 (Thursday, June 16, 2:15 pm – 6:15 pm)*			
SMALL ANIMAL – HEMATOLOGY**			
9:00 am	**HM-1**	**Marc Dhumeaux**	**Comparison of Bone Marrow Sampling in Cats Using Two Different Techniques: Manual and Power Driver-Assisted**
9:15 am	**HM-2**	**Marc Dhumeaux**	**Effects of a Standardized Anesthetic Protocol on Hematologic Parameters in Healthy Cats**
9:30 am	HM-3	Benjamin Brainard	In Vitro Effects of Rivaroxaban on Feline Coagulation Indices
9:45 am	**HM-4**	**Eryn Shipley**	**Comparison of Citrated and Hirudinated Blood on Whole Blood Platelet Aggregation Using the Chrono-Log Impedance Aggregometer**
BREAK			
10:30 am	HM-5	Carolyn Gara-Boivin	Use of Calibrated Automated Thrombogram (CAT) For Monitoring Low Molecular Weight Heparin in Healthy Dogs

Boldface type indicates ACVIM Resident Research Award eligibility.

10:45 am	**HM-6**	**Allison O'Kell**	**The Effects of Prednisone and Prednisone Plus Ultralow-Dose Aspirin on Coagulation Parameters in Healthy Dogs**
11:00 am	HM-7	Ali Dudley	Effects of Low Dose Aspirin on Canine Platelet Function
11:15 am	HM-8	Liliana Marin	Epsilon Aminocaproic Acid Decreases the Prevalence of Postoperative Bleeding in Retired Racing Greyhounds
11:30 am	HM-9	Carrie White	Evaluation of Potential Surrigate Indicators of Hypercoagulability in Dogs with Protein-Losing Nephropathy
11:45 am	HM-10	Catherine Sumner	The Effects of Sample Collection Method on Thromboelastography in Healthy Dogs
12:00 pm	HM-11	Adam Grochowsky	Transfusion of Liquid Plasma to Dogs in the Emergency Setting
12:15 pm	**HM-12**	**Dawn Bachman**	**Prospective Diagnostic Evaluation and Clinical Utilization of an Optimized Assay for Platelet Associated Immunoglobulin in Thrombocytopenic and Non-Thrombocytopenic Dogs**

***Also see Hematology abstracts HM-13 – HM-15 (Thursday, June 16, 4:30 pm – 5:15 pm)*

SMALL ANIMAL – NEPHROLOGY / UROLOGY**

9:00 am	**N/U-1**	**Nicole Smee**	**Investigations into the Effect of Cranberry Extract on the Bacterial Adhesion to Canine Uroepithelial Cells**
9:15 am	**N/U-2**	**Jennifer Wells**	**Association Between Presence of Urovirulence Factors and Antimicrobial Resistance Patterns in 221 Uropathogenic *Escherchia coli* Samples Isolated from Dogs**
9:30 am	N/U-3	Angela Lusby	Prevalence of Asymptomatic Bacterial Urinary Tract Infections in Morbidly Obese Dogs
9:45 am	N/U-4	Rachel Lavoué	Phosphatemia in Healthy Dogs: Within-Day Variation, Effects of Feeding with and without Calcium Carbonate
BREAK			
10:30 am	**N/U-5**	**Vanessa Von Hendy-Willson**	**Rapid Determination of GFR in Dogs with Acute Kidney Injury using a Portable Fluorescence Ratiometric Analyzer**
10:45 am	**N/U-6**	**Elizabeth Lennon**	**Thromboelastography in Non-Azotemic and Azotemic Dogs with Protein-Losing Nephropathy**
11:00 am	N/U-7	Urs Giger	Characterization of Non-Type I Cystinuria in Irish Terriers
11:15 am	**N/U-8**	**Adam Eatroff**	**Long Term Survival of Cats and Dogs Treated with Intermittent Hemodialysis for Acute Kidney Injury**
11:30 am	**N/U-9**	**Emily Klosterman**	**Diagnostic Utility and Concordance with Final Diagnosis of Renal Aspiration in Dogs and Cats**
11:45 am	N/U-10	Nicole Luckschander	Predisposing Factors Associated with Obstruvtive Feline Urinary Tract Disease in Cats
12:00 pm	**N/U-11**	**Catherine Rivara**	**Hyperuricosuria without Alterations in Liver Function is a Risk Factor for Feline Urate Uroliths**
12:15 pm	**N/U-12**	**Rosanne Jepson**	**Use of Surface Enhanced Laser Desorption Ionization Time-of-Flight Mass Spectrometry (SELDI-TOF-MS) as a Urinary Biomarker Discovery Tool in Cats at Risk of Developing Azotemia**

***Also see Nephrology / Urology abstracts N/U-13 – N/U-19 (Thursday, June 16, 4:30 pm – 6:15 pm)*

SMALL ANIMAL – ENDOCRINOLOGY**

9:00 am	EN-1	Tim Williams	Changes in Plasma Fibroblast Growth Factor-23 (FGF-23) Concentrations After Treatment of Hyperthyroidism (HTH) in Cats with Variable Renal Function
9:15 am	EN-2	Samer Al-Murrani	Gene Expression Analysis of Feline Thyroid Tissue and Blood from Cats with Evidence of Mild or Marked Hyperthyroidism Reveals Potential Molecular Causes of the Disease and Identifies Future Routes for Intervention
9:30 am	**EN-3**	**Leyenda Harley**	**IRIS Stages of Chronic Kidney Disease Before and After Treatment with Radioiodine in Cats with Hyperthyroidism**
9:45 am	EN-4	Stijn Niessen	Impact of Home Blood Glucose Monitoring on Quality of Life of Diabetic Cats and Their Owners Recorded Using the Psychometric Tool DIAQoL-Pet
BREAK			
10:30 am	EN-5	Melania Osto	Tissue-Specific Pattern of Insulin Resistance in Glucose-, Lipid- and Lipopolysaccharide-Treated Cats
10:45 am	EN-6	Maria Rodriguez Piñeiro	Image Fusion of Single-Photon Emission Computed Tomography and Computed Tomography Scan in Canine Insulinoma: 10 Cases
11:00 am	**EN-7**	**Miriam Baumstark**	**Evaluation of Aldosterone Concentrations in Dogs with Addison's Disease**
11:15 am	**EN-8**	**Mayank Seth**	**Combining White Blood Cell Counts and the Sodium to Potassium Ratio to Screen for Canine Hypoadrenocorticism**
11:30 am	**EN-9**	**Ruth Gostelow**	**Plasma Free Metanephrine and Normetanephrine Concentrations are Elevated in Dogs with Pheochromocytoma**
11:45 am	EN-10	Ellen Behrend	Assessment by Quantitative PCR of Ability of Sex Hormones to Induce Expression of Classic Glucocorticoid-Induced Genes in Canine Hepatocytes
12:00 pm	EN-11	Ellen Behrend	Evaluation of Measurement of 17-Hydroxy-Progesterone and Estradiol for Diagnosis of Typical and Occult Hyperadrenocorticism
12:15 pm	EN-12	Yaiza Forcada	A Missense Mutation in the Coding Sequence of MC4R (MC4R:C.92 C > T) is Associated with Diabetes Mellitus in DSH Cats (ESVE Award Winner)

***Also see Endocrinology abstracts EN-13 – EN-15 (Thursday, June 16, 5:30 pm – 6:15 pm)*

EQUINE**

9:00 am	**E-1**	**Katarzyna Dembek**	**The Renin-Angiotensin-Aldosterone System (RAAS) and Hypothalamic-Pituitary-Adrenal Axis (HPAA) in Septic Foals**
9:15 am	**E-2**	**Ana Pacheco**	**Evaluation of Coagulation and Blood-Derived Inflammatory Biomarkers in Horses with Acute Onset of Diarrhea During Hospitalization**
9:30 am	**E-3**	**Angelika Schoster**	**Detection and Characterization of *Clostridium difficile* and *Clostridium perfringens* in Gastrointestinal Compartments of Healthy Horses**
9:45 am	**E-4**	**Angelika Schoster**	**Longitudinal Study of *Clostridium difficile*, *Clostridium perfringens*, *Salmonella* and *Escherichia coli* in Healthy Horses**
BREAK			
10:30 am	**E-5**	**Rosie Naylor**	**Influence of Allele Copy Number on Skeletal Muscle Histopathology in Horses with Type 1 Polysaccharide Storage Myopathy**
10:45 am	**E-6**	**Laramie Winfield**	**Factors Associated with Survival in 161 Recumbent Horses**
11:00 am	**E-7**	**Jennifer Read**	**Plasma Peak and Trough Gentamicin Concentrations in Hospitalized Horses Receiving Once Daily Gentamicin**
11:15 am	**E-8**	**Johanne Vanderstock**	**Circulating Neutrophil Phagocytosis and Killing of *Streptococcus equi* Spp. *Equi* and *Zooepidemicus* in Heaves-Affected and Control Horses**
11:30 am	**E-9**	**Heidi Banse**	**Association of Body Condition Score with Metabolic Regulatory Gene Expression and Mitochondrial Content in Skeletal Muscle of Horses**
11:45 am	**E-10**	**Claudia Cruz Villagrán**	**Correlation and Repeatability of Glucose and D-Xylose Intestinal Absorption Tests in Normal Horses**
12:00 pm	E-11	Crystal Hoffman	Prognosis for Racing with Conservative Management of Cervical Vertebral Malformation in Thoroughbreds: 119 Cases (2002 - 2010)
12:15 pm	E-12	Gila Sutton	Two Pain Scales for Horses with Colic: Construction and Validation

***Also see Equine abstracts E-13 – E-26 (Thursday, June 16, 2:15 pm – 6:15 pm) and Equine abstracts E-27 – E-32 (Thursday, June 16, 4:45 pm – 6:15 pm)*

SMALL ANIMAL – CARDIOLOGY**

2:15 pm	C-13	SeungWoo Jung	The Effect of Atenolol on NT-proBNP and Troponin in Asymptomatic Cats with Hypertrophic Cardiomyopathy
2:30 pm	C-14	Sonja Fonfara	Leptin Expression in Blood and Myocardial Samples of Dogs with Cardiac Disease
2:45 pm	**C-15**	**Gretchen Singletary**	**NT-pro-BNP Improves General Practitioners' Accuracy and Confidence of Diagnosis in Cats with Respiratory Signs**
3:00 pm	**C-16**	**Caryn Reynolds**	**NT-proBNP Predicts First Onset of Congestive Heart Failure in Dogs with Degenerative Mitral Valve Disease**
3:15 pm	C-17	Melanie Hezzell	Associations Between N-Terminal Procollagen Type III (PIIINP) and Ventricular Remodelling in Dogs with Mitral Valve Disease
3:30 pm	C-18	Sara Johns	Left Atrial Function in Cats with Left-Sided Cardiac Disease Manifesting Pleural Effusion verses Pulmonary Edema
3:45 pm	**C-19**	**Justin Thomason**	**Presumed Neurally-Mediated Bradycardia and Syncope in Boxers with Normal Echocardiograms and Absence of Ventricular Tachycardia**
BREAK			
4:30 pm	C-20	Eva Oxford	Altered Localization of Beta Catenin and Striatin in Boxer Dogs Afflicted with Arrhythmogenic Right Ventricular Cardiomyopathy Suggests Etiological Involvement of the Canonical Wnt Pathway
4:45 pm	**C-21**	**Daniel Hall**	**Serum Fatty Acid Concentrations in Normal Cats and Cats with Cardiomyopathy**
5:00 pm	**C-22**	**Daniel Hall**	**Application of Platelet Mapping in Normal Cats for Monitoring Antiplatelet Effects of Clopidogrel**
5:15 pm	C-23	Gareth Buckley	The Effect of Using an Impedance Threshold Device (ITD) on Hemodynamic Parameters During Cardiopulmonary Resuscitation in Dogs
5:30 pm	**C-24**	**Emily Olson**	**Tachycardia-Induced Cardiomyopathy in Dogs: Clinical, Holter, and Echocardiographic Findings**
5:45 pm	C-25	Katherine Scollan	Pilot Study Investigation of Antidiuretic Hormone (ADH) Levels in Dogs with Acquired Cardiac Disease
6:00 pm	C-26	Randolph Winter	Aortic Thromboembolism in Dogs: Presentation, Antithrombotic Therapy, and Outcome in 26 Cases

***Also see Cardiology abstracts C-1 – C-12 (Thursday, June 16, 9:00 am – 12:30 pm)*

SMALL ANIMAL – NEUROLOGY / PHARMACOLOGY**

2:15 pm	**N-1**	**Heather Wright**	**Intravenous Fat Emulsion (IFE) for Treatment of Ivermectin Toxicosis in 3 Dogs**
2:30 pm	**N-2**	**Heather Wright**	**Oral Rufinamide in Dogs: Pharmacokinetics and Safety**
2:45 pm	**N-3**	**Courtenay Freeman**	**Preliminary Morphometric Evaluation of Syringomyelia in American Brussels Griffon Dogs**
3:00 pm	**N-4**	**Stephanie Thomovsky**	**Canine Intervertebral Disc Fenestration Using a Novel Tissue Resection Device**
3:15 pm	N-5	Diana Henke	Gliomatosis Cerebri in a Family of Bearded Collies
3:30 pm	N-6	Dennis O'Brien	A Missense Mutation in *HEXA* in Japanese Chin Dogs with GM2 Gangliosidosis
3:45 pm	N-8	Carrie Palm	Therapeutic Apheresis: A Novel Treatment for Myasthenia Gravis in Dogs

BREAK			
4:30 pm	N-7	Elise Craft	Pharmacokinetics and Tolerability of Intravenous Fosphenytoin in Four Healthy Dogs
4:45 pm	N-9	Roberto Jose-Lopez	Spinal Meningiomas in Dogs: Description of Seven Cases, Including a Novel Radiological and Histopathological Presentation
5:00 pm	P-1	Jonathan Mochel	Dynamics of the Renin-Angiotensin-Aldosterone System (RAAS) After Oral Administration of Benazepril in Dogs
5:15 pm	P-2	Ted Whittem	Dose Escalation with Alfaxalone or Propofol in the Dog and Effects on Respiratory Depression
5:30 pm	**P-3**	**Brian Hardy**	**Subcutaneous Administration of Levetiracetam in Healthy Dogs**
5:45 pm	P-4	Butch KuKanich	The Effects of Cytochrome P450 (CYP) Inhibitors on Oral Methadone in Greyhound Dogs
6:00 pm	P-5	Mark Papich	Doxycycline Potency after Storage in a Compounded Formulation for Animals

***Also see Pharmacology abstract P-6 (Thursday, June 16, 3:15 pm – 3:30 pm)*

SMALL ANIMAL – ONCOLOGY / HEMATOLOGY/IMMUNOLOGY/ENDOCRINOLOGY****

2:15 pm	**O-1**	**Sara Allstadt Frazier**	**Phase I Study of Rosiglitazone in Combination with Carboplatin in Cancer Bearing Dogs**
2:30 pm	**O-2**	**Joao Felipe de Brito Galvao**	**Antitumor Effects of Combined Carboplatin and Gemcitabine on Transitional Cell Carcinoma Cells**
2:45 pm	**O-3**	**Jenna Burton**	**Low-Dose Cyclophosphamide Selectively Depletes Regulatory T Cells and Inhibits Angiogenesis in Dogs with Soft Tissue Sarcoma**
3:00 pm	**O-4**	**Carmela Pratt**	**Re-Sensitization of Glucocorticoid-Resistant Canine Lymphoma Cells Using Arsenic Trioxide**
3:15 pm	O-5	Matthew Atherton	Serum Protein Electrophoresis of Dogs with Lymphoma
3:30 pm	O-6	Katja Ruess-Melzer	Evaluation of Aberration in Phenotype as Prognostic Factor in Canine Lymphoblastic Lymphoma
3:45 pm	O-7	Anne Avery	Analysis of Phenotypic Factors Predicting Survival in Canine Patients with B Cell Lymphoma
BREAK			
4:30 pm	HM-13	Sara Zaldivar-Lopez	Investigation of the Haptoglobin Gene in Greyhounds
4:45 pm	HM-14	Sara Zaldivar-Lopez	Molecular Characterization of the Canine Beta-Globin Gene Cluster
5:00 pm	HM-15	Julia Veir	Ease of Procedure and Quality of Bone Marrow Aspirates Obtained from the Costochondral Junction as Compared to Traditional Sites in Healthy Walker Hound Dogs
5:15 pm	IM-1	Claire Fellman	Effects of Cyclosporine and Dexamethasone on Canine T-Cell Expression of IL-2 and IFN-Gamma as Measured by Flow Cytometry and Quantitative RT-PCR
5:30 pm	**EN-13**	**Lauren Reid**	**Comparison of the Effect of Trilostane and Mitotane on Peak Serum Aldosterone Concentrations in Dogs with Pituitary-Dependent Hyperadrenocorticism**
5:45 pm	**EN-14**	**Erinne Branter**	**Antioxidant Status in Hyperthyroid Cats Before and After Radioiodine Treatment**
6:00 pm	EN-15	Susan Gottlieb	Diabetic Cats in Remission Have Mildly Impaired Glucose Tolerance

***Also see Hematology abstracts HM-1 – HM-12 (Thursday, June 16, 9:00 am – 12:30 pm)*
***Also see Endocrinology abstracts EN-1 – EN-12 (Thursday, June 16, 9:00 am – 12:30 pm)*

SMALL ANIMAL – RESPIRATORY / NEPHROLOGY /UROLOGY**

2:15 pm	R-1	Chee-Hoon Chang	Impact of Oral versus Inhaled Glucocorticoids Treatment on Allergen-Specific IgE Testing in Experimentally Asthmatic Cats
2:30 pm	R-2	Elizabeth Rozanski	Serum 25-Hydroxyvitamin D in Cats with Lower Airway Disease
2:45 pm	**R-3**	**Kevin Kumrow**	**Characterization of Interstitial Lung Disease in 13 Dogs: Clinical Course and Computed Tomography**
3:00 pm	R-4	Emilie Krafft	Gene Expression Profiles in Canine Idiopathic Pulmonary Fibrosis
3:15 pm	R-5	Emilie Krafft	Use of Expression Microarray to Identify Potential Biomarkers of Idiopathic Pulmonary Fibrosis in Dogs
3:30 pm	R-6	John Anastasio	Effect of Heliox in Respiratory Parameters in Mesocephalic and Brachycephlic Dogs
3:45 pm	R-7	Marta Planellas	Clinical Evaluation and Biomarkers Determination in Brachycephalic Dogs with Upper Airway Obstruction Before and After Surgical Correction
BREAK			
4:30 pm	N/U-13	Natalie Finch	FGF-23 – Mediator of Renal Secondary Hyperparathyroidism or a Marker of Glomerular Filtration Rate (GFR) in Cats?
4:45 pm	N/U-14	Shubhro Chakrabarti	Urinary Vascular Endothelial Growth Factor as a Marker of Renal Hypoxia in Cats
5:00 pm	N/U-15	Rebecca Geddes	Fibroblast Growth Factor 23 (FGF-23) in Feline Chronic Kidney Disease
5:15 pm	N/U-16	Kevin Le Boedec	Retrospective Evaluation of Pulmonary Abnormalities in Dogs with Renal Azotemia
5:30 pm	N/U-17	Andrew Hanzlicek	The Effects of Rheum Officinale on the Progression of Feline Chronic Kidney Disease
5:45 pm	N/U-18	Mark Rishniw	Azodyl Fails to Reduce Azotemia in Cats with Chronic Kidney Disease (CKD) when Sprinkled Onto Food
6:00 pm	N/U-19	Melissa Daniels	Evaluation of 3 Herbal Compounds Used for Management of Lower Urinary Tract Disease in Cats

***Also see Nephrology / Urology abstracts N/U-1 – N/U-12 (Thursday, June 16, 9:00 am – 12:30 pm)*

SMALL ANIMAL – INFECTIOUS DISEASE**

2:15 pm	ID-1	Melissa Munks	The Use of Anti-Inflammatory Therapy as Adjunctive Treatment in Dogs with Blastomycosis
2:30 pm	ID-2	Audrey Cook	Diagnostic Reliability of an Antigen Assay for Feline Histoplasmosis
2:45 pm	**ID-3**	**Ryan Garcia**	**The Use of a Blood and Urine Galactomannan Antigen Test for the Diagnosis of Canine Systemic Aspergillosis**
3:00 pm	**ID-4**	**Christine Savidge**	***Anaplasma phagocytophilum* in Cats: A Retrospective Analysis and Clinical Evaluation of Fifteen Cases from the North Eastern United States**
3:15 pm	**ID-5**	**Cristina Perez**	**Canine Bartonellosis Mimics Other Tick-Borne Diseases**
3:30 pm	ID-6	Cynda Crawford	Evaluation of a Commercial Blocking ELISA for Detection of Influenza A Nucleoprotein Antibodies in Canine Sera
3:45 pm	ID-7	Catherine McManus	Prevalence of Upper Respiratory Pathogens in Four Management Models for Unowned Cats in the Southeast United States
BREAK			
4:30 pm	ID-8	Susan Little	Sensitivity and Specificity of Quantitative PCR and Virus Isolation for Diagnosis of Feline Immunodeficiency Virus Infection
4:45 pm	ID-9	Susan Little	Seroprevalence of FeLV and FIV in Cats with Oral Disease
5:00 pm	ID-10	Julie Levy	Prevalence and Risk Factors for FELV, FIV, and Heartworm Infection in North American Cats
5:15 pm	ID-11	Cynda Crawford	Use of Antibody Titers and Quantitative PCR as Risk Assessment Tools for Management of an Outbreak of Canine Distemper and Parvovirus
5:30 pm	ID-12	Michael Lappin	Effect of Cyclosporine and Methylprednisolone Acetate on Cats with Chronic Feline Herpesvirus 1 Infection
5:45 pm	ID-13	Laura Andersen	Prevalence of Enteropathogens in Four Management Models for Unowned Cats in the Southeast United States
6:00 pm	ID-14	Michael Lappin	*Toxoplasma gondii* Oocyst Shedding in Normal Cats and Cats Treated with Cyclosporine

***Also see Infectious Disease abstracts ID-15 – ID-18 (Thursday, June 16, 2:15 pm – 3:15 pm)*

SMALL ANIMAL – INFECTIOUS DISEASE/PHARMACOLOGY**/GASTROENTEROLOGY**/HEPATOLOGY/OTHER**

2:15 pm	ID-15	Emily Barker	Whole Shotgun Pyrosequencing of *Mycoplasma haemofelis*: Throwing New Light on an Old Pathogen
2:30 pm	**ID-16**	**Jessie Markovich**	**The Prevalence of Leptospiral Antibodies in Feral Cats in Massachusetts**
2:45 pm	ID-17	Yvonne Fischer	The Rivalta'S Test as a Diagnostic Tool in Cats with Feline Infectious Peritonitis
3:00 pm	ID-18	Jamieson Nichols	Prevalence of Protective Antibody Titers Against Canine Parvovirus (CPV) and Canine Distemper Virus (CDV) and Response to CPV and CDV Vaccination in Shelter Dogs
3:15 pm	P-6	Dawn Boothe	Pharmacokinetics and Time Dependant Killing of Fosfomycin in Multi Drug Resistant-*Escherichia coli* Urinary Tract Infections in Dogs
3:30 pm	GI-19	Romy Heilmann	Development and Analytical Validation of an Enzyme-Linked Immunosorbent Assay for the Quantification of Canine Calprotectin in Serum and Feces from Dogs
3:45 pm	GI-20	Romy Heilmann	Fecal Inflammatory Markers and Microbial Changes in Dogs with Acute Hemorrhagic Gastroenteritis
BREAK			
4:30 pm	GI-21	Tara Bidgood	Pharmacokinetics of Maropitant Citrate Dosed Orally to Dogs at 2 mg/kg and 8 mg/kg Once Daily for 14 Consecutive Days
4:45 pm	GI-22	Niels Grützner	Evaluation of Serum Cobalamin Concentration in Canine Serum Samples Submitted to the Gastrointestinal Laboratory Between 2006-2010
5:00 pm	HP-5	Amanda Mitchell	Sensitivity and Specificity of Tissue Impedance Determination for Correct Veress Needle Placement in Cadaverous Dogs
5:15 pm	**HP-6**	**Sara Hyink**	**Sensitivity and Specificity of Tissue Impedance Determination for Correct Veress Needle Placement in Cadaverous Cats**
5:30 pm	OT-1	Marina Leis	Non-Comparative Study to Evaluate Intraosseous Catheter Insertion with the EZ-IO G3 Device by Novice Participants in a Feline Cadaver Model
5:45 pm	OT-2	Jean-Sebastien Palerme	Causes of Multicavitary Effusion in Dogs: Retrospective Study of 43 Dogs
6:00 pm	OT-3	Maureen Anderson	Observation of Patient and Surgeon Preoperative Preparation in Companion Animal Clinics

***Also see Infectious Disease abstracts ID-1 – ID-14 (Thursday, June 16, 2:15 pm – 6:15 pm)*
***Also see Pharmacology abstracts P-1 – P-5 (Thursday, June 16, 5:00 pm – 6:15 pm)*
***Also see Gasteroenterology abstracts GI-1 – GI-18 (Friday, June 17, 8:00 am – 6:15 pm)*

EQUINE**

2:15 pm	E-13	Gayle Hallowell	Is the Phenotype of Prolapsing Aortic Valve Cusps Different From Normal and Diseased Ones?
2:30 pm	E-14	Colin Schwarzwald	Correlation of Heart Rate and Electrocardiographic Time Intervals to Body Mass in Horses and Ponies
2:45 pm	E-15	Melissa Cordero	Bioactivity of Plasma ACTH from Horses with PPID Compared to Normal Horses
3:00 pm	E-16	John Middleton	Evaluation of Skills and Knowledge Desired By Potential Employers of Entry-Level ACVIM-LAIM Diplomates and Skills and Knowledge Taught During Residency Training

3:15 pm	E-18	Christine Wimer	Infection of PBMC with Neuropathogenic Equine Herpesvirus Type 1 Ab4 Strain Induces Interferon-Alpha and Modulates Interleukin-10 Production
3:30 pm	E-20	Mary Sheats	Myristolated Alanine-Rich C-Kinase Substrate (MARCKS) is an Important Regulator of Equine Neutrophil Migration and Adhesion
3:45 pm	E-21	Emil Olsen	What is the Effect of Head Elevation and Blindfolding on the Gait of Non-Neurologic Horses?
BREAK			
4:30 pm	E-22	Lynne Gardner	Articular Cartilage from Horses Afflicted with Hereditary Equine Regional Dermal Asthenia (HERDA) Exhibits Osteoarthritic Lesions and Isolated Chondrocytes Produce High Levels of the Pro-Inflammatory Mediator Prostaglandin E2
4:45 pm	E-24	Laura Beekman	Corticosteroids Treatment Suppresses the Expression of IL-17 in the Lungs of Horses with Inflammatory Airway Disease (IAD)
5:00 pm	E-25	Joan Norton	The Effect of Clenbuterol on Tracheal Mucociliary Transport in Horses Undergoing Simulated Long-Distance Shipping
5:15 pm	E-26	Sabrina Cummings	Role of Airway Hypertonicity in Exercise Induced Airway Inflammation

***Also see Equine abstracts E-1 – E-12 (Thursday, June 16, 9:00 am – 12:30 pm)*
***Also see Equine abstracts E-27 – E-32 (Thursday, June 16, 4:45 pm – 6:15 pm)*

FOOD ANIMAL / EQUINE**

2:15 pm	FA-1	Niels Grützner	Analytical Validation of a Commercially Available Immunoassay for the Measurement of Serum Cobalamin Concentration in Pigs
2:30 pm	**FA-2**	**Davin Ringen**	**Epidiemiology of Mastitis Pathogens in Heifers on a Grazing Dairy**
2:45 pm	**FA-3**	**Marcio Costa**	**Prevalence of Tetracycline Resistance Genes in *Clostridium difficile* Isolated from Calves Prior and Post Treatment with Oxytetracycline**
3:00 pm	FA-4	Katharine Benedict	Associations Between Antimicrobial Use and Antimicrobial Resistance in *Esherichia coli* Sampled from Individual Feedlot Cattle
3:15 pm	**FA-5**	**Christian Gerspach**	**Premature/Dysmature Syndrome in Cria: A Retrospective Study of 63 Cases (1999-2005)**
3:30 pm	FA-6	Kristen Messenger	The Distribution of Enrofloxacin, and its Active Metabolite Ciprofloxacin, Using an In-Vivo Ultrafiltration Sampling Technique After the Injection of Enrofloxacin to Pigs
3:45 pm	**FA-7**	**Sarai Rivera**	**Pharmacokinetics of Ceftiofur Crystalline Free Acid in Sheep Following Subcutaneous Single-Dose Administration**
BREAK			
4:30 pm	**FA-8**	**Theresa Ollivett**	**The Use of Rapid Thoracic Ultrasonography for Detection of Subclinical and Clinical Pneumonia in Pre-Weaned Dairy Calves**
4:45 pm	E-27	Cris Navas	The Importance of Ultrasonographic Pneumatosis Intestinalis in Equine Neonatal Gastrointestinal Disease
5:00 pm	E-28	Teresa Burns	Distribution of Insulin Receptor in the Digital Laminae of Ponies Fed a High-Carbohydrate Diet: An Immunohistochemical Study
5:15 pm	E-29	Veronique Lacombe	The Novel Role of Glucose Transporter 12 in Insulin-Sensitive Tissues of Healthy and Insulin-Resistant Horses
5:30 pm	E-30	Julie Rapson	Effects of Age and Diet on Glycemic and Insulinemic Responses of Horses to a Concentrate Meal
5:45 pm	E-31	Teresa Burns	Evaluation of the Reponse of the Endocrine Pancreas to Short-Term High-Carbohydrate Feeding in Mixed-Breed Ponies
6:00 pm	E-32	Lisa Tadros	Inflammatory Responses to Intravenous Lipopolysaccharide Infusion in Horses with Equine Metabolic Syndrome

***Also see Equine abstracts E-1 – E-12 (Thursday, June 16, 9:00 am – 12:30 pm)*
***Also see Equine abstracts E-13 – E-26 (Thursday, June 16, 2:15 pm – 5:30 pm)*

ORAL PRESENTATIONS – Friday, June 17

SMALL ANIMAL - NUTRITION / METABOLISM

8:00 am	NM-1	Andrew Specht	Clinical Manifestations of Heterozygous Expression of a Mutation that Causes Glycogen Storage Disease Type Ia in Dogs
8:15 am	NM-2	Thomas Schermerhorn	Analysis of the Hepatic Promoter of the Feline Glucokinase Gene
8:30 am	NM-3	Hervé Lefebvre	Volume of Distribution of Creatinine: An Indicator of Body Water Content in Dogs
8:45 am	NM-4	Jennifer MacLeay	Dietary Alpha Linolenic Acid and EPA Reduce Thromboxane and Improve Mobility in Dogs in a Double Blinded Study
9:00 am	NM-5	Dennis Jewell	Effect of a Medium-Chain Triglyceride:Fish Oil Enriched Diet on Lean Body Mass, Circulating Fatty Acids, and Metabolites in Dogs
9:15 am	NM-6	Aulus Carciofi	Metabolic Adaptations to Weight Loss in Obese Dogs May Require Time to Settle
9:30 am	NM-7	Aulus Carciofi	Methionine Intoxication May be Related to Sulfur Effects on Food Base Excess

SMALL ANIMAL – HEPATOLOGY / GASTROENTEROLOGY**

8:00 am	**HP-1**	**Carrie Kavanagh**	**Canine Acute Liver Failure: Retrospective Review of 50 Cases (1995-2010)**
8:15 am	**HP-2**	**Rae Hutchins**	**The Role of Infectious Agents in Granulomatous Hepatitis in Dogs**
8:30 am	HP-3	Jacqueline Whittemore	Histologic Evaluation of Hepatic Copper and Iron Accumulation in 104 Feline Liver Biopsies
8:45 am	**HP-4**	**Allison Bradley**	**Cyclosporine Therapy for Canine Chronic Hepatitis – A Retrospective Study**
9:00 am	GI-1	Peter Kook	Evaluation of the Safety and Practicability of Endoscopic Ultrasound-Guided Fine Needle Aspiration of the Pancreas in Healthy Dogs
9:15 am	GI-2	Jörg Steiner	Effect of a Low-Fat Diet on Serum Triglyceride, Cholesterol, and Pancreatic Lipase Immunoreactivity Concentrations in Miniature Schnauzers with Hypertriglyceridemia
9:30 am	GI-3	Aarti Kathrani	Polymorphisms in Toll Like Receptor 5 Result in Hyper-Responsiveness to Flagellin
9:45 am	GI-4	Karin Allenspach	Toll-Like Receptor 5 Mutations in German Shepherd Dogs are Hyper-Responsive to Stimulation with Flagellin

***Also see Gasteroenterology abstracts GI-19 – GI-22 (Thursday, June 16, 3:30 pm – 5:00 pm)*

SMALL ANIMAL – GASTROENTEROLOGY**

2:15 pm	GI-5	Olivier Dossin	Effect of Omeprazole on Serum Gastrin Concentrations and Calcium Metabolism in Healthy Dogs
2:30 pm	GI-6	Jose Garcia-Mazcorro	Quantification of Gastric and Duodenal Bacteria Using Fluorescent In Situ Hybridization in Healthy Dogs During Omeprazole Treatment
2:45 pm	**GI-7**	**Jonathan Lidbury**	**Assessment of the Variability Associated with Repeated Measurement of Gastrointestinal Transit Times and Assessment of the Effect of Oral Ranitidine on Gastrointestinal Transit Times Using a Wireless Motility Capsule in Dogs**
3:00 pm	GI-8	Melissa Markel	Evaluation of Bacterial Groups in Dogs with Gastrointestinal Disease using Quantitative PCR Assays
3:15 pm	GI-9	Eric Cassmann	Imbalance in the Composition of Ileal and Colonic Microbiota of Dogs with Chronic Enteropathy
3:30 pm	GI-10	Katie Tolbert	Evaluation of the Role of *Pentatrichomonas hominis* and Enteric Coinfections In Canine Diarrhea
3:45 pm	GI-11	Robert Washabau	Domperidone Augments Cholinergic Contractions in Feline and Rabbit Gastrointestinal Smooth Muscle
BREAK			
4:30 pm	GI-12	Tracy Hill	Co-Administration of NSAIDs and Tramadol Decreases Gastric Mucosal Barrier Function
4:45 pm	GI-13	Nora Berghoff	Effect of the Tight Junction Regulator Larazotide Acetate on Intestinal Permeability and Fecal N-Methylhistamine Concentrations in Soft Coated Wheaten Terriers and Norwegian Lundehunds with Gastrointestinal Disease
5:00 pm	**GI-14**	**Mirjam Equilino**	**Biochemical Markers and Survival in Dogs with Protein-Losing Enteropathy**
5:15 pm	**GI-15**	**Scott Owens**	**Canine Protein Losing Enteropathy: A Retrospective Analysis and Survival Study in 68 Dogs**
5:30 pm	GI-16	Niels Grützner	Evaluation of Serum Methylmalonic Acid Concentrations in Chinese Shar Peis and Dogs of Six Other Breeds with Cobalamin Deficiency
5:45 pm	GI-17	Niels Grützner	Serum Concentrations of Hyaluronic Acid, C-Reactive Protein, and the Calgranulins in Chinese Shar Peis with and without Cobalamin Deficiency
6:00 pm	GI-18	Sabina Lutz	Cobalamin Deficiency in Border Collies

***Also see Gasteroenterology abstracts GI-19 – GI-22 (Thursday, June 16, 3:30 pm – 5:00 pm)*

POSTER PRESENTATIONS

On Display:	Thursday, June 16, 7:00 am – 10:00 pm Friday, June 17, 7:00 am – 10:00 pm Saturday, June 18, 7:00 am – 2:30 pm
Attended by Authors Eligible for ACVIM Resident Research Awards:	Thursday, June 16, 9:50 am – 10:30 am Friday, June 17, 9:50 am – 10:30 am
Attended by ALL Authors – Wine & Cheese Reception	Friday, June 17, 6:00 pm – 7:30 pm

#	**Presenting Author**	**Abstract Title**
SMALL ANIMAL - CARDIOLOGY		
C-27	Mark Oyama	Auscultatory, Echocardiographic, Biochemical, Nutritional, and Environmental Characteristics of Mitral Valve Disease in Norfolk Terriers
C-28	Maria Helena Larsson	Comparative Study of Left Ventricle Measurements by Four Echocardiographic Methods in German Shepherd Dogs

C-29	Arine Pellegrino	Assessment of Left Ventricular Diastolic Function by Color Tissue Doppler Imaging Echocardiography in Maine Coon Cats Tested for MYPBC-AP31 Mutation
C-30	Arine Pellegrino	Assessment of Left Ventricular Sistolic Function by Strain Imaging Echocardiography in Maine Coon Cats Tested for MYPBC-A31P Mutation
C-31	Guilherme Gonçalves Pereira	Systolic Blood Pressure in Healthy Maine Coon Cats
C-32	Maja Spiljak	Advanced Electrocardiography Can Predict Mitral Regurgitation in Cavalier King Charles Spaniels with Myxomatous Mitral Valve Disease
C-33	Rachel Cohen	Exercise Induced Variability of Plasma N-Terminal Pro-BNP in Healthy Dogs
C-34	Denise Schwartz	NT-Pro ANP in Obese Dogs
C-35	**Crystal Hariu**	**Utility of Plasma NT-proBNP for Assessing Hemodynamic Significance of Patent Ductus Arteriosus in Dogs Prior To and Following Transcatheter Ductal Occlusion**
C-36	**Megan McLane**	**Validation of Three Shipping Methodologies for NT-proBNP in a Mixed Cohort of Cats**
C-37	Soojung Lim	Serum Serotonin Concentration in Relation to Severity of Spontaneous Degenerative Mitral Valve Disease in Dogs
C-38	**Neal Peckens**	**Utility of Cardiac Troponin I in Differentiating Between Underlying Etiologies of Pericardial Effusion in the Canine Patient**
C-39	**Gretchen Singletary**	**Plasma Concentrations of ICam-1 and E-Selectin in Dogs with Mitral Valve Disease**
C-40	Denise Schwartz	Chronic Heart Failure Selectively Induces Regional Heterogenity of Glucose Transporter Expression
C-41	Denise Schwartz	Comparison Between Thoracic Computerized Radiography and Echocardiogram in Obese Dogs to Evaluate Heart Enlargement
C-42	Caroline Rasmussen	Heart Rate and Heart Rate Variability in Dogs with Different Degrees of Myxomatous Mitral Valve Disease
C-43	Aparecido Camacho	Bradycardia Induced by Partial Nasal Obstruction by Digital Compression in Non-Sedated Rabbits
C-44	Guilherme Gonçalves Pereira	Tissue Doppler Evaluation in English Cocker Spaniels with Dilated Cardiomyopathy
C-45	Aparecido Camacho	Cardiovascular Function in Obese and Overweight Cats
C-46	Gordon Peddle	Effect of Torsemide vs Furosemide on Clinical, Laboratory, Radiographic, and Quality of Life Variables in Dogs with Heart Failure
C-47	Hiroshi Takano	2D Speckle Tracking Echocardiography in Clinically Healthy Cats and Cats with Hypertrophic Cardiomyopathy
C-48	**April Durtschi**	**Pharmacokinetics and Bioavailability of Carvedilol in Cats**
C-49	Kensuke Nakamura	Effects of Sildenafil Citrate on Dogs with Eisenmenger's Syndrome

SMALL ANIMAL - NEUROLOGY

N-10	**Starr Cameron**	**The Prognostic Value of Admission Blood Gas Parameters in Dogs with Traumatic Brain Injury**
N-11	**Caroline Altenhoefer**	**Serum Lactate Levels in Phenotypical Chondrodystrophic Dogs with Type I Intervertebral Disc Disease**
N-12	Holger Volk	Relationship of Cerebellar Volume to Syringomyelia in the Cavalier King Charles Spaniel
N-13	Holger Volk	Changes Over Time in Caudal Cranial Fossa Volumes and of Cerebellar Herniation in Cavalier King Charles Spaniels with Chiari-Like Malformation
N-14	**James Campbell**	**Assessment of Cord Dorsum Potentials from Caudal Nerves in Clinically Normal Dogs**
N-15	Simon Platt	Expression of Heat Shock Proteins, Ki-67 and E-Cadherin in Canine Intracranial Meningiomas
N-16	Simon Platt	Pharmacokinetic Evaluation of Extended Release Levetiracetam in Dogs
N-17	Simon Platt	Retrospective Evaluation of Repair Techniques for Thoracolumbar and Lumbosacral Vertebral Fractures in Dogs
N-18	Luis Gaitero	Cranial Thoracic Intervertebral Disc Disease in German Shepherd Dogs: A Retrospective Comparative MRI Study
N-19	**Jeanene Harris**	**Signalment, Clinical Features and Magnetic Resonance Imaging Characteristics of 23 Cases of Diskospondylitis in Dogs**
N-20	**Jonathan Huska**	**Comparison of the Access to the Thoracolumbar Spinal Canal Between Hemilaminectomy and Pediculectomy Through Computed Tomography**
N-21	**Cassandra Williams**	**Is Low Level Laser Therapy an Effective Adjunctive Treatment to Hemilaminectomy in Dogs with Acute Onset Paraplegia Secondary to Intervertebral Disc Disease?**
N-22	Byeong-Teck Kang	18F-Fluorodeoxyglucose Uptake of Normal Canine Brain with a High-Resolution Research Tomograph-Positron Emission Tomography and 7 T-Magnetic Resonance Imaging Fusion System
N-23	**Alix Partnow**	**Dysuria in Dogs with Degenerative Lumbosacral Stenosis**

SMALL ANIMAL - ONCOLOGY

O-8	Silvia Lucas	Breast Cancer Resistance Protein (BCRP/ABCG2) Gene Expression by Real Time RT-PCR in Dogs with Multicentric Lymphoma
O-9	Caroline Piskun	High Serum Alkaline Phosphotase (ALP) in Canine Osteosarcoma Patients Correlates with Cellular ALP Expression and Increased Beta-Catenin Expression

O-10	Christine Anderson	Comparison of Abdominal Ultrasound and Magnetic Resonance Imaging for Detection of Abdominal Lymphadenopathy in Dogs with Metastatic Apocrine Gland Adenocarcinoma of the Anal Sac (VCS Award Winner)
O-11	Emma Warry	Pharmokinetics of Cyclophosphamide after Oral or Intravenous Administration to Dogs with Lymphoma (VCS Award Winner)

SMALL ANIMAL - ENDOCRINOLOGY

EN-16	Lynda Melendez	Titration of Dietary Iodine for Reducing Serum Thyroxine Concentrations in Newly Diagnosed Hyperthyroid Cats
EN-17	Lynda Melendez	Titration of Dietary Iodine for Maintaining Normal Serum Thyroxine Concentrations in Hyperthyroid Cats
EN-18	Shiguang Yu	Controlled Level of Dietary Iodine Normalizes Serum Total Thyroxine in Cats with Naturally Occurring Hyperthyroidism
EN-19	**Rachel Oman**	**Outcome of Radioactive Iodine Therapy in Cats Receiving Recent Methimazole Therapy**
EN-20	**Christine Bryan**	**Lack of Cross Reactivity of Feline Insulin with a Commercial Human Insulin Elisa**
EN-21	Kyu-Deok Cho	Comparison of Low- and High-Dose Trilostane Treatment in Dogs with Pituitary-Dependent Hyperadrenocorticism

SMALL ANIMAL - GASTROENTEROLOGY

GI-23	Romy Heilmann	Intra-Individual Variability of Fecal Calprotectin Concentrations in Dogs with Chronic Gastrointestinal Disease
GI-24	Yasushi Minamoto	Prevalence of the *Clostridium perfringens* Enterotoxin Gene and *C. perfringens* Enterotoxin in Fecal Samples from Dogs and Cats
GI-25	**Michael Goldstein**	**Comparison of Culture Methods and Detection of *Clostridium perfringens* from Healthy and Diarrheic Dogs**
GI-26	Katherine Scott	Consistency of Endoscopic Biopsies Obtained from Duodenum and Ileum for Feline Small Cell Lymphoma (SC-LSA) and Inflammatory Bowel Disease (IBD)
GI-27	Panpicha Sattasathuchana	Evaluation of Dental Tartar, Gingivitis, and Subgingival Bacteria in FIV- and FeLV-Negative Cats
GI-28	Aarti Kathrani	Polymorphisms in the Toll Like Receptor 5 Gene Are Significantly Associated with Inflammatory Bowel Disease in a Population Consisting of 38 Different Dog Breeds
GI-29	Aarti Kathrani	Mutational Analysis of the Toll Like Receptor 5 Gene in Boxer Dogs with Inflammatory Bowel Disease Reveals Novel Polymorphism in the Leucine Rich Repeat Domain

SMALL ANIMAL - HEMATOLOGY

HM-16	**Dinaz Naigamwalla**	**Canine Iron-Refractory Iron Deficiency Anemia Caused by Matriptase (*Tmprss6*) Deficiency**
HM-17	**Kenneth Palladino, Jr.**	**A Comparison of Body Weight in Relation to Zinc Toxicity Anemia Secondary to Metalilic Foreign Body**
HM-18	Jennifer Stokes	Retrospective Study of the Safety and Efficacy of Cryopreserved Platelet Concentrate in 26 Dogs
HM-19	Jennifer Stokes	Efficacy and Safety of Fresh Frozen Plasma Stored at 4°C for 5 Days
HM-20	Guldal Inal	Survey of Erythrocytic Pyruvate Kinase Deficiency in Dogs
HM-21	Mitika Hagiwara	Detection of IgG, IgM and C3 on Red Blood Cell Surface of Dogs with Immune-Mediated Hemolytic Anemia
HM-22	Alice Defarges	Sternal Bone Marrow Aspiration in Healthy Cats
HM-23	Anthony Abrams-Ogg	15-Gauge Core Bone Marrow Biopsy in Healthy Cats

SMALL ANIMAL - HEPATOLOGY

HP-7	Sharon Center	Best Predictors of Portosystemic Shunting in 568 Dogs: Post Feeding Serum Bile Acids and Protein-C

SMALL ANIMAL - IMMUNOLOGY

IM-2	Nathalee Prakash	Molecular Signaling Mechanisms of Eosinophilia in Rottweilers
IM-3	Mitika Hagiwara	Proliferative Response of Blood Lymphocytes to Red Blood Cell Membrane in Recently Immunized Dogs

SMALL ANIMAL – INFECTIOUS DISEASE

ID-19	Camille Torres	Role of *Mycoplasma* Spp. in Feline Cat Bite Abscesses
ID-20	Michael Lappin	Effect of Vaccination on Results of Real Time PCR Assays for Adenovirus 2, Parainfluenza, and *Bordetella bronchiseptica* in Puppies
ID-21	Megan Behringer	Ability of a FRET-PCR Assay Targeting *gyrA* to Detect Fluoroquinolone Resistance in Veterinary Clinical Urine Samples Containing *Escherichia coli*
ID-22	Xiaoqiang Liu	Comparative In Vitro Activities of Fluoroquinolone Antimicrobials Against *Escherichia coli* Uropathogens
ID-23	Carlos Larsson	Canine Leproid Granuloma (CLG) in São Paulo: A Retrospective Study of 37 Cases (1990 to 2009)
ID-24	**Michael Kim**	**Amplification of *Anaplasma phagocytophilum* DNA from the Blood and Synovial Fluid of Experimentally Infected Dogs**

ID-25	Melissa Beall	Canine Granulocytic Anaplasmosis and Granulocytic Ehrlichiosis - A Field-Based Comparison
ID-26	**Jeanne Ficociello**	**Comparison of Commercially Available Assays for the Amplification of *Ehrlichia canis* and *Anaplasma phagocytophilum* DNA from the Blood of Naturally Infected Dogs**
ID-27	Pedro Diniz	Prevalence of Tick-Borne Diseases in the Highlands of Peru
ID-28	Michael Lappin	Prevalence of *Bartonella* Spp. DNA in Blood of Naturally Exposed Cats with and without Imidocloprid Treatment
ID-29	Maria Dolores Tabar	Severity of Clinical Signs and Outcome of Dogs Infected by *Wolbachia*, Microfilaria and *Leishmania*
ID-30	Mary Marcondes	Feline Leishmaniasis in the Municipality of Araçatuba, São Pulo, Brazil
ID-31	Mary Marcondes	*Leishmania chagasi* Infection in Cats with Dermatologic Lesions from an Endemic Area of Visceral Leishmaniasis in Brazil
ID-32	**Gavin Olsen**	**Comparison of Radiographic and Urine Antigen Resolution of Clinical Blastomycosis**
ID-33	Valeria Scorza	Prevalence of *Giardia* Spp. and *Cryptosporidium* Spp. in Dogs of the United States
ID-34	Sahatchai Tangtrongsup	Intestinal Parasites of Dogs in Chiang Mai, Thailand
ID-35	Donald Martin	Comparison of 2 Methods of Parasite Recovery from Fecal Specimens of Veterinary Patients: Sedimentation/Concentration vs. Fecalyzer

SMALL ANIMAL – NEPHROLOGY / UROLOGY

N/U-20	Javier Del Angel-Caraza	Canine Urolithiasis in Mexico
N/U-21	**Dinaz Naigamwalla**	**The Use of a Medetomidine-Based Sedation Protocol to Perform Urohydropropulsion and Cystoscopy in the Dog**
N/U-22	**Nicole Smee**	**Effect of Storage, Time, and Temperature on Canine Urine Enzymes**
N/U-23	**Sara Irom**	**Interim Evaluation of the Efficacy and Safety of a High Dose Short Duration Enrofloxacin Treatment Regimen for Urinary Tract Infection in Dogs**
N/U-24	Thomas Daste	Renal Doppler Resistive Index in Dogs with Degenerative Mitral Valve Disease (DMVD)
N/U-25	Chelsea Sonius	Association between Feline Antibody Responses to Alpha-Enolase and Azotemia in Privately-Owned Cats
N/U-26	**Emily Harison**	**Acute Kidney Injury (AKI) as a Predictor of Mortality in Dogs and Cats**
N/U-27	**Jonathan Dear**	**Feline Urate Urolithiasis: 143 Cases (2000-2008)**
N/U-28	Bernard Schmidt	Evaluation of Normal Serum Concentrations of the Putative Uremic Toxins, Indoxyl and p-Cresyl Sulfate, in Healthy Adult Cats
N/U-29	John Kruger	Expression of 15-Hydroxyprostaglandin Dehydrogenase (PGDH) in Urothelium of Cats with Chronic Idiopathic Cystitis
N/U-30	Katherine Hamon	Variation In Kidney Size in a Population of Cats

SMALL ANIMAL – NUTRITION / METABOLISM

NM-8	**Valerie Parker**	**Association Between Body Condition and Survival in Dogs with Acquired Chronic Kidney Disease**
NM-9	Shiguang Yu	Dietary Crude Protein of 28.5% Maintains Long-Term Lean Body Mass in Cats with Impaired Kidney Function
NM-10	Shiguang Yu	Dietary Supplementation of Vitamin B12 Improves Vitamin B12 Status in Geriatric Cats
NM-11	John Bauer	Changes in Serum Essential Fatty Acid Profiles in Dog and Cat Diseases
NM-12	Sungjun Noh	Effect of L-Alanyl-L-Glutamine Supplementation in the Treatment of Canine Parvoviral Enteritis
NM-13	Laura Tonkin	In Vitro Effects of Lipid Emulsion on Platelet Function and Thromboelastography in Healthy Dogs

SMALL ANIMAL - OTHER

OT-4	Marina Leis	Comparison of a New Intraosseous Catheter Insertion Technique Using the EZ-IO G3 Power Driver to a Standard Manual Technique for Intraosseous Catheter Insertion in a Feline Cadaver Model
OT-5	Youngjae Lee	Assessment of Feline Bone Mineral Density: Quantitative Computed Tomography
OT-6	Donghoon Lee	Quantitative Assessment of Aging Effects on Cerebral Blood Flow in Normal Dogs: A Perfusion CT Study
OT-7	Archivaldo Reche, Jr.	Inappropriate Urination in Multi-Cat Households: Are They Really Just Behaving Badly?
OT-8	Archivaldo Reche, Jr.	Living Style, Personality, Owner Subjective Life Quality: What Factors Affect Stress Levels in Domestic Cats (*Felis catus*)?

SMALL ANIMAL - PHARMACOLOGY

P-7	**Kristin Lewis**	**Pharmacokinetics of Diminazene Diaceturate in Healthy Cats**
P-8	Laura Johnston	Determination of Meloxicam Concentration in Synovial Fluid of Inflammed and Non-Inflammed Joints in Dogs
P-9	Kamoltip Thungrat	Trends of Small Animal *Escherichia coli* Antimicrobial Resistance in the United States
P-10	Xiaoqiang Liu	Comparative In Vitro Activities of Fluoroquinolone Antimicrobials Against *Escheria coli* Uropathogens

SMALL ANIMAL - RESPIRATORY

R-8	Aparecido Camacho	Pulmonary Function and Hemogasometric Evaluation of Obese and Overweight Cats

EQUINE

E-33	Marc Kraus	Pharmacokinetics of Intravenously Administered Equine Cardiac Troponin I in Healthy Ponies
E-34	**Elizabeth Finding**	**Flow-Mediated Vasodilation in Healthy Ponies**
E-35	Emil Olsen	How to Obtain Indirect Blood Pressure in the Standing Horse?
E-36	Francisco Mendoza	Thyroid Hormone Cocentrations Differ Between Donkeys and Horses
E-37	**Francois Bertin**	**The Effect of Ovariectomy on Measures of Insulin Sensitivity and Response to Dexamethasone in Mares**
E-38	Nan Huff	Efficacy of Seabuckthorn Berries for Treatment and Prevention of Gastric Ulcers in Horses
E-39	Benjamin Brainard	Evaluation of the VSPro Table-Top Assay for Equine Fibrinogen
E-40	**Seika Hashimoto-Hill**	**Dose-Dependent Effect of *Rhodococcus equi* Infection on Cytokine/Chemokine Responses of Pulmonary Alveolar Macrophages**
E-41	Jennifer Jeske	Predicting Internal Infection in Horses with *Corynebacterium pseudotuberculosis* Using Serum Synergistic Hemolysis Inhibition Test Titers
E-19	Wilson Roberto Fernandes	Immunization of Foals with Attenuated Strain of *Salmonella typhimurium* Carrying Vapa Gene of *Rhodococcus equi*: Clinical and Immunological Evaluation
E-42	Brandy Burgess	Rapid *Salmonella* Detection in Experimentally-Inoculated Equine Feces Using Two Commercially Available Lateral Flow Antigen Detection Systems
E-43	Frank Garza, Jr.	Evaluation of Three Stall Cleaning Methods on the Eradication of *Salmonella* sp. in an Equine Veterinary Teaching Hospital
E-44	Londa Berghaus	Pulmonary Disposition of Gamithromycin in Foals and In Vitro Activity of the Drug Against *Rhodococcus equi* and *Streptococcus equi* Subsp. Zooepidemicus
E-67	**Brent Credille**	**Disposition of Ceftiofur Crystalline Free Acid in Plasma and Pulmonary Epithelial Lining Fluid of Weanling Foals**
E-45	Lutz Goehring	D- Dimer Concentrations and (Experimental) EHV-1 Infection
E-46	Anna Chapman	Identification of Exposure to and Shedding of *Lawsonia intracellularis* in Louisiana Horses Using Serology and Real-Time PCR
E-47	Ashley Boyle	Equine Granulocytic Anaplasmosis: 19 Field Cases in Southeastern Pennsylvania (2005-2010)
E-48	Han van der Kolk	Needle Electromyography of the Esophagus in Friesian Horses with and without Esophageal Dysfunction
E-49	**Karie Vander Werf**	**A Possible Link Between Equine Lymphoma and Equine Herpes Virus 5**
E-50	Martha Hensel	Laminar Inflammatory Events and Epithelial Stress at Obel Grade 3 Lameness in the Carbohydrate Model of Equine Laminitis
E-51	Anna Renier	Oleander Toxicosis in Equids: 30 Cases (1995-2010)
E-52	Henry Staempfli	Validation of the Use of Serum Biochemical Profile for Acid-Base Interpretation in Horses
E-53	Megan Dorris	Increased Metalloproteinase Activity in Chondrocytes from Articular Cartilage of Horses with Hereditary Equine Regional Dermal Asthenia
E-54	Bianca Schwarz	Glycogen Synthase 1 Mutation in the Austrian Haflinger Population
E-55	**Jacquelyn Bowser**	**Tensile Properties of Collagen Rich Tissues of Quarter Horses with Hereditary Equine Regional Dermal Asthenia (HERDA)**
E-56	**Elizabeth Carothers**	**Comparison of a Proprietary Herbal Topical Cream to Silver Sulfadiazene Using an Equine Wound Model**
E-57	Claire Underwood	A Scintigraphic Comparison of 99mTc Labeled IgG, PEG-Liposome and Leukocyte Accumulation in Experimentally Induced Focal Soft Tissue Infection in the Horse
E-59	Carlin Kelly	P38 MAPK Signaling in Equine Laminitis: The Effect of Digital Hypothermia
E-60	Anna Hollis	Does Cefquinome Penetrate the Blood Brain Barrier in the Normal Horse?
E-61	Sherry Cox	Pharmacokinetics of Firocoxib and Its Interaction with Enrofloxacin in Horses
E-62	**Karin Kruger**	**Pharmacokinetics and Pharmacodynamics of Quinapril in Horses**
E-17	Jessica Khodadad	Pharmokinetics and Clinical Safety of Oral Paste Firocoxib, a COX-1 Sparing NSAID, in Foals
E-63	**Natasha Hovanessian**	**Pharmacokinetics of Firocoxib in Neonatal Foals**
E-64	**Johanna Elfenbein**	**A Retrospective Evaluation of Erythrocytosis in Horses with Liver Disease**
E-65	Michele Frazer	Sale Price and Race Earnings in Horses Previously Diagnosed with *Lawsonia intracellularis* Infection

FOOD ANIMAL

FA-9	**Brandon Fraser**	**Lameness in Beef Cattle: Retrospective Study of Foot Disorders (2007-2010)**

Research Report Program of the 2011 ACVIM Forum

Denver, Colorado,
June 15–18, 2011
Index of Abstracts

THURSDAY, JUNE 16

Time	#	Presenting Author	Research Report Title
ONCOLOGY			
9:00 am	7a	Joelle Fenger	Breed-Associated Differential Microrna Expression in Canine Osteosarcoma *(VCS Award Winning Abstract)*
9:25 am	7b	Cecilia Robat	Safety Evaluation of Combination Vinblastine and Toceranib Phosphate (Palladia®) in Dogs: A Phase I Dose-Finding Study *(VCS Award Winning Abstract)*
10:30 am	8a	Timothy Fan	Targeting Tumor Neovasculature with Nanoparticle Therapeutics
10:55 am	8b	Melissa Paoloni	Transitional Cell Carcinoma Comparative Genomics: A Model for Personalized Medicine Approaches to Cancer Care
11:30 am	9a	Gregory Ogilvie	A Prospective Phase I/II, Open Label, Non-Randomized Clinical Trial to Evaluate the Safety and Efficacy of AX-3 Multi-Kinase Inhibitor for the Adjunctive Treatment of Dogs with Spontaneously Occurring Cancers
11:55 am	9b	James Custis	Canine Nasal Tumors: A Stereotactic Radiation Therapy Approach

FRIDAY, JUNE 17

Time	#	Presenting Author	Research Report Title
SMALL ANIMAL INTERNAL MEDICINE			
8:00 am	71a	Rhonda LaFleur	Prevention of Disease and Mortality in Vaccinated Dogs Following Experimental Challenge with Virulent Leptospira
8:25 am	71b	Jody Lulich	Ciprofloxacin Urolithiasis: A Newly Recognized Disease in Dogs
9:00 am	72a	Jody Lulich	Struvite Urolith Dissolution in Cats: A Double Blind Randomized Clinical Trial of 2 Foods
9:25 am	72b	Hasan Albasan	Effects of Formalin on Urolith Composition
8:00 am	73a	Susan Little	Seroprevalence of Three *Ehrlichia* Species in Dogs: A Multi-Institutional Study
8:25 am	73b	Joseph Bartges	Polydimethylsiloxane Urethral Bulking Agent (PDMS UBA) Injection for Treatment of Female Canine Urinary Incontinence – Preliminary Results
9:00 am	74a	Stephanie Smith	Transfusion of Autologous Stored Erythrocyte Concentrates Causes Inflammation and Hypocoagulability
9:25 am	74b	Daniel Hogan	Enhanced Bioconversion of Clopidogrel to its Active Metabolite with Rifampin Induction of Hepatic Isozymes
EQUINE			
8:00 am	85a	Allison Stewart	Identification of Critical Illness Related Corticosteroid Insufficiency in Critically Ill Horses
8:25 am	85b	Dianne McFarlane	Alterations in Autophagy in the Pituitary of Horses with Equine Pituitary Pars Intermedia Dysfunction (PPID)
9:00 am	86a	Lara Maxwell	Efficacy of Delayed Antiviral Therapy Against EHV-1 Challenge
9:25 am	86b	Paul Lunn	Protection from Contemporary Equine Influenza Virus by Commercial Vaccines
NEUROLOGY			
2:10 pm	93a	Simon Platt	Cerebrospinal Fluid Neurotransmitter and Cytokine Concentrations in Canine Degenerative Myelopathy
2:35 pm	93b	Simon Platt	Chiari-Like Malformation, Syringomyelia and Neuropathic Pain: Cerebrospinal Fluid Neurotransmitter and Cytokine Levels

3:10 pm	94a	Steven De Decker	Update on the Conservative Management of Disk Associated Cervical Spondylomyelopathy in Dogs
3:35 pm	94b	Annie Chen	Clinical Evaluation of a Magnetic Resonance Imaging - Guided, Frameless Stereotactic Brain Biopsy System in the Dog
4:25 pm	95a	Dominic Marino	Morphometric Features of the Craniocervical Junction Region in Dogs with Suspected Chiari-Like Malformation
4:50 pm	95b	Dominic Marino	Medical Infrared Imaging (Thermography) of Type I Thoracolumbar Disk Disease in Chondrodystrophic Dogs
5:25 pm	96a	CA Loughlin	Thermographic Imaging (Medical Infrared Imaging) Of Type I Thoracolumbar Disk Disease In Chondrodystrophic Dogs
5:50 pm	96b	Veronika Stein	Spinal Cord Injury Influenced by the Microglia Reaction Profile

ONCOLOGY

4:25 pm	99a	Douglas Thamm	Serum Thymidine Kinase Activity in Canine Splenic Hemangiosarcoma
4:50 pm	99b	Susan Lana	Gene Expression Profiling of Canine T Cell Chronic Lymphocytic Leukemia

SMALL ANIMAL INTERNAL MEDICINE

2:10 pm	105a	Jan Suchodolski	Identification of Fecal Bacterial Dysbiosis in Dogs with Various Gastrointestinal Disorders
2:35 pm	105b	Mark Papich	Oral Absorption of Generic Ciprofloxacin Tablets in Dogs
3:10 pm	106a	Christian Leutenegger	Canine Distemper Virus Quantification by Real-Time PCR Allows to Differentiate Vaccine Virus Interference and Wildtype Infection
3:35 pm	106b	Allyson Berent	The Use of a Subcutaneous Ureteral Bypass (SUB) Device for the Treatment of Feline Ureteral Obstructions
5:25 pm	108a	Katrina Viviano	The Effect of N-acetylcysteine (NAC) in Clinically Ill Dogs: Does it Modulate Oxidative Stress and Improve Clinical Outcome?
5:50 pm	108b	Jeffrey Brockman	Biomarkers Predictive of Insulin Resistance in Cats Improve with Weight Loss/ Healthy Weight Maintenance

ACVIM FOUNDATION RESEARCH

2:10 pm	109a	Nicola Mason	Canine-Derived Antibody Fragments that Target Hemangiosarcoma and its Growth Factors
2:35 pm	109b	Erin Malone	Efficacy of Vitamin D (Calcitriol) for Treatment of Canine Mast Cell Tumors
3:10 pm	110a	Todd Archer	Comprehensive Pharmacodynamics of Cyclosporine in the Dog
3:35 pm	110b	David Panciera	Effect of Hypothyroidism on Reproduction in the Female Dog
4:25 pm	111a	Katherine Williamson	Efficacy of Gastric Acid Suppression for Prevention of Exercise-Induced Gastritis in Racing Alaskan Sled Dogs
4:50 pm	111b	Benjamin Buchanan	Effect of Gallium on Fecal Shedding of *Rhodococcus equi* in Mares
5:25 pm	112a	Marybeth Feutz	Bronchoalveolar Lavage Fluid Proteome in Horses with Recurrent Airway Obstruction
5:50 pm	112b	Lyndi Gilliam	Rattlesnake Envenomation in Horses: Cardiotoxic Effects and Immune Response

EQUINE

4:25 pm	139a	Gillian Perkins	Pyrosequencing of the Gastric Mucosa in Healthy Horses Shows an Abundant Mircrobial Flora Which is Unique in Each Horse and Influenced by Diet and Management
4:50 pm	139b	Beatrice Sponseller	Seasonal Pasture Myopathy in 6 Horses Associated with Multiple Acyl-CoA Dehydrogenase Deficiency (MADD)

FOOD ANIMAL INTERNAL MEDICINE

2:10 pm	145a	Janice Sojka-Kritchevsky	Progression of Parasitemia and Clinical Course in a Splenectomized Alpaca with *Candidatus Mycoplasma Haemolamae* Infection
2:35 pm	145b	Gayle Hallowell	Transabdominal Ultrasonographic Gastro-Intestinal Wall Thickness in Adult Cattle, Sheep and Goats
3:10 pm	146a	Melanie Boileau	Assessment of *Bdellovibrio bacteriovorus* 109J Viability in Bovine Tears
3:35 pm	146b	Henry Staempfli	Hyperchloremia Directly Increases Ionized Calcium Concentration in Bovine Plasma: An Additional Mechanism by Which High Chloride Rations Mitigate Periparturient Hypocalcemia
4:25 pm	147a	John Middleton	A New Look at the Importance of Coagulase Negative Staphylococci in Bovine Mastitis
4:50 pm	147b	Derek Foster	Reduction of Bacterial Contamination of Bovine Colostrum with High Pressure Processing
5:25 pm	148a	TBD	
5:50 pm	148b	Maisie Dawes	Lactoferrin: A Further Investigation into Its Anti-Inflammatory Effects: Intracellular Signaling in Evidence?

SATURDAY, JUNE 18

Time	#	Presenting Author	Research Report Title
CARDIOLOGY			
8:00 am	153a	Amy LeBlanc	Kinetic Assessment of Myocardial 14(R,S)-[18F]fluoro-6-Thia-Heptadecanoic Acid and 18F-Fluorodeoxyglucose Uptake with Positron Emission Tomography/Computed Tomography (PET/CT) in Domestic Cats
8:25 am	153b	Ingrid Ljungvall	Assessment of Global and Regional Left Ventricular Volumes and Geometry Using Real-Time Three-Dimensional Echocardiography in Dogs with Myxomatous Mitral Valve Disease
9:00 am	154a	Mark Rishniw	Heartworm Sex Ratios in Naturally Infected Dogs Differ with Worm Intensities
8:00 am	157a	Masami Uechi	Reverse Remodeling After Mitral Valve Repair Under Cardiopumonary Bypass in Dogs
8:25 am	157b	Herbert Maisenbacher	Effect of Dicholoracetate in Canine Pulmonary Arterial Hypertension: A Pilot Study
NEUROLOGY			
8:00 am	161a	G. Diane Shelton	A Deletion in *ARHGEF 10* is Highly Associated with Early Onset Inherited Polyneuropathy in Leonberger and St. Bernard Dogs
8:25 am	161b	G. Diane Shelton	Startle Disease in Irish Wolfhounds Associated with a Deletion in the Gene Encoding the Glycine Transporter GLYT2
9:00 am	162a	Dennis O'Brien	A Mutation in ATP13A2 Causes Adult-Onset Neuronal Ceroid Lipofuscinosis in Tibetan Terriers
9:25 am	162b	Charles Vite	Intrathecal versus Subcutaneous Therapy of Feline Niemann-Pick Type C Disease
10:30 am	163a	Ronaldo da Costa	A Novel Leukodystrophy in Border Terrier Puppies: The Canine "Rumpshaker"
11:55 am	164b	Susan Taylor	Effects of a Standardized Sheep Herding Strenuous Exercise Protocol on Clinical and Laboratory Parameters in Healthy Border Collies and Dogs with Border Collie Collapse
EQUINE			
8:00 am	197a	Frank Andrews	Diagnostic Frequency, Response to Therapy and Long Term Prognosis Among Horses with Equine Cushing's Disease, 1993–2004
8:25 am	197b	Jenifer Gold	Association of Adrenocorticotropin, Cortisol, and Insulin-Like-Growth Factor-I in Septic and Non-septic Foals
9:00 am	198a	Monica Aleman	Repetitive Stimulation of the Common Peroneal Nerve as a Diagnostic Aid for Botulism in Foals
9:25 am	198b	Nicholas Frank	Effects of Weight Gain on Active Glucagon-Like Peptide-1 Concentrations in Horses with Histories of Obesity and Insulin Resistance
10:30 am	199a	Christopher Fielding	Use of Hypertonic Saline in Endurance Horses: A Randomized Clinical Trial
10:55 am	199b	Modest Vengust	Effect of Inhaled Salbutamol on Pulmonary Circulation Transvascular Fluid Dynamics and Gas Exchange in Exercising Horses
11:30 am	200a	Renaud Leguillette	Inflammatory Airway Disease (IAD) in Horses is Characterized by a Th2 and Th17 mRNA Expression Profile
11:55 am	200b	Virginia Buechner-Maxwell	*In Vitro* Response of Peripheral Blood Lymphocytes From Heavey and Non-Affected Horses To Beta 2-Agonist Stimulation

ABSTRACT C-1

COMPARISON OF SHORT-TERM EFFECTS OF IVABRADINE AND ATENOLOL ON HEART RATE AND ECHOCARDIOGRAPHIC VARIABLES OF LEFT HEART FUNCTION IN HEALTHY CATS. SC Riesen[1,2], KE Schober[1], RM Cervenec[1], JD Bonagura[1]. [1]The Ohio State University, College of Veterinary Medicine, Columbus, Ohio. [2]University of Veterinary Medicine, Vienna, Austria.

Hypertrophic cardiomyopathy (HCM) is the most commonly observed myocardial disease in cats. Beta-blockers and calcium channel inhibitors are frequently administered drugs to cats with preclinical HCM despite the fact that neither drug category has been proven to slow disease progression or improve survival. Ivabradine (Procorolan®, Servier, France) is a novel negative chronotropic agent used in the treatment of ischemic heart disease in people. Little is known about its efficacy and safety in cats. The purpose of this study was to determine the short-term effects of ivabradine on heart rate (HR), blood pressure, left ventricular (LV) systolic and diastolic function, left atrial (LA) performance, and clinical tolerance in healthy cats after repeated oral doses.

Ten healthy laboratory cats were involved in the present study. Physical examination, systolic blood pressure measurement, and transthoracic echocardiography were performed in all cats at baseline and after oral administration (4 weeks each) of ivabradine (0.3 mg/kg, q12 h) and atenolol (6.25 mg/cat, q12h; 1.0–1.7 mg/kg) in a prospective, double-blind, randomized, active-control, fully crossed study. A-priori non-inferiority margins for the effects of ivabradine compared to atenolol were set at 50% ($f = 0.5$) based on predicted clinical relevance, observer measurement variability, and in agreement with FDA guidelines. Variables were compared by use of 2-way repeated measures ANOVA.

Ivabradine was clinically well tolerated with no adverse events observed. HR (ivabradine, $P < 0.001$; atenolol, $P < 0.001$; ivabradine vs. atenolol, $P = 0.721$) and rate-pressure product (ivabradine, $P < 0.001$; atenolol, $P = 0.001$; ivabradine vs. atenolol, $P = 0.847$) were not different between treatments. At the dosages used, ivabradine demonstrated more favorable effects than atenolol on echocardiographic indices of left ventricular (LV) systolic and diastolic function and left atrial performance.

Ivabradine is non-inferior to atenolol with regard to effects on HR, rate-pressure product, LV function, LA performance, and clinical tolerance. Clinical studies in cats with HCM are needed to validate these findings and further assess safety.

ABSTRACT C-2

RANDOMIZED, BLINDED COMPARISON OF EPINEPHRINE AND VASOPRESSIN FOR TREATMENT OF NATURALLY OCCURING CARDIOPULMONARY ARREST (CPA) IN DOGS. Buckley GJ, Rozanski EA, Rush JE. Cummings School of Veterinary Medicine at Tufts University, North Grafton, MA.

The aim of this study was to compare outcome from CPA in dogs following initial administration of either epinephrine or vasopressin during cardiopulmonary resuscitation (CPR).

Dogs having CPA in the ER or ICU of a university hospital were randomized to receive either IV epinephrine (0.01–0.02 mg/kg) or vasopressin (0.5–1.0 U/kg) in a blinded fashion immediately following establishment of IV access and again three minutes later. A standardized CPR protocol was followed. Other vasopressors were not permitted during the six minute study period, at the end of the study period additional CPR interventions were at the discretion of the managing clinician. The primary end point was return of spontaneous circulation (ROSC) within the study period; secondary end points included ROSC at any point, survival to 20 minutes, and survival to one hour.

Sixty dogs completed the study, 31 received epinephrine and 29 received vasopressin. ROSC within six minutes was 53% (13 vasopressin, 19 epinephrine p = 0.2), ROSC at any time was 60% (15 vasopressin, 21 epinephrine p = 0.2). Survival to 20 minutes was 32% (6 vasopressin, 13 epinephrine p = 0.077), survival to one hour was 18% (2 vasopressin, 9 epinephrine p = 0.027). Five dogs survived to 24 hours, one survived to hospital discharge. Of animals dying after ROSC, 13/35 were euthanized and 22/35 rearrested.

No advantage of routine substitution of vasopressin for epinephrine was seen for ROSC, a small survival advantage at one hour was seen in the group receiving epinephrine. The study also demonstrated that prospective clinical CPR research in animals is both possible and practical.

ABSTRACT C-3

SINGLE-DOSE AND APPARENT STEADY STATE PHARMACOKINETICS AND PHARMACODYNAMICS OF AN EXTENDED RELEASE DILTIAZEM FORMULATION IN THE DOG-A PILOT STUDY. SB Leach[1], DM Fine[1], HE Durham[1], MG Papich[2]. [1]University of Missouri, Columbia, MO. [2]North Carolina State University, Raleigh, NC.

Extended release diltiazem formulations (DiltXR) are used in dogs to treat supraventricular tachycardias yet the pharmacokinetic and pharmacodynamic effects have not been evaluated.

Three dogs were evaluated in 3 phases. Phase-1: single-dose DiltXR at approximately 6 mg/kg PO. Phase-2: same dose q12h for 4.5 days. Phase-3: after a 14 day wash-out the single-dose protocol was repeated using cut tablets to assess affect on extended release properties. Blood pressure (BP), 6-lead ECG, echocardiogram, and 24h-ambulatory ECG were performed at baseline, and conclusion of Phase-2. Blood samples and BP was obtained 0, 1, 2, 4, 8, 12, and 24h after final dosing.

Peak median plasma diltiazem concentrations (mcg/ml) measured by HPLC for each phase were 0.0979, 0.016, and 0.07 respectively. Diltiazem concentrations were below the limit of detection in the majority of samples in phase-2. Median diltiazem concentration reached purported therapeutic concentrations (0.05–0.2 mcg/ml) by 2 h post-pill in phase-1 and 1h post-pill in phase-3. Therapeutic concentrations were maintained for 24 h in Phase-1, but only 2h in phase-3. Median BP (mmHg) was 142.7 at baseline and 126.0 at peak concentration in phase-1. Median heart rate (bpm) was 127.1 at baseline. 24h-ambulatory ECG analysis revealed the median hourly heart rate was 101.5 at baseline and 102 during phase-1. Median heart rate at peak concentration in phase-1 was 127.5.

Lack of detectable plasma diltiazem during phase-2 may be due to up-regulation of drug metabolism via P-glycoprotein (ABCB1-1) mutations. Ongoing data collection and analysis will include mutation testing.

ABSTRACT C-4

ADIPONECTIN: A PROTECTIVE ROLE IN DOGS WITH CONGESTIVE HEART FAILURE? RM Wood[1], OL Nelson[1], J Häggström[2], K Höglund[2], I Ljungvall[2], C Kvart[2]. [1]Washington State University College of Veterinary Medicine, Pullman,WA, [2]The Swedish University of Agricultural Sciences,Uppsala, Sweden.

Adiponectin (ADPN) is a cytokine produced by fat cells which has been shown to be correlated with adverse cardiac conditions in humans. In the heart, adiponectin activates several pro-survival reactions, including the AMPK pathway and COX2 receptors, which protect the heart following ischemic injury. Recent studies have shown that higher levels of ADPN influence cardiac remodeling signaling, inhibiting protein synthesis and suppressing pathological cardiac growth. In humans, ADPN plasma levels rise with decreased activity of the sympathetic nervous system and β-adrenergic agonists inhibit ADPN at the level of gene expression. In contrast C-reactive protein (CRP), a marker of systemic inflammation is elevated in humans with congestive heart failure (CHF) and correlates to the severity of disease. First, we hypothesized that dogs with CHF would have reduced ADPN and elevated CRP compared to normal dogs and that cytokine concentrations would predict severity of CHF. Second, we hypothesized that ADPN receptor-1 (R1) and ADPN protein would be elevated in the myocardium of CHF dogs reflecting a compensatory process.

We collected serum from 32 dogs (18 healthy and 18 CHF). Circulating adiponectin and CRP levels were quantified using a Mouse/Rat Adiponectin ELISA and a Canine CRP ELISA. We found lower mean CRP concentrations in normal dogs (3.5 ± 2.3 μg/ml) than dogs with CHF (17.9 ± 17.1 μg/ml), however, the results were not statistically significant due to the large variability seen among the CHF dogs (p = 0.07). We found greater mean ADPN concentrations in normal dogs (11.46 ± 3.3 μg/ml) than CHF dogs (9.2 ± 3.2 μg/ml) (p = 0.05). In general, the greater the severity of the heart failure, the lower the level of serum ADPN. When the 2 tests were combined, CHF dogs (all severity) had significantly lower ADPN:CRP ratios compared to normal dogs ($p < 0.001$) suggesting the ratio has a greater correlation with CHF. In a subset of normal and CHF dogs, myocardial mRNA expression of ADPN R1 was quantified using real-time PCR and ADPN protein by western blot. mRNA expression of ADPN R1 tended to be greater in CHF dogs but not significantly; however, ADPN protein was elevated in myocardium of CHF dogs (25% vs 10%, adj vol) particularly the high molecular weight (active) form. In summary, ADPN was lower in serum but higher in the myocardium of CHF dogs. This may reflect greater utilization and deposition of ADPN in CHF. As suspected in humans, ADPN may play a protective or compensatory role in the pathophysiology of heart failure in the dog.

ABSTRACT C-5

EFFECT OF PIMOBENDAN ON PLATELET AGGREGATION IN DOGS. EA Shipley, DF Hogan, HW Green III, NN Fiakpui, AN Magee, KA Sederquist. Purdue University, School of Veterinary Medicine, West Lafayette, IN.

Pimobendan, a benzimidazole-pyridazinone derivative with vasodilatory and positive inotropic properties, is commonly used for CHF management in dogs. Pimobendan and related molecules can alter platelet aggregation through their phosphodiesterase inhibiting properties. Previous *in vitro* studies in humans, demonstrating a measurable antiplatelet/antithrombotic effect with relatively high pimobendan concentrations, have been cited as a possible beneficial clinical effect of this drug. To the authors' knowledge, there are no studies evaluating the effect of pimobendan on platelet inhibition in dogs.

An *in vitro* study was performed in 10 adult healthy dogs measuring the effect of pimobendan on whole blood platelet aggregation and thromboelastography (TEG). Pimobendan was dissolved in 50% methanol as previously described. Platelet aggregation was performed using the Multiplate® function analyzer and pimobendan was added to hirudinated whole blood for final concentrations of 0, 0.01, 0.1, 1.0, and 10.0 μM. After 5 minutes of incubation, aggregation was induced with either collagen or ADP as an agonist. Thromboelastography was performed on citrated whole blood where pimobendan was added to achieve final concentrations of 0, 0.01, and 10.0 μM. After 30 minutes of incubation with pimobendan, the blood was recalcified and clot formation was induced without an extrinsic activator.

The platelet aggregation area under the curve (AUC) and total aggregation (AU) for collagen were significantly reduced at 10 μM compared to baseline (285.1 ± 72.2, 161.5 ± 37.9 vs. 349 ± 58.4, 196.2 ± 25.8; $p < 0.05$, $p < 0.05$) while the AUC and velocity of platelet aggregation for ADP were significantly reduced at 10 μM compared to baseline (213.4 ± 77.2, 11.4 ± 3.3 vs. 268.5 ± 35.0, 15.3 ± 3.2; $p < 0.05$, $p < 0.05$). There were no significant differences noted in the R, K, α-angle or MA between baseline and any of the pimobendan concentrations on TEG analysis.

Pimobendan exerts a mild *in vitro* inhibitory effect on platelet aggregation induced with collagen and ADP in dogs at a concentration (10.0 μM), which is approximately 1000-fold higher than the reported concentration clinically achievable in dogs (0.01 μM) and similar to concentrations evaluated in previous human *in vitro* studies. Therefore, it would appear that the beneficial clinical profile of pimobendan in dogs with CHF is not related to antiplatelet effects; moreover, there is no evidence that bleeding complications or interactions with other antiplatelet drugs should be included with possible adverse clinical effects of this drug.

ABSTRACT C-6

THE INFLUENCE OF CHRONIC COMBINED ENALAPRIL AND SPIRONOLACTONE ADMINISTRATION ON SERUM ELECTROLYTE CONCENTRATIONS IN DOBERMAN PINSCHERS WITH OCCULT DILATED CARDIOMYOPATHY. Thomason JD, Rapoport G, Fallaw T, and Calvert CA. University of Georgia College of Veterinary Medicine, Athens, GA

The combination of an angiotensin-converting enzyme inhibitor (ACEI) with an aldosterone receptor antagonist can increase serum potassium and magnesium and lower serum sodium concentrations. The objective of this study was to determine whether an ACEI and spironolactone can be co-administered to Doberman pinschers with occult dilated cardiomyopathy without serious adverse influences on serum electrolyte concentrations.

Between 2001 and 2007, 26 client-owned Doberman pinschers were administered enalapril, spironolactone, and carvedilol and followed for at least 6 months. Most dogs were administered mexiletine for ventricular tachyarrhythmia suppression. Dogs were treated with pimobendan when congestive heart failure was imminent.

Baseline and follow-up (3–10 data bases) color-flow Doppler echocardiogram and serum urea nitrogen (SUN), creatinine, sodium, potassium, and magnesium concentration data were retrospectively evaluated and tabulated.

Compared to baseline data, there were no significant changes in serum sodium or serum creatinine concentrations. Serum magnesium ($P = 0.003$), serum potassium ($P = 0.0001$), and SUN ($P = 0.0001$) concentrations increased significantly with time.

Although the combination of enalapril and spironolactone administered to Doberman pinschers was associated with significant increases in magnesium, potassium, and SUN concentrations, these changes were of no apparent clinical relevance. At the dosages used, this combination of drugs appears safe.

ABSTRACT C-7

OCCLUSION DEVICES AND APPROACH IN PATENT DUCTUS ARTERIOSUS: COMPARISON OF OUTCOMES. MK Singh, MD Kittleson, PH Kass, LG Griffiths. William R. Pritchard Veterinary Medical Teaching Hospital, University of California, Davis, CA.

The purpose of this study was to determine if there are any clinically important differences between the approaches (including devices) used in non invasive transvascular (interventional) closure of patent ductus arteriosus (PDA) in dogs in our institution. Initial and follow up records from all dogs (n = 112) that underwent attempted transvascular PDA occlusion from January 2006-December 2009 were examined. Dogs were placed into 4 groups depending on the device and route of vascular access (transvenous or transarterial). Group 1: Amplatz® Canine Ductal Occluder (ACDO) (transarterial) - 36 dogs; Group 2: Gianturco® or MReye Flipper® Detachable Embolization (Flipper) coil (transarterial) - 38 dogs; Group 3: Amplatzer® Vascular Plug (AVP) (transarterial) - 23 dogs; Group 4: Flipper coil (transvenous) - 15 dogs. Statistical comparisons were made using the Kruskal-Wallis test with Mann-Whitney tests to compare pairs of groups when significance was detected. $P < 0.05$ was considered significant.

There was no significant difference in ages between the 4 groups. There was a significant difference in body weight between groups with dogs receiving a coil either transarterially or transvenously (groups 2 and 4) being significantly smaller than dogs receiving an ACDO or AVP. This was by design since the ACDO and AVP cannot be used in small dogs. Overall, the success rate of the total procedure (including vascular access and satisfactory PDA occlusion) was high (92%) with success rates being comparable between groups (87–97%). There was a significant difference in complication rate between groups ($p < 0.0001$) with the ACDO group having a markedly lower complication rate than the remaining groups (3% for ACDO versus 24–33% for the other groups). Total fluoroscopy time ranged from 3–78 minutes (median 8 minutes). Fluoroscopy time for the transvenous method was significantly longer (median 13 minutes; range 10–78 minutes) than in the remaining groups (median 6 minutes; range 3–33 minutes) ($p < 0.0001$). Number of dogs with residual flow immediately following the procedure and 24 hrs later was significantly less in the ACDO group than in the remaining groups (2 dogs from group 2 and 3 from group 3 had moderate persistent flow while 1 dog from group 2 and 1 from group 4 had severe persistent flow 24 hours after the procedure).

The ACDO appears superior in ease of use, complication rate, completeness of occlusion and fluoroscopy time than other devices. The remaining limiting factor with this device is patient size. Until a smaller ACDO device is marketed, coils remain the only choice for interventional closure in very small dogs (< 2.5kg).

Previously presented at the University of California Davis, House Officers Seminar Day.

ABSTRACT C-8

IDENTIFICATION OF TWO DELETION POLYMORPHISMS WITHIN THE CANINE BETA-1 ADRENOCEPTOR GENE.

BA Maran, KM Meurs, SL Lahmers, OL Nelson. Washington State University College of Veterinary Medicine, Pullman, WA.

Beta-adrenergic receptor antagonists (Beta-blockers) are widely utilized for treating a variety of cardiac issues in the dog including subvalvular aortic stenosis, pulmonic valvular stenosis, arrhythmogenic right ventricular cardiomyopathy, and tachyarrhythmias. Recent clinical studies in the dog have also evaluated beta-blocker therapy for possible cardioprotective benefits in early dilated cardiomyopathy and chronic valvular disease. Beta-blockers are widely used for similar indications in human beings as well; however, genetic studies in human beings have identified polymorphisms within the beta-1 adrenoceptor (1-AR) gene that lead to altered receptor function. It has been suggested that these polymorphisms and altered receptor function could affect the dose or type of beta-blocker, or necessitate additional treatment strategies for an individual based on their DNA sequence. The field of pharmacogenomics, which is the influence of genetics on drug response, is very new in the dog and very little is known about breed specific or family variation in the genes that impact cardiac drug function.

The hypothesis of this study is that polymorphisms are present in the canine β1-AR gene, which could result in an altered pharmacologic response to beta-blocker therapy. The objective of the study was to sequence the canine β1-AR gene in unrelated dogs of several different breeds to evaluate for the presence of polymorphisms. The canine β1-AR gene was amplified with specifically designed and optimized primers using the canine β1-AR sequence, standard PCR techniques, and PCR-based genomic sequencing. DNA samples from 40 dogs of five breeds were evaluated. Computer modeling (POLYVIEW3D) was performed to evaluate the structural impact of any observed polymorphisms.

We identified two polymorphisms within the canine β1-AR gene that may alter responsiveness to commonly administered beta-blockers. We sequenced 40 DNA samples from five breeds (Newfoundland, Boxer Dog, Doberman Pinscher, King Charles Cavalier (CKCS), Great Dane) of dogs and identified two separate deletion polymorphisms, with some individuals displaying both deletions. One polymorphism, a six to nine base pair deletion, was identified in five Boxer dogs, six Newfoundlands, seven Great Danes, seven Doberman Pinschers, and one CKCS. This deletion correlates to a loss of two to three aspartic acids. A second polymorphism was identified in four Newfoundlands, four Great Danes, and two CKCS and consists of a 24 base pair deletion. This deletion correlates to a loss of eight amino acids, a repeating strand of glycine and alanine. Computer modeling (POLYVIEW3D) of these deletions determined substantial tertiary structural change within the intracellular domain that has the potential to alter receptor function.

Although the clinical significance of these deletions are unknown, additional investigation into the pharmacogenomics of beta-blockers in dogs is warranted as these polymorphisms may alter the commonly accepted therapeutic strategies for various cardiac diseases in dogs.

ABSTRACT C-9

GENOME WIDE ASSOCIATION ANALYSIS IDENTIFIES LOCATION OF INTEREST ON CHROMOSOME 21 IN GOLDEN RETRIEVERS AND ROTTWEILERS WITH FAMILIAL SUBVALVULAR AORTIC STENOSIS.

JA Stern[1], KM Meurs[1], OL Nelson[1], SM Lahmers[1], LB Lehmkuhl[2], SN White[3,1]. [1]Washington State University, College of Veterinary Medicine, Pullman, WA. [2]MedVet Medical Center for Pets, Worthington, OH. [3]USDA-ARS Animal Disease Research Unit, Pullman, WA.

Subvalvular aortic stenosis (SAS) is one of the most commonly reported canine congenital heart defects and is inherited in Newfoundland dogs and human beings. The golden retriever and Rottweiler are breeds over-represented in dogs with subvalvular aortic stenosis; however, a genetic cause of this disease in these breeds has not been described. We performed genome wide association analysis in both normal and SAS affected Rottweilers and golden retrievers to identify chromosomal regions of interest that could implicate a causative mutation by high density single nucleotide polymorphism (SNP) array.

48 (24 unaffected/24 affected) adult golden retrievers and 48 (20 unaffected/28 affected) adult Rottweilers were included in this study. Criteria for affected included a subcostal continuous-wave Doppler aortic velocity $\geq$ 2.5 m/s and presence of a left basilar systolic ejection murmur; criteria for unaffected included a Doppler aortic velocity $\leq$ 1.8 m/s. DNA samples were obtained from anticoagulated blood. Genotypes were obtained using high density (173,662) SNP arrays, and genome wide association with SAS was evaluated for each breed. Significance cut-off was set at $p = 5 \times 10^{-5}$, and all SNPs meeting this criterion were plotted within each breed and compared across breeds using PLINK.

Affected golden retriever data implicate the most significant region of genetic variation on chromosome 21 at location 27384260 ($p = 1.15 \times 10^{-5}$; odds ratio 7.58) with 11 other significant surrounding SNPs (26093494–28761657). Affected Rottweiler data also implicate the most significant region of genetic variation on chromosome 21 at location 27895300 ($p = 8.98 \times 10^{-6}$; odds ratio 23.39) with 3 other significant surrounding SNPs (25948544–28898527). Other regions of statistical significance were on chromosomes 4 and 7 in the golden retriever and 11 and 38 in the Rottweiler.

Genome wide association with subvalvular aortic stenosis in the golden retriever and Rottweiler implicate overlapping chromosomal regions of interest for causative mutations on chromosome 21. The different secondary chromosomal regions of interest (CHR 4, 7 in golden retrievers and 11, 38 in Rottweilers) supports the known familial nature of this disease within different breeds and may suggest the presence of multiple mutations or breed specific disease modifiers. These data highlight the need for candidate gene evaluation on chromosome 21 in golden retrievers and Rottweilers with SAS.

ABSTRACT C-10

CANINE AND HUMAN DEGENERATIVE MITRAL VALVE DISEASE MIMICS CHONDROGENESIS.

MacLea HB, Lacerda C, Kisiday JD, and Orton EC. College of Veterinary Medicine and Biomedical Sciences, Colorado State University, Fort Collins, CO.

Heart valves share developmental signaling pathways with bone and cartilage. Degenerative aortic valve disease in humans is characterized by valve stenosis and calcification. Recent evidence suggests that degenerative aortic valves are undergoing pathologic processes that mimic osteogenesis. Degenerative mitral valves in dogs and humans are characterized by valve regurgitation, and rarely undergo calcification. We tested the hypothesis that canine and human degenerative mitral valves might be undergoing pathologic processes that mimic chondrogenesis.

To test this hypothesis, expression of bone morphogenic protein 2 (BMP2), a chondrogenic growth factor; Sox 9, a chondrogenic transcription factor; aggrecan, a proteoglycan abundant in cartilage; and type II collagen were evaluated utilizing immunohistochemistry. Normal canine mitral valves, different stages of canine degenerative mitral valves (early, intermediate, and late), and late-stage human degenerative mitral valves were studied.

Canine and human degenerative mitral valves showed focal areas that co-expressed all four markers of chondrogenic signaling and phenotype. Valve interstitial cells and surrounding extracellular matrix in these focal areas adopted a morphologic appearance reminiscent of cartilage. Focal chondrogenesis was present in all stages of canine degenerative mitral valves, but not normal canine mitral valves. Focal areas of chondrogenesis did not coincide with nodular areas of glycosaminoglycan accumulation on the leaflet edge, but rather seemed to occur at points of chordae attachment to leaflets.

In conclusion, canine and human degenerative mitral valves undergo pathologic processes that mimic chondrogenesis. This finding suggests that mitral valve degeneration may be recapitulating developmental signaling pathways shared by heart valves and cartilage. The triggering events for chondrogenesis in mitral valves remain unknown; as does the reason why aortic and mitral valves appear to be undergoing different pathologic processes. The fact that humans exhibit degeneration of both the aortic and mitral valve, and that dogs commonly exhibit only the latter could eventually provide insight into both processes.

ABSTRACT C-11

CHARACTERISTICS OF ARVC BOXERS WITH SUDDEN DEATH. KM Meurs[1], S Lahmers[1], BW Keene[2]. [1]Washington State University College of Veterinary Medicine, Pullman, WA. [2]North Carolina State College of Veterinary Medicine, Raleigh, NC.

Arrhythmogenic right ventricular cardiomyopathy (ARVC) is a familial cardiomyopathy characterized by right ventricular fibrofatty infiltration and ventricular ectopy of left bundle branch block morphology (VPC). A deletion in the striatin gene has been associated with ARVC in at least some boxer families. Syncope and sudden death (SD) occur in some affected dogs, although many affected dogs survive for years.

The objective of this study was to define clinical characteristics of ARVC in boxers that experienced SD, and compare them to those of a contemporaneous group of ARVC boxers that had not died suddenly (NSD). Data for both groups were collected from adult boxers enrolled in a long term prospective study of ARVC in which echocardiograms and 24 hour ambulatory ECG (AECG) are evaluated annually. AECG are quantitated for VPC numbers and arrhythmia grade (1–4). ARVC diagnosis requires at least 100 VPCs/24 hours in the absence of other disease.

Forty three adult boxers that entered the study had died suddenly at the time of analysis (SD defined as the absence of observed clinical signs within 24 hours prior to an unexpected and unexplained death). Striatin genotype was available for 28 of the 43 SD dogs (17 heterozygotes, 11 homozygotes); 19 were female (9 intact) and 24 were male (14 intact). SD occurred at a mean age of 8 years (range, 1–12); 27 SD dogs (63%) had no prior history of syncope. Twelve SD dogs (28%) were on antiarrhythmics at the time of death (metop-p0oooooooooooprolol (1), sotalol (4), amiodarone (1), procainamide (2), mexiletine & atenolol (3), atenolol (1)). Eleven SD dogs (25%) had decreased myocardial systolic function defined by a shortening fraction (%FS) < 25% (range 11–45, mean = 27) on the most recent echocardiogram prior to SD. Median VPCs/24 hours on annual AECG was 7,700 (range 106–91,000) with a median arrhythmia grade of 4 (range 2–4).

Twenty one contemporaneously entered ARVC boxers that had survived to at least the median age of the SD group with NSD were available for comparison; 15/21 were genotyped (10 heterozygous, 1 homozygous, 4 negative), 13 were female (3 intact) and 8 male (2 intact). Twelve NSD dogs (57%) had no prior history of syncope. Median NSD group age was 10 years (range, 8–13); 11/21 (52%) were on an antiarrhythmics (sotalol (9), mexiletine & sotalol (1), mexiletine & atenolol (1)). One NSD dog had decreased %FS (NSD group %FS range 20–42, mean = 32). The NSD median number of VPCs was 5,342 (range 622–62,622), median arrhythmia grade was 4 (range 2–4).

Striatin genotype was significantly associated with SD. No significant differences were found between groups with respect to VPC numbers or arrhythmia grade. Shortening fraction was significantly lower in the SD group ($p < 0.01$). SD in ARVC appears to be associated with the presence of the striatin mutation and reduced % FS, it does not appear to be associated with number of VPCS or arrhythmia grade.

ABSTRACT C-12

BRONCHOMALACIA IN DOGS WITH MYXOMATOUS MITRAL VALVE DEGENERATION. MK Singh, LR Johnson, MD Kittleson, RE Pollard. William R. Pritchard Veterinary Medical Teaching Hospital, University of California, Davis, CA.

Coughing in the small breed dog may be related to cardiac causes associated with myxomatous mitral valve degeneration (MMVD) including pulmonary edema and compression of the mainstem bronchus by a severely enlarged left atrium, or due to respiratory causes such as tracheal and/or bronchial collapse or chronic bronchitis. The purpose of this study was to evaluate the association between left atrial enlargement and large airway collapse in dogs with MMVD and chronic cough. We hypothesized that airway collapse was independent of degree of left atrial enlargement.

Twelve dogs with MMVD and a chronic cough in the absence of congestive heart failure were prospectively evaluated with thoracic and cervical radiography, echocardiography, fluoroscopy, bronchoscopy and bronchoalveolar lavage (BAL). Group 1 dogs (n = 8) had moderate to severe left atrial enlargement based on an echocardiographically calculated left atrial:aortic surface area [LA:Ao(a)] > 6. Group 2 dogs (n = 4) had no to mild left atrial enlargement [LA:Ao(a) ≤ 6]. The site and severity of airway collapse was graded on bronchoscopy and BAL cytology was assessed for evidence of inflammation or infection. The occurrence of bronchoscopic abnormalities was compared between groups using Fisher's Exact Test. $P < 0.05$ was considered significant.

Age and body weight did not differ between groups. Left atrial size was interpreted radiographically as moderately to severely enlarged in 7 of 8 dogs in group 1 and as moderately enlarged in 2 of 4 dogs in group 2. Fluoroscopy revealed variable degrees of airway collapse during normal respiration and induced cough in both groups. Radiography and fluoroscopy were not accurate in identifying site and degree of collapse in either group when compared to bronchoscopy. Cervical tracheal collapse was identified during bronchoscopy in both group 1 (2 of 8) and group 2 (3 of 4) dogs but was subjectively less severe in group 1 dogs. Bronchial collapse > 50% was evident at multiple sites in both groups of dogs with no difference between groups. All dogs had suppurative and/or lymphocytic inflammation on airway cytology. Infection was not present in either group of dogs, although non-specific light bacterial growth was detected in 6 of 8 group 1 dogs and 1 of 4 group 2 dogs (p = 0.22).

Preliminary results failed to identify an association between left atrial enlargement and airway collapse in dogs with MMVD but did suggest that airway inflammation is common in affected dogs. Further studies are needed to identify factors contributing to airway collapse in dogs with and without MMVD.

ABSTRACT C-13

THE EFFECT OF ATENOLOL ON NT-proBNP AND TROPONIN IN ASYMPTOMATIC CATS WITH HYPERTROPHIC CARDIOMYOPATHY. SW Jung, MD Kittleson. William R. Pritchard Veterinary Medical Teaching Hospital, University of California - Davis, Davis, CA.

Atenolol is often used empirically in cats with asymptomatic HCM, even though clinical and experimental evidence of efficacy is lacking. Cardiac biomarkers play a critical role in the early detection of subclinical cardiac disease, in the prediction of long-term prognosis, and in monitoring the response to therapy in humans. We hypothesized that serum concentrations of the biomarkers, N-terminal pro-brain natriuretic peptide (NT-proBNP) and cardiac troponin I (cTnI), would improve following the chronic administration of atenolol PO to asymptomatic cats with HCM.

Six Maine Coon or Maine Coon cross cats with severe HCM from the research colony at UCDavis were administered atenolol (12.5 mg PO twice a day) for 30 days. No cat had severe left ventricular dynamic outflow tract obstruction due to systolic anterior motion of the mitral valve. The concentrations of NT-proBNP and cTnI were assayed prior to drug administration and on the last day of drug administration. There was no statistically significant difference identified in NT-proBNP [median before: 394 pmol/L (range: 71–1500 pmol/L), median after: 439 pmol/L (range: 24–1500 pmol/L); p = 0.63] or cTnI [median before: 0.24 ng/ml (range:0.10–0.97 ng/ml), median after: 0.28 ng/ml (range: 0.09–1.0 ng/ml); p = 0.69] concentrations before and after drug administration using the Wilcoxon matched pairs test.

The cTnI finding suggests that atenolol does not reduce chronic myocyte death in cats with HCM. The lack of improvement in NT-proBNP suggests that atenolol does not improve myocardial wall stress in cats with HCM. A clinical trial is warranted to confirm or refute the findings from this study.

ABSTRACT C-14

LEPTIN EXPRESSION IN BLOOD AND MYOCARDIAL SAMPLES OF DOGS WITH CARDIAC DISEASE. Sonja Fonfara, Udo Hetzel, Simon R. Tew, Joanna Dukes-McEwan, Peter D. Clegg. Dept. of Musculoskeletal Biology and School of Veterinary Sciences, Leahurst Campus, University of Liverpool, Neston, UK.

Leptin, first described as antiobesity hormone, has recently attracted attention due to its effects on the cardiovascular system. Increased leptin is associated with cardiac remodeling, myocardial hypertrophy and dysfunction and poor outcome. Elevated leptin is reported in cases of obesity in veterinary medicine, but the role of leptin in canine cardiac disease is not known.

Therefore, leptin-gene expression was investigated in blood samples of dogs with congestive heart failure (CHF; n = 8) in comparison to dogs presented for cardiac screening (n = 8) without abnormalities. Additionally myocardial samples (interventricular septum, right and left atrium and ventricle) of 10 dogs with no cardiac abnormalities (controls), seven dogs with acquired and three with congenital cardiac diseases were investigated using quantitative RT-PCR.

Leptin blood levels were significantly higher in dogs with CHF in comparison to dogs without diseases (p = 0.013). There was an association with gender with higher myocardial leptin levels in female dogs with cardiac diseases (p = 0.001). Differences between cardiac regions were present (p < 0.05) and cardiac diseases resulted in an increase in atrial leptin levels in both sexes (p = 0.006). Interestingly, a significant reduction of myocardial leptin was present in dogs with congenital cardiac diseases (p = 0.016), whereas acquired cardiac diseases resulted in an increase in leptin (p = 0.035) in comparison to controls.

These results suggest that the heart might be a target of leptin action in the dog and myocardial leptin production might play a role in regulating cardiac function in an auto- and paracrine manner.

ABSTRACT C-15

NT-PRO-BNP IMPROVES GENERAL PRACTITIONERS' ACCURACY AND CONFIDENCE OF DIAGNOSIS IN CATS WITH RESPIRATORY SIGNS. GE Singletary[1], MA Oyama[1], JE Rush[2], PR Fox[3], RL Stepien[4]. [1]University of Pennsylvania, Philadelphia, PA. [2]Tufts University, North Grafton, MA. [3]The Animal Medical Center, New York, NY. [4]University of Wisconsin, Madison, WI.

NT-proBNP can distinguish congestive heart failure (CHF) from primary respiratory disease in cats with respiratory symptoms with a diagnostic accuracy of approximately 90%; however, its additive value to traditional diagnostic testing is unknown. We hypothesized that NT-proBNP would improve the diagnostic accuracy and confidence of general practice veterinarians.

History, physical examination, radiographs, ECG, and routine bloodwork from 10 cats presented to Penn or Tufts for evaluation of respiratory signs were evaluated by 47 general practice veterinarians using an online survey. Participants' median time in practice was between 5–10 yrs. Participants were asked to provide 1) diagnosis of CHF or primary respiratory disease, and 2) level of confidence in their diagnosis (1, lowest-10, highest) prior to and after divulging NT-proBNP results. Diagnosis (CHF, n = 5; primary respiratory, n = 5) was compared to the gold standard as made by consensus opinion of three board-certified cardiologists. In 9 of 10 cases (90%) the NT-proBNP results correctly reflected respiratory or CHF etiology. Prior to NT-proBNP disclosure, correctness of practitioners' diagnosis was 67.4%, and was significantly greater after practitioners were provided NT-proBNP results (85.5%, P = 0.0013). Median practitioner confidence prior to NT-proBNP disclosure was 6 (IQR, 5.5–6.5) and was significantly greater after disclosure (8; IQR, 7–8; P = 0.0039).

These results indicate moderate accuracy and relatively low level of confidence in the diagnosis of feline respiratory signs by general practitioners. NT-proBNP assay, when used in conjunction with traditional evaluation, significantly improved the accuracy and confidence of diagnosis in cats with respiratory signs caused by CHF or respiratory disease.

ABSTRACT C-16

NT-PRO-BNP PREDICTS FIRST ONSET OF CONGESTIVE HEART FAILURE IN DOGS WITH DEGENERATIVE MITRAL VALVE DISEASE. CA Reynolds[1], MA Oyama[1], JE Rush[2], HB Kellihan[3], RL Stepien[3], BK Lefbom[4], J Weidman[4], SG Gordon[5], LB Lehmkuhl[6], TP Nguyenba[6], CK Meier[7], PR Fox[8], DC Brown[1]. [1]University of Pennsylvania, Philadelphia, PA. [2]Tufts University, North Grafton, MA. [3]University of Wisconsin, Madison, WI. [4]Chesapeake Veterinary Cardiology Associates, Vienna VA. [5]Texas A&M University, College Station, TX. [6]MedVet, Worthington, OH. [7]Austin Heart Vet, Austin, TX. [8]The Animal Medical Center, NY.

Predicting risk of CHF in asymptomatic dogs with mitral valve disease (MVD) is challenging. We examined ability of NT-proBNP to identify asymptomatic dogs at high risk for CHF. Dogs with ISACHC-1b (LA:Ao > 1.6) were prospectively recruited; dogs with current or previous CHF or diuretic therapy were excluded. Physical examination, radiography, echocardiography, and NT-proBNP were performed at 2–9 mo intervals for 66 dogs (median follow-up = 276 d, range, 17–1334 d). Thirty-one patients reached a study endpoint of radiographic pulmonary edema; 35 remained asymptomatic. Parameters from the visit immediately previous to onset of CHF (Future-CHF) or prior to the most recent visit without CHF (Remain-Asympt) were analyzed. Median NT-proBNP of Future-CHF (3001pmol/Lpmol/L, IQR 2255-3001) was significantly different from Remain-Asympt (1600 pmol/L, 984–2863; P = 0.0004). Median time to CHF of Future-CHF was 86d (IQR, 35-143). Groups also differed in median LA:Ao (Future-CHF 2.20 [2.00–2.50]; Remain-Asympt 1.96 [1.85–2.13], P = 0.0014); VHS (Future-CHF 12.0 [11.3–12.5]; Remain-Asympt 11.3 [10.75–12.0], P = 0.003); and LVIDd:Ao (Future-CHF 2.82 [2.52–3.25]; Remain-Asympt 2.33 [2.15–2.64], P = 0.0001). ROC analysis to predict if CHF would occur prior to the next visit yielded AUC = 0.753 (95%CI, 0.632–0.851). Sensitivity was 90.3% or 77.4% and specificity 48.6% or 68.6% for NT-proBNP >1490 pmol/L or >2150 pmol/L, respectively. Mean increase in NT-proBNP between penultimate visit and two visits prior to endpoint: Future-CHF = +440.1 pmol/L vs. Remain-Asympt = +10.5 pmol/L. Within 6 mo, 4.5%, 10.7%, 17.4%, and 36.7% of dogs with NT-proBNP < 1000, 1001–2000, 2001–3000, and >3000 pmol/L developed CHF. NT-proBNP and heart size helped assess risk of CHF in asymptomatic MVD. Increasing the assay's upper limit of detection would likely improve utility of NT-proBNP.

ABSTRACT C-17

ASSOCIATIONS BETWEEN N-TERMINAL PROCOLLAGEN TYPE III (PIIINP) AND VENTRICULAR REMODELLING IN DOGS WITH MITRAL VALVE DISEASE. MJ Hezzell, A Boswood, J Elliott. The Royal Veterinary College, London, UK.

PIIINP is a serum biomarker of collagen biosynthesis and is described as a marker of myocardial fibrosis in human patients. We hypothesised that PIIINP concentrations would vary according to the degree of remodelling demonstrable in dogs with naturally-occurring myxomatous mitral valve disease (MMVD).

Serum PIIINP concentrations (μg/ml) were measured in dogs with MMVD and healthy controls using a validated commercially-available radioimmunoassay. Results are reported as (Mean ± SD). Non-normally distributed variables were logarithmically transformed. Comparisons of continuous variables were made between groups using t-tests and one-way ANOVAs with Tukey's post-hoc comparisons. Univariable analyses were used to evaluate associations between PIIINP and clinical characteristics (age, breed [cavalier King Charles spaniel (CKCS) yes/ no], sex, weight, heart rate [measured from ECG], treatment with ACEi [yes/ no], treatment with diuretics [yes/ no] and echocardiographic measurements [LA/Ao, LVEDD/LVFWd, LVEDDN, LVESDN]). Multivariable analysis was initially performed with all dogs included and then repeated excluding all dogs receiving therapy. Dogs with MMVD were divided into those with no cardiomegaly (NC) (LA/Ao < 1.5 and LVEDDN < 1.8), those with cardiomegaly (LA/Ao ≥ 1.5 and/ or LVEDDN ≥ 1.8) but no clinical signs (C) and those dogs with cardiomegaly requiring treatment for congestive heart failure (CHF).

One hundred and fifty-four dogs with MMVD and 23 control dogs were studied. There was no difference in age (P = 0.870) or weight (P = 0.606) between the MMVD and control groups.

There was a significant difference in serum PIIINP (P = 0.034) between normal (11.2 ± 3.66), NC (11.6 ± 4.57), C (10.1 ± 3.44) and CHF (9.4 ± 3.06) groups. Post-hoc comparisons demonstrated a difference between NC and CHF groups (P = 0.038). There was no difference in serum PIIINP between genders (P = 0.228). In the univariable analysis CKCS (yes/ no) (P = 0.016) was positively associated with serum PIIINP. Age (P < 0.0001), Log (LA/Ao) (P = 0.002) and LVEDDN (P = 0.002) were negatively associated with serum PIIINP. In the multivariable model including all dogs, LVEDDN (P < 0.0001, B = −4.04 (95%CI = −6.11 to −1.97)), age (P = 0.006, B = −0.28 (95%CI = −0.47 to −0.08)) and CKCS (yes/ no) (P = 0.003, B = 2.01 (95%CI = 0.67 to 3.34)) were independently associated with serum PIIINP. In the multivariable model including only dogs not receiving therapy (n = 141),

LVEDDN (P = 0.006, B = −4.30 (95%CI = −7.32 to −1.29)), age (P = 0.011, B = −0.31 (95%CI = −0.55 to −0.07)) and CKCS (yes/ no) (P = 0.034, B = 1.75 (95%CI = 0.13 to 3.37)) were independently associated with serum PIIINP. In conclusion, serum PIIINP decreases with age and with increasing LVEDDN. CKCS have higher serum PIIINP measurements independent of age and LVEDDN, which may reflect a difference in collagen turnover in this breed.

ABSTRACT C-18

LEFT ATRIAL FUNCTION IN CATS WITH LEFT-SIDED CARDIAC DISEASE MANIFESTING PLEURAL EFFUSION VERSES PULMONARY EDEMA. SM Johns, OL Nelson. Washington State University College of Veterinary Medicine, Pullman, WA.

Left atrial (LA) chamber dilation and congestive heart failure (CHF) are common consequences of cardiac conditions in cats. In some cases CHF is manifest as right-sided CHF (R-CHF) or pleural effusion, in other cases CHF manifests as left-sided CHF (L-CHF) or pulmonary edema. It is not always readily apparent as to which cats will develop what form of CHF. A general impression has been that LA enlargement is associated with the average burden of elevated filling pressures, but little attention has been paid to the function of the LA chamber itself. Since CHF is classically preceded by abnormal atrial chamber dilation and alterations in atrial chamber function, we want to understand how these changes may help us manage or predict CHF in the cat.

We hypothesized that cats manifesting R-CHF have LA failure with the LA acting primarily as a conduit, resulting in greater pulmonary hypertension, whereas L-CHF cats maintain some booster pump and reservoir function. We measured LA maximum and minimum areas from right parasternal long axis four-chamber views on 2D echo, and LA M-mode dimensions at maximum, minimum, and beginning of atrial contraction. LA area change, fractional shortening, active emptying fraction, and expansion index were calculated from these measurements. Right ventricular internal diastolic diameter was also measured on M-mode views. Preliminary data revealed that maximum left atrial size is not significantly different between R-CHF and L-CHF cats on 2D or M-mode measurements due to high variability. However, total left atrial fractional shortening is significantly reduced in R-CHF cats (11.32% ± 4.9) compared to L-CHF cats (19.7% ± 5.1)(p = 0.001), and R-CHF cats have reduced left atrial active emptying fraction (6.12% ± 3.5) as compared to L-CHF cats (13.7% ± 3)(p < 0.001). Left atrial expansion ability is poorer in R-CHF cats (13.11% ± 6.7) than in L-CHF cats (25.0% ± 8)(p = 0.002). These findings may suggest that atrial stiffness and poorer atrial function is associated with a greater degree of pulmonary venous and thus secondary pulmonary arterial hypertension resulting in pleural effusion (R-CHF). Right ventricular diameter on M-mode was increased in R-CHF cats (4.8 mm ± 2.1) when compared to L-CHF cats (2.89 mm ± 0.8)(p = 0.002) and normal cats (2.74 mm ± 0.8)(p = 0.002), which may also be evidence for a greater degree of pulmonary arterial hypertension in these cats.

ABSTRACT C-19

PRESUMED NEURALLY-MEDIATED BRADYCARDIA AND SYNCOPE IN BOXERS WITH NORMAL ECHOCARDIOGRAMS AND ABSENCE OF VENTRICULAR TACHYCARDIA. Thomason JD, Rapoport G, Fallaw TL, and Calvert CA. University of Georgia College of Veterinary Medicine, Athens, GA.

Episodic weakness and syncope are common in Boxer dogs. Reported causes include rapid ventricular tachycardia (VT) and exertion-excitement triggered neurally-mediated bradycardia (NMB).The purpose of this retrospective study is to describe the features of presumed NMB in Boxers.

To be included in the study, each dog must have been overtly healthy with a history of exertion-excitement triggered syncope or presyncope; had a normal echocardiogram (EC); had absence of VT and fewer than 500 ventricular premature complexes (VPC) on an initial and subsequent 24 hour Holter recordings; and been alive and overtly healthy for at least six months following the initial evaluation.

A total of 27 Boxers were identified. Sixteen were male and 11 were female. Most (90%) dogs were either less than 4 or more than 7 years of age. Most dogs had multiple, but infrequent, episodes and heart rhythm was documented at the time of an episode in only 8 (30%) and only once (bradycardia) on the first Holter recording. Owners were instructed to attempt to precipitate episodes. Bradycardia related episodes were subsequently recorded in 5: during the 2nd (1), 3rd (2) or 4th (1) day of 120 hour Holter recordings and during the 5th day of an event recording (1). Collapse and bradycardia were documented during auscultation in 2 additional dogs.

The heart rate during syncope was never documented in 19 (70%) dogs. A presumptive diagnosis of NMB was based on the absence of initial and follow-up of EC abnormalities and the presence of no or few VPC during extended ECG monitoring. Multiple Holter recordings (48–168 hours) were performed in 8 of 19 (42%) dogs and event monitoring of 5 days (1) and 7 days (1) was performed in 2 additional dogs.

In conclusion, documentation of the heart rhythm during episodes of collapse was difficult, accomplished in only 30% and was unlikely during the first Holter recording. In Boxers with suspected NMB, extended ECG monitoring and implantable loop recorders may be best for HR documentation.

ABSTRACT C-20

ALTERED LOCALIZATION OF BETA CATENIN AND STRIATIN IN BOXER DOGS AFFLICTED WITH ARRHYTHMOGENIC RIGHT VENTRICULAR CARDIOMYOPATHY SUGGESTS ETIOLOGICAL INVOLVEMENT OF THE CANONICAL WNT PATHWAY. Eva M. Oxford, Bruce G. Kornreich and N. Sydney Moïse. Cornell University, College of Veterinary Medicine. Ithaca, NY.

Arrhythmogenic right ventricular cardiomyopathy (ARVC) is an inherited myocardial disease with high prevalence in the boxer dog population, and is associated with sustained monomorphic ventricular tachycardia, sudden cardiac death, and replacement of myocardium with fatty or fibro-fatty tissue. Though several genes have been linked to the disease both in humans and in boxers, the etiology of ARVC is still unclear. Several mechanisms for the development of ARVC have been suggested, including dysfunction of the canonical Wnt pathway, which results in an ARVC phenotype in the mouse. The canonical Wnt pathway has been linked to many cellular functions, including growth and differentiation of adipocytes. With the recent discovery that the gene encoding striatin, a protein involved in Wnt signaling, may be involved in the development of boxer ARVC, we hypothesized that changes in the Wnt pathway may also play a role in the etiology. Here, we show changes in the localization and decreased amount of proteins affiliated with the canonical Wnt pathway.

Afflicted boxers were identified by 24-hour Holter monitoring and histopathological examination of the heart. Samples from the right ventricle RV) of 15 ARVC afflicted boxers, and 5 unafflicted dogs (1 beagle, 2 mongrels, and 2 German shepherds) were collected, fixed in 10% formalin, processed, treated with antibodies recognizing -catenin, striatin, and calnexin, and examined using confocal microscopy. Western blots were performed on 3 unafflicted RV samples, and 2 ARVC afflicted RV samples. Frozen tissue samples were homogenized in Laemmli buffer, and 10 mg of protein was loaded into each well of a 8–16% gradient gel. -catenin, an integral modulator of the Wnt pathway, and striatin were colocalized with the endoplasmic reticulum (ER) marker, calnexin. In the unafflicted animals, -catenin localized at sites of cell-to-cell apposition, and striatin localized in a diffuse intracellular pattern, with no detectable localization in the ER. In contrast, in the 15 ARVC boxers, both -catenin and striatin were colocalized with calnexin in an ER pattern. In the afflicted samples, -catenin and striatin were not visualized to the intercalated disc and intracellular space, respectively. Western blots of striatin and -catenin revealed no changes in the amount of protein. Interestingly, a Western blot for the Wnt protein revealed a decrease in the amount of protein in ARVC samples, compared to unafflicted samples. Our preliminary data suggest that disturbances of the canonical Wnt pathway may play an etiological role in the development of ARVC in the boxer dog.

ABSTRACT C-21
SERUM FATTY ACID CONCENTRATIONS IN NORMAL CATS AND CATS WITH CARDIOMYOPATHY. DJ Hall, LM Freeman, JE Rush. Tufts Cummings School of Veterinary Medicine, North Grafton, MA.

There are numerous benefits of omega-3 fatty acid supplementation in human heart disease, including reduction in arrhythmias, decreased incidence of sudden death, and improved survival in heart failure. Antithrombotic effects of omega-3 fatty acids have been demonstrated in people, which may have particular benefit in cats given their risk of thromboembolic complications with cardiac disease. Benefits also have been found in canine heart disease, and reduced serum concentrations of the omega-3 fatty acids, eicosapentaenoic acid (EPA) and docosahexaenoic acid (DHA) have been found in dogs with congestive heart failure (CHF) secondary to DCM. To the authors' knowledge, no studies to date have investigated fatty acid concentrations in cats with cardiomyopathy. The purpose of this study was to measure serum fatty acid concentrations in normal cats and cats with cardiomyopathy.

Serum fatty acid concentrations were measured in normal cats and cats with cardiomyopathy using gas chromatography. Cats with cardiomyopathy and at least mild left atrial (LA) enlargement (LA to aortic ratio > 1.5 on two-dimensional echocardiography from a right parasternal short axis view) were candidates for study. Normal cats had a normal history, physical examination, echocardiogram, packed cell volume, total solids and platelet count. Cats with evidence of other systemic disease or those receiving anticoagulants were excluded from the study. Normally distributed and skewed data were compared between the cardiomyopathy and control groups with independent t tests or Mann Whitney U tests, respectively. Statistical significance was set at $P < 0.01$.

Thirty cats with cardiomyopathy (23 neutered males and 7 neutered females) and 27 healthy controls (13 neutered males and 14 neutered females) were enrolled. Median age was 8 yr (range, 2–16 yr) in the cardiomyopathy group and 5 yr (range, 1–10 yr) in the control group ($P = .009$). Cats in the cardiomyopathy group were classified in the International Small Animal Cardiac Health Council stage 1b (n = 10), 2 (n = 8), 3a (n = 3) and 3b (n = 9). Compared to control cats, cardiomyopathic cats had higher concentrations of palmitic acid ($P = .002$) and DHA ($P < .001$), and lower concentrations of linoleic acid ($P = .005$). Among cats with cardiomyopathy, there was no significant correlation between any serum fatty acid concentration and left atrial size or age. These findings warrant further investigation into the role of fatty acids in cats with cardiac disease.

ABSTRACT C-22
APPLICATION OF PLATELET MAPPING IN NORMAL CATS FOR MONITORING ANTIPLATELET EFFECTS OF CLOPIDOGREL. DJ Hall, SM Cunningham, JE Rush, SP Shaw. Tufts Cummings School of Veterinary Medicine, North Grafton, MA.

Platelet mapping is an application of thromboelastography that relies on the generation of at least three tracings: $MA_{THROMBIN}$ (maximum platelet activity),MA_{FIBRIN} (fibrin activity only), and$MA_{AA\ or\ ADP}$ (platelet activity not inhibited by arachidonic acid or ADP receptor antagonists, respectively). Using these three tracings, the % inhibition of platelets can be calculated. The purpose of this study was to evaluate the Platelet Mapping assay in normal cats and assess platelet inhibition in cats receiving clopidogrel.

Employee-owned cats with normal history, physical exam, echocardiogram, thromboelastography, packed cell volume, total solids and platelet count were eligible. Clopidogrel (18.75 mg PO Q24h) was administered for 10 days with Platelet Mapping performed on days 0 and 10. Platelet Mapping values were compared using a paired *t* test, with significance set at $P < 0.05$.

Seven cats (4 FS, 3 CM, aged 2–10 years) were enrolled. Compared to day 0, MA_{ADP} ($P < .001$) and MA_{FIBRIN} ($P < .001$) were lower on day 10. The latter unexpected result prompted measurement of fibrinogen concentrations at day 0 and 10 in the last 5 of these 7 cats. Fibrinogen was not different from day 0 to 10 in these 5 cats.

These results suggest that Platelet Mapping may be a simple, outpatient clinical tool to measure antiplatelet activity in cats receiving clopidogrel. This clopidogrel dose resulted in significant platelet inhibition as measured by MA_{ADP} in all cats. Further studies correlating antiplatelet effects measured by Platelet Mapping with clinical outcomes in cats with cardiomyopathy are warranted.

ABSTRACT C-23
THE EFFECT OF USING AN IMPEDANCE THRESHOLD DEVICE (ITD) ON HEMODYNAMIC PARAMETERS DURING CARDIOPULMONARY RESUSCITATION IN DOGS. Buckley GJ[1], Shih A[1], Garcia F[2], Bandt C[1]. [1]University of Florida College of Veterinary Medicine, Gainesville, FL. [2]Michigan State University Veterinary Medical Center, East Lansing, MI.

This study investigated the hemodynamic effects of application of an ITD in a canine model of cardiopulmonary arrest.

8 laboratory beagles which were part of a separate terminal study were anesthetized and instrumented for continuous measurement and recording of right atrial pressure, arterial pressure and carotid blood flow. Following euthanasia, CPR was performed for one minute, then a pause of one minute followed by a second one minute period at a compression rate of 100–120/minute, ventilation with 100% oxygen was delivered at eight breaths/min and 15 ml/kg tidal volume. CPR was performed with the ITD in place (ITD-CPR) and without the ITD (S-CPR) for one period each in a randomized fashion with the rescuer blinded to its application. Baseline, S-CPR and ITD-CPR data were assessed for normality, a Kruskal-Wallis one-way ANOVA on ranks was used (baseline v CPR). When appropriate a pairwise multiple comparison procedure (Dunn's method) was used. Percentage of baseline S-CPR v ITD-CPR was assessed using the Student t-test.

The right atrial diastolic pressure was significantly more negative with the ITD attached than without ($p = 0.02$), the coronary perfusion pressure and carotid blood flow were significantly higher during CPR with the ITD than standard CPR ($p = 0.015$, $p = 0.017$). No significant differences in diastolic, mean or systolic arterial pressure or end tidal CO_2 were seen.

Application of the ITD resulted in significantly improved hemodynamics during CPR in dogs. Clinical evaluation of the device is warranted to examine whether this translates into improved success in resuscitation and survival.

ABSTRACT C-24
TACHYCARDIA-INDUCED CARDIOMYOPATHY IN DOGS: CLINICAL, HOLTER, AND ECHOCARDIOGRAPHIC FINDINGS. EE Olson, M Borgarelli. Kansas State University, College of Veterinary Medicine, Manhattan, KS.

Left ventricular (LV) systolic dysfunction is a common problem in dogs and can be due to a variety of etiologies. One potential etiology for systolic dysfunction is persistent or paroxysmal tachyarrhythmias, leading to tachycardia-induced cardiomyopathy (TICM). In humans, TICM carries a relatively good prognosis in that remodeling may be reversible with normalization of heart rate. Differentiating between primary and secondary tachyarrhythmias in dogs with systolic dysfunction is critical for prognostic purposes as primary tachyarrhythmias may be associated with a better outcome. The goal of our study was to describe a population of dogs with TICM and to determine if treatment of arrhythmias was associated with reversible cardiac remodeling as indicated by standard echocardiographic parameters.

Medical records of dogs referred to the cardiology service of KSU VMTH from 2008 to 2010 were reviewed. TICM was defined as the presence of severe tachyarrhythmias that were reversible with treatment, systolic dysfunction or ventricular enlargement that improved with treatment of the arrhythmia, or dogs with severe tachyarrhythmias and systolic dysfunction of breeds with that are atypical for idiopathic dilated cardiomyopathy. Exclusion criteria were dogs with congenital heart disease, severe mitral regurgitation, and endocarditis. Transthoracic echocardiography, thoracic radiographs and electrocardiography (ECG) were performed in all dogs. Ventricular enlargement and systolic dysfunction were defined according to standard echocardiographic parameters. Arrhythmias were confirmed with a Holter monitor in 7 dogs.

A total of 12 dogs were included in the study. Mean age was 7 years (range 2–12 years) with 8 males (4 intact, 4 castrated) and 4 females (3 spayed, 1 intact). Four dogs had pulmonary venous congestion or pulmonary edema and two dogs had ascites. At initial presentation, the mean±SD values were as follows: heart rate 219±75 bpm, M-mode fractional shortening (FS) 17.29±9.05%, ejection fraction (EF) using the area-length method 30.69±13.45%, and left atrial to aortic root ratio (LA/Ao) 1.92±0.37. Initial mean±SD M-mode derived LV internal dimensions corrected for body weight were as follows: diastolic 1.92±0.37 and systolic 1.49±0.38. At least one of the following tachyarrhythmias were identified in each dog: atrial fibrillation (4), supraventricular tachycardia (5), junctional tachycardia (2), and ventricular arrhythmias (3). Ten dogs were available for follow up. Seven dogs improved in at least one of the following parameters: resolution of tachyarrhythmia (3), improved systolic function (3), or reduced LV dimensions (3).

Improved systolic function, decreased LV size, or resolution of tachyarrhythmia is indicative of TICM, although other causes of reversible myocardial disease cannot be entirely excluded. The ability to accurately diagnose primary tachyarrhythmias could identify dogs with a better prognosis and further substantiate aggressive anti-arrhythmic therapy. Future studies are needed to develop predictive indicators of reversible cardiac remodeling in dogs with tachyarrhythmias.

ABSTRACT C-25

PILOT STUDY INVESTIGATION OF ANTIDIURETIC HORMONE (ADH) LEVELS IN DOGS WITH ACQUIRED CARDIAC DISEASE. KF Scollan, D Sisson. Oregon State University College of Veterinary Medicine, Corvallis, OR.

Antidiuretic hormone (ADH) has been shown to be elevated in humans with congestive heart failure (CHF). Recently, antidiuretic hormone antagonists were successful during investigational treatment of refractory congestive heart failure in humans. ADH levels have been only modestly investigated in dogs with cardiac disease, primarily due to the technical difficulty in measuring ADH levels via radioimmunoassay. Based on the homologous structure of canine and human ADH, we aimed first to determine the feasibility of measuring ADH in dog plasma using a human ELISA kit, and secondly to investigate the level of ADH in dogs with congestive heart failure due to acquired cardiac disease.

ELISA assay kit validation was performed using six healthy dogs with normal clinical and echocardiographic examinations. Pooled canine plasma was spiked with synthetic ADH and intra-assay precision, dilutional parallelism, and linearity were assessed. To address the second aim of the study, samples were collected from normal dogs and dogs with heart failure due to one of two types of acquired cardiac disease: chronic degenerative valve disease (CDVD) or dilated cardiomyopathy (DCM). Patients underwent clinical, radiographic, and echocardiographic examination to confirm diagnosis, assess severity of disease, and determine presence of pulmonary edema. Whole blood was collected into EDTA tubes containing protease inhibitors, cold centrifuged, and plasma was stored at −80° until analysis. Following ether extraction, plasma ADH in each sample was measured in duplicate using a human ELISA kit. Statistical analysis included a D-Agostino and Pearson test for normality; group results were compared using a non-parametric Mann-Whitney test.

ADH was measured in canine plasma using the human ELISA kit with acceptable intra-assay precision, linearity, and dilutional parallelism. Intra-assay coefficient of variation was 11%. Twenty-four dogs were recruited for the second phase of the study. Six normal dogs and twelve dogs with radiographic evidence of pulmonary edema due to either CDVD or DCM were selected for participation. The remaining six dogs were excluded due to lack of definitive radiographic evidence of congestive heart failure. Median ADH values were 3.67 ± .38 pg/mL for the normal group (n = 6) and 6.155 ± .94 pg/mL for the heart failure group (n = 12). Median ADH values for the two groups were statistically different ($p = 0.0057$).

Our preliminary results indicate that measuring canine ADH using a human ELISA kit is feasible and provides results with an acceptable coefficient of variation. We also showed that dogs with congestive heart failure due to CDVD and DCM have elevated ADH levels in comparison to normal dogs. Our findings motivate further investigation to assess the degree of plasma ADH level elevation and the possible use of ADH antagonists as an adjunct treatment for refractory congestive heart failure in dogs.

ABSTRACT C-26

AORTIC THROMBOEMBOLISM IN DOGS: PRESENTATION, ANTITHROMBOTIC THERAPY, AND OUTCOME IN 26 CASES. RL Winter, CD Sedacca, EC Orton. Colorado State University, Fort Collins, CO.

Aortic thromboembolism (ATE) occurs in both cats and dogs. Whereas ATE in cats is strongly associated with structural heart disease and typically has an acute catastrophic presentation; the pathogenesis and presentation of ATE in dogs is less well known or understood. Further, an effective antithrombotic strategy for ATE in dogs has not been reported.

Medical records of dogs diagnosed with ATE between 2003 and 2010 were examined retrospectively. Diagnosis of ATE was based on ultrasonography, Doppler flow studies, and diminished or absent femoral pulses. Dogs were treated with various acute and chronic antithrombotic therapies. The severity of ambulatory dysfunction was graded as none, mild, moderate, severe, or non-ambulatory at presentation and after therapy. A cohort of dogs in this study received a standardized protocol of chronic warfarin therapy with or without antiplatelet drugs. Target international normalized ratio for warfarin therapy was 2 to 3.

Twenty-six dogs were diagnosed with ATE. All had an apparent mural aortic thrombus caudal to the renal arteries with most having evidence of embolization to the iliac and femoral arteries. None had structural heart disease at diagnosis. Twenty dogs (77%) were still ambulatory at diagnosis. The median duration of ambulatory dysfunction prior to presentation was 7.8 weeks (range 1 day – 52 weeks). A majority of dogs (58%) had no concurrent conditions at diagnosis. Nine dogs (33%) had protein-losing nephropathy. Four dogs (15%) were hypothyroid.

Fourteen dogs were treated with a standard warfarin protocol for a median period of 22.9 months (range 0.5–53 months). Eight dogs were treated concurrently with aspirin, 2 dogs were treated concurrently with clopidogrel, and 4 dogs were treated with warfarin only. Ambulatory function improved between 1 and 3 grades in dogs treated with chronic warfarin. The median period until clinical improvement was 13.9 days (range 2–49 days). Two dogs treated with chronic warfarin therapy had documented resolution of ATE, and 5 dogs had complete resolution of ambulatory dysfunction. None of the 14 dogs treated with chronic warfarin became non-ambulatory, died, or underwent euthanasia because of ATE. The median period of freedom from an adverse event was 24.2 months. No serious hemorrhagic events were reported.

Four dogs were treated with tPA. Three of these had an unfavorable outcome. Two dogs were ambulatory before tPA and become non-ambulatory after treatment. Two dogs underwent surgical thrombectomy. One had a favorable outcome until ATE recurred 14 months after surgery.

In conclusion, the pathogenesis of ATE in dogs is not associated with structural heart disease or left atrial thrombus formation. The presentation tends to be for chronic ambulatory dysfunction. Most dogs are still ambulatory at presentation. Warfarin, with or without concurrent antiplatelet therapy, is an effective antithrombotic treatment strategy for dogs with ATE.

ABSTRACT C-27

AUSCULTATORY, ECHOCARDIOGRAPAHIC, BIOCHEMICAL, NUTRITIONAL, AND ENVIRONMENTAL CHARACTERISTICS OF MITRAL VALVE DISEASE IN NORFOLK TERRIERS. MA Oyama[1], LM Freeman[2], BJ Bulmer[2], J MacGregor[3], JE Rush[2], DJ Trafny[1], KM Meurs[4]. [1]University of Pennsylvania, Philadelphia, PA. [2]Tufts University, North Grafton, MA. [3]Massachusetts Veterinary Referral Hospital, Woburn, MA. [4]Washington State University, Pullman, WA.

Degenerative mitral valve disease (DMVD) is the most common cardiac disease of small dogs yet relatively little breed-specific information is known. Through previous work, investigators have encountered Norfolk Terriers (NT) with echocardiographically apparent DMVD in the absence of a heart murmur. In order to more fully understand DMVD in this breed of dog, we sought to characterize findings from the physical and echocardiographic exam, biochemical, biomarker, and nutritional profile, and select environmental variables from a cohort of apparently healthy NT.

Overtly healthy NT $\geq$ 6yrs old were recruited by 3 different veterinary hospitals and underwent historical, physical, ECG, and 2D/color-flow Doppler echocardiographic exam. Anterior mitral valve length, maximal thickness, area, and prolapse were measured from 2D images. Presence of DMVD was defined as thickened, prolapsing, or flail mitral valve leaflets in the presence of color flow Doppler evidence of mitral regurgitation. Blood samples were obtained for serum biochemistry and serotonin, plasma NT-proBNP, amino acid profile, C-reactive protein, and cardiac troponin-I.

Forty-eight dogs were entered into the study (median age, 8yrs IQR [7–10]; gender, 29F, 19M; median BCS, 5.5 [5–6]). Of the 48 dogs, 23 (48%) had murmurs, 2 (4%) had mid-systolic clicks, 11 (23%) had ECG p-pulmonale, and 41 (85%) were deemed to have echocardiographic evidence of DMVD, including 18 NT without a murmur. Seven (15%), 28 (58%), and 13 (27%) dogs were classified as ISACHC 0, 1a, and 1b, respectively. Mean indexed echocardiographic mitral leaflet length ($P < 0.0001$), thickness ($P = 0.019$), prolapse ($P = 0.0005$), and LA:AoD ($P = 0.01$) were significantly different between ISACHC 1a/b vs 0. Between ISACHC 1a/1b and 0, there were no differences in serum amino acids, C-reactive protein, troponin, diet, or environmental factors; however 6 different amino acids (Ala, Gly, Phe, Pro, Trp, Tyr) were significantly higher in ISACHC 1b vs. 1a. Median serum serotonin was increased in dogs with 1a/b vs. 0 ($P = 0.025$). Dogs whose diets contained some canned food ($P = 0.12$) and dogs residing in suburban environments ($P = 0.03$) had higher serotonin concentrations. NT-proBNP tended ($P = 0.07$) to be higher in ISACHC 1a/1b vs. 0. DMVD appears to be relatively common in NT and echocardiographic changes consistent with mild DMVD can be seen in dogs without a heart murmur. The results of this study establish a foundation of useful information upon which additional prospective studies can be developed.

ABSTRACT C-28

COMPARATIVE STUDY OF LEFT VENTRICLE MEASUREMENTS BY FOUR ECHOCARDIOGRAPHIC METHODS IN GERMAN SHEPHERD DOGS. VMC Oliveira, P Chamas, GT Goldfeder, MHMA Larsson. School of Veterinary Medicine and Animal Science - University of São Paulo, São Paulo, SP, Brazil.

Left ventricle (LV) evaluation is one of the most important contributions of echocardiography in the assessment of cardiac function. However, LV analysis can be made from images obtained by different modes and views of the heart. The aim of this study was to compare LV measurements, shortening fraction (SF) and ejection fraction (EF) obtained from four methods: M mode in short-axis and in long-axis, bidimensional mode in short-axis and in long-axis views of the heart.

Forty normal adult German Shepherds were selected. Echocardiographic study of LV of each animal was performed by the four methods described above. ANCOVA test was used to examine the effects of axis, mode, weight and gender over LV measurements.

Isolated effect of the axis was observed for LV end-diastolic diameter (LVEDD), with greater values obtained from short-axis views. There was isolated effect of mode over EF and SF, with greater measurements derived from bidimensional mode methods. Weight correlated with all linear LV measures at least in one method, but not with EF and SF. Weight had positive effect over LV end-systolic diameter and LV end-diastole posterior wall thickness in all methods, except from M mode in short axis in the last one. Gender had isolated effect over LVEDD, males showing greater values than females in bidimensional mode in short and long axis. The combined effect of axis, gender and weight was identified in interventricular septal end-diastolic thickness.

We concluded that normal reference values obtained by different echocardiographic modes and planes should not be used interchangeably.

ABSTRACT C-29

ASSESSMENT OF LEFT VENTRICULAR DIASTOLIC FUNCTION BY COLOR TISSUE DOPPLER IMAGING ECHOCARDIOGRAPHY IN MAINE COON CATS TESTED FOR MYPBC-AP31 MUTATION. A. Pellegrino, A.G.T. Daniel, P.H. Itikawa, F.F. Lima Jr., G.T. Goldfeder, M.H.M.A. Larsson. School of Veterinary Medicine and Animal Science – University of São Paulo, SP, Brazil.

Hypertrophic cardiomyopathy (HCM), characterized by increased cardiac mass and diastolic dysfunction, is the most common feline heart disease. Myocardial analysis by color tissue Doppler imaging (TDI) is more sensitive than conventional echocardiography. This study evaluated diastolic dysfunction in various stages of feline HCM.

Maine Coon cats ($n = 57$) were screened for the MyBPC-A31P mutation and examined with both echocardiography and TDI. Then, were phenotypically classified in: normal ($n = 45$), suspects for HCM ($n = 7$) and HCM group ($n = 5$); and genotypically classified in: negative ($n = 28$), heterozygous ($n = 26$) and homozygous group ($n = 3$). Myocardial velocities, measured in the basal and mild ventricular segment of the interventricular septal wall (IVS), left ventricular free wall (LVW) and in radial segment of left ventricular wall, was compared among different groups.

A significant decreased ($P = 0{,}01$) longitudinal Em/Am at the basal segment of IVS was observed in HCM cats compared with suspects and normal cats. A significant increased ($P = 0{,}01$) longitudinal E/Em at the basal segment of IVS was observed in HCM cats compared with suspects and normal cats. And a significant decreased ($P = 0{,}02$) longitudinal Sm at the basal segment of the LVW was observed in heterozygous cats compared with negative cats, both without hypertrophy. There was a significant positive correlation between summated early and late diastolic velocities (EmAm) and heart rate ($P < 0{,}001$); and a positive correlation between Sm and Em velocities and heart rate ($P < 0{,}01$).

The MYBPC-A31P mutation is not consistently associated with ventricular hypertrophy and negatives cats can also develop HCM. The TDI alone is not able to identify cats with mutation before myocardial hypertrophy.

ABSTRACT C-30

ASSESSMENT OF LEFT VENTRICULAR SISTOLIC FUNCTION BY STRAIN IMAGING ECHOCARDIOGRAPHY IN MAINE COON CATS TESTED FOR MYPBC-A31P MUTATION. A. Pellegrino, G.G. Pereira, P.H. Itikawa, A.G.T. Daniel, F.F. Lima Jr., M.H.M.A. Larsson. School of Veterinary Medicine and animal Science – University of São Paulo, SP, Brazil.

Diastolic dysfunction occurs in many cats with hypertrophic cardiomyopathy (HCM) but less is known about systolic function in various stages of HCM. Myocardial strain analysis by tissue Doppler imaging is a noninvasive echocardiographic method to assess systolic function. This study evaluated systolic function in various stages of feline HCM.

Maine Coon cats ($n = 57$) were screened for the MyBPC-A31P mutation an examined with both echocardiography and strain. Then, were phenotypically classified in: normal ($n = 45$), suspects for HCM ($n = 7$) and HCM group ($n = 5$); and genotypically classified in: negative ($n = 28$), heterozygous ($n = 26$) and homozygous group ($n = 3$). Peak myocardial strain, measured in the basal and mildventricular segment of the interventricular septal wall (IVS), left ventricular free wall (LVW), left ventricular anterior wall (LVAW), left ventricular posterior wall (LVPW) and radial segment of left ventricular wall, was compared among different groups.

Whereas conventional echocardiography demonstrated an apparently normal contractile state based on fractional shortening, myocardial strain (at mildventricular segment of IVS) in HCM cats was significantly decreased compared with normal group ($P = 0{,}01$). Myocardial strain (at basal segment of LVAW) also was significantly decreased in heterozygous cats compared with negative group ($P = 0{,}00$); and was significantly decreased in heterozygous cats compared with negative group, both without ventricular hypertrophy ($P = 0{,}019$). There was a significant negative correlation between strain values and wall thickness ($P < 0{,}05$).

This method allows detection of abnormal systolic deformation in Maine Coons cats with HCM mutation despite apparently normal systolic function. The abnormal systolic deformation already can be present in heterozygous cats without hypertrophy and increased with progressive ventricular hypertrophy.

ABSTRACT C-31

SYSTOLIC BLOOD PRESSURE IN HEALTHY MAINE COON CATS. RM dos Santos, FF Lima Júnior, PH Itikawa, GT Goldfeder, GG Pereira, DAA Costa, CS Oliveira, MHMA Larsson. School of Veterinary Medicine and Animal Sciences, University of Sao Paulo, SP, Brazil.

Systolic blood pressure (SBP) is an important clinical parameter to assess in certain diseases such as hyperthyroidism, chronic renal disease, hyperadrenocorticism and diabetes mellitus. SBP can be measured directly by intra-arterial catheterization or indirectly by Doppler or oscillometric methods. Currently, standardization of SBP in Maine Coon cats (MCC) are not available. This study aimed to provide reference values of SBP in MCC and compare results with those previously described in non Maine Coon cats (NMCC).

SBP were obtained in 39 healthy adult MCC. Animals were screened by physical examination, thoracic radiography, electrocardiography, echocardiography and laboratory tests (CBC, biochemical blood urea, creatinine and total T_4). In face of normal results, SBP was obtained by ultrasonic Doppler method. Cats were gently restrained in right lateral recumbence, and an adequate neonatal cuff was placed around antebrachium. Five consecutive readings were obtained from each cat, with intervals of 20 seconds between measurements. Unpaired *t*-test was employed in order to compare data with those obtained in a previous report of SBP values in a population of NMCC.

Mean SBP obtained in MCC was 113.74 ± 14.0 mmHg. Higher mean SBP was described in a cat population of NMCC (133.6 ± 16.0 mmHg; $p < 0.001$).

It is suggested that SBP values in normal MCC are lower than reference values for NMCC. Hospital environment and restraining apparently did not represent stress factors. Lower values of SBP in MCC may be related to their good nature and giant body size and should be considered when screening for systemic hypertension.

ABSTRACT C-32

ADVANCED ELECTROCARDIOGRAPHY CAN PREDICT MITRAL REGURGITATION IN CAVALIER KING CHARLES SPANIELS WITH MYXOMATOUS MITRAL VALVE DISEASE. M Spiljak[1], AD Petric[2], LH Olsen[3], A Stepancic[4], T Falk[3], CE Rasmussen[3], V Starc[1]. [1]Institute of Physiology, Medical Faculty, University of Ljubljana, Slovenia. [2]Clinic for Surgery and Small Animal Medicine, Veterinary Faculty, University of Ljubljana, Slovenia. [3]Department of Basic Animal and Veterinary Sciences, Faculty of Life Sciences, University of Copenhagen, Denmark. [4]Quality Control Department, Metallurgical and Chemical Industry Cinkarna Celje, INC., Slovenia.

Recently, multiple advanced resting electrocardiographic (ECG) techniques have been applied in humans for detection of cardiac autonomic and repolarisation function. This has improved the diagnostic and/or prognostic value of short-time ECG in detection of common human cardiac diseases even before onset of symptoms or changes in the standard ECG. Therefore, this study investigates, if advanced ECG can predict the severity of mitral regurgitation (MR) in dogs with myxomatous mitral valve disease (MMVD) and thereby improve the diagnostic value of ECG.

The study included 77 privately owned Cavalier King Charles Spaniels (CKCSs) (age 6.0 ± 2.7 years; 30 males and 47 females). All dogs were examined by echocardiography and a short-time (3–5 min) high-fidelity 12-lead ECG, with the dog in a resting position and in sinus rhythm. Dogs were divided into 5 groups according to the degree of MR estimated as the percentage of the left atrium area using color Doppler mapping (0%; 0% < jet ≤ 15%; 15% < jet ≤ 50%; jet > 50%; jet = 100% and with clinical signs of congestive heart failure). ECG recordings were evaluated via custom software programs to calculate 76 different parameters, including heart rate variability (HRV), QT variability (QTV), T-wave complexity, wave morphology and 3-D ECG. One-way ANOVA determined 21 ECG parameters, which were significantly different ($P < 0.05$) between the 5 dog groups. Principal component factor analysis identified a 5-factor model with 83.2% explained variability. QRS dipolar voltage and two repolarization indices of QTV increased significantly with MR severity, whereas total power of the frequency spectrum of RR interval and the standard deviation of QTV decreased significantly with MR severity. For the 5 selected parameters the prediction of MR jet value was tested by multiple linear regression. A correlation coefficient (R) of 0.65 indicated that the prediction value was significant ($P < 0.01$). If age was included in the multiple linear regression the prediction value was further increased ($R = 0.80$).

Our results indicate that for a cut-off criteria of MR ≥ 50% jet the five selected ECG parameters could predict the severity of MR caused by MMVD in CKCSs with sinus rhythm with sensitivity 65% (78% with age inclusion) and specificity 98% (92% with age inclusion) ($P < 0.05$).

ABSTRACT C-33

EXERCISE INDUCED VARIABILITY OF PLASMA N-TERMINAL PRO-BNP IN HEALTHY DOGS. RL Cohen, GE Singletary, CM Michel, MA Oyama. University of Pennsylvania, Philadelphia, PA.

NT-proBNP concentration is increased in canine patients with heart disease. Relatively little is known about the stimuli for release of NT-proBNP in dogs. Physical activity independent of cardiac disease and the stress of being in the hospital could influence NT-proBNP release and affect diagnosis and management of patients. We hypothesized that NT-proBNP concentration in healthy dogs would not exceed the normal reference value (900 pmol/L) following a period of exercise. The goal of this study was to examine whether physical activity could elevate plasma NT-proBNP and cause false positive results in healthy dogs.

The study population included healthy dogs > 1 yr of age without heart murmur or known systemic disease, and normal 2D/color flow echocardiographic exam. Plasma samples for NT-proBNP were obtained before, immediately after, and 1 hour after a standardized 5-minute submaximal exercise regimen. The study included 14 dogs with a median age of 5.3yrs and included 6 females and 8 males. There was no statistical difference in median plasma NT-proBNP concentration across the three time points (baseline median, 617 [IQR, 359–756], immediately post, 582 [314–853], 1-hr post, 476 [328–750]; $P = 0.11$). The average coefficient of variation of NT-proBNP concentration across the exercise regimen was $23.0 \pm 18.9\%$. In 1 of 14 dogs (7.1%), NT-proBNP increased from 790 to 1054 pmol/L immediately after exercise. The results of this study demonstrate that submaximal exercise does not significantly change median NT-proBNP concentration and the incidence of false positive results is low. Further studies should investigate effects of exercise on NT-proBNP concentrations in dogs with heart disease.

ABSTRACT C-34

NT-PRO ANP IN OBESE DOGS. DS Schwartz, PRR Melo, AM Mazini, VMC de Oliveira, GT Goldfeder, MR Azedo. FMVZ-USP (School of Veterinary Medicine, University of São Paulo) São Paulo, SP, Brazil.

Obesity is an increasing problem in veterinary medicine. Obese human patients are shown to present lower levels of natriuretic peptides, regardless of an increased volume and pressure load, what raises the possibility that the natriuretic response is impaired in these individuals. Considering the controversial findings in obese humans, and the lack of studies reported in dogs, this study proposed the evaluation of Nt-proANP concentration in obese dogs.

Nt-proANP concentration was determined prospectively in 39 obese dogs (27 females; 12 males; 6–108 months) and in 23 non-obese dogs (controls; 13 females; 10 males; 12–108 months) from a veterinary hospital population. Obesity was determined by body condition score [2 (4/9); 21 (5/9); 1 (6/9); 3 (7/9); 19 (8/9); 18 (9/9)].

Dogs were excluded if they had any primary cardiac disease, renal insufficiency, endocrine disease, or if they were receiving diuretics, vasodilators, antiepileptic drugs or corticosteroids. Commercial kits were used (Vetsign® Canine Cardio Screen Nt-pro-ANP VC3167 – Guildhay/Biomedica). Mann Whitney test was used for group comparison. Results are presented as median; interval; P25 and P75).

Nt-proANP was significantly lower in obese dogs [413.56fmol/mL (0.0–1287.14); P25 = 228.673; P75 = 595.205] than in controls [647.98fmol/mL (34.99–1596.82); P25 = 490.785; P75 = 957.055]; (P = 0.004).

Results were similar to what has been found in obese humans. Lower levels of natriuretic peptides are also seen in obese heart failure patients.

This study provides important information regarding Nt-proANP concentration in obese dogs, which should be better explored characterizing the behavior of natriuretic peptides after weight loss, and also in obese dogs with primary heart disease.

ABSTRACT C-35

UTILITY OF PLASMA NT-PROBNP FOR ASSESSING HEMODYNAMIC SIGNIFICANCE OF PATENT DUCTUS ARTERIOSUS IN DOGS PRIOR TO AND FOLLOWING TRANSCATHETER DUCTAL OCCLUSION. CD Hariu, AB Saunders, SG Gordon, B Norby, MW Miller. The Michael E. DeBakey Institute, College of Veterinary Medicine and Biomedical Sciences, Texas A&M University, College Station, TX.

Left-to-right shunting patent ductus arteriosus (PDA) is one of the most common canine congenital cardiovascular defects. Human studies have shown that BNP and NT-proBNP concentrations are elevated in patients with PDA, and can be used to detect hemodynamically significant PDA. The purpose of this study was to measure NT-proBNP concentrations in dogs with PDA, and to assess whether additional indicators of hemodynamics correlate with NTproBNP. We hypothesized that NT-proBNP will serve as a simple non-invasive marker of hemodynamic significance in dogs with PDA prior to and following transcatheter ductal occlusion. NT-proBNP was measured in 30 client-owned dogs with echocardiographically normal hearts. Ten dogs with PDA were initially evaluated with thoracic radiographs, transthoracic and transesophageal echocardiography, pulmonary capillary wedge pressure (PCWP) and NT-proBNP. NT-proBNP and echocardiography were repeated at 1 day and 3 months following ductal occlusion. PCWP was repeated at 3 months.

Baseline NT-proBNP was significantly higher in PDA dogs compared to control (1877 ± 2081 pmol/L (mean ± SD), 651 ± 301; $p < 0.0025$). At 1 day and 3 months following ductal occlusion, NT-proBNP was 1347 ± 1256 and 466 ± 380, respectively. The following decreased significantly from baseline: PCWP (11.8 ± 4.7 to 6.4 ± 3.2 mmHg; p = 0.018), and indexed left ventricular internal dimensions in diastole (2.26 ± 0.47 to 1.67 ± 0.19; p = 0.006) but not significantly in systole (1.40 ± 0.31 to 1.20 ± 0.19; p = 0.13). NT-proBNP is elevated in dogs with PDA and transductal closure is associated with a reduction in NT-proBNP, PCWP and left ventricular size.

ABSTRACT C-35

VALIDATION OF THREE SHIPPING METHODOLOGIES FOR NT-proBNP IN A MIXED COHORT OF CATS. M. McLane, M. Baum, J. Allen, G. Farace, S. Ettinger, A. Beardow. California Animal Hospital Veterinary Specialty Group, Los Angeles, CA.

Cardiac biomarkers, particularly NT-proBNP, are becoming more commonly used in dogs and cats as part of a diagnostic work up. Multiple studies already have documented the correlation of this peptide with cardiac disease status and potential clinical implications. In a portion of these reports the manner in which samples were handled was placement of whole blood into an EDTA tube, followed by centrifugation and decanting of the supernatant that was ultimately stored at −20°C or −80°C prior to shipment, either with or without protease inhibition.

Our objective was to compare the NT-proBNP concentrations in feline plasma collected using the previously reported methods to the California Animal Hospital (CAH) collection method using tubes containing a protease inhibitor. This study compared NT-proBNP concentrations using the protease inhibitor tubes vs. EDTA tubes from 18 privately owned feline patients, with confirmed cardiac disease, and 6 control feline patients. For all study participants, we performed a full history and physical examination, a hematology and chemistry panel, thoracic radiographs, ECG, and echocardiogram.

In each study participant, at least 4 ml's of whole blood was drawn from a peripheral vein, and transferred to a plastic EDTA tube. The sample was centrifuged within 1 hour after collection. 1 ml of plasma was then transferred to a tube containing a protease inhibitor, which was stored at 4°C until being shipped within 24 hours of collection. The remaining plasma was placed into 2 separate microtubes, which did not contain a protease inhibitor. One microtube was then stored and shipped as previous studies have reported (−20°C, Styrofoam container, shipment within 24 hours), and the second microtube was frozen at −80°C. All samples were shipped, received and analyzed within 24 hours of collection.

Results of this study showed that no difference was found between the 2 frozen sample methods (663 pmol/L and 650 pmol/L p = 0.40). It was determined that both frozen methods had lower NT-proBNP levels (655 and 646 pmol/L) when compared to plasma samples shipped in protease inhibitor tubes (756 and 794 pmol/L). The findings of this trial demonstrate that the NT-proBNP levels are significantly different between samples placed in EDTA tubes vs. contain protease inhibitor (p = 0.008 and p = 0.0006). Utilizing protease inhibitor tubes allows more accurate measurement of plasma NT-proBNP. As for its relevance for future research and publications, authors should take care to investigate the manner in which blood samples were handled and the conclusion/results of these studies should be taken in light of the methodologies used in collecting, storing, shipping and analyzing the samples.

ABSTRACT C-37

SERUM SEROTONIN CONCENTRATION IN RELATION TO SEVERITY OF SPONTANEOUS DEGENERATIVE MITRAL VALVE DISEASE IN DOGS. [1]S.J. Lim, [1]S.H. Lee, [1]H.J Park, [1]D.W. [1]Jung, [1]H.J. Choi, [2]H.Y Youn, [3]H.M. Park, [1]D.H. Kim, [1]K.H. Song. [1]College of Veterinary Medicine, Chungnam National University, Daejeon, [2]College of Veterinary Medicine, Seoul National University, Seoul, [3]College of Veterinary Medicine, Konkuk University, Seoul, S. Korea.

Degenerative mitral valve disease (DMVD) is one of the most common heart disease and is present approximately 75% of the canine heart disease. Although the high prevalence exists in small dogs, the underlying molecular mechanism of its pathophysiology is rarely known. DMVD is functionally and pathologically similar in humans and dogs, thus, there will be a common pathogenesis in human and dogs with naturally occurring DMVD. Serotonin and serotonin-related mechanisms have been implicated as a cause of valvular disease in human and animals, including spontaneous DMVD in dogs. Increased circulating 5HT concentration as a potential source of heightened 5HT signaling is demonstrated in small dogs with DMVD. The aim of this study was to investigate whether serum 5HT concentrations were associated with severity of naturally occurring DMVD in small dogs and to investigate potential associations of dog characteristics on serum 5HT concentrations in our study population. Forty-eight dogs were included in this study and were classified into control and DMVD groups according to the results of physical and echocardiographic examinations. Based on the LA:Ao ratio, dogs with DMVD were classified as follow: Control (LA:Ao ratio ≤1.5 and no MR), mild (LA:Ao ratio ≤ 1.5 and MR), moderate (1.5 < LA:Ao ratio ≤ 1.8 and MR), Severe (LA:Ao ratio >1.8 and MR).

Serum serotonin concentrations were measured by ELISA. An overall significant difference ($P < .05$) was found among 4 groups and 5HT concentrations (control, 72.38 ng/ml [51.34–95.11]; Mild DMVD, 80.88 ng/ml [65.47–91.41]; moderate DMVD, 93.76 ng/ml [75.71–101.67]; Severe DMVD, 102.13 ng/ml [91.87–113.12]). Significantly higher 5HT concentrations were observed in dogs with moderate ($P < .05$) and severe ($P < .05$) DMVD, compared with concentration in control group. Additionally, 5HT concentration in dogs with moderate DMVD were significantly higher ($P < .05$) than concentration in dogs with mild DMVD. Also, dogs with severe DMVD had significantly higher 5HT concentration than dogs with mild ($P < .05$) and moderate ($P < .05$) DMVD. There was no significant association of age, platelet, and LVIDd, on serum 5HT concentration, however, weak correlation between serum 5HT increased significantly and LA:Ao ratio ($R^2 = .211$, $P < .05$) was observed. The results of this study indicate that serum 5HT concentrations were higher with increasing severity of spontaneous DMVD, which may be the potential cause to advance the progression of DMVD. Further studies should be performed to reveal the role of 5HT in inducing and accelerating spontaneous DMVD and to investigate if lowering serum 5HT concentration could alter the progression of DMVD.

ABSTRACT C-38

UTILITY OF CARDIAC TROPONIN I IN DIFFERENTIATING BETWEEN UNDERLYING ETIOLOGIES OF PERICARDIAL EFFUSION IN THE CANINE PATIENT. N Peckens, M Ferguson, B Lefbom, S Rosenthal, W Tyrrell. Chesapeake Veterinary Cardiology Associates, Vienna, VA.

The objective of this prospective study was to evaluate the utility of cardiac troponin I (cTnI) in differentiating between underlying etiologies of pericardial effusion in the canine patient. Patients were prospectively recruited at time of diagnosis of novel pericardial effusion. Serum samples were collected prior to pericardiocentesis. Patients were evaluated by echocardiography and classified with the diagnosis of hemangiosarcoma (HSA), heart base tumor (HBT), or unknown etiology at the initial evaluation based on established characteristic echocardiographic findings. Idiopathic pericardial effusion (IPE) was defined by histopathology, echonegative for a mass lesion with no recurrence of pericardial effusion >6 months, or symptom free >12 months from time of enrollment. Patients were excluded from analysis if a diagnosis could not be established based on above criteria or concurrent moderate azotemia (Creatinine >3.0 mg/dL) was present at time of sample collection. Serum samples were frozen and analyzed in batches within 60 days of collection by a cTnI assay with a 0.2ng/mL lower limit sensitivity.

Sixty-three patients were recruited over a one year period with 15 patients excluded due to lack of diagnosis (13) or azotemia (2). Median cTnI levels of dogs with HSA (n = 24), HBT (n = 19), and IPE (n = 5) were 14.8 ng/dL (interquartile range (IQR) < 0.2–11.3), 0.23 ng/dL (IQR < 0.2–0.29), and < 0.2 ng/dL (IQR < 0.2- < 0.2) respectively. Concentrations of cTnI differed significantly between dogs with HSA and HBT (p = 0.001) and IPE (p = 0.0029). There was no difference between cTnI concentrations between HBT and IPE dogs (p = 0.911). Receiver operating curve analysis to determine the optimal cutoff for differentiation of dogs affected with HSA and both HBT and IPE revealed a significant (p = < 0.001) area under the curve (0.79). A cut-off point of cTnI of >0.78 yielded a sensitivity of 67% (95% CI, 45–84%) and specificity of 95% (95% CI, 79–99%). Utilizing a higher cut-off point of >3.0 yielded a lower sensitivity of 50% (95% CI, 29–71%), but a higher specificity of 100% (95% CI, 86–100%) which may have more clinical utility given the disparity in prognoses of the etiologies compared.

In conclusion, this study supports the diagnostic utility of cTnI concentrations to delineate between patients with HSA and other etiologies of pericardial effusion, but does not reliably differentiate between dogs with IPE and other neoplastic etiologies.

ABSTRACT C-39

PLASMA CONCENTRATIONS OF ICAM-1 AND E-SELECTIN IN DOGS WITH MITRAL VALVE DISEASE. GE Singletary[1], MA Oyama[1], R Cohen[1], MC Machen[1], J Connolly[2]. [1]University of Pennsylvania, Philadelphia, PA. [2]Children's Hospital of Philadelphia, Philadelphia, PA.

The pathogenesis of degenerative mitral valve disease (DMVD) in dogs remains to be fully elucidated. The high sheer stress caused by mitral regurgitation damages the endothelial surface of the valve, and a previous study demonstrated increased transcription of intercellular adhesion molecule-1 (ICAM-1) and E-selectin in affected mitral leaflet tissue. We hypothesized that this may be responsible for platelet recruitment and adhesion, and initiation of a proliferative cascade, resulting in further myxomatous changes. The goal of this study was to compare plasma levels of ICAM-1 and E-selectin in healthy dogs and those with DMVD.

The study population included dogs with echocardiographic evidence of DMVD and healthy control dogs >1 year old with no heart murmur or known systemic diseases. DMVD dogs underwent 2D/color-flow Doppler echocardiographic exam. Blood samples were obtained for plasma ICAM-1 and E-selectin analysis using commercially available ELISA kits.

The study included 34 dogs, of which 20 had DMVD and 14 were normal. The DMVD group had a median age of 9.5yrs (IQR [8–11]) and included 6 females and 14 males. Two (10%), 13 (65%), 2 (10%) and 3 (15%) dogs were classified as ISACHC 1a, 1b, 2 and 3a, respectively. Of the control dogs, median age was 4.5yrs [2–6.5], with 5 females and 9 males. There was no statistical difference in plasma E-selectin between control dogs (median 2.71 [2.05–7.66]) and those with DMVD (2.46 [1.54–3.55]); P = 0.35. Plasma ICAM-1 concentrations were higher in DMVD dogs (1.58 [1.20–10.58]) than controls (median 1.31 [1.11–1.65], but this difference did not reach statistical significance (P = 0.22). Linear regression analysis showed no significant correlation between ICAM-1 or E-selectin and serum serotonin level, NT-proBNP or echocardiographic measures of DMVD severity (LA:Ao, LVIDd:Ao, LVIDs:Ao). The results of this study demonstrate no significant difference in circulating adhesion molecules ICAM-1 and E-selectin in dogs with DMVD as compared with healthy controls. Further studies investigating adhesion molecules within the mitral valve tissue itself are likely needed if ICAM-1 and E-selectin play a role in the pathophysiology of DMVD.

ABSTRACT C-40

CHRONIC HEART FAILURE SELECTIVELY INDUCES REGIONAL HETEROGENITY OF GLUCOSE TRANSPORTER EXPRESSION. B. Ware[1], Y. Nishijima[1], S. Rogers[3,1], and V.A. Lacombe[1,2]. [1]College of Pharmacy. [2]Department of Veterinary Clinical Sciences, The Ohio State University, Columbus, OH. [3]St. Vincent's Institute of Medical Research, The University of Melbourne, Australia.

The rate of glucose utilization in the heart is greater than in other tissues, and impaired glucose uptake may play a major role in the pathogenesis of heart failure (**HF**). Glucose uptake across the sarcolemma is regulated by a family of membrane proteins called glucose transporters (**GLUTs**), which includes GLUT-4, the major cardiac isoform, and GLUT-12, a recently discovered isoform, the role of which is unknown in the heart. In addition, despite the well-known regional differences in myocardial structure and function, potential regional patterns in glucose transport have not been investigated. Thus, we hypothesized that GLUT-4 and -12 protein and gene expression would be chamber specific in healthy dogs and during chronic HF.

Using a canine model of tachypacing induced chronic HF, GLUT protein and messenger RNA in both ventricles and atria were investigated by immunoblotting and real time PCR.

In control dogs, GLUT-4, but not GLUT-12, protein expression were significantly higher in the atria compared to the ventricles, with the highest content in the right atrium (RA, P < 0.001). GLUT-4 and -12 mRNA were homogeneously expressed in all the cardiac chambers. During chronic HF, GLUT -4 and -12 expression was highest in the left ventricle (LV, by 2.5 and 4.2 fold, respectively, P < 0.01), with a concomitant increase in GLUT-4 and -12 mRNA (P < 0.001). GLUT -4, but not GLUT-12, was decreased in RA during chronic HF (P = 0.001).

Our data suggest that GLUT-4 protein was differentially expressed across the cardiac chambers in the healthy heart. During chronic HF, LV was the primary site dependent on both GLUT4- and GLUT12-mediated glucose transport, which was transcriptionally regulated. In addition, the paradoxical decrease in GLUT4 content in RA may induce perturbations in atrial energy production during chronic HF.

ABSTRACT C-41
COMPARISON BETWEEN THORACIC COMPUTERIZED RADIOGRAPHY AND ECHOCARDIOGRAM IN OBESE DOGS TO EVALUATE HEART ENLARGEMENT. PRR Melo, AM Mazini, VMC de Oliveira, GPR Banon, GT Goldfeder, CN Duarte, ACBCF Pinto, DS Schwartz. FMVZ-USP (School of Veterinary Medicine, University of São Paulo) São Paulo, SP, Brazil.

Some obese dogs are suspected to have cardiac disease because they have enlargement of the heart on thoracic radiograph. It has been reported in cats that the fat increases the cardiac silhouette, while echocardiograms revealed normal cardiac dimensions. The purpose of this study was to determine whether obesity overestimates cardiac dimension in radiographs compared to echocardiographic findings in dogs.

Twenty three obese dogs and 20 controls were included based on a 1–9 body condition scoring (BCS). Computerized radiography was obtained and VHS measurement was performed as previously described. Echocardiographic measurements were interpreted based on reference values according to lean body weight regression equations. Results for echo and VHS measurements were classified in scores as normal, mild, moderate or severe increase. Student's t test was used for comparison of VHS between groups. Mann-Whitney Rank Sum test was used to assess echocardiographic scores between groups. Spearman Rank Order Correlation was used to assess relationships between any pairs of variables between echo and VHS scores, echo vs BCS and VHS vs BCS.

Groups were similar regarding age [obese (69±24); control (62±27); P = 0.368], breeds and gender distribution. Obese dogs had significantly higher VHS and echo scores compared with controls [VHS: (10.58±0.69) vs (9.77±0.54); P < 0.001; Echo score: range (1–4) vs (1–2); P > 0.05]. There were no relationships between any pair of variables analyzed.

These results show that there are changes both in echo and radiographic appearance of the heart in obese dogs, but VHS overestimates cardiac silhouette compared to echo, probably related to pericardial fat accumulation.

ABSTRACT C-42
HEART RATE AND HEART RATE VARIABILITY IN DOGS WITH DIFFERENT DEGREES OF MYXOMATOUS MITRAL VALVE DISEASE. CE Rasmussen[1], T Falk[1], NE Zois[1], SG Moesgaard[1], HD Pedersen[2], J Häggström[3] and LH Olsen[1]. [1]Department of Basic Animal and Veterinary Sciences, Faculty of Life Sciences, University of Copenhagen, Frederiksberg, Denmark. [2]Novo Nordic A/S, Maaloev, Denmark. [3]Department of Clinical Sciences, Swedish University of Agricultural Science, Uppsala, Sweden.

Heart rate variability (HRV) is an indirect measurement of the autonomic modulation of heart rate (HR). Reduced HRV measured from short-time electrocardiography is seen in dogs with heart failure (HF) secondary to myxomatous mitral valve disease (MMVD). However, HRV is suggested to increase with disease severity at early stages of MMVD. The aims of this study were 1) to associate HR and HRV with severity of MMVD in Cavalier King Charles Spaniels (CKCS) and 2) to compare HR and HRV between CKCS and other dog breeds in a group of dogs in HF secondary to MMVD.

One-hundred dogs were examined by echocardiography and 24-hour electrocardiography. The dogs were divided into five groups: 1) CKCS with no/minimal mitral regurgitation (MR) (MR jet ≤ 15% of the left atrial area using color Doppler mapping) and no murmur, 2) CKCS with mild MR (20% < jet ≤ 50%), 3) CKCS with moderate/severe MR (jet > 50%) and no clinical signs of HF, 4) CKCS in HF (HF defined as left atrium to aortic root ratio (LA/Ao) > 1.5, clinical signs of HF and furosemide responsiveness) and 5) non-CKCS in HF. Dogs in HF were allowed HF therapy. Both HR and HRV were analyzed over a 24-hour period, while HRV were also analysed over a 6-hour nightly period. Analyses of variance were performed with HR or HRV as response variables and the explanatory variables dog group and echocardiographic indices of MMVD were included separately. All *P*-values were Bonferroni corrected.

Minimum- and mean HR were significantly higher in CKCS with moderate/severe MR and in HF compared to CKCS with no/minimal and mild MR (all $P < 0.001$). Seven out of 26 HRV variables were significantly decreased in CKCS with moderate/severe MR and in HF compared to CKCS with no/minimal and mild MR (all $P < 0.02$). Another 10 HRV variables showed the same groupwise differences (all $P < 0.02$), except that the difference between CKCS with mild MR and CKCS with moderate/severe MR did not reach statistical significance. Mminimum HR, mean HR and the HRV variables (7 and 10) differing between dog groups, also consistently decreased with increasing MR, LA/Ao and the proximal isovelocity surface area in CKCS. Non-CKCS in HF had a lower minimum HR compared to CKCS in HF ($P = 0.03$) and a higher triangular index measured in both periods (all $P < 0.04$).

In conclusion, HR increased and most HRV variables decreased with increasing severity of MMVD in CKCS, even prior to the development of HF. Other breeds in HF secondary to MMVD had lower minimum HR, but higher triangular index compared to CKCS in HF.

ABSTRACT C-43
BRADYCARDIA INDUCED BY PARTIAL NASAL OBSTRUCTION BY DIGITAL COMPRESSION IN NON-SEDATED RABBITS. E Zacché, FN Gava, EMG Ortiz, T Champion, AA Camacho. Universidade Estadual Paulista – UNESP, Jaboticabal, São Paulo.

Although the cells in the specialized conduction system in the heart are capable of initiating their own impulse, the rate in which those impulses are generated can be influenced by autonomic nervous system. Different types of respiratory patterns can stimulate autonomic nervous system in different manners. Thus, non-sedated rabbits were studied during forced respiration aiming to evaluate the influence of this breathing pattern on heart rate.

Twenty male, one-year-old healthy New Zealand rabbits were enrolled in the study. Animals were set in right lateral recumbency and maintained that way by physical contention. Chemical sedation was not used. Partial nasal obstruction by digital compression was applied to those rabbits for five seconds, eliciting a forced inhaling and exhaling against semi closed nostrils. Heart rate was obtained by measurement of two consecutives RR intervals in the computerized electrocardiography, recorded continuously prior and during the maneuver.

Heart rate before the intervention was 251 ± 34 bpm (mean ± standard deviation). All rabbits submitted to the maneuver showed a dramatic reduction in this parameter. After nasal partial obstruction, heart rate was 142 ± 32 bpm. Data was submitted to statistical analysis by paired Student's t test and a significant difference between the heart rate before and after the maneuver was observed ($p < 0.0001$).

Although the exactly mechanism involved in this response was not elucidated, the presented data support the applicability of this maneuver as an efficient method for non-pharmacological heart rate reduction in rabbits.

ABSTRACT C-44
TISSUE DOPPLER EVALUATION IN ENGLISH COCKER SPANIELS WITH DILATED CARDIOMYOPATHY. GG Pereira, GT Goldfeder, MHMA Larsson. School of Veterinary Medicine and Animal Science, University of São Paulo, SP, Brazil.

Dilated cardiomyopathy (DCM) is recognized by reduced contractility. Although often diagnosed in large breed dogs, English Cocker Spaniels (ECS) are quite predisposed. We hypothesized that a pattern of myocardial dysfunction could be characterized in this breed using tissue Doppler techniques.

Thirty two ECS were studied, being 16 with DCM and 16 healthy for control purpose. Longitudinal tissue Doppler imaging (TDI) of myocardial systolic, early (E) and late (A) diastolic velocities, time from onset of Q wave to peak systolic velocity (QS interval), Doppler derived strain (St) and strain rate (SR) were obtained in mitral annular left ventricular free wall (LVFW) and interventricular septum (IVS). Radial TDI, QS interval, St and SR were obtained in LVFW at papillary muscle level. Systolic myocardial velocity gradients (MVG) were recorded in longitudinal and radial displacements of LVFW. Systolic TDI and St were obtained from right ventricular wall (RVW). Unpaired Student's *t*-test was employed to investigate differences between groups.

LVFW findings in DCM group, compared to control, included lower longitudinal (5.81 ± 1.82 vs. 7.07 ± 1.41 cm/s; $p < 0.05$) and radial (4.69 ± 1.10 vs. 5.92 ± 1.45 cm/s; $p < 0.05$) systolic myocardial velocities, increased longitudinal (114.26 ± 14.26 vs. 93.95 ± 6.42 ms; $p < 0.001$) and radial (115.17 ± 13.78 vs. 92.91 ± 14.53 ms; $p < 0.001$) QS interval, reduced systolic longitudinal (11.21 ± 8.50 vs. 23.8 ± 3.38%; $p < 0.001$) and radial (27.17 ± 8.99 vs. 44.70 ± 6.46%; $p < 0.001$) St and reduced longitudinal (1.71 ± 0.24 vs. 2.87 ± 0.58 s^{-1}; $p < 0.001$) and radial (2.07 ± 0.66 vs. 2.59 ± 0.70 s^{-1}; $p < 0.05$) SR. MVG was lower in LVFW longitudinal (0.90 ± 0.41 vs. 1.44 ± 0.21; $p < 0.001$) and radial (1.11 ± 0.46 vs. 1.66 ± 0.29; $p < 0.01$) displacements in DCM group. Post-systolic contractions were found in IVS of 5/16 dogs with DCM. Reduced systolic St were found in RVW (18.81 ± 5.05 vs. 26.72 ± 5.33%; $p < 0.001$). E/A ratio lower than 1 was the most common diastolic abnormality found in both ventricles.

Lower left ventricular systolic myocardial velocities, electromechanical delay, longitudinal post-systolic contraction, reduced biventricular deformation and relaxation represents main tissue Doppler findings in ECS with DCM.

ABSTRACT C-45

CARDIOVASCULAR FUNCTION IN OBESE AND OVERWEIGHT CATS. T Champion, FN Gava, EMG Ortiz, EZ Pereira, MA Brunetto, AC Carciofi, AA Camacho. Faculty of Agricultural and Veterinary Sciences from São Paulo State University, Jaboticabal, BR.

Obesity can affect cardiac function due to effects on cardiac rhythm, ventricular volume and blood pressure. The purpose of this study was to determine the effects of obesity and overweight on non-invasive systemic blood pressure and Doppler echocardiographic parameters in cats without others causes of cardiac hypertrophy.

The study groups comprised fifteen obese cats with mean body score index (BSI) of 8,8, seven overweight cats (BSI = 6,3) and seven cats with ideal BSI (4,9). The blood pressure was measured by Doppler method and the Doppler echocardiography was performed in conscious animals. The statistical analysis was performed by analysis of variance followed by Tukey's test and Pearson's correlation. The blood pressure values of the obese cats were superior (159,12 ± 11,22 mmHg, $p < 0,0001$) than in overweight (134,45 ± 13,81 mmHg) and normal cats (136,90 ± 13,17 mmHg) and 57% of the obese cats had blood pressure higher than 160 mmHg. There were observed differences on the ratio of early (E) and late (A) left ventricular filling velocity ($p = 0,008$) of obese animals (E/A = 1,07 ± 0,39) compared to overweight (1,68 ± 0,37) and normal cats (1,43 ± 0,24). Seven obese cats (50%) had inversion of E/A compatible with diastolic dysfunction and there were negative correlation ($r = -0,453$, $p = 0,026$) between the E/A ratio and blood pressure values. Other differences observed were increases in left ventricular septum in diastole ($p = 0,002$) and in free wall in diastole ($p = 0,023$) and systole ($p = 0,042$) of the obese animals compared to overweight and normal cats.

These results demonstrate the possibility of cardiovascular effects related to obesity in cats, such as systemic arterial hypertension and secondary diastolic dysfunction.

ABSTRACT C-46

EFFECT OF TORSEMIDE VS FUROSEMIDE ON CLINICAL, LABORATORY, RADIOGRAPHIC, AND QUALITY OF LIFE VARIABLES IN DOGS WITH HEART FAILURE. GD Peddle, MA Oyama, GE Singletary, CA Reynolds, DJ Trafny, MC Machen. University of Pennsylvania, Philadelphia, PA.

Diuretic therapy reduces preload, and relieves congestion secondary to cardiac dysfunction. Torsemide (torasemide) is a loop diuretic with longer duration of action, less diuretic resistance, and adjunctive aldosterone antagonism as compared to furosemide. We hypothesized that torsemide was no less effective than furosemide at diuresis, control of clinical signs, and maintenance of quality of life in dogs with congestive heart failure. A double-blinded, randomized, crossover clinical trial was performed in 7 dogs with stable heart failure receiving BID furosemide and adjunctive medications. Dogs were randomized to their current furosemide dose or torsemide (calculated as 1/10 of the daily furosemide dose divided into BID dosing). Crossover occurred at day 7 and the study ended on day 14. Clinical, laboratory, radiographic, and owner-perceived quality of life variables were evaluated on days 0, 7 and 14.

No dog developed recurrent heart failure during the study. Average furosemide dose on day 0 was 5.13 mg/kg/day (range, 2.8–9.6). Following torsemide treatment, blood urea nitrogen ($P = 0.0028$), albumin ($P = 0.0287$), and albumin:globulin ratio ($P = 0.0012$) were significantly increased, and urine specific gravity ($P = 0.0062$) and chloride ($P = 0.0051$) were significantly decreased as compared to baseline and/or furosemide dosing (one-way ANOVA with Bonferroni correction). No differences in QOL were found.

Results indicate that torsemide is equivalent to furosemide at controlling clinical heart failure in dogs, and might in fact, achieve greater diuresis vs. furosemide. Larger clinical trials evaluating furosemide resistance and/or torsemide as a first-line loop diuretic for congestive heart failure in dogs with heart failure are warranted.

ABSTRACT C-47

2D SPECKLE TRACKING ECHOCARDIOGRAPHY IN CLINICALLY HEALTHY CATS AND CATS WITH HYPERTROPHIC CARDIOMYOPATHY. H. Takano, T. Isogai, T. Aoki, Y. Wakao, Y. Fujii. Azabu University, School of Veterinary Medicine, Kanagawa, Japan.

The purpose of this study was to investigate the feasibility of speckle tracking echocardiography (STE) in healthy cats and to determine whether or not it can detect myocardial dysfunction in cats with diseased heart.

Radial and circumferential strain and strain rate were measured by STE using left ventricle short-axis view in clinically healthy cats. Eighteen cats with HCM whose LV thickness at end-diastole with 6 mm or more were evaluated with STE analysis, and compared with healthy cats. Index of left ventricular synchrony (Trs-SD) was also assessed in cats with HCM, and compared to healthy subjects.

STE resulted in technically adequate images in 100% of the cats. Fusions of early and late diastolic (E and A) wave in strain rate were seen in 3 of 16 cats. Percent errors in analysis with or without simultaneous ECG monitoring were 5.3-14.7% in all parameters. Inter- and intraobserver variability of STE parameters in healthy cats was minimal (4.1-15.6%) except for the systolic circumferential strain rate. Sedation using buprenorphine and acepromazine did not affect any STE parameter. E wave in radial and circumferential strain rate of HCM cats was significantly decreased compared with healthy cats. No significant difference was seen in Trs-SD.

STE analysis was considered clinically feasible to assess cardiac function in cats, and could detect myocardial dysfunction in cats with HCM. Further study is warranted to investigate to assess whether or not STE can differentiate the etiology of left ventricular concentric hypertrophy since it is clinically important.

ABSTRACT C-48

PHARMACOKINETICS AND BIOAVAILABILITY OF CARVEDILOL IN CATS. Durtschi AL, Gompf RE, Martin-Jimenez T, Cox S, Bartges J. The University of Tennessee Veterinary Medical Center, Knoxville, TN.

Carvedilol, a 3rd generation non-selective beta-blocker with ancillary $alpha_1$-blocking and antioxidant properties may have therapeutic implications for multiple diseases in cats; however, pharmacokinetics and bioavailability of commercially prepared oral carvedilol has not been determined. HPLC for carvedilol measurement in feline plasma was validated and standardization curves created. The pharmacokinetics (PK) of carvedilol was evaluated in 5 apparently healthy male neutered adult cats (average weight of 5 kg) following single dose intravenous (IV) of 0.5 mg/kg and single dose oral administration of 1.6 to 2.0 mg/kg. Concentrations of the active parent compound, carvedilol, were detected in plasma using HPLC analysis. Lower limit of quantification was 5 ng/ml.

The mean peak concentration after IV administration of carvedilol was 8639 ng/mL (range, 901 to 8648), elimination half- life was 2.9 hours (range, 2.0 to 5.4), and clearance was 0.35 L/hr/kg. The volume of distribution was 1.18 L/hr. After a single oral administration of carvedilol, the time to peak plasma concentration was 60 minutes (range, 30 to 90 minutes) and the mean residual time was 4.8 hours. The half life was 4.44 hours. Maximal concentration 294 ng/mL and the mean bioavailability was 15.7% with a median of 9.97% (range, 4.7% to 46%).

These data demonstrate a low bioavailability of oral carvedilol and a wide variation in cats. All cats tolerated the oral dose of carvedilol with no major adverse effects. Also, a mean residual time of 4.8 hours would suggest that a more frequent dosing schedule may be required to maintain therapeutic plasma levels. Pharmacodynamic studies investigating beta-adrenergic blockade duration may provide a more accurate dosing interval of carvedilol.

ABSTRACT C-49

EFFECTS OF SILDENAFIL CITRATE ON DOGS WITH EISENMENGER'S SYNDROME. K Nakamura, M Yamasaki, H Ohta, M Takiguchi. Department of Veterinary Clinical Sciences, Graduate School of Veterinary Medicine, Hokkaido University, Sapporo, Hokkaido, Japan.

Sildenafil has shown to be effective for dogs with pulmonary hypertension; however, its efficacy for dogs with Eisenmenger's syndrome (ES) and secondary erythrocytosis has not yet been determined. The objective of this study is to determine the effect of Sildenafil for dogs with Eisenmeger's syndrome and secondary erythrocytosis.

This was a prospective, single arm, open-label study. Five clinical dogs with ES and secondary erythrocytosis were included. New York Heart Association (NYHA) functional class, PCV, and pulmonary artery acceleration time to ejection time ratio (PA AT : ET) were evaluated before and after sildenafil therapy (0.5 mg/kg, bid). NYHA functional class was significantly improved after 1 (median 2; range 1–2, $P = 0.031$) and 3 months (median 2; range 1–2, $P = 0.031$) of sildenafil therapy, compared with the baseline (median 3, range 2–3). PCV was significantly decreased after 1 month ($62.4 \pm 4.7\%$, $P = 0.015$) and 3 months ($59.9 \pm 3.6\%$, $P = 0.01$) of therapy, compared with the baseline ($68.0 \pm 5.6\%$). AT : ET was significantly increased after 1 month of therapy (0.39 ± 0.06, $P = 0.013$) from the baseline (0.28 ± 0.04).

Sildenafil resolved the clinical signs and secondary erythrocytosis in dogs with ES. Sildenafil therapy could be the treatment of choice for dogs with ES.

ABSTRACT E-1

THE RENIN-ANGIOTENSIN-ALDOSTERONE SYSTEM (RAAS) AND HYPOTHALAMIC-PITUITARY-ADRENAL AXIS (HPAA) IN SEPTIC FOALS. K.A. Dembek[1]; K. Onasch[1]; R.J. Barsnick[1]; S.D. Hurcombe[1]; N.M. Slovis[2]; B. Barr[3]; R.E. Toribio[1]. [1]The Ohio State University, College of Veterinary Medicine; Columbus, OH, USA; [2]Hagyard Equine Medical Institute, Lexington, KY, USA; [3]Rood and Riddle Equine Hospital, Lexington, KY, USA.

Sepsis is the number one cause of mortality in neonatal foals. The role of the RAAS and HPAA in systemic inflammation and response to stress is well documented in critically ill human neonates, but limited information exists in foals. We hypothesized that in septic foals the RAAS and HPAA will be activated by systemic inflammation and hypoperfusion and the degree of activation will be associated with severity of sepsis and mortality.

Blood samples were collected on admission from 60 septic (sepsis score > 12), 102 sick non-septic (SNS), and 18 healthy foals of < 7 days of age. Blood concentrations of corticotropin-releasing hormone (CRH), adrenocorticotropin (ACTH), cortisol, aldosterone, angiotensin-II (ANG-II), arginine vasopressin (AVP) and plasma renin activity were determined by radioimmunoassays.

ACTH, cortisol, aldosterone, ANG-II and AVP concentrations were higher while CRH was lower in septic and SNS foals compared to healthy foals ($P < 0.05$). Septic non-survivor foals had higher concentrations of aldosterone, cortisol, ACTH and AVP and lower concentrations of ANG-II and CRH than survivors. AVP was associated with ANG-II in septic, and with ACTH in septic and SNS foals ($P < 0.05$). There was no difference in renin activity and ANG-II concentrations among foal groups. Septic foals had a higher ACTH:aldosterone ratio than healthy foals ($P < 0.001$).

This study shows that in response to sepsis there is RAAS and HPAA activation in critically ill foals. We propose that in sick foals AVP is more important than CRH in regulating ACTH secretion. The increased ACTH:aldosterone ratio further supports relative adrenal insufficiency in septic foals.

ABSTRACT E-2

EVALUATION OF COAGULATION AND BLOOD-DERIVED INFLAMMATORY BIOMARKERS IN HORSES WITH ACUTE ONSET OF DIARRHEA DURING HOSPITALIZATION. A Pacheco, D Bedenice, K Thane, CR Sharp. Cummings School of Veterinary Medicine at Tufts University, North Grafton, MA.

This prospective, cohort study aimed to characterize alterations in coagulation and blood-derived inflammatory biomarkers in adult horses that developed diarrhea during hospitalization. Physical and hematological parameters were evaluated at times 0 (onset of diarrhea), 6, 12, 24 and 48h, then every 48 h until resolution of diarrhea or death. Each hematological analysis included a complete blood count (CBC), thromboelastography (TEG), partial-thromboplastin-time (PTT), prothrombin-time (PT), plasma concentrations of lactate, tumor necrosis factor alpha (TNF-α), interleukin (IL)-1, IL-6, IL-10 and NT-proC-type-natriuretic peptide (pCNP). Horses were categorized into three groups based on the duration of diarrhea and evidence of systemic inflammation. Group 0: Diarrhea < 6 h without systemic inflammation (SI); Group 1 – Diarrhea $\geq$ 6 h without SI; Group 2– Diarrhea with SI. Assessment of vital parameters and CBC established a diagnosis of SI as previously described (Levy, 2003). Descriptive and univariate outcome analyses were based on data normality.

19 horses were enrolled, of which 16 (84.2%) survived to discharge. The mean age was 13.6+/-5.3 years. Eight horses (42.1%) were categorized as group-0, 2 (10.5%) as group-1 and 9 (47.4%) as group-2. Two horses developed thrombophlebitis. Based on the results of TEG, 6/19 (31.6%) were normocoagulable, 7/19(36.8%) were hypocoagulable and 6/19 (31.6%) were hypercoagulable, at one or more time points. Of these, 7/9 (77.8%) group-2 horses were coagulopathic. Additionally, group-2 horses had a significantly lower MA than group-0 horses at baseline (43.6 ± 16.7 vs. 61.9 ± 7.2) and 6h (45.9 ± 15.9 vs. 65.5 ± 5.8). Biomarker analyses are pending.

In conclusion, SI was associated with coagulation disorders in horses with hospital acquired diarrhea.

ABSTRACT E-3

DETECTION AND CHARACTERIZATION OF *CLOSTRIDIUM DIFFICILE* AND *CLOSTRIDIDIUM PERFRINGENS* IN GASTROINTESTINAL COMPARTMENTS OF HEALTHY HORSES. A Schoster, LG Arroyo, HR Staempfli, PE Shewen, JS Weese. Ontario Veterinary College, University of Guelph, Guelph, ON.

Clostridium difficile and *Clostridium perfringens* are commonly associated with colitis and diarrhea in equines but asymptomatic carriers exist. Reported carrier rates of toxigenic *C. difficile* and *C. perfringens* strains in feces range between 0–25% and 0–30% respectively. Toxigenic *C. difficile* has also been isolated from the small intestine of diseased foals and is implicated as etiologic agent of duodenitis/proximal jejunitis in adult horses however scarce information is available on prevalence in gastrointestinal compartments other than feces in healthy horses, and it is unclear whether fecal samples are good predictors of the status of proximal intestinal sites. The objectives of this study were to investigate the presence of *C. difficile* and *C. perfringens* in various intestinal compartments of healthy adult horses and to molecularly characterize isolates.

Intestinal contents were collected from the stomach, duodenum jejunum, ileum, cecum, right dorsal and left ventral colon, small colon and rectum of 10 euthanized horses free of apparent gastrointestinal disease. Enrichment culture was performed for *C. difficile* and *C. perfringens* and *C. difficile* isolates were further characterized via toxin gene detection and ribotyping.

C. difficile was isolated from 9/90 (10%) samples from 5/10 (50%) horses. Between zero and three sites were positive per horse, and multiple sites were positive in three horses. Isolates were recovered from duodenum (n = 1), right dorsal colon (n = 3), small colon (n = 1) and rectum (n = 4). In one horse, the rectal sample was negative but *C. difficile* was isolated from a proximal site, all other horses were positive on the rectal sample if a more proximal compartment was positive. In three horses multiple compartments were positive however different strains were always present within the same horse (n =

2). All isolates possessed genes encoding toxins A and B. Five isolates were ribotype 078 and also possessed genes encoding the binary toxin. The other isolates were ribotype 001 and were negative for the genes encoding the binary toxin. Despite using a method with a detection level as low as 9 cfu/g of feces, no *C. perfringens* was recovered.

Rectal samples were a good predictor of overall *C. difficile* carrier status (4/5 horses), however rectal samples were not always representative for the ribotype carried in more proximal compartments. The presence of variable strains within the same horse suggests transient passage of the bacterium through the gastrointestinal system rather than actual colonization although further study testing multiple colonies per site is needed. The predominance of ribotype 078 is consistent with recent emergence of this strain in this region, as earlier studies found other strains (027, 001) to be more prevalent and a variety of ribotypes were typically recovered from horses. Interestingly ribotype 078 has recently emerged as a hypervirulent strain in humans in our area. The absence of *C. perfringens* is not surprising as several studies have reported similar results for healthy adult horses other than broodmares.

ABSTRACT E-4

LONGITUDINAL STUDY OF *CLOSTRIDIUM DIFFICILE, CLOSTRIDIUM PERFRINGENS, SALMONELLA* AND *ESCHERICHIA COLI* IN HEALTHY HORSES.

A Schoster[1], LG Arroyo[1], HR Staempfli[1], P Shewen[1], LM Sigismondo[1], RJ Reid-Smith[1,2], N Janecko[1,2], JS Weese[1]. [1]Ontario Veterinary College, University of Guelph, Guelph, ON, [2]Public Health Agency of Canada, Guelph, ON.

Clostridium difficile, *Clostridium perfringens* and *Salmonella* are important enteric pathogens in horses, however some healthy animals also harbour these pathogens. Point prevalence studies have reported these carriage rates, but there are no data regarding longitudinal prevalence of these enteric bacteria, information that would be useful to better understand the epidemiology of these pathogens. Additionally, antimicrobial resistance is a pressing concern. Commensal *E.coli* is often used as an indicator organism to evaluate antimicrobial resistance of enteric bacteria, yet there are limited data from horses on farms. The objectives of this study were to longitudinally investigate the above enteric pathogens over the course of one year, molecularly characterize obtained isolates and determine the antibiotic susceptibility profile for *E. coli*.

Fecal samples were collected from 25 adult horses from five farms on a monthly basis over the course of one year. Selective cultures were performed for *C. difficile*, *C. perfringens*, *Salmonella* and *E. coli*. *C. difficile* isolates were characterized via toxin gene PCR and ribotyping. Broth microdilution was performed to assess antimicrobial susceptibility profiles of *E. coli*.

Clostridium difficile was isolated from 15/275 (5.45%) samples from 11/25 (44%) horses. Four horses were positive on more than one occasion, three were positive in two consecutive months. Different ribotypes were found in two of the latter horses. Most isolates were ribotype 078 (n = 6) with ribotype 001 (n = 5) and ribotype C (n = 4) also identified. Ribotypes 078 and C possessed genes encoding toxins A, B and binary toxin, while ribotype 001 only possessed toxin A and B genes. Despite a detection threshold of 9 cfu/g feces, *C. perfringens* was not detected in any samples, nor was *Salmonella*. *E coli* was isolated from 117/225 (52%) samples. Resistance to $\geq$1 antimicrobial was present in only 19/117 (16.9%) isolates. Multidrug resistance ($\geq$ 3 antibiotics) was present in 5/117 (4%). Most commonly, isolates were resistant to sulfisoxazole (17/117) and trimethoprim sulfamethoxazole (16/117).

The overall detection rate for toxigenic *C. difficile* in fecal samples of healthy horses was 5.4% which is consistent with previous studies. The cumulative prevalence of 44% was striking but only one horse shed the same strain for more than one month, indicating *C. difficile* shedding is a transient and dynamic state. The predominant isolation of ribotype 078 is consistent with the suspicion that this strain has emerged and become widely disseminated in this region in recent years. The low prevalence of *C. perfringens* and *Salmonella* is in agreement with some other studies. The low prevalence of antibiotic resistance in commensal *E. coli* was encouraging and suggests that healthy horses on pleasure horse farms are not likely a major reservoir of resistance in enteric bacteria.

ABSTRACT E-5

INFLUENCE OF ALLELE COPY NUMBER ON SKELETAL MUSCLE HISTOPATHOLOGY IN HORSES WITH TYPE 1 POLYSACCHARIDE STORAGE MYOPATHY.

R.J. Naylor[1], S. Johnstone[1], L. Livesey[2], J. Schumacher[2], V. Luis-Fuentes[1], C. Massey[1], N. Henke[2], M. Fernandez-Fuente[1], K. Brock[2] and R.J. Piercy[1]. [1]Comparative Neuromuscular Diseases Laboratory, Royal Veterinary College, London, UK. [2]Auburn University, AL.

Type 1 Polysaccharide Storage Myopathy (PSSM) in horses is associated with a dominant missense mutation (R309H) in the skeletal muscle glycogen synthase gene (*GYS1*). Since disease severity varies between affected horses, we hypothesised that some clinical variability could be accounted for by the underlying genotype. 107 Belgian / Percheron horses were genotyped using a validated restriction fragment length polymorphism assay enabling grouping of horses as homozygotes (HH), heterozygotes(HR) or normal (RR). Subsequently, *semimembranosis* muscle samples were biopsied from each of six matched sedentary horses from each group; one sample was formalin-fixed and one fresh frozen. Sections were stained using haematoxylin and eosin, periodic acid schiff +/- diastase. Anti-dystrophin, nNOS and myosin heavy chain immunohistochemistry was performed to examine sarcolemmal intregrity. There were significant differences in resting CK activity (P = 0.023) (median HH = 364U/l interquartile range(IR) 332–764; HR = 301U/l IR222-377; RR = 260U/l IR216-320) and AST activity (P < 0.001) (AST mean HH = 502U/l SD116; HR = 357U/l SD92; RR = 311U/l SD64) and muscle pathology between the 3 groups, with severity increasing RR < HR < HH. There were significantly more type 2a (P = 0.04) and fewer type 2x fibres (P = 0.02) in homozygotes (2a 55% SD 10.2; 2x 32% SD 18) compared with the other groups (HR 2a 38% SD 10.9, 2x 44% SD 10.8; RR: 2a 36.5% SD 12.4 2x 57% SD 11.5). More type 2a fibres contained polysaccharide inclusions in homozygotes (30% SD 11.1) than in heterozygotes (10.6% SD 6.9) (P < 0.001). Both dystrophin and nNOS expression was normally localised to the sarcolemma in pathologically normal and vacuolated fibres from mutant horses. In conclusion, sedentary homozygotes have more severe skeletal muscle pathology and higher resting plasma CK and AST activities than heterozygotes, and PSSM1 is associated with a fibre type shift towards type 2a. Although subsarcolemmal vacuolation likely disrupts the contractile apparatus's attachment to the sarcolemma, the latter's integrity appeared intact.

ABSTRACT E-6

FACTORS ASSOCIATED WITH SURVIVAL IN 161 RECUMBENT HORSES.

L Winfield, P Kass, K.G. Magdesian, J.E. Madigan, M. Aleman N Pusterla. School of Veterinary Medicine, University of California, Davis, CA.

The recumbent horse presents a logistic, diagnostic, and therapeutic challenge to the equine practitioner. There is currently very little data available on the prognosis and outcome of horses that are recumbent. Therefore, the purpose of this study was to investigate the outcome of hospitalized horses that had been recumbent in the field or in the hospital and the factors affecting their survival.

Records of horses admitted to the School of Veterinary Medicine, University of California Davis from January of 1995 to December of 2010 with a history of recumbency or horses that became recumbent while hospitalized were evaluated. A horse was defined as recumbent if it was unable to stand on its own. The medical record was examined for the following criteria: history pertaining to the current illness including treatment by the rDVM, breed, age, weight, date of presentation, physical and neurological examination findings, CBC and biochemical profile results, initial drugs administered on arrival, time spent recumbent, time spent in a sling, diagnosis, and hospitalization costs. Statistical analysis correlating factors associated with survival was performed using logistic regression.

Overall there were 112 non survivors and 49 survivors. Factors that favored survival included early initiation of treatment in the field by the rDVM, horses that tolerated a sling and spent more time in a sling, increased duration and costs of hospitalization, horses that were recumbent post anesthesia, and those recumbent due to disease of the musculoskeletal system. Factors that increased likelihood of non survival included horses that were ataxic on presentation, horses with increased BUN, horses that spent more time recumbent, those that did not tolerate a sling, and horses diagnosed with botulism and spinal cord disease.

In conclusion, this retrospective study demonstrated that both the cause of recumbency and the ability of horses to tolerate a sling had a direct effect on survival.

ABSTRACT E-7

PLASMA PEAK AND TROUGH GENTAMICIN CONCENTRATIONS IN HOSPITALIZED HORSES RECEIVING ONCE DAILY GENTAMICIN. JR Read[1], PA Wilkins[2], RD Nolen-Walston[1]. [1]University of Pennsylvania, New Bolton Center, Kennett Square, PA. [2]University of Illinois, Champaign-Urbana, IL.

Gentamicin is often used to provide gram negative antimicrobial coverage in horses at 6.6 mg/kg IV every 24 hours. Therapeutic drug monitoring in our hospital suggests larger doses are required in many clinical cases to achieve the desired concentration (8–10× minimum inhibitory concentration) for common bacterial isolates (peak target range 32–40 µg/ml). The aim of this study was to determine the correlation between gentamicin dose and plasma concentration in hospitalized horses receiving gentamicin treatment in order to identify an optimum dose range for this population. Review of records (2002–2010) identified 71 horses ≥ 3 months old receiving once-daily gentamicin with peak and trough assays performed (N = 107 sets). Spearman Rank Correlation coefficient analysis revealed a weak (R = 0.289) but statistically significant correlation (P = 0.003) between gentamicin dose and peak plasma concentration. Horses receiving 7.7–9.7 and > 9.7 mg/kg gentamicin (Groups 2 and 3) had higher median peaks (31 µg/ml) than horses receiving 5.6–7.6 mg/kg (Group 1; 25.7 µg/ml). Higher doses were more likely to result in peaks > 32 µg/ml (42 and 41%, Groups 2 and 3 respectively) than horses receiving 5.6–7.6 mg/kg (Group 1; 16%). All 22 hour post-gentamicin administration trough values were < 2 µg/ml. No correlation was found between dose and change in plasma creatinine during treatment, nor dose and trough level. These data suggest that gentamicin dosage in horses should be individually determined by therapeutic monitoring. Additionally, these data support an initial dose of 7.7–9.7 mg/kg IV every 24 hours in order to achieve desired peak concentration and an appropriately low trough concentration.

ABSTRACT E-8

CIRCULATING NEUTROPHIL PHAGOCYTOSIS AND KILLING OF *STREPTOCOCCUS EQUI* SPP. *EQUI* AND *ZOOEPIDEMICUS* IN HEAVES-AFFECTED AND CONTROL HORSES. J Vanderstock[1], MP Lecours[2], A Lavoie-Lamoureux[1], M Segura[2], M Gottschalk[2], JP Lavoie[1], D Jean[1]. [1]Département de sciences cliniques, Faculté de Médecine Vétérinaire, Université de Montréal, Saint-Hyacinthe, QC, Canada [2]Groupe de recherche sur les maladies infectieuses du porc (GREMIP), Faculté de Médecine Vétérinaire, Université de Montréal, Saint-Hyacinthe, QC, Canada.

Heaves is a common respiratory inflammatory disease, characterized by a pulmonary neutrophilia. This disease is also characterized by an activation of circulating neutrophils after antigen challenge but their specific role in heaves is not well understood. Also, there are anecdotal studies concerning heaves-affected horse to be more susceptible to infection. However, to our knowledge, the antibacterial host defense role mediated by circulating neutrophils was not investigated in heaves-affected horses. The objective of this study was to compare phagocytosis activity and bacterial killing by circulating blood neutrophils of heaves-affected and control horses.

Peripheral neutrophils were isolated from heaves-affected (n = 6) and control (n = 6) horses using a density gradient technique. The killing capacity was assessed by incubating neutrophils with *Streptococcus equi* spp *equi* and spp *zooepidemicus*. After 1 h of bacteria-neutrophil coculture, total viable bacterial cells were measured by quantitative plating. The phagocytosis was evaluated by flow cytometry using fluorescent beads and GFP-transformed *Streptococcus suis* suilysin-negative mutant strain.

Circulating neutrophils from heaves-affected horses showed a significant decrease in their killing capacity toward *S. zooepidemicus* (p = 0.046). A reduced, although not significant (p = 0.1), killing capacity of *S. equi* by these neutrophils was also observed. The phagocytosis activity was not different between groups.

This impairment of blood neutrophil bactericidal activity in heaves-affected horses could contribute to an increase susceptibility to infection.

ABSTRACT E-9

ASSOCIATION OF BODY CONDITION SCORE WITH METABOLIC REGULATORY GENE EXPRESSION AND MITOCHONDRIAL CONTENT IN SKELETAL MUSCLE OF HORSES. HE Banse, D McFarlane. Oklahoma State University Center for Veterinary Health Sciences, Stillwater, OK.

Obesity is a common disorder of the horse, with current prevalence estimated at 30%. In people, obesity is associated with dyslipidemia, insulin resistance, mitochondrial dysfunction and downregulation of lipid and glucose metabolic pathways. In the horse, obesity is similarly associated with insulin resistance and alterations in lipid profiles; however, metabolic regulatory gene expression profiles have not been fully characterized. We hypothesized that obese horses have decreased expression of metabolic regulatory genes and decreased mitochondrial content in skeletal muscle compared with non-obese horses.

Sixteen light breed horses, 2–27 years of age were included. Body condition score (n = 16) and neck circumference (n = 9) were recorded. Post-mortem biopsy samples of the semi-membranosus muscle were obtained. DNA and RNA were isolated. Relative expression of the metabolic genes, peroxisome proliferator activated receptor γ (PPARγ), PPARγ coactivator-1α (PGC-1α), fatty acid translocase (FAT) and estrogen related receptor α (ERRα) was determined by quantitative polymerase chain reaction (qPCR). Mitochondrial content was assessed by determining mitochondrial DNA/nuclear DNA ratio by qPCR, using NADH-dehydrogenase subunit 2 and cytochrome oxidase subunit 2 as mitochondrial genes and beta actin as the nuclear reference gene. Non-normal data was log transformed for analysis and a Pearson coefficient of correlation was calculated for relative gene expression, body condition score and neck circumference. A value of $p < 0.05$ was considered significant. Body condition score was strongly correlated with neck circumference (n = 9, r = 0.72, p = 0.03). Relative expression of ERRα and GLUT-4 increased with body condition score (ERRα: n = 16, r = 0.66, p = 0.005; GLUT-4: n = 16, r = 0.50, p = 0.05). Copy number of the mitochondrial genes (NADH-DH and COX-2) was not related to body condition score or metabolic gene expression. Expression of GLUT-4, ERRα, PPARγ, and FAT were strongly correlated to each other, but not PGC-1α. There was a strong trend towards correlation between PPARγ and PGC-1α in horses with body condition score > 6 (n = 6, r = 0.77, p = 0.07).

In this study, there was no change in mitochondrial content in obese horses. Assessment of mitochondrial function in obese horses and horses with EMS is under way. The strong correlation between PPARγ and PGC-1α observed only in horses with high body condition scores suggests this pathway is activated with obesity. The role of PPARγ and PGC-1α in equine obesity should be further investigated to determine their potential as therapeutic targets. Upregulation of ERRα and GLUT-4 in horses with increasing body condition score is unexpected, and may indicate a compensatory response to dysfunction of a downstream pathway. Further studies to better define the role of metabolic regulatory gene expression in obese horses and those with EMS are ongoing.

Previously presented at the 7th Annual Harold Hamm Diabetes Center Research Retreat, Oklahoma City, OK.

ABSTRACT E-10

CORRELATION AND REPEATABILITY OF GLUCOSE AND D-XYLOSE INTESTINAL ABSORPTION TESTS IN NORMAL HORSES. C. Cruz Villagrán, K. Kalck, S. Elliott, N. Frank, B. Rohrbach, A. Schuver. University of Tennessee College of Veterinary Medicine, Knoxville, TN.

Inflammatory bowel disease is a cause of weight loss, decreased performance, and colic in horses. This condition is difficult to diagnose and clinicians rely upon absorption tests to document malabsorption. The purpose of this study was to compare glucose and xylose absorption tests in normal horses and determine the repeatability of these procedures.

Eight horses received 500 mg/kg dextrose or D-xylose powder mixed as a 10% solution in water or water alone via nasogastric intubation on three different occasions within the same week for three consecutive weeks (9 tests/horse). A crossover design was employed and the order of treatments was randomized. Blood samples were collected at time = 0, 30, 60, 90, 120, 150, and 180 min. Data were analyzed by repeated measures ANOVA and t-tests.

Results showed that the xylose response over time differed significantly from the glucose response over time (test × time; $P < 0.001$). Mean time to maximum concentration differed ($P < 0.001$) between tests (glucose 90 min; xylose 60 min). Within-horse area under the curve, maximum concentration, and time to maximum concentration values for dextrose and xylose did not differ significantly when tests were repeated. Results indicate that glucose and xylose absorption tests are repeatable within the same horse, but plotted curves differ between tests, with peak concentrations occurring at a later time point for the glucose absorption test. We conclude that both tests provide repeatable measures of intestinal absorption, but glucose and xylose appear to differ in their rates of absorption and clearance.

ABSTRACT E-11

PROGNOSIS FOR RACING WITH CONSERVATIVE MANAGEMENT OF CERVICAL VERTEBRAL MALFORMATION IN THOROUGHBREDS: 119 CASES (2002-2010). CJ Hoffman and CK Clark. Peterson & Smith Equine Hospital, Ocala, FL.

The purpose of this study was to examine the records of a population of Thoroughbreds with cervical vertebral malformation (CVM) and to determine which factors have an effect on these horses achieving athletic function.

This was a retrospective case study of 119 Thoroughbreds with CVM treated medically from 2002 to 2010. Forty-one were euthanized after diagnosis, while the remaining 78 were discharged for treatment. Racing records were reviewed to determine which horses raced after treatment. Horses were separated into groups based on whether or not they raced. Medical records were reviewed, and results of neurologic examination, radiographic and laboratory findings, treatments, and outcome were assessed and compared between groups.

Twenty-one of 78 horses treated medically (27%) improved enough to race. Median neurologic grade between groups was significantly different ($p < 0.0001$), with a hind limb grade of 2.0 (range 1–3) for the raced group and 2.5 (range 0.5–4) for the unraced. Intravertebral sagittal ratios measured from standing lateral cervical radiographs were equivocal between groups. Radiographs of all horses were examined for kyphosis, dorsal over-riding arch, caudal epiphysitis, degenerative joint disease, cystic bone lesions, and cranial stenosis of the vertebral canal. Horses with kyphosis ($p = 0.0178$), degenerative joint disease ($p = 0.0497$), or cranial stenosis ($p = 0.0357$) at any site were less likely to return to racing.

Racing prognosis for horses with CVM treated conservatively is equivalent to that of those treated surgically as reported by Rush Moore et al (JAVMA, 1993). Radiographic changes and neurologic grade may help serve as indicators for whether a horse will respond to conservative therapy.

ABSTRACT E-12

TWO PAIN SCALES FOR HORSES WITH COLIC; CONSTRUCTION AND VALIDATION. Gila A Sutton[1], Dan Turner[2], Roee Dahan[1], Yuval Semrik[1], Menashe Soffer[1], Ora Paltiel[3]. [1]Koret School of Veterinary Medicine, The Hebrew University of Jerusalem, Rehovot, Israel. [2]Pediatric Gastroenterology Unit, Shaare Zedek Medical Center, The Hebrew University, of Jerusalem, Israel. [3]Braun School of Public Health, Hadassah-Hebrew University Medical School, Israel.

Since pain assessment is vital for management of colic, a valid, reliable and feasible tool for assessing the severity of acute abdominal pain in horses is urgently needed. Our aim was to construct and validate a behavior-based pain scale by methodology utilized in construction of pain scales in non-verbal humans. The project consisted of four stages. Firstly, behaviors to include in a scale were empirically identified. Thirty equine clinicians noted behaviors in each of 23 random film clips of horses with colic using a checklist. Nine behaviors (e.g. rolling, pawing, and flank watching) demonstrated good inter-observer agreement without bias (multi-rater kappas: 0.5–0.95). Secondly, the clinical judgment of experts was utilized to identify and to weight behaviors. Six expert clinicians independently expressed opinions as to which of 46 behaviors to include and the severity of pain they indicate. Two contending scales (Equine Acute Abdominal Pain Scales (EAAPS) 1 & 2) were constructed based on both the empirical and the judgmental approaches. Each included 12 identical behaviors with a 1–5 point score range; EAAPS-2 with gradations to some of the behaviors and EAAPS-1 without. In the third stage, blood cortisol and lactate levels and heart rate were shown to only approximate pain since they correlate poorly with degree of pain as assessed by visual analog scale (VAS) in 32 horses with colic and 8 controls (Spearman rho; lactate 0.359; cortisol 0.214; heart rate 0.261). Finally, reliability and validity of the pain scales were evaluated including constructs of pain; face validity, convergent and discriminate validity and extreme groups. Thirty of 40 films of horses with colic were randomly presented to 44 expert equine clinicians internationally who were randomly allocated into three groups to score pain; one group by both VAS and numerical rating scale (NRS)), and two groups, each by one of the two EAAPS scales. Inter-observer reliability of both EAAPS scales was excellent (Intraclass Correlation 0.8). Intra-observer reliability based on scores given for identical films demonstrated; 87% and 56% agreement, kappa 0.9 and 0.35, and Spearman's rho = 0.97 and 0.58 for EAAPS-1 and 2, respectively. Both scales varied by ≤ 1 score between observations. Face validity; each group reported their scale to be valid (67% & 81%). Convergent validity; the scales compared favorably with VAS/NRS scores (Spearman's rho: 0.84–0.89). Discriminate validity; correlation to heart rate, lactate and cortisol levels was predictably low (rho = 0.2–0.5). Extreme group validity; colic horses scored significantly higher than control horses; scores of 0.6–0.7 in controls versus 2.6–2.7 in cases. In conclusion, methodology established in human medicine but novel in veterinary medicine was used to construct and validate two clinically feasible equine abdominal pain severity scales that showed excellent reliability and validity. Further refinement of the EAAPS scale is advised prior to introduction into clinical practice.

ABSTRACT E-13

IS THE PHENOTYPE OF PROLAPSING AORTIC VALVE CUSPS DIFFERENT FROM NORMAL AND DISEASED ONES? GD Hallowell and IM Bowen. School of Veterinary Medicine and Science, University of Nottingham, UK.

Aortic valve prolapse (AVP) is a common echocardiographic finding in horses, but when compared with mitral valve prolapse in dogs, little is known about the natural progression of this condition. Previously published data has shown that echocardiographic identification of AVP in horses is reliable, diagnostic criteria have been established and that development occurs with training. The aims of this study were to evaluate the different RNA and protein expressions of smooth muscle actin (SMA), transforming growth factor-β_1 (TGF), nitric oxide synthase (NOS) and the concentrations of elastin and collagen in normal, prolapsing and diseased cusps to evaluate what structural changes may predispose them to prolapse. Valve cusps were harvested and processed from a group of 176 horses at a commercial abattoir following disease classification using echocardiography. Horses were aged 13.7 ± 6.7years, weighing 518 ± 51 Kg and with a median body condition score of 4/9. Cusps were collected in RNAlater® and stored at −80°C prior to processing. cDNA was produced from half a valve using a standard protocoland qRT-PCR performed to assess relative RNA expression of SMA, TGFβ_1, endothelial (eNOS) and inducible NOS (iNOS) and compared with the housekeeping gene 18S. A quarter of cusp was processed using an adapted commercial protocol to evaluate protein expression of SMA, TGF β_1, eNOS and compared to vimentin. Specific antibody binding was assessed with western blotting and protein expression evaluated using dot blots. The remaining quarter cusp was used to measure soluble collagen and elastin concentrations using commercial assays[3]. Statistical analyses included one way ANOVA with post-hoc Bonferoni, paired Student's T-test, linear and logistic regression. There was no effect of gender or age on any of the measurements. Valves from animals with AVP had lower expression of SMA and Elastin compared to normal and diseased valves, increased expression of TGFβ_1 and eNOS, whereas iNOS expression was greater than normal valves (Table 1). Collagen content of valves from horses with AVP was increased compared to normal but lower than horses with valve disease. Prolapsing cusps appear to be a different phenotype from diseased cusps. Further studies will help to elucidate the significance of these findings *in vivo*.

Table [Abstract E-13]: Displaying mean (±SD) RNA expression (displayed as ratio of 18S), protein density (displayed as ratio of vimentin) and soluble elastin and collagen concentrations (μg/mg valve) from equine AoV cusps.

	SMA expression		TGFβ_1 expression		eNOS expression		iNOS	Collagen	Elastin
Cusp	/18S	/vimentin	/18S	/vimentin	/18S	/vimentin	/18S	(μg/mg)	(μg/mg)
Normal	2.3 ± 0.2	1.1 ± 0.1	0.8 ± 0.1	1.4 ± 0.04	1.6 ± 0.3	1.1 ± 0.05	0.7 ± 0.2	66.8 ± 4.9	26.2 ± 1.6
Disease	2.8 ± 0.3*	1.3 ± 0.3*	0.8 ± 0.1	1.5 ± 0.1*	1.6 ± 0.6	1.3 ± 0.04	1.0 ± 0.2*	109.3 ± 7.4*	22.6 ± 1.1*
Prolapse	1.7 ± 0.2*ψ	0.8 ± 0.2*ψ	1.0 ± 0.1*	1.8 ± 0.2*ψ	1.7 ± 0.6*	1.5 ± 0.1*	0.9 ± 0.1*	87.4 ± 6.5*ψ	18.2 ± 1.0*

*denotes differences ($p < 0.05$) from normal cusps and ψdenotes differences from diseased cusps.

ABSTRACT E-14
CORRELATION OF HEART RATE AND ELECTROCARDIOGRAPHIC TIME INTERVALS TO BODY MASS IN HORSES AND PONIES.

CC Schwarzwald[1], M Kedo[2], RL Hamlin[3]. [1]Vetsuisse Faculty of the University of Zurich, Switzerland. [2]Equine Clinic Kottenforst, Wachtberg, Germany. [3]College of Veterinary Medicine, The Ohio State University, Columbus, OH.

A clear association between heart rate (HR) and body mass has been observed across a wide range of mammalian species. Furthermore, it is well known that electrocardiographic (ECG) time intervals vary with heart rate and body mass. Within the equine species, small breeds are generally thought to have higher heart rates than large breeds. However, despite the large differences in size among different equine breeds, there is little information about normal heart rates and normal ECG time intervals in horses and ponies of different body size. Similarly, the relationship between HR and body mass in dogs of various breeds and sizes is still under debate. The goal of this study was to investigate the relationship between heart rate and ECG time intervals to body mass in apparently healthy horses and ponies and to calculate normal ranges for different weight groups.

250 adult horses and ponies at an age of 5.5 (1–30) y [median (range)] and a body weight of 479 (46–1120) kg were included in the study. All animals were considered clinically healthy based on history and physical examination. A standard base-apex ECG was recorded at a speed of 25 (n = 4) or 50 mm/s (n = 246) using a multiparameter monitor (Datascope Passport). During the procedure, the horses were unsedated, standing quiet in a box stall. Mean HR over 15 sec was determined for each recording. The following ECG time intervals were measured in triplicate and averaged for further analyses: PQ interval, QRS duration, QT interval, and difference between QT and QRS (QT-QRS duration). The relationship between HR, ECG time intervals, and body mass was assessed using linear regression analyses. Normal ranges (2.5% to 97.5% percentile) were calculated for 5 different weight groups. The level of significance was $p = 0.05$.

Heart rate was inversely related to body mass ($p < 0.0001$, $R^2 = 0.122$). The PQ interval ($p < 0.0001$, $R^2 = 0.413$), QRS duration ($p < 0.0001$, $R^2 = 0.147$), QT interval ($p < 0.0001$, $R^2 = 0.089$), and QT-QRS duration ($p < 0.0001$, $R^2 = 0.028$) were directly related to body mass. Normal ranges for HR, PQ, QRS, and QT within the different weight groups were 36–64 bpm, 107–230 ms, 60–127 ms, 360–510 ms (< 200 kg); 28–68 bpm, 162–319 ms, 74–155 ms, 362–610 ms (200–399 kg); 28–60 bpm, 211–423 ms, 89–147 ms, 390–581 ms (400–599 kg); 28–54 bpm, 220–380 ms, 87–150 ms, 367–587 ms (600–799 kg); and 24–52 bpm, 240–463 ms, 87–140 ms, 437–533 ms (> 799 kg).

We conclude that in healthy horses there is a significant but weak relationship between body mass and HR and between body mass and ECG time intervals, respectively. This study therefore supports the hypothesis that within the equine species, small breeds have faster heart rates and shorter ECG time intervals than large breeds. Therefore, body mass has to be considered when comparing HR and ECG time intervals to normal ranges in horses.

ABSTRACT E-15
BIOACTIVITY OF PLASMA ACTH FROM HORSES WITH PPID COMPARED TO NORMAL HORSES.

M Cordero, B Shrauner, D McFarlane. Oklahoma State University, Stillwater, OK.

Horses with pituitary pars intermedia dysfunction (PPID) often have elevated plasma ACTH concentrations. However, PPID-affected horses rarely have resting serum cortisol levels above the reference range or adrenal hyperplasia. We hypothesized that this apparent dissociation between plasma ACTH levels and adrenal response in horses with PPID is due to the secretion of ACTH that is less biologically active than that from normal horses.

To test our hypothesis, a bioassay to evaluate ACTH activity was developed. Adrenocortical explants were harvested aseptically from normal horses at euthanasia and stimulated with plasma from healthy (n = 9) and PPID-affected horses (n = 11). The assay was performed three times with explants obtained from different horses. Cortisol secreted by the explants and plasma ACTH levels were measured with commercially available radioimmunoassays. Cortisol secretion stimulated by each sample was standardized to the respective explant protein concentration. Cortisol data was normalized for ACTH concentration in each plasma sample and expressed as a cortisol:protein:ACTH ratio. Ratios from horses with PPID and normal horses were compared by unpaired t-test.

Horses with PPID had significantly lower cortisol:protein:ACTH ratios compared to normal horses. (Assay 1: 0.06 ± 0.04 vs. 0.33 ± 0.11, $P < 0.001$; Assay 2: 0.06 ± 0.04 vs. 0.30 ± 0.10, $P < 0.001$; Assay 3: 0.07 ± 0.06 vs. 0.20 ± 0.13, $P < 0.01$). These results suggest that plasma ACTH from PPID horses is less biologically active than plasma ACTH from normal horses. Our findings give further insight into the pathophysiology of PPID and may aid in the development of novel diagnostic testing protocols.

ABSTRACT E-16
EVALUATION OF SKILLS AND KNOWLEDGE DESIRED BY POTENTIAL EMPLOYERS OF ENTRY-LEVEL ACVIM-LAIM DIPLOMATES AND SKILLS AND KNOWLEDGE TAUGHT DURING RESIDENCY TRAINING.

JR Middleton[1], ACVIM-LAIM Residency Training Committee. [1]University of Missouri, College of Veterinary Medicine, Columbia, MO.

An online survey was conducted to determine perceived needs of potential employers of new ACVIM-LAIM diplomates. The survey was designed as the first step in determining what is needed for success in the various sectors of practice employing ACVIM-LAIM diplomates.

Demographic and background data were collected using questions and drop-down menus on the first page. The survey evaluated 189 skills or concepts in 26 areas of veterinary practice. Participants answered 4 questions about each skill or concept using drop-down ranked lists. Those participants that had completed an ACVIM-LAIM training program were asked 3 additional questions about whether they were taught the skill or concept during their own residency. Data were collated and descriptive statistics calculated. The mean scores or frequencies of use for each skills or concepts were ranked to determine which of the skills or concepts were most important for an entry-level diplomate.

Eighty-eight individuals participated in the survey with 86 respondents being ACVIM diplomates, 1 respondent was not board-certified and 1 respondent was an ACT diplomate. Nineteen respondents were diplomates of ACVIM and an additional specialty. Eighty-three respondents had completed an ACVIM residency. The majority of respondents were in academia (65%) with 30% being in private practice. Equine specialists prevailed (53%) followed by mixed large animal (28%) and then food animal only specialists (13%). The distribution of years post-residency was slightly skewed toward younger diplomates, but overall there was a good distribution of diplomates across years of experience. Most respondents stated that they did not make hiring decisions in their practice. Competency in disciplines other than internal medicine was expected with ultrasonography and radiology being the most desirable followed by theriogenology and lameness. Surgical skills, both equine abdominal (10%) and food animal general surgery (11%) were considered important by some respondents. Thirty-six per cent of respondents thought that a new diplomate should expect to make < $50,000 per annum, while only 13% of respondents thought that a new diplomate should expect to make ≥ $90,000 per annum. Not all respondents answered questions on all skills or concepts. The mean number of skills or concepts evaluated was 86 (SD = 75) with only 18 respondents answering all 189. All skills or concepts evaluated were found to be at least somewhat important, were estimated to be used at least occasionally, were recommended for inclusion in training programs as core or elective, and some level of knowledge was expected. At least some of the respondents were taught each of the skills or concepts during their residency, practiced the skill or concept at least occasionally during their residency, and some degree of competency was expected at the time of completion of their residency. These data will provide a framework for designing future LAIM residency programs.

ABSTRACT E-17

PHARMACOKINETICS AND CLINICAL SAFETY OF ORAL PASTE FIROCOXIB, A COX-1 SPARING NSAID, IN FOALS. McConnico RS, Khodadad JL. Colorado State University College of Veterinary Medicine, Fort Collins, CO.

This study evaluated pharmacokinetics and clinical safety of an oral paste formulation of a commercially available COX1-sparing NSAID in clinically healthy pony foals in a randomized controlled clinical trial. Values for complete blood count, serum chemistry profile, urinalysis, pharmacokinetic assay, and gastric endoscopy were evaluated in eighteen Shetland pony foals treated with firocoxib (0.1 mg/kg, PO, q 24 h) or placebo for 14 days. Foals were divided into 3 treatment groups. Group 1 and 2 foals received firocoxib while a 3rd group was administered an oral placebo. Gastric endoscopy was performed on group 1 and 3 foals prior to treatment and on days 7 and 14 to monitor for the presence of gastric ulcers. Group 2 and 3 foals had blood and urine samples taken sequentially for pharmacokinetic analysis, CBC, serum chemistry evaluation, and urinalysis. Physical examinations were performed prior to treatment and daily for 17 days. Data were analyzed using anova and paired t-tests ($P < 0.05$). None of the foals presented adverse clinical effects. There were no significant changes in CBC, biochemical profiles within groups, or differences between groups. Pretreatment gastric endoscopy scores were not significantly different from evaluations at 7 and 14 days. Firocoxib was quickly absorbed with an observed maximum concentration at 2 hr, the first sampling interval, for the majority of animals. Firocoxib plasma concentrations decreased in a log-linear manner after reaching the maximum concentration and steady state concentrations were achieved by the 7th dose. Based on the sampling times after the final and 14th dose, an average half life of 1.3 days was estimated. Administration of firocoxib did not cause any adverse effects on gastrointestinal, or hematological or serum biochemical variables, appears to have been well tolerated, and follows a predictable pharmacokinetic pattern in 4–6 week old foals.

ABSTRACT E-18

INFECTION OF PBMC WITH NEUROPATHOGENIC EQUINE HERESVIRUS TYPE 1 Ab4 STRAIN INDUCES INTERFERON-ALPHA AND MODULATES INTERLEUKIN-10 PRODUCTION. Bettina Wagner[1], Christine Wimer[1], Heather Freer[1], Armando Damiani[2], Nikolaus Osterrieder[2] and Hollis Erb[1]. [1]Department of Population Medicine and Diagnostic Sciences, College of Veterinary Medicine, Cornell University, Ithaca, NY 14850, USA; [2]Institut für Virologie, Freie Universität Berlin, Philippstrasse 13, 10115 Berlin, Germany.

Equine herpesvirus 1 (EHV-1) is highly prevalent in most horse populations. Horses are routinely vaccinated against EHV-1, and neutralizing antibodies have helped to prevent disease. However, the USDA has recently classified EHV-1 myeloencephalopathy (EHM) as an emerging disease, in response to the apparent increase in incidence, morbidity, and mortality of EHM that suggests a change in virulence of the virus. It has been reported that cellular immune mechanisms, in particular cytotoxic T-cells (CTLs), are important in controlling EHV-1 viremia. Interferon-alpha (IFN-α) has a key function in innate immune regulation by inducing the differentiation and maturation of CTLs. Here, we investigated the influence of abortogenic (RacL11, NY03) and neuropathogenic (Ab4) EHV-1 virus strains on IFN-α, IL-4 and IL-10 secretion in equine PBMC. Equine PBMC were infected with RacL11, NY03 or Ab4 EHV-1 strains or kept in medium for 24 hours. IFN-α, IL-10 and IL-4 secretion was detected in the supernatants by a fluorescent bead-based cytokine assay. The production of IFN-α increased with increasing viral doses and similarly for all three EHV-1 strains. The production of the anti-inflammatory cytokine IL-10 was significantly decreased after Ab4 infection compared to RacL11 and NY03 strains at viral infection doses of MOI 0.3-1. At high doses (MOI 3), IL-10 production was suppressed by all three EHV-1 strains. The results suggested that abortogenic and neuropathogenic EHV-1 strains equally induce anti-viral IFN-α production in equine PBMC. They also illustrated the differences in the ability of EHV-1 strains to modulate anti-inflammatory IL-10. Neuropathogenic Ab4 strain had an increased potential to down-regulate IL-10 production suggesting specific viral mechanisms that interfere with the control of inflammation in the host. The variations in innate IL-10 secretion might influence the development of protective immunity and might offer an explanation why neuropathogenic Ab4 induces more severe disease, including myeloencephalopathy, than abortogenic EHV-1 strains.

Previously presented at a conference of research workers in animal disease.

ABSTRACT E-19

IMMUNIZATION OF FOALS WITH ATTENUATED STRAIN OF *SALMONELLA ENTERIC TYPHIMURIUM* CARRYING *vapA* GENE OF *RHODOCOCCUS EQUI*: CLINICAL AND IMMUNOLOGICAL EVALUATION. A. C. R. C. Porto[1], M. C. Barreira[2], W. R Fernandes[1]. [1]School of Veterinary Medicine and Animal Science, University of São Paulo, Brazil. [2]School of Medicine, University of São Paulo, Ribeirão Preto, Brazil.

Rhodoccocus equi is the major cause of pneumonia in foals during the first six months and control measures are frequently ineffective. Treatment protocols are long, expensive and do not always produce good results. Rhodococcosis prevention through immunization of foals using a safe and efficient vaccine is still a challenge. Recent studies are based on the use of the virulence associated protein A (*vapA*) which has been described as an important inducer of immunity against *R. equi*.

The present study evaluated the clinical and immune response of foals vaccinated with an attenuated strain of *S. enterica typhimurium* expressing *VapA* antigen (test group) or *S. enterica typhimurium* without the *vapA* gene (control group), previous to and following experimental challenge. Two experimental phases were established according to the immunization route: intranasal or oral vaccination up to 12 hrs following birth and at 14 days of age. The experimental and control groups were challenged on day 28 with a virulent stain of *R. equi*. Clinical examination, CBC and image complementary exams were used to evaluate the development of clinical signs. Immune response patterns were evaluated though immunoglobulin dosage, cytokine expression, lymphocyte proliferation assays, isolation of *R. equi* and cytological profiles of TBW.

Clinical manifestation was less intense in the test group during the second experimental phase, and death occurred only in the control group (2/3) and was due to *R. equi* pneumonia. The test group produced a more intense IgGb response when compared to controls however no statistical difference was observed. Lymphoproliferation and Th1 cytokine expression were higher in the test group. In contrast, controls produced an IL-4 response. Local IgA was significantly higher in animals immunized with *Salmonella* carrying *vapA*. Immunization protocols produced no severe toxic effects.

The vaccination of neonatal foals with *S. enterica typhimurium* expressing *VapA* was considered safe, produced efficient modulation of the immune response and is apparently able to protect against experimental *R.equi* infection.

ABSTRACT E-20

MYRISTOLATED ALANINE-RICH C-KINASE SUBSTRATE (MARCKS) IS AN IMPORTANT REGULATOR OF EQUINE NEUTROPHIL MIGRATION AND ADHESION. MK Sheats, KL Carren, EM Hefner, EJ Sung, KB Adler, SL Jones. North Carolina State University College of Veterinary Medicine, Raleigh, NC.

This study was conducted to test the hypothesis that the 32 kD protein, Myristolated Alanine-Rich C-Kinase Substrate (MARCKS), is involved in equine neutrophil migration and adhesion.

In other species, MARCKS phosphorylation and dephosphorylation causes the protein to cycle between the cell membrane and cytosol, respectively. To investigate MARCKS phosphorylation in horses, neutrophils were isolated from whole blood and stimulated with platelet activating factor (PAF), leukotriene B_4 (LTB_4) or phorbol myristate acetate (PMA). Western blot was performed using specific phospho-MARCKS and total MARCKS primary antibodies. These results determined MARCKS phosphorylation is maximal 30 seconds following stimulation and that dephosphorylation occurs within 3 minutes.

To investigate the requirement for MARCKS in equine neutrophil chemotaxis, isolated neutrophils were pre-treated with MANS (a cell permeant peptide identical to the N-terminal 24 amino acids of MARCKS), RNS (a control peptide) or vehicle control (VC) prior to a migration assay toward known neutrophil chemoattractants (LTB_4 or PAF). Pre-treatment of equine neutrophils with MANS significantly inhibited migration while RNS pre-treatment had no effect.

To investigate MARCKS requirement in equine neutrophil adhesion, MANS, RNS or VC treated cells were stimulated to adhere to Immulon 2 plates coated with 5% FBS. Pre-treatment of equine neutrophils with MANS significantly inhibited adhesion while RNS pre-treatment had no effect.

Inhibition of MARCKS using a cell permeant peptide identical to the protein's N-terminus significantly inhibited equine neutrophil adhesion and migration. These results indicate that MARCKS is an important regulator of equine neutrophil chemotaxis and represents a potential target for anti-inflammatory therapy.

ABSTRACT E-21

WHAT IS THE EFFECT OF HEAD ELEVATION AND BLINDFOLDING ON THE GAIT OF NON-NEUROLOGIC HORSES. E. Olsen[1,2], R.J. Piercy[2], P.H. Andersen[1], T. Pfau[2]. [1]Faculty of LIFE sciences, University of Copenhagen, Taastrup, Denmark. [2]Royal Veterinary College, North Mymms, United Kingdom.

Amongst other tests, a thorough neurologic examination of horses may include walking with the head elevated and during blindfolding, in order to help differentiate normal from abnormal and to help with neuroanatomically localising any lesion(s) i.e. in the ataxic horse. Consensus amongst equine neurologists suggests that gait abnormalities associated with these specific tests are often exacerbated in horses with underlying proprioceptive deficits however the effect of these tests on temporal gait characteristics in normal horses has not previously been assessed quantitatively. We hypothesized that head elevation or blindfolding, in comparison with walking in a straight line would result in a compensatory decrease in lateral (left front-on to left hind-on and right front-on to right hind-on) and diagonal coupling intervals (left front-on to right hind-on and right front-on to left hind-on) in normal horses.

Four Thoroughbreds without any history or clinical signs suggestive of neurological disease (age range 3 to 5 years) were included in the study. Retroreflective markers were applied to the withers, to the sacrum and to left and right tuber coxae; for each limb, lateral fetlock markers and dorsal and lateral hoof wall markers were used. A minimum of 3 trials each with 2–4 walk strides for each task were analysed as horses walked across an 8-force-plate runway[i] surrounded by a 12-camera kinematic system.[ii] Force-plate data were processed with semi-automated custom written Matlab[iii] scripts. Data were analysed with a mixed model using the statistical software R. There was a significant fixed effect of normal walk on a straight line and head elevation on left and right lateral coupling intervals ($p < 0.0001$) and of the left and right diagonal coupling intervals ($p < 0.0001$). There was no significant effect of blindfolding on neither lateral nor diagonal coupling intervals. The random effect of horse had no influence on the coupling intervals. The decrease of the lateral coupling intervals indicates a tendency towards a pacing gait during head elevation. We conclude that there is a significant change in temporal gait characteristics of non-neurologic horses when the head is elevated but not during blindfolding compared to normal walking. Current results suggest that pacing and increased variation in foot-placement during head elevation should be interpreted with caution however further work is required to determine whether the change differs between horses with neurological disease and non-neurologic disease.

[i]Kistler Instruments Ltd, Hampshire, UK, [ii]Qualisys AB, Gothenburg, Sweden, [iii] The Mathworks inc., Natick, MA, U.S.A.

ABSTRACT E-22

ARTICULAR CARTILAGE FROM HORSES AFFLICTED WITH HEREDITARY EQUINE REGIONAL DERMAL ASTHENIA (HERDA) EXHIBITS OSTEOARTHRITIC LESIONS AND ISOLATED CHONDROCYTES PRODUCE HIGH LEVELS OF THE PRO-INFLAMMATORY MEDIATOR PROSTAGLANDIN E2. LC Gardner[1], AM Rashmir-Raven[2], LF Heinecke[3], MW Grzanna[3], AC Mrozinski[3], AY Au[3,4], SL Ownby[3], CG Frondoza[3,5,6]. [1]Virginia-Maryland Regional College of Veterinary Medicine, Blacksburg, VA. [2]Michigan State University College of Veterinary Medicine, East Lansing, MI. [3]Nutramax Laboratories, Inc., Edgewood, MD. [4]Syracuse University, Syracuse, NY. [5]Johns Hopkins University, Baltimore, MD. [6]Mississippi State University, Mississippi State, MS.

Hereditary Equine Regional Dermal Asthenia (HERDA) is an autosomal recessive connective tissue disorder associated with a mutation in cyclophillin B that leads to impaired collagen folding, aberrant wound repair, and corneal abnormalities. It affects young Quarter Horses, Appaloosa, and Paints. HERDA shows similarities to the human hereditary connective tissue syndrome Ehlers Danlos (EDS). Many EDS patients suffer from joint pain and osteoarthritis (OA) as adults. The similarity between EDS and HERDA raises the question whether horses suffering from HERDA develop OA. In OA, excess production of inflammatory mediators such as prostaglandin E2 (PGE_2) activate enzymes that degrade cartilage as well as impede wound healing. The present study examined articular cartilage from yearling horses afflicted with HERDA. We hypothesized that chondrocytes from these horses are continually activated to produce inflammatory mediators. To test this hypothesis, articular cartilage from carpal and tarsal joints of HERDA horses were evaluated using histology. PGE_2 production by chondrocyte cultures was measured by ELISA and analyzed by one-way ANOVA, Tukey post-hoc test, $p < 0.05$ significance. We also determined the anti-inflammatory effects of an avocado/soybean unsaponifiables (ASU), glucosamine (GLU), and chondroitin sulfate (CS) mixture (ingredients found in Cosequin® ASU) and phenylbutazone (PBZ) on chondrocytes. Cosequin® ASU and PBZ are used alone or in combination for the management of OA. Chondrocyte cultures were incubated for 24 hrs with control media alone, a clinically relevant concentration of PBZ (4 μg/ml), or the combination of ASU (NMX1000®, 8.3 μg/ml) + GLU (FCHG49®, 11 μg/ml) + CS (TRH122®, 20 μg/ml).

Articular cartilage from joints of five HERDA-afflicted horses showed gross and histologic evidence of osteoarthritic lesions. Chondrocyte cultures from cartilage of horses suffering from HERDA spontaneously produced greater PGE_2 than chondrocytes from normal horses (> 1000-fold). PBZ significantly decreased PGE_2 production by ~90% ($p < 0.001$). The combination of ASU+GLU+CS also significantly reduced PGE_2 production by ~60% ($p < 0.001$). The present study supports anecdotal findings that horses suffering from HERDA are likely to develop OA. The inhibition of PGE_2 synthesis by ASU+GLU+CS suggests that this combination may be beneficial for the management of OA in HERDA.

Research supported by Nutramax Laboratories, Inc.

ABSTRACT E-24

CORTICOSTEROIDS TREATMENT SUPPRESSES THE EXPRESSION OF IL-17 IN THE LUNGS OF HORSES WITH INFLAMMATORY AIRWAY DISEASE (IAD). L Beekman[1], T Tohver[1], R Dardari[2], R Léguillette[1]. [1]Department of Veterinary Clinical and Diagnostic Sciences, Faculty of Veterinary Medicine and [2]Department of Microbiology and Infectious Diseases, Faculty of Medicine, University of Calgary, Calgary, Alberta, Canada.

Equine Inflammatory Airway Disease (IAD) is a common condition often treated empirically with corticosteroids. Gene expression analysis in the bronchoalveolar lavage fluid (BALF) may help understand the effects of corticosteroids in IAD. The first part of the study aimed at identifying reference genes in the BALF of IAD horses treated with corticosteroids. The second part of the study investigated the effects of dexamethasone and fluticasone propionate treatments on the mRNA expression of IL-1β, IL-4, IL-8 and IL-17.

The expression stability of seven candidate reference genes was determined in BALF taken pre- and post-treatment with dexamethasone and fluticasone propionate in horses with IAD. Primers' efficiencies were calculated using LinRegPCR. NormFinder, GeNorm and qBasePlus softwares were used to rank the genes according to their stability. GAPDH was the most stably expressed gene whereas B2M was the least stable gene. In addition, GeNorm analysis revealed that the number of genes required for optimal normalization was four (GAPDH, SDHA, HPRT, RPL32).

In the second part of the study the mRNA expression of IL-1β, IL-4, IL-8 and IL-17 was measured in BALF samples from seven IAD horses treated in a randomized cross-over design with dexamethasone (0.05 mg/kg SID, 15 days) or inhaled fluticasone propionate (3000 mcg BID with Aerohippus, 15 days). The BALF samples were taken at baseline and after each treatment period. There was no significant effect of the corticosteroids treatment on the mRNA expression of IL-1β, IL-4 and IL-8 in the BALF. The mRNA expression of IL-17 was suppressed by dexamethasone and fluticasone propionate treatments.

ABSTRACT E-25

THE EFFECT OF CLENBUTEROL ON TRACHEAL MUCOCILIARY TRANSPORT IN HORSES UNDERGOING SIMULATED LONG-DISTANCE SHIPPING. JL Norton, K Jackson, JW Chen, R Boston, RD Nolen-v. New Bolton Center, University of Pennsylvania School of Veterinary Medicine, Kennett Square, PA.

Pneumonia is observed in horses after long distance transportation in association with confinement of horses' head position leading to a reduction in tracheal mucociliary clearance time (TMCT). We hypothesize that clenbuterol, a beta-2 agonist shown to increase TMCT in the horse, will ameliorate the affect of a fixed head position on large airway contamination and inflammation in a long-distance shipping model. Six adult horses were enrolled in a cross-over design prospective study. Horses were housed with their heads in a fixed position for 48 hours to simulate long distance transport, and treated with clenbuterol (0.8 ug/kg PO q12 h) or a placebo starting 12 hours before simulated shipping. TMCT was measured using a charcoal clearance technique. Data were collected at baseline and 48 hours, and included TMCT, tracheal wash cytology and quantitative culture, rectal temperature, CBC, fibrinogen, and serum TNFα, IL-10 and IL-2 levels. There was a 3-week washout between study arms, and each horse served as its own control. The data was analyzed using regression analysis and Wilcoxon rank-sum tests. No statistically significant difference was seen between treatment and placebo groups for any of the variables investigated. TMCT did not differ after treatment (1.71 ± 0.64 cm/min) versus placebo (1.55 ± 0.82 cm/min; $P > 0.10$), and intratracheal bacterial counts were similar for treatment ($105 \times 10^3 \pm 42 \times 103$ cfu; $P > 0.10$) and placebo ($98 \times 10^3 \pm 41 \times 103$ cfu) groups. A reduction of tracheal β hemolytic *Streptococcus. spp.* after clenbuterol versus placebo was also non-significant (0% versus 33%; $P > 0.10$). In conclusion, treatment with clenbuterol does not appear to combat the deleterious effects of this long-term shipping model.

ABSTRACT E-26

ROLE OF AIRWAY HYPERTONICTY IN EXERCISE INDUCED AIRWAY INFLAMMATION. Sabrina L Cummings[1], Michael S. Davis[1], Katherine K. Williamson[2]. [1]Oklahoma State University Center for Veterinary Health Sciences, Stillwater, OK. [2]Land O Lake Purina Feed, Gray Summit, MO.

Breathing cold air during strenuous exercise is associated with airway inflammation. Under these conditions, warming and humidification of inspired air occurs in the lower respiratory tract resulting in mucosal cooling, desiccation, and hyperosmolarity. The purpose of this research was to test the hypothesis that airway hypertonicity causes inflammatory cell migration and alterations in cytokine expression associated with exercise induced airway inflammation.

Horses ($n = 9$) were examined in a randomized crossover design after exposure to hypertonic aerosols (5 minute nebulization with solutions of either isotonic or hypertonic mannitol). Airway leukocytes were harvested 5 and 24 hours post aerosol challenge via bronchoaveolar lavage, and were used to determine total and differential nucleated cell count and expression of cytokine-specific mRNA.

Hypertonic aerosol challenge resulted in an increase in total number of cells 5 hr after challenge, characterized by increased macrophage ($p = 0.04$) and neutrophil ($p = 0.03$) concentrations, but there was no effect on airway leukocyte concentrations 24 hours after nebulization. No significant changes in the relative quantity of mRNA for airway cytokines were noted at either time point.

These data demonstrate that transient airway hypertonicity can cause airway leukocyte influx and may be responsible for the airway inflammation commonly found in athletes that exercise in cold conditions. However, our data do not support the hypothesis that hypertonicity is the sole initiating cause of changes in cytokine expression secondary to cold weather exercise. It is likely that factors such as airway temperature, shear stress or epithelial damage also play a role in this phenomenon.

ABSTRACT E-27

THE IMPORTANCE OF ULTRASONOGRAPHIC PNEUMATOSIS INTESTINALIS IN EQUINE NEONATAL GASTROINTESTINAL DISEASE. C. Navas de Solis, J.E. Palmer, R.C. Boston, V.B. Reef. New Bolton Center, University of Pennsylvania, Kennet Square, PA.

We studied the importance of abdominal sonograms in neonatal foals suffering from gastrointestinal conditions. We hypothesized that there would be a subgroup of neonates with sonographically detectable pneumatosis intestinalis (PI) as a reflection of a necrotizing component of the disease.

Records of foals ≤ 7 days of age hospitalized between 2005 and 2009 with signs of gastrointestinal disease were evaluated ($N = 89$). The association of sonographic, clinical, pathological and clinicopathological signs with outcome and severity of disease was determined.

Pneumatosis intestinalis was imaged in 19 foals. Twenty-seven foals were classified as having necrotizing gastrointestinal disease based on the presence of gastrointestinal signs (colic, diarrhea, gastric reflux or abdominal distension) and PI detected sonographically (19), surgical (2) or pathological (6) evidence of gastrointestinal necrosis. There was a difference between overall survival rate (58%) and survival rate in foals with necrotizing disease (33%, $P = 0.005$) or foals with PI detected sonographically (37%, $P = 0.02$). Pneumatosis intestinalis was the only sonographic finding associated with outcome. Sonographic abnormalities in peritoneal fluid, stomach, duodenum, jejunum, cecum, umbilicus or the presence of meconium were associated ($P < 0.05$) with surrogates of severity of disease (hospitalization cost or days of hospitalization). Hypoproteinemia was associated with PI ($P = 0.02$). The presence of blood in the feces, reflux and abdominal distension was associated with necrotizing gastrointestinal disease ($P < 0.05$).

Abdominal sonograms have prognostic value in neonatal gastrointestinal disease. Pneumatosis intestinalis was a common sonographic sign that worsened the prognosis. The therapeutic implications of detecting a necrotizing component of the gastrointestinal disease deserve further study.

ABSTRACT E-28

DISTRIBUTION OF INSULIN RECEPTOR IN THE DIGITAL LAMINAE OF PONIES FED A HIGH-CARBOHYDRATE DIET: AN IMMUNOHISTOCHEMICAL STUDY. Burns TA,[1] Watts MR,[1] Geor RJ[2], McCutcheon LJ[2], Belknap JK[1]. [1]College of Veterinary Medicine, The Ohio State University, Columbus, OH. [2]College of Veterinary Medicine, Michigan State University, East Lansing, OH.

The interaction of insulin and the microvascular endothelial insulin receptor (IRc) plays an important role in the normal and insulin resistant (IR) individual. While endothelial IRc signaling is normally vasodilatory, this effect is well-documented to reverse in the IR individual, resulting in vasoconstriction. Although vascular dysfunction has been reported in sepsis-associated equine laminitis, the role of the laminar microvasculature in endocrinopathic laminitis remains poorly characterized. The purpose of this study was to characterize the pattern of IRc expression in digital laminae in ponies subjected to a dietary carbohydrate challenge that mimicked abrupt exposure to pasture rich in nonstructural carbohydrates (NSC). Mixed-breed ponies (body weight 270.9 +/- 74.4 kg) received a diet of hay chop (NSC ~6% on a DM basis) for 4 weeks prior to initiation of the experimental feeding protocol. Following conditioning, ponies either remained on the control diet (n = 11) or received the same diet supplemented with sweet feed and oligofructose (total diet ~42% NSC; n = 11) for a period of 7 days. Serum insulin concentrations were measured prior to and after completion of the feeding protocol. At the end of the feeding protocol, sections of numerous tissues, including dorsal digital laminae, were collected immediately following euthanasia. The samples were formalin-fixed for 48 hours, transferred to 70% ethanol, and paraffin-embedded. Laminar sections were stained immunohistochemically for IRc using a commercially-available antibody (Abcam); the number of IRc (+) cells was quantified in 40x light microscopy fields (n = 10) for each section. The total number of IRc (+) cells was greater in the laminae of challenged ponies than control ponies ($p = 0.0096$), and there was a significant correlation between the change in serum basal insulin concentration and number of laminar IRc (+) endothelial cells ($r = 0.74$; $p < 0.05$). While the number of IRc (+) endothelial cells was significantly greater in the dermal laminae of challenged ponies ($p = 0.0095$), there was no difference in the number of interstitial IRc (+) cells ($p = 0.82$). No epithelial IRc (+) cells were observed in any laminar section, and IRc (+) cells were conspicuously absent from the deep dermal tissue (including vessels). Up-regulation of IRc expression in the laminar vasculature occurs acutely in response to dietary carbohydrate challenge and accompanies hyperinsulinemia in ponies. The dramatic increase in endothelial IRc expression in the laminar microvasculature in nutritionally challenged ponies, with no apparent epithelial IRc present, suggests that hyperinsulinemia associated with exposure to increased dietary NSC may induce laminar injury by causing a similar vasoconstriction in IR equids as described in the microvasculature of IR humans.

ABSTRACT E-29

THE NOVEL ROLE OF GLUCOSE TRANSPORTER 12 IN INSULIN-SENSITIVE TISSUES OF HEALTHY AND INSULIN-RESISTANT HORSES. KR Kohler[1], AP Waller[1], TA Burns[2], MC Mudge[2], JK Belknap[2], VA Lacombe[1,2]. [1]College of Pharmacy. [2]College of Veterinary Medicine, The Ohio State University, Columbus, OH.

Glucose transport from the blood stream into cells, the limiting step in whole-body glucose utilization, is regulated by a family of glucose transporter (GLUT) proteins in insulin-sensitive (i.e., muscle and adipose) tissues. We previously demonstrated that GLUT4, the major isoform, is a key factor in the pathogenesis of equine insulin resistance (IR). While it has been recently demonstrated that GLUT12 (a newly discovered isoform) increases insulin-stimulated glucose transport in human muscle, its role in other tissues, particularly in the setting of IR, is not well characterized in any species. In addition, AS160 has recently emerged as a key downstream signaling molecule regulating translocation of GLUT to the cell surface, the rate-limiting step in glucose uptake. We hypothesized that GLUT12 content would be differentially expressed across tissues and that IR would induce alteration in glucose transport by affecting active cell surface GLUT12.

Biopsies of skeletal muscle, and subcutaneous and visceral adipose tissue were collected from light-breed horses, characterized as either insulin sensitive or compensated IR, based on the results of an insulin-modified frequently-sampled intravenous glucose tolerance test (n = 5/group). We specifically quantified active cell-surface GLUT12 in these biopsies, using an innovative exofacial bis-mannose photolabeled assay, which has not been previously applied to GLUT12. Total GLUT12 protein expression was measured by Western blots, as well as total and phosphorylated (indicating activation of) AS160.

GLUT12 was expressed in all the depots with a significant regional effect. Total GLUT12 protein content was increased ($P < 0.05$) in visceral (omental and mesenteric) compared to subcutaneous (nuchal ligament and tailhead) adipose tissue and skeletal muscle of healthy horses. IR did not induce alterations in active cell-surface and total GLUT12 content nor in total and phosphorylated AS160 in any of the tissues evaluated.

Our data suggests that GLUT12 is abundant in visceral adipose tissue and is therefore likely to play a substantial role in the regulation of glucose transport. However, neither GLUT12 translocation nor AS160 activation are impaired in insulin-sensitive tissues of IR horses. It is concluded that, in contrast with GLUT4, GLUT12 does not appear to contribute to glucose transport alterations during naturally-occurring equine IR.

ABSTRACT E-30

EFFECTS OF AGE AND DIET ON GLYCEMIC AND INSULINEMIC RESPONSES OF HORSES TO A CONCENTRATE MEAL. J. Rapson[1], HC Schott II[1], BD Nielsen[2], LJ McCutcheon[3], P Harris[4], and RJ Geor[1]. [1]Department of Large Animal Clinical Sciences; [2]Department of Animal Science; and [3]Department of Pathobiology and Diagnostic Investigation, Michigan State University, College of Veterinary Medicine, East Lansing, MI; and [4]WALTHAM Centre for Pet Nutrition, Leicestershire, England.

Insulin resistance (IR), characterized by exaggerated glycemic or insulinemic responses to glucose challenge, is a key metabolic disturbance in horses that develop obesity-associated laminitis. In addition to obesity, diet and age have been demonstrated to affect tissue sensitivity to insulin in other species but these factors have received limited investigation in horses. We hypothesized that there would be greater glycemic and insulinemic responses to a sweet feed meal in aged horses, as compared to adult horses, as well as in horses adapted to a forage-only diet. Three diets, grass hay only, grass hay plus sweet feed (starch and sugar-rich, SS), and grass hay plus a fat and fiber (FF) feed, were fed to 17 mares, 8 adult (5–12 yr) and 9 aged (> 19 yr), for a 6-week adaptation period in a randomized design. Glycemic and insulinemic responses to a standardized meal of sweet feed (4 g/kg BW offered for 1 hour) were determined for 6 hours from the onset of feeding. Peak glucose and insulin concentrations and areas under the glucose or insulin vs. time curves (AUC-G, mg/dl/360 min, and AUC-I, μU/ml/360 min) were determined and data were analyzed by one- and two- factor repeated measures ANOVA. There were no differences between age groups in glycemic responses to any of the diets. However, in aged horses peak glucose concentration ($p < 0.03$) and AUC-G ($p < 0.01$) were greater after adaptation to the forage-only diet, as compared to the other two diets. In contrast, aged horses had a greater peak insulin concentration ($p < 0.05$) and AUC-I ($p < 0.03$) than adult horses on all diets but no differences in peak insulin concentration or AUC-I was found between diets within age groups.

As hypothesized, the insulin response, but not the glycemic response, to a sweet feed meal was greater in aged horses, regardless of background diet. Further, the glycemic response was greatest after adaptation to a forage-only diet, but this finding was only significant in aged horses.

Table [Abstract E-30]: Glycemic (AUC-G) and insulinemic (AUC-I) responses to a standardized meal of sweet feed (4 g/kg) in two age groups of horses after adaptation to 3 diets.

Group	Measure	Hay	Hay + FF	Hay + SS	P value
Adult	AUC-G	8360 ± 1440^a	8090 ± 1500^a	4880 ± 745^a	<0.093
Aged	AUC-G	11300 ± 1940^a	7370 ± 765^b	5280 ± 777^b	<0.006
Adult	AUC-I	11800 ± 3160^a	11600 ± 2240^a	7000 ± 1260^a	<0.108
Aged	AUC-I	20900 ± 2950^a	22300 ± 3900^a	21300 ± 5600^a	<0.903

[a]different superscripts across a row indicate significant differences.

ABSTRACT E-31

EVALUATION OF THE RESPONSE OF THE ENDOCRINE PANCREAS TO SHORT-TERM HIGH-CARBOHYDRATE FEEDING IN MIXED-BREED PONIES. Burns TA,[1] Watts MR,[1] Geor RJ[2], McCutcheon LJ[2], Belknap JK[1]. [1]College of Veterinary Medicine, The Ohio State University, Columbus, OH. [2]College of Veterinary Medicine, Michigan State University, East Lansing, OH.

Obesity, insulin resistance, and endocrinopathic laminitis have emerged as critical issues in equine medicine, causing significant morbidity, mortality, and economic loss to the equine industry. In obese humans and rodent models of nutritional obesity, systemic insulin resistance and hyperinsulinemia are followed temporally in a majority of individuals by decreased glucose tolerance, pancreatic β-cell failure, and type II diabetes mellitus. In stark contrast to humans, obese horses and ponies chronically remain in what is termed a "prediabetic" state in human IR, characterized by hyperinsulinemic euglycemia. Few data exist describing the biology of the equine endocrine pancreas in the chronically IR animal that may both: 1) explain this unique equine endocrine physiology and 2) characterize the animal at-risk for hyperinsulinemia-associated laminitis. The purpose of the study reported here was to characterize the morphology and physiology of the equine endocrine pancreas in response to a dietary carbohydrate challenge. Twenty-two mixed-breed ponies (body weight 266.6 ± 170.5 kg) were conditioned to a diet of chopped hay (NSC ~6% on DM basis) for 4 weeks; following conditioning, ponies either remained on the control diet (n = 11), or received the same hay supplemented with sweet feed and oligofructose (total diet ~42% NSC; n = 11) for 7 days. Serum insulin concentrations were measured prior to and after completion of the feeding protocol. At the end of the feeding protocol, sections of numerous tissues, including pancreas, were collected immediately following euthanasia. The samples were formalin-fixed for 48 hours, transferred to 70% ethanol, and paraffin-embedded. Immunohistochemistry was performed on pancreas sections using a commercially-available anti-insulin antibody (Abcam), and measurements of islet surface area and β-cell surface area were performed (n = 10 islets per tissue section) using a commercially-available computer software program (Image J). There was a trend for greater total islet surface area in pancreatic tissue from ponies fed the high NSC diet when compared to the ponies on the hay diet (p = 0.068); however, no difference was noted in β-cell surface area between diet treatments (p = 0.12). The change in serum insulin concentration was significantly greater in the high NSC-fed ponies than in controls (403.8 +/- 317.1 mIU/L vs. 1.00 +/- 4.03 mIU/L; p = 0.002); however, this variable was not correlated with total islet surface area (r = 0.32; p = 0.17) or β-cell surface area (r = 0.25; p = 0.3). Due to the relatively modest changes in pancreatic islet surface area that accompany marked increases in serum insulin concentrations in ponies fed a high NSC diet, it is important to assess both β-cell function and insulin clearance mechanisms in future studies to delineate the mechanism(s) of hyperinsulinemia in this model.

ABSTRACT E-32

INFLAMMATORY RESPONSES TO INTRAVENOUS LIPOPOLYSACCHARIDE INFUSION IN HORSES WITH EQUINE METABOLIC SYNDROME. L Tadros, N Frank. Department of Large Animal Clinical Sciences, College of Veterinary Medicine, University of Tennessee, Knoxville, TN.

Humans that suffer from obesity show exaggerated inflammatory responses and this may be relevant to the association between increased adiposity and laminitis in horses with Equine Metabolic Syndrome (EMS). This study was performed to test the hypothesis that inflammatory responses to endotoxemia differ between healthy horses and those affected by EMS.

Six healthy adult mares and 6 horses with EMS received an intravenous infusion of lipopolysaccharide (LPS; 20 ng/kg in 60 mL sterile saline) or saline alone. A crossover design was employed with a 7-day washout period. Physical examinations were performed hourly for 9h and whole blood was collected at 30, 60, 90, 120, 180, and 240 min for assessment of inflammatory cytokine gene expression. A liver biopsy was performed between 240 and 360 min post-infusion. Data were analyzed using mixed model ANOVA.

Mean rectal temperature, heart rate, and respiratory rate increased following LPS infusion (treatment × time; $P < 0.001$), with higher heart (group × treatment; $P = 0.087$) and respiratory rates (group; $P = 0.017$) detected in EMS horses. Lipopolysaccharide infusion significantly increased whole blood gene expression of tumor necrosis factor α (TNFα), interleukin (IL)-1β ($P < 0.001$), IL-6 ($P < 0.001$), IL-8 ($P < 0.001$), and IL-10 ($P = 0.002$), and hepatic gene expression of IL-6 ($P < 0.001$), IL-8 ($P < 0.001$), and IL-10 ($P = 0.016$). Inflammatory gene expression did not differ significantly between groups, so our hypothesis was not supported. Heart rates tended to be higher when LPS was administered to horses with EMS.

ABSTRACT E-33

PHARMACOKINETICS OF INTRAVENOUSLY ADMINISTERED EQUINE CARDIAC TROPONIN I IN HEALTHY PONIES. MS Kraus[1], BB Kaufer[2], A Damiani[2], N Osterrieder[1,2], M Rishniw[1], W Schwark[1], AR Gelzer[1], TJ Divers[1]. [1]Cornell University, Ithaca, NY, [2]Institut für Virologie, Freie Universität Berlin, Berlin, Germany.

Elevated serum concentration of cardiac troponin I (cTnI) is a biomarker for myocardial damage in horses. Preferred times to test blood for cTnI levels following athletic performance or other events that may cause myocardial injury are not yet established and would be affected by time of release from the myocytes, location of release within the myocytes, duration of release and half-life of cTnI in the horse. This information would be necessary to more accurately and reliably test horses for myocardial injury. The aim of this study was to determine the elimination half-life (T1/2) of equine cTnI. To establish the T1/2 of equine cTnI in horses, cTnI was recombinantly expressed in E.coli. Two healthy ponies received intravenous injections of recombinant equine cTnI and plasma cTnI concentrations were measured with a point-of-care cTnI analyzer at multiple time points after injection. Standard pharmacokinetic analysis was performed to establish the elimination half-life of cTnI. For comparative purposes, data were subjected to pharmacokinetic models describing a single versus biphasic elimination profile. Elimination of recombinant equine cTnI following intravenous administration exhibits a short half-life. Establishing the T1/2 of troponin provides critical information in understanding the clinical application of this cardiac biomarker in clinical practice. This study describes a true biological cTnI T1/2, which has not been documented in any species thus far. Stall-side assessment of this cardiac biomarker in horses should enhance the ability of clinicians to detect myocardial damage and aid in the management and treatment of horses with cardiac disease.

ABSTRACT E-34

FLOW-MEDIATED VASODILATION IN HEALTHY PONIES. EJT Finding, ID Jones, V Luis Fuentes, NJ Menzies-Gow. Royal Veterinary College, Hatfield, UK.

The objective of the study was to evaluate the between-pony, within-pony, between-analyser and within-analyser variation of flow-mediated vasodilation (FMD) measurement in healthy ponies, to investigate the hypothesis that FMD occurs in healthy ponies.

Six healthy, native breed, unrelated pony mares of varying weight (236–406 kg), body condition score (3/9–7/9) and age (14–25 years) were used. The median artery was occluded for 5 minutes. Two-dimensional (2D) ultrasonographic images of the artery were recorded for 30 seconds prior to and for 2 minutes after occlusion. The peak luminal diameter was compared to baseline diameter to calculate the relative percentage increase in luminal size (FMD). Images were obtained from six ponies on one occasion and from one pony on six occasions. Analysis of images was performed by two independent analysers and by one analyser twice.

The mean (SD) FMD in 6 ponies was 12.57% (4.28%) and in 1 pony (6 occasions) was 7.30% (2.11%). Coefficients of variation were 34.09% and 28.84% respectively. Agreement between analysers was fair (ICC = 0.47) and within analyser was poor (ICC = 0.30).

FMD is used to assess endothelial function in humans and has recently been assessed for its use in canine subjects. FMD occurs and measurement is feasible in ponies. FMD could be used to assess endothelial function, in the context of laminitis or other cardiovascular diseases.

ABSTRACT E-35

HOW TO OBTAIN INDIRECT BLOOD PRESSURE IN THE STANDING HORSE. E. Olsen, C. Kronborg, R. Buhl, P.H. Andersen. Faculty of LIFE sciences, University of Copenhagen, Taastrup, Denmark.

Current state-of-the-art technique for measuring blood pressure (BP) in the horse is invasive and involves cannulation of the facial artery. Indirect techniques, such as oscillometry, have proven useful in the anaesthetised horse, but have not become routine in the standing horse. Monitoring BP can be indicated for the diagnosis and treatment of the hypotensive patient (ie. caused by endotoxemia, hypovolemia, Systemic Inflammatory Response Syndrome and cardiac failure) or the hypertensive patient (ie. due to equine metabolic syndrome or pain). The objective of this study was therefore to a) describe the methodology for application of oscillometric BP using a cuff applied to the tail in the standing horse and b) and to determine accuracy and precision of this method applied to the normotensive standing horse.

The oscillometric method is simple to apply in a clinical setting. A pneumatic cuff is snugly applied to the unclipped tail-base with the cuff bladder centered over the middle coccygeal artery. The tail circumference must match the manufacturers description of the cuffs diameter range. The oscillometric apparatus inflates the cuff and obtain systolic, diastolic and mean arterial BP (SAP, DAP and MAP). At least 2 consecutive measurements must be obtained. A correction of 0.7 mmHg/cm vertical distance between cuff and heart level is added to the measurement to correct for hydrostatic pressure difference.

For determination of accuracy and precision of indirect SAP, DAP and MAP, eight healthy horses (age 3 to 16 years), was equipped with an intra-arterial catheter[ii] in the facial artery and a commercial tail-cuff oscillometric apparatus.[i] Measurements were recorded every 2 minutes for 20 minutes. The data were analysed with the statistical software R using a mixed model with repeated measurements and a Bland-Altman analysis corrected for repeated measurements.

Oscillometric BP was accurate and precise for MAP (mean bias, lower confidence level, upper confidence level, variation in difference, all mmHg) (−0.3, −18.5, 19.1, 33.2, respectively) in the conscious horse but not for SAP (−1.5, −19.3, 16.3, 38.2, respectively) and DAP (0.05, 15.9, 16.0, 49, respectively). There was no significant contribution to the statistical model of either horse or measurement number.

All horses tolerated the tail-cuff well and the method was simple to apply. Only MAP could be measured with acceptable accuracy and precision in the normotensive standing horse using the described oscillometric method.

[i]S5 Datex-Ohmeda, GE Healthcare, General Electric Company, Denmark & Datex-Ohmeda division, Finland.

[ii]Intra-arterial cannula with FlowSwitch, 20G, 1.10 mm × 45 mm, Becton, Dickinson and Company, Brøndby, Denmark.

ABSTRACT E-36

THYROID HORMONE CONCENTRATIONS DIFFER BETWEEN DONKEYS AND HORSES.

FJ Mendoza[1], RA Perez-Ecija[1], RE Toribio[2], JC Estepa[1]. [1]Dept. Animal Medicine and Surgery, Cordoba, Spain. [2]The Ohio State University, College of Veterinary Medicine, Columbus, OH, USA.

Reference intervals for thyroid hormone (TH) concentrations have not been established for donkeys. Therefore, clinicians must use reference ranges from horses, potentially leading to misdiagnosis of thyroid diseases. We hypothesized that TH concentrations are different between donkeys and horses. The purposes of this study were: a) to compare TH concentrations between donkeys and horses and, b) to determine whether the age may influence TH concentrations. Thirty-eight healthy donkeys (8.5 ± 0.8 years), mixed breeds, and 20 healthy Andalusian horses (6.4 ± 0.5 years) were used. Donkeys were divided into three groups: < 5 years (n = 13), 5-10 years (n = 12), and >11 years (n = 13). Serum concentrations of total triiodothyronine (tT3), free triiodothyronine (fT3), total thyroxine (tT4), free thyroxine (fT4), reverse triiodothyronine (rT3) and thyroid-stimulating hormone (TSH) were quantified by radioimmunoassay. All blood samples were collected the same day. Neither horses nor donkeys had received any treatment for 30 days before sampling and both farms had similar production conditions. Total T3, fT3, fT4 and tT4 concentrations were higher ($P < 0.01$) in donkeys than horses. In contrast, no statistical differences were found for rT3 and TSH concentrations. Young donkeys (< 5 years) had higher fT4, tT4 and rT3 concentrations compared to other donkey groups ($P < 0.05$). Old donkeys (>11 years) had lower tT3 and fT3 concentrations than both younger donkeys groups ($P < 0.05$). This study shows that there are differences in TH concentrations between donkeys and horses, raising awareness on the possibility of misdiagnosis of thyroid gland dysfunction when using values from horses, being necessary to determine exclusive reference intervals for donkeys.

ABSTRACT E-37

THE EFFECT OF OVARIECTOMY ON MEASURES OF INSULIN SENSITIVITY AND RESPONSE TO DEXAMETHASONE IN MARES.

F.R. Bertin, K. Pader, T.B. Lescun, L. Freeman, J.E. Sojka-Kritchevsky. Purdue University, West-Lafayette, IN.

Ovariectomy is associated with alterations of responses to many hormones, not just those associated with reproductive function. In humans and rats, ovariectomy leads to insulin resistance, increased adiposity and altered fat mobilization. The effects of ovariectomy on energy metabolism have not been reported in horses. Ovariectomized mares have been shown to respond normally to an ACTH stimulation test, but the response to suppression of the hypothalamo-pituitary-adrenal axis has not been previously described. The aim of this study was to evaluate the effect of ovariectomy on insulin response in mares and to determine if mares exhibit alterations in response to dexamethasone administration after ovariectomy.

Six healthy mares underwent an intravenous glucose tolerance test (IVGTT), an insulin sensitivity test (IST) and a dexamethasone suppression test (DST) before and 5 weeks after bilateral ovariectomy. Body weight, cortisol values at baseline, 15 and 24 hours after dexamethasone injection and ACTH values at baseline, 15 and 24 hours after dexamethasone injection, basal insulin/glucose ratio, time to reach a 60% decrease in blood glucose in the IST, time to reach baseline glucose concentration in the IVGTT and area under the curves plotting blood glucose and time to injection of glucose or insulin were compared before and after ovariectomy using a paired *t*-test or an ANOVA for repeated measures. Significance level was $P < 0.05$.

Average body weight was decreased after surgery (6kg [3–48]). The injection of dexamethasone resulted in a serum cortisol concentration of less than 1 µg/dL in all mares before ovariectomy, whereas after ovariectomy, dexamethasone injection resulted in a serum cortisol concentration of less than 1 µg/dL in 5 out of 6 mares. In all cases, ACTH concentration was within the reference range (9–35 pg/mL) before and after ovariectomy. However, ACTH concentrations at t_0 and at t_{15} were significantly higher after ovariectomy. Each mare had a normal IVGTT, both before and after ovariectomy. Additionally, no significant differences were observed in basal blood glucose (84 ± 12 mg/dL before and 85 ± 3 mg/dL after) or in the time to reach glucose baseline (108 ± 66 min before and 99 ± 51 min after). Serum basal insulin concentration and insulin/glucose ratio was not significantly different before or after ovariectomy (22.0 ± 7.9 µIU/mL and 17.5 ± 8.9 µIU/mL and 0.26 ± 0.09 and 0.20 ± 0.10, respectively), nor was the average time to reach a 60% decrease in blood glucose after insulin injection (30 ± 7 min and 25 ± 9 min, respectively).

These findings suggest that, as reported in other species, the short-term effect of ovariectomy may modify dexamethasone response in mares and that, contrary to other species, it may not modify insulin response.

ABSTRACT E-38

EFFICACY OF SEABUCKTHORN BERRIES FOR TREATMENT AND PREVENTION OF GASTRIC ULCERS IN HORSES.

NK Huff[1], AD Auer[2], F Garza[1], M Keowen,[1] B McMullin[3], FM Andrews[1]. [1]Equine Health Studies Program, Department of Veterinary Clinical Sciences, School of Veterinary Medicine, Louisiana State University, Baton Rouge, LA; [2]Purdue University School of Veterinary Medicine, West Lafayette, IN; Seabuck Equine, LLC, Midvale, UT.

Equine Gastric Ulcer Syndrome (EGUS) is a common medical problem in horses. The high prevalence of gastric ulcers, vague clinical signs and negative effect on performance make it a significant clinical and economic problem within the horse industry. Current pharmaceutical treatments are expensive and alter the acidic environment of the stomach. Berries and pulp from the seabuckthorn plant (*Hippophae rhamnoides*) are a rich source of vitamins, trace minerals, amino acids, antioxidants, and other bioactive substances and have been used successfully to treat stomach ulcers in man and rats. The purpose of this study was to evaluate the efficacy of a commercially sold, liquid extract of seabuckthorn berries (SeaBuck™ SBT Gastro-Plus) for treatment and prevention of gastric ulcers in horses.

Eight Thoroughbred and Thoroughbred-cross horses (3–10 years of age, 5 geldings & 3 mares, 380–600 kg) were used in a blinded two-period cross-over study. Treatments consisted of control (untreated) and treatment (SeaBuck™ SBT Gastro-Plus) twice daily mixed with the grain meal. Horses were treated for 5 weeks followed by a 1 week alternating feed-deprivation period to induce or worsen existing ulcers. Gastroscopies were performed on all horses on day 0, week 5, and week 6 (at the end of the alternating feed-deprivation period). Gastric juice was aspirated and pH was measured. During gastroscopy, gastric ulcer scores were assigned to each stomach based on lesion number and severity. Horses acted as their own controls, and between each treatment period the horses had a 2-week washout period. Data was analyzed by ANOVA for repeated measures via the GLM procedure (SAS Inst. Inc., Cary, NC). When significant differences ($P < 0.05$) were observed, a post-hoc Tukey's test was used to determine differences.

Non-glandular gastric ulcer scores significantly increased in all control and SBT-treated horses from Week 5 to Week 6, after the feed-deprivation phase of the study. There was no significant difference in the non-glandular gastric number ($P = 0.84$) and non-glandular gastric severity ($P = 0.51$) scores in SBT-treated horses compared to non-treated controls. Glandular ulcer number ($P = 0.02$) and glandular ulcer severity ($P = 0.02$) was significantly lower in the SBT-treated horses compared to the control horses. There was no significant difference in the pH ($P = 0.06$) in SBT-treated horses compared to non-treated controls. Thus, SeaBuck™ SBT GastroPlus, mixed in the feed twice daily, may be efficacious in controlling the severity of glandular ulcers in horses during stress, without increasing stomach pH.

ABSTRACT E-39

EVALUATION OF THE VSPRO TABLE-TOP ASSAY FOR EQUINE FIBRINOGEN. BM Brainard, KL Epstein. University of Georgia College of Veterinary Medicine, Athens, GA.

The availability of rapid and accurate quantitative fibrinogen measurements may be useful for evaluation of hospitalized equine patients. The Abaxis VSPro analyzer was evaluated for precision using two levels of human fibrinogen controls (300 mg/dL and 150 mg/dL), four different VSPro machines, and two different lots of cartridges, assessed over 5 subsequent days. The coefficients of variation of the assay ranged from 4% (300 mg/dL) to 8% (150 mg/dL). We subsequently evaluated the Abaxis VSPro fibrinogen assay compared to fibrinogen concentration measured using the Beckman Coulter ACL-1000 in 50 equine samples of varying fibrinogen concentrations obtained from horses with gastrointestinal disease. All samples were measured in citrated plasma. Fibrinogen samples measured on the ACL-1000 ranged from 226 to 959 mg/dL (median 501 mg/dL). VSPro samples were run in duplicate, and the mean compared to the ACL values. Pearson correlation coefficient analysis generated an r value of 0.949 ($P < 0.001$). Duplicate measurements on the VSPro were strongly correlated to each other with an r value of 0.9690 ($P < 0.001$). Bland-Altman analysis of these samples for the VSPro compared to the ACL-1000 noted a bias of -84 ± 57 mg/dL The results of this study indicate that the VSPro benchtop fibrinogen analyzer provides accurate and precise fibrinogen data compared to the ACL-1000 reference analyzer.

ABSTRACT E-40

DOSE-DEPENDENT EFFECT OF *RHODOCOCCUS EQUI* (*R. EQUI*) INFECTION ON CYTOKINE/CHEMOKINE RESPONSES OF PULMONARY ALVEOLAR MACROPHAGES. S. Hashimoto-Hill, M. Heller, K. Jackson, J.L. Watson. University of California, Davis, CA.

The immune response of foals to *R. equi* is incompletely understood and believed to be responsible for clinical disease caused by this pulmonary pathogen. In a recent study foals receiving a large inoculum exhibited Th2 skewing with pneumonia and a small inoculum exhibited Th1 skewing without clinical disease. We hypothesized that cytokine/chemokine production by pulmonary alveolar macrophages, *in vitro*, would increase with the infective dose and that the magnitude of the response would differ between foals and adults.

Alveolar macrophages were obtained by bronchoalevolar lavage from 7 healthy mares and their 5-week-old foals. Macrophage cultures were infected with *R. equi* (33701+ or 33701-) at a multiplicity of infection (MOI) of 1 or 100. Total RNA was harvested 4 and 24 hours post- infection, reverse transcribed and used as template for quantitative PCR. The ΔΔCt method was used to calculate relative gene transcripts for IL-6, IL-12p40, TNFα and CXCL10.

Cellular infections at MOI 100 resulted in significantly higher expression of IL-6, IL-12p40 and TNFα mRNA transcripts compared to MOI 1. However, the dose-effect was reversed for CXCL10 with significantly lower expression at the higher MOI. There was no difference in magnitude of cytokine/chemokine responses by the alveolar macrophages between adults and foals.

Dose-dependent responses of alveolar macrophages may represent a novel mechanism by which *R. equi* could modulate immune responses and therefore disease. Significant down-regulation of CXCL10 mRNA transcripts associated with a higher dose is of particular interest as this chemokine plays a role in development of protective Th1 responses.

ABSTRACT E-41

PREDICTING INTERNAL INFECTION IN HORSES WITH *CORYNEBACTERIUM PSEUDOTUBERCULOSIS* USING SERUM SYNERGISTIC HEMOLYSIS INHIBITION TEST TITERS. JM Jeske, SJ Spier, MB Whitcomb, N Pusterla, IA Gardner. University of California at Davis School of Veterinary Medicine, Davis, CA.

The intent of this study was to develop likelihood ratios (LRs) for infection attributable to *Corynebacterium pseudotuberculosis* in horses based on synergistic hemolysis inhibition (SHI) test titers. Medical records for horses presented to the UC Davis Veterinary Teaching Hospital with serum submitted for SHI titer determination were evaluated and 171 cases met study inclusion criteria. These cases were grouped based on evidence of internal and/or external infection attributable to *C. pseudotuberculosis* and likelihood ratios with 95% confidence intervals determined.

Results showed increasing LRs indicating increasing odds for any form of active disease as titer increased with all cases considered. LRs for internal infection were > 1 for titers ≥ 1280 overall and for titers > 160 with external abscess cases excluded. No difference from 1 (and therefore no significant change in pre-test to post-test odds) was seen in any LRs for internal disease when only cases with external disease were examined (external and internal disease vs. external only). Overall, the SHI test results showed usefulness in determining internal *C. pseudotuberculosis* infection in horses with no evidence of external abscessation. Overall, however, higher titers were more indicative of active external or internal disease than internal disease specifically in contrast to previous reports. The SHI test was unable to distinguish internal infection when external abscesses were present.

ABSTRACT E-42

RAPID *SALMONELLA* DETECTION IN EXPERIMENTALLY-INOCULATED EQUINE FECES USING TWO COMMERCIALLY AVAILABLE LATERAL FLOW ANTIGEN DETECTION SYSTEMS. BA Burgess, NR Noyes, DS Bolte, DR Hyatt, DC Van Metre, PS Morley. Animal Population Health Institute, Colorado State University, Fort Collins, CO.

Salmonella enterica is a zoonotic pathogen that has tremendous impact on many different animal production and management systems. Rapid detection of *S. enterica* in fecal samples may facilitate effective infection control practices. Current detection methods require 24–48 hours (Polymerase Chain Reaction or PCR) or 48–72 hours (enriched aerobic culture) to obtain results. Alternatives have been developed, lateral flow antigen detection systems (LFADS), which are currently marketed for *Salmonella* detection related to food safety microbiology. The objective of this study was to evaluate two commercially available rapid *Salmonella* detection systems in equine feces.

Fecal samples collected from repeatedly culture-negative horses were inoculated with known concentrations of *Salmonella enterica* serotype Typhimurium (five uninoculated control samples, and 5 samples of each 10-fold dilution [1.9×10^0–1.9×10^4 cfu/gram of feces]). All samples were aerobically cultured using a standard enrichment technique. In a blinded fashion, samples were tested using two different LFADS as well as plated on agar media for confirmatory testing.

At 24 hours of incubation, using bacterial culture as the reference method, Test 1 was correctly identify 70% of samples (16/25 inoculated samples and 5/5 negative control samples); Test 2 correctly identify 66.7% of samples (15/25 inoculated samples and 5/5 negative control samples). At 42 hours of incubation, both tests correctly identified all samples.

This study demonstrates that both commercially available rapid detection systems evaluated may be valuable diagnostic tools when managing veterinary hospitals and other animal facilities. However, further evaluation of naturally-infected samples is warranted.

ABSTRACT E-43

EVALUATION OF THREE STALL CLEANING METHODS ON THE ERADICATION OF *SALMONELLA* sp. IN AN EQUINE VETERINARY TEACHING HOSPITAL. F Garza, Jr.[1], NK Huff[1], AD Auer[2], SC Eades[1], JR Johnson[1], FM Andrews[1]. [1]Equine Health Studies Program, Department of Veterinary Clinical Sciences, School of Veterinary Medicine, Louisiana State University, Baton Rouge, LA and [2]Purdue University School of Veterinary Medicine, West Lafayette, IN.

Bacterial contamination of stalls with *Salmonella sp.* is a serious problem in equine hospitals. *Salmonella sp.* exposure to horses in the facility can result in nosocomial infections which results in temporary facility closure, until the organism is eradicated. Hospital closure can result in loss of revenue, damage to reputation and interference with patient care. The purpose of this study was to evaluate three stall cleaning methods on eradication of *Salmonella sp.* at an equine veterinary teaching hospital (VTH).

Horses admitted to the VTH were assigned to *Salmonella sp.*-negative stalls within areas of the VTH during the study period (September 2009 - January 2010). When the horses were discharged stalls were randomly assigned to one of three cleaning methods (pressure-washing only [PW], pressure washing and hand scrubbing [PWS], or hand scrubbing only [S]) in a single period, non cross-over design. All stalls were stripped of bedding and surfaces sprayed with tap water and cleaned with a disinfectant quaternary-ammonia solution (Super HDQ Neutral, Spartan Chemical Co., Inc, Maumee, OH). The pressure-washing system (PSC Cleaning Systems, Inc., Toronto, Canada) used, provided a pressure of 3000 psi and a temperature range of 185°–215°F. Following cleaning, each stall was allowed to air dry and within 48 hours, stall surfaces were sampled using three 4″ × 4″ sponges moistened with sterile saline. The person collecting the samples was masked to the method of cleaning. Sponges were submitted to the Louisiana Animal Disease Diagnostic Laboratory (LADDL) for culture of *Salmonella sp.* A Chi-Squared analysis was used to determine significant differences (limit $p < 0.05$) between cleaning methods and *Salmonella sp.* isolation. During the study period, 112 stalls (PW [n = 29]; PWS [n = 50]; S [n = 33] were included. All stalls had negative environmental *salmonella sp.* cultures prior to beginning the study. For PW cleaned stalls, 6/23 (20.7%) were *Salmonella sp.*-positive, for PWS cleaned stalls, 12/38 (24%) were *Salmonella sp.*-positive, and for S cleaned stalls, 4/29 (12.1%) were *Salmonella sp.*-positive. Although, there were fewer *Salmonella sp.*-positive stalls (12.1%) in the hand-scrubbed stalls, cleaning method did not significantly ($p = 0.4057$) affect the isolation of *Salmonella sp.* from the stall environment.

In conclusion, power washing alone, power washing and hand scrubbing, and hand scrubbing alone, using a quaternary- ammonia solution did not significantly affect environmental isolation of *Salmonella sp.* from stalls surfaces in the VTH during this study.

ABSTRACT E-44

PULMONARY DISPOSITION OF GAMITHROMYCIN IN FOALS AND IN VITRO ACTIVITY OF THE DRUG AGAINST *RHODOCOCCUS EQUI* AND *STREPTOCOCCUS EQUI* SUBSP. *ZOOEPIDEMICUS*. LJ Berghaus[1], S Giguère[1], TL Sturgill[1], D Bade[2], TJ Malinski[3], R Huang[3]. [1]University of Georgia, Athens, GA. [2]Microbial Research Incorporated, Fort Collins, CO [3]Merial Limited, Duluth, GA.

The objectives of this study were to determine the plasma and pulmonary disposition of gamithromycin in foals and to investigate the *in vitro* activity of the drug against *Streptococcus equi* subsp. *zooepidemicus* (*S. zooepidemicus*) and *Rhodococcus equi* isolates. A single dose of gamithromycin (6 mg/kg of body weight) was administered intramuscularly. Concentrations of gamithromycin in plasma, pulmonary epithelial lining fluid (PELF), bronchoalveolar lavage (BAL) cells, and blood neutrophils were determined using HPLC with tandem mass spectrometry detection. The minimum inhibitory concentration of gamithromycin required to inhibit growth of 90% of *R. equi* and *S. zooepidemicus* isolates (MIC_{90}) was determined. Additionally, the activity of gamithromycin against intracellular *R. equi* was measured.

Mean peak gamithromycin concentrations were significantly higher in blood neutrophils (8.35 ± 1.77 g/mL) and BAL cells (8.91 ± 1.65 g/mL) compared to PELF (2.15 ± 2.78 g/mL) and plasma (0.33 ± 0.12 g/mL). Mean terminal half-lives in neutrophils (78.6 h), BAL cells (70.3 h), and PELF (63.6 h) were significantly longer than that of plasma (39.1 h). The MIC_{90} of *S. zooepidemicus* isolates was 0.125 g/mL. The MIC_{90} of gamithromycin for macrolide-resistant *R. equi* isolates (128 g/mL) was significantly higher than that of macrolide-susceptible isolates (1.0 g/mL). The activity of gamithromycin against intracellular *R. equi* was similar to that of azithromycin and erythromycin.

Intramuscular administration of gamithromycin at a dosage of 6 mg/kg would maintain PELF concentrations above the MIC_{90} for *S. zooepidemicus* and phagocytic cell concentrations above the MIC_{90} for *R. equi* for approximately 7 days.

ABSTRACT E-45

D-DIMER CONCENTRATIONS AND (EXPERIMENTAL) EHV-1 INFECTION. LS Goehring, G Soboll Hussey, S Hussey, DP Lunn. Colorado State University College of Veterinary Medicine, Fort Collins, CO.

Equine Herpesvirus-associated Myelopathy (EHM) is the result of spinal cord endothelial cell infection with subsequent vasculature pathology. In infection experiments EHM typically occurs in week 2 after nasopharyngeal infection, towards the end of an EHV-1 cell-associated viremia. We hypothesized that horses with endothelial cell infection develop variable stages of a coagulopathy with detectable D-dimer concentrations as a result of cross-linked fibrin degradation.

Eight Western stock yearling horses were infected with EHV-1 (Ab4) by nasopharyngeal instillation. Venous blood samples for collection of plasma were collected in Na-citrate tubes on the day prior to infection (D -1) and on D4 through D11. In addition, clinical data, nasal swabs and peripheral blood mononuclear cells (PBMC) for detection of viremia were collected on the day before infection (D -1) and on D1 through D14 post-infection. D-dimer concentrations were determined in citrated plasma samples using a latex agglutination test (Minutex D-dimer, biopool, Ireland). Viral load in PBMC was determined using quantitative PCR.

All horses showed bi-phasic fevers typical for EHV-1 infections. One horse developed acute EHM on D10 and was euthanized after samples were collected. In all horses D-dimers were undetectable on D -1 and on D4, 5 and 11. In contrast, all horses had increased D-dimer concentrations for 3 to 5 consecutive days starting on day 6 post-infection. D-dimer concentrations in 2 horses increased to 1000 ug/ml and one of these horses was the horse with acute EHM.

Interestingly, mean increased D-dimer concentrations showed timely overlap with the mean fever curve and, delayed by 1 day, with the mean viremia curve. Because plasma samples for D-dimer measurements were not collected during the first 3 days post-infection, which are typically associated with a primary fever, conclusion on the association of D-Dimers with fever of viremia await analysis of a second study currently conducted in our laboratory.

In conclusion our data indicates that during EHV-1 infection with neuropathogenic strains activation of the coagulation cascade and production of cross-linked fibrin is wide-spread; not limited to horses with clinical signs of EHM, and can be expected between days 6 and 10 post-infection.

ABSTRACT E-46

IDENTIFICATION OF EXPOSURE TO AND SHEDDING OF *LAWSONIA INTRACELLULARIS* IN LOUISIANA HORSES USING SEROLOGY AND REAL-TIME PCR. AM Chapman[1], M Buchert[1], CJ Gebhart[2], SC Eades[1]. [1]Louisiana State University, School of Veterinary Medicine, Baton Rouge, LA. [2]University of Minnesota, Department of Veterinary and Biomedical Sciences, St. Paul, MN.

Lawsonia intracellularis is an emerging pathogen in horses and the causative agent in Equine Proliferative Enteropathy (EPE). The goal of this study was to evaluate the exposure of pre-weanling foals and broodmares to *Lawsonia intracelluaris* on several farms in Louisiana with a history of EPE and compare the results to several farms with no known clinical cases of EPE in foals. An additional goal of the study was to identify whether a relationship exists between *Lawsonia intracelluaris* and other gastrointestinal pathogens in foals. Whole blood and fecal samples were collected from 66 mares and 68 foals from four breeding farms in Louisiana. Farms A and B had no known clinical cases of EPE, while Farms C and D had previous know cases of EPE in 2009. Serum samples were examined for the presence of antibodies against *Lawsonia intracellularis* using an immunoperoxidase monolayer assay (IPMA). DNA was extracted from fecal samples using a commercial DNA kit and molecular detection of *Lawsonia intracelluaris* was assayed using real-time PCR. Fecal ova were determined using quantitative sucrose floatation. The presence of fecal *Clostridium difficile* toxin was measured using a commercial enzyme linked immunosorbent assay (ELISA). Three of the 4 farms examined had foals and mares with exposure to *L. intracellularis* as evidenced by serum antibodies against the organism. Of the total population sampled, 6 foals (8.8%) and 14 mares (21.2%) had evidence of antibodies to *L. intracellularis* based on serology. Three foals (4.4%) tested positive for *L. intracellularis* organism by fecal PCR, and all of these foals were located on farm C. Of these, one of the foals was seronegative, while the other two were seropositive. Farm C also had the highest percentage of mares (28.6%) serologically positive for *L. intracellularis*, while Farm A had the highest percentage of foals (14.3%) with antibody titers against *L intracellaris*. Farm C also had the only pairs (n = 3) of serologically positive mares with seropositive foals. While Farm A and B had seropositive mares and/or foals, none of the foals were positive for *L. intracellularis* fecal shedding by PCR. All serum and fecal samples were negative for evidence of *L. intracellaris* on Farm D. Ten foals (14%) had fecal egg counts greater than 200 egg per gram and 2 foals (3%) were positive for *C. difficile* toxin. This study demonstrated evidence of natural exposure to *L intracellularis* on farms both with and without a history of EPE in Louisiana. Further, this study failed to establish a relationship between *L intracellularis* and other gastrointestinal pathogens.

ABSTRACT E-47

EQUINE GRANULOCYTIC ANAPLASMOSIS: 19 FIELD CASES IN SOUTHEASTERN PENNSYLVANIA (2005-2010). A. Boyle, K. Near, H. Aceto. University of Pennsylvania, Clinical Studies New Bolton Center, Kennett Square, PA.

The objective of this study was to examine the clinical, hematological, biochemical, and outcome data from equids infected with Anaplasma phagocytophilum presented to a primary care field setting in southeastern Pennsylvania. Computerized medical records from 19 febrile equids with confirmed Anaplasma phagocytophilum infection were reviewed. Confirmation of Anaplasma phagocytophilum was defined by the presence of granular inclusion bodies seen within leukocytes or eosinophils on microscopic blood smear evaluation and/or a positive polymerase chain reaction (PCR) for Anaplasma phagocytophilum.

18 horses and 1 donkey presented with a mean fever of 104.4°F and mean fever duration of 39 hours. The mean age at presentation was 9.6 years and the mean pack cell volume was 30.8%. 15/19 cases were diagnosed in the months of May to December. Equids ages 5 to 15 years had significantly lower platelet counts. 16/18 cases were positive on blood smear for inclusion bodies and 6/6 cases were positive for Anaplasma phagocytophilum on PCR. Treatments included intravenous oxytetracycline, oral doxycycline, or both. Mean treatment duration was 4.8 days and mean treatment cost was $621. 17/19 cases were normothermic within 48 hours. The treatment used in the two remaining cases was changed from oral doxycycline to intravenous oxtetracycline and was successful.

This is the first case series of equine granulocytic anaplasmosis in the mid-Atlantic states. All cases were examined and treated in the field. In order to make a definitive diagnosis, some cases required PCR. Treatment failures were documented with the use of oral doxycycline alone. 100% of the cases survived.

ABSTRACT E-48

NEEDLE ELECTROMYOGRAPHY OF THE ESOPHAGUS IN FRIESIAN HORSES WITH AND WITHOUT ESOPHAGEAL DYSFUNCTION. JH van der Kolk[1], M Ploeg[1], W Back[1, 3], CM de Bruijn[2], ID Wijnberg[1]. [1]Faculty of Veterinary Medicine, Utrecht University, Utrecht, the Netherlands. [2]Wolvega Equine Clinic, Wolvega, the Netherlands. [3]Faculty of Veterinary Medicine, Ghent University, Ghent, Belgium.

A high incidence of clinical and possibly genetic abnormalities has been reported amongst Friesian horses including dwarfism, hydrocephalus, dissecting aortic aneurism and esophageal dysfunction. The purpose of the current study was to develop a new electromyography (EMG) method to assess neurophysiological function of the esophagus especially for Friesian horses.

Five Friesian horses with esophageal dysfunction were included (ranging in age from 0.5–24 years and comprising 4 mares and a stallion) and two Friesian control horses (a 10- and 12-year-old gelding). All five horses with esophageal dysfunction had a history of recurrent esophageal obstruction and were examined histopathologically post-mortem. Barium contrast radiography was used as the gold standard to distinguish the diseased from the control horses. An endoscopically-guided percutaneous needle EMG procedure (Viking Quest[r]; software version 11.0) was performed just caudal to the larynx and just cranial to the thoracic inlet (to monitor striated and smooth muscle, respectively) to visualize esophageal motility.

Esophageal contractility in both control horses was predominantly reflected by interference patterns associated with longer duration and lower amplitude in smooth muscle compared to striated muscle. Mean ($\pm$ SD) values were 35.1 $\pm$ 19.4 ms and 167.7 $\pm$ 96.7 µV (n = 19 readings) and 10.8 $\pm$ 14.3 ms and 305.8 $\pm$ 233.7 µV (n = 24 readings), respectively. In diseased horses, aperistalsis in smooth muscle was the most remarkable finding suggesting a loss of inhibitory neurogenic input resulting in aperistalsis and thus esophageal dysfunction. Preliminary findings suggest that endoscopically-guided percutaneous needle EMG might become a valuable method in elucidating the pathophysiology of dysfunction of esophageal motility especially in Friesian horses.

ABSTRACT E-49

A POSSIBLE LINK BETWEEN EQUINE LYMPHOMA AND EQUINE HERPES VIRUS 5. KA Vander Werf, EG Davis, K Janardhan, K Williams, S Bolin. Kansas State University School of Veterinary Medicine, Manhattan, KS.

Lymphoma affects horses of all ages. Unlike in humans, no etiologic agent has been discovered. A 9 year old Thoroughbred/Warmblood cross mare presented with signs of upper and lower respiratory disease and was subsequently diagnosed with lymphoma and equine multinodular pulmonary fibrosis (EMPF) and was positive for equine herpes virus 5 (EHV-5) in both the pulmonary tissue and the lymph nodes.

Retrospective polymerase chain reaction (PCR) testing of six lymphoma cases found that 5 of 6 of the cases were positive on PCR for EHV-5 (83.3%, p = 0.0045, RR 5.55). Electron microscopy was performed on one sample and herpes virus particles were identified. Of the samples in which immunohistochemistry was performed (3 of 6), only T-cell rich B-cell lymphoma was identified. Samples of mesenteric or submandibular lymph nodes from 20 clinically healthy horses were submitted for EHV-5 PCR analysis; 15% were positive.

Gamma herpesviruses in humans have been associated with lymphoproliferative diseases such as Kaposi's sarcoma and Burkitt's lymphoma. Equine herpesvirus 5, also a gamma herpesvirus, is found in association with equine lymphoma; although the exact role this virus plays in the initiation or perpetuation of lymphoproliferative neoplasia remains unknown.

ABSTRACT E-50

LAMINAR INFLAMMATORY EVENTS AND EPITHELIAL STRESS AT OBEL GRADE 3 LAMENESS IN THE CARBOHYDRATE OVERLOAD MODEL OF EQUINE LAMINITIS. Hensel, Martha[1], L.A. Fugler[2], B. Leise[1], M. Watts[1], S. Eades[2], and J. Belknap[1]. [1]The Ohio State University College of Veterinary Medicine, Columbus, OH. [2]Louisiana State University School of Veterinary Medicine, Baton Rouge, LA.

Pathologic events reported to occur in the digital laminae in early stages of sepsis-related equine laminitis include leukocyte extravasation into the laminar interstitium, pro-inflammatory cytokine expression, and epithelial stress. While these events have been documented early in the disease process at both a developmental stage and at the onset of Obel Grade 1 (OG1) lameness in the carbohydrate overload (CHO) model of laminitis, the later events occurring at the onset of Obel Grade 3 lameness(OG3, time point at which structural failure of the laminae usually occurs) have not been determined. We hypothesized that the inflammatory events described above are sustained through OG 3 lameness, likely playing an injurious role culminating in laminar failure. Our objectives were to determine pro-inflammatory gene expression, leukocyte extravasation, and epithelial stress at OG 3 induced using the CHO model. Archived laminar tissue samples (snap frozen and paraffin embedded sections) were used from a previous CHO study at Louisiana State University (Control group [n = 6, water], CHO group [n = 7, corn starch]. Calprotectin (CP) immunohistochemistry (IHC) was used to assess both laminar myeloid leukocyte numbers and epithelial stress; RT-qPCR was used to assess inflammatory gene expression. Minimal inflammatory changes were present at OG3 compared to published values at OG1 stage in the CHO lameness model including decreased mRNA concentrations of cytokines (i.e. 20-fold increase in IL-6 at OG3 vs. > 2000-fold increase at OG1, no increase in IL-1β at OG3 vs. 11-fold increase at OG1), chemokines (no change in MCP-1at OG3 vs. > 30 fold increase at OG1, 8-fold increase in IL-8 at OG3 vs. 95-fold increase at OG1) and adhesion molecules (no change in E-selectin at OG3 vs. 10-fold increase at OG1). Laminar leukocyte emigration was also decreased at the onset of OG 3 lameness compared to previously reported leukocyte infiltration at OG 1. Interestingly COX-2, underwent a greater increase at OG3 (approx. 50-fold) compared to that reported at OG1 lameness (35-fold). Finally, epithelial stress at OG3 evidenced by CP IHC did not follow the uniform widespread distribution reported at OG1 lameness, but instead was present in focal areas in which secondary epidermal laminae on either side of a common primary dermal vascular supply demonstrated increased CP signal. Overall, laminar inflammation appears to be subsiding at OG3 lameness, with epithelial stress possibly more dependent on vascular dysregulation instead of inflammatory events. The sustained increase in COX-2, central to the induced production of vasoactive prostanoids in disease processes, may play a role in vascular dysregulation.

ABSTRACT E-51

OLEANDER TOXICOSIS IN EQUIDS: 30 CASES (1995-2010).

AC Renier, P Kass, KG Magdesian, JE Madigan, M Aleman, N Pusterla. School of Veterinary Medicine, University of California, Davis, CA.

This study was conducted to characterize clinical, laboratory and postmortem findings associated with oleander toxicosis in equids and to determine factors predictive of survival in these cases.

Retrospective analysis of medical records from our veterinary medical teaching hospital from January 1, 1995 to July 15, 2010 was completed. Records of equids demonstrating detectable oleandrin in serum, plasma, urine or gastrointestinal fluid samples or detectable serum digoxin in the absence of pharmaceutical cardiac glycoside administration were included. Descriptive statistics were used to evaluate the history, physical examination, and laboratory and postmortem data of affected individuals. Logistic regression analysis was used to detect physical examination and laboratory factors significantly associated with survival.

Thirty equids met inclusion criteria of the study. Three of 30 subjects were dead on arrival or died immediately upon arrival (10%). Of the remaining 27 equids, 85% presented with gastrointestinal abnormalities, 70% were azotemic and 48% had cardiac arrhythmias. Mortality was 50% for all subjects and 44% for those treated. The predominant cause for non-survival was cardiac dysfunction. Factors significantly associated with survival included relatively decreased hematocrit and serum glucose, relatively increased serum chloride, absence of cardiac arrhythmias, and increased duration of hospitalization.

Equids with oleander toxicosis frequently present with gastrointestinal upset and may develop cardiac and renal disturbances. Patients with cardiac arrhythmias and relatively increased hematocrit and serum glucose and decreased serum chloride are significantly less likely to survive. Oleander intoxication is a differential diagnosis for colic in endemic areas, particularly with concurrent azotemia or cardiac dysrhythmia.

ABSTRACT E-52

VALIDATION OF THE USE OF SERUM BIOCHEMICAL PROFILE FOR ACID-BASE INTERPRETATION IN HORSES.

HR Stämpfli[1], A Schoster[1], PD Constable[2]. [1]University of Guelph, Guelph, Ontario, Canada. [2]Purdue University, West Lafayette, IN.

The quantitative physicochemical approach emphasizes the importance of strong ions (Na, K, Cl, lactate), pCO_2, and the plasma protein concentrations in determining plasma pH. Serum concentrations of strong ions, proteins, and total CO_2 are reported on modern biochemical profiles. We hypothesized that the results of serum biochemical analysis can be used for acid-base interpretation in horses. The objective was to determine whether blood pH, anion gap, and strong ion gap could be quantitatively estimated and clinically used based on the results of serum or plasma biochemical analysis.

100 horses (70 adults and 30 foals) presented to the isolation unit of our Veterinary Teaching Hospital for suspected infectious diseases were prospectively enrolled. A venous serum sample was analyzed using a Hitachi 911 or Copas 6000 C501 automated machine. Measured parameters included strong ion difference (SID = {Na+K}-{Cl+lactate}), total protein concentration (TP), and total CO_2 (tCO_2), with lactate being measured by blood gas analyzer. A second venous blood sample was collected into a Na-heparin blood gas syringe and analyzed for pH (pH_m), pCO_2 and concentrations of Na, K, Cl, and lactate using a Radiometer 800 Flex blood gas analyzer; SID was calculated from the measured values, and total solids (TS) were estimated using refractometry. Serum/ plasma pH (pH_{calc}) was calculated using Stewart's 8 factor equation from the results of serum or plasma biochemical analysis, assuming $pCO_2 = 40$ mmHg for serum and pCO_2 accurate for plasma. Anion gap (AG) was calculated as: $AG = (Na+K)-(Cl+tCO_2)$. Strong ion gap (SIG) was calculated as: SIG = 0.21x[total protein, g/L]/$(1+10^{(pKa-pH)})$-AG. Linear regression analysis was used to compare pH_{calc} to pH_m, as well as AG and SIG to blood lactate concentrations.

Measured pH ranged from 7.03 to 7.48 (7.37 ± 0.07). Measured SID from serum biochemistry (SID_{SB}) ranged from 22.3 to 51.4 mEq/L (36.2 ± 4.7 mEq/L) and SID from blood gas analyzer (SID_{BG}) from 14.3 to 46.6 mEq/L (33.3 ± 5.4 mEq/L; $R^2 = 0.59$; $SID_{BG} = 0.919 \times SID_{SB}$). SID_{SB} and SID_{BG} showed small variability in measurements. TP ranged from 18 to 88 g/L (51.0 ± 13.9 g/L) and TS from 20–84 (55.2 ± 13.8 g/L; $R^2 = 0.77$; TS = 1.071 × TP). Using SID_{SB} and tCO_2 values with constant pCO_2, pH_{calc} was poorly associated with pH_m ($R^2 = 0.27$; $pH_{calc} = 0.48 + 3.89$). In contrast, using SID_{BG} with accurate pCO_2, pH_{calc} was closely associated with pH_m ($R^2 = 0.54$; pHcalc = 0.98 + 0.15) and the equation was not different from the line of identity. Anion gap and SIG (mEq/L) calculated were significantly linearly correlated with lactate concentrations (mmol/L); AG = 0.93 × [lactate] + 8.3 (R^2 = 0.39), and SIG = −0.91 × [lactate] + 0.5 (R^2 = 0.41).

We conclude that pH_{calc} using SID_{SB}, tCO_2 and constant pCO_2 values is not accurate. However, variability of measured biochemical parameters between machines was small, permitting use of serum biochemistry for clinical metabolic acid-base abnormalities interpretations of patients. These results reemphasize the importance of strong electrolytes and proteins in acid-base balance.

ABSTRACT E-53

INCREASED METALLOPROTEINASE ACTIVITY IN CHONDROCYTES FROM ARTICULAR CARTILAGE OF HORSES WITH HEREDITARY EQUINE REGIONAL DERMAL ASTHENIA.

ML Dorris[1], AM Rashmir-Raven[2], LF Heinecke[3], MW Grzanna[3], AY Au[3,4], SL Ownby[3], CG Frondoza[1,3,5]. [1]Mississippi State University College of Veterinary Medicine, Mississippi State, MS. [2]Michigan State University College of Veterinary Medicine, East Lansing, MI. [3]Nutramax Laboratories, Inc., Edgewood, MD. [4]Syracuse University, Syracuse, NY. [5]Johns Hopkins University, Baltimore, MD.

Metalloproteinases (MMPs) are critically important in remodeling processes and in wound healing. However, excessive activation of MMPs by pro-inflammatory mediators including cytokines, prostaglandin E2, and nitric oxide lead to tissue breakdown. This is observed in osteoarthritis (OA) which is characterized by erosive lesions in articular cartilage. In Hereditary Equine Regional Dermal Asthenia (HERDA), afflicted horses exhibit collagen abnormalities and can have associated chronic inflammation and aberrant wound repair. HERDA affects horses with Quarter Horse bloodlines and is similar to the human hereditary connective tissue syndrome Ehlers Danlos (EDS). Many adult EDS patients suffer from joint pain and

OA. We hypothesized that chondrocytes from articular cartilage of HERDA horses have increased activity of MMPs. To test this hypothesis, chondrocytes were retrieved from articular cartilage of homozygous HERDA carpal and hock joints. Chondrocytes from normal horses were also obtained for comparison. Chondrocytes were seeded at 5 x 10^6/ml into 6-well plates and incubated at 37°C, 5% CO_2 for up to seven days. Activity of secreted MMPs was determined by zymography using equal amounts of proteins for loading. Secreted MMPs were analyzed by Western blot.

Zymography showed that normal chondrocytes secreted two major bands with gelatinolytic activity observed at 92 and 72 kDa suggestive of the latent form of MMP-2 and MMP-9, respectively. Less intense bands of gelatinolytic activity were observed at about 82 and 62 kDa suggestive of the active form of MMP-2 and MMP-9. Another band of activity was also seen at 240 kDa which is suggestive of a dimer of MMP-9 that has been reported when MMPs are in excess of tissue inhibitors of metalloproteinases (TIMPs). Chondrocyte cultures from homozygous HERDA cartilage showed a similar profile but with decreased activity by 90% at 92 kDa and 10–50% increased activity at 72 kDA compared to normal chondrocytes. Western blot analysis detected MMP-2 and MMP-9 immunoreactivity in chondrocyte culture media of HERDA-afflicted and normal horses. The present study demonstrates for the first time that horses suffering from HERDA have increased MMP activity which may predispose them to the development of lesions in articular cartilage.

Research supported by Nutramax Laboratories, Inc.

ABSTRACT E-54

GLYCOGEN SYNTHASE 1 MUTATION IN THE AUSTRIAN HAFLINGER POPULATION. BC Schwarz, R Ertl, S Zimmer, Y Netzmann, D Klein, I Schwendenwein, R van den Hoven. University of Veterinary Medicine Vienna, Austria.

Equine Polysaccharide Storage Myopathy (PSSM) type 1 is a dominantly inherited glycogenosis caused by a mutation in the gene coding for skeletal muscle glycogen synthase type 1 (GYS-1). The disease has been reported to affect the Haflinger breed but so far its prevalence is unknown. Aim of this preliminary study was to estimate the occurrence of the GYS-1 mutation in Austrian Haflingers and establish which of the seven Haflinger sire lines appear mostly affected.

GYS-1 genotyping of 50 randomly chosen Haflingers was performed with a validated restriction fragment length polymorphism assay. Resting and post-exercise muscle enzyme activities (creatine kinase (CK), aspartate aminotransferase (AST), lacate dehydrogenase (LDH)) and blood lactate concentrations were compared between horses with and without the mutation.

Among the 50 horses 9 were heterozygous (HR) carrier of the mutation. No homozygotes (HH) were identified. All horses with the GYS-1 mutation were descendents of the A- or W-sire lines. The estimated HR prevalence was 18% (95% CI: 8.6–32.4%). CK activity after exercise ($p = 0.022$) was significantly higher in HR horses compared with horses not carrying the mutation (RR). AST activity was significantly higher in the HR group at rest and after exercise ($p < 0.001$). There was no statistically significant difference in resting CK, resting and post exercise LDH activity or blood lactate between HR and RR.

Results suggest that the prevalence of HR in the Austrian Haflinger population is higher than in the overall Quarter Horse population and might be as high as 30%, similar to some Draft horse breeds. Further research is needed to establish the prevalence within the different breeding lines.

ABSTRACT E-55

TENSILE PROPERTIES IN COLLAGEN RICH TISSUES OF QUARTER HORSES WITH HEREDITARY EQUINE REGIONAL DERMAL ASTHENIA (HERDA). JE Bowser[1], S Elder[2], AM Rashmir-Raven[3], JG Grady[1], CE Swiderski[1]. [1]Departments of Clinical Sciences, College of Veterinary Medicine and [2]Agricultural and Biological Engineering, Mississippi State University, Mississippi State, MS. [3]College of Veterinary Medicine, Michigan State University, East Lansing, MI.

Hereditary equine regional dermal asthenia (HERDA) is an autosomal recessive connective tissue disorder affecting Quarter Horse lineages.[1] Although a mutation in the gene encoding cyclophilin B has been genetically linked to HERDA, its causal association with the disease is not yet documented.[2] Previously, we demonstrated reductions in ultimate tensile strength (UTS), modulus of elasticity, and energy to failure (toughness) of skin from many corporal regions of HERDA animals.[3] Given the presumed relationship between HERDA and abnormal collagen structure, and the predominance of Type I collagen in skin, we hypothesized that altered biomechanical properties would be detected in tendons which are rich in Type I collagen. To evaluate this hypothesis we compared the UTS, modulus of elasticity, and energy to failure of forelimb deep digital flexor tendons (DFT) from six HERDA horses to six age-matched controls. Isolated DFT was secured and pulled to failure on an Instron® 1011 Universal Testing Instrument using purpose-built cryogenic clamps. Analysis of Variance was executed using SAS 9.2 PROC GLIMMIX program (SAS Institute, 2009). P-values ≤ 0.05 were identified as significant. UTS and modulus of elasticity were significantly lower in HERDA DFT when compared with controls ($P < 0.0001$); energy to failure did not differ between groups. These findings document abnormal biomechanics in HERDA tendon, leading us to postulate that lower UTS and modulus of elasticity associated with the HERDA defect could convey a competitive advantage in the athletic disciplines in which this defect has segregated. (References on request).

ABSTRACT E-56

COMPARISON OF A PROPRIETARY HERBAL TOPICAL CREAM TO SILVER SULFADIAZINE IN AN EQUINE WOUND MODEL. E Carothers, M Brashier, B Rakauskas, CE Swiderski. Department of Clinical Sciences, College of Veterinary Medicine, Mississippi State, MS.

A proprietary herbal biocontamination product (BIOS) approved for cosmetic use in France, inhibits proliferation of medically relevant bacteria, mold, and viruses. These properties make BIOS potentially useful as a topical wound medication, prompting us to compare BIOS to silver sulfadiazine (SSD) in a distal extremity wound healing model in horses.[1] Using general anesthesia, two 6.25 cm^2 wounds were aseptically created on the dorsomedial aspect of all limbs. For the duration of the study, two contralateral limbs were randomly chosen to be bandaged; the other two limbs were un-bandaged – with one limb of each group being treated with 10% BIOS and the other with SSD. For each limb the most proximal wound served as an untreated control. Every 48 hours wounds were evaluated, digitally photographed, and perimeter and area determined using morphometric software (ImageJ, NIH). Analysis of variance did not identify significant differences between SSD or BIOS treatment for wound perimeter ($p = 0.76$) or area ($p = 0.95$). At individual time points the effect of bandaging was significant when area was evaluated ($p = 0.019$) and trended toward significance for perimeter ($p = 0.084$) comparisons, substantiating published reports that bandaging modifies wound healing. Difference in perimeter and area between control and treatment were highly significant ($p < 0.0001$), substantiating the importance of topical treatment. Over the study duration, effects of bandaging ($p < 0.0001$) and topical treatment (perimeter $p < 0.0001$; area $p = 0.0016$) continued to be highly significant. BIOS performance in the equine distal extremity wound model was equivalent to SSD. Both bandaging and topical treatment significantly impacted wound healing. This effect was compounded when both variables were evaluated over time.

ABSTRACT E-57

A SCINTIGRAPHIC COMPARISON OF 99mTc LABELED IgG, PEG-LIPOSOME AND LEUKOCYTE ACCUMULATION IN EXPERIMENTALLY INDUCED FOCAL SOFT TISSUE INFECTION IN THE HORSE. C Underwood[1], AW van Eps[2], TP Schaer[1], MW Ross[1], G Storm[3], L van Bloois[3], P Laverman[4]. [1]University of Pennsylvania, New Bolton Center, Kennett Square, PA, USA. [2]University of Queensland, School of Veterinary Science, Gatton, Australia [3]Department of Pharmaceutics, Utrecht Institute for Pharmaceutical Sciences, Utrecht University, Utrecht, NL. [4]Radboud University Nijmegen Medical Centre, Nijmegen. NL.

Radiolabeled leukocytes are the only scintigraphic method currently available for identifying sites of infection and/or inflammation in horses; however the clinical applicability of this technique is limited by expense and poor efficacy. This pilot study compares the accumulation of ^{99m}Tc-labeled IgG, PEG-liposomes and leukocytes in an equine muscle abscess model.

Three mixed breed adult horses had 2×10^6 CFU *S. equi zooepidemicus* inoculated into the right semitendonosis to create an abscess. PEG-liposomes were prepared via the film hydration method and labeled using 200 mCi ^{99m}Tc-hexamethyl-propylene-amine-oxime (^{99m}Tc-HMPAO). Autologous leukocytes were obtained from 120 ml whole blood and labelled using 200 mCi ^{99m}Tc-HMPAO. Commercial equine polyclonal IgG was conjugated with the chelator hydrazinonicotinamide (HYNIC) and labelled with 200 mCi ^{99m}Tc. Radiopharmaceutical administration was initiated 24 hours after inoculation. Horses 1 and 2 received 5 mg ^{99m}Tc-IgG, 2.7 μmol/kg ^{99m}Tc-liposomes and ^{99m}Tc-leukocytes, with a 48 hour interval between each radiopharmaceutical. Horse 3 received only ^{99m}Tc-leukocytes. Scintigraphic examinations were performed at 8 and 21 hours post injection (p.i.) with each radiopharmaceutical. After the final study, horses were euthanized and tissue samples collected. The percentage of injected dose per kilogram of tissue (%ID/kg) was calculated for the region of the abscess, normal muscle and multiple organs.

Scintigraphic examinations demonstrated increased radiopharmaceutical in the region of the abscess with all three techniques at both time-points. At 8 hours p.i. abscess-to-background ratio was highest using ^{99m}Tc-IgG (3.7 ± 0.2). At 21 hours p.i. abscess to background ratio was highest using ^{99m}Tc-liposomes (5.9 ± 2). Tissue biodistribution data revealed abscess to muscle ratios of 36 (^{99m}Tc-IgG), 24 (^{99m}Tc-liposomes), and 4.1 (^{99m}Tc-leukocytes).

This preliminary data demonstrates that ^{99m}Tc-liposomes, ^{99m}Tc-IgG and ^{99m}Tc-leukocytes exhibit long circulating characteristics and accumulate at inflammatory/infectious foci after intravenous injection in horses. ^{99m}Tc-IgG and ^{99m}Tc-liposomes appear to be superior to ^{99m}Tc-labelled leukocytes in this model. Due to its low cost and ease of preparation, ^{99m}Tc-IgG has great potential for clinical use where identification of infectious or inflammatory foci is necessary.

ABSTRACT E-59

P38 MAPK SIGNALING IN EQUINE LAMINITIS: THE EFFECT OF DIGITAL HYPOTHERMIA. C. Kelly[1], M. Watts[1], A. Van Eps[2], C. Pollitt[2], J. Belknap[1]. [1]The Ohio State University College of Veterinary Medicine, Columbus, OH. [2]The University of Queensland School of Veterinary Science, St. Lucia, Queensland, Australia.

Digital hypothermia is used clinically to decrease the incidence of sepsis-related equine laminitis, a disease causing structural failure of digital laminae resulting in crippling lameness. Due to the fact that hypothermia was recently reported to effectively decrease laminar expression of inflammatory molecules including pro-inflammatory cytokines, chemokines and COX-2 in equine laminitis, our laboratory is investigating the effect of hypothermia on central upstream signaling cascades which may induce expression of these diverse inflammatory molecules. The p38 MAPK pathway has recently been reported to be a central component of inflammatory signaling in multiple diseases including human sepsis, and is currently being assessed as a therapeutic target. We thus hypothesized that 1) p38 MAPK is upregulated and activated in affected laminae in equine laminitis and 2) digital hypothermia inhibits inflammatory mediator expression by blocking p38 MAPK phosphorylation (indicator of p38 MAPK activation). Western hybridizations using both a total p38 MAPK and a phospho-p38 MAPK antibody were performed on archived pooled laminar samples from black walnut extract (BWE) model (10 control, 2 developmental (DEV) groups [1.5H & 3H post BWE administration] and the onset of Obel Grade 1 lameness (OG1) [n = 5 each]) and carbohydrate overload (CHO) models (CON [n = 8], DEV [n = 6], OG1[n = 6]) of laminitis, and individual laminar samples from two groups of horses from a digital hypothermia (DH) study. In the DH study, one forelimb of each horse was kept at approximately 4°C in ice water and the other at ambient temperature following administration of 10 g/kg oligofructose (OF). Dorsal laminae were harvested for snap freezing at either 24 hours after OF administration (DEV, n = 7) or at the onset of lameness (OG1, n = 6) using protein extracted from treated and untreated digital laminae of each horse. Increased laminar concentrations of phospho-p38 MAPK were present in the developmental periods (1.5H and 3H) in the BWE model, and in both the DEV and OG1 periods in the CHO laminitis models. However, digital hypothermia had no effect on laminar phospho-p38 MAPK concentrations. Thus, p38 MAPK is activated in affected laminae in multiple models of laminitis, but does not appear to be the central signaling cascade through which hypothermia works to block the expression of inflammatory molecules. Therefore, p38 MAPK is not likely to be a viable therapeutic target as a sole source for blocking the multiple inflammatory signaling mechanisms inhibited by local hypothermia.

ABSTRACT E-60

DOES CEFQUINOME PENETRATE THE BLOOD BRAIN BARRIER IN THE NORMAL HORSE? Hollis AR[1] Duggan VE[2] and Corley KTT[3]. [1]Scott Dunn's Equine Clinic, Berkshire, UK; [2]University College Dublin, Dublin, Ireland; [3]Anglesey Lodge Equine Hospital, The Curragh, Ireland.

Meningitis is a rare but serious condition that occurs in both foals and adult horses. There is currently a restricted choice of antimicrobials that are both safe to use in horses and penetrate the blood brain barrier. Cefquinome is a fourth generation cephalosporin that has activity against *Streptococcus*, the most commonly reported causative organism in adult horse meningitis. Therefore, if cefquinome were to achieve therapeutic concentrations in cerebrospinal fluid following routine administration, this would be an exciting advance for the treatment of meningitis in the horse.

5 mature, healthy horses were used on 2 separate occasions, seven days apart, in a crossover design. Each horse was administered either cefquinome (1 mg/kg) or saline (equivalent volume). Cerebrospinal fluid was collected via atlanto-occipital puncture under general anaesthesia 1 and 4 hours after administration of cefquinome or saline placebo. Blood samples were collected prior to, and 1 and 4 hours after administration of cefquinome or placebo. All samples were analysed for the presence of cefquinome by a laboratory masked to treatments administered.

Cefquinome was detectable in the cerebrospinal fluid in all horses 4 hours after intravenous administration, and in 2 horses 1 hour after administration.

Cefquinome penetrates the blood-brain barrier and it is therefore a potential treatment for equine meningitis. Further investigation of the pharmacokinetics and pharmacodynamics of cefquinome in the cerebrospinal fluid is warranted to establish the optimum intravenous dose.

ABSTRACT E-61

PHARMACOKINETICS OF FIROCOXIB AND ITS INTERACTION WITH ENROFLOXACIN IN HORSES. S. Cox, L Dudenbostel, C Sommardahl, J Yarbrough, M Saleh, T. Doherty. University of Tennessee, College of Veterinary Medicine, Knoxville, TN.

The purpose of this study was to determine if enrofloxacin alters the pharmacokinetics of firocoxib in the horse. Firocoxib is a coxib-class nonsteroidal anti-inflammatory drug (NSAID) approved for use in horses to control pain and inflammation associated with osteoarthritis. Dosages of firocoxib are species dependent, with the recommended dose for horses being 0.1 mg/kg as an oral paste every 24 h. The main elimination pathway of firocoxib is hepatic; however the effects of concurrent administration of drugs that may inhibit its metabolism have not been evaluated.

Enrofloxacin is a synthetic antibacterial agent from the flouroquinolone group developed for veterinary use. It is primarily used for gastrointestinal, urogenital, skin and respiratory tract infections in various animals. A well acknowledged problem associated with flouroquinolone usage is their effect on the metabolism of other drugs. Co-administration of multiple drugs can result in unpredictable therapeutic outcomes. Often it is either diminished therapeutic efficacy or increased toxicity of one or more of the administered drugs. Various pharmacokinetic interactions between antimicrobials and NSAIDs have been described.

Six healthy, adult mares were administered 0.1 mg/kg of firocoxib orally. Samples were collected by direct venipuncture of the jugular vein at 0 (control), 15, 30, and 60 min, 2, 4, 8, 12, 16, 24, 48, 72, and 96 h after administration. After a 20 day washout period the six horses were pretreated 3 days with enrofloxacin 5 mg/kg intravenously every 24 h then on the fourth day given 0.1 mg/kg of firocoxib orally. Samples were collected at 0 (control), 15, 30, and 60 min, 2, 4, 8, 12, 16, 24, 48, 72, and 96 h after administration. All samples were stored at −80°C until analysis using a validated HPLC method. The t1/2, C_{max}, T_{max}, AUC_{0-24} and $AUC_{0-\infty}$ after firocoxib administration were 30.66 ± 5.95 h, 91.00 ± 18.44 ng/ml, 0.87 ± 0.62 h, 1057.35 ± 158.63 h*ng/mL and 2366.48 ± 617.19 h*ng/mL respectively. The t1/2, C_{max}, T_{max}, AUC_{0-24} and AUC_{0-} after firocoxib and enrofloxacin administration were 32.38 ± 10.34 h, 78.67 ± 11.62 ng/ml, 1.17 ± 0.41 h, 1031.54 ± 208.17 h*ng/mL and 2658.72 ± 902.10 h*ng/mL respectively. There were no statistically significant differences between the pharmacokinetic parameters indicating that the administration of enrofloxacin at this dose does not impact the pharmacokinetics of firocoxib.

ABSTRACT E-62

PHARMACOKINETICS AND PHARMACODYNAMICS OF QUINAPRIL IN HORSES. K Kruger JL Davis B Breuhaus. North Carolina State University College of Veterinary Medicine, Raleigh, NC.

Angiotensin converting enzyme (ACE) inhibitors improve survival and quality of life in humans and small animals with cardiovascular and renal disease. There is limited information regarding their effects in horses. The purpose of this study was to determine the pharmacokinetics of quinapril and its effects on ACE inhibition in horses.

Six healthy horses were administered quinapril at 120 mg IV, 120 mg PO or 240 mg PO in a 3-way crossover design. Blood was collected at predetermined times for measurement of quinapril and quinaprilat concentrations using high pressure liquid chromatography, as well as ACE concentrations using a radioenzymatic assay. Normally distributed data were analyzed with one way repeated measures analysis of variance (RM-ANOVA) and non-normally distributed data were analyzed using Friedman RM_ANOVA on ranks. Significance was set at $P < 0.05$.

No adverse effects were observed during the study period. Plasma quinapril concentrations were low and rapidly declined after IV administration. Quinaprilat concentrations were below the limit of quantification (0.1 µg/mL). ACE activity was significantly decreased from baseline at 0.5 and 1 hour after IV dosing and at all timepoints after oral dosing. Maximum % ACE inhibition was 72, 53 and 47% with the IV, high and low oral doses, respectively.

These results suggest that, despite low plasma concentrations, quinapril has sufficient oral absorption and results in inhibition of ACE in healthy horses. Controlled studies in clinically affected horses are indicated.

ABSTRACT E-63

PHARMACOKINETICS OF FIROCOXIB IN NEONATAL FOALS. N. Hovanessian[1], M.V. Crisman[1], D.R. Hodgson[1], H.C. McKenzie[2], J.L. Davis[3]. [1]Virginia-Maryland Regional College of Veterinary Medicine, Blacksburg, VA. [2]Marion DuPont Scott Equine Medical Center, Virginia-Maryland Regional College of Veterinary Medicine, Leesburg, VA. [3]North Carolina State University College of Veterinary Medicine, Raleigh, NC.

This study determined the pharmacokinetic profile of firocoxib in healthy neonatal foals. Foals are more sensitive to the side effects of NSAID, primarily due to immature renal clearance mechanisms and ulcerogenic effects on gastric mucosa. Firocoxib, a novel, second generation NSAID, is reported to have reduced side effects due to COX-2 selectivity. The pharmacokinetic profile of firocoxib in neonates has not been established. We hypothesized that firocoxib given PO to neonatal foals would achieve therapeutic concentrations in plasma.

Seven healthy foals of mixed gender were administered 0.1 mg/kg firocoxib PO q24h for nine consecutive days, commencing at 36h old. Blood was collected for firocoxib analysis at 0 (dose #1 only), 0.25, 0.5, 1, 2, 4, 8, 16 and 24h after doses #1, 5 and 9. For all other doses (2, 3, 4, 6, 7 and 8) blood was collected immediately prior to the next dose (24h trough). Elimination samples were collected after dose #9. Plasma was stored at −80°C until analysis. Physical examinations were performed on foals daily and body weight obtained every two days during the sampling period.

Analysis of plasma samples by liquid chromatography-mass spectrometry revealed firocoxib was rapidly absorbed. After the initial dose, a maximum plasma concentration was reached in 30 min, minimal accumulation after repeat dosing occurred and steady state was obtained after approximately four doses. After the final dose, plasma drug concentration decreased in a linear manner with an estimated terminal t1/2 of 11h. Seventy-two hours after the final dose, firocoxib was not detectable (< 2ng/mL).

ABSTRACT E-64

A RETROSPECTIVE EVALUATION OF ERYTHROCYTOSIS IN HORSES WITH LIVER DISEASE. JR Elfenbein[1], AM House[1], N Pusterla[2], EA Carr[3], HE Banse[4], MH Barton[5]. [1]University of Florida College of Veterinary Medicine, Gainesville, FL. [2]University of California-Davis School of Veterinary Medicine, Davis, CA. [3]Michigan State University College of Veterinary Medicine, East Lansing, MI. [4]Oklahoma State University College of Veterinary Medicine, Stillwater, OK. [5]University of Georgia College of Veterinary Medicine, Athens, GA.

Erythrocytosis is reportedly a rare finding associated with hepatocellular carcinoma in horses. The purpose of this study was to determine the relative frequency of erythrocytosis and the clinicopathologic abnormalities and hepatic histopathology associated with erythrocytosis in horses with liver disease.

Ninety-seven horses aged ≥1 year with clinicopathologic or clinical signs of liver disease, a complete blood count (CBC), and hepatic histopathology were included. Information on CBC, biochemical variables, and hepatic histopathology was collected from records. Data from horses with erythrocytosis (packed cell volume > 45%) were compared to those without using the Mann-Whitney rank sum test with significance set at $p < 0.05$.

Thirty-six cases had erythrocytosis (packed cell volume 51.7% [46–61%]). There was significantly greater bilirubin concentration (4.9 mg/dL [0.9–24.4] vs. 3.3 [21.8–0.5]) and bile acids concentration (74 µmol/L [1.2–222] vs. 34 [3–272]) in cases with erythrocytosis. There were no differences between groups in white blood cell count, gamma-glutamyl transferase, sorbitol dehydrogenase, aspartate aminotransferase, and alkaline phosphatase activities, total protein, albumin, globulin, blood urea nitrogen, or glucose concentrations. Fibrosis (64%), biliary hyperplasia (56%), inflammatory infiltrate (50%), megalocytosis (25%), degeneration (25%), necrosis (25%), cholestasis (25%), anisocytosis and anisokaryosis (11%), and lipidosis (3%) were observed in livers of horses with erythrocytosis. Neoplasia (3%) was observed rarely.

This study reports a high frequency of erythrocytosis in horses with liver disease. Erythrocytosis is associated with higher total bilirubin and serum bile acids concentrations. Common histopathologic changes include fibrosis, biliary hyperplasia, and inflammatory infiltrate. Hepatic neoplasia was rare.

ABSTRACT E-65

SALE PRICE AND RACE EARNINGS IN HORSES PREVIOUSLY DIAGNOSED WITH *LAWSONIA INTRACELLULARIS* INFECTION. M. Frazer. Hagyard Equine Medical Institute, Lexington, KY.

This study was performed to determine if horses diagnosed with equine proliferative enteropathy (EPE) from *Lawsonia intracellularis* (LI) infection had long term effects from disease based on their sale price as yearlings and race earnings.

A retrospective review of medical records of Thoroughbred horses that were treated for *Lawsonia intracellularis* infection between January 1, 2002 and January 31, 2008 at Hagyard Equine Medical Institute in Lexington, Kentucky was performed. Three criteria were used for inclusion in this study. First, each horse had presumptively been diagnosed with LI based on physical examination findings such as ventral edema, diarrhea, lethargy, or poor body condition. Second, horses had hypoalbuminemia of less than 2.5 mg/dL (normal reference range: 3.4–4.5 mg/dL). Third, each horse had a positive fecal polymerase chain reaction (PCR) for LI, a positive serum immunoperoxidase monolayer assay (IPMA), or both. An IPMA titer greater than or equal to 60 was considered positive for disease. 116 horses met the initial criteria.

36 of the 116 horses sold at public auction as yearlings. The sale price of these horses was compared to the average sale price of all yearlings by the same sire as the affected horse (control group). 30 of the 116 horses raced in the United States. Their monetary earnings from racing were compared to the average monetary earnings of all progeny by the same sire as the affected horse (control group). Earnings of horses that were between 3 and 7 years of age (23/30 horses) at the conclusion of the study were compared to the lifetime average earnings of the stallion's progeny. Earnings from horses that were two years of age (7/30) at the end of the study were compared to the two year old average earnings of the stallion's progeny. Monetary earnings from all races prior to December 31, 2008 were included in the study. 12 horses both sold at public auction

and raced. As well as being included in the total number of horses that sold and raced, their sale records and monetary earnings were compared to the averages from their respective sire as a separate group.

This retrospective study indicated that yearling horses previously infected with LI do not sell for as much at public auction as their herdmates, but their monetary earnings from racing are not significantly different from other horses. These results should assist practitioners in guiding owners in determing if treatment of horses with EPE is appropriate and it may aid in reassuring owners that despite the poor condition of the horse during and shortly after the course of disease, horse may still have future athletic potential.

This abstract was presented at the AAEP in December 2009.

ABSTRACT E-67

DISPOSITION OF CEFTIOFUR CRYSTALLINE FREE ACID IN PLASMA AND PULMONARY EPITHELIAL LINING FLUID OF WEANLING FOALS. BC Credille[1], S Giguère[1], LJ Berghaus[1], TL Sturgill[1], JM Donecker[2], GS Grover[2], SA Brown[2]. [1]University of Georgia, Athens, GA. [2]Pfizer Animal Health, Kalamazoo, MI.

Bronchopneumonia caused by *Streptococcus equi* subsp. *zooepidemicus* (*S. zooepidemicus*) is one of the most important causes of morbidity in weanling foals. Ceftiofur crystalline free acid (CCFA) is a long acting third-generation cephalosporin antimicrobial recently approved for the treatment of bronchopneumonia associated with *S. zooepidemicus* in adult horses. The objective of the present study was to determine the disposition of CCFA in plasma and pulmonary epithelial lining fluid (PELF) of weanling foals. Six healthy 4- to 5-month-old weanling foals were administered a single intramuscular injection of CCFA at a dose of 6.6 mg/kg of body weight. Concentrations of desfuroylceftiofur acetamide (DCA) and related metabolites were measured by use of ultra-high performance liquid chromatography and tandem mass spectrometry.

Following IM administration, median time to maximum plasma and PELF concentrations was 24 h (12–48 h). Mean (± SD) peak DCA concentration in plasma (1.44 ± 0.46 µg/mL) was significantly higher than that in PELF (0.46 ± 0.03 µg/mL). Terminal half-life of DCA in plasma (74.8 ± 20.9 h) was not significantly different from that of PELF (58.5 ± 11.4 h). Time above the therapeutic target of 0.2 µg/mL was significantly longer in plasma (185 ± 20 h) than in PELF (107 ± 31 h).

Based on the results of the present study, intramuscular administration of CCFA at a dose of 6.6 mg/kg would be appropriate for the treatment of bronchopneumonia caused by *S. zooepidemicus* and other susceptible pathogens in weanling foals.

ABSTRACT EN-1

CHANGES IN PLASMA FIBROBLAST GROWTH FACTOR-23 (FGF-23) CONCENTRATIONS AFTER TREATMENT OF HYPERTHYROIDISM (HTH) IN CATS WITH VARIABLE RENAL FUNCTION. TL Williams, J Elliott, HM Syme. Royal Veterinary College, London, UK.

FGF-23 is secreted by osteocytes and osteoblasts in response to hyperphosphatemia. FGF-23 enhances phosphaturia and is postulated to have a central role in the development of secondary renal hyperparathyroidism. Hyperthyroid cats have elevated plasma phosphate and parathyroid hormone concentrations, which may in part be associated with underlying chronic kidney disease (CKD). The aim of this study was to determine if plasma FGF-23 concentrations were associated with the presence of underlying CKD in hyperthyroid cats, and to investigate the changes in plasma FGF-23 concentrations that occur following treatment of HTH.

Hyperthyroid cats were recruited from two London-based first opinion practices between 1999 and 2009. Cats that were azotemic at diagnosis were excluded. HTH was treated with anti-thyroid medication alone or in combination with thyroidectomy. Cats were included in the study if they had a plasma total thyroxine concentration < 40 nmol/l documented for a six month period following commencement of treatment. Cats were classified as having azotemic CKD if they developed renal azotemia within six months of establishment of euthyroidism. Otherwise cats were deemed to have normal renal function. Stored EDTA plasma samples were assayed for FGF-23 using a recently validated ELISA. The Mann-Whitney U test and the Wilcoxon signed rank test were used to compare between the groups and assess the response to treatment respectively. Results are reported as median [25th, 75th percentiles]. Correlations were made using Spearman's correlation coefficient.

Thirty one cats with HTH (14 azotemic and 17 non-azotemic) were included in the study. Plasma phosphate concentrations decreased following treatment in cats that did not develop azotemia (4.84 [3.91, 5.64] mg/dl vs. 3.91 [3.38, 4.37] mg/dl; n = 13, P = 0.01) whereas plasma phosphate concentrations did not change significantly following treatment in cats that did develop azotemia (4.28 [3.81, 5.64] mg/dl vs. 4.22 [3.10, 5.33] mg/dl; n = 14, P = 0.158). Plasma FGF-23 concentrations were significantly higher in cats that developed azotemia than cats that did not at both pre treatment (211.7 [176.4, 356.3] pg/ml vs. 148.3 [118.8, 274.9] pg/ml; P = 0.039) and post treatment (514.0 [250.2, 800.0] pg/ml vs. 195.1 [160.7, 287.3] pg/ml; P = 0.001) timepoints. Plasma FGF-23 concentrations increased following treatment in both azotemic (P = 0.004) and non-azotemic groups (P = 0.025). Plasma FGF-23 concentrations and plasma phosphate concentrations were not correlated at baseline (r_s = 0.189, P = 0.335) or following treatment (r_s = 0.136, P = 0.472).

Plasma FGF-23 concentrations were higher in pre-azotemic cats than non-azotemic cats and increased following treatment of HTH. The reason that FGF-23 concentrations increased following treatment, particularly in the face of decreasing plasma phosphate concentrations in cats that remain non-azotemic, is unclear but may be related to the decline in glomerular filtration rate.

ABSTRACT EN-2

GENE EXPRESSION ANALYSIS OF FELINE THYROID TISSUE AND BLOOD FROM CATS WITH EVIDENCE OF MILD OR MARKED HYPERTHYROIDISM REVEALS POTENTIAL MOLECULAR CAUSES OF THE DISEASE AND IDENTIFIES FUTURE ROUTES FOR INTERVENTION. X. Gao, S. Malladi, L. Melendez and S. Al-Murrani. Hill's Pet Nutrition, Topeka, KS.

Hyperthyroidism is a disorder resulting from the excessive production and secretion of T4 and T3 by the thyroid gland. Although the disorder and its pathological lesions have been well studied and described the cause remains illusive.

Whole blood and solid tissue samples from non-diseased, severe disease and mild disease cats based on T4 levels and thyroid histology were used in this study. Whole blood samples from 29 non-disease cats, 28 severe disease cats and 17 mild disease cats as well as solid thyroid tissue samples from 30 non-disease cats, 31 severe disease cats and 27 mild disease cats were collected and processed. The resulting total RNA samples were used for GeneChip analysis using our custom feline gene chip designed by Affymetrix. Data analysis was performed using The Partek® GS software for Gene Expression data. The Robust Multichip Average algorithm was used for background adjustment, normalization, and probe-level summarization of the raw data. ANOVA analysis was performed to find significant differentially expressed genes with a minimal False Discovery Rate control of 0.1 and a fold change of 1.3 in each direction.

During the mild disease state, pathways associated with DNA damage and apoptosis are most prominent. At later stages when the histopathological disease is more severe in addition to the aforementioned pathways others associated with TGF-beta signaling, cell adhesion and extracellular matrix remodeling take more prominence.

The analysis of this unique data set generated from the use of our proprietary GeneChip revealed molecular mechanisms that are associated with the transition from non-disease, to mild disease to severe disease, in the thyroid tissue as well as the blood. These mechanisms could provide insights into the causes of the disease and identify potential new therapeutic and diagnostic targets.

ABSTRACT EN-3

IRIS STAGES OF CHRONIC KIDNEY DISEASE BEFORE AND AFTER TREATMENT WITH RADIOIODINE IN CATS WITH HYPERTHYROIDISM. L.S. Harley[1], M.E. Peterson[2], C.E. Langston[1], R. Nichols[2,3]. [1]The Animal Medical Center, New York, NY. [2]Animal Endocrine Clinic, New York, NY. [3]Antech Diagnostics, Lake Success, NY.

Although it is well established that concurrent chronic kidney disease (CKD) develops in about 30% of hyperthyroid cats, no one has reported the use of the IRIS staging system for CKD before and after treatment of these hyperthyroid cats. The purpose of this study was to compare the effects of treatment in hyperthyroid cats with known stage 1 and 2 CKD in order to determine the effects of restoring euthyroidism or inducing hypothyroidism has on the IRIS stage in these cats. We evaluated 36 hyperthyroid cats (median age, 14 years) in this study. One day prior to treatment, serum T_4 concentration, serum chemistry analysis, complete urinalysis, and urine protein-to-creatinine ratio (UPC) were measured. All cats were again evaluated with the same parameters again 3 months after treatment with ^{131}I. Prior to treatment, 26 (72%) of the 36 cats had no evidence of azotemia (serum creatinine < 1.6 mg/dl), whereas 10 cats (28%) had stage 2 CKD (serum creatinine, 1.6–2.8 mg/dl). In the 36 cats, IRIS staging revealed proteinuria in 33 cats (92%), 21 with borderline proteinuria (UPC, 0.2–0.4) and 12 with overt proteinuria (UPC > 0.4). Hyperthyroidism was cured in all 36 cats (median post-T4, 1.3 µg/dl). All cats had a good response to treatment; there were no signs of CKD except for polyuria and polydipsia in some cats. A significant ($P < 0.001$) increase in median values for both serum urea nitrogen (26 mg/dl to 31 mg/dl)) and creatinine (1.1 to 1.7 mg/dl) occurred after treatment. Nine of the 26 cats (34.6%) classified as nonazotemic or IRIS stage 1 prior to ^{131}I progressed to stage 2 CKD after ^{131}I. All 10 cats with stage 2 CKD before treatment remained azotemic after ^{131}I, with 5 cats remaining in stage 2 CKD, and 4 cats progressing to stage 3 CKD (serum creatinine, 3.1–3.7 mg/dl). There was a significant inverse relationship ($P = 0.002$) between pretreatment urine specific gravity (USG) and post-treatment serum creatinine in the 36 cats. Of the 19 cats with post-treatment serum creatinine values > 1.5 mg/dl (stage 2 to 3 CKD), 15 (79%) had pretreatment USG of < 1.040. In contrast, in the 17 cats with post-treatment serum creatinine values < 1.5 mg/dl, only 3 (18%) had pretreatment USG of < 1.040. A significant ($P < 0.001$) decrease in median UPC from 0.3 to 0.1 occurred after treatment, but there was no relationship between degree of proteinuria and IRIS stage in these cats. Two cats developed iatrogenic hypothyroidism after ^{131}I, diagnosed by finding low serum T_4 and high cTSH concentrations. Both hypothyroid cats had progressed from stage 1 before treatment to stage 2 and 3 CKD, respectively, after ^{131}I; after thyroxine replacement, serum creatinine decreased to near pretreatment concentrations in both cats. Conclusions: 1) IRIS stage 2 CKD is common in untreated hyperthyroid cats. 2) Progression to next higher IRIS stage is common after treatment, but most cats with remain relatively asymptomatic for CKD. 3) USG may be helpful in predicting which cat's IRIS stage will progress after ^{131}I. 4) Iatrogenic hypothyroidism worsens azotemia, an effect that appears reversible with replacement therapy.

ABSTRACT EN-4

IMPACT OF HOME BLOOD GLUCOSE MONITORING ON QUALITY OF LIFE OF DIABETIC CATS AND THEIR OWNERS RECORDED USING THE PSYCHOMETRIC TOOL DIAQoL-PET. S. Niessen[1,2], S. Powney[1], J. Guitian[1], A. Niessen[3], P. Pion[4], J. Shaw[2], D. Church[1]. [1]Royal Veterinary College, Hatfield, Herts, UK. [2]Newcastle Medical School, Tyne and Wear, Newcastle, UK. [3]Venlo, Limburg, Netherlands [4]VIN, Davis, CA, USA.

Home blood glucose monitoring (HBGM) of diabetic pets is likely to result in superior glycaemic control, minimizing episodes and impact of dangerous hypoglycaemia and reducing costs. Nevertheless, it has proven difficult to objectively establish a clear benefit of HBGM using biological parameters (clinical signs, blood glucose, fructosamine). The current study aimed to assess the impact of HBGM on owner perceived quality of life (QoL) aspects of diabetes mellitus (DM) treatment, using the recently validated psychometric tool DIAQoL-pet.

Owners of insulin treated diabetic cats were recruited to complete the 29-item tool, evaluating areas affecting the cat's and owner's QoL, including: worry about pet's DM, hypoglycaemia, costs, owner's desire for autonomous control over the pet's DM, etc. Item-weighted-impact-scores (IWIS), reflecting frequency and importance ratings of each item, were calculated, as well as average-weighted-impact-scores (AWIS; average IWIS of all items), as an overall measure of diabetes dependent QoL. Frequencies, IWIS and AWIS were compared between owners practising HBGM and those who did not using Mann Whitney U test (significance $p < 0.05$).

Two hundred and eleven owners of insulin treated diabetic cats completed the DIAQoL-pet; 161 owners practised HBGM, whereas the remaining 50 did not practise any form of home monitoring (including urine glucose). IWIS for 'excessive drinking' and 'owner wanting more control' were significantly different between the HBGM-group (mean+/-standard deviation: −2.01+/−2.4 and −5.07+/−4.8) and the non-HBGM-group (−3.36+/−3.8 and −1.86+/−3.2). There was no significant difference between the groups with regards to the IWIS for other items, including 'worry about hypoglycaemia' or 'worry about pet's DM'. Polydipsia was reported significantly more frequently in the non-HBGM-group and this was the reason for the difference between groups in this item's IWIS as it was considered of equal importance. Frequency and IWIS of reported occurrence of hypoglycaemia signs were not significantly different. AWIS for both groups was not significantly different (HBGM: −1.85+/−1.3; non-HBGM: −1.87+/−1.1).

The current study suggests that HBGM is predominantly practised by owners who desire more autonomous control over their cat's DM. The frequency of polydipsia was lower in the HBGM-group perhaps suggesting superior control. However, HBGM did not detectably affect the impact of the majority of QoL-items, nor the frequency of hypoglycaemic episodes. Overall diabetes dependent QoL of diabetic cat and owner, as measured per DIAQoL-pet, was unaffected by HBGM. These data argue for the use of HBGM in selected pet-owner combinations rather than as part of a practice's standard DM management protocol, although further studies are indicated.

ABSTRACT EN-5

TISSUE-SPECIFIC PATTERN OF INSULIN RESISTANCE IN GLUCOSE-, LIPID- AND LIPOPOLYSACCHARIDE-TREATED CATS. M Osto[1], E Zini[2], M Franchini[3], M Hafner[2], M Ackermann[3], CE Reusch[2], TA Lutz[1]. [1]Institute of Veterinary Physiology; [2]Clinic for Small Animal Internal Medicine; [3]Institute of Virology; Vetsuisse Faculty, University of Zurich, Switzerland.

Insulin resistance is associated with impaired activation of the insulin signaling pathway in peripheral tissues such as skeletal muscle, visceral and subcutaneous (SC) adipose tissue. High plasma glucose, fatty acid and endotoxin levels are three major causes of insulin resistance in feline and human obesity and in type 2 diabetes mellitus. However, the mechanisms by which these factors influence insulin action are still unclear. Therefore, our aim was to investigate the tissue-specific expression of crucial mediators of insulin action such as the insulin-receptor substrate 1 (IRS1), the serine/threonine protein kinase B (PKB/Akt) and of the principal insulin-dependent glucose transporter protein (GLUT4) in feline models of hyperglycemia, hyperlipidemia and subacute endotoxemia.

Healthy cats were infused through the jugular vein with glucose ($n = 5$), lipids ($n = 6$) or lipopolysaccharide (LPS; $n = 5$) for 10 days to clamp their blood concentrations at the approximate level found in untreated feline diabetes (glucose: 25–30 mmol/l; triglycerides: 3–7 mmol/l) or to induce a systemic low-grade inflammation (LPS; rectal temperature: 39.2–40.5°C), respectively. Healthy control cats were infused with saline ($n = 10$). On day 10, specimens were collected from skeletal muscles, visceral and SC fat and processed for IRS1 mRNA expression, total and phosphorylated PKB/Akt and GLUT4 protein expression.

Gene transcripts of IRS1 were not different between the groups. Compared to controls, skeletal muscle PKB/Akt phosphorylation was 91% lower in cats infused with glucose ($P < 0.01$); lipid-infused cats showed a trend for a decrease in PKB/Akt phosphorylation (31% lower than saline) and had decreased GLUT4 expression ($P < 0.05$) in muscle. Total ($p < 0.01$) and phosphorylated ($P < 0.05$) PKB/Akt protein expression were decreased in the SC adipose tissue of LPS-infused cats compared to controls. In these cats, phosphorylation of PKB/Akt protein was also decreased in visceral fat ($p < 0.01$).

Sustained hyperglycemia and, to a lesser extent, hyperlipidemia impaired insulin signaling and glucose transport pathways primarily in skeletal muscle; endotoxemia reduced insulin sensitivity mainly in adipose tissues. Thus, the development of insulin resistance in response to hyperglycemia, hyperlipidemia or endotoxemia might be affected by tissue-specific mechanisms in cats.

ABSTRACT EN-6

IMAGE FUSION OF SINGLE-PHOTON EMISSION COMPUTED TOMOGRAPHY AND COMPUTED TOMOGRAPHY SCAN IN CANINE INSULIMONA: 10 CASES.
Isabel Rodriguez Piñeiro[1,2,3], F Delisle[2,3], Patrick Devauchelle[3], Dan Rosenberg[2], Pauline de Fornel-Thibaud[2,3]. [1]Internal Medicine Unit, VetAgro Sup. Lyon, France. [2]Internal Medicine Unit, National Veterinary School of Alfort, France. [3]Veterinary Anticancer Center, Maisons-Alfort, France.

Separately used, Single Photon Emission Computed Tomography (SPECT) and Computed Tomography (CT) both lack sensitivity and are additionally hampered by a poor anatomical location capacity and a lack of specificity, respectively. These drawbacks suggest an interest in the fusion of images obtained by the 2 techniques. The aim of this study is to test SPECT/CT fusion performance in dogs with insulinoma.

Inclusion criteria were: 1/ a biological diagnosis of insulinoma; 2/ an examination by high resolution CT scan and ^{111}In-pentetreotide SPECT followed by SPECT/CT fusion; 3/ a surgical or post mortem examination completed by histopathological analysis. SPECT examination showing abnormal foci and CT scan showing pancreas, lymph nodes (LN) or liver abnormalities were considered positive. In case of double positivity, presence (IMP+) or absence (IMP-) of superimposition of abnormal images was noted.

Ten dogs were included. In 2/10 dogs, superimposition of abnormalities couldn't be tested. CT scan detected 3 abnormal images [2 pancreatic nodules (PN), 1 enlarged LN (ELN)] while SPECT failed to show any abnormal uptake. Both dogs became euglycemic after removal of PN and LN designed by CT scan. In 6/10, all abnormal images were classified as IMP+ [6 PN, 1 ELN and 1 diffuse hepatic infiltration (DHI)]. Surgery performed on 5/6 resulted in euglycemia in 4; 1 dog remained hypoglycemic after partial removal of 1 PN. PN localization and DHI were confirmed after necropsy in the 6th dog. In 2/10 dogs IMP+ and IMP- images were both recorded. In 1 dog, a DHI was classified as IMP+ but PN localization was IMP-: localized in the left lobe by CT scan and in the corpus by SPECT, the latest localization being confirmed after necropsy. In the other dog PN localization was IMP+ but a diffuse SPECT signal superimposing to the liver considered as normal on CT scan was noted. Hepatic biopsy confirmed SPECT results.

This study confirms an imperfect sensitivity of both CT scan and SPECT. It confirms that CT scan can be associated with unspecific abnormal images. Subject to a confirmation on a larger cohort of dogs, it indicates that IMP+ images provide specific detection and accurate localization of canine insulinomas' primary lesions and metastasis.

ABSTRACT EN-7

EVALUATION OF ALDOSTERONE CONCENTRATIONS IN DOGS WITH ADDISON'S DISEASE.
ME Baumstark, C Mueller, FS Boretti, NS Sieber-Ruckstuhl, M Wenger, CE Reusch. Clinic for Small Animal Internal Medicine, Vetsuisse Faculty, University of Zurich, Zurich, Switzerland.

The majority of dogs with primary hypoadrenocorticism (PH) reveal clinical and laboratory abnormalities of gluco- and mineralocorticoid deficiency. In some of them sodium and potassium levels are normal, a phenomenon currently called atypical Addison's. It has been postulated that in those cases adrenal destruction is confined to the zona fasciculata/reticularis, resulting in isolated glucocorticoid deficiency. However, there are no histological studies confirming a normal zona glomerulosa and in most reported cases diagnosis was based solely on low post-ACTH cortisol levels. The aim of the study was to evaluate aldosterone (ALDO) levels in dogs with PH with and without electrolyte abnormalities.

Seventy dogs with newly diagnosed PH were included. ALDO concentrations (RIA, Coat-A-Count®, Siemens) were measured before and 60 min after administration of 250 μg synthetic ACTH (Synacthen®, Novartis) IV. Results were compared to those of 19 healthy dogs and 17 dogs with diseases mimicking PH. To confirm that peak concentrations were not missed ALDO was additionally measured 15, 30 and 45 min after ACTH in 19 dogs (5 with PH, 14 with PH mimicking diseases). Results were analysed by means of non-parametric statistical methods ($p < 0.05$).

Post-ACTH ALDO was significantly lower in dogs with PH (0–253 pg/ml, median 0 pg/ml) than in healthy dogs (46–602 pg/ml, median 187 pg/ml) and in dogs with PH mimicking diseases (0–639 pg/ml, median 155 pg/ml). Low post-ACTH ALDO was found in 67/70 dogs with PH, in 64/67 of them levels were below the detection limit of the assay. Normal sodium and potassium levels were found in 5/67 dogs (7%), 6/67 dogs (9%) had hyponatremia and normal potassium, 56/67 dogs (84%) had hyponatremia and hyperkalemia. Electrolyte abnormalities ranged from mild to severe. There was no correlation between post-ACTH ALDO and sodium and a weak correlation between post-ACTH ALDO and potassium ($r = -0.28$). ALDO concentrations were not different 30, 45 and 60 min after ACTH.

The results demonstrate that ALDO levels are low in most dogs with PH independent of the degree of electrolyte abnormalities. This implicates that all three zones of the adrenal cortex are compromised and that there are mechanisms which allow maintenance of a normal electrolyte balance without ALDO.

ABSTRACT EN-8

COMBINING WHITE BLOOD CELL COUNTS AND THE SODIUM TO POTASSIUM RATIO TO SCREEN FOR CANINE HYPOADRENOCROTICISM.
M Seth[1], KJ Drobatz[1], DB Church[2], RS Hess[1]. [1]University of Pennsylvania, Philadelphia, PA. [2]Royal Veterinary College, Hatfield, UK.

Definitive diagnosis of canine hypoadrenocorticism (HA) is based on inadequate cortisol secretion following adrenocorticotropic hormone (ACTH) administration. An abnormal serum sodium to potassium (Na:K) ratio can be used to determine whether an ACTH stimulation test is warranted. The aim of this study was to examine the utility of combining the Na:K ratio with white blood cell counts to determine whether an ACTH stimulation test is warranted.

A retrospective review of medical records of dogs examined between 2005 and 2009 was performed. 53 dogs diagnosed with HA and 110 control dogs, in which a diagnosis of HA was excluded during the study period, were included. Inclusion criteria for all 163 dogs were hospitalization with intravenous fluid therapy, a complete blood count, and serum Na and K measurements at the time of initial examination. Dogs were included in the HA group if they also had pre and post ACTH stimulation serum cortisol concentrations ≤ 1.0 μg/dl. Dogs were included in the control group if they had resting or post ACTH stimulation serum cortisol concentration > 2.0 μg/dl. Exclusion criteria were recent administration of glucocorticoids, prior treatment of hyperadrenocorticism, or serum cortisol concentration > 1.0 μg/dl but ≤ 2.0 μg/dl. Continuous variables were compared between groups using the Mann-Whitney U test. Receiver operating characteristic (ROC) curves were produced to assess the sensitivity and specificity of detecting HA with various cutoffs for each variable. Data is presented with 95% confidence intervals (CI) and statistical significance was defined as $p < 0.05$.

The Na:K ratio, neutrophil count and neutrophil:lymphocyte ratio were significantly lower in dogs with HA than in dogs without HA ($p < 0.01$ for each). Lymphocyte and eosinophil counts were significantly higher in dogs with HA compared to dogs without HA ($p < 0.001$ for each). The areas under the curve by ROC analysis were largest for Na:K ratio (0.87, CI:0.81–0.92) and lymphocyte count (0.85, CI:0.78–0.90). A Na:K ratio ≤ 39.6 was 100% sensitive (CI:93–100%) but only 15% specific (CI:9–23%) for detecting HA. A lymphocyte count ≥ 0.77 x10^3 cells/μL was 100% sensitive (CI:93–100%) and 35% specific (CI:27–45%). Conversely a Na:K ratio ≤ 20.1 was 51% sensitive (CI:37–65%) but 100% specific (CI:97–100%) and a lymphocyte count ≥ 6.1 × 10^3 cells/μL was 9% sensitive (CI:3–20%) but 100% specific (CI:97–100%). A Na:K ratio ≤ 35.5 was 96% sensitive (CI:87–100%) and 35% specific (CI:26–44%) for detection of HA and a lymphocyte count ≥ 0.79 × 10^3 cells/μL was 98% sensitive (CI:90–100%) and 36% specific (CI:27–46%) for detection of HA. A combination of this Na:K ratio (≤ 35.5) and lymphocyte count (≥ 0.79 × 10^3 cells/μL) was 96% sensitive (CI:87–100%) and 55% specific (CI:47–66%) for detection of HA.

These results indicate that the combination of lymphocyte count and Na:K ratio results in a better screening test for HA than the use of the Na:K ratio alone.

ABSTRACT EN-9

PLASMA FREE METANEPHRINE AND NORMETANEPHRINE CONCENTRATIONS ARE ELEVATED IN DOGS WITH PHEOCHROMOCYTOMA.
R. Gostelow, H.M. Syme. Royal Veterinary College, London, UK.

Pheochromocytoma is a malignant, catecholamine-producing, adrenomedullary tumor. Clinical signs resulting from excessive catecholamine secretion are typically non-specific, making differentiation from other adrenal tumors a challenge. Elevated plasma concentrations of the catecholamine breakdown products metanephrine (MN) and normetanephrine (NMN) are used to identify pheochromocytoma in humans. This study tested the hypothesis that plasma metanephrine concentrations are greater in dogs with pheochromocytoma than in dogs with other adrenal neoplasms, healthy dogs and dogs with non-adrenal illness.

EDTA plasma was collected from healthy dogs and unwell, hospitalized dogs with non-adrenal illness, pheochromocytoma and cortical tumors between April 2007 and October 2010. Samples were stored at −80°C before measurement of free MN and NMN concentrations using high pressure liquid chromatography at the Central Laboratory for Clinical Chemistry at the University of Groningen (33 samples) or the Mayo Clinic, Rochester, Minnesota (7 samples). Kruskal-Wallis tests followed by Dunn's multiple comparison analysis were used to compare results between groups. Significance was set at $P < 0.05$. Results are reported as median [range].

Eight dogs with pheochromocytoma, 5 healthy dogs, 15 dogs with non-adrenal illness and 12 dogs with cortical tumors were sampled. Pheochromocytoma was diagnosed histologically (7 dogs) or cytologically (1 dog). Cortical tumors were diagnosed histologically (9 dogs) or by response to trilostane treatment after obtaining consistent endocrine test results (3 dogs). MN concentration was significantly higher in dogs with pheochromocytoma (8.3 [1.8–175.2] nmol/l) than in healthy dogs (1.1 [0.9–2.0] nmol/l; $P < 0.05$), dogs with non-adrenal illness (1.9 [0.4–6.6] nmol/l; $P < 0.05$) and dogs with cortical tumors (1.0 [0.3–2.9] nmol/l; $P < 0.001$). NMN concentration was also significantly higher in dogs with pheochromocytoma (56.2 [10.2–190.3] nmol/l) than in healthy dogs (2.5 [2.1–5.0] nmol/l; $P < 0.01$), dogs with non-adrenal illness (3.3 [1.3–10.1] nmol/l; $P < 0.001$) and dogs with cortical tumors (2.8 [0.7–5.4] nmol/l; $P < 0.001$).

Both free MN and NMN concentrations are significantly increased in dogs with pheochromocytoma. Measurement of plasma free metanephrines, particularly NMN, shows promise as a diagnostic test for pheochromocytoma.

ABSTRACT EN-10

ASSESSMENT BY QUANTITATIVE PCR OF ABILITY OF SEX HORMONES TO INDUCE EXPRESSION OF CLASSIC GLUCOCORTICOID-INDUCED GENES IN CANINE HEPATOCYTES. EN Behrend, RJ Kemppainen, RA Kennis, Lee HP. College of Veterinary Medicine, Auburn University, Auburn, AL.

Occult hyperadrenocorticism (HAC) has been theorized to exist in which excess adrenal sex hormone secretion induces the clinical signs and laboratory changes associated with classic HAC. However, the ability of sex hormones to cause such alterations has never been closely evaluated. If sex hormones can cause a syndrome similar to classic HAC, they should be able to induce expression of classic glucocorticoid-induced genes. The purpose of the study was to determine if *in vitro* expression of the gene for corticosteroid-induced ALP (CIALP) could be induced by clinically relevant concentrations of cortisol and sex hormones believed to cause occult HAC.

Canine hepatocytes were purchased from a commercial source (CellzDirect or InVitro) in 6-well plates. Upon arrival (3–4 plates per shipment), the cells were allowed to recover in general media per supplier recommendations. After 4 hrs, media was changed to William's E media (-L-glutamine) containing concentrations of cortisol or sex hormones that have been documented in the literature in dogs with HAC or with purported occult HAC. Each plate was treated with a different hormone (cortisol, 17-hydroxyprogesterone [17OHP], progesterone, estradiol or androstenedione), and each well contained a different concentration (starting with no hormone added as a negative control) to evaluate a dose response. Media was changed daily. After 5 days of hormone exposure, RNA was extracted. Reverse transcription was performed and the product used for quantitative PCR for CIALP and beta-actin (Roche Lightcycler) using a gene-specific fluorescent probe for detection. Standard curves were created for each gene. All samples and standards were run in duplicate. Using the Lightcycler software (vers 4.0), CIALP expression was normalized to that of beta-actin. Fold change in expression was determined relative to the negative control. Each sex hormone was used to treat 3 plates; one plate in each shipment was treated with cortisol as a positive control.

For cortisol, a dose response was seen in expression of the CIALP gene. Compared to no cortisol, 10, 50, 100, 150, 250 and 500 nmol cortisol increased expression 2.8, 4.6, 7.1, 9.3, 9.9 and 9.8 fold, respectively. A 2-fold increase is considered significant (J.Vandesompele et al Genome Biol 2002). Expression of CIALP was not significantly induced in response to any concentration of 17OHP (10 nM maximum), progesterone (5 nM maximum), estradiol (max 400 pM maximum) or androstenedione (100 nM maximum).

We conclude that *in vitro* these sex hormones do not induce expression of the CIALP gene which is classically induced by cortisol *in vivo*; indeed, elevated serum CIALP activity is a hallmark of classic HAC. Thus, the ability of the sex hormones to induce the gene *in vivo* must be questioned and evaluated.

ABSTRACT EN-11

EVALUATION OF MEASUREMENT OF 17-HYDROXY-PROGESTERONE AND ESTRADIOL FOR DIAGNOSIS OF TYPICAL AND OCCULT HYPERADRENOCORTICISM. EN Behrend[1], RJ Kemppainen[1], RA Kennis[1], HP Lee[1], D Bruyette[2], JC Lurye[3]. [1]College of Veterinary Medicine, Auburn University, Auburn, AL; [2]Los Angeles, CA; [3]Pittsburgh, PA.

Measurement of sex hormones has been advocated as an adjunct means for diagnosing typical hyperadrenocorticism (HAC), i.e. disease due to excess cortisol secretion, as well as for diagnosis of atypical HAC, i.e. disease due to excess adrenal sex hormone secretion. However, measurements in either setting have not been widely studied. Therefore, our objectives were: 1. To determine the sensitivity of 17-hydroxy-progesterone (17OHP) and estradiol concentrations pre- and post-ACTH for diagnosis of typical HAC. 2. To determine the specificity of 17OHP and estradiol concentrations pre- and post-ACTH for diagnosis of occult HAC.

Dogs that had PDH (n = 12), dogs that were suspected to have HAC but proven not to (had non-adrenal illness [NAI, n = 89]) or dogs that were healthy (n = 20, used to establish reference ranges [RR]) were enrolled. ACTH stimulation tests were performed (5 mcg/kg cosyntropin IV); blood samples were drawn pre and 60-min post; 17OHP and estradiol were measured by previously validated radioimmunoassays. A Kruskal-Wallis Rank Sum test was used to compare values between the groups. Significance was set at $p < 0.05$.

For basal and ACTH- stimulated 17OHP concentrations, the RR were determined to be 0.03–0.6 ng/mL (mean ± 2 s.d.; range 0.03–0.55) and 0.3–2.2 ng/mL (range 0.31–1.95), respectively. In PDH dogs, 3 and 7 had basal and post-ACTH 17OHP concentrations above the RR, respectively; in the NAI group, 23 and 25 dogs had concentrations above the RR, respectively. Thus, the sensitivity of basal and post-ACTH 17OHP measurement for diagnosis of HAC is 25% and 58%, respectively. Specificity of diagnosis is 72% and 74%, respectively. Post-ACTH 17OHP concentration was significantly different between groups.

For basal and stimulated estradiol concentrations, the RR were determined to be 81–180 pg/mL (range 89–195) and 71–170 pg/mL (range 74–151), respectively. For both basal and stimulated estradiol, 4 PDH dogs (n = 9) had concentrations above the RR; for those with NAI (n = 80), 14 and 15 had concentrations above the RR, respectively. Thus, the sensitivity of estradiol measurement for diagnosis of HAC is 44% for both pre- and post-ACTH. Specificity of estradiol for diagnosis for HAC is 83% and 81% for pre- and post-ACTH, respectively. Overall, 23 dogs with NAI had at least one elevated estradiol concentration (total specificity 70%). Post-ACTH estradiol concentration was not significantly different between groups.

We conclude that use of 17OHP and estradiol concentrations for diagnosis of HAC can be problematic. Sensitivity and specificity are relatively low, potentially leading to misdiagnoses.

ABSTRACT EN-12

A MISSENSE MUTATION IN THE CODING SEQUENCE OF MC4R (MC4R:C.92 C>T) IS ASSOCIATED WITH DIABETES MELLITUS IN DSH CATS. Yaiza Forcada; Angela Holder; Rosanne Jepson; David Church. Brian Catchpole Royal Veterinary College, North Mimms, UK.

Diabetes mellitus is one of the most common feline endocrinopathies and is considered to have a similar pathophysiological basis to human type 2 diabetes. Several studies have identified risk factors for development of diabetes mellitus in cats, which include age, obesity, inappropriate diet and physical inactivity. However, to date, no specific genetic risk factors have been identified. Genome-wide association studies in humans have identified several genes that predispose to obesity and/or diabetes mellitus, one of which is the melanocortin receptor 4 (MC4R) gene. The aim of the current study was to identify polymorphisms (SNPs) in the feline MC4R gene and to use these to perform a case:control study to determine whether these candidate gene SNPs were associated with diabetes mellitus in cats.

Genomic DNA from 10 cats (6 domestic short hair [DSH], 4 Burmese) was initially analysed by PCR and direct sequencing using felMC4R-specific primers, which identified a missense mutation (MC4R:c.92 C > T) in the region encoding the extracellular domain of the receptor protein in DSH cats only. One hundred and nineteen DSH cats were subsequently recruited into the case:control study. Fifty nine cats were obese diabetic (29 male, 30 female), mean age 11.8 years (range 6–18y); mean weight 6.68 kg (range 5.15–10 kg). Sixty lean cats were used as controls (30 male, 30 female), mean age 13.81 years (range 9–19y), mean weight 3.99 kg (range 2.56–5.68 kg). The T to C base change alters a restriction site in the sequence recognized by the enzyme BstOI, such that DNA from cats with the mutant (C) allele can be cut, whereas that from the wild-type (T) allele cannot. Primers were designed that flanked the mutation to allow PCR amplification of this region of MC4R from genomic DNA obtained from EDTA blood. The PCR products were purified and subject to restriction fragment length polymorphism (RFLP) analysis. BstOI digestion products were then analysed by agarose gel electrophoresis. Of the 59 diabetic cats, 32 (54%) were homozygous for the mutation (CC), compared to 21 (35%) of 60 control cats. Statistical analysis (two tailed Fisher's square test) revealed that this difference between groups was statistically significant ($p = 0.0431$).

In conclusion, this pilot study has identified a missense mutation in the coding sequence of MC4R. This could be an important predisposing factor for development of diabetes and/or obesity in DSH cats. Polymorphisms in a similar region of human MC4R predispose to obesity, which in turn is a major risk factor for Type 2 diabetes.

ABSTRACT EN-13

COMPARISON OF THE EFFECT OF TRILOSTANE AND MITOTANE ON PEAK SERUM ALDOSTERONE CONCENTRATIONS IN DOGS WITH PITUITARY-DEPENDENT HYPERADRENOCORTICISM. LE Reid[1], EN Behrend[1], RJ Kemppainen[1], LG Martin[1], TZ Civco[1], JC Lurye[2], CR Ward[3]. [1]College of Veterinary Medicine, Auburn University, Auburn, AL. [2]Pittsburgh, PA. [3]College of Veterinary Medicine, University of Georgia, Athens, GA.

Hyperadrenocorticism (HAC) is one of the most common endocrine disorders of dogs. The two most effective medical treatments are trilostane (Vetoryl®) and mitotane (Lysodren®). Previous studies evaluating the effect of treatment on aldosterone secretion measured the hormone at 60 min post-ACTH administration. However, the optimal sampling time would be at the time of maximal secretion, which occurs 30 minutes after the 5 µg/kg dose commonly used for the test (Carlson et al, JVIM, 2010). Thus, the true effect of either medication on aldosterone secretory capacity is unknown. Our objectives were: 1) to assess and compare the effect of treatment with trilostane and mitotane in dogs with pituitary-dependent HAC (PDH) on aldosterone secretory reserve at 30 min post-ACTH stimulation and 2) to determine if changes in aldosterone concentration at that time correlate with changes in serum sodium and potassium concentrations.

Forty-six dogs being treated for PDH with either mitotane ($n = 30$) or trilostane ($n = 16$) have been enrolled. The dogs could be treated for any length of time. All had ACTH stimulation tests performed (5 mcg/kg cosyntropin IV); blood samples were drawn before and at 30 and 60 min post-ACTH for monitoring of cortisol and aldosterone concentration using previously validated radioimmunoassays. Ten historical normal controls were also included. Serum sodium and potassium concentrations were measured in the basal samples. A Kruskal-Wallis Rank Sum test was used to compare values between normal dogs and those treated with mitotane or trilostane. Linear regression analysis was used to determine if a correlation existed between electrolyte and aldosterone concentrations or between cortisol and aldosterone concentrations. Significance was set at the $p < 0.05$ level.

ACTH-stimulated aldosterone concentrations in mitotane-treated but not trilostane-treated dogs were significantly lower than that in normal dogs at both the 30 and 60 min time points. No difference was detected between aldosterone concentrations at 30 and 60 min after ACTH injection in either treatment group. A positive correlation existed between the 60-min cortisol and 30-min aldosterone concentrations in the trilostane-treated group ($R = 0.813$), i.e. the peak post-ACTH concentration for each hormone, but not in dogs treated with mitotane. Basal serum sodium and potassium concentrations were not correlated with the basal aldosterone concentration in either treatment group.

In conclusion, treatment with mitotane resulted in decreased aldosterone secretory reserve, but this did not correlate with hyperkalemia or hyponatremia. Measurement of aldosterone concentrations is not predictive of electrolyte concentrations.

Previously presented at the Auburn University Phi Zeta Research Emphasis Day, November 10, 2010.

ABSTRACT EN-14

ANTIOXIDANT STATUS IN HYPERTHYROID CATS BEFORE AND AFTER RADIOIODINE TREATMENT. E. Branter, LA Trepanier. Department of Medical Sciences, University of Wisconsin-Madison, Madison, WI.

Antioxidant depletion is documented in humans with hyperthyroidism, and is reversible with treatment. In addition, antioxidant depletion has been shown to increase the risk of methimazole toxicity in rats. The primary aim of this study was to determine whether deficiencies in glutathione (GSH), ascorbate (AA), or vitamin E, along with increases in urinary 8-isoprostanes, were present in hyperthyroid cats, and were reversible after radioiodine treatment. A secondary aim was to determine whether antioxidant abnormalities were associated with a prior history of methimazole toxicity.

Ongoing prospective, controlled, observational study. Otherwise healthy client-owned hyperthyroid cats presenting for radioiodine therapy ($n = 26$ to date) and healthy age-matched controls ($n = 32$ to date) were recruited. All cats were screened with CBC, biochemical panel, urinalysis, and T4, as well as red blood cell (RBC) GSH, plasma AA, plasma vitamin E, and urinary 8-isoprostanes. Hyperthyroid cats were re-evaluated 2 months after radioiodine treatment.

Unlike in humans, median blood antioxidants were not significantly different in hyperthyroid cats (GSH 1.5 mM; AA 11.4 µM, and vitamin E, 19 g/ml) compared to controls (GSH 1.4 mM; AA 12.5 µM, and vitamin E, 17 g/ml). Results for urinary isoprostanes are pending, and associations with methimazole toxicity will be investigated after full recruitment. RBC GSH concentrations did increase significantly (to 1.6 mM; $P = 0.019$) after radioiodine treatment. However, this modest change is unlikely to be clinically significant. Preliminary data do not indicate clinically significant blood GSH, ascorbate, or vitamin E deficiencies in hyperthyroid cats.

ABSTRACT EN-15

DIABETIC CATS IN REMISSION HAVE MILDLY IMPAIRED GLUCOSE TOLERANCE. S Gottlieb[1,2], JS Rand[1], RD Marshall[2]. [1]Centre for Companion Animal Health, University of Queensland, Australia. [2]The Cat Clinic, Brisbane, Australia.

With appropriate insulin therapy and a low carbohydrate diet, up to 90% of newly diagnosed diabetic cats are eventually able to maintain euglycemia without insulin administration, and these cats are considered to have achieved remission. There are currently no published data reporting the glucose tolerance status of cats classified as being in remission, and it is unknown whether these cats are truly in diabetic remission, or should be classified as non-insulin dependent diabetics, or having impaired glucose tolerance, and/or impaired fasting blood glucose. The aim of this study was to determine fasting blood glucose concentrations and glucose tolerance status of cats in remission.

The study was a prospective study in a feline-only clinic. For inclusion, diabetic cats had to have achieved remission through insulin therapy, and insulin withheld for a minimum of two weeks.

Five diabetic cats in remission and five matched non-diabetic cats were enrolled in the study. Blood samples were obtained via the ear vein but where the cat's temperament precluded this, from the jugular.Glucose concentration was measured using a meter calibrated for feline blood (Abbott AlphaTRAK). A simplified glucose tolerance test was performed after food was withheld for 24 hours. A 22G catheter was placed in a cephalic vein three hours before the GTT was commenced, to minimize the effects of stress on blood glucose concentration. Blood glucose concentration was measured at time 0 and then a 1 g/kg dose of glucose was administered slowly via the intravenous catheter. Further blood glucose measurements were made at 2 hours and then hourly until glucose had returned to < 117 mg/dL (< 6.5 mmol/L).

In the control group, all cats had a fasting blood glucose below 117 mg/dL, and following glucose administration, glucose had returned to < 117 mg/dL by 3 hours. Fasting blood glucose in the remission group was < 126 mg/dL (7 mmol/L) in all cats except one, which had fasting blood glucose of 135 mg/dL (7.5 mmol/L). Following glucose administration, all five cats in remission had blood glucose above 117 mg/dL (6.5 mmol/L) at three hours, four were < 117 mg/dL at four hours, and one returned to < 117 mg/dL at five hours. The cat with impaired fasting glucose subsequently became diabetic after steroid administration.

The results of this study show that these cats, while no longer diabetic, have mildly impaired glucose tolerance compared to non-diabetic cats, and a minority have impaired fasting glucose.

ABSTRACT EN-16

TITRATION OF DIETARY IODINE FOR REDUCING SERUM THYROXINE CONCENTRATIONS IN NEWLY DIAGNOSED HYPERTHYROID CATS.

LM. Melendez, RM. Yamka, SD. Forrester, PA. Burris. Hill's Pet Nutrition, Inc. Topeka, KS.

The objective of this study was to determine the role of iodine restriction in the nutritional management of cats with naturally occurring hyperthyroidism. Five domestic shorthair cats ranging in age from 8–17 years were confirmed to have hyperthyroidism based on persistently increased serum total thyroxine concentrations (TT4), palpable thyroid nodule and weight loss. Serum TT4 concentrations ranged from 55- 146 nmol/l (reference range 10–55 nmol/l). The cats were then fed a low iodine containing food (0.47 ppm iodine DMB, as measured by epiboron neutron atomic activation). Serum TT4 concentrations were measured every 3 weeks. Biochemistry parameters were also evaluated at weeks 0, 6 and 9.

At 9 weeks, serum TT4 concentrations had decreased in all cats with 4 of 5 cats (80%) being euthyroid (mean 48 nmol/l; range 41–54 nmol/l). The remaining hyperthyroid cat had an initial serum TT4 of 146 nmol/l, which decreased to 83 nmol/l after being fed the iodine-restricted food. Mean decrease in TT4 for all 5 cats was 26 nmol/l (range 8–63 nmol/l). Renal parameters remained stable in all 5 cats.

These 5 cats along with 4 additional newly diagnosed hyperthyroid cats were transitioned to a similar food that contained less iodine (0.28 ppm DMB). Baseline serum TT4 concentrations in the 4 new cats ranged from 55–73 nmol/l. Serum TT4 and other biochemical parameters were monitored every 3 weeks for 9 weeks, and then every 4 weeks for an additional 8 weeks.

With the 0.28 ppm iodine food the four new cats became euthyroid with a mean TT4 concentration of 41 nmol/ (range 29–50nmol/l). The 4 euthyroid cats from the earlier feeding study had a further decrease in TT4 concentration (mean TT4 = 39 nmol/l, range 38–54nmol/l). The single non-euthyroid cat from the first study had a serum TT4 concentration of 61 nmol/l, a decrease from the baseline concentration of 83 nmol/l. The average decrease in serum TT4 for all 9 cats was 20 nmol/l (range 2–35 nmol/l).

Finally, 8 of the 9 cats were fed a third iodine-restricted food (0.17 ppm DMB) along with one other newly diagnosed hyperthyroid cat (79 nmol/l serum TT4) and evaluated every 4 weeks. All 9 cats in this evaluation were euthyroid (mean TT4 33 nmol/l; range 23–50 nmol/l). This result included the cat whose serum TT4 remained in the hyperthyroid range in the first two evaluations. The average decrease in TT4 was 13 nmol/l (range 0–43 nmol/l). Biochemical features of renal function remained stable and no other biochemical abnormalities were observed. In summary, the results of these three feeding studies demonstrate that feline hyperthyroidism can be managed effectively with dietary iodine restriction.

ABSTRACT EN-17

TITRATION OF DIETARY IODINE FOR MAINTAINING NORMAL SERUM THYROXINE CONCENTRATIONS IN HYPERTHYROID CATS.

LD Melendez, RM Yamka, PA Burris. Hill's Pet Nutrition, Topeka, KS.

We have shown previously that restriction of dietary iodine (I) is a safe and effective method for decreasing serum thyroxine concentrations (TT4) in cats with hyperthyroidism. The objective of this study was to determine the maximum level of iodine in a nutritionally balanced feline mature adult food required to maintain normal serum TT4 concentrations in hyperthyroid cats currently being controlled on a food containing 0.15 ppm I (DMB) as measured by epiboron neutron atomic activation. All cats were previously diagnosed at least 14 months prior to the start of the study and their TT4 concentrations were maintained in the normal range by dietary iodine restriction for a minimum of 10 months (range 10 months- 3 years). Serum TT4 concentrations ranged from 9–42 nmol/l (reference range 10–55 nmol/l) at the beginning of the study.

The cats were divided into two groups each containing 9 cats. Groups were similar in age and gender distribution (mean age = 13.8 years, range 12–18 years). One group (Group A) was placed on a food that was formulated for mature adult cats containing 0.39 ppm I (DMB). The other group (Group B) was placed on a similar food that differed only in that it contained 0.47 ppm I (DMB). Blood was collected from all cats every three weeks and analyzed for serum TT4 concentration. Biochemistry parameters were also evaluated at weeks 0, 6 and 9.

All Group A cats exhibited increases in serum TT4 concentration (mean increase of 25 nmol/l above baseline, range 5–48 nmol/l). Seven of the cats remained in the euthyroid range (mean serum TT4 = 36 nmol/l, range- 27–54 nmol/l). Two cats exceeded the upper limit of the reference range (59 and 76 nmol/l respectively). The cats in Group B also exhibited increases in serum TT4 concentration but to a greater degree than the cats in Group A (mean increase 39 nmol/l, range 20–60nmol/l). Four cats remained in the euthyroid range (mean serum TT4 = 41, range 29–49 nmol/l). The five remaining cats all exceeded the upper limit of the reference range (mean serum TT4 = 76 nmol/l, range- 59–99 nmol/l).

All cats returned to a euthyroid state within 1 month of being returned to a diet containing 0.17 ppm I (DMB). It was determined that serum TT4 concentrations are not ideally controlled in the normal range in hyperthyroid cats fed a food containing ≥ 0.39 ppm I (DMB).

ABSTRACT EN-18

CONTROLLED LEVEL OF DIETARY IODINE NORMALIZES SERUM TOTAL THYROXINE IN CATS WITH NATURALLY OCCURRING HYPERTHYROIDISM.

S Yu, KJ Wedekind, PA Burris, DS Forrester, MF Locniskar. Hill's Pet Nutrition, Inc., Topeka, KS.

Hyperthyroidism is a common disease in old cats. Excessive production of thyroid hormones is the hallmark of the disease. Three main treatments for feline hyperthyroidism include radioactive iodine, thyroidectomy, and antithyroid drugs such as methimazole. Previously we have shown that limiting dietary iodine to or below 0.27 ppm induces euthyroidism in cats with hyperthyroidism compared with a similar diet containing 0.42 ppm iodine. The objective of this study was to test whether dietary iodine at 0.32 ppm would induce euthyroidism in cats with naturally occurring hyperthyroidism.

Fourteen cats with hyperthyroidism confirmed by serum TT_4 and FT_4 measurements were stratified into two groups based on gender and age. One group (control: 4 males and 3 females, age ranged from 11 to 15 years) was given a positive control dry cat food (0.17 ppm iodine) while the other group (test: 3 males and 4 females, age ranged from 12 to 17 years) was fed a commercial dry cat food (1.9 ppm iodine) for at least 6 weeks before the study. Afterwards (week 0), the control cats continued to receive the same food while cats in the test group were given a test food (0.32 ppm iodine) for additional 12 weeks. All cats had free access to their food and deionized water during the study. Blood samples were collected during weeks 0, 3, 6, and 12 of the study.

The control cats maintained euthyroidism during the study. The test food significantly reduced serum TT_4 (72 ± 12, $43 \pm 9^*$, $42 \pm 9^*$, $40 \pm 6^*$ nmol/L in weeks 0, 3, 6 and 12, respectively; *: $p < 0.05$ compared with week 0, Dunnett's t test). It also significantly reduced FT_4 at the end of the study (17 ± 2 vs. 23 ± 2 pmol, week 12 vs. week

0; Dunnett's t test, $p < 0.05$). Serum FT_4 was within the reference range (10–55 pmol/L) in cats in both groups. Serum TT_3, FT_3, and TSH were not affected by the test food and were within the reference ranges (TT_3: 0.6–1.4 nmol/L, FT_3: 1.5–6 pmol/L, and TSH: 0–21 mU/L) in cats of both groups during the study.

This study demonstrates that dietary iodine at or below 0.32 ppm provides an effective and inexpensive therapy for cats with naturally occurring hyperthyroidism.

ABSTRACT EN-19

OUTCOME OF RADIOACTIVE IODINE THERAPY IN CATS RECEIVING RECENT METHIMAZOLE THERAPY. R. Oman, K. F. Lunn. Department of Clinical Sciences, College of Veterinary Medicine and Biomedical Sciences, Colorado State University, Fort Collins, CO.

Radioactive iodine (^{131}I) is a widely used treatment for feline hyperthyroidism. Prior to ^{131}I administration, many cats receive methimazole therapy. It has been suggested that recent withdrawal of methimazole prior to ^{131}I may increase the risk of hypothyroidism, inhibit the response to therapy, or have no effect. To further address this question, a retrospective medical records search was performed to identify hyperthyroid cats that received ^{131}I therapy after methimazole treatment. Inclusion criteria included documentation of the time interval between discontinuation of methimazole and ^{131}I administration, and measurement of thyroxine (T_4) at 7–14 days after ^{131}I. Cats were divided into 2 groups: those receiving ^{131}I within 1 day of stopping methimazole, and those receiving ^{131}I treatment 5 or more days after stopping methimazole.

Sixty cats met the inclusion criteria. Forty received ^{131}I within 1 day of stopping methimazole. Of those, 20 (50%) had a low T4 (< 1.2 mcg/dl), 17 (42.5%) had a normal T4 (1.2–4.8 mcg/dl), and 3 (7.5%) had an elevated T4 (> 4.8 mcg/dl) at 7–14 days after ^{131}I therapy. Fourteen cats received ^{131}I 5 or more days after stopping methimazole: 8 (57%) had a low T_4, 5 (36%) had a normal T_4, and 1 (7%) had an elevated T_4 at 7–14 days after ^{131}I therapy. The results were compared with a Fisher's exact test and there was no difference between the groups ($p = 0.76$). These findings indicate that stopping methimazole therapy within 1 day of ^{131}I therapy does not inhibit the response to therapy.

ABSTRACT EN-20

LACK OF CROSS REACTIVITY OF FELINE INSULIN WITH A COMMERCIAL HUMAN INSULIN ELISA. C Bryan, P Lathan, S Curotto, C Bulla, K Lunsford, S Pruett. Mississippi State University College of Veterinary Medicine, Mississippi State, MS.

Pharmacokinetic studies evaluating synthetic insulin analogs such as glargine necessitate the ability to measure the blood concentrations of glargine without cross-reactivity to endogenous insulin. Although the cross-reactivity between endogenous human insulin assays and synthetic analogs is often known for commercially-available assays, the degree of cross-reactivity of human insulin assays with feline insulin is not. The purpose of this study was to evaluate the cross-reactivity of feline insulin with a commercially-available human insulin ELISA with known cross reactivity to several synthetic analogs.

Pre- and post-prandial blood samples were collected from four healthy cats immediately prior to and approximately 15 minutes following a meal, for a total of 8 samples. Dextrose was added to the meals given to two of the cats. Blood samples were immediately centrifuged and the serum was collected, aliquoted, and stored at −20°C until analysis. Serum insulin levels were determined in parallel with commercially-available feline insulin and human insulin ELISAs. The ELISAs were run in duplicate and according to the manufacturer's instructions.

Concentrations of serum insulin measured by the feline insulin ELISA ranged from 12.7 ng/L to > 700 ng/L. Despite the wide range of concentrations of feline insulin, all 8 samples evaluated with the human insulin ELISA yielded absorbance readings equal to or lower than the absorbance of the negative control, indicating no cross-reactivity between the evaluated human insulin assay and feline insulin. Since this assay is reported to cross-react significantly with glargine, it is a great candidate for determination of serum glargine concentrations in cats.

ABSTRACT EN-21

COMPARISON OF LOW- AND HIGH-DOSE TRILOSTANE TREATMENT IN DOGS WITH PITUITARY-DEPENDENT HYPERADRENOCORTICISM. K Cho, S Noh, D Chang, KJ Na, MP Yang, JH Kang. Chungbuk National University College of Veterinary Medicine, Cheongju, Chungbuk, Republic of Korea.

The aim of this prospective, controlled study was to compare the efficacy of two trilostane protocols for treatment of canine pituitary-dependent hyperadrenocorticism (PDH). Among the 28 client-owned dogs diagnosed with PDH, only the dogs weighing < 5 kg were selected (n = 16). Group A (n = 9; low-dose treatment group) and group B (n = 7; high-dose treatment group) received 0.8 ± 0.3 mg of trilostane/kg orally every 12 hours and 30 mg of trilostane/body orally every 24 hours, respectively. All of the dogs were reassessed at 2, 4, 12, and 24 weeks after the initiation of treatment.

The improvement in post-ACTH stimulation serum cortisol concentration, as well as clinical signs in group A, required more time than group B; however, 2 of 7 dogs in group B had clinical signs and abnormal laboratory findings consistent with hypoadrenocorticism after treatment for 20 weeks. Twenty-four weeks later, all of the dogs of both groups improved the abnormal clinical findings. The present study suggests that twice daily, low-dose administration of trilostane is effective in the management of canine PDH and may be safe without the potential adverse effects of once daily, high-dose treatment. However, because this study involved only a small number of dogs, a population-based control study will be needed to clarify the efficacy of low- compared to high-dose trilostane treatment.

ABSTRACT FA-1

ANALYTICAL VALIDATION OF A COMMERCIALLY AVAILABLE IMMUNOASSAY FOR THE MEASUREMENT OF SERUM COBALAMIN CONCENTRATION IN PIGS. N Grützner[1], D Knabe[2], BD Lawhorn[3], J Kauffold[4], JS Suchodolski[1], JM Steiner[1]. [1]Gastrointestinal Laboratory, [2]Department of Animal Science, [3]Veterinary Large Animal Clinical Sciences, Texas A&M University, College Station, TX, [4]New Bolton Center, University of Pennsylvania, School of Veterinary Medicine, Kennett Square, PA.

Cobalamin is essential for a variety of metabolic processes in many tissues and organs, and has effects on cell growth and peripheral and central nervous system function. Chronic distal small intestinal disease in humans, cats, and dogs has been shown to cause cobalamin deficiency. An immunoassay for the measurement of serum cobalamin concentration in these species is being used in routine practice for the diagnosis of cobalamin deficiency. In pigs, the role of cobalamin has not yet been extensively investigated. Thus, the aim of this study was to analytically validate an immunoassay, labeled for use in humans, for the measurement of cobalamin in porcine serum samples and secondly to determine serum cobalamin concentrations in weaned pigs.

For the analytical validation of the assay, serum cobalamin concentrations were measured using the commercially available IMMULITE® 2000 cobalamin immunoassay (Siemens Healthcare Diagnostics Ltd., Deerfield, IL, USA) in 30 surplus porcine serum samples from a variety of studies. Validation of the assay consisted of determination of dilutional parallelism, spiking recovery, and intra- and inter-assay variability. Additional surplus serum samples from 27 piglets from four litters at a Texas A&M University farm were obtained. Each piglet had been bled twice, the first at weaning (21 days of age) and the second one 12 days later. To investigate results in comparison between age groups, serum cobalamin concentrations were compared using a Wilcoxon matched pairs test. Significance was set at $p < 0.05$.

Observed to expected ratios (O/E) for serial dilutions ranged from 87.3 to 124.9% (mean ± SD: 104.2 ± 14.5%) for four different serum samples at dilutions of 1:1, 1:2, and 1:4, and from 96.8 to 118.3% (mean ± SD: 107.6 ± 15.2%) for one serum sample at dilutions of 1:2, 1:4, and 1:8. O/E for spiking recovery ranged from 87.4 to 116.7% (mean ± SD: 102.0 ± 7.5%) for five different porcine serum samples that had been spiked with each other in a 1:1 dilution. Intra-assay coefficients of variation (%CV) for five different serum samples were 4.3, 5.7, 4.3, 3.7, and 6.1%. Inter-assay %CVs for five different serum samples were 4.9, 7.2, 9.6, 3.5, and 7.6%. Serum cobalamin concentration was significantly lower in piglets post weaning (median: 242 ng/L) compared to those at the time of weaning (median: 324 ng/L; $p = 0.009$).

The IMMULITE® 2000 cobalamin immunoassay labeled for use in humans is linear, accurate, precise, and reproducible for measurement of serum cobalamin concentrations in pigs. This study also showed that piglets that differ in age by only 12 days have significantly different serum cobalamin concentrations. Further investigations of cobalamin concentrations in both sows and piglets at different stages of weaning are warranted.

ABSTRACT FA-2

EPIDIEMIOLOGY OF MASTITIS PATHOGENS IN HEIFERS ON A GRAZING DAIRY. DR Ringen, JR Middleton, D Suntrup, T Markovich. University of Missouri, College of Veterinary Medicine, Columbia, MO.

Primigravid dairy heifers can be infected with mastitis pathogens during the periparturient period. The prevalence of intramammary infection (IMI) ranges from 30–75% of quarters pre-partum and 12–45% at parturition. Some pre-partum infections self-cure before parturition, however a number of these IMIs persist into early lactation. These IMIs may impact milk production and quality and may serve as a reservoir for contagious pathogens. No study has specifically investigated the risk of an IMI persisting from the pre-partum period into early lactation. The objectives of this study were to describe the prevalence of mastitis pathogens in heifers on a grazing dairy before and after parturition and calculate the relative risk (RR) and attributable fraction of population (AFP) for the association between a post-partum and pre-partum IMI.

Two-hundred-ninety-four heifers were systematically assigned to 1 of 3 groups: G1) pre-partum secretions from all mammary quarters (n = 98), G2) no pre-partum secretions collected (n = 98) and G3) pre-partum secretions from two diagonal quarters (n = 98). Group assignments were designed to assess whether pre-partum sampling increased the likelihood of IMI at calving. Mammary quarter secretions were collected for bacterial culture approximately 2 weeks prior to expected calving date. Quarter milk samples were collected for bacterial culture once weekly during the 1st 3-weeks of lactation. Bacterial isolates were classified as staphylococci, non-agalactiae streptococci and Gram-negatives. Mammary quarter samples yielding 2 different bacteria were classified as mixed infections and those yielding $\geq$ 3 bacterial types were classified as contaminated. Bacterial isolates were speciated using gene sequencing methods and strain-typed using pulse-field-gel-electrophorysis to evaluate the relatedness of bacteria isolated from pre-and post-partum samples from the same mammary quarter. Relative risk and AFP were calculated using 2 $\times$ 2 tables.

Forty-five percent of mammary quarters had a pre-partum IMI. During the 1st 3 weeks of lactation the mean prevalence of IMI was 23.3% of quarters. Staphylococci were most frequently isolated bacteria from pre-partum secretions and milk with *S. chromogenes* and *S. aureus* being the most common species. Using data from 228 mammary quarters, the RR and AFP for the association between a post-partum and pre-partum IMI were 11 and 77%, 43 and 86%, and 12 and 68% for all staphylococci, *S. aureus* only and CNS only IMIs, respectively. Mammary quarters sampled pre-partum were no more likely to have a post-partum IMI than those not sampled (Chi-square, P $\geq$ 0.27). These data demonstrate that pre-partum IMIs persist into early lactation and that pre-partum secretion cultures may be a useful, not only in predicting IMI at calving, but also in assessing risk of introducing new contagious mastitis pathogens, e.g., *S. aureus*, into the lactating herd.

ABSTRACT FA-3

PREVALENCE OF TETRACYCLINE RESISTANCE GENES IN *CLOSTRIDIUM DIFFICILE* ISOLATED FROM CALVES PRIOR AND POST TREATMENT WITH OXYTETRACYCLINE. MC Costa, H Staempfli, LG Arroyo, DL Pearl, SJ Weese. University of Guelph, Guelph, ON.

Despite concerns about antimicrobial resistance and *Clostridium difficile* in food animals, there has been little study of the prevalence or mechanisms of resistance. This study evaluated the impact of tetracycline treatment on *C. difficile* shedding in veal calves and the impact on resistance.

Calves arriving on 1 veal farm received oral oxytetracycline for 5 days as per farm protocols. Calves were sampled at arrival and 6 days later. Selective culture for *C. difficile* was performed. Isolates were ribotyped, and tested for tetracycline susceptibility and the presence of tetracycline resistance genes. Multivariable logistic regression models were used to determine the relationship between tetracycline resistance and the presence of tetracycline resistance genes.

Clostridium difficile was isolated from 32% (56/174) and 51% (88/172) calves, at the first and second samples, respectively. The percentage of tetracycline resistant isolates increased from 79% to 93%. Isolates from the second sample were 3 times more likely to be tetracycline resistant (p = 0.016) and 5 times more likely to possess *tet*(M) (p = 0.004). *tet*(M) was detected in 13% (7/53) and 43% (39/91), *tet*(O) in 23% (12/53) and 19% (17/91) and *tet*(W) in 2% (1/53) and 1% (1/91) of isolates from first and second samples, respectively. *tet*(L), *tet*(K) and *tet*(S) were not detected. 54 resistant isolates were not carrying any of the genes investigated.

Routine tetracycline use may have had an impact on both the prevalence of *C. difficile*, as well as the strain distribution and resistance patterns. This is the first report of presence of *tet*(O) in *C. difficile*.

ABSTRACT FA-4

ASSOCIATIONS BETWEEN ANTIMICROBIAL USE AND ANTIMICROBIAL RESISTANCE IN *ESCHERICHIA COLI* SAMPLED FROM INDIVIDUAL FEEDLOT CATTLE. KM Benedict[1], SP Gow[2], CW Booker[3], TA McAllister[4], and PS Morley[1]. [1]Animal Population Health Institute, Colorado State University, Fort Collins, CO. [2]Public Health Agency of Canada, University of Saskatchewan, Saskatoon, SK. [3]Feedlot Health Management Services Ltd., Okotoks, AB. [4]Lethbridge Research Center, University of Lethbridge, Lethbridge, AB.

The objectives of this study were to 1) estimate the prevalence of antimicrobial resistance in the study population and 2) to investigate the associations between exposures to antimicrobial drugs and antimicrobial resistance in fecal non-type specific *E. coli* (NTSEC) recovered from individual feedlot cattle. Two-stage random sampling was used to identify cattle for enrollment at 4 western Canadian feedlots. A fecal sample was collected per rectum from each individual at arrival and in the middle of the feeding period when cattle were rehandled as part of standard feedlot protocol. From samples collected at this second time point, a total of 2,133 NTSEC isolates were tested for susceptibility to antimicrobial drugs by disk diffusion. Parenteral and in-feed exposures to antimicrobial drugs were recorded for each individual enrolled in the study. The least square means estimates and 95% confidence intervals for the prevalence of resistance at each time point were modeled using Poisson regression. Multivariable logistic regression was used to investigate associations between antimicrobial resistance and exposure to antimicrobial drugs. Regression models were adjusted for clustering of observations among individuals and pens.

The most common resistances identified in arrival samples were sulfisoxazole (7.5%; 95%CI: 6.1–9.2), streptomycin (7.7%; 95%CI: 6.3–9.5) and tetracycline (20.0%; 95%CI: 17.7–22.6). At the second sampling point, resistance prevalence was 25.6% (95%CI: 23.5–28.0) for sulfisoxazole, 25.0% (95%CI: 22.8–27.3) for streptomycin, and 72.7% (95%CI: 70.5–75.1) for tetracycline. Logistic regression modeling identified weak associations of exposures to tetracycline and macrolide classes of drugs with antimicrobial resistance at the second time point.

ABSTRACT FA-5

PREMATURE/DYSMATURE SYNDROME IN CRIA: A RETROSPECTIVE STUDY OF 63 CASES (1999-2005). C. Gerspach, D. Anderson. The Ohio State University, Columbus OH.

Prematurity is widely acknowledged as risk factor for subsequent morbidity and mortality in llama and alpaca cria. A review of medical records for premature cria alive at the time of admission to the Veterinary Teaching Hospital between 1999 and 2005 was performed to determine risk factors of prematurity and to report the outcome and related conditions or diseases in affected cria. Medical records for 63 premature or dysmature cria were included in this study. Of these cria, 51 were alpaca and 12 llama, 36 were female and 27 were male. Reasons for referral were prematurity, failure of passive immunity, dyspnoea, weakness and failure to gain weight. Cria were presented at a mean age of 1.4 days and were premature by a mean estimated time of 19.5 days. Overall survival rate was 82.5%, with all llama cria surviving. A multivariate logistic regression model was used to identify risk factors associated with not surviving. Cria receiving camelid colostrum had a significant better outcome than cria receiving no colostrum or colostrum from different species. Dyspnea and tachypnea was associated with a poor outcome. All cria that were able to nurse, without assistance prior to referral, survived. Clinical pathology parameters most commonly associated with death were hyperphosphatemia and acidosis.

ABSTRACT FA-6

THE DISTRIBUTION OF ENROFLOXACIN, AND ITS ACTIVE METABOLITE CIPROFLOXACIN, USING AN IN-VIVO ULTRAFILTRATION SAMPLING TECHNIQUE AFTER THE INJECTION OF ENROFLOXACIN TO PIGS. KM Messenger, AT Blikslager, and MG Papich. North Carolina State University College of Veterinary Medicine, Raleigh, NC.

Enrofloxacin is approved for the treatment of swine respiratory disease, however there are no published studies describing the pharmacokinetics of enrofloxacin at the approved dose and route in pigs (7.5 mg/kg subcutaneously). Furthermore no studies have assessed the unbound concentrations of enrofloxacin at its site of action, the extracellular tissue fluid. Therefore the objective of this study was to use an in-vivo ultrafiltration method to measure the active fraction of enrofloxacin, and the metabolite ciprofloxacin, at 3 tissue sites relevant to pigs, and to compare these concentrations with plasma concentrations collected at similar time points.

Six healthy pigs were used in this study. Pigs were recently weaned and weighed an average 16.3 kg. On the day before the experiment, pigs were anesthetized for the placement of jugular vein sampling catheters and interstitial fluid collection probes. Three ultrafiltration probes were placed in each pig in a subcutaneous site near the right shoulder, an intramuscular site along the epaxial muscles, and in the pleural space of the chest cavity. Each pig received an injection of enrofloxacin (Baytril 100, Bayer Animal Health) at a dose of 7.5 mg/kg subcutaneously behind the left ear. Plasma and interstitial fluid samples were collected at pre-determined time points, and enrofloxacin and ciprofloxacin concentrations were measured using HPLC with fluorescence detection. Protein binding was determined with a microcentrifugation system. Pharmacokinetic data was analyzed using a one compartment model.

The analysis of plasma and ISF showed that only a small fraction of ciprofloxacin was produced in these pigs, therefore ciprofloxacin concentrations were not used in pharmacokinetic measurements. The plasma half-life ($t_{1/2}$), volume of distribution, clearance, and peak concentration (C_{max}) for enrofloxacin was 25.9 hr (± 6.2), 6.29 L/kg (± 1.23), 0.168 L/kg/hr (± 0.076), and 1.07 µg/mL (± 0.28), respectively. The concentrations from each of three tissues were not different in each pig. When pharmacokinetic values from all tissues were combined for the ISF, the $t_{1/2}$ was 23.6 hr (± 4.1) and the C_{max} was 1.26 µg/mL (± 0.10). The enrofloxacin plasma protein binding was 31.1% (± 3.28) and 37.13% (± 16.54) at a high and low concentration, respectively.

This study has demonstrated that the concentration of biologically active enrofloxacin in tissues exceeds the concentration predicted by the unbound fraction of enrofloxacin in pig plasma. The half-life of enrofloxacin is longer in tissues and plasma than has been reported in previous studies. The high tissue concentrations and long half-life produce an AUC/MIC ratio sufficient for the pathogens that cause respiratory infections in pigs.

ABSTRACT FA-7

PHARMACOKINETICS OF CEFTIOFUR CRYSTALLINE FREE ACID IN SHEEP FOLLOWING SUBCUTANEOUS SINGLE-DOSE ADMINISTRATION. S Rivera, JA Angelos, JD Rowe, SE Wetzlich, BA Byrne, LA Tell. School of Veterinary Medicine University of California, Davis, CA.

Ceftiofur crystalline free acid (CCFA), a long-acting ceftiofur formulation labeled for use in cattle, pigs, and horses for treatment of respiratory disease has been used for treatment of ovine respiratory infections in clinical practice. Pharmacokinetic data, however, do not exist for CCFA administered subcutaneously in sheep. The present pharmacokinetic study evaluated the single dose subcutaneous administration of CCFA in sheep (n = 9) at 6.6 mg/kg body weight. Concentrations of ceftiofur free acid equivalents (CFAE) in plasma were measured by high performance liquid chromatography for 14 days following drug administration. Pharmacokinetics of subcutaneous CCFA in sheep were best described using a single compartment model with the following average (± SD) parameters: area under the concentration time curve $_{0-\infty}$ (206.6 h*ug/ml ± 24.8), observed maximum plasma concentration (2.4 ug/ml ± 0.5), and observed time of maximum plasma concentration (23.1 h ± 10.1). No significant adverse drug reactions were observed. Adequate CFAE plasma concentrations were attained to effectively treat respiratory tract pathogens associated with pneumonia in sheep.

ABSTRACT FA-8

THE USE OF RAPID THORACIC ULTRASONOGRAPHY FOR DETECTION OF SUBCLINICAL, AND CLINICAL PNEUMONIA IN DAIRY CALVES. TL Ollivett[1], AJ Burton[2], RC Bicalho[1], and DV Nydam[1]. [1]College of Veterinary Medicine, Cornell University, Ithaca, NY. [2]The University of Georgia, College of Veterinary Medicine, Athens, GA.

The purpose of this study was to assess, using thoracic ultrasonography, the prevalence of lung lesions in pre-weaned dairy calves. Subsequent aims were to describe ultrasonographic changes within the lung, clinical respiratory score, and treatment of respiratory disease.

A longitudinal study was performed using female dairy calves from 6 commercial dairy farms in New York State. Calves were enrolled based on age. Thoracic ultrasound and clinical respiratory scoring were performed on each calf at 2 time points. A standard 5 mHz linear ultrasound probe was utilized to evaluate intercostal spaces 1 through 11 of each hemi-thorax with the calf in lateral recumbency (US1) or standing (US2). Lesion appearance, size, and location were recorded. Respiratory score (RS) was assigned based on a previously published protocol incorporating fever, nasal discharge, cough, ocular discharge and ear droop, with a higher numerical score corresponding to more severe disease. Abnormal lung on ultrasound was defined as one or more areas of ≥1 cm width or depth of non-aerated lung. Farm records were evaluated to identify treated calves. Calves were treated for respiratory disease at the farm manager's discretion, not based upon ultrasound findings or RS. Non-parametric methods were used to evaluate the data.

Ninety-one calves were enrolled into the study, with 6 lost to follow-up. An average of 4 minutes was spent performing the RS and ultrasound on each calf. The median ages at first (US1) and second (US2) examination were 13 (interquartile range 12–15) and 46 (interquartile range 44–47) days, respectively. The majority of calves had a low RS (< 5) and only 3.2% of calves had a RS high enough to warrant treatment based on previous recommendations (RS≥5). The prevalence of calves that had abnormal lungs on ultrasound but a low RS (< 5) was 5.5% (US1) and 16.5% (US2). The prevalence of calves that had abnormal lungs on ultrasound and a high RS (≥ 5) was 0% (US1) and 3.5% (US2). Of the calves that had abnormal lungs on ultrasound but a low RS, 13% were treated with antimicrobials within 7 days of examination. None of the calves with high RS and abnormal lungs on ultrasound were treated with antibiotics within 7 days of examination.

This study demonstrates a high prevalence of abnormal lungs, as detected by thoracic ultrasonography, without significant clinical signs in pre-weaned dairy calves. The relatively low treatment rate in these calves may suggest an area of opportunity for improvement in calf health, welfare, and herd longevity. Further studies and follow up are needed to elucidate the significance of these findings and whether or not treatment is indicated.

ABSTRACT FA-9

LAMENESS IN BEEF CATTLE: RETROSPECTIVE STUDY OF FOOT DISORDERS (2007-2010). B. Fraser, M. Miesner, D. Anderson, S. Laflin, M. Jones. Kansas State University College of Veterinary Medicine, Manhattan, KS.

Literature regarding diseases causing lameness in beef cattle is limited. This retrospective study was undertaken to examine beef cattle presented for lameness. Medical records of beef cattle having a lameness examination done during the period 2007 to 2010 were reviewed and descriptive statistics generated. Lameness was classified based on clinical diagnosis. The medical records of 270 beef cattle were reviewed of which 63.2% were male and 36.8% were female. Beef cattle presented for lameness most often during the summer months (34%) and least during autumn (18%). Causes of lameness were categorized as infectious (44.6%) or non-infectious (55.4%) and infectious lameness subcategorized as either a primary disorder or a secondary infection. All cases of a primary infectious disorder were interdigital phlegmon. Secondary infections diseases included sole abscess (25.9%), septic arthritis (11.1%), tenosynovitis (2.8%), and pedal osteitis (1.3%). Non-infectious lameness included proximal limb lameness (19.6%), foot trauma (14.6%), hoof horn cracks (9.5%), hoof defects (2.5%), interdigital fibromas (1.9%), overgrown hooves (1.9%), sole bruise (1.3%), subclinical laminitis (1.3%), white line disease (0.9%), osteoarthritis (0.9%), heel erosion (0.3%), sole ulcers (0.3%), and sole hemorrhage (0.3%). The most frequently affected claw was the lateral digit of the hind limb (36.4%), followed by the medial digit of the front limb (27.1%), lateral digit of the front limb (23.6%), and the medial digit of the hind limb (12.9%). The findings of this study suggest significant differences in the frequency of disease causing lameness in beef cattle compared to published reports for dairy cattle.

ABSTRACT GI-1

EVALUATION OF THE SAFETY AND PRACTICABILITY OF ENDOSCOPIC ULTRASOUND-GUIDED FINE NEEDLE ASPIRATION OF THE PANCREAS IN HEALTHY DOGS. P.H. Kook[1], P. Baloi[2], M. Ruetten[3], S. Lutz[1], N. Pantchev[4], C.E. Reusch[1], P. Kircher[2]. [1]Clinic for Small Animal Internal Medicine, [2]Unit of Diagnostic Imaging, [3]Institute of Pathology, Vetsuisse Faculty, University of Zurich, Switzerland, [4]IDEXX Vet Med Lab, Ludwigsburg, Germany.

In people, endoscopic ultrasound (EUS) has become the technique of choice for assessing pancreatic disease and EUS-guided fine-needle aspiration (EUS FNA) has proven a useful and safe modality for characterizing pancreatic lesions. Reported complications include infections, bleeding and acute pancreatitis. In dogs, laparoscopic-assisted pancreatic biopsy has been suggested to be a safe procedure, however EUS and EUS FNA have not been evaluated in dogs so far. Thus the aim of the present study was to assess the practicability and safety of EUS examination of the abdominal cavity as well as pancreatic EUS FNA in healthy dogs. This study was approved by the Cantonal Committee for the Authorization of Animal Experimentation, Zurich, Switzerland. The study population consisted of 14 healthy beagle dogs with a median bodyweight of 13.4 kg (10.9–15.7). EUS was performed with an Olympus GF-UC140P-echoendoscope and FNA were performed using 19 G needles (Cook EchoTipUltra). After completion of the EUS-examination of the abdominal cavity from the stomach (liver, gallbladder, bile ducts, kidneys, adrenals, pancreas), the scope was advanced into the duodenum and EUS FNA of the pancreas was performed. FNA tissue acquisition was made applying negative pressure and 6 to 8 needle passes were made. All dogs received 30 mg/kg metimazole IM after EUS FNA and were re-checked ultrasonographically 20 minutes post EUS FNA. Postoperative activity was assessed using a standardized scoring system. A CBC, serum biochemistry, urinalysis and spec cPL® were measured before, as well as 24 and 48 h after EUS FNA.

The EUS examination was complete in 13/14 dogs, the pancreas could not be visualized in 1 dog. The pancreas was hypo- (4/13) to isoechoic (9/13) to the surrounding mesenterium in all cases. In 3/13 dogs parts of the pancreas presented hyperechoic. The mean measured thickness was 0.88 cm. The pancreas was aspirated in 12 dogs using a transgastric approach (3) or transduodenal approach (9). Duodenal transmural puncture was not accomplished in 1 dog where a re-sterilized needle was used. A minimal amount of peripancreatic fluid was observed in 1/12 dogs after EUS FNA. All dogs recovered uneventfully and required no further analgesia. All laboratory results including the spec cPL® measurements were within reference ranges on all three time points. Cytologically, conglomerates of exocrine pancreatic cells were seen in 8/12 cases, duodenal villous epithelial cells were seen in 11/12 cases. In 1 dog the aspirated pancreatic material was sufficient for a histological assessment. The 8 aspirates with exocrine pancreatic cells on cytology were obtained by transgastric (4) and transduodenal (4) aspirations. In conclusion, (1) EUS examination of the abdomen is feasible in medium-sized dogs, (2) the healthy canine pancreas can be difficult to visualize completely, and (3) EUS-guided pancreatic FNA using a 19 G needle is a safe procedure in healthy dogs. Studies evaluating its use in dogs with pancreatic disease are warranted to assess its clinical utility.

ABSTRACT GI-2

EFFECT OF A LOW-FAT DIET ON SERUM TRIGLYCERIDE, CHOLESTEROL, AND PANCREATIC LIPASE IMMUNOREACTIVITY CONCENTRATIONS IN MINIATURE SCHNAUZERS WITH HYPERTRIGLYCERIDEMIA. PG Xenoulis, JS Suchodolski, JM Steiner. Gastrointestinal Laboratory, Texas A&M University, College Station, TX.

Miniature Schnauzers have a high prevalence of idiopathic hyperlipidemia, which is characterized by an increased serum triglyceride (TG) concentration, with or without an increased serum cholesterol (CHOL) concentration. A common initial therapeutic approach for the management of hyperlipidemia is the use of a low-fat diet. Also, it is believed that low-fat diets may be beneficial in the treatment of pancreatitis in dogs. However, the efficacy of this approach has not been evaluated for either condition. The aim of the present study was to evaluate the effect of a commercially available low-fat diet on serum concentrations of TG, CHOL, and canine pancreatic lipase immunoreactivity (cPLI; measured as Spec cPL®) in apparently healthy Miniature Schnauzers with hypertriglyceridemia.

Blood samples were collected from 15 apparently healthy Miniature Schnauzers with hypertriglyceridemia (serum triglyceride concentrations > 108 mg/dL). Common causes of secondary hyperlipidemia were excluded based on historical information, physical examination findings, and the measurement of serum glucose, total T4, and free T4 (by ED) concentrations. The owners of the dogs were asked to switch their dog to the study diet (Royal Canin Gastrointestinal Low Fat®; fat content: 18.6 g/1,000 Kcal) and have a second blood sample collected 8 weeks after their dog had been on the new diet. All blood samples were collected after food had been withheld for 15 hours. Serum TG, CHOL, and Spec cPL concentrations were measured both before and after the diet change. Results were compared between the two time-points using the Wilcoxon signed rank and Fisher's exact tests.

Serum TG concentrations were significantly higher before (median: 432 mg/dL) than after the diet change (median: 178 mg/dL; $p = 0.003$). The proportion of dogs with hypertriglyceridemia was significantly higher before (15/15) than after the diet change (10/15; $p = 0.042$). Also, the proportion of dogs with serum TG > 500 mg/dL was significantly higher before (6/15) than after the diet change (0/15; $p = 0.016$). Serum CHOL concentrations were significantly higher before (median: 296 mg/dL) than after the diet change (median: 258 mg/dL; $p = 0.004$). The proportion of dogs with hypercholesterolemia was significantly higher before (8/15) than after the diet change (0/15; $p = 0.006$). Finally, the difference in serum Spec cPL concentrations before (median: 88 μg/L) and after the diet change (median: 56 μg/L) approached but did not reach significance ($p = 0.052$). Also, the proportion of dogs with high serum Spec cPL concentrations before (4/15) and after the diet change (0/15) was different, but this difference was not significant ($p = 0.099$).

In summary, a commercially available low-fat diet was effective in reducing serum TG and CHOL concentrations in Miniature Schnauzers with hypertriglyceridemia.

ABSTRACT GI-3

POLYMORPHISMS IN TOLL-LIKE RECEPTOR 5 RESULT IN HYPER-RESPONSIVENESS TO FLAGELLIN. A Kathrani[1], D Werling[2], A Holder[2], B Catchpole[2], K Allenspach[1]. [1]Department of Veterinary Clinical Sciences and [2]Department of Pathology and Infectious Diseases, Royal Veterinary College, University of London, UK.

Toll-like receptor 5 (TLR5) is an extracellular pattern recognition receptor which recognizes flagellin present in motile bacteria. We have previously demonstrated a significant association between three non-synonymous single nucleotide polymorphisms (SNPs) in the TLR5 gene (G22A, C100T and T1844C) and inflammatory bowel disease (IBD) in German shepherd dogs (GSDs). Recently, we have confirmed that two of these TLR5 SNPs (C100T and T1844C) are significantly associated with IBD in other canine breeds. To further substantiate the role of TLR5 in canine IBD functional analysis of these polymorphisms would be needed. Therefore the aim of this study was to determine the functional significance of the TLR5 SNPs by transfecting wild-type and mutant receptors in to human embryonic kidney cells (HEK) and carrying out nuclear factor-kappa B (NF-κB) luciferase assay and IL-8 ELISA.

The TLR5 gene containing the risk haplotype for IBD (ACC) and wild-type haplotype (GTT) as determined by the case-control analysis in GSDs with IBD were cloned into plasmids expressing yellow-fluorescent protein (YFP). These were then stably transfected into HEK cells. NF-κB activity was measured by transiently transfecting the cells with NF-κB firefly and HSV-thymidine kinase promoter (pRL-TK) renilla plasmids. The cells were then stimulated with various ligands (0.1 μg/ml flagellin, 0.01 μg/ml flagellin, 1 μg/ml LPS, 1 μg/ml PAM3CSK and media control). Firefly and renilla luciferase activities were measured using the Dual-Glo Luciferase Assay System (Promega, UK) according to the manufacturer's recommendations. The supernatants were harvested and used in an IL-8 ELISA (R&D Systems). Human TLR5 transfected HEK cells (Invivogen) served as positive controls in all experiments.

Independent T-test was used to determine the significance of relative luciferase activity and IL-8 concentration between wild-type and mutated TLR5 cells. Although there was no significant difference between the wild-type and mutated receptor when they were stimulated with 0.01 μg/ml of flagellin ($p = 0.14$), there was a significant increase when the cells with mutated TLR5 were stimulated with 0.1 μg/ml of flagellin compared to the cells expressing wild-type TLR5 ($p = 0.027$). Similarly, there was a significant increase in IL-8 concentration in the supernatants in the cells with the mutated TLR5 receptor when stimulated with 0.1 μg/ml flagellin compared to the wild-type ($p = 0.025$- one-tailed, 0.05-two-tailed) but not with 0.01 μg/ml flagellin ($p = 0.26$).

We show for the first time that polymorphisms associated with IBD are functionally hyper-responsive to flagellin compared to the wild-type receptor. This suggests that TLR5 may play a role in canine IBD and that blocking the hyper-responsive receptor found in susceptible dogs with IBD may alleviate the inappropriate inflammation seen in this disease. However, further in-vivo functional analysis of TLR5, especially at the intestinal mucosal level would be needed to confirm these findings and predict the usefulness of any future therapeutic interventions.

ABSTRACT GI-4

TOLL-LIKE RECEPTOR 5 MUTATIONS IN GERMAN SHEPHERD DOGS ARE HYPER-RESPONSIVE TO STIMULATION WITH FLAGELLIN. K Allenspach[1], A Kathrani[1], A. House[1], A Holder[2], B Catchpole[2], D Werling[2]. [1]Department of Veterinary Clinical Sciences and [2]Department of Pathology and Infectious Diseases, Royal Veterinary College, University of London, UK.

TLR5 has been shown to play a role in the inappropriate inflammation seen in human inflammatory bowel disease (IBD). Similarly, we have recently demonstrated a significant association between three non-synonymous single nucleotide polymorphisms (SNPs) in the canine TLR5 gene (G22A, C100T and T1844C) and inflammatory bowel disease (IBD) in German shepherd dogs (GSDs).

Therefore the aim of this study was to determine the functional significance of the TLR5 SNPs in the breed of GSDs.

The TLR5 gene containing the risk haplotype for IBD (ACC) and wild-type haplotype (GTT) were stably transfected into HEK cells. NF-κB activity was measured by transiently transfecting the cells with NF-κB firefly and HSV-thymidine kinase promoter (pRL-TK) renilla plasmids. The cells were stimulated with various TLR ligands (0.1 μg/ml flagellin, 0.01 μg/ml flagellin, 1 μg/ml LPS, 1 μg/ml PAM3CSK and media control). Firefly and renilla luciferase activities were measured using the Dual-Glo Luciferase Assay System (Promega, UK). The supernatants were harvested and used in an IL-8 ELISA (R&D Systems). Peripheral whole blood from dogs carrying the wild type and mutant TLR5 genes was cultured and stimulated with TLR ligands as above. Canine TNF-alpha was measured in the supernatant by commercially available ELISA (R&D Systems).

T-test was used to determine differences of relative luciferase activity, IL-8 concentration and TNF-alpha concentration between wild-type and mutated TLR5 cells. There was a significant increase in NF-kB activity when the cells with mutated TLR5 were stimulated with 0.1 μg/ml of flagellin compared to the cells expressing wild-type TLR5 ($p = 0.027$), which correlated with IL-8 expression in the supernatant ($p = 0.26$). Similarly, in the whole blood assay the TLR5 risk haplotype for IBD in GSDs (ACC) was significantly hyper-responsive to flagellin at a concentration of 0.1 μg/ml compared to the TLR5 wild-type haplotype (GTT) ($p = 0.001$).

We show for the first time that polymorphisms associated with canine IBD in GSDs are functionally hyper-responsive to flagellin compared to the wild-type receptor. Blocking the hyper-responsive receptor found in susceptible dogs with IBD may alleviate the inappropriate inflammation seen in this disease.

ABSTRACT GI-5

EFFECT OF OMEPRAZOLE ON SERUM GASTRIN CONCENTRATIONS AND CALCIUM METABOLISM IN HEALTHY DOGS. O. Dossin[1], K. Jones[1], M. Ridgway[1], S. Clark-Price[1], JM. Steiner[2], JS. Suchodolski[2]. [1]Department of Veterinary Clinical Medicine University of Illinois, Urbana, IL, USA. [2]Gastrointestinal Laboratory, Texas A&M University, College Station, TX, USA.

Proton pump inhibitors (PPI) are widely used in human and also veterinary medicine. Side-effects of PPI treatment reported in people are atrophic gastritis, gastric and esophageal cancer, and rebound hyperacidity following cessation of treatment, which has been speculated to be due to a sustained increased in circulating gastrin concentration. Moreover, long-term PPI treatment has been associated with an increased risk for osteoporosis in people. Little is known about the effect of PPI treatment on serum gastrin concentration or calcium metabolism in dogs.

Eight healthy adult research dogs (4 males and 4 females) were enrolled into the study. The dogs received an average dose of 1.1 mg/kg of omeprazole orally twice daily for 15 days. Blood samples were collected prior to initiating the treatment and every 3 days during the 15 days of treatment and during the 15 days after discontinuation of treatment for determination of serum gastrin, ionized calcium, PTH, and 25 OH vitamin D3. Gastric fluid was collected via gastroscopy after an overnight fast for measurement of gastric pH prior to, during, and after the omeprazole treatment period. Normally distributed data were compared with a repeated measures ANOVA and post hoc Dunnett's test. Data that were not normally distributed were compared with a Friedman's test and a post-hoc Dunn's test.

Gastric fluid pH was significantly higher ($p < 0.01$) at the end of the treatment period (median: 7.4; range: 4.2–8.1) when compared to pretreatment values (median: 1.7; range: 1.0–6.8). Serum gastrin concentrations increased significantly from a median baseline of 10.0 ng/L (range: 10.0–27.0) to a maximum median of 379.5 ng/L (range: 49.9–566.0) at day 9 of treatment ($p < 0.01$). Serum gastrin remained significantly increased above baseline values from day 6 to day 15 of the treatment, but was not different from pre-treatment values 3 days after the end of the treatment. Omeprazole treatment had no effect on ionized calcium or PTH for the duration of the study. Marginal, but significant changes of 25 OH vitamin D3 were observed at day 15 (end of the treatment period – increased by 13.8%) and day 21 (6 days after the end of the treatment – decreased by 14.7%).

This study shows that treatment with omeprazole for 2 weeks results in a profound and sustained increase in serum gastrin concentration in dogs. This effect is rapidly reversible after cessation of the treatment. No effect on calcium metabolism was observed. However, this study documents only the effect of short-term treatment and it is possible that the effects of long-term administration are different.

ABSTRACT GI-6

QUANTIFICATION OF GASTRIC AND DUODENAL BACTERIA USING FLUORESCENT IN SITU HYBRIDIZATION IN HEALTHY DOGS DURING OMEPRAZOLE TREATMENT. JF Garcia-Mazcorro[1], JS Suchodolski[1], K Jones[2], S Clark-Price[2], JM Steiner[1], O Dossin[2]. [1]Gastrointestinal Laboratory, Texas A&M University, College Station, TX. [2]Department of Veterinary Clinical Medicine, University of Illinois, Champaign, IL.

Omeprazole treatment has been associated with small intestinal bacterial overgrowth and a higher risk for infectious enteropathies in humans. Using a semi-quantitative sequencing approach, we have previously shown that omeprazole treatment may lead to alterations in both duodenal and gastric bacterial populations in healthy dogs (ACVIM 2010). However, a sequencing approach can only estimate relative proportions of genomic bacterial targets. Therefore, significant changes in the total number of bacteria could not be evaluated. The aim of this study was to quantify gastric and duodenal bacterial populations in dogs undergoing omeprazole treatment.

Eight 9 month-old healthy research dogs (4 males and 4 females) were enrolled. The dogs received an average dose of 1.1 mg/kg of omeprazole orally twice a day for 15 days. Endoscopic gastric and duodenal biopsies were harvested 30 and 15 days before starting omeprazole treatment, on the last day of treatment (day 15), and 15 days after the end of treatment (day 30). All biopsies were fixed in 10% formalin for 24 hours, processed, and embedded in paraffin blocks. Fluorescent *in situ* hybridization was used to quantify mucosa-associated bacteria using fluorescently-labeled probes targeting the 16S ribosomal RNA. Statistical analysis aimed to compare changes in *Helicobacter* spp. in gastric biopsies and total bacteria in both gastric and duodenal biopsies using the GLIMMIX and NPAR1WAY procedures in SAS® 9.2.

Bacteria were counted in 1,174 and 989 microscopic fields (63×) obtained from 155 and 132 gastric and duodenal biopsies, respectively. In the stomach, omeprazole treatment led to a decrease in *Helicobacter* spp. (log of average counts ± standard error: 1.98 ± 0.14 at day 15) when compared to the counts 30 (2.43 ± 0.13, p0.0001) and 15 (2.40 ± 0.13, p0.0001) days before treatment. After completion of omeprazole treatment, *Helicobacter* spp. increased and returned to baseline counts (2.45 ± 0.13 at day 30, p0.0001 vs day 15). Also, in the stomach, non-*Helicobacter* spp. bacteria were observed more often during omeprazole treatment (median: 3, range: 0–20) than on days 30 (median: 0, range: 0–3) and 15 (median: 1, range: 0–6) before and 15 days after (median: 0, range: 0–2) omeprazole treatment; however, statistical comparison across time points did not reach significance. In the duodenum, while the median number of bacteria for all time points was zero, non-parametric comparison of median scores (number of points above median) revealed significantly higher numbers of bacteria during omeprazole treatment ($p = 0.0063$).

Our results suggest that omeprazole treatment for 2 weeks leads to a lower abundance of *Helicobacter* spp. organisms in the stomach of healthy dogs. Also, this transient decrease in *Helicobacter* spp. was accompanied by a higher abundance of other bacteria in both the stomach and the proximal duodenum.

ABSTRACT GI-7

ASSESSMENT OF THE VARIABILITY ASSOCIATED WITH REPEATED MEASUREMENT OF GASTROINTESTINAL TRANSIT TIMES AND ASSESSMENT OF THE EFFECT OF ORAL RANITIDINE ON GASTROINTESTINAL TRANSIT TIMES USING A WIRELESS MOTILITY CAPSULE IN DOGS. JA Lidbury, JS Suchodolski, and JM Steiner. Gastrointestinal Laboratory, Texas A&M University, College Station, TX.

The SmartPill pH.P® capsule (The SmartPill Corporation) is a wireless motility capsule that measures pH, pressure, and temperature as it passes through the gastrointestinal (GI) tract. Analysis of this data allows the calculation of gastric emptying time (GET), small and large bowel transit time (SLBTT), and total GI transit time (TGTT). This study evaluated the variability associated with repeated measurement of GI transit times and the effect of oral administration of ranitidine (Zantac®) on GI transit times in dogs using this system. It was hypothesized that ranitidine would reduce GI transit times.

Six privately owned healthy adult dogs weighing between 25.0 kg and 41.0 kg were used. On 3 occasions each dog was fed a standard meal followed by oral administration of a capsule. Data were recorded until the capsule had passed in the dog's feces. On a 4th occasion each dog was given 75 mg of ranitidine *PO* q12 hrs starting 48 hrs prior to testing. The dogs were then fed the test meal and the capsule was administered as above. Ranitidine was given until the capsule had passed in the dog's feces. Proprietary SmartPill software was used to calculate GET, SLBTT, TGTT, and the median gastric pH (MGpH). Mean intra-individual and inter-individual coefficients of variation (CV%) were calculated for GET, SLBTT, and TTT for the first 3 time points. Transit times and gastric pH recorded at all 4 time points were compared using a repeated measures ANOVA. Where significant differences were identified, post-hoc testing was performed using a Bonferroni's multiple comparisons test. Significance was set at $p < 0.05$.

A sharp rise in pH indicating exit of the capsule from the stomach was identified in each experiment. Mean (± SD) GET, SLBTT, and TGTT without ranitidine were 775 ± 144, 1563 ± 614, and 2338 ± 577 min, respectively. Mean GET, SLBTT, and TGTT during treatment with ranitidine were 719 ± 11, 1442 ± 885, and 2155 ± 897 min, respectively. Mean intra-individual CV% before ranitidine for GET, SLBTT, and TGTT were 12.9, 29.7, and 17.8%, respectively. Mean inter-individual CV% before treatment with ranitidine for GET, SLBTT, and TTT were 16.1, 40.5, and 24.7%, respectively. No significant differences in GET, SLBTT, or TGTT were found at any of the 4 time points. The mean MGpH during treatment with ranitidine (pH 2.85) was significantly higher than at all other time points (overall mean pH for the 3 time points: 1.73; $p < 0.05$).

The SmartPill system is an easy to use, ambulatory, non-invasive, non-radioactive method for assessing GI transit times in medium to large breed dogs. Measurements of GI transit times, especially SLBTT, were subject to considerable intra-individual and inter-individual variation. No significant effect of oral ranitidine on GI motility was identified in this group of dogs. However, as expected, oral ranitidine caused a significant increase in gastric pH.

ABSTRACT GI-8

EVALUATION OF BACTERIAL GROUPS IN DOGS WITH GASTROINTESTINAL DISEASE USING QUANTITATIVE PCR ASSAYS. ME Markel, JF Garcia-Mazcorro, JM Steiner, JS Suchodolski. Gastrointestinal Laboratory, Texas A&M University, College Station, TX.

The intestinal microbiota has been implicated in the pathogenesis of various gastrointestinal disorders in both humans and dogs. Recent metagenomic data suggest that specific bacterial groups, including bacteria within the *Clostridium* clusters IV and XIVa (i.e., *Faecalibacterium* spp., Ruminococcaceae, and Lachnospiraceae) and *Bifidobacterium* spp. are decreased, while Proteobacteria are increased in dogs with clinical signs of gastrointestinal disease. The objective of this study was to establish quantitative polymerase chain reaction (qPCR) assays for these specific bacterial groups and evaluate their abundance in healthy dogs and dogs with clinical signs of gastrointestinal disease.

Fecal samples were collected from 21 healthy dogs (14 females and 7 males) and 20 dogs with clinical signs of gastrointestinal disease (11 females and 9 males). Novel quantitative PCR assays were established for *Faecalibacterium* spp., Ruminococcaceae, and Lachnospiraceae by aligning respective group specific sequences against canine specific sequences obtained from 16S rRNA gene clone libraries and sequences available from the Ribosomal Database Project. Primers for *Bifidobacterium* spp. and Proteobacteria were selected from previously published studies. The specificity of the qPCR assays was confirmed by sequencing of obtained qPCR amplicons. The bacterial DNA abundance in fecal samples was compared between healthy dogs and dogs with clinical signs of gastrointestinal disease using a Mann-Whitney *U* test. Significance was set at $p < 0.05$.

A significantly lower abundance of *Faecalibacterium* spp. ($p < 0.01$) and Ruminococcaceae ($p = 0.02$) was observed in dogs with clinical signs of gastrointestinal disease when compared to healthy dogs. Proteobacteria were more abundant in dogs with clinical signs of gastrointestinal disease, but this difference did not reach statistical significance ($p = 0.07$). There was no significant difference in the abundance of Lachnospiraceae ($p = 0.23$) and *Bifidobacterium* spp. ($p = 0.77$) between both groups.

In conclusion, we established novel qPCR assays for *Faecalibacterium* spp., Ruminococcaceae, and Lachnospiraceae. We observed significant decreases in the abundance of *Faecalibacterium* spp. and Ruminococcaceae in dogs with clinical signs of gastrointestinal disease. These bacterial groups are considered major short-chain fatty acid producers and studies are warranted to determine if a decrease in these bacterial groups is associated with decreases in short chain fatty acid production. Further studies are also needed to determine if these bacterial shifts are associated with specific gastrointestinal disorders.

ABSTRACT GI-9

IMBALANCE IN THE COMPOSITION OF ILEAL AND COLONIC MICROBIOTA OF DOGS WITH CHRONIC ENTEROPATHY. AE Jergens[1], E Cassmann, T Atherly[2], S Koda, C Wang[3]. [1]Departments of Veterinary Clinical Sciences and [2]College of Veterinary Medicine, Iowa State University and USDA-ARS, Ames, IA.[3]Veterinary Diagnostic and Production Animal Medicine.

The pathogenesis of chronic enteropathies (CE) in dogs likely involves complex interaction between the mucosal immune system and the intestinal microbiota. While the application of bacterial 16S rDNA sequence-based analysis has shown an association between altered microbial composition and duodenal inflammation in dogs, relatively little is known about alterations in non-invasive mucosal and luminal bacteria seen with diseases involving the ileum and colon. The present study sought to evaluate the relationship of enteric bacteria to type and severity of mucosal inflammation affecting the ileum and colon of dogs with CE.

Eleven client-owned dogs with CE involving both the small and large intestines were prospectively enrolled. CE was diagnosed on the basis of a history of chronic gastrointestinal signs, exclusion of identifiable underlying disorders, and histopathologic evidence of intestinal inflammation. Mucosal bacteria were detected in formalin-fixed ileal and colonic tissue sections with fluorescence in situ hybridization (FISH) using 16S rDNA-targeted probes directed against all bacteria, *Enterobacteriaceae*, *E. coli*, *Eubacterium rectale-Clostridium* coccoides group, *Bacteroides/Prevotella*, and *Helicobacter* spp. Sections were examined by epifluorescence microscopy and the number of bacteria and their spatial distribution (luminal, superficial mucus, epithelial adherent, within mucosa) was determined in ten 60x fields of each section. Microbial composition in CE dogs was compared to the ileal/colonic microbiota of 7 healthy control (HC) dogs using a mixed effect ANOVA model. P values < 0.05 were considered significant.

The final diagnoses for dogs with CE included IBD ($n = 8$) and lymphosarcoma ($n = 3$). When compared to HC dogs, dogs with CE showed regional (ileum versus colon) imbalances in microbiota composition characterized by selective enrichment of mucosa-associated populations. Evaluation of colonic biopsies in dogs with CE showed that the total number of bacteria ($P < 0.04$), *Clostridium* ($P < 0.005$), *Enterobacteriaceae* ($P < 0.04$) and *E. coli* ($P < 0.03$) were increased in the adherent mucus regions of dogs with IBD as compared to HC dogs. Total bacteria ($P < 0.05$) and *E. coli* ($P < 0.02$) were also more numerous in dogs with LSA versus HC and IBD dogs ($P < 0.04$ for *E. coli*). Ileal biopsies from CE dogs similarly showed variable dysbiosis with increased total bacteria ($P < 0.05$) but decreased Helicobacter spp ($P < 0.001$) and Bacteroides ($P < 0.02$) observed within inflamed intestines as compared to HC tissues. The spatial distribution of these bacteria was also appreciably different from HC dogs, with higher numbers of bacteria generally found within the adherent mucus compartment as compared to other ileal regions.

Our data demonstrate that dogs with CE affecting the ileum and colon have altered microbiota composition that may be a cause or consequence of mucosal inflammation. Recognition of these microbiota imbalances may provide new opportunities for therapeutic intervention.

ABSTRACT GI-10

EVALUATION OF THE ROLE OF *PENTATRICHOMONAS HOMINIS* AND ENTERIC COINFECTIONS IN CANINE DIARRHEA. MK Tolbert[1], CM Leutenegger[2], R Lobetti[3], J Birrell[4], JL Gookin[1]. [1]North Carolina State University, Raleigh, NC. [2]IDEXX Reference Laboratories, West Sacramento, CA. [3]Bryanston Veterinary Hospital, Bryanston, South Africa. [4]South African Police Services Veterinary Hospital, Roodeplaat, South Africa.

Trichomonads have been rarely reported in the feces of dogs and their pathogenicity remains uncertain. Although *Pentatrichomonas hominis* (Ph) is considered to be a commensal that may overgrow in dogs with other causes of diarrhea, little is known regarding the history, clinical presentation or prevalence of concurrent GI infections in dogs with trichomonosis. The aim of this study was to determine whether dogs with diarrhea and trichomonosis could be distinguished from dogs having diarrhea without trichomonosis on the basis of clinical signs or presence of concurrent enteric infections.

Fecal samples from 39 dogs were submitted to NCSU from 2007–2010 for *Trichomonas* spp. PCR testing. DNA was extracted using a ZR Fecal DNA mini-prep kit and absence of PCR inhibitors verified by amplification of bacterial 16S rDNA. PCR for Ph and *Tritrichomonas foetus* (Tf) was performed as well as real-time PCR assays for 9 possible concurrent enteric infectious agents. Obtainable medical records were reviewed. All submitted fecal samples were submitted from dogs with diarrhea that was variably described as soft, mucoid, hemorrhagic, or watery.

Mean age of the dogs was 2.33 years (median 1.9; range: 2.5–120 months) and represented a total of 24 breeds. Ph, Tf, or concurrent Ph and Tf were diagnosed in 18, 1, and 1 dogs respectively (Group A). The remaining 19 dogs were negative for Ph and Tf by PCR (Group B). Flagellated organisms were reportedly observed by the RDVM on fecal smear evaluation in 17 of 17 dogs in Group A and 5 of 8 dogs in Group B. Concurrent enteric infections were diagnosed in 12/14 dogs (86%) in Group A and 13/15 dogs (87%) in Group B. Co-infections identified (Group A/B) included canine enteric coronavirus (5/3), *Cryptosporidium* spp. (4/4), *Clostridium perfringens* enterotoxin A (5/9), *Giardia* spp. (4/2), *Salmonella* spp. (1/0), *Clostridium difficile* Toxin B (1/0), and *Campylobacter* spp. (10/8). No dogs were identified as infected with canine distemper virus or parvovirus. Five samples from each group had insufficient quantity or quality of DNA for concurrent infectious disease testing.

In this large study of canine trichomonosis, no differences in age, clinical signs, or prevalence and identity of concurrent enteric infection between diarrheic dogs with or without Ph were identified. Thus, these findings do not appear to support a primary pathogenic role for Ph as a causative agent of diarrhea in dogs.

ABSTRACT GI-11

DOMPERIDONE AUGMENTS CHOLINERGIC CONTRACTIONS IN FELINE AND RABBIT GASTROINTESTINAL SMOOTH MUSCLE. R. Washabau, [1] S. Longhofer, [2] D. Li[1]. [1]College of Veterinary Medicine, University of Minnesota, St. Paul, MN, and [2]Dechra Ltd, Shrewsbury, U.K.

Gastrointestinal motility disorders are a common clinical problem in domestic animals. Many of the G.I. motility disorders have been treated previously with 5-HT_4 agonists although limited availability of drugs in this classification have stimulated interest in the use of new (and old) drug therapies. The dopaminergic antagonists are a group of drugs with well-known anti-emetic effects at central dopamine D_2 receptors, and putative gastrointestinal prokinetic effects at peripheral D receptors. Domperidone has been shown, for example, to reverse gastric relaxation induced by dopamine infusion in the dog. Similar studies have not been reported in the cat or rabbit, two species at risk for distal gastrointestinal motility disorders. Our aim was to study the effects, mechanisms, and sites of action of domperidone in feline colonic and rabbit gastrointestinal smooth muscle contraction.

Portions of stomach (fundus and antrum), intestine (duodenum and ileum), cecum (rabbits only), and colon (ascending and descending) were obtained from healthy cats and rabbits from 9–12 months of age. Longitudinal and circular smooth muscle strips from each site were suspended in physiologic (HEPES) buffer solution, attached to isometric force transducers, and set to optimal muscle length (L_o) using acetylcholine (ACh; 10^{-4} M). Muscle strips were treated with domperidone (D; 10^{-8} to 10^{-4}M) in the presence or absence of ACh (10^{-8} to 10^{-4}M), and maximal force output (P_{max}) was normalized for cross-sectional area (N $= \times 10^4$ Newtons/m^2).

Domperidone (D) had a minor direct effect of inducing feline and rabbit gastric, cecal, and colonic smooth muscle contraction. Direct effects were similar whether in the longitudinal or circular muscle orientation. The direct effect of domperidone was dose-dependent and maximal (feline colon $P_{max} = 0.15$–0.22 N; rabbit colon $P_{max} = 0.10$–0.19 N) at a dose of 10^{-4} M. Domperidone had a much greater indirect effect in augmenting cholinergic (ACh; 10^{-4} M) contractions in feline and rabbit gastric, cecal, and colonic smooth muscle. Domperidone-augmented cholinergic contractions were 157–200% (feline colon $P_{max} = 1.54 \pm 0.24$ N ACh only; feline colon $P_{max} = 2.03 \pm 0.23$N ACh + D) of baseline cholinergic contractions. Domperidone contractions were of a similar magnitude to those induced by cisapride. Domperidone effects were similar in mucosa-intact and mucosa-dissected preparations. Domperidone contractions were unaffected by prazosin (α_1 receptor antagonist), yohimbine (α_2 receptor antagonist), or terbutaline (β_2 receptor agonist), but were somewhat attenuated by dopamine (D_2 receptor agonist) and a non-specific cholinergic antagonist (atropine).

In vitro studies show for the first time that domperidone has minor direct and major indirect effects in augmenting cholinergic contractions of feline and rabbit gastrointestinal (stomach, cecum and colon) smooth muscle.

ABSTRACT GI-12

CO-ADMINISTRATION OF NSAIDS AND TRAMADOL DECREASES GASTRIC MUCOSAL BARRIER FUNCTION. T Hill, A Blikslager. North Carolina State University College of Veterinary Medicine, Raleigh, NC.

As recognition of acute and chronic pain in dogs has increased, so too has the use of non-steroidal anti-inflammatory drugs (NSAIDs) often in conjunction with tramadol. In people and rats, co-administration increases the risk of perforation and gastric injury over NSAIDs alone. Using an *ex vivo* model of acid injury in canine gastric mucosa, we examined the effects of indomethacin and tramadol on gastric permeability and concentrations of gastroprotective prostaglandin E_2 (PGE_2).

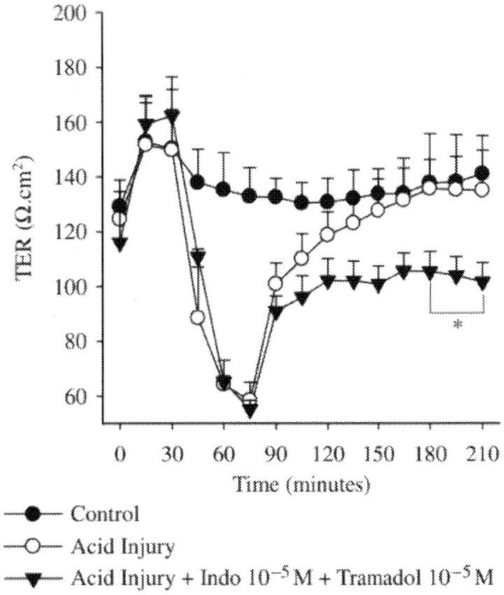

Figure [Abstract GI-12]: Effect of co-administration of indomethacin and tramadol on gastric barrier function, as measured by trans-epithelial resistance. * <0.05, n = 5.

Mucosa from the gastric antrum was harvested from 5 shelter dogs immediately after euthanasia, and mounted on Ussing chambers. The tissues were equilibrated for 30-minutes prior to addition of acidic Ringer's solution (pH, 1.2). After 45-minutes of injury, the acid was replaced with neutral Ringer's and the tissues were treated with indomethacin, tramadol or both. Tissues were maintained for 210-minutes total, during which time permeability was assessed electrically. Prostanoid concentrations were quantified using a commercially available ELISA. Western blots were performed for COX-1 and −2.

Recovery of gastric barrier function after acid injury was inhibited by co-administration of tramadol and indomethacin (Figure 1) but not by tramadol or indomethacin alone (data not shown). Prostaglandin E_2 increased with acid injury. The increase in PGE_2 was inhibited by co-administration of indomethacin and tramadol (in pg/ml: acid injury 681.25 ± 324.88, indo + tramadol 149.11 ± 35.24). There was no significant effect of treatment on COX-1 or −2 expression.

Co-administration of tramadol with a non-selective NSAID inhibits the return of gastric mucosal barrier function after acid injury in canine tissue, suggesting that caution is required in prescribing concurrent use of these drugs in dogs at risk for gastric ulcers. These drugs may exert this effect by decreasing levels of gastroprotective prostanoids. Further study is needed to understand the mechanism of this drug interaction.

ABSTRACT GI-13

EFFECT OF THE TIGHT JUNCTION REGULATOR LARAZOTIDE ACETATE ON INTESTINAL PERMEABILITY AND FECAL N-METHYLHISTAMINE CONCENTRATIONS IN SOFT COATED WHEATEN TERRIERS AND NORWEGIAN LUNDEHUNDS WITH GASTROINTESTINAL DISEASE. N Berghoff, JS Suchodolski, JM Steiner. Gastrointestinal Laboratory, Texas A&M University, College Station, TX.

An increased intestinal permeability (IP) has been suggested to be both cause and consequence of gastrointestinal (GI) disease, such as inflammatory bowel and celiac disease, in people. A novel tight junction regulator, larazotide acetate (Alba Therapeutics, Baltimore, MD) has been shown to significantly decrease IP in rats and in humans with celiac disease. The purpose of this study was to determine if larazotide acetate reduces IP in Soft Coated Wheaten Terriers (SCWT) and Norwegian Lundehunds (NL) with chronic GI disease and ameliorates clinical signs.

Four NL (2 females, 2 males; median age: 2.5 yrs, range: 1.5–4.5 yrs) and 9 SCWT (7 females, 2 males; median age: 5.0 yrs, range: 1.5–9.0 yrs) were enrolled based on presence of clinical signs of GI disease and hypoalbuminemia, increased fecal alpha$_1$-proteinase inhibitor (α_1-PI) concentrations, and/or hypocobalaminemia. SCWT with protein-losing nephropathy were excluded. Dogs were fed q12 hrs and received 0.5 mg (4 NL and 2 SCWT) or 2.0 mg (7 SCWT) of larazotide acetate PO before each meal for 90 days. Prior to start of treatment (day 1) and at the end (day 90), IP was evaluated by calculating the lactulose/rhamnose (L/R)-ratio in serum samples obtained at 30, 60, 90, and 120 min after oral dosing. Also, 3 consecutive fecal samples each were collected prior to day 1 and day 90 for N-methylhistamine (NMH) measurement. Pre- and post-treatment data were compared using a Wilcoxon signed rank test. The 0.5 mg vs. 2.0 mg dose groups were compared using a Mann-Whitney U test. Statistical significance was set at $p < 0.05$.

L/R-ratios (medians) for the 60 min sampling time point were significantly lower on day 90 (0.046) than on day 1 (0.080; $p = 0.018$). Dogs treated with 2.0 mg q12 hrs had significantly lower 60 min L/R ratios on day 90 than dogs treated with 0.5 mg (0.033 vs. 0.094; $p = 0.014$). No difference was found between breeds. Fecal NMH concentrations were not different between time points, treatment groups, or breeds. Fecal α_1-PI concentrations were available for 11 of the 13 dogs and were significantly higher on day 90 compared to day 1 ($p = 0.033$). No differences were found between pre- and post-treatment serum albumin or cobalamin concentrations. Weight gain was seen in all 4 NL. Resolution of diarrhea, vomiting, hyporexia, as well as an increased activity was seen in 1 SCWT. Another SCWT had resolution of diarrhea and a decrease in pruritus. No changes in clinical signs were reported in the remaining 7 SCWT.

This study indicates that larazotide acetate might be able to reduce IP in dogs. This effect may be dose-dependent. However, not all dogs showed an improvement in clinical signs, suggesting that factors other than increased IP might have been responsible for the clinical signs in these dogs. Breed-related effects cannot be ruled out, and further studies are warranted to determine the efficacy of larazotide acetate in dogs of other breeds with GI disease.

ABSTRACT GI-14

BIOCHEMICAL MARKERS AND SURVIVAL IN DOGS WITH PROTEIN-LOSING ENTEROPATHY. M Equilino[1], V Théodoloz[1], M Doherr[2], RM Heilmann[3], JS Suchodolski[3], JM Steiner[3], IA Burgener[1]. [1]Division of Small Animal Internal Medicine and [2]Division of Veterinary Public Health of the University of Bern, Switzerland, [3]Gastrointestinal Laboratory, Department of Small Animal Clinical Sciences, Texas A&M University, College Station, TX, USA.

Protein-losing enteropathy (PLE) is a syndrome which is characterized by excessive loss of protein from the gastrointestinal tract. Commonly reported causes in adult dogs include lymphangiectasia, inflammatory bowel disease (IBD), and alimentary lymphoma. PLE is usually diagnosed by documenting marked hypoalbuminemia and by ruling-out renal protein loss, gastrointestinal bleeding, hepatic insufficiency, and cutaneous protein loss, together with the demonstration of intestinal lesions. The aims of this study were to analyze different biochemical markers, calculate clinical activity scores, and assess survival in dogs with PLE and compare them with those in dogs with food-responsive diarrhea (FRD) without protein loss.

29 dogs with PLE and 18 dogs with FRD, referred to the University of Bern, CH, were enrolled. Selection criteria included a history of chronic diarrhea (> 3 weeks), exclusion of identifiable underlying causes, and histopathologic evidence of intestinal inflammation, but not neoplasia. Underlying disorders were excluded based on CBC, chemistry profile, urinalysis, fecal analysis, trypsin-like immunoreactivity, cobalamin, folate, and transabdominal ultrasound. Also, canine pancreatic lipase immunoreactivity (spec cPL®), C-reactive protein (CRP), calprotectin and alpha$_1$-proteinase inhibitor (α_1-PI) were measured in serum from 18 dogs and compared with 18 dogs with FRD without PLE. All dogs were scored using the canine IBD (CIBDAI) and the canine chronic enteropathy (CCE) clinical activity index (CCECAI).

Total protein, albumin (5–28.1 g/l), and total calcium (1.21–2.32 mmol/l) were decreased in all 29 dogs. Cobalamin was decreased in all but 3 dogs (< 100–490 ng/l). Spec cPL was mildly increased in 3/18 dogs with PLE and normal in 15/18 PLE and all FRD dogs. CRP was normal in 5/18 dogs with PLE (16/18 FRD), mildly increased in 7/18 (1/18 FRD), and moderately increased in 6/18 PLE dogs (1/18 FRD). Calprotectin was slightly higher in dogs with PLE, but all PLE and FRD dogs yielded values in the normal range. Serum α_1-PI was significantly lower in dogs with PLE than in those with FRD (p < 0.001), with 13/18 PLE dogs below the reference range (1/18 FRD). CIBDAI ranged from 4 to 16 and CCECAI from 6 to 19. At the end of the study, 12/29 dogs were still alive with survival times between 26 and 2544 days. 17/29 dogs died with a median survival of 96 days (range 2–874 days). Dogs with mildly increased CRP died earlier than dogs with a normal or moderately increased CRP (p = 0.011), whereas albumin, calcium, Spec cPL, calprotectin, CIBDAI, and CCECAI had no significant impact on outcome and survival.

In conclusion, dogs with PLE have a significantly lower α1-PI in the serum than dogs with FRD. Furthermore, most dogs with PLE have an increased CRP and a decreased cobalamin. A mild increase in CRP appears to be a poor prognostic factor.

ABSTRACT GI-15

CANINE PROTEIN LOSING ENTEROPATHY: A RETROSPECTIVE ANALYSIS AND SURVIVAL STUDY IN 68 DOGS. SL Owens[1], NK Parnell[1], GE Moore[1,2]. [1]Department of Veterinary Clinical Sciences, [2]Department of Comparative Pathobiology, Purdue University School of Veterinary Medicine, West Lafayette, IN.

While hypoalbuminemia is a common finding associated with chronic enteropathies, its impact on survival in this population is poorly defined. The aim of this study was to compare dogs with chronic enteropathies on the basis of their serum albumin concentration at the time of presentation. We hypothesized that dogs with a protein losing enteropathy (PLE) have a significantly shorter survival time compared to dogs with chronic enteropathies which are not hypoalbuminemic (controls).

Information obtained from the medical records included signalment, duration and characteristics of clinical signs, physical examination findings, clinicopathologic data and survival time.

One hundred seventeen cases fit the inclusion criteria; 68 in the PLE group and 49 controls. There was no statistical significance between groups for age (P = 0.12), weight (P = 0.17), weight loss (P = 0.59) and body condition score (P = 0.072). Compared to control dogs, PLE dogs had decreased serum concentrations of cobalamin (P = 0.002), total calcium (P < 0.0001), globulin (P < 0.0001), cholesterol (P < 0.0001) and ionized calcium (P < 0.001). Survival analysis revealed a significantly decreased survival time for PLE dogs (P = 0.0008); median survival was 701 days for PLE dogs and > 3,500 days for controls. While the PLE group did not survive as long, survival was not directly associated with severity of hypoalbuminemia; patients with albumin concentration < 1.3 g/dL survived longer than those with mild hypoalbuminemia (1.6–1.9 g/dL).

This study supports the observation that chronic enteropathy patients have decreased survival time when presented with hypoalbuminemia; however this study suggests the severity of hypoalbuminemia is not a reliable indicator of survival.

ABSTRACT GI-16

EVALUATION OF SERUM METHYLMALONIC ACID CONCENTRATIONS IN CHINESE SHAR PEIS AND DOGS OF SIX OTHER BREEDS WITH COBALAMIN DEFICIENCY. N Grützner, KC Stupka, JS Suchodolski, JM Steiner. Gastrointestinal Laboratory, Texas A&M University, College Station, TX.

Cobalamin (vitamin B_{12}) deficiency in the Chinese Shar Pei (Shar Pei) is suspected to be hereditary. Inherited causes of cobalamin deficiency have been reported in humans and may affect absorption, transport, or cellular processing of cobalamin. Based on human and veterinary studies, an increased serum methylmalonic acid (MMA) concentration has been suggested to reflect cobalamin deficiency at the cellular level. In this context, it has been shown in humans that MMA concentrations are higher in patients with genetic disorders affecting intracellular processing than in patients with genetic defects affecting gastrointestinal processing and extracellular transport of cobalamin. Therefore, the aim of this study was to evaluate serum MMA concentrations in Shar Peis and dogs of six other breeds with cobalamin deficiency.

From 2008 to 2010, 324 serum samples were collected from Shar Peis as well as 6 other dog breeds with undetectable serum cobalamin concentrations (< 150 ng/L), that were tested at the Gastrointestinal Laboratory. Serum MMA concentrations (reference interval: 414.7–1,192.5 nmol/L) were measured using an in-house assay. Serum MMA concentrations were compared between the 7 different dog breeds using a Kruskal-Wallis test with a Dunn's post test. A Mann-Whitney test for non-parametric data sets was used to compare serum MMA concentrations between male and female dogs of each breed. Statistical significance was set at $p < 0.05$.

Thirty-six Shar Peis, 98 German Shepherd dogs, 83 Labrador Retrievers, 47 Yorkshire Terriers, 22 Boxers, 21 Cocker Spaniels, and 17 Beagles with an undetectable serum cobalamin concentration (< 150 ng/L) were included in this study. The median serum MMA concentrations were significantly different among those breeds (Shar Peis: 26,176.7 nmol/L, German Shepherds: 1,986.6 nmol/L, Labrador Retrievers: 1,749.9 nmol/L, Boxers: 2,396.2 nmol/L, Beagles: 1,316.7 nmol/L, Yorkshire Terriers: 1,270.0 nmol/L, and Cocker Spaniels: 1,076.7 nmol/L; $p < 0.0001$). Dunn's post test showed that median serum MMA concentrations in cobalamin-deficient Shar Peis were significantly different when compared to the other 6 dog breeds. Also, serum MMA concentrations were significantly different between Cocker Spaniels and German Shepherds or Boxers ($p < 0.01$ and $p < 0.05$, respectively). Also, serum MMA concentrations were significantly different between the two genders in Beagles and Boxers (medians in Beagles: males 4,268.0 nmol/L and females 1,000.7 nmol/L: $p = 0.027$; medians in Boxers: males 3,041.1 nmol/L and females 2,001.8 nmol/L: $p = 0.015$).

In conclusion, serum cobalamin deficient Shar Peis had a 10 times higher median serum MMA concentration compared to cobalamin deficient dogs of six other dog breeds. Further studies are needed to investigate the intracellular processing of cobalamin in Shar Peis with cobalamin deficiency.

ABSTRACT GI-17

SERUM CONCENTRATIONS OF HYALURONIC ACID, C-REACTIVE PROTEIN, AND THE CALGRANULINS IN CHINESE SHAR PEIS WITH AND WITHOUT COBALAMIN DEFICIENCY. N Grützner, RM Heilmann, SM Cranford, JS Suchodolski, JM Steiner. Gastrointestinal Laboratory, Texas A&M University, College Station, TX.

Chinese Shar Peis (Shar Peis) have a high prevalence of cobalamin deficiency. Two other conditions reported frequently in this breed are Shar Pei fever and cutaneous mucinosis. Shar Pei fever is an autoimmune disorder causing periodic flare-ups and is associated with increased serum concentrations of C-reactive protein (CRP), a non-specific marker of inflammation. Cutaneous mucinosis is characterized by excessive deposition of mucin in the dermis. Also, hyaluronic acid (HA), the main component of mucin, was shown to be significantly higher in serum from Shar Peis with cutaneous mucinosis than in healthy controls. To date, a possible association between Shar Pei fever and/or cutaneous mucinosis on one side and cobalamin deficiency on the other has not been investigated in Shar Peis. Thus, the aim of this study was to compare serum concentrations of HA (an indicator of cutaneous mucinosis) and inflammatory markers (CRP, calprotectin, and S100A12), assumed to be increased in episodes of Shar Pei fever, in Shar Peis with and without cobalamin deficiency.

Serum samples from 40 Shar Peis, collected from 2008 to 2010, were analyzed. Serum HA and CRP (reference interval (RI): 0.0–7.6 mg/L) were quantified by using commercial ELISA kits (Echelon Biosciences, Salt Lake City, UT, USA and Tridelta, Maynooth, Ireland; respectively). Serum calgranulin concentrations were measured using an in-house ELISA (calprotectin; RI: 0.9–11.9 mg/L) and RIA (S100A12; RI: 33.0–233.0 µg/L), respectively. Mann-Whitney *U* tests were used to compare serum HA, CRP, calprotectin, and S100A12 concentrations between Shar Peis with and without cobalamin deficiency. Significance was set at $p < 0.05$.

Fourteen Shar Peis were severely cobalamin deficient, defined by an undetectable serum cobalamin concentration (< 150 ng/L). In the remaining 26 dogs, serum cobalamin concentrations were within the reference interval (251–908 ng/L). Serum concentrations of HA, CRP, calprotectin, and S100A12 were not significantly different between cobalamin deficient Shar Peis (medians: 649.9 ng/ml, 4.9 mg/L, 10.4 mg/L, and 202.4 µg/L, respectively) and Shar Peis with a normal serum cobalamin concentration (medians: 671.5 ng/ml [$p = 0.52$], 1.3 mg/L [$p = 0.11$], 10.8 mg/L [$p = 0.70$], and 190.2 µg/L [$p = 0.61$], respectively). Fifty percent of cobalamin deficient Shar Peis had serum calprotectin concentrations above the upper limit of the reference interval, and 43% had serum S100A12 concentrations above the suggested upper reference limit.

In this study, serum concentrations of HA, CRP, and the calgranulins did not differ between cobalamin deficient Shar Peis and Shar Peis with a normal serum cobalamin concentration. This finding leads us to speculate that increased HA and/or inflammatory markers are not associated with cobalamin deficiency in Shar Peis. Further studies are needed to investigate serum cobalamin concentrations in patients with Shar Pei fever or cutaneous mucinosis.

ABSTRACT GI-18

COBALAMIN DEFICIENCY IN BORDER COLLIES. S.Lutz[1], T.Leeb[2], A. Sewell[3], B. Bigler[4], S. Hartnack[5], B. Riond[6], C.E. Reusch[1], P.H. Kook[1]. [1]Clinic for Small Animal Internal Medicine, Vetsuisse Faculty, University of Zurich, Switzerland; [2]Institute for Genetics, University of Berne, Switzerland; [3]Metabolic Laboratory, Goethe University Frankfurt am Main, Germany; [4]Laupeneck Laboratory, Switzerland; [5]Institute for Epidemiology, [6]Institute for Laboratory Science, University of Zurich, Switzerland.

Cobalamin deficiency (CD) has been associated with gastrointestinal and pancreatic disease in dogs. Hereditary CD has been demonstrated in Giant Schnauzers and single case reports have suggested congenital CD in the Border Collie (BC) breed. Clinicopathologic findings of CD vary and can be unspecific as cobalamin acts as a Co-Factor for a multitude of enzymatic reactions. The two most important reactions concern the conversion of methylmalonyl-CoA to succinyl-CoA and the re-methylation of homocysteine (Hcy). These two metabolites increase when cobalamin is lacking and act as markers for cobalamin availability on a cellular level. Preliminary data from dogs suggested that measurement of methylmalonic acid (MMA) may be a better diagnostic test for CD than serum cobalamin concentration. Therefore the goals of the study were (1) to establish reference values for serum cobalamin, urine MMA and plasma Hcy in healthy pet dogs, (2) to screen a larger BC population from Switzerland for CD, and (3) to perform genomic analyses on BC with CD.

For determination of reference values 35 healthy pet dogs were used. Serum cobalamin was measured using an automated chemiluminescence assay (Immulite 2000), urine MMA was determined using gas chromatography and expressed as a ratio to urine creatinine and plasma Hcy was measured using High Pressure Liquid Chromatography and fluorimetric detection. To calculate reference ranges the 10th and 90th percentile were used. Data were analyzed using non-parametric tests.

Reference ranges for cobalamin, Hcy, and MMA were: cobalamin 279.2–972.8 ng/L; urine MMA 2–4.65 mmol/mol creatinine; and plasma Hcy 4.73–18.34 µmol/L. The screened BC population comprised 113 purebred dogs and 4 BC (median 11.5 months; range 8–41) suffering from congenital CD could be identified. Clinical signs differed and consisted of tiredness (4), stunted growth (4), anemia (3), dysphagia (2) and persistent fever (1). Median (ranges) results for healthy BC and BC with CD were: for cobalamin 592 (142–1855) and 72.00 (30–139) ng/L; for urine MMA 2 (2–360) and 4148 (1800–6665) mmol/mol creatinine; for Hcy 8.5 (2.8–22.4) and 41.00 (40–86.6) µmol/L. Strikingly, healthy BC with cobalamin concentrations well within the reference range had significantly higher urine MMA concentrations compared to control dogs. Under the assumption that the four affected BC are inbred to a single founder animal, first results of genotyping on the 170 k illumina canine_HD SNP chip suggest that mutations in the CUBN and AMN gene can be excluded to cause the observed CD in these dogs. We conclude that CD is a rare familial disease in BC with variable clinical signs. To define the genomic region responsible for CD further genetic analysis is in progress. It remains to be determined why some BC have high urine MMA concentrations despite a serum cobalamin concentration within the reference range.

ABSTRACT GI-19

DEVELOPMENT AND ANALYTICAL VALIDATION OF AN ENZYME-LINKED IMMUNOSORBENT ASSAY FOR THE QUANTIFICATION OF CANINE CALPROTECTIN IN SERUM AND FECES FROM DOGS. RM Heilmann, BC Guard, K Weber, JS Suchodolski, JM Steiner. Gastrointestinal Laboratory, Texas A&M University, College Station, TX.

Calprotectin is a protein complex that plays an important role in the innate immune response. Preliminary data suggest that canine calprotectin (cCP) is a useful marker for the detection of inflammation in dogs. Recently, a radioimmunoassay for the measurement of cCP has been developed and analytically validated, but this test requires the use of a radioactive tracer. Therefore, the aim of this study was to develop and analytically validate an enzyme-linked immunosorbent assay (ELISA) for the quantification of cCP in serum and fecal specimens from dogs.

Canine calprotectin (cCP) was purified, antiserum against purified cCP was raised in rabbits, monospecific antibodies were purified by affinity chromatography, and a sandwich-ELISA was developed. Purified antibodies were used for capturing and, after coupling with horseradish peroxidase (HRP), for reporting. A HRP substrate was used for color development. The assay was analytically validated by determination of analytical sensitivity and specificity, dilutional parallelism, spiking recovery, and intra- and inter-assay variability. Control intervals for serum and fecal cCP were established from 110 and 52 healthy pet dogs, respectively, using the central 95th percentile.

Sensitivity of the assay for serum samples assayed in a 1:400 dilution and for fecal extracts assayed in a 1:4,000 dilution was 0.3 mg/L and 3.2 µg/g, respectively. Over a wide range of the assay, there was no cross-reactivity with cS100A12, the closest structural analogue of cCP available. Observed to expected ratios (O/E) for serial dilutions ranged from 83.2–118.5% (mean ± standard deviation [SD]: 101.3 ± 10.0%) for four different serum samples, and from 81.7–129.1% (mean ± SD: 101.8 ± 14.0%) for five different fecal extracts. O/E for spiking recovery ranged from 87.8–130.4% (mean ± SD: 100.6 ± 6.5%) for four different serum samples and 6 different spiking concentrations, and from 95.9–152.0% (mean ± SD: 104.6 ± 11.6%) for 4 different fecal extracts and 6 different spiking concentrations. Intra-assay coefficients of variation (CV) for 4 different serum samples were 7.8, 5.0, 7.4, and 12.7%, and 10.0, 6.1, 6.2, and 7.0% for 4 different fecal extracts. Inter-assay CV for 4 different serum samples were 17.2, 8.1, 9.9, and 12.6%, and 12.3, 8.3, 7.2, and 9.6% for 4 different fecal extracts. The control intervals for serum and fecal cCP were established as 0.9–11.9 mg/L and 3.2–65.4 µg/g, respectively.

We conclude that this new ELISA for the measurement of cCP is analytically sensitive, linear, accurate, precise, and reproducible, and does not cross-react with canine S100A12. Further studies evaluating the clinical usefulness of measuring serum and/or fecal cCP are currently under way.

ABSTRACT GI-20

FECAL INFLAMMATORY MARKERS AND MICROBIAL CHANGES IN DOGS WITH ACUTE HEMORRHAGIC GASTROENTERITIS. RM Heilmann[1], S Unterer[2], ME Markel[1], JM Steiner[1], JS Suchodolski[1]. [1]Gastrointestinal Laboratory, Texas A&M University, College Station, TX, USA. [2]Ludwig-Maximilians-University, Munich, Germany.

The syndrome of hemorrhagic gastroenteritis (HGE) is characterized by a peracute onset of hemorrhagic diarrhea, vomiting, depression, and anorexia, and can be associated with a high mortality if untreated. The etiology of HGE is unknown, but it is speculated that an abnormal response to bacterial endotoxins, bacteria, or dietary components may play a role. HGE is characterized by an increased vascular/mucosal permeability, thought to represent a type I-hypersensitivity reaction, whereas inflammation and necrosis appear to be rare. However, markers of gastrointestinal (GI) inflammation and changes in the intestinal microbiota have not been studied extensively in dogs with HGE. Therefore, the aim of this study was to evaluate fecal canine calprotectin (CP) and S100A12 (A12), α_1-proteinase inhibitor (α_1-PI, a marker of GI protein loss), and bacterial groups that have previously been shown to be decreased (i.e., *Faecalibacterium* spp., Ruminococcaceae, *Bifidobacterium* spp.) or increased (i.e., Proteobacteria) in fecal samples from dogs with HGE.

Fecal samples from 3 consecutive days were collected from 7 dogs with HGE. Fecal CP, A12, and α_1-PI concentrations were measured by in-house immunoassays. Bacterial DNA was extracted from each fecal sample and was analyzed for *Faecalibacterium* spp., Proteobacteria, Rumino-coccaceae, and *Bifidobacterium* spp. using quantitative PCR assays. Concentrations of fecal CP, A12, and α_1-PI, and the abundance of bacterial DNA were compared using a Friedman test with Dunn's post-hoc tests. Significance was set at $p < 0.05$.

At the time of diagnosis (day 1), fecal CP, A12, and α_1-PI were above the suggested reference intervals in 6, 6, and 5 of the 7 dogs, respectively. Until day 3, this number decreased to 2, 1, and 4, respectively. Decreases in concentrations were significant between days 2 and 3 for A12 ($p = 0.016$), and between days 1 and 3 for α_1-PI ($p = 0.012$), but not for CP despite a trend ($p = 0.085$). No differences in the abundance of *Faecalibacterium* spp. ($p = 0.085$), *Bifidobacterium* spp. ($p = 0.192$), or Proteobacteria ($p = 0.305$) were observed. However, the abundance of Rumino-coccaceae was significantly lower on day 3 when compared to day 2 ($p = 0.008$).

In this study, fecal markers of inflammation and GI protein loss were increased in dogs with HGE. Although the number of patients was small, following initiation of treatment, two of the markers decreased significantly. These results suggest a loss of protein into the GI tract at the onset of HGE. The lack of significant increases of *Faecalibacterium* spp., *Bifidobacterium* spp., and Ruminococcaceae, and decreases in Proteobacteria may suggest GI dysbiosis. Further longitudinal studies are needed and are currently under way to evaluate GI dysbiosis in canine HGE patients.

ABSTRACT GI-21

PHARMACOKINETICS OF MAROPITANT CITRATE DOSED ORALY TO DOGS AT 2 MG/KG AND 8 MG/KG ONCE DAILY FOR 14 CONSECUTIVE DAYS. Steven P. Lesman[1], Joseph F. Boucher[1], G. Scott Grover[1], Jacky Civil[2], Steven R. Cox[1], Tara L. Bidgood[3]. [1]Pfizer Animal Health, Kalamzoo, MI. [2]Pfizer Animal Health, Sandwich, Kent UK. 3-Pfizer Animal Health-Madison, NJ.

The most recent antiemetic approved for use in dogs is maropitant citrate (Cerenia®, Pfizer Animal Health). Maropitant is a selective NK1 receptor antagonist that acts by blocking the binding of substance-P within the emetic center and chemoreceptor trigger zone.

Label dosage recommendations for maropitant citrate are 1 mg/kg SC or 2 mg/kg orally once daily for up to 5 consecutive days (acute emesis) and 8 mg/kg orally once daily for up to 2 consecutive days (motion sickness). The study objective was to determine when steady-state is reached and the pharmacokinetics of maropitant administered at label oral dosages once daily for 14 consecutive days.

Two groups of eight healthy beagles were administered maropitant citrate at 2 or 8 mg/kg orally once daily for 14 days. Concentrations of maropitant and its metabolite were measured in plasma using a LC-MS/MS assay. Pharmacokinetic parameters were estimated using non-compartmental pharmacokinetic techniques and a modeling approach was used to estimate steady-state.

The accumulation ratio for maropitant was 2.46 (AUC0-24) and 2.03 (Cmax) for the 2 mg/kg dose; and 4.81 (AUC0-24) and 2.77 (Cmax) for the 8 mg/kg dose after 14 days. The model estimate for the number of doses required to reach 90% of steady-state was 4.30 for 2 mg/kg and 8.09 for 8 mg/kg. Three dogs experienced a single episode of vomiting.

Dosing maropitant citrate beyond the label duration was well tolerated by healthy dogs. Steady-state was reached after approximately 4 doses for daily 2 mg/kg and 8 doses for daily 8 mg/kg oral dosing.

Previously presented at the Veterinary Cancer Society, November 2010.

ABSTRACT GI-22

EVALUATION OF SERUM COBALAMIN CONCENTRATIONS IN CANINE SERUM SAMPLES SUBMITTED TO THE GASTROINTESTINAL LABORATORY BETWEEN 2006-2010. N Grützner[1], SM Cranford[1], B Norby[2], JS Suchodolski[1], JM Steiner[1]. [1]Gastrointestinal Laboratory, [2]Veterinary Integrative Biosciences, Texas A&M University, College Station, TX.

Cobalamin (vitamin B_{12}) is involved in a variety of metabolic processes. Altered serum cobalamin concentrations have been observed in dogs with gastrointestinal disorders, such as exocrine pancreatic insufficiency (EPI) or severe and longstanding ileal disease. This study was conducted to identify breeds with a higher proportion of a decreased serum cobalamin concentration that were submitted to the Gastrointestinal Laboratory. The study was also aimed at investigating serum trypsin-like immunoreactivity (TLI) concentrations that were diagnostic for EPI in the dogs with a decreased serum cobalamin concentration. Finally, we aimed to compare the number of submissions for cobalamin analysis by breed to the American Kennel Club (AKC) breed ranking list of 2009.

Canine submissions for analysis of serum cobalamin concentration between 2006 and 2010 were analyzed. Results for samples resubmitted from the same dog were excluded. Serum cobalamin concentrations were compared by calculating the odds ratio (OR) and the 95% confidence interval (CI) to find breeds with higher proportions of samples with a cobalamin concentration < 251 ng/L (below the reference interval) or < 149 ng/L (below the assay sensitivity) from dog breeds with a minimum of 30 submissions. Breeds with a significantly higher proportion of samples with a serum cobalamin < 150 ng/L were subsequently investigated for proportions of dogs with a serum TLI diagnostic for EPI (< 2.5 µg/L). Cobalamin submissions were compared to the AKC ranking list of 2009 to find breeds with higher proportions of submissions for serum cobalamin analysis.

28,675 submissions were reviewed. More than 35% of the samples submitted from Chinese Shar Peis (CSP; n = 82, AKC ranking position (rp): 47, OR [CI]: 4.4 [3.2–6.0], $p < 0.0001$), Akita Inus (AI; n = 43, rp: 50, OR [CI]: 2.7 [1.9–4.1], $p < 0.0001$), Greyhounds (GH: n = 174, rp: 140, OR [CI]: 2.5 [2.1–3.1], $p < 0.0001$), and Irish Wolfhounds (n = 14, rp: 79, OR: [CI] 2.1 [1.1–4.1], $p = 0.023$) had serum cobalamin < 251 ng/L. More than 15% of the samples submitted from CSP (n = 63, OR [CI]: 11.4 [8.1–16.1], $p < 0.0001$) and AI (n = 18, OR [CI]: 3.8 [2.3–6.5], $p < 0.0001$) had a serum cobalamin < 150 ng/L. AI with an undetectable serum cobalamin concentration (< 149 ng/L) were 8.6 times [CI: 2.4–31.3] more likely to have a serum TLI consistent with a diagnosis of EPI ($p < 0.001$). GH (n = 110), Parson Russell Terriers (n = 121, rp: 87), and Standard Schnauzers (n = 58, rp: 99) were identified to have the highest rate of samples submitted for serum cobalamin analysis (OR [CI]: 84.8 [65.3–110.0]; 16.1 [13.2–19.5], and 9.0 [CI: 7.2–12.4], respectively; $p < 0.0001$) when comparing submission data to the AKC ranking list (n = 649,677; 164 dog breeds).

Except for CSP, breeds identified here, have not previously been identified to have a higher rate of a decreased serum cobalamin concentration. Also, a possible association between an undetectable serum cobalamin and a decreased serum TLI in AI needs to be further investigated.

ABSTRACT GI-23

INTRA-INDIVIDUAL VARIABILITY OF FECAL CALPROTECTIN CONCENTRATIONS IN DOGS WITH CHRONIC GASTROINTESTINAL DISEASE. RM Heilmann[1], L Toresson[2], N Berghoff[1], N Grützner[1], JS Suchodolski[1], JM Steiner[1]. [1]Gastrointestinal Laboratory, Texas A&M University, College Station, TX, USA. [2]Helsingborg Referral Animal Hospital, Helsingborg, Sweden.

Calprotectin (CP) is a widely used marker for the diagnosis and monitoring of gastrointestinal (GI) inflammation in humans. Studies in humans usually report fecal CP concentrations based on a single stool sample although considerable day-to-day variability of fecal CP was found in patients with GI disease and in healthy controls. Intra-individual variation of canine CP (cCP) was also substantial in a small number of healthy dogs but has not been determined in dogs with chronic GI disease. Thus, the aim of this study was to compare the day-to-day variation of fecal cCP in dogs with chronic GI disease before and during treatment to that in healthy dogs. We hypothesized that fecal cCP would be less variable in patients with chronic GI disease than in healthy controls, and thus collection of a single fecal sample would be sufficient.

Fecal samples from 3 consecutive days were prospectively collected from 15 dogs (group A; median age: 6.1 years) referred for diagnostic work-up of chronic signs of GI disease, from 8 dogs (group B; median age: 5.6 years) with stable GI disease while being treated, and from 44 healthy adult dogs (group C; mean age: 5.4 years). Fecal samples were extracted and cCP was measured by an in-house immunoassay. Mean cCP, standard deviation, coefficient of variation (CV), and difference between maximum and minimum cCP for the 3-day sample collection period were calculated for each dog and were compared among groups using a Kruskal-Wallis test.

Fecal cCP ranged from 2.9–102.7 μg/g (median: 17.6 μg/g) in dogs with GI disease (group A), from 2.9–265.1 μg/g (median: 18.4 μg/g) in dogs of group B, and from 2.9–93.1 μg/g (median: 7.9 μg/g) in healthy controls (group C). CVs were 0–121.3% in group A (median: 20.0%), 2.0–89.4% in group B (median: 47.4%), and 0–145.2% in group C (median: 40.4%), respectively. Patients in group A appeared to have less variable fecal cCP than dogs in group B and C, but this difference was not significant ($p = 0.326$). The difference between maximum and minimum cCP for the 3-day sample collection ranged from 0–31.1 μg/g in group A (median: 7.2 μg/g), from 0.1–240.0 μg/g in group B (median: 12.5 μg/g), and from 0–57.4 μg/g in group C (median: 14.8 μg/g), and were not significantly different between any of the groups ($p = 0.530$).

In this study, considerable day-to-day variation of fecal cCP was found in dogs with chronic GI disease (regardless of treatment) and was comparable to that in healthy dogs. Results of this study suggest that for evaluating fecal cCP in dogs with clinical signs of GI disease, three consecutive fecal samples rather than a single fecal sample should be analyzed. Because we did not intend to evaluate the clinical usefulness of fecal cCP as a marker of GI disease in dogs, disease severity, quality, and location differed among dogs in groups A and B. The diagnostic utility of fecal cCP in dogs with GI disease is currently being investigated.

ABSTRACT GI-24

PREVALENCE OF THE *CLOSTRIDIUM PERFRINGENS* ENTEROTOXIN GENE AND *C. PERFRINGENS* ENTEROTOXIN IN FECAL SAMPLES FROM DOGS AND CATS. Y Minamoto, NM Gossett, JM Steiner, JS Suchodolski. Gastrointestinal Laboratory, Texas A&M University, College Station, TX.

It has been suggested that diagnosis of *Clostridium perfringens* related enteropathy should be based on the detection of the *C. perfringens* enterotoxin gene (*cpe*-gene) by PCR and/or *C. perfringens* enterotoxin (CPE) by ELISA in feces. However, the prevalence of the *cpe*-gene and CPE in dogs and especially cats with gastrointestinal disease has not yet been reported. Also, there is limited information about the stability of CPE in fecal samples at various storage conditions. The aim of this study was to evaluate the prevalence of the *cpe*-gene and CPE and the stability of CPE in fecal samples from dogs and cats.

To evaluate the prevalence of the *cpe*-gene, a total of 481 fecal samples from dogs and cats with clinical signs of gastrointestinal disease (273 dogs and 208 cats) and 109 fecal samples from those without such signs (80 dogs and 29 cats) were examined using PCR. To evaluate the prevalence of CPE, a total of 90 fecal samples from dogs and cats with clinical signs of gastrointestinal disease (31 dogs and 59 cats) and 11 dogs without such signs were evaluated using a commercially available ELISA kit (TechLab, Blacksburg, VA). The results were analyzed using a Fisher's exact test. Significance was set at $p < 0.05$. To evaluate the stability of CPE, 8 fecal samples from dogs and 2 from cats with clinical signs of gastrointestinal disease that were positive for CPE were examined. Also, 5 CPE negative samples from dogs were evaluated as negative controls. Each sample was subdivided into 8 aliquots and evaluated on day 0; on days 2, 5, and 10 after being stored at room temperature (RT) or 4°C; and on day 10 after being stored at −20°C.

The prevalence of the *cpe*-gene was not significantly different between dogs with signs of gastrointestinal disease (99/273; 36.3%) and dogs without (27/80; 33.8%; $p = 0.79$). Also, the prevalence of the *cpe*-gene in cats with signs of gastrointestinal disease (80/208; 38.5%) was not significantly different compared to cats without (6/29; 20.7%; $p = 0.06$). PCR and ELISA results were available for 90 samples. Of the 65 PCR positive samples, only 6 (9.2%) were ELISA positive. Of the 35 PCR negative samples, only 1 (2.9%) was ELISA positive. The prevalence of CPE was not significantly different between dogs with clinical signs of gastrointestinal disease (3/31; 9.7%) and those without (1/11; 9.0%; $p = 1.0$). The prevalence of CPE in cats with signs of gastrointestinal disease was 4/59 (6.8%), but no samples from cats without such signs were available. When evaluating the stability of CPE, results for all aliquots were consistent with the initial result, except for one sample (on day 5, stored at RT, which was initially CPE positive).

These results indicate that only a small proportion of samples that are PCR positive for the *cpe*-gene are also positive for CPE. Studies are warranted to further compare the prevalence of CPE among animals with gastrointestinal disease and those without. Furthermore, the results indicate that CPE is relatively stable in fecal samples at various storage temperatures.

ABSTRACT GI-25

COMPARISON OF CULTURE METHODS AND DETECTION AND CHARACTERIZATION OF *CLOSTRIDIUM PERFRINGENS* FROM HEALTHY AND DIARRHEIC DOGS. M Goldstein, S Kruth, A Bersenas, M Holowaychuk, JS Weese. Ontario Veterinary College, University of Guelph, Guelph, Ontario.

Clostridium perfringens has been implicated as a cause of diarrhea in dogs. The main study objective was to compare two culture methods for the identification of *C. perfringens*. A secondary objective was to evaluate *C. perfringens* toxin genes α, β, β_2, ε, ι and *cpe* from canine isolates using a multiplex PCR and determine their prevalence in a group of normal and diarrheic dogs.

Fecal samples were collected from clinically normal (ND, n = 105) and diarrheic dogs (DD, n = 54) at a primary care veterinary facility. Isolation of *C. perfringens* was performed using direct inoculation of feces onto 5% sheep blood agar (SBA) as well as enrichment of stool in BHI broth followed by inoculation onto SBA. Isolates were tested by multiplex PCR for the presence of α, β, β_2, ε, ι and *cpe* genes.

C. perfringens was isolated from 84% (88/105) of ND fecal samples using direct culture and 87.6% (92/105) with BHI enrichment ($p = 0.79$). In the DD, corresponding isolation rates were 90.7% and 93.8% ($p = 0.45$). All isolates possessed α toxin gene. β, β_2, ε, ι and *cpe* toxin genes were identified in 4.4%, 1.1%, 3.3%, 1.1% and 14.4% of ND isolates, respectively. In the DD group, β and β_2 were identified in 5%, ε and ι were not identified and the *cpe* gene in 16.9% of isolates.

BHI enrichment did not significantly increase the yield of *C. perfringens* compared to SBA but increased time and cost involved. *C. perfringens* ($p = 0.64$) and C. *perfringens* toxin genes were present in equal proportions in ND and DD groups ($p \geq 0.15$). Culture of C. *perfringens* and PCR for toxin genes are of limited diagnostic utility due to the high prevalence of *C.perfringens* in normal dogs and the lack of apparent difference in toxin gene distribution between normal and diarrheic dogs.

ABSTRACT GI-26

CONSISTENCY OF ENDOSCOPIC BIOPSIES OBTAINED FROM DUODENUM AND ILEUM FOR FELINE SMALL CELL LYMPHOMA (SC-LSA) AND INFLAMMATORY BOWEL DISEASE (IBD). KD Scott, DL Zoran, J Mansell, B Norby, MD Willard. Texas A&M University College of Veterinary Medicine and Biomedical Sciences, College Station, TX.

Endoscopic biopsies are a relatively convenient, non-invasive test for feline infiltrative intestinal disorders. Commonly, only the duodenum is examined due to cost, risks and time required to prepare the colon using lavage solutions, cathartics and/or enemas. The purpose of this study was to evaluate the consistency between endoscopic biopsies of the duodenum and ileum in cats.

Endoscopic biopsies from 70 cats which had duodenal and ileal tissue specimens were evaluated retrospectively. All slides were randomized and reviewed by a single pathologist (JM) for quality, number of biopsies, and diagnosis according to WSAVA standards. No information regarding history, clinical signs, endoscopic findings, or previous histological diagnosis was made available to the pathologist. Statistical comparison of the diagnosis of SC-LSA and IBD by intestinal location was conducted using Fisher's exact test ($p < 0.05$ significant).

18 of 70 cats (25.7%) were diagnosed with SC-LSA in the duodenum and/or ileum. Of these 18 cats, 7 (38.9%) were diagnosed with only duodenal SC-LSA, 8 (44.4%) were diagnosed with only

ileal SC-LSA, and 3 (16.7%) had SC-LSA in both duodenum and ileum. In 8 cats with only ileal SC-LSA, 3 had severe IBD in duodenal biopsies, possibly consistent with early SC-LSA. 5 of these 8 had duodenal biopsies without evidence of SC-LSA.

Our results suggest there is a population of cats in which diagnosis of SC-LSA may only be found by evaluating ileal biopsies. Clinicians should consider performing both upper and lower GI endoscopic biopsies in cats with suspected infiltrative small bowel disease.

ABSTRACT GI-27

EVALUATION OF DENTAL TARTAR, GINGIVITIS, AND SUBGINGIVAL BACTERIA IN FIV- AND FELV-NEGATIVE CATS. P Sattasathuchana[1,2], N Thengchaisri[2], P Chongtrakool[3], JS Suchodolski[1], JM Steiner[1]. [1]Gastrointestinal Laboratory, Texas A&M University, College Station, TX, USA. [2]Department of Companion Animal Clinical Sciences, Faculty of Veterinary Medicine, Kasetsart University, Bangkok, Thailand. [3]Department of Pathology, Ramathibodi Hospital, Mahidol University, Bangkok, Thailand.

Periodontitis is one of the most common diseases in cats and is mainly due to the presence of plaque and calculus. In this study, we investigated putative correlations between dental tartar and gingivitis and also between gingivitis and subgingival bacteria in cats.

Twelve cats (median age: 5 years; range: 1–10 years; 6 DSH and 6 Persians; 8 females and 4 males) were enrolled. Dental tartar was obtained during scaling for a dental prophylactic procedure. All cats were negative for FeLV and FIV infection as assessed by a commercial ELISA test (SNAP® FIV/FeLV Combo Test). Severity of gingivitis (scores: 0–3; 0 = normal, 1 = mild, 2 = moderate, and 3 = severe) and dental tartar (scores: 0–3) were scored in each cat. Endodontic paper points were applied for collecting a bacterial sample from the subgingival area and transferred to thioglycollate transporting media for bacterial culture. The relationship between gingivitis and tartar thickness scores was analyzed by spearman correlation. A student's t-test was used to compare the mean differences (gingivitis and tartar thickness scores) between upper and lower teeth. The association between severity of gingivitis and bacterial type was tested by chi square test.

The spearman correlation coefficient for the average gingivitis score and the average tartar thickness score was 0.91 ($p < 0.05$). Interestingly, the average tartar thickness scores from the upper jaw were significantly higher than those from the lower jaw ($p < 0.05$). The highest scores were found for the molar teeth in all cats. Bacterial culture revealed 28.9% anaerobic bacteria species (i.e., *Bacteroides* spp., *Peptostreptococcus anaerobius*, and *Eubacterium aerofaciens*) and 71.1% aerobic bacteria species (i.e., *Pasteurella multocida*, *Streptococcus* spp., *Enterococcus* spp., *Staphylococcus* spp., *Bacillus cereus*, *Escherichia coli*, and *Pseudomonas aeruginosa*). Anaerobic bacteria were found mostly in cats with higher gingivitis scores (2–3; chi square: $p < 0.05$), while *Pasteurella multocida* was found mostly in cats with lower gingivitis scores (0–1; chi square: $p < 0.05$). Antimicrobial sensitivity testing indicated that all of the anaerobic bacteria were sensitive to clindamycin, chloramphenicol, metronidazole, cefoxitin, or tetracycline, 90% were sensitive to erythromycin, and 80% were sensitive to penicillin. The most abundant aerobic bacterial species, *Pasteurella multocida*, was sensitive to cefoxitin in all cases in which it had been cultured.

These results suggest that anaerobic bacteria may be associated in the pathogenesis of severe gingivitis. These data warrant further studies of the prophylactic use of antibiotics in cats undergoing dental prophylactic procedures.

ABSTRACT GI-28

POLYMORPHISMS IN THE TOLL LIKE RECEPTOR 5 GENE ARE SIGNIFICANTLY ASSOCIATED WITH INFLAMMATORY BOWEL DISEASE IN A POPULATION CONSISTING OF 38 DIFFERENT DOG BREEDS. A Kathrani[1], A House[1], D Werling[2], B Catchpole[2], A Murphy[3], K Allenspach[1]. [1]Department of Veterinary Clinical Sciences, [2]Department of Pathology and Infectious Diseasesand [3]Department of Veterinary Basic Sciences, Royal Veterinary College, University of London, UK.

Inflammatory bowel disease is the most common cause of vomiting and diarrhea in dogs. Although it can occur in any canine breed, certain breeds are more susceptible. We have previously shown that polymorphisms in the TLR4 and TLR5 gene are significantly associated with inflammatory bowel disease (IBD) in the German shepherd dog (GSD), a breed at risk of developing this disease. It would be useful to determine if these polymorphisms are significant in other canine breeds as this may allow the development of novel diagnostics and therapeutics to be applied to all canine breeds with IBD. Therefore the aim of this study was to investigate whether polymorphisms in canine TLR 4 and TLR 5 genes are associated with IBD in other non-GSD canine breeds.

Four non-synonymous SNPs in the TLR4 gene; T23C, G1039A, A1572T and G1807A and three non-synonymous SNPs in the TLR5 gene; G22A, C100T and T1844C previously identified in a mutational analysis in GSDs with IBD were evaluated in a case-control study using a SNaPSHOT multiplex reaction. Sequencing information from 85 unrelated dogs with IBD consisting of 38 different non-GSD breeds from the UK were compared to a breed-matched control group consisting of 162 unrelated dogs from patients treated for non-inflammatory disease at the Royal Veterinary College, London, UK.

As in the GSD IBD population the two TLR5 SNPs; C100T and T1844C were found to be significantly protective for IBD in other breeds included in this study ($p = 0.023$ and $p = 0.0195$ respectively).

This study confirms the protective effects of the two TLR5 SNPs (C100T and T1884C) in other canine breeds with IBD. This highlights the importance of TLR5 in the pathogenesis of canine IBD and may represent common pathological pathways of IBD in different canine breeds due to the high degree of haplotype sharing seen among breeds. This may allow for the future expansion of novel diagnostics and therapeutics to be applied to all canine breeds with IBD. Further functional studies looking at the role of TLR5 in the pathogenesis of canine IBD are needed to confirm these findings.

ABSTRACT GI-29

MUTATIONAL ANALYSIS OF THE TLR5 GENE IN BOXER DOGS WITH INFLAMMATORY BOWEL DISEASE REVEALS NOVEL POLYMORPHISM IN THE LEUCINE RICH REPEAT DOMAIN. A Kathrani[1], D Werling[2], K Allenspach[1]. [1]Department of Veterinary Clinical Sciences and [2]Department of Pathology and Infectious Diseases, Royal Veterinary College, University of London, UK.

Toll-like receptor 5 (TLR5) is an extracellular pattern recognition receptor belonging to the innate immune system. We have recently shown that three non-synonymous single nucleotide polymorphisms (SNPs) in the TLR5 gene (G22A, C100T and T1844C) are significantly associated with inflammatory bowel disease (IBD) in German shepherd dogs. In addition, we confirmed that two of these TLR5 SNPs (C100T and T1844C) were significantly associated with IBD in a population consisting of 38 different dog breeds. In order to determine if other novel SNPs exist in the TLR5 gene in addition to the ones identified in the GSD population, mutational analysis was carried out in seven Boxer dogs with IBD.

Polymerase chain reaction was carried out to amplify the TLR5 coding region in the seven dogs with IBD. Sequencing was carried out using sequence based typing with the ABI Prism Sequencing Kit (Applied Biosystems, UK) and analyzed using an ABI3100 automated sequencer (PE Applied Biosystems). Sequencing information from seven Boxer dogs with IBD from the UK were compared to the reference sequence published on the Ensemble webserver (www.ensembl.org/Canis_familiaris).

In addition to the three SNPs identified previously in the TLR5 gene, a novel non-synonymous SNP; T443C was identified in the Boxer dog population with IBD. This SNP has never been reported before and was present as the homozygote genotype in three dogs with IBD and in one dog as the heterozygote genotype. Using the Simple modular architecture research tool (SMART) web server (http://smart.embl.de/) we were able to map the T443C SNP to the leucine rich repeat domain of the TLR5 protein. The leucine rich repeat domain is involved with ligand binding and therefore a change in the amino acid in this region may affect function, especially as the T443C SNP results in a change in the amino acid from non-polar to polar.

Our study further confirms the role of TLR5 in the pathogenesis of canine IBD. Our results suggest that in addition to shared risk polymorphisms amongst breeds, individual breeds may harbor unique SNPs arising after breed formation which may further affect their susceptibility to this disease. However, a case-control study would be needed in the Boxer dog to confirm the significance of the TLR5 T443C SNP and further functional data would be needed to elucidate the exact role of this polymorphism in canine IBD.

ABSTRACT HM-1

COMPARISON OF BONE MARROW SAMPLING IN CATS USING TWO DIFFERENT TECHNIQUES: MANUAL AND POWER DRIVER-ASSISTED. M.P. Dhumeaux, E.C.R. Snead, R.M. Dickinson, M.L. Leis, A.P. Carr, S.M. Taylor, T.Y. Epp. Western College of Veterinary Medicine, Saskatoon, SK, Canada.

An automated power driver device (OnControl, Vidacare) has recently become available for bone marrow aspiration (BMA) and bone marrow biopsy (BMB) in humans. The purpose of our study was to compare this automated technique to the traditional manual technique for bone marrow sampling in cats.

Twelve healthy research cats were anesthetized using a standardized protocol on 2 different occasions, 2 days apart, to have BMAs and BMBs performed by the same operator. On day 1, half of the cats were randomized to have a BMA performed at both the proximal humerus and the iliac crest, and a BMB performed at the iliac wing, using the OnControl device (15-gauge needle for BMA; 11-gauge needle for BMB). The other half of the cats had the same procedures performed using a manual technique (15-gauge Illinois needle for BMA; 11-gauge Jamshidi needle for BMB). On day 3, each cat had BMA performed at the opposite humerus and iliac crest, and a BMB performed at the proximal humerus using the opposite technique from day 1. For each procedure, the operator was given a maximum of 3 attempts to successfully collect a sample. The rate of success, as well as the number of attempts were recorded. The "ease of use" of the device was rated by the operator on a 5-point scale after each procedure. Using previously determined criteria, the macroscopic and microscopic qualities of the BMA and BMB samples were assessed by a board-certified pathologist, blinded to the technique used. The level of pain experienced by each cat was evaluated 6, 12, 18, 24, 36 and 48 hours following each set of procedures, using a previously validated pain scoring system. Two sample t-tests were used to compare the automated technique to the manual technique and to compare the humerus to the iliac crest site for BMAs and the humerus to the iliac wing site for BMBs.

For all procedures, at all sites, the "ease of use" was better for the automated technique than for the manual technique ($P < 0.05$). The duration of the procedure and the number of attempts to collect a sample were significantly lower with the automated technique for BMA at the proximal humerus ($P < 0.05$). There was no significant difference in the level of pain at any time point following each set of procedures with either technique. Performing BMA at the proximal humerus was associated with a higher rate of success ($P < 0.05$), a lower number of attempts ($P < 0.05$), a shorter duration of the procedure ($P < 0.05$), a higher-rated "ease of use" of the technique ($P < 0.05$), and a better quality sample ($P < 0.05$) when compared to sampling from the iliac crest.

In conclusion, we found the automated bone marrow sampling technique suitable for use in adult cats. This technique was easier to use than the manual technique for both BMA and BMB, and reduced the duration of the procedure and the number of attempts for successful BMA at the proximal humerus. Performing BMA at the proximal humerus was faster, easier and allowed collection of better quality samples than at the iliac crest, independently of the technique used.

ABSTRACT HM-2

EFFECTS OF A STANDARDIZED ANESTHETIC PROTOCOL ON HEMATOLOGIC PARAMETERS IN HEALTHY CATS. M.P. Dhumeaux, E.C.R. Snead, S.M. Taylor, A.P. Carr, R.M. Dickinson, T.Y. Epp, M.L. Leis. Western College of Veterinary Medicine, Saskatoon, SK, Canada.

The fractious nature of some feline patients sometimes makes sedation or general anesthesia necessary for routine procedures such as blood collection for hematologic analyses. It has been anecdotally reported that sedation or general anesthesia could induce variations in hematologic parameters in cats, making it important for the clinician to be able to anticipate potential changes on hematologic parameters that could result from chemical restraint.

This study evaluated the effects of a standardizecd anesthetic protocol using ketamine (10 mg/kg, IV), midazolam (0.5 mg/kg, IV) and buprenorphine (10 µg/kg, IM) on the hematologic parameters of 12 healthy adult research cats. Each cat had blood samples collected before and after induction of anesthesia on 2 different occasions, 2 days apart. In total, 24 pairs of complete blood counts were obtained. Analyses were performed at a certified veterinary laboratory. Paired sample t-tests were used to determine whether there were any statistical differences between hematologic parameters before and after induction of general anesthesia, for each cat, on 2 different occasions.

Compared to preanesthetic values there was a significant decrease in red blood cell count, hemoglobin concentration, hematocrit, lymphocyte count and plasma total protein concentration after induction of anesthesia. There was no significant difference in the segmented or band neutrophil, eosinophil, basophil, monocyte and platelet counts between the samples taken before and after induction of anesthesia. On average, there was a 23.7% decrease in the red blood cell count (9.06×10^{12}/L to 6.91×10^{12}/L) ($P < 0.0001$), a 23% decrease in hemoglobin concentration (133.88 g/L to 103.08 g/L) ($P < 0.0001$), a 24.4% decrease in the hematocrit (0.41 L/L to 0.31 L/L) ($P < 0.0001$), a 25.3% decrease in the lymphocyte count (3.68×10^9/L to 2.77×10^9/L) ($P = 0.0023$), and a 12.1% decrease in the plasma total protein concentration (84.79 g/L to 74.54 g/L) ($P < 0.0001$) when samples taken before and after induction of anesthesia were compared.

If only the hematocrit was considered as a marker of anemia, 29% of the samples from these 12 healthy cats, taken while they were under general anesthesia, would have been misinterpreted as belonging to anemic patients (hematocrit < 0.285 L/L), using the reference interval established in our laboratory. None of the cats would have been considered anemic before induction of general anesthesia. In practice, the decrease in lymphocyte count following anesthesia is unlikely to be of clinical relevance, as all the samples except 2 had a lymphocyte count that was within the reference interval for cats established by our laboratory.

This study suggests that complete blood counts performed on blood taken under general anesthesia with this combination of anesthetic drugs in cats should be interpreted cautiously in order not to make a false diagnosis of anemia. The mechanism responsible for the decrease in circulating red blood cell mass following anesthesia induction in cats is unknown and requires further investigation.

ABSTRACT HM-3

IN VITRO EFFECTS OF RIVAROXABAN ON FELINE COAGULATION INDICES. Brainard BM[1], Cathcart CJ[1], Dixon AC[1], Beach WJ[2]. [1]University of Georgia College of Veterinary Medicine, Athens, GA. [2]University of Georgia College of Pharmacy, Athens, GA.

Rivaroxaban is an oral inhibitor of activated coagulation factor X (Xa). It is expected to have similar coagulation effects as low molecular weight heparin, without the need for injection, making it an attractive alternative for long-term anticoagulant therapy in cats.

Citrated blood obtained from five healthy adult cats was exposed in vitro to varying concentrations of rivaroxaban, followed by coagulation testing. The rivaroxaban was extracted from commercially available tablets (Xarelto®) and dissolved in DMSO prior to addition to the blood. Tests performed included kaolin-activated thrombelastography (TEG), prothrombin time (PT), dilute PT (dPT), activated partial thromboplastin time (aPTT), and anti-factor Xa (aXa) activity.

Dose-dependent prolongations were seen in all coagulation parameters. Similar to human data, therapeutic aXa levels (between 0.5–1.0 aXa units) were achieved at in vitro concentrations between 160 and 220 µg/L. At 220 µg/L, dPT measurements were clinically prolonged in all cats (29.2 ± 4 sec vs. 18.5 ± 0.8 sec, P = 0.148), while aPTT values were only mildly prolonged from baseline (21.9 ± 5 sec vs. 15.6 ± 2 sec, P = 0.07). Significant prolongations were seen in dPT at 500 (60.4 ± 42 ec, P = 0.005). TEG R time did not prolong from baseline values until concentrations of 2000 µg/L were reached (16.0 ± 9 min compared to 3.1 ± 0.7 min, P = 0.006).

Rivaroxaban has similar coagulation effects in the cat as in other species and may play a role in feline thromboprophylaxis. Kaolin-activated TEG does not appear to be sensitive to low concentrations of rivaroxaban in the cat.

ABSTRACT HM-4

COMPARISON OF CITRATED AND HIRUDINATED BLOOD ON WHOLE BLOOD PLATELET AGGREGATION USING THE CHRONO-LOG IMPEDANCE AGGREGOMETER. EA Shipley, DF Hogan, HW Green III NN Fiakpui, AN Magee, KA

Sederquist. Purdue University, School of Veterinary Medicine, West Lafayette, IN.

Anticoagulated blood is required for platelet function studies. Sodium citrate, a calcium chelater, is the most commonly used anticoagulant to measure coagulation parameters including platelet aggregation but it may have a negative effect on platelet responsiveness. Dogs are generally considered moderate responders to collagen on platelet aggregation and are notorious for being poor or inconsistent responders to ADP-induced platelet aggregation using citrated whole blood. Hirudin, a selective thrombin inhibitor, can also be used as an anticoagulant for coagulation assays and is the anticoagulant of choice for certain assays including the Multiplate® platelet function analyzer.

Ten adult healthy dogs were used to compare whole blood platelet aggregation between citrated and hirudinated blood samples. Venous blood was collected atraumatically from the external jugular vein directly into tubes containing 3.2% trisodium citrate or hirudin. Whole blood platelet aggregation was performed (Whole-Blood Lumi-aggregometer, Chrono-log Corporation, Havertown, PA, USA) with collagen (5 µg/ml) and ADP (10 mM) as agonists. Maximal platelet aggregation (Ohms) was recorded.

There was a significant increase in collagen-induced platelet aggregation from the hirudinated samples compared to the citrated samples (31.2 ± 6.1 vs. $17.2 \pm 8.6\ \Omega$, $p < 0.0005$). There was also a significant increase in ADP-induced platelet aggregation from the hirudinated samples compared to the citrated samples (15.6 ± 6.0 vs. $2.6 \pm 2.6\ \Omega$, $p = 0.0001$).

The results of this study show a significant difference in platelet responsiveness between citrated and hirudinated whole blood using the Chrono-log impedance aggregometer. While both collagen and ADP-induced platelet aggregation was attenuated from citrated blood samples, this was most notable for ADP-induced aggregation where almost all samples had no objective measurable platelet aggregation. It is suggested from this data that future whole blood platelet aggregation studies performed on the Chrono-log impedance aggregometer should use hirudinated blood samples although new reference limits would need to be established.

ABSTRACT HM-5

USE OF CALIBRATED AUTOMATED THROMBOGRAM (CAT) FOR MONITORING LOW MOLECULAR WEIGHT HEPARIN IN HEALTHY DOGS. C. Gara-Boivin, M. Dunn, C. Thorneloe, C. Bédard. Université de Montréal, Faculté de Médecine Vétérinaire, Saint-Hyacinthe, Quebec.

Low-molecular-weight heparin (LMWH) is now used to prevent thrombotic complications in dogs. A functional assay such as the Calibrated Automated Thrombogram (CAT) may provide a new approach for monitoring LMWH therapy.

We hypothesized that CAT would detect decreased endogenous thrombin potential (ETP) in healthy dogs receiving LMWH (Fragmin®).

Twenty-four healthy adult Beagles were included in this study and divided equally in four groups.

One dose of 50 U/kg, 100 U/kg or 150 U/kg of LMWH were given subcutaneously to healthy dogs and compared to a control group. Platelet poor plasma (PPP) was collected over a 24 hour period.

Using a repeated-measure linear model, effect of LMWH on ETP was time and dose dependent with a significant interaction ($p < 0.0001$). Compared to control dogs, significant differences were obtained for group 50 U/kg at T60 ($p = 0.037$), for group 100 U/kg at T15 ($p = 0.013$) and between T30-T240 minutes ($p < 0.0001$) respectively, and for group 150 U/ at T15 ($p = 0.011$), between T30-T300 minutes ($p < 0.0001$) respectively and at T360 ($p = 0.004$). No significant changes were observed over time in control dogs. At baseline, mean ETP across groups of 50 U/kg, 100U/kg and 150 U/kg were 314.93, 297.05 and 296.17 nM*min respectively, peak effect of ETP was 138.34 ± 20.91, 79.17 ± 27.92 and 55.22 ± 16.08 nM*min respectively, with a decrease from initial ETP value of $54.21 \pm 10.61\%$, $73.66 \pm 7.22\%$ and $81.27 \pm 4.2\%$ respectively.

The CAT assay can be employed to measure the effects of LMWH at different doses in healthy dogs, resulting in significant time and dose-dependent decreases in ETP and warrants further investigation as a tool for monitoring LMWH therapy in dogs.

Figure [Abstract HM-5]

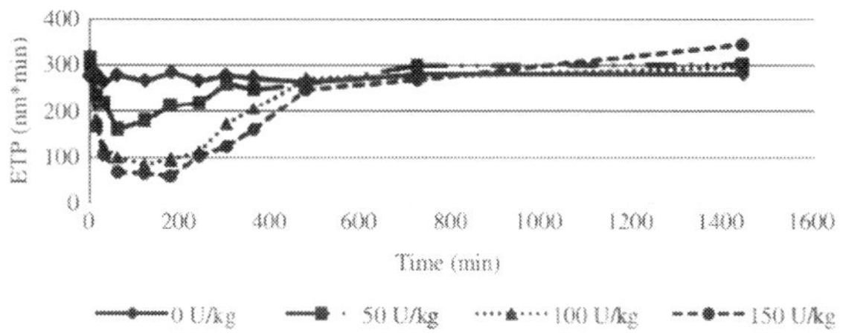

Mean Endogenous thrombin potential over a 24-hour period for each concentration group of Fragmin®.

ABSTRACT HM-6

THE EFFECTS OF PREDNISONE AND PREDNISONE PLUS ULTRALOW-DOSE ASPIRIN ON COAGULATION PARAMETERS IN HEALTHY DOGS. AL O'Kell, DC Grant, DL Panciera, GC Troy, NM Weinstein. Virginia-Maryland Regional College of Veterinary Medicine, Blacksburg, VA.

The purpose of this study was to determine the effects of prednisone and prednisone plus ultralow-dose aspirin on coagulation parameters in healthy dogs, with an emphasis on thromboelastography (TEG).

This was a prospective, randomized, blinded study utilizing fourteen dogs determined to be healthy based on normal physical examination, complete blood count, biochemistry, urinalysis, and fecal floatation. Dogs were evenly divided into either prednisone plus aspirin (PA) or prednisone plus placebo (PP) groups. Baseline values for TEG parameters (R, K, angle, MA, Ly30, Ly60, G, CI) were measured twice two days apart, and thrombin-antithrombin complexes (TAT), and traditional coagulation parameters (prothrombin time, activated partial thromboplastin time, d-dimer, antithrombin (AT), fibrinogen) were measured once. Each dog received 2 mg/kg/day of prednisone, and either 0.5 mg/kg/day of aspirin (PA group) or placebo (PP group) for 14 days. A complete blood count, biochemistry profile, TEG, TAT, and traditional coagulation parameters were then repeated on each dog. Day to day variation was calculated for the TEG parameters using the two baseline measurements. The change from baseline between and within each group were compared using t-tests, or Wilcoxon 2 sample test where appropriate, for TEG, TAT, traditional coagulation parameters, and hematocrit.

Day to day variation in TEG was acceptable ($\leq 10\%$) for MA, G, and angle, unacceptable ($> 10\%$) for R, K, Ly30 and Ly60, and not meaningful for CI. Within both groups, MA, G, CI and fibrinogen significantly increased from baseline ($p < 0.05$). Within both groups, Ly30 and AT significantly decreased from baseline ($p < 0.05$). For the PP group, Ly60 significantly decreased from baseline ($p = 0.03$), and approached significance for the PA group ($p = 0.0504$). All other within group changes from baseline were not statistically significant (p-values > 0.05). For all parameters, there was no difference between groups for change from baseline (p values > 0.05).

Day to day variation in some TEG parameters is high and may preclude their clinical utility. Prednisone causes hypercoagulability in healthy dogs based on increased G, MA, and CI. The addition of ultra-low dose aspirin to prednisone has no effect on the parameters measured in this study.

ABSTRACT HM-7

EFFECTS OF LOW DOSE ASPIRIN ON CANINE PLATELET FUNCTION. A Dudley, J Thomason, J Grady, L Pinchuk, A Mackin, K Lunsford. Mississippi State University College of Veterinary Medicine, Starkville, MS.

'Aspirin resistance' has been identified in people and dogs that develop thrombi despite low dose aspirin therapy. Variability in platelet cyclooxygenase (COX) isoform expression is one proposed mechanism for aspirin resistance in people. Two isoforms (COX-1 and COX-2) have been identified in canine platelets. High (anti-inflammatory) dose aspirin inhibits platelet function and alters expression of both COX isoforms in most dogs. This study evaluated the effects of low dose aspirin on platelet function and COX expression in normal dogs.

Twenty-five healthy client-owned dogs were evaluated before and at two time points (Days 3 and 10) during aspirin therapy (1 mg/kg PO SID). Platelet response to aspirin (Siemens PFA-100®; collagen/epinephrine cartridges), was stratified into one of three groups [aspirin responders (9 dogs), non-responders (8 dogs), or inconsistent responders (8 dogs)].

Flow cytometry identified platelet COX-1 and COX-2 expression. An ELISA was used to measure urine 11-dehydro-thromboxane B_2 (11-dTXB$_2$). There were no significant differences between groups for COX-1, COX-2 or 11-dTXB$_2$ at any time point. When all dogs were considered as a single group, there was a significant increase ($p < 0.0001*$) in COX-1 and COX-2 mean fluorescent intensity (MFI) from baseline to Day 10, 70.1% ± 38.0 (mean ± SD) and 70.8% ± 71.2, respectively. There was a significant decrease in mean urine 11-dTXB$_2$:creatinine on Day 3 and 10 by 23.4% ($p = 0.0044^*$) and 45% ($p < 0.0001^*$).

As with our previous high dose studies, COX-1 expression was increased with aspirin exposure. However, there was a significant increase in COX-2 expression with low dose aspirin in contrast to the decrease seen at higher doses. Our study suggests that levels of platelet COX-1 and COX-2 expression do not influence aspirin response in dogs. Although thromboxane levels decreased in most (23 of 25) dogs on low dose aspirin, platelet function was consistently affected in only 36% of dogs, suggesting that differences in response to thromboxane may play a role in the variable affects of low dose aspirin on canine platelet function.

Figure [Abstract HM-7]

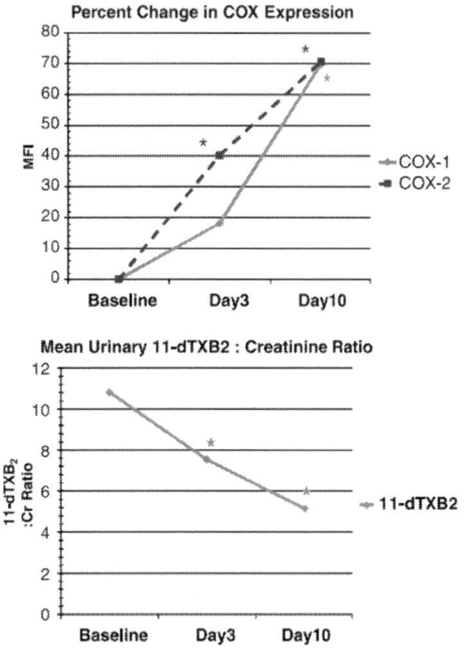

ABSTRACT HM-8

EPSILON AMINOCAPROIC ACID DECREASES THE PREVALENCE OF POSTOPERATIVE BLEEDING IN RETIRED RACING GREYHOUNDS. Marin LM[1], Zaldivar-Lopez S[1], Iazbik MC[2], McLoughlin MA[1], Bednarski R[1], Wittum T[3], Couto CG[1,2,4]. [1]Department of Veterinary Clinical Sciences, [2]Veterinary Medical Center, [3]Department of Veterinary Preventive Medicine; and [4]The OSU Comprehensive Cancer Center, The Ohio State University, Columbus, OH.

Delayed postoperative bleeding is common in retired racing Greyhounds (RRGs), despite normal results of routine hemostasis assays. The excessive postoperative bleeding in the RRGs is not due to primary or secondary hemostatic defects, and may be due to enhanced fibrinolysis or to a clot maintenance dysfunction. Providing a method to prevent or minimize the severity of postoperative bleeding in RRGs will not only have major economic impact for owners, but also will markedly decrease the associated complications of minor or major surgeries in the breed. Epsilon aminocaproic acid (EACA) is a potent inhibitor of fibrinolysis that also supports clot maintenance due to unknown mechanisms.

The objective of this double-blinded, prospective, randomized study was to evaluate the effects of EACA versus placebo on the prevalence of bleeding in RRGs, and to investigate its mechanism of action by using thromboelastography (TEG).

We compared the effects of EACA and placebo in 100 RRGs that underwent elective ovariohysterectomy or orchiectomy at the Veterinary Medical Center, The Ohio State University during 2 years. The main endpoint was bleeding (prevalence and severity); minor endpoints included most TEG parameters.

Thirty percent (15/50) of the RRGs in the placebo group had delayed postoperative bleeding starting 36 to 48 hours after surgery, compared to 10% (5/50) in the EACA group ($P = 0.0124$).

On the TEG parameters, the R time (clot formation time) was significantly different between treatment groups ($P = 0.0321$). The postoperative administration of EACA significantly decreased the prevalence of postoperative bleeding in RRGs.

ABSTRACT HM-9

EVALUATION OF POTENTIAL SURRIGATE INDICATORS OF HYPERCOAGULABILITY IN DOGS WITH PROTEIN-LOSING NEPHROPATHY. CR White, C Langston, AE Hohenhaus, P Fox. Animal Medical Center, New York City, NY.

Thromboembolism associated with protein losing nephropathy (PLN) has been long recognized as a serious and unpredictable complication in dogs, however its prevalence remains unknown. In humans, surrogate indicators are frequently used to assess thromboembolic risk. This study aimed to investigate the prevalence of hypercoagulability in PLN dogs based on thromboelastography (TEG), and to determine whether hypercoagulability in these patients could be predicted by clinical assessments that identify systemic hypertension (systolic blood pressure > 160 mmHg), hypoalbuminemia (serum albumin < 2.7 mg/dl), antithrombin activity (< 70%), and degree of proteinuria (urine protein:creatinine ratio [UPC] ≥ 2). Between March 2009-September 2010, twenty-seven dogs were identified with PLN at the Animal Medical Center. The prevalence of hypercoagulability based on a TEG G-value > 9.6 was 83.3%. There was no statistically significant relationship, either categorically or continuously, in univariate as well as multivariate analyses of all variables. Univariate logistic regression (odds ratio; lower and upper confidence limit; P value) for hypertension was −1.18; 0.201, 6.99; 0.851; for albumin −1.37; 0.228, 8.26; 0.728; and for antithrombin activity −0.73; 0.12, 4.39; 0.728. Thus, in this patient population, in the absence of TEG, prediction of hypercoagulability using abnormalities in commonly measured clinicopathologic variables was not helpful. However, given the documented high prevalence of hypercoagulability in patients with PLN, early institution of prophylactic anti-platelet or anticoagulant therapies should be considered.

ABSTRACT HM-10

THE EFFECTS OF SAMPLE COLLECTION METHOD ON THROMBOELASTOGRAPHY IN HEALTHY DOGS. CM Sumner, SP Shaw, AM deLaforcade. Tufts Veterinary Coagulation Laboratory, Tufts Cummings School of Veterinary Medicine, North Grafton, MA.

Thromboelastography (TEG) is a test of global hemostasis. Due to the effects of extrinsic factors on whole blood coagulation, sample collection method (SCM) may influence results. The purpose was to determine if SCM influenced TEG using kaolin-activated citrated whole blood (WB).

Healthy dogs with normal platelet counts were prospectively enrolled. Three WB samples were obtained from each dog at least 48 hours apart from alternating jugular veins in a randomized order of three methods: 1) vacutainer® into citrated tube (VAC), 2) citrated syringe with transfer into plain tube (CIT), or 3) plain syringe with transfer into citrated tube (PLAIN). Draw time was recorded in seconds. Kaolin-activated TEG was performed, with measurement of reaction time (R), clot formation time (K), maximum amplitude (MA), and alpha angle.

Eleven dogs were enrolled. There were no significant differences in TEG indices between VAC samples and either CIT or PLAIN samples. CIT samples had a significantly higher K value (p = 0.004) and a lower alpha angle (p = 0.004) compared to PLAIN samples. Draw times ranged from 2–10 seconds. A longer draw time was significantly correlated (p = 0.046; R = −0.35) with a shorter R time. Higher platelet count was significantly correlated (p = 0.001; R = 0.575) with a higher MA.

SCM did not have a significant effect on TEG parameters when comparing VAC samples to either CIT or PLAIN samples. Minimizing sample collection time and trauma during venipuncture may be important in minimizing hypercoagulable changes in TEG indices.

ABSTRACT HM-11

TRANSFUSION OF LIQUID PLASMA TO DOGS IN THE EMERGENCY SETTING.

AR Grochowsky, EA Rozanski, JJ Schavone, CR Sharp, AM deLaforcade. Cummings School of Veterinary Medicine at Tufts University, North Grafton, MA.

Liquid plasma (LP) is defined as either plasma collected and refrigerated immediately after collection or fresh frozen plasma (FFP) that is thawed and stored refrigerated until use. Stability studies in people have shown that adequate clotting factor activity is preserved for at least 14 days. LP is transfused in human Level I trauma centers to critically ill people requiring rapid infusion of clotting factors as the time required to defrost FFP is considered prohibitive. The use of LP has not been described in veterinary critical care. The purposes of this study were to 1) determine the length of time required in a water bath for a unit of canine FFP to thaw and 2) describe the use of LP in a busy university emergency room (ER).

For part 1: Six units (250 ml) of canine FFP were individually thawed in a 37°C water bath. The duration of time (in minutes) to thaw was recorded.

For part 2: The transfusion log was reviewed for dogs receiving LP in the last 6 months. The indications and outcome were recorded.

The mean time ± SD thaw time was 34.7 ± 1.38 minutes.

Ten units of LP were transfused to 7 critically ill or injured dogs during the study time. Indications for LP transfusion included hypovolemic shock due to intra-abdominal hemorrhage in 6 dogs (2 traumatic, 4 non-traumatic) and rapid correction of hemorrhage following parenteral tissue plasminogen activator administration in 1 dog. LP volume transfused ranged from 11.1 to 50.5 ml/kg. No transfusion reactions were identified. Effect on coagulation was not consistently evaluated.

Time required to thaw a unit of FFP is greater than 30 minutes which could be detrimental in a bleeding, coagulopathic dog. LP was transfused without incident to critically ill and injured dogs and represents a potential new addition to the armamentarium of treatments in a veterinary ER setting. Further investigation of canine LP is warranted including evaluation of *in vitro* factor stability and *in vivo* efficacy in correcting coagulopathy.

ABSTRACT HM-12

PROSPECTIVE DIAGNOSTIC EVALUATION AND CLINICAL UTILIZATION OF AN OPTIMIZED ASSAY FOR PLATELET ASSOCIATED IMMUNOGLOBULIN IN THROMBOYCTOPENIC AND NON-THROMBOCYTOPENIC DOGS.

DE Bachman[1], MA Forman[2], RA Hostutler[1], SA Martinez[1], L Fulton[1], D Saam[1], L Prior[1], J Ryan[1], DE Prescott[1], J Lin[3], C Deka[3], L Ewing[3], GJ Kociba[4]. [1]MedVet, Medical and Cancer Center for Pets, Worthington, OH. [2]Cornell University Veterinary Specialists, Stamford, CT. [3]IDEXX Laboratory, Westbrook, ME. [4]IDEXX Laboratory, Worthington, OH.

Immune mediated thrombocytopenia (IMT) is associated with increased morbidity and mortality. Large prospective research studies in dog platelet antibodies and clinical utilization of platelet immunoglobulin assays are limited. Potential explanations include limited availability and low specificity due to nonspecific binding. The focus of this study is to evaluate optimized direct and indirect platelet surface associated immunoglobulin (pSAIG) and staining with anti-CD61 antibodies (CD61ab) for the utilization in classifying thrombocytopenic dogs.

One hundred clinically ill and 30 apparently healthy dogs were prospectively evaluated. Data collected included a history of hemorrhage, physical examination evidence of bleeding, complete blood count, and measurement of pSAIG and CD61ab. The pSAIG assay utilized polyvalent antibodies with correction for non-specific binding by subtraction of background fluorescence with control antiserum. Thrombocytopenia was defined as < 164,000/μL and all dogs were clinically classified into 1 of 4 groups (g): g1 IMT, n = 45, g2 Thrombocytopenia from non-immune mediated diseases, n = 37, g3 Ill with normal platelet counts, n = 18, g4 Healthy dogs, n = 30.

Median platelet counts, by groups, were g1, 20,000; g2, 69,000; g3, 324,500; and g4, 212,000/μL. For the direct and indirect pSAIGs in dogs with ITP (g1), more dogs (n = 6 and n = 9) with clinical evidence of bleeding had antibodies compared to those who were not bleeding (n = 3 and n = 6). Considering only direct pSAIG the sensitivity and specificity was 13% and 95%, respectively for the diagnosis of IMT. For indirect pSAIG the sensitivity and specificity was 27% and 97%, respectively, for the diagnosis of IMT. When considering both direct and indirect pSAIG together, the sensitivity was 33% with a specificity of 93%. In g1 interference from high control antiserum background staining was noted in 26.7% of dogs and resulted in a negative direct pSAIG classification. Minimal background interference was noted in g2, g3, or g4. The percentage of platelets stained with CD61ab was significantly less in g1 (median 38, $p = 0.0001$) vs. g2 (median 74, $p = 0.007$) vs. g3 (median 95.1, $p = 0.857$) and g4 (median 95.6).

These findings indicate the optimized platelet surface associated immunoglobulin assay has a high specificity, however poor sensitivity, for the diagnosis of IMT. The decreased CD61 staining in g1 (IMT group) may reflect decreased surface gpIIIa expression, blocking by anti-gpIIIa antibodies or other proteins, clearance by macrophages, or increased non-platelet debris and has potential applications in the diagnosis and treatment of IMT.

ABSTRACT HM-13

INVESTIGATION OF THE HAPTOGLOBIN GENE IN GREYHOUNDS.

S Zaldívar-López[1,2], J Rowell[2], E Fiala[2], CG Couto[1], CE Alvarez[2]. [1]The Ohio State University College of Veterinary Medicine, Columbus, OH. [2]The Research Institute at Nationwide Children's Hospital, Columbus, OH.

Greyhounds have lower serum concentrations of α-globulin than other breeds, explained by negligible levels of haptoglobin (HP) measured using different methods (colorimetric, immunoturbidimetric and protein electrophoresis). The purpose of the present study was to characterize the *HP* gene in Greyhounds.

We isolated DNA and RNA from blood samples of AKC-registered and retired racing Greyhounds (AKCG, RRG), and a German Shepherd Dog (GSD). We sequenced the *HP* exons and splice sites, and conducted array comparative genomic hybridization to identify associated DNA structural variation (custom 1M Agilent oligonucleotide array). Additionally, we tested for the presence of one or multiple haplotypes spanning *HP* in Greyhounds using a high density SNP array (180k Illumina HD). Sequencing results of *HP* in both DNA and cDNA revealed three synonymous SNPs in the racing Greyhound. We did not identify structural variation overlapping or near the *HP* gene. Notably, we detected that the RRG and AKCG do not appear to share a specific haplotype spanning *HP*.

Despite having low or undetectable serum concentrations of HP, we did find that RRG *HP* mRNA is expressed and lacks amino acid variation. This suggests that the clinical absence of the *HP* is attributable to post-transcriptional *HP* effects or to an unknown physiological interaction. Finally, given the existence of distinct RRG and AKCG haplotypes spanning *HP*, it is important to characterize serum levels of HP in AKCG in follow on studies.

ABSTRACT HM-14

MOLECULAR CHARACTERIZATION OF THE CANINE BETA-GLOBIN GENE CLUSTER.

S Zaldívar-López[1,2], J Rowell[2], E Fiala[2], CG Couto[1], CE Alvarez[2]. [1]The Ohio State University College of Veterinary Medicine, Columbus, OH. [2]The Research Institute at Nationwide Children's Hospital, Columbus, OH.

We reported that hemoglobin in retired racing Greyhounds (RRG) has higher oxygen carrying properties and affinity than other breeds. Surprisingly, very little is known about canine hemoglobin genetics. The purpose of this study was to characterize genetics of canine beta globins.

Using computational BLAST analysis of the dog genome, we identified five beta globin genes in a single locus: two human *HBE*-like followed by three *HBB*-like genes. We isolated DNA and RNA from blood of RRGs, AKC registered Greyhounds (AKCG), and German Shepherd Dog (GSD). All beta globin exons and splice sites were sequenced, and the beta globin locus was examined by array comparative genomic hybridization (custom 1M Agilent array). Additionally, we determined the number of common haplotypes that span this locus in RRGs and AKCGs using high density SNP array (180k Illumina HD).

Expression and sequence analysis of cDNA showed all five beta globin genes are actively expressed in adults. CanHBB1 and 2 were created by relatively recent segmental duplication and have identical protein sequence. CanHBB1/2 are abundantly expressed in adults; CanHBB3 is expressed at greatly reduced levels. Sequencing results revealed one rare non-synonymous single nucleotide polymorphism (SNP) in HBE1 of RRGs, but no variation that could explain their abnormal hemoglobin. We did not detect structural variation overlapping or near the beta globin locus. Notably, RRG and AKCG do not share haplotypes spanning the beta globin locus. This is the first characterization of canine hemoglobin genetics, and the first report of canine embryonic hemoglobins and their expression in adults.

ABSTRACT HM-15

EASE OF PROCEDURE AND QUALITY OF BONE MARROW ASPIRATES OBTAINED FROM THE COSTOCHONDRAL JUNCTION AS COMPARED TO TRADITIONAL SITES IN HEALTHY WALKER HOUND DOGS. JK Veir[1], AA Bohn[1], U Lotti[2]. [1]College of Veterinary Medicine, Colorado State University, Fort Collins, CO. [2]Clinica Veterinaria Valdinievole, Monsummano Terme, Italy.

Sampling of the bone marrow in the dog from the costochondral (CC) junction can be performed with minimal to no sedation and readily available equipment but is not in widespread use in the United States. The aim of this study was to compare the number of attempts needed to successfully obtain a sample, the time needed for the procedure, and the sample quality between aspirates obtained from the CC junction and more traditional sites (humerus or femur) in healthy dogs when performed by novice and seasoned practitioners.

Samples were obtained from healthy anesthetized laboratory reared adult dogs after undergoing terminal endoscopic surgery. Paired samples from separate dogs were obtained by each practitioner using either a 22 gauge needle and 6 cc syringe at the CC junction or an 18 gauge Rosenthal needle and 12 cc syringe from either the proximal humerus or femur (clinician preference). Three small animal veterinary interns, one experienced technician and one boarded internist were monitored for number of attempts to success and length of time needed for success of each procedure. Slides were prepared by a single investigator and read by a blinded clinical pathologist. Data were compared using the paired t-test for normally distributed data and Wilcoxen signed rank test for non-Gaussian distributions.

Five pairs of samples from three dogs were evaluated. Two dogs had two pairs drawn from opposite limbs and ribs. Mean number of attempts to success for traditional sampling sites (1.4 +/−0.54) and time to success (8.6 minutes +/−5.2) did not differ significantly from attempts (2.0+/-0.70, p = 0.25) or time (2.7 +/-0.98, 0.06) needed when aspirating from the CC junction. Subjectively, samples were of similar quality with regards to cellularity and number of particles present when compared within practitioners. Myeloid: erythroid ratio and percentage of lymphocytes were also not significantly different between sites (M:E ratio p = 0.37, lymphocyte % p = 0.07) and were within normal limits.

While there were no significant differences between the two sites in terms of number of attempts or time to success, it should be noted that the "seasoned" practitioners had never performed an aspirate at the CC site and had an increased number of attempts compared to the traditional sites. If the number of attempts needed for success decreases with experience, it is likely the time required would decrease as well. Both subjectively and objectively, there were no significant differences in quality or cell populations between the two sampling sites in healthy dogs. Based on this data, bone marrow aspiration from the CC junction appears to be equivalent to more traditional sampling sites in healthy dogs. Larger studies in clinically ill dogs should be performed before routinely using the site in the clinical setting.

ABSTRACT HM-16

CANINE IRON-REFRACTORY IRON DEFICIENCY ANEMIA (IRIDA) CAUSED BY MATRIPTASE (*TMPRSS6*) DEFICIENCY. D Naigamwalla[1], JA Webb[1], P Lee[2], U Giger[3]. [1]Internal Medicine Department, Mississauga-Oakville Veterinary Emergency Hospital, Oakville, ON. [2]Department of Molecular and Experimental Medicine, Scripps Research Institute, La Jolla, CA. [3]Section of Medical Genetics, University of Pennsylvania, Philadelphia, PA.

Recent research on iron homeostasis has elucidated the tightly controlled intestinal uptake of iron. Hepcidin, the major hormone limiting iron absorption and release from macrophages, is down-regulated by matriptase-2, a transmembrane serine protease (*Tmprss6*) produced by the liver. While iron deficiency is commonly caused by chronic blood loss anemia and rarely dietary deficiency or intestinal disorders in dogs and other species, we report here the clinical to molecular investigations of a dog with iron-refractory iron deficiency anemia (IRIDA) caused by a matriptase-2 deficiency homologous to a recently described autosomal recessive disorder in humans.

The proband, a spayed female Cocker Spaniel without any clinical signs except for recent occasional idiopathic seizures, exhibited a life-long history of microcytosis and hypochromasia but not anemia. There was no evidence of any blood loss and the dog was receiving an appropriate meat-based diet. Mean values of complete blood cell counts, performed from 0.5–5 years of age, were: hematocrit 41% (normal reference range 39–56); RBC count $9.6 \times 10^6/\mu L$ ($5.8–8.5 \times 10^6$); MCV 43 fL (62–72); MCHC 31 g/dL (33–36). Serum iron parameters revealed severe iron deficiency with serum iron 34 µg/dL (88–238); Total Iron Binding Capacity 378 µg/dL (246–450); serum iron saturation < 9% (15–50%), and ferritin 410 ng/dL (80–800). Prolonged courses of oral ferrous sulfate supplementation and several short courses of intramuscular (Dextran) injections and intravenous iron infusions did not result in improvement of any red cell or serum iron parameters. However, this dog was never anemic and the partial seizures could not be directly related to the iron deficiency status. No family members were available for further studies.

Genomic DNA was extracted from the proband's EDTA blood and the exons of the *TMPRSS6* gene were amplified with flanking primers and then sequenced. In comparison to the normal canine *TMPRSS6* sequence and that of a sequenced control dog we found a homozygous missense muation, R723H, toward the C-terminal end of the protein in the proband's gene.

In conclusion, the severe microcytosis and hypochromasia, low serum iron parameters and lack of a response to oral and parenteral iron therapy led to the diagnosis of IRIDA. The missense mutation in the matriptase-2 at position 723 from an arginine, which is conserved across all species currently deposited in the Genbank, to a histidine is likely the disease-causing mutation. This is the first report of an IRIDA in the dog with features very similar to those observed in humans. Dogs with naturally-occurring IRIDA may be helpful in developing and assessing novel therapies.

ABSTRACT HM-17

A COMPARISON OF BODY WEIGHT IN RELATION TO ZINC TOXICITY ANEMIA SECONDARY TO METALLIC FOREIGN BODY. K Palladino Jr., J Fondacaro, S Hillock. The Veterinary Medical Center of Long Island, West Islip, NY.

Accidental ingestion of copper-coated zinc pennies minted after 1982 is the most common causes of zinc toxicity anemia in the dog. Zinc toxicity anemia may also be seen with ingestion of zinc from other sources as ingestion of metallic foreign material other than pennies, medicines containing zinc, and zinc supplements. The purpose of this study was to determine if there is a weight below which dogs are more susceptible to zinc toxicity anemia secondary to metallic foreign body ingestion. Records of dogs presented to the Internal Medicine service at the Veterinary Medical Center of Long Island for metallic foreign body ingestion were reviewed for signalment, weight, presenting PCV, and type of metallic foreign body ingested. Eighteen dogs met the inclusion criteria and were compared. Of the 18 dogs, there were 15 cases of coin ingestion (83%), with 11 (73%) involving ingestion of 1 or more pennies. The other 3 cases involved ingestion of a metallic object (1), decorative garland (1), and BB pellets (1).Of the 14 dogs exposed to zinc, 13 (93%) were less than 27 pounds (12.3 kgs). Of those 13 cases 11 (85%) had ingested one or more pennies. Eleven out of the 14 (79%) zinc exposure dogs were anemic at presentation. The average weight of the 11 dogs was 17.5 pounds (8 kg). This study showed that dogs less than 27 pounds appear to be more susceptible to developing anemia secondary to zinc toxicosis, with the majority of cases due to ingestion of pennies minted after 1982. Zinc toxicity anemia secondary to penny ingestion is more commonly seen in small dogs. We suspect larger dogs are able to pass pennies through the pyloric sphincter and thus not develop clinical signs.

ABSTRACT HM-18

RETROSPECTIVE STUDY OF THE SAFETY AND EFFICACY OF CRYOPRESERVED PLATELET CONCENTRATE IN 26 DOGS. L. Alvarez, JE Stokes, J Bartges. College of Veterinary Medicine, University of Tennessee, Knoxville, TN.

Although thrombocytopenia is common in hospitalized dogs, canine cryopreserved platelet concentrate (PC) is used infrequently. Data suggest in vitro efficacy of PC and when administered to research dogs, but efficacy is unknown in clinical patients. Study objectives were to determine clinical characteristics of dogs receiving PC as well as safety and efficacy of PC in thrombocytopenic dogs.

Medical records were evaluated retrospectively to identify dogs that received PC. Information evaluated included patient characteristics, platelet count, acute transfusion reactions, and survival. Twenty six dogs met study criteria. Dogs receiving PC ranged in age from 1–13 years (mean 7.8 years) and 17/26 (65.4%) were spayed or intact females. Hemorrhage was reported in 20/26 dogs (76.9%) prior to PC transfusion. Platelet counts prior to transfusion ranged from 0 to 77 × 10^3/ul (mean 29.9 +/−43.9 × 10^3/ul). Change in platelet count was measured in 23 dogs and the mean change was17.1 +/− 45.5 × 10^3/ul. Dose of PC administered ranged from 4.8 to 40 ml/kg with a mean of 14.5 +/− 8.8 ml/kg. No acute adverse reactions were reported. There was no correlation between transfusion dosage and platelet count change post transfusion. Survival to discharge occurred in 15/26 (57.7%) of dogs. The only variable correlated with survival was age with survivors being younger than non-survivors (6.4 years-old ± 3.9 vs. 10.1 years-old ± 3.1.; p = 0.02).

Efficacy of cryopreserved PC transfusions for improving clinical outcome in dogs with thrombocytopenia is yet to be determined; however, PC is well tolerated in clinical patients.

ABSTRACT HM-19

EFFICACY AND SAFETY OF FRESH FROZEN PLASMA STORED AT 4°C FOR 5 DAYS **CL.** Clement, JE Stokes, JW. Bartges College of Veterinary Medicine, University of Tennessee, Knoxville, TN.

Fresh frozen plasma (FFP) is used to treat coagulopathies in dogs. Current transfusion guidelines recommend that FFP be administered within 4 hours of thawing to avoid decreasing clotting factor function and bacterial contamination. The purpose of this study was to assess clotting factor activity and bacterial contamination of FFP that had been thawed and refrigerated for 5 days.

Blood was collected from 10 client-owned healthy dogs with no known history of coagulopathy or of administration of drugs affecting coagulation. Plasma was separated from whole blood and frozen (−20°C) within 30 minutes of collection. Thawed plasma was maintained at 4°C (+/−2°C). Aerobic and anaerobic bacterial culture, prothrombin time (PT), activated partial thromboplastin time (PTT), and factor II, VII, IX, and X analyses were tested at time of whole blood collection, FFP thaw, 24 hours post-thaw, 72 hours post-thaw, and 120 hours post-thaw.

There were no statistically significant differences in PT and PTT at any of the measured time points. Statistically significant differences occurred between initial measurements of factors II, VII, IX, and X and subsequent time points, but there was no difference in activity levels of the factors once FFP was thawed. One bacterial colony was grown from each of two samples from post-thaw plasma.

Thawed plasma protocols do not significantly decrease the function of factors II, VII, IX, and X or prolong PT and PTT. Bacterial contamination of the plasma supply seems unlikely, but strict aseptic technique should be used when obtaining plasma for patient use.

ABSTRACT HM-20

SURVEY OF ERYTHROCYTIC PYRUVATE KINASE DEFICIENCY IN DOGS. Guldal Inal Gultekin, Karthik Raj, Sarah Lehman, Karen Manhart, and Urs Giger. Sections of Medical Genetics and Pathology, School of Veterinary Medicine, University of Pennsylvania, Philadelphia, PA.

Erythrocyte pyruvate kinase (PK) deficiency is the first and most common erythroenzymopathy described in dogs, cats, and humans. The PK enzyme plays a crucial role in the erythrocyte energy metabolism and its absence causes severe hemolytic anemia, often misdiagnosed as autoimmune hemolytic anemia. The disease is inherited as an autosomal recessive trait and affected dogs also develop osteosclerosis. In dogs, the enzymatic diagnosis is complicated by the anomalous expression of malfunctioning M_2-PK expression, but breed-specific R-PK mutation tests have been reported for Basenjis, West Highland White Terriers (WHWT), and Beagles. We report here on a survey of canine PK deficiency studied at the PennGen Laboratory.

A biased group of samples were received for screening from dog breeds with known mutations as well as from dogs with chronic, prednisone- and antibiotic-resistant hemolytic anemia and their relatives. EDTA blood samples and/or cheek swabs as well as medical record information were received and genomic DNA and/or enzyme activity testing were performed.

Among the 237 WHWTs 7% and 37% were found to be homozygous deficient dogs or carriers, respectively, with a mutant allele frequency of 0.26. The average age at the time of diagnosis was 1.5 years ranging from 2 months to 5 years of age; some samples came from Europe and South America. Of the 67 Beagles studied, 36% were affected and 3% were carriers (mutant allele frequency 0.37). The average age at the time of diagnosis was 2 years ranging from 7 months to 9 years. Surprisingly, very few samples from Basenjis were received for screening, and none showed the mutant allele. While PK-deficient Basenjis lived < 5 years, WHWT and Beagles often show milder signs and can reach 9 years of age.

Several dogs from other breeds were also examined because of chronic regenerative anemia and none had any of the known mutations seen in the other breeds. However, based upon PK enzyme activity studies, Chihuahua, Dachshund, Miniature Poodle, Spitz, Eskimo Toy, and Labrador Retriever dogs were found to be affected; they also had osteosclerosis and at least one Labrador Retriever developed severe hemochromatosis (hepatic iron 37,300 ppm; normal < 1,200 ppm, analyzed on a dry weight basis). Moreover, sequencing of the R-PK cDNA from a PK-deficient Labrador Retriever revealed a new nonsense mutation in exon 6.

In conclusion, PK deficiency appears to be a common cause for hemolytic anemia in certain breeds, and mutation testing makes screening simple. PK deficiency should also be considered in dogs of other breeds which may require the more cumbersome enzyme testing. Studies to identify new mutations will confirm and simplify the diagnosis.

Supported in part by NIH grant RR02512.

ABSTRACT HM-21

DETECTION OF IgG, IgM AND C3 ON RED BLOOD CELL SURFACE OF DOGS WITH IMMUNE-MEDIATED HEMOLYTIC ANEMIA. D Passarelli; SI Miyashiro; ML Franchini; MHS Pelissari; CR Stricagnolo; MK Hagiwara. School of Veterinary Medicine and Zootechny – University of São Paulo, São Paulo, SP, Brazil.

Immune-mediated hemolytic anemia (IMHA) is a common hematological condition observed in dogs. The diagnosis is based on clinical history, presenting signs and hematological evidence of IMHA such as regenerative anemia, leucocytosis and presence of spherocytes. The definitive diagnostic procedure is the Coomb's test (direct antiglobulin test, DAT) which is known to be highly specific but lacks diagnostic sensitivity. Direct flow cytometric assay (FCA) for IgG, IgM or C3 coated red blood cells (RBCs) detection might be more sensitive and thus could be introduced as an alternative diagnostic tool.

To investigate the usefulness of FCA for IMHA diagnosis, evaluation of IgG, IgM or C3 coated RBCs was performed from 15 dogs presented at the Veterinary Hospital at USP that fulfilled clinical and hematological criteria for IMHA. Thirty eight healthy dogs were included as controls.

DAT was performed with polyvalent and monovalent anti-dog sera with twofold serial dilutions of each one, incubated with 2% RBCs suspension at 37°C and 4°C. For FCA, 2% RBCs from anemic and healthy dogs were incubated with FITC anti-dog IgG, anti-dog IgM and anti-dog C3 and submitted to flow cytometry evaluation. Specific software and Mann Whitney U test were used for data analysis.

Five dogs showed positive results for DAT with polyvalent Coombs reagent at 4°C (titer 64 to 4096) and 37°C (titer 64 to 2048) but only three of them had agglutination titer for anti-IgG at 4^0 C (1024 to 8192) and 37°C (512 to 4096). No positive results were observed for anti-IgM and anti-C3 DAT. By FC, percentage of IgG,

IgM and C3 coated RBCs in normal and anemic dogs were, respectively, 1,18% and 17,11% (p < 0,001); 1,15% and 21,29% (p < 0,001); 0,66% and 6,99% (p < 0,001). IgG coated RBCs percentage were higher in dogs showing DAT positive results (min. 33,94%; max. 99,96%; median 54,25%).

Direct flow cytometric erythrocyte immunofluorescence assay is more sensitive than DAT for detection of antibodies coated RBC in anemic dogs and may provide quantitative data useful for laboratorial diagnosis of IMzHA.

ABSTRACT HM-22

STERNAL BONE MARROW ASPIRATION IN HEALTHY CATS. A. Defarges, A. Abrams-Ogg, D. Bienzle. Ontario Veterinary College, University of Guelph, Guelph, Ontario, Canada.

Bone marrow aspirates from cats are typically obtained from the ilium, humerus or femur, but may be difficult to obtain and/or of poor quality. In this study the feasibility, safety, and nature of sternal aspiration in cats was investigated. Under general anesthesia, bone marrow aspirates were obtained in a randomized order by a single investigator from the sternum and ilium of 10 healthy cats weighing 3.4–8.4 kg, with body condition scores of 5–9 (on a scale of 1–9). For sternal aspirates, cats were positioned in sternal recumbency and a 1-inch, 22–23 ga hypodermic needle attached to a 12cc syringe was inserted into the cranial manubrium and directed caudally along the long axis of the sternum. Aspirates were also obtained from the right iliac crest using an 18 ga Illinois needle attached to a 12cc syringe. Difficulty of site localization, needle insertion and advancement, and specimen aspiration, were scored from 1 (easiest) to 5 (hardest). Bone marrow smears were prepared by one investigator and reviewed by a pathologist blinded to aspiration site and cat. Sample quality was scored from 0 (no marrow particles) to 5 (excellent) based on the number of well-smeared marrow particles on the slide. Particle cellularity was scored from 1 (> 75% fat) to 3 (< 25% fat). Post-procedure, cats were treated with tramadol (4–5 mg/kg, PO, q12h) for 3 days, and assessed for post-biopsy pain (Colorado State University Feline Acute Pain Scale, range 0 [no pain] – 4 [maximum]) and site swelling (range 0 [none] – 3 [marked]). Data were analyzed by ANCOVA accounting for effects of weight and body condition score.

Table [Abstract HM-22]: Results [mean ± SE (range)] were (* or ** indicates significant difference, p < 0.05):

Site	Attempt(s)	Time (sec)	Difficulty (4–20)	Quality (1–5)	Pain (at 24 h)	Swelling (at 24 h)
Sternum	1.5 ± 0.15 (1-2)	233 ± 51 (121-345)	6.2 ± 1.2* (4–10)	1.4 ± 0.4** (0–4)	0	0
Ilium	1.2 ± 0.15 (1-2)	205 ± 48 (101-309)	11.7 ± 1.2* (9–14)	3.7 ± 0.4** (0–5)	0	0

Pneumothorax was not identified. It was significantly easier to perform sternal than iliac aspiration, but the quality of the sample was significantly better for iliac than for sternal aspirates. Because of limitations due to sample quality, bone marrow morphology in sternal samples could not be compared to iliac samples in all cats. For samples that could be compared, cellularity was identical for sternal and iliac samples from 1 cat but underestimated in the sternal sample from another cat. Myeloid:erythroid ratios and lymphocyte numbers were the same for sternal and iliac samples in 2 and 3 cats, respectively. Megakaryocyte numbers were the same in one sample, less in sternal samples compared to iliac samples from 2 cats, and overestimated in the sternal sample from 1 cat. Bone marrow cell morphology was normal in all acceptable samples. It was concluded that sternal aspiration of bone marrow using a 22–23 ga hypodermic needle is 1) easier to perform than iliac aspiration; 2) safe; but 3) provides samples of lower quality than iliac aspiration in cats.

ABSTRACT HM-23

15-GAUGE CORE BONE MARROW BIOPSY IN HEALTHY CATS. A. Abrams-Ogg, A. Defarges, D. Bienzle. Ontario Veterinary College, University of Guelph, Guelph, Ontario, Canada.

The diameter of 11–13ga Jamshidi-type needles makes bone marrow core biopsy difficult in cats. In this study, biopsies of the left humeral head were taken under anesthesia using a 1-inch, 15ga needle (EZ-IO® Intraosseous Infusion System, Vidacare) from 10 healthy cats weighing 3.4–8.4 kg with body condition scores of 5–9 (on a scale of 1–9). Humeral biopsies were compared to biopsies taken from the left iliac crest using a 2-inch, 13ga Jamshidi needle. Biopsies were performed in randomized order by one investigator. Biopsy was repeated to a maximum of 3 attempts until a specimen ≥ 5 mm long was obtained. Difficulty of site localization, needle insertion and needle advancement were scored from 1 (easiest) to 5 (hardest). Specimens were wrapped in tissue paper and placed in Davidson's fixative for 15 min and then transferred to formalin. Biopsy sections were reviewed by a pathologist blinded to biopsy site and cat. Biopsy length on the slide was measured, and biopsy quality was scored from 0 (no hematopoietic tissue) to 5 (≥ 5 intertrabecular spaces free of artifact). Post-procedure, cats were treated with tramadol (4–5 mg/kg, PO, q12h) for 3 days, and assessed for post-biopsy pain (Colorado State University Feline Acute Pain Scale, range 0 [no pain] – 4 [maximum]) and swelling of biopsy sites (range 0 [none] – 3 [marked]). Data were analyzed by ANCOVA accounting for effects of weight and body condition score.

There were no significant differences between 15ga and 13ga biopsies except for post-biopsy swelling, and there were no significant effects of body weight and body condition.

Six (60%) of 15 ga and 5 (50%) of 13ga biopsies were considered acceptable specimens for assessment of bone marrow architecture and morphology; all intact spaces in these biopsies had normal hematopoiesis and cell morphology. Comparison of acceptable 15 ga to 13 ga biopsy specimens from 4 cats showed no significant differences for cell density and lymphocytes/plasma cells, while cellularity, assessed as high in 2 of the 13 ga biopsies, was assessed as medium in corresponding 15ga biopsies; and megakaryocytes, assessed as 4–9/low-power field in one 13ga biopsy, were assessed as ≤ 3/low-power field in the 15 ga biopsy. Myeloid:erythroid ratios were greater in 15 ga biopsies compared to 13 ga biopsies in 2 cats, and less in the 15 ga biopsy in one cat. Discordant results between biopsies were not related to differences in quality.

In conclusion, 15 ga bone marrow biopsy of the humerus was as likely to yield a specimen of acceptable quality as was 13 ga biopsy of the ilium, and resulted in less post-biopsy swelling.

Table [Abstract HM-23]: Results [mean ± SE (range)] were (* indicates significant difference, p < 0.05):

Site, needle	Time (min)	Difficulty (3–15)	Quality (0–5)	Length (mm)	Pain (at 24 h)	Swelling (at 24 h)
Humerus, 15 ga	5.1 ± 0.9 (2.2–9.3)	5.6 ± 0.6 (3–9)	2.1 ± 0.5 (0–4)	6.0 ± 0.7 (3–10)	0	0*
Ilium, 13 ga	6.1 ± 0.6 (4.5–9.5)	9.0 ± 0.7 (6–12)	1.9 ± 0.6 (0–5)	7.5 ± 0.8 (5–12)	0.2 ± 0.1 (0–1)	0.9 ± 0.1* (0–1)

ABSTRACT HP-1

CANINE ACUTE LIVER FAILURE: RETROSPECTIVE REVIEW OF 50 CASES (1995-2010). CL Kavanagh, J Cooper, CRL Webster. Tufts Cummings School of Veterinary Medicine, North Grafton, MA.

Reports on canine acute liver failure (ALF) include individual or small case series of animals with a specific diagnosis. The aim of this study was to describe the clinical course, outcome and etiology of ALF in dogs presenting to a referral hospital. Medical records (1995–2010) were reviewed for a diagnosis of ALF (elevated serum bilirubin or icterus with concurrent coagulopathy or hepatic encephalopathy (HE)). Fifty cases were identified representing 22 breeds: 7 Labradors retrievers, 6 Golden retrievers, 3 German shepherds, and 3 Cocker spaniels. Median age was 6 years (1 m to

13 yrs). Presenting signs included anorexia (31/50), vomiting (26/50), polydipsia (9/50) and neurologic signs (6/50). Progressive neurologic impairment compatible with HE occurred in 21/50. Prolongation of PT and aPTT occurred in 36/39 and 33/41, respectively. Concurrent HE and coagulopathy developed in 19 dogs. Common presenting clinicopathologic abnormalities included increased bilirubin (47/48), ALP 45/48, *and* ALT 42/48, thrombocytopenia 36/48, and hypoalbuminemia (27/48). *Definitive etiology was identified in 16/32 dogs with a hepatic biopsy* including neoplasia (12/32), infarction (2/32), congenital PSS (1/32), and hepatic lipidosis (1/32). Etiology was identified in 10 more dogs by serology or history; leptospirosis (6/50), drug toxicity (3/50) and mushroom toxicity (1/50). Secondary complications included ascites (6/50), DIC (6/50), renal failure (5/50), PTE (2/50) and hemoabdomen (2/50). Only 7 dogs survived to discharge. Canine ALF is associated with multiple etiologies and high mortality. Strategies to increase survival in canine ALF are *urgently* needed.

ABSTRACT HP-2

THE ROLE OF INFECTIOUS AGENTS IN GRANULOMATOUS HEPATITIS IN DOGS. RG Hutchins, EB Breitschwerdt, JM Cullen, SA Bissett, JL Gookin. North Carolina State University College of Veterinary Medicine, Raleigh, NC.

Granulomatous hepatitis (GH) is a histopathological diagnosis characterized by focal aggregations of activated macrophages mixed with other inflammatory cells that is usually part of a systemic disease process (WSAVA). Published case reports describe many potential infectious causes, but only one retrospective study involving nine dogs with GH has detailed clinically relevant findings. The aims of this study were to describe the clinical and clinicopathologic findings in dogs with a histopathological diagnosis of GH, and to identify infectious agents using differential staining techniques, PCR, and fluorescence in-situ hybridization (FISH) in archival paraffin-embedded tissues from dogs with GH.

Medical records of dogs with a histopathological diagnosis of GH (n = 22) were reviewed and signalment, historical toxin exposure or evidence of other systemic diseases, clinical signs, physical exam findings, clinicopathologic test results, imaging findings, concurrent diagnoses, treatments administered, and case outcome, when available, were extracted and summarized. Twelve archival formalin-fixed, paraffin-embedded hepatic tissue samples were available for special staining and molecular diagnostic testing. Two of these samples had sufficient tissue for only PCR. The mean age of dogs with GH was 7 years (median 6.5 years; range 2 to 17 years) and included 12 males and 10 females representing 14 different breeds. Common presenting complaints included inappetance or anorexia (n = 12), weight loss (n = 11), lethargy (n = 9), fever (n = 8), and vomiting (n = 8). High mixed liver enzyme activity (14/22) was the most common clinicopathologic abnormality. Leukemia was diagnosed in one dog and copper-associated hepatopathy in 6 dogs. No infectious agents were identified using differential staining techniques. *Bartonella* species DNA was not PCR amplified from the extracted archival tissue. Furthermore, no bacteria were identified by means of FISH using a universal eubacterial probe. These data suggest a possible role for copper accumulation in the genesis of GH in dogs and support further evaluation of dogs with GH for evidence of copper-associated hepatopathy. Future studies should include detailed environmental histories, the collection of adequate sample volumes for quantification of hepatic copper content and the examination of frozen tissues using novel molecular diagnostic platforms.

ABSTRACT HP-3

HISTOLOGIC EVALUATION OF HEPATIC COPPER AND IRON ACCUMULATION IN 104 FELINE LIVER BIOPSIES. Jacqueline C. Whittemore, Kim M. Newkirk, Danielle M. Reel, Ann Reed. University of Tennessee, Knoxville TN.

Hepatocyte copper and iron accumulation contribute to cell loss, inflammation, and fibrosis. The purpose of this study was to compare copper and iron accumulation in feline liver samples with different disease processes.

Liver biopsies (n = 104) submitted between July 1, 2007 and June 30, 2009 were evaluated using WSAVA guidelines and categorized as non-hepatic/normal, congenital, inflammatory/infectious, neoplastic, and other. Copper (by rubeanic acid) and iron (by Prussian blue) accumulation were graded by increasing amounts (0–3) and location (centrilobular = CL, midzonal = MZ, periportal = PP, random = R). Associations between metal scores and diagnosis category were assessed using the Kruskal-Wallis test.

Histologic diagnoses were non-hepatic/normal (n = 12), congenital (n = 6), neoplastic (n = 16), infectious/inflammatory (n = 39), and other (n = 31). Ninety-two samples were negative for copper; remaining samples were graded 1 (n = 5), 2 (n = 6), and 3 (n = 1). Histologic diagnoses (pattern) for positive samples were congenital (1CL), infectious/inflammatory (7: 2CL, 1MZ, 2PP, 2R), neoplastic (2PP), and other (2CL). Iron staining was negative in 18 samples; remaining samples were graded 1 (n = 38), 2 (n = 40), and 3 (n = 8). Distribution was primarily CL (n = 38) or R (n = 33), though MZ (n = 13) and PP (n = 2) distribution occurred. There were no significant differences by Kruskal-Wallis analysis for amount or location of hepatocellular iron or copper for the different disease categories.

In this study, copper accumulation was rare, had variable distribution and occurred primarily in samples with inflammatory/infectious disease. In contrast, iron accumulation was common and did not correlate with disease category. Further prospective evaluation of copper and iron accumulation in feline liver disease and association with outcome may be warranted.

ABSTRACT HP-4

CYCLOSPORINE THERAPY FOR CANINE CHRONIC HEPATITIS – A RETROSPECTIVE STUDY. AM Bradley, DC Twedt. Colorado State University College of Veterinary Medicine and Biomedical Sciences, Fort Collins, CO.

Chronic hepatitis (CH) in dogs is a progressive condition without clearly defined treatment. Glucocorticoids are commonly used to stop progressive inflammation and fibrosis but are associated with significant side effects including a steroid hepatopathy that complicates enzyme monitoring. Cyclosporine is proposed as an alternative therapy, but there are no published reports of its use for canine CH.

Patient records at the CSU Veterinary Teaching Hospital were searched for histologically confirmed cases of CH treated with cyclosporine. Data were compiled on cyclosporine dosing, concurrent medications, clinical course and biochemical parameters.

13 patients over a 50-month period were identified. Serum alanine aminotransferase (ALT) decreased by an average of 71% in 12 dogs. The ALT normalized completely in 6 of 10 dogs treated for > 60 days. In 5 of 6 dogs on > 9 mg/kg/day, the ALT also normalized. Five of the 6 patients that exhibited clinical signs prior to treatment showed measurable improvement (weight gain, fewer gastrointestinal signs). Eight patients had hyperbilirubinemia or ascites prior to treatment; these resolved in 7. Post-treatment histopathology, available in one patient, revealed decreased severity of CH. Five patients exhibited adverse effects including gastrointestinal signs (3), gingival hyperplasia (1), and papillomatosis (1). Cyclosporine was discontinued in 2 dogs with gastrointestinal signs.

Cyclosporine was an effective therapy for many cases of CH and should be considered for patients who are refractory to or cannot tolerate glucocorticoids. Prospective clinical trials with histological documentation are needed to better define cyclosporine's effectiveness in CH.

ABSTRACT HP-5

SENSITIVITY AND SPECIFICITY OF TISSUE IMPEDANCE DETERMINATION FOR CORRECT VERESS NEEDLE PLACEMENT IN CADAVEROUS DOGS. A Mitchell, JC Whittemore, S Hyink, A Reed. University of Tennessee, Knoxville TN.

Insertion of the Veress needle and establishment of pneumoperitoneum is associated with 22 to 57% of all laparoscopic complications in humans. The purpose of this study was to determine the accuracy of interpretation of tissue impedance measurements for Veress needle location.

Two laparoscopists, blinded to impedance measurements, placed reusable Veress needles in 20 cadaverous dogs euthanized for reasons unrelated to the study. Placement order was randomized. A third individual evaluated impedance measurements using a handheld device (SensorMed, Knoxville TN) to determine correct versus incorrect needle placement. Veress needle locations were marked using contrasting colors of India Ink; tissues were dissected to determine ink locations.

Impedance measurement interpretation identified 29/33 correct and 7/7 incorrect placements, respectively. Sensitivity, specificity, accuracy, and precision for correct Veress needle placement are listed below. Agreement was moderate (Kappa 0.50, p = 0.01) for placements by Operator 1 and very high (Kappa = 0.88, p < 0.01) for placements by Operator 2.

Table [Abstract HP-5]

	Sensitivity	Specificity	Accuracy	Precision
Combined	87.9% (29/33)	100% (7/7)	90% (36/40)	100% (29/29)
Operator 1	83.3% (15/18)	100% (2/2)	85% (17/20)	100% (15/15)
Operator 2	93.3% (14/15)	100% (5/5)	95% (19/20)	100% (14/14)

Results for tissue impedance measurement interpretation are superior to published data for currently available tests. Impedance measurements accurately detected all incorrect needle placements. Comparison of needle placement with and without tissue impedance feedback will be necessary to determine whether it increases operator detection of inappropriate Veress needle placement and decreases installment phase complication rates.

ABSTRACT HP-6
SENSITIVITY AND SPECIFICITY OF TISSUE IMPEDANCE DETERMINATION FOR CORRECT VERESS NEEDLE PLACEMENT IN CADAVEROUS CATS.

S Hyink, JC Whittemore, A Mitchell, A Reed. University of Tennessee, Knoxville TN.

Delayed detection of intestinal perforation during Veress needle insertion is associated with high mortality. The purpose of this study was to evaluate the accuracy of tissue impedance measurement interpretation for Veress needle location.

Two laparoscopists, blinded to impedance measurements, placed reusable Veress needles in 24 cadaverous cats. Placement order was randomized. A third individual evaluated impedance measurements (SensorMed, Knoxville, TN) to determine placement location. Needle locations were marked using India Ink; tissues were dissected to determine ink locations.

Impedance measurement interpretation identified 36/38 correct and 2/10 incorrect placements. All 8 undetected incorrect placements were located within the retroperitoneal fat pad. Sensitivity, specificity, accuracy, and precision for correct Veress needle placement are listed below. Correlation was absent (Kappa −0.15, p = 0.34) for placements by Operator 1 and substantial (Kappa 0.78, p < 0.01) for Operator 2. There was no association between correct or incorrect placement and operator on Chi-squared analysis.

Table [Abstract HP-5]

	Sensitivity	Specificity	Accuracy	Precision
Combined	94.7% (36/38)	20% (2/10)	79.2% (38/48)	81% (36/44)
Operator 1	100% (21/21)	66% (2/3)	95.8% (23/24)	95.5% (21/22)
Operator 2	88.2% (15/17)	0% (0/7)	62.5% (15/24)	68.2% (15/22)

Failure of impedance measurements to identify placement in the retroperitoneal fat pad resulted in poor accuracy and discordant Kappa statistics. Small cat size limited the number of appropriate placement sites, perhaps resulting in excessively dorsal placement. Comparison of needle placement with and without tissue impedance feedback will be necessary to determine whether impedance measurements increase detection of inappropriate Veress needle placements or decrease installment phase complication rates.

ABSTRACT HP-7
BEST PREDICTORS OF PORTOSYSTEMIC SHUNTING IN 568 DOGS: POST FEEDING SERUM BILE ACIDS AND PROTEIN-C.

SA Center, JF Randolph, KC Sapa, KL Warner, WE Hornbuckle, AE Yeager, NL Dykes, HJ Harvey, JA Flanders, MB Brooks. College of Veterinary Medicine, Cornell University, Ithaca, NY.

Best clinicopathologic tests detecting portosystemic shunting (PSS) in dogs remains controversial. This retrospective study examined performance of single random "fasting" and paired serum bile acids (SBA; pre- and 2-hr post-feeding) in a large population of non-icteric dogs with confirmed PSS (abdominal ultrasound, colorectal scintigraphy, radiographic or spiral-CT portography, laparotomy, or necropsy). SBA were measured by enzymatic colorimetric method with normal < 25 μmol/L. Dogs meeting inclusion criteria (n = 568) included 467 portosystemic vascular anomalies (PSVA; 368 extrahepatic [E-PSVA], 99 intrahepatic [I-PSVA]), and 101 acquired PSS (APSS). Signalment and laboratory parameters were recorded. Non-parametric statistical analyses were used, two-tailed $P < 0.05$ applied with Bonferroni corrections.

Median age and weight of 74 breeds were 1.2 (0.2–12) yrs and 6.6 (0.2–48) kg, with equal gender distribution. Random "fasting" SBA detected 90% PSVA and 81% APSS, whereas post-feeding SBA detected 100% PSVA and 98% APSS. Low Protein-C (< 70% activity) occurred in 96% PSVA and 74% APSS. Low MCV and creatinine occurred in 82% and 94% of PSVA dogs, respectively; other tests were less helpful. In APSS, post-feeding SBA was superior. Compared to APSS, PSVA had significantly ($P \leq 0.002$) lower MCV, cholesterol, BUN, creatinine, glucose, and Protein-C. Compared to E-PSVA, I-PSVA had significantly ($P \leq 0.009$) lower post-feeding SBA, MCV, albumin, urine specific gravity, and Protein-C but higher cholesterol and glucose.

Post-feeding SBA reflect physiologically provoked bile acid challenge and should be the preferred SBA test in non-icteric dogs for PSS detection. Protein-C assists in identifying PSVA but its utility in APSS may be complicated by concurrent coagulopathies and inflammation.

ABSTRACT ID-1
THE USE OF ANTI-INFLAMMATORY THERAPY AS ADJUNCTIVE TREATMENT IN DOGS WITH BLASTOMYCOSIS.

M. J. Munks, J. K. Byron, S. K. Dowling, N. Szigetvari, M. D. Ridgway, M. Mitchell. University of Illinois College of Veterinary Medicine, Urbana, IL.

This study compared outcomes of treatment with adjunctive nonsteroidal anti-inflammatory drugs (NSAIDs) or anti-inflammatory glucocorticoids in dogs with severe pulmonary blastomycosis. Medical records were reviewed for dogs diagnosed with blastomycosis at the University of Illinois Veterinary Teaching Hospital between 1992 and 2007. Dogs with a presenting PaO_2 of ≤ 80 mmHg, and clinical or radiographic signs of respiratory blastomycosis were included. All dogs were treated with either itraconazole, fluconazole, amphotericin B, or a combination of these. Group 1 (G1) dogs were treated with NSAIDs and Group 2 (G2) dogs were treated with glucocorticoids as anti-inflammatory adjunctive therapy. The following comparisons were made: days of oxygen supplementation, days in hospital, survival to discharge, and long term patient survival. Mann-Whitney U tests and Chi-squared tests were performed on continuous and categorical data, respectively. $P < 0.05$ was considered significant.

Sixty-eight dogs fit the inclusion criteria. G1 consisted of 31 dogs and G2 consisted of 37 dogs. The two groups were found to be similar in weight, age, and sex distribution. There was no significant difference between the two groups with regard to duration of oxygen supplementation, duration of hospitalization, survival to discharge, and patient survival.

There does not appear to be a difference between the clinical course or patient outcomes between groups of dogs with severe pulmonary blastomycosis treated with NSAIDs or anti-inflammatory glucocorticoids. Further studies need to be performed to fully evaluate the impact these adjunct treatments have on prevention of ARDs and additional respiratory complications.

ABSTRACT ID-2
DIAGNOSTIC RELIABILITY OF AN ANTIGEN ASSAY FOR FELINE HISTOPLASMOSIS.

AK Cook[1], LY Cunningham[2], AK Cowell[3], LJ Wheat[4]. [1]Texas A&M College of Veterinary Medicine and Biomedical Sciences, College Station, TX. [2]Oklahoma State University Center for Veterinary Health Sciences, Stillwater, OK. [3]Cat Clinic of Stillwater, Stillwater, OK. [4]MiraVista Diagnostics, Indianapolis, IN.

Diagnosis of feline *Histoplasma capsulatum* infection traditionally relies upon identification of organisms in circulating monocytes or affected organs. In recent years, an antigen assay (AA) was developed for the diagnosis of disseminated histoplasmosis in human patients, but there is little information describing this test in cats. The goal of this study was to determine the sensitivity and specificity of *H. capsulatum* AA in cats with clinical disorders suggestive of histoplasmosis.

Urine and serum *H. capsulatum* AA results for feline patients from 3 veterinary hospitals were evaluated. Medical records were reviewed for confirmatory evidence of histoplasmosis (based on cytological or histopathological findings) or an appropriately supported alternate diagnosis.

AA results were available for 78 cats; initial testing was performed on 72 urine samples, 6 serum samples, and 1 unspecified sample. Of these cats, 17/78 had a definitive diagnosis of histoplasmosis based on organism identification, and 10 had a definitive alternate diagnosis (e.g., neoplasia, other infection) based on necropsy findings (n = 5) or other clinical data (n = 5). An additional 16 cats had a clinical alternate diagnosis with no cytological or histopathological evidence of histoplasmosis in the affected body system(s). The remaining cats had unverified histoplasmosis (n = 9) or an open diagnosis (n = 26). Of the 17 cats with confirmed histoplasmosis, 16 were positive on initial urine AA. One cat (with rectal involvement) was negative, indicating a test sensitivity of 94%. One cat was positive on urine AA but negative on serum AA. All of the 26 cats with definitive or clinical alternate diagnoses had negative results on the AA, suggesting an excellent specificity (100%). However, this result should be interpreted with caution, as the possibility of primary or concurrent histoplasmosis was only definitively excluded in the 5 patients who underwent necropsy examination.

These findings suggest that the AA for *H. capsulatum* is a reliable diagnostic tool in this species. A positive result appears to reliably support the presence of infection, but a small percentage of infected cats may be negative on AA. In addition, tests performed on urine may be more sensitive that those performed on serum.

ABSTRACT ID-3

THE USE OF A BLOOD AND URINE GALACTOMANNAN ANTIGEN TEST FOR THE DIAGNOSIS OF CANINE SYSTEMIC ASPERGILLOSIS. R Garcia[1], J Sykes[1], A Cook[2], L Wheat[3]. [1]Department of Medicine & Epidemiology, University of California, Davis, CA. [2]Department of Small Animal Clinical Sciences, Texas A&M University, College Station, TX. [3]Mira Vista Diagnostics, Indianapolis, IN.

The purpose of this study was to evaluate the sensitivity and specificity of an *Aspergillus* galactomannan antigen enzyme immunoassay (GA-EIA) for the diagnosis of canine systemic aspergillosis. Serum and urine samples were collected from sick dogs at 2 hospitals (UCD and TAMU). Group 1 dogs were diagnosed with systemic aspergillosis using culture (sterile site) or microscopy and culture (non-sterile site). Group 2 dogs had clinical findings suggestive of aspergillosis but an alternate diagnosis was established. Group 3 dogs were not suspected to have aspergillosis. Samples were tested using the GA-EIA and results expressed as a galactomannan index (GMI). GMIs > 0.5 were considered positive. Comparisons were performed using the Mann-Whitney test. There were 10 dogs in Group 1, 22 in Group 2, and 23 in Group 3. Serum was collected from all dogs, and urine from 6, 9, and 10 dogs, respectively. Serum GMIs did not differ from urine GMIs across groups. Serum GMIs of Group 1 dogs were higher than those of Group 2 and Group 3 dogs ($p < 0.0001$). Results from dogs in Group 2 did not differ from those in Group 3 ($p = 0.43$). Two dogs in Group 1 tested negative, but had localized pulmonary infections. One dog in Group 2, which had paecilomycosis, tested positive. Two dogs in Group 3 tested positive. One was being treated with Plasmalyte. The other had a cutaneous opportunistic mycosis. These data support the utility of this assay to aid in the diagnosis of systemic aspergillosis in dogs.

ABSTRACT ID-4

***ANAPLASMA PHAGOCYTOPHILUM* IN CATS: A RETROSPECTIVE ANALYSIS AND CLINICAL EVALUATION OF FIFTEEN CASES FROM THE NORTH EASTERN UNITED STATES.** C Savidge[1], P Ewing [1], J Andrews[2], J Fenrich[3], S Moroff[4]. [1]Angell Animal Medical Center, Boston, MA. [2]Antech Diagnostics, Cary, NCv [3]Antech Diagnostics, Irvine, CA. [4]Antech Diagnostics, Lake Success, NY.

Anaplasma phagocytophilum, an Ixodes tick transmitted rickettsial bacterium has a wide mammalian host range that is not commonly reported in cats. Clinical signs in humans, dogs and cats are often vague and include lethargy, anorexia and malaise.

The purpose of this retrospective study was to describe the clinical signs, laboratory data and response to treatment in cats that tested positive for *A. phagocytophilum* on a commercially available PCR of peripheral blood (FastPanel™). This study describes and reports the appearance of intracellular morulae in feline neutrophils contributing to the diagnosis of *A. phagocytophilum*.

The *A. phagocytophilum* real-time PCR (RT PCR) assay consists of four multiplexed primer systems designed to detect a total of three distinct genes. Amplicons were confirmed as *A. phagocytophilum* by DNA sequencing. Clinicopathologic data was obtained by review of medical records and interview of primary veterinarians. Complete blood counts were available from 13/15 cats and 10/13 blood smears were reviewed.

The cats included in this study were all positive for *A. phagocytophilum* by real-time PCR. The cats ranged from 4 months to 13 years of age with an average age of 3.7 years. Fifteen of 15 cats had a history of tick exposure and lived in the northeastern region of the US, an Ixodes endemic area. All cats presented with lethargy, 13/15 were anorexic and 14/15 had a fever (temperature > 103°F). Other clinical findings included hepatomegaly, splenomegaly, ataxia and ocular changes of conjunctivitis and elevation of the nictitating membrane. Hematologic findings included leukopenia (1/13), neutropenia (1/13) and lymphopenia (4/13). Thrombocytopenia was not noted in any case. Morulae were seen within neutrophils in 2/13 cases. All cases in this report responded to treatment with doxycycline.

This is the first report of the identification of morulae within neutrophils via peripheral blood smear review in cats confirmed by RT PCR to be infected with *Anaplasma phagocytophilum* in North America.

Infection with *Anaplasma phagocytophilum* should be considered in a clinically ill cat with tick exposure, living in an Ixodes endemic area that presents to a veterinarian for lethargy, anorexia and fever.

ABSTRACT ID-5

CANINE BARTONELLOSIS MIMICS OTHER TICK-BORNE DISEASES. C Perez[1], PPVP Diniz[2], RG Maggi[1], EB Breitschwerdt[1]. [1]Intracellular Pathogens Research Laboratory, Center for Comparative Medicine and Translational Research, College of Veterinary Medicine, North Carolina State University (NCSU), Raleigh, NC. [2]Western University of Health Sciences, College of Veterinary Medicine, Pomona, CA.

The spectrum of disease manifestations and the accompanying clinicopathological abnormalities indicative of bartonellosis in dogs have not been thoroughly characterized. The objective of this unmatched case-control study was to compare signalment, clinical and pathologic findings in clinically-ill dogs suspected of a tick-borne disease that were negative for *Bartonella* sp. DNA (controls) as were the dogs diagnosed with bartonellosis by PCR amplification, DNA sequencing and the BAPGM (*Bartonella* alpha Proteobacteria growth medium) enrichment culture approach. Both groups were tested under the same laboratory conditions and in the same time frame. Medical records were reviewed for information regarding signalment, medical history, physical examination findings, clinicopathological abnormalities, microbiological data and treatment.

The study population consisted of 47 *Bartonella*-infected dogs and 93 non-infected dogs. Healthy dogs with no historical illnesses, such as blood donors, were excluded. The following species were amplified: *B. henselae* (n = 28, 59.6%), *B. vinsonii* subsp. *berkhoffii* (n = 20, 42.6%), *B. koehlerae* (n = 3, 6.4%), *B. volans-like* (n = 3, 6.4%), *B. bovis* (n = 1, 2.1%). Nineteen (40.4%) *Bartonella*-infected dogs were febrile and lethargic and ten (21.3%) had neurological signs. Laboratory abnormalities for both groups are summarized below (number of affected dogs provided in parenthesis):

Table [Abstract ID-5]

Group	Anemia	Thrombo-cytopenia	Neutropenia	Mono-cytosis	Lympho-cytosis	Hyperglo-bulinemia	Hypoglo-bulinemia	Proteinuria	Pleural Effusion	Endo-carditis
Bartonella	38% (18/47)	34% (16/47)	15% (7/47)	17% (8/47)	11% (5/47)	27% (12/45)	24% (11/45)	63% (19/30)	13% (6/47)	11% (5/47)
Control	41% (38/93)	33% (31/93)	4% (4/93)	1% (1/93)	0% (0/93)	15% (13/85)	11% (9/85)	60% (41/68)	4% (4/93)	4% (4/93)
Total (n)	140	140	140	140	140	130	130	98	140	140
p-value	0.856	1.000	**0.043**	**< 0.001**	**0.004**	0.159	**0.044**	0.825	0.086	0.1629
Odds ratio	0.898	1.032	**3.894**	**18.87**	-	2.014	**2.732**	1.137	3.256	2.649

Multivariate logistic regression using confounding factors was performed to establish potential associations between specific variables and *Bartonella* sp. infection. There were no differences in signalment, age, sex, body weight and duration of clinical signs between the two groups. Compared to the control population, infection with the genus *Bartonella* was associated with a diagnosis of endocarditis (P = 0.0196, OR = 7.95, 95%CI = 1.39–51.11) and hypoglobulinemia (P = 0.0057, OR = 5.30, 95% CI = 1.62–18.55). Controls were more likely to have joint effusion (P = 0.0059, OR = 5.89, 95% CI = 1.62–27.22) and azotemia (P = 0.0353, OR = 2.93, 95%CI = 1.07–8.86) than were the *Bartonella* sp. infected dogs.

Bartonella was detected in dogs with signs such as fever, anemia, thrombocytopenia, hyperglobulinemia and proteinuria that are typically associated with tick-borne diseases. When endocarditis or hypoglobulinemia are detected, testing for *Bartonella* should be prioritized. Likewise, the detection of *Bartonella* should prompt further testing for endocarditis, if not already investigated.

ABSTRACT ID-6

EVALUATION OF A COMMERCIAL BLOCKING ELISA FOR DETECTION OF INFLUENZA A NUCLEOPROTEIN ANTIBODIES IN CANINE SERA. SA Salomon, TC Anderson, PC Crawford. Maddie's Shelter Medicine Program, College of Veterinary Medicine, University of Florida, Gainesville, FL.

Recent infections of dogs with influenza A viruses of equine, avian, and human origin have raised concerns about the role of dogs in the global ecology of influenza A viruses. These concerns highlight the importance of influenza A virus surveillance in dogs. Surveillance studies in other species depend on detection of antibodies to the highly conserved influenza A nucleoprotein (NP); however, no such antibody detection assay is approved for canine use in the U.S. The purpose of this study was to determine the diagnostic accuracy of a commercial blocking ELISA used for avian species in detecting influenza A NP antibody in dogs. Since the blocking ELISA is not a species-specific or viral subtype-specific format, we hypothesized that it would detect NP antibodies in dogs infected by influenza A virus.

Serum samples from uninfected dogs (n = 204) and dogs naturally infected with canine influenza H3N8 (n = 150) were tested using the IDEXX FlockChek blocking ELISA for influenza A NP antibody according to manufacturer instructions. The sample/negative control (S/N) absorbance ratios for infected dogs ranged from 0.12 to 0.67 compared to 0.53 to 1.40 for uninfected dogs. A receiver operating characteristic (ROC) curve analysis determined optimum diagnostic sensitivity (99.3%) and specificity (99.0%) at a S/N cutoff ratio of 0.647. Using this cutoff ratio, the overall diagnostic accuracy was 99.2%. Coefficients of variation for intra-assay (4.7%) and inter-assay (6.1%) testing demonstrated good repeatability with canine sera.

The excellent diagnostic accuracy of the commercial blocking ELISA makes it a suitable tool for large-scale surveillance of influenza A virus exposure in dogs.

ABSTRACT ID-7

PREVALENCE OF UPPER RESPIRATORY PATHOGENS IN FOUR MANAGEMENT MODELS FOR UNOWNED CATS IN THE SOUTHEAST UNITED STATES. CM McManus[1], JK Levy[1], LA Andersen[1], SP McGorray[2], CM Leutenegger[3]. [1]University of Florida College of Veterinary Medicine, Gainesville, FL. [2]University of Florida College of Medicine, Gainesville, FL. [3]IDEXX Reference Laboratories, West Sacramento, CA.

Upper respiratory disease (URD) can affect a majority of cats in shelters and is one of the leading reasons for euthanasia of otherwise adoptable cats. The purpose of this study was to determine prevalence and risk factors for upper respiratory pathogens in four different models for managing unowned cats: short-term animal shelters (SHEL), long-term sanctuaries (SANC), home-based foster care (FOST), and trap-neuter-return (TNR) programs.

Conjunctival and oropharyngeal swabs were collected from 543 cats, half of which had clinical signs of URD, and tested for feline herpesvirus (FHV), feline calicivirus (FCV), *Chlamydiophila felis*, *Bordetella bronchiseptica* (*Bb*) and *Mycoplasma felis* by real-time PCR. Management model, vaccination, sex, age, body condition, and clinical signs were evaluated as risk factors for infection.

Table [Abstract ID-7]: Prevalence of Upper Respiratory Disease in Unowned Cats

Model	Cat Clinical Status	FCV	FHV	*C. felis*	Bb	*M. felis*
SHEL	Symptomatic (n = 61)	14 (23%)	36 (59%)*	5 (8%)	20 (33%)	23 (38%)
(n = 12)	Asymptomatic (n = 79)	18 (23%)	32 (41%)	2 (3%)	19 (24%)	25 (32%)
SANC	Symptomatic (n = 64)	43 (67%)	12 (19%)	1 (2%)	6 (9%)	54 (84%)
(n = 6)	Asymptomatic (n = 65)	33 (51%)	5 (8%)	4 (6%)	10 (15%)	56 (86%)
FOST	Symptomatic (n = 66)	28 (42%)	15 (23%)*	16 (24%)*	12 (18%)	39 (59%)*
(n = 20)	Asymptomatic (n = 84)	26 (31%)	5 (6%)	8 (10%)	10 (12%)	32 (38%)
TNR	Symptomatic (n = 60)	33 (55%)*	10 (17%)	14 (23%)*	0 (0%)	38 (63%)*
(n = 4)	Asymptomatic (n = 64)	23 (36%)	3 (5%)	3 (5%)	3 (2%)	27 (42%)

* = symptomatic vs asymptomatic significantly different P < 0.05

A majority of cats in all management models carried one or more organisms capable of causing URD. In many cases, prevalence was similar in cats with or without clinical signs. Unlike diseases that can be controlled by segregation of symptomatic animals, the lack of strong correlation between the presence of pathogens with the presence of clinical signs suggests that feline URD control should be managed by vaccination before or at the time of intake ,biosecurity protocols that presume all cats may be shedding pathogens, and minimizing stressful conditions that contribute to disease susceptibility.

ABSTRACT ID-8

SENSITIVITY AND SPECIFICITY OF QUANTITATIVE PCR AND VIRUS ISOLATION FOR DIAGNOSIS OF FELINE IMMUNODEFICIENCY VIRUS INFECTION. M Ammersbach[1], S Little[2], D Bienzle[1]. [1]University of Guelph, Guelph, ON. [2]Bytown Cat Hospital, Ottawa, ON.

Depending on geographical location, sex, age and environment, 2–40% of cats worldwide are infected with the feline immunodeficiency virus (FIV). Knowledge of the FIV status of cats is important to limit the spread of disease and to institute appropriate health management. However, like all lentiviruses, FIV is highly variable in nucleotide sequence, and viral load in cats is variable during different disease stages. Detection of antibodies is the most widely employed diagnostic approach, but does not distinguish FIV-infected from FIV-vaccinated cats. In this study, samples from 30 FIV-seronegative cats, 30 FIV-seropositive cats, and 30 FIV-historically seronegative but vaccinated cats, were analyzed by a commercial quantitative PCR (qPCR) assay and virus isolation. Replicate blood samples were coded, and then submitted for 1) qPCR (Idexx); and 2) mononuclear cell isolation with 7-day culture and viral p24 antigen detection by ELISA. For the p24 antigen ELISA, cutoff absorbance values were established from analysis of 10 FIV-negative samples. FIV infection status was pre-determined based on 2 antibody-ELISA results and vaccination history.

Results indicated that qPCR had a sensitivity of 76% for samples from FIV-seropositive cats, and a specificity of 100% and 94.1% for samples from FIV-seronegative and FIV-vaccinated cats, respectively. At a cutoff value of 3 standard deviations above the mean absorbance for p24 antigen ELISA, results from FIV-negative samples yielded a sensitivity of 69.2% for samples from FIV-seropositive cats, and a specificity of 77.1% and 64.7% for samples from FIV-seronegative and FIV-vaccinated cats, respectively. Conclusions from this study are 1) the commercial FIV qPCR assay has high specificity but limited sensitivity for diagnosis of FIV infection; 2) 7-day virus culture has limited sensitivity and specificity. Hence, detection of antibodies remains the most reliable test for diagnosis of FIV infection, but qPCR may be suitable to rule out infection.

ABSTRACT ID-9

SEROPREVALENCE OF FELV AND FIV IN CATS WITH ORAL DISEASE. Jan Bellows[1], Susan Little[3] Margie Scherk[2]. [1]All Pets Dental Clinic, Weston, FL. Jessica Lachtara, IDEXX Laboratories, Westbrook, ME. [2]catsINK, Vancouver, BC. [3]Bytown Cat Hospital, Ottawa, Ontario.

Oral disease is an important clinical problem in feline medicine and includes common painful conditions such as oropharyngeal inflammation (formerly known as gingivostomatitis) and tooth resorptive lesions. A number of infectious agents have been associated with oropharyngeal inflammation, including feline calicivirus, feline herpesvirus-1, feline immunodeficiency virus (FIV) and feline leukemia virus (FeLV). To date, large scale data on seroprevalence of infectious agents in cats with oral disease are lacking. The purpose of this study was to determine the seroprevalence of FeLV and FIV in a population of cats with oral disease.

Private veterinary clinics in the U.S. were recruited to test feline patients presenting with oral disease. Presenting cases included cats with plaque, calculus, gingivitis, stomatitis, periodontal disease, tooth resorptive lesions and other oral diseases as defined by the practitioner. All cats were tested using a commercially available point-of-care ELISA test (IDEXX Snap Combo). Confirmatory tests were not performed as part of the study. Seroprevalence was calculated as the percentage of positive tests in the study population for each virus.

A total of 11,262 cats were tested. Seroprevalence for FeLV was 6.8% and for FIV was 7.1%. Of these, 119 cats (1.0%) were infected with both viruses. Seroprevalence was higher in cats with inflammatory oral disease than in cats characterized with other types of oral disease. Of 7,805 cats with gingivitis, seroprevalence for FeLV was 7.4% and for FIV was 7.9%, with 1.1% of cats co-infected. Of 1,953 cats with stomatitis, seroprevalence for FeLV was 12.2% and for FIV was 13.9%, with 2.2% of cats co-infected.

The seroprevalence for FeLV and FIV reported in this population of cats with oral disease was higher than in a recent large study where samples from U.S. cats not specifically selected for oral disease were tested (FeLV 2.3%, FIV 2.5%). Results of this study indicate that further investigation of the role of retroviruses in cats with oropharyngeal inflammation is warranted.

ABSTRACT ID-10

PREVALENCE AND RISK FACTORS FOR FELV, FIV, AND HEARTWORM INFECTION IN NORTH AMERICAN CATS. JK Levy[1], HM Scott[2], MM Crandall[1], SJ Tucker[1], EG Wood[3], JD Strout[3]. [1]Maddie's Shelter Medicine Program, University of Florida, Gainesville, FL. [2]Kansas State University, Manhattan, KS, [3]IDEXX Laboratories, Portland, ME.

Reliable tests and preventive vaccines and medications for feline retroviral and heartworm (HW) infections are available, but compliance with protocols to reduce transmission is unknown. No large-scale longitudinal studies evaluating prevalence over time have been reported. The purpose of this study was to determine the prevalence and risk factors for infection compared with a similar study completed for the first time 5 years previously.

Veterinary clinics and animal shelters in the US and Canada submitted results of testing using a point-of-care ELISA for FeLV antigen, FIV antibody, and HW antigen (IDEXX SNAP Triple) and risk factor information for cats tested during March-September 2010. Bivariable and multivariable analyses were used to evaluate risk factors for infections.

Table [Abstract ID-10]

Prevalence of infection by risk factor (unadjusted for other covariates or clinic/shelter effects)

Risk factor	n	FeLV	FIV	HW**
All cats	62,301	3.1	3.6	0.4
Owned	38,741	3.1	3.4	0.4
Unowned	23,560	3.2	3.9*	0.5
Indoor	25,440	2.0	1.7	0.2
Outdoor	36,147	3.9*	5.0*	0.5*
Female	30,741	2.8	2.0	0.4
Male	31,560	3.5*	5.1*	0.4
Juvenile	26,941	2.5	1.6	NA
Adult	35,360	3.6*	5.1*	0.4
Healthy	45,260	1.7	2.0	0.3
Unhealthy	17,041	7.1*	7.7*	0.6*
Respiratory disease	4,889	8.0*	6.5*	1.0*
Oral disease	1,698	5.1*	10.1*	0.8*
Abscess	2,050	5.9*	12.8*	0.9*

*P < 0.05
**adult cats only

A total of 62,301 cats were tested. Only 16% of owned cats were prescribed HW preventive. Risk of retroviral infections was increased by outdoor access, adulthood, and male gender. The most important risk factor associated with all 3 infections was clinical disease; in particular, respiratory and oral diseases and abscesses or bite wounds. Multivariate analysis revealed differences among geodivisions and across infection types.

Feline retroviral and heartworm infections are easily prevented, but difficult to treat. Despite availability of effective management protocols, compliance remains inadequate to reduce the prevalence of these infections. Improved use of preventive care and testing to identify and segregate contagious cats, particularly those at high-risk, is required to reduce the morbidity of these preventable infections.

ABSTRACT ID-11

USE OF ANTIBODY TITERS AND QUANTITATIVE PCR AS RISK ASSESSMENT TOOLS FOR MANAGEMENT OF AN OUTBREAK OF CANINE DISTEMPER AND PARVOVIRUS. PC Crawford[1], JK Levy[1], C Leutenegger[2]. [1]Maddie's Shelter Medicine Program, College of Veterinary Medicine, University of Florida, Gainesville, FL. [2]IDEXX Laboratories Inc., West Sacramento, CA.

Infectious disease outbreaks are common in animal shelters and are frequently managed by depopulation when risk-assessment tools are not available. During a canine distemper virus (CDV) and parvovirus (CPV) outbreak in sheltered dogs, we used a CDV/CPV point-of-care antibody titer ELISA, a CDV quantitative RT-PCR test, and a CPV fecal antigen test as risk assessment tools to guide release of exposed dogs from quarantine and euthanasia of diseased dogs.

Serum samples (for antibody titers) and swabs of the conjunctiva and upper respiratory tract (for CDV PCR) were collected from 111 asymptomatic dogs starting on Day 4 of the outbreak. Dogs with positive CDV PCR tests were retested every 2 weeks until euthanized for progressive disease or released following recovery from infection. Dogs with clinical signs of parvoviral infection were tested using a CPV fecal antigen test.

For dogs ≥ 4 months old, protective antibody titers correlated with resistance to clinical disease, but 10% of dogs shed CDV. Lack of protective CDV antibody titers correlated with susceptibility to clinical infection, but most dogs recovered.

Tables [Abstract ID-11]

Risk assessment and outcome in 60 dogs ≥4 months of age

# Dogs	Day 4 Protective Antibody Titer	Day 4 CDV PCR	Day 14 CDV PCR	Clinical Status	Outcome
36	CDV+/CPV+	Negative	Negative	Healthy	Released on day 15
4	CDV+/CPV+	Negative	Positive	Healthy	All recovered and released
20	CDV−/CPV+	Negative	Positive	Sick	13 recovered, 7 euthanized

Risk assessment and outcome in 51 puppies <4 months of age

Protective Ab Titer	# Dogs	CDV Infected	CDV Died	CPV Infected	CPV Died
CDV+	23	16 (70%)	11 (48%)		
CDV−	28	21 (75%)	16 (57%)		
CPV+	35			3 (9%)	3 (9%)
CPV−	16			6 (38%)	6 (38%)

For puppies < 4 months of age, serology was not predictive of protection against infection and death from CDV, but was predictive for CPV. A total of 31% of puppies < 4 months of age survived and were released for adoption, including 27% of puppies infected with CDV. No dogs that were released from quarantine subsequently developed any signs of CDV or CPV.

Serology and molecular diagnosis were useful risk assessment tools in this disease outbreak, particularly in dogs $\geq$ 4 months of age. This resulted in safe release of 61% of dogs for adoption.

ABSTRACT ID-12

EFFECT OF CYCLOSPORINE AND METHYLPREDNISOLONE ACETATE ON CATS WITH CHRONIC FELINE HERPESVIRUS 1 INFECTION.

MR Lappin[1], L Roycroft[2]. [1]Department of Clinical Sciences, Colorado State University, Ft. Collins, CO. [2]Novartis Animal Health, Greensboro, NC.

Feline herpesvirus 1 (FHV-1) is a common ocular and respiratory pathogen of cats that can have clinical illness exacerbated by stress. Cyclosporine (CsA) is commonly used for the treatment of a number of inflammatory diseases in cats and can induce immune suppression. A small number of cats administered CsA to block renal transplant rejection have developed clinical signs of upper respiratory tract disease that may have been from activated FHV-1.

In this study, young adult cats experimentally inoculated with FHV-1 several months previously were divided into three groups and administered methylprednisolone acetate (8 cats, 5 mg/kg, IM, Day 0 and Day 21), CsA (8 cats, 7.0 mg/kg, PO, daily for 42 days), or a placebo (7 cats, corn syrup; 0.075 mL/kg, PO, daily for 42 days). Each cat was assigned a daily individual clinical score by a trained, masked observer using a standardized score sheet during the initial pre-treatment time period (Day –14 to Day 0) and throughout the 42 day treatment period. Each individual clinical score (conjunctivitis, blepharospasm, ocular discharge, sneezing, nasal discharge, nasal congestion, and body temperature scores), the total clinical score (sum of all parameters), the total ocular score (sum of conjunctivitis, blepharospasm, ocular discharge), and total respiratory score (sum of ocular discharge, sneezing, nasal discharge, nasal congestion) were analyzed using SAS PROC GLIMMIX with 'Treatment', 'Time', and the two-way interaction 'Treatment by Time' all as fixed effects. Statistical significance was defined as $p < 0.05$.

On Day 42 of the study, all of the CsA treated cats had detectable concentrations of CsA in serum (Mean = 406.1 ng/ml; standard deviation = 291.8 ng/ml; Median = 388.5 ng/ml). When group mean values for clinical signs were compared over time as described, significant differences in individual clinical score measurements, in total score, total ocular score, or total respiratory score were not detected over time among any of the treatment groups.

While clinical signs of activated FHV-1 occurred in some cats administered methylprednisolone or CsA, disease was mild and self-limited in most cats and there were no significant CsA side-effects. These results suggest that the CsA protocol described here is unlikely to reactivate latent FHV-1 infection and cause significant clinical illness.

ABSTRACT ID-13

PREVALENCE OF ENTEROPATHOGENS IN FOUR MANAGEMENT MODELS FOR UNOWNED CATS IN THE SOUTHEAST UNITED STATES.

LA Andersen[1], JK Levy[1], CM McManus[1], SP McGorray[2], CM Leutenegger[3]. [1]University of Florida College of Veterinary Medicine, Gainesville, FL. [2]University of Florida College of Medicine, Gainesville, FL. [3]IDEXX Reference Laboratories, West Sacramento, CA.

The purpose of this study was to determine the prevalence and risk factors for enteropathogens in four different models for managing unowned cats: short-term shelter, long-term sanctuary, home-based foster care, and trap-neuter-return (TNR) programs.

Fecal samples were collected from 482 cats, half with diarrhea (D) and half with normal feces (N), and tested for a panel of feline and zoonotic enteropathogens by polymerase chain reaction, antigen, and fecal flotation. Risk factors for infection evaluated include management practices, fecal consistency, and signalment.

Table [Abstract ID-13]

Most common enteropathogens in unowned cats

Model		*Crypto-sporidium*	Coccidia	*Giardia*	*T. foetus*	*C. perf*	FPV	FCoV	Hook-worm	Round-worm
Shelter	D n=60	3 (5%)	14 (23%)	4 (7%)	0	9 (15%)	12 (20%)	28 (47%)	3 (5%)	3 (5%)
n=13	N n=52	5 (10%)	17 (33%)	3 (6%)	0	11 (21%)	5 (10%)	24 (46%)	4 (8%)	3 (65)
Sanctuary	D n=61	9 (15%)	8 (13%)	20 (33%)	16* (26%)	20 (33%)	0	45 (74%)	17 (28%)	18 (30%)
n=10	N n=60	12 (20%)	6 (10%)	13 (22%)	6 (10%)	11 (18%)	0	47 (78%)	9 (15%)	14 (23%)
Foster	D n=60	5 (8%)	9 (15%)	9 (15%)	5 (8%)	16 (27%)	2 (3%)	48* (80%)	5 (8%)	5 (8%)
n=24	N n=62	6 (10%)	8 (13%)	11 (18%)	0	16 (26%)	3 (5%)	36 (58%)	3 (5%)	3 (5%)
TNR	D n=60	5 (8%)	9 (15%)	5 (8%)	0	21 (35%)	0	22 (37%)	39 (65%)	6 (10%)
n=5	N n=67	9 (13%)	7 (10%)	6 (9%)	1 (1%)	21 (31%)	0	16 (24%)	50 (75%)	10 (15%)

*= group with diarrhea significantly different than group with normal feces $P < 0.05$

A majority of cats had at least one enteropathogen of feline or zoonotic importance, regardless of management model or preventive healthcare protocol. For most enteropathogens, the presence or absence of diarrhea did not correlate with infection, the exceptions being *T. foetus* in sanctuary cats and FCoV in foster cats. Prevalence of specific enteropathogens varied between management models, reflecting differences in preventive healthcare and housing conditions.

Management protocols for unowned cats were inadequate for elimination of infections present at the time of intake and for prevention of transmission of enteropathogens among shelter cats. Improved compliance with effective vaccination, deworming, sanitation, and housing protocols is needed to reduce zoonotic and feline health risks.

ABSTRACT ID-14

***TOXOPLASMA GONDII* OOCYST SHEDDING IN NORMAL CATS AND CATS TREATED WITH CYCLOSPORINE.**

MR Lappin, V Scorza. Department of Clinical Sciences, Colorado State University, Ft. Collins, CO.

Several allergic diseases of cats, including atopy and gingivostomatitis, can be resistant to glucocorticoids but responsive to cyclosporine. *Toxoplasma gondii* infection occurs in approximately 30% of cats and the effect cyclosporine therapy has on the *T. gondii* oocyst shedding period is unknown. The objective of this study was to determine whether administration of cyclosporine before or after *T. gondii* infection influences the oocyst shedding period.

The young adult cats were *T. gondii* seronegative when administered 1,000 *T. gondii* tissue cysts orally on day 42. Group 1 cats (n = 10) were never administered cyclosporine; Group 2 cats (n = 10) were administered cyclosporine (7.5 mg/kg, PO) daily on Days 84–126; and Group 3 cats (n = 10) were administered cyclosporine (7.5 mg/kg, PO) daily from days 0–126. Available feces from individual cages were collected daily and fecal flotation by sugar centrifugation was performed for 84 days after *T. gondii* inoculation.

Group 3 shed oocysts for a significantly shorter period than Groups 1 or 2 and had a significantly lower oocyst shedding scores than Groups 1 and 2 on days 5–11 after *T. gondii* inoculation. Group 2 cats had completed the oocyst shedding period prior to being administered cyclosporine and repeat oocyst shedding was not detected during administration of the drug.

Administration of cyclosporine prior to *T. gondii* infection lessened oocyst shedding which is likely from the anti-*T. gondii* effects of the drug. Administration of cyclosporine using this protocol is unlikely to induce repeat *T. gondii* oocyst shedding in client-owned cats.

ABSTRACT ID-15

WHOLE SHOTGUN PYROSEQUENCING OF *MYCOPLASMA HAEMOFELIS*: THROWING NEW LIGHT ON AN OLD PATHOGEN.

EN Barker[1], CR Helps[1], IR Peters[1], AC Darby[2] & S Tasker[1]. [1]School of Veterinary Sciences, University of Bristol, Langford, UK [2]School of Biological Sciences, University of Liverpool, Liverpool, UK.

First described in 1942, *Mycoplasma haemofelis* is a pathogenic hemotropic mycoplasma of cats. Due to it being uncultivable, little is known about its metabolic pathways or mechanism of pathogenicity and whole genome sequencing of feline hemoplasmas has not yet been reported. The aim of this study was to completely sequence the genome of *M. haemofelis* to further characterise this important pathogen.

Mycoplasma haemofelis genomic DNA was purified and subjected to whole shotgun Roche 454 sequencing. Gaps were closed using targeted PCR and amplicon sequencing. Ribosomal genes and potential open reading frames (ORFs) were predicted *in silico*. Putative ORFs were annotated and orthologous groups identified.

Analysis showed a circular genome of 1.15 Mbp with a GC content of 38.9%. Thirty-one transfer RNAs (tRNAs) were identified, accounting for all amino acids, including a tryptophan tRNA for the opal codon (UGA). Of the 1,545 putative proteins identified, 328 (21.2%) matched to proteins from other bacterial species. In common with the *pneumoniae* group of mycoplasmas, the closest phylogenetic relatives of the hemoplasmas, genes involved in carbohydrate metabolism were limited to enzymes of the glycolytic pathway, with glucose appearing to be the sole energy source for *M. haemofelis*. The majority of the pentose phosphate pathway genes present in other cultivatable mycoplasmas appear to be incomplete or absent in *M. haemofelis*, suggesting an alternative mechanism for sourcing purine and pyramidine bases such as scavenging from the host. A gene encoding a glyceraldehyde-3-phosphate dehydrogenase homolog of the immunogenic MSG1 protein of *Mycoplasma suis* was present. Of the uncharacterized hypothetical proteins, 1,115 were arranged in series of orthologous repeats, or comprised fragments there-of, encoding putative proteins of approximately 200 amino acids. The predicted motifs of the majority of these putative proteins were consistent with these proteins being presented on the cell surface; an N' terminal signal peptide or transmembrane region followed by a non-cytoplasmic tail.

These data have provided valuable information as to why this pathogen remains highly fastidious; it lacks some of the metabolic pathways found in cultivatable mycoplasmas. We have also identified a homolog of a known *M. suis* immunogenic protein, and identified a potential mechanism for host immune system evasion by way of highly repetitive, putatively surface-expressed hypothetical proteins with variable sequences.

ABSTRACT ID-16

THE PREVALENCE OF LEPTOSPIRAL ANTIBODIES IN FERAL CATS IN MASSACHUSETTS. J Markovich, L Ross, E McCobb. Tufts University Cummings School of Veterinary Medicine, North Grafton, MA.

Canine leptospirosis has been recognized as a re-emerging disease in the U.S. over the past 20 years, and several serosurveys of the prevalence of leptospiral antibodies in dogs have been published during that time. The role of cats in the epidemiology of leptospirosis has received little attention. Serosurveys of cats for exposure to or infection with leptospires have been published from other geographic areas, but none for cats in the U.S. in the past four decades. The New England states have been found to have a high incidence of canine leptospirosis. The purpose of this pilot study was to determine the prevalence of leptospiral antibodies in a population of feral cats in central Massachusetts.

Blood was collected from 63 sexually intact feral cats presented to a spay and neuter program. Microagglutination titers to *Leptospira* serovars *autumnalis, hardjo, bratislava, icterohaemorrhagiae, canicola, pomona*, and *grippotyphosa* were determined. Three of 63 cats (4.8%) had a positive titer to one or more serovars, with *autumnalis* being the most common. These results are consistent with previously published prevalence rates in feral cats. Further studies are required to determine the role of leptosporosis in clinical disease in the domestic cat.

ABSTRACT ID-17

THE RIVALTA'S TEST AS A DIAGNOSTIC TOOL IN CATS WITH FELINE INFECTIOUS PERITONITIS. Y. Fischer[1], C. Sauter-Louis[2], K. Hartmann[1]. [1]Medizinische Kleintierklinik. [2]Clinic for Ruminants, Ludwig Maximilian University, Munich, Germany.

Since years the Rivalta's test is routinely used in several European countries as a tool to diagnose feline infectious peritonitis (FIP) in cats with effusion. It is inexpensive and easy to perform in private practice. There is, however, only little information about mode of action or its diagnostic value. The objectives of this study were to evaluate sensitivity, specificity, positive (PPV) and negative predictive values of the Rivalta's test to diagnosis of FIP and to examine if there is a correlation with any effusion or blood parameters.

Medical records of 782 cats with effusion in which the Rivalta's test was performed between 1999 and 2010 were reviewed concerning diagnosis, blood and effusion parameters, and survival time. Effusion and blood parameters were compared between Rivalta-positive and -negative effusions using the Mann Whitney U test.

Prevalence of FIP in cats with effusion was 31.7%. The Rivalta's test showed a sensitivity of 91.3%, a specificity of 66.6%, a PPV of 55.9%, and a NPV of 94.3% for the diagnosis of FIP. The PPV improved, when cats with lymphoma or bacterial infection were excluded (PPV 69.2%) and also, when only cats younger than 2 years (PPV 87.5%) or 1 year (PPV 90.8%) of age were included. The most important significantly different parameters between Rivalta-positive and -negative effusions were specific gravity as well as cholesterol, triglyceride, and glucose concentration in the effusion.

The Rivalta's test in general is a useful tool to diagnose FIP, but its sensitivity and specificity are not as high as previously assumed. If the Rivalta's test, however, is performed in young cats or if certain diseases have been ruled-out, its diagnostic value is high. Effusion total protein is not highly correlated with test outcome. Therefore, it is still unclear, which components in the effusion of cats with FIP lead to a positive Rivalta's test.

ABSTRACT ID-18

PREVALENCE OF PROTECTIVE ANTIBODY TITERS AGAINST CANINE PARVOVIRUS (CPV) AND CANINE DISTEMPER VIRUS (CDV) AND RESPONSE TO CPV AND CDV VACCINATION IN SHELTER DOGS. Annette Litster, Jamieson Nichols. Purdue University School of Veterinary Medicine, West Lafayette, IN.

Canine parvovirus (CPV) and canine distemper virus (CDV) infections are relatively common in animal shelters and are important population management issues since the immune status of incoming dogs is usually unknown. Our study aimed to determine the antibody protection status of dogs at the time of admission into an animal shelter (pre-vaccination) and over the following 2 weeks after vaccination.

Serum samples were obtained from 57 incoming shelter dogs aged 4 months and older with no known history of vaccination. Immediately following serum collection, the dogs were vaccinated against CPV and CDV using a modified live vaccine (MLV). CPV and CDV antibody protection status was determined using Synbiotics TiterCHEK. Dogs with unprotective serum antibody levels against CPV and/or CDV were retested at 6–8 days post-vaccination and again at 13–15 days post-vaccination, if antibody levels were still unprotective against CPV and /or CDV. At the conclusion of the study, stored duplicate sera were submitted for batch 'gold standard' testing to determine canine distemper virus serum neutralization and canine parvovirus hemagglutination inhibition antibody titers.

Based on the Synbiotics TiterCHEK results, 43/57 dogs (75.4%) were protected against CPV and 21/57 (36.8%) were protected against CDV at intake. Older incoming dogs were more likely to be protected against CPV ($P < 0.0001$) and CDV ($P = 0.0174$). Dogs that were spayed/neutered were more likely to be protected against CPV on intake than intact animals, although this result was not statistically significant ($P = 0.0627$). The number of dogs with protective titers against CPV/CDV was increased at 6–8 days post-MLV (CPV - 48/56, 85.7%; CDV - 31/52, 59.6%) and further increased at 13–15 days post-MLV (CPV - 54/54, 100.0%; CDV - 47/48, 97.9%).

We conclude that incoming shelter dogs often do not have protective antibody titers against CPV and CDV, but older shelter dogs are more likely to be protected against CPV. Based on this population, we further conclude that a large percentage of dogs develop protective antibody titers to CPV and CDV within 1 to 2 weeks when vaccinated with a MLV.

ABSTRACT ID-19

ROLE OF *MYCOPLASMA* SPP. IN FELINE CAT BITE ABCESSES. C Torres[1], J Hesser[2], M Brewer[1], MR Lappin[1]. [1]Colorado State University, Fort Collins, CO. [2]Alta Vista Veterinary Hospital, Phoenix, AZ.

Mycoplasma spp. are common inhabitants of the feline oral cavity and so likely contaminate many cat bite abscesses. *Mycoplasma* spp. are cell-wall deficient and so do not respond to beta-lactam class antibiotics, the class most commonly use for the treatment of cat bite abscesses. The objectives of this study was to determine whether *Mycoplasma* spp. are common contaminants of cat bite abscesses and are associated with beta-lactam resistant clinical disease.

Privately owned cats with clinical evidence of an acute abscess suspected to be from a cat bite were included in the study. Participants were given a free aerobic and anaerobic culture as well as *Mycoplasma* spp. culture and polymerase chain reaction using *Mycoplasma* genus specific primers. *Mycoplasma* spp. amplicons were sequenced to determine the species. All cats were initially treated with appropriate wound management, were administered an antibiotic in the beta lactam class (amoxicillin-clavulanate or cefovicin), and were rechecked in person or by phone 7 days after beginning treatment.

Of the 26 cats entered into the study to date, *Mycoplasma* spp. were amplified from 4 cats (15.4%). Of the 2 positive samples with adequate DNA for sequencing, one was consistent with *M. felis* and the other was consistent with *M. equigenitalium*. Of the 26 cats, 25 responded by Day 7 to the initial treatment, including 3 of the 4 *Mycoplasma* spp. positive cats. The cat that failed initial treatment was positive for *M. equigenitalium* on both Day 0 and Day 7 and ultimately responded to administration of a fluoroquinolone.

The results suggest that while *Mycoplasma* spp. commonly contaminate cat bite abscesses, routine wound management and antibiotic therapy is adequate for control. However, as *Mycoplasma* spp. infections do not respond to beta lactam class antibiotic therapy, these organisms should be on the differential list for cats with abscesses that fail treatment with this antibiotic class.

ABSTRACT ID-20

EFFECT OF VACCINATION ON RESULTS OF REAL TIME PCR ASSAYS FOR ADENOVIRUS 2, PARAINFLUENZA, AND *BORDETELLA BRONCHISEPTICA* IN PUPPIES.

MR Lappin[1], S Moroff[2]. [1]Colorado State University, Ft. Collins, CO. [2]Antech Diagnostics, Lake Success, NY.

Molecular diagnostic assays are frequently used in clinical practice to aid in the diagnosis of suspected infectious respiratory diseases in dogs. However, most currently available assays cannot distinguish strains of the organisms used in vaccines from naturally occurring strains. Our prior studies demonstrated that previously immune adult dogs are unlikely to shed nucleic acids of vaccine strains of adenovirus 2, parainfluenza, or *Bordetella bronchiseptica*. However, whether this is true for puppies is unknown.

Puppies (n = 8) at a breeding facility were moved into area without other dogs at 6 weeks of age. Swabs of the nasal and pharyngeal mucosa were collected prior to vaccination and on days 0, 1, 2, 3, 4, 5, 6, 7, 10, 14, 17, 21, 24, and 28 after vaccination with an intranasal adenovirus 2, parainfluenza, and *B. bronchiseptica* vaccine (IntraTrac 3, Schering Plough). The swabs were shipped on cold packs by overnight express for DNA/RNA extraction and assay in the FastPanel™ PCR Canine Respiratory Disease Profile at Antech Diagnostics.

All puppies were negative for the infectious agents prior to vaccination. After vaccination, positive assay results for parainfluenza and *B. bronchiseptica* were first detected on Day 1 and on Day 2 for adenovirus 2. By Day 3, DNA or RNA of the agents were amplified from all puppies from both sample sites and most samples were positive for all 3 agents through Day 10. By day 24, only one dog was still positive for *B. bronchiseptica*.

The results indicate that intranasal administration of adenovirus 2, parainfluenza, or *Bordetella bronchiseptica* vaccines commonly leads to positive molecular diagnostic assay results for a short time period after primary vaccination. These findings should be considered when assessing the results of these assays in client-owned puppies with respiratory disease.

ABSTRACT ID-21

ABILITY OF A FRET-PCR ASSAY TARGETING *GYRA* TO DETECT FLUOROQUINOLONE RESISTANCE IN VETERINARY CLINICAL URINE SAMPLES CONTAINING *ESCHERICHIA COLI*.

Megan G. Behringer[1], Kamoltip Thungrat[1], Bashar W. Shaheen[2], and Dawn M. Boothe[1]. [1]Department of Anatomy Physiology and Pharmacology, College of Veterinary Medicine, Auburn University, AL. [2]U.S. Food and Drug Administration, National Center for Toxicological Research, Division of Microbiology, Jefferson, AR.

Antimicrobial resistance in *Escherichia coli* is an increasing concern in both human and veterinary hospitals' patients. The choice drug for treatment in dogs is enrofloxacin, a second generation fluoroquinolone (FQ) whose activity reflects, in part, ciprofloxacin. Among the difficulties in effective *E. coli* treatment is rapid detection of FQ resistance. The purpose of this study was to determine the specificity and sensitivity of a FRET based assay for the rapid detection of urinary tract infections caused by FQ associated multi-drug resistant *E.coli*. 306 clinical *E. coli* isolated from canine urine and 120 clinical veterinary urine samples being examined for *E. coli* were subjected to susceptibility testing for 14 drugs representing 6 drug classes. Pure isolates were designated NDR (no drug resistance), SDR (single drug resistance) and MDR (multi-drug resistant) (n = 101 MDR, 116 SDR and 89 NDR). Minimum inhibitory concentration (MIC) for enrofloxacin ranged from 0.03 µg/ml to 512 µg/ml, with high MIC generally associated with MDR. Extracted DNA from culture and from urine were subjected to FRET-PCR targeting single nucleotide polymorphisms in *gyrA*. The resulting product was sequenced to detect other polymorphisms. Further, to determine the level of detection, microbial free canine urine was inoculated with 10^6 to 10^1 CFU/ml of 7 isolates characterized by variable susceptibility to enrofloxacin (MIC_{Enro} = 0.03, 0.06, 0.15, 1, 64, 128, 256 µg/ml). Of 306 pure isolates, 50 were confirmed positive for enrofloxacin resistance ($MIC_{Enro} > 4$ µg/ml), 43 of which were positively identified by the FRET-PCR assay giving a sensitivity of 86.00%. Only 1 isolate that was resistant was not detected (specificity of 96.66%). However, of the isolates expressing high level resistance (MIC > 8 X breakpoint [64 mcg/ml]), and MDR (n = 34), sensitivity = 97.06%. Of the 120 urine samples 27 contained *E. coli* 7 of which determined to be FQ-resistant by the assay. Colony dilutions of *E. coli* confirmed the assay able to detect enrofloxacin resistance at as low as 101 CFU/ml. The relationship between CFUs and the peak of the –(d/dt) fluorescence of the melting curve was R2 = 0.988. These results conclude that the assay is capable of detecting not only the presence of Escherichia coli in clinical samples, but also detecting severity of fluroquinolone resistance and infection.

ABSTRACT ID-22

COMPARATIVE IN VITRO ACTIVITIES OF FLUOROQUINOLONE ANTIMICROBIALS AGAINST *ESCHERICHIA COLI* UROPATHOGENS.

X.Q. Liu Dawn M.. Boothe Department of Anatomy, Physiology and Pharmacology, College of Veterinary Medicine, Auburn University, Auburn, AL.

The fluoroquinolones (FQs) are a key class of synthetic antimicrobial agents with an established history in both humans and companion animals of efficacy for treatment of urinary tract infections (UTIs) caused by *E. coli*, and fluoroquinolones are common therapy. Among the commonly used FQs in dogs and cats are the 2nd generation drugs, enrofloxacin, marbofloxacin, orbifloxacin (all veterinary approved) and the human drug ciprofloxacin; no 3rd and 4th generation FQ is routinely used. The purpose of this study was to assess the *in vitro* activity of different generation FQs toward *E.coli* uropathogens whose phenotype ranges from no resistance to multidrug resistance. A total number of 51 canine uropathogenic canine or feline *E.coli* isolates had been subjected to susceptibility testing to 6 drugs classes (15 drugs) and phenotyped as to resistance: none (NDR, n = 12), single (SDR, n = 15), or multiple, MDR (resistance to 2–6 drug classes; n = 24). MDR included isolates susceptible (ENR^S-MDR, n = 12) or resistant (ENR^R-MDR) to enrofloxacin. The minimum inhibition concentrations (MICs) for 11 quinolones (1–1st generation, 3-2nd generation, 4–3rd generation and 3–4th generation) were determined for these isolates using broth microdilution methods according to CLSI guidelines (*E. coli* ATCC® 25922 served as a negative control). MIC statistics were generated for each drug among phenotypes. The results showed that companion animal *E. coli* expressing NDR or SDR are largely susceptible to 2nd to 4th generation FQs. However, isolates expressing resistance to 1st or 2nd generation quinolone also express high level resistance based on the MIC $_{90}$ to 3rd and 4th generation FQs. The overall potency (MIC) for the 11 drugs for isolates not expressing ENR resistance (that is, NDR, SDR and ENR^S-MDR) is GAT > MOX, MAR, CIP, LEV, PRA > SAR, ORB, OFL > ENR > NAL. Based on ratio MIC_{90} to susceptible breakpoint MIC with a mean MIC ± SD, the anticipated efficacy for all of isolates would be NAL(4.93 ± 5.13), ENR(30.32 ± 37.47), OFL(52.89 ± 82.47), CIP(128.5 ± 117.5), MOX(201.4 ± 228) ,LEV(180 ± 166.8), GAT(167.6 ± 160.2), PRA(122.5 ± 131.8), MAR(86.47 ± 74.19), SAR(73.54 ± 84.33), ORB(31.83 ± 31,40).

ABSTRACT ID-23

CANINE LEPROID GRANULOMA (CLG) IN SÃO PAULO: A RETROSPECTIVE STUDY OF 37 CASES (1990 TO 2009).

S Maruyama[1], GS Dias[1], M Otsuka[1], CE Larsson Junior[1], R Lucas[1], NS Michalany[2], CE Larsson[1]. [1]School of Veterinary Medicine and Animal Science, University of São Paulo, SP, Brazil. [2]School of Medicine, Federal University of São Paulo, SP, Brasil.

Canine leproid granuloma (CLG) was first reported in Brazil in 1990. Over the past 20 years, 37 cases of CLG were diagnosed in São Paulo, Brazil, and clinical and epidemiological findings were similar to those reported in Australia. All dogs presented with one or more, uni or bilateral, ulcerated or not, papular, nodular or tumoral lesions, mainly observed in the dorsal surface of the ear, site usually more affected. In general, the lesions are painless and confined to the subcutis and skin, and it does not involve regional lymph nodes, nerves or internal organs, and systemic clinical signs frequently are absent. Short-coated breeds show a marked predisposition for this disease.

The definitive diagnosis of CLG was obtained by histological examination of skin biopsies that were stained with acid fast (Ziehl-Neelsen) and DiffQuik®.

Thirty one (83.7%) of the dogs were purebred; in this study the breed pattern comprised 19 (61.3%) Boxers, 2 (6.5%) German Shepherd and Labrador Retriever, 1 (3.2%) Dobermann, 1 (3.2%) Brazilian Terrier, 1 (3.2%) Golden Retriever, 1 (3.2%) Bulldog, 1 (2.8%) American Pitbull, 1 (3.2%) Mastiff, 1 (3.2%) Fila Brasileiro and 1 (3.2%) Cocker spaniel, 6 (16.3%) were of unknown breed. Nineteen (51.4%) of the thirty seven dogs were males. Twenty (54.1%) dogs were 4–6 years old. In most cases, dogs presented with unilateral or bilateral ear lesions, but rarely thoracic, foot and caudal lesions. The animals were successfully treated by use of rifampicin orally ("the brazilian protocol") or enrofloxacin orally and topical rifamicin.

ABSTRACT ID-24

AMPLIFICATION OF *ANAPLASMA PHAGOCYTOPHILUM* DNA FROM THE BLOOD AND SYNOVIAL FLUID OF EXPERIMENTALLY INFECTED DOGS.

MH Kim[1], S Moroff[2], I Piperisova[2], MR Lappin[1]. [1]Colorado State University, Fort Collins, CO. [2]Antech Diagnostics, Lake Success, NY.

Anaplasma phagocytophilum is being recognized more frequently in dogs in endemic areas. Currently, most suspected cases are evaluated for *A. phagocytophilum* antibodies by immunofluorescence assay (IFA) or ELISA. Since *A. phagocytophilum* is an acute disease, detection by antibody measurement may be negative on initial evaluation. It is possible that *A. phagocytophilum* DNA can be amplified from blood or synovial fluid prior to seroconversion.

Wild caught *Ixodes scapularis* adult ticks from Rhode Island were allowed to feed on 18 young adult (2–3 years), mixed sex beagles for up to 7 days. Blood (weekly for 12 weeks), serum (weekly for 12 weeks), and synovial fluid (radiocarpal joint; alternating arthrocentesis weekly for 6 weeks) were collected prior to tick attachment and then weekly after tick attachment. Joint fluid cytology was performed and total DNA was extracted from blood and synovial fluid and assayed in a proprietary real time PCR assay (FastPanel™) that amplifies the DNA of *Anaplasma phagocytophilum, A. platys, Ehrlichia canis, E. chaffeensis*, and *E. ewingii*. Serum was assayed for *A. phagocytophilum* antibodies by IFA. Time to first positive results for serology and PCR were compared by paired Student's t test.

None of the beagles developed clinical evidence of disease, and no major changes in synovial fluid cytology were detected over time. Of the 18 beagles, 15 were positive for *A. phagocytophilum* DNA in blood or synovial fluid or IFA antibodies in at least one sample after tick attachment. Antibody titers appeared in 14 of 15 dogs from weeks 2 to 12 (median to 1st positive = 3 weeks ± 1). Titer magnitude ranged from 1:40 to 1:10,240. *Anaplasma phagocytophilum* DNA was amplified from the blood of 14 of 15 dogs with positive test results ranging from 1 to 12 weeks (median to 1st positive = 2 weeks ± 1.5). *Anaplasma phagocytophilum* DNA was amplified from synovial fluid from 8 of 15 dogs between weeks 1 to 4 (median to 1st positive = 4 weeks ± 1). Of the 8 dogs, 7 were PCR positive for only one week and 1 dog was PCR positive for two consecutive weeks. Of the 15 dogs, 13 were positive for *A. phagocytophilum* in both blood and joints by DNA analysis. *Anaplasma phagocytophilum* DNA was amplified from blood more quickly than seroconversion was detected by IFA antibody titer ($t = -3.4$, $P < 0.01$) or DNA was amplified from synovial fluid ($t = 5.4$, $P < 0.001$).

Anaplasma phagocytophilum DNA can be amplified from the blood prior to development of detectable antibody titers by IFA. Amplification of *A. phagocytophilum* DNA from synovial fluid does not occur in all dogs, appears to be transient in most dogs, and a negative test result does not preclude a diagnosis of *A. phagocytophilum* infection.

ABSTRACT ID-25

CANINE GRANULOCYTIC ANAPLASMOSIS AND GRANULOCYTIC EHRLICHIOSIS – A FIELD-BASED COMPARISON.

MJ Beall[1], B Andrews[2], M Eberts[3], J Sinsabaugh[1], PI Tyrrell[1], R Chandrashekar[1], SE Little[4], EB Breitschwerdt[5]. [1]IDEXX Laboratories, Westbrook, ME. [2]Town and Country Veterinary Hospital, Green Forest, AR. [3]Lakeland Veterinary Hospital, Baxter, MN. [4]Oklahoma State University, Stillwater, OK. [5]North Carolina State University, Raleigh, NC.

Canine granulocytic anaplasmosis and granulocytic ehrlichiosis are tick-transmitted infections caused by *Anaplasma phagocytophilum* (*Aph*) and *Ehrlichia ewingii* (*Eew*), respectively. Both organisms induce an acute clinical disease, frequently accompanied by fever, polyarthropathy and thrombocytopenia. However, *Aph* and *Eew* have different tick vectors, i.e. *Ixodes scapularis* and *Amblyomma americanum*, respectively, with different, but overlapping geographic distributions. In addition, infection outcome may be affected by other regional tick-transmitted pathogens, such as *Borrelia burgdorferi* (MN) or *Ehrlichia chaffeensis* (AR). Therefore, we compared serology and PCR results derived from dogs examined at two private practices located in highly endemic areas for either *Aph* or *Eew*. Serum collected between April-December, 2005 from Minnesota dogs (n = 420) was tested by SNAP® 4Dx® and whole blood was tested by *Aph* PCR. Serum collected from Arkansas dogs (n = 708) for 1 year beginning in August 2009 was tested using microtiter plate ELISAs for antibodies to *Eew*, *E. canis*, and *E. chaffeensis* (*Ech*) while whole blood was tested by *Ehrlichia* PCR. Comparisons were evaluated using Chi Square (*) and Binomial (†) tests with an alpha of 5%.

Table [Abstract ID-25]

Group	Seroprevalence		Evidence of co-exposure	PCR positive	If PCR+, then % seropositive	If seropositive, then % PCR+
MN	*Aph*	*B. burgdorferi*	*Aph/Bb*	*Aph*	*Aph*	*Aph*
(n = 420)	268 (64%)*	162 (39%)	129 (31%)*	35 (8%)†	24 (69%)	9%*
AR	*Eew*	*E. chaffeensis*	*Eew/Ech*	*Eew*	*Eew*	*Eew*
(n = 708)	253 (36%)	80 (11%)	62 (9%)	102 (14%)	93 (91%)	37%

The above results indicated that dogs are frequently exposed to both *Aph* and *Bb* in MN, whereas AR dogs are often exposed to *Eew*, but less frequently to *Ech*. Antibodies to *E. canis* peptides were found infrequently in both MN and AR with only 10 seroreactive dogs detected in both locations. Active *Eew* infection, as determined by PCR, was four times more frequent in AR pet dog seroreactors as compared to active *Aph* infections among *Aph* seroreactors. Although both organisms induce acute disease, the number of *Aph* and *Eew* PCR positive dogs that were also seropositive was relatively high suggesting that both organisms induce persistent infections or that dogs are frequently re-infected, despite the presence of a measurable humoral immune response. Additional studies are needed to determine regional infection profiles in other areas that are endemic for these pathogens.

ABSTRACT ID-26

COMPARISON OF COMMERICALLY AVAILABLE PCR ASSAYS FOR THE AMPLIFICATION OF *EHRLICHIA CANIS* AND *ANAPLASMA PHAGOCYTOPHILUM* DNA FROM THE BLOOD OF NATURALLY INFECTED DOGS.

J Ficociello[1], S Moroff[2], M Brewer[1], MR Lappin[1]. [1]Colorado State University, Fort Collins, CO. [2]Antech Diagnostics, Lake Success, NY.

Anaplasma phagocytophilum and *Ehrlichia canis* are two of the most common vector borne disease agents that infect dogs and cats. While PCR assays that amplify the DNA of these agents from blood are currently available, there is minimal information concerning the performance of these assays in different commercial laboratories that utilize different techniques. The purpose of this study was to compare the *E. canis* and *A. phagocytophilum* results of two different laboratories on the same samples collected from client-owned animals.

Veterinarians in 3 states (AZ, MD, CT) were recruited to participate in the study based on high prevalence rates for *E. canis* or *A. phagocytophilum* infection. Blood in EDTA was collected from dogs or cats with fever, thrombocytopenia, or clinical evidence of polyarthritis and an equal volume of the same blood sample was simultaneously shipped on cold packs by overnight express to Colorado State and to Antech Diagnostics. Standard operating procedures at each laboratory were followed for total DNA extraction and amplification of GAPDH as the DNA control. At Colorado State University, a previously published PCR assay that amplifies the DNA of *Ehrlichia* spp., *Anaplasma* spp., *Neorickettsia* spp., and *Wolbachia* was performed on each sample with positive amplicons sequenced to determine the species. At Antech Diagnostics, a proprietary real time PCR assay (FastPanel™) that amplifies the DNA of *Anaplasma phagocytophilum*, *A. platys*, *Ehrlichia canis*, *E. chaffeensis*, and *E. ewingii* was performed.

In the study to date, samples from 35 animals (30 dogs and 5 cats) have been assayed at both laboratories. DNA of *A. phagocytophilum* (2 cats and 2 dogs) and *E. canis* (1 dog) were amplified at both laboratories with a percentage agreement between laboratories of 100%.

The results to date suggest that the assay results of the two laboratories for *A. phagocytophilum* and *E. canis* are comparable.

ABSTRACT ID-27

PREVALENCE OF TICK-BORNE DISEASES IN THE HIGHLANDS OF PERU. PPVP Diniz[1], B Morton[1], M Kachani[1], T Scott[2], EA Barron[3], CM Gavidia[3]. [1]College of Veterinary Medicine, Western University of Health Sciences, Pomona, CA. [2]Master of Public Health, University of Minnesota. [3]Facultad de Medicina Veterinaria, Universidad Nacional Mayor de San Marcos, Lima, Peru.

Ehrlichiosis and Bartonellosis are zoonotic diseases caused by extremely small, obligate intracellular bacteria that require a mammalian reservoir and a blood sucking arthropod vector. Human ehrlichiosis is present in Peru, with a seroprevalence as high as 23% in the highlands. *Bartonella* species in humans were also identified in Peru since 1905 (*B. bacilliformis*). Recently, a new species (*B. rochalimae*) was isolated from an American woman who became febrile after travelling to Peru. Dogs can become infected with the same *Ehrlichia* species, and the majority of *Bartonella* species that affect human beings. The role of dogs as reservoirs for human infections has not been clearly established, but exposure and/or infection in dogs has been used to monitor human exposure to tick-borne disease (TBD), since they share the same environment. The objective of this study was to determine the serological and molecular prevalence of Anaplasmosis, Ehrlichiosis and Bartonellosis in rural dogs in the highlands of Peru.

A total of 122 healthy adult dogs were enrolled in this study from four communities in the central highlands of Peru: Ondores, Pachacayo, San Juan de Pachayo, and Canchayllo. EDTA-blood samples were collected from 108 dogs, whereas serum samples were available from 110 dogs. Serum samples were tested for Ehrlichia canis, Anaplasma, Borrelia burgdorferi and Dirofilaria immitis infections using a qualitative dot-ELISA (Snap® 4Dx). The EDTA-blood samples were screened by conventional PCR for the groEL gene of the genus *Anaplasma* and *Ehrlichia*, and for the intergenic transcribed spacer of the genus *Bartonella*. Speciation was conducted by nucleotide sequencing.

Bartonella genus DNA was detected from seven of the 108 dogs (6.5%) and *Ehrlichia canis* DNA was detected and sequenced from one dog (0.9%). Four of the *Bartonella* positive samples were identified by DNA sequencing as *B. rochalimae* (GenBank accession numbers HQ185696 and HQ185695). The other three *Bartonella* positive samples were identified as *B. vinsonii* subspecies *berkhoffii*, the causative agent of endocarditis in dogs and humans. No dog was infected with *Anaplasma* species by DNA amplification, but one dog was seroreactive for this genus (0.9%). No specific antibodies against *Ehrlichia canis* and *Borrelia burgdorferi* and no antigens of *Dirofilaria immitis* were detected.

This study expands the current knowledge about TBD in Peru and describes for the first time the infection of *B. rochalimae* in dogs in Peru. The results suggest that dogs may play an important role in the epidemiology of this infection in humans, since they can be asymptomatic but bacteremic.

ABSTRACT ID-28

PREVELANCE OF *BARTONELLA* SPP. DNA IN BLOOD OF NATURALLY EXPOSED CATS WITH AND WITHOUT IMIDOCLOPRID TREATMENT. D Mawby[1], A Stone[2], Julia Wuerz[2], JC Whittemore[1], D Zoran[3], JR Hawley[4], MR Lappin[4]. [1]University of Tennessee, Knoxville TN. [2]University of Florida, Gainesville FL. [3]Texas A & M, College Station, TX. [4]Colorado State University, Fort Collins, CO.

Bartonella spp. DNA is commonly amplified from the blood of cats exposed to *Ctenocephalides felis*. In previous work, it was shown that cats administered imidacloprid and experimentally exposed to *B. henselae* infected cats and *C. felis* did not become PCR positive for *B. henselae* whereas untreated cats all developed infection. The purpose of this study was to determine if administration of imidacloprid to client-owned cats likely to be exposed to *Bartonella* spp. and *C. felis* in the field lessens prevalence of *Bartonella* spp. infection.

Veterinary students in Tennessee and Florida that owned cats that spent at least 20 days per month outside and that were willing to apply imidacloprid to their cats monthly for six months were recruited for the study. Blood for *Bartonella* spp. PCR assay was collected from the cats seven months after starting imidacloprid administration and assayed at Colorado State University. To serve as a control group that was unlikely to have been administered flea control products in the previous 6 months, blood was collected from feral cats during TNR programs in each of the two cities and assayed for *Bartonella* spp. DNA. The *Bartonella* spp. DNA prevalence rates between the groups were compared by chi square analysis with significance defined as $P < 0.05$.

The overall prevalence rates for *Bartonella* spp. DNA in the blood of veterinary student cats (7.4%) and the feral cats (39.1%) were significantly different ($P < 0.0001$). The distribution of results is shown in Table 1.

Table [Abstract ID-28]

	Feral cats	Student cats	All cats	P value (feral vs student cats)
Tennessee	3 of 31 (9.7%)	3 of 45 (6.8%)	6 of 76 (7.9%)	P = 1
Florida	22 of 33 (66.7%)	2 of 23 (8.7%)	24 of 56 (42.9%)	P < 0.0001
All cats	25 of 64 (39.1%)	5 of 68 (7.4%)	30 of 132 (22.7%)	P < 0.0001

The results suggest that Florida feral cats were more commonly exposed to *C. felis* than Tennessee feral cats. While the cats in the groups were not exactly matched, the student cats were allowed outdoors for approximately 20 days per month and lived in the same cities as the feral cats, so *C. felis* exposure rates were likely similar. As previously shown in experimentally-exposed cats, the use of imidacloprid monthly may influence transmission rates of *Bartonella* spp. amongst naturally-exposed cats.

ABSTRACT ID-29

SEVERITY OF CLINICAL SIGNS AND OUTCOME OF DOGS INFECTED BY *WOLBACHIA*, MICROFILARIA AND *LEISHMANIA*. MD. Tabar[1], L. Altet[2], V. Martínez[2], X. Roura[3]. [1]Centro Policlinico Veterinario Raspeig, San Vicente-Alicante (Spain), [2]Molecular Genetics Veterinary Service and [3]Hospital Clinic Veterinari, Universitat Autonoma de Barcelona (Spain).

In an endemic area for leishmaniosis and filariosis, coinfection can occur and immunomodulation produced by *Wolbachia* might influence the clinical signs and progression of both diseases. The aims of the present study were 1) to determine the prevalence of *Wolbachia* in dogs infected with *Dirofilaria immitis (Di)* and other filarial nematodes, 2) to evaluate the level of coinfection of leishmaniosis and filariosis by molecular assays and 3) to evaluate any associations between *Leishmania infantum (Li)* infection, filariosis with or without *Wolbachia* and clinical presentation and outcome. Statistical differences between groups were tested for significance by the Fisher exact test using SPSS v.14.0 software (significance: p-value < 0.05).

One-hundred and eighteen owned dogs from Southeastern Spain presenting for clinical evaluation were included in the study. Criteria for inclusion were infection with *Li* (defined as a high titer of antibodies plus positive results of qPCR and/or *Li* visualization in bone marrow or skin biopsy) and/or filariosis (defined by microfilaremia detected by direct wet smear, plus phosphatase acid stain and/or *Di* antigen determination in some cases). With those criteria, 98 dogs infected with *Li* and 32 microfilaremic dogs were included, and 12 dogs fulfilled both criteria. EDTA-blood samples underwent DNA extraction, and PCR assays were performed to detect *Filaria spp.*, *Wolbachia* and *Li*. The study included 69 males and 49 females (ranging in age from 1 to 14 years) and 36 breeds, the most frequent being the Mongrel (n = 22), German Shepherd (n = 11), Boxer (n = 9) and Cocker (n = 8). Once molecular assays were performed, 98 dogs were infected with *Li*, 49 were infected with filariosis and 29 were coinfected with both of them. In other words, 29.6% (29/98) of dogs infected with *Li* were PCR positive for *Filaria* spp and 59.2% (29/49) of dogs with filariosis were *Li* PCR positive. *Wolbachia* DNA was detected in 30.6% of *Filaria spp* positive dogs (15/49). In dogs infected with filariosis, clinical class of heartworm disease (HWD) before treatment was more severe if they were coinfected with *Li* (p = 0.211) and *Wolbachia* infection was more frequent in dogs not infected with *Li* (p = 0.026). However, filarial infection did not influence disease classification of *Li* infected dogs. There was no relation between outcome and coinfection with *Li* and *Filaria* or presence of *Wolbachia*.

The results of this study highlight the increased sensitivity of PCR for diagnosis of filariosis, confirm the presence of *Wolbachia* in dogs from the Mediterranean basin, show the increased severity of HWD when *Li-Filaria* coinfection is present and suggest that *Wolbachia* could play a protective role for leishmaniosis. *Wolbachia* antigens can stimulate a Th1-type immune response, as has been previously described. However other factors (as treatment with doxycicline) might be responsible for the lower prevalence of *Wolbachia* among filaremic dogs infected with *Li* and further studies must be done to clarify this interaction.

ABSTRACT ID-30

FELINE LEISHMANIASIS IN THE MUNICIPALITY OF ARAÇATUBA, SÃO PAULO, BRAZIL. CN Rossi[1], CM Nunes[1], ESMM Savani[2], VMF Lima[1], MD Laurenti[3], M Marcondes[1]. [1]São Paulo State University, Araçatuba, SP, Brasil. [2]Zoonosis Control Center of São Paulo, SP, Brasil. [3]São Paulo University, São Paulo, SP, Brasil.

The purpose of the present study was investigate the occurrence of leishmaniasis in 200 cats in the municipality of Araçatuba, São Paulo, Brazil, an endemic area for canine visceral leishmaniasis. Animals were evaluated by direct parasitological examination of lymphoid organs and serology for visceral leishmaniosis by immunosorbent assay (ELISA) and indirect immunofluorescence (IFAT). Thirteen (6.5%) out of 200 cats studied were diagnosed with visceral leishmaniasis; eight (4%) by parasitological diagnosis through cytological examination of lymphoid organs, six (3%) were considered positive by ELISA and one (0.5%) by IFAT. Only two (15.38%) out of the thirteen infected cats had clinical signs, characterized by the presence of crusty lesions on the dorsal cervical region and hepatosplenomegaly. Regarding age five cats (38.5%) had between six months and two years, being the others older than 2 years (61.5%). Only one cat (7.7%) was positive for the three employed methods. PCR confirmed *Leishmania sp* infection in nine (69.2%) cats, of which six were diagnosed previously by cytological examination, two by ELISA and one by the three techniques employed.

ABSTRACT ID-31

***LEISHMANIA CHAGASI* INFECTION IN CATS WITH DERMATOLOGIC LESIONS FROM AN ENDEMIC AREA OF VISCERAL LEISHMANIOSIS IN BRAZIL.** JP Vides[1], TF Schwardt, LSV Sobrinho[1], MD Laurenti[2], C Leutenegger[3], M Marcondes[1]. [1]São Paulo State University, Araçatuba, SP, Brasil. [2]São Paulo University, São Paulo, SP, Brasil. [3]IDEXX Laboratories Inc., West Sacramento, CA, USA.

Since its first description in 1912 feline leishmaniosis has been reported in several countries. The purpose of this study was to assess the prevalence of *Leishmania chagasi* infection in cats showing dermatologic lesions from an endemic area for visceral leishmaniasis in Brazil. Animals were evaluated by direct parasitological examination of lymphoid organs, immunohistochemical technique for detection of amastigotes in lesioned skin and serology for visceral leishmaniosis by immunosorbent assay (ELISA) and indirect immunofluorescence (IFAT). Twenty seven (49.1%) out of the 55 cats studied were diagnosed with visceral leishmaniosis. Twelve (44.4%) were positive by parasitological diagnosis; amastigote forms of *Leishmania* sp were identified in lymphoid organs from 10/55 (18.2%) infected cats, and immunohistochemical technique allowed the identification of nine (16.4%) positive animals. The seroprevalence of leishmaniosis was 25.4% (14/55) by ELISA and 10.9% (6/55) by IFAT. FIV specific antibodies were found in 6/55 cats (10.9%), of which 5/6 (83.3%) had leishmaniosis. Real time PCR confirmed *Leishmania chagasi* infection in three cats. Based on the evidence of the high occurrence of leishmaniosis in cats in this study, this disease should be included in the differential diagnosis of skin diseases of felines living in endemic areas.

ABSTRACT ID-32

COMPARISON OF RADIOGRAPHIC AND URINE ANTIGEN RESOLUTION OF CLINICAL BLASTOMYCOSIS. G Olsen, K Deitz, K Miles. Iowa State University College of Veterinary Medicine, Ames, IA.

Blastomyces dermatiditis is a dimorphic fungus that commonly affects large-breed hunting dogs. A recent advancement in diagnosis has come with the advent of a urine antigen screening test that has both high sensitivity and moderately high specificity. Therapy for the disease involves use of antifungal agents, usually itraconazole, and length of treatment is based chiefly on resolution of clinical and radiologic signs. With the new urine antigen test, however, a non-invasive route of monitoring treatment progress is available and could be an adjunct device utilized to determine treatment efficacy and may even reveal a need for prolonged treatment. Therefore, the purpose of this study was to determine if monitoring the *Blastomyces* urine antigen test and comparing to pulmonary radiographic signs would elucidate the necessity for prolonged antifungal therapy, even after resolution of radiologic signs.

To this end, a retrospective case review was performed that identified a series of client-owned animals with naturally occurring blastomycosis. The inclusion criteria were radiographic pulmonary parenchymal signs consistent with fungal disease and urine antigen-confirmed blastomycosis with repeated testing of both radiographs and urine antigen quantification as monitoring parameters until negative results achieved in each. Ideally, intervals between testing dates would be between two and five months. Radiographs were considered negative if all radiographic changes had resolved or if repeated radiographs separated by at least one month were considered static after documented improvement had occurred from original diagnostic radiographs (suspected scarring). Urine antigen testing was considered negative if concentrations were less than 1.0 enzyme immunoassay units, a reference interval set by the testing laboratory.

Preliminary data analysis reveals resolution of radiographic signs of blastomycosis occurred earlier in many of the cases presented than did attaining a negative urine antigen concentration. Ceasing treatment 1 month after radiographic resolution of signs as has been recommended in the past might have resulted in premature discontinuation of therapy in many of the cases.

Monitoring of urine antigen concentrations may be of additional clinical use for determining when cessation of treatment should occur in cases of blastomycosis. Persistent elevation of urine antigen concentrations after radiographic resolution of infection may account for apparent recrudescence of blastomycosis after suspected clinical resolution.

ABSTRACT ID-33

PREVALENCE OF *GIARDIA* SPP. AND *CRYPTOSPORIDIUM* SPP. IN DOGS OF THE UNITED STATES. AV Scorza, MR Lappin. Department of Clinical Sciences, Colorado State University, Ft.Collins, CO.

Giardia spp. and *Cryptosporidium* spp. are both known to cause infections in dogs and humans in the United States. Nevertheless, prevalence rates for dual infection in dogs had not been widely reported.

In this study, fecal samples from dogs housed in a northern Colorado animal shelter (n = 121), dogs owned by veterinary students in northern Colorado (n = 132), and dogs from the Pine Ridge reservation in South Dakota (n = 84) were collected. Each sample was assayed with a commercially available fluorescent antibody assay that detects *Giardia* spp. cysts and *Cryptosporidium* spp. oocysts. Those samples that were positive for *Giardia* spp. or *Cryptosporidium* spp. with adequate DNA available for sequencing were genotyped by the glutamate dehydrogenase [gdh] and by the heat shock protein-70 [HSP-70] genes, respectively. Overall, 45 (13.3%) of the dogs had current evidence of a protozoal infection (Table 1). The dogs from Pine Ridge reservation had the highest prevalence rates for *Giardia* infection and also for dual infections.

Table [Abstract ID-33]: Prevalence of *Giardia* spp. and/or *Cryptosporidium* spp. in dogs

	Giardia alone	*Cryptosporidium* alone	Dual infection	Either infection
Student (n = 132)	3 (2.7%)	0 (0%)	2 (1.5%)	5 (3.8%)
Shelter (n = 121)	6 (5.0%)	2 (1.7%)	4 (3.3%)	12 (9.9%)
Reservation (n = 85)	23 (27%)	2 (2.3%)	3 (3.5%)	28 (32.9%)
Total (n = 338)	32 (9.4%)	4 (1.1%)	9 (2.6%)	45 (13.3%)

From the student dogs, sequencing was successful for the three *Giardia* isolates (assemblage D from 2 dogs; assemblage C from one dog) and one *Cryptosporidium* isolate (*C. canis*). From the reservation dogs, sequencing was successful for nine *Giardia* isolates (assemblage D from 4 dogs; assemblage C from 5 dogs) and one *Cryptosporidium* isolate (*C. canis*).

Cryptosporidium and *Giardia* co-infections are commonly detected in dogs; in this study dual infections were more common than *Cryptosporidium* infections alone. Further studies will be required to determine the clinical importance of this finding. Although the *Giardia* and *Cryptosporidium* isolates that were sequenced were the dog specific assemblages/genotypes, more samples should be analyzed in order to assess the potential for zoonotic transmission of either parasite.

ABSTRACT ID-34

INTESTINAL PARASITES OF DOGS IN CHIANG MAI, THAILAND.

S. Tangtrongsup, AV Scorza, LR Ballweber, JS Reif, MR Lappin, MD Salman. Colorado State University, Fort Collins, CO.

The current study was conducted to determine the prevalence of intestinal parasites in dogs visiting the Veterinary Teaching Hospital, Chiang Mai University, Northern Thailand. Fecal samples (n = 301) were collected and submitted by owners between August 2009 to February 2010. Demographic and geographic data were recorded. Intestinal parasitic infection was diagnosed by both microscopic examination after zinc sulfate centrifugation flotation and commercially available IFA for *Giardia* spp. and *Cryptosporidium* spp. Polymerase chain reaction and DNA sequencing were performed on all *Giardia* and *Cryptosporidium* positive samples to provide genotyic information.

Overall prevalence of intestinal parasitic infection in dogs in Chiang Mai was 38.9%. The most prevalent parasite was *Giardia* spp. (24.9%) followed by *Ancylostoma* spp. (12.0%), *Cryptosporidium* spp. (7.6%), *Cystoisospora* spp. (6.0%), *Toxocara canis* (2.7%), *Trichuris vulpis* (2.0%), coccidian-like (1.7%), *Toxascaris leonina* (0.7%), and *Strongyloides* spp. (0.7%). The prevalence of having at least one parasite in dogs < 1 year, 1–7 years, and > 7 years were 49.3%, 39.8%, and 27.4%, respectively. Of these infected dogs, 59.0%, 34.2%, 5.1%, and 1.7% were infected with one, two, three, and four organisms, respectively. Available DNA sequences from *Giardia* spp. positive samples were shown to be dog specific. Only one adequate DNA sequence was available for *Cryptosporidium* spp., which was shown to be *C. canis*.

The findings suggested that intestinal parasitic infection was common in dogs in Chiang Mai, Thailand. Dogs could be potential source for zoonotic intestinal parasitic infection since dogs in this area are allowed for free roaming. Regular deworming program is indicated to prevent not only transmission among dogs but also to human.

ABSTRACT ID-35

COMPARISON OF 2 METHODS OF PARASITE RECOVERY FROM FECAL SPECIMENS OF VETERINARY PATIENTS: SEDIMENTATION/CONCENTRATION VS. FECALYZER.

D. Martin, Z. Song, J. Goring, and J. Yang. IDEXX Reference Laboratories, Markham, ON.

A retrospective study was conducted on 135 parasite positive fecal specimens consisting of 90 canine, 29 feline, 14 equine and 2 from other host species, comparing recovery of eggs, protozoan cysts and coccidian oocysts using 2 standardized methods of parasite concentration: the formalin/ethyl acetate (F/EA) sedimentation concentration and the commercial Fecalyzer (flotation) kit procedures.

Specimens were processed by each technique either according to manufacturer's instructions or according to standard laboratory procedures. Formalin/ethyl acetate concentrations used at a ratio of 10 ml normal saline to 4 ml ethyl acetate for extraction of lipophilic material from pelleted stool samples, previously fixed in sodium acetate/acetic/acid/formalin (SAF) solution. Flotations with the Fecalyzer kit were performed with concentrated zinc sulfate solution (s.g. 1.22).

The range of parasites recovered from these specimens included flagellate cysts (40 total), coccidian oocysts (28 total), ova and larvae of nematodes (80 total), and ova of trematodes (12 total), and cestodes (16 total).

Recovery rates by Fecalyzer flotation were good for protozoan cysts, coccidian oocysts and nematode eggs and larvae, but very poor for cestode and trematode eggs. Formalin/ethyl acetate concentration showed excellent recovery of all parasites and consistently outperformed Fecalyzer in recovery rates. Recoveries by F/EA concentrations were higher by 27.5% for *Giardia*, by 10.7% for coccidia and by 10.0% for nematode eggs and larvae. With the exception of coccidian oocysts, based on Z-test analyses, recovery rates were significantly higher, at a confidence level of at least 95%, for all parasites, using formalin/ethyl acetate sedimentation concentration.

Although CAPC recommends the use of flotation with centrifugation methods for standard fecal ova and parasite examination for veterinary patients, sedimentation concentration methods are widely and effectively used in human diagnostic parasitology laboratories. These results provide good evidence for the use of F/EA concentration as a preferred method to flotation procedures for stool ova and parasite examinations in veterinary laboratories.

ABSTRACT IM-1

EFFECTS OF CYCLOSPORINE AND DEXAMETHASONE ON CANINE T-CELL EXPRESSION OF IL-2 AND IFN-γ AS MEASURED BY FLOW CYTOMETRY AND QUANTITATIVE RT-PCR.

C Fellman, R Flores, J Stokes, T Archer, K Lunsford, A Mackin. Mississippi State University College of Veterinary Medicine, Starkville, MS.

Cyclosporine and glucocorticoids are powerful immunosuppressive agents used to treat many inflammatory diseases. Cyclosporine inhibits calcineurin-dependent pathways of T-cell activation and the resultant cytokine production, and glucocorticoids directly inhibit genes coding for cytokines. Little work has been done comparing the effects of these agents on cytokine production in dogs. Our study assessed these effects by measuring T-cell cytokine production using flow cytometry, and cytokine gene expression using quantitative reverse transcriptase polymerase chain reaction (qRT-PCR) in activated canine T-cells treated with cyclosporine and dexamethasone.

For flow cytometric assays, peripheral blood mononuclear cells were separated using density gradients and cultured for 12 hours in the presence of cyclosporine (5, 25, or 100 ng/mL), dexamethasone (10^{-7}, 10^{-6}, 10^{-5} M), or cyclosporine plus dexamethasone. For qRT-PCR, whole blood was cultured for 5 hours with the same drugs at the same concentrations, and RNA was then extracted from leukocytes. Expression of cytokines IL-2 and IFN-γ was analyzed in PMA/ionomycin-activated T-cells by flow cytometry, and gene expression for IL-2 and IFN-γ in activated T-cell populations was assessed via qRT-PCR.

Flow cytometry and qRT-PCR both demonstrated inhibition of IL-2 and IFN-γ that was generally dose-dependent in response to both cyclosporine and dexamethasone. Flow cytometry results from the average of samples collected from 3 different dogs are shown in Figure A. Similar results were achieved using qRT-PCR (Figure B).

Figure [Abstract IM-1]

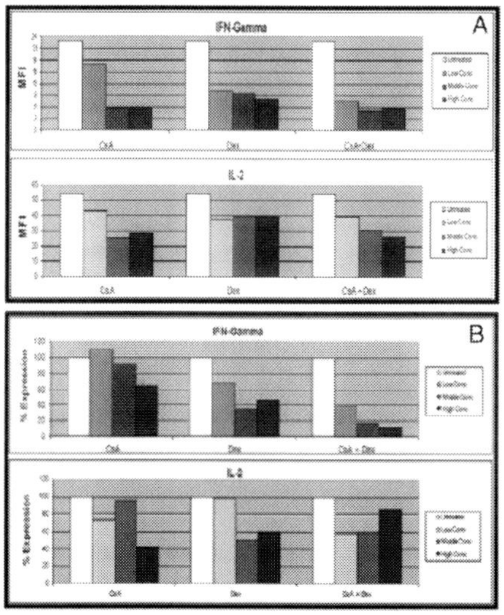

Suppression of IL-2 and IFN-γ in activated T-cells has potential as an indicator of the efficacy of cyclosporine and glucocorticoids in suppressing canine T-cell function *in vivo*, and may therefore be of value for characterizing the immunosuppression induced by these drugs in clinical patients.

ABSTRACT IM-2

MOLECULAR SIGNALING MECHANISMS OF EOSINOPHILIA IN ROTTWEILERS. N Prakash[1], MJ Day[2], P Stumbles[1], IR Peters[2], M Sharman[1], A Paul[1] & C Mansfield[1,3]. [1]Murdoch University, Perth, Australia [2]University of Bristol, Bristol, UK. [3]The University of Melbourne, Melbourne, Australia.

Idiopathic eosinophilic diseases are described in several breeds, but are over represented in Rottweilers. The immunopathogenesis of idiopathic eosinophilic disorders is poorly characterised. Studies in people highlight the importance of cytokines, particularly interleukin-5 (IL-5), in mediating eosinophil maturation, differentiation, egress from the bone marrow, migration and polyclonal expansion. Eotaxin-2 and eotaxin-3 also appear important for induction of chemotaxis and release of reactive oxygen species from eosinophils. The aim of the current study was to establish whether definable differences in specific cytokines associated with mediation of eosinophil production and survival are present between healthy Rottweilers, non-Rottweilers and Rottweilers with non-parasitic eosinophilia. Secondly, by evaluating cytokine profiles the study aimed to improve understanding of the pathophysiology of eosinophilia therefore assisting development of potential molecular treatment options.

Quantitative real-time reverse transcriptase polymerase chain reaction (QRT-PCR) assays were used to quantify messenger RNA (mRNA) encoding cytokines IL-4, IL-5, IL-10, IL-23p19, IL-12p35, IL-12p40, IL-18, interferon gamma (IFN-γ) and chemokines eotaxin-2 and eotaxin-3 from peripheral blood mononuclear cell (PMBC) samples obtained from healthy non-Rottweiler dogs with normal eosinophil counts ($n = 5$) and Rottweilers with normal ($n = 6$), mildly increased ($n = 7$) and high ($n = 3$) eosinophil counts. Quantification of serum IFN-γ was also performed using a commercially available canine-specific ELISA.

All samples were positive for housekeeping genes and all cytokines could be quantified with the exception of eotaxin-2 and -3. Results were normalised using three stably expressed housekeeper genes (RPL13A, SDHA and YWAZ) and a relative copy number was calculated for each sample with the sample with the fewest copies given a value of 1. No significant differences were found between groups but there was a tendency for IFN-γ mRNA expression to be lower in the Rottweilers with moderate to severe eosinophilia versus control dogs ($p = 0.062$). This trend was not seen in the concentration of serum IFN-γ quantified by ELISA as there were no significant differences between normal and diseased animals.

In conclusion, there were no significant differences in cytokine mRNA profiles between normal dogs and Rottweilers with varying degrees of eosinophilia. Additional studies including larger numbers of affected dogs are warranted before any accurate conclusions can be made.

ABSTRACT IM-3

PROLIFERATIVE RESPONSE OF BLOOD LYMPHOCYTES TO RED BLOOD CELL MEMBRANE IN RECENTLY IMMUNIZED DOGS. D Passarelli; SI Miyashiro; MK Hagiwara. School of Veterinary Medicine and Zootechny – University of São Paulo, São Paulo, SP, Brazil.

The presence of large amount of antibody on erythrocyte membrane can accelerate red blood cell (RBC) removal process by the mononuclear phagocyte system. An antigenic stimulus such as the one promoted by vaccines, for example, can induce hypersensitivity reactions and may accelerate RBC destruction. The study objective was to evaluate the erythrocytic membrane potential in inducing lymphocyte proliferative response of recently immunized dogs.

Healthy adult dogs ($n = 17$) were immunized with multiple antigens (commercial vaccine with eight antigens: distemper virus, parvovirus, coronavirus, parainfluenza virus, adenovirus, infectious hepatitis virus and leptospire; and anti-rabies). Blood samples from each animal were collected into EDTA tubes in two moments: PRE (immediately before vaccination) and POS (28 to 35 days after vaccination).

Mononuclear cells were separated by gradient, marked with CFSE-FITC and cultured. The stimuli for lymphocyte proliferation used were autologous erythrocytic membrane (AEM) and Concanavalin A (ConA). AEM was obtained by hypotonic lysis and tested in two concentrations (M1: 0.1ug/100uL; M2: 0.2 ug/100uL).

The proliferation assay was evaluated by flow cytometry and analyzed with specific software. The proliferation index (PI) was calculated dividing the fluorescence intensity of the basal sample by the stimulated one. Statistical analysis was performed using paired t-test for parametric samples and Wilcoxon test for non-parametric samples ($a = 0.05$).

The lymphocyte proliferation index of PRE ConA group (PI: 2.05 ± 0.79) was statistically higher than PRE M1 (PI: 1.03 ± 0.08) ($p < 0.001$) and than PRE M2 (PI: 1.06 ± 0.13) ($p < 0.001$). Within post vaccination group, POS ConA group (PI: 2.18 ± 0.54) was higher than POS M1 (PI: 1.02 ± 0.05) ($p < 0.001$) and than POS M2 (PI: 1.01 ± 0.04) ($p < 0.001$).

For the tested concentrations, autologous erythrocytic membrane does not constitute a stimulus for lymphocyte proliferation *in vitro*, either before or after vaccination procedure. Additionally, there was no evidence of self-reagent lymphocytes to erythrocyte membrane after vaccination.

ABSTRACT N/U-1

INVESTIGATIONS INTO THE EFFECT OF CRANBERRY EXTRACT ON BACTERIAL ADHESION TO CANINE UROEPITHELIAL CELLS. N Smee, GF Grauer, T Schermerhorn. Department of Clinical Sciences, Kansas State University, Manhattan, KS.

E. coli is a common cause of canine urinary tract infection. Current treatment emphasizes eradication of established infection rather than infection prevention but increased antibiotic resistance necessitates strategies to prevent infection. Proanthocyanidins found in cranberry juice inhibit *E. coli* attachment to human uroepithelial cells, impairing bacterial adherence and colonization. We hypothesized that purified cranberry extract (CE) inhibits bacterial adhesion to canine uroepithelial cells.

Five healthy female dogs received an oral CE supplement (Vetoquinol; 100 mg CE/tablet) according to body weight for 30 days. Voided urine collected from each dog before (PRE) and after (30-DAY) completion of the protocol was membrane filtered (22 μm) and stored frozen (-20C). Bacterial adhesion was determined using an *in vitro* assay. Briefly, urine samples were incubated with an uropathogenic *E. coli* strain that had been subcultured to promote fimbriae expression. Urine samples containing *E. coli* were next incubated in 96-well plates containing methanol-fixed Madin-Darby Canine Kidney (MDCK) cells for 1-hr (35C) to permit bacterial attachment. After incubation, plates were washed to remove non-adherent bacteria and fresh media added. Plates were incubated (35C) for 4-hr to grow attached bacteria to detection level. Bacterial concentration in each well was determined using a spectrophotometer (650 nm). Results were analyzed using the Chi-square test.

CE significantly reduced bacterial adhesion by 30% ($n = 5$; $p \leq 0.5$) in 30-DAY urine samples compared with PRE samples. The results show that CE supplementation can reduce adhesion of uropathogenic *E. coli* to canine uroepithelium and suggests one mechanism by which CE might improve urinary tract health.

ABSTRACT N/U-2

ASSOCIATION BETWEEN PRESENCE OF UROVIRULENCE FACTORS AND ANTIMICROBIAL RESISTANCE PATTERNS IN 221 UROPATHOGENIC *ESCHERCHIA COLI* SAMPLES ISOLATED FROM DOGS. J Wells, J Bartges, S Kania, D Bemis. University of Tennessee College of Veterinary Medicine, Knoxville, TN.

The purpose of this study was to determine prevalence of 4 urovirulence factors (UVFs) and antimicrobial resistance in canine uropathogenic *E.* coli (UPEC) and to evaluate associations between UVFs and antimicrobial resistance.

Two hundred and twenty-one UPEC isolates from samples collected from 184 different canine patients submitted to The University of Tennessee Microbiology Laboratory in 2007 were evaluated. A multiplex PCR assay was used to detect *cnf*, *hlyD*, *sfa/foc*, and *papGIII* in DNA lysate. *In vitro* susceptibility was evaluated and if the isolate was resistant to any antimicrobial in a class, it was considered resistant to that class.

Of the 221 samples, the number of UVF expressed per isolate was: 0 = 127/221 (57%), 1 = 27/221 (12%), 2 = 4/221 (2%), 3–22/221 (10%), and 4 = 41/221 (19%). Expression of UVF was *sfa* (33%), *hly* (24%), *cnf* (24%), and *pap* (19%). Presence of 4 UVFs was associated with less resistance ($p < 0.0001$). The combination of *hly*, *cnf*, and *sfa* was associated with less resistance ($p < 0.0001$). When *sfa* was present alone, resistance was less ($p < 0.0001$). Average resistance to antimicrobial class by number of UVFexpressed was: 0 UVF = 4.1 ± 3.3 classes, 1 UVF = 1.2 ± 1.1 classes, 2 UVF = 0.0 ± 0.0 classes, 3 UVF = 0.8 ± 1.6 classes, and 4 UVF = 0.5 ± 1.2 classes.

Urovirulence factors were present in a moderate number of UPEC and correlated negatively with resistance. Neither individual nor combinations of UVFs were associated with increased resistance.

ABSTRACT N/U-3

PREVALENCE OF ASYMPTOMATIC BACTERIAL URINARY TRACT INFECTIONS IN MORBIDLY OBESE DOGS. AL Lusby,[1] CA Kirk,[1] JW Bartges,[1] TD Moyers, PW Toll[2]. [1]University of Tennessee Department of Small Animal Clinical Sciences, Knoxville, TN [2]Hills Pet Nutrition, Topeka, KS.

Obesity is associated with several comorbidities in dogs including pancreatitis, osteoarthritis, oral disease, neoplasia, and lower urinary tract disease. Investigator observations led to the hypothesis that morbidly obese dogs are more likely to have asymptomatic bacterial urinary tract infections (ABUTI) than overweight and moderately obese dogs. Therefore, a pilot study was conducted to screen for ABUTI in obese dogs. Urinalysis with urine culture and dual energy x-ray absorptiometry (DXA) were performed on forty-three dogs with body fat (BF) percentages ranging from 36 to 56%. Following DXA, subjects were categorized as obese (O)(BF = 35–44%, n = 17) and morbidly obese (MO)(BF > 45%, n = 26). No dogs had owner-reported symptoms indicative of UTI. The prevalence of ABUTI in O dogs was 6% (n = 1) and 31% (n = 8) in MO dogs. The dog in the O group with ABUTI was close to being MO with a BF equaling 44.3%. Of the nine dogs with positive cultures, 4 were neutered males and 5 were spayed females. The prevalence ratio of ABUTI in MO dogs was 7.5, indicating dogs with 45% or greater BF are 7.5 times more likely to have the condition then dogs < 45% BF. The results of this pilot study coincide with other surveillance data describing an increased prevalence of lower urinary tract disease in obese dogs. In conclusion, dogs with body fat percentages greater than 45% are at risk for ABUTI, and veterinarians should consider screening all morbidly obese patients for urinary tract infections.

ABSTRACT N/U-4

PHOSPHATEMIA IN HEALTHY DOGS: WITHIN-DAY VARIATION, EFFECTS OF FEEDING WITH AND WITHOUT CALCIUM CARBONATE. R. Lavoué, O. Dossin, B.S. Reynolds, F.A. Granat, R.B. Javard, P.E.S. Charru, H.P. Lefebvre. National Veterinary School of Toulouse, France.

Calcium carbonate (CaC) is recommended to decrease phosphate intake in chronic kidney disease. However, its effect is poorly documented in dogs. Our objectives were to assess within-day, postprandial and CaC effects on phosphatemia variations in healthy dogs.

Phosphatemia was measured every 2 hours for 24 hours in eight adult healthy Beagle dogs in i) fasted condition and ii) a 2 × 2 cross-over design. One group received CaC mixed with maintenance diet (0.8% phosphorus), while the second group received the diet alone. After a 1-week wash-out period, groups were switched. A general linear model was used to test the period, sequence, treatment, dog and time effects on phosphatemia and the area under the phosphatemia versus time curve (AUC_{0-24}).

A significant ($p < 0.001$) circadian variation existed in fasted dogs. The maximum difference (mean: −1.9 mg/dL; 95% C.I.: −2.4 mg/dL; −1.4 mg/dL) was observed between 8 a.m. and midnight. The AUC_{0-24} with CaC (5936 ± 533 mg.min/dL) was mildly but significantly lower ($p = 0.027$) than without CaC (6239 ± 631 mg.min/dL). However, it was similar to the AUC_{0-24} in fasted conditions.

Feeding, with and without CaC, has minor effect on phosphatemia. However, circadian variation of fasted phosphatemia might affect its interpretation.

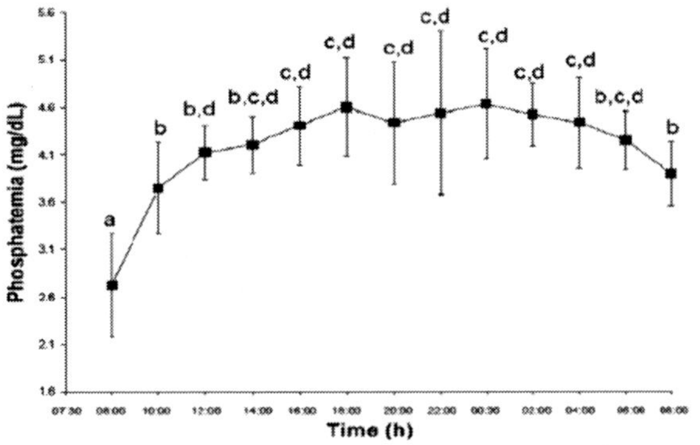

Fig. [Abstract N/U-4]: Circadian variation of phosphatemia in 8 dogs in fasted conditions (when superscripts are different, a statistically significant difference exists).

ABSTRACT N/U-5

RAPID DETERMINATION OF GFR IN DOGS WITH ACUTE KIDNEY INJURY USING A PORTABLE FLUORESCENCE RATIOMETRIC ANALYZER. V. E. Von Hendy-Willson,[1] B. M. Pressler,[1] R. M. Sandoval,[2] D. J. Meier,[3] E. Wang,[2] G. J. Schwartz,[4] B. A. Molitoris[2,3]. [1]School of Veterinary Medicine, Purdue University, West Lafayette, IN. [2]Division of Nephrology, Indiana University School of Medicine, Indianapolis, IN. [3]FAST Diagnostics, Indianapolis, IN. [4]Department of Pediatrics, University of Rochester Medical Center, Rochester, NY.

GFR measurement permits diagnosis of kidney injury prior to development of azotemia, and is the gold standard for kidney function assessment. Accurate and rapid (< 60 min) GFR measurement has been performed in rats by simultaneous transcutaneous assay of two intravascular fluorescently-labeled markers. A recently developed analyzer assays fluorescence via a fiberoptic cable introduced through a peripheral catheter, and thus should also allow rapid GFR determination in larger species. The purpose of this study was to determine correlation and agreement between fluorescent ratiometry (FR) and iohexol plasma clearance (IPC) in dogs over a range of GFRs.

Acute kidney injury (AKI) was induced in 5 female hound-type dogs (10 mg/kg gentamicin IV q8h), and FR and IPC GFR were simultaneously determined on days 0, 3, 6 and 9. A 9-sample, 5-hr protocol was used for IPC; FR was determined following bolus injection of a dextran conjugate mixture (2-sulfohexamine rhodamine-carboxymethyl 150 kD dextran, 5-aminofluorescein-carboxymethyl 5 kD dextran) with fluorescence measured over 60 min. GFR was calculated using 2-compartment model concentration-vs.-time curves for both techniques. Correlation was determined via Spearman's rho; agreement was analyzed via Bland-Altman plots.

IPC GFR and serum creatinine confirmed progressive AKI in all dogs. Correlation between FR and IPC was 0.91 ($p < 0.001$). Bland-Altman plots confirmed good agreement between techniques with slight underestimation of GFR by FR across most observed values.

These results suggest FR is suitable for GFR determination in dogs with AKI. Importantly, the portable analyzer allowed for point-of-care GFR determination in < 60 min using a peripheral vein.

Previously presented at the American Society of Nephrology Renal Week (related but not identical abstract).

ABSTRACT N/U-6
THROMBOELASTOGRAPHY IN NON-AZOTEMIC AND AZOTEMIC DOGS WITH PROTEIN-LOSING NEPHROPATHY. EM Lennon, RM Hanel, SL Vaden. North Carolina State University, Raleigh, NC.

Dogs with protein-losing nephropathy (PLN) are at risk of thromboembolic disease, but the mechanism of hypercoagulability and the population of dogs at risk are unknown. The purpose of this study was to characterize thromboelastography (TEG) in dogs with PLN. Twenty-eight client-owned dogs with PLN (urine protein:creatinine ratio (UPC) > 2.0) and 8 control dogs were enrolled. TEG parameters, antithrombin activity, serum biochemical profiles, and UPC were measured. TEG analyses were run in duplicate with kaolin activation; reaction time (R), clot formation time (K), maximal amplitude (MA), and G (global clot strength) were analyzed. A Wilcoxon Sum Rank Test was used to evaluate differences between groups.

Twelve PLN dogs (42.8%) were azotemic. Nineteen PLN dogs (67.8%) were hypoalbuminemic [serum albumin (SALB) < 3.0 g/dl]; 11 had SALB < 2.5 g/dl.

Dogs with PLN had higher K ($p < 0.01$), MA ($p < 0.005$) and G ($p < 0.005$) than controls. R was similar between the two groups. PLN dogs with SALB < 2.5 g/dl had higher G ($p < 0.05$) values than dogs with SALB > 2.5 g/dl; however, even PLN dogs with normal SALB (> 3.0 g/dl) had significantly higher G values than controls ($p < 0.005$). No significant relationship between UPC and G, SALB and G, antithrombin and G, or SALB and antithrombin was noted using linear regression analysis.

These results indicate that antithrombin, SALB, and UPC cannot be used alone to predict hypercoagulability as assessed by TEG in dogs with PLN. A comprehensive evaluation of the coagulation system in individual patients may be necessary to predict the point at which to initiate anti-thrombotic therapy.

ABSTRACT N/U-7
CHARACTERIZATION OF NON-TYPE I CYSTINURIA IN IRISH TERRIERS. Urs Giger[1], Jessica W. Lee[1], Cait Fitzgerald[1], Junlong Liu[1], Angela Erat[1], Adrian C. Sewell[2], Paula S. Henthorn[1]. [1]Section of Medical Genetics, School of Veterinary Medicine, University of Pennsylvania, Philadelphia, PA. [2]University Children's Hospital, Frankfurt am Main, Germany.

Cystinuria is a hereditary renal tubular reabsorption defect of cystine, ornithine, lysine and arginine (collectively, COLA). The low solubility of cystine in acidic urine predisposes to the formation of uroliths. Type I cystinuria in Newfoundland and Labrador Retriever dogs is an autosomal recessive trait caused by mutations in the SLC3A1 gene, whereas in other breeds, the cause of cystinuria has not yet been determined. We report here on the clinical, biochemical and molecular features of cystinuria in Irish Terriers.

Urine and EDTA blood were collected from 222 Irish Terriers from Europe and Australia. A nitroprusside screening test was used to identify increased cystine in urine. Urinary amino acid concentrations were determined by high-pressure liquid chromatography. Cystinuric dogs were defined as having cystine calculi, a positive nitroprusside result, urinary cystine (> 179 µmol/g creatinine) and/or a COLA concentration of > 700 µmol/g creatinine.

All 83 females tested nitroprusside negative and had normal urinary cystine (< 150 µmol/g creatinine) and COLA (< 500 µmol/g creatinine) concentrations. The 10 intact males that formed calculi as adults exhibited cystine concentrations ranging from 323–1580 and COLA from 1029–4302 µmol/g creatinine. An additional 41 males had similarly high COLA values with cystine levels from 0–1580 µmol/g creatinine. Among the affecteds tested, 75% were nitroprusside positive. The negative nitroprusside results and/or low urinary cystine levels of affecteds may be due to precipitation of cystine in acidic urine. Sequencing the coding regions of the SLC3A1 and SLC7A9 genes from EDTA blood identified no mutations. The mode of inheritance remains undetermined. However, castration appears to lower the urinary cystine and COLA concentrations and to prevent cystine calculi formation, while diet changes have lesser effects.

In conclusion, non-type I cystinuria in Irish Terriers (and several other breeds like Mastiffs and Scottish Deerhounds) is a unique form characterized by increased aminoaciduria only in males, with lower cystine and COLA excretion and fewer and later urolith formation compared to type I cystinuria. Castrating cystinuric Irish Terriers lowers their cystine and COLA excretion and thus their risk for calculi formation.

Supported in part by NIH RR 02512, Canine Health Foundation and LaboKlin, Germany.

ABSTRACT N/U-8
LONG TERM SURVIVAL OF CATS AND DOGS TREATED WITH INTERMITTENT HEMODIALYSIS FOR ACUTE KIDNEY INJURY. A. Eatroff, C. Langston, S. Chalhoub, K. Poeppel, E. Mitelberg. The Animal Medical Center, New York, NY.

Cats and dogs that are diagnosed with acute kidney injury (AKI) and resultant uremia that is not responsive to standard medical therapy are likely to benefit from renal replacement therapies, such as intermittent hemodialysis (IHD). The purpose of this study was to evaluate the long-term outcome of patients with AKI treated with IHD, and to establish whether renal function, as determined by serum or plasma creatinine concentrations, is associated with long-term survival. Medical records of 20 cats and 35 dogs that were diagnosed with AKI, treated with IHD, and survived longer than 30 days following the last IHD treatment were retrospectively analyzed. Standard methods of survival analysis using Kaplan-Meier product limit curves and the log-rank test were performed. For all-cause mortality, the median survival time was 1823 days (95% confidence interval: 841, 4667) for cats and 1049 days (95% confidence interval: 893, 1931) for dogs. When only renal-related causes of death were taken into account, the median survival time was not reached for cats or dogs. Survival time for all-cause mortality was inversely associated with the lowest creatinine concentration within the 30 to 90 day period following the last IHD treatment ($p < 0.0011$ for cats, $p < 0.0104$ for dogs). This study demonstrates that veterinary patients that are diagnosed with AKI, treated with IHD, and survive greater than 30 days after the last IHD treatment have a good long-term prognosis and frequently die from causes that are unrelated to renal impairment.

ABSTRACT N/U-9
DIAGNOSTIC UTILITY AND CONCORDANCE WITH FINAL DIAGNOSIS OF RENAL ASPIRATION IN DOGS AND CATS. ES Klosterman, JY Heo, TA Kurosowa, GE Moore, HG Heng, BM Pressler. Purdue University, West Lafayette, IN.

Renal fine-needle aspiration (R-FNA) is oftentimes attempted during evaluation of dogs and cats with renomegaly, mass lesions, or suspected infiltrative processes. Diagnostic utility of FNA is dependent upon the organ being sampled; additionally, in some organs, certain diagnostic imaging findings are associated with improved concordance of FNA with final diagnosis. Objectives of this study were to evaluate the diagnostic utility of R-FNA and determine whether concordance with final diagnosis is associated with specific clinicopathologic or diagnostic imaging findings. We hypothesized that R-FNA is most useful in patients with diagnostic imaging results suggestive of renal neoplasia (i.e. masses or suspected infiltrative processes).

Dogs and cats that had undergone R-FNA from Jan 1, 1998 to Dec 31, 2008 were identified by database search. Patient signalment, serum creatinine and blood urea nitrogen concentration, urine specific gravity, dipstick protein, R-FNA result, and final diagnosis were recorded. Patients were excluded if abdominal radiographs or sonographic images were not available for review, or if diagnostic test results were insufficient for determination of final diagnosis. A single coauthor blinded to final diagnoses interpreted all abdominal images using a pre-set list of descriptors and grading criteria. Radiographic kidney shape, margin distortion, and ventrodorsal kidney-to-L2 ratio were evaluated. Sonographic kidney margin distortion, cortical echogenicity, and corticomedullary junction distinction were described, and presence of nodules or masses, peri-renal effusion, or a peripheral sonolucent rim was noted. Concordance of R-FNA and final diagnosis was determined, and the chi-squared or Fisher's exact test were used to determine association of concordance with the above variables; $P < 0.05$ was considered significant.

37 dogs and 41 cats (78 animals) met all inclusion criteria. R-FNA results were concordant with the final diagnosis in 43 (55.1%) patients, discordant in 12 (15.4%) patients, and inadequate for cytologic interpretation in 23 (29.5%) patients. Neoplasia or FIP were the final diagnoses in 19 of 43 (44.2%) and 3 of 43 (7.0%) patients with concordant results, respectively. Renal lymphoma ($P = 0.196$), renal carcinoma ($P = 0.364$), and renal neoplasia in general ($P = 0.451$) were not associated with a higher likelihood of R-FNA and final diagnosis concordance. There was no association noted between likelihood of R-FNA and final diagnosis concordance when patients were stratified by species, serum creatinine or blood urea nitrogen concentration, urine specific gravity, dipstick proteinuria, or any diagnostic imaging variables.

This study failed to identify concurrent clinicopathologic or diagnostic imaging findings that enhanced the diagnostic utility of R-FNA. Future studies should use standardized criteria to prospectively identify patients in which R-FNA will be performed, evaluate additional variables that may be associated with increased R-FNA diagnostic utility, and directly compare the utility of R-FNA with that of other diagnostic techniques.

ABSTRACT N/U-10

PREDISPOSING FACTORS; ASSOCIATED WITH OBSTRUCTIVE FELINE URINARY TRACT DISEASE IN CATS.

N. Luckschander[1], V. Wardell[1], R.A. Hirt[1], A. Tichy[2], M. Angerer-Thalhammer[1]. [1]Department for Companion Animals and Horses, University of Veterinary Medicine, Vienna, Austria. [2]plattform biostatistics, department for biomedical sciences, University of Veterinary Medicine, Vienna, Austria.

Feline lower urinary tract disease (FLUTD) is a disease with increasing prevalence in private practices and veterinary teaching hospitals. Although several underlying causes can cause the obstructive form in male cats, the idiopathic form (feline interstitial cystitis) often is diagnosed as underlying reason in cats < 10 years. The goal of this retrospective study was to identify possible predisposing factors in order to optimize the therapy of these patients.

As a study group, 40 cats hospitalized with obstructive FLUTD at the Veterinary University of Vienna were examined during a 2 year period (2008–2010). As a control group 40 cats presented for other reasons were randomly chosen during the same time period. The data were examined concerning the signalment and history. Furthermore, the long-term outcome was evaluated with a questionnaire. Based on assumptions a student's t-test or a chi-square test was used.

There were no significant differences in age and breed. The body weight was significant higher in the FLUTD group than in the control group ($p < 0.01$). We could observe a significant risk for the disease of a weight of > 5 kg ($p < 0.01$). There were significant less cat toilets in the FLUTD group compared to the control group ($p < 0.05$). Furthermore we could observe that in the households of FLUTD cats there was significant less than one toilet per cat ($p < 0.01$) and more cats diseased on FLUTD lived strictly indoor than outdoor ($p = 0.05$).There were no significant differences at the time of hospitalization in age, breed, number of cats per household or season of the year between the two groups.

In summary, we could observe that cats over 5 kg body weight kept indoor with less than one toilet per cat have a significant higher possibility to be affected by obstructive FLUTD. Further studies with an extensive history of animal husbandry are needed to identify risks predispoing cats to this frequent and cost-intensive disease.

ABSTRACT N/U-11

HYPERURICOSURIA WITHOUT ALTERATIONS IN LIVER FUNCTION IS A RISK FACTOR FOR FELINE URATE UROLITHS.

C.M. Rivara, S. Shepard, C.R. Johnson, J.P. Lulich, C.A. Osborne, H Albasan, M Murtaugh. Veterinary Medical Center, University of Minnesota, St. Paul, MN.

Although purine uroliths (ammonium urate, sodium urate, xanthine, uric acid, etc.) represent the third most common stone type in cats, purine uroliths have the highest rate of recurrence (13% in 22 months). In dogs, mutation of the urate transporter (SLC2A9) and portovascular anomalies are common risk factors. However the underlying cause(s) for purine urolith formation in cats is unknown. The purpose of this study was to test the hypothesis that hyperuricosuria without alterations in liver function is common in cats with urate uroliths.

Urine concentrations of purine metabolites were measured by high-performance liquid chromatography in 5 cats with ammonium uroliths (cases), 5 clinically healthy, breed and gender matched cats (negative controls), and 2 cats with naturally occurring xanthine uroliths (positive controls). Prior to urine collection, all cats were fed a standard maintenance food (protein = 7 g/100kcal) for 4 weeks. Urinary xanthine, uric acid, and allantoin concentrations and concentration to creatinine ratios were calculated and compared between groups. Also, serum pre- and post-prandial bile acid concentrations were measured.

When compared to control cats, urinary uric acid concentration was significantly higher in case cats ($P = 0.002$). Xanthine was not detected in the urine of cases or negative controls. A significant difference in fasted and post-prandial serum bile acid concentrations was not detected in cases or controls ($P = 0.197, 0.212$).Hyperuricosuria without increased concentrations of urinary xanthine or allantoin appears to be a risk factor for ammonium urate urolith formation in cats. An association between portovascular shunts and purine urolithiasis was not observed in this population of cats.

ABSTRACT N/U-12

USE OF SURFACE ENHANCED LASER DESORPTION IONIZATION TIME-OF-FLIGHT MASS SPECTROMETRY (SELDI-TOF-MS) AS A URINARY BIOMARKER DISCOVERY TOOL IN CATS AT RISK OF DEVELOPING AZOTEMIA.

RE Jepson, HM Syme, J Elliott. Royal Veterinary College, London, UK.

Studies indicate that proteinuria is predictive, on a population basis, of those cats at risk of developing azotemia. SELDI-TOF-MS is a sensitive, high-throughput, proteomic technique utilising chromatographic surfaces to facilitate separation and detection of proteins and peptides within biological fluids such as urine. Individual low molecular weight (LMW) urinary proteins have been considered as potential biomarkers for renal damage but provide only a limited representation of the urinary proteome; SELDI-TOF-MS may provide a more global assessment.

Normotensive, non-azotemic geriatric cats (> 9 years) were recruited prospectively from two first-opinion clinics for routine health screening. At entry cats received a full physical examination, plasma biochemistry, evaluation of total T4 concentration and urinalysis including urine protein to creatinine ratio. Re-examination was offered at 6 and 12 months. Cats were divided into two groups based on clinical status at the 12 month re-examination (azotemic; creatinine concentration ≥ 2.0 mg/dL and non-azotemic). Optimisation studies were performed to facilitate the automated preparation (Biomek 3000) of CM10 (weak cation exchange) arrays for SELDI-TOF-MS analysis (Ciphergen Enterprise 4000) of urine samples from cats at entry to the study. Results are reported as median [25th, 75th percentile]. Mann Whitney U-test and Wilcoxon signed rank test were used to compare variables between groups and between timepoints, respectively. Ciphergen Express (3.06) software was used to analyse spectral data and a Mann Whitney U-test was used to identify clusters which differed significantly between groups ($p < 0.05$) at entry to the study.

Twenty non-azotemic cats were recruited, of which 10 cats developed azotemia by 12 months. No significant differences in age, body weight, biochemical or urinalysis variables were identified between groups at entry to the study. As might be expected creatinine increased significantly (1.73 mg/dL [1.58, 1.77], 2.17 [2.00, 3.23], p = 0.002) between study entry and 12 months in the cats that developed azotaemia and there was a commensurate increase in phosphate concentration (3.81 mg/dL [3.60, 4.22], 4.65[3.94, 6.48], p = 0.019). Creatinine and phosphorus did not change significantly over time in the cats that did not develop azotaemia. Seven clusters with m/z values of 2822, 10 033, 10 151, 10 234, 11 635, 11 700 were found to differ significantly between groups at entry to the study.

The low protein concentration of feline urine makes the use of proteomic techniques challenging. However, this pilot study indicates that SELDI-TOF-MS can be utilised to examine the feline urinary proteome and that differences in low molecular weight protein patterns may be useful to differentiate those cats which are at risk of the development of azotemia. Further work is necessary to identify these proteins/peptides.

ABSTRACT N/U-13

FGF-23 – MEDIATOR OF RENAL SECONDARY HYPERPARATHYROIDISM OR A MARKER OF GLOMERULAR FILTRATION RATE (GFR) IN CATS? N.C.Finch, R.F.Geddes, H.Syme, J.Elliott. Royal Veterinary College, UK.

Fibroblastic Growth Factor 23 (FGF-23) is a phosphotonin with an important physiological role in the regulation of phosphorous and vitamin D metabolism, and may therefore play a part in the development of renal secondary hyperparathyroidism. Previous studies in cats have shown parathyroid hormone (PTH) to be elevated prior to the development of azotemia. The study objectives were to explore the hypothesis that FGF-23 is a mediator of the development of renal secondary hyperparathyroidism in the non-azotemic stages of feline CKD.

Healthy, non-azotemic (plasma creatinine concentrations (Cr) < 2.0 mg/dl) geriatric cats were recruited into the study prospectively and followed for 12 months. At the study end point cats were categorised into the following 3 groups: group 1 (N = 15)-Cr ≤ 1.58 mg/dl, group 2 (N = 33)–Cr ≥1.58 mg/dl but did not meet the criteria for group 3 and group 3 (N = 14)–Cr > 2.0 mg/dl in association with reduced urine concentrating ability (USG < 1.035) or demonstration of persistent azotemia (Cr > 2.0 mg/dl). Plasma samples were subjected to routine biochemical analysis, intact PTH, calcitriol and intact FGF-23 assay. Variables were compared between the 3 groups at the baseline time point. GFR was measured in an additional group of 19 cats (11 non-azotemic, 4 IRIS stage II, 4 IRIS stage III) using a corrected slope-intercept iohexol clearance method. Relationships were explored using linear regression analysis and determining the coefficient of determination (R^2). Results are presented as median [range].

At the baseline time point FGF-23 concentrations were significantly higher in group 2 (208.1[51.4–814.6], $P = 0.001$) and group 3 (237.6[127.4–908.1], $P = 0.001$) compared to group 1 (126.2[69.4–505.2]). Weak positive relationships were identified between FGF-23 and PTH ($R^2 = 0.126$, $P = 0.005$, N = 62) and FGF-23 and Cr ($R^2 = 0.077$, $P = 0.029$, N = 62). However, the positive relationships between FGF-23 and phosphate ($R^2 = 0.016$, $P = 0.323$, N = 62) and FGF-23 and calcitriol ($R^2 = 0.085$, $P = 0.212$, N = 20) were not significant. The additional group of cats in which GFR measurement was performed there was an inverse relationship between FGF-23 and GFR ($R^2 = 0.208$, $P = 0.040$).

In conclusion, FGF-23 was elevated in cats prior to the development of azotemia. The role of FGF-23 in the development of feline renal secondary hyperparathyroidism remains to be determined and should be explored through interventional studies. However, considering the relationship between FGF-23 and GFR, it cannot be excluded that the phosphotonin is simply a marker of reduced filtration.

ABSTRACT N/U-14

URINARY VASCULAR ENDOTHELIAL GROWTH FACTOR AS A MARKER OF RENAL HYPOXIA IN CATS. S Chakrabarti, H Syme, J Elliott. Royal Veterinary College, London, UK.

Chronic kidney disease (CKD) is common in geriatric cats and hypoxia might contribute to the progression of this disease. The aim of this study was to evaluate urinary vascular endothelial growth factor (VEGF) as a marker of renal hypoxia.

Cats were recruited through geriatric clinics held at two first opinion London practices. VEGF was measured in stored samples using a canine ELISA kit validated for use on feline urine and indexed to creatinine concentration to yield a VEGF to creatinine ratio (VCR). Two studies were undertaken – firstly a cross-sectional analysis of clinical variables associated with VCR in cats with CKD. Diagnosis of CKD was based on concurrent findings of plasma creatinine ≥ 2 mg/dl and USG ≤ 1.035, with persistence of azotemia for ≥ 2 weeks. Only patients receiving no medical therapies were included. Normotensive and (pre-treatment) hypertensive cats were included, but borderline cases (mean systolic blood pressure 160–170 mmHg on the date of sampling) were not. Hyperthyroid cats were also excluded from this cross-sectional study. Associations between VCR and clinical data were initially assessed using the Spearman's coefficient and Mann Whitney Test. Linear regression was then used for multivariate analysis.

The second study used samples from a trial in which hypertensive cats that had been treated with amlodipine for at least 3 months were entered into a randomised cross-over study where they received placebo or benazepril (0.5 to 1 mg/kg daily) for 12 weeks in turn. VCR on placebo was compared with that on benazepril using the Wilcoxon signed ranks test. Cats with well controlled hyperthyroidism were included in this intervention study.

Results are reported as median [25th, 75th percentile]. VCR was higher (49.5[33.3, 74.1] vs. 36.1[25.6, 42.1] fg/g, p = 0.010) in untreated hypertensives (n = 30) than normotensives (n = 63). VCR was correlated with PCV (r = –0.236, p = 0.024, n = 92), UPC (r = 0.444, p < 0.001, n = 93), plasma phosphate (r = 0.286, p = 0.005, n = 93), and USG (r = –0.284, p = 0.006, n = 93), but not plasma creatinine concentration. In the best multivariate model, PCV was associated with VCR independently of UPC ($r^2 = 0.435$, n = 92). VCR was significantly reduced by Benazepril therapy (65.3 [43.1, 92.7] fg/g) compared with placebo (76.0[48.0, 116.7] fg/g; p = 0.031, n = 17) with a reduction seen in 76% of cases.

These results suggest urinary VEGF excretion is associated with proteinuria in cats with CKD and might be a marker of renal hypoxia induced by low PCV. ACE inhibitor therapy might reduce urinary VEGF excretion because angiotensin II causes constriction on efferent arterioles resulting in tubular hypoxia.

ABSTRACT N/U-15

FIBROBLAST GROWTH FACTOR 23 (FGF-23) IN FELINE CHRONIC KIDNEY DISEASE. RF Geddes, NC Finch, J Elliott, HM Syme. Royal Veterinary College, London, UK.

FGF-23 is a phosphaturic hormone. FGF-23 concentrations increase with declining renal function in humans. The objectives of this study were to validate a method for FGF-23 quantification in feline plasma and to assess the association between FGF-23 concentration and plasma creatinine or phosphate concentration in cats with chronic kidney disease (CKD).

Non-azotemic and azotemic (plasma creatinine concentration (Cr) > 2.0 mg/dl) geriatric (> 9yrs) cats were recruited into the cross-sectional study from two London first opinion practices. Cats were excluded from the study if they were fed a phosphate restricted diet, or had evidence of concurrent disease. The cats were categorized, using a modified IRIS staging system, into the following four groups: group 1 (Cr ≤ 1.6 mg/dl), group 2 (Cr 2.0–2.8 mg/dl), group 3 (Cr 2.9–5.0 mg/dl), group 4 (Cr > 5.0 mg/dl). Groups 2 and 3 were further subdivided based on the IRIS targets for plasma phosphate concentration (PO_4): group 2a ($PO_4 \leq 4.5$ mg/dl), group 2b ($PO_4 > 4.5$ mg/dl), group 3a ($PO_4 \leq 5$ mg/dl), group 3b ($PO_4 >$ 5 mg/dl). FGF-23 concentrations were measured in feline EDTA plasma using a human intact FGF-23 ELISA, validated by intra- and inter-assay variability and assessment of dilutional parallelism. Comparisons between groups were made using the Kruskal-Wallis test and Mann-Whitney U test, with statistical significance defined as P < 0.05. Bonferroni correction was applied where appropriate (statistical significance then determined as P < 0.008). Results are reported as median [25th, 75th percentiles]. FGF-23 concentrations ≥ 800pg/ml (upper limit of quantification) were assigned the value of 800pg/ml.

Intra- and inter-assay variability of FGF-23 measurements were < 10.0% and dilutional parallelism between feline samples and the calibration curve were demonstrated. Plasma FGF-23 concentrations increased with increasing creatinine concentrations (group 1:

158 [115, 274], n = 20, group 2: 354 [239, 473], n = 20, group 3: 800 [425, 800], n = 23, group 4: 800 [800, 800], n = 14). FGF-23 measurements were significantly different between all groups (P = 0.005 to < 0.001) except between groups 2 and 3 (P = 0.01). FGF-23 concentrations were significantly higher in cats with higher plasma phosphate concentrations (group 2a: 329 [237, 423], n = 16 vs. group 2b: 576 [374, 793], n = 4; P = 0.047) and (group 3a: 432 [167, 800], n = 10 vs. group 3b: 800 [625, 800], n = 13; P = 0.028).

In conclusion, FGF-23 concentrations were higher in cats with more severe CKD or higher plasma phosphate concentrations as would be predicted from its known biological actions. Further work is warranted to explore the role of FGF-23 in the development of renal secondary hyperparathyroidism by measuring parathyroid hormone (PTH) and calcitriol in cats at different stages of CKD.

ABSTRACT N/U-16

RETROSPECTIVE EVALUATION OF PULMONARY ABNORMALITIES IN DOGS WITH RENAL AZOTEMIA. K. Le Boedec,[1] H.G. Heng,[2] P.W. Snyder,[2] B.M. Pressler[2]. [1]National Veterinary School, Toulouse, France. [2]Purdue University, West Lafayette, IN.

Progressive non-cardiogenic edema and lung dysfunction are common complications of acute kidney injury (AKI) in people. Pulmonary abnormalities have not been systematically reviewed in dogs with renal azotemia, but anecdotal reports of dogs with AKI and concurrent non-cardiogenic pulmonary edema are suggestive of uremic pneumonopathy (UP), a centrally-distributed pulmonary edema syndrome associated with kidney disease in people. We therefore hypothesized that pulmonary-associated clinical signs or thoracic radiograph abnormalities are more common in dogs with renal azotemia than in non-azotemic dogs, and that this association is more likely in dogs with AKI than dogs with chronic renal failure (CRF). Our study objectives were 1) to describe thoracic radiograph and lung histopathologic abnormalities in dogs with renal azotemia, 2) to compare the occurrence of these findings in dogs with AKI, CRF, or non-systemic illness, and 3) to determine if these abnormalities are associated with shorter survival times.

Records of dogs with renal azotemia evaluated from 1/1/2000 to 8/20/2010 were reviewed; dogs which could be classified as having AKI or CRF and which had complete thoracic radiograph studies available for review were included. Dogs with primary intracranial disease and normal serum creatinine and a complete thoracic radiograph study were selected as controls. Signalment, weight, presence of pulmonary-related clinical signs, azotemia duration and severity at time of radiography, and leptospirosis antibody titer were noted. Alveolar, bronchial, interstitial, or nodular lesions were described using a 4-point scale, and lung tissue collected at time of necropsy was reviewed; both the radiologist and pathologist were blinded to final diagnoses. Significance was P < 0.05 for all analyses.

The final study population included 54 AKI, 50 CRF, and 63 control dogs. CRF dogs were older (P < 0.001) than AKI and control dogs. Pulmonary-related clinical signs were more commonly diagnosed at first evaluation in AKI dogs (29/53 dogs, 54.7%) than in CRF (13/50, 26.0%; P = 0.003) or control dogs (9/63, 14.3%; P < 0.001). Presence of an alveolar pattern was the only radiographic finding which differed amongst groups (more common in AKI [n = 8, 14.8%, P = 0.047] and CRF [n = 8, 16%, P = 0.028] dogs than in control dogs [n = 2, 3.2%]). There was no association between presence of an alveolar pattern and any other variable. Alveolar mineralization was the most common lesion in AKI dogs (5/8 dogs; 62.5%), with concurrent alveolar space concretions or mineralization of vessels or bronchioles noted in 1 dog each. Necropsies had not been performed in any of the CRF dogs, but mineralization was not seen in lung tissues from any control dogs (n = 9). Neither pulmonary-associated clinical signs nor alveolar pattern were associated with median number of days from discharge until death in dogs with AKI (P = 0.220 and 0.468, respectively) or CRF (P = 0.280 and 0.253, respectively).

In this group of dogs, presence and type of radiographic pulmonary abnormalities were associated with renal azotemia but not with median time until death. The association between and clinical relevance of alveolar mineralization in AKI dogs were not determined, but both the radiographic and histopathologic abnormalities reported here differ from UP in people.

ABSTRACT N/U-17

THE EFFECTS OF RHEUM OFFICINALE ON THE PROGRESSION OF FELINE CHRONIC KIDNEY DISEASE. AS Hanzlicek[1], MW Sanderson[2], GF Grauer[2]. [1]Oklahoma State University, Stillwater, OK. [2]Kansas State University, Manhattan, KS.

Chronic kidney disease (CKD) is a common cause of morbidity and mortality in cats. The purpose of this study was to investigate the effects of Chinese rhubarb (Rheum officinale) supplementation on the progression of feline CKD.

Cats with stable IRIS stage II or III CKD and without comorbidity were included in the study. Cats were divided into 3 treatment groups and administered rhubarb extract (Group 1, Rubenal®, Vetoquinol, 75 mg tablet PO q 12 h), benazepril as a positive control (Group 2, 0.5 mg/kg PO q 24 h), or both (Group 3). Cats were fed a commercial renal specific diet and enteric phosphate binder as appropriate. Body weight, laboratory data, and blood pressure were recorded every 3 months for up to 34 months.

Variables between groups at enrollment and within groups over visits were compared with ANOVA and repeated measures ANOVA, respectively. A treatment by visit interaction term was included in all repeated measures models. Significance was set at $p \leq 0.05$.

Except for body weight there was no significant differences between treatment groups at enrollment. There was no significant change in body weight, hematocrit (hct), UPC, or creatinine over time as compared to baseline within any group. There was no significant difference between groups over time in regards to change in weight, hct, UPC, or creatinine. The treatment by time interaction was non-significant in all models.

Although there was no benefit associated with combination treatment, the results for rhubarb treatment alone were not different from benazepril treatment.

ABSTRACT N/U-18

AZODYL FAILS TO REDUCE AZOTEMIA IN CATS WITH CHRONIC KIDNEY DISEASE (CKD) WHEN SPRINKLED ONTO FOOD. M. Rishniw[1], S. Wynn[2]. [1]Veterinary Information Network, Davis CA. [2]Georgia Veterinary Specialists, Sandy Springs GA.

Azodyl, an encapsulated, enteric-coated probiotic/prebiotic nutraceutical, is marketed for reduction of azotemia (BUN & Creatinine) in dogs and cats. Cat owners often sprinkle contents onto cat food to facilitate administration. However, exposure to air and stomach acid are thought to inactivate the lyophilized bacteria within the product. Therefore, we examined the ability of food-sprinkled Azodyl to reduce azotemia in cats with CKD.

10 cats with CKD were enrolled in the study and randomized receive Azodyl or placebo. Owners were provided with 3–4 capsules of Azodyl prior to enrollment to ensure compliance with administration. 2 baseline blood samples were obtained 1 month apart, and then 1 & 2 months after beginning therapy. Clinicians and owners were masked as to medication assignment. We hypothesized that a 30% decrease in BUN and/or creat in the Azodyl group would be significant, and set $\alpha = 0.2$. In order to maximize the probability of detecting a difference, we determined the % change as being the difference between the maximal baseline analyte concentration and minimal therapeutic concentration. We compared the % change between groups by Mann-Whitney U Test.

BUN and creatinine did not differ between groups.

Based on these results, Azodyl, applied by sprinkling onto food fails to reduce azotemia in cats with CKD. Whether intact capsule administration reduces azotemia in cats with CKD remains unknown.

ABSTRACT N/U-19

EVALUATION OF 3 HERBAL COMPOUNDS USED FOR MANAGEMENT OF LOWER URINARY TRACT DISEASE IN CATS. M Daniels[1], JW Bartges[1], DM Raditic[1], CA Kirk[1], A Callens[1], S Marsden[2], G Galyon[1]. [1]University of Tennessee, Knoxville, TN. [2]Edmonton, AB, CA.

Lower urinary tract disease (LUTD) occurs commonly in cats, and idiopathic cystitis (FIC) and urolithiasis account for over 80% of cases in cats less than 10 years of age. Although several strategies have been recommended, a common recommendation is to induce dilute urine resulting in more frequent urination and to dilute calculogenic constituents. In addition to conventional therapy using modified diets, traditional Chinese and Western herbs have been recommended, although only one, chorieto, has published data. We evaluated 3 commonly used herbal treatments recommended for use in cats with LUTD including (1) San Ren Tang, (2) Wei Ling Tang, and (3) Alisma. We hypothesized that these 3 Chinese herbal preparations would induce increased urine volume and decreased urine saturation for calcium oxalate and struvite.

Six healthy, spayed female, adult cats were evaluated in a placebo-controlled, randomized, cross-over design study. Cats were randomized to 1 of 4 treatments including placebo (P), San Ren Tang (SRT), Wei Ling Tang (WLT), or Alisma (A). Treatment was for 2 weeks each with a 1 week washout period between treatments. At end of each treatment period, a 24-hour urine sample was collected using modified litter boxes. Urine volume and biochemistries were measured, and urine saturation for struvite and calcium oxalate was estimated using EQUIL 1.5b. Analysis of Variance (ANOVA) was used to analyze data statistically if distributed normally and Kruskal-Wallis was used to analyze data statistically if data were not distributed normally. A $p < 0.05$ was considered significant.

Body weights were not different between treatments. No differences were found in 24-hour urinary analyte excretions, 24-hour urine volume, urine pH, or 24-hour urinary saturation for calcium oxalate or struvite between treatments (Table).

Table [Abstract N/U-19]: Twenty-four hour urine volume and relative supersaturation (RSS) for struvite and calcium oxalate (caox) for 6 adult spayed female domestic short-haired cats receiving placebo (P), alisma (A), Sun Ren Tang (SRT), or Wei Ling Tang (WLT). Data are means ± SD if distributed normally or median (minimum, maximum) if not distributed normally.

Parameter	Units	A	P	SRT	WLT	*p*-value
Volume	ml/kg/d	11.0 ± 4.97	11.5 ± 3.49	14.0 ± 9.16	9.30 ± 3.23	0.7
RSS struvite		0.030 (0.007, 0.148)	0.022 (0.006, 0.074)	0.038 (0.007, 0.921)	0.029 (0.005, 0.058)	0.8
RSS caox		4.89 (2.41, 9.09)	3.62 (2.83, 6.75)	3.95 (2.41, 5.73)	5.50 (2.48, 10.05)	0.8

Results of this study do not support the hypothesis that these 3 Traditional Chinese herbal preparations increase urine volume or decrease urinary saturation for calcium oxalate and struvite.

ABSTRACT N/U-20

CANINE UROLITHIASIS IN MEXICO. J Del Angel Caraza [1], CC Pérez-García[2], I Diez-Prieto[2], MB García-Rodríguez[2], IA Quijano Hernández[1]. [1]Hospital Veterinario para Pequeñas Especies FMVZ-Universidad Autónoma del Estado de México, Toluca, México. [2]Laboratorio de Investigación en Urolitiasis, Universidad de León, España.

Urolithiasis is a multifactorial disease, frequent and recurrent in dogs in the worldwide, in which breed, sex, age, diet, some anatomical abnormalities, urinary tract infection, urine pH and some geographical and hereditary features in the populations studied have been implicated as risk factors. The effective long-term management of urolithiasis depends on identification and control of the pathophysiological mechanisms involved, which, in turn, depend on accurate knowledge of the mineral composition of the uroliths.

The aim of this study was to determine for first occasion the main epidemiological data of canine urolithiasis in Mexico. This study was developed with 491 dogs with urolithiasis from 25 of the 33 states of the country.

Chemical composition of the uroliths was determined by stereoscopic microscopy, infrared spectroscopy, scanning electron microscopy and X-ray microanalysis.

Urolithiasis affected nearly the same number of males and females; with ages ranging from two months to 15 years with a median age of 5 years. Adult animals were the most affected. Breeds more affected were Schnauzer miniature, Poodle, Dalmatian, Yorkshire terrier, Scottish terrier, Chihuahua and Bichon friseé.

Uroliths were found in the lower urinary tract in 97.74% of the cases. Mineral composition of the uroliths was: Struvite 49.69%, followed by calcium oxalate 25.46%, purines 7.13%, silicate 6.72%, others 0.20%, mixed 8.15% and compound uroliths 2.44%. Struvite uroliths affected females in most cases, whereas calcium oxalate, purines and silicate uroliths, were mainly observed in males.

Our results are similar to studies developed in other countries and continents, though we found a higher frequency of uroliths containing silicate, either pure, mixed or compounds uroliths (10.79%); in Mexico City the frequency reached 15%. This high frequency may be due to high consumption of silicate in home-made food or in the groundwater derived from aquifers.

Acknowledgments: This work has been partially supported by a project of Waltham Foundation in Mexico and the Consejo Nacional de Ciencia y Tecnología (CONACyT) of Mexico.

ABSTRACT N/U-21

THE USE OF A MEDETOMIDINE-BASED SEDATION PROTOCOL TO PERFORM UROHYDROPROPULSION AND CYSTOSCOPY IN THE DOG. JA Webb[1], M Rosati[1], D Naigamwalla[1], A Defarges[2]. [1]Mississauga-Oakville Veterinary Emergency Hospital, Oakville, ON. [2]Clinical Studies, Ontario Veterinary College, Guelph, ON.

Voiding urohydropropulsion is a non-invasive method for removing small urocystoliths from the dog, most commonly used in females due to the relatively wider and shorter urethra. This procedure is typically performed under general anesthesia to allow complete relaxation of the urethra, however, anesthesia results in longer procedure times and difficult endotracheal tube stabilization due to the vertical positioning of animals, especially in larger dogs.

The aim of this study was to devise a novel injectable sedation protocol for urohydropropulsion when cystoscopy was not concurrently required. An intravenous catheter was placed, and a combination of medetomidine (10 to 15 µg/kg IV) and hydromorphone (0.025 to 0.05 mg/kg IV) was administered, with the addition of ketamine (2 mg/kg IV) in fractious animals; atipamezole (double volume of medetomidine, administered IM) was used as a reversal agent upon procedure completion. This protocol was considered in cardiovascularly healthy, non-diabetic dogs without evidence of urinary obstruction. Monitoring equipment included electrocardiography, blood pressure measurement, and pulse oximetry, and supplemental flowby oxygen was provided.

Two dogs received the proposed sedation protocol in order to perform urohydropropulsion. Dog one was a 3 year old female spayed Shih Tzu cross, and dog 2 was a 2 year old female spayed Standard Poodle. Ultrasonography revealed a moderate number of urocystoliths present in both dogs, measuring up to 1 mm in dog 1 and 2.3 mm in dog 2. Urohydropropulsion was performed and resulted in retrieval of 15 urocystoliths in dog 1, and approximately 20 urocystoliths in dog 2. Repeat ultrasonography revealed no uroliths present after urohydropropulsion in both dogs. The time from administration of sedation to administration of reversal agent was 6 minutes for dog 1, and 8.5 minutes for dog 2. Records were obtained from 3 dogs that had traditional general anesthetic protocols for urohydropropulsion with cystoscopy for confirmation of urocystolith removal, performed within the last 2 years, and the average anesthetic time was 64 minutes.

Subsequent to the use of medetomine-based sedation protocols for the above dogs, cystoscopy was performed in a 9 year old neutered male Golden Retriever with prostatomegaly. Medetomidine (15 ug/kg IV) and butorphanol (0.2 mg/kg IV) were administered; atipamezole (double volume of medetomidine, administered IM) was used as a reversal agent upon procedure completion. This sedation allowed adequate immobilization for cystoscopy of the urethra and urinary bladder, and endoscopic biopsying of the prostatic urethra and urinary bladder. The time from administration of sedation to administration of reversal agent was 15 minutes for this dog.

In conclusion, a novel sedative protocol for urohydropropulsion is proposed which allows for an appropriate level of sedation along with a short procedure time and rapid recovery. This sedation protocol may also be useful for certain cystoscopic procedures.

ABSTRACT N/U-22

EFFECT OF STORAGE TIME AND TEMPERATURE ON CANINE URINE ENZYMES. N Smee[1], L Pohlman[2], MW Sanderson[1], GF Grauer[1]. [1]Department of Clinical Sciences, [2]Department of Diagnostic Medicine/Pathobiology, Kansas State University, Manhattan, KS.

Increased activity of urinary N-acetyl-β-D-glucosaminidase (NAG) and γ-glutamyl transpeptidase (GGT) can be markers of early tubular cell damage. After acquisition of a urine sample, analysis may be delayed for a variety of reasons, including the need for sample batching within the laboratory or shipping to an outsourced location. Therefore, it is important to know how storage of the sample may affect enzyme activity. We hypothesized that urinary NAG and GGT activity would be affected differently in samples stored by refrigeration vs. freezing.

Thirty-four canine urine samples submitted to the Clinical Pathology Laboratory at Kansas State University were included. Samples were collected from clinical patients with a variety of medical/surgical disorders and were selected based on the day of the week and a minimum volume of 10 ml. A complete urinalysis was performed on each sample; however there were no exclusion criteria based on urinalysis results. NAG and GGT activity in the urine supernatant was assessed by colorimetric assay. Aliquots of each supernatant were refrigerated for 5 days and frozen at −20°C for 5 and 30 days at which time enzyme activity was re-assessed.

Compared to baseline values, enzyme activity for both NAG and GGT were stable after 5 days of refrigeration, however there were significant ($P < 0.01$) declines in GGT and NAG activity when urine supernatants were frozen for 5 and 30 days.

ABSTRACT N/U-23

INTERIM EVALUATION OF THE EFFICACY AND SAFETY OF A HIGH DOSE SHORT DURATION ENROFLOXACIN TREATMENT REGIMEN FOR URINARY TRACT INFECTION IN DOGS. S Irom[1], J Westropp[2], D Chew[3], J Daniels[3]. [1]Great Lakes Veterinary Specialists, Cleveland, OH. [2]University of California Davis School of Veterinary Medicine, Davis, CA. The Ohio State University College of Veterinary Medicine, Columbus, OH.

Treatment for canine urinary tract infections (UTI) typically consists of 7–14 days of antimicrobial drugs in primary care veterinary practice. Compliance with this drug regimen can be difficult for some clients. Enrofloxacin is a veterinary approved fluoroquinolone antimicrobial and is useful for treatment of canine UTI. Fluoroquinolones are often used in human medicine to treat uncomplicated UTIs in women and can be prescribed for as little as 3 days. The primary objective of this study was to determine if dogs with naturally occurring uncomplicated UTI have equivalent microbiologic cure with a high dose short duration protocol of enrofloxacin, compared to a standard antimicrobial protocol.

Client-owned adult dogs with naturally occurring, uncomplicated UTI were prospectively enrolled in a multi-center clinical trial and assigned to 1 of 2 groups in a randomized blinded manner. Group 1 received treatment with 18–20 mg/kg oral enrofloxacin once daily for 3 consecutive days. Group 2 dogs were treated with 13.75–25 mg/kg oral amoxicillin-clavulante twice daily for 14 days. Both groups had urinalyses and urine cultures submitted on day 0, 10, and 21.

At the time of this interim analysis, thirty-six dogs have completed the trial. Bacteriological cure was achieved in 15 dogs (83%) treated with enrofloxacin and 14 dogs (78%) treated with amoxicillin-clavulante, respectively.

These data suggest that the high-dose, short-duration enrofloxacin protocol was equally effective to the standard protocol in treating uncomplicated canine UTI in the sample patient population, and may represent a viable alternative therapeutic regimen for similar patients.

ABSTRACT N/U-24

RENAL DOPPLER RESISTIVE INDEX IN DOGS WITH DEGENERATIVE MITRAL VALVE DISEASE (DMVD). T. Daste[1], V. Chetboul[2], V. Gouni[2], C. Carlos Sampedrano[2], C.J. Layssol-Lamour[1], F. Serres[2], J.L. Pouchelon[2], C. Germain[1], H.P. Lefebvre[1]. [1]Dept. Clinical Sciences, Ecole Nationale Vétérinaire de Toulouse, Toulouse. [2]Cardiology Unit, Ecole Nationale Vétérinaire d'Alfort, Maisons-Alfort cedex, France.

Azotemia is frequent in dogs with DMVD (Nicolle et al; JVIM 2007;21:943–949) and could result from renal hemodynamic alterations. Renal resistive index (RI) allows assessment of renal vascular resistance. The aim of this prospective study was to assess RI in dogs with different DMVD stages.

Fifty-five dogs with DVMD were used (ISACHC class 1 ($n = 28$), 2 ($n = 19$), and 3 ($n = 8$)). Physical examination, renal ultrasonography and echo-Doppler examinations were performed in awake dogs by trained observers. Plasma creatinine, urea and NT-proBNP were measured. Statistical analyses were performed using a general linear model.

Whereas RI of renal and arcuate arteries were unaffected by ISACHC class, left interlobar RI increased ($P < 0.001$) from 0.62 ± 0.05 (mean ± SD) in class 1 to 0.76 ± 0.08 in class 3. Left interlobar RI was also higher ($P < 0.001$) in azotemic (0.74 ± 0.008) than in non azotemic (0.62 ± 0.005) dogs. Similar findings were observed for right interlobar RI. A positive effect of NT-proBNP ($P = 0.002$), urea ($P < 0.001$), creatinine ($P = 0.002$), urea-to-creatinine ratio ($P < 0.001$), left atrium-to-aorta ratio ($P < 0.001$), regurgitation fraction ($P = 0.011$), systolic pulmonary arterial pressure ($P < 0.001$) and shortening fraction ($P = 0.028$) on RI was also observed.

In conclusion, interlobar RI increases with the severity of DMVD and azotemia. A cause-effect relationship remains however to be established.

ABSTRACT N/U-25

ASSOCIATION BETWEEN FELINE ANTIBODY RESPONSES TO ALPHA-ENOLASE AND AZOTEMIA IN PRIVATELY-OWNED CATS. C Sonius[1], JC Whittemore[2], JR Hawley[1], L Schoenberg[3], MR Lappin[1]. [1]Colorado State University, Fort Collins, CO. [2]University of Tennessee, Knoxville, TN. [3]Just Cats Hospital, Cleveland, OH.

Antibodies against alpha-enolase are associated with immune-mediated nephritis in people. It was previously shown that vaccinated cats commonly develop antibodies against alpha-enolase. The purpose of this study was to assess for associations between alpha-enolase antibodies and azotemia in privately-owned cats.

Clinically stable privately owned cats ≥ 10 years of age, with and without azotemia (creatinine > 2 mg/dl), and with an available vaccine history for ≥ 5 years were recruited for the study. Sera were assayed for creatinine concentrations and alpha-enolase antibodies by use of previously validated techniques. Results from cats with and without azotemia were compared by Student's 2-tailed t test or Fisher's exact test with significance defined as $p < 0.05$.

Median ages were 15 years (range: 10–18) and 12 years (range: 10–15) for cats with ($n = 35$) and without azotemia ($n = 27$), respectively. There was no significant difference in vaccine events (number, type, or route of administration) between groups. Azotemic cats (34.3%) were more likely than normal cats (12.5%) to be positive for antibodies against alpha-enolase ($p = 0.016$). In addition, alpha-enolase antibody concentrations were greater ($p = 0.041$) in azotemic cats (mean % ELISA = 62.5%) than cats with normal creatinine concentrations (mean %ELISA = 47.2%).

Results of this study suggest that antibodies against alpha-enolase in cats may be associated with renal disease. Additional prospective evaluation in a larger number of cats is indicated.

ABSTRACT N/U-26

ACUTE KIDNEY INJURY (AKI) AS A PREDICTOR OF MORTALITY IN DOGS AND CATS. E. Harison[1], C Langston[1], D Palma[1], K Lamb[2], P Fox[1]. [1]Animal Medical Center, NY, NY. [2]University of Florida, Gainesville, FL.

AKI is used in human medicine as a predictor of mortality based on the AKIN (Acute Kidney Injury Network) scoring system which utilizes relative increases in creatinine to determine stage. With this scheme, mortality has been shown to increase as the stage of kidney injury (indicated by AKIN score) increases. Accordingly, we hypothesized that this system would improve predicting prognosis in dogs and cats. We retrospectively evaluated 1088 dogs and 856 cats (2008–2009) that had ≥ 2 creatinine measurements within 7 days, and whose first creatinine was < 1.6 mg/dl. Patients were categorized as: Level 0 (no AKI); Level 1 (second creatinine value < 1.6 mg/dL, but creatinine increased ≥ 0.3 mg/dL); or Level 3 (second creatinine > 1.6 mg/dL with a creatinine increase ≥ 0.3 mg/dL). Thirty and 90 day survival for each level was compared to level 0. Adjusted odds ratio (OR) in dogs for 30 day survival was 1.3 for Level 1 (CI 95%, 0.8–2.2) and 3.2 (CI 95%, 1.8–5.5) for Level 2; OR for 90 day survival was 1.3 for Level 1 (CI 95%, 0.8–2.2) and 3.7 (CI 95%, 2.1–6.5) for Level 2. For cats, OR at 30 days was 1.5 (CI 95%, 0.5–4.6) for Level 1 and 3.1 (CI 95%, 1.5–6.7) for Level 2; OR for 90 day survival was 0.9 (CI 95%, 0.3–2.8) for Level 1 and 4.1 (CI 95%,

1.8–9.3) for Level 2. Thus, detecting increasing stage of AKI helps predict mortality in dogs and cats.

ABSTRACT N/U-27

FELINE URATE UROLITHIASIS: 143 CASES (2000-2008). J Dear[1], R Shiraki[2], A Ruby[2], J Westropp[3]. [1]William R Pritchard Veterinary Medical Teaching Hospital, University of California, Davis, CA, [2]Gerald V. Ling Urinary Stone Analysis Laboratory, University of California, Davis, CA and [3]the Department of Veterinary Medicine and Epidemiology, University of California, Davis, CA.

Feline urate urolithiasis accounts for 10% of the feline stones our laboratory analyzes each year; little information is known about this disease, particularly the incidence of those cats with hepatopathies.

The objective of the study was to characterize the signalment, clinicopathologic data, and diagnostic imaging of cats with this disease as well as the salts of uric acid present.

A retrospective analysis of feline urate uroliths submitted to the Stone Lab between January 2000-December 2008 were included. From these data, primary veterinarians were solicited to submit records. Furthermore, all records from cats with urate uroliths from the VMTH were analyzed separately.

143 records were received from the primary care veterinarians.

Sixteen cases were identified from the VMTH. Median values for the CBC and chemistry panels available were within the reference ranges provided, with only a few outliers present. Of the 78 cats with radiographic reports, 70 (90%) had visible evidence of uroliths. Two external cases had confirmed PSS; five cases from the VMTH had a PSS. Cats with urate uroliths and PSS were younger than cats without a documented hepatopathy (2 years vs. 7 years). The Siamese breed was overrepresented. All stones were ammonium hydrogen urate.

The pathogensis of urate uroliths in cats is poorly understood. Most cats were not completely evaluated for PSS, however, there were few clinicopathologic parameters which indicated hepatopathies were present. Further studies are warranted to evaluate genetics and purine metabolism in cats with urate uroliths to help tailor proper management and breeding strategies.

ABSTRACT N/U-28

EVALUATION OF NORMAL SERUM CONCENTRATIONS OF THE PUTATIVE UREMIC TOXINS, INDOXYL AND P-CRESYL SULFATE, IN HEALTHY ADULT CATS. BH Schmidt[1], J Dreisbach[2], R Krebber[3], J Küsters[2], G Beddies[1]. [1]Bayer Animal Health GmbH, Monheim, Germany. [2]Conreso GmbH, Munich, Germany. [3]Bayer CropScience AG, Monheim, Germany.

3-Indoxyl and p-cresyl sulfate (IS, and CS, respectively), small protein-bound molecules derived from gastrointestinal protein metabolism, are among the most important uremic solutes affecting morbidity and mortality in human chronic kidney disease (CKD). In the blood stream, these compounds are predominantly bound to protein, but their debilitating effects on prognosis and quality of life in CKD appear to be driven by the free fraction. The objectives of the present study were to assess the normal, physiological levels of IS and CS in healthy cats and to evaluate the correlation of the respective free and protein-bound levels.

Blood samples were taken from 105 clinically healthy adult cats enrolled at five participating veterinary practices in Germany. After centrifugation, the serum was deep frozen until transport on dry ice to the analytical laboratory. Serum creatinine and urea levels were quantified by VetTest® (IDEXX Laboratories, Inc.). Total and free IS and CS, respectively, were quantified by turbulent flow chromatography coupled with a tandem mass spectrometry detector. Statistical analysis of the results comprised i) a descriptive report of the median with upper and lower bounds of the 95% confidence interval for reference values of IS and CS, ii) a calculation of various Pearson correlation coefficients r, also tested with reference to the null hypothesis of no relationship, and iii) Wilcoxon-Mann-Whitney U-test for an estimation of the effect of hemolysis on serum IS or CS levels. Six animals with serum creatinine or urea levels outside the reference range were excluded from the calculation of reference values.

Median levels of IS in cat serum were 1.19 mg/L with upper and lower bound 95% confidence intervals at 1.46 and 0.99 mg/L, respectively. The corresponding median levels of CS were 2.11 mg/L (median) and 2.46 vs 1.57 mg/L (upper vs lower bound levels, respectively). These values showed a low, non-significant correlation with serum creatinine or urea levels. However, IS and CS serum levels were moderately correlated (total levels $r = 0.4808$, $p < 0001$). Their respective free levels constituted about 5% of the total serum levels ($r \geq 0.8977$, $p < 0.001$). Non-hemolytic samples tended to yield lower values than hemolytic samples. Due to the low number of hemolytic samples ($N = 14$), the group difference could, however, not be statistically confirmed.

The results indicate that it is sufficient to determine total levels of either IS or CS in serum while studying the effects of therapeutic or dietetic interventions on the evolution of these parameters in feline CKD. Reference values are provided for orientation towards clinically relevant changes.

ABSTRACT N/U-29

EXPRESSION OF 15-HYDROXYPROSTAGLANDIN DEHYDROGENASE (PGDH) IN UROTHELIUM OF CATS WITH CHRONIC IDIOPATHIC CYSTITIS. OA Khan[1], JM Kruger[2], SD Fitzgerald[2], M Liebert[1]. [1]University of Michigan, Ann Arbor, MI. [2]Michigan State University, East Lansing, MI.

Disrupted urothelial differentiation has been implicated in the pathogenesis of feline idiopathic cystitis (FIC). Studies of cultured human urothelium have shown that abnormalities in urothelial differentiation and repair may be mediated by persistent 15-hydroxy-prostaglandin dehydrogenase (PGDH) activity and subsequent metabolism of cytoprotective prostaglandins. The goal of this study was to confirm persistent PGDH expression in FIC bladders compared to desmoplakin I+II expression, a marker of urothelial differentiation. Urinary bladder biopsy specimens were obtained by cystotomy from 9 symptomatic cats with chronic FIC. Cats with a history of another major disease, previous cystotomy, or recent treatment with corticosteroids, NSAIDs, antihistamines, antidepressants, or glycosaminoglycans were excluded. Urinary bladder tissue specimens were also obtained from 10 untreated clinically normal specific-pathogen-free cats. Tissue specimens were fixed in buffered 10% formalin and embedded in paraffin. Tissue sections were deparaffinized and subjected to citrate buffer microwave antigen retrieval. Tissues were stained for PGDH using a rabbit anti-PGDH antibody, an isotype negative control or goat anti-desmoplakin I+II and developed using the avidin-biotin peroxidase complex method. All FIC (9/9) and normal (10/10) cat bladder samples showed similar staining of urothelial cytoplasm for PGDH. However, desmoplakin I+II staining, found on the luminal cell surface in 4/4 normal tissues, was disrupted in 6/6 FIC bladder samples. Desmoplakin I+II staining confirmed altered urothelial differentiation in FIC cats. However, PGDH expression remained intact in FIC samples.

We hypothesize that PGDH expression in FIC may contribute to its pathophysiology due to breakdown of prostaglandins essential for urothelial healing. Additional studies will explore this hypothesis.

ABSTRACT N/U-30

VARIATION IN KIDNEY SIZE IN A POPULATION OF CATS. K. Hamon, K. Newkirk, A. Matthews, C. LeBlanc. University of Tennessee College of Veterinary Medicine, Knoxville, TN.

The University of Tennessee College Of Veterinary Medicine's Picture Archiving and Communication System was searched over a 9 month period for cats that had undergone both abdominal radiographs and ultrasound during the same visit. One hundred and three cats were identified (age range < 1 to 18yrs; median 11yrs). Kidney size was determined based on radiographic and ultrasound findings. Of the included cats, 41.8% had two normal sized kidneys, 18.4% had one small and one normal, 15.5% had one large and one normal, 11.7% had two small, 8.7% had two large, and 3.9% had one small and one large kidney. The presence of mineralization, uroliths and hydronephrosis was also noted. Medical records were reviewed for clinical chemistry data and historical information concerning previous urinary disease. No significant differences were found between kidney size and renal function, kidney size and the presence of uroliths, renal mineralization and function or the presence of uroliths and function. The presence of uroliths was significantly associated with hydronephrosis. Of the 30 cats with at least one large kidney, 9 (30%) had hydronephrosis. Of the 23 cats with current or previously diagnosed uroliths, urinary tract infections or other uropathies, 10 (43.5%) had at least one small kidney. Small kidneys were commonly found in older cats, however, this correlation was not statistically significant. Based on these findings, small kidneys are more likely to be the result of urinary disease as opposed to being either congenital or due to aging.

ABSTRACT N-1

INTRAVENOUS FAT EMULSION (IFE) FOR TREATMENT OF IVERMECTIN TOXICOSIS IN 3 DOGS.
HM Wright[1], AV Chen[1], PA Talcott[1], RH Poppenga[2], KL Mealey[1]. [1]Washington State University College of Veterinary Medicine, Pullman, WA. [2]University of California Davis School of Veterinary Medicine, Davis, CA.

This study aimed to evaluate IFE, which has been advocated for treatment of lipid-soluble drug intoxication, in the treatment of clinically-occurring canine ivermectin toxicosis.

One Australian Shepherd and two Miniature Australian Shepherds were included. All three dogs were homozygous for the MDR-1 gene mutation. Two dogs roamed on horse ranches where ivermectin-based deworming products had recently been used. Ivermectin was administered to the third dog (165 µg/kg PO). All three dogs exhibited tremors, ptyalism, and CNS depression, which progressed over several hours to stupor in two dogs, and to a comatose state requiring mechanical ventilation in the remaining dog. A 20% formulation of IFE (Liposyn II, Hospira) was administered as a bolus (1.5 ml/kg) followed by a slow IV infusion (7.5–15 ml/kg over 30 minutes). No change was observed in the neurologic status of any patient. Lipemia visible upon blood sampling persisted for 36 hours in one dog. No other adverse effects were noted. Serum ivermectin levels confirmed ivermectin exposure in each case.

In this study, IFE administration did not result in clinical benefit in cases of ivermectin toxicosis. Brain ivermectin concentrations in MDR1 mutant/mutant genotype dogs may be too high to be overcome by IFE. Additionally, these dogs may lack P-glycoprotein-mediated biliary clearance mechanisms needed for optimal IFE function. Further investigation is needed to determine the utility and optimal dosing of IFE in canine toxicoses, to characterize its safety, and to determine how MDR-1 status may alter the efficacy of IFE in treatment of canine ivermectin intoxication.

ABSTRACT N-2

ORAL RUFINAMIDE IN DOGS: PHARMACOKINETICS AND SAFETY.
HM Wright, AV Chen, SE Martinez, KL Mealey, NM Davies. Washington State University, Pullman, WA.

Rufinamide is a recently approved antiepileptic drug used for the treatment of seizure disorders in human patients. Rufinamide is administered at a dose of 45 mg/kg divided twice daily to achieve therapeutic concentrations of 15 mg/ml. The objective of this study was to determine the pharmacokinetic properties and short-term adverse effects of single-dose oral rufinamide in healthy dogs in preparation for a possible clinical trial evaluating the efficacy of rufinamide in the treatment of canine epilepsy.

Six healthy adult dogs were included. The pharmacokinetics of rufinamide were calculated following administration of a single mean oral dose of 20.0 mg/kg (range 18.6–20.8 mg/kg), extrapolated from the dose used in human patients. Dogs were monitored by repeat physical examinations, electrocardiograms and blood pressure assessments during the course of the study. Plasma rufinamide concentrations were determined using high-performance liquid chromatography. Pharmacokinetic data were analyzed using WinNonlin version 1.0.

No adverse effects were observed. The mean terminal half-life was 9.86 +/− 4.77 hours. The mean maximum plasma concentration was 19.56+/−5.82 mg/ml and the mean time to maximum plasma concentration was 9.33 +/− 4.68 hours. Mean clearance was 1.448 +/− 0.703 L/hr. AUC_{inf} was 410.72 +/− 175.88 mg∗h/ml.

Results of this study suggest that rufinamide given orally at 20 mg/kg twice daily in healthy dogs should result in a plasma concentration and half-life sufficient to achieve the therapeutic level extrapolated from humans without short-term adverse effects. Further investigation into the efficacy and long-term safety of rufinamide in the treatment of canine epilepsy is warranted.

ABSTRACT N-3

PRELIMINARY MORPHOMETRIC EVALUATION OF SYRINGOMYELIA IN AMERICAN BRUSSELS GRIFFON DOGS.
A.C. Freeman, S.R. Platt, M. Kent, E. Huguet. University of Georgia, College of Veterinary Medicine, Athens, GA.

The aims of this study were to investigate the ABG for (i) the prevalence of skull abnormalities; (ii) the prevalence of SM; (iii) an association between lateral ventricular size, cerebellar size and SM; and (iv) associations between SM, skull abnormalities, CSF pleocytosis and clinical signs.

Seventy-six ABGs, recruited as part of a larger epidemiological and genetic study, underwent brain and spinal MRI evaluation (3.0T General Electric Signa HDx, Milwaukee, WI). All dogs were evaluated neurologically, recording deficits and the presence of spinal pain. Sequences acquired included T2W, T1W pre- and post-contrast, and T2W FLAIR, sagittal and transverse. Cervical spinal cord central canal (CC) and or syrinx size and its percent area of spinal cord was measured using OsiriX®. The presence of Chari-like malformation (CM) was assessed by recording the presence of caudal cerebellar deviation and/or foramenal vermal herniation. Lateral ventricle and cerebellar volume was expressed as a percent of the cerebrum and intracranial volume2qa respectively. Forty-five dogs underwent atlanto-occipital cerebrospinal fluid tap at the time of MRI and the white blood cell (WBC) count was recorded. Student's t-tests were used to compare the measured variables between groups with and without skull abnormalities, spinal pain and neurological signs.

The mean age of the 30 males (24 intact) and the 46 females (34 intact) was 50.4 months (range 8–135; median 44 months). Neurological deficits and neck pain were noted in 21 (27%) and 15 (19.7%) of dogs respectively; 5 dogs (6.57%) exhibited both. Cerebellar deviation and vermal herniation were present in 37 (48.68%) and 46 (60.52%) dogs respectively; twenty-three dogs (30.26%) had both. Mean height of the CC was 2.3 mm (0–7.2 mm). Forty (52.63%) CCs were greater than 2 mm in height; the mean length of these lesions was 2.03 vertebrae (0.5–7). Mean CSF WBC count was 4.97/µl (0–39). Syrinx height and extent were significantly higher in dogs with neurological signs (size $p = 0.01$; extent $p = 0.0004$). There were no significant differences in syrinx sizes and extent in dogs with or without skull abnormalities or spinal pain. There were no associations of syrinx height or extent with CSF WBC count or age of dog. Intact females had a significantly lower syrinx extent than intact males ($p = 0.009$). There were no significant differences in presence of spinal pain or neurological signs between dogs with or without skull abnormalities. There was a significant negative association of ventricular percentage and cerebellar percentage ($p < 0.0001$). There was a significant association of ventricular percentage with syrinx percentage ($p = 0.0015$) and height ($p = 0.0007$).

This study suggests that SM and CM are prevalent in ABGs. Syrinx size and extent are associated with neurological signs and ventriculomegaly is associated with both small cerebellar size and large syrinx size. However, SM may not be associated with CM as defined by cerebellar herniation and deviation and is not associated with CSF inflammation.

ABSTRACT N-4

CANINE INTERVERTEBRAL DISC FENESTRATION USING A NOVEL TISSUE RESECTION DEVICE.
S. Thomovsky, R. Packer, N. Lambrechts, G. Moore. Purdue University School of Veterinary Medicine, West Lafayette, IN.

The power tissue resection device (PTRD) is a hand-piece comprised of an outer cannula with motor driven vacuum-assisted inner cutting blade. This device was designed and is marketed for human neurosurgical brain/spinal cord tumor resection. The purpose of this study is to describe the use of the PTRD for intervertebral disc fenestration and to compare the effectiveness of manual fenestration to that of the PTRD.

Fifteen cadaveric lumbar spines were randomly placed into three study groups: group 1 was the control group on which no fenestrations were performed, group 2 was the manual fenestration group and group 3 was the PTRD fenestration group. The effectiveness of fenestration via both manual and PTRD was assessed by calculating the ratio of remaining nuclear weight post fenestration to total nuclear volume. Discs with lower ratios were more effectively fenestrated.

Results showed a smaller ratio of post fenestration remaining nuclear weight to nuclear volume following fenestration with the PTRD (0.23 ± 0.09) as compared to manual fenestration (0.30 ± 0.10). These results did not show statistical significance. When fenestrated samples were compared to control samples (0.39 ± 0.07), there was a statistically significant reduction in ratios.

In conclusion, the PTRD is easy to use and is as effective as the manual technique for canine intervertebral disc fenestration.

ABSTRACT N-5

GLIOMATOSIS CEREBRI IN A FAMILY OF BEARDED COLLIES. D Henke[1], D Gorgas[2], M Vandevelde[1], J Lang[2], F Steffen[3], D York[4], PJ Dickinson[4], T Leeb[5], L Alves[1], A Oevermann[6]. Department of Clinical Veterinary Medicine, [1]Divisions of Clinical Neurologyand [2]Clinical Radiology, [5]Institute of Genetics, [6]Department of Clinical Research and Veterinary Public Health; University of Bern; [3]Department of Clinical Neurology, University of Zürich, Switzerland; and [4]Department of Surgical and Radiological Sciences, School of Veterinary Medicine, University of California, Davis, CA, USA.

According to the human WHO classification gliomatosis cerebri (GC) is a rare astrocytic tumor affecting at least three lobes of the brain with extensive infiltration, but relative preservation of brain architecture. GC has not been reported to occur as a hereditary disease, neither in man nor in animals. Here, we report the temporally clustered occurrence of GC in a family of Bearded collies.

A 7 years old female Bearded collie with forebrain signs was presented. Differentials included inflammatory/ infectious, metabolic/ toxic, and neoplastic diseases. Within a time period of 12 months, 3 offspring of this bitch were presented with similar clinical signs. Two dogs were full siblings (2 males). The remaining female dog originated from a match with a different male dog.

MRI was performed in all 4 dogs and revealed a diffuse and extensive intra-axial lesion with moderate mass effect and midline shift. The ill defined lesion showed mainly a white matter distribution with hyperintense signal in T2-w and FLAIR images and iso- to hypointense signal in T1-w images without contrast enhancement. The lesion was bilateral in all cases, continued along the white matter extending partially into the gray matter with contact to the brain surface.

Neuropathology revealed a diffuse and extensive infiltration of the brain and spinal cord by a neoplastic glial cell population involving white and gray matter of both hemispheres, thalamus, brainstem and cerebellum in all 4 dogs. Based on the cell morphology and immunoexpression of glial fibrillary astrocytic protein by neoplastic cells diagnosis of GC was made.

This is the first report of familial occurrence of GC, which is likely the result of a germ-line mutation. Several human hereditary cancer syndromes are associated with CNS tumors including amongst others the Li-Fraumeni cancer family syndrome (p53 mutation), neurofibromatosis (type 1 and 2) (neurofibromin, merlin mutation), and tuberous sclerosis (hamartin, tuberin mutation). Furthermore, familial clustering of human gliomas unassociated to the known inherited cancer syndromes has been described. In the dog, hereditary CNS tumors are not known.

The exact mode of inheritance and putative gene mutations of GC in this Bearded collie family are currently under investigation. Preliminary results are consistent with a monogenic autosomal dominant mode of inheritance, although a recessive inheritance cannot be completely ruled out at this time. Mutations in the TP53 gene were not found following amplification and sequencing of exons 5–8 in 2 affected dogs.

Previously presented at the ECVN annual meeting in Cambridge, UK.

ABSTRACT N-6

A MISSENSE MUTATION IN *HEXA* IN JAPANESE CHIN DOGS WITH GM_2 GANGLIOSIDOSIS. DN Sanders[1], R Zeng[2], D Wenger[3], GS Johnson[2], ML Katz[1], S Platt[4], DP O'Brien[5]. [1]Mason Eye Institute, University of Missouri School of Medicine, Columbia, MO, USA [2]Department of Veterinary Pathobiology, University of Missouri College of Veterinary Medicine, Columbia, MO, USA [3]Department of Neurology and Neurogenetics, Thomas Jefferson University Jefferson Medical College, Philadelphia, PA, USA [4]Department of Small Animal Medicine & Surgery, University of Georgia, Athens, GA, USA [5]Department of Veterinary Medicine and Surgery, University of Missouri College of Veterinary Medicine, Columbia, MO, USA.

The GM_2 gangliosidoses are characterized by a deficiency of β-hexosaminidase. There are two isoforms: Hex A composed of an α and β subunit encoded by *HEXA* and *HEXB* genes respectively and Hex B with two β subunits. Hex A requires an activator encoded by *GM2A*. Two Japanese chin dogs with confirmed GM_2 gangliosidosis showed elevated total hexosaminidase and normal hexosaminidase A activity, a pattern associated with the AB variant in humans and consistent with prior reports in the breed. This study was performed to identify the mutation responsible using resequencing with an Applied Biosystems 3730xl DNA analyzer as previously described (Awano 2009).

Mutations in *GM2A* cause the AB variant in humans, but resequencing *GM2A* revealed no mutation that could account for the disease. Resequencing *HEXA* and *HEXB* revealed a c.967G > A mutation in *HEXA* which was homozygous in both affected dogs. Sixty-five normal Japanese Chin dogs were screened for the mutant allele; 60 were homozygous for the ancestral allele and 5 heterozygous.

This mutation predicts a p.323E > K substitution affecting one of two primary active-site amino acids that participate in the hydrolysis of GM_2 ganglioside. Substitution of a lysine residue at this site is likely to eliminate subunit A enzymatic activity. The apparently normal levels of hexosaminidase A activity in affected dog samples may be a result of β subunit overexpression. Human Hex B possesses low levels activity against the artificial substrate used to assess Hex A activity, but specificity of activity of the canine enzyme is not known.

Previously presented at the American Society for Neurochemistry: additional data in this abstract.

ABSTRACT N-7

PHARMACOKINETICS AND TOLERABILITY OF INTRAVENOUS FOSPHENYTOIN IN FOUR HEALTHY DOGS. EM Craft[1], EE Patterson[1], LD Coles[2], IE Leppik[2], JC Cloyd[2]. University of Minnesota: [1]Veterinary Medical Center, St. Paul, MN, [2]Center for Orphan Drug Research, Minneapolis MN.

Phenytoin (PHT) is the intravenous drug of choice in humans for seizure emergencies following benzodiazapines. IV fosphenytoin (FOS) is a PHT pro-drug which causes less administration related adverse events. While the short half-life of PHT is not suitable for chronic oral therapy in dogs, IV FOS has not been studied.

Two dogs received 15 mg/kg phenytoin equivalent (PE) and two dogs received 25 mg/kg PE of fosphenytoin intravenously at a rate of 50 mg PE/min. Blood for plasma levels were drawn at 10 time-points over 12 hours; total and unbound drug levels were measured by HPLC. Vital signs including EKG, blood pressure, and neurological examination were monitored.

The half-life of metabolism of FOS to PHT was ~10 min, with > 80% of FOS metabolized to PHT by 30 minutes. Eighty to 84% of PHT was protein-bound during the first 15 minutes after dosing, compared to 90–95% in humans. The elimination half-life for total PHT ranged from 2.8–3.5 hours and for unbound PHT ranged from 1.9–5.4 hours. Dogs receiving 15 mg/kg PE intravenously achieved unbound PHT plasma maximum concentrations of 2.2–2.4ug/ml at 5 minutes, consistent with human loading dose levels. Adverse events observed in some dogs included vomiting, mild ataxia, and short lived tremors, the severity of which appeared dose dependent. All dogs were clinically normal within 30 minutes of all doses.

A 15 mg/kg PE dose of IV FOS appears adequate for production of PHT levels predicted to be effective for the treatment of canine seizure emergencies. Further studies in clinical canine patients are warranted.

ABSTRACT N-8

THERAPEUTIC APHERESIS: A NOVEL TREATMENT FOR MYASTHENIA GRAVIS IN DOGS. CA Palm, LD Cowgill, PJ Dickinson, BK Sturges. UC Davis School of Veterinary Medicine, Davis, CA.

Acquired myasthenia gravis (MG) is caused by antibody-mediated inactivation of the acetylcholine receptor on the neuromuscular endplate causing focal, regional or generalized muscle weakness. Many medical treatments have been reported; however, responses to therapy and outcomes are unpredictable and death often results from aspiration pneumonia.

Therapeutic apheresis is an extracorporeal procedure that separates blood into its components for removal or specific alteration prior to return to the patient. Therapeutic plasma exchange (TPE) is an apheresis treatment in which plasma (containing pathologic antibodies) is removed and exchanged with donor plasma. TPE is used routinely to treat MG in human patients with severe disease or disease unresponsive to conventional therapy.

We report the successful use of TPE to treat 2 large breed dogs with confirmed MG (aceytlcholine receptor antibody concentration: 3.05 and 3.32 nM/L, respectively; normal concentration: < 0.6 nM/L) that was severe and not adequately responsive to traditional therapies. Both dogs were non-ambulatory, recumbent, and demonstrated megaesophagus and aspiration pneumonia.

Three TPE treatments (1 plasma exchange each) were performed over 5 and 7 days, respectively, in each dog without complication. Both dogs became ambulatory within 3 days of starting TPE treatment with subsequent resolution of regurgitation and megaesophagus. Pyridostigmine was continued during TPE sessions and discontinued in both dogs within 3–6 months. Both dogs remain asymptomatic and have had no recurrence of MG during 16 and 4 months of follow-up, respectively.

TPE is a viable treatment option for dogs with MG that have severe disease, life-threatening complications or that remain unresponsive to traditional therapies. TPE may alleviate clinical signs more rapidly, and improve long-term outcomes when compared to historical experiences in patients with comparable disease.

ABSTRACT N-9

SPINAL MENINGIOMAS IN DOGS: DESCRIPTION OF SEVEN CASES INCLUDING A NOVEL RADIOLOGICAL AND HISTOPATHOLOGICAL PRESENTATION. R José-López, C De la Fuente, M Pumarola, S Añor. Facultat de Veterinària, Universitat Autònoma de Barcelona, Spain.

Clinical findings, clinicopathologic data, imaging features, and treatment of canine spinal meningiomas have been described in the veterinary literature, but histological characteristics and tumor grading have less commonly been reported. The aims of this retrospective case series were to describe the clinical, imaging, and histologic features of seven canine spinal meningiomas including a cervical spinal cystic meningioma that had imaging and intraoperative features of a subarachnoid cyst.

Medical records from dogs with a histopathological diagnosis of spinal cord meningioma presented to the Veterinary Teaching Hospital between 2006 and 2010 were reviewed. Signalment, presenting clinical signs, physical and neurologic examination, clinicopathologic data, surgery reports and available images were reviewed. All meningiomas were histologically classified and graded following the international WHO human classification for CNS tumors. Seven dogs were included, 4 males and 3 females. Median age at presentation was 8.7 years (range, 3.5–11.4 years), and median weight was 35 kg (range, 8–45 kg). Median time between onset of clinical signs and diagnosis was 108 days (range, 45 days - 1 year). Cerebrospinal fluid (CSF) analysis was performed in 4 dogs, showing increased protein concentration in 2 cases, and being normal in the other 2. Spinal radiographs revealed vertebral canal widening in one case. Myelography (4/7) showed intradural/extramedullary lesions in three cases, one of them consistent with a CSF-filled subarachnoid cavity, and an extradural lesion in one case. Magnetic resonance imaging (MRI) was performed in all cases and revealed mild to marked hyperintensity on T2W and precontrast T1W images and homogeneous contrast enhancing (CE) intradural/extramedullary masses (4 cervical and 2 thoracic) in six cases, with one of these showing an additional intramedullary CE pattern. A dural tail was identified in two dogs. One dog had a fluid-filled subarachnoid enlargement located dorsally to the spinal cord. This lesion was hyperintense on T2W, hypointense on T1W and FLAIR images, and did not enhance. It was diagnosed as a spinal subarachnoid cyst, but the histopathological study of the surgically resected mass revealed a grade I cystic meningioma. Five other cases underwent cytoreductive surgery, two transitional meningiomas (grade I) that survived 3 (alive at the time of writing) and 7 months; and three anaplastic meningiomas (grade III) that survived 10–16.6 months before neurological deterioration and euthanasia. Another anaplastic meningioma was euthanized right after diagnosis.

There are few reports grading canine spinal meningiomas, with most being grade I or II. Of the few grade III tumors reported, only one had been treated surgically and was euthanized 90 days later because of neurological deterioration. We report four grade III (anaplastic) meningiomas, three of which surgically treated and with longer survival times. Finally, cystic meningioma should be considered in the differential diagnosis of cases with imaging features consistent with arachnoid cyst because of their similar appearance, making histopathological analysis essential for a definitive diagnosis.

ABSTRACT N-10

THE PROGNOSTIC VALUE OF ADMISSION BLOOD GAS PARAMETERS IN DOGS WITH TRAUMATIC BRAIN INJURY. S. Cameron, D. Fletcher. Cornell University College of Veterinary Medicine, Ithaca, NY.

Head trauma is a common veterinary emergency, but few prognostic indicators have been studied in dogs, making it challenging for clinicians to counsel clients about the odds of recovery. A recent meta-analysis showed that higher plasma glucose, lower plasma pH and lower hemoglobin at admission were associated with increased risk of death in human head trauma. The goal of this retrospective study was to investigate the association between admission point of care blood gas parameters and survival to discharge in dogs with head trauma.

Fifty one dogs presenting to the Cornell University Hospital for Animals with head trauma from 2007 to 2010 that had a blood gas analysis done within 1 hour of presentation were eligible for inclusion. Parameters assessed included glucose, base excess (BE), anion gap (AG), pH, hemoglobin, and sodium.

Biochemical data were found to be normally distributed using the Kolmogorov-Smirnov test. T-tests or Welch tests were used to compare parameters between survivors (S,n = 42) and non-survivors (NS, n = 9). Of glucose, BE, AG, pH, hemoglobin, and sodium, only mean glucose (S = 131 mg/dl, NS = 171.4 mg/dl, $p = 0.029$) was significantly different between groups, although there was a trend for a difference in mean BE (S = −3.6, NS = −8.5, $p = 0.055$). Logistic regression analysis showed that of the parameters, only BE was independently associated with outcome (odds ratio 0.79, 95% CI = 0.63–0.98, $p = 0.036$).

These results suggest that two easily measured biochemical parameters (glucose and BE) may yield useful prognostic information in dogs with head trauma, but further studies are needed to further elucidate these findings.

ABSTRACT N-11

SERUM LACTATE LEVELS IN PHENOTYPICAL CHONDRODYSTROPHIC DOGS WITH TYPE I INTERVERTEBRAL DISC DISEASE. CF Altenhoefer, BK Sturges, KM Vernau, MF Knipe, K Hopper. University of California School of Veterinary Medicine, Davis, CA.

Type I intervertebral disc disease (IVDD) commonly affects chondrodystrophic dogs. Neurological recovery and outcome following surgical decompression may be unpredictable due to suspected ischemic neuronal injury. Hyperlactatemia has been associated with spinal cord injury in humans and experimental animals. The purpose of the study was 1) to determine the relationship between serum and CSF lactate levels and 2) to compare lactate levels with neurological outcome following decompressive surgery in dogs with IVDD.

Healthy, chondrodystrophic dogs diagnosed with IVDD localized to the T3-L3 spinal cord were included. Serum lactate levels were obtained at: anaesthetic induction, skin incision, muscle dissection, and extubation. In patients with hyperlacatemia at extubation, additional samples were obtained. CSF was analyzed for lactate concentration. Neurological status was recorded at presentation and multiple times during the recovery period.

31 dogs were included in the study (3–12 years old). 22/31 dogs had normal lactate levels throughout the study. 9/31 dogs had serum hyperlactatemia prior to anaesthetic induction; 6/9 dogs returned to normal during anaesthesia and 3/9 dogs had continued hyperlactatemia until the end of the observation period. Neurological status of the dogs varied similarly between all groups. In 12/14 dogs where CSF lactate levels were measured, initial serum levels were lower than CSF lactate levels; in 5/14 dogs where CSF and serum were collected simultaneously, serum lactate concentration was consistently lower than CSF lactate.

No association between presenting neurological status or neurological outcome and serum or CSF lactate concentration was made. Neither serum nor CSF lactate concentration is useful for predicting neurological outcome in dogs with IVDD.

ABSTRACT N-12

RELATIONSHIP OF CEREBELLAR VOLUME TO SYRINGOMYELIA IN THE CAVALIER KING CHARLES SPANIEL. TA Shaw[1], I McGonnell[1], CJ Driver[1], C Rusbridge[2], HA Volk[1]. [1]The Royal Veterinary College, London, UK. [2]Stone Lion Veterinary Centre, Goddard Veterinary Group, London, UK.

Chiari-like Malformation (CM) has been associated with syringomyelia (SM) in Cavalier King Charles Spaniel (CKCS) and is postulated to result from a mismatch between the volume of the caudal cranial fossa and the brain parenchyma contained within. The objective of this study was to assess the role of cerebellar volume in caudal cranial fossa overcrowding and syringomyelia.

Three dimensional models were created using T2-weighted transverse magnetic resonance images in the commercial software package Mimics ®. Volumes of cerebellar parenchyma were analyzed as percentages of caudal cranial fossa volume (cerebellar caudal cranial fossa percentage) and total brain parenchyma volume (cerebellar brain percentage). Data was assessed for normality and the appropriate statistical test was used to compare means/medians between groups.

Forty-five small breed dogs (SB), 58 CKCS and 31 Labradors (LD) were compared. As SM is thought to be a late onset disease process, two subgroups were formed for comparison: 21 CKCS younger than 2 years with SM (group 1) and 13 CKCS older than 5 years without SM (group 2).

CKCS had a larger cerebellar caudal cranial fossa percentage than the other groups (CKCS 52.19% [40.07–58.76%] vs. SB 47.81% [40.36–62.91%] and LD 41.32% [32.59–52.95%]; $p < 0.001$). The cerebellar brain percentage was also larger in CKCS compared to the other groups (CKCS 8.90% [6.62–11.46%] vs. SB 7.37% [5.25–11.34%] and LD 7.23% [6.36–9.54%]; $p < 0.001$). Group 1 had a significantly larger cerebellar caudal cranial fossa percentage than group 2 (53.71% ± 1.27 vs. 49.31% ± 2.35, $p = 0.001$) and a significantly larger cerebellar brain percentage (9.45% ± 0.43 vs. 8.58% ± 0.55, $p = 0.021$).

Our findings show that the CKCS has a relatively larger cerebellum than small breed dogs and Labradors and there is an association between increased cerebellar volume and SM in CKCS.

ABSTRACT N-13

CHANGES OVER TIME IN CAUDAL CRANIAL FOSSA VOLUMES AND OF CEREBELLAR HERNIATION IN CAVALIER KING CHARLES SPANIELS WITH CHIARI-LIKE MALFORMATION. S. Hamilton[1], L. De Risio[2], C. Rusbridge[3], CJ Driver[1], R. Dennis[2], I McGonnell[1], HA Volk[1]. [1]The Royal Veterinary College, London, UK. [2]The Animal Health Trust, Newmarket, UK, [3]Stone Lion Veterinary Centre, Goddard Veterinary Group, London, UK.

Chiari-like malformation (CM) is nearly omnipresent in the Cavalier King Charles Spaniel (CKCS) breed. The mis-match of the caudal cranial fossa and the parenchyma within is thought to lead to syringomyelia (SM). There is currently a lack of information if the morphological changes seen in CKCS with CM are progressive or non-progressive. In this retrospective study we used established measurements of cerebral volumes, foramen magnum height and cerebellar herniation length to assess if there is a significant difference between subsequent magnetic resonance (MR) imaging of the brain of the same dog.

Electronic patient records were reviewed for CKCS with CM which had two separate MRI scans, which were a minimum of 3 months apart. CKCS with diseases affecting measurements were excluded. For the volumetric measurements three-dimensional models were created using T2-weighted transverse MR images in the medical imaging software (Mimics v12.0, Materialise n.v, 2008). Volumes of the caudal cranial fossa parenchyma were analyzed as percentages of caudal cranial fossa volume and caudal cranial fossa volume was analyzed as a percentage of total cranial cavity volume. The volume of the ventricular system was recorded as a percentage of total parenchymal volume. Data was assessed for normality and the appropriate statistical test was used to compare means/medians.

Twelve CKCS were included with a median scan interval of 9.5 months (3–83 months). The size of the foramen magnum increased significantly between the first and second scan (1.52 ± 0.08cm vs. 1.59 ± 0.09cm; $p = 0.03$), as did the length of cerebellar herniation (0.17 ± 0.05cm vs. 0.22 ± 0.09cm; $p = 0.02$) and the caudal cranial fossa percentage (13.44% [9.9–15.28%] vs. 13.96% [9.9–15.48%]; $p = 0.02$). There was no significant difference noted between the two time points in any of the other volumetric measurements (caudal cranial fossa parenchyma percentage, 87.20 ± 3.09% vs. 88.59 ± 1.97% and ventricle percentage, 3.21% [1.99–14.34%] vs. 2.51% [1.6–17.20%]).

This work could suggest that overcrowding of the caudal cranial fossa in conjunction with the movements of cerebrospinal fluid and cerebellar tissue secondary to pulse pressures created during the cardiac cycle causes pressures on the occipital bone. This leads to a resorption of the bone and therefore an increase in caudal cranial fossa and foramen magnum size allowing cerebellar herniation length to increase.

ABSTRACT N-14

ASSESSMENT OF CORD DORSUM POTENTIALS FROM CAUDAL NERVES IN CLINICALLY NORMAL ADULT DOGS. J Campbell, BDX Lascelles, NJ Olby. North Carolina State University, College of Veterinary Medicine, Raleigh, NC.

The cord dorsum potential (CDP) is a stationary potential arising in dorsal horn interneurons after stimulation of sensory nerves. CDPs have been recorded in normal anesthetized dogs previously, and normal latency values have been determined for tibial and radial nerves. This study was undertaken to determine whether CDPs could be reliably recorded from the caudal nerves in normal dogs, thus allowing electrophysiological assessment of the cauda equina, and whether neuromuscular blockade improved recording quality.

Ten adult dogs weighing from 23.2 to 32.0 kg were anesthetized and cord dorsum recordings were compared before and after administration of atracurium. Recording needles were placed onto the dorsal lamina at intervertebral sites from L7/S1 to L2/3. Stimulations were made on the lateral aspect of the caudal vertebrae approximately 5–8 cm from the tail base. Recordings from 500 stimulations were averaged.

CDPs were recorded successfully in all dogs. Onset latency varied from 2.2 to 4.7 ms. The CDP was largest when recorded closest to the site of entry of the stimulated nerve into the cord, as determined by post-mortem examination immediately after testing in 6 dogs. Administration of atracurium did decrease muscle artifact, and in some cases helped isolate the origin of the CDP.

These data show that CDPs can be readily assessed from the caudal nerves of anesthetized dogs, with or without atracurium. Cord dorsum potentials from caudal nerves may add important information about the integrity of the cauda equina in dogs with suspected degenerative lumbosacral stenosis.

ABSTRACT N-15

EXPRESSION OF HEAT SHOCK PROTEINS, KI-67 AND E-CADHERIN IN CANINE INTRACRANIAL MENINGIOMAS. S.R. Platt, G. Perry, M. Kent, N. Northrup, A.C. Freeman, E.W. Howerth. College of Veterinary Medicine, University of Georgia, Athens, GA.

Canine intracranial glial tumors and many human brain tumors express heat shock proteins (HSPs) associated with their degree of malignancy. The up-regulation of HSPs during tumor cell growth helps keep tumor proteins stable and therefore makes them a reasonable target for therapy. Ki67 expression and EC have been strong indicators of cell proliferation and dedifferentiation, respectively.The aims of this study were to determine (i) if canine meningiomas express HSP 27 and/or HSP 72; (ii) whether the expression of the HSPs was associated with Ki67 and/or E-cadherin (EC) expression; and (iii) whether peritumoral edema was associated with HSP, Ki67 and/or EC expression.

Forty-one formalin-fixed, paraffin-embedded canine intracranial meningiomas underwent immunohistochemical staining using anti-HSP 27, or 72 antibodies. These tumor samples were also immunohistochemically stained for Ki67 and EC expression. Canine mammary carcinoma and squamous cell carcinoma tissues served as the control samples, as both have previously been shown to express HSPs. Skin was used as control for Ki67 and EC. Four non-overlapping high power fields of each stained sample were selected and cell staining was analyzed using a semi-quantitative method for HSPs and Ki67; a qualitative assessment was used for EC. All analyses were performed using SAS V 9.2 (Cary, NC). Descriptive statistics of staining percentages were calculated for all tumors tested. Simple Pearson's correlation was used to test for correlations of EC area with HSP areas and Ki-67 percent positive cells and of EC intensity with HSP intensities and Ki-67 percent positive cells. All hypothesis tests were 2-sided and the significance level was $\alpha = 0.05$. Thirteen meningiomas had MR images quantitatively evaluated for peritumoral edema using T2FLAIR sequences. The edema index (EI) was evaluated for an association with HSP 27, HSP 72, EC and Ki67 expression.

HSP 27 was expressed in 36% (mean 8.7% of cells; range 0–58%), HSP 72 in 52% (mean 5.2% of cells; range 0–26%) and EC in 68% of meningiomas. There was no association demonstrated between either HSP expression variable and EC or Ki-67 expression. There was also no association between the EC expression variables and Ki-67. However, there was a significant negative association between HSP 72 extent ($p = 0.03$) and area ($p = 0.04$) with EI.

In conclusion, HSP 27 and 72 expression was demonstrated in canine intracranial meningiomas but was not associated with Ki-67 or EC expression. This study suggests that HSPs may not have a significant role in the maintenance of canine meningiomas and so do not represent a novel treatment target for this group of tumors unlike canine glial cell tumors. However, HSP 72 may be involved in the pathogenesis of peritumoral edema in meningiomas and warrants further investigation.

ABSTRACT N-16

PHARMACOKINETIC EVALUATION OF EXTENDED RELEASE LEVETIRACETAM IN DOGS. S.R. Platt,[1] M. Kent,[1] A.C. Freeman,[1] I. A. Aljuffali,[2] D. Gullick,[2] R. Arnold[2]. [1]College of Veterinary Medicine, [2]College of Pharmacy, University of Georgia, Athens, GA.

An extended release (XR) formulation of levetiracetam, a second generation antiepileptic drug, was recently approved for human use on a once daily basis. Although levetiracetam is clinically effective for seizure control in dogs, it requires a three times daily administration. The potential benefits of the XR formulation include reduced daily dosing leading to improved compliance and relatively constant plasma concentrations. The aim of this study was to compare the pharmacokinetics of levetiracetam XR tablets with immediate release (IR) tablets following single dosing in dogs.

Five clinically and neurologically normal mixed breed dogs were used in a cross-over design. All dogs (mean body weight 25.9kg; range 23.2–30.5) had normal hematology, serum chemistry and urinalyses. Following a 12 hour fast, each dog was administered oral IR levetiracetam (500 mg; mean dose 19.3 mg/kg; range 16.4–21.6). Heparinized blood for drug analysis was taken from each dog prior to administration and 0.25, 0.5, 0.75, 1, 2, 4 and 8 hours after. Blood was immediately centrifuged and supernatant plasma was stored at −80°C until analysis. After a 4 day wash-out period, each dog was administered 500 mg oral XR levetiracetam and blood samples were taken at identical timings. Plasma samples were thawed at room temperature before preparation by solid phase extraction for HPLC analysis. Reverse phase chromatographic separation was performed. Levetiracetam and an internal standard were detected using ultraviolet spectroscopy at 205 nm. Concentrations of levetiracetam were determined by peak area comparison to the internal standard. Mean data were fit to a one compartment pharmacokinetic model with first order elimination and absorption and included a lag-phase for XR formulation.

No adverse clinical effects were noted in any of the dogs. The AUC associated with XR was 230 hr ug/mL, a 5.14 fold increase over that with IR (44.8 hr ug/mL). The absorption half-life was 3.2 hr with XR and 0.41 hr with IR, a 7.75 fold difference. The elimination half-life was 3.13 hr with XR and 2.19hr with IR, a 1.43 fold difference. The Tmax associated with XR 5.01 hr and 1.22 hr with IR, a 4.21 fold difference. The Cmax associated with XR was 18.5 µg/mL and 9.62 µg/mL with IR, a 1.92 fold difference. The plasma concentration of IR levetiracetam was not detectable at 8hr after administration whereas it was greater than 10 µg/mL at 8hr after XR administration.

Based on the AUC data, there is an approximately 5 fold increase in bioavailability of the XR compared to the IR formulation. The Cmax was approximately 2 times greater following XR administration and a high plasma level in excess of the suggested canine therapeutic concentration (5 µg/mL) for at least 8 hours. Although specific dosing recommendations cannot be made from this data, the favorable pharmacokinetics of XR over IR suggests that single, daily administration could be efficacious.

ABSTRACT N-17

RETROSPECTIVE EVALUATION OF REPAIR TECHNIQUES FOR THORACOLUMBAR AND LUMBOSACRAL VERTEBRAL FRACTURES IN DOGS. S.R. Platt, M. Kraun, M. Kent, A.C. Freeman, J. Eagleson. College of Veterinary Medicine, University of Georgia, Athens, GA.

Thoracic and lumbar vertebrae are frequently affected by fractures and or luxations in dogs following trauma. Surgical repair is part of the emergency treatment described for this disorder but does not guarantee improvement of the associated clinical signs. Multiple surgical repair techniques have been described but have not been compared in terms of their success and the factors associated with a positive outcome. The aims of this study were to retrospectively evaluate the effect of 3 different types of vertebral repair, injury type and injury location on outcome in dogs with thoracolumbar (TL) and lumbosacral (LS) spinal trauma.

Medical records were searched for dogs with radiographic evidence of a TL or LS vertebral fracture and or luxation (1995–2010); signalment, body weight and duration of disease were recorded. Dogs were retrospectively scored neurologically (0–5; normal to plegic with absent pain perception) on admission and at re-evaluation following surgery. Lesion location was classed as T3-L3 and L4-S3; dogs were evaluated as one group and as two separate groups with respect to outcome. A subset of lesions were classed as cord compression or not based on advanced imaging. Three repair techniques were evaluated (i) pins and polymethylmethacrylate (PMMA); (ii) screws and PMMA; and (iii) spinal stapling. Regression analysis was applied to test for an association between the type of surgery and a successful outcome (non-painful and ambulatory). Simple bivariate analyses were performed to investigate for variables predictive of a successful outcome.

Fifty-nine dogs were included. Twenty-eight dogs were classed as T3-L3 and 31 were L4-S3. There were 55 dogs with fractures and 43 with luxations; 23 dogs had both. Thirty-one of 35 dogs evaluated had spinal cord compression. Ten dogs were repaired with IM pins and PMMA, 18 dogs with screws and PMMA and 31 dogs with spinal stapling. Overall, there was a 78.7% success rate; there was no significant difference in outcome between the anatomic sites ($p = 0.5$). All dogs initially graded as 1–3 pre-operation were classed as a successful outcome after at least one week following surgery; 79% of dogs initially graded as 4 (plegic with pain perception) were classed as successful recovery. One dog (12.5%) initially as graded as 5 (plegic with no pain perception) had a successful outcome. A low admission score was statistically predictive of a successful outcome ($p < 0.0001$). Surgery type was not associated with a successful recovery ($p = 0.13$). Signalment, body weight, location of injury, injury type (fracture, luxation or both), presence of compression, and duration of disease did not predict outcome.

From this study, the successful recovery of dogs following surgical fixation is high and is only dependent on the neurological score at the time of admission. The choice of surgical technique does not seem to influence outcome although a prospective study comparing two surgery types is warranted to further investigate this issue the results of which can be confounded by surgeon experience and variable follow-up.

ABSTRACT N-18

CRANIAL THORACIC INTERVERTEBRAL DISC DISEASE IN GERMAN SHEPHERD DOGS: A RETROSPECTIVE COMPARATIVE MRI STUDY. Luis Gaitero[1], Rob Daniel[2], Stephanie Nykamp[1], Gabrielle Monteith[1]. [1]Department of Clinical Studies, Ontario Veterinary College, University of Guelph, Guelph, ON (Canada). [2]Veterinary Emergency Clinic, Toronto, ON (Canada).

Cranial thoracic intervertebral disc disease (IVDD) is extremely rare due to the presence of the intercapital ligament, although anecdotic data suggest German shepherd dogs (GSD) can share some predisposition for this disorder. The objective of the study was to retrospectively evaluate through MRI if cranial thoracic IVDD is significantly more common in GSD compare to other large breed dogs.

A search was done through database of the Ontario Veterinary College. Any GSD were a spinal MRI including T1-T10 spine was performed was recruited. A group of large-breed non-GSD was used as a control. In the midsaggital T2WI plane, three variables were assessed and graded for each intervertebral disc space T1-T9: spinal cord compression (SCC), disc degeneration (DD), and herniation. Wilcoxon sign rank test was used to assess if scores were different between groups. Exact conditional logistic regression was used to determine whether any intervertebral disc space was a risk factor.

22 GSD and 46 large breed non-GSDs were recruited. The GSD group had significantly higher scores than the non-GSD for SCC, and herniation. Regarding the individual intervertebral discs, in the GSD group T2-3, T3-4, T4-5 discs had significantly an increased risk for SCC, and T3-4 for herniation.

The results of this study show that GSD have a higher risk of cranial thoracic disc IVDD than other large breed dogs. That risk was higher in discs T2-T3, T3-4, and T4-5, particularly in T3-4. Genetic and/or conformational factors, such as weakness of the intercapital ligament, may predispose GSD to this lesion.

ABSTRACT N-19

SIGNALMENT, CLINICAL FEATURES AND MAGNETIC RESONANCE IMAGING CHARACTERISTICS OF 23 CASES OF DISKOSPONDYLITIS IN DOGS. JM Harris, AV Chen, RL Tucker, JS Mattoon. Washington State University, Pullman, WA.

Diskospondylitis is a common disease of the canine spine; however, few reports of MR imaging findings in dogs are available. The purpose of this study was to describe the signalment, clinical and MR imaging features in affected dogs.

Twenty-three dogs with a diagnosis of diskospondylitis based on clinical signs, MR imaging, and urine, blood, CSF and/or intervertebral disk cultures were included. Large breed dogs (> 25kg) accounted for 21of the cases. The mean age was 6.8 years with males and females equally represented. Most dogs (15/23) were ambulatory with varying degrees of pain and paresis.

MR imaging characteristics of 27 sites were reviewed. On T2W images, vertebral endplates were of mixed signal intensity (16/27) while the vertebral body was hypointense (11/27). The intervertebral disk space was hyperintense on T2W (16/27) and STIR (14/15) images and mixed signal intensity (7/13) on T1W images. Paravertebral soft tissue hyperintensities were noted on 10/27 T2W and 12/16 STIR images. Contrast enhancement occurred at 12/19 endplates and 15/19 intervertebral disk spaces. Only 4/18 vertebral bodies and 7/18 parvertebral soft tissues contrast enhanced. Intramedullary spinal cord T2W hyperintensity was noted at 10/23 sites.

Spinal cord or cauda equina compression occurred at 22/27 sites. Based on the Spearman correlation coefficient, a significant direct correlation was found between the degree of spinal cord or cauda equina compression and the patient's neurologic status ($P = 0.0011$). The incidence and severity of spinal cord compression in canine diskospondylitis may have prognostic value and may have been previously underestimated using other imaging modalities.

ABSTRACT N-20

COMPARISON OF THE ACCESS TO THE THORACOLUMBAR SPINAL CANAL BETWEEN HEMILAMINECTOMY AND PEDICULECTOMY THROUGH COMPUTED TOMOGRAPHY. J Huska, L Gaitero, B Brisson, S Nykamp, J Thomason, R Poma, W Sears. Ontario Veterinary College, University of Guelph, Guelph, ON (Canada).

Hemilaminectomy and pediculectomy are both well described and commonly utilized techniques to access the spinal canal. These procedures are most often performed to approach a compressive lesion, such as intervertebral disc disease and neoplasia, the goal being adequate visualization of the spinal canal and access to the offending lesion. A proposed benefit of pediculectomy is preservation of the articular facets and thus better maintaining stability of the vertebral column, but at the cost of reduced access to the spinal canal. The purpose of this study was to describe standardized anatomical limits of each technique and report any observed differences that could be considered during pre-surgical planning.

Ten canine cadavers had both procedures performed on opposite sides to access the T11-12, T13-L1, and L2-3 spinal canal. Measurements were obtained after performing a computed tomography study of the spine and recorded from the transverse slice most representative of the defect. The surgical technique, vertebral site, and side of vertebral column were compared with the mean spinal canal and defect height using a covariate model. Dorsal and ventral remnant lamina heights were also compared.

The height of the defect relative to the spinal canal was 76–87% with hemilaminectomy and 58–75% with pediculectomy. The observed difference in defect height of 12–18% ($p < 0.0001$) and varied with spinal canal height. Dorsal remnant lamina height was 0.9–2.4% of spinal canal height with hemilaminectomy and 7–19% with pediculectomy. Ventral remnant lamina height ranged from 1–2% and 0.6–2.3%, respectively, though the difference was not statistically significant.

While a larger defect is expected with a hemilaminectomy procedure, our results demonstrate that this difference increases with increasing spinal canal height. Interestingly, the proportion of exposed spinal canal decreases with increasing canal height for both procedures. The difference in defect height between techniques was due to greater removal of the dorsal spinal canal, possibly making the hemilaminectomy technique better suited for more dorsal lesions, while no statistically difference in access to the ventral canal is observed. No effect of vertebral site was detected. Of note was the involvement of articular facets in half of the pediculectomy defects, involving an average of 22% of the articular facet height. This result questions the suggested benefit for the vertebral stability, but further biomechanical studies would be required.

ABSTRACT N-21

IS LOW LEVEL LASER THERAPY AN EFFECTIVE ADJUNCTIVE TREATMENT TO HEMILAMINECTOMY IN DOGS WITH ACUTE ONSET PARAPLEGLIA SECONDARY TO INTERVERTEBRAL DISC DISEASE? Williams CC, Barone G. Veterinary Medical Center of Long Island, West Islip, NY.

Low level laser therapy (LLLT) is a treatment used in human and veterinary medicine for a variety of clinical syndromes. Some uses in human medicine include acute pain associated with osteoarthritis, rheumatoid arthritis, tendonitis, TMJ disorders, chronic joint disorders, and wound healing. Research is currently on-going to determine the adequate wavelengths to promote effective treatment results with LLLT in these conditions. It is purported that LLLT acts via the mitochondria to increase cellular metabolism promoting wound healing and a decrease in pain and inflammation. In this study, we hypothesized that dogs treated with LLLT in conjunction with hemilaminectomy would display quicker recovery times regardless of the presence or absence of deep pain sensation.

Seventeen dogs (9 Dachshunds, 2 Chihuahuas, 2 French Bulldogs, 2 Lhasa Ahpsos, and 1 each of a Pembroke Welsch Corgi, and a Miniature Poodle) were selected and divided into two groups. The dogs ranged in age from 2 to 11 years old, weighed between 7 and 33 pounds, and underwent hemilaminectamies after acute onset of paraplegia secondary to intervertebral disc disease (surgically confirmed). One group received laser treatments on days 1 through 4 of hospitalization. The second group did not receive LLLT, but followed the same peri-operative medication protocol. The laser used in this study was an Erchonia laser model PL5000 (635nm). The hertz setting was similar for each patient using the previously established protocol for intervertebral disc disease (IVDD) with pulse rate ranging from 9Hz to 1151 Hz. All dogs received advanced imaging pre-operatively with myelogram or MRI.

Results of the study revealed that treatment with LLLT of 635nm wavelength did not shorten or improve recovery times for dogs with acute onset paraplegia secondary to IVDD after hemilaminectomy procedures. Dogs that showed recovery to ambulation at the two week recheck were consistently dogs that were deep pain positive on presentation. A lengthened recovery time or no recovery was seen in the majority of those dogs with absent deep pain on presentation as has been revealed historically in past studies. LLLT did not appear to have an effect on this result.

ABSTRACT N-22

18F-FLUORODEOXYGLUCOSE UPTAKE OF NOMAL CANINE BRAIN WITH A HIGH-REOLUTION RESEARCH TOMOGRAPH-POSITRON EMISSION TOMOGRAPHY AND 7 T-MAGNETIC RESONANCE IMAGING FUSION SYSTEM. BT KANG[1], CY LIM[2], SH GU[2], HM PARK[2]. [1]National Institute of Neurological Disorders and Stroke, National Institutes of Health, Bethesda, MD. [2]Konkuk University College of Veterinary Medicine, Seoul, South Korea.

^{18}F-fluorodeoxyglucose positron emission tomography (FDG-PET) is a useful tool in the arena of clinical neurology by detecting the alteration of glucose utilization shown in various intracranial diseases. However, there are few data describing normal glucose uptake of the canine brain for comparison with suspected or confirmed disease. Thus the purpose of this study was to assess the normal distribution of FDG uptake of canine brain structures using a high-resolution research tomography-PET and 7 T-magnetic resonance imaging (MRI) fusion system.

FDG-PET and T2-weighted MR imaging of the brain were performed on 4 healthy laboratory beagle dogs. Acquired PET and MR images were automatically co-registered by the image analysis software. On MR images, regions of interest (ROI) were manually drawn over 48 intracranial structures, including 6 gross structures (whole brain, telencephalon, diencephalon, mesencephalon, dorsal metencephalon, ventral metencephalon and myelencephalon). A standard uptake value (SUV) and relative SUV ratio (rSUV = SUV of ROI/SUV of whole brain) were calculated for each ROI. 7 T-MR images compensated the low anatomical resolution of PET QJ;by proving good spatial and contrast resolution for the identification of the clinically relevant brain anatomy. Among gross structures, mesencephalon and ventral metencephalon had the highest (SUV: 4.17 ± 0.23; rSUV: 1.12 ± 0.03) and the lowest (SUV: 3.34 ± 0.35; rSUV: 0.90 ± 0.06) FDG uptake respectively. When SUVs were calculated on 42 detailed regions, rostral colliculus and corpus callosum had the highest (SUV: 6.00 ± 0.16; rSUV: 1.62 ± 0.05) and the lowest (SUV: 2.81 ± 0.12; rSUV: 0.76 ± 0.06) value respectively. These data acquired from normal dog brain will be used in clinical neurology to investigate various intracranial diseases such as inflammation, neoplasm and behavioral disorders.

ABSTRACT N-23

DYSURIA IN DOGS WITH DEGENERATIVE LUMBOSACRAL STENOSIS. AB Partnow[1], P Pithua[1], JA Barsanti[2], S Kegge[2], DP O'Brien[1], FA Wininger[1], JR Coates[1]. [1]University of Missouri College of Veterinary Medicine, Columbia, MO. [2]University of Georgia, Athens, GA.

Degenerative lumbosacral stenosis (DLSS) is a multifactorial condition affecting predominantly large breed dogs. The combination of stenosis and compressive neuropathy cause lumbar pain, lameness and neurologic dysfunction. Previous reports describe urinary and fecal incontinence in severely affected dogs. The objectives of this retrospective case series were to describe the clinical signs associated with dysuria and eventual diagnosis of DLSS in dogs, and to describe factors associated with regained micturition following prompt diagnosis and treatment.

Medical records from the University of Georgia and the University of Missouri between 1995 and 2009 of 11 dogs were reviewed. Inclusion required observation of dysuria, urine retention, absence of structural lower urinary tract disease and concurrent presumptive diagnosis of DLSS. Dysuria was defined as inability to initiate or sustain a urine stream. Urine residual volume was evaluated post-voiding. Dysuria was further evaluated using urethral contrast studies, urodynamic testing (urethral profilometry (4) and cystometry (4)), ultrasonography (5), and urine culture (8). Presumptive diagnosis of DLSS was based on imaging using plain radiography and epidurography (8), computed tomography (1) or magnetic resonance imaging (2).

Breeds represented included the German Shepherd Dog (n = 3), Golden Retriever (n = 2), Burnese Mountain Dog (n = 2), and 1 each Labrador Retriever, Weimaraner, Rottweiler and mixed-breed. All dogs were male. 8 were intact at onset of clinical signs. Median body weight was 38.5 kg (range 29.5–46) and median age was 5 years (range 2–10). Median duration of clinical signs prior to admission was 2 months (range 0.25–12). Other pertinent presenting clinical signs included dyschezia (2), fecal incontinence (4), general proprioceptive ataxia (2), weakness (2), and difficulty rising (1). Physical examination findings included pelvic limb muscle atrophy (2) and prostatomegaly (1). Abnormal neurologic examination findings included postural reaction deficits (6), hyporeflexia (4), decreased tail tone (3) and lumbosacral hyperesthesia (6). Neurologic examination was normal in 3 dogs.

Dorsal laminectomy was performed and diagnosis confirmed in 9 dogs; recovery was monitored for a median of 5.5 months (range 0.25–9). Three of the 9 dogs (33%) regained normal micturition within 0.25–1.5 months of surgery. Though not statistically significant, dogs that regained micturition tended to have a shorter duration of clinical signs (median 0.25 months, range 0.25–2) versus dogs that remained dysuric (median 5 months, range 2–12). Two of the 3 dogs that regained micturition were neutered at the onset of clinical signs, but only 1of 6 dogs that remained dysuric was neutered. Signs improved in all dogs with postural reaction deficits and decreased tail tone. Hyperesthesia resolved in 5 of 6 dogs (83%) and fecal continence returned in 2 of 4 dogs (50%).

These findings suggest that following prompt diagnosis and surgical decompression, normal micturition could be regained in DLSS affected dogs presenting with signs of dysuria.

ABSTRACT NM-1

CLINICAL MANIFESTATIONS OF HETEROZYGOUS EXPRESSION OF A MUTATION THAT CAUSES GLYCOGEN STORAGE DISEASE TYPE IA IN DOGS. A Specht[1], L Fiske[2], K Erger[2], T Cossette[2], A Thomas[3], J Verstegen[1], M Campbell-Thompson[2], MB Struck[4], BJ Byrne[2], CE Correia[2], CS Mah[2], DA Weinstein[2], TJ Conlon[2]. [1]University of Florida, College of Veterinary Medicine, [2]College of Medicine, [3]College of Agriculture and Life Sciences, [4]Animal Care Services, Gainesville, FL.

Glycogen storage disease type Ia (GSDIa; von Gierke disease) is an inherited metabolic disorder resulting from a deficiency of glucose 6-phosphatase-α (G6Pase). Previous reports indicate that clinical manifestations of GSDIa occur only in individuals with homozygous expression of a p.I121L mutation. Heterozygote dogs (Het) have been previously reported to exhibit an overall normal outward phenotype. The purpose of this report is to briefly describe some differences that have been observed between Het and homozygous wild type (WT) dogs.

A colony of dogs at the University of Florida contains a mix of affected, WT, and Het individuals. In the course of studies designed to determine the effectiveness of gene therapy for correction of GSDIa in dogs, both WT and Het dogs have been utilized as controls. Available information about body weights, clinical pathology tests, fasting studies, and liver biopsies was retrieved from records for both WT and Het dogs and compared.

Although birth weights are similar, Het dogs have a slower average rate of weight gain than WT dogs and this difference is especially prominent during the first few months of life (Figure 1). In contrast to affected dogs, both WT and Het dogs are able to maintain normal blood glucose concentrations for up to 10–12 hours of fasting, however, after longer fasts of 12–17 hours, Het dogs have lower glucose and higher lactate concentrations (Table 2). In addition, liver biopsy samples from Het dogs had greater apparent levels of glycogen suggested by PAS staining than did samples from WT dogs, and this correlated with the results of proton magnetic resonance spectroscopy which demonstrated 2.9 times greater glycogen content in a liver biopsy sample from a Het dog compared to a sample from a WT dog.

Figure [Abstract NM-1]

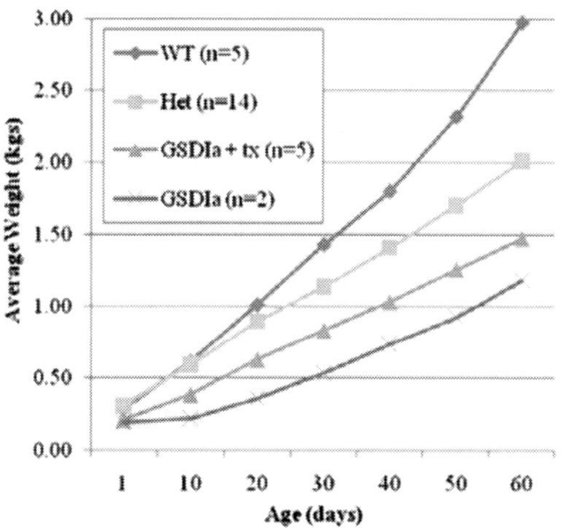

Together, these findings suggest that the level of G6Pase activity in heterozygote dogs does not provide a completely normal physiological, biochemical, or histological phenotype as previously reported.

ABSTRACT NM-2

ANALYSIS OF THE HEPATIC PROMOTER OF THE FELINE GLUCOKINASE GENE. T Schermerhorn, HA Bajaber, JB Black. Department of Clinical Sciences, Kansas State University, Manhattan, KS USA.

The glucokinase gene (GCK) encodes an enzyme involved in cellular glucose-sensing mechanisms in pancreatic beta cells and hepatocytes. GCK mRNA is present in feline pancreas but the gene is not expressed in feline liver. Hepatic GCK expression is abundant in omnivores so its absence may reflect an evolutionary adaptation of strict carnivores, like feline species. We hypothesized species-specific features in the GCK hepatic promoter may underlie the gene expression pattern observed in cats.

Table [Abstract NM-2]: Glucose and lactate levels after 12–17 hours of fasting expressed as mean (± standard deviation)

	WT (n=6)	Het (n=11)
Blood glucose (mg/dL)	78 (±17)	64 (±13)
Blood lactate (mmol/L)	1.1 (±0.5)	1.7 (±1.1)

The putative feline GCK (fGCK) promoter region was located using bioinformatic software to identify homology with human GCK (hGCK). Genomic DNA from a DSH cat was subjected to direct sequencing using a series of PCR reactions with species-specific primers. DNA clones thus obtained were aligned to generate the feline sequence.

Direct sequencing yielded 8.9 kb of genomic DNA sequence with high homology with sequences (ACBE01359231, ACBE01359221) archived in the feline genome project. The feline sequence had six regions homologous with non-coding regions of hGCK; four of these conserved regions are upstream of the putative fGCK start. A 0.8 kb segment immediately upstream of feline hepatic exon 1 is not present in hGCK. The 0.8 kb insert is the reverse complement of a conserved sequence located downstream of exon 1 in feline and human sequences. In conclusion, the putative hepatic promoter of fGCK shares extensive homology with the hGCK promoter but contains a 0.8 kb insert not found in hGCK. Functional studies are needed to confirm the role of the unique insert in regulation of fGCK gene expression.

ABSTRACT NM-3

VOLUME OF DISTRIBUTION OF CREATININE: AN INDICATOR OF BODY WATER CONTENT IN DOGS. HP Lefebvre[1], I Bargues[1], V Biourge[2]. [1]Clinical Sciences, Ecole Nationale Vétérinaire de Toulouse, Toulouse. [2]Royal Canin, Research Center, Aimargues, France.

Deuterium oxide (D_2O) dilution has been proposed for quantifying body water content, but remains difficult to perform routinely. The objective of this study was to assess if the volume of distribution (Vd) of creatinine could be proposed as an alternative in dogs for such a measurement.

Creatinine and D_2O Vd were measured before (C) and after induction (O) of obesity (by giving an hypercaloric diet (6100 kcal/kg) for 6 months) in six healthy adult Beagle dogs. Creatinine (40 mg/kg) and D_2O (200 mg/kg) were simultaneously injected by bolus iv. Blood was collected before administration and then at 5, 10, 30, 60, 120, 180, 240, 360, and 480 min post-injection (creatinine), and 10, 30, 60, 90, 120, 150, 180 and 240 min (D_2O). Plasma concentrations of both markers were determined. Vd was calculated using pharmacokinetic equations.

The body weight increased from 10.7 ± 0.6 (C) (mean ± SD) to 15.8 ± 1.0 kg (O). D_2O Vd decreased from 636 ± 18 (C) to 468 ± 19 (O) mL/kg. Similarly, creatinine Vd decreased from 624 ± 31 (C) to 420 ± 25 (O) mL/kg. The individual difference between creatinine and D_2O Vd (expressed in % of D_2O Vd) ranged from −9.5 to 1.0% (C) and from −17.3 to −7.1% (O).

In conclusion, creatinine Vd provides a good estimate of D_2O Vd in both normal and obese conditions.

ABSTRACT NM-4

DIETARY ALPHA LINOLENIC ACID AND EPA REDUCE THROMBOXANE AND IMPROVE MOBILITY IN DOGS IN A DOUBLE BLINDED STUDY. MacLeay, JM[1]; Frantz, NZ[1]; Baker, SG[2]; Gibson, TW[3]; Moens, NM[3]; Roush, JK[2]; Hahn, K[1]. [1]Hills Pet Nutrition, Topeka, KS, [2]Kansas State University College of Veterinary Medicine, Manhattan, KS, [3]Ontario Veterinary College, University of Guelph, Ontario.

A 16 wk double blinded study was conducted comparing the affect of two foods on mobility in dogs. All work was approved by an IACUC. 53 healthy beagle dogs (7–15 years old, mixed gender) were used. 33 affected (A) and 20 non-affected (NA) dogs were identified based on orthopedic examination and radiography as having or not having evidence of naturally occurring joint pathology (presence of osteophytes, dysplasia, effusion, pain on manipulation etc) in one or more joints. A and NA dogs were evenly distributed between two locations.

Foods had nutrient profiles adequate for maintenance according to the 2009 AAFCO Official Publication. The Test food contained greater amounts of methionine, manganese, carnitine, vit. E., vit. C, alpha linolenic acid (ALA), and eicosapentaenoic (EPA) acid: the food provided 388 mg n3 fatty acids and 847 mg n6 fatty acids per 100kcal. All dogs were fed the Control food for 4 wks followed by a 12 wk feeding period where 17 A and 10 NA dogs consumed the Test food and 16 A and 10 NA dogs the Control. Blood and urine were collected at weeks 0, 4, 8 and 12 and analyzed for serum fatty acids and urine thromboxane:creatinine ratios were determined.

Evaluators in this study were different than those making the original diagnosis and so were blinded as to treatment and diagnosis. Orthopedic exams were performed by two veterinary surgeons at each site on weeks 0, 4, 8 and 12. The same two evaluators examined the same dogs throughout. The data was evaluated for the difference between A and NA dogs and between foods with age, gender and location as covariates. Body weight, disease status, age and gender were blocked. Analysis included ANOVA repeated measures mixed procedure (SAS version 9.0) to determine treatment effects over time.Serum EPA was greater and arachidonic acid lower at weeks 4, 8 and 12 in the Test food fed dogs ($p < 0.05$). Urine thromboxane:creatinine ratios were decreased in the A dogs fed the Test food compared to the A dogs fed the Control food at 12 wks ($p < 0.05$). Lameness score was significantly improved ($p < 0.05$) within and between groups of dogs fed the Test food. A significantly greater proportion of A dogs fed the Test food had improvement in total joint score, lameness, functional disability and overall assessment score at 12 wks compared to A dogs fed the Control food. 69% of A dogs had an improved Overall Assessment score on the Test food after 4 wks and at 12 wks compared to 40% at 4 wks and 27% at 12 wks of A dogs consuming the Control food. This study shows that a food with moderate amounts of added linolenic acid and EPA can have a positive impact on systemic inflammation and mobility in 4–12 weeks.

A similar abstract will be presented at the Orthopedic Research Society meeting in January 2011 to an audience largely of orthopedic researchers interested in human orthopedics.

ABSTRACT NM-5

EFFECT OF A MEDIUM-CHAIN TRIGLYCERIDE:FISH OIL ENRICHED DIET ON LEAN BODY MASS, CIRCULATING FATTY ACIDS, AND METABOLITES IN DOGS. DE Jewell[1], JL Greitl[1], DK Joshi[1], JA Hall[2]. [1]Pet Nutrition Center, Hill's Pet Nutrition, Inc, Topeka, KS; [2]Dept. of Biomed. Sci., Coll. Vet. Med., Oregon State University, Corvallis, OR.

Fat is an important dietary component, serving both as a source of energy and as a supplier of essential fatty acids (FA). Medium-chain triglycerides (MCT) contain intermediate length FA that do not rely on L-carnitine for transport across the inner mitochondrial membrane, bypassing this rate-limiting step in FA oxidation. Long-chain (n-3) polyunsaturated fatty acids (PUFA) from fish oil (FO), and in particular eicosanoids derived from eicosapentaenoic acid (EPA), may protect against excessive inflammatory reactions, which may be exacerbated by eicosanoids derived from (n-6) arachidonic acid (AA). This study investigated the effects of adding MCT:FO and L-carnitine to a control diet (Prescription Diet® k/d®) on lean body mass, and serum FA and metabolites. Forty healthy Beagles (3.1 to 14.8 y) were fed one of three foods (n = 13 to 14 dogs each) for 6 mo. The study protocol was reviewed and approved by IACUC, Hill's Pet Nutrition, Inc. All foods were complete, balanced, and sufficient for maintenance of adult dogs; and had similar concentrations of moisture, protein, and fat (approx. 7.4%, 14.0%, 18.1%, respectively).

Table [Abstract NM-5]: Food composition of three treatment foods (all values are % of food as fed)

Prescription Diet® k/d®	Treatment 1	Treatment 2	Treatment 3
Added L-carnitine	–	300 mg/kg	300 mg/kg
Added fish oil	–	0.6%	1.5%
Added MCT (animal FA)	–	–	+
Lauric, 12:0	0.01	0.01	0.87
Myristic, 14:0	0.17	0.18	0.46
EPA, 20:5 (n-3)	0.01	0.10	0.23
DHA, 22:6 (n-3)	0.00	0.07	0.17

Composition of serum FA was determined by gas chromatography of FA methyl esters.

Metabolomic profiles of serum samples were determined from extracted supernatants that were split and run on GC/MS and LC/MS/MS platforms, for identification and relative quantification of small metabolites. Body composition was determined by dual energy x-ray absorptiometry.

Serum concentrations of lauric and myristic FA increased; EPA and DHA increased in a dose-dependent manner; and AA decreased in dogs fed treatment food 3 (proc-mixed procedure in SAS; all $P \leq 0.05$) when compared to dogs fed treatment foods 1 or 2. Serum concentrations of acetylcarnitine and succinylcarnitine increased, indicating L-carnitine incorporation, in dogs fed treatment foods 2 and 3. Thus, a diet enriched with MCT:FO significantly altered serum FA composition, enriching (n-3) PUFA and lowering AA concentrations. There was no change in lean body mass for any of the diets compared to baseline values, and no difference between treatments, showing that all three treatment foods met protein requirements.

ABSTRACT NM-6

METABOLIC ADAPTATIONS TO WEIGH LOSS IN OBESE DOGS MAY REQUIRE TIME TO SETTLE. AC Carciofi, MA Brunetto, MOS Gomes, MC Peixoto, FC Sá, SP Nogueira, JT Jeremias. Veterinary Medicine and Surgery Department, UNESP - São Paulo State University, Jaboticabal, SP, Brazil.

Ten owned dogs, obese for more than 12 months (body condition score [BCS] of 9; fat mass [FM] = 45.7 ± 1.51%) were studied. These dogs had their weight reduced by 20% (BCS = 8; FM = 33.5 ± 1.92%; $P < 0.001$) being designated weight reduced (WR) group and then were fed to maintain constant body weight during 150 days (BCS = 8.2; $P = 0.6$), designated maintenance (MAIN) group. A control (CT) group of 10 beagles was also included (BCS = 4.5; FM = 18.3 ± 1.38%; $P < 0.01$). In all groups the glucose postprandial response test was performed after 12 hours of fasting. Blood samples were taken pre-feeding and after 5, 10, 15, 30, 45, 60, 120, 180, 240, 300 and 360 minutes of the consumption of cooked rice enough to the ingestion of 6g of starch/kg body weight. TNF-α and IL-6 were dosed in MILLIPLEX™MAP panel, insulin and leptin by radioimmunoassay. Statistical analysis included paired or non-paired T-tests and Wilcoxon ($P < 0.05$). The regimen normalized meal glucose response, the area under the curve (AUC) of glucose for WR was lower than for obese ($P < 0.05$) and similar to MAIN and CT ($P > 0.05$). Insulin secretion did not normalize immediately, as obese and WR exhibited similar AUC of insulin and higher values than for CT ($P < 0.05$). MAIN, however, presented similar AUC of insulin than CT, with lower values than obese and WR ($P < 0.01$), suggesting that dogs require some time to adapt their metabolism. Leptin, TNF-α, and IL-6 presented significant reductions after weight loss ($P < 0.05$), without differences between WR, MAIN and CT ($P > 0.05$), suggesting an improvement of the pro-inflammatory state consequent to obesity.

ABSTRACT NM-7

METHIONINE INTOXICATION MAY BE RELATED TO SULFUR EFFECTS ON FOOD BASE EXCESS. AC Carciofi, SP Nogueira, JT Jeremias, APJ Maria, MA Brunetto, MOS Gomes. Veterinary Medicine and Surgery Department, UNESP - São Paulo State University, Jaboticabal, SP, Brazil.

Studying food base excess (BE) modification, methionine intoxication was described. In a basal kibble dog diet (BE = 305 mEq/kg; 1.98 g/kg of S) two dosages of ammonium sulphate and Methionine was added, resulting in diets with BE of 135 mEq/kg (4.26 g/kg of S) and −51 mEq/kg (6.65 g/kg of S), or BE of 142 mEq/kg (4.68 g/kg of S) and 4 mEq/kg (7.49 g/kg of S), respectively. A 2 × 2 factorial plus a control diet design, resulting in five treatments, and 32 adult health beagle dogs were used, in a completed randomized design with six dogs per diet. A 5-d adaptation phase followed 3-d of total urine collection (in bottles with 100 mg of thymol). Urine were pooled by dog and analyzed for density, volume and pH. Food macroelements were determined by standards methods (AOAC, 1995) and used for BE calculation. Dog's acid-basic status was studied by blood gas analysis of venous blood, at 8:00h (pre feeding) and 6 hours after meal. A dose-dependent reduction of urinary pH was verified for both compounds ($P < 0.01$). Blood bicarbonate ($r = 0.98$; $P < 0.01$), and blood base excess ($r = 0.75$; $P < 0.01$) were highly correlated with food BE. Acidemia and reduced blood BE were verified in diets with BE close to zero (higher dose of both compounds, or > 6.6 g/kg of S), resulting in daily or each other day vomiting episodes in the dogs. Ataxia, seizures, and vomiting were previously describe in dogs fed 47 g/kg of methionine, but our results suggest that a much lower value (27.7 g/kg) was toxic and that the safe upper limit should be between this value and 15.7 g/kg (the lower evaluated dose).

ABSTRACT NM-8

ASSOCIATION BETWEEN BODY CONDITION AND SURVIVAL IN DOGS WITH ACQUIRED CHRONIC KIDNEY DISEASE. VJ Parker,[1,2] LM Freeman[2]. [1]Iowa State University College of Veterinary Medicine, Ames, IA. [2]Tufts Cummings School of Veterinary Medicine, North Grafton, MA.

In people with chronic kidney disease and heart failure, obesity is associated with longer survival times. This association, called the "obesity paradox," also has been recognized in dogs and cats with heart failure. Excess weight appears to modulate the serious deleterious effects of muscle loss in these diseases. The purpose of this study was to determine the effects of body condition and body weight changes in dogs with naturally-occurring chronic kidney disease (CKD).

Dogs diagnosed with ≥ IRIS stage II CKD between 2008 and 2009 at Iowa State University and Tufts Cummings School of Veterinary Medicine were eligible for the study. Dogs < 1 year of age and those with acute renal failure or suspected congenital renal diseases were excluded. Medical records were reviewed using a standardized data form, and data were collected for initial body weight and body condition score (BCS, 1–9 scale), clinicopathologic values, changes in body weight and BCS, comorbidities, and treatments. Dogs were classified as underweight (BCS = 1–3), moderate weight (BCS = 4–6), or overweight (BCS = 7–9). A change in body weight was defined as > 0.2 kg. Survival times were determined for all dogs that were discharged from the hospital and lived > 1 day. Associations between survival and BCS or body weight changes were analyzed using Cox proportional hazards models.

One hundred two dogs were enrolled in the study. At the time of diagnosis, 21 dogs were classified as IRIS Stage II, 57 dogs were Stage III and 24 dogs were Stage IV. Median body weight at baseline was 18.7 kg (range, 2.4–50.1 kg). For dogs with body condition scores recorded (n = 74), 13 were underweight (18%), 51 were moderate (69%), and 10 were overweight (14%). For dogs that had at least two body weights recorded over the course of their disease, 22 gained weight, 46 lost weight, and 10 had no change in weight. Changes in body weight were not associated with survival; however, BCS at the time of diagnosis was significantly associated with survival. Dogs classified as underweight had a significantly shorter survival time compared to both moderate (P < 0.001) and overweight dogs (P < 0.001).

These results suggest that body condition is an important consideration in dogs with acquired chronic kidney disease. Further studies are warranted to evaluate the relationship between obesity and longer survival in dogs with CKD.

ABSTRACT NM-9

DIETARY CRUDE PROTEIN OF 28.5% MAINTAINS LONG-TERM LEAN BODY MASS IN CATS WITH IMPAIRED KIDNEY FUNCTION. S Yu, R Velliquette, R Yamka, D Jewell. Hill's Pet Nutrition, Inc., Topeka, KS.

Protein restriction is the cornerstone of dietary management of kidney disease. The National Research Council recommends 20% crude protein and the American Association of Feed Control Officials (AAFCO) recommends a minimum of 26% crude protein for maintenance for healthy adult cats. Protein requirement is unknown for adult cats with kidney disease. Most commercially produced cat foods for adult maintenance contains 30% or more crude protein on a dry matter basis. A typical therapeutic food for cats with kidney disease contains about 28% crude protein. The objective of the present study was to investigate whether dietary crude protein at 28.5% would be adequate for the maintenance for adult cats with impaired kidney function.

Seven adult cats, 3 female and 4 male, with age ranging from 5 to 13.7 years old (mean: 8.1 years) were used in the study. All cats had elevated serum creatinine concentration (> 1.6 mg/dl, range: 1.61–2.10 mg/dl) and reduced glomerula filtration rate (mean: 32% reduction; range: 12–60% reduction) during the study. They did not have other systematic diseases, e.g., hyperthyroidism, at the beginning of the study. Cats were fed an expanded dry food made with ingredients commonly used in commercial dry cat foods. The food contained 28.5% crude protein (chemical analysis) and 4366 kcal/kg (calculated) on a dry matter basis, or 65.3 g protein/1000 kcal. Each essential amino acid in the food was at least 130% of that recommended by AAFCO. Other nutrients in the food also exceeded AAFCO's recommendations for maintenance for adult cats. Cats were fed the food for 30 weeks. Lean body mass (dual x-ray absorptionmetry; Hologic, Hologic, Inc, MA) and serum albumin concentration were measured periodically to monitor protein status of cats.

The average lean body mass (mean ± SD) was 3.63 ± 0.82 kg, 3.72 ± 0.85 kg, 3.74 ± 0.84 kg, and 3.75 ± 0.89 kg in weeks 3, 11, 17, and 30 of the study, respectively. Paired t-test did not detect statistical difference (p > 0.05) when comparing the lean body mass in weeks 3 versus weeks 11, 17, and 30, respectively. Serum albumin concentration were within the normal reference range during the study (mean ± SD: 3.04 ± 0.35%, 2.80 ± 0.35%, 3.04 ± 0.32%, and 3.07 ± 0.40% in weeks 3, 11, 17, and 30, respectively). These data show that 28.5% dietary crude protein in a dry food with 4366 kcal/kg on a dry matter basis, or 65.3 g protein/1000 kcal, is adequate for maintenance for cats with impaired kidney function.

ABSTRACT NM-10

DIETARY SUPPLEMENTATION OF VITAMIN B_{12} IMPROVES VITAMIN B_{12} STATUS IN GERIATRIC CATS. S Yu, L Melendez, N Ahle, P Toll, W Carter. Hill's Pet Nutrition, Inc., Topeka, KS.

Cats with gastrointestinal diseases sometimes were reported to have impaired vitamin B_{12} status. Recently, we have observed marginal vitamin B_{12} deficiency in apparently healthy old cats. The objective of this study was to investigate whether dietary supplementation of vitamin B_{12} would improve vitamin B_{12} status in geriatric cats (> 10 years). Healthy geriatric cats (n = 46, male: 10, female: 36) were stratified based on gender and age into two groups, 23 cats per group. One group was fed a nutritionally complete and balanced dry cat food (control: vitamin B_{12} = 11 mcg/100g) while the other group was given the same food with additional vitamin B_{12} supplementation (treatment: vitamin B_{12} = 49.3 mcg/100 g) for 90 days. Blood samples were taken on days 0, 45 and 90 for serum vitamin B_{12}, methylmalonic acids (MMA), and folate measurements. Dietary vitamin B_{12} supplementation significantly (p < 0.05, repeated measures) increased serum vitamin B_{12} and decreased serum MMA. Serum vitamin B_{12} was 586, 676, 749 ng/l and serum MMA was 499, 426, 452 nmol/l in the treatment group on days 0, 45, and 90, respectively; in the control group, serum vitamin B_{12} was 681, 632, 652 ng/l and serum MMA was 434, 443, 483 nmol/l on days 0, 45, and 90, respectively. Serum folate concentration was not affected by the diets. This study suggests that geriatric cats may have higher dietary requirement for vitamin B_{12} and shows that dietary supplementation of vitamin B_{12} improves vitamin B_{12} status of geriatric cats.

ABSTRACT NM-11

CHANGES IN SERUM ESSENTIAL FATTY ACID PROFILES IN DOG AND CAT DISEASES. E Phenix, KE Bigley, JE Bauer. Companion Animal Nutrition Laboratory, and Faculty of Nutrition, Texas A&M Univ., College Station, TX, USA.

In humans, several disease conditions exist that involve abnormal patterns of polyunsaturated fatty acids and similar abnormalities may be present in companion animals. Indeed there have been reports of decreased plasma arachidonic acid and reduced delta-6 desaturase activities in dogs with atopy and other skin disorders. The present study investigated serum fatty acid profiles in dogs and cats presented to the Texas A&M University Veterinary Teaching Hospital, Clinical Pathology Laboratory over the past one year period. Results were compared with normative data generated among dogs and cats from earlier feeding studies. Sera used were residual samples submitted to the laboratory for other diagnostic procedures and stored frozen for no more than 2 months after collection. The samples were grouped according to presenting disorders involving liver, kidney, digestive, and cardiac diseases. Total lipids were extracted using chloroform:methanol (2:1 v/v) and fatty acid methyl esters were prepared for capillary gas chromatography. Relative percentage distribution of individual serum fatty acids for each animal were then compared with average normative serum phospholipid fatty acid values (dogs, n = 44; cats, n = 29) by calculating the ratio of the value in the diseased individual to the normal mean value and used as an index of normalcy. Normalcy ratios were then plotted on a logarithmic scale with normal at 1.0. The ratio was then compared to changes greater than 3, 2 and 1 standard deviations of the normal mean values. In this way a graphical presentation of resultant values was obtained. Although the animals had been fed various commercial diets and some home-prepared foods, a number of noteworthy patterns emerged from this analysis. Dogs showed increased linoleic acid, decreased arachidonic acid, increased total monounsaturated and decreased saturated fatty acids at P < 0.001; oleic acid was increased at p < 0.01. Remarkably, these findings were similar for all canine disease categories evaluated (n = 30, heart: n

= 17, kidney; n = 28, liver, and n = 28 digestive disorders). In cats, a slight decrease in arachidonic acid and large decrease in 22:0 was observed but only in heart disorders. By contrast, modest elevations of arachidonate were observed in kidney, liver, and digestive disease groups but at $p < 0.01$. Sample sizes of the feline sera were considerably smaller (range of 4-13 per group). A limitation of this analysis is that variability of normal data may exist depending on diet fed making comparisons less reliable However, these preliminary data suggest that metabolic diseases of dogs may depress plasma arachidonic acid independent of diet fed suggesting either reduced conversion from linoleic acid or increased utilization of arachidonate for eicosanoid production during times of metabolic stress. Conversely, in cats, increases in arachidonic acid may be associated with diet arachidonate or other mechanisms. Additional studies to verify these findings are warranted.

ABSTRACT NM-12

EFFECT OF L-ALANYL-L-GLUTAMINE SUPPLEMENTATION IN THE TREATMENT OF CANINE PARVOVIRAL ENTERITIS. JH Kang[1], S Noh[1], K Cho[1], S Lee[2], D Chang[1], MP Yang[1]. [1]Chungbuk National University College of Veterinary Medicine, Cheongju, Chungbuk, Republic of Korea. [2]Illmare Animal Medical Center, Seoul, Republic of Korea.

The objective of this study was to determine whether or not L-alanyl-L-glutamine (Ala-Gln) supplementation in dogs with parvoviral enteritis improves the survival rate and ameliorates clinical signs without side effects. This randomized, double-blinded, placebo-controlled clinical trial included 39 client-owned dogs. The dogs were randomly assigned into two groups and administered Ala-Gln solution (Dipeptiven; 0.4 g/kg) or an equivalent volume of placebo orally twice a day. All of the dogs (Ala-Gln group [n = 20] and placebo group [n = 19]) received standard treatment while hospitalized and were monitored daily according to a clinical scoring system and diagnostic evaluation for 11 days. Among the 39 dogs, 17 (Ala-Gln-treated group [n = 8] and placebo group [n = 9]) were vaccinated and 22 (Ala-Gln-treated group [n = 12] and placebo group [n = 10]) were not vaccinated. The population consisted of 29 purebreds and 10 mixed breed dogs, with a mean age of 12.2 ± 2.2 weeks. The survival data were compared statistically by means of a log-rank test for the Kaplan-Meier survival curves.

The clinical scores of Ala-Gln-treated dogs improved significantly relative to the placebo group. There was a significant difference between the two groups in the survival distribution ($P = 0.038$); specifically, 3 of the Ala-Gln-treated dogs (15.0%) died, whereas 8 of the dogs in the placebo group (42.1%) died. No side effects were associated with the administration of Ala-Gln. These results suggest that the oral administration of Ala-Gln is effective in improving clinical signs and survival rate in dogs with parvoviral enteritis.

ABSTRACT NM-13

IN VITRO EFFECTS OF LIPID EMULSION ON PLATELET FUNCTION AND THROMBOELASTOGRAPHY IN HEALTHY DOGS. LR Tonkin, NK Parnell, DF Hogan. Purdue University, School of Veterinary Medicine, West Lafayette, IN.

Bleeding disorders, thrombocytopenia and alterations in platelet function have been documented in humans receiving lipid-containing parenteral nutrition formulations. Despite a lack of evidence in the veterinary literature, it is believed that parenteral lipids are contraindicated in critical illness when the development of bleeding disorders is likely. The objective of this study was to determine if there is an in vitro effect on platelet function and thromboelastography (TEG) in normal dogs with varying concentrations of a 20% soybean oil emulsion (Intralipid®).

Twelve clinically healthy dogs were used for this study. Whole blood platelet aggregation, using ADP and collagen agonists, was measured using multiple electrode aggregometry in hirudinated blood with final lipid concentrations of 0, 1, 10, and 30 mg/ml. The TEG parameters R, K, α-angle, and maximum amplitude (MA) were evaluated from citrated whole blood with equivalent final lipid concentrations as platelet aggregation.

There was no significant difference between groups with collagen-induced platelet aggregation. There was a significant increase in the area under the curve (AUC) with ADP-induced aggregation at a lipid concentration of 30 mg/ml ($p = 0.027$). The MA was significantly reduced at both the 10 mg/ml ($p < 0.001$) and 30 mg/ml ($p < 0.001$) lipid concentration. There was no statistical difference between groups evaluating the other TEG parameters.

While platelet aggregation appeared enhanced at the highest concentration evaluated, this concentration is not clinically relevant. The reduction in MA seems discordant but both fibrinogen and platelets contribute to the MA. Therefore the higher lipid concentrations may be interfering with fibrinogen kinetics or fibrinogen-platelet interaction. In vivo studies are indicated to determine if any of these changes are clinically significant.

ABSTRACT O-1

PHASE I STUDY OF ROSIGLITAZONE IN COMBINATION WITH CARBOPLATIN IN CANCER BEARING DOGS. SD Allstadt Frazier[1], DS McKemie[2], TA Guerrero[1], KA Skorupski[1], CO Rodriguez, Jr[1]. University of California-Davis, [1]Department of Surgical and Radiological Sciences and [2]Department of Molecular Biosciences, Davis, CA.

Rosiglitazone is a peroxisome proliferator-activated receptor gamma (PPARγ) agonist and an FDA-approved anti-diabetic agent in humans that has been investigated for its ability to reduce tumor cell growth. Specifically, the combination of rosiglitazone and carboplatin has demonstrated enhanced tumor control.

The purpose of this study was to determine the peak plasma concentrations and side effect profile of rosiglitazone after oral administration in dogs with spontaneously occurring cancer. All dogs received carboplatin intravenously concurrently with oral rosiglitazone. Ten cancer-bearing dogs with normal pre-treatment hepatic and renal function were enrolled. Complete pre-treatment hematological and biochemical parameters were available in ten dogs and post-treatment parameters in nine dogs.

Peak plasma concentrations varied with dose and ranged from 150.3–959.4 ng/mL and occurred between 30 minutes and 4 hours post administration and rapidly declined after the peak. The dose limiting toxicity was hepatic at a dose of 9 mg/m^2. There was one Grade III, two Grade I ALT, and one Grade III AST elevations noted. No changes in total bilirubin, alkaline phosphatase, or GGT values were noted. Blood glucose values remained within normal limits. Mild, self-limiting gastrointestinal and hematologic toxicities were observed when rosiglitazone was administered in combination with carboplatin.

Based on this study, the recommended dose of rosiglitazone in cancer-bearing dogs with normal hepatic function is 6 mg/m^2 orally once daily. Side effects of the combination appear similar to side effects noted with carboplatin alone. Further study is needed to determine efficacy of this combination and if more frequent dosing is required to maintain plasma concentrations.

ABSTRACT O-2

ANTITUMOR EFFECTS OF COMBINED CARBOPLATIN AND GEMCITABINE ON CANINE TRANSITIONAL CELL CARCINOMA CELLS. JF de Brito Galvao, S Murahari, WC Kisseberth, DJ Chew, N Inpanbutr. The Ohio State University, Columbus, OH.

Carboplatin has shown little activity as a single agent for the treatment of canine transitional cell carcinoma (TCC). However, gemcitabine has shown synergism with carboplatin in human cell lines. The purpose of this study was to evaluate the activity of gemcitabine against canine TCC cell lines alone or in combination with carboplatin. We hypothesized that gemcitabine in combination with carboplatin would have synergistic effects *in vitro*. The results of this study could provide a rationale for treatment of canine TCC with the combination of these drugs.

TCC cell lines TCC-Kiss, TCC-Knapp-JS, TCC-AXA, TCC-HXC, and TCC-SH were treated with gemcitabine, carboplatin, or the combination. Cell proliferation was assessed using CyQUANT assay, cell cycle was evaluated using propidium iodide staining, and apoptosis was assessed by measuring caspase-3/7 activation. Synergy was quantified by combination index analysis using Compusyn software.

Treatment of canine TCC cell lines with carboplatin or gemcitabine decreased cell proliferation, induced cell cycle arrest, and apoptosis. When TCC cell lines were treated with gemcitabine and carboplatin in combination at a therapeutically relevant concentration (gemcitabine < 100uM, carboplatin < 250uM), a significant decrease in cell proliferation was observed compared to gemcitabine or carboplatin alone, and the drug combination was synergistic in 3 of 5 cell lines, and additive in the remaining 2 lines.

Gemcitabine exhibits biologic activity against canine TCC cell lines and carboplatin combined with gemcitabine exhibits synergistic activity at biologically relevant concentrations. Our results support further evaluation of these drugs in dogs with TCC to determine the clinical efficacy of this combination.

ABSTRACT O-3

LOW-DOSE CYCLOPHOSPHAMIDE SELECTIVELY DEPLETES REGULATORY T CELLS AND INHIBITS ANGIOGENESIS IN DOGS WITH SOFT TISSUE SARCOMA. JH Burton, L Mitchell, DH Thamm, SW Dow, BJ Biller. Colorado State University, Fort Collins, CO.

Metronomic chemotherapy has been shown in murine models and humans to improve tumor control by inhibiting tumor angiogenesis and suppressing regulatory T cells (Treg). Treg are a subset of T lymphocytes demonstrated to be increased in humans and dogs with cancer and are thought to suppress cellular immune responses against tumors. The purpose of this study was to determine whether metronomic cyclophosphamide therapy depletes Treg and/or exhibits antiangiogenic activity in dogs with soft tissue sarcoma.

Client owned dogs with histologically confirmed grade I or II soft tissue sarcoma were administered cyclophosphamide at $12.5\,mg/m^2$ or $15\,mg/m^2$ orally once daily for 28 days. Whole blood and tumor biopsies were obtained on days 0, 14, and 28. Flow cytometric analysis of blood was performed to assess changes in T lymphocyte subsets, including $CD4^+$ and $CD8^+$ cells as well as $CD4^+FoxP3^+$ Treg. Tumor microvessel density (MVD) was assessed by performing immunohistochemistry for CD146.

Five dogs were enrolled in the $12.5\,mg/m^2/day$ dose cohort and six dogs were enrolled in the $15.0\,mg/m^2/day$ dose cohort. In patients that received cyclophosphamide at $12.5\,mg/m^2/day$, the mean number of Treg decreased from day 0 to 28 but there was no change in the mean percentage of Treg or MVD. For patients that received $15.0\,mg/m^2/day$, both the mean number and percent of Treg as well as MVD decreased over the 28 day time period.

Cyclophosphamide at $15.0\,mg/m^2/day$ or greater selectively depletes Treg and inhibits angiogenesis in dogs with soft tissue sarcoma.

ABSTRACT O-4

RE-SENSITIZATION OF GLUCOCORTICOID-RESISTANT CANINE LYMPHOMA CELLS USING ARSENIC TRIOXIDE. CL Pratt, RK Sellon, JN Bryan, J Fidel. Washington State University College of Veterinary Medicine, Pullman, WA.

Arsenic trioxide (ATO) is used to treat leukemias, multiple myeloma, and relapsed lymphoid malignancies in humans; its use has not been explored in veterinary oncology. Prior therapy with glucocorticoids decreases likelihood and duration of remission for dogs with lymphoma treated with chemotherapy. We hypothesized that ATO will re-sensitize glucocorticoid-resistant canine lymphoma cells to glucocorticoid-induced apoptotic death.

The OSW canine lymphoma cell line was cultured with 200 uM dexamethasone. Remaining viable dividing cells were considered resistant. Resistant cells were exposed to 0.01 uM and 0.02 uM of ATO without dexamethasone, after which cells were washed and re-exposed to 200 uM dexamethasone. After 24, 48 and 72 hours of dexamethasone exposure, cells were counted using Trypan blue stain. Apoptosis was assessed by TUNEL assays on cytospin preparations collected at 24, 48, and 72 hours from ATO-exposed and control groups. Statistical analysis was performed using 1 way ANOVA and Tukey's test.

The proportion of dead cells increased over time in both 0.01 uM and 0.02 uM ATO exposed groups. The proportion of dead cells was greater for 0.01 uM ATO ($p < 0.001$) and 0.02 uM ($p < 0.001$) groups compared to control. Apoptosis increased with increasing ATO concentration and duration of dexamethasone exposure compared to control.

These results support the effectiveness of ATO at re-sensitizing glucocorticoid-resistant canine lymphoma cells to apoptotic death following re-exposure to glucocorticoids. Ongoing gene expression studies aim to elucidate this mechanism. Additional studies to determine if this effect is seen with other chemotherapeutic agents are warranted.

ABSTRACT O-5

SERUM PROTEIN ELECTROPHORESIS OF DOGS WITH LYMPHOMA. M.J. Atherton, J.S. Morris, P.D. Eckersall, J. Harvie. School of Veterinary Medicine, College of Medical, Veterinary and Life Sciences, University of Glasgow, Glasgow, Scotland, UK.

Lymphoma is the most common hematopoietic tumor of dogs. Protein disturbances may be associated with this disease including monoclonal gammopathies in a low percentage of cases. Serum protein electrophoresis (SPE) is routinely used to aid diagnosis of various canine diseases including lymphoma when total protein concentration is elevated. The purpose of this study was to compare SPE changes in lymphoma patients without elevated total proteins with a population of healthy dogs.

Agarose gel electrophoresis was performed on residual serum from 17 healthy control dogs and 21 untreated dogs with multicentric lymphoma (stage III -V) after measuring total protein (TP) using the biuret method. Densitometric traces of the protein bands were obtained using computer software (Totallab 100) and the albumin, alpha-1, alpha-2, beta-2 and gamma globulin subfractions were identified by visual inspection. The total protein concentration, the number of subfractions and the relative and absolute protein subfraction concentrations were then compared statistically between the two populations.

In lymphoma dogs, TP, absolute albumin, beta-2 and gamma globulin concentrations and both relative and absolute concentrations of the alpha-1 globulins were significantly lower however relative and absolute alpha-2 globulin concentrations were significantly elevated. No monoclonal gammopathies were identified in any of the dogs and not every patient with lymphoma had the above changes in their electrophoretogram.

This study has demonstrated that significant changes occur in the albumin and globulin fractions of canine lymphoma patients despite no obvious increase in TP. Further investigation is required to identify the proteins responsible for these changes.

ABSTRACT O-6

EVALUATION OF ABERRATION IN PHENOTYPE AS PROGNOSTIC FACTOR IN CANINE HIGH-GRADE LYMPHOMA. Chiara Leo, Katja Ruess-Melzer, Laura Marconato, Paola Valenti, Stefano Comazzi, Maria Elena Gelain, Barbara Kaser-Hotz. Animal Oncology and Imaging Center, Hünenberg, Switzerland.

It is well known that immunophenotype has a prognostic value for the outcome of canine lymphoma, with T-cell lymphomas having a worse prognosis than B-cell lymphomas. The recent advent of flow-cytometric techniques allowed easy detection of many different markers on lymphoma cells and therefore, not only distinguish between T and B cells, but also estimate possible aberration on immunophenotype. In human oncology, although some controversy persists, it seems that non-Hodgkins lymphoma and acute leukemia carrying aberrations have a worse prognosis.

The aim of this study was to evaluate the role of immunophenotype aberration in canine high-grade lymphoma considering outcome and time span to achieve complete response under chemotherapy. Samples of bone marrow, blood and lymph node suspensions from twenty-three dogs were evaluated with flow-cytometry.

Eleven dogs had aberrant expression of neoplastic lymphocytes and twelve were non-aberrant. The most common aberrations found were: positivity to CD34, biphenotypes, double expression of T-antigens (CD4+, CD8+), diminished expression of CD45. All dogs were treated with a CHOP-based protocol. There was a significant difference for the time to achieve response to chemotherapy (partial or complete). 12/12 non aberrant lymphomas went into CR or PR after the first treatment (L-Asparaginase), while aberrant lymphomas needed more than 2 treatment to reach CR or PR. There was a trend for a prolonged disease free interval with non-aberrant versus aberrant, although it was not statistically significant.

Aberration of immunophenotype may be a prognostic factor for canine lymphomas, but further studies with larger groups are needed.

ABSTRACT O-7

ANALYSIS OF PHENOTYPIC FACTORS PREDICTING SURVIVAL IN CANINE PATIENTS WITH B CELL LYMPHOMA. S. Rao[1], S. Lana[1,2], J. Eickhoff[4], E. Marcus[3], P.R. Avery[3], P.S. Morley[1], and A.C. Avery[2,3]. [1]Department of Clinical Sciences, [2]Flint Animal Cancer Center and [3]Department of Microbiology, Immunology and Pathology, College of Veterinary Medicine and Biomedical Sciences, Colorado State University, Fort Collins, CO. [4]Department Department of Biostatistics & Medical Informatics, University of Wisconsin, Madison, WI.

Class II major histocompatibility expression is a significant and independent predictor of prognosis in human B cell lymphoma. Low class II MHC is consistently associated with poorer outcome. The mechanism underlying this relationship is not clear, but one hypothesis is that high class II MHC allows for better antigen presentation and tumor–specific immune responses. In the this study, we investigated whether that class II MHC expression in canine B cell lymphoma was associated with remission and survival times. A total of 160 patients were categorized by level of class II MHC,expression of CD34 and cell size for on neoplastic B cells. Multivariable Cox-proportional hazard analysis was used investigate this research question using a randomly selected subset of the data, and the predictive ability of this model was validated on the remaining 1/3 of patient data. Results suggested that low class II MHC expression was associated with decreased times to relapse and death as is seen in human B cell lymphoma, and that large neoplastic cells were associated with decreased survival time. CD34 expression was not associated with patient outcomes. These findings have implications for the use of dogs to model human lymphomas, for the study of tumor vaccines, and for prediction of mortality in dogs with B cell lymphoma with a high level of specificity.

ABSTRACT O-8

BREAST CANCER RESISTANCE PROTEIN (BCRP/ABCG2) GENE EXPRESSION BY REAL TIME RT-PCR IN DOGS WITH MULTICENTRIC LYMPHOMA. LCS Rodrigues[1], D Levy[2], SP Bydlowski[2], SRR Lucas[1]. [1]School of Veterinary Medicine and Animal Science - University of São Paulo-Brazil [2]Laboratory of Genetic and Molecular Hematology (LIM31) Faculty of Medical Science - University of São Paulo-Brazil.

One of the reasons for the failure of canine lymphoma treatment is related to the resistance of tumor cells against chemotherapy drugs. The major form of this resistance is provide by multidrug resistance ABC transporters. ABC transporters proteins comprise a large superfamily of transmembrane proteins, ATP-dependent, that extrude a large variety of drugs from the cells. Multidrug resistance phenotype in cancer cells is associated with overexpression of these transmembrane proteins. ABCG2, also known as BCRP, is a 655 residue half-transporter protein that protect hematopoetic stem cells against toxic compounds. The aim of this study was to investigate the expression of BCRP (ABCG2) in canine multicentric lymphoma. Samples were collected by fine needle aspiration of an enlarged lymph nodes, from 25 dogs with multicentric lymphoma (stage III to V) at diagnosis, and 8 normal lymph nodes (control). Dogs that were previously treated with prednisone or chemotherapy were excluded from the study. Quantitative RT-PCR was used to measure the mRNA expression level of BCRP and FLNB expression as a endogenous reference canine gene. A widely range expression value for ABCG2 expression was found for canine multicentric lymphoma. High gene expression was observed in 52% (13/25) canine lymphoma, but 48% of dogs had a lower expression when compared with normal lymph node. Gene expression was not associated with clinical staging, complete or partial remission, relapse and survival time. In conclusion, ABCG2 was expressed in canine lymph node and canine multicentric lymphoma at the diagnosis, and it was not correlated with clinical response.

ABSTRACT O-9

HIGH SERUM ALKALINE PHOSPHOTASE (ALP) IN CANINE OSTEOSARCOMA PATIENTS CORRELATES WITH CELLULAR ALP EXPRESSION AND INCREASED β-CATENIN EXPRESSION. C. Piskun, M. Huelsmeyer, V. Thompson, T. Stein. University of Wisconsin-Madison School of Veterinary Medicine, Madison, WI.

Osteosarcoma (OSA), the most frequent primary malignant bone tumor of dogs, is both locally aggressive and highly metastatic. Prognostic factors for canine OSA include tumor location, distant metastatic disease, and serum alkaline phosphatase (ALP) concentration. An increased serum ALP concentration is associated with poor prognosis; however the mechanisms underlying this phenomenon are currently unclear. During normal bone development ALP may be used as a marker for osteoblasts. Additionally, ALP is a downstream target of activated canonical Wnt/β-catenin signaling. Therefore, we hypothesized that increased serum ALP would be associated with increased expression of β-catenin in canine OSA. The goals of this study were: (1) characterize and compare cellular ALP expression in OSA tissue from patients with normal and high serum ALP; and (2) assess β-catenin expression in those same patient populations.

We used frozen OSA samples collected from patients with either high ALP ($n = 3$) or normal ALP ($n = 3$). Total RNA was isolated from the frozen tissue, converted to cDNA, and analyzed using quantitative reverse-transcriptase polymerase chain reaction (qRT-PCR) with either target gene ALP (aim 1), or target gene β-catenin (aim 2). Additionally, β-catenin expression was analyzed by Western blot. QPCR data for β-catenin and ALP expression were normalized to 18S, and relative expression was calculated by the $\Delta\Delta Ct$ method.

The relative expression of cellular ALP was higher in high serum ALP samples compared to normal serum ALP samples: 44.74 ± 22.04 (mean relative expression $\pm$ standard deviation; $p < 0.001$). Further, the relative expression of β-catenin was also increased; β-catenin expression of high serum ALP samples relative to low serum ALP samples was 24.57 ± 11.41 ($p < 0.001$), which is also seen by Western blot.

This study begins to clarify the mechanism behind high serum ALP in canine OSA, and suggests the Wnt signaling pathway may be active in this population of patients. Further work will focus on elucidating the role active Wnt signaling plays in the biology of OSA. In the future, serum ALP status of OSA patients may help identify patients that would benefit from therapies targeting this pathway.

ABSTRACT O-10

COMPARISON OF ABDOMINAL ULTRASOUND AND MAGNETIC RESONANCE IMAGING FOR DETECTION OF ABDOMINAL LYMPHADENOPATHY IN DOGS WITH METASTATIC APOCRINE GLAND ADENOCARCINOMA OF THE ANAL SAC. Christine L. Anderson, Janean Fidel, Rance Sellon, Jeffrey N. Bryan and Chelsea D. Tripp. Washington State University College of Veterinary Medicine, Pullman, WA.

Accurate assessment of abdominal lymph node status is of vital importance for appropriate treatment planning and determining prognosis in dogs with apocrine gland adenocarcinoma of the anal sac (AGAAS). Pretreatment knowledge of lymph node status is helpful for determining prognosis and planning the optimal extent of lymphadenectomy. In addition, pretreatment knowledge of lymph node status may help in selecting patients who might benefit from adjuvant chemotherapy and radiation therapy. Abdominal ultrasound is currently the most commonly employed test to screen for abdominal lymphadenopathy in dogs with AGAAS. Imaging studies in people indicate that magnetic resonance imaging (MRI) is a more sensitive technique than abdominal ultrasound for the detection of abdominal lymphadenopathy. Systematic studies comparing abdominal MRI and abdominal ultrasound techniques in dogs with AGAAS are absent. The aim of this retrospective study was to directly compare the efficacy of abdominal ultrasound and abdominal MRI techniques in detecting abdominal lymphadenopathy in dogs with AGAAS. Sixteen dogs with histologically confirmed AGAAS had abdominal ultrasound and abdominal MRI. Abdominal ultrasound detected lymphadenopathy in 8 out of 16 dogs. Abdominal MRI detected lymphadenopathy in 14 out of 16 dogs. 50% of abdominal MRI detected lymphadenopathy was identified on abdominal ultrasound. This data indicates that MRI is significantly more sensitive than ultrasound for detecting lymphadenopathy in dogs with AGAAS. As such, MRI should be considered in any patient with AGAAS to more reliably stage the disease, particularly when accurate characterization of the extent and distribution affects therapeutic planning.

Previously presented at the Veterinary Cancer Society Annual Conference, October 2010(VCS Award Winner).

ABSTRACT O-11
PHARMACOKINETICS OF CYCLOPHOSPHAMIDE AFTER ORAL OR INTRAVENOUS ADMINISTRATION TO DOGS WITH LYMPHOMA. Emma Warry, Ryan Hansen, Dan Gusterfson, Susan Lana. Colorado State University Veterinary Teaching Hospital, Animal Cancer Center, Ft. Collins, CO.

To determine and compare the plasma concentration of cyclophosphamide and its metabolite 4-OHCP, within the plasma of lymphoma bearing dogs being treated with either oral or intravenous cyclophosphamide.

In this prospective study, patients were randomly assigned to either receive oral or intravenous cyclophosphamide, at a dose of 250 mg/m2. Based on a priori power calculation eight patients per treatment group were enrolled. Plasma was obtained at times 0, 15, 30, 60 minutes, and then at 2, 4, 6, 8, 24 hours post administration for evaluation of 4-OHCP concentrations by liquid chromatography-dual mass spectrometry (LC/MS/MS).

Average values were obtained for both cyclophosphamide and 4-OHCP concentrations within the plasma of both groups. The following values were obtained, half life (HL), time to maximum concentration (Tmax), maximum concentration (Cmax), and area under the curve (AUC). The Mann-Whitney statistical test was used to compare the groups. The AUC for cyclophosphamide was statistically significant ($P < 0.05$) when compared between the two groups. The AUC for 4-OHCP was not statistically significant between the groups. The difference between Cmax for cyclophosphamide and 4-OHCP was statistically significantly ($P < 0.05$) between the groups.

Although the AUC for cyclophosphamide was statistically significant between the two groups, the AUC for the active metabolite 4-OHCP was not different when administered intravenously or orally. Thus drug exposure to the active metabolite of cyclophosphamide is the same when administered intravenously or orally.

Previously presented at the Veterinary Cancer Society Annual Conference, October 2010. (VCS Award Winner)

ABSTRACT OT-1
NON-COMPARATIVE STUDY TO EVALUATE INTRAOSSEOUS CATHETER INSERTION WITH THE EZ-IO G3 DEVICE BY NOVICE PARTICIPANTS IN A FELINE CADAVER MODEL. ML Leis, ECR Snead, MP Dhumeaux, TY Epp, AP Carr. Western College of Veterinary Medicine, Saskatoon, SK.

The percentage of successful intraosseous (IO) catheter insertions, insertion times, and "ease of use" scores using the EZ-IO G3 power driver by a wide spectrum of novice participants in feline cadavers were evaluated. Novice users' mean IO catheter insertion time using the EZ-IO G3 driver was also compared to the mean IV catheter insertion time in normovolemic feline and canine patients presented to the Western College of Veterinary Medicine (WCVM) Small Animal Hospital.

Novice users included 40 WCVM personnel (8 technicians, 11 veterinary students, 5 interns, 8 residents, 8 clinicians). After watching a 5-minute EZ-IO G3 training video, each participant inserted 3 IO catheters using the EZ-IO G3 driver. Site (proximal humerus or trochanteric fossa of the femur) and side of cat (right or left) were randomized for each attempt for each participant. A 15 gauge x 15 mm long needle and a 15 gauge x 25 mm needle were used for IO catheter insertion in the humerus and femur, respectively. Participants then graded the "ease of use" of the EZ-IO G3 device on a Visual Analog Scale (VAS) that was converted to a 5-point scale. Twenty-six IV catheter insertions in normovolemic feline and canine patients performed by WCVM Small Animal Hospital personnel (6 technicians, 16 veterinary students, 1 intern, 1 resident, 2 clinicians) were then timed and compared to the mean IO catheter insertion time in feline cadavers by study participants using the EZ-IO G3 device.

The IO catheter was inserted correctly on every attempt by 70% (28/40) of participants. No difference was found between participant groups for mean IO catheter placement confirmation percentage ($p = 0.72$). Percentage of IO catheter "slippage off the bone" at the time of placement did not vary across participant groups ($p = 0.12$). Mean IO catheter insertion times were all less than 10 seconds and did not differ significantly as a function of attempt number ($p = 0.93$) or as a function of participant group ($p = 0.78$). Participants rated the EZ-IO G3's "ease of use" favorably and subjective scores did not differ across participant groups with varying levels of clinical experience ($p > 0.2$). Compared to the mean insertion time for IV catheterization (188 sec), mean IO catheter insertion by participants using the EZ-IO G3 (8 sec) was significantly faster ($p = 0.001$).

Regardless of their level of clinical experience, participants rated the EZ-IO G3 device favorably in terms of its "ease of use" and their willingness to use the device in the future. Regardless of their level of clinical experience, study participants successfully placed IO catheters using the EZ-IO G3 device and did so significantly faster than the reported IV catheter insertion time in normovolemic feline and canine patients in the WCVM Small Animal Hospital. Intraosseous catheterization using the EZ-IO G3 has the potential to provide very rapid vascular access and is a skill that can be easily learned.

Previously presented at the Western College of Veterinary Medicine Undergraduate Poster Competition.

ABSTRACT OT-2
CAUSES OF MULTICAVITARY EFFUSION IN DOGS: RETROSPECTIVE STUDY OF 43 DOGS. Palerme JS, Goldstein M, Kruth SA. Ontario Veterinary College, Guelph, Canada.

Multicavitary effusion is a common cause of presentation for dogs to emergency medical centers. The goal of this study was to identify common underlying causes of multicavitary effusion as well as determine their relative importance. A retrospective analysis of 43 cases of multicavitary effusion admitted to the ICU of a tertiary referral center (Ontario Veterinary College) from 2007 to 2010 was performed. Twenty-three different breeds, with Golden and Labrador retrievers (25.6% and 14%, respectively) being most commonly seen, were included in the study. Ages ranged from 2 to 14 years with a median age of 9 years and a mean of 8.1 years. 69.8% of cases were males (30/43 cases). Most common presenting signs included lethargy (48.8%), anorexia (32.6%), vomiting (21%) and dyspnea (21%). Cavitary effusion was detected by either ultrasonography (pericardial, pleural or abdominal) or radiographs (pleural). Bicavitary effusion was present in 28 cases (65.1%) whereas 15 cases (34.9%) had tricavitary effusion. Neoplasia was found to be the most common underlying cause overall (51.2%), with hemangiosarcoma being the leading type (40.9% of neoplasia cases), followed by congestive heart failure (9.3%), gastrointestinal lymphangectasia (4.7%), peritonitis/pancreatitis (4.7%), cirrhotic liver disease (2.3%) and acute renal failure (2.3%). In 11 cases (25.6%), no underlying cause could be found. Of these, 4 (9.3% of all cases) were diagnosed as having idiopathic pericardial effusion. Taken together, these findings suggest a strong association between multicavitary effusion and diseases carrying a guarded prognosis in dogs.

ABSTRACT OT-3
OBSERVATION OF PATIENT AND SURGEON PREOPERATIVE PREPARATION IN COMPANION ANIMAL CLINICS. MEC Anderson, BA Foster, JS Weese. Ontario Veterinary College, University of Guelph, Guelph, ON, Canada.

Infection control practices in veterinary clinics and hospitals are becoming increasingly important, with rising client expectations, growing concern about the spread of antimicrobial-resistant pathogens, and the potential for zoonotic transmission of disease. Surgical patients are at increased risk of developing infections, and can serve as sources of these pathogens for other animals and people with whom they have contact within and outside the clinic. Taking all reasonable precautions to reduce the risk of surgical site infections, beginning with preoperative preparation of the surgeon and patient, is therefore an important part of any infection control program. While guidelines are available for preoperative preparation procedures, there has been no objective investigation of compliance with these guidelines in veterinary practices. The objectives of this pilot study were to describe a range of preoperative hand scrub and surgical site preparation practices in veterinary clinics, and to determine if there were any areas that consistently require improvement.

Observation of preparation practices was performed in each of ten clinics over 9–14 days using 2–3 small wireless surveillance cameras. Data was coded for 148 surgical patients, and 31 surgeons performing a total of 190 hand scrubs. Patient hair removal was most commonly performed after induction of the animal (129/140, 92%) and using clippers (114/137, 83%). Steps in surgical site aseptic preparation ranged from 1–4. Contact time with soap ranged from 18–369s (mean 82s, median 53s), and with alcohol from 3–220s (mean 41s, median 30.5s). Application of alcohol or antiseptic using a "cleanest to dirtiest" pattern was infrequent (29/66 (44%) and 11/83 (13%), respectively). Potential contamination of the surgical site occurred most frequently when the animal was moved to the surgery table after initial preparation (40/58, 69%). Preoperative alcohol hand rub was used in 2/10 facilities, but soap and water hand scrub was still more commonly used even at these clinics. Proximal-to-distal scrubbing was noted in 95/142 (67%) of soap and water scrubs. Contact time during surgeon hand preparation ranged from 7–529s (mean 144s, median 124s) for soap and water and from 4–123s (mean 34s, median 25s) for alcohol-based hand rub. Approximately 75% of the variation in contact time was due to inter-surgeon variation. No significant changes in practices were identified over the course of the observation period.

Some preoperative preparation practices were fairly consistent between clinics in this study, while others varied considerably. Contact times with preparatory solutions were often far shorter than recommended, and there was a high frequency of non-sterile contact with the surgical site during movement of patients to the surgical suite. The camera system used to perform this study did not have a significant time-dependent effect on the behavior of participants, and could be useful for performing similar field-based observational studies in the future.

ABSTRACT OT-4

COMPARISON OF A NEW INTRAOSSEOUS CATHETER INSERTION TECHNIQUE USING THE EZ-IO G3 POWER DRIVER TO A STANDARD MANUAL TECHNIQUE FOR INTRAOSSEOUS CATHETER INSERTION IN A FELINE CADAVER MODEL. ML Leis, ECR Snead, MP Dhumeaux, TY Epp, AP Carr. Western College of Veterinary Medicine, Saskatoon, SK.

This prospective randomized study compared the percentage of successful intraosseous (IO) catheter insertions, insertion times, and "ease of use" scores using the EZ-IO G3 power driver to manual IO catheterization in feline cadavers. The IO catheter insertion time in cadavers using the EZ-IO G3 device was also compared to IV catheter insertion time in normovolemic feline and canine patients.

After a purposely limited training period, a preclinical veterinary student was timed and video-recorded as she performed 20 IO catheter placements in feline cadavers (10 IO insertions by placing an Illinois needle manually and 10 IO insertions using the EZ-IO G3). Order of technique (manual or EZ-IO G3), site of IO placement (proximal humerus or trochanteric fossa of the femur), and side of cat (right or left) were randomized for each attempt. When using the EZ-IO G3, a 15 gauge x 15 mm long needle and a 15 gauge x 25 mm needle were used for IO catheter insertion in the humerus and femur, respectively. After each attempt, the student graded the "ease of use" of each technique on a visual analog scale (VAS) that was converted to a 5-point scale. Twenty-six IV catheter insertions in normovolemic feline and canine patients performed by Western College of Veterinary Medicine (WCVM) Small Animal Hospital personnel (6 technicians, 16 veterinary students, 1 intern, 1 resident, 2 clinicians) were then timed and compared to the student's mean IO catheter insertion time using the EZ-IO G3.Median IO catheter insertion times for the 2 techniques were significantly different (manual IO technique 24 sec; EZ-IO G3 5 sec) ($p < 0.001$); the manual method took 48 seconds longer (95% confidence interval of 28 to 67 seconds) than the EZ-IO G3 method. Insertion time was more variable for the manual technique than for the EZ-IO G3. Percentage of catheter "slippage off the bone" and extravasation around the inserted catheter were significantly higher for placement of the manual IO catheter compared with placement of the EZ-IO G3 catheter ($p < 0.001$). Student's subjective ratings were more favorable and more consistent for the EZ-IO G3 technique compared to the manual technique for IO catheter insertion. Compared to the mean insertion time for IV catheterization in the WCVM Small Animal Hospital, IO catheter insertion by the student using the EZ-IO G3 was significantly faster (IV catheter 188 sec; EZ-IO G3 IO catheter 7 sec) ($p < 0.001$).

Intraosseous catheter insertion using the EZ-IO G3 can be said to be significantly faster, less traumatic, more user-friendly, and as effective as IO catheter placement using the manual technique. Vascular access via IO catheter insertion using the EZ-IO G3 device may be suggested to be faster than IV catheter insertion.

Previously presented at the Western College of Veterinary Medicine Undergraduate Poster Competition.

ABSTRACT OT-5

ASSESSMENT OF FELINE BONE MINERAL DENSITY: QUANTITATIVE COMPUTED TOMOGRAPHY. Y Lee, H Cheon, D Lee, MP Yang, J Chang, JH Kang, D Chang. Chungbuk National University College of Veterinary Medicine, Cheongju, Chungbuk, Republic of Korea.

Computed tomography (CT) has been widely investigated and applied as a means for non-invasive quantitative bone mineral determination in human medicine. The aim of this study was to assess age-related changes and anatomic variation in bone mineral density (BMD) using quantitative CT in normal cats. Seventeen normal cats were included in this study and divided into the following 3 age groups: < 1 year ($n = 4$); 2–5 years ($n = 10$); and > 6 years ($n = 3$). A computed tomographic scan of each vertebra from the 12th thoracic to the 7th lumbar spine, and the pelvis, was performed with a bone-density phantom (50, 100, and 150 mg/cm^3, calcium hydroxyapatite, CIRS phantom®). On the central transverse section, the elliptical region of interest (ROI) was drawn to measure the mean Hounsfield unit value. Those values were converted to equivalent BMD by use of the bone-density phantom and linear regression analysis ($r^2 > 0.95$).

The mean BMD value of the thoracic vertebrae (651.3 ± 100.4 mg/cm^3) was significantly higher than of the lumbar vertebrae (520 ± 119.5 mg/cm^3). The maximum BMD occurred at the T12, T13, and L1 levels in all age groups. There was a statistically significant difference in the mean BMD value among the 3 age groups at the T12 ($P < 0.001$), T13 ($P < 0.001$), and L4 levels ($P = 0.013$), respectively. In addition, there was no significant difference between the mean BMD value of the left and right iliac bodies (485.1 ± 140.4 mg/cm^3 and 482.1 ± 148.2 mg/cm^3, respectively). The present study suggests that age-related changes and anatomic variation in BMD values should be considered when assessing BMD using quantitative CT in cats with bone disorders.

ABSTRACT OT-6

QUANTITATIVE ASSESSMENT OF AGING EFFECTS ON CEREBRAL BLOOD FLOW IN NORMAL DOGS: A PERFUSION CT STUDY. D Lee, H Cho, Y Lee, MP Yang, J Chang, JH Kang, D Chang. Chungbuk National University College of Veterinary Medicine, Cheongju, Chungbuk, Republic of Korea.

Dynamic contrast-enhanced computed tomography (DCE-CT) is a rapid and widely available method of cerebral perfusion imaging. However, there is no established reference value of cerebral blood flow (CBF) measured by DCE-CT according to a dog's age. The purpose of this study was to identify the correlation between regional CBF and aging in clinically normal dogs using DCE-CT. Fourteen dogs with no evidence of hemodynamic disorders and central nervous system dysfunction were included in this study. Dogs were assigned to the following 3 age groups: < 1 year (group 1); 3–6 years (group 2); and < 10 years (group 3). DCE-CT scans were performed at the level of the third ventricle and mesencephalic aqueduct. CBF in the gray and white matter was calculated using STROKETOOL-CT® software.

The overall mean $\pm$ standard deviation quantitative estimate for regional CBF in clinically normal dogs was 67.8 ± 12.9 ml/min/100 g, 44.7 ± 4.1 ml/min/100 g, and 31.0 ± 3.4 ml/min/100 g in groups 1, 2, and 3, respectively. There was no significant regional CBF difference between the right and left sides of the brain in each group. Also, a statistically significant difference in the regional CBF was observed between groups 2 and 3 ($P < 0.001$). Thus, aging affects the regional CBF in normal dogs and the values should be considered assessing the results of DCE-CT.

ABSTRACT OT-7

INAPPROPRIATE URINATION IN MULTI-CAT HOUSEHOLDS: ARE THEY REALLY JUST BEHAVING BADLY?

D. Ramos[1], A. Reche Junior[1], D.S. Mills[2], P.L. Fragoso[1], A.G.T. Daniel[1], M.F. Freitas[1]. [1]Department of Medical Clinics, Faculty of Veterinary Medicine and Animal Science, University of São Paulo, Brazil; [2]Faculty of Agriculture, Food and Animal Sciences, University of Lincoln, UK.

According to several clinical behavior guidelines, "toileting" type inappropriate urination (i.e. large amounts of urine deposited in horizontal surfaces) can arise in cats suffering from a medical problem (typically lower urinary tract disease). By contrast, "spraying" type behaviour (i.e. possibly smaller amounts of urine deposited on vertical areas) is more typically associated with anxiety brought about by a threat to local resources, arising from either a change in the physical environment or threat to these resources from another cat. However, there is some evidence that "sprayers" may also be presented with a medical problem, which might be linked to the disease (e.g. painful voiding associated with crystalluria may lead to a standing posture being adopted and small amounts eliminated at a given time). This might be associated with an apprehensive state or simply a co-morbid state.

As part of a larger research project aimed at investigating behavioral and physical aspects of cats presented with inappropriate urination, owners of 14 "spraying" and 18 "toileting" cats with appropriate control subjects from the same households were recruited throughout local media coverage and the internet. The case-control dyads were brought by the owners to the veterinary hospital of the University of São Paulo, at the same time, for a medical work-up (i.e. physical examination, complete blood count, biochemical profile, urinalysis, urine culture and abdominal ultrasound).

No significant differences between the "sprayers" and "toileters" regarding the occurrence of medical problems were found. Both groups had a similar proportion of cats affected by medical illnesses (sprayers: 35.7%, toileters: 44.4%; Chi^2, $p = 0.618$), directly or indirectly relating to the urinary system (e.g. diabetes, chronic kidney disease). In both groups, control cats also had a relatively high occurrence of medical concerns (21.4% and 27.8%, respectively for each control group). These results emphasize the importance of careful medical evaluation of cats presented for a urinary house-soiling problem. The relatively high prevalence of medical concerns among apparently healthy cats in multi-cat households may have arisen, at least in part, as a result of an inability/failure of owners to monitor individuals, thus allowing some early signs to pass unnoticed. The way in which medical and behavioral elements are linked (if at all) remain unknown but deserve further investigation.

ABSTRACT OT-8

LIVING STYLE, PERSONALITY, OWNER SUBJECTIVE LIFE QUALITY: WHAT FACTORS AFFECT STRESS LEVELS IN DOMESTIC CATS (*FELIS CATUS*)?

D. Ramos, M.N. Arena, A. Reche Junior, A.G.T. Daniel, M.V.C. Albino, A.S. Vasconcellos, P. Viau, C.O. Alvarenga. Faculty of Veterinary Medicine and Animal Science, University of São Paulo, Brazil.

Considered as a semi-social species, domestic cats appear to be highly sensitive to the effects of social stress, especially when living in high density populations. Cats are capable of adapting to live in-group; nonetheless, they do not appreciate living in close proximity with others as result of an environment lacking of great opportunities of escaping and hiding.

This study aimed at testing the following hypotheses: (A) owners' perceived quality of life affects cats' global levels of stress; (B) cats' global levels of stress are influenced by cats' personality; (C) cats' living style (single housing versus large group housing) does affect stress levels in cats. To our knowledge, this is the first study investigating stress levels of domestic owned cats, under natural conditions, throughout measurement of faecal glucocorticoids metabolites concentration, and taking into consideration cat personality, cat living style and owner's subjective life quality.

In this study, adrenocortical activity, as a valuable physiological indicator of emotional stress, was evaluated throughout the measurement of faecal glucocorticoids metabolites in fourteen single and sixteen in-group housed cats. Cat personality as well as owners life quality was evaluated by self reported questionnaires given to the owners to answer.

Significant differences in mean glucocorticoids metabolites concentrations (mGCM) between the two populations (i.e. single versus in-group cats) were not detected (Random effect model, $p = 0.178$). However, when mGCM were taken as a function of cat personality, there were differences regarding single cats- *timid* cats showed higher levels in comparison to *easy-going* (Random effect model, $p = 0.018$) and *bossy* (Random effect model, $p = 0.020$) cats. As to owner subjective life quality, a direct association between the scores given by the owners to the *social* dimension and mGCM was found for single cats only (i.e. the better the owner felt itself social wise the higher the mGCM of the cat; Random effect model, $p = 0.002$).

Social stratification may compensate the stress resulting from spatial restriction in large in-group living cats. Other underexplored factors such as feline personality and owner life style seem to play an equally important role in domestic cats' day to day levels of stress, especially in the cats kept as single pets.

ABSTRACT P-1

DYNAMICS OF THE RENIN-ANGIOTENSIN-ALDOSTERONE SYSTEM (RAAS) AFTER ORAL ADMINISTRATION OF BENAZEPRIL IN DOGS.

J Mochel[1], G Strehlau[2], R Mohamed[1], M Peyrou[1]. [1]Novartis Centre de Recherche Santé Animale SA, St-Aubin, Switzerland. [2]Novartis Animal Health Inc., Basel, Switzerland.

In dogs, RAAS activation is a major feature of congestive heart failure (CHF). Benazepril (Fortekor®) is a potent ACE inhibitor with well-documented effectiveness in canine CHF. Although ACE activity (ACE_A) has been used in preclinical studies as a surrogate marker of efficacy, some authors have reported a poor correlation between plasma ACE_A and changes in angiotensin II (AII) or aldosterone (AL). The purpose of this study was to investigate the effect of benazepril on canine plasma renin activity/concentration (PRA/PRC), angiotensin I (AI), AII, AL, and fractional excretion of potassium (UfeK), sodium (UfeNa) and aldosterone (UfeAL). Sixteen beagle dogs were fed a low-sodium diet and dosed with placebo or benazepril tablets (10 mg PO, q24h) for 5 days. Blood and urine samples were collected on day 1 (D1) and day 5 (D5) over 24-hour periods. Data were analyzed by repeated measures ANOVA of baseline corrected values, and ANOVA of $AUC_{24hours}$. Compared with placebo, benazepril induced a significant increase in PRA and AI at D1 (p-value $_{[PRA]}$:0.001, p-value $_{[AI]}$:0.002) and D5 (p-value $_{[PRA]}$:0.001, p-value $_{[AI]}$ < 0.001). No differences in PRC were noticed. Based on $AUC_{24hours}$, AII levels were 34% lower in the benazepril group at D5 (p-value $_{[AII]}$:0.01). UfeAL and AL decreased by up to 27% and 24% at D1 and D5, respectively, though differences did not reach statistical significance. Benazepril markedly influences RAAS dynamics in dogs. Decreased exposure to AII and AL are likely to be the key events required to counteract pathological remodeling of the heart in CHF.

Figure [Abstract P-1]: Difference from placebo (expressed as a percentage) [geometric mean + 90% confidence interval]. ** Indicates statistically significant differences ($p < 0.01$).

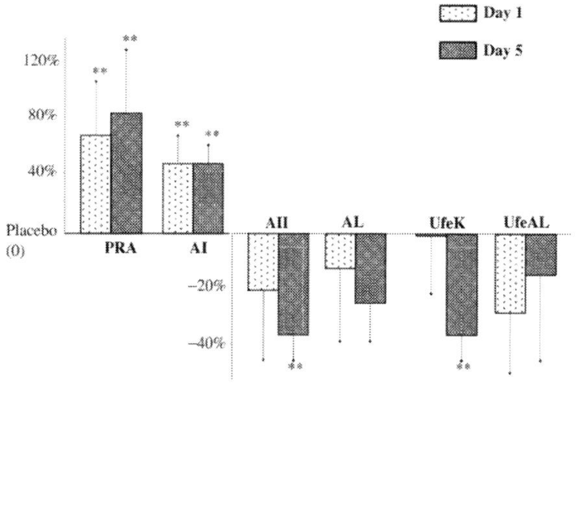

ABSTRACT P-2

DOSE ESCALATION WITH ALFAXALONE OR PROPOFOL IN THE DOG AND EFFECTS ON RESPIRATORY DEPRESSION. T Whittem[1], H Keates[2]. [1]University of Melbourne Faculty of Veterinary Science, Werribee, Victoria, Australia. [2]University of Queensland School of Veterinary Science, Gatton, Queensland, Australia.

This study compared two intravenous anesthetic agents, alfaxalone (ALF) (Alfaxan®, Jurox Pty. Ltd.) and propofol (PPF) (Rapinovet®, Schering Plough Animal Health) and their effects on spontaneous ventilation after induction of anesthesia in dogs at various doses.

This randomized, crossover, dose-escalation study used six dogs in weight and gender-matched pairs (3M-3F). For each drug, each dog was dosed incrementally at 1, 2, 5, 10 and 20 times the labeled anesthetic induction dose rate (ALF 2 mg/kg, PPF 6.5 mg/kg) or until a dose was reached that rendered the dog apneic. A minimum of three days was allowed between doses. For each dose administration, the entire calculated dose was delivered constantly over 1 min.

The primary variable was apnea, defined as an absence of spontaneous ventilation for 1 minute. Apneic dogs were manually ventilated with oxygen until they resumed adequate spontaneous ventilation. Once the apneic dose was determined for an individual dog for one drug, the dog began incremental doses with the alternate drug. For each anesthetic episode times were recorded from completion of induction dose to; removal of endotracheal tube, dog lifting head, dog attaining sternal recumbency and dog standing. Pulse rate, respiratory rate, SpO_2 and $ETCO_2$ were each measured every 5 min. Within-dog comparisons were made using the paired Student's t-test.

For both ALF and PPF all 6 dogs respired voluntarily at the labeled (1 X) dose. For PPF at 2 and 5 X doses, 4 and 0 dogs respired voluntarily respectively. For ALF at 2, 5 and 10 X doses, all 6, 4 and 1 dog respired voluntarily respectively. For all six dogs to become apneic required 5 X dose of PPF and 20 X dose of ALF. The mean no observable adverse effect dose (NOAEL) expressed as a multiple of the labeled dose was higher for ALF (4.8 X) than for PPF (1.7 X) ($p = 0.035$). There were no significant differences between times to extubation, head lift or attaining sternal recumbency after ALF and PPF at 1, 2 and 5 X doses. At the 2 X dose, dogs took longer to stand after ALF (29.0 ± 7.0 min) than PPF (25.0 ± 8.3 min).

We concluded that based on anaesthetic duration, the manufacturer's labeled dose rates of 2 mg/kg for ALF and 6.5 mg/kg for PPF were equivalent. However, based on the dose escalation, the number of dogs becoming apneic at each dose-multiple is consistent with PPF having a narrower safety margin, i.e., PPF caused more respiratory depression than ALF.

ABSTRACT P-3

SUBCUTANEOUS ADMINISTRATION OF LEVETIRACETAM IN HEALTHY DOGS. BT Hardy[1], EE Patterson[1], JM Cloyd[2]. [1]Veterinary Clinical Sciences, University of Minnesota, St Paul, MN. [2]Experimental and Clinical Pharmacology, University of Minnesota, Minneapolis, MN.

Parenteral levetiracetam (LEV) has been shown to rapidly attain therapeutic levels in dogs when given IV or IM, and has been used off-label for the treatment of seizure emergencies. The purpose of this study was to determine the safety and pharmacokinetics of subcutaneously administered levetiracetam in healthy dogs. Potential application of these results would be use of SQ LEV instead of or in addition to rectal diazepam for the treatment of cluster seizures at home.

LEV was administered SQ between the shoulder blades to 4 healthy, purpose-bred hound dogs, at a dose of 60 mg/kg (undiluted). Blood samples were collected at 15, 120 and 420 minutes after LEV administration via jugular venipuncture. Plasma LEV concentrations were measured by high pressure liquid chromatography.

None of the dogs became sedated, nor was there pain evident on palpation of the injection site. Mean (standard deviation) LEV concentration was 65.2 (29.5), 114.5 (10.5) and 84.9 (20.6) µg/ml at 15, 120 and 420 minutes, respectively.

Administration of SQ LEV was well tolerated, and exceeded the suggested therapeutic range (5–45 µg/ml) within 15 minutes of administration, and remained above the range for at least 7 hours. These data indicate that SQ LEV administration may be an alternative for the at-home treatment of cluster seizures in dogs, and prospective studies in epileptic dogs are warranted.

Figure [Abstract P-3]

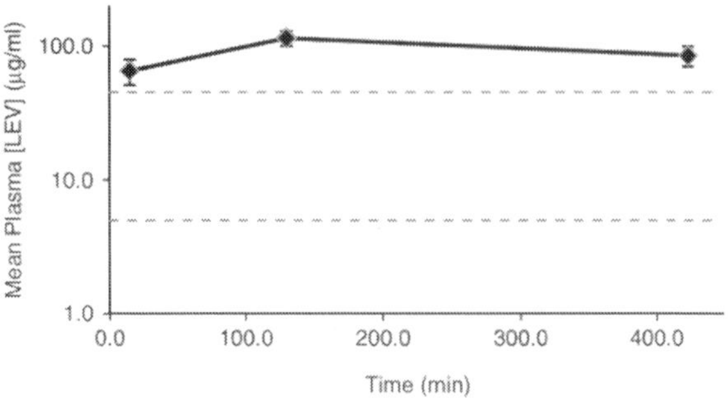

ABSTRACT P-4

THE EFFECTS OF CYTOCHROME P450 (CYP) INHIBITORS ON ORAL METHADONE IN GREYHOUND DOGS. Butch KuKanich, Kate Stenske KuKanich, Jessica Rodriguez. Kansas State University, College of Veterinary Medicine, Manhattan, KS.

The purpose of this study was to assess the effects of CYP inhibitors (ketoconazole, chloramphenicol, fluoxetine, trimethoprim, cimetidine, and medetomidine) in varying combinations on the bioavailability of oral methadone in healthy Greyhound dogs. The IACUC approved this study.

CYP inhibitors were administered PO for 48 hours prior to methadone administration. Methadone hydrochloride was administered PO at a targeted dose of 1 mg/kg. Blood was obtained for the determination of methadone plasma concentrations by mass spectrometry. The area under the curve (AUC) of methadone for each treatment group was compared statistically to the AUC of methadone administered without inhibitors using the Mann-Whitney Rank Sum Test.

Significant increases ($P < 0.001$) in the methadone AUC occurred in all treatment groups which included chloramphenicol, including chloramphenicol as the only inhibitor. The magnitude of increase was at least 50 fold. Mean concentrations of methadone exceeded 10 ng/mL for at least 10 hours in all groups administered concurrent chloramphenicol. No significant increases in the AUC occurred in any of the groups which did not include chloramphenicol.

In conclusion, chloramphenicol significantly inhibits the metabolism of methadone in Greyhound dogs. As a result, the oral bioavailability of methadone is significantly increased and plasma concentrations are achieved that are reported to be effective in humans for 10–36 hours after a single oral administration.

ABSTRACT P-5

DOXYCYCLINE POTENCY AFTER STORAGE IN A COMPOUNDED FORMULATION FOR ANIMALS. Mark G. Papich, Daria DiGiovanni, and Gigi Davidson. North Carolina State University, College of Veterinary Medicine, Raleigh, NC.

Doxycycline hyclate is used frequently in small animals, horses and exotic animals for treatment of a wide variety of infections. Because doxycycline hyclate tablets may not be suitable for oral administration in some animals, particularly horses and cats, it has been compounded into liquid suspensions. The commercially available doxycycline calcium 10 mg/ml oral suspension, Vibramycin® (Pfizer) is not suitable for use in animals due to its low concentration and flavoring that animals find unpalatable. Because of the known inherent instability of doxycline in aqueous vehicles under storage, this study was conducted to determine the potency of two formulations stored in dark and light conditions. A high pressure liquid chromatography (HPLC) assay with UV absorption at 350 nm was developed for analyzing doxycycline in formulations, in comparison to a reference standard from the United States Pharmacopeia (USP). Doxycycline hyclate 100 mg tablets were first tested for potency. The tablets were then crushed and mixed with a pharmaceutical vehicle to make two concentrations: 33.3 mg/mL and 166.7 mg/mL. The vehicle used was a 50:50 mixture of a Vehicle for Oral Solution (Ora-Sweet, USP-NF) and Vehicle for Oral Suspension (Ora-Plus, USP-NF). The suspensions were prepared in replicates of 3. Each replicate was divided, with one aliquot stored at room temperature in lighted conditions, and the other aliquot stored at room temperature in the dark. Doxycycline was extracted from the formulations and measured by HPLC at Day 0, 1, 4, 7, 14, 21, and 28. Each replicate was tested and the potency reported as the percent doxycycline relative to the USP reference standard. On Day 0, 1, 4, and 7, the potency of each formulation was within 90–110% of the reference standard (range 93.4–109%). This value is within the accepted range cited in USP < 795 > on Pharmaceutical Compounding-Non-Sterile Preparations. However, starting at Day 14, the potency declined dramatically and remained low for the tests performed on Day 21 and 28. The potency on Day 14, 21, and 28 was below 20% of the reference standard (range 14–18%). There was also a noticeable change in the quality of the formulation starting on Day 14, and a change in the color of the formulation to a dark brown. These results indicate that when doxycycline hyclate tablets are compounded as a suspension in an aqueous vehicle as described in this study, at 33.3 and 166.7 mg/mL under the storage conditions used in this study, potency of the formulation cannot be assured beyond 7 days. We recommend a beyond-use-day (BUD) of 7 days for formulations prepared and stored at room temperature in light or dark conditions.

ABSTRACT P-6

PHARMACOKINETICS AND TIME DEPENDENT KILLING OF FOSFOMYCIN IN MULTI DRUG RESISTANT-*ESCHERICHIA COLI* URINARY TRACT INFECTIONS IN DOGS. D.M. Boothe, P. Hubka. Department of Anatomy, Physiology, and Pharmacology, College of Veterinary Medicine Auburn University, AL.

Therapeutic options for multidrug resistance (MDR) *Escherichia coli* urinary tract infections (UTI) are limited. Fosfomycin (FOS) tromethamine is an oral, broad-spectrum, cell-wall active, bactericidal drug approved for treatment of uncomplicated UTI in humans. The purpose of this study was to determine time dependency of FOS and the disposition of FOS tromethamine in dogs. Using a randomized, double crossover design, 12 client-owned dogs received FOS sodium IV (40 mg/kg) and FOS tromethamine (PO, 80 mg/kg) either with (n = 6) or without food (n = 6). Serum and urine were collected for 24 hr; FOS was quantitated with a bioassay (ATCC *E. coli* 25922, serum or ATCC *Proteus vulgaris* 13315, urine). In-vitro killing curves (cell counts through 24 hours) were performed at 0 (control),0.5,1,2,8,16 and 32 x MIC for MDR *E. coli* canine FOS susceptible (E-test®) uropathogens. Killing curves indicated FOS to be time dependent. After IV administration, clearance (ml*kg/hr), volume of distribution (L/kg), elimination half-life (HL; hr) and mean residence time (MRT, hr) were (mean ± sd): 210 ± 104, 0.23 ± 0.15, 0.36 ± 0.19, 1.14 ± 0.35 and 1.7 ± 0.4, respectively. For PO, C_{max}, HL and MRT were 66 ± 21, 2.5 ± 1.09 and 5.1 ± 1.7, respectively. Serum FOS exceeded the MIC_{90} reported for multidrug resistant (MDR) *E. coli* (1.5 μg/ml) for 7hr (IV; 2.5 μg/ml) and 12hr (PO, 9 μg/ml). Oral bioavailability was greater with food (109 ± 31%) compared to fasting (66 ± 16%). Adverse events were limited to diarrhea (5/12). This study shows serum FOS exceeds the MIC $_{90}$ of MDR *E.coli* uropathogens for at least 12 hr after administration of 80 mg/kg PO FOS tromethamine.

ABSTRACT P-7

PHARMACOKINETICS OF DIMINAZENE DIACETURATE IN HEALTHY CATS. KM Lewis[1], LA Cohn[1], MG Papich[2]. [1]University of Missouri College of Veterinary Medicine, Columbia, MO. [2]North Carolina State University College of Veterinary Medicine, Raleigh, NC.

Diminazene is an aromatic diamidine, anti-protozoal drug that has shown promise in a small number of cases of cytauxzoonosis. In a non-controlled case series, 5 of 6 cats with clinical cytauxzoonosis given 3 mg/kg of diminazene aceturate survived infection. Dosage frequency was two intramuscular injections given one week apart. Commercial formulations contain the diminazene diaceturate salt. The active base is diminazene with the salt consisting of two aceturate molecules. Currently there is no data available on the pharmacokinetics of either diminazene compound in cats. The objective of this study was to determine the pharmacokinetics of diminazene diaceturate in healthy cats.

Four purpose bred cats with normal physical examination, CBC, chemistry and urinalysis were used. A powdered commercial drug formulation (Veriben®, Ceva Sanet Animale) was freshly reconstituted with sterile water to a concentration of 7 mg/mL prior to administration and sterile filtered solution. Heparinized blood samples were collected just before (hour 0) or 0.5, 1, 2, 4, 8, 12, 18, 24, 36, 48, 72, 120, and 168 hours after intramuscular administration of 3 mg/kg (1.68 mg/kg of diminazene base) diminazene diaceturate. The plasma was separated by centrifugation within 30 minutes of collection and frozen (−80°C) until analysis. Concentrations of diminazene were measured by HPLC analysis using UV absorption and ion-pairing conditions. The pharmacokinetic profile was analyzed using a simple one-compartment model. In these cats, diminazene had a mean terminal half life ($T_{1/2}$) of 1.70 (+/- 0.29) hrs and mean peak plasma concentration (C_{MAX}) 0.51 (+/− 0.11) μg/mL. The mean residence time (MRT) of diminazene was 2.45 hrs (+/−0.42). Systemic clearance (CL/F) was 1.38 (+/− 0.26) L/kg/hr. The volume of distribution per fraction absorbed (VD/F) was 3.36 (+/− 0.72) L/kg.

A single intramuscular dose of diminazene diaceturate was well tolerated by all 4 cats. Without knowing the concentration required to inhibit or kill *Cytauxzoon felis*, it is not yet possible to make suggestions regarding optimum dosing schedules for this drug. Additional toxicology data and studies to assess clinical efficacy for the treatment of cytauxzoonosis are indicated before routine clinical use can be considered.

ABSTRACT P-8

DETERMINATION OF MELOXICAM CONCENTRATION IN SYNOVIAL FLUID OF INFLAMED AND NON-INFLAMED JOINTS IN DOGS. Johnston L[1], Ingwersen W[2], Narbe R[1]. [1]Boehringer Ingelheim Animal Health, Ingelheim, Germany [2]Boehringer Ingelheim Vetmedica Burlington, Canada. [3]Boehringer Ingelheim Vetmedica, Ingelheim, Germany.

Meloxicam has been shown to accumulate in areas of inflammation in both the rat and human. The objective of this study was to investigate the concentration of meloxicam in synovial fluid of inflamed joints versus that of non-inflamed joints in dogs.

Eight male dogs were treated with 0.2 mg/kg of meloxicam on Day One and 0.1 mg/kg of meloxicam on Day Two. All treatments were administered orally. On Day Three reversible acute synovitis was induced in one stifle by aseptic, intra-articular administration of 1 ml sodium urate crystal suspension (10 mg/ml). In four dogs synovitis was induced in the L stifle and in four dogs the same procedure was used in the R stifle. In each dog the stifle without induction of synovitis served as the "normal" joint sample. A synovial fluid sample was collected from both the R and L stifle of each dog. Sample collection occurred eight hours after administration of sodium urate and twenty four hours after the last administration of meloxicam.

Synovial meloxicam concentration was analysed using High Performance Liquid Chromatography-Mass Spectrometry (HPLC/MS-MS). The concentration of meloxicam in the inflamed versus non-inflamed joint in each dog was compared using the paired t-test.

Table [Abstract P-8]

Mean meloxicam concentration in normal joint ng/mL	246 (SEM 35.3)
Mean meloxicam concentration in inflamed joint ng/mL	391 (SEM 43.4)
p-value of paired t-test	p = 0.002

The results indicate that meloxicam preferentially accumulates in inflamed joints in the dog as meloxicam concentrations are statistically significantly higher in inflamed joints than in non-inflamed joints.

ABSTRACT P-9

TRENDS OF SMALL ANIMAL *ESCHERICHIA COLI* ANTIMICROBIAL RESISTANCE IN THE UNITED STATES. K Thungrat[1], P Hubka[1], J Sofge[1], DM Carpenter[2], DM Boothe[1]. [1]College of Veterinary Medicine, [2]Department of Mathematics and Statistics, Auburn University, Auburn, AL.

No national surveillance system exists for monitoring emergent resistance in companion animals. However, *E. coli* resistance is an increasing therapeutic and public health concern in these in dogs and cats. The purpose of this study was to describe current resistance patterns of canine and feline pathogenic *E. coli* throughout the United States and identify risk factors of antimicrobial resistance.

Isolates (n = 1512) of clinical *E. coli* collected from dogs or cats from May 2008 through May 2010 located in 6 different regions. Susceptibility was determined to 15 drugs (6 drug classes) by broth microdilution methods. Pharmacodyamaic statistics were described regionally. Phenotypes were determined and type of resistance was based on the number of drug classes to which resistance was expressed: none (NDR), single (SDR) and multi (MDR).

The majority of isolates were from urinary tract (71.5%) and dogs (75.7%). The proportion of resistance type for each drug was: NDR (17.5%), SDR (56.4%) and MDR (26.12%). The proportion of MDR was greatest in the Southwest (25.29%) and least in the Northwest (9.19%) ($P < 0.05$). For all regions, the proportion of resistance was: cephalothin (CPH, 65.2%) > amoxicillin-clavulanic acid (AMX, 59.4%), ampicillin (AMP, 52.9%), tricarcillin-clavulanic acid (TCX, 21.8%), doxycyline (DXY, 15.9%) > cefoxitin (CFX, 15.2%), cefpodoxime (CPX, 13.7%), chloramphenicol (CHP, 13.2%), enrofloxacin (ENR, 13.0%), ciprofloxacin (CIF,12.3%), trimethoprim-sulfamethoxazole (TMX, 10.7%), ceftazidime(CFZ, 10.0%), gentamicin (GTM, 10.0%), cefotaxime (CFT, 9.2%) > meropenem (1.1%) ($P < 0.05$). The MIC_{90} exceeded the resistant breakpoint for AMP, AMX, CPX, CPH, CIF, CFX, DXY and ENR whereas MIC_{50} did not surpass the susceptible breakpoint. Beta-lactams (96.37%) was the most and aminoglycosides the least (0.12%) SDR. The drug class most frequently involved in MDR was beta-lactams (97.72%) and least, GEN (30.13%).

Resistance differs regionally, being greatest in the Southwest. CPH is the most and meropenam is the drug least associated with resistance; these patterns are consistent with current drugs used by veterinarians.

ABSTRACT P-10

COMPARATIVE IN VITRO ACTIVITIES OF FLUOROQUINOLONE ANTIMICROBIALS AGAINST *ESCHERICHIA COLI* UROPATHOGENS. X.Q. Liu, Dawn M. Boothe. Department of Anatomy, Physiology and Pharmacology, College of Veterinary Medicine, Auburn University, Auburn, AL.

The fluoroquinolones (FQs) are common choices for treatment of *E. coli* urinary tract infections (UTIs) in animals and humans. 2nd generation drugs approved in animals include enrofloxacin (ENR), marbofloxacin (MAR), orbifloxacin (ORB); human drugs include ciprofloxacin (CIP). 3rd and 4th generation FQ for humans include moxifloxacin (MOX), gatifloxacin (GAT) and ofloxacin, (OFL]), its L-isoform levofloxacin (LEV). For animals, pradofloxacin (PRA) is approved for use in Europe. The purpose of this study was to assess the *in vitro* activity of 1st (Naladixic acid [NAL] through 4th generation FQs (n = 11) toward dog or cat *E.coli* uropathogens (n = 51). Isolates were subjected to susceptibility testing to 6 drugs classes (15 drugs). Isolate phenotypes included no (NDR; n = 12), single (SDR; n = 15) or multidrug (to more than 2 drug classes; MDR; n = 24) resistance (including ENR resistant [ENR^R-MDR; n = 12] or ENR susceptible (ENR^S-MDR, n = 12). The minimum inhibition concentrations (MICs) were determined for each isolates using broth microdilution (*E. coli* ATCC® 25922 served as a negative control). MIC statistics were generated for each drug among phenotypes. The overall potency (MIC_{90}) for all ENR susceptible isolates (NDR, SDR and ENR^S-MDR) was GAT > PRA, MOX, MAR, LEV, CIP > SAR, ORB, OFL > ENR > NAL. Each *E. coli* isolate expressing NDR or SDR was susceptible to all FQ. However, isolates expressing resistance to 1st or 2nd generation FQ were also resistance to later generation drugs. The ratio of MIC_{90} breakpoint MIC (MIC ± SD), was NAL(4.93 ± 5.13), ENR(30.32 ± 37.47), ORB(31.83 ± 31,40), OFL(52.89 ± 82.47), SAR(73.54 ± 84.33), MAR(86.47 ± 74.19), CIP(128.5 ± 117.5), LEV(180 ± 166.8), GAT(167.6 ± 160.2), MOX(201.4 ± 228), PRA(122.5 ± 131.8).

ABSTRACT R-1

IMPACT OF ORAL VERSUS INHALED GLUCOCORTICOIDS TREATMENT ON ALLERGEN-SPECIFIC IgE TESTING IN EXPERIMENTALLY ASTHMATIC CATS. CH Chang[1], K Robinson[1], TM Lee-Fowler[2], AE DeClue[1], LA Cohn[1], CR Reinero[1]. [1]University of Missouri College of Veterinary Medicine, Columbia, MO. [2]Alabama Veterinary Specialists, Birmingham, AL.

Glucocorticoids (GC) are standard therapy for allergic asthma but do not reverse the underlying type I hypersensitivity. Allergen-specific immunotherapy (ASIT), a process of "desensitization", is potentially curative but requires identification of offending allergens. The purpose of this study was to determine if oral or inhaled GC administered at routinely used dosages would interfere with allergen identification. We hypothesized that oral but not inhaled GC would interfere with accurate identification of allergen-specific IgE using skin and serum testing in experimentally asthmatic cats.

Asthma was induced in eighteen cats using Bermuda grass allergen (BGA). Cats (n = 6/group) were randomized to receive oral GC (10 mg prednisolone q 24 hr PO), inhaled GC (600 ug budesonide q 24hr) or placebo (gelatin capsule q 24hr PO) for one month. Intradermal skin testing (IDST) and BGA-specific IgE amounts were measured prior to, during (weeks one and four) and every two weeks after treatment until both tests were positive. A paired T test was used to compare serum IgE among groups pre- and post-treatment ($P < 0.05$ significant).

IDST reactivity was eliminated in 4/6 cats on oral GC, 3/6 on inhaled GC, and 1/6 placebo-treated cats. Within two weeks after stopping treatment, IDST was again positive in all cats. Contrary to our hypothesis, serum IgE reactivity to BGA was not significantly diminished by any treatment.

In conclusion, a two week withdrawal from GCs is adequate for IDST identification of allergen but no withdrawal is required prior to serum IgE testing to identify the sensitizing allergens.

Previously presented at the Veterinary Comparative Respiratory Society Symposium, Raleigh, North Carolina, September 28-30, 2010.

ABSTRACT R-2

SERUM 25-HYDROXYVITAMIN D IN CATS WITH LOWER AIRWAY DISEASE. EA Rozanski[1], CR Sharp[1], SL Beyerlien,[2] PA Schenck[2]. [1]Cummings School of Veterinary Medicine at Tufts University, North Grafton, MA. [2]Diagnostic Center for Population and Animal Health, Michigan State University, East Lansing, MI.

In people, increasing severity of asthma is associated with low serum concentrations of 25-hydroxyvitamin D (25-OH-D). 25-OH-D is thought to ameliorate lower airway inflammation primarily by decreasing the production of pro-inflammatory mediators, and by increasing the production of the anti-inflammatory cytokine IL-10. In people, serum 25-OH-D concentration is associated with sunlight exposure as well as dietary intake. Cats do not rely on sunlight for vitamin D synthesis; all vitamin D comes from dietary intake. Cats have a naturally occurring lower airway disease syndrome (LAD) that shares many features with human asthma. The goal of this study was to evaluate serum 25-OH-D concentrations in cats with LAD.

Cats with naturally developing LAD were enrolled. Criteria for a diagnosis of LAD included a history of cough, wheeze or respiratory distress, radiographic evidence of a bronchial pattern and hyperinflation, negative heartworm antigen and antibody test, and a resolution of clinical signs in response to glucocorticoids. Dietary history was obtained. 25-OH-D concentrations were determined on serum samples by a commercial laboratory.

Twelve cats with LAD were enrolled. All cats ate commercial cat food. The median 25-OH-D concentration was 112 nmol/l with a range of 65–176 nmol/l which is within the reported reference range of 65–170 nmol/l.

In contrast to human asthma, lower airway disease in cats is not associated with low serum concentrations of 25-OH-D.

ABSTRACT R-3

CHARACTERIZATION OF INTERSTITIAL LUNG DISEASE IN 13 DOGS: CLINICAL COURSE AND COMPUTED TOMOGRAPHY. K. Kumrow, EA Rozanski, J Sutherland-Smith, JE Rush. Cummings School of Veterinary Medicine at Tufts University, North Grafton, MA.

Interstitial lung diseases (ILD) are uncommon in dogs, with the most commonly recognized ILD idiopathic pulmonary fibrosis ("Westie fibrosis"). In human medicine, ILD represent a large umbrella of pulmonary diseases, with IPF only a subset. Other, more treatable, ILDs are also identified, and may respond to either the removal of a stimulus (hypersensitivity) or steroid therapy. The goal of this report is to describe the clinical course, including outcome, computed tomography and histopathology of dogs affected with an ILD.

The computed tomography (CT) log was reviewed for dogs that underwent thoracic CT scanning for evaluation of respiratory signs, and had changes consistent with ILD as the primary abnormality, including the presence of diffuse disease in all lobes, and at least 2 of the following: reticulation, ground glass opacity, consolidation, or traction bronchiectasis. Survival time from CT date was calculated. The presence of moderate pulmonary hypertension [pHTN]

(> 50 mmHg) as estimated by tricuspid regurgitant jet, was also reported and survival times were compared with a Mann-Whitney rank sum with $P < 0.05$ considered significant.

Thirteen dogs were identified. Terriers and Chihuahuas were the most commonly affected breeds. Two dogs were adolescents, the remaining dogs ranged from 7–16 years, with a median of 11 years. Histopathology results (n = 5), including moderate to severe interstitial fibrosis (3) alveolar proteinosis with fibrosis (1), and interstitial eosinophilic pneumonia (1). One had suspected cryptogenic organizing pneumonia and had a good response to glucocorticoids. Eight dogs died of respiratory failure, with a median post CT survival time of 42 days (range 2–365), two dogs died of non-pulmonary disease, 2 dogs had severe lower respiratory infections as puppies with persistent respiratory signs, and both are still alive at > 2 years since diagnosis, 1 terrier is alive at 19 months and 1 was lost to follow up. 5 dogs had PHTN, with a median survival of 60 days (42–590), while the 5 dogs without had a survival of 180 days (range 21–730). [p = 0.3].

Interstitial lung disease in dogs is not just idiopathic pulmonary fibrosis. Following respiratory infection, young dogs may develop an ILD with a relatively indolent course and rare ILD is steroid responsive. CT is useful to identify ILD but further research correlated with echocardiography and histopathology is advised to use it to prognosticate.

ABSTRACT R-4

GENE EXPRESSION PROFILES IN CANINE IDIOPATHIC PULMONARY FIBROSIS. E. Krafft[1], HP. Heikkilä[2], M.Vanherberghen[1], I.Peters[3], D. Peeters[1], MM. Rajamäki[2], C. Clercx[1]. [1]Department of Equine and Companion Animal Clinical Sciences, Faculty of Veterinary Medicine, University of Liège, Belgium; [2]Department of Equine and Small Animal Medicine, Faculty of Veterinary Medicine, University of Helsinki, Finland; [3]School of Clinical Veterinary Science, University of Bristol, Bristol, UK.

Idiopathic pulmonary fibrosis (IPF) is an interstitial pulmonary disease, mainly described in West Highland white terriers (WHWT). Identification of molecular pathways important in the pathogenesis of IPF would improve our understanding of this disease and may help identify therapeutic targets. The aim of the present study was to investigate gene expression in lungs of WHWT with IPF using oligonucleotide microarray.

Total RNA was extracted from post-mortem pulmonary samples from five WHWT with IPF and five control dogs (CTRL) without pulmonary disease. The RNA was pooled from each group (IPF and CTRL) and analysed using the canine specific Affymetrix microarray technology. Genes with a minimum of a two-fold difference in expression between the two groups were selected for further analysis. The most significant biological functions for these genes were identified using Ingenuity Pathways Analysis.

More than 1000 genes were identified as having greater than two-fold difference in expression. The significant biological functions associated with these genes were related to cellular movement, cellular proliferation and apoptosis. Most notable among these were genes encoding the leukocyte chemotactic proteins: CCL2 (fold change +4.93), CCL17 (+2.74) and IL8 (+4.32); the proteins involved in fibroblast migration; and the matrix metalloproteinases (MMPs) involved in matrix degradation: MMP7 (−17.55), MMP9 (-3.42), MMP1 (−2.76).

This study has identified genes which may be important in pathogenesis of IPF, e.g. proteins involved in leukocytes chemotaxis, fibroblast recruitment and activation, regulation of apoptosis, and extracellular-matrix turn-over. However, real-time quantitative RT-PCR studies are needed to confirm these results before any definitive conclusions can be drawn.

ABSTRACT R-5

USE OF EXPRESSION MICROARRAY TO IDENTIFY POTENTIAL BIOMARKERS OF IDIOPATHIC PULMONARY FIBROSIS IN DOGS. E. Krafft[1], HP. Heikkilä[2], M. Vanherberghen[1], D. Peeters[1], MM. Rajamäki[2], C. Clercx[1]. [1]Department of Companion Animal and Equine Clinical Sciences, Faculty of Veterinary Medicine, University of Liège, Belgium; [2]Department of Equine and Small Animal Medicine, Faculty of Veterinary Medicine, University of Helsinki, Finland.

Idiopathic pulmonary fibrosis (IPF) is an interstitial disease, mainly described in West Highland white terriers (WHWT). Definitive diagnosis ultimately relies on lung histopathology. Identification of specific biomarkers would be very helpful. Expression microarray is a powerful screening tool to study local gene expression in a disease state. The aim of the present study was to measure gene expression profiles in lungs of WHWT with IPF to identify potential blood or bronchoalveolar lavage fluid (BALF) biomarkers.

Total RNA was extracted from post-mortem pulmonary samples from five WHWT with histopathologically confirmed IPF and five control dogs (CTRL) without pulmonary disease. The RNA was pooled from each group (IPF and CTRL) and analysed using the canine specific Affymetrix microarray technology. IPA-Biomarkers Analysis (Ingenuity System) was used to filter and prioritize biomarkers candidates using the three following criteria: a minimum of a two-fold difference in expression between IPF and CTRL; expression of the gene in lung tissue; possible detection of the protein in blood or in BALF.

Fifty-four molecules met all the criteria. Based on difference in expression, promising proteins included CCL7 (fold change +7.8), α3-Actinin (+5.7), CCL2 (+4.9), Serum Amyloid A1 (+4.4), IL8 (+4.3), PLUNC (−25.1), MMP7 (−17.6). Some are well-known biomarkers of IPF in humans either for diagnosis (MMP7, IL8) or prognosis (CCL2).

These results provide novel potential biomarkers of canine IPF. Measurement of these proteins in blood and BALF of healthy dogs, dogs with IPF and with other respiratory diseases is needed to assess their use as biomarkers of canine IPF.

ABSTRACT R-6

EFFECT OF HELIOX ON RESPIRATORY PARAMETERS IN MESOCEPHALIC AND BRACHYCEPHLIC DOGS. J Anastasio, EA Rozanski, D Bedenice. Lung Function Laboratory, Cummings School of Veterinary Medicine at Tufts University, North Grafton, MA.

Heliox is a mixture of helium and oxygen that has been used therapeutically in human medicine for treatment of airway obstruction. Helium's low density and other physical properties have been shown to reduce the work of breathing by limiting turbulence. The purpose of this study was, therefore, to evaluate respiratory parameters in response to inhaled heliox in dogs with meso- and brachycephalic conformation.

Eleven healthy dogs were recruited, five were mesocephalic and six were brachycephalic. Flow-volume loops were collected using commercial software (Buxco©) while breathing 70:30 Helium: oxygen (Heliox) and 70:30 Nitrogen:oxygen (Nitrox) in a randomized order via a low dead-space face mask. Due to the intrinsic gas properties, gas flow rates and volumes were corrected *in-vitro* by a conversion factor for the effect of helium on the pneumotachograph. Respiratory rate, tidal volume (ml), minute ventilation (L), inspiratory time (Ti), expiratory time (Te), peak inspiratory flow (PIF) and peak expiratory flow (PEF) were recorded while breathing heliox or nitrox. Values were compared using a paired sample t-test, with $P < 0.05$ considered significant.

All dogs cooperated with testing. There was no significant difference in respiratory rate, tidal volume, minute ventilation, inspiratory or expiratory times, or peak inspiratory flow. Peak expiratory flow was significantly higher ($P = 0.01$) while breathing heliox than when breathing nitrox in brachycephalics but not in mesocephalics ($P = 0.22$).

Heliox is well-tolerated in healthy dogs and results in an increased expiratory flow rate in brachycephalic dogs. Further investigation of heliox is warranted in dogs with airway obstruction.

ABSTRACT R-7

CLINICAL EVALUATION AND BIOMARKERS DETERMINATION IN BRACHYCEPHALIC DOGS WITH UPPER AIRWAY OBSTRUCTION BEFORE AND AFTER SURGICAL CORRECTION. M. Planellas[1], R Cuenca[1], MD. Tabar[2], C. Bertolani[3], C. Poncet[3], JM. Closa[4], J. Lorente[5], J. Pastor[1]. [1]Universitat Autònoma de Barcelona, Spain [2]Centro Policlínico Veterinario Raspeig, Alicante, Spain [3]entre Hopitalier Vètèrinaire Fregis, Paris, France, [4]Hospital Veterinari ARS, Barcelona, Spain, [5]Hospital General Universitari de la Vall d'Hebron, Barcelona, Spain.

Brachycephalic dogs are characterized by abnormal anatomy of upper airways (UA) that leads to an obstructive syndrome, named Brachycephalic Obstructive Airway Syndrome (BOAS).The purpose of this prospective multicentric study is to assess the effects that surgical correction has on the severity of clinical signs and levels of acute phase proteins (C-reactive protein [CRP], Haptoglobin [Hp]) and cardiac troponin I (cTnI). Thirty three brachycephalic dogs with BOAS were included and evaluated before and, approximately two months, after surgical correction. The most common components of BOAS found were elongated soft palate (33/33; 100%), stenotic nares (31/33; 94%) and everted laryngeal saccules (13/33; 39.4%). Staphylectomy was performed by means of two different surgical techniques: laser (n = 12) or electrical scalpel (n = 21). There were significant differences between dogs depending on the surgical technique used, with a higher reduction of respiratory signs ($p < 0.002$) and a better postsurgical improvement ($p < 0.015$) with the use of laser. The levels of CRP, Hp and cTnI were categorized into normal or elevated. Before surgical treatment three (9.1%), six (18.2%) and thirteen (39.4%) dogs had elevated values of CRP, Hp and cTnI, respectively. Two months after surgical correction, five (15.1%), eleven (33.3%) and fourteen (42.4%) dogs had elevated values of CRP, Hp and cTnI, respectively. There were no statistical differences between values of CRP and cTnI before and after surgical correction but the levels of Hp increased significantly after surgical treatment ($p < 0.008$), probably due to postsurgical treatment with corticosteroids. As previously suggested by others, there was a statistically significant reduction of respiratory and gastrointestinal signs in dogs with BOAS submitted to surgical correction ($p < 0{,}001$).

According to the results obtained in the present study, the determination of CRP, Hp and cTnI before and two months after surgical treatment do not have a prognostic value in dogs with BOAS. Even though, near half of the dogs studied had elevated levels of cTnI (39.4%) that persisted after surgical treatment (42.4%), suggesting some degree of myocardial damage is present. Further studies are needed considering the influence of breed and age. To the authors' knowledge, this is the first description of CRP, Hp and cTnI determination in dogs with BOAS.

ABSTRACT R-8

PULMONARY FUNCTION AND HEMOGASOMETRIC EVALUATION OF OBESE AND OVERWEIGHT CATS. T Champion, AP Gering, E Zacché, EMG Ortiz, FN Gava, MA Brunetto, AC Carciofi, AA Camacho. Faculty of Agricultural and Veterinary Sciences from São Paulo State University, Jaboticabal, BR.

Overweight and obesity are common conditions that lead to alterations in respiratory mechanics, airway resistance, pattern of breathing and gas exchange in humans. The objective of the present study was to investigate if there are significant differences on respiratory parameters and arterial gas analysis of obese and overweight cats, in conscious state and under general anesthesia.

Twenty nine adult cats were arranged in three groups: obese (n = 15), overweight (n = 7) and with ideal body score index (BSI) (n = 7). Mean of BSI in the groups were: 8,8 (obese), 6,3 (overweight) and 4,9 (ideal BSI). Cats did not had respiratory, cardiac or others systemic diseases.

The respiratory parameters were evaluated with a ventilometer equipment coupled to facemasks in conscious cats and directly to the endotracheal tube in anesthetized cats under spontaneous respiration. The anesthesia was performed with propofol ($3 \pm 1{,}2$ ml/kg) and the cats were maintained in the same anesthetic plan. The three groups were compared by analysis of variance followed by Tukey's test and conscious and anesthetized cats were compared by Student's T test, with a 5% significance level.

There were not observed differences on the respiratory parameters evaluated on ventilometry (tidal volume, expiratory and inspiratory times and peak pressures, respiratory rate and partial pressure of end tidal CO_2 ($PETCO_2$)) and on arterial gas parameters (PaO_2 e $PaCO_2$) in the three groups. The PaO_2 of cats with ideal BSI was $88{,}1 \pm 13{,}5$ mmHg, although was not significantly different ($p = 0{,}06$) from overweight ($72{,}0 \pm 11{,}9$ mmHg) and obese cats ($72{,}6 \pm 14{,}6$ mmHg). Comparison of anesthetized to conscious cats, it was detected decreases in tidal volume, expiratory and inspiratory times and peak pressures and increase in $PETCO_2$ in respiratory rate in the anesthetized cats. Only $PETCO_2$, inspiratory time and respiratory rate in overweight cats did not differ in anesthetized cats.

These results suggest that obesity and overweight did not result in impairment of respiratory function in cats and propofol induced respiratory depression.

ACVIM FORUM SESSION 7A

BREED-ASSOCIATED DIFFERENTIAL MICRORNA EXPRESSION IN CANINE OSTEOSARCOMA. Fenger, J.M.[1], Jaworski, L.N.[1], Ozer, H.G.[3], Sarver, A.L.[5], Subramanian, S.[4,5], Breen, M.[7], Modiano, J.F.[5,6], London, C.A.[2,3], Kisseberth, W.C.[1,3]. [1]Department of Veterinary Clinical Sciences, College of Veterinary Medicine, The Ohio State University, Columbus, OH, USA. [2]Department of Veterinary Biosciences, College of Veterinary Medicine, The Ohio State University, Columbus, OH, USA. [3]Ohio State University Comprehensive Cancer Center, The Ohio State University, Columbus, OH, USA. [4]Department of Laboratory Medicine and Pathology, School of Medicine, University of Minnesota, Minneapolis, MN, USA. [5]Masonic Cancer Center, University of Minnesota, Minneapolis, MN, USA. [6]Department of Veterinary Clinical Sciences, College of Veterinary Medicine, University of Minnesota, St. Paul, MN, USA. [7]Department of Molecular Biomedical Sciences, North Carolina State University and Center for Comparative Medicine and Translational Research, North Carolina State University, Raleigh, NC, USA; Lineberger Comprehensive Cancer Center, UNC-Chapel Hill, NC, USA.

Osteosarcoma (OSA) is the most common bone tumor in dogs, however, little is known regarding the mechanisms underlying malignant transformation in these tumors. Breeds such as Rottweilers and Greyhounds are at higher risk for developing OSA, suggesting that heritable factors play a role in this disease. MiRNAs have tumor/tissue specific roles in regulating gene expression and dysregulated miRNA expression is found frequently in cancer. We hypothesize that canine OSA is characterized by a unique miRNA expression profile(s) with dysregulation of some miRNAs being associated with specific breeds.

MiRNA expression profiling of primary OSA tumors from 6 Greyhounds and 6 Rottweilers was performed using the Nanostring Technologies nCounter miRNA Expression Assay Kit, interrogating the miRNA expression profile of 651 human miRNAs, 168 of whose mature sequences are 100% conserved between human and dog.

17 miRNAs were differentially expressed in Greyhound versus Rottweiler tumors ($p < 0.05$), suggesting that breed-specific dysregulation of miRNAs may contribute to the development and progression of spontaneous OSA.

Hierarchical clustering revealed distinct miRNA expression signatures in Greyhound OSA tumors as compared to Rottweilers. Based on these preliminary results, we are evaluating a larger cohort of OSA tumor samples including Greyhounds, Rottweilers, Golden Retrievers, and a mixed population of other breeds. Statistical analysis will be performed to determine the association of miRNA transcript levels with specific breeds and overall outcome. Characterization of miRNA expression in canine OSA will facilitate our understanding the biology of this disease and has the potential to identify targets for therapeutic intervention.

Originally presented at VCS, October 29 – November 1, 2010, San Diego, CA. (VCS Award Winner)

ACVIM FORUM SESSION 7B

SAFETY EVALUATION OF COMBINATION VINBLASTINE AND TOCERANIB PHOSPHATE (PALLADIA®) IN DOGS: A PHASE I DOSE-FINDING STUDY. Cecilia S. Robat Dr Vet.Med, Cheryl A. London DVM, Ph.D., DACVIM(Onc), Kim A. Selting DVM, MS, DACVIM(Onc), David M.Vail DVM, DACVIM(Onc). Madison, WI.

Combination therapies using drugs with documented single-agent activity and lack of overlapping toxicities could potentially improve outcome. The hypothesis intended to be tested is that Palladia® can be safely administered concurrently with a standard weekly protocol of vinblastine (VBL), at dosages known to have activity against mast cell tumors.

Dogs with histologically confirmed measurable mast cell tumors were evaluated for eligibility to enter a standard phase I dose-finding trial (3+3 cohort), at a starting dose of 2.3 mg/m2 IV VBL (weekly for a total of 4 treatments) and 2.25 mg/kg PO Palladia® EOD, concurrently. Dose escalation of Palladia® was scheduled in 0.25 mg/kg increments until MTD was established or FDA label dose completed (3.25 mg/kg). Safety evaluation was performed weekly throughout the 4 week study period. Dose-limiting toxicities were described following established VCOG-CTCAE(v1.0) criteria. While antitumor response is not a primary endpoint of phase I trials, activity was documented prior to VBL treatments 2-4, and monthly thereafter, based on RECIST criteria.

Nine dogs have been enrolled; cohort 3 is filled and approaching completion of the evaluation period. Hematologic dose limiting toxicity led to 2 de-escalations of VBL. The current safe combination appears to include VBL at 1.6 mg/m2 every other week and Palladia® at 2.25 mg/kg EOD. Response was seen in all but one dog.

Without head to head trials comparing efficacy of bi-weekly VBL combined with Palladia® and VBL alone, choice of therapy should remain at the clinician's discretion.

Originally presented at VCS, October 29 – November 1, 2010, San Diego, CA. (VCS Award Winner)

ACVIM FORUM SESSION 8A

TARGETING TUMOR NEOVASCULATURE WITH NANO-PARTICLE THERAPEUTICS. V Coyle[1], R Tong[2], L Tang[2], J Cheng[2], TMFan[1]. [1]University of Illinois Department of Veterinary Clinical Medicine, Urbana, IL. [2]University of Illinois, Department of Chemistry, Materials Science and Engineering, Urbana, IL.

Prostate specific membrane antigen (PSMA) is a transmembrane protein expressed by tumor-associated neovasculature, but not normal blood vessels. Based upon its selective expression in endothelial cells associated with cancer, PSMA may serve as a conserved angiogenic target shared by macroscopic solid tumors of various histologies.

To investigate the feasibility of targeting a homogenous population of PSMA-expressing endothelial cells as a novel anticancer strategy, we have investigated PSMA expression in several canine hemangiosarcoma (cHSA) cell lines, and subsequently developed self-assembling nanoparticles containing diagnostic (near infrared dyes) and therapeutic (doxorubicin) cargo which selectively bind to PSMA by means of the A10 aptamer, a commercially-available oligonucleotide.

The expression of PSMA by cHSA cells was confirmed transcriptionally and translationally by real time PCR and immunohistochemistry, respectively. Selective binding and endocytosis of A10 decorated nanoparticles was studied by fluorescent microscopy. The ability of A10 decorated nanoparticles encapsulating doxorubicin to exert *in vitro* cytotoxic effects in cHSA cells was assessed by colony forming assays. Using a cHSA xenograft murine tumor model, clinically-relevant anticancer effects of A10 decorated nanoparticles encapsulating doxorubicin were tested.

All cHSA cell lines expressed PSMA mRNA and protein. A10 decorated nanoparticles were selectively endocytosed by PSMA-expressing cells, and when these nanoparticles encapsulated doxorubicin, significant cytotoxic effects were exerted *in vitro*. Finally, A10 decorated nanoparticles encapsulating doxorubicin significantly reduced the size of macroscopic cHSA tumor burdens in transplanted mice.

Diagnostic and therapeutic nanoparticles can be targeted to PSMA-expressing endothelial cells, and cHSA provides a comparative model for the future study of nanoparticle therapeutics.

ACVIM FORUM SESSION 8B

TRANSITIONAL CELL CARCINOMA COMPARATIVE GENOMICS: A MODEL FOR PERSONALIZED MEDICINE APPROACHES TO CANCER CARE. M Paoloni[1], D Dhawan[2], J Stewart[2], K Nuckolls[1], J Lee[3], D Theodorescu[4], C Khanna[1], D Knapp[2]. [1]National Cancer Institute, Bethesda, MD. [2]Purdue University School of Veterinary Medicine, West Lafayette, IN. [3]University of Virginia, Charlottesville, VA. [4]University of Colorado, Denver, CO.

Canine transitional cell carcinoma (TCC) is the most common tumor of the urinary tract, and is similar to human invasive TCC in histopathologic characteristics, molecular features, sites of metastasis, and response to medical therapy. Prevalence is increasing, and novel therapies and strategies are needed to effectively treat this aggressive form of cancer in both species.

Personalized medicine techniques intend to improve treatment outcome by using patient tumor profiling to identify potential and individualized therapeutic targets. A genomic algorithm has been developed termed "coexpression extrapolation", or COXEN, that aims to use expression microarray data to predict drug activity in patient TCC samples. The utility of this predictive methodology has been established in other types of cancer *in vitro*, however its clinical utility has not yet been determined.

Validation studies of COXEN in 10 canine TCC cell lines were conducted. The goal was to determine the value of COXEN in predicting baseline sensitivity of canine TCC to 5 chemotherapy agents (gemcitabine, mitoxantrone, carboplatin, vinblastine and cisplatin) that would then be used in a proposed clinical trial. Additionally, expression data from 25 canine treatment-naïve primary tumor samples were generated on an Affymetrix array platform (Canine Genome v 2.0). Both the expression data and TCC cell line data (anti-proliferative effects, 50% growth inhibition or GC_{50}) were used to establish a canine specific predictive COXEN algorithm. COXEN scores for canine TCC cell-line drug activity were then analyzed. Scores predicted the activity of cisplatin, gemcitibine, and mitoxantrone in all 10 cell lines, and of carboplatin in 6 cell lines. Because all of the cell lines were sensitive to vinblastine ($GI_{50} < 0.05\,\mu M$), the COXEN score was not predictive of its potency. Interestingly, COXEN fails to predict vinblastine response in human TCC cell line data as well. In concurrent work, comparative genomic studies to define and compare the gene expression signatures of TCC in dogs and humans provides further evidence that canine TCC is a valuable genomic model of the human disease. Current studies involve testing the chemo-predictivity of this derived canine COXEN algorithm in additional canine TCC cell lines.

Canine TCC offers an excellent model for *in vitro* and *in vivo* studies of the COXEN approach. This preclinical work will be used to guide the feasibility of future COXEN clinical trials in dogs and humans with TCC.

ACVIM FORUM SESSION 9A

A PROSPECTIVE PHASE I/II OPEN LABEL, NON-RANDOMIZED CLINICAL TRIAL TO EVALUATE THE SAFETY AND EFFICACY OF AMINOACT MULTI-KINASE INHIBITOR FOR THE ADJUVANT TREATMENT OF DOGS WITH SPONTANEOUSLY OCCURING CANCER. GK Ogilvie*, HY Chen, O Mollstedt, DM Proulx*, L Van Horn*, MD Price, MH Tsai. Angel Care Cancer Center*, California Veterinary Specialists, Carlsbad CA; OncoNutrition, Riverside, CA.

A small molecule complex (Aminoact) isolated from bovine milk is a natural peptide mixture with multi-kinase inhibitory effects against epidermal growth factor receptor (EGFR) and insulin-like growth factor receptor-1 (IGFR-1). Ingestion of Aminoact in people with cancer results in lower serum TNF-alpha, an increase in antioxidant superoxide dismutase (SOD) enzyme activity, and subjects' blood serum causes apopsotis in cancer cell lines. This study was designed to first assess safety and secondly the efficacy of three dosage levels of AX-3 in sustaining progression free survival (PFS) for dogs with refractory advanced and/or metastatic cancer.

The prospective, open label study included dogs of different breeds with naturally occurring histologically confirmed malignancies. The first 13 dogs received Aminoact at $1 g/m^2$; the second group of 7 dogs subsequently received the same dosage + 350 mg of Aminoact; and the third group of dogs subsequently received $2 g/m^2$. Each dog was treated orally daily for six weeks along with 550 mg betaine HCl, that aids in peptide absorption. All patients were evaluated for toxicity using the VCOG-CTCAE and efficacy using the RECIST criteria via assessment of clinical parameters, blood work and client questionnaires.

No toxicity other than mild, transient (grade I) nausea was noted, nor were there any changes in hemograms or biochemical profiles in any patient. Dogs with tumors that were confirmed as responders (> 50% reduction in size) include pulmonary adenocarcinoma, mast cell tumor, trichoepithelioma and soft tissue sarcoma. It appears in limited studies that the response rate may be more durable at higher dosages.

The response to Aminoact is dose dependent and only transient mild toxicity was observed, which suggest the maximum effective dosage has not been reached. Further clinical studies will be valuable in determining the effective dosage and response duration. Treating cancer in dogs with Aminoact offers a unique opportunity as a model for human cancer biology and translational cancer therapeutics.

ACVIM FORUM SESSION 9B

CANINE NASAL TUMORS: A STEREOTACTIC RADIATION THERAPY APPROACH. JT Custis, JF Harmon, SD Ryan, SL Kraft, SM LaRue. Colorado State University Animal Cancer Center (CSU), Fort Collins, CO.

Stereotactic radiation therapy (SRT) combines patient immobilization, image guidance, and intensity modulated delivery to achieve ablative radiation doses within the tumor, while preferentially sparing surrounding normal tissues. The purpose of this study was to evaluate the efficacy of SRT as a means of achieving local tumor control for canine nasal tumors.

Retrospective analysis was performed on dogs with a nasal tumor confirmed by histopathology and computed tomography, no previous surgical or radiation therapy, at least six months of follow-up, and completion of three fractions of SRT at CSU.SRT was administered via the Varian Trilogy linear accelerator once daily for three consecutive days. The Varian Eclipse treatment planwas reviewed to determine the planned target volume (PTV) and dose to 95% of the PTV. Kaplan-Meir survival analysis was performed for disease free interval (DFI) and overall survival (OS).

Sixteen patients with nasal tumors (8 adenocarcinoma/carcinomas, 2 squamous cell carcinomas, 3 chondrosarcomas, 2 osteosarcomas, and 1 undifferentiated sarcoma) were treated with SRT. A median dose of 29.1Gy was administered to 95% PTV with a median PTV of 75.3cc. SRT was well tolerated by the normal tissues with minimal, manageable side effects. To date, the median DFI is 270 days, while the median OS is 367 days.

Based upon the initial clinical experience, stereotactic radiation therapy is an emerging modality in the management of canine nasal tumors.

ACVIM FORUM SESSION 71A
PREVENTION OF DISEASE AND MORTALITY IN VACCINATED DOGS FOLLOWING EXPERIMENTAL CHALLENGE WITH VIRULENT LEPTOSPIRA. R LaFleur, J Dant, T Wasmoen. Intervet/Schering Plough Animal Health, Elkhorn, NE.

Canine Leptospirosis can vary from subclinical infection to illness that ranges from mild to severe, including death, depending on the susceptibility of the dog, virulence of the organism, and route and degree of infection. The objective of this study was to evaluate the ability of a canine Leptospira bacterin to prevent infection and disease following challenge with virulent *Leptospira canicola*, *L. pomona*, *L. grippotyphosa*, or *L. icterohaemorrhagiae*. Groups of 8-week-old beagles were vaccinated (day 0) and boosted (day 21) with placebo ($n = 10$) or the 4-way bacterin ($n \geq 20$) and subsequently challenged with each serovar. The results demonstrated that blood and various tissue samples from placebo-recipients became reliably infected, and the dogs developed typical clinical signs of Leptospirosis including loss of appetite, ocular congestion, depression, dehydration, jaundice, hematuria, melena, vomiting, petechiae, and death. In addition, placebo-recipients developed kidney and liver dysfunction. In contrast, some vaccine-recipients became infected, but the organisms were cleared quickly from the blood. Vaccinated dogs failed to develop severe clinical disease requiring medical intervention, and no animals died ($p \geq 0.001$). A few of the vaccinated dogs developed clinical abnormalities, but the clinical signs remained mild and were self-limiting ($p < 0.0001$ for each serovar). Administration of the bacterin also prevented thrombocytopenia ($p < 0.0001$), kidney complications caused by *L. canicola* ($p < 0.0001$), *L. icterohaemorrhagiae* ($p < 0.0001$), and *L. pomona* ($p = 0.012$), and liver dysfunction caused by *L. pomona* ($p < 0.0001$) and *L. grippotyphosa* ($p < 0.0001$). The results therefore confirmed that vaccinating dogs with the 4-way Leptospira bacterin provided a high degree of protection (99.5%–100%) against the clinical signs of Leptospirosis including mortality.

ACVIM FORUM SESSION 71B
CIPROFLOXACIN UROLITHIASIS: A NEWLY RECOGNIZED DISEASE IN DOGS. J Lulich, C Osborne, A Cokley, S Kemp, L Ulrich, M Buettner, S Davidson, L Koehler, P Ashton, L Swanson, H Albasan, E Nwaokorie, S Leach. University of Minnesota, St Paul, MN.

Ciprofloxacin, a synthetic fluoroquinolone antimicrobic, is not FDA-approved for veterinary use. However, due to recent availability of less expensive generic formulations, extra-label use of ciprofloxacin by veterinarians appears more common. Although ciprofloxacin crystalluria and uroliths have been reported in humans, we are unaware of any published reports in dogs. This is surprising since mean urine ciprofloxacin concentration (0.36 mg/ml) in dogs following a modest IV dose (13 mg/kg) was 3 times higher than the solubility of ciprofloxacin in water (0.12 mg/ml). To identify the occurrence of ciprofloxacin uroliths in dogs, records from the Minnesota Urolith Center were reviewed. Between January 2001 and December 2009, ciprofloxacin was identified in uroliths from 58 dogs; uroliths were composed of 100% ciprofloxacin in 10, mixed uroliths containing ciprofloxacin were identified in 6, a shell of ciprofloxacin was observed in 21, and ciprofloxacin surface crystals were identified in 21.

Based on an experimental study in which 83% of human volunteers consuming 1000 mg of ciprofloxacin with $NaHCO_3$ exhibited ciprofloxacin crystalluria (urine pH > 7.3), while no volunteers consuming 1000 mg of ciprofloxacin and NH_4Cl to acidify urine formed crystals; we postulated that ciprofloxacin uroliths could be dissolved in acidic urine. To test this hypothesis, canine uroliths composed of 100% ciprofloxacin from a single source (6-yr-old male, English bulldog receiving 24 mg/kg of ciprofloxacin PO, q12 hr to manage superficial pyoderma; turbulent flow chromatography/tandem mass spectrometry detected 517 mg of ciprofloxacin/g of urolith) were incubated in urine at selected pH's and monitored for dissolution. Urine obtained from multiple dogs not receiving fluoroquinolones, was pooled and divided into 5 aliquots. Aliquots were adjusted with HCl or NaOH to a pH of 3, 5, 6, 7, or 8. Aliquots were capped and preserved by refrigeration; pH was monitored and readjusted weekly. Ten uroliths of approximately equal weight were randomly assigned to individual flasks containing 10 mls of urine. Flasks were constantly agitated and maintained at 38°C. Every 24 hours, urine was discarded and replaced with 10 mls of urine of identical pH until stone dissolution was complete.

Ciprofloxacin urolith dissolution times at each urine pH are reported below.

Table ACVIM Forum Session 71B

Urine pH	3	5	6	7	8
Mean Initial Urolith Weight (mg)	80.5	66	76.5	75	78.5
Mean Dissolution Time (day)	7	15	23	18.5	14.5

Ciprofloxacin uroliths are a newly recognized disease and a potential adverse effect of ciprofloxacin administration in dogs. In vitro dissolution of ciprofloxacin uroliths was achieved in canine urine, supporting the premise that in vivo dissolution is possible. Urolith dissolution times were shortest at lower and higher pH's, which is consistent with the pKa (6.0 and 8.8) of this amphiprotic antimicrobic (more soluble at pH below the acidic pKa and above the alkaline pKa).

ACVIM FORUM SESSION 72A
STRUVITE UROLITH DISSOLUTION IN CATS: A DOUBLE BLIND RANDOMIZED CLINICAL TRIAL OF 2 FOODS. J. Lulich[1], J. Kruger[2], J. Merrills[2], J. MacLeay[3], I. Paetau-Robinson[3], C. Koenigsknecht[2], S. Davidson[1], C. Osborne[1]. [1]University of Minnesota, St. Paul, MN. [2]Michigan State University, E. Lansing, MI. [3]Hills Pet Nutrition, Topeka, KS.

Foods designed to promote struvite urolith dissolution may be designed for short term feeding facilitating rapid dissolution or may be formulated with a more moderate target urine pH to allow for dissolution and then life-long maintenance feeding minimizing recurrence. The purpose of this study was to compare the efficacy and rate of dissolution of a maintenance food with a struvite dissolution food.

Sixteen client-owned adult cats (13 FS, 3 MC) with naturally occurring struvite urocystoliths (mineral composition based on history, radiographs, urinalysis, urine culture and physical examination) were randomized to either a dry maintenance food (Test) or a dry food known to dissolve struvite uroliths (Control). The clinical care team and owner were blinded to treatment assignment. The Test food was formulated to provide 0.06% Mg (DM), 0.65% P, 35% protein, and a calculated target urine pH value (UpH) of 6.2–6.4. The Control food was formulated to provide 0.06% Mg (DM), 0.77% P, 35% protein, and a targeted urine pH of 5.9–6.1. Owners were advised to feed the assigned diet exclusively in an amount to maintain body condition. After diet assignment radiographs were performed at eight weekly intervals until there was no evidence of uroliths or until there was evidence that the uroliths were the same size or larger. A physical examination, complete blood count, serum chemistry profile, urinalysis and urine culture were repeated at the conclusion of the study. Statistical analysis was by ANOVA.

All uroliths dissolved in all cats and both foods were palatable. Radiographs of cats fed the Control food indicated the uroliths dissolved in a significantly shorter time (mean ± Std Dev of 1.6 ± 0.7 weeks) compared to cats consuming the Test food (mean 3.9 ± 2 weeks; $P<0.05$).). Cats in the Control group finished the study at 1 (n = 4), 2 (n = 3) and 3 weeks. Cats in the Test group finished the study 1, 2, 3 (n = 2), 4, 5, 6, and 7 weeks.

All cats had similar UpH values at study start (Control 6.516 ± 1.013, Test 6.638 ± 0.432). Cats consuming the Control food had significantly lower UpH values at the time of dissolution compared to cats consuming the Test food: mean UpH values for the Control group was 5.949 ± 0.243 compared to 6.337 ± 0.163 ($p < 0.05$) for the Test group. Mean urine specific gravity values at baseline and at dissolution were not different between or within groups. Adverse events were not observed in either group of cats. This study provides evidence that clinicians may choose either food to facilitate struvite dissolution based on clinical assessment of individual cases and owner preferences.

ACVIM FORUM SESSION 72B

EFFECTS OF FORMALIN ON UROLITH COMPOSITION. H Albasan[1,2], CA Osborne[1], JP Lulich[1], LK Ulrich[1], LA Koehler[1]. [1]University of Minnesota, College of Veterinary Medicine, Minnesota Urolith Center, St. Paul, MN. [2]Erciyes University, Faculty of Veterinary Medicine, Department of Internal Medicine, Kayseri, Turkey.

The Minnesota Urolith Center occasionally receives uroliths for analysis that are immersed in formalin. Results of quantitative analysis of these uroliths revealed that some submitted in formalin consisted of newberyite (magnesium hydrogen phosphate trihydrate). Because newberyite is uncommonly found in uroliths formed by cats and dogs, we hypothesized that this mineral was an in vitro artifact caused by exposure of struvite (magnesium ammonium phosphate hexahydrate) to formalin. The purpose of this study was to determine if formalin alters the mineral composition of uroliths.

Urolith submissions containing stones of either 100% struvite (n = 5 dogs and 5 cats), 100% calcium oxalate (n = 5 dogs and 5 cats), 100% calcium phosphate apatite (n = 5 dogs and 5 cats), 100% cystine (n = 5 dogs and 5 cats), 100% ammonium urate (n = 5 dogs and 5 cats), and 100% silica (n = 5 dogs) preserved by only air drying were tested. One urolith from each submission was quantitatively analyzed by polarized light microscopy or infrared spectroscopy. A subsequent urolith from the same submission was immersed in 1 ml of 10% buffered formalin for 48 hours at room temperature. Uroliths were then air dried for 30 minutes and the analysis was repeated.

After exposure to formalin, portions of all 10 struvite uroliths were transformed into newberyite. Three (1 dog and 2 cats) of 10 ammonium urate uroliths were completely dissolved. Newberyite was not detected in any of the remaining uroliths. Likewise quantitative mineral analysis of non-struvite uroliths remained unchanged.

To avoid misdiagnosis of mineral composition, uroliths should not be immersed in formalin prior to analysis.

ACVIM FORUM SESSION 73A

SEROPREVALENCE OF THREE *EHRLICHIA* SPECIES IN DOGS – A MULTI-INSTITUTIONAL STUDY. MJ Beall[1], EB Breitschwerdt[2], LA Cohn[3], CG Couto[4], MW Dryden[5], S Eddlestone[6], L Guptill[7], WE Hoffmann[8], MA Kennedy[9], P Lathan[10], SE Little[11], A Roy[6], KA Sayler[12], J Sinsabaugh[1], K Snowden[13], BA Stillman[1], EG Welles[14], MJ Yabsley[15]. [1]IDEXX Laboratories, Westbrook, ME. [2]North Carolina State Univ., Raleigh, NC. [3]Univ. of Missouri, Columbia, MO. [4]The Ohio State Univ., Columbus, OH. [5]Kansas State Univ., Manhattan, KS. [6]Louisiana State Univ., Baton Rouge, LA. [7]Purdue Univ., West Lafayette, IN. [8]Univ. of Illinois, Champaign, IL. [9]Univ. of Tennessee, Knoxville, TN. [10]Mississippi State Univ., Mississippi State, MS. [11]Oklahoma State Univ., Stillwater, OK. [12]Univ. of Florida, Gainesville, FL. [13]Texas A&M Univ., College Station, TX. [14]Auburn Univ., Auburn, AL. [15]Univ. of Georgia, Athens, GA.

The geographic range of the lone star tick (*Amblyomma americanum*) is expanding together with the disease agents it vectors, including *Ehrlichia chaffeensis* and *Ehrlichia ewingii* which cause both canine and human ehrlichiosis. This study examined the frequency with which dogs are exposed to *E. chaffeensis* and *E. ewingii* relative to *E. canis*, which is transmitted by the more ubiquitously distributed brown dog tick (*Rhipicephalus sanguineus*). A total of 6,512 canine serum samples, ranging from 182 to 614 from each of the 14 participating institutions, collected at random from clinical accessions, diagnostic laboratories and/or shelters were evaluated. All serum samples were tested by three microtiter plate ELISAs using species-specific peptides for antibodies to *E. canis*, *E. chaffeensis* and *E. ewingii*. Zip code information for sample origin was provided by the collaborator and was used to assess seroprevalence by region. Comparisons were evaluated using the Chi-square test. Seroreactivity for at least 2 of 3 *Ehrlichia* spp was found in samples from every institution. On an average, seroprevalence of *E. canis* (0.9%) was significantly ($p < 0.001$) lower than *E. chaffeensis* (2.8%) or *E. ewingii* (5.2%). Samples from Texas had the highest *E. canis* seroprevalence (2.9%), while samples from Missouri had the highest seroprevalence of *E. chaffeensis* (12.2%) and *E. ewingii* (21.3%; $p < 0.001$). Both Mississippi and Oklahoma had greater than a 7% *E. ewingii* seroprevalence, while North Carolina had an *E. chaffeensis* seroprevalence of 7.7%. Samples from Ohio had the lowest aggregate seroprevalence (1.0%) with only 4 dogs *E. canis* seropositive, one *E. ewingii* seropositive and no *E. chaffeensis* seroreactors. The geospatial pattern of *E. chaffeensis* and *E. ewingii* seropositive samples was similar to that previously reported based on modeling seroreactivity to *E. chaffeensis* in white-tailed deer as well as the distribution of human monocytic ehrlichiosis (HME) cases reported by the CDC. This study provides the first large scale regional documentation of canine exposure to these three *Ehrlichia* spp., highlighting where infections most commonly occur and thus identifying areas where heightened awareness about these emerging vector-borne disease agents by both veterinarians and public health agencies is needed.

ACVIM FORUM SESSION 73B

POLYDIMETHYLSILOXANE URETHRAL BULKING AGENT (PDMS UBA) INJECTION FOR TREATMENT OF FEMALE CANINE URINARY INCONTINENCE – PRELIMINARY RESULTS. JW Bartges, A Callens. University of Tennessee, Knoxville, TN.

Urinary incontinence (UI) occurs in approximately 20% of spayed female dogs. The most common cause is urethral sphincter mechanism incompetency (USMI). Pharmacological agents are effective, however, not all dogs respond, and dogs may become refractory to treatment over time. Urethral bulking, where a compound is injected submucosally in the urethra, has been used in women and in female dogs with urinary incontinence. New synthetic compounds have been used in human medicine; the most promising is Polydimethylsiloxane (PDMS), which has been shown to be more effective than glutaraldehyde cross-linked collagen. The purpose of this descriptive clinical trial is to evaluate the safety and effectiveness of PDMS urethral bulking agent (PDMS UBA) in client-owned, spayed female dogs with naturally-occurring UI due to USMI.

Twenty-two, spayed female dogs were included. Dogs had a median age of 6 years (2 to 11 years). Eighteen dog breeds were represented, and dogs weighed a median of 29.9 kg (7.3 to 62.7 kg). Average length of time of UI was 2.5 ± 2.3 years; 18/22 dogs had been treated medically, of which 2/18 were continent, 14/18 were improved, and 2/18 had no improvement. Dogs were deemed healthy based on results of physical examination, complete blood cell counts, plasma biochemical analysis, and urinalysis; urine cultures were negative.

Dogs were anesthetized, positioned in dorsal recumbency, and cystoscopy performed using a 2.7 mm, 0- or 30- degree, 18 cm rigid cystoscope. Urethral bulking was performed with PDMS UBA. On average, 2.5 ± 0.9 ml were injected in 3 to 5 locations approximately 1 to 1.5 cm distal to the trigone submucosally in the proximal urethra. Good coaptation was achieved in all dogs. The procedure took on average 15.9 ± 4.3 minutes. One dog experienced urethral obstruction after the procedure; a Foley catheter was inserted for approximately 12 hours and removed at which time she urinated normally and was continent. Three dogs experienced an acute allergic reaction characterized by blepharedema and urticaria treated successfully with diphenhydramine. Dogs were discharged on day of procedure except for the one dog that experienced urethral obstruction. All dogs were treated with Meloxicam (0.1 mg/kg PO q24h for 3 days).

Owners were contacted on day after discharge and 21/22 dogs were continent; 1/22 dogs was improved. Dogs were re-evaluated 1 week after discharge and 21/22 dogs were continent and 1/22 dogs were improved; urine cultures were negative for all dogs at that time. At one month after discharge, 17/22 dogs were continent, 2/22 dogs had experienced 1 to 2 episodes of incontinence but were continent at this recheck, 3/22 dogs were improved. Continent dogs were not receiving medications for UI. Preliminary results support use of PDMS UBA for female dogs with UI that have either failed medical therapy or as a first line therapy.

ACVIM FORUM SESSION 74A

TRANSFUSION OF AUTOLOGOUS STORED ERYTHROCYTE CONCENTRATES CAUSES INFLAMMATION AND HYPOCOAGULABILITY.

SA Smith[1], M McMichael[2], JM Herring[2], T Ngwenyama[2], R. Corsi,[2] A Galligan[2], AN Beloshapka[3], P Deng[3], K Swanson[2,3]. [1]College of Medicine, [2]College of Veterinary Medicine, or[3]Department of Animal Sciences, University of Illinois, Urbana, IL.

We previously reported that transfusion to normal dogs of autologous erythrocyte concentrates (pRBCs) that had been stored for 21 days causes a profound inflammatory response (2X increase in leucocyte count and fibrinogen, 60X increase in C-Reactive Protein). We speculated that inflammation was due to cytokines produced during the storage period, and hypothesized that transfusion of fresh (F) pRBCs would elicit less inflammation than would stored (S) pRBCs. A whole blood unit was collected from healthy dogs (n = 10) for pRBCs on day 0, then again on day 32. On day 35 dogs received an autologous transfusion of pRBCs stored for either 35 days (S, n = 5) or 3 days (F, n = 5). CBCs and in-tem thromboelastometry (CT:coagulation time, CFT:clot formation time, α:alpha, MCF:maximum clot firmness) were evaluated on blood samples collected at 0 (pre) and 5,9,24,48, and 72 hours after transfusion.

Fresh pRBCs did not elicit any change in leucocytes, platelets, or thromboelastometry. Stored pRBCs elicited a degenerative left shift (5 hr) followed by a regenerative left shift (9–48 hr), thrombocytopenia (36% decrease at 5 hr), and marked hypocoagulability characterized by prolonged CT (5,9,24 hr) and CFT (5,9 hr), and decreased α (5,9 hr) and MCF (5,9, 24 hr).

Table [Abstract 74A]

		0	5	9	24	48	72
Segs	F	4.8(1.7)	5.0(1.4)	3.3(2.1)^a	3.8(0.7)^a	3.9(1.3)^a	4.0(1.5)
(10^3/μl)	S	6.1(3.8)	5.4(7.1)	16.0(10.6)^b	24.8(12.6)^b	19.2(5.5)^b	10.1(2.6)
CT	F	164(32)	158(12)^a	161(28)^a	161(12)^a	152(22)	161(38)
(s)	S	162(38)	225(48)^b	212(15)^b	221(24)^b	173(15)	147(30)
CFT	F	99(17)	102(25)^a	86(21)^a	103(20)	101(22)	105(27)^a
(s)	S	139(49)	279(47)^b	256(63)^b	121(21)	97(15)	75(10)^b
α	F	71(3)	72(3)^a	73(5)^a	72(2)	72(2)	72(4)^a
(°)	S	67(6)	54(5)^b	58(6)^b	67(4)	71(3)	75(2)
MCF	F	59(6)	58(6)^a	60(6)^a	62(6)	57(2)^a	57(6)^a
(mm)	S	53(7)	49(2)^b	48(5)^b	59(2)^b	65(3)^b	66(2)^b

Data are mean(SD). a: $p < 0.05$ between groups F and S by t test. b: $p < 0.05$ compared to "0" by RM ANOVA.

Transfusion of autologous stored pRBCs elicits a greater inflammatory response than fresh pRBCs, and results in hypocoagulability on thromboelastometry.

ACVIM FORUM SESSION 74B

ENHANCED BIOCONVERSION OF CLOPIDOGREL TO ITS ACTIVE METABOLITE WITH RIFAMPIN INDUCTION OF HEPATIC ISOZYMES.

DF Hogan. Purdue University, School of Veterinary Medicine,West Lafayette, IN.

Clopidogrel is a potent antiplatelet drug that is gaining popularity in veterinary medicine for antithrombotic therapy. The parent molecule is an inactive prodrug that must be converted by hepatic isozymes to an active metabolite. The majority of the parent molecule is directed to the formation of inactive metabolites with only an extremely small proportion of parent molecule directed to the formation of the active metabolite. There are multiple hepatic isozymes responsible for the formation of the active metabolite. A non-specific hepatic isozyme inducer such as rifampin could increase the formation of the active metabolite of clopidogrel thereby increasing the pharmacodynamic response which may allow a reduced drug dose to achieve a clinical effect. We have previously presented data supporting the increased pharmacodynamic response of clopdiogrel after rifampin therapy. The goal of this study was to demonstrate an increased pharmacokinetic response of clopidogrel after rifampin induction of hepatic isozymes.

Six healthy, purpose-bred dogs were used for this study. The pharmacokinetics of clopidogrel were determined by measuring the parent molecule, primary inactive metabolite and active metabolite through LC/MS/MS. The pharmacodynamics of clopidogrel were determined by measuring collagen-induced whole blood aggregation. Blood samples were collected prior to clopidogrel administration (baseline), after 7 days of 2 mg/kg clopidogrel PO q 24 hrs, and after 7 days of 2 mg/kg clopidogrel PO q 24 hrs + 10 mg/kg PO q 12 hrs rifampin.

Given the absence of a known standard for the active metabolite, only a semi-quantitative assessment of active metabolite concentration can be made. There was no identifiable active metabolite peak noted at baseline or after clopidogrel treatment. However, with clopidogrel and rifampin combined administration there was an active metabolite peak identified in all dogs with a mean area of 41.7 ± 23.5. The development of the active metabolite peak was associated with an increase in the pharmacodyamic response of clopidogrel in the dogs.

This is the first study in any species to document the increased formation of the active metabolite of clopidogrel in response to a strong, non-specific hepatic isozyme inducer. This increased pharmacokinetic response was associated with an increased pharmacodynamic response of clopidogrel. This data provides supportive evidence to develop therapeutic protocols to improve the pharmacodynamic response to clopidogrel in dogs that may reduce dosing requirements or correct subtherapeutic pharmacodynamic response.

ACVIM FORUM SESSION 85A

IDENTIFICATION OF CRITICAL ILLNESS RELATED CORTICOSTEROID INSUFFICIENCY IN CRITICALLY ILL HORSES.

AJ Stewart[1], E Hackett[2], TJ Towns[1], H Weaver[1], B Johnson[1], AA Wooldridge[1], AS Munsterman[1], Zhong Q[1]. College of Veterinary Medicine, Auburn University, Auburn, AL. [2]College of Veterinary Medicine & Biomedical Sciences, Colorado State University, Fort Collins, CO.

Critical illness-related corticosteroid insufficiency (CIRCI) has been identified in humans, foals, dogs and cats with lower-than-expected circulating cortisol concentrations, and/or by a blunted cortisol response to ACTH stimulation. Our purpose was to determine if CIRCI exists in critically ill horses.

Endogenous plasma ACTH and serum cortisol concentrations, and cortisol at T = 0 and T = 30 min after 0.1 μg/kg cosyntropin, were measured by radioimmunoassay from horses with colic or systemic illness on admission, and days 2, 4 and 6 of hospitilization. Horses were divided into mild, moderate, or severe illness groups based on clinicopathologic data. Inappropriately low cortisol was defined as endogenous cortisol < mean-1SD achieved after administration of 0.1 μg/kg cosyntropin to normal horses (< 269 nmol/L). Inadequate delta cortisol was defined as < mean delta cortisol in normal horses after 0.1 μg/kg cosyntropin (< 159 nmol/L). Cortisol, ACTH and delta cortisol were compared using ANOVA between groups, with $P < 0.05$ considered significant. Fifty-eight horses classified as having mild (11), moderate (30) and severe (17) disease at admission had survival rates of 100%, 97% and 35% respectively. Admission ACTH and cortisol concentrations were highest in severely ill horses (93 ± 198pg/mL, 361 ± 137 nmol/L) compared to moderate (31 ± 36, 279 ± 137) and mildly ill horses (18.0 ± 12.0, 237 ± 133). Admission cortisol concentrations were higher overall in severely ill horses ($P = 0.016$), but were low in 24% (4/17). Admission delta cortisol was low in 85% (11/13) of severely ill horses, and was associated with marked adrenal hemorrhage in non-survivors.

Severely ill horses have high cortisol and ACTH, but low cortisol and delta cortisol may indicate CIRCI secondary to adrenal hemorrhage.

ACVIM FORUM SESSION 85B
ALTERATIONS IN AUTOPHAGY IN THE PITUITARY OF HORSES WITH EQUINE PITUITARY PARS INTERMEDIA DYSFUNCTION (PPID). Dianne McFarlane, Kristen McDaniel, Kim Hill, Melissa Cordero. Oklahoma State University.

Equine pituitary pars intermedia dysfunction (PPID) is a common endocrinopathy of aged horses that results from neurodegeneration of the dopaminergic periventricular neurons that innervate the intermediate lobe of the pituitary. Factors that initiate spontaneous dopaminergic neurodegenerative disease remain elusive, however accumulation of misfolded α-synuclein protein and dysfunctional protein clearance have been implicated. Misfolded protein accumulation occurs due to increased protein production or decreased clearance of damaged macromolecules through the process of autophagy. While have previously demonstrated that horses with PPID have increased α-synuclein in the periventricular neurons compared to controls, it remains unknown whether the protein accumulates due to increased production or decreased clearance. We hypothesized that autophagy is decreased in the pituitary neurointermediate lobe from horses with PPID compared to controls.

Neurointermediate lobe pituitary tissue was from collected from horses with PPID (n = 12) and healthy horses (n = 37, 2–35 years). Realtime PCR was used to determine the relative expression of autophagy genes (mTOR, Beclin1, ATG12, ATG7, ATG5, PINK, LAMP2) and α-synuclein Relative gene expression from horses with PPID were compared to healthy horses by t-test following log transformation. A Pearson coefficient of correlation was calculated comparing α-synuclein expression with autophagy gene expression.

The expression of α-synuclein, autophagy-related genes (ATG12, Beclin, Lamp2), and mTOR was greater in horses with PPID than in healthy horses. Age was not correlated to α-synuclein or autophagy gene expression. There was a significant positive correlation between expression of α-synuclein and Beclin1, ATG12, ATG7, ATG5, and PINK, but not mTOR expression. Accumulation of α-synuclein protein in horses with PPID may result from increased α-synuclein expression. Autophagy genes are upregulated in horses with PPID, suggesting a compensatory response, although these findings need to be confirmed by demonstrating an increased functional response. α-synuclein expression was positively correlated to expression of autophagy genes except mTOR, suggesting α-synuclein may stimulate autophagy in an mTOR independent manner.

ACVIM FORUM SESSION 86A
EFFICACY OF DELAYED ANTIVIRAL THERAPY AGAINST EHV-1 CHALLENGE. LK Maxwell[1], LL Gilliam[1], N Pusterla[2], R Carmichael[1], RW Eberle[1], JW Ritchey[1], TC Holbrook[1], T Gull[1], GB Rezabek[1], D McFarlane[1], CG MacAllister[1]. [1]Oklahoma State University, Stillwater, OK. [2]University of California, Davis, CA.

Equine herpes virus type-1 (EHV-1) outbreaks are often not recognized until exposed horses are at immediate risk for developing equine herpes myeloencephalopathy (EHM). The objective of this study was to determine whether delayed therapy with the antiviral drugs valacyclovir or ganciclovir could protect those horses most at risk for EHM. Eighteen aged (> 20 years) mares were randomized to treatment: no therapy (control), oral valacyclovir therapy, or intravenous ganciclovir therapy. Drug administration was initiated at the onset of the second febrile phase, between days 4–6 after EHV-1 inoculation (PI), and continued for one week. Neurological examinations were performed prior to the study and for three weeks PI. One horse was excluded from the study for failure to become febrile. Body temperature was significantly lower in the ganciclovir-therapy horses as compared to control horses on days 6–8 PI ($P < 0.05$), whereas valacyclovir-therapy horses did not differ from control horses. Viremia in whole blood, as determined by PCR, was also lower in the ganciclovir-therapy horses on days 7–10 PI and on day 7 PI in the valacyclovir-therapy horses ($P < 0.05$). Although antiviral drug administration did not reduce the risk of ataxia ($P = 0.06$) or nasal shedding, ganciclovir therapy did decrease the severity of ataxia ($P < 0.05$) as compared to valacyclovir-therapy and control horses, where 0/6, 2/5, and 4/6 horses, respectively, developed at least a two grade change in ataxia. In summary, ganciclovir administration provided better protection against EHM than did valacyclovir when therapy was initiated just prior to the onset of neurological disease.

ACVIM FORUM SESSION 86B
PROTECTION FROM CONTEMPORARY EQUINE INFLUENZA VIRUS BY COMMERCIAL VACCINES. A Sellers[1], G. Landolt[1], HGG Townsend[2], J. Tuttle[3], RM Stenbom[3], DP Lunn[1]. [1]CSU, Fort Collins, CO. [2]WCVM, Saskatoon, SK. [3]BIV, St. Joseph, MS.

Equine vaccination is amongst the most important method of prophylaxis against equine influenza virus (EIV), a pathogen in which continuous antigenic drift can lead to vaccine failure. A 6-month duration of immunity (DOI) challenge infection study was conducted using commercial inactivated vaccines containing different strains of A/equine/2/influenza virus's, including Innovator™, containing Kentucky/97 (Pfizer Animal Health, New York, NY), and Calvenza, containing a combination of Ohio/2003, Kentucky/95, and Newmarket93 (Boehringer Ingelheim Vetmedica, St. Joseph, MS). The challenge virus strain was Colorado/07, the most contemporary challenge strain currently in use.

The study design was a blinded, randomized challenge trial. Three groups of 10 yearling ponies, with no history or serological evidence of EIV infection were established. Each group received one of three treatments: vaccination with Innovator™; vaccination with Calvenza™; or injection with a saline placebo. Each treatment was administered 3 times, at intervals of 1 month between the first two treatments, and 3 months between the second and third treatments. All ponies were challenged by nasal nebulization of $5X10^7$ EID_{50} influenza virus A/eq/2/Colorado/07 6 months after the third treatment. Clinical signs of disease, including rectal temperature, nasal discharge, anorexia, coughing, and depression, were recorded daily for 2 days prior to challenge infection, and 14 days post-challenge. Nasal shedding of EIV was measured on the same days, using a real-time PCR test procedure. EIV-specific antibody responses were measured by ELISA. Differences between groups were analyzed by non-parametric repeated measures ANOVA, and differences were declared significant when $P < 0.05$.

All control group ponies demonstrated clinical signs of disease consistent with EIV infection post-challenge infection, including pyrexia, nasal discharge, inappetance and partial anorexia. These signs were significantly lower in both vaccine groups; mean body temperature was elevated (> 101.5°F) for 8 days in controls, but only 2 days in vaccine groups. Nasal shedding of EIV was detected in all ponies in all groups; over the duration of the study the Calvenza group shed significantly less virus than Innovator and Control. Over time antibody titers were significantly higher in the Calvenza than the Innovator group, and both were significantly greater than controls.

This study demonstrated that both current commercial inactivated EIV vaccines have a duration of clinical protection of at least 6 months after a highly pathogenic challenge with a recent EIV isolate. Both antibody responses and virological protection differed between the vaccines. Formulation difference between the vaccines, including the EIV antigens employed, may have contributed to this performance difference.

ACVIM FORUM SESSION 93A
CEREBROSPINAL FLUID NEUROTRANSMITTER AND CYTOKINE CONCENTRATIONS IN CANINE DEGENERATIVE MYELOPATHY. S. Platt,[1] R. Barber,[1] M. Kent,[1] S. Kanazono,[2] A.C. Freeman,[1] G. Edwards,[1] J.R. Coates[2]. [1]College of Veterinary Medicine, University of Georgia, Athens, GA. [2]College of Veterinary Medicine, University of Missouri, Columbia, MO.

Degenerative myelopathy (DM) may be homologous to a form of amyotrophic lateral sclerosis in humans which has excitotoxic and immunologic pathogeneses described. The aims of this study were to determine (i) presence or absence of abnormalities in concentrations of CSF amino acid (AA) neurotransmitters (glutamate, glycine and γ–aminobutyric acid (GABA)) and cytokines in dogs with DM and if present (ii) investigate associations with disease severity.

Twenty-two dogs histopathologically confirmed for DM and 21 dogs with suspected DM based on thorough diagnostic investigations and 42 clinically normal age-matched control dogs were included in the study. The neurological severity of the DM dogs was graded (1–4) using an established scale. CSF was evaluated for presence of glutamate, glycine and GABA by high performance liquid chromatography and for GM-CSF, IFN-γ, IL-2, IL-4, IL-6, IL-7, IL-8, IL-10, IL-15, IL-18, IP-10, KC (keratinocyte chemoattractant), MCP-1 (monocyte chemotactic protein-1) and TNF-α using a commercially available, canine multiplex immunoassay (Millipore, Billerica, MA). All data analyses were performed using SAS V 9.2 (Cary, NC). Analyte levels were compared between DM confirmed, DM suspected and control dogs by an analysis of variance (ANOVA). Spearman correlation was used to test for correlations of analyte levels and neurological grades. All hypothesis tests were 2-sided with $\alpha = 0.05$.

There were no significant differences between individual CSF analytes in DM confirmed and DM suspected dogs. Glutamate levels were not significantly different between DM affected (mean = 0.09 µg/ml; range = 0.01–0.29; SD = 0.069) and control dogs (mean = 0.07 µg/ml; range = 0.01–0.29; SD = 0.056). Control dogs (mean = 0.86 µg/ml; range = 0.08–1.16; SD = 0.213) had significantly higher levels of GABA ($p < 0.0001$) than DM dogs (mean = 0.12 µg/ml; range = 0.07–0.79; SD = 0.12). Control dogs (mean = 2.14 µg/ml; range = 0.01–3.83; SD = 0.76) also had significantly higher glycine concentrations ($p < 0.0001$) than DM dogs (mean = 0.45 µg/ml; range = 0.01–0.98; SD = 0.24). DM-affected dogs also had significantly higher levels of IL-2 (p = 0.03), KC ($p < 0.0001$) and MCP-1 (p = 0.005) than control dogs. Neurotransmitter levels were not significantly associated with neurological grade. KC levels were significantly higher in the least affected dogs (p = 0.0015). There were no associations with disease severity and analyte concentrations.

DM affected dogs have an imbalance of CSF AA concentrations creating a relatively excitotoxic environment. Reports in human ALS confirm an imbalance between CSF excitatory and inhibitory AAs suggesting a pathogenic role for excitotoxicity in ALS. It also appears that DM affected dogs have increases in CSF cytokines and chemokines suggestive of an immunologic component to the pathogenesis as is similar to ALS. Further prospective analysis of DM is warranted to evaluate the role of treatment on CSF variables.

ACVIM FORUM SESSION 93B

CHIARI-LIKE MALFORMATION, SYRINGOMYELIA AND NEUROPATHIC PAIN: CEREBROSPINAL FLUID NEUROTRANSMITTER AND CYTOKINE LEVELS. S.R. Platt, A.C. Freeman, M. Kent, E. Huguet, G. Edwards, R. Barber. College of Veterinary Medicine, University of Georgia, Athens, GA.

The pathogenesis of neuropathic pain (NP) and syringomyelia (SM) in association with Chiari-like malformation (CLM) in dogs has focused on the anatomical anomalies and secondary cerebrospinal fluid (CSF) flow abnormalities. Neuropathic pain in humans has been associated with abnormalities of neurotransmitters such as glutamate and serotonin as well as immunologic mechanisms. The aim of this study was to investigate the CSF neurotransmitter and cytokine levels in Brussels Griffon dogs (BGs) with CLM, SM and NP.

As part of an MRI study investigating the prevalence of SM in BGs, atlanto-occipital CSF was acquired from 46 dogs and stored at -80C until analysis. All dogs underwent a neurologic exam prior to MRI; Osirix® software was used to measure SM and the presence of cerebellar herniation and deviation were recorded. Deproteinized CSF samples were analysed for presence of serotonin (ng/ml), glutamate, glycine and GABA (µg/ml) by high performance liquid chromatography. All CSF samples were evaluated simultaneously for GM-CSF, IFN-γ, IL-2, IL-4, IL-6, IL-7, IL-8, IL-10, IL-15, IL-18, IP-10, KC, MCP-1 and TNF-α. A commercially available, canine multiplex immunoassay (Millipore, Billerica, MA) was used for the cytokine analysis (pg/ml). Student's t-tests were used to compare the means of neurotransmitter and cytokine values between groups with and without skull abnormalities or spinal pain. Simple Pearson's correlation was used to test for correlations of neurotransmitter and cytokine values with syrinx dimensions and correlations of neurotransmitter with cytokine values. All hypothesis tests were 2-sided and the significance level was $\alpha = 0.05$.

NP was detected in 8 dogs (17%); SM was present 24 dogs (52%); and CM was detected in 24 dogs (52%). IFN-γ levels were significantly lower in dogs with NP than without (p = 0.036). There were significant positive correlations between syrinx size and IL-8 (p = 0.017), KC (p = 0.025) and MCP-1 (p = 0.003). There were significant negative correlations between IFN-γ and syrinx height (p = 0.025) and extent (p = 0.042). There was a significant negative correlation between IL-2 and syrinx height (p = 0.042). Neurotransmitter levels were not associated with skull abnormalities or spinal pain, but there was a positive correlation of glycine with IL-2 (p = 0.004) and MCP-1 with glutamate (p = 0.0147) and serotonin (p = 0.0059).

The size of the syrinx in BGs with SM is associated with several cytokine elevations but only a decrease of IFN-γ was associated with NP. Based on this study it does not appear that excitotoxicity plays a role in either SM development or NP. Further work is justified on the role of the immune system in CM, SM and NP.

ACVIM FORUM SESSION 94A

UPDATE ON THE CONSERVATIVE MANAGEMENT OF DISK ASSOCIATED CERVICAL SPONDYLOMYELOPATHY IN DOGS. S. De Decker[1,2], I. Gielen[1], L. Duchateau[1], A. Oevermann[3], L. Van Ham[1]. [1]Faculty of Veterinary Medicine, Ghent University, Merelbeke, Belgium. [2]The Royal Veterinary College, London, United Kingdom. [3]Neurocenter, University Bern, Vetsuisse Faculty, Switzerland.

Current knowledge about the conservative management of disk associated cervical spondylomyelopathy (DA-CSM) is rather limited and mainly based on retrospectively retrieved data. The goals of this study were to prospectively evaluate the evolution of clinical signs in dogs treated conservatively for DA-CSM. Additionally, several potential prognostic parameters and the correlation of initial clinical signs with magnetic resonance imaging (MRI) and transcranial magnetic stimulation (TMS) were investigated.

Twenty-one dogs were included. After neurological evaluation, neurological status was graded from 0 (= normal) to 6 (= tetraplegia). All animals underwent low-field MRI and TMS with measurement of onset latencies and peak-to-peak amplitudes from the extensor carpi radialis and cranial tibial muscles. From the MR images, the following dimensions were calculated: remaining spinal cord area; compression ratio; vertebral occupying ratio of the spinal cord; canal height to body height ratio (CBR); canal height to body length ratio (CBLR); and the canal compromise ratio. Intraparenchymal intensity (ISI) changes were graded from 0 to 3. All dogs were reevaluated by the same person after 1, 3, 6, 12, and 24 months.

Eight of 21 dogs (38%) experienced a positive clinical evolution with improvement of clinical signs or stabilization of mild clinical signs. All dogs with a negative clinical evolution 1 month after diagnosis experienced a further progression of clinical signs resulting in a poor outcome. The opposite was true for all dogs with a positive clinical evolution after 1 month. Outcome was further significantly associated by the remaining spinal cord area and the vertebral canal compromise ratio. Prognosis was not significantly affected by clinical presentation or TMS. Progression of clinical signs, in unsuccessfully treated dogs, was generally characterized by a rapid and dramatic deterioration of neurological status. There were no significant correlations between clinical presentation, MRI and TMS. Two dogs underwent necropsy and histopathological examination. This revealed in both cases chronic Wallerian degeneration and segmental myelomalacia.

The results of this study suggest that conservative treatment of DA-CSM is associated with a rather guarded prognosis. Clinical evolution 1 month after diagnosis and selected MRI parameters can be considered as prognostic indicators. The lack of correlation between clinical presentation and outcome, medical imaging and electrophysiological evaluation is disturbing and warrants further investigation.

ACVIM FORUM SESSION 94B

CLINICAL EVALUATION OF A MAGNETIC RESONANCE IMAGING (MRI) - GUIDED, FRAMELESS STEREOTACTIC BRAIN BIOPSY SYSTEM IN THE DOG. AV Chen[1], RL Tucker[1], GJ Haldorson[1], SA Thomovsky[1], AR Schneider[1], RS Bagley[2]. [1]Washington State University College of Veterinary Medicine, Pullman, WA. [2]Iowa State University, College of Veterinary Medicine, Ames, IA.

A MRI–guided stereotactic brain biopsy system has not been clinically evaluated in dogs. The purpose of this study was to determine the ability of the Brainsight™ system to obtain histologically diagnostic samples and access the impact of this procedure on neurologic status for 72 hours after the biopsy.

Five dogs with MRI definable lesions in the brain have been enrolled. Breeds included a Pitbull mix, Pembroke Welsh Corgi, French Bulldog, Border Terrier and West Highland White Terrier. Age ranged from 5–11 years. Weight ranged from 8.5–18.8 kg. Dogs presented with seizures (n = 5), ambulatory paresis(n = 4), unilateral blindness(n = 2) and head tilt(n = 1). One dog had a normal neurologic exam. Lesions chosen for biopsy were in the olfactory and/or frontal lobes (n = 3), parietal lobe(n = 1), and pyriform lobe(n = 1). Lesions were between 14–22 mm in diameter. All lesions were well-circumscribed and contrast enhancing except for one. Histologic diagnosis of meningioma(n = 3) and granulomatous meningoencephalitis(n = 1) were made. The poorly-circumscribed, non-contrast enhancing frontal mass yielded non-specific necrosis. Following biopsy, three dogs returned to pre-biopsy neurologic status within 12 hours. The French Bulldog took 48 hours to return to previous neurologic status due to brachycephalic syndrome that required oxygen support. One dog had acute respiratory arrest 16 hours post-biopsy. Necropsy is pending.

These results suggest that this MRI-guided biopsy system can provide an accurate histologic diagnosis of brain lesions. Biopsies of poorly-circumscribed and non-contrast enhancing brain lesions may be less diagnostic. Further evaluation is on-going to determine the true diagnostic yield and complication rate of this procedure.

ACVIM FORUM SESSION 95A

MORPHOMETRIC FEATURES OF THE CRANIOCERVICAL JUNCTION REGION IN DOGS WITH SUSPECTED CHIARI-LIKE MALFORMATION: 274 CASES (2007–2010). Dominic J. Marino. The Canine Chiari Institute at Long Island Veterinary Specialists Plainview, NY.

Concurrent malformations of the craniocervical junction are commonly identified in humans with Chiari type I malformation. Recent evidence suggests such craniocervical junction abnormalities (CJAs) also occur in dogs suspected of having Chiari-like malformation (CLM). The purpose of this study was to objectively describe morphometric features of the craniocervical junction region of dogs with suspected CLM and to investigate for associations between these features and the occurrence of other malformations in this region.

Magnetic resonance (MR) and computed tomographic (CT) images from 274 dogs with CLM were evaluated. Three regions of neural tissue compression were assessed: cerebellar compression (CC); ventral compression at the C1/C2 articulation, termed "medullary kinking" (MK); and dorsal compression (DC) at the C1/C2 articulation. A compression index (CI) was calculated for all abnormal regions for each dog. Multiple logistic regression analysis was performed ($p < 0.05$) to ascertain whether CI values for the different regions of compression were associated with the incidence of other craniocervical junction abnormalities.

68% of dogs had MK and 38% of dogs had DC. 28% of dogs also had evidence of atlanto-occipital overlapping (AOO). Breed and severity of CC were the only significant predictors of AOO ($p < 0.0001$ and $p < 0.0092$). 28% of suspected CLM cases have AOO as the anatomic abnormality responsible for CC. Compression index values may help differentiated subtypes of CJAs in dogs.

ACVIM FORUM SESSION 95B

MEDICAL INFRARED IMAGING (THERMOGRAPHY) OF TYPE I THORACOLUMBAR DISK DISEASE IN CHONDRODYSTROPHIC DOGS. CA Loughin, DJ Marino. The Canine Chiari Institute at Long Island Veterinary Specialists, Plainview, NY.

Medical infrared imaging (MII) is a non-invasive diagnostic imaging technique that measures skin surface temperature and generates thermal pattern maps based on predetermined color scales. Because skin temperature, dependent on regional perfusion, is under direct control of the sympathetic nervous system, MII provides information about the function of the autonomic nervous system. Because of recent advances in technology and lack of sedation needed to image patients, MII has potential use as a screening test for a variety of conditions that may result in autonomic dysregulation like Chiari-like malformation in dogs (CLM). The purposes of this study were to establish a MII protocol for dogs suspected of having CLM, to identify thermal imaging patterns for various regions of interest (ROI), to evaluate changes in thermal patterns and compare the results to those of MRI findings, considered the standard for diagnosing CLM in dogs. One hundred and five Cavalier King Charles Spaniel dogs with clinical signs attributable to CLM and confirmed CLM with MRI were evaluated with a complete blood count and chemistry profile, examination by a board certified surgeon/neurologist, multidetector CT scan of the cranio-cervical junction, whole body MRI and MII. The protocol for thermal imaging included cranial and caudal views of the body, full lateral right and left body views, dorsal views of the head and body, and right and left lateral views of the head. Thermal patterns were assessed with custom image recognition software. After each dog was imaged awake, general anesthesia was administered and the dogs re-imaged using the same protocol. MRI findings in dogs with severe or moderate cerebellar compression and cerebellar herniation were compared with MII results. The top of head and front of head ROI were 89.2% and 97.3% successful in identifying dogs with CLM. Based on these preliminary findings, MII may be a viable screening tool to detect CLM in dogs.

ACVIM FORUM SESSION 96A

THERMOGRAPHIC IMAGING (MEDICAL INFRARED IMAGING) OF TYPE I THORACOLUMBAR DISK DISEASE IN CHONDRODYSTROPHIC DOGS. CA Loughin, DJ Marino. The Canine Chiari Institute at Long Island Veterinary Specialists, Plainview, NY.

Medical infrared imaging (MII) is an imaging technique that measures skin surface temperature derived from cutaneous perfusion and generates thermal pattern maps based on color scales. MII has been used as a test for a variety of conditions that cause autonomic dysregulation resulting in altered cutaneous perfusion. Acute thoracolumbar intervertebral disk disease (TLIVDD) is common in dogs. The purpose of this study was to: 1) determine the success of MII in identifying dogs with TLIVDD, 2) compare the MII localization with MRI results and surgical findings 3) determine if the MII pattern returns to that of normal dogs following decompression surgery. 72 small breed chondodystrophic dogs with TLIVDD confirmed with MRI and 14 dogs with no TLIVDD were evaluated with MRI and MII. Regions correlating with the intervertebral disk spaces were analyzed for average temperatures and thermographic patterns. Thermal patterns were assessed with computer recognition pattern analysis (CRPA) software. 21 dogs were re-evaluated 8 weeks after surgery using the same protocol. When analyzing temperature averages over a region, no significant difference was found between control and affected dogs. CRPA was 90% successful in differentiating normal from affected dogs. CRPA was 97% successful in identifying the intervertebral disk space when compared with MRI and surgical findings. Based on these findings, MII may be a viable screening tool to detect TLIVDD in dogs.

ACVIM FORUM SESSION 96B

SPINAL CORD INJURY INFLUENCED BY THE MICROGLIA REACTION PROFILE. V.M. Stein[1], T.M.A. Boekhoff[1], E.-M. Ensinger[1], R. Carlson[1], P. Bock[2], W. Baumgärtner[2], K. Rohn[3], A. Tipold[1]. [1]Department of Small Animal Medicine and Surgery, [2]Department of Pathology, [3]Institute of Biometry, Epidemiology, and Information Processing, University of Veterinary Medicine, Hannover, Germany.

Microglia physiologically shows regional topographical differences in immunophenotype and function within the central nervous system indicating the endowment for a prompt response to pathological stimuli such as trauma. Spinal cord injuries (SCI) consist of a primary injury encompassing the mechanical impact and the "secondary wave" of injury occurring minutes to weeks later and comprising various consecutive effects such as increased production of free radicals, excessive release of excitatory neurotransmitters and inflammatory reactions. Activated microglia has the potential to perform some of these reactions, their contribution to the secondary wave is therefore controversially discussed. It has to be considered a double-edged sword as both, beneficial and deleterious effects have been attributed to these cells.

The purpose of the presented study was to assess microglial involvement, particularly in the "secondary wave" following SCI. Microglia from 15 dogs with SCI was isolated and characterized ex vivo in terms of morphology, immunophenotype, and function by flow cytometry. The results were compared to region-specific findings obtained from healthy control dogs (n = 30).

The histopathological exam confirmed the diagnosis of SCI in the cervical (n = 5) and thoracolumbar (n = 10) spinal cord, and revealed a significant activation of microglia/ macrophages and upregulation of myelinophagia in dogs with SCI 5 days or longer prior to euthanasia. Microglial ex vivo examination showed significantly increased expressions of B7-1, B7-2, MHC II, CD1c, ICAM-1, CD14, CD44, and CD45, and significantly enhanced phagocytosis and generation of reactive oxygen species (ROS) in SCI compared to healthy controls.

Microglial cells seem to be highly activated following SCI with an immunophenotype indicating their active role in co-stimulation of T cells, in leukocyte adhesion and aggregation, and in lipid and glycolipid presentation. Microglial phagocytosis might play a pivotal role in removal of injured or damaged cells and initialize subsequent healing processes. However, as ROS can be directly neurotoxic an enhanced microglial generation might lead to bystander damage of the traumatized spinal cord and might therefore add to the deleterious effects of the secondary wave. Modulating the microglial response in SCI might be a valuable novel therapeutic strategy alleviating further damage to the spinal cord.

ACVIM FORUM SESSION 99A

SERUM THYMIDINE KINASE ACTIVITY IN CANINE SPLENIC HEMANGIOSARCOMA.

DH Thamm,[1] DS Kamstock,[1] CR Sharp,[2] S Johnson,[3] E Mazzaferro,[4] LV Herold,[5] S Barnes,[6] K Winkler,[7] K Selting[8]. [1]Colorado State University, Fort Collins, CO., [2]Tufts University, North Grafton, MA., Emergency Animal Hospital of NW Austin, Austin, TX., [4]Wheat Ridge Veterinary Specialists, Wheat Ridge, CO., [5]DoveLewis Emergency Animal Hospital, Portland, OR., [6]Animal Emergency Critical Care, The LifeCentre, Leesburg, VA., [7]Georgia Veterinary Specialists, Atlanta, GA., [8]University of Missouri, Columbia, MO.

Thymidine kinase (TK) is a soluble biomarker present in S-phase of a salvage pathway for DNA synthesis, and can be measured in serum. TK activity correlates with stage, prognosis, and relapse in canine and human lymphoma. We previously reported the results of a pilot study evaluating TK activity in archived canine osteosarcoma, transitional cell carcinoma, and hemangiosarcoma (HSA) sera, and found elevated TK activity in 80% of canine HSA sera evaluated. The purpose of this study was to prospectively evaluate serum TK activity in a large number of dogs presenting to emergency clinics with hemoabdomen and a splenic mass, to determine if TK activity could be used as a noninvasive means to distinguish HSA versus benign conditions in this population.

Dogs presenting with hemoabdomen and a splenic mass identified on ultrasound examination were studied. Serum was collected prior to anesthesia, euthanasia or surgical intervention and frozen until batch analysis. Tissue from all patients was evaluated histologically by a single pathologist. Sera from age-matched normal dogs comprised a control population. An ELISA using azidothymidine as a TK1 substrate was used. Comparisons between groups were made using 2-tailed student T-tests, and receiver-operator characteristic (ROC) curves were generated.

Sixty-two patients and 39 normal controls were studied. There were 35 dogs with HSA, 10 dogs with other splenic neoplasia, and 17 dogs with benign diseases. Using a training set of 24 normal dogs, a cutoff of 6.55 U/L was established from the ROC curve. TK activity was significantly higher ($p < 0.0001$) in dogs with HSA than in the validation set of 15 normal dogs (mean+/−SD = 17.71+/−4.5 and 2.01+/−0.6 respectively), but not between dogs with HSA and benign splenic disease (mean+/−SD = 7.02+/−3.7, p = 0.13). Using a cutoff of 6.55 U/L, TK activity demonstrated a sensitivity of 0.54, specificity of 0.76, positive predictive value of 0.83 and negative predictive value of 0.45 for distinguishing HSA versus benign splenic disease. When interval thresholds of < 1.55 and >7.95 U/L were used together, diagnostic utility was markedly increased for distinguishing both HSA versus normal and HSA versus benign disease.

In conclusion, serum TK evaluation may assist in detection of canine HSA, and may also discriminate between benign disease and HSA in dogs with hemoabdomen and a splenic mass.

ACVIM FORUM SESSION 99B

GENE EXPRESSION PROFILING OF CANINE T CELL CHRONIC LYMPHOCYTIC LEUKEMIA.

S. Lana[1], R Burnett[1], J. Modiano[2], A Avery[1]. [1]Colorado State University, Fort Collins CO. [2]University of Minnesota, Minneapolis, MN.

T cell chronic lymphocytic leukemia (CLL) is a heterogeneous disease that affects a number of dog breeds. CLL patients have variable disease outcomes. The objectives of this study were to use gene expression profiling of CD8 T cell leukemias with variable outcomes in order to identify markers that can be used in routine diagnostic tests to distinguish good from poor prognosis disease, and to identify potential targets for novel therapy.

Gene expression profiling of 12 CD8 T cell leukemias (7 good, 5 poor prognosis) was conducted. Samples from 6 normal dogs were also profiled. Several differentially expressed genes were found including CD9, CD94, and CD 25. These were selected for further study using flow cytometry to determine expression of protein on the cell surface. Seventy nine cases of CD8+ T cell leukemia were screened for CD 9 expression. Forty seven had associated outcome information. Based on analysis to date, CD9 expression as assessed by flow cytometry does not appear to provide prognostic information.

A monoclonal antibody to CD25 was recently made available. To date 33 patients with CD8 T cell leukemia have been profiled. CD25 is variably expressed on T cell leukemias compared to normal CD8 T cells.

CD25 is the receptor for interleukin 2. Cyclosporin, a commonly used immunosuppressive drug, inhibits IL-2 production, and has been used to treat a subset of T cell leukemias in people. Thus, the finding that CD25 is up regulated on T cell leukemias compared with normal T cells suggests a possible new therapeutic avenue.

ACVIM FORUM SESSION 105A

IDENTIFICATION OF FECAL BACTERIAL DYSBIOSIS IN DOGS WITH VARIOUS GASTROINTESTINAL DISORDERS.

JS Suchodolski[1], RM Heilmann[1], SE Dowd[2], L Toresson[3], S Unterer[4], EM Dillman[1], AK Cook[1], JM Steiner[1]. [1]Gastrointestinal Laboratory, Texas A&M University, College Station, TX, USA; [2]Research and Testing Laboratory, Lubbock, TX, USA; [3]Helsingborg Referral Animal Hospital, Helsingborg, Sweden; [4]Ludwig-Maximilians-University, Munich, Germany.

Recent molecular studies have revealed a highly complex bacterial microbiota in the intestine of dogs. There is mounting evidence that microbes play an important role in the pathogenesis of acute and chronic enteropathies of dogs, including idiopathic inflammatory bowel disease (IBD). Similarly, compositional changes of the intestinal bacterial ecosystem have been associated with IBD in humans. The aim of this study was to characterize the bacterial microbiota in dogs with various gastrointestinal disorders using a next generation sequencing technique.

Fecal samples were obtained from healthy dogs (n = 31), dogs with acute uncomplicated diarrhea (n = 7), dogs with acute hemorrhagic diarrhea (AHD; n = 13), and dogs with active (n = 8) and therapeutically controlled IBD (n = 10). The bacterial composition was analyzed by massive parallel 16S rRNA gene 454-pyrosequencing. Differences between groups were analyzed using Mann-Whitney *U* tests and Kruskal-Wallis tests followed by Dunn's multiple comparison tests. Statistical significance was set at $p < 0.05$.

Significant differences in the proportions of several bacterial groups were identified between healthy and diseased dogs. Dogs with gastrointestinal disease had significantly higher proportions of Proteobacteria ($p < 0.01$). Proportions of Firmicutes were lower in diseased dogs, but this difference did not reach significance (p = 0.06). Within the Firmicutes the most notable findings were decreases in bacterial groups belonging to *Clostridium* clusters IV and XIVa (i.e., *Ruminococcus, Dorea,* and *Faecalibacterium* spp.; $p < 0.01$ for all). Dogs with AHD had the most profound changes of the microbiota, followed by dogs with acute uncomplicated diarrhea, and dogs with active IBD. *Faecalibacterium* spp. was the bacterial group most prominently depleted in dogs with active IBD, but was not significantly different between healthy dogs and dogs with therapeutically controlled IBD (p = 0.66).

Results of this study revealed bacterial dysbiosis in fecal samples of dogs with various GI disorders. Bacterial changes were more profound in dogs with severe disease, but were not identified in dogs with therapeutically controlled IBD, suggesting that the microbiota is stable in non-active disease. The bacterial groups identified are considered to be important short chain fatty acid producers and may serve as candidates for the diagnosis or therapeutic monitoring of GI disease. Future studies are necessary to determine if these microbial changes correlate with functional changes in the intestinal microbiota.

ACVIM FORUM SESSION 105B
ORAL ABSORPTION OF GENERIC CIPROFLOXACIN TABLETS IN DOGS. Mark G. Papich, and Daria DiGiovanni. North Carolina State University, College of Veterinary Medicine, Raleigh, NC.

Ciprofloxacin oral tablets, available in a generic formulation for people, are widely used for treatment in dogs. Oral absorption data for ciprofloxacin in dogs has been variable, and too limited to guide accurate dosing. Subsequently, published doses for dogs in veterinary formularies have varied from 5 to 25 mg/kg. This study was undertaken to explore the factors that may affect oral absorption of generic ciprofloxacin in dogs, and to derive a pharmacokinetic-based dose for treating susceptible bacteria. Six healthy adult beagle dogs were used for the study (11.2 kg mean weight). After placing jugular vein catheters for collecting blood samples, these dogs were administered either a single oral dose of ciprofloxacin (250 mg tablet; mean dose 23 mg/kg), or an intravenous (IV) dose (10 mg/kg; 2 mg/mL solution). A randomized crossover design was used with a washout time between treatments. Blood was collected for plasma drug analysis for 24 hours. Ciprofloxacin concentration in plasma was analyzed using high pressure liquid chromatography (HPLC) and pharmacokinetics analyzed using a computer program. Oral absorption was also evaluated via deconvolution analysis. The oral dose was well-tolerated, but the IV dose produced transient vomiting and depression in some dogs. After the oral dose, the peak plasma concentration (C_{MAX}) was 4.4 µg/mL (CV 55.9%), terminal half-life (t1/2) 2.6 hr (CV 10.8%), AUC 22.5 µg · hr/mL (CV 62.3%), and systemic absorption (F) 63.7% (CV 41.6%). After the IV dose, the t1/2 was 3.7 hr (CV 52.3%), systemic clearance 0.587 L/kg/hr (CV 33.9%), and volume of distribution 2.39 L/kg (CV 23.7%). After examining the pharmacokinetic results from the oral dose, it was apparent that oral ciprofloxacin was absorbed well in some dogs (approximately 80%), but poorly in others (approximately 30%). To explore the factors that may have affected oral absorption, two high absorbers and two low absorbers were administered an additional oral dose as a 10 mg/mL solution (250 mg total dose) via gastric tube. After administration of the oral solution, the plasma concentrations were more uniform and consistent among dogs. Absorption of the oral solution of ciprofloxacin was 71.0% (CV 7%) with a t1/2 of 3.1 (CV 18.6%) hr and C_{MAX} of 4.67 µg/mL (CV 17.6%). Therefore, it appears that inconsistent oral absorption of ciprofloxacin in some dogs may be formulation-dependent, and affected by tablet dissolution in the canine small intestine. Doses were calculated using the data for oral tablets in these dogs. The pharmacokinetic-pharmacodynamic (PK-PD) target was an AUC/MIC ratio of 100. Because of the wide range in oral absorption of tablets, a dose to reach the PK-PD target ranged from 12–75 mg/kg, with a mean dose of 32 mg/kg, once daily for bacteria with a MIC value $\leq$ 0.25 µg/mL. Some bacteria may have higher MIC values because the CLSI breakpoint for susceptible bacteria is $\leq$ 1 µg/mL.

ACVIM FORUM SESSION 106A
CANINE DISTEMPER VIRUS QUANTIFICATION BY REAL-TIME PCR ALLOWS TO DIFFERENTIATE VACCINE VIRUS INTERFERRENCE AND WILDTYPE INFECTION. CM Leutenegger, C Crawford, J Levy, M Estrada. IDEXX Laboratories, Inc., West Sacramento, CA.

Canine distemper (CD) is a highly contagious, acute or subacute systemic viral disease of dogs and other carnivores which can be controlled efficiently by the use of modified live-virus (MLV) vaccines. However, MLV strains do cross-react with molecular diagnostic tests and cause significant confusion for clinicians.

The purpose of this study was to use quantitative real-time PCR viral load information to differentiate between vaccine virus used in MLV vaccines and wildtype infections in dogs.

A real-time PCR test for CD virus (CDV) based on the P gene for phosphoprotein was used to determine viral loads in vaccinated and wildtype infected animals. A total of 158 respiratory mucosal swab samples from MLV vaccinated and asymptomatic dogs were obtained within the first 3 weeks after MLV vaccination. Based on the viral load in vaccinated animals, a cutoff value was established for the differentiation of dogs with clinical signs of respiratory distress and presumably infected with a wildtype strain of CDV. Two hundred clinical cases with known clinical and vaccination histories were analyzed to validate the cutoff value.

The CDV real-time PCR proved to be of high analytical and diagnostic sensitivity: A standard curve was established using known numbers of CDV molecules to allow absolute quantitative CDV viral load data. The limit of detection was in the single molecule range while the limit of quantitation was established at around 10 molecules per PCR reaction. A comparison to IFA showed real-time PCR to be 30% more sensitive. The CDV viral load in vaccinated animals averaged 26,738 viral particles per swab. A cutoff value of 107,903 viral particles was calculated by adding 3 standard deviations to the average value. This cutoff value correctly detected 95.2% of the vaccinated samples. Acutely infected dogs with CDV compatible clinical signs have high viral loads normally several logs higher than the cutoff value. In dogs with clinical distress, recent CDV MLV vaccination but viral loads below the cutoff value, other infectious agents were detected by using a panel of real-time PCR tests. Testing additional infectious agents in clinical settings is important in order to explain clinical signs when viral loads below cutoff values indicate that CDV is not the cause of clinical signs.

In conclusion, quantitative real-time PCR is a sensitive, rapid and reliable test regardless of recent vaccination. The use of a cutoff value will be of significant help to discriminate between vaccine interference and wildtype infection in clinical settings.

ACVIM FORUM SESSION 106B
THE USE OF A SUBCUTANEOUS URETERAL BYPASS (SUB) DEVICE FOR URETERAL OBSTRUCTIONS IN CATS. A Berent[1], C Weisse[1], M Wright,[1] K Todd[2], D Bagley[3]. [1]The Animal Medical Center, New York, NY. [2]The Veterinary Hospital of the University of Pennsylvania [3]Thomas Jefferson University Hospital, Philadelphia, PA.

Feline ureteral obstructions are a common urinary dilemma and traditional therapy is associated with substantial morbidity/mortality. Feline nephrostomy tubes are reported as being effective when pelvic drainage is required. The biggest limitation is externalized drainage, requiring careful management to prevent infection/dislodgement. The development of an indwelling ureteral bypass using a combination locking-loop nephrostomy/cystostomy tube was modified from humans, resulting in permanent indwelling drainage, reduced complications, and improved quality of life. The objective is to describe the technical and clinical outcome using a novel device called a subcutaneous ureteral bypass (SUB) in cats with ureteral obstructions.

Fifteen cats (16 kidneys) had a SUB placed for: ureterolithiasis (9), ureteral stricture (+/− stones) (6), and ureteral stent rejection (2). The median pre- and post-procedure creatinine was 6 mg/dL (range: 2.8–13) and 2.5 mg/dL (range: 2.4–9), respectively. The median pelvis diameter pre and post-procedure were 15 (range: 7–28) and 5 mm (range 3.4–6), respectively. Six French tubes were placed in 14, and 5 Fr. in 2. The bypass remained indwelling for a median of >240 days (range 64->450). There were 3 major complications resulting in nephrostomy tube dislodgement (2) and port leakage (1) 5 days after surgery. One patient with severe coagulopathy developed a clot which resolved with tPA infusion through the port. No SUB got occluded/obstructed long-term.

Overall, the use of a SUB for cats with ureteral obstructions can be considered a functional option when other therapies have failed or are contraindicated, but shtime.

Figures [ACVIM Forum Session 106B]

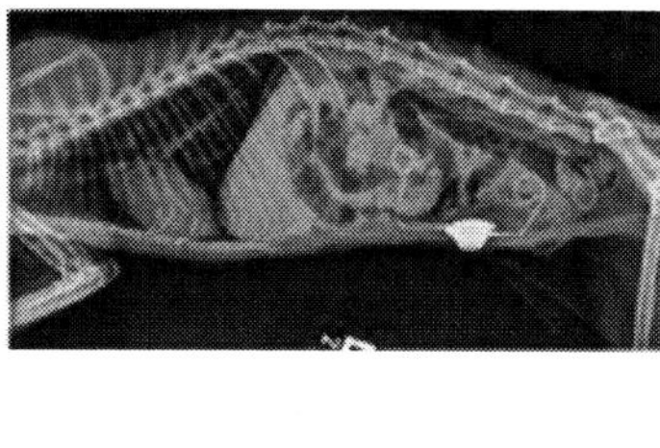

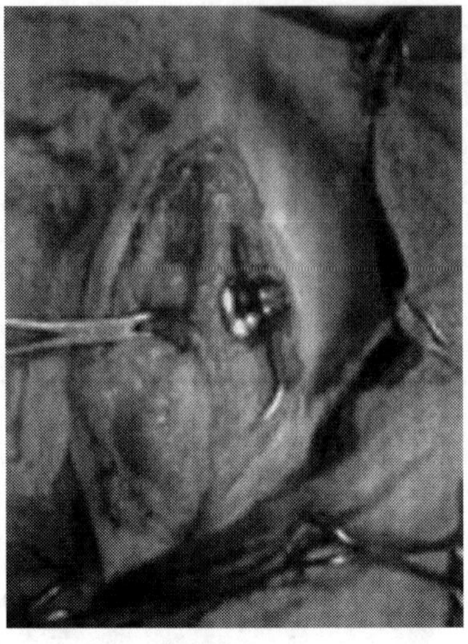

ACVIM FORUM SESSION 108A

THE EFFECT OF N-ACETYLCYSTEINE (NAC) IN CLINICALLY ILL DOGS: DOES IT MODULATE OXIDATIVE STRESS AND IMPROVE CLINICAL OUTCOME? Katrina R Viviano and Beth Vander Wielen. Department of Medical Sciences, University of Wisconsin-Madison, School of Veterinary Medicine, Madison, WI.

Oxidative stress is considered central to the pathogenesis of many systemic diseases. In humans, biomarkers of oxidative stress, antioxidant depletion and lipid peroxidation, have been correlated with disease severity and associated with poor clinical outcomes. Therapeutic antioxidant supplementation with NAC in glutathione (GSH)-deficient patients has shown clinical benefits, including repletion of intracellular GSH levels. We have shown that clinically ill dogs are GSH-deficient, and that GSH deficiency correlates with mortality, but it is not clear whether there are direct benefits of antioxidant intervention in these patients. The purpose of this randomized, investigator-blinded, placebo-controlled, prospective study was to evaluate the effect of NAC to normalize blood antioxidants (RBC reduced GSH (RBC GSH), plasma cysteine (CYS), serum vitamin E (vit E), and whole blood selenium (Se)), reduce lipid peroxidation (urine isoprostane/creatinine ratio (U I/C)), and improve illness scores (SPI2) and outcome (survival to discharge) in clinically ill dogs.

Clinically ill client-owned dogs, admitted to the UW Veterinary Medical Teaching Hospital that did not receive blood transfusions, TPN, vitamins, or antioxidants were eligible for the study. Dogs enrolled in the study were randomized to receive IV infusions q. 6 h. of either NAC (1 × 140 mg/kg and 6 × 70 mg/kg) or equal volumes of 5% dextrose (placebo) over 48 hours. At the time of enrollment, and 2 hours following the final 48 hour infusion, blood and urine were collected to quantify RBC GSH, CYS, vit E, and Se concentrations; U I/C ratios; and calculate SPI2 scores. RBC GSH and CYS concentrations were quantified by HPLC. Commercially available HPLC, atomic absorption spectroscopy, and EIA were used to quantify vit E, Se, and U I/C ratios, respectively. Nonparametric statistical analyses were used, with results reported as medians and P < 0.05 considered significant.

Sixty-one ill dogs were randomized to either NAC (n = 30) or placebo (n = 31). Overall this group of ill dogs had significantly decreased RBC GSH (1.50 vs. 1.91 mM; P = 0.0013), vit E (27 vs. 56 µg/mL; P = 0.0002), and Se (0.37 vs. 0.55 µg/mL; P = 0.0308) levels and elevated U I/C ratios (969 vs. 398 pg/mg; P = 0.0005) in comparison to healthy control dogs. Dogs in the placebo group showed a significant further decrease in RBC GSH over the next 48 hours (1.48 to 1.42; *P* = 0.035). NAC supplementation significantly increased plasma CYS levels (9.1 to 15.1 µM; P < 0.0001), and prevented a further decline in RBC GSH (1.55 to 1.58 mM; P = 0.174). However, serum vit E (30 vs. 28 µg/mL), Se (0.40 vs. 0.35 µg/mL), U I/C ratios (946 vs. 806 pg/mg), SPI2 scores (0.74 vs. 0.75), and outcome (81% vs. 76%) were not significantly different between the NAC and placebo groups after treatment.

The results of this study further support that clinically ill dogs experience oxidative stress, and suggest that antioxidant supplementation with NAC within the first 48 hours of hospitalization prevents further RBC GSH depletion. Further studies are necessary to investigate whether longer duration or combined antioxidant supplementation normalizes the redox state and impacts long-term outcome.

ACVIM FORUM SESSION 108B

BIOMARKERS PREDICTIVE OF INSULIN RESISTANCE IN CATS IMPROVE WITH WEIGHT LOSS/ HEALTHY WEIGHT MAINTENANCE. J. Brockman and R. Yamka. Hills Pet Nutrition, Topeka, KS.

Diabetes Mellitus in cats is very similar to Type II Diabetes in Humans, preceded by a period of insulin resistance. Evaluating insulin resistance in a cat is a time consuming, expensive, and difficult procedure. There is a need for a simple biomarker based test predictive of insulin resistance. There is a biomarker based assay predicative of insulin resistance in humans. The purpose of this study was to evaluate the utility of this assay in overweight cats and show improvement in insulin sensitivity following weight loss and weight maintenance.

The insulin resistance assay is based on the quantitative analysis of 5 metabolites (2-hydroxybuterate, creatine, palmitate, decanoyl-carnitine, and oleoyl-LPC). A proprietary algorithm (Metabolon, Inc, Durham NC) was used to generate a predictive RD (Rate of Disposal) value (normal range in cats 6.5–10.5). Individuals with an RD value less than 6 will have a greater than 50% chance of being insulin resistant and an RD value less than 3 will have a greater than 90% chance of being insulin resistant.

Initial studies demonstrated that the RD values indicating insulin resistance in cats correlated with age, obesity and severity of diabetes as determined by histopathology and blood glucose levels. In a feeding study of 40 cats (> 39% vs. < 25% body fat) RD values improved from 6.19 ± 1.23 to 6.84+1.38 (p = 0.029). During weight maintenance, 19% body fat for 4 months, further improvement was observed (RD, 8.30 + 1.08 (p = 9.2E-12)). These results demonstrate that long term weight maintenance following weight loss is critical for increasing insulin sensitivity in cats.

ACVIM FORUM SESSION 109A

CANINE-DERIVED ANTIBODY FRAGMENTS THAT TARGET HEMANGIOSARCOMA AND ITS GROWTH FACTORS. A Braganza, B Clerc-Renaud, S Huang, NJ Mason. University of Pennsylvania School of Veterinary Medicine, Philadelphia, PA.

The use of monoclonal antibodies and antibody fragments to directly target tumor antigens and neutralize their growth factors has shown promising results in human clinical trials. However, these targeted approaches have not been possible in dogs since specific tumor antigens have not been identified, monoclonal antibodies of canine origin are not available and the efficacy of xenogeneic antibodies in the dog is limited by neutralizing antibody responses. To overcome these obstacles, we have generated canine antibody phage display libraries from canine splenocytes. These libraries consist of single chain variable fragments (scFv) comprised of canine variable heavy (VH) and variable light (VL) immunoglobulin chains displayed on the surface of bacteriophage (Fig. 1). The antigen specificity within these libraries is diverse and recapitulates the antigen-experienced immunoglobulin repertoire of the dog. We can now use simple panning techniques to isolate scFv of canine origin that bind to either known targets or unknown targets which can then be identified using standard molecular techniques.

Canine HSA is a highly aggressive malignancy of vascular endothelial cells that affects large breed dogs. Although there are no confirmed immunological targets for HSA, serum levels of Vascular Endothelial Growth Factor (VEGF) are elevated in these patients and, as in many human cancers, VEGF may represent an important therapeutic target for neutralization. We used simple panning techniques to screen canine scFv libraries generated from the spleens of dogs with HSA against canine VEGF and successfully isolated 3 scFv clones that bind and neutralize canine VEGF *in vitro*. These scFvs are now being taken into a murine model of canine HSA to determine whether they can inhibit tumor growth and metastases. In addition, we have panned the same antigen-experienced scFv phage display libraries against allogeneic primary canine HSA cells of low passage number to isolate canine-derived antibody fragments that can target malignant endothelial cell surface molecules. Early results demonstrate enrichment of scFv phage libraries for malignant endothelial cell binders. These scFv can be readily linked to chemotherapeutic agents or other toxins and used to deliver high doses directly to the malignant cell. This novel approach aims to reduce side effects of systemic chemotherapy and augment therapeutic response.

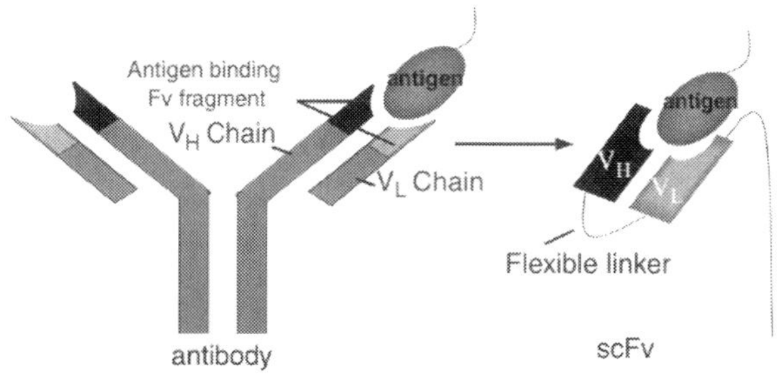

Fig [ACVIM Forum Session 109A]: Cartoon of a single chain fragment variable (scFv) comprised of the VH and VL regions of immunoglobulin linked by a flexible linker. scFvs are the smallest antibody fragments that maintain antigen-binding specificity and affinity.

ACVIM FORUM SESSION 109B
EFFICACY OF VITAMIN D (CALCITRIOL) FOR CANINE MAST CELL TUMORS.

EK Malone[1], KM Rassnick[1], JJ Wakshlag[1], DS Russell[1], R Al-Sarraf[2], DM Ruslander[3], CS Johnson[4], and DL Trump[4]. [1]Cornell University College of Veterinary Medicine, Ithaca, NY. [2]Animal Emergency and Referral Associates, Fairfield, NJ. [3]Veterinary Specialty Hospital of the Carolinas, Cary, NC. [4]Roswell Park Cancer Institute, Buffalo, NY.

Calcitriol, (vitamin D_3), has antineoplastic activity and acts synergistically to potentiate the antitumor activity of a diverse array of chemotherapeutics. CCNU, vinblastine, corticosteroids, and tyrosine kinase inhibitors, are used to treat canine mast cell tumors (MCT). Vitamin D receptor is expressed in the majority of canine MCTs, suggesting a role for calcitriol in the management of dogs with these tumors. The purpose of our study was to examine the in vitro effects of calcitriol in combination with CCNU, vinblastine, imatinib, or toceranib on canine mastocytoma C2 cells. Also, we evaluated the antitumor activity of DN101, a highly concentrated oral formulation of calcitriol, as single-agent treatment in dogs with naturally occurring MCTs.

C2 cells were incubated with serial dilutions of calcitriol (0.1–25 nM). Twenty-four hours later, cells were then treated with vehicle control or serial dilutions of CCNU (2.5–20 uM), vinblastine (1.25–10 nM), imatinib (1.67–12.5 nM), or toceranib (3.13–25 nM). Cell viability was assessed with an MTT assay after 48 hours and data was used to derive a combination index (CI: values < 1, 1, > 1 indicate synergism, additivity, antagonism, respectively). In the phase II clinical trial, dogs were eligible if they had at least 1 measurable, histologically confirmed, MCT. Calcitriol was administered orally. RECIST criteria were used to assess tumor response.

Calcitriol, CCNU, vinblastine, imatinib, and toceranib each suppressed C2 cell viability in a dose-dependent manner. CI values < 1 were obtained for calcitriol (0.1–6.25 nM) combined with CCNU (5 and 10uM), vinblastine (2.5 and 5 nM), imatinib (1.67–12.5 nM) and toceranib (0.1–12.5 nM). Due to the occurrence of toxicity (vomiting, anorexia, hypercalcemia), the phase II trial was terminated early; only 10 of 20 planned patients were treated. One dog with a metastatic muzzle MCT had a complete response that lasted 89 days. Three dogs achieved partial response lasting from 74–90 days.

In summary, our in vitro data demonstrate that calcitriol combined with CCNU, vinblastine, imatinib or toceranib has synergistic effects on C2 mastocytoma cells. Antitumor responses were observed in dogs with spontaneously occurring MCTs treated orally with single-agent calcitriol, but the frequency of adverse effects was high. Together these results suggest calcitriol combination therapies might have significant clinical utility in the treatment of canine MCTs but refinement of the calcitriol-dosing regimen must be done.

ACVIM FORUM SESSION 110A
ACVIM FOUNDATION RESEARCH REPORT: DEVELOPMENT OF ASSAYS FOR THE PHARMACODYNAMIC MONITORING OF CYCLOSPORINE IN THE DOG.

T Archer, C Fellman, J Stokes, K Lunsford, A Mackin. Mississippi State University, Starkville, MS.

Cyclosporine is a potent immunosuppressive agent used to treat many canine inflammatory and immune-mediated diseases. Cyclosporine has gained popularity as an immunosuppressive agent because of a favorable toxicity profile compared to many other immunosuppressive agents. Optimal dosing regimens for cyclosporine in the dog remain unclear, primarily because standard methods that monitor effectiveness of immunosuppression have not been established. Pharmacokinetic testing is currently used during treatment with oral cyclosporine to adjust doses based on measurement of blood drug levels. Individual patients, however, often demonstrate marked variations in blood drug levels while on similar oral doses of cyclosporine, and can also demonstrate different clinical responses even at comparable drug levels, making correlation of blood cyclosporine levels and degree of disease control extremely difficult. Pharmacodynamic testing offers an alternative method for regulating cyclosporine dosing by objectively measuring the effects of cyclosporine on T-cells, the drug's main cellular target in the body. Our ACVIM Foundation-funded research has focused on developing and evaluating a comprehensive panel of biomarkers of immunosuppression that can be utilized for pharmacodynamic monitoring during treatment with cyclosporine and other immunosuppressive agents that affect T-cell function.

We have completed several studies using flow cytometry to evaluate activated T-cell expression of surface molecules (CD25 & CD95) and cytokines (IL-2, IFN-γ & IL-4) as potential biomarkers. Our first study was an *in vitro* study evaluating expression of surface molecules and cytokines in canine T-cells exposed to varying concentrations of cyclosporine. This study established consistent drug-associated suppression of the cytokines IL-2, IFN-γ and IL-4. Our second study was an *in vivo* study in normal dogs evaluating the effects of two doses of oral cyclosporine, a high dose considered to be reliably immunosuppressive (starting dose 10 mg/kg bid, titrated upwards as needed to attain trough drug blood levels of at least 600 ng/mL) and a lower dose used to treat atopy (5 mg/kg sid), on T-cell expression of these three cytokines. Significant suppression of IL-2 and IFN-γ expression was seen at the high cyclosporine dose, while at the lower dose only IFN-γ expression was suppressed. Because T-cell expression of IL-4 was not significantly suppressed at the high cyclosporine dose, IL-4 was not evaluated at the lower drug dose.

Because of specialized sample handling requirements, flow cytometry is not as practitioner friendly as other assays (such as PCR) for routine use in pharmacodynamic testing. We have therefore conducted an *in vitro* study comparing the effects of cyclosporine on activated T-cell expression of IL-2 and IFN-γ using flow cytometry and qRT-PCR, and demonstrated dose dependent and comparable suppression of IL-2 and IFN-γ using either methodology. We are currently evaluating, using qRT-PCR, the effects of oral cyclosporine on T-cell expression of IL-2 and IFN-γ in normal dogs prior to moving on to pharmacodynamic trials in our clinic patients.

ACVIM FORUM SESSION 110B
EFFECT OF HYPOTHYROIDISM ON REPRODUCTION IN BITCHES.

DL Panciera[1], BJ Purswell[2], KA Kolster[2], SR Werre[3]. Departments of [1]Small Animal Clinical Sciences, [2]Large Animal Clinical Sciences, and[3]Laboratory for Study Design and Data Analysis, Virginia-Maryland Regional College of Veterinary Medicine, Virginia Tech, Blacksburg, VA.

Numerous reproductive abnormalities, including irregular interestrous period, anestrus, and infertility have been attributed to hypothyroidism. We previously documented reduced fertility and lower birth weight and increased periparturient mortality in pups born to bitches with experimentally-induced hypothyroidism for a median duration of 56 weeks. The purpose of this study was to evaluate reproductive function in these same bitches after hypothyroidism was treated with a replacement dose of levothyroxine.

Twelve multiparous bitches were studied. Hypothyroidism was induced in 6 dogs by administration of 1 mCi/kg ^{131}I. Hypothyroidism was confirmed by finding serum T4 concentrations before and 4 hours after IV administration of human recombinant TSH that were

< 5 nmol/L. Levothyroxine (0.02 mg/kd q 24 h) was administered to all hypothyroid bitches. Six bitches served as euthyroid, untreated controls. Dogs were evaluated daily for signs of estrus and were bred by 1 of 2 males when serum progesterone was ≥5 ng/ml. Interestrous interval, gestation length, strength and duration of contractions during whelping, time between pups, number of live pups and stillbirths, viability of pups at birth, weight of pups, and periparturient mortality were recorded. The Student's t-test and ANOVA were used to compare differences between control and hypothyroid bitches for continuous, normally distributed data. The Wilcoxon rank sum test was used to analyze data between groups that was not normally distributed.

The mean duration of hypothyroidism prior to levothyroxine administration was 102 ± 8.1 weeks. Breeding took place after levothyroxine treatment for 46 ± 12.6 weeks in the hypothyroid group. All 6 dogs in the hypothyroid group and 5/6 control dogs were pregnant, while 4/8 hypothyroid and all 6 control bitches became pregnant prior to levothyroxine administration. No difference in interestrus interval or gestation length was noted between groups. During whelping, no difference in strength of contractions, contraction duration, interval between pups, or viability scores of pups was found between groups. Litter size, birth weight and peirparturient mortality were similar between groups.

Levothyroxine administration reverses the detrimental effects of hypothyroidism on fertility and neonatal health.

ACVIM FORUM SESSION 111A

EFFICACY OF GASTRIC ACID SUPPRESSION FOR PREVENTION OF EXERCISE-INDUCED GASTRITIS IN RACING ALASKAN SLED DOGS. KK Williamson[1], MD Willard[2], MS Davis[1], ME Payton[1]. [1]Oklahoma State University, Stillwater, OK. [2]Texas A&M University, College Station, TX.

Racing sled dogs have a high prevalence of exercise-induced gastric erosions/ulcers, with reports ranging from 50–67% of dogs running at least 100 miles in a day or less. Omeprazole reduces the severity of, but does not completely prevent, gastritis under racing conditions, and can be difficult to administer under these conditions. Famotidine can be administered in food, but has only demonstrated efficacy under less intense training conditions. The purpose of these studies was to evaluate different acid suppression strategies under racing conditions for the prevention of exercise-induced gastritis.

Experiment #1 was a randomized placebo-controlled study using 36 sled dogs (3–8 years) competing in a 330 mile race over 50–60 h. Treatment groups were famotidine (approx 1 mg/kg qd) or no treatment, beginning 2 days prior to the start of the race and proceeding until gastroscopy was performed 24h after the race. Experiment #2 was a randomized positive-control study using 52 sled dogs (2–8 years) running a mock race of 300 miles in 50h. Dogs were divided into omeprazole (approx 1 mg/kg qd, administered 30 min prior to a meal) or famotidine (approx 2 mg/kg bid) groups beginning 2 days prior to the exercise challenge and continuing for 24h after completion. Gastroscopy was performed immediately prior to the start of dosing and 24h after completion of the exercise. In all cases, mucosal appearance during gastroscopy was blindly scored using previously described scoring system.

Famotidine (1 mg/kg qd) reduced the prevalence of clinically-relevant, exercise-induced gastric lesions compared to no treatment (7/16 vs 11/16, $p = 0.031$). Compared to famotidine at 2 mg/kg bid, omeprazole significantly decreased the severity (0.4 vs 1.2, $p = 0.0002$) and prevalence (2/23 vs 7/21, $p = 0.049$) of gastric lesions.

Although famotidine provides some benefit in the prevention of exercise-induced gastric lesions, neither the recommended dose nor the higher dose were considered acceptable in the prevention of exercise-induced gastritis as between 33–44% of the dogs receiving famotidine had clinically significant lesions. A previous study examining omeprazole under racing conditions, but without careful administration on an empty stomach, resulted in a 22% prevalence of clinically significant gastric lesions. However, the bioavailability of omeprazole is reduced in the presence of food, and when the daily administration of the drug is carefully scheduled to coincide with an empty stomach, the resulting prevalence of clinically significant lesions induced by racing-intensity exercise is reduced to just over 10%. The conclusions of these studies are that omeprazole is superior to famotidine in preventing gastritis in racing sled dogs during competition. Routine administration of omeprazole is recommended to prevent stress-associated gastric disease in exercising and racing Alaskan sled dogs.

ACVIM FORUM SESSION 111B

EFFECT OF ORAL GALLIUM NITRATE ON FECAL SHEDDING OF *RHODOCOCCUS EQUI* IN MARES. BR Buchanan[1], K Kuskie[3], N Pusterla[2], N Cohen[3]. [1]Brazos Valley Equine Hospital, Navasota, TX. [2]University of California Davis SVM, Davis, CA. [3]Texas A&M University CVM, College Station, TX.

Mares may be an important source of environmental contamination with *Rhodococcus equi* on breeding farms. Attempts to reduce fecal shedding of *R equi* by the mare and the effects of the mare's fecal *R equi* concentration on airborne concentrations in the foaling stall have not been previously reported. Twenty-one Arabian mares were treated daily with either oral gallium nitrate or placebo in a randomized double-blind study. Fecal samples were collected at day 320 of gestation (time 1), the week before foaling (time 2), and the week after foaling (time 3). Airborne concentration of *R equi* were measured in the stall within 6 hours post foaling using a microbial air sampling system into which standard (100-mm) culture plates with a media selective for *R. equi* have been loaded.

Concentration of total *R equi* were determined by morphological characteristics. The concentration of virulent *R equi* was determined using a modified colony immunoblot method. Concentrations of total and virulent *R equi* were compared among mares to examine effects of treatment, time, and treatment by time interaction.

There were significant ($P < 0.05$) effects of treatment that depended on time of sample collection. At sample times 1 and 2 there were no significant differences between groups in the fecal concentration of virulent *R equi*. At time 3 concentrations of virulent *R equi* were significantly lower among mares in the treatment group ($P < 0.05$) compared to control. Effects of time depended significantly on groups: for the control group, there were no significant effects of time. For the treatment group, concentrations tended to decrease over time, and concentrations at time 3 were significantly ($P < 0.05$) lower than those at time 1. No other differences among times for concentrations in the treatment group were statistically significant. There were no significant effects of treatment, sample time, or their interaction on the concentration of total *R equi* between groups; however, the pattern for these data was similar to that observed for the virulent isolates. No significant differences were determined between treatment groups for airborne concentrations of virulent or total *R equi*.

Treatment of mares with oral gallium nitrate significantly reduced the fecal concentrations of virulent *R equi* over time, but had no impact on the airborne concentration of *R equi* shortly after foaling.

ACVIM FORUM SESSION 112A

BRONCHOALVEOLAR LAVAGE FLUID PROTEOME IN HORSES WITH RECURRENT AIRWAY OBSTRUCTION. Marybeth Miskovic Feutz,[1] C. Paige Riley,[2] Jiri Adamec,[3] Laurent L. Couetil[1]. [1]Purdue University School of Veterinary Medicine, West Lafayette, IN. [2]Bindley Bioscience Center, Purdue University, West Lafayette, IN. [3]University of Nebraska, Lincoln, NE.

The purpose of this study was to evaluate the protein profile of bronchoalveolar lavage fluid (BALF) in horses affected with recurrent airway obstruction (RAO) and in control horses using proteomics and Western blot techniques. RAO-affected ($n = 5$) and control horses ($n = 6$) were subjected to an experimental exposure trial; when the RAO-affected horses showed clinical signs of disease, BALF was collected from all horses. BALF was also collected from client-owned RAO-affected horses ($n = 15$) with naturally-occurring clinical signs of disease and client-owned control horses ($n = 12$) from the same environments. The BALF from the experimental exposure trial horses was subjected to trypsin digestion and proteomics analysis with mass spectrometry (MS). Peaks detected with MS were identified using tandem MS analysis and database searches. Western blot was used to confirm the identity and expression levels of two proteins identified using proteomics techniques in the BALF of all horses. Data from MS experiments were analyzed with the Student's t-test to compare peak intensity between RAO-affected and control horses. Western blot band density data was analyzed with the Kruskal-Wallis ANOVA for comparison between groups of horses. Significance level was set at $P < 0.05$.

With MS proteomic analysis of the BALF from the experimental exposure trial horses, 2049 total peaks (peptides) were identified. Of these peaks, 100 were differentially expressed between the RAO-affected (24 over-expressed) and control horses (76 over-expressed). Identifications were made for 250 BALF proteins. Transferrin and secretoglobin were chosen for validation with Western blot. Proteomics indicated that secretoglobin was not differentially expressed between the experimental exposure trial group; this was confirmed with Western blot analysis. Western blot also showed that client-owned RAO-affected horses had lower secretoglobin expression than client-owned control horses and control horses before experimental exposure. According to the proteomics data, transferrin was over-expressed in control horses after experimental exposure compared to RAO-affected horses. While the Western blot analysis did not show a statistically significant difference in this comparison, transferrin was significantly over-expressed in control horses before experimental exposure compared to client-owned RAO-affected horses. In addition, both secretoglobin and transferrin band densities on Western blot were negatively correlated with airway obstruction and neutrophilic pulmonary inflammation.

This study demonstrates that proteomics techniques can be used in the investigation of equine BALF proteins. The proteins identified as differentially expressed between RAO-affected and control horses in this study including, but not limited to, secretoglobin and transferrin should undergo further evaluation for their use as biomarkers of RAO, and as potential targets of new therapeutic agents for RAO.

ACVIM FORUM SESSION 112B
RATTLESNAKE ENVENOMATION IN HORSES: CARDIOTOXIC EFFECTS AND IMMUNE RESPONSE. Lyndi L. Gilliam[1], TC. Holbrook[1], ME Payton[1], SM Martin[2], R Carmichael[1], JM Taylor[3], AA Giulioli[4], K Levis[2], CL Ownby[1], D McFarlane[1]. [1]Oklahoma State University, Stillwater, OK. [2]Hansford County Veterinary Clinic, Spearman, TX. [3]Red Rock Biologics, Woodland, CA. [4]Tri-County Veterinary Clinic, Anna, TX.

Cardiotoxic effects of rattlesnake venom in the horse are not well defined. The first aim of this study was to document cardiac damage in naturally envenomated horses. Twenty horses with clinical diagnosis of snake bite were included. A snake venom ELISA was utilized to confirm envenomation when possible. Serum and plasma were collected at selected intervals. Plasma was assayed for cardiac troponin I (cTnI) using a flurometric assay (Stratus CS®, Dade Behring). Holter monitors (Zymed®, Philips) were placed at admission, 1 week and 1 month post presentation. Echocardiography was performed on available horses 5–7 months after envenomation.

The second aim of this study was to investigate potential mechanisms of the cardiac damage. Serum samples were assayed for TNFalpha using a commercial assay (Endogen). Antibody titers to *Crotalus atrox* venom were measured at admission, 1 week and 1 month after natural envenomation and compared to titers in vaccinated horses (Crotalus Atrox Toxoid, Red Rock Biologics).

A significant number of horses showed elevations in cTnI ($p < 0.05$) at one or more time point indicating myocardial damage. Holter readings revealed the presence of arrhythmias or persistent tachycardia in 19 horses. Five of twenty horses were available for echocardiography; no abnormalities were noted. Horses with increased cTnI tended to have greater TNFalpha concentrations compared with horses without increased cTnI. Peak venom titers in bitten horses were significantly higher than peak titers in vaccinated horses ($p < 0.05$).

Rattlesnake envenomation was associated with evidence of cardiac damage in a significant proportion of bitten horses. Further studies are needed to determine the cause as well as mechanisms to treat and/or prevent its occurrence.

ACVIM FORUM SESSION 139A
PYROSEQUENCING OF THE GASTRIC MUCOSA IN HEALTHY HORSES SHOWS AN ABUNDANT MIRCROBIAL FLORA WHICH IS UNIQUE IN EACH HORSE AND INFLUENCED BY DIET AND MANAGEMENT. G.A. Perkins, H.C. den Bakker, A.B. Burton, H.N. Erb, S.P. McDonough, P.L. McDonough, J. Parker, R. Rosenthal, M. Wiedmann, S. Dowd, and K.W. Simpson. Cornell University, College of Veterinary Medicine, Ithaca, NY.

Little is known about the gastric mucosal flora in healthy horses and its role in gastric disease has not been critically examined. Our laboratory previously reported that a diverse microbial flora with a predominance of *Streptococcus spp.* and *Lactobacillus spp.* exists in healthy horses using Fluorescence *in situ* hybridization (FISH). The present study sought to further characterize the gastric mucosal flora of healthy horses using massive parallel 16SrRNA bacterial tag encoded FLX-titanium amplicon pyrosequencing (bTEFAP). Biopsies of the squamous, glandular, antral and any ulcerated mucosa were obtained from 4 healthy horses via gastroscopy after a 12-hour fast and 2 horses immediately post-mortem. DNA was extracted from the mucosal biopsies and bTEFAP and data processing was performed. Hierarchical cluster analysis based on relative abundance data on the genus level were performed to look for trends in bacterial diversity among the individual horses.

Pyrosequencing yielded between 4,500 and 13,000 reads per horse with 9238, 16587, 16587 reads in the antrum, squamous and glandular regions, respectively. The microbiome segregated into two distinct clusters: Cluster 1 comprised of 2 horses that were stabled, fed hay and sampled at post-mortem and Cluster 2 consisted of 4 horses that were pastured on grass, fed hay and biopsied gastroscopically after a 12-hour fast. Samples from different antomic regions clustered by horse rather than region. Despite being very similar at the higher taxonomic level (phyla) differences in the distribution of bacteria were seen at the genus and species level. The dominant bacteria in Cluster 1 horses were Firmicutes (>83% reads/sample) consisting of mainly *Streptococcus spp.*, *Lactobacillus jensenii*, *L. fornicalis* and *Sarcina maxima*. Cluster 2 had more diversity with a predominance of Proteobacteria, Bacteroidetes and Firmicutes and 51 genera identified such as *Streptococcus spp.*, *Moraxella spp.*, *Actinobacillus spp.*, and others. Though the relative abundance of the individual taxonomic groups was significantly different between individual horses, no significant differences in the overall diversity could be found (as assesed by Shanon Weaver, ACE and Choa I diversity indices). *Helicobacter spp.* sequences were not identified in any sample (out of 58,891 reads). The ulcerated mucosa from Horse 3 (group 1) had lower diversity and higher numbers of bacteria predominated by *Lactobacillus equigenerosi*.

This data shows that the equine gastric mucosa harbors an abundant and diverse microbiome which is unique to each individual and differs by sampling method, fasting prior to sampling and diet.

ACVIM FORUM SESSION 139B
SEASONAL PASTURE MYOPATHY IN SIX HORSES ASSOCIATED WITH MULTIPLE ACYL-COA DEHYDROGENASE DEFICIENCY (MADD). BT Sponseller[1], SJ Valberg[2], NE Schultz[2], HE Bedford[3]. [1]Iowa State University College of Veterinary Medicine, Ames, IA. [2]University of Minnesota College of Veterinary Medicine, St. Paul, MN. [3]Performance Equine Vets, Aiken, SC.

Seasonal pasture myopathy (SPM; atypical myopathy [AM] in Europe), typified by nonexertional rhabdomyolysis, occurs in pastured horses during autumn or spring. Clinical signs rapidly progress from muscular weakness to recumbency and frequently death. Extensive myonecrosis and intramyofiber lipid storage occur in highly oxidative respiratory and postural muscles. Recently, a defect of lipid metabolism called MADD has been identified in European horses with AM. This report documents the first cases of equine MADD in the United States.

Six Midwestern US horses suspected of having SPM in the spring or fall of 2009 were evaluated for MADD by urine organic acids, plasma acylcarnitines and/or muscle carnitine and histopathology. Five horses had clinical signs and clinicopathologic data consistent with severe rhabdomyolysis. One horse was found dead on pasture after 2 days of rear limb stiffness and inappetance. Urinary organic acid profiles revealed markedly elevated ethylmalonic and methylsuccinic acids, butyrylglycine, isovalerylglycine, and hexanoylglycine, consistent with equine MADD. Plasma acylcarnitine profiles from 2 horses had marked elevations of short chain acylcarnitines, while the third horse and only survivor had minor elevations of short chain acylcarnitines. Affected muscle showed extensive degeneration with intramyofiber lipid accumulation, a marked decrease in free carnitine, and high levels of carnitine esters.

SPM appears to be a highly fatal emerging disease of pastured horses in the US characterized by weakness, colic-like signs and myoglobinuria. The disease is associated with a defect in muscular lipid metabolism that can be diagnosed by performing lipid staining of muscle samples and urine organic acid profiles.

ACVIM FORUM SESSION 145A
PROGRESSION OF PARASITEMIA AND CLINICAL COURSE IN A SPLENECTOMIZED ALPACA WITH *CANDIDATUS MYCOPLASMA HAEMOLAMAE* INFECTION.

JE Sojka-Kritchevsky, B Toth, JB Messick, AN Baird. Purdue University, West Lafayette, IN.

Candidatus Mycoplasma haemolamae (*C*Mhl) is a common red blood cell parasite of new world camelids. The high degree of parasitemia that develops in an infected splenectomized animal allows for the efficient collection of parasitic DNA. This DNA can then be used in the development of genetically-derived tools such as PCR and in-situ hybridization. Thus, one splenectomized animal can replace many immunologically intact animals within a research setting. The purpose of this study was to track the natural progression of *C*Mhl parasitemia and associated clinical signs in a splenectomized alpaca.

An intact, 9-month-old, 39.1 kg male alpaca was used in this study. He had tested positive via PCR for *C*Mhl on three different occasions, although no organisms were seen on peripheral blood smears. The alpaca was placed under general anesthesia and a ventral midline incision was made. The spleen was located, the vessels ligated, and the organ removed. Buprenorphine and flunixin meglumine were given for 2 and 4 days after surgery respectively. Body weight, attitude, rectal temperature, blood glucose, and PCV were recorded daily. In addition, a peripheral blood smear was examined daily and the percent of red blood cells that were infected with Mycoplasma organisms was determined.

The alpaca was not parasitemic prior to surgery. One percent of the RBC's contained mycoplasma on days 2 and 3 after splenectomy. Parasitic bloom developed on day 4 with 85% of the red blood cells infected, and over 70% containing 3 or more organisms. The alpaca was treated with 20 mg/kg Oxytetracycline I.V. on day 4. On post-operative day 5 no parasites were seen in the peripheral blood. The peripheral blood remained free of parasites for 11 days. On the morning of the 12th day, 3% of the peripheral red blood cells contained Mycoplasma. By late that afternoon, 75% of the observed RBC's contained 3–4 organisms. The alpaca again received oxytetracycline. There were no more parasites observed from that time until the alpaca was euthanized 5 days later.

The alpaca lost 3.7 kg between days −1 and +2 after surgery. His weight fluctuated between 34.9 and 35.4 kg for the remainder of the study period. Blood glucose ranged between 87 and 163 mg/dL There was no major change in PCV (range 21–29%), a finding that was expected as the spleen was not available to remove infected red blood cells. Body temperature ranged between 38.1 and 39 degrees Celsius except for days 4 and 16 when more than 70% of red blood cells contained parasites. On those days rectal temperature reached 39.4 and 39.1 degrees respectively.

This study confirmed that a non-parasitemic, yet PCR positive alpaca did indeed harbor *C*Mhl. The time from splenectomy to parasitic bloom was shorter, and the length of oxytetracycline suppression longer than has been observed in other species.

ACVIM FORUM SESSION 145B
TRANSABDOMINAL ULTRASONOGRAPHIC GASTRO-INTESTINAL WALL THICKNESS IN ADULT CATTLE, SHEEP AND GOATS.

GD Hallowell[1] IM Bowen[1] and TJ Potter[2]. [1]School of Veterinary Medicine and Science, University of Nottingham, UK [2]Westpoint Veterinary Group, West Sussex, UK.

Gastro-intestinal (GI) disease frequently results in increased wall thickness in many species. Identification of changes in GI wall thickness using ultrasound has proved to be a useful diagnostic tool and is widely used in human patients, small animals and horses. Although GI motility has been evaluated in cattle, normal reference ranges for wall thickness has not been reported in ruminants. The aims of this study were to report normal values for wall thickness of various GI structures and to assess the repeatability of this technique in adult dairy cows, sheep and goats. Eight healthy adult Holstein Friesian (HF) cattle (656 ± 11 Kg), eight Jersey (J) cattle (458 ± 56 kg), thirteen adult sheep (79 ± 11 Kg) and eleven adult goats (43.5 ± 8 Kg) were recruited and examined on three consecutive days. Ultrasonographic images were optimised for the structure of interest. Structures were identified based upon appearance and anatomical position. A minimum of three cineloops were obtained of the abdominal organs per intercostal space (ICS) and three along the ventral midline in each ICS; images were analysed offline. Data were analysed using ANOVA and post-hoc Bonferoni, Student's T-test and intra-class correlation coefficients. Each structure was measured per ICS per species; if no differences were noted for structures in different ICS, then measurements were pooled. No differences were noted between HF and J cattle so data were pooled. Data are displayed in table 1. Good repeatability (ICC > 0.91) was obtained for all measurements and no differences were noted between animals of the same species or between days. These measurements for assessment of normal GI thickness are repeatable and may allow valuable additional information to be gained from ruminants with GI disease.

Table [ACVIM Forum Session 145B]: Transabdominal ultrasonographic GI measurements in ruminants. *represents differences ($p < 0.05$) in wall thickness when compared with cattle and ø if different from goats.

	Reticulum (cm)	Ruminal blind sac (cm)	Rumen (dorsal and ventral; cm)	Ruminal groove (cm)	Omasum (cm)	Abomasum (cm)	Small intestine (cm)	Caecum (cm)
Cattle	0.33±0.08	0.38±0.07ø	0.41±0.09ψ	1.45±0.52ψ	0.41±0.07ψ	0.27±0.05ψ	0.25±0.06	0.31±0.03ψ
Goats	0.32±0.04	0.32±0.03*	0.30±0.10*	0.67±0.20*	0.31±0.04*	0.21±0.05*	0.22±0.04	0.26±0.06*
Sheep	0.26±0.06*ψ	0.24±0.11*ψ	0.29±0.11*	0.85±0.26*ψ	0.30±0.05*	0.21±0.04*	0.23±0.04	0.26±0.05*

ACVIM FORUM SESSION 146A
ASSESSMENT OF *BDELLOVIBRIO BACTERIOVORUS* 109J VIABILITY IN BOVINE TEARS.

MJ Boileau, KD Clinkenbeard. Oklahoma State University, Stillwater, OK.

Ocular infections with the infectious bovine keratoconjunctivitis (IBK) agent *Moraxella bovis* (*M. bovis*) are associated with significant economic loss in the cattle industry.Although antibiotic therapy is the treatment of choice for IBK, treatment failures are common and current vaccines are not optimally effective mainly due to antigenic variation. As a result, our laboratory has been actively investigating the therapeutic potential of *Bdellovibrio bacteriovorus* 109J (*B. bacteriovorus*); a predatory bacterium capable of attacking and inducing lysis of Gram-negative bacteria, as a new treatment for IBK.

We have previously shown that *B. bacteriovorus* can reduce the number of *M. bovis* attached to bovine epithelial cells in an *in vitro* model of IBK and that *B. bacteriovorus* can be trained to kill *M. bovis* as effectively as *E. coli* using serial passages. In this study, we hypothesized that *B. bacteriovorus* can remain viable in bovine tears without its prey for up to 24 hours. This hypothesis was addressed by incubating inocula of active *B. bacteriovorus* in its preferred media peptone yeast extract (PYE) and comparing *B. bacteriovorus* viability in bovine tears or phosphate buffered saline (PBS) at time 0, 2, 4, 12, and 24 hours. Using a plaque assay to quantify the mean amount of plaque forming units (PFUs) of *B. bacteriovorus* exposed to each treatment, it was determined that viability of *B. bacteriovorus* over time was comparable between treatment groups.

Overall, the results supported that *B. bacteriovorus* can remain viable in tears for up to 24 hours in the absence of prey bacteria. Further studies are needed to determine the therapeutic potential of *B. bacteriovorus* in an *in vivo* model of IBK.

Table 1 [ACVIM Forum Session 146A]. Comparison of *B. bacteriovorus* (*Bb*) viability in PYE and bovine tears over time.

Time (hour)	Bb in PYE ($\times 10^9$ PFU/mL)	Bb in tears ($\times 10^9$ PFU/mL)
0	5.2 ± 1.89	4.3 ± 1.04
2	4.7 ± 1.52	4.7 ± 1.04
4	3.3 ± 0.29	3.8 ± 1.6

Table 2 [ACVIM Forum Session 146A]: Comparison of *B. bacteriorovorus* viability in PYE and PBS over time.

Time (hour)	Bb in PYE ($\times 10^9$ PFU/mL)	Bb in PBS ($\times 10^9$ PFU/mL)
0	5.8 ± 4.93	5.6 ± 3.32
2	5.3 ± 2.47	4.2 ± 0.57
4	4.8 ± 1.44	3.3 ± 1.53

Table 3 [ACVIM Forum Session 146A]: Comparison of *B. bacteriovorus* viability in PYE and bovine tears over a 24-hour period

Time (hour)	*Bb* in PYE ($\times 10^9$ PFU/mL)	*Bb* in tears ($\times 10^9$ PFU/mL)
0	5.1 ± 0.65	5.1 ± 2.10
12	2.6 ± 1.78	2.7 ± 0.70
24	$5.7. \pm 0.55$	6.3 ± 0.74

ACVIM FORUM SESSION 146B

HYPERCHLOREMIA DIRECTLY INCREASES IONIZED CALCIUM CONCENTRATION IN BOVINE PLASMA: AN ADDITIONAL MECHANISM BY WHICH HIGH CHLORIDE RATIONS MITIGATE PERIPARTURIENT HYPOCALCEMIA.

PD Constable[1], HR Stämpfli[2]. [1]Department of Veterinary Clinical Sciences, Purdue University, West Lafayette, IN, USA;[2]Department of Clinical Studies, University of Guelph, Guelph, ON, Canada.

Correction of the measured ionized calcium concentration (cCa^{2+}) to a pH = 7.40 is routinely applied in experimental studies in order to assist in the interpretation of measured values relative to a reference range. The equation most commonly used for pH correction in bovine plasma is: $cCa^{2+}_{pH = 7.40} = cCa^{2+} \times 10^{(-0.24 \times (7.40 - pH))}$. The validity of this equation for bovine plasma is unknown. Accordingly, our first objective was to characterize the in vitro relationship between cCa^{2+} and pH for bovine plasma. Feeding rations with a low dietary cation-anion difference (DCAD) during late gestation mitigates periparturient hypocalcemia in dairy cows, particularly when chloride containing acidodgenic salts are fed. The mechanism for this beneficial effect remains unclear. Our second objective was to determine whether hyperchloremia displaces calcium from binding sites to albumin, thereby increasing cCa^{2+}.

The *in vitro* relationship between plasma log(cCa^{2+}) and pH in was investigated using lithium heparin anticoagulated blood from 10 healthy Holstein-Friesian calves. Plasma was harvested and tonometered with CO_2 at 37°C over a pH range of 7.10–7.70. Plasma chloride concentration (cCl^-) was altered by equivolume dilution of plasma with 3 electrolyte solutions of varying cCl^- (97 ± 1, 110 ± 1, and 123 ± 1 mEq/L; mean $\pm$ SD).

The slope of the linear regression equation relating log(cCa^{2+}) to pH for 112 tonometered plasma samples from the 10 calves was -0.24 ± 0.05 at normal values for cCa^{2+} (2.63 ± 0.09 mEq/L), albumin concentration (30.9 ± 2.0 g/L), and cCl^- (99.1 ± 2.2 mEq/L). The experimentally-determined value for the slope for bovine plasma was identical to that determined previously for human plasma. The formula for correcting cCa^{2+} in bovine plasma for change in pH from 7.40 is therefore: $cCa^{2+}_{pH = 7.40} = cCa^{2+} \times 10^{(-0.24 \times (7.40 - pH))}$. This equation is only valid at normal concentrations of albumin and chloride in plasma. Equivolume dilution of plasma by electrolyte solutions of varying cCl^- indicated that $cCa^{2+}_{pH = 7.40}$ increased by 0.007 mEq for every 1 mEq/L increase in cCl^-. In other words, plasma cCa^{2+} at a given pH increases directly in response to an increase in plasma cCl^-, presumably because the additional chloride displaces calcium that is electrostatically bound to albumin. Furthermore, the increase in cCa^{2+} is independent to the change in plasma pH induced by an increase in cCl^- and decrease in plasma strong ion difference.

Our finding that hyperchloremia directly increases plasma cCa^{2+} provides an additional mechanism by which ingestion of high chloride (acidogenic) rations prevents the clinical signs of periparturient paresis. Our finding is consistent with the results of other studies that indicate acidogenic salts that contain chloride as the predominant anion (ie, NH_4Cl, $CaCl_2$) are more effective in increasing cCa^{2+} than equimolar quantities of acidogenic salts such as $MgSO_4$.

ACVIM FORUM SESSION 147A

A NEW LOOK AT THE IMPORTANCE OF COAGULASE NEGATIVE STAPHYLOCOCCI IN BOVINE MASTITIS.

JR Middleton[1], J Perry[1], RN Zadoks[2], S Dufour[3], ME Paradis[3], DT Scholl[3]. [1]University of Missouri, Columbia, MO. [2]Moredun Research Institute, Edinburgh, Scotland. [3]University of Montreal, Quebec, Canada.

Coagulase negative staphylococci (CNS) are among the most common bacteria isolated from the bovine mammary gland. Historically, these bacteria were lumped together as minor mastitis pathogens. Modern molecular techniques have allowed accurate speciation and fingerprinting of the CNS species. These methodologies have recently been applied to the study of CNS in bovine mastitis. The aim of the studies presented here was to evaluate the role of individual CNS species on milk somatic cell count (SCC) and duration of intramammary infection (IMI).

In the first study, mammary quarter foremilk samples were aseptically collected from all lactating cattle (~180 head) at the University of Missouri dairy research center once monthly for 17 months for bacterial culture and milk SCC. Staphylococcal isolates were speciated by sequencing the *rpo*B gene and strain-typed using pulsed-field gel electrophoresis (PFGE). Using species and fingerprint data along with published definitions for staphylococcal IMI, 91 CNS IMIs were identified. Overall, 11 species of CNS were identified with *Staphylococcus chromogenes*, *S. cohnii*, *S. epidermidis*, and *S. simulans* being most prevalent. Duration of IMI and SCC data were analyzed using regression models accounting for repeated measures. Mean milk SCC and duration of IMI were found to differ between CNS species ($P < 0.05$). Although most IMIs were of short duration (1 month), *Staphylococcus capitis* and *S. chromogenes* IMIs had longer mean durations of infection than 3 or more of the other species isolated. Mean SCCs were under 350,000 cells/ml in most cases. However, *Staphylococcus simulans* and *S. xylosus* IMIs were more inflammatory (mean > 600,000 cells/ml) and had a higher mean SCC than *S. cohnii*, *S. epidermidis*, and *S. haemolyticus*.

To examine the relationship between CNS IMI and milk SCC in a larger population of cattle, CNS isolates from the Canadian Bovine Mastitis Research Network (CBMRN) culture collection were obtained for speciation. Speciation and fingerprinting were performed as above. Isolates were from subclinical IMI from before and after the dry period and from subclinical IMI during lactation. Data associated with each isolate were obtained from the CBMRN database. Nine-hundred-thirty-eight isolates from 696 mammary quarters in 89 herds were successfully speciated. Twenty-two different species of CNS were identified. *Staphylococcus chromogenes* was the most frequent species identified accounting for 40% of the infections. Three species, *S. chromogenes*, *S. xylosus*, and *S. simulans* accounted for >75% of all infections. Data were analyzed using a linear hierarchical repeated measures mixed model. Differences in mean SCC were found between some CNS species and culture negative control quarters and also between different species of CNS ($P < 0.05$). Overall, our data demonstrate potential differences in pathogenicity between strains of CNS that cause bovine mastitis.

ACVIM FORUM SESSION 147B

REDUCTION OF BACTERIAL CONTAMINATION OF BOVINE COLOSTRUM WITH HIGH PRESSURE PROCESSING.

KP Poulsen[1], DM Foster[2], BE Farkas[3], SM McGuirk[1], GW Smith[2], CJ Czuprynski[1]. [1]University of Wisconsin School of Veterinary Medicine, Madison, WI. [2]North Carolina State University College of Veterinary Medicine, Raleigh, NC. [3]North Carolina State University Department of Food, Bioprocessing & Nutrition Sciences.

Passive transfer of maternally derived antibodies via ingestion of good-quality colostrum within the first 24 hours of life is crucial for the health and future productivity of dairy calves. However, infectious diseases can be transmitted via colostrum feeding, which may require use of a colostrum replacement product or pasteurization to decrease disease transmission. While pasteurization of colostrum is effective for sterilization, heating during pasteurization can alter the viscosity of colostrum, destroys important nutritional biomolecules, and has been shown to decrease colostral IgG concentrations. The purpose of this study was to investigate the effect of high pressure processing (HPP) on the viscosity, IgG concentration, and bacterial contamination of bovine colostrum.

First milking colostrum samples were collected from 18 cows from 3 different farms, and 50 ml aliquots of each sample were pooled for analysis. Pooled colostrum was processed in triplicate using an isostatic press at 400 MPa (60,000 psig) for 0, 5, 15, 30, and 45 minutes. Samples were tested for the effects of HPP on the viscosity, bacterial load (CFU/ml), and IgG concentration.

There was a significant decrease ($p < 0.05$) in bacterial load at each time point when compared to time 0. No significant difference in IgG concentration was found between any time points. Subjectively, the colostrum viscosity appeared to increase with the processing time, though the rheologic assessment has not been completed at this time.

HPP appears to be an effective method to decrease bacterial contamination of colostrum while maintaining appropriate IgG concentrations. Minimizing the processing time or pressure may be necessary to maintain an acceptable viscosity of the colostrum. Based on these results, additional studies are justified in order to determine the optimum combination of processing time and pressure and the effect of HPP on specific bovine pathogens.

ACVIM FORUM SESSION 148A
TBA

ACVIM FORUM SESSION 148B
LACTOFERRIN: A FURTHER INVESTIGATION INTO ITS ANTI-INFLAMMATORY EFFECTS; INTRACELLULAR SIGNALING IN EVIDENCE? Maisie E Dawes. Western University of Health Sciences College of Veterinary Medicine, Pomona, CA.

The heme-associated iron-binding apoprotein lactoferrin (LF) is known for its, anti-inflammatory, anti-parasitic, antimicrobial and bactericidal effects. Lactoferrin demonstrates ubiquity throughout mammalian host biological fluids: Saliva; tears; mammary secretions, as well as at mucosal surfaces. It is also released from immune cells under pathogenic stimulation. The purpose of this study which has been approved by Western University's Institutional Animal Care and Use Committee is to further characterize the mechanisms through which LF modulates inflammation in the face of bacterial endotoxin. It was hypothesized that LF would inhibit p38 phosphorylation. Numerous studies speak to the ability of LF to alter leukocyte function, inhibit cytokine production, and bind lipopolysaccharide (LPS); mechanisms through which it is believed to achieve its anti-inflammatory effects. Recently, investigators demonstrated its ability to interact with host DNA while others describe regulation of granulocyte adhesion and motility; elucidating its roles in the apoptotic signaling.

In earlier studies, Dawes ME, et al. demonstrated lactoferrin's ability to limit the expression of inducible cyclooxygenase-2 and the gelatinase, matrix metalloproteinase -9 by LPS-induced macrophages. The generation of these inflammatory mediators is modulated by pro- inflammatory cytokines such as interleukin-1 β (IL-1 β) and tumor necrosis factor-alpha (TNF-α), the production of both being dependent on signaling through the p38 mitogen-activated protein kinase (MAPK) pathway.

Peripheral mononuclear cells (5×10^6)isolated from buffy coat cells of healthy neonatal to 4-month old Holstein calves were cultured in the presence and absence of LF (200 ng/ml), LPS (1 μg/ml), anisomycin (25 μg/ml), a known p38 activator - the positive control, and 50 mM of SB203580, a known p38 inhibitor - the negative control. Sample lysates obtained post culture was subjected to immunoprecipitation and kinase reactions. Reactions were terminated under reducing conditions and evaluated using western immunoblotting. Phosphorylation of activated transcription factor-2 (ATF-2) by phosphorylated p38 served as the marker of investigation. Immunologically reactive ATF-2 expression by LPS and anisomycin-treated cells was compatible with a prominent band at 40 kD. Evidence of LF-induced inhibition of LPS-induced p-38 activation was observed in lanes representative of co-cultures of LF + LPS; LF + anisomycin; and anisomycin + SB203580, which was demonstrated by decreased immunological reactivity at 40kD. The findings here, suggest that LF interferes with LPS-induced p-38 activation of transcription factor ATF-2, *in vitro*. This serves as additional proof of its potential use in attenuating the systemic effects of LPS.

ACVIM FORUM SESSION 153A
KINETIC ASSESSMENT OF MYOCARDIAL 14(R,S)-[^{18}F]FLUORO-6-THIA-HEPTADECANOIC ACID AND ^{18}F-FLUORODEOXYGLUCOSE UPTAKE WITH POSITRON EMISSION TOMOGRAPHY/COMPUTED TOMOGRAPHY (PET/CT) IN DOMESTIC CATS. AK LeBlanc [1,2], MR Akula[2], EB Martin[2], RE Gompf[1], GD Galyon[1], TD Moyers[1], AC Stuckey[2], MJ Long[2], SJ Kennel, GW Kabalka, MC Besozzi[2,3], JS Wall[2]. [1]University of Tennessee College of Veterinary Medicine; [2]University of Tennessee Graduate School of Medicine, Molecular Imaging and Translational Research Program; [3]University of Tennessee Medical Center; Knoxville, TN.

Fatty acid oxidation (FAO) is an important myocardial biochemical pathway with therapeutic and diagnostic applications. The domestic cat provides a robust and relevant model to study myocardial fatty acid metabolism with 14(R,S)-[^{18}F]Fluoro-6-thia-heptadecanoic acid ([^{18}F]FTHA) as a powerful tracer for cardiac PET/CT imaging.

Six (6) clinically normal, purpose-bred cats of similar age and body condition were imaged with [^{18}F] fluorodeoxyglucose ([^{18}F]FDG) and [^{18}F]FTHA by using dynamic cardiac-gated fused PET/CT for kinetic assessment of myocardial glucose and fatty acid uptake and metabolism, respectively. Kinetic tracer uptake within the myocardium was achieved by initiating image data acquisition simultaneously with tracer injection. PET data were acquired over a 1 hour period with the heart in the center of the scanner field of view.

Regions of interest were drawn in the left ventricular wall and thoracic aorta for the purpose of measuring the kinetics of tracer redistribution. Serial blood samples were also taken during PET imaging for comparison with image data. The equilibrium biodistribution of both tracers was documented 1 hour post-injection in a whole body PET/CT image. Standard echocardiographic examination of cardiac structures was also performed.

Both radiotracers remained in the plasma fraction; however, [^{18}F]FTHA was cleared from the more rapidly than [^{18}F]FDG ($t_{1/2}$ ~ 2 and ~ 20 min, respectively). The tracers were readily visualized within the feline myocardium in dynamic PET images and analysis of the blood pool clearance from the kinetic image data agreed with blood sampling data. Myocardial uptake of each tracer was best described by a double exponential analysis and was rapid but variable among animals (range 1–30 Bq/cc/min), although blood glucose levels were similar in all cats during image acquisition. Physiologic [^{18}F]FDG was observed in the brain, salivary tissue, gastrointestinal tract, renal pelves and urinary bladder, with [^{18}F]FTHA seen in the myocardium, liver and renal cortex. All cats were normotensive with normal echocardiographic parameters.

This study demonstrates the utility of kinetic imaging using [^{18}F]FDG and [^{18}F]FTHA to study feline myocardial metabolism in the fasted state using PET/CT imaging. This work further provides validation of [^{18}F]FTHAin a relevant *in vivo* model of myocardial metabolism to support future clinical trials in animals with myocardial disease. Application of [^{18}F]FTHA uptake to the study of feline hepatic and renal disease should also be considered.

ACVIM FORUM SESSION 153B

ASSESSMENT OF GLOBAL AND REGIONAL LEFT VENTRICULAR VOLUMES AND GEOMETRY USING REAL-TIME THREE-DIMENSIONAL ECHOCARDIOGRAPHY IN DOGS WITH MYXOMATOUS MITRAL VALVE DISEASE.
I. Ljungvall[1], K. Höglund[1], C. Carnabuci[1], A. Tidholm [2], J. Häggström[1]. [1]Swedish University of Agricultural Sciences, Uppsala, Sweden. [2]Albano Animal Hospital, Danderyd, Sweden.

The left ventricle (LV) shape has been suggested to change from elliptical to more globular in response to chronic volume overload. Real-time three-dimensional echocardiography (RT3DE) offers new modalities for LV assessment. The aim of the study was to investigate left ventricular changes in shape and volume occurring in response to different severities of naturally acquired myxomatous mitral valve disease (MMVD) in dogs using RT3DE.

Privately owned dogs were classified by standard echocardiography into: Healthy (20), mild (20), moderate (8) and severe MMVD (17). A LV cast was obtained using semi-automated endocardial border tracking from RT3DE dataset, from which global and regional (automatically acquired basal, mid, and apical segments based on LV long-axis dimension) end-diastolic (EDV) and end-systolic volumes (ESV), LV Long-axis dimension and RT3DE Sphericity index, were derived.

Global and regional EDV and ESV increased significantly with increasing MMVD severity, assessed by MMVD group-wise comparisons and linear regression analyses using left atrial to aortic root ratio, and LV end-diastolic and end-systolic dimensions. All three segments contributed to the overall increased global volumes, but the mid EDV segment was strongest associated with increasing LV end diastolic dimension ($P = 0.048$). Furthermore, LV Long axis distance and LV Sphericity index increased with increasing MMVD severity. The basal and apical EDV segments were strongest associated with Sphericity index ($P < 0.0001$).

In conclusion, this RT3DE study showed that increased LV EDV, primarily in the mid segment, leads to rounding of LV apical and basal segments in response to increasing MMVD severity in dogs.

ACVIM FORUM SESSION 154A

HEARTWORM SEX RATIOS IN NATURALLY INFECTED DOGS DIFFER WITH WORM INTENSITIES.
M. Rishniw[1], Y. Schukken[1], E.Greiner[2]. [1]Cornell University, Ithaca, NY. [2]University of Florida, Gainesville, FL.

Male:female (M:F) ratios in various biological systems often differ from 1. Determinants of distorted M:F include genetic, epigenetic and environmental factors. Wolbachia infection can distort sex ratios of nematodes. Whether *Dirofilaria immitis* (DI) maintain equal M:F in naturally infected dogs is unknown.

84 dogs from shelters in Florida with naturally acquired DI infection were euthanized and necropsied. All adult DI in each dog were sexed using morphological features. Total worm burdens and numbers of males and females were recorded. No other information was available for any dog. All data, raw and transformed, were examined visually and descriptively. Raw numerical data were further examined by a paired T-test; log-odds transformed data were examined by logistic regression. We also conducted a binomial distribution goodness of fit analysis assuming a null hypothesis of a M:F = 1.0.

Worm intensities ranged from 1 to 143 DI per dog. Eight dogs had unisex infections: 7/8 had all-female infections. Dogs with low-intensity dual-sex infections were more likely to have greater numbers of female DI. Overall, sex ratios were equal (paired t-test, P = 0.7). However, logistic regression demonstrated that the probability of being female is strongly affected by the total worm intensity, with lower intensities increasing the probability of having a predominance of female worms.

Our data show that DI sex ratios in naturally-infected dogs equal 1 when examining the entire dog population, but deviate to favor female worms at low worm intensities. These data could impact adulticide treatment strategies. The reasons for sex ratio distortion in DI are unknown.

ACVIM FORUM SESSION 157A

REVERSE REMODELING AFTER MITRAL VALVE REPAIR UNDER CARDIOPUMONARY BYPASS IN DOGS.
Masami Uechi. Nihon University, Veterinary Cardiovascular Medicine and Surgery, Kanagawa, Japan.

We evaluated cardiac reverse remodeling after mitral valve repair under cardiopulmonary bypass (CPB) for mitral regurgitation in small breed dogs. Fifty dogs (body weight 1.8–9.3 kg, age 5–14 years) with mitral regurgitation were treated between August 2006 and November 2009. The cardiac murmur was grade 4/6–6/6. The preoperative chest X-rays showed cardiac enlargement (vertebral heart scale (VHS) 11.0–13.1). Echocardiography showed severe mitral regurgitation and left atrium enlargement (LA/Ao 2.0–4.2). After inducing anesthesia, a thoracotomy was performed in the fifth intercostal space. CPB was started by using a CPB circuit connected to carotid artery and jugular vein catheters. After inducing cardiac arrest, the left atrium was sectioned and chordae tendineae rupture confirmed. The chordae tendineae were replaced with expanded polytetrafluoroethylene. A mitral annulus plasty was also done, and the left atrium was closed. After de-clamping for restarting the heart, the chest was closed. Heart rate decreased from 118–164 bpm to 75–138 bpm. The grade of cardiac murmur was reduced to 0/6–3/6 three months postoperatively, and the heart shadow was reduced (VHS 9.8–11.5) in the chest x-rays. Echocardiography confirmed the marked reduction in mitral regurgitation and the left atrial dimensions (LA/Ao 1.2–2.2). Mitral valve repair reduced enlarged cardiac size by reduction of regurgitant rate.

ACVIM FORUM SESSION 157B

EFFECT OF DICHLOROACETATE IN CANINE PULMONARY ARTERIAL HYPERTENSION: A PILOT STUDY.
HW Maisenbacher III[1], PW Stacpoole[2], AH Estrada[1], BJ Pogue[1]. [1]University of Florida, College of Veterinary Medicine, Gainesville, FL. [2]University of Florida, College of Medicine, Gainesville, FL.

Pulmonary arterial hypertension (PAH) is a well recognized condition in dogs leading to considerable morbidity and mortality. The majority of therapeutics has focused on endothelial dysfunction causing reduced production of vasodilators, such as nitric oxide and prostacyclin, coupled with overproduction of vasoconstrictors, such as endothelin-1. More recently, it has been shown that the mitochondria play an important role in the development of PAH as oxygen sensors and regulators of cellular proliferation. In PAH, pulmonary artery smooth muscle cells undergo a metabolic shift from oxidative phosphorylation in the mitochondria to glycolysis in the cytoplasm as the major energy source and this leads to suppression of apoptosis and increased proliferation. Dichloroacetate (DCA) inhibits pyruvate dehydrogenase kinase to activate pyruvate dehydrogenase which catalyzes the rate limiting step for entry of pyruvate into the Krebs cycle, thus increasing mitochondrial respiration. In three different rat models of PAH, DCA has been shown prevent and reverse PAH by normalizing molecular pathology, stimulating apoptosis of pulmonary artery smooth muscle cells, and reducing pulmonary artery hypertrophy. DCA has known toxic effects, including reversible hepatotoxicity and peripheral neuropathy, and has not been studied in any species with naturally occurring PAH. The objective of this open label pilot study is to evaluate the therapeutic and toxic effects of DCA in naturally occurring canine PAH.

Three dogs with PAH diagnosed by Doppler echocardiography and no correctable underlying cause are enrolled in the study. Dogs are orally administered 25 mg/kg of DCA divided daily for 2 weeks, and then 12.5 mg/kg of DCA divided daily for the remainder of the study. At baseline, 2, 4, 8, and 12 weeks, an echocardiogram, CBC, serum chemistry profile, urinalysis, NT-proBNP, blood uric acid, blood lactate, noninvasive blood pressure, nerve conduction study, and trough DCA level (12 hr post-dose) are obtained. The measured echocardiographic parameters include peak and mean tricuspid regurgitant flow velocity and pressure gradient, peak and end-diastolic pulmonary regurgitant flow velocity and pressure gradient, pulmonary valve flow velocity acceleration time and ejection time, pulmonary valve flow velocity time integral, right ventricular myocardial performance index, tricuspid annular plane systolic excursion, and systolic tricuspid annular tissue velocity. Variables are inspected for normalcy and equality of variances and a two-sided paired t-test is used to compare the variables before and after treatment at each evaluation time.

The basis for the role of the mitochondria in PAH and the results of this pilot study will be presented to determine if DCA warrants further study as a therapy for dogs with PAH.

ACVIM FORUM SESSION 161A

A DELETION IN *ARHGEF10* IS HIGHLY ASSOCIATED WITH EARLY ONSET INHERITED POLYNEUROPATHY IN LEONBERGER AND ST. BERNARD DOGS. G.D. Shelton[1], K.J. Ekenstedt[2], C. Drgemüller[3], K.M. Minor[4], D. Becker[3], T. Leeb[3], N. Patterson[2], J. Mickelson[4]. [1]Department of Pathology, School of Medicine, University of California-San Diego, La Jolla, CA, [2]Department of Veterinary Clinical Sciences, College of Veterinary Medicine, University of Minnesota, St. Paul, MN, [3]Institute of Genetics, University of Berne, Bern, Switzerland, [4]Department of Veterinary and Biomedical Sciences, University of Minnesota, St. Paul, MN.

Polyneuropathy in large breed dogs is a relatively common clinical problem for which the genetic basis is generally unknown. The first cases of polyneuropathy in the Leonberger breed (Leonberger polyneuropathy or LPN) were identified in 1999 by one of the authors (GDS) and a report published in 2003 (MuscleNerve 27:471–477). In this report a spontaneous, distal and symmetrical polyneuropathy with onset between 1 to 9 years of age was described and characterized clinically, electrophysiologically, histologically and morphometrically. There were striking similarities between LPN and the Charcot-Marie-Tooth group of human inherited sensory and motor polyneuropathies, which have many known genetic mutations.

A genome-wide case-control association study for LPN was performed with 53 cases and 42 controls on high-density 170K canine SNP arrays and revealed a significantly associated region on CFA 16 ($P_{raw} = 2.36 \times 10^{-10}$ $P_{genome} = 9.99 \times 10^{-5}$). A clear association of an approximately 1 Mb CFA16 haplotype with cases ($P = 1.71 \times 10^{-8}$) was observed, particularly with those cases that were affected more severely and at a younger age ($P = 2.55 \times 10^{-11}$). A positional candidate gene, *ARHGEF10*, which has previously been associated with peripheral nerve abnormalities in humans, was sequenced, revealing a deletion that results in a frame shift and premature stop codon. Of all Leonbergers with young onset LPN (before 4 years), 48.5% (32 of 66) have two copies of this deletion, and, of all young onset Leonbergers that are nerve biopsy positive for LPN, 59.4% (19 of 32) have two copies of this deletion. Importantly, nearly all dogs carrying two copies of the deletion (32 of 34 or 94.1%) are affected with LPN by the age of 4 years.

The Leonberger breed was generated from crossing several breeds, including the St. Bernard, and a polyneuropathy clinically and histologically similar to LPN occurs in this breed. To determine if the *ARHGEF10* mutation was associated with polyneuropathy in the St. Bernard, DNA was extracted from archived frozen muscle biopsy specimens from clinical cases (n = 3). The identical *ARHGEF10* mutation was identified in all tested St. Bernard dogs affected with early onset polyneuropathy suggesting the mutation may have been passed through the St. Bernard to the Leonberger.

ACVIM FORUM SESSION 161B

STARTLE DISEASE IN IRISH WOLFHOUNDS ASSOCIATED WITH A DELETION IN THE GENE ENCODING THE GLYCINE TRANSPORTER GLYT2. G.D. Shelton[1], J.L. Gill[2], D. Capper[3], R.J. Higgins[4], R.J. Harvey[2]. [1]Department of Pathology, School of Medicine, University of California-San Diego, La Jolla, CA, USA; [2]Department of Pharmacology, The School of Pharmacy, London, UK; [3]Halsey East Animal Clinic, Portland, OR, USA; [4]Department of Pathology, Microbiology and Immunology, School of Veterinary Medicine, University of California-Davis, Davis, CA, USA.

Startle disease or hyperekplexia is caused by defects in mammalian glycinergic neurotransmission resulting in an exaggerated startle reflex and extensor hypertonia triggered by noise or touch. In humans and animals, startle disease is typically caused by mutations in one of three genes (*GLRA1*, *GLRB*, and *SLC6A5*) encoding postsynaptic glycine receptor subunits (α1 and β) or a presynaptic glycine transporter (GlyT2). A litter of seven Irish Wolfhounds was recently identified in which two puppies developed muscle stiffness and tremor beginning at 5–7 days of age post-partum. Signs were dramatic when the puppies were handled and resolved when the puppies were relaxed or sleeping. Both puppies were euthanized due to ongoing stiffness, tremor and breathing difficulties. Necropsies were performed, but no microscopic pathological abnormalities were identified in the peripheral or central nervous system.

Based on the clinical signs, exons from the three candidate genes were amplified from genomic DNA isolated using PCR and directly sequenced. No deleterious polymorphisms were identified in either *GLRA1* or *GLRB*. However, difficulties were experienced in amplifying *SLC6A5* exons 2 and 3 from affected animals, although control samples were positive, suggesting that the PCR primer designs and conditions were not at fault. Further PCRs revealed that the reason for this anomaly was the presence of a homozygous 4.2 kb deletion encompassing exons 2 and 3 of the GlyT2 gene in both affected animals. This deletion is predicted to result in the loss of part of the large cytoplasmic N-terminus that is vital for trafficking of GlyT2 to synaptic sites, and a loss of all subsequent transmembrane domains via a frameshift. This genetic lesion was confirmed by defining the deletion breakpoint, Southern blotting and Multiplex Ligation-dependent Probe Amplification (MLPA). This analysis enabled the development of a rapid genotyping test that revealed heterozygosity for the deletion in the dam and sire and three other siblings, suggesting recessive inheritance of this disorder. Wider testing of related animals has identified a total of 18 carriers of the *SLC6A5* deletion and enabled the identification of non-carrier animals to guide future breeding strategies.

ACVIM FORUM SESSION 162A

A MUTATION IN ATP13A2 CAUSES ADULT-ONSET NEURONAL CEROID LIPOFUSCINOSIS IN TIBETAN TERRIERS. Fabiana H.G. Farias, Gary S. Johnson, Dennis P. O'Brien, Fred A. Wininger, Rong Zeng, Stephanie D. McKay, Robert D. Schnabel, Jeremy F. Taylor, Claire Wade, Kerstin Lindblad-Toh, Hannes Lohi, Corinne S. Mann, and Martin L. Katz. From the Department of Veterinary Pathobiology (Farias, Johnson, Zeng, Katz), Division of Animal Sciences (McKay, Schnabel, Taylor), Mason Eye Institute (Mann, Katz) and Department of Veterinary Medicine and Surgery (O'Brien, Coates), University of Missouri, Columbia, Missouri. Faculty of Veterinary Science, The University of Sydney, NSW 2006 Australia (Wade). Department of Medical Biochemistry and Microbiology, Uppsala University, Uppsala Sweden (Lindblad-Toh). Department of Molecular Genetics, the Folkhälsan Institute of Genetics, Finland. (Lohi).

The neuronal ceroid lipofuscinoses (NCLs) are a group of neurodegenerative lysosomal storage diseases defined by the lysosomal accumulation of autofluorescent proteinaceous material with characteristic ultrastructural appearances. This study was performed to identify the mutation responsible for the adult-onset NCL in Tibetan Terriers. A 19-case $\times$ 15-control genome wide association study produced the strongest associations between the NCL phenotype and CFA2 markers. All 19 NCL-affected Tibetan Terriers were homozygous for the same haplotype which extended for 22 consecutive SNPs spanning 0.95 Mb. None of the 20 annotated genes within this target region had previously been associated with human or rodent NCL. We used DNA from NCL-affected Tibetan Terriers to resequence the coding regions and intron-exon borders of several genes harbored within the target region and found a single base pair deletion, c.1623delG, in exon 16 of positional candidate ATP13A2. This deletion produces a frame shift and a predicted premature termination codon. We genotyped all 454 Tibetan Terrier DNA samples in our collection and found all 45 NCL-affected Tibetan Terriers to be homozygous for the c.1623delG allele. Eleven additional c.1623delG homozygotes were either less than 5 years old, or lost to follow up. There were no known cases of NCL in the remaining 398 Tibetan Terriers which were either heterozygous (n = 149) or homozygous for the ancestral allele (n = 249). ATP13A2 is a member of group of ion transport genes and has been associated with lysosomes. Mutations in human ATP13A2 cause Kufor-Rakeb syndrome (KRS), a rare neurodegenerative disorder with clinical features that include parkinsonism plus spasticity, supranuclear upgaze paresis, and dementia. Post-mortem findings in KRS have not been reported. We conclude that NCL in Tibetan terriers is caused by a mutation in ATP13A2. Our results suggest that KRS may be a form of adult onset NCL in humans.

ACVIM FORUM SESSION 162B

INTRATHECAL VERSUS SUBCUTANEOUS THERAPY OF FELINE NIEMANN-PICK TYPE C DISEASE.

C.H. Vite,[1] E. Mauldin,[1] S. Ward,[1] V. Stein,[2] M.E. Haskins,[1] R. Stratton,[3] L.M. Kao,[4] D. Ory,[5] S.U. Walkley,[6] M.T. Vanier[7]. [1]University of Pennsylvania, Philadelphia, PA. [2]Stiftung Tierärztliche Hochschule, Hannover, Germany. [3]Cyclodextrin Technologies Development, Inc, High Springs, FL. [4]Johnson and Johnson Pharmaceutical Research and Development, Raritan, NJ. [5]Washington University, St Louis, MO. [6]Albert Einstein College of Medicine, Bronx, NY. [7]Institut National de la Santé et de la Recherche Médicale (INSERM), Lyon, France.

Niemann-Pick type C (NPC) disease is a progressive neurological disorder characterized by dementia and ataxia, hepatic and pulmonary disease, and death typically within the first or second decade. Despite the identification of causative mutations, the pathogenesis is not clear and therapies to successfully treat NPC disease have been ineffective to date. The recent use of intravenously administered 2-hydroxypropyl-beta-cyclodextrin (HPBCD), an FDA-designated orphan drug (May 2010), in a small number of children with NPC disease is based on favorable treatment outcome data in subcutaneously treated mouse and cat models. To rigorously evaluate the mechanistic, pharmacologic, and toxicity issues associated with HPBCD therapy in NPC disease, we have utilized the spontaneous feline NPC model harboring a missense mutation in NPC1 (pC955S), orthologous to the most common mutation in juvenile-onset patients. The feline NPC model has clinical, neuropathological and biochemical abnormalities similar to those present in juvenile-onset patients making this model homologous to the most common disease form seen in human patients. We identified that intrathecal administration of HPBCD ameliorated all clinical aspects of neurological disease at least up to 24 weeks of age (an age when untreated cats die) but had no effect on hepatic disease. We identified that while subcutaneous therapy with HPBCD at all doses ameliorated liver disease, only 8000 mg/kg substantially affected neurological disease but also resulted in early death due to pulmonary toxicity. Finally, we identified a dose-related toxic effect of HPBCD on hearing function that had not been described in any other species.

ACVIM FORUM SESSION 163A

A NOVEL LEUKODYSTROPHY IN BORDER TERRIER PUPPIES – THE CANINE "RUMPSHAKER.".

Ronaldo C. da Costa,[1] Paula Martin-Vaquero,[1] Jessica K. Simmons,[1] GillianL. Beamer,[1] Karin Jaderlung,[2] Michael J. Oglesbee[1]. [1]College of Veterinary Medicine, The Ohio State University, Columbus, OH. [2]Norwegian School of Veterinary Science, Oslo, Norway.

Leukodystrophies are disorders of myelin synthesis and maintenance that affect CNS myelin. They are subdivided as leukodystrophies, hypomyelinating disorders and spongy degenerations. Although infrequently seen, several forms have been described in various dog breeds. We present a novel form of complex leukodystrophy consisting of hypomyelination and spongy degeneration that presents primarily with hind end tremors in Border terrier puppies.

Three Border terriers from two different litters (and lineages) are described here that presented with a history of shaking movements. The youngest dog was a 3-week old male. It was the only dog affected in the litter. The other two dogs were 6-week old female littermates. There were two unaffected males in the same litter. Physical examination revealed no abnormalities. On neurological examination, the affected dogs displayed severe hind end tremors, with a characteristic swinging side-to-side movement (best described as "rumpshaker"). The tremors also involved the head and thoracic limbs but to a lesser degree, and disappeared when the dogs were asleep or at rest. Severe cerebellar ataxia was observed when the dogs ambulated. Proprioceptive positioning was delayed in the pelvic limbs of all 3 dogs. Spinal reflexes and nociception appeared normal. Necropsy was performed in all 3 puppies. No macroscopic changes were observed. Histologic evaluation of the CNS revealed spongy degeneration and hypomyelination in all funiculi of the cervical and thoracic spinal cord. White matter of the frontal, temporal and parietal cortices had mild multifocal spongy degeneration and hypomyelination, whereas white matter of the cerebellum, medulla and pons showed severe diffuse spongy degeneration and hypomyelination with gliosis. The combination of reduced myelin formation combined with spongiform white matter changes in the absence of microglial responses suggest a complex pathogenesis affecting both oligodendrocytes' capacity to synthesize myelin and the stability of the myelin that was formed. The number of oligodendrocytes and axons appeared subjectively normal indicating a primarily hypomyelinating process.

The clinical and pathological features of this disease have not been described in any other canine leukodystrophy. The primary and most striking clinical feature is the presence of severe tremors in the hind end, causing the "rumpshaker" pheynotype. Genetic studies are underway to determine if the disease is inherited and the inheritance mode.

ACVIM FORUM SESSION 164B

EFFECTS OF A STANDARDIZED SHEEP HERDING STRENUOUS EXERCISE PROTOCOL ON CLINICAL AND LABORATORY PARAMETERS IN HEALTHY BORDER COLLIES AND DOGS WITH BORDER COLLIE COLLAPSE.

S.M. Taylor[1], C.L. Shmon[1], L. Su[1], K. Minor[2], E.E. Patterson[2], J.M. Mickelson[2], G.D. Shelton[3]. [1]Western College of Veterinary Medicine, Saskatoon, SK. [2]College of Veterinary Medicine University of Minnesota, St. Paul, MN. [3]School of Medicine, University of California-SD, La Jolla, CA.

A syndrome of Border collie collapse (BCC) appears to be common in dogs used for working stock. This syndrome has also been called malignant hyperthermia, heat intolerance, exercise-induced collapse and "wobbles". A presumptive diagnosis of BCC can only be made by eliminating other causes of exercise intolerance and weakness. The purpose of this study was to describe the clinical features of collapse in affected dogs and determine if there were characteristic clinical or laboratory features at rest or after exercise that could aid in diagnosis.

Seven adult Border collies with a history of collapse during sheep herding (affected) and 5 adult Border collies regularly used for sheep herding but showing no signs of exercise intolerance (normal) were evaluated before and after participating in a videotaped 10 minute exercise protocol consisting ofa series of continuous short outruns and fetches of three sheep in an outdoor pen. Exercise was halted at 10 minutes or earlier if there were signs of gait or mentation abnormalities. Pre-exercise evaluation included physical examination, orthopedic and neurological exam. Pre and immediate post exercise rectal temperature, pulse and respiration, patellar reflexes, ECG, CBC, serum biochemistry profile, cortisol, arterial blood gas and plasma lactate and pyruvate concentrations were measured. Clinical parameters (gait, temperature, reflexes) and lactate and pyruvate concentrations were evaluated at intervals up to 120 minutes after exercise. Additional testing in affected dogs included measurement of acetylcholine receptor antibodies (AChRab) and DNA testing for dynamin-associated exercise induced collapse (dEIC) and the ryanodine receptor mutation associated with canine malignant hyperthermia(MH). One week after exercise affected dogs had thoracic radiographs and echocardiography performed and were anesthetized for EMG and muscle biopsies.

There were no significant differences in temperature, pulse, respiration, or any laboratory parameter at any time point between normal and affected dogs. No arrhythmias were detected. Affected dogs were negative for the DNA mutations tested and for AChR ab. Thoracic radiographs, echocardiograms, EMGs and muscle biopsies were normal.

The 5 normal dogs had no alterations in mentation or gait during or after exercise. Three of the affected dogs had exercise halted early (6 min–9 min) because of altered gait or mentation. All 7 of the affected dogs were abnormal in the 15 minutes following exercise. Abnormalities seen in all dogs included disorientation, dull mentation, swaying, falling to the side, exaggerated lifting of limbs each step, choppy gait, delayed limb protraction, scuffing of rear and/or forelegs, and crossing legs when turning. All dogs returned to normal by 30 minutes.

BCC appears to be an episodic nervous system disorder that can be triggered by exercise. Genetic testing excluded dEIC and the described canine MH mutation. Common causes of exercise intolerance were eliminated, but the cause of collapse in BCC was not determined and no clinical or biochemical marker to aid diagnosis was established.

ACVIM FORUM SESSION 197A
DIAGNOSTIC FREQUENCY, RESPONSE TO THERAPY AND LONG TERM PROGNOSIS AMONG HORSES WITH EQUINE CUSHING'S DISEASE, 1993–2004.

Rohrbach BW[1], Stafford JR[2], Willis RS[1], Reed SM[3], Schott HC[4], Andrews FM[5]. [1]Department of Comparative Medicine, College of Veterinary Medicine, University of Tennessee, Knoxville, TN; [2]VCA Veterinary Referral Associates, Gaithersburg, MD; [3]Rood & Riddle Equine Hospital, Lexington, KY; and [4]Department of Large Animal Clinical Sciences, College of Veterinary Medicine, Michigan State University, East Lansing MI. [5]Equine Health Studies Program, Department of Veterinary Clinical Sciences, School of Veterinary Medicine, Louisiana State University, Baton Rouge, LA.

Equine Cushing's disease (ECD) is common in older horses. The purpose of this study was to determine the frequency of diagnosis, identify prognostic factors and assess owner satisfaction with treatment. The study was a retrospective cohort design evaluating equine accessions reported to the Veterinary Medical Data Base (VMDB) and the Ohio State University from 1993–2004. Proportional accessions, annual incidence and demographic characteristics of horses with ECD were compared with all accessions in the VMDB. Medical records for a subset of horses were extracted and owners contacted to obtain long-term follow up information.

Two hundred seventeen new cases of ECD were reported to the VMDB. Incidence increased from 0.25/1,000 in 1993 to 3.72/1,000 in 2002. Eighty-one percent of horses were $\geq$ 15 years of age. Average delay from onset of signs to diagnosis was 180 days (range 1 to 1,824 days). Hirsutism (84%) and laminitis (50%) were the most common clinical signs. Improvement in one or more signs 2 months after diagnosis was reported by 9/22 (41%) of horse owners. None of the clinical or laboratory data were associated with survival and, 50% of horses were alive, 4.6 years after diagnosis. 17/20 (85%) of horses were euthanized and 13/17 (76%) were euthanized due to conditions associated with ECD. Twenty-eight of 29 (97%) of horse owners said they would treat a second horse for ECD.

ECD is becoming a more frequent diagnosis. Fifty percent of horses survived 4.5 years after diagnosis and owners were satisfied with the horse's quality of life.

Supported by Centers of Excellence in Livestock Diseases and Human Health, College of Veterinary Medicine, University of Tennessee.

ACVIM FORUM SESSION 197B
ASSOCIATION OF ADRENOCORTICOTROPIN, CORTISOL AND INSULIN-LIKE GROWTH FACTOR-I IN SEPTIC AND NON-SEPTIC FOALS.

JR Gold[1], ND Cohen[1], TH Welsh[2], LC Caldwell[2] RL Belgrave[3]. [1]Texas A&M University, College of Veterinary Medicine, College Station, TX. [2]Texas A&M University, College Station, TX. [3]Mid-Atlantic Equine Hospital, Ringoes, NJ.

The role of the hypothalamic-pituitary-adrenal (HPA) axis in sepsis has been a subject of a great deal of research. The role that the somatogenic axis plays in sepsis is less well understood and how these two axes interact during critical illness is not clear. The purpose of this study was to assess inter-relationships of adrenocorticotropin (ACTH), cortisol, and insulin-like growth factor-I (IGF-I), in septic and non-septic term foals.

Blood samples were obtained from term septic foals less than 7 days of age (n = 20) admitted to Texas A&M University Veterinary Medical Teaching Hospital or Mid-Atlantic Equine Hospital. The foals were classified as septic by a sepsis score $\geq$ 11 and/ or a positive blood culture. Non-septic term foals less than 7 days of age (n = 8) and having a sepsis score < 11 and a negative blood culture, were obtained from Texas A&M University Veterinary Medical Teaching Hospital and Mid-Atlantic Equine Hospital. Plasma and serum were processed from whole blood collected by jugular venipuncture upon admission, at 24 hours post admission and at 5 days post admission or at the time of discharge.

Plasma concentrations of ACTH, and serum concentrations of cortisol and IGF-I were determined by specific RIAs. Data were analyzed using linear mixed-effects modeling with foal modeled as a random effect and day of admission modeled as an ordered categorical variable; post-hoc testing of pair-wise comparisons was made using the method of Sidak. Significance was set at $P < 0.05$, and analyses were performed using S-PLUS software (TIBCO, Inc., Seattle, WA).

Plasma concentrations of ACTH were not significantly different between septic and non-septic foals whereas septic foals had greater serum cortisol (37 ± 8 ng/mL vs 25 ± 7 ng/mL) but lower serum IGF-I (116 ± 14 ng/mL vs 152 ± 15 ng/mL) relative to non-septic foals pooled overall sampling times. The positive association of the peripheral blood concentrations of ACTH and cortisol depended on disease status of the foals. Specifically, cortisol and ACTH were positively correlated for the septic foals ($P = 0.027$) but not significantly correlated in the non-septic foals. Peripheral concentrations of ACTH and IGF-I were not significantly correlated whether data were pooled overall or stratified by sepsis status. However, peripheral concentrations of cortisol and IGF-I were negatively associated ($P = 0.019$); disease status did not influence this association, although it appeared to be a stronger association for the septic than the non-septic foals. The negative correlation between serum concentrations of the adrenal axis steroid cortisol and the somatogenic axis peptide IGF-I may reflect interactions of these homeorhetic hormones. Further studies of these and other metabolic hormones in a greater number of foals are warranted to better understand how these factors contribute to survival or non-survival of critically ill foals.

ACVIM FORUM SESSION 198A
REPETITIVE STIMULATION OF THE COMMON PERONEAL NERVE AS A DIAGNOSTIC AID FOR BOTULISM IN FOALS.

M Aleman, DC Williams, JE Nieto, KG Magdesian, RJ Brosnan, D Feary, H Hilton, T Kozikowski, J Higgins, JE Madigan, RA LeCouteur. University of California, Davis, CA.

Botulism is a potentially fatal paralytic disorder which definitive diagnosis is difficult. The purpose of this study was to investigate if repetitive stimulation of the common peroneal nerve will aid in the diagnosis of suspected botulism in foals.

Four healthy foals were used for its comparison with 3 foals with suspected botulism. Controls were anesthetized and affected foals were sedated to avoid risks of anesthesia. The common peroneal nerve was chosen for its superficial location and easy access. Stimulating electrodes were placed along the common peroneal nerve. For recording, the active and reference electrodes were positioned over the midpoint and distal end of the extensor digitorum longus muscle, respectively. Repeated supramaximal stimulation of the nerve was performed utilizing a range of frequencies (1 to 50 Hz). Amplitude, area under the curve and percentages of decrement or increment for each M wave over subsequent potentials for each set of stimuli were analyzed.

Baseline M waves were decreased in affected foals compared to controls. A decremental response was seen at all frequencies in control foals. Decremental responses were also observed in affected foals at low frequencies. However, an incremental response in amplitude and area under the curve was seen in all affected foals at 50 Hz.

Reduced baseline M waves with incremental responses at high rates are supportive of a presynaptic neuromuscular disorder which botulism was the most likely cause in these foals. Repetitive nerve stimulation is a safe, simple, fast, and non-invasive technique that can aid in the diagnosis of suspected botulism in foals.

ACVIM FORUM SESSION 198B

EFFECTS OF WEIGHT GAIN ON ACTIVE GLUCAGON-LIKE PEPTIDE-1 CONCENTRATIONS IN HORSES WITH HISTORIES OF OBESITY AND INSULIN RESISTANCE. K. Chameroy, N. Frank, S. Elliott, A. Schuver. Department of Large Animal Clinical Sciences, College of Veterinary Medicine, University of Tennessee, Knoxville, TN.

Insulin resistance (IR), obesity, and type 2 diabetes affect glucagon-like peptide 1 (GLP-1) concentrations in humans and rodents, but this incretin hormone has not been examined in horses. We therefore hypothesized that GLP-1 concentrations would change in horses as obesity and IR were induced or exacerbated by overfeeding. Six horses previously diagnosed with Equine Metabolic Syndrome were provided with twice the amount of digestible energy required for maintenance as sweet feed and hay for 8 weeks. Intravenous and oral glucose tolerance tests (OGTTs) were performed at 0 and 8 weeks. Effects of time and period (0 and 8 weeks) were assessed by repeated measures ANOVA.

Mean body weight increased from 438 ± 61 kg (range, 381 to 533 kg) to 464 ± 61 kg (range, 394 to 550 kg) over 8 weeks, with individual horse weight gain varying from 2 to 10%. Mean body condition score increased ($P = 0.006$) from 6 ± 2 (range, 4 to 8.5) to 8 ± 1 (range, 7 to 9). Three horses developed mild laminitis. Glucagon-like peptide 1 concentrations increased over time during OGTTs ($P = 0.023$), but the period × time effect was not significant ($P = 0.141$). Area under the GLP-1 curve remained unaffected by weight gain, whereas area under the insulin curve increased ($P = 0.003$) over time, indicating a reduction in insulin sensitivity. Obesity and IR were induced or exacerbated when horses previously diagnosed with EMS were overfed, but GLP-1 concentrations did not change as a result.

ACVIM FORUM SESSION 199A

USE OF HYPERTONIC SALINE IN ENDURANCE HORSES: A RANDOMIZED CLINICAL TRIAL. CL Fielding[1] and KG Magdesian[2]. [1]Loomis Basin Equine Medical Center, Loomis, CA. [2]School of Veterinary Medicine, University of California, Davis, CA.

Hypertonic saline solution (7.2%) (HSS) is an intravenous fluid used for the emergency treatment of intravascular volume deficits. The use of this fluid in horses with severe dehydration is controversial. The purpose of this study was to compare the use of HSS and isotonic saline solution (0.9%) (ISS) for the emergency treatment of endurance horses.

Endurance horses eliminated from competition and requiring intravenous fluid therapy were eligible for enrollment in the study. Twenty-two horses were randomly assigned to receive 4 ml/kg of either HSS or ISS along with 5 L lactate Ringer's solution (LRS). Following this bolus, all horses were treated with an additional 10L of LRS. Blood and urine samples were collected before, during and after treatment. Data was compared using two-way ANOVA with repeated measures.

As compared to ISS, HSS horses showed a greater decrease in PCV ($p = 0.04$), total protein ($p = 0.01$), albumin ($p = 0.01$), and globulin ($p = 0.02$). HSS horses showed a greater increase in sodium and chloride ($p < 0.001$) as compared to ISS horses. Horses receiving HSS had a shorter time to urination ($p = 0.03$) and lower specific gravity ($p < 0.001$) than those receiving ISS.

Results of this study indicate that HSS may provide faster restoration of intravascular volume deficits than ISS in endurance horses receiving emergency medical treatment. More profound electrolyte changes should be expected with HSS however.

ACVIM FORUM SESSION 199B

EFFECT OF INHALED SALBUTAMOL ON PULMONARY CIRCULATION TRANSVASCULAR FLUID DYNAMICS AND GAS EXCHANGE IN EXERCISING HORSES. M. Vengust[1], G. Heigenhauser[2], C. Kerr[3], H. Staempfli[3], J. Pringle[4], L. Viel[3]. [1]University of Ljubljana, Ljubljana, Slovenia; [2]McMaster University, Hamilton, Canada; [3]University of Guelph, Guelph, Canada; [4]Swedish University of Agricultural Sciences, Uppsala, Sweden.

β_2-adrenergic receptor agonists have been shown to increase erythrocyte carbonic anhydrase activity, which may stimulate the Jacobs-Stewart cycle and increase pulmonary circulation transvascular fluid fluxes during exercise. Increase in pulmonary transvascular fluid fluxes ($J_{V\text{-}A}$) and consequent increase in the pulmonary interstitial fluid would be detrimental for alveolar O_2 exchange during the fast erythrocyte transition time across the pulmonary capillaries. Therefore, we hypothesised that treatment with inhaled β_2-adrenergic receptor agonist will increase $J_{V\text{-}A}$ and the alveolar-arterial PO_2 difference ($AaDO_2$) during exercise.

Six STB horses were exercised on a high-speed treadmill at 80% VO_2 peak until fatigue. Horses were randomly assigned to treatment with salbutamol (Sal: 500mcg) or placebo (Control: Con) inhalation via Aeromask* 60 min prior to exercise, with cross over treatment used at the repeated exercise test (8 days later). Arterial and mixed-venous blood, as well as CO_2 elimination and O_2 uptake, were sampled simultaneously at rest, during exercise at 60 sec intervals until fatigue, and into recovery. Blood gases were analyzed. $AaDO_2$ was calculated using the inspired PO_2 (149 mmHg), and blood partial pressure of O_2 and CO_2. Blood volume (%) changes across the lung were calculated from changes in hemoglobin and hematocrit values in venous and arterial blood. Cardiac output (Q) was calculated using the Fick equation. $J_{V\text{-}A}$ was calculated using Q and blood volume changes across the lung. Variables were analyzed using two-way repeated-measures ANOVA ($P < 0.05$).

The duration of exercise to fatigue was 4.3 ± 0.3 min and 4.4 ± 0.4 min in both Con and Sal, respectively. At rest Sal had no effect on $J_{V\text{-}A}$, oxygen consumption (VO_2), blood oxygen saturation (SO_2) or $AaDO_2$ ($P > 0.05$). At the onset of exercise $J_{V\text{-}A}$ increased in Con and Sal ($P < 0.0001$) and at fatigue reached 10.0 ± 2.4 L/min and 10.0 ± 1.6 L/min, respectively. Treatment with Sal had no effect on $J_{V\text{-}A}$ during exercise ($P = 0.9$). At the onset of exercise SO_2 and VO_2 increased in Con and Sal ($P < 0.0001$). Treatment with Sal had no effect on SO_2 or VO_2 during exercise ($P > 0.05$). $AaDO_2$ increased during exercise in Con and Sal ($P < 0.0001$) and at fatigue reached 19.4 ± 2.3 mmHg and 18.1 ± 2.4 mmHg, respectively. Treatment with Sal had no effect on $AaDO_2$ during exercise ($P = 0.3$).

Inhaled β_2-adrenergic receptor agonist salbutamol at the dose of 500mcg given 60 min before exercise did not affect the duration of exercise to fatigue, $J_{V\text{-}A}$, VO_2, SO_2 or $AaDO_2$. Therefore, it had no detrimental effect on alveolar-capillary diffusion distance and the ventilation/perfusion mismatch in exercising horses.

ACVIM FORUM SESSION 200A

INFLAMMATORY AIRWAY DISEASE (IAD) IN HORSES IS CHARACTERIZED BY A TH2 AND A TH17 mRNA EXPRESSION PROFILE. R Léguillette, L Beekman. Department of Veterinary Clinical and Diagnostic Sciences, Faculty of Veterinary Medicine, University of Calgary, Calgary, Alberta, Canada.

Inflammatory Airway Disease (IAD) and recurrent airway obstruction (RAO) represent two classes of equine lung inflammatory diseases that may share some similar immunologic mechanisms. There is evidence that Th2 cytokines and IL-17 play some role in RAO. IAD is a common condition in horses, but its pathophysiology is still not understood. The aim of the present study was therefore to determine the mRNA expression of Th1, Th2 and Th17 inflammatory cytokines, to understand the immunological mechanisms of IAD.

The mRNA expression of ten inflammatory cytokines and chemokines was measured in the bronchoalveolar fluid (BALF) of seventeen horses with IAD and compared with ten control horses. The horses were selected based on 1- their clinical signs, 2- the inflammatory cells count in the BALF, 3- their physical examination and 4- their medical history. The mRNA expression of IL-5, IL-1β, IL-6, IL-8 and IL-10 was significantly up-regulated in BALF from horses with IAD.

Furthermore, the BALF samples were subdivided in two groups based on the differential cells count 1- BALF with increased mast cells (IAD-Mast) and 2- BALF with increased neutrophils (IAD-Neutro). IL-4 was significantly down-regulated in the IAD-Neutro group compared to the IAD-Mast group. IL-17, IL-5 and IL-8 were significantly up-regulated in the IAD-Neutro group compared to the IAD-Mast group.

The present study shows that IAD in horses is characterized by a Th2 and a Th17 mRNA inflammatory expression profile and that different immunological mechanisms are involved in mast cells or neutrophils accumulation in the BALF of horses with IAD.

ACVIM FORUM SESSION 200B

***IN VITRO* RESPONSE OF PERIPHERAL BLOOD LYMPHOCYTES FROM HEAVEY AND NON-AFFECTED HORSES TO β2-AGONIST STIMULATION.** M. Werner, M.F. Ehrich, M.R. Prater, P. Eyre, S. Werre, V.A. Buechner-Maxwell. Virginia-Maryland Regional College of Veterinary Medicine, Virginia Tech, Blacksburg, VA.

β_2-adrenoreceptor (β_2-AR) agonists are a class of medications that promote smooth muscle relaxation and bronchodilation in horses and humans with airway disease. Activated human peripheral blood lymphocytes (PBLs) also respond to β_2-AR agonist stimulation by attenuating the production of cytokines associated with the pathogenesis of asthma and Recurrent Airway Obstruction (RAO). The aim of this study was to develop an *in vitro* technique for measuring the response of equine PBLs to stimulation with Salbutamol, a β_2-AR agonist. This method was then used to compare the response of PBLs from RAO-affected and non-affected horses to β_2-AR agonist stimulation. PBLs from 4 RAO and 4 non-affected horses were cultured ($4x10^6$/ml) in RPMI complete media with Concanavalin A (ConA, 2ug/10^6cells) for 0, 1, or 2 days then stimulated with Salbutamol (30 minutes). Using flow cytometric techniques, response was measured by detecting protein kinase A phosphorylation of vasodilator stimulated phosphoprotein (VASP). Results were verified by western blot analysis. Activated PBLs were then incubated with ConA for one day were pre-incubated with β_2 or β-adrenoreceptor antagonist (ICI 118,551, Sigma®; Atenolol, Sigma®) for 15 minutes, followed by 30 minutes Salbutamol (500 nM) stimulation. Results were analyzed by ANOVA or ANCOVA and differences were considered significant when $p < 0.05$.

Response to β-antagonist was only observed in activated PBLs (pre-cultured with Con A) and was greater in cells from RAO horses as compared to cells from non-affected horses. The addition of β-antagonist attenuated the response of PBLs to Salbutamol while the addition of a β_1-antagonist had no effect. These findings indicate that activated PBLs from RAO-affected horses have a greater response to Salbutamol as compared to PBLs from non-affected horses, and this response is mediated mainly through the β_2-AR.

Human β_2-AR are known be polymorphic and this polymorphism results in a variable response to β_2-agonist binding that affects long term outcome in human asthmatics. Further studies are required to determine if the difference in response of PBLs from RAO affected as compared to non-affected horses is due to genetic polymorphism in the equine β_2-AR, and whether this difference is associated with a propensity for horses to develop equine RAO.

ABSTRACT DISCLOSURE INDEX

Abstract	Presenter	Disclosure
C-7	Rita Singh	Previously presented at the University of California Davis, House Officers Seminar Day.
C-15	Gretchen Singletary	Drs. Oyama, Rush, and Fox serve on a veterinary advisory board for IDEXX Laboratories, Westbrook ME, makers of the BNP assay. Dr. Stepien performs ad hoc consulting for IDEXX Laboratories. The study was independently planned, executed, analyzed, and reported by the authors with no input from IDEXX. IDEXX provided BNP assays for study patients free of charge. No IDEXX employees are authors or coauthors on this study and abstract.
C-27	Mark Oyama	Drs. Oyama and Rush serve on a veterinary advisory board for IDEXX Laboratories, Westbrook ME.
C-33	Rachel Cohen	Dr. Oyama consults as part of a veterinary advisory board for IDEXX Laboratories, maker of the NT-proBNP assay
E-9	Heidi Banse	Previously presented at the 7th Annual Harold Hamm Diabetes Center Research Retreat, Oklahoma City, OK.
E-16	John Middleton	Industry grants and contracts: Delaval Manufacturing Microdermis (formerly DRSS Global), LLC Government grants: USDA-AFRI, USDA-AES, USDA-AHFF Other grants: University of Missouri College of Veterinary Medicine Committee on Research
E-17	Jessica Khodadad	I was asked and agreed to participate as an advisory council member (1 of 23 total from across the US) for Merial regarding Equioxx. This present project was already underway when I was asked to participate. Merial provided some funding for the study and covered the costs of PK studies.
E-18	Christine Wimer	Previously presented at a conference of research workers in animal disease.
E-22	Lynne Gardner	Lowella Heinecke, Mark Grzanna, Abigayle Mrozinski, Angela Au, Stacy Ownby, and Dr. Carmelita Frondoza are employed by Nutramax Laboratories, Inc. who funded this research.
E-39	Benjamin Brainard	This study was funded by Abaxis, Inc.
E-44	Londa Berghaus	Roser Huang and Thomas Malinski work for Merial. The drug studied was donated by Merial but the project was not funded by Merial. Steeve Giguere has acted as consultant for Merial on a different project in the past.
E-53	Megan Dorris	Lowella Heinecke, Mark Grzanna, Angela Au, Stacy Ownby, and Carmelita Frondoza are all employed by Nutramax Laboratories, Inc. who funded the research.

E-65	Michele Frazer	This abstract was presented at the AAEP in December, 2009.
EN-10	Ellen Behrend	In relation to #4, no animals were used by us.
EN-13	Lauren Reid	Previously presented at the Auburn University Phi Zeta Research Emphasis Day, November 10, 2010.
EN-16	Lynda Melendez	All authors are employees of Hill's Pet Nutrition, Inc.
EN-17	Lynda Melendez	All authors are employees of Hill's Pet Nutrition, Inc.
GI-3	Aarti Kathrani	Maybe presented at Digestive Disease Week 2011 in Chicago, will be submitted, not yet accepted.
GI-7	Jonathan Lidbury	The dog food used for this study was provided by Nestle Purina.
GI-11	Robert Washabau	The research project, from which data were generated and described in the abstract, was financially sponsored by Dechra Ltd. This was outlined as bullet #7 ("research by the speaker sponsored by the Company") in the ACVIM Policy on Disclosure of Conflict of Interest.
GI-13	Nora Berghoff	This study was partially funded by Alba Therapeutics, the manufacturer of the drug investigated in this study (larazotide acetate). Alba Therapeutics also provided additional funds for enrolling dogs into the study (owner incentive). None of the authors are affiliated with Alba Therapeutics or have any investments or other financial interest in this company.
GI-19	Romy Heilmann	The authors work at the Gastrointestinal Laboratory at Texas A&M University, a non-profit organization where analysis of samples using in-house immunoassays is performed on a service basis.
GI-20	Romy Heilmann	Dr. Heilmann, Ms. Markel, Dr. Steiner, and Dr. Suchodolski work at the Gastrointestinal Laboratory at Texas A&M University, a non-profit organization where analysis of samples using in-house immunoassays is performed on a service basis.
GI-21	Tara Bidgood	Previously presented at the Veterinary Cancer Society, November 2010. All authors are employed at Pfizer Animal Heath and the study was conducted at Pfizer.
GI-23	Romy Heilmann	Dr. Heilmann, Dr. Berghoff, Dr. Grützner, Dr. Suchodolski, and Dr. Steiner work at the Gastrointestinal Laboratory at Texas A&M University, a non-profit organization where analysis of samples using in-house immunoassays is performed on a service basis.
HM-12	Dawn Bachman	This study was funded by IDEXX lab, Inc.
ID-2	Audrey Cook	One of the authors (LJ Wheat) markets the antigen assay test evaluated in this study. He helped provided information regarding patient test results but did not have any input on case selection, categorization or data analysis.
ID-10	Julie Levy	Erin Wood and Jessie Strout are employees of IDEXX Laboratories, the manufacturer of the test that was used in this study.
ID-11	Cynda Crawford	Dr. Leutenegger is employed by IDEXX Laboratories which performed the PCR diagnostic testing

ID-12	Michael Lappin	L. Roycroft is employed by Novartis Animal Health
ID-24	Michael Kim	Drs. Moroff and Piperisova are employees of Antech Diagnostics. The testing methodology employed in our study was performed by utilizing their proprietary FastPanel (TM) PCR technique.
ID-25	Melissa Beall	Dr. Melissa Beall is an employee of IDEXX Laboratories, Inc.
N/U-1	Nicole Smee	The product tested in this study and the funding was by Vetoquinol.
N/U-3	Angela Lusby	Dr. Lusby and Kirk have research sponsored by Hills Pet Nutrition and Dr. Toll is employed by Hill's Pet Nutrition
N/U-5	Vaness Von Hendy-Willson	Previously presented at the American Society of Nephrology Renal Week (related but not identical abstract).
N/U-15	Rebecca Geddes	The presenter's PhD Studentship is sponsored by Royal Canin, but since this abstract does not involve any data on dietary therapy it is not considered to be a conflict of interest.
N/U-17	Andrew Hanzlicek	This study is funded by Vetoquinol which is the maker of Rubenal the rhubarb supplementation used in this study.
N/U-20	Javier Del Angel-Caraza	This work has been partially supported by a project of Waltham Foundation in Mexico and the Consejo Nacional de Ciencia y Tecnologia (CONACyT) of Mexico.
N/U-23	Sara Irom	This clinical trial was sponsored and funded by Bayer Healthcare.
N/U-28	Bernard Schmidt	BS and GB are representatives of the sponsor of the study, i.e., Bayer Animal Health GmbH; JD, JK and RK received financial support from the sponsor for running their parts of the study
N-5	Diana Henke	Previously presented at the ECVN annual meeting in Cambridge, UK.
N-6	Dennis O'Brien	Previously presented at the American Society for Neurochemistry: additional data in this abstract.
NM-3	Hervé Lefebvre	Research sponsored in part by Royal Canin
NM-4	Jennifer MacLeay	A similar abstract will be presented at the Orthopedic Research Society meeting in January 2011 to an audience largely of orthopedic researchers interested in human orthopedics. Drs. MacLeay and Hahn are current employees of Hills Pet Nutrition, Inc.. Dr. Frantz was an employee of Hills Pet Nutrition, Inc. at the time the work was done. Drs. Moens, Gibson, Baker and Roush were paid for their services on this project by Hills Pet Nutrition, Inc.
NM-5	Dennis Jewell	Presenting author, second, and third authors are employed by Hill's Pet Nutrition.
OT-1	Marina Leis	Previously presented at the Western College of Veterinary Medicine Undergraduate Poster Competition.
OT-4	Marina Leis	Previously presented at the Western College of Veterinary Medicine Undergraduate Poster Competition.
P-2	Ted Whittem	Presenting author is past employee of Sponsor. No current conflict of interest.

P-8	Laura Johnston	All authors work for Boehringer Ingelheim, manufacturers of meloxicam
R-1	Chee-Hoon Chang	Previously presented at the Veterinary Comparative Respiratory Society Symposium, Raleigh, North Carolina, September 28-30, 2010.

WELCOME TO THE 2011 ACVIM FORUM IN DENVER, CO

Welcome to Denver! I am pleased and honored that you have chosen to attend the 2011 ACVIM Forum, the premier veterinary internal medicine continuing education meeting! Much of the credit for this distinction goes to the ACVIM Forum Program Committee as well as the many specialists who deliver state-of-the-art clinical science and veterinary internal medicine advances to our attendees.

This year's agenda is full, with a core program of approximately 280 scientific lectures, more than 350 research abstracts, and a technician lecture series. All sessions of the core program are open to every attendee. Thursday through Saturday you'll find a variety of special interest group (SIG) sessions (workshops, case discussions, and labs) that address diverse topics in veterinary internal medicine. Many of these sessions are available in the evenings. Veterinary technicians will be pleased to see a scientific program and workshops designed specifically for them. Veterinary technicians also are welcome to attend other scientific sessions.

Make plans to attend the Opening Night Reception, sponsored by Hill's Pet Nutrition and the ACVIM on Thursday evening. On Friday morning, all attendees are invited to the Keynote Address: Extra-Ordinary Veterinarian: Dr. David Sherman, DVM, DACVIM *(Nominated by ACVIM Diplomates to inspire attendees with the story of his unusual career path)*. After the Keynote address, please stay for the Small Animal or Large Animal Consensus Statement DRAFT session.

Research posters are on display throughout the meeting. Content may be discussed with the authors on Friday evening during the Wine and Cheese Reception sponsored by Nestlé Purina.

There is an abundance of quality session topics on the daily scientific sessions grid. However, if you are unable attend a session of interest the Proceedings will help to fill in the gaps.

Make sure to leave plenty of time to meet with our many exhibitors in the Exhibit Hall. You will be surprised by the number and diversity of companies that come to bring you the very latest in veterinary technology, pharmaceutical advances, superior nutritional products, educational materials, and career opportunities.

The ACVIM is indebted to the 2011 ACVIM Forum Program Committee (chaired by Dr. Marlene Hauck) for the incredible effort it takes to put together a program of this magnitude. We also want to reiterate our thanks to our many speakers, facilitators, and moderators who are willing to share their knowledge, talent, and time. We are very grateful to our sponsors for their recognition, presence, and support of our efforts to provide an unequaled educational experience for all attendees. The ACVIM Forum continues to grow substantially in both size and quality due to a wide range of sponsor support.

The dedicated ACVIM staff members have handled all of the organizational details needed to bring this meeting to you, and are available at the registration desk to help you with any special needs or requests. Please take a few minutes to personally thank them for a job well done.

The members of the ACVIM Board of Regents realize that you have choices regarding where to spend your continuing education dollars, and we appreciate the fact that you have chosen to join us in Denver. We hope that you have an outstanding educational experience, and that your attendance at the 2011 ACVIM Forum will facilitate your continued commitment to the highest quality of service in veterinary practice, academia, or industry. Mark you calendars now to join us for the 2012 ACVIM Forum, May 30 – June 2, in New Orleans, Louisiana.

Wishing you a stimulating and productive meeting,

Dr. Jean A. Hall, DVM, MS, PhD, DACVIM

President, ACVIM Board of Regents

2011 ACVIM FORUM SPONSORS

The ACVIM appreciates the generous support provided by our sponsoring companies.

PLATINUM

Hill's

GOLD

Nestlé PURINA

HESKA Smarter, Together

Pfizer Animal Health

INFINITI MEDICAL Advanced Technology for Vets and Pets

SILVER

Elanco

Dechra Veterinary Products

ROYAL CANIN

Abbott Animal Health

BRONZE

STORZ KARL STORZ—ENDOSKOPE

Boehringer Ingelheim

Intervet Schering-Plough Animal Health

VAC Therapy KCI

IDEXX LABORATORIES

MERIAL

Bayer HealthCare Animal Health

Universal Ultrasound See the difference

GENERAL PROGRAM

Abaxis
AnimalScan
Butler Schein Animal Health
CareCredit
Digicare Biomedical Technology
Live Oak Bank
Mila International
Novartis
Rogue Research
Vetel Diagnostics
Virbac Animal Health

MEDIA

International Veterinary Information Service (IVIS)
Journal of Veterinary Internal Medicine (JVIM)
Veterinary Cancer Society (VCS)
VetLearn

Keynote Session

2011 ACVIM Award Winners & Honorees

2011 ACVIM Forum Sponsor Recognition

Keynote Address

Nominated by ACVIM Diplomates, Dr. David Sherman – Dr. Sherman's career is grounded in internal medicine but he has not followed a traditional path. During his residency he became interested in the diseases of goats and gradually became expert in that field. With goats being far more important outside the United States than within, Dr. Sherman found himself responding to requests for consultations on goat health and production worldwide. His travels took him mainly to Asia and Africa where his interests broadened to include the role of livestock in poverty alleviation, providing veterinary services to the rural poor, the interactions of livestock with wildlife, the environmental impacts of livestock production, and the application of a one health approach in our global society. His various interests culminated in the publication of the textbook "Tending Animals in the Global Village: A Guide to International Veterinary Medicine."

ENDOCARDIAL MAPPING AND RADIOFREQUENCY CATHETER ABLATION OF ATRIAL TACHYCARDIAS IN DOGS

Roberto Santilli, DVM, PhD, DECVIM-CA (Cardiology)
Samarate, Varese, Italy

INTRODUCTION

The generic term atrial tachycardia refers to a group of arrhythmias that originate in the atria and then propagate to the ventricles with variable atrioventricular (AV) conduction ratio depending on the electrophysiologic properties of the AV node. According to human guidelines two different groups of atrial tachycardias can be identified: focal atrial tachycardia (FAT) and macro reentrant atrial tachycardia or atrial flutter (AFL)[1]. Focal atrial tachycardia is defined as an atrial activation starting rhythmically at a small area (focus) from which it spreads out centrifugally and without activation over a significant portion of the cycle length (CL)[2]. Focal atrial tachycardias manifest an atrial rate between 100 and 250 bpm and rarely at 300 bpm [2]. Despite the fact that FATs exhibit three different electrogenic mechanisms (abnormal automaticity, micro-reentry and triggered activity), all of them are grouped in the same category. According to the response of the electrophysiologic tests FATs caused by triggered activity and micro-reentry can also be called non-automatic FAT[3-5].

Prevalence of FATs in people is low (0.34 – 0.46 %), FATs account for 10 to 23 % of supraventricular tachycardias (SVT) in children with spontaneous remission often reported (24-63 %)[6]. Automatic FATs tend to be more common in young people and they are present in 80 % of patients with tachycardia-induced cardiomyopathy (TICM)[7]. Sixty-tree percent of patients with FAT have left ventricular dysfunction most commonly caused by a TICM[7]. Focal atrial tachycardias have been described also in dogs of different breeds, with a mean body weight of 33 $\pm$ 17.4 Kg and a median age of 51 months (range 5-156 months)[8]. Dogs are usually presented for episodic weakness or for congestive heart failure compatible with TICM (19%)[8]. Focal atrial tachycardia resulted incessant in 19 % of dogs, paroxysmal in 56 % of dogs and deteriorated into persistent atrial fibrillation in 25 %[8]. Mean atrial CL reported in dogs is 238.2 $\pm$ 69.2 msec; it is conducted in 1/3 of the dogs with different degree of AV nodal block with a mean ventricular CL of 292.7 $\pm$ 72.5 msec [8].

In people FATs have a characteristic anatomic distribution with typical clustering around the crista terminalis, coronary sinus, para-hissian region, tricuspid and mitral valve annulus, right and left interatrial septum, right and left appendages, pulmonary veins, superior vena cava and non coronary aortic cusp [9-17]. In people 63 % of FATs are right-sided, while 37% arise from the left atrial wall or pulmonary veins[9]. One-half to 2/3 of right atrial FATs and 21 % of total FATs arise from the crista terminalis an area with marked anisotrophy and rich of automatic tissue, factors that promote respectively reentry and abnormal automaticity[9]. Coronary sinus ostium and body account for approximately 10 % of FATs in people, the region of Thebesian valve presents an abrupt change in myocardial fiber orientation that facilitates reentry[9]. Coronary sinus musculature can be also a source of abnormal automatism foci[18]. Interatrial septum and perinodal area accounts for 34 % of FATs while tricuspid valve annulus for 13-22 %[9,12]. Pulmonary veins tachycardias comprise 46 % of left atrial FATs and 24 % of all FATs [9,15,19]. Left and right superior pulmonary veins are the ones most commonly affected [15]. Mitral valve annulus is the second most common location of left FATs and accounts for 28-36 % of total FATs [15]. Four to 17 % of human patients with FATs have more than one focus in the right free wall (29.5%) or in the right free wall and right septal area [20]. Patients with multiple ectopic foci presented a cardiovascular co-morbidity, shortest tachycardia CL and lower radiofrequency catheter ablation (RFCA) success rate [21]. According to detailed endocardial mapping 63% of the dogs with FATs presented right sided foci (crista terminalis 41%; triangle of Koch 25%; tricuspid valve annulus 16 %; right interatrial septum 9% and right auricle 9%) and 37% presented FATs arising from the pulmonary veins (right superior pulmonary veins 57%; left superior pulmonary veins 28%; right inferior pulmonary veins 15%). Twelve percent of dogs presented multiple ectopic foci [8].

The reported success rate of RFCA of FAT in people ranges between 69 and 100 %, with higher success rate in the right FATs compared to left FATs and septal tachycardias (80%, 72% and 52% respectively)[22-24]. The recurrence rate ranges from 0 to 33 %. Positive predictor for successful RFCA of atrial foci are right atrial location, while the negative ones are male gender, older age, multiple foci and repetitive forms [22-24]. Recurrence rate reported is approximately 8%, complication rate reported is around 5% and the most common sites of damage are the AV node and the phrenic nerve during RF energy delivery to ablate the cristal tachycardias[22-24].

Macro-reentrant atrial tachycardia is a reentrant activation around a large central obstacle (normal or abnormal structures, fixed, functional or a combination) in at least one of its dimension[25]. There no part of atrial tissue outside the circuit that are activated. Atrial flutter is a macro reentrant atrial tachycardia characterized by an organized atrial rhythm with a rate typically between 250 and 350 bpm in people[1,25-26]and usually more than 300 bpm in dogs. [27-28] According to the location of protected isthmus, an area of slow conduction which is part of the anatomic circuit of the arrhythmia, AFL are classified in typical and atypical[1,25-26]. The most common anatomic circuit of AFL includes the cavo-tricuspid isthmus (CTI, an area of the low posterior right atrium bounded by the inferior vena cava and the Eustachian ridge posteriorly and the tricuspid valve annulus anteriorly.[29] According to the rotation pattern around the tricuspid valve annulus, two forms of CTI-dependent AFL have been identified: typical and reverse typical.[26] During typical AFL the activation wavefront turns counterclockwise around the tricuspid valve annulus, while during reverse typical AFL it turns clockwise.[25-26] Other forms of CTI-dependent AFL reported in people include the double-wave and lower-loop reentry[1]. A further type of AFL, for some authors classifiable as atypical, is the left atrial macroreentrant atrial tachycardia, a stable arrhythmia that can originate in the left atrium[25]. A variety of left atrium circuits associated with large silent areas over the posterior left atrium has been described[31]. Last type of AFL reported in people is the left septal atrial flutter, an arrhythmia that shows a slowing of the electrical conduction in the left atrial septum caused by antiarrhythmic treatments and/or atrial myopathy[32]. We reported two cases of endocardial mapping and radiofrequency catheter ablation of CTI-dependent AFL typical and typical reverse in two dogs with episodic weakness and documented SVT[28]. Atrial CL found were respectively 165 msec and 130 msec. Electrocardiographic pattern of typical AFL was continued succession of depolarizing segments and electrical deflections that conferred a classical saw-tooth pattern with predominant negative polarity in inferior leads (F waves). Reverse typical atrial flutter waves appeared as positive deflections in inferior leads with isoelectric lines between them[28].

RFCA is considered the treatment of choice for AFL in people with a long term success rate after induction of bidirectional block in the protected isthmus area of 90-100%, a recurrence rate of 7% and a complication rate of 1.34% [1,33-34].

ENDOCARDIAL MAPPING AND RADIOFREQUENCY CATHETER ABLATION TECNIQUES

Radiofrequency catheter ablation of either FAT or AFL should be performed after a detailed endocardial mapping to assess the arrhythmogenic mechanism and the location of targets for ablation, respectively ectopic foci and isthmus. In dog electrophysiological studies are performed in general anesthesia as previously described, after discontinuing antiarrhythmic drugs for at least 5 drug half-lives. The animals are placed in dorsal recumbence. Two multipolar electrode catheters and one 7-F ablation catheter are inserted in the venous system, with a modified Seldinger technique, using fluoroscopic and intra-cardiac ECG guidance [8-28]. A decapolar catheter is inserted through the right external jugular vein into the coronary sinus, a second decapolar electrode catheter is inserted through the right femoral vein and placed around the tricuspid valve annulus to record with the proximal electrodes pairs the potentials of the antero-lateral aspect of the crista terminalis (HRA), with the middle pairs the potentials of the middle lateral right atrium (MRA) and with the distal pairs the potentials of the low postero-lateral right atrium (LRA). A 7-F ablation catheter, 8 mm, with deflectable curve is advanced through the right femoral vein and placed alternatively at the His bundle, right ventricle, different area of right and left atrial wall and CTI to map endocavitary atrial electrograms and to perform programmed stimulation [8,28]. To map left atrial chamber or pulmonary veins ostium and bodies, in case a patent foramen ovale is not present, a transeptal puncture should be performed using the described technique [35]. From the right femoral vein, under fluoroscopic guidance, a specially designed guide wire is first inserted into a transeptal needle and pushed across the fossa ovalis into the left chamber. A jump of the tip of the needle from the thicker muscular septum to the thin wall of the fossa ovalis is used to localize fossa ovalis. Along the wire positioned into the left atrium the transeptal needle is pushed across the interatrial septum to enlarge the hole, and then an introducer is left in place and used to move the mapping and the ablation catheters. Lasso catheters are special electrodes positioned in a circular fashion that are used to map pulmonary vein ostium, during ablation procedures. In some cases left atrial tachycardias arising from right superior pulmonary vein, can be identified mapping high posterior right atrium [36]. The analysis of double potentials in this area can help to establish the location of the ectopic foci, if the amplitude of the first potential is greater than that of the second potential, this indicated a right posterior atrial focus, and in the opposite case a right superior pulmonary vein focus should be suspected. All intra-cardiac electrograms, 12-lead surface ECG are displayed and analyzed on an EP recorder.

Focal atrial tachycardia

Programmed atrial and ventricular stimulation, bursts atrial pacing and isoproterenol hydrochloride 0.04-0.1 µg/kg/min IV constant rate infusion are usually used to test inducibility of documented and no documented arrhythmias and to verify electrogenic mechanism. Automatic FAT can be initiated only with isoproterenol infusion; there is no effect of programmed stimulation neither on initiation nor on termination of the tachycardia. Non automatic FAT could be ruled in case the initiation of the tachycardia is reproducible with incremental atrial pacing, programmed atrial pacing or burst atrial pacing[37]. Reported automatic FATs in people ranges between 7% and 16%, while non automatic from 4% to 20%[3]. In dogs 63% of FATs are automatic 21% are non automatic[8].

Endocardial mapping permits to evaluate the origin of activation defined as the site of earliest pre-systolic activity relative to the onset of the P wave during tachycardia (usually from 30 to 100 ms) where sharp and negative unipolar recording, with QS pattern, appeared [38]. In dogs mean pre-systolic activities reported is -39,9 $\pm$ 17.7[8]. Other tools to evaluate the location of atrial ectopic foci are mechanical interruption of the tachycardia caused by application of pressure to the area and the presence of fragmented, long duration, low amplitude electrograms. Fractioned potential consisted of three or more consecutive negative deflections with low peak-to-peak amplitude [39]. In the dog 57% of atrial potentials of the focus area are fractionated and 31% are low-amplitude. During FATs endocardial mapping show s a significant portion of the CL without recorded activity even when recording the entire right and left atrial chambers and the coronary sinus. In some cases intra-atrial conduction disturbances prolong the atrial activation over a large portion of the CL and conduction spread can follow a circular pattern. In these cases entrainment of the tachycardia circuit is used to differentiate reciprocating tachycardia from focal atrial tachycardia. After overdrive atrial pacing from the coronary sinus area during tachycardia the presence of variable (more than 10 ms from baseline) ventricular-atrial interval after pacing is considered diagnostic for FAT [40]. Reentrant atrial tachycardias are diagnosed according to the presence of entrainment with concealed fusion. According to this electrophysiologic test pacing at a site with pre-systolic activity or mid-diastolic potential with a CL 10-20 ms faster than the atrial tachycardia should entrain a reentrant tachycardia to the pacing rate without changing the morphology of P wave or the intracardiac electrograms sequence.[41] As reported in people also in dogs FATs can trigger atrial fibrillation particularly in case of pulmonary veins foci and multiple foci [8,19].

Regardless of the underlining electrogenic mechanism FATs are ablated by targeting the foci where the atrial activation starts according to the presence of the earlier pre-systolic activities with fragmented and low-amplitude potential and sharp and negative unipolar recording. 7-F thermo-coupled electrode is usually used to ablate ectopic foci. Radiofrequency current is generated by conventional generator which monitors the catheter tip temperature, power output and impedance. The procedure is performed in controlled temperature during tachycardia with a maximum of 65° and 75 watts. Maximum ablation time for a particular target is 60 sec., and the radiofrequency delivery should be interrupted in case of catheter displacement of increase of the impedance. The procedure is considered successful in case of tachycardia interruption during radiofrequency delivery with no re-induction after 45 minutes post-ablation. In our EP laboratory acute success rate of RFCA of focal atrial tachycardia is 42.1 %, since ¼ of FATs cannot be induced with programmed or incremental atrial stimulation, atrial burst or isoproterenol infusion. Our complication rate is around 15, 7% and includes two cases of anesthesia- related death, one minor stroke with spontaneous resolution during RFCA of a pulmonary vein ostium automatic focus and one programmed AV block in a dog with triangle of Koch tachycardia unresponsive to the pharmacological treatment.

Macro-reentrant atrial tachycardia

During macro-reentrant atrial tachycardia the reentrant activation is continuous and should be present of the entire CL of the tachycardia, and therefore also during the isoelectric line of the ECG. In case of AFL line of block with double potentials can be recorded in the area of functional or anatomic barriers [25]. In the isthmus area or in sites with local conduction disturbances fractioned electrograms can be recorded [25]. Transient entrainment of the tachycardia is present in most cases and a return CL equal to 20 ms of the tachycardia CL at the site of pacing identifies that the area is part of the circuit. In case the return CL is present in at least 2 pacing sites separated by at least 2 cm a diagnosis of macro-reentry can be done[41]. CTI-atrial flutter is diagnosed when a concealed entrainment during LRA pacing can be shown[42]. Typical AFL presents a counterclockwise reentry descending along the anterior and lateral walls and ascending along septal and posterior wall of right atrium. A reverse direction of rotation is observed during typical reverse AFL with an ascending direction along the anterior wall and descending along the posterior and septal walls [1,25-26]. The circuits of left atrial flutter and of septal flutter are less commonly studied. The criteria to diagnose the macro-reentrant circuit and the mapping technique of the protected isthmus are although similar. In our EP laboratory we identified beyond, several cases of CTI-dependent atrial flutter, two cases of right septal atrial flutter with the isthmus area located in the right interatrial septum.

Radiofrequency catheter ablation of AFL circuits is performed in controlled temperature with a thermocouple-tipped steerable 7-F catheter. Maximal catheter tip temperature and power output are usually set respectively at 65 °C and 100 W. Radiofrequency delivery for each target lasts at least 45 seconds. Persistence of block should be verified for at least 20-30 minutes. The target sites for ablation to cure AFL are the protected isthmus. Radiofrequency catheter ablation of the CTI is performed using both anatomical and an electrophysiological approach using the right lateral fluoroscopic view and the presence of concealed entrainment at the target site. Radiofrequency delivery can be performed during tachycardia or sinus rhythm. The catheter should be then gradually withdrawn towards the caudal vena cava to produce a linear lesion. The endpoint of ablation is the induction of bidirectional conduction block at the CTI demonstrated by a collision of waveform, while pacing the coronary sinus, with a supero-inferior activation of the right free wall post-ablation, by the presence of a slow conduction from low right free wall and coronary sinus during pacing of controlateral site, and by recording double potential at the site of linear ablation at the CTI [29, 33, 43]. Reported success rate of RFCA of CTI-dependent AFL and our experience of RFCA of right septal flutter in dogs are promising.

REFERENCES

[1]Blomstrom-Lundqvist C, et al. ACC/AHA/ESC Guidelines. **J Am Coll Cardiol** 2003;42:1493.. [2] Saoudi N et al. **Eur Heart J** 2001;22:1162. [3]Chen SA et al. **Circulation** 1994;90:1262. [4] De Bakker JM et al. **J Cardiovasc Electrophysiol** 1994;5:335. [5]Kammeraad JA et al. **Pacing Clin Electrophysiol** 2003;26:736. [6]Poutiainen AM et al. **Eur Heart J** 1999;20:694. [7] Chen SA, et al. **J Cardiovasc Electrophysiol** 1998;9:355. [8] Santilli RA et al. **J Vet Intern Med** 2010;24:539. [9] Kistler PM et al. **J Am Coll Cardiol** 2006;48:1010. [10] Badhwar N et al. **J Am Coll Cardiol** 2005;46:1921. [11] Chang KC et al. **Jpn Circ J** 2001;65:1034. [12] Chen CC et al. **J Cardiovasc Electrophysiol** 2000;11:744. [13] Das S et al. **J Cardiovasc Electrophysiol** 2008;19:231. [14] Freixa X et al. **Europace** 2008;10:105. [15] Hachiya H et al. **Circ J** 2005;69:205. [16] Kalman JM et al. **J Am Coll Cardiol** 1998;31:451. [17] Kistler PM et al. **Circulation** 2003:108:1968. [18] Tritto M et al. **J Cardiovasc Electrophysiol** 2001;12:1187. [19] Haisseguerre M et al. **N Engl J Med** 1998;339:659. [20] Hillock RJ et al. **Heart Rhythm** 2007;4:435. [21] Hu YF et al. **Heart Rhythm** 2009;6:198. [22] Anguera I et al. **Am J Cardiol** 2001;87:886. [23] Hindricks G. **Eur Heart J** 1993;14:1644. [24] Kay GN et al. **J Am Coll Cardiol** 1993;21:481. [25] Saoudi N et al. **Eur Heart J** 2001;22:1162. [26] Cosio FG. **Card Electrophysiol Rev** 2002;6:356. [27]Schoels W et al. **Circ Res** 1990;67:35. [28] Santilli RA et al. **J Vet Cardiol** 2010;12:59. [29] Olgin JE et al. **J Cardiovasc Electrophysiol** 1996;7:1112. [30] Morrouche F et al. **Circulation** 2004;109:2440. [31] Jais P et al. **Circulation** 2000;101:2928. [32] Marrouche NF et al. **Circulation** 2004;109:2440. [33] Chen SA et al. **J Am Coll Cardiol** 1996;27:860. [34] Kay GN, et al. **J Cardiovasc Electrophysiol** 1993;4:371, [35] Brockenbrough E et al. **Am J Cardiol** 1960;6:219. [36] Yamada T et al. **J Cardiovasc Electrophysiol** 2004;15:745. [37] Lin YJ et al. **J Cardiovasc Electrophysiol** 2005;16:163. [38] Tank K et al. **Chin Med J Engl** 2003;116:1455. [39]De Groot NM et al **J Cardiovasc Electrophysiol** 2006;17:1086. [40] Colombowala IK et al. **J Cardiovasc Electrophysiol** 2008;19:142. [41] Stevenson WG et al **J Cardiovasc Electrophysiol** 2005;16:173. [42] Saoudi N et al. **PACE** 1998;21:2105. [43] Calkins H et al **Am J Cardiol** 1994;73:353.

INTERVENTIONAL MANAGEMENT OF AV BYPASS TRACT MEDIATED TACHYCARDIA IN THE DOG

Kathy Wright, DVM, DACVIM
Cincinnati, OH

INTRODUCTION

One form of tachycardia in the dog involves a macroreentrant circuit that includes the atrial myocardium, atrioventricular node, ventricular myocardium, and an accessory pathway or AV bypass tract. When this circuit conducts from the atrial myocardium anterograde through the AV node to the ventricular myocardium, then retrograde up the AV bypass tract to the atrial myocardium, it produces a narrow QRS complex tachycardia known as orthodromic atrioventricular reciprocating tachycardia (OAVRT) . On the surface ECG, retrograde P waves can sometimes be distinguished and will typically be found in the ST segment of these narrow complex tachyarrhythmias, i.e. a short RP tachycardia. A special form of orthodromic atrioventricular reciprocating tachycardia is known as permanent junctional reciprocating tachycardia or PJRT. PJRT involves a decrementally-conducting, retrograde-only accessory pathway resulting in a long RP tachycardia.

Far less common in the dog is antidromic reciprocating tachycardia. The impulse in this form travels from the atrial myocardium, anterograde over the AV bypass tract to the ventricular myocardium, then retrograde up the AV node to the atrial myocardium to complete the circuit. This produces a wide and bizarre complex tachycardia that can be mistaken for ventricular tachycardia. There are two reasons why this is less common in the dog. First, typical canine accessory pathways do not conduct nearly as well in the anterograde direction as they do in the retrograde direction. Second, the typical canine AV node does not conduct well retrograde.

AV bypass tracts can also participate in tachyarrhythmias in which they are not essential to the circuit (unlike orthodromic or antidromic reciprocating tachycardia), but are bystanders, albeit sometimes deadly ones. Examples would include atrial fibrillation with anterograde conduction over an accessory pathway to the ventricles. If the 1:1 conduction cycle length of the anterograde AV bypass tract is short, then ventricular fibrillation can result.

Tachycardia-induced cardiomyopathy or tachycardiomyopathy is secondary ventricular myocardial dysfunction that results from incessant of frequent tachyarrhythmias. Clinically and echocardiographically, it can be difficult to distinguish this from idiopathic dilated cardiomyopathy. The caveat with this disease, is that the myocardial dysfunction and congestive heart failure can resolve completely or substantially if the arrhythmia is well-controlled or eliminated.

ELECTROPHYSIOLOGY STUDY

The dog is prepared by weaning them from the oral antiarrhythmic drugs currently being administered. This requires admission to the hospital and careful monitoring, as incessant atrioventricular reciprocating tachycardia will almost always result. Ideally, this is done for five half-lives of the particular drug(s) the animal is taking, although we have typically gone to three half-lives for owner convenience/cost and safety reasons. Intravenous diltiazem is used to control sustained orthodromic reciprocating tachycardia, and several dogs have been taken to the electrophysiology laboratory on a constant rate infusion of this antiarrhythmic drug, just to keep them out of incessant OAVRT and thus minimize morbidity/mortality at the time of anesthetic induction and early maintenance. One must realize that this will influence AV nodal measurements and your ability to induce AV reciprocating tachycardia at the time of study. Induction of OAVRT is not necessary to map the accessory pathway and assess its characteristics in dogs, however, as the retrograde pathway can be mapped using ventricular pacing.

Following anesthetic induction, the dog is placed in dorsal recumbency on the fluoroscopic table. Both groin regions and the ventral cervical regions are sterilely prepared. Catheter introducer sheaths are introduced into the right femoral vein (2 in a larger dog), both external jugular veins, and the left femoral artery (the latter for continuous blood pressure monitoring). Multipolar electrode catheters are placed in the His bundle region, coronary sinus, high right atrium, and right ventricular apex. Atrial train and extrastimulus testing is performed, followed by ventricular train and extrastimulus testing. The AV nodal anterograde Wenckebach point, shortest 1:1 atrial pacing cycle length that conducts anterograde over the accessory pathway, ventricular activation sequence, and induction of tachyarrhythmias that are dependent on critical AV nodal delay are determined with incremental atrial pacing. Atrial extrastimulus testing allows definition of the AV nodal effective and functional refractory periods, anterograde accessory pathway effective and functional refractory periods, and atrial effective refractory period. It also allows the induction of certain tachyarrhythmias. Incremental ventricular pacing is performed to determine the shortest cycle length at which one-to-one conduction occurs over a retrograde accessory pathway and the retrograde AV node. The retrograde atrial activation sequence is also determined, which demonstrates the general location of retrograde AV bypass tract insertion into the ventricular myocardium. Ventricular extrastimulus testing allows one to determine the effective and functional refractory periods of any retrograde accessory pathways and the retrograde AV node. To prove that an AV bypass tract is present and forming the retrograde limb of a tachycardia circuit, a ventricular premature impulse is delivered during tachycardia at the time of His bundle refractoriness. If the impulse is able to preexcite the atrium with the same activation sequence as during the tachyarrhythmia, then an accessory pathway is present and OAVRT is very likely. If that impulse in fact terminates the narrow complex tachyarrhythmia without reaching the atrium, then OAVRT is proven.

Once basic electrophysiologic testing has been completed, and it has been determined that an accessory pathway is present, then detailed mapping of that pathway must occur. Detailed mapping involves maneuvering the multipolar ablation catheter in the general vicinity of the AV bypass tract that was determined at baseline study. The goal is to find the shortest VA interval possible in that dog, as that will be the site of the atrial insertion point of the accessory pathway. This is done either during ventricular pacing or during orthodromic atrioventricular reciprocating tachycardia. Each method has its pros and cons. In dogs that exhibit manifest anterograde accessory pathway conduction, mapping can be performed during sinus rhythm. The challenges with this approach include determining the exact onset of the delta wave, and small variations in its determination can introduce error into delta-V time measurements.

CATHETER ABLATION

Standard Radiofrequency Technique

The standard radiofrequency ablation unit we use is a temperature-controlled generator that allows the operator to set the target temperature (to the tip of the catheter), duration of energy delivery, and maximum wattage that can be delivered (50 watts in this case). Impedance, wattage, temperature, and time are continuously displayed. The catheter tip temperature rises rapidly at the beginning of an application. It typically approaches the target temperature within about 2-3 seconds. The radiofrequency generator then adjusts the delivered power to maintain the target temperature until the application is ended. Several factors have been shown to affect the cardiac tissue temperatures and thus the lesion size achieved during radiofrequency catheter ablation. Some of these factors can be controlled by the choice of catheter and programmable settings on the radiofrequency generator, such as electrode tip size, duration of energy delivery, irrigation rate (if irrigated tip catheters are being used, and power or temperature setting. Other factors are either difficult or impossible to control, such as convective cooling of the electrode tip by intracavitary blood flow, electrode–tissue contact pressure, and electrode tip orientation against the endocardium. These latter factors, however, have also been shown to have major impact on cardiac tissue temperatures and hence lesion size.

It is radiofrequency power delivered to the tissue that determines lesion size and the catheter tip temperature is poorly correlated to lesion size in vivo. Limits of standard temperature-controlled ablation include: 1) sites of poor blood flow in which the catheter tip is not convectively cooled and target temperature is quickly reached without adequate power being delivered to the tissue to destroy it permanently. 2) sites of unstable tissue contact with significant convective loss of energy to the surrounding blood pool and inadequate tissue heating.

If at a successful endocardial site for radiofrequency catheter ablation in the typical dog, accessory pathway conduction should be terminated within 5-7 seconds. This does not apply to epicardial AV bypass tracts; however, these are less common. If no retrograde AV nodal conduction is present in a particular dog, then when ventricular pacing you will see VA dissociation once retrograde AP conduction has terminated. If retrograde AV nodal conduction is present, you will see a shift in the retrograde activation sequence from the AP to the AV node and usually a significant increase in VA conduction time. If ablating during orthodromic reciprocating AV tachycardia, the tachycardia will terminate. If ablating during preexcited sinus rhythm, the delta wave will disappear and standard anterograde AV nodal conduction will become apparent. We will review various cases to illustrate these points.

Cases that can present unique challenges

1. Accessory pathways associated with coronary sinus diverticula or the middle cardiac vein
2. Right free wall accessory pathways
3. Midseptal accessory pathways
4. Multiple accessory pathways
5. Epicardial accessory pathways

REFERENCES

Vaksmann G et al. Heart 2006;92:101. Calo et al. J Cardiovasc Med 2007: 893):138. Wright KN et. al. J Vet Cardiol 2006;8:95. Santilli RA et. al. J Vet Cardiol 2000; 2(1):25. Wright KN et. al. J Vet Int Med 1999;13(4):361. Kulig J et al. Circ 2010;122;e480. Eick O et al. Ind Pacing Electrophysiol 2002:2(3):66. Santilli RA et al. J Am Vet Med Assoc 2007;231(3):393.

CURRENT PACING PRACTICES AND FUTURE DIRECTIONS

Daniel F. Hogan, DVM, DACVIM (Cardiology)
West Lafayette, IN

HISTORICAL PERSPECTIVE AND TIMELINE FOR PROGRESSION OF PACEMAKER THERAPY

The first reported implantation of an artificial pacing system in a dog was more than 40 years ago. During these early years, most reports were individual case reports or small case series. The pacing equipment available for veterinary clinical use at that time was relatively basic in complexity, had limited availability, and was not specifically designed for the dog. However, even with these limitations the clinical response was considered quite good and pacemaker therapy began its solid foundation as the standard of care for symptomatic bradycardia in veterinary medicine.

Over the ensuing decades the equipment has dramatically increased in complexity and availability which has had an impact on the technique of implantation as well as the type of pacing system. There have been episodic case reports, case series, and retrospective case studies during the past 40 years which act as "snap shots" in time with regard to pacing practices.

1968 (Buchanan et al, JAVMA): The first published report of a pacemaker implanted in a dog with complete heart block. Two unipolar, epicardial leads were implanted via thoracotomy. The pacing system was an asynchronous, fixed-rate system (VOO). There were no complications and the therapy allowed resolution of CHF.

1976 (Musselman et al., J Small Anim Pract): The first published report of a transvenous, endocardial, bipolar lead implanted in a dog with complete AV block.

1981 (Yoshioka et al., JAAHA): Retrospective study on 12 dogs that received a sutureless epicardial lead. Pacing systems included asynchronous VOO (5/11) and demand VVI (6/11).

1981 (Lombard et al., JAAHA): Retrospective study on 31 dogs where 30 received epicardial leads and 1 received a transvenous endocardial lead. There was an equal split of asynchronous (VOO) and demand (VVI) pacing systems.

1983 (Bonagura et al., JAVMA) Retrospective study of 11 dogs receiving epicardial pacemakers where the objective description of complications were reported to determine how these could be improved.

1985 (Darke et al., JSAP) First report of a transvenous, dual-chambered pacemaker implanted in a dog with complete AV block.

1986 (Fox et al., JAVMA): Transdiaphragmatic approach for epicardial lead placement described.

1989 (Darke et al., JSAP): European retrospective study of 19 dogs and 1 cat where all received transvenous endocardial leads. This was the first report of a pacemaker in the cat. All of the cases were demand pacing systems with the majority receiving VVI while a minority received dual-chambered systems.

1991 (Sisson et al., JVIM): Retrospective study of 40 dogs that received single chamber, demand pacemakers. Descriptive population summaries were objectively reported as well as questions regarding the most appropriate type of lead fixation to avoid lead dislodgement.

1992 (Roberts et al., Vet Rec): First report of a rate-adaptive ventricular demand pacemaker (QT-interval).

1999 (Flanders et al., JAVMA): Pilot study evaluating the use of the costocervical vein in lieu of the external jugular vein in dogs to avoid the issue of lead dislodgement.

2001 (Oyama et al., JVIM): Retrospective study of 154 dogs through CANPACERS. Essentially all were fixed-rate ventricular demand pacing systems. There was an even mixture of pacing leads including epicardial and transvenous, unipolar and bipolar, as well as active and passive fixation. Complications and yearly survival were critically evaluated and related to operator experience. Client satisfaction was also critically evaluated.

2005 (Santilli et al., Med Sci Monit): Retrospective study of 42 dogs all receiving transvenous, ventricular demand pacemakers. This was the first report of a case series of dogs (12) with a rate-adaptive ventricular demand pacemaker (VVIR).

2005 (Ferasin et al., JSAP): Pilot study evaluating the appropriate rate response in exercising dogs with rate-responsive, ventricular demand pacemakers.

2006 (Green et al., JVIM): First published report (abstract) of a case series of dogs (25) that received transvenous, endocardial dual-chambered pacemakers.

2006 (Bulmer et al., JVIM): First report and prospective study of single lead, dual-chambered pacing systems in 10 dogs.

2006 (Wess et al., JVIM): Retrospective study of 105 dogs where increased success and reduced complication rates were concluded to be due to operator experience and supervision.

2007 (Johnson et al., JSAP): Retrospective study of 104 dogs with a majority receiving a rate-adaptive, ventricular demand pacemaker. First published report of pacemaker therapy for treatment of vasovagal syncope.

2009 (Hildebrandt et al., JVIM): Retrospective study of 33 dogs that received dual-chambered pacing systems. Concluded that dual-chamber pacing in dogs is feasible and does not have higher complication rates than single-chamber systems.

2009 (Maisenbacher et al., AJVR): Prospective study in 10 normal dogs demonstrating that LV performance and ventricular synchronization was best achieved with right atrial pacing andwas worse with RV apical pacing although this was improved with simultaneous LV and RV (BiV) pacing.

2009 (Estrada et al., JVC): Prospective study in 7 dogs with naturally occurring complete AV block demonstrating that cardiac performance was increased with BiV pacing compared to RV apical pacing.

CURRENT PACING PRACTICES

Given the time delay between gathering of clinical cases and date of publication and the frequency of retrospective pacemaker studies within the literature, it would appear that veterinary medicine is ready for a new study that would reflect the current pacing practices. An online survey was performed in preparation for this discussion and the data will be presented during the conference session.

In the absence of this current data, reviewing the last few published retrospective studies can give us an idea of where we are today. The vast majority of pacing systems utilize transvenous endocardial leads. Regardless of the underlying etiology for bradycardia (AVB, SSS, atrial standstill), the majority of systems implanted are single-chambered and equally split or slightly in favor of rate-adaptive ventricular demand systems (VVIR) over fixed-rate ventricular demand systems (VVI). A smaller population of dogs receive single lead physiologic pacing systems [VDD(R)] or dual-chamber physiologic pacing systems [DDD(R) or VDD(R)]. Lastly, a very small minority of cases with sinus node dysfunction and normal AV nodal conduction receive a single chamber atrial demand system [AAI(R)]. However, this is dependent upon operator and institutional experience and preference and witnessed by the majority of cases from the author's institution receiving dual-chambered pacing systems.

Complications occur with all procedures and pacemaker implantation is no exception. The frequency of reported major and minor complications range from 20%-40% and 14% in the 1980s; 55% and 13% in the 1990s; and 13%-33% and 11%-36% in the 2000s. The most common reported complication is lead dislodgement with reported frequencies of 10% in the 1980s, 59% in the 1990s, and 22%-50% in the 2000s.

FUTURE DIRECTIONS

It is difficult in some ways to give insight to where we might be going with regards to pacemaker therapy in veterinary medicine. Much of this is because none of the equipment is manufactured with our patients in mind so we must evaluate the newer human pacing products as they come to market, decide if they would be beneficial to our patients, and then wait for them to become available. However, we are already witnessing some of the future today when we look at pacing practices such as physiologic AV synchronous pacing (single or dual-chambered), rate-adaptive pacemakers, and biventricular pacing. I expect we will see these practices more frequently performed in the next 5-10 years.

ALTERNATIVE PACING SITES

Ventricle

Right ventricular apical pacing has been the standard site for lead location from the beginning of transvenous pacemaker therapy. However, RV apical pacing has been shown to result in ventricular dysynchrony, reduced left ventricular function, and increased risk for the development of CHF and atrial fibrillation. Alternative pacing sites include His-bundle pacing, *para*-Hisian pacing, and RVOT pacing. These sites appear to improve ventricular function and reduce myocardial deterioration but the results are not uniform between sites and these sites present challenges to lead attachment and myocardial capture.

Atrioventricular septal synchronous pacing

The atrioventricular septum is a muscular portion of the IVS that separates the RA and LV where the myocardium is predominantly LV in origin and continuous with fibers in the LV free wall. It has been demonstrated that deep penetration of this area by intramyocardial leads within the RA can appropriately sense and result in relative ventricular synchrony.

LEADLESS PACING SYSTEMS

In an attempt to avoid lead related issues such as infection, lead fracture, and dislodgement, experimental leadless systems have been evaluated in animal studies. A small receiver (leadless lead) is implanted into the chamber that is to be paced and a transmitter (generator) is inserted in proximity to the heart. The transmitter can use ultrasound or magnetic field waves to activate the receiver. This work is still very experimental and not ready currently for field testing. There are many issues to work out including durability of the receivers and transmitters, accurate sensing and capturing, as well as interference from outside signals.

CONCLUSION

Veterinary medicine has seen dramatic changes and improvements to pacemaker therapy in the past 40 years since it was first described. While there are expected changes and improvements to equipment in the next 10-20 years, the slope of advancement is likely to not be as steep. Focusing on improvement to current implantation techniques to reduce complication rates would appear to result in the most dramatic advancement for pacemaker therapy in veterinary medicine.

REFERENCES

Buchanan et al., JAVMA 1968;152:1099-1109. Musselman et al., JSAP 1976;17:149-162. Yoshioka et al., JAAHA 1981;17:746-750. Lombard et al., JAAHA 1981;17:751-758. Bonagura et al., JAVMA 1983;182:149-155. Darke et al., JSAP 1985;26:295-303. Fox et al., JAVMA 1986;189:1303-1308. Darke et al., JSAP 1989;30:491-499. Sisson et al., JVIM 1991;5:322-331. Roberts et al., Vet Rec 1992;130:471-472. Flanders et al., JAVMA 1999;215:46-48. Oyama et al., JVIM 2001;15:229-239. Santilli et al., Med Sci Monit 2005;11:BR168-175. Ferasin et al., JSAP 2005;46:115-120. Green et al., JVIM 2006;20:731. Bulmer et al., JVIM 2006;20:257-271. Wess et al., JVIM 2006;20:877-884. Johnson et al., JSAP 2007;48:4-11. Hildebrandt et al., JVIM 2009;23:31-38. Maisenbacher et al., AJVR 2009;70:455-463. Estrada et al., JVC 2009;11:79-88.

CARDIOVERSION OF ATRIAL FIBRILLATION: IS IT BETTER THAN RATE CONTROL?

Janice M. Bright, BSN, MS, DVM, DACVIM
Fort Collins, CO

INTRODUCTION

Atrial fibrillation (AF) and chronic heart failure (HF) are common and interrelated disorders, and the coexistence of these conditions is associated with a poor prognosis. Heart failure causes structural and electrical atrial remodeling thereby creating the basis for development and perpetuation of AF; and AF frequently results in hemodynamic deterioration and the worsening of heart failure.

In dogs, the goals of management of AF are to reduce or eliminate clinical signs and to prevent long term complications such as tachycardia-mediated cardiomyopathy and arterial thromboembolism. The question becomes whether these goals are best achieved by allowing the patient to remain in AF and controlling the ventricular rate (rate control) or by cardioversion to sinus rhythm (rhythm control). Because AF has deleterious effects on cardiac output, ventricular filling pressures and myocardial oxygen delivery, patients with HF may theoretically benefit more from restoration of sinus rhythm than dogs with lone AF.

HUMAN CLINICAL TRIALS

Various human clinical trials have been done in an attempt to compare rhythm control to rate control in patients with AF. Initial clinical trials including AFFIRM, PIAF, RACE, STAF and HOT-CAFÉ showed no survival benefit of rhythm control.[1-4] Because few of the patients in these trials had heart failure or sustained AF, the more recent AF-CHF study focused on patients in this subgroup of patients. Nonetheless, this study also failed to document a favorable effect of rhythm control on mortality.[6]

Several subsequent studies, however, including SCAF and CAFÉ-II have shown benefits of sinus rhythm including improved quality of life, improved left ventricular function, and improved long-term prognosis compared to HF patients managed with rate control.[7, 8] In both of these studies significant improvement was noted only if patients with no relapse of AF were compared to patients with adequate rate control. Similarly, retrospective analyses of patients with heart failure in the RACE study showed improvement in quality of life for those patients who could be maintained in sinus rhythm.[9]

CANINE STUDIES

What data are available from dogs with AF and HF to guide our AF management decision? There is certainly a dearth of information in the veterinary literature at this time to confirm benefits of restoration of sinus rhythm in patients with AF and HF. However, it has been shown in a canine model of progressive heart failure induced by sequential coronary microembolizations that the dogs developing AF had a more rapid progression of failure and higher plasma norepinephrine concentrations than dogs with equally severe HF remaining in sinus rhythm.[10] We have looked retrospectively at the severity of heart failure in dogs converted to sinus rhythm at our hospital using the International Small Animal Cardiac Health Council Heart Failure Classification to define the severity of clinical signs before and after cardioversion in 22 symptomatic dogs. Restoration of sinus rhythm was associated with either a reduction in severity in or a complete resolution of clinical signs in 18 of these 22 dogs (81.8%).[11]

Currently, the "Achilles heel" of a rhythm control strategy for management of AF and HF, in people and dogs is arrhythmia recurrence. Patients with HF are prone to a higher rate of recurrence of AF after cardioversion because they have structural heart disease and dilated atria, both of which are stimuli for AF relapse.[12 - 14] Nonetheless, many dogs with structural heart disease will remain in sinus rhythm after electrical cardioversion long enough to benefit from the risks and costs of the procedure.[12, 13] Our data allow us to predict with 95% confidence that for dogs with structural heart disease if the estimated chronicity of AF is 31 days of less, 70% will remain in sinus rhythm for more than 28 days after cardioversion.[13]

MANAGEMENT GUIDELINES

For human patients with AF and HF the ACC/AHA/ESC management guidelines (including a recent focused update) continue to advocate that the choice of rate versus rhythm control be dictated by the clinical scenario keeping in mind the clinical signs related to the arrhythmia and the ability of the patient to tolerate the rate and rhythm control medications.[15, 16] The best management choice for dogs with AF and HF (heart rate control or rhythm control) should similarly be individualized based on whether the dog is symptomatic primarily from an uncontrolled ventricular rate or from the AF itself (loss of AV synchrony & atrial contraction). In many dogs with HF it is difficult to make this determination, and a trial of aggressive rate control or a trial of cardioversion may be necessary for differentiation. However, because the duration of AF adversely affects maintenance of sinus rhythm after cardioversion, time spent on trials of rate control should be minimized.

It is important for veterinarians to bear in mind the fact that although currently available human and canine clinical studies do not irrefutably document the superiority of rhythm control over rate control, neither do these studies disprove the hypothesis that restoration of sinus rhythm is preferable to the continuation of AF, particularly when rate control fails to restore adequate quality of life.[17]

REFERENCES

[1]Wyse DG et al. New Engl J Med 2002;347:1825. [2]Hohnloser SH et al. Lancet 2000;356:1789. [3]Carlsson J et al. J Am Coll Cardiol 2003;41:1690. [4]Van Gelder TC et al. New Engl J Med 2002;347:1834. [5]Opolski G et al. Chest 2004;126:475. [6]Roy D et al. New Engl J Med 2008;358:2667. [7]Friberg L et al. Heart 2009;95:1000 . [8]Shelton RJ, et al. Heart 2009;95:924. [9]Hagens VE et al. Heart Rhythm 2005;149:1106. [10]Tisdale JE et al. J Card Failure 2006;12:747. [11]Bright JM et al. Kirk's Current Veterinary Therapy XIV 2009;739. [12]Bright JM et al. J Vet Cardiol 2005;7:85. [13]Bright JM et al. J Vet Intern Med 2008;22:114. [14]Deedwania PC et al. Am J Med 2010;123:198. [15]Fuster V, et al. 2006; JACC 48:854. [16]Wann LS et al. Heart Rhythm 2011;8:157. [17]Reiffel JA. J Atrial Fib 2008;1:31.

ASSESSMENT OF DIASTOLIC FUNCTION: PRINCIPLES & ECHOCARDIOGRAPHIC CORRELATES

Dr. Karsten E Schober, Dr med vet, PhD, DECVIM (Cardiology)
Columbus, OH

INTRODUCTION

The heart spends more than half its time in diastole (first relaxing and then filling to prepare for the next ejection). Although abnormalities of diastolic function are well recognized in clinical practice, in particular in cats with cardiomyopathy, its relative role in clinical heart failure and therapeutic decision-making remains unclear.[1-4] By definition, diastolic dysfunction refers to abnormalities in 1) active ventricular *relaxation* (delayed or incomplete return of myofibrils to pre-systolic length and tone) and 2) passive *compliance* (decreased LV diastolic distensibility or decreased chamber volume relative to any change in pressure, dV/dP) or its reciprocal, diastolic *stiffness* (abnormally increased chamber pressure induced by change in a given volume, dP/dV) leading to 3) abnormal *filling* (decreased end-diastolic volume or adequate end-diastolic volume only at the expense of increased filling pressure [FP]).[5] Diastolic suction closely related to ventricular untwisting, pericardial restraint, right ventricular pressure, heart rate, rhythm, afterload, and atrial function are additional factors influencing diastolic filling. An increase in left atrial pressure (LAP) to a level that creates a LA-LV pressure gradient great enough to provide adequate ventricular filling is a pathophysiologic hallmark of diastolic dysfunction[5-8] linking LV diastolic dysfunction to clinical signs. Ultimately, this elevation of LAP will be transmitted into the pulmonary vascular system leading to venous congestion and CHF. Although epidemiologic data on the incidence and importance of diastolic heart failure in small animals are sparse, evidence indicates that diastolic abnormalities in small animals are common.[9,10] Diastolic dysfunction is frequently encountered with feline cardiomyopathy associated with myocardial remodeling, injury, and repair, specifically with ischemic, hypertrophic, and restrictive heart disease, infiltrative cardiomyopathy, LV hypertrophy and fibrosis due to aortic stenosis or systemic hypertension, pericardial disease, tachyarrhythmia, and normal aging.

Conceptually, LV diastole can be divided into 4 distinct phases: isovolumic relaxation, rapid filling, slow filling, and atrial contraction. Isovolumic relaxation begins during midsystole (phase of reduced ejection), is energy dependent, and does not contribute to ventricular filling. In normal resting animals, the rapid/early filling phase driven by passive (instantaneous LA-LV pressure gradient) and active (suction and untwisting) forces provides most of the end-diastolic volume whereas slow filling or diastasis provides only 5% of the total filling capacity. The final phase, atrial contraction (or atrial "booster"), contributes the remaining LV volume. During exercise, tachycardia, and in situations in which early LV filling is compromised, the LA contribution to LV filling becomes crucial in maintaining cardiac output. Clinical features of LV diastolic dysfunction are nonspecific and may include gallop sounds, exercise intolerance, and CHF.[4] Isolated diastolic heart failure (CHF with normal EF [HFNEF] does exist in small animals.[9,10]

BIOCHEMICAL AND BIOMECHANICAL EVENTS

The lusitropic state of the LV (lusitropy, the ability of the cardiac myocytes to relax) is affected by biochemical events within the cardiomyocyte and biomechanical events in the LV (active relaxation) as well as biophysical properties of the heart (passive stiffness).[11,12] The major processes within cardiac myocytes that initiate diastole are cross-bridge detachment and Ca^{2+} reuptake by the sarcoplasmatic reticulum. From a hemodynamic point of view, the timing of the transition from systole to diastole in normal animals corresponds to the moment at which 81% to 84% of peak LV pressure has developed, but may occur even before aortic valve opening in CHF with poor systolic function.[11] Increasing cytosolic cAMP due to β-adrenergic stimulation phosphorylates phospholamban (a reversible inhibitor of SERCA pump activity) and thus, enhances the rate of relaxation. In contrast, calcium handling during diastole is altered in failing hearts leading to intracellular calcium overload and delayed or incomplete relaxation. It should be noted that relaxation delay per se is not necessarily associated with increased late-diastolic stiffness. However, the combination of delayed relaxation and myocardial hypertrophy, tachycardia, exercise, and increased afterload causes decreased filling and increased LV end-diastolic pressures (upward shift of the diastolic pressure-volume relationship).[3,11] Myocardial relaxation is also influenced by external factors such as the inherent viscoelastic properties of the myocardium (elastic recoil or passive restoring forces) determined by the extracellular matrix that consists primarily of connective tissue. In addition, titin and switched titin isoforms seem to contribute significantly to altered myocardial stiffness in heart failure. Abnormal collagen and myocardial mass are key determinants of myocardial stiffness and, in excess, cause LV filling abnormalities and elevated end-diastolic pressures. In addition, ventricular relaxation is inherently related to volume status, afterload, non-uniformity, and inotropic state.

DIAGNOSIS OF DIASTOLIC FUNCTION

Assessment of LV diastole using echocardiographic variables is difficult for several reasons, depends on many assumptions, and has become a controversial clinical subject: **1)** There are a high number of Doppler indices to assess LV diastolic function, but experts in the field do not uniformly agree on what constitutes appropriate measurement of diastolic function. In addition, interpretation of commonly used variables reflects integrative LV properties that lack diastolic specificity. Moreover, the predictive value of abnormal diastolic indices in small animals is not well known. **2)** There are differences between diastolic abnormalities (e.g., an abnormal Doppler E:A ratio), diastolic dysfunction (e.g., an abnormal LV pressure-volume loop), and diastolic heart failure (a clinical syndrome),[13,14] all of which are incompletely defined in small animals. **3)** Studies comparing invasive with non-invasive measures of LV diastolic function are rare in animals with naturally occurring disease. However, there may be important differences between Doppler measures of filling and indices of true intrinsic diastolic function and chamber properties. **4)** There is still a lack of echo reference values for diastolic function in dogs and cats owing to the complexity of diastolic events. Therefore, diagnostic thresholds are often arbitrarily chosen making the objective assessment of diastolic alterations impossible.

Invasive Assessment of Diastolic Function.

Two events are commonly assessed in the determination of LV diastolic properties: active relaxation and passive stiffness.[2,3,11,12] Relaxation may be characterized by the time constant of LV pressure decay during isovolumic relaxation tau (τ, equals the time required for the LV pressure to fall from its initial value to $1/e$ of that value), isovolumic relaxation time (IVRT; time between closure of the aortic valve and opening of the mitral valve), relaxation half time ($t_{1/2}$), average rate of relaxation (R_{avg}), or the peak negative rate of LV pressure fall in early diastole ($-dP/dt_{max}$), whereas ventricular passive stiffness (or operating chamber compliance) is related to the slope (Kc) of the LV pressure-volume curve (end-diastolic P-V or length-tension relationship) all requiring cardiac catheterization for determination. Tau and $-dP/dt_{max}$ may also be measured by Doppler aortic regurgitant velocity analysis, however, determination is time-consuming and thus not practically feasible. Other diagnostic imaging methods such as nuclear medicine radionuclide techniques, MRI, and ultra-fast CT may be applied to measure variables of LV diastolic function but are not widely available in veterinary medicine.

Echocardiographic Assessment of Diastolic Function.

LA enlargement may be an early manifestation of abnormal LV diastolic function and fluid retention. In addition, LV hypertrophy, excessive moderator bands, "sparkling" appearance of the myocardium (consistent with an infiltrative process), or pericardial thickening or tamponade make the presence of a diastolic abnormalities likely. As a general rule, systolic ventricular dysfunction is almost uniformly associated with diastolic dysfunction whereas isolated diastolic dysfunction (in the presence of unchanged or even exaggerated variables of global LV function) may exist. The most commonly used methods quantifying LV diastolic function employ echocardiography, using IVRT and transmitral and pulmonary vein Doppler flow patterns.[1,8,14] The diagnostic value of such Doppler-derived indices is, however, somewhat limited by the strong influence of heart rate, ventricular load, and age. More recently, new techniques such as color M-mode velocity propagation[12] and pulsed wave tissue Doppler imaging (TDI)[6,17] have been used clinically to assess LV diastolic function. Unlike mitral inflow and pulmonary vein Doppler flow patterns, TDI and color M-mode variables are less load-dependent, as shown in experimental animals with abnormal LV diastolic function.[6] Strain rate imaging including speckle tracking echocardiography and related variables[18] measuring ventricular diastolic deformation and myocardial stretch as well as myocardial untwisting are promising new tools to assess LV diastolic function, however, clinically useful applications of such techniques are still waiting for discovery.

1. Assessment of LV Relaxation.[12,15-19]

The isovolumic relaxation time (IVRT), peak early diastolic filling rate, early diastolic LV flow propagation velocity (Vp), tissue Doppler early diastolic wall motion velocity (Ea), and tissue Doppler isovolumic deceleration velocity are variables most frequently used to assess LV relaxation. In clinical terms, each of those variables has diagnostic advantages and limitations, and the most suitable single variable for clinical assessment of LV relaxation has not yet been determined. The IVRT (range, 37-55 ms in normal cats and 40-80 ms in normal dogs)[20,21] commonly measured from simultaneous Doppler recordings of transmitral flow and aortic outflow reflects LV relaxation, but also depends, among many, on LV load, heart rate, sympathetic tone, and age. Diastolic dysfunction cannot be excluded on the basis of normal values of IVRT. Prolongation of IVRT is suggestive of delayed or incomplete relaxation with normal or only slightly increased LVFP, whereas shortening of IVRT suggests relaxation delay with abnormally increased LVFP (or exaggerated relaxation as seen with sympathetic stimulation). Exceptions to the latter may occur, particularly in situations characterized by severe LV hypertrophy and fibrosis. Color M-mode early diastolic transmitral flow propagation velocity (Vp) may also be used to estimate LV relaxation. Vp has been shown to be related to invasively obtained *tau*,[22] and has been found to be less preload and heart rate dependent than IVRT and other Doppler-derived indices of transmitral flow. Vp has been reported in normal dogs (range, 45-70 cm/s)[20] and cats (range, 55-111 cm/s).[21] As with transmitral and pulmonary vein Doppler flow variables, Vp is markedly influenced by age. Technical problems to adequately record and measure Vp and its dependency on systolic LV function and chamber geometry causes high measurement variation that limits its practical use. Tissue Doppler imaging (or tissue Doppler) of the mitral annulus has also been used to assess LV relaxation. In the absence of regional ischemia and wall motion abnormalities, the early diastolic wall motion velocity (Ea or E') as well as wall motion indices of isovolumic relaxation (peak isovolumic deceleration velocity and delay, pDEC, and IVRT) have all been shown to be useful, relatively preload independent surrogates of *tau* and $-dP/dt_{max}$ and thus, LV relaxation.

2. Assessment of LV Compliance.[14,23]

Echocardiographic measures of LV compliance include the deceleration time of early diastolic flow (DT_E) and the ratio of late transmitral flow duration (A wave) to atrial reversal (AR wave) of pulmonary vein flow duration (A duration:AR duration), representing measures of early and late diastolic LV compliance, respectively. The diagnostic and prognostic value of DT_E is limited to cardiac abnormalities with concurrently reduced LV systolic function whereas the A duration:AR duration ratio may be applied independently of LV systolic function.

3. Estimation of LV Filling Pressures.[1,7,8,14,22,2-27]

Increased E velocity, an E:A ratio above 2.0 (2.2 in young, growing animals), an abbreviated DT_E, an increased E:IVRT ratio, and a decreased systolic fraction of pulmonary vein flow have all been reported in dogs and people as indicators of increased LV filling pressure (> 15 to 20 mm Hg), particularly if systolic dysfunction was present. However, similar echocardiographic findings may be seen in animals with left atrial dysfunction, markedly increased LV wall stiffness, and tachycardia with normal or only mildly elevated filling pressures. Moreover, if LV systolic performance is normal and myocardial stiffness is severely elevated such as found in cats with hypertrophic cardiomyopathy, or in the absence of atrial contraction such as with atrial fibrillation, the E:A ratio and the DT_E may not reliably be used to estimate LV filling pressure. In addition, different disease conditions (e.g., degenerative mitral valve disease – "volume overload-dominant influence" on LV filling; dilated cardiomyopathy – "systolic dysfunction-dominant influence" of LV filling; and hypertrophic cardiomyopathy – "relaxation abnormality-dominant influence" on LV filling) make the use of disease-specific reference values necessary, making the clinical use of DE variables in the assessment of diastolic chamber function more challenging. Recent studies in dogs and cats revealed that the simultaneous consideration of the peak E wave velocity divided by either the peak velocity of the tissue Doppler mitral annulus Ea wave (E:Ea, normal < 9 in canine DCM, < 13 in canine mitral valve disease, < 12 in cats with HCM) or IVRT (E:IVRT, normal < 1.88 in dogs with DCM and < 2.50 in dogs with mitral valve disease; value in cats unknown)[26,27] can provide a more accurate assessment of LAP. Those methods "adjust" the E velocity for the effects of relaxation and in essence yield estimates of LVFP. Of particular importance is that the relationship between those variables to LV filling pressures remains valid even in patients with preserved LV systolic function, fused transmitral flow waves, or atrial fibrillation. It needs to be proven, however, if those ratios become the diastolic equivalent to ejection fraction in the echocardiographic assessment of LV diastolic function.

Tissue Doppler-derived myocardial velocity gradients have also been used to assess segmental LV diastolic function and synchrony. Another novel, promising tool to study regional LV diastolic function is Strain Rate Imaging (SRI)[18] a technique relatively immune to cardiac translational motion and tethering. SRI depicts myocardial deformation properties and enables quantification of regional myocardial function with high temporal and spatial resolution. This is important when studying diastole, as very high amplitude, short-lived deformations occur during this time period (especially during the isovolumic period). Indices such as the time-to-onset of regional relaxation (T_R) and early diastolic strain (ε) and strain rate (SR) have been reported in people and have clinical potential to detect ischemic wall segments, investigate myocardial viability, and characterize cardiac function in cardiomyopathies.

NATURAL HISTORY AND STAGEING OF DIASTOLIC DYSFUNCTION

Doppler transmitral flow patterns are traditionally used to describe the natural history and document the severity of global LV diastolic dysfunction and obtain information on prognosis. Complimentary modalities such as pulmonary vein flow and TDI are needed for interpretation of equivocal findings and incomplete transmitral flow recordings. Diastolic dysfunction has been graded as follows: **Grade 1** – Delayed relaxation pattern (mild diastolic dysfunction), **Grade 2** – Pseudonormal flow pattern (moderate diastolic dysfunction), **Grade 3** – Reversible restrictive pattern, and **Grade 4** – Irreversible (fixed) restrictive pattern (both severe diastolic dysfunction). It has been shown in people, dogs,[9,26] and cats[10] that LV diastolic dysfunction often follows a progression through these categories with increasing severity of cardiac disease. In asymptomatic animals, normal LA size, IVRT, and E:A ratio are used to document *Normal diastolic function*. Relaxation delay is the earliest abnormality[11] associated with diastolic dysfunction leading to prolonged IVRT, E:A reversal (E:A < 1.0), and prolonged DT_E ("*Delayed relaxation*"). Early diastolic LV filling is reduced and compensated for by increased late diastolic filling flow. At this stage, LVFP is usually normal or only mildly elevated (particularly during exercise). Patients are asymptomatic at rest. As diastolic failure progresses, LV relaxation remains abnormal, LV compliance gets progressively worse, and LAP increases. This leads to a shortening of IVRT and DT_E and a temporary "normalization" of a previously abnormal E:A ratio ("*Pseudonormal flow*", intermediate pattern). That is, a pseudonormal flow pattern may indicate worsening of underlying diastolic abnormalities with the appearance of "normal" diastolic function on Doppler echocardiograms. Indices used to unmask pseudonormality of transmitral flow include decreased A duration:AR duration ratio, TDI early-to-late diastolic wall motion velocity ratio, and increased E:Ea or E:IVRT.[1,7,8,26,27] Pseudonormal mitral inflow due to markedly elevated LVFP is most often observed in animals with abnormal systolic function. If diastolic dysfunction is the primary pathophysiologic mechanism such as with hypertrophic cardiomyopathy and preserved systolic performance, the transmitral flow pattern may indicate "pure" relaxation abnormality for an extended time period despite markedly elevated LVFP. This is due to the impaired relaxation-dominant influence on transmitral flow that overshadows the effect of increased filling pressures. A further increase in LVFP and a severe decrease in LV compliance may lead to "*Restricticve filling*", characterized by significantly shortened IVRT and DT_E, increased E and decreased A wave velocities (E:A > 2.0). and increased E:IVRT. The LA is dilated and hypocontractile. Patients with restrictive filling are symptomatic and have a poor prognosis (irreversible) unless restrictive filling can be reversed by treatment (reversible). More recently, "*Pseudorestrictive flow*" has been identified in situations where there is volume-overload dominant influence on LV filling (e.g., in dogs with moderate and severe mitral valve disease) characterized by elevated E:A (>> 2.0) and elevated Ea with often normal E:Ea and E:IVRT.

Table 1: Commonly used echocardiographic variables to assess LV diastolic function in dogs and cats

Diastolic Component	Echocardiographic Variable
Relaxation	*IVRT (ms), *Ea (cm/s), Vp(cm/s), *E:A ratio, pDEC delay (ms) and velocity (cm/s)
Filling	*E:A ratio, S:D ratio, end-diastolic volume index (ml/m^2)
Compliance/stiffness	*LA size, *DT_E (ms; early diastolic compliance), *A-duration:AR-duration ratio (late diastolic compliance)
LV filling pressure	*E:IVRT[1,*] ratio, *E:A ratio, *E:Ea ratio[1], E:Vp ratio[1], DT_E

*Variables commonly preferred by the author; [1]applicable also with fused E and A waves and with atrial fibrillation.

REFERENCES

1. Ommen SR. **Curr Opin Cardiol** 2001;16:240; 2. DeMaria AN, et.al. **J Am Coll Cardiol** 199;34:1659; 3. Banerjee A. **Progr Pediatr Cardiol** 1999;10:65; 4. Kass DA, et.al. **Circ Res** 2004;94:1533; 5. Marwick TH. **J Am Coll Cardiol** 2009;54:47; 6. Nagueh SF, et.al. **J Am Coll Cardiol** 2001;37:278; 7. Nagueh SF, et.al. **Circulation** 1999;99:254; 8. Firstenberg MS, et.al. **Am J Cardiol** 2000;36:1664; 9. Schober K, et al. **J Vet Cardiol** 2002;4:7; 10. Schober K, et al. **J Vet Intern Med** 2008;22:1469; 11. Gillebert TC, et.al. **Acta Cardiol** 1997; 52:223; 12. Constable P, et.al. **J Vet Intern Med** 1999;13:5; 13. Zile MR, et.al. **N Engl J Med** 2004;350:1953; 14. Gilman et al. **J Am Soc Echocardiogr** 2007;20:199; 15. Aurigemma GP, et.al. **Am Heart J** 2004;148:e12; 16. Brun P, et.al. **J Am Coll Cardiol** 1992;20:420; 17. Garcia MJ, et.al. **J Am Coll Cardiol** 2000;35:201; 18. Abraham TP, et. al. **J Am Coll Cardiol** 2002;39:1531; 19. Hashimoto I, et.al. **J Am Coll Cardiol 2004;44:1459**; 20. Schober K, et.al. **Tieraerztl Prax** 2002;30:417; 21. Schober K **Proc 7th Meet FECAVA** 2001:6; 22. Little WC, et.al. **Circulation** 1995;92:1933; 23. Schober K, et.al. **Am J Vet Res** 2003;64:93; 24. Oyama MA, et.al. **J Vet Intern Med** 2004;18:667; 25. Schwammenthal E, et.al. **Am J Cardiol** 2004;94:488; 26. Schober K, et al. **J Vet Intern Med** 2010;24:1358; 27. Schober K, et al. **J Vet Intern Med** 2008;22:578.

INSIGHTS INTO ANTIARRHYTHMIC MODALITIES FOR THE TERMINATION OF ATRIAL FIBRILLATION

Bruce G. Kornreich, DVM, PhD, DACVIM (Cardiology)
Ithaca, NY

Atrial fibrillation (AF) is characterized by either intermittent (paradoxical) or permanent disorganized atrial electrical activity that overrides normal sinoatrial node pacemaker function and that is variably conducted through the atrioventricular node to the ventricles. AF leads to loss of atrioventricular synchrony; and the rapid, irregular depolarization of the atria caused by AF decreases ventricular preload as the fibrillating atrium is unable to move blood forward in a normal fashion. Combined with the rapid ventricular rate that occurs in response to AF, clinically significant tachycardia may predispose to decreased cardiac output, the development of systolic failure, and to increased mortality in patients with AF. In humans, patients with AF are at an increased risk for the development of thromboembolic disease. This tendency has not been identified in veterinary patients with AF, in which AF is most commonly permanent.

AF is the most common supraventricular arrhythmia diagnosed in both human and veterinary patients, and its etiology remains controversial. In small animals, AF is most commonly diagnosed in patients with atrial dilation, presumably due to increased surface area of affected atria in addition to altered atrial myocardial fibrosis, cell cycle control, and cellular differentiation, all of which may promote the maintenance of a sufficient number of reentrant circuits for the maintenance of AF. Abnormalities of the autonomic nervous system have also been implicated in some individuals with AF. Dogs with structural heart disease combined with AF have decreased survival times when compared to dogs with AF in the absence of structural heart disease (lone AF). Lone AF is a common cause of poor performance in horses, perhaps because the large surface and high vagal tone (which decreases action potential duration) in equine atrial tissue predisposes to the maintenance of a critical number/pattern of spiral waves. Lone AF is common in humans, where it has most commonly been shown to originate at the junction of the pulmonary veins and the left atrium. The origin of lone AF in veterinary patients with AF has not been definitively identified in clinical patients, although studies suggesting a role for the pulmonary vein autonomic ganglia in vagally mediated lone AF have been carried out in both *in vitro* and *in vivo* canine atrial preparations.

Various models and techniques have been used to study the origin and maintenance of AF over the past few decades. These include *in vivo* electrophysiologic studies, *ex vivo* electrophysiologic studies, single channel and whole cell patch clamp studies, molecular biologic and proteomic studies, optical mapping, and mathematical/*in silico* modeling. These studies have provided valuable information about the mechanism of AF in both human and animals, and have shed light on both conserved and unique aspects of AF arrhythmogenesis across species.

The ideal treatment goal in patients with AF is termination of AF and restoration of normal sinus rhythm (cardioversion). Successful cardioversion is less likely in patients with structural heart disease and/or with long standing AF, likely due to electrical and structural remodeling that occurs at the molecular and protein levels in the atria of affected individuals. Available modalities for cardioversion can be roughly divided into pharmacologic, electrical, and surgical interventions, with pharmacologic and electrical defibrillation being the methods commonly available in veterinary patients. Each of these modalities carries a risk:benefit ratio that varies with patient condition, and this must be considered along with availability of technology and financial considerations when making clinical decisions regarding a given patient.

Recent studies have identified novel modalities for the termination of AF in clinical patients. Among these are new drugs including late sodium channel blockers, atrial selective ion channel blockers, antifibrotic agents, and stretch receptor blockers. Cell and gene-based therapies for termination of AF and for ventricular rate control in AF patients have also been investigated, as has low voltage electrical cardioversion. These potential treatment modalities hold the promise of exciting new therapeutic options for the treatment of AF in clinical patients. This seminar will discuss some of the recent advances in pharmacologic, electrical, and gene-based therapies for patients with AF.

REFERENCES

Ambrosi CM et.al. **Heart Rhythm** 2011;8(1):101, Fenton FH et.al. **J Vet Cardiol** 2008;10(2):87, Ehrlich JR, et.al. **Drugs** 2009;69(7):757, Lu y, et.al. **Circulation** 2010;122(23):2378, Menaut P, et.al. **J Vet. Cardiol** 2005;7(2):75.

SINUS NODE MESSAGING TO THE REST OF THE HEART: LINKING THE ECG TO THE ELECTROPHYSIOLOGY

N. Sydney Moise, DVM, MS, DACVIM (Cardiology & Small Animal Internal Medicine)
Ithaca, NY

INTRODUCTION

Recently we discovered a unique phenomenon about the way the heart of the dog beats.[1] In the dog we have found that certain heart rates occur very infrequently, such that over the range of heart rates the probability for certain frequencies is not equal.[1] Based on 24-hour electrocardiographic recordings of hundreds of dogs (data from our Holter Lab), a consistent range of heart rates are in low density or in other words, avoided. An analogy of this peculiar observation might be the height of people. In a randomly selected population it would be expected that the height of persons would cover all possible heights over the range possible; however, what if in the measurement of 140,000 people (this is the number of times that the heart of a dog beats in 24 hours), no person measured between 67 to 70 inches. In the dog, between the heart rates of 60 and 80 beats/minute the beat density is very low. We named this range of heart rates the Zone of Avoidance. This pattern of the beating canine heart has not been previously described or the mechanism ascribed. We have identified the Zone of Avoidance using methods to examine over time the beat to beat intervals and density of beats at the respective rates.

RR INTERVALS

In electrophysiology the heart rate is most commonly identified at its instantaneous rate which is related to the time period between each beat. Because this is measured on the electrocardiogram using the time distance between R waves (when heart block is absent), the term RR interval is used. Thus, the instantaneous beat rate is given as the RR interval duration in msec rather than the clinical term of heart rate which is given as beats per minute. Herein, we use the term 'RR interval' as we discuss the Zone of Avoidance.

AUTONOMIC INFLUENCE ON RHYTHM

Normally the dominant pacemaker that regulates the heart rate and rhythm is the sinus node.[2,3] The ion channels in the sinus node are responsible for spontaneous discharge of electrical activity to stimulate the heart to beat.[3,4] Once these cells discharge an electrical impulse must exit the sinus node region and spread a wave of depolarization throughout the heart.[2] In addition to the inherent control by ion channels of the sinus discharge rate, the sinus node is influenced by the autonomic nervous system.[5-10] The sinus node responds to such neurological input to increase or decrease the spontaneous rate of discharge. During high physical activity or excitement the heart rate is fast, whereas during sleep the heart rate is slow. Different animal species have varying influence from the balance of autonomic tone.[5-10] As a species, the dog has strong influence from the parasympathetic nervous system and thus, the sinus node of the dog does not have a constant discharge rate, but instead sinus arrhythmia is prevalent.[5-10] While the mechanism behind the variation in sinus rhythm is complex and still disputed,[10] it is probably a composite of influences from the dorsolateral pons, features of breathing (rate, tidal volume, expiratory/inspiratory ratio) and impact of the baroreceptor response.[5-10]

TACHOGRAMS, HISTOGRAMS AND POINCARE PLOTS

To evaluate the beat to beat intervals and heart rate densities we are examining three visual methods (24-hour tachograms with hourly displays, 24-hour histograms with hourly displays, and hourly Poincaré plots) for which we are developing means of quantification. We obtain these representations of RR intervals from our Holter analysis system (Forest Medical, Syracuse, NY).

A beat to beat interval tachogram gives the RR interval data over the course of time. The time of day (X-axis) is plotted against the RR interval (Y-axis).

An RR interval histogram provides insight into the frequency of RR intervals. Over a given time period a long RR interval series is used to construct the histogram such that the RR interval (X-axis) is plotted against the number or occurrences (Y-axis) of that interval. If the heart rate over the course of a day increases and decreases over the full range of RR interval possibilities, with slowing and speeding of equal probability, the histogram might be expected to have bell-shaped curve or some degree of skewness, depending on the distribution of the rate.

A Poincaré plot (also known as a Lorenz plot) is a two-dimensional RR interval scattergram of beat to beat variability.[11] By convention, each RR interval is plotted (X-axis, RRn interval) against the following RR interval (Y-axis, RRn + 1 interval). Because the heart rhythm and RR interval is not random the resulting structure of this plot has shape.[12,13]

SPECIES AND HEART RHYTHM

Other species, including humans and horses have sinus arrhythmia due to the influence of autonomic tone; yet, only the dog has the Zone of Avoidance. In Figure 1 the Poincare plots of the human, horse, and cat are shown. In Figure 2 the tachograms of a horse and cat are displayed. These show no Zone of Avoidance. In contrast, the tachogram of the dog as shown in Figure 3 shows a 'white strip' of low density beats and the corresponding Poincare of one hour in this time frame is shown in Figure 4 of a dog. However, the Zone of Avoidance is most prominent during periods of high parasympathetic tone and absent when the heart rate is rapid.[1] The question posed is whether it is the impact of the parasympathetic tone that causes the Zone of Avoidance, or is it simply a slow heart rate such that other influences that decrease rate would also cause this phenomenon.

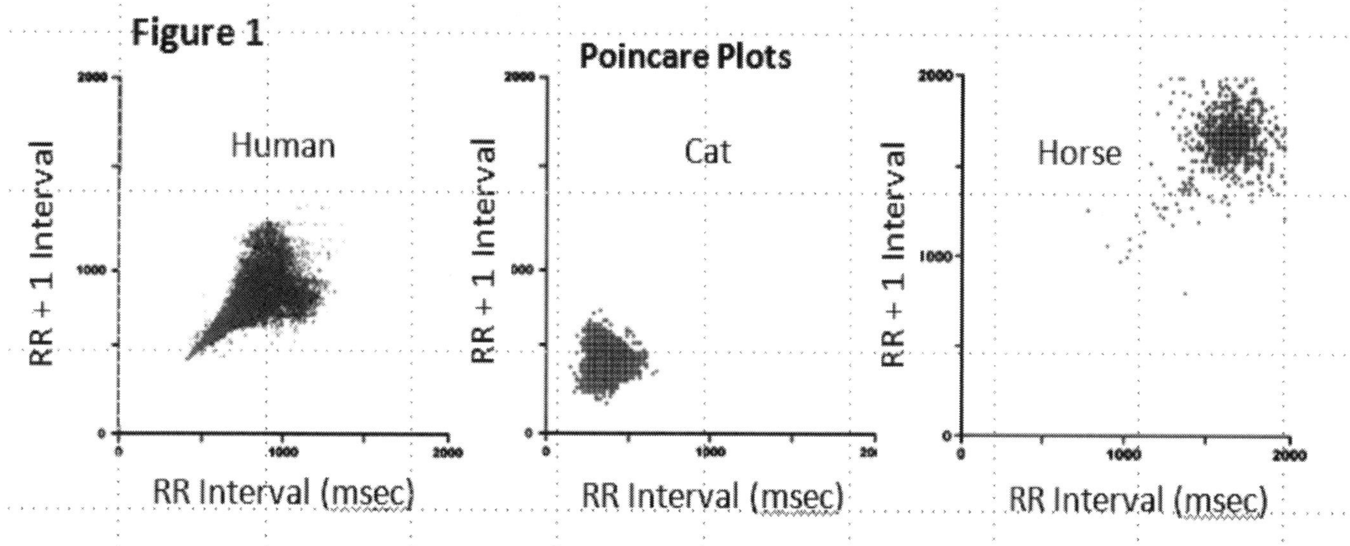

Figure 1 Poincare plots show a comparison amongst a person, cat and horse. The positioning is reflective of the respective heart rates.

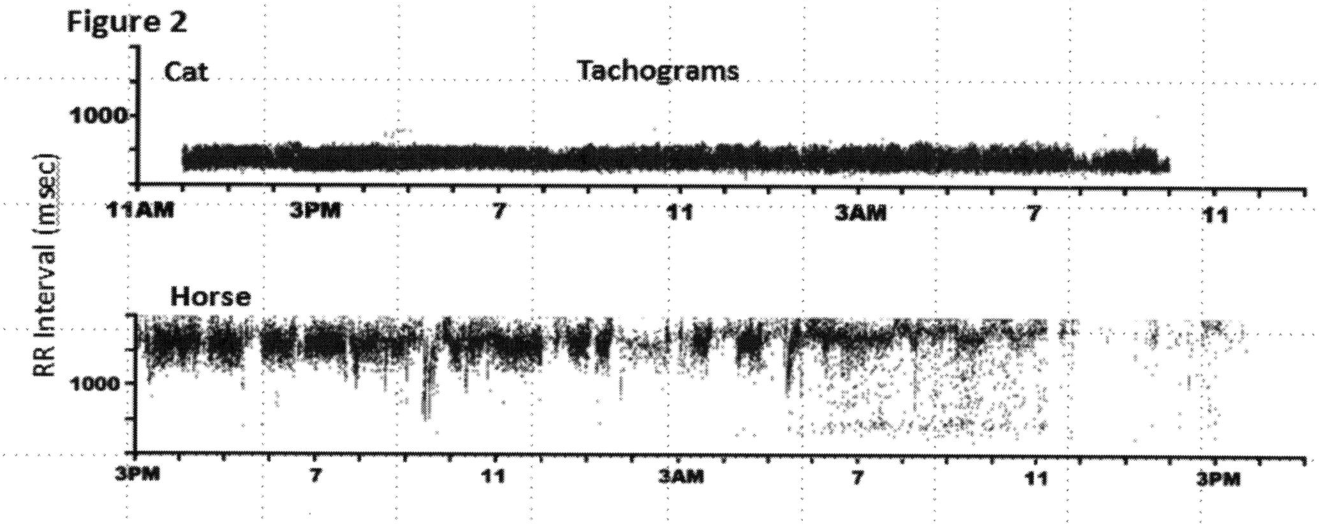

Figure 2 The tachogram of the cat and horse show the faster and slower rates of these two species.

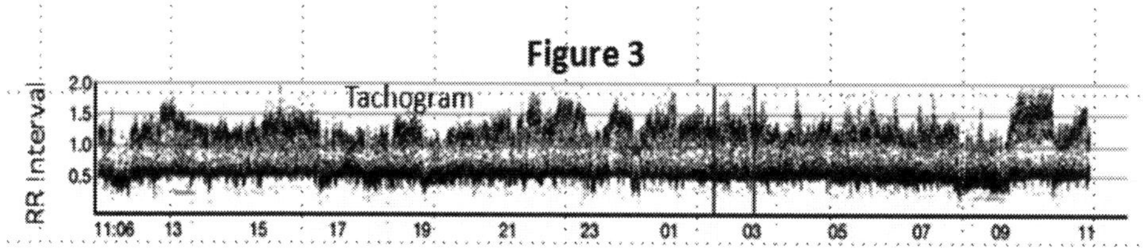

Figure 3 Tachogram over 24 hours from a dog showing a 'white strip' whereby RR intervals are 'avoided'.

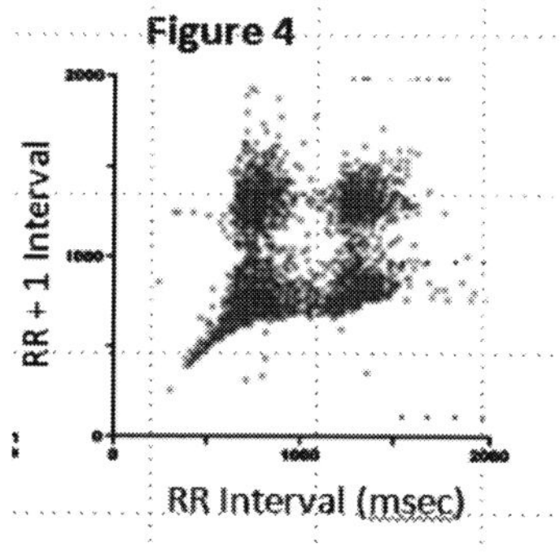

Figure 4 Poincare plot from a dog showing the corresponding low density of certain RR intervals in the dog. This single hour corresponds to a single hour of the tachogram.

REFERENCES

1.Moïse NS, et al. J Vet Cardiol. 12(3):191-196, 2010. 2. Fedorov VV, et al.. *Circ Res* 2009;104:915-923. 3. Boineau JP, et al. *Circ Res* 1984;54:332-347. 4. Bucchi A, et al. *J Physiol* 2006;572:335-346. 5. Horner RL, et al. J Appl Physiol. 1995 Jun;78(6):2003-13. 6. Miyashita Y, et al. J Cardiovasc Electrophysiol. 1999;10(8):1066-76. 7. Dick TE, et al. Respir Physiol Neurobiol. 2009;168(1-2):76-85. 8. Eckberg DL. J Appl Physiol 106: 1740–1744, 2009. 9. Strauss-Blasche G, et al. Clinical and Experimental Pharmacology and Physiology (2000) 27, 601–606. 10. Bucchi A, et al. J Am Coll Cardiol.43(3):483-490,2004. 11. Climent AM, et al. IEEE Transactions on Biomed Engin 2009;56:433-442. 12. Esperer HD, et al. A.N.E. 2008;13:44–60. 13. Nakajima K, et al. J Cardiovasc Electrophysiol.2002;13:896-901.

EQUINE REFERRAL GUIDELINES

Debra C. Sellon, DVM, PhD, DACVIM
Pullman, WA and Athens, GA

INTRODUCTION

In December 2009, representatives of the American College of Veterinary Internal Medicine (ACVIM), American College of Veterinary Surgery (ACVS), American College of Theriogenology (ACT), American College of Veterinary Emergency and Critical Care Medicine (ACVECC), and the American Association of Equine Practitioners (AAEP) met in Las Vegas, NV during the annual meeting of the AAEP. The purpose of the discussion was to identify areas of potential collaboration between board-certified veterinary specialists and general practitioners as they relate to equine veterinary practice. One topic of discussion at this meeting was the possibility of developing a joint statement with guidelines for referral of equine patients between veterinarians. After this meeting, Dr. Michelle Barton, President of the Specialty of Large Animal Internal Medicine in the ACVIM worked to confirm the collaboration of all participating groups and identify a mutually agreeable timeline for development of referral guidelines. This document summarizes the organization and timeline for this effort. The accompanying presentation will outline the contents of the referral guidelines and provide a forum for ACVIM Diplomate input on the document.

PROCESS AND TIMELINE

Senior leadership from each specialty college and from the AAEP were engaged in initial conversations to obtain a consensus on the scope of the project, appropriate procedures and timeline, nomination of task force members, and development of a mutually-agreeable reporting structure. The intent of this effort is to provide clear information to help the public understand the definition of a specialist, how to find or request a specialist consultation or examination for their horse, and assist veterinarians in understanding how they may work together to optimize their relationship for the benefit of their equine patients. Dr. Debra Sellon was chosen to chair the task force. The participating specialty colleges and AAEP leadership indicated their willingness to participate in the process by October 1, 2010. After negotiation of details, all colleges and the AAEP agreed on procedures and timelines by November 1, 2010. Panel members representing each organization were nominated and their participation confirmed. The panel members signed contracts and conflict of interest statements in early December 2010. The task force consists of two members from each participating specialty college and four members from the AAEP. All specialty college representatives on the task force are concurrent members of the AAEP. Efforts were made to include representation from private referral hospitals, academic referral hospitals, specialists in ambulatory practice, and veterinarians working in a wide variety of non-specialty practices. This group of individuals is referred to as the Equine Referral Ethics Task Force; their names and professional affiliations are shown in Table 1. Each organization was encouraged to develop their own methods and procedures for obtaining additional member input into the drafting of the guidelines if so desired.

Table 1. Members of the Equine Referral Ethics Task Force

Affiliation	Name
ACVIM	Debra Sellon (Task Force Chair)
ACVIM	Nathan Slovis
ACT	Claire Card
ACT	Ben Espy
ACVECC	Pam Wilkins
ACVECC	Fairfield Bain
ACVS	Eric Mueller
ACVS	Rolf Embertson
AAEP	Reynolds Cowles
AAEP	Matt Brokken
AAEP	Melinda Freckleton
AAEP	Rich Forfa

The Referral Ethics Task Force convened an initial meeting at the 2010 annual meeting of the AAEP in Baltimore, MD. The first draft statement of their work is due to be submitted to the AAEP Board of Directors by May 15, 2011. The first public presentation of this draft will occur at the ACVIM Forum in Denver, Colorado in June 2011. The information will subsequently be presented at the annual meetings of the ACT, ACVECC, ACVS, and AAEP between August and December 2011. At each public presentation comments will be solicited to guide editing of the document. A second draft of the manuscript is due no later than February 1, 2012. This draft will be made available to all members of the participating organizations for additional comment. A final written draft of the guidelines are due to each organization by May 1, 2012. Approval of the final draft is requested from the participating organizations no later than June 1, 2010 with a goal of submitting the final statement for website and journal publication by July 1, 2012. Individuals wishing to comment on this task force and its efforts are encouraged to contact Dr. Sellon or any member of the task force.

HOW TO MAINTAIN OR INCREASE FA CASE LOAD IN VETERINARY SCHOOLS: 10 RULES WE FEEL HELP US OVER THE YEARS

Gilles Fecteau, DMV, DACVIM and André Desrochers, DMV, MS, DACVS, DECBHM
St-Hyacinthe, Québec

HISTORICAL BACKGROUND INFORMATION

The farm animal service at St-Hyacinthe Veterinary School is a referring center for Eastern Canada bovine practitioners. Approximatively 1000 cases are referred each year to the hospital. Amongst the clinician team, there are 3 Diplomates of the ACVIM, 2 Diplomates of the ACVS and 2 Diplomates of the ECBHM. The classical patient is an Holstein dairy cow from one of the 2 following population: a commercial cow from a client near the Veterinary school or a purebred cow from a anywhere in the province of Québec, Ontario or the Maritimes. The small ruminant case load is proportional to its economical importance in Québec.

(1) Complementary expertise and team effort

One important aspect is that the different clinicians should have complementary expertise. All can perform most tasks and correctly manage any type of cases, however we feel that natural interest for particular subject and type of cases is useful to broad the service offer to the clientele. By diversifying the interests, the group ensures that the literature search and body of knowledge remains current. In fact, not only the level of expertise is current but the development of new technique or new approach will emerge in some specific fields of interest. The achievement of the common goal has to be a team effort. This means that the individual goals may be a bit behind the actual collective objective. Always remember the reason we are there; teach veterinary medicine while treating true patients…

(2) Define clearly who your clients are

As the team develops and evolves (new hiring and retirement), it is extremely important to remember our mission. The primary goal of our work and as a consequence, the primary client is the veterinary student. The society needs you to graduate enough veterinarians each year to maintain the steady state between retirement and new graduates in the profession. Therefore, our primary and most important client is the veterinary student. A secondary and also extremely important client is the referral veterinarian. Since our policy is to accept only referrals, we would loose our clientele if the contact with the referral was inadequate. Of course, another important client is the producer, owner of our patient. Remember that they play an important role in supporting the veterinary school through funding that becomes available. Of course, the patient is also your client and patient care has to be amongst the most important criteria that directs the decision of your services. The interns and residents also clients and future colleagues, deserve time and attention. They are close from your staff and students and often offer valuable suggestions to improve the delivery of services.

Refuse case direct from the client

It can be very tempting to have client directly calling you instead of having the referral veterinarians involved. We systematically refuse that approach and found that in the long run, the service is winning. If a client calls us directly, we remain polite and ask him to confirm with his primary veterinarian that the patient be referred to us. In other words, we explain that we work in team with his veterinarian and that we need to talk and discuss the case prior to the hospitalisation. The type of case evolved over the years and very few cases are now considered routine case.

(3) Today's farm animal students become tomorrow's referral

It is important that students interested by farm animal practice realise and understand how a referred case is managed and what services can be offered to bovine practitioners. They often think naively that referral veterinarians are practitioners with fewer competencies and in fact they will do better once they graduate. It is our role to demonstrate that it is in fact a proof of competency to refer a case for care that cannot be provided at the farm. Teach them their limits and what we can do for them as a referral veterinarian. We also explain in rounds that it is always easier to criticize a case while you come later. The reality of the situation in which the primary veterinarians was working cannot be described and understood easily after the case is referred to us. The condition may also evolve after the first examination was performed. Treat them professionally because they will be your referral for the next 20 years.

(4) Availability: phone (pager)

Referral veterinarians need you at any time of the day or night. Most phone calls will NOT result in a case coming through your clinics. However, if you are not answering all phone calls, you will miss the phone call related to a referral case. We must accept that most phone calls are in fact free consult or free continuing education. It is the price to pay to see interesting cases and maintain clinical activities. We are extremely easy to reach during the day and multiple team members are in fact available to answer the phone. There is a price to pay and how many publications or grant are lost while you are on the phone over the years is difficult to estimate?

(5) Referral letters and phone calls

One extremely important factor for referral veterinarians is that they need to be kept informed. This is actually even more important if the primary diagnosis changes after your assessment of the case. This could be misinterpreted by the owner as a mistake by the referral veterinarians. For each referral, a summary (very brief) is sent by fax or e-mail the day of discharge. Our hospital software system does not allow any possibility to easily share information with veterinarians. The summary must be sent quickly even if it means that it is not as complete or perfect or informative, as you would like it to be. Many producers will call the referring veterinarians as soon as the cow returns home. It is very uncomfortable for a veterinarian to answer questions from his clients if he does not know the details of the case. Cases where the producer appears unhappy with the care provided by his referring veterinarians will be approached with caution. Phone contact will be made with the referring veterinarian to explain the situation, what was said and not said. We do not believe that long and fastidious referral letters are appreciated by our referral veterinarians if they arrive late.

(6) Be on site for emergency

Emergency situations are always problematic situations. There is a lot of tension in the air. Moreover, your personal is reduced (AHT, barn crew, students, colleagues). It is probably the worst time for a junior Faculty or residents to be initiated to manage a case by himself. "Be on site" to take the pressure off the shoulder of the residents; we can share responsibilities and alternate roles during the next emergency. The senior staff takes care of client communication (often a bit more tricky) or technical aspects of surgery. We tend not to let the residents manage emergencies by themselves until the final year. We feel that the client often makes their idea of our quality of service based on the level of care they received after hours. It is difficult for us to let an inexperienced veterinarian in charge of emergencies because of they are all referral and complicated cases.

(7) Monitor the billing policy and ensure possibilities of academic reductions of fees

I know this is easier said than done but in order to maintain some cases you need some flexibility with billing. All cases cannot blindly bill in the same manner. In some situations extraordinary case needs extraordinary cares and extraordinary bills as well. The subject of billing is always prone to conflict with administration and this should be addressed with tenacity. It is the yin and yan theory. They want us to charge more; we want them to charge less… It is a constant battle. The only thing we can say is if you loose this battle, you will be left without a caseload and it will likely cost more to the veterinary school to achieve its primary goal: train veterinary students how to practice! However, the billing reductions need to be monitored and rules needs to be established. We don't want a clinician to have a reputation to charge less than others. We also feel that some opportunities are good to increase the income.

Seek opportunities and accept the challenges

Often, our service is challenge by unusual type of cases. This forces us out of our comfort zone and at first the temptation can be to say No! Think twice since some of those challenges can be tomorrow's case load. As examples, the down cows and tank did great for us; the cloned calves also. The secret is to do it in a way that creates a win-win situation. We charge for what it is worth and adapt our clinics and time to feel good about the services provided.

(8) Support for government agencies (epidemio-surveillance)

We feel that support from government agencies is very important. As a referral center, we could provide sentinel surveillance to detect strange cases or exotic diseases. Referral veterinarians tend to call us if they see cases with unusual presentation. This is in fact a service that some government agencies should be offering? Collaboration with government agencies must be encouraged. Sometimes funding becomes available to establish surveillance and monitoring of specific diseases.

(9) Contact with referral veterinarians through CE and publications

The proximity with referral veterinarians can be maintained by CE and workshops. In our experience, in the week following a conference or a workshop, some of the veterinarians who attended the session refer cases. Almost as if they were reminded that that possibility exists. It is important to present challenging cases but not to become too esoteric. Cases series or retrospectives studies with a good clinical description and some ideas in percentage of the prognosis are excellent. Practitioners appreciate presentation of all the mistakes that you did yourself before diagnosing the problem and realising that you were facing a new disease or clinical entity.

(10) Match clinical needs and academic objectives

As Faculty member, it is important that you prepare your summary for promotion as well as making sure the clinics work. It is an equilibrium and what we found works best is to try to match those 2 objectives instead of making them in opposition. That would mean clinical cases become teaching material, CE material and your research material. We try to stay away from research grants that keep you away from the clinics unless it is your professional objective. Make it a mission or a section statement so the administration will acknowledge it and most likely consider it in the process of promotions. At the same time it will be more rewarding and encourage others to do the same.

Conclusion

There is no secret recipe to maintain a case load. It needs time and commitment. The time in clinics should be a priority. Recruiting a good team with a common goal, means you're on the road for success. If you are looking for big grant and zillions of publications in prestigious journal, you may be disappointed with a heavy commitment in clinics. However, do not think that your life will be boring.

A SHORTAGE OF LARGE ANIMAL VETERINARIANS: WHAT THE ECONOMICS TELL US

Jonathan M. Naylor DACVIM, DACVN, PhD
University of Ross School of Veterinary Medicine, St Kitts and Nevis.

INTRODUCTION

There is considerable interest in the supply of food supply veterinarians (FSVM). A number of recent reports suggest that there is currently a shortage. A Delphi study found that experts predicted a need for an additional 12.5% more FSVM for the period from 2004 to 2016 (Prince et al. 2006). An invited opinion suggested that it is difficult to interest students in a career in FSVM because practitioners are too busy to encourage high school students to take up this career path and because there is a paucity of sufficient food animal experience in veterinary schools to capture a student's interest (Sterner 2006). However, a study of livestock owners in Tennessee did not find evidence that producer's perceived a wide scale shortage of FSVM veterinarians (Jensen et al. 2009). Another approach to the question of veterinary supply has been to use economic indicators, principally the change in cost of veterinary services and change in veterinary salaries as an indicator of relative changes in supply and demand (Brown and Silverman 1999). This paper reviews the literature and reports new data on economic indicators and the supply of veterinary services over the last 10 years.

METHODS

Data on U.S. population were retrieved from the U.S. census bureau (www.census.gov) using the American Fact Finder information sheets for the years 2000-2009 and DataFerrett for 1999. Population data for the month of July of each year were used in the analysis. Numbers of beef cows, milk cows, hogs and pigs, and sheep including lambs, were retrieved from the USDA/NASS website http://quickstats.nass.usda.gov/. January 1st figures were used for Sheep and Lambs, and Cattle Inventory; June 1st values for Hog Inventory. Chicken numbers were measured as commercial chickens, F1, slaughtered. Numbers of owned cats and dogs were provided by the American Pet Products Association (Greenwich, CT) and are derived from surveys conducted every two years.

The number of veterinarians in the U.S. was measured using 3 sources. The Bureau of Labor Statistics generates Occupational Employment Statistics using an annual mail survey measuring occupational employment and wage rates for wage and salary workers in nonfarm establishments, by industry. This survey measures employment of veterinarians in all walks of life including private industry, and local, state, and federal government agencies. The survey does not measure self employed veterinarians. Statistics were downloaded from www.bls.gov/oes/. The AVMA maintains a database and publishes membership data. Membership is voluntary; voting members must be graduates of a Veterinary Medical College and either be a member of a constituent veterinary organization or be sponsored by 2 AVMA members. The third source was the AVMA database of actively employed US veterinarians residing in the US; this includes AVMA members and non-members. Non-members are defined as non-AVMA member veterinarians born after 1937, who received their veterinary degree after 1961-3. Periodically, the number of veterinarians in this database is published (Shepherd et al. 2004, 2005, 2006, 2007, 2008, 2009, Wise et al, 2000, 2001, 2002, 2003, 2004).

The employment of new graduates was estimated from published surveys of initial career choice of new graduates. The numbers were adjusted for the survey response rate and the number of schools surveyed to give an estimated total number of new graduates taking each initial career choice. Data were analyzed from 2000 to 2009 as there was insufficient published data for 1999 to be included in the results.

Details on the Consumer Price Index (CPI) for all products and for veterinary services to pets in urban areas were obtained from the U.S. government, bureau of labor statistics using the searchable database at http://data.bls.gov:8080/PDQ/outside.jsp?survey=cu.

To investigate the relationship between starting salary and field of initial employment, all salaries were converted to 1999 dollars using the appropriate change in CPI for all items. The number of veterinarians entering each field was then compared to starting salary using linear regression over the range of years from 2000 to 2009 inclusive.

RESULTS AND DISCUSSION

Changes in Size of the Veterinary Workforce

In 1999 a survey predicted an overall 7% increase in the demand for veterinarians needed to satisfy all demands between 1999 and 2009 (Brown and Sliverman 1999). In reality, from 1999 to 2009 the number of AVMA members increased by 24%, the number of working veterinarians who are not self employed by 27% and the estimated number of total graduates from US veterinary schools increased by 11%.

Another change in employment pattern is the increase in the number of new graduates opting for advanced studies as internships, residencies, or other forms of further education.

Changes in Animal Numbers

From 1999 to 2009 the number of beef cows declined by 6%, the number of sheep declined by 21%, while there were increases in the number of milking cows by 2%, the number of hogs and pigs by 10%, and the number of F1 hybrid chickens slaughtered by 4%. A major change has been an increase in the number of owned pets. Between 1998 and 2008, the number of owned dogs increased by 23% and the number of owned cats by 22%. The increase in the number of owned pets parallels the increase in size of the veterinary workforce.

Changes in Cost of Veterinary Services and Annual Salary

A 1999 paper predicted that the supply of veterinarians would exceed demand until 2009 and that there would be downward pressure on real incomes for veterinarians (Brown and Silverman 1999). However, salary data for the last 10 years show that real income of veterinarians rose and that the increase is faster than that of the general population. From 2001 to 2009, the mean annual salary of all employed U.S. workers who were not self employed increased by a compound rate of 3.11% annually, for all veterinarians who were not self employed by 3.35%, the starting salary for small animal exclusive veterinarians in private practice by 5.28%, and the starting salary for large animal exclusive veterinarians by 6.53%. Mean student debt increased by 6.84% annually over the same time period. From 1999 to 2007, surveys of AVMA members indicated that the salary of all small animal exclusive veterinarians rose by 5.18% and of all food animal exclusive veterinarians by 7.28% compounded annually. Similarly, from 1999 to 2009, the cost of veterinary services to pets in urban environments has risen by 79%, which is more rapid increase than the 29% increase in the Consumer Price Index for all items.

Growth in the number of veterinarians, salary, and cost of veterinary services, would be an expected response to increased demand for veterinary services and a shortage of labor in response to the rapid growth in the numbers of owned pets.

Correlations between Starting Salary and Career Path of New Graduates

The correlations between the number of new graduates seeking animal employment and selected variables over the period 2000 – 2009 are shown in Table 1. A decline in the number of graduates entering large animal practice was associated with an increase in starting salaries, expressed in 1999 dollars. Similar trends exist for employment as small animal veterinarians. Instead progressively more new graduates have opted for advanced studies at reduced rates of compensation compared to those available in practice. There is also a high positive correlation between starting salary and the number of students undertaking advanced studies. This may indicate that high starting salaries are an attempt to lure students away from advanced study. Surveys of new graduates repeatedly show progressiveness of the practice and mentorship as being more important than salary in graduate's choice of early career path, so enhanced mentorship opportunities in internships and other forms of advanced training may explain initial preference for advanced studies (Jelinski et al. 2009, 2009a). Another possibility is a decline in how confident new graduates feel about their ability to perform in private practice. First opinion case load in veterinary teaching hospitals has been declining while concerns over liability and welfare may limit student experience, increasing the attractiveness of mentored training as an intern or similar form of advanced study.

Table 1. Correlations between starting salary and the estimated number of veterinarians accepting initial employment in food animal predominant private practice between 2000 and 2009. Starting salary was adjusted to a 1999 dollar value.

Item Description	Large Animal Salary	Small Animal Salary	Small Animal Employment	Total Number of Graduates
Large Animal Employment	-0.712	-0.780	0.791	-0.564

SUMMARY

The economic data in this report suggest an unanticipated increase in the size of the veterinary profession in the last decade. There has been only a modest increase in the number of graduates from U.S. schools during this period. The unanticipated increase in size of the profession is probably mainly due to U.S. citizens obtaining veterinary training at out of country locations. Ross University school of Veterinary Medicine, for example is the largest single supplier of veterinarians into the U.S. market. An increase in starting salaries at rates much greater than those of the general work force may have made it easier for students to accommodate the additional debt typically associated with out of state education. Despite the increase in the size of the profession a variety of economic indicator are consistent with a shortage of veterinarians between 2000 and 2009. These include an increase in starting salaries at rates much greater than those for the overall population and an above average increases in the cost of veterinary services for pets in an urban setting. Despite the increasing financial rewards for large animal veterinary practice, the number of new graduates from U.S. schools opting for initial employment in food animal practice continues to be in decline. In 2010 an average of only one new U.S. veterinary school graduate per state entered food animal predominant or food animal exclusive veterinary practice (Shepherd and Majchrzak 2010). An encouraging trend is the widespread political interest in addressing this issue and the deployment of new initiatives such as the USDA's Veterinary Medical Loan Repayment Program and the AVMA/AVMF Veterinary School Loan Repayment Program. Many schools are increasing their enrollment which may help satisfy pent up demand in both large and small animal practice.

REFERENCES

Brown JP and Silverman JD. J Am Vet Med Assoc 1999; 215:161. Jensen KL et al. J Vet Med Educ 2009;36:30. Jelinski MD, et al. Can Vet J 2009; 50:943. Jelinski MD, et al. Can Vet J 2008; 49:161. Prince, JB et al. J Am Vet Med Assoc 2006;229: 57. Shepherd AJ. J Am Vet Med Assoc 2004; 224:1677. Shepherd AJ. J Am Vet Med Assoc 2005; 227:1084. Shepherd AJ. J Am Vet Med Assoc 2006; 229:1087. Shepherd AJ and Majchrzak S. J Am Vet Med Assoc 2007; 231:1813. Shepherd AJ. J Am Vet Med Assoc 2008; 233:1067. Shepherd AJ and Majchrzak S. J Am Vet Med Assoc 2009;235:523. Shepherd AJ and Majchrzak S. J Am Vet Med Assoc 2010;237, 795. Wise JK and Adams C-L. J Am Vet Med Assoc 2000;216:184. Wise JK and Lloyd JW. J Am Vet Med Assoc 2001;218:710. Wise JK and Gonzalez ML. J Am Vet Med Assoc 2002; 220, 179. Wise JK and Lloyd JW. J Am Vet Med Assoc 2003;222:312. Wise JK and Shepherd AJ. J Am Vet Med Assoc 2004;224:213.

CURRENT METHODS IN INFECTIOUS DISEASE RESEARCH

Paul J. Plummer, DVM PhD DACVIM (LAIM)
Ames, IA

INTRODUCTION

The development of modern molecular techniques and the rapidly decline cost of genetic sequencing has resulted in a plethora of molecular information regarding infectious disease agents. Furthermore, over the last decade there has been an increasing awareness that many disease processes are the result of multiple infectious disease agents working in synergy to result in physical manifestation of the disease. Perhaps one of the first examples of such a disease was periodontal disease in humans where it was demonstrated that there was the necessity of synergistic growth of two or three bacterial species (*Treponema* sp., *Porphorymonas* sp. and *Bacteroides* sp. in what is known as the "red complex"). It is now commonly known that the presence of one of these organisms in the absence of at least one other does not result in significant disease. Further research demonstrated that not only was there the necessity for synergistic growth of these organisms but there was also an association of organisms with a decline in other bacterial flora resulting a large scale population shift in the normal flora of the gingival pocket. These findings, along with other similar disease processes identified in humans, demonstrated that the traditional view of Koch's postulates were perhaps too simplistic and opened up a new area of disease research focused on understanding infectious disease from a broader population structure perspective. In order to facilitate this change there has been a rapid development of new molecular techniques and research methods that allow studying infectious disease from a more global perspective and these techniques are rapidly augmenting or in some cases replacing traditional molecular techniques such as the polymerase chain reaction. As an increasing number of papers are published utilizing these new techniques, internist are more often faced with the need to translate research findings into clinical relevant knowledge. A basic understanding of the techniques used to obtain the data is critically necessary to effective translation. This review will attempt to summarize some of the more prevalent methods and approaches to this research with the goal for making the reader aware of the limitations and applications of the method. Additionally, a better understanding of what techniques are available would be expected to result in the development of further translational research collaboration between internal medicine clinicians and molecular microbiologist with the ultimate goal of increasing our knowledge of disease and therapeutic interventions.

TECHNIQUES FOCUSED ON PURE CULTURE

Although there is an increasing awareness of the role of synergistic polymicrobial disease processes in veterinary medicine, many diseases are still associated with single agent infections. Additionally, detailed study of synergistic associations necessitates an understanding of the molecular interactions between organisms and requires complete genetic information regarding each of the species of interest. As a consequence there is still a significant need for information derived from pure culture growth conditions of single organisms. This information acts as a foundation for studying disease pathogenesis and allows for the detailed study of interactions between organisms on the molecular scale. The following techniques are focused on improving our understanding of the molecular architecture of single organisms and allow for the application of the community level studies that will be described later in this manuscript.

Full Genome Sequencing

As with all biology, the genetic sequence of an infectious disease organism ultimately determines its molecular repertoire with which to cause disease. Improving our understanding of the genetic sequences of infectious disease agents not only allows us to study gene expression and molecular regulation but also plays an important role in populating our genetic databases with the information necessary to utilize the community based research methods that will be discussed later. In the early days of genetic sequencing the process was very time consuming and laborious, involving months of labor to sequence even small portions of a bacterial genome. The time and effort involved in this process necessitated a very high price tag for such work and limited its application to a limited number of medically important organisms – mostly associated with human disease. A major milestone in genetic sequencing technology came with the development of new sequencing by synthesis methods in the mid 1990s. Prior to that time, sequencing was performed by the traditional "Sanger method" which required the incorporation of dideoxynucleotides that resulted in chain termination and a band that could be observed in an agarose gel. The new sequencing by synthesis methods rely on monitoring for the release of pyrophosphate during DNA chain elongation. As a consequence the sequencing method can be easily automated. This method was then paired to a platform that allowed for monitoring these reactions on a nano scale and a new generation of massively parallel sequencing methods were born. These techniques allow for large numbers of DNA strands to be simultaneously sequenced in relatively short periods of time. The two best known platforms for performing these techniques are the Roche 454 GS FLX platform and the Illumina platform, both of which are widely available in academic settings. The GS FLX platform currently offers the ability to sequence 400-600 million base pairs (in 400-500 bp reads) in a single 10 hour sequencing reaction. The company expects to increase their read length to 1000 bp per read in the near future. Current cost of a single run is between $5000-7000 which allows for full genome sequencing of most bacteria for roughly $10,000. This number is less than half the price of the first bacterial genome that we sequenced using this technology in 2006 and the price is expected to continue to drop. The other platform, Illumina, can generate 25 million sequences of up to 150 bp long per lane for an approximate cost of $2500. The obvious price advantage of the Illumina platform must be balanced with the shorter base read sequences which makes de novo sequencing using Illumina more difficult. However, if a good genome backbone exist (i.e. a genetically similar organism is already sequenced to be used as a backbone for sequence assembly) we can get a full genome sequence for under $2500. Recently a new sequencing platform marketed by Pacific Biosystems was released. This platform allows for even longer base reads then the 454 platform at a cost similar to the Illumina. Currently these sequencers are not commonly available, however their use will likely increase as availability issues are addressed. In short, these new methods have now made full genome sequencing of bacterial species and even specific strains a common place event in microbiology laboratories. Using this sequencing data in association with comparative genomic techniques that rely on comparing genome sequences to predict virulence or potential virulence associated genes have allowed for rapid improvements in our understanding of the molecular basis for disease induction by infectious organisms.

Transcriptomics

Once the full genome sequence of an organism is known we can utilize several methods for monitoring mRNA synthesis and gene transcription. The focus of these studies would be to determine which genes are integrally involved in pathogenesis, survival in a given environment or synergistic interactions. In principle the idea would be to compare the gene expression in two samples of interest to see how transcription changes in response to the "treatment" tested. The two samples could represent a variety of different specimens such as a virulent strain vs. a non-virulent strain, an *in vivo* sample obtained from a site of infection vs. *in vitro* growth, or a strain grown in the presence of a second bacterial species (testing for synergism) vs. pure culture. Essentially, transcriptomics measure the relative amount of mRNA in the sample for each gene of interest compared to the baseline sample. There are two commonly used platforms for transcriptomics – solid-phase microarray probes and Illumina sequencing. In the case of microarrays, small DNA probes are designed to hybridize to cDNA (DNA complementary to mRNA, used because of propensity of RNA to degrade) from each of the genes of the organism. These probes are immobilized on a microscope slide in an "array" that allows the researcher to know that a given spot on the array matches a given gene probe. Each of the cDNA sample is then labeled with a fluorescent dye hence the color of the spot (i.e. fluorescence of the color matching the sample labels) gives a relative ratio of the expression of that gene. If sample A is labeled "green" and the other sample B is labeled "red", then a green spot on the array is highly expressed in sample A and not in sample B while a yellow sample is equally expressed in both. An alternate method of monitoring transcription is the use of the Illumina sequencing platform described above. In this application the mRNA of the sample is collected and made into cDNA as for the microarrays. However, instead of hybridizing the cDNA to the solid array the cDNA is subjected to sequencing using the Illumina short read method. Since 25 million sequences can be obtained the actual number of sequences for each open reading frame (i.e. gene, based on gene sequencing matching) can be determined and compared to the reference sample. Both techniques have been widely applied to study the molecular basis of infectious disease in a variety of processes.

Proteomics

Although less frequently used, new techniques to monitor the synthesis of proteins on the cellular scale have been developed and are commercially available. This technique has the advantage over transcriptomics of being able to determine if post-transcriptional modification or regulation alters the amount of protein present. The most widely utilized platform at this time is the Matrix assisted laser depolarization/ionization time of flight methodologies (MALDI-TOF). Using this technique the individual identities of a complex mixture of proteins can be deciphered. Alternatively, fluorescently labeled 2 dimensional protein electrophoresis can be performed and used to determine differences in protein expression. This technique however has the disadvantage of having a lower sensitivity and requires additional testing to determine the identity of altered proteins.

COMMUNITY BASED CULTURE INDEPENDENT METHODS

In cases where a disease process involves either the presence of multiple synergistic organisms or a change in microbial community structure (i.e. enteritis associated with loss of normal flora) it is necessary to be able to determine a local "census" of organisms present in a sample. In such a case, both the types and relative number of organisms is important to this process. At the foundation of this technique is the ability to use one of a number of molecular techniques to compare amplified gene sequences to a database of known gene sequences – hence the absolute necessity of continued genome sequencing to build the database. Many of these procedures are termed metagenomics which is the study of genomes directly isolated from some specific environmental sample. The techniques used to collect the data; the depth (i.e. how much data) and the analysis vary from procedure to procedure and are associated with significant differences in cost.

Culture Independence

One common theme of many of these techniques is that they rely on culture independent methods to avoid the bias imposed by bacterial or viral culture. It is estimated that less than 10% of the bacteria residing in most environments can be cultured using known laboratory media. As a consequence, even culture with a wide variety of medias in multiple environments (aerobic, anaerobic and microaerophilic) is likely to only result in isolation of 10% of the organisms present in the sample. Culture independence relies of isolation of total DNA from an environmental sample and subsequent molecular techniques to determine what is present in the sample. Thus, it is limited only by the ability to collect DNA from all of the organisms and by our ability to see what DNA is there using molecular techniques.

Terminal Restriction Fragment Length Polymorphism (TRFLP)

Readers may be aware of a technique called RFLP which relies on PCR amplification of a specific gene from a sample and the subsequent digestion of the PCR product with a restriction enzyme to result in fragments of different lengths. This technique has been widely used in molecular epidemiology studies to estimate the genetic relatedness of a variety of samples of the same genus and species. For community studies this procedure has been modified in several specific ways. First, the gene used for PCR amplification is the 16S ribosomal gene conserved in all bacterial species. Thus regardless of the bacterial genus the primers are likely to amplify a product. Second, the PCR primers are terminally labeled with a fluorescent marker that allows for automated size determination using capillary electrophoresis. This technique is applied to community based studies as follows. First total DNA would be collected from the sample of interest and purified using commercially available kits. Then terminally-tagged universal 16S PCR primers are used to amplify a portion of the 16S gene of all bacteria present in the sample. This amplified DNA now becomes our target of interest and the number of each type of 16S gene sequences is determined by the number of original bacteria of that species (and hence the number of DNA strands with that sequence). In TRFLP we then digest those amplicons using specific restriction enzymes to result in the formation of a variety of different length terminal fragments (labeled with a fluorescent dye due to primer labeling) each coinciding with a specific genus of bacteria (the 16S region is conserved in its presence but highly variable in its sequence thus meaning that each genus has a different sequence that will result in different cut-sites for the restriction enzyme). Following digestion the sample is run on a capillary electrophoresis unit and the sizes of each terminal fragment are recorded. These sizes can be compared to a database that contains over 60,000 bacterial species to determine the identity of the sample. Furthermore, the fluorescent intensity of a given fragment will directly relate to the number of bacteria represented by that size fragment. Thus, both the diversity and the population structure of the sample can be determined. The limitations of this technology are that some bacteria have some overlap in sizing and a finite number of bacteria can easily be identified using the electrophoresis. Furthermore, each sample needs to be processed and analyzed separately. Analysis is fairly straight forward and in relatively simple environments the results can be very useful. Total cost for sample processing and analysis is roughly $25 per sample.

Metagenomics by sequencing

Perhaps the best platform for performing metagenomic studies is the use of the massively parallel sequencing platforms described above for full genome sequencing. Using these platforms we can acquire large amounts of sequence data derived from total DNA in a sample and then compare those sequences with known DNA sequences in databases to determine the origin. This allows the identification of not only bacterial components but also viral and fungal if applied appropriately. The number of sequences obtained for each type of organism can then be used to directly determine the relative ratios of the various isolates in the sample. This allows for the development of pie graphs and bar graphs that can readily demonstrate community population structure changes over time or with disease. If the primary focus of the project is on bacterial species, then a 16S PCR amplification can be performed prior to sequencing and allows one to maximize their depth of coverage on bacterial species. The addition of barcodes to samples allow for pooling of sample into a single sequencing reaction to improve cost efficiency. In one recent sample submitted by our laboratory we pooled 15 samples together that were barcoded for a single run. We received just over 1.5 million usable reads (average 350 bp long) with a range of 50,000-72,000 sequences per pooled sample. This means that for each sample we can essentially count and divide a minimum of 50,000 bacterial sequences into their genus and species giving us a good picture of what is there and the relative concentrations.

CONCLUSIONS

The techniques described in this manuscript have the potential to greatly impact the study of infectious disease in veterinary medicine. Our laboratory is already applying most of the techniques to a variety of disease processes of interest. With an improved understanding of the limitations that traditional culture methods have imposed on our research these new techniques allow us to pursue culture independent approaches with much broader global community based studies.

SPINAL CORD INJURY CLINICAL TRIALS UPDATES

Natasha Olby, Vet MB, PhD, DACVIM (Neurology)
Raleigh, NC

INTRODUCTION

Acute spinal cord injury remains one of the most common complaints for dogs presenting to veterinary neurologists, and while understanding of the pathophysiology of traumatic spinal cord injury has improved over the last 20 years (1), there remains a lot of controversy over the optimal therapy of such injuries. There is a need for carefully performed, randomized, blinded clinical trials to evaluate potential therapies. There is evidence dating back to the 1960s that surgical decompression of acute intervertebral disc herniations improves the outcome of dogs that have suffered thoracolumbar disc herniations causing paraplegia with loss of nociception (2). This evidence was not generated in blinded, randomized trials, but the outcomes of this grade of injury with and without surgery are so different as to provide compelling evidence for surgery. While the majority of clinical reports of different therapies of spinal cord injury in dogs are anecdotal or descriptions of case series, the number of carefully designed studies is growing. A blinded, placebo controlled trial has been performed to evaluate the efficacy of an oscillating field stimulator at improving outcome (3) and this trial showed some beneficial effect on recovery of sensory function. Another blinded trial evaluated the effect of N-acetylcysteine, and failed to show an effect (4). There are also interesting reports of phase 1 trials describing therapies such as polyethylene glycol (5) and olfactory nerve ensheathing cell transplants (6).

While blinded, placebo controlled trials are held up as the gold standard for determining the optimal therapy of a condition, medical history is littered with reports of trials that are either misleading or plain wrong. Ultimately, the trial is only as good as its design, and simply adding a placebo will not correct for design errors. Appropriate inclusion criteria and outcome measures are extremely important, and the course of recovery from the disease without additional intervention must be clearly described to allow appropriate statistical design. The course of recovery from acute spinal cord injury due to thoracolumbar intervertebral disc herniations in dogs has been described in detail (7) and two problems emerge. The first is that the recovery of dogs with clinically incomplete lesions is good. The vast majority of these dogs recover even from paraplegia to achieve a level of motor and autonomic recovery that is very acceptable to the owner and difficult to improve upon. This renders the majority of dogs that present to veterinary neurologists unacceptable for the purposes of a clinical trial, other than perhaps to evaluate whether speed of recovery could be ameliorated. This leaves dogs that present with clinically complete lesions: such lesions cause paraplegia with loss of nociception. The natural recovery of these dogs following decompressive surgery varies from a near complete recovery to permanent paralysis, making this group far more attractive to the investigator. As a group, the recovery of these dogs is incomplete and can potentially be improved, but the inability to identify *a priori* which of these dogs is most severely injured means that group sizes needed to detect a treatment effect are large. This is compounded by the relative scarcity of dogs with this grade of injury and thus the need for large multicenter collaboration. In this presentation, the development of new outcome measures will be described and the interim results of two different clinical trials will be discussed.

OUTCOME MEASURES IN CANINE SPINAL CORD INJURY

The outcome of acute, thoracolumbar spinal cord injury in dogs has largely focused on pelvic limb function, and most score function on scales modified from Frankel based on the presence or lack thereof of nociception and motor function. Groups have generated combined scores of motor function, nociception and proprioceptive placing (3, 8), but it is important to understand the limits of evaluating these rather subjective parameters in dogs on ordinal scales. More sophisticated scores evaluate the lateral position of the pelvic limb paw and the variability of strides using kinematic analysis (9, 10).

Gait combines both motor and sensory function, and an ordinal gait scale, known as the BBB scale, is used universally in experimental studies on rodents. At the Canine Spinal Cord Injury Program at North Carolina State University, we developed and tested an expanded open field gait scale as a modification of the BBB scale for use in dogs (11) and used this gait scale to describe the outcome of surgically treated acute disc herniations (7). However, this scale generates ordinal data, and so in an effort to complement the open field gait scale with more objective continuous data, we developed two different scores of dogs walking on the treadmill. We determined that normal dogs walk with a particular foot sequence on the treadmill more than 99% of the time, and therefore defined normal coordination according to this foot sequence. The next problem we encountered when trying to develop an objective scoring system is that paraparetic dogs can go from apparent paralysis to effective ambulation and back to paralysis in a short space of time. By scoring extended sequences of treadmill walking, we determined the minimum number of consecutive footsteps that had to be captured and scored to accurately represent the dog's full abilities was 100 thoracic limb steps. We generate two scores from dogs walking on the treadmill. The first is a simple representation of the number of times the patient takes a step with the pelvic limbs. A total of 100 consecutive thoracic limb steps are videotaped, and the number of pelvic limb steps taken over the same period is counted. The stepping score is calculated from the number of pelvic limb steps divided by 100 and expressed as a percentage. Additional information on the coordination of thoracic and pelvic limb steps is gained from the coordination score in which the number of pelvic limb steps that are coordinated with the thoracic limbs is counted and again expressed as a percentage. In order to make this scoring consistent, we have strict definitions of what constitutes a footstep that have to be fulfilled in order to count a step. We also complement our gait score, stepping score and coordination score with ordinal scales based on owner perception of changes in function over a period of a week. We have established a working group of veterinary neurologists interested in spinal cord injury, and plan to develop recommendations on scoring systems to measure the outcome of dogs with acute spinal cord injuries.

POLYETHYLENE GLCYOL(PEG) AND METHYLPREDNISOLONE SODIUM SUCCINATE (MPSS) IN CANINE SPINAL CORD INJURY: A MULTICENTER CLINICAL TRIAL

There is experimental evidence that both PEG and MPSS can improve outcome of acute spinal cord injury, although clinical evidence is more limited. The NASCIS trials evaluating MPSS in the 1980s and 1990s are examples of trials that had some design limitations and led to more confusion than answers. In a multicenter trial funded by Morris Animal Foundation, we aim to recruit 145 dogs in the first 24 hours after acute intervertebral disc herniations to be randomized to one of three treatment groups: MPSS, PEG and placebo. All dogs are surgically decompressed. In order to be included, dogs have to be paraplegic and nociception negative. Dogs are evaluated at 0, 24 hours, hospital discharge, 2, 4 8 and 12 weeks by videotaping neurological examinations and gait, and by completion of standard neurological examination sheets by veterinarians. The videotapes are scored by two blinded investigators at North Carolina State University at a later date. Twenty centers have been recruited including both private specialty practices and academic institutions, surgeons and neurosurgeons (http://cvm.ncsu.edu/vth/clinical_services/neuro/acute_disc.html) and all are sent blinded drug packages and receive training in the trial procedures. Recruitment to the trial has been negatively impacted by the recession, but at this point 61 cases have completed the trial and an interim analysis is ongoing. Early analysis of the outcomes has revealed some interesting findings. The first is a lack of adverse side effects across the population evaluated. At least one third of the cases have received high doses of MPSS, but to date, adverse consequences of this drug regime have not been reported to us. The second, somewhat alarming finding is the rate of ascending myelomalacia. Just over 16% of all cases recruited developed this fatal complication. While we still do not know the identity of dogs in each treatment group, the study safety monitor did ensure that this complication was evenly distributed across all three treatment groups and is unlikely to be a consequence of therapy. The incidence of this complication seems high, but matches early reports of 'the ascending syndrome' from Funquist (2). The overall recovery rate of independent ambulation at 12 weeks was only 49% across all three groups, with final breakdown of the outcomes in each group pending statistical analysis by independent statisticians. It is unlikely that we will have sufficient power to detect modest treatment effects at this time, and will continue to recruit cases to the trial for two additional years.

4-AMINOPYRIDINE (4AP) AND ITS DERIVATIVES IN CHRONIC CANINE SPINAL CORD INJURY: A BLINDED, DOUBLE CROSS OVER TRIAL

Blockade of rapidly activated voltage gated potassium channels can restore conduction to demyelinated axons in addition to enhancing synaptic neurotransmission and muscle contraction strength. 4AP has been used for its potassium channel blocking effects since the 1970s and was recently shown to improve walking speed in patients with multiple sclerosis in a blinded clinical trial. 4AP does have potentially beneficial effects in dogs with spinal cord injury (12) but unfortunately is somewhat limited by adverse side effects such as tremors and seizures. Derivatives of 4AP have been developed (13) and shown to be safe in dogs (14), and the most promising of these, TBOC, is being evaluated in a blinded clinical trial on chronically paraplegic dogs in a trial funded by Neurometrix. Dogs recruited to this trial must have suffered an acute spinal cord injury that has left them nociception negative at least three months after injury. The study design is shown below and ensures that each dog receives both drugs with a two-week washout period in between therapies.

Weeks 1 and 2	Weeks 3 and 4	Weeks 5 and 6	Weeks 7 and 8	Weeks 9 and 10
Placebo or drug 1	Placebo or drug 1	Washout	Placebo or drug 2	Placebo or drug 2

Once a week, patients are videotaped when walking along a carpeted floor for open field gait scoring, and when walking on the treadmill for stepping and coordination scores. Owners complete a questionnaire on a weekly basis. The study aims to recruit 20 cases, and currently 19 dogs have either completed or are completing the study protocol. Early analysis of the first five cases showed that owners could identify the weeks their pet was receiving 4AP, but not TBOC. The initial drug dosing rate for TBOC was based on guinea pig and *in vitro* work and was found to be inadequate. Escalation of the dose of TBOC to a comparable dose of 4AP (approximately 1mg/kg TID for TBOC and BID for 4AP) had a noticeably beneficial effect and so this dose was adopted for the rest of the study. An interim analysis after eight dogs had completed the protocol showed a trend for both 4AP and TBOC to improve stepping, gait and coordination scores. This trend was not significant, but was consistent, with both 4AP and TBOC having similar effects. Owners were clearly able to identify the weeks their pets were on placebo and several opted to receive 4AP in the long term (due to the lack of commercial availability of TBOC at this time). One patient had an adverse response to 4AP, developing anxiety, tremors and seizures. No patients had adverse responses to TBOC.

ADDITIONAL CLINICAL TRIALS

We have three additional clinical trials running at the Canine Spinal Cord Injury Program at North Carolina State University. These include a blinded trial evaluating TBOC in dogs that are ambulatory paraparetic, evaluation of the effect of cranberry extract on the incidence of urinary tract infections following acute spinal cord injury (in collaboration with Gulf Coast Veterinary Specialists), and evaluation of percutaneous transplantation of autologous adipose derived stromal cells and Schwann cells with inosine infusion in chronically paraplegic dogs. All are blinded, placebo controlled trials designed to detect modest treatment effects and all are funded by Morris Animal Foundation.

CONCLUSIONS

There are a growing number of veterinary neurologists taking advantage of this common, naturally occurring problem in dogs to investigate improvements in therapy of spinal cord injury in a manner that will help to guide our treatment choices in the future. There is a need for some consensus on the most appropriate methods of scoring outcomes, and continued collaboration to generate adequate case numbers for suitably powered clinical trials.

REFERENCES

1. Olby NJ . **Vet Clin N Am Sm Anim Pract** 2010;40:791-807. 2. Funquist, **Acta Orthop Scand Suppl.** 1962;56:1-50. 3. Borgens et al. **J Neurotrauma** 1999;16:639-57. 4. Baltzer WI, et al. **Spine** 2008;33:1397-402. 5. Laverty et al. **J Neurotrauma** 2004;21:1767-77. 6. Jeffery et al. **J Neurotrauma** 2005;22:1282-93. 7. Olby NJ et al. **J Neurotrauma** 2004;21:49-59. 8. Levine et al. **Prev Vet Med** 2009;89:121-7. 9. Hamilton et al. **BMC Neurosci.** 2007;8:77. 10. Hamilton et al. **BMC Vet Res**. 2008;4:47. 11. Olby NJ et al. **Am J Vet Res** 2001;62:1624-1628. 12. Blight et al. **J Neurotrauma** 1991;8:103-119. 13. McBride et al. **Eur J Pharm Sci**. 2006;27:237-42. 14. Olby NJ et al. **J Vet Pharmacol Ther** 2009;32(5):485-91.

Canine Brain Tumor Clinical Trials at the University of Minnesota

G. Elizabeth Pluhar, DVM, PhD, DACVS
Saint Paul, MN

Intracranial neoplasia occurs more frequently in dogs than in humans (14.5 per year per 100,000 canine populations at risk compared with 4–5 per 100,000 humans) (Frenier 1990, Kornegay 1986). Canine glioma reportedly occurs most commonly in brachycephalic breeds such as Boxers and Boston terriers (Snyder 2006). No sex predilection has been reported and brain tumors are recognized with greater incidence in animals over six years of age. Gliomas can arise in almost any location in the brain, but are more commonly supratentorial. We and others have found many similarities between human and canine GBM such as: overexpression of the epidermal growth factor receptor and mutation of the p53 tumor suppressor gene, extensive invasion into normal brain, peritumoral edema and necrosis, hemorrhage, compression, herniation, and obstructive hydrocephalus. Similar to humans, the prognosis for dogs with brain tumors in general is poor regardless of therapeutic intervention. However, much less is known about canine glioma treatment outcomes because only a small number of studies with few dogs have been reported. The median survival time for dogs with glioma that did not receive any type of treatment ranges between 6 and 13 days (Foster 1988, Heidner 1991), and between 60 and 80 days in dogs that receive palliative therapy only (Turrel 1986, LaRue 2001). Radiation therapy may have increased survival time in one dog with glioma (176 days) as compared to corticosteroid therapy in three dogs with glioma (18, 40 and 64 days) (Turrel 1986). We calculated the survival data for 9 dogs recently diagnosed with glioma based on MRI characteristics of an intra-axial mass at our institution. The median survival time of these dogs that did not receive any therapy other than corticosteroids and anticonvulsants was 29 days, range 1-128 days.

Canine meningioma is reported to be the most commonly occurring intracranial neoplasm in dogs, comprising ~ 40% of all primary spontaneous tumors diagnosed at necropsy. Dolichocephalic breeds (Braund 1986) and Golden retrievers (McGrath 1962, Andrews 1973, Greco 2006) may be at risk for developing meningiomas; and most tumors are detected in dogs > 7 years old (Axlund 2002, Greco 2006). Although human meningiomas usually occur in middle-aged females (Altinores 2004), no sex predilection has been reported in dogs.

Therapeutic treatment goals for canine meningiomas are complete excision or cytoreduction of the tumor. Current treatments for dogs include palliative therapy, surgery, and radiation therapy. Survival times for dogs given no or palliative therapy ranged from 59 to 81 days. The median survival time for dogs treated with either surgical excision or radiation therapy alone is to 6 to 7.5 months (Axlund 2002, Niebauer 1991). Combining radiation therapy with surgical excision increased median survival to 16.5 months in 13 dogs (Axlund 2002). More complete excision using a surgical aspirator prolonged the median survival to 1254 days in 16 dogs with grade I meningiomas; but the only dog with anaplastic meningioma in this series did not survive surgery (Greco 2006). It seems obvious that complete surgical removal of these tumors is rarely, if ever, attained. This may be due to anatomic location of the tumor, inability to distinguish tumor from normal brain tissue or meninges, tumor infiltration into healthy brain, and tumor friability.

Canine meningiomas are very similar to human meningiomas in neuroimaging characteristics, gross and histological appearance, and expression of growth factors and cell surface receptors (Dickinson 2009, Kraft 1997, Platt 2006, Thomas 1996). Recently, a more detailed histological classification of canine meningiomas has been promoted that mirrors the World Health Organization system used in people where 3 histological grades correlate with biological behavior (Sturges 2008). The new grading system for canine meningiomas includes benign (Grade I), atypical (Grade II), and anaplastic (Grade III) tumors. It seems there is a higher incidence of Grade II that have a more aggressive behavior and a higher potential for recurrence in dogs as compared to humans. This may account for the difference in survival times between the two species. The 5-year disease-free rates in people are 90% or greater, which correlates with the percentage of tumors that are WHO grade I (Mirimanoff 1985). The prognosis for people with recurrent meningioma is much more guarded and resembles that for dogs.

In 2005, we began work to initiate a canine glioma clinical trials program at the University of Minnesota (UMN) Veterinary Medical Center (VMC). Our team currently has six publications showing the similarity to human glioma histologically and conducting *in vivo* studies to determine the safety and efficacy of adenoviral and nonviral gene transfer into the canine brain. We are collaborating with a group at UCLA to determine the safety and efficacy of Flt3L and HSV-tk gene therapy in pet dogs with glioma, and with a small local biotechnology company to test a novel catheter to infuse therapeutic agents into gliomas in dogs. We also have three funded clinical trials to test the efficacy of combinations of autologous tumor lysate/adjuvant and IFNγ gene transfer given after surgical debulking of glial tumors and meningiomas in dogs.

Inclusion criteria for the gene therapy and vaccine therapy are: (i) client-owned dog with an intracranial mass on magnetic resonance imaging (MRI) consistent with a diagnosis of glioma; (ii) mass must be amenable to surgical resection; (iii) must have informed consent provided by the dog's owner; (iv) histologic confirmation of grade III-IV glioma. Exclusion criteria for the trial are: (i) severe neurological dysfunction that would preclude completion of the trial; (ii) tumor not amenable to surgical resection; (iii) existence of concomitant disease that would affect the dog's prognosis and preclude completion of therapy; (iv) metastatic brain tumors.

STUDY 1: AMERICAN CANCER SOCIETY VACCINE STUDY FOR GLIOMAS

Treatment groups: 1) surgery + CpG (adjuvant)/lysate vaccination, 2) surgery + IFNγ gene therapy + CpG(adjuvant)/lysate. Each dog has all grossly visible tumor removed via the appropriate craniotomy approach. The dogs in group 2 that receive IFNγ gene therapy have the gene vector injected into the brain around the resection cavity. All dogs have an immediate postoperative MRI to assess for any residual tumor/completeness of resection. The tumor lysate is produced from cultures the dog's own tumor cells grown *in vitro* and are given intradermally every 2 weeks for a total of 6 vaccines starting 2 weeks after surgery. Dogs are monitored by neurologic and MRI examinations 3, 6, and 12 months after surgery. Clinical response is considered as CR (complete response) = disappearance of all target lesions, PR (partial response) = 30% decrease in the longest diameter of tumor, PD (progressive disease) = 25% increase in the longest diameter of tumor, or SD (stable disease) = small changes that do not meet above criteria. Any toxicities are assessed using a standard veterinary scale of grade I-V events, determined by regular blood chemistry and neurological examinations.

As of February 1, 2011, we have recruited 15 dogs in to the ACS study: 5 grade II, 1 grade III, and 2 grade IV astrocytomas; 2 grade II and 2 grade III oligodendrogliomas, 2 primitive neuroectodermal tumors; and 1 neuroblastoma. Mean survival time for all dogs is 267 days and median survival time is 241.5 days. Four dogs were censored that died of causes unrelated to the brain tumor and 4 dogs are still alive (112-387 days).

STUDY 2: NIH NATIONAL INSTITUTE OF NEUROLOGICAL DISORDERS AND STROKE GENE THERAPY STUDY FOR GLIOMAS

Treatment groups: 1) surgery + chemotherapy with Temozolomide + Adenovirus w/reporter gene, 2) surgery + chemotherapy with Temozolomide + Adenovirus w/ Flt3L & HSV-tk genes. Each dog has all grossly visible tumor removed via the appropriate craniotomy approach. The Adenoviral vectors with either the active or reporter genes are injected into the brain around the resection cavity. All dogs have an immediate postoperative MRI to assess for any residual tumor/completeness of resection. All dogs receive an oral chemotherapy drug that is effective in human glioma patients called temozolomide daily for 6 weeks starting 2 days after surgery. Dogs are monitored by neurologic and MRI examinations 2, 6, and 12 months after surgery. Clinical response will be considered as CR (complete response) = disappearance of all target lesions, PR (partial response) = 30% decrease in the longest diameter of tumor, PD (progressive disease) = 25% increase in the longest diameter of tumor, or SD (stable disease) = small changes that do not meet above criteria. Any toxicities are assessed using a standard veterinary scale of grade I-V events, determined by regular blood chemistry and neurological examinations.

As of February 1, 2011, we have recruited 8 dogs in to the NINDS combined gene therapy study: 3 grade III and 3 grade IV astrocytomas; 2 grade III oligodendrogliomas. Median progression free time for all dogs is 198 days. Two were censored that died of causes unrelated to the brain tumor and 2 dogs are still alive and progression free (73 and 237 days).

STUDY 3: AMERICAN BRAIN TUMOR ASSOCIATION VACCINE STUDY FOR MENINGIOMAS

Treatment groups: 1) surgery + vaccination w/autologous lysate + CpG as adjuvant, 2) surgery + vaccination w/autologous lysate + imiquimod as adjuvant

Each dog has all grossly visible tumor removed via the appropriate craniotomy approach. All dogs have an immediate postoperative MRI to assess for any residual tumor/completeness of resection. The tumor lysate is produced from cultures the dog's own tumor cells grown *in vitro* and are given intradermally every 2 weeks for a total of 6 vaccines starting 2 weeks after surgery. Dogs in group 1 have CpG oligodeoxynucleotide (ODN), 18-25 base pair oligonucleotides of cytosine (C) and guanine (G) on a phosphodiester backbone (p) that bind toll-like receptor-9 (TLR9), as an immunostimulant. Dogs in group 2 have imiquimod, a topical cream that activates dendritic cells in the skin by binding TLR7 and used to treat sarcoids, as an immunostimulant. Dogs are monitored by neurologic and MRI examinations 3, 6, and 12 months after surgery. Clinical response is considered as CR (complete response) = disappearance of all target lesions, PR (partial response) = 30% decrease in the longest diameter of tumor, PD (progressive disease) = 25% increase in the longest diameter of tumor, or SD (stable disease) = small changes that do not meet above criteria. Any toxicity is assessed using a standard veterinary scale of grade I-V events, determined by regular blood chemistry and neurological examinations.

As of February 1, 2011, we have recruited 12 dogs in to the ABTA meningioma study: 6 grade I (subtypes: 2 meningothelial, 2 psammomatous, 2 transitional), 3 grade II – atypical, and 3 grade III (subtypes: 2 papillary, 1 rhabdoid). Five dogs are dead, only one with grade III tumor from disease progression at 670 days. Mean survival time for the censored dogs is 247 days. Seven dogs are still alive and progression free, and range from 145 to 600 days since treatment.

STUDY 4: NIH NATIONAL INSTITUTE OF NEUROLOGICAL DISORDERS AND STROKE VACCINE STUDY FOR GLIOMAS

Treatment groups: 1) surgery + chemotherapy with temozolomide, 2) surgery + chemotherapy with temozolomide + lysate from cells grown in 20% O_2, 3) surgery + chemotherapy with temozolomide + lysate from cells grown in 5% O_2. Each dog has all grossly visible tumor removed via the appropriate craniotomy approach. All dogs have an immediate postoperative MRI to assess for any residual tumor/completeness of resection. The tumor lysate is produced from cultures the dog's own tumor cells grown *in vitro* are given intradermally every 2 weeks for a total of 6 vaccines starting 2 weeks after surgery. It appears that the culture conditions, such as oxygen tension, affects the phenotype of the tumor cells that results in preferential activation of either a cell-mediated or a humoral immune response. Dogs are monitored by neurologic and MRI examinations 3, 6, and 12 months after surgery. Clinical response is considered as CR (complete response) = disappearance of all target lesions, PR (partial response) = 30% decrease in the longest diameter of tumor, PD (progressive disease) = 25% increase in the longest diameter of tumor, or SD (stable disease) = small changes that do not meet above criteria. Any toxicity is assessed using a standard veterinary scale of grade I-V events, determined by regular blood chemistry and neurological examinations.

As of February 1, 2011, we have recruited 2 dogs in to the NINDS vaccine therapy study. One dog with an anaplastic astrocytoma that died from a heart base tumor 96 days after surgery and one dog with a PNET that is alive and progression free 86 days after surgery.

STUDY 5: NIH SBIR/STTR CONVECTION ENHANCED DELIVERY OF CARBOPLATIN FOR GLIOMAS

The final study involves a new catheter that provides convection-enhanced delivery (CED) of a chemotherapeutic agent into brain tumors in dogs. Four dogs diagnosed with a solitary intra-axial brain mass suspected to be a glioma will be inducted into the trial. All dogs will have MRIs of the brain performed before catheter placement and after CED. The dogs will be placed under general anesthesia for the entire procedure. The first MRI will be obtained with skin fiducials to determine the tumor volume and characteristics and the coordinates for stereotactic biopsy and subsequent catheter placement projected into the center of the tumor mass. The CED catheter will be placed in the tract made by the biopsy needle. CED will be done by infusing a 10-mg/ml:0.004-mmol/ml carboplatin:Gad-DPTA cocktail solution at ~6 µl/min to deliver a volume appropriate to cover the tumor (1-2 ml of infusate is expected). After CED is completed a second MRI will be done to show the distribution of the carboplatin:Gad-DPTA solution, the volume of tumor covered by infusate, and the degree of reflux and backflow by comparing the pre- and post-CED MR images. We will conduct follow up MRI at 1 and 3 months after CED to assess tumor volume/response to the infusion.

A summary of the results and findings for each study up to the date of the conference will be presented. The results will include tumor types and grades treated, progression-free survival times, effects of surgical removal of glial tumors, median survival times, cause of death, and side effects of the various therapies.

REFERENCES

Altinörs N, et al. (2004) **J Invest Surg** 17:283–289. Andrews EJ. (1973) **J Am Vet Med Assoc** 163:151–157. Axlund TW, et al. (2002) **J Am Vet Med Assoc** 221(11):1597-600. Braund KG, et al. (1986) **Comp Contin Educ** 8:241–248. Dickinson PJ, et al. (2009) **Vet Pathol** 46(5):884-92. Foster ES, et al. (1988) **J Vet Intern Med** 2:71–74. Frenier LS, et al. (1990) **Comp Contin Educ** 12(10):1422–1433. Greco JJ, et al. (2006) **J Am Vet Med Assoc** 229:394–400. Heidner GL, et al. (1991) **J Vet Intern Med** 5:219–226. Kornegay J. **Contemporary Issue in Small Animal Practice: Neurologic Disorders.** 1986:79-108. Kraft SL, et al. (1997) **J Vet Intern Med** 11:218–225. LaRue SM, et al. **Small animal clinical oncology** 2001:119-137. McGrath JT. (1962) **Acta Neuropathol** Suppl 1:3–4. Mirimanoff RO, et al. (1985) **J Neurosurg** 62:18. Niebauer GW, et al. (1991) **J Am Vet Med Assoc** 198:89-95. Platt SR, et al. (2006) **J Vet Intern Med** 20(3):663-8. Snyder JM, et al. (2006) **J Vet Intern Med** 20(3):669-75. Sturges BK, et al. (2008) **J Vet Intern Med** 22:586–595. Thomas WB, et al. (1996) **Vet Radiol Ultrasound** 37:20–27. Turrel JM, et al. (1986) **J Am Vet Med Assoc** 188(8):851–856.

TRANSLATIONAL LESSONS LEARNED FROM A CANINE MODEL OF DUCHENNE MUSCULAR DYSTROPHY

Joe N. Kornegay, DVM, PhD, DACVIM (Neurology)
Chapel Hill, NC

INTRODUCTION

Duchenne muscular dystrophy (DMD) is an X-linked recessive disorder affecting approximately 1 of 3,500 newborn human males in whom absence of the protein dystrophin causes progressive degeneration of skeletal and cardiac muscle.[1] No treatment halts or reverses progression of DMD. Although several therapies are promising, key questions must first be addressed in relevant animal models. Spontaneous forms of X-linked muscular dystrophy due to dystrophin deficiency have been identified in mice, multiple dog breeds, and cats. Unlike the dystrophin-deficient mdx mouse, which remains relatively normal clinically,[2] affected dogs develop progressive, fatal disease strikingly similar to the human condition.[3-9] Accordingly, studies in canine dystrophin-deficient models such as golden retriever muscular dystrophy (GRMD) may be more likely than those in mdx mice to predict pathogenesis and outcome of treatment in DMD. Potential treatments for DMD may be broadly categorized as molecular, cellular, or pharmacologic.[10] In this review, we will briefly discuss the role that the GRMD model has played in advancing our understanding of two therapeutic approaches, namely myostatin inhibition and AAV-based gene therapy.

BACKGROUND, RATIONALE FOR MYOSTATIN INHIBITION, AND HYPERTROPHY IN DMD AND GRMD

Se-Jin Lee and colleagues have shown that the myostatin gene is a key negative regulator of muscle growth.[11-14] Humans,[15] cattle,[16] sheep,[17] and dogs[18] with myostatin mutations have dramatic muscle hypertrophy. Dystrophin-deficient mdx mice in which myostatin is knocked out[19] or inhibited postnatally[20] also have a less severe phenotype. Based on these findings, there has been increasing interest in treatments to inhibit myostatin and thus promote muscle growth.[21,22] However, results from other dystrophic murine models in which myostatin was knocked-out have varied,[23,24] with some mice showing increased morbidity.[25] There are questions about potential senescence of myostatin-deficient cells that have undergone multiple divisions.[24] Moreover, abnormalities have been identified in muscle tendons from myostatin-null mice.[26] Considering these inconsistencies, additional studies in a more relevant large animal model are needed.

Phenotypic Variation in DMD and GRMD

Effects of dystrophin deficiency vary among species, individuals, and muscles. Reasons for phenotypic variation are poorly understood and raise questions about primary versus secondary effects of dystrophin deficiency.[27]

There are two distinctive clinical alleles in humans with dystrophin gene mutations. Those in DMD patients disrupt the reading frame and cause a severe skeletal muscle phenotype.[28] In contrast, with Becker MD, the reading frame is maintained and signs of skeletal muscle disease are less pronounced. Even among DMD patients, who all lack dystrophin except for rare *revertant fibers* produced through alternative splicing, symptoms can vary markedly.[29] GRMD dogs also demonstrate pronounced phenotypic variation, with some dogs unable to walk within the first 6 months of life and others living well into adulthood.[30,31]

Extensor muscles that undergo eccentric muscle contraction (e.g. quadriceps femoris) are particularly vulnerable in DMD.[32] In contrast, the extraocular muscles are largely spared.[33] Most muscles in mdx mice undergo early necrosis, regenerate, and remain relatively normal until late in life.[34,35] However, the extraocular muscles are spared during even the initial necrotic phase, while lesions in the diaphragm occur early and progress.[36] Studies have shown that flexor muscles are generally more severely involved than extensors in neonatal GRMD dogs,[37] potentially due to their role in crawling. Presumably, as dogs begin to walk, weight-bearing extensor muscles are predisposed to injury, just as in DMD.

Muscle Hypertrophy in DMD and GRMD

DMD muscles generally undergo profound atrophy *but some enlarge*.[38,39] Initial studies emphasized that deposition of fat and connective tissue led to muscle enlargement (pseudohypertrophy).[38] More recently, ultrasound and MRI imaging have shown increased contractile tissue (true hypertrophy).[39,40] Better definition of the mechanisms involved could provide insight into DMD's pathogenesis and the effects of myostatin inhibition. While most GRMD muscles also atrophy, some hypertrophy.[41] Our data have shown that the cranial sartorius undergoes early *true hypertrophy* followed by loss of muscle mass with an associated increase in endomysial space.[41]

The Role of Muscle Hypertrophy in Postural Instability and Contractures in DMD and GRMD

Both muscle weakness and joint contractures contribute to postural instability and ultimate loss of ambulation in DMD.[42] Debilitating distal joint contractures have been attributed to one of two mechanisms, namely a local imbalance of agonist and antagonist muscle strength[43] or postural changes due to *selective sparing* of proximal flexor muscles.[42,44] Weakness of antagonist extensor muscles in DMD correlates highly with flexor contracture severity.[43] As opposing extensor muscles weaken, flexor contractures worsen. This imbalance may be exacerbated by *selective hypertrophy* of muscles that oppose atrophied muscles. We have shown that GRMD extensor and flexor muscles operating at the tibiotarsal joint (TTJ) are differentially affected, with extensors being proportionally weaker than flexors.[45] Moreover, the degree of *CS hypertrophy* in GRMD has a strong *negative correlation* with the TTJ angle,[41] suggesting that the CS could play a role analogous to iliotibial band tightening in DMD.[46] Taken together, these data indicate that some dystrophin-deficient muscles are already prone to hypertrophy and that disproportionate muscle enlargement may exacerbate debilitating contractures.

Cross-Breeding GRMD and Myostatin-Deficient Dogs

While murine models have been powerful tools in defining disease mechanisms,[47] results do not necessarily extrapolate to humans, presumably due to differences between murine and human size and physiology.[48] These shortcomings are partially obviated with large animal (mostly dogs and pigs) models. With mice, the role of specific genes and their interactions can be studied using knockout and transgenic models. Unfortunately, transgenic technology in the dog is very cumbersome, essentially precluding use of this approach.[49] Thus, it is not possible to conduct myostatin knock out studies in dystrophic dogs. Fortuitously, a spontaneous deletion has been identified in the myostatin gene of the Whippet breed.[18] Dogs that are homozygous for the myostatin-null allele (so called *bully whippets)* have gross enlargement of muscle, while those that are heterozygous have intermediate muscle mass and are significantly faster than their myostatin-normal counterparts.

To provide further justification for use of myostatin blockade as a treatment for DMD, we have initiated a set of experiments in which dogs carrying dystrophin and myostatin gene mutations are crossbred. Our initial hypothesis was that dystrophic myostatin-null heterozygote or homozygote dogs would have a less severe phenotype. But, in our initial litter produced from mating an obligate GRMD carrier with a myostatin-null homozygote Whippet, the single myostatin-null heterozygote GRMD dog actually developed a more severe phenotype in association with contractures, seemingly due to an imbalance between selectively hypertrophied flexor muscles and weakened extensors. The phenotype of this dog has been documented with magnetic resonance imaging (MRI), functional tests, and quantitative histopathologic analysis. Similar results have been seen in a second litter of dogs produced by mating a GRMD male and a double dystrophin/myostatin-null heterozygote. Reasons for this differential muscle involvement are not clear but could relate to greater expression of the myostatin activin type IIB receptor in flexor muscles.

In summary, while myostatin inhibition offers a logical approach for treatment of DMD and other muscle-wasting disorders, the potential for preexisting muscle hypertrophy to exaggerate postural instability and joint contractures in both DMD and GRMD raises questions about deleterious consequences. We were originally motivated to cross breed dystrophin and myostatin-deficient dogs to corroborate studies in mdx mice. But, instead, our preliminary data suggest that myostatin inhibition in dystrophin-deficient individuals can cause phenotypic deterioration. Muscle is equipped with a complex set of regulatory genes that establish a balance between forces that would otherwise lead to either muscle atrophy or hypertrophy.[50] In fact, not surprisingly, our preliminary mRNA microarray data on muscle from GRMD dogs suggest that myostatin is already down-regulated in dystrophin-deficient muscles. Clearly, data derived from animals in which myostatin has been down-regulated from conception do not fully extrapolate to a clinical situation in which treatment is applied postnatally as the disease progresses. However, our results from dystrophic, myostatin-null heterozygote dogs suggest that down-regulation of myostatin can potentially tip the balance between agonist and antagonist muscles, exacerbating contractures.

ADENO-ASSOCIATED VIRUS (AAV)-MEDIATED MINI-DYSTROPHIN THERAPY

The two most common molecular approaches for treatment of DMD are *gene therapy*, whereby the dystrophin gene is introduced into muscle either locally or through the vasculature, generally by plasmids or viral vectors,[51-54] and *gene correction*, which involves introduction of oligonucleotides (chimeric or antisense) to induce either inherent repair mechanisms or exon skipping to reestablish the correct nucleotide sequence (reading frame).[55-57]

Use of AAV in Dystrophin Gene Therapy

Variables to be considered in choosing a viral vector for gene therapy include potential pathogenicity, maximum insert size, route of delivery, degree of integration into the genome, ability to transduce non-dividing cells, duration of expression in vivo, and the potential for immune rejection.[52] Properties of AAV-based vectors generally compare favorably with other commonly used vectors such as retroviral, lentiviral, and adenoviral vectors. This small DNA virus is largely non-pathogenic, integrates into non-dividing cells, and has relatively long-term expression with minimal immune response. One limitation relates to the small insert size of ~ 4.5 kb, which is inadequate to carry the ~ 14 kb cDNA of the dystrophin gene. This problem has been circumvented, in part, by deleting non-essential components of the rod domain within the dystrophin gene, creating so-called mini- or micro-dystrophins of ~ 4.5 kb that can be carried by AAV.[58,59] Several AAV serotypes, distinguished by capsid proteins and variable tissue tropism, have been identified. AAV serotypes 2, 6, 8, and 9 have primarily been used in dystrophic dogs.[53,54,60]

Use of AAV-Mini-Dystrophin Constructs in the GRMD Model

Studies of intramuscular AAV-mediated dystrophin mini- or micro- gene therapy in GRMD dogs have documented a marked immune response to either components of the transgene (to include dystrophin)[61] and/or viral capsid proteins.[62,63] Working in parallel with these other groups, we have collaborated with Xiao Xiao, first while he was at Pittsburgh and now at UNC-CH to replicate his seminal mdx mouse studies[59,64] in GRMD dogs. Our studies initially focused on intramuscular injection of AAV-dystrophin mini-gene constructs and, more recently, have entailed regional limb and systemic vascular delivery. We have been particularly interested in the degree of immunologic reaction to constructs. In keeping with other published studies, preliminary data from direct muscle injection showed a greater immune response than that encountered in the mdx mouse. The severity of this reaction was markedly reduced when canine versus human mini-dystrophin was used. These studies have been extended to an afferent transvenular retrograde extravasation method[65,66] for somatic gene transfer to muscle. This method allows perfusion of viral constructs via retrograde saphenous or cephalic vein infusion in conjunction with occlusion of limb vasculature with a tourniquet applied at the inquinal or axillary area. Strong, near-uniform expression of canine mini-dystrophin gene and other components of the glycoprotein complex have been seen in some dogs with minimal immune response. We are continuing these studies, in tandem with a phase 1 human clinical trial in which only saline is infused in adult (limb girdle and Becker) muscular dystrophy patients, to define key variables such as perfusion rate and pressure.

Because of the relative immaturity of the neonatal immune system, animals treated soon after birth may mount a less robust immune response. Working with the Duan laboratory at the University of Missouri, we achieved widespread muscular transduction in normal, neonatal dogs with a human placental alkaline phosphatase reporter gene after systemic intravenous delivery of an AAV9 construct.[67] We have subsequently demonstrated long-term expression of mini-dystrophin in neonatal GRMD dogs in collaboration with Xiao Xiao.[60] Despite the prominent gene expression, an apparent, dramatic, early immune response was seen with MRI in these dogs and they developed severe joint contractures similar to those seen with *Neospora caninum* infection.[68] Factors that have led to an immune response, with minimal persistent dystrophin expression, in some dogs, while others have had long-term transduction without immune suppression are not clear. Localized treatment may lead to a more pronounced response due to the more concentrated viral load. While AAV capsid proteins are well conserved across serotypes,[69] dogs treated with either AAV2 or 6 generally have generally had a greater immune response than those in which AAV8 or 9 were used. Dystrophin, itself, may serve as a neoantigen and induce an immune response. Use of codon-optimized mini-dystrophin in our neonatal GRMD dogs may have increased the antigen load, aggravating the immune response.

Use of AAV-Mini-Dystrophin Constructs in DMD

An initial phase I trial of an AAV-mini-dystrophin construct similar to the one used in our GRMD dogs has been completed in six DMD patients at 9 years of age or older.[70] The biceps brachii muscle was injected with the construct and biopsied at either 42 or 95 days. Mini-dystrophin was seen at the early time point but not later. T-cells that reacted specifically to the injected dystrophin were identified. Moreover, dystrophin-specific T cells were seen in patients prior to treatment, suggesting that so-called *revertant fibers* may sensitize the immune system. This was surprising because most scientists have assumed that truncated, partially functional dystrophin protein produced from alternatively spliced transcripts would be seen as self and protect against an immune response to the injected dystrophin.

In summary, AAV-mini- or micro- dystrophin constructs have shown promise first in the mdx mouse and, more recently, in the GRMD model. Studies initially involved localized (intramuscular) injections and have been extended to include regional limb and systemic delivery. In some GRMD studies, a marked immune response has been seen in muscle injected with these constructs, either because of a reaction to AAV capsid proteins or the dystrophin gene, itself. On the other hand, in our hands, long term dystrophin expression has been seen in GRMD dogs without immune suppression. Because of the promising results from animal models, a phase 1 clinical trial in which DMD patients were treated with similar AAV-mini-dystrophin constructs has been completed. In addition, proof-of-principle studies of regional limb delivery of saline are being conducted in adult muscular dystrophy patients. It will be critical to continue animal studies while human trials are evolving.

REFERENCES

[1]Hoffman EP, et al. **Cell** 1987;51:919. [2]Bulfield G, et al. **Proc Natl Acad Sci USA** 1984;81:1189. [3]Kornegay JN, et al. **Muscle Nerve** 1988;11:1056. [4]Jones BR, et al. **J Neurol Sci** 2004;217:143. [5]Baltzer WI, et al. **J Am Anim Hosp Assoc** 2007;43:227. [6]Sharp NJH, et al. **Genomics** 1991;13:115. [7]Cooper BJ, et al. **Nature** 1988;334:154. [8]Valentine BA, et al. **Am J Med Gen** 1992;42:352. [9]Schatzberg SJ, et al. **Neuromuscul Disord** 1999;9:289. [10]Chakkalakal JV, et al. **FASEB J** 2005;19:880. [11]Lee S-J. **Annu Rev Cell Dev Biol** 2004;20:61. [12]McPherron AC, et al. **Nature** 1997;387:83. [13]Lee S-J and McPherron AC. **Curr Opin Genet Dev** 1999;9:604. [14]McPherron AC and Lee S-J: **J Clin Invest** 2002;109:595. [15]Schuelke M, et al. **N Engl J Med** 2007;350:2682. [16]McPherron AC and Lee S-J. **Proc Natl Acad Sci USA** 1997;94:12457. [17]Clop A, et al: **Nat Genet** 2006;38:813. [18]Mosher DS, et al. **PLoS Genet** 2007;3:779. [19]Wagner KR, et al. **Ann Neurol** 2002; 52:832. [20]Wagner KR, et al. **Proc Natl Acad Sci USA** 2005;102:2519. [21]Bradley L, et al. **Cell Mol Life Sci** 2008;65:2119. [22]Patel K and Amthor H. **Neuromuscul Disord** 2005;15:117. [23]Mendias CL, et al. **J Appl Physiol** 2006;101:898. [24]Parsons SA, et al. **Am J Pathol** 2006;168:1975. [25]Li ZF, et al. **Am J Pathol** 2005;166:491.[26]Mendias CL, et al. **Proc Natl Acad Sci USA** 2008;105:388. [27]Hoffman EP and Gorospe JRM. In: **Ordering the Membrane-Cytoskeleton Trilayer. Current Topics in Membranes**, Vol. 38. Eds. Morrow J, Mooseker M. Academic Press, New York, NY, 1991: 113. [28]Koenig M, et al: **Am J Hum Genet** 1989;45:498. [29]Heydemann A, et al. **Biochim Biophysi Acta** 2007;1772:216. [30]Valentine BA, et al. **J Neurol Sci** 1988;88:69. [31]Ambro'sio CE, et al. **Neuromuscul Disord** 2008;18:892. [32]Edwards RHT, et al. **Lancet** 1984;1:548. [33]Karpati G, et al. **Am J Med Genet** 1986;25:653. [34]Pastoret C and Sebille A. **J Neurol Sci** 1995;129:97. [35]Karpati G, et al. **Muscle Nerve** 1988;11:795. [36]Stedman HH, et al. **Nature** 1991;352:536. [37]Valentine BA and Cooper BJ. **Neuromuscul Disord** 1991;1:31. [38]Jones DA, et al. **J Neurol Sci** 1983;60:307. [39]Cros D, et al. **J Neurol** 1989;236:43. [40]Marden FA, et al. **Skeletal Radiol** 2005;34:140. [41]Kornegay JN, et al. **Neuromuscul Disord** 2003;13:493. [42]Vignos PJ, Jr, et al. **JAMA** 1963;184:89. [43]Brooke MH, et al. **Muscle Nerve** 1983;6:91. [44]Siegel IM, et al. **J Bone Joint Surg** 1968;50A:1437. [45]Kornegay JN, et al. **J Neurol Sci** 1999;166:115. [46]Rideau Y. In: **Neuromuscular Disease**. Eds. Serratrice G, Cros D, Desnuelle C, Raven, New York, NY, 1984: 557. [47]Capecchi MR. **Nat Rev Genet** 2005;6:507. [48]Sacca R, et al. **Methods Mol Biol** 2010;602:37. [49]Hong SG, et al. **Genesis** 2009;47:314. [50] Sandri M, **Physiol** 2007;23:160. [51]Xiao X, et al. **J Virol** 1996;70:8098. [52]Takeda S: **Neurosci News** 2000; 3, 51. [53]Miyagoe Y and Takeda S: **Exp Cell Res** 2010;316:3087. [54]Wang Z, et al. **ILAR J** 2009;50:87. [55]Wilton SD and Fletcher S: **Curr Pharm Design** 2010;16:988. [56]Aartsma-Rus A and van Ommen G-JB. **Eur J Hum Gen** 2010;18:146. [57]Le Roy F, et al. **Trends Mol Med** 2009;15:580. [58]Chamberlain JS. **Hum Mol Genet** 2002;11:2355. [59]Watchko J, et al. **Hum Gene Ther** 2002;13:1451. [60]Kornegay JN, et al. **Mol Ther** 2010;18:1501.[61]Yuasa K, et al. **Gene Ther** 2007;14:1249. [62]Wang Z, et al. **Hum Gene Ther** 2007;18:18. [63]Wang Z, et al. **Mol Ther** 2010;18:617. [64]Wang B, et al. **J Orthop Res** 2009;27:421. [65]Su LT, et al. **Circulation** 2005;112:1780. [66]Arruda VR, et al. **Blood** 2005;105:3458. [67]Yue Y, et al. **Mol Ther** 2008;16:944. [68]Knowler C and Wheeler SJ. **J Sm Anim Pract** 1995;36:172. [69]Gao GP, et al. **Proc Natl Acad Sci USA** 2002;99:11854. [70]Mendell JR, et al. **New Eng J Med** 2010;363:1429.

CYTOSAR: ITS USE AND MISUSE IN THE TREATMENT OF NEUROLOGIC DISEASE

Carlos O. Rodriguez, Jr., DVM PhD DACVIM (Oncology)
Davis, CA

ABSTRACT

Cytarabine (cytosar, ara-C) has demonstrated clinical utility in and is most frequently used for the treatment of lymphoid malignacies[1-2]. The demonstrated efficacy is dependent on its metabolism, specifically to its monophosphate, diphosphate, and triphosphate (ara-CTP) metabolites[3-4]. The rate-limiting step in the intracellular accumulation of the triphosphate is the phosphorylation of the drug to its monophosphate, which is catalyzed in the first step by deoxycytidine kinase (dCK)[5]. The triphosphate moiety then competes with native cytosine deoxyriboside (**Figure 1**) for incorporation into DNA and incorporation (and consequently response to therapy) is directly proportional to concentrations of intracellular ara-CTP x time of exposure[6]. Therefore, maximizing the accumulation of the triphosphate is dependent on achieving plasma cytarabine concentrations that do not exceed the phosphorylating capacity of dCK. Interestingly, the optimization of ara-C delivery for leukemia is still being sought despite more than 30 years of clinical use.

In veterinary neurology, ara-C has been used in the management of inflammatory brain disease[7-10]. The dose that is most commonly used in conjunction with prednisone is 50mg/m2 administered subcutaneously twice daily for 2 consecutive days every 3 to 4 weeks. Ara-C has been shown to accumulate in the CSF at about 20-40% of that found in the plasma[11-12]. When administered subcutaneously, peak plasma levels are reached in about 20-60 minutes and decrease below steady state levels after about 100 minutes[13]. While higher doses could be administered, there is the potential for increased neurotoxicity[14-17]. The purpose of this talk is to discuss the biochemistry of dCK and to describe pharmacologically-directed dosing strategies that may maximize ara-C in the plasma, CSF, and thus increase the utility of ara-C in veterinary neurology.

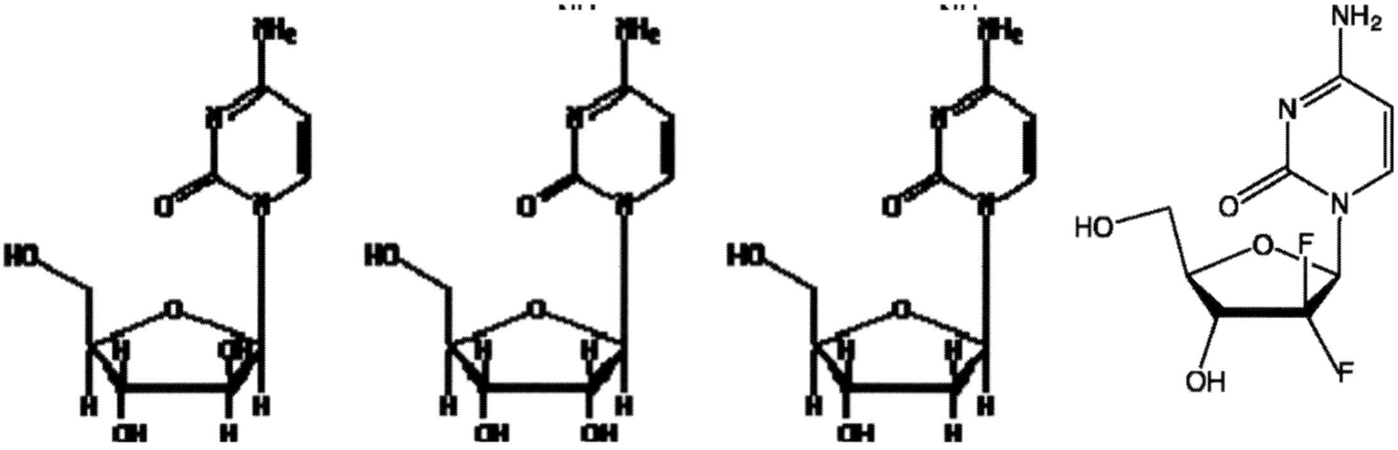

Cytosine Arabinoside | Cytosine Riboside | Cytosine Deoxyriboside | Gemcitabine

Figure 1. Structures of ara-C, native nucleosides, and gemcitabine.

REFERENCES

[1]Jabbour E et.al. **Cancer.** 2007;109 (11):2171. [2]Shah, M and Agarawal B. **Ind J Pediat**. 2008;75(8):831. [3]Plageman PW et al **Cancer Res** 1978;38:978. [4]Wiley JS et. al. **J. Clin Invest** 1982;69:479. [5]Wiley JS et. al. **J. Clin Invest** 1985;75:632. [6]Kufe D, et.al. **Blood** 1984; 64:54 [6]Chou T-C, et. al. **Cancer Res.** 1977 37:3561. [7]Behr S, et.al. **Vet Rec**. 2009 16;164(20):627. [8]Menaut P et.al **Vet Rec**. 2008 23;162(8):241. [9]Zarfoss M, et.al . **J Small Anim Pract**. 2006;47(10):588. [10]Nuhsbaum MT, et.al **Vet Ophthalmol**. 2002;5(1):29. [11]Dennis R, et.al **Brain Research** 2000 8(56):281. [12]Ho DHW and Frei E III. **Clin Pharmacol Ther** 1971;12:944. [13]Slevin, ML, et.al. **Cancer Chemother Pharmacol** 1983;10(2):112. [14]Lazarus H et.al. **Cancer** 1981;48:2577. [15]Herzig RH et.al. **Semin Oncol** 1987; 14(Supp 1);21. [16]Nand S et.al. **J Clin Oncol** 1986;4:571. [17]Grossman L et.al. **Med Pediatr Oncol** 1983;11:246

A REVIEW OF THE RESEARCH AND INTERESTS OF DR. ROBERTO POMA, DIPLOMATE ACVIM (NEUROLOGY)

Fiona M.K. James, DVM, MSc, DVSc, DACVIM (Neurology); Robbie J. Daniel, BSc, DVM
Guelph, ON, Canada

Dr. Roberto Poma was an Assistant Professor at the Ontario Veterinary College (OVC), University of Guelph, Guelph, Ontario, Canada. He was a father of two and friend to many. His passion was the use of electroencephalography (EEG) for the diagnosis and treatment of canine epilepsy. His objective was to characterize epileptic syndromes in dogs, with the additional benefit of supporting the investigation of human epilepsy. In support of this, he won a large grant from the Canadian Foundation for Innovation in 2006 to develop a comparative epilepsy program. The canine epilepsy research program under his guidance had multiple parallel projects.

He started with describing the semiology of canine idiopathic epileptics. This unpublished research was funded through the OVC Summer Student Research Assistantship. Using a specifically designed questionnaire for clients living with epileptic dogs, he recruited 38 dogs with generalized seizures and 21 dogs with focal seizures, describing the physical signs of each seizure type, the frequency with which the pre-ictal phase could be identified, and the length of each phase. He hoped that by characterizing the pre-ictal phase, further investigation into pre-emptive therapy would be possible. Additionally, he planned to investigate the correlation between clinical signs and type of seizures.

To characterize specific breed-related epilepsy syndromes, he worked with the Golden Retriever, the Australian Shepherd, and the Nova Scotia Duck Tolling Retriever. He collaborated with Dr. Berge Minassian, a specialist in epilepsy and genetics at the Toronto Hospital for Sick Children, to find the links between genes and epilepsy in both canines and humans. Dr. Minassian's laboratory had discovered the first canine epilepsy gene, epm2b, the same gene later implicated in Lafora progressive myoclonus epilepsy in humans. This research was part of a broader collaboration with international groups including Dr. Hannes Lohi in Finland, Dr. Gary Johnson in Missouri, USA, and Drs Ned Patterson and Jim Mickleson in Minnesota, USA. The hope, for canines, was to devise a diagnostic test to help to eradicate epilepsy from affected breeds.

For each patient, along with the standard neurologic examination and magnetic resonance imaging (MRI) of the brain, he recorded EEG. Initially, he correlated the EEG montage with the anatomic substrate. Preliminary research involved using previously reported anatomic landmarks for electrode placement and determining the specific areas/lobes underneath these electrodes. Through electrode placement, MRI imaging and anatomic dissection, this study revealed the struggle with varying breed morphology and the difficulties implicit in finding abnormal waveforms and localizing abnormalities to specific areas of the brain. He had also developed canine protocols for two types of EEG systems – a routine EEG unit for outpatient recordings, and a human ambulatory system adapted for the home environment. Dr. Poma additionally designed a canine epilepsy monitoring unit within the intensive care unit at the Veterinary Teaching Hospital of the OVC. All the EEG systems provided synchronized video monitoring, allowing correlation of observable clinical signs with the abnormal brain electrical activity. This culminated in his recent publication describing absence seizures with myoclonic features in a juvenile Chihuahua, a first for veterinary neurology.

Mapping the epileptic seizure focus via EEG, Dr. Poma hoped to advance therapeutic intervention for canine epileptics. Via sophisticated software 3-D reconstruction of the brain MRI, the goal was to eventually target the epileptic seizure focus via transcranial magnetic stimulation (TMS) or surgical excision, similar to treatment for refractory human epileptic patients. Dr. Poma wanted to know if TMS might become a non-surgical therapy for refractory epilepsy. TMS thresholds required to stimulate motor evoked potentials (MEPs) from the tibialis cranialis muscles were measured in a group of 9 refractory canine epileptics and 5 control (healthy) dogs in a pilot study funded by the Pet Trust Foundation, OVC, University of Guelph. Following the epileptics for 60 days before and after TMS treatment, he observed that the epileptic group had decreased TMS excitability compared to the controls. Also, he observed a decrease in the number of seizures and more prolonged interictal period in the post-treatment period vs. pre-treatment period for the refractory epileptics. He had hoped to pursue this promising early result further, testing applicability in a larger treatment group.

Dr. Poma had a wide-ranging vision for research into canine and human epilepsy. His thorough approach was to have detailed description of the condition in canines, while elucidating underlying genetic causes and forecasting future avenues for potential treatment.

REFERENCES

Lohi H, et al. **Science** 2005;7(307):5706. Pellegrino FC, et al. **Clin Neurophysiol** 2004;115(2):477. Poma, R. et al. **Epileptic Disorders** 2010;12(2):138-41.

DRUG MOVEMENT AND THE BRAIN: THE BATTLE OF THE BARRIERS

Dawn Merton Boothe, DVM, PhD, DACVIM, DACVCP
Auburn, AL

General Mechanisms of Drug Movement

Drugs move through the body through two major mechanisms: bulk flow and passive diffusion. The cardiovascular system is the primary determinant of bulk flow; glomerular filtration is one of the more important specific examples. The chemical nature of the drug does not affect bulk flow. Most drugs are characterized by a molecular weight (MW of 350 or less, ensuring movement of unbound drug betweenendothelial cells despite the presence of protein filters in the endothelial gaps. Exceptions are made for those tissues whose capillaries are not fenestrated because endothelial cell junctions are tight. Drug movement for these tissues must occur across cell membranes. Drugs can move through cells directly through the lipid layers of the cell (e.g., passive diffusion), through aqueous pores formed by aquaporins (proteins) that span the width of the membrane, by combination with transmembrane carrier proteins, and by pinocytosis. Of these, transmembrane (passive) diffusion and carrier-mediated movement are most important.

Passive diffusion occurs independent of any other mechanism of drug movement and requires no energy. The single most important factor influencing passive diffusion through cell membranes is concentration of diffusible drug, which is most easily manipulated by increasing the dose. Host determinants of passive diffusion are subject to change and include thickness of the membrane to be traversed (inversely proportional; e.g., edematous compared with normal tissues), surface area (directly proportional; e.g., small intestine versus stomach), environmental pH (see ionization), and temperature (directly proportional). In contrast, drug characteristics influencing passive diffusion generally influence lipid solubility are not amenable to change[2,3] Lipid solubility is influenced by the inherent chemical structure of a drug, such as molecular weight (smaller molecules diffuse more easily) and **partition coefficient** (PC) or ratio which refers to the relative distribution of the drug into an organic solvent (e.g., octanyl) versus water. A ratio greater than 1 suggests greater distribution to the organic phase, indicating lipid solubility. Benzene rings, carbon double bonds and methyl groups tend to make a drug more lipid soluble, whereas polar compounds such as amine or hydroxyl groups contribute to drug water solubility. Because only un-inonized drug effectively distributes, drug pKa and environmental (host) pH also influence drug distribution. If a drug is sufficiently lipid soluble to cross cell membranes, a steady-state equilibrium theortetically may be reached when the amount of drug moving from one compartment (ie, into the cell) area equals the amount moving in the opposite direction (ie, out of the cell); the concentration of drug will be equivalent on either side of a membrane. However, in its ionized form, a drug cannot traverse lipid membranes and becomes trapped and thus "partitioned" by the pH in its surrounding environment. Thus, the total concentration on either side may differ with pH. A drug is trapped when it is in an "unlike" environment (e.g., acid in a basic pH). Although less common, drug movements other than passive diffusion and bulk flow influence PDC.

Carrier-mediated transport includes both facilitated diffusion (nonactive) or active transport. An example important transport system is the **P-glycoprotein** system, an MDR-1 gene product present in several tissues in the body, and in portals of entry (gastrointestinal tract) or exit (renal tubular brush borders, bile canniculi) and sanctuaries (brain, eye, testicle). [8] Pinocytosis is a rare drug movement exemplified by the uptake of vitamin B_{12} in the ileum and aminoglycosides by renal tubular cells. Distribution is influenced by the multidrug transport protein P-gp. Polymorphism has been demonstrated in the Collie, yielding differences in P-gp content. Two P-gp transporting proteins are encoded by ABCB1 (previously MDR1 or PGY1) and MDR3 (also named *MDR2* and *PGY3)*; only the MDR1 gene product is thought to significantly influence drug metabolism. P-gp acts as an efflux pump by translocating drugs from the intracellular to extracellular compartments. The protein transports a large number of drugs that are chemically divergent; further, these drugs are associated with a specific CYP450 responsible for metabolism of the drugs that have been transported. Examples drug substrates include: antimicrobial Drugs (macrolides, imidazole antifungals, selected fluoroquinolones); anticancer drugs (doxorubicin, vincristine, vinblastine, mitoxantrone), anthelmintics (avermectins), cardiac drugs (digoxin, quinidine, diltiazem, verapamil),CNS active drugs (selected morphine derivatives, behavior modifying drugs, phenothiazines), steroidal hormones, and endogenous substrates such as bilirubin and bile acids. Inhibitors of P-gp include macrolides, cyclosporine, imidazole antifungals, quinidine, verapamil, and selected morphine derivatives. Inducers include dexamethasone, morphine, rifampin, phenothiazines, St. John's wort and others.

Polymorphism of the MDR1 gene and P-gp have been reported in humans and are associated with altered drug disposition and thus susceptibility to adverse drug events. Interestingly, polymorphism also has been associated with an increased risk of certain illnesses (e.g., refractory seizures, Parkinson's disease, inflammatory bowel disease). Polymorphism reflecting a mutation deletion of MDR1 that causes nonfunctional P-gp has been documented in Collie and related working-breed dogs. The incidence of the deletion is impressively high: In the U.S., in one study, 35% of Collies were homozygous and another 42% heterozygous for the mutation deletion.[84] A similarly high incidence was found in dogs in France: 20% of Collies and related breeds were found to be homozygous for the normal allele, 32% heterozygous for the deletion (carrier), and 48% homozygous for the mutant allele (affected dogs).[85] The impact of the mutation on drug safety in afflicted animals can be profound. Substrate specificity for P-pg appears to be similar among species, suggesting that human data can be used to predict which drugs might be more likely to cause adverse effects in these breeds.[86]

Drug Movements: Plasma – Tissue Drug concentrations

After administration of a fixed dose of a drug, several drug movements act in concert to determine PDC; they are largely dependent on passive diffusion of the drug and include absorption (A) from the site of administration to systemic circulation, defined as the major vessels and well-perfused organs; distribution (D) of the drug from systemic circulation to tissues (target and nontarget) and back again; and elimination of the drug from the body by metabolism (M) and excretion (E). These drug movements are dynamic, occurring simultaneously, and their net effects determine PDC at any time during the dosing interval after administration of a fixed dose. Because passive diffusion is the major determinant of each drug movement, each in turn is influenced by a number of other factors. The impact of these factors on drug movements and the time course of drug in the body is modeled mathematically, leading to the study of pharmacokinetics. Each drug movement is influenced by host and drug factors that may affect therapeutic success.

Absorption

The percentage of an administered dose of drug that reaches systemic circulation and is thus able to induce a response is referred to as **bioavailability** (F).[2,3,9,10] Absolute bioavailability is the actual bioavailability and can be determined only after intravenous administration of a drug. In contrast, relative bioavailability of two different preparations or by two routes of administration of the same drug can be evaluated by comparing their area under the curve (AUC).[10] The area under the concentration versus time curve (AUC) describes the entire time course of the drug. Pharmacologic response of two the same drug given by different routes may vary even if equally bioavailable if the rate of absorption differs. Most absorption after oral administration occurs in the small intestine. Factors impacting absorption include drug pKa, lipid and water solubility, and surface area. However, general, the intestinal surface area is so large that changes seldom are of sufficient magnitude to effect absorption. Enterocytes can decrease oral bioavailability of a drug by causing its efflux from enterocytes, in part because of the presence of active transport proteins located on enterocyte cell membranes[13] including the efflux transporter, P-glyocoprotein which is characterized by broad substrate specificity. Hepatic metabolism also can profoundly affect the PDC of an orally administered if it undergoes first pass metabolism due to high extraction (>70%) with each pass through the liver.

Distribution

Once a drug reaches the systemic circulation, it must be distributed from the central (blood) compartment to peripheral tissues to impart a pharmacologic effect and then back again so that it can be eliminated. The major factors that determine drug distribution to and from tissues include drug lipid solubility and its ability to penetrate cell membranes, the degree to which the drug is bound to plasma or tissue proteins, and regional (organ) blood flow. The presence of transport proteins (p-glycoprotein, OATPs, and POTs) also influence drug distribution, particularly in "sanctuary" tissues characterized by tissue–blood barriers (e.g., brain, cerebrospinal fluid, placenta).[13] Binding of lipid-soluble drugs to plasma proteins facilitates their circulation but prevents drug distribution and activity. Binding is considered significant (high) if ≥ 80% of the drug is bound to protein (usually albumin). The clinical relevance of displacement (ie, due to competition among drugs or low serum albumin) and increased concentrations of unbound drug may be avoided by increased renal and hepatic clearance as long as function is normal.[16] In addition to its ability to reach target tissues, distribution of a drug is important because the extent of distribution proportionately but inversely determines PDC and directly prolongs drug elimination half-life. The amount of tissue to which a drug is distributed, often estimated by the theoretical parameter, (apparent) **volume of distribution** (Vd) of the drug directly determines dose (Dose = Vd*target drug concentration); Vd is mathematically determined by the drug concentration (Co) that is measured immediately after distribution of an intravenously administered dose (Vd=Dose/Co). In most tissues, capillary are fenestrated and present no barrier to movement of drug into interstitial fluid (comprising the majority of extracellular fluid, ECF). Drugs which are limited to ECF in distribution are characterized by a Vd of ≤ 0.3 L/kg. Capillaries of selected organs, including the brain, cerebrospinal fluid, eye, testis, and prostate, are not fenestrated,[18,19] and drugs must diffuse through the capillary endothelium to penetrate these tissues. For treatment of such organs, lipid-soluble drugs (i.e., Vd ≥ 0.6 l/kg) are more likely than water-soluble drugs to penetrate the endothelial barrier and achieve therapeutic concentrations. Such drugs also are more likely to penetrate cell membranes and thus are distributed to total body water (TBW). Documenting the extent to which a drug distributes to a specific site is difficult because techniques are required that allow collection of interstitial versus intracellular tissue. Measurement of drug in interstitial fluid or ICF might confirm the site of drug distribution.. Several techniques have been developed to collect fluid representativeof ECF. Examples include tissue cage models; "wick" methods; and more recently and perhaps most appropriately, the use of microfiltration systems that collect directly from ECF using a microprobe. A less representative method of reporting tissue concentrations is measurements of drug in homogenized tissues; these measure both extracellular and intracellular drug. Drugs bound (trapped) inside a cell (e.g., as a result of binding to proteins, or ionization) may markedly increase drug concentrations in homogenized tissue, but the drug may not be pharmacologically active or available to interstitial tissues

The term **distribution equilibrium,** or steady-state equilibrium, is often used to refer to the state in which the amount of drug leaving plasma for tissues is equivalent to the amount of drug leaving tissues for plasma. The reality is that distribution equilibrium may not ever be achieved because drugs are characterized by an "open" system—that is, drug is continually being eliminated. Distribution equilibrium is less likely to be achieved for drugs with a short half-life, particularly when administered at a long dosing interval, but may be achieved if the drug is administered at an interval that is very small compared to the elimination half-life (such as occurs with constant rate infusions). Drugs that distribute to TBW may take longer than distribution to ECF. Clinically, peak PDC (C_{max} can not be assessed until distribution has reached equilibrium [or more appropriately, pseudoequilibrium]. If PDCs are being monitored in a patient, blood samples for peak drug concentrations should not be collected until this time. For orally administered drugs, distribution occurs during absorption and sampling should occur after absorption is complete, generally within 1 to 2 hours after administration. The rate at which a drug is excreted from the body is the final determinant of PDC. Most drugs are eliminated by hepatic metabolism, renal excretion, or both.[2,3]

Metabolism

Lipid-soluble drugs require conversion to a water-soluble form before they can be eliminated by the kidney; otherwise, they passive resorbed in the renal tubules would preclude excretion. Metabolism can occur in two phases. Drugs or their metabolites may undergo both phases or only phase II metabolism. In general, phase I metabolism chemically changes the drug so that it is more water soluble and thus more susceptible to phase II metabolism; phase II metabolism generally renders the drug sufficiently water soluble to be renally excreted. The majority of, but not all, drug metabolism occurs in the liver via *cytochrome P450 enzymes (CYP450* or simply *CYP)*. identified genetically as a superfamily containing several families of enzymes.[20-23] The most important the enzymes and proportion of drugs metabolized are 2C (18%), 2E1 (7%), and 3A4 (30%). The sequelae of phase I metabolism commonly is drug inactivation but phase I metabolites often are equally or more (e.g., prodrugs) active or toxic. With phase II metabolism, a large water-soluble molecule is to either the parent drug or its phase I metabolite. Glucuronidation is the most common although the glutathione transferase system is important for several different roles, including detoxification of radical metabolites.

Excretion

Excretion is the irreversible removal of the drug (either as parent or metabolite) from the body. Renal excretion is the most important route of drug elimination for both parent (especially water soluble) drugs and their metabolites [2,3,26,27] Host factors that determine renal excretion include renal blood flow (including glomerular filtration rate), active tubular secretion, and tubular (generally passive) reabsorption. The kidney is also capable of metabolizing some drugs (e.g., imipenem), although this capacity is only occasionally of clinical importance. Glomerular filtration is a passive process. Drugs enter the glomerulus by bulk flow, being excluded if too large (> 60,000 MW) as occurs for highly protein-bound drug if the glomerulus is healthy. In contrast, active transport of drugs in the proximal tubules is very efficient, rapid, and insensitive to protein binding. Reabsorption of drugs from renal tubules into peritubular capillaries slows renal excretion; the extent depends on its lipid solubility and its ionization. Drugs excreted in the urine that do not undergo passive reabsorption will be progressively concentrated in the renal tubule. Tubular cells may be exposed to higher concentrations of drugs, which may increase the risk of nephrotoxicity. In contrast to renal excretion, biliary excretion is very slow and much less clinically. Drugs eliminated in the are generally >600 MW, and are excreted by one of three classes of transport proteins. Drugs excreted in the bile are in greater contact with the intestine and its flora and may undergo enterohepatic circulation if bacteria unconjugate the drug such that drug reabsorption may occur.

Stereoisomers are chemical compounds with the same molecular formula that differ only by the orientation of functional groups in space around the central atom (e.g., carbon). Enantiomers are stereoisomers that differ owing to orientation of four different groups around a single *chiral* carbon (or phosphorous or nitrogen) atom but are non- superimposable mirror images of one another. The center atom around which the asymmetry occurs is referred to as the *chiral molecule.* The enantiomers are differentiated as to either D (dextrorotatory) or L (levorotatory). More commonly, they are defined by the direction in which they rotate light. As such, they are "optical isomers": if one enantiomer rotates light to the left (S, or sinister [counter clockwise]), the other will rotate it to the right (R, rectus [clockwise]) the same number of degrees. The only physical difference between enantiomers is the direction in which they rotate light. However, although the physical characteristics of enantiomers are similar, the pharmacodynamics (including both potency and efficacy) and pharmacokinetics (all phases of disposition) may be profoundly different for each isomer. Most enantiomeric drugs are sold as racemic mixtures (generally 50:50), and administration of such drugs represents polypharmacy.[25] Some drugs are prepared only as one isomer, usually identified by prefixes, e.g., "**lev**iteracetam" , "**es**molol" or "**dex**tromethorphan" in the name. Further complicating enantiomer disposition is the potential for each enantiomer to be converted to the other after administration. Species differences are likely to be profound.

CNS Barriers To Drug Movement

The blood–brain barrier and blood–cerebrospinal fluid (CSF) barrier form a permeability boundary, rendering the brain a sanctuary by limiting penetration of macromolecules. They also act as selective barriers to inflow and outflow of small, charged molecules, including drugs, neurotransmitters, and their precursors or metabolites. The blood–brain barrier is absent in selected locations, thus allowing the brain to monitor chemicals. These areas include but are not limited to the area postrema, median eminence of hypothalamus, and the posterior pituitary and pineal glands. Cerebral ischemia and inflammation modify the integrity and function of the barriers. The blood–brain barrier comprises blood capillaries that differ in structure compared to other tissues in that they lack fenestrae and are joined by tight junctions similar to the epithelium of other tissues. Microvessels comprise about 95% of the total surface area of the barrier. Astrocyte foot processes encapsulate the capillaries, forming the tight junctions. Because pinocytosis also is absent, compounds can enter the brain only by passive diffusion, limiting penetration to small, lipid-soluble, non-ionized molecules. In addition to the physical barrier, function mechanisms exist to preclude central nervous system (CNS) penetration. These include degradative enzymes, which target not only chemicals but also many peptides, and high concentrations of efflux transporters. These include adenosine triphosphate (ATP)–binding cassette proteins such as P-glycoprotein (among the most abundant and most well characterized) and others, resulting in a combined broad substrate specificity that targets many drugs.[2] The substrate specificity of the P-glycoprotein transporter is large; accordingly, it impact drugs safety (see above). Interactions involving P-glycoprotein (or other efflux transporters) are among the drug interactions described at the level of the blood–brain barrier.[3] In the dog this protein is regulated by only one gene (ABCB1).[4] The blood–CSF barrier prevents compounds moving from interstitial fluid into brain parenchyma.[5] This barrier is located in the epithelium of the choroid plexus, limiting movement of compounds from the blood into the CSF. The choroid plexus consists of the highly vascularized pia mater and cells with microvilli on the CSF side. The plexus dips into pockets formed by folding of ependymal cells onto themselves, forming a double layer structure between the dura and pia (i.e., the arachnoid membrane). This double layer contains the subarachnoid space. Although the capillaries of the choroid plexus are fenestrated, the adjacent choroidal epithelial cells form tight junctions, precluding movement of most macromolecules. Like the blood–brain barrier, the blood–CSF barrier contains organic acid transporters that preclude penetration of many therapeutic agents.[6] The role of efflux pumps and P-glycoprotein in particular is discussed later. The formidable barrier presented to chemical penetration of the brain contributes to therapeutic failure when treating dysfunction of the CNS. A number of strategies have been described to enhance drug delivery to the brain.[6] These include manipulation of the drugs such that they are chemically more likely to penetrate the barrier, administration of prodrugs, the combination of drugs with compounds that decrease drug efflux, the design of vectors or other carriers that are able to penetrate, and temporary disruption of the barriers themselves. The use of nanotechnology offers potentially viable options.[7] Finally, alternative routes of delivery include intraventricular, intrathecal, olfactory, and direct interstitial routes. None thus far has proved consistently safe or relevant to all therapeutic agents.

REFERENCES

All cited references can be found in Ch1. Principles of Pharmacology; Ch. 2 Factors Affecting Drug Disposition, and Ch. 27, Anticonvulsant and other CNS Active Drugs, in Boothe, DM, Small Animal Clinical Pharmacology and Therapeutics, 2nd Ed, Elsevier.

ENHANCING THERAPEUTIC SUCCESS IN THE NEUROLOGIC PATIENT WITH DRUG MONITORING:

Dawn Merton Boothe, DVM, PhD, DACVIM, DACVCP
Auburn, AL

The success of any fixed dosing regimen most often is based on the patient's clinical response to the drug. Fixed dosing regimens are designed to generate plasma drug concentrations (PDC) within a therapeutic range, ie, achieve the desired effect while avoiding toxicity. However, a therapeutic range (C_{min} and C_{max}) is a population parameter that describes the range between which 95% of the animals might respond. Monitoring determines the individual patient's therapeutic range. Response below the therapeutic range does not necessarily indicate therapy is not needed; likewise, failure should not be considered only if the drug is above the maximum range. For some patients, the maximum range will need to be exceeded and should be considered if the drug is safe and respon. Thus, the absence of seizures in a dog with subtherapeutic concentrations is not justification for discontinuing the drug. On the other hand, a very small proportion of animals respond at concentrations higher than the recommended maximum and risk-benefit considerations should determine the need to add a second drug.

Marked inter-individual variability in physiology, response to disease and response to drugs results in variability in dose-response relationships which might be detected through TDM. All factors which determine drug disposition (absorption, distribution, metabolism and excretion) are amenable to change due to physiologic, pathologic and pharmacologic factors. The most recent examples are Collie breeds with the MDR gene mutation, and drug interactions involving CYP3A4 or P-glycoprotein. Changes in drug metabolism and excretion induced by age, sex, disease or drug interactions are among the more important factors which can cause PDC to become higher or lower than expected. Recommended dosing regimens are sometimes designed to compensate for the effects of some of these factors. Unfortunately, the effects of many factors are unpredictable and cannot be anticipated in the individual patient, despite innovated dosing calculations.

Therapeutic drug monitoring replaces the trial and error approach to dosing regimen designs that may prove costly both financially and to patient health. Monitoring is indicated in clinical situations in which an expected therapeutic effect of a drug has not been observed, or in cases where drug toxicity related due to high toxic PDC is suspected. In addition, TDM can be used to establish whether or not optimum therapeutic drug concentrations have been achieved for drugs characterized by a response that is difficult to detect or in which the manifestations of disease are life threatening and the trial and error approach to modification of dosing regimen is unacceptable. In situations in which chronic drug administration is expected, TDM can be used to define the effective target PDC in the patient. The target PDC can then be used if pharmacokinetics change in the patient over the course of chronic drug administration due to disease, environmental changes, age or drug [or diet] interactions. Drug monitoring has also been useful in identifying owner noncompliance as a cause of therapeutic failure or adverse reactions. As such, drugs for which TDM is most useful are characterized by one or more of the following: 1) serious toxicity coupled with a poorly defined or difficult to detect clinical endpoint (eg, antibimicrobials, anticonvulsants and cyclosporine); 2) a steep dose-response curve for which a small increase in dose can result in a marked increase in desired or undesired; 3) a narrow therapeutic range; 4) marked inter-individual pharmacokinetic variability which increases the variability in the relationship between dose and PDC (eg, PB, cyclosporine, zonisamide, levetiracetam); 5) non-linear pharmacokinetics which may lead to rapid accumulation of drugs to toxic concentrations (eg, phenytoin or, in cats, PB); and an unexpected toxicity due to drug interactions (eg, enrofloxacin induced TPH toxicity or chloramphenicol or clorazepate induced PB toxicity). In addition, TDM is indicated when a drug is used chronically, and thus is more likely to induce toxicity or changes in pharmacokinetics (ie, anticonvulsants), or in life-threatening situations in which a timely response is critical to the patient (eg, epilepsy or bacterial sepsis). Drugs for which TDM might not be indicated include those characterized by a wide therapeutic index which are seldom toxic even if PDC are higher than recommended and therapeutic failure does not put the patient or risk, or those for which response can be easily monitored by clinical signs.

Not all drugs can be monitored by TDM; certain criteria must be met. Patient response to the drug must correlate with (ie, parallel) PDC. Drugs whose metabolites (eg, diazepam) or for which one of two enantiomers comprise a large proportion of the desired pharmacologic response cannot be as effectively monitored by measuring the parent drug. Rather, all active metabolites and/or the parent drug should be measured. For cyclosporine (CsA), for which parent and *some* metabolites are active, HPLC measures only the parent whereas immunoassays measure parent and some metabolites. Species differences vary in the production of these metabolites. An effective therapeutic range must have been identified for the drug in the species and for the disease being treated. For many drugs, recommended therapeutic ranges in animals have been extrapolated from those determined in humans, but care must be taken for this approach (eg, bromide and procainamide). The drug must be detectable in a relatively small serum sample size, and analytical methods must be available to rapidly and accurately detect the drug in plasma. Cost of the method must be reasonable.

Implementation and response to TDM requires an understanding of the relationships between PDC, interval (T) and drug elimination half-life [t½]. In general, TDM should not be implemented until PDC have reached steady state in the patient. Steady-state PDC occur at the point when drug input and drug elimination (ie, distribution, metabolism and/or excretion) are equilibrated. Although PDC change to some degree during the dosing interval, they remain constant between intervals at steady-state (note that "steady-state is not actually reached with drugs whose half-life [t½] is substantially shorter than the dosing interval). With multiple drug dosing at the *same* regimen, PDC will reach 50, 75 and 87.5% of steady-state concentration at one, two or three half-lives, respectively (and so on) regardless of the drug. The same time period (ie, 3-5 drug half-lives) must elapse prior to monitoring if any portion of the original dosing range (ie, dose, frequency or route) is changed. For drugs with a long t½, compared to the dosing interval, drug accumulation can be very dramatic (ie, the drug concentrations following the first dose (PDC_{first}) are much lower than drug concentrations at steady state (PDC_{ss}). The dosing regimen of such drugs is designed such that drug concentrations will be in the.therapeutic range, but only when steady state concentrations have been achieved.. The amount that the drug accumulates depends on how much shorter the interval is compared to the t½ (ratio of T: t½): a T: t½ ratio of 1 results in an accumulation ratio (PDC_{first}/PDC_{ss}) of 2, 2(2.4), 3(5) and 4 (6.3 fold ; for bromide [t ½ = 21 d], a 12 h interval will result in a an accumulation ratio of 72).). For drugs characterized by a long t½, TDM can be implemented by measuring concentrations at approximately one drug t½ at which time PDC will be approximately 50% of PDC_{ss}. A third alternative to proactive monitoring is available for patients for whom steady-state concentrations must be reached immediately. A loading dose an be administered to rapidly achieve therapeutic PDC. The loading dose needed to achieve a known therapeutic concentration of a drug depends upon the Vd of that drug in the patient and the target (ie, the therapeutic concentration:). If the drug is orally administered, the bioavailability (F) must also be taken into account when the dose is determined. After a loading dose is administered, the maintenance dose should be "just right" to maintain PDC achieved after loading. If not, a problem may not become obvious until steady-state occurs (ie, 3 to 5 t½; for bromide, this would be 3 months). However, monitoring can be used to pro-actively evaluate the proper maintenance dose (see Box). In general, monitoring of a drug with a long half-life requires only one sample. Generally, for consistency's sake, we suggest collection of a trough (before the next dose). The trough PDC is the lowest drug concentration that develops during a dosing interval and it generally should not drop below the C_{min} , particularly for anticonvulsants.

Table 1. Therapeutic Drug Monitoring Data for Selected Anticonvulsants

Drug	Dose[1] mg/kg	Route	Interval	Half-life	Time to SS[2] SS	MW	Therapeutic Range (3) Low	High	Units	Low	High	Units
(+)Bromide	25-30	PO	12-24	21-24 d (D) 14 d (C)	2-3 mos	79	1405	3937	mmol/L	0.8-1[10,11]	3[11]	mg/ml
Clorazepate	0.5-1	PO	8 hrs	4-6 (d); 5-10	<24 hrs	409	367	978	nmol/L	150[3]	400	ng/ml
Clonazepam										70		
Felbamate	15	PO	divided	5-8 (d)	<24 hrs	238.2	125	250	micromol/L	30	60	mcg/ml
Gabapentin	10-30	PO	8-12	< 14 hours	2 – 3 days	171.25	70	120	micromol/L	12	21	mcg/ml
Leviteracetam				2-3		170	32	120	micromol/L	6	21	mcg/ml
Phenobarbital	2	PO	12 hrs	45-90 (D); 45 (C)	2 to 3 wks[7]	232.2	86	194	micromol/L	20	45	mcg/ml
	3-6	IM		(9-12; d)	2-3 days							
	3-16[6] total	IV inf	60 min									
	6-12[8]	IV slow										
Phenytoin				2.5 after multiple dosing	<24 hrs	252	40	80	micromol/L	10	20	mcg/ml
Pregabalin	3-4	PO	8-12	7	<24 hrs	159				3	9.5	mcg/ml
Topiramate						212.2	15	60	micromol/L	2	13	mcg/ml
Zonisamide	4-6	PO	8-12	16-44*	2-3 days	339	35	180	micromol/L	10	40	mcg/ml

Many drugs are characterized by half-lives that are much shorter than the dosing interval. For these drugs, no to little accumulation occurs, the concept of "steady-state" is perhaps irrelevant, and response can be evaluated with the first dose (or as soon as the disease has had time to respond). The amount PDC declines during a dosing interval, that is, the fluctuation between C_{max}/C_{min}) depends, again on the relationship between t½ and interval. If the interval is 1, 2, 3, or 4 times the t½, PDC will decrease 50, 75 and 87.5%, respectively during the dosing interval . This fluctuation may be unacceptable (eg, antiepilpetics, some cardiac drugs, potentially cyclosporine). Detection of this fluctuation will require collection of both a peak and trough sample. The peak PDC (C_{max}) is the maximum concentration achieved after a dose is administered and presumably it should not exceed the recommended C_{max} if the drug is not safe. Timing of peak sample collection can be difficult to predict; ideally, absorption and distribution should be complete. The route of drug administration can influence the time at which peak PDC occur, which will vary among drugs. For orally administered drugs, absorption is slower (1-2 hours) and distribution is often complete by the time peak PDC have been achieved. However, the absorption rate can vary widely due to factors such as product preparation, the effect of food or patient variability. Because food can slow the absorption of many drugs, fasting is generally indicated (if safe) prior to therapeutic drug monitoring. Generally, peak PDC occur 2-4 hours after oral administration. Some drugs are simply absorbed more slowly than others (eg, PB) and the time of peak PDC sample collection is longer (eg, 2 to 5 hours for PB). For drugs administered intravenously, TDM must take place after distribution. For some IM and SC administrations, absorption occurs rapidly (ie, 30-60 minutes), but, again, drug distribution may take longer. Thus, PDC generally are measured 1-2 hours after administration after parenteral drug administration.

Collection of peak and trough sample is particularly important for drugs characterized by a narrow therapeutic range and a short half-life. For such drugs, calculating the t½ might be useful for determined an appropriate dosing interval, although both a peak and trough sample should be collected (t½ =0.693/ kel, where kel = $LN(C1/C2]/t_2-t_1$, where C and t are the concentration and time point of the 1st (peak) and 2nd (trough) sample, respectively (Table 1). This can be easily calculated using Microsoft Excel). In contrast to drugs with a short t½, peak and trough concentrations will not differ substantially for drugs whose t½ is much longer than the dosing interval (eg, bromide, and for some patients, PB) and a single sample is generally sufficient for such drugs. Single samples might also be indicated for slow release products (eg, levetiracetam) since constant drug absorption mitigates a detectable difference between peak and trough concentrations. Single samples also can be collected following a loading dose (ie, bromide) or at the first t½ (ie, 3 to 4 weeks) in a patient that has just begun bromide therapy. Finally, if the question to be answered by TDM is one of toxicity, a single peak sample may answer the question. The impact of drug interactions (eg, induction [eg PB] or inhibition [eg, cyclosporine and ketaconazole]), or disease [eg, cardiac disease an beta blockers or digoxin] on drug clearance may cause a drug to shift from a short half-life to a long half-life or vice versa. For example, we have measured half-lives as short as 12 hrs for PB and less than 1 hr for levetiracetam, or as short as 5 hrs and as long as 150 hrs for cyclosporine. For such drugs, a prudent approach would be monitoring at baseline and then at proposed steady-state once the second drug has begun, or the disease changes. Peak and trough samples might be collected before and after the change has been implemented such that time to steady-state might actually be predicted based on half-life (eg, if ketaconazole is added to cyclosporine). Peak and trough samples might be collected in any patient that is not responding well to therapy with any drug which may have a short t½ compared to the dosing interval. The most accurate kinetic information is generated from patients receiving an IV dose since the volume of distribution can be estimated along with drug elimination t½. For oral doses, only the rate of elimination and drug t½ can be obtained.

The minimum information necessary for interpretation of PDCs includes the following: 1) The total daily dose of drug which will be correlated with the patient's measured PDC. 2) Time intervals of drug administration are particularly important for drugs with short half-lives (eg, levetiracetam, gabapentin). Provision of this information assures the clinical pharmacologist that blood samples contain the actual trough and/or peak drug concentrations. From this data, a drug t½ can be calculated and a proper dosing interval can be determined. 3) The patient's clinical status is important because both acute and chronic diseases can dramatically alter drug disposition patterns. This is particularly true for patients with renal, liver or cardiac disease. If this information is lacking, disease-induced changes in drug disposition cannot be distinguished from other causes such as non-compliance or drug interactions. 4) Concurrently administered drugs may alter drug disposition patterns and thus contribute to individual differences in drug disposition. Frequency, dose, amount and the actual times of all drugs given to the patient must be known in order to recognize or predict potential drug interactions. 5) Physiologic characteristics such as patient species, breed and age are often important to the interpretation of PDC because known or predictable differences they may induce drug disposition, or because of known differences in pharmacodynamic responses. Weight must be provided in order to determine Vd. 6) The reason for TDM should be given, ie, has the patient failed therapy or is the patient exhibiting signs of toxicity?

Once results are received, either the dose or interval of a drug might be modified (or if all is well, not changed). If patient PDC is to high or low, and particularly if the t½ is long, then the dose can be change in proportion to the desired change in PDC. Thus, if the PB is 20 µg/ml and the target is 25 µg/ml, then the old dose should be increased (or decreased) by 25-20/20 or 25%. This approach can be repeated until the maximum (or minimum) end of the therapeutic range is reached. If the t½ is short, decreasing the dosing interval may be more cost effective. Note that for each t½ to be added to the dose interval, the dose must be doubled (to add 2 t½, the dose must be quadrupled, for 3- t½, the dose must be increased 8 fold, etc).

BOX-1: BASICS OF THERAPEUTIC DRUG MONITORING IN THE EPILEPTIC PATIENT

Start-up: To establish base-line concentrations, and ideally the minimum effective concentration for the patient. This is particularly important in the patient that subsequently has break-through seizures or a sudden change in clinical signs suggestive of side effects or adverse events. Start up monitoring should be designed to establish a minimum effective dose such that the patient is not exposed to higher than necessary concentrations. Start-up is also indicated to establish when a uncontrolled patient has exceeded the maximum therapeutic range and further increases in therapy are not likely to contribute significantly to seizure control. At that point (and ideally only at that point) is combination therapy implemented. Start-up monitoring also should be implemented within a day of a loading dose.

Follow-up: Every time a new dosing regimen is implemented the patient should be remonitored at steady-state (this might be withheld until response is realized). A new dosing regimen is defined by any change in dose or interval, or the appearance of a patient factor that might change drug disposition. This includes transitioning from a loading dose to a maintenance dose or adjusting a maintenance dose. Monitoring also should be implemented if a change is anticipated in the patient that might alter drug half-life and thus drug concentrations. Examples include the addition (or discontinuation) of one or more drugs (antiepileptic drugs or nonantiepileptic drugs that can influence the disposition of the antiepileptic drug), a change in the diet (e.g., for bromide), or a change in patient health such that disposition might change as disease progresses or responds to therapy. Determining the most appropriate time point for monitoring will be complicated in patients in which drug half-life has changed and is no longer predicted by sample population statistics. For such patients more frequent monitoring is suggested.

Check-up: Monitoring proactively is particularly important for patients at risk for therapeutic failure. Monitoring is also important as patients age, are placed on diets, etc. The interval of routine monitoring should be individually determined, ranging from 3 months in patients with proven history of variability in drug concentrations to yearly as part of the annual physical examination.

What's-up? Monitoring should be implemented if a previously controlled patient begins to experience seizures or clinical signs emerge suggested of side adverse events. Availability of baseline greatly facilitates interpretation. Monitoring helps identify factors which might cause drug concentrations to change (e.g., drug interactions, change in diet, poor owner compliance). Monitoring should also be implemented to help clarify the role of disease in clinical signs.

Loading dose: If a loading dose is used, drug concentrations should be monitored after a loading dose, one drug elimination half-life later and, to establish a new steady state, three to five drug half-lives later. The sample collected at one drug elimination half-life should be equivalent to the loading dose sample; if not, the maintenance dose must be adjusted accordingly. The collection of a sample one half-life after a loading dose is minimally useful unless a postload sample has been collected; because the patient is not yet at steady state with the maintenance dose, plasma drug concentrations are still in a state of flux. ***No loading dose:*** If a loading dose is not administered, and the drug is characterized by a long half-life (i.e., several weeks to months until steady state is reached), a sample can be collected at one half-life to allow proactive evaluation of the appropriateness of a chosen dosing regimen: doubling the concentration measured at one half-life predicts baseline steady-state concentrations. If the predicted concentrations are not on target, the maintenance dose can be adjusted. Otherwise, for most drugs, monitoring can begin with a baseline sample collected at three to five drug half-lives.

Number of samples: The need for peak, trough, or both samples depends on the half-life of the drug. If the half-life in the patient is substantially longer than the dosing interval (we recommend greater than 2 to 3 times), peak and trough concentrations will be very similar, and a single trough sample collected just before a dose is sufficient (e.g., bromide, and for the majority of patients, phenobarbital; zonisamide). The trough sample is preferred: not only does it reveal the lowest concentration to which the patient is exposed during a dosing interval, it also is not impact by absorption and distribution and thus is most likely to be consistent for across-time comparisons. For drugs that might fluctuate markedly during a dosing interval (e.g., levetiracetam, gabapentin, pregabalin, and zonisamide in some pateints), both a peak and trough sample should be taken to determine patient half-life. If the half-live is revealed to be long, then subsequent single trough samples are indicated.

Other considerations: Animals should be fasted before samples are collected. Special handling preparations generally are unusual, although each laboratory should be called to confirm special handling procedures. Serum separator tubes are contraindicated because the silicon separator can bind anticonvulsant drugs and falsely decrease serum drug concentrations. If used, serum should be immediately harvested after centrifugation.[149] Interpretation of therapeutic drug monitoring results can be facilitated by a clinical pharmacologist.

REFERENCES

All cited references can be found in Ch5. Therapeutic Drug Monitoring, in Boothe, DM, Small Animal Clinical Pharmacology and Therapeutics, 2nd Ed, Elsevier

IMMUNOSUPPRESSION AND THE CNS

Dawn Merton Boothe, DVM, PhD, DACVIM, DACVCP
Auburn, AL

The Immune Response

The innate systems are nonspecific in their response and are best exemplified by barriers such as the intimate vasculature brain, White blood cells also represent innate immunity. These include neutrophils, eosinophils, basophils, macrophages, and dendritic cells, and unconventional T-cell subsets bearing T-cell receptors (TCR), natural killer (NK; CD1d) cells and gamma/delta (γδ) T-cells. The adaptive immune system provides the host an ability to recognize and, with subsequent presentation, remember specific pathogens (B and T memory cells), and increasingly stronger responses each time the pathogen is presented. The antigen is exposed to an antigen-processing or presenting cell (APC), which includes dendritic cells (the principle APC), macrophages, and activated B cells.[4] The antigen is identified by the APC as foreign and is subsequently phagocytized by the APC. The APC "processes" the antigen and "exhibits" it on the cell surface in a groove made by the MHC molecule. The APC migrates to the T cell area of a lymph node and presents the antigen to naïve CD4 T cells. If the APC comes into contact with a CD4 T cell with a TCR that recognizes the antigen associated with the MHC, it becomes activated. Interleukin-1 (IL-1) release by the APC and interleukin-2 (IL-2) from the activated CD4 cell itself amplifies the sequelae of CD4 activation. The activated CD4 cell differentiates, proliferates, and produces a number of cytokines which results in recruitment of other leukocytes, initiation of B cell production of immunoglobulins (Igs), and formation of other T cell colonies, including "memory" cells.

Cellular components

T lymphocytes provide the primary regulation of the immune response. T cell activity begins with specific antigen recognition by a receptor on the surface of the cell.[4] T cells are further subdivided into several populations of cells depending on their role in immunoregulation. Helper cell (CD4 T cells; T_h cells) receptors recognize and bind to the peptide–MHC class II complex of APC cells. In response to IL-1, CD4 cells consequently proliferate and become primed as either T_h1 or T_h2 CD4 cells, which modulate further responses in both the cell-mediated and humoral arms. Migration of activated lymphocytes from lymph nodes to tissues is facilitated by adhesion molecules expressed by endothelial cells in tissues. Whether a T cell becomes T_h1 or T_h2 reflects the stimulating cytokine: IL-12 activates signal transducer and activator of transcription 4 (STAT-4), which regulates T_h1 differentiation, whereas IL-4 activates STAT-6 and T_h2 differentiation.[5] The T_h1 subsets of CD4 produce primarily proinflammatory cytokines such as IFN-γ (IFN-γ) and tumor necrosis factor-alpha (TNF-α) (cachexin). As such, T_h1 cells regulate signals that promote cell-mediated immunity and control intracellular pathogens. T_h1 response is paramount to successful resistance to most microbial pathogens, including bacteria, intracellular protozoa, and fungal organisms. Additionally, T_h1 cells mediate organ-specific autoimmunity and as such are crucial to the pathogenesis of autoimmune diseases. Activated CD4 cells also yield, in response to IL-2, helper T cells that contain the glycoprotein CD8 and are responsible for the T cell–mediated cytotoxic response. Receptors of the CD8 T cells recognize peptide–MHC class I complexes (generally associated with endogenous peptides present on all nucleated cells) on APCs. Recognition of the CD8 receptor and subsequent interaction with CD4 cells result in the generation of cytolytic (CTL) or cytotoxic T cells (T_C; killer T cells). Cytotoxic T cells are capable of directly (without further interaction with CD4 cells) causing lysis of cells expressing the targeted specific peptide–MHC complex.[4] This includes normal but otherwise damaged (include virally infected or cells damaged by TNF-β or lymphotoxin) cells. CD8 cells release several different cytotoxins which cause pores in the target cell membrane. Cytotoxic T cells also undergo clonal expansion, resulting in prepared effector cells. In contrast to T_h1 cells, T_h2 cells also secrete *antiinflammatory* cytokines (IL-4, IL-5, IL-9, IL-10, IL-13). Because of their antiinflammatory effects, T_h2 cytokines can decrease autoimmune diseases associated with cell-mediated immunity; however, an imbalance toward T_h2 cells have been implicated in the pathogenesis of asthma and allergy (see below).[5] Dysregulation of the influence of STAT-4 and STAT-6—which have contrary effects—has been associated with immune-mediated diseases.[5] In contrast to immune-mediated diseases, in which T_h1 cells play key roles, key effector cells in all chronic allergic diseases include eosinophils, basophils, and T_h2 cells.[6] Their involvement occurs through release of preformed, or formed in situ, granule proteins and cytokines.[7] Cysteinyl leukotrienes appear to have an important role in the regulation of human eosinophil hematopoiesis, recruitment, and activation.[7] An imbalance in the activities of CD4 T_h1 and T_h2 subsets may be responsible for the onset or exacerbation of immune-mediated diseases. Increased concentrations of T_h1 (and decreased T_h2) are associated with response to viral and fungal infections, whereas increased concentrations of T_h2 (and decreased T_h1) are associated with increased production of IgE and IgA antibodies. Autoimmune diseases are associated with a predominance (reflecting an imbalance) of T_h1. Imbalances also have been associated with resistance to infectious disease and malignancy.[15] In contrast, allergic inflammatory diseases (eg., atopy, inflammatory bowel disease, asthma) may reflect an imbalance that favors T_h2.

Soluble components

These include cytokines previously described, immunoglobulins, and the complement cascade. TNFα induces a broad spectrum of activity. Cytotoxicity of multiple cell types, including tumors, reflects both apoptosis and necrosis. A large number of proteins are targeted, along with other central mediators of the inflammatory process and immune activation. Examples include nuclear factors (e.g., NFκB), nitric oxide synthetase (NOS), cell-surface molecules, MHC classes I and II, and secreted proteins such as IL (e.g., -1, -6, -8), IFN, granulocyte-macrophage colony-stimulating factor, platelet-derived growth factor, urokinase plasminogen activator, and TNF-α itself. The complement cascade is a major effector mechanism of the immune response. The cascade results in highly amplified events that interact with other physiologic cascades, including the coagulation pathway, kinin formation, and fibrinolysis.[11] Consequences of complement activation include opsonization, the release of biologically vasoactive peptides, and cellular lysis. Cell-mediated cytotoxicity is implemented by cytotoxic T lymphocytes, NK cells (and antibody-dependent cell-mediated cytotoxicity (ADCC), mediated by a variety of cells that express surface Fc receptors. Effector cells of ADCC include monocytes, neutrophils, eosinophils, selected cytotoxic T cells, and NK cells.

Inflammation in the CNS

The body minimizes destruction that often accompanies inflammation by restricting the passage of immune cells from systemic circulation into the central nervous system; white cells are generally present in nerve tissue only when blood vessels are damaged by trauma or disease. The brain lacks a lymphatic system, limiting localized immune response and the blood brain barrier precludes antibody entry. In the brain, microglial cells serve as macrophages, scavenging for damaged neurons, plaques and infectioius materials. The ability of microglial cells to detect signals is heighted compared to other tissues because of the need for rapid and effective response such that inflammation is minimized. They lack many cell surface proteins associate with APC, including MHC I and II antigens. Proliferation and replacement generally occurs only upon activation. During states of severe infection, disruption of the blood brain barrier may allow bone marrow derived cells access to the CNS. Perivascular microglial cells facilitate repair of damaged endothelial cells. Microglial cells are characterized by plasticity that accommodates their varied function. Activation occurs through a variety of signals; morphologic changes are accompanied by protein uptake and release of cytotoxic signals as well as recruitment of other inflammatory cells. Cytokinse (IL-1alpha and beta and TNF) are released. T-cells are activated. Microglial cells assume the traditional functions of macrophages during acute inflammation (ie. Neuronal damage), but may contribute to negative sequelae associated with chronic (neuro) inflammation during which persistent release of inflammatory mediates may result in neuronal death. The, mechanisms by which T cells access the CNS are not clear. Investigators have recently identified, in association with toxoplasmosis infection,CD8+ cell influx along a reticular type of tissue pathway. Investigators interpret this to suggest that in the brain, "specialized structures are induced by inflammation that guide migration of T cells in this immune-privileged environment and allow them to perform a search-and-destroy type of mission required to find abnormal cells or microbes with the brain."

Glucocorticoids as Immunomodulators

Mechanism of Action

Two primary types of corticosteroid receptors exist. Type I, or mineralocorticoid receptors (MRs), bind both endogenous GLCs and aldosterone. Type II, or GRs, bind endogenous and exogenous GLCs but have a poor affinity for mineralocorticoids. The GRs and MRs are sufficiently similar that drugs may bind to both, causing similar responses. Every cell appears to have GRs, although the liver is the primary target. The type and concentration of GRs varies between species and tissue. Type II receptors are more ubiquitous in the brain. The degree to which GLCs bind to each receptor type varies with the circulating concentration. At basal levels type I receptors are preferred, but as cortisol concentrations increase (e.g., during stress), type II receptors increasingly are activated.[6] The molecular description of the GR as it relates to its mechanism of interaction has been described.[11-14] The GR consists of at least two isoforms.[11] The α isoform binds to GLCs, DNA, and transcription factors and thus modulates transcription activity. The α isoform may also act through nongenomic mechanisms. In contrast, although the β isoform binds to DNA, it does not bind other ligands, fails to activate transcription, and appears to be able to interfere with the activities of α isoforms.[12] Corticosteroid receptors are located in the cytoplasm of the target cell complexed with heat shock proteins (Hsp 70 and Hsp 90) that act as chaperones, proteins which in non-functional activities (e.g., folding, assembly) of macromolecules, and an immunophilin, an intracellular protein that binds other immunosuppressive compounds.[9,11] The complexed receptor is inactive until bound to a steroid ligand.[9] Steroids enter the cells by passive diffusion, although a rate-limited active transport mechanism may also exist (Figure 30-2) Once in the cell, the GLC binds to the receptor, causing the heat shock protein and other molecules to dissociate. The GCL diffuses into the nucleus where it binds to regulatory proteins of target genes. The short DNA sequences recognized by the activated GR are referred to as *glucocorticoid responsive elements* (GREs). It is through these GREs that specificity of GLC-modulated gene transcription is controlled.[4] Transcription of the gene and subsequent formation of the targeted protein is either induced (transactivated; positive GRE or pGRE) or inhibited (transrepressed; negative GRE or nGRE). Thus, intracellular GLCs impart their effects in one of three ways: The steroid–receptor complex moves into the nucleus and (1) binds to specific DNA sequences causing the regulation of genes (for example, lipocortin or genes responsible for gluconeogenesis) (referred to as a "cis" or element, resulting in transactivation), (2) interacts with other transcription factors (e.g., AP-1, NFκB), generally at lower concentrations, that prevent the transcription of targeted genes (e.g., TNFα, IL-1B, IL-4 and 5; referred to as a "trans element" resulting in transrepression); or (3) through non-genomic mechanisms involving membrane-associated GR and secondary messenger systems.[2, 12a] Non-genomic responses are rapid in onset, and includes vascular responses (eNOS activation), lymphocytolysis, and possibly, induction of inflammatory prostaglandins. "Permissive" effects of GLC may reflect non-genomic responses. The receptor and GLC are eventually metabolized (the exact location or timing is not documented). The cellular half-life of the activated complex is about 10 hours.[15] The GR generally has a lower affinity for nGRE than pGRE. Transrepression involves protein–protein coupling to other transcription factors (e.g., nuclear factor kappa-B [NFκB], activator protein-1 [AP-1], STAT-5 (signal transducer and activator of transcription) or nuclear factor of activated T-cells [NFAT]) such that their activity is modulated.

The importance of these different GR interactions results in differential physiologic (and pharmacologic) effects and ultimately provides a basis for directed drug therapy. The largely undesirable metabolic effects of long-term GLC therapy appear to be mediated by pGREinteraction (i.e., transactivation gene expression). In contrast, the desirable antiinflammatory effects appear to reflect modulation of transcription factors (i.e., transrepression through interaction with nGREs).[16,17] Trans (protein-coupling) effects, and repression in particular, appear to occur at concentrations lower than that necessary for cis (transactivation) effects,[18] providing another mechanism of directed therapy. Designer GLC might be developed such that potency for protein coupling (trans) repression is greater than DNA interaction, leading to cis or trans activation. Dexamethasone and prednisolone are examples of a "symmetric" GLC, characterized by equal binding affinity for both actions. Interestingly, medroxyprogesterone acetate exhibits a preponderance of transrepression activity rather than transactivation.[16] Newer GLCs are likely to have a preponderance of activity for either activation or repression. A term that has emerged is "Dissociated GLCs" which target GLC-dependent transactivation or alternatively act as inhibitors of NFκB-dependent transcription, thus retaining their in vivo antiinflammatory activity but avoiding their undesirable metabolic effects.[19] Differences in GR affinity for nGREs versus pGREs may also offer a mechanism for differential targeted drug response based on dose (or concentration). Treatment with low, rather than high, doses of GLCs appears to minimize undesirable side effects.[20] Using a retrospective analysis of published literature (meta analysis precluded by method variability) coupled with a prospective study, researchers found that most adverse effects to GLCs in humans were associated with a high, rather than low, doses (7.5 mg prednisolone equivalent in humans [approximately 0.1 mg/kg])of GLC (see the section on adverse events

Resistance to glucocorticoids

Failure to respond to steroid therapy is observed in human patients being treated for a variety of diseases. Poor compliance and low bioavailability must be ruled out before other factors are considered. The documented phenomenon of GLC resistance (both familial and iatrogenic or drug induced forms have been reported in humans) may reflect, among other causes, reduced receptor numbers or affinity for GLC.[21] Cellular response to GLCs has been directly correlated with receptor numbers. Reversible downregulation of receptor numbers and a subsequent decrease in biological effect are documented sequelae of GLC treatment.[7] Such effects have been demonstrated on lymphocyte receptors.[22] GLC failure in patients receiving long-term GLC therapy for treatment of host-versus-graft rejection has been attributed to downregulation of cytoplasmic GRs in T lymphocytes. GLC doses subsequently must be increased.[23] Pulsing of high doses of GLCs may overcome this relative resistance (see the section on therapeutic use).[24] The relative balance of the α and β isoforms of GR appear to influence cell sensitivity to GLCs; higher concentrations of β contribute to GLC resistance.[12] Some human patients with severe disease are nonresponsive to even high doses of GLCs. Lack of response has been associated with poor antiproliferative effects on blood T lymphocytes, in contrast to near-complete inhibition of responders. A similar relationship has been demonstrated for other chronic allergy-based diseases, such as asthma or rheumatoid arthritis, and renal allograft rejection. A relatively fast in vitro T lymphocyte proliferation assay has been useful in identifying nonresponders before treatment.[25] Several molecular mechanisms have been proposed to account for poor control of T cell proliferation by GLCs. These include overexpression of the multidrug resistance gene (MDR1 polymorphism), resulting in increased P-glycoprotein–mediated efflux of cytoplasmic (see relevant discussion in the section on neoplasia). This mechanism might be reversible by co-administration of cyclosporine, an inhibitor of P-glycoprotein. Other mechanisms include impaired GLC signaling by the GR and increased activation by epithelial cells of proinflammatory mediators that directly inhibit GR transcription. This latter mechanism (postreceptor failure) is supported by the presence of clinical signs consistent with Cushing's disease in humans with poor inflammatory control.[25] Differences in activation patterns of NFκB have been associated with GLC resistance in human patients with chronic inflammatory bowel disease.[19] Ultimately, differences in response at the molecular level may also be shown .

Immunmodulatory effects of GLCs are most frequently used in clinical medicine for their antiinflammatory and immunosuppressive actions (see Table 30-1).[31] Because the antiinflammatory and immunosuppressive effects of GLCs reflect specific actions on white blood cells, these effects are inextricably linked. They generally occur only when the amounts of steroid are present in concentrations greater than those found in the normal physiologic state (i.e., pharmacologic concentrations). GLCs also profoundly affect white blood cell function. Ultimately, both the humoral and cell-mediated arms of the immune response are affected. The antiinflammatory effects of GLCs reflect all three mechanisms of GLC action (genomic transcription and transactivation; nongenomic membrane receptor signaling).[12] GLCs upregulate or downregulate as many as 2000 genes involved in the regulation of the immune response GLCs inhibit early and late phases of the inflammation. Responses that are inhibited include edema formation, fibrin deposition, leukocyte migration, phagocytic activity, collagen deposition, and capillary and fibroblast proliferation. Many of these processes involve lymphokines and other soluble mediators of inflammation, and it is through these mediators that GLCs exert their antiinflammatory actions. GLCs induce annexin I (also known as lipocortin), which inhibits phospholipase 2, thus blocking the release of arachidonic acid and its subsequent conversion to eicosanoids. GLCs also preferentially inhibit transcription of cyclooxygenase-2 (COX-2), the inducible form of cyclooxygenase.[12,32,33] Interestingly, recent evidence indicates that under some circumstances, GLCs may actually induce COX-2.[34] The net effect of specificity to inhibit COX-2 may be inhibition of inflammatory prostaglandins without negatively affecting the protective effects of prostaglandins in other body systems (e.g., gastrointestinal, renal, hemostasis). GLCs induce the protein MAPK phosphatase 1, which through various actions inactivates a number of proteins important in the signaling of cytokines.[12] Through physical interaction, transcription of NFκB is inhibited by GLCs. Many of the antiinflammatory effects of GLCs occur rapidly, apparently independently of changes in gene expression (nongenomic). Examples include activation of endothelial nitric oxide synthetase,[12] offering an alternative mechanism of action of a new class of more selective antiinflammatory drugs.[35] Among the sequelae of these mechanisms, GLCs inhibit release of TNFα and IL-2 from activated macrophages. TNFα induces cytotoxicity and can enhance neutrophil and eosinophil function.[27] The release of platelet-activating factor from leukocytes and mast cells is inhibited by GLCs.

The immunosuppressive actions of GLCs, like their antiinflammatory actions, involve disruption of intercellular communication of leukocytes through interference with lymphokine production, biological action, or both. GLCs inhibit the synthesis and release of IL-1 by macrophages, thereby suppressing the activation of T cells. GLCs also inhibit IL-2 synthesis by activated T cells. IL-2 plays a critical role in amplification of cell-mediated immunity.. GLCs block the effects of the migration inhibitory factor-γ (causing macrophages to migrate away from the affected area)and interferon-γ (IFN-γ) on macrophages.[27] IFN-γ, which is released from activated T cells, plays an important role in facilitating antigen processing by macrophages. Additionally, GLCs suppress the bactericidal and fungicidal actions of macrophages. GLCs also alter synthesis of and biologic response to collagenase, lipase, and plasminogen activator..The antiinflammatory effects of GLCs also may reflect inhibition of the inducible form of nitric oxide synthase (iNOS).[37] Synovial macrophage nitric oxide production and iNOS synthesis are inhibited by dexamethasone. The inhibitory effect appears to be mediated by lipocortin.The immunosuppressive actions of GLCs are more pronounced on the cellular arm than the humoral arm of the immune system. GLCs have minimal effects on plasma immunoglobulin concentrations but can modulate immunoglobulin function. For example, opsonization of bacteria is inhibited. Therapeutic doses of GLCs do not significantly decrease an animal's antibody response to antigenic challenge (e.g., vaccinations). The immunosuppressive effects of GLCs may also reflect actions on the hypothalamic–pituitary–adrenal axis. Multiple cytokines appear to regulate this axis. Specifically, IL-1 appears to stimulate the release of CRH and directly increase the release of ACTH, and it may cause the adrenals to release GLCs. These interactions appear to be important to modulation of stress and thus maintenance of homeostasis.[9]

Glucocorticoid effects in the CNS

In the central nervous system (CNS), the highest concentration of MRs is in the limbic system, whereas GRs are more ubiquitous in the brain.[6] Indirectly, GLCs maintain adequate plasma concentrations of glucose for cerebral functions, maintain cerebral blood flow, and influence electrolyte balance in the CNS. GLCs decrease formation of cerebrospinal fluid, which results in a reduction of intracranial pressure. In human beings GLCs are believed to influence mood (e.g., producing euphoria), behavior, and brain excitability.[9] The euphoric effect commonly recognized in dogs also is likely to reflect differences in GRs. Steroids, including GLCs, also appear to regulate neuronal excitation, with differential effects occurring for some drugs. For example, dexamethasone inhibits microglial cell proliferation; cortisone also will inhibit proliferation at higher doses through GRs, but stimulate proliferation through MRs.[26] GLCs appear to have a protective effect in the CNS through induction of glutamine synthetase in both the CNS and peripheral nervous system; the enzyme adds an amine to and thus removes glutamate.[43] Increased glutamate has been associated with CNS pathology, although the relationship between the two remains controversial.[44,45] However, efficacy reflects regulation through transcription, which takes at least 24 hours. Glutamine has several roles in the CNS; in the presence of hyperammonia (e.g., liver disease) increases formation of glutamine in brain cells and the potenetial for cerebral edema. [45b] The use of glucocorticoids for treatment of cerebral edema might be avoided in situations in which increased glutamine increases the risk of cerebral edema. Interestingly, GLC concentrations appear to be directly related to the severity of memory impairment in animals, including humans.[6] Learning is impaired in animals exposed to prolonged reductions in circulating concentrations of GLCs; conversely, memory also appears to be impaired if GLC concentrations are too high.

The effects of GLCs are generally recognized to be dose dependent. Indeed, GLCs can be classified on the basis of their potency, generally relative to hydrocortisone. For example, dexamethasone is 30 times and prednisolone 4 times as potent as hydrocortisone in impairing glucose metabolism (Table 30-2).[20]). Daily doses of glucocorticoids used therapeutically have been variably referred to as physiologic, which generally reflects 0.1 to 0.25 mg/kg predniosolone or equivalent, and a pharmacologic or supraphysiologic, ranging, for prednisolone, 0.8 mg/kg IV, to 1 to 2 mg/kg/day, and a high dose of approximately 10 mg/kg orally (or 4 to 60 mg/m^2/day). Some references refer to very high supraphysiologic doses (in dogs: hydrocortisone at 2-4 mg/kg, IV every 6 to 8 hrs, 4 to 20 mg/kg prednisolone IV every 2 to 6 hrs, 0.5 to 2.0 mg/kg dexamethasone, or 30 to 100 mg/kg methylprednisolone), which also might be referred to as a "shock" dose.

Drugs

Prednisone and prednisolone generally are considered equivalent in terms of therapeutic dosing in both dogs and cats. However, the two are not necessarily equivalent, as has been demonstrated and previously demonstrated in dogs and cats. Colburn and coworkers[63] demonstrated in Beagles that relative bioavailability is greater for prednisolone. After 5 mg orally of either drug, the C_{max} was twice as high and the area under the curve 40% higher for prednisolone compared with prednisone. That the two are not equivalent in cats was demonstrated by Graham-Mize and Rosser.[59] Disposition of prednisolone in cats after oral administration was as follows: area under the curve 3230.55 ng/mL/hr and C_{max} 1400.81 ng/mL, with a half-life for excretion of 1 hour. This compares to an area under the curve of 672.63 ng/mL/hr and C_{max} of 122.18 ng/mL after oral administration of prednisone. The half-life of prednisone was longer at 2.46 hours. According to these data, the dose of prednisone must be at threefold to fivefold higher than that of prednisolone to achieve equivalent activity. **Methylprednisolone** possesses lipid antioxidant activity that has been shown to be beneficial in the treatment of experimental spinal cord trauma in cats and experimentally induced *Escherichia coli* bacteremia.[65,66] In contrast, methylprednisolone did not appear clinically beneficial in a canine model of spinal trauma, although the model may have caused insufficient damage for effective evaluation.[67] Although some other GLCs (dexamethasone, prednisolone) are efficacious as lipid antioxidants, methylprednisolone is the most potent. Methylprednisolone has an intermediate duration of action (12 to 36 hours) and is also a good candidate for alternate-day administration. **Dexamethasone** is a highly potent GLC, but it has virtually no mineralocorticoid activity. It possesses some lipid antioxidant activity. The disposition of dexamethasone in dogs exhibits dose dependency. The prolonged duration of action (biological half-life approximately 48 hours) of dexamethasone makes it inappropriate for alternate-day administration. **Tirilazad mesylate** is a novel, non-GLC, 21-aminosteroid (lazaroid) that possesses potent antioxidant activity (i.e., it protects against oxygen-derived free radicals). Unlike the GLCs, this agent does not inhibit phospholipase A_2, but it does inhibit lipid peroxidation-induced arachidonic acid release.[73] The mechanism appears to reflect, in part, insertion into the cell membrane.

REFERENCES

All cited references can be found in Ch. 30. Glucocorticoids and Ch 31. Immunomodulators and Biological Response Modifiers in Boothe, DM, Small Animal Clinical Pharmacology and Therapeutics, 2nd Ed, Elsevier.

ONCOLOGY ACVIM FORUM UPDATE: WHERE DO WE GO FROM HERE?

Susan Lana, DVM, DACVIM (Oncology)
Ft Collins, CO

INTRODUCTION

Approximately 25% or less of boarded oncologists have attended the ACVIM Forum each of the last 3 years. This is the lowest percentage attendance of all of the ACVIM specialty groups. We understand that unlike other Specialties, the Veterinary Cancer Society annual meeting does provide an alternative to the Forum. The ACVIM Forum Program Committee was interested in finding out why oncologist attendance has been low and how we might modify the Forum program to better meet the needs of oncologists. Pursuant to this goal, the Oncology Forum Program Committee generated a survey to build an understanding of how to improve the ACVIM Forum for Oncologists.

The goal of the survey was to query boarded ACVIM oncologists to find out how they perceive the ACVIM forum with the intent to increase attendance by providing content that meets the needs of the oncologist. The survey was completed on line and participation was solicited via the list serve and links sent directly to oncology diplomats. We also made announcements at the fall VCS meeting to encourage more participation.

METHODS AND RESULTS

The survey had a total of 20 questions. There were a total of 103 respondents which is approximately 40% of all active oncology diplomats. When asked if they had attended the forum in the last 10 years 93% of respondents had. All but 20 had attended in the last 3 years. The average number of times was 2.6 with a median of 5 times. When asked how frequently they attended the meeting, 24% said yearly, 14% every other year, 46% felt their attendance was dependant on location, and 16% felt their attendance depended on cost.

The respondents were then asked to pick the top three reasons for attending the forum from a list provided. There was also a write in option. When considering the number one reason, the most common reason listed was Oncology Learning Opportunities (31 respondents rated this reason number one for why they attend) followed by Visiting with Colleagues (22 number one ratings). When evaluating which of the 4 choices received the highest number of 1, 2, or 3 ratings, Visiting with Colleagues had 79 top three responses while Oncology Learning Opportunities had 66 top three responses. When asked if they attended more than one CE meeting per year 67% reported yes and 33% no. When asked if they could attend only one which would they choose: VCS or ACVIM, 92% chose VCS. Other meetings respondents listed as a primary choice included ECVIM, ACVR, AACR, ACVS, and a variety of human scientific meetings. The survey then asked "if you choose VCS, which of the following factors impact that choice?" Respondents were allowed to put more than one answer. Ninety three individuals responded and 87 of those listed content as one of the reasons for choosing VCS. Social interaction was mentioned 52 times.

When asked about the new forum format, 66% were familiar with the change. When asked how they felt about the change, 51% were positive, 27% negative, and 22% were neutral. When asked what type of presentations they wanted to see, respondents were given 4 choices and could choose more than one. Clinically applicable updates was selected 77 times, discussion panels on "Current Standards of Treatment" was selected 65 times, state of the art topics selected 59 times and short scientific abstracts was selected 54 time. This generally reflects room attendance based on topics at last year's forum. When asked to rate the forum with respect to oncology presentations and content, respondents gave good to excellent ratings for quality of the speakers 76% of the time; quality of the information 59% of the time; quality of the topics 52% of the time. When asked to rate the pre forum day using the same criteria the respondents gave good to excellent ratings for quality of the speakers 84% of the time although 30% of respondents had no opinion; quality of the information 78% of the time; quality of the topics 77% of the time. The general perception of the pre-forum format was excellent for 21%, good for 39%, average for 12%, poor for 4% and 24% had no opinion. Finally when asked what their overall satisfaction with the ACVIM forum was 91 oncology dips responded. Eight (9 %) were very satisfied, 46 (50 %) were satisfied, 29 (32 %) were somewhat satisfied and 8 (9 %) were not satisfied. When asked if they attended the Oncology business meeting, only 33% did.

When asked if they encourage their residents to attend ACVIM only 31% of those respondents with residents did, the majority did not. Another way to encourage attendance and involvement in the forum is to target or encourage specialty candidates to stay for the meeting after taking their exams. Cost to register for the forum has often been voiced as a concern by candidates. In the years preceding board certification, specialty resident candidates incur the following expenses: registration of residency with ACVIM- $100; General exam- $640; Credentials- $335; specialty exam- $950. The total cost paid in fees to ACVIM is $2025 which does not include travel expenses to get to the exams. The resident registration fee to attend ACVIM forum is $315.

CONCLUSION

In conclusion, it will be difficult to increase attendance at the ACVIM forum as long as people are reluctant to attend more than one meeting and the VCS meeting remains highly popular for both content and social interaction.

HDACs AS A THERAPEUTIC TARGET

Douglas H. Thamm, VMD, DACVIM (Oncology)
Fort Collins, CO

THE HISTONE DEACETYLASE ENZYMES

It is easy to forget that, in the cell, DNA is not a linear molecule; rather, it adapts secondary and tertiary structure and assembles itself into complexes with a variety of structural proteins. The core unit of structure of DNA is the **histone**, a phylogenetically conserved octamer of protein subunits around which the DNA strand associates. Histones undergo a variety of posttranslational modifications, including methylation, acetylation, ubiquitination, and phosphorylation among others. It is this combination of modifications that is thought to comprise the "histone code", which governs diverse functions such as transcriptional regulation, DNA repair, and chromosome condensation in mitosis.[1,2]

The differential acetylation of histones, controlled by various histone acetyltransferases and ***histone deacetylases*** (HDACs) is important in governing the structure of chromatin and thereby plays a role in modulating expression of genes associated with cellular proliferation, differentiation and survival.[2] The acetylation of histones by HDACs results in formation of heterochromatin (compact chromatin) and transcriptional repression. This transcriptional repression can negatively regulate the expression of a variety of developmental and tumor suppressor genes.[2]

To date, 18 mammalian HDAC enzymes have been reported. They are generally divided into 2 classes, based on sequence homology to yeast counterparts. **Class I HDACs** are expressed in most cell types, and are found almost exclusively in the nucleus. **Class II HDACs** have a more restricted expression pattern, and are capable of both nuclear and cytoplasmic localization in response to different cellular signals.[3]

The precise mechanisms of action of all HDAC enzymes are not yet well defined. The primary function of HDACs is to affect repression of transcription by altering chromatin condensation, selectively altering gene expression.[2] However, HDACs also are capable of differentially acetylating a variety of non-histone protein substrates, which can further regulate gene transcription, as well as regulating protein stability and protein-protein interactions. Well-characterized non-histone targets of HDACs include p53, tubulin, estrogen and androgen receptors, the E2F1 and HIF1-alpha transcription factors, c-MYC, NF-kappaB, and the HSP90 chaperone protein.[4] It is likely that there are many other non-histone targets that have yet to be elucidated.

The pharmacologic inhibition of HDAC is capable of antitumor effects in a wide variety of human cancer models. Importantly, normal cells are more resistant to the cell-death inducing properties of HDAC inhibitors than transformed cells.[5] Growth inhibition is commonly observed, at least in part as a result of induction of expression of the cell cycle checkpoint regulator p21 (CDKN1A). The pathways identified in mediating HDAC inhibitor-induced transformed cell death include apoptosis by the intrinsic and extrinsic pathways, mitotic catastrophe/cell death, autophagic cell death, senescence, and reactive oxygen species facilitated cell death.[2]

In addition to direct effects on tumor cell proliferation and survival, inhibitors of HDAC can mediate a variety of other putative antineoplastic effects, which are listed in **Table 1**.

HISTONE DEACETYLASE INHIBITORS

A number of compounds that inhibit HDACs have progressed rapidly through clinical trials.[2] For example, suberoylanilide hydroxamic acid (SAHA, also known as Vorinostat and marketed as Zolinza®; Merck & Co., Inc., NJ, USA) was approved recently for treatment of cutaneous T cell lymphoma. Approval was granted as part of the FDA orphan drug program, following a priority review. A variety of other natural, semi-synthetic, and synthetic inhibitors of HDAC are in various stages of clinical development (**Table 2**). There has been encouraging antitumor activity in a variety of tumor types in these early studies, including hematopoietic and solid tumors.

Table 1: *Putative Antineoplastic Mechanisms Associated with HDAC Inhibition*

Mechanism	Target(s)
Angiogenesis inhibition	Reduction in activity of Hif-1/VHL/VEGF axis: Decreased angiopoietin-2 synthesis: Decreased endothelial commitment of progenitor cells: Decreased endothelial NOS: Decreased bFGF synthesis: Inhibition of *in vitro* endothelial proliferation and tube formation: Induction of endothelial cell apoptosis and growth inhibition
Decreased bone loss	Induction of osteoclast apoptosis: Promotion of osteoblast maturation: Enhanced bony differentiation of mesenchymal stem cells
Immunomodulation	Upregulation of MHC class I and class II antigen expression: Upregulation of costimulatory molecule expression (B7.1, B7.1, CD40, ICAM-1): Re-expression of repressed tumor antigens: Alterations in regulatory T cell function
Telomerase inhibition	Decreases in TERT expression, telomerase activity, and telomerase length
Anti-inflammatory effects	COX-2 inhibition: Decreased Toll-like receptor mediated proinflammatory gene transcription: Decreased IL-6 and IL-1beta production

Table 2. *HDAC Inhibitors in Human Clinical Development*

Phase of Development	Compound	Company / Sponsor
Phase-I	AR-42	Arno Therapeutics
	CUDC-101	Curis
Phase-II	LAQ824	Novartis
	Sulforaphane	Various
	CI-994	Pfizer
	MGCD0103	MethylGene / Pharmion
	PCI-24781	Pharmacyclics
	SB939	S*BIO
	MS-275	NCI
	ITF2357	Italfarmaco
	4SC-210 / resminostat	4SC
	Pivaloxymethyl butyrate / AN-9 / Pivanex	Titan
	PDX101 / belinostat	CuraGen / TopoTarget
Phase-III	LBH589 (CTCL)	Novartis
Approved	SAHA / vorinostat / Zolinza® (approved for cutaneous T cell lymphoma)	Merck
	FK228 / depsipeptide / romidepsin / Istodax® (approved for T cell lymphoma)	Celgene
	Phenylbutyrate / Buphenyl® (approved for urea cycle disorders)	Ucyclyd
	Sodium valproate / Depakote® (approved for epilepsy)	Abbott

There is tremendous structural heterogeneity among the HDAC subclasses and even among the individual isoforms. Increasingly, this heterogeneity is being shown to give rise to distinct functions.[3] Despite these emerging differences, many of the known HDAC inhibitors are class indiscriminate with respect to the HDACs they inhibit. For example, SAHA is a non-specific agent, inhibiting the activity of class I and II HDAC enzymes equally. Therefore, much of the activity of HDAC inhibitors has been broadly attributed to all HDAC subclasses. However, with respect to cancer therapy it is emerging that class I HDAC enzymes may be more clinically relevant targets.[6] This is still very controversial and not absolute, as illustrated by findings that inhibition of HDAC6, a class II HDAC, leads to acetylation and disruption of the chaperone function of heat-shock 90 protein (Hsp90) in leukemic cells.[7] Nevertheless, the current evidence is swaying towards class I HDACs as the important cancer targets and this may call the therapeutic potential of broad-spectrum inhibitors such as SAHA into question.

HDAC Inhibitors in Veterinary Oncology

To date, there is little to no information available regarding the use of HDAC inhibitors in domestic animals. A single case report described the postoperative administration of SAHA to a dog with hemangiosarcoma with a favorable outcome;[8] however, the dose and schedule were empirically chosen, and no pharmacokinetic or pharmacodynamic evaluation was performed to demonstrate biologic activity. A recent abstract reported on the administration of SAHA to a slightly larger cohort of dogs with hemangiosarcoma.[9] Gastrointestinal toxicity was dose-limiting, and the study was terminated prematurely as a result of toxicity. Again, pharmacodynamic data were not reported. The astronomical cost of SAHA will undoubtedly limit its broad use in veterinary patients for the next several years.

The HDAC inhibitors phenylbutyrate and sodium valproate (VPA) are FDA approved for other indications, and are comparatively inexpensive versus SAHA. VPA is marketed as an anticonvulsant and is available in both oral and injectable forms, and preliminary pharmacokinetic information in dogs is available from studies performed in the 1980's. VPA is rarely used as an anticonvulsant in dogs due to its very short half-life. Sustained-release oral formulations of VPA are now available. Human clinical trials have evaluated VPA as a single-agent, as well as in combination with demethylating agents, retinoids, or conventional cytotoxic agents. It appears that VPA concentrations capable of modulating HDAC activity are achievable *in vivo* in humans.

HDAC INHIBITORS AND TRADITIONAL CANCER THERAPY

As with most novel, targeted therapies, they may find their greatest utility when combined with standard forms of therapy such as chemotherapy, radiation therapy and surgery. Preclinical data support the concept that HDAC inhibition can potentiate the efficacy of both chemotherapy and radiation therapy.

Following pretreatment with HDAC inhibitors, enhanced sensitivity to many classes of chemotherapeutic agent has been demonstrated. These include topoisomerase inhibitors such as doxorubicin, epirubicin and etoposide, antimetabolites such as 5-fluorouracil, gemcitabine, fludarabine and cytosine arabinoside, antimicrotubule agents such as paclitaxel and docetaxel, and platinum drugs, as well as many other novel/targeted agents. In many cases, the mechanisms of chemosensitization are not completely understood. One mechanism responsible for enhancement of sensitivity to DNA-interacting agents by HDAC inhibitors may be through induction of a conformational change in chromatin structure, allowing better access of the drug to the DNA molecule.[10] Hyperacetylation of tubulin may alter the sensitivity of cells to microtubule-interacting agents such as vinca alkaloids and taxanes.[11] Double-strand DNA breaks may be repaired inefficiently as a result Ku70 hyperacetylation, leading to decreased DNA binding affinity.[12] Additionally, alterations in the expression of pro- and anti-apoptotic proteins may enhance apoptosis in response to diverse chemotherapeutic agents. In many, but not all cases, these synergistic effects are sequence-specific and most profound when cells are pretreated with the HDAC inhibitor. Interestingly, at least one study has suggested that HDAC inhibitors may protect against the cardiotoxic effects of doxorubicin as well.[13] However, there are potential adverse effects associated with HDAC inhibition and chemotherapy: some studies have suggested that continuous exposure to HDAC inhibitors may induce the expression of P-glycoprotein, an ATP-dependent efflux pump which may confer resistance to a variety of antitumor natural products.[14] Importantly, these data suggest that short-term HDAC inhibitor treatment around the time of chemotherapy administration may be effective, and even preferable to continuous treatment. This schedule of treatment may reduce toxicity and cost.

Similarly, enhancement of radiation sensitivity has been observed with HDAC inhibitor treatment. Interestingly, the topical application of HDAC inhibitors was shown in one study to mitigate acute cutaneous effects from radiation therapy and to accelerate radiation-induced cutaneous wound healing.[15]

HDAC INHIBITORS – CURRENT AND FUTURE VETERINARY STUDIES

Our laboratory recently demonstrated that pretreatment of canine and human osteosarcoma cells with clinically relevant concentrations of VPA enhances the antiproliferative and pro-apoptotic effects of doxorubicin, and significantly inhibits tumor growth in a murine subcutaneous xenograft model. This appears to be due, at least in part, to increased nuclear accumulation of doxorubicin by the tumor cells following VPA exposure.[16] We have recently completed accrual to a phase-I clinical trial in tumor-bearing dogs of escalating VPA followed by a standard dose of doxorubicin; No toxicity was observed at the highest doses evaluated, and increased histone acetylation in peripheral blood mononuclear cells increased linearly with dose. Increased histone acetylation was documented in tumor tissue from 4 of 8 evaluable patients.[17]

REFERENCES

[1]Jones PA, et al. **Cell** 2007;128(4):683. [2] Xu WS, et al. **Oncogene** 2007;26:5541. [3]de Ruijter AJM, et al. **Biochem J** 2003;370:737. [4]Glozak MA, et al. **Gene** 2005;363:15. [5]Kelly WK, Marks PA. **Nat Pract Clin Oncol** 2005;2:150. [6]Karagiannis TC, El-Osta A. **Leukemia** 2005;21:61. [7]Bali P, et al. **J Biol Chem** 2005;280:26729. [8]Cohen LA, et al. **Vet Comp Oncol** 2004;2:243. [9]Overley B, et al. **Proc Vet Cancer Soc** 2007:84. [10]Marchion DC, et al. **J Cell Biochem** 2004;92:223. [11]Catalano MG, et al. **Endocr Relat Cancer** 2007;14(3):839. [12]Chen C-S, et al. **Cancer Res** 2007;67(11):5318. [13]Daosukho C, et al. **Free Radical Biol Med** 2007;42:1818. [14]Tabe Y, et al. **Blood** 2006;107(4):1546. [15]Chung, Y-L, et al. **Mol Cancer Ther** 2004;3(3):317. [16]Wittenburg LA, et al. **Cancer Chemother Pharmacol** 2011;67(1):83. [17]Wittenburg LA, et al. **Clin Cancer Res** 2010;16(19):4832.

MACROPHAGE DEPLETION THERAPY TO STIMULATE CANCER IMMUNITY

Dr. Amanda M. Guth, DVM, PhD and Steven W. Dow, DVM, PhD, DACVIM
Fort Collins, CO

INTRODUCTION

Tumor cells evade destruction by the immune system through various mechanisms. Recently, it was discovered that one of the main ways tumor cells avoid immune destruction is by recruiting populations of cells that inhibit the immune system through various mechanisms. Among these cells are myeloid-derived suppressor cells (MDSCs), made up of immature monocytes and neutrophils, and tumor-associated macrophages (TAMs), which produce cytokines and other factors that directly block the anti-tumor response by T helper cells, cytolytic T cells and natural killer (NK) cells. Studies in mouse tumor models and in humans with cancer demonstrate that increased numbers of these cells systemically and in the tumor microenvironment correlate with a poor disease outcome.

The heterogeneous MDSC population of cells is characterized based on their ability to suppress the immune response of T cells and NK cells. Recent data demonstrate that MDSC can block T cell responses by depleting L-arginine (via the production of arginase I), depleting cystine and cysteine[1], releasing reactive oxygen and nitrogen species, and producing peroxide and nitric oxide. These molecules directly affect T cell signaling following antigen stimulation. Moreover, these suppressive cells produce cytokines that can inhibit immune cellular functions, i.e. TNFa-mediated inhibition of NK cell function, or that recruit other immunosuppressive cells such as T regulatory cells and inhibitory dendritic cells (DCs). In addition, MDSCs and TAMs in the tumor microenvironment produce factors that enhance tumor growth. Among these factors are VEGF, which plays a role in blood vessel formation, and various matrix metalloproteinases (MMPs), which help break down tissue matrix and allow for the invasion of the tumor cells into normal, surrounding tissues. Lastly, it has been suggested that these immunosuppressive cells could play an important role in establishing the "pre-metastatic niche," thus contributing to the systemic spread of disease.[2]

LIPOSOMAL CLODRONATE

Given the nature of these immunosuppressive cells, it is becoming more apparent as to why most tumor vaccines have thus far been clinically unsuccessful. Therefore, it is imperative to develop ways in which these cells can either be eliminated or manipulated to make them less immunosuppressive in order to accelerate the success of various immunotherapies. Our lab has been investigating the use of mannosylated liposomal clodronate (LC, Figure 1) as a novel therapeutic to enhance the efficacy of tumor vaccines. Clodronate is a non-nitrogen bisphosphonate, that when metabolized, acts as an analog of ATP, thus leading to apoptosis of susceptible cells. When administered *i.v.*, clodronate is quickly taken up by bone and, thus, has been used to treat bone diseases such as osteosarcoma. However, when it is encapsulated in a liposome, it is taken up only by phagocytic cells in the bloodstream and the spleen.[3,4] In our hands, the addition of the mannosylated residues increases its uptake by these cells. Studies from our lab and others demonstrate that injection of LC *i.v.*, in mice, leads to depletion of monocytes (and occasionally neutrophils) in the bloodstream around 18-24 hours post-injection.[3] Moreover, there is also a depletion of macrophages and dendritic cells in the spleen observed 24 hours after LC treatment.[4]

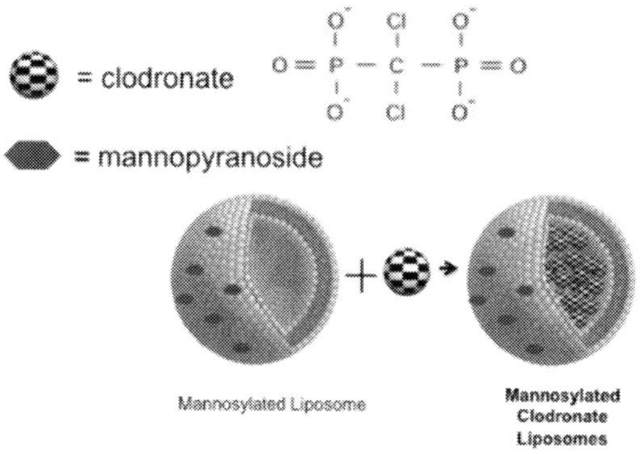

Figure 1: Mannosylated liposomal clodronate

LIPOSOMAL CLODRONATE AND TUMORS

Since LC can deplete splenic macrophages, we and others have proposed that treatment of tumor-bearing mice with LC would lead to a decrease in the percentages of TAMs and thus possibly reverse some of the immune suppression brought on by these cells. In fact, our current data and other published data in the literature utilizing numerous different mouse tumor models show tumor regression following LC treatment.[5-11] In these published studies, it was speculated that the tumor regression was due mostly to the decreased angiogenesis seen when TAMs were depleted in the tumor. However, in our hands using various mouse tumor models, we observed that not only were TAMs depleted in the mice treated with LC, but that MDSCs were reduced following treatment. Further studies revealed that this depletion of both TAMs and MDSCs (as observed in both the blood and the spleen) led to a reversal of the T cell immune suppression. This conclusion is based on the observations that: 1) the effects of LC on inhibiting tumor-growth in CD4 and CD8 knockout mice was eliminated and 2) we were able to detect increased production of IFNg and cytolytic activity of spleen-derived T cells from tumor-bearing mice that were treated with LC as compared to untreated and liposomal alone treated mice. Thus, we hypothesize that LC is effective at reversing T cell immune suppression in tumor-bearing mice.

CANINE CANCER AND LIPOSOMAL CLODRONATE

Utilizing our previous findings in mice and safety data on administering LC in dogs[12], we established a clinical trial in which dogs bearing measureable soft tissue sarcomas were treated once weekly with LC *i.v.* The dogs enrolled were monitored for toxicity in addition to changes in tumor size. Moreover, CBC data and blood samples were collected and analyzed by flow cytometry. Tumor biopsies were obtained prior to treatment, mid-way through treatment and at the conclusion of the study, which ended after the dogs received 6 weekly treatments. Of the dogs enrolled in the study that received at least treatments, 77% demonstrated either a partial response or stable disease. Analyzing tumor biopsies obtained prior to treatment and halfway through treatment, we observed a trend towards decreased numbers of monocytes/macrophages in the tumor and decreased vessel formation as analyzed via immunohistochemistry (IHC). These preliminary findings demonstrate that LC is safe to use in dogs with cancer and our results suggest that it may have a similar mechanism of action on TAMs and MDSCs in the dogs as it does in our mouse models of disease.

Due to LC's ability to eliminate myeloid/phagocytic cells, we also examined the effectiveness of LC as a treatment for malignant histiocytosis (MH), a disease of myeloid lineage cells. *In vitro* data demonstrated that LC was effective at directly killing canine MH cells. A pilot study of five pet dogs with MH revealed that 40% of the dogs had significant tumor regression while receiving the LC treatment.[13] Given that MH is often resistant to most conventional chemotherapies, our results suggest that LC, and perhaps other liposome-encapsulated bisphophonates, could be used as an effective treatment of MH is dogs.

FUTURE DIRECTIONS

Our results suggest that we can reverse tumor-mediated immune suppression by reducing the numbers of TAMs and MDSCs present in the tumor and systemically. Logically, we predict that administration of LC along with a tumor vaccine would enhance the efficacy of the tumor vaccine. Moreover, TAMs, and possibly MDSCs, may play a role in protecting cells from chemotherapy, so it is likely that administration of LC with conventional chemotherapy may enhance the effects of the chemotherapy. Finally, a recent publication highlights that elimination of macrophages in tumors enhances the anti-tumor effects of ionizing radiation.[14] Currently, in the United States, clodronate is not available for use clinically. However, it is possible that other newer and more potent bisphosphonates such as zoledronate and pamidronate may function more effectively. In addition, inhibitors of MDSC recruitment or agents that induce MDSC maturation (eg, all-trans retinoic acid) may also be effective methods for reversing the immune suppressive effects of MDSC.

CONCLUSION

In conclusion, our data demonstrate that LC can be used to reduce the number of immunosuppressive cells and thus reverse T cell immune suppression-leading to increased IFNγ production and increased cytolytic activity. It is safe, well-tolerated and easily administered in cancer bearing animals and can result in partial responses and/or stable disease. Given our findings, we believe that the addition of LC to conventional and novel tumor vaccines and immunotherapies may enhance their efficacy.

REFERENCES

[1]Srivastava et.al. **Cancer Res**. 2010:70(1):68, [2]Ye et.al. **J Genet Genomics** 2010:37(7):423, [3]Tacke et.al. **J Exp Med**. 2006:203(3):583, [4]Van Rooijen et.al. **J Immunol Methods** 1994:174(1):83, [5]Oosterling et.al. **J Pathol.** 2005:207(2):147, [6]Zeisberger et.al. **Br J Cancer** 2006:95(3):272, [7]Gazzaniga et. al. **J Invest Dermatol** 2007:127(8):2031, [8]Hiraoka et.al. **Cancer Sci**. 2008:99(8):1595, [9]Miselis et.al. **Mol Cancer Ther**. 2008:7(4):788, [10]Halin et.al. **Neoplasia** 2009:11(2):177, 11Lepique et.al. **Clin Cancer Res.** 2009:15(13):4391, [12]Mathes et.al. **Exp Hematol** 2006:34(10):1393, [13]Hafeman et. al. **Cancer Immunol Immunother.** 2009;59(3):441, [14]Meng et.al. **Cancer Res.** 2010:70(4):1534.

MASITINIB IN CANINE MAST CELL DISEASE

Albert Ahn, DVM
Short Hills, NJ

INTRODUCTION

Since their introduction about 10 years ago, tyrosine kinase inhibitors (TKI) have revolutionized the treatment of certain diseases in human medicine and represent a cutting-edge technology in cancer therapeutics. Today, armed with positive findings from two phase III randomized controlled trials [1, 2], including approximately 6 years long-term follow-up data in canine mast cell tumor [3], 2 years of European post-market experience, and recently acquired regulatory approval from the US Food and Drug Administration (FDA), the TKI known as masitinib is set to do likewise for veterinary medicine. Drugs of this class inhibit protein tyrosine kinases, enzymes involved in cellular signaling pathways that regulate key cell functions such as proliferation, differentiation, migration, activation and survival. Activation of these enzymes, through mechanisms such as point mutations or over expression, can lead to various forms of cancer as well as non-oncological disorders. The first oncological disorder targeted by this drug has been canine mast cell tumors (MCT). This cancer is the most common cutaneous malignant neoplasm in dogs. However, masitinib has therapeutic potential beyond the field of chemotherapeutics, its highly targeted mechanism of action generating treatment possibilities in a large number of immune-mediated and neurological conditions [4]. This diversity is exemplified via a successful phase III study that showed significant reduction in the symptoms of canine atopic dermatitis following treatment with masitinib.

DEVELOPMENT OF MASITINIB

Masitinib is an oral tyrosine kinase inhibitor (TKI), which is being developed as a targeted therapy for numerous indications in veterinary and human medicine. Although masitinib has been selectively available in the US under a personal importation program for veterinarians since June 2009, it has only recently become openly available to US veterinarians following its conditional approval by the FDA. Marketed under the trade name Kinavet CA-1®, it is indicated for treatment of recurrent (post-surgery) or non-resectable grade II/III cutaneous mast cell tumors in dogs that have not previously received radiotherapy and/or chemotherapy except corticosteroids. In fact, masitinib was the first ever approved drug for veterinary oncology, having been registered by the European Medicines Agency (EMA) in November 2008 for the treatment of non-resectable MCTs grade II/III, and subsequently marketed in Europe under the trade name Masivet®.

Masitinib has been designed specifically to optimize its selectivity against just a few key tyrosine kinases that are implicated in various cancers; primarily KIT, platelet-derived growth factor receptors (PDGFR α and β) and Lyn, while not inhibiting, at therapeutic doses, those tyrosine kinases attributed to possible toxicity [4]. This atypically high specificity permits masitinib to be administered at higher doses for a more potent therapeutic effect, whilst maintaining an acceptable level of tolerability; i.e. it can be expected to exhibit a better safety profile compared to multi-targeting TKIs. Masitinib's main cellular target is the mast cell and therefore the origin of MCTs, an important distinction from cytotoxic chemotherapies that block multiplication of all cells, including healthy ones. Evidence also indicates that recruitment of inflammatory cells, especially infiltration by mast cells, facilitates the growth and spread of cancers by producing molecules that enhance tumor invasiveness and metastasis. Therefore, reduction of the mast cell burden may prove to be therapeutically beneficial in restraining the growth of numerous cancers, even those without a direct association with mast cell proliferation.

Complementing this direct anti proliferative and pro-apoptosis action on mast cells, masitinib is also capable of regulating the activation of mast cells through its targeting of Lyn, Fyn and Lck [4]. The subsequent down regulation of mast cell activity will reduce the array of mediators released, including vascular endothelial growth factor and large amounts of pro-inflammatory mediators, which in turn is likely to disrupt as yet unidentified downstream signaling pathways and angiogenesis processes. Additionally, it is thought that inhibition of Lyn kinase activity also has potential benefits in defense against metastasis and drug-resistance, as well as masitinib's inhibition of PDGFR-β possibly reducing angiogenesis and enhancing chemotherapy sensitivity and availability at the tumor site. This latter aspect may be one mechanism through which masitinib acts as a potentiator of chemotherapies [5], with the implication that it could hold potential therapeutic benefit for a varied range of cancers.

Table 1. Summary of study results for masitinib treatment of MCT.

	Masitinib	Placebo	p-value
Best response rate (first 6 months)	50%	29%	0.020
Overall TTP for non-resectable tumors	173 days	75 days	0.001
TTP (non-resectable, non-mutated KIT)	140 days	75 days	0.027
TTP (non-resectable, mutated KIT)	241 days	83 days	0.002
12-month survival rate (non-resectable)	62%	36%	0.024
24-month survival rate (non-resectable)	40%	15%	0.040
Median overall survival (first-line, non-resectable)	803 days	322 days	0.045

MASITINIB FOR TREATMENT OF MAST CELL TUMORS

Registration of masitinib was based upon the successful outcome of a phase III randomized clinical trial, which included 202 dogs with grade II/III cutaneous MCTs, recruited from 25 veterinary centers in the USA and France [1]. This study reported that masitinib significantly delayed time-to-progression (TTP), i.e. increased overall median TTP compared to placebo, with the effect being more pronounced when used as first-line therapy and regardless of whether the tumors expressed mutant or wild-type KIT (Table 1).

To further assess the long-term impact of masitinib on the survival of dogs with MCT, those dogs benefiting from treatment on the phase III study were enrolled into a compassionate-use program. Findings from the follow-up study revealed that masitinib can achieve the primary goal of any anti-cancer treatment; namely, tumor control leading to prolonged survival and ideally a complete (curative) response. More specifically, a significant increase of survival rates at 12 and 24 months was seen in dogs with non-resectable tumors regardless of the MCT's KIT mutation status (Table 1) [3].

These conclusions, as well as claims of masitinib representing a curative therapy for some dogs, are supported further by a number of case studies from dogs with complete response that have remained on this follow-up study for over 4.5 years [6] (Table 2). Moreover, such cases also exemplify the possible long-term tolerability of masitinib, with none of these cases reporting adverse reactions of any significance over the duration of treatment.

Table 2. Case studies exemplify the long-term survival, even curative, benefits of masitinib

Case	1	2	3
Breed	Jack Russell Terrier	Labrador Retriever	Mixed Breed
Treatment	First-line	First-line	Third-line
KIT Mutation	Exon 8	Exon 11	Wild-type
Survival	**56 months**	**60 months**	**58 months**

Analysis from the follow-up study also revealed that control of disease was highly correlated with long-term survival, indicating that tumor stabilization also provides important clinical benefits. This is in contrast to the conventional approach of short-term chemotherapy, for which the aim is to achieve a disappearance of the tumor before too much collateral damage occurs in healthy tissue. Indeed, if this chemotherapeutic goal is not attained then the survival time is worsened. The targeted nature of masitinib and its ability to stop tumor progression allows for a long-term therapeutic approach of tumor control, effectively making MCT a chronic rather acute disease.

Other key points from these studies included:

- For dogs with non-resectable tumors and regardless of the presence of KIT mutations, masitinib induced: long-lasting tumor control and significantly improved long-term survival.
- 50% of the overall population treated by masitinib had their tumor decrease in size by ≥50% during the first 6 months, vs. 29% for placebo.
- Complete tumor response at 12 and 24 months was reported in 15% and 9% of observed cases, respectively, vs. 0% with placebo.
- Survival at 12 months in non-resectable MCT was 62.1% vs. 36% ($p=0.024$); and at 24 months it was 39.8% vs. 15% ($p=0.04$).
- Masitinib demonstrated a significant anti-metastatic potential in preventing the emergence of metastasis ($p=0.006$). Additionally, complete remission in cases of metastatic MCT have been reported in post-registration cases.
- Masitinib introduced a new paradigm in oncology treatment for animals, with control of disease at 6 months being highly predictive of long-term survival, whereas short-term response at 6 weeks was not. This indicates that disease control over 6 months is a valuable treatment objective and recommended minimum treatment duration.
- Unlike many chemotherapy treatments, masitinib is delivered orally in tablet form and can be administered everyday directly by the owner at the point-of-care.
- Common side effects typical associated with masitinib were diarrhea and vomiting. These reactions are usually mild to moderate in severity, typically short-lived, and with less frequent occurrence after the first 3 months of treatment [2]. All adverse reactions resolved spontaneously upon cessation of treatment with no withdrawal symptoms.
- Masitinib showed an increased risk of reversible protein loss syndrome (1.6% of dogs following the implementation of a management plan) with regular surveillance of albumin and proteinuria allowing for treatment discontinuation while the dog is still clinically asymptomatic [2]. Thus, dogs being treated with masitinib must be regularly monitored by the veterinarian (bi-weekly during the first 3 months and then at least monthly).

CONCLUSION

Masitinib can significantly improve long-term survival and prevent tumor recurrence in dogs with recurrent or non-resectable grade II/III cutaneous MCT. Owing to its novel mechanism of action, masitinib is likely to be of therapeutic benefit to a large number of oncological and immune-mediated conditions of relevance to veterinary medicine, including canine atopic dermatitis and feline asthma.

ACKNOWLEDGEMENTS

Colin Mansfield, PhD provided medical writing services on behalf of AB Science. Albert Ahn is an employee of AB Science.

REFERENCES

[1] Hahn KA, et al.; **J Vet Intern Med** 2008; 22:1301-1309. [2] Cadot P, et al.; **Vet Dermatol** 2010; 21. [3] Hahn KA, et al.; **Am J Vet Res** 2010;71:1354-1361. [4] Dubreuil P, et al.; **PLoS One** 2009; 4:e7258. [5] Thamm D, et al.; **Vet J** 2011, DOI:10.1016/j.tvjl.2011.01.001. [6] Phillips B, et al.; **Proc 30th VCS Conf** 2010; Oct 29-Nov 1.

PHARMACOLOGICALLY DIRECTED DOSING OF NUCLEOSIDE ANALOGS: TARGETING AN ENZYME

Carlos O. Rodriguez, Jr., DVM PhD DACVIM (Oncology)
Davis, CA

ABSTRACT

Gemcitabine (dFdC) has demonstrated clinical utility in the treatment of diverse malignancies[1-7]. The demonstrated efficacy is dependent on its metabolism, specifically to its monophosphate, diphosphate, and triphosphate (dFdCTP) metabolites. The rate-limiting step in the intracellular accumulation is the phosphorylation of the drug to its monophosphate, which is catalyzed in the first step by deoxycytidine kinase (dCK)[8-9]. Therefore, maximizing the accumulation of the triphosphate is dependent on achieving plasma gemcitabine concentrations that do not exceed the phosphorylating capacity of dCK.

In vitro data suggest that the rate of dCK-mediated phosphorylation is expected to become saturated at gemcitabine concentrations that exceed 20µmol/L with an inhibitory effect occurring at higher levels. In vivo in humans, the rate of dFdCTP accumulation was saturated at 10 to 20 µM gemcitabine in plasma; this optimal concentration is achieved in plasma when gemcitabine is infused at a dose rate of 8 to 10 $mg/m^2/min$[10-14] (**Figure 1**).

Gemcitabine is the quintessential chemotherapeutic agent because it possesses such a strong association between the MTD and the infusion schedule[9]. When administered frequently (ie, daily for 5 days), the MTD was found to be only 9 $mg/m^2/d$ when infused over 30 minutes[15]. This is in sharp contrast to the MTD of more than 4,000 $mg/m^2/d$ when the drug is administered using a biweekly schedule[16]. Additionally, doses ranging from 700 to 2,600 $mg/m^2/d$ have been successfully administered to patients as a 30-minute infusion on a weekly (for 3 weeks) schedule [1-5]. The common denominator in these clinical trials is the duration of infusion, which by convention has been 30 minutes. Indeed, this has become a standard infusion duration[17], even though it is known that such administration schedules produce plasma concentrations of gemcitabine that are unlikely to be used completely for intracellular conversion to gemcitabine triphosphate[6, 18-21]. Therefore, under current administration practices, it is not possible for the target cells to use a substantial portion of the infused drug, which is metabolically cleared by deamination and excreted in the urine[21].

When compared to trials in humans, relatively fewer studies have evaluated gemcitabine in veterinary medicine[22-31]. To date only two pharmacokinetic studies in normal dogs have been performed[30-31] and all published studies have utilized the 30 minute infusion. While minimal toxicity has identified over a broad range of doses, the veterinary oncology studies have failed to demonstrate the anti-tumor efficacy observed in the human studies. The human in vivo data would suggest that this is due to suboptimal plasma gemcitabine (**Table 1**). The purpose of this talk is to discuss the biochemistry of dCK and to describe pharmacologically-directed dosing strategies (**Figure 1**) that may maximize plasma gemcitabine and thus increase the intracellular accumulation of dFdCTP in veterinary species.

Table 1 Sample of gemcitabine doses reported in literature.

mg/m2/30 min	**mg/m2/min**	infusion schedule	Ref
9	**0.3**	q24h x 5	15
25	**0.8**	weekly as radiosensitizer	22
27	**0.9**	weekly x 3	1-3, 5, 7
50	**1.7**	weekly as radiosensitizer	22
87	**2.9**	weekly x 3	1-3, 5, 7
300	**10.0**	EOW	25
400	**13.3**	weekly x 3	26
675	**22.5**	EOW	25, 26
700	**23.3**	weekly x 3	1-5
800	**26.7**	weekly x 4	27
2600	**86.7**	weekly x 3	1-5
4000	**133.3**	EOW	17
2 mg/kg	**2**	with carboplatin	28, 29

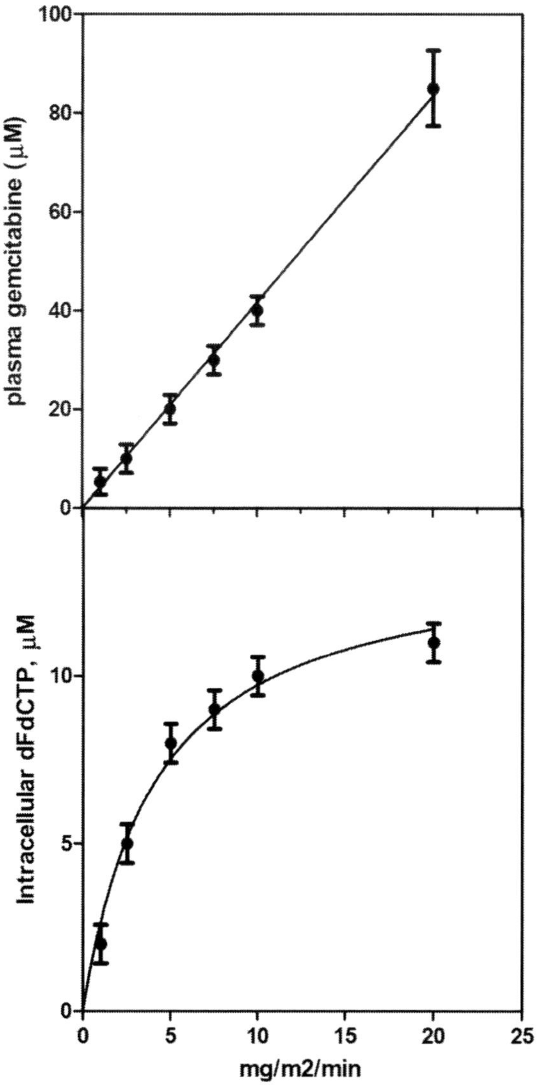

Figure 1. Theoretical fixed-dose rate vs plasma gemcitabine (upper panel) or interacellular gemcitabine triphosphate (lower panel).

REFERENCES

[1]Abratt RP, et. al. **J Clin Oncol** 1994;12:1535. [2]Anderson H, et. al. **J Clin Oncol** 1994;12:1821. [3]Perez-Manga G, et. al. **J Clin Oncol** 2000;18:2545-2552. [4]Santoro A, et. Al. **J Clin Oncol** 2000;18:2615. [5]Zinzani PL, et. al. **J Clin Oncol** 2000;18:2603. [6]Grunewald R, et. al. **J Clin Oncol** 1992;10:406. [7]Rizzieri DA, et. al. **J Clin Oncol** 2002;20:674. [8]Boven E, et. al. **Br J Cancer** 1993;68:52. [9]Kaye SB. **J Clin Oncol** 1994;12:1527. [10]Gandhi V and Plunkett W. **Cancer Res** 1990;50:3675. [11]Plunkett W, et. al **Nucleosides Nucleotides** 1989;8:775. [12]Shewach DS, et. al. **Cancer Res** 1994;54:3218. [13]Grunewald R, et. al. **Cancer Chemother Pharmacol** 1991;27:258. [14]Grunewald R, et. al. **Cancer Res** ;50:6823. [15]O'Rourke T, et. al. **Investigational New Drugs** 1989;10:165. [16]Shewach DS, et. al. **Cancer Res** 1994;54:3218.[17]Brown TD, et. al. **Cancer Chemother Biol Response Modif** 1991;12:111. [18]Martin C and Pollera CF. **Int J Clin Pharmacol Res** 1996;16:9. [19]Gandhi V, et. al. **J Clin Oncol** 2002;20:665. [20]Patel SR, et.al. **J Clin Oncol** 2001;19:3483.[21]Abbruzzese JL, et al. **J Clin Oncol** 1991;9:491.[22]LeBlanc AK, et. al. **Vet Radiol Ultrasound** 2004;45:466. [25]Kosarek CE, et. al. **J Vet Intern Med** 2005;19:81. [26]Turner AI, **J Vet Intern Med** 2006;20:1384. [27]Marconato L, et. al. **Vet Comp Oncol** 2008;6:90. [28]Dominguez PA, et. al. **J Vet Intern Med** 2009;23:130. [29]Martinez-Ruzafa I, et. al. **J Vet Intern Med** 2009;23:570. [30]Freise KJ and Martin-Jimenez T. **J Vet Pharmacol Ther** 2006;29:147. [31]Freise KJ and Martin-Jimenez T. **J Vet Pharmacol Ther** 2006;29:137.

INCRETIN HORMONES PART 1: PHYSIOLOGY

Chen Gilor, DVM, PhD, DACVIM
Wetherby, United Kingdom

INTRODUCTION

In the past 20 years, incretin physiology has transformed from an esoteric branch of gastrointestinal physiology into a dominant area of research and development of new drugs for treatment of diabetes and obesity. Incretin-based therapies are based on administration of incretin analogs and mimetics as well as on manipulations of other components of the incretin system. Thus, it is important to understand the basic concepts of incretin physiology and the role of incretin hormones in health and disease.

The physiological role of incretin-secreting cells (K and L cells) is to sense the type and quantity of digested nutrients in the gut. They then secrete incretin hormones as preparatory signals to other remote organs (e.g. brain, pancreas etc.). In the pancreas, the incretin signal is translated to increased sensitivity to the stimulatory effect of glucose. This effect is responsible for the observed difference in insulin secretion between oral and intravenous glucose and is defined as the "incretin effect"; oral glucose leads to much greater insulin secretion compared to intravenous glucose even when blood glucose concentrations are equal.[1,2] The incretin effect is thought to be exclusively mediated by 2 peptide hormones: Glucagon-like peptide-1 (GLP-1) which is secreted from L cells, and glucose-dependent insulinotropic polypeptide (GIP) which is secreted from K cells. Incretins also participate in regulation of pancreatic beta-cell differentiation, proliferation and survival; incretins affect glucagon secretion, slow the rate of gastric emptying and increase satiety. Thus, incretins play a major role in glucose homeostasis. [3,4] In people, incretin hormones play a crucial role in the pathogenesis and treatment of type 2 diabetes. Diabetes mellitus in cats shares many similarities with type 2 diabetes in people but in cats, little is known about incretin hormones and their role in the diabetes. In the first part of this presentation, comparative aspects of incretin physiology will be discussed, including stimulation of GIP and GLP-1 secretion, their effect of on the pancreas and other organs, and their degradation and elimination. In the second part of this presentation, current incretin-based therapies will be discussed.

BIOLOGY OF K AND L CELLS

K and L cells are interspersed in the epithelium of the gut. They are described as open-type cells, reaching the lumen of the gut via a slender apical process. On their basal side they are located in close proximity to the vasculature and neurons of the lamina propria.[5] The main stimulus for secretion of hormones from K and L cells is the presence of nutrients in the lumen of the gut (but not by nutrients in the blood). Hormone secretion takes place in the basolateral side (to the blood and nerve endings). K and L cells sense the type and quantity of digested nutrients via at least 2 mechanisms. In people and rodents simple sugars are sensed via a specific G protein-coupled "sweet taste receptor" that belongs to the same family of taste receptors in the tongue.[6] Cats lack this sweet taste receptor and they do not seem to sense sweet tastes.[7] Whether or not cats have a different G protein-coupled taste receptor or a different mechanism of sensing carbohydrates by intestinal enteroendocrine cells is unknown. Some studies questioned the importance of GPCR taste receptors in stimulation of secretion from K and L cells.[8,9] In these studies, glucose stimulated GIP and GLP-1 secretion via glucose metabolism and K_{ATP} channel closure in a mechanism similar to stimulation of insulin secretion in pancreatic beta cells. Nonmetabolizable sugars were still able to promote GLP-1 release via a sodium-glucose cotransporter-dependent mechanism.[9,10] Lipids are sensed by specific GPCR's (GPR40, GPR120, GPR119).[9] The mechanism by which amino acids are sensed is less clear. GPCR's for amino acids have not been described on K and L cells. In dogs there is some evidence that stimulation of K cells by protein is mediated by pH changes in the duodenum that are secondary to protein digestion in the stomach.[11] Amino acids can also activate secretion of incretins by inducing sodium-dependent inward currents when they enter the cells via specific transporters.[12] The species-specific profile of receptor expression and distribution of K and L cells along the gut determine in part the magnitude response to different dietary components. Amino acids are weak stimulators of GIP secretion in people but strong stimulators of it in dogs, cats and rats.[5,13] In most species, fat and carbohydrates are strong stimulators of GIP secretion. Fat is more potent than carbohydrates in stimulating GIP secretion in people and dogs but the opposite is true in rodents and pigs.[5] In cats, fat is more potent than amino acids in stimulating GIP secretion, but glucose does not stimulate GIP secretion.[13] Post-prandial GLP-1 secretion is typically biphasic and is stimulated by ingested lipid, carbohydrate and protein. Monounsaturated fatty acids stimulate GLP-1 secretion more than polyunsaturated fatty acids or saturated fatty acids.[14] In dogs and rodents, but not in people, GLP-1 secretion is also stimulated by GIP.[15]

Quantitative immunohistochemical studies in rats, pigs and people have demonstrated that the majority of L cells are located in the ileum and colon, with small numbers in the distal jejunum. In contrast, K cells are concentrated in the duodenum and proximal jejunum. A population of K/L cells that contain both GIP and GLP-1 is also present in rats, pigs and people. In the mid-jejunum, about two thirds of K and L cells are actually K/L cells.[4] In dogs, the density of L cells increases from the duodenum aborally to the ileum as in other species, however, quantitative analysis revealed that over 70% of L cells are found in the jejunum. Interestingly, K cells in the dog are also concentrated in the jejunum with lesser concentrations in the duodenum. Importantly, over 30% of L cells in the jejunum of the dog are found in close proximity to K cells, suggesting the existence of a paracrine interaction between the two.[15] K/L cells were not described in dogs.

DEGREDATION AND ELIMINATION OF INCRETIN HORMONES

Active GIP (1-42) and active GLP-1 (7-37 and 7-36 amide) are degraded by the enzyme dipeptidyl peptidase-4 (DPP-4) into inactive forms: GIP (3-42), GLP-1 (9-37) and GLP-1 (9-36) amide. DPP-4, also known as CD26, is a serine protease that specifically cleaves dipeptides from the amino terminus of oligopeptides or proteins that contain an alanine or proline residue in position 2, thereby modifying or inhibiting their activity.[5] Active GLP-1 (but not active GIP) is also degraded by the enzyme neutral endopeptidase 24.11 (NEP-24.11), a membrane-bound zinc metallopeptidase. DPP-4 and NEP are ubiquitous in tissues, and DPP-4 is also present in a soluble form in the blood. Notably, DPP-4 is present on the surface of endothelial cells, including those lining blood vessels that drain the intestinal mucosa, which are positioned directly adjacent to the sites of GIP and GLP-1 secretion. Consequently, more than 75% of the GLP-1 that enters the portal circulation already has been inactivated. Similarly, about 50% of the remaining active GLP-1 is degraded in the liver. Therefore, only 10-15% of secreted GLP-1 reaches the systemic circulation in its active form. It is therefore widely accepted that for estimation of GLP-1 secretion, measuring the active form is inadequate and the total (active and inactive forms) should be measured.[16] Measuring total GLP-1 also better estimates the full effect of the active peptide, including its neurally mediated effects within the gut and portal veins, as well as its direct effect on the pancreas. For estimation of the action of GLP-1 directly on the pancreas, it is accepted that active GLP-1 should be measured.[16]

When injected intravenously, active GLP-1 has a half-life of about 1-2 minutes. The inactive GLP-1 is quickly cleared by the kidneys and has a half-life of about 4-5 minutes. Clearance by the kidneys is not affected by disease states such as obesity and diabetes. The half-life of active GIP in the blood is similarly short (about 2 minutes in rodents and about 5 minutes in healthy people). Like GLP-1, GIP is also cleared in the kidneys. In contrast to active GLP-1, the kidneys are a major degradation site for active GIP and thus, more active GIP escapes the portal system and can have systemic effects.[3]

THE EFFECTS OF INCRETIN HORMONES IN PANCREATIC BETA CELLS

The primary function of pancreatic beta cells is the secretion of insulin in response to changing concentrations of blood glucose. The primary functions of incretin hormones are stimulation of insulin biosynthesis, stimulation of glucose-dependent insulin secretion and regulation of beta cell mass.[5] GIP and GLP-1 are ligands of the G protein-coupled receptors GIPR and GLP-1R, respectively. Both receptors are present on pancreatic beta cells. The effects of GIP and GLP-1 on insulin secretion are similar and (in sub-maximal concentrations) additive. It is suspected that their final pathway to insulin secretion is a common one. On a molar basis, however, GLP-1 is a more potent insulinotropic agent than GIP in at least 2 orders of magnitude.[3]

Activation of GLP-1R and GIPR leads to activation of adenylate cyclase and production of cAMP. Subsequently, cAMP stimulates insulin secretion via PKA-dependent phosphorylation of downstream targets and PKA-independent activation of Epac2. The effects downstream include the following: (1) inhibition of K_{ATP} channels, which leads to cell membrane depolarization; (2) increases in intracellular calcium caused by influx of extracellular Ca^{2+} through voltage-gated Ca^{2+} channels, activation of nonselective cation channels, and mobilization of intracellular Ca^{2+} stores; (3) increases in mitochondrial ATP synthesis, which lead to further membrane depolarization; (4) closure of voltage-dependent K^+ (Kv) channels and consequent reductions in Kv currents, thereby preventing cell membrane repolarization; and (5) direct effects on the protein machinery of exocytosis (that occur distal to increases in ATP and intracellular Ca^{2+}), including induction of docking, priming and fusion of vesicles.[3] These effects are responsible for the potentiation of insulin secretion in response to an increase in blood glucose. The cAMP signal transduction system is also responsible for 2 other major effects in the beta cell: (1) the synergistic effect of incretins with glucose to promote insulin gene transcription, mRNA stability, and biosynthesis (2) enhancement of the glucose-sensing apparatus by increasing the expression of its components: GLUT-2 and glucokinase. Thus incretins have the potential to replenish beta cell insulin stores, prevent exhaustion of beta cell reserves and increase the sensitivity of beta cells to glucose.[3]

For the treatment of diabetes, expansion of beta cells mass is probably the most unique and exciting effect of incretins. Incretins potentiate glucose induced proliferation and differentiation of beta cells. GLP-1R agonists also promote conversion of pancreatic ductal cells of the exocrine pancreas into insulin-secreting, glucose-responsive endocrine-like cells. Incretins protect beta cells from apoptosis induced by various cytotoxic agents or dexamethasone. They also prevent apoptosis secondary to glucolipotoxicity.[3]

OTHER EFFECTS OF INCRETIN HORMONES

Glucagon secretion

GIP increases glucagon secretion in a glucose-dependent manner by stimulating pancreatic alpha cells directly. This effect is typically masked during hyperglycemia when GIP-stimulated insulin secretion decreases glucagon secretion. In contrast to GIP, GLP-1 inhibits glucagon secretion in a glucose-dependent manner, thus reducing hepatic glucose output especially in hyperglycemia. The glucagonostatic effect of GLP-1 is lost when blood glucose concentrations decrease below euglycemia, thus minimizing the risk of severe hypoglycemia. The effect GLP-1 on alpha cells is likely indirect, mediated at least in part by activation of secretion from beta cells and a paracrine effect of insulin and amylin on glucagon secretion.[3]

Extra-pancreatic effects of incretin hormones

GLP-1 is a regulator of appetite, food intake and body weight. Studies in rodents show that GLP-1R agonists reduce short-term food intake when injected peripherally or directly into the CNS. Repeated injections inhibit weight gain. These effects are observed also in lean, obese, and diabetic people. GLP-1 crosses the blood brain barrier and exerts its effect on the CNS directly but it also affects it via stimulation of the vagus nerve in the intestines and in the portal system.[3]

Physiologic concentrations of GIP inhibit gastric acid secretion in dogs but only supra-physiologic concentrations achieve a similar response in people. This correlates with the fact that GIP secretion is stimulated by acid in the duodenum in dogs but not in people as mentioned before. GLP-1 is a potent inhibitor of gastric emptying and gastric acid secretion, thus it slows the entry of nutrients into the circulation and improves post-prandial hyperglycemia in healthy as well as diabetic patients (type 1 and type 2). The effect of GLP-1 on the stomach is mostly mediated by the vagus nerve, but some evidence exists for a direct effect.[3]

GIPRs are expressed on adipocytes and GIP seems to play a role as an anabolic regulator of fat metabolism. The anabolic effects of GIP in fat include stimulation of fatty acid synthesis and re-esterification, enhancement of insulin-stimulated incorporation of fatty acids into triglycerides, up-regulation of lipoprotein lipase synthesis, and reduction of glucagon-stimulated lipolysis. GIPR is not present in liver or muscle tissue and thus increased GIP concentrations preferentially increase FFA uptake and oxidation in adipose tissue.

GLP-1R expression is detected in canine muscle and adipose tissues and in some experiments, GLP-1 stimulated glucose incorporation into glycogen in muscle and liver as well as lipolysis in adipose tissue.[17]

GLP-1, GIP, and their receptors are expressed in other tissues including thyroid C cells (secreting calcitonin), bone, heart, blood vessels, brain, and others. Their effects in those tissues do not seem related to glucose homeostasis or diabetes and therefore will not be discussed here.[5]

INCRETIN HORMONES IN CATS

Little is known about incretin hormones in cats. Oral glucose stimulates GLP-1 secretion in cats but in contrast to other species, it does not stimulate GIP secretion.[13] The concentrations and the pattern of secretion of GIP and GLP-1 correlate with the concentrations and pattern of secretion of insulin but overall, the incretin effect (the observed difference in insulin secretion between oral and intravenous glucose) is smaller in cats compared to people although.[13] In one study, injection of a DPP-4 inhibitor lead to greater inhibition of glucagon secretion than stimulation of insulin secretion in cats.[18] In cats, baseline GIP concentrations are about a fifth of the basal concentration in people, but there is a much higher increase in response to lipids and amino acids.[13] Also in contrast to people but similar to dogs, amino acids are strong stimulators of GIP secretion in cats (although not as strong as lipids). Basal GLP-1 concentrations are similar in cats and people, with similar post-prandial fold increases.[13]

REFERENCES

1. McIntyre N, et al. J Clin Endocrinol Metab. 1965;25:1317-24. 2. Nauck MA, et al. J Clin Endocrinol Metab 1986;63:492-498. 3. Kim W, Egan JM. Pharmacol Rev 2008;60:470-512. 4. Holst JJ, et al. Mol Cell Endocrinol. 2009;297:127-36. 5. Baggio LL, Drucker DJ. 2007;132:2131-57. 6. Jang HJ, et al. Proc Natl Acad Sci U S A. 2007;104:15069-74. 7. Li X, et al. J Nutr. 2006;136:1932S-34S. 8. Reimann F, et al. Cell Metab. 2008;8:532-9. 9. Parker HE, et al. Diabetologia. 2009;52:289-98. 10. Gribble FM, et al. Diabetes. 2003;52:1147-54. 11. Deacon CF. Regul Pept. 2005;128:117-24. 12. Reimann F, et al. Diabetologia. 2004;47:1592–601. 13. Gilor C, et al. Domest Anim Endocrinol 2011(accepted for publication). 14. Beysen C, et al. Diabetologia. 2002;45:1533-41. 15. Damholt AB, et al. Cell Tissue Res. 1999;298:287-93. 16. Deacon CF, Holst JJ. Best Pract Res Clin Endocrinol Metab. 2009;23:425-32. 17. Sandhu H, et al. Diabetes. 1999;48:1045-53. 18. Furrer D, et al. Vet J. 2010;183:355-7.

INCRETIN HORMONES PART 2: INCRETIN-BASED THERAPY FOR DIABETES

Chen Gilor, DVM, PhD, DACVIM
Wetherby, United Kingdom

INTRODUCTION

Traditional treatments for diabetes are either just partially effective (diet change, oral drugs) or associated with significant side effects (insulin therapy). Owner compliance is a major problem because twice-a-day injections are required with most commonly used insulin preparations. Weight gain may indicate a good response to therapy initially but can eventually become a problem. Hypoglycemia is a common complication of insulin therapy and can be life-threatening. These side effects have been significantly reduced in human medicine with the introduction of incretin-based treatments. Incretin hormones (GLP-1 and GIP) stimulate glucose-dependent insulin secretion. Their effect is blunted in euglycemia and therefore they are less likely to cause hypoglycemia. GLP-1 also promotes weight loss. GLP-1 analogs are not only effective in controlling blood glucose in diabetics but they are also associated with weight loss in diabetic and non-diabetic obese people. One major advantage of incretin-based therapies is their potential to reverse the course of the disease. Incretins increase the proliferation and survival of pancreatic beta cells and decrease the rate of apoptosis. Incretin-based treatments in diabetic people correct not only hyperglycemia but also all the major markers of beta-cell dysfunction.

Incretin hormones are peptides that have a very short half after injection. They are inactivated by ubiquitous enzymes in the blood and in tissues (DPP-4 and NEP-24.11). Thus, their use for treatment of diabetes is limited. There are numerous treatment strategies that overcome this problem and take advantage of incretin-hormone physiology. In this lecture, incretin-based therapies that are in clinical use in human medicine will be emphasized. Data from experiments in healthy cats will be presented as well.

INCRETIN HORMONES IN DIABETES AND OBESITY

The incretin effect is severely reduced or absent in people with type 2 DM, contributing to glucose intolerance and post-prandial hyperglycemia. In diabetics, the secretion of GIP is normal or slightly reduced but its effect on the pancreas is markedly decreased. In contrast, GLP-1 retains its insulinotropic effects in type 2 DM (at least in supraphysiologic concentrations), but its secretion is decreased.[1] Whether a blunted incretin response is a primary process in type 2 DM or a secondary process caused by diabetes, is debatable. A blunted incretin response is seen in other types of diabetes, suggesting it is a secondary process[1] but recently it was shown that intensified treatment of type 2 diabetes, resulting in near normal blood glucose concentrations, did not restore the normal secretion of incretin hormones in response to a meal.[2] More evidence for the role of incretins in development of diabetes is the involvement of the Wnt signaling pathway in diabetes. Stimulation of beta cell proliferation by incretins, as well as the effects of incretins on insulin biosynthesis and secretion, is mediated in part by PKA activation of the bipartite transcription factor beta-catenin/T cell transcription factor (β-cat/TCF). β-cat/TCF is also crucial in induction of GLP-1 biosynthesis. β-cat/TCF is the ultimate effector of the Wnt signaling pathway and it is encoded by the TCF7L2 gene.[3] Polymorphisms in the TCF7L2 gene are strongly associated with reduced incretin action and with a strong risk of type 2 diabetes.[4,5]

There is some evidence that obesity contributes to attenuation of the incretin effect independently of diabetes although the mechanisms of this phenomenon are unknown and the effect of obesity is variable. In a recent study, obesity was associated with decreased incretin effect including decreased beta-cell response to GLP-1 but also decreased GLP-1 secretion.[6] In some studies, GLP-1 secretion was normal in obese people but GIP secretion was increased during fasting and early after a meal.[7]

Insulin resistance in itself, regardless of obesity or diabetes, has been associated with abnormal secretion of GIP and GLP-1.[8] Increased insulin concentrations in the insulin-resistant subjects might down-regulate incretin secretion. Decreased incretin secretion might be the limiting factor on appropriate compensatory responses of pancreatic beta cells to insulin resistance, leading to an insufficient increase in insulin secretion and development of glucose intolerance in early diabetes. The possibility that abnormalities in GLP-1 and GIP secretion do not necessarily result from obesity, but rather contribute to its development, should also be considered because of their roles in appetite control (GLP-1) and increasing insulin sensitivity in adipose tissue (GIP).[9]

In cats, glucose-stimulated GLP-1 plasma concentrations were higher in lean *vs.* obese cats.[10] After gastric glucose administration, GLP-1 concentrations did not return to baseline within 5 hours in lean cats but they did return to baseline in obese cats within 3 hours. GLP-1 concentrations were also a lot more variable in lean compared to obese in that study.[10]

INCRETIN-BASED THERAPY STRATEGIES

There are several treatment modalities that overcome the problems associated with direct administration of incretin hormones. Long-acting, DPP-4-resistant, synthetic GLP-1 mimetic peptides are most commonly used. Available formulations are all for SQ injections but other delivery systems are being developed (including enteral, pulmonary, or sublingual). Also under development are non-peptidic GLP-1 receptor agonists.[11] Oral DPP-4 inhibitors that prolong the half-life of incretin hormones are also commercially available.[12] Non-peptidic drugs that activate free fatty acid receptors in L cells and stimulate GLP-1 secretion are being developed (GRP120 and GRP119 agonists).[13,14,15] Diet manipulation should be investigated. Finally, GLP-1 mimetics have been used successfully to increase the survival of transplanted beta cells in people with type 1 diabetes.[16]

Drugs that are commercially available or that are currently in advanced clinical trials are described below.

GLP-1 MIMETICS AND ANALOGS

Exenatide

The peptide exendin-4 was first isolated from the poisonous venom of the Gila Monster (*Heloderma suspectum*). Exendin-4 is a 39-amino acid peptide that shares only a 53% sequence homology with GLP-1 but its affinity for the GLP-1 receptor is 1000 times greater than the affinity of GLP-1. Unlike GLP-1, exendin-4 is not a substrate for DPP-4 and NEP.[1] Exenatide is a synthetic exendin-4. Resistant to degradation, exenatide is eliminated by the kidneys and has a half-life of 3-4 hours in people. Its biological effect lasts about 8 hours after subcutaneous injection and it can be detected in the plasma for up to 15 hours.[17] Multiple studies, both *in vitro* and *in vivo*, have shown that, in general, exendin-4 has the same physiologic effects as GLP-1 in the pancreas, GI tract, and brain.[12] Exenatide lowers blood glucose through multiple mechanisms, including enhancement of glucose-dependent insulin secretion, suppression of excess glucagon secretion, reduction of food intake, and slowing of gastric emptying. Exenatide is associated with improvement in some of the earliest and most fundamental abnormalities of type 2 diabetes: diminished "first-phase insulin response" and proinsulin/insulin ratio. Acute administration of exenatide in type 2 diabetic patients corrects the abnormal insulin secretion pattern after an IV glucose bolus (first phase and second phase insulin responses) and restores the ability of beta cells to respond to rapid changes in blood glucose concentrations.[12] Exenatide also improves proinsulin/insulin ratio after 30 weeks of treatment.[18] Exenatide has been shown to be as effective as insulin glargine in the treatment of DM but with less side effects (e.g. hypoglycemia and weight gain).[19] In a 2-year follow-up of patients receiving exenatide, patients achieved sustained and significant reductions in glycosylated hemoglobin, accompanied by significant weight loss (instead of weight gain commonly seen in diabetics receiving insulin) and improvement in serum liver enzyme activity and blood pressure. Most importantly, treatment with exenatide improved beta cell function as measured by homeostasis model assessment of beta cell function (HOMA-B).[20]

Exenatide has minimal side effects in people. It is mostly associated with nausea and less frequently with vomiting. Infrequently, it might cause hypoglycemia. Severe hypoglycemia (requiring assistance) was reported rarely (only 5 of 2781 patients) and only in patients who also received sulfonylurea drugs. The rate of hypoglycemic events decreased over time. Interestingly, the risk of hypoglycemia was not different in exenatide *vs.* insulin treatment. Antibodies to exenatide developed in 67% of patients but this did not affect outcome and was not associated with side effects (Amori et al., 2007).[21]

When first discovered, exendin-4 was shown to potentiate amylase release from rat acinar pancreatic cells in response to other hormones such as cholecystokinin. This was shown *ex vivo* and in high doses. Although exendin-4 was implicated as a cause of pancreatitis in one case report, a cause and effect has not been proven and exenatide is still considered a safe drug.[21]

In healthy cats, exenatide was quickly absorbed after a SQ injection and caused glucose-dependent insulin secretion. Increased glucose tolerance, however, was not observed after a single SQ injection.[22] Recently, a study in rats revealed an unexpected consequence of acute exendin-4 administration (Perez-Tilve et al., 2010).[23] Exendin-4, whether administered into the circulation or into the CNS, caused stimulation of the sympathetic nervous system, leading to hyperglycemia. This effect was independent of the glucose-dependent insulinotropic effect of exendin-4 and it waned with more chronic administration (after 6 days). Similar activation of the sympathetic system might explain the acute effects of exenatide in cats. At a dose of 1.0 mcg/kg SQ (about 10 times the dose that is used in diabetic people), exenatide injection did not cause any side effects in healthy cats, except for hypoglycemia in 1 out of 9 cats.[22]

Exenatide is commercially available in the USA under the trade name Byetta®.

Exenatide OW

Exenatide is a peptide and thus it has to be injected to retain efficacy. A long-acting sustained-release formulation of exenatide has recently been described (Exenatide OW or Exenatide LAR). It consists of injectable microspheres of exenatide and poly(D,L-lactic-co-glycolic acid), a common biodegradable medical polymer with established use in absorbable sutures and extended-release pharmaceuticals, that allows gradual drug delivery at controlled rates. This drug is administered in people subcutaneously once weekly. It has been shown in a recent clinical study to be more effective than twice-a-day exenatide in achieving glycemic control with no increased risk of hypoglycemia and with similar reductions in body weight.[24] If this drug can be used in cats as effectively, it will revolutionize the treatment of feline diabetes. Studies of exenatide in cats have not been reported to date.

Liraglutide

Liraglutide is a human GLP-1 synthetic analog with 2 amino acid substitutions and a fatty acid acyl group that enables noncovalent binding to albumin, thereby extending the pharmacokinetic profile of the GLP-1 molecule. Liraglutide exhibits a prolonged pharmacokinetic profile after a single injection, and exhibits all of the actions of native GLP-1.[1] Liraglutide (once-a-day) was recently compared to exenatide (twice-a-day). Liraglutide provided significantly greater improvements in glycemic control than did exenatide and was generally better tolerated.[25] Liraglutide has been used successfully to treat obesity in non-diabetic patients.[26]

Liraglutide is commercially available in the USA and Europe under the trade name Victoza®.

Albugon (Albiglutide, Naliglutide)

Albugon is a recombinant GLP-1-albumin fusion protein that exhibits a reduced affinity for the GLP-1R, but displays a broad spectrum of GLP-1R-dependent actions in preclinical studies, including inhibition of food intake and gastric emptying and reduction of glycemia excursion after meal ingestion. Importantly, it has the potential to be used as a very long-acting drug: It has been investigated in people for use as a weekly, biweekly and monthly injection.[27]

DPP-4 INHIBITORS

DPP-4 inhibitors (e.g. sitagliptin, vildagliptin) are administered orally. In people, vildagliptin and sitagliptin are well-tolerated and not associated with hypoglycemia when used alone. Both agents increase plasma concentrations of GLP-1 and GIP after meal ingestion, enhance glucose-stimulated insulin secretion and reduce ratios of proinsulin:insulin, consistent with an improvement in beta cell function. They are, however, less potent than other oral hypoglycemic drugs.[12] In contrast to GLP-1 analogs, DPP-4 inhibitors are not associated with nausea or vomiting in people, but they are associated with weight gain. DPP-4 inhibitors are also associated with increased risk of nasopharyngitis, urinary tract infections and headaches.[21] Increased risk of infections might be related to the action of DPP-4 in T-cells as a co-stimulatory molecule (CD26).

DPP-4 inhibitors are commercially available alone or in combination with other oral hypoglycemic drugs (e.g. sitagliptin is marketed alone as Januvia® and in combination with metformin as Janumet®).

A DPP-4 inhibitor has been used experimentally in cats and was effective in enhancing insulin secretion and inhibiting glucagon secretion after an intravenous glucose challenge. Glucagon inhibition was also observed after a meal challenge.[28] In that study, the drug was administered subcutaneously and not orally.

REFERENCES

1. Holst JJ, et al. Mol Cell Endocrinol. 2009;297:127-36. 2. Højberg PV, et al. Diabet Med. 2008;25:1268-75. 3. Jin T. Diabetologia. 2008;51:1771-80. 4. Lyssenko V. Curr Opin Clin Nutr Metab Care. 2008;11:385-92. 5. Pilgaard K, et al. Diabetologia. 2009;52:1298-307. 6. Muscelli E, et al. Diabetes 2008;57:1340-8. 7. Vilsboll T, et al. J Clin Endocrinol Metab. 2005;88:2706-13. 8. Rask E, et al. Diabetes Care. 2001;24:1640-5. 9. Kieffer TJ. Trends Pharmacol Sci. 2003;24:110-2. 10. Hoenig M, et al. Domest Anim Endocrinol. 2010;38:95-102. 11. Sloop KW, et al. Diabetes, 2010 Dec;59(12):3099-107. 12. Kim W, Egan JM. Pharmacol Rev 2008;60:470-512. 13. Adachi T, et al. Biol Pharm Bull. 2008;31:1021-3. 14. Morishita M, et al. J Control Release. 2008;132:99-104. 15. Overton HA, et al. Br J Pharmacol. 2008;153:S76-81. 16. Vaithilingam V, et al. Curr Opin Organ Transplant. 2008 Dec;13(6):633-8. 17. Kolterman OG, et al. Am J Health Syst Pharm. 2005;62:173-81. 18. DeFronzo RA, et al. Diabetes Care. 2005;28:1092-100. 19. Glass LC, et al. Curr Med Res Opin. 2008;24:639-44. 20. Buse JB, et al. Clin Ther. 2007;29:139-53. 21. Amori RE, et al. J Am Med Assoc. 2007;298:194-206. 22. Gilor C, et al. J Vet Int Med. 2010;24(6):1567 (abst.). 23. Perez-Tilve D, et al. Am J Physiol Endocrinol Metab. 2010 May;298(5):E1088-96. 24. Buse JB, et al. Diabetes Care. 2010 Jun;33(6):1255-61. 25. Buse JB, et al. Lancet. 2009 Jul 4;374(9683):39-47. 26. Astrup A, et al. Lancet. 2009;374:1606-16. 27. Rosenstock J, et al. Diabetes Care. 2009 Oct;32(10):1880-6. 28. Furrer D, et al. Vet J. 2010;183:355-7.

ENDOSCOPIC, BIOPSY, AND HISTOPATHOLOGIC GUIDELINES IN THE DIAGNOSIS OF GASTROINTESTINAL INFLAMMATION

Robert J. Washabau, VMD, PhD, DACVIM on behalf of the WSAVA Gastrointestinal Standardization Group*
St. Paul, MN

INTRODUCTION

Diagnosis and treatment of companion animal gastrointestinal tract disorders have long been complicated by the absence of clinical, diagnostic, histopathologic, and therapeutic standards. The World Small Animal Veterinary Association (WSAVA) International Gastrointestinal (GI) Standardization Group was convened in 2004 for the purpose of developing standards for history taking, physical examination, laboratory diagnostic tests, imaging procedures and reports, endoscopic procedures and reports, biopsy procedures and reports, histopathologic interpretation, immunohistochemistry, treatment trials, and patient response and outcome in dogs and cats with gastrointestinal disease. A summary of Phase I studies was presented at the WSAVA Congress in Dublin in 2008. During Phase I (2004-2008), the GI Standardization Group published standards for endoscopy, biopsy, and histopathological evaluation of inflammation in endoscopic biopsies of the gastrointestinal tract of dogs and cats. In 2008, the GI Standardization Group developed an ACVIM Consensus Statement on "Endoscopic, Biopsy, and Histopathologic Guidelines for the Evaluation of Gastrointestinal Inflammation in Companion Animals" for presentation at the 26th Annual ACVIM Forum in San Antonio, Texas, and a Consensus Statement was published in the Journal of Veterinary Internal Medicine in 2010.

INFLAMMATORY BOWEL DISEASE - SCOPE OF THE PROBLEM

Inflammatory bowel disease (IBD) broadly refers to a group of idiopathic, chronic gastrointestinal disorders characterized by mucosal inflammation. Although the prevalence of IBD is unknown, it is arguably the most common histopathologic diagnosis obtained in dogs and cats with chronic vomiting or diarrhea. Despite perceived importance, the accuracy of these diagnoses has been the subject of some contention. IBD may represent one or more forms of chronic enteropathy that are distinguished from food-responsive and antibiotic-associated causes by their therapeutic responsiveness to immunosuppressive agents but not to dietary or antibiotic therapy alone. Although the underlying cause of IBD remains unknown, accumulating evidence in animal models suggests that intestinal inflammation results from altered interaction between gut microbes and the mucosal immune system in a susceptible host. Diagnosis of IBD currently is defined by: (1) chronic (i.e., > 3 weeks) persistent or recurrent gastrointestinal signs; (2) histopathologic evidence of mucosal inflammation; (3) inability to document other causes of gastrointestinal inflammation; (4) inadequate response to dietary, antibiotic, and anthelmintic therapies alone; and (5) clinical response to anti-inflammatory or immunosuppressive agents. Clinical signs (e.g., vomiting, small bowel diarrhea, large bowel diarrhea, weight loss, alterations in appetite) are attributed to mucosal cellular infiltrates, inflammatory mediators, and inflammation-associated enterocyte dysfunction and intestinal dysmotility. Histopathologic evaluation of biopsy specimens is required for definitive diagnosis, but no standard microscopic grading system of IBD lesions has been universally accepted.

LIMITATIONS OF HISTOPATHOLOGY

Over the past 2 decades, several independent groups have developed and applied classification systems for characterizing the nature and severity of gastrointestinal inflammatory changes. In most of these studies, the nature of gastrointestinal inflammation is portrayed primarily by the dominant population of inflammatory cells (e.g. lymphoplasmacytic, eosinophilic, pyogranulomatous) within the lamina propria. Such populations however may overlap and occur in various combinations and patterns. In many instances, the morphologic or cytoarchitectural changes of the epithelium and mucosa have been inappropriately under-emphasized. The severity of gastrointestinal inflammation usually has been graded using a simple 4 point scale (i.e., normal, mild, moderate, marked or severe). Although this approach appears logical, the specific criteria defined by various groups have differed so that it is impossible to conclusively compare the histopathological changes described in different studies. Even when specific criteria are applied, substantial variation may occur among pathologists' interpretations of changes in gastrointestinal tissue samples. Willard et al., for example, reported lack of uniformity in the assessment of 50% of biopsy samples examined by 5 veterinary pathologists. This interpretive variation poses problems for the routine diagnosis of gastrointestinal disease as well as for monitoring the progress of patients undergoing post-therapeutic endoscopy. Moreover, multi-center diagnostic and therapeutic clinical trials are not possible with such variation. With this background, a GI Standardization Group was convened with the support of the WSAVA, with the purpose of developing standards for the diagnosis and treatment of gastrointestinal diseases in the dog and cat. One of the first tasks of this group was to develop a consensus on the normal histology of the gastrointestinal tract with the subsequent aim of developing a set of histopathological standards for the nature and severity of mucosal inflammatory and associated morphological changes.

NORMAL HISTOLOGY OF THE GASTROINTESTINAL TRACT

The normal histology of the canine and feline gastrointestinal tract is affected by variables such as developmental stage (e.g., age of the animal dietary history, medication history) and therefore remains the subject of considerable controversy. Lack of agreement on normal histology has been one reason for erroneous diagnosis of gastrointestinal inflammation in many veterinary patients. Lack of agreement on standards for normal histology also has limited universal acceptance of grading systems in the evaluation of IBD. The GI Standardization Group used an evidence-based medicine approach to establish a reference range for normal histologic findings in the gastrointestinal tract of dogs and cats. Several examples of Class II and III evidence-based data were found in the Group's review of the scientific literature. Most of the studies employed microscopic evaluation of hematoxylin and eosin (HE) stained tissues, whereas others used immunohistochemistry (IHC) to label and count leukocyte populations. Representative examples from the GI Standardization Group's archives have been published already.

DEVELOPMENT OF STANDARDS FOR DIAGNOSIS OF GASTROINTESTINAL INFLAMMATION

In response to perceived limitations, the GI Standardization Group developed guidelines for the standardized interpretation of inflammatory change in the gastrointestinal mucosa of the dog and cat. These recently published guidelines provide a simple visual and textual description of the major inflammatory changes in the gastric body and antrum, duodenum and colon, and define what constitutes mild, moderate and severe pathological change. The guidelines are applicable to tissues from both dogs and cats; the only distinction between the species being with respect to the numbers of duodenal intra-epithelial lymphocytes (IEL) which are greater in cats compared with dogs. The guidelines are designed to be used 'microscope-side' by veterinary pathologists and define changes at the level of the 40x microscope objective which is considered to be the magnification at which most pathologists will refine their morphological diagnosis. Morphologic and inflammatory changes typical at each of the 4 anatomic sites were outlined in Table format.

The guidelines adopted by the GI Standardization Group are accompanied by a set of standard reporting forms, which encourage pathologists to evaluate biopsies and record findings in a consistent fashion. The forms could serve as the basis for numerical scoring of inflammatory changes as would be undertaken in research investigations. Pathologists are encouraged to report the total number of tissue samples present on the microscope slide and document the quality of these samples using descriptions of 'adequate', 'marginal' and 'inadequate' as defined by the GI Standardization Group. If such information is not included in a biopsy report, the group recommends that clinicians specifically request that it be included in the final report.

The GI Standardization Group pathologists have used the guidelines to evaluate a large slide set, derived from both dogs and cats from 9 referral institutions in 6 different countries. The Group currently is using these guidelines to identify factors affecting inter-pathologist variation and histologic lesions associated with hypoalbuminemia. Although the interpretation of endoscopically-obtained biopsies of gastrointestinal mucosa will remain a diagnostic challenge, acceptance and refinement of the GI Standardization Group's guidelines should help address current problems related to lack of standardization. Additional studies will be needed to evaluate the relative importance of each criterion and whether a weighted or non-weighted cumulative score is appropriate. The ultimate value of any grading system will be determined by its ability to accurately diagnose disease, direct therapy, and predict outcome.

GUIDELINES FOR ENDOSCOPIC EXAMINATION AND BIOPSY

Representative tissue samples containing lesions of interest are crucial for the diagnosis of most gastrointestinal tract diseases. There are 3 means of obtaining such biopsies: flexible endoscopy, laparoscopy, and surgery. Flexible endoscopy has 5 advantages. (1) Endoscopy permits the operator to see mucosal changes that cannot be visualized by the serosal approach of the surgeon. This in turn permits directed biopsy at these sites. (2) Endoscopy also permits the collection of multiple tissue biopsies (e.g., 10 or more, if necessary) from each site, which is potentially important because some diseases may have a multifocal distribution, even within 1 section of the intestine. (3) In some instances, endoscopy permits diagnosis of selected lesions without the need for tissue biopsy (e.g., ulceration, erosion, lymphangiectasia). (4) Endoscopic procedures have minimal risk of perforation and septic peritonitis, compared to surgical biopsy. (5) The procedure is quicker, less stressful and less invasive to the patient, and may be less expensive than surgery.

Flexible endoscopy has some disadvantages. Standard duodenoscopy cannot access the entire gastrointestinal tract (although enteroscopy could), and duodenoscopy alone in animals with severe gastrointestinal tract disease might not permit detection of the most important lesions. It is very easy to obtain inadequate tissue samples (i.e., mostly tips of villi) that do not readily permit diagnosis. Finally, even well-trained endoscopists cannot reliably sample duodenal muscularis mucosa or dense, submucosal infiltrative lesions with flexible endoscopic forceps.

Endoscopy is not necessarily appropriate for every animal with chronic gastrointestinal disease. It is impossible to make an all encompassing list of when to do and when not to do gastrointestinal endoscopy. Substantial latitude must be given to the clinician who continually must weigh the specifics of the case, client expectations, monetary concerns, risk to the patient and other factors. Nonetheless, certain general principles can be stated. First, endoscopy seldom benefits patients with acute diarrhea (i.e., < 3 weeks in duration) unless the disease is particularly severe or a specific disease needs to be quickly diagnosed or eliminated (e.g. histiocytic ulcerative colitis, histoplasmosis, neoplasia). Second, clinical assessment usually seems more useful and appropriate than endoscopic biopsy in determining response of inflammatory diseases to therapy. Third, endoscopy is primarily of value in diagnosing infiltrative, erosive or other anatomic (e.g., lacteal dilatation, foreign body) problems. It seldom allows diagnosis of dietary-responsive enteropathy, antibiotic-responsive diarrhea, or gastrointestinal motility disorders. The healthier the patient (i.e., modest to no weight loss, relatively good body condition score, normal serum albumin concentration, not lethargic, not anorexic, no ultrasonographic evidence of infiltrative disease), the more consideration should be given to therapeutic trials (e.g., dietary, antibiotic, anthelmintic, or probiotic trials) instead of endoscopic biopsy, at least initially. Conversely, the more clinically ill the patient (e.g., severe weight loss, very poor body condition score, hypoalbuminemia, anorexia, ultrasonographic evidence of infiltrative disease), the more reasonable it usually is to perform endoscopic biopsy before therapeutic trials. Fourth, if the clients allow, it is generally helpful to image the abdomen ultrasonographically before endoscopy in an attempt to ensure that infiltrative lesions out of reach of the endoscope (e.g., mid-jejunum) are not present.

When endoscopy is performed, careful and thorough examination of the stomach, small intestine, and large intestine is the first step. Standardized endoscopic report forms have been developed that require a systematic, rigorous, and complete examination of the gastrointestinal tract. Good endoscopy forms have several features. They include patient identification and date, reason for procedure, specific equipment used (i.e., endoscopes, biopsy forceps, foreign body retrieval devices, etc), complications encountered, extent of examination (i.e., how far the endoscope was advanced), generation of images, specific lesions, and final recommendations. Check boxes are strongly recommended in such report forms so as to document whether specific lesions were or were not seen and whether specific problems did or did not occur. They also are useful in helping to ensure that examinations are complete.

Good tissue quality is as important as good endoscopic technique because poor quality tissue samples may not be interpretable by microscopy. There are many objective studies and recommendations regarding optimal technique and technology for endoscopic biopsy of the human gastrointestinal tract, but similar data are not as readily available for the dog and cat. Some generalizations may be made. It seems intuitively obvious that larger (e.g., 2.8 mm) biopsy forceps procure larger, and perhaps better quality, tissue samples than smaller (e.g., 2.2 mm) forceps. That said, overall tissue quality still depends upon mucosal thickness. Willard et al. reported that duodenal biopsy quality was equivalent between dogs and cats, despite the assumption that feline intestinal biopsies probably involved the use of smaller diameter endoscopes and biopsy forceps. The thinner feline duodenal mucosa may be more readily sampled, even into the muscularis mucosa, than the thicker canine duodenal mucosa. This observation also may explain why the thinner ileal mucosa is more readily sampled than the thicker duodenal mucosa.

Variability in tissue quality may result from variability in sample submission technique as well as tissue processing in the diagnostic laboratory. Therefore, tissue samples should be carefully removed from the biopsy forceps and submitted in such a manner as to avoid artifacts and to permit optimal tissue orientation in the laboratory. The clinician must communicate and work with the laboratory to assure proper tissue orientation on glass slides because the approach may vary from laboratory to laboratory. Techniques for handling and mounting tissue samples have been described previously. If samples are to be submitted free-floating in formalin, the only way to ensure they will be properly oriented is for the histopathology technician to examine each piece of tissue with a dissecting microscope as they are embedded.

Substantially fewer biopsy samples are needed to establish a diagnosis as the quality of the tissue increases from inadequate to marginal to adequate, although there are some important differences between the dog and the cat. Approximately 6 marginal or adequate tissue samples from the feline stomach or duodenum are sufficient to diagnose villus atrophy and mild to moderate cellular infiltration. In the dog however approximately 6 to 7 adequate or 10 to 15 marginal gastric or duodenal tissue samples are required to reliably diagnose villus atrophy, lymphangiectasia, and mild or moderate cellular infiltrates. Canine duodenal crypt lesions are seemingly more difficult to diagnose reliably, and approximately 13 adequate or 28 marginal samples may be required. The actual numbers probably will change as more studies evaluate this issue, but superior quality samples enhance the diagnostic sensitivity of the biopsies. Therefore, the total number of tissue samples that should be taken during a procedure will depend upon the skill of the endoscopist. In general, skilled endoscopists must take fewer samples than less skilled endoscopists to achieve the same number of adequate samples on the histology slide.

The dependence of diagnosis on the quality of the tissue samples supports the notion that clinicians should insist upon pathology reports including both the total number of tissue samples submitted and the quality of these samples (i.e., inadequate, marginal, adequate), to determine the level of confidence in the reported histological diagnosis. If most of the samples are inadequate or marginal, the clinician should re-assess his or her technique for procedural error. If uncertain, the clinician could request a second opinion on the slides to assess their quality.

*WSAVA Gastrointestinal Standardization Group Membership - T. Bilzer, M.J. Day, E.J. Hall, A.E. Jergens, J. Mansell, T. Minami, R.J. Washabau, B. Wilcock, M.D. Willard

REFERENCES

(1) Gastrointestinal Endoscopy Standards. http://www.wsava.org/standardizationgroup.htm. (2) Willard MD, Mansell J, Fosgate GT, et al. Effect of sample quality on the sensitivity of endoscopic biopsy for detecting gastric and duodenal lesions in dogs and cats. J Vet Intern Med, 2008; 22: 1084-1089. (3) Day, M.J., Bilzer T., Mansell J., et al. Histopathological standards for the diagnosis of gastrointestinal inflammation in endoscopic biopsy samples from the dog and cat: a report from the World Small Animal Veterinary Association Gastrointestinal Standardization Group. J Comp Path 2008; 137: S1-S43. (4) Washabau RJ, Day MJ, Willard MD et al. ACVIM Consensus Statement on Gastrointestinal Inflammation. J Vet Intern Med 2010; 24: 10-26.

LEUKEMIA: FACTOIDS, FUN WITH FLUID AND CONFOUNDING CASES

Tracy Stokol, BVSc, PhD, DACVP
Ithaca, NY

LEUKEMIA CONCEPTS

The syndromes of myeloid (non-lymphoid) hematopoietic neoplasia, though variable in their manifestion and clinical course, are all clonal diseases arising from a hematopoietic stem cell (HSC). Somatic mutations in an HSC can result in an expanding clone of progeny that are not subject to normal regulation of proliferation and/or maturation. Inhibition of normal hematopoiesis is concurrent with expansion of the neoplastic clone, leading to deficiencies of normal mature cells (and cytopenias in peripheral blood, e.g. anemia, neutropenia, thrombocytopenia). Dependent on the mutation or combination of mutations, the disease can be expressed as ineffective dysplastic hematopoiesis (the myelodysplastic syndromes), proliferation of immature cells (the acute myeloid leukemias [AML]), or accumulation of mature cells of one type (chronic myeloid leukemias). Both MDS and chronic myeloid leukemias can progress to AML.

The syndromes of lymphoid neoplasia are also clonal but evolve from cells at any stage of lymphocyte ontogeny. The acute lymphoid leukemias arise in marrow (usually) from cells at the earliest stages of lymphoid maturation (pro-T, pro-B or NK cells), whereas chronic lymphoid leukemias involve the more mature counterparts. The lymphomas are clonal proliferations of lymphocytes (T, B or NK cells) that arise in extramedullary tissue from anatomic compartments (e.g., follicle center, mantle zone, paracortical zone, etc.) of the lymph node or extranodal lymphoid tissue. The cells of different lymphomas share many features with their normal counterparts, which can influence the distribution, behavior, morphologic characteristics, and immunophenotypic profile of a given lymphoma population. However, lymphoma cells are rarely exact replicas of normal counterparts and may display abnormal phenotypes.

The concept that leukemias and lymphomas are clonal neoplasms has direct relevance to diagnosis. For example, recognition (either by morphologic features, immunophenotype or tests of clonality) that the lymphoid cells in blood, an enlarged lymph node or other tissues are of one type is a key part of the diagnosis of lymphoma.

TYPES OF LEUKEMIA

The term leukemia indicates neoplastic cells of hematopoietic origin in the bone marrow and/or circulation. For simplicity, leukemia can be categorized into three general types: acute, chronic and lymphoma-associated leukemia.

Acute leukemia

These leukemias arise primarily in the bone marrow and infiltrate extramedullary sites secondarily. They consist of immature cells, with morphologic features of blasts. Because normal hematopoietic tissue is replaced by the neoplastic cells (called myelophthisis), affected animals usually present with severe cytopenias and readily identifiable circulating blasts. In some cases, very low numbers or no circulating blasts may be observed in peripheral blood smears. These latter patients are usually bi- or pancytopenic and a bone marrow aspirate is needed to identify the underlying cause for the cytopenias. An acute leukemia is defined as >20% blasts in the bone marrow, i.e. a bone marrow aspirate is essential for diagnosis. Acute leukemias can be myeloid (AML) or lymphoid (ALL) in origin. Rare variants are mixed lineage (expressing markers of both myeloid and lymphoid cells) or undifferentiated (only expressing stem cell markers) acute leukemias. Although a diagnosis of acute leukemia can be made from blood and bone marrow aspirate smears, differentiation of AML from ALL requires more than morphologic assessment of blasts (since blasts of different origin can look alike). Additional diagnostic tests, specifically immunophenotyping and cytochemistry, are required for classification of acute leukemias. This may be very important as ALL may be more responsive to aggressive chemotherapy than AML, which has a uniformly poor prognosis. AML is the most frequent type of acute leukemia in dogs, followed by B-ALL[1]. Similarly, AML is more common than ALL in cats. Acute leukemias are quite rare in horses, but ALL seems to outnumber AML in this species.

Chronic leukemia

These leukemias are characterized by the proliferation of mature cells (myeloid or lymphoid), resulting in a persistently elevated count of the involved cell line in peripheral blood. There are several different types of chronic myeloid leukemia, which are based on the involved cell lineage, e.g. polycythemia vera, chronic neutrophilic leukemia. The chronic lymphocytic leukemias (CLL) consist of B, T or NK CLL. Chronic leukemias usually have a protracted clinical course and are often diagnosed as an incidental finding in older animals presenting for other complaints[1,2]. These leukemias must be distinguished from extramedullary causes of persistently elevated mature hematopoietic cells, e.g. chronic antigenic stimulation for CLL, cytokine-induced leukocytosis for chronic neutrophilic leukemia. Bone marrow aspiration is generally not useful for differentiation as the marrow shows hyperplasia of the associated cell lines in both neoplastic or non-neoplastic mature cell proliferations, i.e. a myeloid hyperplasia is observed in marrow in both chronic neutrophilic leukemia and a leukocytosis that is cytokine-driven from a neoplasm (i.e. paraneoplastic). The diagnosis of chronic leukemia requires documentation of a persistently elevated cell line (e.g. neutrophilia, lymphocytosis) without an identifiable underlying cause (e.g. occult neoplasia)[1,3]. In the case of a chronic lymphocytosis, assessment for clonality or immunophenotyping is useful for distinguishing between reactive and neoplastic causes[1-4]. The most common chronic leukemia is CLL, which is usually of T cell origin in both dogs and cats[1,2]. T-CLL typically arises in the spleen and may infiltrate marrow secondarily, whereas B-CLL arises in the marrow.

Lymphoma-associated leukemia

Circulating lymphoma cells can be observed in the peripheral blood of animals with lymphoma (with or without marrow involvement), producing what is termed lymphoma-associated leukemia. This is also defined as stage V disease. These cells will usually resemble the primary tumor, e.g. they will be lymphoblasts if the primary tumor is a large cell lymphoma or will be small lymphocytes if the primary tumor is a small cell lymphoma. The number of circulating lymphoma cells varies from very low (i.e., not included in the differential cell count) to very high. The latter cases (many lymphoma cells in blood, particularly when they are large or blasts) can mimic an acute leukemia. In these cases, additional diagnostic tests (including immunophenotyping and cytochemistry) are needed to distinguish between acute leukemia and lymphoma. This is obviously important, because these two disorders have very different prognoses. Fortunately, these severe lymphomas are infrequent in animals. Circulating lymphoma cells are more commonly seen in dogs than cats; in fact, this is the most common of the three types of leukemia in the dog. Circulating lymphoma cells can be seen in cats with thymic, hepatic, and splenic lymphoma. Cows with lymphoma due to bovine leukemia virus can also have circulating lymphoma cells (of B cell origin). Lymphoma rarely manifests with circulating neoplastic cells in the horse.

TOOLS FOR DIFFERENTIATING BETWEEN LEUKEMIA

These tools include immunophenotyping, cytochemistry and clonality assessment and are used for: 1) Classification of acute leukemia (immunophenotyping and cytochemistry); 2) Differentiation of lymphoma from acute leukemia (immunophenotyping and cytochemistry) and 3) Differentiation of reactive from neoplastic lymphocytosis (immunophenotyping, e.g. abnormal or uniform phenotype, and clonality). They should be used as adjunctive tests and should always be interpreted with results from other clinical pathologic (hemogram, chemistry, bone marrow or other aspirates) and diagnostic (e.g. imaging) tests. They should never be used alone to make a diagnosis of hematopoietic neoplasia.

Immunophenotyping

This involves the use of antibodies to detect specific antigens on the surface or in the cytoplasm or nucleus of cells. Different hematopoietic cells express unique or specific combinations of antigens, allowing their identification with this technique. Antibodies can be applied to cells on cytology or histology slides (immunocyto- or immunohistochemistry) or suspended cells in blood, bone marrow or tissue aspirates (immunophenotyping with flow cytometry). In veterinary medicine, we have far more antibodies against antigens of lymphoid (T cells especially) cells, but fewer are available for myeloid and B cells. More antibodies are also available for the dog compared to the cat. Common antigens for different cell types are: CD21, pax-5, and CD79a for B cells; CD3, CD5, CD4 (helper) and CD8 (cytotoxic) for T cells, CD4, CD11b, CD11c, myeloperoxidase for neutrophils, CD11b, CD11c, CD1c, CD11d for monocytes and CD34 for stem cells[1,5].

A few facts about immunophenotyping: 1) These assays are technically demanding and labor intensive. 2) Interpretation of flow cytometry is both science and art. It requires experience AND taking into account the hematologic data and morphologic features of the neoplastic cells. Flow data can be readily misinterpreted, particularly if appropriate controls or analysis (e.g. gating strategy) is not used. Thus, scatter and histogram plots should be provided along with the numerical results and the interpretation. 3) All laboratories that offer immunophenotyping for leukemia should optimize assay procedures and test each antibody in relevant samples from healthy animals for specific and non-specific binding (which can yield false positive results). 4) Formalin-fixation destroys antigens (e.g. CD4, CD8), therefore more antibodies can be applied to cytologic or fluid specimens, increasing the versatility of these specimens for immunophenotyping (particularly fluid samples using flow cytometry). 5) Non-specific binding is known to occur with antibodies against specific antigens, e.g. CD79a is a membrane antigen. If the nucleus of the cell in question is positive for CD79a, this is a false positive result. 6) Flow cytometry is usually performed on unfixed, intact cells so the results will be positive only if the antigens are surface-expressed. Intracellular antigens (e.g. CD79a and CD20 – both B cell markers) can be detected by flow cytometry, but the cells need to be permeabilized first, which alters their size and complexity with flow (which can hinder interpretation). With permeabilization, you cannot see the location of the antibody on the cell (e.g. you may get false positive results with CD79a if it is binding to the cell nucleus – see point above). 7) Demonstrating a uniform phenotype in lymphoid cells (e.g. all lymphocytes are CD3+CD4+ helper T cells) is supportive (not definitive) evidence of neoplasia. Expansion of a uniform population of cells can occur in reactive conditions. 8) Demonstrating an abnormal phenotype is very supportive of a diagnosis of neoplasia, e.g. B cells do not normally express CD1c (a T cell marker), but B cell lymphomas and B-CLL can express this marker. 9) CD34 is a very useful stem cell marker for differentiating between lymphoma and acute leukemia, i.e. acute myeloid leukemias are usually positive, whereas most lymphomas are negative, for CD34. However, T-ALL are frequently negative for CD34 and some large B cell lymphomas can be positive for this stem cell marker[1,5]. 10) Drugs can affect antigen expression on lymphoid cells, e.g. glucocorticoids affect expression of commonly used antigens[6].

Cytochemistry

This involves the detection of enzymes within cells by applying substrates for the enzymes to the cells on venous blood or cytology smears. If the cells contain the enzyme, they will cleave the substrate forming a specific color that can be detected by light microscopy. Since myeloid (except erythroid and megakaryocytic) cells contain more enzymes than lymphoid cells, this is useful for identifying leukemias of myeloid origin. It is less useful for lymphoid leukemias because they will often be negative for all enzymes. Common enzyme substrates are myeloperoxidase (granulocytes and monocytes), sudan black (mostly granulocytes), chloroacetate esterase (granulocytes), non-specific esterase (monocytes, megakaryocytes) and alkaline phosphatase (monoblasts in the dog).

Some facts about cytochemistry: 1) This can only performed on cytology smears (blood and bone marrow). Sufficient numbers of neoplastic cells must be present to reliably distinguish them from normal cells, since cell morphologic features (crucial for identification of cells) are difficult to discern with these stains. 2) Cytochemistry is time-sensitive. Enzyme activity decreases in stored samples, so the fresher the sample, the better. 3) There are species-specific differences in the enzyme content of individual cells. For example, canine eosinophils are strongly positive but feline eosinophils are negative for myeloperoxidase. 4) Negative cytochemiical results are expected in most lymphoid cells and erythroid cells. 5) Like immunophenotyping, this technique is labor intensive and technically demanding. 6) Using a combination of test results, rather than a single test result, to phenotype a leukemia is optimal. We usually use cytochemistry in CONJUNCTION with immunophenotyping to distinguish between AML and ALL (or lymphoma).

Clonality

These tests are based on Polymerase chain reaction assays for Antigen Receptor Rearrangements (PARRs) in lymphoid cells, i.e. the T cell receptor gene (for T cell neoplasia) or the immunoglobulin M heavy chain gene (for B cell neoplasia). Lymphoid cells from animals with clonal or neoplastic conditions usually have a single gene rearrangement that will be amplified by PCR as a single band. In contrast, animals with reactive lymphoid conditions will have many rearranged genes and multiple amplified bands (coming from multiple clones of lymphocytes), usually producing a smear on a PCR gel. PCR tests for clonality are excellent for confirming an expanded lymphoid population in blood, bone marrow or tissues is neoplastic[1,3,4].

Some facts about clonality testing: 1) False positive and negative reactions occur for a variety of reasons[3,4]. PCR requires scrupulous attention to detail and good technique (to prevent DNA contamination from other sources). Results of PARR testing should ideally be interpreted along with the other clinical and laboratory findings (NOT in isolation). Specific infectious agents and other antigens can cause clonal expansions of lymphoid cells, for instance, *Ehrlichia canis* can result in a clonal expansion of cytotoxic T cells in dogs[3,4]. Clonality tests should not be used to distinguish AML from ALL, because animal patients with AML may have clonally re-arranged B or T cell receptors.

REFERENCES

[1]Vernau and Moore, **Vet Immunol Immunopathol** 1999; 69:145, [2]Workman and Vernau **Vet Clin North Amer Sm Anim Pract** 2003; 33:1379, [3]Avery and Avery **Vet Clin North Amer Sm Anim Pract** 2007; 37: 267, [4]Burnett et.al. **Vet Clin Pathol** 2003;32, [5]Wilkerson et.al. **Vet Immunol Immunopathol** 2005; 106:179, Ammersbach et.al. **J Vet Intern Med** 2006, 20:1166.

EXPERIENCE WITH AN ELECTRONIC ONLINE DATABASE IN A MULTICENTER CLINICAL TRIAL

George E. Lees, DVM, MS, DACVIM
College Station, TX

Beginning early in 2005, the International Veterinary Renal Pathology Initiative was conceived and subsequently organized under the auspices of the of the World Small Animal Veterinary Association (WSAVA) to improve standards for diagnosis of renal diseases in animals by integrating pathologic findings from histologic, ultrastructural, and immunostaining examinations of kidney samples with clinical information about each affected animal's illness.

From the outset, the organizers recognized several important challenges. Chief among these was the fundamental nature of nephropathology, which is a sub-specialty of anatomic pathology that is unique because of its routine use of ultrastructural (electron microscopic) examinations along with histologic (light microscopic) findings and immunostaining (immunofluorescence microscopy) for the diagnosis of common diseases, especially those affecting glomeruli. Moreover, the renal pathologic findings must be carefully integrated with information derived from the clinical investigation of the patient's illness in order to formulate the correct diagnosis.

Additionally, veterinary nephropathology confronted several further challenges in 2005. Well-established standards for diagnosis of kidney diseases in animals (particularly in dogs and cats) were lacking, especially for glomerular disorders. Most of the veterinary laboratories that did pathologic evaluations of kidney never performed ultrastructural examinations or immunostaining. Indeed, thorough histologic evaluations (ie, routinely using appropriately thin sections, special stains, etc.) often were not performed. The clinical and pathologic features of the disorders rarely were adequately integrated either by the pathologist or the clinician when renal diseases were diagnosed. Pathologists evaluating renal biopsies rarely were given access to detailed clinical information about the affected patient, and clinicians rarely had access to adequately detailed pathologic observations (including TEM and IF findings). And finally, individuals with special expertise in veterinary nephropathology were very few in number and widely scattered. In order to surmount these challenges, the organizers formulated and implemented a plan that had the following key elements:

- Establish a few "Diagnostic Renal Pathology Centers" (DRPC) having the capability to routinely perform comprehensive pathologic evaluations, including light microscopy (LM) transmission electron microscopy (TEM), and immunofluorescence (IF) staining, of renal specimens from animals.
- Prospectively study a large number of renal biopsy cases gathered from geographically diverse origins; the international scope of the project was a key element from the outset.
- Initially focus on developing improved standards for glomerular diseases in dogs, and then extend the effort to include other topics (eg, feline renal disease).
- Build an extensible hardware/software infrastructure to facilitate operation of the DRPCs and connect them with one another by means of a shared database capable of supporting both near term (ie, the 'renal standardization study' re: diagnosis of dog glomerular disease) and more distant goals (eg, eventual prospective therapeutic trials).
- Routinely collect appropriate clinical information in concert with the pathologic evaluation of every case and maintain that information in the database along with the pathologic information.
- Make all tissue sections (LM, TEM, and IF) available in digital formats having sufficient detail (ie, resolution) for diagnostic assessment on readily accessible (ie, web-based) viewing platforms.
- Enable the investigators to review and discuss the clinical and pathologic features of individual cases in detail during online (ie, web-based) meetings.
- Ultimately build consensus among the investigators about disease classification, diagnostic criteria, and nomenclature, etc. (ie, establish the improved standards).

To accomplish these goals, a commercial provider of electronic data capture services for clinical trials (Prelude Dynamics, LLC, Austin, TX) was contracted to build a custom edition of their software (called, 'Vision') to manage the clinical and pathologic; however, that robust software platform was not capable of managing the digital pathology component of the project. Therefore, we evaluated several commercially available digital pathology hardware/software systems, and selected one made by leader in this field (Aperio Technologies, Inc., Vista, CA). The digital pathology system consists of a slide scanner (ScanScope) that makes whole-slide digital images of light microscopic sections and a server/software platform (called, 'Spectrum') that stores/manages the digital image files and associated data. The digital pathology platform not only manages the digital image files created by the ScanScope (ie, the LM image files), it also stores and manages all of the digital pathology image files that are generated by the TEM and IF evaluations. Thus, the selected digital pathology solution works equally well for all modalities (LM,

TEM, and IF) of pathologic examination required for the project. Finally, to make the user experience as efficient and effective as possible, we worked with the two software providers (ie, Prelude Dynamics and Aperio) to develop a software interface between the 'Vision' system and the 'Spectrum' system that permits individuals using the 'Vision' system to see (as thumbnail images) all the digital image files are available for each case, as well as to deploy and examine any or all of those images (which are seamlessly delivered from 'Spectrum', the image server platform, to any internet-connected computer anywhere in the world) without exiting the 'Vision' system (and thus maintaining full access to all the clinical and pathologic information stored in the 'Vision' database while the pathologic sections are being examined).

MULTICENTER CLINICAL TRIALS: THE PARTICIPANT EXPERIENCE

Philip R. Fox, DVM, DACVIM/DECVIM (Cardiology), DACVECC
New York, NY

Comprehensive information and resources about clinical trials can be downloaded from many web sites including: *http://www.ClinicalTrials.gov; http://www.cancer.gov/clinicaltrials/search;* http://clinicalstudies.info.nih.gov; http://www.ncbi.nlm.nih.gov; http://www.americanheart.org; http://www.acc.org/clinical/guidelines/failure/hf_index.htm

The Multicenter Clinical Trial

Clinical trials are research studies that investigate naturally occurring disease using pre-defined study protocols. Interventional studies allocate a treatment and the investigator records outcome. Observational studies focus upon observations that are measured or evaluated.

All clinical trials use guidelines (research protocols) that specify inclusion and exclusion criteria, identifies which pets can and cannot be enrolled, keeps them safe, and helps ensure that researchers will be able to produce reliable results. Some studies include pets with illness or conditions to be studied; others need healthy participants. All studies require written and signed informed client consent forms.

Clinical trials are conducted in phases which have different objectives:

Phase I trials test an experimental drug or treatment in a small group for the first time to evaluate its safety, determine a safe dosage range, and identify side effects.

Phase II trials study a drug or treatment in a larger cohort to assess if it is effective, and determine whether further evaluate its safety.

Phase III trials involve large patient groups to confirm drug or treatment effectiveness, monitor side effects, compare the study treatment to commonly used therapies, and collect safety information.

Phase IV trials involve post marketing studies intended to identify additional information regarding drug risks, benefits, and optimal use.

Why Participate in a Clinical Trial?

The well constructed clinical trial benefits the study sponsor, pet owner, and investigator. It is imperative for all parties to read the fine print to assure that all expectations are met.

Clients- take an active role in their pet's health care, obtain access to new or advanced therapies, help by contributing to medical research that could help other pets, and often have certain medical costs paid for by the study trial sponsor.

Sponsors- pay for studies in order to generate data intended for regulatory purposes (CVM drug approval) or post label marketing.

Investigators- can participate in cutting edge studies, gain access to therapies that are not yet commercially available, and collaborate with investigators to publish findings. Some studies generate research revenue.

Challenges to the Participant of Clinical Trials

Participants should not overlook the fact that clinical trials are work intensive, require meticulous record keeping, and usually need technical support. Participants should also be prepared to spend time with study monitors for periodic assessments of forms, data sheets, and medical records. In addition, study folders, medical records and related diagnostic information are often required to be retained for many years and be subjected to FDA audits. In cases where the participant is the primary investigator, the time commitment can become substantial.

Survey of Board Certified Cardiologists Regarding Multicenter Clinical Trials

Eighty ACVIM board certified cardiologists responded to a survey designed to identify preferences and perceptions regarding multicenter cardiac clinical trials. While this survey was conducted several years ago, the responses reflect current attitudes regarding multicenter trials amongst cardiologists.

Fox PR, 2007
AMC

Multicenter Clinical Trials
ACVIM Cardiology Specialist Survey, 2007

Q: **Have you** *directly participated* **in clinical trial design?**

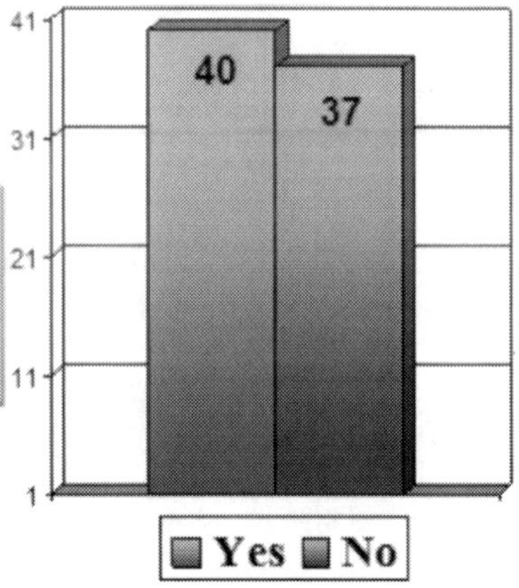

Fox PR, 2007
AMC

Multicenter Clinical Trials
ACVIM Cardiology Specialist Survey, 2007

Q: **Have you** *directly participated* **in clinical trial design?**

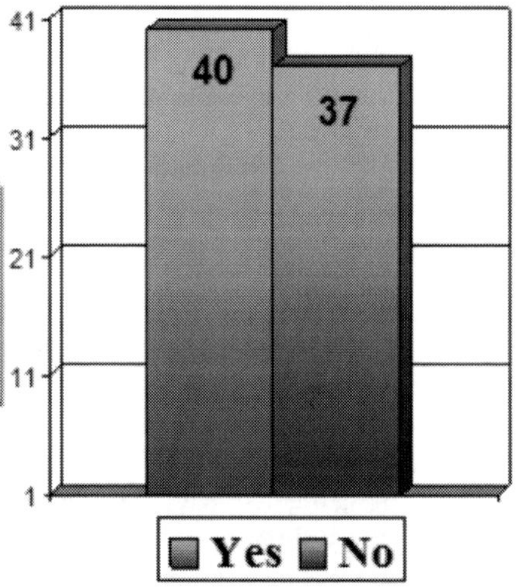

Fox PR, 2007
AMC

Multicenter Clinical Trials
ACVIM Cardiology Specialist Survey, 2007

Q: **Have you been a** *PI or Co-PI* **of a clinical trial?**

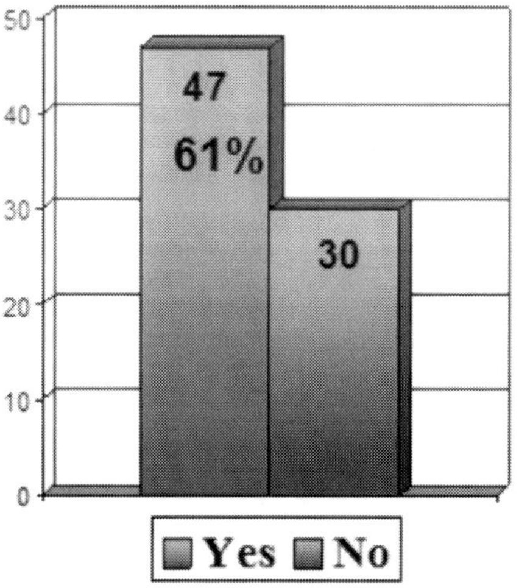

Fox PR, 2007
AMC

Multicenter Clinical Trials
ACVIM Cardiology Specialist Survey, 2007

Q: **Have you** *directly contributed cases* **to clinical trials?**

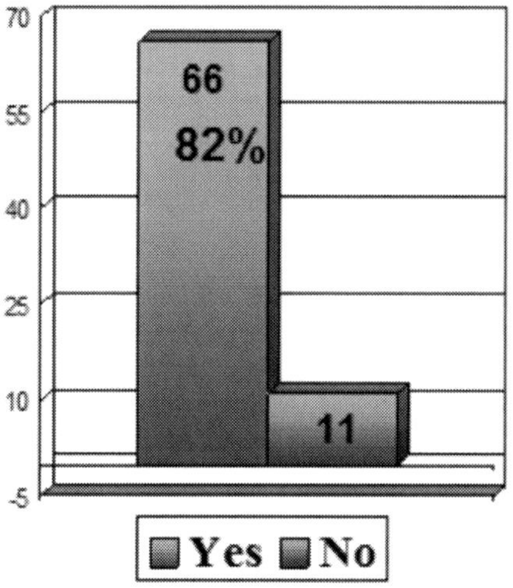

Fox PR, 2007
AMC

Multicenter Clinical Trials

ACVIM Cardiology Specialist Survey, 2007

Q: **How important is** ***participation in study conception/design*** **in** ***your decision*** **to enter a clinical trial?**

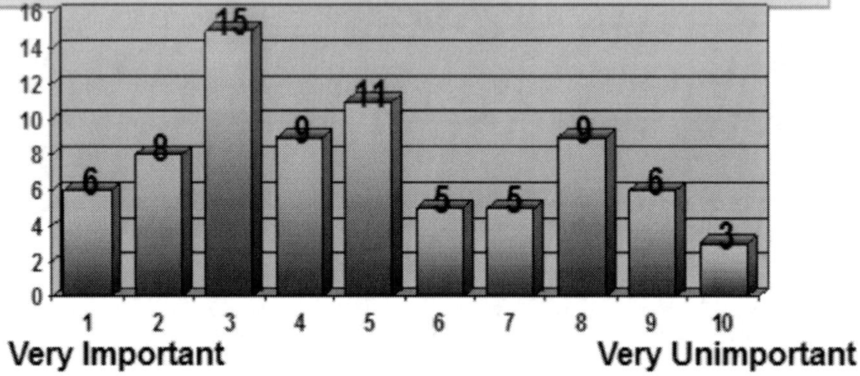

Fox PR, 2007
AMC

Multicenter Clinical Trials

ACVIM Cardiology Specialist Survey, 2007

Q: **How important is** ***authorship*** **to your decision to participate?**

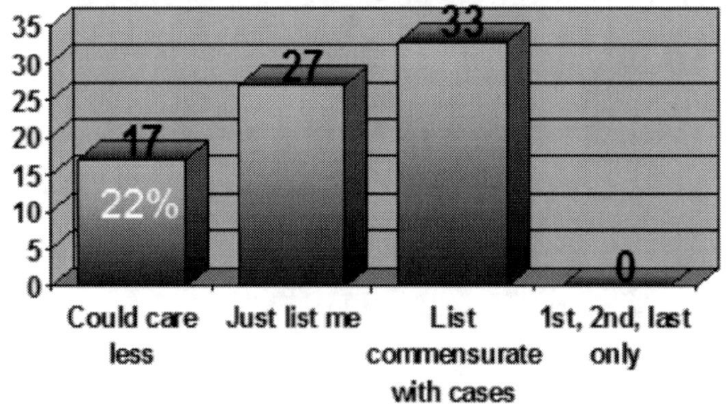

Fox PR, 2007
AMC

Multicenter Clinical Trials
ACVIM Cardiology Specialist Survey, 2007

Q: How important is *recovery* of *indirect* study *expenses* to participation?

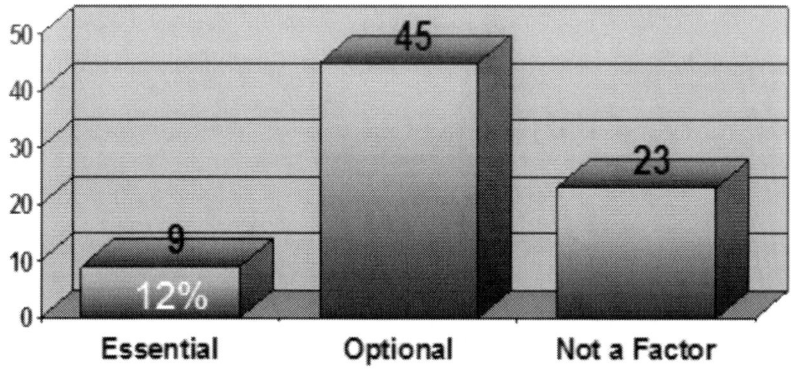

Fox PR, 2007
AMC

Multicenter Clinical Trials
ACVIM Cardiology Specialist Survey, 2007

Q: How important is *recovery* of *direct* study *expenses* to participation?

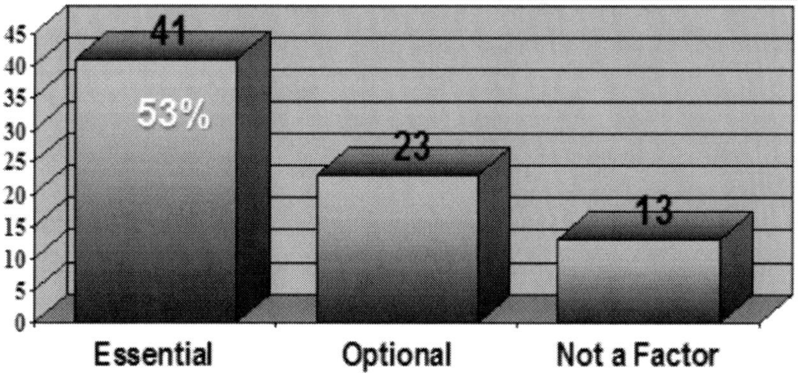

Fox PR, 2007
AMC

Multicenter Clinical Trials
ACVIM Cardiology Specialist Survey, 2007

Q: **How important is** *money for client incentives* **for your participation**

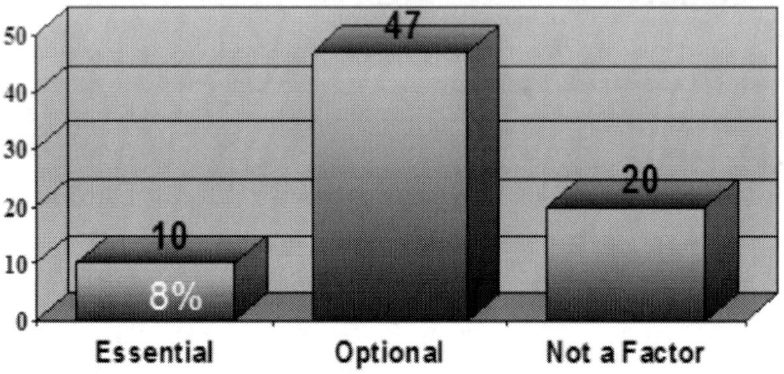

Fox PR, 2007
AMC

Multicenter Clinical Trials
ACVIM Cardiology Specialist Survey, 2007

Q: **How Motivated are you to** *encourage your residents/faculty contribution* **to other studies?**

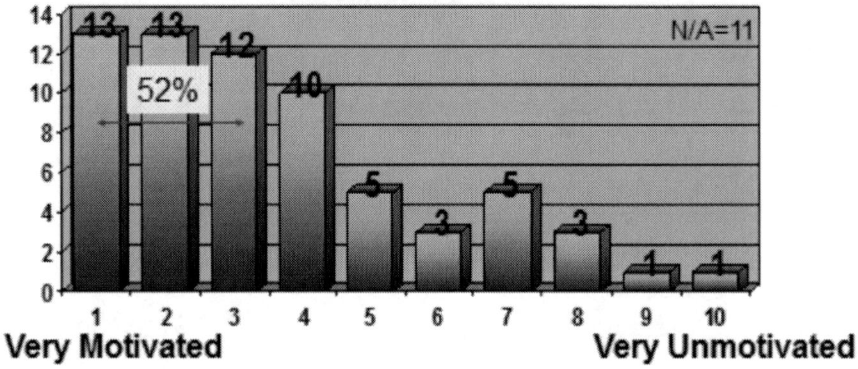

Fox PR, 2007
AMC

Multicenter Clinical Trials
ACVIM Cardiology Specialist Survey, 2007

Q: **How Motivated are <u>you</u> to *contribute cases <u>yourself</u>* to others' studies?**

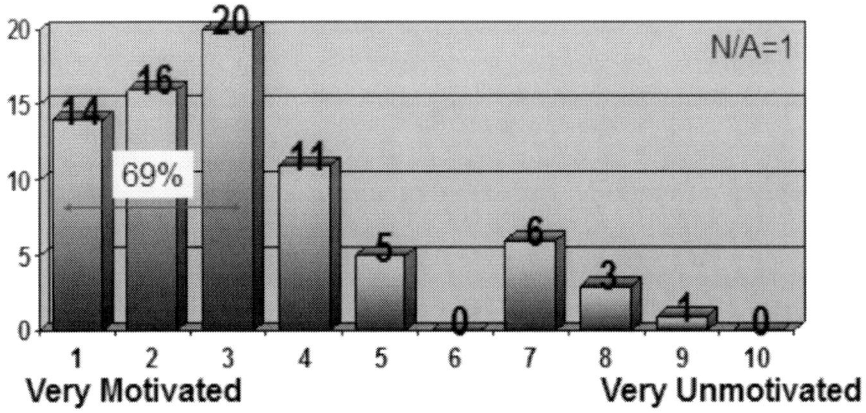

Fox PR, 2007
AMC

Multicenter Clinical Trials
ACVIM Cardiology Specialist Survey, 2007

Q: **What are** your *major hurdles* ***for you*** to *organize* or *lead* **a multicenter trial?**

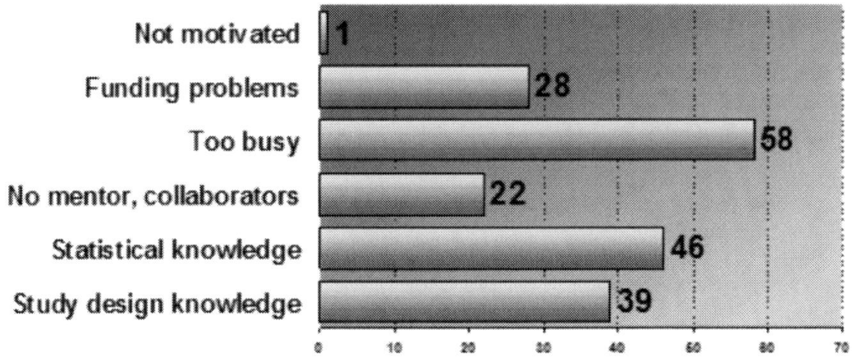

Fox PR, 2007
AMC

Multicenter Clinical Trials

ACVIM Cardiology Specialist Survey, 2007

Q: How Comfortable or reluctant to are you regarding *industry sponsored* studies?

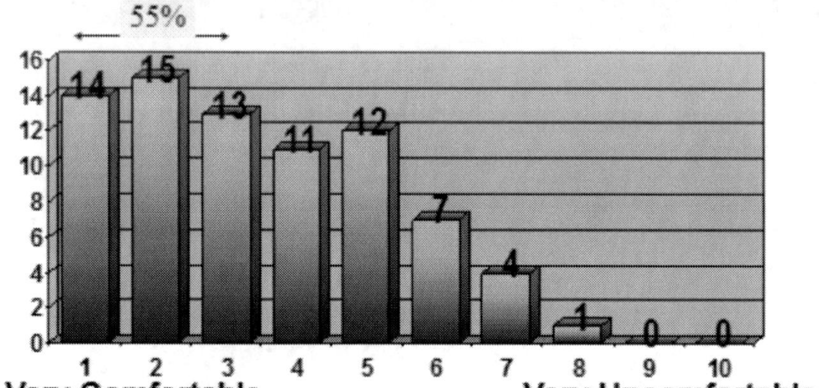

Fox PR, 2007
AMC

Multicenter Clinical Trials Survey

ACVIM Cardiology Specialist Wish List- 2007

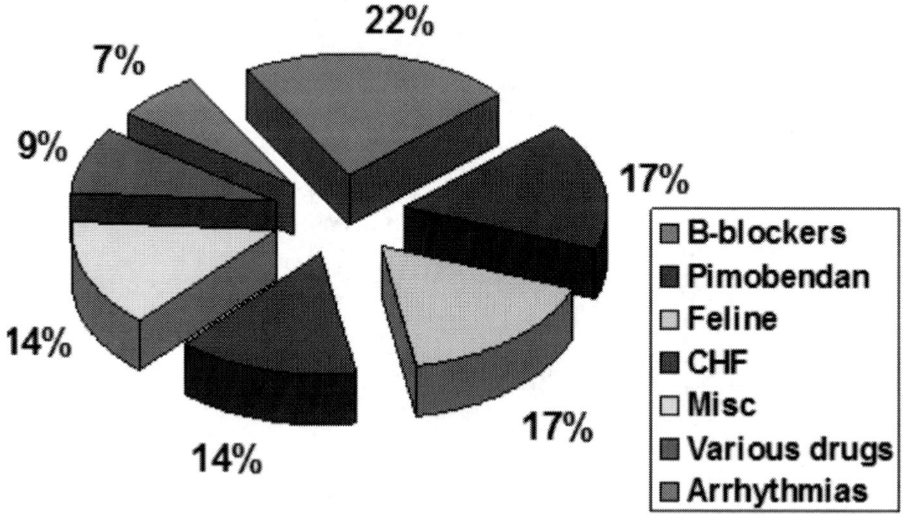

PAIN MANAGEMENT IN FOALS: BEYOND THE NSAIDS

Debra C. Sellon, DVM, PhD, DACVIM
Pullman, WA

INTRODUCTION

Pain is an inevitable and essential sensation in life. Without pain, impending or actual tissue trauma is unrecognized until significant and possibly irreversible tissue damage has occurred. The inability to recognize and respond to pain is therefore a life-threatening condition. Chronic unmanaged pain however, provides few, if any, physiologic benefits but may have significant adverse effects on function of body systems. For many years the adverse effects of uncontrolled pain on patient recovery and well-being were rarely acknowledged by physicians or veterinarians. Analgesic therapy was considered primarily a humane consideration or a necessity to facilitate painful diagnostic and therapeutic procedures such as surgery. These attitudes were exemplified and compounded by depictions of the "tough it out" attitude of the traditional 19th century American West. In the last few decades, however, it has become apparent that failure to adequately manage pain in human and animal patients can have long-standing and severe consequences independent of the inciting cause of the pain. This recognition has resulted in major changes in analgesic practices in human and companion animal care. Changes in veterinary attitudes and practices towards analgesic medication for large animal patients have been adopted much more slowly.

Only in the last decade has peer-reviewed veterinary literature shown a significant increase in publication of manuscripts related to recognition and treatment of pain in horses. A PubMed search using the terms "horse" and "pain" yields 1564 articles of which half were published since the turn of the century. Articles published prior to 1980 were predominantly related to treatment of acute abdominal pain. Most pain-related research prior to 1990 investigated effective methods for animal restraint or control of painful behaviors to facilitate examination and treatment (e.g. relief of acute colic pain) or on recognition of adverse effects of analgesic medications (e.g. gastrointestinal or renal effects of non-steroidal anti-inflammatory medications [NSAIDs]). Evaluations of the pharmacokinetic and pharmacodynamic properties of analgesic agents used in the horse became more prevalent in the mid-1980s when alpha-2 agonists began to be used regularly in equine practice. Studies of the comparative efficacy of various analgesic regimens in postoperative equine patients first appeared in the mid-1990s. The first publication in this search list that related to recognition and objective evaluation of acute pain in horses appeared in a German journal in 1996.[1]

As recently as 2001, a letter to the editor of *Veterinary Record* from a highly experienced equine practitioner suggested that horses experience little or no pain after castration and questioned the routine administration of postoperative analgesics for all equine castrations.[2] This conclusion was based on the lack of behavioral signs of inguinal or abdominal pain (e.g. sweating, rolling, stretching) after castration. Rebuttal letters from equally experienced equine practitioners argued that horses do experience significant pain after castration but that pain is manifest in more subtle behavioral alterations.[3-5] The varied opinions expressed in these letters are not unique to British veterinarians. Surveys of equine veterinarians from the United Kingdom, the United States, Canada, and New Zealand have similarly reported wide ranges of estimates in the degree of pain experienced by horses after routine castration.[6-8]

Peer-reviewed information about pain management in foals appeared even later in the veterinary literature than similar information related to adult horses. A PubMed search of the terms "pain" and "foal" returns 48 references, only 2 of which were published prior to 1980. The few published studies that exist focus almost entirely on pharmacokinetics and adverse effects of specific drugs, especially NSAIDs, in neonatal foals. Arguably, the earliest references of significant clinical interest were manuscripts on phenylbutazone toxicosis[9] and gastroduodenal ulceration in foals[10] published in 1983 and 1986, respectively. Neither of the excellent recent editions of Veterinary Clinics of North America – Equine Practice focusing on pain management (published in 2002[11] and 2010[12]) included chapters on pain management in foals.

Neonatal medicine is one of the most intensively practiced aspect of equine internal medicine and much has been written about intensive care treatment for neonatal foals. The apparent discrepancy between the interest in equine neonatal critical care medicine practice and the nearly complete absence of peer-reviewed literature related to pain or analgesic use in neonatal foals suggests one of two possibilities: (1) equine internists are providing analgesic medications to foals without adequate evidence upon which to base their therapeutic decisions; or, (2) sick neonatal foals are not receiving analgesic medications. The author's personal experience suggests that both possibilities are real. This manuscript seeks to review the available evidence for analgesic therapies in foals with the goal of highlighting areas where additional information is needed.

RECOGNITION OF PAIN IN FOALS

This author was unable to identify any peer-reviewed literature addressing recognition or objective assessment of pain in neonatal foals. This literature is sparse for adult horses but appears to be nonexistent for foals. The two most widely read texts on equine neonatal medicine do not explicitly address assessment of pain in foals and have minimal discussion of analgesic regimens for these patients. The conclusion is that there is almost no evidence on which practitioners can base their assessment of degree of pain in equine neonatal patients.

The paucity of information related to assessment of pain in foals is not surprising if the history of human neonatal pain assessment is considered. Recognition of the clinical importance of pain in human neonates was delayed by several factors including a belief that infants do not feel pain or are less sensitive to it because of an immature central nervous system; belief that there are no long-term adverse outcomes related to infant pain; lack of accurate assessment methods; and lack of evidence for safety and efficacy of management approaches.[13,14] Clearly the same factors impact equine neonatal pain management. The belief that newborns lacked sufficient neurologic pain and therefore perinatal pain would have no lasting effect on behavior and development persisted until the 1980s. Prior to that time behavioral responses to potentially painful stimuli were often considered to be reflex responses and not indicative of pain.[15] As late as the 1970s surgery was performed in some infants with only neuromuscular blocking agents. In 1987, researchers demonstrated that the stressful response to surgery in preterm infants was significantly decreased if an analgesic was used and that infants receiving analgesics had fewer post-operative complications.[16] Only in 2006 was it demonstrated that nociceptive responses are associated with cerebrocortical responses even in preterm infants, indicating that true pain perception exists from an early age.[17] Today, it is generally accepted that events that are considered painful in adult humans should be considered painful for infants and appropriate analgesic treatment is indicated regardless of the age of the patient.[18]

Because pain is a subjective and very personal experience, it is very difficult to accurately assess pain even in a conscious adult human. This difficulty is only compounded when working with a nonverbal patient whether animal or human. Health care professionals for these patients must rely on behavioral and physiological indicators in their attempts to assess pain and distress. In human infants these types of responses are less vigorous and more variable in preterm infants than in full-term infants. The same is likely to be true for our equine patients. Multi-dimensional tools that combine behavioral and physiological information are the preferred tools. In human infants, physiologic indicators of pain that are included in such tools may include changes in heart rate, respiratory rate, blood pressure, oxygen saturation, vagal tone, palmar sweating, and plasma cortisol or catecholamine concentrations.[18] Behavioral indicators may include changes in facial expressions, body movements, and vocalizations but these may be absent in some neonates.[18] When pain is prolonged some neonates may become quite passive with few body movements, decreased heart rate, and decreased oxygen consumption.[19]

The central and peripheral nervous system anatomy and function of neonatal foals, to the best of our knowledge, is sufficiently mature at birth to accommodate nociception and perception of pain in the same ways it occurs in adults. In the absence of objective tools for the assessment of pain in neonatal foals, this aspect of patient evaluation becomes heavily dependent on the individual perceptions and experiences of the attending veterinarian. These interpretations appear to vary depending on the veterinarian's age, education, beliefs, personal bias, and professional experiences.[7,8,20]

At this time veterinarians have no evidence on which to base their interpretations of the degree of pain experienced by equine neonates. Anecdotally, most veterinarians would appear to recognize moderate to severe pain in the form of overt lameness or "colic" behavior; beyond these situations there is tremendous variability in the assessment of degree of pain. Assessment is especially difficult in recumbent, dysmature infants with multiple organ dysfunction, sepsis, hypoxic-ischemic encephalopathy, or other severe metabolic derangements. A simple and reliable tool for assessment of pain in neonatal foals is essential if veterinary medicine is to advance the quality of humane care for equine neonates.

FOAL PHYSIOLOGY AND PHARMACOLOGY

There are a number of physiologic and pharmacologic differences between foals and adult horses that may significantly impact the disposition and effect of analgesic drugs. Newborn foals have a relatively high heart rate and cardiac output as compared to adults. This impacts their ability to compensate for pharmacologically induced bradycardia resulting from administration of alpha-2 agonists.[21] Conversely, vascular resistance in foals is lower than that of adults resulting in a lower systemic blood pressure. This lower vasomotor tone is thought to persist until at least a month of age.[21] Because foals have a large surface area to body mass ratio, they have significant heat loss and a high metabolic rate. Their oxygen consumption per kg of body weight is two to three times greater than in adults.[21] Total body water content in foals is approximately 70-75% of body weight compared to 50-75% in adult horses, primarily due to a considerably larger extracellular fluid volume.[22,23] These differences in total body water balance result in a much higher apparent volume of distribution for many drugs given to foals as compared to the same drug administered to an adult horse and possibly higher concentrations of drug at the site of action. The relative ECF volume decreases progressively through the first few months of life until it reaches adult levels. Although foals have significantly less body fat than adult horses (2-3% and 5%, respectively), this difference is not thought to have a major impact on drug disposition for most drugs.[22] Neonates of most species have an immature and more permeable blood-brain barrier as compared to adults. This may explain the apparently more rapid and profound responses to sedative agents such as diazepam and xylazine that are observed in foals as compared to adult horses.[23] Other characteristics of neonatal animals that may impact drug disposition and effect include better absorption from the gastrointestinal tract, lower binding to plasma proteins, slower clearance of many drugs, and a decreased hepatic microsomal activity. Because of the immaturity of some drug metabolism pathways, lipid-soluble drugs often have longer half-lives in the neonate and may accumulate with repeated dosing unless the dosing interval is lengthened. Studies of the pharmacokinetics of chloramphenicol, a drug that undergoes hepatic metabolism, suggest that hepatic metabolic processes may be immature in the first 1-3 days of life but likely mature rapidly in the first weeks.[24,25] Full-term neonatal foals at 3 to 4 days of age appear to have a glomerular filtration rate and effective renal plasma flow that is similar to adult horses.[26] The many confirmed and likely physiologic differences between neonatal foals and adult horses probably account for the differences in pharmacokinetics and pharmacodynamics observed for many drugs requiring that foal dosage regimens be altered to avoid either subtherapeutic or toxic blood concentrations.[23,24,27-32]

ALPHA-2 ADRENERGIC RECEPTOR AGONISTS IN FOALS

Alpha-2 adrenergic receptor agonists such as xylazine and detomidine produce dose-dependent sedation, analgesia, and muscle relaxation with significant cardiovascular and respiratory depression. Although widely used in healthy foals, there are concerns associated with the use of these drugs in very sick foals because of the profound negative cardiovascular and respiratory effects. Severe hypotension and bradycardia may result. Foals with upper airway obstruction or respiratory disease may have very significant upper airway obstruction because of laryngeal and pharyngeal relaxation after administration of these drugs. These drugs should be administered with caution and at lower than adult doses on a mg/kg basis with some recommendations that xylazine administration not exceed 0.2 to 0.3 mg/kg IV.[21] Deaths have been reported in neonatal foals with doses of 0.55 mg/kg.[30] Contrary to reports in adult horses, xylazine administration to neonatal foals is reported to have no significant effect on blood glucose or plasma insulin concentrations.[32] In a comparison of xylazine and detomidine in young foals, authors concluded that detomidine was less desirable than xylazine because of its arrhythmogenic effects.[31]

NSAIDS IN FOALS

Phenylbutazone, flunixin meglumine, and ketoprofen are the NSAIDs that are most commonly used by equine veterinarians.[7] The pharmacokinetics and clinical effects of these drugs and ibuprofen have been studied in healthy neonatal foals.[9,28,29,33-37] For these NSAIDs, there tends to be a higher volume of distribution and prolonged clearance compared to adult horses, resulting in a longer elimination half-life for each drug. Dosing recommendations are that the initial dose may be at the high end of the range and that subsequent dosing intervals may be increased in order to achieve plasma concentrations similar to those of adult horses. Dose-related toxicities including gastrointestinal ulceration and renal papillary necrosis have been described, especially with phenylbutazone and to a lesser extent with flunixin meglumine. The availability of an oral and intravenous formulation of firocoxib, a selective COX-2 inhibitor in horses, may provide clinicians with an NSAID alternative that is less toxic to the gastrointestinal tract of foals.

OPIOIDS IN FOALS

In humans, age and gender can alter clinical efficacy of opioid analgesic drugs. Age is the most important factor affecting morphine requirements for analgesia in infants after surgery. In general, opioid analgesic potency tends to be greater in neonatal animals. These differences may be due to receptor density, signaling, and blood brain permeability differences. Importantly, morphine plasma concentrations are not clearly correlated with degree of analgesia in human neonatal patients. There are also clear differences in behavioral (sedative or stimulatory) responses to these drugs between adults and neonates.

Of the opioid analgesics available to veterinarians, butorphanol has been most studied and most used in neonatal foals. It is primarily a kappa agonist. Administration of butorphanol to neonatal foals at 0.05 mg/kg IV results in plasma concentrations that are below those predicted to be analgesic in adult horses.[27] Treated foals, however, exhibited profound sedation, decreased responsiveness to human interaction, and decreased physical activity as compared to untreated foals. This is in sharp contrast to the mild locomotory stimulation that is typically observed in adult horses treated with butorphanol. In addition, butorphanol administration stimulated suckling behavior in foals and resulted in mild decreases in respiratory rate, body temperature, and gastrointestinal sounds.[27]

Transdermal fentanyl has been recommended for use in neonatal foals. Transdermal patches are easy to apply to the medial muscles of the antebrachium in foals but result in variable plasma concentrations.[38] Anecdotally, transdermal fentanyl appears to provide analgesia in many foals but objective pharmacodynamic studies have not been performed.

OTHER ANALGESIC OPTIONS IN FOALS

There are a number of emerging analgesic drug options in foals for which little or no evidence is available beyond anecdotal reports of their safety and efficacy. Drugs that may prove beneficial with additional study include continuous rate infusions of lidocaine or ketamine, intramuscular ketamine, and oral gabapentin or tramadol. Much additional information is needed before these drugs can be safely recommended for use in neonatal foals.

ACKNOWLEDGEMENTS

The author acknowledges the substantial contributions of Marta Gonzalez Arguedas, DVM, MA, DACVIM to this work.

REFERENCES

1. Zierz J, et al. *Tierarztl Prax* 1996;24:108. 2. Green P. *Vet Rec* 2001;149:160. 3. Capner C. *Vet Rec* 2001;149:252. 4. Harris R. *Vet Rec* 2001;149:279. 5. Jones R. *Vet Rec* 2001;149:252. 6. Hewson CJ, et al. *Can Vet J* 2007;48:155. 7. Price J, et al. *Vet Rec* 2002;151:570-575. 8. Waran N, et al. *N Z Vet J*;58:274-280. 9. Traub JL, et al. *Am J Vet Res* 1983;44:1410. 10. Becht JL, et al. *Equine Vet J* 1986;18:307. 11. Mama KR, et al. *Veterinary Clinics of N Amer Equine Pract* 2002;18. 12. Muir WW, 3rd. *Veterinary Clinics of N Amer Equine Pract* 2010;26. 13. Anand KJ. *Arch Pediatr Adolesc Med* 2001;155:173. 14. McClain BC, et al. *Pediatrics* 2005;115:1073. 15. Lee BH. *J Am Vet Med Assoc* 2002;221:233. 16. Anand KJ, et al.. *Lancet* 1987;1:243. 17. Slater R, et al. *J Neurosci* 2006;26:3662. 18. Coleman MM, et al. *Adv Neonatal Care* 2002;2:123. 19. Batton DG, et al. *Pediatrics* 2006;118:2231. 20. Dohoo SE, et al. Postoperative use of analgesics in dogs and cats by Canadian veterinarians. *Can Vet J* 1996;37:546-551. 21. Dunlop CI. *Vet Clin North Am Equine Pract* 1994;10:67. 22. Caprile KA, et al. *Vet Clin North Am Equine Pract* 1987;3:123. 23. Baggot JD, et al. *Equine Vet J* 1984;16:364. 24. Adamson PJ, et al. *Am J Vet Res* 1991;52:426. 25. Norman WM, et al. *Am J Vet Res* 1997;58:878. 26. Brewer BD, et al. *J Vet Intern Med* 1990;4:301. 27. Arguedas MG, et al. *J Vet Intern Med* 2008;22:1416. 28. Breuhaus BA, et al. *J Vet Res* 1999;60:1066. 29. Carrick JB, Papich MG, Middleton DM. *Can J Vet Res* 1989;53:195. 30. Carter SW, et al. *Equine Vet J* 1990;22:384. 31. Oijala M, et al. *Equine Vet J* 1988;20:327. 32. Robertson SA, et al. *Equine Vet J* 1990;22:43. 33. Crisman MV, et al. *Am J Vet Res* 1996;57:1759. 34. Crisman MV, et al. *J Vet Pharmacol Ther* 1991;14:330. 35. Semrad SD, et al. *Am J Vet Res* 1993;54:2083. 36. Wilcke JR, et al. *Am J Vet Res* 1993;54:2064. 37. Wilcke JR, et al. *Am J Vet Res* 1998;59:290. 38. Eberspacher E, et al. *Vet Anaesth Analg* 2008;35:249.

NEONATAL FOAL DIARRHEA: PATHOPHYSIOLOGY, DIAGNOSIS, & TREATMENT

K. Gary Magdesian, DVM, DACVIM, DACVECC, DACVCP
Davis, CA

INTRODUCTION

Diarrhea is one of the most common causes of morbidity in foals, especially in the neonatal period. The spectrum of disease varies considerably, with signs ranging from mild diarrhea to those consistent with severe enterocolitis. In many cases, diarrhea may be part of the normal gastrointestinal (GI) maturational events. It is critical to differentiate this process from those with pathological etiologies, as any delay in recognition or treatment of enteritis can affect outcome. This lecture will review the pathophysiology of diarrhea in foals, as well as the specific etiologies, diagnostics, and treatment.

PATHOPHYSIOLOGY

Diarrhea may develop as a result of GI inflammation, dysmotility, hypersecretion, luminal hyperosmolarity, and hyperpermeability. It is likely that most or all of these mechanisms contribute to the diarrhea observed with enterocolitis. Risk factors for the development of diarrhea in foals have been examined. These include use of shavings as bedding, the prophylactic administration of antibiotics, and treatment with oral vitamins. Foals born to visiting rather than resident mares were at higher risk of developing diarrhea.[1] Protective factors include disinfection of foaling stalls, maternal udder and tail washing, and pasture foaling.[2]

Diarrhea in foals varies widely in terms of etiology and severity. These include foal heat diarrhea, and infectious and noninfectious causes. Foal heat diarrhea is incompletely understood. What is known is that it represents maturation of the GI luminal flora during the neonatal period. Previously thought to be due to hormonal changes in the dam during the first postpartum estrus, hence the name 'foal heat', it was later shown that orphan foals also develop this type of diarrhea. Changes that have been documented to occur with foal heat diarrhea include mild intestinal hypersecretion, with an increase in luminal electrolyte concentrations and pH. These are thought to coincide with the foal's initial intake of feed and coprophagia. Affected foals should be otherwise healthy, with self limiting diarrhea that resolves within a week.

INFECTIOUS CAUSES OF DIARRHEA

Bacterial Etiologies:

Salmonella

Salmonellosis is one of the most severe forms of enterocolitis that affect foals. Signs can vary from subclinical to highly fatal hemorrhagic diarrhea and a severe systemic inflammatory response syndrome. Colic, fever, and anorexia are common. Several serotypes of salmonellae can affect foals, but serotype Typhimurium is among the most virulent. Septicemia is a concern with salmonellosis, with localized infections such as septic arthritis occurring as a sequel. Systemic antimicrobials should be administered to prevent bacteremia and extra-intestinal infection. Foals of any age can be affected with salmonellosis. A complete blood count often reveals leukopenia with neutropenia, and a left shift.

Clostridium

Clostridial enteritides are similar to that of salmonellosis, in that foals can be subclinical carriers or severely affected. The most recognized clostridia associated with enterocolitis in foals include *Clostridium difficile* and *Clostridium perfringens*. Spores are encountered in the environment, and may be shed in horse or foal feces. Infection may occur in individual foals or as outbreaks. *Clostridium difficile* can affect horses of any age, but is a primary pathogen only in young foals. *C. perfringens* most commonly affects foals in the neonatal period. Both microbes produce toxins that are part of the pathogenesis of enterocolitis. Types A and C are the most common types of *C. perfringens* isolates affecting young foals. Type A isolates produce alpha toxin, whereas type C produce both alpha and beta toxins. Beta-2 toxins may also play a role in disease. Type C infections tend to be more severe and can cause hemorrhagic diarrhea. One report described a mortality rate of 54 % in foals with *C. perfringens*.[3] *Clostridium difficile* produces toxins A and B, which are important in the pathophysiology of disease and also aid in diagnosis of the disease. Foals with clostridiosis have hematologic changes consistent with systemic inflammation, including neutropenia, immature neutrophilia, hemoconcentration, and often signs of organ dysfunction on serum biochemistry analysis.

Clinical signs of clostridiosis include colic and abdominal distention. Death can precede development of diarrhea, so early recognition, even when presumptive, is key. Clostridiosis should be high on the differential diagnosis list for any neonatal foal with colic and fever or leukopenia, and specific treatment should be instituted until proven otherwise. Anorexia and watery to hemorrhagic diarrhea may accompany these signs.

Other Bacteria

Bacteria that have been associated with diarrhea, though less commonly than the above pathogens, include attaching/effacing *E. coli, Aeromonas hydrophila, Streptococcus durans, Campylobacter jejuni, Clostridium sordelli*, and *Bacteroides fragilis*.

Viral Diarrhea

Viral diarrheas are very common in neonatal foals. Rotavirus is considered the most common viral cause of enteritis in foals, and coronavirus is receiving renewed attention. Rotavirus is highly contagious, and infections usually occur as outbreaks on breeding farms. Most foals are 5-35 days old; however rotavirus has a short incubation period and can affect foals as young as 2 days and up to

6 months old. Hemorrhagic feces are not a common feature of rotaviral diarrhea, and are more likely due to a bacterial cause. Rotavirus causes significant blunting of the small intestinal villi, leading to secondary lactose intolerance. A commercial vaccine, used as a three dose series during gestation, is available for mares on endemic farms. Studies have demonstrated a variable reduction in morbidity, length of diarrhea, and degree of shedding of viral particles in foals born to vaccinated dams.

Coronavirus previously was thought to primarily affect immunocompromised foals. However, recent reports have described infections in immunocompetent foals, suggesting it is a primary pathogen. There is some preliminary evidence to suggest coronavirus may be involved with co-infections with other pathogens involved in foal diarrhea.[4] The role of adeno-, parvo-, breda, and other viruses in foal diarrhea remains to be studied.

Protozoal Diarrhea

Cryptosporidium parvum is increasingly recognized as a cause of diarrhea in individual foals as well as herd outbreaks. Affected foals are 4–21 days old, with a ≥ 3 day incubation period. As with coronavirus, it was previously thought to only affect immunocompromised foals. However, it is now known to cause disease in normal foals. *Cryptosporidium* has zoonotic potential, and affected foals should be handled with strict biosecurity protocols. Recovering foals can serve to infect other foals, highlighting the importance of isolation.

Giardia requires further study in foals, as concrete data associating fecal shedding with clinical disease are lacking. There have been isolated cases of foals with diarrhea and high fecal *Giardia* counts, and these foals have responded to metronidazole.

Parasitic Diarrhea

Strongyloides westeri (threadworm) infests foals through their dam's milk. Whereas light loads likely do not cause clinical disease, heavy infestations may cause mild diarrhea. There is an association between *S. westeri* and diarrhea when >2000 eggs per gram of feces are present. Prophylactic deworming of the dam 4 weeks prior to foaling is effective.

NONINFECTIOUS CAUSES OF DIARRHEA

Most severe cases of diarrhea are associated with infectious agents. However, occasionally noninfectious etiologies result in diarrhea of similar severity. Extensive pica is one of those examples, particularly when foals ingest sand or dirt. Dietary causes of diarrhea include overfeeding of orphan and premature foals. This also includes foals that are intolerant of enteral nutrition, such as those with ileus. Milk replacers can cause diarrhea if they contain excessive whey or lactose amounts, leading to osmotic diarrhea.

Lactose intolerance is another cause of diarrhea in foals and can represent a primary or secondary process. Secondary lactose intolerance is more common, and can occur after rotaviral or clostridial enteritis due to diffuse small intestinal damage, where lactase is synthesized. A recent report described suspected primary lactose intolerance in two foals; these foals responded well to administration of lactase enzyme.[6] One of these was an 8-day old Quarter Horse, while the second was a 6-day old Thoroughbred-Irish Draught cross.

DIAGNOSIS

There are many diagnostic tools currently available. It should be noted that the finding of a positive diagnostic test does not necessarily point to that agent as the only or even primary pathogen. Many of these agents can be shed subclinically. In addition, potential pathogens are often present as mixed infections[4], and some are opportunistic as secondary pathogens. Therefore, a thorough and comprehensive evaluation should be performed, especially during outbreaks when finding the causative agent is critical to prevention. In addition, the laboratory should be contacted prior to submission of fecal samples, so that storage and shipping instructions can be followed. Some of these agents are anaerobic or fragile, so it is important to properly handle the samples during shipping. Finally, one negative result should not rule out an infectious cause. Many of the following diagnostic tests lack 100% sensitivity, and therefore a repeat test may be indicated if clinical suspicion is high.

Diagnosis of salmonellosis rests with fecal culture or PCR, the latter of which has higher sensitivity, especially when performed on fecal samples that have undergone overnight enrichment in selective broth. Fecal culture should also include enrichment to enhance sensitivity.

Diagnosis of *C. difficile* is more complicated, because culture of the microbe alone does not necessarily make a diagnosis. Instead, diagnosis is currently regarded as contingent on identification of both the microorganism and its toxins. Identification of the microorganism can be made through culture or immunoassays that identify glutamate dehydrogenase (GDH) antigen of the microbe. Toxin identification is made through either stool cell culture cytotoxin assays, which have the highest sensitivity, or fecal immunoassays for toxins A and B. Immunoassays are the most readily available tests in most commercial labs, but unfortunately they have variable sensitivities. Therefore, if a negative result is obtained in the face of high clinical suspicion, the test should be repeated.

Diagnosis of *C. perfringens* is challenging, because the microbe frequently can be cultured from healthy foals. Therefore, identification of toxigenicity, while ruling out other causes, is key. This is best done by submitting fresh feces to a laboratory that runs fecal immunoassays for *C. perfringens* toxins, including alpha, beta, and other toxins. Fecal assays for *C. perfringens* enterotoxin (CPE) are more readily available, but lack sensitivity because only a minority of isolates produce CPE. The alternative is to submit cultured isolates for PCR analysis for toxin gene sequences. Gram stains of feces, though neither highly sensitive nor specific, can give a rapid indication of clostridial overgrowth if a predominance of Gram positive rods is present.

Rotavirus or coronavirus diarrhea can be diagnosed with fecal antigen tests (enzyme immunoassay), electron microscopy, or PCR. The use of PCR for these diseases is in its early stages and may be the most difficult to interpret if positive results are obtained. Fecal immunoassays have a rapid turnaround time and high accuracy.

Cryptosporidial diarrhea is diagnosed through identification of the agent in feces. While it shows up on routine fecal flotation analysis, this technique lacks sensitivity because the oocysts are very small and many labs lack expertise in this area. Therefore, more sensitive assays should be requested, including acid fast stains or immunofluorescent antibody (IFA) tests. *Giardia* can also be detected on fecal flotation, enzyme immunoassays, and IFA tests

.*Strongyloides westeri* is detected on routine fecal flotation. Quantitative analyses can be useful in trying to determine the significance of finding embryonated ova on fecal floats.

In addition to the above specific diagnostic tests, blood cultures should be obtained on any neonatal foal with enterocolitis. Approximately 50 % of foals with acute diarrhea are bacteremic.[7] In addition, plasma or blood IgG concentration should be measured to ensure adequate passive transfer of colostral antibodies. Complete blood counts, serum biochemistry profile, and blood gas analysis should be obtained. Foals with enterocolitis often exhibit leukopenia, immature neutrophilia (left shift), hemoconcentration, hypoproteinemia, metabolic acidemia, electrolyte disturbances (sodium, chloride, and potassium), hyperlactatemia, and signs of organ dysfunction. The latter frequently consists of azotemia and increased liver enzymes. Urinalyses are useful to monitor renal function.

Imaging is also useful in diagnosis and monitoring of foals with diarrhea. Abdominal ultrasonography is indicated for monitoring intestinal thickness, luminal content echogenicity, and qualitative assessment of peritoneal fluid volume. Abdominal radiographs allow for assessment of GI distention, as well as for the presence of sand or other foreign material.

TREATMENT

Treatment of enterocolitis and acute diarrhea in foals includes: 1. specific therapy for agents; 2. hemodynamic support, and 3.general therapy and supportive care.

There is no specific therapy for salmonellosis. Antibiotics do not alter the course of diarrhea, although they are important in preventing bacteremia. Antibiotics for preventing septicemia are listed below. Specific therapy for clostridiosis consists of metronidazole (10-15 mg/kg PO or IV, q 8-12 h with lower dose or longer interval for foals < 5 days of age). Viral enteritides are primarily treated with supportive care. *Cryptosporidium* is usually not treated with specific therapy, although there are new treatments being used in humans with cryptosporidiosis that potentially could be applied to foals. These include nitazoxanide (NZT), paromomycin, and azithromycin. Giardia is treated with fenbendazole or metronidazole. A new treatment for some causes of diarrhea, including rotavirus, cryptosporidiosis and clostridiosis, includes NZT. Though research for efficacy and safety is still required in foals, there is some evidence for their use in human medicine. A note of caution is that NZT itself can cause diarrhea.

Hemodynamic support is primarily accomplished through IV fluids. This can be accomplished through a number of methods; however the author prefers the fluid challenge method. This consists of a bolus of isotonic replacement fluid, such as PlasmaLyte 148 or A (Baxter), Normosol R (Abbott) or lactated Ringer's solution (LRS), at a dose of 10-20 mL/kg over 20-30 minutes, with subsequent reassessment. Most foals with moderate to severe diarrhea need 2-4 of these challenges, but the subsequent boluses can be given more slowly, over 30-60 minutes depending on the severity of hypovolemic shock. Clinical indicators of adequate fluid resuscitation include urine production (as per ultrasonographic evidence), warming of extremities, improvement in capillary refill time and mucous membrane color, and an increase in level of awareness. Measurement of arterial blood pressure and plasma lactate can aid in determining when to stop boluses. At this point, fluid therapy should be decreased to a twice maintenance rate, consisting of 8-10 ml/kg/h, with adjustments made based on clinical status. Foals that are in septic or endotoxic shock may require inotrope (dobutamine) or pressor (norepinephrine) therapy if they fail to respond to fluid therapy in terms of resolution of signs of shock. This will require hospitalization and intensive care. Acid-base and electrolyte derangements should also be addressed.

Systemic antimicrobials should be administered in foals with enterocolitis because of the risk of bacteremia. Broad spectrum antimicrobials such as a combination of ceftiofur (5 mg/kg IV, q 12 h) and amikacin (25 mg/kg IV, q 12h), or amikacin and ampicillin (22 mg/kg IV, q 6-8 h) should be used. Ceftiofur alone should be administered if the foal is azotemic. If amikacin is used, serial serum creatinine and urinalyses should be monitored, as should therapeutic drug levels. General supportive care of foals with diarrhea includes nutritional support, gastroprotectants, lactase enzyme supplementation, and anti-endotoxic therapies. For foals with abdominal distention, reflux or signs of colic, the author withholds nursing or any enteral feeding for 12-24 h in order to allow for gut rest. Nutritional support initially consists of dextrose supplementation (4 mg/kg/min) along with B vitamins. If anorexia or intolerance of enteral feeding persists beyond 24 h, parenteral nutrition is indicated. Omeprazole (4 mg/kg, PO q 24 h) or pantoprazole (1.5 mg/kg, IV diluted, q 24 h for foals with reflux) is indicated as foals with diarrhea are at high risk of gastric ulcer formation. Lactase enzyme (120 IU/kg q 3-8 h, PO) is helpful to aid in digestion of lactose and minimizing osmotic contribution to diarrhea. Oral di-trioctahedral smectite (Bio-sponge, Platinum Performance) binds clostridial toxins in vitro[8] and may adsorb endotoxins. Plasma transfusions are beneficial for foals with significant hypoproteinemia, and though concrete evidence is lacking, they may aid in binding endotoxins or clostridial toxins. Bismuth subsalicylate may also be administered. Nonsteroidal anti-inflammatory drugs such as flunixin meglumine should be used judiciously, due to potential adverse effects on the gut and kidneys.

Any foal with acute diarrhea should be handled with strict biosecurity protocols and housed in isolation facilities, due to the highly contagious nature of many of the associated pathogens. Many are also zoonotic. Disinfectants should be selected based on the individual pathogen, such as phenolic compounds for rotavirus and peroxygen compounds for *Clostridium* or *Cryptosporidium*.

REFERENCES

[1]Traub-Dargatz, JL, et. al. **J Am Vet Med Assoc** 1988;192:1553-6. [2]Cohen ND, et al. **J Am Vet Med Assoc** 1994;204:1644-51. [3]East LM et al, **J Am Vet Med Assoc** 1998;212:1751-6 [4]Slovis NM, et al. **Proc AAEP**, 2010;56;262-265. [5]Netherwood T, et al. **Epidemiol Infect** 1996;117;375-383. [6]Roberts, VLH, et al, **Equine Vet Educ** 2008;20;249-251. [7]Hollis AR, et al. J Vet Int Med 2008;22;1203-9. [8]Lawler JB, et al. **Am J Vet Res** 2008;69;233-9.

PERINATAL ASPHYXIA SYNDROME: A COMPARATIVE OVERVIEW OF EQUINE & HUMAN NEONATAL NEUROCRITICAL CARE MEDICINE

Samuel D. Hurcombe, BSc, BVMS, MS, DACVIM (LAIM), DACVECC
Columbus, OH

INTRODUCTION

Perinatal asphyxia syndrome (PAS) remains a poorly documented yet clinically relevant condition of the newborn foal. PAS is also known as neonatal encephalopathy (NE), hypoxic ischemic encephalopathy (HIE), neonatal maladjustment syndrome, neonatal maladaptation disorder, barker foals, and wanderer foals. There is no real consensus among perinatologists in naming this condition because the underlying pathophysiological basis for the condition is poorly understood.

In human newborn babies an equivalent condition associated with perinatal hypoxia and/or ischemia *in utero* or during parturition is referred as neonatal encephalopathy; this represents common causes of mortality or mental disability in infant populations. PAS and resulting HIE occur in 1-3 per 1000 births in the United States. In contrast, PAS is estimated to affect up to 2% of all foal births.[1]

Depending on the severity of injury, deficits in people are characterized as those related to generalized motor deficits, neurocognitive/behavioural disorders and neurodevelopmental disabilities.

It is evident from the clinical presentation of these foals, that in addition to the CNS, organs with high oxygen consumption / blood supply can be affected. Thus, encephalopathy, enteropathy, and nephropathy are part of this syndrome.

PATHOPHYSIOLOGY

There is limited evidence based information regarding the cause, risk factors, specific diagnostic approach, pathophysiology, and treatment of PAS in foals. Most clinicians know of PAS and its manifestation in the neonatal foal and have treated many successfully with supportive care as well as targeted neuroprotective strategies based on clinical and laboratory findings.

PAS has been associated with conditions affecting the mare, the foal or both. Pre-partum diseases in the mare include placentitis, premature placental separation, maternal illness (hypoxemia, hypotension, malnutrition), dystocia and cesarean section have been implicated to increase the risk of PASin foals.[1,2] Pulmonary insufficiency and reversal to fetal circulation have been found in foals with PAS where authors speculate that the degree of pulmonary pathology or shunting causes significant hypoxemia to cause neurologic disease.[3]

While several organs may be affected, CNS dysfunction is often pronounced, severe and due to adverse effects on energy metabolism in the brain. It is speculated that either *in utero* or during parturition, uterine blood flow leads to fetal sympathetic nervous system activation and redistribution of cardiac output to central organs. If the reduced oxygen delivery is sustained or severe enough, fetal cardiac output declines leading to cerebral ischemia and decreased oxygen availability in metabolically active tissues, particularly neuronal grey matter.[4]

In the absence of oxygen, the concentration of lactate rises sharply as glucose is metabolized anaerobically. Free radical production or reactive oxygen species (ROS) during ischemic/hypoxic episodes lead to cell membrane lipid peroxidation and disruption of cellular integrity. With dysoxia and limited aerobic capacity, production and utilization of ATP for ionic pumps (e.g. Na-K ATPase) is decreased, leading to ionic shifts, notably calcium through voltage-gated channels and membrane disruption. Intracellular accumulation of ROS, calcium in concert with excitatory neurotransmitter generation such as glutamate contributes to neuronal cell death. Intracellular calcium (ICa) activates calcium-dependent proteases, lipases and endonucleases, resulting in cell membrane damage and apoptosis. Reperfusion of affected tissues leads to a second wave of free ROS production, nitric oxide production, inflammatory mediators and the cycle of neuroexcitatory cellular destruction continues. ICa triggers glutamate release, which activates N-methyl-D-aspartate (NMDA) receptors. In turn, activated NMDA receptors allow additional sodium and calcium entry, leading to further neuroexcitation and cell death. Intracellular ion shifts and accumulation results in intracellular edema and neuronal dysfunction. This cascade of injury is thought to continue days beyond the initial ischemic episode.

DIAGNOSTICS

PAS foals have variable CNS disturbance with a varying time to onset of clinical signs. Foals believed to have sustained hypoxemia and/or ischemia *in utero*, may exhibit clinical signs from birth, while others who sustain hypoxemia and/or ischemia during parturition, may have a delayed onset of signs, accounting for the time for neuronal ATP depletion and altered cerebral energy metabolism. Delayed onset disease is most commonly seen a few hours to a few days following parturition.

Mild signs include poor suckle reflex, indiscriminant suckling on inanimate objects, difficulty finding the mare's teats, aimless wandering, disorientation and poor affinity for the dam. More severe signs can reflect severe neuroexcitatory activity including focal, generalized and refractory seizure activity occasionally with secondary associated cranial trauma. Many foals exhibit cranial nerve deficits, particularly optic and oculomotor (nystagmus, mydriasis) as well as brainstem deficits, i.e. respiratory center dysfunction such as apneustic breathing and hypoventilation. Many specific neurodiagnostic techniques in foals are targeted to rule in other diseases including infectious conditions (meningitis, brain abscess, encephalitis), congenital malformations, metabolic disturbances (glucose and electrolytes) and trauma. PAS is often a diagnosis of exclusion, provided history and clinical findings are also supportive.

Biomarkers: Recently plasma biomarkers of neuronal injury were evaluated in 33 foals with suspected PAS as well as control foals. Specifically, ubiquitin C-terminal hydrolase 1 (UCHL1) a biomarker believed to reflect increased blood brain barrier permeability and neuronal destruction, and phosphorylated axonal forms of neurofilament H (pNF-H), a biomarker believed to reflect axonal injury, were measured by ELISA. UCHL1 as a potential biomarker out-performed pNF-H and was increased in foals with

PAS.[5] In humans, additional biochemical markers investigated have included adrenomedullin, S100B, activin A, neuronal specific enolase (NSE), and glial fibrillary acid protein (GFAP).[6]

Bloodwork abnormalities are poorly specific for the diagnosis of PAS. Previous studies have evaluated neuronal biomarkers[5], endocrinologic variables including pregnenolone and pregnenediol[7] in affected foals with variable discriminatory success. Other findings are indicative of co-morbid disease such as sepsis, pulmonary insufficiency[3], and failure of transfer of passive immunity. *Cerebrospinal fluid (CSF)* is often normal in PAS foals and mostly used to assist in diagnosing meningitis; a common differential diagnosis.

Imaging: Magnetic resonance imaging (MRI) can be used to detect anatomical abnormalities such as hemorrhage and edema. The sequences, timing and type of imaging studies in affected foals have not been established. Advanced sequences such as diffusion-weighted imaging (DWI) and apparent diffusion coefficient (ACD) measurements have been associated with outcomes in humans.[8]

Necropsy: CNS lesions in foals with PAS are variable. Multifocal neuronal necrosis is a consistent finding, supporting cerebral ischemia in the pathogenesis. Gyri of the cerebrum appear flat, sulci are shallow and the cortical white matter-grey matter junction exhibits auto-fluorescence shown by an apple-green color (365 nm ultraviolet light) indicating necrosis.[9] The grey matter of the cerebral cortex, caudate nuclei, thalamus, hippocampus, cerebellar cortex, and medulla oblongata appear to be affected mostly with hemorrhage frequently observed.[5,10] Other lesions have included periventricular leukoencephalomalacia detectable by MRI.

TREATMENTS

Treatment for the neurological manifestation of PAS is somewhat controversial in terms of evidence based medicine and what truly works. The therapeutic window for pharmacologic intervention may be limited, at a time when brain damage is already occurring.[6] One difficulty in treating equine PAS is our limited pathophysiologic understanding of this condition, in particular in the CNS, and therefore appropriate therapeutic interventions remain somewhat empirical as prospective evaluation of targeted therapies is lacking. Key points in the treatment of the neurological presentation include: enhance blood/O_2 supply, limit additional cell injury, reduce cerebral edema and control seizure activity. Again, treatment is largely symptomatic (seizure control, fluid support, and infection control) as well as theoretical. Many practitioners use hyperosmolar agents for to treat cerebral edema, which clinically sometimes work, however, if we believe that equine PAS is similar to human PAS, edema is largely intracellular and considered to be poorly responsive to agents such as mannitol and hypertonic saline.

Seizure activity must be treated as prolonged excitatory activity increases cerebral oxygen consumption. The use of benzodiazepines and barbiturates as drugs that potentiate the effect of inhibitory gamma-aminobutyric acid (GABA), and enhanced neuronal hyperpolarizing chloride conductance are frequently employed.

NMDA antagonists (magnesium and ketamine) are in vogue, and in human neonatal critical care, clinical trials assessing the utility of $MgSO_4$ have proven inconclusive. Some studies evaluating $MgSO_4$ show a positive benefit in cognitive and motor function in mild to moderately affected children, and no effect in severely affected patients.[11] A similar lack of improvement has been seen in acute head trauma patients, presumably due to an intact blood brain barrier at the time of $MgSO_4$ administration.

An important concept in treatment is to maintain cerebral perfusion. Crystalloid and colloid administration is an important adjunct in treating affected foals being cautious to avoid hyperglycemia. Hypercapnia is to be avoided as increased CO_2 leads to vasodilation in unaffected tissues causing a preferential shunting to these regions while depriving affected areas and worsening the ischemic insult. Foals are treated with respiratory stimulants such as caffeine (phosphodiesterase) and doxapram where recent studies indicate doxapram is effective to treat hypoventilation (hypercapnia) in PAS foals.[12] Other treatments used in infants and foals have included naloxone, thiamine, antioxidants, erythropoietin, allopurinol, non-steroidal antiinflammatories and glucocorticoids.

Non-pharmacological strategies employed by human criticalists' for the treatment of affected babies include hypothermic neuroprotective strategies to reduce cerebral metabolic rate and tissue damage.[13]

OUTCOMES

The prognosis for PAS in foals is generally considered good provided CNS deficits are mild, seizures are amenable to pharmacologic control and foals are metabolically and nutritionally supported. Reported survival rates for PAS foals vary between 72%-80% after hospitalization and supportive care.[5,14] Most surviving foals become athletes with similar performance outcomes when compared to age-matched unaffected horse.[15,16] In previous studies, poor prognostic factors associated with mortality include having severe deficits at the time of birth, concurrent sepsis, failure to improve neurologically in the first 5 days after birth and a persistent comatose state.[17]

REFERENCES

[1] Bernard WV, Reimer JM, Cudd T. Historical factors, clinicopathologic findings, clinical features, and outcome of equine neonates presenting with or developing signs of central nervous system disease. **Proc Am Assoc Equine Pract** 1995;41:222-224 [2] Furr M. Perinatal asphyxia in foals. **Compend Cont Educ Pract Vet** 1996;18:1342-1351 [3] Anonymous. The role of pulmonary insufficiency in maladjusted foals: 6 case examples. **Equine Vet J Suppl** 2005;35:62-72 [4] Rudolph AM. The fetal circulation and its responses to stress. **J Dev Physiol** 1984;46:276-284 [5] Ringger NC, Giguere S, Morresey PR et al. Biomarkers of brain injury in foals with hypoxic-ischemic encephalopathy. **J Vet Intern Med** 2010; [6] Florio P, Abella R, Marioni E, et al. Biochemical markers of perinatal brain damage. **Front Biosci** 2010;2:47-72 [7] Houghton E, Holtan D, Grainger et al. Plasma progestagen concentrations in the normal and dysmature newborn foal. **J Reprod Fertil Suppl.** 1991;44:609-617 [8] Liauw L, van Wezel-Meijler G, Veen S, et al. Do apparent diffusion coefficient measurements predict outcome in children with neonatal hypoxic-ischemic encephalopathy?. **Am J Neuroradiol** 2009;30:264-270 [9] Wilcox AL, Calise DV, Chapman SE et al. Hypoxic/ischemic encephalopathy associated with placental insufficiency in a cloned foal. **Vet Pathol** 2009;46:75-79 [10] Palmer AC and Rossdale PD. Neuropathological changes associated with the neonatal maladjustment syndrome in the thoroughbred foal. **Res Vet Sci** 1976;20:267-275 [11] Vink R, Cook NL, van den Heuvel C. Magnesium in acute and chronic brain injury: an update. **Magnes Res** 2009;22:158-162 [12] Giguere S, Slade JK, Sanchez LC. Retrospective comparison of caffeine and doxapram for the treatment of hypercapnia in foals with hypoxic-ischemic encephalopathy. **J Vet Intern Med** 2008;22:401-405 [13] Shankaran S. Neonatal encephalopathy: treatment with hypothermia. **J Neurotrauma** 2009;26:437-443 [14] Katz L. Perinatal asphyxia syndrome in a Quarter horse foal. **Vet Clin North Am Equine Pract** 2006;22:193-208 [15] Axon J, Palmer J, Wilkins PA. Short-term and long-term athletic outcome of neonatal intensive care unit survivors. **Proc Am Assoc Equine Pract** 1999;45:224-225 [16] Bryant JE,

Bernard W, Wilson WD et al. Race earnings as an indicator of future performance in neonatal foals treated for neurologic disorders. **Proc Am Assoc Equine Pract** 1994;40:197-198 [17] Clement SF, Behavioral alterations and neonatal maladjustment syndrome in the foal. **Proc Am Assoc Eq Pract** 1985;31:145-148

CAUSES AND MANAGEMENT OF AZOTEMIA IN FOALS

Harold C. Schott II, DVM, PhD, DACVIM
East Lansing, MI

INTRODUCTION

Azotemia in neonatal foals ($\leq$7 days of age) may be an indicator of prerenal failure, acute kidney injury (AKI), obstructive disease, congenital renal disorders, or disruption of the collecting system leading to uroperitoneum. Because clinical signs of renal failure may be similar to those with septicemia or asphyxia (weakness, recumbency, or poor nursing vigor), performing a serum biochemical analysis in compromised neonates is an essential part of the minimum database to detect azotemia. Spurious hypercreatininemia may also be a transient finding in asphyxiated foals or foals delivered from mares with placentitis.[1] Neonates with spurious hypercreatininemia have normal serum electrolyte concentrations, supporting normal renal function, and creatinine concentration, which may exceed 20 mg/dL, typically drops by 50% or more over the first 1-2 days of life, as long as foals are well hydrated and nursing well. In older foals, azotemia and renal failure are almost always a consequence of another primary disease process although limited renal function with congenital anomalies may delay onset of renal failure for until foals are several years old.

NEONATAL RENAL DEVELOPMENT AND PHYSIOLOGY

Nephrogenesis, in terms of nephron numbers, is essentially complete at birth and as foals grow the kidneys increase in size through the first 1-2 years of life.[2] Foals are born with about 10 million glomeruli in each kidney and these enlarge in size as the kidneys grow. Of interest, low birth weight and premature infants may be born with fewer nephrons and there is mounting evidence that this reduced renal endowment may increase the risk of developing hypertension and chronic kidney disease (CKD) in later life.[3]

Normal term foals typically first urinate between 5 and 10 hours of age, with fillies tending to urinate 1-2 hours later than colts.[4] Initial urine may be dilute or concentrated (specific gravity up to 1.040) but hyposthenuria develops by 24 hours of age in all foals that are nursing well. Urine pH is nearly neutral and significant proteinuria (2-3+ on reagent strips) is commonly observed from 24-48 hours of age in foals that received good passive transfer of colostral antibodies.[5] Serum urea nitrogen (BUN) and creatinine (Cr) concentrations are variable during the initial 24 hours of life with values in the range of 15-30 mg/dL and 2-4 mg/dL, respectively. BUN generally drops below the lower limit of the adult reference range by 24 hours of age while Cr may not drop below 2 mg/dL until 48 h of age. By 1-2 weeks of age BUN is typically below 10 mg/dL and Cr may fall below 1 mg/dL. These low values can be attributed to the large volume of fluid intake as milk (five-fold greater fluid intake as compared to adult horses on a ml/kg basis) and associated diuresis that may persist for the first 3-4 months of life, as long as foals continue nursing their dams.

Renal function of neonatal foals, as assessed by glomerular filtration rate (GFR), is similar to that of adult horses.[6,7] Further, kidneys of foals are capable of excreting a sodium load, provided fluid administration,[8] and fractional electrolyte clearance values are similar to those of adult horses, supporting that renal tubular function in foals is similar to that of adult horses.[5]

PRERENAL FAILURE

Renal hypoperfusion leading to prerenal azotemia is common in compromised neonates. Several compensatory mechanisms accompany decreased renal perfusion to maintain renal blood flow (RBF) and GFR over a range of perfusion pressures (autoregulation). These include a decrease in afferent arteriole tone and intrarenal production of vasodilatory prostanoids (PGE2 and PGI2) to maintain blood flow, particularly in the medulla. Renal hypoperfusion increases the risk of nonsteroidal anti-inflammatory drug (NSAID) associated acute renal injury (AKI); for example, nearly 40% of premature infants that receive indomethacin to speed closure of the ductus arteriosus develop alterations in renal function.[9] When the more selective COX-2 NSAIDs were initially introduced, there was hope that they would be more sparing of renal function than non-selective COX inhibitors. However, it is now recognized that these newer NSAIDs provide little benefit of limiting renal damage; thus, all NSAIDs should be used sparingly in compromised foals.

Renal tubular function remains essentially normal with prerenal failure. Thus, serum electrolyte concentrations are typically normal and urine sodium concentration is <20 mEq/L. Depending on hydration status, urine specific gravity may be variable but the classic finding that supports prerenal azotemia is a specific gravity $\geq$1.025 and a urine osmolality (U_{osm}) exceeding 500 mOsm/kg.

INTRINSIC RENAL FAILURE

Intrinsic renal disease and acute renal failure (ARF) can develop as a consequence of hypoxic-ischemic injury associated with birth asphyxia, septicemia (*Actinobacillus equuli* appears to have a predilection for neonatal kidneys) or leptospirosis, or treatment with nephrotoxic medications (aminoglycoside antibiotics and NSAIDs).[9] It is often difficult to separate the effects of these factors because many compromised foals are routinely treated with nephrotoxic medications. Over the past decade there has been a shift away from routine use of penicillin and an aminoglycoside for broad spectrum antimicrobial therapy of neonatal foals (largely replaced by ceftiofur) and this change has likely decreased the incidence of aminoglycoside nephrotoxicity in foals. Further, the shift away from multiple daily doses to once daily dosing regimens for aminoglycoside antibiotics has been well documented to decrease the risk of nephrotoxicity in infants.[10] Unfortunately, during the same time period there has been increased use of oxytetracycline in foals with contracted tendons and tetracycline-associated ARF may go undetected in compromised neonates unless a serum chemistry profile is submitted as part of case management. The incidence of ARF in foals in neonatal intensive care units is not well documented but human neonatal intensive care units have reported incidences ranging from 6-24% with cardiac surgery and severe asphyxia being recognized risk factors. Of interest, ARF was nonoliguric, oliguric, and anuric in 60, 25, and 15% of patients, respectively.[9]

Although there are no specific values for BUN and Cr that separate prerenal failure from intrinsic renal failure, azotemia is generally greater with intrinsic renal failure. Further, there is not a sudden transition from prerenal failure to intrinsic renal failure, rather these are overlapping conditions. To emphasize the point that some intrinsic renal damage likely occurs with simple renal hypoperfusion, the term acute kidney injury, characterized by a 50% increase in Cr, has been introduced to describe early damage with intrinsic renal disease. This predominantly tubular damage may be characterized by proteinuria, microscopic hematuria, glucosuria, increased urine sodium and chloride concentrations and excretion, increased urinary enzyme activity, and detection of novel biomarkers of tubular epithelial damage in urine (kidney injury molecule 1 and others). Examination of urine sediment may also reveal hyaline or granular casts. These urinary changes can develop a few days before onset of significant azotemia.

With both prerenal and renal azotemia, acute changes are better assessed by the comparatively greater increase in Cr than BUN, due to the larger size and polarity of Cr that limits diffusion of Cr out of the extracellular fluid (ECF) space. Thus, following changes in Cr over time is more reliable for assessment of renal function than monitoring BUN. However, it warrants emphasis that increases in Cr are insensitive for detection of early changes in renal function (GFR must decrease by 75% or more before azotemia develops). An occasional foal may have a more marked increase in BUN than Cr and in such cases upper intestinal ulceration with blood loss into the intestine should be considered as a potential cause for the increase in BUN. Next, another simple question to ask is whether or not acid-base status and serum electrolyte concentrations are normal. These parameters typically remain normal with prerenal azotemia while alterations are common with intrinsic renal failure (hyponatremia, hypochloremia, hyperkalemia, and metabolic acidosis). A final hallmark of tubular damage associated with AKI and intrinsic renal failure is loss of concentrating ability. Urine of affected foals is typically hyposthenuric (specific gravity$\leq$1.008 and $U_{osm}\leq$300 mOsm/kg).

With AKI and intrinsic renal failure, it is important to determine whether or not affected neonates are oliguric (urine output <0.5-1 ml/kg/h) because oliguric AKI carries a more guarded prognosis than nonoliguric AKI. However, most cases in neonates are nonoliguric and these may only be detected by finding a progressive increase in Cr with serial serum biochemical analyses. Fortunately, neonatal kidneys are forgiving and treatment can sometimes be successful with anuria/oliguria lasting up to 72 hours. Dialysis, either peritoneal or hemodialysis, is also feasible in foals, as compared to full sized adult horses, when anuria/oliguria persists.[11-13] However, recent reports of peritoneal dialysis as a treatment for oliguric renal failure in horses are not without controversy as careful review of the reports reveals that urine output was not well documented[12] and that Cr had already begun to decline before dialysis was initiated.[13] The main challenge of managing patients with anuric/oliguric renal failure is not to control Cr, rather it is to maintain acid-base balance and serum electrolyte concentrations within reasonable ranges while renal tissue is repaired and urine production returns. Of interest, recent evidence in management of ARF in people provides further support for cautious intravenous fluid support of compromised foals (so-called "running them dry") as excessive fluid replacement may increase intrarenal pressure and further compromise restoration of renal function.[14]

POSTRENAL FAILURE

Azotemia arising from disruption of the urinary tract typically occurs in the face of nearly normal kidney function but failure to eliminate urine leads to progressive abdominal distension and electrolyte alterations. Hyponatremia, hypochloremia, and hyperkalemia develop with "mixing" or "dialysis" of the expanded peritoneal fluid compartment with the remainder of the ECF compartment, including plasma.[15] Again, because Cr does not diffuse out of the ECF compartment as rapidly as urea, finding a peritoneal fluid Cr to serum Cr ratio $\geq$2 is the clinicopathologic test of choice to confirm uroperitoneum. However, this test is being performed less than in the past because detection of a large volume of hypoechic fluid within the abdomen via transabdominal ultrasonography is essentially a pathognomonic finding in foals with electrolyte alterations expected with uroabdomen.

The classic case of uroperitoneum occurs in 2-3 day old colts that develop abdominal distension consequent to a dorsal bladder wall tear sustained as they pass through the pelvic canal with a full bladder.[15] However, silent uroabdomen can also develop in compromised neonates in intensive care units, especially when they are unable to stand and void urine normally. Uroabdomen in these foals may develop as a consequence of bladder distension and leakage/rupture or can be due to a small leak in association with sepsis of the urachus.[16,17] Uroabdomen in these foals may not become a problem until 4-7 days of age and is often first detected as an increase in free abdominal fluid during routine transabdominal ultrasonographic monitoring of sick neonates to assess intestinal motility and bladder size. At initial suspicion of uroabdomen, enough time mat not have elapsed for electrolyte alterations to develop but uroperitoneum can be confirmed by comparing peritoneal fluid Cr to serum Cr. Finally, another cause of "late onset" uroperitoneum can be a ruptured ureter. Most ureteral leaks occur near the kidney and may be associated with blunt abdominal trauma during parturition or after birth. The author has seen several cases of ureteral leakage in foals with multiple rib fractures supporting birth trauma and possible delivery with the mare in a standing position. Urine initially accumulates more slowly in the retroperitoneal space until it ruptures into the peritoneal cavity. Further, with unilateral ureteral rupture, the bladder continues to fill from the contralateral side and normal urination may still be observed.

With uroabdomen, hyperkalemia ($\geq$7 mEq/L) may cause life-threatening cardiac arrhythmias and requires immediate intervention by administration of potassium-free intravenous fluids supplemented with calcium and dextrose, and possibly administration of β_2-agonist medications.[18] It is also necessary to drain urine that has accumulated in the abdomen and this may be the most important step in overall management of hyperkalemia in neonatal foals. Once the patient has been stabilized and serum potassium concentration is $\leq$5.5 mEq/L, surgical correction of urinary tract disruption is generally the treatment of choice, although small bladder rents may heal with placement of an indwelling bladder catheter alone.

CONGENITAL DISORDERS

Anomalies of renal development reported in foals include agenesis/aplasia, hypoplasia, dysplasia, and polycystic kidney disease (PCKD).[19] Although present at birth, anomalies may not lead to azotemia and renal failure until later in life, especially renal hypoplasia or PCKD. For example, the author diagnosed PCKD as the cause of CKD in a 19-year-old Arabian mare. Anomalies of the collecting system such as ureteral ectopia do not typically cause renal failure and azotemia, unless complicated by obstruction or ascending urinary tract infection. Megavesica, or a markedly enlarged bladder, has been described in several foals in association with suspected urachal obstruction and can cause postrenal failure and abdominal distension (due to the large bladder) without actual disruption of the urinary tract.

EVALUATION AND MANAGEMENT OF AZOTEMIA

When azotemia is detected in a neonatal foal (≤7 days of age), repeating the physical examination may detect subtle findings that may not have been appreciated during the initial exam. For example, mild edema or "looseness" of the skin in the axillary and inguinal regions may be appreciated with oliguria and foals with hyperkalemia may have fine muscle tremors that may be easier to feel than to see. Evidence of incontinence may support ectopic ureter and moistness of the umbilicus may support localized sepsis that may lead to subcutaneous urine accumulation and/or uroperitoneum. In addition to physical examination and a clinicopathologic database, abdominal ultrasonography should be a standard part of the work-up of neonatal foals. With regard to the urinary system, the first question to answer is whether or not the foal actually has two kidneys that are of normal shape and size (about 5 x 10 cm for a 50 kg foal). Foals that have nursed well or that have been started on intravenous fluid support and are producing adequate urine often have mild distension (to 1-1.5 cm) of the renal pelvis, but this should be a symmetrical finding in both kidneys. Next, the bladder should be imaged to determine whether it is small (supporting oliguria) or possibly larger than normal (supporting intrauterine urachal obstruction or lack of detrusor tone).

If the bladder is larger than 10 cm in diameter and urination is not observed within 30-60 minutes after this finding, a catheter should be passed to collect a urine sample and to empty the bladder. Indwelling bladder catheters with closed collection systems are being used more frequently in recumbent neonates and are certainly helpful in documenting rate of urine production. However, even well managed catheters are commonly colonized with multidrug resistance bacterial spp.; thus, short-term use (≤3 days) is the goal and prophylactic antimicrobial treatment is warranted. With uroabdomen, drainage of accumulated urine is indicated but can be challenging. The author prefers using a small trocar chest tube (16 French), placed abaxial to the linea alba, over a teat cannula or bitch catheter in order to allow more rapid removal of urine from the abdomen.

One of the hallmark treatments of azotemia is supportive fluid therapy. However, the approach should error on the side of caution. For example, with prerenal azotemia and spurious hypercreatininemia, fluid support can often be limited to enteral milk feeding or close observation to ensure adequate nursing. With milk alone provided at 10% or more of body weight daily, foals will be receiving a fluid intake at a rate more than adequate for maintaining hydration. Although adequate nutritional support with enteral milk feeding can be accomplished in all neonates that tolerate milk feedings, it is important to remember that foals will produce dilute urine (specific gravity ≤1.005) when fed either 10 or 20% of body weight daily as milk. However, they will produce a greater amount of dilute urine when they are fed more. Thus, assessment of body weight gain daily is a superior method of assessing adequacy of either nursing or enteral feeding, as compared to only measuring urine specific gravity. When enteral feeding is not tolerated well and/or when acid-base balance or serum electrolyte concentrations are altered, judicious intravenous fluid therapy support may also be required. Potassium free intravenous fluid (0.9% saline solution) should be used when hyperkalemia (≥5.5 mEq/L) is present while a polyionic replacement fluid can be used when potassium concentration is not elevated. The goal should be to limit fluid replacement to maintenance to 1.5-times maintenance requirements and patients on intravenous fluids should be monitored twice daily for development of edema and body weight change. Ideally, urine production will also be monitored with an indwelling collection system and central venous pressure can be monitored via the jugular venous catheter placed for fluid administration (as the catheter end is often near or in the right atrium). Ideally, CVP should be maintained below 10 cm H_2O. Further, the enteral route, when it is working, can also be used for administration of sodium chloride, sodium bicarbonate, potassium chloride, or potassium bicarbonate. Finally, intravenous fluid therapy does not need to be continued until azotemia has fully resolved. Foals that are nursing adequately or tolerating enteral feeding well will have a fluid intake that may be 3-5-times maintenance requirements. The resultant diuresis will be adequate for resolution of azotemia, as long as damaged renal tissue undergoes repair. What is difficult (impossible) to know at the start of therapy is what the final Cr will be after repair of damaged renal tissue. Many cases will have Cr return to values within the reference range in 5-10 days while others will be discharged from care with persistent elevations in Cr. In the latter cases, Cr should be monitored at monthly intervals and may still return to normal values over 1-3 months.

In foals with oliguric ARF, judicious fluid therapy is even more critical to prevent volume overload. Although efficacy remains unproven, furosemide (1-3 mg/kg, q 2h IV, or a continuous rate infusion) may be used in an attempt to increase urine flow rate. If no effect is observed, mannitol (1 g/kg IV) has also been used as an osmotic diuretic but, again, supportive data is lacking in foals. Finally, the current consensus is that use of a dopamine infusion in an attempt to increase RBF is no longer recommended as potential development of arrhythmias outweighs any potential benefits of this treatment. Finally, peritoneal dialysis can be considered with refractory oliguria to partly correct electrolyte alterations and azotemia but a more guarded prognosis should be issued for owner consideration before this more invasive treatment is pursued.

REFERENCES

1. Chaney KP, et al. **J Vet Emerg Crit Care** 2010;20:244. 2. Beech DJ, et al. **Equine Vet J** 2001;33:438. 3. Puddu M, et al. **Am J Nephrol** 2009;30:162. 4. Knottenbelt DC. **Equine Neonatology Medicine and Surgery** 2004 Saunders, p. 6. 5. Edwards **Aust Vet J** 1990;67:251. 6. Brewer BD, et al. **J Vet Int Med** 1990;4:301. 7. Holdstock NB, et al. **Equine Vet J** 1998;30:335. 8. Buchanan BR, et al. **J Am Vet Med Assoc** 2005;227:1123. 9. Andreoli SP. **Pediatr Nephrol** 2009;24:253. 10. Contopoulos-Ioannidis DG, et al. **Pediactrics** 2004;114:111. 11. Vivrette S, et al. **J Am Vet Med Assoc** 1983;203:105. 12. Gallatin LL, et al. **J Am Vet Med Assoc** 2005;226:756. 13. Han JH, et al. **Equine Vet Educ** 2008;20:256. 14. Prowle JR, et al. **Nature Rev: Nephrology** 2010 6:107. 15. Richardson DW, et al. **J Am Vet Med Assoc** 1983;182:267. 16. Kablack KA, et al. **Equine Vet J** 2000;32:505. 17. Dunkel B, et al. **J Vet Intern Med** 2005;19:889. 18. Weisberg LS. **Crit Care Med** 2008;36:3246. 19. Chaney KP. **Vet Clin North Amer:Equine Pract** 2007;23:691.

DIAGNOSTIC TESTING FOR FELINE HYPER- AND HYPOTHYROIDISM

Mark E. Peterson, DVM, DACVIM
New York, NY

INTRODUCTION

The diagnosis of hyperthyroidism and hypothyroidism in cats is primarily based on a constellation of typical clinical features, including the cat's signalment, history, clinical signs, and physical examination findings (eg, palpation of thyroid nodule). Since many nonthyroidal diseases can mimic the signs of thyroid disease in cats, a complete database (e.g., CBC, serum chemistry profile, urinalysis) must always be evaluated to help exclude other illness. After reviewing this data, the next step is to then use thyroid function tests to confirm the diagnosis of thyroid dysfunction.

DIAGNOSING CATS WITH HYPERTHYROIDISM

The diagnosis of hyperthyroidism is usually straightforward, as 90% of hyperthyroid cats will have a serum total T_4 concentration that is clearly high.[1-3] Serum total T_3 is not recommended as a screening test for hyperthyroidism because 25-30% of hyperthyroid cats have values that remain within normal, reference range limits (ie, not diagnostic for hyperthyroidism).[2-4]

Practitioners should be aware of the assay techniques being used by their commercial or in-house laboratories. Radioimmunoassay (RIA) has long been considered to be the gold standard technique, but the regulations regarding radioactivity has resulted in a search for alternative methods.[5-8] Chemiluminescent enzyme immunoassays (Immulite®) utilize the same type of antibody testing as RIA; but instead of measuring a radioactive isotope bound to the protein in question, they use a photomultiplier tube that counts light emissions have also been validated for feline serum.[5,6] Many commercial laboratories prefer this technique as it is more automated. An enzymatic chemistry method (DRI® thyroxine assay, Microgenics Corporation) is also now being used by many laboratories. This technique has the advantage for the laboratory of being fully automated, which provides the advantage for the practitioner and patient that results can be available sooner.[7] Finally, in-house enzyme-linked immunosorbent assay (ELISA) test kits are also available.[8] It is important to realize that all of these assays can provide serum T4 values that are falsely high or falsely low; no assay will provide perfect test sensitivity and specificity.

Difficulties arise in the routine diagnosis of hyperthyroidism in three general scenarios: 1) when a cat is hyperthyroid but the serum T_4 concentration is within normal, reference range limits; 2) when a cat has a palpably enlarged thyroid nodule (s) but does not have other clinical or biochemical evidence of hyperthyroidism; and finally, 3) when a euthyroid cat is screened and is misdiagnosed as hyperthyroid based on a falsely-high serum T_4 (or free T_4) concentration.

Hyperthyroid Cat but Normal Serum T_4 Concentration

Although serum total T_4 is preferable as a screening test for hyperthyroidism, approximately 10% of all hyperthyroid cats (and 40% of cats with early or mild hyperthyroidism) have serum T_4 within the reference range limits.[2,3] Such T_4 values are usually within the mid- to high-end of the reference range. Thus, finding a single reference range T_4 value does not preclude such a diagnosis. In general, there are 2 possible reasons why a hyperthyroid cat could have a normal T_4 value. (1) In early or mildly affected cats, serum total T_4 concentrations can fluctuate in and out of the reference range.[9] Such fluctuation of thyroid hormones occurs in all hyperthyroid cats but the degree of fluctuation is of little diagnostic significance in cats with markedly elevated T_4 concentrations. In early or mildly hyperthyroid cats (no concurrent illnesses), serum total T_4 concentrations will eventually increase into the diagnostic thyrotoxic range upon retesting a few weeks later. (2) Severe non-thyroidal illness is also capable of suppressing serum total T_4 concentrations to below the reference range in euthyroid cats.[10,11] Similarly, marginally elevated serum total T_4 concentrations may be suppressed to the mid- to high-end of the reference range in cats with concurrent mild hyperthyroidism and concurrent moderate to severe non-thyroidal disease.

If hyperthyroidism is suspected based upon history, clinical signs (weight loss despite a good appetite), and examination findings (tachycardia, palpable thyroid nodule) but the serum T_4 remains with the upper half of the normal range, hyperthyroidism is still possible. There is no definitive approach to diagnose hyperthyroidism in this scenario but we have several options:

(1) **Repeat total T_4:** In cats in which overt, manageable underlying disease is identified, such concurrent disease should be managed first, before proceeding with further thyroid testing. Once concurrent disease is resolved, most hyperthyroid cats will develop a clearly high T_4, confirming the diagnosis. On the other hand, in cats without overt underlying disease, simply repeating the serum T_4 concentration in 2 weeks may be diagnostic if the T_4 is fluctuating in and out of the reference range.[9]

(2) **Scintigraphy:** Thyroid scintigraphy is a nuclear medicine procedure that produces a visual display of functional thyroid tissue based on the selective uptake of various radionuclides by thyroid tissue.[12] Thyroid scintigraphy is able to identify thyroid disease and define the degree of that thyroid disease relatively unaffected by the presence of concurrent nonthyroidal illness. Because thyroid scintigraphy directly visualizes functional thyroid tissue, thyroid imaging can diagnose hyperthyroidism before laboratory tests are abnormal. Thyroid scanning can also prevent misdiagnosis of hyperthyroidism in cats with falsely high serum T_4 values.

With thyroid imaging (scintigraphy), hyperthyroid cats usually exhibit increased thyroidal uptake of radioisotope administered, either radioactive iodine (^{123}I or ^{131}I) or technetium-99M as pertechnetate ($^{99m}TcO_4^-$). Percentage uptake or increased thyroid:salivary ratio may be calculated and both are strongly correlated with circulating T_4 concentration and provide a sensitive means of diagnosing hyperthyroidism.[12,13] However, apart from expense and the difficulties in dealing with radioisotopes, few veterinarians have access to the equipment needed to obtain thyroid images or perform thyroid uptake determinations.

3) **Free T_4 concentrations**. In cats with mild hyperthyroidism and normal T_4 values, free T_4 concentrations can aid in diagnosis. In support of that, serum free T_4 concentrations measured by equilibrium dialysis were more consistently elevated in hyperthyroid cats (over 98% of cases).[2] Although the free T_4 was more sensitive than the total T_4 for diagnosing hyperthyroidism, the test specificity for free T_4 was poor, with up to 20% of sick euthyroid cats having false-positive free T_4 results.[2,14] These sick cats generally have corresponding total T_4 values in the lower half or below the reference range.

Caution is therefore advised in using serum free T_4 measurements by equilibrium dialysis as the sole diagnostic test for hyperthyroidism. It is more reliable if interpreted with a corresponding total T_4 measurement. A T_4 value within the mid- to high-normal reference range combined with a high free T_4 concentration is consistent with hyperthyroidism, whereas a low or low-normal T_4 with a high free T_4 is usually associated with non-thyroidal illness. One must always combine these lab results with the cat's clinical features and the presence of a palpable thyroid nodule to make the diagnosis.

To complicate the free T4 assay situation further, unpublished results indicate that the present free T_4 by dialysis assay is not the same assay (originally by Nichols Diagnostics) that was used in all of the published studies evaluating the free T_4 assay in cats with thyroid disease. Nichols Diagnostics sold the distribution rights for the assay a few years ago, and the assay performance appears to have since deteriorated with many more false positive and false negative test results. In addition, commercial diagnostic labs are now using many other free T_4 assays, most of which do not have the equilibrium dialysis step that we believed was important. At this time, it is unclear how much additional information is truly gained by present free T_4 assays over the use total T_4 estimations alone.

4) **Thyroid stimulating hormone (Thyrotropin or TSH).** In human patients, measurement of circulating thyroid stimulating hormone (TSH) concentration is usually used as a first-line discriminatory test of thyroid function. A species-specific feline TSH assay has not yet been developed; however, assays for measuring canine TSH (cTSH) are widely available, and it has been suggested that its measurement may provide some diagnostic information in cats with suspected hyperthyroidism.[15,16] Theoretically, as in people, it could be expected that TSH levels should be low in early stages of hyperthyroidism before T_4 is elevated. Caution is advised in over interpreting TSH values in cats. The current canine assays only detect approximately 35% of recombinant feline TSH, making it difficult to distinguish normal values from the suppressed values expected in cats with hyperthyroidism. Perhaps the only use for such TSH measurement would be to exclude hyperthyroidism, i.e., finding a normal rather than suppressed cTSH value.

5) **Dynamic testing.** In the majority of hyperthyroid cats with normal total T_4 concentrations, identification of concurrent disease, repeat total T_4 analysis, or simultaneous measurement of free T_4 allows confirmation of the diagnosis. Further diagnostic tests are rarely required. Dynamic thyroid function tests (T_3 Suppression, TRH Stimulation) have been recommended in the past as helpful in confirming a diagnosis of hyperthyroidism.[1,3] Nowadays, use of these tests are only considered when repeated total T_4 concentration remains within reference range, when the free T_4 concentration is equivocal, or when thyroid scintigraphy is unavailable.

Enlarged Thyroid Gland but Not Hyperthyroid

Non-functional enlargement of thyroid glands (goiter) has been recognized since the 1960's but has taken on new significance since functional hyperthyroidism arose as an entity in the late 1970's. Non-functional goiter was 're-recognized' about a decade ago.[18] Many believe that clinical hyperthyroidism has a prodromal period (also called 'subclinical hyperthyroidism' or 'pre-hyperthyroidism'). However, it is not clear whether all goiters are indicative that the cat will develop hyperthyroidism. Surgical removal of non-functional goiters has been proposed as a preventative measure, but there is no evidence to support this approach.

Misdiagnosed as Hyperthyroid based on Falsely-High Serum T4 Concentration

A high circulating total T_4 concentration is the biochemical hallmark of hyperthyroidism and is extremely specific for its diagnosis. False positive results (ie, a high T_4 in a cat without hyperthyroidism) are rare but are being seen with increasing frequency. The reason for this is unclear but may be related to the increasing use of screening T4s as part of the cat's annual "wellness" program.

If a high total T_4 concentration is measured in a cat without the characteristic signs of hyperthyroidism, one should always verify the T_4 concentration. It's extremely important to remember that hyperthyroidism is a clinical diagnosis and should never be based on a serum T_4, free T_4, or TSH concentration alone. One MUST combine these lab results with the cat's clinical features <u>and</u> the presence of a palpable thyroid nodule to make the diagnosis.

DIAGNOSING CATS WITH HYPOTHYROIDISM

Diagnosing feline hypothyroidism can be challenging, regardless of its etiology. This syndrome rarely develops spontaneously. Most commonly, feline hypothyroidism is iatrogenic, secondary to overtreatment of hyperthyroidism. A presumptive diagnosis is based on a combination of the history, clinical signs, physical examination findings, and routine laboratory tests.[20,21]

However, a number of factors can make the diagnosis of hypothyroidism difficult to confirm. First, concurrent diseases can result in an "euthyroid sick syndrome," characterized by falsely-low serum thyroid hormone concentrations.[10,11] Second, one should expect weight gain and a decreasing activity level after successfully treating a cat with hyperthyroidism. Therefore, the clinical signs of iatrogenic hypothyroidism and the expected return to a euthyroid status can overlap. Finally, many cats develop a marked transient decrease in total T_4, within the first month of therapy with ^{131}I. This transient hypothyroid state is followed by a return to euthyroidism over the next 3 to 6 months as the remaining normal thyroid tissue recovers and starts to function once again.[21]

In most cats, we recommend waiting 3 months before making a definitive diagnosis of iatrogenic hypothyroidism after ^{131}I treatment, especially if the cat is not presenting with typical clinical features of hypothyroidism. However, the veterinarian should diagnose or exclude hypothyroidism as soon as possible in cats with renal disease because hypothyroidism, treatment for hyperthyroidism, and chronic renal disease all lower the glomerular filtration rate (GFR).[22,23] The combined effect of these 3 factors can lead to severe azotemia or even total renal failure. Treating the hypothyroidism can raise the GFR to an acceptable level.[24,25]

Serum T_4 concentration

By definition, cats with hypothyroidism have deficient thyroid hormone secretion. Therefore, finding a low serum T_4 concentration is key in diagnosing feline hypothyroidism. A normal T_4 concentration virtually excludes hypothyroidism. While important, a subnormal T_4 concentration alone is not definitive. The serum T_4 concentration can also be low in cats with nonthyroidal illness.[2,10,11] In general, the severity of the illness correlates inversely with the serum T_4 concentration (i.e., sicker cats have lower serum T_4 concentrations). Because multiple diseases and other factors can falsely lower the serum T_4 concentration in cats, the veterinarian must first rule out nonthyroidal disease before considering a diagnosis of hypothyroidism.

Cats with suspected hypothyroidism and a low T_4 concentration still require additional testing before the veterinarian can make a definitive diagnosis. Further tests such as serum free T_4 concentration, cTSH concentration, recombinant human TSH (rhTSH) stimulation test, or thyroid scintigraphy are indicated to confirm the disease.

Serum free T_4 concentration

Free T_4 is the non-protein bound fraction of circulating T_4 that can enter cells, producing the biologic effect of thyroid hormone and regulating the pituitary feedback mechanism. Free T_4 accounts for less than 1% of circulating T_4. Because only the free T_4 is biologically active, measuring free T_4 is a more sensitive test for diagnosing hypothyroidism. In addition, nonthyroidal illness influences free T_4 less than it influences the total T_4.[2] Therefore, free T_4 is better at distinguishing a euthyroid cat with nonthyroidal disease from a hypothyroid cat. Although measuring free T_4 concentration by equilibrium dialysis is a more accurate stand-alone test than total T_4 concentration, free T_4 is far from perfect for confirming feline hypothyroidism. First, moderate to severe nonthyroidal illness can falsely lower the free T_4 concentration, although to a lesser degree than seen with total T_4.[2] Second, up to 20% of cats with non-thyroidal disease develop a falsely high free T_4 concentration, further confusing the interpretation.[2,14]

Serum TSH concentrations

In dogs with hypothyroidism, a high serum TSH concentration confirms that the disease is primary (located within the thyroid gland). A specific assay for feline TSH is not yet available, but the commercially available cTSH assay cross-reacts with feline TSH enough to enable its use as a diagnostic test for hypothyroid cats. In one of the reported adult cats with spontaneous hypothyroidism, the serum TSH concentration was high when measured with the cTSH assay.[26] Similarly, many cats with iatrogenic hypothyroidism will also develop high serum TSH concentration as measured by the cTSH assay.[19]

Normal cats and cats with nonthyroidal illness generally maintain normal values for serum TSH. Therefore, the finding of a low total or free T_4 in combination with a high TSH concentration greatly improves the diagnostic sensitivity for hypothyroidism.

Thyroid stimulating hormone (TSH) response test

The TSH stimulation test provides important information for diagnosing hypothyroidism because it directly tests the thyroid's secretory reserve. In normal cats, administering exogenous TSH produces a consistent rise in serum T_4 concentration. In contrast, hypothyroid cats show little, if any, rise in the low basal serum T_4 concentrations after TSH stimulation.[19,20] In the past, bovine TSH was the preferred preparation for TSH stimulation testing in cats. However, bovine TSH is no longer available. An rhTSH preparation has been validated for TSH stimulation testing in cats.[27,28] The testing protocol involves collecting samples for serum T_4 concentration before and 6 hours after IV administration of 25-200 µg of rhTSH (Thyrogen). Administrating rhTSH to clinically normal cats generally increases the basal T_4 concentration by at least 2-fold. Further studies are needed to validate the use of this test for diagnosing feline hypothyroidism, but one would expect that these cats would experience little to no rise in serum T_4. The major disadvantage of this test is that rhTSH is extremely expensive, making it cost prohibitive for most owners.

Thyroid scintigraphy

Thyroid scintigraphy is considered the best imaging technique for dogs with suspected hypothyroidism because it can distinguish between hypothyroid dogs and dogs with a falsely low serum T_4 concentration. In hypothyroid dogs, thyroid imaging typically reveals decreased or even absent radionuclide uptake (thyroid gland is not visible on the scan). In contrast, dogs with a falsely low serum thyroid hormone concentration secondary to illness or drug therapy will have a normal thyroid image. Thyroid scintigraphy is also a powerful tool for diagnosing adult hypothyroid cats (both spontaneous and iatrogenic). As in hypothyroid dogs, thyroid imaging shows minimal to no uptake of the radionuclide.[26]

REFERENCES

[1]Graves TK, et al. *Vet Clin North Am Small Anim Pract* 1994;24:567. [2]Peterson ME, et al. *J Am Vet Med Assoc* 2001;218:529. [3]Peterson ME. *Clin Tech Small Anim Pract* 2006;21:2. [4]Broussard JD, et al. *J Am Vet Med Assoc* 1995; 206:302. [5] Kemppainen RJ, et al. *Am J Vet Res* 2006; 67:259. [6]Singh A, et al. *J Vet Diag Invest* 1997;9:261. [7]Horney BS, et al. *Vet Clin Pathol* 1999; 28:20. [8]Lurye JC, et al. *J Am Vet Med Assoc* 2002; 221:243. [9]Peterson ME, et al. *J Vet Intern Med* 1987;1:142. [10]Peterson ME, et al. *J Am Vet Med Assoc* 1990;197:1203. [11]McLoughlin MA, et al. *J Am Anim Hosp Assoc* 1993;29:227. [12]Broome MR. *Clin Tech Small Anim Pract* 2006; 21:10, 2006. [13]Daniel GB, et al. *Vet Radiol Ultrasound* 2002; 43:374. [14]Mooney CT, et al. *J Am Vet Med Assoc* 1996;208:2004. [15]Wakeling J, et al. *J Small Anim Pract* 2008; 49:287. [16]Wakeling J. *Can Vet J* 51:33, 2010 [18]Norsworthy G, et al. *J Fel Med Surg* 2002;4:145. [19]Greco DS. *Clin Tech Small Anim Pract* 21:40, 2006. [20]Peterson ME. *Current Veterinary Therapy X*. 1989,1000. [21]Peterson ME, et al. *J Am Vet Med Assoc* 1995;207:1422. [22]van Hoek I, et al. *Domest Anim Endocrinol* 2009:36:45. [23]Williams TL, et al. *J Vet Intern Med* 2010;24:863. [24]van Hoek IM, et al. *J Feline Med Surg* 2010;12:117. [25]Gommeren K, et al. *J Vet Intern Med* 23:844, 2009. [26]Blois SL, et al. *J Feline Med Surg* 2010;12:156. [27]Stegeman JR, et al. *Am J Vet Res* 2003;64:149. [28]van Hoek IM, et al. *J Feline Med Surg* 2009;11:309. [29]Diaz Espineira MM, et al. *J Vet Intern Med* 2007;21:25.

METHIMAZOLE, CARBIMAZOLE & ALTERNATIVE MEDICAL THERAPIES FOR FELINE HYPERTHYROIDISM

Mark E. Peterson, DVM, DACVIM
New York, NY

INTRODUCTION

The thioureylene class of antithyroid drugs, which includes methimazole and carbimazole, are most commonly used drugs for treatment of cats with hyperthyroidism.[1-3] However a variety of other drugs are also available for medical treatment of hyperthyroid cats, some of which are not familiar to many practicing veterinarians. Occasionally, such alternative medical therapies are required because of adverse reactions to the standard antithyroid drugs or for other specific reasons (usually cat owner preference for a "natural" treatment method).

ANTITHYROID DRUGS

Methimazole and carbimazole are two antithyroid drugs that can be used in cats for pre-operative control and long-term management of hyperthyroidism.[1-5] Both have a potent and reliable effect on suppressing thyroid hormone production. A related drug, propylthiouracil, often used in human medicine, is not recommended for cats because of a high incidence of serious adverse reactions (immune-mediated hemolytic anemia and thrombocytopenia).[6]

All of these drugs act by inhibiting thyroid peroxidase, an enzyme that normally oxidizes the iodide anion (I^-) to iodine.[7] This oxidation of iodide is a necessary step that results in either a mono-iodinated tyrosine (MIT) or di-iodinated tyrosine (DIT) being incorporated into thyroglobulin. The thioureylene drugs also effectively inhibit the coupling reaction, where the DIT/MIT residues on thyroglobulin are coupled together to yield thyroxine (T4) and of triiodothyronine (T3).

These drugs do not inhibit the action of the sodium-dependent iodide transporter located on follicular cells' basolateral membranes so thyroid uptake is not effected.[7] In addition, these drugs do not stop thyroid tumor growth, so thyroid goiters generally become larger over time with despite therapy. Also, antithyroid drugs do not prevent the transformation of thyroid adenomatous hyperplasia/adenoma to carcinoma that occurs in some cats on long-term antithyroid drug treatment.[8]

Methimazole Tablets

Methimazole is specifically licensed for treatment of feline hyperthyroidism both in the USA and Europe where the drug is supplied as 2.5- and 5-mg tablets (Felimazole®, Dechra Veterinary Products). It is also available as a brand name drug for human use (Tapazole® and Northyx®), as well as generic formulations (5- and 10-mg tablets).

The generic methimazole costs about a third of brand name Tapazole. Although no studies of cats comparing the efficacy of brand name vs. generic methimazole have been reported, their efficacies appear to be equivalent.

Toxicity of the human and generic methimazole develops, at least in part, because of the bitter taste of the methimazole tablet; this presumably occurs frequently in cats that may bite or crush the tablet after oral administration.[1] With the methimazole licensed for veterinary use (Felimazole®), the tablet is sugar coated, thereby avoided the bitter taste and lessening gastrointestinal side effects in most cats.

Carbimazole Tablets

Carbimazole is available for human use in many European countries and Japan, and may be available in the US soon. It exerts its antithyroid effect through immediate conversion to methimazole when administered orally.[9]

In addition to regular carbimazole,[5] a once daily controlled-release formulation (10- or 15-mg tablets) was recently licensed for cats in Europe (Vidalta®, Intervet Schering Plough).[10,11] Pharmacokinetic studies of this controlled-release formulation have shown no pronounced concentration peak and a sustained presence of methimazole in plasma (> 24 hours) with an apparent terminal half-life of approximately nine hours after oral administration.[10] Based on relative bioavailability and conversion it is estimated that 15 mg of this preparation is equivalent to approximately 7.5 mg of conventional methimazole.[11] It has been shown that administration of this drug with food significantly enhances its absorption.[10]

Compounded Methimazole/Carbimazole Preparations

While antithyroid dugs are routinely administered orally, compliance can be problematic particularly in fractious cats or in those that develop GI side effects from the drug. One alternative means of administering antithyroid drugs is vs. transdermal administration.[12-16] The most common transdermal vehicle used for antithyroid drug administration is pluronic lecithin organogel (PLO). Another vehicle available is Lipoderm® which is less greasy than PLO and can be refrigerated. Lipoderm also appears to cause less skin irritation and may have a better ability to penetrate with drug (at least in humans, but is be slightly more expensive than PLO.

The transdermal gel is applied in a thin layer to the non-haired portion of the inner pinnae using a concentration approximating 5 mg/0.05-0.1ml (50-100 mg/ml). Owners are instructed to wear exam gloves or finger cots, to apply the gel to alternate ears, and to wipe away any crusted material prior to the next dose, which prevents excess vehicle build up. Transdermal methimazole is associated with fewer GI side effects than oral therapy, but some cats resent manipulation of the ear and crusting can occur between doses leading to erythema.[12-16]

Custom transdermal formulation increases expense of antithyroid drug therapy. In addition, the efficacy and long-term stability of transdermal products can never be guaranteed. In addition to transdermal preparations, compounding pharmacies will custom supply methimazole in a number of other preparations, including chewable tablets, oral suspensions, and oral capsules.

Overview of Antithyroid Drug Preparations

When choosing an antithyroid drug formulation, it is important to consider the efficacy, shelf life, adverse effect, and cost of the product used. All of these formulations, of course, have advantages and disadvantages.

Efficacy of the drug formulation. All formulations of methimazole and carbimazole appear to be efficacious in most cats, depending on owner and cat compliance. However, the tablets licensed for use by the FDA are considered to be the most reliable, since the company must guarantee the concentration of drug in each tablet. This is not the case with compounded formulations, where wide variability in drug concentration is likely.

Adverse effects of drug formulation. Side effects of the medication, especially methimazole, can be lessened in many cats by switching to brand name Felimazole® or using compounded transdermal or flavored oral products (which avoid the bitter taste of the drug).

Cost of drug formulation. Generic brands are by far the cheapest, followed by brand-name tablets, then compounded oral suspensions and pastes, and finally transdermal preparations. Part of this cost is related to shorter shelf life with some of the compounded products.

ALTERNATIVE MEDICAL TREATMENTS FOR HYPERTHYROIDISM

Occasionally, alternative medical therapies are required in cats with hyperthyroidism because of adverse reactions to methimazole or for other specific reasons (usually cat owner preference for a "natural" treatment method). For the most part, these alternative therapies should be used for the short-term and only recommended prior to a more permanent treatment option.

β-Adrenoceptor Blockers (Propranolol and Atenolol)

Propranolol and atenolol are the most frequently used β-adrenoceptor blocking agents in hyperthyroid cats.[1-3,17,18] They are used to symptomatically control the tachycardia, tachypnea, hypertension and hyperexcitability associated with hyperthyroidism. Although traditionally considered to have no discernable effect on serum thyroid hormone concentrations, propranolol may inhibit peripheral conversion of T_4 to T_3, as the drug does in human patients.[19]

These β-adrenoceptor blocking drugs are recommended when rapid control of clinical signs is desirable and may be used either in combination with stable iodine[17] or methimazole. Alone, they are a useful treatment option for cats awaiting radiation therapy or in those cases in which there is a delayed return to euthyroidism after treatment because neither have any direct effect on the thyroid gland. Propranolol is a non-selective β-adrenoceptor blocker and is contraindicated in cats with pre-existing uncontrolled asthma or congestive cardiac failure. Atenolol is often preferred because it is a selective β_1-adrenoceptor blocking agent.

Stable Iodine

Large doses of stable iodine (I-127) acutely decrease the rate of thyroid hormone synthesis (Wolff-Chaikoff effect) and release although these effects are erratic, inconsistent, short-lived and escape from inhibition can occur.[20] In addition stable iodine is contraindicated prior to treatment radioactive iodine, since stable iodine will result in a lowered uptake of ^{131}I by the thyroid tumor (and will thereby decrease the efficacy of radioiodine treatment).[21] The drug is associated with a high incidence of adverse reactions (excessive salivation and partial to complete anorexia) purportedly because of its brassy taste.[1] Placing the dose in a gelatin capsule can reduce the prevalence of these side effects.

For these reasons, it is not used as sole therapy but usually in association with β-adrenoceptor blocking agents in a staged regimen for short-term pre-operative treatment.[17] Using this combination of drugs, serum T_3 concentrations normalize in most cats resulting in clinical improvement, although serum total T_4 concentrations only decrease in approximately one third of cases. A reported advantage of stable iodine before thyroidectomy is its apparent effect on decreasing the vascularity and friability of the thyroid gland, but this effect is controversial.

Calcium Ipodate and Iopanoic Acid

A number of oral cholecystographic agents (e.g. calcium ipodate) decrease T_4 production, an effect presumably mediated by release of iodine as the drug is metabolized, and also acutely inhibit peripheral T_4 to T_3 conversion. The latter effect has been clearly demonstrated in hyperthyroid cats where administration of calcium ipodate was associated with clinical improvement and normalization of serum total T_3 concentrations in over 60% of cases.[22] Waning of the effect is possible after 3 months of therapy.

Therefore, calcium ipodate is only likely to serve as an alternative to stable iodine in the short-term preparation for surgery. Calcium ipodate is no longer widely available. Similar agents such as iopanoic acid and diatrizoate meglumine have been used at the doses described.[23] All of these drugs are quite expensive to use on a long-term basis and are of limited effectiveness.

L-Carnitine

Carnitine is an amino acid that can ameliorate hyperthyroid symptoms in human patients, and I have used this as an adjunct therapy in hyperthyroid cats as well.[24] Daily doses up to 250 to 500 mg per day appear safe in cats, with gastrointestinal side effects being most likely. It is extremely **important** not to use D-carnitine, as the D-isomer is potentially toxic.

The primary effect of L-carnitine appears to be on peripheral tissues, not on the thyroid gland itself. The drug's effects are related to its ability to inhibit the entry of both T_4 and T_3 into the cell nucleus.[24] Although L-carnitine acts to reduce symptoms of hyperthyroidism, it does not change the circulating T_4 and T_3 concentrations, reduce thyroid tumor volume or size, or prevent adenomatous transformation into thyroid carcinoma.

Alternative/Complementary Therapies

More and more cat owners are asking veterinarians for a "natural" treatment alternative for hyperthyroidism. Some of the alternative medicine approaches that have been claimed to be effective include the following:[25,26]

1. Traditional Chinese Medicine, including herbs and acupuncture;
2. Bugleweed/Lycopus;
3. Lemon balm/Melissa officinalis;
4. Homeopathic drugs.

Resthyro is one common herbal therapy marketed (primarily on the internet) as a natural product for cat hyperthyroidism *support*. This popular product is loaded with a powerful combination of seven different herbal extracts. Together, it is claimed through many testimonials that this they herbs are very effective in helping with the symptoms of feline hyperthyroidism. An "all-natural product," the company asserts that Resthyro has no known side effects and is "gentle" enough for long-term use. The ingredients and actions are as follows:

1. *Eleutherococus Senticosus*: formerly known as Siberian ginseng, this is a rugged shrub that is native to eastern Russia and parts of northern Japan, China and Korea. The herb is believed to help control the overgrowth of the thyroid, while stimulating gland function. It is also thought to aid in metabolism stimulation.
2. *Bugleweed*: member of the mint family. In the past, the leaves and flowers of the plant were used medicinally as a sedative and to help treat coughs. More recently, it was discovered that the herb may somehow help reduce the levels of thyroid hormone in the body.
3. *Lemon Balm*: member of the mint family, native to Mediterranean and Southern European regions. Acts to block thyroid stimulating activity; support normal digestion; and calm nervous tension and excitability.
4. *Hawthorn*: white-flowering shrub and member of the rose family. Heart and vascular tonic, helps maintain cardiovascular structures and functions.
5. *Skullcap*: North American herb long-regarded for its relaxing and detoxifying qualities. A natural sedative; moderates nerve transmissions; supports kidney function.
6. *Valerian*: native to Europe and some parts of Asia. Helps control inflammation, relaxes spasms and is a sedative.
7. *Gotu Kola:* a small green herb used medicinally for centuries in countries like China, India and Indonesia. Known for its cleansing and diuretic effects, and for its antibacterial and anti-inflammatory properties.

Personally, I have never seen a cat treated with one of these natural alternative medications that did well, at least on a long-term basis. Most of the active ingredients in the herbs act by supposedly lowering TSH secretion; however, that could not explain the "effectiveness" of these agents in our patients, since circulating TSH is already low in these cats with hyperthyroidism. It is also important to remember that cats with hyperthyroidism usually have one or more thyroid adenomas and, less commonly, thyroid carcinoma. No one has ever documented that these alternative medications reduce the size of the thyroid tumor or normalize the high serum T_4 concentrations associated with feline hyperthyroidism.

Herbs and homeopathic drugs are vastly powerful entities. Holistic practitioners may tell you that some of these therapies work well. However, this is based on an improvement in the cat's clinical signs alone. It is not based a dramatic lowering or normalization of the cat's serum T_4 value or a reduction in thyroid tumor size. To my knowledge, no cases have ever been published in a referred journal that document that these therapies cure or even results in significant improvement in cats with hyperthyroidism.

REFERENCES

[1]Mooney CT, et al. *Kirk's Current Veterinary Therapy XIII*. 2000;333. [2]Trepanier LA. *Clin Tech Small Anim Pract* 2006;21:22. [3]Trepanier LA. *Vet Clin North Am Small Anim Pract* 2007;37:775. [4]Peterson ME, et al. *J Vet Intern Med* 1988;2:150. [5]Mooney CT, et al. *J Small Animal Pract* 1992;33:228. [6]Peterson ME, et al. *J Am Vet Med Assoc* 1984;184:806. [7]Cooper DS. Antithyroid drugs. *N Engl J Med* 1005; 352:905. [8]Hibbert A, et al. *J Feline Med Surg* 2009;11:116, 2009. [9]Peterson ME. *Res Vet Sci.* 1993;54:351. [10]Frenais R, et al. *J Vet Pharmacol Ther* 2008;31:213. [11]Frenais R, et al. *J Small Anim Pract* 2009;50:510. [12]Hoffman SB, et al. *J Vet Pharmacol Ther* 2002;25:189. [13]Hoffmann G, et al. *J Feline Med Surg* 2003;5:77. [14]Sartor LL, et al. *J Vet Intern Med* 2004;18:651. [15]Buijtels JJ, et al. *Tijdschr Diergeneeskd* 2006;131:478. [16]Lecuyer M, et al. *Can Vet J* 2006;47:131. [17]Foster DJ, et al. *J Small Anim Pract* 1999;40:307. [18]Henik RA, et al. *J Feline Med Surg* 2008;10:577. [19]Wiersinga WM. *Thyroid* 1991;1:273. [20]Burgi H. *Best Pract Res Clin Endocrinol Metab* 2010;24:107. [21]Peremans K, et al. *J Feline Med Surg* 2008;10:460. [22]Murray LA, et al. *J Am Vet Med Assoc* 1997;211:63. [23]Gallagher AE, et al. *J Feline Med Surg* 2009;11:69. [24]Benvenga S, et al. *J Clin Endocrinol Metab* 2001;86:3579. [25]Pitcarin RH, et al. Complete Guide to Natural Health for Dogs and Cats. 2005; 402. [26]Dobias P. World Congress WSAVA/ FECAVA/CSAVA 2006;124. [27]Wakeling J. *Can Vet J* 2010;51:33.

TREATMENT OF HYPERTHYROIDISM AND CONCURRENT RENAL DISEASE: IS THE "TAPAZOLE TRIAL" NECESSARY?

Mark E. Peterson, DVM, DACVIM
New York, NY

INTRODUCTION

Hyperthyroidism and chronic kidney disease (CKD) are both common disorders in older cats. Therefore, it should not be surprising that both disorders frequently develop together in the same cat. The prevalence of concurrent renal disease in cats with hyperthyroidism is estimated to be approximately 30-35%.[1,2]

Hyperthyroidism tends to "artificially" increase the renal blood flow (RBF) as well as the glomerular filtration rate (GFR).[1-4] When the GFR is increased in a hyperthyroid cat with underlying CKD, it can mask renal insufficiency; serum concentrations of urea nitrogen and creatinine may be normal despite mild to moderate kidney disease.[1-7] Decreased muscle mass, which is a common feature of hyperthyroidism, also contributes to the lowered serum creatinine concentration in these cats (since creatinine is derived from muscle tissue).[1,2]

Treating hyperthyroidism restores the high serum T_4 concentration to normal and, in cats *without* CKD, also returns the high GFR back to normal values. In cats with CKD, however, the GFR will fall to the low-normal or subnormal levels expected with moderate renal dysfunction. Therefore, this decrease in GFR can result in the apparent worsening of the serum kidney function tests or the *development* of renal disease. It is important to remember, however, that treating the hyperthyroidism itself does not cause the CKD in these cats. The renal disease was already present before treatment but was masked by the hyperdynamic state of the hyperthyroidism.

PHYSIOLOGICAL INTERACTIONS BETWEEN THYROID HORMONES AND RENAL FUNCTION

Hyperthyroidism decreases peripheral vascular resistance by dilating the arterioles of the peripheral circulation. Because of this decrease in systemic vascular resistance, the effective circulating volume decreases, stimulating the renin-angiotensin-aldosterone system. This leads to renal sodium retention with a resultant increase in blood volume. Cardiac output may increase dramatically in cats with hyperthyroidism. In addition to the decrease in systemic vascular resistance, an increase in heart rate, increases in left ventricular contractility and ejection fraction, and increase in blood volume all contribute to this increase in cardiac output.[1,2]

These systemic hemodynamic factors (i.e., the increased cardiac output) combined with intrarenal vasodilation lead to increases in renal blood flow (RBF), glomerular hydrostatic pressure, and glomerular filtration rate (GFR). Thyroid hormones also influence renal tubular function including electrolyte handling. With time, hyperthyroidism can lead to renal changes including glomerular hypertension, glomerulosclerosis, proteinuria, and hyperplasia and hypertrophy of the renal tubules.[1,2]

HYPERTHYROIDISM ITSELF MAY CONTIBUTE TO CHRONIC KIDNEY DISEASE

Recent research provides evidence that hyperthyroidism may contribute to the development or progression of CKD in cats. First of all, a number of recent reports indicate that many untreated hyperthyroid cats develop proteinuria, which resolves within 4 weeks of successful treatment.[8,9] This proteinuria, which reverses after treatment, could be a reflection of glomerular hypertension and hyperfiltration, changes in tubular protein handling, or a change in the structure of the glomerular barrier.[1] Whatever the cause of the proteinuria, no treatment is generally needed other than treatment of the hyperthyroid state itself.

Secondly, cats with untreated hyperthyroidism have high levels of retinol binding protein (RBP), a urinary marker for tubular dysfunction or damage.[10,11] This high urinary RBP excretion may reflect tubular damage or dysfunction resulting from the thyroid-induced hypertrophy and hyperplasia of the tubular cells. After treatment, these high urinary RBP levels fall in cats without azotemia but may remain slightly high in cats with pre-existing CKD. This too suggests that hyperthyroidism can cause reversible renal dysfunction; however, the renal tubular changes may become irreversible with time as CKD progresses.

Thirdly, many cats with untreated hyperthyroidism have high values for urinary N-acetyl-b-D-glucosaminidase (NAG), a lysosomal glycosidase found primarily in epithelial cells of the proximal convoluted tubule.[12] Like RBP, NAG is a specific marker of active proximal tubular damage. After treatment, these high urinary NAG levels decrease, again suggesting that these renal changes can be reversed, at least in cats without pre-existing CKD.

Overall, these studies all suggest that that leaving a hyperthyroid cat untreated (or poorly regulated with methimazole) may be detrimental to long-term kidney function. Treating and curing hyperthyroidism may help to both reverse renal damage and preserve remaining kidney function.

CLINICAL IMPLICATIONS IN HYPERTHYROID CATS

Decision-making with regard to treating cats with hyperthyroidism and CKD can be difficult. Clinicians can be confronted with two different scenarios concerning hyperthyroid cats with concurrent CKD.

First of all, about 10% of hyperthyroid cats have known pre-existing CKD at time of diagnosis. These cats have obvious clinical and biochemical evidence of mild to moderate kidney disease, and do not present a diagnostic dilemma. However, hyperthyroid cats with pre-existing azotemia are more difficult to treat successfully, at least on a long-term basis. They have a worse prognosis than do cats that are not azotemic prior to treatment for hyperthyroidism.

In the second scenario, cats are initially not azotemic but develop high serum concentrations of urea nitrogen or creatinine only after they have been treated for hyperthyroidism. These cats, which represent 20% to 25% of all hyperthyroid cats, are much more of a diagnostic dilemma because one may not even suspect that they have concurrent renal disease.

Pre-existing Azotemia in a Newly Diagnosed Hyperthyroid Cat

In cats with overt CKD, the diagnosis of hyperthyroidism can sometimes be complicated by suppression of circulating T_4 and T_3 concentrations to within the reference range limits. Although most of these cats will have serum T_4 concentrations that are clearly above reference range limits, some cats with moderate to severe CKD have T_4 concentrations that fall into the mid- to high-end of the normal range. Since any nonthyroidal illness would be expected to decrease serum thyroid hormone concentrations into the low to undetectable range in sick cats without concurrent hyperthyroidism, concomitant hyperthyroidism should be suspected in any middle- to old-aged cat with severe CKD and normal serum T_4 and T_3 concentrations, especially if signs of hyperthyroidism are also present.

Because we can predict that the GFR will fall once the euthyroidism is restored, it is generally advised to try medical management prior to a more definitive treatment in cats with concurrent hyperthyroidism and pre-existing CKD. A low starting dose (i.e., 1.25 mg orally once daily) of methimazole with gradual dose escalation is prudent when starting a cat with CKD and hyperthyroidism on medical therapy. Cats should be monitored every 2 weeks with a CBC, biochemical profile, urinalysis, and serum T_4 concentration. Because the initial decline in GFR stabilizes after a month of successful resolution of the hyperthyroidism, one can decide at that time whether or not the cat's renal function is stable or worsening.

Concurrent management of CKD is also extremely important these cats. If the renal function remains stable after a euthyroid state is established, a more definitive treatment, such as ^{131}I, should be strongly considered. If, on the other hand, renal function declines dramatically after antithyroid drug treatment, especially if accompanied by marked clinical deterioration of the cat's renal failure (to IRIS Stage 3 or 4), it may be best to maintain the cat on a reversible anti-thyroid therapy, decrease the dose, or stop it all together if need be.

In some of these cats with severe CKD, maintenance of a mild hyperthyroid state may give the best short-term clinical result. However, this course of action is far from ideal, and these cats have a very guarded to poor prognosis. Except in these extreme circumstances, the validity of maintaining a cat in a mildly hyperthyroid state is questionable given that uncontrolled hyperthyroidism, in itself, appears to be damaging to renal function.[1,8,10-12]

Non-Azotemic Before Treatment of Hyperthyroidism; Development of Post-Treatment Azotemia

Hyperthyroidism is known to increase GFR, decrease circulating creatinine concentrations, and mask underlying renal disease. About 20% to 25% of hyperthyroid cats without known CKD develop azotemia after successful treatment of hyperthyroidism, irrespective of therapeutic modality (methimazole/carbimazole, surgical thyroidectomy, or radioiodine).[1]

Predicting which hyperthyroid cats will develop overt azotemia after treatment of hyperthyroidism can be difficult to impossible. The determination of GFR is clearly the best predictor of post-treatment CKD, with a low to low-normal GFR indicating that a hyperthyroid cat is at increased risk for post-treatment azotemia. However, techniques for assessment of GFR are not widely used in practice, and even GFR determinations are not a 100% perfect predictor of CKD. Routine pre-treatment parameters such as serum urea or creatinine concentrations, and urine specific gravity are certainly useful, but they cannot consistently predict impending azotemia.[13]

Should Methimazole Trials be Performed in all Hyperthyroid Cats?

Again, determining which untreated hyperthyroid cats have clinically significant underlying CKD can sometimes be difficult. Use of methimazole or carbimazole can provide a "preview" of how the cat will be after curing hyperthyroidism. Thus, many veterinarians attempt trial therapy with methimazole or carbimazole to help test what renal function might remain after treating the hyperthyroidism. If no marked deterioration occurs, then a more permanent therapeutic option for hyperthyroidism may be recommended.

Except for advanced (IRIS Stage 3-4) CKD, the necessity of this approach in cats without pretreatment azotemia is questionable, given that treatment for the hyperthyroidism is strongly recommended whatever the outcome. In support of this reasoning, the survival of cats that do develop **azotemia** is not shorter than those that do not develop azotemia after treatment of hyperthyroidism. In one study, the **median** survival time of cats that developed azotemia (595 days) was similar to that in cats that remained non-azotemic (584 days) after treatment.[14]

Preventing Hypothyroidism after Treatment for Hyperthyroidism

Whatever treatment option for hyperthyroidism is considered, it is important to avoid hypothyroidism as it may have its own detrimental effects on GFR. Hypothyroidism in both humans and dogs has been showed to reduce GFR,[15,16] and it is likely that similar changes occur in cats that develop iatrogenic hypothyroidism.

In a recent study by Williams *et al*,[17] cats with iatrogenic hypothyroidism were more likely to develop azotemia in the 6 months after treatment than cats that remained euthyroid. Hypothyroid cats with azotemia also had shorter survival times than nonazotemic cats, whereas no difference in survival between euthyroid cats with or without azotemia could be detected. This suggests that the development iatrogenic hypothyroidism contributes to the development of azotemia, at least in cats with mild underlying CKD (IRIS Stage I or II).[17,18] More importantly, the hypothyroidism may shorten survival after treatment of hyperthyroidism.

If a cat with post-treatment azotemia develops a low T_4 concentration, a serum TSH level should be measured to help exclude hypothyroidism.[19] The finding of a high serum TSH concentration confirms hypothyroidism. A specific assay for feline TSH is not yet available. However, the commercially available canine TSH assay cross-reacts with feline TSH enough to enable its use as a diagnostic test for hypothyroid cats.[19,20]

If iatrogenic hypothyroidism is diagnosed, treatment with L-thyroxine (0.1 mg once to twice daily) is indicated.[20,21] The dosage should be adjusted based on post-pill serum T4 and cTSH determinations. Most cats treated will show improvement in their azotemia as the hypothyroidism resolves and euthyroidism is restored.[18]

Long-Term Renal Function in Cats that Develop Azotemia after Treatment

In most cats that develop post-treatment azotemia, the CKD is not that severe or life threatening. It is also unusual to see a jump of more than one IRIS stage after treatment. In other words, hyperthyroid cats with IRIS Stage I-II CKD may develop overt azotemia after treatment, but one would not expect those cats' CKD to advance to more than IRIS Stage II-III after treatment.

In addition, the decline in GFR after successful treatment of a cat's hyperthyroidism is not very progressive. Rather, the fall in GFR is detectable within 1 month but then remains stable at this level for months thereafter.[1,6] The rise in serum urea nitrogen and creatinine values in cats with CKD follow the decrease in GFR, so that azotemia, when it does occur, would be expect to develop within 1 month of treatment but remain relatively stable over many months.

SUMMARY

Hyperthyroidism is associated with an increased GFR and decreased serum creatinine concentration relative to euthyroidism. An increase in cardiac output and renal perfusion leads to an increase in GFR. However, an increase in glomerular capillary pressure (glomerular hypertension), glomerulosclerosis, and proteinuria may contribute to progression of renal disease in cats with hyperthyroidism. There does not appear to be a clinically practical method of predicting post-treatment renal failure, although pre-treatment GFR measurement seems to be useful. About 30% of all hyperthyroid cats can be expected to develop azotemia within a month of successful treatment of hyperthyroidism. In most cats, however, their azotemia tends to remain stable for prolong**ed** periods and does not significantly affect survival.

REFERENCES

[1]Syme HM. *Vet Clin North Am Small Anim Pract* 2007;37:723. [2]Langston CE. *Clin Tech Small Anim Pract* 2006;21:17. [3]Graves TK, et al. *Am J Vet Res* 1994;55:1745-1749. [3]Adams WH, et al. *Can J Vet Res* 1997;61:53. [4]Adams WH, et al. *Vet Radiol Ultrasound* 1997;38:231. [5]Becker TJ, et al. *J Am Anim Hosp Assoc* 2000;36:215. [6]Boag AK, et al. *Vet Rec* 2007;161:711. [7]DiBartola SP, et al. *J Am Vet Med Assoc* 1996;208:875. [8]van Hoek I, et al. *Domest Anim Endocrinol* 2009;36:45. [9]Williams TL, et al. *J Vet Intern Med* 2010;24:863. [10]van Hoek I, et al. *J Immunol Methods* 2008;329:208. [11]van Hoek I, et al. *J Vet Intern Med* 2009;23:1031. [12]Lapointe C, et al. *J Vet Intern Med* 2008;22:1103. [13]Riensche MR, et al. *J Feline Med Surg* 2008;10:160. [14]Wakeling J, et al. *J Vet Intern Med* 2006;20:1523. [15] Iglesias P, et al. *Eur J Endocrinol* 2009;160:503. [16]Gommeren K, et al. *J Vet Intern Med* 2009;23:844. [17]Williams TL, et al. *J Vet Intern Med* 2010;24:1086. [18]van Hoek IM, et al. *J Feline Med Surg* 2010;12:117. [19]Wakeling J. *Can Vet J* 2010;51:33. [20] Greco DS. *Clin Tech Small Anim Pract* 2006;21:40. [21]Peterson ME. *Current Veterinary Therapy X*. 1989,1000.

TREATMENT OF SEVERE, UNRESPONSIVE, OR RECURRENT HYPERTHYROIDISM IN CATS

Mark E. Peterson, DVM, DACVIM
New York, NY

INTRODUCTION

Most hyperthyroid cats, at least in the early stages of their disease, can be readily controlled with antithyroid drugs.[1-3] Similarly, most cats with mild to moderate hyperthyroidism are cured quite easily with standard "low" doses of radioiodine or by use of surgical thyroidectomy.[4-6] Occasionally, the practicing veterinarian will see hyperthyroid cats that become unresponsive to antithyroid drug treatment, generally after months of satisfactory control. Other cats treated with either thyroidectomy or radioiodine remain persistently hyperthyroid or experience relapse shortly after treatment. Such problematic hyperthyroid cats can be frustrating to manage, especially as their disease becomes more severe and they develop complications of advanced and/or poorly controlled hyperthyroidism.[7]

PERSISTENT OR RECURRENT HYPERTHYROIDISM AFTER THYROIDECTOMY

Because most hyperthyroid cats have involvement of both thyroid lobes, bilateral thyroidectomy is generally indicated if long-term cure is the goal. The two major techniques for bilateral thyroidectomy include the intracapsular and extracapsular methods.[5-8] The major problem with the intracapsular technique for thyroidectomy is that it can be difficult to remove the entire thyroid capsule (and therefore all abnormal thyroid tissue) while concurrently preserving parathyroid function. Small remnants of thyroid tissue that remain attached to the capsule may regenerate and produce recurrent hyperthyroidism.[5-8] With the extracapsular technique, the incidence of relapse is much less than because the entire thyroid capsule is removed at time of surgery.

When recurrent hyperthyroid does occur after bilateral thyroidectomy, it generally takes many months to years for the serum T_4 to increase or clinical signs of hyperthyroidism to develop.[5-8] In contrast, we occasionally see cats that remain hyperthyroid or develop relapse very shortly after surgery; in these cats, either ectopic thyroid tissue or thyroid carcinoma should be suspected.[7,9]

Although disease is bilateral in 70% cats, some cats have asymmetrical thyroid enlargement, with one thyroid lobe being very large and the other being only minimally enlarged.[7] In these cats, unilateral thyroidectomy generally restores euthyroidism, at least for a few weeks, and it may take many months for the remaining lobe to grow to a size for hyperthyroidism to recur. However, the client must be warned of the possibility of persistent or recurrent hyperthyroidism whenever unilateral thyroidectomy is contemplated.

Overall, because of the potential for recurrence of hyperthyroidism, all cats should have their serum T_4 concentration monitored once or twice a year after unilateral or even bilateral thyroidectomy.

FAILURE OF STANDARD ^{131}I TREATMENT OR RELAPSE AFTER ^{131}I TREATMENT

Most hyperthyroid cats treated with radioactive iodine can be cured with administration of a single dose (2 to 6 mCi).[4,9,10] Approximately 5% of cats, however, fail to respond completely and remain mildly hyperthyroid after treatment with these standard doses of radioiodine. Almost all of these hyperthyroid cats can be cured by a follow-up ^{131}I treatment, although a slightly higher dose of ^{131}I may be needed to achieve euthyroidism.[4] On the other hand, if little or no response to ^{131}I is observed, especially after retreatment with higher ^{131}I doses, thyroid carcinoma must always be considered; in such cases, thyroid scintigraphy and/or thyroid biopsy is strongly recommended.[4,9-12]

Relapse of hyperthyroidism after ^{131}I is possible but rare. When it does occur, relapse generally develops between 2 to 4 years after treatment. In these cats, it is unclear whether such relapse is due to regrowth of the original thyroid tumor tissue or the development of an entirely new thyroid nodule (not related to the original goiter). On the other hand, if relapse develops quickly after initial ^{131}I therapy, thyroid carcinoma should be excluded and thyroid scintigraphy and/or thyroid biopsy strongly recommended.[4,9-12]

Thyroid carcinomas are more resistant to the effect of ^{131}I than benign thyroid disease (ie, adenomas or adenomatous hyperplasia) and the size of thyroid carcinomas is usually much larger. Therefore, extremely high doses of radioiodine (30 mCi) are almost always needed for destruction of this malignant tissue and cure of the cat's hyperthyroid state.[9,12]

POOR RESPONSE TO LONG-TERM TREATMENT WITH ANTITHYROID DRUGS

This is a relatively common problem seen in hyperthyroid cats treated with long-term methimazole or carbimazole. Because these antithyroid drugs do not slow or stop thyroid tumor growth, the thyroid goiter in these cats tends to grow progressively larger with time. After 1-3 years or more, many of these cats will develop very large, palpable goiters and will become difficult to regulate, even with high daily doses of oral or transdermal methimazole. Eventually, it may be difficult to impossible to raise the antithyroid drug dose to one high enough to keep maintain circulating thyroid values even close to the normal reference range.

With enough time and as the disease progresses, the thyroid adenoma/adenomatous hyperplasia in some hyperthyroid cats will also transform into malignant thyroid carcinoma. Again, methimazole or other antithyroid drug therapy does nothing to the thyroid tumor pathology and cannot stop this from happening.

All practicing veterinarians should always explain to all owners of cats with hyperthyroidism that the disease is caused by a thyroid tumor, and that this tumor will gradually grow progressively larger with time. If methimazole (carbimazole) is used as initially treatment, they should be told that the antithyroid drug may eventually stop working and that other treatments may be needed. However, both surgery and radioiodine can be more difficult to achieve a cure in cats with extremely large goiters, even if the thyroid tumors are benign, so early definitive therapy is always best. In cats that are young to middle-aged, it is not logical (at least to me) to treat with methimazole (carbimazole) on a long-term basis unless you do not believe that the cat will live for more than a year or two or the owner refuses definitive therapy.

As someone who does thyroid imaging and radioiodine treatment, I see a couple of cats each month that have been treated for years with antithyroid drugs and now have very large to huge thyroid tumors. In about 10% of these cats with severe, long-standing hyperthyroidism, the diagnosis is thyroid carcinoma with extension of tumor or metastasis into the chest. It's quite likely that transformation of the tumor from adenoma to carcinoma has occurred in these cats. This may be the natural evolution of the disease; however, antithyroid drugs have been shown to be carcinogenic in rodents, it is possible that the drug treatment is playing a role.[15,16]

Cats with huge benign or malignant tumors require much larger doses of radioiodine to ablate (10 to 30 mCi) the thyroid tumor(s) than do the typical recently diagnosed cats with mild to moderately hyperthyroidism (2 to 6 mCi). Almost all of these cat owners comment to me that "only if their veterinarian has told them about this treatment before." Yes, I can generally help these cats with severe, advanced thyroid disease, but it is much more difficult to cure them with a single dose. These cats are also older and have many more complications, mainly because of the concurrent disease (eg, renal disease) that is common in older cats.

Unfortunately, in a newly diagnosed hyperthyroid cat, we have no way to determine how fast the thyroid tumor is going to grow or its potential for malignant transformation. If you have a relatively young hyperthyroid cat (<13 years of age) with no concurrent problems, I would certainly recommend surgery or radioiodine since those cats will likely live for quite a few more years.

THYROID CARCINOMA

Thyroid carcinoma must always be suspected if the palpable goiter is extremely large, serum total T_4 is particularly high, or if hyperthyroidism persists despite high doses of methimazole or carbimazole. However, each of these criteria may also apply to benign hyperthyroidism. Lack of response to increasing doses of antithyroid drugs, persistent hyperthyroidism or rapid recurrence following routine bilateral thyroidectomy or radioiodine therapy should all should alert the clinician of the possibility of malignancy.[14] While thyroid carcinoma in cats appears to be rare, many cases appear to go undiagnosed, at least in the earlier stages of disease. Diagnosing thyroid carcinoma is very important because standard treatment strategies for hyperthyroidism almost always fail to manage this malignant condition adequately.

Clinical Features of Thyroid Carcinoma in Cats

Most thyroid carcinoma in cats are functional and are associated with typical clinical and biochemical features of hyperthyroidism, such as polyphagia, weight loss, and restlessness. Non-secretory carcinomas also occur in cats and may represent up to a third of cats with thyroid carcinoma; such cats are euthyroid and usually present because of the incidental discovery of a moderate-sized to large cervical mass.

Whether the cat is hyperthyroid or euthyroid cannot, however, determine the carcinoma's capacity to take up and concentrate radionuclide. With thyroid scintigraphy, about half of euthyroid cats with thyroid carcinoma show some degree of radioisotope uptake, indicating that radioiodine may still be a useful treatment in a subset of these cats.

How to Distinguish Thyroid Carcinoma from Adenoma (Adenomatous Hyperplasia)

The clinical features of hyperthyroid cats with thyroid carcinoma are similar those of hyperthyroidism cats with benign adenoma (or adenomatous hyperplasia). Features that should increase one's suspicion for thyroid carcinoma include the following:

- Larger-than-expected goiter; sometimes multiple, nodular thyroid masses
- Long history of hyperthyroidism (> 2 years)
- Extremely high serum T_4 concentration (> 5 to 10-times the upper limit of normal)
- Lack of response to high doses of methimazole or carbimazole (> 15-20 mg/day)
- Failure or recurrence of hyperthyroidism after thyroidectomy or radioiodine therapy
- Locally invasive or highly vascular thyroid tumor (usually noted at surgery)
- Multifocal, irregular, and heterogeneous areas of increased radionuclide uptake with thyroid imaging
- Signs of distant metastases on radiography or thyroid imaging (usually pulmonary)

In a few cats with thyroid carcinoma, however, very few of the above features will apply. Alternatively, it is possible for a cat with advanced benign disease to show some of the same features (e.g., large goiter, long history of hyperthyroidism, very high serum T_4 value). Thyroid carcinoma can be extremely variable on palpation (if in the cervical region), ranging from discrete solitary mobile nodules, to chains of tissue extending along the neck or fixed firm masses. Both benign and malignant thyroid tumors may become extremely large if cystic changes develop. Gross appearance at surgery, however, may be more informative since thyroid carcinomas frequently infiltrate local tissues and are highly vascular.

An inadequate response to conventional treatment in any hyperthyroid cat should raise the question of a possible thyroid carcinoma. For example, if hyperthyroidism is not controlled or quickly recurs after bilateral thyroidectomy or standard doses of radioiodine fail to cure the cat's hyperthyroidism, thyroid carcinoma must be considered. Many cats eventually diagnosed with thyroid carcinoma had initially been controlled *fairly* well with antithyroid drugs. Although these drugs are typically effective in suppressing serum T_4 concentration for variable periods, the daily dose requirement needed to control the hyperthyroid state gradually and progressively increases in these cats. Eventually, the thyroid tumor will grow to such a size that antithyroid drug therapy, no matter what the dose, no longer blocks T_4 secretion from the large volume of carcinomatous tissue.

Scintigraphic features that can help distinguish malignant from benign thyroid disease include multiple and extensive areas of hyperfunctional thyroid tissue, a heterogeneous pattern of radioisotope uptake with irregular margins, linear multifocal patterns (suggestive of tumor extension along fascial planes), and uptake of masses within the cranial mediastinum. A recent review of scintigraphic findings in hyperthyroid cats, however, found that similar imaging changes are occasionally observed in cats with benign disease.[11] The identification of metastatic disease in thyroid carcinoma is complicated by the existence of ectopic thyroid tissue in up to 25% of cats. Hence, multiple areas of radioisotope uptake in the cervical or intra-thoracic locations (or both) could potentially reflect metastases or multiple primary tumors in ectopic sites.

Despite its limitations, thyroid scintigraphy remains a valuable and essential tool in evaluating a cat with suspected thyroid carcinoma. Thyroid imaging is the best means we have to assess tumor functionality and determine thyroid volume to assist in treatment planning with surgery or high-dose radioiodine.

Definitive Diagnosis of Thyroid Carcinoma

The only definitive way to confirm a diagnosis of thyroid carcinoma is by histopathologic examination. Pathological criteria for malignancy include extracapsular invasion, vascular invasion, cellular pleomorphism, and increased mitotic activity.

Definitive identification of thyroid carcinoma on histopathology generally necessitates excisional biopsy. Routine histopathology of the thyroid tumor should always be done after any thyroid surgery. Because many thyroid carcinomas are highly vascular, excision may be difficult and hemorrhage may occur; in these cats, it may not be possible to do a thyroidectomy, but one should at least attempt to obtain a biopsy for diagnosis.

Scintigraphy can be used adjunctively, especially in recognizing regional and distance metastasis. But again, scintigraphy alone cannot reliably distinguish whether the thyroid tissue is malignant since ectopic thyroid tissue commonly is found in cats with benign thyroid disease.

Interestingly, mixed adenoma and carcinoma tissue has been documented in some cats with thyroid carcinoma. This strongly suggests that, at least in some cats, thyroid carcinomas appear to arise from transformation of benign into malignant neoplasia.

Treatment of Cats with Thyroid Carcinoma

Radioiodine is considered the gold standard treatment for hyperthyroidism. It has the advantages of treating all hyperfunctional tissue, regardless of site (including ectopic thyroid tissue). In addition, there is no need for general anesthesia and treatment complications are minimal with radioiodine therapy. The efficacy of radioiodine is well established in cats with benign disease, with success rates of up to 94% reported.[4]

The use of radioiodine for treatment has also been reported in a limited number of cats with thyroid carcinoma. Higher ^{131}I doses are required for cats with thyroid carcinoma (3-10 x higher than for benign disease), primarily because of the much larger tumor volume associated with carcinoma. In addition, at least in euthyroid cats with thyroid carcinoma, the malignant tissue appears less efficient at taking up and concentrating radioiodine. Therefore, higher administered ^{131}I doses are needed in order to have enough ^{131}I taken up by the tumor to destroy the carcinomatous tissue.

Most studies of cats with thyroid carcinoma have used an empirical high ^{131}I dose of 30 mCi. With this protocol, long-term prognosis is good with extended survival commonly achieved.[12-14] In one recent study, 12 hyperthyroid cats with hypersecretory thyroid carcinomas were treated with high-dose radioiodine. Of these, treatment was successful in 11 cats, with complete resolution of hyperthyroidism and extended survival times (median 1,065 days). The remaining cat had a partial response.

The combination of surgical debulking followed by administration of high-dose radioactive iodine also has been reported to be successful in cats with thyroid carcinoma. This allows one to obtain thyroid tissue for histopathologic examination in order to confirm thyroid carcinoma. Because the total tumor volume is lessened by the debulking procedure, a lower ^{131}I dose is generally needed to destroy the remaining tumor tissue. However, if excision biopsy is not possible, higher doses of ^{131}I will be needed, necessitating a longer hospital stay. Unfortunately, many of these cats are not surgical or anesthetic candidates because of their severe, advanced, and uncontrolled hyperthyroidism.

In conclusion, thyroid carcinoma is a relatively rare (but increasingly common) cause of hyperthyroidism. Although there are many clinical and imaging features that suggest thyroid carcinoma, diagnosis can only be made by histopathologic examination of tissue. Treatment with high-dose radioiodine is clearly the best therapeutic option. Successful response to radioiodine is associated with prolonged survival times, comparable to those achieved in hyperthyroid cats with benign disease.

REFERENCES

[1]Trepanier LA. *Clin Tech Small Anim Pract* 2006;21:22. [2]Trepanier LA. *Vet Clin North Am Small Anim Pract* 2007;37:775. [3] Peterson ME, et al. *J Vet Intern Med* 1988;2:150. [4]Peterson ME, et al. *J Am Vet Med Assoc* 1995;207:1422. [5]Welches CD, et al. *Vet Surg* 1989; 18:392, 1989. [6]Naan EC, et al. *Vet Surg* 2006;35:287. [7]Mooney CT, et al: Manual of Canine and Feline Endocrinology, 2004;95. [8]Birchard SJ. *Clin Tech Small Anim Pract* 2006;21:29. [9]Broome MR. *Clin Tech Small Anim Pract* 2006; 21:10. [10]Chun R, et al. *Vet Radiol Ultrasound* 2002;43:587. [11]Harvey AM, et al. *J Feline Med Surg* 2009;11:96. [12]Hibbert A, et al. *J Feline Med Surg* 2009;11:116. [13]Guptill L, et al. *J Am Vet Med Assoc* 1995;207:1055. [14]Turrel JM, et al. *J Am Vet Med Assoc* 1988;193:359. [15]Owen NV, et al. *Food Cosmet Toxicol* 1973;11:649. [16]Todd GC. *Vet Pathol* 1986; 23:110.

CYP 450: FAMILIES, PHARMACOGENETICS, & IMPACT OF HEALTH & DISEASE

Butch KuKanich, DVM, PhD, DACVCP
Manhattan, KS

INTRODUCTION & BACKGROUND

Metabolism is a life essential process that occurs whether or not drugs are administered to an animal. Metabolism of both endogenous and exogenous compounds is a process that enhances elimination of a compound from the body by altering the chemical structure to increase the compound's hydrophilicity. Increasing the hydrophilicity of the compound enhances the elimination of the compound in the urine and bile. Metabolism is a basic feature of our patients as they are constantly exposed to hormones, plant and animal chemicals, chemicals in drinking water, and metabolic waste products among other routes of exposure. Metabolism provides a means to enhance elimination of these chemicals otherwise they would continuously accumulate in the animal until toxicity occurred. Although we often term these metabolism enzymes as "drug metabolizing enzymes," it is important to realize drugs we administer are only a small part of the chemicals these enzymes act upon.

The liver is the major organ of metabolism in the body. However metabolic enzymes are located throughout the body and can contribute significantly to drug elimination. The intestines, kidneys, spleen, and skin all contain metabolizing enzymes that can contribute to drug elimination. As an example, dogs undergoing experimental hepatic devascularization had a decrease of morphine systemic clearance from 51.5 mL/min/kg to 27.3 mL/min/kg, less than a 50% reduction in the morphine clearance despite complete loss of hepatic function (Jacqz, et al, 1986).

Two major metabolic pathways occur, Phase I and Phase II reactions. Phase I reactions result in formation of functional groups on the molecule either by addition of the functional group or by removal of a non-functional group subsequently exposing a functional group. The cytochrome P450 (CYP) mixed-function oxidase system is a group of enzymes that is most commonly associated with Phase I reactions, but other enzymes such as alcohol dehydrogenase also catalyze Phase I reactions. The major Phase I reactions are: oxidation, reduction, hydrolysis, and hydration. A rate limiting process of CYP metabolism is the function / activity of the CYP enzyme itself. Increases in CYP content or activation of inactive enzymes will result in increased drug metabolism capability. In contrast, destruction of CYP or inactivation of CYP by competitive or noncompetitive inhibitors will result in decreased capacity for drug metabolism.

Phase II reactions are conjugation reactions. Conjugation reactions add large polar groups to a functional group on the molecule, in which the functional group is either already present or is exposed following a Phase I reaction. The major conjugation reactions are glucuronide, acetyl, sulfate, and amino acid conjugation. Cats are deficient in glucuronide conjugation reactions. Dogs are deficient in acetyl conjugation reactions. Drug interactions occur less frequently with Phase II reactions compared to Phase I reactions.

Metabolism substrates such as drugs can also be classified as high hepatic extraction or low hepatic extraction substrates. Drugs that are almost completely eliminated (>70%) from the blood as it flows through the liver are considered high extraction drugs. For example, fentanyl is a high hepatic extraction drug, because >95% of the drug is eliminated from the blood as it flows through the liver. High hepatic extraction drugs typically have a systemic clearance approaching hepatic blood flow, suggesting the rate-limiting step for high hepatic extraction drugs is the rate of delivery of the drug to the liver by blood flow. These drugs are also termed "flow-dependent" drugs as the elimination is often proportional to blood flow. For example, the systemic clearance of fentanyl in dogs is ~ 30 mL/min/kg, similar to hepatic blood flow in dogs, ~ 30 mL/min/kg suggesting it is a high extraction ratio drug (KuKanich & Hubin, 2010). The metabolism enzymes are essentially in "excess" of the drug concentrations therefore mild to moderate changes in enzyme function have minor effects on the systemic clearance of a high hepatic extraction drug. However moderate to severe changes in hepatic metabolizing enzyme activity will alter systemic drug clearance. Alterations in hepatic blood flow will have proportional changes in systemic clearance of the high hepatic extraction drug. For example, decreases in cardiac output and subsequently hepatic blood flow will result in proportional decreases in the systemic clearance of a high hepatic extraction drug such as fentanyl. Therefore fentanyl clearance is more likely to be affected by uncompensated cardiac disease than mild-moderate hepatic disease.

Low hepatic extraction drugs are often termed "flow independent" drugs, as systemic drug clearance is independent of mild to moderate changes in hepatic blood flow. However the systemic clearance of low hepatic extraction drugs is proportional to the activity of the hepatic metabolizing enzyme. Theophylline is considered a low hepatic extraction drug with a systemic clearance of 1.3 mL/min/kg compared to hepatic blood flow ~ 30 mL/min/kg. Administration of enrofloxacin, a CYP inhibitor, significantly decreased systemic clearance to 0.75 mL/min/kg (Intorre, et al, 1995), resulting in significantly higher maximum plasma drug concentrations.

The effects of metabolism on a drug's activity can be variable. Many times drug metabolites are inactive. For example tramadol is rapidly and predominantly metabolized in dogs to the inactive metabolite N-desmethyltramadol (KuKanich & Papich, 2011). However loss of a drug's activity is not always the case after metabolism. Hydrocodone is metabolized in part to hydromorphone in dogs, with the metabolite (hydromorphone) being approximately equipotent to the parent drug (hydrocodone) (KuKanich & Paul, 2010). Tramadol, which has little opioid effects, is metabolized to O-desmethyltramadol in cats, which has substantial opioid effects (Pypendop & Ilkiw, 2008). Metabolism may also increase the toxicity of some drugs, with the metabolism of acetaminophen to n-acetyl-para-benzoquinonimine (NAPQI) and the subsequent toxicity of NAPQI resulting in methemoglobinemia and hepatic injury (Hinson, et al 2010). Therefore the notion that metabolism inactivates or detoxifies all drugs is incorrect.

CYTOCHROME P450'S

Cytochrome P450 is a superfamily of enzymes that is divided into families in which at least 40% homology occurs and are labeled with numbers. For example CYP1, CYP2, CYP3, etc., are CYP families. The families are subdivided into subfamilies in which at least 55% homology occurs and are labeled with capital letters, for example CYP2B, CYP 2C, CYP2D, etc. The specific enzyme is identified with an additional number, for example CYP2B6 (human), CYP2B11 (dog), CYP2B17 (monkey). The substrate specificity is specific for each enzyme, but as expected much overlap occurs between species and enzymes. Therefore many generalizations are made in dogs, cats, and horses based on enzymatic activity of specific probe substrates, but without molecular identification of the species specific CYPs, some generalizations may be incorrect. For example, S-mephenytoin is selectively metabolized (4'-hydroxylation) by human CYP2C19, but not by the homologous canine cloned enzymes, CYP2C21 and CYP2C41 (Shou, et al, 2003). However it was extensively metabolized by dog liver microsomes suggesting another CYP in dogs actively metabolizes S-mephenytoin.

The CYP1A1 and CYP1A2 appear to be conserved across humans, cats and dogs with approximately 81% protein homology between human and canine (Martignoni, et al, 2006; Tanaka, et al, 2006; Shou et al 2003). CYP2E1 also appears to be well conserved in dogs and humans. CYP2B6 is the human CYP2B enzyme most associated with drug metabolism while CYP2B11 is the dog enzyme with 78% protein homology. CYP2D6 is the human CYP2D enzyme most associated with drug metabolism while CYP2D15 is the analogous CYP2D in dogs with 75% protein homology. CYP2C9 and CYP2C19 are human CYP2C enzymes active in drug metabolism while CYP2C21 and CYP2C41 are the analogous dog enzymes. The protein homology of CYP2C9 with CYP 2C21 and CYP2C41 is 69 and 70%, respectively. The protein homology between CYP2C19 with CYP2C21 and CYP2C41 is 75 and 74%, respectively. CYP3A4/5 are the primary CYP3A enzymes in humans while CYP3A12 and CYP3A26 are the canine isoforms. The protein homology of CYP3A12 with CYP3A4 and CYP3A5 is 79 and 78%, respectively.

CYTOCHROME P450 SPECIES DIFFERENCES

Cytochrome P450 expression and activity is dependent on the CYP isoform and species of interest (Chauret et al, 1997; Shimada 1997). It is also important to realize the reported comparative CYP activities will also be dependent on the CYP probe used and assay as species differences in CYP homology may result in different specificity of the probe for the enzyme. *In vitro* and *in vivo* activity determination of CYP will also affect the estimation of species differences in the various isoforms. In general, CYP3A4 substrate metabolism is greatest in humans, similar in dogs and cats, and markedly lower in horses. CYP2B6 substrate metabolism is greatest in cats, followed by dogs and is much lower in humans. CYP2D6 substrate metabolism is greatest in the horse, followed by the cat, then human, and lowest in the dog. CYP2C9 substrate metabolism is greatest in humans and horses and significantly (~10-20 fold) less in dogs and cats. CYP2C19 substrate metabolism is greatest in horses, followed by humans and dogs, and least in cats. CYP1A2 is similar in humans and dogs and less activity occurs in cats and horses.

CYTOCHROME P450 PHARMACOGENETICS

Several CYP polymorphisms have been identified in dogs. A CYP2D15 polymorphism was identified and characterized for celecoxib metabolism in Beagles where approximately half of the Beagles were poor metabolizers (half-life~5.1 hr) and half were rapid metabolizers (half-life ~1.3 hr) (Paulson et al, 1999). The extent of this polymorphism in the mixed-breed or purebred dog populations has not been characterized, but Greyhounds may be more similar to the poor metabolizers with a half-life of 4 hours (Hunter et al, 2005). Additionally, the extent of this polymorphism for other drug substrates (i.e. tramadol to O-desmethyltramadol) has not been reported.

A CYP1A2 polymorphism has been also identified in Beagles and later identified in other dog breeds including: Irish Wolfhound, Whippet, Dalmation, and Australian Shepherd (Aretz & Geyer, 2011). However the clinical relevance of this polymorphism remains unclear as no clinically used drug has been identified as a CYP1A2 substrate in dogs.

A CYP2C41 polymorphism has also been identified in Beagles and mixed-breed dogs (Blaisdell,et al, 1998). Clinically used drugs associated with CYP2C41 metabolism have not been specifically identified in dogs. It is also unclear if a functional CYP2C21 would override clinically relevant effects of the mutation.

Researchers have speculated that Greyhounds are poor metabolizers of some drugs. The activity of cytochrome P450 2B11 (CYP2B11), as determined by propofol metabolism, was significantly slower in Greyhounds as compared to Beagles, but was not different than mixed breed dogs (Court, et al, 1999). Therefore it is unclear if Greyhounds are poor metabolizers or if Beagles are ultra-metabolizers of propofol. Similarly, the plasma clearance of antipyrine, a nonspecific substrate for CYP activity, was significantly slower in Greyhounds compared to Beagles (KuKanich, et al, 2007). Again it is unclear from the study if Greyhounds are poor metabolizers of antipyrine or Beagles are ultra-metabolizers of antipyrine. The recovery of Greyhounds from thiopental anesthesia is prolonged compared to mixed-breed dogs and has been hypothesized to be delayed due to decreased drug elimination (Sams, et al, 1985). It is unclear as to the extent of any metabolism associated differences in Greyhounds, but metabolism difference do not appear to be extensive as most drugs have similar clearance in Greyhounds as other dog breeds or mixed-breed dogs.

CYTOCHROME P450 AND AGE

The activity of CYP does appear to vary with age as assessed *in vivo*. Activity is the least at 1 week of age for caffeine clearance in dogs (purported to be a CYP1A2 substrate). Significant increases (2.5 fold) in caffeine clearance occur by 3 weeks of age and by 5 weeks of age it is the most rapid, after which it progressively decreases back to the rate at 3 weeks of age when the animal is an adult (Tanaka, et al, 1998). Trimethadione clearance (purported to be a CYP2C, CYP2E, and CYP3A substrate) is again slowest at 1 week of age and significantly increases by 3 weeks of age and peaks at 10 weeks of age and then progressively decreases by 150 weeks of age (adult) to the same rate at 1 week of age (Tanaka, et al, 1998). Therefore puppies between 3-30 weeks are expected to have increased metabolism of CYP 1A, and potentially at least one of CYP2C, CYP2E, and CYP3A, relative to adult dogs. This study suggests a very rapid development of CYP drug clearance, similar to adult rates within 2-3 weeks of age for CYP1A2 and within 1 week of age for some other CYP enzymes.

THE EFFECTS OF HEPATIC DISEASE ON CYTOCHROME P450

The effects of hepatic disease on drug metabolism by CYP will depend on the severity of the disease and drug. High hepatic extraction drugs will be expected to have decreased clearance only in moderate to severe hepatic impairment (see introduction / background). The clearance of low hepatic extraction drugs will be proportional to the degree of hepatic impairment (see introduction / background).

DISCLAIMEER

The information is accurate to the best of the author's knowledge. However recommendations change as new data become available and errors are possible. The author recommends double-checking the accuracy of all information including dosages.

REFERENCES:

Aretz & Geyer. **J Vet Pharmacol Ther**. 2011. Blaisdell, et al. **Drug Metab Dispos**. 1998;26(3):278. Chauret, et al. **Drug Metab Dispos.** 1997;25(10):1130.Court, et al. **Drug Metab Dispos**. 1999;27(11):1293. Hinson, et al. **Handb Exp Pharmacol**. 2010;(196):369. Hunter, et al. **Am J Vet Res.** 2005;66(8):1441. Intorre, et al. **J Vet Pharmacol Ther**. 1995;18(5):352. Jacqz, et al. **Drug Metab Dispos**. 1986;14(6):627. KuKanich & Hubin. **J Vet Pharmacol Ther**. 2010;33(1):42. KuKanich & Papich. **Am J Vet Res.** 2011. KuKanich & Paul. **J Vet Intern Med**. 2010; 24(3):775. KuKanich, et al. **J Vet Pharmacol Ther**. 2007;30(4):314. Martignoni, et al. **Expert Opin Drug Metab Toxicol**. 2006;2(6):875. Paulson , et al. **Drug Metab Dispos**. 1999;27(10):1133. Pypendop & Ilkiw. **J Vet Pharmacol Ther**. 2008;31(1):52. Sams, et al. **Am J Vet Res**. 1985;46(8):1677-83. Shimada, et al. **Arch Toxicol**. 1997;71(6):401. Shou, et al. **Drug Metab Dispos**. 2003;31(9):1161. Tanaka, et al. **Life Sci**. 2006;79(26):2463.

GENETIC VARIABILITY IN PHASE II ENZYMES: IDENTIFYING THE RISK OF DRUG TOXICITY

Lauren A. Trepanier, DVM, PhD, DACVIM, DACVCP
Madison, WI

BACKGROUND

So called "Phase II" enzymes are defined by their role in conjugating drugs and other compounds to small molecules. These reactions include glucuronidation by UDP-glucuronosyltransferases (UGTs), *N*-acetylation by *N*-acetyltransferases, sulfonation by sulfotranferases, methylation by thiopurine methyltransferase (TPMT) or catecholamine-*O*-methyltransferase (COMT), and glutathione conjugation by glutathione-*S*-transferases (GSTs) (Table 1). Conjugation may follow Phase I metabolism (i.e. by cytochrome P450 and others), or may act directly on parent drugs. Phase II reactions typically inactivate compounds, increase water solubility, and promote drug excretion. Genetic variability in Phase II enzymes is well documented in human patients, and is associated with altered drug responses. This area is just beginning to be explored in veterinary patients.

UDP-GLUCURONOSYLTRANSFERASES

UDP-glucuronosyltransferases (UGTs) are expressed in the biliary tree, liver, kidney, colon, and other tissues. There are multiple UGT families that catalyze the glucuronidation of a wide range of drugs, to include propofol, morphine, buprenorphine, doxorubicin, acetaminophen, and oxazepam, as well as endogenous bilirubin, thyroxine, corticosteroids, sex hormones, and bile acids. [1] A promoter polymorphism in UGT1A1 leads to hereditary hyperbilirubinemia (Gilbert's syndrome) in humans. [2] Another UGT1A1 promoter polymorphism (UGT1A1*28, found in 10% of Americans) is associated with myelosuppression and GI toxicity from the antineoplastic drug irinotecan (Camptosar®); this association is strong enough to warrant drug label recommendation for UGT1A1*28 genotyping. [3] UGTs also inactivate dietary and tobacco carcinogens; consistent with this, defective UGT alleles have been associated with an increased risk of gastroesophageal and colorectal cancers. [4,5]

Dogs glucuronidate a variety of drugs, to include acetaminophen, carprofen, morphine (to the inactive morphine-3-glucuronide), [6] codeine and dihydrocodeine, coumarin, omeprazole, dobutamine, and prazocin. Unlike humans, dogs do not glucuronidate propofol [7] or t produce the active morphine-6-glucuronide metabolite. [8] A canine UGT has been cloned, [9] but genetic variability in canine UGTs has not yet been explored.

It is well known that cats are deficient in glucuronidation of some xenobiotics; for example, UDP-glucuronosyltranferase (UGT) activity for acetaminophen is 10-fold lower in cats compared to dogs and humans. [10] This is due to a nonfunctional feline pseudogene for UGT1A6, [11] the primary UGT isoform that metabolizes acetaminophen in humans. Because of this, the threshold for acetaminophen toxicity is 3 to 4-fold lower in cats compared to dogs and humans. [12] In addition, cats have prolonged elimination half-lives for some NSAIDs that are metabolized by UGTs, including aspirin and carprofen; [13] this translates to increased clinical susceptibility to aspirin and carprofen toxicity. However, cats are able to normally glucuronidate endogenous compounds such as thyroxine [14] and bilirubin, [15] as well as some drugs, to include lorazepam [16] and morphine, [17] which are catalyzed by UGT2B7 rather than UGT1A6 in humans. Consistent with this, the elimination half-life of morphine in cats (1-1.5 h) [18] is similar to that in dogs (1.2 h). [8]

N-ACETYLTRANSFERASES

N-acetyltransferases (NAT1 and NAT2) inactivate drugs such as sulfamethoxazole, sulfasalazine, dapsone, hydralazine, procainamide, and isoniazid. High NAT2 activity leads to rapid clearance of isoniazid and may warrant dosage increases in patients with tuberculosis. [19] About half of Caucasians and African Americans have slow NAT2 alleles; slow *N*-acetylation increases the risk of sulfonamide hypersensitivity, [20,21] dapsone methemoglobinemia, and gastrointestinal upset from sulfasalazine. [22]

In dogs, *N*-acetylation activity is absent due to a genome-wide loss of both NAT genes. [23] This may contribute to the prevalence of sulfonamide hypersensitivity in dogs. In cats, hepatic *N*-acetylation is measurable but low, due to the presence of only one NAT1 gene. [24] NAT detoxifies *p*-aminophenol, the acetaminophen metabolite that causes methemoglobinemia. Dogs completely lack this pathway, and cats have very low activity, which may contribute to the development of methemoglobinemia with acetaminophen overdose in both dogs and cats. [25]

SULFOTRANSFERASES

Sulfotransferases (SULTs) conjugate a hydrophilic SO_3- group to endogenous molecules such as dopamine, estradiol, thyroid hormones, and bile salts. SULT1A1, which predominates in human liver and kidney, acts upon drug substrates such as acetaminophen, diethylstilbestrol (DES), and tamoxifen. SULTs can also bioactivate tobacco carcinogens, and high activity SULT1A1 has been associated with an increased risk of bladder cancer in smokers. [26]

In dogs, sulfate conjugates have been found for propranolol and acetaminophen. Several canine SULTs have been cloned. Canine SULT1A is expressed in liver, kidney, lung, and intestine, with high affinity for the prototypical phenolic compound *p*-nitrophenol (which is structurally similar to acetaminophen). [27] Other canine SULTs catalyze the sulfonation of dopamine (SULT1D1)

[28] and the thyroid hormones T3 and rT3 (SULT1B1). [29] SULT polymorphisms have yet to be explored in dogs and SULTs have not yet been cloned in cats.

METHYLTRANSFERASES (TPMT, COMT)

Thiopurine methyltransferase (TPMT) catalyzes several steps in the disposition of azathioprine, with a net overall effect of detoxification. About 5% of human subjects are homozygous for genetic polymorphisms in the TPMT gene that lead to impaired activity. These patients are at high risk for myelosuppression from azathioprine and 6-mercaptopurine, and require adjustments to less than 10% of standard dosages to avoid toxicity. [30] Because of this, measurement of red blood cell TPMT activity has been adopted by many oncologists prior to the use of azathioprine or 6-mercaptopurine for the treatment of human leukemias. In fact, 6-mercaptopruine was the first human drug to have pharmacogenetic testing recommended on its label.

In dogs, we have found wide variability in TPMT activity as well, with about 40% of this variability attributable to coding sequence polymorphisms in the canine TPMT gene. [31] However, we have not established a relationship between TPMT polymorphisms and azathioprine response in dogs; several dogs with azathioprine-associated bone marrow suppression that we have evaluated have had normal TPMT activities (unpublished observations).

Cats overall have TPMT activities that are about one-fifth of canine levels. [32-34] This may explain why cats as a species are especially sensitive to myelosuppression from azathioprine. [35] Further individual variability in TPMT among cats (more than 9-fold) can be attributed to genetic polymorphisms in the feline gene. [33]

Another Phase II enzyme, catecholamine-O-methyltransferase (COMT), metabolizes epinephrine, norepinephrine, and dopamine. Low activity human COMT alleles have been linked to increased pain sensitivity in fibromyalgia [36] and during migraines. [37] These low-activity alleles, which presumably lead to higher receptor availability of catecholamines, have also been associated with enhanced responses to propranolol in patients with temporomandibular pain. [38] Conversely, high activity COMT predicts a good response to methylphenidate, with fewer side effects, in children with ADHD. [39]

COMT has been cloned in dogs and one polymorphism has been identified to date (482 G>A). [40] However, canine COMT has not been functionally characterized or studied in relation to drug response in dogs. COMT activity is present in cat liver, but is less than 10% of that in human liver, possibly due to the absence of one of the two COMT isoforms; [41] however, the molecular mechanism for this is as yet unknown.

GLUTATHIONE-S-TRANSFERASES

Glutathione-S-transferases (GSTs) are Phase II biotransformation enzymes that detoxify many xenobiotics, including environmental carcinogens, via conjugation to glutathione. There are multiple families of GSTs in humans, to include, among others, GST-mu (GSTM), GST-pi (GSTP), and GST-theta (GSTT). These families include various isoforms, and exhibit considerable inter-individual variability in expression and activity. Both GSTM1 and GSTT1 show a "null" genotype in humans, in that the respective genes are partially missing at the genomic level, and expression and activity are absent. [42] These GSTM1 and GSTT1 null alleles are present in about 50 and 20% of Americans, respectively. Defective GST polymorphisms have been associated with the risk of lymphomas in humans; [43,44] this increased risk is thought to be due to impaired detoxification of environmental pollutants. For example, the chlorinated hydrocarbon dichloromethane is a GSTT1 substrate, and exposure to chlorinated hydrocarbons is significantly associated with lymphoma in humans. [45] GSTs also inactivate many antineoplastic drugs, including cisplatin, chlorambucil, melphalan, and the cyclophosphamide metabolite acrolein. This may explain why low activity GST polymorphisms have also been associated with enhanced response to, and toxicity from, chemotherapy in humans with leukemias and non-Hodgkin's lymphoma. [43,46]

In dogs, hepatic GSTT activities vary more than 45-fold, [47] with 12% of beagles lacking GSTT activity and protein expression. [47,48] Other GST isoforms have not been evaluated. Our laboratory is working to characterize the genetic basis for this wide GSTT variability in dogs, and to determine its significance in both drug toxicity and lymphoma risk.

REFERENCES

1. Bock KW. *Biochem Pharmacol*;80:771-777. 2. Clarke DJ, et al. *Clin Chim Acta* 1997;266:63-74. 3. Perera MA, et al. *Pharmacotherapy* 2008;28:755-768. 4. Tang KS, et al. *World J Gastroenterol* 2005;11:3250-3254. 5. Vogel A, et al. *Z Gastroenterol* 2002;40:497-502. 6. Merrell WJ, et al. *Anesthesiology* 1990;72:308-314. 7. Liang SC, et al. *J Pharm Biomed Anal*;54:236-241. 8. KuKanich B, et al. *J Vet Pharmacol Ther* 2005;28:371-376. 9. Soars MG, et al. *Biochem Pharmacol* 2003;65:1251-1259. 10. Court MH, et al. *Biochem Pharmacol* 1997;53:1041-1047. 11. Court MH, et al. *Pharmacogenetics* 2000;10:355-369. 12. Savides MC, et al. *Toxicol Appl Pharmacol* 1984;74:26-34. 13. Parton K, et al. *J Vet Pharmacol Ther* 2000;23:73-79. 14. Hays MT, et al. *Endocr Res* 1988;14:51-58. 15. Spivak W, et al. *Biochem J* 1985;225:787-805. 16. Schillings RT, et al. *Drug Metab Dispos* 1975;3:85-88. 17. Krishnaswamy S, et al. *Drug Metab Dispos* 2004;32:862-869. 18. Taylor PM, et al. *J Vet Pharmacol Ther* 2001;24:391-398. 19. Kinzig-Schippers M, et al. *Antimicrob Agents Chemother* 2005;49:1733-1738. 20. Carr A, et al. *AIDS* 1994;8:333-337. 21. Shear NH, et al. *Annals of Internal Medicine* 1986;105:179-184. 22. Pullar T, et al. *Ann Rheum Dis* 1985;44:831-837. 23. Trepanier LA, et al. *Biochemical Pharmacology* 1997;54:73-80. 24. Trepanier LA, et al. *Pharmacogenetics* 1998;8:169-179. 25. McConkey SE, et al. *J Vet Pharmacol Ther* 2009;32:585-595. 26. Wang YH, et al. *Int J Urol* 2008;15:1029-1034. 27. Tsoi C, et al. *Arch Biochem Biophys* 2002;401:125-133. 28. Tsoi C, et al. *Biochem J* 2001;356:891-897. 29. Tsoi C, et al. *Arch Biochem Biophys* 2001;390:87-92. 30. Wang L, et al. *Oncogene* 2006;25:1629-1638. 31. Salavaggione OE, et al. *Pharmacogenetics* 2002;12:713-724. 32. Foster AP, et al. *J Vet Intern Med* 2000;14:552-554. 33. Salavaggione OE, et al. *J Pharmacol Exp Ther* 2004;308:617-626. 34. White SD, et al. *J Vet Intern Med* 2000;14:499-502. 35. Beale KM, et al. *Am J Vet Res* 1992;53:1236-1240. 36. Barbosa FR, et al. *Rheumatol Int.* 37. Park JW, et al. *J Clin Neurol*

2007;3:24-30. 38. Tchivileva IE, et al. *Pharmacogenet Genomics*;20:239-248. 39. Kereszturi E, et al. *Am J Med Genet B Neuropsychiatr Genet* 2008;147B:1431-1435. 40. Dias Pereira P, et al. *Vet Pathol* 2008;45:124-130. 41. Huh MM, et al. *J Biol Chem* 1979;254:299-308. 42. Hayes JD, et al. *Annu Rev Pharmacol Toxicol* 2005;45:51-88. 43. Hohaus S, et al. *Clin Cancer Res* 2003;9:3435-3440. 44. Kerridge I, et al. *Br J Haematol* 2002;118:477-481. 45. Seidler A, et al. *J Occup Med Toxicol* 2007;2:2. 46. Hohaus S, et al. *Clin Cancer Res* 2005;11:2175-2179. 47. Watanabe T, et al. *Arch Toxicol* 2004;78:218-225. 48. Watanabe T, et al. *Arch Toxicol* 2006;80:250-257.

Table 1: Phase II conjugating enzymes: substrates and known polymorphisms in humans, dogs, and cats

Phase II enzyme	**Species**	**Substrates in humans**	**Polymorphisms**
UDP-glucuronosyltransferases	Human	Depends on isoform: propofol, morphine, buprenorphine, doxorubicin, acetaminophen, oxazepam, thyroxine, endogenous bilirubin, corticosteroids, sex hormones, and bile acids Known substrates in dogs: morphine (inactivation), hydromorphone, carprofen, codeine, coumarin, omeprazole, dobutamine, prazocin	UGT1A1*28, found in 10% of Americans, leads to myelosuppression from the antineoplastic drug irinotecan
	Dog		None characterized
	Cat		Cats have pseudogene for UGT1A6; lack glucuronidation for acetaminophen and some other NSAIDs
N-acetyltransferases	Human	NAT2: potentiated sulfonamides, dapsone, hydralazine, sulphasalazine, procainamide	About half of all Caucasians and African Americans have defective *NAT2* activity; confers increased risk of sulfonamide hypersensitivity, dapsone methemoglobinemia, and sulphasalazine toxicity
	Dog		Both *NAT1* and *NAT2* genes absent in dogs and all canids; dogs lack all *N*-acetylation activity
	Cat		Cats and other felids lack *NAT2* but express *NAT1*; increased susceptibility to hematologic toxicity from the acetaminophen metabolite *p*-aminophenol
Sulfotransferases	Human	Primarily endogenous compounds (dopamine, estradiol, thyroid hormones, and bile salts) SULT1A1 acts upon acetaminophen, diethylstilbestrol (DES), tamoxifen	Rapid SULT1A1 leads to bioactivation of tobacco carcinogens and increases risk of bladder cancer in smokers
	Dog		No canine polymorphisms characterized
	Cat		Feline SULTs not characterized
Thiopurine methyltransferase	Human	Azathioprine metabolites	Patients with defective TPMT alleles have a high risk of azathioprine myelosuppression Patients with high activity alleles have poorer remission rates
	Dog		Dogs have wide variation in TPMT activity. Low activity seen in some Giant Schnauzers; high activity seen in some Malamutes Clinical significance not yet established
	Cat		Feline TPMT activity is about 80% lower than in dogs; may explain feline risk of myelosuppression from azathioprine
Catecholamine-*O*-methyltransferase	Human	Epinephrine, norepinephrine, and dopamine	High activity COMT alleles associated with ADHD drug response
	Dog		One canine COMT SNP identified to date, but not characterized
	Cat		COMT activity in feline liver is only 10% of that in human liver
Glutathione-S-transferases	Human	Chlorambucil, melphalan, cisplatin, acrolein, industrial solvents (dichloromethane)	GSTM1 and GSTT1 “null” alleles (partial gene deletion) present in 50 and 20% of humans, respectively Confers increased risk for lymphoma and other cancers
	Dog		Wide range of GSTT1 activity in dog liver, with 12% of beagles lacking GSTT1 activity; substrates yet uncharacterized
	Cat		Feline GST genes not characterized

GENETIC VARIABILITY IN DRUG TRANSPORTERS: IMPACT ON HEALTH AND DISEASE

Katrina L. Mealey DVM, PhD, DACVIM, DACVCP
Pullman, WA

INTRODUCTION

Once a drug is administered to a patient, its fate is dictated by its physiochemical characteristics (i.e., size, lipophilicity, charge). These properties affect drug absorption, distribution, metabolism and excretion, all of which are dependent on drug movement through cell membranes. Movement of a drug molecule through cell membranes can occur by simple diffusion down a concentration gradient. Drug molecules can also be transported against a concentration gradient by drug transport pumps. One of the most important drug transport families is the ATP-Binding Cassette (ABC) family. ABC transporters are expressed on cell membranes where they serve primarily as drug efflux pumps. The ABC transporter family contains over 40 members which have been subdivided into 7 subfamilies (designated ABC 'A' through ABC 'G'). Thus far, only 4 members of the ABC transporter family are known to be responsible for transporting drugs that are commonly used in dogs and cats: ABCB1 (P-glycoproein), ABCG2 (Breast Cancer Resistance Protein), ABCC1 (Multidrug Resistance Related Protein-1) and ABCC2 (Multidrug Resistance Related Protein-2). Among these, the most well-characterized in terms of clinically relevant polymorphisms in both human and veterinary medicine are ABCB1 and ABCG2. Polymorphisms in these drug transporters can contribute to variability in drug pharmacokinetics and pharmacodynamics. However, because of overlap in substrate specificity, a defect in function of one transporter may be "covered" by another transporter (Table 1). Information gleaned from research in humans and rodent models has, to date, been applicable to veterinary patients. Specifically, substrates for a particular ABC transporter in humans are also substrates for the canine and feline orthologues. Similarly, compounds that inhibit a particular ABC transporter in humans also appear to inhibit the orthologous canine and feline transporters. What does not cross species lines is the particular genetic polymorphisms that have been described for each of these transporters. In two important genes, ABCB1 and ABCG2, the polymorphisms identified in veterinary species have had greater impact on drug disposition than polymorphisms described in human patients. This article will provide an overview of the pharmacogenetics of selected ABC drug transporters, namely ABCB1 (P-gp) and ABCG2 (BCRP) because these have been characterized (at least to some degree) in dogs and cats, respectively.

Table 1. Partial listing of drugs/drug classes that are substrates for four ABC Transporters

Drug Class	ABCB1	ABCG2	ABCC1	ABCC2
Fluoroquinolones	some	most	some	some
Anthracyclines	most	most	some	some
Vinca Alkaloids	all	none	some	some
Macrocyclic Lactones	all	none	none	none
Immunosuppressants	some	some	none	none
Loperamide	yes	no	no	no
Digoxin	yes	no	no	no

ABCG2 (BREAST CANCER RESISTANCE PROTEIN OR BCRP)

BCRP is expressed on enterocytes, the canalicular membrane of hepatocytes, proximal renal tubule cells, erythrocytes, stem calls and functions as an important component of the blood-brain barrier and the blood-retina barrier[1]. At these locations, BCRP functions in a protective capacity excreting substrate drugs into the intestinal lumen, bile, and urine, or restricting access of substrate drugs to sensitive brain and retinal tissues. Additionally, BCRP is expressed in mammary tissue where it transports substrates into milk[2]. A number of polymorphism in ABCG2 have been described in humans. The most well-characterized polymorphism, ABCG2 421C>A results in a glutamine (polar) to lysine (nonpolar) amino acid change[3]. Affected individuals have decreased protein expression and lower ABCG2 transport capacity compared to individuals with the wildtype allele.

While canine ABCG2 has not been well-characterized, characterization of feline ABCG2 was recently described[4]. Feline ABCG2 was sequenced and the consensus amino acid sequence was compared to that of 10 other mammalian species. Four feline-specific amino acid changes in conserved regions of ABCG2 were identified. The effect of these feline-specific amino acid changes on ABCG2 function was assessed using transfection experiments. cDNA constructs for feline and human ABCG2 were inserted in a pcDNA3 expression vector and expressed in HEK

293 cells. ABCG2 transport function was assessed using the two classic ABCG2 substrates (mitoxantrone and BODIPY-prazosin). The ability of feline ABCG2 to transport either substrate was defective. These results have a number of important clinical consequences.

Fluoroquinolone antimicrobials have been extensively used in veterinary medicine for approximately 20 years. Domestic cats experience an unusual, and apparently species-specific, adverse drug reaction associated with fluoroquinolones—acute retinal degeneration and blindness[5,6]. This toxicity was first documented with enrofloxacin, which is structurally similar to the human-approved fluoroquinolone ciprofloxacin. Enrofloxacin received FDA approval for use in cats in 1990 at a dosage of 2.5 mg/kg/day per os. In 1997, based on results of safety studies in cats, a flexible dosing label for enrofloxacin was approved, consisting of 5 to 20 mg/kg, per os, once daily or as a divided daily dose. Shortly after introduction of the flexible dosing label, veterinarians reported blindness in cats receiving enrofloxacin, particularly after once-daily administration of doses at the high end of the dose range. In a prospective study, enrofloxacin was administered at a dose of 50 mg/kg to 12 cats while 12 cats received saline control. All enrofloxacin-treated cats developed retinal degeneration as early as 3 days after enrofloxacin administration, while none of the control cats developed retinal lesions[5]. Histopathologic changes included severe vacuolization progressing to necrosis in the photoreceptor layer, outer nuclear layer and outer plexiform layer of the retina. Thus, high doses of enrofloxacin are acutely toxic to the retina of cats. Orbifloxacin, another fluoroquinolone approved for use in cats and dogs, also causes retinal damage in cats (www.fda.gov freedom of information summary NADA 141-081).

In human patients, cutaneous photosensitivity is described as one of the more common adverse reactions to fluoroquinolones[7]. When exposed to light, fluoroquinolones generate reactive oxygen species (ROS) that attack cellular lipid membranes causing tissue damage. Excepting the skin, the organ most susceptible to phototoxicity is the eye, including the retina[8]. Extensive phototoxic damage to the retina can lead to retinal degeneration and blindness. Under normal conditions, the retina is protected from photoreactive compounds by the blood-retina barrier. The blood-retinal barrier consists of both capillary endothelial cell tight junctions and retinal pigment epithelial cell tight junctions as well as important transporters including ABCG2 that function to prevent distribution of drugs to the retina[9,10]. ABCG2 is strategically positioned on the luminal membrane of retinal capillary endothelial cells where it actively transports substrates from the endothelial cell back into the capillary lumen. Because fluoroquinolones are substrates for ABCG2[11,12] their distribution to retinal tissues is normally restricted by ABCG2.

Dysfunction of feline ABCG2 at the blood retinal barrier results in accumulation of photoreactive fluoroquinolones in feline retina. Exposure of the retina to light then generates ROS that cause the characteristic retinal degeneration and blindness documented in cats receiving high doses of fluoroquinolones. Defective BCRP function in cats may also contribute to their sensitivity to acetaminophen. In dogs and humans, acetaminophen is metabolized primarily by glucuronidation and sulfation to nontoxic metabolites. Cytochrome P450 enzymes, which generate a toxic metabolite (NAPQI), are not involved. Relative to other species, cats have low hepatic levels of UDP-glucuronyl transferase and are unable to use glucuronidation pathways. Thus, acetaminophen metabolism is shifted to the sulfation pathway. In most species ABCG2 mediates the biliary excretion of acetaminophen sulfate. However, in *abcg2* knockout mice, biliary excretion of acetaminophen sulfate was negligible[13]. Because feline ABCG2 is also dysfunctional, a similar phenomenon would be expected to occur in cats. This would result in shunting of acetaminophen metabolism away from sulfation in favor of cytochrome P450-mediated pathways where the toxic metabolite NAPQI is generated. Thus, defective ABCG2 function in cats may be a significant contributing factor to the high susceptibility of cats to acetaminophen toxicity.

A long list of substrates and inhibitors of ABCG2 already exists, and the list continues to expand. With regard to cancer therapy, important substrates of BCRP include mitoxantrone, camptothecin derivatives, etoposide and methotrexate. Of particular interest, due to the recent FDA approval of toceranib for treatment of canine mast cell tumors, many tyrosine kinase inhibitors have been shown to be substrates for ABCG2[1]. In fact tyrosine kinase inhibitors are reported to be both substrates for and inhibitors of ABCG2. Whether or not defective ABCG2 function in cats will affect efficacy and/or toxicity of tyrosine kinase inhibitors has yet to be determined.

ABCB1 (P-GLYCOPROTEIN; P-GP)

P-gp pharmacogenetics has been fairly well characterized in dogs and is described in more detail elsewhere[14]. P-gp substrates include many anticancer drugs (anthracyclines, vinca alkaloids, epipodophyllotoxins), macrocyclic lactones (ivermectin, selamectin, milbemycin, etc), loperamide, and other drugs. A number of studies have addressed P-gp's role in drug absorption, distribution and excretion and how these pharmacokinetic parameters affect drug safety and efficacy. Much of this information has been generated from studies and clinical observations of dogs affected by a mutation in the ABCB1 gene. The ABCB1 polymorphism in dogs consists of a 4 base-pair deletion mutation[15]. This deletion results in a shift of the reading frame that generates several premature stop codons. Because protein synthesis is terminated before even 10% of the protein product is synthesized, dogs with two mutant alleles exhibit a P-gp null phenotype, similar to mdr1 (-/-) knockout mice. Affected dogs include many herding breeds. For example, roughly 75% of collies in the United States, France, and Australia have at least one mutant allele.

Drug distribution can be dramatically affected by alterations in P-gp function. Because P-gp is a component of the blood-brain barrier, the blood-testes barrier, and the placenta, distribution of P-gp substrate drugs to these tissues is greatly enhanced in ABCB1(mut/mut) dogs. ABCB1(mut/mut) dogs experience adverse neurologic effects after a single dose of ivermectin (>120 μg/kg). Affected dogs also appear to have increased susceptibility to neurologic adverse effects of other avermectins (milbemycin, selamectin, and moxidectin) as well as loperamide. Loperamide is an opioid that is generally devoid of CNS activity because it is excluded from the brain by P-gp. At routinely used therapeutic doses, loperamide causes profound CNS depression in ABCB1 mutant/mutant dogs[16]. Less information is available regarding P-gp and the blood brain barrier in cats. The author has received anecdotal reports of ivermectin toxicity in cats after standard doses, but whether or not the underlying cause is a result of altered P-gp expression or function is not currently known.

P-gp can also affect the elimination of drugs, including anticancer agents, from the body. P-gp is expressed on renal tubular cells and biliary canalicular cells, suggesting that it plays an important role in drug excretion. Concurrent administration of a P-gp-inhibitor decreases the biliary and renal clearance of doxorubicin in rats. Studies comparing biliary excretion of the P-gp substrate ^{99m}Tc-sestamibi in ABCB1 WT/WT and ABCB1 mut/mut dogs demonstrate that ABCB1 mut/mut dogs do not excrete this drug into bile. ^{99m}Tc-sestamibi is essentially undetectable in gallbladders of ABCB1 mutant/mutant dogs but is highly concentrated in gallbladders of ABCB1 WT/WT dogs[17]. Diminished biliary drug excretion may play a role in the apparent increased sensitivity of herding breeds to chemotherapeutic drugs that are P-gp substrates. For example, ABCB1(mut/mut) and ABCB1(WT/mut) dogs are significantly more likely to develop hematologic toxicity (neutropenia and thrombocytopenia) after treatment with the P-gp substrate vincristine than ABCB1 (WT/WT) dogs[18]. However, these dogs tolerate cyclophosphamide (a drug that is not a P-gp substrate) at full doses. In dogs, vincristine is eliminated primarily via biliary excretion of parent drug with some urinary excretion of parent drug and metabolites. Therefore, reduced doses of P-gp substrate chemotherapeutic agents are recommended in ABCB1 mut/mut and ABCB1 WT/mut dogs.

INHIBITORS OF ABC TRANSPORTERS CAN MIMIC PHARMACOGENETICALLY MEDIATED ADVERSE REACTIONS

Because of the evidence that P-gp, and ABCG2 contribute to multidrug resistance in cancer patients, numerous drugs have been developed to inhibit these transporters in an effort to improve outcome in patients treated with chemotherapy. Clinical trials involving inhibitors of ABC transporters have resulted in unexpected and undesired pharmacokinetic interactions between the inhibiting agents and the chemotherapeutic drugs used for treating cancer patients. For example, inhibiting agents often block renal and/or biliary excretion of anti-cancer drugs by the ABC transporter resulting in their accumulation in plasma which then increases their toxicity. Inhibiting ABC transporters not only blocks excretion of drugs, but alters their distribution as well. Inhibiting P-gp has been shown to enhance brain penetration of substrate drugs in dogs (mimicking what is seen in MDR1 mutant/mutant dogs). For example ketoconazole increases brain penetration of ivermectin in MDR1 WT/WT dogs causing neurological toxicity. Inhibiting ABCG2 is likely to enhance retinal accumulation of substrate drugs. Thus, concurrent administration of a fluoroquinolones and an ABCG2-inhibitor has the potential to cause retinal degeneration and blindness in any species.

REFERENCES

1. Cusatis and Sparreboom, Pharmacogenomics;9:2008. 2. Jonker et al, Nat Med;11:2005. 3. Tamura et al, J Exp Ther Oncol;6:2006. 4. Ramirez et al, Pharmacogenomics: 21:2011,66. 5. Ford et al, Am J Vet Res;68:2007. 6. Wiebe and Hamilton, J Am Vet Med Assoc;221:2002. 7. Melhorn and Brown, Ann Pharmacother;41:2007. 8. Roberts, J Photochem Photobiol;64:2001. 9. Asashima et al, Pharm Res;23:2006. 10. Tagami et al, Brain Res 1283 :2009. 11. Alvarez et al, J Pharm Sci ;97 :2008. 12. Merino et al, Drug Metab Dispos ;34 :2006. 13. Zamek-Gliszczynski et al, Mol Pharmacol ;70 :2006. 14. Cascorbi, Pharmacol Ther ;112 :2006. 15. Mealey et, Pharmacogentics, 2001;11:727-733. 16. Mealey et al, Drug Metab Disp, 2008;36:1073-1079. 17. Coelho et al J Vet Pharm Ther;32:2009. 18. Mealey et al, J Vet Int Med, 2008;22:996-1000.

TARGETING MYOCARDIAL REMODELING: THE ROLE OF CARDIOACTIVE DRUGS

AR Dillon DVM, MS, MBA, DACVIM (SAIM)
Auburn University, AL

INTRODUCTION

Cardiac remodeling involves cellular, molecular, and interstitial changes that are evaluated globally via imaging for shape and function, and clinically via changes in physical signs. Although in human literature, "cardiac remodeling" was first used to describe changes after myocardial infarction, other injury induces similar but different changes associated with pressure overload (hypertension, aortic stenosis), idiopathic dilated cardiomyopathy, myocarditis, and volume overloads. Because of the limitation of space, this presentation will address the most common heart disease of dogs, mitral insufficiency, and concentrate on the early phases of remodeling.

Cardiac hypertrophy is consistent with cardiac remodeling, which occurs with elevated work loads as a means of reducing ventricular wall and septal stress; and is classified as "physiological" when associated with exercise and "pathological" in response to cardiac disease. The early dysfunctional remodeling in mitral regurgitation (MR) has the ultimate goal of reducing wall stress, even in volume overloads, but the ultimate result is loss of ventricular function and heart failure. This presentation will selectively consider drugs considered to affect remodeling.

EFFECTS OF EARLY REMODELING IN CANINE MR

The majority of the emphasis of clinical therapy for MR in dogs has focused on the various treatment modalities once the dog has become "clinical", typically from preload associated with enlarged left atria or pulmonary edema. The struggle for quality of life during obvious heart failure has been frustrating in the inability to make major strides in the extension of life. Auburn University, in collaboration with the University of Alabama at Birmingham Medical Center, has for the last 15 years been involved in studies of early cardiac remodeling. Models have included ischemic cardiomyopathy (coronary artery occlusion in dogs and humans), diabetic heart disease in humans, pressure overloads (aortic banding in dogs), mitral regurgitation (MR) in humans, and MR in dogs induced by mitral valve chordal rupture. Early cardiac remodeling associated with volume overload is distinctively different to that of pressure overloads of dogs and humans, idiopathic cardiomyopathies, and ischemic cardiomyopathies.

The data currently suggest that early in the course of MR of dogs, before the first clinical disease or even global change in cardiac MRI or echocardiographic measures can be detected, significant cellular and molecular remodeling have been initiated. This very early remodeling, within the first month of mild experimental MR in dogs, includes myocyte growth without proliferation, altered excitation-contraction proteins, abnormal Ca flux, expression of fetal genes, and alterations of myocyte contractility. To emphasize the point, before any global changes in the heart can be determined, there is significant elongation of cardiomyocytes, loss of extracellular matrix, and annular ring changes, and early intracardiac dyssynergy.

Attempts at early intervention in this remodeling with current medications (Angiotensin converting enzymes, Angiotensin II receptor blockers, mast cell inhibitors, diuretics, etc) have not slowed the progression of the remodeling or delayed the projected onset of clinic disease (a non-specific target in spontaneous disease). All of these various drugs have obvious justified uses at various stages in this progressive deterioration of cardiac function. Although early intervention of some of these agents inappropriately may in fact accelerate the dysfunction. The later stages of MR resulting in very elongated myocytes, poor myocardial contractility, abnormal adhesions myocyte molecules (FAT and tinin), loss of cardiac collagen and matrix, chamber dilation, and systolic and diastolic dysfunction. To emphasize the point, based on cardiac catheterization, cardiac MRI, and echocardiography, despite all the early and late phases of remodeling of the heart in canine MR, the forward cardiac output (stroke volume/ beat via the aorta) never equals 75% of pre MR, even as the total stroke volume (volume/beat via aorta and atrium) may be 2X the pre MR volume.

The heart simply cannot overcome the pathology of the real cause of heart failure which is the abnormal mitral valve apparatus. Attempts to alter the course of the primary disease or slow remodeling in humans has had limited success, thus MR in man is a surgically disease with minimal documented therapy to delay the time to surgery.

TARGETS OF EARLY REMODELING

Support of Cardiac Output via Afterload Reduction: Blocking experiments RAS

Dilation of the LV is associated with myocyte stretch, annular stretch, loss of collagen, and loss of myocyte to myocyte connection for systolic function. Early observations of an increase in the mast cell numbers was directly correlated with the increased intracellular concentrations of Chymase, via in vivo microdialysis techniques, and the loss of collagen in MR dogs in the first phases of canine MR. The resultant lengthening of cardiomyocytes was further detected in these initial experiments. With the discovery that dogs and humans have Chymase which converts Angiotensin I to Ang II much faster than ACE and is present in the LV interstitium in high concentrations in early MR, it was perceived as a potential target.

Early attempts to decrease afterload via ACE-inhibitors or ANG II receptor blockers in dogs with MR demonstrated that there was no beneficial affect on forward cardiac output or remodeling. Even when systemic vascular resistance was returned to normal in early MR dogs, there was no demonstrated hemodynamic effect and myocardial remodeling and collagen loss were not attenuated, and in fact collagen loss and myocyte stretch were more severe than untreated controls. These data led to an accompanying cautionary editorial comment in JACC to human MDs on the lack of support and implied danger of ACE-inhibitor therapy early in volume overload diseases in humans. Because of the lack of any demonstrated beneficial affect from RAS blockade in early MR dogs and the attending increase in loss of interstitial collagen and myocyte stretch in treated dogs compared to controls, the proposed hypothesis of afterload reduction attenuate cardiac remodeling, chamber dilation or to increase cardiac output was not been supported in dogs with early MR. Because the myocardial interstitial conversion of ANGI to ANGII in dogs is controlled by chymase ACE was not effective in myocardium, even in the presence of extreme ANGII receptor blockade, based on mRNA and AT1 receptor densities, the cardiomyocytes simply adjust to the blockade by increasing receptors. The effects of ANG II on peripheral arterial resistance is ACE dependent, and the effects of ACE Inhibitors late in heart failure, after majority of remodeling has already occurred, has been documented.

Support of Myocardial Collagen and Matrix:

Further attempts to attenuate Chymase activity was via mast cell blockade which did slow the cardiac LV enlargement, did attenuate the collagen loss, and clinically the heart "size" appeared better that other treatments, but the drug was toxic to cardiomyocytes and the result was a globally less enlarged heart with a rigid hypocontractile myocardium. The loss of collagen as early as 2 and 4 weeks after minimal MR in dogs and the appearance of mast cells suggest a cause and effect relationship. With the ability of chymase to active MMPs and initiate destruction of the matrix, the relatively quick presence of mast cells in response to physical LV wall stress is impressive. With the evidence that MR in dogs is associated with collagen loss, the use of aldosterone as an antifibrotic agent (as in hypertension and ischemic disease of man) would not have support in MR of dogs.

Extracellular Matrix: Genes activation in dog MR is anti-matrix

Using comparative gene array 42,000 genes, 659 were altered by MR in the LV of dogs, including 217 upregulated and 442 downregulated. Western blot were used to demonstrate gene expression of key genes. Basically, the MMPs and genes associated with collagen destruction were upregulated and the noncollagen genes important in extracellular matrix structure were downregulated (decorin, fibulin 1, fibrillin 1, connective tissue growth factor). Further multiple genes in transforming growth factor–beta pathway were decreased and decreased activity confirmed. Basically, with the initiation of MR in dogs, the dog simply fails to replace interstitial collagen in the heart with the inevitable result of LV dilation and myocardial cell stretch and stress. Gene pathway analysis demonstrated the multiple pathways in the dog heart with MR is closely associated with known inflammatory, immune mediated, and cardiac processes of humans.

Specific Chymase Inhibition:

With the down regulation of ECM scaffolding genes and TGF-B signaling, the Chymase associated with mast cells was targeted with a specific Chymase inhibitor to "protect" the collagen from the activation of MMPs by the Chymase. In dogs with experimental mild MR, Chymase inhibition not only normalized the anticipated decrease in fibronectin and FAK phyosphorylation, but prevented cardiomyocyte myofibrillar degeneration. Total tinin and stiffer isoform paralleled the preservation of matrix. The result was an increase in the torsion angle of the LV during contraction demonstrated via tagged MRI, supporting the suggestion that conservation of matrix participates in myocardial function and perhaps a tinin-dependent cardiomyocyte shortening in MR of dogs. Because experiments to block Chymase activation of ANG II with ACE inhibitors and ANG II receptor blockers early in MR did not preserve collagen or myocyte function, these data suggest that direct role of Chymase may be to be a target in early MR to assist both the matrix and myocytes. A 12X increase in LV bradykinin in MR dogs was also normalized by Chymase inhibition, again theory supporting the protection of matrix because bradykinin in some models will reduce fibronectin mRNA. However, despite these early results, the loss of collagen and eventual dilation of LV was not attenuated by Chymase inhibition alone, as the downregulation of matrix production is probably unaffected.

Targeting an "Acute" Protective Sympathetic System

Blocking the systemic sympathetic neuroendocrine or especially the directly "wired" portion of the sympathetic NS on myocardial function has been extensively examined in various models. Central to understanding is the distinctly different myocardial environment of the myocytes early in MR vs late in MR. Using the microdialysis techniques on MR dog LV, excessive norepinephrine (NE) release into the interstititium in early MR was demonstrated which is directly via stellate ganglion adrenergic stimulation which is augmented by Angiotensin II receptor activation. Continued activation of neuronal catecholamine release into the interstitial fluid (ISF) may be necessary in sustaining LV function initially; however, there is an attendant risk of functional deterioration. Early in MR of dogs, marked increases (5X) in concentrations of NE in myocardial ISF are associated with toxic myocyte death, receptor desensitization, and diminished functional responses to sympathetics. Increased ISF NE causes oxidative metabolites which increases MMPs and is toxic to nerves and myocytes. The result of constant stellate activation on the heart is a decrease in the mRNA for NE and Epi myocardial receptors and a failure of myocytes to respond to activation. Initial use of beta blockade alone had a slight transient decrease on myocyte function, but continued use of beta2 specific drugs was associated with the return of beta2 receptors on myocytes and an increased responsiveness to adrenergic stimulation to normal in these dogs with early MR. The extended trials in dogs early in MR did result in an attenuation of myocyte stretch and an improvement in cardiac output compared to control MR dogs but with the anticipated result that the matrix loss was less attenuated that the cardiomyocyte support. The isolated myocytes had a return to normal in Ca flux and improved contractility. Important to remember that these results reflected changes early in volume overload (pre-clinical signs), not later in MR when systolic dysfunction can be demonstrated. However, the utility to use selective beta blockers early in MR when only a murmur is detected to delay remodeling is being evaluated.

Support the Myocyte

Calcineurin and Calmodulin-dependent Kinase II pathyways, antihypertrophic and/or alterations of ANP production and receptors, L and T type calcium channel modulators, oxidative stress modulators, etc all represent a multitude of targets which more directly influence the cardiomyocyte. The phosphodiesterase superfamily expressed in the myocardium has been targets in humans and dogs for a variety of vascular and cardiac issues. The data set on PDE inhibitors in early stages of MR in dogs is incomplete. The wide class of inotropes and inodilators (levosimendan) each with different mechanisms, and some with effects outside of the initial anticipated activity, as in Ca sensitizers of tropopin C by pimobendan. The discussion of specifics is too broad for this venue. Use of clinical trials are difficult where entry into studies do not control for prior drug administration and clinical disease is advanced beyond reversible remodeling. Support of cardiomyocytes in systolic failure is a different question than in early remodeling in MR. The open question to any direct myocardial supportive drug is can we anticipate at what point in MR the myocytes are starting into systolic dysfunction, and would early intervention in MR be advantageous to the cardiomyocyte or damage the valves apparatus or increase the regurgitant volume in a clinically compensated stage of the disease.

FUTURE TARGETS

First the caveat that "it is not the ventricles fault" in routine MR of dogs, blame the phenotypic expression of the genetic disease of the valvular apparatus. Drugs directed to alter valvular changes may be breed specific as there may be different genotypic diseases of the valves. However the myocardial response to the valvular disease, very early in MR before the dog has any clinical disease or minimally detectable chamber enlargement, is extensive collagen loss, severe cardiomyocyte elongation and wall stress, and unhealthy cardiomyocytes with decreased contractility. Further, the genetic response in the LV is to not rebuild the supporting matrix of the heart and this predicts the outcome.

The dysfunctional remodeling which develops in this programmed response to wall stress eventually leads to preload disease (pulmonary edema) and eventually systolic dysfunction. The majority of the molecular remodeling occurs while the dog is subclinical with the unloaded LV being able to appear hypercontractile.

At what point in this early stage of MR should the veterinarian intervene with pharmacologic drugs (many of which are indicated later in the disease)? Successful therapies derived from pressure overloads, ischemic heart disease, cardiomyopathies, or even later stages of mitral valve disease may have little value in this debate. The variable penetration of different drugs from the same class into the cardiac interstitium (ACE inhibitors), differences in lipid solubility, pharmacokinetics of drugs, and especially the varied receptor binding affinity of a group of drugs (that are often discussed as if they were one drug, ie beta blockers) should constantly entered into each debate. Data supporting potential candidate drugs are being investigated, but until empirically demonstrated or justified by clinical trials, drug therapy based on considered opinions should be carefully weighed.

REFERENCES

Killingsworth C, **AHA** resabst 2007, Hankes GH et al. **Am J Physiol Heart Circ** 2006;291, H147. Dillon AR, et al. **JVIM** 2005; 19, 417. Tallaj J et al. **Circulation** 1994; 108, 225. Stewart JA, et al. **J Mol Cell Cardiol** 2003;35, 311. Perry GJ et al. **J Am Coll Cardiol** 2002 ; 39, 1374. Mehta RH et al. **Am Heart J**, 2003 ;145, 1078. Pat G et al. **Circulation** 2010 ; 122:1488. Dell'Italia LJ et al. **Am J Physio** 1997;273:H961. Pat B et al. **Am J Physio Heart Circ Physio** 2008;295:H2321. Dillon AR et al. ACVIM Proc 2010. Zheng J et a **Circulation**. 2009:119:2086. Ahmed M et al. **J AM Coll Cardiol** 2010;55:671. Kruger M et al **J Mol Cell Card** 2009; 46:490. Tsutsui H et al. **J Clin Invest** 1994; 93:2639. Sabri A et al. **Circ Res** 2008;102:1127. Chetboul et al **V *J Vet Intern Med*** 2007; 21:742. Haggstrom J et al ***J Vet Intern Med*** 2008; 22:1124.

ACE-INHIBITORS, EMERGING & ESTABLISHED ROLES IN THE TREAMENT OF HEART FAILURE

Clarke E. Atkins, DVM, DACVIM (SAIM & Cardiology)
Raleigh, NC

In landmark veterinary studies of enalapril in NYHA phase III and IV heart disease (moderate to severe heart failure), due to mitral regurgitation (MR) and dilated cardiomyopathy (DCM), enalapril improved survival by >100% as well as reducing pulmonary edema and, improving quality of life scores. Exercise capacity is also improved in dogs with experimental MR. Benazepril has likewise been shown to improve survival and imidapril was shown to be equal to enalapril in survival studies of dogs in heart failure due to MR and DCM. ACE-I have proven to provide additional benefits in human patients by blocking pathological remodeling, presumably slowing progression of heart disease, by blunting electrical remodeling and by normalizing serum electrolyte concentrations. Today, ACE-I represent the cornerstone in the chronic management of CHF. They are indicated in virtually all cases of systolic heart failure in which they are tolerated. In a survey of ACVIM-boarded cardiologists, ACE-I were utilized in the management of heart failure due to MR by >90%%[1] and the ACVIM Consensus Panel on Diagnosis and Management of Canine Chronic Valvular Heart Disease unanimously recommended their use in dogs suffering from heart failure, with a majority recommending their use in asymptomatic dogs with cardiomegaly.[1a] Published data support the use of ACE-I in DCM prior to the onset of heart failure and, although controversial, in asymptomatic, moderate to severe MR, as well.

EVIDENCE FOR THE USE OF ACE-I

There is a large body of evidence that ACE-I, coupled with other modalities (digoxin, furosemide, salt restriction) can improve symptoms and prolong life in humans with CHF due to a variety of causes, including dilated cardiomyopathy, coronary artery disease (CAD), and hypertensive heart disease.[1b]

Indeed, one of the best-studied areas in veterinary pharmacologic efficacy is that of ACE-I in the treatment of CHF in dogs. The IMPROVE study, a randomized, double-blind, placebo-controlled trial (RPCT), demonstrated clinical improvement and exercise, reduced heart rate, mean pulmonary artery pressure, and pulmonary edema score, and improved overall evaluation in congestive CHF (NYHA class III, IV).[2] The COVE study, a RPCT, evaluated dogs in NYHA class III and IV CHF found that the use of enalapril with other treatments for CHF improved NYHA class, pulmonary edema score, cough/dyspnea, exercise, and attitude and fewer dogs withdrew from the study due to death or worsening of CHF, as compared to placebo plus other therapies.[3] The LIVE study, another RPCT of enalapril, demonstrated >100% improvement in survival time or dropout due to worsening of signs in CHF due to both mitral regurgitation and dilated cardiomyopathy.[4] A European RPCT demonstrated a 62% reduction in mortality over a 56 day period in dogs with CHF receiving benazepril.[5] Finally, imidapril has been shown to be equal to enalapril in survival studies of dogs in heart failure due to MR and DCM.[5a] Hamlin, et al. demonstrated significant improvement in exercise capacity (and non-significant increase in survival time) after 7 weeks of enalapril therapy in dogs with CHF due to experimental MR.[6] ACE-I are useful in improving quality of life and survival in CHF.

There are a number of studies which have evaluated ACE-I vs placebo or other therapies earlier in the course of cardiac disease. The SOLVD study prevention arm (NYHA class I, II dilated cardiomyopathy and CAD) evaluated human patients with no or mild symptoms and showed a reduction in overt CHF and a non-significant 8% reduction in deaths.[7] The RAAS was shown to be activated. The V-HEFT II study, involving humans, many of which had mild CHF (87% NYHA class II, III) due to hypertension and CAD, demonstrated activation of the RAAS (but only in those receiving concomitant diuretic therapy) and a 28% reduction in mortality in patients treated with enalapril as compared with the combination of hydralazine and isorbide dinitrate.[8] The SAVE study of post-myocardial infarction patients demonstrated a 37% reduction in cardiovascular deaths with captopril and there was evidence of activation of the RAAS.[9] These studies show reduced mortality in mild/asympotmatic heart failure and early activation of the RAAS in most cases. Importantly, however, these are all human studies; the majority of patients suffered from diseases not commonly seen in animals, none of the studies evaluated MR, none were monotherapy trials, and many different ACE-I were used.

There are some important veterinary or animal model studies as well. In an ischemic microsphere model of ventricular disease, enalapril prevented progression of left ventricular dysfunction.[10] Pederson, et al. evaluated plasmal renin activity (PRA) and aldosterone concentration (PAC) in Cavilier King Charles Spaniels (CKCS) with mitral regurgitation, NYHA class I and II, and compared them to normal controls.[11] There was statistically significant elevation in PRA and PAC in these mildly affected dogs, indicating activation of RAAS; however there was a good deal of overlap between groups with many affected dogs falling within the normal range for PRA and PAC. A retrospective and as yet unpublished, study demonstrated a substantial and significant (601 vs 314 days) increase in survival duration in asymptomatic (occult) Doberman Pinschers which received one of 3 ACE-I (captopril, enalapril, or lisinopril), as compared to those which did not.[12] This has been followed in 2009 with a second retroprospective study of 91 dogs with occult DCM with 34 not receiving ACE-I (presumably using some of the dogs from the previous report) and 57 receiving benazepril. Though less dramatic, time to death or onset of heart failure was significantly greater in treated dogs (425 vs 339 days).[13]

EVIDENCE AGAINST THE USE OF ACE-I

The RAAS is abnormally activated in CHF, but its role may vary depending on the initial cardiac insult. Haggstrom and colleagues studied CKCS with NYHA class III CHF due to MR and demonstrated no activation in RAAS (PAC and Ag II normal) at baseline.[14] Monotherapy with enalapril did, however, produce reductions in ACE, PAC, and Ag II. The enalapril-treated dogs demonstrated a slight but significantly greater increase in left ventricular dimensions than did the group receiving hydralazine at 6 months; both groups received concurrent furosemide therapy after week 3. These investigators also evaluated CKCS with compensated MR (NYHA class I or II) every 6 months until decompensation occured.[13] The PAC and Ag II fell significantly from 4.6

months prior to decompensation until decompensation, suggesting that the RAAS was not activated at the time of CHF onset. In a study of experimental MR, ramipril failed to prevent pathological hypertrophy, possibly because of up-regulation of cardiac AT1 receptors.[15] In addition, more recent studies from this laboratory demonstrated that blocking the RAAS system using an angiotensin-receptor-blocker did not prevent remodeling or the loss of the myocardial collagen weave in this very acute model of heart failure.[15a] Lastly, there is some evidence that while quality of life improves in ACE-I treated human patients with CHF, this benefit is not demonstrable in patients treated prior to the onset of CHF.[16]

EARLY USE OF ACE-I

To the authors' knowledge there are 2 prospective trials specifically addressing the utility of ACE-I in asymptomatic, naturally acquired cardiac disease (MR) in dogs. First, Kvart, et al. published the results of the SVEP (Scandanavian Veterinary Enalapril Project), a multicenter (13) PCRT of asymptomatic MR in CKCS receiving either enalapril or placebo.[16a] This 4-year study evaluated 229 dogs, with 98 dogs reaching the endpoint of CHF. Dogs, most with mild disease, receiving enalapril (average .38 mg/kg) could not be distinguished from those receiving placebo, based on time to CHF (median 1,150 vs 1,130 days, respectively). The second study, centered at North Carolina State University, is likewise a multicenter (9) U.S. PCRT of dogs with asymptomatic, moderate to severe MR, randomized to receive either enalapril (average 0.46 mg/kg) or placebo.[16b] This study ended October 1, 2001, 5 years after its commencement. Kaplan-Meier estimates of the effect of enalapril on the primary endpoint (time to CHF in 124 dogs with a 60-day run-in period) failed to demonstrate a significant benefit (P=0.06), though the median heart-failure-free survival was extended by ~4 months. Chronic enalapril administration did have a significant benefit on the combined endpoint of CHF/all-cause mortality (benefit was 317 days or 10.6 months vs placebo). In addition, dogs were significantly more likely to be CHF-free at day 500 (64% vs 44.6%) and at study end (22% vs 9.2%). Chronic enalapril treatment (mean dose, 0.46 mg/kg) of dogs with naturally-occurring, moderate to severe MR significantly delayed the onset of CHF, compared to placebo-treated dogs, based on number of CHF-free days, number of dogs CHF-free at days 500 and study end, and increased time to a combined secondary endpoint of CHF/all-cause mortality. Improvement measured in the primary endpoint, Kaplan-Meier-estimated CHF-free survival (median benefit, ~4 months), was not significant. Taken together, however, these results suggest that enalapril modestly delays the onset of CHF in dogs with moderate to severe MR.

SAFETY

There was early concern regarding the renal safety of these compounds[17-19] and all ACE-I, which have enjoyed extensive clinical use, have been associated with renal dysfunction, usually temporary.[20] There is evidence that, at very high doses (180x the clinical dosage), ACE-I have direct nephrotoxic effects but it is generally felt that the major impact of ACE-I on the kidney, with clinically relevant dosages, is through production of hypotension, with reduced renal perfusion pressure and resulting in worsening of azotemia.[17] To date, veterinary clinicians have had experience with enalapril, captopril, benazepril, imidapril, ramipril and lisinopril. A long-term study of enalapril in elderly dogs with compensated MR showed no evidence of renal damage over 2 years, as compared to the placebo group.[21] The active metabolite of benazepril is known to be excreted both in the bile and in the urine so that lower serum concentrations are evident in experimental renal disease.[22] The clinical relevance of this is unclear. Over 25 years of veterinary clinical experience with ACE-I (mainly captopril, benazepril and enalapril) have taught us that their impact on kidney function is minimal even in the face of severe heart failure. When azotemia is observed, ACE-I are almost always being used in conjunction with diuretics and sodium restriction and hypotension results. Typically, cessation of diuretic therapy or reduction in the dosage results in the reversal of azotemia.[20]

EMERGING USES FOR ACE-I IN HEART FAILURE

Future uses of ACE-I include the use of other drugs concurrently in the advent of aldosterone escape. Reciprocally speaking, the use of ACE-I to suppress the activation of RAAS as is seen with amlodipine and furosemide will become more important as it is better understood. The discovery and better understanding of tissue RAAS activators will help us to suppress the tissue RAAS which makes up 90-99% of the body's RAAS. Electronic remodeling during arrhythmias can be blocked in part with ACE-I, so this drug group will play a role in management of arrhythmias in the future. Lastly, ACE-I will likely find greater use in combination with pimobendan (which have to date only been studied in designs comparing their efficacy, without concern for additive benefit) and beta-blockers which show promise in treatment of heart failure and slowing the progression of asymptomatic heart disease.

REFERENCES

1. Bonagura J. Current Clinical Practices Survey. *ACVIM Heart Failure Symposium* 2010; Oral presentation. 1a. Atkins C, Bonagura J, Ettinger S, Keene, B (chair) et al. Guidelines for the diagnosis and treatment of canine chronic valvular heart disease. *J Vet Intern Med* 2009;23:1142-50. 1b. CONSENSUS Trial Group: Effects enalapril on mortality in severe congestive heart failure. New Eng J Med 1987; 316:1429-1435. 2. IMPROVE Study Group: Acute and short-term hemodynamic, echocardiographic, and clinical effects of enalapril maleate in dogs with naturally acquired heart failure: Results of invasive multicenter prosepective veterinary evaluation of enalapril study. J Vet Intern Med 1995; 9:234-242. 3. Cove Study Group: Controlled clinical evaluation of enalapril in dogs with heart failure: Results of the cooperative veterinary enalapril study group. J Vet Intern Med 1995; 243-252. 4. LIVE Study Group: Effects of enalapril on survival in dogs with naturally acquired heart disease: Results of the long-term investigation of veterinary enalapril (LIVE) study group. *J Am Vet Med Assoc* 1998;213:1573–1577.5. The BENCH Study Group. The effect of benazepril on survival times and clinical signs of dogs with congestive heart failure: results of a multi center, prospective, randomized, double-blinded, placebo-controlled, long-term clinical trial. *J Vet Cardiol* 1999;1:5–18. 5a. Amberger C, Chetboul V, Bomassi E, et al. Comparison of the effects of imidapril and enalapril in a prospective, multicentric randomized trial in dogs with naturally acquired heart failure. *J Vet Cardiol* 2004;6:9-16. 6. Hamlin, RL, Benitz, AM, Ericsson, GF, et al.: Effects of enalapril on exercise tolerance and longevity in dogs with heart failure produced by iatrogenic mitral regurgitation.J Vet Intern Med 1996; 10:85-87. 7. SOLVD Investigators: Effect of enalapril on mortality and the development of heart failure in asymptomatic patients with reduced left

ventricular ejection fractions. New Eng J Med 1992; 327:685-691. 8. Cohn JN, Johnson G, Ziesche S, et al.: A comparison of enalpril with hydralzine-isosorbide dinitrate in the treatment of chronic congestive heart failure. New Eng J Med 1991; 325:303-310. 9. Pfeifer, MA, Braunwald, E, Moye, LA, et al.: Effect of captopril on mortality and morbidity in patients with left ventricular dysfunction after myocardial infarction: Results of the survival and ventricular enlargement trial (SAVE). New Eng J Med 1992; 327:669-677. 10. Sabah, H, Shimoyama, H, Kono, T: effects of long-term monotherapy with enalapril, metoprolol, and digoxin on the progression of left ventricular dysfunction and dilation in dogs with reduced ejection fraction. Circulation 1994; 89:2852-2859. 11. Pederson, HD, Koch, J, Poulsen, K, et al. Activation of renin-angiotensin system in dogs with asymptomatic and mildly symptomatic mitral valvular insufficiency. J Vet Intern Med 1995; 9:328-331. 12. OGrady, M, Horne, R, Gordon, SG: Does angiotensin enzyme inhibitor therapy delay the onset of congestive heart failure or sudden death in Doberman Pinschers with occult dilated cardiomyopathy (abst). Jour Vet Intern Med 1997; 11:138. 13. O'Grady MR, O'Sullivan ML, Minors SL, Horne R. Efficacy of benazepril hydrochloride to delay the progression of occult dilated cardiomyopathy in Doberman Pinschers. J Vet Intern Med. 2009 Sep-Oct;23(5):977-83. 14.Haagstrom, J, Hansson, K, Kvart, C, et al.: Effects of naturally acquired decompensated mitral valve regurgitation on the renin-angiotensin-aldosterone system and atrial naturietic peptide concentration in dogs. Amer J Vet Res 1997; 58:77-82.14. Haagstrom, J, Hansson, K, Bengt, E, et al.: Effects of long-term treatment with enalapril or hydralazine on the renin-angiotensin-aldosterone system and fluid balance in dogs with naturally acquired mitral valve regurgitation. Amer J Vet Res 1996; 57:1645-1652. 15. Dell'Itallia, LJ, Balcells, E, Meng, QC, et al.: Volume overload left ventricular hypertrophy is unaffected by ACE inhibitor treatment in dogs. Amer J Physiol 1997; 273:H961-970. 15a. Perry GJ, Wei CC, Hankes GH, et al. Angiotensin II receptor blockade does not improve left ventricular function and remodeling in subacute mitral regurgitation in the dog. *J Am Coll Cardiol* 2002;39:1374-1379. 16. SOLVD Investigators: Quality of life among 5,025 patients with left ventricular dysfunction randomized between placebo and enalapril: The studies of left ventricular dysfunction. J Amer Coll Cardiol 1994; 23:393-400.16a. Kvart C, Haggstrom J, Pedersen HD, et al. Efficacy of enalapril for prevention of congestive heart failure in dogs with myxomatous valve disease and asymptomatic mitral regurgitation. *J Vet Intern Med* 2002;16:80-8. 16b. Atkins CE, Keene BW, Brown WA, et al. Results of the veterinary enalapril trial to prove reduction in onset of heart failure in dogs chronically treated with enalapril alone for compensated, naturally occurring mitral valve insufficiency. *J Am Vet Med Assoc* 2007;231:1061-9. 17. MacDonald JS, Bagdon AJ, Peter CP, et al. Renal effects of enalapril in dogs. *Kidney International* 1987:S20:148–153. 18. Packer M, Leen WH, Medina N, et al. Functional renal insufficiency during long-term therapy with captopril and enalapril in severe, chronic, heart failure. *Ann Intern Med* 1987:106:346–352. 19. Schlessinger DP, Rubin SI. Potential adverse effects of angiotensin-converting enzyme inhibitors in the treatment of congestive heart failure. *Compend Contin Educ Pract Vet* 1994;16:275–283. 20.Bakris GL, Weir MR. Angiotensin-converting enzyme inhibitor-associated elevations in serum creatinine. *Arch Intern Med* 2000;160:685–693. 21. Atkins CE, Brown WA, Coats JR, et al. Effects of long-term administration of enalapril on clinical indicators of renal function in dogs with compensated mitral regurgitation. *J Am Vet Med Assoc* 2002;221:654-658. 22. Hervé P. Lefebvre, Valérie Laroute, Didier Concordet, Pierre-Louis Toutain. Effects of Renal Impairment on the Disposition of Orally Administered Enalapril, Benazepril, and Their Active Metabolites, J Vet Intern Med 1999; 13:21-27.

BETA BLOCKERS: EMERGING & ESTABLISHED ROLES IN THE TREATMENT OF HEART FAILURE

Jonathan A. Abbott, DVM, DACVIM (Cardiology)
Blacksburg, VA

INTRODUCTION

The beta-blockers or, beta-adrenergic antagonists (BAA), constitute a class of drugs that are competitive antagonists of the catecholamines, epinephrine and norepinephrine. They have important effects on the cardiovascular system and the indications for their use are diverse. The BAA have been used in the management of cardiac arrhythmias, for the treatment of myocardial ischemia and as a means to address the unique pathophysiology of hypertrophic cardiomyopathy. In human beings, heart failure (HF) due to systolic myocardial dysfunction is now an established indication for beta-blockade. There are experimental data that suggest that this approach may have promise in the management of canine HF. This review will briefly address the clinically relevant pharmacology of the BAA. The current and potential role of beta-blockade in the management of veterinary patients will be reviewed with emphasis on the use of beta-blockers in canine patients with systolic dysfunction.

PHARMACOLOGY

The neurotransmitters of the sympathetic, or adrenergic, nervous system (ANS) are epinephrine and norepinephrine. Both are synthesized from the amino acid, tyrosine, and are ligands for the adrenergic receptors. Adrenergic receptors are categorized based on their affinity for agonists.[1] While α-adrenergic receptors are expressed in myocardium, it is the ß-1 and ß-2 receptors that have the greatest relevance to cardiac function. Binding of epinephrine – secreted by the adrenal glands – or norepinephrine, which is secreted by adrenergic nerve endings as well as the adrenal glands, results in a conformational change in the receptor. The effect of this conformational change is mediated by the stimulatory G-proteins and results in activation of adenylate cyclase. Beta-agonism therefore increases the concentration of the intracellular second messenger, cyclic AMP. Phophorylation of regulatory proteins including phospholamban, as well as the voltage gated calcium channels, is responsible for inotropic, lusitropic, chronotropic and dromotropic effects of adrenergic activation.

The BAA are structurally similar to the endogenous catecholamines and bind reversibly to receptors; they function therefore as competitive adrenergic antagonists.[2] The many BAA differ in terms of receptor selectivity, intrinsic sympathomimetic activity, pharmacokinetics as well as properties that cause ancillary biological effects including vasodilation. Propranolol, a non-selective antagonist of ß-1 and ß-2 receptors is the prototype BAA. This drug and other non-selective agents have been classified as 1st generation BAA. The second generation BAA include metoprolol, atenolol and others that exhibit relative selectivity for ß -1 receptors. Carvedilol and other third generation BAA have ancillary effects that include, in the case of carvedilol, vasodilation resulting from α-1 antagonism.

The cardiovascular effects of BAA generally can be predicted based on the physiologic responses to adrenergic activation. Beta-blockade slows heart rate, decreases the inotropic state of the myocardium and slows conduction velocity. However, it is relevant that these effects are, in fact, mediated through receptor antagonism. Therefore it is more accurate to state that BAA prevent effects, such as the rise in heart rate, that result from adrenergic activation. Because of this, the effect of BAA in any individual depends not only drug concentration but also on prevailing autonomic tone. The response to beta-blockade is also determined by the association-dissociation kinetics of drug and receptor. For these reasons, the relationship between drug plasma concentration and drug effect is complex. For example, in contrast to metoprolol, carvedilol inhibits catecholamine response after elimination from plasma.[3]

AUTONOMIC PATHOPHYSIOLOGY IN HEART FAILURE

The development of systolic dysfunction is associated with activation of specific neuroendocrine pathways that include the adrenergic nervous system (ANS) and the renin-angiotensin-aldosterone system (RAAS).[4] In the short-term, these responses serve to preserve systemic arterial blood pressure and cardiac output. Indeed, similar responses are evident in acute, non-cardiac, circulatory dysfunction; hypovolemic shock also is associated with release of renin and adrenergic activation. However, there is now abundant evidence that supports the contention that persistent neuroendocrine activation in chronic cardiac dysfunction is maladaptive. Raised concentrations of neuroendocrine mediators such as norepinephrine and renin are associated with poor prognosis in people with HF yet pharmacologic blunting of these systems with angiotensin converting enzyme inhibitors (ACE-I) and BAA improve prognosis.[5] The conceptual model in which the syndrome of HF results simply from diminished pump function has been replaced by one in which activation of the ANS and RAAS contribute to progressive myocardial dysfunction. Specifically with regard to activation of the ANS, baroreceptor insensitivity is believed to importantly contribute to a state of "adrenergic dysregulation" in which adrenergic activation and raised concentrations of circulating catecholamines are unopposed by parasympathetic restraint.[6] No doubt other factors are also responsible as there are experimental and clinical data that demonstrate that some characteristics of abnormal adrenergic function depend on local, not systemic, factors. Specifically, the effects of increased adrenergic drive are apparent in the right, but not left, ventricles of hearts explanted from human beings with isolated right HF caused by pulmonary hypertension.[7] Regardless, it is established that the HF state is associated with increases in adrenergic activity. This increase in adrenergic activity is associated with myocardial norepinephrine depletion, a finding that is seemingly paradoxical but reflects increases in interstitial norepinephrine concentration and diminished re-uptake of norepinephrine.[8] The way in which increases in norepinephrine harm the myocardium has not been resolved. Chronic adrenergic activation is associated with uncoupling of beta-receptors from G proteins and potentially, destruction of the receptors.[9] This phenomenon, known as beta-receptor down-regulation may be important to pathogenesis of myocardial dysfunction. More simply, it might be that norepinephrine is simply cardiotoxic - an effect that is evident in experimental models.[10]

BETA-BLOCKADE IN HEART FAILURE

BAA are negatively inotropic and as a result, seemingly pose a threat to patients that have HF caused by systolic dysfunction. Indeed, early reports of myocardial norepinephrine depletion were interpreted to reflect a casual deficit of adrenergic support for the failing heart. However, beginning in the 1970's innovative physicians began testing the hypothesis that beta-blockade might improve outcome in human patients with systolic HF.[11] Early investigations were non-randomized and enrolled relatively small numbers of patients. However, encouraging initial results were the impetus for later, systematic investigation of the role of BAA in HF. Several randomized, placebo-controlled double-blinded clinical trials (RCT) have been published and with few exceptions, results of these investigations have been positive. Separate clinical trials have demonstrated mortality reduction when bisoprolol, carvedilol and metoprolol were compared with placebo in people with HF receiving background therapy of ACE-I and diuretics with or without digoxin.[12-15]

In human beings with myocardial dysfunction, chronic beta-blockade is associated with increases in ventricular ejection fraction.[16] The results of a trial intended to compare the effects of ACE inhibition, beta-blockade and their combination on echocardiographic indices of systolic function in people with mild HF has been published. In this study, the combination of an ACE-I and BAA reversed ventricular remodeling and supports the use of BAA early in human HF.[17]

The results of RCT that have evaluated BAA in human beings with HF due to systolic dysfunction have been compelling. Positive effects were evident in all classes of HF and the positive clinical response was associated with favorable effects on ventricular remodeling and myocardial function. Furthermore, available data suggest that quality of life is preserved or improved. Indeed, the use of BAA in the management of HF due to systolic dysfunction is now considered standard therapy; the most recent iteration of the American Heart Association Guidelines includes the recommendation of BAA for all patients with HF.[18]

Beta-blockade in Experimental Models

The effects of beta-blockade have been evaluated using a number of different experimental models and some of the resulting data are relevant to a consideration of the use of BAA in veterinary patients with spontaneous cardiac disease. A number of investigators have independently evaluated beta-blockade in the context of mitral regurgitation (MR) induced by chordal rupture. Using this model, 3 months treatment with atenolol improves myocardial function.[19] Another group of investigators, using a similar model of MR, determined that 4 months administration of metoprolol was associated with improved cardiomyocyte function. In this study, favorable ventricular remodeling was not evident but, recognizing that the experiment was not specifically designed to test the effect of drug on survival, relative to the treatment group, fewer experimental subjects in the control group survived the 4 months experiment.[20] The effects of BAA also have been investigated in cardiomyopathy induced by sequential coronary embolisation of microspheres. Using this model, studies have demonstrated that 3 months administration of metoprolol attenuates the process of remodeling and progression of myocardial dysfunction, decreases apoptosis and improves ventricular filling.[21,22]

There is persisting uncertainty with regard to the mechanisms by which BAA exert favorable effects in HF. Up-regulation of beta-receptors might be partly responsible. However, up-regulation of receptors is observed in association with the administration of some, but not all, BAA. Carvedilol for example, does not result in up-regulation of beta receptors despite the wealth of clinical trial data that support its use in people with HF.[23] NE is myo-toxic and protection from the adverse effects of chronic beta-stimulation might explain the favorable effect of beta-blockade. And of course, BAA also decrease heart rate and therefore reduce myocardial oxygen demand. The use of electronic pacing to maintain heart rate in canine HF due to induced MR prevents the improvement in myocardial function that is otherwise associated with beta blockade.[24]

Beta-blockade in Spontaneous Disease of Veterinary Patients

Investigations of beta-blockade in spontaneous disease of veterinary patients have provided mixed results. Rush et al retrospectively evaluated the effect of metoprolol on dogs with MR or dilated cardiomyopathy (DCM).[25] The patient population consisted of dogs with degenerative mitral valve disease or dilated cardiomyopathy. Because of the retrospective study design it was not possible to evaluate the effect of metorprolol on survival. However, this study provided evidence that administration of metorpolol was tolerated and apparently safe. In another retrospective investigation, median survival of dogs with DCM receiving propranolol in addition to digoxin and furosemide was 126 days; the study did not include a control group but the administration of propranolol was apparently tolerated.[26] More recently, a randomized placebo controlled trial evaluated the effect of 0.3mg/kg BID carvedilol on canine DCM; there was no effect on remodeling or neuroendocrine variables.[27] Possibly, the dose used in this study was inadequate as the trial was designed and initiated prior to publication of reports on the pharmacodynamic effects of oral carvedilol in healthy dogs.[28,29] Another RCT evaluated the effect of carvedilol on echocardiogaphic indices of myocardial function and on indices of quality of life in dogs with spontaneous MR.[30] In this study, relative to placebo, carvedilol improved quality of life.

The efficacy of BAA in heart failure has been unequivocally demonstrated in people with HF and experimental evidence favors the use of BAAs in dogs with systolic dysfunction. Based on this, the cautious use of BAA in veterinary patients can be justified. However, beta-blockade in canine patients with systolic dysfunction is currently an unproven therapy. Furthermore, BAA can have catastrophic effects in patients with severe myocardial dysfunction who may be critically dependent on a limited inotropic reserve in order to maintain adequate cardiac output. The author considers the use of BAA, generally metoprolol or carvedilol, in patients with echocardiographic evidence of myocardial dysfunction who have not developed congestive signs or in those in which congestive signs have resolved with therapy. Unwillingness on the part of the pet-owner to contend with frequent dosage adjustments or accept the potential for adverse effects is a relative contraindication. The initial dose is very low and, if tolerated, is increased at 10-14 day intervals until a target dose is reached or adverse effects are evident. An initial carvedilol dose of 0.1 mg/kg (not to exceed 3.125 mg/patient) q12h is probably reasonable. The dose is gradually increased, if tolerated, to 0.5-1 mg/kg PO q 12 h over 6-8 weeks. The patient should be carefully observed for evidence of decompensation characterized by congestive signs, clinically evident bradycardia and/or hypotension. If these signs are observed, the dose is decreased to the one that was most recently tolerated. Published reports of clinical trials and anecdotal evidence suggest that the key to patient toleraability is gradual dose titration. Anecdotally, even small doses of a BAA may be associated with decompensation and worsening of congestive signs, although it is important to recognize that HF is a progressive disorder in which clinical deterioration may occur independent of therapy.

OTHER INDICATIONS FOR BETA-BLOCKADE

Feline Hypertrophic Cardiomopathy (HCM)

In feline HCM, beta-blockade can decrease or abolish dynamic left ventricular outflow tract obstruction caused by systolic anterior motion of the mitral valve (SAM) and may prevent tachycardia thereby indirectly improving diastolic function. Unfortunately, published evidence regarding the efficacy of BAA, or for that matter, other agents in feline HCM is sparse. The results of a multicenter, randomized, placebo-controlled trial that had been designed to evaluate the relative efficacy of atenolol, diltiazem and enalapril in feline patients with diastolic heart failure have been reported but not yet published. In that trial, relative to placebo, administration of atenolol increased the risk of recurrence of pulmonary edema.[31] This trial did not address the therapeutic management of subclinical HCM and it is possible that decreases in myocardial oxygen demand, decreases in heart rate and abolition of SAM associated with BAA might be favorable in that scenario.

Canine Subvalvular Aortic Stenosis (SAS)

Canine patients with severe SAS are at high risk for sudden, unexpected cardiac death. The use of BAA in this disorder is predicated on the belief that a decrease in myocardial oxygen demand associated with the negatively inotropic and chronotropic effects of BAA is beneficial. Additionally, blunting of the chonotropic response to exercise may improve coronary perfusion in patients with severe SAS because coronary flow in this disorder occurs primarily during diastole.[32] It is also possible that the effects of beta-blockade reduce the frequency and potentially fatal outcome of arrhythmias in SAS. Systematic evaluation of the effect of beta-blockade in relation to a suitable control group has not been reported.

Arrhythmias

Adrenergic activation is relevant to the pathogenesis of many forms of pathologic tachyarrhythmia. In clinical trials that have enrolled people with cardiac disease, beta-blockade has been associated with a decrease in the risk of sudden arrhythmic death. Comparable data have not been published in the veterinary literature. However, the combination of mexiletine and atenolol decreases the frequency of ventricular tachyarrhythmia in boxer dogs with arrhytyhmogenic right ventricular cardiomyopathy.[33]

REFERENCES

1. Ahlquist RP. Am J Phys 1948;153:586. 2. Hoffman BL et al. In: Hardman JG LL, ed. Goodman & Gilman's The Pharmacological Basis of Therapeutics, 9th ed; 1996:199. 3. Kindermann M et al. Circ 2004;109:3182. 4. Katz AM. Med Clin North Am 2003;87:303. 5. Cohn JN et al. N Engl J Med 1984;311:819. 6. Zucker IH et al. Basic Res Cardiol 1991;86 Suppl 3:133. 7. Bristow MR et al. J Clin Invest 1992;89:803. 8. Delehanty et al. Am J Phys 1994;266:H930. 9. Chakraborti S et al. Cell Signal 2000;12:499. 10. Mann DL et al. Circ 1992;85:790. 11. Waagstein F et al. Br Heart J 1975;37:1022. 12. CIBIS Investigators and Committees. Circ 1994;90:1765. 13. Packer M et al. N Engl J Med 1996;334:1349. 14. Packer M et al. N Engl J Med 2001;344:1651. 15. Lancet 1999;353:2001. 16. Bristow MR et al. Circ 1996;94:2807. 17. Komajda M et al. Eur J of Heart Fail 2004;6:467. 18. Hunt SA et al. 2009 J Am Coll Cardiol 2009;53:e1. 19. Tsutsui Het al. J Clin Invest 1994;93:2639. 20. Pat B et al. Am J Physiol 2008;295:H2321. 21. Sabbah HN et al. J Am Coll Cardiol 2000;36:1698. 22. Sabbah HN et al. Circ1994;89:2852. 23. Flesch M et al. Cardiovasc Res 2001;49:371-380. 24. Nagatsu M et al. Circ 2000;101:653. 25. Rush et al. J Vet Cardiol 2002;4:23. 26. Tidholm A. J Vet Cardiol 2006;8:41. 27. Oyama MA et al. J Vet Int Med 2007;21:1272. 28. Abbott JA et al. H Am J Vet Res 2005;66:637-641. 29. Gordon SG et al. J Vet Int Med 2006;20:297. 30. Marcondes-Santos Met al. J Vet Int Med 2007;21:996. 31. Fox PR. J Vet Int Med2003;17:398. 32. Pyle RL et al. Circ Res 1973;33:34. 33. Meurs KM et al. J Am Vet Med Assoc 2002;221:522.

INODILATORS: EMERGING & ESTABLISHED ROLES IN THE TREATMENT OF HEART DISEASE

Sonya Gordon, DVM, DVSc, DACVIM
College Station, TX

INTRODUCTION

This class of agents is characterized by a combination of both positive inotropic and vasodilatory properties and were labeled inodilators as a reflection of the term 'inodilation' which was coined by Lionel H Opie in 1989. Examples of agents in this class that have been used with some frequency in veterinary medicine include; milrinone, pimobendan, and levosimendan. By definition digoxin can also be considered a member of this class but its relative hemodynamic potency makes it a member in name only. The mechanism of action by which each of these agents lead to augmented inotropy and vasodilation differ as do ancillary properties making each agent relatively unique. Availability and dosage formulations also impact the potential clinical utility of individual agents in this class. Currently, pimobendan is available as an oral formulation and milrinone and levosimendan are available as intravenous (IV) preparations. Prototypical agents in this class such as milrinone and similar agents were evaluated for the treatment of chronic human heart failure and reportedly increased mortality. Following the outcome of these studies further development and testing of inodilators for the treatment of chronic human heart failure was not actively pursued. Current acceptable human usage is for short-term IV support in advanced heart failure (levosimendan, milrinone). However, there is renewed interest in the potential use of specific agents in this class (pimobendan) as part of a multimodal treatment strategy for pulmonary hypertension (PH) as well as its chronic use in combination with a beta-blocker for the treatment of chronic heart failure (HF) characterized by systolic dysfunction. Historically, all positive inotropes have relied on the same final common pathway to exert their myocardial effects. Increased intracellular calcium leads to increased contractility but calcium loading of myocytes has the potential disadvantage of increasing myocardial oxygen consumption, risk of arrhythmias, cell injury and cell death. Sensitizers do not increase myocardial oxygen consumption and have the potential to reverse contractile dysfunction under pathologic conditions. Thus calcium sensitizers can theoretically overcome the disadvantages of traditional inotropes and may represent the only 'safe' inotropic mechanism. However pure sensitizing agents may have a risk of impairing relaxation thus the specific mode of calcium sensitization and balance of other properties may determine the overall clinical utility of individual agents.

Milrinone

Mirlinones main mechanism of action is inhibition of phosphodiesterase III (PDE-III). Phosphodiesterase III is an intracellular enzyme that specifically hydrolyzes cAMP in myocardial and vascular tissue. When PDE-III is inhibited, intracellular cAMP concentration increases leading to an increase cytosolic calcium. This mechanism of action has the same type of inotropic effect in the myocardium as is produced by sympathomimetics (e.g. dobutamine). The major difference is that PDE-III inhibitors 'bypass' the beta-receptors and so there is no decrement in inotropic effect over time or in the setting of chronic heart failure when beta-receptors may be down regulated. Milrinone also produces veno- and arteriolar dilation which is mediated by PDE-III inhibition in vascular smooth muscle. Lastly, agents that inhibit myocardial PDE-III improve relaxation, a positive lusitropic effect. Inhibition of myocardial PDE-III leads to increased cytosolic cAMP and enhanced phosphorylation of sarcoplasmic proteins that enhance the rate of reabsorption of cytosolic calcium after contraction which speeds the rate of relaxation. Cardiovascular effects of milrinone are species dependent. Myocardial contractility increases to a similar degree as is observed following the administration of a beta-agonist in dogs and cats (i.e. approximately 100% above baseline). Because of this marked species difference, data obtained from human patients given milrinone cannot be extrapolated to dogs or cats. The clinical veterinary indication for milrinone is for acute cardiovascular support in dogs with HF, however in most cases dobutamine is the preferred agent for this indication. No clinical studies of the effects of intravenous milrinone administration for acute myocardial failure in dogs or cats have been performed. Clinical studies of the effects of chronic oral administration have been performed but this form of the drug has not been approved for veterinary use and is not available for human use. Milrinone lactate is supplied in 10 ml and 20 ml single-dose vials containing 1 mg/ml. It can be diluted in 0.45% and 0.9% sodium chloride and 5% dextrose in water. Single canine IV injection 30–300 μg/kg (effects reduced to 50% by 10 min and gone in 30 min), constant-rate intravenous infusions (1–10 ug/kg/min) with peak effect in 10–30 min. Milrinone is chemically incompatible with furosemide and thus should not be administered in the same intravenous line without flushing adequately in between.[1]

Pimobendan (Vetmedin®)

Pimobendan exerts its positive inotropic effects primarily through sensitization of the cardiac contractile apparatus to intracellular calcium. It is also a PDE-III and thus can potentially increase intracellular calcium concentration and increase myocardial oxygen consumption. However, the cardiac myocyte PDE-III inhibitory effects of pimobendan are reportedly minimal at pharmacologic doses in dogs with heart disease, which is a major advantage relative to pure PDE-III inhibitors such as milrinone.[2,3] Pimobendan's calcium sensitization of the contractile apparatus is achieved by enhancement of the interaction between calcium and the troponin C complex resulting in a positive inotropic effect that does not increase myocardial oxygen consumption.[4] Overall, pimobendan enhances systolic function by improving the efficiency of contraction, limiting the arrhythmogenic side effects of positive inotropes whose sole mechanism of action is to increase myocardial intracellular calcium. Pimobendan inhibits PDE-III in vascular smooth muscle resulting in dilation of both arteries and veins leading to a reduction of both cardiac preload and afterload. Pimobendan has also been demonstrated to have additional vasodilatory properties that are reportedly endothelium mediated (nitric oxide) and may involve inhibition of PDE-V. This form of vasodilation may have particular benefit in the setting of pulmonary hypertension. The significance of alterations in pro-inflammatory cytokine concentrations such as tumor necrosis factor-α, interleukins 1-β and 6 on the progression of HF has been documented in many etiologies of heart disease. Alterations in these cytokine concentrations are associated with increased morbidity and mortality. Pimobendan has demonstrated beneficial modulation of a several such cytokines in various models of HF, representing an additional potential benefit.[5,6] Pimobendan has been shown to reduce collage induced platelet aggregation in humans. The mechanism is thought to be related to inhibition of platelet PDE-III. The role of platelet aggregation in canine heart disease (chronic valvular disease, dilated cardiomyopathy, pulmonary hypertension) is not clearly elucidated to date but cats with most forms of cardiomyopathy are recognized to have a pro-thrombotic risk. The potential ancillary anti-platelet effect of pimobendan, if demonstrated in dogs and cats, may have implications with respect to its clinical utility.[7] Lastly, agents that inhibit myocardi PDE-III have the potential to improve relaxation, a positive lusiotropic effect and pimobendan has some ability to inhibit myocardial PDE-III The clinical significance of this effect will be related to relative amount of PDE-III (may be reduced in failing myocardial cells) and the potency of its inhibition by pimobendan. These potential ancillary properties may be related to pimobendans proven clinical efficacy in canin congestive heart failure due to chronic valvular disease (CVD) and dilated cardiomyopathy (DCM). In veterinary medicine pimobendan has been proven to prolong survival and reduce morbidity in dogs with congestive heart failure due to CVD and DCM and is recognized as a cornerstone in the therapy of these conditions.[8-10] Pimobendan is available in the United States as a licensed oral formulation for the dog (chewable tablet) which is manufactured by Boehringer Ingelheim Vetmedica® as Vetmedin®. Some other countries have Vetmedin® capsule formulations. This product in not currently available as an IV preparation however the oral preparation is rapidly absorbed with peak effects i 2-4 hours. The recommended canine dose is 0.25-0.3 mg/kg twice daily. In later stages of heart failure or when clinical signs become refractory the dose (0.25-0.3 mg/kg) is frequently increased to three times daily. There are no published studies to date on the pharmacokinetics and pharmacodynamics of pimobendan in cats but published abstracts report cat dosages that are similar to dog dosages.[11-1] Generic formulations of this product have not been clinically evaluated and should be avoided in lieu in the licensed product. In addition reformulation (suspensions) should be avoided since this product in known to have limited stability.

Levosimendan

Levosimendan exerts its positive inotropy predominantly through calcium sensitization. It appears to increase myofilament calcium sensitivity by binding to cardiac troponin C in a calcium-dependant manner. This stabilizes the calcium-induced conformational change of troponin C thereby; changing the actin-myosin cross bridge kinetics apparently without increasing the cycling rate of the cross-bridges or myocardial ATP consumption, increasing the effects of calcium on cardiac myofilaments during systole and improving the contraction at low energy cost. Calcium concentration and therefore sensitization decline during diastole, allowing normal or improved relaxation. Levosimendan leads to vasodilation through the opening of ATP-sensitive potassium channels in vascular smooth muscle leading to smooth muscle relaxation. Levosimendan also has a selective PDE-III inhibitory action that may contribute to the inotropic effects. It has been suggested that that its calcium sensitization effects may predominate at lower concentrations and like pimobendan when it is used in the setting of myocaridal failure myocardial PDE-III levels may be reduced favoring the sensitization mechanism. Lastly levosimendan may open mitochondrial (ATP)-sensitive potassium channels in cardiomyocytes which could exert a cardioprotective effect. This agent is available in an IV formulation for human use in many countries. It is used for the acute/short-term support of decompensated heart failure. An oral formulation of levosimendan was recently developed for evaluation in the treatment of canine congestive heart failure but to date no clinical trial results have been reported.

THERAPEUTIC RATIONAL FOR INODILATORS

Tolerable reductions in cardiac preload and afterload through balanced venous and arterial dilation are desirable in patients with symptomatic heart failure secondary to the majority of structural heart diseases. Relative exceptions include diseases characterized by obstruction to flow such as sub-aortic stenosis, systolic anterior motion of the mitral valve. Augmentation or support of inotopy in diseases characterized by reductions in myocardial contractility would likewise have potential value. The majority of structural heart diseases are characterized by systolic dysfunction at some time during their course. In many diseases systolic dysfunction is only present in advanced stages. Thus, some diseases and some stages of disease that are characterized by preserved inotropy would not be expected to benefit from these agents. Agent specific ancillary properties may also contribute to potential indications for specific agents in this class.

ESTABLISHED ROLES FOR THE INODILATOR PIMOBENDAN

Dog

Pimobendan has been proven to prolong survival and reduce clinical signs in dogs with congestive heart failure secondary to both CVD and DCM.[8-10] According to the ACVIM consensus statement guidelines for the diagnosis and treatment of CVD pimobendan should be considered a cornerstone of multimodal therapy in dogs with Stage C and D heart failure due to CVD.[15]

Cat

None

EMERGING ROLES FOR THE INODILATOR PIMOBENDAN

Dog

ACVIM Stage B2 heart failure is characterized by the absence of any current or historical signs or symptoms of heart failure and substantive cardiac remodeling. No medication to date has been definitively proven to delay the onset of heart failure in Stage B2 CVD and DCM however angiotensin converting enzyme inhibition (ACEI) may have efficacy in some sub-populations. Stage B2 DCM and CVD have significant morbidity and mortality risks and thus investigation of agents that have the potential to delay the onset of heart failure are needed. Pimobendan is currently under investigation in prospective placebo controlled clinical trials for this indication in both Stage B2 CVD and DCM.

Pimobendan is currently used off label for the treatment of both right and left heart failure due to a variety of underlying etiologies. It is used as a rescue agent on a case by case basis in patients with symptomatic heart failure when conventional heart failure therapy is failing and/or when the underlying cardiac disease (based on echocardiogram) is characterized by ventricular systolic dysfunction or myocardial failure.

There is substantial interest in the potential benefit of beta-blockade in many etiologies of heart disease and heart failure characterized by myocardial systolic dysfunction such as CVD and DCM. However, particularly in symptomatic heart failure (Stage C and D) the use of beta-blockers may be difficult due to the relative risk of decompensation as a result of their short-term negative effects on inotropy. This relative risk may be minimized or eliminated by the combination of a beta-blocker with pimobendan and the potential cumulative benefits of both agents may then be realized.

Pimobendan may have a role to play in the treatment of pulmonary hypertension particularly when used in combination with PDE-V inhibitors (e.g. sildenafil). In PH PDE-III is relatively high in large and resistance pulmonary arteries and PDE-V is increased primarily in the large pulmonary arteries. PDE-III inhibitors such as pimobendan have a greater affinity for cAMP over cGMP and PDE-V inhibitors such as sildenafil have a greater afginity for cGMP over cAMP and thus the combination of these two agents may be superior for the treatment of PH. This may be especially relevant in the setting of clinical right heart failure, when the echocardiogram suggests right ventricular myocardial failure is present or when the underlying etiology for the PH is related to advanced CVD.[16,17]

Pimobendan has recognized use in the emergency treatment of congestive heart failure but this indication could be better evaluated for this indication if an IV formulation were available. The availability of an IV formulation would likely lead to its evaluation for short-term support of the cardiovascular system in dogs with recognized heart disease (e.g. Stage B2 CVD and DCM) that require general anesthesia.

Cat

Feline heart disease leading to congestive HF is characterized predominantly by diastolic dysfunction. Chronic oral treatment has conventionally emphasized control of clinical signs associated with volume overload (e.g. furosemide with or without an angiotensin converting enzyme inhibitor), arterial thromboembolism prophylaxis (e.g. clopidogrel with or without aspirin), and medications that may directly or indirectly improve ventricular relaxation (e.g. diltiazem or a beta-blocker). However, there are a number of feline heart diseases that are characterized by echocardiographic evidence of systolic ventricular dysfunction (myocardial failure). These diseases include many forms of unclassified cardiomyopathy (CM), dilated CM, arrhythmogenic right ventricular CM, some forms of hypertrophic CM and some severe forms of congenital and other acquired feline heart disease. Cats with congestive HF due to systolic (myocardial) failure have a very poor long term prognosis with a reported median survival time of 13 days with conventional therapy[18]. Pimobendan has been used as a rescue agent in cats for this indication and appears to be well tolerated.[11,12,14] However due to pimobendans ancillary properties (positive lusiotropy, anti-platelet effects, cytokine modulation) there may be rationale for its use in congestive HF due to all cardiomyopathies.[13] However, its safety in CM characterized by obstruction to outflow (hypertrophic obstructive cardiomyopathy) would need to be better established.[12-14] Ultimately the potential utility of pimobendan for these indications will need to be evaluated in well designed prospective placebo controlled clinical trials.

REFERERENCES

Sonya G Gordon and Mark D Kittelson. In Small Animal Clinical Pharmacology 2nd edition. Saunders Elsevier, Edinburgh 2008, pp.380[1], Sato N, et.al. Am J Physiol 1999;276:H1699[2], Verdouw PD, et.al. Eur J Pharmacol 1986;126:21[3], Ruegg JC. Circulation 1986;73: III-78[4], Starling M, Proceedings. Valves in the heart of the big apple II: Evaluation and management of valvular heart disease. 2002[5], Iwasaki A et.al. J Am Coll Cardiol 1999;33:1400[6], Saniabadi AR et al. Cardiovas Res 1989;23:184[7], Lombard CW, et.al. J Am Anim Hosp Assoc 2006; 42:249[8], Lombard CW, et.al. J Am Anim Hosp Assoc 2006[9], Fuentes VL, et.al. J Vet Intern Med 2002;16:255[10], Sturgess CP et al. J Vet Intern Med 2007;21:1423 (abst)[11], Roland RM et al. J Vet Intern Med 2008;22:1457 (abst)[12], MacGregor JM et al. J Vet Intern Med 2010;24:692 (abst)[13], Gordon SG et al. 2010 ACVIM Forum Research Report 2010 ACVIM Forum Proceedings. PP158[14], Atkins C et.al. J Vet Intern Med 2009;23:1142[15], Watanabe et al. Cardiovasc Drug Therapy 2003;17:375[16], Sahara et al. Cardiovasc Drugs Ther 2006;20:377[17], Fox et.al. J Vet Intern Med 1997;11:141 (abst)[18].

THE TRIANGLE OF CARE: REFERRING & REFERRAL PROCESSES

Katherine Dobbs, RVT, CVPM, PHR
Appleton, WI

There is a Triangle of Care within veterinary medicine, where one point is the referring or general practitioner, another point is the emergency clinician or specialist, and the third point is the client or pet owner. In the middle of this triangle we have the patient, the beloved pet…and these relationships work in collaboration to keep that pet healthy and happy. Relating to this Triangle of Care, the AVMA Ethics and the AAHA Referral Guidelines can help us develop a Code of Conduct to help our teams within the referral practice deliver the type of communication and compassionate care that will help us build these relationships. The first section of the code focuses on the relationship between the two practices who are working to help the family and the pet. The second section focuses on the relationship between our pet owners and our referral practices, and finally the last section involves the relationship between our practice and the pet at the center of this triangle. The goal is to balance these relationships to keep pets healthy, and clients and referring practices happy.

TRIANGLE CODE OF CONDUCT

The Referral Relationship

- Recognize that a Triangle of Care exists between the general practice, referral practice, and pet owner—with the pet's best interest central to all we do. Inherently we know that the pet's best interest is first and foremost, but it becomes easy to let this slip to the back of our mind when faced with other concerns within the relationships of this triangle. We must also assume that the other two points, the pet owner and the general practitioner, are acting on the behalf of the pet, unless determined otherwise by our own observation or investigation.
- Honor the veterinarian-client-patient-relationships (VCPR) that exist in this Triangle of Care. There are two distinct VCPR's that exist in this triangle, which can become complicating when dealing with follow-up care, new problems discovered or reported, and disagreements between points.
- Nurture the relationship between practices by providing friendly service and responsive communication to all members of the referring community. It is not only a relationship between practitioners, but instead a relationship between all members of the referring community including the referring practice's front office personnel, technicians, managers, and associates. If any member of the referring community is treated poorly, we risk our relationship with that entire referring practice.
- Demonstrate respect for the referring community through words and actions, both in private and within the practice team and in public to our mutual clients. Always give the referring community the benefit of the doubt. As emergency and specialist practices, we are often required or asked to deal with an issue or perception that can reflect poorly on the referring practice. We must handle these situations carefully, and encourage our team to maintain an attitude of support rather than superiority.
- Represent accurate credentials of all veterinarians on staff, demonstrating an understanding of professional designations, initials, titles, and recognized credentials. This is important from a legal and ethical standpoint, but we must look beyond the rules and realize that we also have the opportunity to educate both of our client bases by illustrating these titles and credentials. We must also ensure that all members of our team understand and can relay these designations.
- Provide accurate estimates of fees and availability of services within the scope of designated authority. Your technician may be the one who explains the entire financial estimate to the pet owner once they are in the referral practice, but there may only be certain fees your support staff is authorized to give over the phone to a pet owner or referring practice. All members of the referral practice must also be aware of what services the practice provides, including those services provided on an out-patient basis.

The Pet Owner Relationship

- Realize that the way we treat the referring community will be interpreted as the way we treat our mutual clients. While the referring practice will hear things, both good and bad, from our mutual clients who come through our doors, the referring practice also will interpret the way that we treat them as individuals as the way we will or have treated the pet owners they send to us for care.
- Recognize that new callers are potential new clients for our practice, are generally more interested in our level of care and compassion than our prices, and are only committed to the examination fee to seek our opinion. While we have relatively few "phone shoppers" in referral practice, there are those who want to inquire about our prices before they schedule an appointment. It is most important to share our compassion and begin to build a relationship, even when a price is eventually discussed on the call.
- Recognize that both pets and pet owners need triage, and while in our practice their comfort and needs are our primary responsibility. The referral practice typically has protocols in place to handle urgent needs of patients that come through the door. However, often the pet owner is in a more critical emotional state than their pet. We must also develop protocols designed to comfort and care for our pet owners during this difficult time for them.
- Acknowledge that communication is largely non-verbal, and remain aware of our tone of voice, body language, and facial expressions at all times. A relatively small portion of our message is the words we choose; even more important is our tone of voice and our body language. We must monitor our own body language during interactions, and monitor that of our pet owner to determine if they are receiving the message and understand our words.

- Never judge clients' decisions about their pet's care, choice of humane euthanasia, or ability to afford the recommended treatment. We are often in very difficult situations involving tough choices for our pet owners. The ability to separate our own judgment from the choice of the pet owner is difficult, but needs to be in place for us to deliver the support needed by our pet owners and maintain our own emotional health as well.
- Provide compassionate service to all clients regardless of their emotional state, realizing that complaints often result from a pet owner's loss of control during an emotional or traumatic event in the life of their beloved pet. In referral practice, our pet owners are often facing a very difficult time in the life of their pet, and it is important that we not take their behavior personally whether they express sorrow, frustration, or anger. Empathy can help us separate from these emotional discussions, and understanding the predicament they are currently facing.
- Handle issues between the pet owner and referring practice with grace and respect, remaining impartial and neutral, but displaying appropriate concern for the pet. It is vital that while offering support to the pet owner, we maintain the reputation and relationship of our practice with the referring community. Support moving forward with the pet's care, rather than comment on care provided up to that point.

The Patient Relationship

- Realize that the way we treat the pet owner will be interpreted as the way we treat their pet. Pet owners typically have no way to judge our medical expertise. They will, however, make assumptions about our medical care from the way that they are treated by members of our team. If we deliver compassion to the person, they will assume we are compassionate to their pet.
- Recognize the role of each person's position on the healthcare team, and take part in medical training and continuing education to enhance our ability to provide both client and patient care within our area and scope of authority. Each person no matter their position is important in the care that is delivered by the referral practice. The front office is the center of communication, the assistant is responsible for the patient's basic needs, and the technicians deliver the medical care; all of them will benefit from staying current with the trends and technologies in veterinary medicine, and in turn your patients will benefit from the best medical care.
- Strive to earn the respect of your colleagues on the practice team and display respect in return, while actively pursuing ways to facilitate and enhance this mutual respect. Respect is critical to the relationships within your team, and the health of these relationships become paramount to the care delivered to the pet.
- Recognize each person's unique and vital position in the practice, including your own. The more we understand about each other's positions in the referral practice, the more we realize their important role in the medical care provided to our patients and compassionate care delivered to our clients.
- Acknowledge your responsibility to communicate the need for policy and protocol changes, additions, or deletions. Each team member should feel empowered to make suggestions regarding policies and protocols within the practice. Encourage initiative by involving the team whenever possible, and pursue suggestions they bring to the table.
- Seek advice from management as needed, realizing it is more important to ask then to know. It is important that we display confidence to our clients, but at the same time we must realize when we need to ask for more information. It is this ability to "know what you don't know" that becomes critical to the care of the patient and client.
- Seek assistance from other team members as needed, realizing it is more important to request help then to fail alone. It is the nature of veterinary medicine that care is delivered by a team of professionals working together.
- Remain aware of potentially dangerous situations, exercise caution, and follow safety protocols to protect the health and property of all involved, including yourself. There are occupational hazards in our profession, and we must maintain the proper attitude and obtain training on a regular basis to stay safe and keep others around us safe, including the patients.
- Recognize your own personal limits, and actively monitor your emotional health. One of those occupational hazards is the emotional toll that this profession can take on us as individuals. It is important to practice self-care so that you are able to provide the best care within your ability to patients, pet owners, and the referring community.

While we can develop a Code of Conduct for the referring practice as well, there would be no likely way we can implement or reinforce adherence to this Code. However, there are ways in which we can influence the "reverse relationships" within the Triangle of Care. The most important is communication. Yet communication takes place in a variety of ways which we can take advantage of and use to reinforce these reverse relationships. Information that we want relayed from the referring practice to our mutual clients can be initiated and affected by the materials that we provide the referring practices. For example, our written materials and website should contain information about the credentials of veterinarians on staff, with enough explanation to educate the pet owner and the referring community. The fees and services we can provide are communicated to the referring practice, and then passed along to the pet owner. We may encourage the referring practice to have a pet owner call us if they have concerns about a first visit, so we can begin to build that relationship. When a difficult situation exists where the referral practice receives complaints about a referring practice from the pet owner, the way in which this situation is handled can make or break the referral relationship in the future.

We often feel that the actions of the referring practice are beyond our control. A specific example may be the absence of a formal referral form or history provided on a pet by the referring practice. We reinforce the need for these items on our communications with the referring practice by providing the form on our website, and asking for adequate patient history prior to the referral appointment. Yet it may continue to be a problem. This is when it's time to get creative. Perhaps you mail the practice a stack of forms, even though they could have accessed them on the website, to eliminate that one step that might be getting in their way.

Perhaps you develop a newsletter communication with the referring community, and include a list of helpful things they can do to benefit the pet's appointment. In this type of communication, perhaps you include an interesting case report and stress how helpful it was to have adequate information before the pet even arrived. It may help to actually set up a face-to-face meeting with the manager or practice owner to discuss the referral process and any issues they may have, while including a pitch for them to send the information needed. Thus when we think we may not have any control over how the referring community behaves, we may just need to see where we can intersect the situation ourselves.

The Triangle of Care is essential to the continued well being and recovery of pets. The stronger the relationships that exist in this triangle, the more pets we have the ability to help and heal. We must continually evaluate the strength of these relationships and develop ways to improve and enhance them so that we can continue to provide the care we are called upon to provide.

THE VETERINARIAN'S ROLE IN THE TRIANGLE OF CARE

Katherine Dobbs, RVT, CVPM, PHR
Appleton, WI

There is a Triangle of Care within veterinary medicine, where one point is the referring or general practitioner, another point is the emergency clinician or specialist, and the third point is the client or pet owner. In the middle of this triangle we have the patient, the beloved pet…and these relationships work in collaboration to keep that pet healthy and happy. From this Triangle of Care, a Code of Conduct can be developed to help our teams within the referral practice deliver the type of communication and compassionate care that will help us build these relationships. Each person on the referral practice team has a role in ensuring these relationships remain balanced. The emergency or specialist veterinarian is essential to the two veterinary-client-patient-relationships (VCPR's) that exist in the Triangle of Care. Many of their responsibilities as veterinarians, practice owners, or associate leaders in the referral practice revolve around the education of the team around them to make a positive impact on these Triangle of Care relationships.

TRIANGLE CODE OF CONDUCT

The Referral Relationship

- Recognize that a Triangle of Care exists between the general practice, referral practice, and pet owner—with the pet's best interest central to all we do.
    - As a member of the leadership in a referral practice, the veterinarian should ensure that the entire practice team understands this concept, and the importance of each of the relationships involved. This can be accomplished by instituting a Code of Conduct that is presented to new hires, revisiting with existing team members, reinforced in coaching team members, and referred to when discipline or termination is necessary.
- Honor the veterinarian-client-patient-relationships (VCPR) that exist in this Triangle of Care.
    - As a role model for behavior in the practice, the veterinarian should exhibit a positive attitude regarding their VCPR with the pet owners in the referral practice, but most importantly demonstrate how the referring practice VCPR needs to be respected as the primary means of continued pet healthcare and well being.
- Nurture the relationship between practices by providing friendly service and responsive communication to all members of the referring community.
    - As a representative of the referral practice, the veterinarian should illustrate how all members of the referring practice teams should receive great client service. This includes when a team member from a referring practice comes to the referral practice on behalf of their own personal pet, and in communications regarding mutual clients of both practices.
- Demonstrate respect for the referring community through words and actions, both in private and within the practice team and in public to our mutual clients. Always give the referring community the benefit of the doubt.
    - As a leader and role model, the veterinarian should illustrate the proper attitude in respect to the referring community. Often we are faced with the results of an unfortunate incident that happened at the referring practice, or a diagnosis that was inadequate or incorrect. The way in which these incidents are handled by the attending veterinarian in the referral practice will be a representation of the attitude the entire referral practice team should exhibit in regards to the referring practice.
- Represent accurate credentials of all veterinarians on staff, demonstrating an understanding of professional designations, initials, titles, and recognized credentials.
    - Both the existing veterinarians and newly hired veterinarians should ensure that the entire practice teams understand the credentials and their meanings, so they can adequately represent these to the referring community. The specific terminology is important when working with residents, interns, and residency-trained ("board ready") specialists, so that the client understands the level of expertise and experience they are being provided by the attending doctor. The team members should understand the process of becoming a specialist and each stage, for example that specialists are fully qualified DVM's before they pursue specialization.
- Provide accurate estimates of fees and availability of services within the scope of designated authority.
    - The veterinarian must ensure that the team members understand which fees they can discuss with different people they come in contact with, such as referring team members and pet owners. The team members need to understand and be able to explain the process for pets that need "outpatient" care so the referring community can communicate to the pet owner, and the pet owner will be aware of the process involved in reporting results or findings.

The Pet Owner Relationship

- Realize that the way we treat the referring community will be interpreted as the way we treat our mutual clients.
    - As leader and role model, the veterinarian should demonstrate great client service to our pet owners, and explain to the team members that the way we treat pet owners will be a direct reflection of our patient care and client service in the eyes of the referring community.
- Recognize that new callers are potential new clients for our practice, are generally more interested in our level of care and compassion than our prices, and are only committed to the examination fee to seek our opinion.

- As a member of the leadership team, the veterinarian should ensure that there is a specific policy in place for those handling these type of phone calls. The policy should include what services can be discussed, what fees can be passed along, and how to create a conversation that begins to build a relationship, without implying a specific diagnosis or treatment plan until the pet is examined by the veterinarian.

➢ Recognize that both pets and pet owners need triage, and while in our practice their comfort and needs are our primary responsibility.
  - While the veterinarian is primarily involved in establishing an adequate triage protocol to attend to the immediate needs of the patient, they also need to be an example of how to take the client's feelings into consideration, including coaching the support staff to maintain adequate continuing communication and updates as the veterinarian attends to the patient.

➢ Acknowledge that communication is largely non-verbal, and remain aware of our tone of voice, body language, and facial expressions at all times.
  - As a role model for the type of communication that should occur between team members and with the two client bases, demonstrate the proper and effective non-verbal communication. When sending a message through a team member, ensure that the proper body language accompanies that message, so it is relayed to the client in the proper manner as well.

➢ Never judge clients' decisions about their pet's care, choice of humane euthanasia, or ability to afford the recommended treatment.
  - As a leader in the practice, and the highest qualified to understand and explain medical circumstances, help to demonstrate a respect for client choices without judgment. Assist the team members in understanding the medical necessities for the recommended treatment, and the choices selected by the client, so the team members can have a better understanding of difficult choices made by the clients that can affect the team member's emotional state.

➢ Provide compassionate service to all clients regardless of their emotional state, realizing that complaints often result from a pet owner's loss of control during an emotional or traumatic event in the life of their beloved pet.
  - As a role model for how to handle difficult clients and delicate situations, illustrate the proper way to be compassionate in the face of highly emotional situations. Help the team member to understand the basis for the client's action or reaction, so that the team member can decrease the likelihood of taking the client's behavior personally as a reflection of that team member's actions.

➢ Handle issues between the pet owner and referring practice with grace and respect, remaining impartial and neutral, but displaying appropriate concern for the pet.
  - The veterinarian is the primary person who should be communicating with a referring veterinarian if a pet owner expresses frustration, disappointment, or outrage regarding the care of their patient by the referring practice. The veterinarian should obtain information from the pet owner, and then move forward by having a discussing regarding this information with the referring veterinarian. If a manager or other support team member is involved in these situations as well, the veterinarian should provide coaching and support.

The Patient Relationship

➢ Realize that the way we treat the pet owner will be interpreted as the way we treat their pet.
  - As a role model, the veterinarian sets the tone of the practice, and the way in which the team members should communicate with pet owners. They need to explain that despite their advanced qualifications to provide the highest quality medical care, the reputation of the practice depends on the way in which clients are treated by every team member in the practice.

➢ Recognize the role of each person's position on the healthcare team, and take part in medical training and continuing education to enhance our ability to provide both client and patient care within our area and scope of authority.
  - As a role model, the veterinarian should participate in continuing education both outside and within the practice, to demonstrate their commitment to continued advancement. The veterinarian should also help the team seek continuing education by passing along information about CE events, suggesting topics, and even presenting in-house CE seminars to the practice team.

➢ Strive to earn the respect of your colleagues on the practice team and display respect in return, while actively pursuing ways to facilitate and enhance this mutual respect.
  - As the highest credentialed medical professional, the veterinarian typically expects respect from the support team. However, respect is gained more by your actions than the initials following your name. The veterinarian needs to earn the respect of the team by showing interest in the opinions of team members and the way in which they communicate. The veterinarian should demonstrate respect for every position on the practice team.

➢ Recognize each person's unique and vital position in the practice, including your own.
  - As a leader in the practice, the veterinarian should have personal recognition of the importance of each position, but they should also demonstrate this importance through their actions and words in front of other team members.

➢ Acknowledge your responsibility to communicate the need for policy and protocol changes, additions, or deletions.
  - As part of the practice leadership, the veterinarian is in a prime place to observe the policies and protocols that affect the patients, pet owners, referring community, and internally within the practice team. They should communicate to the proper person with suggestions for positive change.

- Seek advice from management as needed, realizing it is more important to ask then to know.
  - As a credentialed medical professional, the veterinarian has an expanded knowledge of the medical aspect of veterinary practice. However, they also need to demonstrate an understanding of the policies and protocols that affect situations other than patient care. They should use the management team to seek advice on how to handle certain situations, and be open to admitting when they need assistance.
- Seek assistance from other team members as needed, realizing it is more important to request help then to fail alone.
  - While the veterinarian is the primary medical professional, they still need the support and assistance of the entire team to deliver adequate care to patients and clients. Ensure that the requests for assistance are presented in a respectful way, asking rather than telling the support staff to pitch in.
- Remain aware of potentially dangerous situations, exercise caution, and follow safety protocols to protect the health and property of all involved, including yourself.
  - As a role model, the veterinarian will demonstrate how to remain safe during the care of patients and clients. The veterinarian is in the position to affect the safety of the team when it comes to dealing with fractious animals or unsafe medications, zoonotic diagnoses, or risky procedures.
- Recognize your own personal limits, and actively monitor your emotional health.
  - The training received by the veterinarian focuses mainly on medical knowledge and expertise, yet the veterinarian needs to be prepared for the emotional toll that will be part of providing veterinary care. While the veterinarian may acknowledge the emotional hazard for the support team and how burnout and compassion fatigue may be affecting those around him/her, the veterinarian needs to be aware that their position does not make them immune to those same hazards. As a leader and role model, the veterinarian can demonstrate adequate self-care by participating in those activities that improve the emotional health of the team.

The AVMA Principles of Veterinary Medical Ethics are directed at the veterinarians, yet the entire team must understand the concepts presented in order to participate in the operation of ethical veterinary medicine practice. The AAHA Referral Guidelines are directed at the veterinarians as well, yet the entire team must understand the protocols and theories to see that the practice operates in a way that supports a healthy referral relationship. For example, the Ethics support the accurate representation of credentials, which must be communicated by the entire team. The AAHA guidelines outline how often the practices should communicate with each other, yet this communication is often initiated by the medical support staff and delivered by the front office personnel. It is the veterinarians' responsibility to educate and enforce the actions that will support the Ethics and Guidelines.

Veterinarians also have a powerful influence on the actions and attitudes of the practice team, whether they are practice owners or associates. This power is not to be taken lightly; the team is constantly watching the veterinarians and will be affected by their behaviors, whether positive or negative. Each veterinarian should receive the information and training to become a positive influence on the entire practice.

THE TECHNICIAN'S ROLE IN THE TRIANGLE OF CARE

Katherine Dobbs, RVT, CVPM, PHR
Appleton, WI

There is a Triangle of Care within veterinary medicine, where one point is the referring or general practitioner, another point is the emergency clinician or specialist, and the third point is the client or pet owner. In the middle of this triangle we have the patient, the beloved pet…and these relationships work in collaboration to keep that pet healthy and happy. From this Triangle of Care, a Code of Conduct can be developed to help our teams within the referral practice deliver the type of communication and compassionate care that will help us build these relationships. Each person on the referral practice team has a role in ensuring that these relationships remain balanced. The technician is essential to the delivery of medical care to the patient, and interacting with the client along the way. Their interactions will reflect the level of medical care and compassion that is being delivered to the pet owner and referring community.

TRIANGLE CODE OF CONDUCT

The Referral Relationship

- Recognize that a Triangle of Care exists between the general practice, referral practice, and pet owner—with the pet's best interest central to all we do.
  - The technician is trained to deliver medical nursing care to animal patients. Those who have acquired a degree from an institute of higher learning also possess the knowledge and understanding behind the tasks they perform. It is assumed that the technician has the pet's best interest in mind at the conclusion of this training or experience. Yet they also need to understand the importance of the relationships that support the continued care of that pet.
- Honor the veterinarian-client-patient-relationships (VCPR) that exist in this Triangle of Care.
  - The importance of a valid and current VCPR between the patient and the referral practitioner must be understood when it comes to tasks such as refilling medications or offering consultative advice. Just as important is the VCPR that must be maintained between the patient and the referring veterinarian. The technician must support the health of this long term relationship as well, no matter how long the patient is receiving treatment by the referral team.
- Nurture the relationship between practices by providing friendly service and responsive communication to all members of the referring community.
  - The technician is often in a position to come in contact with the medical staff of the referring practices, and must remember that they represent their practice when they are in the community. They can help take this relationship to another level by initiating contact with technicians in the referring community, whether it's mingling at a seminar or hosting CE at the referral practice for the teams of the referring practices. Both parties should remain open to what they can learn from each other.
- Demonstrate respect for the referring community through words and actions, both in private and within the practice team and in public to our mutual clients. Always give the referring community the benefit of the doubt.
  - With the medical knowledge to realize the result of care given to the patient by the referring veterinarian, the technician may face the difficult situation of questionable care. The technician must not judge the referring practitioner, and instead support the best interest of the pet moving forward.
- Represent accurate credentials of all veterinarians on staff, demonstrating an understanding of professional designations, initials, titles, and recognized credentials.
  - The technician often understands the career ladder of the veterinarian, but this is not to be taken for granted. They need to be educated as well, so they can accurately represent the practitioners. This becomes important, for example, when a client is scheduled to see a resident and did not fully comprehend the difference between a resident and a boarded specialist.
- Provide accurate estimates of fees and availability of services within the scope of designated authority.
  - The technician may be the one who discusses fees and services most often with the referring community and pet owners, so they must be fully trained to represent these accurately. Fees should be delivered with an emphasis on the care provided, and services must be explained in a way that is either friendly to the pet owner, or medically advanced for the referring community. Out-patient services need to be explained to both client bases to prevent misunderstanding.

The Pet Owner Relationship

- Realize that the way we treat the referring community will be interpreted as the way we treat our mutual clients.
  - The technician may interact with members of the referring community during patient admission, treatment, or discharge. They may also interact during veterinary events such as CE and meetings. The referring community will assume that the way they are treated by referral practice technicians is the way that our mutual clients are treated during a referral.
- Recognize that new callers are potential new clients for our practice, are generally more interested in our level of care and compassion than our prices, and are only committed to the examination fee to seek our opinion.

- It may be the technician's duty to handle calls when a pet owner is seeking advice or fees for possible future treatment. It is important that the technician knows how much they can presume to know the patient's problem, and that the pet owner knows that a second opinion is within reach for the cost of the examination.

➢ Recognize that both pets and pet owners need triaged, and while in our practice their comfort and needs are our primary responsibility.
  - The technician is primarily focused on the immediate medical needs of the patient which they were trained to help. However their interaction with the pet owner often sets the tone of the visit and the level of trust that family will have which will allow the technician to move forward with care for the patient. The technician needs to remain aware of the pet owner's emotional state so that the information needed can be obtained to move forward with treatment.

➢ Acknowledge that communication is largely non-verbal, and remain aware of our tone of voice, body language, and facial expressions at all times.
  - The technician often is skilled at communicating with animals of all sorts, but their ability to communicate with people will determine their success in the referral practice. This may be more difficult for those technicians who entered the profession more for the pet, than the person. Communication training should be an ongoing emphasis in the referral practice.

➢ Never judge clients' decisions about their pet's care, choice of humane euthanasia, or ability to afford the recommended treatment.
  - With the pet's best interest foremost in their minds, technicians may have a difficult time understanding the choices pet owners are forced to make. Many of the conclusions that pet owners must choose go against the nature of the technician trained to heal animals. This is when the reduction of suffering might be a hard concept for the technician to support. This is a topic that requires continued discussion within the referral practice team.

➢ Provide compassionate service to all clients regardless of their emotional state, realizing that complaints often result from a pet owner's loss of control during an emotional or traumatic event in the life of their beloved pet.
  - The more emotional the pet owner, the more difficult it may be to obtain the patient for treatment away from the family, and obtain history from a family that is suffering with fierce emotions. The technician needs to understand how to provide care for clients in a variety of emotional states, and avoid taking it personally when these emotions spill out in a way that affects the technician.

➢ Handle issues between the pet owner and referring practice with grace and respect, remaining impartial and neutral, but displaying appropriate concern for the pet.
  - The technician may be in the position to hear "horror stories" from pet owners regarding the referring practice. The technician must understand that their role is to encourage care for the pet from this point forward, while supporting the pet owner but not commenting on the nature of the client's feelings toward the referring practice or veterinarian.

The Patient Relationship

➢ Realize that the way we treat the pet owner will be interpreted as the way we treat their pet.
  - The technician can only acquire the patient for treatment if the family feels the technician can be trusted. This requires attention to the communication between the technician and pet owner. If the pet owner is treated without obvious compassion or regard, they will assume their pet will be treated in the exact same way.

➢ Recognize the role of each person's position on the healthcare team, and take part in medical training and continuing education to enhance our ability to provide both client and patient care within our area and scope of authority.
  - The technician who is credentialed will seek continuing education to meet their CE requirements, but even those on the medial team that are not credentialed need to have the opportunity and possess the initiative to continue advancing their medical knowledge.

➢ Strive to earn the respect of your colleagues on the practice team and display respect in return, while actively pursuing ways to facilitate and enhance this mutual respect.
  - The technician should be aware of the "chasm" between "the front" and "the back" that is often felt within the veterinary practice and work to achieve mutual respect by appreciating the jobs of the other team members.

➢ Recognize each person's unique and vital position in the practice, including your own.
  - The technician who is trained and/or experienced must work to demonstrate appreciation and admiration for the other jobs that help make the practice operate. This is done by giving compliments, verbally recognizing accomplishments, and using "good manners" such as asking for help and thanking those who do respond.

➢ Acknowledge your responsibility to communicate the need for policy and protocol changes, additions, or deletions.
  - The technician has an up close view of the patient care delivered by the practice, and should take part in suggesting changes as they see necessary. The technician should play an active part in contributing to the development of new policies and protocols, both for patient care and client service.

➢ Seek advice from management as needed, realizing it is more important to ask then to know.
  - Even with education and training, the technician is not necessarily expected to know it all. The most important attribute is often the ability to "know what you do not know" in order to ask for clarification. With patients' lives on the line, and the reputation of the practice in the eyes of the clients, seeking the right information is vital.

➢ Seek assistance from other team members as needed, realizing it is more important to request help then to fail alone.

- The technician cannot provide patient care alone, it requires the help of assistants and the involvement of the front office team. Asking for help, and being ready to help others, is the key to teamwork and the success of the practice.

➢ Remain aware of potentially dangerous situations, exercise caution, and follow safety protocols to protect the health and property of all involved, including yourself.
    - Whether working together or alone, the technician must be aware to remain safe. Whether this involves heavy patients, the potential for patients to be aggressive, or working with dangerous substances, the technician is expected to know and minimize occupational hazards.

➢ Recognize your own personal limits, and actively monitor your emotional health.
    - Technicians receive great satisfaction from the work they do caring for animals, but it can come with a cost. They often handle emotional clients, difficult life-and-death situations, and long-term relationships that end with the death of a pet. The technician needs to remain aware of their need for self-care and maintenance of their state of mind.

The technician is typically one step below the veterinarian when it comes to medical credentials, and with these credentials comes the responsibility of being a leader and role model within the practice even when the position title doesn't necessarily indicate this role. Other medical staff such as non-credentialed patient care personnel provide life saving and health retaining medical care while assisting the other members of the medical team. The medical team overall affects the referral relationships by interacting with pet owners and the referring community in a variety of ways, both inside and outside of the practice. They should set the tone for creating positive referral relationships during the delivery of medical patient care.

THE RECEPTIONIST'S ROLE IN THE TRIANGLE OF CARE

Katherine Dobbs, RVT, CVPM, PHR
Appleton, WI

There is a Triangle of Care within veterinary medicine, where one point is the referring or general practitioner, another point is the emergency clinician or specialist, and the third point is the client or pet owner. In the middle of this triangle we have the patient, the beloved pet…and these relationships work in collaboration to keep that pet healthy and happy. From this Triangle of Care, a Code of Conduct can be developed to help our teams within the referral practice deliver the type of communication and compassionate care that will help us build these relationships. Each person on the referral practice team has a role in ensuring that these relationships remain balanced. The receptionist is essential to these relationships because they are positioned at the center of all communications that go in and out of the referral practice. Whether they are checking out a pet owner, or receiving a referral call from a referring practitioner, the receptionist reflects the referral practice's commitment to the health of these relationships within the Triangle of Care.

TRIANGLE CODE OF CONDUCT

The Referral Relationship

- Recognize that a Triangle of Care exists between the general practice, referral practice, and pet owner—with the pet's best interest central to all we do.
    - The front office personnel are figuratively and literally in the middle of these relationships, enabling communication to flow back and forth along the sides of this triangle. While they may spend the least amount of time interacting with the animals, they should realize they are a part of the healthcare team that delivers care to pets.
- Honor the veterinarian-client-patient-relationships (VCPR) that exist in this Triangle of Care.
    - The front office personnel manage the referral practice VCPR using data input and creation and maintenance of patient charts. They should remain aware that the VCPR with the referring practice also exists, so they will need to be actively participating in communication back and forth with the client and these two VCPR's.
- Nurture the relationship between practices by providing friendly service and responsive communication to all members of the referring community.
    - The front office personnel will be given the opportunity to speak with many members of the referring community, including the referring practice front office team and the referring practitioners. Regardless of who is on the other end of the line, the front office personnel should demonstrate respect and friendly service with any member of the referring community, even when having to repeat information or requests that should have already been provided.
- Demonstrate respect for the referring community through words and actions, both in private and within the practice team and in public to our mutual clients. Always give the referring community the benefit of the doubt.
    - At the front desk, the personnel may have the opportunity to hear from the pet owners regarding their experience at the referring practice. While remaining neutral and avoiding comment about the referring practice, the front office team should focus on what their practice can do for the pet now and in the future, rather than discussing the past.
- Represent accurate credentials of all veterinarians on staff, demonstrating an understanding of professional designations, initials, titles, and recognized credentials.
    - It is imperative that the front office personnel understand and use the correct credentials for every veterinarian on staff. This usually involves staff training upon hire, even if the team member has worked in general practice. The qualifications can be quite different when working with specialists, and particularly practices that offer residency programs. Ensure that the front office is accurately explaining these qualifications to both client bases.
- Provide accurate estimates of fees and availability of services within the scope of designated authority.
    - Depending on the practice, the front office may have very few fees they can discuss, but they should receive training on how best to present them. They will need to understand the protocol for scheduling out-patient procedures or tests to adequately explain to pet owners and referring practices what outcome to expect for all involved.

The Pet Owner Relationship

- Realize that the way we treat the referring community will be interpreted as the way we treat our mutual clients.
    - Each time a member of the front office talks with the referring community, they are representing the referral practice's overall level of client service. Polite enthusiasm and respect should be demonstrated regardless of the nature of the conversation or interaction.
- Recognize that new callers are potential new clients for our practice, are generally more interested in our level of care and compassion than our prices, and are only committed to the examination fee to seek our opinion.
    - The practice will determine who will field these types of phone calls. If it is the front office personnel, they should receive training that will best represent the practice's policy and compassion. If these messages are to be passed along, the front office personnel should ensure that they are forwarded to the proper person with the proper amount of information so that calls can be prioritized.
- Recognize that both pets and pet owners need triage, and while in our practice their comfort and needs are our primary responsibility.

- This is especially important for the front office team as they are on the front lines of comforting clients who are often times in worse shape than their pet. While the client remains in the front area, their comfort and needs are priority for the front team. The front office should also be trained to recognize visible signs of a pet's decline so they can seek immediate care if needed.

➢ Acknowledge that communication is largely non-verbal, and remain aware of our tone of voice, body language, and facial expressions at all times.
  - The front office team is the "face" of the practice for clients coming, and going, and interacting over the phone as well. Body language should be used in a way that reinforces relationships, and team members should constantly monitor their body language for the proper message.

➢ Never judge clients' decisions about their pet's care, choice of humane euthanasia, or ability to afford the recommended treatment.
  - Being on the front lines also means that the front office will witness some difficult choices being made by pet owners that they may not personally agree with, but have been made for the pet's or family's best interest. The front office team members should resist voicing their disagreement, except in predetermined places and times communicated by the manager or practice owner.

➢ Provide compassionate service to all clients regardless of their emotional state, realizing that complaints often result from a pet owner's loss of control during an emotional or traumatic event in the life of their beloved pet.
  - Being on the front lines also means that the front office team is the first one to receive the brunt of the client's emotions, including some emotions that are terribly negative. It will help if team members can relate from a place of empathy with the client, rather than judge their client's choices. This will also help to preserve their faith in human beings in general.

➢ Handle issues between the pet owner and referring practice with grace and respect, remaining impartial and neutral, but displaying appropriate concern for the pet.
  - It is difficult when the front office team member is faced with inquiring as to the history of the patient, when perhaps the client is not pleased about the level of care they received at the referring practice. The front office team is in a prime position to hear these complaints, yet they must remain neutral and focus on the future care of the pet and what the referral practice can do to make the situation better for the pet and the family.

The Patient Relationship

➢ Realize that the way we treat the pet owner will be interpreted as the way we treat their pet.
  - The front office team spends more time with the clients than with the pets. Even so, the client will assume that if they are treated well by the front office team, their pet is likely treated well by the medical team up front. Therefore client service has an impact beyond that person's interpretation of how they themselves are treated.

➢ Recognize the role of each person's position on the healthcare team, and take part in medical training and continuing education to enhance our ability to provide both client and patient care within our area and scope of authority.
  - While the front office team does not deliver patient care, they are involved in conversations regarding pet healthcare with pet owners and the referring community. The front office team should be knowledgeable on how to spell, pronounce, explain, and apply medical terminology, laboratory tests, common procedures, medications, etc.

➢ Strive to earn the respect of your colleagues on the practice team and display respect in return, while actively pursuing ways to facilitate and enhance this mutual respect.
  - The "chasm" between "the front" and "the back" in many veterinary practices can be bridged by the enhancement of mutual respect. The front office team should strive to gain an understanding of the positions in the patient care portion of the practice, and educate those positions in the front office's importance to the flow of information within the practice.

➢ Recognize each person's unique and vital position in the practice, including your own.
  - Each front office team member should recognize the importance of each position in the practice, and most importantly, they need to realize their importance on the team. Self respect should be an ongoing goal, even in the face of difficult clients, frustrated referring practice team members, and stressed team members throughout the referral practice.

➢ Acknowledge your responsibility to communicate the need for policy and protocol changes, additions, or deletions.
  - The front office team is responsible for managing and organization "the office" of the veterinary practice. Each member should feel free to suggest new ways of organization or implementation of new procedures that can help the front office overall. It is each person's responsibility to contribute to the success of their primary area in the practice.

➢ Seek advice from management as needed, realizing it is more important to ask then to know.
  - The front office is on the front lines of the numerous phone calls that come in asking for advice rather than services. There is a line which the front office team member must not cross, such as diagnosing a problem or implying a certain service needs to be rendered. If in doubt, the front office team member must feel compelled to ask their management team for help.

➢ Seek assistance from other team members as needed, realizing it is more important to request help then to fail alone.
  - At any moment, changes can occur up front with each ring of the phone or swing of the door. Front office team members must be constantly looking out for their team, and jump into situations that require their assistance. Rather

than staying narrowly focused on their own task at the moment, the team member must see the big picture and be flexible in helping others when situations suddenly arise.

- Remain aware of potentially dangerous situations, exercise caution, and follow safety protocols to protect the health and property of all involved, including yourself.
  - Even at the front desk, there are occupational hazards within the veterinary practice. Exposure to pets, cleaning materials, and some office supplies require that the front office team member remain aware of the dangers and importance of safety training.
- Recognize your own personal limits, and actively monitor your emotional health.
  - One of the occupational hazards for the front office team member lies in the proximity to the difficult client emotions when emergency cases arrive, difficult financial decisions must be made, and unwelcome news must be delivered. Each team member should actively monitor their emotional state, realizing that they are not immune to the difficulties of being involved in veterinary healthcare.

The receptionist or front office team member is vital to the veterinary practice, maintaining the office functions and flow of communication within the relationships of the Triangle of Care. They are in the position of being right where these relationships intersect, where the patient's history must meet the client's visit must meet up with the appointment in the referral practice. Without the proper flow of communication, it would be impossible to deliver the best medical care to pets. The front office team members need to remain aware of the importance of their position on the healthcare team.

5 KEY MARKETING QUESTIONS FOR SPECIALISTS IN PRIVATE PRACTICE

Karyn Gavzer, MBA, CVPM
Springboro, OH

Veterinary specialists do not become specialists because they love selling. They love medicine and pets, not selling. Unfortunately, many specialists equate marketing with selling and fail to understand that it is so much more than that!

Marketing is a master discipline. Advertising, pricing, brand-building, public relations (online and off), distribution, packaging and – yes, selling -- tuck under it. The practice's marketing plan should set strategy and drive your business revenue. It should also make marketing choices on where and how much money to spend easier. Keep in mind that every marketing choice you make should contribute to your brand (image) in its appearance and messaging and either promote awareness or attract business for your specialty practice.

Marketing is a basic business tool that all specialists need to understand and use wisely and well. It isn't simple or easy to know what to do, however, especially in the face of the many new marketing choices that are available today. The following five questions address key marketing questions that need to be decided to help your practice go and grow, especially in today's challenging economic environment. I hope they will be helpful to you in making thoughtful, smart choice for your specialty practice:

1. WHAT IS THE MOST IMPORTANT MARKETING DECISION I CAN MAKE FOR MY SPECIALTY PRACTICE?

Location, location, location! Specialists need a certain number of referrals per month to pay the bills and make enough money to meet their business goals for their practices. The number of referrals required will be different, according to the type and/or combination of specialty services offered and the size of the practice.

The three primary things to look for in a location are: 1) Population density sufficient to support the number of referrals needed to make the practice's finances work; this is co-dependant on the number of general veterinary practices located within the referring area that can send you cases.. 2) Demographics that support specialist-level care, e.g., income and education. 3) Affordable lease or building terms that won't overly burden the practice. And a bonus factor would be a practice location that is a highly visible and accessible.

2. WHAT IS THE MOST EFFECTIVE MARKETING TO BUILD MY PRACTICE?

Build word-of-mouth recommendations! Referring veterinarians that recommend you to their patients and colleagues is the best marketing of all. It takes time to build trust and rapport with referring veterinarians but once you reach the "tipping point," word-of-mouth can't be beat.

Word-of-mouth marketing requires a clear referral policy and consistency in following it on cases. Personal contact is also critically important because veterinarians refer to who they know and trust. If they've never met you and know little about you, it may deter referrals. Make sure that you give veterinarians in your referral area a chance to meet you – hold CE at your hospital, give talks to local veterinary groups and take them to lunch to make this happen.

New marketing tools that can also help build word-of-mouth go far beyond phone calls, case reports, mailings and newsletters. Social media is proving itself to be a strong marketing channel for increasing awareness, educating, communicating and building relationships with referring veterinarians and clients. It's one more tool that should be in everyone's marketing toolbox, especially because it is rapidly replacing other, less personable forms of advertising.

3. WHAT DO I NEED TO DO, BESIDES OFFER EXCELLENT, EXPERT MEDICAL SERVICES TO ATTRACT MORE REFERRALS?

The most important thing that you can do is to understand your "customers," in this case, we are talking about veterinarians who refer to you. This is easily done in a small veterinary practice by keeping in personal contact with referring veterinarians and responding to what they tell you or to changes in their referring patterns. It is also good, even in small specialty practices, to survey referring DVMS at least annually to find out how you are doing.

In general, referring veterinarians want:

- To look good in the eyes of their clients and their hospital teams
- To know personally the specialist to whom they are referring their patients
- To receive consistent, courteous, professional treatment
- To be shown respect
- To receive timely information about the cases they send you
- To learn from the cases they send you
- To have their clients and their patients returned as soon as possible
- To know that you will do no unnecessary work, such as repeating tests or radiographs that they have just done; or do any general work, such as vaccinating pets

4. MARKETING COSTS TIME AND MONEY. HOW DO I PRIORITIZE MARKETING DECISIONS WHEN I CAN'T DO EVERYTHING?

Your marketing priorities will naturally grow out of your business goals. Do you want to build more referrals for a new service e.g., cancer treatment, cardiac cases? Do you want to increase the number of local veterinary practices that refer to you? Do you want to increase referrals from veterinarians who are already referring to you? Do you want to increase community awareness?

In addition, don't overlook the basics! Appearances count online and off. When was the last time you "remodeled" your website? Are your grounds, facility and signage in good repair or worn and dated in appearance? Eighty percent of the impression that people take away about you is impacted by what they see. What does your practice's appearance, online and off, say about you?

When new marketing options evolve, always do a "T" appraisal to determine if they are a good bet for you. Think of a "T" as a scale. Under the left side of the "T," list all of the pros of the decision. On the right side of the "T," list all of the negatives. Use this to inform your decision. Finally, there are some things that we can only learn by doing; so if you decide to go ahead with a new marketing activity, make sure to track and monitor it to determine if you want to continue it, e.g., frequent Facebook postings.

5. WHAT IS A PROVEN, CHEAP AND EFFECTIVE MARKETING CHOICE FOR MY SPECIALITY PRACTICE?

There is a proven, cheap, effective marketing choice, but it is commonly undervalued and overlooked. It's your employees.

Think about it, the most powerful communication and the best opportunity you have to bond with rDVMs and clients is through one-on-one personal interactions. These happen every day on the phone and inside and outside of your practice. They are high-touch moments that need to be managed as part of your overall marketing plan.

Your employees are your brand and their interactions in the exam room, reception room and on the telephone "trump" any other form of marketing you can do. Turn your employees into brand ambassadors by taking care of them so that they feel supported and energized to take care of referring veterinarians and clients. Share your mission and core values with them so that they know what they are part of and provide training in communication skills to help them connect and bond with clients.

HOW TO SURVIVE IN A "YELLOW LIGHT" ECONOMY

Karyn Gavzer, MBA, CVPM
Springboro, OH

Consumers are feeling the pain of higher prices for food and gasoline. They are uncertain about the economy and some are afraid of losing their jobs and homes. Everyone wants to believe the economy will get better soon … but when? And what can veterinary practitioners do in the meantime to ensure their own survival?

The good news is that there are favorable trends to help veterinarians during tough times. There is also reason to believe that things won't get worse and that gives us psychological stability – we're still standing. It feels like the recovery is starting. It may be a slow road back but we are at least headed in the right direction and there are things you can do to help your practice survive and thrive in a sluggish economy.

Maslow's Hierarchy of Needs[1] provides a useful framework to help predict people's behavior in different economic circumstances. A quick review can yield insight and understanding into your clients' and your staff's behaviors in different scenarios. It can help you gauge where they are at in the hierarchy and help you decide on the best strategies for working with them.

Maslow's Hierarchy of Need

It is also worthwhile to look at practice implications and at time-proven business strategies successfully used in the past to get through economic downturns. Leadership is mission critical during times of economic instability. What is the practice's mission and vision? Long-term, where is the practice going?

Keep your perspective and stay focused on the future, even as you wrestle with the challenges of today. False bravado is not needed. Calm confidence in yourself, your team, your practice and the value you provide is. Fear is paralyzing. Choosing confidence over fear resets your mind and allows you to think clearly and make good choices.

Use this time to build teamwork and set new goals. Talk about the practice's vision and values. Ask your team to help you make the hard choices to get through today and prepare for tomorrow. Winning strategies to consider include:

- Identifying ideas to *creatively* control expenses:
  - Labor costs
  - Inventory
  - Energy costs
  - Other?
- Identifying new opportunities:
  - Establish new alliances through networking
  - Professionally promote your services through social media
  - Fee strategies to make help clients fund advance care
  - Improve service to clients and rDVMs

- Use the equipment you have better
- Pioneer new services

Above all else, it is important to work on yourself so that you can be a role model to others. Keep a long-term perspective and a good attitude. Take inspiration from the words of Rev. Robert Schuller and remember –

"Tough times never last. Tough people do!"

REFERENCES

1. Huitt, W. (2004). Maslow's hierarchy of needs. *Educational Psychology Interactive*. Valdosta, GA: Valdosta State University. From, http://chiron.valdosta.edu/whuitt/col/regsys/maslow.html.

SOCIAL MEDIA TIPS TO HELP YOU TWEET, BLOG, FRIEND, CONNECT & GROW

Karyn Gavzer, MBA, CVPM
Springboro, OH

OVERVIEW

Even though 2011 has started better than 2010, patient visits to specialists are still down compared to the pre-recession years. Unemployment is still high and that is a factor that affects pet care decisions. For discussion purposes, let's estimate the negative impact of unemployment on referrals at about 10%. Chances are your specialty practice is down even more than that in patient volume so unemployment alone doesn't explain the low case counts. The question to ask is: Where are the patients?

According to the new Bayer-Brakke-NCVEI study* (BBN study) that was released in January 2011, declining patient visits were first observed in general veterinary practices years before the 2008 recession began; in earlier years, however, general practitioners were able to increase their fees sufficiently to continue to grow and didn't notice the gradual decline. In earlier years, general veterinary practitioners referred cases to specialists and your practices grew even more rapidly. The BBN study concludes that the recession did not cause this trend but it did accelerate it. Almost overnight, general practitioners became aware that they were seeing fewer pets as their patient numbers dramatically dropped in 2009. Another change they experience was clients saying no to recommendations, including seeking specialty care and your referred case counts dropped as well.

The new study blames the drop in patient visits on the lack of value that pet owners put on wellness care, which is the primary driver of visits to veterinarians. The study also said that clients were turning to the internet for information and advice when their pets had problems, rather than call their veterinarians as they had in the past. This last is good news because it gives us something with which to work. Clearly the BBN study tells us where our clients are – and clearly where veterinary specialists we need to be -- and that is online!

A GOOD FIT

The skillful use of Social Media can help specialists bring their patient counts back up. We know from research that the people we want to reach are online and Facebook, in particular, has the volume to reach anyone you want to reach. Facebook demographics are a good match with pet owners. Published Facebook statistics tell us:

- Facebook has more than 500 million active users; 200 million of whom access Facebook through their mobiles
- 50% of Facebook users log on every day
- An average user has 130 friends
- Well over 70% of Generation X (Age 27- 43) and Boomers (Age 44 – 62) are Facebook users

The average veterinary client tend to be a woman, age 34 – 54, and this tells us the Facebook especially offers the veterinary specialist a forum in which to reach pet owners with information that can drive them to their general practice veterinarians and on to a referral to you.

MAKE SMART USE of SOCIAL MEDIA TO BUILD BUSINESS

- Social Media can take up a lot of time, if you let it. To avoid time-wasting, hire a technician to set up things for you and link your Social Media platforms to your website. This will also have a beneficial effect on your search engine optimization.
- Post often, but don't "sell" – new, novel, helpful stories work. Pet owners love to be in the know and what better expert than you to speak on your specialty?
- Feature posts about pets that you have helped and include mention of the rDVMs in your story so that pet owners see the teamwork involved between you and them to help their pets. rDVMs will see it too and they will see how you value your relationship with them.
- Seek expertise to help you set a strategy for using Facebook and other social media platforms so that you present a consistent branding message for your practice.
- Remember always that you have two audiences – the pet owner and rDVMs. Your message should be engaging for both groups.
- Post at least once a week, if you can.

- Adapt your message for different Social Media platforms – start with your blog and link it to your website; then shorten the message again for Facebook and link to your blog and website; adapt it for Twitter and link it to your blog and website. (These links can be set up automatically by your technology person, but you will need to rephrase the message for each platform.)
- Start a blog – they are easy to change and will also help to optimize your website in internet searches. Blogs are taking on a life of their own and because they can look like your website and are so easy to use they may soon become more popular place for pet owners, clients and rDVMs to go. Websites, on the other hand, are useful for "evergreen" information about your practice. Another way to think of websites is as the new "business card" of the internet.

SUMMARY

Clients and rDVMs are online and you need to be where your clients are! It's easy, practical and smart to engage them in conversations about pets as of way of educating them about the many ways you can help their pets. Use the tips and ideas mentioned to take advantage of the new social media platforms, make new friends for the practice, and build your business.

* For a copy of the preliminary findings on the Bayer-Brakke-NCVEI study that was presented at the NAVC 2011, go to www.NCVEI.org.

THE 3RD LEVEL OF CLIENT ENGAGEMENT

Karyn Gavzer, MBA, CVPM
Springboro, OH

OVERVIEW

If you think that client relationships are built solely on successful diagnoses and treatments, think again! New research from the emerging field of neuroscience is yielding deeper understanding about the way people *really* think and the dominant role that emotions play in all human interactions, including the ones that play out in your practice. This talk discusses new findings and provides applications to use to build more successful and rewarding relationships with clients at the third level of client engagement.

Please note that this talk is geared to client communications but the same techniques work equally well when used to build stronger relationships with referring veterinarians.

THE RATIONALITY OF IRRATIONAL BEHAVIOR

Research published in 2008 by the AVMA* shows that pet owners are more compliant, more loyal and more accepting of fees when they have a good relationship with you. Information published by the Gallup Press ** and other sources confirm the important role that feelings play in human behavior. Taken together, these data suggest that success in practice is based on clients' feelings about you more than on your medical expertise. This may seem irrational, but it isn't once you understand the new findings about how people think and process information.

It has only been in the last couple of decades that the emerging field of neuroeconomics – a hybrid of neurology, psychology and economics – has begun to unlock the secret of how emotions work as a function of biology. Scientists are now using advanced technology to see how the brain responds to feelings and correlating brain activity with observable behavior. This has provided new insights into how the brain is wired. It also helps explains why even intelligent people make irrational choices and do things that make no logical sense.

Feelings drive human behavior in ways that we are not aware of and sometimes, in spite of our better judgments. Neuroscience has shown that the basic connection between feelings and actions is so primary and deep that it easily overrides more rational thoughts. For instance, folk wisdom cautions, "Count to 10 before you loose your temper!" It turns out that this is scientifically sound advice. Counting to 10 gives the weaker, slower, rational thought process a chance to catch up with the lightening bolt response emotions evoke. These "lightening bolts" occur when we first meet people, fall in love, or see something we want; as well as when we are frightened, experience loss, disappointment, or anger and other human emotions.

Most of us like to think that we are rational human beings. New research indicates, however, that we are first and foremost products of our biology. We are emotionally-driven creatures, as are our clients. The fact is that emotions exert such a strong influence on perceptions and behaviors that they must be attended to in order to successfully communicate and build trusting relationships.

In the past, client service initiatives were mostly directed at meeting common, first level satisfaction criteria based on quality patient care. In other words, meeting the basic clients' expectations for their pets' care was enough – after all, isn't that what they came for? Today, we know that quality patient care is only the starting point of client relationships that are strengthened through emotional engagement and trust-building.

The following chart is adapted from the work of John Fleming, Ph.D. and Jim Asplund.** It shows how emotional engagement with clients can be broken down into four distinct levels, starting with Level 1 and building up to Level 4:

HIERARCHY OF EMOTIONAL ENGAGEMENT **

Think of a pyramid with the lowest level of client engagement at the bottom and the highest and rarest level of client engagement at the top. Most practices today are at the second level of client engagement, but it isn't enough to create trusting, rewarding relationships. Level three is the standard of client care and it requires new skills and communication . The 4 levels of the Hierarch of Emotional Engagement are:

4. **Passion** Can't imagine going anywhere else! Like to recommend you and tell others about your practice.
3. **Fulfillment** Feel good about their choice to come to you and your referral practice.
2. **Integrity** Feel you will treat them fairly and that they can trust you to act in their best interests.
1. **Confidence** This is the foundation piece = trust in your competency and the ability to provide good service.

Using the above chart, what activities might you undertake in practice to consistently achieve engagement at the third level of client engagement?

AUTOMATED RESPONSE –VS – AUTHENTIC CONNECTION

If human beings were machines, we could automate processes to the point where we could program communications to achieve almost perfect and predictable results every time we interacted with another. When we try to automate responses with people, however, just the opposite usually happens and it backfires. For instance, if we try to substitute automated answering machines for people and we frustrate our clients – so much so that they quickly circumvent the system to get to talk to a real person. It also backfires when we try to automate human behavior by asking receptionists to read from a script in response to client questions; or if the our doctors are directed to automatically shake clients' hands when they enter exam rooms for the first time. Even though these seem like reasonable things to do, in practice they prevent individualized and authentic people-to-people connections. Sometimes, they create negative impressions when others see the "automated" behaviors as insulting and insincere – just the opposite of your intentions!

If you are ready to communicate at a more effective level, you'll have to give up on automated responses and focus instead on doing things differently to create authentic connections.

The "soft stuff" of relationship-building communication requires attention to body language and tone of voice as well as to the words that you say to others. It requires you to read the person and the situation and incorporate both factual as well as emotional components into your communications.

SUMMARY

Clients have changed! They are willing to spend money on their pets and they appreciate the advanced medical options that you provide to help their pets. Clients are also more educated, sophisticated and demanding. They are quick to tell others about their experiences, which are based on their feelings, not facts. Today, it is necessary to incorporate both informational and emotional components into your client communications. This will help you make a positive emotional connection and build a solid foundation for trusting, satisfying relationships.

REFERENCES

* Pet Owner Price Sensitivity and Attitude Study, American Veterinary Medical Association 2007, Schaumburg, IL, BNResearch, Portland, OR. ** Human Sigma Managing the Employee-Customer Encounter, 2007, Gallup Press, New York, NY, ISBN: 978-1-59562-016-3 .

HYPERTROPHIC CARDIOMYOPATHY SCREENING IN CATS

Virginia Luis Fuentes, VetMB, PhD, CertVR, DVC, MRCVS, DACVIM (Cardiology), DECVIM-CA (Cardiology)
Hatfield, UK

HCM- BACKGROUND

HCM is probably the most prevalent cardiac disease in cats, both in apparently healthy cats[1,2] and in cats presenting with clinical signs due to heart disease.[3] Furthermore, HCM is the most common cardiac disease in both young and old cats, and for both pedigree and non-pedigree cats. HCM is the most common inherited cardiac disease in people, occurring in approximately 0.2% of the population, yet the prevalence appears to be even higher in cats. In one necropsy study of nearly 5000 cats, the prevalence of HCM was reported as 5.1%.[4] Two recent reports suggest the prevalence of HCM in asymptomatic cats (using current echocardiographic criteria) is around 15%, even in non-pedigree cats.[1,2] If accurate, this is an incredibly high prevalence.

It might not matter much that the HCM prevalence is so high, were it not for the dire consequences in some affected cats. Pulmonary edema and/or pleural effusion may result from severe HCM in some cats, and arterial thromboembolism (ATE) is a catastrophic complication in other affected cats. Sudden death is a third (and probably under-reported) effect of HCM. Sudden death is widely recognized in human HCM, but we have few mechanisms in place to document the prevalence of sudden death in cats with HCM. Veterinary cardiologists are probably the least likely people to identify such cats. Evidently not all cats with HCM will develop these serious complications, although the exact proportion of cats that progress to clinical signs is not known. Without doubt, HCM represents an important health problem in cats.

DEFINING HCM

Hypertrophic cardiomyopathy (HCM) is defined in people as 'left ventricular hypertrophy (LVH) associated with a nondilated chamber in the absence of another cardiac or systemic disease capable of producing the magnitude of hypertrophy evident.'[5] In practical terms, the usual human clinical diagnostic criterion is a maximal left ventricular (LV) diastolic wall thickness greater than or equal to 15mm.[5] In cats, similar clinical diagnostic criteria are used; commonly (but not always) using a maximal LV diastolic wall thickness greater than or equal to 6mm.[2] Pathological definitions include an increase in cardiac mass to bodyweight ratio and the presence of myocytes disarray, but a very heterogeneous spectrum of phenotypes is recognized in human HCM, which includes a variable phenotype even for individuals with the same genotype. Feline HCM exhibits a similarly variable phenotype. A molecular definition of HCM also poses problems, as hundreds of different mutations can result in a human HCM phenotype. The majority of HCM mutations affect one of 10 genes encoding sarcomeric proteins, including myosin binding protein C (MyBPC). Of the 2 mutations identified so far in feline HCM, both affect the MyBPC gene.[6,7]

DIAGNOSING HCM

Unlike chronic mitral valve disease in dogs, it is difficult to rule out feline HCM using physical examination alone. Some physical exam findings (such as a gallop sound) are strongly associated with a diagnosis of HCM (or at least significant myocardial disease). The negative predictive value of physical exam is low, however, and a definitive diagnosis requires echocardiography performed by a trained and experienced operator. While this is assured with a board-certified cardiologist, the standard of echocardiography among non-cardiologists is very variable. Echocardiography is more challenging in cats than dogs, yet it plays a more important role in diagnosis of feline cardiac disease compared with canine cardiac disease. With echocardiography, a diagnosis of 'affected' versus 'normal' can depend on correct placement of a measurement calliper to within less than a millimeter. This is a big problem in general practice, as many small animal veterinarians are unable to make a definitive diagnosis of one of the most common life-threatening feline diseases.

REASONS FOR SCREENING FOR HCM IN CATS

This combination of high prevalence, potentially serious consequences, and diagnostic uncertainty means that there is a critical need for HCM screening. There are many possible reasons why owners or veterinarians might wish to identify cats with HCM before they demonstrate clinical signs. Breeders of pedigree cats of breeds predisposed to HCM often seek screening for their breeding stock to prevent inadvertent breeding from affected cats. Similarly, pet-owners may wish to screen their pet cats for HCM if a related cat is diagnosed with, or dies from, HCM, and many owners will request more information on prognosis when their veterinary suggests their cat may have heart disease. Not unreasonably, many veterinarians will wish to identify cats at risk of congestive heart failure (CHF) prior to procedures that may increase its risk (general anesthesia, or enthusiastic intravenous fluid therapy). Veterinarians may also wish to identify cats at risk of CHF, ATE or sudden death in order to institute therapy that might reduce this risk, although evidence supporting such therapies is still lacking. Animal shelters may also wish to screen for heart disease to avoid rehoming cats with serious HCM and a poor prognosis. The steps necessary for effective screening will vary, depending on the reason for screening.

HCM screening for breeders

Breeders seek HCM screening because they want to breed responsibly and avoid producing affected offspring, although this is sometimes prompted because of a known family history of HCM or because screening for certain diseases can become fashionable. This is the most challenging type of HCM screening by far. It is important to detect even mildly affected cats in this context, as it is known in human HCM that phenotypic expression may vary widely even among family members with the same mutation. As there may be little echocardiographic difference between normal cats and mildly affected cats with HCM, clearly defining the criteria for 'affected' is vital. Individual cats may be classified differently according to the echocardiographic technique used, so that values obtained using M-mode echocardiography are not interchangeable with those obtained using two-dimensional (2D) echocardiography, even when the maximal measured wall thickness is recorded rather than a mean of several measurements. The maximal focal thickness is recorded by some, versus the wall thickness affecting at least 50% of a wall segment by others. If 2D views are used, the precise planes (and number of planes) should be specified. Whether or not the endocardial surfaces are included should also be specified, as this can lead to a difference of almost a millimeter. Great care must be taken to avoid inclusion of false tendons, which can also increase the recorded wall thickness if inadvertently included. The significance of other findings such as dynamic outflow tract obstruction (including systolic anterior motion of the mitral valve) and papillary muscle hypertrophy should also be defined.

The lack of a gold standard for defining feline HCM prevents appropriate validation of the above criteria for distinguishing affected and normal, but this is a difficult concept for cardiologists, let alone cat breeders. One of the problems associated with breeder-driven schemes is that the screener can feel coerced into making judgements for which we have insufficient information. More information on the natural history and progression from normal to a classic HCM phenotype would help this.

The ability to identify HCM-affected cats using a genetic blood test (or even a cheek swab) would be highly desirable in terms of availability and ease of interpretation. Unfortunately, interpretation of genetic testing is also fraught with difficulties. In Maine coon cats, the penetrance of the identified MyBPC mutation is incomplete, so that echocardiographic testing is still necessary to identify phenotype in individual cats.[8,9] In addition, wild-type mutations may exist alongside the previously documented HCM mutations. The lack of a gold standard for defining feline HCM also hampers progress in identifying abnormal genotypes. Identification of a positive HCM phenotype is straightforward in severely affected cats, but there is a broad 'equivocal' category of asymptomatic cats and age-related penetrance exacerbates the problem. The potential advantage of screening pedigree cats over nonpedigree cats is that the number of HCM mutations might be expected to be smaller in the more closed gene pools of purebred cats in comparison with cats in the general population.

HCM screening to identify cats at risk of CHF, ATE

Although it can be difficult for the general practitioner to make a definitive diagnosis of HCM, this is rarely necessary outside of screening breeding cats. Most veterinarians (and owners) are more concerned with identifying cats at risk of the complications associated with HCM, rather than the HCM itself. If the priority is to recognize cats at risk of CHF or ATE, then there are a number of more widely available options. Some physical exam findings can be helpful, such as an audible gallop or arrhythmia. Presence or absence of a murmur seems to be relatively unhelpful.[2] Presence of some arrhythmias (such as atrial fibrillation) may be an ominous sign. The role of biomarkers in stratifying risk of CHF in asymptomatic cats with HCM has still to be defined, but NT-proBNP shows promise. Left atrial (LA) enlargement is a very important sign associated with increased risk of CHF and ATE.[10] LA enlargement can be identified relatively easily with echocardiography, and to a lesser extent, with radiography. Identification of myocardial changes becomes easier as disease severity worsens, so the cats at highest risk of CHF or ATE should also be the easiest to identify using echocardiography. Qualitative changes on 2D echocardiography are often sufficient to identify such cats. The presence of spontaneous echocontrast is believed to be an indicator of risk of thromboembolism, and is readily imaged in standard views. Although we are still waiting for evidence that therapy can prevent CHF in high risk cats, it is at least possible to warn owners about what to expect in cats with severe changes, and antithrombotic therapy is at least available for cats at risk of ATE.

It is more difficult to determine the risk of interventions (such as anesthesia or IVFT) in moderately affected cats, and these cats can be difficult to identify using echocardiography. HCM fulfils certain criteria for diseases that should be screened for, in that there are potentially serious consequences if HCM is missed, action may be taken if HCM is identified, and there are tests available. There is hope that biomarkers may offer one option for initial risk stratification in cats undergoing potentially hazardous procedures.

For cat shelters, presence or absence of a heart murmur is a poor indicator of future risk of cardiac complications. For veterinarians involved in shelter medicine, an accurate biomarker would provide the ideal solution for mass screening and prevention of rehoming apparently healthy cats that unexpectedly develop CHF or ATE after moving to a new home.

FUTURE DIRECTIONS

A major priority is to obtain more information on the natural history of HCM, so that we can identify reliable indicators of future progression to a malignant HCM phenotype. HCM is a very heterogeneous disease in cats, and it would be invaluable to be able to single out cats at high risk of future complications. This would not only help with providing better prognostic information, but would also allow clinical trials to target the most appropriate patient groups. A second major priority is to standardise the diagnostic criteria for HCM. With the many different measurement techniques in use, it is difficult to compare 'like with like', and this will impede future progress in identifying both new mutations and other factors which modify gene expression.

REFERENCES

1. Paige CF, et al. **JAVMA** 2009;234:1398. 2. Wagner T, et al. **J Vet Cardiol** 2010;12:171. 3. Ferasin L, et al. Feline **J Fel Med & Surg** 2003;5:151. 4. Liu SK. **Vet Clin N Am** 1977;7:323. 5. Maron BJ, et al. **J Am Coll Cardiol** 2003;42:1687. 6. Meurs KM, et al. **Human Molecular Genetics** 2005;14:3587. 7. Meurs KM, et al. **Genomics** 2007;90:261. 8. Wess G, et al. **J Vet Intern Med** 2010;24:527. 9. Mary J, et al. **J Vet Cardiol** 2010;12:155. 10. Payne J, et al. **J Small Anim Pract** 2010;51:540.

DIAGNOSIS & PREVALENCE OF DILATED CARDIOMYOPATHY IN DOBERMAN PINSCHERS

Gerhard Wess, DVM, Dr. med vet., PhD, DACVIM (Cardiology), DEVCIM (Internal Medicine & Cardiology)
Munich, Germany

INTRODUCTION

Dilated cardiomyopathy (DCM) is an important cause of cardiac morbidity and mortality in dogs and is the most commonly acquired cardiac disorder in large and giant breed dogs.[1-2] Dobermans Pinschers are one of the most commonly affected breeds, and DCM in this breed is an inherited, slowly progressive primary myocardial disease.[1,3] The occult stage of the disease is characterized by evidence of morphologic or electrical derangement in the absence of clinical signs of heart disease.[1] The morphologic abnormality consists of left ventricular (LV) enlargement in systole and later in diastole. Ventricular premature contractions (VPCs) are a common finding in the occult phase of DCM in Doberman Pinschers.[3] Sudden death, caused by ventricular tachycardia-fibrillation, occurs during the occult phase in at least 25 to 30% of affected dogs.[4] These abnormalities, morphologic or electrical, can coexist or can be of predominantly one form at any time during this occult stage.[1,4]

PREVALENCE

Existing prevalence information on cardiomyopathy in Doberman Pinschers is from dogs in the US or Canada where prevalence ranges between 45 and 63%.[5] A recent study from Europe showed a similar prevalence with 58% of the dogs being affected.[6] This European study showed that 37% of the dogs had only VPCs without echocardiographic changes and that arrhythmias often were the first abnormality detected. Only a few dogs (13%) presented with only echocardiographic changes and no arrhythmias on Holter examination. Another interesting finding was that although there was no overall difference in the occurrence of cardiomyopathy between male and female dogs, there was a difference between sex concerning disease manifestation.[6] Female dogs had significantly more VPCs without echocardiographic changes than male dogs and this difference became more apparent with increasing age. On the other hand, male dogs developed earlier echocardiographic changes than did female dogs. Some older studies concerning Doberman Pinscher cardiomyopathy found a higher prevalence of the disease in male dogs. However, the prevalence study from Europe found an equal sex distribution, which would support the suspected autosomal dominant mode of inheritance.[6] The different findings concerning the sex distribution may be explained by the results of the study that showed an equal sex distribution but different disease progression between male and female dogs. Female dogs seem to experience a more slowly progressive disease with VPCs as the only abnormality found even in the older age groups. Male dogs, on the contrary, showed echocardiographic changes earlier than did female dogs. These changes are easier to detect because no 24-hour-ambulatory-ECG (Holter) examination is necessary, and male dogs therefore are also more likely to develop CHF at an earlier age than female dogs and also die earlier from their disease. From a clinical perspective, it is important that despite the fact that female dogs are more likely to show only VPCs as an abnormality, a Holter examination also should be performed in every male dog, because the risk of sudden death is high in all dogs with VPCs.

DIAGNOSIS

For diagnostic, prognostic, therapeutic, and breeding purposes it is essential to detect the occult phase of the disease. Diagnosis primarily depends on evidence of ventricular arrhythmias, detected by Holter and echocardiographic evaluation of left ventricular dimensions and function. Established and newer tests such as biomarkers or new echocardiographic methods have been recently be evaluated in their value to diagnose DCM in Doberman Pinschers and will be presented below.

5-Minute-ECG versus 24-hour-ECG

Ventricular premature contractions (VPCs) are a cardinal finding in the occult phase of DCM in Doberman Pinschers. The number of VPCs usually increases over time and eventually systolic dysfunction develops and dogs start to show clinical signs. The gold standard for detecting VPCs is a Holter examination.[7] Currently several cut-off values (> 50 or >100 VPCs) are considered diagnostic for DCM in Doberman Pinschers, when systemic causes for arrhythmias are excluded. Holter examinations, however, are not as readily available to the veterinarian, and are more expensive and time consuming to perform than in-house ECGs. Therefore, a recent study evaluated the use of an in-house ECG and compared the results to the Holter recordings as a gold standard for detecting arrhythmias in Doberman Pinschers.[7] This study showed a low sensitivity but high specificity of a 5-minute ECG to predict > 100 VPCs/24 hours, if at least 1 VPC is detected within 5 minutes. The absence of VPCs in the 5-minute ECG should not lead to the assumption that the dog is healthy, because false negative cases were found in 35.8 % of the examinations.[7] At least 1 VPC in a 5-minute ECG strongly warrants further examination, because specificity (96.7 %) and positive predictive value (85.6 %) are very high. Sensitivity increased with higher VPC numbers detected by the Holter examination. Therefore, a 5-minute ECG cannot replace a 24-hour ECG examination for the purpose of detecting the occult phase of DCM in Doberman Pinschers.[7]

Echocardiography

M-mode echocardiography used to be considered the gold-standard to diagnose echocardiographic changes in Doberman Pinschers with DCM.[8] A recent study in Doberman Pinschers compared several established with new calculated M-Mode reference values using a longitudinal follow-up study.[9] A multivariate analysis including weight, age, and gender revealed that only gender had a significant influence on M-mode values. Therefore, separate M-mode cut-off values should be used for male and female dogs. The best discriminating values to differentiate healthy dogs from DCM were for the left ventricular diastolic diameter, LVIDd >48mm for males and >46 mm for female Doberman Pinschers. The best cut-off value for the systolic left ventricular diameter, LVIDs was >36 mm and can be used for both, male and female dogs.[9] However, M-mode is a 1-dimensional method that relies on geometric assumptions that may assess geometric changes occurring in diseases such as DCM less accurately. Accordingly, in humans it is now recommended to use the Simpson's method of disks (SMOD) for 2D volume measurements.[10] Using SMOD, heart volume is measured as the summation of parallel cylinders, whose diameters are derived from endocardial border tracing performed on 1 or 2 orthogonal LV apical views. The principle underlying this method is that the total LV volume is calculated from the summation of a stack of elliptical disks. The SMOD method was recently also evaluated in Doberman Pinschers using a longitudinal follow-up study and compared to M-Mode measurements.[11] The ability of the new SMOD cut-off values, normalized to body surface area (BSA), for left ventricular end-diastolic volume (LVEDV/BSA >95ml/m^2) and end-systolic volume (LVESV/BSA >55ml/m^2) to detect echocardiographic changes in Doberman Pinschers with DCM was compared to currently recommended M-mode values. Dogs with elevated SMOD values but normal M-mode measurements were followed-up using a prospective longitudinal study design. The study showed that the SMOD method is superior to conventional M-mode measurements in its ability to detect early echocardiographic changes in Doberman Pinschers with DCM. SMOD detected correctly early echocardiographic abnormalities in Doberman Pinschers that only later also became abnormal using M-Mode. Therefore, SMOD could be considered as a new gold-standard in the diagnosis of echocardiographic changes in Doberman Pinschers.[11] New methods such as tissue Doppler may a in the future useful in the diagnosis of DCM in Doberman as well.[12] A new study established new reference values for tissue velocity imaging (TVI), Strain and Strain Rate in healthy Doberman Pinschers; theses parameters might be used for a supportive diagnostic approach.[13]

Biomarker

Screening for occult disease stages is one of the most promising areas of blood sample-based biomarker research. Blood-based testing to detect cardiomyopathy is attractive because of its minimally invasive nature, ease of sample collection, potential for widespread availability, quantitative nature, and theoretic costs efficiency. Cardiac biomarkers, such as cardiac troponin I (cTnI) or natriuretic peptides are all elaborated, presumably along a continuum concurrent with disease severity. Therefore, they may be found in patients with early stages of nonclinical disease. Circulating cTnI has been demonstrated to be a highly specific and sensitive marker for myocardial cellular damage in humans and animals. Recently, cTnI is has been also evaluated in various stages of DCM in Doberman Pinschers.[14] A cut-off value of >0.22 ng/ml had a sensitivity of 79.5 % and a specificity of 84.4 % to detect all forms of cardiomyopathy (including the stage with only VPCs or only echocardiographic changes). Although cTnI should not replace Holter or echocardiography, cTnI measurement is a valuable additional test that even can detect cardiomyopathy in cases, where conventional tests are still unremarkable, i.e. in cases that became only abnormal using conventional tests within 1.5 years. N-terminal pro-B-type natriuretic peptide (NT-proBNP) is a valuable biomarker which can be used to distinguish cardiac causes of dyspnea from noncardiac causes, and high circulating concentrations of the peptide have been reported in humans and dogs with DCM, myxomatous mitral valve degeneration, hypertrophic cardiomyopathy, and various congenital heart diseases. Recently, NT-proBNP concentrations were evaluated in various stages of dilated cardiomyopathy (DCM) in Doberman Pinschers.[15] that study showed, that similar to cTnI, plasma concentrations of NT-proBNP were increased in dogs with evidence of DCM and in apparently healthy dogs that developed DCM within 1.5 years after samples were obtained.[15] Sensitivity and specificity of NT-proBNP to detect all stages of DCM with a cutoff value of > 400 pmol/L were 81.1% and 75.0%, respectively. Sensitivity was 90.0% and specificity 75.0% to predict echocardiographic changes. Specificity to detect echocardiographic changes reached 90.4% using a cut-off value of > 550 pmol/L. It is important to realize that the cutoff values for Doberman Pinschers established in the present study are only valid for the NT-proBNP assay (Guildhay) that was used and for samples collected according to the described method. At least 1 other test for NT-proBNP evaluation is commercially available for use in dogs, for which sample collection and test characteristics are likely to be different from those described in this report. Cutoff values may also be different for such tests.

Genetic Tests

Two different research groups, one from the US and one from Europe have performed genome wide association studies in order to find the underlying genetic defect for DCM in Doberman Pinschers. Whereas one group found significant associations on chromosome 14, the other group found in their study significant associations on chromosome 5. Therefore it might be possible that either different forms of DCM in Doberman Pinschers have different mutations, or that there might be other factors leading to different results. Future study will evaluate these differences.

REFERENCES

1. O'GRADY MR, ET AL. VET CLIN NORTH AM SMALL ANIM PRACT 2004;34:1187. 2. TIDHOLM A, ET AL. VET J 2001;162:92. 3. CALVERT CA, ET AL. J AM ANIM HOSP ASSOC 2004;40:24 . 4. CALVERT CA, ET AL. J AM VET MED ASSOC 2000;217:1328. 5. O'GRADY M R, ET AL. J VET INTERN MED 1998;12:199. 6. WESS G, ET AL. J VET INTERN MED 2010;24:533. 7. WESS G, ET AL. J VET INTERN MED 2010;24:367. 8. CALVERT CA, ET AL. J AM VET MED ASSOC 1986;189:293. 9. WESS G, ET AL. ACVIM-PROCEEDINGS, ANAHEIM 2010;232. 10. LANG RM, ET AL. EUR J ECHOCARDIOGR 2006;7:79. 11. WESS G, ET AL. J VET INTERN MED 2010;24:1069. 12. O'SULLIVAN ML, ET AL. J VET INTERN MED 2007;21:81. 13. SIMAK J, ET AL. JVC 2011 (IN PRESS). 14. WESS G, ET AL. J VET INTERN MED 2010;24:843. 15. WESS G, ET AL. AJVR 2011 (IN PRESS).

PRE-CLINICAL CHRONIC DEGENERATIVE MITRAL VALVE DISEASE IN THE DOG

Michele Borgarelli, DVM, PhD, DECVIM-CA
Manhattan, KS

INTRODUCTION

Chronic degenerative mitral valve disease as a result of myxomatous degeneration (MMVD) represents the most common acquired cardiovascular disease in the dog. The disease is characterized by a long pre-clinical period and many dogs die for other reasons and do not progress to heart failure. This pre-clinical group of dogs represent a very heterogeneous group with some animal affected by a mild disease (stage B1 accordingly to ACVIM consensus) and others that have a more advanced status likely leading to develop clinical signs of heart failure (stage B2 ACVIM). This consideration underscores the importance of identifying the sub-population of pre-clinical dogs at higher risk for cardiac death or progression to a more severe class of heart failure, as such as sub-population would benefit from a closer monitoring activity or medical treatment or in the future from surgical repair.

NEUROHORMONAL ACTIVATION

There are conflicting data about neurohormonal activation in dog with pre-clinical MMVD. Haggstrom and others reported that plasma concentrations of angiotensin II (A II) and aldosterone were not elevated in dogs with stage B2 MMVD. The not significant activation of the circulating renin-angiotensin-aldosterone system (RAAS) was also reported by Pedersen and collaborators in two studies. Other studies have shown that activation of local RAAS may play a more important role. Dell'Italia reported that cardiac ACE activity, chymase like activity and cardiac A II levels were significantly higher in the ventricles of dogs with experimentally induced mitral regurgitation (MR) compare to controls. The same study suggested that in dogs the majority of A II forming activity was from chymase-like activity rather than from ACE activity in both normal and MR dogs. A recent study from Fujii and others confirmed that in a similar animal model cardiac ACE activity was elevated despite normal levels of plasma A II and aldosterone concentration. Interestingly, results from this study, in contrast to what has been reported by Dell'Italia, show that cardiac chymase-activity was significant decrease compare to control. We found significantly increased plasma concentration of aldosterone in a group of dogs with advanced pre-clinical MMVD compare to normal dogs. All together these data seem reflecting the clinical heterogeneity of dogs affected by pre-clinical MMVD and that there are differences between natural occurring diseases and experimentally induced MR.

NATURAL HISTORY

Myxomatous degeneration of mitral valve appears to be a relative benign condition for many dogs. We have previously reported that more than 70% of asymptomatic dogs were alive at the end of 6.6 years of observation. Preliminary data of a study from our group aimed at evaluating natural history and risk factors for progression of MMVD stage B1 and B2 in 256 dogs, demonstrates a median survival time of 27.6 months. During the observation the overall mortality was 27%, whereas the cardiac deaths represented 11%. The same study show that 13% of dogs progressed to a more advance stage of MMVD during the observation period. Another recent study reported that 82% of pre-clinical dogs were still asymptomatic at 12 months from inclusion of the study. The VETPROOF trial reported a median survival time for placebo dogs of 778 days and similar results were reported for the SVEP trial. The mortality and progression rate for dogs is similar to what has been reported in human beings affected by asymptomatic MR due to MMVD.

PROGNOSTIC INDICATORS

Several prognostic indicators have been proposed for MMVD. These include age, gender, intensity of heart murmur, degree of MR, left atrial enlargement, rate of changes of cardiac enlargement in radiographics or echocardiographic examinations and increased N-terminal fragment of proBNP (NT-proBNP). Our preliminary data suggest that left atrial enlargement, left ventricle enlargement and increased E wave to A wave ratio of transmitral flow represent risk factors for cardiac death. Whereas, left atrial enlargement, increased peak velocity of E wave of the transmitral flow and presence of cough represent the most significant risk factor for progression to a most advanced stage of heart failure.

In people, NT-proBNP has been showed of being an excellent biomarker for identifying patients with cardiovascular disease that are at higher risk of heart failure or death. Brain natriuretic peptide has been demonstrated to correlate with severity of MMVD in dogs. However, in veterinary literature there are few studies aimed to establish the prognostic value of NT-proBNP in dogs with pre-clinical MMVD. A recent study from Chetboul and others suggests that a cutoff of 466 ρmol/L had 80% sensitivity and 76% specificity for predicting 12 months cardiac death or progression to heart failure. In another study Moonarmart and others report that a 100 ρmol/L increase in NT-proBNP was associated with an increased hazard ratio of all cause mortality by 7%.

Recently Ljungvall and others reported that plasma cardiac troponin I (cTnI) measured by a high sensitivity test was associated with MMVD severity, suggesting this test can be used as biomarker of overall cardiac remodeling. This study however, is not providing any data about the prognostic value of cTnI to identify pre-clinical dogs at higher risk of progression to heart failure or death. So further larger prospective studies are needed to investigate the prognostic value of cTnI and BNP in dogs with pre-clinical MMVD.

TREATMENT

Treatment of dogs with asymptomatic MMVD has been subject of controversy. The ACVIM consensus group did not recommend treating dogs with MMVD in stage B1. However, the same group did not reach a consensus for dogs with cardiac signs of remodeling (i.e. left atrial or left ventricular enlargement). In experimentally induced MR ACE-I treatment did not attenuated volume-overload hypertrophy, suggesting that this treatment is not able to block cardiac remodeling process. This can explain why the VETPROOF and the SVEP trial have failed to demonstrate a significant effect of ACE-I treatment on the primary outcome variable represented by time of inclusion to the onset of heart failure. On the other hand, the heterogeneity of the population of dogs with pre-clinical MMVD can be responsible for this failure too. Although both studies were rigorously well designed multicenter double-blinded trials, the lack of clinical repeatable criteria for assessing severity of the disease makes it difficult to have a homogeneous group of patients. Therefore, it is possible that dogs with MMVD and more severe cardiac enlargement may benefit from medical treatment. At this moment there are two ongoing multicenter double-blinded prospective studies testing the effect of pimobendan (EPIC) and association of spironolactone with benazepril (DELAY) on onset of heart failure in dogs with advanced pre-clinical MMVD.

CONCLUSIONS

Dogs with pre-clinical MMVD represent a heterogeneous population. Because the high prevalence of this disease in this species and the relative benign history for many affected dogs, it is important that future studies will actively seek for useful clinical marker able to identify dogs at higher risk of progression and death, as they could benefit from a more watchful monitoring or treatment. This can be of particular importance if surgical repair of mitral valve will become a feasible option also in dogs.

REFERENCES

[1.] Atkins C. et al., **J Vet Int Med** 2009; 23: 1142 [2.] Atkins C. et al., **J Am Vet Med Assoc** 2007; 231: 1061 [3.] Borgarelli M et al., **J Vet Int Med** 2008; 22: 120. [4.] Borgarelli M et al., **Vet Clin North Am Small Anim Pract** 2010; 40: 651 [5.] Chetboul V. et al. **J Vet Int Med** 2009; 23: 984 [6.] Dell'Italia LJ. et al, **Am J Physiol** (Heart Circ Physiol) 1995; 38: H2065 [7.] Dell'Italia LJ et al, **Am J Physiol** (Heart Circ Physiolol) 1997; 42: H961 [8.] Fujii Y. et al., **Am J Vet Res** 2007; 68: 1045 [9.] Haggstrom J et al., **Am J Vet Res** 1997; 58: 77. [10] Grigioni F et al. **J Am Coll Cardiol Img** 2008; 1: 133 [11.] Kvart C. et al. **J Vet Int Med** 2002; 16: 80 [12] Ling LH et al. **N Engl J Med** 1996; 335: 1417 [13.] Ljungvall I et al., **J Vet Int Med** 2010; 24: 153 [14.] Moonarmart W et al., **J Small Anim Pract** 2010; 51: 84 [15.] Pedersen HD et al., **Res Vet Sc** 1998; 66: 11 [16] Pedersen HD et al., **J Vet Int Med** 1995; 9: 331

TRANSESOPHAGEAL ECHOCARDIOGRAPHY: TIPS, TECHNIQUES & CLINICAL APPLICATION

Ashley B. Saunders, DVM, DACVIM (CARDIOLOGY)
College Station, TX

"After some initial trepidation, it is now appreciated that tranesophageal echocardiography is a procedure that is relatively easy to perform, uncomplicated, and capable of providing unique insight into cardiothoracic structures" - Jae K Oh

INTRODUCTION

Transesophageal echocardiography (TEE) is an established imaging modality for interventional cardiac procedures in humans where it was first introduced in the late 1970's.[1] Early probes were linear scanners with large foot prints that provided limited image quality and patient use. Over the years, multiplane phased array transducers and miniaturization have led to high frequency, small probes and improved real time 2D and 3D imaging.[2] Transesophageal echocardiography is a practical and often essential tool for the diagnosis and treatment of congenital and acquired cardiovascular diseases.

IMAGE ACQUISITION

Transesophageal echocardiography is performed using an ultrasound transducer mounted on a flexible endoscopic in small animals or a colonoscope in horses. Probes are available in a variety of sizes (adult, pediatric, micro and 3D) depending on the imaging system. Small, pediatric probes are easier to place in small animals (< 3kg) and offer more room to maneuver; however pediatric probes may not maintain adequate contact in large dogs, do not image as well in the far field, and often have no provision for lateral angulation. Lateral angulation allows the operator to avoid intrathoracic structures that negatively affect image quality. Alternatively, adult size probes are too large to use in very small animals (typically < 3kg). The author and colleagues have used an adult probe to image dogs ranging from 2.5 to 52kg. For TEE studies, veterinary patients require a balanced anesthesia protocol appropriate for the patient's cardiovascular status. A mouth guard placed prior to insertion of the probe into the esophagus prevents damage to the probe. Three standard imaging positions (cranial, middle and caudal) are used to display the cardiac structures in a variety of longitudinal and transverse planes.[3] Transgastric imaging does not routinely provide valuable imaging windows in dogs but may provide limited views in cats.[3,4] The TEE probe can be manipulated in a variety of ways including withdrawal and advancement, clockwise or counter clockwise rotation, retroflexion or anteflexion, and lateral angulation. Mechanical wheels and electronic controls provide a continuum of imaging possibilities.

CLINICAL APPLICATION

Transesophageal echoardiography has been utilized to image a variety of species including dogs, cats, horses, and birds.[3-6] A fundamental understanding of cardiac anatomy is vital for successful TEE imaging. Transesophageal echocardiography provides exceptional anatomic information, assessment of ventricular function, and is particularly useful for evaluating congenital defect morphology, determining whether surgical or interventional therapy can be performed and predicting the type and size of devices to be deployed.[7,8] Acquiring accurate morphologic characteristics and dimensions is invaluable for minimizing procedural complications. Additionally, continuous imaging during device placement may identify improperly sized or positioned devices prior to deployment and release. Successful transcatheter occlusion of patent ductus arteriosus (PDA) and atrial septal defect (ASD) necessitates acquisition of accurate anatomic dimensions to select the most appropriate device type and size to achieve complete closure and minimize the risk of systemic or pulmonary embolization. Transesophageal echocardiography provides critical intra-operative spatial orientation for device sizing, positioning and deployment. Once devices are deployed, TEE provides intra-operative assessment of residual flow or degree of occlusion, and documentation of appropriate positioning prior to final release. An ASD must have a sufficient circumferential tissue rim with adequate margins between the defect and adjacent cardiac structures for successful device occlusion. Appropriate angulation of the ASD occlusion device and delivery catheter relative to the defect is crucial to optimize positioning and successful deployment. The orientation of the device relative to the defect can be continuously monitored and redirected based on TEE guidance. Although, fluoroscopy allows the operator to visualize the device, it cannot provide the guidance needed to appropriately position the septal occlusion device.

Transesophageal echocardiography allows continuous monitoring during heartworm extraction.[9] Because heartworms cannot be visualized fluoroscopically, it can be challenging to determine if all worms have been removed without using echocardiography. Although transthoracic echocardiography (TTE) can identify and locate worms, its use necessitates interruption of fluoroscopy and risks breaks in sterile technique.

Coronary artery anomalies in dogs with pulmonic stenosis may be suspected during TTE but cannot always be confirmed without angiography. If a coronary artery anomaly is suspected but cannot be confirmed using TTE, TEE can be performed for confirmation and might preclude the need for selective coronary angiography and help determine if medical, interventional or surgical therapy is the next step. In addition to coronary artery anomalies, TEE has been utilized to image dogs with aortic aneurysm, cranial vena cava syndrome and pulmonary thromboembolism.[10-12] Potential advantages of TEE include a reduction in fluoroscopic and procedural time, volume of contrast administered, and reduced exposure to radiation for the operator and patient.[8]

POTENTIAL COMPLICATIONS AND LIMITATIONS

Reported complications associated with TEE in humans are uncommon and include thermal esophageal burns, gastrointestinal perforation, bacteremia and transient bradycardia associated with placement of the TEE probe.[13] Complications have not been reported in the veterinary literature. Experimentally, prolonged exposure to the TEE probe did not reveal macroscopic or microscopic esophageal damage in dogs.[14] Partial obstruction to pulmonary venous flow can occur at the caudal esophageal position in cats and could theoretically occur in small dogs.[4]

Limitations include the need for a specialized TEE probe, and the requirement for an experienced operator during the procedure. Patients may be too small for insertion of an adult TEE probe, while in large dogs the pediatric probes may not maintain adequate contact with the esophagus to produce high quality images. Acoustic shadows from implanted devices may affect image quality and visualization of color Doppler flow. Infrequently, the TEE probe will compress cardiac structures or overlie the area of interest on fluoroscopic images and the probe will have to be removed to continue the procedure.[8]

Techniques for successful image acquisition and clinical applications for diagnostic and therapeutic utilization of TEE related to interventional cardiology will be reviewed in this session.

REFERENCES

[1]Shanewise JS, et al. Anesth Analg 1999;89:870. [2]Djoa KK, et al. Ultrasound Med Biol 1996;22:1. [3]Loyer C, Thomas WP. Vet Radiol Ultrasound 1995;36:212. [4]Kienle RD et al. Vet Radiol Ultras 1997;38:288. [5]Raisis et al. Am J Vet Res 2000;61:1282. [6]Beaufrere et al. J Am Vet Med Assoc 2010;236:540. [7]Saunders AB, et al. J Vet Intern Med 2007;21:68. [8]Saunders AB, et al. J Vet Intern Med 2010;24:1407. [9]Small M, et al. J Am Vet Med Assoc 2008;233:1441. [10]Chetboul V, et al. J Am Vet Med Assoc 2003;223:1159. [11]Mulz J, et al. J Vet Cardiol 2010;12:217. [12]Venco, et al. Vet Radiol Ultras 1998;39:564. [13]Augoustides JGT, et al. J Cardiothorac Vasc Anesth 2006;20:379. [14]O'Shea JP, et al. J Am Coll Cardiol 1991;17:1426.

IMPLANTABLE CARDIOVERTER-DEFIBRILLATORS IN DOGS: HOW CLOSE TO CLINICAL REALITY?

Romain Pariaut, DVM, DACVIM (Cardiology), DECVIM-CA (Cardiology)
Baton Rouge, LA

INTRODUCTION

Rapid sustained ventricular tachycardia causes signs of weakness and syncope because of low cardiac output secondary to decreased ventricular filling in diastole. In addition, ventricular tachycardia can degenerate into ventricular fibrillation. Ventricular fibrillation leads to rapid loss of consciousness and sudden death unless treatment is given within minutes following its initiation. In dogs, common cardiac conditions that are associated with a risk of sudden death from arrhythmias include arrhythmogenic right ventricular cardiomyopathy, severe subaortic stenosis, dilated cardiomyopathy in Dobermans, inherited ventricular arrhythmias in German Shepherd dogs and Chagas myocarditis.

Pharmacological treatment of ventricular tachycardia is usually effective at decreasing the number and complexity of ventricular arrhythmias and preventing signs of weakness and collapse. However, it may not decrease the risk of sudden death and the effect may dissipate over time as the underlying cardiac disease progresses.

HISTORY AND EVOLUTION OF IMPLANTABLE CARDIOVERTER-DEFIBRILLATORS (ICDs)

The first experimental model of ICDs was developed in 1969, followed by the first human implant in 1980. ICDs were approved by the FDA in 1985. Current ICDs are small multi-programmable devices, which are able to deliver therapies tailored to the characteristics of the arrhythmia and include pacemaker features. ICD generators are connected to an electrode that is secured within the right heart chambers. A right ventricular defibrillation coil (RV coil) at the extremity of the electrode, and a superior vena cava coil (SVC coil) positioned 15 to 20 cm more proximally, allow to select various directions for the vector of defibrillation; high-energy shocks can be delivered between the 2 coils, between the RV coil and the generator or between the 2 coils and the generator.[1] The cost of the devices ($50000) has limited their availability to veterinarians. In dogs, the first ICD implantation was reported in 2006.[2]

ICD FEATURES

ICDs must detect, diagnose, and treat ventricular arrhythmias; the time from arrhythmia detection to treatment delivery must be short; they must not fail to detect potentially life-threatening ventricular tachyarrhythmias, but at the same time should not be fooled by supraventricular tachycardias into delivering inappropriate therapy.

Treatment features of current multi-programmable ICDs include antitachycardia pacing for overdrive suppression of slower monomorphic VT by delivering rapid sequences of pacing pulses, cardioversion for synchronized high-energy termination of rapid monomorphic VT (FVT), and biphasic defibrillation for treatment of VF. The programmable parameters used for tachyarrhythmia detection and discrimination include rate and duration. This feature allows selecting up to 3 zones of detection (VT, FVT and VF zones), defined by incremental ranges of heart rate, which trigger different treatment responses. Ventricular pacing for brady-arrhythmia therapy is also a programmable option, in the event of temporary post-defibrillation shock asystole.[1,2]

ICD PLACEMENT

ICD positioning is critical because the shock of defibrillation must travel through most of the left ventricular muscle to be effective. The generator cannot be placed in the cervical region as it is usually done with pacemakers. Indeed, the position of the SCV coil in the cranial vena cava does not allow the vector of the current of defibrillation to simultaneously depolarize a sufficiently large portion of the LV for successful therapy. It is believed that successful defibrillation requires concurrent depolarization of 75 to 90% of the myocardium. In dogs, the SVC coil cannot be used to deliver shocks. Defibrillation has to be programmed between the RV coil and the generator, that is placed subcutaneously on the left side of the thorax.[2,3]

ICD TESTING

At the time of implantation, it is important to test the ability of the device to detect and convert ventricular fibrillation. Ventricular fibrillation is induced by a shock on the T wave or 50-Hz burst pacing. After detection of ventricular fibrillation, conversion is attempted with a high-energy shock. Because defibrillation success is probabilistic and defibrillation threshold varies over time, ICDs are usually programmed to deliver up to 6 consecutive shocks. If it is unsuccessful, external defibrillation must be available to terminate the arrhythmia. It is best achieved by placing self-adhesive patch electrodes connected to a biphasic external defibrillator on both sides of the dog's thorax at the beginning of the procedure. The lowest energy that defibrillates the myocardium, or defibrillation threshold, is determined. It is preferable to obtain successful defibrillation with an energy at least 10 J lower than the maximum programmable energy.[4]

ICD PROGRAMMING

Arrhythmia detection is programmed based on the characteristics of the ventricular arrhythmia occurring in the patient. Rate is the major criterion used to differentiate sinus rhythm from ventricular tachycardia. Zones of heart rate corresponding to sinus rhythm, slow ventricular tachycardia and rapid ventricular tachycardia can be selected. The maximum heart rate for ventricular tachycardia detection is usually 250 bpm, with some units allowing 300 bpm. Duration of the arrhythmia can also be programmed. A long detection time will decrease the risk of erroneous detection of arrhythmias. Other discriminators have been developed to decrease the risk of inappropriate therapy. A "sudden onset" criterion differentiates between sinus tachycardia, which shows a progressive increase in heart rate from ventricular tachycardia which has a sudden onset. An "interval stability" criterion differentiates between atrial fibrillation, which is characterized by an irregular heart rate from other tachyarrhythmias, which are usually regular. The "morphology" criterion evaluates the amplitude and duration of the QRS complexes to clearly identify the wide QRS complexes of tachycardia.[5]

ICD COMPLICATIONS

In ICDs, detection algorithms for VT or VF and the decision to defibrillate are based on heart rate sensed from the electrode tip and duration of the tachycardia. Maximum programmable rate for VF detection is usually 250 bpm, but sinus rhythm may reach above 250 bpm in normal dog. In addition, inappropriate shock can occur as a result of double-counting of the T waves, that have large amplitude in dogs.

Sepsis is a risk after implantation. The duration of the procedure to implant the ICD may increase the infectious risk. Infection rates associated with pacemaker implantation in dogs have been reported between 0.95% and 5%. In children, ICD implantation is associated with an infection rate of 3 to 5 %. Antibiotic therapy does not usually eliminate the infection. However, partial response to antibiotics may allow delaying the removal of the ICD. It has been shown that infection usually does not settle if the lead and generator are not extracted. Removal of the lead is associated with the risk of vessel or right ventricular rupture if the implant stays in place for more than 1 year.[1,6,7]

Electrical interference may also affect appropriate ICD functioning.

CHALLENGES OF ICDs IN DOGS

One of the challenges of the future is to identify electrocardiographic and clinical parameters that are predictors of the risk of sudden death from ventricular fibrillation in dogs. In people, ICD implantation is currently indicated combined with an antiarrhythmic treatment, which may help decrease the duration, rate and complexity of arrhythmias, and therefore the number of shocks. Observational studies have allowed identification of patients at increased risk for sudden death, who will benefit from ICD implantation. However, this information is not available about dogs with ventricular arrhythmias.

A second challenge is to limit the incidence of inappropriate high-energy therapy. Inappropriate defibrillation is still frequent in people with ICDs. It has been reported in 5 to 10% of ICD patients and adequate programming is critical to limit their occurrence. In dogs, the risk is mainly associated to T wave oversensing and elevated heart rate that can reach a pre-programmed treatment zone under physiologic conditions. However, it appears that T wave oversensing can be prevented by decreasing detection sensitivity. In addition, beta-blocker therapy can be used to blunt maximum rate of sinus tachycardia and prevent it from entering the VF zone. Finally, innovative programmable features available on recently developed ICDs eliminate heart rate as the only parameter triggering therapy.

A third challenge is to develop the use of less aggressive therapies available on current ICDs. Antitachycardia pacing may be used to overdrive suppress re-entrant tachycardias. Low-energy cardioversion may terminate more rapid ventricular tachycardia, and limit the need for high-energy shocks.

CONCLUSION

In conclusion, ICD implantation is associated with technical challenges and a high risk of complications. However, it may represent a treatment option for dogs thought to be at high risk for sudden death or those that are poorly respond to antiarrhythmics. Multi-programmable ICDs offer new programming options that may further facilitate their use in dogs.

REFERENCES

[1] Silka MJ, Bar-Cohen Y. **Heart Rhythm** 2006;3:1360-1366. [2] Nelson OL, Lahmers S, et al. **J Vet Intern Med** 2006;20:1232-1237. [3] Walcott GP, Ideker RE. Clinical cardiac pacing, defibrillation, and resynchronization therapy. 3rd ed. Philadelphia, WB Saunders; 2000. 75-161. [4] Wathen MS, DeGroot PJ, et al. **Circulation** 2004;110:2591-2596. [5] Dorian P, Philippon F, et al. **Heart Rhythm** 2004;1:540-547. [6] Oyama MA, Sisson DD, et al. **J Vet Intern Med** 2001;3:229-239. [7] Henrickson CA, Brinker JA. **Heart Rhythm** 2008;5:1083-1087.

ACCESSORY PATHWAYS IN UK LABRADOR RETRIEVERS: PRESENTATION, MANAGEMENT & GENETICS

Geoff J. Culshaw, BVMS, DVC; Kim M. Summers, BSc (Hons), PhD; Anne T. French, DECVIM, PhD
Edinburgh, UK

INTRODUCTION

Accessory pathways consist of electrically conducting cardiac muscle fibres that cross the electrically insulating annulus fibrosus of the heart, thus allowing electrical communication between the atria and ventricles other than through the atrioventricular node (AV node). Their presence can potentially lead to an electrical short-circuit or "macro re-entrant" circuit consisting of atrial tissue, AV node, ventricular tissue and accessory pathway and consequently the possibility of a supraventricular tachycardia (SVT) called orthodromic atrioventricular reciprocating tachycardia (OAVRT).

Accessory pathways usually lack the rate-modifying properties of the AV node and therefore conduct more rapidly but take longer to repolarise. Some can only conduct in one direction, from the ventricles to the atria, but some can conduct in both directions. Unidirectional conducting pathways are termed "concealed".

OAVRT can occur in a rapid onset-offset manner resulting in rapid reductions in cardiac output. Where it is sustained for a period of time, it can result in congestive heart failure due to tachycardia-induced cardiomyopathy (TICM), which may be very difficult to distinguish from dilated cardiomyopathy (DCM).

PRESENTATION

Signalment

Previously published reports on dogs with OAVRT have suggested a predisposition in Labrador retrievers, boxers and in male dogs[1-4]. In the UK, the canine breed most represented with OAVRT is the Labrador retriever and it appears to be predominantly in chocolate and black coated members of this breed. OAVRT is seen less frequently in golden retrievers and other breeds, predominantly gundogs. The age range for initial detection of a tachycardia by a veterinary surgeon is 3 months – 9 years although most dogs first present within the first 18 months of life.

History and Clinical Findings

Dogs not in congestive heart failure typically present with a history of insidious onset lethargy, or with rapid onset exercise intolerance. Many owners notice an extremely rapid, vigorous apex beat. Syncope does not appear to be a feature in dogs but gastrointestinal signs at the time of SVT are common and a rapid tachycardia in a weak dog with poor cardiac output and GI signs can be mistaken for an abdominal catastrophe. If in congestive heart failure, dogs present with signs of either left, right or biventricular failure. For both groups of dogs, the tachycardia may not be present at the time of initial examination.

When in tachycardia, heart rates tend to be extremely high. It has been reported that rates with OAVRT are lower in comparison to another form of SVT called focal atrial tachycardia (FAT)[4] however in UK Labrador retrievers, in common with other reports,[1, 2, 5] very high rates have been recorded, ranging from 280-400 beats/min.

Diagnosis

Definitive diagnosis requires an electrophysiological study (EPS) but there are certain ECG findings that give a high index of suspicion and differentiate OAVRT from FAT. These include ventricular pre-excitation, rapid onset-offset of SVT, initiation or termination of the SVT by an atrial premature complex (APC) or ventricular premature complex (VPC) [6], echo beats and the appearance of a P′ wave in the ST segment of SVT[4].

Pre-excitation either on resting ECG or 24 hour Holter ECG either before or after starting therapy occurs in approximately two thirds of affected UK Labradors. It occurs in pathways with bidirectional conduction when a normal impulse generated from the sinoatrial node is conducted through the pathway to the ventricles rather than through the AV node. It appears as a short PQ interval with, typically, an abnormal QRS and a deflection called a delta wave.

Because OAVRT is due to a macro re-entrant circuit, appropriately-timed stimulation at any point along that circuit by an APC or VPC can either trigger or terminate an SVT. Macro re-entrant circuits tend to either be "on" or "off" for this reason and so the gradual slowing of SVT before conversion into sinus rhythm observed in FAT does not occur.

Analysis of the termination of the SVT is often the most useful way of diagnosing OAVRT. Spontaneous termination of the SVT is usually due to refractoriness of the AV node breaking the macro re-entrant circuit. This is immediately preceded by depolarisation of the atria, observed as a P′ wave in the ST segment of the final QRST of the SVT. The P′ wave is usually negative in polarity[4] and its presence detected by comparison of this ST segment with the ST segment of the first normal sinus beat. The ST segment of the first normal sinus beat is less negative or even positive. Additionally, the P′ wave in OAVRT is usually positive on lead aVR[4]. With FAT, on the other hand, the P′ wave during SVT is not within the ST segment but lies much closer to the QRS that follows it[4, 6]. Echo beats are observed where an SVT almost occurred but was stopped by the AV node. On the ECG there is a sinus rhythm with occasional more negative ST segments (the P′ echo beat) followed by a non compensatory pause.

MANAGEMENT

Dogs in congestive heart failure should be stabilised first using conventional therapy such as furosemide, pimobendan and an ACE inhibitor. Once congestion is under control or if congestive failure is not present, then medical cardioversion can be attempted.

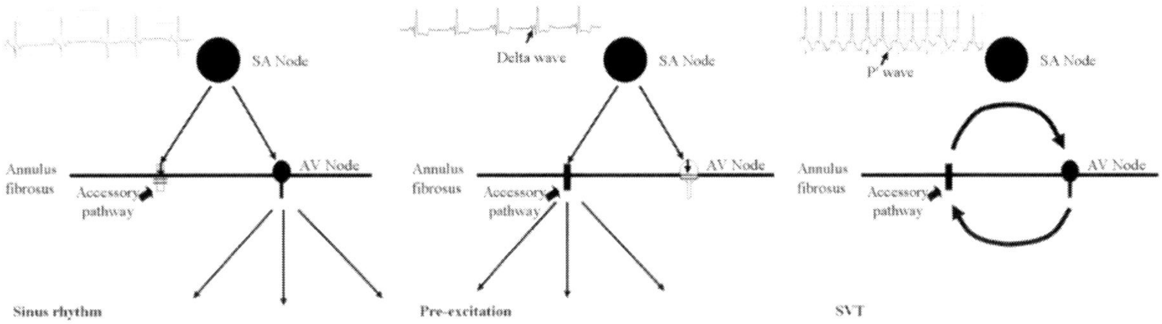

Figure 1. Schematic representations of accessory pathways with sinus rhythm, pre-excitation and SVT.

Drug Management

Dogs presented to the Royal (Dick) School of Veterinary Studies (R(D)SVS) have been treated with a range of anti-arrhythmic medications prior or subsequent to referral which reflects the variable success of drug management, the development of drug refractoriness and the multiple therapeutic target sites within the macro re-entrant circuit. For rapid cardioversion, intravenous medications include lidocaine, verapamil and esmolol. Approximately 50% of Labradors at the R(D)SVS have responded to intravenous lidocaine despite the fact that lidocaine theoretically should not affect SVTs[5]. When lidocaine is ineffective, slow intravenous verapamil is usually successful although several doses may be required. For oral medication, mexiletine is usually effective if lidocaine has been efficacious, and diltiazem is usually effective if the SVT has responded to verapamil. Another useful drug is sotalol because of its multiple effects on the macro re-entrant circuit but in our experience it is only effective in approximately 50% of cases. Very often combination therapy is employed such as mexiletine plus diltiazem, or sotalol plus diltiazem. It should be remembered though that use of these medications is not without risk. They are usually negatively inotropic so caution should be exercised when administered to patients with myocardial failure. Drugs such as diltiazem and sotalol should be titrated to effect. In our experience vagal manoeuvres are ineffective in these dogs and drug efficacy is reduced where congestive heart failure is present, presumably due to enhanced conduction through the AV node caused by elevated sympathetic tone.

Long term (possibly lifelong) medical management is often problematic because of development of drug refractoriness and the use of polypharmacy with associated drug side effects and cost. Labradors tend to be indiscriminate foragers with gastrointestinal upsets commonly resulting in reduced drug absorption and relapse to SVT. Monitoring control of SVT may be difficult as it may only occur at night or because many Labradors may not show overt clinical signs when in SVT and so frequent 24 hour Holter ECG monitoring is required. For many patients, permanent cure is desirable and this can be achieved through radiofrequency ablation (RFA).

Radiofrequency Ablation

RFA consists of passing an alternating electrical current through a small, well defined area of cardiac tissue (in this case, the accessory pathway) via a special ablating catheter inserted transvenously. The resistive properties of the cardiac muscle result in heating of the area to the point of thermal destruction of the tissue[7]. There are 4 stages to RFA:

- Electrophysiological testing to establish AV node health and refractoriness
- Pacing tests to induce SVT and determine the presence of OAVRT or FAT
- Mapping to identify the location of the accessory pathway
- Ablation of the accessory pathway followed by repeat testing 30-40 minutes later[5].

Multi-polar electrode catheters are inserted into the femoral and jugular veins, advanced into the heart and located in the right atrium, coronary sinus, bundle of His and the right ventricle. These catheters are also capable of pacing. Catheter placement is guided both fluoroscopically and by intracardiac ECG.

On the intracardiac electrogram, atrial impulses are denoted as A and ventricular impulses as V. Therefore 3 normal sinus beats would appear as AV AV AV deflections while OAVRT appears as VA VA VA. RFA can be performed either during SVT where termination of the VA cycle on V confirms the pathway has been damaged, or during ventricular pacing when VA dissociation confirms the pathway has been damaged or during pre-excitation when loss of pre-excitation should confirm the correct area has been targeted. Ablation during SVT is preferable as it allows assessment of the AV nodal conduction during ablation. During ablation attempts, the power of the current and the temperature of the catheter are closely monitored. Evidence of accessory pathway damage should appear within 10-15 seconds. If it does then the ablation is continued for one minute to permanently damage the pathway.

Although left-sided and multiple pathways have been occasionally described in dogs[3, 6], all of the Labradors that have undergone RFA at the R(D)SVS have had a single right-sided accessory pathway, usually on the right posteroseptal wall of the atrioventricular groove.

Minor complications of RFA include induction of atrial fibrillation which is usually self-limiting, skin burns if the dog is not electrically earthed and wound seroma formation. The major potential complication is inadvertent damage of the AV node by ablation leading to temporary or permanent AV block. This risk is reduced by careful mapping of the location of the pathway and the bundle of His, lower power ablation and ablation during SVT if possible. Consent should be obtained beforehand for pacemaker implantation if required so that this can be performed at the time of the procedure if AV nodal damage occurs.

Follow-up

Follow-up echocardiography after RFA usually shows complete or partial reversal of TICM as previously reported[2, 8]. If RFA is successful, there is usually no requirement for long term cardiac medication.

GENETICS

The R(D)SVS is currently the only centre in the UK performing RFA in dogs and cases have been referred to the R(D)SVS from throughout the UK. In this referred population, the male, darker coated Labrador retrievers appear predisposed to OAVRT and it was therefore decided to undertake a genetic study of this population of dogs.

In humans, accessory pathways that cause pre-excitation and SVT result in Wolff-Parkinson-White (WPW) Syndrome. WPW Syndrome is termed "sporadic" with an incidence of 0.15-0.25%. However, children of affected individuals have a higher incidence of up to 3.4%[9]. There is also a male predisposition in humans (2:1). WPW Syndrome may also be observed in combination with HCM for which a genetic mutation on chromosome 7 has been identified[10, 11] and as part of multisystemic disorders[9]. It is also seen in humans with Ebstein's anomaly which is of interest as Labrador retrievers are predisposed to tricuspid valve dysplasia.

Pedigree Analysis

We have acquired pedigrees and DNA of Labradors diagnosed with OAVRT at the R(D)SVS or by other veterinary cardiologists within the UK. Common ancestry has been found in all the dogs where we have sufficient pedigree information. This may reflect the narrow gene pool of coloured Labradors within the UK but could indicate a founder effect leading to inheritance of a common mutation for accessory pathways. A small number of lines appear more than once in most affected dogs and on both the sides of the sire and dam. Affected dogs can be classified into 2 closely related groups. No direct link between the 2 groups closer than 5 generations back has been found as yet but some of the pedigrees are incomplete. European and American ancestry is also within these groups. Pedigree analysis suggests either an autosomal recessive or autosomal dominant with reduced penetrance pattern of inheritance. X-linked recessive inheritance is also possible, given the preponderance of male patients.

Determination of Linkage with Coat Colour

The hypothesis that OAVRT is a recessive trait linked to coat colour is being tested. In Labradors there are 2 main genes responsible for coat colour. They are *MC1R* and *TYRP1*[12]. *MC1R* encodes an "on/off" switch for pigment production. Yellow Labradors (and most pale coloured dogs) are homozygous for a recessive mutation in this gene and produce only yellow or red phaeomelanin pigment. Black and chocolate Labradors are either heterozygous or homozygous for the original dominant form of *MC1R* so produce dark (eumelanin) pigment in their coat. *TYRP1* determines the colour and amount of dark pigment produced. It too has dominant and recessive forms. If homozygous or heterozygous for the original dominant form of *TYRP1*, the dog is black. If homozygous for the recessive form, the dog is chocolate.

If the mutation for accessory pathways is recessive and linked to coat colour, then affected dogs must be homozygous for the gene it is linked to. This means that it cannot be linked to any form of *TYRP1* as both black and chocolate dogs are affected. Also, if the trait is linked to *MC1R*, affected dogs must be homozygous for the dominant form of the gene. Sequencing of the MC1R genes in known affected dogs is currently under way.

The hypothesis that OAVRT is a trait that is autosomal dominant with reduced penetrance (that is, not all animals carrying the mutation show the condition) linked to coat colour is also being tested. In order for this to be valid, the disease mutation must be linked to the recessive allele of *TYRP1* so all affected black dogs would be heterozygous for *TYRP1*. Sequencing of the *TYRP1* genes in known affected dogs is also under way.

It is possible that there is only a chance association without linkage between accessory pathways and coat colour in Labradors. Breeders of black and chocolate Labrador retrievers tend not to use yellow Labrador breeding stock and a random mutation within black and chocolate Labrador lines would result in such an association.

Further Investigation

A screening programme for Labrador retrievers within the UK has been established. This will attempt to identify subclinically affected dogs through identification of SVT, pre-excitation or echo beats on 24 hour Holter ECG. The aim is to acquire pedigrees and DNA samples from more known affected dogs and likely unaffected dogs for further pedigree and genetic analysis. A Genome Wide Association Study (GWAS) will be undertaken to look for common single nucleotide polymorphisms (SNPs) where specific alleles are associated with disease status. Candidate genes closely linked to these SNPs can then be sequenced to determine whether significant mutations are present

REFERENCES

[1]Atkins C, et.al. **J Vet Intern Med** 1995;9:43. [2]Wright K, et.al. **J Vet Intern Med** 1999;13:361. [3]Santilli R, et.al. **J Am Vet Med Assoc** 2007;231:393. [4]Santilli R, et.al. **J Vet Intern Med** 2008;22:915. [5]Johnson M, et.al. **J Vet Intern Med** 2006;20:272. [6]Wright K, et.al. **J Vet Cardiol** 2006;8:95. [7]Haissaguerre M, et.al. **J Cardiovasc Electrophysiol** 1994;5:532. [8]Packer D, et.al. **Am J Cardiol** 1986;57:563. [9]Ehtisham J, et.al. **J Cardiovasc Electrophysiol** 2005;16:1258. [10]MacRae C, et. al. **J Clin Invest** 1995;96:1216. [11]Gollob M, et.al. **N Engl J Med** 2001;344:1823. [12]Schmutz S, et.al. **Mamm Genome** 2002;13:380.

EFFICACY OF SPIRONOLACTONE TO PREVENT SYMPTOMATIC STAGES IN MVD

Gerard Le Bobinnec, DVM, DECVIM (Cardiology)
Bordeaux, France

Step 1 : What is the role of Aldosterone in heart failure (HF)?

Its role is very deleterious, causing increased volume with sodium retention, and progressive myocardial and vascular (including coronary) fibrosis (1).

Step 2 : What is the role of Mineralocorticoid Receptor Blockade (anti-aldosterone: Spironolactone & Eplerenone) in the treatment of human HF ?

Rales study proved beneficial effects in terms of hospitalizations and duration of life (2)

Step 3 : How are they currently being used in human medicine?

Essentially in classes NYHA III & IV of HF (evaluated stages) (2)

Step 4 : What is the future(prospective in human cardiology)?

The efficiency has recently been proved in class NYHA II (4,5) , and the next step is class I (3).

Step 5 : Comparison with canine MVD,: are the lesions similar?

In MVD there is deleterious remodeling including progressive myocardial and vascular (& coronary) fibrosis (6,7).

Step 6 : What is the role of Mineralocorticoid Receptor Blockade (anti-aldosterone : Spironolactone & Eplerenone) in the treatment of canine MVD ?

As in human medicine, to prolong duration of life by Aldosterone receptor Blockade (8)

Step 7 : How is it currently be used in Veterinary Medicine?

Spironolactone is marketed for symptomatic stages, in addition of the conventional treatment (8)

Step 8 : Do we have arguments to use it in asymptomatic stages ? Yes

- the first one is the example of human cardiology (3,4,5)
- the second one is that in asymptomatic stages, there is no increased level of aldosterone EXCEPT in case of feeding with a low sodium diet (9,10) or in case of left ventricular dilation (11).
- the third one is that ACE Inhibitors failed in this indication, and also seem to be inefficient to control increasing aldosterone levels (12).

Step 9 : The French trial :

This open study started in September 2004 and is now including 419 MVD. The main objective is to detect beneficial effects of Spironolactone given in class 1A-1B to delay symptomatic stages. The definitive result and statistics will be discussed at the time of the presentation.

Step 10 : The Italian trial :

The objective is exactly the same as the French trial, but it is a randomized, double blind, placeno controlled study.. It is schedule d to conclude in 2015.

REFERENCES

1. Weber KT, Brilla CG. Pathological hypertrophy and cardiac interstitium. Fibrosis and renin-angiotensin-aldosterone system. Circulation 1991; 83 : 1849-1865. 2. Pitt B et al. The effect of Spironolactone on morbidity and mortality in patient with severe heart failure (RALES study). N Engl J Med 1999 ; 341 : 709-717. 3. Amstrong PW . Aldosterone Antagonist- Last Man Standing ?. N Engl J Med 2010 ; 10.1056/NEJMe1012547. 4. Zannad F et al . Rationale and design of Eplerenone in Mild Patients Hospitalisation And SurvIval Study in Heart Failure (EMPHASIS-HF). Eur J Heart Fail 2010 ; 12 : 617-622. 5. Zannad F et al . Eplerenone in Patients with Systolic heart failure and Mild Symptoms. N Engl J Med 2010 ; 10.1056/NEJMoa1009492. 6. Falk T et al. Arteriosclerotic changes in myocardium, lung and kidney in dogs with congestive heart failure and myxomatous mital valve disease.J Vet Int Med 2005 ; 19 : 927-950 (abstract n°5). 7. Falk T et al ; Correlation of cardiac pathology and clinical findings in dogs with naturally occuring congestive heart failure. J Vet Int Med 2007 ; 21 : 1420-1445 (absract n°75). 8. Bernay F et al. Efficacy of Spironolactone on survival in dogs with naturally occuring mitral regurgitation caused by myxomatous mitral valve disease. J Vet Int Med 2010 ; 24 : 331-341. 9. Rush J, Freeman L et al .Clinical, echocardiographic, and neurohormonal effects of a sodium-restricted diet in dogs with heart failure. J vet Int Med 2000 ; 14 : 513-520. 10. Freeman l, Rush J, Markwell P. Effects of dietary modification in dogs with early chronic valvular disease. J Vet Int Med 2006 ; 20 : 1116-1126. 11. Hezzel MJ, Boswood A, Elliot J. Relationships between serum and urinary aldosterone, ventricular remodelling, and outcome in dogs with mitral valve disease. J Vet Int Med 2010 ; 24 : 660-795 (abstract n°7). 12. Lantis AC et al . Aldosterone escape in furosemide-activated circulating renin-angiotensin-aldosterone system (RAAS) in normal dogs.J Vet Int Med 2010 ; 24 : 660-795 (abstract n°6).

SLEEPING AND RESTING RESPIRATORY RATES IN DOGS WITH SUBCLINICAL HEART DISEASE. M. Rishniw[1], D. Ohad[2], I. Ljungvall[3], F Porciello[4], J. Haggstrom[3]. [1]Cornell University, Ithaca NY, [2]Hebrew University of Jerusalem, Rehovot, Israel, [3]University of Agricultural Science, Uppsala, Sweden, [4]University of Perugia, Perugia, Italy.

Reference intervals for at-home sleeping respiratory rates (SRR) in client-owned healthy dogs exist. Whether SRR is similar in dogs with subclinical heart disease, or increases with disease progression prior to CHF remains unknown.

We acquired at-home SRR and resting respiratory rates (RRR) from client-owned dogs with subclinical volume-overload heart disease. All dogs underwent echocardiographic evaluation, and the severity of the subclinical disease was staged based on left atrial size. Owners collected 10 SRR and 10 RRR measurements over the following 1-2 months.

Seventeen participants provided data on 67 dogs (age 125±44 mo; median weight 9 kg). 60 dogs had mitral valve disease,7 had other diseases. 21/67 were receiving cardiac medications. LA size varied: subjectively, 41 dogs had no/mild LA enlargement, 19 had moderate LA enlargement, 7 had severe LA enlargement. Within-dog SRR_{mean} was 16.5±5.2 breath/min, 2 dogs with severe disease had SRR_{mean} >30. Within-dog RRR_{mean} was 21.3±6.2 breath/min, 7 dogs had RRR_{mean} >30. SRR_{mean} was significantly lower than RRR_{mean} ($p<0.0001$). When classified subjectively on LA size, SRR did not differ between groups ($p=0.092$),and did not correlate with subjective LA size (Spearman Rho= 0.11, $p=0.1$). There was no relationship between SRR_{mean} and bodyweight or age. SRR_{mean} of dogs with subclinical heart disease differed from SRR_{mean} of healthy dogs ($p<0.001$).

These data suggest that dogs with subclinical heart disease have SRR_{mean} that falls within the range of healthy dog SRR_{mean} unless they have severe subclinical heart disease. SRR might be useful in identifying dogs with impending CHF or severe subclinical disease.

PEFORMANCE COMPARISON OF A COMMERCIALLY AVAILABLE POCKET ECHOCARDIOGRAPHY DEVICE AND THE VIVD E9 FOR 2D AND COLOR IMAGING IN DOGS WHEN PERFOMRED BY EXPERIENCES EXAMINERS. SG Gordon, AB Saunders, MW Miller, SE Achen, C Hariu, R Fries, and J Carlson. The Michael E. DeBakey Institute, College of Veterinary Medicine and Biomedical Sciences, Texas A&M University, College Station, TX.

The new echocardiography device, Vscan™, by GE Healthcare® weighs 390g, has a 3.5 inch display with 240 x 320 pixel resolution and fits into your pocket. It comes with one broad-bandwidth, 1.7-3.8 MHz, phased array probe offering 2D and color Doppler imaging with distance and area measurement capabilities. We hypothesized that the Vscan 2D and color Doppler imaging capabilities were clinically comparable to a high quality, stand alone echocardiography system by the same manufacturer (Vivid E9). Twenty dogs of various weights with a variety of cardiovascular abnormalities were imaged by a cardiologist; first with the Vscan and subsequently with the Vivid E9. Measurements and clinical conclusions derived from the Vscan study were completed before the patient was imaged with the Vivid E9. Studies emphasized subjective and objective evaluation of chamber morphology and size, left ventricular contractility, valve morphology and motion, identification of any valvular insufficiencies and semi-quantitative estimation of severity when present. Agreement between studies was evaluated statistically.

MULTI-CENTER EVALUATION OF PLASMA N-TERMINAL PRO-BRAIN NATRIURETIC PEPTIDE (NT-PROBNP) TO DETECT ASYMPTOMATIC (OCCULT) FELINE CARDIOMYOPATHY.

PR.Fox,[1] MA.Oyama,[2] CA.Reynolds,[2] JE.Rush,[3] TC.DeFrancesco,[4] BW.Keene,[4] CE.Atkins,[4] SG.Gordon,[5] KE.Schober,[6] JD.Bonagura,[6] RL.Stepien,[7] HB.Kellihan,[7] KA.MacDonald,[8] LB.Lehmkuhl,[9] TP.Nguyenba,[9] NS.Moise,[10] BK.Lefbom,[11] DF.Hogan[12].
[1]Animal Medical Center, NY, NY; [2]University Pennsylvania, Philadelphia PA; [3]Tufts University Cummings School Veterinary Medicine, North Grafton, MA; [4]North Carolina State University, Raleigh, NC; [5]Texas A&M University, College Station, TX; [6]Ohio State University, Columbus, OH; [7]University Wisconsin, Madison, WI; [8]Animal Care Center, Rohnert Park, CA; [9]MedVet Associates, Worthington, OH; [10]Cornell University, Ithaca, NY; [11]Chesapeake Veterinary Cardiology Associates, Springfield, VA; [12]Purdue University, West Lafayette, IN.

The utility of plasma NT-proBNP concentration (BNP) to detect occult cardiomyopathy (OCM) is unresolved. We intended to determine whether BNP can discriminate OCM from normal cats, and whether BNP correlated with clinical, biochemical, and echocardiographic parameters.

This prospective, case-controlled study included 114 normal and 113 OCM cats. Diagnosis was determined from history, physical examination, and M-mode/2D/Doppler echocardiography. Optimal cut-off values were derived from ROC curve analysis.

BNP (median, IQR) was higher in: OCM (186pmol/L; 79, 478.2pmol/L) vs. normal (24pmol/L; 24, 32pmol/L) (P<0.001); and in HOCM (396pmol/L; 205.2, 685.5pmol/L) vs. HCM (112pmol/L; 48.5, 318.5pmol/L) (P<0.001). BNP in OCM cats correlated 1) positively with LVPWd (ρ= 0.228; P=0.015), LA/Ao ratio (ρ= 0.309; P=0.001), LVs (ρ= 0.328; P<0.001), and troponin-I (ρ= 0.637; P<0.001), and 2) negatively with %FS (ρ= - 0.272; P=0.004). Area under ROC curve was 0.92; >46 pmol/L cut-off distinguished normal from OCM (91.2% specificity, 85.8% sensitivity; >99 pmol/L was 100% specific, 70.8% sensitive.

BNP reliably discriminated normal from OCM cats. Further studies must assess test performance in general feline populations.

UTILITY OF NTproBNP TO ASSESS THE PRESENCE OF CARDIAC DISEASE, DISEASE SEVERITY AND SURVIVAL IN A POPULATION OF 388 CATS. A Beardow[2], K Yee[1], J Allen[1], G Farace[2], W Schoellkopf[2], M Beall[2], SJ Ettinger[1], 1. California Animal Hospital Veterinary Specialty Group, Los Angeles, CA. 2. IDEXX Laboratories, Westbrook, ME.

In smaller studies, NT-proBNP has demonstrated utility to detect and stage heart disease in cats. NTproBNP levels have also been shown to be predictive of survival in humans and dogs.

388 client-owned cats were evaluated at a mixed general/specialty hospital over a three year period. Evaluation included a full medical history, complete physical examination including BCS, chest radiographs, Doppler and 2-D echocardiographic studies, Doppler BP, USG, serum BUN, Creatinine, TT4, and plasma NT-proBNP concentrations. Cats were classified according to the previously reported CAH scheme. Survival information was recorded for 385 cats.

NTproBNP levels were correlated with CAH class. NTproBNP concentrations in cats classified as CAH grade 2 or higher were significantly different from those cats classified as grade 0.

Echocardiographic parameters that correlated most closely with NTproBNP concentration were LA/Ao. IVSd and LVPWd.

Controlling for age, an increasing BCS at entry was associated with a decreased rate of death ($p=0.0029$), with a hazard ratio of 0.795 (95% CI = 0.683-0.924). Increasing NTproBNP levels at entry are associated with an increased rate of death ($p<0.0001$) with a hazard ratio of 1.107 (95% CI = 1.071-1.144) implying rate of death increases by 10.7% for every 100pmol/L increase in NTproBNP concentration.

These results suggest that NTproBNP is useful for identifying the presence of disease, assessing disease severity and identifying those cats likely to succumb to their disease. Interestingly, an increased BCS on diagnosis is associated with a decreased risk of death.

I_f CURRENT INHIBITION IN CATS WITH HYPERTROPHIC CARDIOMYOPATHY

Karsten E. Schober, Dr. Med. Vet., PhD, DECVIM (Cardiology)
Columbus, OH

INTRODUCTION

Hypertrophic cardiomyopathy (HCM) is the most common cardiac disease of cats and is associated with significant mortality.[1-3] Histomorphologic hallmarks of the disease include myofiber hypertrophy and disarray; narrowing of intramural coronary arteries leading to decreased myocardial perfusion, potential for myocardial oxygen demand-supply mismatch especially during tachycardic episodes or exercise, myocardial ischemia and necrosis; and replacement fibrosis.[2] Left ventricular (LV) hypertrophy also reduces LV chamber size, impairs ventricular relaxation, and increases wall stiffness, all of which impede filling. As a consequence, LV filling pressures increase leading to left atrial enlargement, arrhythmias, congestion, blood stasis, neurohormonal activation, and finally decompensation. More than half of the cats with HCM also develop dynamic obstruction of the LV outflow tract. The latter increases LV wall stress and thus myocardial oxygen demand, and may worsen LV hypertrophy, aggravate diastolic dysfunction, and lead to atrial pathology.

The progression of HCM is variable with most cats having a long asymptomatic period that may end with the abrupt development of congestive heart failure (CHF), arterial thromboembolism, or sudden cardiac death. Several risk factors have been identified including severity of LV hypertrophy, diastolic functional class, LA size, and heart rate (HR).[1,4] In addition to wall stress and systolic function, HR is a major determinant of myocardial oxygen consumption.[5,6] Tachycardia is associated with decreased myocardial perfusion by limiting diastolic coronary blood flow and thus may induce myocardial ischemia along with higher myocardial oxygen demand,[5] further decreasing diastolic chamber filling, and worsening of LV outflow tract obstruction. It is estimated from studies in awake dogs that a 1% increase in diastolic time fraction (the ratio between the duration of diastole to the length of the entire cardiac cycle) increases sub-endocardial blood flow by 3% to 6% in the normal heart.[7] Although the general association between progression of HCM and HR has not yet been fully elucidated, and it remains to be determined if HR itself has prognostic importance or is only an epiphenomenon. Several studies in cats have shown that advanced disease is accompanied by elevated HR. In a recent retrospective study completed at our institution in 43 healthy cats, 49 cats with asymptomatic HCM, and 49 cats with symptomatic HCM, mean HRs taken during physical examination were 175/min, 191/min, and 201/min, respectively ($P < 0.05$). A study done with 56 cats with HCM found HR to be a prognostic indicator with a medium survival time after diagnosis of 1,830 days if HR was < 200/min and 152 days if HR was ≥ 200/min.[1] In another study with 27 normal cats and 29 cats with cardiomyopathy and CHF, significant elevation of HR in the latter group (mean 182/min versus 217/min, $P < 0.05$) was reported.[8] Human studies using logistic regression analysis found HR to be an independent variable of outcome.[9]

Evidence of myocardial ischemia in cats with HCM has rarely been reported owing to the difficulty of diagnosing this condition. ST segment elevation or depression on the ECG suggestive of myocardial ischemia has specific diagnostic merit but lacks sensitivity. Serum cardiac troponin I has been used as a surrogate marker of myocardial ischemia and subsequent cardiomyocyte injury in people[10] as well as in cats.[11] Circulating cTnI is elevated in most cats with asymptomatic HCM, and is even more elevated in cats with CHF. Although the pathogenesis of acute decompensation of otherwise clinically stable cats with HCM is not fully understood, anecdotal evidence exists that stressful events leading to tachycardia including but not limited to routine examinations at a veterinary hospital, sudden changes in the cat's home environment, pain of any source, anesthesia, or elective surgery all can trigger cats with well compensated HCM to suddenly develop CHF.[3,12] In people, physical or emotional stress may lead to unwanted tachyarrhythmias that are known triggers for acute decompensation of recently stable ischemic heart disease.[9] It is also conceivable that stress-induced tachycardia leads to hyperdynamic LV outflow and reduced LV dimension, favoring systolic anterior motion in cats with HCM. Also, HR not only is important in the development of ischemia, but also influences whether or not ischemic episodes trigger malignant ventricular arrhythmias.[9] Therefore, pharmacological modulation of HR aimed at lowering resting HR and preventing sudden periods of uncontrolled tachycardia may reduce the risk of acute decompensation in cats with HCM, life-threatening arrhythmias during ischemic periods, sudden cardiac death, dynamic outflow obstruction, and myocardial fibrosis.

In the past, treatment of cats with HCM has been directed at relieving obstruction of the LV outflow tract, management of CHF, arrhythmia control, and prevention and management of ATE. The most common medications used include bradycardic agents such as beta adrenergic blockers and calcium channel antagonists; CHF drugs including angiotensin-converting enzyme inhibitors, venodilators, diuretics; and anti-platelet drugs. Beta blockers and calcium channel blockers are the most frequently used agents in preclinical feline HCM although their general use remains controversial. Adverse effects induced by such drugs have been reported including weakness, lethargy, salivation, weight loss, and reduced left atrial function. Beta receptor blockade may be contraindicated if CHF, systolic dysfunction, conduction system abnormalities, reactive airway disease, hypotension, and thromboembolic disease are present. Moreover, studies in animals and people revealed that the beneficial long-term effects of β-adrenoreceptor blockade can be attributed, to a large degree, to the resultant HR reduction.[13-15] Negative inotropism without concomitant HR reduction is indeed suggested to be detrimental to the ischemic myocardium.[15] Therefore, selective negative chronotropic effects without negative inotropy has been a therapeutic goal for the treatment of ischemic heart disease in people. Whether or not this is applicable to cats with HCM is currently unknown. However, the fact that ischemia is integrated part in the pathophysiology of HCM and that unwanted tachycardia may trigger decompensation makes this approach at least appealing.

A novel class of drugs, the selective inhibitors of the "funny" inward pacemaker current (I_f), has recently gained attention in people with ischemic heart disease and angina and with inappropriate sinus tachycardia. I_f inhibitors effectively lower HR, relief symptoms of angina, and improve exercise tolerance.[16] Recent large epidemiological studies have shown resting HR to be an independent predictor of cardiovascular and all-cause mortality in people with and without cardiac disease and thus, may obtain importance as a potential therapeutic target.[13,14] Moreover, in addition to cardiac I_f expression in the sinoatrial node (SAN), pathologic myocyte hypertrophy may lead to re-expression and thus upregulation of fetal I_f isoforms in working myocardium.[17] As previously shown[17] the resulting changes in I_f density are correlated to the type of disease, with I_f over-expression being of greater magnitude in ischemic than in dilated cardiomyopathy. From a clinical point of view, myocardial over-expression of I_f may contribute to calcium overload and may predispose ventricular myocytes to enhanced automaticity, a condition associated with an increased risk for sudden cardiac death. Thus, in addition to HR control, selective I_f inhibitors may exert additional beneficial (possibly anti-arrhythmic, anti-remodeling, and anti-injury) effects beyond simple HR lowering.[18] Specific sinus node inhibitors (SNIs) have been studied in experimental dogs[7,9-21] and cats[22-26] and cats with naturally-occurring HCM[23] and will be subject to further discussions as outlined below.

FUNNY INWARD CURRENT (I_f) INHIBITORS

The hyperpolarization-activated, autonomic nervous system-modulated, cyclic nucleotide-gated (HCN), inward mixed cation (K^+ and Na^+) current termed I_f (or I_h) is a prominent depolarizing current in a variety of tissues that show spontaneous electrical activity including neuronal (retina) and cardiac cells. Cardiac I_f is predominantly found in the conduction system (the sinoatrial pacemaker cells and less prominent in Purkinje fibers) where I_f is believed to mediate HR via direct modulation by cAMP. The amplitude of the I_f current determines the slope of spontaneous diastolic depolarization.[27] As far as currently known, four HCN channels underlie the I_f current to produce a tissue and cell-specific I_f throughout the organism. In the SA node, HCN4 is most abundant in all species,[28] whereas HCN1, HCN2, and HCN3 are expressed at lower quantity in and less specificity for the SAN compared to the working myocardium. Several SNIs have been studied in experimental animals (cats, dogs, rats, mice, pig, rabbits, sheep, and guinea pigs) and people including zatebradine, alinidine, cilobradine, and ivabradine all of which produce a use- and voltage-dependent blockade of the I_f current, resulting in a slow onset of current particularly during early phase-4 spontaneous depolarization.[29,30] SNIs bind to HCN from inside the cell, and thus, must first diffuse into the cell to produce their effects (i.e., the funny channel must be open for SNIs to reach their binding sites). All SNIs have been shown to have no or only negligible direct effects on myocardial inotropy and lusitropy, systemic blood pressure, systemic vascular resistance, and intracardiac conduction.[13,27] Although the therapeutic index of SNIs is high and ivabradine is well tolerated in people, dose-dependent bradycardia, temporary visual disturbances (luminous phenomena),[31,32] and headaches have been reported as (albeit rare) side effects.

Ivabradine is a highly selective I_f inhibitor that has gained particular attention in recent years. It has been approved in Europe for use in people with ischemic heart disease and sinus rhythm that do not tolerate or do not respond sufficiently to beta adrenergic blockade. In addition to its bradycardic effects, ivabradine has been shown to preserve cardiac output, increase coronary blood flow, improve myocardial oxygen demand-supply mismatch, and prevent the loss of coronary vessels in failing myocardium in people and experimental animals.[33-36] Auxiliary effects such as modifications in the extracellular matrix including decreased myocardial collagen accumulation and increased LV capillary density were also reported.[35,37] Ivabradine has been shown in people to be non-inferior to atenolol and diltiazem for anti-anti-anginal and anti-ischemic effects.[22,32] Ivabradine does not cross the blood-brain barrier and, therefore, has no effect on the I_h current in central nervous system neurons. Due to ivabradine's unique selective HR-reducing properties it may become a new therapeutic option in cats with HCM. It should prolong LV diastolic filling and coronary perfusion time, decrease myocardial oxygen consumption, and reduce myocardial ischemia and its chronic consequence, fibrosis. It may also attenuate or relieve LV outflow tract obstruction due to improved LV filling making SAM less likely, and most importantly reduce tachycardia-induced crisis events.

STUDIES IN CATS

The following studies were executed in the author's laboratory, and results are briefly summarized:

1. Hemodynamic effects of intravenous cilobradine in anesthetized, healthy cats:[22] A single dose of cilobradine (0.2 mg/kg, IV) to healthy anesthetized cats led to a consistent and lasting (several hours) reduction of HR (mean±SD, 115±17 min^{-1}) starting within a few minutes of administration. Cilobradine slightly (mean, +4 ms) increased tau and decreased dP/dt_{max} (mean, -350 mmHg/s) but did not affect LV filling pressures. Co-administration of dobutamine (5 µg/kg/min, CRI) did not override the negative chronotropic effect of cilobradine (no change of HR), but significantly and relevantly influenced tau (mean -15 ms) and dP/dt_{max} (+1,035 mmHg/s).

2. Pharmacokinetics of oral ivabradine in healthy cats:[24] Ivabradine and its major metabolite, S-18982, were analyzed on day 1 and 4 after oral administration of ivabradine (0.3 mg/kg) to healthy cats. Concentration-time plots for ivabradine indicated biphasic profiles for all cats on days 1 and 4, and profiles for S-18982 indicated monophasic profiles for all cats on both days. Peak plasma concentrations of the parent compound and S-18982 were reached within 1 hour of administration. Plasma half life (t1/2) was 4-6 h for ivabradine and 3-4 h for S-18982. Both analytes were below the LLOQ after 12- 24 h in 7 (88%) cats. A 4-compartmental model was fitted characterizing pharmacokinetics of combined ivabradine and S-18982 and suggested twice daily dosing for clinical use in cats. Ivabradine was well tolerated in all cats and adverse effects were not observed.

3. Pharmacodynamic effects of oral ivabradine in healthy cats:[25] The pharmacodynamic effects of 3 single doses of oral ivabradine (mg/kg; 0.1, 0.3, and 0.5) were evaluated using a single-blind, placebo-controlled, randomized, fully-crossed study design and 8 chronically-instrumented healthy cats. HR, ECG, and systemic blood pressure were continuously monitored via radio-telemetry for 18 to 24 h after drug administration. HR decreased significantly in a dose-dependent manner with peak negative chronotropic effects observed 2-3 hours after ivabradine (median; 145 min^{-1} after placebo, 135 min^{-1} after ivabradine 0.1 mg/kg, 115 min^{-1} after ivabradine 0.3 mg/kg, and 103 min^{-1} after ivabradine 0.5 mg/kg). HR was still reduced at 12 hours after ivabradine (0.3 mg/kg and 0.5 mg/kg) compared to placebo. The tachycardic response to stress applied via acoustic startle was blunted at all three doses of ivabradine. No effect of ivabradine on systolic, diastolic, and mean blood pressure was identified. There was a total number of 14 VPCs observed in all cats and all recordings with no difference between placebo and ivabradine. Clinically discernable side effects were not observed.

4. Hemodynamic effects of intravenous ivabradine in healthy cats and cats with HCM:[23] Six healthy cats and 6 cats with preclinical HCM were studied. Anesthetized cats underwent cardiac catheterization and were studied over a range of hemodynamic states induced by treatments with esmolol; esmolol and dobutamine; ivabradine (0.3mg/kg, IV); and ivabradine and dobutamine. LV function, cardiac output, and LA function were studied by catheter-based methods and simultaneous echocardiography. Treatment with ivabradine resulted in a significant and consistent reduction of HR (mean, -48 min^{-1}), rate-pressure product (mean, -3,711 mmHg/min), and LV contractile function (mean, -613 mmHg/s dP/dt_{max}) and a slight increase in LVEDP (mean, +4.1 mmHg), LV end-diastolic wall stress, and tau (mean, +4 ms) in cats with HCM with differences noted compared to healthy cats. Concurrent administration of ivabradine and dobutamine resulted in an increase of LV dP/dt_{max} and tau, with blunted chronotropic effects of catecholamines. Systemic blood pressure was not affected. LA performance improved after ivabradine in cats with HCM but decreased with esmolol.

5. Effects of oral ivabradine of echocardiographic variables of left heart function in healthy awake cats:[26,38]
a) *Single oral dose:*[38] Ten healthy, non-sedated cats underwent transthoracic echocardiography at baseline and 3 hours after oral administration of a single dose of ivabradine (0.3 mg/kg) and atenolol (6.25 mg/cat) in a prospective, double-blind, randomized, active control, and fully-crossed study. Echocardiographic variables of chamber size and LV diastolic function were determined. Three hours after ivabradine and atenolol, HR was reduced (mean±SD, 58±21 min^{-1} and 40±23 min^{-1}, respectively). At baseline and after ivabradine, 5/10 cats and 0/10 had fusion of transmitral filling waves, respectively. In cats with non-fused filling waves at baseline (5/10), ivabradine did not change transmitral flow velocities and diastolic classification. In contrast to atenolol, ivabradine did not change IVRT, Peak E, Peak A, and E:A. Left atrial and LV dimensions were equally increased after both drugs. In summary, ivabradine consistently leads to separation of fused transmitral flow waves with minimal effects on echocardiographic variables characterizing LV diastolic function. However, left atrial and LV size increase relevantly after a single dose ivabradine. The apparent usefulness of ivabradine in the echocardiographic assessment of LV diastole needs to be confirmed in cats with cardiomyopathy.
b) *Repeated oral doses:*[26] Ten healthy laboratory cats underwent physical examination, systolic blood pressure measurement, and transthoracic echocardiography at baseline and after oral administration (4 weeks each) of ivabradine (0.3 mg/kg, q12h) and atenolol (6.25 mg/cat, q12h) in a prospective, double-blind, randomized, active-control, fully-crossed study. A-priori non-inferiority margins for the effects of ivabradine compared to atenolol were set at 50% (f=0.5) based on predicted clinical relevance, observer measurement variability, and in agreement with FDA guidelines. Ivabradine was clinically well tolerated with no adverse events observed. HR (ivabradine, P<0.001; atenolol, P<0.001; ivabradine vs. atenolol, P=0.721) and rate-pressure product (ivabradine, P<0.001; atenolol, P=0.001; ivabradine vs. atenolol, P=0.847) were not different between treatments. At the dosages used, ivabradine demonstrated more favorable effects than atenolol on echocardiographic indices of LV systolic and diastolic function and left atrial performance.

6. Myocardial expression of HCN proteins in healthy cats and cats with HCM: Myocardial samples of the right atrium, LV, and right ventricle of 20 control cats and 8 cats with HCM were collected. Expression of feline HCN was determined by immunoblot analysis using antibodies against HCN2 and HCN4. If cell lysate proteins from control cats and cats with HCM were analyzed simultaneously on the same nitrocellulose membranes, optical densities were compared directly. For all other analyses, optical densities of Western blot RV and LV bands were expressed quantitatively by normalization to their right atrial band on the same lane and then compared subsequently. HCN4 expression was increased in LV myocardial samples of cats with HCM (P=0.036) compared to control cats, whereas in RV samples only a trend was observed (P=0.055). Age had no effect on feline myocardial HCN4 expression. Results indicate that myocardial HCN4 expression is upregulated in LV myocardial tissue of cats with HCM. The pathophysiological importance of myocardial HCN overexpression with regard to myocyte function, altered automaticity, and subsequent arrhythmogenicity deserves further study.

7. Effects of oral ivabradine on LV outflow tract obstruction in cats with HCM: Based on our previous observations in cats that ivabradine slightly increases LV volume and decreases LV systolic function it is expected that ivabradine has a reducing effect of LV outflow tract obstruction in cats with HCM. Studies are currently underway to test this hypothesis.

REFERENCES

[1]Atkins CE, et al. **J Am Vet Med Assoc** 1992;201:613; [2]Liu SK, et al. **Am J Cardiol** 1993;72:944; [3]Rush JE, et al. **J Am Vet Med Assoc** 2002;220:202; [4]Schober KE, et al. **J Vet Intern Med** 2006;20:120; [5]Colin P, et al. **J Pharmacol Exper Ther** 2004;308:236; [6]Colin P, et al. **Am J Physiol Heart Circ Physiol** 2003;284:H676; [7]Bache RJ, et al. **Circ Res** 1977;41:648; [8]Hamlin RL. **J Am Anim Hosp Assoc** 1989;25:284; [9]Fox K, et al. **J Am Coll Cardiol** 2007;50:823; [10]Adams III J. **Curr Opin Cardiol** 1999;14:310; [11]Connolly DJ, et al. **J Fel Med Surg** 2003;5:209; [12]Smith SA, et al. **Intern J Appl Res Vet Med** 2004; 2:159; [13]Fox K, et al. **J Am Coll Cardiol** 2007;50:823; [14]Boehm M, et al. **Lancet** 2010;376:886; [15]Colin P, et al. **Am J Physiol Heart Circ Physiol** 2002;282:H672; [16]Borer JS, et al **Circulation** 2003;107:817; [17]Cerbai I, et al. **J Mol cell Cardiol** 2001,33:441; [18]Tavazzi L, et al. **Dialogues Cardiovasc Med** 2006;11:30; [19]Cheng Y, et al. **J Pharmacol Exper Ther** 2007;321:469; [20]Kalman JM, et al. **J Pharmacol Exper Ther** 1995;272:85; [21]Pagel PS, et al. **J Pharmacol Exper Ther** 1995;275:127; [22]Schober KE, et al. **Am J Vet Res** 2003;64:93; [23]Riesen S, et al. **Am J Vet Res** 2011 (in press); [24]Riesen S, et al. **J Vet Pharmacol Ther** 2010 (ePub, Dec); [25]Cober R, et al. **J Vet Cardiol** 2011 (submitted); [26]Riesen S, et al. **J Vet Intern Med** 2011 (in press); [27]Riccioni G. **Adv Ther** 2010;27:160; [28]Chandler NJ, et al. **Circulation** 2009;119:1562; [29]Breall JA, et al. **J Am Coll Cardiol** 1993;21:471; [30]Van Bogaert PP, et al. **Eur J Pharmacol** 2003;478:161; [31]Gargini C, et al. **Vision Res** 1999;39:1767; [32]Tardif JC, et al. **Eur Heart J** 2005;26:2529; [33]Simon L, et al. **J Pharmacol Exper Ther** 1995;275:659; [34]Tardif JC, et al. **Eur Heart J** 2005;26:2529; [35]Mulder P, et al. **Circulation** 2004;109:1674; [36]Boldt A, et al. **Pharmacology** 2010;86:249; [37]Couvreur N, et al. **Eur Heart J** 2009;ePub554; [38]Schober K, et al **J Vet Intern Med** 2011;25 (abstract).

KINETIC ASSESSMENT OF MYOCARDIAL 14(R,S)-[^{18}F]FLUORO-6-THIA-HEPTADECANOIC ACID AND ^{18}F-FLUORODEOXYGLUCOSE UPTAKE WITH POSITRON EMISSION TOMOGRAPHY/COMPUTED TOMOGRAPHY (PET/CT) IN DOMESTIC CATS.

AK LeBlanc [1,2], MR Akula[2], EB Martin[2], RE Gompf[1], GD Galyon[1], TD Moyers[1], AC Stuckey[2], MJ Long[2], SJ Kennel, GW Kabalka, MC Besozzi[2,3], JS Wall[2]. [1]University of Tennessee College of Veterinary Medicine; [2]University of Tennessee Graduate School of Medicine, Molecular Imaging and Translational Research Program; [3]University of Tennessee Medical Center; Knoxville, TN.

Fatty acid oxidation (FAO) is an important myocardial biochemical pathway with therapeutic and diagnostic applications. The domestic cat provides a robust and relevant model to study myocardial fatty acid metabolism with 14(R,S)-[^{18}F]Fluoro-6-thia-heptadecanoic acid ([^{18}F]FTHA) as a powerful tracer for cardiac PET/CT imaging.

Six (6) clinically normal, purpose-bred cats of similar age and body condition were imaged with [^{18}F] fluorodeoxyglucose ([^{18}F]FDG) and [^{18}F]FTHA by using dynamic cardiac-gated fused PET/CT for kinetic assessment of myocardial glucose and fatty acid uptake and metabolism, respectively. Kinetic tracer uptake within the myocardium was achieved by initiating image data acquisition simultaneously with tracer injection. PET data were acquired over a 1 hour period with the heart in the center of the scanner field of view. Regions of interest were drawn in the left ventricular wall and thoracic aorta for the purpose of measuring the kinetics of tracer redistribution. Serial blood samples were also taken during PET imaging for comparison with image data. The equilibrium biodistribution of both tracers was documented 1 hour post-injection in a whole body PET/CT image. Standard echocardiographic examination of cardiac structures was also performed.

Both radiotracers remained in the plasma fraction; however, [^{18}F]FTHA was cleared from the more rapidly than [^{18}F]FDG ($t_{1/2}$ ~ 2 and ~ 20 min, respectively). The tracers were readily visualized within the feline myocardium in dynamic PET images and analysis of the blood pool clearance from the kinetic image data agreed with blood sampling data. Myocardial uptake of each tracer was best described by a double exponential analysis and was rapid but variable among animals (range 1 – 30 Bq/cc/min), although blood glucose levels were similar in all cats during image acquisition. Physiologic [^{18}F]FDG was observed in the brain, salivary tissue, gastrointestinal tract, renal pelves and urinary bladder, with [^{18}F]FTHA seen in the myocardium, liver and renal cortex. All cats were normotensive with normal echocardiographic parameters.

This study demonstrates the utility of kinetic imaging using [^{18}F]FDG and [^{18}F]FTHA to study feline myocardial metabolism in the fasted state using PET/CT imaging. This work further provides validation of [^{18}F]FTHA in a relevant *in vivo* model of myocardial metabolism to support future clinical trials in animals with myocardial disease. Application of [^{18}F]FTHA uptake to the study of feline hepatic and renal disease should also be considered.

ASSESSMENT OF GLOBAL AND REGIONAL LEFT VENTRICULAR VOLUMES AND GEOMETRY USING REAL-TIME THREE-DIMENSIONAL ECHOCARDIOGRAPHY IN DOGS WITH MYXOMATOUS MITRAL VALVE DISEASE. I. Ljungvall[1], K. Höglund[1], C. Carnabuci[1], A. Tidholm[2], J. Häggström[1]. [1]Swedish University of Agricultural Sciences, Uppsala, Sweden, [2]Albano Animal Hospital, Danderyd, Sweden.

The left ventricle (LV) shape has been suggested to change from elliptical to more globular in response to chronic volume overload. Real-time three-dimensional echocardiography (RT3DE) offers new modalities for LV assessment. The aim of the study was to investigate left ventricular changes in shape and volume occurring in response to different severities of naturally acquired myxomatous mitral valve disease (MMVD) in dogs using RT3DE.

Privately owned dogs were classified by standard echocardiography into: Healthy (20), mild (20), moderate (8) and severe MMVD (17). A LV cast was obtained using semi-automated endocardial border tracking from RT3DE dataset, from which global and regional (automatically acquired basal, mid, and apical segments based on LV long-axis dimension) end-diastolic (EDV) and end-systolic volumes (ESV), LV Long-axis dimension and RT3DE Sphericity index, were derived.

Global and regional EDV and ESV increased significantly with increasing MMVD severity, assessed by MMVD group-wise comparisons and linear regression analyses using left atrial to aortic root ratio, and LV end-diastolic and end-systolic dimensions. All three segments contributed to the overall increased global volumes, but the mid EDV segment was strongest associated with increasing LV end diastolic dimension ($P = 0.048$). Furthermore, LV Long axis distance and LV Sphericity index increased with increasing MMVD severity. The basal and apical EDV segments were strongest associated with Sphericity index ($P < 0.0001$).

In conclusion, this RT3DE study showed that increased LV EDV, primarily in the mid segment, leads to rounding of LV apical and basal segments in response to increasing MMVD severity in dogs.

HEARTWORM SEX RATIOS IN NATURALLY INFECTED DOGS DIFFER WITH WORM INTENSITIES. M. Rishniw[1], Y. Schukken[1], E.Greiner[2]. [1]Cornell University, Ithaca NY, [2]University of Florida, Gainesville FL.

Male:female (M:F) ratios in various biological systems often differ from 1. Determinants of distorted M:F include genetic, epigenetic and environmental factors. Wolbachia infection can distort sex ratios of nematodes. Whether *Dirofilaria immitis* (DI) maintain equal M:F in naturally infected dogs is unknown.

84 dogs from shelters in Florida with naturally acquired DI infection were euthanized and necropsied. All adult DI in each dog were sexed using morphological features. Total worm burdens and numbers of males and females were recorded. No other information was available for any dog. All data, raw and transformed, were examined visually and descriptively. Raw numerical data were further examined by a paired T-test; log-odds transformed data were examined by logistic regression. We also conducted a binomial distribution goodness of fit analysis assuming a null hypothesis of a M:F=1.0.

Worm intensities ranged from 1 to 143 DI per dog. Eight dogs had unisex infections: 7/8 had all-female infections. Dogs with low-intensity dual-sex infections were more likely to have greater numbers of female DI. Overall, sex ratios were equal (paired t-test, P=0.7). However, logistic regression demonstrated that the probability of being female is strongly affected by the total worm intensity, with lower intensities increasing the probability of having a predominance of female worms.

Our data show that DI sex ratios in naturally-infected dogs equal 1 when examining the entire dog population, but deviate to favor female worms at low worm intensities. These data could impact adulticide treatment strategies. The reasons for sex ratio distortion in DI are unknown.

FREQUENCY OF RIGHT-TO-LEFT SHUNTING IN DOGS WITH CONGENITAL CARDIAC DISEASE UNDERGOING INTERVENTIONAL CARDIOVASCULAR PROCEDURES. DF Hogan, L Meltzer. Purdue University, School of Veterinary Medicine, West Lafayette, IN.

Paradoxic embolism (PE) from deep venous thrombi can occur in humans with intracardiac connections usually associated with elevations in right heart pressures. An unrecognized ASD, usually in the form of patent foramen ovale (PFO), has been shown to be one of the most common causes of cryptogenic stroke. Dynamic echocardiographic techniques including Valsalva maneuvers are used in an attempt to identify these defects. We have experienced at least two documented PE events in animals associated with right heart interventional procedures. Our objective was to evaluate animals with congenital heart defects (CHD) where an interventional procedure was planned in an attempt to identify right-to-left (R-L) shunting not clearly elucidated on color Doppler echocardiography as a possible risk factor for PE.

During a period of 6 years (2005-2011) 45 dogs had an agitated saline study performed prior to an interventional procedure. The vast majority of dogs (40/45, 89%) had evidence or suspicion of elevated right heart pressures on two-dimensional, color- or spectral Doppler echocardiography. Of those dogs, 21/40 had evidence of R-L shunting with 14/21 (67%) (14/40, 35% total population) having evidence of R-L shunting at the atrial level only, presumably through a previously unrecognized PFO. In the dogs with no evidence for elevated right heart pressures, 2/5 (40%) had evidence of R-L at the atrial level. Interestingly, 2/40 dogs (5%) with evidence for elevated right heart pressures and 2/5 (40%) without evidence of elevated right heart pressures had evidence of very late R-L shunting at the atrial level, presumably through pulmonary arterial to venous connections. Additionally, 1 dog with evidence for elevated right heart pressures that did not have evidence of R-L shunting exhibited evidence of R-L shunting during a breath-hold technique in an attempt to approximate Valsalva physiology.

There was no attempt at risk stratification for PE in this study and there was no clinical evidence of PE in any of the patients that underwent interventional procedures. While one could conclude that this means there is no clear clinical risk with such R-L intracardiac shunting, it should be noted that a positive study simply meant that contrast was seen within the systemic circulation and has not been currently graded in severity of shunting. However, one of the benefits of agitated-saline contrast studies is the rapid and objective assessment of severity through the density of contrast within the systemic circulation. Other benefits include minimal invasiveness, large safety profile, high sensitivity, and lack of required sedation or anesthetic protocols.

We conclude that unrecognized intracardiac defects such as PFO are relatively common in dogs undergoing interventional procedures. These defects could result in PE and should be considered prior to interventional procedures for possible prophylactic measures such as antithrombotic therapy.

GENE-NUTRIENT INTERACTIONS IN HYPERTROPHIC CARDIOMYOPATHY IN MAINE COON CATS. LM Freeman,[1] JE Rush, KE Meurs,[2] BJ Bulmer,[1] SM Cunningham.[1] [1]Tufts Cummings School of Veterinary Medicine, North Grafton, MA; [2]Washington State University, Pullman, WA.

While a mutation in the myosin binding protein C (MYBPC) gene has been identified as a cause of hypertrophic cardiomyopathy (HCM) in the Maine Coon cat, phenotypic expression of the disease varies widely and cannot be accounted for only by the mutation. The cause for this phenotypic disparity is unknown but may be the result of environmental modifiers, such as nutrition. Nutrition and growth patterns in utero or in early life may interact with a genetic predisposition for HCM and increase the risk for the development of cardiovascular disease. The purpose of this study was to characterize associations between nutritional factors, MYBPC status, and phenotypic expression of HCM in Maine Coon cats.

Maine Coon cats >2 year of age and without other major diseases were enrolled in the study. Echocardiography, blood pressure assessment (Doppler), and radiography were performed. In addition, blood was collected for biochemistry profile, T4 (if >7 years), and assessment of insulin, glucose, insulin-like growth factor (IGF)-1, IGF binding proteins, and DNA testing for the MYBPC mutation.

Fifty-six cats have been enrolled in the study. Fifteen cats had HCM [$MYBPC^+$ (n=2) and $MYBPC^-$(n=13)] and 41 had echocardiographically normal hearts [$MYBPC^+$(n=13) and $MYBPC^-$ (n=28)]. Cats with HCM were older (P<0.001), heavier (P=0.003), had a higher body condition score (P=0.02), a larger abdominal circumference (P=0.001), and longer humerus length (P=0.006) than cats without HCM. Cats with HCM also had significantly higher serum glucose concentrations (P=0.02). Historical findings included a larger size at 6 months of age (P=0.03) and heavier body weight at one year (P=0.04) in cats with HCM compared to the cats without HCM. The only significant difference found between cats with and without the MYBPC mutation was a lower serum glucose concentration in the $MYBPC^+$ group compared to $MYBPC^-$ group (P=0.02).

Early growth, larger body size, and obesity may be environmental modifiers of a genetic predisposition for HCM. Additional studies are required to further investigate the relationship between growth rate, body size, and MYBPC mutations in the pathogenesis of this common feline disease.

REGULATION OF CARDIAC β-1, β-2, and β-3 ADRENERGIC RECEPTORS IN CANINE VALVULAR HEART DISEASE. OL Nelson[1], J Häggström[2], K Höglund[2], I Ljungvall[2], C Kvart[2]. [1]Washington State University College of Veterinary Medicine, Pullman,WA, [2]The Swedish University of Agricultural Sciences, Uppsala, Sweden.

Use of β -receptor blocking drugs in dogs with heart failure has been disappointing as compared to results in people. Dogs with overt congestive failure do not tolerate the hemodynamic drug effects and clinical trials in subclinical heart disease have not shown clear benefit in survival or disease progression. It is therefore logical to hypothesize that the canine species may express β-adrenergic receptors in different proportions than humans, and these receptor proportions may regulate differently in canine heart disease. We also hypothesized that dogs with CHF would have upregulation of β1 & 3 adrenergic receptors compared to healthy dogs.

We quantified mRNA expression of β1, 2, and 3 adrenergic receptors in atrial and ventricular myocardium of 13 dogs that were euthanized at various stages of degenerative mitral valvular disease (DMVD) and congestive heart failure (CHF). The mean age of DMVD dogs was 10.1 (6 males, 7 females). Breeds of dogs with DMVD were 1 border collie, 1 beagle, 1 miniature poodle and 10 Cavalier King Charles Spaniels (CKCS). Six healthy mix breed dogs (mean age, 5.5years; mean weight, 15.9kg; 3 males, 3 females) without cardiac disease were used as controls. Normal canine LV myocardium expressed similar proportions of β1 and β2 adrenergic receptors, whereas β3 receptors were slightly less. Δ Ct values did not differ in LV myocardium between healthy dogs and CHF dogs for RPS5 (housekeeping gene), β1 or β3 adrenergic receptors. β2 adrenergic receptor was significantly upregulated in LV myocardium of CHF dogs. The LA myocardium results were very similar to the LV myocardium. Six CKCS did not register β1 receptor signal for LA or LV myocardium.

These results may suggest that upregulation of β2 adrenergic receptors are a predominant finding in the dog, which could explain the differences in tolerance of beta-blocking drugs in dogs compared to humans (which predominately upregulate β1). We also postulate that some CKCS may have polymorphisms of their β1 adrenergic receptor to account for lack of amplification. These results could be further investigated by quantifying β adrenergic receptor protein in myocardium and sequencing the β1 receptor in the CKCS. Understanding regulation of β-adrenergic receptors in various stages of canine heart disease will elucidate answers to these common questions regarding the use of β -receptor blocking drugs for heart failure therapy in canine patients.

MITOCHONDRIAL COMPLEX I AND COMPLEX IV ACTIVITY IN CANINE MYOCARDIUM UNDER NORMAL, ANESTHETIZED, AND CARDIOMYOPATHIC CONDITIONS. Sleeper MM, Bansal S, Rosato B, Avadhani N. Veterinary school of the University of Pennsylvania; Philadelphia, PA.

Mitochondrial diseases are well described in the human literature with most causing dysfunction of the multiple organs which rely particularly heavily on mitochondrial energy production. Therefore, mitochondrially induced cardiomyopathy may be heart specific or associated with other abnormalities (neurologic, skeletal muscle, etc.). It has also previously been demonstrated that volatile anesthetic agents depress mitochondrial function. This study was undertaken to compare mitochondrial complex I and complex IV activity in normal left ventricular myocardium samples which had been exposed to volatile anesthesia (isoflurane), normal myocardium (which had not been exposed to volatile anesthetics), and samples from a group of dogs affected with DCM. Both complexes were assayed spectrophotometrically on isolated mitochondria from left ventricular tissue from a control group, a group with dilated cardiomyopathy (DCM), and a group that was under volatile anesthesia prior to euthanasia The activity level of complex I was 24.6% reduced in DCM samples compared to the control groupand 36.5% decreased in the anestitized group. The activity level of complex IV was not significantly different between the groups. Similar results were found in previous studies of complex I activity and suggest that volatile anesthesia and DCM disrupt the ETC in the dog, potentially leading to an increase in reactive oxygen species. The studies on complex IV did not show a significant difference. A larger scale study is warranted.

REVERSE REMODELING AFTER MITRAL VALVE REPAIR UNDER CARDIOPUMONARY BYPASS IN DOGS. Masami Uechi. Nihon University, Veterinary Cardiovascular Medicine and Surgery, Kanagawa, Japan.

We evaluated cardiac reverse remodeling after mitral valve repair under cardiopulmonary bypass (CPB) for mitral regurgitation in small breed dogs. Fifty dogs (body weight 1.8–9.3 kg, age 5–14 years) with mitral regurgitation were treated between August 2006 and November 2009. The cardiac murmur was grade 4/6¬–6/6. The preoperative chest X-rays showed cardiac enlargement (vertebral heart scale (VHS) 11.0–13.1). Echocardiography showed severe mitral regurgitation and left atrium enlargement (LA/Ao 2.0–4.2). After inducing anesthesia, a thoracotomy was performed in the fifth intercostal space. CPB was started by using a CPB circuit connected to carotid artery and jugular vein catheters. After inducing cardiac arrest, the left atrium was sectioned and chordae tendineae rupture confirmed. The chordae tendineae were replaced with expanded polytetrafluoroethylene. A mitral annulus plasty was also done, and the left atrium was closed. After de-clamping for restarting the heart, the chest was closed. Heart rate decreased from 118¬–164 bpm to 75–138 bpm. The grade of cardiac murmur was reduced to 0/6–3/6 three months postoperatively, and the heart shadow was reduced (VHS 9.8–11.5) in the chest x-rays. Echocardiography confirmed the marked reduction in mitral regurgitation and the left atrial dimensions (LA/Ao 1.2–2.2). Mitral valve repair reduced enlarged cardiac size by reduction of regurgitant rate.

EFFECT OF DICHLOROACETATE IN CANINE PULMONARY ARTERIAL HYPERTENSION: A PILOT STUDY. HW Maisenbacher III[1], PW Stacpoole[2], AH Estrada[1], BJ Pogue[1]. 1. University of Florida College of Veterinary Medicine, Gainesville, FL. 2. University of Florida College of Medicine, Gainesville, FL.

Pulmonary arterial hypertension (PAH) is a well recognized condition in dogs leading to considerable morbidity and mortality. The majority of therapeutics has focused on endothelial dysfunction causing reduced production of vasodilators, such as nitric oxide and prostacyclin, coupled with overproduction of vasoconstrictors, such as endothelin-1. More recently, it has been shown that the mitochondria play an important role in the development of PAH as oxygen sensors and regulators of cellular proliferation. In PAH, pulmonary artery smooth muscle cells undergo a metabolic shift from oxidative phosphorylation in the mitochondria to glycolysis in the cytoplasm as the major energy source and this leads to suppression of apoptosis and increased proliferation. Dichloroacetate (DCA) inhibits pyruvate dehydrogenase kinase to activate pyruvate dehydrogenase which catalyzes the rate limiting step for entry of pyruvate into the Krebs cycle, thus increasing mitochondrial respiration. In three different rat models of PAH, DCA has been shown prevent and reverse PAH by normalizing molecular pathology, stimulating apoptosis of pulmonary artery smooth muscle cells, and reducing pulmonary artery hypertrophy. DCA has known toxic effects, including reversible hepatotoxicity and peripheral neuropathy, and has not been studied in any species with naturally occurring PAH. The objective of this open label pilot study is to evaluate the therapeutic and toxic effects of DCA in naturally occurring canine PAH.

Three dogs with PAH diagnosed by Doppler echocardiography and no correctable underlying cause are enrolled in the study. Dogs are orally administered 25 mg/kg of DCA divided daily for 2 weeks, and then 12.5 mg/kg of DCA divided daily for the remainder of the study. At baseline, 2, 4, 8, and 12 weeks, an echocardiogram, CBC, serum chemistry profile, urinalysis, NT-proBNP, blood uric acid, blood lactate, noninvasive blood pressure, nerve conduction study, and trough DCA level (12 hr post-dose) are obtained. The measured echocardiographic parameters include peak and mean tricuspid regurgitant flow velocity and pressure gradient, peak and end-diastolic pulmonary regurgitant flow velocity and pressure gradient, pulmonary valve flow velocity acceleration time and ejection time, pulmonary valve flow velocity time integral, right ventricular myocardial performance index, tricuspid annular plane systolic excursion, and systolic tricuspid annular tissue velocity. Variables are inspected for normalcy and equality of variances and a two-sided paired t-test is used to compare the variables before and after treatment at each evaluation time.

The basis for the role of the mitochondria in PAH and the results of this pilot study will be presented to determine if DCA warrants further study as a therapy for dogs with PAH.

ASSESSMENT OF SYSTOLIC AND DIASTOLIC FUNCTION VIA TISSUE DOPPLER ECHOCARDIOGRAPHY IN BOXERS AFFECTED WITH ARRHYTHMOGENIC RIGHT VENTRICULAR CARDIOMYOPATHY COMPARED TO NORMAL BOXERS AND NON-BOXER CONTROL DOGS. J Orr[1], ML O'Sullivan[2], M O'Grady[2], R Hamilton[3]. 1. Aurora, IL. 2. Ontario Veterinary College, Guelph, ON. 3. The Hospital for Sick Children, Toronto, ON.

Arrhythmogenic right ventricular cardiomyopathy (ARVC) is the most common acquired cardiac disease reported in the Boxer breed and shares many similarities with ARVC in humans. It is characterized by ventricular arrhythmia of right ventricular (RV) origin, resulting in clinical signs of syncope, exercise intolerance, congestive heart failure and sudden death. Diagnosis of the disease currently relies on the use of Holter monitors, however given the inherent variability associated with Holters, false negatives can occur. Therefore, adjunctive diagnostic modalities are needed to aid in the diagnosis of ARVC-suspected dogs that have inconclusive Holter recordings. Furthermore, RV function has not been well characterized in ARVC in Boxers, and newer modalities are now available to quantitate RV function, which has historically been a challenge with standard echocardiographic methods. The purpose of our study was to assess both systolic and diastolic function using Tissue Doppler imaging (TDI) measured at the medial and lateral mitral valve annulus and the lateral tricuspid valve annulus in Boxers affected with ARVC compared to non-affected normal Boxers and non-Boxer control dogs. Given that TDI is abnormal in humans with ARVC and the similarities of the disease in Boxers, we hypothesized that TDI would be abnormal in Boxers with ARVC.

70 dogs were evaluated and classified into five groups using Holter monitors and measurement of frequency of ventricular arrhythmia (VPCs/hr): 15 Boxers with ≥ 1000 VPCs/24 hrs, 10 Boxers with 200-999 VPCs/24 hrs, 15 Boxers with 25-199 VPCs/24 hrs, 15 Boxers with ≤ 24 VPCs/24 hrs, and 15 non-Boxer controls dogs (with ≤ 24 VPCs/24 hrs). Two-dimensional echocardiography and TDI was performed in all dogs. Signal averaged ECG (SAECG) was also performed and serum troponin I concentrations were measured.

When comparing all groups, there were no significant differences in the TDI parameters measured between groups. This finding suggests that ventricular dysfunction may be uncommon in affected Boxer dogs, or that ventricular function is maintained until the later stages of the disease. Furthermore, no significant correlations between TDI and SAECG variables were found. Significant differences in troponin I concentrations were found between groups of dogs, namely Boxers and the non-Boxer control dogs.

M-MODE MEASUREMENT OF LEFT VENTRICULAR WALL THICKNESS IN A LARGE POPULATION OF SCREENED CATS WITH OR WITHOUT HYPERTROPHIC CARDIOMYOPATHY. J. Häggström [1], T. Falk [2], L. Nilsfors [3], U. Olsson [4], A. Dajm[4], J.G. Kresken [5], I. Ljungvall [1], K. Höglund [1], A. Tidholm [6]. [1]Swedish University of Agricultural Sciences, Uppsala, Sweden, [2] Faculty of Life Science, University of Copenhagen, Frederiksberg, Denmark, [3] Vallentuna, Sweden, [4]PawPeds, Gagnef, Sweden, [5] Clinic for Small Animals Kaiserberg, Duisburg, Germany, [6]Albano Animal Hospital, Danderyd, Sweden.

Pure-bred cats are currently screened for presence of hypertrophic cardiomyopathy (HCM) using echocardiography. Measurement of left ventricular (LV) wall thickness is an important part of the examination. Currently used reference ranges are based on comparably old studies including a limited number of presumably normal cats. The present study was aimed at investigating M-Mode derived measurements of diastolic LV wall thickness in a large population of pure-bred cats.

8476 privately owned pure-bred cats, grouped into 11 breed groups, underwent echocardiographic screening between 2003 and 2009. Signalment data, age, results from physical and echocardiographic examinations were entered into a database. Effects of age, body weight, breed-group, heart rate, year of examination and screener were studied using linear and multiple regression analyses.

Mean age and body weight were 2.3±1.7 years and 4.4±1.2 kg, respectively. 7578 cats were considered normal, 463 cats were considered border-line cases, 326 were diagnosed with HCM, and 109 with other diagnoses. Overall, mean septal and lateral wall thicknesses were 3.97±0.77 mm and 3.91±0.80 mm, respectively. Both lateral and septal diastolic wall thicknesses increased with age, body weight and year of examination, were higher in males, varied between breed groups, and between screeners (all $P<0.001$). Including all cats, the upper 90% confidence range was <5 mm in cats <6 kg. Including only healthy cats, the weight range increased to <7kg.

In conclusion, several factors influence M-mode measurement of LV wall thickness. A diastolic LV wall thickness >5 mm in cats <6 kg is considered outside the normal variation.

USE OF THE VAGAL MANEUVER IN THE DIAGNOSIS, MANAGEMENT, AND TREATMENT OF ARRHYTHMIAS

Gerard Le Bobinnec, DVM, DECVIM (Cardiology)
Bordeaux, France

Vagal maneuvers involve stimulation (by carotid sinus compression) or inhibition (by atropine or glycopyrrolate injection) of the parasympathetic system. They are helpful in the electrocardiographic differentiation and diagnosis of arrhythmias.

The autonomic nervous system includes the sympathetic cardiostimulating system and the parasympathetic inhibitory system.The cardiac anatomic division of the nerve fibers of these two antagonist systemes are quite different : the vagal parasympathetic innervation is primarily at the supraventricular level (sinus node, atrial tissue and atrio-ventricular (junctional) node, whereas the sympathetic innervation involves the ventricular level as well. Physiologic responses identified on the surface ECG include, for the sympathetic system, a positive chronotropic effect (increased frequency), positive dromotropic effect (increased speed of conduction), and positive ventricular bathmotropic effect (increased ventricular excitability). For the parasympathetic , there are a negative chronotropic effect (decreased frequency), negative dromotropic effect (decreased speed of conduction), and positive atrial bathmotropic effect (increased atrial excitability).

Carotid Sinus Compression (CSC) involves noninvasive stimulation of the vagal nerve to enhance vagal tone. The technique consists of gently squeezing the thumb and index fingers progressively along the neck in the anatomic natural depression located between the atlas, the angle of the mandible, and the larynx. This direct manual stimulation of vagal baroreceptors located at the bifurcation of the carotid arteries triggers the cardioinhibitory center of the medulla in the brain.

The cholinergic effects recorded on the ECG are negative chronotropism, negative dromotropism, and shortening of the atrial refractory period. The consequent parasympathetic cardiac effects are : reduced heart frequency, possible second degree AV block, and cardiac standstill with escape complexes. When pressure is slackened, a short reflex sinus tachycardia appears because of the feedback release of epinephrine ; this indirect sympathetic stimulation may reveale concealed ventricular excitability.

CSC can be used on pre-existing arrhythmias, or to reveale latent (concealed) arrhythmias. On pre-existing arrhythmias, CSC may be helpful in the differential diagnosis of tachycardias, and may also sometimes bring explanations about the mechanism of an arrhythmia.

About differentiating tachycardias, a classical exemple is the false interpretation of ventricular tachycardia arising from the combination of atrial fibrillation and right Bundle Branch Block : this classic pitfall in ECG analysis is avoided by CSC which may transiently slow the tachycardia and yet would have no effect on a true ventricular tachycardia.

On Supraventricular Tachycardias (SVT), CSC may be helpful both for diagnosis, to explain the mechanism, and even to interrupt the SVT : AV node dependant SVT are quite all sensitive to CSC ; with AV node independant SVT, it depends of mechanism : for example, in Ectopic Atrial Tachycardia by neoautomatism , CSC decreases only ventricular frequency (functional AV block), whereas the same ectopic AT by triggered activity (early or delayed afterdepolarisations) can be broken with a return in sinus rhythm.

CSC can help to differentiate ventricular reentry from parasystole : after CSC with ventricular reentry, there is a negative dromotopic effect that (often, not always) breaks the reentrant loop and then decreases the frequency of VPCs ; in Parasystole, the frequency of ventricular premature beat increases, because the sinus rhythm is slowed and this decreases the refractory periods of the ventricular tissue, opening the way for the parasystolic ectopic focus.

For latent arrhythmias, CSC can reveale a concealed conduction trouble, essentialy through its dromotropic negative effect. It can also sometimes replace a Holter recording by revealing concealed excitability. Concealed excitability may be revealed by a direct effect of CSC on increasing supraventricular sensitivity and also indirectly by the trigerring of the sympathetic reflex occuring after cessation of the maneuver.

CSC is very simple to use and is safe in dogs : after 30 years, we never had any accident, in spite of some frightening situations (trigerred Ventricular Flutter), even on digitalized patients, as it's described in human medicine.

REFERENCES

Collet M, Le Bobinnec G. Les manœuvres vagales et paravagales. In : Electrocardiographie et Rythmologie canine, Ed Point Veterinaire, Maisons-Alfort, 2001 ; 199-222. Ettinger SJ, Le Bobinnec G, Côté E ; Electrocardiography. *In* Ettinger SJ, Feldman EC. Textbook of Veterinary Internal Medicine, 5th Ed, Philadelphia, , Saunders, 2000, chapter 114 : 800-833.

DIFFERENTIATING BETWEEN RESPIRATORY AND CARDIAC CAUSES OF DYSPNEA

John M. MacGregor, DVM, DACVIM (Cardiology)
Woburn, MA

INTRODUCTION

When confronted with a dyspneic dog or cat, one of the most immediate and important differentiations that a practitioner can make is to determine the cause for the patient's respiratory distress. This task is made more problematic by the fact that many patients are at risk for two or even three different causes of breathing difficulty. Making the correct assessment as the cause of the respiratory embarrassment is important because choosing the wrong cause delays proper treatment and treatments for some causes of dyspnea may be harmful in patients for whom it is the inappropriate treatment. Decision making is aided but potentially further complicated by the wide array of diagnostic testing modalities available for diagnosis causes of respiratory distress. This talk will evaluate the various means that have been proposed as methods by which the cause for dyspnea can be elucidated including physical examination, electrocardiograms, echocardiograms, biomarkers and thoracic radiographs.

HISTORY AND PHYSICAL EXAMINATION

A complete history and comprehensive physical examination are crucial to best assessing causes of dyspnea. Clinicians should have a list of potentially differential diagnoses in their minds based on the signalment and historical information before even examining the patient. Physical examination, including thorough thoracic auscultation can help refine the differential list and by the time confirmatory testing occurs, a working hypothesis as to the cause of the dyspnea should be in hand. Factors such as presence or absence of a murmur or gallop, heart rate, respiratory pattern and sounds, and body temperature can all help localize findings. The hypothesis generated by the physical examination may or may not be supported by diagnostic testing and one should not be chained to the original supposition.

DIAGNOSTIC TESTING

Electrocardiogram

While an electrocardiogram can potentially provide information about cardiac size, it is neither sensitive nor specific for such findings. Useful supportive information can be obtained if a cardiac arrhythmia is present that usually accompanies significant heart disease, but treatment for dyspnea should NEVER be initiated based solely on an ECG.

Echocardiogram

An echocardiogram is a very useful test for confirmation of the presence or absence of cardiac disease. However, routine echocardiograms are rarely useful in establishing the exact cause for a patient's dyspnea. Dogs and cats can have severe cardiac disease yet their respiratory signs may be caused by other conditions, leading to faulty conclusions. This is not to say that echocardiograms are not useful – routine echocardiograms are useful generally as negative screening tests. Echocardiograms can be particularly useful in cats with pleural effusion as this can be due to congestive heart failure but can also be due to a variety other causes. In general, if an animal has an echocardiographically normal heart or mild cardiac changes, there is very little chance that dyspnea is due to cardiac causes. Finally, evidence will be presented that some Doppler echocardiographic variables are actually good predictors of the presence or absence of congestive heart failure.

Biomarkers

Biomarkers such as brain naturietic peptide (BNP) and cardiac Troponin I (cTnI) have gained attention as a potentially useful diagnostic test to determine whether dyspnea is due to cardiac or respiratory causes. Biomarkers are considered useful in humans to determine the cause of respiratory difficulty and to help guide treatment. A commercially available test for the amino terminus portion of the pro-BNP peptide (NT- proBNP) has been shown to have high sensitivity and specificity for differentiating between respiratory and cardiac causes of dyspnea. Studies evaluating levels of cTnI and atrial naturietic peptide (ANP) in dyspneic cats and dogs are few in number and there is currently insufficient evidence to support their use. A rational approach to the use of biomarkers in determining the cause of dyspnea in dogs and cats will be presented, including potential pitfalls. In short, treatment should NEVER be initiated based on biomarker levels alone.

Radiography

Thoracic radiographs are both sensitive and specific in determining the cause of respiratory distress and have long been considered the gold standard in making that determination. Thoracic radiographs are still the most definitive test for determining the presence or absence of congestive heart failure when high quality radiographs are interpreted by experienced radiologists or cardiologists. A recent study demonstrated that most clinicians could distinguish the more clear-cut cases of congestive heart failure but accuracy improved in difficult cases with experience/training of the evaluator. Evaluation of indicators that allow differentiation of cardiac disease such as pulmonary edema (with emphasis on typical peri-hilar location of that fluid in dogs), pulmonary venous congestion and cardiomegaly (particularly increased left atrial size) are paramount to determining whether dyspnea is due to cardiac or other causes.

THERAPEUTIC RESPONSE

The determination as to the cause of dyspnea does not stop with arriving at an original diagnosis. Response to therapy, both in the short term and the long term, should be part of the determination process as well. All cases should always be re-examined based on ongoing information being gathered with particular attention being paid to cases that were challenging to diagnose initially, cases that respond with very low doses of therapy, cases that don't seem to respond to appropriate therapy, and cases that respond in a manner that is unexpected.

SUMMARY

Differentiation between respiratory and cardiac causes for dyspnea can be challenging, particularly in cases where multiple potential causes for respiratory distress are present in one patient. The use of more than one diagnostic modality and integration of all results is the best method for determining the cause of your patient's dyspnea.

REFERENCES

1.Oyama, M et al. J Am Vet Med Assoc .2009; 235(11):1319. 2. Fine D et al J Am Vet Med Assoc. 2008;232(11):1674. 3.Kellihan HB J Vet Cardiol. 2009; Suppl 1:S93. 4. Raffan E J Vet Intern Med 2009; 23(6):1184. 5. Rush et al ECVIM (abs), 2010. 6. Connolly DJ et al. J Vet Cardiol. 2009; Suppl 1:S41. 7. Fox et al J Vet Cardiol. 2009; Suppl 1:S51. 8. Oyama et al Vet Clin North Am Small Anim Pract. 2010 40(4):545. 10. Connolly DJ Vet Clin North Am Small Anim Pract. 2010;40(4):559. 11. Boswood A J Vet Cardiol. 2009;Suppl 1:S23. 11. Porcello F et al Aust Vet J. 2008;86(10):390. 12. Hansson K J et al Small Anim Pract. 2009; Suppl 1:44. 13. Boswood A et al J Small Anim Pract. 2003;44(3):104. 14. Schober K et al. J Vet Intern Med. 2010;24(6):1358.

JUST WHEN MY DAY WAS GOING WELL: DIAGNOSING THE CONFUSING CARDIOPULMONARY PATIENT

Rebecca L. Stepien, DVM, DACVIM (Cardiology), Heidi B. Kellihan, DVM, DACVIM (Cardiology)
Madison, WI

Diagnosing cardiovascular diseases in geriatric patients is challenging due to the possible presence of concurrent diseases that resemble cardiovascular disease, concurrent diseases that cause complications of cardiovascular disease and the possibility of one or more chronic concurrent medications obscuring organ function. In addition, physical findings typically used to diagnose cardiac disease in patients with clinical signs suspicious for heart failure (heart murmurs, irregular heart rhythms, weak pulses) may be found in many geriatric patients with aging changes of their cardiovascular system or pre-clinical disease.

DISEASES THAT LOOK LIKE HEART FAILURE BUT ARE NOT

1. Bronchopulmonary Disease

Chronic bronchopulmonary disease (BPD) is the most common clinical diagnosis mistaken for heart disease or heart failure. Overlapping clinical signs (consistent with both BPD and heart disease/CHF) include chronic non-productive cough, exercise intolerance, dyspnea and gradual loss of body condition. Older animals commonly have heart disease *with* BPD, so the presence of a heart murmur does not necessarily imply that the heart is the cause of the clinical signs.

Dogs:
BPD commonly confused with CHF: bronchitis, collapsing airway syndromes, pulmonary interstitial disease or pulmonary fibrosis ± pulmonary hypertension (PH).

Table 1. Physical findings compared:

Physical or Diagnostic Finding	Bronchopulmonary Disease	Heart Disease or Heart Failure
Non-productive cough elicited by tracheal palpation	Yes	Yes (left atrial enlargement)
Pulmonary sounds	Increased airway sounds, inspiratory/expiratory crackles, wheezes	Increased airway sounds, inspiratory crackles
Heart rate	Normal to low with respiratory sinus arrhythmia	Normal to elevated, no respiratory sinus arrhythmia, arrhythmias possible
Heart murmur	Concurrent disease or TR due to pulmonary hypertension	Usually
Jugular distension	If PH is present (i.e. not with bronchitis)	If right-sided heart failure is present (often due to PH)

How can we differentiate these problems?

1. Thoracic radiographs: Most valuable and easiest, documents collapsing airways or bronchiectasis, cardiac size and shape with pulmonary fields and vasculature

- Left-sided CHF:
  - Symmetrical distribution of pulmonary infiltrates, especially in caudal lung lobes
  - Left-sided or generalized heart enlargement with left atrial (LA) enlargement
  - Pulmonary venous engorgement
- Chronic pulmonary disease with pulmonary hypertension
  - Patchy interstitial to alveolar infiltrates, widely distributed
  - Right-sided or generalized cardiomegaly (LA enlargement may be present if concurrent MR)
  - Pulmonary arterial enlargement (main PA or peripheral pulmonary arteries)
  - Enlargement of caudal vena cava
- Chronic bronchitis
  - Thickened bronchial walls (doughnuts, tramlines)
  - Bronchiectasis

2. Echocardiography or Doppler-echocardiography: Cannot diagnose heart failure but find cardiac changes compatible with heart failure or exclude severe left-sided changes, excellent for diagnosis of pulmonary hypertension

- Left-sided CHF:
  - Marked LA enlargement for any reason, wide pulmonary veins may be visible as they enter the LA
  - Left ventricular dilation is supportive
- Chronic pulmonary disease with pulmonary hypertension

- Right atrial and right ventricular enlargement
- High velocity tricuspid insufficiency
- High velocity pulmonary insufficiency
- Pulmonary artery wider than aortic root at same level

3. Endotracheal wash or bronchoscopy: Examination of bronchial secretions and anatomy may be helpful in the diagnosis of infectious or inflammatory bronchitis or pulmonary disease or allow diagnosis of dynamic airway collapse. Positive aerobic or Mycoplasma spp. cultures or the presence of large amounts of mucus or inflammatory cells are supportive of airway disease.

Cats:
BPD commonly confused with CHF: feline asthma, feline bronchial disease (inflammatory diseases of the bronchial tree and pulmonary parenchyma)

Table 2. Physical findings compared:

Physical or Diagnostic Finding	Feline Bronchial Disease	Heart Disease or Heart Failure
Cough	Yes	No
Pulmonary sounds	Increased airway sounds, inspiratory/expiratory crackles, wheezes	Increased airway sounds, inspiratory crackles
Heart rate	Normal	Normal to elevated, occasionally bradycardic with severe CHF, gallop or arrhythmias possible
Heart murmur	Concurrent disease	Often but not invariably

How can we differentiate these problems?
1. Thoracic radiographs: Documents bronchial pattern supportive of asthma, cardiac size and shape with pulmonary fields and vasculature, presence of pleural effusion

- Left-sided CHF:
  - Variable distribution of pulmonary infiltrates, any pattern possible
  - Left-sided or generalized heart enlargement with possible left atrial (LA) enlargement
  - Pulmonary venous engorgement
  - If biventricular failure is present, pleural effusion (DDx: chylothorax, pyothorax)
- Feline asthma or feline bronchial disease
  - Patchy interstitial to alveolar infiltrates, widely distributed
  - Normal looking cardiac silhouette
  - May see pronounced bronchial pattern

2. Echocardiography or Doppler-echocardiography: Cannot diagnose heart failure but find cardiac changes compatible with heart failure or exclude severe left-sided disease

- Left-sided CHF:
  - Marked LA enlargement for any reason, wide pulmonary veins may be visible as they enter the LA
  - Left ventricular dilation or hypertrophy is supportive IF severe LA enlargement is present
- Chronic pulmonary disease
  - Variable right atrial and right ventricular enlargement
  - Normal LA and LV measurements or mild LA and LV changes

3. Endotracheal wash or bronchoscopy: Examination of bronchial secretions and anatomy may be helpful in the diagnosis of infectious or inflammatory bronchial disease. Significant numbers of inflammatory cells (neutrophils or eosinophils) are NOT supportive of CHF.

2. Body Cavity Effusions
Causes of effusions that may be confused with CHF: neoplasia (carcinomatosis, mesothelioma), obstruction or damage to the lymphatics (chylous effusions), severe hypoproteinemia, inflammatory diseases (e.g. FIP), infections (pyothorax, peritonitis).[1]

Table 3. Physical findings compared:

Physical or Diagnostic Finding	Non-Cardiac Cavity Effusion	Heart Failure-Related Effusions
Abdominal Palpation	Fluid wave, possibly mass lesion	Fluid wave, may palpate hepatomegaly
Pulmonary sounds	Muffled	Muffled
Heart rate	Normal	Normal to elevated, occasionally bradycardic with severe CHF, gallop or arrhythmias possible
Heart murmur	Concurrent disease	Often but not invariably
Jugular distention	No	Yes

How can we differentiate these problems?

1. Thoracic radiographs: Presence of pleural effusion, heart size if visible
- If CHF is the cause:
  - Cardiomegaly is often noted, may be obscured by fluid.
  - Pulmonary infiltrates may be present, any pattern possible
  - Right-sided or generalized heart enlargement with possible RA enlargement
  - Caudal vena cava enlargement
  - Globoid cardiomegaly may indicate pericardial effusion
- Other causes of effusion:
  - Normal cardiac size and shape if visible
  - Normal caudal vena cava size if visible

2. Echocardiography or Doppler-echocardiography: Used to identify heart disease if present
- CHF:
  - Identify atrial enlargement (either side) to identify elevated atrial pressures
- Other effusions:
  - Normal right atrial and right ventricular or mild changes
  - Normal LA and LV measurements or mild LA and LV changes

3. Fluid analysis:
- CHF:
  - Total protein in fluid usually < 4.5 g/dl
  - Often serosanguinous, PCV should be < 10%
  - Cytology shows mixed cell population (neutrophils, eosinophils, macrophages) with low total cell counts and no neoplastic cells or mitotic figures identified.
  - Fluid may have some gross and cytologic features of chylous effusion but usually not overtly "creamy"
- Other causes:
  - Total protein < 2 g/dl (suspect hypoproteinemic) or > 4.5 g/dl (suspect neoplastic)
  - Cytology may show inflammation, presence of bacteria, neoplastic cells
  - PCV > 10% may reflect hemothorax or hemoabdomen (compare to peripheral PCV)
  - Effusion triglycerides > peripheral triglycerides may indicate chylous effusion.

4. Body cavity ultrasounds may reveal mass lesions or anatomic abnormalities (e.g. lung lobe torsion)

3. Arrhythmias

Cardiac arrhythmias are frequently documented in elderly dogs and to a lesser extent, cats. These arrhythmias raise questions about the presence of cardiac disease and the necessity for antiarrhythmic therapy.

Causes of arrhythmias in elderly patients

There is a natural concern that previously undetected primary cardiac disease may be present in a geriatric patient with newly discovered arrhythmias. A series of cardiologic diagnostic testing allows diagnosis of primary cardiac disease if present, but other causes of arrhythmias should be considered:

1. Primary cardiac disease:

Ventricular arrhythmias are often associated with myocardial disease (dilated, hypertrophic or restrictive) or valvular disease when the presence of CHF contributes to hypoxia. Severe atrial enlargement commonly results in atrial premature complexes, atrial or supraventricular tachycardia or atrial fibrillation. Some arrhythmias (primarily diseases of the conduction system such as AV block or accessory pathways) are detected via ECG but not associated with detectable anatomic changes.

2. Hypoxia:

Temporary or chronic hypoxia may contribute to development or potentiation of arrhythmias. Local myocardial hypoxia may occur due to (difficult to detect) myocardial infarctions or generalized systemic hypotension and generalized hypoxia may be secondary to anesthetic mishaps, pulmonary edema or pleural effusion of any cause. Cats are particularly susceptible to CHF-related hypoxia and even severe ventricular arrhythmias may resolve with successful therapy for edema or pleural effusion.

3. Sepsis:

The circulating pro-inflammatory cytokines, along with severe hypotension that is typical of septic shock are thought to contribute to the occurrence of arrhythmias in these patients. Rapid therapy of hypotension and infection along with supportive care often leads to resolution of arrhythmias, but temporary antiarrhythmic support may be needed.

4. Metabolic disease:

This category includes diseases that are often more common in geriatric animals. Systemic diseases associated with hypertension (e.g. hyperadrenocorticism, renal failure) may lead to LV hypertrophy and arrhythmias. Other diseases (e.g. hyperthyroidism or hypoadrenocorticism) cause neurohormonal or electrolyte changes that directly cause arrhythmias. Neurologic diseases often cause ectopy by mechanisms that remain unclear, but in some cases, may involve derangement of the sympathetic nervous system modulation.

5. Drugs and toxins:

Only a few commonly-encountered toxins cause arrhythmias as a primary sign (e.g. naturally-occurring digitalis glycosides in foxglove plants, certain snake venoms), many cause systemic abnormalities (i.e. GI upset) that can lead to electrolyte-related arrhythmias. Arrhythmias due to drug administration or toxicity are extremely commonly and range from mild bradycardias and atropine-responsive AV blocks during anesthesia to severe arrhythmias following administration of cardiotoxic doses of doxorubicin or digoxin. Significantly, many antiarrhythmic medications cause arrhythmia at non-standard doses.

6. Autonomic causes of arrhythmias:

This category is considered to be a diagnosis of exclusion, commonly suspected when all other causes of diseases have been ruled out. Autonomically-mediated arrhythmias such as accelerated idioventricular rhythms are common in sedated or anesthetized patients when systemic disease is present, therefore, they are commonly diagnosed in patients with significant but non-cardiac disease.

Diagnostic Testing

If the cause of the arrhythmia is not obvious, further testing may help to find or rule out various causes (always treat before investigation if arrhythmia is life-threatening).

1. Generalized cardiac screening:

- Auscultation: define murmurs, arrhythmias, pulse strength, other signs of CHF
- ECG: define the arrhythmia and document severity
- Thoracic radiographs: cardiomegaly and evidence of CHF
- Echocardiography: evidence of significant myocardial or valvular disease, severe chamber dilation, mass lesions or evidence of myocardial hypoxia
- Blood pressure: rule out hypertension, monitor if hypotension is part of clinical syndrome

2. Systemic screening:

- Physical exam: presence of abdominal masses or signs of systemic/neurologic disease
- Screening blood tests (CBC, biochemistry, urinalysis) for metabolic disease
- Specific testing if endocrine disease is suspected
- Additional imaging (e.g. abdominal ultrasound): if suspicious findings are present
- Review of current drugs/exposure to toxins/recent sedation or anesthesia

Common Diseases of the Elderly Affecting the Heart

Canine hypothyroidism:

Hypothyroidism is a disease of aging dogs associated with decreased exercise tolerance, lethargy and weight gain. Long-standing severe hypothyroidism can result in severe myocardial systolic dysfunction without dilation, congestive heart failure and atrial fibrillation.[2] In affected dogs, recognition of underlying hypothyroidism results in long-term survival without aggressive CHF therapy, an outcome that is unlikely if the metabolic abnormalities are not addressed.

Recognizing CHF secondary to chronic hypothyroidism:

- Left-sided heart failure may be present
- Atrial fibrillation with slow ventricular response even when untreated (120-140 bpm) is common
- Echo findings: normal to increased LV diameter, decreased contractile function, no major valvular changes (looks like DCM except relatively less LV dilation)
- Physical findings suggestive of hypothyroidism: weight gain, dry hair coat, hair loss over tail and bridge of nose
- Biochemical findings: 80% of hypothyroid dogs have hypercholesterolemia

Therapeutic Tips:

- Treat CHF as usual until stable normal thyroxine concentrations are obtained via supplementation
- Begin thyroxine supplementation at low doses, increase about every 3-4 weeks until mid-range therapeutic concentrations are achieved
- Expect increased ventricular response to AF when thyroxine is supplemented; may require additional medications to keep < 160 bpm (does not usually resolve)
- Taper furosemide and then discontinue when endocrine situation is stable
- Survival times > 4 years are expected if no other cardiac complications are present

Feline hyperthyroidism:

Hyperthyroidism is an extremely common disease of aging cats and may be complicated by concurrent renal disease and hypertension. Long-standing and severe hyperthyroidism may result in overt CHF, typified by severe pleural effusions. Less severe cardiac effects of hyperthyroidism include tachycardia and tachyarrhythmias, gallop rhythms, open-mouthed breathing when excited/stressed, characteristic echocardiographic changes and systemic hypertension.

Recognizing CHF secondary to chronic hyperthyroidism:

- Signs of CHF are accompanied by a palpable thyroid mass (goiter) in a feline patient most commonly > 10 years of age
- Serum thyroxine concentrations are elevated, may see mild ALT elevations
- If CHF is not present, thyrotoxic heart disease signs include tachycardia/arrhythmias, gallop rhythms and variable heart murmurs
- Echocardiographic findings: LV diameter normal to enlarged and hypercontractile. All four cardiac chambers are typically mildly dilated. LA enlargement may predominate. LV walls are normal to mildly, symmetrically thickened. No evidence of significant valvular disease or asymmetry in cardiac chamber sizes.

Therapeutic Tips:

- Hyperthyroidism can be treated medically, surgically or via radiation as available
- IF CHF is present, it should be treated as usual (furosemide, ACEI) until endocrine status is stable
- Systolic BP> 180 mmHg should be treated with antihypertensives until hyperthyroidism is successfully controlled
- If HR > 240 bpm, beta-blocker therapy may be helpful until resolution of hyperthyroidism

REFERENCES

[1]Beatty J, et. al. **J Feline Med Surg.** 2010 Sep;12(9):693. [2]Gerritsen RJ, et.al. **Vet Q.** 1996 Jun;18(2):49.

SALVAGING THE CARDIOVASCULAR PATIENT WITH UNEXPECTED DEVELOPMENTS

Rebecca L. Stepien DVM, DACVIM (Cardiology), Heidi B. Kellihan DVM, DACVIM (Cardiology)
Madison, WI

INTRODUCTION

Unexpected developments in a previously stable cardiovascular patient may be due to worsening of the cardiac disease, development of complications of that disease, or development of a concurrent, unrelated illness that is interacting with the cardiac disease. The assessment of patients with any of these causes of unexpected developments is similar, beginning with reassessment of cardiovascular status, assessment for new cardiovascular complications, and a larger look at the whole patient for other concurrent diseases.

UNEXPECTED DEVELOPMENT #1: RECURRANCE OF CONGESTIVE HEART FAILURE OR UNEXPECTED DEVEOLPMENT #2: LACK OF RESPONSE OF CHF TO REASONABLE THERAPY

Heart failure that doesn't respond to typical aggressive therapy is often referred to as "refractory" heart failure. This term may also be used to describe sequential recurrence of heart failure signs that increase in severity over time as the heart disease progresses, necessitating step-wise increases in medical therapy.

Treatment of refractory HF represents a dilemma for the veterinarian. Altering or adding further medications is tempting but hazardous because heart failure patients are often hemodynamically and metabolically fragile. The goals of therapy in refractory patients are the same as in "naïve" HF patients: improving quality of life and extending life.

The most common reasons for recurrent or refractory congestive heart failure include:

- Diagnosis of CHF was not correct, or not *quite* correct
- Development of new systemic abnormalities
- Progression of underlying abnormalities not previously recognized
- Medication-related problems (cardiac or other medications)
- Complication of current cardiac disease
- Progression of cardiac disease without "new" complications

General Clinical Approach to the Patient

History:

Any recent changes in the patient's life should be investigated. Have there been changes in routine? Kenneling? Medications? Weight? Geriatric patients are at increased risk for stress during holidays that include new visitors, changes in daily routines and access to new or unusual food. This is also a good moment to check the *owner's* report of medication doses and schedule – it is often NOT what it says in the record.

Physical examination:

Physical examination of the patient may reveal evidence of concurrent systemic illness. Are the signs of heart failure the same as previously or are they new signs? Is an arrhythmia present or has the patient's arrhythmia changed?

Recommended diagnostic approach for the chronic heart failure patient with new illness:

The diagnostic testing choices for newly decompensated chronic heart failure patients or those with concurrent illness are based on what type of problem the animal is exhibiting. In most cases where a new disease is suspected, blood testing can be very helpful and usually includes hematology, chemistry and urinalysis to detect new diseases or new complications. Thoracic radiographs are used to assess heart failure status and in some cases, can be helpful to identify dehydration. If a new or different arrhythmia is present, and ECG is warranted. Additional testing that can be performed on patients in which it is indicated are serum digoxin concentration and baseline or post-pill thyroxine concentration. An echocardiogram should be considered in cases where the manifestation of heart failure is unexpected (e.g. a dog with chronic left-sided heart failure now has ascites and pulmonary hypertension is suspected), if a cause for new signs may be found (e.g. intracardiac thrombus in a cat with signs of peripheral embolism, or a ruptured chordae tendineae in a dog with acute, fulminant pulmonary edema). In general, if findings may change therapy, echocardiography is justified.

Once initial testing has been performed, a prioritized problem list should be developed. Is the new problem a magnified version of the historical disease, or is it completely different? Which conditions are life-threatening and require immediate assistance? What effects might a suspected new disease have on the heart (e.g. development of systemic hypertension may make left-sided heart failure more difficult to resolve).

Development of a new therapeutic plan:

A new therapeutic plan must take into account the systemic illnesses or abnormalities that may be present as well as the underlying chronic heart disease itself. Rebalancing the decompensated patient may involve short term lifestyle change, short or long term changes in medications or a change in the way a patient's medications are administered.

Systemic illness:

Temporary adjustments of cardiac medications may be necessary during the acute pre-treated stages of systemic illness (e.g. brief discontinuation of diuretics in severely hypotensive dehydrated animals). As the problem is resolved, medications will likely be restarted and gradually need to be increased to previous dose levels. If systemic illness occurs BECAUSE of cardiac medications, there may be permanent changes in dose (e.g. electrolyte imbalances, azotemia, inappetence).

New cardiac signs:

New cardiac signs may develop if new complications of old diseases are present, such as ruptured chordae tendineae or left atrium, new arrhythmias or pulmonary hypertension from left-sided failure leading to right-sided failure. In addition, new signs may develop due to distant complications or new cardiovascular disease, such as thromboembolism (systemic or pulmonary), dirofilariasis or cardiac tumors. These new complications are often uncovered with standard diagnostic tests, including radiographs, echocardiography and directed laboratory testing.

Example: The Fluid Retainer

The problem I: The patient had pulmonary edema that was controlled and is now back.

Causes: Typical precipitating causes of recurrence of fluid retention problems in chronic heart failure patients include natural progression of disease (degree of neurohormonal activation exceeds therapeutic adjustments), sudden change in disease status (e.g. development of pulmonary hypertension secondary to left heart failure causes right heart failure), addition of outside medications or change in environmental or care circumstances or development of a new arrhythmia.

Presenting signs: Recurrent pulmonary edema in previously stable patient is usually due to progression of disease or occurrence of a precipitating cause (e.g. new arrhythmia). The diagnosis typically involves review of history and both cardiac and noncardiac medications, physical examination and thoracic radiography to confirm the presence of edema, an ECG if indicated by physical examination and laboratory tests to exclude other problems (e.g. hypoproteinemia, anemia). Therapy includes discontinuation of any offending medications (e.g. corticosteroids), and a temporary increase in the aggressiveness of therapy (e.g. change of route of furosemide administration to parenteral administration, addition of temporary catecholamine administration). Chronically, additional vasodilators and positive inotropes like pimobendan can be added, neurohormonal blockade should be maximized (e.g. ACEI and spironolactone ± digoxin). Any new significant arrhythmias should be treated with acute and then chronic anti-arrhythmic therapy.

The problem II: Canine: Long term "cougher" due to left atrial enlargement and left-sided heart failure develops ascites or starts fainting OR **Feline:** Pleural effusion previously controlled with medication is now recurring.

Cause: If apparently "left-sided" disease leads to right-sided failure, look for evidence of pulmonary hypertension (PH). Left sided HF (elevated left atrial pressure) leads to increased pulmonary arterial pressures through "back up" of pressure. In cats, this is manifest as pleural effusion occurring even though disease is left-sided. In dogs, when the right heart fails due to PH, left atrial pressure decreases and the dog's previous problem with coughing often eases (at the cost of ascites development). In dogs, chronic lung disease may contribute to development of PH.

Diagnosis: Physical diagnosis of PH may include detection of an accentuated (loud) second heart sound ("lub-DUP"). On thoracic radiographs, enlargement of pulmonary veins AND arteries may be noted. Echocardiography is required to make a definitive diagnosis and judge the severity of PH. Typical echocardiographic findings include the expected evidence of previously-diagnosed left-sided disease PLUS right atrial and ventricular dilation ± apparent RV thickening, high (> 3 m/sec) velocity tricuspid insufficiency, high velocity (> 2.5 m/sec) pulmonary insufficiency and pulmonary arterial enlargement.[1]

Therapy: Mechanical removal of fluid should be performed for safety and comfort of patient. In the case of pleural effusion, most or all of the effusion can be safely removed. In the case of ascites, approximately 75% of the fluid should be removed at the maximum to avoid hypotension. Overall congestive heart failure management should be optimized, with maximal neurohormonal blockade, addition of pimobendan if not already on the medication schedule, and the heart rate control in atrial fibrillation. When these maneuvers have been instituted and pulmonary hypertension persists, therapy with sildenafil can be initiated as a direct pulmonary arterial dilator. This medication is not typically required in cats (in which pimobendan is quite effective vs. pleural effusion caused by heart failure) but in dogs, can be used at a dose of 1 mg/kg orally twice to three times daily.[2] Normal systemic blood pressure should be confirmed prior to therapy with sildenafil.

UNEXPECTED DEVEOPLMENT #3: HEART FAILURE RESOLVED, PATIENT HAS OTHER PROBLEMS

New problems in patients in whom heart failure is recently resolved may include complications of medical therapy (electrolyte disturbances), effects of "overtreatment" (dehydration, hypotension) or reflect a completely different disease process. The diagnostic approach to these patients involves careful analysis of history and PE findings. Careful discussion of the patient's recent history may reveal a change in cough quality that may indicate a bronchopulmonary problem has arisen, or reveal changes in the household or medication routine that has resulted in decreased or absent appetite and resulting dehydration. A full physical examination is required and may reveal findings overlooked when a short "cardiac" physical is performed in a familiar patient. Some key points that may arise with a full physical examination include:

- Body weight – weight loss over time leads to overdoses of medications that are dosed on a per kg basis
- Body temperature: fever may indicate infection is present

- Heart rate, respiratory rate: If these are within normal range, CHF is less likely
- Careful abdominal palpation: may reveal a mass lesion
- Lymph node palpation
- Retinal exam: may reveal findings that reflect systemic hypertension or systemic infectious disease

Example: Hypotension

Hypotension is a common presenting clinical sign in treated cardiovascular patients. The problem usually involves a previously balanced HF patient is presented with client complaints of weakness, lethargy, generally doing poorly or a sudden deterioration of clinical condition. The history often includes…"he hasn't been eating but he has been getting ALL of his medications…"

Dehydration is the most common cause of hypotension in cardiovascular patients, followed by unintentional medication overdose in a patient that has lost weight but continues to receive "original weight" medication doses. Dehydration may involve a lack of access (new living situation, accidentally empty water bowl) or a new problem may have developed that keeps the pet from getting to the water source (blindness, orthopedic problems etc.), or may result from general inappetence that is drug-related (toxic concentrations or idiosyncratic reaction, drug effect-related (e.g. prerenal azotaemia due to diuretic therapy) or related to another, non-cardiac disease (e.g. patients vomiting or with diarrhea from any cause).

Physical examination:

The physical examination of these patients reveals a patient that is persistently weak and lethargic (not with episodic collapse), inappetent or anorexic, and often thin/muscle wasted. Weak pulses, tacky mucous membranes and hypothermia are common. In cats, bradycardia may be present.

Diagnostic testing:

With a history and physical examination that is content with hypotension, other diagnostic testing confirms clinical suspicions. Blood pressure measurement frequently returns values "too low to measure" by Doppler or oscillometric means. A direct BP measurement via an arterial catheter commonly shows systolic BP < 100 mmHg. Blood tests are important to identify evidence of dehydration or electrolyte imbalance and importantly, to screen for other disease. Thoracic radiograph and echocardiography commonly reveal a small heart and blood vessels and apparent contractility is decreased due to decreased ventricular loading. It is very important to screen patients with thoracic radiographs even if congestive heart failure signs are not present – masses, infections and thoracic problems (trauma, hemo- or pneumothorax) may be revealed. Severe electrolyte disturbances can result in arrhythmias, and an ECG should be recorded on any animal with a heart rate above or below normal range or with an irregular rhythm.

Initial therapy:

Initial therapy of the hypotensive, dehydrated patient includes temporary discontinuation of diuretics and most vasodilators (in some cases, drugs with multiple effects like pimobendan may be continued). Cautious fluid administration is indicated, because these patients are unlikely to take in enough fluid orally to rehydrate. The intravenous route of fluid administration is best for close control of amounts and rate of administration and because administration can be stopped immediately if needed (as opposed to the subcutaneous route). Warmed fluids are helpful, particularly in cats, and low sodium fluids are preferred. If the patient's heart rate is low (< approximately 120 bpm in the dog and < approximately 160 bpm in cats), any medications that slow heart rate (e.g. diltiazem, digoxin) should be discontinued (if beta blockers are on board, these must be decreased in dose for 24-36 hours before discontinuation. Temporary catecholamine administration may provide both systolic and diastolic support for hypotensive patients, and dobutamine is preferred to avoid excessive tachycardia. Any electrolyte abnormalities (hypokalemia and hypomagnesemia are common) should be rectified by supplementation either in fluid therapy or by oral therapy.

Table 1. Potassium and Magnesium Supplementation

	Intravenous	Oral	Other
Potassium	15-60 mEq/l in IV fluids (based on level of deficiency), administer at a rate no higher than 0.5 mEql/kg/hour	Potassium gluconate: 2 mEq per 4.5 kg body weight in food q 12 hrs	Encourage oral intake
Magnesium	0.75-1 mEq/kg/day of $MgCl_2$ or $MgSO_4$ in D5W for rapid repletion if hypomagnesaemia is present, then 0.3-0.5 mEq/kg/day for next 3-5 days[3] Alternate dosing: 20-40 mg/kg of $MgSO_4$ slow IV bolus, the 20-40 mg/kg over the next 24 hours *Note: incompatible with calcium or bicarbonate!*	1-2 mEq/kg/day, magnesium oxide or magnesium hydroxide	Decrease loss by adjusting loop diuretic dose Encourage oral intake

Acute monitoring of rehydration of the hypotensive patient:

Careful observation of the patient's urinary output will help judge the adequacy of rehydration efforts. This need not (and usually should not) involve invasive monitoring. If the patient is allowed the opportunity to urinate and does so after fluid administration, some efficacy can be assumed. Alternatively, bladder filling can be assessed by simple palpation or even "quick" ultrasound checks. Respiratory rate is a valuable indication of "overhydration" – when a patient's resting respiratory rate (non-panting) climbs to > 30-32 breaths per minute during fluid therapy, fluid therapy rate should be reassessed. Return to normal temperature, heart rate and appetite are indications of successful therapy but the patient may require 2-3 days of careful support before recovery is evident. An important component of care at this stage is frank discussion with the owner regarding the delicacy of the treatment protocol and the importance of not rushing the therapy in these fragile patients.

When appetite returns, medications may gradually reinstated AS NEEDED. Potassium and magnesium should be monitored every 24 hours if heavy supplementation is used or every 48 if less aggressive supplementation is used. The patient may be discharged when eating and breathing easily (e.g. creatinine does not have to be normal). After discharge, a 3 day phone call to check at-home respiratory rate and appetite is helpful, and these patients benefit from a recheck approximately one week post-discharge to check physical examination finings and drug and electrolyte concentrations as needed.

REFERENCES

[1]Kellihan HB, et.al. **Vet Clin North Am Small Anim Pract.** 2010 Jul;40(4):623. [2]Kellum HB, et. al. **J Vet Intern Med.** 2007 Nov-Dec;21(6):1258.
[3]Dhupa N. Magnesium Therapy, In: **CVT XII**. 1995, p. 132.

CANINE-DERIVED ANTIBODY FRAGMENTS THAT TARGET HEMANGIOSARCOMA AND ITS GROWTH FACTORS. A Braganza, B Clerc-Renaud, S Huang, NJ Mason, University of Pennsylvania School of Veterinary Medicine, Philadelphia, Pennsylvania.

The use of monoclonal antibodies and antibody fragments to directly target tumor antigens and neutralize their growth factors has shown promising results in human clinical trials. However, these targeted approaches have not been possible in dogs since specific tumor antigens have not been identified, monoclonal antibodies of canine origin are not available and the efficacy of xenogeneic antibodies in the dog is limited by neutralizing antibody responses. To overcome these obstacles, we have generated canine antibody phage display libraries from canine splenocytes. These libraries consist of single chain variable fragments (scFv) comprised of canine variable heavy (VH) and variable light (VL) immunoglobulin chains displayed on the surface of bacteriophage (Fig. 1). The antigen specificity within these libraries is diverse and recapitulates the antigen-experienced immunoglobulin repertoire of the dog. We can now use simple panning techniques to isolate scFv of canine origin that bind to either known targets or unknown targets which can then be identified using standard molecular techniques.

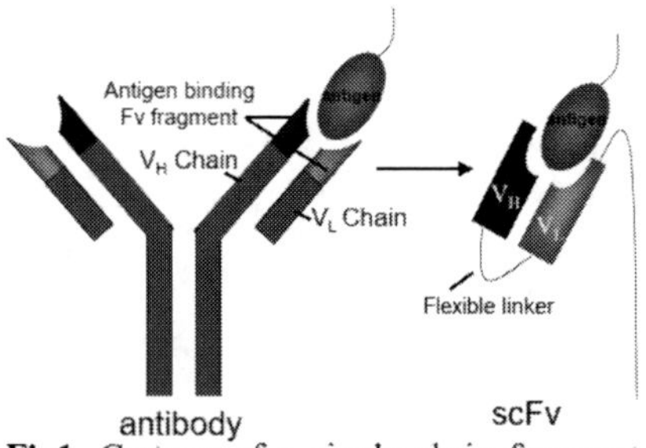

Fig1. Cartoon of a single chain fragment variable (scFv) comprised of the VH and VL regions of immunoglobulin linked by a flexible linker. scFvs are the smallest antibody fragments that maintain antigen-binding specificity and affinity.

Canine HSA is a highly aggressive malignancy of vascular endothelial cells that affects large breed dogs. Although there are no confirmed immunological targets for HSA, serum levels of Vascular Endothelial Growth Factor (VEGF) are elevated in these patients and, as in many human cancers, VEGF may represent an important therapeutic target for neutralization. We used simple panning techniques to screen canine scFv libraries generated from the spleens of dogs with HSA against canine VEGF and successfully isolated 3 scFv clones that bind and neutralize canine VEGF *in vitro*. These scFvs are now being taken into a murine model of canine HSA to determine whether they can inhibit tumor growth and metastases. In addition, we have panned the same antigen-experienced scFv phage display libraries against allogeneic primary canine HSA cells of low passage number to isolate canine-derived antibody fragments that can target malignant endothelial cell surface molecules. Early results demonstrate enrichment of scFv phage libraries for malignant endothelial cell binders. These scFv can be readily linked to chemotherapeutic agents or other toxins and used to deliver high doses directly to the malignant cell. This novel approach aims to reduce side effects of systemic chemotherapy and augment therapeutic response.

EFFICACY OF VITAMIN D (CALCITRIOL) FOR CANINE MAST CELL TUMORS. EK Malone[1], KM Rassnick[1], JJ Wakshlag[1], DS Russell[1], R Al-Sarraf[2], DM Ruslander[3], CS Johnson[4], and DL Trump[4]. 1. Cornell University College of Veterinary Medicine, Ithaca, NY. 2. Animal Emergency and Referral Associates, Fairfield, NJ. 3. Veterinary Specialty Hospital of the Carolinas, Cary, NC. 4. Roswell Park Cancer Institute, Buffalo, NY.

Calcitriol, (vitamin D_3), has antineoplastic activity and acts synergistically to potentiate the antitumor activity of a diverse array of chemotherapeutics. CCNU, vinblastine, corticosteroids, and tyrosine kinase inhibitors, are used to treat canine mast cell tumors (MCT). Vitamin D receptor is expressed in the majority of canine MCTs, suggesting a role for calcitriol in the management of dogs with these tumors. The purpose of our study was to examine the in vitro effects of calcitriol in combination with CCNU, vinblastine, imatinib, or toceranib on canine mastocytoma C2 cells. Also, we evaluated the antitumor activity of DN101, a highly concentrated oral formulation of calcitriol, as single-agent treatment in dogs with naturally occurring MCTs.

C2 cells were incubated with serial dilutions of calcitriol (0.1 - 25nM). Twenty-four hours later, cells were then treated with vehicle control or serial dilutions of CCNU (2.5-20uM), vinblastine (1.25-10nM), imatinib (1.67-12.5nM), or toceranib (3.13-25nM). Cell viability was assessed with an MTT assay after 48 hours and data was used to derive a combination index (CI: values <1, 1, >1 indicate synergism, additivity, antagonism, respectively). In the phase II clinical trial, dogs were eligible if they had at least 1 measurable, histologically confirmed, MCT. Calcitriol was administered orally. RECIST criteria were used to assess tumor response.

Calcitriol, CCNU, vinblastine, imatinib, and toceranib each suppressed C2 cell viability in a dose-dependent manner. CI values <1 were obtained for calcitriol (0.1-6.25nM) combined with CCNU (5 and 10uM), vinblastine (2.5 and 5nM), imatinib (1.67-12.5nM) and toceranib (0.1-12.5nM). Due to the occurrence of toxicity (vomiting, anorexia, hypercalcemia), the phase II trial was terminated early; only 10 of 20 planned patients were treated. One dog with a metastatic muzzle MCT had a complete response that lasted 89 days. Three dogs achieved partial response lasting from 74-90 days.

In summary, our in vitro data demonstrate that calcitriol combined with CCNU, vinblastine, imatinib or toceranib has synergistic effects on C2 mastocytoma cells. Antitumor responses were observed in dogs with spontaneously occurring MCTs treated orally with single-agent calcitriol, but the frequency of adverse effects was high. Together these results suggest calcitriol combination therapies might have significant clinical utility in the treatment of canine MCTs but refinement of the calcitriol-dosing regimen must be done.

ACVIM FOUNDATION RESEARCH REPORT: DEVELOPMENT OF ASSAYS FOR THE PHARMACODYNAMIC MONITORING OF CYCLOSPORINE IN THE DOG. T Archer, C Fellman, J Stokes, K Lunsford, A Mackin. Mississippi State University, Starkville, MS

Cyclosporine is a potent immunosuppressive agent used to treat many canine inflammatory and immune-mediated diseases. Cyclosporine has gained popularity as an immunosuppressive agent because of a favorable toxicity profile compared to many other immunosuppressive agents. Optimal dosing regimens for cyclosporine in the dog remain unclear, primarily because standard methods that monitor effectiveness of immunosuppression have not been established. Pharmacokinetic testing is currently used during treatment with oral cyclosporine to adjust doses based on measurement of blood drug levels. Individual patients, however, often demonstrate marked variations in blood drug levels while on similar oral doses of cyclosporine, and can also demonstrate different clinical responses even at comparable drug levels, making correlation of blood cyclosporine levels and degree of disease control extremely difficult. Pharmacodynamic testing offers an alternative method for regulating cyclosporine dosing by objectively measuring the effects of cyclosporine on T-cells, the drug's main cellular target in the body. Our ACVIM Foundation-funded research has focused on developing and evaluating a comprehensive panel of biomarkers of immunosuppression that can be utilized for pharmacodynamic monitoring during treatment with cyclosporine and other immunosuppressive agents that affect T-cell function.

We have completed several studies using flow cytometry to evaluate activated T-cell expression of surface molecules (CD25 & CD95) and cytokines (IL-2, IFN-γ & IL-4) as potential biomarkers. Our first study was an *in vitro* study evaluating expression of surface molecules and cytokines in canine T-cells exposed to varying concentrations of cyclosporine. This study established consistent drug-associated suppression of the cytokines IL-2, IFN-γ and IL-4. Our second study was an *in vivo* study in normal dogs evaluating the effects of two doses of oral cyclosporine, a high dose considered to be reliably immunosuppressive (starting dose 10 mg/kg bid, titrated upwards as needed to attain trough drug blood levels of at least 600 ng/mL) and a lower dose used to treat atopy (5 mg/kg sid), on T-cell expression of these three cytokines. Significant suppression of IL-2 and IFN-γ expression was seen at the high cyclosporine dose, while at the lower dose only IFN-γ expression was suppressed. Because T-cell expression of IL-4 was not significantly suppressed at the high cyclosporine dose, IL-4 was not evaluated at the lower drug dose.

Because of specialized sample handling requirements, flow cytometry is not as practitioner friendly as other assays (such as PCR) for routine use in pharmacodynamic testing. We have therefore conducted an *in vitro* study comparing the effects of cyclosporine on activated T-cell expression of IL-2 and IFN-γ using flow cytometry and qRT-PCR, and demonstrated dose dependent and comparable suppression of IL-2 and IFN-γ using either methodology. We are currently evaluating, using qRT-PCR, the effects of oral cyclosporine on T-cell expression of IL-2 and IFN-γ in normal dogs prior to moving on to pharmacodynamic trials in our clinic patients.

EFFECT OF HYPOTHYROIDISM ON REPRODUCTION IN BITCHES. DL Panciera[1], BJ Purswell[2], KA Kolster[2], SR Werre[3]. Departments of [1]Small Animal Clinical Sciences, [2]Large Animal Clinical Sciences, and [3]Laboratory for Study Design and Data Analysis, Virginia-Maryland Regional College of Veterinary Medicine, Virginia Tech, Blacksburg, VA.

Numerous reproductive abnormalities, including irregular interestrous period, anestrus, and infertility have been attributed to hypothyroidism. We previously documented reduced fertility and lower birth weight and increased periparturient mortality in pups born to bitches with experimentally-induced hypothyroidism for a median duration of 56 weeks. The purpose of this study was to evaluate reproductive function in these same bitches after hypothyroidism was treated with a replacement dose of levothyroxine.

Twelve multiparous bitches were studied. Hypothyroidism was induced in 6 dogs by administration of 1 mCi/kg ^{131}I. Hypothyroidism was confirmed by finding serum T4 concentrations before and 4 hours after IV administration of human recombinant TSH that were < 5 nmol/L. Levothyroxine (0.02 mg/kd q 24 h) was administered to all hypothyroid bitches. Six bitches served as euthyroid, untreated controls. Dogs were evaluated daily for signs of estrus and were bred by 1 of 2 males when serum progesterone was ≥5 ng/ml. Interestrous interval, gestation length, strength and duration of contractions during whelping, time between pups, number of live pups and stillbirths, viability of pups at birth, weight of pups, and periparturient mortality were recorded. The Student's t-test and ANOVA were used to compare differences between control and hypothyroid bitches for continuous, normally distributed data. The Wilcoxon rank sum test was used to analyze data between groups that was not normally distributed.

The mean duration of hypothyroidism prior to levothyroxine administration was 102 ± 8.1 weeks. Breeding took place after levothyroxine treatment for 46 ± 12.6 weeks in the hypothyroid group. All 6 dogs in the hypothyroid group and 5/6 control dogs were pregnant, while 4/8 hypothyroid and all 6 control bitches became pregnant prior to levothyroxine administration. No difference in interestrus interval or gestation length was noted between groups. During whelping, no difference in strength of contractions, contraction duration, interval between pups, or viability scores of pups was found between groups. Litter size, birth weight and peirparturient mortality were similar between groups.

Levothyroxine administration reverses the detrimental effects of hypothyroidism on fertility and neonatal health.

EFFICACY OF GASTRIC ACID SUPPRESSION FOR PREVENTION OF EXERCISE-INDUCED GASTRITIS IN RACING ALASKAN SLED DOGS. KK Williamson[1], MD Willard[2], MS Davis[1], ME Payton[1]. [1]Oklahoma State University, Stillwater, OK, [2]Texas A&M University, College Station, TX.

Racing sled dogs have a high prevalence of exercise-induced gastric erosions/ulcers, with reports ranging from 50-67% of dogs running at least 100 miles in a day or less. Omeprazole reduces the severity of, but does not completely prevent, gastritis under racing conditions, and can be difficult to administer under these conditions. Famotidine can be administered in food, but has only demonstrated efficacy under less intense training conditions. The purpose of these studies was to evaluate different acid suppression strategies under racing conditions for the prevention of exercise-induced gastritis.

Experiment #1 was a randomized placebo-controlled study using 36 sled dogs (3-8 years) competing in a 330 mile race over 50-60h. Treatment groups were famotidine (approx 1 mg/kg qd) or no treatment, beginning 2 days prior to the start of the race and proceeding until gastroscopy was performed 24h after the race. Experiment #2 was a randomized positive-control study using 52 sled dogs (2-8 years) running a mock race of 300 miles in 50h. Dogs were divided into omeprazole (approx 1 mg/kg qd, administered 30 min prior to a meal) or famotidine (approx 2 mg/kg bid) groups beginning 2 days prior to the exercise challenge and continuing for 24h after completion. Gastroscopy was performed immediately prior to the start of dosing and 24h after completion of the exercise. In all cases, mucosal appearance during gastroscopy was blindly scored using previously described scoring system.

Famotidine (1 mg/kg qd) reduced the prevalence of clinically-relevant, exercise-induced gastric lesions compared to no treatment (7/16 vs 11/16, $p = 0.031$). Compared to famotidine at 2 mg/kg bid, omeprazole significantly decreased the severity (0.4 vs 1.2, $p = 0.0002$) and prevalence (2/23 vs 7/21, $p = 0.049$) of gastric lesions.

Although famotidine provides some benefit in the prevention of exercise-induced gastric lesions, neither the recommended dose nor the higher dose were considered acceptable in the prevention of exercise-induced gastritis as between 33-44% of the dogs receiving famotidine had clinically significant lesions. A previous study examining omeprazole under racing conditions, but without careful administration on an empty stomach, resulted in a 22% prevalence of clinically significant gastric lesions. However, the bioavailability of omeprazole is reduced in the presence of food, and when the daily administration of the drug is carefully scheduled to coincide with an empty stomach, the resulting prevalence of clinically significant lesions induced by racing-intensity exercise is reduced to just over 10%. The conclusions of these studies are that omeprazole is superior to famotidine in preventing gastritis in racing sled dogs during competition. Routine administration of omeprazole is recommended to prevent stress-associated gastric disease in exercising and racing Alaskan sled dogs.

EFFECT OF ORAL GALLIUM NITRATE ON FECAL SHEDDING OF RHODOCOCCUS EQUI IN MARES. BR Buchanan[1], K Kuskie[3], N Pusterla[2], N Cohen[3]. 1. Brazos Valley Equine Hospital, Navasota, TX. 2. University of California Davis SVM, Davis, California. 3. Texas A&M University CVM, College Station, TX.

Mares may be an important source of environmental contamination with *Rhodococcus equi* on breeding farms. Attempts to reduce fecal shedding of *R equi* by the mare and the effects of the mare's fecal *R equi* concentration on airborne concentrations in the foaling stall have not been previously reported. Twenty-one Arabian mares were treated daily with either oral gallium nitrate or placebo in a randomized double-blind study. Fecal samples were collected at day 320 of gestation (time 1), the week before foaling (time 2), and the week after foaling (time 3). Airborne concentration of *R equi* were measured in the stall within 6 hours post foaling using a microbial air sampling system into which standard (100-mm) culture plates with a media selective for *R. equi* have been loaded.

Concentration of total *R equi* were determined by morphological characteristics. The concentration of virulent *R equi* was determined using a modified colony immunoblot method. Concentrations of total and virulent *R equi* were compared among mares to examine effects of treatment, time, and treatment by time interaction.

There were significant ($P<0.05$) effects of treatment that depended on time of sample collection. At sample times 1 and 2 there were no significant differences between groups in the fecal concentration of virulent *R equi*. At time 3 concentrations of virulent *R equi* were significantly lower among mares in the treatment group ($P<0.05$) compared to control. Effects of time depended significantly on groups: for the control group, there were no significant effects of time. For the treatment group, concentrations tended to decrease over time, and concentrations at time 3 were significantly ($P<0.05$) lower than those at time 1. No other differences among times for concentrations in the treatment group were statistically significant. There were no significant effects of treatment, sample time, or their interaction on the concentration of total *R equi* between groups; however, the pattern for these data was similar to that observed for the virulent isolates. No significant differences were determined between treatment groups for airborne concentrations of virulent or total *R equi*.

Treatment of mares with oral gallium nitrate significantly reduced the fecal concentrations of virulent *R equi* over time, but had no impact on the airborne concentration of *R equi* shortly after foaling.

BRONCHOALVEOLAR LAVAGE FLUID PROTEOME IN HORSES WITH RECURRENT AIRWAY OBSTRUCTION. Marybeth Miskovic Feutz,[1] C. Paige Riley,[2] Jiri Adamec,[3] Laurent L. Couetil.[1] 1. Purdue University School of Veterinary Medicine, West Lafayette, IN. 2. Bindley Bioscience Center, Purdue University, West Lafayette, IN. 3. University of Nebraska, Lincoln, NE.

The purpose of this study was to evaluate the protein profile of bronchoalveolar lavage fluid (BALF) in horses affected with recurrent airway obstruction (RAO) and in control horses using proteomics and Western blot techniques. RAO-affected (n=5) and control horses (n=6) were subjected to an experimental exposure trial; when the RAO-affected horses showed clinical signs of disease, BALF was collected from all horses. BALF was also collected from client-owned RAO-affected horses (n=15) with naturally-occurring clinical signs of disease and client-owned control horses (n=12) from the same environments. The BALF from the experimental exposure trial horses was subjected to trypsin digestion and proteomics analysis with mass spectrometry (MS). Peaks detected with MS were identified using tandem MS analysis and database searches. Western blot was used to confirm the identity and expression levels of two proteins identified using proteomics techniques in the BALF of all horses. Data from MS experiments were analyzed with the Student's t-test to compare peak intensity between RAO-affected and control horses. Western blot band density data was analyzed with the Kruskal-Wallis ANOVA for comparison between groups of horses. Significance level was set at $P<0.05$.

With MS proteomic analysis of the BALF from the experimental exposure trial horses, 2049 total peaks (peptides) were identified. Of these peaks, 100 were differentially expressed between the RAO-affected (24 over-expressed) and control horses (76 over-expressed). Identifications were made for 250 BALF proteins. Transferrin and secretoglobin were chosen for validation with Western blot. Proteomics indicated that secretoglobin was not differentially expressed between the experimental exposure trial group; this was confirmed with Western blot analysis. Western blot also showed that client-owned RAO-affected horses had lower secretoglobin expression than client-owned control horses and control horses before experimental exposure. According to the proteomics data, transferrin was over-expressed in control horses after experimental exposure compared to RAO-affected horses. While the Western blot analysis did not show a statistically significant difference in this comparison, transferrin was significantly over-expressed in control horses before experimental exposure compared to client-owned RAO-affected horses. In addition, both secretoglobin and transferrin band densities on Western blot were negatively correlated with airway obstruction and neutrophilic pulmonary inflammation.

This study demonstrates that proteomics techniques can be used in the investigation of equine BALF proteins. The proteins identified as differentially expressed between RAO-affected and control horses in this study including, but not limited to, secretoglobin and transferrin should undergo further evaluation for their use as biomarkers of RAO, and as potential targets of new therapeutic agents for RAO.

RATTLESNAKE ENVENOMATION IN HORSES: CARDIOTOXIC EFFECTS AND IMMUNE RESPONSE.

Lyndi L. Gilliam[1], TC. Holbrook[1], ME Payton[1], S M Martin[2], R Carmichael[1], JM Taylor[3], AA Giulioli[4], K Levis[2], CL Ownby[1], D McFarlane[1]. 1.Oklahoma State University, Stillwater, OK. 2. Hansford County Veterinary Clinic, Spearman, TX. 3. Red Rock Biologics, Woodland, CA. 4.Tri-County Veterinary Clinic, Anna, TX.

Cardiotoxic effects of rattlesnake venom in the horse are not well defined. The first aim of this study was to document cardiac damage in naturally envenomated horses. Twenty horses with clinical diagnosis of snake bite were included. A snake venom ELISA was utilized to confirm envenomation when possible. Serum and plasma were collected at selected intervals. Plasma was assayed for cardiac troponin I (cTnI) using a flurometric assay (Stratus CS®, Dade Behring). Holter monitors (Zymed®, Philips) were placed at admission, 1 week and 1 month post presentation. Echocardiography was performed on available horses 5-7 months after envenomation.

The second aim of this study was to investigate potential mechanisms of the cardiac damage. Serum samples were assayed for TNFalpha using a commercial assay (Endogen). Antibody titers to *Crotalus atrox* venom were measured at admission, 1 week and 1 month after natural envenomation and compared to titers in vaccinated horses (Crotalus Atrox Toxoid, Red Rock Biologics).

A significant number of horses showed elevations in cTnI ($p<0.05$) at one or more time point indicating myocardial damage. Holter readings revealed the presence of arrhythmias or persistent tachycardia in 19 horses. Five of twenty horses were available for echocardiography; no abnormalities were noted. Horses with increased cTnI tended to have greater TNFalpha concentrations compared with horses without increased cTnI. Peak venom titers in bitten horses were significantly higher than peak titers in vaccinated horses ($p<0.05$).

Rattlesnake envenomation was associated with evidence of cardiac damage in a significant proportion of bitten horses. Further studies are needed to determine the cause as well as mechanisms to treat and/or prevent its occurrence.

EARLY GOAL-DIRECTED THERAPY: IS EQUINE MEDICINE READY?

Rose D. Nolen-Walston, DVM, DACVIM (LAIM)
Kennett Square, PA

INTRODUCTION

The term Early Goal-Directed Therapy (EGDT) came to popular attention after publication of a seminal study in the New England Journal of Medicine in 2001 which has now been cited over a thousand times.[1] This paper describes a randomized clinical trial using 288 patients with septic shock presenting to an academic tertiary care hospital emergency room. Criteria for inclusion included showing 2/4 criteria of systemic inflammatory response syndrome, a systolic blood pressure < 90 mmHg after a 20-30 ml/kg crystalloid bolus, or a blood lactate concentration > 4 mmol/L. Patients were randomly allocated to two groups, receiving either standard treatment or resuscitation following an Early Goal-Directed protocol (figure 1) for up to six hours before admission to the intensive care unit. All patients received broad spectrum antibiotics within the first hour of admission, and supplemental oxygen and ventilatory support as needed. The experimental protocol included the normalization of mean arterial pressure (MAP), central venous oxygen saturation ($ScvO_2$), and urine output over the first 6 hours of hospitalization by the use of fluid boluses, vasopressors, and transfused red blood cells administered to preset goals. Specific adjustments of cardiac preload, afterload, and contractility were used to balance oxygen delivery with oxygen demand, and aimed to reduce the progression of reversible into irreversible cellular dysfunction.

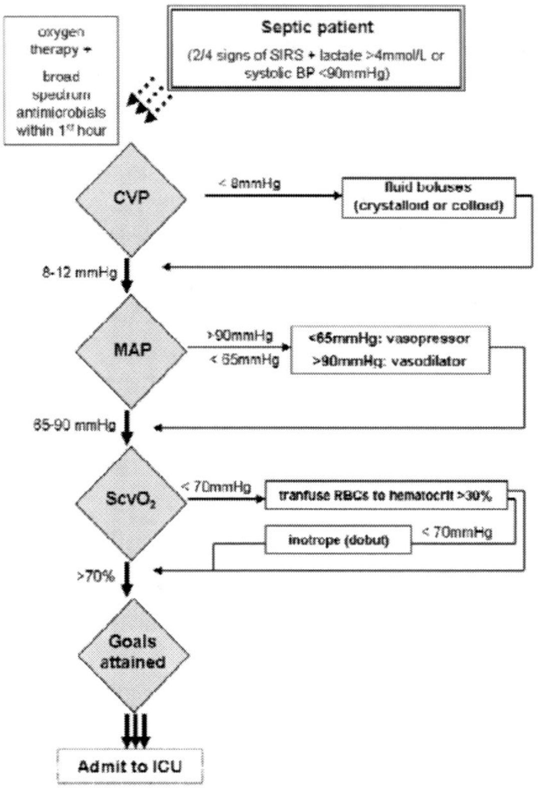

Figure 1: EGDT flowchart (adapted from Rivers E, et al. N Engl J Med 2001;345:1368)

The study reported that EGDT resulted in an absolute reduction of mortality of 16% when compared to standard care. Additionally, there was a 14% reduction in vasopressor use during the first 3 days of hospitalization, the length-of-stay decreased by 5 days, and hospital charges were lower by $5,882 per patient. These outcomes were considered extraordinary, and have received significant discussion and scrutiny over the decade since publication. In 2008, the Surviving Sepsis Campaign published the "International Guidelines for Management of Severe Sepsis and Septic Shock", and included EGDT interventions as a level 1 recommendation, meaning that the intervention's desirable effects clearly outweigh its undesirable effects.[2]

APPLICATIONS OF EGDT IN EQUINE MEDICINE

Septic shock is well described in equine patients of all ages, but is probably most common in neonatal foals, and adult horses with colitis or recovering from large colon volvulus and other post-surgical bowel ischemia. These animals frequently fulfill the Rivers criteria for inclusion, and are generally considered to have a guarded prognosis. The body of literature concerning the treatment of clinical sepsis is considerably larger in foals than adult horses, probably for several reasons. First, sepsis is a very common reason for hospitalization of neonates, and thus the caseload available for study is often significant. Secondly, due to their small size, much instrumentation designed for the human market can often be used without modification. And thirdly, again due to their small size, interventions that may be cost-prohibitive in a 500kg adult are plausible in a 50kg neonate. But even if these limitations disappeared, it is possible that the parameters used for human EGDT would not be optimal for the treatment of equine sepsis. The following discusses our current state of knowledge regarding the three main components of EGDT as they relate to equine medicine, and discusses their potential applicability to the formulation of an EGDT schema for horses and foals.

Central Venous Pressure

Central venous pressure (CVP) is used in many species to monitor right-sided intravascular volume status and optimize cardiac preload, especially in critical care medicine. Measured at either the cranial vena cava or the right atrium, the two determinants of CVP are the cardiac function, which is dependent on the Starling forces of the length-tension relation, and cardiac return of blood from the venous reservoir. As a result, changes in CVP can be seen associated with changes in contractility, venous tone, and circulating blood volume. In EGDT, achieving an adequate CVP is one of the earliest steps in resuscitation of patients presenting to the emergency service with sepsis.

The healthy, standing adult horse has a normal CVP of 7-12 cmH_2O, with higher values seen under anesthesia, especially in lateral recumbency compared to dorsal decubitus.[3] The effect of various manipulations on CVP in horses has been documented, including increase in CVP after intravenous hypertonic solution, bovine hemoglobin blood substitute, exercise, and high intraabdominal pressure, and decreased after administration of xylazine and aceytlpromazine.[3] With regards to hypovolemia,

Magdesian showed that a decrease in CVP is associated with reduction of intravascular volume after acute blood loss in adult horses.[4] The New Bolton Center sepsis study group recently performed a series of experiments to further characterize CVP in adult horses. Using a fureosemide/water restriction model of volume depletion, we found a significant linear relationship between dehydration and CVP, with a 2.2cm$H_2$0 decrease in CVP for every 1% increase in dehydration as measured by change in bodyweight. This change in hydration resulted in only minimal alterations in PCV, total protein, and lactate (figure 2), despite 2/10 horses developing severe neurological abnormalities at 6.5-7% dehydration.[3] Additionally, we evaluated the repeatability and reproducibility of CVP measurements in healthy adult horses and found excellent repeatability of measurements obtained in triplicate at any given timepoint.[5] We also compared the use of electronic monitors to water manometry, and found that both electronic monitors systematically provide measurements that are approximately 2 cmH_2O lower than water manometry; but the differences between the two electronic devices are small enough (less than 0.5 cmH_2O) to be considered clinically insignificant.[6] The effect of head position on CVP was significant, with an elevated head position decreasing CVP by -2.0 ± 6.5cmH_2O, while the lowered head position increased CVP by 3.7 ± 5.5cmH_2O.[5] These finding offer clinicians specific tools to confidently acquire and interpret CVP readings in un-anesthetized adult horses.

However, the use of CVP to determine volume replacement has come under considerable scrutiny in the human literature. A systematic review of the human literature cites Magdesian's study in horses as being the only example of a study demonstrating a clear relationship between CVP and volume.[7] Recent human studies have indicated that arterial pulse pressure variation and stroke volume variation derived from pulse contour analysis,[8] vena caval diameter in patients undergoing positive-pressure ventilation,[9] and even $ScvO_2$[10] are superior to CVP for predicting fluid responsiveness. However, neither of these techniques is likely to be practical in the population of hospitalized adult equids.

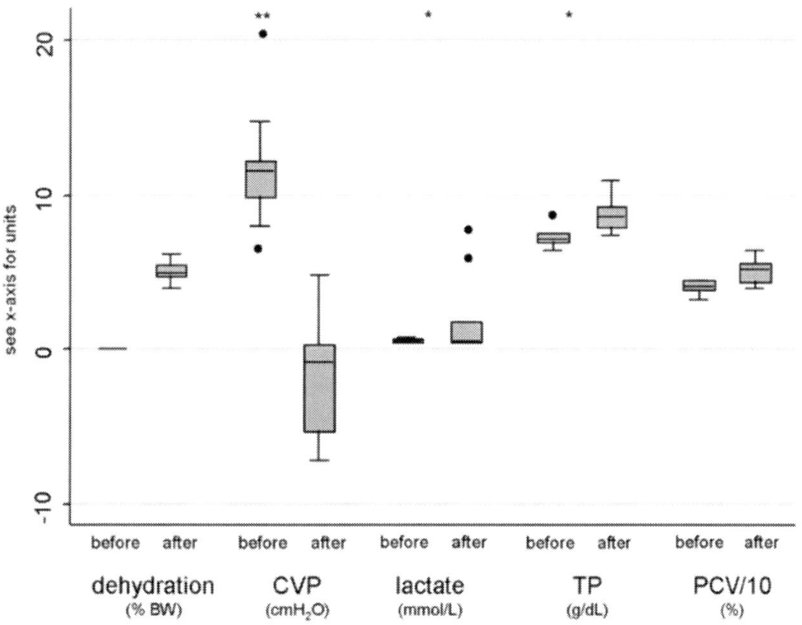

Figure 2. The effect of dehydration on selected variables over time. Each box-and-whisker pair shows the median, 25th and 75th quartiles, 95% confidence interval, and outliers, both at baseline (before) and after dehydration (after). ** indicates $P<0.001$, * indicates $P<0.05$. Y-axis units are specified by variable on the x-axis. (From Nolen-Walston, R., et al. **J Vet Internal Med**, no. doi: 10.1111/j.1939-1676.2010.0641.x)

Mean Arterial Pressure

The main impediment to the use of MAP as a resuscitation endpoint in the adult, awake horse with sepsis lies in the difficulty in obtaining reliable measurements. Although direct blood pressure is obtained as a standard of care in most hospitals for cases undergoing general anesthesia, placement and especially maintenance of an arterial catheter in adult horses and ambulatory foals is often difficult. In foals, blood pressure measurement using indirect oscillometry with a cuff placed over the coccygeal artery or dorsal metatarsal artery results in acceptable accuracy, [11] but in adults, methods of indirect blood pressure management are typically less reliable, although may be useful for measuring trends. However, no direct comparison of direct versus indirect sphygmomanometry in hemodynamically deranged horses has been performed, and thus clinicians struggle to obtain and interpret MAP data in even critically ill, septic adult horses, thus impeding our ability to enact appropriate interventions to support blood pressure and perfusion. A continued debate in human medicine, the choice of first-line and subsequent vasopressors continues to be deliberated in equine medicine. Using humans as our experimental model, some extrapolation can probably be made from the level 1 recommendation of vasopressor preference for norepinephrine or dopamine to maintain target MAP $\geq$ 65mmHg,[2] but especially in adult horses, it would seem prudent to select agents that preferentially spare the splanchnic circulation

Central Venous Oxygen Saturation

Central venous oxygen saturation ($ScvO_2$) is the percent saturation of oxyhemoglobin in a central vein, typically the superior vena cava, and it is used as a substitute for the more invasive mixed oxygen saturation (SvO_2) obtained from the pulmonary artery. These measurements provide a determination of the ratio oxygen consumption (VO_2) to oxygen delivery (DO_2), where low venous oxygen saturation represents excess oxygen extraction at the tissue level. In shock, this is generally thought to be caused by an interaction of increased tissue demand (high VO_2) due to high metabolic needs combined with a decrease on DO_2 associated with reduced cardiac output and/or arterial oxygenation. The EGDT guidelines recommend resuscitation to an ScvO2 ≥70% or SvO_2 ≥65%.[1] This is achieved using two strategies: the first is to ensure that the patient is volume repleted to a CVP of ≥8mmHg using bolus crystalloids, and that the hematocrit is corrected to ≥30% using packed red cells. By maintaining adequate circulating volume and appropriate arterial oxygen content, the increase in DO_2 to the tissue beds may correct the oxygen extraction imbalance. If these are unsuccessful in remediating the low $ScvO_2$, inotropes (generally dobutamine) are used to improve DO_2 via an increase in cardiac output, which may be reduced due to the myocardial dysfunction of sepsis. However, $ScvO_2$ obtained from the cranial vena cava reflects oxygen balance from the brain and cranial half of the body, and in humans is ~10mmHg lower than $ScvO_2$ obtained from the inferior vena cava. While in septic humans this may not be an important distinction, in the adult horse sepsis is almost universally associated with severe gastrointestinal disease, and thus cranial $ScvO_2$ levels may underestimate profound abnormalities of oxygen extraction in the abdominal viscera. Although in horses, blood transfusion is rare due to the short-half life of exogenous erythrocytes, improvement of $ScvO_2$ by means of inotropes may be achievable, but has yet to be demonstrated in either horses or foals. In humans, $ScvO_2$ is measured continuously using an oximetry probe integral to the catheter; although these catheters are of an appropriate size for foals, they are expensive to buy, and obtaining $ScvO_2$ from aliquots submitted for blood gas analysis is likewise to expensive to obtain at appropriately frequent intervals.

EGDT IN EQUINE MEDICINE

Investigation into the ideal treatment of sepsis should remain a priority for equine internists and intensivists. Despite extensive human data regarding the optimal management of patients with septic shock, equine medicine lags due to a lack in both clinical caseload as well as funding compared to human medicine. Nonetheless, as veterinary clinician-scientists, we have the advantage of being able to create experimental models using our specific species, and the paucity of hospital cases can be balanced by our ability to design and employ highly repeatable models of sepsis to clarify outcome differences in therapeutic interventions. It is unlikely that the trio of CVP, MAP, and $ScvO_2$ will be ideal for our use based mainly on the difficulty in the measurement all of these parameters, or the cost in obtaining them. However, since the publication of the Rivers paper, many other parameters have been investigated,[12] of which several show promise for the equine patient. CVP is an easily obtained in both large and small equine patients, and seems to be more associated with intravascular volume status than it is in humans, although optimizing CVP has not been investigated in equine sepsis *per se*. The use of venous lactate concentration as a surrogate for $ScvO_2$ to measure severe tissue hypoperfusion is already used extensively in equine medicine, although can be confounded by Type B hyperlactatemia in some disease states. It is likely that frequent lactate readings accompanied by occasional $ScvO_2$ or $PcvO_2$ (partial pressure of oxygen measurements to evaluate oxygen extraction are more plausible than real-time $ScvO_2$ analysis as is done in human medicine. The use of arterial blood pressure remains problematic; it is interesting to note that many large colon volvulus patients have continuous blood pressure monitoring with extensive pressor and inotropic support throughout the intra-operative period, but are then placed in the recovery stall and never have another blood pressure reading obtained despite severe ongoing hemodynamic compromise.

Another feature of the EGDT protocol was the use of broad spectrum antimicrobials in all patients within 1 hour of admission to the emergency department. The used of antimicrobials in foals with signs of sepsis is universal, but it remains controversial in adult horses with severe GI disease despite findings that a large percentage of these cases are blood culture positive, and may have a worse prognosis than those who have negative blood cultures.[13] The criteria for use of antibiotics in horses with signs of SIRS, sepsis, and compromised gastrointestinal barriers deserves considerable further attention. All patients in the Rivers protocol also received physiologic (APACHE) scoring to describe illness severity. Several scoring systems have been proposed for equine patients, but none have been universally adopted, thus making comparison of interventions across institutions and patient groups more difficult.

In conclusion, EGDT remains an intriguing goal on the horizon of equine hospital care, but enough preliminary data exist that investigation of the components and protocols of EGDT for both adult horses and foals should be considered both desirable and feasible.

REFERENCES

[1]Rivers E, et al. **N Engl J Med** 2001;345:1368; [2]Dellinger RP, et al. **Crit Care Med**. 2008 Jan;36(1):296; [3]Nolen-Walston RD, et al. **J Vet Intern Med**. 2010 Nov 23. doi: 10.1111/j.1939-1676.2010.0641; [4]Magdesian KG, et al. **J Am Vet Med Assoc** 2006;229:1458; [5]Norton JL, et al. **J Vet Intern Med** 2009;23:777; [6]Norton JL, et al. **J Vet Intern Med**. 2011 Jan 31. doi: 10.1111/j.1939-1676.2010.0670; [7]Marik PE, et al. **Chest** 2008;134:172-178.[8]Marik PE, et al. **Crit Care Med** 2009;37:2642; [9]Feissel M, et al. **Intensive Care Med** 2004;30:1834; [10]Giraud R et al. **J Trauma.** 2010 Aug 27. [Epub ahead of print]; [11]Giguère S et al. **J Vet Intern Med**. 2005 Jul-Aug;19(4):571; [12]Otero RM, et al. **Chest** 2006;130: 1579; [13]Johns I et al. **Equine Vet. J.** (2009) 41 (2) 160.

SMALL VOLUME RESUSCITATION IN EQUIDS

Lucas Pantaleon, DVM, MS, DACVIM
Versailles, KY

INTRODUCTION

Diverse etiologies can initiate shock, ultimately resulting in tissue hypoperfusion.[1] Circulatory failure leads to a decrease in oxygen delivery and a decline in partial pressure of oxygen at the cellular level. [2] When a critical partial pressure of oxygen is reached, a shift from aerobic to anaerobic metabolism ensues. [2] This is defined as dysoxia, a state in which inadequate tissue oxygenation produces cellular injury.[2] If shock is not reversed, then irreversible cellular and tissue damage occurs, leading to organ dysfunction and ultimately failure. [1] Shock often, but not always, results from circulatory failure and decreased oxygen delivery. [2]

Systemic inflammatory response syndrome is the body's inflammatory response to an insult (shock). During this syndrome, a complex interaction between pathologic vasodilatation, relative and absolute hypovolemia, myocardial dysfunction, abnormal blood flow distribution, decreased capillary recruitment, intravascular coagulation, increased blood cell adhesion, endothelial edema and mitochondrial dysfunction (cytopathic hypoxia) ensues, making resuscitation challenging. [1-3]

Shock can be quantitative (when the oxygen delivery is decreased) or distributive(when the oxygen extraction ratio is decreased). In turn, the decline in oxygen delivery can be caused by two distinct etiologies. First, a decrease in flow, originated by either a decrease in circulatory volume (relative or absolute hypovolemia) or a failure of the heart as a pump can cause a decrease in oxygen delivery. [2] Secondly, a drop in the arterial oxygen content, as occurs in an acute respiratory disorder, will decrease oxygen delivery to the tissues. Absolute hypovolemia arises from severe dehydration or hemorrhage. Septic or endotoxemic shock are forms of distributive shock, in which relative hypovolemia occurs due to the loss of vascular tone and leaky vasculature. Thus, relative hypovolemia is associated with altered redistribution of blood flow among and within organs.[2] In addition, absolute hypovolemia due to diarrhea, gastric reflux, volume loss by drains or sequestration also occurs in septic or endotoxic patients. [4]

It is important to keep in mind that shock can result from a mixture of quantitative and distributive features and absolute and relative hypovolemia. [2] Moreover, in septic or endotoxic patients, due to persistent microvascular dysfunction, organ dysfunction can occur in the absence of global hypoperfusion and despite volume repletion and hemodynamic normalization.[5]

In adult horses, sepsis or endotoxemia is originated by a number of different etiologies, more commonly associated with intestinal diseases such as colitis, enteritis or strangulation.[6,7] The mortality rate in horses with strangulating lesions remains high. [8] The main cause for the high mortality is due to the severe systemic inflammation caused by absorption of endotoxin and bacteria through the compromised gut – blood barrier and associated hemodynamic alterations. [8] Moreover the gastrointestinal tract is one of the first organs affected by global hypoperfusion and postulated to be the primary trigger for multiple organ dysfunction. [9]

RESUSCITATION

The main goal in critically ill patients is to resuscitate early and aggressively with the aim of maintaining a reasonable mean arterial pressure and cardiac output. Thus the goals of fluid resuscitation are to restore oxygen delivery, improve tissue oxygen level, maintain an aerobic metabolism and preserve organ system perfusion and function. [1] The ultimate aim is to prevent organ dysfunction and enhance survival, by stabilizing macro-hemodynamics and exerting beneficial effects on the microcirculation.

Resuscitation therapy should be guided by parameters, measured at different time points during the resuscitation process, which reflect adequacy of tissue and organ perfusion. [1] Some of the parameters that are most commonly used in veterinary medicine are mentation status, urine output, heart rate, capillary refill time, rectal and extremities temperature, mean arterial blood pressure, blood gases, central venous oxygen saturation, lactate concentration, serum chemistry profile and coagulation profile.

FLUID RESUSCITATION

Fluid resuscitation is a mainstay for management of critically ill patients.[10] The major question regarding fluid therapy in critically ill patients has been what solution to use: crystalloids, colloids (which colloids), hypertonic saline or combinations; and which of these regimens is best in terms of safety and effectiveness.[11-14] Despite years of human and animal research, the optimal fluid for resuscitation in a given clinical situation remains unanswered in both horses and humans.[11,15,16] Moreover, opinions on resuscitation strategies in humans patients are also unsettled and vary from "running the patient dry" (hypovolemic) to supra-physiologic resuscitation strategies (hypervolemic) and from "scoop and run" to "stay and play" pre-hospital resuscitation techniques.[17,18]

In a recent human meta-analysis, evidence of improved survival in critically ill patients treated with crystalloid or colloids fluid resuscitation was not found. [10]

Crystalloids

The most commonly type of fluid therapy used for resuscitation in humans and equids are polyionic isotonic crystalloid fluids (Lactated Ringers and 0.9 % sodium chloride).[1,7]

Resuscitation of endotoxemic horses requires large volumes of crystalloids (i.e. 80 ml/kg). [19] Evidence supports that large volumes of crystalloid fluids are associated with cardiac and pulmonary complications, gastrointestinal dysmotility and bacterial translocation, coagulation disturbances, and immunological and inflammatory mediator disorders.[17] Homeostasis of cellular volume is paramount for normal cell function, thus the importance of tight regulatory mechanisms. Cellular swelling activates phospholipase A_2

resulting in the increased production of prostaglandins, leukotrienes and lipooxygenases.[17] Moreover increased macrophage volume leads to the production of tumor necrosis factor α (TNF-α), an important inflammatory mediator of sepsis.[17] TNF-α is elevated in approximately 20 % of the horses with colic and it has been associated with high mortality.[8]

Resuscitation of endotoxic horses with 60 ml/kg of crystalloids was associated with a marked increase in the central venous pressure (peaking at 37 mmHg), an indication of volume overload.[20] Therefore, more conservative crystalloid fluid resuscitation protocols or small volume resuscitation might be a better way to manage horses in need of resuscitation.

Prompt resuscitation is paramount in critically ill patients. The large volume of crystalloids recommended for resuscitation of horses with shock would be very difficult to administer in a timely manner. Time is crucial in horses with intestinal strangulation since delaying surgery worsens the prognosis. [21] These horses can also exhibit intractable pain, thus making the administration of large volumes of crystalloid difficult and often impossible.

Small Volume Resuscitation

Small volume resuscitation is defined as a rapid infusion of 7.2 – 7.5 % hypertonic saline (HS), at a dose between 3 to 6 ml/kg, with or without addition of colloids.[5,22,23] In horses, most commonly dosages of 5 ml/kg of HS and 5 to 10 ml/kg of hydroxyethyl starch (HES) are used. In order to prolong the intravascular volume effects of HS, it is combined with a colloid containing high water binding capacity (i.e. HES).[5,22] This combination has synergistic effects by increasing plasma osmolality (due to mobilization of intracellular water) and providing high plasma colloid osmotic pressure for conservation of volume expansion effects.[5,22,24]

During shock states due to the release of inflammatory mediators and dysfunction of the sodium – potassium pump, erythrocytes and endothelial cells become edematous. [24] The main sources of volume expansion for HS are the red blood cells and endothelial cells, which lose 8 % of their volume to the intravascular compartment. [24] HS also increases blood flow by reducing vascular resistance due to arteriolar vasodilatation. The beneficial hemodynamic effects of HS are two fold: intravascular volume expansion and improvement of the microcirculation.

More recently the inmmunomodulatory effects of small volume resuscitation have being recognized. HS increases levels of antiinflammatory cytokines (Interleukin 1 receptor antagonist and Interleukin 10), these effects last over 24 hours, far beyond the rise in plasma osmolality. [25] HS and HES are associated with a decrease in inflammatory cytokines and attenuation of endothelial adhesion molecules expression, when compared to crystalloids fluids. [17]

Small volume resuscitation simultaneously allows for rapid improvement of systemic hemodynamics with less administered fluids and decreases the post-injury edema at the microcirculation level, thus improving tissue perfusion. Moreover, the antiinflammatory effects could provide beneficial effects beyond hemodynamic improvement.

REFERENCES

[1] Hollenberg SM, Ahrens TS, Annane D, et al. Practice parameters for hemodynamic support of sepsis in adult patients: 2004 update. **Crit Care Med** 2004;32:1928-1948. [2] Vallet B. Resuscitation from Circulatory Shock. In: Fink M, et al, ed. **Textbook of Critical Care**, Fifth edElsevier Saunders; 2005:905-910. [3] Marino P. Tissue Oxygenation. In: Marino P, ed. **The ICU Book**, Third ed. Philadelphia: Lippincoott Williams & Wilkins; 2007:193-207. [4] Marx G. Fluid therapy in sepsis with capillary leakage. **Eur J Anaesthesiol** 2003;20:429-442. [5] Oliveira RP, Velasco I, Soriano F, et al. Clinical review: Hypertonic saline resuscitation in sepsis. **Crit Care** 2002;6:418-423. [6] Roy MF. Sepsis in adults and foals. **Vet Clin North Am Equine Pract** 2004;20:41-61. [7] Shuster R, Traub-Dargatz J, Baxter G. Survey of diplomates of the American College of Veterinary Internal Medicine and the American College of Veterinary Surgeons regarding clinical aspects and treatment of endotoxemia in horses. **J Am Vet Med Assoc** 1997;210:87-92. [8] Weiss DJ, Evanson OA. Evaluation of activated neutrophils in the blood of horses with colic. **Am J Vet Res** 2003;64:1364-1368. [9] Barros JM, do Nascimento P, Jr., Marinello JL, et al. The Effects of 6% Hydroxyethyl Starch-Hypertonic Saline in Resuscitation of Dogs with Hemorrhagic Shock. **Anesth Analg**. [10] Perel P, Roberts I. Colloids versus crystalloids for fluid resuscitation in critically ill patients. **Cochrane Database Syst Rev** 2007:CD000567. [11] Choi PT, Yip G, Quinonez LG, et al. Crystalloids vs. colloids in fluid resuscitation: a systematic review. **Crit Care Med** 1999;27:200-210. [12] Boldt J, Mueller M, Menges T, et al. Influence of different volume therapy regimens on regulators of the circulation in the critically ill. **Br J Anaesth** 1996;77:480-487. [13] Khandelwal P, Bohn D, Carcillo JA, et al. Pro/con clinical debate: do colloids have advantages over crystalloids in paediatric sepsis? **Crit Care** 2002;6:286-288. [14] Oliveira RP, Weingartner R, Ribas EO, et al. Acute haemodynamic effects of a hypertonic saline/dextran solution in stable patients with severe sepsis. **Intensive Care Med** 2002;28:1574-1581. [15] Bedenice D. Evidence-based medicine in equine critical care. **Vet Clin North Am Equine Pract** 2007;23:293-316. [16] Marik PE, Iglesias J. Would the colloid detractors please sit down! **Crit Care Med** 2000;28:2652-2654. [17] Cotton BA, Guy JS, Morris JA, Jr., et al. The cellular, metabolic, and systemic consequences of aggressive fluid resuscitation strategies. **Shock** 2006;26:115-121. [18] Cotton BA, Jerome R, Collier BR, et al. Guidelines for prehospital fluid resuscitation in the injured patient. **J Trauma** 2009;67:389-402. [19] McFarlane D. Hetastarch: A Synthetic Colloid with Potential in Equine Patients. **Compendium** 1999;21:867-877. [20] Pantaleon LG, Furr MO, McKenzie HC, 2nd, et al. Cardiovascular and pulmonary effects of hetastarch plus hypertonic saline solutions during experimental endotoxemia in anesthetized horses. **J Vet Intern Med** 2006;20:1422-1428. [21] van der Linden MA, Laffont CM, Sloet van Oldruitenborgh-Oosterbaan MM. Prognosis in equine medical and surgical colic. **J Vet Intern Med** 2003;17:343-348. [22] Kreimeier U, Thiel M, Peter K, et al. Small-volume hyperosmolar resuscitation. **Acta Anaesthesiol Scand Suppl** 1997;111:302-306. [23] Oi Y, Aneman A, Svensson M, et al. Hypertonic saline-dextran improves intestinal perfusion and survival in porcine endotoxin shock. **Crit Care Med** 2000;28:2843-2850. [24] Rocha-e-Silva M, Poli de Figueiredo LF. Small volume hypertonic resuscitation of circulatory shock. **Clinics** 2005;60:159-172. [25] Duchesne JC, McSwain NE, Jr., Cotton BA, et al. Damage control resuscitation: the new face of damage control. **J Trauma** 69:976-990.

SODIUM DISORDERS IN LARGE ANIMAL PATIENTS

Allison Stewart BVSc (Hons), MS, DACVIM (Large Animal Internal Medicine), DACVECC
Auburn, AL

INTRODUCTION

Severe hypo- and hypernatremia or the consequences of inappropriate correction can result in neurologic manifestations which are not uncommonly fatal. Homeostatic mechanisms are designed to keep serum sodium concentrations [Na^+] within a narrow range as [Na^+] strongly influences extracellular fluid (ECF) tonicity and regulates water balance.

WATER HOMEOSTASIS

Total body water (TBW) is ~60% of body weight (0.6 x BW) in adult horses (79% in foals), and is distributed between the intracellular compartment (2/3) and the extracellular compartment (1/3). The extracellular fluid (ECF) is divided between the interstitial fluid compartment (IFC) (3/4) and the plasma compartment (1/4). Concentration of body fluid solutes, as reflected by the ECF [Na+] is tightly regulated. Precise control is achieved by maintenance of water balance, which compensates for vast fluctuations in daily solute and water intake. Water intake is determined by availability and regulated by thirst. Water excretion is controlled by vasopressin (antidiuretic hormone [ADH]) and its target tissue- the renal collecting ducts.

Osmolality is a physical property of the solution, and is determined by the concentrations (number of particles) of all ECF solutes, both permeant and impermeant. Particles that cannot cross cellular membranes generate ***osmotic pressure.*** Lipid soluble solutes that are regularly present in body fluids, most notably urea, permeate cell membranes as freely as water. These solutes contribute to osmolality but have no impact on tonicity. ***Tonicity*** is the ***effective osmolality*** of a solution, and is determined by osmotically active solutes that are impermeable across cellular membranes. Only solutes that cannot cross the membrane separating two compartments generate an ***effective osmotic pressure***. Water flows from the more dilute (hypotonic) to the more concentrated (hypertonic) solution. Tonicity determines the flux of water movement. Clinical conditions associated with ***hyper- or hypo tonicity*** often have neurological sequelae (especially after rapid correction), whereas those associated with ***hyperosmolality*** in the absence of hypertonicity are rarely associated with neurological signs, therefore the distinction is important.

Plasma osmolality (mOsm/kg water) = (2 x [Na^+] in mEq/L) + ([BUN] in mg/dL/2.8) + ([glucose] in mg/dL/18)

Role of thirst

Hypertonicity is the most potent stimulus for thirst. If the serum osmolality rises 2-3% above normal, osmoreceptors are stimulated, thirst is appreciated and ADH is released. Tonicity sensors are located in the anterior hypothalamus close to the supraoptic nuclei. A second neuronal area important in controlling osmolarity and ADH secretion is located along the anteroventral region of the **third ventricle (AV3V).** The AV3V is external to the blood brain barrier and therefore diffusion of most electrolytes into the interstitium of the region readily occurs, resulting in rapid osmoreceptor response. Sustained hypertonicity and hypernatremia occur only when thirst or access to water is impaired.

Role of vasopressin and the kidney

Vasopressin is synthesized in the hypothalamus and released from the posterior pituitary gland in response to increased tonicity and shrinkage of osmoreceptor cells or decreased effective arterial blood volume. Sodium and its accompanying anions largely determine ECF tonicity, therefore sodium is the primary stimulus for vasopressin release in a hyperosmolar state. Glucose promotes ADH release only with diabetes mellitus.

Vasopressin acts on the principal cells of the collecting ducts binding to V2 receptors on the basolateral membrane. This results in insertion of aquaporin 2 on the luminal membrane, increasing water permeability and water reabsorption, resulting in excretion of concentrated urine. In high concentrations, vasopressin causes vasoconstriction and elevates blood pressure. The opposite occurs when ECF becomes hypo-osmotic. When vasopressin is absent, collecting ducts remain impermeable to water and dilute urine (<100mOsmol/kg) is excreted. Kidneys can dilute or concentrate urine such that osmolality ranges from 50 to 1200 mOsmol/kg H_2O.

CLINICAL DISORDERS OF WATER HOMEOSTASIS

Hyponatremia is caused by an increase in water relative to Na^+ (ECF water excess), and hypernatremia results from a decrease in water relative to Na^+ (ECF water deficit). In normal animals, NaCl excess transiently increases tonicity; the resulting stimulation of thirst and ADH secretion causes water retention, and prevents sustained hypernatremia. Inhibition of ADH secretion with secondary increase in water excretion prevents sustained hyponatremia.

HYPONATREMIA

Hyponatremia (<130mEq/L) is usually associated with hypo-osmolality, but plasma osmolality can be normal or high. Hyponatremic patients may have decreased, normal or increased total body Na.

Translocational hyponatremia occurs when water moves from the intracellular space to the extracellular space in response to an osmotically active solute confined within the extracellular compartment. Hyperglycemia is the most common cause. There is no change in TBW. Although the serum [Na^+] is reduced, serum osmolality is increased, resulting in cellular dehydration. For each 100 mg/dL increase in serum glucose, the serum Na^+ will decrease: ***Corrected serum*** [Na^+] = [Na^+] + ***1.6 ([serum [glucose] – 100] / 100)***
Infusion of sodium free isotonic solutions should cease. Hyperglycemia should be controlled with insulin.

Hypovolemic hyponatremia occurs when there is both total body Na^+ and water deficits, with Na^+ deficit exceeding the water deficit. There is a nonosmotic release of vasopressin in response to hypovolemia. Urine [Na^+] can be used to differentiate extrarenal (urine [Na^+] < 20 mEq/L: appropriate renal salt conservation in response to volume contraction) and renal (urine [Na^+] > 20 mEq/L) losses. When volume depletion is of extrarenal origin (secretory diarrhea, ileus, reflux, sweat, blood loss and 3rd space sequestration (peritonitis, uroadomen)) and the excessive loss of water and solute is accompanied by free water or hypotonic fluid intake, hyponatremia results. Renal loss of Na^+ and water include the use of diuretics, osmotic diuresis (glucosuria, mannitol or urea), bicarbonaturia (renal tubular acidosis) and ketonuria. Fluid losses are hypotonic compared to the ECF, but replacement with relatively more hypotonic fluid results in increased thirst and nonosmotic release of vasopressin which accompanies ECF volume depletion.

Euvolemic/isovolemic hyponatremia is the most common dysnatremia in hospitalized humans. These patients have increased TBW, but normal total body Na^+. Causes include drugs (oxytocin, cyclophosphamide, NSAIDs, fluphenazine, etc), hypothyroidism, glucocorticoid deficiency and stress. The most common cause is the syndrome of inappropriate antidiuretic hormone (SIADH) release, where the serum [Na^+] falls as the kidneys inappropriately retain water. The patient's normal total body Na^+ is diluted by the retained water. There is an impaired suppression of vasopressin release relative to the degree of hypotonicity. Therefore, there is increased ADH secretion or increased renal response to ADH. Patients have impaired free water excretion, with normal excretion of Na^+. Urine is inappropriately concentrated (>100mOsmol/kg), urine [Na^+] is elevated (>20-30mEq/L), yet plasma osmolality is <270 mOsmol/kg. SIADH is diagnosed after ruling out diuretics, pituitary, adrenal, thyroid disease or renal insufficiency.

Hypervolemic hyponatremia occurs when total body Na^+ is increased but TBW is increased to an even greater degree, causing hyponatremia and edema formation. These syndromes include congestive heart failure (CHF), cirrhosis and acute and chronic renal failure. They all are associated with impaired water and salt excretion. In CHF, decreased effective arterial blood volume activates aortic and carotid baroreceptors leading to vasopressin release. Water excretion is also inhibited by actions of the renin-angiotensin-aldosterone system and the sympathetic nervous system. There is low cardiac output, reduced glomerular filtration rate, increased angiotension II production, thirst stimulation and further development of hyponatremia.

Cerebral Effects of Chronic Hyponatremia

It is crucial the brain be protected against changes in plasma tonicity as an increase in brain water of >10% is fatal. Brain adaption occurs within minutes after the development of hypotonicity. Water gain causes swelling and decreased brain osmolality. Increased hydrostatic pressure in the interstitial compartment of the brain immediately forces Na^+ containing ECF into the CSF. Movement of water out of the brain occurs within minutes and limits the change in brain water content to less than expected based on osmotic calculations alone. Potassium also moves out of cells within 24h of hypotonicity. Normalization of brain volume is complete within a few days through loss of organic osmolytes (idiogenic osmoles such as amino acids (taurine, glutamate, glutamine), phosphocreatine and myoinositol) from brain cells. Low brain osmolality persists despite normalization of brain volume. Careful (slow) correction of serum hypotonicity re-establishes normal osmolality without risking brain damage. Aggressive correction of hyponatremia that proceeds faster than the recovery of lost organic osmolytes leads to irreversible brain damage (osmotic demyelination) as the brain is now hypotonic compared to the ECF, resulting in fluid loss and brain shrinkage.

Clinical Signs of Hyponatremia

The symptoms of early hyponatremia (>125mEq/L) in humans are often nonspecific, possibly causing nausea and vomiting. In humans, once [Na^+] is <125mEq/L progressive cerebral edema occurs with headache, lethargy, ataxia, seizures, coma and death. If untreated, mortality rates in humans patients with severe hyponatremia approach 50%, with severe cerebral edema resulting in tentor-herniation. Cases of hyponatremia in foals will be presented.[1-3]

Treatment of Hyponatremia. (Careful monitoring is ESSENTIAL)

Symptomatic Hyponatremia		**Asymptomatic Hyponatremia**
Acute (duration < 48 hr)	Chronic (duration > 48 hrs, or unknown)	
Emergency correction until clinical signs relieved	Limited immediate correction during the initial few hours, then very slow final correction over several days	Chronic (>48 hrs) No immediate correction required
3% hypertonic saline: 1-2 ml/kg/h with furosemide	0.9% saline or 3% hypertonic saline: 1-2 ml/kg/h with furosemide. Commence water restriction when [Na^+] increases by 10%, or when patient is asymptomatic	
	Risk of osmotic demyelination with too rapid correction, frequent monitoring of serum and urine electrolytes recommended	**Do not raise serum [Na^+] by >0.5 mEq/L/hr or > 12 mEq/L within 24 hrs.**
Do not raise serum [Na^+] by >1-2 mEq/L/hr	Long term management: identify and correct predisposing etiologies. In humans demeclocycline, urea or V_2 receptor antagonists are used	

For severe acute (< 48 hrs) symptomatic hyponatremia the sodium required should be calculated to estimate replacement needs:
Sodium required (mEq) = TBW x (125 –serum [Na$^+$]) where ***TBW (L) = BW (kg) x 0.6***

One may estimate the anticipated change in serum [Na^+] after IV of 1 L of x% administered hypertonic saline using the following formula: ***Change in serum sodium concentration = (infusate [Na$^+$] – serum [Na$^+$])/ TBW + 1***

(Infusate concentration: 0.9% normal saline = 154 mEq/L, 3% hypertonic saline = 513 mEq/L, 7.5% hypertonic saline = 1232 mEq/L) Note: concentration of NaCl to be used is VERY controversial. Some clinicians recommend normal saline, and believe hypertonic saline is contraindicated. With ECF volume contraction (dehydration) then isotonic saline is recommended to expand ECF volume, decrease vasopressin release and help correct hyponatremia. However, hypertonic saline allows administration of smaller volumes. If serum [Na^+] is >125mEq/L, then 50% of the deficit can be safely administered over 24 hrs, with the remainder over the next 24-48 hrs. For chronic severe hyponatremia DO NOT raise serum [Na^+]>0.5mEq/L/hr or >12 mEq/L/day. This figure is obtained from clinical experience and the fact that cerebral water increases by about 10% during severe hyponatremia, therefore it is safe to increase serum [Na^+] by 10% (~10 mEq/L). Serum [Na^+] should initially be monitored every 2-4h, than every 4–8h.

With severe (symptomatic or asymptomatic) hyponatremia, it is tempting to fully correct the dysnatremia, but even Normosol R™ (with [Na^+]=140mEq/L), at 60-120mls/kg/day in a dehydrated animal, can dramatically raise [Na^+] over a 24 hrs, sometimes with fatal results. It is unnecessary to raise serum [Na^+] past the lower limit of normality (132 mEq/L). Initial rapid, but carefully monitored partial correction of [Na^+] is warranted in patients with acute, symptomatic hyponatremia, as the risk of cerebral edema outweighs the risk of iatrogenic osmotic demyelination. Furosemide administration enhances free water excretion and prevents volume expansion from the administered NaCl. Urine output and urine [Na^+] should be monitored. If a patient is excreting large quantities of hypotonic urine, and once the desired serum [Na^+] is obtained, hypotonic fluids may be required to prevent further increase in [Na^+].

Central Pontine Myelinolysis (osmotic demyelination)

Overly rapid correction of chronic hyponatremia can induce neurologic complications, including central pontine and extrapontine myelinolysis from osmotic demyelination. There is gradual onset of neurologic alterations (behavioral changes, cranial nerve palsies, mouth champing, ear twitching, quadriplegia, seizures, coma and death) after 1-6 days. During chronic hyponatremia the brain limits swelling by extruding electrolytes and organic osmolytes. Following rapid correction there is an intracellular to extracellular shift in water resulting in cellular dehydration, with shrinkage of endothelial cells and disruption of tight junctions. Circulating immune mediators cause oligodendroglial injury and demyelination.

In humans MRI is diagnostic, but typical lesions may take 2 weeks to be seen. Treatment involves lowing serum [Na^+] to within 12mEq/L of the initial precorrection concentration. Glucocorticoids regulate BBB permeability and prevent BBB disruption induced by hypertension and hyperosmolality. Humans and dogs have survived after several weeks of intensive care.

HYPERNATREMIA

Hypernatremia (serum [Na^+]>145 mEq/L) reflects water deficit relative to total body [Na^+] resulting in hypertonic serum. Total body Na^+ may be increased, decreased, or normal. Hypernatremia develops with thirst or renal concentration malfunction; with water restriction; or iatrogenically after hypertonic or long-term replacement fluids. Classification requires volume status assessment.

Hypovolemic hypernatremia reflects loss of sodium and water, but relatively more water, with signs of volume depletion due to hypotonic water losses from the kidney (urine [Na^+] ≤20mEq) or extrarenal sources such as the GIT (urine [Na^+]>20mEq/L).

Euvolemic/isovolemic hypernatremia reflects water losses accompanied by inadequate water intake, and no change in total body [Na^+]. Water losses can be extrarenal (respiratory tract or skin) which results in elevated urine osmolality, or renal from decreased vasopressin production or collecting tubule response. Urine [Na^+] is variable depending on water intake. Hypodispsia can occur in animals with a neurologic deficit that impairs thirst perception. Isovolemic hypernatremia is typically seen in patients with diabetes insipidus (DI). The relative water deficit should be calculated to guide treatment, and sodium restriction is warranted. 5% dextrose should be administered to balance urine losses. Treatment of central DI includes an ADH analog (desmopressin acetate).

Hypervolemic hypernatremia usually results from hypertonic saline or bicarbonate administration, leading to increased total body [Na^+] that exceeds increases in TBW. Patients may have concurrent ongoing losses of hypotonic body fluids (GI losses). Neonatal foals are designed to be efficient at Na conservation, and if given large volumes of replacement fluids over several days, they often become hypernatremic and volume overloaded. A normal foal drinking 10 to 20% of its body weight in milk ingests only 0.9 to 2.4 mEq/kg/d of sodium, based on mare's milk [Na^+] of 9 to 12 mEq/L. Foals are therefore designed to thrive on a low-Na diet. Administering a mix of isotonic fluids diluted with 5% dextrose can prevent development of hypernatremia. Treatment of hypervolemic hypernatremia requires Na^+ restriction and if severe edema develops loop diuretics.

Clinical manifestation of hypernatremia

Similar to hyponatremia, clinical signs can be nonspecific, and most are related to changes in serum osmolality in the CNS. Signs include lethargy, restlessness, thirst, muscle irritability, spasticity, hyperreflexia, seizures, coma and death.[2,3] Mortality in humans is 40-70%, with acute serum [Na^+] >160 mEq/l associated with a 75% mortality rate. Mortality rate is often due to the underlying disease process or inappropriate treatment. Treatment decisions depend on the characterization of hypernatremia (hypo-hyper- or isovolemic), its magnitude, duration and severity of symptoms. Hypernatremia and the attendant hypertonicity cause water extraction from the brain resulting in its shrinkage, which can induce vascular rupture with cerebral bleeding and subarachnoid hemorrhages.

Treatment of hypernatremia

If hypovolemic, isotonic 0.9% saline can be used to correct the hemodynamic instability. Otherwise, 5 % dextrose or hypotonic (0.45%) saline can be used for mild hypernatremia. Symptomatic or severely hypernatremic patients should have their serum [Na^+] monitored every 2 to 4 hrs. Sometimes fluid [Na^+] concentrations need to be elevated to equal that of the patient's serum [Na^+] concentration to allow rehydration without rapid changes in serum [Na^+]. For acute hypernatremia, rapid correction is usually well tolerated, however serum [Na^+] should not fall >1-2 mEq/L/hr. For chronic hypernatremia serum [Na^+] should not fall >0.5 mEq/L/hr (or <10-12 mEq/L/d) to a target of 145mEq/L. This requires careful administration of hypertonic fluids. An estimate of relative water deficit associated with hypernatremia is: ***Water deficit (in liters) = TBW x ([serum*** $[Na^+]$ ***/reference [Na+] – 1).*** Half of the water deficit should be replaced over the first 24 hrs, and remainder over the next 24 to 72h. The anticipated change in serum $[Na^+]$ after IV infusion of 1L of any fluid is: ***Change in serum*** $[Na^+]$ ***= (infusate*** $[Na^+]$ ***– serum***$[Na^+]$***)/ (TBW + 1)***

Effects of Hypernatremia and Correction on the Brain

Within minutes, loss of water from brain cells causes increased osmolality and cellular shrinkage. Partial restitution of brain volume occurs within hours as electrolytes enter (rapid adaptation). Brain volume normalization is complete within several days due to intracellular accumulation of organic osmolytes (slow adaptation). With reversal of systemic hypernatremia, it takes ~2-3 days for removal of ***idiogenic osmoles*** from the brain. Intracellular brain tissue is hypertonic relative to the ECF and water rushes down its concentration gradient resulting in cerebral edema. Salt poisoning of pigs fed a high sodium diet occurs after water restriction results in hypernatremia, and return of water supply induces acute cerebral edema.

REFERENCES

[1]Lakritz et al. J Am Vet Med Assoc 1992:200(8):1114. [2]Wong DM et al. J Vet Emerg Crit Care 2007;17(3):275 [3]Arroyo LG et al. Can Vet J 2008;49(5):494. [4]Angelos et al. J Am Vet Med Assoc 1999;214(9):1364. [5]Heath et al J Am Vet Med Assoc 1995;207(6):738.

EQUINE PROTOZOAL MYELOENCEPHALITIS: DIAGNOSTIC TESTING

Frank M. Andrews, DVM, MS, DACVIM
Baton Rouge, LA

INTRODUCTION

Equine Protozoal Myeloencephalitis (EPM) is a debilitating neurologic disease of horses that has focal or multifocal distribution involving the spinal cord, brain, and brainstem[1]. The primary cause of EPM is *Sarcocystis neurona* (SN), although *Neospora hughesi* (NH), a related protozoan parasite, has been isolated from horses with EPM [2,3]. Clinical signs depend on the number of organisms and their location within the nervous system.

The life cycle of SN involves the definitive host, the North American opossum (*Didelphis virginiana*) and the South American opossum (*Didelphis albiventris*), both of which eat skeletal muscle from intermediate hosts (i.e. striped skunk, raccoon, nine-banded armadillo, and cat)[2]. The parasite is found in the sarcocyst stage in skeletal muscle of these intermediate hosts and the opossum must ingest this muscle to complete the life cycle. A recent report from Michigan State University, College of Veterinary Medicine, identified SN sarcocysts in the tongue of a 4-month-old cross-bred colt that was euthanized due to severe neurologic signs[4]. The characteristic protozoal organisms were found in the brain and spinal cord of this colt. The presence of sarcocysts in the tongue muscle of this colt suggests that horses have the potential to be an intermediate host for SN. However, additional studies need to be completed to determine the frequency of occurrence. It is important to remember that infected horses do not shed the SN and cannot transmit the parasite to healthy horses.

Clinical signs depend on lesion localization and there is no breed, age, or sex predilection. However, in a recent retrospective study, Thoroughbreds, Standardbreds and Tennessee Walking Horses were found to have a higher proportional morbidity rate compared to ponies and draft breeds[5]. Also, in that same study, horses from 5-15 years of age were most affected. In another study, adult horses (mean age 6.5 years) were most commonly affected.[6] Clinical signs may be triggered by a stressful event such as transport, pregnancy, other diseases, or administration of corticosteroids. Signs may be subtle or severe, and the duration of onset may be acute or chronic. Most horses suffer from chronic insidious onset neurologic disease, which often starts as an obscure lameness that does not respond to peripheral nerve blocks. Signs slowly progress to neurologic gait deficits, such as weakness, ataxia, and spasticity. Head tilt or other cranial nerve signs occur in 12% of horses with EPM and muscle atrophy can occur in any muscle, however the gluteal muscle is most often (57%) affected when compared to other muscles[7].

Differential diagnoses should include trauma, Cervical Stenotic Myelopathy, Equine Herpes Virus I Myeloencephalitis, Polyneuritis Equi (Cauda equina neuritis), Equine Motor Neuron Disease (EMND), Degenerative Myelopathy, occipitoatlantoaxial (A/O) malformation and botulism. In horses that have acute onset of neurologic disease, Rabies, West Nile Virus, and Eastern or Western Encephalitis should also be considered[6].

DIAGNOSIS

The antemortem diagnosis of EPM is difficult due to nonspecific clinical signs and its similarity to other neurologic diseases and musculoskeletal lameness. Prior to 1993, the diagnosis of EPM was made based on the presence of asymmetric neurologic signs with multifocal distribution, ruling out other causes of neurologic disease, and post-mortem findings of mononuclear cells and/or the presence of protozoal organisms in CNS neurons. In 1993, the Western blot test (WBT) was developed for the detection of SN) specific antibodies to surface antigens (SAGs) in serum and cerebrospinal fluid (CSF), and was widely used with, and without a clinical examination to diagnose EPM[8]. Also, the WBT was used to screen "clinically" normal horses suspected of having subclinical disease or as part of the "prior to purchase examination." The presence of antibodies to SN in the CSF was considered positive for active disease and treatment was frequently based on the presence of specific SN antibodies alone. A report showed the presence of CSF antibodies to a foreign protein, ovalbumin, in horse CSF, sixty days after IM administration [9]. Also, CSF antibodies to SN have been found in horses following the administration of a killed SN whole merizoite vaccine[10]. This recent information on the presence of antibodies in CSF and serum in normal horses, has challenged our assumption that the presence of antibodies to EPM in CSF is correlated with clinical disease. Since the WBT, several other tests have been made available for veterinarians to diagnose EPM. These include Western blot tests with various modifications, an ELISA test using SnSAG 1, an indirect immunofluorescent test (IFAT), and a new unique modified ELISA using rSnSAG4/2 chimera (domain 1 and SnSAG4 fused to SnSAG2)[8,11,12]. However, all of these tests are based on antibody production by the horse in response to SnSAG proteins. Thus, presence of antibodies in the CSF to SnSAG antigens, in the absence of clinical disease, has brought into question the "validity" of an antibody-based test for the diagnosis of EPM.

Test Validity

Several criteria have been established to determine the validity of diagnostic tests, such as the immunodiagnostic testing for EPM. These criteria include the sensitivity and specificity of the test, its reliability (repeatability), and its yield (the amount of disease detected in the population)[13]. Also, a diagnostic test should be innocuous, rapid, inexpensive and simple to perform.. Furthermore, a valid diagnostic test should accurately identify which horses have disease and which ones do not have disease. In the context of EPM, sensitivity is defined as:

$$\frac{\text{No. of true positive}}{\text{No. of true positive + No. of false negatives}}$$

Specificity is defined as:

$$\frac{\text{No. of true negatives}}{\text{No. of true negatives + No. of false positives}}$$

Although, sensitivity and specificity give some idea about the test validity, clinicians need to know what the probability is that a horse with a positive immunodiagnostic test truly has EPM and the probability that a horse with a negative test does not have EPM. These probabilities are defined as the positive (PPV) and negative predictive value (NPV) and are based on the sensitivity and specificity of the test for the disease in question (see above), and disease prevalence (number of horses in a given area that have EPM at a given time). The lower the disease prevalence in an area, the greater the likelihood that a horse with a positive immunodiagnostic test will not truly have EPM. In a recent study, disease prevalence in horses with neurologic disease in California was calculated to be 23% (23% of horses with neurologic disease had EPM). Disease prevalence for EPM in horses with neurologic disease in the Midwest and Southern US (Tennessee, Ohio, and Florida) has been estimated to be 50% (50% of horses with neurologic disease have EPM). Thus, horses with a positive immunodiagnostic test in the Midwest or Southern US are more likely to have EPM than horses with a positive immunodiagnostic test in California. The following example illustrates how prevalence affects PPV and NPV. We consider a population of 100 horses with neurologic disease presented to each of these regions of the country. According to prevalence figures, 23 horses in California would have EPM and 50 horses in the Eastern US would have EPM (true positives). Utilizing the sensitivity and specificity of Western blot immunodiagnostic test, 89% and 89%[8], respectively, 20 of the 23 horses with EPM would test positive in California (Figure 1), whereas 45 of 50 horses would test positive in the Midwestern and Southern US (Figure 2).

Prevalence = 23%

	Test +	Test -	
EPM +	20	3	23
EPM -	7	70	77
	27	73	100

Figure 1.

Prevalence = 50%

	Test +	Test -	
EPM +	45	5	50
EPM +	5	45	50
	50	50	100

Figure 2.

There would be more false positive tests (horses with positive immunodiagnostic test but do not have EPM) in California (7/27, 25%) when compared to horses in the Midwest or Southern US (5/50, 10%). The PPV of a positive immunodiagnostic test in California would be 20/27 or 75%, whereas, the PPV of a positive immunodiagnostic test in the Midwest and Southern US would be 45/50 or 90%. In California, 3/4 horses with neurologic disease and a positive immunodiagnostic test would truly have EPM, whereas in the Midwest or Southern US 9/10 horses with neurologic disease and positive immunodiagnostic test would have EPM. Also, the NPV (the likelihood that a horse with a negative immunodiagnostic test does not have disease) of the immunodiagnostic test is 70/73 or 97% in California and 45/50 or 90% in the Midwest or Southern US. Regardless the region of the country, a negative immunodiagnostic test is strong evidence (≥ 90%) a horse does not have EPM.

It is reasonable to assume that as long as an antibody-based test is used for diagnosis of EPM there will be controversy surrounding the results. However, from the results above, regardless the diagnostic testing used, positive predictive values vary from 75% to 90% depending on the region of the country and prevalence of disease in that area. In response to the controversy of diagnostic testing the ACVIM developed a consensus statement in 2003 to address this very issue.[14] Several considerations are key to using and interpreting an antibody-based immunodiagnostic test of EPM and they include:

1. The immunodiagnostic test should be performed in conjunction with thorough clinical and neurologic examinations.
2. The immunodiagnostic test should not be performed on normal horses.
3. The immunodiagnostic test on CSF is slightly better than serum to diagnose EPM.
4. It is likely that a horse with a negative immunodiagnostic test does not have EPM (≥ 90% NPV).
5. A weak positive WBT result is considered equivocal and test should be repeated if EPM is suspected.
6. A test may be negative in acute disease, so repeating the immunodiagnostic test in 1 to 2 weeks may be helpful[15].

REFERENCES

1. Dubey JP, Lindsay DS, Saville WJA, et al. A review of *Sarcocystis neurona* and equine protozoal myeloencephalitis (EPM). Vet Parasitol 2001;95:89-131. 2. Cheadlem MA, Lindsay DS, Rowe S, et al. 1999. Prevalence of antibodies to *Neospora* sp. in horses from Alabama and characterization of an isolate recovered from a naturally infected horse. Int J Parasitol. 1999;29;1537–1543. 3. Marsh AE, Barr BC, Madigan J, et al 1996a. Neosporosis as a cause of equine protozoal myeloencephalitis. J Am Vet Med Assoc 1996;209:1907–1913. 4. Mullaney T, Murphy AJ, Kiupel M, et al. Evidence to support horses as natural intermediate hosts for *Sarcocystis neurona. Vet Parasitol* 2005;133:27–36. 5. Odoi A, Witonsky S, Andrews F. Are cases of Equine Protozoal Myeloencephalitis (EPM) in North America on the rise? In Proceedings 56th Annual Am Assoc Equine Practitnrs 2010;56:342. 6. Furr MO. Equine Protozoal Myeloencephalitis. In Equine Neurology. Eds: M. Furr and S. Reed. Blackwell Publishing, Ames, Iowa. 2008;195-212. 7. Mayhew IG, De Lahunta A, Whitlock RH, et al. 1977 Equine protozoal myeloencephalitis. Proc. Annu Meet Am Assoc Eq Pract 22:107-114. 8. Granstrom DE, Dubey JP, Davis SW, et al. 1993. Equine protozoal myeloencephalitis: antigen analysis of cultured *Sarcocystis neurona* merozoites. J Vet Diagn Invest. 1993;5:88–90. 9. Furr M. Antigen-specific antibodies in cerebrospinal fluid after intramuscular injection of ovalbunim in horses. J Vet Intern Med 2002;16:588-592. 10. Witonsky S, Morrow JK, Leger C, et al. *Sacrocystis neurona*-specific immunoglobulin G in the serum and cerebrospinal fluid of horses administered *S. neurona* vaccine. J Vet Intern Med 2004;18:98-103. 11. Yeargan MR, Howe DK. Modification of enzyme-linked immunosorbent assays based on Sarcocystis neurona surface antigens (SNSAGS) for improved equine antibody detection. 2010 EPM Society SIG, ACVIM Forum, Anaheim, CA. 12. Duarte PC, Daft BM, Conrad PA et al. 2003 Comparison of a serum indirect fluorescent antibody test with two Western blot tests for the diagnosis of equine protozoal myeloencephalitis. J Vet Diagn Invest 15:8-13. 13. Mausner JS, Bahn AK. Epidemiology: An introductory text. WB Saunders Publishing. 1985:237-263. 14. Furr M, MacKay R, Granstrom D, et al. Clinical diagnosis of equine protozoal myeloencephalitis (EPM). J Vet Intern Med 2002;16:618-621. 15. Johnson AL, Burton AJ, Sweeney RW. 2010 Utility of 2 immunological tests for antemortem diagnosis of equine protozoal myeloencephalitis (*Sarcocystis neurona*) infection in naturally occurring cases. J Vet Intern Med 1184-1189.

EQUINE PROTOZOAL MYELOENCEPHALITIS: TREATMENT OPTIONS

Frank M. Andrews, DVM, MS, DACVIM
Baton Rouge, LA

INTRODUCTION

Equine Protozoal Myeloencephalitis (EPM) is a debilitating neurologic disease of horses that has focal or multifocal distribution involving the spinal cord, brain, and brainstem.[1] The primary cause of EPM is *Sarcocystis neurona* (SN), although *Neospora hughesi* (NH), a related protozoan parasite, has also been isolated from horses with EPM[2,3]. Clinical signs depend on the number of organisms and their location within the nervous system.

The life cycle of SN involves the definitive host, the North American opossum (*Didelphis virginiana*) and the South American opossum (*Didelphis albiventris*), both of which eat skeletal muscle from intermediate hosts (i.e. striped skunk, raccoon, nine-banded armadillo, and cat)[2]. The parasite is found in the sarcocyst stage in skeletal muscle of these intermediate hosts and the opossum must ingest this skeletal muscle to complete the life cycle. A recent report from Michigan State University, College of Veterinary Medicine, identified SN sarcocysts in the tongue of a 4-month-old cross-bred colt that was euthanized due to severe neurologic signs[4]. The characteristic protozoal organisms were also was found in the brain and spinal cord of this colt. The presence of sarcocysts in the tongue muscle of this colt suggests that horses have the potential to be an intermediate host for SN, However, additional studies need to be completed to determine the frequency of occurrence. It is important to remember that infected horses do not shed parasite and cannot transmit the parasite to healthy horses.

Clinical signs depend on lesion localization and there is no breed, age, or sex predilection. However, in a recent retrospective study, Thoroughbreds, Standardbred and Tennessee Walking Horses were found to have a higher proportional morbidity rate compared to ponies and draft breeds.[5] Also, in that same study horses from age 2-15+ years of age were most affected[5]. In another study, adult horses (mean age 6.5 years) were most commonly affected. Clinical signs may be triggered by a stressful event such as transport, pregnancy, other diseases, or administration of corticosteroids. Signs may be subtle or severe and the duration of onset may be acute or chronic. Most horses suffer from a chronic insidious onset neurologic disease, which often starts as an obscure lameness, that does not respond to peripheral nerve blocks). Signs slowly progress to neurologic gait deficits, such as weakness, ataxia, and spasticity. Head tilt or other cranial nerve signs occur in 12% of horses with EPM and muscle atrophy can occur in any muscle, however the gluteal muscle is most often (57%) affected when compared to other muscles[6].

Diagnosis is based on the presence of clinical neurologic signs and a positive immunodiagnostic test on CSF or in some cases serum. Every effort should be made to rule out other neurologic diseases such as trauma, Cervical Stenotic Myelopathy, Equine Herpes Virus I Myeloencephalitis, Polyneuritis Equi (Cauda equina neuritis), Equine Motor Neuron Disease (EMND), Degenerative Myelopathy, ocipitoatlanoaxial (A/O) malformation and botulism. Also, in horses that have acute onset of neurologic disease Rabies, West Nile Virus, and Eastern or Western Encephalitis should also be considered[7]. An accurate diagnosis is the goal as treatment is expensive and may require a long duration of administration.

TREATMENT

Early and effective treatment of horses with EPM is essential to minimize residual neurologic deficits. Antiprotozoal agents are the cornerstone of treatment for EPM[7]. The principles of treatment for EPM include reduction of inflammation, reduction of brain and spinal cord edema and elimination of the causative agent. To accomplish these goals, non-specific and specific agents can be used in the treatment of EPM in horses.

Non-Specific Treatment

Dimethyl Sulfoxide (DMSO)

General medical therapy consists of administration of Dimethyl Sulfoxide (DMSO, 1 ml/kg [in a 10% solution or 500 ml in 5 liters] of saline, lactated Ringer's solution (LRS), Multisol®, or Normosol®), IV or PO, q12h). DMSO is an industrial solvent that reduces inflammation by scavenging free oxygen radicals. DMSO has also been shown to maintain blood flow to tissues by inhibiting prostaglandins and thromboxane and prevents platelet aggregation and vasoconstriction. DMSO may be administered with corticosteroids and is synergistic.

Corticosteroids

Corticosteroids may be administered to decrease inflammation. Dexamethasone (0.1 – 0.2 mg/kg, IV, q24h) or methyl prednisolone (1.0 – 2.0 mg/kg, IV, q24h) may be used aswell, but are controversial in horses with EPM!!! Horses with acute life-threatening neurologic signs may benefit from single dose corticosteroid administration. Corticosteroid administration to horse with chronic EPM may cause immunosuppression and potentiate spread of organisms throughout the nervous system. However, in a previous study, experimentally infected horses only showed a mild increase in clinical neurologic signs when corticosteroids were administered at typical anti-inflammatory doses. The potential side effects of corticosteroids include laminitis, immunosuppression, diarrhea, and adrenal insufficiency, and the efficacy of corticosteroids for treatment of inflammation in neurologic disease has been questioned. I believe corticosteroids have a place in treatment EPM by stabilizing neuronal cell membranes and suppressing the acute inflammation which lead to observed clinical signs. Corticosteroids, in high doses, should not be administered for more than 5 days in duration. However, if high doses are administered for longer durations, the horse should be withdrawn from the drug using a progressive halving of the dose over a 2 to 3 week period!!

Specific Treatment

Ponazuril (Toltrazuril Sulfone, Marquis®)

Specific medical treatment for EPM includes the FDA-approved antiprotozoal medication, Marquis® (ponazuril (toltrazuril sulfone), 5 mg/kg, PO, q24h, Bayer Animal Health, Shawnee Mission, KS). This pharmaceutical agent is a low molecular weight triazine compound approved for treatment of EPM. Marquis® exceeds minimal inhibitory concentrations for *S. neurona* in CSF and inhibits the respiratory chain enzymes of the plastinoid or apicoplast body and mitochondria. The apicoplast and mitochondria are important in organism reproduction so the drug is likely protozoalstatic. This drug has been shown to have 60% efficacy in the treatment of EPM. No major side effects have been noted with use of Ponazuril. A recent study was undertaken by Dr. R. J. MacKay at the University of Florida to study the effect of intermittent ponazuril administration on immunoconversion against *S. neurona* in horses. Horses immunoblot-negative for *S. neurona* IgG (5/group) were given 5 x 10^5 *S. neurona* sporocysts by nasogastric tube, then given ponazuril at 20 mg/kg every 7 or 14 d for 84 d. Control horses (5/group) were either not treated or not challenged. There was significant ($P < 0.05$) effect of treatment on immunoconversion (detected by immunoblot) in CSF but not in serum. All challenged horses became serum-positive within 56 d. Only 2/5 horses treated every 7 d had antibodies in CSF; all other *S neurona*-challenged horses became CSF-positive within 84 d. Magnitude of the anti-17-kD antibody response in CSF ("relative quantity CSF") was significantly reduced by 7-d but not by 14-d treatments. Consistent neurologic or other clinical abnormalities were not seen in any group. Likewise, histologic examination of CNS did not reveal either protozoa or consistent inflammatory changes. Although ponazuril treatment every 7 d reduced immunoconversion in CSF, this effect was not absolute. Protocols using ponazuril may have application to prevention of EPM.

The bioavailability of ponazuril was reported to be 32% in horses. The low bioavailability of ponazuril is because it is relatively insoluble in GI fluids and has limited contact with intestinal mucosa. The low bioavailability of ponazuril is likely the reason for the compound to require 8 days to reach steady state MIC. In a recent study, ponazuril (2.2 mg/kg, orally) was mixed with DMSO to improve its bioavailability[8]. In this study, DMSO increased the oral bioavailability of toltrazuril sulfone threefold, from 32% to 71%. This improved bioavailability of toltrazuril sufone would allow clinicians to administer a loading dose in acute cases of EPM or to reach MIC faster. The authors' this paper suggested that the toltrazuril sulfone in DMSO could be mixed with feed and given to horses without exceeding the FDA recommendations for levels of DMSO permissible in pharmacological preparations.

Diclazuril (Protozil®)

Diclazuril (Protozil®, Intervet Schering Plough Animal Health, Roseland, NJ), a low molecular weight lipid soluble triazine compound, was recently launched as a FDA-approved treatment for EPM in horses[9]. Protozil® (1 mg/kg, orally, Q24h, 28 days) is sold has a top-dress alfalfa-based pelleted formulation. The approval studies utilized 214 mares, stallions, and geldings (various breeds; 9.6 mos - 30 yrs of age) that were confirmed to have EPM based on neurologic examination, positive Western blot test (WBT) and ruling out other diseases such as cervical stenotic myelopathy, Equine Herpes Virus 1 myeloencephalopathy, and West Nile Virus encephalopathy. Diclazuril 1.56% antiprotozoal pellets were administered to horses as a top-dress on their daily grain ration at 1mg/kg, 5mg/kg, or 10mg/kg once daily for 28 days. Horses in all the treatment groups had no adverse reactions and there was no difference in treatment successes between groups, thus horses receiving 1mg/kg (n=42) were evaluated for efficacy. Clinical investigator success was defined as a horse that improved ≥1 grade on CNS exams performed on day 48 vs day 0. WBT success was defined as a horse with a (-) CSF WB by day 48, regardless of the horse's neurologic status. Based on the number horses that seroconverted to (-) CSF WB status by day 48, 28/42 horses (67%) were classified as successes. Only 2 adverse events were observed. Diclazuril pellets administered as a top-dress at 1 mg/kg BW once daily for 28 days were considered safe and effective treatment for EPM in naturally infected horses under field conditions. According to label, Protozil (1.56% diclazuril) is indicated for the treatment of EPM caused by *S. neurona* in horses.

Sulfadiazine/Pyrimethamine and NTZ

Other FDA-approved treatments exist including Re-Balance® (Phoenix Scientific) a folic acid antagonist, containing sulfadiazine and pyrimethamine and Navigator® (32% Nitazoxanide [NTZ] oral paste, IDEXX Pharmaceuticals, Inc.). Neither of these medications are currently available commercially.

Sulfadiazine/Pyrimethamine Compounded

The combination of sulfadiazine (20 mg/kg, PO q24h) and pyrimethamine (1 mg/kg, PO q24h) is still available for treatment of EPM. This combined antiprotozoal therapy is available through compounding pharmacies but is not FDA-approved. This combination requires at least 6 months of treatment. Toxicity included anemia, fetal loss, diarrhea, and fetal abnormalities. The advantage of this combination is the lower cost and can be used for 30 days in horses where the diagnosis is in question. There is some evidence that sulfadiazine/pyrimethamine are synergistic with the triazine compounds[7]. This is empirical, but may be used in horses that have recurrence or refractory disease[7].

Immunostimulants

Immunostimulants have been recommended as an ancillary treatment for hoses with EPM. Levamisole (1 mg/kg, orally, daily), Eqstim (5 ml IM on days 1, 3, and 7, then monthly), or Equimune IV (1.5 ml, IV, weekly for 3 weeks) have been advocated. However there is no evidence that these immune modulators have an effect on the course of EPM.

Prevention

Prevention of EPM consists primarily of management changes (keeping lids tightly on grain barrels, disposing of dead wildlife on premises, and keeping wildlife out of the feed areas). A provisionally approved, killed merozoite vaccine is no longer available (Ft. Dodge Animal Health, S. neurona EPM Vaccine®). The vaccine has been shown to stimulate serum neutralizing antibodies that block the infection of cell cultures in vitro. However, the ability to prevent clinical disease in the horse has not been demonstrated. A multi-center efficacy study failed to enroll enough horses that were vaccinated to determine the efficacy of the vaccine in prevention of EPM[10]. Some clinicians have advocated vaccination of horses on farms with a high incidence or prevalence of EPM and some veterinarians advocate the vaccination of horses with chronic recurrent EPM that responds poorly to treatment; however, the effectiveness of these practices have not been established.

REFERENCES

1. Dubey JP, Lindsay DS, Saville WJA, et al. A review of *Sarcocystis neurona* and equine protozoal myeloencephalitis (EPM). Vet Parasitol 2001;95:89-131. 2. Cheadlem MA, Lindsay DS, Rowe S, et al. 1999. Prevalence of antibodies to *Neospora* sp. in horses from Alabama and characterization of an isolate recovered from a naturally infected horse. Int J Parasitol. 1999;29:1537–1543. 3. Marsh AE, Barr BC, Madigan J, et al 1996a. Neosporosis as a cause of equine protozoal myeloencephalitis. J Am Vet Med Assoc 1996;209:1907–1913. 4. Mullaney T, Murphy AJ, Kiupel M, et al. Evidence to support horses as natural intermediate hosts for *Sarcocystis neurona. Vet Parasitol* 2005;133:27–36. 5. Odoi A, Witonsky S, Andrews F. Are cases of Equine Protozoal Myeloencephalitis (EPM) in North America on the rise? In Proceedings 56th Annual Am Assoc Equine Practitnrs 2010;56:342. 6. Mayhew IG, De Lahunta A, Whitlock RH, et al. 1977 Equine protozoal myeloencephalitis. Proc. Annu Meet Am Assoc Eq Pract 22:107-114. 7.Furr MO. Equine Protozoal Myeloencephalitis. In Equine Neurology. Eds: M. Furr and S. Reed. lackwell Publishing, Ames, Iowa. 2008;195-212. 8. Dirikolu L, Darpiesiuk W, Lehner AF, et al. 2008 Synthesis and detection of toltrazuril sulfone and its pharmacokinetics in horses following administration of dimethylsulfoxide. J Vet Pharmacol Therap 32:368-378. 9. Reed S, Andrews F, MacKay R, et al. Safety and efficacy of diclazuril 1.56% pellets for treatment of horses naturally infected with equine protozoal myeloencephalitis. Presented at the EPM Society Special Interest Group, ACVIM Forum Anaheim, CA 2010 (Abstract). 10. Cohen ND, MacKay RJ, Toby E, et al. 2007 A multicenter case-control study of risk factors for equine protozoal myeloencephalitis. J Am Vet Med Assoc 231:1857-1863.

LAMINITIS AS A CONSEQUENCE OF SEPSIS: PATHOPHYSIOLOGY

Andrew W. van Eps, BVSc, PhD, MACVSc, DACVIM
Queensland, Australia

INTRODUCTION

Laminitis is a common and frustrating complication of a variety of primary diseases in the horse. In recent years, researchers have come to realise that there are important differences in the pathophysiology of laminitis depending on the inciting cause. This has divided laminitis into essentially three forms, based upon the purported basic mechanisms of disease. Laminitis that occurs in horses with pituitary *pars intermedia* dysfunction and in horses with equine metabolic syndrome has been linked to insulin resistance/ hyperinsulinemia and derangements of glucose regulation[1-4]. Laminitis that occurs secondary to prolonged, continuous/excessive weight bearing (supporting limb laminitis) is widely regarded as being primarily due to reduced lamellar perfusion[5]. Laminitis secondary to diseases characterised by the systemic inflammatory response syndrome (SIRS) and in particular sepsis (SIRS caused by bacterial infection/.bacterial products), has been the most extensively studied form of laminitis, and perhaps has the most complex pathophysiology. Recent evidence suggests that this form of laminitis has much in common with the end organ dysfunction that occurs in human sepsis[6]. A similar, complex interaction of bioenergetic failure, microvascular dysfunction, inflammation and enzymatic degradation appears to contribute to lamellar failure. In septic horses, the severity (and clinical importance) of damage to the lamellae over other "end organs" is likely due to the addition of mechanical distraction (caused by weight bearing) in this tissue. Experimental laminitis models utilising alimentary overload with carbohydrate (starch or oligofructose) produce profound sepsis as a result of death and proliferation of different microbial populations in the hind gut and the liberation and absorption of microbial products through damaged intestinal mucosa[7-11]. Alimentary dosing with Black Walnut extract (BWE), although widely considered mechanistically unique, also causes profound local and systemic inflammation[12]. Much of the information gained from these experimental laminitis induction models is therefore relevant in the discussion of sepsis related laminitis here.

THE ACUTE LAMINITIS LESION

Laminitis is ultimately a separation of epidermis from dermis, and therefore hoof wall from distal phalanx. Prior to the last decade, researchers had focused on studying the lesion in horses with acute (clinically evident) laminitis. Attempting to identify the inciting events at this stage of the disease is akin to looking for a penny on the track in the aftermath of a train derailment. In more recent years, focus has shifted to studying events occurring much earlier in the development of laminitis, prior to any clinical signs and prior to much of the secondary damage that occurs as a result of mechanical distraction. A combination of cellular injury and extracellular matrix degradation has been identified early in the course of laminitis induction using experimental models that reproduce SIRS/sepsis.

The primary epidermal and dermal lamellae divide into secondary epidermal lamellae (SEL) and secondary dermal lamellae (SDL) respectively, interdigitating to form a normally very strong connection between the hoof wall and the distal phalanx. At the interface between the dermis and the epidermal basal cells (EBCs) lies the basement membrane (BM)[13]. Lysis of the BM, as well as its detachment from the underlying EBCs is a key event in the development of laminitis[14]. The disappearance of the BM from around the bases of the SELs, and its detachment from the tips of the SELs was first observed using light microscopy during the acute phase of laminitis[14]. This degradation and detachment of the BM, however, was considered only a consequence of epidermal cellular damage by some authors[15,16], due to the detection of other cellular pathology preceding any detectable BM damage. Recently, the use of more sensitive techniques has enabled lamellar BM changes to be detected far earlier in the developmental period of experimentally induced laminitis (only 12 h after alimentary oligofructose dosing)[17]. This study showed evidence of BM disruption (loss of immunostaining for collagen type IV and laminin-332 in the region of the BM) approximately 12 hours or more before even the earliest clinical signs of laminitis would be expected with this model[17]. There appeared to be not only degradation and lysis of the BM, but also separation of intact BM from the underlying EBCs: likely due to failure/dissolution of hemidesmosomes (HD), which are responsible for anchoring the BM to the cytoskeleton of the EBCs. This is consistent with the previously documented finding of reduced HD density noted later in the developmental and acute phases of experimentally-induced laminitis.[18,19]

Architectural changes observed in the SELs and EBCs themselves have been variably described during the development of acute laminitis. A consistent early finding is thinning and lengthening of the SELs observed with light microscopy, occurring in conjunction with early separation of the BM[14,15]. The EBC nuclei become rounded instead of oval and take an abnormal position in the cytoplasm of the cell, reflecting the underlying disruption of cytoskeletal attachments to the BM. Swelling, vacuolation, pyknosis and displacement of epithelial cells have also been described early in acute laminitis[15]. Capillaries of the lamellar dermis become compressed as the SELs continue to deform under distractive load after the loss of BM integrity[14]. Microthrombi, once primarily implicated in the pathogenesis of acute laminitis[20] are not usually observed in sections from developmental and early acute laminitis [14,15,17] Leukocyte emigration into the lamellar tissue is a consistently documented phenomenon in developmental and early acute laminitis.[14,16,17,21]

Progression of the laminitis lesion from these early acute changes depends largely on the initial extent of lamellar damage (and subsequent degree of destabilisation of the lamellar attachments of the digit as a whole) as well as the presence/absence of an ongoing lamellar insult. If a large area of lamellar tissue is affected at once and there is an overwhelming loss of mechanical integrity of the lamellar interface, the distal phalanx will displace distally within the hoof capsule ("sinking") causing additional separation of the epidermis from the dermis, haemorrhage, contusion and tearing of tissues and mechanical occlusion of the lamellar vasculature[22-24]. Less widespread and less severe lamellar damage results in varying amounts of distraction and resultant disorganisation of the lamellar tissues. In the absence of total destabilisation, healing processes begin within the lamellar tissue over the ensuing days after the development of acute laminitis. Proliferation of EBCs and repopulation and repair of BM occurs; however the original architecture of the lamellar interface is lost and the resultant disorganised structure lacks the surface area and strength of the original attachment.[25]

MECHANISMS OF ACUTE LAMELLAR DAMAGE

The exact mechanisms that cause the lesions of acute laminitis remain unclear. Researchers studying sepsis-related laminitis have largely departed from the traditional hypothesis of ischemia and reperfusion injury in favour of pursuing inflammation and enzymatic activation/degradation as key early pathophysiological events. These processes, together with cellular energy failure (due either to microvascular failure or cellular bioenergetic failure in the face of adequate nutrient supply) parallel the purported mechanisms of end organ damage in human sepsis.[6]

Inflammation

The central role of inflammation in the pathogenesis of acute laminitis has been highlighted experimentally. [7,16,21,26-32] Endothelial activation, cytokine and chemokine upregulation, and leukocyte emigration into the lamellar tissue occur early during the development of experimentally induced laminitis. Much of this evidence comes from experimental laminitis induction utilising BWE, a unique model characterised by rapid, multisystemic, neutrophil margination and activation[12]. There is, however, recent evidence that inflammation also plays a key role in laminitis induced by alimentary carbohydrate (CHO) overload, an experimental model that appears to more closely mimic naturally occurring sepsis in the horse. Similar lamellar inflammatory events, including inflammatory signalling and leukocyte accumulation, occur during CHO laminitis induction albeit somewhat delayed and closer to the onset of lameness[7,33]. Once they have emigrated into the lamellar tissue, leukocytes (particularly neutrophils) produce reactive oxygen species, enzymes and proinflammatory cytokines that contribute to inflammation and tissue damage.[34] With respect to determining the contribution of leukocyte infiltration to the pathophysiology of sepsis-related laminitis, the timing of leukocyte emigration into the lamellar tissue is critical. If leukocytes infiltrate the lamellar tissue prior to structural BM damage, then it is likely that they play a critical role in the initial injury. Immunostaining with calprotectin (a stain that highlights both leukocytes and stressed epithelium) in recent studies examined this timing: in the BWE model leukocyte infiltration clearly preceded epithelial stress, however in the CHO model leukocyte accumulation and epithelial stress appeared to occur simultaneously and it was not clear whether ultrastructural BM damage had already occurred[16,21]. Very early evidence of BM damage has been documented recently using electron microscopy[19] and immunostaining techniques[17], and it appears from these studies that leukocyte infiltration in CHO laminitis may be occurring in response to early BM damage rather than prior to it.

The initiators of systemic and local inflammation in laminitis models and naturally occurring sepsis related laminitis have not been identified, however a range of bacterial products including endotoxin have been suggested. Endotoxin has long been implicated in the development of laminitis; however infusion of endotoxin alone has not resulted in laminitis experimentally. Clinical signs consistent with endotoxemia were the major risk factor and the common clinical finding in hospitalized horses that developed laminitis in an epidemiological study.[35] In a recent study utilising the oligofructose CHO model, Bailey and colleagues demonstrated the presence of endotoxemia, platelet activation and TNF alpha within the blood early in the developmental period. [10] In a separate study, pre-treatment with endotoxin also increased the severity of laminitis induced with the OF CHO model[36]. Endotoxin and other bacterial products undoubtedly contribute to initiation of the early inflammatory changes observed in experimental models of acute laminitis, and also the pathogenesis of naturally occurring laminitis due to sepsis.

Enzymatic degradation of BM components

Enzymatic degradation of components of the dermo-epidermal junction was first suggested by Pollitt as a potential pathogenesis for laminitis.[37] Matrix metalloproteinase enzymes (MMPs) were originally implicated. Zymography of lamellar tissue from clinical cases of acute and chronic laminitis showed increased activation of MMPs.[38] Marked upregulation of MMP-2 transcription has been demonstrated in the acute phase of carbohydrate-induced laminitis[39]. The modest increase in MMP-9 observed in laminitic tissue[38,40] is likely to be due the influx of PMNs[41]. Incubation of lamellar hoof explants with an activator of MMPs causes a laminitis-like separation at the dermo-epidermal junction[40]. Although MMPs undoubtedly play a role in perpetuating lamellar damage, their role as initiators of laminitis has been recently challenged: new evidence places MMP-2 and MMP-9 production and activation many hours later than other molecular events in the oligofructose CHO model.[17] Recently, ADAMTS-4 (a disintegrin and metalloproteinase with thrombospondin motifs) has been identified as an enzyme capable of processing and destabilising the proteoglycans and glycoproteins of the lamellar BM. A marked increase in the expression of ADAMTS-4 mRNA has been identified early in experimental laminitis development, as well as in natural cases of the disease[17,42,43]. This enzyme may play a central role in initiating the laminitis lesion. The expression, production and activation of protease enzymes (including MMPs and ADAMs), and inhibition of their inhibitors (tissue inhibitors of metalloproteinase; TIMPs) may occur via multiple pathways, including as a consequence of inflammation, ischemia, or the direct action of endogenous molecules or bacterial toxins. It is interesting to note that disruption of BM occurs in epithelial tissues from other organs (skin, stomach) during the development of experimentally induced laminitis[44]. It is important to consider that whilst universal BM degradation may occur, it may only cause catastrophic consequences in the lamellar tissue as a result of the unique mechanical stress placed on this tissue.

Vascular and metabolic derangements

The traditional notion that ischemia is an initiator and a key event in the development of sepsis-related laminitis is no longer accepted by most research groups. There is no doubt, however, that changes in blood flow to the lamellae occur in association with the development and progression of laminitis. The fundamental question of whether lamellar ischemia is actually occurring as an initial event in the pathogenesis has been addressed directly by Loftus et al. [45] in the BWE model: the authors were unable to demonstrate molecular biochemical or metabolic evidence of ischemia during the developmental phase. This work, however, has not been repeated in the CHO model or in naturally occurring laminitis. There is evidence that derangements in microcirculatory function do contribute to the laminitis lesion. The digital circulation is highly sensitive to endogenous vasoactive substances [46] and the digital vasculature appears to be predisposed to venoconstriction[47], potentially disturbing lamellar perfusion during the development of laminitis. Vasoactive substances (including prostaglandins and thromboxane), are released by activated platelets, endothelial cells and leukocytes, contributing to both inflammation and dysregulation of the microvasculature.[10,48] In addition to reduced blood flow, disturbances of mitochondrial function and cellular energy metabolism may occur in the presence of adequate blood flow/nutrient delivery, as is the case in human sepsis related organ dysfunction [6]. The contribution of disturbed glucose metabolism and indeed the contribution of insulin to sepsis-related laminitis has not been investigated; however increases in serum insulin levels during the development of OF-induced laminitis were minor and not considered biologically significant [49]. Endotoxin has been shown to reduce insulin sensitivity after experimental infusion.[50]

PAIN

It is now clear that some major components of the initial acute laminitis lesion (BM degradation, leukocyte infiltration) occur well before clinical evidence of foot pain is apparent. The precise source of foot pain in acute laminitis is not clear, however the reliable analgesic response noted after phenylbutazone administration in acute laminitis implies that inflammation is at least partially responsible[25]. Other potential origins of pain in laminitis have been suggested, including increased submural pressure, tearing of submural tissues, ischemia, excessive contact between the sole and the distal phalanx and neuropathic mechanisms[51,52]. Driessen and colleagues noted that the sensory nerve terminals within the lamellae are located primarily at the base of the dermal lamellae and therefore the earliest laminitis pathology (cocnetrated around the BM and SELs) was unlikely to result in direct nociception[53]. In addition, inflammation at this early stage may be too distant and lack the intensity to result in nociception via neurochemical signalling. For these reasons, the developmental phase of laminitis proceeds without causing pain and goes unnoticed in naturally occurring laminitis cases. Progression of the laminitis lesion to involve tension and shearing of the dermal tissue (including connective tissue and capillaries) is likely to result in direct activation of perivascular nerve terminals together with local inflammation[53]. The first signs of pain in laminitis (incessant weight shifting) are then noted. This initial pain often coincides with increased digital pulse amplitude and increased hoof temperature. Although these findings have been interpreted as indicative of inflammation, they also likely represent a change in the nature of the foot circulation, with blood bypassing the capillary bed (due to initial compression of lamellar capillaries in the SDL) through dilated arteriovenous shunts[54], as well as increased post-capillary resistance.[55]

CONCLUSION

The utilisation of modern research tools is beginning to uncover molecular and ultrastructural events during the development of laminitis that occur far earlier than originally thought only a decade ago. These events occur long before the onset of clinical signs in experimental cases, and this is also likely to be the case in naturally occurring cases of laminitis. Further characterisation of these events: enzymatic activation, BM degradation, inflammation, microvascular dysregulation and bioenergetic failure; will eventually lead to improved early recognition and prevention of sepsis-related laminitis.

REFERENCES

1. Johnson PJ, Wiedmeyer CE, LaCarrubba A, et al. Laminitis and the equine metabolic syndrome. Vet Clin North Am Equine Pract 2010; 26:239-255. 2. McGowan CM. Endocrinopathic laminitis. Vet Clin North Am Equine Pract 2010; 26:233-237. 3. Asplin KE, Sillence MN, Pollitt CC, et al. Induction of laminitis by prolonged hyperinsulinaemia in clinically normal ponies. Vet J 2007;174:530-535. 4. Treiber KH, Kronfeld DS, Geor RJ. Insulin resistance in equids: possible role in laminitis. J Nutr 2006;136:2094S-2098S. 5. van Eps A, Collins SN, Pollitt CC. Supporting limb laminitis. Vet Clin North Am Equine Pract 2010; 26:287-302. 6. Belknap JK, Moore JN, Crouser EC. Sepsis-From human organ failure to laminar failure. Vet Immunol Immunopathol 2009;129:155-157. 7. Belknap JK, Giguere S, Pettigrew A, et al. Lamellar pro-inflammatory cytokine expression patterns in laminitis at the developmental stage and at the onset of lameness: innate vs. adaptive immune response. Equine Vet J 2007;39:42-47. 8. Pollitt CC, Visser MB. Carbohydrate alimentary overload laminitis. Vet Clin North Am Equine Pract 2010; 26:65-78. 9. Milinovich GJ, Trott DJ, Burrell PC, et al. Changes in equine hindgut bacterial populations during oligofructose-induced laminitis. Environ Microbiol 2006;8:885-898. 10. Bailey SR, Adair HS, Reinemeyer CR, et al. Plasma concentrations of endotoxin and platelet activation in the developmental stage of oligofructose-induced laminitis. Vet Immunol Immunopathol 2009;129:167-173. 11. Sprouse RF, Garner HE, Green EM. Plasma endotoxin levels in horses subjected to carbohydrate induced laminitis. Equine veterinary journal 1987;19:25-28. 12. Belknap JK. Black walnut extract: an inflammatory model. Vet Clin North Am Equine Pract 2010; 26:95-101. 13. Pollitt CC. The basement membrane at the equine hoof dermal epidermal junction. Equine Vet J 1994;26:399-407. 14. Pollitt CC. Basement membrane pathology: a feature of acute equine laminitis. Equine veterinary journal 1996;28:38-46. 15. Morgan SJ, Hood DM, Wagner IP, et al. Submural histopathologic changes attributable to peracute laminitis in horses. American journal of veterinary research 2003;64:829-834. 16. Faleiros RR, Johnson PJ, Nuovo GJ, et al. Laminar leukocyte accumulation in horses with carbohydrate overload-induced laminitis. Journal of veterinary internal medicine American College of Veterinary Internal Medicine 25:107-115. 17. Visser MB. Investigation of proteolysis of the basement membrane during the development of equine laminitis. . In: School of Veterinary Science. Brisbane: The University of Queensland; 2009. 18. French KR, Pollitt CC. Equine laminitis: loss of hemidesmosomes in hoof secondary epidermal lamellae correlates to dose in an oligofructose induction model: an ultrastructural study. Equine veterinary journal 2004;36:230-235. 19. Nourian AR, Baldwin GI, van Eps AW, et al. Equine laminitis: ultrastructural lesions detected 24-30 hours after induction with oligofructose. Equine Vet J 2007;39:360-364. 20. Weiss DJ, Geor RJ, Johnston G, et al. Microvascular thrombosis associated with onset of acute laminitis in ponies. American journal of veterinary research 1994;55:606-612. 21. Faleiros RR, Nuovo GJ, Belknap JK. Calprotectin in myeloid and epithelial cells of laminae from horses with black walnut extract-induced laminitis. Journal of veterinary internal medicine American College of Veterinary Internal Medicine 2009;23:174-181. 22. Ackerman N, Garner HE, Coffman JR, et al. Angiographic appearance of the normal equine foot and alterations in chronic laminitis. Journal of the American Veterinary Medical Association 1975;166:58-62. 23. Hood DM, Grosenbaugh DA, Slater MR. Vascular perfusion in horses with chronic laminitis. Equine veterinary journal 1994;26:191-196. 24. Hood DM. The mechanisms and consequences of structural failure of the foot. Vet Clin North Am Equine Pract 1999;15:437-461. 25. Van Eps AW, Pollitt CC. Equine laminitis model: lamellar histopathology seven days after induction with oligofructose. Equine veterinary journal 2009;41:735-740. 26. Blikslager AT, Yin C, Cochran AM, et al. Cyclooxygenase expression in the early stages of equine laminitis: a cytologic study. J Vet Intern Med 2006;20:1191-1196. 27. Black SJ, Lunn DP, Yin C, et al. Leukocyte emigration in the early stages of laminitis. Vet Immunol Immunopathol 2006;109:161-166. 28. Black SJ. Extracellular matrix, leukocyte migration and laminitis. Vet Immunol Immunopathol 2009;129:161-163. 29. Waguespack RW, Kemppainen RJ, Cochran A, et al. Increased expression of MAIL, a cytokine-associated nuclear protein, in the prodromal stage of black walnut-induced laminitis. Equine veterinary journal 2004;36:285-291. 30. Waguespack RW, Cochran A, Belknap JK. Expression of the cyclooxygenase isoforms in the prodromal stage of black walnut-induced laminitis in horses. Am J Vet Res 2004;65:1724-1729. 31. Fontaine GL, Belknap JK, Allen D, et al. Expression of interleukin-1beta in the digital laminae of horses in the prodromal stage of experimentally induced laminitis. American journal of veterinary research 2001;62:714-720. 32. Faleiros RR, Leise BB, Westerman T, et al. In vivo and in vitro evidence of the involvement of CXCL1, a keratinocyte-derived chemokine, in equine laminitis. Journal of veterinary internal medicine American College of Veterinary Internal Medicine 2009;23:1086-1096. 33. Leise BS, Faleiros RR, Watts MW,

et al. Laminar inflammatory gene expression in the carbohydrate overload model of equine laminitis. Equine veterinary journal 2010. 34. Belknap JK. Pathogenesis of laminitis. In: Robinson NE, Sprayberry KA, eds. Current Therapy in Equine Medicine, 6 ed. Philadelphia: Saunders; 2008:541-543. 35. Parsons CS, Orsini JA, Krafty R, et al. Risk factors for development of acute laminitis in horses during hospitalization: 73 cases (1997-2004). J Am Vet Med Assoc 2007;230:885-889. 36. Toth F, Frank N, Chameroy KA, et al. Effects of endotoxaemia and carbohydrate overload on glucose and insulin dynamics and the development of laminitis in horses. Equine veterinary journal 2009;41:852-858. 37. Pollitt CC. Equine laminitis: a revised pathophysiology. In: Proceedings of the 45th Annual Convention of the American Association of Equine Practitioners, Albuquerque, New Mexico 1999;188-192. 38. Johnson PJ, Tyagi SC, Katwa LC, et al. Activation of extracellular matrix metalloproteinases in equine laminitis. Veterinary record 1998;142:392-396. 39. Kyaw Tanner M, Pollitt CC. Equine laminitis: increased transcription of matrix metalloproteinase-2 (MMP-2) occurs during the developmental phase. Equine veterinary journal 2004;36:221-225. 40. Pollitt CC, Pass MA, Pollitt S. Batimastat (BB-94) inhibits matrix metalloproteinases of equine laminitis. Equine veterinary journal Supplement 1998;26:119-124. 41. Loftus JP, Belknap JK, Black SJ. Matrix metalloproteinase-9 in laminae of black walnut extract treated horses correlates with neutrophil abundance. Vet Immunol Immunopathol 2006;113:267-276. 42. Coyne MJ, Cousin H, Loftus JP, et al. Cloning and expression of ADAM-related metalloproteases in equine laminitis. Vet Immunol Immunopathol 2009;129:231-241. 43. Budak MT, Orsini JA, Pollitt CC, et al. Gene expression in the lamellar dermis-epidermis during the developmental phase of carbohydrate overload-induced laminitis in the horse. Vet Immunol Immunopathol 2009;131:86-96. 44. Visser MB, Pollitt CC. Immunohistochemical Distribution of Laminin-332 and Collagen Type IV in the Basement Membrane of Normal Horses and Horses with Induced Laminitis. Journal of comparative pathology. 45. Loftus JP, Belknap JK, Stankiewicz KM, et al. Laminar xanthine oxidase, superoxide dismutase and catalase activities in the prodromal stage of black-walnut induced equine laminitis. Equine Vet J 2007;39:48-53. 46. Baxter GM, Laskey RE, Tackett RL, et al. In vitro reactivity of digital arteries and veins to vasoconstrictive mediators in healthy horses and in horses with early laminitis. American journal of veterinary research 1989;50:508-517. 47. Peroni JF, Moore JN, Noschka E, et al. Predisposition for venoconstriction in the equine laminar dermis: implications in equine laminitis. J Appl Physiol 2006;100:759-763. 48. Robertson TP, Bailey SR, Peroni JF. Equine laminitis: a journey to the dark side of venous. Vet Immunol Immunopathol 2009;129:164-166. 49. van Eps AW, Pollitt CC. Equine laminitis induced with oligofructose. Equine Vet J 2006;38:203-208. 50. Toth F, Frank N, Elliott SB, et al. Effects of an intravenous endotoxin challenge on glucose and insulin dynamics in horses. Am J Vet Res 2008;69:82-88. 51. Jones E, Vinuela-Fernandez I, Eager RA, et al. Neuropathic changes in equine laminitis pain. Pain 2007;132:321-331. 52. Morgan SJ, Grosenbaugh DA, Hood DM. The pathophysiology of chronic laminitis. Pain and anatomic pathology. Vet Clin North Am Equine Pract 1999;15:395-417, vii. 53. Driessen B, Bauquier SH, Zarucco L. Neuropathic pain management in chronic laminitis. Veterinary clinics of North America Equine practice 26:315-337. 54. Molyneux GS, Haller CJ, Mogg K, et al. The structure, innervation and location of arteriovenous anastomoses in the equine foot. Equine veterinary journal 1994;26:305-312. 55. Allen D, Jr., Clark ES, Moore JN, et al. Evaluation of equine digital Starling forces and hemodynamics during early laminitis. American journal of veterinary research 1990;51:1930-1934.

LAMINITIS AS A CONSEQUENCE OF SEPSIS: PREVENTION AND TREATMENT

Andrew W. van Eps, BVSc, PhD, MACVSc, DACVIM
Queensland, Australia

INTRODUCTION

Although the practice of true evidence-based medicine is difficult in the veterinary field, in any case it is important to base therapeutic decisions upon the available scientific evidence. A summary of available information relating to the pathogenesis of sepsis-related laminitis is presented in the accompanying article. The key to successful prevention (and treatment) of sepsis-related laminitis may be early recognition of a horse that is at risk. In the absence of a biomarker for laminitis, horses with clinical signs or clinical laboratory evidence of systemic inflammation (endotoxemia/sepsis; the major risk factor for laminitis development in hospitalised horses[1]) must be identified as early as possible, so that laminitis prophylaxis as well as aggressive treatment of the primary disease can be initiated.

PHARMACOLOGIC THERAPY

Anti-inflammatory drugs

Evidence from experimental models and clinical cases suggests that there is a significant contribution of inflammation to the pathogenesis of developmental and acute laminitis[2-6]. High doses of flunixin meglumine have been suggested to have potential benefit in the septic horse for laminitis prophylaxis[7] however there is no experimental evidence to support this. Corticosteroids are potent inhibitors of both the COX and lipoxygenase pathways of inflammation. Their link to the development of laminitis under certain conditions[8] has generally precluded their use in clinical cases of acute laminitis. In the horse with sepsis, corticosteroids may be considered too immunosuppressive when given systemically. Targeted delivery of potent anti-inflammatory drugs (via regional perfusion or nanoparticle encapsulation) warrants future investigation in the prophylaxis/treatment of laminitis due to sepsis. Pentoxifylline exerts anti-inflammatory effects via inhibition of inflammatory mediators (particularly TNF-α) and recent evidence suggests a possible beneficial effect in experimentally induced laminitis[9]. Lidocaine infusion failed to have an effect on lamellar inflammation in a recent study of experimental laminitis induction[10].

Endotoxin neutralisation

Although endotoxin alone does not appear to be capable of causing laminitis, and laminitis cannot be prevented by binding endotoxin alone during its development[11] , there is no doubt that endotoxin contributes to systemic inflammation in the septic horse. Therefore it may be beneficial to attempt to directly neutralise circulating endotoxin through the administration of intravenous polymixin B and hyperimmune plasma [12-14].

Protease inhibitors

Enzymatic degradation of epidermal-dermal attachments by proteases (including MMPs and ADAMs) contributes to lamellar failure in acute laminitis[15-21]. Specific protease inhibitors are currently prohibitively expensive for systemic use, although local or targeted delivery may be possible in the future[22]. Recently Eades and colleagues tested a range of readily available drugs that have MMP inhibitory actions: only pentoxifylline ameliorated clinical signs in an experimental carbohydrate overload model of acute laminitis, though its anti-inflammatory actions may have contributed to the favourable outcome[9].

Drugs that alter microvascular perfusion

Treatments aimed at increasing lamellar blood flow have not resulted in empirical or experimental evidence of successful prevention nor treatment of sepsis-related laminitis. Isoxuprine[23] and topical nitroglycerine[24]have since been shown to have no effect on digital perfusion. Although acepromazine effectively increases total blood flow to the digit[25], there is little evidence to support this as being beneficial in the prevention or treatment of sepsis-related laminitis. A platelet aggregation inhibitor appeared to reduce the incidence of laminitis in a single study of experimental laminitis induction [26]. A recent report showed an apparent reduction in laminitis incidence after colic surgery in horses administered low molecular weight heparin[27] however the control group in this study was historical and the total incidence of laminitis low.

Analgesia

In most cases of sepsis-induced laminitis, initial analgesia is successfully provided by NSAIDs (phenylbutazone or flunixin meglumine). In the author's experience, the more COX 2-selective NSAIDs (firocoxib, meloxicam) are inferior to non COX-selective NSAIDs for acute laminitis pain. Adjunctive analgesia may be provided through the use of opioids, lidocaine, ketamine and alpha-2 agonist drugs alone or in combination: analgesic medications for laminitis have been extensively reviewed elsewhere[28]. The demonstration of a neuropathic component to chronic laminitis pain[29] has prompted the investigation of gabapentin as an analgesic for laminitis[30]. Gabapentin (or perhaps pregabalin) may be useful for inhibition of temporal summation and spinal cord "wind-up" [31,32] in acute cases of laminitis due to sepsis.

DIGITAL HYPOTHERMIA

Continuous distal limb cryotherapy effectively ameliorates experimentally-induced laminitis when applied continuously during the developmental phase[33,34]. This experimental evidence is most directly applicable to the septic horse at high risk of laminitis development. There is currently no data on the efficacy of its use during acute laminitis, however many of the same processes (inflammation and enzymatic activation) are known to continue after clinical signs have developed. Hypothermia also appears to provide some digital analgesia. Possible protective mechanisms of digital hypothermia in the face of sepsis include inhibition of inflammation (including cytokine production and leukocyte infiltration), inhibition of enzymatic degradation of the lamellar extracellular matrix and a reduction in cellular metabolic requirements in the face of microvascular failure or non-ischemic cellular bioenergetic failure. Recently studies in rodent models of sepsis have demonstrated the profound anti-inflammatory effect of hypothermia on end-organ damage [35-38]. Pre-emptive hypothermia (10˚C below normal) markedly reduced the severity of acute lung injury by reducing neutrophil emigration, inhibiting pro-inflammatory cytokine activity and increasing anti-inflammatory cytokine activity[36]. Less profound hypothermia (5˚C below normal) applied even after the lung was primed with neutrophilic inflammation, also decreased the severity of acute lung injury, suggesting a therapeutic role for hypothermia beyond its preventive effect [38]. Digital hypothermia appears to have similar anti-inflammatory effects in lamellar tissue during experimental laminitis induction with carbohydrate[39]. Methods and protocols for digital hypothermia have been reviewed elsewhere[40].

OTHER MANAGEMENT

Provision of deep bedding, removal of shoes and provision of frog and caudal sole support are advisable both in patients with acute laminitis and in the septic horse at risk. The (presumably) profound contribution of mechanical stress (through weight bearing) to the acute laminitis lesion behoves us to identify an effective means of countering it, particularly in the septic horse where the primary disease (and laminitis stimulus) is temporary. Absolute reduction of weight bearing on multiple limbs can be achieved only through encouraging/enforcing recumbency, or providing sling support for the entire body. Reducing weight bearing by the use of a sling is technically difficult and not well tolerated for more than short periods with existing sling technology. Forced recumbency was well tolerated for up to 72 hours in ponies anesthetised and allowed to recover in a stall with a lowered roof [41]. This technique warrants further investigation and evaluation as an adjunct therapy/prophylaxis for laminitis.

REFERENCES

1. Parsons CS, Orsini JA, Krafty R, et al. Risk factors for development of acute laminitis in horses during hospitalization: 73 cases (1997-2004). J Am Vet Med Assoc 2007;230:885-889. 2. Faleiros RR, Nuovo GJ, Belknap JK. Calprotectin in myeloid and epithelial cells of laminae from horses with black walnut extract-induced laminitis. Journal of veterinary internal medicine American College of Veterinary Internal Medicine 2009;23:174-181. 3. Black SJ. Extracellular matrix, leukocyte migration and laminitis. Vet Immunol Immunopathol 2009;129:161-163. 4. Belknap JK, Giguere S, Pettigrew A, et al. Lamellar pro-inflammatory cytokine expression patterns in laminitis at the developmental stage and at the onset of lameness: innate vs. adaptive immune response. Equine Vet J 2007;39:42-47. 5. Blikslager AT, Yin C, Cochran AM, et al. Cyclooxygenase expression in the early stages of equine laminitis: a cytologic study. J Vet Intern Med 2006;20:1191-1196. 6. Fontaine GL, Belknap JK, Allen D, et al. Expression of interleukin-1beta in the digital laminae of horses in the prodromal stage of experimentally induced laminitis. American journal of veterinary research 2001;62:714-720. 7. Belknap JK. The pharmacologic basis for the treatment of developmental and acute laminitis. Vet Clin North Am Equine Pract 2010 26:115-124. 8. Bailey SR, Elliott J. The corticosteroid laminitis story: 2. Science of if, when and how. Equine Vet J 2007;39:7-11. 9. Eades SC. Clinical and matrix metalloproteinase inhibitory effects of pentoxifylline on carbohydrate overload laminitis: preliminary results In: Proceedings of the 2nd AAEP Foundation Equine Laminitis Research Workshop, West Palm Beach 2009;44. 10. Williams JM, Lin YJ, Loftus JP, et al. Effect of intravenous lidocaine administration on laminar inflammation in the black walnut extract model of laminitis. Equine veterinary journal 42:261-269. 11. Raisbeck MF, Garner HE, Osweiler GD. Effects of polymyxin B on selected features of equine carbohydrate overload. Veterinary and human toxicology 1989;31:422-426. 12. Sykes BW, Furr MO. Equine endotoxaemia--a state-of-the-art review of therapy. Australian veterinary journal 2005;83:45-50. 13. Morris DD, Whitlock RH, Corbeil LB. Endotoxemia in horses: protection provided by antiserum to core lipopolysaccharide. American journal of veterinary research 1986;47:544-550. 14. Morresey PR, Mackay RJ. Endotoxin-neutralizing activity of polymyxin B in blood after IV administration in horses. American journal of veterinary research 2006;67:642-647. 15. Pollitt CC. Equine laminitis: a revised pathophysiology. In: Proceedings of the 45th Annual Convention of the American Association of Equine Practitioners, Albuquerque, New Mexico 1999;188-192. 16. Johnson PJ, Tyagi SC, Katwa LC, et al. Activation of extracellular matrix metalloproteinases in equine laminitis. Veterinary record 1998;142:392-396. 17. Kyaw-Tanner MT, Wattle O, van Eps AW, et al. Equine laminitis: Membrane type matrix metalloproteinase-1 (MMP-14) is involved in acute phase onset. Equine Vet J 2008. 18. Kyaw Tanner M, Pollitt CC. Equine laminitis: increased transcription of matrix metalloproteinase-2 (MMP-2) occurs during the developmental phase. Equine veterinary journal 2004;36:221-225. 19. Coyne MJ, Cousin H, Loftus JP, et al. Cloning and expression of ADAM-related metalloproteases in equine laminitis. Vet Immunol Immunopathol 2009;129:231-241. 20. Budak MT, Orsini JA, Pollitt CC, et al. Gene expression in the lamellar dermis-epidermis during the developmental phase of carbohydrate overload-induced laminitis in the horse. Vet Immunol Immunopathol 2009;131:86-96. 21. Visser MB. Investigation of proteolysis of the basement membrane during the development of equine laminitis. . In: School of Veterinary Science. Brisbane: The University of Queensland; 2009. 22. Nourian AR, Mills PC, Pollitt CC. Development of intraosseous infusion of the distal phalanx to access the foot lamellar circulation in the standing, conscious horse. Veterinary journal London, England 1997 2009. 23. Ingle-Fehr JE, Baxter GM. The effect of oral isoxsuprine and pentoxifylline on digital and laminar blood flow in healthy horses. Veterinary surgery VS the official journal of the American College of Veterinary Surgeons 1999;28:154-160. 24. Gilhooly MH, Eades SC, Stokes AM, et al. Effects of topical nitroglycerine patches and ointment on digital venous plasma nitric oxide concentrations and digital blood flow in healthy conscious horses. Veterinary surgery VS the official journal of the American College of Veterinary Surgeons 2005;34:604-609. 25. Leise BS, Fugler LA, Stokes AM, et al. Effects of intramuscular administration of acepromazine on palmar digital blood flow, palmar digital arterial pressure, transverse facial arterial pressure, and packed cell volume in clinically healthy, conscious horses. Veterinary surgery VS the official journal of the American College of Veterinary Surgeons 2007;36:717-723. 26. Weiss DJ, Evanson OA, McClenahan D, et al. Effect of a competitive inhibitor of platelet aggregation on experimentally induced laminitis in ponies. American journal of veterinary research 1998;59:814-817. 27. de la Rebiere de Pouyade G, Grulke S, Detilleux J, et al. Evaluation of low-molecular-weight heparin for the prevention of equine laminitis after colic surgery. J Vet Emerg Crit Care (San Antonio) 2009;19:113-119. 28. Driessen B, Bauquier SH, Zarucco L. Neuropathic pain management in chronic laminitis. Veterinary clinics of North America Equine practice 26:315-337. 29. Jones E, Vinuela-Fernandez I, Eager RA, et al. Neuropathic changes in equine laminitis pain. Pain 2007;132:321-331. 30. Terry RL, McDonnell SM, van Eps AW, et al. Pharmacokinetic profile and behavioral effects of gabapentin in the horse. Journal of veterinary pharmacology and therapeutics Accepted. 31. Harding LM, Kristensen JD, Baranowski AP. Differential effects of neuropathic analgesics on wind-up-like pain and somatosensory function in healthy volunteers. Clin J Pain 2005;21:127-132. 32. Arendt-Nielsen L, Frokjaer JB, Staahl C, et al. Effects of gabapentin on experimental somatic pain and temporal summation. Reg Anesth Pain Med 2007;32:382-388. 33. van Eps AW, pollitt CC. Equine laminitis model: Cryotherapy reduces the severity of lesions evaluated seven days after induction with oligofructose. Equine veterinary journal 2009;41:741-746. 34. van Eps AW, Pollitt CC. Equine laminitis: cryotherapy reduces the severity of the acute lesion. Equine Veterinary Journal 2004;36:255-260. 35. Scumpia PO, Sarcia PJ, Kelly KM, et al. Hypothermia induces anti-inflammatory cytokines and inhibits nitric oxide and myeloperoxidase-mediated damage in the hearts of endotoxemic rats. Chest 2004;125:1483-1491. 36. Lim CM, Kim MS, Ahn JJ, et al. Hypothermia protects against endotoxin-induced acute lung injury in rats. Intensive Care Med 2003;29:453-459. 37. Fujimoto K, Fujita M, Tsuruta R, et al. Early induction of moderate hypothermia suppresses systemic inflammatory cytokines and intracellular adhesion molecule-1 in rats with caerulein-induced pancreatitis and endotoxemia. Pancreas 2008;37:176-181. 38. Chin JY, Koh Y, Kim MJ, et al. The effects of hypothermia on endotoxin-primed lung. Anesth Analg 2007;104:1171-1178, tables of contents. 39. Van Eps AW, Leise BB, Watts CS, et al. Digital hypothermia inhibits early laminar inflammatory signaling in the oligofructose laminitis model. In: American College of Veterinary Internal Medicine Forum, Anaheim, CA 2010. 40. van Eps AW. Therapeutic hypothermia (cryotherapy) to prevent and treat acute laminitis. Vet Clin North Am Equine Pract 2010; 26:125-133. 41. Wattle O, Ekfalck A, Funkquist B, et al. Behavioural studies in healthy ponies subjected to short-term forced recumbency aiming at an adjunctive treatment in an acute attack of laminitis. Zentralblatt fur Veterinarmedizin Reihe A 1995;42:62-68.

ASSESSMENT OF AORTIC REGURGITATION IN HORSES: WHAT IS THE EVIDENCE-BASED MEDICINE?

Dr Gayle D. Hallowell, MA, VetMB, PhD, CertVA, DACVIM, MRCVS
Leicestershire, United Kingdom

INTRODUCTION

Aortic regurgitation (AR) is a disease of the older horse and is thought to be a slowly progressive degenerative condition[1]. The development of valvular disease appears to be affected by both training and age[2-4]. Aortic regurgitation manifests itself as a disease entity in the horse in many different ways. It can be identified as a diastolic murmur during routine physical examination with no other signs associated with cardiac disease. Alternatively, it can cause ventricular enlargement with or without associated dysrhythmias, which can lead to poor performance and ultimately congestive cardiac failure. Aortic valve disease can also predispose to aortic root (AoR) rupture and sudden death. The prevalence of AR is likely dependent upon the age, breed and activity of the group evaluated, but is estimated to be approximately 1-9% in racehorses in training[3,5,6] and 6% in a large study of the (older) general horse population[7].

DIAGNOSIS

Aortic regurgitation was first reported in horses over 40 years ago[8-10] and is diagnosed from the presence of a holodiastolic, decrescendo murmur loudest over the left (LICS4) and/or right heart base (RICS4) that radiates caudoventrally [11] with bounding or 'waterhammer' pulses in moderate to severe cases. This diagnosis is now confirmed and severity assessed using 2D and Doppler echocardiography. Echocardiographic changes seen include thickening and nodules present on the aortic velocity (AoV) (usually the left coronary cusp), increased left ventricular chamber size, increased aortic sinus diameter, increased fractional shortening and decreased left ventricular free wall thickness, particularly notable in end-diastole and vibrations of the septal leaflet of the mitral valve and occasionally of the AoV[11].

SIGNIFICANCE

Despite AR being regarded as a serious and progressive cardiac abnormality, one large, longitudinal survey found that only 8% of deaths in middle aged and older horses? were attributed to cardiac disease[7]. The presence of cardiac murmurs or valvular regurgitation have not been associated with poor performance[3]. It is proposed that left ventricular enlargement and associated ventricular dysrhythmias make affected animals unsafe to be ridden.

Ventricular dysrhythmias and exercise

It has been noted that horses free of cardiac disease have ventricular (VPD) and supraventricular premature depolarisations (SVPD) during exercise, particularly in the recovery period[13,17,18]. It is proposed that any dysrhythmia at peak exercise is abnormal and indicative of intrinsic cardiac disease, hypoxia, or electrolyte abnormalities that are having direct effects on the heart. It is currently recommended that if >2 isolated VPD's are detected during peak exercise, or if >5 pairs or paroxysms of VPD's are detected immediately after exercise they are significant[19]. These numbers are however based upon expert opinion rather than studies/evidence based medicine?. A recent study evaluating a relatively large group of normal racehorses in training were shown to have VPD's during all phases of exercise with 12% in the warm-up phase, 3.0% during maximal exercise, 15 % in the recovery period with a total of 29% of horses affected in at least one phase[20], which brings into question current guidelines.

Sudden death at exercise

A small number of horses die, without preceding clinical signs, during exercise. One study reported 0.4 fatalities per 1000 flat starts and 8.9 per 1000 starts over fences, of which 19% (flat), and 3.5% (fences), were reported to have been sudden deaths[12]. Ventricular dysrhythmias are assumed to be the main causative factor for sudden death[12-15], although there are no post mortem or electrocardiographic data to support or refute this assumption. The most common post mortem findings in animals that suddenly die were acute pulmonary edema, congestion and hemorrhage[15,16], which could be secondary to cardiac abnormalities. The importance of VPDs detected in horses with AR is commonly associated with an increased perceived risk of sudden death.

Ventricular dysrhythmias in human athletes

Sudden death occurs in 1 in 9,000 to 1 in 300,000[21-25] athletes. The most common causes are related to undiagnosed structural or electrical cardiac abnormalities[22,23,26], with inherited/congenital cardiac disease, such as cardiomyopathies, congenital coronary anomalies and ion channelopathies being the most frequently identified causes[27]. The relative risk of death is 2.5 times higher in athletes when compared with aged-matched, non-athletic individuals[28]. There is controversy in human medicine regarding the value of pre-competition resting electrocardiograms[29,30], despite it being more sensitive for the diagnosis of cardiomyopathies than history and physical examination[28-30], and the fact that there has been a marked reduction in deaths in certain countries where this technique has been implemented[31]. Resting electrocardiograms are used to identify changes in cardiac structure and size, and the practice of performing exercising electrocardiograms is less common in this group. This makes comparisons between equine and human athletes more challenging as the underlying etiology of the cause of sudden death is different. Frequent VPD's at rest in athletes may or may not be a marker of 'athlete's heart' and deconditioning has no effect on prevalence of this dysrhythmia[32]. The presence of frequent VPDs or non-sustained ventricular tachycardia may be a hallmark of underlying pathology and increased risk of death. As such, European Society of Cardiology recommends further evaluation[33], although this is not universally undertaken.

CONSENSUS ON THE ASSESSMENT OF SAFETY IN HORSES WITH AR

A group (n=18) of equine clinicians were invited to respond to a questionnaire regarding how horses with AR are commonly assessed and how the information gleaned is interpreted. A reply was obtained from 69% of those approached. 85% of respondents assessed left ventricular size using rounding of the apex and M-mode measurements, with a further 38% considering e-point to septal separation (EPSS), 23% assessing inter-ventricular septum (IVS) shape and movement, and a minority assessing left ventricle (LV) volume and measuring long axis views of the LV. 46% of respondents use their own normal reference ranges and 61% used published reference ranges (43% of these using ranges in textbooks and 71% using original papers). Majority of respondents (77%) adapted the normal reference ranges using (their?) experience for those animals in full training, and for those of non-standard breeds. 33% of respondents compared either the LV to the left atrium (LA) or right ventricle (RV) in breeds where there were no published ranges. Majority of respondents did not regard cardiac size as being an important variable on its own in terms regarding safety (69%). Majority of respondents performed exercise tests on animals with AR; 31% always performed them, 38% only if AR is deemed moderate to severe, and 8% only if AR is mild to moderate. Majority of respondents performed an exercise test that reflected the horse's normal exercise regime (69%), 15% performed a standardized exercise test and 8% performed a test that exceeds the exercise regime that horses normally undertake. Majority of respondents did not routinely perform 24-hour Holter monitoring in cases with AR (69%). There were significant differences regarding what was deemed abnormal and unsafe on both exercising ECGs and Holter recordings. Some respondents felt that any VPDs at rest or exercise were unacceptable (23%). Others felt that multiple single VPDs, particularly if were multiform, doublets or triplets of VPD's, R on T or ventricular tachycardia at peak exercise were unacceptable (69%) or as horses increased pace or during recovery phases (31%). Other findings that were deemed important regarding severity of AR included pulse quality (23%), presence of mitral regurgitation (31%) and indirect blood pressure measurements (15%). Many respondents (38%) felt reassessment was important in order to assess progression and safety. There were a number of respondents that commented on the significance of dysrhythmias at exercise and also the value of administering pharmacologic agents in these cases (e.g. quinalapril). Some respondents (31%) commented that it is easy to over-interpret results such that the animal is no longer fit for its purpose.

CONCLUSIONS

The aim of this talk was to evaluate the evidence of how horses with AR are diagnosed and managed. The diagnosis of AR is well established and undertaking exercise tests widely performed. However, the interpretation of any dysrhythmias noted at rest or exercise varies. It is widely accepted that normal horses have dysrhythmias before, during and after exercise, but what is not known in many instances is when will these dysrhythmias lead to collapse and death. Nor is it known how common dysrhythmias result in sudden death, nor how many horses with AR die due to their cardiac condition. More work is needed to try and establish the significance of some of the findings that are commonly seen in horses with AR.

ACKNOWLEDGEMENTS

I thank the following clinicians whom have contributed significant information and data for this presentation: Drs Helene Amory, Michelle Barton, Mark Bowen, Rikke Buhl, Mary Durando, Sophy Jesty, John Keen, Gunther Van Loon, Celia Marr, Johanna Reimer, Colin Schwarzwald, JoAnn Slack and Lesley Young.

REFERENCES

1.Else RW, Holmes JR. Cardiac pathology in the horse. 1. Gross pathology. *Equine Vet J* 1972;4:1-8; 2.Buhl R, Ersboll AK, Eriksen L, et al. Changes over time in echocardiographic measurements in young Standardbred racehorses undergoing training and racing and association with racing performance. *J Am Vet Med Assoc* 2005;226:1881-1887; 3.Young LE, Rogers K, Wood JLN. Heart murmurs and valvular regurgitation in thoroughbred racehorses: epidemiology and associations with athletic performance. *J Vet Intern Med* 2008;22:418-426; 4.Hallowell GD. Aortic Valve Prolapse in the Horse. *School of Veterinary Medicine and Science*: Nottingham, 2010; 5.Zucca E, Ferrucci F, Stancari G, et al. The prevalence of cardiac murmurs among standardbred racehorses presented with poor performance. *J Vet Med Sci* 2010;72:781-785; 6. Buhl R, Ersboll AK, Eriksen L, et al. Use of color Doppler echocardiography to assess the development of valvular regurgitation in Standardbred trotters. *J Am Vet Med Assoc* 2005;227:1630-1635; 7.Stevens KB, Marr CM, Horn JN, et al. Effect of left-sided valvular regurgitation on mortality and causes of death among a population of middle-aged and older horses. *Vet Rec* 2009;164:6-10; 8.Bishop SP, Cole CR, Smetzer DL. Functional and morphologic pathology of equine aortic insufficiency. *Pathol Vet* 1966;3:137-158; 9.Smetzer DL, Bishop S, Smith CR. Diastolic murmur of equine aortic insufficiency. *Am Heart J* 1966;72:488-497; 10.Sporri H. Two clinical types of aortic insufficiency in horses. *Ann N Y Acad Sci* 1965;127:358-363; 11. Reef VB, Spencer P. Echocardiographic evaluation of equine aortic insufficiency. *Am J Vet Res* 1987;48:904-909; 12. Boden LA, Anderson GA, Charles JA, et al. Risk of fatality and causes of death of Thoroughbred horses associated with racing in Victoria, Australia: 1989-2004. *Equine Vet J* 2006;38:312-318; 13.Physick-Sheard PW, McGurrin MK. Ventricular arrhythmias during race recovery in Standardbred Racehorses and associations with autonomic activity. *J Vet Intern Med* 2010;24:1158-1166; 14.Brown CM, Kaneene JB, Taylor RF. Sudden and unexpected death in horses and ponies: an analysis of 200 cases. *Equine Vet J* 1988;20:99-103; 15. Gelberg HB, Zachary JF, Everitt JI, et al. Sudden death in training and racing Thoroughbred horses. *J Am Vet Med Assoc* 1985;187:1354-1356; 16. Boden LA, Charles JA, Slocombe RF, et al. Sudden death in racing Thoroughbreds in Victoria, Australia. *Equine Vet J* 2005;37:269-271; 17.Jose-Cunilleras E, Young LE, Newton JR, et al. Cardiac arrhythmias during and after treadmill exercise in poorly performing thoroughbred racehorses. *Equine Vet J Suppl* 2006:163-170; 18.Scheffer CW, Robben JH, Sloet van Oldruitenborgh-Oosterbaan MM. Continuous monitoring of ECG in horses at rest and during exercise. *Vet Rec* 1995;137:371-374; 19. Martin BB, Jr., Reef VB, Parente EJ, et al. Causes of poor performance of horses during training, racing, or showing: 348 cases (1992-1996). *J Am Vet Med Assoc* 2000;216:554-558; 20.Ryan N, Marr CM, McGladdery AJ. Survey of cardiac arrhythmias during submaximal and maximal exercise in Thoroughbred racehorses. *Equine Vet J* 2005;37:265-268; 21.Van Camp SP, Bloor CM, Mueller FO, et al. Nontraumatic sports death in high school and college athletes. *Med Sci Sports Exerc* 1995;27:641-647; 22.Maron BJ, Doerer JJ, Haas TS, et al. Sudden deaths in young competitive athletes: analysis of 1866 deaths in the United States, 1980-2006. *Circulation* 2009;119:1085-1092; 23. Maron BJ, Gohman TE, Aeppli D. Prevalence of sudden cardiac death during competitive sports activities in Minnesota high school athletes. *J Am Coll Cardiol* 1998;32:1881-1884; 24.Maron BJ, Haas TS, Doerer JJ, et al. Comparison of U.S. and Italian experiences with sudden cardiac deaths in young competitive athletes and implications for preparticipation screening strategies. *Am J Cardiol* 2009;104:276-280; 25.Eckart RE, Scoville SL, Campbell CL, et al. Sudden death in young adults: a 25-year review of autopsies in military recruits. *Ann Intern Med* 2004;141:829-834; 26.Maron BJ. Sudden death in young athletes. *N Engl J Med* 2003;349:1064-1075; 27.Borjesson M, Pelliccia A. Incidence and aetiology of sudden cardiac death in young athletes: an international perspective. *Br J Sports Med* 2009;43:644-648; 28.Corrado D, Basso C, Rizzoli G, et al. Does sports activity enhance the risk of sudden death in adolescents and young adults? *J Am Coll Cardiol* 2003;42:1959-1963;

29.Drezner J, Corrado D. Is there evidence for recommending electrocardiogram as part of the pre-participation examination? *Clin J Sport Med* 2010;21:18-24; 30.Borjesson M, Dellborg M. Is there evidence for mandating electrocardiogram as part of the pre-participation examination? *Clin J Sport Med* 2010;21:13-17; 31.Pelliccia A, Corrado D. Can electrocardiographic screening prevent sudden death in athletes? Yes. *Bmj* 2010;341:c4923; 32.Delise P, Lanari E, Sitta N, et al. Influence of training on the number and complexity of frequent VPBs in healthy athletes. *J Cardiovasc Med (Hagerstown)* 2010; 33. Heidbuchel H, Panhuyzen-Goedkoop N, Corrado D, et al. Recommendations for participation in leisure-time physical activity and competitive sports in patients with arrhythmias and potentially arrhythmogenic conditions Part I: Supraventricular arrhythmias and pacemakers. *Eur J Cardiovasc Prev Rehabil* 2006;13:475-484.

EVALUATION OF SEIZURES IN HORSES

Véronique A. Lacombe, DVM, PhD, DACVIM; Stephen Reed, DVM, DACVIM
Columbus, OH; Lexington, KY

Seizures (from the Latin sacire, "to take possession of") were already recognized during the antiquity (with the first description around 500 - 700 B.C), and remain very prevalent today in both human and veterinary medicine. A seizure is characterized by non-specific, paroxysmal, abnormal and often catastrophic event. A variety of terms have been used to describe equine seizure disorders, which may be referred to as fits, attacks, strokes, convulsions, or epilepsy. Epileptic seizures (ES), which result from abnormal and excessive discharge of a set of neurons in the brain, present with a wide variety of non-specific clinical manifestations.[1,2] Therefore, diagnosing seizures in horses is challenging because seizures refer to a specific clinical event regardless of the underlying cause, which can be difficult to determine. In addition, evaluating seizures in horses is difficult because of the rapid return of clinical signs without persistent neurologic deficits combined with the difficulty in imaging the equine brain. Finally, little common agreement exists on the definition of epilepsy in horses. Therefore, establishment of a standardized classification in equine epileptology could enhance communication among clinicians and researchers. In addition, it could facilitate a logical clinical approach for the clinician to establish a diagnosis, a therapeutic plan and a prognosis for seizures. Therefore, the first goal of this presentation is to provide a classification of ES in horses, according to the most recent and commonly accepted definitions in human and small animal epileptology. The second goal is to review the diagnostic utility of the ancillary tests used in equine medicine.

CLASSIFICATION OF EPILEPTICAL SEIZURES

The commission on classification and terminology of the International league against Epilepsy (ILAE) first created a comprehensive classification for epilepsy and epileptic syndromes ~ 25 years ago.[3] This classification is a dynamic concept and has been revised based on the advances in neurodiagnostic technology. A standardized classification has numerous advantages, including the foundation for a coherent and systematic approach in the diagnosis and treatment of epilepsy, as well as for a common mode of communication among clinicians. Although epilepsy classification has been established in dogs and cats, such classification has not been attempted in horses.[1,4-5] Therefore we propose to use the following classification, accordingly to previous veterinary studies and current guidelines of the ILAE, which is based on a comprehensive diagnostic scheme organized in 5 axes.[1,3-6]

Ictal Phenomenon (axis 1)

Axis 1 consists of a description of the ictal semiology, including the "pre-ictal", "post-ictal" and "interictal" period, during which the behavior of the horse should be observed. Please note that the inter-ictal physical and neurological examinations may be within normal limits. The postictal phase can range from mild depression, to stupor and blindness that can last for hours. Status epilepticus is defined as more than 30 minutes of continuous seizure activity or 2 or more sequential seizures with no complete recovery of consciousness between seizures. It is uncommon in adult horses.[2]

Classification by Seizure Type (axis 2)

Partial seizures, which are caused by a localized (focal) abnormal discharge, result in localized motor signs or sensations, the first clinical signs being highly indicative of the epileptic focus localization.[6] Partial seizures are further categorized as simple, if alertness and normal mentation are maintained, or complex, if impairment of consciousness is reported. Tremors/fasciculations involving the facial muscle, abnormal tonus/clonus in one limb, or paroxysmal behavior changes (including self-mutilation) could be suggestive of partial seizures. Generalized ES involve the entire cerebral cortex, and result in generalized bilateral motor activity over the whole body, including convulsive seizures (formerly referred to "grand-mal"), or non-convulsive seizures (formerly referred to "petit-mal") or myoclonic seizures. Consciousness may be impaired and this impairment may be the initial manifestation. Generalized seizures can originate from both cerebral hemispheres from the onset (primary generalized seizures), or progress from partial seizures (secondary generalized seizures).[1-3,4,7] In human and small animal classification, the majority of the seizures described were partial seizures with or without secondary generalization.[4] Similar findings were recently reported in adult horses.[7]

Classification by Etiology (axis 4)

Seizures are classified by etiology based on the results of complementary diagnostic tests. Reactive ES are seizures occurring from the reaction of a healthy brain to a temporary systemic disease (e.g., endogenous metabolic disturbance). Epilepsy is a group of diseases with diverse etiologies and electrophysiologic patterns that predispose the affected patient to recurrent unprovoked ES (i.e. 2 or more ES). Epilepsy is further classified as symptomatic, cryptogenic, and idiopathic (i.e., suspected genetic predisposition).[3]

Recurrent ES secondary to an identifiable structural brain disorder (e.g., inflammatory, neoplastic, and traumatic), are classified as symptomatic epilepsy. When an intracranial origin of seizures is suspected despite unremarkable results of neurodiagnostic tests, recurrent ES are classified as cryptogenic epilepsy (Greek cryptos =

hidden), which is also referred as "probable symptomatic" epilepsy.[5] When no underlying brain abnormality could be identified combined with a response to anti-epileptic drugs and a suspected genetic predisposition (i.e., Arabian horses under the age of one year), a diagnosis of "idiopathic epilepsy" is made.[8]

One of the main difficulties with the proposed classification is to differentiate symptomatic epilepsy from recurrent symptomatic seizures due to on going brain insult. Another difficulty is the differentiation of cryptogenic epilepsy from idiopathic epilepsy (i.e., no underlying structural disorders combined with a genetic predisposition), due to the limitations of the neurodiagnostic tests available in equine medicine.

Other Axes

Axis 3: Epilepsy syndrome, which is characterized by a cluster of specific symptoms, has been not been well characterized in veterinary medicine. It has been reported during genetic disorders such as idiopathic juvenile Arabian and lavender foal syndrome.[8,9]

Axis 5: An optional designation of the degree of impairment caused by the epileptic condition.[6]

DIAGNOSTIC UTILITY OF ANCILLARY TESTS

Blood work

A complete serum biochemistry profile should be performed to rule out metabolic abnormalities that may trigger seizures (i.e., *reactive ES*). Test for bile acids and ammonia should be performed if hepatic encephalopathy is suspected. Presence of hypoglycemia may indicate a cause for seizures, although the likelihood of hypoglycemia-induced seizures in adult horses is rare. Renal disease and hyperlipidemia should be ruled out as well.

Cerebrospinal fluid (CSF)

CSF analysis is a valuable tool in determining the underlying etiology of seizures caused by an active structural cerebral abnormality (i.e., *symptomatic ES*). In general, collection of CSF should be performed close to the site of the suspected neuroanatomic lesion to increase the likelihood of finding abnormalities in the spinal fluid cytology, and therefore determining the etiologic agent. In most instances, the site of choice is the atlanto-occipital cistern. If this site in not an option, because of the risk (and/or cost) associated with the procedure and general anesthesia, the other alternative site is the lumbosacral space, providing this technique can be performed safely in the standing and sedated animal.

Electroencephalography (EEG)

EEG is defined as the graphic recording of the rhythmic bioelectrical activity arising predominantly from the cerebral cortex. It is the most important paraclinical test in humans suffering from epilepsy. As a non-invasive method of determining abnormal cerebral electrical activity, EEG is most useful in the investigation and management of human patients with epilepsy, in particular in establishing probability of recurrence, need for treatment, and prognosis. In addition, EEG has been used in small and large animals as a screening procedure in patients with seizures, because of its excellent sensitivity.[10-12] The ictal EEG reflects bilateral neuronal discharge widespread in both hemispheres during generalized seizures.[3,10,13] Similar to reports in humans and small animals, the presence of epileptiform paroxysmal activity on the EEG recording, which is defined as abnormal paroxysmal transient events (such as spikes, sharp waves, and spike-and-wave discharges), supports the diagnosis of epilepsy in horses.[3,10,13] However, normal interictal EEG recordings do not exclude the diagnosis of epilepsy since the use of tranquilizers may increase the threshold for seizures, and since the window of recording may not have been long enough to record interictal paroxysmal epileptiform discharges.[11] In addition, changes are not pathognomonic of a disease, but rather reflect the kind of process occurring (e.g., acute vs. chronic; focal vs. diffuse; inflammatory vs. degenerative).

Advanced imaging diagnostic techniques

Magnetic resonance imaging (MRI) and computed tomography (CT) should be used to detect any structural abnormalities in the brain and the skull (e.g., tumor, hemorrhage, hydrocephalus, fractures). A 2007 evidence-based review from the American Academy of Neurology and the American Epilepsy Society recommends an electroencephalogram combined with a CT or MRI scan in the work-up of seizures.[14]

Similar to conventional X-ray imaging, CT of the head measures the attenuation of X-ray photon beams through different tissues, except that radiation passes successively through tissues from multiple directions.[15] It allows visualization of the anatomy of bony structures of the head in great detail, although small fractures may be missed if appropriate window parameters are not chosen. Thus, CT should be primarily conducted in cases of seizures secondary to head trauma (i.e., symptomatic ES), and is warranted to rule out structural cerebral lesions as an etiological diagnosis, especially if associated with other neurologic signs such as cranial nerve deficits or altered mentation.[16] With the exception of subarachnoid hemorrhage and mineralized lesions, CT showed limited sensitivity in identifying inflammatory disorders and small soft tissue lesions in the equine brain, which were not further detectable after contrast administration.[17] In addition, the presence of epilepsy was significantly negatively associated with the presence of intracranial lesions detectable by CT imaging of the head.[16]

The high tissue contrast and anatomical detail provided by exploiting differences in magnetic properties of hydrogen within tissue water and lipids make MRI ideal for examination of the brain parenchyma.[18] Therefore, MRI is more sensitive than CT in detecting CNS lesions such as demyelinating and other white matter diseases, cerebral neoplasms, degenerative diseases and non-acute hemorrhage.[15,17] MRI has become the gold standard test to diagnose structural intracranial abnormalities in small animals, and MRI is more sensitive in first, apparently unprovoked seizures in humans. Unfortunately, most referral centers do not have MRI equipment to accommodate an equine adult head.

Finally, results from these diagnostic tests, but also information obtained from the history, physical and neurologic examinations, as well as progression of clinical signs and response to treatment, remain essential in the diagnostic work-up of horses with ES.[17] In conclusion, diagnosis of seizures remains a challenge in equine medicine and there is a need for common agreement on the terminology use in equine epileptology. Therefore, the use of a standardized classification could help equine clinicians in the establishment of a diagnosis, a therapeutic plan and a prognosis for ES.

REFERENCES

[1]Podell M. Proc SEVC Barcelona, 2009. [2]Lacombe VA, et. al. In: Equine Neurology. Ed: SM Reed and M Furr. pp 77-93, 2007. [3]ILAE Commission on the classification and terminology of the International League Against Epilepsy. Epilepsia 1989: 30:389-399.. [4]Berendt M, et. al. J Vet Intern Med 1999; 13:14-20. [5]Schriefl S, et.al. JAVMA 2008; 233:1591-1597. [6]Engel J. Epilepsia 2001; 42:796-803. [7]Mosseri S, et.al. J. Vet Intern Med 2010; 24 (3): 789-790. [8]Aleman M, et.al. J Vet Intern Med 2006;20:1443-9. [9]Page P, et.al. J Vet Intern Med 2006;20:1491-4. [10]Berendt M, et.al. Acta Neurol Scand 1999; 99:276–283. [11]Lacombe VA, et.al. In Equine Neurology. Ed: SM Reed and M. Furr. pp 127-148, 2007. [12]Lacombe VA, et.al. J Vet Intern Med 2001; 15:385-393. [13]Podell M. Vet Clinic North Am: Small animal practice 1996; 26:779-809. [14]Krumholz A., et.al. Neurology 2007; 69 (21): 1996–2007. [15]Gilman S. N Engl J Med 1998;338, 812–820. [16]Sogaro-Robinson C, et.al. JAVMA 2009; 235:176-183. [17]Lacombe VA, et.al. Equine Vet J 2010; 42 (5) 393-399. [18]Jose Cunilleras, et al. Equine vet Educ 2007; 19 (4) 179-181.

IDENTIFICATION OF CRITICAL ILLNESS RELATED CORTICOSTEROID INSUFFICIENCY IN CRITICALLY ILL HORSES. **AJ Stewart[1], E Hackett[2], TJ Towns[1], H Weaver[1], B Johnson[1], AA Wooldridge[1], AS Munsterman[1], Zhong Q[1]. College of Veterinary Medicine, Auburn University, Auburn AL. [2]College of Veterinary Medicine & Biomedical Sciences, Colorado State University, Fort Collins, CO.**

Critical illness-related corticosteroid insufficiency (CIRCI) has been identified in humans, foals, dogs and cats with lower-than-expected circulating cortisol concentrations, and/or by a blunted cortisol response to ACTH stimulation. Our purpose was to determine if CIRCI exists in critically ill horses.

Endogenous plasma ACTH and serum cortisol concentrations, and cortisol at T=0 and T=30 min after 0.1µg/kg cosyntropin, were measured by radioimmunoassay from horses with colic or systemic illness on admission, and days 2, 4 and 6 of hospitilization. Horses were divided into mild, moderate, or severe illness groups based on clinicopathologic data. Inappropriately low cortisol was defined as endogenous cortisol <mean-1SD achieved after administration of 0.1µg/kg cosyntropin to normal horses (<269nmol/L). Inadequate delta cortisol was defined as <mean delta cortisol in normal horses after 0.1µg/kg cosyntropin (<159nmol/L). Cortisol, ACTH and delta cortisol were compared using ANOVA between groups, with $P < 0.05$ considered significant.

Fifty-eight horses classified as having mild (11), moderate (30) and severe (17) disease at admission had survival rates of 100%, 97% and 35% respectively. Admission ACTH and cortisol concentrations were highest in severely ill horses (93±198pg/mL, 361±137nmol/L) compared to moderate (31±36, 279±137) and mildly ill horses (18.0±12.0, 237±133). Admission cortisol concentrations were higher overall in severely ill horses (P=0.016), but were low in 24% (4/17). Admission delta cortisol was low in 85% (11/13) of severely ill horses, and was associated with marked adrenal hemorrhage in non-survivors.

Severely ill horses have high cortisol and ACTH, but low cortisol and delta cortisol may indicate CIRCI secondary to adrenal hemorrhage.

ALTERATIONS IN AUTOPHAGY IN THE PITUITARY OF HORSES WITH EQUINE PITUITARY PARS INTERMEDIA DYSFUNCTION (PPID). Dianne McFarlane, Kristen McDaniel, Kim Hill, Melissa Cordero. Oklahoma State University.

Equine pituitary pars intermedia dysfunction (PPID) is a common endocrinopathy of aged horses that results from neurodegeneration of the dopaminergic periventricular neurons that innervate the intermediate lobe of the pituitary. Factors that initiate spontaneous dopaminergic neurodegenerative disease remain elusive, however accumulation of misfolded α-synuclein protein and dysfunctional protein clearance have been implicated. Misfolded protein accumulation occurs due to increased protein production or decreased clearance of damaged macromolecules through the process of autophagy. While have previously demonstrated that horses with PPID have increased α-synuclein in the periventricular neurons compared to controls, it remains unknown whether the protein accumulates due to increased production or decreased clearance. We hypothesized that autophagy is decreased in the pituitary neurointermediate lobe from horses with PPID compared to controls.

Neurointermediate lobe pituitary tissue was from collected from horses with PPID (n=12) and healthy horses (n= 37, 2-35 years). Realtime PCR was used to determine the relative expression of autophagy genes (mTOR, Beclin1, ATG12, ATG7, ATG5, PINK, LAMP2) and α-synuclein Relative gene expression from horses with PPID were compared to healthy horses by t-test following log transformation. A Pearson coefficient of correlation was calculated comparing α-synuclein expression with autophagy gene expression.

The expression of α-synuclein, autophagy-related genes (ATG12, Beclin, Lamp2), and mTOR was greater in horses with PPID than in healthy horses. Age was not correlated to α-synuclein or autophagy gene expression. There was a significant positive correlation between expression of α-synuclein and Beclin1, ATG12, ATG7, ATG5, and PINK, but not mTOR expression. Accumulation of α-synuclein protein in horses with PPID may result from increased α-synuclein expression. Autophagy genes are upregulated in horses with PPID, suggesting a compensatory response, although these findings need to be confirmed by demonstrating an increased functional response. α-synuclein expression was positively correlated to expression of autophagy genes except mTOR, suggesting α-synuclein may stimulate autophagy in an mTOR independent manner.

EFFICACY OF DELAYED ANTIVIRAL THERAPY AGAINST EHV-1 CHALLENGE. LK Maxwell[1], LL Gilliam[1], N Pusterla[2], R Carmichael[1], RW Eberle[1], JW Ritchey[1], TC Holbrook[1], T Gull[1], GB Rezabek[1], D McFarlane[1], CG MacAllister[1]. 1. Oklahoma State University, Stillwater, OK. 2. University of California, Davis, CA.

Equine herpes virus type-1 (EHV-1) outbreaks are often not recognized until exposed horses are at immediate risk for developing equine herpes myeloencephalopathy (EHM). The objective of this study was to determine whether delayed therapy with the antiviral drugs valacyclovir or ganciclovir could protect those horses most at risk for EHM. Eighteen aged (> 20 years) mares were randomized to treatment: no therapy (control), oral valacyclovir therapy, or intravenous ganciclovir therapy. Drug administration was initiated at the onset of the second febrile phase, between days 4-6 after EHV-1 inoculation (PI), and continued for one week. Neurological examinations were performed prior to the study and for three weeks PI. One horse was excluded from the study for failure to become febrile. Body temperature was significantly lower in the ganciclovir-therapy horses as compared to control horses on days 6-8 PI ($P<0.05$), whereas valacyclovir-therapy horses did not differ from control horses. Viremia in whole blood, as determined by PCR, was also lower in the ganciclovir-therapy horses on days 7-10 PI and on day 7 PI in the valacyclovir-therapy horses ($P<0.05$). Although antiviral drug administration did not reduce the risk of ataxia ($P=0.06$) or nasal shedding, ganciclovir therapy did decrease the severity of ataxia ($P<0.05$) as compared to valacyclovir-therapy and control horses, where 0/6, 2/5, and 4/6 horses, respectively, developed at least a two grade change in ataxia. In summary, ganciclovir administration provided better protection against EHM than did valacyclovir when therapy was initiated just prior to the onset of neurological disease.

PROTECTION FROM COMTEMPORARY EQUINE INFLUENZA VIRUS BY COMMERCIAL VACCINES. A Sellers[1], G. Landolt[1], HGG Townsend[2], J. Tuttle[3], RM Stenbom[3], DP Lunn[1]: [1]CSU, Fort Collins, CO, [2]WCVM, Saskatoon, SK, [3]BIV, St. Joseph, MS.

Equine vaccination is amongst the most important method of prophylaxis against equine influenza virus (EIV), a pathogen in which continuous antigenic drift can lead to vaccine failure. A 6-month duration of immunity (DOI) challenge infection study was conducted using commercial inactivated vaccines containing different strains of A/equine/2/influenza virus's, including Innovator™, containing Kentucky/97 (Pfizer Animal Health, New York, NY), and Calvenza, containing a combination of Ohio/2003, Kentucky/95, and Newmarket93 (Boehringer Ingelheim Vetmedica, St. Joseph, MS). The challenge virus strain was Colorado/07, the most contemporary challenge strain currently in use.

The study design was a blinded, randomized challenge trial. Three groups of 10 yearling ponies, with no history or serological evidence of EIV infection were established. Each group received one of three treatments: vaccination with Innovator™; vaccination with Calvenza™; or injection with a saline placebo. Each treatment was administered 3 times, at intervals of 1 month between the first two treatments, and 3 months between the second and third treatments. All ponies were challenged by nasal nebulization of $5X10^7$ EID_{50} influenza virus A/eq/2/Colorado/07 6 months after the third treatment. Clinical signs of disease, including rectal temperature, nasal discharge, anorexia, coughing, and depression, were recorded daily for 2 days prior to challenge infection, and 14 days post-challenge. Nasal shedding of EIV was measured on the same days, using a real-time PCR test procedure. EIV-specific antibody responses were measured by ELISA. Differences between groups were analyzed by non-parametric repeated measures ANOVA, and differences were declared significant when $P < 0.05$.

All control group ponies demonstrated clinical signs of disease consistent with EIV infection post-challenge infection, including pyrexia, nasal discharge, inappetance and partial anorexia. These signs were significantly lower in both vaccine groups; mean body temperature was elevated (>101.5°F) for 8 days in controls, but only 2 days in vaccine groups. Nasal shedding of EIV was detected in all ponies in all groups: over the duration of the study the Calvenza group shed significantly less virus than Innovator and Control. Over time antibody titers were significantly higher in the Calvenza than the Innovator group, and both were significantly greater than controls.

This study demonstrated that both current commercial inactivated EIV vaccines have a duration of clinical protection of at least 6 months after a highly pathogenic challenge with a recent EIV isolate. Both antibody responses and virological protection differed between the vaccines. Formulation difference between the vaccines, including the EIV antigens employed, may have contributed to this performance difference.

THE EVER CHANGING NEMATODE PARASITE GENOME AND THE EFFECTS ON CONTROL PROGRAMS

Louis C. Gasbarre, MS, PhD
Buffalo, WY

INTRODUCTION

A prerequisite to the adoption of a parasitic life cycle is the ability to maintain high reproductive rates. All parasites must find a suitable host to survive and a primary survival strategy is to produce very high numbers of offspring to insure continuation of the species. Because of these very high reproductive rates the genome of any given parasitic species has a high potential rate of changes. As such, any control practice that exerts a strong selective pressure on the parasitic genome can rapidly select for alleles that favor survival of the parasite under the selecting pressure.

The discovery of the avermectin anthelmintics markedly changed animal husbandry production in the US. Because these drugs have large margins of safety, high efficacy and broad range of activity the manner and frequency in which anthelmintics are used, as well as livestock raising practices changed dramatically. The safety and efficacy of these drugs led to the development of strategic treatment regimens which had as their target optimal animal health and productivity through the reduction of parasite transmission resulting from clearing pastures of infectious larvae. The combination of highly efficacious, broad spectrum drugs with treatment programs that apply the drugs in such a way to maximize reductions in the parasite populations led to a number of alterations in production practices including higher stocking rates and reductions in the use of alternative control practices.

While these programs have proven to be profitable for many livestock raisers in the US, there has been concern that programs which rely strictly upon drug administration without regard for good pasture parasite management will lead to more rapid selection of drug resistance in the nematode populations. This has been well documented in small ruminant species, in which GI nematodes resistant to all classes of anthelmintics has been well documented (see Waller; 1997). Until recently, there has been a scarcity of well documented reports of drug resistance in cattle and horse nematodes. In cattle this has led to the suggestion that in cattle the immune system seem to handle GI nematode infections better than those of sheep and goats, and as such anthelmintic resistance was unlikely to arise due to lower selective pressure because of lower drug usage (Waller, 1994). In horses, where the number of treatments can be very high, the absence of well documented anthelmintis resistance is perplexing. One idea has been that this is a result of the complexity of the parasite fauna found in equines, and to the fact that the selective pressure may not be as severe because the drugs are not effective against tissue dwelling larval forms thus leading to a natural "refugia". Recent studies in cattle have shown profound changes in the cattle nematode populations in the US, and we will focus on these changes to demonstrate how quickly this can lead to problems in any host-parasite system.

WHAT HAS CHANGED OVER THE LAST 5-6 YEARS IN US CATTLE NEMATODE POPULATIONS

In the US in the early part of the last decade there were no documented cases of macrocyclic lactone resistance in US cattle nematodes. At this time there were increasing anecdotal reports of lowered anthelmintic efficacy in cattle, and well documented reports of drug resistance to benzimidazoles by *Cooperia* sp. in New Zealand (Vermunt et al., 1995; Hosking et al. 1996), to ivermectin by *Cooperia* sp. in Great Britain (Stafford and Coles 1999), and to both macrocyclic lactones and benzimidazoles by *Haemonchus* sp. and to the macrocyclic lactones by *Cooperia* sp. in Argentina (Anziani et al. 2004).

The initial demonstration of macrocyclic lactone resistance was in a stocker operation in Wisconsin. This producer was an early adopter of the concept of strategic deworming and had been using this technology for 17 years on his operation. At the end of each grazing season he took a cursory sample to assess the effect of his control program. In the previous year he had found a few parasite eggs, but not enough for concern. But near the end of the grazing season in 2002 he had very poor condition in his animals, and had a couple of deaths. One animal was sent to the STATE laboratory and the results came back as " parasitic gastroenteritis". In 2003 we began a study involving female or castrated male beef or beef cross calves weighing 80-100 kg that were purchased in the winter of 2002-2003 at auction barns in the southeastern United States. Upon assembly, all were dewormed with Dectomax® injectable, and then placed with various contract grazers in Alabama and Mississippi. While on pasture, the animals were dewormed once or twice with either Ivomec® or Cydectin®. During the first week of May 2003, animals were re-assembled from the grazers and dewormed with Ivomec Plus® injectable, after which they were shipped to Wisconsin. There, the 152 lightest animals were placed on the pasture used in this study for the summer 2003. The pastures had been grazed by only cattle for the previous 40 yr. Since approximately 1980, all animals grazed on this pasture had been dewormed at turnout with either fenbendazole or one of the avermectins, then strategically twice more during the grazing season. To test the efficacy of current anthelmintics on parasites in the herd, the animals were fecal sampled, treated with either one of the brand name macrocyclic lactones, a benzimidizole, or left untreated. Fourteen days later a second fecal sample was taken, and the level of reduction in fecal egg count was calculated. The reduction in mean fecal EPG, at 14 days post treatment, for all groups tested was < 85%. The highest reduction (82%) was seen in the moxidectin group and the lowest reduction (42%) in the eprinomectin group. Lower than expected efficacy was seen in all treatment groups, with few animals in any group showing reductions ≥ 90%.

A subset of 3 animals was randomly selected from each treatment group and euthanized for recovery of nematode parasites. At necropsy, all treated groups contained worms. The fewest total worms were found in the albendazole group, while the greatest numbers of worms were found in the ivermectin group. All groups contained similar numbers of parasites in their abomasa, but significant differences were seen in the number of parasites recovered from the small intestine. The smallest number of parasites was seen in the albendazole treated cattle, while all avermectin and moxidectin treated cattle harbored several thousand parasites. The predominant species recovered from the abomasum was *Haemonchus placei*, accounting for >half the parasites recovered in all groups (range 55 - 99%). Only in the benzimidazole group did *H. contortus* account for all parasites found in the organ. *O. ostertagi* found were in the untreated group, the eprinomectin group and the albendazole group, but these accounted for < 0.5% of the total abomasal parasites for each group. The most commonly found species in the small intestine of all groups was *Cooperia punctata* (range 85-100%), with small numbers of *C. oncophora* and *C. spatulata*.

The study was repeated the next year with the following changes: 1. monitor the animals over the course of the grazing season; 2. increase the number of animals slaughtered at the end of the season to minimize the effect of potential outlying observations; 3. increase the time interval between treatment and slaughter to ensure sufficient time for drug action; 4. utilize repeated fecal sampling in the fecal egg reduction test to minimize procedural variation; and 5. increase the number of drugs tested . The results of this study were in agreement with those of the previous study and demonstrated that during the course of the grazing season the pastures harbored significant numbers of parasites that were refractory to avermectins, milbemycin, and a benzimidazole at the label recommended doses. As seen previously, *Haemonchus contortus* resistant to all these anthelmintics were present over the course of the study period. In contrast, *Cooperia* sp., mainly *punctata*, and *H. placei* were resistant only to the macrocyclic lactones. There was no apparent resistance against the older anthelmintic levamisole, which had not been used for > 20 yr in the operation. However, animals treated with this drug continued to harbor small but measurable numbers of *Ostertagia ostertagi.*

While this study was the first conclusive demonstration of anthelmintic resistance in cattle nematodes questions remained as to the general applicability of this information especially because of the high stocking rates and frequent anthelmintic use by the producers. About this time we were contacted by the West Virginia University Extension Service of the West Virginia Department of Agriculture, who with the West Virginia Cattleman's Association conduct the Southern Bull and Replacement Heifer Test Program. In 2006, researchers observed, that in spite of a strict vaccination program on arrival, test cattle developed more health problems than expected. Fecal samples were found to have high egg counts in many cases, even though the animals had been treated multiple times with cattle wormers. In 2007, fecal samples were collected from the rectum from a total of 86 heifers from 9 different producers and 103 bulls from 13 producers on delivery to the test station. At this time, the bulls were treated with an oral benzimidazole (fenbendazole) and the heifers with a pour-on endectocide (ivermectin). Fecal samples were taken from the animals for a second time, 14 days after treatment. Initial egg counts did not differ between the bulls and heifers (49.4 and 42.4 respectively). In contrast, endectocide treatment reduced fecal eggs counts by only 45 %, whereas treatment with an oral benzimidazole reduced the mean fecal egg count by 95% (26.9 and 2.2 respectively). In the heifers the effectiveness of the treatment varied based upon the source of the animals, but in all cases the FECRT showed less than an 85% reduction (range of 83 to 26%).

Subsequently studies were carried out at University managed herds at both North Carolina State University and Clemson University. In both cases, the herds had a history of the use of macrocyclic lactones as the only or predominant means of nematode control for the herds. In both herds, the use of a macrocyclic lactone reduced fecal egg counts by less than 85%. The parasites remaining after treatment at both locations were members of the genus *Cooperia* as determined by PCR performed on eggs recovered from the feces after treatment.

To date studies in our laboratory and others have shown less than expected efficacy using macrocyclic lactones in 15 different States in all geographic areas parts of the US. Reduced efficacy was most often seen when pour-on formulations were used, but have also been seen when injectable products were applied. Reduced efficacy was seen with both brand name and generic products. Follow-up studies have in 5 States have demonstrated that the less than expected reduction in fecal egg counts was due to a low efficacy of the drug and not simply to an improper application of the drug. The 5 States were Wisconsin, West Virginia, North Carolina, South Carolina, Oklahoma and Kansas. To date all demonstrated instances of macrocyclic lactone resistance in US cattle have been in members of the genus *Cooperia* with the most common species of parasite demonstrating macrocyclic lactone resistance being *Cooperia punctata.*

Another change that has taken place is an increased incidence and intensity of infection by the intestinal worm, *Nematodirus helvetianus* . *Nematodirus* has traditionally been considered to be a problem only in young animals, as older animals seem to be effectively immune to the parasite. Additionally the parasite is not usually seen in high numbers and it has been assumed that this was because the parasite has a much lower reproduction rate than the other nematode parasites found in cattle. *Nematodirus* also differs in that eggs from this species will lie for long periods of time on the pastures and only hatch in warm weather. *Nematodirus* does not seem to have yet acquired resistance to the drugs commonly used. The apparent increase in this parasite seems to be the result of two factors. The first is that strategic programs are usually based upon *Ostertagia ostertagi* a colder weather species. As such by the time the eggs hatch, the drugs are gone. Additionally, the heavy use of the drugs is reducing the normal complex parasite fauna to either monoculture or at most two species. This seems to favor the establishment of large numbers of *Nematodirus* in the host.

WHAT IS THE IMPORTANCE OF THESE FINDINGS

The economic impacts of macrocyclic lactone resistance in US cattle parasites remain to be accurately assessed. Current anthelmintic treatment programs are based upon the assumptions that pastured animals contain a mixture of different species of parasites, with the most pathogenic being the medium brown stomach worm, *Ostertagia ostertagi.* As a result optimal treatment times are based upon transmission optima's for *Ostertagia,* with little regard being paid to the other parasites. The appearance of resistant species has changed this paradigm. Treatment programs are producing animals that have monospecific infections of *Cooperia* or *Nematodirus.* And often the numbers of these species are much higher than the numbers seen in normal mixed species infections. While the precise reason for these numbers has yet to be demonstrated. it is likely due to several factors including the removal of parasites that are in direct competition for preferred sites in the gut, differences in the effectiveness and nature of the immune responses elicited by monospecific versus polyspecific infections, and differences in the optimal time to apply anthelmintics for the different parasite species. A second confounding factor in determining the economic impact of the drug-resistant parasites is the relative paucity of information available on the true pathogenicity of either *Cooperia punctata* or *Nematodirus helvetianus.* While there have been numerous studies of the pathogenicity of *Ostertagia* and to a lesser degree *Cooperia oncophora*, very few studies have assessed the true economic impact of *Cooperia punctata*. A recent study performed on cattle in an experimental feedlot demonstrated that drug resistant *C. punctata* at a level comparable to that seen under normal transmission levels significantly reduced both the average daily gain and feed consumption in cattle. Additionally, the intestine showed classic changes associated with intestinal nematodes, and there was marked involvement of the gut immune system.

It is very interesting to note that in horses the large strongyles have essentially disappeared due to the repeated use of macrocyclic lactones and benzimidizoles. What were once considered to be the dominant pathogens in horses have now been replaced by the cyathosomes (Kaplan, 2002), much like the replacement of *Osteragia* by *Cooperia/Nematodirus* in cattle. In the case of the small strongyles drug resistance against the benzimidizoles is widespread, while resistance to the macrocyclic lactones has not yet been conclusively demonstrated. The impact of this change on overall equine health and the functioning of the gut immune system has yet to be determined.

RECOMMENDATIONS

The results generated over the last 2-3 years indicate that the repeated and almost exclusive use of macrocyclic lactones to control nematodes in cattle in the US over the last 40 years has altered the genetic make-up of the parasites. In cattle, the parasites of today appear to be less susceptible to the drugs and the dominant parasite in today's herds appears to be *Cooperia* sp. In horses, the widespread use of suppressive parasite control programs has made cyathosomes the dominant parasitic species. It is important that producers and Veterinarians no longer assume that treatment with a any anthelmintic has achieved the desired level of parasite control. Testing of the effectiveness of treatment should be performed on a routine basis. Control programs should not rely on a single class of anthelmintic, and rotation of drug classes or combinational therapies should be considered. It is important to define the transmission patterns of the nematodes found at a given location and to adjust the timing of drug application to address the species on hand, More is not always better. Drug therapy should be designed to alleviate health or production problems, not to eradicate the parasites. Any program built solely on the repeated administration of chemotherapeutic agents against nematode parasites is doomed to failure. Any successful program must include good management practices and appropriate and effective chemotherapeutic agents. (For Review in horses see Nielsen et al. 2010).

REFERENCES

Anziani, O.S., Suarez, V., Guglielmone, A.A., Warnke, O., Grande, H., Coles, G.C. 2004. Resistance to benzimidazoles and macrocyclic lactone anthelmintics in cattle nematodes in Argentina. Vet. Parasitol. 122: 303-306. Hosking, B.C., Watson, T.G., Leathwick, D.M., 1996. Multigeneric resistance to oxfendazole by nematodes in cattle. Vet. Rec. 138, 67–68. Kaplan, R.M. 2002. Anthelmintic resistance in nematodes of horses. Vet. Res. 491-507. Nielsen MK, Fritzen B, Duncan JL, Guillot J, Eysker M, Dorchies P, Laugier C, Beugnet F, Meana A, Lussot-Kervern I, von Samson-Himmelstjerna G. 2010. Practical aspects of equine parasite control: a review based upon a workshop discussion consensus. Equine Vet J. 42(5):460-468. Stafford, K., Coles, G.C., 1999. Nematode control practices and anthelmintic resistance in dairy calves in the southwest of England. Vet. Rec. 144, 659–661. Vermunt, J.J., West, D.M., Pomroy, W.E., 1995. Multiple resistances to ivermectin and oxfendazole in *Cooperia* species of cattle in New Zealand. Vet. Rec. 137, 43–45. Waller, P.J. 1994. The development of anthelmintic resistance in ruminant livestock. Acta Trop. 56, 233-243. Waller, P.J., 1997. Anthelmintic resistance. Vet. Parasitol. 72, 391–412.

USING NEW TECHNOLOGIES TO DIAGNOSIS AND CONTROL NEMATODE PARASITES

Louis C. Gasbarre, MS, PhD
Buffalo, WY

INTRODUCTION

Gastrointestinal nematodes severely hinder the health and well-being of pastured animals throughout the world. Disease caused by these parasites can include stunted growth, tissue edema, and severe diarrhea. Even in systems where broad spectrum anthelmintics are routinely used the parasites still hinder optimal growth, health, and productivity. Livestock raisers have increasingly depended on anthelmintic drugs and have increased both the frequency in which the drugs are used and the percentage of animals treated. These factors have resulted in the wide-spread appearance of anthelmintic-resistant parasites in small ruminants throughout the world (Waller, 1977), and the appearance of anthelmintic resistance in cattle nematodes in New Zealand, Great Britain and most recently in the US, (Gasbarre, et.al, 2009). The combination of frequent anthelmintic adminstation and the subsequent appearance of drug resistant parasites have resulted in dramatic changes in nematode parasite populations throughout the US.

These changes have necessitated the development of new and more accurate methods to diagnosis the presence and composition of the parasite fauna found in the host and the development of new non-chemical methods to limit parasite transmission. Armed with these tools it is possible to develop control programs that are less dependent on anthelmintics, and that are more precise in the application of the anthelmintics.

NEW DIAGNOSTIC TOOLS

The diagnosis of parasitic nematode infections in livestock has classically relied on the demonstration and enumeration of parasite eggs in the feces. This usually involves emulsification of feces in a solution of high specific gravity followed by flotation of the eggs for counting. There has been much discussion over the years as to the accuracy and precision of this procedure in gauging the parasite burden in the host. To assess the variation associated with this procedure we performed a study in calves over a four year period. The study involved about 200 animals per year with each animal being sampled for 3 consecutive days at 3 different times during the year (spring, summer, fall). These data were analyzed to determine the sources of variation in the data set. In order the sources were: between animal variation, sample variation, season, and day of sampling. The day of sampling accounted for 2-3% of the total variation meaning that the day to day herd mean was very precise. Conversely the sample variation for the same animal day to day was large (about 30%), meaning that a single sample from a given animal is not a good estimator of it's parasite burden. In simple terms this means that in cattle the fecal egg count is a good measure of the parasite load of a given group of animals, but if you want an adequate assessment of an individual animal you need to sample that animal multiple times. The logical questions that arise from this are: how many animals do you need in the group?, or how many times do you need to sample an individual ? Because parasites are not normally distributed in a herd, sampling must account for the skewed distribution, To adequately assess the worm burden in a group you must sample 17-20 animals, in groups less than 17 you would sample all the animals (Gasbarre et.al.1996). Sampling an individual requires a very large number of replicates but you receive maximum effect for effort with 3 consecutive samplings. Samplings in excess of this result in only small gains in accuracy.

After enumeration of the parasite eggs the next question usually asked is "What are the species of parasite present?" With the exception of a few nematode species, egg identification by visual inspection is extremely difficult if not impossible. An alternative has been to collect the eggs place them in an appropriate culture medium and then identify the larvae after a period of development. The problem with this approach is that optimal culture conditions are not the same for all species. As such, changes in temperature, moisture, and time will result in different ratios of parasite species. Thus, at best culture is qualitative not quantitative, and can be prone to false negatives. Over the past several years a number of laboratories have begun to develop molecular probes that accurately identify parasite DNA at the genus an species levels. These probes offer the advantages of being extraordinarily precise and sensitive. In the case of cattle nematodes they have been used to identify the genera of parasites in a given individual or group by isolation of eggs followed by DNA isolation and PCR for distinct genomic regions (Zarlenga et al 2001). In horses a similar technology has been used of eggs recovered from feces to identify species of small strongyles after drug treatment (Traversa et. Al. 2009, Ionita et.al. 2010). The major problem to be resolved with this technology lies in the areas of making the assays quanitative and in developing means to easily isolate high quality parasite DNA from the feces and means to preserve it during shipment to the laboratory. The problem of quanitation may be solved by using sequencing to identify the parasite species. As the cost of sequencing come down, mass sequencing using a metagenomics approach in the identification of parasite DNA may prove superior to PCR amplification.

USING HOST GENOMICS TO CONTROL TRANSMISSION

An increasingly attractive alternative for control of the parasites is the use of the diversity of the host genome to reduce parasite transmission patterns. This has been best described in cattle. The first step in the development of such programs is the identification of regions of the host genome that contain genes that influence acquired or innate resistance to the parasites. Once identified such genomic regions, commonly referred to as QTL, could be used in a selection program to reduce parasite transmission within cattle populations. In a larger context, the discovery of QTL affecting parasite resistance is the first step in the identification of genes that are responsible for resistance to the parasites.

Resistance to infection with nematodes in cattle can be manifested in a number of ways. Most parasites will stimulate an effective level of protective immunity in most individuals in a herd after several months on pasture. Reinfection of these animals results in a significant reduction in the number of parasites that become established. Other parasites, most notably *Ostertagia ostertagi* requires many months, and possibly years, before animals demonstrate sufficient immunity to reduce development of newly acquired parasites. This prolonged susceptibility is a major reason that this parasite remains the most economically important cattle nematode throughout temperate regions of the world. Because different parasite species may or may not be susceptible to the same types of immune responses, the identification of QTL affecting parasite resistance requires very thorough characterization of individual host resistance patterns. In addition, such an effort requires: reagents capable of probing the entire host genome, a designed host population in which gene segregation can be followed, statistical expertise to discern phenotypic and genotypic relationships, and high throughput genomic and bioinformatic capacity.

Previous studies herd demonstrated that the number of nematode eggs/gram (EPG) in feces of pastured cattle is strongly influenced by host genetics (Leighton et al., 1989),and that the heritability of this trait was approximately 0.30 (Gasbarre et al., 1990). In addition, EPG values were not distributed normally, and a small percentage of the herd was responsible for the majority of parasite transmission (Gasbarre et al., 1990). This "overdispersion" of EPG values was first described by Crofton (1971a,b) and has been reported in other cattle populations. In this distribution, most individuals have relatively low fecal EPG values, and a small percentage of animals, estimated to be between 15 and 25% of the total population (Anderson and May, 1985), exhibit high EPG values. This pattern strongly suggests that genetic management of a small percentage of the herd could considerably reduce overall parasite transmission. Results also indicated that the odds of certain bulls producing susceptible (high EPG) calves were 20 times greater than other bulls (Gasbarre et al., 1995).

To further investigate this genetic relationship and experimental Angus population was developed. Analyses of the cattle produced in the subsequent breeding program demonstrated a number of additional characteristics of GI nematode infections in cattle. The best indicator of level of *Cooperia* infection was the EPG value ($r = 0.6$), while *Ostertagia* was best measured by serum pepsinogen levels ($r = 0.7$), weight gain ($r = -0.5$) and measures of anemia ($r = 0.5$). Other phenotypic measures were poor predictors of parasite numbers (Gasbarre, 1997). These relationships agree with reports of similar parasite-host systems). In contrast, studies in tropical areas have indicated significant correlations between fecal EPG values and total worm numbers.

In addition, calves could be assigned to one of three types: (1) Type I never demonstrate high EPG values; (2) Type II show rises in EPG values through the first 2 months of test then fall and remain at levels associated with the Type I calves; and (3) Type III calves maintain high EPG levels throughout the test. The approximate percentage of these calves in the first generation of breeding was 25:50:25, respectively. A secondary infection experiment showed that EPG values for Type I and II calves remained low, while Type III calves continued to shed high numbers of eggs. These phenotypes can be categorized as innately immune, acquired immune, and immunologically non-responsive. Control of parasite transmission and parasite-induced loss should be targeted in the non-responsive animals.

Results indicate selection has been effective in increasing the proportions of innately immune and immunologically responsive calves. In addition, the range of EPD in the herd is about half the mean EPG value for all calves tested to date (unpublished data). These results provide further evidence that genetics of the host play an important role in parasite transmission.

Genotypying of animals using both microsatellite markers and the new bovine SNP (single nucleotide polymorphism) chip that can interrogate about 50,000 individual genetic polymorphisms has been completed on more than 700 animals, including animals on test and their ancestors (over 70 historical sires). . Variables analyzed using this model have included mean and peak numbers of nematode EPG, mean serum pepsinogen level, and EPG type. These analyses have identified QTL on several chromosomes. Current studies are aimed at the validation of these putative QTL and fine mapping of the loci.

At the same time, both a candidate gene and a global microarray have been used to discern differences in gene expression patterns in tissues taken from resistant and susceptible individuals. In these studies, the following tissues have been analyzed: abomasal lymph nodes, abomasal mucosa, mesenteric lymph nodes, and jejunal mucosa. Preliminary results to date have indicated marked differences between the different tissue sources and apparent differences between the responses elicited in the abomasum versus the jejuneum. In general, resistant animals seem to have a higher level of activation of genes involved in immune effector mechanisms and tissue repair, while susceptible cattle seem to have higher levels of activation in genes associated with specific recognition of parasite antigens. While these results are very preliminary and do not address cause and effect relationship, this approach coupled with both QTL analysis and comparative genomics using the available human genome information will greatly speed the identification of relevant bovine genes.

An important question remains regarding how we will use this data. In most cases it would not be wise to implement a breeding program based on resistance to a single disease entity. Resistance to GI nematodes is a complex mixture of many types of immune responses and recent work indicates that protection against the different parasites may involve different types of immune responses . It is also becoming evident that different types of immune responses, i.e. the so called Th1 and Th2 response, are counter-regulatory. Thus it is possible that animals that are superior in some types of responses will be inferior in others. This coupled with the fact that many individuals already question the loss of diversity in agricultural species makes it unlikely there will be major programs to breed for helminth resistance in cattle except in the areas of extremely high parasite pressure.

Even though it seems unlikely that there will be major breeding programs for helminth resistance in cattle, there are several ways host genetics could be used to greatly increase the efficiency of cattle and dairy production. The first is based on the fact that nematodiasis is a "quantitative" disease. As parasite burdens increase so does loss, and it is probable that very low parasitemias induce little or no economic effect in otherwise well-managed animals. With this in mind, the goal of modern parasite control programs is to keep transmission rates down, and not to eradicate the parasites, which is probably impossible. The second important factor is the "overdispersed" nature of nematode distributions. This pattern implies that a few members of the herd are responsible for most of the parasite transmission. In an elegant analysis Anderson and May (1985) calculated that the treatment/removal/vaccination of the susceptibles would be almost as effective in controlling disease as would be treatment of the entire population. From an economic standpoint rapid and inexpensive identification of these "susceptible/non-responders" could reduce treatment costs substantially. Such an approach will also lessen the selective pressures placed on parasite populations, and should slow the development of drug resistance in the parasites. The key is the identification of reliable markers for susceptibility/ resistance.

In a similar manner, identification of the genes involved in protection should provide producers with the information to make informed breeding decisions. The past few years have seen cattle producers become more aware of selection for a variety of production traits, and have seen a movement away from selection for a single trait. Cattle producers today judge the genetic potential of bulls for many traits. The use of breeding values is a common subject of cattle producers. Providing producers with reliable information on the performance of their cattle under parasite pressure will allow them to make informed breeding decisions. Producers in areas of high parasite pressure could chose not to use a "susceptible" bull, while those in areas of less parasite transmission could use those bulls if they provided other traits important in the overall breeding program.

Finally, as our ability to manipulate the genome becomes more refined. It may be possible to genetically engineer cattle to meet the specific needs of a given producer. The ability to do this with parasite resistance will depend upon the exact mechanisms involved, the number of important genes, and the precise regulatory processes entailed. In any of these future schemes, it will be important to first identify the gene loci involved in resistance, and to find means to rapidly and reliably identify the resistance phenotype of individual animals.

REFERENCES

Anderson, R.M., May, R.M., 1985. Herd immunity to helminth infection and implications for parasite control. Nature 315, 493–496. Crofton, H.D., 1971a. A quantitative approach to parasitism. Parasitologia 62, 179–193. Crofton, H.D., 1971b. A model of host–parasite relationships. Parasitologia 63, 343–364. Gasbarre, L.C., Leighton, E.A., and Bryant, D. 1996. Reliability of a single fecal egg per gram determination as a measure of individual and herd values for trichostrongyle nematodes of cattle. Am. J. Vet. Res. 57: 168-171. Gasbarre, L.C., 1997. Do fecal egg counts accurately reflect worm numbers in cattle? Vet. Proc. 11, 900–903. Gasbarre, L.C., Leighton, E.A., Bryant, D., 1995. Genetically controlled resistance to gastrointestinal nematode infections of cattle. Proc. Beef Improv. Fed., 212–217. Gasbarre, L.C., Leighton, E.A., Davies, C.J., 1990. Genetic control of immunity to gastrointestinal nematodes of cattle. Vet. Parasitol. 37, 257–272. Gasbarre, L. C., Smith, L.L., Lichtenfels, J.R., Pillit, P.A. 2009.The development and identification of cattle nematode parasites resistant to multiple classes of anthelmintics in a commercial cattle population in the US. Vet Parasitol.166:281-285. Ionita M, Howe DK, Lyons ET, Tolliver SC, Kaplan RM, Mitrea IL, Yeargan M . 2010. Use of a reverse line blot assay to survey small strongyle (Strongylida: Cyathostominae) populations in horses before and after treatment with ivermectin. Vet Parasitol. 168(3-4):332-7. Leighton, E.A., Murrell, K.D., Gasbarre, L.C., 1989. Evidence for genetic control of egg-shedding rates in calves. J. Parasitol. 75, 498–504. Traversa D, Iorio R, Otranto D, Giangaspero A, Milillo P, Klei TR. 2009.Species-specific identification of equine cyathostomes resistant to fenbendazole and susceptible to oxibendazole and moxidectin by macroarray probing. Exp Parasitol. 121(1):92-5. Waller, P.J., 1977. Anthelmintic resistance. Vet. Parasitol. 72, 391–412. Zarlenga, D.S. Chute, M.B., Gasbarre, L.C., and Boyd, P. 2001. A multiplex PCR for differentiating economically important gastrointestinal nematodes of cattle. Vet. Parasitol. 97: 199-209.

PYROSEQUENCING OF THE GASTRIC MUCOSA IN HEALTHY HORSES SHOWS AN ABUNDANT MIRCROBIAL FLORA WHICH IS UNIQUE IN EACH HORSE AND INFLUENCED BY DIET AND MANAGEMENT. G.A. Perkins, H. C. den Bakker, A.B. Burton, H.N. Erb, S.P. McDonough, P.L. McDonough, J. Parker, R. Rosenthal, M. Wiedmann, S. Dowd, and K.W. Simpson. Cornell University, College of Veterinary Medicine, Ithaca, NY.

Little is known about the gastric mucosal flora in healthy horses and its role in gastric disease has not been critically examined. Our laboratory previously reported that a diverse microbial flora with a predominance of *Streptococcus spp.* and *Lactobacillus spp.* exists in healthy horses using Fluorescence *in situ* hybridization (FISH). The present study sought to further characterize the gastric mucosal flora of healthy horses using massive parallel 16SrRNA bacterial tag encoded FLX-titanium amplicon pyrosequencing (bTEFAP). Biopsies of the squamous, glandular, antral and any ulcerated mucosa were obtained from 4 healthy horses via gastroscopy after a 12-hour fast and 2 horses immediately post-mortem. DNA was extracted from the mucosal biopsies and bTEFAP and data processing was performed. Hierarchical cluster analysis based on relative abundance data on the genus level were performed to look for trends in bacterial diversity among the individual horses.

Pyrosequencing yielded between 4,500 and 13,000 reads per horse with 9238, 16587, 16587 reads in the antrum, squamous and glandular regions, respectively. The microbiome segregated into two distinct clusters: Cluster 1 comprised of 2 horses that were stabled, fed hay and sampled at post-mortem and Cluster 2 consisted of 4 horses that were pastured on grass, fed hay and biopsied gastroscopically after a 12-hour fast. Samples from different antomic regions clustered by horse rather than region. Despite being very similar at the higher taxonomic level (phyla) differences in the distribution of bacteria were seen at the genus and species level. The dominant bacteria in Cluster 1 horses were Firmicutes (>83% reads/sample) consisting of mainly *Streptococcus spp., Lactobacillus jensenii, L. fornicalis* and *Sarcina maxima*. Cluster 2 had more diversity with a predominance of Proteobacteria, Bacteroidetes and Firmicutes and 51 genera identified such as *Streptococcus spp., Moraxella spp., Actinobacillus spp.*, and others. Though the relative abundance of the individual taxonomic groups was significantly different between individual horses, no significant differences in the overall diversity could be found (as assesed by Shanon Weaver, ACE and Choa I diversity indices). H*elicobacter spp.* sequences were not identified in any sample (out of 58,891 reads). The ulcerated mucosa from Horse 3 (group 1) had lower diversity and higher numbers of bacteria predominated by *Lactobacillus equigenerosi*.

This data shows that the equine gastric mucosa harbors an abundant and diverse microbiome which is unique to each individual and differs by sampling method, fasting prior to sampling and diet.

SEASONAL PASTURE MYOPATHY IN SIX HORSES ASSOCIATED WITH MULTIPLE ACYL-COA DEHYDROGENASE DEFICIENCY (MADD). BT Sponseller[1], SJ Valberg[2], NE Schultz[2], HE Bedford[3]. 1. Iowa State University College of Veterinary Medicine, Ames, IA. 2. University of Minnesota College of Veterinary Medicine, St. Paul, MN. 3. Performance Equine Vets, Aiken, SC.

Seasonal pasture myopathy (SPM; atypical myopathy [AM] in Europe), typified by nonexertional rhabdomyolysis, occurs in pastured horses during autumn or spring. Clinical signs rapidly progress from muscular weakness to recumbency and frequently death. Extensive myonecrosis and intramyofiber lipid storage occur in highly oxidative respiratory and postural muscles. Recently, a defect of lipid metabolism called MADD has been identified in European horses with AM. This report documents the first cases of equine MADD in the United States.

Six Midwestern US horses suspected of having SPM in the spring or fall of 2009 were evaluated for MADD by urine organic acids, plasma acylcarnitines and/or muscle carnitine and histopathology. Five horses had clinical signs and clinicopathologic data consistent with severe rhabdomyolysis. One horse was found dead on pasture after 2 days of rear limb stiffness and inappetance. Urinary organic acid profiles revealed markedly elevated ethylmalonic and methylsuccinic acids, butyrylglycine, isovalerylglycine, and hexanoylglycine, consistent with equine MADD. Plasma acylcarnitine profiles from 2 horses had marked elevations of short chain acylcarnitines, while the third horse and only survivor had minor elevations of short chain acylcarnitines. Affected muscle showed extensive degeneration with intramyofiber lipid accumulation, a marked decrease in free carnitine, and high levels of carnitine esters.

SPM appears to be a highly fatal emerging disease of pastured horses in the US characterized by weakness, colic-like signs and myoglobinuria. The disease is associated with a defect in muscular lipid metabolism that can be diagnosed by performing lipid staining of muscle samples and urine organic acid profiles.

ASSESSMENT OF SERUM AMYLOID A (SAA) TESTING AND ITS CLINICAL APPLICATION IN A SPECIALIZED EQUINE PRACTICE. Rodney L. Belgrave[1], Meranda Dickey[2], Kristopher L. Arheart[3], Carolyn Cray[2]. 1. Mid-Atlantic Equine Medical Center, Ringoes, NJ; 2. Division of Comparative Pathology, Department of Pathology, University of Miami Miller School of Medicine, Miami, FL; and 3. Department of Epidemiology and Public Health, University of Miami Miller School of Medicine, Miami, FL.

Serum amyloid A (SAA) has been determined to be a major acute phase protein in the equine species. In the current study, SAA levels were quantitated using an automated assay and compared with the traditional assessments of inflammation and infection such as total WBC count, fibrinogen, and serum protein changes as determined by electrophoresis in both healthy and clinically ill animals. Significant differences were observed in SAA levels, A:G ratio, and total WBC count between clinically normal and abnormal patients. There was no significant difference with respect to fibrinogen between the two groups. Marked variation was apparent with respect to mean SAA levels in clinically abnormal horses whereas only mild variation was present in the mean A:G ratio, total WBC count and fibrinogen of the same group. Poor correlation was observed between SAA and WBC count and a SAA and fibrinogen. A weak negative correlation was noted between SAA and A:G ratio. SAA was the analyte most consistently increased in all clinically abnormal horses indicating it can provide valuable information regarding the clinical state of the patient. Repeated measures in clinically ill animals demonstrated a smoother gradual return to normal levels versus fibrinogen in clinically ill patients. In some cases, fibrinogen levels remained elevated despite normalization of the SAA and resolution of the clinical disease and total WBC count was unaffected throughout the treatment period. The marked variation in SAA levels in clinically ill horses and rapid decrease in elevated levels seen in response to successful therapy, allow SAA to be more useful as a monitoring tool and prognostic indicator than traditional markers of inflammation.

THE 'SECOND GENOME' OF THE HORSE: METAGENOMIC INVESTIGATION OF THE EQUINE INTESTINAL MICROBIOME. JS Weese, M Costa, A Sturgeon, H Staempfli. University of Guelph, Guelph, ON, Canada.

The intestinal tract houses one of the richest and most complex microbial populations on the planet. In humans, the gastrointestinal microflora (microbiome) is believed to consist of ~1000 bacterial species at up to 1 trillion bacteria/g, and has been referred to as our 2nd genome. It is increasingly evident that this microbiome plays a critical role in health and a wide range of diseases. Presumably, the situation is comparable in horses, yet there has been limited study. The objective of this study was to use next generation sequencing to investigate the equine gut microbiome in a non-culture dependent, metagenomic fashion.

Fresh feces were collected from 6 healthy horses. Following DNA extraction and purification, 16s rRNA gene PCR was performed. Emulsion PCR and next generation sequencing were then performed and sequences were evaluated using the MG-RAST server.

Between 6579 and 21600 high quality, approximately 500bp sequences were obtained per horse, with overall evaluation of 40,483,773 bp. In all horses, the Firmicutes Phylum predominated, ranging from 49-73% (mean 65%, SD 12%) of sequences. Bacteriodetes organisms were the next most abundant, ranging from 17-34% (mean 23%, SD 7.4%), followed by Actinobacteria at 6.6-19.6% (mean 11%, SD 5.5%) of sequences. Proteobacteria were highly variable, ranging from 0.3-32% (mean 12%, SD 15%). The only 2 horses with less than 70% Firmicutes were co-housed, and they had the most similar microbiomes based on hierarchical cluster analysis of relative abundance at the Phylum level. They also had a marked increase in Proteobacteria (particularly *Acinetobacter* spp) compared to the other horses; 31% vs 2.1%. A large percentage of sequences were not identified to the Genus (or even higher) level despite being of high quality and length, indicative of organisms that have not yet been characterized. At Class level, Clostridia were dominant, ranging from 30-63% of sequences, with Bacilli (7-32%) 2nd most abundant. 100-200 Genus level identifications were present per horse, with much broader species diversity.

The species richness and number of unidentified sequences indicate the complexity of the equine intestinal microbiome. The predominance of Clostridia and related organisms is in contrast to many other species, and demonstrates the importance of this much-maligned group of bacteria. The finding that co-housed (but not directly co-mingled) horses had the most similar microbiomes indicates an effect of farm or management, and suggests that modification of the microbiome of a healthy horse may be feasible, something that could be targeted for preventive purposes. Ongoing comparison of healthy horses with horses with enteric disease, particularly colitis, will provide significant insight into the role of this microbiome in health and disease, and could lead to better diagnostic, therapeutic and preventive measures.

DIAGNOSTIC FREQUENCY, RESPONSE TO THERAPY, AND LONG TERM PROGNOSIS AMONG HORSES WITH EQUINE CUSHING'S DISEASE: 1993-2004

Rohrbach BW[1], Stafford JR[2], Willis RS[1], Reed SM[3], Schott HC[4], Andrews FM[5]
[1]Knoxville, TN
[2]Gaithersburg, MD
[3]Lexington, KY
[4]East Lansing MI
[5]Baton Rouge, LA

Equine Cushing's disease (ECD) is common in older horses. The purpose of this study was to determine the frequency of diagnosis, identify prognostic factors and assess owner satisfaction with treatment. The study was a retrospective cohort design evaluating equine accessions reported to the Veterinary Medical Data Base (VMDB) and the Ohio State University from 1993-2004. Proportional accessions, annual incidence and demographic characteristics of horses with ECD were compared with all accessions in the VMDB. Medical records for a subset of horses were extracted and owners contacted to obtain long-term follow up information.

Two hundred seventeen new cases of ECD were reported to the VMDB. Incidence increased from 0.25/1,000 in 1993 to 3.72/1,000 in 2002. Eighty-one percent of horses were $\geq$ 15 years of age. Average delay from onset of signs to diagnosis was 180 days (range 1 to 1,824 days). Hirsutism (84%) and laminitis (50%) were the most common clinical signs. Improvement in one or more signs 2 months after diagnosis was reported by 9/22 (41%) of horse owners. None of the clinical or laboratory data were associated with survival and, 50% of horses were alive, 4.6 years after diagnosis. 17/20 (85%) of horses were euthanatized and 13/17 (76%) were euthanatized due to conditions associated with ECD. Twenty-eight of 29 (97%) of horse owners said they would treat a second horse for ECD.

ECD is becoming a more frequent diagnosis. Fifty percent of horses survived 4.5 years after diagnosis and owners were satisfied with the horse's quality of life.

Supported by Centers of Excellence in Livestock Diseases and Human Health, College of Veterinary Medicine, University of Tennessee.

ASSOCIATION OF ADRENOCORTICOTROPIN, CORTISOL AND INSULIN-LIKE-GROWTH FACTOR IN SEPTIC AND NON-SEPTIC FOALS

Jenifer R. Gold, DVM, DACVIM; Noah D. Cohen, VMD, DACVIM; Lisa Caldwell, MS;
Rodney L. Belgrave, DVM, DACVIM; Thomas Welsh, PhD
College Station, TX

INTRODUCTION

The role of the hypothalamic-pituitary-adrenal (HPA) axis in critical illness and sepsis has been the subject of a great deal of research.[1-6] The importance of an intact HPA axis for metabolic and immunologic homeostasis during critical illness has also been well recognized. [1,2] The role that the somatotrophic axis plays in critical illness is less well understood. In the normal physiologic state, growth hormone (GH) is released from somatotropes in a pulsatile fashion.[7] Synthesis and secretion of GH is regulated by the integrative actions of 2 hypothalamic neurohormones; specifically the stimulatory actions of growth hormone-releasing hormone (GHRH) and the inhibitory actions of somatostatin (SS). [8] Critical illness such as sepsis causes serious metabolic derangements.[9] Several studies have linked the critically ill catabolic state and suppression of the somatotrophic axis (e.g., the relationships among GH, insulin-like growth factor I, IGF-I, and IGF-I binding protein 3 (IGFBP-3).[10,11,12] During critical illness the normal profile of GH is altered such that GH pulse frequency and peak concentration magnitude are increased.[7,10,12] It is still unclear which factor ultimately controls GH in response to critical illness.[14,15]

All these factors have been interpreted in animal and human studies as peripheral resistance to GH under times of stress and critical illness.[14-19] The peripheral resistance to GH is likely brought about by inflammatory cytokines such as tumor necrosis factor (TNFα), interleukin-1 (IL-1) and interleukin-6 (IL-6).[17,18] The hypothesis is that reduced GH-receptor expression and thus low circulating levels of IGF-I are driven by inflammatory cytokines which in turn through decreased negative feedback inhibition, increases release of GH during times of stress and critical illness.[17-19] The abundance of GH exerts lipolytic, insulin-antagonizing and immune stimulation, while the IGF-I mediated somatrophic effects are reduced.[18-19] None the less, the cellular effects of IGF-I in the peripheral tissues (indirect GH action) during critical illness are complicated and much remains to be understood.

The role of the HPA axis in critical illness in neonatal foals has become better understood in recent years.[20-23] The role the somatotrophic axis plays in septic foals is poorly described, and how the HPA and somatotrophic axis interact during critical illness is not known. Thus the purpose of this study was to assess inter-relationships of adrenocorticotropin (ACTH), cortisol, and IGF-I in septic and non-septic foals. The hypotheses were that septic foals would have a decrease in their IGF-I concentrations compared to non-septic foals and that an association would exist between the HPA and somatotrophic axes.

MATERIALS AND METHODS

Blood samples were obtained from term septic foals less than 7 days of age (n=20) admitted to Texas A&M University Veterinary Medical Teaching Hospital or Mid-Atlantic Equine Medical Center. The foals were classified as septic if they either had a sepsis score[24] $\geq$ 11 at admission or a positive blood culture. Non-septic term foals (i.e., sepsis score <11 with negative results of blood culture) <7 days of age (n=8) were obtained from the same study centers. Plasma and serum were processed from whole blood collected by jugular venipuncture upon admission, at 24 hours post admission and at 5 days post admission or at the time of discharge.

Plasma concentrations of ACTH, and serum concentrations of cortisol and IGF-I were determined by specific RIAs. Data were analyzed using linear mixed-effects modeling with foal modeled as a random effect and day of admission modeled as an ordered categorical variable; post-hoc testing of pair-wise comparisons was made using the method of Sidak. Significance was set at $P < 0.05$, and analyses were performed using S-PLUS software (TIBCO, Inc., Seattle, WA).

RESULTS

Plasma concentrations of ACTH were not significantly different between septic and non-septic foals whereas septic foals had greater serum cortisol (37 ± 8 ng/mL vs 25 ± 7 ng/mL) but lower serum IGF-I (116 ± 14 ng/mL vs 152 ± 15 ng/mL) relative to non-septic foals pooled overall sampling times. The positive association of the peripheral blood concentrations of ACTH and cortisol depended on disease status of the foals. Specifically, cortisol and ACTH were positively correlated for the septic foals ($P=0.027$) but not significantly correlated in the non-septic foals. Peripheral concentrations of ACTH and IGF-I were not significantly correlated whether data were pooled across disease status or stratified by sepsis status. However, peripheral concentrations of cortisol and IGF-I were negatively associated ($P = 0.019$); disease status did not influence this association, although the association appeared stronger among the septic than the non-septic foals.

CONCLUSION

The negative correlation between serum concentrations of the adrenal axis steroid cortisol and the somatotrophic axis peptide IGF-I may reflect interactions of these homeostatic hormones. Further studies of these and other metabolic hormones in a greater number of foals are warranted to better understand how these factors contribute to survival or non-survival of critically ill foals.

REFERENCES

1. Marik P, Zaloga G. Adrenal insufficiency in the critically ill: A new look at an old problem. Chest 2002;122:1784-1796. 2. Gibson S, Hartman D, Schenk J. The endocrine response to critical illness: Update and implication for emergency medicine. Emerg. Med Clin. North Am 205;23:909-929. 3. Annane D, Maxime V, Ibrahim F, et al. Diagnosis of adrenal insufficiency in severe sepsis and septic shock. Am J Resp Crit Care Med 2006;174:1319-1326. 4. Vanhorebeek I, Van den Berghe G. The Neuroendocrine response to critical illness is a dynamic process. Crit Care Clin 2006;22:1-15. 5. Gonzalex H, Nardi O, Annane D. Relative adrenal failure in the ICU: An identifiable problem requiring treatment. Crit Care Clin 2006;22:105-118. 6. Beishuizen A, Thijs L, The immunoendocrine axis in critical illness:

Beneficial adaptation or neuroendocrine exhaustion? Curr Opin Crit Care 2004;10:461-467. 7. Van Den Berghe G, De Zegher F, Bouillon R. Acute and prolonged critical illness as different neuroendocrine paradigms. J Clin Endocrin and Metab 1998;83:1827-1834. 8. Mesotten D, Van den Berghe G. Changes within the GH/IGF-IGFBP axis in critical illness. Crit Care Clin 2006;22:17-28. 9. Wolfe RR, Martini WZ. Changes in intermediary metabolism in severe surgical illness. World J Surg 2000;24:639-647. 10. Ross R, Miell J, Freeman E, et al. Critically ill patients have high basal growth hormone levels with attenuated oscillatory activity associated with low levels of insulin-like growth factor-I. Clin Endocrinol (Oxf) 1991;35:47-54. 11. Van de Berghe G, de Zegher F, Lauwer P, et al. Growth hormone secretion in critical illness: Effect of dopamine. J Clin Endocrin Metab 1994;79:1141-1146. 12. Van de Berghe. Novel insights into the neuroendocrinology of critical illness. Euro J Endocrinol 2000;143:1-13. 13. Voerman HJ, Strack van Schijndel RJM, de Boer H, et al. Growth hormone: secretion and administration in catabolic adult patients, with emphasis on critically ill patient. Netherlands J Medicine 1992;41:229-244. 14. Hermansson M, Wickelgren RB, Hammerqvist F. et al. Measurement of human growth hormone receptor messenger ribonucleic acid by quantitative polymerase chain reaction-based assay: demonstration of reduced expression after elective surgery. J of Clin Endocrinol Metabol 1997;82:421-428. 15. Baxter RC, Hawker FH, To C et al. Thirty day monitoring of insulin-like growth factors and their binding proteins in intensive care patients. Growth Hormone and IGF Research 1998;8:455-463. 16. Rodriguez-Arnao J, Yarwood Y, Ferguson C. et al. Reduction in circulating IGF-I and hepatic IGF-I mRNA levels after cecal ligation and puncture associated with differential regulation of hepatic IGF-binding-1, 2 and-3 mRNA levels. J Endocrin 1996;151:287-292. 17. Bentham J, Rodriquez-Arnao J, Ross RJ. Acquired growth hormone resistance in patients with hypercatabolism. Hormone Research 1993;40:87-91. 18. Timmins AC, Cotterill AM, Cwyfan Hughes SC, et al. Critical illness is associated with low circulating concentrations of insulin-like growth factors I, and II alterations in insulin-like growth factor binding proteins and induction of an insulin like growth factor binding protease. Crit Care Med 1996;24:1460-1466.
19. Gibson FA, Hinds CJ. Growth hormone and insulin-like growth factors in critical illness. Intensive Care Med 1997;23:369-378. 20. Gold J, Divers T, Barton M, et al. Plasma adrenocorticotropin, cortisol and adrenocorticotropin/cortisol ratios in septic and normal-term foals. J Vet Intern Med 2007;21:791-796. 21. Hurcombe S, Toribio R, Slovis N, et al. Blood arginine vasopressin, adrenocorticotropin hormone, and cortisol concentrations at admission in septic and critically ill foals and their association with survival. J Vet Intern Med 2008;22:639-647. 22. Costagnetti C, Rametta M, Tudor R, et al. Plasma ACTH and cortisol in normal and critically ill neonatal foals. Vet Res Commun 2008;32(Suppl 1)S127-S129. 23. Hart KA, Slovis NM, Barton MH. Hypothalamic-Pituitary-Adrenal Axis Dysfunction in Hospitalized Neonatal Foals. J Vet Intern Med 2009;23:901-912. 24. Brewer B, Koterba A. Development of a scoring system for the early diagnosis of equine neonatal sepsis. Equine Vet J 1988;20:89-95.

REPETITIVE STIMULATION OF THE COMMON PERONEAL NERVE AS A DIAGNOSTIC AID FOR BOTULISM IN FOALS. M Aleman, DC Williams, JE Nieto, KG Magdesian, RJ Brosnan, D Feary, H Hilton, T Kozikowski, J Higgins, JE Madigan, RA LeCouteur, University of California, Davis, CA.

Botulism is a potentially fatal paralytic disorder which definitive diagnosis is difficult. The purpose of this study was to investigate if repetitive stimulation of the common peroneal nerve will aid in the diagnosis of suspected botulism in foals.

Four healthy foals were used for its comparison with 3 foals with suspected botulism. Controls were anesthetized and affected foals were sedated to avoid risks of anesthesia. The common peroneal nerve was chosen for its superficial location and easy access. Stimulating electrodes were placed along the common peroneal nerve. For recording, the active and reference electrodes were positioned over the midpoint and distal end of the extensor digitorum longus muscle, respectively. Repeated supramaximal stimulation of the nerve was performed utilizing a range of frequencies (1 to 50 Hz). Amplitude, area under the curve and percentages of decrement or increment for each M wave over subsequent potentials for each set of stimuli were analyzed.

Baseline M waves were decreased in affected foals compared to controls. A decremental response was seen at all frequencies in control foals. Decremental responses were also observed in affected foals at low frequencies. However, an incremental response in amplitude and area under the curve was seen in all affected foals at 50 Hz.

Reduced baseline M waves with incremental responses at high rates are supportive of a presynaptic neuromuscular disorder which botulism was the most likely cause in these foals. Repetitive nerve stimulation is a safe, simple, fast, and non-invasive technique that can aid in the diagnosis of suspected botulism in foals.

EFFECTS OF WEIGHT GAIN ON ACTIVE GLUCAGON-LIKE PEPTIDE-1 CONCENTRATIONS IN HORSES WITH HISTORIES OF OBESITY AND INSULIN RESISTANCE. K. Chameroy, N. Frank, S. Elliott, A. Schuver. Department of Large Animal Clinical Sciences, College of Veterinary Medicine, University of Tennessee, Knoxville, TN.

Insulin resistance (IR), obesity, and type 2 diabetes affect glucagon-like peptide 1 (GLP-1) concentrations in humans and rodents, but this incretin hormone has not been examined in horses. We therefore hypothesized that GLP-1 concentrations would change in horses as obesity and IR were induced or exacerbated by overfeeding. Six horses previously diagnosed with Equine Metabolic Syndrome were provided with twice the amount of digestible energy required for maintenance as sweet feed and hay for 8 weeks. Intravenous and oral glucose tolerance tests (OGTTs) were performed at 0 and 8 weeks. Effects of time and period (0 and 8 weeks) were assessed by repeated measures ANOVA.

Mean body weight increased from 438 ± 61 kg (range, 381 to 533 kg) to 464 ± 61 kg (range, 394 to 550 kg) over 8 weeks, with individual horse weight gain varying from 2 to 10%. Mean body condition score increased ($P = 0.006$) from 6 ± 2 (range, 4 to 8.5) to 8 ± 1 (range, 7 to 9). Three horses developed mild laminitis. Glucagon-like peptide 1 concentrations increased over time during OGTTs ($P = 0.023$), but the period × time effect was not significant ($P = 0.141$). Area under the GLP-1 curve remained unaffected by weight gain, whereas area under the insulin curve increased ($P = 0.003$) over time, indicating a reduction in insulin sensitivity. Obesity and IR were induced or exacerbated when horses previously diagnosed with EMS were overfed, but GLP-1 concentrations did not change as a result.

USE OF HYPERTONIC SALINE IN ENDURANCE HORSES: A RANDOMIZED CLINICAL TRIAL. CL Fielding[1] and KG Magdesian[2]. 1. Loomis Basin Equine Medical Center, Loomis, CA. 2. School of Veterinary Medicine, University of California, Davis, CA.

Hypertonic saline solution (7.2%) (HSS) is an intravenous fluid used for the emergency treatment of intravascular volume deficits. The use of this fluid in horses with severe dehydration is controversial. The purpose of this study was to compare the use of HSS and isotonic saline solution (0.9%) (ISS) for the emergency treatment of endurance horses.

Endurance horses eliminated from competition and requiring intravenous fluid therapy were eligible for enrollment in the study. Twenty-two horses were randomly assigned to receive 4ml/kg of either HSS or ISS along with 5 L lactate Ringer's solution (LRS). Following this bolus, all horses were treated with an additional 10L of LRS. Blood and urine samples were collected before, during and after treatment. Data was compared using two-way ANOVA with repeated measures.

As compared to ISS, HSS horses showed a greater decrease in PCV ($p=0.04$), total protein ($p=0.01$), albumin ($p=0.01$), and globulin ($p=0.02$). HSS horses showed a greater increase in sodium and chloride ($p<0.001$) as compared to ISS horses. Horses receiving HSS had a shorter time to urination ($p=0.03$) and lower specific gravity ($p<0.001$) than those receiving ISS.

Results of this study indicate that HSS may provide faster restoration of intravascular volume deficits than ISS in endurance horses receiving emergency medical treatment. More profound electrolyte changes should be expected with HSS however.

EFFECT OF INHALED SALBUTAMOL ON PULMONARY CIRCULATION TRANSVASCULAR FLUID DYNAMICS AND GAS EXCHANGE IN EXERCISING HORSES

M. Vengust[1], G. Heigenhauser[2], C. Kerr[3], H. Staempfli[3], J. Pringle[4], L. Viel[3]. [1]University of Ljubljana, Ljubljana, Slovenia; [2]McMaster University, Hamilton, Canada; [3]University of Guelph, Guelph, Canada; [4]Swedish University of Agricultural Sciences, Uppsala, Sweden.

β_2-adrenergic receptor agonists have been shown to increase erythrocyte carbonic anhydrase activity, which may stimulate the Jacobs-Stewart cycle and increase pulmonary circulation transvascular fluid fluxes during exercise. Increase in pulmonary transvascular fluid fluxes ($J_{V\text{-}A}$) and consequent increase in the pulmonary interstitial fluid would be detrimental for alveolar O_2 exchange during the fast erythrocyte transition time across the pulmonary capillaries. Therefore, we hypothesised that treatment with inhaled β_2-adrenergic receptor agonist will increase $J_{V\text{-}A}$ and the alveolar-arterial PO_2 difference ($AaDO_2$) during exercise.

Six STB horses were exercised on a high-speed treadmill at 80% VO_2 peak until fatigue. Horses were randomly assigned to treatment with salbutamol (Sal: 500mcg) or placebo (Control: Con) inhalation via Aeromask* 60min prior to exercise, with cross over treatment used at the repeated exercise test (8 days later). Arterial and mixed-venous blood, as well as CO_2 elimination and O_2 uptake, were sampled simultaneously at rest, during exercise at 60sec intervals until fatigue, and into recovery. Blood gases were analyzed. $AaDO_2$ was calculated using the inspired PO_2 (149 mmHg), and blood partial pressure of O_2 and CO_2. Blood volume (%) changes across the lung were calculated from changes in hemoglobin and hematocrit values in venous and arterial blood. Cardiac output (Q) was calculated using the Fick equation. $J_{V\text{-}A}$ was calculated using Q and blood volume changes across the lung. Variables were analyzed using two-way repeated-measures ANOVA ($P<0.05$).

The duration of exercise to fatigue was 4.3±0.3 min and 4.4±0.4 min in both Con and Sal, respectively. At rest Sal had no effect on $J_{V\text{-}A}$, oxygen consumption (VO_2), blood oxygen saturation (SO_2) or $AaDO_2$ ($P>0.05$). At the onset of exercise $J_{V\text{-}A}$ increased in Con and Sal ($P<0.0001$) and at fatigue reached 10.0±2.4 L/min and 10.0±1.6 L/min, respectively. Treatment with Sal had no effect on $J_{V\text{-}A}$ during exercise ($P=0.9$). At the onset of exercise SO_2 and VO_2 increased in Con and Sal ($P<0.0001$). Treatment with Sal had no effect on SO_2 or VO_2 during exercise ($P>0.05$). $AaDO_2$ increased during exercise in Con and Sal ($P<0.0001$) and at fatigue reached 19.4±2.3 mmHg and 18.1±2.4 mmHg, respectively. Treatment with Sal had no effect on $AaDO_2$ during exercise ($P=0.3$).

Inhaled β_2-adrenergic receptor agonist salbutamol at the dose of 500mcg given 60min before exercise did not affect the duration of exercise to fatigue, $J_{V\text{-}A}$, VO_2, SO_2 or $AaDO_2$. Therefore, it had no detrimental effect on alveolar-capillary diffusion distance and the ventilation/perfusion mismatch in exercising horses.

INFLAMMATORY AIRWAY DISEASE (IAD) IN HORSES IS CHARACTERIZED BY A TH2 AND A TH17 mRNA EXPRESSION PROFILE. R Léguillette[1], L Beekman[1]. 1. Department of Veterinary Clinical and Diagnostic Sciences, Faculty of Veterinary Medicine, University of Calgary, Calgary, Alberta, Canada.

Inflammatory Airway Disease (IAD) and recurrent airway obstruction (RAO) represent two classes of equine lung inflammatory diseases that may share some similar immunologic mechanisms. There is evidence that Th2 cytokines and IL-17 play some role in RAO. IAD is a common condition in horses, but its pathophysiology is still not understood. The aim of the present study was therefore to determine the mRNA expression of Th1, Th2 and Th17 inflammatory cytokines, to understand the immunological mechanisms of IAD.

The mRNA expression of ten inflammatory cytokines and chemokines was measured in the bronchoalveolar fluid (BALF) of seventeen horses with IAD and compared with ten control horses. The horses were selected based on 1- their clinical signs, 2- the inflammatory cells count in the BALF, 3- their physical examination and 4- their medical history. The mRNA expression of IL-5, IL-1β, IL-6, IL-8 and IL-10 was significantly up-regulated in BALF from horses with IAD.

Furthermore, the BALF samples were subdivided in two groups based on the differential cells count 1- BALF with increased mast cells (IAD-Mast) and 2- BALF with increased neutrophils (IAD-Neutro). IL-4 was significantly down-regulated in the IAD-Neutro group compared to the IAD-Mast group. IL-17, IL-5 and IL-8 were significantly up-regulated in the IAD-Neutro group compared to the IAD-Mast group.

The present study shows that IAD in horses is characterized by a Th2 and a Th17 mRNA inflammatory expression profile and that different immunological mechanisms are involved in mast cells or neutrophils accumulation in the BALF of horses with IAD.

IN VITRO RESPONSE OF PERIPHERAL BLOOD LYMPHOCYTES FROM HEAVEY AND NON-AFFECTED HORSES TO β2-AGONIST STIMULATION. M. Werner, M.F. Ehrich, M.R. Prater, P. Eyre, S. Werre, V.A. Buechner-Maxwell. Virginia-Maryland Regional College of Veterinary Medicine. Virginia Tech, Blacksburg, Virginia.

β_2-adrenoreceptor (β_2-AR) agonists are a class of medications that promote smooth muscle relaxation and bronchodilation in horses and humans with airway disease. Activated human peripheral blood lymphocytes (PBLs) also respond to β_2-AR agonist stimulation by attenuating the production of cytokines associated with the pathogenesis of asthma and Recurrent Airway Obstruction (RAO). The aim of this study was to develop an *in vitro* technique for measuring the response of equine PBLs to stimulation with Salbutamol, a β_2-AR agonist. This method was then used to compare the response of PBLs from RAO-affected and non-affected horses to β_2-AR agonist stimulation.

PBLs from 4 RAO and 4 non-affected horses were cultured ($4x10^6$/ml) in RPMI complete media with Concanavalin A (ConA, 2ug/10^6cells) for 0, 1, or 2 days then stimulated with Salbutamol (30 minutes). Using flow cytometric techniques, response was measured by detecting protein kinase A phosphorylation of vasodilator stimulated phosphoprotein (VASP). Results were verified by western blot analysis. Activated PBLs were then incubated with ConA for one day were pre-incubated with β_2 or β_1-adrenoreceptor antagonist (ICI 118,551, Sigma®; Atenolol, Sigma®) for 15 minutes, followed by 30 minutes Salbutamol (500nM) stimulation. Results were analyzed by ANOVA or ANCOVA and differences were considered significant when $p<0.05$.

Response to β_2-antagonist was only observed in activated PBLs (pre-cultured with Con A) and was greater in cells from RAO horses as compared to cells from non-affected horses. The addition of β_2-antagonist attenuated the response of PBLs to Salbutamol while the addition of a β_1-antagonist had no effect. These findings indicate that activated PBLs from RAO-affected horses have a greater response to Salbutamol as compared to PBLs from non-affected horses, and this response is mediated mainly through the β_2-AR.

Human β_2-AR are known be polymorphic and this polymorphism results in a variable response to β_2-agonist binding that affects long term outcome in human asthmatics. Further studies are required to determine if the difference in response of PBLs from RAO affected as compared to non-affected horses is due to genetic polymorphism in the equine β_2-AR, and whether this difference is associated with a propensity for horses to develop equine RAO.

CLOSTRIDIUM DIFFICILE: UPDATE ON SPORADIC DISEASE & OUTBREAKS

J. Scott Weese, DVM, DVSc, DACVIM
Guelph, Ontario, Canada

INTRODUCTION

Clostridium difficile is a Gram positive anaerobic sporeforming bacterium that is an important pathogen in many species, including horses. First identified in healthy human infants in the early 20th Century, it was regarded as a commensal gastrointestinal bacterium and essentially ignored until the 1970s, when it was implicated as an important cause of antimicrobial associated diarrhea and essentially the sole cause of pseudomembranous colitis. *Clostridium difficile* is now known to be a critically important pathogen in humans, as the leading cause of antimicrobial- and hospital-associated diarrhea, and an emerging cause of disease in people in the general population. *Clostridium difficile* can also be isolated from a diverse range of animals, with variable clinical impacts. In horses, it is the most commonly diagnosed cause of colitis in some regions, and a potential cause of diarrhea outbreaks on breeding farms.

EPIDEMIOLOGY

Clostridium difficile infection (CDI) is a leading cause of colitis in horses in some regions, but uncommon (or at least uncommonly diagnosed) in others. Reasons for the apparent geographic variation are unknown, but may relate to regional variability in the commensal (and poorly characterized) gastrointestinal microflora. *Clostridium difficile* can cause disease in both adults and foals, ranging from mild diarrhea to rapidly fatal hemorrhagic enteritis.[1-4] All ages can be affected, including foals less than 24 hours of age. It has also been implicated as a cause of duodenitis/proximal jejunitis,[5] however this has only been reported in one region and broader study is needed.

In humans, *C. difficile* is most commonly diagnosed in hospitalized and otherwise high risk individuals, but in recent years there has been increasing recognition of the role of *C. difficile* in community-associated diarrhea, including in individuals without classical risk factors such as antimicrobial administration.[6,7] In horses, community-associated infections have predominated and most infections likely occur in horses outside of hospitals and with no recent history of hospitalization. Similarly, CDI has classically been considered an antimicrobial-associated disease in humans, but recent evidence indicates that antimicrobial therapy is not required for development of CDI.[8] While antimicrobial administration has been associated with CDI in horses[9-11] and undoubtedly plays a role in disease in many cases, there is far from an absolute relationship between CDI and antimicrobial administration. Some studies have reported that a majority of affected horses had received antimicrobials,[3,10] while in others a minority of horses with CDI had a history of antimicrobial administration.[12] In one study, there was no difference in antimicrobial exposure history for horses with CDI versus other types of colitis,[12] and the role of antimicrobials may be more as a risk factor for colitis of any sort rather than a risk factor for CDI versus other types of infectious colitis (e.g. salmonellosis).

Despite the clear role of *C. difficile* in equine colitis, it can also be found in healthy horses. Comparison of prevalence studies should be done cautiously because of differences in study populations and culture methodology, however, *C. difficile* can be found in 0-33% of healthy adult horses (usually <10%) and 1-44% of healthy foals, with higher rates in antimicrobial-treated individuals.[4,10,13-15] There is an effect of age, with younger foals more likely to harbour *C. difficile*, as has been reported in other animal species.

There can be significant diversity in the types of *C. difficile* found in horses,[16-19] including strains that are important causes of disease in humans. In general, strains found in horses are often indistinguishable from those found in people. As an example, in one study of race horses and broodmares from Ontario, Canada, the most common strain was ribotype 027/NAP1/toxinotype III.(C Medina et al, unpublished data) This is an international outbreak strain that occurs in humans in many countries, in both endemic cases and in outbreaks associated with high morbidity and high mortality. It is also the 3rd most common strain found in people in Ontario hospitals.[20] An important aspect of this strain is the presence of a deletion and mutation in *tcdC,* a gene that down regulates toxin A and B production. This results in markedly increased toxin production *in vitro*, and it may play a role in the apparent enhanced virulence that is evident clinically, at least in humans. The 2nd most common strain in horses was ribotype 001, which is the most common strain in people in the province. Ribotype 078/NAP8/toxinotype V, a strain often associated with food animals, only accounted for 5.8% of isolates, a similar prevalence to an earlier study of horses (predominantly from Canada).[18] This strain is of concern because it appears to be an emerging cause of community-associated CDI in humans and it also possesses a mutation and deletion in *tcdC*. These are different than in ribotype 027, but this strain still produces more toxins A and B than 'normal' *C. difficile* strains. While ribotype 078 has been uncommon in horses, it is possible that there is ongoing evolution of strain distribution. More recent study in the same region has identified ribotype 078 as a dominant ribotype,(A Schoster, unpublished data) suggesting that this strain may be emerging in the equine population. Further study is needed to determine whether there is any difference in the likelihood or severity of disease associated with different strains in horses. Considering the apparent over-representation of variant toxinotypes (toxinotypes III and V) and the presence of alterations in *tcdC* in ribotypes 027 and 078 that result in excessive *in vitro* toxin production, this could be an area of relevance. Another variant type, a toxin A negative/toxin B positive genotype (typically ribotype 017) has been reported in horses, sometimes at higher frequencies than typically encountered in humans.[15,17]

DIAGNOSIS

Diagnosis of CDI in horses has not been adequately investigated. The clinical standard in humans has been detection of toxins in feces. The Gold Standard is the cell cytotoxicity assay, a test that detects toxin B, but this test is expensive, time consuming, technically demanding and of limited availability. ELISAs have been developed with reasonable sensitivity and specificity in humans, and these are the mainstay of clinical diagnosis. It is accepted that there may be limitations in sensitivity of ELISAs (as well as the Gold Standard), but they are the most commonly used diagnostic tests. Unsurprisingly, these tests are often cross-applied to other species, including horses, but it is important to remember that there may be differences in test performance in samples from different species, so validation of tests is desired for use in horses. Recently, one commercial toxin A/B ELISA was shown to have a sensitivity of 84% and a specificity of 96% compared to the Gold Standard.[21] Use of this test for diagnosis of CDI in horses is therefore reasonable. While it is likely that other commercial assays would perform well in equine feces, this is not guaranteed. Since a validated assay is available for horses, it is difficult to justify the use of other tests that have not been validated. Another concern about test selection is the target. Some assays only detect toxin A, while others detect toxins A and/or B. Because of the high prevalence of toxin A negative/toxin B positive strains reported in some equine studies, the use of a test that only targets toxin A is hard to justify.

Culture is of limited usefulness for clinical diagnosis. It is time consuming and of limited availability, but the main problem is the high rate of colonization in some horse populations. Since a potentially large percentage of healthy horses carry *C. difficile*, interpretation of positive results is difficult. Further, some strains are non-toxigenic and cannot cause disease. Isolation of *C. difficile* with subsequent testing for toxin production or toxin genes can eliminate that problem but not the issue of the baseline colonization rate. The main clinical use of culture is to rule out CDI, as long as a highly sensitive method is used. Failure to isolate *C. difficile* using enrichment culture in a lab with experience isolating *C. difficile* should provide a reasonable assurance that CDI is not present.

ELISAs directed against 'common antigen' are also available. These rapid and inexpensive tests as often used in humans as screening tools because of their very high sensitivity. Accordingly, negative results quickly and confidently rule out CDI, while a positive result leads to further testing of the sample (usually by toxin ELISA). This type of approach is appealing in horses, but there is currently inadequate information supporting this approach. Comparison of one commercial antigen ELISA to direct and enrichment culture in diarrheic horses only reported sensitivities of 93% and 68%, respectively.[3] Considering enrichment culture is a better standard than direct culture, the sensitivity of this particular test is inadequate for a screening tool.

Real time PCR assays are increasingly being used for detection of *C. difficile* in humans, particularly hospitalized individuals, however there is currently no evidence supporting their use in horses. Since PCR only detects organism DNA, it cannot differentiate horses with CDI from horses that are colonized with *C. difficile* but have diarrhea of another etiology. If the baseline colonization prevalence of the population was very low, PCR could be a potentially useful tool. However, the baseline colonization rates that exist in many horse populations presumably have a significant impact on the positive predictive value of a positive PCR result. PCR might be useful to rule out CDI, if the assay was shown to be highly sensitive *in horse feces* and therefore had a high corresponding negative predictive value.

TREATMENT

Unfortunately, there is little new objective information regarding treatment of CDI in horses. Treatment is focused on supportive therapy, typically with inclusion of metronidazole. Di-tri-octahedral smectite has been shown to effectively bind *C. difficile* toxins *in vitro*[22] but clinical trials involving CDI are needed. Limited additional information can be obtained from human medicine. One area that might be of interest for CDI (and other types of colitis) is fecal bacteriotherapy (stool transplantation/transfaunation…). This 'old' technique has received much attention in the past few years because of its apparent success for the treatment of recurrent CDI in humans.[23] Small stool transplantation studies have reported astoundingly high cure rates for people with recurrent CDI, a potentially debilitating and difficult to treat condition. This may rekindle interest in both the study and use of fecal bacteriotherapy in horses. Objective research is needed to evaluate efficacy and define optimal treatment regimens. In humans, both oral and enema routes are used, and it is unclear whether enema administration in horses would allow for subsequent colonization of proximal regions of the intestinal tract. However, it is a route worth investigating because it bypasses the potential bactericidal effects of gastric acid, bile and other inhibitory substances in the proximal GI tract. Other adjunctive therapies that are sporadically used in humans include intravenous immunoglobulin and polystyrene toxin binders, but these have not been investigated in horses.

OUTBREAKS

Equine Hospitals

While CDI is the leading cause of hospital-associated diarrhea in humans, hospital-associated CDI is curiously uncommon in horses, despite their clear susceptibility to this pathogen and the occurrence of hospital-associated diarrhea outbreaks of other etiologies. There is only one published report of a CDI outbreak in an equine hospital.[24] In that outbreak, fecal toxin testing was not performed and 6 different strains were identified among 9 isolates, suggesting that if an outbreak was actually present, it was not analogous to typical enteropathogen (i.e. *Salmonella*) outbreaks in equine hospitals that involve widespread dissemination of a single strain. The limited reports of hospital-associated CDI outbreaks should not be taken as an indication that there is minimal risk. Since horses with diarrhea of any etiology are typically isolated and handled with enhanced infection control precautions, the lack of reported outbreaks may indicate that typical management practices are largely effective for containment of *C. difficile* in equine hospitals. However, the potential that outbreaks, particularly small outbreaks, could be undiagnosed or unreported cannot be dismissed, and there are periodic anecdotal reports of *C. difficile* outbreaks in facilities that are never published.

Farms

While there is limited published information, anecdotal evidence indicates that CDI outbreaks are not uncommon on farms in some regions, particularly amongst foals on large breeding farms. A common pattern involves sporadic cases of mild CDI in foals early in the foaling season, with a steady increase in the incidence of diarrhea, often to the point where virtually all foals born on the farm by the end of the season develop disease. Often, a variety of different strains are involved in single farm outbreaks, suggesting that farm-associated risk factors (e.g. antimicrobial use, hygiene practices) are more important than introduction or propagation of specific virulent strains.

Farm outbreaks can be very difficult to control. This could be the result of high levels of colonization and environmental contamination by the time the outbreak is recognized. Various methods of control have been attempted during outbreaks, including changes in foaling practices (e.g. foaling location, hygiene practices), comprehensive cleaning and disinfection, restriction of antimicrobials, prophylactic treatment of foals, and similar measures, with little evidence that any work. It is likely that once a large outbreak is underway, prevention of exposure of neonatal foals to *C. difficile* is very difficult. Despite the ability of *C. difficile* spores to persist in the environment for years, outbreaks uncommonly occur in subsequent years, although this certainly can occur. In addition to any measures implemented during an outbreak, a review of practices and potential changes should be made after the foaling season to reduce the risk of recurrence. Depopulation of barns with thorough cleaning and disinfection (as thorough as can be achieved on a farm), along with comprehensive assessment of areas such as routine antimicrobial use, stocking density, turnout and intermingling patterns, and routine infection control practices, is indicated.

In contrast to foals, outbreaks of CDI in adult horses appear to be very rare on farms. Regional variation in CDI susceptibility must be considered, however. This is highlighted by the phenomenon reported from Sweden, whereby mares of foals being treated for *R. equi* pneumonia were dying of CDI.[9] Subsequently, low dose erythromycin exposure was shown to cause fatal CDI in Swedish mares.[11] This effect has not been reported elsewhere but there must be consideration of regional variation in susceptibility and drug use. Further, since outbreaks of CDI in adult horses appear to be rare, any apparent CDI outbreak involving adults should trigger an investigation of antimicrobial exposure, including antimicrobial contamination of feed.

ZOONOTIC DISEASE RISK

The risk of zoonotic infection is unclear. *Clostridium difficile* is an important human pathogen and strains that affect horses are typically the same as those that cause disease in humans. Independent evolution of these different strains in humans and animals is exceedingly unlikely, so there must be some linkage between human and animal infections, be it human-animal, animal-human or common source environmental infection (from contamination that would have ultimately been human or animal). The relevance of this to human or horse health is not known. The risk of CDI from exposure to an infected or colonized horse is presumably quite low if basic infection control and hygiene practices are used. Risks may be higher in certain people (e.g. people treated with antimicrobials, proton pump inhibitors or chemotherapeutic agents, the elderly, people going into hospital) and from certain horses (e.g. diarrheic horses, foals, antibiotic-treated horses), but objective data are lacking. In lieu of convincing evidence to the contrary, *C. difficile* should be considered a zoonotic infection that can be acquired from horses, and general infection control and hygiene practices should be used, particularly with diarrheic horses.

REFERENCES

[1] Donaldson MT, et al. **Journal of the American Veterinary Medical Association** 1999;215:358-361.[2] Jones RL, et al. **J Clin Microbiol** 1987;25:1225-1227. [3] Ruby R, et al. **Journal of the American Veterinary Medical Association** 2009;234:777-784.[4] Weese JS, et al. **Equine Vet J** 2001;33:403-409.[5] Arroyo L, et al. **J Med Microbiol** 2006;55:605-608. [6] Centers for Disease Control and Prevention. **MMWR Morb Mortal Wkly Rep** 2005;54:1201-1205. [7] Centers for Disease Control and Prevention (CDC). **MMWR Morb Mortal Wkly Rep** 2008;57:340-343. [8] Wilcox MH, et al. **J Antimicrob Chemother** 2008;62:388-396. [9] Båverud V, et al. **Equine Veterinary Journal** 1998;30:482-488. [10] Båverud V, et al. **Equine Veterinary Journal** 1997;29:279-284. [11] Gustafsson A, et al. **Equine Veterinary Journal** 1997;29:314-318. [12] Weese J, et al. **Equine Vet J** 2006;38:185-188. [13] Båverud V, et al. **Equine Veterinary Journal** 2003;35:465-471. [14] McNamara SE, et al. **Zoonoses and Public Health** 2010. [15] Ossiprandi MC, et al. **Comp Immunol Microbiol Infect Dis** 2009. [16] Arroyo L, et al. **J Med Microbiol** 2005;54:163-166. [17] Arroyo L, et al. **Vet Microbiol** 2007;120:179-183. [18] Keel K, et al. **J Clin Microbiol** 2007;45:1963-1964. [19] Magdesian K, et al. **J Am Vet Med Assoc** 2006;228:751-755. [20] Martin H, et al. **J Clin Microbiol** 2008;46:2999-3004. [21] Medina-Torres CE, et al. **J Vet Intern Med** 2010. [22] Weese JS, et al. **Equine Vet J** 2003;35:638-641. [23] Rohlke F, et al. **J Clin Gastroenterol** 2010;44:567-570. [24] Madewell BR, et al. **J Vet Diagn Invest** 1995;7:343-346.

COMPLICATED *CORYNEBACTERIUM PSEUDOTUBERCULOSIS* INFECTIONS IN HORSES

Lutz S. Goehring, DVM, MS, PhD, DACVIM
Fort Collins, CO

Corynebacterium pseudotuberculosis (Cptbc) is a pleomorphic gram-positive, intracellular, facultative anaerobe with a worldwide distribution. It has the ability to cause disease in many species; however, horses and small ruminants are most commonly affected by the disease. Horses and small ruminants, each, are infected by different biovars with a difference in their ability to reduce nitrate.[1-3]

Common names for a *Cptbc*-infection in horses are 'Pigeon Fever', 'Dryland Distemper', or 'False Strangles', and an infection has basically 3 main clinical manifestations: i) external, (deep) subcutaneous abscesses; ii) internalized infection, and iii) ulcerative lymphangitis.[4] The rarely occurring *Cptbc*-associated ulcerative lymphangitis, which typically involves a single limb and is the result of a primary infection, needs to be distinguished from the even more infrequent presentation of a *Cbtbc*-associated purpura hemorrhagica, a vasculitis and thrombocytopenia that is similar to the pathogenesis of *Streptococcus equi subsp equi*-infections, resulting in symmetrical limb (or head) swelling and ventral edema as a result of immune-mediated, complement-activated immune complex disease.[5]

External, subcutaneous abscesses with often extremely thick fibrous capsules are most common, and include about 90% of all infections. The popular name 'Pigeon Fever' originates from the high prevalence of subcutaneous abscesses in the pectoral region of affected horses, but external lesions are also reported elsewhere on the body including the ventral abdomen, sheath, mammary gland, head and limb musculature.[6]

Internal abscesses and internal organ involvement occur infrequently (<10%), and are more insidious in onset and nature.[7] Renal involvement has been described in approx. 25% of internal abscess development, but other organs, single or in combination, can be involved: liver, lungs, spleen. An infection can also to lead to abortion and fetal infection.[4,8-10] 'Pigeon Fever' is common in the western United States with epidemics reported in Colorado during 1997 and 2009. [11,12]

C. pseudotuberculosis can persist for extended periods in the environment on infected fomites and in the soil. 'Pigeon Fever' is typically a disease of summer and fall. Although its epidemiology has not been entirely elucidated, it is thought that flies (*Haematobia irritans, Musca domestica, Stomoxys calcitrans*) function as mechanical vectors to transmit infectious organism from horse to horse through feedings on skin abrasions. Contact of a wound with contaminated surfaces or soil are other possibilities of infection.[9,11] Periodic reoccurrence of outbreaks with epidemic proportions very likely have a multi-factorial background involving previous winter and spring conditions favoring insect (vector) survival and replication, existing equine herd immunity, relative humidity, and others.

Internalized disease commonly presents with lethargy, pyrexia, weight loss, abdominal pain and/or respiratory signs depending on abscess location.[7] Complete blood count and biochemistry panel findings are most consistent with chronic-active inflammation: leukocytosis caused by a mature neutrophilia, anemia, hyperfibrinogenemia and hyperglobulinemia. A specific diagnosis of a *Cptbc*- infection is corroborated by a serum synergistic hemolysis inhibition (SHI) test result of a titer ≥ 512 or by culture. Biomarkers of specific organ system dysfunction such as azotemia or increased liver enzyme activity are helpful in further localizing the disease; however, imaging modalities and further testing of a variety of body fluids are often necessary to conclusively determine the abscess location.[7] In a number of cases the only indicator for a putative abscess location in the horse's abdomen remains an increased nucleated cell count and protein increase in the abdominal fluid. Clinical outcome is largely dependent on the location and size of an abscess. Case fatality rates are much higher in horses with internal abscesses (40%) relative to horses with only external lesions (0.8%).[4] However, case fatality rates can approach 100% in horses with internal abscesses that are not treated with antimicrobials. Clinical resolution can be achieved, with prolonged antimicrobial therapy, and a success rate of approximately 70% is reported.[7] A favorable outcome is more likely to be accomplished if an abscess can be either drained or surgically removed in addition to antimicrobial therapy. *Per os* administered potentiated sulfonamides, chloramphenicol, doxycycline or doxycycline in combination with rifampin have empirically shown efficacy; however, these antimicrobials require a long-term administration and owner compliance. An internal *Streptococcus equi subsp equi* abscess, 'Bastard Strangles', a foreign body abscess, or abscess otherwise should always be considered as differential diagnoses.

REFERENCES

1. Biberstein EL, Knight HD, Jang S. Two biotypes of Corynebacterium pseudotuberculosis. Vet Rec 1971;89:691-692. 2. Costa LR, Spier SJ, Hirsh DC. Comparative molecular characterization of Corynebacterium pseudotuberculosis of different origin. Vet Microbiol 1998;62:135-143. 3. Songer JG, Beckenbach K, Marshall MM, et al. Biochemical and genetic characterization of Corynebacterium pseudotuberculosis. Am J Vet Res 1988;49:223-226. 4. Aleman M, Spier SJ, Wilson WD, et al. Corynebacterium pseudotuberculosis infection in horses: 538 cases (1982-1993). J Am Vet Med Assoc 1996;209:804-809. 5. Pusterla N, Watson JL, Affolter VK, et al. Purpura haemorrhagica in 53 horses. Vet Rec 2003;153:118-121. 6. Farstvedt EG, Hendrickson DA, Dickenson CE, et al. Treatment of suppurative facial cellulitis and panniculitis caused by Corynebacterium pseudotuberculosis in two horses. J Am Vet Med Assoc 2004;224:1139-1142, 1112. 7. Pratt SM, Spier SJ, Carroll SP, et al. Evaluation of clinical characteristics, diagnostic test results, and outcome in horses with internal infection caused by Corynebacterium pseudotuberculosis: 30 cases (1995-2003). J Am Vet Med Assoc 2005;227:441-448. 8. Miers KC, Ley WB. Corynebacterium pseudotuberculosis infection in the horse: study of 117 clinical cases and consideration of etiopathogenesis. J Am Vet Med Assoc 1980;177:250-253. 9. Perkins SL, Magdesian KG, Thomas WP, et al. Pericarditis and pleuritis caused by Corynebacterium pseudotuberculosis in a horse. J Am Vet Med Assoc 2004;224:1133-1138, 1112.10. Poonacha KB, Donahue JM. Abortion in a mare associated with Corynebacterium pseudotuberculosis infection. J Vet Diagn Invest 1995;7:563-564. 11. Hall K, McCluskey BJ, Cunningham W. Corynebacterium pseudotuberculosis infections (Pigeon Fever) in horses in western Colorado: an epidemiological investigation. J Eq Vet Sci 2001;21:284-286. 12. anonymous: www.news.colostate.edu/Release/2692.

NEW PERSPECTIVES ON SALMONELLA: NOSOCOMIAL & COMMUNITY-ACQUIRED

Jill R. Johnson, DVM, MS, DABVP, DACVIM
Baton Rouge, LA

INTRODUCTION

Nosocomial *Salmonella* infections are a major concern in veterinary hospitals. Outbreaks of salmonellosis have dictated closure of our equine hospital on several occasions in the past decade and prompted implementation of stringent biosecurity measures. Expanded surveillance culturing of hospitalized patients revealed that the proportion of horses culturing positive for *Salmonella* exceeded expectations. This retrospective analysis of information is an attempt to identify factors associated with this observation.

MATERIALS AND METHODS

Fecal samples were collected for salmonella culture from hospitalized equine patients at the Louisiana State University Veterinary Teaching Hospital between 2000 and May 2010 for diagnostic and surveillance purposes. From 2000 – 2004, selected groups considered to be 'high risk' (colics, patients on IV fluids) were sampled. From 2004 until May 2010, all hospitalized patients were sampled daily for the first 5 days of hospitalization then weekly. Cultures were performed by the Louisiana Animal Disease Diagnostic Laboratory. Serotyping of *Salmonella* isolates (one isolate per admission) was performed by National Veterinary Services Laboratory. Cultures were designated as 'diagnostic' if the sample was submitted because the horse exhibited diarrhea or 'surveillance' if the sample was part of the routine hospital *Salmonella* surveillance program. Outpatients were not sampled.

Hospital admissions were categorized as salmonella-positive if one or more cultures were positive during hospitalization. The day of hospitalization for which the first positive culture was obtained was calculated from the admission date and culture submission date. Positive admissions were grouped into community-acquired or hospital-acquired based on the day of first positive culture. A range of cutoff points ranging from 1-5 days was examined.

The stall location of the patient during hospitalization was identified from hospital records. Using a geographic information system (GIS) (ArcGIS 9.3, ESRI), a map of the hospital floor plan was constructed and various spatial feature layers were developed. Using Tracking Analyst® extension, admissions were temporospatially mapped in the hospital.

Based on zip code of the owner, trainer or referring veterinarian recorded in the medical record, an approximate geographic location of the horse prior to admission was determined. Using a GIS, locations of patients prior to admission were mapped by zip code area.(Golden R and Schell JD, 2008)

Timelines were constructed showing occurrence of positive admissions in relation to total hospital admissions.

FINDINGS

The number of cultures and the number of admissions positive for *Salmonella* are shown in Table 1. Eleven horses shed more than one serotype during a single admission. Four-hundred twelve patients were admitted more than once (range 2-6 hospitalizations). Of these, 34 (8.2%) were cultured positive on more than one admission.

Table 1. Number of cultures and admissions positive for *Salmonella*.

	Total	# *Salmonella*-positive	% *Salmonella*-positive
All Cultures	14,784	1,438	9.7%
All Admissions	4,469	699	15.6%

Fifty-four serotypes of *Salmonella* were identified. Of these, 32 'orphan' serotypes were identified in only one patient each. The most commonly identified serotypes occurring in at least 12 positive admissions were *S. javiana, S. anatum , S. newport, S. braenderup, S. typhimurium*, and *S. rubislaw*. (Table 2)

Table 2. Most common serotypes identified among positive admissions

Serotype	# Admissions	% of positive admissions
All positive admissions	699	
S. aviana-positive admissions	336	48.0
S. anatum -positive admissions	60	8.5
S. newport-positive admissions	45	6.0
S. braenderup-positive admissions	19	2.7
S. typhimurium-positive admissions	17	2.4
S. rubislaw-positive admissions	12	1.7
Salmonella not serotyped	123	17.5
'Orphan' serotypes	32	4.5

Five hundred thirty four of 4,469 admissions were cultured because of diarrhea, 133 of which were *Salmonella*-positive (24.9%). Of 3,944 admissions sampled for surveillance, 577 were positive for *Salmonella* (14.6%). (Table 3)

Table 3. Number of admissions positive for any serotype of *Salmonella* based on reason for culturing

Sampling Reason	# Admissions in group	# *Salmonella* positive	% *Salmonella* positive
Diagnostic (diarrhea)	534	133	24.9%
Surveillance	3,944	577	14.6%

Two major temporal clusters of cases, one involving *S. anatum* (2002) and the other *S. javiana* (2006-2010), were observed. Temporal clusters of other serotypes were observed (mini-outbreaks) on several occasions. Spatially the positive admissions were widely distributed in the hospital.

Depending on the cutoff point selected to distinguish community- vs hospital-acquired, a substantial numbers of cases were assigned as community-acquired. (Table 4)

Table 4. Cumulative percentage of admissions culturing positive by hospitalization day.

Group	Hosp Day 1 %	Hosp Day 2 %	Hosp Day 3 %	Hosp Day 4 %	Hosp Day 5 %
All *Salmonella*	14.3	31.5	54.2	73.1	84.6
S. javiana (all)	9.5	25.9	51.2	74.1	87.2
S. javiana (dx)	16.7	41.7	66.7	83.3	91.7
S. javiana (survey)	8.3	23.3	48.6	72.6	86.5
Orphans	28.6	50.0	53.6	75.0	82.1
Surveillance	12.0	29.3	53.4	71.9	84.2
Diagnostic (diarrhea)	24.8	41.3	57.9	78.2	86.5

The geographic origin of all horses presented to the hospital was evenly dispersed regionally. The distribution of 'orphan' serotypes (arguably community-acquired) did not appear to cluster spatially. Horses with community-acquired *S. javiana* infection (based on a 2 day cutoff) predominantly originated from southern Louisiana and Mississippi.

DISCUSSION

The aggregate rate of isolation of *Salmonella* during the period of study was unexpectedly high (15.6% of admissions) and triggered aggressive biosecurity measures in the hospital. However, when *Salmonella* culture results were examined by serotype, it was clear that the problem was not so much a salmonella problem as multiple *Salmonella* problems. Without serotyping the scope of the problem would have been misinterpreted.

The rate of *Salmonella* isolation was higher in diagnostic samples from horses exhibiting diarrhea (24.9% vs. 14.6% for surveillance samples) but the rate of shedding in surveillance samples was still unexpectedly high.

Epidemiologically, the cutoff between community-acquired and hospital-acquired is often set at 2 or 3 days. Using this criterion, 30-50% of hospital admissions from which salmonella was cultured could be considered to be community-acquired infections. Clinically, 5 cultures over 5 days is often used as the basis for ruling out infection and shedding. (Hyatt DR and Weese JS, 2004) Applying this criterion, as many as 85% of *Salmonella*-positive horses could potentially be community-acquired infections. In support of this, some of the orphan serotypes were not cultured until day 7 of hospitalization. There is a paucity of information with regard to experimental salmonellosis in horses that would indicate the infectious dose or time course of events post-inoculation with various serotypes.

The serotype responsible for our most recent major outbreak, *S. javiana*, is a reptile-associated strain of *Salmonella* which was only sporadically detected in horses in our population until 2006. Geographically it occurs primarily in the southeastern part of the United States. This strain is one of the most common serotypes reported in people since 2000 and association with reptiles has been identified as a risk factor (Srikantiah P et al. 2004). A change in the herpetofaunal diversity and abundance in various habitats following hurricanes Ivan (2004) and Katrina and Rita (2005) resulted in displacement of certain reptiles to higher ground (Schriever et al) potentially increasing exposure of other inhabitants of the shared environment to their pathogens. Additionally, following hurricanes Katrina and Rita, transient water contamination with fecal-indicator organisms was documented (Demcheck et al, 2007). *S. javiana*-positive horses classified as community-acquired (based on positive culture within the first 2 days of hospitalization) primarily originated from geographic locations in the southern part of Louisiana and Mississippi that were heavily impacted by the hurricanes.

In a practical sense for biosecurity, a high rate of community-acquired infection strongly suggests that one should assume that <u>all</u> patients present a risk and take appropriate precautions to avoid nosocomial transmission. Environmental decontamination and screening have become a significant part of our hospital biosecurity protocol.

ACKNOWLEGEMENTS

This project would not have been possible without the assistance of Dr.Susan Eades, Dr. Alma Roy, Keith Neal, Genuine Parker, Chris Truman and Kerry Hawthorne.

REFERENCES

Demcheck DK et al. in Farris et al. eds. Science and the Storms: the USGS Response to the Hurricanes of 2005. U.S. Geological Survey Circular 1306. : 239, 2007. Golden R and Schell JD. Environmental Health Perspectives 116:a18, 2008. (zip code and GIS) Hyatt DR and Weese JS. Vet Clin Equine 20:577, 2004. Schriever TA et al. Wetland 29:112, 2009. Srikantiah P et al. Epidemiol. Infect. 132:273, 2004

ANTIMICROBIAL RESISTANCE: IS IT RELEVANT TO EQUINE VETERINARIANS?

Imogen Johns, BVSc, DACVIM, MRCVS
North Mymms, UK

INTRODUCTION

Antimicrobial drugs (AMD) are essential for human and animal health and welfare. Since the advent of antimicrobial use, morbidity and mortality rates for infectious diseases in both humans and animals have decreased substantially. However, within several years of the appearance of these drugs, resistance of bacteria to previously effective drugs was observed. Antimicrobial resistance (AMR) is defined as the ability of a micro-organism to withstand the effect of an AMD. Although the ability of bacteria to resist the action of antimicrobials pre-dates the advent of these drugs, exposure to antimicrobials can promote the emergence of resistant strains of bacteria, either by promoting the survival of inherently resistant strains or by inducing the expression of existing AMR genes. Since the initial observations of resistant isolates, AMR in human pathogens has become recognised as one of the key developing health concerns globally, the fear being that resistance within certain disease-causing organisms will render them untreatable with currently available drugs.[1] The situation is compounded by the fact that it appears unlikely that new classes of antimicrobials will be developed in the near future.

Resistance to an antimicrobial may be inherent/intrinsic to bacteria, such as resistance of gram-negative bacteria to macrolides due to an inability to cross the cell wall, or it may be acquired. Acquired antimicrobial resistance typically occurs via acquisition of genes from other bacteria that confer resistance either to a single AMD or in some instances, to multiple AMD. Resistance genes can be located either on the bacterial chromosome, or most commonly, on mobile genetic elements such as plasmids. Conjugation, the process by which plasmids move from one bacterium to another, allows transfer of resistance genes even between unrelated bacteria. These resistant genes can also be transferred via transformation and transduction, resulting in potential widespread dissemination of resistance genes.[2] The ease with which these resistance genes can be transferred between bacteria accelerates the risk of spread of resistance.

Antimicrobial resistance, in particular multi-drug resistant isolates, has traditionally been associated with hospital infections. For example, methicillin-resistant *Staphylococcus aureus* (MRSA) is one of the most important causes of hospital associated infection. Recently strains of MRSA have been identified as causes of community acquired infections.[3] The widespread dissemination of resistant isolates highlights not only the ease with which AMR can be spread, but also the relevance the subject has to all practitioners, hospital based or not.

ANTIMICROBIAL RESISTANCE IN ANIMALS

Antibiotics are widely used in animals, both as therapeutic agents and, in some areas of the world, as growth promoters in food animal species. As in human medicine, AMR is of major concern in animals, not only because of the potential for treatment failure, but also because of the potential for zoonotic transfer of resistant isolates or genes. Despite considerable attention, research and speculation regarding the role which antimicrobial use in food animal species has in the development of antimicrobial resistance in human pathogens, the link remains uncertain. Antimicrobial resistance is a well recognised phenomenon in bacteria isolated from food animals, but the extent to which these isolates are then transferred to humans is unclear. [4,5]Despite this, because of the importance to human health, regulatory bodies are placing increasing pressure on veterinarians to limit antimicrobial use. As an example, in 2009, the Chief Medical Officer in the UK called for cephalosporins and fluoroquinolones to be banned from veterinary use.

ANTIMICROBIAL RESISTANCE IN HORSES

In comparison with humans and food producing animals, relatively few studies have addressed the emergence of resistance in horses. The development of AMR in bacteria that reside and cause disease in predominantlyequine species is likely, however, to have implications for both human and horse health and welfare. As in other species, the emergence of AMR in equine pathogens has the potential to limit the efficacy with which infectious diseases in horses can be treated. Widespread AMR in important equine pathogens would result in increased morbidity and mortality, increased cost of treating infections and may result in the use of antimicrobials which potentially should be reserved for treating humans.

In addition, although horses are considered food producing animals in some parts of the world, for the most part they are considered companion animals. The close contact that exists between horses and their handlers raises the issue that AMR bacteria could also be transmitted, not via the food chain, but via direct contact. Although specific studies addressing this possibility have not been performed in horses, there is evidence in other animal species.[6,7] For example, an increase in the proportion of tetracycline resistant enteric commensals was identified in both poultry and poultry farm workers following introduction of tetracycline supplemented feed on the farm.[7]

COMMENSAL BACTERIA AND ANTIBIOTIC RESISTANCE

During treatment with AMD, resistance can develop in both the targeted pathogen, as well as in commensal bacteria such as enteric *E coli*.[8] Monitoring for AMR can thus be performed by surveying clinical isolates for resistance, or by evaluating resistance patterns of commensal isolates. The information regarding AMR of commensal organisms is relevant not only because they have the potential to become pathogens, especially in immunocompromised hosts, but more importantly because they act as a vast reservoir of mobile resistance genes which can potentially be transferred between animals and humans.[9]

ANTIBIOTIC RESISTANCE IN SELECT EQUINE PATHOGENS

Salmonella

Nosocomial infections of *Salmonella*, including multi-drug resistant serotypes *S. typhimurium* and *S. newport* have been reported in several large animal hospitals in recent years.[10,11] Whilst multi-drug resistance (MDR) is not typically associated with increased virulence, it does complicate treatment if required, by limiting the spectrum of appropriate antimicrobials. Outbreaks of *Salmonella* can have major impacts on veterinary hospitals, ranging from patient morbidity and mortality through to disruption of clinical services and loss of client confidence. The emergence of MDR equine isolates is also of concern because *Salmonella* represents a significant zoonotic pathogen. In humans, *Salmonella* is a leading cause of food borne infections, with infections increasingly being caused by drug resistant isolates. Whilst the link between *Salmonella* infection in horses and infection in people is not as clearly established as in other food animals, an outbreak of MDR *Salmonella newport* in humans in France was attributed to the consumption of contaminated horse meat.[12] In addition, the close physical contact between owners/handlers and their horses suggests that direct transmission may occur, as has been shown in small animals.[6]

MRSA

Methicillin-resistant *Staphylococcus aureus* has for some time been a major cause of nosocomial infections in human hospitals. There are also increasingly frequent reports of severe community-associated MRSA infections in people.[3] MRSA possess an altered penicillin binding protein that renders the bacteria resistant to all β-lactam antimicrobials, and frequently to a wide range of other antimicrobials. In recent years, MRSA has also been identified in horses, in both community and hospital settings.[13,14] In horses, MRSA infection is most commonly associated with surgical site infections and joint infections, and, as in humans, can also be identified as nasal colonizers in asymptomatic horses. MRSA infections in horses are typically associated with a good outcome, but can be difficult to treat because of the limited number of effective antimicrobials. Both horses with nasal carriage of MRSA and those with clinical infections can play a role in the zoonotic transfer of the bacteria. Although transmission of equine-origin MRSA to humans has been documented, the potential role of horses in human MRSA colonization and infection requires further investigation.[14] It seems most likely that, based on low colonization rates in healthy animals, horses act as contaminated vectors rather than as true reservoirs of the organism.

Rhodococcus equi

A recent study evaluated the prevalence of resistance to macrolides and rifampin in clinical isolates of *R equi*, and its impact on foal survival.[15] An overall prevalence of 3.7% was identified, with foals infected with resistant isolates being significantly less likely to survive. Although antimicrobial resistance is not uniformly associated with poor outcome, this study highlights the challenges that equine clinicians face when patients are infected with resistant bacterial isolates.

CHANGING RESISTANCE PATTERNS IN EQUINE PATHOGENS

Because of the lack of effective surveillance schemes, little is known about changes in AMR over time in equine pathogens. In one study, bacterial isolates from horses with ulcerative keratitis over 10 years were collated, and sensitivity patterns compared between isolates collected earlier in the decade to those later.[16] A statistically significant increase in resistance of *Pseudomonas aeruginosa* isolates to gentamicin and tobramycin was found between the isolates from 1992 to 1998 and those from 1999 to 2000. An increase in resistance of *Streptococcus equi* subspecies *zooepidemicus* to gentamicin was found between the isolates from 1993 to 1997 and those from 1998 to 2000. From this study, and from information acquired via human surveillance systems, it is evident that antimicrobial resistance patterns are fluid and can change depending on various factors, including selection pressure. Ongoing surveillance and dissemination of resistance patterns is thus a vital element in the control of AMR.

ANTIBIOTIC RESISTANCE IN EQUINE COMMENSAL BACTERIA

Several studies, both within Europe and the USA have investigated the occurrence of antimicrobial resistant commensal bacteria in hospitalized horses and those treated with antimicrobials.[17-22] One study evaluated the effect of peri-operative penicillin use on skin commensal *Staphylococci*.[19] At hospital admission, horses harbored isolates that were susceptible to antibiotics or resistant to only one group of drugs. After 3 days, the percentage of *Staphylococcus* isolates displaying antibiotic resistance, as well as the number of resistance genes per isolate, increased moderately in hospitalized horses without surgery or penicillin treatment but dramatically in hospitalized horses after colic surgery as well as penicillin treatment

In a survey of fecal *E coli* isolates collected from both hospitalized horses and those housed in livery premises in the UK, the proportion of antibiotic resistant isolates was significantly higher in hospitalized horses (48% v 12%; P<0.0001).[20] In this study, MDR isolates (resistant to 4 or more antibiotic classes) were identified in 38.8% of horses. The effect of antibiotic administration on commensal fecal *E coli* has been evaluated in several other studies.[17,18,21] A recent study performed in Ireland indicated that resistance patterns in these bacteria can change within three days of antibiotic administration in hospitalized horses.[21] A similar trend was also detected in hospitalized horses that were not treated with antibiotics, suggesting that hospitalization alone increases the chance of identifying resistant isolates in feces.

Whilst the information regarding the development of these resistance patterns is of value, whether or not the resistant isolates persist for a long period of time may be more relevant, as this could increase the potential for dissemination of AMR. Studies in humans, cattle, pigs and poultry have shown that resistance develops in commensal isolates during treatment with AMD, and have also determined that these resistant isolates can still be present for up to several months after AMD discontinuation.[8,23-25] There is however, no information available regarding persistence of resistant isolates in horses. If, as in other species, resistant isolates persist in the feces of treated horses, these isolates could potentially be transferred to in-contact horses and other animals (or humans) even after treatment has ceased, increasing the potential for dissemination of AMR.

The results of a prospective study investigating the development and persistence of antimicrobial resistance in commensal enteric *E coli* in horses treated with antibiotics will be discussed.

REFERENCES

[1] Anon 2001. www.who.int/drugresistance. [2] Tenover. **CID** 2001;33:S108. [3] Boucher, et al. **Clin Infect Dis** 2010;51:S183. [4] Angulo, et al. **J Vet Med** 2004;51:374. [5] Phillips, et al. **J Antimicrob Chemother** 2004;53:24. [6] Wright, et al. **Emerg Infect Dis** 2005;11:1235. [7] Levy, et al. **NEJM** 1976;295:583. [8] Lowrance, et al. **Am J Vet Res** 2007;68:501. [9] Johnson, et al. **Infect Immun** 2001;69:1306. [10] Dallap Schaer, et al. **J Vet Intern Med** 2010;24:1138. [11] Tillotson, et al. **JAVMA** 1997;211:1554. [12] Espiè, et al. **Euro Surveill** 2003;27:2252. [13] Weese, et al. **J Vet Intern Med** 2006;20:182. [14] Weese, et al. **JAVMA** 2005;226:580. [15] Giguère, et al. **JAVMA** 2010;237:74. [16] Sauer, et al. **Vet Ophthalmol** 2003;6:309. [17] Williams, et al. **J Vet Intern Med** 2009;23:427. [18] Dunowoska, et al. **JAVMA** 2006;228:1909. [19] Schellmann, et al. **J Clin Microbiol** 2006;44:4444. [20] Ahmed, et al. **Ann Clin Microbiol Antimicrob** 2010;9:12. [21] Bryan, et al. **Irish vet J**.2010;63:373. [22] Grønvold, et al. **Vet Microbiol** 2010;145:366. [23] Lautenbach, et al. **Antimicrob Agents Chemother** 2006;50:3939. [24] Delsol, et al. **J Appl Microbiol** 2003;95:1226. [25] McDermott, et al. **Appl Environ Microbiol** 2005;71:4986

STANDARD AND NOVEL TREATMENTS FOR SOLID TUMORS IN EQUIDS

Jeffrey Phillips, DVM, MSpVM, PhD, DACVIM (Oncology)
Knoxville, TN

INTRODUCTION

In this lecture I will present some of the more commonly seen tumors in horses. The list is certainly NOT all-inclusive; only the most clinically applicable tumors are presented. The presentation follows an organ-system based approach where all the relevant tumors within an organ-system are presented concurrently. Standard and novel treatments are presented for each tumor type.

Skin and Subcutaneous Tumors

Skin and subcutaneous tumors are the most commonly diagnosed neoplasms in equids. These tumors have both benign and malignant variants. The main treatment option is of course surgical resection. Alternate treatment options are presented.

Equine Sarcoid

Sarcoids are locally aggressive, fibroblastic skin tumors that are the most common skin tumor in horses comprising ~90% of skin tumors (~20% of all tumors)

- Signalment:
  - There is no apparent age or sex predilection (but see below) but the majority of cases are seen in horses <4 yrs
  - Predilection for thoroughbreds and geldings has been described
- Etiology:
  - Unknown etiology but strong evidence to suggest that bovine papillomavirus (types 1 and 2) are associated with pathogenesis
  - Mode of transmission unknown
  - No evidence of heritability
- Behavior:
  - Tumors can appear as benign wart-like lesions that can spontaneously regress (rare cases) to large ulcerated fibrous growths
  - Can be found anywhere on the body but more common on the head, lips, neck, around the eyes, legs; rarely found on the back. Also seen at the sites of previous trauma
  - Six clinically recognized forms based on their appearance and behavior:
    - Occult, nodular, verrucose (warty), fibroblastic (proud-flesh), mixed (warty and proud-flesh), and malevolent
    - Names are based on clinical behavior and histopath appearance
    - Fibroblastic, mixed, and malevolent types are more aggressive; histologically these can be diagnosed as soft tissue sarcomas when there is no obvious skin involvement of the tumor.
  - Lower-grade do not metastasize but higher-grade tumors can (i.e. those "called" soft tissue sarcomas
  - May spontaneously regress months to years after diagnosis
- Histopathology:
  - All tumors contain both epidermal and dermal components. The more aggressive variants may invade the deeper tissues. Again, if there is no skin (dermal) involvement, the more aggressive variants are usually diagnosed as soft tissue sarcomas (e.g. fibrosarcoma, etc).
  - Dermal components consist of invasive fibroblasts and connective tissue components
- Diagnosis:
  - Typically made based on clinical appearance and location
  - Histopathology is required to differentiate between other benign and malignant tumors
- Prognosis:
  - Generally good to excellent
  - Periocular lesions are often more difficult to treat due to anatomic constraints
  - Recurrent lesions tend to be more aggressive than original tumors

Hemangiosarcoma

Malignant tumors of endothelial cell origin. Major features of these tumors in horses are listed below:

- Signalment:
  - Seen in horses of all ages but more common in middle aged to older horses
  - Thoroughbreds are overrepresented
  - No gender predilection
- Behavior:
  - Highly aggressive both locally and systemically.
  - Common metastatic/primary sites include lung and pleura (77%), skeletal muscle (46%), and spleen (43%)
  - Primary cutaneous tumors comprise <10% of all tumor sites

- Clinical signs:
  - Pend on tumor location but include epistaxis, anemia, increased heart and respiratory rates, SQ swellings, lameness, leukocytosis, thrombocytopenia, fever, depression, joint effusion, etc.
  - Hemorrhage into SQ or body cavities
- Prognosis:
  - Generally guarded to grave for most cases; early cases may be effectively treated with surgery. Actinic (purely cutaneous with no SQ involvement) tumors are generally considered less aggressive than other sites.

Mast Cell Tumors

These tumors are much less common in horses than in cats or dogs. Behavior is as follows:

- Signalment: No age (seen as young as at birth), sex (some support to males being more frequently affected than females but numbers are small), or breed predilection. Important to rule out parasitic infection +/- ivermectin therapy.
- Behavior:
  - Most frequently involve the head, neck (similar to cats), and legs but other sites have been described.
  - Usually develop in the skin or SQ but visceral sites are also possible
  - Two distinct types are noted: single cutaneous nodule and the disseminated form where multiple lesions are noted
  - Metastasis (visceral) has not been reported; however, the multiple cutaneous form may represent disseminated disease
  - Usually considered benign tumors and occasionally regress spontaneously
- Histopathology:
  - Lesions are characterized by the presence of well-differentiated mast cells (most commonly) with an extreme eosinophilic infiltrate. These lesions must be differentiated from parasitic abscesses such as those caused by onchocerca.
  - Focal areas of necrosis are common within the tumors
  - No established grading system has been described
- Clinical Signs:
  - Typically raised cutaneous swellings which may appear bruised or reddened. May be solitary or multiple
- Prognosis:
  - Generally good to excellent. Disseminated cases have a guarded prognosis but should rule out parasitic involvement.

Melanocytic Tumors

Melanocytic tumors are seen in virtually all species. In large animals they are most common in horses (up to 15% of all skin tumors); specifically grey horses.

- Signalment: Seen in both grey and non-grey horses but more commonly grey. No age or sex predilection. Some thought that malignant versions are more common in non-grey horses but little proof. Incidence increases with age up to 80% (for grey horses) by age 15.
- Etiology: In most horses, the origin is NOT thought to be related to sun exposure. In grey horses, the gene associated with "graying" leads to increased risk of tumor formation. Furthermore, there are specific SUBSETS of grey horses that have **mutations** in the agouti gene (gene responsible for bay colored horses) that further increase risk for tumor formation and lead to the generalized melanomatosis presentation. These mutations are inherited in an autosomal dominant fashion and can be tested for.
- Behavior:
  - More than 90% are benign at initial presentation (nevi or dermal melanomas), but up to 2/3 can progress to malignancy.
  - There are four distinct clinical syndromes in horses:
    - Melanocytic nevus are solitary superficial masses seen in both grey and nongrey. Surgical excision is curative. These are benign as name suggests.
    - Dermal melanoma are typically seen only in grey horses and can occur singly or with multiple lesions. Incidence is extremely high in grey horses (numbers as high as 50% noted). They are locally invasive and can spread to visceral sites but metastatic rate is low.
    - Dermal melanomatosis this is the more aggressive version of the melanoma and is only clinically different than above tumor. Multiple cutaneous tumors and visceral involvement is noted. Only noted in grey horses. Essentially, this is the malignant version of dermal melanoma in grey horses
    - Anaplastic malignant melanoma can occur in any coat color horse and are highly aggressive. They typically consist of solitary skin lesions with visceral involvement. Histologically they differ from dermal melanomas by being anaplastic. Clinically they are highly metastatic
- Histopathology:
  - Nevus are well differentiated tumors that show minimal infiltrative behavior
  - Dermal melanomas and melanomatosis are identical histologically and vary only in clinical behavior

- Clinical signs:
  - Most tumors arise in skin of perineal region, ventral tail, external genitalia, seen in guttural pouch, other areas
  - Oral site relatively common after above sites
  - See pigmented masses
  - Non-specific signs for internal lesions
- Diagnosis:
  - Signalment (grey horse) and tumor appearance usually characteristic
  - Biopsy or cytology in questionable cases
- Prognosis: Good to excellent for benign variants. Guarded to grave for malignant

Lymphoproliferative Diseases in Horses

Lymphoproliferative diseases are much less common in Equids than in small animals. The most common type seen in large animals is lymphoma. Some of you may recall several well known horses that were diagnosed with lymphoma including the Thoroughbred "Lost in the Fog" (Eclipse winning sprinter).

Lymphoma in the Horse

Unknown incidence but thought to account for ~3% of all tumors in horses. One study noted that intestinal lymphosarcoma was the most commonly diagnosed neoplasia of the intestines (38%).

- Signalment: No breed or sex predilection. Age ranges from few years to >10 years.
- Clinical Signs:
  - Depend on site of involvement but include lymphomegaly, weight loss (common), ventral edema (common), abdominal masses, mediastinal mass.
  - Lost in the Fog was noted to have a retroperitoneal mass and splenic masses
  - Can also see cutaneous masses that are either proliferative or relatively flat alopecic lesions.
  - Weight loss and ventral edema are most common signs
- Behavior:
  - Typically aggressive disease similar to presentation in dogs. Multicentric form is most common. Atypical sites much less common than in cattle or cats.
  - T-cell lymphoma accounts for majority of cases (most cells are small to medium in size); B-cell tumors often have profound T-cell infiltrates (non-neoplastic) and thus are called T-cell rich B-cell lymphoma
- Diagnosis:
  - Similar to small animals (cytology, bloodwork, biopsy, etc)
  - Abnormal lymphocytes are seen in circulation in 25-50% of cases aiding the diagnosis.
  - Serum IgM levels have been used in an attempt to provide non-invasive diagnostic modalities. Decreased IgM level is often seen in horses with lymphoma. These levels can be measured; however, the sensitivity is poor (~28%) while specificity is fairly good (~88%).
- Prognosis:
  - Few animals are treated so accurate figures are unknown. Durable remissions have been reported.

Urogenital tumors

The most common urogenital tumors include granulosa cell tumors and squamous cell carcinoma.

Granulosa cell tumors

This is the most common tumor of the equine ovary (~5-6% of all tumors in the mare). Diagnosis is typically made following appearance of clinical signs consistent with hyperestrogenism. Common features include:

- Signalment: Typically middle-aged to older animals
- Clinical signs: Signs of hyperestrogenism including heat cycles, pyometra, etc. May also see aggressive behavior due to elevations in testosterone. In rare cases these tumors can obtain massive proportions and result in abdominal distension.
- Behavior: Metastases are unusual. Clinically usually results in cystic lesions in the ovary
- Diagnosis: History, clinical signs, transrectal palpation, ultrasound, and hormonal assays are used for diagnosis. Histopath confirms the diagnosis.
- Prognosis: Good to excellent.

Large granulosa cell tumor

Squamous Cell Carcinoma of the Horse

Two different variants will be discussed: those that involve the skin or mucocutaneous junctions and those that involve the external genitalia.

Cutaneous SCC

These occur primarily around the eye and at mucocutaneous junctions. Features include:

- Signalment: Most common in middle to older age but no gender predilection. Lightly pigmented horses are more commonly affected. Horses at increased risk include Appaloosas, Clydesdales, shires, and Belgians.
- Etiology: Thought to be solar induced. Also can occur in animals that have ingested plants with photosensitizing agents. Some have suggested a viral etiology. Finally, can also be seen in previously traumatized skin (burned etc).
- Behavior: Locally invasive but rarely metastasize
- Clinical signs: Ulcerative or proliferative lesions, conjunctivitis for eyelid lesions, other signs pend site. Disease is often multifocal in any given horse.
- Diagnosis: Location, breed, appearance, histopathology in questionable cases

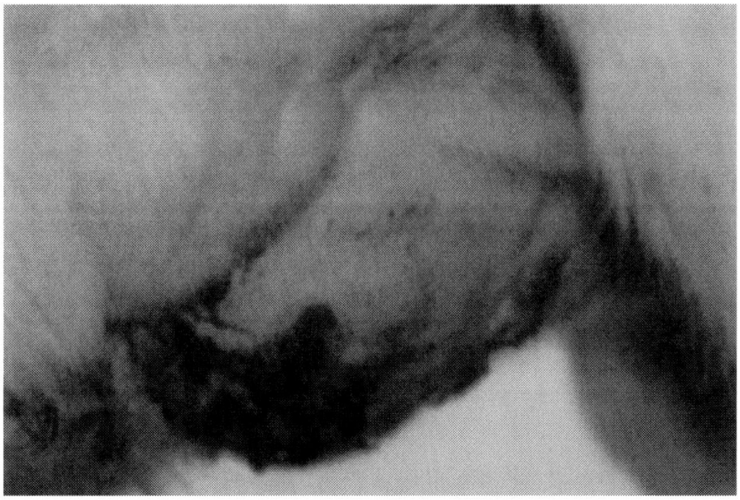

Large ulcerative preputial SCC in a horse

Urogenital SCC

This tumor is most commonly seen on the penis or sheath of intact or geldings. Features include:

- Signalment: Most common in stallions or geldings. More commonly seen in same breeds at risk for cutaneous SCC
- Etiology: Solar exposure or perhaps irritation from smega may play a role. There is some initial evidence suggesting the involvement or association of a novel equine papilloma virus and tumor formation.
- Clinical signs: Foul odor, ulceration, bleeding, masses, failure to drop
- Behavior: Very aggressive locally, metastasis can occur but are typically late in disease course
    - Lesions may be ulcerative or proliferative and can affect any portion of the penis but typically the non-pigmented portion, the glans, and the inner sheath
    - Metastasis tends to be to the draining lymph nodes
    - Local recurrence is extremely common following marginal resection
    - "Field cancerization" is also commonly seen where multiple independent lesions develop over time. In other words, if they get one lesion they are at risk for developing others in the future
- Diagnosis: Clinical signs, breed, location, biopsy
- Prognosis: Typically good but local recurrence is common. Adjuvant therapies (topical 5-FU, cryotherapy, piroxicam) may help lower the incidence of local recurrence. Concerning features such as high mitotic index or vascular invasion on biopsy may denote a more guarded prognosis.

TREATMENT

Standard Options

Surgery: Surgical resection has long been the most common treatment modality for solid tumors. Complete resection is curative but some of the more common locations are associated with a high incidence of local recurrence (periocular, facial, perirectal). Additional therapies are typically required in these select locations.

Cryotherapy: Has been used as an adjuvant therapy to surgery to sterilize residual disease after surgical resection. Must use a thermocouple of most locations. Typically requires 1-3 treatment cycles.

Immunotherapy: Commonly used and many techniques have been described including:

- Intralesional BCG is associated with ~60-70% success rates with sarcoids, unknown with other tumor types. Both modified live and killed cell wall preparations have been used. Successes are typically seen only with the minimally invasive, superficial lesions.
- Autogenous vaccines have been described for sarcoids and SCC. Studies have failed to demonstrate a significant benefit.

- Novel agents such as topically applied Imiquimod have shown response rates as high as 70-80% for sarcoids; although this is most likely to work on the superficial, minimally invasive lesions.

Chemotherapy: Two common approaches may be considered: intralesional and systemic. Drugs include:

- 5-Fluorouracil: Can be injected intralesionally at doses of 1-3 ml per tumor site (pending size of tumor). Epinephrine is typically adding to the injection at ratio of 1 part : 10 parts 5-FU. Can also be mixed with sterile sesame seed oil (1:1) immediately prior to usage then injected. Generally successful for resolution of small sarcoids and SCC. Often ineffective for melanoma. Maximum systemic dosage ~750 mg for average horse.
- Cisplatin: This has been used in a similar fashion to 5-FU. However, it is currently only available in an aqueous form (at 1 mg/ml). Previously, it was available in a lyophilized form that could be diluted to desired concentration. Technique and dilutions are same as 5-FU. Maximum systemic dosage is ~100 mg. Impregnated beads also available at 1.6 mg/bead.
- Carboplatin: Commonly used in place of cisplatin due to availability of a generic form. The platinum drugs have much higher response rates than 5-FU. Generally injected at a dosage of 1 ml/cm^3 of tumor (all drugs use this ratio of drug to tumor) while staying below maximum systemic dosage. Maximum systemic dosage ~1000 mg for average horse.
- Piroxicam: Has been used empirically to treat and/or prevent SCC at doses of 0.15 mg/kg/day. Appears well tolerated based on anecdotal observations. Responses have been seen in ~10-20% of patients.
- Multicentric or disseminated lymphoma is typically treated with chemotherapy. Virtually all chemotherapeutics have been tried in horses including:
  - Doxorobucin at 35 mg/m2 (or potentially more)
  - Vincristine at 0.5-0.7 mg/m2
  - Cytoxan 200 mg/m2
  - Prednisolone 1 mg/kg, tapering dose

Radiation therapy: May be considered for solid tumors. Cutaneous lesions are typically treated with brachytherapy (Iridium-192) and success rates reach 100% for small SCC and sarcoids. For larger or inaccessible lesions, external beam irradiation may be considered but has not been thoroughly evaluated. In general, due to the cost and limited availability radiation therapy is rarely pursued.

Hyperthermia: Involves heating the tumor and surrounding tissue to temperatures above 41 degrees Celsius. Methodologies include ultrasound, radiant, and electromagnetic energy. Electromagnetic energy is typically supplied as either radiofrequency ablation or microwave hyperthermia. Radiofrequency techniques are limited to small, superficial lesions. Microwave hyperthermia can be used for larger, deeper tumors. New methodology allows for the delivery of tumoricidal doses of heat to depths of >8 cm. This heat is often combined with chemotherapy in order to increase local response. We will present data illustrating the effectiveness of this modality for large, non-resectable tumors.

REFERENCES

1. Sundberg JP, Burnstein T, Page EH, et al: Neoplasms of equidae. JAVMA 170:150, 1977. 2. MacGillivray KC, Sweeney RW, Del Piero F. Metastatic melanoma in horses. J Vet Intern Med. 2002 Jul-Aug;16(4):452-6. 3. Johnson PJ. Dermatologic tumors (excluding sarcoids). Vet Clin North Am Equine Pract. 1998 Dec;14(3):625-58, viii. 4. Valentine BA. Equine melanocytic tumors: A retrospective study of 53 horses (1988–1991). J Vet Intern Med 1995;9:291–297. 5. Smith SH, Goldschmidt MH, McManus PM. A comparative review of melanocytic neoplasms. Vet Pathol. 2002, Nov;39(6):651-78. 6. Saulez MN, Schlipf JW Jr, Cebra CK, McDonough SP, Bird KE. Use of chemotherapy for treatment of a mixed-cell thoracic lymphoma in a horse. JAVMA 2004. 224(5):733-8. 7. Perkins GA, Nydam DV, Flaminio MJ, Ainsworth DM. Serum IgM concentrations in normal, fit horses and horses with lymphoma or other medical conditions. J. Vet Intern Med. 2003;17(3):337-42. 8. Martens A, De Moor A, Vlaminck L, Pille F, Steenhaut M. Evaluation of excision, cryosurgery, and local BCG vaccination for the treatment of equine sarcoids. Vet Rec. 2001;149(22):665-9. 9. McCue PM, Roser JF, Munro CJ, Liu IK, Lasley BL. Granulose cell tumors of the equine ovary. Vet Clin North Am Equine Pract. 2006;22(3):799-817.

NOVEL TREATMENTS FOR EQUINE MELANOMA

Jeffrey Phillips, DVM, MSpVM, PhD, DACVIM (Oncology)
Knoxville, TN

INTRODUCTION

In this lecture I will build on the concepts presented in the previous lecture while focusing specifically on melanoma. Cutaneous melanocytic tumors are among the most common tumors as previously noted. Surgical resection is considered the mainstay of therapy; but due to tumor location, curative surgery is rarely feasible. Other treatment options that can be considered include radiation therapy, intralesional cytotoxic chemotherapy, and immune modulatory therapy. Prognosis is determined by initial tumor staging, histopathology, and available treatment options. Because most horses present for treatment with locally advanced, non-resectable tumors, effective non-surgical therapies are clearly needed to improve survival in these patients. Ideally, these therapies would include drugs that have not only local but also systemic activity to either treat or prevent disease spread.

Several studies have been undertaken to elucidate the molecular basis of equine melanoma as a comparative model for human melanocytic tumors. For years, veterinarians have been aware of the increased risk for tumor formation associated with the loss of coat color due to graying. Recent work has identified the genetic basis for this premature graying as a 4.6-kb duplication in intron 6 of the STX17 (syntaxin-17) gene. This duplication appears to function as a cis-acting regulatory mutation leading to the overexpression of the surrounding genes STX17 and NR4A3. Although overexpression of both STX17 and NR4A3 were found in melanomas of gray horses, neither have been directly associated with the development of melanocytic tumors in horses. Changes in melanocortin-1 receptor signaling have also been widely studied to determine their role in melanocytic tumor development. It has long been known that signaling through the melanocortin-1 receptor is a key component in regulating skin pigmentation. Activation of the receptor leads to the upregulation of a variety of genes important for melanogenesis including microphthalmia-associated transcription factor (MITF), tyrosinase (TYR), tyrosinase-related proteins (TRYP), and others.

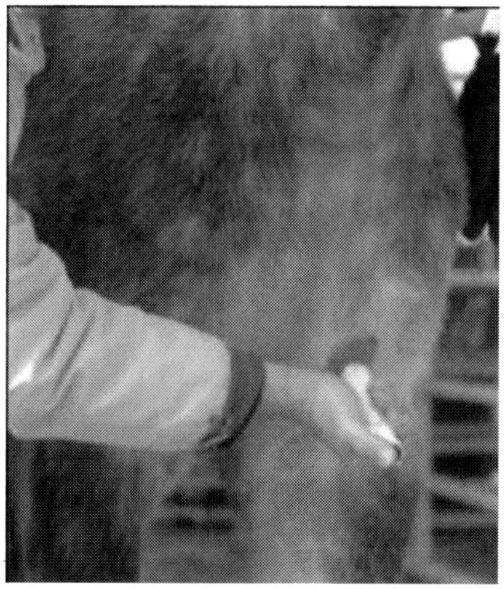

Use of needle-free injector for vaccination with HuTyr DNA

Understanding the role of melanogenic proteins in tumor formation allows for the development of molecularly targeted therapies to treat equine patients with melanocytic tumors or in some cases for prevention in those at risk for their development. A logical molecular target is the protein tyrosinase. In general, tyrosinase expression is tightly controlled both spatially and temporally. In tumor tissue, however, tyrosinase expression appears to constitutively elevated. Gray horses at increased risk of tumor formation would be expected to have elevated tyrosinase expression through enhanced signaling from the MC1R pathway.

TREATMENT

A novel methodology used to target tyrosinase in both humans and veterinary patients is DNA vaccines encoding tyrosinase. In veterinary medicine, a USDA licensed xenogenic DNA vaccine encoding *human* tyrosinase (HuTyr) is available for the treatment of canine melanoma. This xenogenic vaccine exploits the close homology of human and canine tyrosinase (92%) to generate a tyrosinase-specific anti-tumor response. In comparison, the equine tyrosinase sequence shares 90% homology to the human sequence; based on this sequence homology, cross reactivity of HuTyr DNA vaccine in the horse would be expected. Herein, we describe the use of the HuTyr xenogenic DNA vaccine in a cohort of normal horses and the development of immunologic assays to document their cell-mediated and humoral responses to vaccination. We further provide comparative immunologic results in normal dogs vaccinated with HuTyr DNA. Based on these results, this therapy may be considered for the treatment and (potentially) prevention of melanocytic tumors.

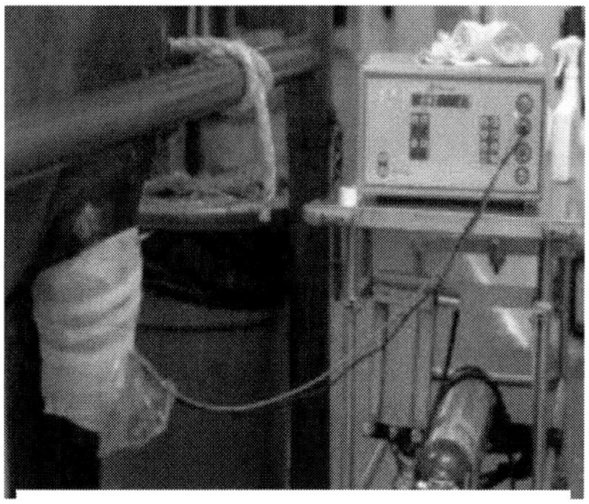

Local hyperthermia for cutaneous tumor in horse

In addition to immunologic therapies, many horses with locally-advanced melanocytic tumors require more immediate palliative therapy to reduce tumor-related symptoms. Horses with peri-rectal tumors may present with ulcerative to necrotic tumors or with compressive lesions. Surgical options for these patients are limited and immunologic approaches would be expected to produce minimal improvement. For the past two years, we have been utilizing a form of local therapy termed thermochemotherapy. In this approach, we combine intralesional doses of chemotherapy (carboplatin) with local hyperthermia of the tumor. Hyperthermia is produced by a microwave frequency applicator that can effectively heat the tumor to depths >8 cm. Treatment temperatures range from 41-45 degrees Celsius within the tumor. By combining these two treatment options, we have been able to demonstrate exceptionally high response rates and improved palliative options for advanced tumors.

REFERENCES

1. Sundberg JP, Burnstein T, Page EH, et al: Neoplasms of equidae. JAVMA 170:150, 1977. 2. MacGillivray KC, Sweeney RW, Del Piero F. Metastatic melanoma in horses. J Vet Intern Med. 2002 Jul-Aug;16(4):452-6. 3. Johnson PJ. Dermatologic tumors (excluding sarcoids). Vet Clin North Am Equine Pract. 1998 Dec;14(3):625-58, viii. 4. Valentine BA. Equine melanocytic tumors: A retrospective study of 53 horses (1988–1991). J Vet Intern Med 1995;9:291–297. 5. Smith SH, Goldschmidt MH, McManus PM. A comparative review of melanocytic neoplasms. Vet Pathol. 2002, Nov;39(6):651-78. 6. Fleury, C. Bérard F, Leblond A, Faure C, et al. The study of cutaneous melanomas in Camargue-type gray-skinned horses (2): epidemiological survey. Pigment Cell Res. 13, 47–51 (2000). 7. Seltenhammer MH, Simhofer H, Scherzer S, et al. Equine melanoma in a population of 296 grey Lipizzaner horses. Equine Vet J. 2003 Mar;35(2):153-7. 8. Rosengren Pielberg G, Golovko A, Sundström E, et al. A cis-acting regulatory mutation causes premature hair graying and susceptibility to melanoma in the horse. Nat Genet. 2008 Aug;40(8):1004-9.

EQUINE PIROPLASMOSIS: OVERVIEW AND CURRENT STATUS IN THE U.S.A.

Josie L. Traub-Dargatz, DVM, MS, DACVIM; Angela M. Pelzel, DVM; Donald P. Knowles, DVM, PhD, DACVP; Robert H. Mealey, DVM, PhD, DACVIM
Fort Collins, CO

INTRODUCTION

Equine piroplasmosis is a tick-borne disease that affects horses, donkeys, mules, and zebras. There are two distinctive EP causative agents, *Babesia caballi* and *Babesia (Theileria) equi*. It has been proposed that *Babesia equi* be given a taxonomic classification of *Theileria equi*[1] while other researchers have suggested that a more accurate classification for this parasite falls between *Babesia* and *Theileria* based on genomic analysis.[2,a] Until additional studies are available that determine a final taxonomic classification, for this proceeding paper we will refer to the parasite as *Babesia (Theileria) equi* or *B. equi*.

TRANSMISSION

B. caballi and *B. equi* are transmitted by many of the same tick vectors, with multiple types of ticks implicated in the transmission of EP agents on a global perspective. Ticks serve as a reservoir of *B. caballi* because the organism persists in the ticks through several generations, with transtadial and transovarial transmission occurring in some types of ticks.[3] In contrast, horses are the primary reservoir of *B. equi* with no transovarial transmission demonstrated to date.[3] Few countries are considered free from native transmission of EP agents via ticks and the prevalence of EP infections is consistent with the distribution of known competent tick vectors.[3]

Dermacentor (Anocentor) nitens, the tropical horse tick, was reported to be the primary vector of *B. caballi* transmission during the 1960's outbreak of EP in Florida. A joint USDA-Animal and Plant Health Inspection Services (APHIS) and State of Florida eradication program for *B. caballi* brought the outbreak under control through a multifaceted approach that emphasized tick control on horses and equine premises.[4] Prior to 2009, no tick transmission of *B. equi* was recognized in the U.S.A.

The EP agents can also be transmitted by iatrogenic means, for example, the reuse of equipment, such as needles and syringes that have become contaminated with blood, has been implicated in transmission of EP agents.[5] In addition, the use of horses that are carriers of the EP agent as a source of blood for transfusion could result in transmission.

CLINICAL SIGNS AND DIFFERENTIAL DIAGNOSES

Horses infected with either agent have similar clinical signs. Clinical signs of EP can include fever, anemia, icterus and anorexia.[3] Digestive tract signs can occur including colic, constipation or diarrhea.[3] It has been reported that *Babesia equi* can be vertically transmitted, but how often this occurs is not well documented.[6]

In countries where EP is endemic, foals born to infected mares maybe protected from clinical disease through ingestion of protective colostral antibodies.[3] Thus, in some regions of the world where infection is common, little or no clinical disease maybe observed in native horses. However, disease is frequently observed in adult horses suddenly introduced into areas with large numbers of infected ticks.

It is important to recognize that inapparent carriers represent the majority of infected horses. Because these horses appear clinically normal, diagnosis of infection relies on laboratory testing that will be described in a separate section of these proceedings. Clinical signs can alert the veterinarian to the possibility of EP. However, the clinical signs of EP are consistent with other diseases including equine infectious anemia (EIA), purpura hemorrhagica, idiopathic immune mediated anemia, and intoxications.

DIAGNOSIS

In cases having clinical signs consistent with EP, examination of blood smears can assist in diagnosis. Giemsa staining of blood smears followed by careful microscopic examination can reveal the intraerythrocytic parasites in acute cases. *B. caballi* can appear pyriform shaped and occurs in pairs (Figure 1) while *B. equi* appears as four pyriform parasites in a Maltese cross formation (Figure 2).[3] Because the parasitemia can be very low in horses that have recovered from clinical disease, the examination of direct blood smears may not allow for detection of the infection in chronic carriers.

To confirm the diagnosis in clinical cases and to detect infection in inapparent chronic carriers of EP agents, serologic tests have been used. There are three types of serologic tests that are used to detect antibodies to *B. caballi* and *B. equi*. These include the complement fixation test (CFT), the competitive inhibition enzyme-linked immunoabsorbant assay (cELISA) test and the indirect immunofluorescent antibody (IFA) test. The cELISA tests are available as kits[b], one test kit for detection of antibodies to *B. caballi* and a second kit for detection of antibodies to *B. equi* licensed by the USDA:APHIS:VS, Centers for Veterinary Biologics. Currently the recognized tests for purposes of equine importation by the World Organization for Animal Health (OIE) are the IFA and cELISA tests. Through comparison testing of sera using multiple test methods it has been determined that the cELISA test is more sensitive at detection specific antibody in carrier horses than is the CFT, however recent clinical cases of EP have illustrated that in acute infection the CFT may be positive while the cELISA test is negative until several weeks after clinical disease. Thus, within an outbreak, multiple tests may be necessary to accurately determine a horse's infection status.

Several types of polymerase chain reaction (PCR) tests have been developed and are used for research purposes. What role the PCR test will play in the future for determining a horse's infection status is still being determined.

Until 2010 all testing for EP has been conducted at the USDA:APHIS:VS National Veterinary Services Laboratories (NVSL). As of February 1, 2011 there are diagnostic laboratories in 8 states (TX, NM, CO, OK, KY, FL, LA and AR) which are approved to perform the cELISA test on domestic horses for movement purposes within the U.S.A. All import testing, testing of suspect clinical cases of EP in domestic horses, and official regulatory traceout testing is to be conducted at NVSL. In addition all positive cELISA tests at regional laboratories are to be confirmed at NVSL.

CURRENT EP SITUATION

Outbreak investigations since 2008 and surveillance findings since 2009 have revealed horses infected with EP agents are in three sectors of the U.S. equine population: 1) among horses imported prior to 2005; 2) among horses associated with an index premises in south Texas; 3) among a subpopulation of horses involved in both sanctioned and non-sanctioned racing, primarily in Quarter Horses.

Since 1970 when a test for EP infection became available, horses imported into the U.S.A. were tested using the complement fixation test until the summer of 2005. In 2005, the NVSL switched from the CFT to the cELISA test. Some horses imported prior to 2005 have subsequently bee n tested with the cELISA test for purposes of movement either interstate or into events requiring testing. Some of these horses that were previously legally imported based on a negative CFT have tested positive on the cELISA test.

A horse with clinical signs consistent with EP was confirmed to be infected with *B. equi* in October of 2009. This case prompted a regulatory response and investigation including testing of all horses on this index premises with 292 of the 360 horses on the index premises test positive for *B. equi* . There were over 2500 horses tested as part of the investigation which included traceout of the horses that had left the index premises, dangerous contacts and horses on ranches in proximity to the index premises. Based on the investigation there were 73 additional horses positive for *B. equi* located in 15 states. The epidemiologic investigation revealed there was tick transmission on the index premises and 4 other epidemiologically linked premises located nearby. Subsequent testing of ticks recovered from horses on the index premises revealed that *Amblyomma cajennense* and, to a lesser degree *Dermacentor variabilis* , were involved. No tick transmission outside of south Texas occurred.

There have been two EP outbreaks, one in Florida in 2008 and one in Missouri/Kansas, that occurred among Quarter Horses used in non-sanctioned racing. Transmission in these outbreaks appeared to be due to iatrogenic means including reuse of needles and the practice of blood boosting.

Based on the EP outbreak in Texas and the outbreaks among race horses in Florida and Missouri/Kansas, race tracks in multiple states (NM, TX, OK, CO, IA, LA, AR, FL, KY, and IN) have implemented entry testing. In addition multiple states have requirements for interstate movement of horses from areas that they perceive to pose a risk for introduction of EP agents. More than 60,000 horses have been tested for EP infection at NVSL and several thousand more have been tested at recently approved regional diagnostic laboratories. As of December 2010, based on movement testing or traceout, there have been 112 Quarter Horse racehorses in 9 states and 10 Thoroughbred race horses in 4 states detected as EP test positive.

All EP test positive horses have been quarantined, euthanatized, exported from the U.S.A. or enrolled in the USDA research initiative. The treatment of EP infected horses is not allowed in the U.S.A. other than those horses that have become officially enrolled in the recently initiated USDA treatment research project that is the focus of the second proceedings paper on EP by Dr. Robert Mealey.

Education to heighten awareness of the clinical signs of EP and potential routes of transmission of EP agents for equine veterinary practitioners and owners and trainers will be important in the control of EP in the U.S.A. There is a recently posted extensive literature review on EP[7] and also an owner education information sheet regarding protecting your horse against EP in both English and Spanish [8]available on the USDA website.

ACKNOWLEDGEMENT

Portions of this proceedings appeared previously in the American Association of Equine Practitioners 56th Annual Convention Proceedings, pages 1-7, 2010.

FIGURES

Figure 1: *B. caballi* merizoites appear pyriform shaped and occur in pairs within erythrocytes. Photograph courtesy of Dr. Donald Knowles.

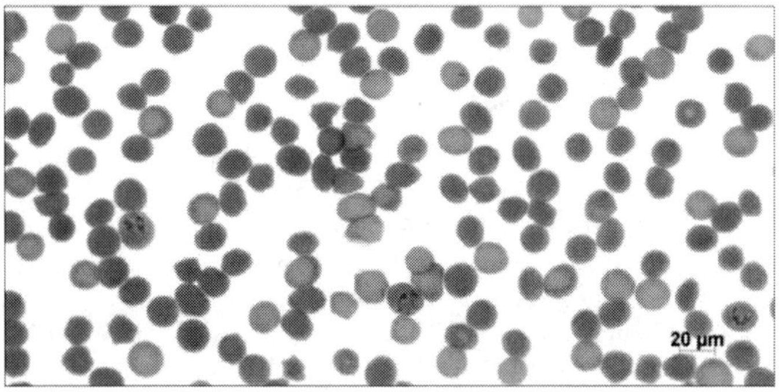

Figure 2: *B. equi* merizoites appear as four pyriform parasites in a Maltese cross formation within erythrocytes. Photograph courtesy of Dr. Donald Knowles.

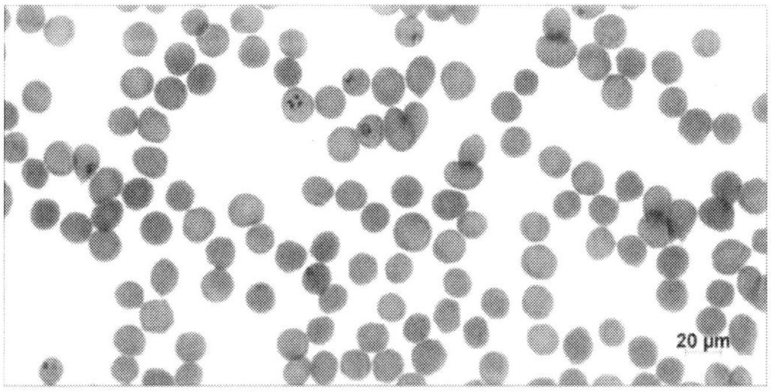

FOOTNOTES

[a] Knowles, Donald, Personal Communication on March 17, 2010.[b] Veterinary Medical Research Diagnostics, Pullman, WA, USA.

REFERENCES

[1] Mehlorn H, et al, Parasitol Res 1998, 84:467. [2] Allsopp MT et al Parasitology., 1992 108 (pt 2):147. [3] Rothschild CM et al,. In: Sellon DC and Long MT, ed *Equine Infectious Diseases*, Saunders Elsevier, St. Louis, 1st ed, 2007:465 [4] USDA:APHIS:VS. APHIS Factsheet Equine Piroplasmosis, December 2009, accessed May 4, 2010 http://www.aphis.usda.gov/publications/animal_health/content/printable_version/fs_equine_piro.pdf. [5] Gerstenberg C et al, in *Proceedings* Equine Infectious Diseases VIII, 8th International Conference, Dubai, 1998, Wernery U et al Editors, R and W Publications (Newmarket): Newmarket UK,1999: 217.[6] Donnellan CMB et al, , In Mair TS and Hutchinson RE ed *Infectious Diseases of the Horse*, Equine Veterinary Journal Ltd, 2009:333. [7] USDA:APHIS:VS. Veterinary Services Equine Piroplasmosis Literature Review, September 2010, accessed January 31, 2011 http://www.aphis.usda.gov/animal_health/animal_diseases/piroplasmosis/downloads/ep_literature_review_september_2010.pdf [8] USDA:APHIS:VS. Veterinary Services Infosheet, Protecting your horses against Equine Piroplasmosis, April, 2010, accessed January 31, 2011 http://www.aphis.usda.gov/animal_health/animal_diseases/piroplasmosis/downloads/ep_protect_your_horses_en_sp.pdf

EQUINE PIROPLASMOSIS: CONCEPTS FOR TREATMENT AND PREVENTION

Robert H. Mealey, DVM, PhD, DACVIM; Massaro W. Ueti, DVM, PhD; Josie L. Traub-Dargatz, DVM, MS, DACVIM; Angela M. Pelzel, DVM; Donald P. Knowles, DVM, PhD, DACVP
Pullman, WA

INTRODUCTION

Babesia equi is a tick-transmitted apicomplexan hemoprotozoan parasite of equids, and along with *Babesia caballi*, is an etiologic agent of equine piroplasmosis (EP). *B. equi* is endemic in tropical, subtropical, and some temperate regions, and is thought to be more widespread than *B. caballi*.[1-3] It is estimated that only 10% of the world's horses reside in regions free of EP.[1] The United States, Canada, Australia, Japan, England, Ireland, and Iceland are not considered endemic regions. EP is reportable to the World Organization for Animal Health (OIE).[4] Clinical *B. equi* infections can be acute, characterized by hemolytic disease of varying severity. Most horses in endemic regions become infected within the first year of life, with case fatality rates of 5-10% in naïve horses. Case fatality rates can exceed 50% in previously unexposed mature horses when introduced into endemic regions.[2] Many horses become subclinically infected with low levels of parasitemia and no obvious clinical signs. Importantly, horses infected with *B. equi* are persistently infected for life,[1-3] and despite low levels of parasitemia serve as reservoirs for transmission to naïve horses.[1-3,5]

Babesia equi is the cause of most of the EP cases currently in the U.S., and therefore this presentation and proceedings will focus on *B. equi*. Based on published observations in the vertebrate host and tick vector, including the ability of sporozoites to infect equine lymphocytes and form schizonts (see below), and the absence of transstadial transmission in the tick, it has been proposed that *B. equi* be reclassified as *Theileria equi*.[6] Recent genomic analysis places the organism somewhere between the *Babesia* and *Theileria* genera (unpublished data), and for the purposes of this discussion the organism will be referred to as *B. equi*.

OVERVIEW OF *B. EQUI* LIFE CYCLE

Infected ticks feed on horses and inject *B. equi* sporozoites with their saliva. Based on published work, sporozoites initially infect lymphocytes within the new equine host.[7-9] Ongoing research is beginning to corroborate these observations (unpublished data). Schizonts form within these cells and give rise to merozoites, which are released and invade erythrocytes. Asexual division occurs within erythrocytes giving rise to four pear-shaped stages which appear as a "Maltese cross" within the parasitized erythrocyte. After a tick ingests infected erythrocytes during acquisition feeding, sexual reproduction within the tick midgut results in zygotes, which penetrate the midgut and give rise to kinetes which enter the salivary glands via the hemolymph. Finally, sporozoites form in the tick salivary glands, and *B. equi* can then be transmitted to another susceptible horse via subsequent tick feeding.

CHEMOTHERAPEUTIC AGENTS AGAINST *B. EQUI*

Several different antiprotozoal drugs and treatment protocols have been used to treat *B. equi* infections in horses. These drugs include the anti-babesial compound imidocarb dipropionate, the anti-theilerial compounds parvaquone and buparvaquone, and various drugs in combination.[2,10-13] Variable results are likely related to drug dosage and treatment regimens, differences in susceptibility and/or resistance patterns among the different *B. equi* strains evaluated, and the methods used to define treatment success. It is thought that horses in endemic regions, where the prevalence of *B. equi* infection is high, are protected against subsequent infection and disease by premunition. Premunition is protective immunity induced by persistent subclinical infection. Thus, the goal of treatment for horses in endemic regions has been to reduce clinical disease, not clear infection. However, immune responses in horses that have been cleared of infection have not been completely evaluated and it is currently not known whether these immune responses could be protective against re-infection (see below). In contrast, the goal of treatment for horses in non-endemic regions (like the U.S.) is to clear infection and eliminate the risk of transmission. Imidocarb dipropionate appears to be effective against the *B. equi* strains currently causing infections in horses in the U.S., and is likely the drug of choice for this purpose.[14] It should be noted here that EP is a regulated disease in the U.S. and treatment is only allowed under the supervision of state and federal animal health authorities.

PHARMACOKINETICS AND SAFETY OF IMIDOCARB DIPROPIONATE IN HORSES

Imidocarb dipropionate (Imizol®, Intervet/Schering-Plough Animal Health) is a diaminide of the carbanilide series of antiprotozoal compounds and is labeled for the treatment of babesiosis in dogs. The mechanism of action is not precisely known. In horses injected with a single intramuscular (IM) dose of 2.4 mg/kg, imidocarb dipropionate distributes rapidly and is not detected in plasma 12 hours after administration.[15] The drug is eliminated more slowly in horses however, with urine concentrations similar to peak plasma concentrations for up to 36 hours after treatment and fecal drug concentrations detectable out to 10 days post treatment.[15] These pharmacokinetic data suggest that imidocarb is rapidly sequestered in tissues and can remain in the body for a prolonged period. Studies in sheep injected with a single IM dose of 3 mg/kg and in pigs injected with 2.0 mg/kg (IM, daily for 3 days) found that imidocarb dipropionate was detected at the highest concentrations in the liver and kidneys.[16,17] A 30-day storage period to peak concentrations was observed in the liver of sheep,[16] while in pigs, imidocarb was detected in the liver and kidneys out to 56 days.[17] Taken together, these findings suggest that current imidocarb dipropionate treatment protocols in the horse could result in high levels in storage tissues such as the liver and kidneys, leading to a prolonged antiprotozoal effect. Alternatively, the release of sub-lethal levels of drug from these tissue reservoirs might promote resistance in *B. equi* organisms that are not eliminated.

Toxicity has been evaluated in horses after two daily IM injections of 2, 4, 8, 16, or 32 mg/kg of imidocarb dipropionate. The LD_{50} was 16 mg/kg, with death occurring within six days of the first injection.[18] Dose-dependent nephrotoxicity and hepatotoxicity was observed, and death was attributed to acute renal cortical tubular necrosis and acute periportal hepatic necrosis induced by the 16 and 32 mg/kg doses.[18]

The protocol for treating *B. equi* infection in horses (with the goal being elimination of the parasite) consists of four IM injections of 4 mg/kg imidocarb dipropionate administered at 72 hour intervals.[1,2,11,19] Adverse effects are common but usually transient, and are primarily due to anticholinesterase effects. These include spasmodic colic, diarrhea, and inappetence.[2,19] To minimize these side effects, pretreatment with atropine sulfate has been suggested,[19] while others have suggested the use of N-butylscopolammonium bromide (Buscopan®, Boehringer Ingelheim) when colic signs become severe and/or prolonged (Dr. Benjamin Espy, personal communication). Local injection site reactions can also occur.[19] It has been concluded that this treatment regimen has no clinically significant deleterious effects on hepatic function based on the lack of elevations in serum GGT and bile acids in experimental ponies.[19] Although transient, elevations in urine GGT:creatinine ratios occur and mild azotemia can develop during treatment in experimental ponies, indicating that changes in renal function are possible and should be considered.[19]

EFFICACY OF IMIDOCARB DIPROPIONATE FOR CLEARANCE OF *B. CABALLI* AND *B. EQUI*

The treatment protocol described above (four IM injections of 4 mg/kg imidocarb dipropionate administered at 72 hour intervals) has recently been shown to clear persistent *B. caballi* infection in experimental horses and remove transmission risk.[20] In that study, horses were inoculated with the Puerto Rico strain of *B. caballi* and allowed to progress to the persistent phase of infection. After 70 days, horses were treated with imidocarb using the protocol above. Within five days of treatment, parasite levels in the blood of the treated horses were no longer detectable by quantitative polymerase chain reaction (PCR) or the more sensitive nested PCR.[20] Treated horses remained negative by PCR over a nine month period. Untreated controls remained positive by PCR throughout the entire period. Treated horses became seronegative by complement fixation test (CFT) within 56 days and seronegative by competitive ELISA (cELISA) within 201 days. Untreated controls remained antibody positive. Inoculation of 100 ml blood obtained from each of the treated horses three months after treatment did not transmit *B. caballi* to naïve recipient horses, while transfer of the same amount of blood from the untreated controls readily transmitted *B. caballi* to naïve recipient horses. Finally, uninfected transmission-competent *Dermacenter nitens* ticks acquisition fed on the two treated horses failed to transmit *B. caballi* when subsequently transmission fed on naïve horses, while ticks acquisition fed on the untreated controls successfully transmitted *B. caballi* to naïve horses within 12 days after transmission feeding.[20]

Based on evidence that experimental *B. caballi* infection was cleared with imidocarb with subsequent removal of transmission risk, numerous naturally-infected horses involved in the current Texas *B. equi* outbreak, as well as several other naturally-infected *B. equi* positive horses from other regions, have been treated with the same imidocarb dipropionate protocol. Preliminary results show that a pilot group of treated horses have become negative by quantitative PCR and nested PCR. Importantly, transmission risk has been determined for these treated horses by transfer of 100 ml blood. For each treated horse tested thus far, blood transfer has failed to transmit *B. equi* when inoculated into a highly susceptible naïve splenectomized horse. Although tick transmission studies are ongoing, the results of this pilot project are encouraging in terms of clearance of *B. equi* infection which is critical in future control strategies. Interestingly, and unlike the experimental *B. caballi* infected horses in the study above, many of the *B. equi* infected horses treated with imidocarb dipropionate have remained seropositive on the cELISA despite parasite clearance based on PCR results and challenge transmission studies described above. Although persistence of specific antibody complicates the assessment of imidocarb efficacy for the complete elimination of *B. equi*, antigen-independent models have been proposed to explain the persistence of long-term antibody titers. These models include memory B lymphocytes with special "memory" qualities that need fewer signals to mature to plasma cells, and/or the presence of long-lived antibody producing plasma cells. A clear understanding of the meaning of the presence of specific antibody in a treated PCR negative horse must be defined. It may be that persisting antibody is the result of the length of the infection prior to treatment with subsequent persistence of long lived memory B lymphocytes/plasma cells. Monitoring post-treatment antibody titers in these horses for additional lengths of time may reveal that they eventually become antibody negative. In contrast, it has been hypothesized that in PCR negative/ antibody positive horses, parasites could be sequestered in capillary beds or the bone marrow, and, with time, could emerge in the peripheral blood. Such a horse would reconvert to PCR positive and remain a transmission risk. However, it is our opinion that the collective data available thus far indicate that imidocarb treatment for complete removal of *B. equi* (and *B. caballi*) is worthy of further investigation for treatment of infected horses in non-endemic countries.

CONCEPTS FOR IMMUNOLOGIC PROTECTION AGAINST *B. EQUI*

Vaccination of at-risk horses against *B. equi* represents a promising means of control, but no effective vaccines are available. Moreover, the determinants of protective immunity are not known. During natural transmission, infected ticks feeding on equine hosts inject *B. equi* sporozoites with their saliva. Based on previously published work, these sporozoites can invade lymphocytes and form schizonts, which give rise to approximately 200 merozoites per cell.[7,9] Ongoing work is beginning to corroborate these observations (unpublished data). Merozoites then invade and replicate within erythrocytes, leading to clinical disease. Although horses infected with *B. equi* can be resistant to disease associated with re-infection, the mechanisms of this protective effect are not known. In *Theileria annulata*, a related apicomplexan of cattle with similar lymphocyte and erythrocyte stages of infection,[21] resistance to infection is associated with schizont-specific cellular immune responses.[22] In *Theileria parva*, another related apicomplexan of cattle, there is direct evidence that CD8+ cytotoxic T lymphocytes (CTL) can control infection.[23] Moreover, infected cattle develop antibodies that neutralize infectivity of *T. parva* sporozoites,[24] and a monoclonal antibody has been described that can block different strains of *T. parva* sporozoite infectivity in vitro and in vivo.[25,26]

Although the early infection events represent potential targets for immune intervention, few studies have characterized the lymphocyte stages of *B. equi* infection. Lymphocyte infection by sporozoites leads to transformation into lymphoblastoid cells in vitro.[8,9] These cells are capable of stimulating memory lymphocyte proliferation and cytostatic activity,[27] suggesting that *B. equi* schizont-containing lymphocytes are targets for cell-mediated immune responses. However, no studies to date have evaluated the ability of sporozoite/schizont-specific humoral or cell-mediated responses to protect against *B. equi* infection nor have these responses been analyzed in detail. Since sporozoites are the infective forms entering the equine host, immune responses directed against them (and the earliest schizont lymphocyte stages) could prevent infection. Relevant humoral responses would include antibodies capable of blocking sporozoite entry into lymphocytes, and antibodies capable of eliciting antibody-dependent cellular cytotoxicity (ADCC) of schizont-containing lymphocytes. Important cell-mediated responses would include CD8+ CTL that eliminate schizont-containing lymphocyte target cells in an MHC class I-restricted manner, and CD4+ helper T lymphocytes that promote IgG class switching, activation of macrophages, and enhancement of CTL killing.

More is known about immune responses directed against the erythrocyte stage of *B. equi* infection. The spleen has a critical role since horses with an intact spleen often control *B. equi* parasitemia and survive while splenectomized horses do not.[28,29] However, the spleen is not sufficient for control in the absence of adaptive immunity because in contrast to immunocompetent foals, foals with severe combined immunodeficiency (SCID), which have intact spleens and innate immune responses, rapidly develop high levels of parasitemia and severe clinical disease following inoculation with blood stabilate containing *B. equi* merozoite-parasitized erythrocytes.[30] Because SCID foals completely lack functional B and T lymphocytes,[31-33] these results indicate that adaptive immune responses are required for control. Although the precise mechanisms of adaptive immune control of *B. equi* erythrocyte stages are not known, both humoral and cell-mediated responses are thought to be important. In immunocompetent horses, the development of merozoite-specific IgGa (IgG1) and IgGb (IgG4,7) during acute *B. equi* infection correlates with control of parasitemia, while merozoite-specific IgG(T) (IgG3,5) appears only after resolution of parasitemia.[34] Since these isotypes bind complement and interact with Fc receptors,[35] complement activation and opsonization by *B. equi* merozoite-specific antibodies likely play important roles in resolution of acute parasitemia and maintenance of long term control.[34] Work by others has shown that immunization with killed merozoites results in reduced parasitemia and clinical disease in donkeys undergoing lethal *B. equi* challenge.[36] Protective effects are associated with high titers of whole merozoite antigen-specific antibodies and merozoite antigen-specific lymphocyte proliferative responses,[36] suggesting that both antibody and cell-mediated responses contribute to *B. equi* immune control. To further dissect the role of antibodies, *B. equi* immune plasma has been adoptively transferred into SCID foals before and after *B. equi* stabilate challenge. No protection occurred in five foals despite high merozoite-specific serum antibody titers. Parasitemia was delayed in one foal with moderate titers, suggesting that although merozoite-specific antibodies contribute to protection, antibody titers alone do not predict protective effects (unpublished data). Clearly, additional work is needed to define the determinants of immune protection against *B. equi* infection.

AREAS OF ONGOING AND FUTURE RESEARCH

The dilemma for equine health and international commerce is the stark contrast in management protocols between regions of the world endemic and non-endemic for *B. equi* infection. The equine industry in the U. S. has indicated a desire to continue their non-endemic status with continued import controls and to find methods to remove transmission risk from infected domestic horses. In regions of the world where EP is endemic, the treatment goal is resolution of clinical signs by reduction in the level of acute parasitemia but with persistence of low-level infection and presumed immunity against superinfection. Should our current chemotherapeutic research prove successful in clearing *B. equi* infection in horses and thus remove transmission risk, data mostly derived from other persistent parasitic infections indicate cleared horses would again be susceptible to re-infection and clinical disease. Our preliminary data indicate that imidocarb diproprionate removes transmission risk from *B. equi* infected horses, at least with the U. S. isolates tested thus far. An emerging issue though is the long-term (years) persistence of specific antibody in cleared horses, limiting the use of current serological methods to confirm parasite clearance. Therefore successful global management of *B. equi* infections to protect equine health and promote commerce requires the resources to complete the following research objectives: (1) methods to clear persistent *B. equi* infection and remove transmission risk; (2) methods to diagnostically confirm clearance, and (3) methods to protect cleared horses from clinical disease associated with re-infection and therefore promote international trade and equine health. Although research is ongoing and/or planned in all three areas, a current focus is understanding immunity against the early *B. equi* lymphocyte stages because these have the potential to prevent infection. Of special interest are the immune responses generated in naturally-infected horses successfully cleared by imidocarb dipropionate treatment, particularly sporozoite-specific neutralizing antibody responses and schizont-specific T cell responses, especially CTL.

REFERENCES

[1]D. T. De Waal, *Br.Vet.J.* 148, 6-14 (1992). [2]C. M. Rothschild et al., in *Equine Infectious Diseases* (2007). [3]E. Schein, in *Babesiosis of Domestic Animals and Man* (1988). [4]OIE Listed Diseases. http://www.oie.int/eng/maladies/en_classification.htm (2006). [5]M. W. Ueti et al., *J.Clin.Microbiol.* 43, 3755-3759 (2005). [6]H.Mehlhorn et al., *Parasitol.Res.* 84, 467-475 (1998). [7]U. G. Moltmann et al., *J.Parasitol.* 69, 111-120 (1983). [8]G. Rehbein et al., *Z.Parasitenkd.* 67, 125-127 (1982). [9]E. Schein et al., *Tropenmed.Parasitol.* 32, 223-227 (1981). [10]S. Kumar et al., *J.Vet.Med.Sci.* 65, 1171-1177 (2003). [11]K. L. Kuttler et al., *Am.J.Vet.Res.* 48, 1613-1616 (1987). [12]J. L. Zaugg et al., *Am.J.Vet.Res.* 50, 782-785 (1989). [13]J. L. Zaugg et al., *Am.J.Vet.Res.* 53, 1396-1399 (1992). [14]T. Bunn, http://www.usaha.org/committees/reports/2009/report-hd-2009.pdf (2009). [15]C. Belloli et al., *Equine Vet.J.* 34, 625-629 (2002). [16]O. Lai et al., *Vet.Hum.Toxicol.* 44, 79-83 (2002). [17]Z. Wang et al., *J.Agric.Food Chem.* 57, 2324-2328 (2009). [18]L. G. Adams, *Res.Vet.Sci.* 31, 54-61 (1981). [19]C. Meyer et al., *J.S.Afr.Vet.Assoc.* 76, 26-32 (2005). [20]O. N. Schwint et al., *Antimicrob.Agents Chemother.* 53, 4327-4332 (2009). [21]H. Mehlhorn et al., *Adv.Parasitol.* 23, 37-103 (1984). [22]U. Seitzer et al., *Vaccine* 26 Suppl 6, G24-G28 (2008). [23]D. J. McKeever et al., *Proc.Natl.Acad.Sci.U.S.A* 91, 1959-1963 (1994). [24]A. J. Musoke et al., *Immunology* 45, 663-668 (1982). [25]D. A. Dobbelaere et al., *Parasite Immunol.* 6, 361-370 (1984). [26]D. A. Dobbelaere et al., *Proc.Natl.Acad.Sci.U.S.A* 82, 1771-1775 (1985). [27]E. Zweygarth et al., *Zentralbl.Bakteriol.Mikrobiol.Hyg.A* 254, 281-289 (1983). [28]W. M. Frerichs et al., *Am.J.Vet.Res.* 30, 1333-1336 (1969). [29]K. L. Kuttler et al., *Am.J.Vet.Res.* 47, 1668-1670 (1986). [30]D. P. Knowles, Jr. et al., *Infect.Immun.* 62, 1909-1913 (1994). [31]T. C. McGuire et al., *Clin.Immunol.Immunopathol.* 3, 555-566 (1975). [32]E. K. Shin et al., *J.Immunol.* 158, 3565-3569 (1997). [33]R. Wiler et al., *Proc.Natl.Acad.Sci.U.S.A.* 92, 11485-11489 (1995). [34]C. W. Cunha et al., *Clin.Vaccine Immunol.* 13, 297-300 (2006). [35]M. J. Lewis et al., *Mol.Immunol.* 45, 818-827 (2008). [36]S. Kumar et al., *Vet.Parasitol.* 106, 19-33 (2002).

PHARMACOTHERAPY FOR CERVID SPECIES

James M. Jensen, DVM, DACZM
San Antonio, TX

INTRODUCTION

Medical therapy for nontraditional animals has always been compromised by the lack of species-specific drug dosage information. This situation has improved in recent years due to the publication of pharmacological research results and clinical reports of successful dosages in a wide variety of exotic animals. Drug formularies have been created for exotic pets, camelids, and chemical immobilization of wildlife.

A definitive pharmacotherapy text for deer species has not been produced. This is likely due to the fact that drug dosage information for cervids is somewhat scarce and is diffusely placed in a variety of publications. Correct drug dosing in deer requires several considerations that will be discussed in this presentation. These include drug delivery systems, body condition scoring, metabolic scaling/allometry, and proper drug selection.

White tail deer and North American elk are the prominent species of native deer that are produced in the United States. However it is important to recognize that other cervid species are farmed and ranched in good numbers. Species that originate from temperate climates include mule deer, red deer, fallow deer, sika deer, and Pere David deer. Axis deer are the most prominent tropical species and their production is concentrated in the southern states.

DRUG DELIVERY

Intravenous infusion of therapeutics and fluids is often necessary in cervid medicine. Certain products are very beneficial for these administrations. Polyurethane intravenous catheters (Jorgenson Labs, Terumo, et al) are valuable for long-term administration because they reduce the incidence of thrombosis and infection. They can be used for several days without replacement, and are more comfortable to patients than Teflon catheters because of their flexibility. They are normally placed in the jugular vein. External catheter extensions are very useful to extend the catheter system to the antlers or the withers of a patient. This allows better human access. These can be attached to the deer with skin sutures of various types such as loops, stents, and Chinese finger cuffs.

When deer are stalled for intensive therapy, spin-lock (coiled) catheter extensions suspended above the patient allow the animal to be mobile without damaging the drug/fluid delivery system. Novel efforts have also been made to create infusion devices that are portable and attached to the animal. One of the most successful attempts utilized is the Provider 5500[1], a portable infusion system manufactured by Abbott Laboratories, Hospital Products Division. This was attached to various zoo hoof stock with refractory bacterial infections via stint sutures and skin glue. It provided continuous antibiotic infusion over protracted time periods.

Oral administration of systemic drugs has limitations in deer because they are ruminants. Many therapeutic compounds are catabolized by rumen microflora and are rendered ineffective. Others may have reduced effect because of dilution and slow passage. Based heavily on domestic ruminant research and clinical evidence in exotic ruminants, the following drugs survive rumen function and can reach therapeutic systemic concentrations: sulfonamides, quinolones, benzimidazoles, haloperidol, phenothiazines, benzodiazepines, avermectins and milbemycin. It is noteworthy that several families of antibiotics are not represented on this list.

Cutaneous application of drugs on cervid species is a questionable practice. This route is affected by variables such as skin thickness, subcutaneous fat, and repellency of the hair coat. These variables in deer species can be quite different from those of cattle.

BODY CONDITION SCORING

Body condition scoring refers to a subjective evaluation of the body mass of an animal. In a 9 point system, 1 point represents extreme emaciation, 5 represents optimum body mass and 9 represents morbid obesity. Veterinarians make this assessment of patients several times a day, whether or not they assign a score to each case. Evaluating the body condition of a cervid can provide a fair estimate of the actual mass (weight) of the animal when getting an exact weight is impossible. Scoring the body condition of an animal also alerts the practitioner to discrepancies in the animal's condition that may require drug dosage adjustments. Texts that provide normal weight ranges for deer species include Exotics on the Range (Texas A&M University Press), Grzimek's Animal Life Encyclopedia Vol. 13 Mammals (Van Nostrand, Reinhold Co.) and Walker's Mammals of the World, Vol. 2, 6th ed. (The Johns Hopkins University Press).

METABOLIC DRUG SCALING

Pharmacokinetic research has been used extensively to derive safe and effective drug dosages for a wide variety of exotic species. It is best described as research that entails drug administration to a target species followed by analytical sampling of blood and/or tissue to identify therapeutic dosage regimens. This is the most definitive way to create drug doses and intervals. Pharmacokinetic research has been conducted in several deer species and but many drug regimens are extrapolated rather that scientifically proven.

Veterinarians should be very cautious about using linear extrapolation of drug dosages from domestic animals to deer species. Linear extrapolation is only effective when the domestic specie is taxonomically similar to the deer specie, the mass of the two species are similar, and the drug has wide therapeutic and safety margins.

Metabolic drug scaling is based on equations that take metabolic energy and animal mass difference into account. The known dose of certain ruminant specie can be used to create a dosage for a specific animal of another specie by a multi-step mathematical process. The mathematics requires that each specie or animal family has a research derived minimum energy cost

(MEC) equation. The following list of MEC values reveals metabolic rates of animals in a declining order. Most deer species likely have MEC values in the lower range of this abbreviated table: (In these equations W stands for the body mass in kilograms.)

Dairy cattle = $80(W^{.75})$

Beef cattle = $77(W^{.75})$

Placental Mammals (general) = $70(W^{.75})$

Llama = $59(W^{.75})$

Sheep = $56(W^{.75})$

Dromedary camel = $50.9(W^{.75})$

For various reasons drugs are either good or poor candidates for mathematical scaling. Drugs that require specific receptor sites such as narcotics and reproductive hormones are not considered scalable. Drugs that have direct flow dependent excretion with minimal hepatic metabolism seem to be the best candidates for metabolic scaling. The following drugs are considered good examples of scalable compounds: ampicillin, apramycin, carbenicillin, cephapirin, chlortetracycline, diazepam, erythromycin, gentamycin, oxytetracycline, prednisone, tetracycline, and thiamphenicol.

EXTRA-LABEL DRUG USE IN DEER

The American Veterinary Medical Association (AVMA) produced an extra-label drug use (ELDU) information outline that summarizes and clarifies the American Drug Use Act (AMDUCA) of 1994. This outline can be obtained through the AVMA website (www.avma.org). Veterinarians are well advised to comply not only with the AMDUCA guidelines but also with the U.S. Department of Agriculture's guideline for drug withdrawal times. It is wise to apply domestic cattle withdrawal times for individual drugs in deer species even if a faster clearance time is expected.

PHARMACOTHERAPY

Antimicrobials

Several injectable antibiotics have good therapeutic value in deer species. These include tulathromycin (Draxxin), ceftiofur (Excenel, Naxcel) and florfenicol (Nuflor). They are particularly effective against *Pasteurella* and *Mannheimia* species that cause significant soft tissue (i.e. pulmonary) infection in cervid species. All three drugs come in products with extended dosing intervals. This is very beneficial due to the fractious nature of deer and the stress that frequent medication can cause. The following antimicrobial dosages have been documented in literature:

Ceftiofur sodium[2] – 2.2 mg/kg, IM, q12h

Chlortetracycline[3] (elk) – 10 mg/kg, IV, q24h for 5 days to treat *Listeria monocytogenes*

Florfenicol[4] (elk) – 40 mg/kg, SC, q24h

Oxytetracycline[3] (elk) – 10 mg/kg, IV, q24h for 3 days to treat *Pasteurella spp.*

Oxytetracycline long-acting[3] (elk) – 20 mg/kg, IM to treat *Pasteurella spp.*

Penicillin G[3] (elk) – 20,000 to 44,000 IU/kg, IM, q24h for 7-14 days

Tilmicosin[5] (elk) – 10 mg/kg, SC

Oral antibiotic therapy is limited in deer species. Sulfonamides and enrofloxacin can reach appropriate systemic levels through oral administration in many ruminant species. In contrast aminoglycosides such as neomycin and spectinomycin are virtually unabsorbed by the oral route but are valuable for treating gastrointestinal infections.

Anthelmintics

Parasite control is a major preventive medicine principle for deer farms. Control programs and anthelmintics are very similar to those of domestic livestock. Benzimidazole compounds such as fenbendazole, albendazole and oxfendazole have a wide range of efficacy including, strongyles, hookworms, roundworms and tapeworms. They also have an effect on flukes. Macrocytic lactones are also considered broad spectrum de-wormers. This group includes avermectins and milbemycins. They have a wide effect on nematode parasites but no effect on tapeworms or flukes. Levamisole is the most commonly used imidothiazole compound. It has no effect against flukes or tapeworms but does treat intestinal nematodes and lungworms. The following anthelmintic doses have been recorded in literature:

Albendazole[6] (white tail deer) – 26mg/kg, PO for treatment of F*ascioloides magna*

Clorsulon[6] (white tail deer) – 24 mg/kg, PO, for treatment of *Fascioloides magn*

Febantel (elk) – 7.5 mg/kg, PO for treatment of pulmonary nematodes

Fenbendazole (red deer, fallow deer, roe deer) – 10mg/kg, PO, for treatment of pulmonary nematodes

Ivermectin[7] (roe deer) – 0.3-0.4 mg/kg, PO, q7d

Oxfendazole – 4.5 mg/kg, PO

Rafoxanide[8] (red deer) – 15mg/kg, PO for treatment of *Fascioloides magna*

Parasite drug resistance is a growing concern in all ruminants. *Haemonchus contortus* and other strongyle types are particularly problematic. When resistance is evident it may be necessary to combine drugs from different pharmacological families to improve efficacy. A distinct treatment advantage can be obtained by using a larval sensitivity test (Drenchrite) to identify the best drug or drug combination to use. It may be valuable to consider historic organophosphate anthelmintics when dealing with resistant parasites since new drug families are not emerging rapidly. Products such as haloxon and dichlorvos have not been significantly used in ruminants for several years. For this reason significant parasite resistance to them would not be expected. Proper de-worming schedules and pasture rotation are very important when dealing with such resistance.

Certain food substances have been identified as being helpful in parasite control on red deer farms in New Zealand. Chicory[9,10] and bird's foot trefoil[11] are two forage plants used on these farms. They both contain condensed tannins and chicory also has crude sesquiterpene lactones that suppress nematode larval motility. Research has documented the value they add to an effective parasite control program.

REFERENCES

[1]Spelman L, et.al. **Proc. ARAV** 1994, 279-280. [2]Drew M, et.al. **J Vet Pharm Therap** 2004:27(1):7-11. [3]Haigh J, et.al., **Farming Wapiti and Red Deer** 1993, Mosby. [4]Alcorn J, et.al. **J Vet Pharm Therap** 2004;27(5):289-292. [5]Clark C, et.al. **J Vet Pharm Therap** 2004; 27(6):385-387. [6]Foreyt W. **J Am Vet Med Assoc** 1985;187(11):1187-8. [7]Kutzer E, et.al. **Berliner und Munchener Tierarztliche Wochenschrift** 2000:113(4):149-151. [8]Rajsky D, et.al. **Helminthologia** 2002:39(2):67-70. [9]Molan A, et.al. **Parasit Internat** 2003;52(3):209-210. [10]Hoskin S, et.al. **Proc NZ Soc An Prod** 2003;63:269-272. [11]Ramirez-Restrepo C, et.al.**An Feed Sci Tech** 2004;117(1-2):85-105.

SPECIAL MEDICAL CONSIDERATIONS RELATED TO DEER PRODUCTION

James M. Jensen, DVM, DACZM
San Antonio, TX

INTRODUCTION

Deer production often creates challenges that are somewhat unique in comparison to domestic livestock. Cervid species are generally less tractable and require remote medication and chemical immobilization capabilities. Their fractious nature also contributes to high incidences of capture myopathy, rumenitis/gastritis, and enterotoxemia. These factors encourage veterinarians to place strong emphasis on animal conditioning and challenge us to develop new therapeutic approaches.

REMOTE DRUG DELIVERY SYSTEMS

Drug therapy is ineffective without a proper and exact delivery system. Physical restraint greatly facilitates correct drug therapy because it allows hands-on administration of therapeutics. Capture chutes, solid box stalls, squeeze gates, rotundas and physical restraint techniques often benefit pharmacotherapy[1]. Remote injection syringes are also valuable for drug delivery. Several manufacturers (e.g. Tel-Inject, Pneu-Dart, Dan-Inject) produce capture darts and weapon systems that can deliver fluid volumes from 0.5 to 15 ml. Most modern delivery systems use pneumatic propulsion of the dart as compared to powder charges that drove older systems. Likewise synthetic dart syringes have replaced older metal syringes. This has significantly improved the accuracy of darting and has reduced the incidence of dart-induced tissue trauma. The maximum effective range is realistically 60 meters for small syringes. However 20 meters or less is a far more reasonable darting distance for cervid species.

CHEMICAL IMMOBILIZATION

A large number of immobilization drugs and drug combinations have been used on deer species. Discussion of each of these is beyond the intent of this presentation. However it is useful to evaluate representative examples of immobilizing protocols.

A simple and reliable combination of ketamine and medetomidine has proven to be effective not only in cervid species but also in a wide variety of exotic mammals. This product is made by mixing 4.5 ml of Domitor (1 mg/ml medetomidine) with 1g of powdered (or lyophilized) ketamine. Ketamine dissolves well in Domitor with some warming. The resulting 5 ml mixture has a concentration of 200 mg ketamine and 0.9 mg medetomidine per ml. It will immobilize small cervids at a dose of 1 ml per 100 lb of body weight. Large deer species are dosed at 1 ml per 125 lb. When the ketamine has been metabolized (~30 minutes after darting) atipamezole can be given SC or IM to antagonize the medetomidine.

A relatively new immobilization drug mixture goes by the acronym BAM. This stands for butorphanol, azaperone, and medetomidine. This combination was pioneered by the specialized veterinary pharmacy, ZooPharm. This pharmacy provides the products separately by prescription. They are then mixed in a ration of 28 mg of butorphanol, 18 mg of azaperone, and 11 mg of medetomidine. It is administered at 1 ml per 100 lb of body weight. It is reversible with tolazoline/atipamezole/naltrexone given intramuscularly

Diazepam is a somewhat forgotten drug with regard to deer immobilization. However it is a safe and inexpensive product that provides good sedation and muscle relaxation to immobilization combinations. Years ago it was successfully used orally in a variety of ruminant species at the Metro Zoo in Toronto to provide deep sedation prior to restraint and shipment. It was administered in feed at dose ranges of 5-8 mg/kg. The animals were allowed 30-45 minutes induction time after eating. They became very approachable under the diazepam influence and were assisted to their feet for examination and crating. The effects lasted 3-4 hours. This is an example of innovative drug therapy that can be used in the cervid industry. It is noteworthy that the previously mentioned powdered ketamine and powdered diazepam can be purchased inexpensively through compounding pharmacies such as Gallipot and Spectrum.

These additional drugs and drug combinations represent the variety of immobilization options for deer species[2]:

Ketamine 3-4 mg/kg combined with xylazine 2 mg/kg, IM in white tail and mule deer
Ketamine 3-4 mg/kg combined with xylazine 1 mg/kg, IM in elk
Thiafentanil oxalate (A3080) 0.005 - 0.02 mg/ kg, IM in elk
Naltrexone (narcotic antagonist) 1 mg/kg, IV in elk and white tail deer
Yohimbine (xylazine antagonist) 0.1–0.2 mg/kg, divided ½ IV and ½ IM
Tolazoline (xylazine antagonist) 2-4 mg/kg, divided ½ IV and ½ IM

Constant rate infusion of analgesic compounds is often used to control pain in domestic species. When circumstances call for continuous, intensive analgesia in captive deer, a combination of buprenorphine, ketamine and lidocaine can be employed. The dosages for each drug are listed below:

Butorphanol, 0.013 mg/kg/h

Ketamine, 2.4 mg/kg/h

Lidocaine, 3 mg/kg/h

When the fluid infusion rate is calculated for the patient, the doses of the three analgesics can then be calculated and added to the administration bag. Cervids on this regimen should be monitored for proper flow rate and to observe for any indications of excessive narcosis.

One cervid specific activity that requires analgesia is velvet antler removal. The growing antler has a remarkable amount of sensory innervation and it is important to block pain during the removal process. Amino-amide local anesthesia agents such as lidocaine, mepivicaine and bupivicaine can be used to deaden the skin surrounding the antler pedicle[2]. A combination of equal parts of 2% lidocaine and 2.5% bupivicaine gives rapid induction and prolonged effect (several hours). This can be given as a ring block injected at a dose of 1 ml per linear centimeter. It is not advisable to add epinephrine to this mixture because of its potential to cause avascular necrosis of the pedicle.

LONG-ACTING TRANQUILIZATION

Long-acting tranquilizers are slowly finding acceptance in exotic hoof stock management in the United States. Many of these products have been used historically on wild ungulates to modify behavior and facilitate taming after capture. They have also been valuable in reducing stress and self-destruction during transport and habituation in new environments. Long-acting tranquilizers predominantly come from the phenothiazine and thioxanthene drug groups. They can be combined to take advantage of overlapping lengths of activity. The following drugs and drug combinations have been useful in deer species:

Azaperone[3] (white tail deer, elk) 0.2 mg/kg, IM post reversal – 6 hours of tranquilization

Haloperidol 0.25 – 0.5 mg/kg – 12-24 hours

Perphenazine enanthate[3] (red deer) 1 mg/kg, IM – 7-10 days

Zuclopenthixol acetate[4,5] (elk, white tail deer, red deer) 1 mg/kg, IM – up to 4 days

Zuclopenthixol acetate[6] (elk) 1-1.6 mg/kg, IM

HEALTH CONDITIONING

Conditioning is a concept that veterinarians should embrace whenever possible. This implies addressing an animal's physical, nutritional and psychological conditions in order to ensure successful treatment and survival. It also entails addressing any concurrent disease processes. Physical conditioning may include factors such as shade, shelter, cooling mechanisms, warmth and comfortable confinement. Nutritional conditioning implies that ample levels of appropriate diet are available. Potential deficiencies must also be addressed during nutritional conditioning (i.e. vitamin E/selenium, vitamin A, vitamin B_1). Psychological conditioning refers to the well-being of the patient. Stress can often be minimized in cervid patients by maintaining a comfortable social surrounding of conspecific animals. Reduced visual light can also have a calming effect.

Elimination of concurrent illnesses is also very important when trying to maximize the effect of good pharmacotherapy. Deer may have primary diseases that are impacted by secondary problems such as rumenitis, gastritis, endoparasites or organ specific diseases. It is imperative to simultaneously treat these complications in order to have successful recovery.

COMMON MEDICAL CHALLENGES OF DEER

Capture Myopathy

The acute stage of capture myopathy requires immediate therapeutic action to avoid death of the animal and to reduce the likelihood of permanent muscle disability. The following drug therapy should be considered:

Sodium bicarbonate, 5 meq/kg by slow IV (5minutes) – treatment of acidosis

Dexamethasone SP, 1 mg/kg, IV – shock treatment, gluconeogenesis, stabilization of cell membranes, controls pulmonary edema

Flunixin meglumine, 2.2 mg/kg, IV, q8h – anti-inflammatory, antipyretic, endotoxemia treatment

Methocarbamol, 22 mg/kg, IV, q8-12h – muscle relaxation

or

Diazepam, 0.25-0.5 mg/kg, q6-8h, IV – muscle relaxation, sedation

Vitamin E/Se – not immediately effective but is necessary for long-term recovery

Buprenorphine, 0.005-0.02 mg/kg, IV or IM, q6-12h for pain relief

Rumenitis/Gastritis

These syndromes can be primary or secondary medical issues. Rumenitis is often precipitated by a high amount of concentrated feed in the diet. It is usually characterized by rumen atony, acidosis, loss of digestive organisms (i.e. large and small protozoans) and inappetence.

Gastritis involves inflammation and ulceration of the abomasum. This produces abdominal pain and inappetance. Changes occur in the patient's hemogram and acute phase blood proteins (fibrinogen, haptoglobin) increase dramatically. Rumenitis and gastritis are often concurrent problems. Drug therapy may include:

Rumenitis

5% $NaHCO_3$ solution, 10 ml/kg, IV over 30 minutes
Magnesium hydroxide, 1 g/kg, PO via gavage in warm water
Transfaunation with a large amount of healthy rumen effluent via gavage

Gastritis

Omeprazole, 1mg/kg, PO, q24h for 7-14 days
Appropriate fluid and antibiotic therapy
Buprenorphine (previous dose)

Endotoxemia

Acute gastroenteritis is often compounded by the absorption of bacterial endotoxins through the inflamed intestinal mucosa. This is particularly threatening in young deer but all ages can be affected. The treatment regimen may include:

Aggressive isotonic fluid therapy – 1.5 to 2 times the maintenance rate (70 mg/kg); monitor electrolytes
Flunixin meglumine (previous dose) – anti-inflammatory; binds endotoxins
Polymixin B, 6 mg/kg, IV, q8h until signs improve - binds endotoxins and prevents their attachment to cell walls
Dexamethasone SP (previous dose)
Enteric antibiotics – metronidazole, 22 mg/kg, q12h, PO: neomycin 10 mg/kg, q12h, PO

REFERENCES

[1]Fowler M, et.al. **Restraint and Handling of Wild and Domestic Animals**, 2nd ed., 1996. [2]Walsh V, et.al. **New Zealand Vet J** 2002; 50(6):237-243. [3]Caulkett N, et.al. **Zoological Restraint and Anesthesia, IVIS**. [4]Read M. **Zoological Restraint and Anesthesia, IVIS**. [5]Divierio S, et.al. **Appl Anim Behav Sci** 1996;49:83-88. [6]Read M, et.al. **J Wildl Dis** 2000;36:450-459

HEMORRHAGIC DISEASE OF DEER: NATURAL HISTORY, PATHOGENESIS, CLINICAL PRESENTATION, AND DIAGNOSIS

Mark G. Ruder, DVM, David E. Stallknecht, PhD, Elizabeth W. Howerth, DVM, PhD, DACVP
Athens, GA

INTRODUCTION

Hemorrhagic disease (HD) is caused by epizootic hemorrhagic disease (EHD) viruses (EHDV) and bluetongue (BT) viruses (BTV), and has long been recognized as one of the most significant infectious diseases of white-tailed deer (WTD; *Odocoileus virginianus*) in North America.[29,19] Reports of large-scale epizootic mortality among WTD in the southeastern United States (US) date back to the late 1800s and early 1900s, but it wasn't until 1955 when the viral etiology of HD was first confirmed after EHDV-1 was isolated from WTD during an epizootic in New Jersey.[35] By 1966, the BT viruses were included as causative agents of HD following isolations from WTD and bighorn sheep (*Ovis canadensis*).[31,38] Since that time, HD has occurred regularly in the US.

The EHDV and BTV serogroups are in the genus *Orbivirus* of the family *Reoviridae*. Orbiviruses are non-enveloped, icosohedral particles that contain a genome consisting of 10 double-stranded RNA segments[32], which encode a total of 11 viral proteins, seven structural proteins (VP1-7) and four nonstructural proteins (NS1-3 and 3A). The outer shell is composed of VP2 and VP5 and together determine serotype specificity, with VP2 playing the more significant role.[32] Currently, the EHDV serogroup is divided into seven different serotypes, 1, 2, and 4 through 8.[2] Worldwide, there are currently 24 distinct serotypes of BTV.[32]

NATURAL HISTORY

Host Range and Distribution

Both EHDV and BTV are primarily arthropod-borne, infecting a variety of wild and domestic ruminants and are widely distributed throughout much of the world. Their geographic ranges have been historically limited to latitudes (approximately 40-50 °N and 35 °S) with temperate and tropical climates that supply habitat and conditions suitable for the *Culicoides* spp. vector;[27] however, the recent emergence of BTV-8 throughout northern Europe suggests these historic climatic barriers to orbiviral spread may no longer be intact. Despite the widespread distribution of viruses and the frequent exposure of wild ungulates throughout the world, clinical HD has only been reported in wildlife in North America, where it is primarily considered a disease of WTD.[19] Although morbidity and mortality is occasionally reported in mule deer (*Odocoileus hemionus*), American pronghorn (*Antilocarpa americana*), elk (*Cervus elaphus*) and bighorn sheep. This variation in susceptibility of wild ungulates to infection and clinical disease with EHDV and/or BTV is similar to that observed in domestic ruminants. For instance, historically, the most significant morbidity and mortality has been largely confined to sheep infected with BTV and WTD infected with EHDV or BTV. Although infection of cattle with these viruses is common, disease is often considered to be mild with BTV infection and absent in EHDV infections. However, in recent years there has been an apparent increase in cases of clinical disease in cattle associated with EHDV throughout the world. For instance, since 2003, clinical EHDV infections in cattle have occurred in the French islands of Réunion and Guadeloupe, Algeria, Morocco, Israel, Turkey, and Tunisia.[6,44,46]

Historically, only EHDV serotypes 1 and 2, and BTV serotypes 2, 10, 11, 13, and 17 were known to circulate in North America, but this has drastically changed over the last 12 years. In 2006, EHDV-6, an exotic serotype, was initially isolated from a WTD in Indiana.[1] However, since the initial discovery, EHDV-6 (Indiana) has been isolated in subsequent years from deer in Texas, Kansas, Missouri, Illinois, Michigan, and Arkansas, thus is now considered endemic in the US.[33] Interestingly, these EHDV-6 isolates represent a novel EHDV reassortment of gene segments from serotypes 2 and 6.[1] In addition to the detection of an exotic EHDV, 10 non-endemic BTV serotypes (1, 3, 5, 6, 9, 12, 14, 19, 22, and 24) have been isolated between 1998 and 2010 in the southeastern US, primarily from asymptomatic cattle.[20] Many of these serotypes are known or suspected to be endemic to the Caribbean and Central America,[25] where these incursions may have originated from.

Transmission and Epidemiology

Both EHDV and BTV are transmitted by *Culicoides* species, of which there are approximately 1400 species worldwide; however, only a small portion of these are proven vectors.[27] Known competent vectors of BTV in North America include *C. sonorensis* and *C. insignis*; however, many other species are suspected vectors based on experimental infections or isolations from field specimens.[11,43] The only proven biological vector of EHDV in North America is *C. sonorensis*; however, additional species are likely to be competent.[11] For instance, previous studies have demonstrated viral replication in *C. lahellei* and *C. venustus*.[37,21]

Culicoides are small (1 to 3 mm), primarily hematophagous flies in the family Ceratopogonidae that breed in a variety of semi-aquatic habitats and are distributed throughout the temperate and tropical climatic regions of the world.[27] The duration of the life cycle varies among species and climatic conditions, ranging from as little as seven days in tropical climates to seven months or more in temperate regions, where midges generally overwinter as third or fourth instar larvae.[5] Thus, when environmental conditions are optimal, populations can rapidly reach a high density, which likely plays a critical role in spawning outbreaks.[27] These seasonal peaks in abundance are often associated with warm, overly wet periods.[26] During these seasonal peaks, attack rates on preferred hosts are highest on warm summer nights lacking significant winds.[26] For instance, over 20,000 *C. lahillei* have been collected from a single white-tailed deer during a morning collection of midges.[36] Although adult midges are short-lived and do not disperse far from productive breeding sites via active flight, they can passively disperse great distances via wind currents, which may be involved in introducing infected midges into new areas and facilitating outbreaks in susceptible host populations.[28,34]

Hemorrhagic disease is seasonal throughout most of North America, with activity peaking in late summer and early fall. In the United States, HD exists in both enzootic and epizootic cycles and whether HD presents as epizootic mortality or as enzootic likely depends on 1) the virus serotype, strain, and dosage; 2) vector competency and abundance; and 3) host related factors.[29] Examining the US landscape via physiographic regions helps to elucidate whether epizootic or enzootic cycles exist, as well as how the disease will clinically manifest in WTD. The driving forces behind this variation among physiographic regions likely results from a combination of factors, such as vector activity, and host acquired and innate resistance. In enzootic regions of the coastal plain in the southeastern US where midge activity can persist much of the year, HD activity occurs annually and most cases are subclinical or occur as the chronic form of the disease. Moving inland into the Piedmont and Appalachian Mountains in the southeast and in portions of the Midwest, epizootics with high mortality generally occur every 3-5 years. Epizootics are even more infrequent in more northern latitudes and result in significant mortality when they occur.[40] However, these physiographic delineations and time periods are general trends, and small isolated epizootics involving small numbers of animals likely occur in both enzootic and epizootic regions regularly. In addition to epizootic and enzootic cycles, it has been suggested that in certain regions of the US, such as Texas, and in portions of the Midwest and Florida, WTD and orbiviruses co-exist in a state of enzootic stability.[42] In these regions, seroprevalence among WTD is near 100%; however, reports of clinical disease are extremely rare.[12,40,42] This enzootic stability is likely the result of acquired immunity via frequent exposure to the viruses, high subsequent passive immunity in fawns, and mechanisms of innate immunity via co-evolution of the host and pathogen.[40]

Evidence to support these mechanisms has been demonstrated during previous controlled experiments. Acquired immunity with EHDV in white-tailed deer has been demonstrated in earlier studies in which previous infection protected deer from clinical disease when subsequently challenged with a homologous EHDV,[32,35] but only partial protection following challenge with other serotypes.[15] It does not appear that previous EHDV infection protects against subsequent challenge with BTV and vice versa.[17,30] Passively acquired maternal antibodies persist for 18 – 23 weeks in WTD and likely offer some protection for fawns during their first transmission season.[13] In addition, there is evidence of an innate resistance to clinical disease caused by EHDV-1 and EHDV-2 infection in a southern subspecies of WTD (*O. v. texanus*) when compared to a northern subspecies (*O. v. borealis*).[14] These mechanisms of immunity are potentially lost when penned deer from non-enzootic regions are imported to enzootic regions, possibly explaining the high rates of morbidity and mortality encountered among some captive herds.

PATHOGENESIS

Once the virus enters the skin of a susceptible host via the bite of an infected midge, the virus undergoes primary replication in regional lymph nodes.[3,4] Evidence suggests that conventional dendritic cells are responsible for transporting the virus from the skin to the lymph nodes.[16] Following primary replication in a lymph node, the virions enter systemic circulation and disseminate to a variety of tissues throughout the body for secondary replication in endothelium, phagocytic mononuclear cells, lymphocytes, and potentially others.[3,4,8] The highly cell-associated viremia in WTD generally peaks between day 5 – 10 post-infection and is prolonged, but not persistent. The primary viremia is frequently associated with peripheral blood mononuclear cells and erythrocytes, as has been shown for BTV-infected calves,[24] as well as EHDV-infected deer.[39] Following secondary replication, a high titer secondary viremia ensues that is primarily erythrocyte-associated,[24,39] which can last for up to 49 days in BTV-infected calves[24] and up to 56 days in experimentally infected WTD.[30] The virions reside within invaginations of erythrocytes during this prolonged viremia, which protects the virus from recognition by co-circulating neutralizing antibodies. Serotype-specific neutralizing antibodies in WTD can be detected as early as day 8 post-infection and generally peak in 2-3 weeks.

The widespread viral replication in endothelial cells and subsequent endothelial damage in susceptible species, such as WTD infected with EHDV or BTV and sheep infected with BTV,[18,19] exposes the underlying collagen and basement membrane, resulting in massive activation of the coagulation system and a prothrombotic state. This differing susceptibility of endothelium to EHDV/BTV infection and replication may be one of the factors that play a role in variations observed in susceptibility to clinical disease between ruminant species.[8,24] Subsequent to microvascular injury, a consumptive coagulopathy with concurrent fibrinolysis results in DIC.[18] The ensuing hemorrhagic diathesis generally culminates in death, and is what gave hemorrhagic disease its name.

CLINICAL PRESENTATION

In WTD, clinical disease and pathological findings caused by EHDV are indistinguishable from those caused by BTV, which is why viruses from both serogroups are etiologic agents of HD. Observations made during both experimental trials and field investigations have clearly elucidated the clinical signs associated with HD in white-tailed deer. The disease is highly variable, ranging from subclinical infections to sudden death without premonitory signs of disease. Other animals may recover after a mild clinical course, whereas some may succumb months after acute disease due to chronic sequelae. The myriad of clinical signs and pathologic changes potentially observed with HD is refined when the disease is categorized into one of three forms: peracute, acute, or chronic.[7] The peracute form is primarily characterized by edema and cases generally present as sudden death with or without premonitory signs. In addition to edema, congestion, hemorrhage, and ulcerative lesions manifest in acute HD, often considered the classical form. Following recovery of the acute form of the disease, the chronic form is typified by hoof abnormalities, such as growth interruption or sloughing, ruminal scarring, and subsequent emaciation in the winter months. It is important to note that the chronic form does not represent active disease, but rather sequelae of HD.

Most of the prominent clinical signs of HD are a direct or indirect consequence of the virus-mediated vascular injury described above. Common clinical signs initially include fever, loss of appetite, rough hair coat, lethargy, hyperemia of mucous membranes, erythema of nonhaired regions of the skin, and ptyalism.[30,35] As documented during experimental infections, fever generally peaks between days 5-7 post-inoculation, which coincides with peak viremia. The body temperature then returns to baseline, although animals that succumb to the disease may have subsequent hypothermia.[30] Additional clinical signs include hemorrhages in the oral cavity eventually progressing to erosions and ulcerations, hemorrhages in the skin and coronary bands with or without lameness, bleeding tendencies, respiratory difficulty, bloody diarrhea, recumbency and terminal convulsions.[29] For example, in an individual animal that becomes moderately to severely ill, clinical signs typical of acute HD have the following progression over a two week period: decreased activity and inappetence, hyperemia and erythema → facial and cervical swelling and varying degrees of lameness → oral hemorrhages and bleeding tendencies → oral erosions and ulcerations. Death typically may occur between day 4-15 post-infection.

Leukopenia and lymphopenia are consistent clinicopathologic findings during the first week of infections in WTD with EHDV and BTV, generally reaching a nadir by day 5-7 post inoculation before returning to baseline by day 10.[19] Lymphopenia is often profound (<1,000 cells/ml) in WTD that have severe disease in an experimental setting.[30] Hypoproteinemia is also a common finding during the first week in clinically affected WTD. Between day 7 and 10 post-infection, activated partial thromboplastin time and prothrombin time become progressively prolonged and animals are frequently thrombocytopenic.[14,18]

Postmortem Findings

Gross postmortem findings are highly variable and can be very subtle in peracute cases. Common early findings include pleural, pericardial, and peritoneal effusions; pulmonary edema of varying severity; and subcutaneous edema, which is especially prominent in the fascial planes of the head, neck, and ventrum.[9,35] As the disease progresses and there is a consumptive insult to the coagulation system, widespread petechial, ecchymotic, and suffusive hemorrhages are common in many tissues, especially the gastrointestinal tract and heart.[9,35] Hemorrhagic gastrointestinal tract content is occasionally observed.[19] Testicular hemorrhage is commonly encountered in males. When present, two nearly pathognomonic gross findings of HD in white-tailed deer are hemorrhage in the tunica media at the base of the pulmonary artery and hemorrhage over the serosal surface of the pylorus.[19] If the animal survives the acute phase of the infection, erosions and ulcerations are frequently observed on the gingiva, palate, dental pad, tongue, buccal papillae, and the forestomachs.[9,35] Animals that survive can manifest with lesions of chronic disease, including scaring of the tongue and rumen mucosa and cracked or sloughing hooves subsequent to interruption of growth of the hoof wall.[19] These animals may succumb to malnutrition or other complications during the coming winter and spring.

DIAGNOSIS

Epizootic morbidity and/or mortality among one or more individuals in a herd during the transmission season should lead to the suspicion of HD. Depending on whether the animal presents in the peracute or acute form, differential diagnoses may include heartwater, sepsis, clostridial enterotoxemia, anthrax, malignant catarrhal fever, foot-and-mouth disease, bovine viral diarrhea and vesicular stomatitis. In mule deer, adenovirus infection should also be considered.[45]

Diagnosis of HD should be based on virus isolation coupled with RT-PCR from blood or tissues.[1,30] Preferred diagnostic samples include fresh (not frozen) spleen, lung, lymph node, or unclotted blood (EDTA, heparin, or citrate). Both EHDV and BTV are readily isolated on a number of cell lines, although cattle pulmonary artery endothelial (CPAE) cells and baby hamster kidney (BHK) cells (ATCC, Rockville, Maryland, USA) are among the most sensitive. Since many deer are subclinically infected, positive test results must be supported by consistent clinical or postmortem findings. Given the recent detection of novel or exotic EHD and BT viruses in the US, the importance of serotype-specific molecular diagnostics cannot be overemphasized. Numerous serogroup- and serotype-specific RT-PCR protocols to detect BTV and EHDV RNA have been published.[22,23] The chronic form of HD is simply a consequence of previous infection and infectious virus is not present in these animals. However, a presumptive diagnosis can be made based on postmortem findings and examination of the hoof walls. Given that the hoof wall of a WTD grows approximately 5 mm per month, measuring the distance from the coronary band to the line of growth interruption can estimate when the animal was infected.[19]

Circulating antibodies against EHDV and BTV are common in WTD throughout much of the US, thus serology should not be relied upon for diagnosis of infection unless comparison of acute and convalescent titers can be made or seroconversion documented. Available serologic assays include AGID and a cELISA (Veterinary Diagnostic Technology, Inc., Wheatridge, Colorado, USA), and serum neutralization (SN).[41] The AGID and cELISA detect antibodies at the serogroup level, although the commercially available cELISA is only for BTV antibodies. Currently, SN is the only serotype-specific serologic assay available for diagnosis.

REFERENCES

[1]Allison A, et al. **Gen Virol** 2010;430; [2]Anthony S, et al. **Virus Res** 2009;145:200; [3]Barrat-Boyes S, et al. **Vet Micro** 1994;40:361; [4]Barrat-Boyes S, et al. **Vet Immuno and Immunopathol** 1995;45:73; [5]Braverman Y. 1994 **Rev Sci Tech OIE**;13:1175; [6]Breard E, et al. **Vet Rec** 2004;155:422; [7]Couvillion C et al. **Proc USAHA** 1981:522; [8]Darpel K, et al. **Bluetongue** 2009:265; [9]Fletch A, et al. **Can J Comp Med** 1971;35:274; [10]Foster N, et al. **Am J Vet Res** 1963;24:1195; [11]Foster N, et al. **J Wildl Dis** 1977;13:9; [12]Flacke G, et al. **J Wildl Dis** 2004;40:288; [13]Gaydos J, et al. **J Wildl Dis** 2002;38:253; [14]Gaydos J, et al. **J Wildl Dis** 2002;38:713; [15]Gaydos J, et al. **J Wildl Dis** 2002;38:720; [16]Hemati B, et al. **J Virol** 2009;83(17):8789; [17]Hoff G, et al. **J Wildl Dis** 1974;10:25; [18]Howerth E, et al. **Am J Vet Res** 1988;49:1906; [19]Howerth E, et al. **Infectious Diseases of Wild Mammals** 2001:77; [20]Johnson D, et al. **Proc AAVLD** 2007:118; [21]Jones R, et al. **Mosq News** 1983;42(3):184; [22]LeBlanc N, et al. **J Virol Methods** 2010;167(2):165; [23]Maan N, et al. **PLoS ONE** 2010;5(9):9524; [24]MacLachlan NJ, et al. **Vet Pathol** 1990;27:223; [25]MacLachlan NJ, et al. **Vet Res** 2010;41:35; [26]Meiswinkel R, et al. **Infectious Diseases of Livestock** 2004;1:93; [27]Mellor P, et al. **Ann Rev Entomol** 2000;45:307; [28]Mullen G. **Medical and Veterinary Entomology** 2009;169; [29]Nettles V, et al. **Trans NA Wildl and Nat Res Conf** 1992:499; [30]Quist C, et al. **J Wildl Dis** 1997;33:584; [31]Robinson R, et al. **J Wildl Manag** 1967;31:165; [32]Roy P. **Fields Virology** 2007:1975; [33]SCWDS, unpublished data; [34]Sellers R. **BT, AHS, and Related Orbiviruses** 1992:284; [35]Shope R, et al. **J Exp Med** 1960;111:155; [36]Smith K, et al. **J Med Entomol** 1996;33:507; [37]Smith K, et al. **J Med Entomol** 1996;33:117; [38]Stair E, et al. **Pathol Veteraria** 1968;5:164; [39]Stallknecht D, et al. **J Wildl Dis** 1997;33:574; [40]Stallknecht D, et al. **Trans NA Wildl and Nat Res Conf** 2002:75; [41]Stallknecht D, et al. **J Wildl Dis** 1995;31:331; [42]Stallknecht D, et al. **J Wildl Dis** 1996;32:695; [43]Tanya V, et al. **Vet Micro** 1992;32:1; [44]Temizel E, et al. **Emerg Infect Dis** 2009;15:317; [45]Woods L, et al. **Vet Pathol** 1996;33:125; [46]Yadin H, et al. **Vet Rec** 2008;162:53

ANATOMIC AND BIOCHEMICAL CHARACTERIZATION OF FETLOCK HYPEREXTENSION IN LLAMAS/MOLECULAR EXPRESSIONS & EPIDEMIOLOGICAL STUDIES OF THE FETLOCK HYPEREXTENSION IN LLAMAS

Shannon K. Reed, DVM, MS, DACVS; Stacy Semevolos
Columbia, MO

INTRODUCTION

This set of presentations is intended to summarize research [1,2,3] into the condition of idiopathic fetlock hyperextension and suspensory apparatus breakdown in llamas. This condition is poorly understood in the llama. The suspensory apparatus supports the distal limb and allows the joints of normal animals to maintain relatively upright anatomical positions (figure 1). The apparatus consists of the superficial digital flexor tendon, deep digital flexor tendon, the suspensory ligament, and the proximal sesamoid bones and associated tendons. Idiopathic hyperextension of the fetlock is anecdotally reported to affect llamas frequently. Llama breeders refer to the condition as "down in the pasterns" or "down in the fetlocks." The condition can result in debilitating lameness, most likely due to mineralization of soft tissue as well as osteoarthritis in the distal limb. Affected llamas have a characteristic appearance with a sudden rapid onset of hyperextension of the fetlock joint and the first phalanx almost parallel to the ground (figure 2) in severely affected animals. Similar conditions have been reported in other species, with the most widely publicized being degenerative suspensory ligament desmitis (DSLD) in horses[4,5,6,7]. This disease is characterized by widespread, progressive degeneration of the collagen in the suspensory ligaments-particularly the branches, and occurs in multiple limbs. Although the condition occurs in many breeds, the Peruvian Paso breed is frequently diagnosed with this disease. These horses demonstrate ultrasonographic changes to the branches of the suspensory ligament including diffuse loss of echogenicity, enlarged suspensory branches, and irregular fiber pattern[8]. Pathologically, DSLD is consistently characterized by degeneration and swelling of collagen bundles, clusters of fibrocytes in affected collagen with associated degeneration, necrosis, and chondroid hyperplasia[4,5]. We have performed and published three studies to date and three are in progress evaluating llamas affected by fetlock hyperextension.

ANATOMICAL STUDY

No previous description of the suspensory apparatus structures, origins, and insertions exists for the llama. Six pairs of forelimbs and six pairs of hindlimbs were dissected and described. The full descriptions are available in the International Journal of Morphology[9,10]. When comparing llamas to other domestic species, noteworthy anatomic variations include lack of distal sesamoid (navicular) bones, multiple branches to the suspensory ligament to accommodate 2 pairs of proximal sesamoid bones per limb (one per digit), a prominent lumbricalis muscle at the level of the branching of the suspensory ligament, and weight bearing on both the third and second phalanx.

INITIAL ADULT LLAMA STUDIES

The initial study evaluated morphologic and biochemical characterization of fetlock hyperextension in llamas[1]. In that study, 6 affected animals and 6 age and sex matched controls were evaluated. Lameness evaluations revealed an obvious lameness at a walk in 2/6 affected llamas. Radiographs evaluating pastern to ground axis revealed a significant difference between affected and control llamas with control llamas having a mean angle of 52° (figure 3) in the forelimbs and affected 12 ° (figure 4). The hindlimbs were similar with 57 ° in the controls and 30 ° for the affected. One severely affected animal in this group had osteoarthritis of the both metacarpophalangeal joints as well as mineralization of soft tissue structures. Ultrasound was performed in all animals. Because none of the llamas would tolerate standing ultrasonography, the llamas were sedated and placed in lateral recumbency, which may have induced some artifact. Two affected animals in this group had lesions detected ultrasonographically, although the extent of damage was not similar to that reported in horses with DSLD. Complete blood counts and serum biochemistry were performed on all animals with no significant findings. Serum samples were obtained for trace mineral analysis. Following euthanasia, multiple samples were collected from the tendons and ligaments of all animals. The samples were either placed in formalin or snap frozen in liquid nitrogen. Liver samples were collected and submitted for trace liver analysis. Formalin samples were prepared for immunohistochemistry with antibodies directed against collagen type I and collagen type III. Formalin samples were also processed for H&E staining as well. For lysyl oxidase activity assays, frozen samples from each component (DDFT, SDFT, and suspensory ligament) were processed for lysyl oxidase activity assessment by modification of a fluorometric assay[11,12,13].

This study group revealed some interesting findings. Animals in the affected group were significantly lower in liver copper levels animals (55 +/- 35.6 ppm) than control animals (102.3 +/- 38.09 ppm) (P=0.01). Additionally, serum zinc was significantly higher in affected animals than controls (0.3683 +/- 0.06 µg/dl) than in age matched controls (0.2850 +/- 0.03 µg/dl) (p=0.03). Copper deficiency has been implicated in tendon and ligament breakdown in other species[12,13]. As copper is essential for organic cofactor formation in amine oxidases such as lysyl oxidase[12,13], a change in copper levels would be expected to decrease functional activity of lysyl oxidase. However, despite the low liver copper levels, lysyl oxidase activity was not significantly different between groups. Additionally, as measured by immunohistochemistry scoring, Collagen types I and III were expressed similarly between affected and unaffected animals in this group, although this in contrast to what was expected with damaged connective tissue undergoing repair. The superficial digital flexor tendon sections displayed a moderate number of tenocytes spread throughout the tendon matrix compared to the decreased cellularity of the deep digital flexor tendon The tenocytes of the DDF were much smaller than the SDF or suspensory ligament, and tendon matrix predominated in the DDF samples. The suspensory ligament contained bundles of rudimentary muscle which were similarly distributed in all animals. In general, the suspensory ligaments were more vascular than the SDF or DDF.

The minimal ultrasound and gross changes apparent in these llamas suggested a molecular basis for the disease, but initial testing did not clarify the disease. Therefore, additional studies were performed on banked tissue from this initial group to attempt to further characterize the nature of fetlock hyperextension in the llama through molecular and histological techniques that identify specific gene expression alterations and connective tissue matrix changes. Histologic sections of the superficial digital flexor tendon, deep digital flexor tendon, and suspensory ligament were evaluated for collagen, elastin, copper, and proteogylcan content by Picrosirius red[o], Masson's trichrome stain, Verhoeff's silver stain, rhodanine stain, and alcian blue respectively, using standard laboratory staining protocols. Masson's trichrome stain differentiates between collagen and muscle, with the nuclei staining black, the cytoplasm, keratin, and muscle fibers staining red and the collagen staining blue. Verhoeff's silver stain differentiates elastic fibers using hematoxylin, ferric chloride and iodine. Rhodanine stain shows excess copper (400ppm) in tissue as a bright red to red-yellow color. Proteoglycan levels were evaluated using alcian blue stain in a 0.1 N hydrochloric acid solution. Picrosirius red stained sections were evaluated using both polarized and non-polarized light microscopy for short and long collagen fibers. Sections were evaluated by two evaluators (SKR and SAS) blinded to animal identification. The tenocytes, matrix, paratenon and vasculature were evaluated for staining patterns and compared between control and affected animals. In addition to the further histologic evaluation, frozen tissue samples were processed for quantitative real-time PCR evaluation of gene expression of Collagen type I, Collagen Type III, lysyl oxidase, β-actin, and MMP-13.

No significant difference was found between affected and control animals in this group for mRNA expression of collagen type I ($p=0.22$, collagen type III ($p=0.58$), or lysyl oxidase ($p=0.35$). There was a trend for decreased MMP-13 expression in this group of affected llamas ($p=0.08$). Based on Alcian blue staining, proteoglycan content differed by region and animal. In general, there was mild proteoglycan staining near vessels and septa within tendon/ligament bundles. Also, there was moderate proteoglycan staining in the fibrous connection between the superficial digital flexor tendon (SDFT) and the suspensory ligament (SL). In two of the affected llamas, there was marked proteoglycan accumulation associated with lesions in the SL. In two of the control llamas, there was moderate proteoglycan staining in discrete areas of the SL associated with altered fiber patterns and increased cellularity. The remaining animals showed sporadic mild proteoglycan staining in the SL matrix.

In the tissue specimens stained with Verhoeff's silver stain, elastin content was found to be very low (around 1-2%) throughout the tendon and ligament samples. Elastin tended to be oriented longitudinally in the tendon matrix and was primarily adjacent to vascular components. No difference was observed between affected and control animal specimens. Collagen content and orientation were evaluated by the Picrosirius red and trichrome stains. Long chain collagen (type I) was apparent throughout the tendon and ligament matrix. No short chain collagens were observed in the matrix of any of the samples. Short chain collagens were seen in the loose connective tissue in the peritendinous region. No differences were appreciated between affected and control animals in collagen content. The rhodanine staining for copper accumulation revealed no evidence of excess copper in any of the samples.

The lack of difference in lysyl oxidase mRNA expression between the affected and control groups was similar to our previous study. This lack of difference suggests an etiology other than the proposed copper deficiency. However, it is reported that even when dietary copper or lysyl oxidase activity is low, normal cross-linking can occur and the reduction in cross-linking has to be severe before significant changes in viscoelastic properties occur in a tissue in which collagen fibers are highly aligned, such as in bone (Opsahl). We had also hypothesized that matrix metalloproteinase (MMP) expression would be increased in llamas with fetlock hyperextension. MMPs are a family of proteinases that are central to degradation of the extracellular matrix. MMP-13 is a collagenase that is increased in cartilage degeneration with joint disease, synovial inflammation in osteoarthritis, and has been shown to be up-regulated by interleukin-1β in damaged tendon cells in humans[15]. MMP-13 was also up-regulated in tissue harvested from naturally occurring tendonitis in the horse[16] and other studies have shown an increase in MMP-13 in induced tendonitis[17]. The trend for decreased MMP-13 expression in affected llamas in our study was unexpected and implied a non-degenerative, non-inflammatory etiology for this condition. Collagen type I is predominantly expressed in normal tendons and ligaments, while collagen type III shows increased expression in damaged tendons and ligaments in horses[18]. In horses with experimental collagenase-induced damage to the superficial digital flexor tendon, expression of collagen type III was highest at 4 weeks post injury and decreased thereafter.[18] We therefore hypothesized that collagen type I expression would be lower and collagen type III expression would be higher in affected animals. However, we found no difference in collagen type I and III expression between the affected and control animals. This may be due to chronicity of disease, as the history and length of disease process of these particular animals is not known. It may also reflect a mechanism of breakdown that is not yet understood and differs from that of the horse. Horses with degenerative suspensory ligament desmitis have demonstrated a systemic disorder of the connective tissue with abnormal accumulation of proteoglycans between collagen and elastic fibers of many tissues, including the suspensory ligament, SDFT, DDFT, patellar and nuchal ligaments, cardiovascular system, and sclera[7]. However in this study, only the suspensory ligament was noted to have increased proteoglycan content associated with lesions in a few of the affected animals. The lack of obvious accumulation in the other affected animals may be due to sampling from non-lesional sites. If more areas within the suspensory ligament were evaluated, perhaps lesional proteoglycan accumulation would have been appreciated. Additionally, proteoglycan accumulation was seen in two control animals with altered fiber patterns and increased cellularity.

JUVENILLE STUDY

Because of the potential that chronicity of disease was altering the properties of the affected tissue, a new study was instituted using 6 age and sex matched juvenile llamas (<4 years) in an effort to decrease the time from initiation of disease to study evaluation[3]. The objective of this study was to determine trace mineral levels, gene expression alterations, and connective tissue matrix composition changes in affected juvenile llamas. The hypothesis was that collagen type III and MMP-13 gene expression and PG content would be increased and lysyl oxidase and collagen type I gene expression would be decreased in the SL of affected juvenile llamas. It was also predicted that liver copper levels would be decreased in affected llamas when compared to age-matched controls. These animals underwent initial evaluation similar to the adult group: lameness examination, radiographs, ultrasound, CBC, serum chemistry, serum trace mineral analysis and post euthanasia collection of liver as well as tendons and ligaments of the suspensory apparatus. Histologic sections of the superficial digital flexor tendon, deep digital flexor tendon, and suspensory ligament were evaluated for collagen, elastin, copper, and proteogylcan content by Picrosirius red°, Masson's trichrome stain, Verhoeff's silver stain, rhodanine stain, and alcian blue respectively, using standard laboratory staining protocols. Formalin samples were prepared for immunohistochemistry with antibodies directed against collagen type I and collagen type III. Frozen tissue samples were processed for quantitative real-time PCR evaluation of gene expression of Collagen type I, Collagen Type III, lysyl oxidase, β-actin, and MMP-13. Lysyl oxidase activity through fluorometric assay was not measured in this group.

No significant findings were seen on CBC or serum chemistry. On ultrasonographic evaluation of the suspensory apparatus, lesions were noted in 2 of the affected llamas and none of the control llamas, which was not statistically significant ($p=0.15$). On the basis of measurements obtained from radiographs, the mean P1-to-ground angle of the forelimbs (35°; range 17-55 °) and hindlimbs (40°; range 34-48 °) of affected llamas was significantly less ($p<0.05$) than the mean angle of the forelimbs (65°; range 59-70 °) and hindlimbs (67°; range 57-74 °) of the control llamas. This was similar to the adult group. On assessment of radiographic abnormalities, more affected animals had distal displacement of the proximal sesamoid bones (5/6 affected, 0/6 controls; $p<0.05$) and osteoarthritis of the fetlock joint (3/6 affected, 0/6 controls; $p<0.05$) than normal control animals.

There was significantly increased gene expression of collagen type III ($p<0.05$) in the SL of affected juvenile llamas compared to control llamas. This is a variation from the adult animals. No significant difference was found between affected and control animals in gene expression of collagen type I ($p=0.22$), lysyl oxidase ($p=0.28$), or MMP-13 ($p=0.28$). Juvenile llamas affected with metacarpophalangeal hyperextension had significantly increased serum molybdenum ($p<0.05$) and decreased liver cobalt ($p<0.05$) compared to control llamas, but no other significant differences were found. Upon evaluation of trace mineral levels lying outside established laboratory reference ranges, there were more affected juvenile llamas having clinically low liver cobalt ($p<0.05$) compared to control llamas. No other significant differences were found.

Based on the total SL scores of Alcian blue stained sections, there was significantly increased PG content ($p<0.05$) in the SL of affected llamas compared to their age- and sex-matched controls. There was strong PG staining in the SL of 4 affected juvenile llamas, compared to only 1 control llama. In the 4 affected llamas, there was strong PG staining associated with increased blood vessels, septa, and muscle content within the SL ($p<0.05$). In 2 of these affected llamas, marked PG accumulation was found within lesions of the SL. Another affected llama also showed general accumulation of PG in many areas of the SL matrix. One control llama had strong PG staining in some areas of the SL matrix. In animals not having strong PG staining of the SL, there was generally mild PG staining near blood vessels and septa within ligament bundles, as well as irregular areas of mild PG staining within the SL matrix. No differences were found in PG content of the SDFT ($p=0.21$) and DDFT ($p=0.12$) of affected and control llamas.

Collagen type III immunostaining was found primarily in the epitenon, paratenon, loose connective tissue, vessels, and septa of the SDFT, DDFT, and SL. Mild expression was evident in some tenocytes and ligamentocytes, but with minimal associated matrix staining. No difference was found in collagen type III immunostaining of the SDFT ($p=0.09$), DDFT ($p=0.21$) or SL ($p=0.21$) between affected and control llamas.

Based on the picrosirius red and Masson trichrome staining of sections, collagen content of the SDFT, DDFT, and SL was comparable between affected juvenile animals and normal controls. Similarly, Verhoeff elastin staining did not reveal any differences in elastin content or orientation between affected and control llamas.

Increased collagen type III expression and PG content in SL of juvenile llamas having metacarpophalangeal and metatarsophalangeal hyperextension provides evidence for ongoing matrix repair, possibly in response to injury or degeneration[7,18]. The trace mineral differences between affected and control llamas may have reflected dietary mineral imbalances in affected animals.

Our previous study showed that llamas affected by metacarpophalangeal hyperextension had increased serum zinc and decreased liver copper compared to control animals[1]. In that study, there were 5 adult llamas and 1 juvenile llama included in the group of affected animals. However, in the current study, affected juvenile llamas did not show the expected decrease in liver copper levels compared to controls. Copper deficiency is known to decrease the activity of lysyl oxidase[14,19], an enzyme necessary for proper cross linking of collagen fibers. However, when lysyl oxidase activity and gene expression were compared in previous studies[2,3], no significant differences were found between affected and control animals. Similarly, the current study did not reveal a significant difference in lysyl oxidase gene expression within the SL between affected and control llamas.

Collagen type I is predominantly expressed in normal tendons and ligaments, while collagen type III is expressed in damaged tendons and ligaments. The previous study[2] in adult llamas did not show differences in gene expression of collagen types I and III in affected animals, possibly reflecting the chronicity of the condition in older animals. While the current study confirmed the lack of collagen type I expression changes in younger llamas, increased collagen type III gene expression in affected juvenile llamas supported SL injury as a potential inciting factor in fetlock hyperextension. However, collagen type III protein was not apparent in histologic lesions within the SL of affected llamas. In addition, in all of the llamas examined, collagen type III was mainly limited to loose connective tissue in septa surrounding vessels and nerves within the SL, plus the epitenon and paratenon. Therefore, increased gene expression of collagen type III in affected llamas may have been due to a greater proportion of septa within the SL rather than increased expression by SL tenocytes and matrix.

Based on the radiographic findings of distally displaced sesamoid bones in most of the affected llamas, we concluded that fetlock hyperextension was the result of SL elongation, rather than involving the SDF, DDF or distal sesamoidean ligaments. This elongation may be due to inherent SL weakness, injury or degeneration. The lack of concurrent widespread ultrasonographic SL lesions in all affected llamas made injury or degeneration seem less likely. Further clarification of this syndrome is necessary and should include evaluation of young and acutely affected animals. Studies of the physical characteristics of collagen and its cross links would be possible with electron microscopy and high performance liquid chromatography (HPLC). Additionally, biomechanical testing of the suspensory apparatus comparing affected llamas to normal llamas could be performed.

ONGOING NEAR COMPLETED STUDIES (AS OF 2/2011)

Studies we are currently completing include evaluation of serum trace mineral analysis in approximately 90 camelids comparing findings in affected animals to age/sex matched controls. Additionally, we have almost completed an epidemiological survey of camelids in the Northwest United States. There is a disease prevalence rate of 17% in llamas and 0.7% of alpacas in the herds surveyed (including close to 1000 animals.)

REFERENCES

1. Reed S, Semevolos SA. **Am J Vet Res** , 2007. 2. Reed SK, Semevolos SA. **Am J Vet Res,** 2010. 3. Semevolos SA, Reed SK. **Am J Vet Res** 2011. 4. Mero J, Scarlett J.. **J Eq Vet Sc** 2005; 224-228. 5. Mero J, Pool R. **AAEP proceedings.** 2002 (pp. 329-334). 6. Gibson K, Steel CM. **Eq Vet Ed** 2002;4:50-64. 7. Halper J, Byoungjae K. Ahrar K, et al. ***BMC Vet Res*** 2006; 2:12. 8. Dyson S, Arthur R. Palmer SR. Suspensory Ligament Desmitis. ***Vet Cl N Am***1995;11:172-215. 9. Constantinescu G, Reed S, ***Int J Morph*** . 2007. 10. Constantinescu G, Reed S, ***Int J Morph***. 2007. 11. Palamakumbura AH, Trackman PC. **Anal Biochem** 2002. 12. Rucker RB, et al.**J Nutr** 1999;129:2143-2146. 13. Rucker RB, et al. **Am J Clin Nutr** 1998;67(suppl):996S-1002S. 14. Opsahl W et al. **J of Nut** 1982Reed SK, , et al. *Am J Vet Res* 2007. 15. Corps A, et al. **Rheumatology** 2005. 16. Nomura et al. **J Vet Med Sci** 2007. 17. Dudhia J, et al. **Aging Cell** 2007. 18. Dahlgren L, et al. **Am J Vet Res** 2005; 66(2):266-270. 19. Chou et al. **J Biol Chem** 1969;21:5785-5789.

GRAPHICS

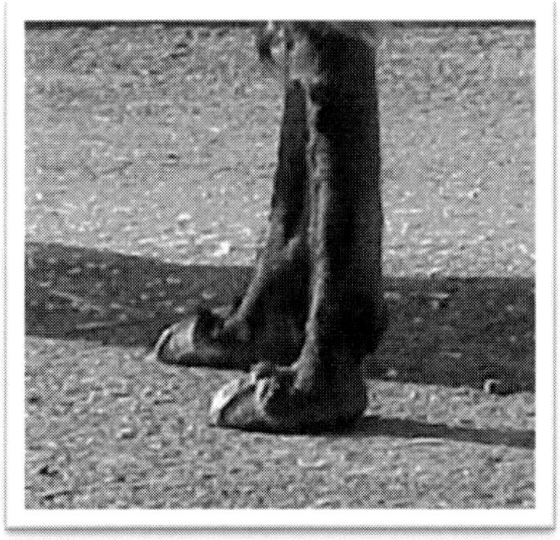

Figure 1: Normal fetlocks in adult llama

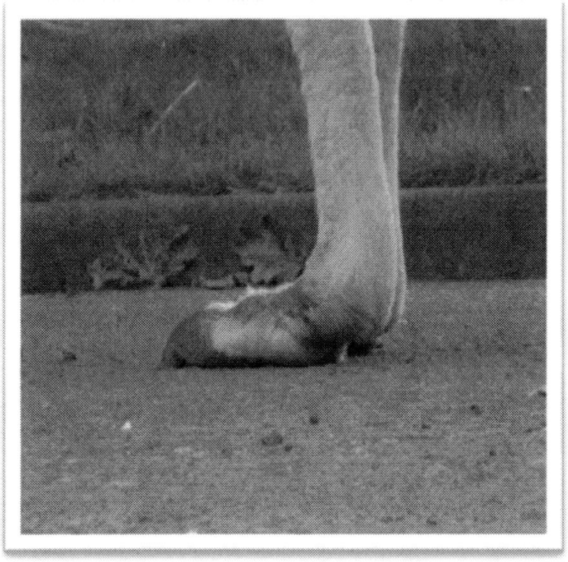

Figure 2: Severely affected llama

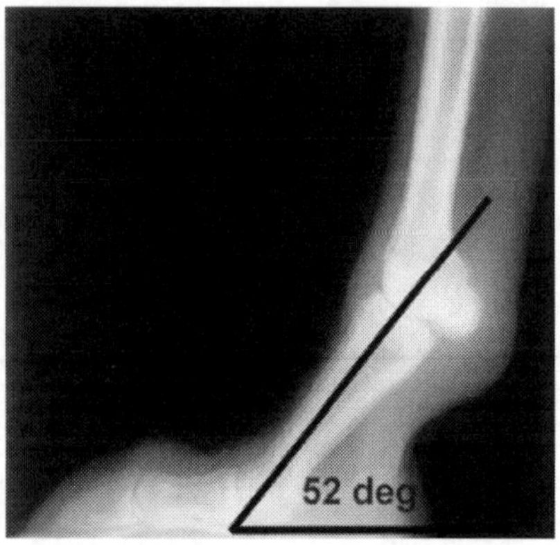

Figure 3: Radiograph of normal llama

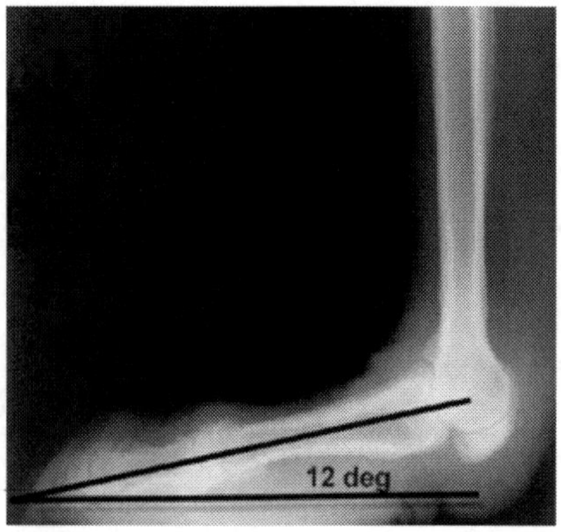

Figure 4: Radiograph of severely affected llama

ANATOMIC AND BIOCHEMICAL CHARACTERIZATION OF FETLOCK HYPEREXTENSION IN LLAMAS/MOLECULAR EXPRESSIONS & EPIDEMIOLOGICAL STUDIES OF THE FETLOCK HYPEREXTENSION IN LLAMAS

Shannon K. Reed, DVM, MS, DACVS; Stacy Semevolos
Columbia, MO

INTRODUCTION

This set of presentations is intended to summarize research [1,2,3] into the condition of idiopathic fetlock hyperextension and suspensory apparatus breakdown in llamas. This condition is poorly understood in the llama. The suspensory apparatus supports the distal limb and allows the joints of normal animals to maintain relatively upright anatomical positions (figure 1). The apparatus consists of the superficial digital flexor tendon, deep digital flexor tendon, the suspensory ligament, and the proximal sesamoid bones and associated tendons. Idiopathic hyperextension of the fetlock is anecdotally reported to affect llamas frequently. Llama breeders refer to the condition as "down in the pasterns" or "down in the fetlocks." The condition can result in debilitating lameness, most likely due to mineralization of soft tissue as well as osteoarthritis in the distal limb. Affected llamas have a characteristic appearance with a sudden rapid onset of hyperextension of the fetlock joint and the first phalanx almost parallel to the ground (figure 2) in severely affected animals. Similar conditions have been reported in other species, with the most widely publicized being degenerative suspensory ligament desmitis (DSLD) in horses[4,5,6,7]. This disease is characterized by widespread, progressive degeneration of the collagen in the suspensory ligaments-particularly the branches, and occurs in multiple limbs. Although the condition occurs in many breeds, the Peruvian Paso breed is frequently diagnosed with this disease. These horses demonstrate ultrasonographic changes to the branches of the suspensory ligament including diffuse loss of echogenicity, enlarged suspensory branches, and irregular fiber pattern[8]. Pathologically, DSLD is consistently characterized by degeneration and swelling of collagen bundles, clusters of fibrocytes in affected collagen with associated degeneration, necrosis, and chondroid hyperplasia[4,5]. We have performed and published three studies to date and three are in progress evaluating llamas affected by fetlock hyperextension.

ANATOMICAL STUDY

No previous description of the suspensory apparatus structures, origins, and insertions exists for the llama. Six pairs of forelimbs and six pairs of hindlimbs were dissected and described. The full descriptions are available in the International Journal of Morphology[9,10]. When comparing llamas to other domestic species, noteworthy anatomic variations include lack of distal sesamoid (navicular) bones, multiple branches to the suspensory ligament to accommodate 2 pairs of proximal sesamoid bones per limb (one per digit), a prominent lumbricalis muscle at the level of the branching of the suspensory ligament, and weight bearing on both the third and second phalanx.

INITIAL ADULT LLAMA STUDIES

The initial study evaluated morphologic and biochemical characterization of fetlock hyperextension in llamas[1]. In that study, 6 affected animals and 6 age and sex matched controls were evaluated. Lameness evaluations revealed an obvious lameness at a walk in 2/6 affected llamas. Radiographs evaluating pastern to ground axis revealed a significant difference between affected and control llamas with control llamas having a mean angle of 52° (figure 3) in the forelimbs and affected 12 ° (figure 4). The hindlimbs were similar with 57 ° in the controls and 30 ° for the affected. One severely affected animal in this group had osteoarthritis of the both metacarpophalangeal joints as well as mineralization of soft tissue structures. Ultrasound was performed in all animals. Because none of the llamas would tolerate standing ultrasonography, the llamas were sedated and placed in lateral recumbency, which may have induced some artifact. Two affected animals in this group had lesions detected ultrasonographically, although the extent of damage was not similar to that reported in horses with DSLD. Complete blood counts and serum biochemistry were performed on all animals with no significant findings. Serum samples were obtained for trace mineral analysis. Following euthanasia, multiple samples were collected from the tendons and ligaments of all animals. The samples were either placed in formalin or snap frozen in liquid nitrogen. Liver samples were collected and submitted for trace liver analysis. Formalin samples were prepared for immunohistochemistry with antibodies directed against collagen type I and collagen type III. Formalin samples were also processed for H&E staining as well. For lysyl oxidase activity assays, frozen samples from each component (DDFT, SDFT, and suspensory ligament) were processed for lysyl oxidase activity assessment by modification of a fluorometric assay[11,12,13].

This study group revealed some interesting findings. Animals in the affected group were significantly lower in liver copper levels animals (55 +/- 35.6 ppm) than control animals (102.3 +/- 38.09 ppm) (P=0.01). Additionally, serum zinc was significantly higher in affected animals than controls (0.3683 +/- 0.06 μg/dl) than in age matched controls (0.2850 +/- 0.03 μg/dl) (p=0.03). Copper deficiency has been implicated in tendon and ligament breakdown in other species[12,13]. As copper is essential for organic cofactor formation in amine oxidases such as lysyl oxidase[12,13], a change in copper levels would be expected to decrease functional activity of lysyl oxidase. However, despite the low liver copper levels, lysyl oxidase activity was not significantly different between groups. Additionally, as measured by immunohistochemistry scoring, Collagen types I and III were expressed similarly between affected and unaffected animals in this group, although this in contrast to what was expected with damaged connective tissue undergoing repair. The superficial digital flexor tendon sections displayed a moderate number of tenocytes spread throughout the tendon matrix compared to the decreased cellularity of the deep digital flexor tendon The tenocytes of the DDF were much smaller than the SDF or suspensory ligament, and tendon matrix predominated in the DDF samples. The suspensory ligament contained bundles of rudimentary muscle which were similarly distributed in all animals. In general, the suspensory ligaments were more vascular than the SDF or DDF.

The minimal ultrasound and gross changes apparent in these llamas suggested a molecular basis for the disease, but initial testing did not clarify the disease. Therefore, additional studies were performed on banked tissue from this initial group to attempt to further characterize the nature of fetlock hyperextension in the llama through molecular and histological techniques that identify specific gene expression alterations and connective tissue matrix changes. Histologic sections of the superficial digital flexor tendon, deep digital flexor tendon, and suspensory ligament were evaluated for collagen, elastin, copper, and proteogylcan content by Picrosirius red°, Masson's trichrome stain, Verhoeff's silver stain, rhodanine stain, and alcian blue respectively, using standard laboratory staining protocols. Masson's trichrome stain differentiates between collagen and muscle, with the nuclei staining black, the cytoplasm, keratin, and muscle fibers staining red and the collagen staining blue. Verhoeff's silver stain differentiates elastic fibers using hematoxylin, ferric chloride and iodine. Rhodanine stain shows excess copper (400ppm) in tissue as a bright red to red-yellow color. Proteoglycan levels were evaluated using alcian blue stain in a 0.1 N hydrochloric acid solution. Picrosirius red stained sections were evaluated using both polarized and non-polarized light microscopy for short and long collagen fibers. Sections were evaluated by two evaluators (SKR and SAS) blinded to animal identification. The tenocytes, matrix, paratenon and vasculature were evaluated for staining patterns and compared between control and affected animals. In addition to the further histologic evaluation, frozen tissue samples were processed for quantitative real-time PCR evaluation of gene expression of Collagen type I, Collagen Type III, lysyl oxidase, β-actin, and MMP-13.

No significant difference was found between affected and control animals in this group for mRNA expression of collagen type I (p=0.22, collagen type III (p=0.58), or lysyl oxidase (p=0.35). There was a trend for decreased MMP-13 expression in this group of affected llamas (p=0.08). Based on Alcian blue staining, proteoglycan content differed by region and animal. In general, there was mild proteoglycan staining near vessels and septa within tendon/ligament bundles. Also, there was moderate proteoglycan staining in the fibrous connection between the superficial digital flexor tendon (SDFT) and the suspensory ligament (SL). In two of the affected llamas, there was marked proteoglycan accumulation associated with lesions in the SL. In two of the control llamas, there was moderate proteoglycan staining in discrete areas of the SL associated with altered fiber patterns and increased cellularity. The remaining animals showed sporadic mild proteoglycan staining in the SL matrix.

In the tissue specimens stained with Verhoeff's silver stain, elastin content was found to be very low (around 1-2%) throughout the tendon and ligament samples. Elastin tended to be oriented longitudinally in the tendon matrix and was primarily adjacent to vascular components. No difference was observed between affected and control animal specimens. Collagen content and orientation were evaluated by the Picrosirius red and trichrome stains. Long chain collagen (type I) was apparent throughout the tendon and ligament matrix. No short chain collagens were observed in the matrix of any of the samples. Short chain collagens were seen in the loose connective tissue in the peritendinous region. No differences were appreciated between affected and control animals in collagen content. The rhodanine staining for copper accumulation revealed no evidence of excess copper in any of the samples.

The lack of difference in lysyl oxidase mRNA expression between the affected and control groups was similar to our previous study. This lack of difference suggests an etiology other than the proposed copper deficiency. However, it is reported that even when dietary copper or lysyl oxidase activity is low, normal cross-linking can occur and the reduction in cross-linking has to be severe before significant changes in viscoelastic properties occur in a tissue in which collagen fibers are highly aligned, such as in bone (Opsahl). We had also hypothesized that matrix metalloproteinase (MMP) expression would be increased in llamas with fetlock hyperextension. MMPs are a family of proteinases that are central to degradation of the extracellular matrix. MMP-13 is a collagenase that is increased in cartilage degeneration with joint disease, synovial inflammation in osteoarthritis, and has been shown to be up-regulated by interleukin-1β in damaged tendon cells in humans[15]. MMP-13 was also up-regulated in tissue harvested from naturally occurring tendonitis in the horse[16] and other studies have shown an increase in MMP-13 in induced tendonitis[17]. The trend for decreased MMP-13 expression in affected llamas in our study was unexpected and implied a non-degenerative, non-inflammatory etiology for this condition. Collagen type I is predominantly expressed in normal tendons and ligaments, while collagen type III shows increased expression in damaged tendons and ligaments in horses[18]. In horses with experimental collagenase-induced damage to the superficial digital flexor tendon, expression of collagen type III was highest at 4 weeks post injury and decreased thereafter.[18] We therefore hypothesized that collagen type I expression would be lower and collagen type III expression would be higher in affected animals. However, we found no difference in collagen type I and III expression between the affected and control animals. This may be due to chronicity of disease, as the history and length of disease process of these particular animals is not known. It may also reflect a mechanism of breakdown that is not yet understood and differs from that of the horse. Horses with degenerative suspensory ligament desmitis have demonstrated a systemic disorder of the connective tissue with abnormal accumulation of proteoglycans between collagen and elastic fibers of many tissues, including the suspensory ligament, SDFT, DDFT, patellar and nuchal ligaments, cardiovascular system, and sclera[7]. However in this study, only the suspensory ligament was noted to have increased proteoglycan content associated with lesions in a few of the affected animals. The lack of obvious accumulation in the other affected animals may be due to sampling from non-lesional sites. If more areas within the suspensory ligament were evaluated, perhaps lesional proteoglycan accumulation would have been appreciated. Additionally, proteoglycan accumulation was seen in two control animals with altered fiber patterns and increased cellularity.

JUVENILLE STUDY

Because of the potential that chronicity of disease was altering the properties of the affected tissue, a new study was instituted using 6 age and sex matched juvenile llamas (<4 years) in an effort to decrease the time from initiation of disease to study evaluation[3]. The objective of this study was to determine trace mineral levels, gene expression alterations, and connective tissue matrix composition changes in affected juvenile llamas. The hypothesis was that collagen type III and MMP-13 gene expression and PG content would be increased and lysyl oxidase and collagen type I gene expression would be decreased in the SL of affected juvenile llamas. It was also predicted that liver copper levels would be decreased in affected llamas when compared to age-matched controls. These animals underwent initial evaluation similar to the adult group: lameness examination, radiographs, ultrasound, CBC, serum chemistry, serum trace mineral analysis and post euthanasia collection of liver as well as tendons and ligaments of the suspensory apparatus. Histologic sections of the superficial digital flexor tendon, deep digital flexor tendon, and suspensory ligament were evaluated for collagen, elastin, copper, and proteogylcan content by Picrosirius red°, Masson's trichrome stain, Verhoeff's silver stain, rhodanine stain, and alcian blue respectively, using standard laboratory staining protocols. Formalin samples were prepared for immunohistochemistry with antibodies directed against collagen type I and collagen type III. Frozen tissue samples were processed for quantitative real-time PCR evaluation of gene expression of Collagen type I, Collagen Type III, lysyl oxidase, β-actin, and MMP-13. Lysyl oxidase activity through fluorometric assay was not measured in this group.

No significant findings were seen on CBC or serum chemistry. On ultrasonographic evaluation of the suspensory apparatus, lesions were noted in 2 of the affected llamas and none of the control llamas, which was not statistically significant ($p=0.15$). On the basis of measurements obtained from radiographs, the mean P1-to-ground angle of the forelimbs (35°; range 17-55 °) and hindlimbs (40°; range 34-48 °) of affected llamas was significantly less ($p<0.05$) than the mean angle of the forelimbs (65°; range 59-70 °) and hindlimbs (67°; range 57-74 °) of the control llamas. This was similar to the adult group. On assessment of radiographic abnormalities, more affected animals had distal displacement of the proximal sesamoid bones (5/6 affected, 0/6 controls; $p<0.05$) and osteoarthritis of the fetlock joint (3/6 affected, 0/6 controls; $p<0.05$) than normal control animals.

There was significantly increased gene expression of collagen type III ($p<0.05$) in the SL of affected juvenile llamas compared to control llamas. This is a variation from the adult animals. No significant difference was found between affected and control animals in gene expression of collagen type I ($p=0.22$), lysyl oxidase ($p=0.28$), or MMP-13 ($p=0.28$). Juvenile llamas affected with metacarpophalangeal hyperextension had significantly increased serum molybdenum ($p<0.05$) and decreased liver cobalt ($p<0.05$) compared to control llamas, but no other significant differences were found. Upon evaluation of trace mineral levels lying outside established laboratory reference ranges, there were more affected juvenile llamas having clinically low liver cobalt ($p<0.05$) compared to control llamas. No other significant differences were found.

Based on the total SL scores of Alcian blue stained sections, there was significantly increased PG content ($p<0.05$) in the SL of affected llamas compared to their age- and sex-matched controls. There was strong PG staining in the SL of 4 affected juvenile llamas, compared to only 1 control llama. In the 4 affected llamas, there was strong PG staining associated with increased blood vessels, septa, and muscle content within the SL ($p<0.05$). In 2 of these affected llamas, marked PG accumulation was found within lesions of the SL. Another affected llama also showed general accumulation of PG in many areas of the SL matrix. One control llama had strong PG staining in some areas of the SL matrix. In animals not having strong PG staining of the SL, there was generally mild PG staining near blood vessels and septa within ligament bundles, as well as irregular areas of mild PG staining within the SL matrix. No differences were found in PG content of the SDFT ($p=0.21$) and DDFT ($p=0.12$) of affected and control llamas.

Collagen type III immunostaining was found primarily in the epitenon, paratenon, loose connective tissue, vessels, and septa of the SDFT, DDFT, and SL. Mild expression was evident in some tenocytes and ligamentocytes, but with minimal associated matrix staining. No difference was found in collagen type III immunostaining of the SDFT ($p=0.09$), DDFT ($p=0.21$) or SL ($p=0.21$) between affected and control llamas.

Based on the picrosirius red and Masson trichrome staining of sections, collagen content of the SDFT, DDFT, and SL was comparable between affected juvenile animals and normal controls. Similarly, Verhoeff elastin staining did not reveal any differences in elastin content or orientation between affected and control llamas.

Increased collagen type III expression and PG content in SL of juvenile llamas having metacarpophalangeal and metatarsophalangeal hyperextension provides evidence for ongoing matrix repair, possibly in response to injury or degeneration[7,18]. The trace mineral differences between affected and control llamas may have reflected dietary mineral imbalances in affected animals.

Our previous study showed that llamas affected by metacarpophalangeal hyperextension had increased serum zinc and decreased liver copper compared to control animals[1]. In that study, there were 5 adult llamas and 1 juvenile llama included in the group of affected animals. However, in the current study, affected juvenile llamas did not show the expected decrease in liver copper levels compared to controls. Copper deficiency is known to decrease the activity of lysyl oxidase[14,19], an enzyme necessary for proper cross linking of collagen fibers. However, when lysyl oxidase activity and gene expression were compared in previous studies[2,3], no significant differences were found between affected and control animals. Similarly, the current study did not reveal a significant difference in lysyl oxidase gene expression within the SL between affected and control llamas.

Collagen type I is predominantly expressed in normal tendons and ligaments, while collagen type III is expressed in damaged tendons and ligaments. The previous study[2] in adult llamas did not show differences in gene expression of collagen types I and III in affected animals, possibly reflecting the chronicity of the condition in older animals. While the current study confirmed the lack of collagen type I expression changes in younger llamas, increased collagen type III gene expression in affected juvenile llamas supported SL injury as a potential inciting factor in fetlock hyperextension. However, collagen type III protein was not apparent in histologic lesions within the SL of affected llamas. In addition, in all of the llamas examined, collagen type III was mainly limited to loose connective tissue in septa surrounding vessels and nerves within the SL, plus the epitenon and paratenon. Therefore, increased gene expression of collagen type III in affected llamas may have been due to a greater proportion of septa within the SL rather than increased expression by SL tenocytes and matrix.

Based on the radiographic findings of distally displaced sesamoid bones in most of the affected llamas, we concluded that fetlock hyperextension was the result of SL elongation, rather than involving the SDF, DDF or distal sesamoidean ligaments. This elongation may be due to inherent SL weakness, injury or degeneration. The lack of concurrent widespread ultrasonographic SL lesions in all affected llamas made injury or degeneration seem less likely. Further clarification of this syndrome is necessary and should include evaluation of young and acutely affected animals. Studies of the physical characteristics of collagen and its cross links would be possible with electron microscopy and high performance liquid chromatography (HPLC). Additionally, biomechanical testing of the suspensory apparatus comparing affected llamas to normal llamas could be performed.

ONGOING NEAR COMPLETED STUDIES (AS OF 2/2011)

Studies we are currently completing include evaluation of serum trace mineral analysis in approximately 90 camelids comparing findings in affected animals to age/sex matched controls. Additionally, we have almost completed an epidemiological survey of camelids in the Northwest United States. There is a disease prevalence rate of 17% in llamas and 0.7% of alpacas in the herds surveyed (including close to 1000 animals.)

REFERENCES

1. Reed S, Semevolos SA. **Am J Vet Res** , 2007. 2. Reed SK, Semevolos SA. **Am J Vet Res,** 2010. 3. Semevolos SA, Reed SK. **Am J Vet Res** 2011. 4. Mero J, Scarlett J.. **J Eq Vet Sc** 2005; 224-228. 5. Mero J, Pool R. **AAEP proceedings.** 2002 (pp. 329-334). 6. Gibson K, Steel CM. **Eq Vet Ed** 2002;4:50-64. 7. Halper J, Byoungjae K, Ahrar K, et al. ***BMC Vet Res*** 2006; 2:12. 8. Dyson S, Arthur R, Palmer SR. Suspensory Ligament Desmitis. ***Vet Cl N Am***1995;11:172-215. 9. Constantinescu G, Reed S, ***Int J Morph*** . 2007. 10. Constantinescu G, Reed S, ***Int J Morph***. 2007. 11. Palamakumbura AH, Trackman PC. **Anal Biochem** 2002. 12. Rucker RB, et al.**J Nutr** 1999;129:2143-2146. 13. Rucker RB, et al. **Am J Clin Nutr** 1998;67(suppl):996S-1002S. 14. Opsahl W et al. **J of Nut** 1982Reed SK, , et al. *Am J Vet Res* 2007. 15. Corps A, et al. **Rheumatology** 2005. 16. Nomura et al. **J Vet Med Sci** 2007. 17. Dudhia J, et al. **Aging Cell** 2007. 18. Dahlgren L, et al. **Am J Vet Res** 2005; 66(2):266-270. 19. Chou et al. **J Biol Chem** 1969;21:5785-5789.

GRAPHICS

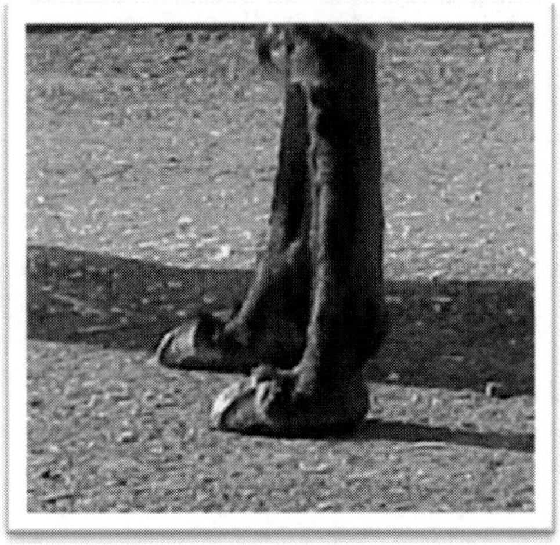

Figure 1: Normal fetlocks in adult llama

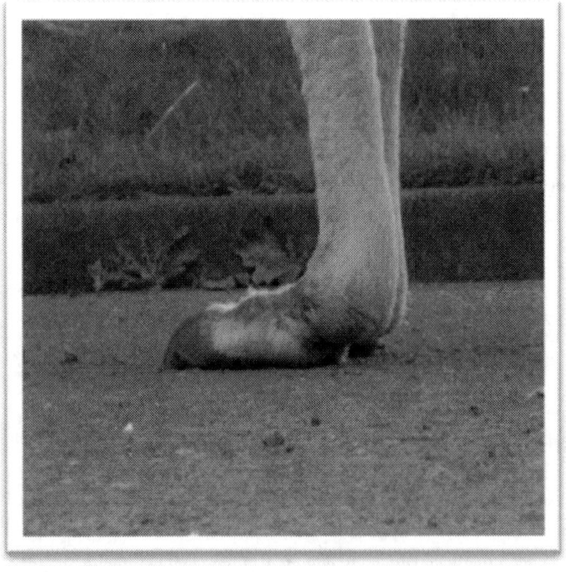

Figure 2: Severely affected llama

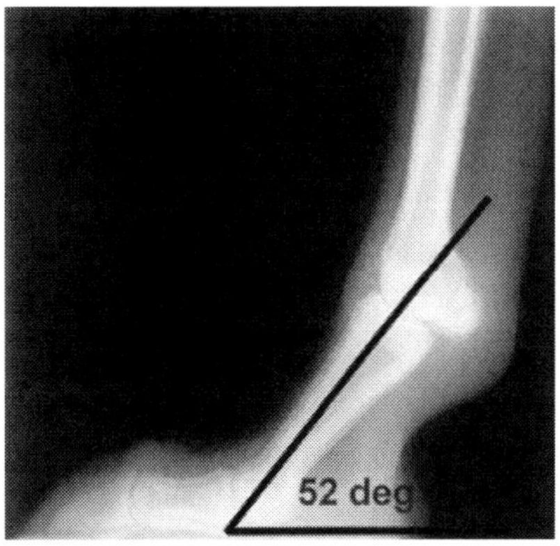

Figure 3: Radiograph of normal llama

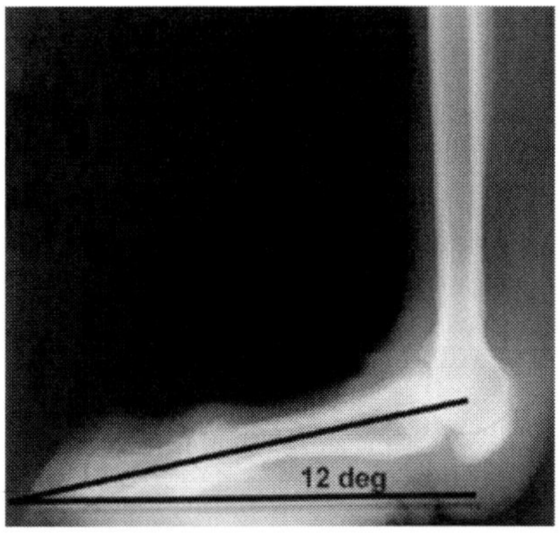

Figure 4: Radiograph of severely affected llama

USE OF COMPUTED TOMOGRPAHY FOR GASTROINTESTINAL DISEASE IN CAMELIDS

Susanne M. Stieger-Vanegas, DVM, PhD, DECVDI
Corvallis, OR

INTRODUCTION

In llamas gastro-intestinal disorders are difficult to diagnose based on medical history and physical examination. In the llama colic symptoms can be obscure and the cause of the symptoms can be difficult to localize due to the limited availability of diagnostic techniques evaluating the abdomen. Colic symptoms caused by bacterial enteritis or enteritis caused by Eimeria macusaniensis can be very difficult to differentiate from colic symptoms caused by intestinal strangulation, entrapments, or intraluminal obstruction.[1,2] For comparison in horses, colic symptoms are more easily detected as horses appear to be more sensitive to abdominal pain. In horses with colic symptoms a rectal examination is routinely performed in addition to complete blood count, abdominocentesis and evaluation of nasogastric reflux. In some cases ultrasound and abdominal radiographs are included in the diagnostic work-up of the colic. In young llamas and alpacas, the rectal examination is limited due to the small size of the animals. In adult llamas dependent on the size of the hands of the examiner often only the most caudal aspect of the rectum can be evaluated during the examination. Other diagnostic techniques, such as radiography are limited due to the size of the animals and may provide only limited information due to the overlap of abdominal organs and the ability of only obtaining a lateral view. Ultrasound, which is often readily available, is only able to evaluate relatively superficial structures and might further be limited due to the presence of gas in the intestinal tract.[3] Laboratory evaluation of the blood might point towards a gastro-intestinal disorder, but might not allow differentiating between a non-surgical case such as enteritis or a surgical case such as intestinal entrapment, strangulation, neoplasia or intraluminal obstruction.[4,5,6] Intraluminal obstruction in New World camelids can be caused by bezoars. Bezoars are gastric or small intestinal foreign bodies that either contain predominantly hair (trichobezoar) or plant fibers (phytobezoar). Phytobezoars are generally formed in the stomach and may be passed into the small intestine. Bleeding and ulceration may occur, however intestinal obstruction is the most common clinical problem.

Computed tomography (CT) is a well-established abdominal imaging technique in humans and small animals, which is often used to further investigate the extent and origin of a lesion diagnosed on radiographs or ultrasound. The fast technology advancement in CT in the last few years, especially in the area of multi-detector scanning has also lead to a range of new possibilities and applications for veterinary patients including llamas and alpacas. In veterinary medicine the development of multi-detector scanners with much shorter scan times has allowed to scan patients using only sedation instead of full anesthesia, which can be especially useful in diseased animals, where long anesthesia times are not warranted due to the status of the patient. The newer scanner designs also allow performing helical CT scans to acquire volumetric datasets, which then can be used to create two-dimensional and three-dimensional images of varying thickness and in various planes anywhere within the scan volume.

Often adequate images of the abdomen including the gastro-intestinal tract can be obtained without using contrast; however the use of contrast should be encouraged whenever possible to differentiate between vascular and non-vascular structures. In addition, the enhancement characteristics might help to differentiate between certain lesions. Unenhanced CT is very good to evaluate the size and location of abdominal organs, and to detect and localize stones and mineralizations. In veterinary medicine most commonly intravenous iodinated contrast agents are used to evaluate the vascularity of abdominal abnormalities. However, especially in smaller sized large animals such as llamas and alpacas, where the radiographic or ultrasonographic evaluation of the abdomen is limited an abdominal CT post oral contrast application might be used to evaluate the gastro-intestinal tract. In combination with oral contrast, CT can identify and differentiate the intestine from surrounding soft tissue. In humans it has been demonstrated that oral contrast agent distend the bowel lumen and allows therefore evaluating the wall thickness.[7] CT has also successfully been used for identification of the intestinal tract of healthy adult llamas. In one reported study barium sulfate was used as an oral contrast agent to differentiate between the gastro-intestinal tract and the soft tissue, however the imaging study was delayed for 18 hours to allow the barium sulfate to move through the gastro-intestinal tract.[2] Barium sulfate moves relatively slowly through the gastro-intestinal tract and therefore time delays are necessary to allow the barium sulfate to be transported through the entire gastro-intestinal tract. In humans also oral iodinated contrast agents are used for evaluation of the gastro-intestinal tract. The faster transport of non-ionic contrast agents through the intestinal tract, allows working with much shorter time delays, which might be more feasible way of examining clinical patients with colic symptoms.

Independent of the contrast application, timing of the contrast is very important. Intraoral contrast might have to be given several hours before the abdominal CT is performed allowing the contrast to pass through the gastrointestinal tract. The delay time chosen for intravenous contrast agents is dependent on the area of interest and the suspected disease. Routinely at least one post contrast study with an approximately 60 sec delay should be included in a CT study of the abdomen. If a vascular abnormality such as a portal shunt is suspected, the arrival of contrast agent in the area of interest has to be considered and correct timing of the scan has to be performed. This can be done with either a small intravenous injection of a test bolus or a bolus-tracking technique.

REFERENCES

1. Cebra CK, et al. *Eimeria macusaniensis* infection in 15 llamas and 34 alpacas. JAVMA, 2007; 230:94-100. 2. Van Hoogmoed L, et al. Use of Computed tomography to evaluate the intestinal tract of adult llamas. Vet Radiol Ultrasound, 1998; 39:2, 117-122. 3. Cebra CK, et al. Transabdominal ultrasonographic appearance of the gastrointestinal viscera of healthy llamas and alpacas. Vet Radiol Ultrasound, 2002; 43(4):359-66.4. Cebra CK, et al. Acute gastrointestinal disease in 27 New World camelids: clinical and surgical findings. Vet Surg ,1998; 112-121.5. Fowler M. In: Medicine and surgery of the South American camelids. Ames:Iowa State University Press, 1989; 224-262.6. Belknap, E. Medical Problems of llamas. Vet Clin North Am (Food Anim Pract), 1994; 10:291-298.7. Megibow AJ. In: Haaga JR. Computed tomography and magnetic resonance imaging of the whole body. St. Luis: CV Mosby CO, 1990;855-895.

PROGRESSION OF PARASITEMIA AND CLINICAL COURSE IN A SPLENECTOMIZED ALPACA WITH *CANDIDATUS MYCOPLASMA HAEMOLAMAE* INFECTION. JE Sojka-Kritchevsky , B Toth, JB Messick, AN Baird. Purdue University, West Lafayette, IN 47907.

Candidatus Mycoplasma haemolamae (CMhl) is a common red blood cell parasite of new world camelids. The high degree of parasitemia that develops in an infected splenectomized animal allows for the efficient collection of parasitic DNA. This DNA can then be used in the development of genetically-derived tools such as PCR and in-situ hybridization. Thus, one splenectomized animal can replace many immunologically intact animals within a research setting. The purpose of this study was to track the natural progression of CMhl parasitemia and associated clinical signs in a splenectomized alpaca.

An intact, 9-month-old, 39.1 kg male alpaca was used in this study. He had tested positive via PCR for CMhl on three different occasions, although no organisms were seen on peripheral blood smears. The alpaca was placed under general anesthesia and a ventral midline incision was made. The spleen was located, the vessels ligated, and the organ removed. Buprenorphine and flunixin meglumine were given for 2 and 4 days after surgery respectively. Body weight, attitude, rectal temperature, blood glucose, and PCV were recorded daily. In addition, a peripheral blood smear was examined daily and the percent of red blood cells that were infected with Mycoplasma organisms was determined.

The alpaca was not parasitemic prior to surgery. One percent of the RBC's contained mycoplasma on days 2 and 3 after splenectomy. Parasitic bloom developed on day 4 with 85% of the red blood cells infected, and over 70% containing 3 or more organisms. The alpaca was treated with 20 mg/kg Oxytetracycline I.V. on day 4. On post-operative day 5 no parasites were seen in the peripheral blood. The peripheral blood remained free of parasites for 11 days. On the morning of the 12th day, 3% of the peripheral red blood cells contained Mycoplasma. By late that afternoon, 75% of the observed RBC's contained 3-4 organisms. The alpaca again received oxytetracycline. There were no more parasites observed from that time until the alpaca was euthanized 5 days later.

The alpaca lost 3.7 kg between days -1 and +2 after surgery. His weight fluctuated between 34.9 and 35.4 kg for the remainder of the study period. Blood glucose ranged between 87 and 163 mg/dL There was no major change in PCV (range 21 – 29%), a finding that was expected as the spleen was not available to remove infected red blood cells. Body temperature ranged between 38.1 and 39 degrees Celsius except for days 4 and 16 when more than 70% of red blood cells contained parasites. On those days rectal temperature reached 39.4 and 39.1 degrees respectively.

This study confirmed that a non-parasitemic, yet PCR positive alpaca did indeed harbor CMhl. The time from splenectomy to parasitic bloom was shorter, and the length of oxytetracycline suppression longer than has been observed in other species.

ASSESSMENT OF *BDELLOVIBRIO BACTERIOVORUS* 109J VIABILITY IN BOVINE TEARS

Melanie J Boileau, DVM, MS, DACVIM; Kenneth D Clinkenbeard, DVM, PhD
Stillwater, OK

INTRODUCTION

Ocular infections with the infectious bovine keratoconjunctivitis (IBK) agent *Moraxella bovis* (*M. bovis*) are associated with significant economic loss in the cattle industry. The clinical manifestations of IBK range from mild unilateral or bilateral conjunctivitis to central corneal ulceration and perforation. Administration of antibiotics such as oxytetracycline,[1,2] florfenicol,[3,4] ceftiofur crystalline free acid[5] and tulathromycin[6] have been shown effective in the treatment of experimentally induced or naturally occurring IBK in cattle however treatment failures are common and current commercially available vaccines are not optimally effective[7,8] mainly due to *M. bovis* antigenic strain variation. As a result, our laboratory has been actively investigating the therapeutic potential of *Bdellovibrio bacteriovorus* 109J (*B. bacteriovorus*); a predatory bacterium capable of attacking and inducing lysis of Gram-negative bacteria, as an effective non-chemotherapeutic alternative treatment for IBK; delivered topically on affected eyes.

We have previously shown that *B. bacteriovorus* can reduce the number of *M. bovis* attached to bovine epithelial cells in an *in vitro* model of IBK and that *B. bacteriovorus* can be trained to kill *M. bovis* as effectively as *E. coli* using serial passages.[9] However, the specific/non-specific antimicrobial action of the tear film proteins, immunoglobulins (IgA) and enzymes[10] on *B. bacteriovorus* viability has not been investigated. The purpose of the study was therefore to determine whether *B. bacteriovorus* can remain viable in bovine tears without its prey for up to 24 hours.

MATERIAL AND METHODS

Tears were collected from both eyes of one clinically healthy steer by placing Wek-cell cellulose sponges in the lower medial conjunctival fornix until saturated. Tears were extracted from the sponges by centrifugation (4000 rpm x 10 min), filtered twice sequentially through 0.8 μm, then 0.45 μm membranes, and frozen at -20 °C for later use.

Using a plaque assay to quantify the mean amount (± SD) of plaque forming units (PFUs) of *B. bacteriovorus,* the viability of active *B. bacteriovorus* inocula incubated in its preferred media peptone yeast extract (PYE) was compared to *B. bacteriovorus* viability in bovine tears or phosphate buffered saline (PBS) at time 0, 2, 4, 12, and 24 hours (Table 1-3).

DISCUSSION & CONCLUSION

Based on the PFUs of *B. bacteriovorus* exposed to each treatment, it was determined that viability of *B. bacteriovorus* over time was comparable between treatment groups. Overall, the results supported that *B. bacteriovorus* can remain viable in tears for up to 24 hours in the absence of prey bacteria. Further studies are needed to determine the therapeutic potential of *B. bacteriovorus* in an *in vivo* model of IBK.

Table 1. Comparison of *B. bacteriovorus* viability in PYE and bovine tears over time

Time (hr)	*B. bdellovibrio* in PYE ($x\ 10^9$ PFU/mL)	*B. bdellovibrio* in tears ($x\ 10^9$ PFU/mL)
0	5.2 ± 1.89	4.3 ± 1.04
2	4.7 ± 1.52	4.7 ± 1.04
4	3.3 ± 0.29	3.8 ± 1.6

Table 2. Comparison of *B. bacteriovorus* viability in PYE and PBS over time

Time (hr)	*B. bdellovibrio* in PYE ($x\ 10^9$ PFU/mL)	*B. bdellovibrio* in PBS ($x\ 10^9$ PFU/mL)
0	5.8 ± 4.93	5.6 ± 3.32
2	5.3 ± 2.47	4.2 ± 0.57
4	4.8 ± 1.44	3.3 ± 1.53

Table 3. Comparison of *B. bacteriovorus* viability in PYE and bovine tears over a 24-hour period

Time (hr)	*B. bdellovibrio* in PYE ($x\ 10^9$ PFU/mL)	*B. bdellovibrio* in tears ($x\ 10^9$ PFU/mL)
0	5.1 ± 0.65	5.1 ± 2.10
12	2.6 ± 1.78	2.7 ± 0.70
24	5.7. ± 0.55	6.3 ± 0.74

REFERENCES

[1] George LW, et. al. **J Am Vet Med Assoc** 1988;192(10):1415. [2] Eastman TG, et. al. **J Am Vet Med Assoc** 1998;121(4):560. [3] Angelos JA, et. al. **J Am Vet Med Assoc** 2000;216(1):62. [4] Dueger EL, et. al. **Am J Vet Res** 1999;60(8):960. [5] Dueger EL, et. al. **Am J Vet Res** 2004;65(9):1185. [6] Lane VM, et al. **J Am Vet Med Assoc** 2006;229(4):557. [7] Davidson HJ, et. al. **Can Vet J** 2003;44:577. [8] Smith PG, et. al. **Am J Vet Res** 1990;51(7):1147. [9] Boileau MJ, et al. **Can J Vet Res** 2011: In Press. [10] Gionfriddo JR. **Am J Vet Res** 2000;61:1289.

HYPERCHLOREMIA DIRECTLY INCREASES IONIZED CALCIUM CONCENTRATION IN BOVINE PLASMA: AN ADDITIONAL MECHANISM BY WHICH HIGH CHLORIDE RATIONS MITIGATE PERIPARTURIENT HYPOCALCEMIA. **PD Constable[1], HR Stämpfli[2]. [1]Department of Veterinary Clinical Sciences, Purdue University, West Lafayette, IN, USA; [2]Department of Clinical Studies, University of Guelph, Guelph, ON, Canada.**

Correction of the measured ionized calcium concentration (cCa^{2+}) to a pH = 7.40 is routinely applied in experimental studies in order to assist in the interpretation of measured values relative to a reference range. The equation most commonly used for pH correction in bovine plasma is: $cCa^{2+}{}_{pH=7.40} = cCa^{2+}\times 10^{(-0.24\times\{7.40 - pH\})}$. The validity of this equation for bovine plasma is unknown. Accordingly, our first objective was to characterize the in vitro relationship between cCa^{2+} and pH for bovine plasma. Feeding rations with a low dietary cation-anion difference (DCAD) during late gestation mitigates periparturient hypocalcemia in dairy cows, particularly when chloride containing acidodgenic salts are fed. The mechanism for this beneficial effect remains unclear. Our second objective was to determine whether hyperchloremia displaces calcium from binding sites to albumin, thereby increasing cCa^{2+}.

The *in vitro* relationship between plasma log(cCa^{2+}) and pH was investigated using lithium heparin anticoagulated blood from 10 healthy Holstein-Friesian calves. Plasma was harvested and tonometered with CO_2 at 37°C over a pH range of 7.10-7.70. Plasma chloride concentration (cCl^-) was altered by equivolume dilution of plasma with 3 electrolyte solutions of varying cCl^- (97±1, 110±1, and 123±1 mEq/L; mean±SD).

The slope of the linear regression equation relating log(cCa^{2+}) to pH for 112 tonometered plasma samples from the 10 calves was -0.24 ± 0.05 at normal values for cCa^{2+} (2.63 ± 0.09 mEq/L), albumin concentration (30.9 ± 2.0 g/L), and cCl^- (99.1 ± 2.2 mEq/L). The experimentally-determined value for the slope for bovine plasma was identical to that determined previously for human plasma. The formula for correcting cCa^{2+} in bovine plasma for change in pH from 7.40 is therefore: $cCa^{2+}{}_{pH=7.40} = cCa^{2+}x10^{(-0.24x\{7.40-pH\})}$. This equation is only valid at normal concentrations of albumin and chloride in plasma. Equivolume dilution of plasma by electrolyte solutions of varying cCl^- indicated that $cCa^{2+}{}_{pH=7.40}$ increased by 0.007 mEq for every 1 mEq/L increase in cCl^-. In other words, plasma cCa^{2+} at a given pH increases directly in response to an increase in plasma cCl^-, presumably because the additional chloride displaces calcium that is electrostatically bound to albumin. Furthermore, the increase in cCa^{2+} is independent to the change in plasma pH induced by an increase in cCl^- and decrease in plasma strong ion difference.

Our finding that hyperchloremia directly increases plasma cCa^{2+} provides an additional mechanism by which ingestion of high chloride (acidogenic) rations prevents the clinical signs of periparturient paresis. Our finding is consistent with the results of other studies that indicate acidogenic salts that contain chloride as the predominant anion (ie, NH_4Cl, $CaCl_2$) are more effective in increasing cCa^{2+} than equimolar quantities of acidogenic salts such as $MgSO_4$.

A NEW LOOK AT THE IMPORTANCE OF COAGULASE NEGATIVE STAPHYLOCOCCI IN BOVINE MASTITIS. JR Middleton[1], J Perry[1], RN Zadoks[2], S Dufour[3], ME Paradis[3], DT Scholl[3]. 1. University of Missouri, Columbia, MO. 2. Moredun Research Institute, Edinburgh, Scotland. 3. University of Montreal, Quebec, Canada.

Coagulase negative staphylococci (CNS) are among the most common bacteria isolated from the bovine mammary gland. Historically, these bacteria were lumped together as minor mastitis pathogens. Modern molecular techniques have allowed accurate speciation and fingerprinting of the CNS species. These methodologies have recently been applied to the study of CNS in bovine mastitis. The aim of the studies presented here was to evaluate the role of individual CNS species on milk somatic cell count (SCC) and duration of intramammary infection (IMI).

In the first study, mammary quarter foremilk samples were aseptically collected from all lactating cattle (~180 head) at the University of Missouri dairy research center once monthly for 17 months for bacterial culture and milk SCC. Staphylococcal isolates were speciated by sequencing the *rpo*B gene and strain-typed using pulsed-field gel electrophoresis (PFGE). Using species and fingerprint data along with published definitions for staphylococcal IMI, 91 CNS IMIs were identified. Overall, 11 species of CNS were identified with *Staphylococcus chromogenes*, *S. cohnii*, *S. epidermidis*, and *S. simulans* being most prevalent. Duration of IMI and SCC data were analyzed using regression models accounting for repeated measures. Mean milk SCC and duration of IMI were found to differ between CNS species ($P < 0.05$). Although most IMIs were of short duration (1 month), *Staphylococcus capitis* and *S. chromogenes* IMIs had longer mean durations of infection than 3 or more of the other species isolated. Mean SCCs were under 350,000 cells/ml in most cases. However, *Staphylococcus simulans* and *S. xylosus* IMIs were more inflammatory (mean > 600,000 cells/ml) and had a higher mean SCC than *S. cohnii*, *S. epidermidis*, and *S. haemolyticus*.

To examine the relationship between CNS IMI and milk SCC in a larger population of cattle, CNS isolates from the Canadian Bovine Mastitis Research Network (CBMRN) culture collection were obtained for speciation. Speciation and fingerprinting were performed as above. Isolates were from subclinical IMI from before and after the dry period and from subclinical IMI during lactation. Data associated with each isolate were obtained from the CBMRN database. Nine-hundred-thirty-eight isolates from 696 mammary quarters in 89 herds were successfully speciated. Twenty-two different species of CNS were identified. *Staphylococcus chromogenes* was the most frequent species identified accounting for 40% of the infections. Three species, *S. chromogenes*, *S. xylosus*, and *S. simulans* accounted for >75% of all infections. Data were analyzed using a linear hierarchical repeated measures mixed model. Differences in mean SCC were found between some CNS species and culture negative control quarters and also between different species of CNS ($P < 0.05$). Overall, our data demonstrate potential differences in pathogenicity between strains of CNS that cause bovine mastitis.

REDUCTION OF BACTERIAL CONTAMINATION OF BOVINE COLOSTRUM WITH HIGH PRESSURE PROCESSING. KP Poulsen[1], DM Foster[2], BE Farkas[3], SM McGuirk[1], GW Smith[2], CJ Czuprynski[1] 1. University of Wisconsin School of Veterinary Medicine, Madison, WI 2. North Carolina State University College of Veterinary Medicine, Raleigh, NC, 3. North Carolina State University Department of Food, Bioprocessing & Nutrition Sciences.

Passive transfer of maternally derived antibodies via ingestion of good-quality colostrum within the first 24 hours of life is crucial for the health and future productivity of dairy calves. However, infectious diseases can be transmitted via colostrum feeding, which may require use of a colostrum replacement product or pasteurization to decrease disease transmission. While pasteurization of colostrum is effective for sterilization, heating during pasteurization can alter the viscosity of colostrum, destroys important nutritional biomolecules, and has been shown to decrease colostral IgG concentrations. The purpose of this study was to investigate the effect of high pressure processing (HPP) on the viscosity, IgG concentration, and bacterial contamination of bovine colostrum.

First milking colostrum samples were collected from 18 cows from 3 different farms, and 50ml aliquots of each sample were pooled for analysis. Pooled colostrum was processed in triplicate using an isostatic press at 400 MPa (60,000 psig) for 0, 5, 15, 30, and 45 minutes. Samples were tested for the effects of HPP on the viscosity, bacterial load (CFU/ml), and IgG concentration.

There was a significant decrease ($p<0.05$) in bacterial load at each time point when compared to time 0. No significant difference in IgG concentration was found between any time points. Subjectively, the colostrum viscosity appeared to increase with the processing time, though the rheologic assessment has not been completed at this time.

HPP appears to be an effective method to decrease bacterial contamination of colostrum while maintaining appropriate IgG concentrations. Minimizing the processing time or pressure may be necessary to maintain an acceptable viscosity of the colostrum. Based on these results, additional studies are justified in order to determine the optimum combination of processing time and pressure and the effect of HPP on specific bovine pathogens.

EFFICACY OF A NOVEL NON-ANTIBIOTIC FOOT BATH SOLUTION FOR TREATING DIGITAL DERMATITIS IN LACTATING DAIRY COWS. TS Markovich[1], PD Constable[2], [1]Department of Veterinary Clinical Science, University of Missouri, Columbia MS; [2]Department of Veterinary Clinical Sciences, Purdue University, West Lafayette, IN.

Digital dermatitis (DD) is a contagious claw disease that commonly causes lameness in lactating dairy cattle in North America and Europe. Treatment protocols for DD include foot baths and topical application of antibiotics. Control measures for DD include decreasing moisture where cattle are standing, disinfecting foot trimming equipment, and foot baths. The objective of this study was to determine whether application of a novel foot bath solution (Accelerator™) was superior to, or as effective as, a 5% copper sulfate foot bath to control DD in lactating dairy cows. The novel foot bath solution does not contain antibiotics or copper and requires less frequent application than a copper sulfate foot bath.

Lactating Holstein-Friesian cows from a herd in central Missouri were examined for evidence of DD and thirty-four cows with visible DD lesions (hairy heels warts) in at least one hind limb and a lameness score of $\geq 2/5$ were randomly assigned to one of two treatment groups. During a 28 day study period, 17 cows walked through the Accelerator™ foot bath once weekly, while the other 17 cows walked through a 5% copper sulfate solution foot bath once daily five days a week (Monday through Friday). The hind feet were evaluated on day 0 (immediately before starting the foot bath), and on days 7, 14, 21 and 28. Data collection included lameness score (1-5), pain score (0-2) via application of a water jet to the distal hind limb, lesion score (M0, M1, M2, M3, and M4), and digital image of each hoof that permitted measurement of lesion area using a freely available software program (ImageJ 1.42i). Changes in DD lesion area and lameness score over time were evaluated using repeated measures analysis of variance. The number of limbs cured and the incidence of new DD infections were compared between treatments using Chi-Square test. $P < 0.05$ was considered significant.

The number of limbs with DD in the Accelerator™ (65%, 22/34) and $CuSO_4$ (71%, 24/34) groups were similar at the start of the study (Day 0). However, a lower percentage of limbs in the Accelerator™ group (76%) were affected by DD over the 28 day study period than in the 5% copper sulfate group (88%, $P = 0.0036$). The Accelerator™ foot bath treatment also resulted in a clinically and statistically ($P < 0.05$) significant reduction in DD lesion area and lower lameness scores than cows walking through the 5% copper sulfate foot bath.

In conclusion, the Accelerator™ footbath shows promise for controlling DD in dairy herds that wish to use a less labor intensive footbath regimen than that commonly used with a 5% copper sulfate foot bath. Moreover, the Accelerator™ footbath offers an alternative foot bath for producers interested in decreasing environmental contamination with copper or antibiotics.

LACTOFERRIN: A FURTHER INVESTIGATION INTO ITS ANTI-INFLAMMATORY EFFECTS; INTRACELLULAR SIGNALING IN EVIDENCE? Maisie E Dawes

The heme-associated iron-binding apoprotein lactoferrin (LF) is known for its, anti-inflammatory, anti-parasitic, antimicrobial and bactericidal effects. Lactoferrin demonstrates ubiquity throughout mammalian host biological fluids: Saliva; tears; mammary secretions, as well as at mucosal surfaces. It is also released from immune cells under pathogenic stimulation. The purpose of this study which has been approved by Western University's Institutional Animal Care and Use Committee is to further characterize the mechanisms through which LF modulates inflammation in the face of bacterial endotoxin. It was hypothesized that LF would inhibit p38 phosphorylation. Numerous studies speak to the ability of LF to alter leukocyte function, inhibit cytokine production, and bind lipopolysaccharide (LPS); mechanisms through which it is believed to achieve its anti-inflammatory effects. Recently, investigators demonstrated its ability to interact with host DNA while others describe regulation of granulocyte adhesion and motility; elucidating its roles in the apoptotic signaling.

In earlier studies, Dawes ME, et al. demonstrated lactoferrin's ability to limit the expression of inducible cyclooxygenase-2 and the gelatinase, matrix metalloproteinase -9 by LPS-induced macrophages. The generation of these inflammatory mediators is modulated by pro- inflammatory cytokines such as interleukin-1 β (IL-1 β) and tumor necrosis factor-alpha (TNF-α), the production of both being dependent on signaling through the p38 mitogen-activated protein kinase (MAPK) pathway.

Peripheral mononuclear cells (5 x 10^6) isolated from buffy coat cells of healthy neonatal to 4-month old Holstein calves were cultured in the presence and absence of LF (200 ng/ml), LPS (1μg/ml), anisomycin (25 μg/ml), a known p38 activator - the positive control, and 50 mM of SB203580, a known p38 inhibitor - the negative control. Sample lysates obtained post culture was subjected to immunoprecipitation and kinase reactions. Reactions were terminated under reducing conditions and evaluated using western immunoblotting. Phosphorylation of activated transcription factor-2 (ATF-2) by phosphorylated p38 served as the marker of investigation. Immunologically reactive ATF-2 expression by LPS and anisomycin-treated cells was compatible with a prominent band at 40 kD. Evidence of LF-induced inhibition of LPS-induced p-38 activation was observed in lanes representative of co-cultures of LF + LPS; LF + anisomycin; and anisomycin + SB203580, which was demonstrated by decreased immunological reactivity at 40kD. The findings here, suggest that LF interferes with LPS-induced p-38 activation of transcription factor ATF-2, *in vitro*. This serves as additional proof of its potential use in attenuating the systemic effects of LPS.

This page intentionally left blank.

PATHOLOGY AND PATHOGENESIS OF BOVINE TUBERCULOSIS

Randall J. Basaraba, DVM, PhD, DACVP
Fort Collins, CO

INTRODUCTION

The purpose of this presentation is to describe the pathologic features of tuberculosis in cattle and other species and to introduce new concepts related to how bacilli evade host immunity to establish latent infections. Animals or humans with latent infections are clinically normal and often lack detectable lesions ante-mortem, yet have evidence of infection based on a positive tuberculin skin tests or other immunological assays. The ability of bacilli to persist for long periods of time without causing clinical disease and the cross reactivity with closely related environmental bacteria are among the greatest challenges in the development of accurate and reliable diagnosis of mycobacterial infections. The complex "life cycle" and the expression of various virulence mechanisms by mycobacteria accounts for why vaccines that confer sterile immunity for bovine or human tuberculosis, remain elusive. A better understanding of the molecular pathogenesis of mycobacterium and how the host responds to infection may aid in improved methods for diagnosis, prevention and treatment of mycobacterial infections in animals and humans.

ETIOLOGY OF TUBERCULOSIS

Bovine tuberculosis (BTB) is caused by *Mycobacterium bovis*, which is closely related to the classical tubercle bacillus of humans, *Mycobacterium tuberculosis* (MTB). Not only are the two agents morphologically indistinguishable but they share identity over 99% at the whole genome. For the purpose of this presentation, another important similarity is that *M. bovis* and MTB infections produce similar pulmonary and extra-pulmonary pathology in humans and animals. In standard microbiology text books, pathogenic mycobacteria including *M. bovis* and MTB are classified as intracellular, non-spore forming, Gram positive bacilli. In addition, both *M. bovis* and MTB are considered obligate aerobic bacteria and thus require oxygen to actively replicate. These microbiological characteristics are not only important for laboratory diagnosis but have important implications in the molecular pathogenesis of mycobacterial infections. Recent data, primarily from MTB, has called into question the majority of these fundamental physiologic and phenotypic characteristics including the ability of mycobacteria to from spores (1). Even the most characteristic feature of these pathogens, their ability to resist decoloration by inorganic acids following staining with carbol-fuchsin or acid-fast staining characteristics has been challenged. Even though classified as Gram positive, mycobacteria are not easily stained with simple bacterial stains due to the high lipid content of the complex cell wall. While the chemical reaction that confers acid-fastness is still poorly understood, it is generally thought to be dependent on the spatial arrangement of the mycolic acids and their esters in the bacterial cell wall. Recent studies however, using a combination of the fluorescent "acid-fast" stains rhodamine and auramine, have demonstrated that within infected tissues, there exists multiple populations of bacilli that show variability in acid-fast staining reactions such that many bacilli stain acid-fast negative (2). This feature has important implications not only in the diagnosis of tuberculosis but also in drug treatment. Particularly important in human patients infected with BTB or MTB, the mode of action of one of the most important first line anti-tuberculosis drugs isoniazid, is the inhibition of mycolic acid synthesis. These studies suggest that under certain physiologic conditions such as hypoxia, the cell wall and possibly the mycolic acid structure is altered such that it renders the bacilli drug resistant or drug tolerant.

TRANSMISSION OF MYCOBACTERIA

Many host and pathogen factors influence the pathogenesis and thus the pathology of mycobacterial infections. These include the resistance of the host, the virulence of the challenge strain, the mode of transmission and others. The transmission of *M. bovis* and MTB in their natural hosts is primarily through aerosol spread of droplet nuclei during close contact with infected individuals. Human infections with *M. bovis* are well documented and are on the rise, particularly in the southwest and south central United States that border Mexico (3-6). The ingestion of unpasteurized dairy products contaminated with *M. bovis* is an important source of human infection which leads primarily to extra-pulmonary infections including cervical lymphadenitis and involvement of the gastrointestinal tract and associated mesenteric lymph nodes (7). Oral infections generally require a higher dose of bacilli compared to airborne exposure. The significance of extra-pulmonary tuberculosis especially caused by *M. bovis* in people is that bacilli are less responsive to drug therapy and extra-pulmonary sites are more likely to serve as a site of persistent infections (5). Transplacental transmission in cattle is due to endometrial involvement and can manifest as infection of the fetal liver and lymph nodes associated with the portal circulation. Other but less common routes of infection include percutaneous inoculation, genital transmission and mammary infections resulting from infusion of contaminated needles or pharmaceutical preparations.

PATHOLOGY OF TUBERCULOSIS GRANULOMAS

As mentioned previously, BTB and MTB are classified as intracellular pathogens and in fact, the ability of mycobacteria to survive within macrophages and other cell types is central to the pathogenesis of tuberculosis. The ability to survive intra-cellularly is not only a strategy used by the bacteria to evade host immunity but is the main stimulus for granuloma formation. The localized focus of chronic inflammation incited by mycobacteria is referred to as a granuloma. A granuloma is defined as a localized aggregate of mixed inflammatory cells with predominately mononuclear and multinucleated macrophages (giant cells), lymphocytes and plasma cells and fewer granulocytes. Granuloma formation associated with mycobacterial infections reflects the progressive cellular response to a chronic infection but can occur in response to non-infectious stimuli as well. Mycobacterial infection however, remains one of the top differential diagnoses when granulomatous inflammation is confirmed grossly or histologically. Other features of tuberculous granulomas are the development of central caseous necrosis, calcification, encapsulation with fibrous connective tissue and commonly in MTB lesions in humans and experimentally infected rabbits, lesion cavitation. The "post-lesion" modifications of granulomata are often a reflection of the chronicity of the disease and the presence or absence of unknown stimuli that incite a shift in the inflammatory response. The most common but poorly understood is the massive infiltration of neutrophils that are associated with cavitary disease and abscess formation in chronic mycobacterial infections.

ROLE OF NEUTROPHILS IN THE PATHOGENESIS OF TUBERCULOSIS

The role neutrophils play in the pathogenesis of tuberculosis is poorly understood. Mycobacterium can survive phagocytosis and intracellular killing by neutrophils and have been shown to be among the first cells to engulf and transport bacilli from the site of experimental injection to regional lymph nodes. Recently, it has been shown that neutrophils modulate the immune response to Bacillus Calmette-Guerin (BCG) vaccination through a myeloid suppressor cell (8). The infiltration of neutrophils during mycobacterial infections while providing antimicrobial activity can be detrimental to the host through the generation of tissue damaging reactive oxygen species (ROS) and nitrogen species (RNS) or free radicals. The progressive infiltration of neutrophils into lesions and subsequent necrosis of granulomas following experimental infection of experimental animals with MTB as well as M. bovis infection of cattle and deer can be abrogated by prior vaccination with BCG. Mycobacteria have evolved several defense mechanisms to resist killing by ROS and RNS, which enables bacilli to survive intra-cellularly within phagocytic cells. Mycobacteria have recently been shown to induce and resist killing by neutrophils through a newly described antimicrobial killing mechanism of referred to as Extracellular Traps (ET) formation (9). ET formation is an important innate immune defense mechanism that leads to extra-cellular rather than intra-cellular bacterial killing by not only neutrophils but macrophages and other inflammatory cells. ETs consist of host cell derived extra-cellular DNA with attached antimicrobial peptides and enzymes that trap and kill extra-cellular bacteria (10). The significance of mycobacteria and other bacteria being able to resist killing by ETs is that the extrusion and entrapment of bacteria in extracellular DNA provides and additional survival strategy for the bacteria through the formation of complex, drug-tolerant microbial communities referred to as biofilms as described below.

PATHOGENESIS OF GRANULOMA NECROSIS

One of the most important features of tuberculosis in cattle and humans is the formation of caseous necrosis in granulomas. Granuloma necrosis represents an adverse host response to infection that leads to irreversible tissue damage and thus a loss of the structural and functional integrity of the lung or extra-pulmonary tissue parenchyma. The term caseous is derived from the dry, crumbly appearance (cheese like) of the cellular exudate in the center of necrotic granulomas. The development of lesion necrosis is in is a complex and poorly understood response to the cellular and humoral immune response incited by mycobacteria infection. Both host and pathogen factors contribute to the formation of granuloma necrosis. Mycobacteria and mycobacteria cell wall extracts stimulate a strong delayed-type hypersensitivity (DTH) resulting in a positive response to the tuberculin skin test. A localized inflammatory response at the site of injection of purified protein derivative (PPD), the tuberculin skin test (TST) remains the most wildly used ante-mortem diagnostic test worldwide for the diagnosis of both bovine and human tuberculosis. It has been long assumed that the DTH reaction stimulated by mycobacterial infection accounts for necrosis of the pulmonary and extra-pulmonary granuloma. This model links the necrosis of granulomas to the lysis of mycobacteria infected macrophages by cytotoxic lymphocytes. While the pathogenesis of granuloma necrosis includes an immune mediated process, several lines of evidence argue against infected macrophages being the target of the immune-mediated cytolytic response. The spatial organization of lymphocytes at the periphery of the lesion argues against the target cell being infected macrophages but rather endothelial cells since the initiation of necrosis is consistently within the lesion center. Our laboratory has studied the possibility that endothelial cells of the lung vasculature that supplies the affected lung are the targets of immune-mediated and oxidative stress mediated damage. The loss of lesion blood supply and thus ischemia as the lesions expand in size, explains the consistent appearance of central necrosis within the first 2 to 3 weeks of infection. This model is supported by immunohistochemistry data that shows early and extensive loss and then remodeling of the blood vasculature that supplies the affected lung parenchyma with new vessel formation as the inflammation subside and granulomas begin to heal. Besides the irreversible tissue damage, granuloma necrosis is thought to be the primary process that promotes the hematogensous dissemination of bacilli from the primary site of infection in the lung to essentially any part of the body (11).

CONSEQUENCE OF GRANULOMA NECROSIS

As previously mentioned, the consequence of infected host cells dying of necrosis in *M. bovis* and MTB infections is that there is irreversible tissue damage that can lead to a loss of function especially if extensive. The other consequence is that foci of necrosis can serve as site of bacterial persistence even following aggressive antimicrobial drug therapy. Using the guinea pig model of MTB infection we have shown that it is primarily the extra-cellular population of bacilli entrapped within the necrotic cellular debris of necrotic granulomas that persists following drug treatment (12). Besides the lung, necrosis of extra-pulmonary lesions can harbor persistent bacilli and may explain why these lesions are less responsive to drug therapy (13). While it is well accepted that mycobacteria are primarily intracellular pathogens, bacilli can survive in an extra-cellular microenvironment with necrotic granulomas or lesions with central cavitation. Within these lesions are remnants of necrotic host cells as well as large molecular weight plasma proteins that leak from the blood vasculature into the extra-vascular space (14). With the disruption of granuloma blood supply and the leakage plasma proteins into the extracellular space, necrotic granulomas represent a nutrient rich environment that harbors extracellular bacilli. Even within the lung, the avascular granuloma is deprived of oxygen and therefore, extra-cellular bacteria within these lesions can survive. In this anaerobic or hypoxic microenvironment, there is minimal or very slow replication which contributes to the loss of sensitivity of drug therapy (15, 16). Moreover, with the lack of blood supply and thus no means for immune cells to efficiently traffic between the circulation and the lesion, bacilli sequestered within necrotic granulomas are also sheltered from the cellular immune system which is necessary for eradication of mycobacteria (17, 18).

BACTERIAL BIOFILMS

In response to stressful environmental conditions, bacteria and other microorganisms adopt a variety of genotypic and phenotypic adaptations to give them a survival advantage. As previously mentioned, simply slowing the rate of replication in response to a low oxygen microenvironment of an obligate aerobic bacteria like mycobacteria, can significantly reduce the susceptibility to drug therapy. Another mechanism that microbes use to survive adverse conditions is to establish complex microbial communities composed of individuals or groups of bacteria that are genotypically or phenotypically heterogenous. The term biofilm has been given to the "community lifestyle" of bacteria in the environment or in vivo associated with a wide variety of infections in humans and animals. Most of what we know about the mechanisms and physiology of biofilms has been extrapolated from in vitro studies or the study of environmental bacteria. To complicate matters further, not only do single species of bacteria differ but biofilms can be made up of a number of different types of bacteria. There is increasing evidence that different types of bacteria within a polymicrobial community live synergistically to enhance the survival of each other (19). There are multiple stages in the formation of bacterial biofilms. The first is the transition between free living or planktonic to attached. The ability to attach to a biotic or abiotic surface is critical for the establishment of early biofilms. The next stage is the production of a pathogen derived extracellular matrix that encases the community as the biofilm matures. The extracellular matrix is important not only for holding the community together but it is used to acquire and concentrate micronutrients but also protects the community from environmental insults including drug therapy and host immune responses. The last stage of biofilm formation is dispersion of bacteria from the community that when released, can disseminate to reestablish infection in other parts of the body. As previously mentioned, most of what we know about biofilm formation has been derived from in vitro studies where attachment is to abiotic surfaces rather than to host derived macromolecules that one would expect to find associated with in vivo infections. Recently, however it has been shown that lysed human neutrophils enhance *in vitro* biofilm formation by *Pseudomonas aeruginosa* one of the most extensively studied biofilm forming bacteria of humans and animals (20, 21). The importance of macromolecules derived from necrotic host cells is that bacteria can utilize proteins, nucleic acids and lipids as an attachment matrix in the initiation and early maturation of biofilm communities.

BIOFILM FORMATION BY MYCOBACTERIA

Both environmental and pathogenic mycobacteria have been shown to form biofilms *in vitro* and in their natural environment. Recently, *Mycobacterium avium subsp. paratuberculosis*, the causative agent of Johne's disease in cattle and other ruminant and non-ruminant species, has been shown to survive within the environment through the formation of biofilms in water (22). While other pathogenic mycobacteria including the opportunistic respiratory pathogen *Mycobacterium abscessus* have been shown to form biofilms, adoption of this mode of growth as a survival by *M. bovis* and MTB remains controversial (23, 24). Recently, *in vitro* studies have demonstrated that MTB can form *in vitro* drug-resistant communities when grown as a pellicle at the liquid, air interface of artificial growth media (25, 26). Therefore, rather than complex polysaccharides, MTB may use free mycolic acids as an extra-celluar matrix in the initiation and stabilization of biofilm communities. Similar to the studies with *Pseudomonas*, data from our laboratory show that MTB and *M. bovis* BCG can utilize DNA and other macromolecules derived from necrotic host cells to initiate the formation of drug-tolerant biofilm-like communities *in vitro*. Pathogen-derived DNA originating from non-viable bacteria has been shown to be important as an attachment matrix in the initiation of biofilm formation. The focus of our laboratory is to show that host-derived DNA from necrotic cells or from cells undergoing ET formation represents an even more abundant source for the attachment and maturation of in vivo biofilms. Preliminary data suggests that digestion of extracellular DNA with DNase potentiates the activity of antimicrobial drugs in an *in vitro* model of mycobacterial biofilms.

CONCLUSION

The interaction between mycobacteria and the host is complex. There is a fine balance between the ability of the host response to control or eliminate infection without creating irreversible tissue damage in the form of lesion necrosis or cavitation. Mycobacteria have evolved survival strategies to not only survive the intracellular stage of the life cycle in the early stages of infection by also the extra-cellular stage of infections when sequestered within necrotic granulomas. The pathogens my utilize remnants of necrotic host cells and plasma proteins to initiate the formation of microbial communities to not only resist killing by phagocytic cells but to survive drug therapy as well. Strategies the prevent or disrupt the formation of biofilm communities are being explored as a way to potentiate the efficacy of current antimicrobial drugs that are increasingly being rendered ineffective through the development of drug resistance. Understanding this basic biology can also enhance the development of new vaccines that target antigens that are important in the initiation and maturation of bacterial biofilms (27).

REFERENCES

[1]Ghosh, J., et al. *Proc Natl Acad Sci U S A*2009 106:10781, [2]Ryan, G. J. et al. *PLoS One* 2010. 5:e11108. [3] Dankner, W. M *Pediatrics,* 2000 105:E79. [4]LoBue, P. A. *Int J Tuberc Lung Dis* 2003 7:180. [5]LoBue, P. A. 2005 *Int J Tuberc Lung Dis* 9:333. [6]de Kantor, I. N. 2010 *Int J Tuberc Lung Dis* 14:1369. [7]Thoen, C. 2006 *Vet Microbiol* 112:339. [8]Martino, A., E. 2010 *J Immunol* 184:2038. [9]Ramos-Kichik, V. 2009 *Tuberculosis (Edinb)* 89:29. [10]Wartha, F 2008 *Sci Signal* 1:pe25. [11]Hernandez Pando, R. 2010 *Tuberculosis (Edinb)* 90:268. [12]Lenaerts, A. 2007 *Antimicrob Agents Chemother* 51:3338. [13]Basaraba, R. J. 2006. *Tuberculosis (Edinb)* 86:386. [14]Turner, O. C. 2003. *Infect Immun* 71:864. [15]Aly, S. 2006. *J Pathol* 210:298. [16]Tsai, M. C. 2006. *Cell Microbiol* 8:218. [17]Klinkenberg, L. G. 2008. *J Infect Dis* 198:275. [18]Converse, P. J. 2009. *Infect Immun* 77:1230. [19]Al-Mutairi, D. 2011 *Curr Opin Allergy Clin Immunol* 11:18. [20]Walker, T. S. 2005. *Infect Immun* 73:3693. [21]Parks, Q. M. 2009. *J Med Microbiol* 58:492. [22]Beumer, A. 2010. *Appl Environ Microbiol* 76:7367. [23]Greendyke, R. 2008. *Antimicrob Agents Chemother* 52:2019. [24]Williams, M. M. 2009. *Appl Environ Microbiol* 75:2091. [25]Ojha, A. K 2010. *J Biol Chem* 285:17380. [26]Ojha, A. K 2008. *Mol Microbiol* 69:164. [27]Harro, J. M 2010. *FEMS Immunol Med Microbiol* 59:306.

VETERINARY PERSPECTIVE ON THE RAW MILK INDUSTRY

Franklyn Garry, DVM, MS, DACVIM
Fort Collins, CO

INTRODUCTION

Consumers in the United States have benefited from heat pasteurization of dairy products since the early 1900s. There is no question that pasteurization is an effective method to improve the microbiological safety of milk. National disease eradication programs that have decreased the prevalence of bovine tuberculosis and bovine brucellosis since the beginning of the 20th century have certainly decreased risks of these zoonotic pathogens in dairy products even in raw (unpasteurized) milk. There are numerous other pathogens shed by asymptomatic cows that can contaminate milk during milking or handling. Even with processes that improve milk hygiene, the risk for transmission of pathogens to humans via milk and other dairy products is decreased by pasteurization. Therefore it is interesting that controversy continues about the sale of raw milk in the United States. It appears that debate about regulations formulated to restrict access to raw milk has intensified over recent years.

MILK PASTEURIZATION IN THE UNITED STATES

In the early 1890s a philanthropist named Nathan Strauss began distributing pasteurized milk in New York City and other large cities throughout the United States. The voluntary use of pasteurized milk substantially lowered infant mortality rates and numerous cities developed ordinances requiring that commercial milk sold in the city be pasteurized. In 1909 research findings were published by the US Public Health and Marine Hospital Service describing the thermal death curves for a variety of microorganisms. In 1924 the US Public Health Service developed the "Standard Milk Ordinance" as a model for voluntary adoption by state and local milk control agencies. Controversy about the necessity of pasteurizing milk as a safety procedure and about the potential deleterious effects of pasteurization on milk taste and quality were particularly strong in the early 1900s. Gradually these arguments subsided, but concerns about the effects of pasteurization have continued and intensified in recent years.[1]

The federal standards first established in 1924 have been upgraded and revised periodically since that time with the development of new information and are known as the Grade A Pasteurized Milk Ordinance (PMO). All 50 states have voluntarily adopted these guidelines which cover aspects of production, handling, transportation, processing, testing, allowable bacterial limits, and sale of milk. Specific details of required time and temperature for pasteurization of fluid milk are included in the PMO.[2] The US Food and Drug Administration (FDA) oversees these standards but has no jurisdiction to enforce the standards within state borders. Individual states are responsible for establishing their own regulations and use the PMO recommendations as guidelines. The FDA oversees interstate shipments and prohibited interstate shipment of raw milk for human consumption in 1987. There is substantial variation between states in the regulation of raw milk sale. In some states the commercial sale of raw milk is illegal but raw milk producers can sell "cow shares" so that consumers have a legal investment in the animals producing milk and therefore have access to raw milk outside of normal commercial channels. Other states allow commercial sale of raw milk under a variety of different restrictions.[3]

RAW MILK

Throughout the time since the PMO was originally promulgated, raw milk consumption has continued to be common among farm families because of its taste and convenience. Additionally a small percentage of the general public consumes raw milk.[3] The notion that pasteurization of the commercial milk supply has improved public health by decreasing the risk of milkborne disease outbreaks seems indisputable. In 1938 milk borne outbreaks in the US represented about 25% of all disease outbreaks associated with food and water. This compares with less than 1% of all such outbreaks by 2000.[2] Despite the fact that only about 4% of the populace drinks raw milk, in calendar year 2010 there were 11 confirmed milk borne disease outbreaks associated with raw milk consumption reported from eight states, and a total of 98 people with confirmed infections.[4] The microbial agents identified in these cases included Campylobacter, Salmonella, and E. coli 0157 – H7. A published report of a 2008 milk borne outbreak of E. coli 0157 suggests that 70% of reported outbreaks associated with dairy products occur from raw milk, although only 1 to 3% of the public consumes such products. This report from Connecticut demonstrates that the outbreak occurred even though the farm of origin appeared to have acceptable milking standards and sanitation procedures.[5]

Those who support the availability and consumption of raw milk allege that there are a number of benefits to raw milk consumption and disadvantages of the pasteurization process. At the extreme, raw milk advocates suggest that unpasteurized products are not only completely safe, but also confer a number of health benefits that are removed by pasteurization. These advocates assert that heat treatment destroys a number of enzymes and vitamins that prevent or treat a wide spectrum of disease problems. A recent review by Lejeune at al presents a list of the nutrients that fit this description and also provides documentation of the fallacy of these arguments.[3] Although some of these substances are affected by heat treatment, that effect is minor and for some of these nutrients milk is only a minor source while other components of a balanced diet provide much greater quantity.

From a scientific point of view it seems inarguable that pasteurization reduces the risk of microbial contamination of milk products. The primary nutrients expected to come from milk products are unaffected by pasteurization. The minor nutrients present in raw milk are either not substantially affected by pasteurization, or are expected to be derived from other food sources in much greater quantity. If the argument about the benefits of raw milk hinged solely on scientific reasoning then consumers should be better served by consuming pasteurized rather than raw milk. As Lejeune points out, scientific reasoning is ineffective in convincing raw milk advocates that pasteurized milk products are superior.[3]

Given that a portion of the population will continue to desire raw milk, and that there are means to obtain the product, some attempts have been made to promote regulations that decrease risks associated with the product. Following a well-publicized disease outbreak associated with raw milk, attempts were made in Connecticut to legislate new rules regarding raw milk sale. These included requirements for prominently displaying labels that included detailed warnings about the potential for infectious agents being present in the product. The legislation also recommended increased testing frequency and limiting sale of the product to farm premises.[5] This legislation was not passed and it is reasonable to argue that the positive impact of such regulations would not be very high. In particular, testing is a poor way to ensure safety. Microbial contamination can occur sporadically and be missed by testing. Such contamination may be unevenly distributed in a product and may be low at the time of testing but with the potential to proliferate over time. Furthermore susceptible individuals may be infected even with very low doses.[3]

POLICY AND THE ROLE OF SCIENCE

I believe the most fascinating aspect of the controversy about raw milk consumption and regulations to prevent it, lies not in the scientific arguments about the risks of raw milk, but rather about some sociological elements of the debate. The FDA, attempting to fulfill its role as a protector of public health, has clearly defined an advocacy role against raw milk sale and consumption. For over a century consumers have looked to the FDA for standards that will keep the public healthy and safe. One might assume that the FDA would present the scientific bases for regulations and then follow a dispassionate course of action in enforcing those regulations. This assumption however would be incorrect. In fact the FDA goes a step beyond and in the context of education, takes on an advocacy role. This is most clearly characterized in a PowerPoint presentation from 2005 entitled "On the Safety of Raw Milk (with a word about pasteurization)", written by John Sheehan, director of the Division of Dairy and Egg Safety. http://www.fda.gov/downloads/Food/FoodSafety/Product-SpecificInformation/MilkSafety/ConsumerInformationAboutMilkSafety/UCM166069.pdf

The Weston A. Price Foundation has taken offense at this presentation and has chosen to refute it by giving a slide by slide rebuttal along with a complete listing of references on their website www.realmilk.com. The foundation argues that the FDA has made a grave mistake by considering the safety of raw milk in its own category, outside of the context of general food safety. They also argue against the FDA statement that pasteurization is the only way to make milk safe. Perhaps the most compelling critique offered by the foundation notes that raw milk seems to be singled out as if it has unique risks for the consuming public, although realistically all raw food products present similar risks. Their argument questions why raw milk is identified by FDA as a unique public health threat while raw produce can also harbor infectious organisms and numerous cases of food borne infection have resulted from consumption of such food. Indeed raw milk harbors similar types of risks as eating raw meat, vegetables and salads, or working with food producing animals, and yet none of these are the focus of 'demonizing' efforts intended to restrict access.

What has developed appears to be a reflection of the general polarization of viewpoints occurring on many fronts within the United States. There appears to be little middle ground or compromise. Advocates become extreme on either side of the argument. Several of the sources used in the FDA presentation did not support the FDA position, suggesting they were not carefully read, or that the authors willfully ignored the real content of the manuscript. For example one citation is used to implicate raw milk as the cause of death in 22 patients. However the authors of that citation noted that the cause of death in most cases was due to another underlying disease and stated that "sources of infection were not identified in 61% of our cases". The authors went on to cite six different plausible sources of infection, of which one was raw milk.[6] Reviewing the supporting literature for the presentation, for many of the case studies no clear causal relationship was found between disease outbreak and the consumption of raw milk. These errors provide opportunities for the rebuttal to suggest that the presentation is in fact biased. For their part the Weston A. Price Foundation presents a rebuttal that is more or less scientifically sound, but fails to present compelling scientific information to support their belief that raw milk has properties superior to those of pasteurized milk.

The issue of consumer choice in a free society is important. The notion that consumers buying food from a commercial marketplace should be able to assume that the food is safe seems reasonable, and provides a specific justification for the development of regulations about the food sold in the marketplace. Sale of specialty foods that do not meet specific food safety standards should logically require that vendors assure that consumers of these foods recognize that they are not covered by the same assurances of food safety that apply to the commercial food supply. What should be the role of scientists and regulatory agencies in overseeing the production and distribution of these specialty food items? It appears that the raw milk advocates reject the hand of regulatory agencies and see the regulations as attempts to impinge upon their options to choose. There are many examples of situations where the public has the option to make choices that contain inherent risks. Oppositional advocacy stances tend to become extreme such that one argument exaggerates the risk, while the other argument minimizes the risk, making it more difficult for the consumer to establish a balanced sense of the real risk.

REFERENCES

[1]Steele JH. **JAVMA** 2000;217(2):175-177. [2] US Food and Drug Administration, Center for Food Safety and Applied Nutrition. **Grade "A" Pasteurized Milk Ordinance** 2005 http://www.cfsan.fda.gov/~ear/pmo05toc.html [3] LeJeune JT, et al. **Clin Infect Dis** 2009; 48:93-100. [4] Centers for Disease Control and Prevention. Food Safety 2010 http://www.cdc.gov/foodsafety/rawmilk/raw-milk-resources.html [5]Guh A, et al. **Clin Infect Dis** 2010;51(12):1411-1417. [6] Werner SB, et al. **Br Med J** 1979;2(6184):238-241

ANIMAL & PUBLIC HEALTH CONCERNS WITH *CLOSTRIDIUM DIFFICILE* & MRSA IN FOOD ANIMALS

J. Scott Weese, DVM, DVSc, DACVIM
Guelph, Ontario, Canada

CLOSTRIDIUM DIFFICILE

Clostridium difficile is a Gram positive anaerobic bacterium that is an important pathogen in many species, and which can also be found in a wide range of healthy animals. First identified in healthy human infants in the early 20th Century, it was regarded as a commensal and essentially ignored until the 1970s, when it was implicated as the cause of antimicrobial associated diarrhea and pseudomembranous colitis in humans. *Clostridium difficile* is now known to be a critically important pathogen in humans, as the leading cause of antimicrobial- and hospital-associated diarrhea and an emerging cause of disease in people in the general population. *Clostridium difficile* can also be isolated from various food animals, with different clinical implications. Further, identification of *C. difficile* in food animals has led to concern that food animals and food might be sources of human infection.

Clostridium difficile in food animals

The role of *C. difficile* in disease in food animals varies with species, with a clear role in enteric disease in young piglets [1,2] and less convincing evidence for calves.[3-5] In piglets, *C. difficile* infection (CDI) is characterized by enteritis with moderate to severe mesocolonic edema in 1-7 day old piglets.[6] Diarrhea is most common, however constipation or obstipation may be present. It is a leading cause of neonatal diarrhea in death in some regions but rare in others, for unknown reasons. Disease appears to be rare to non-existent in older pigs. Thus, while *C. difficile* can have a major impact on pig health, it is essentially confined to a neonatal period.

The role of *C. difficile* in disease in cattle is unclear. There is no evidence that it is a cause of disease in older cattle, and if it is a pathogen, it is probably a cause of neonatal diarrhea. However, data regarding calves are conflicting. An association between diarrhea and the presence of *C. difficile* toxins in feces (the typical clinical standard for diagnosis) was reported in one study,[4] while an inverse association was reported in another.[7] Curiously, despite the toxin results, an inverse association between isolation of *C. difficile* and the presence of diarrhea was identified in both studies, with *C. difficile* isolated more commonly from healthy calves. Further, disease could not be experimentally induced in neonatal, colostrum-deprived cattle.[5] Thus, it is currently unclear whether CDI is an important cause of neonatal calf diarrhea, a geographically relevant problem or a non-entity.

Clostridium difficile can be commonly found in healthy pigs and cattle, with much higher rates in young piglets and calves compared to older animals.[8-10] Reported shedding rates vary between studies but in most, *C. difficile* can be found in the majority of young piglets, and very high (close to 100%) cumulative prevalence can be found with longitudinal study.[8] Colonization occurs very early, with the peak colonization rate (94%) on in 2 day old piglets in one longitudinal study.[8] The prevalence of colonization decreases quickly as piglets age, and by the time of slaughter, the prevalence is much lower (<7%).[9,11](JS Weese, unpublished data)

There has been less study in cattle, but the same general pattern seems to be present, with higher colonization (10-38%) rates in young calves[4,5,7,12] compared to slaughter age or breeding cattle, where rates are low.[11](JS Weese, unpublished data)

In many studies of food animals, one strain, ribotype 078/toxinotype V, has predominated. In some studies, it has accounted for all *C. difficile* isolates, while other studies have reported more variability.[4,12-16] This strain is also found in humans, where it has historically been a relatively uncommon cause of disease. However, recently, the incidence of CDI caused by ribotype 078 has increased, and this clone appears to be an important cause of community-associated CDI.[17] But, despite its association with food animals, ribotype 078 is not the only strain found in food animals. Another important strain that has been reported is ribotype 027/NAP1/toxinotype III or related strains. This is a human epidemic clone that is associated both with endemic disease and outbreaks of high morbidity and mortality. Reports of ribotype 027 in food animals have raised significant concerns because of the importance of this strain in human disease, including large outbreaks.

Clostridium difficile in retail meat

Not surprisingly, identification of *C. difficile* in food animals led to studies of retail meat.[18] Studies performed in various countries have identified *C. difficile* contamination in a range of meat products, with prevalence ranging from 0-42%.[18-24] The strains found in meat tend to be the same as those found in food animals, however the association may not be as clear as expected. While one should be hesitant to draw many conclusions based on a limited number of studies, the prevalence of ribotype 027/NAP1 appears disproportionately high in food compared to food animals, often higher than ribotype 078. Whether this is truly the case is not known and explanations are lacking, however this strain is particularly adept at persisting in the hospital environment based on its high sporulation rate.[25] It is plausible, that high sporulation of ribotype 027 could also be an important component of environmental persistence in slaughterhouses and meat processing environments.

An important aspect that needs to be considered is the amount of *C. difficile* in meat. Prevalence studies have typically used enrichment culture, which is optimal for recovery of *C. difficile* but cannot determine the amount of *C. difficile* that is present. Limited quantitative study has been performed. In one study of ground beef and pork, spore counts ranged 20 to 60 spores/g in pork and 20 to 240 spores/g in beef in samples that were quantifiable, with the majority of samples being positive only on enrichment culture. [23] This suggests that while contamination may be relatively common, spore numbers tend to be low. The relevance of dose is not known, but it is logical to assume that low levels of spore contamination are less of a concern than high levels, particularly to the low-risk, general population. The inability of recommended cooking temperatures to kill *C. difficile* spores[26] raises more questions and concerns compared to non-sporeforming bacteria such as *E. coli,* where proper cooking will be effective at killing the organism.

Public Health Concerns

There is currently no objective evidence indicating that *C. difficile* is a foodborne pathogen, but no studies have reported concerted (or any) efforts to investigate this area. Determination of the role of food in human CDI may be rather difficult. If ingestion of food contaminated with an appropriate infectious dose leads shortly thereafter to signs of enteric disease, then epidemiologic studies may be successful in identifying an association. However, this may not represent the pathophysiology of CDI, where foodborne exposure may only be of concern for selected, at-risk populations and/or when ingestion of *C. difficile* might lead to colonization with subsequent development of CDI at a later date. This potentially indirect or delayed onset of disease could certainly hamper objective determination of foodborne risks as identification of the role of food in disease would be very difficult if the point of exposure may have been days, weeks or months in the past. While studies investigating the role of food in CDI are certainly warranted, definitive determination of the role of food in CDI may be very difficult.

There is also concern that zoonotic transmission of *C. difficile* could result from exposure while working with animals or their environment. The risk of this is completely unclear at this time. It is reasonable to suspect that people working with high prevalence groups (e.g. young piglets and calves) would be at greater risk of exposure. Whether that actually corresponds to increased risk of *C. difficile* colonization or disease remains to be determined. The use of good general hygiene practices around food animals, particularly diarrheic animals and neonates, is a reasonable preventive measure for *C. difficile*, among other potential pathogens.

METHICILLIN-RESISTANT *STAPHYLOCOCCUS AUREUS* (MRSA)

MRSA is a critically important human pathogen that causes tremendous numbers of infections every year. Historically, MRSA has been a leading cause of hospital-associated infection, but since the 1990s, community-associated MRSA has emerged and in some regions, MRSA is an important of community-associated infections. MRSA has predominantly been considered a human pathogen, with infections identified initially in animals thought to be 'spillover' from humans. However, in the past 5 years, there has been an apparent dramatic change in the epidemiology of MRSA in some regions and evidence that food animals may play an important role in human infection.

Epidemiology

Despite the importance of *S. aureus* as a cause of bovine mastitis and periodic reports of MRSA from milk, MRSA in food animals received little attention until 2005, when a unique type of MRSA, sequence type 398 (ST398) was identified as a cause of "unexpected" MRSA infections in people in the Netherlands who had been in contact with pigs.[27] Exceptionally high MRSA colonization rate of pig farmers, 760 times that of the general Dutch population, were identified, exclusively involving ST398. This strain was then found in pigs in the Netherlands, and subsequently elsewhere in Europe, as well as Asia and North America. Some studies found up to 100% of healthy pigs were carrying ST398 in their nares or intestinal tract. ST398 is now referred to as 'livestock-associated MRSA' and it has become a significant human health concern in some regions. However, the situation appears to be somewhat different outside of Europe. In China, a different clone, ST9, has been found.[28,29] In Canada, ST398 is found, but so are other strains, particularly a very common human epidemic clone (CMRSA-2/USA100). The Canadian situation highlights in interesting concept. Presumably, ST398 MRSA originated in pigs, probably imported from Europe where it evolved from methicillin-susceptible ST398, a porcine commensal. ST398 MRSA was subsequently transmitted to pig farmers and veterinarians, as discussed further below. However, the presence of human epidemic MRSA indicates that human-pig transmission has also likely occurred. Thus, MRSA in pigs is probably the result of multiple instances of movement of MRSA between pigs, from pigs to people and from people to pigs.

Risk factors for MRSA colonization of pigs have not been adequately investigated. Tetracycline resistance is typical in ST398 and has led to the assumption that the emergence of this strain resulted from heavy tetracycline use in pig production.[31] However, such conclusions may be misleading since tetracycline resistance is also very common in equine ST398 and CMRSA-5/USA500 isolates[32,33] despite the sparing use of tetracycline in horses. MRSA is also present on antibiotic-free swine farms.[34] Recently, attention has been given the presence of heavy metal resistance genes in ST398 MRSA, and the potential that zinc supplementation could be a factor in co-selection for MRSA in pigs.[35]

The evolution of MRSA appears to have been different in cattle. Following the initial report of MRSA in a dairy cow in 1972, sporadic cases of MRSA mastitis in dairy cattle were described, typically at a low prevalence. Notwithstanding more recent reports, the incidence of MRSA mastitis and the prevalence of methicillin resistance among bovine *S. aureus* isolates appears to be quite low, so MRSA does not appear to be a common or important bovine mastitis pathogen at this time.

Minimal information is available about MRSA colonization of healthy cattle. A recent Dutch study found 28% of veal calves colonized with MRSA,[36] but in Switzerland the pathogen was isolated from only 1.3% of calves and 0.4% of adult cows[37] and in Canada from none of 491 Canadian feedlot cattle.[38] Clearly, more study of healthy cattle is required.

Food Animal Health Concerns

MRSA is of minimal animal health concern. Despite the high prevalence of colonization in pigs, clinical infections are rare, with only sporadic reports of exudative dermatitis, urinary tract infection, and mastitis-metritis-agalactia syndrome.[39,40] Similarly, MRSA mastitis has been reported in cattle, but this appears to be quite uncommon. Indeed, considering the importance of *S. aureus* as a cause of bovine mastitis and the amount of beta-lactam antimicrobial exposure for the treatment of mastitis, it is rather surprising that MRSA is not a common cause of mastitis. While MRSA can certainly cause opportunistic infections of various types in cattle and pigs, it is of limited consequences overall. The main food animal concerns are with respect to public health and public relations.

Zoonotic Transmission of MRSA

Livestock-associated MRSA has emerged as a pressing concern in Europe. Since the first reports of human infections associated with MRSA in pigs, subsequent studies have confirmed that pig contact is a risk factor for MRSA infection and colonization.[41,42] There have been dramatic increases in human ST398 infection and colonization in the Netherlands, where the prevalence of *smaI* PFGE nontypable (and therefore presumed ST398) MRSA isolates recovered from humans rose from 0% in 2002 to more than 21% in 2006.[43] Changes in surveillance likely accounted for some of this increase, but it is clear that ST398 rates are rising and are to a large degree associated with animal contact. Colonization rates have been particularly high in various groups with livestock contact, such as veterinarians, farmers, and people living on farms.[42-45] Studies have identified colonization in 32% of people with veal calf contact in the Netherlands[36] and in 32% of hospitalized people who had contact with pigs and veal calves.[42] These rates themselves are remarkable, but they are astounding considering that the prevalence of colonization in the general Dutch population is less than 1%.

A recent Dutch study indicated that the annual incidence of human MRSA increased more than threefold from 2001 to 2006 and that this increase was entirely attributable to animal-related MRSA.[46] According to the study, animals accounted for 26% of the cases in people who were identified as colonized upon hospital admission. Because of the risk associated with pig contact, Dutch hospitals have adopted strict MRSA control measures, including isolation of people with pig contact at the time of their hospital admission until MRSA screening results are available.

Farm-based risk factors for MRSA acquisition have been only minimally investigated. Not surprisingly, working in barns with colonized pigs is a risk factor,[47] and more intensive and longer contact is probably an important factor, as evidenced by transient MRSA carriage in people who visit farms to collect samples versus higher rates of colonization in workers with more intensive or longer contact.

The geographic variation in the prevalence of ST398 colonization and incidence of infection in humans in quite interesting. Although it is a leading cause of community-associated infection in some European countries, ST398 MRSA infections are currently rare in North America, despite the presence of ST398 in livestock.[48] The reason for this is unclear. Various possibilities exist, including differences in direct and indirect contact with food animals, much lower population density in North American pig-rearing regions, and the more common presence of other (perhaps competing) MRSA strains in people in the general population. Another possibility is that this is an emerging disease in North America, with a later onset than in Europe.

MRSA in Food

Considering the increasing evidence of MRSA in food animals, it is logical that concerns would emerge about MRSA contamination of food, and food as a source of infection or colonization (from ingestion, handling or environmental contamination). Indeed, MRSA can be found in retail with prevalence ranging from 0.4% to 12%.[49-52] As with *C. difficile,* most studies have used enrichment culture, and a study evaluating MRSA levels in meat found low level contamination in most positive samples.[51] ST398 MRSA has predominated in some, but not all, reports of MRSA in meat, raising questions about the origin of MRSA contamination. Studies have reported the presence in meat of strains more commonly found in people, including strains not yet reported in food animals, or clearly disproportionate rates of human epidemic clones in meat compared to food animals. For example, a Canadian study found MRSA in retail ground beef at a similar prevalence to pork,[51] despite the fact that MRSA was not found in a large study of Canadian feedlot cattle. While this may just represent inadequate testing of cattle, the fact that a human epidemic clone predominated in meat suggests that humans could play a role in meat contamination. Clearly, more information is required, necessitating research beyond the cross-sectional study of retail food contamination. There is no question that food can be contaminated with MRSA, and a role in human disease can be hypothesized, but no objective data are available to support or refute such a hypothesis. More intensive surveillance and the inclusion of queries about food exposure (ingestion and contact) in community-associated MRSA infections investigations are necessary to elucidate the true role of food in human disease. Even so, determination of an association may be difficult, since, if infection occurs, it probably does so only in certain situations and often well after exposure.

REFERENCES

[1] Songer J, et al. **Vet Microbiol** 2007;124:358-361.[2] Yaeger MJ, et al. **J Vet Diagn Invest** 2007;19:52-59. [3] Hammitt M, et al. **Vet Microbiol** 2007;127.[4] Rodriguez-Palacios A, et al. **Emerging Infect Dis** 2006;12:1730-1736. [5] Rodriguez-Palacios A, et al. **Vet Microbiol** 2007;124:166-172. [6] Songer J. **Clin Microbiol Rev** 1996;9:216-234. [7] Hammitt, et al. **Vet Microbiol** 2007. [8] Weese JS, et al. **Anaerobe** 2010;16:501-504. [9] Norman K, et al. **Anaerobe** 2009. [10] Norman KN, et al. **Anaerobe** 2009;15:256-260. [11] Hoffer E, et al. **J Food Prot** 2010;73:973-975. [12] Avbersek J, et al. **Anaerobe** 2009;15:252-255. [13] Rodriguez-Palacios A, et al. **Vet Microbiol** 2007;124:166-172. [14]Costa M, et al. European Conference of Clinical Microbiology and Infectious Diseases. Helsinki, Finland, 2009. [15] Keel K, et al. **J Clin Microbiol** 2007;45:1963-1964. [16] Pirs T, et al. **J Med Microbiol** 2008;57:790-792. [17] Goorhuis A, et al. **Clin Infect Dis** 2008;47:1162-1170. [18] Rodriguez-Palacios A, et al. **Emerging Infect Dis** 2007;13:485-487. [19] Bouttier S, et al. **Emerging Infect Dis** 2010;16:733-735. [20] Rodriguez-Palacios A, et al. **Emerging Infect Dis** 2009;15:802-805. [21] Von Abercron SM, et al. **J Food Prot** 2009;72:1732-1734. [22] Indra A, et al. **Wien Klin Wochenschr** 2009;121:91-95. [23] Weese J, et al. **Appl Environ Microbiol** 2009. [24] Weese JS, et al. **Lett Appl Microbiol** 2010. [25] Akerlund T, et al. **J Clin Microbiol** 2008;46:1530-1533. [26] Rodriguez-Palacios A, et al. **Anaerobe** 2010. [27] Voss A, et al. **Emerging Infect Dis** 2005;11:1965-1966. [28] Cui S, et al. **J Antimicrob Chemother** 2009;64:680-683. [29] Wagenaar J, et al. **Vet Microbiol** 2009. [30] Smith TC, et al. **PLoS ONE** 2008;4:e4258. [31] de Neeling A, et al. **Vet Microbiol** 2007;122:366-372. [32] Van den Eede A, et al. **Veterinary Microbiology** 2009;133:138-144. [33] Weese J, et al. **Emerging Infect Dis** 2005;11:430-435. [34] Weese JS, et al. **Zoonoses and Public Health** 2010;ePub ahead of print. [35] Cavaco LM, et al. **Antimicrob Agents Chemother** 2010;54:3605-3608. [36] Graveland H, et al. **PLoS ONE** 2010;5:e10990. [37] Huber H, et al. **Euro Surveill** 2010;15. [38]Weese JS, et al. Proc ASM-ESCMID Conference on methicillin-resistant staphylococci in animals 2009;47. [39] Schwarz, et al. **J Antimicrob Chemother** 2007. [40] van Duijkeren E, et al. **Emerging Infect Dis** 2007;13:1408-1410. [41] Lewis HC, et al. **Emerging Infect Dis** 2008;14:1383-1389. [42] van Rijen MM, et al. **Clin Infect Dis** 2008;46:261-263. [43] van Loo I, et al. **Emerging Infect Dis** 2007;13:1834-1839. [44] Köck R, et al. **Eur J Clin Microbiol Infect Dis** 2009. [45] Vandenbroucke-Grauls CM, et al. **Nederlands tijdschrift voor geneeskunde** 2006;150:1710-1712. [46] van Rijen MM, et al. **J Hosp Infect** 2009;72:299-306. [47] van den Broek IV, et al. **Epidemiol Infect** 2008:1-9. [48] Golding GR, et al. **Emerging Infect Dis** 2010;16:587-594. [49] de Boer E, et al. **Int J Food Microbiol** 2009;134:52-56. [50] Pu S, et al. **Appl Environ Microbiol** 2009;75:265-267. [51] Weese JS, et al. **Lett Appl Microbiol** 2010;51:338-342. [52] Weese JS, et al. **Can Vet J** 2010;51:749-752.

SMALL RUMINANT ABORTIONS: NEW RESEARCH ON DISEASE, MANAGEMENT AND ZOONOSIS

Paul J. Plummer, DVM, PhD, DACVIM (LAIM)
Ames, IA

INTRODUCTION

Clinical abortions are one of the most common infectious disease outbreaks observed in sheep and goat operations in the United States. Although a variety of etiologic agents can be responsible for these outbreaks, a common theme amongst all of them is the potential for zoonotic infections associated with either the environment or animals that are aborting. For this reason it is imperative that veterinary practitioners be aware of the potential for zoonotic spread both in terms of protecting themselves but also to allow appropriate recommendations to the producers concerning their health. In this presentation we will discuss the most important abortion pathogens with a focus on recent findings regarding their ability to cause disease, their treatment and their potential as zoonotic pathogens.

CAMPYLOBACTER SPP.

Based on the sheep NAHMS studies, *Campylobacter* spp. associated abortions are the most common cause of small ruminant abortion outbreaks in the United States.[1] Although *C. fetus* subsp. *fetus* was traditionally considered the primary pathogen associated with small ruminant abortions, the proportion of *C. fetus* isolations has drastically dropped over the last decade.[2,3] Instead, *C. jejuni* has emerged as the predominate cause of small ruminant abortions in the Unites States.[3] This finding is in sharp contrast to that observed in other sheep producing areas of the world (Australia and UK) where *C. fetus* still predominates. Furthermore, genetic testing of the variability of the *C. jejuni* isolates obtained from a large number of sheep abortions in the Unites States has demonstrated that there is a convergence to a single highly virulent genotype associated with sheep abortions. Although this organism predominantly causes abortions in sheep, we do have several goat and cattle isolates from abortion outbreaks that match this genotype. In terms of clinical medicine, these finding have several significant implications. First, all of the isolates that match this genotype are highly tetracycline resistant. Since tetracycline therapy is often the first line of treatment in sheep abortion outbreaks clinicians should be aware that the majority of *Campylobacter* associated abortion outbreaks in the US are likely not to respond to that therapy. Several other drugs commonly used in food animal medicine are effective *in vitro* including macrolides and fluorfenicol. We do not, however, have any data regarding their frequency of dosing or clinical efficacy at this time. Second, this genotype is highly virulent with a very strong propensity to quickly move from the intestinal lumen to the uterus via the blood. In comparison to other sheep fecal isolates and non-type abortion isolates the ability to induce abortions is significantly higher when this genotype is the causative organism. Third, there appears to be significant differences in terms of vaccine efficacy between this isolate and the two commercially available *Campylobacter* bacterins. Data from our laboratory that is currently in press demonstrates that one vaccine performed equivalent to placebo while the other performed similar to a strain specific bacterin when accessed in a guinea pig abortion model commonly used in vaccine research.[4] Finally, we have demonstrated that this genotype is associated with significant zoonotic potential and has been isolated from a number of human patients in this country (manuscript in preparation). In terms of diagnostics, dark field microscopy on fetal stomach contents and microaerophilic culture reliably isolate the organism. Genotyping and antimicrobial sensitivities are likely only available from specialty labs.

CHLAMYDOPHILA

Chlamydophila abortus, previously known as *Chlamydia,* is one of the top three abortion agents in both sheep and goats in our experience. The organism is associated with abortion outbreaks that cannot be distinguished from any of the other abortion agents based purely on clinical signs. Due to the fact that the organism is an obligate intracellular organism it cannot be cultured on routine bacteriological culture of aborted fetus or tissue. The organism is believed to enter herds/flocks through poor biosecurity practices, however recent unpublished data from our laboratory suggest that the majority of farms are positive for the organism based on real-time PCR. There is an increasing interest in the role of stress on development of abortion outbreaks associated with this organism and it may be a situation where the presence of a normally non-pathogenic isolate can result in significant disease when stress induces immuno-compromise. This is an area that warrants further research before sound scientific evidence can be used to make conclusions.

Current diagnostics rely on histopathologic identification of lesions consistent with the organism or on real-time PCR for *Chlamydophila* sp. At least one commercially available PCR test has the ability to distinguish *C. abortus* from *C. pecorum* which is more commonly associated with small ruminant ophthalmic disease. A commercial vaccine is available but good evidence based medicine regarding efficacy of it's use is missing. Anecdotally, there are reports of poor protection in certain herds/flocks. The organism is reportedly susceptible to tetracycline prompting the common use of feed-grade tetracycline products at doses that exceed the labeled dose. Such use would be considered extra label and clinicians should consider the implications of feed grade delivery in light of this issue. Although this organism has been associated with zoonotic risk, disease is much more likely with some of the other *Chlamydophila* spp. then *C. abortus*. Precautions should be taken since definitive diagnosis is often not made until several weeks after the initial abortions.

TOXOPLASMOSIS

T. gondii is also one of the top three causes of both sheep and goat abortions in our experience. Disease occurs only when the animal is exposed to the organism for the first time during pregnancy. If the animal is exposed prior to pregnancy a protective immunity develops and subsequent exposure typically does not result in abortion. Infective oocyst are shed in the environment by infected felines with higher and more prolonged shedding occurring in young kittens. The organism is maintained in the environment in rodent populations where it can be passed vertically to offspring, however rodents do not shed infective oocyst into the animal environment. Transmission between infected sheep or goats is typically not a concern although there is some anecdotal evidence of vertical transmission in utero, through milk and following consumption of infected placenta. The primary zoonotic risk to humans would be exposure to infective oocyst in the same environment as the small ruminants although direct transmission of the disease from a small ruminant to a human is rare (potentially possible via milk). Overseas, there is a commercially available vaccine; however it is a live vaccine with a very short shelf life and is not marketed or available in the United States. In some countries decoquinate is labeled for control of toxoplasma at a dose higher then generally used in the United States for control of coccidia. Diagnosis is generally based on identification of characteristic histopathologic change, serology performed on fetal thoracic fluid or paired IgM/IgG toxo titers that are commercially available.

Q-FEVER (COXIELLA BURNETTI)

In our experience *C. burnetti* is in the top three causes of goat abortions as a result of decreased rates of *Campylobacter* in goats compared to sheep. *C. burnetti* is also commonly associated with abortions in sheep, however *Campylobacter, Chlamydophila and* toxoplasmosis exceed it in number. The traditional wisdom concerning the disease process and abortions associated with this agent have been recently questioned in light of several large human disease outbreaks, most recently in the Netherlands associated with dairy goat operations. The organism occurs in two predominate variants, a large cell metabolically active form (called the large cell variant or phase I) and a small cell form that is metabolically inactive and functions as a spore (although it is not truly a spore). The small cell variant is highly resistant to environmental degradation and can be easily aerosolized. The aerosol route of transmission is considered to be the primary route of transmission to both animals and humans. Recent epidemiologic data from the Netherlands suggest that the organism is capable of aerosolizing up to 5 km in appropriate environmental conditions. The organism is apparently more prevalent in US ruminant populations then previously believed. Two recent studies in dairy cattle bulk tank milk (one using serology and the other using PCR) demonstrated upwards of 90% herd level prevalence in US dairy cattle operations. Work in our laboratory and others (unpublished, personal communication) has demonstrated significant herd/flock level prevalence in sheep and goat operations although this prevalence is likely somewhat more seasonal then that of cattle due to shedding characteristics. Based on experimental infection models and observation of infected animals following abortion storm the following trends appear to occur. Dairy cattle shed the organism more in their milk then vaginal fluids (perhaps why more zoonotic risk associated with attending small ruminant abortions), however they also shed the organism in their milk for prolonged periods of time relative to small ruminants. In contrast, vaginal shedding of the organism is very high in small ruminants with lesser amounts in the milk. Shedding in small ruminants also appears to be much more prevalent in the periparturient period in comparison to cattle. It is also very clear that small ruminants can shed the organism in the absence of serological positive status. Furthermore, in infected herds/flocks most shedding occurs in the complete absence of clinical signs such as abortion. The consequence of this finding is that the zoonotic risk associated with this organism is not confined to periods of high abortion rates. In fact, in the Netherlands outbreak the majority of source farms for human infection did not have any clinical signs of disease.

Diagnosis of Q-fever typically relies on the identification of characteristic histopathological changes although the organism is capable of being shed in animals without histopathologic evidence, In these cases it is unlikely that the organism is the primary agent associated with the abortion, however the samples should be treated with appropriate precautions to prevent zoonotic transmission. There are several commercial PCR tests available and validated for use on vaginal fluids, milk, feces, urine and aborted tissues. The test are highly sensistive and are likely the best option to testing shedding status. Serologic testing has traditionally relied on the complement fixation test, however an increasing body of evidence suggest that this test lacks sensitivity in the clinical setting. ELISA and IFA test are rapidly surpassing the CF test as the preferred serologic assays. In human medicine the IFA is the gold standard serologic test since it allows for determination of independent phase I and phase II responses that can assist with determining the chronicity of infection in human patients. Our laboratory has recently developed and validated an IFA assay for sheep, goats and cattle. The ELISA performs similar to the IFA assuming that an ELISA specific for both phase I and II antigens is utilized, however it is unable to distinguish between the two responses and simply provides a positive/negative result. The utility of determining the phase response with regards to clinical disease is well validated in humans but has not been rigorously tested in ruminant species. Although the organism is typically susceptible to tetracyclines, use of these drugs in the face of abortion outbreaks does not appear to limit shedding.

This organism poses a significant zoonotic risk to veterinarians, farmers and possibly individuals in aerosol distance from infected farms. A recent serologic survey of veterinarians attending the AVMA conference showed that roughly 25% of veterinarians were seropositive for exposure. Individuals that performed large animal veterinary services or had significant surface water exposure (ie. swim in lakes, ponds etc) were at increased risk of being seropositive. The cause for an association with surface water is unclear although the organism has been show to survive in amoebas for up to six months. In the Netherlands humans that lived within 2 km of an infected dairy goat operation (ie. bulk tank positive by PCR) were more then 30 times more likely to develop Q-fever the individuals that lived more the 5 km from those same operations. The Dutch government initially instituted a vaccination strategy along with a ban on manure spreading but the human case incidence continued to rise. Under intense pressure to address the issue in 2010 they elected to euthanize all pregnant goats on farms that had at least one positive bulk tank PCR, a decision that has been questioned by many experts. A breeding ban was placed on all non-pregnant animals from those same farms and the human case rate appears to have dropped significantly.

OTHER INFECTIOUS CAUSES

Although less common, placental infection with *Listeria, E. coli, Salmonella* and *Leptospirosis* have also been associated with significant abortion outbreaks in sheep and goats. Viral agents such as border disease, bluetongue virus, caprine herpes virus, BVDV and cache valley fever are also possible. While most of the viral agents are unlikely to be associated with zoonotic risk the other bacterial species all pose the potential for zoonotic transmission.

CONCLUSIONS

A variety of different infectious agents are associated and commonly observed in small ruminant abortion outbreaks. The majority of these organisms carry at least some risk of zoonotic transmission and thus appropriate biosecurity precautions should be taken to protect both veterinarians and producers.

REFERENCES

[1] USDA. Part II: Reference of sheep health in the United States. Fort Collins, CO: USDA, APHIS, VS, NAHMS, 2001. [2] Delong WJ, et.al. Am J Vet Res 1996;57:163-167. [3] Sahin O, et al. J Clin Microbiol 2008;46:1663-1671. [4] Burroughs, E et.al. AJVR – in press.

DECISION MAKING FOR SURGICAL & MEDICAL MANAGEMENT OF ACUTE ABDOMEN IN CALVES

Mireille Meylan, Dr.med.vet. FVH, MS, PhD, Dr habil., DACVIM, DECBHM
Berne, Switzerland

INTRODUCTION

Whereas some forms of acute abdomen such as intestinal volvulus are encountered at similar frequencies in young animals and in adult cattle, other causes are seen either exclusively (e.g. intestinal atresia) or distinctly more often in calves than in adults (e.g. intraluminal obstruction due to a phytobezoar). The course of disease may be more acute in young than in older animals (e.g. faster devitalization of the intestine after volvulus because lack of mesentarial fat reduces the protection of the vessels at the location of the twist). In young animals as in adults with acute abdomen, it is important that a complete physical examination be performed, in order to reach a tentative diagnosis or a list of differential diagnoses that are as precise as possible. The objective findings of a thorough examination of the patient and the results of further diagnostic investigations allow the best decision making for appropriate treatment (e.g. surgical vs. medical) or to recommend salvage of the animal if the prognosis is very poor, thus avoiding further suffering for the affected animal and unnecessary costs for the owner.[1,2,3]

PHYSICAL EXAMINATION OF CALVES WITH SIGNS OF ABDOMINAL PAIN

Signs of abdominal pain, such as abnormal posture, kicking, restlessness, or enlargement of the abdomen can be caused by disturbances not only in the digestive system, but also in other organs (e.g. umbilicus, kidneys). Diseases unrelated to abdominal pain can also bear similarities upon presentation, such as one calf that was presented to our clinic during emergency hours because of severe colic manifested by restlessness and kicking to the abdomen, the cause of which turned out to be infestation with biting lice.

Signalement and history

The age of the patient and the course of disease are relevant, as knowledge of these factors may already limit the list of differential diagnoses to consider. For example, increasing signs of colic and progressive abdominal enlargement in a 3-day-old calf is strongly suggestive of intestinal atresia, whereas the first differential for peracute severe unrelenting colic with pronounced abdominal distention in a 2-month-old calf will be intestinal volvulus.

A detailed history should thus include the signs observed by the patient's owner (colic, depressed appetite, abnormal demeanor, bloat, decreased or absent fecal output, straining to defecate or urinate), and information about the onset, duration and progression of clinical signs. In intestinal obstructions characterized by tissue compromise such as intussusception or volvulus, abdominal pain caused by dilatation and tension on the mesentery diminishes in the course of devitalization of the affected tissues through reduced perfusion and hypoxia. A phase of depression due to increasing toxemia follows. Thus, the episode of colic is of short duration and can easily be missed. In contrast, in simple obstipation, e.g. due to obstruction with a phytobezoar, the intensity of colic symptoms increases during the course of disease because of progressive distention of the intestine. Volvulus of the root of the mesentery is associated with severe colic symptoms and rapid progression to shock due to vascular compromise and ischemic necrosis of the intestine.[1,2,4,5,6] Assessment of the feeding regimen and water supply also conveys important information, as unbalanced feed can lead to digestive disturbances (e.g. risk of ruminal drinking associated with inadequate milk quantity and/or quality). Finally, the owner is best qualified to notice and report changes from normal behavior and demeanor of his/her animal(s).[1,2]

Physical examination

As mentioned above, a complete and detailed physical examination is required for any calf presenting with abdominal pain and/or an enlarged abdomen. The animal is first observed from a distance for assessment of demeanor, signs of colic (looking or kicking at the abdomen, lying down and standing up, straining). Abnormal shape or enlargement of the abdomen can be noticed at the same time. Abnormal posture, e.g. with the limbs extended away from the body or with an arched back, can reflect attempts to reduce intraabdominal pressure or painful stimuli in the abdomen.[1,2,3]

Of the vital parameters (heart and respiratory rates, temperature), the heart rate provides the most useful information in patients with abdominal pain: constant tachycardia (in calves > 140 bpm) is indicative of serious illness, especially if it does not respond to treatment with spasmolytics or antiiflammatory drugs (e.g. flunixin meglumine). Body temperature can be normal, elevated in cases involving inflammation and /or toxemia (e.g. peritonitis), or below normal in advanced cases with circulatory and/or toxic shock. The respiratory rate can be elevated as a sign of stress or pain. Acid-base imbalances such as metabolic acidosis also influence respiratory patterns. Careful examination of the circulatory and respiratory systems allows one to ascertain the patient's general condition. Circulatory and/or toxic shock develops rapidly in the case of strangulating lesions or volvulus (abomasum, intestine, root of the mesentery).[1,2,3,7]

Abdominal shape and contours must be assessed in regard to the age, body condition, and feeding history of the animal. In cases with abdominal enlargement, the localization of increased volume (left side, right side, both sides, ventral or dorsal) helps to define possible causes: distention of the rumen as in ruminal drinking or gas tympany lead primarily to distention of the left flank, with gas accumulating dorsally whereas solid or liquid contents cause rather ventral enlargement. Left-sided displacement of the abomasum (LDA) can also lead to visible enlargement of the left flank (the dilated displaced abomasums may be visible dorsally, caudally to the last rib). Ventral enlargement of the abdomen on both sides can be caused by impaction of the rumen or abomasum, by intestinal dilatation, or by the accumulation of free abdominal fluid (peritonitis, ascites). Causes of distention of the right flank are numerous, including right displacement of the abomasum (RDA) or abomasal volvulus (AV), intestinal dilatation or intestinal volvulus. In small calves, distended organs can often be visualized and palpated through the thin body wall.[1,2]

Findings at rumen auscultation must be evaluated in relation to the age and feeding regimen of the calf, as the onset of rumen development depends on the consumption of solid feed. The forestomachs are normally not fully developed until the age of 4 months.[1] Intestinal borborygmi can be normal, decreased or increased in abdominal disease. Intestinal atony occurs in metabolic disturbances, e.g. toxemia, but also as a consequence of obstructive lesions where distention of an intestinal segment leads to reflectory atony of the entire gut. Simultaneous percussion and auscultation, as well as succussion and auscultation of both flanks are an important part of the examination of young bovine patients with abdominal enlargement and/or colic. A high-pitched tone ("ping") heard upon percussion/auscultation and splashing sounds elicited by succussion of the body wall indicate the presence of a distended organ filled with gas and fluid. Precise localization of these findings is important in identifying their cause: in the left flank, a "ping" and positive succussion are mainly due to LDA, but may also be present with rumen tympany. The list of potential causes for a "ping" and positive succussion in the right flank is longer, including RDA and AV (large round-shaped "ping" reaching cranially under the last ribs), cecal dilatation ("ping" located more dorsally and extending caudally to the pelvis), dilatation of the small intestine ("ping" of small diameter, location not always consistent over time -"wandering ping"- or several smaller areas with positive percussion/auscultation). Pneumoperitoneum caused by peritonitis, in most cases due to a perforating abomasal ulcer in calves, produces a positive percussion/auscultation dorsally on both sides but a negative result at succusion.[8] In calves, deep palpation of the abdomen replaces the pain tests for traumatic reticuloperitonitis performed in adult cattle. Bimanual palpation allows for evaluating the tension of the body wall and the presence of pain elicited upon manipulation. In many cases, dilated organs (abomasum, loops of small or large intestine) can be identified through the thin body wall of young calves. Beside peritonitis due to a perforating abomasal ulcer, abomasal or intestinal distention and umbilical diseases are the most important differential diagnoses for abdominal pain in calves. Careful palpation of the external umbilicus in relation to the age of the calf and deep palpation of the internal umbilical structures (umbilical vein, arteries and urachus) belong to the basics of clinical examination in calves.[9]

Rectal examination, one of the most important elements of the clinical evaluation of adult bovine patients with acute abdomen, is not possible in calves due to their small size. However, digital rectal examination permits the assessment of feces (amount, consistency, color). In neonates, obvious malformations such as atresia ani can be ruled out. Assessment of fecal output, of the color, consistency, odor of the feces, and of the presence of blood, fibrin or fragments of intestinal mucosa are particularly important if an obstructive lesion or hemorrhage is suspected in the gastrointestinal tract (GIT). Caution must be used in the appreciation of fecal output, however, because it can be maintained for several hours after occurrence of an obstructive lesion in the proximal segments of the intestine or be strongly reduced or even absent without mechanical obstruction.[10] The color and consistency of fecal material is influenced by the feeding regimen, especially whether the calf is on a milk diet or already weaned.

Ascending infections of one or both kidneys as in pyelonephritis caused by *Corynebacterium renale* in cows is uncommon in calves. Few cases of malformation such as polycystic kidneys or ectopic ureters have been reported in calves.[1] Painful micturition is mostly associated with pathology of the umbilical structures closely associated with the urinary tract (urachus, umbilical arteries).[11,12]

Further diagnostic investigations

A complete blood count is not the ancillary test of first choice in calves with signs of abdominal pain, as a red blood cell count is mostly only indicative of the degree of dehydration (fluid sequestration in the GIT, shock) or anemia (bleeding abomasal ulcer). The leukogram reveals the severity of inflammation or toxemia e.g. due to tissue necrosis or peritonitis. Acute inflammation results in elevated serum fibrinogen levels. A blood chemistry panel typically reveals hypokaiemia, hypochloremia and metabolic alkalosis in acute cases of abomasal disease or proximal intestinal obstruction, reflecting reflux of abomasal contents and failure of electrolyte re-absorption in the duodenum and proximal jejunum. These changes are less pronounced to absent with obstructive lesions located in the distal small intestine or in the large intestine because electrolytes and water can be absorbed normally in the proximal gut, orally to the obstruction. In protracted cases, metabolic acidosis is superimposed on the original alkalosis in the course of developing cardiovascular insufficiency and tissue hypoxemia; the switch from alkalosis to acidosis is considered to be indicative of a poor prognosis in calves with acute abdominal disease.[1,10,13] Reflux of abomasal contents into the rumen results in elevated concentrations of chloride ions in ruminal fluid, however this parameter is unreliable in calves because the normal values established for adult ruminants (< 25-30 mmol/L) may not be valid for unweaned calves. Passage of an orogastric tube, however, is recommended to confirm patency of the esophagus and cardia, and to assess the nature of rumen contents (e.g. milky and malodorant with a low pH in ruminal drinkers). Abdominal tap may be useful if peritonitis is suspected, an increased volume of serosanguinous peritoneal fluid indicates the presence of devitalized abdominal organs (abomasum, intestine after volvulus or incarceration). Urinalysis may be indicated if pollakiuria or stranguria are observed and when abnormalities are evident at macroscopic evaluation of the urine.[1,2,10,13]

Ultrasonographic examination of the abdomen is in many cases the most useful further diagnostic investigation in calves with acute abdomen.[14] Its results will directly influence the clinical management of these patients, especially the decision to take them to surgery or not. Thanks to the size of the animals, ultrasonography allows for assessment of nearly the entire abdominal cavity in small calves, and thus is very helpful in the assessment of patients with colic and suspected intestinal obstruction. Evidence of dilated loops of intestine without propulsive motility at ultrasonography of the right flank is an indication for immediate laparotomy. The exact cause of intestinal distention, however, can only be directly visualized in rare cases, e.g. intussusception in form of a "bulls-eye"-shaped structure.[14] The size and location of the abomasum can be assessed by ultrasonography. Perforating ulcers are normally not visualized, but the resulting peritonitis can be identified (free abdominal fluid with or without fibrin tags). Ultrasonographic evaluation of intraabdominal umbilical remnants (umbilical vein, arteries and urachus) allows not only a precise preoperative diagnosis but also a prognosis, especially if omphalophlebitis is present. The prognosis for a calf with infection of the umbilical vein extending deep into the liver parenchyma is grave because uncomplicated surgical removal of the abscess is not possible. It is therefore important that the liver be assessed carefully if omphalophlebitis is present. Ultrasonographic assessment of the umbilicus is further helpful when incarceration of abdominal organs (abomasum, intestine, omentum) in an umbilical hernia is suspected.[11,14]

In many cases, a definitive causative diagnosis cannot be reached despite extensive clinical work-up. When a surgical lesion is suspected, exploratory laparotomy should be considered as a part of the diagnostic investigations and will at the same time allow for correction e.g. of an intestinal obstruction.[1,2]

PATIENT MANAGEMENT

The decision to proceed to surgery with a calf showing signs of abdominal pain is based on a tentative diagnosis which should be as precise as possible. Some diagnoses, such as RDA, AV or intestinal obstruction, imply immediate surgical treatment, but decision-making is distinctly less obvious in other cases. Careful clinical evaluation must address the possibility of disease in organ systems outside of the GIT causing signs of abdominal pain, in young animals especially in relation with the umbilicus.

Primary forestomach disturbances requiring immediate surgical intervention are uncommon in calves. In case of life-threatening tympany that cannot be relieved by stomach tube in calves, careful examination of the abdomen should ensure that rumen tympany is, in fact, the main problem. Indeed, acute pronounced abdominal distention is more often caused by abomasal or intestinal tympany than by ruminal bloat in calves.[2]

Abomasal displacement is best corrected surgically, allowing for fixation (by abomaso- or omentopexy) and prevention of recurrence. Surgery for RDA and AV must be considered an emergency due to vascular compromise associated with AV and potential impending torsion in cases of RDA.[2] Laparotomy may be indicated if peritonitis caused by perforating abomasal ulcers is suspected, allowing for confirmation of the diagnosis, evaluation of the prognosis based on the degree of abdominal contamination, and in some cases for repair of the abomasal defect(s).[15,16]

Decision-making about surgical vs. conservative treatment is often less obvious in colic of suspected intestinal origin because the nature of the primary lesion (partial or complete intestinal volvulus, intussusception, incarceration, impaction, luminal compression through adherences or other abnormal abdominal structures outside of the intestine, functional ileus) mostly cannot be determined at physical examination. Acute abomasal and/or intestinal tympany may resolve without surgical intervention, but it cannot be differentiated clinically from abomasal or intestinal volvulus, which have a lethal course if not corrected surgically as rapidly as possible.[1,2] Criteria for decision-making include onset and progression of colic signs, heart rate, findings at abdominal auscultation, fecal output, evidence of dilated intestines at ultrasonography, and response to therapy with spasmolytics and fluids. Exploratory laparotomy is warranted with ongoing tachycardia (> 140 bpm) and abdominal pain unresponsive to conservative treatment, no or strongly reduced fecal output, and evidence of serious intestinal distention in the absence of obvious alternative causes for ileus. Even if the exploration of the abdomen does not reveal a mechanical obstruction, it provides important information for the further management of the case: prokinetic drugs (e.g. neostigmin) and purgatives (mineral oil, Epsom salts) should not be administered as long as an obstructive lesion has not been ruled out. Thus, evidence of "functional ileus only" at surgery allows for aggressive post-operative medical treatment.[1,2]

CONCLUSION

Decision making in the clinical management of calves with signs of abdominal pain is directed by the results of a complete physical examination combined with appropriate further diagnostic investigations. The decision to take an animal to surgery is based either on the diagnosis of a condition for which surgical treatment is the therapy of choice, or on the severity of clinical signs (circulatory shock, "ping" in the right flank, distended intestines at ultrasonography, absent or strongly reduced fecal output), possibly combined with a lack of response to conservative therapy. An exploratory laparotomy should be considered as a constitutive part of the diagnostic work-up of such cases.

REFERENCES

[1]Mulon P.Y. and Desrochers A. **Vet Clin N Am Food Anim Pract** 2005;21:101. [2]Van Metre D.C.et al. **Vet Clin N Am Food Anim Pract** 2005;21:655. [3]Naylor J.M. and Bailey J.V., **Can Vet J** 1987;28:657. [4]AbutarbushS.M. et al. **J Am Vet Med Assoc** 2006;229:1627. [5]Constable P.D. et al. **J Am Vet Med Assoc** 1997;210:531. [6]Anderson D.E. et al. **J Am Vet Med Assoc** 1993;203:1178. [7]Huskamp B. **Dtsch Tierärztl Wochenschr** 1969:76:33. [8]Marshall T.S. **Vet Clin N Am Food Anim Pract** 2009;25:209. [9]Rademacher G. **Tierärztl Umschau** 2006;61:3. [10]Smith D.F. **Vet Clin N Am Food Anim Pract** 1990;6:449. [11]Steiner A. and Lejeune B. **Clin N Am Food Anim Pract** 2009;25:781. [12]Baird A.N. **Clin N Am Food Anim Pract** 2008;24:467. [13]Anderson D.E. **Clin N Am Food Anim Pract** 2008;24:383. [14]Streeter R.N. and Step D.L. **Vet Clin N Am Food Anim Pract** 2007;23:541. [15]Katchuik R. **Can Vet J** 1992;33:459. [16]Tulleners E.P. and Hamilton G.F. **Can Vet J** 1980;21:262.

CLINICAL MANAGEMENT & ANCILLARY DIAGNOSTICS OF ABDOMINAL EMERGENCIES IN CALVES

Mireille Meylan, Dr.med.vet. FVH, MS, PhD, Dr habil., DACVIM, DECBHM
Berne, Switzerland

INTRODUCTION

The complete physical examination and recommended diagnostic procedures in calves with acute abdomen have been discussed in the previous paper (# 205). In this contribution, the pathogenic mechanisms relevant to case management, diagnostic procedures and therapeutic measures will be reviewed for specific causes of abdominal emergencies in calves.

The main indications for immediate surgery in calves with acute abdomen include strong suspicion of a strangulating obstruction (e.g. abomasal volvulus, intussuception, mesenteric volvulus), progressive abdominal distention, persistent severe abdominal pain, absence of feces (especially if no feces are passed after fluid resuscitation), evidence of left displacement of the abomasum (prompt surgery is recommended because of possible development of abomasal ulcers, a common complication in calves with left-displaced abomasum), and abomasal tympany.[3] In all cases, aggressive medical treatment prior to, during and after surgery greatly influences the chances of survival of young animals with acute abdominal disease. Medical treatment includes fluid therapy to support the cardiovascular system and correct electrolyte and acid-base imbalances, antiiflammatory drugs, antibiotics to prevent post-operative complications, and specific treatment depending on the final diagnosis and the presence of concurrent diseases.[1] Abdominal surgery in adult cattle is mostly performed with the animal standing, but the size of young animals and their tendency to lay down in response to pain mostly preclude from standing surgery in calves. As a consequence, the positioning of the calves in dorsal recumbency (e.g. for umbilical surgery) or in lateral recumbency (e.g. right flank laparotomy if intestinal obstruction is suspected) leads to decreased gas exchange in the lungs and thus reduced tissue oxygenation in critically ill patients. Except for this aspect, no negative effects on the prognosis have been identified in calves with acute abdomen undergoing exploratory laparotomy, even if a non-surgical lesion that would have been amenable to medical treatment was eventually identified.[1,2]

ABDOMINAL EMERGENCIES IN CALVES ASSOCIATED WITH RUMINAL DISEASES

Tympany of the rumen in calves is mostly related to ruminal drinking (moderate recurrent bloat), indigestion associated with weaning (intermittent tympany associated with reduced weight gain) or obstruction of the cardia e.g. by trichobezoars (recurrent acute tympany).[1] Whereas ruminal drinking and indigestion are best treated with dietary measures, intraruminal foreign bodies such as a hairballs or phytobezoars must be removed surgically. Rumenotomy from the left flank with the calf in right lateral recumbency is performed according to a similar technique as in adults animals. It is however, important to remember that no indication is normally given for such an intervention prior to the onset of ruminal development, i.e. before the calf is approximately 2 months old. The surgical technique for rumenotomy has been described in detail. Rumenotomy is basically the only indication for left flank laparotomy in calves (with the rare exception of nephrectomy of the left kidney).[1,4]

ABDOMINAL EMERGENCIES IN CALVES ASSOCIATED WITH ABOMASAL DISEASES

Abomasal disorders are common in calves, they include acute abomasal tympany, abomasal ulcers, abomasal displacement to the left or to the right side of the abdomen, and abomasal volvulus.

Acute abomasal tympany

The etiology of abomasal tympany, a syndrome in young calves characterized by anorexia, abdominal distention, bloat and often death in 6 to 48 hours, remains unclear. Necropsy of such cases generally shows abomasal tympany, forestomach and abomasal edema, hemorrhage, mucosal necrosis, and occasionally mural emphysema. The disease has been associated with the feeding of milk replacers and fermentation by lactobacilli as observed in lambs, but a direct link has not been demonstrated. Furthermore, the fact that the abomasal tympany is also encountered in calves suckling their dam points at other causative factors.[2] However, a predisposition for the disease in the presence of large amounts of nutrients (e.g. after ingestion of an excessive volume of milk) is generally accepted as part of the pathogenesis of acute abomasal tympany. The broadly admitted practice of fasting affected calves for 12-24 hours after an episode of abomasal tympany also supports this hypothesis.[2,5] A role of *Clostridium perfringens*, especially type A, has been suspected, but the disease could not be reproduced experimentally in its full extent, thus suggesting a multifactorial pathogenesis of the disease.[6,7,8] Abomasocentesis under visual control, i.e. during right flank laparotomy, is the best surgical option for the relief of abomasal tympany, as opposed to blind transabdominal puncture which is associated with a high risk of peritonitis. Laparotomy also allows abomasotomy to remove putrefying milk or trichobezoars if necessary.[3] A method of ventral puncture of the tympanic abomasum between the xyphoid and the umbilicus with the calf in dorsal recumbency has been described, which appears to be associated with minimal risk of leakage of abomasal contents in the abdominal cavity, however this method is not adequate for cases where tympany is secondary to abomasal volvulus.[9] As both conditions (uncomplicated abomasal tympany and volvulus) cannot be differentiated clinically, laparotomy is recommended if the best possible therapy is to be provided; ventral puncture can be considered for animals of low economic value. Antibiotic treatment with penicillin has also been advocated in regard to a possible role of clostridial organisms in the pathogenesis of the disease.[5]

Abomasal ulcers

Four main factors have been associated with the development of abomasal ulcers: trace mineral deficiency, bacteria (especially *Cl. perfringens*), stress and the presence of abrasive agents (e.g. hairballs). The fact that calves with above average milk intake often suffer from abomasal ulcers has led to the hypothesis that spill-over of milk in the reticulo-rumen may cause bacterial fermentation and result in colonization of the abomasum. A role of *Clostridium perfringens*, especially type A, has been discussed, but could not be confirmed. Thus, the etiopathogenesis of abomasal ulcers in calves remains unclear to date. Prevalence of 0.2-5.7% in beef calves and 32-76% in healthy veal calves have been reported. The pre-ruminant calves (< 3 weeks) and calves during transitional phase (3-8 weeks) are at highest risk of developing abomasal ulcers. Abomasal ulcers are classified in 4 types: I (non-perforating, limited bleeding), II (non-perforating with severe hemorrhage after erosion of a vessel in the abomasal wall), III (perforating, causing local peritonitis), and IV (perforating with generalized peritonitis). Non-perforating ulcers (types I and II) causing chronic ill-thrift in calves are difficult to identify, especially type I, because they are not associated with typical clinical signs. Fecal occult blood tests may be positive (increased sensitivity with repeated testing) and packed cell volume may be low, especially in case of type II ulcers. Even at surgery, a good visualization of abomasal ulcers is not possible.[1,5] A technique has been described to expose the entire abomasal mucosa after exteriorization through an incision in the large curvature, allowing for identification and resection of ulcers.[10] The prognosis is guarded to poor in calves with type IV ulcers, survival rates from 0 to 40% can be found in the literature.[10,11] Histamine receptor type 2 antagonists (cimetidine, ranitidine) have been shown to increase the mean pH in the abomasum of healthy neonatal calves, however, the therapeutic efficacy of these drugs has not been evaluated in ill calves.[12]

Abomasal displacement and volvulus

Abomasal displacement to the left and to the right (with or without volvulus) has been described in calves as early as in the 1960s. The surgical principles for reposition and fixation of the abomasum via omentopexy or abomasopexy are similar in calves and in adults. Because the wall of abdominal organs is thinner in calves and because a lesser amount of fat is present to stabilize the mesentery and the omentum, it is advised to manipulate these structures with particular care in young animals.[1,13] Dirksen has reported survival rates of 69% after surgery for LDA and 33% for abomasal volvulus in calves.[14]

ABDOMINAL EMERGENCIES IN CALVES ASSOCIATED WITH INTESTINAL DISEASES

Colic of intestinal origin in calves is mostly associated with the different forms of intestinal obstruction described below, however, in rare occasions, enteritis can also cause symptoms of abdominal pain even before the onset of diarrhea. In such cases, diarrhea may appear after fluid resuscitation of the affected calves.[3] However, colic due to intestinal ileus is definitively more common. The main causes of intestinal obstruction in calves include intussuception, volvulus, incarceration and entrapment, intraluminal obstruction, and malformation.[3] The associated clinical signs and treatment options are shortly reviewed below.

Intussuception

In most species intussuception is primarily a disease of the young; this is likely to be true also in cattle. Most intussuceptions occur in the distal jejunum and the proximal ileum, where the mesentery is longest and most mobile (mesenteric flange). Ileocecal or ileocolic intussuceptions are not seen in adult cattle because of the stability afforded by the ileocecal ligament, but they do occur in young calves, as the mesentery is very thin (less mesenteric fat than in adults) and does not provide the same degree of support as in older animals. In adults, masses or tumors are often found at the leading edge of the intussuception. In calves, postulated causative factors include dietary changes as well as viral and bacterial enteritis, i.e. conditions that might cause asynchronous peristalsis or hypermotility and thus predispose to intussuception. Occurrence of intestinal intussuception is characterized by abrupt onset of clinical signs with severe abdominal pain due to the tension on the mesentery of the intussuceptum. This phase lasts approximately 6-12 hours and is followed by a period of decreasing signs of colic as necrosis of the intestine and mesenterium progress, leading to increasing depression caused by peritonitis and toxemia.[15,16] Mild hyponatremia, hypochloremia, hypocalcemia, azotemia, hyperglycemia and hemoconcentration are mostly present in calves with intussuception, but no differences in hematology or blood chemistry parameters were found between survivors and non-survivors.[17] The intussuception itself cannot always be visualized at ultrasonographic examination even if signs of intestinal dilatation are evident, and the definitive diagnosis is often made at surgery.[15,16] The best approach for surgery of the intestine in calves is through the right flank. A ventral midline approach with the calves in dorsal recumbency is not recommended because of the difficulty in exposing the small intestine for resection and anastomosis.[17] The technique for resection of intestinal intussuception has been described in detail.[15,16] Manual reposition of the intussuception through traction is not recommended because of the risk of rupture and contamination of the abdominal cavity. Furthermore, resection of up to 30 cm of the small intestine orad to the intussuception is advocated due to difficulty in differentiating vital and non-vital intestinal wall visually in the immediate vicinity of the intussuception.[16]

Intestinal volvulus

Segmental volvulus most often occurs in the jejunoileum, i.e. in the distal-most aspect of the mesenteric flange, with twisting of 180 to 360°. Cattle of all ages can be affected with the sudden onset of severe colic. The pain then decreases with ischemic necrosis of the affected intestinal segment. Metabolic acidosis rapidly becomes superimposed on initial metabolic alkalosis as the condition progresses. In case of volvulus of the intestine about the mesenteric root, the jejunun, ileum, cecum and the majority of the ascending colon are included in the torsion (only the duodenum and parts of the colon located within the dorsal parts of the mesentery are spared). This is a most dramatic and rapidly fatal condition because of the large amount of intestine undergoing acute vascular compromise. Volvulus of the mesenteric root is characterized clinically by peracute onset of severe unrelenting pain and associated with a poor prognosis. Even if rapid surgical correction of the mesenteric root torsion can be performed, the risk of death through toxemia and peritonitis due to tissue devitalization remains high.[15,16,18]

Incarceration and entrapment of the small intestine

Incarceration of the small intestine is typically seen in the distal flange of the jejunoileum, but is possible throughout the jejunum and ileum. Incarceration of the intestine occurs mostly in a rent in the mesentery which may have developed spontaneously or after intestinal resection.[15,16] Incarceration of intestine in body wall defects around the umbilicus are discussed below.

Remnants of embryonic structures can persist as thin, tough ligamentous bands fixed at both ends (often between umbilicus and liver -as the round ligament of the liver-, umbilicus and ileum -as the vitelloumbilical band-, or between umbilicus and urinary bladder -as urachal remnant-). Entrapment causes compromise of vascular supply and cessation of intestinal transit. Surgical correction with removal of the abnormal structures is warranted, resection of the intestine at the location of entrapment may be necessary.[15,16]

Intraluminal obstruction

Intraluminal intestinal obstruction in calves is mostly associated with trichobezoars (hairballs) or phytobezoars (consisting of fibrous plant material). Young animals are particularly at risk for trichobezoars because they are more often affected with ectoparasites (lice, mange) causing pruritus, thus increased licking and ingestion of hair. Intraluminal obstruction is characterized clinically by slow onset of disease with colic of increasing intensity and progressive abdominal enlargement, as the intestine orad to the obstruction becomes dilated with ingesta. Therapy consists of enterotomy for removal of the obstructing foreign body.[15,16]

Intestinal atresia

Atresia coli is most common in calves, followed by atresia of the jejunum, rectum and anus. Whereas atresia ani can be readily identified and corrected surgically in neonatal calves, the diagnosis of atresia of more oral parts of the intestine mostly requires exploratory laparotomy. The different presentations of intestinal atresia have been recently reviewed.[16,19] Because intestinal atresia is often associated with other malformation, e.g. of the innervation of the intestinal wall or of the heart, and because a genetic component has not been ruled out, further breeding with such animals is not recommended. At our clinic, a tentative diagnosis of intestinal atresia is based on the history of a calf born bright and alert and showing increasing signs of colic, no fecal output and progressive abdominal enlargement from the second day of life. The suspicion is confirmed by evidence of dilated loops of intestine at ultrasonography of the right flank. If the diagnosis is confirmed at exploratory laparotomy, the calf is euthanized. Salvage procedures such as colostromy or cecostomy are not performed for considerations of animal welfare, neither is intestinal anastomosis because of the high rate of complications due to concomitant malformations (e.g. of the neural plexus in the abdominal wall), and to prevent further dissemination of possibly defective genes in the population.[19]

ABDOMINAL EMERGENCIES IN CALVES ASSOCIATED WITH UMBILICAL DISEASE

The most common presentation of umbilical disease as an emergency occurs in the case of incarceration of abdominal organs in an umbilical hernia. Open hernia with extra abdominal prolapse of the jejunum through the umbilical ring in newborn calves obviously also requires immediate surgical repair. The incidence of umbilical hernias has been reported to be 0.65 - 1.04% in calves. Herniated abdominal structures include in decreasing order of frequency the omentum, abomasum, rumen and small intestine.[1] Entrapment of loops of the small intestine or of the pyloric region of the abomasum can cause acute interruption of ingesta passage and severe signs of colic. The umbilical region is enlarged and painful upon palpation, ultrasonographic examination allows the identification of the incarcerated structures in most cases.[20] Therapy for incarcerated umbilical hernia consists of immediate resection of the umbilical ring from ventral with the calf in dorsal recumbency, with or without resection of the structures entrapped in the umbilical ring. Although sedation and epidural anesthesia with lidocaine-xylazine has been described for umbilical surgery, general anesthesia with intubation and maintenance with anesthetic gas vaporized in oxygen is the preferred method for complicated umbilical interventions, especially if intestinal anastomosis or abomasotomy may have to be performed.[21]

Infectious affections of umbilical remnants occur frequently in calves, with urachitis being more common than omphaloarteriitis and –phlebitis, but these forms of umbilical disease rarely represent surgical emergencies,[1,21] except for infected periarterial hematomas in neonatal calves which can rapidly lead to peritonitis, sepsis and death. Such infected hematomas must be removed as soon as possible, i.e. in the first days of the calf's life. Clinical symptoms and specific clinical features of that form of umbilical disease have been described in detail.[22]

REFERENCES

[1]Mulon P.Y. and Desrochers A. **Vet Clin N Am Food Anim Pract** 2005;21:101. [2]Naylor J.M. and Bailey J.V., **Can Vet J** 1987;28:657. [3]Van Metre D.C.et al. **Vet Clin N Am Food Anim Pract** 2005;21:655. [4]Niehaus A.J. **Vet Clin N Am Food Anim Pract** 2008;24:341. [5]Marshall T.S. **Vet Clin N Am Food Anim Pract** 2009;25:209. [6]Panciera R.J. et al. **J Vet Diagn Invest** 2007;19:392. [7]Songer J.G. and Miskimins D.W. **Anaerobe** 2005;11:290. [8]Roeder B.L. et al. **Am J Vet Res** 1988;49:201. [9]Kümper H. **Tierärztl Prax** 1994;22:25. [10]Katchuik R. **Can Vet J** 1992;33:459. [11]Tulleners E.P. and Hamilton G.F. **Can Vet J** 1980;21:262. [12]Ahmed A.F. et al. **Am J Vet Res** 2001;62:1531. [13]Niehaus A.J. **Vet Clin N Am Food Anim Pract** 2008;24:349. [14]Dirksen G. **Bov Pract** 1994;28:120. [15]Smith D.F. **Vet Clin N Am Food Anim Pract** 1990;6:449. [16]Anderson D.E. **Clin N Am Food Anim Pract** 2008;24:383. [17]Constable P.D. et al. **J Am Vet Med Assoc** 1997;210:531. [18]Anderson D.E. et al. **J Am Vet Med Assoc** 1993;203:1178. [19]Meylan M. **Clin N Am Food Anim Pract** 2008;24:479. [20]Steiner A. and Lejeune B. **Clin N Am Food Anim Pract** 2009;25:781. [21]Baird A.N. **Clin N Am Food Anim Pract** 2008;24:467. [22]Rademacher G. **Tierärztl Umschau** 2006;61:3.

ARE INTRAMAMMARY ANTIBIOTICS EFFECTIVE FOR THE TREATMENT OF MASTITIS?

Ted Whittem, BVSc, PhD, DACVCP, FACVSc
Melbourne, Australia

INTRODUCTION

The focus of this discussion is acute mastitis and its treatment using intramammary antibiotics in dairy cattle. Acute mastitis is bacterial infection of the udder and may be localized (mild) or include systemic signs (severe) with or without bacteraemia. Acute mastitis occurring during lactation is the indication for administration of lactating-cow intramammary antibiotics (LCIA).

Despite many decades of comprehensive mastitis control programs including LCIA use, acute mastitis remains a significant cost to the dairy industry. The cost can be indirect from lost production, changed managerial practices and implementation of control programs. In addition, direct costs are incurred including the cost of replacing culled animals and treatment with antibiotics as LCIA, parenteral injectables or both. Dairymen spend dollars on treatment of acute mastitis using LCIA with the expectation that the products are effective.

EFFICACY

To different people "efficacy" has different meanings. Although the most important effect of antimicrobials is to minimize economic impact of mastitis on the production unit, it is difficult to evaluate the effect of individual cow treatment on overall production profitability.

Measures of efficacy at the individual cow level are needed for evaluation of the outcome from LCIA. Such measures may include remission of clinical signs or elimination of the pathogen. Appropriate definitions of these measures allow the declaration of cases as 'cured' or 'failed-to-cure'. Efficacy is often expressed as the percentage of cases cured from those treated. It should be noted that a proportion of these case may have been self-cured without treatment and that the stated efficacy percentage usually includes self-cured cases.

Clinical Cure

Clinical signs of acute mastitis vary with the pathogen and severity of the infection. Clinical signs include; watery or bloody or clotted milk, swollen or hot or hard quarters, loss of appetite and systemic signs of infection or shock with localization of some signs in the udder. Remission of clinical signs without recrudescence prior to expiry of the milk discard time is a typical definition of clinical cure. Declaration of clinical cure is a subjective evaluation the accuracy of which depends on the skill of the observer.

Bacteriological Cure

Many authorities, especially drug regulating authorities, define the failure to identify the same pathogen (as initially identified prior to treatment) in 2 of 3 consecutive post treatment samples as evidence of a bacterial 'cure'. There are many pitfalls of this definition because of potential false positive pretreatment samples and resulting false positive diagnoses, and false negative samples post treatment. Reports of efficacy in research publications need to be reviewed critically.

Efficacy of Lactating Cow Intramammary Antibiotics

There are objective prospective data describing efficacy of LCIA, but mostly without untreated controls. A retrospective study of sub-clinical mastitis in 9007 cows in North America helps to identify the expected "cure rate" for untreated cases; the spontaneous bacteriological cure rate was 65%. In that study, some antibiotics improved this cure rate to 75% and some did not significantly affect the rate.[1] Bacteriological 'response' measured at day 14 post treatment of acute mastitis to LCIA with penethemate was merely 38%, not different from the response to infusion with *Lactococcus lactis*. Unfortunately, this study used a variable for bacteriological response which allowed persistence of large numbers of colony forming units of the pathogen, making the data not comparable to other studies since its primary variable is inconsistent with the usual definition of bacteriological 'cure'.[2] With experimentally induced subclinical staphylococcal mastitis LCIA with amoxicillin was bacteriological effective in only 7/23 cows (30%), but in this study the numbers are very small and therefore difficult to rely upon.[3] That the causative pathogen can markedly affect the cure rates was well demonstrate in an early study of LCIA with erythromycin; bacteriological cures ranged from 74.4 to 100% depending on the bacterium identified. In cases including both subclinical and clinical mastitis, the overall cure rate for 102 quarters was 81.5%.[4] After LCIA with penicillin/novobiocin in experimentally induced bacterial mastitis, bacteriological cure at 28 days was achieved in 70 and 35% cases of staphylococcal mastitis of <2 or > 4 weeks duration respectively; duration of infection affected cure rates. The bacteriological cure rate also varied widely between causative pathogens.[5] Bacteriological cure rates in the UK and Europe for naturally acquired clinical mastitis in 434 cows after LCIA with cefaperazone were just 69%. Notably, this was after only a single-dose treatment. Clinical cure rates in the same study were higher, at 82% but again, no untreated controls were available.[6] The clinical and bacteriological cure rates in a French study for LCIA with amoxicillin/cloxacillin were 92 and 56.7 % at 22 days post treatment.[7] The bacteriological cure rates in a New Zealand study for LCIA with two dissimilar products were both 77 % at 21 days post treatment.[8] An Australian study of approximately 400 cases showed overall bacterial and clinical cure rates at 70 and 81.7 % respectively for two LCIA, cefuroxime and cloxacillin. The two products did not differ.[9] Some studies have evaluated the efficacy of LCIA with and without systemic antibiotics versus supportive only treatment (oxytocin and milking out) by comparing economic variables such as milk yield loss. In one study of 124 cases, antibiotic use reduced milk yield loss.[10] However, in another study using a similar approach differences of treatment cost between oxytocin and amoxicillin LCIA were not demonstrated whereas LCIA with cephapirin was more expensive.[11]

Comparison between these studies is confounded by their different methodology and different production and treatment systems. Nevertheless, the picture emerges of some efficacy of LCIA. It seems that acute clinical mastitis, like many infections, can be self-limiting and self-curing, possibly depending on the pathogen, severity and/or duration of infection. Using LCIA with some of the existing proprietary products seems likely in a few productions systems to be of commercial advantage by increasing the clinical cure rate, and perhaps increasing the bacteriological cure rate from approximately 25% self-cure to approximately 50% cure.[12] Some pathogens seem to be less readily addressed by LCIA.

Factors affecting efficacy of Lactating Cow Intramammary Antibiotics

Antimicrobial efficacy is influenced by factors such as the choice of antibiotic, its spectrum of activity, its access to the site of infection, its concentration and persistence at the site of infection and its activity in the medium at the site of infection. Disease factors may also alter expected efficacy in LCIA for acute mastitis; the causative organism, the pathophysiological response to the infection or to the product formulation.

The chosen antibiotic must be active against the infecting pathogen. Unfortunately *in vitro* sensitivity testing is not useful for predicting efficacy in resolution of acute lactational mastitis.[12] This lack of correlation is probably due to factors such as different rates of bacterial replication, differences in the growth medium, effects of formulation excipients on the drug release and the presence of physical or functional barriers to antimicrobial drug distribution. Fundamentally, the antimicrobial drug must have adequate access to the bacteria in sufficient concentrations with an appropriate concentration-time profile. Appropriate profiles are discussed elsewhere, as the pharmacokinetic-pharmacodynamic correlates to efficacy differ between antimicrobial drug families.

For non invasive pathogens, e.g., *Streptococcus agalactiae* and *Streptococcus dysgalactiae*[13] the antibiotic must simply distribute adequately from the formulation infusion site through the milk, as these pathogens reside in the milk. The rate and extent of dissolution and dissemination of the active antimicrobial drug from the formulation into the milk can be markedly affected by variations of formulation. For this reason it is possible that two formulations with the same mass of the same antibiotic, but manufactured by different companies, may have distinctly different efficacy for non invasive pathogens.

In contrast, for invasive bacteria such as *Streptococcus uberis* and *Escherichia coli* (or other enterobacteriaciae) [13] the concentration reached and persistence in the mammary interstitial tissue of active drug is more important than its concentration and persistence in the milk. While formulation factors can influence the distribution of drug to tissues, the inherent properties of the active molecule itself are likely to be more important: factors such as molecular size, lipid solubility, ionization state and protein binding. For these infections then, the differences in efficacy could be more marked between different actives even though their *in vitro* activity against the pathogen might be quite similar.

Finally, infections caused by *Staphylococcus aureus* have pathophysiology that results in changes in the form and function of the infected tissue.[13] Inflammation, fibrin, collagen, cellular infiltration, and potential intracellular invasion by bacteria all occur. The host cells containing intracellular staphylococci can be either interstitial or in the milk. The access of the antibiotic, therefore, to the 'site of infection' changes as a function of the severity, duration and location of the disease process, meaning that efficacy is likely to alter as a function of time (even without the development of resistance).

To better understand drug efficacy within the bovine udder, a complete understanding of milk-drug concentrations in the different compartments of the mammary gland is needed. A physiologically-based pharmacokinetic model of the udder would be of use.

PROPOSED PHARMACOKINETIC CONSIDERATIONS FOR EFFICACY OF LCIA

Antimicrobial efficacy is directly related to the concentration of the drug at the site of infection, and its persistence at sufficient concentration (above MIC). Modeling the elimination of drugs from the udder in the milk is a way of predicting the concentration of the drug in the milk at any time after dosing. Although elimination of drug from the udder is traditionally analyzed using standard first-order pharmacokinetics, these models are clearly incorrect. There are several acknowledged difficulties with these traditional models. Collecting milk samples only at well-spaced milking intervals provides data that are too sparse and too few to make accurate estimations of elimination rates, but by collecting milk samples more frequently the pharmacokinetics are changed because collection directly increases the drug clearance. Further, these models treat the udder as a single well mixed compartment of constant volume; both these assumptions are incorrect.

A simplified physiological model of milk compartments in the bovine udder has been described, containing (at least) two distinct anatomical compartments; the cisternal compartment and the secretory compartment (alveoli and ductules).[14,15] In the bovine udder, both compartments begin filling directly after milking but the cisternal compartment fills more slowly, especially in the first hours post milking. The milk production curve varies based on milking frequency.[14-16]

Milk being produced and entering the two holding compartments between milkings dilutes drugs infused in to the udder. Dilution or changes in concentration not only affect the efficacy of drugs against pathogens in the milk, but also alter the rates of diffusion of drug into interstitial tissue according to Fick's first law of diffusion.

The inter-milking interval affects the extent to which the secretory compartment is filled. Filling, especially overfilling, stretches the tissue and leads to changes in epithelial permeability which in turn may alter the rates of diffusion of drug into interstitial tissue. Efficacy may be affected by changes in the milking routine and may be correlated to production characteristics of individual animals.

The udder is 'emptied' once every milking. However, residual milk not harvested and remaining in the udder contains residual drug. The mass of residual drug remaining in the udder will, therefore, be affected by the efficiency of milking. The residual volume of the bovine udder has been determined experimentally to be in the range of 5 to 14% total udder volume.[17-19] The estimates vary widely because of different methodology, but the estimates provide a starting place to evaluate the effect of under or over milking.

CLINICAL RELEVANCE

Despite the lack of overwhelming evidence for efficacy of LCIA, prudent rational LCIA therapy both persists as common practice and is recommended by 'authorities'. It is common practice to use LCIA, at least in some parts of the world. There is a need to identify and further research factors that may influence the efficacy of LCIA, since increases in efficacy would be of value to the dairy industry.

Improvements in formulations may provide an opportunity for increasing efficacy to targeted pathogens, noting that prevalence of mastitis pathogens varies between different production systems and therefore use-patterns for particular formulations probably should also vary.

REFERENCES

1. Wilson DJ, Gonzalez RN, Case KL, et al. J Dairy Sci 1999;82:1664-1670. 2. Klostermann K, Crispie F, Flynn J, et al. J Dairy Res 2008;75:365-373. 3. Owens WE, Watts JL, Boddie RL, et al. J Dairy Sci 1988;71:3143-3147. 4. Schultz EJ. J Am Vet Med Assoc 1968;152:376-379. 5. Owens WE, Ray CH, Watts JL, et al. J Dairy Sci 1997;80:313-317. 6. Wilson CD, Agger N, Gilbert GA, et al. Vet Rec 1986;118:17-19. 7. Serieys F, Raguet Y, Goby L, et al. J Dairy Sci 2005;88:93-99. 8. McDougall S. Efficacy New Zealand Veterinary Journal 1998;46:226-232. 9. Wraight MD. New Zealand Veterinary Journal 2003;51:26-32. 10. Shim EH, Shanks RD, Morin DE. J Dairy Sci 2004;87:2702-2708. 11. Van Eenennaam AL, Gardner IA, Holmes J, et al. J Dairy Sci 1995;78:2086-2095. 12. Sandholm M, Kaartinen L, Pyorala S. J Vet Pharmacol Ther 1990;13:248-260. 13. Erskine RJ, Wagner S, DeGraves FJ. Vet Clin North Am Food Anim Pract 2003;19:109-138, vi. 14. Davis SR, Farr VC, Copeman PJ, et al. J Dairy Res 1998;65:1-8. 15. Stelwagen K, Hopster H, Van Der Werf JT, et al. J Dairy Sci 2000;83:48-51. 16. Vetharaniam I, Davis SR, Soboleva TK, et al. J Dairy Sci 2003;86:1987-1996. 17. Carruthers VR, Davis SR, Bryant AM, et al. Journal of Dairy Research 1993;60:1-11. 18. Isaksson A, Arnarp L. Acta Vet Scand 1988;29:259-262. 19. Knight CH, Hirst D, Dewhurst RJ. J Dairy Res 1994;61:167-177.

UPDATE ON BOVINE HEMATOLOGY

Susan Tornquist, DVM, PhD, DACVP
Corvallis, OR

INTRODUCTION

Blood cells and other hematologic features of cattle have a few unique and interesting characteristics and some long-held assumptions regarding bovine hematology have recently been challenged. This review will focus on these differences and new information.

ERYTHROCYTES

Bovine erythrocytes are larger than those of some other ruminants, such as sheep and goats, but are smaller than those of dogs and the area of central pallor isn't prominent in bovine RBCs. Poikilocytosis, which includes the presence of echinocytes, schistocytes and acanthocytes is seen fairly often in calves, even those that appear clinically healthy. The reason for this is not known, though it could be related to the shift from fetal hemoglobin (HgbF) to the adult form of hemoglobin (HgbA) as the erythrocytes produced in the calf *in utero* are replaced by new red blood cells in the neonatal period. The lifespan of bovine erythrocytes is approximately 150 days, which is relatively long among domestic animals.

Reported reference intervals for cattle RBC parameters don't tend to vary markedly based on breed, though beef cattle generally have higher RBC counts and hemoglobin concentrations than dairy cattle and a similar difference is seen in bulls as compared to cows. Cattle that are moved to high altitude for some parts of the year can have a 10-15% change in packed cell volume (PCV).

Cattle typically show morphologic changes in circulating erythrocytes as part of a regenerative response to anemia. These changes include polychromasia, macrocytosis, anisocytosis and sometimes basophilic stippling (due to aggregated ribosomes) , which is more commonly seen in cattle, than in most other species as part of a regenerative response. Nucleated RBCs may also be present in the peripheral blood in a vigorous regenerative response. Increased reticulocytes are also indicative of a regenerative anemia, and they may begin to appear in the peripheral blood as early as 3-4 days after acute blood loss or hemolysis with a peak at 7-14 days.

Regenerative anemia in cattle occurs, as it does in other species, in the face of blood loss and hemolysis. Causes of hemolysis that are seen in cattle include Heinz body anemia, associated with hypophosphatemia (postpaturient hemoblobinuria), ingestion of kale, cabbage, onions, and copper and selenium toxicity. Infectious agents associated with hemolytic anemia in cattle are *Anaplasma marginale*, *Babesia bigemina*, *Theileria* spp., *Leptospira* spp, *Clostridum haemolyticum* and *C. novyii* type D, *Mycoplasma wenyonii* and Trypanosome spp. though the trypansomes found in cattle in the United States (*T. theileri*) are not considered pathogenic.

The most common cause of non-regenerative anemia in cattle is probably the mild to moderate normocytic, normochromic anemia of chronic disease or inflammation. Iron deficiency will also result in non-regenerative anemia, but is more often microcytic and hypochromic. This may be seen in cattle with chronic gastrointestinal parasitism or sometimes in milk-fed calves. Copper and cobalt deficiencies have also been associated with non-regenerative anemias.

More unusual causes of non-regenerative anemia include congenital dyserythropoiesis associated with macrocytic, normochromic anemia in Polled Hereford calves and bracken fern toxicity which may result in a generalized bone marrow suppression.

Cattle have at least 11 blood group systems and over 700 known blood group factors.

Bovine hemoglobin has been used to make a cell-free oxygen-carrying blood substitute for other species, including humans.

LEUKOCYTES

Neutrophils of cattle have a third type of neutrophil granule which gives the bovine neutrophil its characteristic pinkish cytoplasmic color. This is a large granule with high antimicrobial activity. Bovine lymphocytes are also somewhat unique in that there is considerably more variability in the size of normal circulating bovine lymphocytes, compared with other species. They include small lymphocytes that are about 8 microns in diameter with a relatively condensed chromatin pattern to the round nucleus. These have a morphologic appearance similar to those of other species. Medium-sized bovine lymphocytes have a larger, less-condensed nucleus and more cytoplasm that may entirely surround the nucleus. Nuclei may be round to oval to indented. The large lymphocytes of cattle are about 15 microns in diameter with an often indented nucleus with more dispersed chromatin, even more cytoplasm than the medium-sized lymphocyte, and often look very similar to monocytes. Large granular lymphocytes, that contain a few small magenta-colored granules, are relatively common in cattle.

Cattle have a smaller bone marrow reserve population of neutrophils than other species, leading to a unique pattern of granulocytic response to acute, suppurative inflammation. This relatively smaller bone marrow reserve can't be mobilized quickly and the initial response to acute inflammation in cattle is often a marked neutropenia. By 24 hours following an acute inflammatory insult, a vigorous left shift response is often noted, with the neutropenia still present. This pattern would be considered a degenerative left shift with a poor prognosis in other species, however cattle will eventually overcome their sluggish neutrophil production and develop a neutrophilia by about 4-5 days. When inflammation becomes chronic, cattle may have either a neutrophilia or normal neutrophil numbers. Since the response is somewhat unpredictable, the use of plasma fibrinogen concentration to detect inflammation in cattle is common. Fibrinogen concentrations in cattle may be as high as 1.5 g/dl in marked inflammatory conditions.

The idea that a shift in neutrophil to lymphocyte ratio (N:L) to > 1 is a reliable indicator of inflammation in cattle has recently been challenged by a report by George et al (2010) suggesting that commonly-used hematology reference intervals for cattle may no longer be appropriate. She generated new reference intervals for dairy cattle, using clinically-healthy, California herds that were similar in age to those sampled in 1957 when Schalm collected bovine blood to calculate the reference intervals in 1965's textbook Schalm's Hematology. In comparing the 1957 and 2001 data, it appeared that neutrophil counts were now significantly higher and lymphocyte counts, as well as monocytes and eosinophils were significantly lower than they were in the 1957 populations. The mean N:L ratio in current, healthy dairy cattle was 1.17 in contrast to the most frequently reported ratio of approximately 0.5. Reasons for this change are not entirely clear, but may include the influence of changed management practices, diet, genetic selection, and the decreased prevalence of bovine viral diarrhea virus (BVDV) infections as BVD can result in mean lower neutrophil counts in infected animals, even those that are clinically healthy. Although cattle carrying the gene for bovine leukocyte adhesion deficit (BLAD), including heterozygotes, have lower neutrophil counts than those without this gene, the presence of BLAD carriers did not appear to affect the California-generated data. These carriers, could, however, affect the mean neutrophil percentage in other bovine populations. The presence of absence of bovine leukemia virus (BLV) infected cattle can also affect leukocyte reference intervals and the cows included in the California study were all certified BLV free.

Leukocyte morphologic abnormalities may be seen in acquired Pelger –Huet anomaly, in the face of inflammation, the hereditary Chediak-Higashi syndrome, reported in Hereford cattle and in GM_1 gangliosidosis, a hereditory disorder of Friesian cattle. BLAD is associated with persistent moderate to marked neutrophilia and neutrophil functional abnormalities, but morphologic abnormalities are not seen in neutrophils from affected cattle.

Infectious agents that may rarely be found in bovine leukocytes include *Anaplasma phagocytophilum* and other bacteria during septicemia.

PLATELETS

Bovine platelets have an interesting ultrastructure in that they lack an open canalicular system, but this doesn't affect their function or numbers. Thrombocytopenia may occur in conditions similar to those in other species, as well as BVDV infection and grazing of bracken fern, which can result in either pancytopenia or more selective hematopoietic cell deficiencies.

BONE MARROW

The unique feature of bovine bone marrow is the low myeloid to erythroid (M:E) ratio compared to other species. This correlates with the depletion of neutrophils seen early in acute inflammation.

PLASMA PROTEINS

As mentioned earlier, fibrinogen is often used as an indicator of acute inflammation and will typically increase within 2 days, often to very high concentrations. As it can easily (though roughly) be assayed via heat precipitation and refractometer, it is the most commonly evaluated acute phase protein in cattle. Haptoglobin and serum amyloid A are other acute phase proteins that increases in inflammation in cattle, but are not as easily measured, so are not as widely used.

REFERENCES

Bertram TA. Neutrophilic leukocytes structure and function in domestic animals. Adv Vet Sci Comp Med 1985; 30:91-129.George JW, et al. Comparison of bovine hematology reference intervals from 1957 to 2006. Vet Clin Path 2010;39(2):138.Jones JL, et al. 2007. Evaluation of the ruminant complete blood cell count. *Vet Clin N Am Food Anim Prac* 23: 377. Kehrli ME ,e t al. 1990. Molecular definition of the bovine granulocytopathy syndrome: Identification of deficiency of the Mac-1 (CD11b/CD18) glycoprotein. *Am J Vet Res* 51: 1826.Keller SL,et al. 2006. Effects of noncytopathic type 2 Bovine viral diarrhea virus on the proliferation of bone marrow progenitor cells. *Can J Vet Res* 70: 20.Lumsden JH, et al. 1974. The Piromen test as an assay of bone marrow granulocyte reserves in the calf: Studies on bone marrow and peripheral blood leukocytes. *Can J Comp Med* 38:56. McNair J, et al. Evluation of a competitive immunoassay for the detection of bovine haptoglobin. Res Vet Sci 1997; 63:145. Osborn BI, et al. 1968 Acquired Pelger-Huet anomaly in cattle. *J Am Vet Med Assoc* 152: 11.Paape MJ,et al, . 2003. The bovine neutrophil: Structure and function in blood and milk. *Vet Res* 34: 597.Steffen DJ, et al. Congenital dyserythropoiesis and progressive alopecia in Polled Hereford calves: hematologic, biochemical, bone marrow cytologic, electrophoretic, and flow cytometric findings. J Vet Diagn Invest 1992: 4(1): 31Stockham SL and Scott MA. Fundamentals of Veterinary Clinical Pathology, 2nd edition. Ames, Iowa. Blackwell Publishing, 2008. Tornquist SJ, et al. Interpretation of ruminant leukocyte responses. In : Weiss and Wardrop, eds, Schalm's Veterinary Hematology, 6th ed., Ames, Iowa: Wiley-Blackwell; 2010:307.Wood D, Quiroz-Rocha GF. Normal hematology of cattle. In: Weiss and Wardrop, eds, Schalm's Veterinary Hematology, 6th ed., Ames, Iowa: Wiley-Blackwell; 2010:829.

ULTRASOUND AS A VALUABLE DIAGNOSTIC TOOL IN CAMELID MEDICINE

Claire E. Whitehead, BVM&S, MS, DACVIM, MRCVS
London, United Kingdom

INTRODUCTION

The diagnostic importance of ultrasonography in camelid medicine cannot be emphasised enough. In the author's practice, abdominal ultrasound is used as an extension of the physical examination in all sick camelids but especially in adult animals. Ultrasonography may also be used in evaluation of externally palpable masses prior to aspiration, as a guide for centesis of the abdominal or thoracic cavities, or to guide biopsy of internal organs, as well as for evaluation of the thorax in certain cases and for echocardiography. Clearly, ultrasound is also useful in reproductive applications but will not be discussed here. Surprisingly, a review of the available literature reveals limited use of abdominal ultrasonography in evaluation of sick camelids, although several authors point out that it may have been useful to have performed it in certain case reports. The normal ultrasonographic appearance of gastrointestinal viscera in healthy llamas and alpacas has been described.[1]

APPROACH TO ULTRASONOGRAPHIC EVALUATION IN CAMELIDS

Portable ultrasound machines have been available for many years and their availability greatly facilitates proper examination in the field. The diagnostic quality of these portable machines has improved enormously over recent years and portable machines are now able to offer Doppler technologies. However, Doppler is not required for most cases unless the operator has a particular need to evaluate vasculature or wishes to perform echocardiography. For most purposes, probes in the range of 5-9 MHz frequencies are ideal for use in camelids and variable frequency probes in these ranges are readily available. However, a higher frequency probe may be useful for evaluation of very superficial structures such as tendons and masses and a lower frequency probe for a deeper (but less detailed) look inside the abdomen. The type of probe that you choose depends on what you wish to look at - and a little bit on the behavior of the subject. A microconvex probe is perfect for transabdominal use, especially in neonatal patients, but many camelids will adopt sternal recumbency when the probe is placed on the inguinal area making it challenging to fit both the probe and the operator's hand in the small space that remains. In these cases, it makes sense to switch to a linear rectal probe which may be much better tolerated and allow the operator to work "with" rather than against the patient. Some linear probes can be too wide to fit snugly between the ribs and allow evaluation of the liver and thorax: in these cases, either a slimmer linear probe or a microconvex probe will usually work better. The linear rectal probe may also be used for evaluation of the reproductive tract and caudal urinary tract structures. When performing rectal ultrasound, it is vital that adequate restraint is achieved either by physical or chemical means since rectal tears may easily occur when an alpaca or llama resents examination. Additionally, the rectum must be evacuated of fecal material and adequate lubrication used. For transabdominal ultrasound, it is usually not necessary to clip the fleece for a survey examination. If it becomes necessary to follow a particular structure, or if biopsy is required, then clipping will facilitate this. However, most owners will resent their animals being clipped unless this is absolutely necessary and is not a great idea in winter. Since alpaca fleece is so fine, usually it is possible to acquire excellent images by parting the fleece and using alcohol as the contact medium. The author does not find that clipping and use of ultrasound gel is superior: in fact alcohol provides a better contact medium and has the advantage of not requiring removal.

When performing abdominal ultrasound, it usually does not matter whether the patient is standing up or sitting in sternal recumbency. A standing or sternal recumbent position is preferable to lateral recumbency as this allows a more normal anatomical relationship of the internal organs. However, ventral midline structures may require the patient to remain standing or to be positioned laterally. Sedation is rarely required. When performing the examination, it helps to adopt a consistent approach in order to ensure that the entire abdomen is evaluated, including the gastrointestinal tract, liver, spleen, the bladder and uterus (+/- fetus) where relevant, and both kidneys in the retroperitoneal space. It is important to realise that it may not be possible to evaluate the entire abdomen in adult animals since ultrasound will only penetrate part way through the abdomen depending on the frequency of probe being used.

For echocardiography, it is usually possible to position the patient in lateral recumbency as one would for small animal patients, taking particular care to restrain the head and neck. Sedation is not normally required although may be required for adults. For adults it is particularly important to ensure that the hind legs are not free to be kicked: ensuring that the legs are tied to the examination table using thick ropes placed above the fetlocks is usually appreciated by the operator (and the owner of the machine)! Echocardiography can be performed in the standing animal if necessary although it can be difficult to get sufficient access behind the elbows as most animals will not easily allow their legs to be pulled cranially.

ABDOMINAL ULTRASONOGRAPHY

Evaluation of the abdomen by ultrasonography is a vital part of the examination of any sick camelid patient. This is because sick camelids often have very similar presentations, physical examination findings are often not helpful at differentiating various diseases, and neither abdominal palpation nor rectal palpation are particularly helpful for evaluation of the abdomen of these species. Although palpation may be useful in neonates, the proportion of space available caudal to the ribs for abdominal palpation is smaller in camelids than in small animals, and the presence of the large first stomach compartment that is always full of feed material in non-neonates limits this as a useful examination method. The small size of most camelids also limits the use of rectal palpation and only the reproductive tract and other intrapelvic structures can be evaluated via this route by a practitioner with small enough hands and forearms. Transabdominal ultrasound must include evaluation of the main organs (liver, kidneys, spleen) as well as the different areas of the gastrointestinal tract (forestomachs and intestines) and the bladder and uterus where relevant.

Liver

A good place to start the examination is with the liver. Find the liver by placing the probe in the second to last rib space on the right side, about midway down the body wall. This area will typically allow you to see a triangular view of the liver, with the portal vein on the medial surface. The portal vein is the largest vessel visible within the liver and has a hyperechoic border. The hepatic veins are also visible and these can be differentiated from the portal veins since they do not have a hyperechoic border. Hepatic arteries are not typically visible unless distended. Camelids have no gall bladder and the bile ducts are only observed if distended or hyperplastic as in liver fluke infestation.[2] Liver parenchyma is homogeneous and more hypoechoic than the spleen. It is quite common for camelids imported from South America to have hyperechoic spots throughout the liver parenchyma and these are thought to be migration tracts of larval stages of the nematode *Lamanema chavezi*. These are usually considered to be incidental findings. If the operator becomes familiar with the normal appearance of the liver, it is possible to get some idea of whether the liver is larger or smaller than normal, displaced, or whether there are changes in echogenicity. The liver may become diffusely more hyperechoic or enlarged if there is fatty infiltration as a result of excessive fat mobilization. The liver is a primary site for development of abscesses secondary to bacteremic episodes: these may be solitary or multiple, and vary in size hugely. Multiple granulomas were observed by ultrasound in the livers of all camelids with tuberculosis caused by *Mycobacterium microti* in one report and in another case report describing the clinical findings in an alpaca with *Mycobacterium kansasii* infection.[3,4] Abscesses less than a few millimeters diameter are unlikely to be seen, or at least to be differentiated clearly as abscesses. The appearance of abscesses varies, but they are typically of mixed echogenicity and more chronic abscesses will have a capsule. Findings of hyperfibrinogenemia and/or hyperglobulinemia on blood work may be compatible with the presence of abscesses within the liver or anywhere else. It is worthwhile to spend time moving the ultrasound probe dorsally and ventrally as well as cranially and caudally in order to evaluate the liver as fully as possible. The liver is also a good location for finding neoplastic lesions. The most common tumour in camelids is lymphoma or lymphosarcoma and this may result in multifocal circular lesions, or diffuse enlargement of the liver. Aspiration of liver cells or liver biopsy is easily guided by the use of ultrasound. Directly medial to and ventral to the liver is the third stomach compartment (C3). In the camelid with a normally positioned and sized liver, the wall of C3 next to the liver in the 2nd to last rib space is a good site to inspect for the presence of ulcers. This site is a common site for ulcers to develop and these may be suspected by the finding of a thickened, edematous wall.

Urinary Tract

The kidneys in llamas and alpacas are ovoid and nonlobulated, each measuring approx 5 x 9 cm.[5] The left kidney is found immediately caudal to the spleen in the dorsal left paralumbar fossa beneath the transverse processes of the 5th-7th lumbar vertebrae while the right kidney is slight more cranial and observed through the right paralumbar fossa. The renal pelvis appears hyperechoic while the medulla is anechoic/hypoechoic and the cortex contrasts with this being more echogenic. Loss of corticomedullary contrast may reflect dehydration or renal disease. The renal cortex should appear homogeneous. Occasionally anechoic renal cysts may be observed but these are usually incidental findings. However, in some cases, renal cysts may be very large (normally a congenital abnormality) and necessitate nephrectomy. The renal artery and vein may be seen medially and the aorta and caudal vena cava may be seen medial to this, particularly in neonates. Lymphoma affecting the kidneys may appear as solitary or multiple circular lesions. Unilateral and bilateral renal agenesis has been reported in camelids[6,7] as well as other urinary tract abnormalities. Ureters may be distended due to ectopia or obstruction but are not usually easy to find. Pyelonephritis is not common in camelids. Bladder distension may occur in males secondary to urethral obstruction. The bladder is normally found by placing the probe in the caudal/most ventral abdomen, directly in front of the pelvis, or mammary gland in females.

Spleen

The spleen is located in the dorsal left paralumbar fossa, immediately caudal to the first stomach compartment (C1). It usually appears triangular and has homogeneous echogenicity, containing many small anechoic blood vessels. It may be enlarged in camelids having severe or chronic anemia, as the spleen may be a site for extramedullary hematopoiesis. Splenomegaly may also occur during infections or secondary to congestive heart failure or liver disease.

Gastrointestinal Tract

The normal ultrasonographic appearance of the gastrointestinal tract has been described.[1] Knowledge of the normal appearance allows the clinician to determine abnormalities of significance, especially in the case of colic presentation, so that the most appropriate treatment decisions can be made. C1 fills most of the left side of the abdomen and may extend into the right side ventrally: dorsally, a curved hyperechoic line caudal to the last rib marks the caudodorsal extent. The cranial and caudal saccular regions may be identified ventrally by visualisation of series of semi-circular shaped hyperechoic saccules measuring 1-2cm diameter.[1] Normally the presence of gas prevents examination of the contents of C1 by ultrasound: if fluid is seen in C1, this may reflect accumulation of fluid in the forestomachs as a result of upper GI obstruction or ileus. In neonates, this occurs more often in bottle-fed crias if overfeeding has occurred. In these cases, milk mixed with saliva is clearly seen swirling around in C1. This is an abnormal finding since milk should be transported directly through to C3 in the esophageal groove. C3 is best seen on the right side ventrolaterally, caudal to the last rib. It may be difficult to see the contents of C3 due to the presence of gas although more easily than for C1. Thickening or edema of the walls of C3 may reflect ulceration. The best site to evaluate for the presence of ulcers is just medial or ventral to the liver. The small and large intestines will be seen mostly on the right of midline, from ventrum to paralumbar fossa: however, during pregnancy they will be pushed over to the left side. When observing intestines by ultrasound, it is normal to see regular movement due to peristaltic contractions and occasional fluid-filled segments will be observed. Small intestine and spiral colon segments are normally not more than about 1.5 cm in diameter but the proximal loop of the ascending colon can measure up to 7cm diameter and is likely to have more hyperechoic content than that of the small intestines.[1] Fluid distension of bowel loops with increasing length and number of observable fluid-filled segments may indicate enteritis, ileus or obstruction. In enteritis, there is likely to be intestinal activity observed, with auscultable sounds and fecal material present in the rectum whereas obstructions or ileus are more likely to have atonic fluid-distended bowel, and lack of intestinal mixing sounds or feces. Thickening of the walls of the intestines may occur in chronic inflammatory bowel disease, neoplasia, Johne's disease and enteritis. Changes in wall thickness are usually subtle. The one caveat with ultrasonography of the gastrointestinal tract is that some impactions may be difficult to visualise due to the presence of gas or fibre, for example large trichobezoars in C3. To visualise these, radiography may be required.

Free Peritoneal Fluid

Increased peritoneal fluid may occur in ascites, uroperitoneum, peritonitis or with strangulating obstructions. Collection of this fluid allows cytological differentiation. Camelids rarely have much free peritoneal fluid and blind ventral or paramedian taps are often not successful. An ultrasound-guided, paracostal approach on the right side is recommended.[1] *Streptococcus equi* subsp *zooepidemicus* has been associated with polyserositis in alpacas and llamas.[8-10] Clinical cases tend to have accumulation of large volume of free peritoneal fluid and sometimes thickened pleura evident on ultrasound: early on in the disease process there may be no increase in the volume of peritoneal fluid, although abdominocentesis reveals increased protein concentration and high white cell count. Cytology and culture of this fluid may be diagnostic.

Umbilical Structures

The umbilical structures can be evaluated in young crias where concern arises about umbilical infections. In the author's experience, umbilical abscessation is not nearly as common as it is in calves. Crias are more likely to develop sepsis before this and treatment of sepsis is likely to prevent abscesses from occurring. In normal crias, the umbilical vein (cranial to the umbilical stump), urachus and umbilical arteries (caudal to the umbilical stump) should shrink over the first week of life such that they will be hard to identify. Enlargement may indicate infection or inflammation and structures can be followed cranially or caudally to determine the extent of infection. Umbilical abscessation is more likely once crias are 2 weeks old or more.

THORACIC ULTRASONOGRAPHY

Ultrasonographic evaluation of the thorax is limited by anatomical considerations (bones of the ribs, vertebrae and sternum) and acoustics since ultrasound is reflected by air. Microconvex probes are likely to be superior in camelids as they will fit between the ribs more easily. Thoracic ultrasonography is most useful for detection of pleural disease and evaluation of areas of collapsed lung or superficial lung lesions/masses. The latter will often have been identified/suspected on radiographs and subsequent ultrasonographic evaluation allows visual assessment of the mass if next to the pleura. Ultrasound-guided aspiration of the mass will then allow cytology ± culture where applicable. Pleural effusion can be evaluated by ultrasound which then may be used to guide thoracocentesis. Normally the pleural membranes are thin, appearing as thin echogenic layers, and the visceral and parietal layers should glide smoothly over each other with each inspiration and expiration. Movement of the probe should be slow enough to visualise these movements during each respiration. In pleuritis, the pleural membranes may appear thickened and characteristic "comet tails" may be seen – these are reverberation artefacts due to the roughened surface. Adhesions between the pleural membranes will prevent movement of the layers against each other and reflects more chronic disease. Camelids affected with *Streptococcus equi* subsp *zooepidemicus* may develop pleuritis in addition to peritonitis and thickened pleural membranes have been identified in some cases where performed.[8,10]

REFERENCES

[1]Cebra CK, et al. **Vet Radiol Ultrasound** 2002;43:359. [2]Hamir AN, et al. **Vet Pathol** 2002;39:592. [3]Zanolari P, et al. **J Vet Intern Med** 2009;23:1266. [4]Braun U, et al. **Schweiz Arch Tierheilkd** 2009;151:287. [5]Fowler M. In: **Medicine & Surgery of South American Camelids**, 2nd ed. 1998:430. [6]Hardefeldt LY, et al. **Aust Vet J** 2007;85:185. [7]Poulsen KP, et al. **Can Vet J** 2006;47:159. [8]Cebra CK, et al. **Am J Vet Res** 2000;61:1525. [9]Hewson J, Cebra CK. **Can Vet J** 2001;42:465. [10]Jones M, et al. **J Vet Intern Med** 2009;23:220.

BIOSECURITY IN CAMELID OPERATIONS & QUALITY ASSURANCE IN THE UK

Claire E. Whitehead, BVM&S, MS, DACVIM, MRCVS
London, United Kingdom

INTRODUCTION

Recently the emergence of several infectious diseases in llamas and alpacas, including Bovine Viral Diarrhea (BVD) Virus, coronavirus and tuberculosis, have caused an increase in the level of concern among owners and breeders and their veterinarians regarding the importance of biosecurity practiced on camelid operations. For many of these diseases, the pathogenesis and clinical effects in these species may be largely unknown or extrapolated from other species, and the potential for these species to act as carriers of disease remains questionable due to the lack of expensive experimental infection studies. While these headline diseases have raised the level of awareness, there are many other more common diseases that may have just as serious an effect on their hosts (e.g. gastrointestinal parasites, mange) that could be easily controlled by the use of appropriate biosecurity arrangements. Many camelid owners and breeders have very little prior livestock experience so that some fairly basic on-farm biosecurity measures may be lacking and education, particularly from their local veterinarians, is key in the approach towards helping protect these animals from the spread of disease. Instituting adequate biosecurity for the camelid industry may mean changing some existing husbandry practices: however the benefits of avoiding contact with disease and the financial and emotional costs that may ensue should far outweigh any inconveniences faced.

The aim of "biosecurity" is to prevent the introduction and spread of disease in susceptible populations or herds of animals.[1] This concerns mainly viruses, bacteria, protozoa and parasites. Table 1 shows a non-exhaustive list of potential diseases that have been diagnosed in llamas or alpacas and organisms that may be transmitted from animal to animal for which appropriate biosecurity arrangements may be able to minimise or prevent transmission.

Table 1. Potential diseases and organisms that may be transmitted from animal to animal, and their potential sources for camelids.

Disease-causing Group	Examples	Source
Gastrointestinal parasites	Nematodes, trematodes, cestodes	Other camelids, sheep cattle
Protozoa	Coccidia *Giardia, Cryptosporidium*	Other camelids Other camelids, livestock
Ectoparasites	*Sarcoptes, Chorioptes, Psoroptes* mites	Other camelids, foxes?
Bacteria	*Strep equi* subsp *zooepidemicus* *Salmonella*	Horses Horses, cattle, other camelids, poultry
Mycobacteria	*M. bovis* (TB) *M. avium* subsp *paratuberculosis* (Johne's Disease)	Cattle, badgers (UK), infected camelids Cattle, infected camelids
Viruses	Llama adenovirus, coronavirus, respiratory viruses, West Nile Virus, equine herpesvirus-1 (EHV-1), BVD, bluetongue, Foot and Mouth Disease (FMD)?	Other camelids, cattle, horses, sheep, pigs, birds…

For some of the viral diseases, BVD and bluetongue in particular, the pathogenesis of disease in camelids and the effects of these viruses in terms of causing clinical disease in our species is largely unknown. Despite the fact that the Office International des Épizooties (OIE) lists South American Camelids as being susceptible to FMD, there is no clear documented evidence of South American Camelids showing clinical disease although they have been shown to produce antibodies when directly exposed in experimental studies.[2] Further research is needed for many of these diseases, but in the meantime, it would be prudent to include them in biosecurity considerations to try and minimise their potential effects.

Essentially, biosecurity means protecting existing herds from exposure to disease that may be brought in by new animals, and minimizing the risk of the spread of disease amongst newly introduced animals, while also minimizing exposure to potential pathogens (disease-causing agents) that already exist on the farm. It is important to understand that animals may be "clinically normal" and showing no signs of disease while actually either incubating disease or being subclinically affected – for example, shedding coccidia although not having diarrhea. Therefore, it is not adequate merely to exclude animals that appear to be unhealthy as those that appear healthy (and may have passed veterinary health examinations) may still introduce disease onto a farm. A veterinary health certificate only shows that the animal appeared to be healthy *at the time of the examination* and does not certify the animal as free from disease. It is also important to realize that the stress of moving animals from one farm to another, attending shows, changes in environmental conditions and social status can cause immune system suppression such that an animal that was previously healthy may be more susceptible to development of disease or may start to excrete the virus, bacterium or parasite. For example, it is possible for an animal to have a negative fecal evaluation for parasites prior to transportation, and then several weeks later to find that the same animal has a clinically significant parasite burden. Commingling of animals at shows or while visiting breeding farms increases the potential exposure of animals to disease while also causing a certain degree of stress-induced immunosuppression. For comparison, imagine a night spent on an aircraft travelling back from holiday – the effects of sleep deprivation (a stressor) suppress your immune system such that your body is less able to prevent the development of a cold from exposure to a cold virus breathed out by someone on the plane nearby.

BIOSECURITY RECOMMENDATIONS

The following practices are suggested for biosecurity on alpaca farms.

1. Avoid introducing any animals that do not appear healthy to the rest of the herd.
2. Know the disease status of any herd from which animals are being bought, or where animals are being temporarily housed. It is important to ask questions about current health status on the farm, but also previous problems with diseases such as gastrointestinal problems (diarrhea, parasites…), respiratory disease, abortion, ill-thrift or failure of

passive transfer. Knowing that a particular farm had a problem with diarrhea for example is not necessarily a problem: knowing that the problem was identified, diagnosed and correct treatment and prevention measures employed should be reassuring and also shows that the farm itself has good health control practices and potentially good biosecurity as well!

3. Insist on the pre-movement testing and treatment of animals prior to their arrival. This is to show that they are at least as healthy as the farm's own animals, assuming that optimum standards of health screening and disease prevention are employed on the existing farm. There is little point in demanding testing of incoming animals unless the animals pre-existing in the home herd are healthy! Testing might involve blood testing for BVD antigen (not antibody), and fecal testing. *Appropriate* treatment for endoparasites should be performed prior to arrival if required, and animals retested if concern exists regarding parasite resistance to helminthicides or anticoccidials. No farm without an obvious mange problem should accept a new animal with active lesions.
4. Ideally all new animals brought to a farm should be isolated and quarantined before introducing them to the herd. This includes animals that have been to shows or off-farm for breeding purposes since they have been commingling with other animals. If solitary animals are being introduced, it may be advisable to place herd animals in the quarantine area to reduce the separation stress: these animals are then treated as isolated like the animal just arriving on farm – potentially less valuable animals could be chosen for this purpose such as geldings, but avoid using breeding or lactating females.
    - Recommended quarantine period: 6 weeks. This would be the gold standard. Six weeks will allow any previously prepatent *Eimeria macusaniensis* to be detected in fecal tests. The prepatent period for *E macusaniensis* is approximately 34 days.[3] Other parasites and viruses are likely to be apparent on fecal tests or due to clinical disease within this time period.
    - Operate an all-in, all-out policy such that all animals housed in the quarantine area leave quarantine at the same time. The quarantine period only starts from the time the last animal entered. Do not have animals moving in and out of the quarantine area on an individual basis.
    - The location of the quarantine site should be physically separate from the main herd as well as its handling facilities and housing. Ideally this should be several hundred yards away if possible and positioned such that it is downwind from the main herd, but at the very least, there should be a double fence-line between the groups in order to prevent direct contact. This may be unrealistic for most situations but would be desired for a quarantine area for imported animals. For regular animal movements, a separate paddock area should be allocated and ideally separated from the home herd by a double fence providing at least 3m separation.
    - On a larger facility, have different staff working with the quarantine group and do not have the same staff return to the main herd. If this is not possible, tend to the main herd *before* tending to the quarantine group.
    - Wear clothing and footwear that is dedicated to the quarantine group when tending to this group – must be washable (e.g. rubber boots okay, suede or fabric work boots are not). Color coding of overalls is an obvious and visible reminder.
    - Animals in quarantine should be monitored on a daily basis. Monitor attitude, appetite, water consumption, urination and defecation as well as for any signs of coughing, discharge from the eyes or nose, diarrhea etc. Animals showing any signs of disease should be further separated from the rest of the animals in quarantine and evaluated by a veterinarian.
    - Evaluate a fecal sample on entry to the facility and 4 weeks later. Ideally fecal samples should be collected every 3 days and stored in the fridge.[4] If there is a subsequent problem with parasite-related illness or excretion then the source of the infestation can be determined. Treatment should be done if animals are found to have significant parasite burdens and their feces retested 14-21 days later.
5. Minimise exposure of the herd to human visitors. If visitors come to the farm, they should be questioned about exposure to other livestock and where they have come from and make sure that they come to the farm wearing clean clothes and boots. Provide visitors with overalls and either rubber boots or overboots kept on the farm for this purpose. Make sure that hand-washing facilities are available and that they are used.
6. Maintain optimum condition of animals in the herd. This will make them less susceptible to disease.
    - Monitor and record body condition on a monthly basis – over *and* under conditioning can cause problems.
    - Ensure optimal nutrition in terms of the feed and minerals that are provided. Mineral deficiencies can be responsible for disease problems in herds. Also, ensure that young growing stock receive sufficient vitamin D to prevent rickets.
    - Provide unlimited access to fresh clean water.
    - Provide protection from the elements – shade in summer and shelter from rain, wind, cold and snow in the winter. Appropriate bedding (e.g. straw) may also be necessary for protection from cold.
7. General cleaning and disinfection.
    - Personal hygiene is vital to prevent transmission of pathogens between animals and from animals to humans (zoonotic disease).
        - Frequent hand washing with hot water and soap.
        - Cleaning and disinfection of boots with appropriate disinfectants.
        - Thorough washing of clothing used when caring for animals.
    - Disinfection requires surfaces to be clean. All organic matter (feces, urine, milk, spit) must first be removed before disinfection can work on fences, floors, walls, clothing or hands. Disinfection may be difficult for

certain organisms such as Cryptosporidia (which can also cause disease in humans) so is best avoided in the first place by following optimal biosecurity on the farm.

QUALITY ASSURANCE SCHEME IN THE UNITED KINGDOM

There are currently approximately 25,000 registered alpacas and an estimated 5,000 llamas in the UK. 85% of owners have fewer than 20 animals on their holding and most keep them mainly for non-commercial reasons. However, regardless of herd size, most owners move their animals about the country for breeding, showing and agistment and there is no regulation of the movement of llamas and alpacas unlike the situation in other livestock industries in the UK. Recent problems with the emergence of BVD and tuberculosis in parts of the UK have led to owners becoming extremely mistrustful of others at shows and when buying or selling animals. The lack of a worthwhile antemortem screening tool for camelids amplifies this concern when herds that have previously had disease are declared "free of disease" on the basis of intradermal skin testing, the current global standard for other livestock. In addition to these biosecurity concerns are concerns about the level of knowledge among some owners of some very basic husbandry. Therefore, it was thought that optimization and standardization in the three key areas of health, welfare and biosecurity would be welcomed by the industry leading to improved education and consumer confidence for those showing, buying or selling, breeding or agisting animals. These are the objectives of the proposed Quality Assurance Scheme (QAS).

Initially a number of breeders were consulted regarding their concerns, especially those who had experienced problems. The membership of the British Alpaca Society was invited to complete an online questionnaire in order to assess current health and husbandry practices, owner attitudes and also their receptivity to such a scheme. Administration routes needed to be researched and a certification body identified. A workshop was organised with veterinarians working with camelids in order to discuss how a QAS could be run and what factors were regarded as important for inclusion in any inspection criteria that would allow membership of the scheme.

There were some interesting findings from the online questionnaire. There were 240 responses out of a possible approximately 1000 owner/breeders. Most owner/breeders thought that there were problems with biosecurity within the alpaca industry in the UK – although nearly three quarters of respondents believed that biosecurity on their own holdings was adequate. Most owner/breeders were practicing some form of precautions when bringing new animals into their herd which was encouraging, although the specific precautions being taken varied hugely. Less than half of owner/breeders taking their animals to shows were quarantining them on return to the farm. And over one-third of owner/breeders responding to the survey took animals away to other farms for "drive-by" breeding(s) or received "drive-by" visitors for breeding(s) on their own farms. Approximately 50 respondents reported fence-line contact with other species on neighbouring farms and these were mostly sheep or cattle and some horses while approximately half had other species on their own farm, including sheep, cattle, pigs, goats and llamas (in order of decreasing importance). These species often shared the same pasture. Fortunately most respondents had some spare pasture not being grazed at all. Responses to questions about parasite control suggested that parasite control strategies were not often especially tailored to their own premises. Very few farms had regular veterinary visits and many expressed concern at the level of knowledge about camelids. Nearly all respondents reported having had problems with mange, while some had had TB, and very few had had problems with either BVD or Johne's disease. On the whole, most owner/breeders were in favour of the establishment of a QAS, although they were concerned about costs, and that it shouldn't introduce unnecessary bureaucracy that burdens other livestock keepers in the UK. They were especially keen that it be a worthwhile tool used in combination with education to enable higher welfare and biosecurity standards nationally and to help control or eradicate problems.

It was decided that different tiers of scheme membership could be established, in order not to exclude the smaller breeder who wouldn't necessarily want to go to great added expenses associated with quarantine areas etc. In this way, an entry level standard would be available as well as a "premier" standard. In addition, layers could be added later on with specific disease testing showing herd freedom from certain diseases such as BVD. Some requirements for the scheme would be enforced only for the "premier" level members, and these include farm borders maintaining at least 3m separation from any neighbouring livestock as well as a 6 week quarantine period. Some requirements would be essential in both entry and premier levels and these included: microchip identification (for movement tracing), accurate movement records, and recording of deaths along with cause of death and number of animals not necropsied (and why). For many areas of health and husbandry it was decided that inspections should include evaluation of standard operating procedures for certain things in order to show that the owner or breeder is aware of a particular recommendation and shows that planning has taken place appropriate to that specific farm: for example, parasite control strategy, vaccination program, body condition scoring and nutrition. The farm inspection could include a spot-check of body condition to ensure that the owners are adequately skilled in performing this husbandry task.

One area that caused a great deal of discussion among the vets attending the workshop was the issue of drive-by breeding(s). All agreed that this was a potentially hazardous practice with respect to disease transmission, but it seemed likely that trying to police this under a QAS was likely to doom it to failure. Therefore a compromise was reached under which the vets were given a list of all potential practices relating to breeding, and they assigned a risk score (0-5) to each activity relating to the risk of transmission of TB and also for general disease transmission. The consensus score could then be provided to owners and breeders so that they would be able to see what practices constituted relatively most risk and make their own decisions accordingly. Under the scheme, owners/breeders would be required to show what activities they practice so that others may make decisions on the basis of which farms they would like to come into contact with.

This scheme is currently only in the planning stages, and it is hoped that funding will be obtained to make it a reality in the near future.

REFERENCES

[1]Barrington GM, et al. **Small Rumin Res** 2006;61:217. [2]Wernery U, et al. **Vet J** 2004;168:134. [3]Foreyt WJ. **Vet Parasitology Ref Manual**; 2001. [4]Cebra C. International Camelid Conference, 2007.

DIAGNOSIS AND EPIDEMIOLOGY OF HEMOPARASITES OF CAMELIDS

Susan Tornquist, DVM, PhD, DACVP
Corvallis, OR

INTRODUCTION

Organisms that have been reported to affect blood cells of New World camelids include '*Candidatus* Mycoplasma haemolamae', commonly known as *Mycoplasma haemolamae* and *Anaplasma phagocytophilum*, the agent of granulocytic ehrlichiosis.

REVIEW OF *M. HAEMOLAMAE*

Organism Characteristics

M. haemolamae is a hemotropic mycoplasma that attaches to red blood cell membranes of llamas and alpacas. The organism was first described in 1990 in llamas in two different geographical locations and reported to be an *Eperythrozoon*-like organism based on light and scanning microscope morphology. In 2002, it was subsequently re-classified as a mycoplasma based on the 16S rRNA sequence which was most similar to *Mycoplasma haemosuis* and *Mycoplasma wenyonii,* mycoplasmas that infect the erythrocytes of swine and cattle respectively. The new name '*Candidatus* Mycoplasma haemolamae' was proposed for this organism based on morphology, genomic and structural characteristics. The organism has not yet been cultured, precluding an official species designation, but it is often referred to as *M. haemolamae*.

M. haemolamae organisms are small (0.4-0.6 micron) bacteria that lack a cell wall. They may appear as coccoid, linear or ring-shaped and most often appear to be on the edge of the erythrocyte or free in the background, especially if blood smears aren't made soon after blood is collected. The organisms do not actually enter the erythrocyte, though they may cause a depression in the cell membrane.

Clinical Signs

Llamas and alpacas infected with *M. haemolamae* may have mild to severe anemia that variably appears to be regenerative. Lethargy, depression and fever are sometimes present in infected camelids. Deaths have occurred in heavily infected camelids. However, it appears that most infected camelids do not show any clinical nor laboratory abnormalities associated with the infection. In many cases, clinical disease is seen when animals are immune-suppressed, stressed, or concurrently infected with other organisms, such as gastrointestinal parasites.

Diagnosis and Experimental Infections

In 2001, a polymerase chain reaction (PCR) - based assay was developed to provide a more sensitive and specific diagnostic method than the previously used blood smear examination. The assay was used in experimental infections and showed that many camelids are subclinical chronic carriers. Some of the experimentally-infected camelids were transiently anemic, while others were not and they did not develop fever, depression, nor hypoglycemia. Using the PCR assay, multiple experimental infections in camelids were performed to study the efficacy of treatment regimes on infection. Healthy adult llamas and alpacas nearly always became infected when transfused with blood from a chronically-infected alpaca.

Treatment of experimentally-infected camelids with a number of drugs including oxytetracycline, florfenicol, enrofloxacin did not effectively clear infection, as shown by detection of the organism following immune suppression of animals who had been previously treated and thought to be negative. Overall, infection and treatment studies showed that most infected camelids are chronic carriers though organisms may most often be at undetectable levels unless the animals are immune-suppressed.

Transmission

Transmission by biting insect vectors and vertical transmission are suspected but not proven. A few efforts to amplify *M. haemolamae* from biting insects found on infected camelids have not resulted in positive amplification.

Transplacental transmission has been hypothesized and is supported by the finding of organisms in blood of crias as young as 24 hours old.

A recent report showed that crias born to dams that are positive for infection at the time of birth may be negative by PCR assay both prior to and after nursing on the infected dam. Other crias born to infected dams are positive before and after ingestion of colostrum and remain positive up to 6.5 months. *M. haemolamae* has not been amplified from colostrum from dams that tested positive at the time of parturition.

Prevalence

Using the PCR assay to test 6,466 llama and alpaca blood samples from all of the United States, as well as Canada, Australia and the U.K., the prevalence rate appears to be approximately 30%. An additional prospective study of apparently healthy herds in the Pacific Northwest included 534 llamas and alpacas. Results of these retrospective and prospective studies including demographic and seasonal occurrence will be published.

Samples from smaller numbers of llamas and alpacas in selected herds in Peru and Chile have shown a lower prevalence i.e. between 10 to 20% positive. No association with anemia or clinical signs was found in these animals.

In a comprehensive review of anemia in camelids, 3,662 complete blood counts were reviewed. The PCR assay for *M. haemolamae* was performed on 941 of these camelids. There was a significantly higher percentage of anemic animals that were positive for *M. haemolamae* compared to the anemic animals that were negative.

REVIEW OF EHRLICHIOSIS IN CAMELIDS

Ehrlichiosis in llamas and alpacas has been infrequently reported but is known to occur on the West coast as well as New England. The agent identified is now classified as *Anaplasma phagocytophilum*, which causes granulocytic ehrlichiosis in a wide range of mammalian host species and appears to have world-wide distribution. It is carried and transmitted by *Ixodes* ticks.

In the two case reports of this organism causing clinical signs in camelids, the diagnosis was initially made on detection of morulae in neutrophils and eosinophils in a blood smear, and later confirmed by molecular methods. Both camelids presented for non-specific signs that included partial anorexia and lethargy with slight to progressive ataxia also noted. Other laboratory findings were also non-specific and did not include the anemia, leukopenia and thrombocytopenia that may be seen in acute ehrlichial infection. In one of the cases, there was a concurrent infection with *M. haemolamae*.

A. phagocytophilum organisms may be seen in leukocytes on blood smears for only a brief time period during the infection. Because the clinical signs are non-specific and the associated hematologic findings are inconsistent, it is likely that there are infections in camelids that go undiagnosed and untreated. Oxytetracycline is the most common treatment for ehrlichiosis and was used successfully, along with supportive care, in both of the clinical cases reported.

REFERENCES

Barlough JE, et al. **J Clin Microbiol** 1997;35:1005. Lascola K, et al. **J Vet Intern Med** 2009;23:379. McLaughlin BD, et al. **J Am Vet Med Assoc** 1990; 197:1170. Messick JB, et al. **Intl J of Syst and Evolu Micro** 2002;52:693. Reagan WJ, et al. **Vet Pathol** 1990;27:426. Tornquist SJ, et al. (Abs) **Vet Path** 2002;39: 616. Tornquist SJ, et al. **Am J Vet Res** 2009;70:1102. Tornquist SJ, et al. **J Vet Diagn Invest** 2010;22:766.

PATHOPHYSIOLOGY AND DIAGNOSIS OF COMPLEX CONGENITAL HEART DISEASE IN NORTH AMERICAN CAMELIDS: BEYOND VSDS

Brian A Scansen DVM, MS, DACVIM (Cardiology)
Columbus, Ohio

INTRODUCTION

Congenital heart disease (CHD) is seen with considerable frequency in North American camelids, with an estimated incidence of 2.6% in the llama.[1] As such, it is important that practitioners who treat these species be familiar with the variety of lesions that may be present. While few forms of CHD are curable, some diseases are well-tolerated and the prognosis for others may be fair for production. An understanding of the potential diagnoses and the pathophysiology underlying their circulatory derangements will allow camelid veterinarians to inform clients and help make management recommendations. Within the context of CHD, the author will provide a review of common and rare CHD with case examples of camelids afflicted with these defects, focusing on the diagnosis and pathophysiology. Potential treatment options will be presented in brief.

LEFT-TO-RIGHT SHUNTING LESIONS

A shunt is defined as an abnormal mixing of blood in the heart or great vessels. The consequences of mixing between the left and right heart (shunting) vary depending on their magnitude and direction. In a left-to-right shunt (meaning from the systemic circulation to the pulmonary circulation) then the signs are principally of volume-overload to the left heart because the normal amount of blood flowing to the lungs is augmented by the amount added from the shunt before returning to the left heart. Conversely, right-to-left shunts (where blood is diverted from the venous system to the systemic arterial system) result in poor oxygenation and cyanosis.

Ventricular Septal Defects

Ventricular septal defects (VSD) are often the first form of CHD camelid producers think of when heart disease is suspected in a young animal. Although large-scale epidemiologic studies have not been performed in both llamas and alpacas, it does appear that VSDs are the most common CHD in llamas.[1,2] The pathophysiology of VSD is such that small defects (restrictive) may be well-tolerated and not result in overt signs of cardiac dysfunction. Larger defects, however, result in the shunting of blood from left ventricle to the pulmonary circulation. The result of this increased pulmonary flow is damage to the lung vasculature and volume overload to the left heart. VSDs can occur throughout the interventricular septum and numerous classification schemes have been developed to characterize the location of the defect. A slight modification of the classification scheme by Soto and colleagues[3] is preferred by the author. Briefly, the septum is divided into inlet, muscular/trabecular, outlet, and membranous sections. Perimembranous (also called membranous or infracristal) defects occur in the membranous septum, which is positioned within the left ventricle (LV) outflow tract just below the aortic valve, adjacent to the septal tricuspid leaflet and caudal to the supraventricular crest of the right ventricle. Muscular defects are surrounded by muscular septum and the trabeculations of the right heart and are positioned more apically than the membranous type. Doubly-committed juxtaarterial defects (also called subarterial, supracristal, subpulmonary, or infundibular) lie immediately below the cusps of the pulmonic valve on the right and below the aortic valve on the left. The diagnosis of a VSD is suspected from a loud systolic murmur on the right chest wall and confirmed with Doppler echocardiography. Exact confirmation of the size of the defect and secondary cardiac changes (e.g., aortic insufficiency or left ventricular enlargement) carry prognostic information for the client; additionally, some defects may be in a position amenable to catheter-based closure. Examples of each of these forms of VSD have been reported in camelids[4], with perimembranous seemingly the most common.[1]

Atrial Septal Defects

As with a VSD, an atrial septal defect (ASD) is a hole in the intracardiac septum separating the left and right heart. ASD is a rare defect in camelids, but is seen in association with other defects. The degree of shunting depends on the size of the defect and the relative resistance of the pulmonary versus the systemic circulation, typically from the left atrium across the ASD to the right thereby increasing blood flow through the lungs. With profound shunting, pulmonary hypertension and shunt reversal (Eisenmenger's physiology) may occur. There are three forms of ASD depending on their location in the atrial septum – ostium primum defects (in the ventral portion), ostium secundum defects (in the mid-atrial septum in the area of the fossa ovalis), and sinus venosus defects (high in dorsal atrial septum, adjacent to the cranial or caudal vena cava). Flow across the atrial septum may also develop if right atrial pressure becomes elevated, even in the presence of a normally formed atrial septum. This develops because the atrial septum is composed of two membranes that are closed during the first week of life as left atrial pressure exceeds right atrial pressure. If the patient has right-sided heart disease that elevates right atrial pressure (pulmonic stenosis, tricuspid dysplasia, tetralogy of Fallot), then the membranes may separate and allow flow across the atrial septum from right to left, referred to as a patent foramen ovale (PFO). In the setting of a large ASD or concurrent defects, presenting signs will be comparable to VSD – pulmonary overcirculation and predominately respiratory symptoms or weakness, lethargy, and cyanosis associated with pulmonary hypertension and shunt reversal. Echocardiography confirms the diagnosis and allows for evaluation of any concurrent defects. ASDs have been reported in camelids, most commonly as a component of an atrioventricular septal defect (see below), tricuspid atresia,[5] or other complex CHD.[4]

Atrioventricular Septal Defects

Defects of the atrioventricular septum (AVSD) occur in which the patient has an ASD above the tricuspid valve (a primum defect), a VSD just below the tricuspid valve (inlet membranous defect), and often malformations of the atrioventricular valves. These defects were formally referred to as endocardial cushion defects or defects in the atrioventricular canal. Signs vary depending on the relative size of each defect, but both congestive heart failure and cyanosis are common. There is a report of 3 llamas with this defect, of which 2 were related.[6] The author has also seen this defect in a 2 month old alpaca. Unfortunately, AVSD carries a poor prognosis, though survival of up to 2 years has been reported.[6] The diagnosis is made by demonstration of a primum ASD, a VSD in the dorsal septum, and a common atrioventricular valve. There are also partial forms of AVSD recognized, depending on the presence/absence of the ASD and VSD as well as the degree of development of distinct mitral and tricuspid annuli.[7]

Patent Ductus Arteriosus

The ductus arteriosus is a normal structure in the fetus derived from the left sixth aortic arch and allows for the diversion of blood away from the lungs in utero when they are not expanded by air. Within the first 12-24 hours after birth, the ductus begins to constrict in response to increased oxygen tension and is closed within the first week of life. If the ductus remains open, pulmonary vascular resistance will fall and blood will flow from the aorta to the pulmonary artery. Auscultation of a patient with PDA displays a continuous machinery heart murmur at the dorsal left heart base. In patients with pulmonary hypertension, or in very young patients with elevated pulmonary vascular resistance, only a systolic component may be apparent. The pulse is strong because the pulse-pressure (difference between systolic and diastolic pressure) is wide; this occurs because of diastolic run-off from the aorta across the ductus to the pulmonary artery resulting in a low diastolic aortic pressure. Chest films are suggestive and echocardiography confirms the ductus as turbulent flow entering the main pulmonary artery. The author has seen PDA in camelids most commonly when imaging neonates in the first 1-2 days of life (considered normal at that age). However, the author has also seen PDA in camelids in the setting of obstructive pulmonary outflow (tricuspid atresia, pulmonary atresia) and once as an isolated defect in a 7 week old alpaca cria. Definitive therapy is an option for animals with PDA – either via interventional closure or surgical ligation. Surgical ligation has been reported in a 3 month old alpaca cria with a right-sided PDA as an addendum to a report of the condition in dogs.[8]

Coronary Cameral Fistulae

The author has seen a case of fistulous communication between the right coronary artery and the right ventricle in a 6 day old llama with a continuous heart murmur. Such communications are very rare in animals, but do occur in humans as a congenital anomaly or after trauma.[9] The pathophysiology of this rare defect is similar to the other left-to-right shunts as arterial blood is diverted to the right heart (although these fistula may enter into any cardiac chamber), setting the stage for pulmonary overcirculation. In addition, coronary steal syndrome may develop in which the myocardium is deprived of coronary blood flow because of the shunt. The diagnosis in the llama cria was suspected by echocardiography and confirmed on postmortem examination.

OTHER MALFORMATIONS OF THE GREAT VESSELS

Coarctation of the Aorta

Aortic coarctation is a common affliction of children, but rarely seen in veterinary species. The coarctation is a ridge of tissue at the aortic isthmus – the junction of the arch and the descending aorta, right where the ductus/ligamentum attaches. Causes of coarctation are unclear, but it has been theorized to occur secondary to ectopic ductal tissue or abnormal preductal flow. If severe, the coarctation can cause poor perfusion to the descending aorta and hypertension in the cranial limbs. Left ventricular hypertrophy occurs and severe collateral vessels may develop. There are three forms of coarcation – a discrete ridge (coarctation), tubular hypoplasia (segmental narrowing of the aortic arch), and aortic interruption (considered the most extreme form of coarctation where the descending aorta is anatomically separate from the arch). The author has seen a postmortem specimen of a young camelid with tubular hypoplasia, with the descending aortic flow supplied through the ductus arteriosus.

Vascular Ring Anomalies

Several malformations of the arterial system have been described in animals; the most common clinical presentation is a pediatric animal with dysphagia. The basic pathophysiology of these malformations is an embryologic defect resulting from abnormal persistence or regression of fetal vascular structures (aortic arches), ultimately leading to an abnormal spatial arrangement of the major intrathoracic arteries relative to the trachea and esophagus. Clinical signs are the direct result of extraluminal compression of the esophagus (resulting in megaesophagus, regurgitation, or aspiration pneumonia) and/or trachea (resulting in inspiratory stridor, coughing, or dyspnea). There are several vascular rings which may be present in camelids, with reports of persistent right aortic arch with left ligamentum in a llama,[10] and a right ductus arteriosus with left aortic arch in an alpaca.[8] At least four other cases of vascular ring anomalies in camelids have been diagnosed.[4] The diagnosis may be confirmed with cross-sectional imaging (computed tomographic angiography), non-selective or selective angiographic studies, or surgical exploration. Therapy is directed at surgical decompression and alleviation of the offending structure, though esophageal dysfunction may persist if not corrected at a young age.

CYANOTIC HEART DISEASE

Cyanotic heart disease is not uncommon in camelids; in the author's experience, it is seen more in these veterinary species than any other. Crias typically present in the first weeks of life with failure to thrive, dyspnea, and cyanotic membranes. A useful tool to recall the common cyanotic heart defects can be shown with the use of one's hand – called the 5 'T' mnemonic. Consider the thumb to represent a truncus arteriosus, the index and middle finger intertwined to represent transposition of the great arteries, 3 fingers represent tricuspid atresia, 4 represent the tetralogy of Fallot, and 5 digits represent each letter of total anomalous pulmonary venous connection. These 5 'Ts' of cyanotic CHD can then be easily considered in a differential for a cria with cyanosis and suspected CHD.

Truncus Arteriosus Communis

Truncus arteriosus communis, or common arterial trunk, is a congenital abnormality in which only one great vessel arises from the heart. It develops when there is a lack of septation within the embryologic truncus arteriosus and therefore no delineation between aorta and pulmonary artery. The left and right pulmonary arteries arise off the trunk at variable locations, a VSD is common, and coronary artery anomalies are also seen. There are few reports of this defect, though the author has seen one case in a young camelid and there is a report of this anomaly in a guanaco.[11] The diagnosis is confirmed by demonstrating the origin of the pulmonary arteries from the trunk – by echocardiography, cross-sectional imaging, or postmortem. In addition to truncus arteriosus communis, the author recently diagnosed a case of what has been termed "hemitruncus" in a 7 day alpaca cria, where the right pulmonary artery arose from the aortic arch, and the left pulmonary artery was intact arising from a small main pulmonary artery.

Transposition of the Great Arteries

Transposition of the great arteries, also called ventriculoarterial discordance, refers to a condition in which the aorta arises from the right ventricle and the pulmonary artery arises from the left ventricle. As can be seen, the pathophysiology of this defect is the development of two parallel circulations and is incompatible with post-fetal life unless a mixing pathway is present such as a VSD, ASD, or PDA. This defect has been observed in a 4 day old alpaca.[4]

Tricuspid Atresia

Tricuspid atresia refers to a CHD in which the right atrioventricular valve fails to develop. Systemic venous return, therefore, has no exit from the right atrium unless the atrial septum is perforate and these animals therefore typically have an ASD or a PFO with right-to-left shunting of all venous return. Pulmonary blood flow, therefore, requires a ductus arteriosus or a VSD. The author has seen this condition in a 6 day old alpaca and it has also been reported in an 8 day old female alpaca.[5]

Tetralogy of Fallot

Fallot's tetralogy is the most common cyanotic CHD of humans and is also seen in veterinary species. The four components of the tetralogy are right ventricular outflow obstruction, right ventricular hypertrophy, VSD, and dextroposition of the aorta. The underlying pathophysiology of this defect is believed to result from incomplete rotation and faulty partitioning of the conotruncus (comprising the distal portions of the outflow tracts and the proximal portion of the great vessels). Malrotation of the truncal-bulbar ridges results in misalignment of the outlet and trabecular septum and consequent straddling of the aorta over the malaligned VSD. To put this simply, the aorta wins at the expense of the pulmonary artery and the ventricular septum cannot close because the aorta is deviated to the right. The result of right ventricular outflow obstruction with a VSD is that de-oxygenated blood is diverted to the systemic circulation causing cyanosis, weakness, and collapse. Longstanding sequelae include polycythemia and hyperviscosity due to increased erythropoiesis. There are palliative surgeries that may be considered, but these have not been pursued in camelids. There are few reports of this condition in camelids in the literature,[4] though the author has seen 3 alpaca crias with this affliction.

Pulmonary Atresia

Pulmonary atresia, the condition formerly referred to as pseudotruncus, should be considered as a severe form of the tetralogy of Fallot. Indeed, in many human medical texts the two are considered the same defect and characterized as either Tet-PS or Tet-PA depending upon the presence of stenosis or atresia of the pulmonary valve, respectfully. The pathophysiology of pulmonary atresia can be considered comparable to Tet-PS with severe right-to-left shunting across the VSD, though in Tet-PA the pulmonary flow is entirely dependent on the presence of a patent arterial duct. As such, these animals typically become cyanotic and present around day of life 3 to 10 as this is the natural time for ductal closure to occur. The author has seen severe cases of pulmonary atresia in crias, and it has been described in a recent review of camelid cardiology.[4]

Total Anomalous Pulmonary Venous Connection

Total anomalous pulmonary venous connection is listed here as the 5th differential for cyanotic heart disease, given the above mnemonic. In this condition, the pulmonary veins drain into the systemic venous circulation – typically at an atrial level, though occasionally into vena cava. To the author's knowledge, this defect has not yet been reported in camelids.

SUMMARY

In summary, numerous forms of CHD are seen in young camelids. Given the wide differential diagnosis, knowledge of these lesions is useful to inform management decisions, to begin to search for trends and genetic factors, and to potentially serve as a model for further study of these conditions in human CHD.

REFERENCES

[1] Boon JA, et al. **Vet Clin North Am Food Anim Pract** 1994;10:353. [2] Leipold HW, et al. **Vet Clin North Am Food Anim Pract 1994;**10:401. [3] Soto B, et al. **Br Heart J** 1980;43:332. [4] Margiocco ML, et al. **Vet Clin North Am Food Anim Pract** 2009;25:423. [5] Slack J, et al. **Vet Radiol Ultrasound** 2008;49:309. [6] Cebra ML, et al. **J Zoo Wildl Med** 1998;29:225. [7] Rudolph AM. **Congenital Diseases of the Heart.** Wiley-Blackwell, West Sussex, UK, ed. 3rd, 2009, pp. 203-224. [8] Holt D, et al. **Vet Surg** 2000;29:264. [9] Bruckheimer E, et al. **Catheter. Cardiovasc Interv** 2010;**75**:850. [10] Butt TD, et al. **Vet Rec** 2001;148:118. [11] Michaelsson M, et al. **Congenital Heart Malformations in Mammals: An Illustrated Text.** Imperial College Press, London, UK, 2000, pp. 113-118.

RUMINANT CASE WORKSHOP & DISCUSSION

Kevin E. Washburn, DVM, DABVP, DACVIM
College Station, Texas

CASE 1

3 year-old bull presented for weight loss, chronic diarrhea of about one month's duration and some blood in the stool for the past week. The bull was on pasture with cows and had maintained a good appetite throughout. He had been dewormed one month prior to presentation. Vaccinations were current. The pasture was heavily wooded. No other animals affected. The bull was raised from a calf on the farm. Physical examination abnormalities included bloody, watery diarrhea containing black flecks of a hard material, oral and preputial erosions and a roughened, inflamed rectal mucosa upon palpation.

Any other questions?

Differentials?

BVDV, oak toxicosis, BTV, abomasal ulcer, enzootic lymphosarcoma, others?

Diagnostics?

BVDV ELISA, BLV serology, CBC/electrolytes, chemistry panel, fecal floatation exam, abdominal ultrasound, others?

Results

BVDV negative, BLV negative, fecal floatation exam negative, normal abdominal ultrasound, mildly decreased A/G ratio

Diagnosis

Treatment

CASE 2

3 year-old intact Boer goat presented for lameness of right rear limb of two weeks duration that was worsening. The referring veterinarian radiographed the right stifle and discovered effusion. He was placed on phenylbutazone with no response. Physical examination abnormalities included pain upon manipulation of the right stifle and right coxofemoral joints. The right stifle was palpably slightly larger than left but not warm to the touch. Some crepitus was appreciated over the right coxofemoral joint. All other findings were within normal limits.

Any other questions?

Differentials?

Trauma (fracture of the head of the femur, ruptured cruciate ligament, subluxation of coxofemoral joint), others?

Diagnostics

Radiographs of stifle and coxofemoral joints

Results

Effusion of right stifle, mild. Lucency of right caudal acetabulum, increased joint space and fracture of femoral neck. Suggestive of pathologic fracture from infection.

Further diagnostics: CT scan

Diagnosis

Treatment

CASE 3

5 year-old Brahman female presented for recumbency of 36 hours duration. She had calved 6 days prior to presentation. Two weeks prior to presentation, she was presented to the referring veterinarian for udder edema and vulvar edema. She was treated with isoflupredone and a diuretic (furosamide) for one week until she calved. The owner noted the edema went away completely following calving. Calving was uneventful. Three days after she calved, the cow was noted to be down and reluctant to stand. She progressively got weaker up to the point of presentation. She had continued to eat and drink normally throughout. Physical abnormalities included recumbency, moderate dehydration and her head and neck were flexed to the left toward her flank. Neurologic examination revealed depressed reflexes in all limbs and generalized decreased muscle tone and was interpreted as generalized paresis.

Any other questions?

Differentials?

Metabolic derangement (Ca, K, P), trauma, lymphosarcoma,

Diagnostics?

CBC, chemistry with electrolytes, cervical radiographs, BLV gp 51, others?

Results

Elevated CK, K of 2.5, metabolic alkalosis, cervical radiographs – no abnormalities, BLV negative,

Treatment

CASE 4

Three 3 to 4 year-old cross breed does presented for recumbency and ataxia with weakness for three days duration. These were 3 does from a group of 15 that were kept in a 20 acre pasture. This pasture was one of 4 similar sized pastures and the owner rotated these goats with two other groups of about 15 in each (always had one pasture open). All 3 of these were from the same pasture and none of the others in the other two pastures were affected. One other goat in the same pasture was left as she had become recumbent after a three day period of ataxia and weakness. Physical examination abnormalities were: two of the does were recumbent and one was ambulatory but had significant deficits. All three were bright, alert, eating, drinking and normal otherwise. Neurologic examination abnormalities were: the two down does had no cranial nerve deficits, however, one doe had markedly depressed withdrawal reflexes and patellar reflexes while the other had a depressed patellar reflex of the left rear limb. The video depicts the deficits in the ambulatory goat.

Lesion localization

Lower motor neuron disease with a lesion or lesions in the spinal cord, asymmetrical distribution suggesting possible multiple lesions.

Any other questions?

Differentials?

Delayed OP toxicity, trauma?, meningeal worm?, others?

Diagnostics

CSF fluid analysis

Results

eosinophilic pleocytosis

Diagnosis

Treatment.

CASE 5

2.5 year-old Angus bull presented for weight loss, diarrhea and anorexia of approximately 2 weeks duration. He was a show bull that had been weaned off full feed and had been receiving 14% protein ration and alfalfa hay. When the diarrhea started, he was placed on higher quality forage. Over the last few days he just picked at the hay and had become more lethargic. He drank well and no problems were noted with urination. No other animals were affected and he was from a well vaccinated herd. Physical examination abnormalities included a 2/9 BCS, moderate dehydration and fecal staining over the perineum and back legs.

Any other questions?

Differentials?

Parasitism, peritonitis, Johnes, ruminal acidosis, BVD, others?

Diagnostics

Abdominal ultrasound, CBC, chemistry panel/electrolytes, fecal with fluke finder, rumen fluid analysis, Johnes ELISA, others?

Results

Multiple large hypoechoic regions in the liver surrounded by hyperechoic zones suggestive of abscessation. Inflammatory leukogram with a low A/G ratio, hypoproteinemia and hypoalbuminemia, fecal and fluke finder negative, rumen fluid analysis normal, Johnes ELISA negative

Diagnosis

Treatment

CASE 6

2 year-old intact male Hampshire ram presented for breathing heavily and lack of urination over the last 24 hours. The ram had been anorexic for about the last 3 days and depressed. Upon physical examination, the patient was depressed and walked stiffly in the rear legs. He had bilateral serous nasal discharge with increased respiration rate and effort. Crackles were present in the right dorsal lung field. No signs of urination were observed.

Differentials?

Pneumonia, urolithiasis (partial or complete), ruptured urinary bladder

Diagnostics: Radiographs; serum chemistry, abdominal ultrasound, urinalysis

Results

Bronchopneumonia of ventral lung fields, azotemia, distended urinary bladder, occult blood in urine

Diagnosis

Treatment.

Ram developed red urine.

Further diagnostics

urine copper

Results

0.6 ppm copper (normal < 0.3)

Diagnosis part two

Treatment

CASE 7

15 month-old Angus bull presented for lethargy, partial anorexia and distended abdomen. The lethargy started about 1 month previously, partial anorexia about 2 weeks and abdominal distension about 10 days prior to presentation. The bull had been out on pasture with cows. No prior history of illness. Urination and defecation not noted by owner. Physical examination abnormalities included mildly pale mucous membranes, tachycardia, tachypnea and marked bilateral ventral abdominal distention. Rectal

examination revealed the rumen was small and collapsed, there seemed to be a lot of "space" and the caudal pole of the kidney was barely reachable. No feces were noted in the rectum. The bull did not respond to withers pinch and actually bristled up to the stimuli.
Any other questions?
Differentials?
Intestinal obstruction, uroabdomen, urolithiasis, hardware, others?
Diagnostics?
CBC/electrolytes, abdominal ultrasound, abdominocentesis, chemistry, rumen chloride, others?
Results
Azotemia, free fluid in the abdomen, hyponatremia, hypochloremia, normal rumen chloride, uroabdomen
Diagnosis
Treatment

CASE 8

5-year old Angus cow presented for stumbling/knuckling in both rear limbs and ataxia that was progressively becoming worse over the last week. She calved 3 weeks prior to presentation without assistance and they noticed some ataxia about ten day's post-partum. They thought she had lost weight since calving. She was reportedly febrile at the referring veterinarian where she had been treated with flunixin meglumine and florfenicol. Physical examination abnormalities included tachycardia, 103.5 F and generalized lymphadenopathy. Neurologic examination revealed buckling of the rear limbs and a rear limb ataxia with weakness. Tail and anal tone were normal. She had a normal appetite and drank water. Within 36 hours of presentation, she became recumbent.
Lesion localization?
L4 to L6
Any more questions?
Differentials?
Trauma, enzootic lymphosarcoma, others?
Diagnostics?
BLV gp 51 serology, BLV p24 antigen, peripheral lymph node aspirate, lumbar radiographs, others?
Results
BLV gp 51 and p24 antigen positive, lumbar radiographs – no notable abnormalities, cytology – predominately large lymphoblasts with multiple nucleoli and mitotic figures.
Diagnosis
Treatment

CASE 9

12 month-old Longhorn heifer clone presented for a 3 week history of diarrhea. The heifer had lost a considerable amount of weight. She was dewormed when the diarrhea started with ivermectin. She was eating 12% pelleted feed, alfalfa hay and free choice Bermuda grass hay. Her appetite had been normal throughout. Physical examination abnormalities included dehydration, fecal staining and matting over the hind legs and tail, a 3/9 BCS, decreased rumen contractions and a few small oral erosions.
Any other questions?
Differentials?
BVD, parasitism, vesicular stomatitis, Salmonella, Johnes?, chronic hardware, others?
Diagnostics?
BVDV ELISA, fecal, abdominal ultrasound, CBC, chemistry with electrolytes, Salmonella screening, others?
Results
BVDV negative, abdominal ultrasound within normal limits, hypoproteinemia and albuminemia, Salmonella negative, fecal = over 8,000 Strongyle eggs per gram
Diagnosis
Treatment

THE BEST PAPERS OF 2010

Peter Constable, BVSc(Hons), MS, PhD, DACVIM
West Lafayette, IN

THE OTHER BEST PAPERS OF 2010

Allen Roussel, DVM, MS, DACVIM
College Station, TX

Peer-reviewed food animal medicine articles published in 2010 were examined. English-language articles published in top tier veterinary journals and a smaller number of animal agricultural journals based on 5-year impact factor (ISI Web of Knowledge) and articles published in other journals based on recommendations from colleagues were considered. The title and abstract of eligible articles were reviewed and articles selected for further consideration. Ten evidenced-based articles that dealt with an important topic were subsequently selected for presentation and discussion. Criteria used for selection included diversity, excellence in study design and statistical analysis, and a perceived likelihood that the results of the study will impact the way we practice food animal medicine. The papers selected contained a clearly stated rationale and hypothesis, and a well described materials and methods section that will permit anyone, anywhere, to reproduce the study.

Study designs used in food animal medicine include descriptive studies (case reports, case series, surveys) and analytic studies (observational studies such as case-control, cross-sectional, and cohort studies; experimental studies such as laboratory studies and randomized clinical trials; Dohoo et al 2003). The 10 papers selected used an appropriate method of randomization (where appropriate) and had appropriate statistical power as well as internal validity and external validity (generalizability). In general terms, large multi-site studies have greater external validity and small controlled laboratory or experimental station studies have greater internal validity (Lean et al 2009). Although the ideal study has excellent internal and external validity, completion of the ideal study usually requires extensive resources which are usually unavailable. Guidelines for materials and methods section for randomized controlled trials were developed in 2010 (the REFLECT statement; O'Connor et al 2010) and should be adhered to whenever possible.

The intent of limiting the list to 10 studies was to focus attention and discussion on a small number of evidenced-based papers that are likely to become "game changers". A similar approach was used to select papers from 2008 and 2009 for discussion at the 2009 and 2010 ACVIM Forum, respectively (Constable 2009; Constable and Roussel 2010). The selected papers for discussion at the 2011 ACVIM Forum increase our understanding of the pathophysiology of disease or change the way we diagnose, treat, and control disease of food animals. Many excellent papers were not able to be listed because of the desire for diversity in species and organ system(s) affected, age of animal, causative agent, and geographic location. The strong emphasis on diversity meant that the 10 selected papers were based on 3 food animal species (cattle, sheep, goats) and were published in 7 different journals. The 5-year impact factor rank (and impact factor) for the Veterinary Sciences subject category was: 12 (2.26) for J Vet Intern Med; 19 (2.01) for J Am Vet Med Assoc; 35 (1.50) for Vet Rec; and 72 (0.81) for Berl Munch Tierarztl Wochenschr. The 5-year impact factor rank (and impact factor) for the Agriculture, Dairy and Animal subject category was: 1 (2.94) for J Dairy Sci; and 3 (2.76) for J Anim Sci. The 10 papers originated from 6 countries (Canada, Germany, Switzerland, Turkey, United Kingdom, USA) and are listed below in alphabetical order:

1. Braun U, Forster E, Steininger K, Irmer M, Gautschi A, Previtali M, Gerspach C, Nuss K. Ultrasonographic findings in 63 cows with haemorrhagic bowel syndrome. *Vet Rec* 2010;166:79-81.
2. Chigerwe M, Tyler JW. Serum IgG concentrations after intravenous serum transfusion in a randomized clinical trial in dairy calves with inadequate transfer of colostral immunoglobulins. *J Vet Intern Med* 2010;24:231-234.
3. Coskun A, Sen I, Guzelbektes H, Ok M, Turgut K, Canikli S. Comparison of the effects of intravenous administration of isotonic and hypertonic sodium bicarbonate solutions on venous acid-base status in dehydrated calves with strong ion acidosis. *J Am Vet Med Assoc* 236;1098-1103.
4. Kaler J, Daniels SLS, Wright JL, Green LE. Randomized clinical trial of long acting oxytetracycline, foot trimming, and flunixin meglumine on time to recovery in sheep with footrot. *J Vet Intern Med* 2010;24:420-425.
5. Kappe EC, Halami MY, Schade B, Alex M, Hoffman D, Gangl A, Meyer K, Dekant W, Schwarz BA, Johne R, Buitkamp J, Böttcher J, Müller H. Bone marrow depletion with haemorrhagic diathesis in calves in Germany: characterization of the disease and preliminary investigations on its aetiology. *Berl Munch Tierarztl Wochenschr* 2010;23:31-41.
6. Oevermann A, Di Palma S, Doherr MG, Abril C, Zurbriggen A, Vandevelde M. Neuropathogenesis of naturally occurring encephalitis caused by *Listeria monocytogenes* in ruminants. *Brain Pathology* 2010;20:378-390.
7. Rollin E, Berghaus RD, Rapnicki P, Godden SM, Overton MW. The effect of injectable butaphosphan and cyanocobalamin on postpartum serum β-hydroxybutyrate, calcium, and phosphorus concentrations in dairy cattle. *J. Dairy Sci.* 93:978–987.
8. Speijers MHM, Baird LG, Finney GA, McBride J, Kilpatrick DJ, Logue DN, O'Connell NE. Effectiveness of different footbath solutions in the treatment of digital dermatitis in dairy cows. *J Dairy Sci* 2010;93:5782-5791.
9. Todd CG,.Millman ST, McKnight DR, Duffield TF, Leslie KE. Nonsteroidal anti-inflammatory drug therapy for neonatal calf diarrhea complex: Effects on calf performance. *J Anim Sci* 2010. 88:2019–2028.
10. Wittek T, Grosche A, Locher LF, Fürll M. Diagnostic accuracy of D-dimer and other peritoneal fluid analysis measurements in dairy cows with peritonitis. *J Vet Intern Med* 2010;24:1211-1217.

REFERENCES

Constable PD. Top ten evidenced-based papers in food animal medicine and surgery for 2008. Proceedings of the 27th ACVIM Forum, June 2009. Constable PD & Roussel AJ. Five of the best food animal internal medicine journal articles published in 2009. The other five best food animal internal medicine journal articles published in 2009. Proceedings of the 28th ACVIM Forum, June 2010. Dohoo IR, Martin W, Stryhm H. Introduction to observational studies. In Veterinary Epidemiologic research. AVC Inc, Charlottetown, Prince Edward Island, Canada, 2003, pp 139-150. Lean IJ, Rabiee AR, Duffield TF, Dohoo IR. Invited review: Use of meta-analysis in animal health and reproduction: Methods and applications. *J Dairy Sci* 2009; 92:3545-3565. O'Connor AM, Sargeant JM, Gardner IA, Dickson JS, Torrence ME and consensus meeting participants. The REFLECT statement: methods and processes of creating reporting guidelines for randomized controlled trials for livestock and food safety. J Vet Intern Med 2010;24:57-64.

ASSESSMENT OF OUTCOME AFTER INTRASPINAL OLFACTORY ENSHEATHING CELL TRANSPLANTATION FOR SEVERE SPINAL CORD INJURY IN DOGS

NICK D. JEFFERY, BVSC, PHD, DECVS, DECVN, FRCVS
AMES, IA

INTRODUCTION

Olfactory ensheathing cell (OEC) transplantation came to prominence as a potential regenerative therapy for spinal cord injury (SCI) during the late 1990s and early 2000s[1,2]. Experiments in rodents demonstrated that intraspinal transplantation of OECs was associated with improved functional outcome after SCI and was accompanied by axon regeneration. Such findings prompted a surge of optimism that their application in human patients might provide relief of the disabilities associated with SCI. Subsequent experience has highlighted how progress from laboratory to clinic is slow and unpredictable; despite the elapse of more than a decade, the first human clinical trials to measure efficacy of OECs for SCI have yet to be carried out.

As part of the method by which safety and efficacy of a putative intervention can be measured, we proposed the use of naturally-occurring severe SCI in pet dogs as a model in which to assess the clinical value of OEC transplantation. In the first part of this investigation we determined that intraspinal OEC transplantation was safe[3]. In the second phase, which is currently underway, we needed to quantify in detail the effects of OEC transplants in order to measure their efficacy in reversing the functional deficits associated with SCI.

We used three methods to measure deficits that follow thoracolumbar SCI in dogs: gait analysis, electrophysiological function and assessment of bladder function. Each of these has direct parallels with assessments that are currently used for human SCI patients:

1. Two aspects of quadripedal gait are abnormal following SCI: i) the coordination of thoracic and pelvic limb motion; and, ii) lateral stability (coordination of pelvic limb paw position with the center of gravity). We considered that documentation and analysis of the kinematics of limb motion would be a promising method to determine the magnitude of effect of SCI on these two variables. We also hypothesized that it was possible that intraspinal OEC transplants may differ in their effects on each of these two aspects of gait coordination.

2. Although electrophysiological measurements can be viewed as surrogate measures of outcome (since they measure a parameter that is not directly a problem to the patient) it is complementary to gait analysis. As a secondary outcome measure it can be used to provide a mechanistic explanation for observed changes in gait.

3. Reversal of deficits in continence and sexual function has greater priority for paraplegic human patients than restoration of locomotion[4]. Evaluation of sexual function in canine patients is problematical but bladder function can be conveniently assessed through urodynamics and there are distinct parallels between the responses in human and canine spinal patients, enabling direct interspecies comparisons of treatment effects.

METHODS

Gait analysis

Digital motion capture equipment provides an ideal non-invasive method for capturing the kinematics of limb and paw movement. Reflective markers were placed on prominent bony landmarks on both pelvic and thoracic limbs and their motion as each dog walked in on a treadmill was recorded for a period of at least 20 seconds. This data was subsequently analyzed using a custom-written Matlab script to extract points of interest such as the stance length, time delay between thoracic and pelvic paw placement on the treadmill and paw position on weight-bearing.

Subsequent analysis was used to determine the temporal relationship between thoracic and pelvic paw placement (a measure of coordination between these limbs) and the variability in lateral distance between the pelvic limb paws (a measure of the reproducibility of paw placement in the lateral plane)[5,6].

Initially we quantified these aspects of gait in normal dogs and compared with those with chronic spinal cord injury. Subsequently we analyzed the change in these parameters during recovery from mild SCI and their inter-measurement variability.

Electrophysiology

We measured both sensory evoked potentials (SSEPs) and transcranial magnetic motor evoked potentials (TMMEPs) to provide data on both ascending and descending pathways. The procedures were similar to those published previously[7,8].

Urodynamics

A dedicated urodynamic system was used to measure the response of the detrusor to gradually increased pressure resulting from a constant-rate infusion of saline into the urinary bladder. In normal patients the bladder is able to tolerate increases in volume without increased pressure up to a set limit at which detrusor contractions occur. The ability to accommodate the increasing volume and the frequency of contractions both form measurable end-points indicative of abnormal bladder function and coordination that typically result from 'upper motor neuron' spinal cord lesions.

RESULTS

Gait

Our results show a change in gait pattern as dogs recover from mild SCI; we were also able to document changes in gait pattern following our intervention (although the relationship between intervention type and outcome has not yet been analyzed because the trial is still 'blinded'). Using the currently available data we can document a statistically significant change (overall improvement in coordination) with time following intervention.

Electrophysiology

Surprisingly, we were able to record SSEPs and / or TMMEPs in many of our cases before they were treated, despite there being no evidence of 'deep pain sensation' in any case. We have also documented reduction in latencies with time following the intervention, perhaps suggesting that recovery in spinal cord conduction might play a role in recovery of more normal gait patterns.

Urodynamics

Our data has shown a change in bladder compliance with time in some of our treatment cohort. Thus there is an increased ability to accommodate fluid without an increase in bladder pressure. This is also reflected in the increased time to first bladder contraction. Both of these measures suggest improved supraspinal control of bladder function.

CONCLUSIONS

Our studies demonstrate that the detrimental effects on function caused by clinical SCI in domestic dogs can be evaluated in an objective and comprehensive manner. The use of these analysis tools now permit detailed evaluation of the effects of novel interventions which, if beneficial, would provide a solid basis for recommending specific interventions for more widespread use in dogs or human patients. The establishment of pet dogs as 'models' for human SCI investigation implies the possibility of many further investigations, since at least 100 interventions have been shown to improve outcome after SCI in experimental rodents[9].

REFERENCES

[1]Li Y, et al. **Science** 1997;277(5334):2000. [2]Ramon-Cueto A, et al. **Neuron** 2000;25(2):425. [3]Jeffery N, et al. **J Neurotrauma** 2005;22(11):1282. [4]Anderson K, **J Neurotrauma** 2004;21(10):1371. [5]Hamilton L, et al. **BMC Neurosci** 2007;8:77. [6]Hamilton L, et al. **BMC Vet Res** 2008;4:47. [7]Holliday T, **Vet Clin N Am** 1992;22(4):833. [8]Nollet H, et al. **Vet J** 2003;166(1):28. [9]Kwon B, et al. **J Neurotrauma** 2010:Jun16[Epub].

UPDATE ON THE DIAGNOSIS OF DISK ASSOCIATED CERVICAL SPONDYLOMYELOPATHY

Steven De Decker, DVM, PhD, MRCVS
Merelbeke, Belgium

INTRODUCTION

Disk associated cervical spondylomyelopathy (DA-CSM) is probably the most common and typical cause of cervical spondylomyelopathy (CSM) in dogs. This complex and incompletely understood neurological syndrome occurs in adult to older large breed dogs. The adult Doberman Pinscher is overrepresented. In DA-CSM, caudal cervical spinal cord compression is typically caused by protrusion of the intervertebral disk between the sixth and seventh cervical vertebrae (C6-C7) and/or between the fifth and sixth cervical vertebrae (C5-C6), sometimes in combination with dorsal compression resulting from ligamentum flavum hypertrophy. Other abnormalities that can be seen include rather mild vertebral malformations consisting of flattening of the ventrocranial border of the vertebral body, craniodorsal tilting of the vertebral body into the vertebral canal, spondylosis deformans, a funnel-shaped caudal vertebral canal with a narrowed cranial orifice, and intervertebral foraminal stenosis.[1] Although articular facet hypertrophy is not a common cause of spinal cord compression in DA-CSM, articular facet abnormalities can be seen in Doberman Pinschers with CSM.[2] This disorder can be diagnosed by a variety of imaging modalities such as myelography, postmyelographic computed tomography (CTM) and magnetic resonance imaging (MRI). The last decades, MRI has gained popularity in veterinary medicine and has largely replaced the more invasive techniques, such as myelography and CTM, to establish a diagnosis of DA-CSM. The last years, several studies have demonstrated new insights into the diagnosis of DA-CSM. Therefore, the aim of this presentation is to provide a summary of recent developments in the diagnosis of DA-CSM.

MAGNETIC RESONANCE IMAGING IN CLINICALLY NORMAL DOBERMAN PINSCHERS

The main advantages of MRI are its non-invasive nature, that it involves no radiation, is not known to be associated with any major side effects, obtains images in multiple anatomic planes without the need for reconstruction, and provides superior resolution particularly of soft-tissue structures. This high sensitivity of MRI can become disadvantageous as clinically unimportant lesions can become apparent. This can lead to overinterpretation, which may lead to false positive results. This limitation of MRI has been recognised in human neuroradiology for several decades. A recent veterinary study has demonstrated a high prevalence of cervical MRI abnormalities in clinically normal Doberman Pinschers.[2] These abnormalities included spinal cord compression in 25% of the investigated dogs and disk degeneration, intervertebral foraminal stenosis and mild disk herniation in the majority of cases.

Another study evaluated caudal cervical low-field MRI studies of clinically normal Doberman Pinschers and English Foxhounds.[3] The English Foxhounds represented a breed with comparable body conformation as Doberman Pinschers, but no known breed predisposition to any neurological syndromes affecting the cervical spinal cord. This study confirmed the high prevalence of MRI abnormalities in clinically normal dogs. Disk associated spinal cord compression was demonstrated in 30% of all included cases, complete intervertebral disk degeneration in 38% of all included cases, and vertebral body abnormalities in 35% of the studied Doberman Pinschers. The occurrence of vertebral body abnormalities was the only variable significantly associated with the Doberman Pinschers breed. The majority of demonstrated abnormalities were significantly influenced by higher age and the location of the intervertebral disk spaces with the more caudally located intervertebral disk spaces being more severely affected. The majority of these dogs were available for a follow-up period of 16 to 18 months after the MRI study. None of these dogs developed clinical signs suggestive for a cervical spinal cord disorder. These findings suggest that MRI abnormalities are commonly seen in the caudal cervical vertebral column of clinically normal older Doberman Pinschers and English Foxhounds. Such lesions are probably part of the common aging process of spinal degeneration and are not necessarily associated with the development of clinical signs further in life.

INFLUENCE OF THESE MRI FINDINGS ON CLINICAL INTERPRETATION

The occurrence of irrelevant abnormalities on cervical MRI studies of older Doberman Pinschers and Foxhounds questions the reliability of clinical MRI interpretations. This is of major importance because several large breed dogs are predisposed to neurological syndromes affecting the caudal cervical vertebral column. Therefore, a randomized blinded study was performed to determine the intra –and interobserver agreement for results of cervical low-field MRI in dogs with and without clinical signs of DA-CSM.[4] Additionally, this study evaluated the occurrence of false positive and false negative clinical assessments. Most of the assessed variables were associated with a very good intraobserver, but only a moderate interobserver agreement. Seventeen percent of the included clinically normal dogs were erroneously categorized as clinically affected and 9.5% of the clinically affected dogs were erroneously categorized as clinically normal. In 6.8% of the studied dogs, no consensus interpretation was reached between the different observers. This resulted in an overall sensitivity of 0.85 and overall specificity of 0.71for MRI to differentiate between dogs with and without clinical signs of DA-CSM. Several of the clinically normal dogs demonstrated rather severe MRI abnormalities, while in some of the obviously affected dogs only subtle MRI abnormalities were seen. This indicates that imaging findings do not always correlate with clinical signs.

DIFFERENTIATION BETWEEN CLINICALLY RELEVANT AND IRRELEVANT MRI FINDINGS

Several methods have recently been investigated to differentiate between clinically relevant and irrelevant MRI findings in dogs with and without clinical signs of DA-CSM. These methods can be subdivided into morphological MRI parameters, morphometric MRI parameters, and electrophysiological evaluation of the cervical spinal cord by transcranial magnetic stimulation (TMS).

Two independent veterinary studies have suggested that the occurrence of hyperintense intraspinal intensity changes can be considered the most reliable morphologic MRI variable to differentiate between clinically relevant and clinically irrelevant spinal cord compressions.[2,4]

Several morphometric MRI variables have been investigated to differentiate between clinically relevant and irrelevant spinal cord compression.[5] Results indicate that a certain degree of spinal cord compression, expressed by the remaining spinal cord area, should be reached before clinical signs will occur. Although a reliable threshold value was identified to differentiate between clinically affected and clinically normal dogs, less than reliable interobserver agreement limits the application of this value in individual dogs. Presence of clinical signs was not associated by degree of dorsoventral flattening of the spinal cord and degree of vertebral canal compromise by protruded intervertebral disk. Results further suggested that clinically affected Doberman Pinschers have a relatively narrower vertebral canal compared to clinically normal English Foxhounds. This finding supports the results of previous studies [2,6], suggesting a contributing role of relative vertebral canal stenosis in the development of DA-CSM.

Transcranial magnetic stimulation is a non-invasive, painless and sensitive technique for stimulating the cerebral cortex in order to evaluate the functional integrity of the descending motor pathways of the spinal cord. Previous studies have evaluated the use of TMS in Doberman pinschers with and without clinical signs of CSM and in dogs with other cervical spinal cord disorders.[7,8] Recently, it has been demonstrated that TMS can be used as a diagnostic tool to differentiate clinically relevant from irrelevant spinal cord compressions seen on MRI.[9] Clinically normal Doberman Pinschers with clinically irrelevant spinal cord compressions demonstrated similar TMS-values as clinically normal Doberman Pinschers without spinal cord compression and significantly different TMS-values compared to clinically affected Doberman Pinschers. Furthermore, reliable threshold values with combined high sensitivity and high specificity were identified to differentiate between clinically normal and clinically affected Doberman Pinschers.

VERTEBRAL RATIOS IN DISK ASSOCIATED CERVICAL SPONDYLOMYELOPATHY

Although several possibilities have been proposed, little is known about the etiology or predisposing risk factors for the development of DA-CSM. A multifactorial etiology is considered most likely. Relative vertebral canal stenosis with or without a funnel-shaped caudal vertebral canal, leading to a narrowed cranial orifice, have been suggested as contributing risk factors for the development of this disease. In relative stenosis, the diameter of the vertebral canal is less than that expected for an individual animal, but does not cause neural compression in itself. This implies decreased available space between the spinal cord and the vertebral canal and carries an increased risk of becoming clinically relevant on the development of space-occupying conditions of the vertebral canal, such as age related intervertebral disk degeneration and protrusion. Absolute vertebral canal measurements obtained from survey radiographs are influenced by radiographic magnification. This variability, owing to magnification errors, can be resolved by the use of ratios of different measurements. Vertebral ratios have been proposed as a screening or diagnostic tool in humans [10] and horses [11] with cervical spondylopathy or cervical stenotic myelopathy, respectively. Although controversial, it has been suggested that the ratio of the vertebral canal height to the vertebral body height quantifies vertebral canal stenosis.[10]

Recent studies have investigated the use of several vertebral ratios in DA-CSM. In two studies, the vertebral canal height to body height ratio (CBR), vertebral canal height to body length ratio (CBLR), caudal canal height to cranial canal height ratio (CCHR), and vertebral body length to vertebral body height ratio (BLHR) were calculated. [12,13] It was suggested that CBR and CBLR represented vertebral canal size, CCHR vertebral canal shape, and BLHR vertebral body shape. For all evaluated vertebral ratios, significant differences were demonstrated between Doberman Pinschers with DA-CSM and clinically normal Doberman Pinschers [12,13] and clinically normal English Foxhounds.[13] No differences were noted between both groups of clinically normal dogs.[13] These results suggest that Doberman Pinschers with DA-CSM have a narrower midsagittal vertebral canal height, a more funnel shaped caudal vertebral canal, and more square shaped vertebral bodies compared to clinically normal Doberman Pinschers and English Foxhounds. Therefore, it can be argued that altered vertebral column dimension may play a potential contributing role in the pathogenesis of DA-CSM in dogs. However, considerable overlap existed between values of individual dogs of different groups, no reliable threshold values could be identified to differentiate between clinically affected and clinically normal dogs, and limits of agreement for interobserver agreement were wider than the differences between the mean values of both groups.[12] Therefore, it is currently not indicated to use these vertebral ratios in individual animals as a screening or diagnostic tool for DA-CSM. Significant differences for the vertebral body length to vertebral body height ratio (BLHR) suggest that the shape and/or size of the vertebral body can vary between dogs.[13] This limits reliable interpretation of the vertebral canal height to vertebral body height (CBR) and vertebral canal height to vertebral body length (CBLR) ratios. This is further demonstrated by the lack of correlation between both ratios (CBR and CBLR) and relative vertebral canal stenosis, expressed by the vertebral occupying ratio of the spinal cord.[13] Therefore, it is suggested that CBR and CBLR do not quantify relative vertebral canal stenosis in dogs.

COMPARISON OF AVAILABLE DIAGNOSTIC TECHNIQUES

Disk associated cervical spondylomyelopathy can be diagnosed by myelography, CTM or MRI. Each of these techniques is associated with specific well known advantages and disadvantages. Although several studies have established the use of these different diagnostic modalities to obtain a diagnosis of DA-CSM, little is known about the relative contributions and limitations of these individual techniques for the assessment of this disorder. A previous study has compared the use of myelography and MRI in 18 Doberman Pinschers with CSM.[14] Results suggested that considerable variation exists between the interpretation of myelography and MRI. Furthermore, abnormalities were more commonly seen on MRI and 5 of 18 dogs developed postmyelographic seizures.

A recent study has evaluated the intraobserver, interobserver, and intermethod agreement of myelography, CTM, and low-field MRI in 22 dogs with DA-CSM.[15] The following variables were assessed by all three techniques: number, site and direction of spinal cord compressions; narrowed intervertebral disk spaces; vertebral body abnormalities; spondylosis deformans; and abnormal articular facets. Intervertebral foraminal stenosis was assessed on CT-M and MRI. Site and direction of worst spinal cord compression were the most reliably assessed variables with always very good or good intraobserver, interobserver or intermethod agreement for the different imaging modalities. Assessment of abnormal articular facets and intervertebral foraminal stenosis were the least reliably assessed variables with poor interobserver agreement for all evaluated imaging modalities. Myelography demonstrated very good to good intraobserver agreement for several of the assessed variables. However, 6 of 22 dogs developed postmyelographic seizures and 3 of 22 dogs experienced postmyelographic neurological deterioration. Furthermore, the lack of transverse imaging on myelography limited reliable interpretation of articular facet abnormalities and intervertebral foraminal stenosis. Postmyelographic computed tomography and low-field MRI demonstrated only moderate intraobserver agreement for most assessed variables. The three diagnostic techniques demonstrated moderate to fair interobserver and intermethod agreement for most assessed variables. Postmyelographic computed tomography was considered the superior technique for evaluation of articular facet abnormalities and intervertebral foraminal stenosis. Magnetic resonance imaging was considered the least reliable technique for interpretation of spondylosis deformans and narrowed intervertebral disk spaces. Spinal cord compressions, craniodorsally tilted vertebral bodies, narrowed intervertebral disk spaces, abnormally shaped vertebral bodies, and intervertebral foraminal stenosis were significantly more often diagnosed on MRI compared to CTM and myelography.

There is considerable variation in image interpretation between different observers and imaging modalities in dogs with DA-CSM. It seems currently inappropriate to consider one of the evaluated diagnostic techniques as the imaging modality of choice. In contrast, they should rather be considered as complementary to each other.

REFERENCES

[1]Van Gundy TE. **Vet Clin North Am Small Anim Pract** 1988; 18(3):667. [2]da Costa RC, et al. **Am J Vet Res** 2006; 67(9):1601. [3]De Decker S, et al. **Am J Vet Res** 2010;71(4):428. [4]De Decker S, et al. **Am J Vet Med Assoc** 2011; 238(1):74. [5]De Decker S, et al. **Vet J** 2011, doi 10.1016/j.tvjl.2010.12.017. [6]Breit S and Kunzel W. **J Anat** 2001; 199 (5):527. [7]da Costa RC, et al. **Am J Vet Res** 2006; 67(9):1613. [8]Poma R, et al. **J Am Vet Med Assoc** 2002; 221(1):60. [9]De Decker S, et al. **Am J Vet Med Assoc** 2011; 238(1):81. [10]Pavlov H, et al. **Radiology** 1987; 164(3):771. [11]Moore BR, et al. **Am J Vet Res** 1994; 55(1):5. [12]De Decker S, et al. **Am J Vet Res** 2011a, In Press. [13]De Decker S, et al. **Am J Vet Res** 2011b, In Press. [14]da Costa RC, et al. **Vet Radiol Ultrasound** 2006; 47(6):523. [15]De Decker S, et al. **J Am Vet Med Assoc** 2011, In Press.

RECENT DEVELOPMENTS IN THE SURGICAL TREATMENT OF THE CAUDAL CERVICAL SPONDYLOMYELOPATHY

P. Filippo Adamo, DVM, DECVN
Madison, WI

INTRODUCTION

Wobbler syndrome in dogs refers to a collection of disorders of the cervical vertebrae and intervertebral discs of large breed dogs resulting in cervical canal stenosis and spinal cord compression. All these different clinical entities result in the same clinical signs, mainly characterized by a typical wobbling gait (predominantly affecting the hind limbs), paresis and cervical pain. The most typical and predominant syndrome is the disc associated Wobbler syndrome (DAWS). DAWS, predominantly affects middle to older age, large-breed nonchondrodystrophoid dogs, particularly Doberman Pinschers. Doberman Pinschers may be predisposed to clinical cervical spinal cord compression by congenital relative vertebral canal stenosis with a loss of reserve space. Many other breeds have also been reported as being affected by this condition (including small dogs), and in one study of 90 dogs, where Doberman Pinschers were excluded, thirty-two breeds were represented, of which Labrador retrievers (13), Dalmatians (13) and Rottweilers (12) were the most commonly affected.[4] The main factor in DAWS is the underlying chronic disc degenerative disease. The C5-C6 and C6-C7 disc spaces are most commonly affected, with lesion in both interspaces present in about 20% of affected dogs.[1]

PATHOGENESIS

Although the pathogenesis of DAWS is not well understood, it is thought to be multifactorial including primary developmental abnormalities and secondary degenerative changes that lead to vertebral canal stenosis and spinal cord compression.[5] Chronic degenerative disc disease seems to be such an important factor that the term "DAWS" has been suggested. Spinal cord compression in DAWS is often dynamic and secondary to a combination of degenerative disc diseases (leading to protrusion of the intervertebral disc) and the relative redundancy, relative hypertrophy, or "in folding" of the ligamentous structures (dorsal longitudinal ligament, dorsal annulus, interarcuate ligament and joint capsule) because of collapse of the disc space. The relative redundancy of the dorsal longitudinal ligament and the dorsal annulus results more often in ventrally located spinal cord compression. The relative redundancy of the interarcuate ligaments and joint capsules may result in dorsally located spinal cord compression, with or without osteoarthrosis of the overloaded facets, or articular processes. Simultaneous narrowing of the intervertebral foramina because of disc space collapse may cause neurovascular compression at the foramina.[1,2,3,5] Generally in a traction responsive lesion, the dorsal and/or ventral spinal cord compression may be reduced by the application of linear traction to the cervical spine which re-establishes disc width, flattens the relatively redundant soft tissue structures, and opens the narrowed foramina. In most dogs, 2-3 mm of distraction restores a normal disc width of 4-6 mm.[2,6]

DIAGNOSIS

1. Clinical signs

Animals affected with DAWS are usually 4-8 years of age. The most common presentation of DAWS is ataxia (lack of gait coordination), which is most severe in the pelvic limbs, cervical pain and low carriage of the head. Clinical signs may range from only cervical pain (5-10% of patients) to tetraplegia.[1,7] The owners commonly report a gradual onset, although the symptoms can sometimes occur or be exacerbated more acutely. A slowly progressing pelvic limb ataxia and/or paresis is usually noted. Progression to thoracic limb involvement with a short stilted gait can also occur. A broad-based stance can be noticed in the hind limbs. Affected dogs often show a characteristic "disconnected" gait, where the thoracic limbs seem to advance at different rates. The disease is usually chronic progressive if left untreated. Prognosis is usually worse for dogs with chronic clinical signs and severe tetraparetics.

2. Survey radiography

Survey radiography may be suggestive of DAWS but they are not conclusive. They are used as preliminary screening before more specific advanced imaging diagnostic techniques such as myelography, computed tomography (CT) and magnetic resonance imaging (MRI). Survey radiographs are useful to rule out potential other causes of cervical diseases such as vertebral fractures, subluxations, vertebral tumor, and discospondylitis. General anesthesia or sedation is necessary to obtain correct positioning of the dog. Mild deformity of the cervical vertebral body and spondylosis deformans ventral to the intervertebral space may be seen, and narrowing of the intervertebral disc space is frequently seen. However, abnormalities on survey radiographs do not always correlate with spinal cord compression. Some dogs with severe radiographic abnormalities will have no spinal cord compression on myelography or MRI and demonstrate no clinical signs. Survey radiographs may be normal in some dogs with DAWS.

3. Advanced Imaging (Myelography, CT and MRI)

Diagnosis of DAWS can be made by myelography with traction views (stress myelography) or MRI. In DAWS the extent of cord compression can vary with flexion, extension and linear traction (distraction). Owing to the risk of neurologic deterioration after these cervical manipulations during general anesthesia, only linear traction myelography or linear traction MRI views areroutinely used. Cross sectional computed tomography (CT) combined with myelography may also help to better visualize spinal cord compression and spinal cord atrophy.[3] Recently, MRI was shown to be more accurate and more informative than cervical myelography or CT-myelography. MRI better defines the site, severity and nature of spinal cord compression, and allows visualization and characterization of intraparenchymal spinal lesions. Associations between spinal cord MRI findings and histologic abnormalities have been documented.[1,8] The possible complications associated with myelography are not seen with MRI, because itdoes not require the injection of a contrast medium into the subarachnoid space.

4. Static versus Dynamic lesion

The concept of static and dynamic lesions was first established in 1982; determination of the "dynamic" or "traction responsive" feature, in contrast to "static" lesions (such as extruded disc material, malformed facets, or deformed vertebral arches), has been suggested as being essential to determine the most appropriate operative technique. However, distinction between dynamic and static lesions is currently unclear, subjective, and dependent on personal opinion.[1] However it appears that 20% of the dog's weight is enough to produce sufficient traction; it is unnecessary to use traction forces higher then 25% of the patient's weight.[6] Generally with traction responsive lesions, the dorsal and/or ventral spinal cord compression may be reduced by the application of linear traction to the cervical spine which re-establishes disc width, flattens the relatively redundant soft tissue structures, and opens the narrowed foramina. In most dogs, 2-3 mm of distraction restores a normal disc width of 4-6 mm.[2]

TTREATMENT

1. **Medical treatment**

Medical treatment with activity restriction and corticosteroids may be indicated in a normal dog with a first episode of neurologic deficits following minor trauma; otherwise, surgery is the treatment of choice.[3] Conservatively treated dogs should have restricted activity for at least two months. In one study, where medical treatment was used, a successful outcome was achieved in 45% of the patients.[9] This study concluded that conservative treatment for DAWS is associated with a guarded prognosis. In another study, that compared medical versus surgical treatment, it was found that 81% of dogs surgically treated (via ventral slot, dorsal laminectomy or distraction fusion) improved, compared to the 54% of dogs that improved with medical treatment.[10]

2. Surgical treatment

There are many surgical techniques described to treat DAWS, which can be broadly divided into two categories: direct decompressive surgeries and distraction-stabilization surgeries.[11,12] Directdecompressive surgeries involve removal of the hypertrophied annulus fibrosus and dorsal longitudinal ligament via ventral slot or dorsal decompression via dorsal laminectomy.

a) Dorsal laminectomy: dorsal decompression is usually recommended for dogs with single dorsal lesions that do not respond to traction as well as do patients with multiple dorsal lesions.[13] The major disadvantages are that extensive soft tissue dissection is needed, the ventrally located disk material cannot be removed, and there is significant, short term morbidity with deterioration of neurological status, which can cause considerable nursing problems in giant-breed dogs. In one study, fourteen of twenty dogs undergoing dorsal laminectomy had immediate decline in neurologic grade.[13] Although dorsal laminectomy should not cause domino lesions, recurrence of clinical signs has been reported in 10% of dogs secondary to restrictive fibrosis.[13]

b) Ventral slot: ventral decompression alone, can be technically challenging and exacerbates instability, particularly with cervical extension. Ventral decompression may also result in further collapse of the disc space exacerbating spinal cord compression with additional infolding of the ligamentous structures and joint capsules. Bleeding from the venous plexus is also a major complication of ventral decompression, which in rare cases requires a blood transfusion. In over 20% of patients treated with ventral slot alone, failure to respond to treatment was associated with inadequate removal of disc material, which increased spinal cord compression as the intervertebral space collapsed at the treated site.[14] Collapse of the intervertebral space can also compress the nerve roots in the intervertebral foramen, which in turn may cause cervical hyperesthesia and focal spinal cord ischemia.[1] In the long term this technique seems to be clinically effective and instability, if present, may be alleviated because after ventral decompression many of these interspaces will fuse, even without cancellous autografting.

c) Distraction-stabilization: distraction-stabilization-fusion techniques distract the vertebrae to stretch the hypertrophied tissue and relieve spinal cord compression; the vertebrae are then stabilized with appropriate implants and fusion is promoted with bone autografts (cancellous, cortical or corticocancellous) and cancellous bone allografts. Various techniques have been used to maintain distraction and/or graft retention to allow for bony fusion of the affected interspace. These include screws and washers, smooth pins, threaded pins or bone screws and polymethylmethacrylate (PMMA) bridges, interbody or intervertebral cement plugs, modified K-wire spacers with bone screws and PMMA bridge, metallic plates (stainless steel and titanium), plastic plates with unicortical and bicortical screws, and metallic plates with locking screws.[3,15-17] Early implant failure with loss of distraction before fusion has been the most common cause of failure in distraction-stabilization techniques. Common complications include interspace collapse because ventral or dorsal migration of the graft, penetration of bicortical pins or screws into the vertebral canal with spinal cord damage or transverse foramen with vertebral artery or nerve root compromise, fracture of bone cement bridges, pins, screws and plates, and various soft tissue complications such as esophageal erosion because of ventral hardware or PMMA prominence.[3,18] These disastrous complications have caused some surgeons to rely only on direct ventral decompression for both static and traction-responsive lesions. Many authors using either direct or indirect decompression claim 70 - 90% success rate.

DOMINO LESION

Recurrence of clinical signs secondary to a "domino" lesion may occur as a late postoperative complication with any of these techniques. [11,12,19] Recurrence can be caused by compression at the original site or by a domino lesion at an adjacent site. Domino lesions or adjacent segment disease are believed to be the result, at least in part, of abnormal stresses imposed on one intervertebral space by fixation of an interspace adjacent to it.[20-24] These stresses can exacerbate any pre-existing subclinical instability, and produce either disc extrusion or hypertrophy of annular or ligamentous structures. Recurrence of paraparesis to tetraparesis occurs in up to one third of dogs after either ventral decompression or metal implant and bone cement fixation. It usually occurs between six months and four years after the original surgery, with a mean recurrence of around two years.[20-24] A recent meta-analysis examining surgically treated DAWS dogs found an 80% short term success rate with about 20% of successful surgeries having significant recurrence after long term follow-up.[7] The type of surgery performed (decompression versus distraction fusion), did not influence outcome. Given the high rate of surgical failure and long term recurrence, new methods are continually investigated for the treatment of DAWS in dogs.

CERVICAL ARTHROPLASTY: Ventral decompression and placement of artificial disc.

The goal of cervical arthroplasty is to preserve motion after neural decompression while providing distraction and stability. Ventral slot followed by the implantation of a cervical disc prosthesis in dogs with DAWS, has the potential to achieve the optimal goal of spinal decompression, restoration of the biomechanics at the surgical treated sites with sparing the adjacent vertebral motor units from the alterations in loading associated with ventral slot and distraction-fusion surgical techniques, which may eventually prevent the occurrence of a domino lesion. Additionally, advantages of cervical arthroplasty over the standard ventral slot is that the artificial disc acts as a spacer preventing the early collapse of the vertebral space usually associated with ventral slot and/or vertebral subluxation at the treated site. Advantages of artificial disc implantation over pins or screws and PMMA distraction fixation techniques include the elimination of the potential complication associated with pin impingement on neurovascular structures.

In a preliminary *in vitro* study in dogs[25] it was concluded that cervical spine specimens with the an implanted prosthesis specifically designed and manufactured for the dog's cervical spine (Table 1), have biomechanical behavior more similar to an intact spine compared to spinal specimens treated with ventral slot and PMMA procedures. In a recent clinical pilot study[26,27] an equivalent Titanium artificial disc was implanted in two dogs clinically affected by DAWS. The artificial disc used in this study is relatively easy to implant, it is cost-effective (its cost may be equivalent to the cost of the pins and PMMA), it doesn't require special instrumentation, and it could be applied to multiple sites if needed. In both dogs a good distraction at the treated site was achieved at the time of surgery, the implant was well tolerated and there were no signs of implant infection on serial post-operative radiographs. Both dogs had an excellent short and long term clinical outcome. The cervical prosthesis provided immediate distraction at the treated site. This immediately and effectively relieves the mechanical compression caused by the redundant dorsal annulus and/or ligamentum flavum, reopens the narrowed intervertebral foramina relieving the potential spinal cord ischemia caused by compression of the vertebral spinal artery, and reduces the cervical pain caused by nerve roots compression. Both dogs were re-assessed with MRI (18 and 24 months post-surgery respectively). In both dogs all remaining cervical discs showed no signs of degeneration, and overall the MRI showed no significant changes relative to the MRI at the time of diagnosis. The possibility of performing MRI studies after implantation of this prosthesis, without interfering with spinal cord visibility, is an advantage over any other distraction stabilization techniques reported at this time. This allows, in the event of a recurrence of clinical signs, an accurate re-evaluation of the spinal cord at the same site as well as at adjacent locations, for an early detection of domino lesions or any other spinal cord disease. This preliminary clinical study proved the principle that cervical disc prosthesis in dogs could be a valuable method to treat DAWS in dogs.

- Sizes of the prosthesis.

Disc	A (mm)	B (mm)	C (mm)	D (mm)	E (mm)
sm 1	9.0	10.5	7.4	6.3	4.5
sm 2	10.5	10.5	7.4	7.0	5.2
sm 3	12	10.5	7.4	7.8	5.9

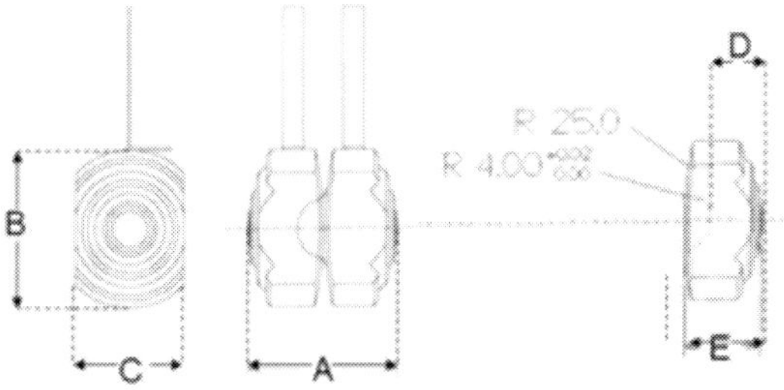

Table 1. Canine cervical prosthesis: The external surface of each end plate is convex to prevent migration. The concavity and the convexity in the center area of each surface acts like a ball and socket. Each surface has concentric grooves to allow bone in growth into the implant. In the latest prototypes. the two breakable fins attached to each endplate to facilitate handling during implantation, have been replaced by two screwable octagonal pins and the prosthesis is hold together by a barrel holder.

CONCLUSIONS

Disc associated Wobbler syndrome is a relatively common cause of chronic spinal cord compression in adult large breed dogs. This disease can be very challenging, both for the referring veterinarians and even for the specialists. DAWS is considered a surgical disease but the ideal surgical procedure still doesn't exist. Further studies are needed to improve the outcome of this devastating disease in dogs.

REFERENCES

1. Da Costa RC, **Vet Clinic Small** 2010, 40:881-913. 2. Trotter E: **Vet Surgery** 2009, 38:705-718. 3. Sharp N et al: in Sharp W (ed): Small animal spinal disorders (ed 2nd), 2005, 93-120. 4. Downes C et al: **British Small An Vet Congr** 2009. 5. Seim H et al: **JAAHA** 1982, 18:241-251. 6. Penderis J: **Vet Radiol Ultras** 2004 45:216-219,. 7. Jefferey N et al: **J. Sm Anim Pract** 2001, 42:574-581. 8. Oshio I et al: **Spine** 1993, 18:1140-1149. 9. De Decker S et al: **J Small Anim Pract** 2009, 50:136-142. 10. Da Costa RC et al: **J Am Vet Med Assoc:** 2008, 1284-1290. 11. Dixon B, et al: **J Am Vet Med Assoc** 1996, 208:61-68. 13. De Risio L et al: **Vet Surgery** 2002, 31:418-427. 14. Chambers J et al: **J Am Anim Hosps Assoc** 1986, 22:775-778. 15. McKee W at al: **J Small Anim Pract** 1999, 40:465-472. 16. Shamir M et al: **Vet Surgery** 2008, 37:186-192. 17. Bergman R et al: **Vet Surgery** 2008, 37:530-536. 18. Koehler C et al: **Am J Vet Surg** 2005, 66:678-687. 19. Wilson E et al: **J Am Vet Med Assoc** 1994, 205:1297-1299. 20. Rusbridge C et al: **J Small Anim Pract** 1998, 39:425-431. 21. Brucker K et al: **J Am An Hosp Ass** 1989, 25:677-683. 22. Brucker K et al: 1989, **Vet Surg** 18:197-203. 23. Chambers J et al: **JAVMA** 1982, 180:410-414. 24. McKee W et al: **J Small Anim Pract** 1990, 31:22-27. 25. Adamo PF et al: **Vet Surgery** 2007, 36:729-741. 26. Adamo PF et al: **Annual ACVIM Forum, Montreal, CA,** 2009; 27. Adamo PF **JAVMA** 2011 *in press*.

JUVENILE EPILEPSY: AN UPDATE FROM THE WORLD OF PEDIATRIC NEUROLOGY

Sam Long, BVSc, PhD, DECVN
Melbourne, Australia

CLASSIFICATION SCHEMES: SPLITTING HAIRS?

In the world of human neurology, seizures have been classified in several ways historically. These include:

1) according to supposed etiology (e.g idiopathic/primary or symptomatic/secondary)
2) according to site of origin
3) according to clinical form (or phenomenology) – e.g. generalized or focal
4) according to frequency (isolated, cyclic, repetitive, status epilepticus)
5) according to electroencephalographic appearance

However, a distinction must be made between the appearance of seizures in a patient, and the classification of epilepsy syndromes which are essentially disease constellations, most of which manifest several seizure types.

A classification scheme to describe **seizures** themselves was first described in 1970 by Gastaut, and this was then refined by the Commission on Classification and Terminology of the International League Against Epilepsy (the ILAE) in 1981. A similar scheme was then introduced in 1989 to describe the **epilepsy syndromes**. Overall this is probably the most widely recognised and adopted classification system used in human neurology internationally. Subsequently various authors have argued for changes to the scheme since methods did not exist in 1989 to further characterize the basis of either seizures or epilepsy (i.e. the availability of genetic sequencing and identification of the genetic basis for many of the epilepsies, along with the widespread use of sophisticated EEG techniques). The most recent significant update to the 1989 scheme was been proposed by the ILAE in 2001, and minor modifications subsequently applied (the most recent iteration of the scheme appeared in 2010). These updates have argued for the description of epilepsy along 5 'axes' which essentially support an increasing level of understanding of the underlying etiology of the seizure. These axes are:

1) Ictal phenomenology (what does the seizure look like?)
2) Seizure type (does the seizure description fit a known pattern?)
3) Syndrome (does the seizure type suggest one of the seizure syndromes?)
4) Etiology (does the seizure syndrome have a known underlying cause?)
5) Impairment (What is the disease course in the patient?)

Note that the term '**syndrome**' is commonly interchanged with '**epilepsy**' (and that any patient having had more than a single seizure can be described as having epilepsy). In addition, within the human field there is some debate about the definition of these axes (with some authors arguing that Axis 2 has essentially turned into Axis 3 in many situations and that the 2 should be merged). Importantly, it has been argued that identification of a syndrome is useful in that it is more likely to convey diagnostic and prognostic information than a seizure type. However, the use of sophisticated diagnostic techniques (interictal EEG, video-ictal EEG, MRI, inter- and ictal SPECT and MEG) allows for identification of a seizure syndrome in, at best, 62% of human patients (Seino, 2004).

Although these axes are perhaps poorly defined, and may not be definable at all in some patients, the concept of seizure types and epilepsy syndromes are useful ones, and the seizure classification scheme as currently applied, is similar to that commonly adopted by most veterinary neurologists (see Tables 1 and 2).

Table 1: Classification of seizures

GENERALIZED SEIZURES	FOCAL SEIZURES
Tonic clonic (in any combination) Absence * Typical * Atypical * Absence with special features Myoclonic absence Eyelid myoclonia Myoclonic * Myoclonic * Myoclonic atonic * Myoclonic tonic Clonic Tonic Atonic	**Descriptors of focal seizures according to degree of impairment during seizure:** * Without impairment of consciousness/responsiveness + With observable motor or autonomic components (roughly corresponds to the concept of "simple partial seizure) + Involving subjective sensory or psychic phenomena only (corresponds to the concept of aura) * With impairment of consciousness/responsiveness (roughly corresponds to the concept of complex partial seizure) * Evolving to a bilateral, convulsive seizure (involving tonic, clonic or tonic and clonic components; replaces the term secondarily generalized seizure)
MAY BE FOCAL, GENERALIZED, OR UNCLEAR	
Epileptic Spasm	

Table 2: Electro-clinical syndromes and other epilepsies arranged by age at onset *

Neonatal period	**Adolescence - Adult**
Early myoclonic encephalopathy (EME) Benign familial neonatal seizures (BFNS) Ohtahara syndrome	Juvenile absence epilepsy (JAE) Juvenile myoclonic epilepsy (JME) Epilepsy with generalized tonic-clonic seizures alone Progressive myoclonus epilepsies (PME) Autosomal dominant partial epilepsy with auditory features (ADPEAF) Other familial temporal lobe epilepsies
Infancy	**Less Specific Age Relationship** *
Migrating partial seizures of infancy West syndrome Myoclonic epilepsy in infancy (MEI) Benign infantile seizures Benign familial infantile seizures Dravet syndrome Myoclonic encephalopathy in nonprogressive disorders	Familial focal epilepsy with variable foci (childhood to adult) Reflex epilepsies
Childhood	**Distinctive Constellations**
Febrile seizures plus (FS+) (can start in infancy) Early onset benign childhood occipital epilepsy (Panayiotopoulos type) Epilepsy with myoclonic atonic (previously astatic) seizures Benign epilepsy with centrotemporal spikes (BECTS) Autosomal-dominant nocturnal frontal lobe epilepsy (ADNFLE) Late onset childhood occipital epilepsy (Gastaut type) Epilepsy with myoclonic absences Lennox-Gastaut syndrome Epileptic encephalopathy with continuous spike-and-wave during sleep (CSWS) including: Landau-Kleffner syndrome (LKS) Childhood absence epilepsy (CAE)	Mesial temporal lobe epilepsy with hippocampal sclerosis (MTLE with HS) Rasmussen syndrome Gelastic seizures with hypothalamic hamartoma
Epilepsies attributed to and organized by structural-metabolic causes	**Conditions with epileptic seizures that are traditionally not diagnosed as a form of epilepsy *per se*.**
Malformations of Cortical development (hemimeganencephaly, hetertopias etc) Neurocutaneous syndromes (Tuberous sclerosis complex, Sturge-Weber, etc) Tumor Infection Trauma Angioma Peri-natal insults Stroke	Benign neonatal seizures (BNS) Febrile seizures (FS)

*This arrangement does not reflect etiology

SEIZURE TYPES IN INFANTS, CHILDREN AND ADOLESCENTS

Of all the seizure **types** mentioned, two forms are significantly overrepresented in childhood – **absence** and **myoclonic** seizures are both much more common in this age group as compared with adults. Absence seizures constitute the most characteristic epilepsy of childhood. Children with this form of epilepsy usually have their first seizure after four years of age, and may have several hundred seizures per day. In people these seizures are brief and devoid of motor activity – sometimes so brief that the patient does not perceive their occurrence. There is no obvious prelude or warning, and the patient stares and briefly stops talking or ceases to respond. In most there is a very brief burst of clonic movement of the eyelids, facials muscles or fingers, the frequency of which corresponds to that of the EEG abnormality which takes the form of a generalized 3Hz spike and waver pattern. Postural tone may be decreased or increased, but as a rule patients do not fall and may even continue complex movements such as walking or riding a bicycle. After 2 to 10s the patient then reeastablishes contact with the environment and resumes activity. These seizures often related to periods of inattention and may appear in the classroom when the child is sitting quietly. They may occur frequently enough to disturb attention and thinking to the point where performance at school is impaired and can occur for hours with no interval of normal mental activity between them. This is **absence status**. Absence seizures may be the only seizure type in childhood epilepsy - the attacks tend to diminish in frequency in adolescence and then disappear, although they are then often replaced by major generalized seizures.

Myoclonic jerks are characterized by brief muscular contractions, some being so small as to involve only one muscle (or even a part) with others large enough to involve a limb, both sides of the body, or the entire trunk. They are brief, lasting 50-100ms and may occur intermittently and unpredictably or present as a single jerk or a brief salvo. The appearance of several small, rhythmic myoclonic jerks as part of absence seizures or as isolated events in patients (of any age) with generalized tonic clonic seizures are usually benign and respond well to medication. However, disseminated myoclonus appearing in childhood is a more serious condition that is often the result of acute viral encephalitis, drug toxicities (particularly lithium), or a condition known as subacute sclerosing panencephalitis. If it is progressive and associated with dementia, the prospect of metabolic and other degenerative conditions such as juvenile lipidosis, Lofara-type familial myoclonic epilepsy, mitochondrial disorders or other chronic familial degenerative diseases becomes more likely.

THE JUVENILE EPILEPSIES: WHAT ARE THEY?

Lennox-Gastaut Syndrome

Lenox-Gastaut syndrome is a condition characterized by the onset of atonic seizures between 2 and 6 years of age. This is often succeeded by combinations of motor, tonic-clonic and partial seizures, and progressive intellectual impairment together with a distinctive slow (1-2 Hz) spike and wave EEG pattern. This early onset of atonic seizures and associated abnormalities usually is associated with serious neurologic disease – the most common underlying causes include prematurity, preinatal injury, and metabolic diseases of infancy.

Juvenile myoclonic epilepsy

This is the most common form of idiopathic generalized epilepsy in older children and young adults. It begins typically in adolescence (typically at 15 years of age but spanning all the teenage years). Classically the first presentation is either with a generalized seizure or because of myoclonic jerks involving the whole body. Absence seizures may be prominent, and myoclonic jerks may be more common in the morning. Myoclonic jerks of the arm and upper trunk are common and become more prominent with fatigue, during early stages of sleep, or after alcohol ingestion. EEG is characterized by bursts of 4-6Hz irregular polyspike activity and a linkage has been established in some patients to chromosome 6 (although no mendelian inheritance has been established). The disease does not impair intelligence and tends not to be progressive although infrequent seizures tend to occur through life. Valproic acid is very effective in eliminating these seizures and should be continued indefinitely.

Benign childhood epilepsy with centrotemporal spikes and epilepsy with occipital spikes

This is a type of focal motor epilepsy that tends to be a self-limited disorder, transmitted in families as an autosomal dominant condition. The age of onset is between 5 and 9 years of age, and usually the first seizure is a tonic-clonic seizure with focal onset that occurs at night. Following this come clonic contractions of one side of the face, less often of one arm or leg. These seizures are readily controlled with a single anticonvulsant and gradually disappear during adolescence.

Infantile spasms (West Syndrome)

Infantile spasms are a term applied to a particular form of epilepsy of infancy and early childhood. This seizure disorder appears during the first year of life and is characterized by recurrent brief episodes of gross flexion movements of the trunk and limbs (less frequently extension movements; these have also been termed salaam or jackknife seizures). Most of these patients show severe EEG abnormalities consisting of continuous multifocal spikes and large amplitude slow waves, although this is not specific for infantile spasms. As the patient matures the seizures diminish, usually disappearing by the fourth or fifth year. The underlying pathology is most commonly cortical dysgenesis, and the seizures may respond dramatically to treatment with ACTH, corticosteroids, or benzodiazepines (clonazepam is most commonly used). However most patients are left mentally impaired once the seizures depart. Infantile spasms my also be part of the Lennox-Gastaut syndrome.

Febrile seizures

Febrile seizures are commonly not referred to as epilepsy. These seizures are peculiar to children and children between 6 months and 5 years of age, with a strong tendency to be inherited, and are generally regarded as benign. Most commonly this condition involves a single generalized motor seizure that occurs as the temperature rises or reaches its peak. The seizure seldom lasts for more than a few minutes, and EEG (by the time it is obtained) is normal. The risk of epilepsy later in life is minimal (slightly greater than that of the general population at 2.4%) and in some families a gene defect by linkage analysis has been implicated.

A more serious condition is that of seizures as a result of encephalitis accompanied by febrile illness. In this situation focal or prolonged seizures, EEG abnormalities and repeated episodes of febrile seizures with the same or other diseases occur (known as complicated febrile seizures). A reasonable percentage of these may be complicated by atypical petit mal, atonic and tonic seizures, together with mental retardation. On long term (18 year follow-up) study followed children with complicated febrile seizures (focal, prolonged or repeated episodes of febrile seizures) and found the long term risk of unprovoked seizures in later life to be increased by 8, 17 or 49 percent respectively.

REFERENCES

Adams and Victor's Principles of Neurology, 8th Edition; Ropper AH, Samuels MA eds. McGraw-Hill, 2005. Seino, M. Epilepsy Res 2006; 70(S):S27-S33. Engel, JE Jr. Epilepsy Res 2006; 70(S): S5-S10.

THE EMERGING ROLE OF PARENTERAL LEVETIRACETAM FOR THE EMERGENCY TREATMENT OF SEIZURES

Edward (Ned) Patterson DVM, PhD, DACVIM (SAIM) and Brian Hardy DVM
St. Paul, MN

INTRODUCTION

Status epilepticus [SE] and acute repetitive seizures [ARS] - are common reasons for presentation to veterinary emergency clinics. ARS is also commonly referred to as cluster seizures in veterinary medicine. Morbidity and mortality from SE and ARS is high in both humans and dogs. In the largest studies to date in veterinary medicine, the mortality rate was 25.3 - 38.5% (Bateman 1999; Zimmerman 2009). In humans with SE and ARS, mortality estimates typically range from 3-40% (Chin 2004), depending on the population evaluated, but overall mortality is approximately 22% (DeLorenzo 1996), similar to that seen in dogs. The optimal treatment for SE is unknown. To date, there are not any controlled studies for the emergency treatment of seizures in dogs, and there have been no significant studies in human status epilepticus since the double blind study of diazepam compared to lorazepam by Leppik et al. in 1983 and a comparison of phenytoin, phenobarbital, diazepam and lorazepam by Trieman et al. in 1998.

Levetiracetam (LEV) is the S-enantiomer of α-ethyl-2-oxo-1-pyrollidine acetamide. Oral Levetiracetam (LEV) was approved in 1999 for adjunctive treatment of adult persons with epilepsy (partial seizures). It has a unique purported mechanism of action binding to a synaptic vesicle protein - SV2A (Lynch 2004). Its pharmacokinetic profile is very favorable with absence of hepatic metabolism and low protein binding, and LEV has the highest safety margin of any anti-epileptic drug (AED) in animal models (Klitgaard 1998). LEV is very soluble in water and an IV formulation became available in 2006 which is supplied in single dose 5 mL vials (100 mg/mL). It is FDA approved for humans as bridge therapy when oral administration is not possible. The package label indicates it to be infused over 15 min. Reported side effects of LEV in humans include headache, somnolence, respiratory infections, and anxiety (Cereghino 2000). Benzodiazepines have been used for decades as first line therapy for the emergency treatment of seizures in humans and dogs (Thomas, 2000). There are few other IV anti-epileptic drugs available for use in emergencies and of those even fewer can be given IM - phenobarbital, fosphenytoin, and some benzodiazepams. There is an IV formulation of valproic acid but an attempt to use it IM but it caused muscle necrosis in dogs (Gallo 1997). Phenytoin was developed in the early 1950's for IM as its initial indication, however, the injection is extremely painful and studies in the 1970's demonstrated that it was erratically and poorly absorbed (Serrano 1974). Because LEV is highly water soluble, it has the clinical potential to be safe and well tolerated for parenteral use, either as an IM dose or IV even with inadvertent extravasation (Patterson 2008). The serum half-life is 3-4 hours in dogs. The typical starting oral dose in dogs is 20mg/kg TID (60mg/kg/day). In repeated oral dose studies of 300mg/kg/day, 1 of 8 dogs developed an unsteady gait. In single dose studies, dogs receiving 1200mg/kg exhibited some salivation, vomiting, tachycardia, and restlessness (UCB Pharmacia internal data). Reported therapeutic plasma level for people is 5-45 *u*g/ml.

LEV's unique spectrum of anticonvulsant activity, very high therapeutic index, and potential neuroprotective properties, make it a potentially interesting agent in the treatment of SE/ARS.

RECENT EXPERIENCE WITH PARENTERAL LEVETIRACETAM IN RODENTS, HUMANS, AND DOGS

IV LEV in laboratory animals with induced seizures:

One recent study showed a synergistic effect of diazeapam and levetiracetam administered together in an experimental electrically induced rodent model of self-sustaining status epilepticus (SSSS). Treatment with LEV during the maintenance phase of SSSE diminished (at 200 mg/kg) or aborted seizures (in doses of 500 or 1000 mg/kg). Addition of LEV significantly enhanced the anticonvulsant effects of diazepam (DZP), even when both drugs where given in doses below their therapeutic level. (Mazarati, 2004).

Other studies in rodents have had more mixed results. In one, the effect of intravenous LEV in the rat lithium-pilocarpine model of status epilepticus was evaluated. Ten or 30 min after the onset of behavioral SE, animals were treated with LEV (200-1200 mg/kg i.v.) administered in a single bolus, which resulted in attenuation of behavioral manifestations of seizures and in reduction of neuronal injury but it did not significantly alter ictal discharge recorded by EEG (Zheng 2010).

IV and IM LEV in normal dogs

A number of studies have shown that LEV is safe in healthy dogs, and in doses of up to 60mg/kg IV (Dewey 2008). In our initial study in 6 normal dogs using LEV, we gave IV doses of 20-25mg/kg. We monitored ECG activity and blood pressure during the IV injection - undiluted over 5 minutes. We also gave the same dogs similar doses of LEV IM and observed minimal to no tissue injury (less than from saline injection) and minimal discomfort. The half-life of IV and IM LEV in these dogs was 3.0 hours and in addition blood levels indicated that IV LEV might have anti-epileptic effects in dogs for up to 12 hours after a single injection (Patterson 2008).

IV LEV for status epilepticus and acute repetitive seizures in people

Levetiracetam,, valproate, and topiramate have recently become available for intravenous use, and the use of ketamine and of other agents (topiramate, felbamate, etc.) have seen renewed interest. A rapidly increasing but largely anecdotal body of literature reports success in seizure control at the price of relatively few complications with the clinical use of those agents in refractory SE. (Westerlan 2008). There are, however, conflicting reports as to efficacy of IV LEV and none to date are randomized clinical trials. In one study, efficacy and effectiveness of IV LEV in SE were assessed in an observational, multicentric and retrospective study. IV LEV was used as add-on therapy (2nd or 3rd line). IV LEV was effective in approximately half of the patients (57.5%), The conclusion was that IV LEV was an effective antiepileptic drug for SE, but the efficacy depends on the timing of its administration, being more effective when used as early treatment, and less effective as add-on treatment. (Aiguabella 2011).

In a different study, a retrospective review was performed for children who received intravenous levetiracetam within 30 minutes of any seizure, which could have been a solitary seizure, ARS or SE . Patients received a mean levetiracetam dose of 29.4 mg/kg. Most children (67%) needed additional antiepileptic drugs to abort their seizure. Most patients (71%) were placed on maintenance levetiracetam within 24 hours of their loading dose. The predictive ability of patient and drug regimen variables in outcomes was poor. Only the number of concomitant antiepileptic drugs consistently predicted outcomes. Levetiracetam was well tolerated at the doses studied, and appeared most effective in single seizure events rather than ARS or SE. (Reiter 2010).

A prospective study in 9 elderly human patients with concomitant medical conditions (arrhythmias/respiratory distress/hepatic diseases) was performed with IV LEV, as the traditional therapy for SE was considered unsafe. IV LEV was used as first-line therapy (loading dose of 1500 mg/100 ml/15 min, mean maintenance daily dose of 2500 mg/24 h) and administered during video-EEG monitoring. In all the patients but one, IV LEV was effective in the treatment of SE and determined either the disappearance of (7/8), or significant reduction in (1/8), epileptic activity; no patient relapsed in the subsequent 24 h. No adverse events or changes in the ECG/laboratory parameters were observed (Fattouch 2010).

A controlled study of IV LEV in epileptic dogs with SE/ARS

A randomized, placebo-controlled, double-masked study was recently performed at the University of Minnesota Veterinary Medical Center in 19 dogs presenting for SE or cluster seizures (ARS) comparing IV diazepam quickly followed by IV LEV to IV diazepam quickly followed by IV saline (placebo control) as the initial treatment. The IV LEV was well tolerated. Fifty six percent of the dogs receiving LEV had no further seizures for 24 hours versus ten percent of dogs receiving saline control. The full details of this study have been submitted for publication.

Subcutaneous (SQ) LEV in normal dogs and at home for epileptic dogs

In a pilot study at the University of Minnesota College of Veterinary Medicine, SQ LEV was administered to 4 normal dogs, at a dose of 60mg/kg. No adverse effects were seen from SQ LEV administration at the injection site or behaviorally. LEV plasma concentrations ranged from 60 to 120 ug/ml over 7 hours exceeding the suggested human therapeutic range of 5-45μg/ml. Additional data from this study is being presented in a neurology abstract at this meeting.

Based on this pilot data in normal dogs, clinicians from our neurology service have dispensed SQ LEV for possible use by owners at home for epileptic dogs who do not always respond sufficiently to diazepam per rectum and/or intranasally. To date this has been in a very limited number of dogs, and more cases are needed to be able to make any possible conclusions about the effectiveness of at home SQ LEV.

CONCLUSIONS

With very few if any exceptions, parenteral levetiracetam has a wide therapeutic index and it is exceptionally well tolerated at varying dosages in a number of species including the dog. Results regarding efficacy to date, have been someone mixed, but it appears that in general the earlier it is administered the more effective it might be. This might be related to how quickly it crosses the blood brain barrier since it is highly water soluble, and also what levels are maintained in the CSF and its pharmacodynamic interactions in the brain. Well designed, larger scale controlled clinical trials of IV, SQ (in dogs) and possibly IM LEV would help resolve how to best use parenteral LEV in timing, dose, and route and the resulting efficacy. Additional well designed, studies could also help determine what plasma level is optimal for the treatment of SE and/or ARS.

REFERENCES

[1] Aiguabella 2011 et al. (2011) **Seizure** 20(1):60-4 [2] Bateman et al. (1999) **JAVMA**. 215(10):1463-8 [3] Cereghino, et al. (2000) **Neurology** 55, 236–242. [4] Chin et al. (2004) **Eur J Neurol**. 11(12):800-10 [5] Delorenzo et al. (1996) **Neurol**. 46(4):1029-35 [6] Dewey et. al (2008). **JVECCS** 18(2): 153-7 [7] Fattouch et al. (2010) **Acta Neurol Scan**. 121(6):418-21 [8] Gallo et al. (1997) **Epilepsy Research** 28, 11–15. [9] Klitgaard et al. (1998) **Eur J Pharmacol**. 353(2-3):191-206 [10] Leppik et al. (1983) **JAMA**. 249(11):1452-4 [11] Leppik et al. (2001) **Epilepsia**, 42 (Suppl. 4), 44–45 [12] Lynch et al. (2004) **PNAS**, 101, 9861–9866. [13] Mazarati et al. (2004) **Epil Res**. 58 (2-3):167-74 [14] Patterson et al. (2008) **J Vet Pharmacol Ther**. 31(3):253-8 [15] Reiter et al. (2010) **Pediat Neurol** 43(2):117-21. [16] Serrano et al. (1974) **Archives of Neurology**, 31, 276–278 [17] Thomas, W.B. (2000) **Vet. Clin of N Am**. 30, 183–186 [18] Treiman et al. (1998) **N Engl J Med**. 339(12):792-8 [19] Westerlan et al. **Epilepsia** (2008), 49, Suppl 9:63. [20] Zheng et al. (2010) **Neuropharmacology** 58 (4-5):793-8

USE OF ELECTROENCEPHALOGRAPHY IN EQUINE SPECIES

Monica Aleman, MVZ, PhD, DACVIM
Davis, CA

SUMMARY

Electroencephalography is a useful diagnostic aid in the investigation of neurological disorders in horses with particular attention to sleep disorders, paroxysmal episodes such as seizures, seizure-like, altered behavior or mentation, monitoring of the critical neurological compromised patient, and determination of brain death. This lecture is aimed at highlighting the value of this underutilized diagnostic modality in equine medicine.

INTRODUCTION

Electroencephalography (EEG) is the graphic representation of the difference in voltage between two different locations within the cerebral cortex plotted over time. During the electroencephalogram, multiple graphs of voltage differences between pairs of electrodes plotted over time are recorded. Electroencephalography is a well established diagnostic technique in human medicine. The first peer reviewed reports of the use of EEG to study the human brain date to the 1930's by Hans Berger and Loomis et al.[1] Since then over 82,500 studies have been published. The first EEG studies reported in the literature in horses are from the 1960's on which states of arousal and sleep were investigated.[2] Since then most of the studies have evaluated the effects of intravenous drugs and anesthetics on the EEG, evaluation of sleep, CNS disease, and depth of anesthesia and nociception.

In general, neuroelectrodiagnostic testing has been more challenging in the horse due to its size, physiological differences, client compliance, financial constraints, and clinician selection of a diagnostic work up. Part of the problem is that electrodiagnostic testing has not been given enough importance in equine medicine, and therefore this useful diagnostic modality is not being offered to owners. This lack of consideration of EEG studies is reflected in the limited number of scientific publications and case reports. Baseline EEG studies in healthy horses and foals are still lacking (e.g. small sample size, not systematically performed in foals). Therefore variations of electrical brain activity among healthy individuals cannot be fully established and may result in misinterpretation. To this limitation, we have to add the effects of drugs and anesthetics that may alter or suppress brain electrical activity.

EEG IN EQUINE MEDICINE

EEG has been usually associated with the investigation of seizures; however there are other applications for its use in horses along with continuous/simultaneous video recording, and includes:

1) Paroxysmal activity (seizures, seizure-like, fine localized tremors, myoclonus-like).
2) Diagnostic refinement of epileptic disorders (e.g. classification: generalized, focal).
3) Drug therapy monitoring (antiseizure medication).
4) Sleep and sleep disorders (restraint-induced collapse in foals, sleep deprivation versus sleep disorder [narcolepsy with and without cataplexy, hypersomnia, abnormal rapid eye movement-sleep]).
5) Alterations in behavior and mentation (may be associated with paroxysmal discharges).
6) Assessment of nociception during surgical procedures.
7) Patient monitoring in the intensive care unit (head trauma, status epilepticus, comatose [is there brain death?]).
8) Welfare aspects (assess methods of humane euthanasia [immediate versus delayed brain death]).

SEIZURES

Both seizures and epilepsy occur in horses though less common compared to other species. According to Sengoku, epilepsy refers to a group of chronic heterogenous disorders characterized by a tendency to have recurrent seizures.[3] The International League Against Epilepsy (ILAE) used to classify epilepsies into primary (idiopathic with a genetic basis) and secondary (symptomatic and cryptogenic).[4] This classification is no longer in favor and was changed to: 1) genetic on which a known or presumed genetic disorder results in seizures as the core symptom of the disorder; 2) structural/metabolic on which a known acquired condition is associated with a high risk of seizures (e.g. traumatic brain injury); and 3) unknown on which the cause is unknown or has not been recognized. Epilepsy occurs in both, foals and adult horses and can be classified under the previous categories (unpublished data). Currently, the only published presumed genetic epileptic disorder in horses is juvenile idiopathic epilepsy of Egyptian Arabian foals.[5] The disorder has an early onset of seizures (median 2 months, range 2 days to 6 months of age) and affects both females and males. Physical and neurological examinations in between seizures are normal except if complications occur such as aspiration pneumonia and trauma (the 2 most common). Postictal blindness occurs in all cases and can last from a few seconds to 3 weeks. Minimum data base, diagnostic imaging and CSF are normal. Seizures are usually easily manageable with phenobarbital and in a few cases addition of potassium bromide may be necessary. This disorder appears to be self limiting since seizures appear to resolve by 1 to 2 years of age. Families of diseased horses tend to have affected foals and pedigree analysis supports a dominant mode of inheritance.[5]

Epileptiform activity consisting of spikes (duration < 70 ms), sharp waves (70-200 ms), spike-and-wave discharges, or multiple spike complexes are detected in these foals (Fig.1).[5] Epileptiform activity appears to be multifocal, although voltage maximum is frequently encountered on midline in the central and parietal regions. Many foals were observed to have transients of sleep such as sleep spindles and K complexes, indicating non-rapid eye movement (REM) sleep. Photic, and sound, and tactile stimulation at the end of EEG recordings did not elicit paroxysmal activity.

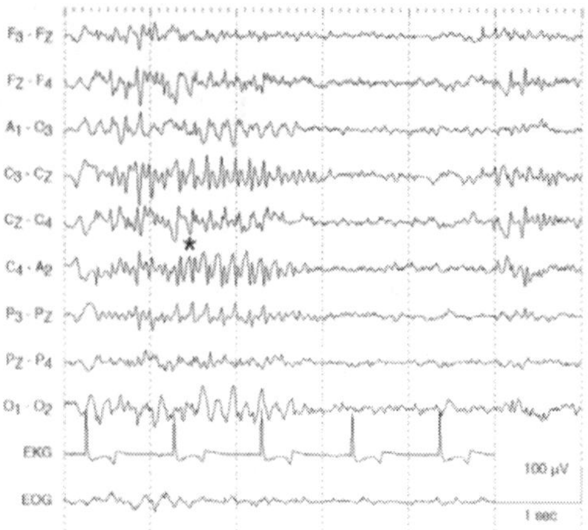

Burst of spike-and-wave discharges (*) at Cz/C4 in a 4-week old Arabian colt. From Aleman et al. JVIM 2006;20:1443-1449.

It is accepted that diagnosis of epilepsy is made primarily based on clinical grounds in human medicine, and that a normal EEG does not exclude seizures. However, EEG in patients with epilepsy can contribute to the diagnostic refinement, identification of subclinical seizures, and classification by location (generalized versus local).[6] In human medicine, EEG can aid in the classification of different epileptic syndromes and direct therapy more accurately. Other suspected familial epileptic disorders have been seen by the author and are currently under study (unpublished).

Acquired epilepsy in horses can occur as a permanent sequela of traumatic brain injury.[7] Epilepsy of unknown etiology is also seen in horses on which complete work ups including full body postmortem evaluation have failed to identify a cause. Seizures have been reported in adult horses due to intracranial and extracranial causes as follows in no particular order: brain tumors (uncommon), infection (viral, bacterial, parasitic encephalitis), hepatoencephalopathy, leukoencephalomalacia, intermittent hypoglycemia (e.g. associated with renal cell carcinoma releasing insulin-like growth factors), hypocalcemia, toxicity (e.g. metaldehyde, organophosphates), pregnancy, estral cycle (last 2 unpublished), among others. Seizures in foals can be caused by septicemia, hypoxic ischemic encephalopathy, metabolic disturbances, bacterial meningitis, abscess, viral encephalitis, liver failure, trauma, and congenital abnormalities.

SLEEP

Similar to humans, sleep is important for the overall health of our equine companions. Normal sleep and sleep disorders are also poorly understood in horses as reflected in the limited literature available. Limited information and lack of understanding of sleep and associated disorders may lead to inaccurate diagnosis and management. The use of EEG redefined the stages of sleep and contributed to the classification of multiple sleep disorders in people. Most of the studies about sleep in the horse have been observational. Horses sleep an average of 3 to 4 hours a day. Intermittent episodes of sleep occur throughout the day with most of the sleep occurring at night. Foals have more periods of sleep throughout the day and of longer duration. Attempts at studying and defining stages of sleep in the horse based on physical observation and continuous EEG recording have been limited to "in doors" studies.[8] Lack of telemetric units with recording distances that allow free range of motion without producing movement artifacts, and electrodes that will stay for long periods of time are some of the challenges when attempting to study natural sleep in the horse.

In people stages of sleep include wake stage, drowsiness (stage 1), light sleep (stage 2), slow wave sleep or delta sleep (stages 3 and 4), and rapid eye movement sleep (REM-sleep). Stages 1 through 4 are collectively termed as non-rapid eye movement sleep (NREM-sleep). In horses, definition of stages of sleep is less sophisticated. Wakefulness in people has a prominent alpha rhythm (8-13 Hz), and the transition from this stage to drowsiness is noted when alpha rhythm diminishes to less than 50% of the recording and is replaced by theta activity (4-<8Hz). Based on our studies, a true alpha rhythm has not been recognized in horses, although some background activity falls in the alpha frequency range. Therefore, the transition from wakefulness to drowsiness in the horse solely based on EEG is difficult to distinguish. For this reason, terms such as relaxed wakefulness, drowsiness, and somnolence have been used for these stages in horses.

Normal transients of sleep such as sleep spindles and K-complexes as those observed in light sleep in humans have also been identified in EEGs from normal horses. Slow wave sleep characterized by high amplitude delta activity is also seen in horses. This stage can occur while the horse is standing with one limb primed and the head held at withers height. However, if the surrounding environment is safe or horses are in a group, this stage of sleep will occur in sternal recumbency. Unique to the horse is the development of secondary AV block based on an ECG while on this stage of sleep.[8] REM-sleep characterized by low voltage and mixed frequencies with episodic fast eye movements, and lack of EMG activity is observed in horses. Horses in this stage of sleep are in lateral recumbency and less commonly in sternal recumbency with their head supported in the ground. REM-sleep in horses is about 15% of their total sleep time. Reflexive movement such as blinking, facial twitching, and paddling may be observed and commonly confused with seizures. Loss of muscle tone precludes standing REM-sleep.

SLEEP DISORDERS

Pure sleep deprivation is not a disorder and simply means lack of sleep. Sleep deprivation is relatively not uncommon in horses, and it can be induced after a few days (5-7) of loss or incomplete periods of sleep or lack of REM-sleep. Examples of situations that may result in lack of sleep in the horse include environmental stress (excessive noise, extreme temperatures, unfamiliar or unsafe conditions such as presence of wildlife), constant traveling, rank in the herd, introduction or removal of horses in the herd, illnesses that may prevent the horse from laying down resulting in lack of REM-sleep such as orthopedic and neurological disease, and inadequate bedding, among others.[9] Hospitalized patients, particularly those in intensive care units on where constant traffic of personnel, horses and equipment prevent complete cycles of sleep, can have profound lack of sleep.[9] Sleep deprivation should be considered in the list of possible causes for horses with collapsing episodes. Sleep deprivation can resolve if the primary cause is properly addressed. Sleep deprivation is commonly confused with narcolepsy and cataplexy.

Hypersomnia, narcolepsy with and without cataplexy, and other less defined disorders (abnormal REM-sleep) have been reported or suspected rarely in horses. Hypersomnia means excessive amount of sleep despite apparently "normal or complete cycles of sleep". This disorder may be primary or secondary to endocrinopathies and neurological disease.[9] Narcolepsy is a lifelong disabling sleep disorder characterized by excessive daytime sleepiness, disturbed nocturnal sleep, and abnormal REM sleep in people.[10] Manifestations of the disorder include cataplexy (sudden loss of muscle tone), sleep paralysis, and hypnagogic (sleep-onset) hallucinations (humans).[10] The disorder is triggered by strong emotions such as humor, surprise, and anger in people; and excitement, play and feeding in dogs.[11] The disease in humans is a genetically complex disorder and an autoimmune basis is suggested by the presence of HLA DQBi*0602 alleles in affected individuals resulting in early loss of neurons in the hypothalamus that produce hypocretin.[11] In the dog, both familial and sporadic narcolepsy occurs. Mutations in the hypocretin-2 gene transmitted as an autosomal recessive trait is responsible for familial cases in Dobermann, Labrador Retriever, and Dachshund breeds.[12] Sporadic or acquired cases have been reported in over 17 breeds and are associated with a loss of production of hypocretin.[12] Hypocretin concentrations can be measured in CSF.[12] Similar to humans and small animals, familial and acquired narcolepsy have been documented in horses. Familial narcolepsy has been reported or suspected in Miniature horses, Suffolks, Shetland Pony, and Fell Pony.[13] Hypocretin levels in CSF have been measured and found low in horses with pituitary pars intermedia dysfunction, and in one Icelandic foal with narcolepsy.[14] Sleep disorders have been observed by the author in horses with CNS disease such as West Nile virus and Equine protozoal myelitis.

Observational studies are very important in the evaluation of sleep disorders. However, identification of alterations in sleep and determination of the stage at which they occur are not possible without the use of simultaneous EEG recording. Limited information and lack of understanding of sleep and associated disorders may lead to inaccurate diagnosis and management. Abrasions or scars in the dorsal aspect of the fetlocks and carpi may be the result of collapsing episodes resulting from either, sleep deprivation or disorders in horses. However, other causes of collapse must be ruled out (e.g. syncope). Continuous EEG recording with simultaneous video monitoring and documentation of the sleep behavior are essential in the diagnostic workup of horses with suspected alterations in sleep.

REFERENCES

[1]Loomis AL, et al. **J Exp Psychol** 1936;19:249-279.[2]Ruckebusch Y. **CR Soc Biol** 1963;157:840-844.[3]Sengoku A. **Epilepsia** 2002;43:6-8. [4]www.ilae-epilepsy.org.[5]Aleman M, et al. **J Vet Intern Med** 2006;20:1443-1449.[6]Koutroumanidis M, et al. **Epilepsia** 2005;46:96-107.[7]Feary DJ, et al. **J Am Vet Med Assoc** 2007;231:259-266.[8]Williams DC, et al. **J Vet Intern Med** 2008;22:630-638.[9]Aleman M, et al. Proc 54th Ann Conv AAEP 2008;.[10]Dauvilliers Y, et al. **Lancet** 2007;369:499-511.[11]Mignot E. **Sleep Medicine** 2004;5:S2-S8.[12]Tonokura M, et al.**Vet Rec** 2007;161:375-380.[13]Lunn DP, et al. **Equine Vet J** 1993;25:483-487.[14]Bathen-Nothen A, et al. **J Vet Intern Med** 2009;23:1299-1302.

CERVICAL HEMILAMINECTOMY: A VALID ALTERNATIVE TO THE VENTRAL SLOT PROCEDURE Cervical

Dominik Faissler, Dr. med. vet., DECVN
North Grafton, MA

INTRODUCTION

Intervertebral disk disease (IVDD) is a common disorder. Cervical IVDD is less common than thoracolumbar disk disease but still accounts for 14% or more of all disk extrusions. A large number of chondrodystophic and non-chondrodystrophic breeds can be affected with Dachshunds and Beagles being the most common small breeds and Labrador Retrievers, German shepherds, and Rottweilers being the most common large breeds reported. In most of the studies median or mean age at diagnosis is close to 7 years with a wide range from 1 to 16 years. Clinical signs can vary from neck pain only to ambulatory or non-ambulatory tetraparesis to tetraplegia. Respiratory function can be compromised in rare cases. Any of the cervical intervertebral disk spaces can be affected with a trend of disk herniations occurring most commonly in small breed dogs at C2-3 and in larger breed dogs at C6-7. Diagnostic tools include myelography, computed tomography (CT), CT myelography and MRI. Advanced imaging modalities offer better 3-dimensional resolution with the surgeon receiving more information about the specific location of spinal cord compression. Since at our institution all dogs with a cervical localization of the neurological signs undergo either MRI or CT myelography to obtain a final diagnosis, more variation in the direction of disk extrusion is observed. Options include disk material at midline ventral to the spinal cord, disk material lateralized often under or around nerve roots, extruded nucleous polposus lateral to dorsolateral of the spinal cord, and rarely extrusion of disk material into the lateral foramen.

VENTRAL SLOT

A ventral slot procedure is the most common surgical technique to correct cervical disk extrusions. The basic idea is to enter the spinal canal ventrally through a small, midline opening and evacuate extruded disk material from the ventral aspect of the spinal canal. The size of the bone window is limited and should not exceed 33% of the vertebral diameter and should be no longer than 33% of each vertebral length to preserve stability of the motion unit. Vertebral subluxations are reported after ventral slot procedures. The advantage of the ventral slot procedure is the minimal soft tissue involvement, short procedure time and fast recovery. The major disadvantages are the small visual field, the difficulty removing lateralized disc material and the impossibility of visualizing the nerve roots. Severe bleeding from the venous sinuses makes visualization of spinal canal structures even more difficult, may lead to abortion of the procedure or may cause iatrogenic trauma if the procedure is continued despite obvious venous sinus bleeding.

CERVICAL HEMILAMINECTOMY

An alternative procedure to enter to the spinal canal is the cervical hemilaminectomy. The basic idea for this approach is to get access to the lateral aspect of the spinal canal by removing the dorsal lamina, articular facets and pedicules on one side. The major advantage is good visualization of lateral and ventral spinal cord as well as the space under and around the nerve roots. Disk material extruded laterally, dorsolateral or trapped under the nerve roots can be removed even when more chronic and adhered to the surrounding tissue. Bleeding from the venous sinuses or the interarcuate branch of the venous sinus is still a potential complication but can be much better handled and stopped than in a ventral slot because of the larger bone window. The major disadvantage is the more extensive dorsal soft tissue approach down to the spine. Different ways of muscle dissection are reported: a) the "classical" dorsal approach through midline structures down to the spine, b) a modified dorsal approach to the lateral cervical spine, and c) a modified lateral approach to canine cervical spine. In large breed dogs all the different ways of muscle dissection described above are more intensive whereas in small and toy breed dogs the soft tissue approach appears to be more limited. Potential complications of a cervical hemilaminectomy include nerve root damage, muscular swelling and subfascial seroma in large breed dogs resulting in transitional worsening in the neurological condition after the procedure, higher level of surgical pain and prolonged recovery time. Another potential complication, damage to vertebral artery has not been personally observed by the author.

CONCLUSIONS

Limited information is available on the long-term outcomes of cervical hemilaminectomies in dogs. One study evaluating the effect of cervical hemilaminectomy in 18 small breed dogs with symmetrical and asymmetrical disc herninations reported an excellent outcome in 15/18 (83%) of the patients and a good outcome in 3/18 (17%) patients. A second study of 6 dogs with IVDD undergoing cervical hemilaminectomy were assessed to have an excellent outcome in 4/6 (66.6%) and a good outcome in the remaining two cases. At our institution cervical hemilaminectomy is the standard procedure for dogs with lateralized cervical disk extrusions. The recovery time is only slightly longer compared to the ventral slot procedure. In the majority of the cases the outcome of our patients undergoing cervical hemilaminectomies are excellent to good.

REFERENCES

Cherrone KL, et al. J Am **Anim Hosp Assoc** 2004;40:316-329. Dallman MJ, et al. **J Am Vet Med Assoc** 1992;200(12):2009-2011. Fauber AE, Bergman RL. **ESVN Proceedings Cambridge, UK,** 2010:60. Felts JF, Prata RG. **J Am Anim Hosp Assoc** 1983;19(5):755-760. Gill PJ, et al**. J Am Anim Hosp Assoc** 1996;32(1):77-80. Hoerlein BF. **WB Saunders Company**, 3rd ed. 1978:530-541. Lemarie RJ, et al. **J Am Anim Hosp Assoc** 2000:36(4):348-358. Lipsitz D, Bailey CS. **Prog Vet Neurol** 1995;6(2);60-65. Rossmeisl JH, et al. Vet Surg 2005;34:436-444. Seim HB, Prata RG. **J Am Anim Hosp Assoc** 1982;18(2):233-240. Sharp NJH, Wheeler J. **Elsevier Mosby**, 3rd ed. 2005:93-120. Tanaka H, et al. **J Vet Med Sci** 2005;67(5):679-683.

INTRACTABLE PARTIAL EPILEPSY

Gregory A. Worrell, MD, PhD
Rochester, MN

INTRACTABLE PARTIAL EPILEPSY

The World Health Organization estimates that epilepsy is the cause of 1% of the global burden of disease based on productive years lost to disability or premature death [13] [4]. Partial, or localization-related epilepsy, is the most common seizure disorder and 20% - 40% of patients will experience physically and socially disabling drug resistant seizures[14] [5, 15, 16]. In fact, the majority of studies show that patients who fail to respond to initial first-line anti-epileptic drug (AED) therapy trials have a low *probability of being seizure-free with continued drug trials [5, 15].* The most wdely quoted study found less than 3% of patients are rendered seizure free after failure (due to lack of efficacy) of only 2 antiepileptic drugs [5, 15]. This study showed that 47% of patients respond to the first AED, 13% respond to the second AED, and only 3% respond to a third drug or multiple drugs. Of those patients who did not respond to the first AED because of lack of efficacy, only 11% subsequently became seizure free. These results suggest that a patient who has adequate trials of 2 AEDs, failed because of efficacy and not side effects, the probability of pharmacological seizure remission are low. Generally, after failing 2 or 3 AED individuals are considered to have drug resistant epilepsy (DRE) [5, 15]. It should be noted, however, that recent studies have shown higher seizure remission rates in some populations [17, 18]. French et al [17] reported that 32 (16%) of 198 patients with poorly controlled partial epilepsy achieved a 6 month seizure remission during a three year period. Notably, 21 of these patients also had surgery. Nonetheless, the prediction and classification of DRE remains a challenge [4, 17, 18].

EPILEPSY SURGERY

Epilepsy surgery is an effective and safe therapy for carefully selected patients with drug resistant partial epilepsy [19-22] [23]. While only one study represents class I evidence [24] there is now widespread consensus (Neurology consensus practice parameter[25]) that epilepsy surgery should be considered for patients with drug resistant localization related partial epilepsy. In particular, many patients with medial temporal lobe epilepsy or lesional epilepsy have surgically remediable epilepsy syndromes[4, 7]. The majority of these patients will experience a significant reduction in seizure frequency following surgical resection of the region of epileptic brain [7, 20, 25-30]. A recent study using a decision analysis model showed that on average anterior temporal lobe resection provides substantial gains in life expectancy and quality adjusted life expectancy for surgically eligible patients with DRE compared with medical management [23].

Patients with medial temporal lobe epilepsy and lesional epileptic syndromes are generally the most favorable candidates for surgical resection. The frequency of seizure activity, type of seizure(s), co-morbidity, and underlying pathology are important variables when evaluating a patient's candidacy for surgical treatment. The impact of chronic epilepsy on psychosocial, economic, and quality of life should be taken into account for all individuals[16]. The goals of epilepsy surgery are to eliminate seizures and improve their quality of life, e.g. allow individuals to work, attend school, live independently, and drive [23, 31, 32].

The most common operative procedures are resection of epileptic brain tissue, e.g., focal corticectomy, and anterior temporal lobectomy [7, 25]. The rationale for surgical treatment is that removing the epileptogenic zone, i.e., site of seizure onset and initial seizure propagation [21, 29, 33] will eliminate seizures. The preoperative evaluation is directed at identifying the site of seizure onset and the location of functional eloquent cortex. Most patients undergo video EEG monitoring to record their habitual seizures, MRI to identify possible structural abnormalities, neuropsychological evaluation, visual field testing, and WADA testing (Table 2). The high diagnostic yield of MRI to identify the common pathological substrates, e.g., post-traumatic encephalomalcia, vascular malformation, tumor, malformations of cortical development, mesial temporal sclerosis (MTS), is well established in patients with DRE (Table 1). Selected patients will require functional neuroimaging, i.e., PET, SPECT, functional MRI, and possibly chronic intracranial EEG monitoring to localize the region of seizure onset [20, 21, 29, 33-38].

The outcome of epilepsy surgery is dependent on many variables including the age of seizure onset, location of the epileptic brain tissue, underlying pathology, and surgical strategy. Favorable prognostic indicators include an early age of seizure onset, medial temporal lobe seizure onset, and a completely resectable structural lesion on MRI, e.g. MTS, cavernous malformation, low-grade neoplasms [26, 30, 39]. Approximately 80% to 90% of patients with medial temporal lobe epilepsy and the above lesional epilepsy syndromes may be rendered seizure-free or near seizure-free following a total excision of the epileptogenic zone. Less favorable operative candidates include patients with malformations of cortical development (MCD) and non-lesional epilepsy [35, 40-43]. In particular non-lesional extratemporal partial epilepsy remains a significant challenge [30, 44, 45] [46] [47, 48].

The cost of the presurgical evaluation and treatment is considerable, ranging from approximately $30,000 to $100,000. This cost, however, must be compared within the context of intractable epilepsy that includes antiepileptic drug therapy, office visits, laboratory studies, unemployment, and lost productivity[4, 13, 23]. In addition to the psychosocial benefits for individuals the large-scale economic benefits to society are now well recognized.

LOCALIZATION OF THE EPILEPTOGENIC ZONE

The presurgical evaluation for epilepsy surgery is focused on localization of epileptic brain, i.e. the region of brain generating the patients habitual seizures. The cornerstones of the presurgical evaluation are the 1.) Clinical history and Neurological exam 2.) Scalp video-EEG to record the patients habitual seizures 3.) Seizure protocol MRI, and for some patients 4.) functional neuroimaging (SPECT, PET, MRS, fMRI), and chronic intracranial EEG monitoring (iEEG). The surgical treatment is based on the concept that seizures begin in a discrete region of brain, the *seizure onset zone (SOZ)*, and then propagate to a critical volume of "susceptible tissue" to generate seizures. In order to obtain seizure freedom, the SOZ and the surrounding *epileptogenic zone (EZ)* must be resected[33]. Unfortunately, the EZ does not have an anatomical, imaging, or electrophysiological definition, and is basically a theoretical construct acknowledging that resection of the SOZ does not guarantee seizure free outcomes. In patients with a focal lesion on MRI, e.g. low grade tumor or a vascular malformation, if the SOZ localizes to the MRI lesion then *complete* resection of the MRI lesion will most often eliminate the seizures[7]. The rate of seizure free outcomes in these selected patients can approach 80-90%, ***demonstrating the efficacy of epilepsy surgery*** [4, 7, 21, 29, 30] [39]. However, many patients have a normal MRI, or lesions that extend beyond what is evident on MRI, e.g. malformation of cortical development [43], or into functionally eloquent brain that cannot be safely removed. These patients make up at least 30% of evaluations at major epilepsy centers. For patients with normal MRI scans only 30-50% of those finally deemed surgical candidates will achieve seizure freedom [30, 44, 45] [46] [47, 48], ***demonstrating the current limitation of epilepsy surgery.***

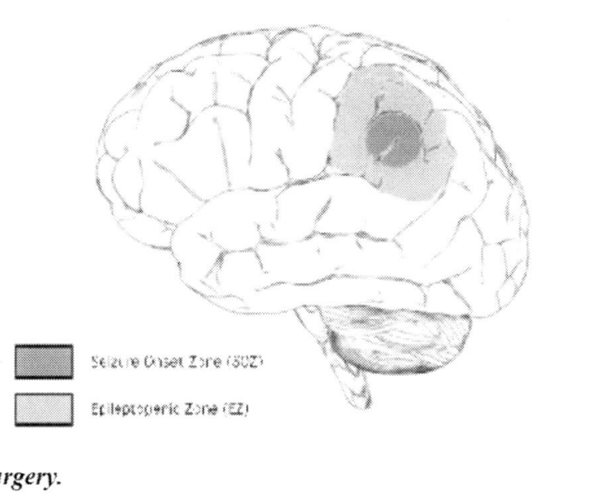

Figure 1: The epileptogenic zone is defined as the region that must be resected for the patient to be seizure free. The seizure onset zone is the region of seizure onset determined by iEEG.

Many patients with DRE are not candidates for epilepsy surgery. Fortunately, there are emerging therapeutic options, including responsive brain stimulation, chronic deep brain stimulation, focal cooling, and seizure prediction devices. In this talk DRE will be reviewed, and non-pharmacological treatments including epilepsy surgery and emerging therapies will be discussed.

REFERENCES

1.Bell, G.S., et al., Drowning in people with epilepsy: how great is the risk? Neurology, 2008. 71(8): p. 578-82. 2. Hitiris, N., et al., *Mortality in epilepsy.* Epilepsy Behav, 2007. 10(3): p. 363-76. 3. So, E.L., et al., Report of the American Epilepsy Society and the Epilepsy Foundation Joint Task Force on Sudden Unexplained Death in Epilepsy. Epilepsia, 2008. 4. Engel, J., Jr., Surgical treatment for epilepsy: too little, too late? Jama, 2008. 300(21): p. 2548-50. 5. Kwan, P. and M.J. Brodie, *Early identification of refractory epilepsy.* N Engl J Med, 2000. 342(5): p. 314-9. 6. Szaflarski, J.P., et al., Seizure control in patients with epilepsy: the physician vs. medication factors. BMC Health Serv Res, 2008. 8(1): p. 264. 7. Cascino, G., P. Boon, and D. Fish, *Surgically remedial lesional syndromes.*, in *Surgical treatment of the epilepsies*, J.J. Engel, Editor. 1993, Raven Press: New York. p. 77-86. 8. Cascino, G.D., Commentary: how has neuroimaging improved patient care? Epilepsia, 1994. 35: p. S103-104. 9. Oommen, J., M. Morrell, and R.S. Fisher, *Experimental Electrical Stimulation Therapy for Epilepsy.* Curr Treat Options Neurol, 2005. 7(4): p. 261-271. 10. Theodore, W.H. and R. Fisher, *Brain stimulation for epilepsy.* Acta Neurochir Suppl, 2007. 97(Pt 2): p. 261-72. 11. Theodore, W.H. and R.S. Fisher, *Brain stimulation for epilepsy.* Lancet Neurol, 2004. 3(2): p. 111-8. 12. Cascino, G.D., *When drugs and surgery don't work.* Epilepsia, 2008. 49 Suppl 9: p. 79-84. 13. Murray, G. and A. Lopez, Global Comparative Assessments in the Health Sector: Disease Burden, Expenditure, Intervention Packages. 1994, World Health Organization: Geneva, Switzerland. 14. Berg, A.T., Understanding the delay before epilepsy surgery: who develops intractable focal epilepsy and when? CNS Spectr, 2004. 9(2): p. 136-44. 15. Kwan, P. and M.J. Brodie, Drug treatment of epilepsy: when does it fail and how to optimize its use? CNS Spectr, 2004. 9(2): p. 110-9. 16. Fisher, R.S., et al., The impact of epilepsy from the patient's perspective I. Descriptions and subjective perceptions. Epilepsy Res, 2000. 41(1): p. 39-51. 17. Callaghan, B.C., et al., Likelihood of seizure remission in an adult population with refractory epilepsy. Ann Neurol, 2007. 62(4): p. 382-9. 18. French, J.A., *Refractory epilepsy: one size does not fit all.* Epilepsy Curr, 2006. 6(6): p. 177-80. 19. Cascino, G., P. Boon, and D. Fish, *Surgically remediable lesional syndromes.*, in *Surgical Treatment of the Epilepsies*, J.J. Engel, Editor. 1993, Raven Press: New York. p. 77-86. 20. Cascino, G.D., *Surgical treatment for epilepsy.* Epilepsy Research, 2004. 60(2-3): p. 179-86. 21. Engel, J.J. and T.A. Pedley, *Epilepsy: a comprehensive textbook.* 1997, Philadelphia, PA: Lippincott-Raven Publishers. 22. Luders, J.J. and Y. Comair, eds. *Epilepsy Surgery.* 2001, Lippincott Williams & Wilkins: Philadelphia. 23. Choi, H., et al., Epilepsy surgery for pharmacoresistant temporal lobe epilepsy: a decision analysis. Jama, 2008. 300(21): p. 2497-505. 24. Wiebe, S., et al., A randomized, controlled trial of surgery for temporal-lobe epilepsy.[see comment]. New England Journal of Medicine, 2001. 345(5): p. 311-8. 25. Engel, J., Jr., et al., Practice parameter: temporal lobe and localized neocortical resections for epilepsy: report of the Quality Standards Subcommittee of the American Academy of Neurology, in association with the American Epilepsy Society and the American Association of Neurological Surgeons. Neurology, 2003. 60(4): p. 538-47. 26. Cohen-Gadol, A.A., et al., Long-term outcome of epilepsy surgery among 399 patients with nonlesional seizure foci including mesial temporal lobe sclerosis. Journal of Neurosurgery, 2006. 104(4): p. 513-24. 27. Engel J, H.-G.W., Spencer D, *Overview: Surgical Therapy*, in *Epilepsy: A Comprehensive Textbook*, P.T. Engel J, Editor. 1997, Lippincott-Raven Publishers: Philadelphia. p. 1673-76. 28. Engel, J.J., *Principles of epilepsy surgery*, in *The treatment of epilepsy*, S. Shorvon, F. Dreifuss, and D. Fish, Editors. 1996, Blackwell: Oxford. p. 519-529. 29. Luders, H., *Epilepsy Surgery.* 1992, New York: Raven Press. 30. Radhakrishnan, K., et al., Predictors of outcome of anterior temporal lobectomy for intractable epilepsy: a multivariate study. Neurology, 1998. 51(2): p. 465-71. 31. Spencer, S. and L. Huh, *Outcomes of epilepsy surgery in adults and children.* Lancet Neurol, 2008. 7(6): p. 525-37. 32. Tellez-Zenteno, J.F. and S. Wiebe, *Long-term seizure and psychosocial outcomes of epilepsy surgery.* Curr Treat Options Neurol, 2008. 10(4): p. 253-9. 33. Luders, H. and I. Awad, *Conceptual Considerations*, in *Epilepsy Surgery*, H. Luders, Editor. 1992, Raven Press: New York. p. 51-62. 34. Cascino, G.D., et al., *Ictal SPECT in nonlesional extratemporal epilepsy.* Epilepsia, 2004. 45 Suppl 4: p. 32-4. 35. Cascino, G.D. and T.J. O'Brien, *Resection for Epilepsy in the setting of a nonlocalizing MRI*, in *Treatment of Epilepsy*, E. Wyllie, Editor. 2001, Lippincott Williams and Williams: Philadelphia. p. 1135-1145. 36. So, E.L., Integration of EEG, MRI, and SPECT in localizing the seizure focus for epilepsy surgery. Epilepsia, 2000. 41(Suppl 3): p. S48-54. 37. So, E.L., Role of neuroimaging in the management of seizure disorders. Mayo Clin Proc, 2002. 77(11): p. 1251-64. 38. Spencer, S., M. Sperling, and D. Shewmon, *Intracranial Electrodes*, in *Epilepsy: a comprehensive textbook*, J. Engel, Jr. and T. Pedley, Editors. 1997, Lippincott-Raven Publishers: Philadelphia, PA. p. 1719-1747. 39. Awad, I.A., et al., Intractable epilepsy and structural lesions of the brain: mapping, resection strategies, and seizure outcome. 40. Epilepsia, 1991. 32(2): p. 179-86. 41. Alonso-Nanclares, L., et al., *Microanatomy of the dysplastic neocortex from epileptic patients.* Brain, 2005. 128(Pt 1): p. 158-73. 42. Fauser, S., et al., Clinical characteristics in focal cortical dysplasia: a retrospective evaluation in a series of 120 patients. Brain, 2006. 129(Pt 7): p. 1907-16. 43. Kim, D.W., et al., Predictors of surgical outcome and pathologic considerations in focal cortical dysplasia.

Neurology, 2008. 44. Palmini, A., et al., Focal neuronal migration disorders and intractable partial epilepsy: a study of 30 patients. Ann Neurol, 1991. 30(6): p. 741-9. 45. Chapman, K., et al., *Seizure outcome after epilepsy surgery in patients with normal preoperative MRI.* Journal of Neurology, Neurosurgery & Psychiatry, 2005. 76(5): p. 710-3. 46. Mosewich, R.K., et al., Factors predictive of the outcome of frontal lobe epilepsy surgery. Epilepsia, 2000. 41(7): p. 843-9. 47. Smith, J.R., et al., *Results of lesional vs. nonlesional frontal lobe epilepsy surgery.* Stereotact Funct Neurosurg, 1997. 69(1-4 Pt 2): p. 202-9. 46. Schomer, D.L. and P.M. Black, A 24-year-old woman with intractable seizures: review of surgery for epilepsy. Jama, 2008. 300(21): p. 2527-38. 48. Jeha, L.E., et al., Surgical outcome and prognostic factors of frontal lobe epilepsy surgery. Brain, 2007. 130(Pt 2): p. 574-84.

BEYOND DRUGS: SURGERY & DEVICES FOR INTRACTABLE EPILEPSY

Gregory A. Worrell, MD, PhD
Rochester, MN

SURGICAL ADVANCES

The use of gamma-knife for treatment of DRE is currently under investigation. While the us of gamma-knife for treatment of arterial venous malformations is widely used, the role of gamma-knife for treatment of substrate directed epilepsy remains investigational. The current data suggest slightly reduced and delayed effacicy compared to resective surgery for mesial temporal epilepsy. The psot pocedure morbidity and mortality are higher, and currently there is not convincing evidence that there are cognitive benefits.

Whether gamma-knife has a role in treatment of seizures that co-localize with eloqent brain or regions that are not safely accessible for resection remains unexplored.

DEVICES FOR THE TREATMENT OF EPILEPSY

Advances in neural engineering are poised to deliver exciting new technologies to the treatment of brain disease (neurological and psychiatric diseases). One of the most active areas has been the development of devices that may predict, prevent, and abort seizures.

Electrical stimulation has a long history in experimental neuroscience. The clinical use of electrical stimulation was first reported by Frisch and Hitzig (1870) to localize eloquent cortex. Krause (1931) and Foerster (1926) later used electrical stimulation during operations on patients with epilepsy, and further explored localization of brain function. The seminal work of Penfield and Jasper greatly extended the use of electrical stimulation for localization of cortical function, and reported the first description of a seizure resulting from cortical stimulation.

I.S. Cooper first proposed the use of electrical stimulation for the treatment of epilepsy in the 1970's and he reported a reduction in seizure frequency with chronic cerebellar stimulation. However, later controlled trials did not confirm the efficacy of cerebellar stimulation for the treatment of epilepsy. This early failure underscores the need for well designed clinical trials to establish the efficacy of electrical stimulation before these methods can be recommended in clinical practice.

Vagus Nerve Stimulation

VNS is an adjunctive treatment for patients with medically refractory partial epilepsy. The device was approved by the US Food and Drug Administration (FDA) for partial epilepsy in patients 12 years of age or older in 1997. It delivers intermittent electrical stimulation to the left vagus nerve. The device is uses a helical electrode that wraps around the vagus nerve and is powered by an implanted programmable pulse generator. Currently, more than 50,000 patients are treated with VNS. This is a palliative therapeutic modality with a response rate similar to AED therapy, with approximately 50% reduction in seizure activity in approximately one-third to one-half of individuals. Very rarely will patients achieve a seizure remission with this device.

Neuropace Responsive Neurostimulation

The RNS trial was a multi-center, double-blind, randomized trial of responsive neural stimulation applied to the region of brain generating seizures. In this study from 32 epilepsy centers, 191 participants were implanted with the Neuropace RNS system. In the pivotal trial phase 93 patients randomized to sham stimulation and 96 patients received responsive stimulation delivered on detection of epileptiform activity. Unadjusted median declines at the end of the blinded phase were 41% reduction in seizures in the group receiving stimulation compared to a 9% reduction in the sham stimulation group. Using a generalized estimating equation there was a 21% greater reduction in the group receiving stimulation compared to the control (sham) group not receiving stimulation After the blinded phase all patients received stimulation, and over the most recent 3 months period 51% of patients had a seizure reduction of at least 50%.

The RNS trial showed evidence of efficacy and modest complication rates. During the course of the trial five deaths occurred, but none were believed to be related to implantation or stimulation.

STIMULATION OF ANTERIOR NUCLEUS OF THALAMUS

The SANTE trial recently reported by Fisher et al. (Fisher R, et al. Epilepsia 51: 899-908, 2010). new reference was a multi-center, double-blind, randomized trial of bilateral stimulation of the anterior nuclei of thalamus for partial epilepsy. In this study from 17 epilepsy centers, 110 participants were implanted with the Medtronic DBS device, and half randomized to sham stimulation and half to duty cycle stimulation. Baseline monthly median seizure frequency was 19.5. In the last month of the blinded phase the stimulated group had a 29% greater reduction in seizures compared with the control group as estimated by a generalized estimating equations (GEE) model (p=0.002). Unadjusted median declines at the end of the blinded phase were 14.5% in the sham stimulation (control group) and 40.4% in the group receiving stimulation. After the blinded phase all patients received stimulation, and after two years 54% of patients had a seizure reduction of at least 50%.

The SANTE trial showed evidence of efficacy and modest complication rates. During the course of the trial five deaths occurred, but none were believed to be related to implantation or stimulation.

CONCLUSIONS

In carefully selected patients patients who do not respond to medication epilepsy surgery remains the best chance for eliminating seizures. The results of well-designed, but relatively small trials investigating the benefit of Adkins and modified Adkins diets for adults are very encouraging. The safety and feasibility of deep brain, hippocampus, and neocortical stimulation for the treatment of epilepsy is established. The evidence for efficacy is emerging. A well designed multi-center trial (Medtronics, SANTE) has demonstrated the feasibility, safety and efficacy of anterior nucleus of thalamus stimulation. A multi-center trial investigating responsive stimulation of hippocampal and neocortical seizure is underway. The trial (Neuropace, RNS) has already demonstrated feasibility and safety, and results for proof of efficacy should be coming soon.

Well designed clinical trials are leading the way!

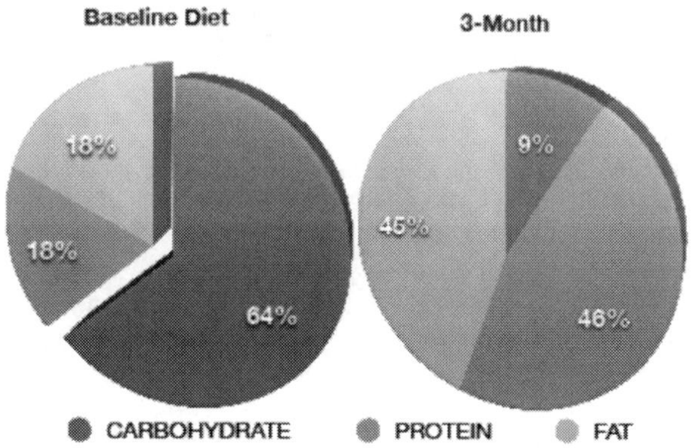

REFERENCES

1.Bell, G.S., et al., Drowning in people with epilepsy: how great is the risk? Neurology, 2008. 71(8): p. 578-82. 2.Hitiris, N., et al., Mortality in epilepsy. Epilepsy Behav, 2007. 10(3): p. 363-76. 3. So, E.L., et al., Report of the American Epilepsy Society and the Epilepsy Foundation Joint Task Force on Sudden Unexplained Death in Epilepsy. Epilepsia, 2008. 4. Engel, J., Jr., Surgical treatment for epilepsy: too little, too late? Jama, 2008. 300(21): p. 2548-50. 5. Kwan, P. and M.J. Brodie, Early identification of refractory epilepsy. N Engl J Med, 2000. 342(5): p. 314-9. 6. Szaflarski, J.P., et al., Seizure control in patients with epilepsy: the physician vs. medication factors. BMC Health Serv Res, 2008. 8(1): p. 264. 7. Cascino, G., P. Boon, and D. Fish, Surgically remedial lesional syndromes., in Surgical treatment of the epilepsies, J.J. Engel, Editor. 1993, Raven Press: New York. p. 77-86. 8. Cascino, G.D., Commentary: how has neuroimaging improved patient care? Epilepsia, 1994. 35: p. S103-104. 9. Oommen, J., M. Morrell, and R.S. Fisher, Experimental Electrical Stimulation Therapy for Epilepsy. Curr Treat Options Neurol, 2005. 7(4): p. 261-271. 10. Theodore, W.H. and R. Fisher, Brain stimulation for epilepsy. Acta Neurochir Suppl, 2007. 97(Pt 2): p. 261-72. 11. Theodore, W.H. and R.S. Fisher, Brain stimulation for epilepsy. Lancet Neurol, 2004. 3(2): p. 111-8. 12. Cascino, G.D., When drugs and surgery don't work. Epilepsia, 2008. 49 Suppl 9: p. 79-84. 13. Murray, G. and A. Lopez, Global Comparative Assessments in the Health Sector: Disease Burden, Expenditure, Intervention Packages. 1994, World Health Organization: Geneva, Switzerland. 14. Berg, A.T., Understanding the delay before epilepsy surgery: who develops intractable focal epilepsy and when? CNS Spectr, 2004. 9(2): p. 136-44. 15. Kwan, P. and M.J. Brodie, Drug treatment of epilepsy: when does it fail and how to optimize its use? CNS Spectr, 2004. 9(2): p. 110-9. 16. Fisher, R.S., et al., The impact of epilepsy from the patient's perspective I. Descriptions and subjective perceptions. Epilepsy Res, 2000. 41(1): p. 39-51. 17. Callaghan, B.C., et al., Likelihood of seizure remission in an adult population with refractory epilepsy. Ann Neurol, 2007. 62(4): p. 382-9. 18. French, J.A., Refractory epilepsy: one size does not fit all. Epilepsy Curr, 2006. 6(6): p. 177-80. 19. Cascino, G., P. Boon, and D. Fish, Surgically remediable lesional syndromes., in Surgical Treatment of the Epilepsies, J.J. Engel, Editor. 1993, Raven Press: New York. p. 77-86. 20. Cascino, G.D., Surgical treatment for epilepsy. Epilepsy Research, 2004. 60(2-3): p. 179-86. 21. Engel, J.J. and T.A. Pedley, Epilepsy: a comprehensive textbook. 1997, Philadelphia, PA: Lippincott-Raven Publishers. 22. Luders, J.J. and Y. Comair, eds. Epilepsy Surgery. 2001, Lippincott Williams & Wilkins: Philadelphia. 23. Choi, H., et al., Epilepsy surgery for pharmacoresistant temporal lobe epilepsy: a decision analysis. Jama, 2008. 300(21): p. 2497-505. 24. Wiebe, S., et al., A randomized, controlled trial of surgery for temporal-lobe epilepsy.[see comment]. New England Journal of Medicine, 2001. 345(5): p. 311-8. 25. Engel, J., Jr., et al., Practice parameter: temporal lobe and localized neocortical resections for epilepsy: report of the Quality Standards Subcommittee of the American Academy of Neurology, in association with the American Epilepsy Society and the American Association of Neurological Surgeons. Neurology, 2003. 60(4): p. 538-47. 26. Cohen-Gadol, A.A., et al., Long-term outcome of epilepsy surgery among 399 patients with nonlesional seizure foci including mesial temporal lobe sclerosis. Journal of Neurosurgery, 2006. 104(4): p. 513-24. 27. Engel J, H.-G.W., Spencer D, Overview: Surgical Therapy, in Epilepsy: A Comprehensive Textbook, P.T. Engel J, Editor. 1997, Lippincott-Raven Publishers: Philadelphia. p. 1673-76. 28. Engel, J.J., Principles of epilepsy surgery, in The treatment of epilepsy, S. Shorvon, F. Dreifuss, and D. Fish, Editors. 1996, Blackwell: Oxford. p. 519-529. 29.Luders, H., Epilepsy Surgery. 1992, New York: Raven Press. 30. Radhakrishnan, K., et al., Predictors of outcome of anterior temporal lobectomy for intractable epilepsy: a multivariate study. Neurology, 1998. 51(2): p. 465-71. 31. Spencer, S. and L. Huh, Outcomes of epilepsy surgery in adults and children. Lancet Neurol, 2008. 7(6): p. 525-37. 32. Tellez-Zenteno, J.F. and S. Wiebe, Long-term seizure and psychosocial outcomes of epilepsy surgery. Curr Treat Options Neurol, 2008. 10(4): p. 253-9. 33. Luders, H. and I. Awad, Conceptual Considerations, in Epilepsy Surgery, H. Luders, Editor. 1992, Raven Press: New York. p. 51-62. 34. Cascino, G.D., et al., Ictal SPECT in nonlesional extratemporal epilepsy. Epilepsia, 2004. 45 Suppl 4: p. 32-4. 35. Cascino, G.D. and T.J. O'Brien, Resection for Epilepsy in the setting of a nonlocalizing MRI, in Treatment of Epilepsy, E. Wyllie, Editor. 2001, Lippincott Williams and Williams: Philadelphia. p. 1135-1145. 36. So, E.L., Integration of EEG, MRI, and SPECT in localizing the seizure focus for epilepsy surgery. Epilepsia, 2000. 41(Suppl 3): p. S48-54. 37. So, E.L., Role of neuroimaging in the management of seizure disorders. Mayo Clin Proc, 2002. 77(11): p. 1251-64. 38. Spencer, S., M. Sperling, and D. Shewmon, Intracranial Electrodes, in Epilepsy: a comprehensive textbook, J. Engel, Jr. and T. Pedley, Editors. 1997, Lippincott-Raven Publishers: Philadelphia, PA. p. 1719-1747. 39. Awad, I.A., et al., Intractable epilepsy and structural lesions of the brain: mapping, resection strategies, and seizure outcome. Epilepsia, 1991. 32(2): p. 179-86. 40. Alonso-Nanclares, L., et al., Microanatomy of the dysplastic neocortex from epileptic patients. Brain, 2005. 128(Pt 1): p. 158-73. 41. Fauser, S., et al., Clinical characteristics in focal cortical dysplasia: a retrospective evaluation in a series of 120 patients. Brain, 2006. 129(Pt 7): p. 1907-16. 42. Kim, D.W., et al., Predictors of surgical outcome and pathologic considerations in focal cortical dysplasia. Neurology, 2008. 43. Palmini, A., et al., Focal neuronal migration disorders and intractable partial epilepsy: a study of 30 patients. Ann Neurol, 1991. 30(6): p. 741-9. 44. Chapman, K., et al., Seizure outcome after epilepsy surgery in patients with normal preoperative MRI. Journal of Neurology, Neurosurgery & Psychiatry, 005. 76(5): p. 710-3. 45. Mosewich, R.K., et al., Factors predictive of the outcome of frontal lobe epilepsy surgery. Epilepsia, 2000. 41(7): p. 843-9. 46. Smith, J.R., et al., Results of lesional vs. nonlesional frontal lobe epilepsy surgery. Stereotact Funct Neurosurg, 1997. 69(1-4 Pt 2): p. 202-9. 47. Schomer, D.L. and P.M. Black, A 24-year-old woman with intractable seizures: review of surgery for epilepsy. Jama, 2008. 300(21): p. 2527-38. 48. Jeha, L.E., et al., Surgical outcome and prognostic factors of frontal lobe epilepsy surgery. Brain, 2007. 130(Pt 2): p. 574-84. 49. Cascino, G.D., A. Hopkins, and S.D. Shorvon, eds. Epilepsy. 1995, Lippincott Williams & Wilkins: Philadelphia. 50. Malow, B.A., et al., Treating obstructive sleep apnea in adults with epilepsy: a randomized pilot trial. Neurology, 2008. 71(8): p. 572-7. 51. Chen, D.K., et al., Sensitivity and specificity of video alone versus electroencephalography alone for the diagnosis of partial seizures. Epilepsy Behav, 2008. 13(1): p. 115-8. 52. So, E.L., Value and limitations of seizure semiology in localizing seizure onset. J Clin Neurophysiol, 2006. 23(4): p. 353-7. 53. Curio, G., Ain't no rhythm fast enough: EEG bands beyond beta. J Clin Neurophysiol, 2000. 17(4): p. 339-40. 54. Vanhatalo, S., J. Voipio, and K. Kaila, Full-band EEG (FbEEG): an emerging standard in electroencephalography. Clin Neurophysiol, 2005. 116(1): p. 1-8. 55. Jack, C.R., Jr., Magnetic resonance imaging. Neuroimaging and anatomy. Neuroimaging Clin N Am, 1995. 5(4): p. 597-622. 56. Jack, C.R., Jr., Magnetic resonance imaging in epilepsy. Mayo Clinic Proceedings, 1996. 71(7): P. 695-711. 57. Berkovic, S., et al., Preoperative MRI predicts outcome of temporal lobectomy: an actuarial analysis. Neurology, 1995. 45: p. 1358-1363. 58. Kuzniecky, R.I., et al., Multimodality MRI in mesial temporal sclerosis: relative sensitivity and specificity. Neurology, 1997. 49(3): p. 774-8. 59. Yun, C.H., et al., Prognostic factors in neocortical epilepsy surgery: multivariate analysis. Epilepsia, 2006. 47(3): p. 574-9. 60. Glikmann-Johnston, Y., et al., Structural and functional correlates of unilateral mesial temporal lobe spatial memory impairment. Brain, 2008. 131(Pt 11): p. 3006-18. 61. Kanner, A.M., Can fMRI replace the Wada test in predicting postsurgical deterioration of verbal memory? Nat Clin Pract Neurol, 2008. 4(7): p. 364-5. 62. Krsek, P., et al., Incomplete resection of focal cortical dysplasia is the main predictor of poor postsurgical outcome. Neurology, 2008. 63.Carne, R.P., et al., 'MRI-negative PET-positive' temporal lobe epilepsy (TLE) and mesial TLE differ with quantitative MRI and PET: a case control study. BMC Neurol, 2007. 7: p. 16. 64.Dunn, J.F., et al., Functional brain mapping at 9.4T using a new MRI-compatible electrode chronically implanted in rats. Magn Reson Med, 2009. 61(1): p. 222-8. 65.Madhavan, D. and R. Kuzniecky, Temporal lobe surgery in patients with normal MRI. Curr Opin Neurol, 2007. 20(2): p. 203-7. 66. Jack, C.R., Jr., et al., Mesial temporal sclerosis: diagnosis with fluid-attenuated inversion-recovery versus spin-echo MR imaging.[see comment]. Radiology, 1996. 199(2): p. 367-73. 67. Andermann, L.F., et al., Psychosis after resection of ganglioglioma or DNET: evidence for an association. Epilepsia, 1999. 40(1): p. 83-7. 68. Burneo, J.G., et al., Adult-onset epilepsy associated with dysembryoplastic neuroepithelial tumors. Seizure, 2008. 17(6): p. 498-504. 69. Schramm, J. and A.F. Aliashkevich, Surgery for temporal mediobasal tumors: experience based on a series of 235 patients. Neurosurgery, 2008. 62(6 Suppl 3): p. 1272-82. 70. Stavrou, I., et al., Long-term seizure control after resection of supratentorial cavernomas: a retrospective single-center study in 53 patients. Neurosurgery, 2008. 63(5): p. 888-96; discussion 897. 71. Dodick, D.W., G.D. Cascino, and F.B. Meyer, Vascular malformations and intractable epilepsy: outcome after surgical treatment. Mayo Clin Proc, 1994. 69(8): p. 741-5. 72. Kraemer, D.L. and I.A. Awad, Vascular malformations and epilepsy: clinical considerations and basic mechanisms. Epilepsia, 1994. 35 Suppl 6: p. S30-43. 73. Lowe, A.J., et al., Epilepsy surgery for pathologically proven hippocampal sclerosis provides long-term seizure control and improved quality of life.[see comment]. Epilepsia, 2004. 45(3): p. 237-42. 74. McIntosh, A.M., et al., Temporal lobectomy: long-term seizure outcome, late recurrence and risks for seizure recurrence. Brain, 2004. 127(Pt 9): p. 2018-30. 75. Jeha, L.E., et al., Predictors of outcome after temporal lobectomy for the treatment of intractable epilepsy. Neurology, 2006. 66(12): p. 1938-40. 76. Janszky, J., et al., Temporal lobe epilepsy with hippocampal sclerosis: predictors for long-term surgical outcome. Brain, 2005. 128(Pt 2): p. 395-404. 77. Quesney, L.F., Intracranial EEG investigation in neocortical epilepsy. Adv Neurol, 2000. 84: p. 253-74. 78. Carne, R.P., et al., MRI-negative PET-positive temporal lobe epilepsy: a distinct surgically remediable syndrome. Brain, 2004. 127(Pt 10): p. 2276-85. 79. O'Brien, T.J., et al., Comparative study of 99mTc-ECD and 99mTc-HMPAO for peri-ictal SPECT: qualitative and quantitative analysis. J Neurol Neurosurg Psychiatry, 1999. 66(3): p. 331-9. 80. O'Brien, T.J., et al., Subtraction SPECT coregistered to MRI in focal malformations of cortical development: localization of the epileptogenic zone in epilepsy surgery candidates. Epilepsia, 2004. 45(4): p. 367-76. 81. O'Brien, T.J., et al., Subtraction peri-ictal SPECT is predictive of extratemporal epilepsy surgery outcome. Neurology, 2000. 55(11): p. 1668-77. 82. O'Brien, T.J., et al., Subtraction ictal SPECT co-registered to MRI improves clinical usefulness of SPECT in localizing the surgical seizure focus. Neurology, 1998. 50(2): p. 445-54. 83. O'Brien, T.J., et al., Subtraction SPECT co-registered to MRI improves postictal SPECT localization of seizure foci. Neurology, 1999. 52(1): p. 137-46. 84. Theodore, W.H., et al., FDG-positron emission tomography and invasive EEG: seizure focus detection and surgical outcome. Epilepsia, 1997. 38(1): p. 81-6. 85. Salamon, N., et al., FDG-PET/MRI coregistration improves detection of cortical dysplasia in patients with epilepsy. Neurology, 2008. 71(20): p. 1594-601. 86. Henry, T.R., et al., Hippocampal neuronal loss and regional hypometabolism in temporal lobe epilepsy. Ann Neurol, 1994. 36(6): p. 925-7. 87. Ho, S.S., et al., Comparison of ictal SPECT and interictal PET in the presurgical evaluation of temporal lobe epilepsy. Ann Neurol, 1995. 37(6): p. 738-45. 88. Theodore, W., ed. Positron Emission Tomography in the Evaluation of Epilepsy., ed. J.C.J. Cascino GD. 1996, Butterworth-Heinemann: Boston. 165 – 175. 89. Spencer, S.S., The relative contributions of MRI, SPECT, and PET imaging in epilepsy. Epilepsia, 1994. 35 Suppl 6: p. S72-89. 90. Sata, Y., et al., Quantitative analysis of benzodiazepine receptor in temporal lobe epilepsy: [(125)I]iomazenil autoradiographic study of surgically resected specimens. Epilepsia, 2002. 43(9): p. 1039-48. 91. Juhasz, C., et al., Alpha-methyl-L-tryptophan PET detects epileptogenic cortex in children with intractable epilepsy. Neurology, 2003. 60(6): p. 960-8. 92. Suhy, J., et al., 1H MRSI predicts surgical outcome in MRI-negative temporal lobe epilepsy. Neurology, 2002. 58(5): p. 821-3. 93. Kuzniecky, R., Magnetic resonance spectroscopy in focal epilepsy: 31P and 1H spectroscopy. Rev Neurol (Paris), 1999. 155(6-7): p. 495-8. 94. Vermathen, P., et al., Hippocampal N-acetylaspartate in neocortical epilepsy and mesial temporal lobe epilepsy. Ann Neurol, 1997. 42(2): p. 194-9. 95. Duncan, J.S., Imaging and epilepsy. Brain, 1997. 120 (Pt 2): p. 339-77. 96. Duncan, J.S., Magnetic resonance spectroscopy. Epilepsia, 1996. 37(7): p. 598-605. 97. Cendes, F., et al., Proton magnetic resonance spectroscopic imaging and magnetic resonance imaging volumetry in the lateralization of temporal lobe epilepsy: a series of 100 patients. Ann Neurol, 1997. 42(5): p. 737-46. 98. Kutsy, R.L., D.F. Farrell, and G.A. Ojemann, Ictal patterns of neocortical seizures monitored with intracranial electrodes: correlation with surgical outcome. Epilepsia, 1999. 40(3): p. 257-66. 99. Tan, K.M., et al., Influence of subtraction ictal SPECT on surgical management in focal epilepsy of indeterminate localization: A prospective study. Epilepsy Res, 2008. 82(2-3): p. 190-3. 100. Quesney, L.F., et al., Presurgical EEG investigation in frontal lobe epilepsy. Epilepsy Res Suppl, 1992. 5: p. 55-69. 101. Lee, S.A., D.D. Spencer, and S.S. Spencer, Intracranial EEG seizure-onset patterns in neocortical epilepsy. Epilepsia, 2000. 41(3): p. 297-307. 102. Wetjen, N.M., et al., Intracranial electroencephalography seizure onset patterns and surgical outcomes in nonlesional extratemporal epilepsy. J Neurosurg, 2008. 103. Worrell, G., et al., High Frequency Oscillations and Ictogenesis in Neocortical Epilepsy. Submitted, 2002. 104. Bragin, A., et al., High-frequency oscillations in human brain. Hippocampus, 1999. 9(2): p. 137-42. 105. Buzsaki, G., Large-scale recording of neuronal ensembles. Nat Neurosci, 2004. 7(5): p. 446-51.106. Bragin, A., et al., Interictal high-frequency oscillations (80-500 Hz) in the human epileptic brain: entorhinal cortex. Ann Neurol, 2002. 52(4): p. 407-15. 107.Engel Jr, J., et al., High-frequency oscillations: What is normal and what is not? Epilepsia, 2008. 108. Worrell, G.A., et al., High-frequency oscillations in human temporal lobe: simultaneous microwire and clinical macroelectrode recordings. Brain, 2008. 131(Pt 4): p. 928-37. 109. Jirsch, J.D., et al., High-frequency oscillations during human focal seizures. Brain, 2006. 129(Pt 6): p. 1593-608. 110. Urrestarazu, E., et al., Interictal high-frequency oscillations (100-500 Hz) in the intracerebral EEG of epileptic patients. Brain, 2007. 130(Pt 9): p. 2354-66. 111. Sander, J.W., et al., National General Practice Study of Epilepsy: newly diagnosed epileptic seizures in a general population. Lancet, 1990. 336(8726): p. 1267-71. 112. Semah, F., et al., Is the underlying cause of epilepsy a major prognostic factor for recurrence?[see comment]. Neurology, 1998. 51(5): p. 1256-62. 113. Wiebe, S., Blume, W.T., Girvin, J.P., Eliasziw, M., A randomized, controlled trial of surgery for temporal-lobe epilepsy. New England Journal of Medicine, 2001. 345(5): p. 311-318. 114. Jay, V. and L.E. Becker, Surgical pathology of epilepsy: a review. Pediatr Pathol, 1994. 14(4): p. 731-50. 115. Cascino, G.D., et al., Routine EEG and temporal lobe epilepsy: relation to long-term EEG monitoring, quantitative MRI, and operative outcome. Epilepsia, 1996. 37(7): p. 651-6. 116. Pringle, C., et al., Pathogenesis of mesial temporal sclerosis. Can J Neurol Sci, 1993. 20: p. 184-193. 117. O'Brien, T.J., et al., Temporal lobe epilepsy caused by mesial temporal sclerosis and temporal neocortical lesions. A clinical and electroencephalographic study of 46 pathologically proven cases. Brain, 1996. 119(Pt 6): p. 2133-41. 118. Cascino, G., et al., Routine EEG and temporal lobe epilepsy: Relation to long-term EEG monitoring, quantitative MRI, and operative outcome. Epilepsia, 1996. 37: p. 651-656. 119.Trenerry, M.R., et al., MRI hippocampal volumes and memory function before and after temporal lobectomy. Neurology, 1993. 43: p. 1800-1805. 120. White, J.R., et al., Predictors of postoperative memory function after left anterior temporal lobectomy. Epilepsy Behav, 2002. 3(4): p. 383-389. 121. Cascino, G.D., et al., MRI in the presurgical evaluation of patients with frontal lobe epilepsy and children with temporal lobe epilepsy: pathologic correlation and prognostic importance. Epilepsy Res, 1992. 11(1): p. 51-9. 122. Diaz-Arrastia, R., et al., Neurophysiologic and neuroradiologic features of intractable epilepsy after traumatic brain injury in adults. Arch Neurol, 2000. 57(11): p. 1611-6. 123. Dunoyer, C., et al., The use of stereotactic radiosurgery to treat intractable childhood partial epilepsy. Epilepsia, 2002. 43(3): p. 292-300. 124. Kraemer, D.L., et al., Surgical outcome in patients with epilepsy with occult vascular malformations treated with lesionectomy. Epilepsia, 1998. 39(6): p. 600-7. 125. Luhmann, H.J., et al., Characterization of neuronal migration disorders in neocortical structures: extracellular in vitro recordings. Eur J Neurosci, 1998. 10(10): p. 3085-94. 126. Palmini, A., et al., Intrinsic epileptogenicity of human dysplastic cortex as suggested by corticography and surgical results. Ann Neurol, 1995. 37: p. 476-487. 127. Andermann, F., Cortical dysplasias and epilepsy: a review of the architectonic, clinical, and seizure patterns. Adv Neurol, 2000. 84: p. 479-96. 128. Kuzniecky, R., ed. Magnetic resonance imaging in cerebral developmental malformations and epilepsy. Neuroimaging in epilepsy: principles and practice, ed. G. Cascino and C.J. Jack. 1996, Butterworth-Heinemann: Boston. 51-63. 129. Spencer, S.S., W.H. Theodore, and S.F. Berkovic, Clinical applications: MRI, SPECT, and PET. Magnetic Resonance Imaging, 1995. 13(8): p. 1119-24. 130. Jack, J., CR. , et al., Intractable nonlesional epilepsy of temporal lobe origin: lateralization by interictal SPECT versus MRI. Neurology, 1994. 44(5): p. 829-36. 131. Mosewich, R., et al., Outcome of lesional vs. nonlesional frontal lobectomy for intractable epilepsy. Electroencephalogr Clin Neurophysiol, 1997. 103: p. 136. 132. Jack, C.J., et al., Intractable nonlesional epilepsy of temporal lobe origin: lateralization by interictal SPECT versus MRI. Neurology, 1994. 44: p. 829-836. 133. Holmes, M.D., et al., Outcome after surgery in patients with refractory temporal lobe epilepsy and normal MRI. Seizure, 2000. 9(6): p. 407-11. 134. Blume, W.T., et al., Indices of resective surgery effectiveness for intractable nonlesional focal epilepsy. Epilepsia, 2004. 45(1): p. 46-53. 135. Wieser, H.G., et al., Long-term seizure outcomes following amygdalohippocampectomy. J Neurosurg, 2003. 98(4): p. 751-63. 136. Mosewich, R., et al., Factors Predictive of the Outcome of Frontal Lobe Epilepsy Surgery. Epilepsia, 2000. 41: p. 843-849. 137. Sylaja, P.N., et al., Seizure outcome after anterior temporal lobectomy and its predictors in patients with apparent temporal lobe epilepsy and normal MRI. Epilepsia, 2004. 45(7): p. 803-8. 138. Berkovic SF, et al., Preoperative MRI predicts outcome of temporal lobectomy: An actuarial analysis. Neurology, 1995. 45(1358-1363). 139. Cohen-Gadol, A.A., et al., Normal magnetic resonance imaging and medial temporal lobe epilepsy: the clinical syndrome of paradoxical temporal lobe epilepsy.[see comment]. Journal of Neurosurgery, 2005. 102(5): p. 902-9. 140. Theodore, W., et al., Temporal lobectomy for uncontrolled seizures: the role for positron emission tomography. Ann Neurol, 1992. 32: p. 789-794. 141. Bartolomei, F., et al., Long-term efficacy of gamma knife radiosurgery in mesial temporal lobe epilepsy. Neurology, 2008. 70(19): p. 1658-63. 142. Spencer, S.S., Gamma knife radiosurgery for refractory medial temporal lobe epilepsy: too little, too late? Neurology, 2008. 70(19): p. 1654-5. 143. Wheless, J.W., History of the ketogenic diet. Epilepsia, 2008. 49 Suppl 8: p. 3-5. 144. Sirven, J., et al., The ketogenic diet for intractable epilepsy in adults: preliminary results. Epilepsia, 1999. 40(12): p. 1721-6. 145. Kossoff, E.H. and J.L. Dorward, The modified Atkins diet. Epilepsia, 2008. 49 Suppl 8: p. 37-41. 146. Penfield, W., Erickson, T.C., Epilepsy and cerebral localization. 1941, Springfield, IL: Charles C. Thomas. 147. Cooper, I.S., I. Amin, and S. Gilman, The effect of chronic cerebellar stimulation upon epilepsy in man. Transactions of the American Neurological Association, 1973. 98: p. 192-6. 148. Cooper, I.S., et al., Chronic cerebellar stimulation in epilepsy. Clinical and anatomical studies. Archives of Neurology, 1976. 33(8): p. 559-70. 149. Cooper, I.S., et al., Safety and efficacy of chronic cerebellar stimulation. Applied Neurophysiology, 1977. 40(2-4): p. 124-34. 150. Cooper, I.S., A.R. Upton, and I. Amin, Chronic cerebellar stimulation (CCS) and deep brain stimulation (DBS) in involuntary movement disorders. Applied Neurophysiology, 1982. 45(3): p. 209-17. 151. Ben-Menachem, E., et al., Evaluation of refractory epilepsy treated with vagus nerve stimulation for up to 5 years. Neurology, 1999. 52(6): p. 1265-7. 152. Schachter, S.C., Vagus nerve stimulation therapy summary: five years after FDA approval. Neurology, 2002. 59(6 Suppl 4): p. S15-20. 153. Group, T.V.N.S.S., A randomized controlled trial of chronic vagus nerve stimulation for treatment of medically intractable seizures. The Vagus Nerve Stimulation Study Group. Neurology, 1995. 45(2): p. 224-30. 154. Cramer, J.A., E. Ben Menachem, and J. French, Review of treatment options for refractory epilepsy: new medications and vagal nerve stimulation. Epilepsy Res, 2001. 47(1-2): p. 17-25. 155. Cramer, J.A., Exploration of Changes in Health-Related Quality of Life after 3 Months of Vagus Nerve Stimulation. Epilepsy Behav, 2001. 2(5): p. 460-465. 156. Schachter, S.C. and C.B. Saper, Vagus nerve stimulation. Epilepsia, 1998. 39(7): p. 677-86. 157. Wright, G.D., D.L. McLellan, and J.G. Brice, A double-blind trial of chronic cerebellar stimulation in twelve patients with severe epilepsy. Journal of Neurology, Neurosurgery & Psychiatry, 1984. 47(8): p. 769-74. 158. Chkhenkeli, S.A. and I.S. Chkhenkeli, Effects of therapeutic stimulation of nucleus caudatus on epileptic electrical activity of brain in patients with intractable epilepsy. Stereotact Funct Neurosurg, 1997. 69(1-4(Pt 2)): p. 221-4. 159. Mirski, M.A. and R.S. Fisher, Electrical stimulation of the mammillary nuclei increases seizure threshold to pentylenetetrazol in rats. Epilepsia, 1994. 35(6): p. 1309-16. 160. Mirski, M.A., et al., Anticonvulsant effect of anterior thalamic high frequency electrical stimulation in the rat. Epilepsy Res, 1997. 28(2): p. 89-100. 161. Velasco, A.L., et al., Electrical stimulation of the hippocampal epileptic foci for seizure control: a double-blind, long-term follow-up study. Epilepsia, 2007. 48(10): p. 1895-903. 162. Velasco, F., et al., Effect of chronic electrical stimulation of the centromedian thalamic nuclei on various intractable seizure patterns: I. Clinical seizures and paroxysmal EEG activity. Epilepsia, 1993. 34(6): p. 1052-64. 163. Velasco, F., et al., Electrical stimulation of the centromedian thalamic nucleus in control of seizures: long-term studies. Epilepsia, 1995. 36(1): p. 63-71. 164. Velasco, M., et al., Effect of chronic electrical stimulation of the centromedian thalamic nuclei on various intractable seizure patterns: II. Psychological performance and background EEG activity. Epilepsia, 1993. 34(6): p. 1065-74. 165. Fisher, R.S., et al., Placebo-controlled pilot study of centromedian thalamic stimulation in treatment of intractable seizures. Epilepsia, 1992. 33(5): p. 841-51. 166. Kerrigan, J.F., et al., Electrical stimulation of the anterior nucleus of the thalamus for the treatment of intractable epilepsy. Epilepsia, 2004. 45(4): p. 346-54. 167. Graves, N.M. and R.S. Fisher, Neurostimulation for epilepsy, including a pilot study of anterior nucleus stimulation. Clin Neurosurg, 2005. 52: p. 127-34. 168. Hodaie, M., et al., Chronic anterior thalamus stimulation for intractable epilepsy. Epilepsia, 2002. 43(6): p. 603-8. 169. Benabid, A.L., et al., Therapeutic electrical stimulation of the central nervous system. C R Biol, 2005. 328(2): p. 177-86. 170. Loddenkemper, T., et al., Deep brain stimulation in epilepsy. J Clin Neurophysiol, 2001. 18(6): p. 514-32. 171. Vonck, K., et al., Neurostimulation for refractory epilepsy. Acta Neurol Belg, 2003. 103(4): p. 213-7. 172. Vonck, K., et al., Long-term amygdalohippocampal stimulation for refractory temporal lobe epilepsy. Ann Neurol, 2002. 52(5): p. 556-65. 173. Tellez-Zenteno, J.F., et al., Hippocampal electrical stimulation in mesial temporal lobe epilepsy. Neurology, 2006. 66(10): p. 1490-4. 174. Lesser, R.P., et al., Brief bursts of pulse stimulation terminate afterdischarges caused by cortical stimulation. Neurology, 1999. 53(9): p. 2073-81. 175. Engel, J.J., P. Van Ness, and T. Rasmussen, Outcome with respect to epileptic seizures, in Surgical treatment of the epilepsies, J. Engel, Jr., Editor. 1993, Raven Press: New York. p. 609-621. 176. Radhakrishnan, K., et al., Predictors of outcome of anterior temporal lobectomy for intractable epilepsy. A multivariate study. Neurology, 1998. 51: p. 465-471.

CEREBROSPINAL FLUID NEUROTRANSMITTER AND CYTOKINE CONCENTRATIONS IN CANINE DEGENERATIVE MYELOPATHY S. Platt,[1] R. Barber,[1] M. Kent,[1] S. Kanazono,[2] A.C. Freeman,[1] G. Edwards,[1] J.R. Coates.[2] College of Veterinary Medicine, University of Georgia, Athens, GA[1] & College of Veterinary Medicine, University of Missouri, Columbia, MO.[2]

Degenerative myelopathy (DM) may be homologous to a form of amyotrophic lateral sclerosis in humans which has excitotoxic and immunologic pathogeneses described. The aims of this study were to determine (i) presence or absence of abnormalities in concentrations of CSF amino acid (AA) neurotransmitters (glutamate, glycine and γ–aminobutyric acid (GABA)) and cytokines in dogs with DM and if present (ii) investigate associations with disease severity.

Twenty-two dogs histopathologically confirmed for DM and 21 dogs with suspected DM based on thorough diagnostic investigations and 42 clinically normal age-matched control dogs were included in the study. The neurological severity of the DM dogs was graded (1-4) using an established scale. CSF was evaluated for presence of glutamate, glycine and GABA by high performance liquid chromatography and for GM-CSF, IFN-γ, IL-2, IL-4, IL-6, IL-7, IL-8, IL-10, IL-15, IL-18, IP-10, KC (keratinocyte chemoattractant), MCP-1 (monocyte chemotactic protein-1) and TNF-α using a commercially available, canine multiplex immunoassay (Millipore, Billerica, MA). All data analyses were performed using SAS V 9.2 (Cary, NC). Analyte levels were compared between DM confirmed, DM suspected and control dogs by an analysis of variance (ANOVA). Spearman correlation was used to test for correlations of analyte levels and neurological grades. All hypothesis tests were 2-sided with $\alpha=0.05$.

There were no significant differences between individual CSF analytes in DM confirmed and DM suspected dogs. Glutamate levels were not significantly different between DM affected (mean=0.09µg/ml; range=0.01-0.29; SD=0.069) and control dogs (mean=0.07µg/ml; range=0.01-0.29; SD=0.056). Control dogs (mean=0.86µg/ml; range=0.08-1.16; SD=0.213) had significantly higher levels of GABA ($p<0.0001$) than DM dogs (mean=0.12µg/ml; range=0.07-0.79; SD=0.12). Control dogs (mean=2.14µg/ml; range=0.01-3.83; SD=0.76) also had significantly higher glycine concentrations ($p<0.0001$) than DM dogs (mean=0.45µg/ml; range=0.01-0.98; SD=0.24). DM-affected dogs also had significantly higher levels of IL-2 ($p=0.03$), KC ($p<0.0001$) and MCP-1 ($p=0.005$) than control dogs. Neurotransmitter levels were not significantly associated with neurological grade. KC levels were significantly higher in the least affected dogs ($p=0.0015$). There were no associations with disease severity and analyte concentrations.

DM affected dogs have an imbalance of CSF AA concentrations creating a relatively excitotoxic environment. Reports in human ALS confirm an imbalance between CSF excitatory and inhibitory AAs suggesting a pathogenic role for excitotoxicity in ALS. It also appears that DM affected dogs have increases in CSF cytokines and chemokines suggestive of an immunologic component to the pathogenesis as is similar to ALS. Further prospective analysis of DM is warranted to evaluate the role of treatment on CSF variables.

CHIARI-LIKE MALFORMATION, SYRINGOMYELIA AND NEUROPATHIC PAIN: CEREBROSPINAL FLUID NEUROTRANSMITTER AND CYTOKINE LEVELS.

S.R. Platt, A.C. Freeman, M. Kent, E. Huguet, G. Edwards, R. Barber. College of Veterinary Medicine, University of Georgia, Athens, GA.

The pathogenesis of neuropathic pain (NP) and syringomyelia (SM) in association with Chiari-like malformation (CLM) in dogs has focused on the anatomical anomalies and secondary cerebrospinal fluid (CSF) flow abnormalities. Neuropathic pain in humans has been associated with abnormalities of neurotransmitters such as glutamate and serotonin as well as immunologic mechanisms. The aim of this study was to investigate the CSF neurotransmitter and cytokine levels in Brussels Griffon dogs (BGs) with CLM, SM and NP.

As part of an MRI study investigating the prevalence of SM in BGs, atlanto-occipital CSF was acquired from 46 dogs and stored at -80°C until analysis. All dogs underwent a neurologic exam prior to MRI; Osirix® software was used to measure SM and the presence of cerebellar herniation and deviation were recorded. Deproteinized CSF samples were analysed for presence of serotonin (ng/ml), glutamate, glycine and GABA (μg/ml) by high performance liquid chromatography. All CSF samples were evaluated simultaneously for GM-CSF, IFN-γ, IL-2, IL-4, IL-6, IL-7, IL-8, IL-10, IL-15, IL-18, IP-10, KC, MCP-1 and TNF-α. A commercially available, canine multiplex immunoassay (Millipore, Billerica, MA) was used for the cytokine analysis (pg/ml). Student's t-tests were used to compare the means of neurotransmitter and cytokine values between groups with and without skull abnormalities or spinal pain. Simple Pearson's correlation was used to test for correlations of neurotransmitter and cytokine values with syrinx dimensions and correlations of neurotransmitter with cytokine values. All hypothesis tests were 2-sided and the significance level was $\alpha = 0.05$.

NP was detected in 8 dogs (17%); SM was present 24 dogs (52%); and CM was detected in 24 dogs (52%). IFN-γ levels were significantly lower in dogs with NP than without (p=0.036). There were significant positive correlations between syrinx size and IL-8 (p=0.017), KC (p=0.025) and MCP-1 (p=0.003). There were significant negative correlations between IFN-γ and syrinx height (p=0.025) and extent (p=0.042). There was a significant negative correlation between IL-2 and syrinx height (p=0.042). Neurotransmitter levels were not associated with skull abnormalities or spinal pain, but there was a positive correlation of glycine with IL-2 (p=0.004) and MCP-1 with glutamate (p=0.0147) and serotonin (p=0.0059).

The size of the syrinx in BGs with SM is associated with several cytokine elevations but only a decrease of IFN-γ was associated with NP. Based on this study it does not appear that excitotoxicity plays a role in either SM development or NP. Further work is justified on the role of the immune system in CM, SM and NP.

UPDATE ON THE CONSERVATIVE MANAGEMENT OF DISK ASSOCIATED CERVICAL SPONDYLOMYELOPATHY IN DOGS. S. De Decker [1,2], I. Gielen [1], L. Duchateau [1], A. Oevermann [3], L. Van Ham [1]. [1] Faculty of Veterinary Medicine, Ghent University, Merelbeke, Belgium. [2] The Royal Veterinary College, London, United Kingdom. [3] Neurocenter, University Bern, Vetsuisse Faculty, Switzerland.

Current knowledge about the conservative management of disk associated cervical spondylomyelopathy (DA-CSM) is rather limited and mainly based on retrospectively retrieved data. The goals of this study were to prospectively evaluate the evolution of clinical signs in dogs treated conservatively for DA-CSM. Additionally, several potential prognostic parameters and the correlation of initial clinical signs with magnetic resonance imaging (MRI) and transcranial magnetic stimulation (TMS) were investigated.

Twenty-one dogs were included. After neurological evaluation, neurological status was graded from 0 (= normal) to 6 (= tetraplegia). All animals underwent low-field MRI and TMS with measurement of onset latencies and peak-to-peak amplitudes from the extensor carpi radialis and cranial tibial muscles. From the MR images, the following dimensions were calculated: remaining spinal cord area; compression ratio; vertebral occupying ratio of the spinal cord; canal height to body height ratio (CBR); canal height to body length ratio (CBLR); and the canal compromise ratio. Intraparenchymal intensity (ISI) changes were graded from 0 to 3. All dogs were reevaluated by the same person after 1, 3, 6, 12, and 24 months.

Eight of 21 dogs (38%) experienced a positive clinical evolution with improvement of clinical signs or stabilization of mild clinical signs. All dogs with a negative clinical evolution 1 month after diagnosis experienced a further progression of clinical signs resulting in a poor outcome. The opposite was true for all dogs with a positive clinical evolution after 1 month. Outcome was further significantly associated by the remaining spinal cord area and the vertebral canal compromise ratio. Prognosis was not significantly affected by clinical presentation or TMS. Progression of clinical signs, in unsuccessfully treated dogs, was generally characterized by a rapid and dramatic deterioration of neurological status. There were no significant correlations between clinical presentation, MRI and TMS. Two dogs underwent necropsy and histopathological examination. This revealed in both cases chronic Wallerian degeneration and segmental myelomalacia.

The results of this study suggest that conservative treatment of DA-CSM is associated with a rather guarded prognosis. Clinical evolution 1 month after diagnosis and selected MRI parameters can be considered as prognostic indicators. The lack of correlation between clinical presentation and outcome, medical imaging and electrophysiological evaluation is disturbing and warrants further investigation.

CLINICAL EVALUATION OF A MAGNETIC RESONANCE IMAGING (MRI) - GUIDED, FRAMELESS STEREOTACTIC BRAIN BIOPSY SYSTEM IN THE DOG.

AV Chen[1], RL Tucker[1], GJ Haldorson[1], SA Thomovsky[1], AR Schneider[1], RS Bagley[2]. 1. Washington State University College of Veterinary Medicine, Pullman, WA. 2. Iowa State University, College of Veterinary Medicine, Ames, IA.

A MRI–guided stereotactic brain biopsy system has not been clinically evaluated in dogs. The purpose of this study was to determine the ability of the Brainsight™ system to obtain histologically diagnostic samples and access the impact of this procedure on neurologic status for 72 hours after the biopsy.

Five dogs with MRI definable lesions in the brain have been enrolled. Breeds included a Pitbull mix, Pembroke Welsh Corgi, French Bulldog, Border Terrier and West Highland White Terrier. Age ranged from 5-11 years. Weight ranged from 8.5-18.8kg. Dogs presented with seizures (n=5), ambulatory paresis(n=4), unilateral blindness(n=2) and head tilt(n=1). One dog had a normal neurologic exam. Lesions chosen for biopsy were in the olfactory and/or frontal lobes (n=3), parietal lobe(n=1), and pyriform lobe(n=1). Lesions were between 14-22mm in diameter. All lesions were well-circumscribed and contrast enhancing except for one. Histologic diagnosis of meningioma(n=3) and granulomatous meningoencephalitis(n=1) were made. The poorly-circumscribed, non-contrast enhancing frontal mass yielded non-specific necrosis. Following biopsy, three dogs returned to pre-biopsy neurologic status within 12 hours. The French Bulldog took 48 hours to return to previous neurologic status due to brachycephalic syndrome that required oxygen support. One dog had acute respiratory arrest 16 hours post-biopsy. Necropsy is pending.

These results suggest that this MRI-guided biopsy system can provide an accurate histologic diagnosis of brain lesions. Biopsies of poorly-circumscribed and non-contrast enhancing brain lesions may be less diagnostic. Further evaluation is on-going to determine the true diagnostic yield and complication rate of this procedure.

MORPHOMETRIC FEATURES OF THE CRANIOCERVICAL JUNCTION REGION IN DOGS WITH SUSPECTED CHIARI-LIKE MALFORMATION: 274 CASES (2007-2010).

Dominic J. Marino, CA Loughin, CW Dewey. The Canine Chiari Institute at Long Island Veterinary Specialists Plainview, NY.

Concurrent malformations of the craniocervical junction are commonly identified in humans with Chiari type I malformation. Recent evidence suggests such craniocervical junction abnormalities (CJAs) also occur in dogs suspected of having Chiari-like malformation (CLM). The purpose of this study was to objectively describe morphometric features of the craniocervical junction region of dogs with suspected CLM and to investigate for associations between these features and the occurrence of other malformations in this region.

Magnetic resonance (MR) and computed tomographic (CT) images from 274 dogs with CLM were evaluated. Three regions of neural tissue compression were assessed: cerebellar compression (CC); ventral compression at the C1/C2 articulation, termed "medullary kinking" (MK); and dorsal compression (DC) at the C1/C2 articulation. A compression index (CI) was calculated for all abnormal regions for each dog. Multiple logistic regression analysis was performed ($p<0.05$) to ascertain whether CI values for the different regions of compression were associated with the incidence of other craniocervical junction abnormalities.

68% of dogs had MK and 38% of dogs had DC. 28% of dogs also had evidence of atlanto-occipital overlapping (AOO). Breed and severity of CC were the only significant predictors of AOO ($p<0.0001$ and $p< 0.0092$). 28% of suspected CLM cases have AOO as the anatomic abnormality responsible for CC. Compression index values may help differentiated subtypes of CJAs in dogs.

MEDICAL INFRARED IMAGING (THERMOGRAPHY) AS A SCREENING MODALITY IN DOGS WITH CHIARI-LIKE MALFORMATION. CA Loughin, DJ Marino, The Canine Chiari Institute at Long Island Veterinary Specialists, Plainview, NY.

Medical infrared imaging (MII) is a non-invasive diagnostic imaging technique that measures skin surface temperature and generates thermal pattern maps based on predetermined color scales. Because skin temperature, dependent on regional perfusion, is under direct control of the sympathetic nervous system, MII provides information about the function of the autonomic nervous system. Because of recent advances in technology and lack of sedation needed to image patients, MII has potential use as a screening test for a variety of conditions that may result in autonomic dysregulation like Chiari-like malformation in dogs (CLM). The purposes of this study were to establish a MII protocol for dogs suspected of having CLM, to identify thermal imaging patterns for various regions of interest (ROI), to evaluate changes in thermal patterns and compare the results to those of MRI findings, considered the standard for diagnosing CLM in dogs. One hundred and five Cavalier King Charles Spaniel dogs with clinical signs attributable to CLM and confirmed CLM with MRI were evaluated with a complete blood count and chemistry profile, examination by a board certified surgeon/neurologist, multidetector CT scan of the craniocervical junction, whole body MRI and MII. The protocol for thermal imaging included cranial and caudal views of the body, full lateral right and left body views, dorsal views of the head and body, and right and left lateral views of the head. Thermal patterns were assessed with custom image recognition software. After each dog was imaged awake, general anesthesia was administered and the dogs re-imaged using the same protocol. MRI findings in dogs with severe or moderate cerebellar compression and cerebellar herniation were compared with MII results. The top of head and front of head ROI were 89.2% and 97.3% successful in identifying dogs with CLM. Based on these preliminary findings, MII may be a viable screening tool to detect CLM in dogs.

MEDICAL INFRARED IMAGING (THERMOGRAPHY) OF TYPE I THORACOLUMBAR DISK DISEASE IN CHONDRODYSTROPHIC DOGS. CA Loughin, DJ Marino, BP Grossbard. The Canine Chiari Institute at Long Island Veterinary Specialists, Plainview, NY.

Medical infrared imaging (MII) is an imaging technique that measures skin surface temperature derived from cutaneous perfusion and generates thermal pattern maps based on color scales. MII has been used as a test for a variety of conditions that cause autonomic dysregulation resulting in altered cutaneous perfusion. Acute thoracolumbar intervertebral disk disease (TLIVDD) is common in dogs. The purpose of this study was to: 1) determine the success of MII in identifying dogs with TLIVDD, 2) compare the MII localization with MRI results and surgical findings 3) determine if the MII pattern returns to that of normal dogs following decompression surgery. 72 small breed chondodystrophic dogs with TLIVDD confirmed with MRI and 14 dogs with no TLIVDD were evaluated with MRI and MII. Regions correlating with the intervertebral disk spaces were analyzed for average temperatures and thermographic patterns. Thermal patterns were assessed with computer recognition pattern analysis (CRPA) software. 21 dogs were re-evaluated 8 weeks after surgery using the same protocol. When analyzing temperature averages over a region, no significant difference was found between control and affected dogs. CRPA was 90% successful in differentiating normal from affected dogs. CRPA was 97% successful in identifying the intervertebral disk space when compared with MRI and surgical findings. Based on these findings, MII may be a viable screening tool to detect TLIVDD in dogs.

SPINAL CORD INJURY INFLUENCED BY THE MICROGLIA REACTION PROFILE. V.M. Stein[1], T.M.A. Boekhoff[1], E.-M. Ensinger[1], R. Carlson[1], P. Bock[2], W. Baumgärtner[2], K. Rohn[3], A. Tipold[1]. [1]Department of Small Animal Medicine and Surgery, [2]Department of Pathology, [3]Institute of Biometry, Epidemiology, and Information Processing, University of Veterinary Medicine, Hannover, Germany.

Microglia physiologically shows regional topographical differences in immunophenotype and function within the central nervous system indicating the endowment for a prompt response to pathological stimuli such as trauma. Spinal cord injuries (SCI) consist of a primary injury encompassing the mechanical impact and the "secondary wave" of injury occurring minutes to weeks later and comprising various consecutive effects such as increased production of free radicals, excessive release of excitatory neurotransmitters and inflammatory reactions. Activated microglia has the potential to perform some of these reactions, their contribution to the secondary wave is therefore controversially discussed. It has to be considered a double-edged sword as both, beneficial and deleterious effects have been attributed to these cells.

The purpose of the presented study was to assess microglial involvement, particularly in the "secondary wave" following SCI. Microglia from 15 dogs with SCI was isolated and characterized ex vivo in terms of morphology, immunophenotype, and function by flow cytometry. The results were compared to region-specific findings obtained from healthy control dogs (n = 30).

The histopathological exam confirmed the diagnosis of SCI in the cervical (n = 5) and thoracolumbar (n = 10) spinal cord, and revealed a significant activation of microglia/ macrophages and upregulation of myelinophagia in dogs with SCI 5 days or longer prior to euthanasia. Microglial ex vivo examination showed significantly increased expressions of B7-1, B7-2, MHC II, CD1c, ICAM-1, CD14, CD44, and CD45, and significantly enhanced phagocytosis and generation of reactive oxygen species (ROS) in SCI compared to healthy controls.

Microglial cells seem to be highly activated following SCI with an immunophenotype indicating their active role in co-stimulation of T cells, in leukocyte adhesion and aggregation, and in lipid and glycolipid presentation. Microglial phagocytosis might play a pivotal role in removal of injured or damaged cells and initialize subsequent healing processes. However, as ROS can be directly neurotoxic an enhanced microglial generation might lead to bystander damage of the traumatized spinal cord and might therefore add to the deleterious effects of the secondary wave. Modulating the microglial response in SCI might be a valuable novel therapeutic strategy alleviating further damage to the spinal cord.

A DELETION IN *ARHGEF10* IS HIGHLY ASSOCIATED WITH EARLY ONSET INHERITED POLYNEUROPATHY IN LEONBERGER AND ST. BERNARD DOGS. G.D. Shelton[1], K.J. Ekenstedt[2], C. Dr□gemüller[3], K.M. Minor[4], D. Becker[3], T. Leeb[3], N. Patterson[2], J. Mickelson[4]. 1. Department of Pathology, School of Medicine, University of California-San Diego, La Jolla, CA, 2. Department of Veterinary Clinical Sciences, College of Veterinary Medicine, University of Minnesota, St. Paul, MN, 3. Institute of Genetics, University of Berne, Bern, Switzerland, 4. Department of Veterinary and Biomedical Sciences, University of Minnesota, St. Paul, MN.

Polyneuropathy in large breed dogs is a relatively common clinical problem for which the genetic basis is generally unknown. The first cases of polyneuropathy in the Leonberger breed (Leonberger polyneuropathy or LPN) were identified in 1999 by one of the authors (GDS) and a report published in 2003 (MuscleNerve 27:471-477). In this report a spontaneous, distal and symmetrical polyneuropathy with onset between 1 to 9 years of age was described and characterized clinically, electrophysiologically, histologically and morphometrically. There were striking similarities between LPN and the Charcot-Marie-Tooth group of human inherited sensory and motor polyneuropathies, which have many known genetic mutations.

A genome-wide case-control association study for LPN was performed with 53 cases and 42 controls on high-density 170K canine SNP arrays and revealed a significantly associated region on CFA 16 (P_{raw} = 2.36 x $10^{-10,}$ P_{genome} = 9.99 x 10^{-5}). A clear association of an approximately 1 Mb CFA16 haplotype with cases (P=1.71 x 10^{-8}) was observed, particularly with those cases that were affected more severely and at a younger age (P=2.55 x 10^{-11}). A positional candidate gene, *ARHGEF10*, which has previously been associated with peripheral nerve abnormalities in humans, was sequenced, revealing a deletion that results in a frame shift and premature stop codon. Of all Leonbergers with young onset LPN (before 4 years), 48.5% (32 of 66) have two copies of this deletion, and, of all young onset Leonbergers that are nerve biopsy positive for LPN, 59.4% (19 of 32) have two copies of this deletion. Importantly, nearly all dogs carrying two copies of the deletion (32 of 34 or 94.1%) are affected with LPN by the age of 4 years.

The Leonberger breed was generated from crossing several breeds, including the St. Bernard, and a polyneuropathy clinically and histologically similar to LPN occurs in this breed. To determine if the *ARHGEF10* mutation was associated with polyneuropathy in the St. Bernard, DNA was extracted from archived frozen muscle biopsy specimens from clinical cases (n=3). The identical *ARHGEF10* mutation was identified in all tested St. Bernard dogs affected with early onset polyneuropathy suggesting the mutation may have been passed through the St. Bernard to the Leonberger.

STARTLE DISEASE IN IRISH WOLFHOUNDS ASSOCIATED WITH A DELETION IN THE GENE ENCODING THE GLYCINE TRANSPORTER GLYT2. G.D. Shelton[1], J.L. Gill[2], D. Capper[3], R.J. Higgins[4], R.J. Harvey[2]. 1. Department of Pathology, School of Medicine, University of California-San Diego, La Jolla, CA, USA; 2. Department of Pharmacology, The School of Pharmacy, London, UK; 3. Halsey East Animal Clinic, Portland, OR, USA; 4. Department of Pathology, Microbiology and Immunology, School of Veterinary Medicine, University of California-Davis, Davis, CA, USA.

Startle disease or hyperekplexia is caused by defects in mammalian glycinergic neurotransmission resulting in an exaggerated startle reflex and extensor hypertonia triggered by noise or touch. In humans and animals, startle disease is typically caused by mutations in one of three genes (*GLRA1*, *GLRB*, and *SLC6A5*) encoding postsynaptic glycine receptor subunits (α1 and β) or a presynaptic glycine transporter (GlyT2). A litter of seven Irish Wolfhounds was recently identified in which two puppies developed muscle stiffness and tremor beginning at 5-7 days of age post-partum. Signs were dramatic when the puppies were handled and resolved when the puppies were relaxed or sleeping. Both puppies were euthanized due to ongoing stiffness, tremor and breathing difficulties. Necropsies were performed, but no microscopic pathological abnormalities were identified in the peripheral or central nervous system.

Based on the clinical signs, exons from the three candidate genes were amplified from genomic DNA isolated using PCR and directly sequenced. No deleterious polymorphisms were identified in either *GLRA1* or *GLRB*. However, difficulties were experienced in amplifying *SLC6A5* exons 2 and 3 from affected animals, although control samples were positive, suggesting that the PCR primer designs and conditions were not at fault. Further PCRs revealed that the reason for this anomaly was the presence of a homozygous 4.2 kb deletion encompassing exons 2 and 3 of the GlyT2 gene in both affected animals. This deletion is predicted to result in the loss of part of the large cytoplasmic N-terminus that is vital for trafficking of GlyT2 to synaptic sites, and a loss of all subsequent transmembrane domains via a frameshift. This genetic lesion was confirmed by defining the deletion breakpoint, Southern blotting and Multiplex Ligation-dependent Probe Amplification (MLPA). This analysis enabled the development of a rapid genotyping test that revealed heterozygosity for the deletion in the dam and sire and three other siblings, suggesting recessive inheritance of this disorder. Wider testing of related animals has identified a total of 18 carriers of the *SLC6A5* deletion and enabled the identification of non-carrier animals to guide future breeding strategies.

A MUTATION IN ATP13A2 CAUSES ADULT-ONSET NEURONAL CEROID LIPOFUSCINOSIS IN TIBETAN TERRIERS. Fabiana H.G. Farias, Gary S. Johnson, Dennis P. O'Brien, Fred A. Wininger, Rong Zeng, Stephanie D. McKay, Robert D. Schnabel, Jeremy F. Taylor, Claire Wade, Kerstin Lindblad-Toh, Hannes Lohi, Corinne S. Mann, and Martin L. Katz. From the Department of Veterinary Pathobiology (Farias, Johnson, Zeng, Katz), Division of Animal Sciences (McKay, Schnabel, Taylor), Mason Eye Institute (Mann, Katz) and Department of Veterinary Medicine and Surgery (O'Brien, Coates), University of Missouri, Columbia, Missouri. Faculty of Veterinary Science, The University of Sydney, NSW 2006 Australia (Wade). Department of Medical Biochemistry and Microbiology, Uppsala University, Uppsala Sweden (Lindblad-Toh). Department of Molecular Genetics, the Folkhälsan Institute of Genetics, Finland. (Lohi).

The neuronal ceroid lipofuscinoses (NCLs) are a group of neurodegenerative lysosomal storage diseases defined by the lysosomal accumulation of autofluorescent proteinaceous material with characteristic ultrastructural appearances. This study was performed to identify the mutation responsible for the adult-onset NCL in Tibetan Terriers. A 19-case × 15-control genome wide association study produced the strongest associations between the NCL phenotype and CFA2 markers. All 19 NCL-affected Tibetan Terriers were homozygous for the same haplotype which extended for 22 consecutive SNPs spanning 0.95 Mb. None of the 20 annotated genes within this target region had previously been associated with human or rodent NCL. We used DNA from NCL-affected Tibetan Terriers to resequence the coding regions and intron-exon borders of several genes harbored within the target region and found a single base pair deletion, c.1623delG, in exon 16 of positional candidate ATP13A2. This deletion produces a frame shift and a predicted premature termination codon. We genotyped all 454 Tibetan Terrier DNA samples in our collection and found all 45 NCL-affected Tibetan Terriers to be homozygous for the c.1623delG allele. Eleven additional c.1623delG homozygotes were either less than 5 years old, or lost to follow up. There were no known cases of NCL in the remaining 398 Tibetan Terriers which were either heterozygous (n = 149) or homozygous for the ancestral allele (n = 249). ATP13A2 is a member of group of ion transport genes and has been associated with lysosomes. Mutations in human ATP13A2 cause Kufor-Rakeb syndrome (KRS), a rare neurodegenerative disorder with clinical features that include parkinsonism plus spasticity, supranuclear upgaze paresis, and dementia. Post-mortem findings in KRS have not been reported. We conclude that NCL in Tibetan terriers is caused by a mutation in ATP13A2. Our results suggest that KRS may be a form of adult onset NCL in humans.

INTRATHECAL VERSUS SUBCUTANEOUS THERAPY OF FELINE NIEMANN-PICK TYPE C DISEASE.

C.H. Vite,[1] E. Mauldin,[1] S. Ward,[1] V. Stein,[2] M.E. Haskins,[1] R. Stratton,[3] L.M. Kao,[4] D. Ory,[5] S.U. Walkley,[6] M.T. Vanier[7] 1. University of Pennsylvania, Philadelphia, PA. 2. Stiftung Tierärztliche Hochschule, Hannover. Germany. 3. Cyclodextrin Technologies Development, Inc, High Springs, FL. 4. Johnson and Johnson Pharmaceutical Research and Development, Raritan, NJ. 5. Washington University, St Louis, MO. 6. Albert Einstein College of Medicine, Bronx NY. 7. Institut National de la Santé et de la Recherche Médicale (INSERM), Lyon, France.

Niemann-Pick type C (NPC) disease is a progressive neurological disorder characterized by dementia and ataxia, hepatic and pulmonary disease, and death typically within the first or second decade. Despite the identification of causative mutations, the pathogenesis is not clear and therapies to successfully treat NPC disease have been ineffective to date. The recent use of intravenously administered 2-hydroxypropyl-beta-cyclodextrin (HPBCD), an FDA-designated orphan drug (May 2010), in a small number of children with NPC disease is based on favorable treatment outcome data in subcutaneously treated mouse and cat models. To rigorously evaluate the mechanistic, pharmacologic, and toxicity issues associated with HPBCD therapy in NPC disease, we have utilized the spontaneous feline NPC model harboring a missense mutation in NPC1 (pC955S), orthologous to the most common mutation in juvenile-onset patients. The feline NPC model has clinical, neuropathological and biochemical abnormalities similar to those present in juvenile-onset patients making this model homologous to the most common disease form seen in human patients. We identified that intrathecal administration of HPBCD ameliorated all clinical aspects of neurological disease at least up to 24 weeks of age (an age when untreated cats die) but had no effect on hepatic disease. We identified that while subcutaneous therapy with HPBCD at all doses ameliorated liver disease, only 8000 mg/kg substantially affected neurological disease but also resulted in early death due to pulmonary toxicity. Finally, we identified a dose-related toxic effect of HPBCD on hearing function that had not been described in any other species.

A NOVEL LEUKODYSTROPHY IN BORDER TERRIER PUPPIES – THE CANINE "RUMPSHAKER."

Ronaldo C. da Costa,[1] Paula Martin-Vaquero,[1] Jessica K. Simmons,[1] Gillian L. Beamer,[1] Karin Jaderlung,[2] Michael J. Oglesbee.[1] [1] College of Veterinary Medicine, The Ohio State University, Columbus, OH. [2] Norwegian School of Veterinary Science, Oslo, Norway.

Leukodystrophies are disorders of myelin synthesis and maintenance that affect CNS myelin. They are subdivided as leukodystrophies, hypomyelinating disorders and spongy degenerations. Although infrequently seen, several forms have been described in various dog breeds. We present a novel form of complex leukodystrophy consisting of hypomyelination and spongy degeneration that presents primarily with hind end tremors in Border terrier puppies.

Three Border terriers from two different litters (and lineages) are described here that presented with a history of shaking movements. The youngest dog was a 3-week old male. It was the only dog affected in the litter. The other two dogs were 6-week old female littermates. There were two unaffected males in the same litter. Physical examination revealed no abnormalities. On neurological examination, the affected dogs displayed severe hind end tremors, with a characteristic swinging side-to-side movement (best described as "rumpshaker"). The tremors also involved the head and thoracic limbs but to a lesser degree, and disappeared when the dogs were asleep or at rest. Severe cerebellar ataxia was observed when the dogs ambulated. Proprioceptive positioning was delayed in the pelvic limbs of all 3 dogs. Spinal reflexes and nociception appeared normal. Necropsy was performed in all 3 puppies. No macroscopic changes were observed. Histologic evaluation of the CNS revealed spongy degeneration and hypomyelination in all funiculi of the cervical and thoracic spinal cord. White matter of the frontal, temporal and parietal cortices had mild multifocal spongy degeneration and hypomyelination, whereas white matter of the cerebellum, medulla and pons showed severe diffuse spongy degeneration and hypomyelination with gliosis. The combination of reduced myelin formation combined with spongiform white matter changes in the absence of microglial responses suggest a complex pathogenesis affecting both oligodendrocytes' capacity to synthesize myelin and the stability of the myelin that was formed. The number of oligodendrocytes and axons appeared subjectively normal indicating a primarily hypomyelinating process

The clinical and pathological features of this disease have not been described in any other canine leukodystrophy. The primary and most striking clinical feature is the presence of severe tremors in the hind end, causing the "rumpshaker" pheynotype. Genetic studies are underway to determine if the disease is inherited and the inheritance mode.

LOW LEVEL LASER THERAPY AS AN ADJUNCTIVE THERAPY TO THORACOLUMBAR DECOMPRESSION FOR CANINE INTERVERTEBRAL DISK DISEASE. WE Draper, TA Schubert. University of Florida, College of Veterinary Medicine, Gainesville, FL.

The purpose of our study was to determine if low level laser therapy and surgical decompression of the spinal cord encourages a more rapid clinical resolution than surgical decompression alone.

Dogs presented for acute paraparesis/paraplegia secondary to an acute thoracolumbar disk herniation were given a modified Frankel score (MFS). 36 dogs scoring 0-3 were included in this prospective study. Based on alternating order of presentation dogs were assigned to 1 of 2 treatment groups (19 in the no laser treatment control group and 17 in the laser treatment group). All dogs underwent surgery (hemilaminectomy +/- pediculectomy) for a herniated thoracolumbar intervertebral disk as determined by MRI or CT. Group 2 was treated with transcutaneous low power laser therapy (LLLT) for 5 days, or until they achieved an MFS of 4. We used a 5 x 200mW 810nm cluster array to deliver $25W/cm^2$ to the skin. Based on penetrance data, this delivered an appropriate energy density to the spinal cord. All treatments were done by the investigators, and there were no sham treatments. Dogs in both treatment groups were scored daily by the investigators using the MFS system until they achieved a score of 4.

The mean time to achieve an MFS of 4 in each group was analyzed using the pooled Student's t-test. An MFS of 4 in the laser treatment group was realized in 5.97 days. This was a significantly shorter time ($p = 0.004$) than the control group, which was 12.97 days. 2 of the dogs in the control group did not finish the study as they never achieved an MFS of 4. These dogs were not included in the statistical analysis.

Dogs receiving low level laser therapy and decompressive surgery for thoracolumbar intervertebral disk herniation have a decreased time to ambulation compared to dogs that received only decompressive surgery. Decreasing the time to ambulation could be important with regards to decreasing morbidity associated with a non-ambulatory dog (pneumonia, urinary tract infections, decubital ulcers, etc.) as well as decreased economic impact on our clients.

CANINE CENTRONUCLEAR MYOPATHY – NEW INSIGHTS INTO THE ROLE OF PTPLA IN MYOGENESIS. GL Walmsley[1], M Maurer[2], IM McGonnell[1], SC Brown[1], CA Sewry[3], L Tiret[2] and RJ Piercy[1]. 1 RVC, London, UK, 2: INRA, Maisons-Alfort, France, 3 ICH, London, UK.

Centronuclear myopathies (CNM) are a group of inherited disorders characterized clinically by weakness and skeletal muscle hypotrophy, and histologically by abnormal centralization of nuclei within myofibres. Identification of several genetic causes of human CNM suggests the disorders may be caused by dysfunctional membrane trafficking and defective T-tubule structure or function in skeletal muscle. A myopathy with similar pathological features, caused by a SINE insertion in the *PTPLA* gene, occurs in Labrador Retrievers but the associated protein's function in normal muscle and the mutation's effect on T-tubule structure and function are unknown. We aim to validate canine CNM as a model for human CNM and investigate the role of PTPLA in muscle development and maintenance.

Skeletal muscle from CNM-affected and age matched control dogs was evaluated by immuno-histochemistry and electron microscopy. Staining for T-tubule markers (BIN1 and DHPR) was disorganised in affected animals from 2 years of age and ultrastructural examination revealed some dilated T-tubules with altered morphology in young (10mo) and aged (12yo) CNM dogs when compared with controls. The relative expression and cellular localisation of the principal PTPLA splice variants (*PTPLAfl* and *PTPLAd5*) were investigated during muscle differentiation. Fluorescence microscopy of murine myoblasts and myofibres transfected with expression plasmids containing each of the above tagged-PTPLA isoforms revealed localisation to the endoplasmic/sarcoplasmic reticulum. Conventional and quantitative RT-PCR revealed that myoblast differentiation to myotubes involves an increase in expression of *PTPLAfl* in comparison with the *PTPLAd5* transcript. In murine embryos (E14.5), *PTPLAd5* is the main expressed variant; in adult mice, alternative splicing specifically occurred in striated muscles (which mainly expressed *PTPLAfl*). RT-PCR using primers designed to amplify a putative zebrafish *PTPLA* homologue, also revealed several splice variants and a similar switch in the relative expression of *PTPLAd5* and *PTPLAfl* during development from gastrula/segmentation stages to larvae and adult fish. Expression patterns of *PTPLA* variants in zebrafish embryos evaluated by in situ hybridization, were similar to those seen in mammals.

In conclusion, dogs deficient in PTPLA have similar changes to tubulo-reticular structures to those seen in human autosomal forms of CNM. Skeletal muscle PTPLA variants appear to localise to the endoplasmic reticulum and an evolutionarily conserved relative change in expression of PTPLA isoforms occurs during development. This information contributes to our understanding of myofibre differentiation. In future, studies on primary cells and strategies to knockdown, or knockout *PTPLA* in cell lines, mouse and zebrafish models may provide novel insight into the pathogenesis of canine centronuclear myopathy and related human diseases and, more generally, may help determine the functional roles of PTPLA in mammals.

EFFECTS OF A STANDARDIZED SHEEP HERDING STRENUOUS EXERCISE PROTOCOL ON CLINICAL AND LABORATORY PARAMETERS IN HEALTHY BORDER COLLIES AND DOGS WITH BORDER COLLIE COLLAPSE. S.M.Taylor[1], C.L.Shmon[1], L.Su[1], K.Minor[2], E.E.Patterson[2], J.M.Mickelson[2], G.D.Shelton[3]. 1.Western College of Veterinary Medicine, Saskatoon, SK. 2.College of Veterinary Medicine University of Minnesota, St.Paul,MN.3.School of Medicine, University of California-SD, La Jolla, CA.

A syndrome of Border collie collapse (BCC) appears to be common in dogs used for working stock. This syndrome has also been called malignant hyperthermia, heat intolerance, exercise-induced collapse and "wobbles". A presumptive diagnosis of BCC can only be made by eliminating other causes of exercise intolerance and weakness. The purpose of this study was to describe the clinical features of collapse in affected dogs and determine if there were characteristic clinical or laboratory features at rest or after exercise that could aid in diagnosis.

Seven adult Border collies with a history of collapse during sheep herding (affected) and 5 adult Border collies regularly used for sheep herding but showing no signs of exercise intolerance (normal) were evaluated before and after participating in a videotaped 10 minute exercise protocol consisting ofa series of continuous short outruns and fetches of three sheep in an outdoor pen. Exercise was halted at 10 minutes or earlier if there were signs of gait or mentation abnormalities. Pre-exercise evaluation included physical examination, orthopedic and neurological exam. Pre and immediate post exercise rectal temperature, pulse and respiration, patellar reflexes, ECG, CBC, serum biochemistry profile, cortisol, arterial blood gas and plasma lactate and pyruvate concentrations were measured. Clinical parameters (gait, temperature, reflexes) and lactate and pyruvate concentrations were evaluated at intervals up to 120 minutes after exercise. Additional testing in affected dogs included measurement of acetylcholine receptor antibodies (AChRab) and DNA testing for dynamin-associated exercise induced collapse (dEIC) and the ryanodine receptor mutation associated with canine malignant hyperthermia(MH). One week after exercise affected dogs had thoracic radiographs and echocardiography performed and were anesthetized for EMG and muscle biopsies.

There were no significant differences in temperature, pulse, respiration, or any laboratory parameter at any time point between normal and affected dogs. No arrhythmias were detected. Affected dogs were negative for the DNA mutations tested and for AChR ab. Thoracic radiographs, echocardiograms, EMGs and muscle biopsies were normal.

The 5 normal dogs had no alterations in mentation or gait during or after exercise. Three of the affected dogs had exercise halted early (6min-9min) because of altered gait or mentation. All 7 of the affected dogs were abnormal in the 15 minutes following exercise. Abnormalities seen in all dogs included disorientation, dull mentation, swaying, falling to the side, exaggerated lifting of limbs each step, choppy gait, delayed limb protraction, scuffing of rear and/or forelegs, and crossing legs when turning. All dogs returned to normal by 30 minutes.

BCC appears to be an episodic nervous system disorder that can be triggered by exercise. Genetic testing excluded dEIC and the described canine MH mutation. Common causes of exercise intolerance were eliminated, but the cause of collapse in BCC was not determined and no clinical or biochemical marker to aid diagnosis was established.

MOLECULARLY TARGETED CYTOTOXIC AND ELECTROPORATION-BASED THERAPIES FOR CANINE INTRACRANIALGLIOMA

John H. Rossmeisl, Jr., DVM, MS, DACVIM (SAIM and Neurology)
Blacksburg, VA

INTRODUCTION

Malignant gliomas (MG), and in particular glioblastoma multiforme (GBM), represent some of the most biologically aggressive and treatment refractory malignancies recognized in humans and dogs. Despite several decades of intensive research focused on MG/GBM, significant progress towards improving the survival of humans with GBM has been lacking. Currently, the average survival of a human with GBM is approximately 15 months. Only 5% of humans with GBM survive for 5 years, which is a statistic that has remained largely unchanged since 1959. Recently, multiple independent investigators have demonstrated that the clinical manifestations, natural course of disease, pathology, molecular etiopathogenesis, and genetic aberrations of canine and human brain tumors have been shown to be similar, which makes dogs an ideal translational model for the study of brain neoplasia. This presentation will provide background theory, an overview of pre-clinical data, and preliminary results of two translational canine intracranial glioma clinical trials.

MOLECULARLY TARGETED CYTOTOXINS

Several recent human studies have demonstrated the potential value of the clinical usage of molecularly targeted cytotoxins for the treatment of GBM. Investigators in our collaborative research group have demonstrated that the majority of human GBMs express the monomeric from of the IL-13 receptor (IL-13Rα2) on their surface, which is not expressed in normal brain. This discovery led to the development of first generation IL-13 based, Pseudomonal exotoxin (PE) derived cytotoxins. Subsequently, IL-13-PE cytotoxins have been shown to be extremely potent anti-glioma agents, and Phase III clinical trials demonstrated that the progression free survival of humans with recurrent GBM treated with IL-13-PE was superior to those receiving standard of care (carmustine wafers).[1] IL-13 based cytotoxins have been one of the few compounds in the last two decades to have to have been shown to have demonstrable clinical efficacy in the setting of recurrent GBM.[1,2] However, these trials have also highlighted the obstacles researchers and clinicians are faced with when designing and applying molecularly targeted cytotoxins. Ultimately, the clinical utility of any molecularly targeted cytotoxin will depend on three fundamental factors: 1) the target specificity, 2) the relative percentage of the patient population that can be predicted to respond to the cytotoxin, and 3) effective and efficient delivery of the cytotoxin to the tumor.

In an effort to improve upon previous results, ongoing research in our laboratory has resulted in the identification of an additional molecular target for GBM, the tyrosine kinase receptor Eph2A. Overexpression of Eph2A receptor has been demonstrated in many human GBM specimens, as well as various other types of human malignancies, including breast, ovarian, and pancreatic carcinoma. Physiologically, activated Eph2A negatively regulates integrin-mediated cellular adhesion, cell spreading, and migration. However, in most solid tumors studied to date, Eph2A is present in its inactive (non-phosphorylated) state and thus cells on which it is expressed may have tendencies to be more motile, faster-growing, and invasive, features which are characteristic of the phenotype of tumors on which Eph2A overexpression have been documented. However, most interesting is that results to date indicate that overexpression of Il-13Rα2 and/or Eph2A together can be documented in nearly 100% of human GBM, thus indicating that predictable anti-tumor efficacy should be nearly universal if a combinatorial treatment approach using both of these molecular cytotoxins is employed.[1,2]

As justification for performance of a combinatorial molecularly targeted cytotoxin clinical trial in dogs with MG, we have further shown that the molecular fingerprints of canine high-grade gliomas and GBM are homologous to that of humans. Investigations by our collaborative group have demonstrated that IL-13Rα2 and Eph2A overexpression also occur in the majority of canine high-grade gliomas, but not in normal brain tissue. Thus, we provide additional evidence that the dog is an excellent model for translational brain tumor research and the molecular signature of IL-13Rα2 and Eph2A overexpression represents a unique, specific and predictable target for human and canine glioma treatment.[1] In preparation for canine clinical trials, new generation IL-13 based cytotoxic mutants have been designed and shown to have potent and more specific anti-glioma activity than previous cytotoxins. A prototype recombinant Eph2A based cytotoxin has also been conceived, manufactured, and is currently undergoing testing in our laboratory.

Due to the inherent limitations of the blood-brain-barrier, macromolecules such as the targeted cytotoxic proteins previously discussed are preferentially delivered locally into brain tissue. To accomplish this task in our clinical trial, we will be using a convection enhanced delivery (CED) technique. CED is essentially a low-pressure continuous infusion process that occurs over hours to days that has been shown to be capable of achieving substantial tissue concentrations of macromolecules over large areas of the brain without producing tissue injury, substantially increasing intracranial pressure, or decreasing blood flow. Several previous attempts at CED techniques in the brain have been troubled by unpredictable, undesirable, or unknown delivery of agents due to technical issues with catheter design and placement. Building on the pioneering CED work done in dogs by Dickinson et al.,[3] several new innovations relating to CED catheter design and intraoperative neuroimaging have now made it possible to efficiently and effectively monitor delivery of macromolecules into the brain using CED. We will carry these studies further by using real time PET based imaging techniques with several tracers to monitor catheter placement, drug delivery and distribution, while preventing shortcomings related to MRI techniques dependent on infusion of standard gadolinium based contrast agents, which can linger when injected locally into tumor beds.

In summary, our pre-clinical work has identified specific and predictable molecular signatures within canine and human gliomas that can serve as therapeutic targets for bacterial based cytotoxins with potent activity against gliomas. In conjunction with recent improvements in CED techniques, we are well poised to embark on a clinical trial using spontaneous canine gliomas to further evaluate the *in vivo* efficacy of these molecularly targeted cytotoxic compounds.

ELECTROPORATION-BASED THERAPIES

Electroporation, which results in an increase in the permeability of the cell membrane, is initiated by exposing cells or tissues to electric pulses.[4] As a function of the induced transmembrane potential the electroporation pulse can result in: no effect on the cell membrane, reversibly permeabilize the cell after which cells can survive (reversible electroporation), or permeabilize the cell membrane in a manner that leads to cell death (irreversible electroporation [IRE]) presumably through a loss of homeostasis if not from other superimposed damage modes. This increase in transmembrane potential is dependent on a variety of conditions such as tissue type, cell size, and pulse parameters including pulse shape, duration, number, and repetition rate. However, for a specific tissue type and set of pulse conditions, the primary parameter determining the extent of electroporation is the electric field to which the tissue is exposed.[4] Reversible electroporation has been used for years to facilitate genetic modifications of cell function via delivery of plasmid DNA, mRNA and viruses through the reversible creation of nanopores within the cell membrane. Reversible electroporation can be used to increase permeability in order to allow intracellular delivery of otherwise impermeable cytotoxic macromolecules for targeted cell death in a process known as electrochemotherapy (ECT). IRE technology has been applied for decades within *in vitro* cellular systems, in particular for the sterilization and preprocessing of foodstuffs and for bactericidal effects during water purification.

The destruction of tumors and other undesirable tissue through focal ablation has become an important, minimally invasive alternative to open surgery. Methods of tissue destruction include phototherapy, radiofrequency lesioning, cryoablation, microwave ablation, high-intensity focused ultrasound (HIFU), and laser interstitial thermotherapy (LITT). Mechanisms of tissue destruction include changes in local temperature resulting in coagulation necrosis (radiofrequency lesioning, LITT, and ultrasound) and cell membrane rupture (cryoablation), generation of free-radicals (phototherapy) and *in situ* tissue fixation (microwaves). The confines of the cranial vault and the precision necessary for treatment have limited the use of some of these tools for intracranial disorders. When the electric field strength is above a critical value, IRE is achieved, resulting in cell death within the treated volume. IRE has been reported in conjunction with tissue thermal damage caused by electrical Joule heating. Joule heating is the dissipation of energy when an electrical current flows through a medium with resistance to that current. However, it is possible at certain electric parameters to produce IRE without thermal damage to the tissue in a process referred to as non-thermal IRE.[5,6]

In 1993, Salford et al.[7] first reported on the combined use of ECT with bleomycin in a rodent glioma model, demonstrating a doubling in survival time compared to animals treated with bleomycin alone. The rats showed no immediate adverse effects from the delivery of the pulsed electric fields. It should be noted that the effects of electric pulses alone were never investigated. Since this time, ECT has evolved as a promising new anticancer therapy for the treatment of cutaneous and other systemic tumors, but investigations of its use for the treatment of brain disorders were largely abandoned until recently.[6]

IRE may offer some advantages over more traditional surgical resection methods for selected brain tumors. The small electrode size makes the procedure minimally invasive and adaptable to virtually any neuroanatomic location with existing stereotactic guidance systems. It also offers biologic advantages (minimal heat generation during treatment and sparing of major blood vessels) that may make it attractive as a complimentary technique to tumor resection. IRE creates a sharply delineated volume of ablated tissue with sub-millimeter resolution.[5,6] Moreover, outside the zone of tissue ablation, there is a penumbra of non-destructive increase in membrane permeability. The cells outside the zone of ablation but within this penumbra are more susceptible to exogenous agents such as chemotherapy. IRE also offers the advantage of speed, as typical treatment durations range from 5-15 minutes, and numerous commercial entities manufacture relatively inexpensive electrical generators capable of pulse delivery within the realm of IRE.[6] Finally, preliminary studies also suggest that the mechanism by which IRE induces cell death is remains effective in neoplasms that have developed chemo- and radioresistance. Currently, several centers worldwide are conducting clinical trials in humans using various IRE and electrochemotherapy approaches for hepatic, renal, pancreatic, and pulmonary neoplasms.

In vitro investigations in our laboratory have demonstrated the cytotoxic effects of IRE against canine, rodent, and human glioma cell lines; identified and capitalized on the effects of various pulse paradigms to maximize anti-neoplastic field effects while simultaneously preserving major vascular structures in the treated volume and preventing formation of vasogenic edema; and provided insight into the mechanism of cellular death induced by IRE which has implications for selecting combinatorial treatment approaches. Additional studies have indicated that tumor cytotoxicity can be enhanced within the reversible penumbra of electroporation using various chemotherapeutic agents with known efficacy against gliomas at doses 10-100 fold less than those recommended for systemic use, thus minimizing any systemic dose-related toxicosis.[7] Recent progress in device development has also yielded promising advancements in electrode design and pulse generator capabilities that may allow for future delivery of IRE without the need for administration of neuromuscular blocking agents, as well as completely non-invasive pulse delivery via the transcranial route similar to techniques currently in use for HIFU based brain therapies.

Additionally, we have shown *in vivo* that IRE ablation of brain tissue: 1) can be achieved safely and predictably in rodent and canine models; 2) can be monitored using routine clinical procedures such as ultrasound, CT, or MRI; 3) occurs without any contribution from thermal damage; and 4) is capable of submillimeter resolution between ablated and normal cells.[6,8] Our *in vivo* work has also shown that normal and neoplastic brain tissue can be safely and controllably ablated using electrical charges lower than those typically delivered to the human brain during electroconvulsive therapy.[6,8,9] Although an intact immune system is not required to achieve the desired effects of IRE, studies in brain and extraneural tumors suggest that IRE treatment may be associated with induction of host tumor infiltrating lymphocytes, which has been shown in some other spontaneous neoplasms to have a positive correlation with therapeutic response and survival. Preliminary work has also indicated that IRE may also have a radiosensitizing effect in the brain, which may offer additional advantages for treating intracranial neoplasms with adjunctive radiotherapy.[9]

We are currently conducting an IRE-based, funded canine glioma clinical trial for dogs. All dogs with intra-axial prosencephalic lesions on MRI consistent with glial-origin neoplasia that are free of other significant concurrent systemic disease are candidates for inclusion. Dogs will be randomized into treatment groups to receive either stereotactic IRE ablation + placebo or stereotactic IRE ablation + ECT.

REFERENCES

[1]Debinski W, et al. **Clin Cancer Res** 1999; 5: 985-990; [2]Wyksoky J, et al. **Clin Cancer Res 2008**; 14: 199-208; [3]Dickinson PJ, et al. **J Neurosurg 2008**; 108:989-998; [4]Edd JF, et al. **IEEE Trans Biomed Eng 2006;** 53:1409-1415;[5] Davalos RV, et al. **Ann Biomed Eng 2005**; 33: 223-231; [6]Ellis TL, et al. **J Neurosurg 2010**; doi:10.3171/2010.5.JNS091448; [7]Salford LG, et al. **Biochem Biophys Res Comm 1993** 194:938-943; [8]Garcia PA, et al. **J Mem Biol 2010**; 236: 127-136; [9]Garcia PA, et al. **Technol Cancer Res Treat 2011**; 10:73-83.

NANOPARTICLES IN BRAIN TUMOR THERAPY

Rebecca A. Packer, MS, DVM, DACVIM (Neurology)
West Lafayette, IN

INTRODUCTION

The primary cause of disease progression and death in glioma patients is not spread of the tumor to distant locations, but rather local recurrence at the initial tumor site. Recurrence results because of tumor cell resistance to treatment. For this reason, vast amounts of funding and effort are being invested to identify improved treatments for gliomas. Therapeutic research using nanoparticles has become an expanding theme in this field. The field of nanotherapy is too large to present a comprehensive review in this session. Instead, literature and preliminary data applicable to clinical research and clinical treatment will be presented. Though such therapies are applicable to other types of brain tumor, this discussion will focus on canine gliomas due to their use as a translational model.

CURRENT APPLICATIONS IN BRAIN TUMOR MANAGEMENT

The rationale for developing and studying nanoparticles for cancer therapy include the following: 1) reduced toxicity of drug cargo, 2) ability to penetrate membranes (e.g., the blood brain barrier), 3) delivery vehicle for drugs/genes/etc, 4) evasion of the mononuclear phagocytic system (MPS)/reticuloendothelial system (RES), and 5) enhanced permeability and retention (EPR) effects.

By definition, nanoparticles are less than 1 micron (e.g., submicron) in size in at least one dimension. These constructs are most often spherical in geometry, with diameters ranging from >10 nm to nearly 1 micron, however, constructs with different shapes, compositions, and fabrication methods have been reported.[1] These include spherical nanoparticles such as liposomes (solid or fluid shell) and polymer micelles (solid core);[1] rod-like ("string of pearls") nanoparticles such as polyrotaxanes, in which the drug or cargo are threaded on a linear polymer core;[2,3] cylindrical or elliptical nanoparticles that are fabricated from templates or stretched polymers (fabricated nanoparticles);[4,5] and "chip"-like nanoparticles which are under development by several laboratories and may be available for future use that include biosensors and/or reporters.[6] One of the most common strategies used in cancer therapy is drug delivery by attaching or incorporating drugs into the particles.

Pertaining to reduced toxicity, although there are reports of nanoparticle and liposomal drug formulations minimizing systemic side effects by limiting drug diffusion into normal cells,[7,8] this may vary with the chosen carrier and its own biodistribution profile.[8] There are also opposing reports of increased toxicity and increased drug deposition in such organs as bone marrow (and enhanced myelosuppression), liver, spleen, and lungs, when drug was delivered via specific nanoparticle carriers compared to free drug.[9] Liver, spleen, kidney and lung are the common organs where nanoparticle accumulation occurs, and where one would anticipate deposition of carrier drug. Based on these conflicting studies, such hypotheses that nanoparticle delivery vehicles will empirically reduce drug toxicity below that of free drug should be investigated carefully.

Nanoparticles are a unique class of drug carrier due to their ability to diffuse through tissues with tight intercellular spaces, and to be constructed with both diagnostic and therapeutic functional characteristics. Nanoparticles can be conjugated not only with drugs to serve as therapeutic ***drug carriers***, but also with cell surface antibodies as a means of ***cellular targeting***, and with imaging contrast agents (iron oxide or gadolinium-based agents) to allow their diffusion and penetration to be monitored via ***diagnostic imaging modalities***. Imaging of nanoparticles has been demonstrated using paramagnetic particles for MRI, iodinated surfaces for computer tomography, or photoemitters for fluorescence microscopy or scanning.[8,9] Initial results using paramagnetic nanoparticle contrast agents for different tumor models, for example breast cancer cells in mice, demonstrate a precise and prolonged delineation of the tumor margin or lymph node metastasis.[8,9,10,11] In a spontaneous canine glioma model, real-time MRI confirmed accurate administration of gadoteridol-carrying liposomes containing a topoisomerase-inhibitor delivered via convection-enhanced interstitial infusion.[12]

Magnetic resonance imaging (MRI) contrast agents (e.g. gadolinium diethylene-triamine-pentacetic acid [Gd-DTPA]) are typically injected systemically, and function passively and non-specifically to enhance the signal of certain tissues of interest. These contrast agents rely on breakdown of the blood-brain barrier in order to passively enhance the tissue of interest. Targeted contrast agents, however, are modified so that they localize to a specific cell type or tissue through a passive or active mechanism.[13] Active targeting involves the use of biomarkers such as antibodies or small peptides to encourage specific binding. Novel biomolecular strategies have been reviewed in the literature, and data indicate that targeting via actively-targeted nanoparticles conjugated with contrast-agents could be of diagnostic value.[13-17] Initial efforts to create targeted contrast agents for MRI involved direct conjugation of a contrast agent (typically, Gd-DTPA) onto an antibody. However, this method did not prove to be very successful owing to the relatively low sensitivity of MRI and the low concentration of most cellular targets. More recent efforts have focused on photo-emission imaging (e.g., PET, NMR spectroscopy, or fluoroscopy).[18]

FUNCTIONAL CONSIDERATIONS

Treatment failures in glioma patients result mainly from local tumor recurrence, rather than distant metastasis or novel sites of tumor growth.[19] Local recurrence and survival were not significantly different between high-grade glioma patients receiving local irradiation compared to whole brain irradiation.[19] Outcomes were also not significantly different between patients who underwent extensive compared to less extensive surgical resection.[20] These studies suggest that tumor recurrence is not due to failure to recognize tumor tissue during treatment approaches; but rather it is due to tumor cell resistance. Such outcomes illustrate the need for improved drug delivery mechanisms and improved ability to penetrate local tissues. To this end, nanoparticles can be designed with optimal characteristics to diffuse through the target tissue (e.g., brain tissue in this study) and maintain an optimal residence time (e.g., maintain therapeutic concentrations of drug locally), with minimal systemic side effects.

Effects of size. The effect of nanoparticle shape and size significantly affects residence time, cellular uptake, and tissue diffusion.[21-24] Although diffusion is affected by a complex set of interactions including size, shape, and surface chemistry; smaller particles have been shown to diffuse more rapidly than larger particles. It is also important to consider the likelihood of agglomeration of nanoparticles when performing diffusion studies. Higher concentrations of particles, as well as smaller particles, tend to agglomerate moreso than larger particles, forming large aggregates.[24] Agglomeration then slows or even halts diffusion contributing to accumulation of nanoparticles in tissues. Of note, iron oxide nanoparticles, a very common particle type, are generally not stable and tend to agglomerate fairly readily,[24] whereas a flash precipitation method has been described in which nanoparticle size is controlled by varying the concentrations of two solutions rapidly mixed to form self-assembled nanoparticles which, under certain conditions, are stable against aggregation.[25] Size also has been shown to affect clearance of nanoparticles in the body, and internalization of nanoparticles into cells.[26,27] In clearance studies, smaller particles have been shown to more effectively evade uptake by splenic filtration and the MPS system, and diffuse through membranes such as vessel walls or the blood brain barrier.[26] Specific nanoparticle dimensions that allow for these features are dependent on the relative size of membrane fenestrations and cell-cell spacing. However, although smaller particles may evade clearance by the MPS system (< 200 nm), particles that are too small (< 10 nm) will be rapidly filtered by the kidney.[27]

Effects of shape. The effect of nanoparticle shape has been shown in many studies to significantly affect the pharmacokinetic properties of nanoparticles.[2,3,5,21-24,27,28] Rod-like or filamentous 'stealth' nanoparticles have unique characteristics that allow evasion of clearance by macrophages, improved residence time, and improved diffusion through tissue.[2,3,21,22] Worm micelles administered intravenously can remain in circulation for days to weeks.[21] The common theme appears to be the aspect ratio, with higher aspect ratios allowing more effective evasion of the MPS system.[5,21,28] Increased residence time directly equates to longer drug exposure times and improved opportunities for cellular uptake. Flexible rod-type polymer micelle carrier systems display exceptionally long plasma circulation times (exceeding 7 days in some cases) due to flow alignment in blood that minimizes contact of the vehicle with and clearance by macrophages,[21] and enhances their extravasation into tumor tissues via the "enhanced permeation and retention effect" (EPR).[29,30] Polyrotaxanes are such rod-like nanoparticles, and fall within the applicable size range for EPR effect. As a result, they should afford an advantageous pharmacokinetic profile for use as drug carriers. The effect of shape on cellular uptake has been studied independent of size, and particles with larger aspect ratios (e.g., cylindrical or filamentous particles) have shown increased cellular binding and internalization of nanoparticles.[22,23,28]

Effects of targeting ligands. The efficacy of nanoparticles as diagnostic and therapeutic agents is enhanced by both passive and active targeting.[31,32] Chelating the nanoparticles with antibody to a surface antigen allows for specific targeting toward cell lines of interest. In pharmacokinetic studies, although targeted nanoparticles have prolonged residence time and greater cellular uptake compared to nontargeted particles, optimal target and drug-loading densities are not consistent across applications.[31,33,34] Also, diffusion distance may be inversely related to the density of targets (the "binding site barrier" effect).[31] Thus, optimal target-loading densities must be established for each tissue application, particle type and targeting ligand to achieve optimal residence time and diffusion distance in the tissue of interest.

Effects of surface chemistry. While nanoparticle size and shape have been studied extensively, the effect of surface chemistry is not as well understood. As with conventional pharmacokinetics, surface chemistry (e.g., amphiphilicity) has certainly been implicated in affecting the membrane permeability, cellular uptake and residence time of nanoparticles.[21,27] PEGylation, for example, has been shown to prolong residence time in circulation, but may impair interactions between the nanoparticle and target of interest, and also may promote particle adhesion to red blood cells.[27,35,36] As evidenced by these examples, nanoparticle pharmacokinetics and pharmacodynamics are a complex interaction of size, shape, and surface chemistry, which will additionally vary with each tissue type of interest, targeting ligands, and cargo. Thus, strict extrapolation of nanoparticle behavior from one application to another will not be reliable.

PRELIMINARY RESEARCH AND FUTURE DIRECTIONS

Preliminary research at Purdue has focused on optimizing the construct size and shape for optimal diffusion and residence time in brain tissue, as well as optimal densities of drug and targeting ligands. Preliminary data is briefly outlined below, and updated results will be presented. The long-term goal is to develop and lead a multi-center treatment trial on clinical cases of dogs with naturally-occurring brain tumors. Colleagues who are interested in being involved in potential future collaborations in a clinical trial should contact Dr. Packer.

Shape. The Thompson laboratory (Department of Chemistry, Purdue University; collaborator) has prepared rod-like polyrotaxanes,[2,3] polymer micelles, and fabricated polymer nanoparticles that can be produced in cylindrical (and other) shapes. We have conducted preliminary studies comparing passive diffusion rates of various sizes of spherical liposomes, polymer micelles, and polyrotaxanes, and cylindrical fabricated nanoparticles in a previously validated gel phantom model.[37]

Targeting. Targeted nanoparticles have been shown to have increased local residence time and internalization of drug compared to non-targeted nanoparticles.[31,33,34] In our preliminary studies, we have validated binding of a specific cell surface target to all three glioma cell lines tested in our laboratory. Targeting this surface protein of brain tumor cells should increase drug exposure and internalization of drug. Validation of binding to the canine cell line also provides evidence for anticipated use of a spontaneous canine model of glioma for pre-clinical trials following pharmacokinetic and pharmacodynamic studies.

CLOSING REMARKS

A critical component of translational research with nanoparticles is the continued investigation into the relationships between the physicochemical properties of nanostructures and their behavior in vivo, which is currently poorly understood. Although advances have been made in some aspects, greater insight into the mechanisms dictating the pharmacokinetics and fate of nanoparticles in vivo is needed. Because nanoparticle behavior and physiochemical properties vary between tissue types, such studies are needed for each particle construct and tissue type of interest. By incorporating advances in nanoscale engineering, molecular imaging, and novel therapeutics, targeted carriers have the potential to enable physicians to diagnose and treat oncologic diseases with greater effectiveness than ever before. Additional considerations to improve nanoparticle stability, uniformity in construction, and the ability to produce mass quantities sufficient for large-scale therapeutics are warranted.

ACKNOWLEDGEMENTS

We thank Dr. Karen Pollock at the Indiana University School of Medicine for providing the SF767 and U87-MG human glioma cell lines, and Dr. John Ohlfest at the University of Minnesota for providing the canine astrocytoma cell line.

REFERENCES

[1]Tong R, et al. **Polym Rev** 2007;47:345-381. [2]Loethen S, et al. **J Macromol Sci C** 2007;47:383-418. [3]Loethen S, et al. **Biomacromolecules** 2006;7:2501-2506. [4]Canelas DA, et al. **Wiley Interdiscip Rev Nanomed** Nanobiotechnol 2009;1(4):391-404. [5]Champion JA, et al. **Pharm Res** 2009;26(1):244-249. [6]Kricka LJ. **Clin Chem Acta** 2001;307(1-2):219-223. [7]Huang G, et al. **Int J Pharm** 2008;355(1-2):314-320. [8]Brigger I, et al. **Adv Drug Del Rev** 2002;54:631-651. [9]Couvreur P, et al. **J Pharm Sci** 1982;71:790-792. [9]Gallego O, et al. **Clin Transl Oncol** 2006;8(11):788-795. [10]Xhu W, et al. **Magn Reson Med** 2008;59(4):679–685. [11]Harisinghani MG, et al. **N Engl J Med** 2003;348:2491-2499. [12]Dickinson PJ, et al. **Neuro-Oncology** 2010;928-940. [13]Morawski AM, et al. **Curr Opin Biotechnol** 2005;16(1):89-92. [14]Laurent S, et al. **Bioconjug Chem** 2004;15(1):99-103. [15]Artemov D, et al. **Cancer Res** 2003;63(11):2723-2727. [16]Artemov D, et al. **Magn Reson Med** 2003;49(3):403-408. [17]Botnar RM, et al. **Circulation** 2004; 110(11):1463-1466. [18]Hahn MA, et al. **Anal Bioanal Chem** 2011;399(1):3-27. [19]Sharma RR, et al. **Neurol India** 2003;51:512-517. [20]Sanai N, et al. **Neurosurgery** 2008;62:753-764. [21]Geng Y, et al. **Nat Nanotechnol** 2007;2:249-255. [22]Huang X, et al. **Biomaterials** 2010;31:438-448. [23]Gratton SEA, et al. **PNAS** 2008;105(33):11613-11618. [24]Stark WJ. **Angew Chem Int** Ed 2011;50:2-21. [25]Valencia PM, et al. **ACS Nano** 2010;4(3):1671-1679. [26]McNeil SE. **J Leuk Biol** 2005;78:585-594. [27]Yoo JW, et al. **Curr Pharm Des** 2010;16(21):2298-2307. [28]Sharma G, et al. **J Control Release** 2010;147:408-412. [29]Maeda H. **Advan Enzyme Regul** 2001;41:189-207. [30]Maeda H, et al. **Eur J Pharm Biopharm** 2009;71:409-419. [31]Lee H, et al. **Mol Pharm** 2010;1195-1208. [32]Wang PP, et al. **Adv Drug Deliv Rev** 2002;54:987-1013. [33]Kirpotin DB, et al. **Cancer Res** 2006;6732-6740. [34]Drummond DC, et al. **J Control Release** 2010;13-21. [35]Bazile D, et al. **J Pharm Sci** 1995;84(4):493-498. [36]Mishra S, et al. **Eur J Cell Biol** 2004;83(3):97-111. [37]Chen ZJ, et al. **J Neurosurg** 2004;101:314-322.

DRUG RESISTANCE – DOES IT MATTER IN HUMAN & CANINE BRAIN TUMORS ?

Kaspar Matiasek, DVM, DrMedVet, DrMedVetHabil
Kentford, Suffolk, UK

INTRODUCTION

Tumor location, growth pattern, vascularity, its intrinsic sensitivity/resistance to radiotherapy and chemotherapeutics as well as the dose-limiting toxicity of radiation and pharmacological agents are the main determinants for the success and efficacy of tumor therapy [1]. Primary intraaxial tumors and those penetrating the pial membrane are prone to incomplete resection on brain surgery and therefore require adjunct treatment for delay of recurrence. Failure of physical removal is one of the major concerns, driving veterinary surgeons and pet owners against invasive treatments even if the animal may benefit from debulking cytoreductive surgery for reasons such as decrease of intracranial pressure. If available, radiation therapy is given favour amongst non-invasive therapies since it appears to slow down the rate of tumor cell proliferation considerably without being dependent on the permeability of the blood-brain-barrier (BBB) [2,3].

The paradigm of the tight BBB for some reasons has not been questioned even though experimental studies have shown that the BBB in solid brain tumours is replaced by simple fenestrated capillaries during neo-angiogenesis. It therefore was not until lipophilic nitrosourea derivates lomustine and carmustine and nowadays temozolomide became available that chemotherapy was appreciated as a considerable treatment option. According to the very few trials, however, the outcome of chemotherapy as single treatment in canine brain tumours is less than promising [4,5].

This contrasts somehow to response rates of up to 70% in human oligodendrogliomas and oligoastrocytomas and the success of polychemotherapy in tumors with a significant oligodendrocyte component, in general [6]. The selective failure of chemotherapy may involve different mechanisms such as up-regulation of anti-apoptotic pathways, effective DNA-repair mechanisms, prevention of drug uptake, and metabolic inactivation and elimination of the agent[7]. In the human brain, ATP-dependent drug efflux pumps – socalled multidrug resistance proteins (MDRP) - have been associated with excretion of cytotoxic, antiviral and anticonvulsive agents, especially those that resemble nucleoside analogues or organic anions conjugated with sulfate, glucuronate or glutathione [7].

MDRP-EXPRESSION IN THE BRAIN AND BRAIN TUMORS

Several MDRP of the ATP-binding cassette superfamily are constitutively expressed in brain microvessels, the ependyma and choroid plexus epithelium. Most abundantly the luminal endothelial surface in humans and other mammals, including dogs, express p-glycoprotein (P-GP) and breast cancer resistance protein (BCRP), while the multidrug resistance protein 1 (MRP1) contributes significantly to the blood-CSF barrier at the basolateral aspects of the plexus epithelium [8]. P-GP and BCRP show substrate specificity for a set of lipophilic agents that are prevented from entering the neuroparenchyma. Thus, lipophilia of an anticancer drug does not guarantee for effective CNS uptake. Moreover, the expression can be essentially upregulated under various structural and functional disturbances.

Consequently, a set of studies has focussed on these interface molecules and their specific expression pattern and load in intratumoral blood vessels, perivascular astrocytes and tumor cells from human glioblastoma, astrocytoma, oligodendroglioma, oligoastrocytoma, ependymoma, neuroblastoma and meningioma [7,10,11]. In summary, MDRP are expressed in all entities but with some essential differences regarding the topology. Most intraaxial tumours appear to express P-GP and BCRP strongly in endothelial cells. MRP1 in contrast, was seen mainly in both reactive and neoplastic astrocytes [7,11] and the latter also are capable of de novo expression of P-GP amongst different other members of the ATP-binding cassette superfamily [7]. Other gliomas only occasionally show a tumor cellular MDRP expression [7,11]. P-GP also was strongly expressed in neuroblastoma cell lines [11]. Similarly, meningioma cells reveal high loads of P-GP plus MRP1 [11].

In veterinary medicine, MDRP have been mainly studied in dogs suffering from seizures [12]. The above mentioned unpopularity of chemotherapy of brain tumours in this species plus the omission of biopsies and follow-up studies hitherto has precluded an elaboration of correlative data. In a multicenter investigation, we have recently assessed the immunohistochemical expression pattern in more than 90 canine gliomas and meningiomas. The majority of gliomas show a similar P-GP and BCRP pattern like the human cases with strong endothelial and rare to absent tumor cell labelling. MRP1 expression is prerequisite for astrocytes and occasionally stains intratumoral blood vessels. P-GP also is the predominant membrane pump in the vascular tree of the meningiomas. On the other hand, it is nearly absent from tumor cells where both other molecules appear to be significantly expressed.

FUNCTIONAL SIGNIFICANCE OF MDRP-EXPRESSION

Speculations on the relevance of MDRP in brain tumors have been based on the expression pattern, the mRNA and protein content and in vitro experiments, employing transporter antagonists [11]. Comparison of fresh biopsy slides, extracted cell lines and organotypic cultures, however, demonstrate a significant plasticity of MDRP expression in vitro. In particular, the P-GP value appears to decrease in tumor cell culture and therefore may be underestimated if not assessed via quick sections. The MRP1 pattern, on the other hand, is considered to be rather stabile and its functionality can be proven in culture by daunomycin cytotoxicity after antagonist application [11]. For the less well preserved P-GP, immediate investigation of the cellular expression is recommended. Furthermore, the distribution amongst the tissue and subcellular compartments should be recorded. Strong endothelial BCRP and P-GP expression in ependymomas has been associated with significant chemoresistance [10]. Membrane labelling is likely to reflect the functional integration with more accuracy than a cytoplasmic signal. Likewise, the circumference of the immunopositive endothelial lining should be evaluated for possible gaps. Drugs that have successfully passed the endothelial-pericytic barrier may be uptaken by astrocytes and expelled from their perivascular end feet that appear to enrich MDRP in various tumors. If the drug has gained access to the tumor, the tumor cellular expression may reflect the degree of protection. According to both human and canine data, low expression in most oligodendrogliomas would suggest a negligible excretion by the molecules tested. Meningioma cells on the other hand, for many instances show a strong immunopositivity for multiple efflux pumps. The variability between the cases in most studies underlines the need for individual testing and anti-MDRP strategies if chemotherapy is the only treatment option. Depending on the substrate specificity of the transporter, the type of the applicable drug may be chosen individually. Further adjunct treatments with calcium blockers, cyclosporin analogues and other MDRP modulators, including cyclooxygenase inhibitors, as well as via RNA intereference may increase the effective intracellular dose of the cytotoxic agent within the brain tumor and thereby improve the outcome [1].

REFERENCES

[1]MacDiarmid, et al. **Nature Biotec** 2009;27(7):643. [2]Spugnini, et al. **Vet Radiol Ultrasound** 2000; 41:377. [2] Bley CR, et al. **J Vet Intern Med** 2005;19:849. [4]Jeffery & Brearley. **J Small Anim Pract** 1993;34:367. [5]Fulton & Steinberg. **Semin Vet Med Surg (Small Anim)** 1990; 5:241. 127. [6]Carpentier. **Lancet Neurol** 2005;4:4. [7]Bronger, et al. **Cancer Res** 2005;65(24)11419. [8]Leslie, et al. **Toxicol Appl Pharmacol** 2005;204:216. [9]Conrad, et al. **Toxicology** 2001;156:81. [10]Ginguene, et al. **Brain Pathol** 2010:20:926. [11]Spiegl-Kreinecker, et al. **J Neuro-Oncol** 2002;57:27. [12]Pekcec, et al. **Epilepsy Res** 2009;83(2-3):144.

CANINE INTRACRANIAL NEOPLASIA AS A TRANSLATIONAL BRAIN TUMOR MODEL

R. Timothy Bentley, BVSc, DACVIM (Neurology)
West Lafayette, IN

INTRODUCTION

Rodent xenograft models of human brain cancer have provided many insights into human glioma and glioblastoma, and are the most common *in vivo* model utilized to study this disease. Flank and intracranial xenografts are typically used. However, xenograft models have not always accurately recapitulated the human disease, and the translational of therapies from conventional rodent models has been fraught with difficulties.[1]

An estimated 61,414 Americans were diagnosed with a primary brain tumor in 2009[2], and gliomas represent 80% of malignant brain tumors and 33% of all brain tumors[3]. The most recent 5 year survival percentages for malignant brain tumors are only 31-36%, with minimal increase from a figure of 27-30% in the 1980's.[2-4] An improved model of this common and devastating disease must be found.

Spontaneous intracranial glioma is common in dogs, and shares a remarkable number of clinical, pathological, radiological, molecular and prognostic features and also cell culture properties with the human disease. Additionally, the dog has been suggested to be more generally an improved translational model of human cancer, being closer to humans in genetic and molecular biology critical to tumor development, such as telomere biology, cancer stem cell biology, genetic mutations including those affecting p53, and molecular therapeutic targets[5,6].

THE EPIDEMIOLGICAL AND CLINICAL BASIS FOR THE CANINE TRANSLATIONAL MODEL

Compared to a human incidence of around 6 malignant brain tumors per 100,000 per year[4], dogs develop 14.5-26 primary CNS tumors per 100,000 per year[7,8], and 36% of canine brain tumors are glial in origin[9]. As in human patients (but unlike rodents utilized in models), canine patients show variation in the tumor type, tumor molecular biology, age at onset, concurrent disease, neurological impairment and seizures, and the tumor location within the cerebral hemispheres or occasionally occur elsewhere.[9,10] Possibly due to the compressed lifespans of dogs with respect to humans, and the strong founder effect in brachycephalic breeds such as the Boxer, dogs develop glioma more frequently than in people, but with similarly devastating outcomes. The median survival for human glioblastoma, following surgery, radiotherapy and corticosteroids is 10-12 months, or 14-20 months with the addition of temozolomide.[11-14] For canine glioma, the median survival following radiotherapy, corticosteroids +/- surgery is around 9.3-10.1 months[15], and the use of temozolomide for canine glioma is currently being studied. In both species, gliomas are typically focal, intracranial masses that are locally invasive but rarely if ever undergo hematagenous metastasis.

THE MOLECULAR BASIS FOR THE CANINE TRANSLATIONAL MODEL

The suitability of spontaneous canine cancer as a model of human disease has been thoroughly reviewed.[6,16] p53 is heavily related to the development of malignancy in both dogs and humans, and the dog has been established as a model for the p53 family; the canine p53 gene shares 87% sequence similarity.[17] p53 is the most commonly mutated gene in human cancer, and a similar incidence of mutation has been observed between canine and human cancers such as osteosarcoma and breast carcinoma. Significantly, p53 alterations were seen in 35% of canine astrocytomas, a figure to be compared with 30-40% of human astrocytomas.[10,18] p53 plays a pivotal role in cancer biology, and p53 has been described as the master guardian and executioner of apoptosis. p53 is a tumor suppressor gene (oncogene) and is required for p21 expression The p53 protein is upregulated in response to a multitude of events induced by cancer and its treatments, such as DNA damage, hypoxia and cellular stress, and the targets of p53 include cell cycle arrest genes (such as p21), regulators of apoptosis (including BAX and PUMA), anti-angiogenic proteins (including thrombospondin) and DNA repair genes. As such, p53 is the single common link between both DNA mutations leading to cancer and DNA-damaging therapy and a host of responses. This critical role of p53 in both normal and cancerous states is highly conserved between human and canine subjects. Additionally, in both species, p53 is rapidly turned over by HDM2 (*syn.* MDM2), and p53 has been shown to accumulate in canine cancer cells in response to both DNA damage and HDM2 inhibition by nutlin3.[17] Critically, p53 production is not necessarily reduced by cancerous mutations; rather, abnormal p53 may accumulate, inhibiting the function of wild-type p53, and inducing the production of HDM2 and the subsequent ubiquitylation of the wild-type. Finally, HDM2 is overproduced in many tumors, leading to suppression of p53 function without any direct mutation of the p53 gene.[19]

Given the potential shown by HDM2 antagonists, such as nutlin3, in the treatment of human glioma, the shared molecular tumor biology between human and canine tumors is critical in establishing the dogs as a translational model. Additional features of shared tumor biology between dogs and humans have been thoroughly reviewed, such as similarity in telomere length (15-20kb in humans and dogs; >60kb in mice), sites of telomerase activity (tumors, stem cells, gonadal tissue and certain white blood cells in humans and dogs; ubiquitous in mice), histology, tumor behavior, and molecular targets such as receptor tyrosine kinases[6].

THE CANINE RADIOLOGICAL MODEL AND ADVANCES IN CANINE NEURO-ONCOLOGIC RADIOLOGY

Marked radiological similarities exist between common human and canine primary intracranial tumors. This may facilitate translational clinical trials, supporting the use of advanced imaging in the diagnosis and management of brain tumors. In both species, meningiomas are typically severely and homogenously contrast-enhancing, discrete, broad-based, extra-axial masses that may contain cystic regions. Gliomas, including glioblastoma, are poorly demarcated, variably contrast-enhancing, intra-axial masses.

As the field of human neuro-oncologic radiology advances, the ability to predict the type of glioma based on MRI features has been assessed. Grade II astrocytomas are typically non-enhancing expansions of affected brain, Grade III astrocytomas (anaplastic astrocytomas) typically show patchy enhancement and T2 hyperintensity, and Grade IV astrocytomas (glioblastoma) are irregularly enhancing, but centrally non-enhancing, with marked mass effect secondary to vasogenic edema. Increasing tumor grade has been related to increasing mass effect, cyst formation or necrosis, border definition, edema, heterogeneity and hemorrhage and tumors that cross midline.[20]

The pronounced radiological similarities between human and canine primary intra-axial brain tumors have been discussed.[21-24] However, while canine glioma has been repeatedly reported to be similar in appearance to human glioma, the ability to differentiate between types of canine glioma with MRI has been minimally investigated to date.

Preliminary Findings

The MRI features of 24 histopathologically confirmed, primary intra-axial canine brain tumors were compared with tumor grade and type. Nine oligodendrogliomas, 12 astrocytomas (including 4 glioblastomas), 2 primitive neuroectodermal tumors and 1 gliomatosis cerebri were evaluated by one investigator blinded to the histopathology. Increasing severity of contrast enhancement ($p=0.0021$), and the formation of cysts or intra-tumoral fluid accumulations ($p=0.0354$) were associated with increasing tumor grade. A potentially important relationship identified was the possible association between increasing distribution of gradient echo signal voids and increasing tumor grade ($p=0.0594$). These findings provide further support that canine intra-axial brain tumors represent a solid radiological model of human brain tumors, and show surprising similarities in the relationship between certain MRI features and increasing malignancy between these two species.

The MRI interpretation scheme has been improved following this preliminary study, and is now being applied to a larger number of tumors by 5 investigators at 2 different institutions. Further results will be submitted for publication in the future.

CANINE GLIOMA CANCER STEM CELLS

Current therapies for malignant brain tumors, such as glioblastoma, successfully decrease tumor mass, but typically do *not* eradicate all tumor-initiating cells, the result of which is that tumor recurrence may be inevitable. Such conventional therapies are used under the assumption that all tumor cells have the same capability to initiate and maintain tumors. Recently, the "Cancer Stem Cell" theory has challenged prior thinking, and states that stem cells that are inherently immortal and capable of long-term division accumulate mutations and produce the progenitor and somatic cells that make up the mass of the tumor.[6] Cancer Stem Cells or Cancer Stem-like Cells (CSCs) make up less than 1% of the tumor mass, but can be highly resistant to therapy and may be the primary tumor initiating cells. Isolation of human glioblastoma CSCs is based on the stem cell-associated protein CD133.[25,26] Such CD133+ glioma cells are resistant to radiotherapy, the mainstay of conventional treatment, through DNA repair and preferential activation of the DNA damage checkpoint response.[27] Further study of this critically important population, and therapies targeted against cancer stem cells, may revolutionize the management of a broad range of cancers that are resistant to conventional therapies, including glioblastoma.

Cancer stem cells have recently been successfully isolated from an intracranial glioblastoma in a Boxer dog.[21] This ground-breaking development has significantly furthered the development of canine glioma as a translational model. As with human glioblastoma CSC, these cells were CD133 and nestin positive, and displayed all the defining properties of glioma CSC: 1) self-renewal; 2) initiation of brain tumors upon orthotopic implantation; 3) multipotent capacity to differentiate into neuronal, astrocytic or oliogodendroglial cell phenotypes.

Preliminary Findings

The results of preliminary work with canine intracranial glioma cancer stem cells, at the Purdue School of Veterinary Medicine and Bindley Bioscience Center, will be discussed.

DISEASE BURDEN ESTIMATES AND PROGNOSTIC MARKERS FOR CANINE NEOPLASIA

A potential limitation of the translational canine intracranial neoplasia model is found in the evaluation of treatment response. Conventional assessment of brain tumor disease burden is performed via advanced imaging, such as MRI, a procedure which in dogs is invasive (in the requirement for lengthy anesthesia episodes) and expensive. Furthermore, differentiating tumor progression from other MRI abnormalities such as post-operative changes and radiation-therapy induced alterations, including radiation necrosis, can be challenging.

The evaluation of both disease burden and treatment response could be improved by the development of a biomarker which is: 1) non-invasive; 2) cost-effective; 3) applicable to any type of intracranial neoplasia (rather than limited to one tumor type). The ideal biomarker would differentiate brain tumors from non-neoplastic space-occupying lesions, aid in differentiating a host of MRI abnormalities (treatment-induced changes as above, post-ictal changes, concurrent disease etc.) from tumor recurrence or progression, and also have prognostic value. Peripheral blood biomarkers of angiogenesis hold promise as a disease burden estimate that is not specific to any one tumor type, and may potentially be of great value in the canine translational brain tumor model by replacing or complementing MRI in the evaluation of treatment response.

Circulating cells which contribute to angiogenesis were previously thought to originate from the endothelium itself and were described as circulating endothelial progenitor cells or EPCs. Recent ground-breaking work at the Indiana University School of Medicine using polychromatic flow cytometry has defined a subset of cells ($CD34^{+}AC133^{+}CD45^{dim}CD31^{+}$) which originate from the bone marrow and contribute to angiogenesis.[28,29] These cells exist at an elevated level in a number of human cancers, including those which are common in dogs such as breast carcinoma and osteosarcoma. Infusion of these cells was associated with significantly faster tumor growth in a rodent xenograft model. The discovery of this population of cells, named Circulating Progenitor Cells (CPCs), has resulted in a paradigm-shift in our understanding of neo-angiogenesis, in addition to providing a new biomarker that holds promise in the evaluation of treatment response, and a potential new treatment target.

Glioblastoma is characterized by extensive necrosis and hypoxia, which results in the production of HIF-1α and the recruitment of bone marrow derived cells.[30] Markers of angiogenesis may therefore have particular utility in studies of malignant brain tumors, and preliminary results from the Indiana University School of Medicine show that the percentages of these cells are elevated in patients with glioblastoma over normal human subjects.

The biology of angiogenesis in canine brain tumors is remarkably similar to that in human subjects. Angiogenesis within various solid canine tumors including glioma mimics the human disease in various respects.[31-34] The angiogenic features shared between the two species were expression of VEGF and VEGF-Receptor, the process of "vasculogenic mimicry", the genetics of angiogenesis, and alterations in cerebral blood flow and permeability induced by brain tumors.

Preliminary Findings

The results of preliminary work with canine CPCs by the Purdue School of Veterinary Medicine and the Indiana University School of Medicine will be discussed.

ACKNOWLEDGEMENTS

The following acknowledgements must be made, for assistance without which this presentation would not have been possible. Dr. Packer (Purdue University); Drs. Ohlfest and Pluhar (University of Minnesota) and the support of American Cancer Society RSG-09-189-01-LIB; Drs. Ober, Feeney and Anderson (University of Minnesota); Drs. Rickus and Clase and Miss Osbourne (Purdue University) and the support of the Showalter Trust; Drs. Pollok and Case (Indiana University) and the support of the Indiana CTSI-Project Development Team.

REFERENCES

1. Rainov N, et al. *Journal of Neuropathology and Experimental Neurology*. 2000;59(7):607-613. 2. Surveillance, Epidemiology, and End Results Program. 2008. Analyzed July 2009 by CBTRUS, published on ABTA website (www.abta.org). 3. CBTRUS Statistical Report (2009): Primary Brain and Central Nervous System Tumors Diagnosed in the United States in 2004-2005. website: www.cbtrus.org. 4. Deorah S, et al. *Neurosurgical Focus*. 2006;20(4):E1-7. 5. Rubinstein L. *NCI Monograph*. 1977;46:43-49. 6. Pang LY, Argyle DJ. *Biochimica et Biophysica Acta*. 2009;1792:380-391. 7. Dobson J, et al. *Journal of Small Animal Practice*. 2002;43:240-246. 8. Vandevelde M. *Proceedings. Brain Tumors in Man and Animals, Research Triangle Park, NC, Sept 5-6.*; 1984. 9. Snyder JM, et al. *Journal of Veterinary Internal Medicine*. 2006;78(June 2004):669-675. 10. Stoica G, et al. *Veterinary Pathology*. 2004;41:10-19. 11. Laws E, et al. *Journal of Neurosurgery*. 2003;99:467-473. 12. Stupp R, et al. *The Oncologist*. 2006;11:165-180. 13. Back F, et al. *Ann Acad Med Singapore*. 2007;36:338-342. 14. Parlato C, et al. *Front Biosci*. 2006;1(11):1280-3. 15. Brearley MJ, et al. *Journal of Veterinary Internal Medicine*. 1999;13:408-412. 16. Paoloni MC, Khanna C. *Nature Rev Cancer*. 2008;8(2):147-56. 17. Zhang J, et al. *Molecular Cancer Research*. 2009;7(1):67-78. 18. Nozaki M, et al. *Neuro-Oncology*. 1999;1(2):124-137. 19. Vassilev LT. *Trends in molecular medicine*. 2007;13(1):23-31. 20. Dean BL, et al. *Radiology*. 1990;174:411-415. 21. Stoica G, et al. *Veterinary Pathology*. 2009;46:391-406. 22. Lipsitz D, et al. *Veterinary Pathology*. 2003;40:659-669. 23. Turrel J, et al. *Journal of the American Veterinary Medical Association*. 1986;188:851-856. 24. Frenier S, et al. *The Small Animal Compendium*. 1990;12:1422-1433. 25. Singh S, et al. *Nature*. 2004;432(7015):396-401. 26. Galli R, et al. *Cancer Research*. 2004;64(19):7011-7021. 27. Bao S, et al. *Nature*. 2006;444(7120):756-60. 28. Estes ML, et al. *Current protocols in cytometry*. 2010;Chapter 9:Unit 9.33.1-11. 29. Estes ML, et al. *Cytometry A*. 2010;77(9):831-9. 30. Du R, et al. *Cancer cell*. 2008;13(3):206-20. 31. MacLeod A, et al. *Acad Radiol*. 16(10):1187-95. 32. Clemente M, et al. *Veterinary Pathology*. 2010;47:265-274. 33. Al-Dissi AN, et al. *J Comp Pathol*. 2009;141:229-236. 34. Kodama A, et al. *J Comp Pathol*. 2009;141:199-203.

BASIC GENOMICS

Dennis O'Brien, DVM, PhD, DACVIM (Neurology)
Columbia, MO

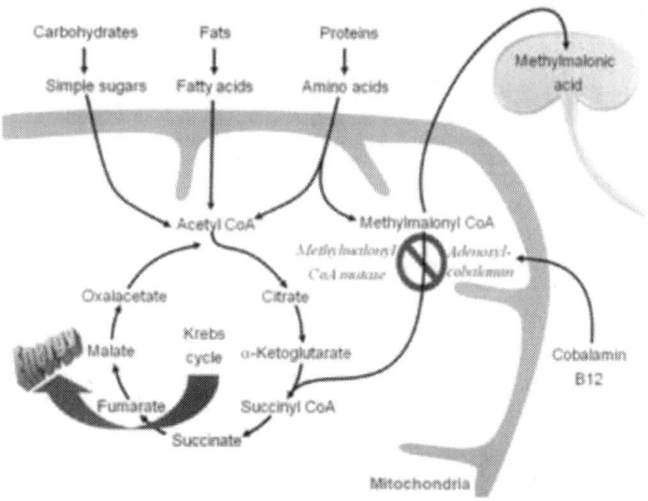

Figure 1: Inborn errors of metabolism, such as methylmalonyl CoA mutase deficiency, can lead to excretion of an abnormal metabolite in the urine. Acquired disease such as cobalamin deficiency would also need to be considered.

The advent of modern molecular genomics and the publication of the canine genome reference sequence has revolutionized our ability to tease out the contribution of genetics to disease. The unique evolutionary history of the dog make it an ideal species for such studies, and as the genome of other species are sequenced, similar advances can be expected. This talk will discuss the characteristics of a neurologic disease that should raise suspicion of a hereditary disease, and review the basic principles of genomics. The second will discuss the strategies for identifying a mutant gene and the use of DNA testing.

RECOGNIZING GENETIC DISEASE

Breed predilection & the popular sire effect

One of the main clues to a hereditary disease is a high incidence of the disease in certain breeds or inbred lines, but this does not necessary reflect poor breeding practices. With rare exceptions (e.g. airway problems associated with brachycephalic breeds), the high incidence of a hereditary disease in a breed is not the result of conscious decisions on the part of breeder. Inbreeding does not cause genetic disease per se, but rather makes the expression of recessive traits more likely. These can either be desirable traits such as a particular coat texture, or a disease causing mutation.

Recessive mutations can become widespread in a breed due to the popular sire effect. A dog wins in shows or field trials because they have the characteristics that the fanciers of that breed find desirable. Thus it is only natural that other breeds will want those characteristics in their line. With the advent of shipping of frozen semen, a winning male can sire offspring throughout the world. It has been estimated that even in a highly out-bred population like humans, every individual carries ~5 disease causing mutations in their genome. Thus even the best dog is likely to carry a similar number of mutant alleles. Unless a mutant allele is already widespread throughout a population, the first generation of offspring from the popular sire will not have a major problem, but 50% of his offspring will carry the mutant alleles (Figure 1-left). When the next generation starts to breed, however, carriers of the mutant allele will mate and produce affected offspring. The disease may not be immediately recognized as hereditary or concealed, and then the mutant allele can become even more widespread in the breed.

Each hereditary disease will be unique, but some general characteristics of a disease can raise suspicion of a genetic basis (Table 1). Many have a neonatal or young age onset, though some such as degenerative myelopathy can be delayed until old age. Congenital malformation will usually show static signs. Imaging may show the malformation though care must be taken in interpreting MRI in neonates.[3] Inborn errors of metabolism can lead to the accumulation of abnormal metabolites in the lysosome in storage diseases or in the urine in organic acidurias. In the latter case, urine organic acid screens can often detect the metabolite (Figure 1), but in storage diseases histopathology is often necessary to make a diagnosis.

Table 1: General characteristics of some hereditary neurologic diseas

Congenital malformations	Neonatal or young onset Static or progressive signs Diagnosed on imaging or necropsy
Organic acidurias	Neonatal or young onset Waxing and waning encephalopathy Dietary influences Ketonuria, acidosis or anion gap, hypoglycemia, hyerammonemia Diagnosed on urine organic acid screens
Lysosomal storage diseases	Neonatal to old age onset Varying signs of ataxia, blindness, weakness dementia, seizures Inexorably progressive course Diagnosed on necropsy or occasionally liver biopsy or leukocytes
Channelopathies	Neonatal to adult onset Altered excitability of muscle or nerves Signs of myopathy, ataxia or seizures May be episodic Diagnosed on clinical signs or electrodiagnostics
Neurodegenerative diseases	Young to old age onset Selective or diffuse degeneration of neurons, myelin, or muscle May see selective atrophy on imaging

BASIC GENETICS

The ABCs of DNA

All of the information needed to produce an individual animal is encoded in their genome. The basic building block of the genome is the DNA nucleotides or base pairs. Only 4 nucleic acids make up the "alphabet" that is used to create a genome. These nucleotides are always paired which is what allow the DNA to replicate an exact copy of itself during mitosis or meiosis and allows transcription of the precise information to RNA. Groups of three nucleotides are codons which code for a single amino acid in a peptide chain or signal the end of a gene (a stop codon). Within a gene, there will be stretches of DNA that do not contribute to the final product (introns) interspersed between the stretches that code for the actual peptide sequence (exons). During transcription, the introns are spliced out and only the exons (the coding sequence) ultimately generate the protein. In many genes, different forms of a protein can be formed by altering the way the exons are spliced together. Chromosomes are physical structures visible on light microscopy during cell division which are just somewhat arbitrary collections of individual genes. Genomics studies not only the individual genes, but how the entirety of the genome contributes to what the phenotype (the physical makeup) of an individual is.

Polymorphism and mutant alleles

Chromosomes and hence genes come in pairs. The separate forms of a pair of genes are the alleles and the combination in an individual is their genotype. During mitosis, an exact duplicate of both chromosomes gets transmitted to the daughter cell. During meiosis, however, only one of the chromosomes gets transmitted to the egg or sperm. Thus an offspring receives one chromosome and hence one allele from each parent. The alleles can either be the same (homozygous) or they can be different (heterozygous). Mutations can arise from a variety of causes such as ionizing radiation, mutagenic toxins, retroviral like elements in the genome that insert themselves into genes, or errors in replication. Mutations give rise to a different allele, and different forms of an allele are called polymorphisms. Most mutations are of no consequence to the final protein structure, but they can be used as markers to track a particular allele on a given chromosome through a family or population. Some mutations, of course, will affect the function of a protein and cause disease. A mutant allele that causes disease can either be dominant or recessive. In a dominant trait, one copy of the mutant allele is enough to cause disease. Though the normal allele may produce a normal protein, the mutant allele may produce an abnormal protein that interferes with the function of the other. In a recessive trait, the animal must be homozygous for the mutant allele to show disease. If for example the mutant allele does not produce protein, having one copy of the normal disease may produce enough normal protein to function. With two mutant alleles, howver, no normal protein can be formed and disease results. The ultimate objective in genetic studies is to identify the mutation that produces a given trait so that it can be selected against if it is disease-causing or selected for if it produces a desirable trait.

SEGREGATION ANALYSIS

If a hereditary disease is suspected, an extended family history should be collected. Breeders can be defensive when approached about a potentially genetic disease, so it is important to assure them that we are working as a team to find a solution to the problem. As breeders have seen the success with control of diseases such a progressive retinal atrophy, they are becoming more willing to help with research to find the genes responsible. The number, sex and phenotype of prior or subsequent litters from the breeding pair should be obtained. The anticipated percentage of affected offspring can be calculated using Punnet squares (Figure 2). Remember that these are only probabilities and in any give litter, any combination is possible.

Various factors can alter the actual number of seen. There will be an ascertainment bias when assessing clinical populations. Breeding two carriers may not produce any affected offspring simply by chance. This breeding will then not be recognized as one that should have been counted towards the % of normal offspring, skewing the observed number towards more affected offspring. This bias can be corrected using the Davie correction[2]. The number of probands (the first affected animal that brings a family to the attention of the clinician) is used in the calculation, and thus must be noted when collecting histories. Some dominant traits, such as Manx tailless or Chinese crested dog hairless, are lethal when homozygous. Thus breeding two affected animals will produce a greater than expected % affected offspring because the offspring homozygous for the trait die in utero or as neonates.

Not all hereditary diseases are fully penetrant, that is an animal can have the genotype that should cause the disease but be clinically normal. Presumably other genetic or environmental factors affect which animals succumb to the disease. While penetrance is an all-or-none function, expressivity is the degree of expression of a gene. In Manx cats for example, the tailless trait can vary from cats with a shorter than normal tail, to those with sacral agenesis and neurologic signs. Variable expression can also occur in female carriers of an x-linked trait due to x-inactivation or Lyonization. In any give cell of a female, only one X chromosome is expressed and the other is inactivated. Thus there will be a mosaic of expression of an X-linked gene depending on which chromosome is inactivated. The simplest example of this is the mosaic of orange coat color (an X-linked trait) in the female tortoise-shell cat. A similar mosaic of expression of mutant myelin proteins in female "shaker" pups can produce mild clinical signs[1].

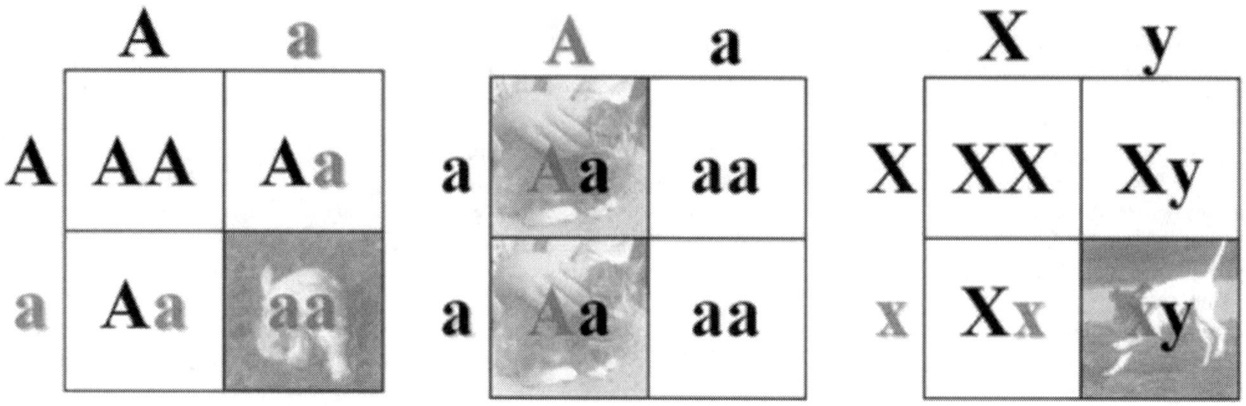

Figure 2: Punnet squares are used to calculate the expected percentage of affected and carriers of a trait when the genotypes of the parents are know. (Left) Breeding two carriers of a recessive trait (a) such as neonatal ataxia in Coton de Tulears will produce 25% affected pups and 50% carriers. Note that since the affected pups are easily identified 66% of the normal pups are carriers. (Middle) Breeding an animal with a dominant trait (A) such as Manx tailess will produce 50% affected and 50% normal kittens. (Right) Breeding a normal male to a female carrier of an X-linked recessive trait (x) such as hereditary ataxia in English Pointers will produce 50% affected males and 50% carrier females.

Mitochondria contain their own small genome which is separate from the nuclear DNA. The mitochondrial genome codes for some of the enzymes involved in energy production in the mitochondria. Thus mutations in the mitochondrial disease can produce signs of encephalopathy or myopathy due to inadequate ATP production. The sperm do not contribute any mitochondria during fertilization, so all the mitochondria and their genomes are transmitted through the egg. As a result mitochondrial diseases follow a maternal inheritance pattern, where even an affected male will not transmit the mutant gene to his offspring. The female will have a mix of mitochondria with and without the mutation in her cells. Whether an offspring is affected or not will depend on the "dose" of mutant versus normal mitochondria that particular egg receives.

REFERENCES

[1]Cuddon PA et al. **Annals of Neurology**1998;44(5):771; [2]Davie AM. **Annals of Human Genetics** 1979:42:507; [3]Gross B et al. **Veterinary Radiology and Ultrasound** 2010;52(4):361.

GENE DISCOVERY & DNA TESTING

Dennis O'Brien, DVM, PhD, DACVIM (Neurology)
Columbia, MO

When a hereditary disease is recognized, the ultimate goal will be to identify the mutation responsible for the disease. The techniques available to discover disease causing genes are rapidly evolving, but currently consist of candidate gene sequencing, linkage mapping, and genome wide association studies. Identifying the mutation will permit the development of a DNA test that can be used to diagnose the disease as well as identify carriers of the mutation so that breeders can take steps to eliminate the disease from the breed. Interpretation of DNA test results, however, requires knowledge of the nature of the tests.

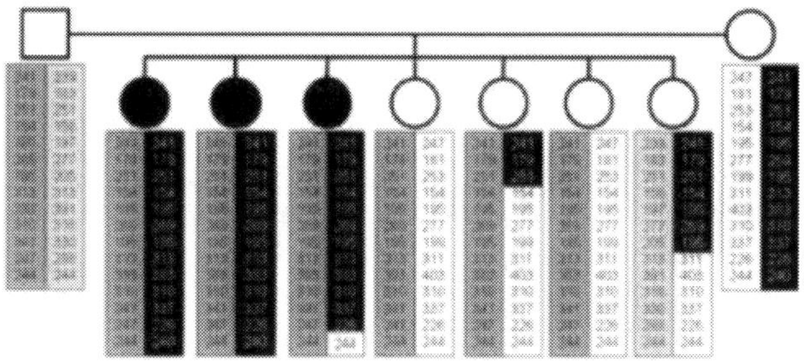

Figure 1 A: Markers are polymorphic genes (A1 & A2 and B1 & B2). If they are on the same chromosome as the gene containing the disease-causing mutation (d), they will segregate with the disease. B: During meiosis, recombinations can occur. The closer the marker is to the mutant allele, the less likely it is that there will be a recombination such as occurred between d and B2.

GENE DISCOVERY STRATEGIES

Candidate genes

A gene becomes a candidate for sequencing when there is a very high index of suspicion that a mutation in that particular gene might be responsible for the disease. The candidate gene may already be associated with the disease in another species. Typically this is humans or mice, but genes are increasingly being identified first in domestic species. The more accurate and molecular the original diagnosis, the better the list of candidate genes will be. For example, when a necropsy was performed on a dachshund with a suspected lysosomal storage disease, we demonstrated storage material in neurons that autofluoresced suggesting that the condition was one of the neuronal ceroid lipofuscinoses (NCL). There are nine genes associated with NCL in humans, mice, sheep and dogs which then became prime candidates for the dachshund disease. On electron microscopy, the inclusion material formed curvilinear bodies which are characteristic of the form of NCL in humans caused by mutations in the tripeptidyl peptidase 1 gene (***TPP1***). Sequencing the coding regions of ***TPP1*** gene from the affected dog showed a deletion that predicted a premature stop codon and a severely truncated protein[2]. Thus having sufficient details on the disease permitted us to identify the mutation when studying just a single affected individual. Alternatively, a specific biochemical clue such as the presence of an abnormal organic acid in the urine such as L-2 hydroxyglutaric acid[9], could point toward a specific enzyme as the candidate, or a specific electrophysiologic finding such as myotonia[3], could point toward a specific ion channel.

Often however, the candidate gene approach is not sufficient. There may be too many candidates such as in seizure disorders or progressive retinal atrophy where dozens if not hundreds of candidates exist. Species difference may mask the relationship. For example mutations in the ***SOD1*** gene in dogs cause degenerative myelopathy which is characterized by initial upper motor neuron paralysis and sensory loss[1]. In humans, ***SOD1*** mutations cause familial amyotrophic lateral sclerosis, which presents as a lower motor neuron paralysis without sensory involvement. Finally, the cause of the disease being studied may not have been discovered yet in any species. In these situations, gene mapping or association studies are needed to identify the location of the disease causing mutation in the genome.

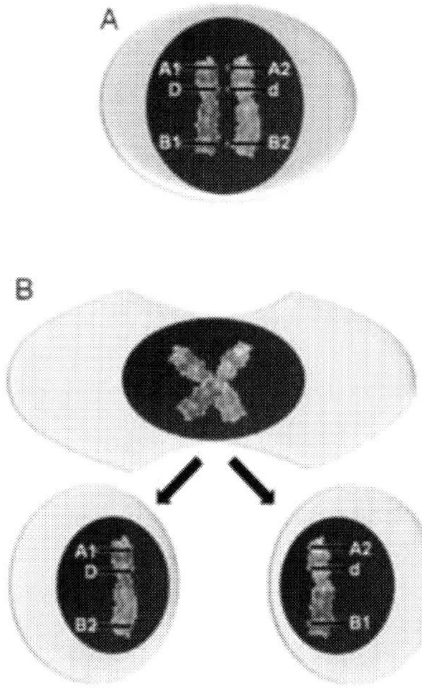

Linkage mapping

Some genes will be polymorphic, that is there is variation in the DNA sequence that can be used to distinguish different forms of the gene. These can be either single nucleotide polymorphisms (SNPs discussed in more detail below) where a single base pair is substituted, or microsatellites. Microsatellites are repeated nucleotides usually found in the non-coding regions of the genome. When microsatellite regions are being duplicated, there can be slippage resulting in variations in the number of repeats. Since the number of repeats can vary widely, there can be many polymorphisms with microsatellites. The polymorphisms are used as markers to trace the

Figure 2: The haplotype of this family containing the mutant allele for neonatal seizures[4] is shown in black in the dam and dark grey in the sire. Numbers represent the different alleles for the markers. Recombinations in the 3rd affected dog and the 2nd normal dogs define the borders of the locus since the mutant allele must be in the black haplotype in the former and cannot be in the black haplotype in the latter. The 4th normal dog is not informative because she inherited the normal haplotype from the sire.

inheritance of an area of the genome through a family[5].

A marker is linked to the gene with disease-causing mutation (Figure 1) if it is close enough that there is unlikely to be a recombination between the marker and the mutant allele. The degree of linkage is expressed as a log of odds (LOD) score and a score of >3.2 is generally considered significant.[6] When a significant association is identified it defines a locus, a region where the mutant allele is thought to reside. A group of alleles on a chromosome that are transmitted together in a family are a haplotype. By analyzing the haplotype of affected versus normal dogs in a family, recombinations can be identified that will define the borders of the locus where the disease allele resides (Figure 2). Fine mapping can then be performed using additional markers to further define the locus.

Genome wide association studies

While microsatellite markers are more polymorphic, SNPs have the advantage of being very numerous and lending themselves to automation in "SNP chips" which contain the chemistry to perform thousands of SNP discriminations on a DNA sample[7]. Genome wide association studies (GWAS) calculate the association between the mutant allele and specific SNPs and the results are displayed in a "Manhattan plot" (Figure 3). A significant association will again suggest the disease locus. Since GWAS compares populations, they can be performed when sampling DNA from extended families is not plausible, such as in old-age onset disease. GWAS must be controlled for population stratification. The line or family where the disease is segregating, may also have other characteristics, such as a particular set to the ears or texture to the coat, that is also segregating in the family and could give a spurious association. Fine mapping and haplotype analysis are again used to further define the locus.

Figure 3: In a Manhattan plot, the individual chromosomes are indicated by shades along the y axis and the probability of association with the disease on the x axis. This plot shows a significant association with markers on chromosome 2.

From locus to specific mutation

Once the locus has been identified, the genes within the locus become candidates. They are prioritized for sequencing based on criteria such as whether they have been associated with the disease in other breeds or species, whether the function of the gene suggests it may be a likely candidate (e.g. a transcription factor such as ATF2 in a developmental disease like neonatal encephalopathy with seizures[4]), or where and when during development the gene is expressed (e.g. as was done for Labrador Retriever myopathy[8]. These candidates are sequenced and if a potential mutation is identified, the entire family or population being studied is genotyped to see if the allele segregates as expected. If it does, then a large number (typically about 100) of dogs from breeds where the disease has not been reported are screened for the mutant allele. If a dog is identified that is homozygous for the mutant allele but does not have the disease is identified, then it must be questioned whether the causative mutation has been identified.

Some mutations, such as a deletion which produces a premature stop codon and a severely truncated protein, are self-evident. A single nucleotide polymorphism, however, that changes a single amino acid of the protein could be significant or it could be an inconsequential SNP, thus deserves closer scrutiny. An amino acid switch that changes the charge of the protein is more likely to cause a conformational change in the final protein. Areas that are critical to the normal function of a protein are likely to be conserved amongst various species, while variations can occur in areas that are not as critical. Ultimately, it may be necessary to demonstrate that the mutation identified disrupts function of the protein. Even if the mutation does not produce a premature stop, some will affect transcription or translocation of the protein. Immunohistochemistry may demonstrate a lack of normal protein in the tissues, though care must be taken since commercial antibodies developed for human or laboratory rodents may not recognize canine protein. Alternatively, biochemical assays on tissue, leukocytes or fibroblast cultures may be needed to determine function. This would be especially important when a candidate gene approach was used with few affected dogs (e.g. the *TPP1* in dachshunds[2]).

DNA TESTING

DNA based tests for hereditary diseases can be used to diagnose the disease and to identify carriers of the mutant allele for breeding purposes. A current listing of where DNA tests can be obtained for various diseases and breeds can be found at the Orthopedic Foundation for Animals website (www.offa.org/dna_alltest.html). Different types of tests for hereditary disease are available. They vary in their sensitivity and specificity and how they have to be interpreted.

When a locus is first identified using linkage analysis, the linked markers can be used to assess the risk of a dog being affected with the disease or being a carrier of the mutant allele. Because recombinations could occur between the linked marker and the disease allele, false positives/negatives are possible. These errors, however, can be valuable in identifying haplotype borders that further narrow the locus. The linked markers are most useful when assessing individuals in a family where the disease is known to be segregating. Though the marker allele by definition is segregating with the disease in a family, the same marker will be in the general population and may not be associated with the disease allele in some lines. Thus linked markers will have higher false positive rates when randomly testing an individual's genotype. Fortunately, identification of the linkage typically soon leads to a specific DNA test for the mutation.

Specific DNA tests are highly sensitive and specific for identifying the mutation they are designed to detect. The same disease phenotype, however, can be caused by different mutations in the same gene, or sometime by mutations in other genes. The

DNA tests are only validated for specific breeds and even then, there can be variability. For example, mutations in two different genes have been associated with neuronal ceroid lipofuscinosis in dachshunds[2,10]. Testing for one of the mutations would yield a "false negative" result if the disease were caused by the other mutation in that individual. A dog who is homozygous for the mutant allele of an autosomal recessive disease clearly has the mutation and is at risk for the disease, but variable penetrance or expressivity can yield a false positive result. For example, a mutation in the ***SOD1*** gene has been identified in a number of breeds with degenerative myelopathy, but dogs that are homozygous for the mutation have lived to an old age without developing signs[1]. Maybe if they lived long enough they would develop signs, but none-the-less, there is at the very least an age related variable penetrance. As the disease is studied further we will be able to say what the degree of penetrance is and have a better idea about the sensitivity of the test.

A DNA test will be highly specific for the mutation it tests for. An old German Shepherd who is homozygous for the ***SOD1*** mutation could, however, have concurrent disease such as a type II disk herniation or a spinal tumor producing similar signs. Thus depending on the disease, it may still be necessary to rule out other causes. Surgery for a herniated disk could significantly improve the quality of life in the example above, but the owners would also need to be informed of the risk of the signs progressing due to degenerative myelopathy even if the surgery is successful.

The other use for DNA testing is to guide breeding decisions. The goal should be to never produce an affected animal while maintaining the best characteristics of the breed and desirable genetic diversity. Over time, the frequency of the mutant allele can be decreased as well. A dog with a dominant trait should not be bred since they will produce 50% affected offspring. These dogs would typically be identified by phenotype unless the disease has a delayed onset, when a DNA test may be necessary. If the mutant allele for a recessive disease is rare enough in the breed and/or the breeding population large enough, then it may be acceptable to simply remove carriers of a mutant allele from breeding. More commonly, however, the frequency of the mutant allele is high, or the breeding pool is small as in new or less popular breeds. Then it is wise to continue to breed carriers in order to maintain desirable genetic diversity or other traits, but they should be bred to a dog who is clear of the mutant allele so that affected dogs are not produced. When selecting for the next generation of breeding stock, the genotype of the dogs should be an important factor in deciding which dogs to breed, but can be secondary to other considerations so long as the above principles are applied.

CONCLUSIONS

The era of molecular genetics will permit the diagnosis and elimination of a wide variety of diseases. Already gene therapies, enzyme replacement therapies, and stem cell therapies are being investigated to treat animals affected with these conditions. As our knowledge in this area expands we will be able to determine the role of genetics in complex traits such as behavior or diseases where a number of different genes are involved in determining the risk of disease. The identification of the gene responsible for a hereditary disease can also shed light on the pathogenesis of similar acquired diseases by identify important molecular pathways that may be involved. Ultimately, regulation of gene expression through therapies such as interfering RNAs may become as commonplace as drug therapies today.

REFERENCES

[1]Awano T et al. **PNAS** 2009;106(8)2794; [2]Awano T et al **Molecular Genetics and Metabolism** 2006;89(3):254; [3]Bhalero DP et al. **American Journal of Veterinary Research** 2002;63(10):1442; [4]Chen X et al. **Neurogenetics** 2008;9:41; [5]Clark LA et al. **Genomics** 2004;84(3):550; [6]Gordon D et al. **Journal of Heredity** 2003;94(1):1; [7]Karlsson EK et al. **Nature Genetics** 2007;39(11):2321; [8]Pele M et al. **Human Molecular Genetics** 2005;14(11):1417.

GENETICS OF DOMESTIC CAT DISEASES & HEALTH

Leslie A. Lyons, PhD
Davis, CA

INTRODUCTION

The previous two lectures in this session have provided a wealth of background genetics that will be expanded in the following two lectures. Firstly, the nuances of the domestic cat will be explored. Each of our companion animals have different population dynamics that create unique genetic aspects that must be considered when examining their genomes for inherited traits and diseases. Secondly, the genetic tests available for the domestic cat will be discussed, including how cat breeders use this information to assist breeding programs and promote the health of the individual and the populations. The discussions of the cat tests will be followed with a closer examination of why genetic tests are not 100% accurate and how both aspects about the disease and the assay methods affect testing accuracy in the second lecture. Indirect genetic tests will be considered, which includes linkage and haplotype-based tests and tests focusing on quantitative traits, such as behavior, production traits, and morphology. Finally, the highly experimental approaches for improving health will be examined, such as gene therapies, transgenics, and personalized medicine.

FELINE GENETICS

Nuances of cat populations

Other than the South African sub-species of African wildcat, Felis lybica cafra, most species of wildcat and their associated sub-species may be progenitors to domestic cat populations[1,2], Felis lybica having the most scientific support. In the Old World, domestic cats have overlapping ranges with their wildcat progenitors and gene flow is still proceeding. Domestic cats likely actively participated in their own domestication, both humans and felines developing a symbiotic, mutual tolerance as early civilizations developed between 8,000 – 3,000 B.C., year ago. Via human migrations, cats have migrated around the world, generally to work as vermin control on boats and farms, thus their population management has generally been unrestricted. Arguably, the domestication of the cat can be considered incomplete and in flux, since abandoned cats and feral populations can quickly adapt to a "wildcat" lifestyle. In the United States, there are 88.3 million cats compared to 74.8 million dogs in households, implying roughly 13.5 million more cats than dogs. However, the proportion of pedigreed versus random bred cats is still fairly low, only 10 - 15% of feline patients at the University of California - Davis Veterinary Medicine Teaching Hospital is represented by pedigreed cats[3].

The creation of cat breeds has been a significantly different process than breed development in other companion animal and agricultural species[4,5]. Within the early 1900's, several cat fancy registries were established and stricter breeding practices for cats were developed. Various modern encyclopedic volumes pertaining to the domestic cat list approximately 50 - 80 cat breeds, worldwide [6,7]. However, a majority of breeds developed within the past fifty years, and, many listed breeds did not develop into viable populations, some even lost to posterity. Cats naturally performed their required function, vermin control. Dog breeds, however, have been selected for morphologies that support diverse functions and behaviors. The breeds of agricultural species are selected for meat and milk quality and quantity. These behavioral and production traits are complex interactions between many genes, suggesting more selection has been applied to the genome of dogs, cattle, pigs and chickens, than for the genome of the domestic cat. Less selection likely is resulting in more genetic diversity for cats, implying potentially comparatively less health issues in cat breeds as compared to other species.

Although not all cats produced by breeders are registered, perhaps only 20 – 30%, the CFA, one of the largest cat registries worldwide, generally registers approximately 40,000 pedigreed cats annually[8]. Approximately 16,000 – 20,000 are Persians and approximately 3,000 are Exotics, thus, the Persian group of cats represents over 50% of the cat fancy population. Common breeds that generally have at least one thousand annual registrants are Abyssinians, Maine Coons and Siamese. Other popular breeds include the Birman and Burmese, which are more prevalent in other areas, such as the United Kingdom. A more recently developed cat breed, the Bengal[9], which is a hybrid between the Asian Leopard cat, Prionailurus bengalensis, and the domestic cat, has gained significant popularity throughout the world, even though some registries currently do not recognize the breed. These hybrid cats may also have allelic incompatibilities for a given gene, as the genes between the two species, leopard cat and domestic cat, have millions of years of evolutionary divergence, which allows differences at the DNA sequence level of a gene. Hence, an accumulation of different genetic variants that are functional within the species, but non-functional across the felids, are likely present in some Bengal cats. Thus, hybrid cat breeds may have unexpected health problems and infertility, creating a challenge to both genetic tests and health care.

Nuances of the feline genome

Chromosomes of the domestic cat are easily distinguishable, consisting of 18 autosomal chromosomes and the XY sex chromosome pair, resulting in 38 chromosomes for the cat genome. Karyotypic studies of male tortoiseshell cats have shown that they are often mosaics, or chimeras, being XX/XY in all or some tissues. The minor chromosomal differences that are cytogenetically detectable between a domestic cat and an Asian leopard cat are likely the cause of fertile problems in the Bengal cat breed. The high conservation of chromosomal arrangement of cat to humans has greatly facilitated candidate gene approaches since the location of particular genes could be anticipated in cats from comparison with the genetic map of humans.

The cat's importance in human health, comparative genomics, and evolutionary studies supported the National Institutes of Health – National Human Genomics Research Institute (NIH-NHGRI) decisions to produce of a low coverage, 2x, sequence of the cat genome. Because of the small scale sequencing effort, only approximately 65% of the euchromatin (gene-coding) sequence was identified. The sequence assembly suggested the identification of 20,285 feline genes that have counter parts (orthologs) in the human genome. Recently, an additional approximate 14x genome coverage has been completed for the cat and the work is in preparation for publication. The more in depth sequencing, which will provide a deeper coverage draft sequence of the cat, will improve the euchromatin coverage to ~90 – 95%. The better coverage implies that more cat-specific genetic sequence will be known for any gene of interest. Mutation screening methods will be more efficient, leading to the identification of causative mutations more rapidly and more efficiently. An important by-product of the DNA sequencing effort is the identification of the normal genetic variation in the cat genome, SNPs. Hill's has also provided funding to support feline genome resource development, thus a cat SNP-based DNA array was commissioned and is now available. Each DNA chip has 12 regions, each region is used to test one cat. Each region has the assays for ~74,000 SNPs. The major benefit of the arrays is the ability to assess the entire genome via GWAS.

FELINE GENETIC TESTS

Approximately thirty-three genes contain fifty mutations that cause feline health problems or alterations in the cat's appearance. A variety of commercial laboratories can now perform cat genetic diagnostics, allowing both the veterinary clinician and the private owner to obtain DNA test results. DNA is easily obtained from a cat via a buccal swab using a standard cotton bud or cytological brush, allowing DNA samples to be easily sent to any laboratory in the world. The DNA test results identify carriers of the traits, predict the incidence of traits from breeding programs, and influence medical prognoses and treatments. Thus, genetic testing is an effective preventative medicine and a potential ultimate cure.

The first mutations identified in cats were for lipid and lysosomal storage diseases in the early and mid-1990's[10,11], as these diseases have well defined phenotypes and known genes with mutations that were as found in humans (See reviews[12,13]). Most of the common diseases, coat colors, and coat types have been deciphered in the cat following the same candidate gene approach, implying finding a replicate trait in another species and checking the same gene for causative mutations. Presented in Table 1 are the commercially available simple genetic traits and diseases with known mutations in the cat. To date, other than the muscular dystrophy mutation[10], all other mutations in the cat are autosomal, not found on the X or Y chromosomes. The coat color mutations are common to all cats and are effective for genetic typing in all breeds and populations. However, even though long fur is common in breeds and random bred cats, long fur is an exception because four different mutations in *FGF5* can cause a cat to have long fur[14,15]. Most diseases are rare in breeds, however, polycystic kidney disease (PKD) and hypertrophic cardiomyopathy (HCM), are prevalent. PKD in Persians is estimated at 30 – 38% worldwide[16-18]. Because of cross breeding with Persians, many other breeds, such as British Shorthairs, American Shorthairs, and Scottish Folds, also need to be screened for PKD[19-21]. Thus, veterinarians need to be aware of cross-breeding practices, which differ in different cat registry organizations, so that genetic tests can be placed as high or low priority for differentials and diagnostics.

Breeding Recommendations

Cat breeders are very knowledgeable in regards to weighing different factors to produce healthy cats that are of good type and temperament. Many genetic tests help a breeder make a clearer, more educated decision. Cats with a positive genetic test for diseases should be screened by other diagnostics, such as ultrasound in the case of HCM and PKD, to determine disease status and this overall information used in breeding decisions. Other health, type, and behavioral attributes should certainly be considered in the overall breeding program. However, breeders need to work hard to reduce the risks with any health issue. With the HCM A31P mutation, every cat that has the mutation is at risk for developing HCM and every cat with the mutation will pass it on to some or all of its offspring. Cats that are homozygous for the A31P mutation will definitely pass the mutation to their offspring. The homozygous cats are at high risk of developing severe HCM. Cats that are heterozygous for the mutation should not be bred unless they have other qualities that are either highly beneficial or necessary to the breed. Kittens that test negative for the mutation should be used to replace them in the gene pool. A slow eradication of disease is recommended for highly prevalent diseases, such as HCM and PKD, as quick elimination of such a high number of cats could lead to other effects of inbreeding depression. Breeds with very low populations sizes, such as Korats, have learned to manage the gangliosidoses in their breed, never breeding carriers together. Hopefully, all the disease mutations can be eventually eradicated, but good breeding decisions and balancing population diversity must be considered.

REFERENCES

[1]Driscoll CA et al. **Science** 2007;317:519-523. [2]Lipinski MJ et al. **Genomics** 2008;91:12-21. [3]Louwerens M et al. **J Vet Intern Med** 2005;19:329-335. [4]Bowling AT. **Genetics of the Horse.** New York: CABI; 2000;25-52. [5]Wayne R. **Genetics of the Dog**. New York: CABI; 2001;43-60. [6]Morris D. **Cat breeds of the world**. New York: Penguin Books; 1999. [7]Morris D. **Cat Breeds of the World: A Complete Illustrated Encyclopedia.** New York: Viking Penquin; 1999;10-12. [8]Cat Fanciers' Almanac 2004;20:72-86. [9]Johnson G. **The Bengal Cat**. Greenwell Springs, LA: Gogees Cattery; 1991. [10]Winand NJ et al. Neuromuscul Disord 1994;4:433-445. [11]Yogalingam G et al. **J Biol Chem** 1996;271:27259-27265. [12]Banks G, Chamberlain J. **Curr Top Dev Biol** 2008;84:431-453. [13]Valayannopoulos V et al. **Orphanet J Rare Dis** 2010;5:1-20. [14]Drogemuller C et al. **Anim Genet** 2007;38:218-221. [15]Kehler JS et al. **J Hered** 2007;98:555-566. [16]Cannon MJ et al. **Vet Rec** 2001;149:409-411. [17]Barrs VR et al. **Aust Vet J** 2001;79:257-259. [18]Barthez PY et al. **J Feline Med Surg** 2003;5:345-347. [19]Biller DS et al. **J Am Vet Med Assoc** 1990;196:1288-1290. [20]Eaton KA et al. **Vet Pathol** 1997;34:117-126. [21]Lyons LA et al. **J Am Soc Nephrol** 2004;15:2548-2555.

Table 1. Common Commercialized DNA Tests for Domestic Cats.

Disease / Trait	MOI[‡]	Phenotype	Breeds	Gene	Mutation
Agouti	AR	Banded fur to solid	All breeds	*ASIP*	del122-123
Amber	AR	Brown color variant	Norwegian Forest	*MC1R*	G250A
Brown	AR	Brown, light brown color variants	All breeds	*TYRP1*	b = -5IVS5, b^l = C298T
Color	AR	Burmese, Siamese color pattern, full albino	All breeds	*TYR*	c^b = G715T, c^s = G940A, c = C975del
Dilution	AR	Black to grey / blue, Orange to cream	All breeds	*MLPH*	T83del
Gloves	AR	White feet	Birman	*KIT*	c.1035_1036delinsCA
Hairless (Naked)	AR	Atrichia	Sphynx	*KRT71*	c.816_1G_A
Long fur	AR	long fur	All breeds[§]	*FGF5*	c.356insT, C406T, c.474delT, A475C
Rexing (curly fur)	AR	Curly hair coat	Devon Rex	*KRT71*	c.1108-4_1184del, c.1184_1185insAGTTGGAG
AB Blood Type	AR	Determines Type B	All breeds	*CMAH*	18indel-53, G139A
Gangliosidosis 1	AR	Lipid storage disorder	Korat, Siamese	*GBL1*	G1457C
Gangliosidosis 2	AR	Lipid storage disorder	Burmese	*HEXB*	15bp del (intron)
Gangliosidosis 2	AR	Lipid storage disorder	Korat	*HEXB*	C39del
Glycogen Storage Dis. IV	AR	Glycogen storage disorder	Norwegian Forest	*GBE1*	230bp ins 5' - 6kb del
Hypertrophic Cardiomyopathy	AD	Cardiac disease	Maine Coon	*MYBPC3*	G93C
Hypertrophic Cardiomyopathy	AD	Cardiac Disease	Ragdoll	*MYBPC3*	C2460T
Progressive Retinal Atropy	AR	Late onset blindness	Abyssinian	*CEP290*	IVS50 + 9T>G
Progressive Retinal Atropy	AD	Early onset blindness	Abyssinian	*CRX*	n.546delC
Polycystic Kidney Disease	AD	Kidney cysts	Persian	*PKD1*	C10063A
Pyruvate Kinase Def.[†]	AR	Hemopathy	Abyssinian	*PKLR*	13bp del in exon 6
Spinal Muscular Atrophy	AR	Muscular atrophy	Maine Coon	*LIX1-LNPEP*	140kb del, exons 4-6

† Unpublished test, presented only as abstract. ‡ Mode of inheritance of the non-wildtype variant, § Long fur variants are more or less common depending on the breed. References can be found in Lyons LA. **Top Companion Anim Med** 2010;25(4):203-12.

GENOMICS IN FUTURE PET HEALTH CARE

Leslie A. Lyons, PhD
Davis, CA

LIMITATIONS OF GENETIC TESTING

To date, most cat genetic tests have been for traits that have nearly complete penetrance, have little variability in expression, and are early in onset. However, some imperfect examples do exist in cats that have not caused as much controversy as the HCM test. The *CEP290* PRA mutation in Abyssinians has a late age of onset and some cats with subclinical disease have been identified[1]. Some cats with the pyruvate kinase deficiency can have very mild and subclinical presentations[2]. The interplay of various coat color genes often muddle the determination of the true coat color of cats. As true in humans with cardiac disease, the finding that not all cats with the A31P mutation in *MYBPC3* get HCM is actually usual in the field of HCM genetic testing. Thus, disease or trait causing mutations may not be 100% penetrant, thus, they do not always cause clinically detectable disease. Presence of clinical disease in an individual cat and the severity of disease expression is likely affected by the known genetic aspects presented below.

Incomplete Penetrance

For some traits and diseases, even though a known causative mutation has been identified, an individual with that mutation does not present with the condition. Incomplete penetrance is an extreme of variable expression (see below). In general, the reason as to why a condition would not present is unknown, but other genetic, biological, and environmental interactions certainly play a role in the overall appearance and health of an individual and its organs. The sensitivity of clinical diagnostics may also influence the determination of penetrance. In the case of HCM, echocardiography (cardiac ultrasound) can be considered an insensitive tool for detecting mild forms of HCM in cats, thus, many cats with mild do not clinically appear to have cardiac disease. Experience and bias also play a role in diagnosis. For example, individuals who do not have expertise with ultrasonic examinations for HCM or PKD are less likely to be able to provide an accurate diagnosis for these diseases.

Age of Onset (age-related penetrance)

Some diseases have a slow progression and may not present until later in life. In humans, HCM due to *MYBPC* mutations is clearly a disease that has slow progression and commonly does not express until the individual is over 50 years of age. HCM in Maine Coon cats can also develop in older cats, especially in cats that are heterozygous for the mutation (carry only one copy of the mutated gene) and, for some unknown reason, in females. Often, an autosomal dominant disease may be more severe if two copies of the risk mutation are present in an individual, leading to earlier and more severe disease, which appears to be the case with the A31P mutation. The definitive age as to when a cat is clear of developing HCM is not precisely determined.

Variable Expression

Most traits and diseases have some amount of variable expression depending on the individual. For example, not all cats with the mutation for blue dilution have the same color of blue / gray. Obviously, the background genetics and environment of the individual influence the overall presentations of traits and diseases. Thus, the level of presentation can be variable in regard to left ventricular wall thickness in cats with HCM. Cats can have mild, moderate or severe HCM. Only those cats with severe HCM show clinical signs although a few cats with lesser severity of disease may die suddenly. Cats with HCM may fall in the "equivocal" range for wall thickness, thus, definitive affected status is difficult to declare. These equivocal cats may progress to more severe disease with time, or the equivocal status may be as severe as the disease gets. Some cats with PKD have only a few cysts and never progress to renal failure, others have severe and fast progression of disease and succumb renal failure in a few years.

Disease Heterogeneity

Often, more than one mutation in the same gene, or mutations in different related genes can cause the same disease. Genetic heterogeneity for HCM in humans is well established, thus, there is no reason to not think the same situation is true for cats. Currently over 1000 mutations in over 10 genes are known to cause HCM in humans. Only two mutations have been identified that cause HCM in cats, the A31P mutation in Maine Coon cats and the R820W mutation in Ragdolls[3,4], which also causes disease in humans[5]. Both mutations are in *MYBPC3*, the most commonly mutated gene in humans with HCM (see review[6]). Other breeds of cats including Bengals, Siberians, Devon Rex, and Sphynx and mixed breed cats either do not have or have an extremely low prevalence of the A31P or the R820W mutation. However, there are Maine Coon cats that have HCM that do not have the A31P mutation and so there has to be at least one more cause of HCM, most likely another mutation, in this breed. The long fur mutations in the cat are examples of trait heterogeneity.

Genetic Testing Accuracy

Even though a specific genetic mutation may be identified for a genetic trait or disease, research laboratories use different methods to assay for the mutation. Errors in genetic assays may produce inaccurate DNA results, leading to the confusion of genetic test interpretation. Direct DNA sequencing is considered the most robust method, "the Gold Standard," but also one of the more costly methods of analysis. Because DNA primers must bind to the DNA sequence flanking a specific mutation, other, unimportant mutations may be in the areas where the primers bind, causing poor or no amplification of one or both alleles for a given individual. This condition is known as allelic drop-out and all testing laboratories are aware of this potential source of error for a genetic test. Even direct DNA sequencing can suffer from allelic drop-out, but because a larger portion of the gene which may have other DNA variants is generally amplified, a higher likelihood of detecting allelic drop-out is available. Laboratories will place PCR primers in different locations surrounding the mutation of interest, which is often proprietary information, in attempts to lower the risk of allelic drop-out. Thus, some laboratories have better assays than others, even if they are doing the same assay method and testing for the same mutation. The different DNA assay methods are usually developed to reduce cost and to fit the laboratory's expertise and instrumentation. But, some assays may have, in general, some increased risk of test failure. Different common methods for DNA testing include real-time PCR (TaqMan), restriction fragment length polymorphism (RFLP), allele-specific oligos (ASO), or even now mass spectroscopy-based methods. Just as a veterinarian may want to know if a FIV test is performed by a SNAP test, versus an ELISA, versus a Western Blot because each method has different sensitivities and specificities, so too is true for DNA testing methods. Veterinarians will need to become failure with the different genetic testing approaches and not hesitate to ask a testing laboratory about their methods and sensitivity and specificity for their approaches.

Inaccurate Clinical Diagnosis

Ultrasonic examination of the heart, echocardiography, is the most common and currently the only useful method for detecting cardiac disease in cats. Several studies have evaluated HCM presence in domestic shorthair and Maine Coon cats[7-9]. Not all cardiac disease is HCM and even the definition of HCM can be debated. A consistent definition for HCM is not always used by all cardiologists, thus, there is some difficulty with correlating a genetic test result with an ultrasound report, especially if detailed diagnostic criteria are not presented in the report. Misinterpretations in ultrasound examinations may lead to different interpretations with disease status.

Other sources of error

Often, breeders will insist that a genetic test is incorrect because tested parents were clear of the mutation in question. Several issues need to be considered to resolve these problems. Firstly, the sample needs to be confirmed to be the cat in question. Samples can be unintentionally confused before being sent to the laboratory, or even at the laboratory. Many laboratories will re-evaluate a sample if demonstratively questioned. Gender testing and DNA identification testing can help determine the identity of the sample. Some laboratories request that samples are sent only by veterinarians who have also documented the chip number for a cat. Other times, the parentage has proven to be inaccurate for a cat. Laboratories may offer a parentage test, refunding the costs if parentage and the mutation status proves to be accurate but the test in question inaccurate. Having different laboratories perform tests can often prove accuracy, but often leads to confusing when different results are obtained. Testing laboratories strive to obtain consistent results by testing controls via workshops at the International Society of Animal Genetics. Overall, the reputation of the laboratory is correlated with the accuracy of the tests. Laboratories associated with universities often providing more through service with deciphering problems.

COMPLEX TRAITS

All the mutations influencing a disease may not be identified at any given time, thus, usually the mutations that influence a condition the most, have the highest heritability, are identified first. Because multiple mutations may act additively to cause a disease, each mutation may be said to confer a "risk" for disease development. Thus, some mutations may be considered risk factors, predisposing an individual to health problem. These risk conferring mutations are neither necessary nor sufficient for causing disease. An excellent example of mutations that confer a risk are the DNA variants associated with cardiac disease in cats. GWAS studies using the developed DNA array for the cat will also facilitate the detecting of genes effecting complex traits.

The major benefit of the arrays is the ability to assess the entire genome, via the genome-wide association studies (GWAS). Because the SNPs are at such a high density, the cats used for a GWAS can be from a population, not direct relatives. Thus, individual cases of diseases or traits can be examined from a population, or across breeds and populations. To date, genetic studies have required large extended families, now "cases (cats with the trait) and controls (cats without the trait)" from the families and the population can be examined with the arrays. In addition, there is less concern for the mode of inheritance of the trait, GWAS can be performed with dominant, recessive, X-linked, and even traits that may have complex inheritance but a high heritability or relative risk in a population. Less cases are required for a recessive trait, more for a dominant trait, and more for relative risk studies – the more cases the lower the relative risk. The case-control design also supports the interaction with the private practice veterinarian and researchers as case and controls can come from populations in diverse areas.

FUTURE OF COMPANION ANIMAL HEALTH CARE

An overall goal of identifying these genetic mutations is the correction of the defect via gene therapies and designer drug therapies. Several inborn errors of metabolism in cats are already exploring gene therapy approaches and viral vector-mediated gene therapy has return vision in a canine model of progressive retinal atrophy[18,19]. Thus, genetic testing is an effective preventative medicine and a potential ultimate cure. Genetics is also moving cats towards the veterinary version of designer medicine termed "P4 Medicine", which implies using genetics to predict disease in the individual, prevent the occurrence of disease by controlled breeding, personalize drug selection based on genotype, and encourage active participation in pet and population health by the owner / breeders. Once causative mutations for heritable conditions are identified in the cat, felines become a more important asset to human health.

Even with limited genetic resources, the domestic cat has already become the focus of research technologies that use reproductive methods to manipulate the cat genome. Although gene knock-outs for cat allergens were proposed, the currently available "hypoallergenic" cats have been developed by natural breeding methods (http://www.lifestylepets.com/). The scientific support for these hypoallergenic cats is currently unpublished. However, cloning of the cat by either nuclear or chromatin transfer has been successful and even commercialized[10,11]. At least one company used client provided materials to clone a pet as a commercial adventure, however, this commercial endeavor has been terminated. Clients are now asking veterinarians to collect viable cells from their cats and send them to cloning companies for the eventual use as the sample to clone the donor cat. Besides several domestic cats, two exotic felids, the African wildcat and the Sand Cat, have also been successfully cloned[12,13]. The cat cloning has exemplified important aspects regarding this highly controversial topic. The first cloned cat, Cc, was cloned from a calico cat, which is a female cat (Rainbow). Rainbow has white spotting and different alleles affecting coat color (normal versus orange-hue) at the *Orange* locus on the X chromosome. During X-inactivation only the color allele on the expressed X is active, thus, producing patches of black (normal) or orange coloration. The cloned cat, Cc, did not express the orange allelic variants, hence, apparently X-inactivation did not get reprogrammed during the cloning process[10]. This outcome highlights two important concepts; 1) researchers still do not recognize or understand all the normal biological processes during embryo development, 2) cloned animals can definitely look different that the original, donor individual. The successful cloning of the African wildcat potentially opens avenues for the reintroduction of the genetics of individuals that have been lost to conservation breeding programs, such as animals that have been sterilized or are too old or unhealthy to breed. However, domestic cat eggs were used for the cloning of the wildcats, hence, their mtDNA is not representative of the endangered species. But overall, cloning may provide unexpected opportunities in conservation, or perhaps in the replication of individuals with certain phenotypes of interest in research. Representing the ultimate of an inbred species, a series of clones from the same individual could foster interesting studies of nature versus nurture. The recent successes with the development of red and green fluorescent protein transgenic cats[14] also support felines as models for novel diseases that are not currently recognized in the cat populations at large, improve the cats role in biomedical research.

The deeper sequencing of the cat genome and the investigation of variation by resequencing in different cat breeds will allow the cat to leap forward in genetics, from the analysis of single gene traits to the investigation of more complex traits, such as asthma, diabetes, and obesity. Cats are notorious for viral infections, hence the identification of susceptibility and resistance genes for acquired diseases should become a mainstay in cat research. The costs of genome sequencing has dropped from millions of dollars to tens of thousands of dollars. The intention is to be able to sequence a genome for $1,000 or less in the near future. Already, several individual humans have been sequenced in efforts to find rare mutations that cause diseases in only a few individuals. The mutations are so rare, the case – control design can't be used. Thus, as these technologies have already been applied to other species, the direct sequencing of a genome for a companion animal should be feasible within our lifetime, if not within the next decade. Already, tumors are genotyped to identify markers that can facilitate the use of more targeted chemotherapies, the same will become common for different types of heart and renal failures.

CONCLUSION

The available genetic resources for the cat now places the research bottleneck for felines back to the acquisition of appropriate patients with sufficient cases and controls. Hence, the primary care veterinarian, the veterinary specialists and veterinary researchers need to join forces to properly characterize diseases and routinely collect research materials so that patients are not lost to important studies and health investigations. The development of the DNA tests for parentage and identification (http://www.isag.org.uk/), coat colors and the prominent diseases, such as PKD and HCM, has encouraged cat breeders to explore genetic research in a more positive manner and has encouraged their participation in research studies. Forensic applications [15-17] and mass disasters have brought to the forefront the DNA profiling of loved ones, including our pets. For these reasons, more cat breeders are banking DNA sources on their animals or providing DNA to service and research laboratories. Many veterinary hospitals and large clinical conglomerates are developing electronic database systems that could facilitate the identification of proper patients, cases and controls. Combined with DNA banking and specialty health care, the veterinary world stands to enhance the possibilities of complex disease research in the cat by leaps and bounds. Even though feline origins remain a mystery and "domesticated" may not be the most appropriate term for the domestic cat, the cat is unlocking its genetic secrets that explain the cat's form and function.

REFERENCES

[1]Menotti-Raymond M et al. **Vet J** 2010; 186(1):32-8. [2]Kohn B, Fumi C. **J Feline Med Surg** 2008;10(2):145-53. [3]Meurs KM et al. **Genomics 2007**;90:261-264. [4]Meurs KM et al. **Hum Mol Genet** 2005;14:3587-3593. [5]Ripoll Vera T et al. **Int J Cardiol** 2010; 145(2):405-7. [6]Tsoutsman T et al. **Clin Exp Pharmacol Physiol** 2008;35:1349-1357. [7]Fries R et al. **J Vet Intern Med** 2008;22:893-896. [8]Gundler S et al. **Acta Vet Scand** 2008;50:22. [9]Paige CF et al. **J Am Vet Med Assoc** 2009;234:1398-1403. [10]Shin T et al. **Nature** 2002;415:859. [11]Yin XJ et al. **Reproduction** 2005;129:245-249. [12]Gomez MC et al. **Cloning Stem Cells** 2004;6:247-258. [13]Gomez MC et al. **Cloning Stem Cells** 2008;10(4):469-83 [14]Gómez M et al. **Cloning Stem Cells** 2009;11(1):167-76. [15]Menotti-Raymond M et al. **J Forensic Sci** 1997;42:1039-1051. [16]Menotti-Raymond MA et al. **Nature** 1997;386:774. [17]Menotti-Raymond MA et al. **J Forensic Sci** 2005;50:1061-1070. [18]Komáromy AM et al. **Hum Mol Genet** 2010 Jul 1;19(13):2581-93. [19]Acland GM et al. **Mol Ther** 2005;12(6):1072-82.

BREED-ASSOCIATED DIFFERENTIAL MICRORNA EXPRESSION IN CANINE OSTEOSARCOMA. Fenger, J.M.[1], Jaworski, L.N.[1], Ozer, H.G.[3], Sarver, A.L.[5], Subramanian, S.[4,5], Breen, M.[7], Modiano, J.F.[5,6], London, C.A.[2,3], Kisseberth, W.C.[1,3] [1]Department of Veterinary Clinical Sciences, College of Veterinary Medicine, The Ohio State University, Columbus, OH, USA. [2]Department of Veterinary Biosciences, College of Veterinary Medicine, The Ohio State University, Columbus, OH, USA. [3]Ohio State University Comprehensive Cancer Center, The Ohio State University, Columbus, OH, USA. [4]Department of Laboratory Medicine and Pathology, School of Medicine, University of Minnesota, Minneapolis, MN, USA. [5]Masonic Cancer Center, University of Minnesota, Minneapolis, MN, USA. [6]Department of Veterinary Clinical Sciences, College of Veterinary Medicine, University of Minnesota, St. Paul, MN, USA.[7] Department of Molecular Biomedical Sciences, North Carolina State University and Center for Comparative Medicine and Translational Research, North Carolina State University, Raleigh, NC, USA; Lineberger Comprehensive Cancer Center, UNC-Chapel Hill, NC, USA.

Osteosarcoma (OSA) is the most common bone tumor in dogs, however, little is known regarding the mechanisms underlying malignant transformation in these tumors. Breeds such as Rottweilers and Greyhounds are at higher risk for developing OSA, suggesting that heritable factors play a role in this disease. MiRNAs have tumor/tissue specific roles in regulating gene expression and dysregulated miRNA expression is found frequently in cancer. We hypothesize that canine OSA is characterized by a unique miRNA expression profile(s) with dysregulation of some miRNAs being associated with specific breeds.

MiRNA expression profiling of primary OSA tumors from 6 Greyhounds and 6 Rottweilers was performed using the Nanostring Technologies nCounter miRNA Expression Assay Kit, interrogating the miRNA expression profile of 651 human miRNAs, 168 of whose mature sequences are 100% conserved between human and dog.

17 miRNAs were differentially expressed in Greyhound versus Rottweiler tumors ($p<0.05$), suggesting that breed-specific dysregulation of miRNAs may contribute to the development and progression of spontaneous OSA.

Hierarchical clustering revealed distinct miRNA expression signatures in Greyhound OSA tumors as compared to Rottweilers. Based on these preliminary results, we are evaluating a larger cohort of OSA tumor samples including Greyhounds, Rottweilers, Golden Retrievers, and a mixed population of other breeds. Statistical analysis will be performed to determine the association of miRNA transcript levels with specific breeds and overall outcome. Characterization of miRNA expression in canine OSA will facilitate our understanding the biology of this disease and has the potential to identify targets for therapeutic intervention.

Originally presented at VCS, October 29 – November 1, 2020, San Diego, CA.

SAFETY EVALUATION OF COMBINATION VINBLASTINE AND TOCERANIB PHOSPHATE (PALLADIA®) IN DOGS: A PHASE I DOSE-FINDING STUDY. Cecilia S. Robat Dr Vet.Med, Cheryl A. London DVM, Ph.D., DACVIM(Onc), Kim A. Selting DVM, MS, DACVIM(Onc), David M.Vail DVM, DACVIM(Onc), Madison, WI.

Combination therapies using drugs with documented single-agent activity and lack of overlapping toxicities could potentially improve outcome. The hypothesis intended to be tested is that Palladia® can be safely administered concurrently with a standard weekly protocol of vinblastine (VBL), at dosages known to have activity against mast cell tumors.

Dogs with histologically confirmed measurable mast cell tumors were evaluated for eligibility to enter a standard phase I dose-finding trial (3+3 cohort), at a starting dose of 2.3mg/m2 IV VBL (weekly for a total of 4 treatments) and 2.25mg/kg PO Palladia® EOD, concurrently. Dose escalation of Palladia® was scheduled in 0.25mg/kg increments until MTD was established or FDA label dose completed (3.25mg/kg). Safety evaluation was performed weekly throughout the 4 week study period. Dose-limiting toxicities were described following established VCOG-CTCAE(v1.0) criteria. While antitumor response is not a primary endpoint of phase I trials, activity was documented prior to VBL treatments 2-4, and monthly thereafter, based on RECIST criteria.

Nine dogs have been enrolled; cohort 3 is filled and approaching completion of the evaluation period. Hematologic dose limiting toxicity led to 2 de-escalations of VBL. The current safe combination appears to include VBL at 1.6mg/m2 every other week and Palladia® at 2.25mg/kg EOD. Response was seen in all but one dog.

Without head to head trials comparing efficacy of bi-weekly VBL combined with Palladia® and VBL alone, choice of therapy should remain at the clinician's discretion.

TARGETING TUMOR NEOVASCULATURE WITH NANOPARTICLE THERAPEUTICS. V Coyle[1], R Tong[2], L Tang[2], J Cheng[2], TM Fan[1]. 1. University of Illinois Department of Veterinary Clinical Medicine, Urbana, IL. 2. University of Illinois, Department of Chemistry, Materials Science and Engineering, Urbana, IL.

Prostate specific membrane antigen (PSMA) is a transmembrane protein expressed by tumor-associated neovasculature, but not normal blood vessels. Based upon its selective expression in endothelial cells associated with cancer, PSMA may serve as a conserved angiogenic target shared by macroscopic solid tumors of various histologies.

To investigate the feasibility of targeting a homogenous population of PSMA-expressing endothelial cells as a novel anticancer strategy, we have investigated PSMA expression in several canine hemangiosarcoma (cHSA) cell lines, and subsequently developed self-assembling nanoparticles containing diagnostic (near infrared dyes) and therapeutic (doxorubicin) cargo which selectively bind to PSMA by means of the A10 aptamer, a commercially-available oligonucleotide.

The expression of PSMA by cHSA cells was confirmed transcriptionally and translationally by real time PCR and immunohistochemistry, respectively. Selective binding and endocytosis of A10 decorated nanoparticles was studied by fluorescent microscopy. The ability of A10 decorated nanoparticles encapsulating doxorubicin to exert *in vitro* cytotoxic effects in cHSA cells was assessed by colony forming assays. Using a cHSA xenograft murine tumor model, clinically-relevant anticancer effects of A10 decorated nanoparticles encapsulating doxorubicin were tested.

All cHSA cell lines expressed PSMA mRNA and protein. A10 decorated nanoparticles were selectively endocytosed by PSMA-expressing cells, and when these nanoparticles encapsulated doxorubicin, significant cytotoxic effects were exerted *in vitro*. Finally, A10 decorated nanoparticles encapsulating doxorubicin significantly reduced the size of macroscopic cHSA tumor burdens in transplanted mice.

Diagnostic and therapeutic nanoparticles can be targeted to PSMA-expressing endothelial cells, and cHSA provides a comparative model for the future study of nanoparticle therapeutics.

TRANSITIONAL CELL CARCINOMA COMPARATIVE GENOMICS: A MODEL FOR PERSONALIZED MEDICINE APPROACHES TO CANCER CARE. M Paoloni[1], D Dhawan[2], J Stewart[2], K Nuckolls[1], J Lee[3], D Theodorescu[4], C Khanna[1], D Knapp[2]. 1. National Cancer Institute, Bethesda, MD. 2. Purdue University School of Veterinary Medicine, West Lafayette, IN. 3. University of Virginia, Charlottesville, VA. 4. University of Colorado, Denver, CO.

Canine transitional cell carcinoma (TCC) is the most common tumor of the urinary tract, and is similar to human invasive TCC in histopathologic characteristics, molecular features, sites of metastasis, and response to medical therapy. Prevalence is increasing, and novel therapies and strategies are needed to effectively treat this aggressive form of cancer in both species.

Personalized medicine techniques intend to improve treatment outcome by using patient tumor profiling to identify potential and individualized therapeutic targets. A genomic algorithm has been developed termed "coexpression extrapolation", or COXEN, that aims to use expression microarray data to predict drug activity in patient TCC samples. The utility of this predictive methodology has been established in other types of cancer *in vitro,* however its clinical utility has not yet been determined.

Validation studies of COXEN in 10 canine TCC cell lines were conducted. The goal was to determine the value of COXEN in predicting baseline sensitivity of canine TCC to 5 chemotherapy agents (gemcitabine, mitoxantrone, carboplatin, vinblastine and cisplatin) that would then be used in a proposed clinical trial. Additionally, expression data from 25 canine treatment-naïve primary tumor samples were generated on an Affymetrix array platform (Canine Genome v 2.0). Both the expression data and TCC cell line data (anti-proliferative effects, 50% growth inhibition or GC_{50}) were used to establish a canine specific predictive COXEN algorithm. COXEN scores for canine TCC cell-line drug activity were then analyzed. Scores predicted the activity of cisplatin, gemcitibine, and mitoxantrone in all 10 cell lines, and of carboplatin in 6 cell lines. Because all of the cell lines were sensitive to vinblastine (GI_{50} <0.05μM), the COXEN score was not predictive of its potency. Interestingly, COXEN fails to predict vinblastine response in human TCC cell line data as well. In concurrent work, comparative genomic studies to define and compare the gene expression signatures of TCC in dogs and humans provides further evidence that canine TCC is a valuable genomic model of the human disease. Current studies involve testing the chemo-predictivity of this derived canine COXEN algorithm in additional canine TCC cell lines.

Canine TCC offers an excellent model for *in vitro* and *in vivo* studies of the COXEN approach. This preclinical work will be used to guide the feasibility of future COXEN clinical trials in dogs and humans with TCC.

A PROSPECTIVE PHASE I/II OPEN LABEL, NON-RANDOMIZED CLINICAL TRIAL TO EVALUATE THE SAFETY AND EFFICACY OF AMINOACT MULTI-KINASE INHIBITOR FOR THE ADJUVANT TREATMENT OF DOGS WITH SPONTANEOUSLY OCCURING CANCER. GK Ogilvie *, HY Chen, O Mollstedt, DM Proulx*, L Van Horn*, MD Price, MH Tsai. Angel Care Cancer Center*, California Veterinary Specialists, Carlsbad CA; OncoNutrition, Riverside, CA.

A small molecule complex (Aminoact) isolated from bovine milk is a natural peptide mixture with multi-kinase inhibitory effects against epidermal growth factor receptor (EGFR) and insulin-like growth factor receptor-1 (IGFR-1). Ingestion of Aminoact in people with cancer results in lower serum TNF-alpha, an increase in antioxidant superoxide dismutase (SOD) enzyme activity, and subjects' blood serum causes apopsotis in cancer cell lines. This study was designed to first assess safety and secondly the efficacy of three dosage levels of AX-3 in sustaining progression free survival (PFS) for dogs with refractory advanced and/or metastatic cancer.

The prospective, open label study included dogs of different breeds with naturally occurring histologically confirmed malignancies. The first 13 dogs received Aminoact at $1g/m^2$; the second group of 7 dogs subsequently received the same dosage + 350 mg of Aminoact; and the third group of dogs subsequently received $2g/m^2$. Each dog was treated orally daily for six weeks along with 550 mg betaine HCl, that aids in peptide absorption. All patients were evaluated for toxicity using the VCOG-CTCAE and efficacy using the RECIST criteria via assessment of clinical parameters, blood work and client questionnaires.

No toxicity other than mild, transient (grade I) nausea was noted, nor were there any changes in hemograms or biochemical profiles in any patient. Dogs with tumors that were confirmed as responders (>50% reduction in size) include pulmonary adenocarcinoma, mast cell tumor, trichoepithelioma and soft tissue sarcoma. It appears in limited studies that the response rate may be more durable at higher dosages.

The response to Aminoact is dose dependent and only transient mild toxicity was observed, which suggest the maximum effective dosage has not been reached. Further clinical studies will be valuable in determining the effective dosage and response duration. Treating cancer in dogs with Aminoact offers a unique opportunity as a model for human cancer biology and translational cancer therapeutics.

CANINE NASAL TUMORS: A STEREOTACTIC RADIATION THERAPY APPROACH. JT Custis, JF Harmon, SD Ryan, SL Kraft, SM LaRue. Colorado State University Animal Cancer Center (CSU), Fort Collins, CO.

Stereotactic radiation therapy (SRT) combines patient immobilization, image guidance, and intensity modulated delivery to achieve ablative radiation doses within the tumor, while preferentially sparing surrounding normal tissues. The purpose of this study was to evaluate the efficacy of SRT as a means of achieving local tumor control for canine nasal tumors.

Retrospective analysis was performed on dogs with a nasal tumor confirmed by histopathology and computed tomography, no previous surgical or radiation therapy, at least six months of follow-up, and completion of three fractions of SRT at CSU.SRT was administered via the Varian Trilogy linear accelerator once daily for three consecutive days. The Varian Eclipse treatment planwas reviewed to determine the planned target volume (PTV) and dose to 95% of the PTV. Kaplan-Meir survival analysis was performed for disease free interval (DFI) and overall survival (OS).

Sixteen patients with nasal tumors (8 adenocarcinoma / carcinomas, 2 squamous cell carcinomas, 3 chondrosarcomas, 2 osteosarcomas, and 1 undifferentiated sarcoma) were treated with SRT. A median dose of 29.1Gy was administered to 95% PTV with a median PTV of 75.3cc. SRT was well tolerated by the normal tissues with minimal, manageable side effects. To date, the median DFI is 270 days, while the median OS is 367 days.

Based upon the initial clinical experience, stereotactic radiation therapy is an emerging modality in the management of canine nasal tumors.

TEACHING T CELLS TO TARGET TUMORS: TOWARDS THE DESIGN OF MORE EFFECTIVE CANCER VACCINES

Barb Biller, DVM, PhD, DACVIM and Jill Slansky, PhD
Fort Collins, CO.

INTRODUCTION

Tumor vaccines are a great idea but usually don't work very well

Treating cancer with a vaccine is an attractive strategy because of the potential to specifically target the tumor itself, mobilizing an immune response that kills tumor cells, spares normal host tissues and results in immunologic memory that can protect against the development of metastatic disease. There are a number of approaches to vaccine design that illustrate the strategies used to focus the immune system on tumor targets. Whole cell or tumor lysate vaccines are usually patient and/or species specific and have the advantage that specific tumor antigens do not need to be isolated or defined in order to generate the vaccine. DNA vaccines usually incorporate syngeneic or xenogeneic plasmid DNA chosen to stimulate different aspects of the adaptive immune response. An example of this is the recently developed xenogeneic DNA vaccine for canine melanoma that uses a human DNA plasmid to target a tumor antigen frequently expressed on malignant melanoma cells. Other approaches, such as dendritic cell vaccines and viral vector-based products are designed to deliver genes or other immunostimulatory products to tumor tissues. Peptide vaccines are also of significant interest because of their ability to selectively target defined tumor antigens, particularly antigens which are common to multiple tumor types.

Despite their collective potential, however, the results of human clinical trials using tumor vaccines have been disappointing (1, 2). One prominent reason for their failure is that most cancer vaccines do not adequately stimulate tumor-specific T cells in patients because they engage T cells with low affinity interactions (3). This outcome is perhaps not unexpected, since most T cells do not identify tumor antigens as foreign, identifying them instead as "self", and T cells that react strongly against self tissues are removed from the immune system as it develops to prevent autoimmune diseases. In addition, the typically immunosuppressive environment associated with tumor progression dampens the development of effective antitumor immunity through multiple mechanisms including the production of immunosuppressive cytokines such as TGF-β and IL-10 and recruitment of inhibitory cells such as regulatory T lymphocytes (Treg), tumor-associated macrophages and myeloid-derived suppressor cells.

Using the T cells that infiltrate and target tumor tissues to make a vaccine is novel approach

One method to boost the T cell response to tumors is to vaccinate with *mimotopes* or mimics of tumor epitopes, which stimulate T cells that recognize the tumor with much greater affinity (4, 5). In general, mimotopes are engineered to improve the binding of peptide tumor antigens to both the antigen presenting machinery (MHC) molecules and their cognate T cell receptors (TCR). Several studies in mouse tumor models show the superior ability of mimotope vaccines to significantly increase the number and cytotoxic function of tumor-specific T cells compared to traditional peptide vaccines (6-8).

An exciting approach in designing mimotope vaccines is to use the T cells that infiltrate tumors to select those mimotopes that activate tumor-specific T cells with high efficiency. We are currently developing a method in which the T cells from the tumor are removed and introduced to a wide variety of mimotopes genetically encoded in large peptide/MHC libraries (9). The mimotopes are then selected based on the strength of their binding interaction to the T cells.

Mimotopes elicit effective antitumor immunity in the CT26 colon carcinoma model

We have been studying the immune response to mimotope vaccines in the CT26 colon carcinoma model of mice. Similar to human tumor antigens, the CD8- T cell response to this tumor is directed against a non-mutated self-antigen (known as AH1) presented by an MHC class I molecule. We have found that tumor-specific T cells can be isolated from CT26 tumors and used to design peptide mimotope vaccines that protect mice from tumor challenge (Fig 1). In contrast, vaccination with the native AH1 peptide fails to confer protection. We also found that the tumor-specific T cells in mimotope vaccinated mice proliferate extensively in vivo and fully develop into effective cytotoxic T lymphocytes (CTLs) compared to mice vaccinated with the native peptide (6-9). These results support both the feasibility and superiority of using T-cell selected mimotope vaccines to stimulate effective antitumor immunity in an aggressive murine model of colon carcinoma.

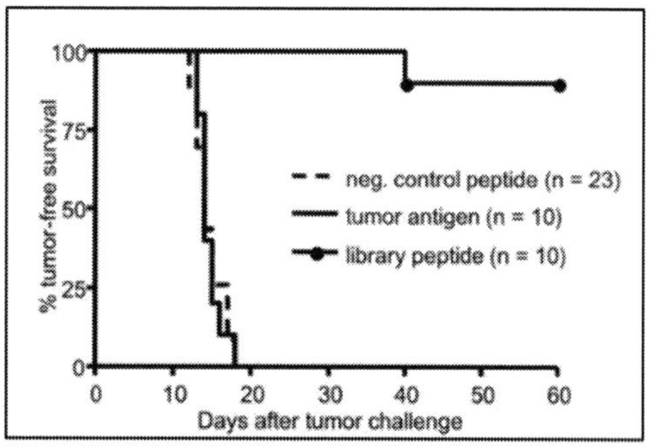

Figure 1

TRANSLATING PROMISING RESEARCH IN MICE TO DOGS WITH CANCER

Building a tool box to design a mimotope vaccine for dogs: Step 1 - Understanding MHC Class I alleles

Clinical trials using peptide vaccines in humans are frequently designed to identify individuals with a compatible Human Leukocyte Antigen (HLA) phenotype since the tumor antigens are presented to T cells through the MHC Class I system. For example, studies evaluating melanoma tumor antigen vaccines show that T-cell presentation of the NY-ESO-1 antigen is restricted by the HLA-A*0201 phenotype, which is present in about 40% of humans (10). Identifying the subgroup of people most likely to respond strongly to a vaccine can confer a significant survival advantage. To determine if dogs can be similarly selected requires additional understanding of the canine MHC class I molecules. Although the canine DLA molecules have not been fully characterized, there is evidence for dominant alleles, particularly within the DLA-88 loci. Of the known class I loci in dogs, DLA-88 is by far the most polymorphic (11, 12). In humans, class I genes that are more polymorphic (such as HLA-A or HLA-B) tend to have higher cell surface expression than less polymorphic genes, correlating with increased function as antigen presenting molecules. It is likely therefore, that DLA-88 plays the greatest role in presentation of MHC class I restricted peptides in dogs. Therefore we have recently begun sequencing the DLA-88 loci using an oligonucleotide primer pair that binds in critical regions of the gene and covers the corresponding TCR contact region. An update from this effort will be discussed briefly in this session.

Step 2: Evaluating tumor infiltrating lymphocytes (TIL) in canine tumors

The immune cells that infiltrate tumors typically include key immunostimulatory cells such as dendritic cells and CD8+ cytotoxic T cells (CTL) as well as immunosuppressive cells such as tumor-associated macrophages, myeloid-derived suppressor cells, Treg and sometimes CD4+ T helper cells. In some human malignancies such as melanoma, lymphoma and carcinomas of breast and ovaries, the ratio of "good" to "bad" lymphocytes within tumor tissues correlates with prognosis and can sometimes be a better predictor of outcome than tumor grade or stage of disease. The prognostic value of TIL is less clear in veterinary patients, however, in a recent study of dogs with malignant mammary gland tumors an increased lymphocytic infiltrate and an increased ratio of CD4+ to CD8+ T cells within the TIL was associated with a significantly poorer prognosis (13). The ratio of CD8+ T cells to Treg has also been linked to outcome in dogs with osteosarcoma and malignant melanoma (14, 15).

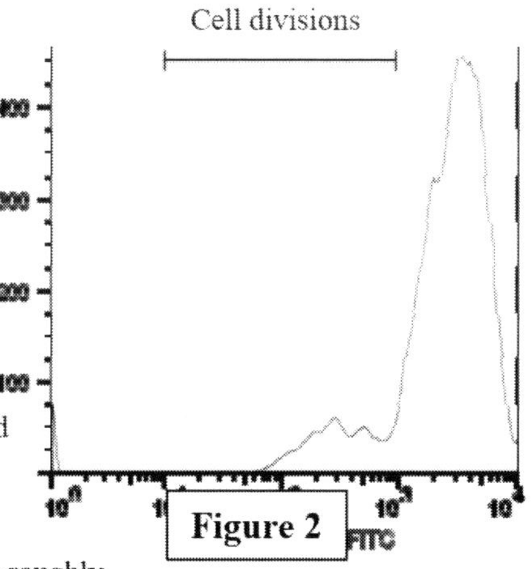

Figure 2

We are currently evaluating the lymphocyte subsets (CD4+, CD8+ and Treg) that infiltrate malignant lymph nodes in dogs with B cell lymphoma. In addition to phenotype of the TIL, we are also interested in their function. Our preliminary data shows that, while roughly 78% of the lymph node contains malignant B cells, the remaining lymphocytic infiltrate consists of approximately 8% non-regulatory CD4+ T lymphocytes, 2% CD4+FoxP3+ (Treg) T cells and 2% CD8+ T cells. Despite their small numbers, most of these CD8+ T cells proliferate vigorously in response to stimulation with conA (a strong mitogen) as shown in **Fig. 2**. Proliferation in response to autologous tumor cells and correlation of TIL with outcome is still being investigated.

CONCLUSIONS AND FUTURE DIRECTIONS

The results presented here are early steps towards the goal of building a more effective tumor vaccine. Future work will include identification of canine MHC anchor residues and production of a peptide library that can be used to select effective mimotopes using CD8+ T cells within the TIL population in dogs with B cell lymphoma. This information will permit construction of an autologous vaccine and followed by testing in a Phase I/II clinical trial. While initial vaccines will be patient-specific, an understanding of the rules governing effective mimotope vaccine design will ultimately enable testing of a disease specific rather than a patient specific product. Because clinical testing can be done in relevant large animal models rather than moved directly from mice to humans, products that are effective in mice but ineffective in humans are much more likely to be detected relatively early in the vaccine development process. We anticipate that by selecting T cells that vigorously respond to the tumor to design the vaccine, some of the limitations of current vaccine approaches will be resolved. However, as with many novel therapies, the best use of tumor vaccines will be in combination with traditional treatment modalities such as surgery, chemotherapy and radiation therapy. Therefore, the challenges associated with developing effective combinatorial therapies will likely be the focus of cancer research for many years to come.

REFERENCES

[1.] Khazaie K, et al. **Curr Opin Oncol** 2009;21:524. [2.] Lero M, et al. **Cancer Immunol Immunother** 2009;58:1159. [3.] McWilliams JA, et al. **Vaccine** 2008;26:1863. [4.] De Visser K, et al. **J Immunol** 2001167:3818. [5.] Sharav T, et al. **Vaccine** 2007;25:3032. [6.] Jordan K, et al. **Proc Natl Acad Sci** 107;4652. [7.] Le Gai F, et al. **J Immunother** 2005;28:252. [8.] McMahan R, et al. **J Clin Invest** 2006;116:2543. [9.] Jordan K, et al. **J Immunol** 2008;28:188. [10.] Khong H, et al. **J Immunother** 2004;27:472. [11.] Graumann M, et al. **Tissue Antigens** 1998;51:374. [12.] Wagner J, et al. **Tissue Antigens** 2002;59:205. [13.] Estrela-Lima A, et al. **BMC Cancer** 2010;10:256. [14.] Biller B, et al. **JVIM** 2010:24:1118. [15.] Tominaga M, et al. **J Vet Diagn Invest** 2010;22:438.

MOLECULAR MECHANISMS OF LUNG TUMORIGENESIS & IMPLICATIONS FOR TARGETED THERAPIES

Kelly R. Hume, DVM, DACVIM (Oncology)
Ithaca, NY

INTRODUCTION

Lung cancer remains the leading cause of cancer deaths worldwide with over one million deaths annually.[1] In the United States, approximately 200,000 new cases are diagnosed each year.[2] At its current rate, lung cancer is expected to move from the tenth most common cause of death in the world to the fifth most common cause.[3] The reported relative 5-year survival rate is only 50% for patients with localized disease and an abysmal 3% for patients with distant metastatic disease.[2] While the incidence of primary lung cancer in our veterinary patients is significantly lower than that in people, the prognosis for dogs and cats with advanced disease is equally poor. In comparative fashion, this review summarizes established molecular mechanisms of lung tumorigenesis and associated targeted therapies. Current lung cancer research in veterinary medicine is also discussed.

CLASSIFICATION OF LUNG CANCER

The two main categories of primary pulmonary cancer are non-small cell lung cancer (NSCLC) and small cell lung cancer (SCLC). In humans, about 85% of all lung cancers are NSCLC. The most common subtypes of NSCLC are squamous cell carcinoma (SCC) and adenocarcinoma (ADC).[4] ADC of the lung can have a papillary, nodular, alveolar, or broncho-alveolar (BAC) histopathologic pattern. ADC most commonly occurs in smokers, but is also the most common subtype among never-smokers, especially women.[5] The majority of patients present with advanced disease as there are no cancer-specific symptoms to look for at an early stage. The 5 year survivorship for stage I NSCLC is 58-73% after surgery; however, a significant number of patients will relapse and die of both local recurrence and distant metastatic disease. Approximately 20-30% of patients with NSCLC are amenable to radical surgery, but only 40-50% are cured. The 5 year survival rate of this disease has not changed over the last 30 years.[3] SCLC is characterized by neuroendocrine differentiation. About 20% of patients with limited stage SCLC can be cured.[6] Unfortunately, over 60% of patients have radiographic evidence of metastasis at the time of diagnosis. Permutations in the standard regimen of platinum and etoposide chemotherapy have not improved survival. 5 yr survival for limited stage disease is 15-25%; with extensive stage disease median survival time is only 10 months. Less than 10% of patients survive two years.

ADC is the most common histologic type of lung cancer in dogs and cats. While dogs and cats with grade I disease can have extended survival, survival for animals with grade III disease or lymph node involvement typically ranges from 5 – 75 days.[7]

GENETIC ABNORMALITIES AND MOLECULAR MECHANISMS

Germ-line mutations in p53 and Rb (tumor suppressor genes) are associated with increased lung cancer susceptibility and risk, as is germ-line mutation in EGFR and reduced DNA repair capacity resulting from germ-line alterations in nucleotide excision repair genes such as ERCC1.[4] Genetic changes common to all types of lung cancer include loss of heterozygosity in chromosome 3p (detected in up to 80% of cases independent of histological subtype), mutations in p53 (up to 50% of NSCLC and 70% of SCLC), and inactivation of Rb (80-100% of high grade SCLC, only 15-35% of NSCLC).[5] p53 mutation or overexpression is inversely related to prognosis in cases of ADC but not SCC. This data is supported by a mouse model of ADC where loss of p53 hastens lung tumor progression following oncogenic Kras activation.[8] Loss of Rb is predictive of poor survival in SCC.[5] TTF-1 (thyroid transcription factor-1) is frequently expressed in well-differentiated adenocarcinoma, but BAC are often negative. High grade tumors (low differentiation), vascular infiltration, increased mitotic activity, necrosis are reported as adverse prognostic markers for ADC. ERCC1 expression is associated with resistance to cisplatin; RRM1 (a subunit of the enzyme ribonucleotide reductase) expression is associated with resistance to gemcitabine.[5] Work with mouse cancer models has identified that rapid repair of DNA damage mediates resistance to cisplatin chemotherapy, suggesting that inhibiting DNA repair mechanisms may be one strategy to overcome resistance to DNA damaging agents like cisplatin.[9]

Approximately 10-15% of all lung cancers arise in never-smokers.[10] Risk factors for lung cancer for this group include secondhand smoke, radon exposure, indoor air pollution, and exposure to occupational agents such as asbestos and silica. In general, lung cancer in never-smokers peaks at a younger age than in smokers, is more likely to arise in women, and is more likely to be ADC; never-smokers tend to fare better with treatment than smokers.[11] In clinical trials of gefitinib and erlotinib never-smokers were found to have a better clinical outcome; this ultimately led to the determination that mutations in the ATP binding site of the tyrosine kinase domain of EGFR are strongly associated with response to EGFR tyrosine kinase inhibitors. EML4-ALK fusion genes (discussed below) have also been found in never-smokers. Lung cancers in never-smokers have a lower frequency of p53 and Kras mutations.

Kras and EGFR in NSCLC

Kras mutations are noted in up to 30% of ADC, but are rare in SCC and SCLC. EGFR mutations or amplifications are relatively common in both SCC and ADC, but are rare in SCLC. EGFR mutations seem to occur early in the development of ADC unrelated to smoking, whereas Kras mutations occur early in the development of smoking-related ADC. EGFR mutation is associated with an improved prognosis for NSCLC, even when treated with cytotoxic chemotherapy. Targeted agents against EGFR include the small molecule tyrosine kinase inhibitors gefitinib and erlotinib and the monoclonal antibody cetuximab. EGFR mutations are detected more often in women, never-smokers, patients of Asian origin, and in ADC histology. EGFR mutations include deletions, missense point mutations, and insertional mutations. In contrast to EGFR mutations, higher EGFR copy number is associated with a worse prognosis.[5] Amplification of the oncogene MET is present in about 20% of both ADC and SCC and is associated with resistance to EGFR inhibitors and a poor prognosis.[4]

SCLC

Genetic differences between NSCLC and SCLC include expression of Kras and Cox2 in NSCLC and c-myc amplification in SCLC. In terms of loss of cell cycle control, Rb is inactivated and E2F is overexpressed in SCLC, whereas $p16^{INK4}$ inactivation occurs in NSCLC. As noted above, p53 mutations are more common in SCLC than in NSCLC.[5] Increased expression of the anti-apoptotic gene Bcl-2, reduced expression of MMP inhibitors, and increased expression of vascular growth factors have all been identified in SCLC. However, to date, no histopathologic or genetic change has been correlated with prognosis.[5] Initial investigations targeting bcl-2 were with an antisense compound and were unsuccessful. However, the compound may not have been hitting and downregulating its target. Currently, small molecule inhibitors of bcl-2 are in phase I and II trials.[6] Also under investigation are a monoclonal antibody (BB-10901) against CD56 (member of the neural cell adhesion molecule family), a Hedgehog pathway inhibitor (with hopes of affecting tumor stem cells), dual c-Src and c-Abl inhibition (AZD0530), and FAK inhibition.[6,9] Temsirolimus, an inhibitor of mTOR, shows promise in improving survival of extensive stage SCLC patients when given as maintenance therapy following chemotherapy-induced remission.[12]

Other targets under investigation

The current impact of targeted therapies in lung cancer has been quite modest, in contrast to that achieved in chronic myelogenous leukemia, gastrointestinal stromal tumor, and a subset of breast cancers.[13] While targeted therapy may be used to help patients who do not tolerate standard cytotoxic therapy, the ultimate goal is to use targeted therapy only in patients who will benefit from the treatment. (In contrast to standard cytotoxic chemotherapy, where all patients are treated for a minority to experience a benefit.) Gefitinib resulted in prolonged progression-free survival in selected patients with activating EGFR mutations in their tumors; cetuximab improved survival compared to chemotherapy alone in patients with advanced EGFR-expressing NSCLC. Concurrent use of a tyrosine kinase inhibitor with chemotherapy did not improve outcome, but use of a tyrosine kinase inhibitor as maintenance therapy did improve outcome for patients with NSCLC. Challenges for use of targeted therapies include proper analysis of biopsy specimens and standardization of handling and analysis protocols to identify patients that would benefit from therapy, cost, and determining how to use targeted agents in combination protocols with multimodality therapy.[3]

The monoclonal antibody bevacizumab, targeting VEGF, is an FDA approved treatment for NSCLC. Factors associated with increased tumor angiogenesis correlate with the development and prognosis of lung cancer. Mouse models show that mice with stabilized Hif2a expression, in concert with oncogenic Kras, develop larger tumors, have an increased tumor burden and decreased survival.[14] Bevacizumab improved progression free survival when added to palliative chemotherapy in selected patients with advanced non-squamous cell lung cancer.[3]

New clinical targets include EML4-ALK (fusion gene echinoderm microtubule-associated protein-like 4 and anaplastic lymphoma kinase) and IGF1R. EML4-ALK is validated in mouse models. Treatment of patients with an ALK/c-Met inhibitor has resulted in partial responses even in heavily pre-treated NSCLC patients. Testing for the fusion protein is challenging however, as different sites can fuse. Multiple antibodies and small molecule inhibitors targeting IGF1R are in development with many in clinical trials. Antibodies are less likely to affect the insulin receptor. A possible advance in screening involves detecting early markers of epigenetic change, such as covalent histone modifications. The zinc-finger transcription factor Snail is a repressor of E-cadherin and its overexpression in lung cancer and premalignant lesions can lead to epithelial-mesenchymal transition, angiogenesis, and tumor growth. Studies of epithelial-mesenchymal transition have identified lung cancer self-renewal cells with elevated CD133 and ALDH levels, which may serve as therapeutic targets.[9] p16 methylation occurs early in the development of SCC but only rarely in ADC. Agents that reverse epigenetic changes may thus have a role in the treatment of lung cancer.[15] Patients with strong staining for COX-2 via in situ hybridization with a digoxgenin-labeled antisense riboprobe had a median survival time of 1 year, in contrast to 8.5 years in patients with weak staining (stage I NSCLC). While data achieved significance, individual patient numbers were small.[16]

Lung stem cells may not control steady-state epithelial cell homeostasis, but appear critically important for repair following injury. It is hypothesized that the longevity of lung stem cells means they have a lengthy exposure to potential toxic insults and potentially accumulate genetic mutations making them ideal candidates for tumor initiation. Cancer specific stem cells seem to have a common property of resistance to therapeutics due to expression of at least one member of the ATP binding cassette (ABC) transporter protein family.[3] Putative bronchioalveolar stem cells have been identified that expand in response to oncogenic Kras in culture and in precursors of lung tumors *in vivo*.[17] If the majority of lung cancers are ever to be cured, it is likely that these stem cells will have to be targeted.

COMPARATIVE ASPECTS WITH VETERINARY MEDICINE

While it has been difficult to prove a causal link between environmental smoke exposure and other carcinogens in spontaneous canine lung tumors, higher levels of anthracosis have been associated with increased risk for primary lung cancer.[18] Anthracosis is the deposition of black dusty material within lung macrophages, an indicator for lung exposure to air pollutants such as polycyclic aromatic hydrocarbons like benzopyrene and trace metals like arsenic. Lung tumor growth rates in dogs are similar to that reported in humans with equivalent histologic types.[19] A detailed study of a feline bronchioalveolar lung carcinoma revealed nuclear TTF-1 immunoreactivity, aneuploidy, and a p53 mutation similar to that found in lung cancers of smokers, although the cat was not known to have been exposed to mainstream or second-hand smoke.[20] Nuclear staining for TTF-1 was detected in 59-64% of canine primary pulmonary epithelial tumors.[21,22] Metastatic tumors were always negative.[22]

Greater EGFR binding has been observed in neoplastic canine lung, as compared to normal lung, due to a difference in higher concentration of EGFR sites.[23] Another study found that 47% of canine lung tumors had moderate to intense immunoreactivity for EGFR (spontaneous and plutonium-induced); this reactivity was associated with specific histologic subtypes: highest in SCC and nonexistent in BAC.[24] Kras mutations have been detected in 16 – 25% of canine NSCLC; presence of mutation was also associated with a higher incidence of metastasis (60% vs 32% if Kras wildtype).[25,26] Whether these genetic abnormalities can be used to direct therapy in animals has not yet been investigated.

In 30 cases of primary canine lung cancer evaluated at Cornell University, median body weight was 21.6 kg and median age at diagnosis was 11 years. Cough was the presenting complaint in 57% of dogs, whereas 40% were diagnosed incidentally. The most prevalent histologic subtype was BAC (50%). In a large, multi-institutional study (344 cases; 17 institutions), Dobermans, Irish setters, and Boxers were identified as at risk breeds; median age was 11.5 y, and overall survival was 200 days. Mitotic index appeared not to be predictive of outcome and there was an extremely preliminary suggestion that dogs receiving platinum/gemcitabine chemotherapy fared better than dogs receiving vinorelbine/doxorubicin chemotherapy.[27]

The author is currently determining if loss of the DNA damage checkpoint protein Hus1 can halt the progression of lung cancer and is also working to develop multiphoton microscopy as a diagnostic tool for lung cancer.

REFERENCES

[1]Youlden DR, et al. **J Thorac Oncol** 2008;3:819. [2]Jemal A, et al. **CA Cancer J Clin** 2009;59:225. [3]Spiro SG, et al. **Respirology** 2010;15:44. [4]Herbst RS, et al. **N Engl J Med** 2008;359:1367. [5]Mountzios G, et al. **Crit Rev Oncol Hematol** 2010;75:94. [6]Dowell JE. **Am J Med Sci** 2010;339:68. [7]Rissetto KC, et al. **Vet Med** 2008;Mar:154. [8]Jackson EL, et al. **Cancer Res** 2005;65:10280. [9]Haura EB, et al. **J Thorac Oncol** 2010;5:S207. [10]Samet JM, et al. **Clin Cancer Res** 2009;15:5626. [11]Rudin CM, et al. **Clin Cancer Res** 2009;15:5646. [12]Pandya, KJ et al. **J Clin Oncol** 2005;23:622s. [13]Camidge R, et al. **J Thorac Oncol** 2009;4:930. [14]Kim WY, et al. **J Clin Invest** 2009;119:2160. [15]Belinsky SA, et al. **Cancer Res** 2003;63:7089. [16]Khuri FR, et al. **Clin Cancer Res** 2001;7:861. [17]Kim CFB, et al. **Cell** 2005;121:823. [18]Bettini G, et al. **Vet J** 2010;186:364. [19]Perry RE, et al. **Am J Vet Res** 1992;53:1740. [20]Grossman DA, et al. **Cancer Res** 2002;62:3826. [21]Ramos-Vara JA, et al. **Vet Pathol** 2005;42:315. [22]Bettini G, et al. **Vet Comp Oncol** 2009;7:28. [23]Leung FC, et al. **Proc Soc Exp Biol Med** 1991;196:385. [24]Gillett NA, et al. **Vet Pathol** 1992;29:46. [25]Griffey SM, et al. **Carcinogenesis** 1998;19:959. [26]Kraegel SA, et al. **Cancer Res** 1992;52:4724. [27]Selting K, et al. VCS Proceedings 2010 and personal communication, with permission.

ANTI-VACUOLAR ATPase DRUGS: A NOVEL ANTICANCER STRATEGY

Enrico P. Spugnini, DVM, Ph D, DACVIM (Oncology), DECVIM-CA (Oncology)
Rome, Italy

INTRODUCTION

Tumor cells rely on H^+ exchangers to relieve themselves from the dangerous protons byproduct of cancer metabolism that could trigger a cascade of lytic enzymes that ultimately would lead to self-digestion. Among these mechanisms, one of the most investigated are the vacuolar H^+-ATPases (V-ATPases).[1-4] V-ATPases are ATP dependant H^+ transporters that utilize the energy freed by the hydrolysis of ATP with the active transport of protons from the cytoplasm to the lumen if intracellular compartments or, if located within the cytoplasmic membrane, the extracellular compartment. Structurally speaking, the V-ATPases are composed of a peripheral domain (V_1) that carries out ATP hydrolysis and an integral domain (V_0) responsible for exchanging protons.[5] The regulatory mechanisms of V-ATPases are neutralized in malignant tumor cells that overexpress lysosomal proteins on the cell surface.[6,7] The acidic tumor environment is a consequence of anaerobic glucose metabolism with secondary production of lactates byproducts through the upregulation of hypoxia-inducible factor 1α or can be due to inadequate tumor perfusion, tumor hypoxia or enhanced transmembrane pH regulation.[8,9] These pumps, coupled with other ion exchangers, play a crucial role in the establishment and maintenance of malignant tumor environment and promote the selection of more aggressive cell phenotypes able to survive in this highly selective environment.

ROLE OF V-ATPASES IN TUMOR SPREAD

V-ATPases play a critical role in the maintenance of an appropriate relatively neutral intracellular pH, an acidic luminal pH, and an acidic extracellular pH by actively pumping protons either through ion exchange mechanisms or by segregating H^+ within cytoplasmic organelles that are subsequently expelled.[10] It is hypotesized that the low extracellular pH of tumors might trigger proteases, leading to the dissolution of extracellular matrix. This phenomenon, as is well known, significantly contributes to tumor invasion and dissemination.[11-12] In fact, tumor invasion depends on tumor acidifying ability that ultimately leads to secretion and activation of several classes of proteases. [13-14] The most commonly triggered enzymes are MMP-2, MMP-9, cathepsin B, and cathepsin L. Low pH can also up-regulate or activate proangiogenic factors such as VEGF-A and IL-8.[15-16]

ROLE OF V-ATPASES IN CHEMORESISTANCE

The acidic microenvironment caused by changes in the pH gradient between the intracellular and the extracellular compartments as well as the pH gradient between the cytoplasm and the intracellular organelles can be significantly involved in the mechanism of drug resistance.[17,18] There are several mechanisms involved in this phenomen, including decreased uptake or neutralization of weakly basic drugs by the acidic tumor microenvironment or the sequestration of chemotherapy drugs within lysosomal vescicles.[18,19] An accelerated turn over of acidic vesicles may represent an additional tumor strategy of drug resistance based on counteracting current transportation. Several investigators developed new approaches to better characterize tumor pH in animal models mostly through imaging systems in order to identify novel targets.

RESULTS IN PRECLINICAL STUDIES

Several studies in murine models have shown that the neutralization of the H^+ transporters could result in increased tumor control and extended survival. A recent study showed that blocking V-ATPases thorugh RNA interference led to prevention of cancer metastases in a murine model. As a result, new approaches have been developed to modulate drug efficacy within the low pH tumor milieu including the use of RNA interference, bicarbonates or the induction of metabolic alcalosis.[20,21] Finally, two recently published articles describe the chemosensitizing action of proton pump inhibitors in a murine model of orthotopic cutaneous melanoma, a well known chemo-refractory neoplasm, opening a novel field of investigation.[22,23]

CURRENT CLINICAL APPLICATIONS IN HUMANS AND PETS

Several phase II trials are ongoing in humans with advanced oral carcinoma, lymphoma and melanoma. The preliminary results appear very promising. A large phase II trial is currently being conducted to identify the best drug(s), their dose and schedule in dogs and cats with advanced tumors

FUTURE DIRECTIONS

Novel vacuolar ATP-ase inhibitors are being made and evaluated in cancer patients, in combination with different chemotherapy agents in order to maximize tumor control. Finally, besides being a potential target of anticancer drugs, it is conceivable that V-ATPases might become a predictive factor of tumor behavior and final outcome through the immunohistochemical evaluation of their expression and cellular distribution in tumor biopsies

REFERENCES

[1]Spugnini et al. **J Exp Clin Cancer Res** 2010; 29: 44. [2]Forgac. **Nat Rev Mol Cell Biol** 2007 ; 8: 917. [3]Cipriano et al. **Biochim Biophys Acta** 2008; 1777: 599. [4]Jefferies et al. **Arch Biochem Biophys** 2008; 476: 33. [5]Arai et al. **J Biol Chem** 1988; 263: 8796. [6]Saitoh et al. **J Biol Chem** 1992; 267:5700. [7]Glunde et al. **Neoplasia** 2003; 5: 533. [8]Gatenby et al. **Nat Rev Cancer** 2004; 4: 891. [9]Fais et al. **Cancer Res** 2007; 67: 10627. [10]Nishi et al. **Nat Rev Mol Cell Biol** 2002; 3: 94. [11]Martinez-Zaguilan et al. **Am J Physiol** 1993; 265: 1015. [12]Martínez-Zaguilán et al. **Clin Exp Metastasis** 1996; 14: 176. [13]Razaq et al. **Eur Spine J** 2003 ; 12: 341. [14]Webb et al. **Novartis Found Symp** 2001; 240:169. [15]Koukourakis et al. **Br J Cancer** 2003; 89: 877. [16]Rofstad et al. **Cancer Res** 2006; 66: 6699. [17]Mahoney et al. **Biochem Pharmacol** 2003; 66: 1207. [18]Simon et al. **Proc Nat Acad Sci USA** 1993; 91: 1128. [19]Raghunand et al. **Biochem Pharmacol** 1999; 57: 1047. [20]Raghunand et al. **Biochem Pharmacol** 1999; 57: 309. [21]Raghunand N. **Methods Mol Med** 2006; 124: 347. [22]Luciani et al. **J Natl Cancer Inst** 2004; 96:1702. [23]De Milito et al. **Int J Cancer** 2010; 127: 207.

DOES DNA METHYLATION PLAY A ROLE IN CANINE LYMPHOMA?

Jeffrey N. Bryan, DVM, MS, PhD, DACVIM (Oncology)
Pullman, WA

INTRODUCTION

Epigenetic alterations are modifications of chromatin that alter gene expression. These alterations include methylation of DNA and histones as well as acetylation of histones.[1,2] Enzymes add or remove methyl or acetyl groups from specific DNA bases or amino acid residues of histones. During this process, no change occurs in the sequence of the DNA. Thus, the code of the genes is unaltered, but expression of the gene product is modulated. The best studied form of DNA methylation is the addition of a methyl group at the 5th carbon position of cytosine in a C followed by G (CpG) dinucleotide.[3,4] This is accomplished by DNA methyltransferases (DNMT) using S-adenosyl methionine (SAMe) as the methyl donor. The CpG dinucleotide is statistically rare in the genome. CpG dinucleotides which reside outside of CpG islands tend to be methylated. The presence of methylcytosine identifies DNA as being of self rather than pathogen origin, denotes the parent strand of DNA during replication for purposes of base mismatch repair, and contributes to the overall stability of DNA. Loss of normal levels of methylation (genomic hypomethylation) is one of the earliest molecular changes of cancer.[5] There are regions of DNA that are highly GC-rich with a higher than expected abundance of CpG dinucleotides, termed CpG islands. These islands are defined as regions of DNA spanning at least 500 bases with a G + C content > 55% and an observed/expected ratio of CG > 0.6.[6] These typically reside at the 5' end of genes and include promoter and first exon regions.[6] When the cytosines in these CpG islands are largely unmethylated, an open chromatin pattern results and transcription can proceed.[1] However, when the cytosines are largely methylated (hypermethylation), chromatin is more compact, and transcription factors, such as SP-1, Thyrotropin Receptor, and Nuclear Factor kappa B, are inhibited from binding, potentially resulting in the silencing of the gene.[1] This occurs normally in imprinting of genes of the X chromosome, which must be expressed monoallelically.[7] Silencing of tumor suppressor genes has been identified in neoplastic cells, apparently contributing to their cancerous phenotype.[1] A number of individual genes have been reported silenced in non-Hodgkin's Lymphoma (NHL) with known roles in a variety of cellular processes including cell cycle progression, signal transduction, apoptosis, and gene regulation, suggesting that CpG island methylation is important in the genesis and clinical behavior of these malignancies.[8,9,10] DNA methylation marks are stable and heritable from one cell generation to the next and across generations of animals.[7,11]

The *DLC1* gene possesses tumor suppressor function.[12,13] The protein is a Rho-GTPase Activating Protein (RhoGAP) that counteracts the signaling of RhoA and Cdc42 Ras signaling proteins.[14] Loss of this function causes unconstrained growth signaling from the surface of the cell to the nucleus, alters cell mobility through increased ROCK-mediated events, and extracellular signaling.[12,14-16] The tumor suppressor function has been confirmed in a mouse model.[13] Transfection of *DLC1* reverses the malignant behavior of non-small cell lung carcinoma (NSCLC) cells and breast carcinoma cells *in vitro* and *in vivo*.[17-19]

The human DLC1 protein contains three functional domains: a sterile α motif (SAM), a RhoGAP, and a steroidogenic acute regulatory-related lipid transfer (START) domain.[20] The START and RhoGAP domains are required for tumor-suppressor function of the protein.[21] *DLC1* silencing has been demonstrated to be a significant contributor to many human cancers. This gene and its protein have only recently been characterized in the dog in our laboratory.

The presence of hypermethylation in the promoter region of the *DLC1* gene has been demonstrated in humans with NHL.[16,22] The *DLC1* gene has been found to be hypermethylated in six NHL cell lines, with concomitant silencing of transcription, reversible by treatment with a combination of a demethylating agent and a histone deacetylase inhibitor.[22] Seventy-five human NHL patient samples were examined for hypermethylation of several candidate genes, including *DLC1*. Of these, 87% demonstrated hypermethylation of *DLC1*.[22] Overall, expression of mRNA for this gene was significantly downregulated in tumor tissue compared to normal tissue.[22] Only recently have epigenetic mechanisms begun to be examined in dogs.[23] Initial work by our group has identified strong evidence that the tumor suppressor gene *DLC1* is frequently hypermethylated in canine NHL, as it is in human NHL.[22,24]

EXPERIMENTAL DESIGN

A region of the canine genome corresponding to the sequence surrounding the promoter region of the human *DLC1* gene (NM_006094) was identified as previously described.[24] A list of transcription factor motifs were compared with the canine and human *DLC1* DNA sequences and the start and stop sites were recorded for all perfect matches. The sequence of canine *DLC1* was translated with an online Open Reading Frame Finder (http://www.ncbi.nlm.nih.gov/gorf/gorf.html). The sequences were compared to the human *DLC1* sequence in Gen-Bank using ClustalW (http://www.ebi.ac.uk/clustalw/index.html).[25] Protein functional group analysis was performed using http://www.ebi.ac.uk/Tools/InterProScan/.[26] Lymphoma samples were collected from dogs presenting to the University of Missouri-Columbia Veterinary Medical Teaching Hospital. The normal lymphocytes used for this study were from dogs used for a concurrent protocol[27], blood samples of volunteer dogs, or node samples harvested after euthanasia of donated animals without lymphoma. Peripheral blood mononuclear cells were isolated over a ficoll-hypaque (Sigma-Aldrich, St. Louis, MO) gradient. All available samples were reviewed by a pathologist to assign a Working Formulation diagnosis. DNA and RNA were extracted from each sample. Aliquots of each DNA sample were treated with bisulfite for analysis. The Ambion FirstChoice RLM-RACE (rapid amplification of cDNA ends) (Ambion, Inc. Austin Texas) kit was used to isolate mRNA and perform 5' and 3' RACE. The MethPrimer website (http://www.urogene.org/methprimer/ index1/html) was used design primers for combined bisulfite restriction analysis (COBRA) and methylation-specific PCR (MSP).[28] Results were analyzed with a gel documentation and analysis system (Kodak, Rochester, NY). Proportion of methylation for COBRA samples was calculated as the intensity of the cut bands divided by the total intensity of all bands in the lane. For MSP, a ratio of the intensity of the methylated band to the unmethylated band was calculated. Bisulfite sequencing was performed using five clones of each purified plasmid DNA using a standard dideoxynucleotide sequencing technique. CpG dinucleotides within the sequences were mapped and labeled as methylated if cytosine remained unconverted, and unmethylated if a thymidine was in a cytosine position. Expression analysis was performed using primers that reside in two adjacent exons, spanning an intron, near the 3' end of the mRNA. Assays were run in triplicate. Data were evaluated using the $\Delta\Delta Ct$ method. The relationship between methylation and expression and methylation and age were evaluated using a linear correlation model. Methylation density was compared to expression of the mRNA relative to normal lymphoid tissue with a Mann-Whitney rank sum test. Association between methylation at each restriction enzyme site or MSP site and the neoplastic phenotype or lymphoma subtype result was evaluated using the Fisher's Exact Test. Differences in survival were calculated for dogs which did or did not have methylation at each restriction site and MSP analysis using the Kaplan Meier Log-Rank Analysis. Significance was considered to be $P<0.05$.[29]

RESULTS

Canine *DLC1* cDNA (GenBank FJ602870) sequence shares 88% identity with the human sequence. The canine and human peptide sequences share 89% identity. All of the major human exons are present in the canine sequence. Transcription factor binding sites are highly similar between the species, particularly Sp1 sites. The COBRA demonstrated that the gene was methylated in seven of 21 cases at the BstuI cut-site (0 of 7 normal) (P=0.141), 19 of 21 cases at the TaqaI cut-site (2 of 7 normal) (P=0.009), and in every case and four of seven normal samples at the HpyCh4IV cut-site (P=1.00). MSP demonstrated 17 of 21 cases and no normal samples to be positive. Bisulfite sequencing showed significantly more methylation in the lymphoma samples compared to normal samples, with little methylation present in the 3' end of the promoter region. A non-significant positive relationship between age and methylation was identified. Methylation did not result in silencing of the gene in the dogs in this series.[29] Working Formulation classification was not associated with any particular pattern of methylation in this gene.

DISCUSSION

Both human and canine sequences contain a large CpG island, the same number of exons, and protein functional groups. The similarities in the canine and human sequences and in transcription factor sites suggest highly common functions. The COBRA and MSP results demonstrate a pattern of greater density of methylation edges of the promoter with lower density of methylation near the core region where, presumably, transcription is controlled. This is supported by the bisulfite sequencing data. In human cancers, methylation of genes has been identified that was not associated with silencing. A hypomethylated region has been identified around the transcription start site (TSS) of the *hTERT* gene that permits transcription in spite of heavy methylation upstream.[30] Two of these cases were T cell lymphomas and demonstrated methylation, a phenomenon identified in human T cell lymphoproliferative disease.[31] Genomic hypomethylation and accumulation of CpG island hypermethylation occurs with increasing age in humans, but has not been demonstrated in dogs.[32] In this series of lymphoma samples, there was not a statistically significant association between age and the degree of methylation. There was also no association between the classification of lymphoma and patterns of methylation, something that has been observed in humans. It is significant that this gene is aberrantly methylated in canine lymphoma. Our lab is currently evaluating a microarray tool for evaluating whole genome CpG island methylation to elucidate the role of methylation in development of canine non-Hodgkin lymphoma.

CLINICAL IMPLICATIONS OF DNA METHYLATION

Evaluation of the DNA methylation changes associated with NHL has allowed the subclassification of disease, the identification of minimal residual disease, and differentiation from normal and hyperplastic lymphoid tissue.[33-35] Methylated DNA in blood has shown promise for tumor identification in lung and other cancers.[36,37] The marks present in lymphoma of dogs could be useful in the ongoing effort to subtype these tumors with clinical significance. The stability of DNA in circulation could also offer readily accessible biomarkers that could be inexpensively identified.[38] Drug therapies that alter DNA methylation have been shown to be beneficial in hemolymphatic diseases in mouse models and humans.[39,40] Further elucidation of the alteration of DNA methylation in canine lymphoma will be necessary to fully exploit these diagnostic and therapeutic strategies.

REFERENCES

1. Herman JG, Baylin SB. N Engl J Med 2003;349:2042-2054. **2**. Bachman KE, et al. Cancer Cell 2003;3:89-95. **3**. Robertson KD, Jones PA. Carcinogenesis 2000;21:461-467. **4**. Singal R, Ginder GD. Blood 1999;93:4059-4070. **5**. Feinberg AP, et al. Nat Rev Genet 2006;7:21-33. **6**. Takai D, Jones PA. Proc Natl Acad Sci U S A 2002;99:3740-3745. **7**. Jones PA, Baylin SB. Nat Rev Genet 2002;3:415-428. **8**. Katzenellenbogen RA, et al. Blood 1999;93:4347-4353. **9**. Kawano S, et al. Blood 1999;94:1113-1120. **10**. Baur AS, et al. Blood 1999;94:1773-1781. **11**. Guerrero-Bosagna C, et al. PLoS ONE 2010;5:e13100. **12**. Yuan BZ, et al. Cancer Res 1998;58:2196-2199. **13**. Xue W, et al. Genes Dev 2008;22:1439-1444. **14**. Wong CM, et al. Cancer Res 2003;63:7646-7651. **15**. Sahai E, Marshall CJ. Nat Rev Cancer 2002;2:133-142. **16**. Ullmannova-Benson V, et al. Leukemia 2009;23:383-390. **17**. Yuan BZ, et al. Oncogene 2004;23:1405-1411. **18**. Goodison S, et al. Cancer Res 2005;65:6042-6053. **19**. Wu PP, et al. Ann Clin Lab Sci 2009;39:263-269. **20**. Liao YC, et al. J Cell Biol 2007;176:43-49. **21**. Wong CM, et al. Cancer Res 2005;65:8861-8868. **22**. Shi H, et al. Carcinogenesis 2006;28:60-70. **23**. Pelham JT, et al. Res Vet Sci 2003;74:101-104. **24**. Bryan JN, et al. Cancer Ther. 2008;6:137-148. **25**. Larkin MA, et al. Bioinformatics 2007;23:2947-2948. **26**. Zdobnov EM, Apweiler R. Bioinformatics 2001;17:847-848. **27**. McCaw DL, et al. Clin Cancer Res 2007;13:2496-2503. **28**. Li LC, Dahiya R. Bioinformatics 2002;18:1427-1431. **29**. Bryan JN, et al. BMC Genet 2009;10:73. **30**. Zinn RL, et al. Cancer Res 2007;67:194-201. **31**. Taylor KH, et al. Cancer Res 2007;67:2617-2625. **32**. Fraga MF, et al. Ann N Y Acad Sci 2007;1100:60-74. **33**. Hagiwara K, et al. Leuk Res 2010;34:50-54. **34**. Bennett LB, et al. Genes Chromosomes Cancer 2009;48:828-841. **35**. Rahmatpanah FB, et al. Leukemia 2006;20:1855-1862. **36**. Anglim PP, et al. Mol Cancer 2008;7:81.:81. **37**. Anglim PP, et al. Mol Cancer 2008;7:62.:62. **38**. Schaefer DM, et al. Vet Comp Oncol 2007;5:145-155. **39**. Issa JP, et al. J Clin Oncol 2005;23:3948-3956. **40**. Appleton K, et al. J Clin Oncol 2007;25:4603-4609.

ELECTROCHEMOTHERAPY: PROTOCOL CONSIDERATIONS & RESEARCH SUMMARY

Enrico P Spugnini, DVM, Ph. D., DACVIM (Oncology), DECVIM-CA (Oncology)
Rome, Italy

INTRODUCTION

Chemotherapy drugs are effective provided that they reach their site of action. The primary obstacle for chemotherapy agents is the cytoplasmic membrane, especially for lipophobic agents such as bleomycin. Electrochemotherapy (ECT) couples the administration of an anticancer agent (usually bleomycin or cisplatin), to the delivery of electric pulses having appropriate waveforms. The application of permeabilizing pulses leads to perturbation of the cell membrane, thus resulting in increased uptake of chemotherapeutic drugs, ultimately leading to cell apoptotic death.[1] Clinical trials evaluating humans treated with ECT for cutaneous neoplasms were first reported in the early 90s, and have been followed by many trials that confirm its potential for the treatment or palliation of dermatological tumors.[2-4] Recently, a standard operating procedure has been proposed.[5]

ELECTROCHEMOTHERAPY EQUIPMENTS

Different waveforms have been adopted by investigators: exponentially decaying, square, rectangular or biphasic. In general the number of pulses delivered is set in 8 single pulses applied per cm of tumor area at a voltage of 1300 V/cm (800 V/cm for intraoperative use), with a duration of 100 μs and a frequency of 1 Hz. Treatments are repeated until the whole tumor area is covered.

RESULTS IN VETERINARY MEDICINE USING SQUARE PULSES BASED ECT

The first clinical trial in veterinary patients involved the treatment of feline sarcomas recurring after radiation therapy using bleomycin and Interleukin 2 producing cells coupled with square pulses.[6] The authors did not report any clinical response but claimed prolonged control for cats undergoing ECT compared to untreated patients. More recently, CDDP or bleomycin based ECT has been reported to treat cutanous neoplasms in dogs and cats as well as sarcoids in horses.[7-9]

EVALUATION OF BIPHASIC PULSES BASED ECT IN PETS

Experimental design

Our group investigated in the past 8 years the feasibility of electrochemotherapy in companion animals affected by different neoplasms. We adopted trains of biphasic pulses delivered as a burst instead of single impulses. The electric waveform were generated by a *Chemopulse* equipment. This device is built up by a toroidal core transformer generating a roughly rectangular pulse which is split in two halves that are sequentially driven to obtain a biphasic pulse. Pulses are not singularly produced but are created in bursts. The equipment allows choosing electric field intensity, number of pulses and pulse duration. The standard wave train used for this experiment was set to 8 pulses of 50 + 50 μs (biphasic pulse), and the electric field intensity was 1250 V/cm. The pulse repetition frequency is 1 Hz while the frequency of burst repetition is 1 kHz, resulting in a total burst duration of 7.1 ms. Bleomycin was locally injected at the tumor site and in the tumor's surrounding tissues. Five minutes after the injection, trains of biphasic pulses were administered through caliper electrodes as above described. Adherence of the caliper electrodes to the lesions was maximized using an electrophoresis gel. The treatment was repeated after 1 week.

RESULTS

In the first trial, ECT was used to directly attack neoplasms.[10] Overall response rate was 80%, with 40% long lasting remissions (in excess of 1 year). This phase I/II study led to the development of several electrodes specific for the different body districts.[11] On the basis of the preliminary studies, cohorts of dogs and cats affected by melanoma, soft tissue sarcoma mast cell tumor and nasal squamous cell carcinoma were enrolled in phase II studies.

Melanoma

Ten patients with malignant melanoma (MM) of the oral cavity were enrolled in a phase II study. The overall response rate was 80% (median time to recurrence 6 months), with a 50% of patients in remission after 1 year.[12] Of interest, most of the long term responders experienced a vitiligo like discoloration at the treatment site, potentially suggestive of recruitment of the immune system after uncovering of deep antigens.

Soft tissue sarcoma

A total of 72 cats affected by soft tissue sarcoma were assigned to three different groups: 1) surgery alone, 2) surgery coupled with intraoperative ECT, 3) surgery coupled with post operative ECT.[13] The median control times were respectively 4, 12 and 19 months. In this study were identified some prognostic factors such as previous treatment and tumor size. A cat with a recurring hemangiopericytoma was also referred at the time of the study, and is still in remission 2.5 years after the completion of the ECT treatment.[14] The only side effects were tumor dehiscence in 2 patients and focal inflammation in 2 patients. Systemic side effects were not reported among the enrolled patients.

In another study, adjuvant ECT was adopted to increase local control in a group of 22 dogs, obtaining a mean time to recurrence of 730 days, with 50% of the dogs still disease free at the time of writing.[15] In one dog with a large, unresectable sarcoma, ECT has been used to decrease tumor size and allow surgical excision.[16]

Mast cell tumor

Adjuvant bleomycin-based ECT was used in 28 dogs with incompletely excised mast cell tumors. Overall response rate was 85%. The median time to recurrence was not reached. Estimated time to recurrence was 52.76±6.5 months.[17] One dog had wound dehiscence. Another study, enrolling 37 dogs, evaluated the efficacy of cisplatin for the treatment of incompletely excised canine mast cell tumors obtaining a mean disease-free interval of 1218 days with 78% of the patients experiencing local control at different times.[18]

Squamous cell carcinoma

A small cohort of 9 cats with sun induced squamous cell carcinoma of the nasal planum were treated with intralesional bleomycin coupled with ECT with a rate of complete responses of 77%.[19]

CURRENT PROTOCOLS

Electrochemotherapy is mostly used as a adjuvant therapy, locally injecting the chemotherapy drugs bleomycin or cisplatin and then following with the application of permeabilizing pulses. For larger tumors drugs can be administered locally[16] or systemically[20], mostly on the basis of tumor volume, firmness or resistance to insertional electrodes. Caution should be exerted when using electrochemotherapy in patients previously treated with radiation therapy due to the risk of radiation recall.[21]

CONCLUSION

Electrochemotherapy is a safe and efficacious approach to veterinary soft tissue and cutaneous neoplasms. Its low cost and ease of administration make it a valuable addition to the currently available oncological therapies. One advantage of this technique is the possibility of repeated treatments in case of local recurrence and the recruitment of the patients' immune system ,as suggested by the low incidence of metastatic disease among our treated patients and by the tumor selection observed among some pets with recurring disease.

FUTURE DIRECTIONS

Ongoing studies are evaluating the efficacy of ECT when other drugs than bleomycin are used, in particular anthracyclines and their analogues (mitoxantrone) and platinum compounds such as CDDP and carboplatin. Studies are being performed to characterize the membrane alterations induced by trains of biphasic pulses in order to improve the efficacy of the technique.[22] Novel electrochemotherapy equipments adopting different electrical parameters are currently being developed. Finally, our group is assessing the potentials for electrogenetherapy using electric pulses to drive chemosensitizing plasmids within tumor cells.

REFERENCES

[1]Gothelf et al. **Cancer Treat Rev** 2003; 29: 371. [2]Belehradek et a. **Cancer** 1993; 72: 3694. [3]Daskalov et al.**IEEE Eng Med Biol** 1999 ; 18 : 62. [4]Gehl et al. **Melanoma Res** 2000; 10: 585. [5]Mir et al. **Eur J Cancer** 2006 ; 4(S) : 14. [6]Mir et al. **Br J Cancer** 1997 ; 76 : 16. [7]Tozon et al. **Anticancer Res** 2001; 21: 2483. [8]Rols et al. **Bioelectrochem** 2002 ; 55 : 101. [9]Tozon et al. **Anticancer Res** 2005 ; 25 : 839. [10]Spugnini et al. **J Exp Clin Cancer Res** 2003 ; 22 : 571. [11]Spugnini et al. **J Exp Clin Cancer Res** 2005; 24 : 245. [12]Spugnini et al. **Melanoma Res** 2006; 16 : 23. [13]Spugnini et al. **Cancer Chemoter Pharmacol** Epub 2006; 2007; 59 : 375. [14]Spugnini et al. **Vet Rec** 2006; 159 : 598. [15]Spugnini et al. **In Vivo** 2007; 21: 819.[16]Spugnini et al. **Vet Rec** 2008; 162: 186. [17] Spugnini et al. **Anticancer Res** 2006; 26: 4585.[18] Spugnini et al. **J Vet Intern Med** In Press.[19] Spugnini et al. **Vet J** 2009; 179: 117.[20] Spugnini et al. **In Vivo** 2008; 22: 51. [21] Spugnini et al. **In Vivo** 2008; 22: 751. [22] Spugnini et al. **Microsc Res Tech** 2007; 70: 1041.

ELECTRONIC BRACHYTHERAPY IN VETERINARY MEDICINE

Dominic J. Marino, DVM, DACVS, DACCT, CCRP
Plainview, NY

INTRODUCTION

The word brachytherapy is derived from the Greek word "brachys" meaning short distance and refers to the therapeutic use of radiation within or close to a tumor. The use of brachytherapy is not a recent development, on the contrary, it's place in the history of radiation therapy is well documented.[1] Recognized as one of the greatest discoveries in the field of medicine, Henri Becquerel first discovered natural radioactivity in 1896 and two years later, Marie and Pierre Curie, working in Becquerel's laboratory, identified the radioactive substances Polonium and Radium. Their work ushered in the age of radiation medicine for which they received the Nobel Prize for Physics in 1903. Soon after, both Pierre Curie and Alexander Graham Bell, working independently, proposed radium be inserted into a tumor within a glass tube revealing that radiation caused cancers to shrink and thus launching the age of brachytherapy.[1] \Brachytherapy originally began with the treatment of head and neck and gynecologic cancers utilizing interstitial and surface applications. Ward and Smith developed a radon seed gun in 1933, followed by a special gold seed gun developed at the Royal Marsden Hospital broadening the use of interstitial therapy. During the 1930's, the Holt Radium Institute in Manchester, England became the leading center for brachytherapy in the world pioneering a low dose treatment method. Afterloading techniques were developed in the 1960's based on the work of Ulrich Henchkein in the 1950's describing the placement of hollow tubes permitting manual and eventually remote afterloading of interstitial implants. Application techniques like remote afterloading significantly improved safety for both the patient and the operator resulting in the more widespread use of brachytherapy techniques.[1]

The advent of teletherapy for deeper tumors and the reports of complications associated with radiation exposure from high-energy radionuclides led to a decline in the use of brachytherapy towards the middle of last century. However, over the past decades, there has been renewed interest in the use of brachytherapy for a number of reasons. The development of more elegant low dose rate (LDR), medium dose rate (MDR), and high dose rate (HDR) techniques and equipment, the identification of various radionuclides exhibiting a greater diversity of radioactive properties, sophisticated imaging techniques and the computerized treatment plans have resulted in better dose distribution and improved sparing of normal tissues. The benefits of brachytherapy including decreased total treatment volume because of the close proximity of the radioactive source to the tumor volume and shortened treatment times were not realized until recently in veterinary medicine. Reports of using interstitial brachytherapy in veterinary medicine involve treating equine squamous cell carcinoma (SCC)[2-4] or sarcomas [5], and cattle [3]and can be found only as early as the 1970s.

PERMANENT SEED THERAPY

Additional benefits of seed implantation include the continuous low-dose rate delivery of radiation to the treatment volume. Radioisotopes historically used in brachytherapy therapy as permanent implants include: radon-222, iodine-125, gold-198, and as temporary implants: cesium-137, iridium-192, and radium-226. Permanent implantation involves using radioisotopes with shorter half lives, lower levels of radioactivity and is associated with lower exposure levels for both patients and medical personnel. Because they are generally available as seeds, they can be implanted in a wide variety of tumor shapes. Disadvantages of permanent implantation include imprecise dose distribution because of variations in seed application techniques. Additionally, once seeds are placed, corrections are difficult if not impossible. Since permanent implantation seeds continuously emit radiation, precautions must be taken to ensure the safety of those in close proximity of the patient for pre-determined periods of time after seed application. These restrictions are dependent on health department regulations regarding the specific radioisotope used. Potential liability exists if radioactive seeds are lost or the patient dies outside the control of the hospital to ensure proper seed disposal post mortem.

TEMPORARY SEED THERAPY

Temporary radioisotope implantation involves the use of radioisotopes with relatively longer half lives, but are removed when the time needed to reach the radiation dose to the treatment volume has been reached. Radioisotopes can be utilized as needles of fixed lengths and radioactivity or as "ribbons" with seeds spaced at predetermined intervals. Needle implantation has several disadvantages namely the difficulty applying rigid needles around curved structures, shielding requirements for inventory storage and a substantial financial commitment to maintain an adequate array of needles.[6] Using a "ribbon" method for temporary seed delivery has several advantages including more flexibility afforded the clinician in treatment planning. Ribbons can be cut and modified to be placed using an afterloading techniques into preplaced hollow tubes, thus minimizing exposure to the clinician and removed when the desired dose has been delivered.[7]. Klueter and coworkers described an iridium-192 HDR treatment protocol for intranasal tumors in dogs that was well tolerated and found to have results consistent with previous reports of alternative forms of radiation therapy.[8]

RECENT BRACHYTHERAPY INNOVATIONS

The development of a balloon catheter system in human medicine for the delivery of iodine-125 LDR brachytherapy (BCLDR) in cavities left after brain tumor removal began with studies designed in dogs.[9, 10] Existing methods like interstitial "seed' therapy involve placement of a stereotactic frame, multiple incisions, drilling of multiple portals, while utilizing teletherapy, doses necessary for tumor control result in unacceptable toxicity to surrounding normal tissue.[11]

BCLDR therapy involves placement of the inflatable balloon end of the catheter in the tumor cavity after tumor removal and the infusion port in a superficial subcutaneous location in the case of brain tumors or externally in the case of breast cancer patients. According to treatment planning guidelines, the iodine-125 is infused in the subcutaneous port allowing the balloon to serve as a volumetric LDR radiation source with exact conformance to the to the resection margin of the tumor bed.[12, 13] Preliminary reports in treating dog brain tumors with BCLDR therapy cited risk of seizures, infection, and development of new neurologic deficits as disadvantages of this form of therapy. Large post resection voids and substantially irregular defects may preclude the use of this technique because of balloon size after inflation. Additionally, dogs were found to have small amounts of detectable radioactivity in their urine. Although diffusion of minute quantities of radioactivity through the walls of the silicone balloon walls was expected, other mechanisms were suspected including: inadvertent administration during port infusion or needle leakage after infusion. Overt catheter or port failure was not suspected because of reliable manufacturing methods, however, facilities employing this techniques should be prepared to handle appropriate patient restrictions and handling of urine and other patient secretions in accordance with health department regulations in the event of a catheter failure.[10]

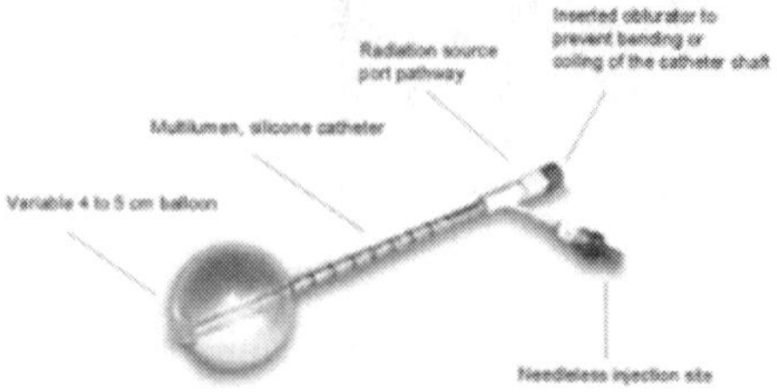

Figure 1. MammoSite® brachytherapy catheter with multiple lumen ports.

BCLDR therapy has been in clinical use in human medicine for the treatment of mammary cancer affording the same tissue sparing benefits in breast surgery patients post surgical lumpectomy. [14-18] The MammoSite® brachytherapy (MSB) system (Hologic, Marlborough, MA) applicator was developed to be more easily applied and better tolerated in breast therapy patients. The MSB consists of a silicone balloon connected to a double-lumen catheter (Figure 1).

The catheter has both a small inflation channel and a channel for the passage of an Ir-192 HDR brachytherapy source. The balloon is inflated with saline solution mixed with a small amount of contrast material to aid visualization. The balloon is inflated to a size that would completely fill the lumpectomy cavity and ensures conformance of the tissue to the balloon (Figure 2). An Ir-192 radioactive source, connected to a computer-controlled HDR remote after-loader, is inserted through the catheter into the balloon to deliver the prescription radiation dose.[14, 18] The MSB radiation therapy device generally delivers 34 Gy over 10 fractions (3.4 Gy per fraction, twice daily (BID)) with the prescription point at 1 cm from the balloon surface.[14, 18]

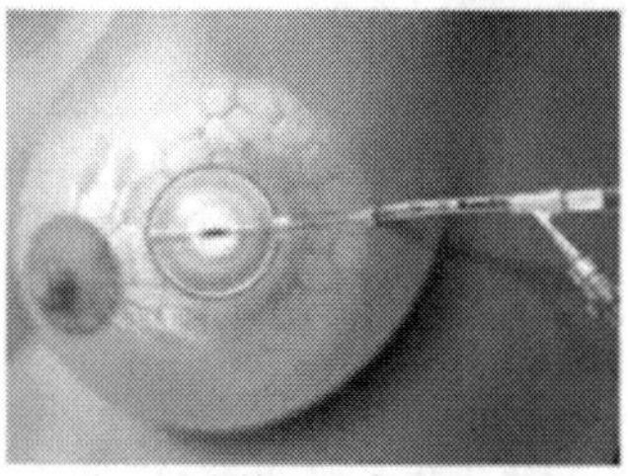
Figure 2. MammoSite® brachytherapy expanded to fill the lumpectomy cavity and ensures conformance of the tissue to the balloon.

ELECTRONIC BRACHYTHERAPY

An electronic brachytherapy (EB) system (Accent® , Xoft, Fremont, CA) is a modified form of balloon-based brachytherapy that it uses an electronic 50 kilo-voltage x-ray source rather than an iridium-192 HDR source. Xoft has developed a system utilizing X-rays to mimic the treatment parameters of iridium-192 and replicate the benefits of radiation therapy without the drawbacks of an iridium-192 HDR radioactivity source. The EB system consists of a controller, flexible X-Ray source, balloon applicators and miscellaneous accessories (Figure 3 &4). Treatment is delivered by a disposable, miniaturized X-Ray tube assembly called the Axxent® HDR X-Ray Source (Figure 5). The X-ray source consists of a water cooled miniaturized x-ray tube that is inserted into the balloon catheter or a closed end treatment cannula and delivers the X rays to the planning tumor volume (PTV) with an output approximating 1Gy/min at 1cm from the source. Using this intra-cavity approach, the Xoft EB device can potentially reduce the toxic radiation load to adjacent normal tissue while shortening total radiation therapy time.

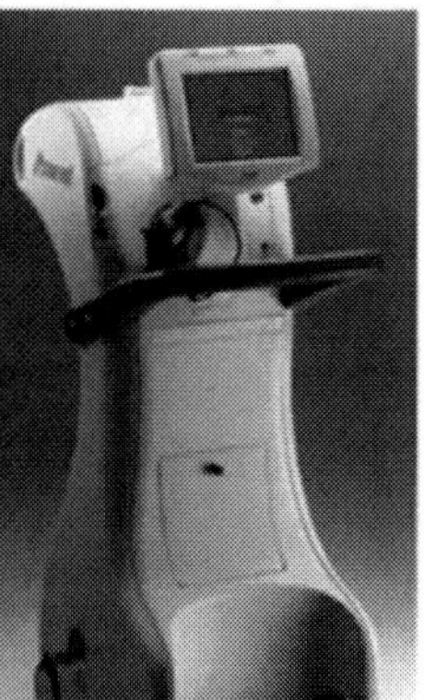
Figure 3. The Axxent® controller is a mobile unit, consisting of a digital touch -screen for the input of treatment data, self calibration and to monitor treatment progress.

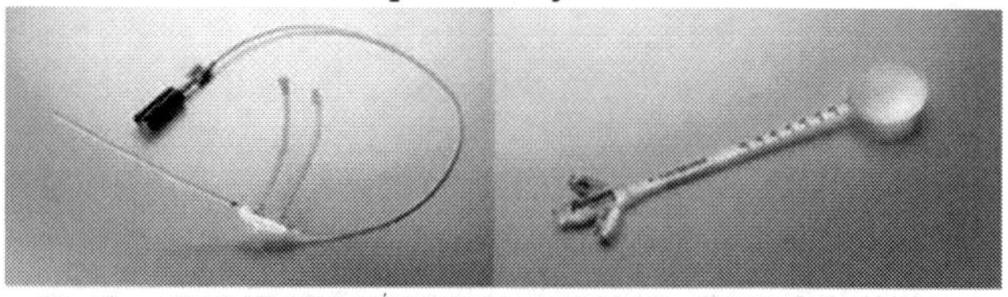
Figure 4. The Axxent® system utilizes a flexible wand attached to the X-Ray source that is placed within the channel of the balloon applicator.

The EB system is similar to MSB system, consisting of a balloon catheter that is intended to be inserted into the PTV by means of a percutaneous approach. The catheter for both systems presents similar limitations because of large size of balloon relative to veterinary patient size and cost and therefore is not regularly used by the authors when treating with the EB system.

A smaller diameter closed end plastic cannula prototype is under clinical investigation by the authors. By sacrificing the multiport system deemed unnecessary by the authors, a smaller treatment cannula has been found to be successful in preliminary tests (Figure 6). One inherent problem with the MSB technique is the high skin dose when treatment is near the skin surface that can result in late effect skin toxicity. The EB source exhibits pronounced anisotropy resulting in decreased dose at the proximal portion of the source.[16] In treatment plans near the skin, this property can be exploited by placing the proximal end of the source near the skin.

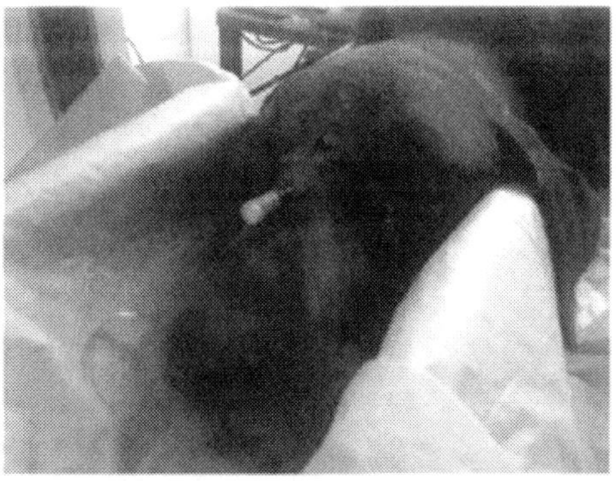

Figure 6. A prototype closed end plastic cannula with smaller diameter under clinical investigation by the authors placed through the maxillary recess and across the ethmoid bone to treat the resection bed of an olfactory meningioma. The smaller diameter was achieved by sacrificing the multiport system.

Another difference to be considered is the relative biologic effectiveness (RBE) related to the lower energy of the photons emitted by the EB versus the high energy gamma rays emitted by the MSB source. It is well established that the RBE of low-energy photons is large compared with high energy gamma rays, because of the dominance of photoelectric absorption at low energies.[19] The RBE for a 40 kVp source similar to the EB source discussed here has been calculated to be 1.28 greater than an 192-Ir source and should be considered when comparing the two.[19]

The EB controller is a mobile unit, consisting of a digital touch-screen for the input of treatment data, self calibration and to monitor treatment progress. Because the EB system does not require a shielded radiation room or an HDR afterloader unit, both of which are necessary for treatment using Ir-192 as a source, the cost savings are substantial. Because of the decreased costs and regulatory burden associated with the EB system, a wider application of this technology can be contemplated, including surface and intraoperative use.

REFERENCES

1. Mould RF, B.J., Martinez BL, Radium Brachytherapy: Historical Review Brachytherapy from radium to optimization, ed. M. RF. 1994, The Netherlands: Nucletron International. 2. Silver, I. and D. Cater, Radiotherapy and chemotherapy for domestic animals. Acta Radiol Ther Phys Biol, 1964. 2: p. 222-82. 3. Banks, W.C. and R.B. England, Radioactive gold in the treatment of ocular squamous cell carcinoma of cattle. J Am Vet Med Assoc, 1973. 163(7): p. 745-8. 4. Gavin, P. and E. Gillette, Interstitial Radiation Therapy of Equine Squamous Cell Carcinomas. Vet Rad, 1978. 19(4): p. 138-141. 5. Frauenfelder, H.C., W.E. Blevins, and E.H. Page, Rn for treatment of periocular fibrous connective tissue sarcomas in the horse. J Am Vet Med Assoc, 1982. 180(3): p. 310-2. 6. Turrell, J.M. and P.D. Koblik, Techniques of afterloading iridium-192 interstitial brachytherapy in veterinary medicine. Veterinary Radiology, 1983. 24(6): p. 278-283. 7. Northrup, N.C., et al., Iridium-192 interstitial brachytherapy as adjunctive treatment for canine cutaneous mast cell tumors. J Am Anim Hosp Assoc, 2004. 40(4): p. 309-15. 8. Klueter, S., et al., High-dose-rate brachytherapy for intranasal tumours in dogs: results of a pilot study. Vet Comp Oncol, 2006. 4(4): p. 218-31. 9. Dempsey, J.F., et al., Dosimetric properties of a novel brachytherapy balloon applicator for the treatment of malignant brain-tumor resection-cavity margins. Int J Radiat Oncol Biol Phys, 1998. 42(2): p. 421-9. 10. Stubbs, J.B., et al., Preclinical evaluation of a novel device for delivering brachytherapy to the margins of resected brain tumor cavities. J Neurosurg, 2002. 96(2): p. 335-43. 11. Salcman, M., et al., Technique and preliminary results of interstitial irradiation for primary brain tumors. J Neurooncol, 1986. 4(2): p. 141-9. 12. Tatter, S.B., et al., An inflatable balloon catheter and liquid 125I radiation source (GliaSite Radiation Therapy System) for treatment of recurrent malignant glioma: multicenter safety and feasibility trial. J Neurosurg, 2003. 99(2): p. 297-303. 13. Chan, T.A., et al., Treatment of recurrent glioblastoma multiforme with GliaSite brachytherapy. Int J Radiat Oncol Biol Phys, 2005. 62(4): p. 1133-9. 14. Dickler, A., et al., Treatment volume and dose optimization of MammoSite breast brachytherapy applicator. Int J Radiat Oncol Biol Phys, 2004. 59(2): p. 469-74. 15. Niehoff, P., et al., Clinical experience with the MammoSite radiation therapy system for brachytherapy of breast cancer: results from an international phase II trial. Radiother Oncol, 2006. 79(3): p. 316-20. 16. Shah, N.M., et al., MammoSite and interstitial brachytherapy for accelerated partial breast irradiation: factors that affect toxicity and cosmesis. Cancer, 2004. 101(4): p. 727-34. 17. Zannis, V., et al., Descriptions and outcomes of insertion techniques of a breast brachytherapy balloon catheter in 1403 patients enrolled in the American Society of Breast Surgeons MammoSite breast brachytherapy registry trial. Am J Surg, 2005. 190(4): p. 530-8. 18. Dickler, A., et al., The MammoSite breast brachytherapy applicator: a review of technique and outcomes. Brachytherapy, 2005. 4(2): p. 130-6. 19. Brenner DJ, C.-S.L., Beatty JF, Clinical relative biological effectiveness of low-energy x-rays emitted by miniature x-ray devices Physics in Medicine and Biology 1999. 44(2).

SERUM THYMIDINE KINASE ACTIVITY IN CANINE SPLENIC HEMANGIOSARCOMA.

DH Thamm,[1] DS Kamstock,[1] CR Sharp,[2] S Johnson,[3] E Mazzaferro,[4] LV Herold,[5] S Barnes,[6] K Winkler,[7] K Selting[8]. [1]Colorado State University, Fort Collins, CO., [2]Tufts University, North Grafton, MA., Emergency Animal Hospital of NW Austin, Austin, TX., [4]Wheat Ridge Veterinary Specialists, Wheat Ridge, CO., [5]DoveLewis Emergency Animal Hospital, Portland, OR., [6]Animal Emergency Critical Care, The LifeCentre, Leesburg, VA., [7]Georgia Veterinary Specialists, Atlanta, GA., [8]University of Missouri, Columbia, MO.

Thymidine kinase (TK) is a soluble biomarker present in S-phase of a salvage pathway for DNA synthesis, and can be measured in serum. TK activity correlates with stage, prognosis, and relapse in canine and human lymphoma. We previously reported the results of a pilot study evaluating TK activity in archived canine osteosarcoma, transitional cell carcinoma, and hemangiosarcoma (HSA) sera, and found elevated TK activity in 80% of canine HSA sera evaluated. The purpose of this study was to prospectively evaluate serum TK activity in a large number of dogs presenting to emergency clinics with hemoabdomen and a splenic mass, to determine if TK activity could be used as a noninvasive means to distinguish HSA versus benign conditions in this population.

Dogs presenting with hemoabdomen and a splenic mass identified on ultrasound examination were studied. Serum was collected prior to anesthesia, euthanasia or surgical intervention and frozen until batch analysis. Tissue from all patients was evaluated histologically by a single pathologist. Sera from age-matched normal dogs comprised a control population. An ELISA using azidothymidine as a TK1 substrate was used. Comparisons between groups were made using 2-tailed student T-tests, and receiver-operator characteristic (ROC) curves were generated..

Sixty-two patients and 39 normal controls were studied. There were 35 dogs with HSA, 10 dogs with other splenic neoplasia, and 17 dogs with benign diseases. Using a training set of 24 normal dogs, a cutoff of 6.55 U/L was established from the ROC curve. TK activity was significantly higher ($p < 0.0001$) in dogs with HSA than in the validation set of 15 normal dogs (mean+/-SD = 17.71+/-4.5 and 2.01+/-0.6 respectively), but not between dogs with HSA and benign splenic disease (mean+/-SD = 7.02+/-3.7, $p = 0.13$). Using a cutoff of 6.55 U/L, TK activity demonstrated a sensitivity of 0.54, specificity of 0.76, positive predictive value of 0.83 and negative predictive value of 0.45 for distinguishing HSA versus benign splenic disease. When interval thresholds of <1.55 and >7.95 U/L were used together, diagnostic utility was markedly increased for distinguishing both HSA versus normal and HSA versus benign disease.

In conclusion, serum TK evaluation may assist in detection of canine HSA, and may also discriminate between benign disease and HSA in dogs with hemoabdomen and a splenic mass.

GENE EXPRESSION PROFILING OF CANINE T CELL CHRONIC LYMPHOCYTIC LEUKEMIA. S. Lana[1], R Burnett[1], J. Modiano[2], A Avery[1]. 1Colorado State University, Fort Collins CO, 2University of Minnesota, Minneapolis, MN.

T cell chronic lymphocytic leukemia (CLL) is a heterogeneous disease that affects a number of dog breeds. CLL patients have variable disease outcomes. The objectives of this study were to use gene expression profiling of CD8 T cell leukemias with variable outcomes in order to identify markers that can be used in routine diagnostic tests to distinguish good from poor prognosis disease, and to identify potential targets for novel therapy.

Gene expression profiling of 12 CD8 T cell leukemias (7 good, 5 poor prognosis) was conducted. Samples from 6 normal dogs were also profiled. Several differentially expressed genes were found including CD9, CD94, and CD 25. These were selected for further study using flow cytometry to determine expression of protein on the cell surface. Seventy nine cases of CD8+ T cell leukemia were screened for CD 9 expression. Forty seven had associated outcome information. Based on analysis to date, CD9 expression as assessed by flow cytometry does not appear to provide prognostic information.

A monoclonal antibody to CD25 was recently made available. To date 33 patients with CD8 T cell leukemia have been profiled. CD25 is variably expressed on T cell leukemias compared to normal CD8 T cells.

CD25 is the receptor for interleukin 2. Cyclosporin, a commonly used immunosuppressive drug, inhibits IL-2 production, and has been used to treat a subset of T cell leukemias in people. Thus, the finding that CD25 is up regulated on T cell leukemias compared with normal T cells suggests a possible new therapeutic avenue.

INTERMITTENT SINGLE-AGENT DOXORUBICIN FOR TREATMENT OF CANINE B CELL LYMPHOMA. ML Higginbotham, DL McCaw, JC Nietfeld, MJ Wilkerson, KB Reeds, DB Burr. Kansas State University, Manhattan, KS.

The treatment protocol used for individual lymphoma cases is dependent upon overall patient health, toxicity risk, cost, owner time commitment and comfort of the administering veterinarian. The objective of this study was to prospectively evaluate intermittent single-agent doxorubicin for treatment of canine B cell lymphoma in an attempt to decrease travel, cost, and work loss for the owners.

Patients were entered with a diagnosis of B cell lymphoma on cytology or histopathology. Immunophenotype was confirmed by flow cytometry of a lymph node aspirate or immunohistochemistry. Lymphoma morphology was determined when histologic specimens were available. All patients were staged with CBC, biochemical profile, urinalysis, abdominal ultrasound and bone marrow aspirate. Patients were treated with $30mg/m^2$ or 1mg/kg of doxorubicin every 2 weeks for 3 treatments. When progressive disease occurred, patients were treated with an additional doxorubicin treatment as necessary. Patients were eligible to receive other treatments when non-responsive to doxorubicin.

Eighteen dogs were enrolled. The median number of doxorubicin doses was 4 (range 1-9). The median first remission time was 68.5 days (range 0-335 days), median doxorubicin remission time was 80.5 days (range 0-335 days), and the overall median survival time was 169.5 days (range 0-865 days). Two dogs are alive at 335 and 865 days.

Based on these findings, intermittent doxorubicin is an acceptable treatment protocol for canine B cell lymphoma.

SERUM ANTIBODY REACTIVITY PROFILING FOR CANINE LYMPHOMA DETECTION.

DH Thamm,[1] JB Legutki,[2] P Carrigan,[2] A Roesler,[2] SA Johnston.[2] 1. Colorado State University, Fort Collins, CO. 2. Arizona State University, Tempe, AZ.

Identifying effective disease-specific biomarkers can be challenging. The physical features of antibodies make them ideal biomarkers. Furthermore, antibodies may be elicited early in the ontogeny of different chronic diseases, including cancer. Autoantibodies have been identified for specific diseases, but as yet there is no systematic, simple method to identify informative antibodies. Recently, protein arrays have facilitated this process by immobilizing most of the proteins from a pathogen or human onto a glass slide. We have created arrays with 10,000 random sequence 20-mer peptides spotted in an addressable, machine-readable fashion. The random sequences allow an unbiased display of antibody binding; the 20-mer length provides many possible epitope positions per peptide and (potentially) allows for some amount of structural complexity. In addition, the arrays are inexpensive and can be used to assay large numbers of samples quickly. Lymphoma is one of the most common canine cancers, for which novel diagnostic and monitoring methods would be useful. As proof-of-concept, we sought to determine whether patterns of antibody reactivity could be used to distinguish canine lymphoma from clinically normal dogs.

The CIM10K Array consists of 10,000 random peptides spotted through a maleimide linkage onto activated commercially produced aminosilane slides. Each peptide sequence is derived from a random number generator that places any of 19 amino acids (excluding cysteine) in 17 positions, and no peptides demonstrate homology with known mammalian proteins. The serum samples were diluted 1:500 in aqueous buffer and applied to the array. Bound antibody was detected using a biotinylated rabbit anti-canine IgG followed by fluorescently labeled streptavidin. The array was then washed, dried and then scanned in a conventional microarray laser scanner. Differentially recognized peptides were identified using ANOVA, followed by refinement of the discriminatory capability using classification algorithms including distance based clustering and Principal Components Analysis.

We evaluated sera from 10 dogs with B cell lymphoma, 5 dogs with T cell lymphoma, and 10 age-matched healthy dogs. An ANOVA test identified 83 peptides capable of segregating lymphoma versus normal dogs. When this 83-peptide "subarray" was generated *in silico* and used to probe the same dataset, 12 of 15 lymphoma sera segregated together, and furthermore, B cell and T cell lymphomas were correctly discriminated in these same 12 samples.

In conclusion, these preliminary studies demonstrate that a simple and powerful antibody profiling technique is capable of discriminating between lymphoma and normal canine sera. Follow-on studies are validating this 83-peptide "subarray" on an independent set of canine sera, and future studies will address the utility of serum antibody reactivity profiling for monitoring disease status in lymphoma patients, as well as the diagnostic utility of this technique for other important canine cancers.

DIAGNOSIS AND MANAGEMENT OF ACUTE AND CHRONIC LEUKEMIAS IN DOGS

Joanne L Intile, MS, DVM, DACVIM (Oncology)
Rochester, NY

INTRODUCTION

The term leukemia refers to a group of diseases characterized by the clonal proliferation of neoplastic hematopoietic cells within bone marrow. Leukemias are classified as being either of myeloid or lymphoid derivation, which is determined based upon identifying the cell of origin. Myeloid leukemias are further classified into being of granulocytic, monocytic, erythroid, or megakaryocytic origin, whereas lymphoid leukemias are classified into being of T-cell, B-cell, or natural killer (NK) cell origin. Myeloid and lymphoid leukemias are also classified as being acute or chronic based on upon the degree of cell differentiation shown by the neoplastic cell population, and the typical clinical progression of disease.

Classification of leukemias in dogs historically was made based upon examination of morphological features of cells from peripheral blood smears and bone marrow aspirates, and cytochemical staining. Immunophenotyping by flow cytometry, which is considered the gold standard for classification of leukemias in people, is becoming more routine in veterinary patients. Ultimately, distinguishing the different forms of leukemia from each other requires combining the results of the patient's history, physical exam findings, and diagnostic imaging results, as well as results of clinicopathological data (CBC, cytology of tissue aspirates, bone marrow cytology, immunophenotyping, cytochemical staining, and presence or absence of clonal gene rearrangements.)

Studies of the clinical and hematological features of leukemia are lacking in veterinary medicine, with most information coming from research reports with small patient numbers or individual cases. Another important consideration is that many studies are outdated, and results must be interpreted carefully following the more recent recognition that the classification of leukemias using morphological features and/or cytochemical staining alone is exceedingly inaccurate.

ACUTE LEUKEMIAS

In acute leukemias, neoplastic cells are blasts, which are immature, undifferentiated cells that cannot be distinguished from each other based on morphological assessment alone. The term acute also implies duration of clinical signs (short) and prognosis (poor). Acute leukemias are further divided into being of lymphoid or myeloid origin.

Acute leukemias have typically been thought to affect younger dogs, however a more recent study indicated middle-aged dogs are most commonly affected. No consistent breed or gender predilection is recorded. Dogs with acute leukemia are generally ill at the time of presentation. The most common clinical signs reported are lethargy, vomiting, diarrhea, poor appetite, pyrexia, weight loss, and shifting leg lameness. Physical examination typically reveals hepatosplenomegaly and lymphadenopathy (if present) is usually mild. Hematology can be variable, ranging from a leukocytosis with circulating blasts to leukopenia with few blasts in circulation to complete pancytopenia. Cytopenias are thought to arise from infiltration of bone marrow with neoplastic cells that causes a depression of production of hematopoietic elements secondary to competition for space and nutrients (myelophthisis). Other factors that are thought to contribute to cytopenias include the production of inhibitory substances by the neoplastic cells, failure of trophic factor production by bone marrow stroma, and immune-mediated destruction. Cytopenias can also arise due to peripheral consumption resulting from infection or hemorrhage. Serum chemistry panel results are variable. Bone marrow cytology classically will show >20% blasts, however more typically 80-100% of marrow elements are replaced by neoplastic cells. Further staging tests such as thoracic radiographs and abdominal ultrasound are indicated to examine for extent of disease.

Acute myeloid leukemia (AML) is caused by the aberrant proliferation of myeloid progenitor cells within the bone marrow and is subclassified as being M0-M7 (Table 1). Other forms of myeloid neoplasia include myelodysplastic syndromes and chronic myeloproliferative diseases. Between 2-8% of people with AML will present with concurrent myeloid sarcoma, which is a solid tumor of myeloblasts and other granulocytic precursors occurring within an extramedullary site or bone, and these tumors are frequently found within the mediastinum. Interestingly, mediastinal masses are commonly observed in dogs with AML. There is no consistent recommended treatment protocol for AML in dogs. Combination protocols using cytosine arabinoside and doxorubicin with or without cyclophosphamide are most commonly utilized. Remissions are uncommon and median survival time ranges from 2-4 weeks.

Acute lymphoid leukemia (ALL) represents aberrant proliferation of lymphocytes within bone marrow and is classified into being of T-cell, B-cell (most common), or natural killer (NK) cell origin. Historically, AML is thought to occur more frequently than ALL, however there are conflicting results in the literature regarding which form of acute leukemia is most common. Treatment generally consists of CHOP based chemotherapy protocols as used for multicentric lymphoma. Complete remissions are achieved in 30-45% of cases, and median survival times range between 1-3 months.

CHRONIC LEUKEMIAS

Chronic leukemias are also classified into being of lymphoid or myeloid origin. In this form of leukemia, neoplastic cells arise from a clonal proliferation of morphologically appearing mature cells. Chronic also implies duration of clinical signs (long) and prognosis (good). The diagnosis is made upon the presence of persistent and unexplained leukocytosis where a reactive or cytokine-mediated source of hyperplasia has been ruled out.

Chronic lymphoid leukemia (CLL) is the most common form of chronic leukemia. CLL arises either from the bone marrow (myeloid, B-cell) or spleen (T-cell). CLL typically affects middle aged to older dogs, with a median age of about 10 years. No consistent breed or gender predilection is reported. CLL can often be an incidental diagnosis, with a mature lymphocytosis being detected on routine geriatric labwork. Clinical signs, if present, are again non-specific, and include weight loss, lethargy, reduced appetite, among many others. A mild generalized lymphadenopathy with or without concurrent hepatosplenomegaly may be seen. The most common laboratory finding is a leukocytosis, characterized by a mature lymphocytosis, which can vary from being just above the reference range to > 1,000,000 cells/uL. Anemia and thrombocytopenia are seen in 50% and 25-50% of dogs, respectively. Monoclonal gammopathies are seen in 2/3rds of dogs with CLL, as well as hemolytic anemia, pure red cell aplasia, and hypercalcemia. CLL generally follows an indolent clinical course in dogs, therefore the timing of when to initiate treatment is controversial. Many dogs can simply be monitored with monthly physical examinations and complete blood counts. Treatment with prednisone and chlorambucil is generally recommended for dogs showing clinical signs, significant organomegaly, cytopenias, and/or lymphocyte counts greater than 60,000 cells/uL. Uncomplicated cases (lymphocytosis without concurrent cytopenias and minimal or no other organ involvement) can have an excellent prognosis with treatment, with survival times exceeding 3 years. However dogs with cytopenias and significant organ involvement and/or poor clinical response to therapy, and dogs with a B-cell phenotype have a poorer prognosis. Additionally, there are reports of blast transformation; therefore CLL should be considered a relatively heterogenous disease with realistic expected survival times to be in the range of 1-3 years.

Chronic myeloid leukemias (CML) are exceedingly rare in dogs, and as such, not well characterized. Examples include polycythemia vera, chronic myelogenous leukemia, chronic neutrophilic, eosinophilic, or baophilic leukemia, and essential thrombocythemia. Clinical signs and exam findings are similar to dogs with CLL. Laboratory findings are poorly documented and vary depending upon subtype of CML and can include a leukocytosis with a left shift, anemia, thrombocytopenia or thrombocytosis as possible concurrent findings. There are no consistent recommended treatment protocols for dogs with CML. In general, the prognosis for dogs with CML is thought to be poorer than for dogs with CLL, with survival times ranging from 4 – 15 months. In people, CML is characterized by a specific chromosomal aberration known as the Philadelphia chromosome (translocation between the *c-abl* oncogene and the breakpoint cluster region (BCR)). The discovery of this mutation along with the development of tyrosine kinase inhibitor drugs that target the mutated *c-able* receptor has revolutionized the treatment of CML in people. There is a report of a homologous translocation to the Philadelphia chromosome in dogs with spontaneously occurring CML. As such, the use of tyrosine kinase inhibitors in the treatment of canine CML may prove promising.

DIAGNOSING ACUTE AND CHRONIC LEUKEMIAS

Achieving a confirmative diagnosis of leukemia is challenging. Although the neoplastic cells in chronic leukemias are morphologically mature and therefore more readily distinguishable, determining if increased numbers of such cells are present due to neoplastic proliferation versus a reactive condition to a distant source or inflammation or infection can be difficult. As stated above, in acute leukemias the neoplastic cells are immature blasts, and therefore morphology cannot be used to determine origin. Also, differentiating an acute leukemia from lymphoma is difficult, as clinical signs, physical exam findings, and clinicopathological test results can be identical. The following advanced diagnostic tests are recommended in cases where leukemia is suspected:

1) Immunophenotyping: Immunophenotyping involves the detection of cell-surface antigens using antibodies against lineage-specific markers. Each hematopoietic cell will express unique markers during differentiation and maturation. Antibodies can be generated against these markers that can be used to identify the lineage of a particular cell. Antibodies can be applied to fixed (acetone for cytology or formalin for histology) or fresh tissue (for flow cytometry) samples. In veterinary medicine, there are many more antibodies available for identifying lymphoid cells than for myeloid cells. Common antigens used for confirming a B-cell lineage are CD21, pax-5, and BLA-36. Common antigens for confirming T-cell lineage are CD3, CD4, and CD8. CD34 is generally considered a marker for hematoipoietic stem cells and therefore, can be useful for differentiating an acute leukemia (which arises within the bone marrow) from lymphoma (which arises in extramedullary sites). Neoplastic cells from dogs with acute leuekmias will generally be positive for CD34, where as neoplastic lymphocytes from dogs with lymphoma are typically negative.

2) Cytochemistry: Cytochemistry involves detecting the presence or absence of intracellular enzymes via the application of substrates that react with the enzyme to produce color changes that are detected using light microscopy. Cytochemistry is performed on cytology smears from blood, bone marrow, or tissue samples. As myeloid cells contain more enzymes than lymphoid cells, cytochemical stains are considered more valuable for identifying cells of myeloid origin, whereas lymphoid cells are typically negative for all cytochemical markers. Common cytochemical stains used in veterinary medicine include peroxidase for identifying granulocytes and monocytes, chloroacetate esterase for identifying granulocytes, non-specific esterase for identifying monocytes and megakaryocytes, and alkaline phosphatase for identifying monoblasts.

3) Clonality testing: Clonality testing attempts to differentiate reactive expansions of cell lines from neoplastic proliferations by examining gene rearrangements. Specifically, a polymerase chain reaction (PCR) assay is used to detect rearrangements in the B-cells receptor (surface Ig) or T-cell (TCR) genes. Individual B and T-lymphocytes have unique cell surface antigen-binding receptors that arise from rearrangement events between V, D, and J gene segments. In PCR, DNA is extracted from cells and gene sequences for the T-cell or B-cell receptors are amplified. If the population of lymphocytes is clonal, as seen with neoplastic conditions, a single receptor will be amplified, and results will show a single band on the PCR assay. If the population of lymphocytes is polyclonal, as seen in reactive conditions, multiple receptors will be amplified, and results will show multiple bands on the PCR assay. False positive single bands can be obtained in cases of infections diseases (e.g. Ehrlichia canis and Borrelia). Clonality is not useful for differentiating AML from ALL or an acute leukemia from a stage V lymphoma. PCR can be performed on fixed or fresh tissue samples.

Clinical cases will be presented as correlates

Table 1. Classification of subtypes of AML in Veterinary Medicine

AML Subtype	Type of Leukemia
AML M0	Acute myeloblastic leukemia with minimal differentiation
AML M1	Acute myeloblastic leukemia without maturation
AML M2	Acute myeloblastic leukemia with maturation
AML M3	Acute promyelocytic leukemia
AML M4	Acute myelomonocytic leukemia
AML M5a	Acute monoblastic leukemia
AML M5b	Acute monocytic leukemia
AML M6a	Acute erythroleukemia
AML M6b	Acute erythroleukemia, erythroid predominance
AML M7	Acute megakaryoblastic leukemia

REFERENCES

1. Adam F et al. Vet Comp Onc 2009;7(3):181. 2. Antagnoni MT et al. Vet Res Comm 2003;27(1)367. 3. McManus PM. Vet Clin Path 2005;34(3):189. 4. Modiano JF. Hematopoietic Neoplasia. In Weiss DJ and Wardrop KJ. Schalm's Veterinary Hematology 6th ed. Ames: Wiley-Blackwell:2010: 421. 5. Jain NC et al. Vet Clin Path1991;20(3); 63. 6. Juopperi TA et al. Vet Path 2011;48(I):182. 7. Presley RH et al. Compendium 2006:28(12):831. 8. Ramasamy K et al. Leuk and Lymphoma 2007:48(2):290. 9. Tasca S, et al. Vet Clin Path 2009;38(1):2. 10. Vernau W and Moore PF. Vet Path and Immunopath 1999;69(2-4):145. 11. Villers E et al. Vet Clin Path 2006;35(1):55.

AUTHOR INDEX

AUTHOR INDEX

AUTHOR INDEX

AUTHOR INDEX

AUTHOR INDEX

AUTHOR INDEX

American College of Veterinary Internal Medicine (ACVIM) Forum 2011

Denver, Colorado, USA
15-18 June 2011

Volume 2 of 2

Editors:

Dianne McFarlane **Jeff Toll**

ISBN: 978-1-61839-030-1

Printed from e-media with permission by:

Curran Associates, Inc.
57 Morehouse Lane
Red Hook, NY 12571

CURRAN ASSOCIATES INC. proceedings.com

Some format issues inherent in the e-media version may also appear in this print version.

Printed by Curran Associates, Inc. (2011)

American College of Veterinary Internal Medicine
1997 Wadsworth Boulevard Suite A
Lakewood CO 80214-5293

Phone: (800) 245-9081
Fax: (303) 231-0880

acvim@acvim.org

Additional copies of this publication are available from:

Curran Associates, Inc.
57 Morehouse Lane
Red Hook, NY 12571 USA
Phone: 845-758-0400
Fax: 845-758-2634
Email: curran@proceedings.com
Web: www.proceedings.com

ACVIM Specialty Symposium (Pre-Forum) Proceedings

Cardiology

Large Animal Internal Medicine - Equine

Large Animal Internal Medicine - Food Animal

Neurology

Oncology

Small Animal Internal Medicine

Veterinary Scientific Program Proceedings

AAEP

ABVP

ACVCP/AAVPT

ACVIM Business Management

ACVIM Cardiology

ACVIM Foundation Grant Presentations

ACVIM Large Animal Internal Medicine - Equine

ACVIM Large Animal Internal Medicine - Food Animal

ACVIM Neurology

ACVIM Oncology

ACVIM One Medicine

ACVIM Small Animal Internal Medicine

ASVCP

AVHTM

CGS

Food Animal Internal Medicine - AAVN/ACVN

ISCAID

Liver Study Group

SCE

SVN/U

VCRS (LA)

VCRS (SA)

Session# **VECCS** **Page#**

VECCS

VUS

Technician Program Proceedings

Technician

Neurology

Title / Author

TOCERANIB PHOSPHATE FOR THE TREATMENT OF SOLID TUMORS IN DOGS AND CATS

Cheryl London, DVM, PhD, DACVIM (Oncology)
Columbus, OH

INTRODUCTION

Toceranib phosphate (Palladia) is an orally bioavailable multi-targeted receptor tyrosine kinase (RTK) inhibitor active against several members of the split-kinase RTK family including VEGFR, PDGFR and Kit. Toceranib is closely related to the small molecule inhibitor sunitinib (Sutent) that is active against VEGFRs 1, 2, and 3, PGFDRα and β, Flt-3, Kit, CSF-1, and Ret. Sunitinib and toceranib were developed as anti-angiogenic agents based on their inhibition of VEGFR and PDGFR, two receptor families known to play an important role in tumor driven angiogenesis. Toceranib was approved in June 2009 for the treatment of dogs with grade 2 and 3 mast cell tumors (MCTs) based on a clinical field study demonstrating a 43% objective response rate. Toceranib's activity against MCTs is largely mediated by its ability to inhibit signaling mediated by Kit, an RTK often dysregulated through activating mutation in this disease. The objective response rate of MCTs possessing a Kit mutation to toceranib is 70-80% and is predicted based on the reliance of MCTs with such mutations on constitutively activated Kit for growth and survival. Interestingly, the objective response rate of MCTs without Kit mutation to toceranib is 20-37%. While some MCTs may rely at least partly on Kit signaling pathways for sustained survival, it is also possible that toceranib's activity in these tumors is anti-angiogenic in nature.

TOCERANIB IN CANINE SOLID TUMORS

Given that toceranib was developed as a small molecule inhibitor targeting new blood vessel growth, biologic activity in a broad range of tumors without obvious driver mutations might be expected. In support of this, sunitinib has demonstrated single agent activity in gastrointestinal stromal tumors, renal cell carcinoma, thyroid carcinoma, insulinoma, angiosarcoma and hepatocellular carcinoma. Furthermore, in the phase I study of toceranib, partial responses and stable disease were noted in a variety of cancers including metastatic carcinoma, transitional cell carcinoma, multiple myeloma, melanoma, and metastatic sarcoma. These data suggest that toceranib may have anti-tumor activity beyond primarily Kit driven MCTs. Following approval in 2009, toceranib was used off label to treat a variety of canine cancers, typically in the setting of failed primary or standard of care treatments. A retrospective analysis was performed to provide preliminary information regarding the potential biologic activity of toceranib in canine anal gland anal sac adenocarcinoma (AGASACA), metastatic osteosarcoma (OSA), thyroid carcinoma, head and neck carcinomas (squamous cell carcinoma, SCC; ceruminous gland carcinoma) and nasal carcinoma. A secondary end point of this study was to survey the toceranib dose and dosing regimen used in the treatment of dogs with these solid tumors. The OncoListserve was used to solicit for information regarding cases in which toceranib was used to treat dogs for the tumor types described above.

Anal sac anal gland adenocarcinoma

Of the 32 dogs with ASAGACA treated with toceranib, there were 9 Labrador retrievers and 2 Labrador mixed breed dogs suggesting that his breed may be overrepresented. The median age of all dogs was 11 years; 13 were spayed females and 19 were neutered males. Prior therapy (surgery, chemotherapy, radiation therapy, or a combination of these) had been given to 25/32 dogs and 7 were hypercalcemic secondary to disease. Metastasis was present in most dogs (n=28) including sublumbar lymph nodes (n=25), lungs (n=9), liver (n=2) and other sites (n=11). The median toceranib dose used was 2.81 mg/kg (range 2.2-3.25 mg/kg). Most of the dogs (22/32, 68.8%) were treated on a Monday, Wednesday, Friday (MWF) schedule with the remainder receiving drug every other day (EOD, n=8) or twice per week (n=2). Of the 32 dogs treated, 8 (25%) experienced partial response (PR) and 20 (62.5%) experienced stable disease (SD) for a clinical benefit (CB) rate of 87.5%. The median duration for PR was 22 weeks (range 10-40 weeks) and the median duration for SD was 30.5 weeks (range 10-47). The median duration of treatment for all 32 dogs treated with toceranib was 25 weeks (range, 0-47+ weeks).

Metastatic osteosarcoma

To be included in this retrospective analysis, dogs must have had radiographic evidence of pulmonary metastatic disease as it was not considered possible to accurately assess primary bone lesions for response to therapy. Of the 23 dogs with OSA treated with toceranib, there were 4 mixed breeds, 4 Golden retrievers, 3 Labrador retrievers, 3 Greyhounds, with the remainder a variety of other primarily large breed dogs. The median age of all dogs was 9 years: 13 were spayed females, and 10 were neutered males. Nearly all dogs (21/23) had undergone surgery to remove the primary tumor and received chemotherapy prior to the development of metastasis. The median toceranib dose used was 2.7 mg/kg (range 2.3-3.25 mg/kg). Approximately half of the dogs (13/23, 56.5%) were treated on a MWF schedule with the remainder receiving drug EOD. Of the 23 dogs treated, 1 (4.3%) experienced PR and 10 (43.5%) experienced SD for a CB rate of 47.8%. The median duration of treatment for the 11 dogs that experienced CB was 24 weeks (range 10-42+ weeks).

Thyroid carcinoma

Of the 15 dogs with thyroid carcinoma treated with toceranib, there were 4 mixed breeds, 2 Labrador retrievers, 2 beagles, with the remainder a variety of other breeds. The median age of all dogs was 10 years; 9 were spayed females, 2 were intact males, and 4 were neutered males. Prior therapy had been given to 10/15 dogs including surgery (n=4), chemotherapy (n=7) and radiation therapy (n=3). The primary tumor was present in 13 dogs, 9 dogs had metastatic disease to the lungs and 1 dog had metastatic disease to the liver and spleen. The median toceranib dose used was 2.75 mg/kg (range 2.47-3.25 mg/kg). More than half of the dogs (9/15,

60%) were treated on a MWF schedule with the remainder receiving drug EOD (n=5) or every third day (n=1). Of the 15 dogs treated, 4 (26.7%) experienced PR and 8 (53.3%) experienced SD for a CB rate of 80%. The median duration of treatment for the 12 dogs that experienced CB was 24.5 weeks (range 11-50+ weeks).

Head and neck carcinoma

There were 8 dogs with head and neck carcinomas including 3 squamous cell carcinomas (SCC) of the nasal planum, 4 SCC of the oral cavity (mandible, maxilla, lingual) and one dog with a ceruminous gland adenocarcinoma. There were 3 mixed breeds, 2 Labrador retrievers, and 2 Golden retrievers, and one Boxer. The median age was 11.5 years; there were 3 spayed females, 2 intact males, and 3 neutered males. The majority (n=5) had received no prior treatment, while 2 had undergone surgery and 1 had completed radiation therapy. Metastasis to the mandibular nodes was present in 2 dogs and to the lungs in one dog. The median toceranib dose used was 2.87 mg/kg (range 2.5-3.2 mg/kg). Half of the dogs were treated on a MWF schedule and half received drug EOD (n=5). Of the 8 dogs treated, 1 (12.5%) experienced CR, 5 experience PR (62.5%), 1 experienced SD although this was not of sufficient duration to meet criteria for inclusion, and 1 experienced PD for a CB of 75%. The median duration of treatment for the 6 dogs that experienced CB was 19.5 weeks (range 4-48+ weeks).

Nasal carcinoma

There were 7 dogs with nasal carcinoma eligible for inclusion in the analysis. The median age was 10 years; there were 3 spayed females and 4 neutered males. The majority (n=4) had received prior radiation therapy, and metastasis to the lymph nodes was present in 2 dogs and to the lungs in 1 dog. The median toceranib dose used was 2.67 mg/kg (range 2.4-3.2 mg/kg). Six of the dogs were treated on a MWF schedule and only 1 received drug EOD. Responses in the nasal tumors are difficult to assess without serial CT and only 2/7 dogs had imaging performed following initiation of toceranib. Therefore, for those dogs with no repeat CT, evidence of biologic activity was considered positive if clinical signs (epistaxis, nasal obstruction) improved and these dogs were considered to have experienced CB. Of the 2 dogs that were re-imaged, 1 had a CR and 1 had SD; an additional 3 dogs experienced CB for an overall CB rate of 71.4%. The median duration of treatment for dogs with CB was 18 weeks (range 12-20+ weeks).

TOCERANIB DOSING IN DOGS WITH SOLID TUMORS

The toceranib label indicates dosing should be initiated at 3.25 mg/kg every other day (EOD) and dose reduced as needed based on clinical side effects; the lowest recommended dose is 2.2 mg/kg. While this has been established as the maximum tolerated dose, evidence exists that good biologic activity occurs when doses are initiated below the 3.25 mg/kg rate and flexible dosing schedules may also be effective. For example, in the Phase I study, the overall biologic activity in the 2.5 mg/kg group was 68% compared to 60% in the 3.25 mg/kg group. Based on these data, it is reasonable to assume that for many dogs lower doses of toceranib result in significant clinical activity. This is important as the recent clinical experience with toceranib in dogs suggests that dosing at 2.5-2.75 mg/kg EOD is better tolerated than the higher dose, resulting in less toxicity, better owner compliance, and importantly, fewer drug holidays. Data regarding toceranib dose and regiment for the 85 dogs used in the OncoListserve analysis are summarized in the table below. Clinical benefit (CB) was noted in 63 (74%) cases (2CR, 18PR, 43SD) and the median dose for all 85 dogs was 2.8mg/kg (range 2.2-3.25). Of the 63 dogs that experienced CB, 37 (59%) were dosed on a MWF basis, 46 (73%) received less than 3 mg/kg, and 47 (75%) were on therapy for 4 months or longer. The median dose was 2.8 mg/kg for dogs that experienced SD and the median dose was 2.8 mg/kg for dogs with PR/CR. These data did not differ from the 20 dogs that experienced PD: the median dose of toceranib was 2.8 mg/kg, 13 (59%) were dosed on a MWF basis and 14 (64%) received less than 3 mg/kg.

	All	CR/PR	SD	CR/PR/SD	PD
Number of Dogs	85	20 (24%)	43 (51%)	63 (74%)	22 (26%)
Age	10 (3-18)	10 (7.5-13)	10 (3-18)	10 (3-18)	10 (4-13)
Gender					
FS	41	8	25	33	8
M	4	3	0	3	1
MC	40	9	18	27	13
Dose (mg/kg)	2.8 (2.2-3.25)	2.8 (2.48-3.25)	2.8 (2.2-3.25)	2.8 (2.2-3.25)	2.8 (2.2-3.25)
≥3 mg/kg	25 (29%)	6 (30%)	11 (26%)	17 (27%)	8 (36%)
<3 mg/kg	60 (71%)	14 (70%)	32 (74%)	46 (73%)	14 (64%)
MWF Schedule	50 (59%)	7 (35%)	30 (70%)	37 (59%)	13 (59%)
Metronomic Con Meds					
NSAID	17	6	5	11	6
Low-dose CTX	14	3	9	12	2
NSAID/CTX	12	3	7	10	2
Total	43 (51%)	12 (60%)	21 (49%)	33 (52%)	10 (45%)
Duration of Tx (wks)	N/A	22 (4-48+)	25 (10-50+)	23 (4-50+)	N/A

TOCERANIB IN CATS

A retrospective analysis evaluating the activity of Toceranib in feline tumors was performed. Oncologists were asked to retrospectively provide data on cats administered toceranib phosphate for any tumor type. Measurable objective response rate was defined as the percentage of cats with complete (CR) or partial remission (PR); biological response rate (BRR) referred to CR, PR, or stable disease (SD) >30 days. Four-two cats were treated with a median dose of 2.8 mg/kg of toceranib phosphate 3-4 times weekly (range 1.67-4.0 mg/kg). Tumor types included 13 squamous cell carcinomas (SCC), 8 presumptive injections site sarcomas (VAS), 6 mast cell tumors, 6 carcinomas, 4 mammary gland carcinomas, 3 sarcomas and 1 multiple myeloma. Tumor response was adequately assessed for 13 SCC, 7 VAS, and 4 MCT. For the SCC, 8% of cats had an objective response and the BRR was 54% including one CR and 6 SD. The objective response rate for VAS was 42.9% and BRR was 71% (3PR and 2SD). Response was assessed in 4 cats with MCT. The objective response rate was 33% and BRR was 100%. Adverse events in the 15 cats included anorexia and weight loss (6), vomiting (5), hiding (2), tumor necrosis/ulceration (2), and one each with elevated liver enzymes, lameness, fever, and diarrhea. These data suggest that toceranib may have biological activity against feline SCC, VAS and MCT. Future clinical trials will be necessary to define its true activity in these disease settings.

TOCERANIB AND PIROXICAM: RESULTS OF A PHASE I STUDY

There is significant interest in the use of non-steroidal anti-inflammatory medications (NSAIDs) for the treatment of cancer in dogs and cats, not only for the management of cancer related pain, but as part of metronomic chemotherapy protocols. One of the more common NSAIDs used in canine metronomic protocols is piroxicam, a mixed COX-1/COX-2 inhibitor. Piroxicam has demonstrated single-agent activity in some carcinomas and is routinely used in combination with low-dose continuous cyclophosphamide. Given the fact that both toceranib and NSAIDs exhibit dose limiting gastrointestinal toxicity, it would be predicted that combining these drugs would result in an enhanced side effect profile, thereby precluding their use together. A phase I trial was performed in tumor-bearing (non-mast cell) dogs to establish the safety of the combination. Five dose-escalating cohorts, up to and including approved label dosage for toceranib and standard dosage for piroxicam, were completed without observing a frequency of dose-limiting toxicity. Therefore, the combination of standard dosages of both drugs (toceranib, 3.25 mg/kg, every other day; piroxicam, 0.3 mg/kg daily) was generally found to be safe. Several anti-tumor responses were observed in this clinical trial as well. Based on these data, piroxicam is now being used in combination with toceranib and low dose cyclophosphamide for the treatment of several tumor types in dogs at the institution of the author (OSU). In general, the toceranib (2.5-2.75 mg/kg) is alternated every other day with the piroxicam (0.3 mg/kg) and cyclophosphamide (10 mg/m2) which are given together.

TOCERANIB AND VINBLASTINE: RESULTS OF A PHASE I STUDY

There is increasing interest in combining Toceranib with chemotherapy agents. A phase I clinical trial was recently performed to identify an appropriate dosing regimen combining Toceranib with vinblastine in canine mast cell tumors. These two drugs were chosen for combination because both exhibit single agent activity in mast cell tumors, and they have non-overlapping primary toxicities (gastrointestinal for toceranib, myelotoxicity for vinblastine). The dose-limiting toxicity for the simultaneous combination was neutropenia and the maximally tolerated dose was vinblastine (1.6 mg/m^2 every other week) concurrent with toceranib (3.25 mg/kg PO, every other day). The 50% reduction in dose intensity for vinblastine was necessitated based on enhanced myelosuppression, when combined with toceranib. Despite the reduction in vinblastine, 71% of dogs experienced an objective response to therapy to the drug combination. The enhanced myelosuppression and significant biologic activity suggest the possibility of additive or synergistic activity, although future prospective studies are necessary to appropriately define this activity.

TOCERANIB AND COARSE FRACTIONATED RADIATION: RESULTS OF A TOLERABILITY STUDY

Both radiation therapy and toceranib have single agent activity in the treatment of mast cell tumors. A prospective clinical trial was performed in dogs with unresectable mast cell tumors to determine the efficacy, tolerability, and adverse event profile of combined treatment with toceranib and hypofractionated radiation therapy. Dogs (n=17) were enrolled in this study. All had measurable mast cell tumors and all received prednisone, omeprazole, and diphenhydramine72 hrs prior to starting the toceranib which was then administered at a dose rate of 2.75 mg/kg on MWF for 1 week prior to starting the radiation therapy. The objective response rate was 78% (53% CR, 24% PR) and another 12% of dogs had stable disease for a CB rate of 90%. The most common toxicity was gastrointestinal (grade I). The hypofractionated radiation/toceranib protocol was well-tolerated in most dogs and demonstrated short-term efficacy in a large number of treated patients. Future clinical trials will be necessary to determine the long-term activity of combined radiation/toceranib in dogs with mast cell tumors.

SUMMARY AND ONGOING/FUTURE WORK

Toceranib has now been used in a variety of settings to treat cancer in both dogs and cats. Several clinical trials are underway to determine the utility of Toceranib in canine hemangiosarcoma, transitional cell carcinoma, and osteosarcoma, among others. The role of toceranib in combination with radiation therapy for the treatment of canine nasal carcinomas and feline oral squamous cell carcinomas is also being explored. Lastly, the pharmacokinetics of toceranib at below label doses is being explored to assess the affects on pharmacodynamic end points such as serum VEGF. Data from these clinical efforts should provide important new information that will help to guide appropriate use of this novel agent to treat cancer in companion animals.

REFERENCES

London CA et al., Phase I dose-escalating study of SU11654, a small molecule receptor tyrosine kinase inhibitor, in dogs with spontaneous malignancies. Clin Cancer Res 9, 2755; 2003. London CA et al., Multi-center, placebo-controlled, double-blind, randomized study of oral toceranib phosphate (SU11654), a receptor tyrosine kinase inhibitor, for the treatment of dogs with recurrent (either local or distant) mast cell tumor following surgical excision. Clin Cancer Res 15, 3856; 2009. Mena AC, Pulido EG, Guillén-Ponce C. Understanding the molecular-based mechanism of action of the tyrosine kinase inhibitor: Sunitinib. Anticancer Drugs 21 Suppl 1:S3-S11; 2010. Chow LQ, Eckhardt SG. Sunitinib: From rational design to clinical efficacy. J Clin Oncol 25:884-896; 2007. London CA et al. Preliminary evidence for biologic activity of toceranib phosphate (Palladia) in solid tumors. Submitted, Vet Comp Oncol, 2011. Hohenhaus A et al. Biologic activity and adverse event profile in cats treated with toceranib phosphate. Presented at the 30th Annual Veterinary Cancer Society Meeting, October 29-November 1, 2010. Hon E et al. Phase I safety evaluation of toceranib phosphate (Palladia) combined with piroxicam in dogs. Accepted for publication, Vet Comp Oncol, 2011. Robat C et al. Safety evaluation of combination vinblastine and toceranib phosphate (Palladia) in dogs with mast cell tumors: A phase I dose finding study. Accepted for publication, Vet Comp Oncol, 2011. Carlsten K et al. Hypofractionated radiation therapy plus toceranib for unresectable canine MCTs. Submitted, Vet Comp Oncol, 2011.

FELINE LYMPHOMA: TIME FOR A NEW PARADIGM?

Antony S. Moore, BVSc, MVSc, MACVSc, DACVIM (Oncology)
Wauchope, NSW, Australia

INTRODUCTION

One of the most difficult aspects of treating lymphoma in cats is the lack of prognostic factors (especially when compared with their canine counterparts). Also multiple studies have looked for such factors; the heterogeneous nature of this disease entity defies broad, sweeping categorization.

PROGNOSTIC FACTORS

To date those studies that have found prognostic significance are listed below.

Response to treatment has been shown to be one of the few prognostic factors for cats with lymphoma. In one large study, cats achieving a complete remission had median response duration of 7 months and median survival of 8.5 months compared with 3 weeks and 7 weeks respectively if they achieved a partial remission. The same factor has been repeated in multiple studies, but since this requires treating the patient it cannot be truly used to advise owners as to the suitability of their pet for therapy.

Cats that are "constitutionally unwell" (sick) are classified as having substage "b" lymphoma. Substage "a" was applied to cats with lymphoma that are otherwise feeling well. In one study, cats with substage "b" had a median survival of 3.5 months compared with a 9.5 month median survival for cats with substage "a" lymphoma. This same effect of illness has been seen in other studies and underscores the importance of supportive care for these cats while chemotherapy is being administered.

Stage of disease, as defined by Mooney[1] was significantly related to response in one study where cats with stage I lymphoma had higher response rates than those in stages IV to V (93% compared with 40 to 60%) and cats with stage I and II lymphoma had longer survival times (7.6 months compared with 3 months). This five-stage system is more complex than the one used for dogs, and the difficulties in applying it may explain why it has not shown prognostic value in other studies.

As further studies are published it appears that the true "break point" in staging where prognosis is strongly *clinically* affected is for cats with stage 1 lymphoma. Stage 1 lymphoma occurs primarily in cats as nasal lymphoma which have been shown to have longer remissions and survival times.

A cat with FeLV antigenemia is less likely to have a long survival, possibly due to development of other diseases as well as the lymphoma. Both remission duration (4 weeks vs. 5 months) and survival (5 weeks vs. 6 months) were shorter for cats with FeLV. In contradiction to these findings, FeLV test status was not an independent predictor of survival in another study. FeLV is an uncommon disease, and the majority of cats with lymphoma are unaffected, so the importance of this factor is comparatively low.

Histologic grade of lymphoma is important in prognosis. Most low-grade lymphoma has been described in the intestinal tract, but any site can be involved. In one study cats with intestinal well differentiated lymphoma had no better response to COP chemotherapy than cats with high grade intestinal lymphoma. In contrast, a series of 29 cats with well differentiated lymphoma were treated with prednisone and chlorambucil. Twenty cats (69%) achieved a CR for up to 49 months (median 16 months). The median survival for all 29 cats regardless of response was 17 months, while complete responders lived a median of 23 months. Interestingly, 11 cats with high grade lymphoma in the same study received multiagent chemotherapy and only 2 (11%) had a CR. In a recently published Australian study of cats with low-grade intestinal lymphoma, 75% had complete clinical remission with a median remission time of 19 months. Absence of lethargy and complete response to chemotherapy were positive prognostic indicators in that study. This study and two others[2,3,4], have shown that cats with diffuse low grade alimentary lymphoma can have durable remissions when treated with oral prednisone in combination with chlorambucil. The dosage of chlorambucil does not seem to be important; the studies above reported high-dose pulse (15 mg/m^2 daily for 4 consecutive days every 3 weeks), or 6 mg/m^2 every day for 2 weeks, then every alternate day. Response rates and durations were similar. High dose pulse chlorambucil in cats has rarely been associated with liver toxicity, bone marrow suppression, and very rarely, seizures; but all these resolve with treatment cessation and this protocol appears easier for owners to manage as it reduces the number of negative interactions (pill giving) with their pet. The recent literature has convinced me that a relatively conservative chemotherapy protocol is the best choice with cats with this disease. In many affected cats the chlorambucil protocol is a more attractive choice as well because they are older, and often have low-grade renal disease. However; if the patient fails to respond, or relapses, then I would recommend a more aggressive protocol; one study reported a 100% response rate in cats experiencing relapse of low grade GI lymphoma.

In other studies, individual factors have been assessed to affect outcome in cats with lymphoma. These include; cats that received doxorubicin as part of their treatment had longer remissions; and cats with a high proliferative fraction of their tumor cells had longer survival times. One study[5] found that elevations in serum LDH were associated with poor prognosis, which mirrors the same effect in human lymphoma patients, but this is rarely included in serum chemistry profiles any more.

TREATMENT

Regardless of the primary site of lymphoma, survival times in untreated cats tend to be very short. The majority of cats in one study that were not treated died within 2 weeks, although cats that were FeLV negative appeared to live a little longer. In another group of 39 untreated cats, only two cats lived longer than 4 months. Cats with lymphoma treated with supportive care only had an average survival of 5 days.

Single agent treatment of lymphoma in cats is not generally recommended due to the low chance for lasting response. Cyclophosphamide, vincristine, and prednisone form the basis for most chemotherapy protocols. The individual drugs had been shown, albeit anecdotally, to have significant activity in the treatment of feline lymphoma, sometimes resulting in long-term control. In these early reports, vinblastine was also found to be useful while L-asparaginase had some limited activity and chlorambucil had questionable efficacy.

The COP (cyclophosphamide, vincristine, prednisone) protocol most often used involves a 4 week induction period followed by maintenance therapy every 3 weeks until 1 year at which time all treatment is ceased. Cats that were treated with the 4 week induction period alone had a median remission of less than 2 months, as compared to median remissions ranging between 4.5 and 28 months for cats that received the complete COP protocol, implying that maintenance therapy was important. 38 cats were treated with COP and maintenance therapy for various anatomic forms of lymphoma. The overall complete response rate was 30 of 38 cats (74%). The duration of complete responses ranged from 2 to 42 months. The CR rate varied from 100% for multicentric lymphoma to 50% for extranodal lymphoma. Similarly, the median duration of remission ranged from 4.5 months for alimentary lymphoma to 28 months for cats with peripheral nodal lymphoma. All groups contained individuals that lived longer than a year.

Further reports of the use of COP protocol have not found the same efficacy. A further 38 cats treated in the 1990's differed from the earlier cats in being less likely to be FeLV positive or have mediastinal lymphoma. Eighteen of the 38 cats (47%) had a complete response. The median response duration was less than 3 months, and no cat had a remission longer than 6 months. All cats with mediastinal, nasal and peripheral nerve lymphoma responded (a total of five cats), and six of eleven cats with multicentric, five of twelve with alimentary, and two of nine with renal lymphoma responded. To further complicate the issue a recent European study found that COP gave a similar response rate (about 70%) to that seen in earlier U.S. studies in a population of cats that was mainly young purebred and with a low proportion of GI lymphoma.

Interestingly, two studies that investigated the efficacy of doxorubicin as an induction agent found that fewer than 30% of cats had a complete response. Similarly doxorubicin appears to be a poor "rescue" agent for feline lymphoma. This is in contrast to dogs, where doxorubicin is a very effective single agent in the treatment of lymphoma. A further recent study found that doxorubicin was an ineffective *rescue* agent in cats with lymphoma that failed to respond or relapsed after chemotherapy with other drugs. It appears then, that doxorubicin is best used in combination with other chemotherapy agents.

Further evidence of the utility for doxorubicin as a component of multiagent chemotherapy came from a study in which cats that had achieved a complete response to COP were randomized to either continue COP maintenance, or to receive doxorubicin alone. Cats that received only COP had a median remission of 3 months while cats receiving doxorubicin had a median remission of 9.5 months. The only long term survivors in the study received both doxorubicin and COP. This protocol also used a total of 6 months of chemotherapy, which is a much shorter maintenance phase than most other protocols that use one or more years of treatment. This implies that shorter protocols, if multidrug combinations are used, may be adequate to treat feline lymphoma, if effective multi-drug combinations are used. These findings were supported by a multi-institutional study which found that cats receiving doxorubicin as part of their chemotherapy protocol were likely to have a longer remission duration that those receiving non-doxorubicin containing protocols (9 months vs. 3 months).

COP plus doxorubicin, L-asparaginase, and methotrexate was used to treat 132 cats with lymphoma of all stages and anatomic sites. In this group only 11 cats were FeLV positive and 4 cats were FIV positive. Most cats (100 cats or 76%) were clinically ill, and most had gastrointestinal lymphoma. Complete remissions were seen in 88 cats (67%). Remission times for all cats ranged from 0 days to over 5 years with a median remission of 5 months. Response to therapy and whether the cat was sick or not appeared to be the only factors to influence survival. Another similar protocol was used to treat Australian cats with lymphoma, however, the median survival was only about 6 months in this group of FeLV negative, but often FIV positive, cats.

MOPP (Mustargen, Vincristine, Prednisone, Procarbazine) is a protocol that combines 2 alkylating agents with vincristine and prednisone. A 28 day cycling MOPP protocol that used mustargen, vincristine, procarbazine and prednisone, was reported to be effective when other chemotherapy failed to give a remission. Twenty-three cats, most of which had gastrointestinal lymphoma, were treated with MOPP when a combination chemotherapy protocol failed. 10 cats had a complete response and 4 cats had a partial response, the median remission duration for these 14 cats was 166 days. Overall it was well tolerated. Mild anorexia was seen in two cats, and five cats showed significant leukopenia or anemia. Anorexia with MOPP protocol is most likely to resolve when procarbazine is reduced to every alternate day administration, or given concurrently with metoclopramide.

In another study, MOPP was administered with doxorubicin and cyclophosphamide as first line therapy. In this study, thirty-five cats with lymphoma were categorized as having multicentric (15), extranodal (12), GI (6) or mediastinal (1) lymphoma. 52% had a CR with a median survival of 195 days with 7 cats still in CR up to 796 days after starting treatment. 20% had a PR with a median survival of 115 days. Median survival for 27% of cats that did not respond to treatment was 21 days. 6-month and 1-year survival was 36% and 21%, respectively.

Radiation has limited application in the treatment of feline lymphoma. The main indications for radiation would be an apparently localized (usually extranodal) lymphoma, or to cause rapid reduction in tumor bulk for comfort (for example with a large mediastinal lymphoma causing dyspnea). Radiation therapy can occasionally cause long remissions alone, but should be used in conjunction with chemotherapy as lymphoma should always be considered to be a systemic disease. Our limited experience at Tufts between 1998 and 2002 with 10 cats with alimentary lymphoma treated with multimodality chemotherapy, and radiotherapy (8 Gy, given in 2 Gy fractions on 4 consecutive days), and recent publication of cats treated with chemotherapy and abdominal radiotherapy[6] (10 daily fractions of 1.5 Gy) and cats with resistant lymphoma treated with radiotherapy alone[7] (8 Gy, given in 4 Gy fractions on 2 consecutive days) indicate there may be role for localized radiotherapy to improve outcome in this difficult group of patients.

SUMMARY

Outcomes for feline lymphoma patients cannot be easily predicted using current staging schemes, although clearly cats with stage 1 disease have a better outcome, cats with more advanced disease may be better characterized by location and stage rather than either alone. Histologic grading should also be considered, with low-grade lymphomas much more responsive than non-low grade lymphoma. Chemotherapy is still the best treatment modality, but the addition of radiotherapy may improve outcomes, particularly if used in first line treatment

REFERENCES

[1] Mooney SC, et al. **J Am Vet Med Assoc** 1989;194:696. [2] Lingard AE, et al. **J Feline Med Surg** 2009;11:692. [3] Kiselow MA, et al. **J Am Vet Med Assoc** 2008;232:405. [4] Stein TJ, et al. **J Am Anim Hosp Assoc** 2010;46:413. [5] Hadden AG, et al. **J Vet Internal Med** 2008:22:153. [6] Williams LE, et al. **Vet Radiol Ultrasound**. 2010;51(6):681. [7] Parshley DL, et al. **J Feline Med Surg**. 2011 E-pub

DEVIL FACIAL TUMOR DISEASE: BIOLOGY, EPIDEMIOLOGY & TREATMENT OUTCOMES

Antony Moore, BVSc, MVSc, DACVIM, David Phalen DVM, PhD, Dipl. ABVP, Angela Frimberger VMD, DACVIM, Stephen Pyecroft BVSc, PhD MACVSc, Sarah Peck BVetMed, Colette Harmsen BVSc, Suzanneth Lola, DVM.
Wauchope and Camden, NSW and Kings Meadow, TAS, Australia

INTRODUCTION

Devil facial tumor disease (DFTD) is an aggressive, transmissible and commonly fatal, malignancy for which no treatment has been reported; it is one of only two known naturally occuring clonally transmissible cancers (the other being transmissible venereal tumor of canids).[1] It was first described in 1996 and by 2006, it was estimated that 59% of Tasmania was affected by the disease, accounting for population declines of more than 80% in some areas.[2] Since first being described there has been an 89% decline in spotlighting sightings in the north-east of Tasmania and a 41% decline in average sightings state-wide.[2] Older animals have declined as a proportion of the affected population, and the age at breeding has compensatorily declined leading to a shift in life-history of this species.[3] The Tasmanian devil was listed as endangered by IUCN in 2009.

CLINICAL PRESENTATION AND PATHOLOGY

Clinically the disease appears as multiple, firm, raised soft-tissue nodules and masses, which are often centrally ulcerated and necrotic. As they progress, they ultimately become large space occupying lesions. The anatomic distribution of DFTD lesions suggests that transmission of transplantable tumor cells occurs during "jaw-wrestling" a common interaction among devils.[4] There are usually multiple lesions initially arising on the face (especially oral areas) and neck, and it is progressive growth of these lesions that lead to the death of most of the affected animals. While the cause of death for most affected Tasmanian devils is thought to be starvation, metastatic disease occurs rapidly and widely. In a survey of wild caught animals with DFTD 45% had metastasis, most commonly to the lungs (32%) and to the regional lymph node (25%). Other affected sites included liver, spleen, kidney, pouch, abdominal and thoracic musculature and fat. When affected animals were kept in captivity for an extended period, there were additional metastatic sites identified (including to the brain, spine, heart and adrenal glands). Our observations of affected devils show that when protected in a captive situation, untreated devils live an average of 3 months from capture.

While normal Tasmanian devils have 14 chromosomes, DFTD cells in all affected animals studied contain only 13 chromosomes with deletions and unidentified marker chromosomes present similar to those seen in canine transmissible venereal tumor (CTVT), another transplantable cancer.[4] Additional genetic studies confirm that DFTD is a monophyletic clonally transmissible tumor.

Cytologically, the tumor cells appear as round cells, and histologically it is characterized as a multi-nodular tumor, densely cellular with round cells arranged in sheets, packets and cords separated by fine fibrous stroma. There is an infiltrative pattern of growth at margins with pseudocapsule formation in some areas. There is often a marked neutrophilic infiltrate, but rarely lymphoid infiltration. In the original immunohistochemistry studies, tumor cells stained positively for vimentin, S100, neuron-specific enolase, chromogranin A, and synaptophysin markers; which was suggestive of neuroectodermal origin.[5,6] More recently, a genetic analysis of DFTD showed that the gene with the highest expression in DFTD encodes myelin basic protein, normally produced by CNS oligodendrocytes and peripheral Schwann cells. Further supporting a Schwann cell derivation was the finding that all DFTD stain positively with periaxin, an antibody against Schwann cell–specific myelin protein.[7]

MANAGEMENT STRATEGIES

Current management strategies for this disease include isolation of populations of devils by fencing or establishment of isolated peninsulas, and capture and removal of affected devils. Additional "insurance" populations of Tasmanian devils have been established in breeding colonies around mainland Australia. With such reliance on captive populations for survival of the species, investigation into treatment strategies would appear to be important; however none have been reported to date.

CHEMOTHERAPY TRIAL

Canine transmissible venereal tumor (CTVT) is readily transmitted during coitus by transfer of cells from diseased to healthy dogs, but it also may be spread by social behaviour, such as sniffing or licking. Tumor behaviour was studied through 40 generations of dogs and implantation was found to occur in 68% of exposed animals. CTVT is an aggressive disease, capable of invasive growth and metastasis, but with appropriate treatment the cure rate is high. Vincristine, administered weekly to 201 dogs with CTVT resulted in a complete remission (CR: 100% tumor reduction) in 197 dogs and a partial remission (PR: >50% tumor reduction but less than CR) in one dog; only one dog relapsed in a period of 12 months. Similar to DFTD, the chromosomal complement of CTVT is constant worldwide, with 58 to 59 chromosomes found in the tumor (compared with 78 for cells of normal dogs).[1] Unlike DFTD, metastasis of CTVT is uncommon and occurs in less than 5% of reported cases, mainly in immunosuppressed dogs. The inguinal lymph nodes are the most likely site of lymphatic metastasis, although distant metastases have been seen in the abdominal viscera, skin, and central nervous system. The disease homology between DFTD and CTVT suggested to us that attempting a treatment known to be highly successful against CTVT may be worthwhile and successful. Vincristine and other chemotherapy may have similar efficacy in DFTD.

We designed a trial to evaluate the feasibility of treating affected captive devils, to assess toxicities associated with chemotherapy, to identify maximally tolerated dosages for the drugs used, and to evaluate tumor response in affected animals. The drugs chosen for evaluation were vincristine (due to the impressive cure rate seen in CTVT), as well as doxorubicin and carboplatin, as both are considered broad spectrum chemotherapeutics from different chemotherapy classes.

Tasmanian devils captured as part of the management strategies noted above, that were affected but still estimated to have a survival time untreated of more than 6 weeks, were entered into this trial. All animals were housed in large outdoor enclosures, and anaesthetized using isoflurane for all intravenous treatments.

All chemotherapy agents were administered at increasing dosages during the trial in order to establish the maximally tolerated dosage level for Tasmanian devils. This escalation was necessary, as treatment of devils with an arbitrary and non-toxic dosage level may have been ineffective, and led to an invalid assumption that the drug itself was ineffective. If there was no response seen in devils at the maximally tolerated dosage then it would be assumed that it would not be a helpful agent in the treatment of this disease. The plan was that at each dosage level, three devils would be treated and evaluated for adverse events. If there was no grade 3 or higher toxicity, then the dosage for the next cohort of three animals would be increased according to Fibonacci's Modified scheme.

To evaluate toxicity to patients during this trial, previously published (VCOG) criteria used to evaluate adverse events during chemotherapy in companion animals (specifically dogs and cats), were modified. These adverse events were graded as to their severity 1 through 5.

While the established VCOG criteria were very helpful in identifying systems that needed to be evaluated for adverse events, and while many of the criteria used to grade the events were directly transferable from companion animals to Tasmanian devils, the grading of some toxicities was probably not ideal. Specifically, gastrointestinal toxicities that would be considered as grade 2 (moderate) or 3 (severe) in dogs and cats, become life-threatening (grade 4) in captive wild devils due to restrictions on handling, hospitalization, intravenous fluid administration and continued medications. While no animal died as a result of these toxicities, subcutaneous fluid support and aggressive antibiotic use were needed in 3 animals, and diet changes, as well as medications to reduce nausea were used as needed to keep animals eating during some portions of the trial. Evaluation of the severity of gastrointestinal (GI) toxicities was further complicated by the feeding habits of Tasmanian devils, and their adjustment to captive living and feeding. Whether reduction in appetite and weight loss should be considered an adverse event of chemotherapy was difficult to evaluate in this trial; many devils lost weight in the period between capture and starting chemotherapy and continued to lose weight during chemotherapy. All animals were allowed a period to "acclimatize" to captive living, but that needed to be kept brief (usually 1 to 2 weeks) due to the rapid progression of tumor growth, and the need to start treatment before the tumors themselves impacted on their quality of life. Tumor progression and development of visceral metastases often causes signs of appetite loss, and other GI signs; this also complicated the interpretation of such signs in devils undergoing treatment.

Doxorubicin, vincristine and carboplatin were each administered and efficacy measured by physical tumor measurement as well as monitoring for metastatic disease. Pharmacokinetic data was also collected to measure vincristine plasma levels.

Tasmanian devils appear to tolerate higher dosages of some chemotherapy drugs than domestic carnivore companion animals. On the other hand, this is not a generalizable finding, and Tasmanian devils tolerate doses of other chemotherapy at similar dosage levels to dogs and cats. An unexpected finding from this trial is that despite a lower basal metabolic rate, the clinical effect and toxicity of drugs could not be predicted without careful evaluation of each drug at different dosage levels. This has implications for treatment of Tasmanian devils with other drugs that are not chemotherapeutics. This finding may also translate to other marsupial species.

ACKNOWLEDGEMENTS

This trial was funded through a Tasmanian Government Project grant and completed as a joint collaboration between the Wildlife Health and Conservation Centre at the University of Sydney, Veterinary Oncology Consultants, and the Tasmanian Department of Primary Industries and Water. Additional support was provided by Carmel Pharma (donation of PhaSeal products), Pfizer Animal Health Australia (donation of maropitant) and Hills Pet Nutrition Pty Ltd (donation of Hill's a/d).

REFERENCES

1 Murchison EP. **Oncogene** 2009;27:S19. 2 Hawkins CE. et al.**Biol Conserv** 2006;131:307. 3 Lachish S. et al. **J Anim Ecol** 2009:78: 427. 4 Pearse AM, et al. **Nature** 2006 439: 549. 5 Loh R, et al. **Vet Pathol** 2006; 43:890. 6 Loh R, et al. **Vet Pathol** 2006; 43:896. 7 Murchison EP, et al. **Science** 2010;327:84.

EVALUATING METRONOMIC CHEMOTHERAPY USING RELEVANT TUMOR BIOMARKERS

Barb Biller, DVM, PhD, DACVIM
Fort Collins, CO

INTRODUCTION

The survival benefits of conventional maximum tolerated dose (MTD) chemotherapy are modest for the treatment of many canine and feline malignancies. In addition, conventional chemotherapy may induce significant acute and chronic toxicities. Recent studies have demonstrated that metronomic chemotherapy, which is defined as the uninterrupted administration of low doses of cytotoxic drugs at regular and frequent intervals, may be at least as effective as MTD therapy and is associated with substantially less toxicity and expense (1-3). In comparison to conventional MTD chemotherapy, the critical difference of metronomic therapy appears to be the elimination of long break periods between treatments. Elimination of treatment gaps also eliminates, or at least substantially reduces, the ability of tumor cells to undergo repopulation and damage repair and to alter their microenvironment (3). Even in the setting of drug-resistant disease, metronomic dosing protocols have been found to elicit significant and durable treatment responses in people with heavily pre-treated, metastatic malignancies (3-5).

How Metronomic Chemotherapy Works

In contrast to conventional chemotherapy protocols, which target rapidly dividing tumor cells, a key target of metronomic chemotherapy is the tumor-associated vasculature. Tumor angiogenesis is thought to occur at two levels in the cancer patient: (1) locally within the tumor microenvironment whereby circulating endothelial cells (CECs) are stimulated to proliferate and (2) via systemic effects on bone marrow-derived circulating endothelial progenitor cells (CEPs). Continuous administration of low doses of a surprisingly wide range of drugs (such as cyclophosphamide, methotrexate, vinblastine and paclitaxel) is cytotoxic to both CECs and CEPs but has no effect on non-endothelial cells such as tumor cells, epithelial cells and leukocytes. Because of their relative genetic stability, endothelial cells are inherently less susceptible to the development of drug resistance than are tumor cells (1, 6 – 7). Even when maximally tolerated doses of chemotherapy drugs are no longer effective, significant inhibition of tumor growth and sparing of normal tissue can sometimes be achieved by simply changing to a metronomic dosing regimen.

Another target of metronomic chemotherapy is the regulatory T cell (Treg), a subset of the CD4+ T-lymphocyte population that facilitates tumor survival in cancer patients through immunosuppressive effects. The selective toxicity of metronomic chemotherapy for Treg has been best studied for the alkylating agent cyclophosphamide (CYC). In mice and in humans, administration of low doses of CYC leads to a transient but profound decrease in the number of circulating Treg and directly inhibits Treg function (4, 8 – 9). Several recent studies show that it may also be possible to decrease Treg through the administration of cyclooxygenase (COX) inhibitors. For example, administration of COX inhibitors to mice with lung cancer resulted in suppression of Treg function and a decrease Treg numbers, in part through inhibition of tumor-induced expression of COX-2 and prostaglandin E2 (10). The immunostimulatory effects of low dose CYC on dendritic cells has also been reported (11).

THE APPLICATION OF METRONOMIC CHEMOTHERAPY TO VETERINARY PATIENTS

Although the results of many clinical trials investigating metronomic chemotherapy in human cancer patients appear promising, a significant limitation has been considerable empiricism in the establishment of optimal metronomic dosing schedules and in evaluating therapeutic activity (3). Several studies have recently shown, however, that the optimal biologic dose of metronomic chemotherapy (defined as the dose eliciting maximum reduction in tumor volume without inducing significant systemic toxicity) corresponds closely to the nadir in CEP levels in the blood (12). Endothelial cell numbers also appear to correlate well with other standard measures of tumor angiogenesis such as tumor microvessel density (MVD) assessment. Circulating endothelial cells have tentatively been identified in dogs using flow cytometry and, as in people, may be useful in helping to identify optimal drug doses and schedules for metronomic regimens (13). Although Treg levels are known to decrease in human patients receiving low doses of CYC, the validity of using Treg numbers to monitor metronomic dosing protocols awaits further investigation.

To date, the application of metronomic chemotherapy to veterinary patients has been based on anecdotal reports and extrapolation from human studies. Evaluation of metronomic drug dosing has recently been described in two small clinical trials (14, 15). In one of these studies, dogs with splenic hemangiosarcoma received a combination of 3 drugs (CYC, etoposide and piroxicam) (14). Study design did not permit determination of single drug efficacy or optimal drug dosing, nor were tumor biomarkers assessed. In the other study a significant treatment effect from administration of low-dose continuous CYC and piroxicam therapy was described in dogs with incompletely resected soft tissue sarcoma (15). As in the canine hemangiosarcoma trial, the relative contribution of each drug could not be evaluated and drug dosages were empirically determined based on human clinical trials. Several unpublished clinical studies have recently described tolerability of metronomic dosing of chlorambucil and lomustine in tumor-bearing dogs; assessment of antiangiogenic markers to determine drug dosages were not reported (Proceedings of VCS Annual Meeting, 2010).

How do we know if metronomic chemotherapy is working the way we think it is?

The low risk for toxicity, reasonable cost, ease of administration and potential for significant therapeutic benefit makes metronomic chemotherapy a particularly attractive treatment option for companion animals with cancer. There is clearly a need, however, to conduct basic studies that enable the identification of effective protocols using relevant tumor biomarkers in addition to measuring clinical parameters such as disease-free interval and survival time. Towards this end, we investigated the effectiveness of metronomic dosing of single-agent oral CYC in dogs with soft tissue sarcoma (STS) to begin to identify the effects of individual agents in these multi-drug protocols. The primary goals of this trial were to determine whether low-dose CYC therapy decreases Treg and/or exhibits antiangiogenic activity.

Low-dose CYC decreases Treg and tumor microvessel density in dogs with soft tissue sarcoma

To determine the effects of metronomic CYC therapy on numbers of Treg in circulation, blood samples were collected from client-owned dogs with measurable grade I or II soft tissue sarcoma on days 0, 14, and 28 after beginning oral CYC. Two cohorts were evaluated based on preliminary studies, with one group of dogs receiving daily oral CYC at a target dosage of 12.5 mg/m2 daily and the other at a target dosage of 15 mg/m2 daily. For dogs with STS receiving CYC 12.5 mg/m^2/day, there was no significant change in the mean percentage of Treg over the 28 day period (Fig. 1A). A significant decrease in the mean absolute number of Treg was seen, however, from day 0 to 28 ($P = 0.001$; Fig 1B). Dogs receiving CYC at a target dosage of 15.0 mg/m^2/day had significant decreases in both the absolute number and percentage of Treg in the peripheral blood during the 28 day study period (Fig 2). Despite the decrease of Treg at the higher dosage of CYC, there were no changes in the absolute numbers or percentages of either $CD4^+$ or $CD8^+$ lymphocytes during the course of the clinical trial, suggesting that the effects of CYC were selective for the Treg subset of T-lymphocytes.

Figure 1

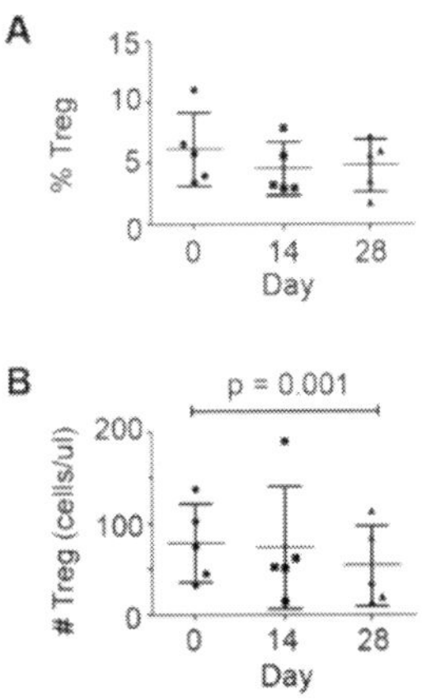

Figure 2

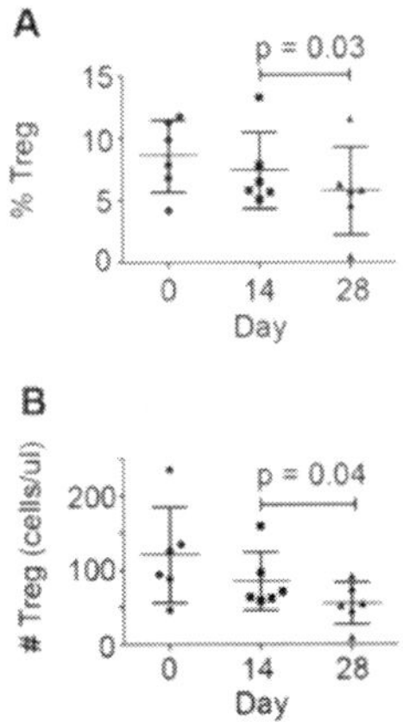

Tumor MVD was also assessed in these patients by measuring the percentage of $CD146^+$ cells using immunohistochemistry from serial tumor biopsies from dogs with STS receiving oral CYC. There were no significant changes in the mean percentages of $CD146^+$ cells between the three study time points in the tumors of dogs treated with 12.5 mg/m^2/day CYC (Fig 3A). However, in the dogs receiving CYC at 15.0 mg/m^2/day, the percent of $CD146^+$ cells significantly decreased from the pretreatment mean. This decrease persisted throughout the 28 day study period and was also significantly decreased from the MVD prior to therapy ($P = 0.004$, Fig 3B). Together, these results suggest that both antiangiogenic and immunomodulatory effects of CYC occur when administered at 15 mg/m2/day.

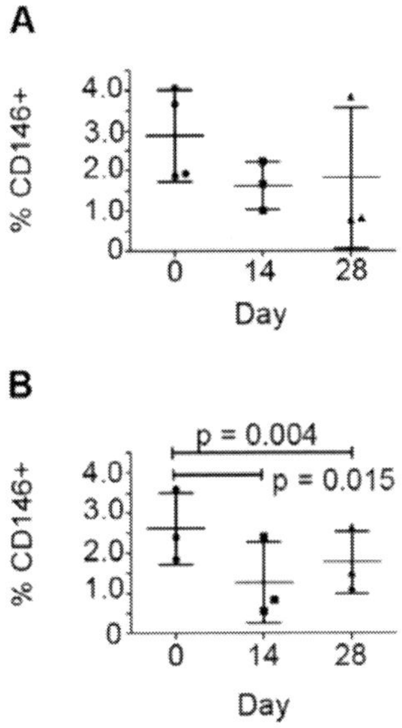

Figure 3

CONCLUSIONS AND FUTURE DIRECTIONS

Although this small study provides a starting point for metronomic dosing of CYC, many factors such as the pharmacokinetics of low-dose CYC therapy and effects of drug combinations await investigation. It is also not clear whether changes in circulating numbers of Treg corresponds to antiangiogenic effects for drugs other than CYC or whether Treg analysis could serve as a surrogate for tumor MVD and CEC assessment. At this time, assessment of canine Treg and particularly of CECs remains technically challenging; neither assay is available for widespread use. Other potential biomarkers for antiangiogenic therapy, such as gene expression of VEGFR-2 and VE-Cadherin in peripheral blood, are also being evaluated in human patients and may be applicable to monitoring of veterinary patients as well (16). In conclusion, more clinical research on safety, efficacy and validation of surrogate markers of angiogenesis should be performed before the widespread use of chemotherapy drugs in a metronomic schedule makes its way into standard oncology practice.

REFERENCES

[1] Browder T, et al. **Cancer Res** 2000;60:1878. [2] Emmenegger U, et al. **Cancer Res** 2004;64:3994. [3] Kerbel RS, et al. **Nat Rev** 2004;4:423. [4] Ghiringhelli F, et al. **Cancer Immunol Immunother** 2007;56:641. [5] Orlando L, et al. **Anti-cancer drugs** 2006;17:961. [6] Bocci G, et al. **Cancer Res** 2002;62:6938. [7] Miller KD, et al. **J Clin Oncol** 2001;19:1195. [8] Lutsiak ME, et al. **Blood** 2005;105:2862. [9] Man S, et al. **Cancer Res** 2002;62:273. [10] Sharma, et al. **Cancer Res** 2005;65:5211. [11] Nakahara, et al. **Blood** 2010;115:4384. [12] Shaked Y, et al. **Blood** 2005;106:3058. [13] Rusk A, et al. Clin Cancer Res 2006;12:7444. [14] Lana S, et al. **J Vet Intern Med** 2007;21:764. [15] Elmslie R, et al. **J Vet Intern Med** 2008;22:1373. [16] Rabascio C, et al. **Cancer Res** 2004;64:4373.

THE ROLE OF METABOLIC MANIPULATION IN CANCER THERAPY

Angela E. Frimberger, VMD, MACVSc, DACVIM (Oncology)
Wauchope, NSW, Australia

INTRODUCTION

Established chemotherapy has provided massive gains in cancer survival but has limitations, and ongoing research strives to identify new therapies. One possible therapeutic target is based on common metabolic alterations in cancers. This developing targeted therapy is not to be confused with the complementary and alternative cancer management strategy known as metabolic therapy, also known as Kelley's treatment, Gonzalez treatment, Issels whole body therapy, or Gerson therapy, which variably uses a combination of special diets, enzymes, nutritional supplements, and other measures (sometimes including laetrile) in an attempt to remove "toxins" form the body and body's defenses against disease, and for which there is no convincing scientific evidence.

TUMOR CELL METABOLISM AND ITS POTENTIAL FOR THERAPEUTIC MANIPULATION

Tumor cells are often highly reliant on glucose as an energy source, and commonly this is via activation of glycolysis even in the presence of adequate oxygen (aerobic glycolysis or Warburg effect), despite the fact that this metabolic pathway is a much less efficient process for production of energy from glucose than oxidative phosphorylation. The reason for this metabolic switch remains unclear, but proposed mechanisms include mitochondrial defects and malfunction, adaptation to hypoxic tumor microenvironment, oncogenic signaling, and abnormal expression of metabolic enzymes. This metabolic change is one of the fundamental metabolic alterations occurring in malignant transformation, and its result is that the glucose requirements of rapidly proliferating tumors are often very high.[1] This increased dependence on the glycolytic pathway could serve as a biochemical target for anticancer therapy, especially in tumors with mitochondrial defects or under hypoxic conditions (a very common situation and one which contributes to resistance to other therapies) and several glycolytic inhibitors are under study for their anticancer potential.

Very rapidly proliferating tumors with high bioenergetic requirements are unable to downregulate metabolism in response to starvation. This means that tumor cells may be vulnerable to metabolic stress which might be induced pharmacologically, and this could provide a means of inducing relatively cancer-cell specific toxicity with non-cytotoxic drugs.[2, 3] Potential triggers for this metabolic stress would include inhibitors of glycolysis, angiogenesis inhibition and nutrient deprivation, or forcing mitochondrial respiration. Several strategies and drugs that may interfere with the glycolytic metabolism of cancer cells are under study and may lead to the development of anticancer approaches with clinical relevance.[4, 5]

CASE IN POINT: DIABETES MELLITUS & CANCER

It is known that people with type 2 diabetes mellitus are at increased risk for various cancers, including hepatic, colon, bladder, breast, and endometrial,[6] and that their prognosis is worsened relative to nondiabetic cancer patients.[7] Several different possible mechanisms to explain this risk have been postulated, and include diabetes-related metabolic abnormalities such as obesity, hyperlipidemia, and oxidative stress, and additional effects of exogenous insulin administration such as stimulation of insulin-like growth factor receptors.[8]

A metabolic drug with anticancer potential: Metformin

Epidemiologic studies

Epidemiologic evidence began to emerge in the early 2000's suggesting that diabetics who receive the biguanide metformin, a commonly prescribed oral hypoglycemic agent in people, for control of hyperglycemia have a reduction in their diabetes-related excess cancer risk. In the last 3 years a raft of epidemiological studies have strengthened that suggestion, and shown that the benefit is limited to metformin and not other antidiabetic medications. Further, in addition to showing a reduced risk of cancer occurrence, recent studies have indicated an improvement in the prognosis of diabetics receiving metformin who develop cancer, relative to diabetic cancer patients that do not receive metformin.[6, 9 - 15] A November 2010 meta-analysis of 11 studies including over 4,000 cancer events and over 500 cancer deaths found a 31% reduction in overall summary relative risk (0.69; 95% confidence interval, 0.61-0.79) in diabetics taking metformin compared with other antidiabetic drugs.[15] These epidemiologic studies have spurred investigations into possible mechanisms of action of this apparent protective effect of metformin and preclinical drug studies.

Mechanistic and preclinical studies

The mechanism of this apparent protective effect of metformin is not yet entirely clear, but a number of potential mechanisms have been postulated, and some valuable evidence has so far been published. It is clear that metformin activates AMP kinase (AMPK), which has tumor suppressor functions in addition to its central role in energy homeostasis. AMPK is also activated by the product of the tumor suppressor gene LKB1.[6] In addition, because metformin reduces blood glucose level it may cause metabolic stress to tumor cells committed to aerobic glycolysis, as discussed above. Further, by reducing blood glucose level and insulin resistance, blood insulin level is reduced, so reduced stimulation of insulin / insulin-like growth factor (IGF) receptor may play a role.[6]

Metformin acts as an inhibitor of the mammalian target of rapamycin (mTOR) via AMPK signaling, which has antitumor effects. However, while mTORC1 inhibition by metformin is directed through AMPK, it has also been shown to occur in the absence of AMPK, therefore at least another pathway also exists. [16] Further, metformin's stimulation of AMPK may result in reduced production of estrogen in breast tissue by inhibiting aromatase expression.[17]

Metformin is also commonly used in the treatment of polycystic ovary syndrome (PCOS). In a study that used motility and migration assays as a marker for invasive and metastatic capacity in a human endometrial carcinoma cell line, the presence of sera from women receiving metformin for PCOS resulted in significantly reduced in vitro invasion compared to sera from matched control women with PCOS not receiving metformin, suggesting potential antimetastatic and / or anti-invasive properties.[18]

A number of other potential beneficial effects of metformin have been raised. CD24, an adhesion molecule associated with increased metastatic potential was shown to be inhibited on a breast cancer cell line in the presence of metformin.[19] In another study, a breast cancer cell line incubated with metformin had significantly reduced MDR1 expression.[20] A recent study that used an in vivo model of tobacco carcinogen-induced lung cancer reported that tumor burden was decreased in animals that were administered metformin. [21] Finally, another study examined the effects of AMPK activation in human lung, prostate, and breast cancer cells irradiated in vitro and found that cancer cells irradiated in the presence of metformin had increased levels of AMPK activation and a lower surviving fraction than cells irradiated without metformin.[22]

Evidence has also been reported that metformin can be toxic to tumor stem cells – at least in cell lines - and may act synergistically with conventional chemotherapy in this way.[23] Investigators treated mice implanted with transformed mammary epithelial cells with metformin alone, metformin plus doxorubicin, and doxorubicin alone. Metformin alone had no effect, doxorubicin alone shrank the tumors but they then regrew, and the combination caused complete responses and no recurrence.[24]

Clinical trials

Encouraging findings in epidemiologic and preclinical studies have led to the suggestion that metformin warrants investigation as an anticancer therapeutic and / or chemopreventive drug in its own right, separate from its role in diabetes management. [25] At this time, attention is especially focused on metformin's therapeutic potential in breast [26 - 28] and lung cancer. [29]

In a recent phase 1 trial, patients with advanced solid tumors who had exhausted standard treatment options received a combination daily metformin and weekly tensirolimus (another mTOR inhibitor). At the starting dose, dose-limiting toxicity was observed in all patients, and at the next (reduced dose), 25% of patients experienced dose-limiting toxicity; 1 out of 8 patients experienced a partial response and 5 out of 8 patients had stable disease. [30]

Some investigators have argued that metformin's long established history of clinical use in people and its relatively moderate and well understood toxicity profile mean that the drug could potentially be brought into clinical trials as an anticancer agent more quickly. In fact, a phase III adjuvant trial (NCIC MA.32) has now been initiated. [31]

Related drugs

Two biguanide drugs related to metformin, phenformin and buformin, were previously used as oral antidiabetic agents but withdrawn in the 1970s because their greater toxicity risk compared to metformin. However, the intriguing evidence for metformin's potential value in chemoprevention and cancer treatment has led to the suggestion that these two drugs should be re-evaluated for potential utility in cancer management and that other related compounds should also be sought and investigated. [6]

SUMMARY

Retrospective studies strongly suggest that diabetics treated with metformin have a substantially reduced cancer burden compared with other diabetics, although the precise mechanism remains unclear, as does the relevance to people without diabetes. However, laboratory models have provided strong evidence for the activity of metformin and other biguanides in both cancer treatment and chemoprevention. Investigations of the mechanism of action of these drugs have shown that the mechanism is complex, and that much remains to be learned in order to use these drugs optimally. [33]

Most oncologists would readily acknowledge that successful cancer management in patients almost always requires a multifaceted treatment approach, and that the availability of new agents or approaches with some efficacy, even if incomplete, is welcome. A number of drugs that are not cytotoxic agents and were originally intended for other purposes have been found to have utility in the management of cancer, including NSAIDs, thalidomide, retinoic acid, and others. In some ways their non-cytotoxicity contributes to their utility because they can be combined with cytotoxic treatments in novel ways to increase the efficacy of therapy. It is possible that metformin and related drugs may join the ranks of these drugs to become an adjunctive / supportive drug in cancer management; or, it may be that the findings for metformin discussed here ultimately become just a clue towards understanding a target in cancer cells that could be exploited with greater potency through a potential novel treatment approach.

REFERENCES

[1] Pelicano H, et al. **Oncogene**. 2006;25:4633. [2] Karantza-Wadsworth V, et al. in DeVita VT et al, **Cancer PPO** 2008. [3] Jin S, et al. **J Cell Sci** 2007;120:379. [4] López-Lázaro M. **Anticancer Agents Med Chem** 2008;8:305. [5] Sattler UG, et al. **Curr Med Chem** 2010;17:96. [6] Chong CR, et al. **Oncol** 2009;14:1178. [7] Barone BB, et al. **JAMA** 2008;300:2754. [8] Vigneri P, et al. **Endocr Relat Cancer** 2009;16:1103. [9] Lee MS, **BMC Cancer** 2011;11:20. [10] Yang X, et al. **Diabetes Care** 2011 Jan 7. [Epub ahead of print] [11] Monami M, et al. **Diabetes Care** 2011;34:129-31. [12] Baur DM, et al. **Metabolism** 2010 Nov 15. [Epub ahead of print] [13] McFarland MS, et al. **Pharmacotherapy** 2010;30:1159. [14] Papanas N, et al. **Expert Opin Investig Drugs** 2010 Jun 18 . [Epub ahead of print]. [15] Decensi A, et al. **Cancer Prev Res (Phila)** 2010;3:1451. [16] Kalender A, et al. **Cell Metab** 2010;11:390. [17] Brown KA, et al. **Breast Cancer Res Treat** 2010 Mar 19. [Epub ahead of print] [18] Tan BK, et al. **J Clin Endocrinol Metab** 2010 Dec 29. [Epub ahead of print] [19] Vazquez-Martin A, et al. **Oncol Rep** 2011;25:135. [20] Kim HG, et al. **Br J Pharmacol** 2010 Nov 3. [Epub ahead of print] [21] Memmott RM, et al. **Cancer Prev Res (Phila)** 2010;3:1066. [22] Sanli T, et al. **Int J Radiot Oncol Biol Phys** 2010;78:221. [23] Vazquez-Martin A, et al. **Breast Cancer Res Treat** 2010 May 11. [Epub ahead of print] [24] Hirsh HA et al. **Cancer Res** 2009;69:7507. [25] Ben Sahra I, et al. **Mol Cancer Ther** 2010;9:1092. [26] Gonzalez-Angulo AM, et al. **Clin Cancer Res** 2010;15:1695. [27] Cassniga M, et al. **Cancer Epidemiol Biomarkers Prev** 2009;18:701. [28] Muti P, et al. **Cell Cycle** 2009;8:2661. [29] Antonoff MB, et al. **Semin Thorac Cardiovasc Surg** 2010;22:195. [30] Mackenzie MJ, et al. **Invest New Drugs** 2010 Oct 27. [Epub ahead of print] [31] Goodwin PJ, et al. **Breast Cancer Res Treat** 2011;126:215. [32] Pollak M. **Cancer Prev Res (Phila)** 2010 Sep;3:1060.

CHEMOTHERAPY PREPARATION & ADMINISTRATION SAFETY ISSUES

Angela E. Frimberger, VMD, MACVSc, DACVIM (Oncology)
Wauchope, NSW, Australia

INTRODUCTION: WHY ALL THE FUSS?

Most chemotherapeutic agents are both toxic and mutagenic. Alkylating agents have been associated with the highest risks to handlers. Organ damage and increased risk of fetal loss have been reported in persons handling and administering chemotherapy with inadequate attention to personal safety. This was first suspected in the 1970's when chemotherapy drugs had already been in clinical use for 2 decades, and unfortunately (as in many other industries) many healthcare workers were exposed to hazardous substances that led to illness and even deaths among those workers.[1]

In the US, the Occupational Safety and Health Administration (OSHA) investigated chemotherapy preparation practices in the early 1980's and in1986 published guidelines describing the equipment, garments, and work practices appropriate to protect pharmacists and nurses from exposure to cytotoxic agents.[2]

After the OSHA guidelines were implemented, concern about safety in handling chemotherapy was again reduced; however although the magnitude of risk was certainly greatly reduced by following the guidelines, new studies were published in the 1990's showing that drug residues on work surfaces were a potential significant source of worker exposure.[3] One review pointed out that despite the use of precautions cyclophosphamide was present in the urine of health care workers in 11 of 12 studies.[4] The presence of contamination outside of biological safety cabinets demonstrates that their protection is imperfect, and that manipulations such as reconstitution, transfer between containers, spiking and unspiking IV containers, priming IV tubing, and connecting or disconnecting syringes from injection ports cause leakage and contamination. In addition the contamination of the working environment is persistent, and other workers who are not directly involved in handling hazardous substances may be exposed.

REGULATORY CONSIDERATIONS

Regulatory guidelines for employers whose workers are handling hazardous drugs including cytotoxic chemotherapeutics have been discussed and determined to some extent around the world, however in large part they are not legally binding. Overall, precautions are similar in any geographic location, but veterinarians are encouraged to check the specific guidelines and regulations for their location to ensure they are in compliance with any relevant regulations.

In the US, the relevant regulatory instruction is to be found in the OSHA Technical Manual, Section VI: Health-Care Facilities, Chapter 2. *Controlling Occupational Exposure to Hazardous Drugs*; which can be found online at: http://www.osha.gov/dts/osta/otm/otm_vi/otm_vi_2.html#5. While the OSHA guidelines are not in themselves law, they can be enforced under the general duty clause in the Code of Federal Regulations (CFR) which requires that employers provide a safe or healthful employment and places of employment, should OSHA be called in investigate a complaint. In addition, the Hazard Communication Standard is a federal regulation that requires employers to inform employees of the risks of all hazardous materials in their workplace; and the Access to Employee Exposure and Medical Records standard requires that employees who work with toxic substances have access to workplace monitoring records as well as their employee medical records.

Additional extremely useful and more recent – but voluntary - guidelines can be found in NIOSH Publication No. 2004-165: *Preventing Occupational Exposure to Antineoplastic and Other Hazardous Drugs in Health Care Settings*, to be found at: http://www.cdc.gov/niosh/docs/2004-165/.

While other specific requirements may exist, the OSHA and NIOSH publications can also provide useful guidance for veterinary practitioners in other countries.

In addition, it is important to ensure that any cytotoxic-contaminated waste is disposed of according to any regulations. In the US, the Environmental Protection Agency (EPA) regulates the management of hazardous wastes, including several chemotherapy drugs according to the Resource Conservation and Recovery Act (RCRA). Two relevant EPA publications are *Managing Hazardous Waste: A Guide for Small Businesses* [EPA 2001] and *RCRA Hazardous Waste Regulations* [40 CFR Parts 260–279].

DEVELOPING SAFE CHEMOTHERAPY HANDLING POLICIES IN VETERINARY PRACTICE

Precautions should be taken during every phase of chemotherapy preparation, administration and disposal.

Employers should:

- Develop policies and procedures for the safe storage, preparation, administration, and disposal of cytotoxic agents.
- Seek input of workers on all policies and procedures that will affect them.
- Provide safety training for employees, including how to recognize and evaluate possible risks.
- Whenever possible, provide a work area that is designated for chemotherapy and is limited to authorized personnel.
- Whenever possible, provide a biological safety cabinet (below).
- Identify the hazardous drugs used in the practice and keep a written inventory up to date.
- Make Material Safety Data Sheets (MSDS) and other relevant information available to employees.
- Provide workers with proper personal protective equipment PPE (can include correct chemotherapy gloves, disposable gowns and sleeve covers, eye and face protection, and NIOSH-certified respirators (surgical masks do not provide adequate respiratory protection)) (these are generally available from chemotherapy drug suppliers), train workers in its use and require workers to do so.

- Prohibit eating, drinking, etc. in areas where cytotoxic agents are handled.
- Develop a spill management procedure.
- Develop a plan for medical surveillance of personnel handling hazardous drugs.
- Develop a policy regarding chemotherapy handling during pregnancy, conception, and breastfeeding. The Oncology Nursing Society recommends that "employers provide alternate duty to employees who request other assignments due to pregnancy, the desire to conceive, or breast-feeding".[1]
- Provide syringes and IV sets with Luer-Lok fittings for preparing and administering chemotherapy drugs.
- Consider using closed-system drug transfer devices and needleless systems (below).
- Monitor compliance with safety procedures.

Workers should:

- Participate in training before taking on chemotherapy duties.
- Review guidelines and other documents for more information when needed.
- Consistently use personal protective equipment and follow recommended procedures.
- Wash hands after chemotherapy-related duties.
- Disposing of any contaminated waste in designated containers.
- Clean up any spills according to set procedures (above).
- Report and follow up on any accidental exposures.

SAFE CHEMOTHERAPY HANDLING

There are 3 basic systems for safe chemotherapy handling:

The traditional gold standard is to use a vertical laminar flow biological safety cabinet (BSC) (or "hood") for drug preparation. A BSC has vertical airflow that moves away from the worker, as opposed to horizontal airflow that moves away from the product toward the worker (which is used in cabinets designed to keep the contents of the cabinet sterile). However, as noted above, BSCs provide imperfect protection and should not be relied on solely. "Chemo-pins" (or the PhaSeal system) (both below) are used for preparing injectable chemotherapeutics in the hood. Both can be used as a multi-dispensing pin. Further, a disposable gown with closed-cuff sleeves and chemotherapy gloves should be worn.

A newer approach that appears to offer equivalent (if not better) safety and better convenience is to use a closed system drug transfer device, as suggested by the NIOSH guidelines published in 2004, discussed above. Several commercial systems are now on the market but the best developed and documented is the PhaSeal system, a component-based closed system drug transfer device that mechanically prohibits the escape of drug aerosols.[4, 5, 6] Although a hood is not necessary, gloves should be worn. The components are relatively costly compared to regular syringes and needles, but much less costly than installing a hood. A major safety advantage over using a hood with traditional needles and syringes is that it provides protection during administration as well as preparation. As well as being safer, this also allows most clients (unless pregnant, see below) to remain with their pets during chemotherapy administration, which usually greatly reduces anxieties!

A third alternative is to prepare and administer drugs in a quiet, low traffic area using protective eyewear, a respirator mask, a disposable gown with closed-cuff sleeves and gloves during both preparation and administration. Further, when preparing injectable chemotherapeutics, it is preferable to use "Chemo Dispensing Pin" hydrophobic filters that insert into drug vials and prevent aerosolisation. If a filter is not used, alcohol moistened gauze should be wrapped around the vial top and needle when the needle is withdrawn. *This option is only appropriate in situations where chemotherapy is being handled very infrequently.*

Further chemotherapy safety procedures include:

- Whenever chemotherapy is to be administered, have ready a "SPILL KIT" containing personal protective equipment, disposable gown, over-shoes, chemotherapy gloves, protective eyewear, NIOSH-approved respirator; disposable absorbent material cleaner such as strong alkaline detergent, disposable mop head, approved container for cytotoxic waste and sharps, possibly a disposable brush and shovel should be available in case a glass container is broken, and potentially spill pillows for larger spills.
- Gloves should be powder free, nitrile or heavy latex (not vinyl), not exam or surgical gloves. If exam or surgical gloves must be used, they should be double-gloved.
- Gowns should be disposable, low-permeability, with closed front and cuffs, intended for single use.
- A plastic-backed absorbent disposable pad should always be used ("incontinence pads" are suitable).
- Luer lock syringes should always be used.
- Chemotherapy vials and syringes should be stored and carried in zipper-lock plastic bags.
- For drugs that are kept refrigerated, a designated small refrigerator is preferred; they at least MUST be kept in a separate refrigerator from that used to store food.
- Pills should never be split or broken.
- Chemotherapy waste (e.g. used syringes, gloves, empty bottles) should be managed separately from the main waste stream. Most hospital waste collectors can provide suitable containers.

- For practice staff handling body fluids (e.g. cleaning cages) from chemotherapy patients, routine protection should be taken against contact with excreta. Cages and runs must never be pressure-hosed.

Premixed chemotherapy drugs

Some veterinarians find it convenient to order custom-prepared individual chemotherapy doses; which can be ordered with PhaSeal components attached. This is more costly compared to purchasing multi-dose vials, but for practices that use chemotherapy infrequently it is more cost effective by saving on inventory, drug wastage and equipment costs. One disadvantage is that if chemotherapy administration is delayed the drug dose may not be able to be stored in the syringe, so is wasted.

Oral medications

Most chemotherapy safety guidelines focus on injectable medications, and guidelines for handling orally administered chemotherapeutics safely are sparse. For oral chemotherapy in smaller patients, it is often necessary to have drugs reformulated into smaller size capsules. This must only be done by a compounding pharmacist equipped for handling cytotoxic medications and must never be done in the veterinary practice.

PREGNANCY AND CHEMOTHERAPY

In the author's opinion, despite all safety measures, veterinary staff who are, or are trying to become, pregnant or are breastfeeding, should not assist in chemotherapy preparation or administration, or clean cages of chemotherapy patients.

Pregnant pet owners should consult their obstetricians regarding whether they should have their pet undergo chemotherapy and how to manage aftercare. A general recommendation is that pregnant clients not be in the room during drug preparation and administration, and that aftercare be performed by another member of the family; however social contact such as petting is safe.

CHEMOTHERAPY SAFETY AT HOME

The main chemotherapy safety effort should be directed at the level of preparing and administering drugs, where higher levels of exposure to the drug occur, and often involve the same people repeatedly. However it is important to remember that clients will be taking chemotherapy drugs home - if not in bottles, then at least in their pets' bodies – and will require instructions on safe handling.

If oral chemotherapy is dispensed for administration at home, medications should be placed in medication bottles (not paper envelopes or plastic bags) marked "CYTOTOXIC" and chemotherapy gloves should be provided.

Usually there is no significant risk to the owner of a pet receiving chemotherapy when routine hygiene is used. However, it is important to be aware of basic considerations involved. The two times pet owners need to take precautions are when administering medications, or when handling bodily wastes.

Many veterinary clients (and practice staff) are very concerned about handling excreta of chemotherapy patients. While it is appropriate to employ safety precautions, normal hygiene is largely adequate. Simply avoiding direct contact with urine and feces prevents the majority of exposure and most pet owners do this anyway. Weather conditions and contact with soil inactivate most active substances (remember how carefully most chemotherapy drugs must be stored to maintain potency) so pets should eliminate in an area where urine will drain into the soil. If solid wastes need to picked up, gloves or an inside-out plastic bag should be worn and they should be put in a plastic bag and disposed of in the trash. If bedding is soiled it is preferable to dispose of than launder it if possible. The concentration of active substances in excreta is usually low so a single accidental exposure followed by washing should not be cause for panic.

These precautions should be followed for approximately 48 to 72 hours following administration. However, specific guidelines for handling excreta are not well defined for most drugs in animals, so it is best to err on the conservative side. Recent reports showed that low levels of drug residues of some chemotherapeutics were detectable in the urine of dogs for days and in some cases weeks after therapy.[7, 8] For pets receiving frequent chemotherapy treatment (for example, on metronomic chemotherapy), the timing guidelines obviously need to be adjusted to the individual situation.

Table 1. An example of the type of instructions that should be given to pet owners following chemotherapy.

Most drugs (including chemotherapy) are excreted to some extent in the urine or feces as the actual drug, or metabolites which may be active, for approximately 24 to 48 hours. In particular, the platinum drugs (carboplatin and cisplatin) are highly excreted in the urine for about 48 hours after administration. So as a general rule, direct contact with these body fluids should be avoided for about 3 days to be on the safe side. Most people do this anyway. Here are some procedures for managing body wastes after chemotherapy:

- Any time you need to handle bodily wastes, wear latex gloves, "peel" the gloves off afterwards so they come off inside out and dispose of in the trash, and wash your hands.
- For cats that use a litter pan, consider using clumping litter. Wear latex gloves when scooping and wash your hands afterwards, peel the gloves off afterwards and dispose of in the rubbish, and wash your hands. Place waste in a plastic bag for disposal.
- If other cats in the household can be encouraged to use a different litter pan that is ideal, but if not possible then the next best arrangement is to keep the pan scooped as frequently as possible so the other cats are not continually getting the patients' wastes on their feet and then grooming afterwards.
- If your pet relieves him / herself outdoors, try to have them do it in a quiet area, on soil where urine will drain away quickly, and drug and metabolites will be inactivated by sun and rain.
- If you need to pick up solid waste, you should wear latex gloves; put the waste in a plastic bag, and dispose of in the trash.
- At home, wastes can be put either in a plastic bag in the trash, or place in the toilet, close the lid, and double flush.
- If you need to clean up an "accident" in the home, wear gloves and use disposable absorbent materials like paper towels (or old towels that you don't mind throwing away) and put them in a plastic bag in the trash. Bleach is the most effective cleaner for chemotherapy drugs but if that is not an option on the surface, other soaps or cleaners can also be effective.
- Any soiled bedclothes should be discarded rather than washed if possible.
- Other animals should not be allowed to consume the patient's wastes (sorry, but it happens).
- Finally, if you, a family member, or another pet accidentally contacts body wastes; don't panic, just wash it off with soap and plenty of water. Even with the drugs that are excreted in the highest concentration, it is over about 48 hours, so the amount in any single urination will be a fraction of the dose your pet received, and the amount that actually gets on your skin is a small fraction of that; and only a very small percentage could possibly penetrate the skin.

If you have chemotherapy tablets or capsules to give to your pet at home:

- Remember to keep bottles and waste well out of reach of children.
- Always wear latex gloves when handling these medications. Your veterinarian can provide these.
- The medications should not be handled in the kitchen; it should be done in another room such as a laundry room or bathroom (away from the toothbrushes). If medications need to be refrigerated, only place the tightly closed bottle in the fridge, and carry the bottle to another room before opening.
- Pills should not be split or broken, and capsules should not be opened.
- If you need to place the medications directly on a surface, use something disposable such a paper towel or piece of kitchen foil, and then dispose of it afterwards.
- Afterwards, "peel" your gloves off so they come off inside out, dispose in the trash, and wash your hands.

In addition to the procedures listed above, women who are or may be pregnant should:

- **not** be present in the room when chemotherapy is being administered
- **never** handle any chemotherapy medications that are dispensed to be given at home
- **never** handle the urine or faeces of a pet that has received chemotherapy within the past 3 days – these jobs should be delegated to another family member.

Very young children should also not be present when chemotherapy is being administered. There are no clear guidelines on whether it is safe for older children to be present when chemotherapy is being given to a pet, and this decision should be left to the judgment of the veterinarian and the parent.

These procedures may seem very detailed, but this is only because many people caring for pets receiving chemotherapy have lots of questions about safety. We want to be sure that all your concerns are addressed and prepare you for any eventuality. You'll soon see that the times you need to be careful are really quite limited; most of the things you need to do are more or less the same basic hygiene that you normally use; and the additional things you need to do will become a simple routine.

REFERENCES

[1] Polovich M. **Online J Iss Nurs** 2004;9:5.E-pub. [2] McDevitt JJ, et al. **J Occup Med** 1993;35:57. [3] Sessink PJM, et al. **Drug Safety** 1999;20:347. [4] Harrison BR, et al. **Am J Health Syst Pharm** 2006;63:1736. [5] Sessink PJ, et al. **J Oncol Pharm Pract** 2010;Epub ahead of print [6] Siderov J, et al. **J Oncol Pharm Pract** 2010;Epub ahead of print. [7] Knobloch A et al. **J Vet Int Med** 2010;24:379. [8] Knobloch A et al. **J Vet Int Med** 2010;24:384.

Do NSAIDS make a difference in cancer?

Rowan J. Milner, BVSc (Hons), MMedVet (Med), DACVIM (Oncology), DECVIM (Int Med & Oncology)
Gainesville, Florida

INTRODUCTION

The medicinal benefits of a powder made from the leaves and bark of the willow tree have been known for over 3500 years. Hippocrates left historical records showing the medical benefits of willow tee extracts for headaches, pains and fever. In the 1800's chemists isolated the active ingredient, acetylsalicylic acid (aspirin), which lead to the commercial production of the drug. The annual modern day consumption of the drug is estimated to be 40,000 to metric tons.[1] Apart from the past uses of aspirin it is now commonly used to help prevent heart attacks, strokes and blood clot formation in high risk individuals.[2] More recently, the benefits of aspirin and other NSAIDS in the treatment of cancer have been reported; for a detailed review of past human and animal studies see references.[3,4] Because of the plethora of information out there on NSAIDS and cancer, this review seeks to provide a summary of the mechanism of action of NSAIDS, newly described pathways, and published veterinary literature that underpins the use of NSAIDS in cancer treatment.

Mechanism of Action

Traditionally the primary targets of NSAIDS are the enzymes responsible for the enzymatic catabolism of arachidonic acid following phosphilase-A_2 enzymatic activity on the cellular phospholipid bilayer of the cell. It is worth noting that corticosteroids exert their anti-inflammatory effect by inhibiting phospholipase-A_2, thus potentiating the negative side effects of NSAIDS. The resultant metabolites include prostaglandins, eicosanoids and leukotrienes. More specifically, certain NSAIDS e.g. aspirin or piroxicam are nonspecific, targeting both cyclooxygenase (COX-1 and COX-2) enzymes, and are known as nonselective NSAIDS. Others like tepoxalin target not only cyclooxygenases but also lipoxygenase. More recently developed NSAIDS or more selective and target COX-2 at lower concentrations than COX-1. This differential in dose is called the COX-1: COX-2 ratio and will be discussed in detail later. However it is important to emphasize that selective COX-2 inhibitors, if given at high enough concentrations will inhibit COX-1. The inhibition of COX-1 by NSAIDS is a double-edged sword since it is constitutively expressed in nearly all cell types at a constant level and is responsible for the cytoprotective prostaglandins of the stomach and in the maintaining normal renal blood flow. In addition, it is found in high concentrations in platelets, where it produces the eicosanoid, thromboxane that is needed for primary plug formation. Its location on the luminal side of the smooth endoplasmic reticulum of the cell is common to most enzymes. COX-2 is almost identical in structure to COX-1 and is primarily an inducible enzyme that can accept a wide range of fatty acids as substrates for prostaglandin production. It is only found constitutively expressed in the brain, kidney and reproductive tract. Its intracellular location in the endoplasmic reticulum and nuclear membrane may allow it to act as transcription factor for DNA. COX-2 can be induced by lipopolysaccharide, cytokines IL-1, IL-2, TNF, oncogenes, growth factors, and tumor promoters leading to increases in inflammatory prostaglandins within hours. In cancer, COX-2 is found to be overexpressed in certain preneoplastic and neoplastic tumors and its metabolite, PGE_2 is linked to inhibition of apoptosis, increased angiogenesis, increased proliferation, increased invasiveness, immunosuppression and chronic inflammation. This is particularly true of high-grade carcinomas, with an inflammatory component in which inflammation is, in part, responsible for the pro-malignant transformation known as mesenchymal-epithelial transition (MET). Recently an enzyme variant of the COX-1 gene, COX-3 has been described.[5] Interestingly the enzyme in the dog is similar to COX-1 and thus capable of enzymatic action on arachidonic acid, however the gene product in the mouse is not.[6] COX-3 in the brain is postulated to be the target of acetaminophen and responsible for its antipyretic properties.[5]

Current discoveries on inflammation have identified Nuclear Factor- B (NF- B) as a COX-independent mechanism of inflammation and thus a target for NSAIDS. When activated in the cytoplasm, NK- B translocates to nucleus and acts as a pro-inflammatory and proliferation transcription factor. Aberrant activation of NF- B is frequently observed in many cancers and suppression of NF- B limits the proliferation of cancer cells. High doses of NSAIDS do inhibit NF- B by interfering with its activation pathway.

Pharmacology

The reader is referred to a thorough review of NSAID pharmacology by Papich (2008).[6] However, what is key to understanding NSAID pharmacology is the reversible or irreversible inhibition of COX by NSAIDS, selectivity COX-1: COX-2 ratio of some NSAIDS and that different forms of the same drug exist. By definition, the IC_{50} of an NSAID is defined as the drug concentration needed to inhibit the enzyme activity by 50%. The selectivity of COX-2 inhibitors is based on the IC_{50} ratio of COX-1 inhibition relative to COX-2 inhibition. Thus, a COX-2 selective drug like meloxicam has a higher COX-1:COX-2 ratio than ketoprofen that is nonselective. See Table-1 for details on published COX-1:COX-2 ratios.

It important to also note that some drugs (carprofen, ketoprofen) may exist as *racemic mixtures* i.e. the commercial drugs are in a 50:50 ratio of two mirror-image enantiomeric forms, R and S.[7] The implication of this finding is that the enantiomers do not have the same pharmacological properties in the body leading to differences in pharmacokinetics and pharmacodynamic. Since not all drugs are created equal, this may be of particular importance when using generic drugs.[7]

Table 1: Published IC50 ratios in dogs for selective and nonselective COX-2 inhibitory NSAIDS

Drug	COX-2 selectivity	COX-1:COX-2 ratio [8]
Deracoxib	Yes	1275
Carprofen	Yes	16.8 (1.75) [9]
Meloxicam	Yes	2.72 (12.3) [9]
Piroxicam	No	2
Etodolac	No	0.53
Aspirin	No	0.39
Ketoprofen	No	0.17 (0.36) [9]

Non-steroidal anti-inflammatory drugs and Cancer

The current lay press and scientific literature abounds with references on NSAIDS in research and clinical medicine. Epidemiologic studies with aspirin have shown an inverse relationship with prolonged usage and colon cancer, resulting in a decreased risk of colon cancer by 30-50%. In addition, aspirin may also reduce risk of esophageal, stomach, breast, and ovarian cancer.[3] Normal gut epithelium has no COX-2 expression but adenomatous polyps and colorectal cancer show between 40 and 80% expression respectively. COX 2 expression in human colon cancer is associated with larger tumor size, deeper invasion but not with metastasis.

In a canine study by Mohammed SI et al [10] the authors examined the expression of COX-1 and -2 using immunohistochemistry in Canine lymphoma (LSA), prostatic carcinoma (PCA), osteosarcoma (OSA), oral melanoma (MEL), oral squamous cell carcinoma (SCC), oral fibrosarcoma (FSA), mammary carcinoma (MCA), and normal tissues. They found COX-2 expression in 65% of SCC, 61% of MCA, 55% of PCA and in 60% of oral melanoma cases. COX-2 was generally absent in normal tissues. COX-2 expression was minimal or absent in mesenchymal tumors and canine lymphoma. As expected COX-1 was expressed in normal epithelial tissues and in some osteoclast and osteoblast in bone, but was not in normal lymph node. The same authors examined the expression of PGE_2 in the similar tumors and found increased levels mirroring COX-2 expression.[11] Several of other studies have found similar COX-2 expression including 100% expression in transitional cell carcinomas (TCC).

In a landmark 1992 Phase-1 study by Knapp et al[12] in 62 dogs with naturally occurring tumors piroxicam, several partial remissions were documented in 30% of TCC, 60% of SCC and 30% of MCA. This study established the once a day dose of 0.3 mg/kg piroxicam as an acceptable dose. In a subsequent study by Knapp et al[13] using 0.3mg/kg q 24 hours in 34 TCC dogs: 45% of dogs had a complete remission (CR), 12% a partial remission (PR), 58% stable disease (SD), and 25% progressive disease. Crucially the median survival of the dogs without surgery was 181 days which was a significant improvement in survival. Other findings included a fall in PGE_2 levels in circulating monocytes with decreasing tumor size. Interestingly piroxicam had little antitumor effect in an in-vitro study of 4 canine cell lines. Indeed this was a common finding through subsequent in-vitro trials with NSAIDS. Where cancer cell toxicity did occur it was at unobtainable serum levels. In an effort to improve efficacy piroxicam has been combined with cisplatin, [14, 15] mitoxantrone, [16] and carboplatin. [17] Piroxicam combined with cisplatin was found to be too nephrotoxic in both studies and cannot be recommended at this time. The combination of carboplatin with piroxicam did not result in a significant increase in survival when compared to piroxicam alone. The current gold standard is mitoxantrone (5 mg/m^2 IV every 3 weeks for 4 cycles) combined with piroxicam (0.3mg/kg q 24hours). The median survival was found to be 350 days, however the study reported a median time to failure of 291 days. This begs the question, Is chemotherapy an effective adjunct to piroxicam since the median survival of piroxicam alone is 181 days? Nonetheless, the anecdotal experience at UF oncology clinic is that some cases have prolonged survivals which we did not see with piroxicam alone.

Not surprisingly NSAID therapy is being used in other tumors, principally carcinomas, and has been the focus of a number studies. The use of piroxicam in canine oral SCC was reported by Schmidt et al.[18] Responses were seen with 5% showing a CR, 11% PR, 29% SD and 55% PD. However the addition of carboplatin in 7 dogs with oral SCC resulted in a 57% CR rate with no surgery and a median follow-up of 534 days.[19] This seems to be a promising combination and early results at UF seem to be supportive of the use of carboplatin with piroxicam in oral SCC. In a study piroxicam and carprofen were used in canine prostatic carcinoma with promising results.[20] The median survival was 6.9 months compared to 0.7 months. Responses have also been seen in nasal adenocarcinoma, mammary carcinoma. Non-steroidal anti-inflammatory drugs also form part of the "metronomic therapy", which is a combination of doxycycline, cyclophosphamide and piroxicam.

In conclusion, the mechanism of action of NSAIDS now includes inhibition of COX-1, COX-2, lipoxygenase, and NF- B resulting in the suppression of inflammation, angiogenesis and tumor proliferation. Clinical research has explored the use of NSAIDS in numerous carcinomas in canines. While clinical research has focused mainly on piroxicam, research on meloxicam, deracoxib, carprofen, and tepoxalin is likely in the near future.

REFERENCES

1. Warner TD, Warner, T. D. Cyclooxygenase-3 (COX-3): filling in the gaps toward a COX continuum? *Proc Natl Acad Sci USA*. 2002;99(21):13371–3. 2. Lewis HD, Davis JW, Archibald DG, et al. Protective effects of aspirin against acute myocardial infarction and death in men with unstable angina. Results of a Veterans Administration Cooperative Study. *N. Engl. J. Med.* 1983;309(7):396-403. 3. Harris RE, Beebe-Donk J, Doss H, Burr Doss D. Aspirin, ibuprofen, and other non-steroidal anti-inflammatory drugs in cancer prevention: a critical review of non-selective COX-2 blockade (review). *Oncol. Rep.* 2005;13(4):559-583. 4. Baek SJ, McEntee MF, Legendre AM. Review paper: Cancer chemopreventive compounds and canine cancer. *Vet. Pathol.* 2009;46(4):576-588. 5. Chandrasekharan NV, Dai H, Roos KLT, et al. COX-3, a cyclooxygenase-1 variant inhibited by acetaminophen and other analgesic/antipyretic drugs: cloning, structure, and expression. *Proc. Natl. Acad. Sci. U.S.A.* 2002;99(21):13926-13931. 6. Papich MG. An Update on Nonsteroidal Anti-Inflammatory Drugs (NSAIDs) in Small Animals. *Veterinary Clinics of North America: Small Animal Practice.* 2008;38(6):1243-1266. 7. Lees P. Analgesic, Antiinflammatory, Antipyretic Drugs. In: *Veterinary Pharmacology and Therapeutics*. 9th ed. Ames, Iowa: Wiley-Blackwell; 2009:457-486. 8. Streppa HK, Jones CJ, Budsberg SC. Cyclooxygenase selectivity of nonsteroidal anti-inflammatory drugs in canine blood. *Am. J. Vet. Res.* 2002;63(1):91-94. 9. Kay-Mugford PA, Benn SJ, LaMarre J, Conlon PD. Cyclooxygenase expression in canine platelets and Madin-Darby canine kidney cells. *Am. J. Vet. Res.* 2000;61(12):1512-1516. 10. Mohammed SI, Khan KNM, Sellers RS, et al. Expression of cyclooxygenase-1 and 2 in naturally-occurring canine cancer. *Prostaglandins Leukot. Essent. Fatty Acids.* 2004;70(5):479-483. 11. Mohammed SI, Coffman K, Glickman NW, et al. Prostaglandin E2 concentrations in naturally occurring canine cancer. *Prostaglandins Leukot. Essent. Fatty Acids.* 2001;64(1):1-4. 12. Knapp DW, Richardson RC, Bottoms GD, Teclaw R, Chan TC. Phase I trial of piroxicam in 62 dogs bearing naturally occurring tumors. *Cancer Chemother. Pharmacol.* 1992;29(3):214-218. 13. Knapp DW, Richardson RC, Chan TC, et al. Piroxicam therapy in 34 dogs with transitional cell carcinoma of the urinary bladder. *J. Vet. Intern. Med.* 1994;8(4):273-278. 14. Chun R, Knapp DW, Widmer WR, et al. Cisplatin treatment of transitional cell carcinoma of the urinary bladder in dogs: 18 cases (1983-1993). *J. Am. Vet. Med. Assoc.* 1996;209(9):1588-1591. 15. Greene SN, Lucroy MD, Greenberg CB, Bonney PL, Knapp DW. Evaluation of cisplatin administered with piroxicam in dogs with transitional cell carcinoma of the urinary bladder. *J. Am. Vet. Med. Assoc.* 2007;231(7):1056-1060. 16. Henry CJ, McCaw DL, Turnquist SE, et al. Clinical evaluation of mitoxantrone and piroxicam in a canine model of human invasive urinary bladder carcinoma. *Clin. Cancer Res.* 2003;9(2):906-911. 17. Boria PA, Glickman NW, Schmidt BR, et al. Carboplatin and piroxicam therapy in 31 dogs with transitional cell carcinoma of the urinary bladder. *Vet Comp Oncol.* 2005;3(2):73-80. 18. Schmidt BR, Glickman NW, DeNicola DB, de Gortari AE, Knapp DW. Evaluation of piroxicam for the treatment of oral squamous cell carcinoma in dogs. *J. Am. Vet. Med. Assoc.* 2001;218(11):1783-1786. 19. de Vos JP, Burm AGD, Focker AP, et al. Piroxicam and carboplatin as a combination treatment of canine oral non-tonsillar squamous cell carcinoma: a pilot study and a literature review of a canine model of human head and neck squamous cell carcinoma. *Vet Comp Oncol.* 2005;3(1):16-24. 20. Sorenmo et al., 2004.

NON-SURGICAL PAIN MANAGEMENT STRATEGIES FOR CANINE OSTEOSARCOMA

Timothy M. Fan, DVM, PhD, DACVIM
Urbana, IL

INTRODUCTION

Osteosarcoma (OSA) is by far the most common primary bone tumor in dogs, and it is estimated that over 10,000 dogs per year will be diagnosed with OSA in the United States.[1] Canine patients diagnosed with OSA are often middle-aged to older dogs of large or giant breeds, with definite overrepresentation of certain breeds such as the Rottweiler, Saint-Bernard, Golden Retriever, Irish Wolfhound, and Greyhound. The standard therapy for canine OSA includes radical surgical resection of the affected limb followed by adjuvant systemic chemotherapy, and provides a median survival time approaching 10-12 months.[1]

The most common and striking presenting sign of dogs presented with OSA is lameness associated with severe pain of the affected bone.[1] The mechanisms of bone pain have been well characterized. The link between osteolytic bone metastasis and pain is provided by the concomitant activation of osteoclasts and nociceptors by prostaglandins and cytokines.[2] In fact, osteolytic destruction of bone is mediated by several humoral mechanisms in addition to prostaglandins and include parathormone, cytokines, and growth factors that stimulate osteoclast activity.[2] It is suspected that similar mechanisms take place in pain associated with primary bone tumors such as canine osteosarcoma. Since both malignant osteoblasts and resident osteoclasts appear to play a preponderant role in the genesis of bone pain associated with malignant osteolysis, therapies that inhibit malignant osteoblast proliferation and survival, as well as concurrently minimizing pathologic osteoclastic activities should be effective in significantly decreasing bone pain in canine osteosarcoma patients that are not candidate for radical surgery.

PALLIATIVE THERAPIES

Standard-of-care therapy, defined as the treatment option that results in the longest median survival times, is surgical resection of the primary tumor followed by 3 to 6 cycles of either a platinum- or doxorubicin-based chemotherapy protocol. Unfortunately, not all dogs with OSA are considered good candidates for amputation, and alternative palliative treatment options for controlling bone pain should be considered. Reported survival times for canine patients treated with palliative intent therapy ranges from 3 to 10 months. With the commercial boom of pharmacologic pain medications approved for use in dogs and cats, the general practioner is now offered a plethora of novel analgesics that may provide some moderate relief for chronic osteolytic pain associated with appendicular OSA. In addition to the administration of conventional nonsteroidal anti-inflammatory drugs (NSAIDs) or the application of transdermal opioids, newer analgesics such as tramadol and gabapentine may also alleviate cancer-related pain.

Palliative Radiation Therapy

Radiation therapy (RT) is considered the most effective treatment modality in human cancer patients for the management of osteolytic bone pain. Unfortunately, in human cancer patients, the incidence of multifocal or diffuse skeletal metastases secondary to primary carcinoma progression often precludes the efficient implementation of megavoltage radiation therapy. Luckily, this is the not case in dogs with OSA, as most often these patients only experience severe bone pain associated with a focal primary tumor. In this clinical setting when only one or few lesions are involved, the effective management of malignant bone pain is typically achieved by administering coarse fractions of 8 to 10 Gy of megavoltage irradiation.[3-5] Palliative RT reportedly improves limb function and quality of life in about 75% of patients, and for a median of 2-3 months duration. Although megavoltage radiation therapy appears effective when used as a single agent treatment option, it remains to be definitively demonstrated that the concurrent administration of systemic chemotherapy along with palliative radiation enhances analgesic response rates and durations. However, given that canine OSA causes localized bone destruction AND metastases, combining radiation therapy with systemic chemotherapy makes biologic sense for achieving pain alleviation and delaying the onset of distant metastases.

^{153}Samarium-EDTMP

The use of a therapeutic radionuclide called ^{153}Samarium-EDTMP has been described for both appendicular and axial OSA in dogs, and provides pain relief in many treated patients.[6-7] ^{153}Samarium-EDTMP is commercially available for the treatment of skeletal metastases in human cancer patient, and is sold by the name Quadramet™. In human cancer patients, Quadramet™ is effective for alleviating bone pain associated with diffuse or multifocal skeletal lesions, and therefore has the advantage over palliative radiation therapy which is limited for treating only a few lesions. By means of delivery concentrated radiation doses to the site of active bone remodeling, ^{153}Samarium-EDTMP administration is capable of providing significant and meaningful palliation of bone pain in dogs suffering from appendicular OSA. ^{153}Samarium-EDTMP therapy is well tolerated and alleviates osteolytic bone pain in the majority of dogs treated. Side effects associated with treatment include transient decreases in platelet and white blood cell counts. The combination of ^{153}Samarium-EDTMP with systemic carboplatin therapy has been recently investigated for the management of bone pain associated with appendicular OSA, and appears well-tolerated but does carry the risk for grade III or IV neutrophil nadirs.

Stereotactic radiosurgery (SRS)

Although the delivery of radiation therapy is effective for managing a variety of tumor histologies, there always remains the concern for excessive normal tissue damage. For the treatment of brain tumors in human cancer patients, excessive normal tissue damage is unacceptable, and prompted the development of stereotactic radiosurgery, as coined the term "gamma knife". Stereotactic radiosurgery involves the precise delivery of a single large dose of radiation to a designated tumor target; achieving by multiple arrays of overlapping radiation beams. Similar to its utility for treating brain tumors in people, stereotactic radiosurgery has also been evaluated for alleviating bone pain associated with appendicular OSA. Recent results suggest that the use of stereotactic radiosurgery in dogs with OSA can provide pain alleviation, long-term local tumor control and improvement in limb function.[8] Similar to palliative radiation therapy, combining systemic chemotherapy with stereotactic radiosurgery may improve long term outcomes for controlling localized bone pain and delaying the onset of metastases.

Aminobisphosphonates

Bisphosphonates are synthetic analogs of inorganic pyrophosphate (PP_i) that were initially utilized in the detergent industry as demineralizing agents, and then for diagnostic purposes in bone scanning, based on their ability to adsorb to bone mineral. The pharmaceutical use of bisphosphonates has now gained wide acceptance in human non-neoplastic bone disorders such as osteoporosis and Paget's disease. In the last decade, bisphosphonates have been intensely investigated as novel antineoplastic agents.

The effective treatment of bone disorders by bisphosphonates is attributed to their differential effect on bone resorption and bone mineralization. At low concentrations bisphosphonates inhibit bone resorption without inhibiting the process of bone mineralization. This results in stabilization and even enhancement of bone mineral density. Bisphosphonates directly inhibit bone resorption by binding to hydroxyapatite crystals. Once incorporated into the hydroxyapatite matrix of bone, bisphosphonates inhibit further calcium and phosphorous mineral dissolution. Perhaps more importantly, bisphosphonates impede osteoclast activity and induce osteoclast apoptosis; both mechanisms result in inhibition of bone resorption.

Given the prevalence of primary and metastatic bone tumors that affect veterinary patients, potent aminobisphosphonates (NBPs), such as pamidronate, may have a role in an adjuvant setting for the management of cancer-induced bone pain. To date, there have been several studies evaluating the pain alleviating and bone biologic effects of intravenous pamidronate in companion animals suffering from focal bone tumors.

In one prospective study principally evaluating the safety of IV pamidronate was conducted in 33 dogs diagnosed with primary and secondary skeletal tumors.[9] Intravenous pamidronate (1.0 mg/kg diluted with 0.9% sodium chloride to a total volume of 250 mls) as a 2-hour constant rate infusion (CRI) every 28 days was well-tolerated. In a subset of osteosarcoma-bearing dogs, bone biologic and clinically-relevant therapeutic effects were documented as significant reductions in urine N-telopeptide concentrations, increases in relative primary tumor bone mineral density, and subjective pain alleviation.[9]

Following the established safety of IV pamidronate in dogs with primary and secondary skeletal tumors, a second study consisting of 43 dogs with appendicular osteosarcoma treated with IV pamidronate (comparing 1.0 mg/kg versus 2.0 mg/kg) was conducted.[10] While the two different dosages of pamidronate did not demonstrate differences for pain alleviation, reductions in urine N-telopeptide concentrations, or increases in relative bone mineral density, there was a strong (although not significant) biologic trend for higher doses of pamidronate (2.0 mg/kg) to exert greater bone biologic effects as reflected by larger absolute reductions in urine N-telopeptide concentrations. Overall, 12/43 (28%) osteosarcoma-bearing dogs treated with single-agent IV pamidronate achieved pain alleviation for greater than 4 months. In addition to the subjective analgesic effects of pamidronate reported by pet owners, changes in urine N-telopeptide concentrations and relative bone mineral density of the primary tumor correlated with therapeutic response. These findings are highly significant, as they validate the use of biochemical and radiological surrogate indices of bone turnover for monitoring response to pamidronate therapy in bone cancer-bearing dogs.

Finally, the adjuvant use of IV pamidronate in combination with palliative radiation therapy and intravenous doxorubicin for managing malignant osteolytic pain in dogs with OSA has also been investigated. For OSA-bearing dogs receiving palliative therapy, achievement of effective and durable analgesia remains of utmost importance to owners seeking to improve or maintain their pets' quality of life. Given the documented palliative effectiveness of ionizing radiation and IV pamidronate as monotherapy for dogs with appendicular OSA, combining these 2 treatment modalities appears clinically justified. In a double-blinded, placebo-controlled study of 50 OSA-bearing dogs, the adjuvant use of pamidronate in combination with radiation and doxorubicin did not improve owner's perception of pain alleviation in their pets when compared to radiation and doxorubicin alone. The median pain-free interval for dogs treated with radiation and doxorubicin alone or in conjunction with pamidronate, was 75 days and 76 days, respectively.[11] Despite the apparent lack of improved analgesia, dogs that received adjuvant pamidronate demonstrated improved quality-of-life scores, and importantly superior bone biologic effects representative of decreased malignant bone resorption at the level of the primary tumor. Collectively the findings from this clinical trial suggest that adjuvant pamidronate may not subjectively improve analgesia in dogs already being treated with megavoltage radiation and IV doxorubicin, but still may be indicated in OSA-bearing dogs given the beneficial bone biologic effects within the bone tumor microenvironment.

REFERENCES

[1]Dernell WS et al. **Small Animal Clinical Oncology**, Third edition. Philadelphia, WB Saunders, 2001, pp.378-417.[2]Payne R. **Cancer** 1997;80(8 Suppl):1608-13.[3]Coomer A. **Vet Comp Oncol.** 2009;7(1):15-27.[4]Mayer MN. **Can Vet J**. 2006;47(7):707-9.[5]Green EM.**J Am Anim Hosp Assoc**. 2002;38(5):445-51.[6]Barnard SM. J **Am Vet Med Assoc.** 2007;230(12):1877-81.[7]Aas M. Clin Cancer Res. 1999;5(10 Suppl):3148s-3152s.[8]Farese J.**J Am Vet Med Assoc.** 2004;225(10):1567-72.[9]Fan TM. **J Vet Intern Med**. 2005;19(1):74-80.[10]Fan TM. **J Vet Intern Med.** 2007;21(3):431-9.[11]Fan TM. **J Vet Intern Med.** 2009;23(1):152-60.

LIMB SPARING OPTIONS: SURGICAL & STEREOTACTIC RADIOTHERAPY

Stewart D. Ryan, BVSc (Hons), MS, DACVS
Fort Collins, CO

INTRODUCTION

Osteosarcoma (OSA) is the most common primary bone tumor in dogs, accounting for 85% of all bone tumors. It is a biologically aggressive tumor causing local bone destruction and has a high metastatic rate. Osteosarcoma occurs more frequently in the appendicular skeleton (75%) compared to the axial skeleton (25%). Morbidity due to local disease is attributed to tumor-associated osteolysis that leads to bone pain, lameness and frequently pathologic fracture. The current standard of care therapy for dogs with appendicular osteosarcoma is removal of the local tumor by means of amputation or limb salvage followed by adjuvant chemotherapy. The reported median disease free interval and median survival time with this therapy ranges from 202 to 257 days and 258 to 321 days respectively [1-4]. Surgery has a role in many aspects of therapy for osteosarcoma, ranging from diagnosis, amputation, limb salvage and metastatectomy. This presentation will focus on surgical and stereotactic radiotherapy limb salvage options.

AMPUTATION

Limb amputation is the gold standard for local management of primary bone tumors in dogs. The technique for forelimb and hindlimb amputation is well described in various veterinary surgical textbooks. Forelimb amputation should include removal of the scapula, for local tumor control and cosmetic reasons, particularly for dogs with proximal humeral osteosarcoma. Coxofemoral disarticulation is the preferred amputation method for hind limbs affected by osteosarcoma distal to the proximal femur. Dogs with proximal femoral osteosarcoma should receive either en-bloc acetabulectomy or subtotal hemipelvectomy to achieve adequate oncologic resection margins.

The majority of dogs, even large and giant breeds, patients with mild to moderate osteoarthritis or heavy body weight, will function well after amputation. [5, 6] However, limb salvage may be chosen as an alternative to amputation due to owner preference or concurrent orthopedic or neurological disease.

LIMB SALVAGE

The goals of limb salvage surgery are complete resection of the primary tumour and preservation of limb function and, in combination with adjuvant chemotherapy, median survival time equivalent to amputation and chemotherapy. Good candidates for surgical limb salvage are dogs with osteosarcoma confined to the bone, with minimal extension into adjacent soft tissue and involving less than 50% of the bone length. [7-9] Dogs must be free of metastasis, free of infection, and healthy enough to undergo anaesthesia, major surgery and post-operative rehabilitation. Pathological fracture is a relative contraindication for limb spare surgery as tumor cells can be locally disseminated in the soft tissues and hematoma, so local recurrence could be more likely.

There are a wide variety of limb salvage techniques that can be used as an alternative to amputation with good functional results. The most common methods of surgical limb salvage involve resection of the local tumor with a 3-5cm margin of normal bone and replacement of the resultant bone defect with a cortical allograft filled with polymethylmethacrylate (PMMA) or a metal endoprosthesis. Other reported limb salvage methods include: bone transport osteogenesis, pasteurization, autoclaving or intraoperative irradiation of tumoral autograft, ulna rollover and manus translation. Marginal resection of the soft tissue component of the bone tumor is common to all techniques. The extent of tumor involvement of bone is based on radiographic findings; however, these changes may underestimate the true tumor margins. Findings of nuclear scintigraphy, CT or MRI scans can be used to more accurately assess tumor margins, especially marrow extension.

Surgical limb salvage with allograft bone has been attempted in a variety of anatomic sites, but can only reliably be performed in the distal radius location in conjunction with fusion of the radiocarpal joint. Limb function and use was reported as good to excellent in 80% of cases in a case series of 220 dogs treated with allograft limb salvage of the distal radius. [7] Allograft limb salvage of the humerus was reported in 17 cases and due to the high complication rate and poor functional outcome, cannot be recommended. [10] Ulnectomy is a suitable limb salvage option for lesions confined to the ulna.

Surgical limb salvage is associated with a high complication rate. The most common complications are: infection, local tumor recurrence and implant related complications. Infection is the most common complication (26-44% of cases). The high infection rate is most likely due to lack of early blood supply, poor local recipient environment, poor soft tissue coverage, preoperative radiation and adjuvant chemotherapy. Infection is managed with long term systemic antimicrobials or locally implanted antibiotic impregnated PMMA beads. Interestingly, dogs that have an infection have a significantly longer survival time compared to those that do not develop infection. [11] Local recurrence rates are reported between 15-28%. Local recurrence can be treated with amputation or a second limb salvage surgery and same survival as dogs treated without a local recurrence is observed. Implant loosening or failure can be observed in 11-20%) of cases. Less commonly reported complications include allograft rejection which can lead to failure, resorption or fracture.

Surgical limb salvage can provide good to excellent results in 90% of dogs without compromising survival, but is associated with a high complication rate.

STEREOTACTIC RADIATION THERAPY (SRT) LIMB SALVAGE

Stereotactic radiosurgery (SRS) was first described as a curative-intent limb salvage technique for appendicular osteosarcoma in 11 dogs by Farese *et. al.* in 2004. [12] Clinical outcome was good to excellent and adverse effects related to radiation were minimal. Pathologic fracture occurred in 4 dogs and surgical stabilization was possible in those cases. Median survival time was 363 days.

A Varian Trilogy linear accelerator was installed at Colorado State University in November 2008. The Trilogy is able to deliver image guided, stereotactic radiation therapy with sub-millimeter precision. The Trilogy has an on board imaging system that can acquire megavoltage portal images, kVp radiographic images and cone bean computer tomography (CBCT) images. These time-of-therapy imaging modalities permit verification of accurate patient positioning and reference to the radiation treatment planning CT.

We have developed a SRT and adjuvant carboplatin chemotherapy protocol for treatment of extremity osteosarcoma in dogs with the hypothesis that this protocol can provide a non-surgical limb salvage option in a variety of anatomic sites. Over 50 dogs have been treated with this protocol with osteosarcoma lesions located in the humerus, radius, ulna, femur and tibial anatomic sites. The mean radiation dose to 95% of the prescribed tumor volume (PTV) for all patients was 36.0 Gy. Eight dogs received SBRT in 2 fractions (mean 95% PTV dose = 31.2Gy) and the remaining dogs received SBRT in 3 fractions (mean 95% PTV dose = 37.3Gy)

Clinical outcome has been good to excellent based on subjective assessment of increased limb use and durable pain palliation. Local tumor control has been excellent based on serial radiography, repeat scintigraphy in 3 cases, decreasing urine NTx levels and radiographic follow up and histology on retrieved limbs. There has been one case of an out of field tumor recurrence documented by radiography and histology. Overall median survival time is 275 days with a range of 66 to 723 days which is comparable to survival times achieved with amputation and adjuvant chemotherapy. The majority of dogs died due to development of metastatic disease. Limbs have been retrieved in 16 cases and histology has not demonstrated viable tumor in the bone within the irradiated field to date.

Post treatment complications included moderate to severe acute radiation effects to the skin in the first three cases. The radiation dose prescription was modified and only mild grade I or no acute radiation skin effects have been observed subsequently. Pathologic fractures occurred in 13 cases (16%). These fractures were treated with amputation, surgical stabilization or external coaptation. Nine fractures occurred early (mean 40 days post SRS) and 4 cases occurred late (mean 312 days post SRS). With improved case selection criteria and pre-emptive surgical stabilization for lytic tumors, only three cases have sustained fracture in the last 17 cases (18%) compared to 10 out the first 19 cases (53%). A retrospective analysis of pre-therapy imaging parameters in SRT cases that went on to fracture identified criteria that can be used to identify patients at high risk of post irradiation fracture. In patients at high risk for fracture, pre-emptive surgical stabilization with plates and screws or interlocking nails has been offered. These implants are placed using a minimally invasive technique where possible immediately after the third fraction of SRT.

The SRT limb salvage protocol has been used for tumors in the proximal humerus, distal femur and distal tibia regions, anatomic sites that are not amenable to successful surgical allograft limb salvage procedures. The multimodality approach of stereotactic radiotherapy combined with neoadjuvant bisphosphonate therapy and adjuvant chemotherapy provides a functional non-surgical limb salvage option in a variety of anatomic locations compared to surgical limb salvage, which can only reliably be used in the distal radius anatomic site.

REFERENCES

1. Ranney, D., Thoracic outlet: an anatomical redefinition that makes clinical sense. Clin Anat, 1996. 9(1): p. 50-2. 2. Muhammed, N., P. Campbell, and I.S. Smith, Peripheral nerve entrapment syndromes: diagnosis and management. Br J Hosp Med, 1995. 53(4): p. 141-6. 3.Kent, M.S., et al., Alternating carboplatin and doxorubicin as adjunctive chemotherapy to amputation or limb-sparing surgery in the treatment of appendicular osteosarcoma in dogs. J Vet Intern Med, 2004. 18(4): p. 540-4. 4. Bergman, P.J., et al., Amputation and carboplatin for treatment of dogs with osteosarcoma: 48 cases (1991 to 1993). J Vet Intern Med, 1996. 10(2): p. 76-81. 5. Kirpensteijn, J., et al., Ground reaction force analysis of large breed dogs when walking after the amputation of a limb. Vet Rec, 2000. 146(6): p. 155-9. 6. Kirpensteijn, J., R. van den Bos, and N. Endenburg, Adaptation of dogs to the amputation of a limb and their owners' satisfaction with the procedure. Vet Rec, 1999. 144(5): p. 115-8. 7. Straw, R.C. and S.J. Withrow, Limb-sparing surgery versus amputation for dogs with bone tumors. Vet Clin North Am Small Anim Pract, 1996. 26(1): p. 135-43. 8. Liptak, J.M., et al., Cortical allograft and endoprosthesis for limb-sparing surgery in dogs with distal radial osteosarcoma: a prospective clinical comparison of two different limb-sparing techniques. Vet Surg, 2006. 35(6): p. 518-33. 9. LaRue, S.M., et al., Limb-sparing treatment for osteosarcoma in dogs. J Am Vet Med Assoc, 1989. 195(12): p. 1734-1744. 10. Kuntz, C.A., et al., Limb salvage surgery for osteosarcoma of the proximal humerus: outcome in 17 dogs. Vet Surg, 1998. 27(5): p. 417-22. 11. Lascelles, B.D., et al., Improved survival associated with postoperative wound infection in dogs treated with limb-salvage surgery for osteosarcoma. Ann Surg Oncol, 2005. 12(12): p. 1073-83. 12. Farese, J.P., et al., Stereotactic radiosurgery for treatment of osteosarcomas involving the distal portions of the limbs in dogs. J Am Vet Med Assoc, 2004. 225(10): p. 1567-72, 1548.

CHEMOTHERAPY FOR OSA: PROTOCOL PROS & CONS

Timothy M. Fan, DVM, PhD, DACVIM
Urbana, IL

INTRODUCTION

Osteosarcoma (OSA) is the most common primary bone tumor diagnosed in dogs. Anatomically, OSA may develop anywhere osteoblasts normally reside, however, the metaphyseal regions of appendicular bones is most commonly affected, accounting for 80% of all OSA diagnoses and typically seen in large to giant breeds of dogs with males being over-represented in many reports.[1] Our understanding of the risk factors and genetic changes in canine OSA patients is growing but has yet to lead to specific innovative therapies. Similarly, despite identifying several intracellular signaling pathways which may be deranged in OSA, preferential inhibition of these mitogenic or survival pathways has not yet clinically materialized as a "silver bullet" for curbing the highly metastatic tendencies of OSA.[2-6] As such, the management of OSA metastases continues to rely solely on the administration of systemic chemotherapy. Although definitive clinical evidence exists for the positive benefit derived from adjuvant systemic chemotherapy, a "gold standard" chemotherapy protocol has yet to be identified and implemented universally.

METASTATIC CASCADE

By convention, metastasis requires the ability of a tumor cell to successfully (1) disseminate through epithelial-mesenchymal transition, (2) invade and migrate through extracellular matrix (ECM), (3) intravasate into blood or lymph vessels, (4) survive circulatory transport in blood or lymphatic fluid, (5) extravasate into distant target organs, (6) establish a metastatic niche, and (7) grow and form tumor nodules. Although the "metastatic cascade" is viewed as a continuous process, the efficiency at each stage of metastases is highly variable. The early steps of tumor cell invasion, migration, intravasation and extravasation are considered to be highly efficient, while the events that take place post-extravasation are highly inefficient. As such, systemic therapies which target the latter inefficient steps of metastases are likely to exert greater biologic anticancer effects, which may translate into improved survival times in dogs with OSA. Current systemic chemotherapy protocols are thought to kill or enforce dormancy of micrometastases within the lung parenchyma. Because it is believed that dogs with OSA at diagnosis already have pulmonary micrometastases, the institution of systemic chemotherapy very likely targets "metastatic cascade" events which take place post-extravasation. Based upon this logic, systemic chemotherapy is effective because it makes the latter inefficient events of the "metastatic cascade" even more inefficient.

CHEMOTHERAPY BASICS

Chemotherapy maybe viewed as a form of controlled "poison", which when administered properly and rationale, provides improved quality-of-life and survival times for the cancer patient. Because chemotherapy is not specific for tumor cells, it becomes the responsibility of the veterinary practioner to be well informed and educated on the appropriate principles of chemotherapy handling and administration. To gain a mastery of using chemotherapy, it remains of paramount importance for the administrator to understand the basic chemistry and biology behind the "cytotoxic" effects of everyday chemotherapeutic agents.

Most chemotherapeutic agents induce cytotoxicity by either damaging DNA or interfering with necessary substrates or enzymes required for the faithful synthesis and replication of genetic material. Any "bump-in-the-road" during DNA replication may potentially result in irreparable damage, subsequently resulting in cellular death in both normal and neoplastic cells. Fortunately, most chemotherapeutic agents exert the majority of their cytotoxic properties on rapidly dividing cells, which are most often inclusive of rapidly growing and dividing cancer cells. However, because many normal cells also undergo relatively quick replicative cycles (bone marrow stem cells and gastrointestinal epithelial cells), adverse side effects in the cancer patient should be anticipated.

Understanding the Cell Cycle

All cells, regardless if they are normal or neoplastic, go through an orderly and stepwise division process, and is termed the cell cycle. This divisional cycle can be roughly divided into five stages:

1. G0: Quiescent phase where cells are not actively dividing, but maintain the ability to be recruited back into active division should the appropriate stimulus be applied.
2. G1: "Gap 1" phase where proteins and enzymes required for DNA synthesis are produced as a prelude to active DNA synthesis.
3. S: This phase is when DNA is actually synthesized, requiring the activity of various proteins including DNA polymerases, which are responsible for the faithful replication of single-stranded DNA (leading and lagging strands).
4. G2: "Gap 2" phase where proteins and enzymes required for protein synthesis, RNA synthesis, and mitotic spindle formation are produced.
5. M: "Mitosis" phase where duplicated DNA is divided equally between two daughter cells.

The time it takes for a cell to complete the cell cycle is called doubling-time. Among various cell types, most of the cell cycle phases are relatively similar, except for the length of the G1 phase. The G1 phase can be highly variable, thereby explaining the faster or slower replicative rates of various cell types. "Stop Signs" or "Checkpoints" regulates transition through each phase of the cell cycle. All necessary machinery in the form of proteins or enzymes must be in place prior to a cell commencing through a checkpoint. Checkpoints serve as safeguards to ensure appropriate conditions for cell growth and division. Many times, cancer cells no longer obey these regulatory checkpoints, thereby allowing them to have a growth advantage over normal resident cells. In addition to unregulated growth, a cancer cell's disrespect for regulatory checkpoints may allow for the acquisition of additional genetic mutations, which might favor more aggressive phenotypes.

Understanding the relationship of the cell cycle and chemotherapy

The extent and effect of cytotoxic chemotherapy is intricately linked to the cell cycle and the type of anticancer agent. Broadly, most chemotherapeutic agents may be categorized into one of five groups:

1. Alkylating Agents
2. Antimetabolites
3. Antineoplastic Antibiotics
4. Microtubule Spindle Poisons
5. Miscellaneous Agents

Cytotoxicity of alkylating agents is a direct result of DNA adducts formation. DNA adducts will interfere with the replication of DNA, potentially resulting in cellular apoptosis. Because DNA is present in every phase of the cell cycle, alkylating agents are able to affect all phases of cellular division, also termed cell-cycle non-specific cytotoxicity. The action of antimetabolites is generally through inhibition of enzymes or proteins required for the synthesis of DNA. Failure to synthesize DNA results in halting of the cell cycle, and ultimately cell death. Because DNA synthesis is restricted to predominantly one phase of the cell cycle (the "S" phase), the cytotoxicity of antimetabolites is termed cell-cycle specific cytotoxicity. Antineoplastic antibiotics make up a family of natural compounds capable of inducing cellular death via several mechanisms including free radical generation that can damage DNA, intercalation of DNA that prevents the synthesis of DNA, and inhibition of critical enzymes necessary for the unwinding of DNA. Because antineoplastic antibiotics primarily exert their cytotoxic effects on DNA, all phases of cellular division may be adversely affected. Antineoplastic antibiotics are cell-cycle non-specific agents, similar to alkylating agents. Microtubule spindle poisons are agents that interfere with either the polymerization (formation) or depolymerization (destruction) of structures that are formed by the protein called tubulin. One of the major structures that is formed from tubulin is the mitotic spindle, which is responsible for the equal division of DNA material between two daughter cells. Interference with either spindle formation or spindle deconstruction will result halting of the cell cycle and subsequent cell death. Because the mitotic spindle apparatus is only formed during the "M" phase of the cell cycle, microtubule spindle poisons are considered cell-cycle specific agents. Although there are many miscellaneous chemotherapeutic agents available that do not "neatly" fall into the above four categories, most often their mechanism of action is similar to alkylating agents, thereby their associated cytotoxicity is often cell-cycle non-specific.

OSA CHEMOTHERAPY

The use of systemic chemotherapy is known to improve survival from an average of 4-5 months, with surgery alone, to 10-14 months. The drugs most commonly used are the platinum agents (cisplatin or carboplatin) and doxorubicin. Reports of combination or alternating protocols with these drugs have resulted in variable survival times, but in general so far, no such combination has shown clear and repeatable improvement over single-agent therapy.[7-8] Dogs typically receive 4 to 5 cycles of chemotherapy.

Once metastatic disease is clinically detectable, chemotherapy is usually ineffective. Pulmonary metastatectomy has been described and can contribute to significantly prolonged survival in selected patients. Clinical guidelines for determining patient eligibility for metastatectomy include: primary tumor in complete and prolonged remission (ideally for >300 days); one to two radiographically detectable nodules only; no evidence of distant metastatic disease at non-pulmonary sites; no evidence of life-limiting concurrent disease; and a long tumor doubling time (>30 days) with no new lesions observed during this time period.

REFERENCES

[1]Dernell WS et al. **Small Animal Clinical Oncology**, Third edition. Philadelphia, WB Saunders, 2001, pp.378-417.[2]Khanna C **Clin Cancer Res**. 2002;8(7):2406-12. [3]Paoloni MC **PLoS One.** 2010 Jun 8;5(6):e11013.[4]Liao AT. **Anim Genet**. 2006;37(3):248-52.[5]MacEwen EG. **J Cell Biochem.** 2004;92(1):77-91.[6]MacEwen EG. **Clin Exp Metastasis**. 2003;20(5):421-30.[7]Kent MS. **J Vet Intern Med**. 2004;18(4):540-4.[8]Bailey D. **J Vet Intern Med**. 2003 Mar-Apr;17(2):199-205.

BEST TREATMENT OPTIONS FOR CANINE OSTEOSARCOMA

Stewart Ryan, BVSc, MS, DACVS and Timothy M. Fan, DVM, PhD, DACVIM
Fort Collins, CO and Urbana, IL

INTRODUCTION

Osteosarcoma (OSA) is the most common primary bone tumor in dogs, accounting for 85% of all bone tumors. It is a biologically aggressive tumor causing local bone destruction and has a high metastatic rate. Systemic dissemination is expected to be present microscopically in the vast majority (90-95%) of dogs diagnosed with the disease. The pulmonary parenchyma is the most common metastatic site. Lymph nodes appear to be an uncommon site of metastases, with a reported rate of 6.1% to 37%. With the more common use of adjuvant chemotherapy, skeletal and soft-tissue metastases are increasingly diagnosed. Osteosarcoma occurs more frequently in the appendicular skeleton (75%) compared to the axial skeleton (25%). Morbidity due to local disease is attributed to tumor-associated osteolysis that leads to bone pain, lameness and frequently pathologic fracture. The current standard of care therapy for dogs with appendicular osteosarcoma is removal of the local tumor by means of amputation or limb salvage followed by adjuvant chemotherapy. The reported median disease free interval and median survival time with this therapy ranges from 202 to 257 days and 258 to 321 days respectively.[1-4]

SIGNALMENT

Canine appendicular osteosarcoma is most frequently diagnosed in large to giant breeds. The age distribution at presentation is bimodal with most dogs 7 to 9 years of age at diagnosis and a smaller number of dogs 1 to 2 years of age at diagnosis. Osteosarcoma most frequently arises in the metaphyseal region of long bones with the distal radius and proximal humerus the most common locations and the proximal and distal aspects of the femur and tibia affected less frequently. Lameness and localized limb swelling are the most common presenting clinical signs. It is not an uncommon history in osteosarcoma cases for a presumptive diagnosis of and orthopedic pathology (e.g. osteoarthritis, cruciate disease, hip dysplasia, ligament sprain) to be made and to observe an initial favorable response to rest and anti-inflammatory therapy before progression of clinical signs.

DIAGNOSIS

A minimum database consisting of a complete blood count, serum chemistry profile and urinalysis is recommended to screen for concurrent disease, and to aid in establishing a prognosis by quantitating the serum alkaline phosphatase concentration. Certain therapeutic decisions might also be altered based of the results of such tests. For example, a patient with marginal renal function might not be an ideal candidate for cisplatin therapy, a known nephrotoxic agent, and could require dose reductions of a renally excreted drug, such as carboplatin.

Diagnostic imaging plays an important role in staging dogs with OSA. Radiographs of the affected bone have a characteristic radiographic appearance ranging from lytic to blastic but most commonly a mixed lytic-blastic pattern. Other radiographic include cortical bone loss starting on the endosteal surface, periosteal proliferation, palisading cortical bone (sunburst effect), periosteal elevation (Codman's triangle), loss of the fine trabecular pattern in metaphyseal bone, and pathologic fracture. Thoracic radiographs reveal gross metastatic disease in less than 10% of dogs at presentation. However, it is well established that >90% of these dogs will go on to develop pulmonary metastatic disease if treated by surgery alone. Pulmonary metastatic lesions appear as discrete soft-tissue nodules, and multiple lesions are common. Nuclear scintigraphy is more sensitive than radiography for imaging skeletal metastases. However, the specificity of this imaging modality is relatively low, and any scintigraphic lesions beyond the primary tumor should ideally be further evaluated with radiographs and/or biopsy to verify their significance.

While biopsy remains the gold standard for diagnosis of OSA, fine needle aspirate via an 18 gauge needle may provide a diagnosis by less invasive means. Cells from lytic OSA lesions will readily exfoliate. Aspirates often reveal large, immature mesenchymal cells that may have intracytoplasmic or extracellular osteoid, which appears as an amorphous pink matrix. In addition to malignant osteoblasts, benign osteoblasts and osteoclasts may also be present. Fine needle aspiration of lytic bone lesions may help to rule out fungal or bacterial osteomyelitis, where inflammatory cells or organisms may be observed. A tissue biopsy can also be used to provide preliminary cytologic information by touching or rolling the sample on a slide.

PROGNOSTIC FACTORS

While breed and sex have not been recognized to have prognostic importance, young dogs with OSA appear to have shorter survival times and biologically more aggressive disease. The presence of detectable metastatic lesions at the time of diagnosis is a recognized poor prognostic factor, standard chemotherapy being ineffective at improving survival in that setting. Elevated alkaline phosphatase levels (total, bone or liver isoenzymes) are also negative prognostic factors.[5-6] Animals with elevated levels have shorter survival times by ~50% (approximately 170d vs. >400d), even when treated aggressively with surgery and chemotherapy. Percent tumor necrosis, induced by radiation therapy or chemotherapy prior to a surgery, is predictive of local tumor control when compared to dogs treated with amputation alone.[7] Quantitative bone scintigraphy of the primary tumor (correlates with tumor volume) demonstrated a relationship between larger tumors and earlier metastasis. Recently a new histologic grading system for OSA provided prognostic information (higher grade = shorter survival), while a previous study had shown increased tumor microvascular density in patients presenting with detectable pulmonary metastases, and recognized it as a negative prognostic indicator.[8] Finally and interestingly, although not being a prognostic factor found at initial diagnosis, it has been reported that dogs with infections of their allografts, after limb-sparing surgery, were twice as likely to survive longer, compared to dogs without infection. A non-specific activation of the immune system has been advanced as a potential explanation.[9]

TREATMENT OPTIONS

Treatment options for osteosarcoma can be broadly classified in to palliative and curative intent treatment options, and recommendations for either palliative or curative disease management is dictated by 2 principle factors, being pet owner- and patient-dependent. Palliative treatment options for appendicular OSA include analgesia, radiation therapy, limb amputation, and metronomic chemotherapy. The goal of palliative therapy is to control pain and lameness associated with the primary tumor. Palliative therapy does not attempt to modify metastatic disease progression or improve survival time, but rather emphasized quality of life scores. Palliative therapy options are appropriate for dogs with metastatic disease or owners that do not want to pursue curative intent therapy. Curative-intent treatment options include removal of the primary tumor by amputation or limb salvage therapy followed by adjuvant chemotherapy.

The use of systemic chemotherapy is known to improve survival from an average of 4-5 months, with surgery alone, to 10-14 months. The drugs most commonly used are the platinum agents (cisplatin or carboplatin) and doxorubicin. Reports of combination or alternating protocols with these drugs have resulted in variable survival times, but in general so far, no such combination has shown clear and repeatable improvement over single-agent therapy. Dogs typically receive 4 to 5 cycles of chemotherapy.

Once metastatic disease is clinically detectable, chemotherapy is usually ineffective. Pulmonary metastatectomy has been described and can contribute to significantly prolonged survival in selected patients. Clinical guidelines for determining patient eligibility for metastatectomy include: primary tumor in complete and prolonged remission (ideally for >300 days); one to two radiographically detectable nodules only; no evidence of distant metastatic disease at non-pulmonary sites; no evidence of life-limiting concurrent disease; and a long tumor doubling time (>30 days) with no new lesions observed during this time period.

REFERENCES

[1]Ranney D. **Clin Anat.** 1996. **9**(1): p. 50-2. [2]Muhammed N. **Br J Hosp Med**, 1995. **53**(4): p. 141-6.[3]Kent MS. **J Vet Intern Med** 2004. 18(4): p. 540-4.[4]Bergman PJ. **J Vet Intern Med**, 1996. **10**(2): p. 76-81.[5]Ehrhart N. **J Am Vet Med Assoc**. 1998;213(7):1002-6.[6]Garzotto CK. **J Vet Intern Med**. 2000;14(6):587-92.[7]Powers BE. **Cancer.** 1991 Jan 1;67(1):126-34.[8]Coomber BL. **Can J Vet Res.** 1998;62(3):199-204.[9]Lascelles BD. **Ann Surg Oncol**. 2005 Dec;12(12):1073-83.

MORE THAN A FRIEND: HOW CLINICAL STUDIES IN PET DOGS AND CATS ADVANCE HUMAN MEDICINE

Steven W. Dow, DVM, PhD, DACVIM
Fort Collins, CO

DUAL ROLE OF PETS IN SOCIETY

The One Medicine concept suggests that companion animals can play an important role in society beyond their role as companions and family members. For example, companion animals (dogs and cats) are physiologically, immunologically, and genetically more closely related to humans than the rodents (mice, rats) that are frequently used in research. In addition, companion animals also live in close association with humans, sharing shelter and sometimes food. They are also treated with many of the same medications used to treat humans. Thus, studies of new therapeutics conducted in companion animals with spontaneous diseases are often considered by regulatory agencies and pharmaceutical companies to be more relevant to human clinical study outcomes than studies of experimental diseases in mice.

At present, most clinical studies conducted by veterinarians are intended to improve the health of the species being studied. However, there are several important diseases of companion animals where studies conducted in companion animals may have direct relevance to both animals and humans. These diseases include cancer, chronic degenerative diseases (joints, kidneys), and inherited diseases with known genetic defects. Within each disease category, there are opportunities to conduct clinical studies investigating new treatment approaches that will have immediate relevance to human health. Broadly, the greatest opportunities can be found in evaluating the safety and efficacy of new drugs, the effectiveness of new gene therapy and delivery approaches, and the application of new stem cell therapy technologies.

ANIMAL MODELS OF HUMAN DISEASE: WHERE CAN CLINICAL STUDIES IN COMPANION ANIMALS HAVE THE BIGGEST IMPACT ON HUMAN HEALTH?

For clinical studies in companion animals to be relevant to human health, it is critical that the disease in dogs, cats, or horses be similar in numerous respects to the disease in humans. Thus, diseases that are relatively unique to dogs (eg, hip dysplasia, granulomatous meningoencephalomyelitis, mast cell tumor) are not strong models for human disease. On the other hand, several canine spontaneous disease models are highly relevant to human health studies, including certain cancers of dogs, especially melanoma, lymphoma, and osteosarcoma. The canine cancer model is the best established companion animal spontaneous disease model and is particularly valuable for investigating new cancer drugs and the effectiveness of immune-based therapies. Inherited, single-gene defect diseases in dogs and cats are very valuable as models for investigating gene replacement approaches, which has been exploited in both ocular and hematologic disease models. In the case of cats, relevant models for human diseases include retroviral-induced immune suppression (eg, FIV infection), certain inborn errors of metabolism, feline allergic bronchitis/asthma, and potentially feline end stage renal disease. In the case of horses, degenerative joint disease and tendon and ligament injuries are valuable models with direct relevance to humans. Below, specific examples where clinical studies in companion animal disease models have already had an impact on human health, or where there is a strong potential for such an impact, will be discussed.

BONE CANCER IN DOGS AS A MODEL FOR HUMAN BONE CANCER

Osteosarcoma in dogs is a particularly valuable animal model for studying new therapeutic approaches to bone cancer in humans. As summarized in Table 1, there are a number of tumor features that are shared between dogs and humans that make osteosarcoma in dogs a particularly valuable animal model. The fact that there are few existing therapeutics (other than conventional chemotherapy) to treat osteosarcoma also make it easier to test new approaches to treatment in this model.

Table 1. Characteristics of bone cancer in dogs and humans

Tumor characteristic	Dog	Human
Mean age at diagnosis	7 yrs	14 yrs
Patient gender	Female>Male	Female>male
Body size	Large breeds	Heavy
Site of primary tumor	Appendicular	Appendicular
Metastases present at diagnosis	80-90%	80-90%
Percent high-grade tumors	95%	85-95%
Metastases with no chemotherap	90%	80%
Site of metastases	Lung, bone, soft tissu	Lung, bone, soft tissu
Chemotherapy effective?	Improved survival	Improved survival

USE OF L-MTP-PE AS ADJUVANT THERAPY FOR BONE CANCER IN DOGS AND HUMANS

The use of L-MTP-PE (liposomal muramyl tripeptide) for adjuvant treatment of osteosarcoma represents one of the best examples to date of the dual benefit that companion animal studies can have for humans. Muramyl tripeptide is a bacterial-derived peptide that activates innate immunity, and when delivered with liposomes specifically targets lung macrophages for activation [1]. In clinical trials in dogs, L-MTP-PE was evaluated in over 150 animals with osteosarcoma [2-4]. Based on these studies, an effective and safe dose of L-MTP-PE was established. Importantly, it was also shown that L-MTP-PE when administered concurrently with chemotherapy significantly improved survival in dogs with osteosarcoma [4]. The studies in canine osteosarcoma helped pave the way for studies in human osteosarcoma. A large, randomized, placebo-controlled trial of adjuvant L-MTP-PE in humans with osteosarcoma was based on the design of the original canine trials [5-6]. The clinical trial in humans found that disease-free intervals and overall survival times in humans treated with chemotherapy and L-MTP-PE were significantly prolonged compared to patients treated with chemotherapy alone, as was predicted by the earlier trials in dogs.

INTRAVENOUS LIPOSOME-DNA COMPLEXES (CLDC) FOR TREATMENT OF ADVANCED CANCER IN DOGS AND HUMANS

A second cancer immunotherapy study clearly illustrates the progression of investigations from rodents to dogs to humans. A potent immune stimulant comprised of non-coding plasmid DNA complexed to liposomes (CLDC) was found originally to activate antitumor immunity in mice [7]. Subsequent studies in dogs with advanced, metastatic osteosarcoma found that i.v. administration of CLDC significantly increased overall survival times in dogs, compared to historical control animals [8]. In a second study in dogs, CLDC administration was also found to inhibit tumor angiogenesis in animals with soft tissue sarcoma [9]. Based in large part on the results of canine studies, clinical trials of intravenous CLDC immunotherapy have been initiated in humans with advanced, inoperable cancer or with chemotherapy-resistant acute myelogenous leukemia.

CANINE MELANOMA MODEL FOR EVALUATING CANCER IMMUNOTHERAPY AND VACCINES

Melanoma in dogs is also an effective model for human cancer, based on similar biological behavior of the tumor and similar tumor genetic abnormalities [10]. Studies in mice originally demonstrated that vaccines prepared using inactivated whole melanoma cells could be used to generate antitumor immunity. Subsequently, a number of human clinical trials followed, building on murine and canine clinical studies. For example, studies of allogeneic melanoma vaccines in dogs demonstrated safety and evidence of antitumor efficacy [11-12]. From these studies and others, the first approved cancer vaccine for cancer in dogs or humans was developed for immunization against canine melanoma [13-14]. In a second approach, intra-tumoral gene therapy was shown to be effective for treatment of canine melanoma [15]. These studies in dogs led directly to the first human cancer gene therapy studies performed at the University of Colorado, which were done in human advanced melanoma patients, and were designed directly from the original dog studies [16].

STEM CELL THERAPY: A UNIQUE OPPORTUNITY FOR COMPANION ANIMAL RESEARCH

A second major opportunity for studies in companion animals to directly support and advance human clinical trials is in the area of stem cell therapy. Though a great deal of hype currently surrounds the stem cell therapy field, there are nonetheless important and scientifically valid advances being made in the field. The challenge of course is distinguishing hype from reality. In many cases, convincing proof-of-principle pre-clinical studies can only be done in naturally-occurring disease models of dogs, cats, and horses.

STEM CELL THERAPY FOR TENDON AND JOINT INJURIES IN DOGS AND HORSES

Stem cell therapy in veterinary medicine at present generally involves the administration of mesenchymal stem cells (MSC) comprised of stromal cells with multi-lineage potential derived from bone marrow or adipose tissues [17-20]. These MSC have the potential to differentiate into connective tissues such as bone, tendon, or cartilage when injected into injury sites under appropriate growth and anchorage conditions. However, it is now appreciated that MSC also exert potent immune suppressive effects, and these effects may in fact account for much of the treatment efficacy associated with MSC therapy [21].

Stem cell therapy has been reported to stimulate joint repair in experimental injuries in horses [22]. Already stem cell studies performed in equine patients at Colorado State University and several other veterinary teaching hospitals are widely cited as early examples of successful stem cell treatments for ligament and tendon injuries. In addition, intra-lesional injection of MSC may stimulate healing of chronic tendinosis in horses, though the results of randomized trials have not yet been reported [23-24]. Similarly, injection of autologous adipose-derived MSC in dogs with osteoarthritis of the coxofemoral joint was reported to be significantly effective in improving lameness, pain and range-of-motion scores in a randomized clinical trial [25]. While MSC therapy trials are getting underway in humans, the results from ongoing trials in horses and companion animals are likely to have a significant impact on stem cell therapy study design and implementation.

OTHER AVENUES FOR EXPLORING STEM CELL THERAPY IN COMPANION ANIMALS

Several other disease models in companion animals are particularly promising in terms of relevance to human health. For one, treatment of traumatic neurological injury with stem cells holds considerable promise as a model for investigating the effectiveness of such approaches in human spinal cord injury patients. Indeed, studies in dogs have already demonstrated the potential effectiveness of stem cell therapy for ameliorating spinal cord injury [26]. It may also be possible to harness the immune modulatory effects of MSC for treatment of immune-mediated diseases of companion animals, including inflammatory bowel disease, autoimmune diseases, and feline allergic airway disease. The feline spontaneous allergic airway disease (asthma) model is particularly promising, as the disease in cats resembles in many respects asthma in humans [27-28]. Moreover, our preliminary studies in rodent asthma models, as well as the work of others, indicate that MSC therapy can be very effective for ameliorating the inflammatory and physiological changes associated with allergic airway disease [29]. Thus, clinical evaluation of MSC treatment in the feline asthma model has considerable promise as a new stem cell therapy model.

STEM CELL THERAPY AND THE FELINE CHRONIC KIDNEY DISEASE MODEL

Finally, there is evidence that the feline chronic kidney (CKD) disease model may be an effective model for study of human end stage renal disease (ESRD). End-stage renal disease affects more than 26 million adults in the US and is estimated to result in more than 65,000 deaths annually. While the initiating causes of ESRD and CKD in humans and cats are often very different, the final common pathways (eg, chronic, tubulointerstitial nephritis) appear to converge. Since ESRD is very difficult to model effectively in rodents, the feline CKD model appears to be an effective model in many regards for human ESRD.

Our laboratory has been investigating the use MSC therapy for feline CKD for several years now. While it is theoretically possible that MSC may help stimulate renal regeneration, we believe that the anti-inflammatory effects of MSC are likely to be the predominant mechanism of action. Therefore, we hypothesize that MSC administration will suppress intra-renal inflammation in feline CKD, and thereby stabilize renal function. In this model, it is not necessary to have large numbers of MSC engrafting in the kidney to exert an effect, as suggested by rodent studies [30]. In our pilot MSC therapy studies for feline CKD, the MSC were injected intra-renally and unilaterally, allowing us to monitor the untreated kidney as a control. These studies demonstrated that the approach could be implemented safely and might exert some benefit on renal function [31]. However, the multiple sedations required for the intra-renal stem cell therapy approach appeared to compromise patient management. Therefore, the intra-renal MSC injection approach was judged to be largely unworkable. We have therefore now instituted new studies with intravenously administered MSC in cats with CKD. Preliminary results indicate that the i.v. administration protocol is much better tolerated by cats and may possibly provide greater clinical benefit, though the studies are still ongoing.

SUMMARY

Our interactions with companion animals undoubtedly already offer us great rewards on an emotional level. However, our companion animals may offer even more, by serving as participants in clinical studies designed to benefit both animals and man. As veterinary specialists interested in clinical research, the opportunity to perform such clinical studies can offer double rewards. We must also however continue to serve as strong advocates for the well-being of our patient "volunteers" who are serving in such trials.

REFERENCES

1. Fidler IJ, et al. . **Invasion Metastasis** 1989;9:75-88. 2. Kurzman ID, et al. **Clin Cancer Res** 1995;1:1595-1601. 3. MacEwen EG, et al. **J Natl Cancer Inst** 1989;81:935-938. 4. MacEwen EG, et al. **Clin Cancer Res** 1999;5:4249-4258. 5. Meyers PA, et al. **J Clin Oncol** 2005;23:2004-2011. 6. Meyers PA,et al. . **J Clin Oncol** 2008;26:633-638. 7. Dow SW, et al. . **J Immunol** 1999;163:1552-1561. 8. Dow S, et al. . **Hum Gene Ther** 2005;16:937-946. 9. Kamstock D, et al. . **Cancer Gene Ther** 2006;13:306-317. 10. MacEwen EG, et al. **Cancer Invest** 1986;4:397-402. 11. Hogge, et al. **Hum Gene Ther** 1998;9:1851-1861. 12. Hogge GS, et al. . **Cancer Gene Ther** 1999;6:26-36. 13. Bergman PJ, et al. **Vaccine** 2006;24:4582-4585. 14. Bergman PJ et al. **Clin Cancer Res** 2003;9:1284-1290. 15. Dow SW, et al. . **J Clin Invest** 1998;101:2406-2414. 16. Walsh P, et al. . **Hum Gene Ther** 2000;11:1355-1368. 17. Koch TG, **Can Vet J** 2009;50:155-165. 18. Arthur A, et al. . **J Cell Physiol** 2009;218:237-245. 19. Richardson LE, et al. **Trends Biotechnol** 2007;25:409-416. 20. Taylor SE, et al **Equine Vet J** 2007;39:172-180. 21. Nauta AJ . **Blood** 2007;110:3499-3506. 22. Frisbie DD, **Equine Vet J** 2010;42:86-89. 23. Lacitignola L,et al **Vet Res Commun** 2008;32 Suppl 1:S33-38. 24. Chong AK, et al . **Front Biosci** 2009;14:4598-4605. 25. Black LL, et al. **Vet Ther** 2007;8:272-284. 26. Lee SH, et al. . **Neurol Res** 2009;31:996-1002. 27. Kirschvink N, **Curr Drug Targets** 2008;9:470-484. 28. Day JH,et al **Ann Allergy Asthma Immunol** 2006;96:263-277; 277-268, 315. 29. Nemeth K., et al. **Proc Natl Acad Sci** U S A 2010;107:5652-5657. 30. Semedo P, et al. . **Stem Cells** 2009;27:3063-3073. 31. Quimby J, et al. **J Fel Med Surg,** 2011., in press.

FELINE AND HUMAN ASTHMA: BI-DIRECTIONAL COMPARATIVE MEDICINE

Carol Norris Reinero, DVM, DACVIM, PhD
Columbia, MO

OVERVIEW

The similarities in the anatomic, physiologic and immunologic manifestations of allergic asthma between humans and cats allows for a rich opportunity for bi-directional comparative research. Additionally, having an experimental model of feline allergic asthma with rigid control of type and amount of allergen exposure, environmental conditions and to some extent, genetic contributions to the asthmatic phenotype provides a means by which controlled pilot studies can be performed with relatively small numbers of animals. Results from these pre-clinical studies can guide larger clinical trials in both species, allowing for investigations into underlying immunologic and physiologic mechanism of disease, diagnostic tests and therapeutics.

DEFINITION, INCIDENCE AND IMPACT OF ALLERGIC ASTHMA

Hallmark features of asthma include airway inflammation, airway hyperresponsiveness, airflow limitation and airway remodeling. In humans, asthma is considered one of the most common chronic illnesses and is estimated to affect up to 10% of the population in developed countries, being especially important in the pediatric population. In cats, asthma is one of the most common respiratory diseases, estimated to affect up to 5% of the pet cat population. In both species it is responsible for substantial morbidity and occasional mortality. Despite the considerable impact on quality of life and economic burden of this disease, there have been surprisingly few effective novel diagnostics for early diagnosis, severity stratification, exacerbation and control of asthma which are non- or minimally invasive. Similarly, there have been few novel therapeutics which have been widely beneficial for the majority of asthmatics.

DO CATS HAVE MORE TO OFFER TO ASTHMA THAN BEING A TRIGGER?

The vast majority of biomedical publications using search terms "cat" and "asthma" implicate the feline species as a cause for human allergic asthma. However, it is important to note that of all the animal species, only the cat naturally and commonly develops a syndrome of allergic asthma which replicates all the major features of human asthma including eosinophilic airway inflammation, airflow limitation and airway remodeling. Spontaneously developing feline asthma has been recognized for over 100 years, and more recent data supports allergy as the most common cause, similar to what is noted in humans. Humans and their pet cats share the same environment, so it is not surprising that when affected individuals were screened for allergen-specific IgE, positive reactions to many of the same allergens (e.g., house dust mite allergen (HDMA), Bermuda grass allergen (BGA)) occurs in both species. Overall, this makes study of naturally developing feline asthma appealing. Additionally, feline models of induced asthma have been beneficial for research (see below).

While cats are blamed frequently as triggers for human asthma, ironically pet cats have tested positive for human dander.

EXPERIMENTAL FELINE ASTHMA

There are still inherent challenges to making a definitive diagnosis of feline asthma in pet cats, including discriminating it from chronic bronchitis, minimally invasive testing, and very rare use of pulmonary mechanics measurements which are key for asthma diagnosis/discrimination in humans. Additionally, as is appreciated in humans, there are differing triggers, amounts and timing of allergen exposure, other environmental influences and genetic variations within pet cats with asthma making clinical studies problematic. Because cats are a species which naturally develops asthma (unlike mice, rats, guinea pigs, sheep and non-human primates which have been used for many basic science discoveries) it makes sense to use them as an experimental model of disease. That is, the feline species "is set up" to be able to respond to allergen by developing important immunologic and physiologic features seen in humans with asthma.

Historically, several models of feline asthma have been used. The earliest used ovalbumin, a very common antigen used in rodent models of asthma. Ovalbumin is not an allergen cats are exposed to naturally in their environments. The next model used *Ascaris suum* antigen; while cats are naturally exposed to ascarids, this parasite normally is ingested in the wild causing gastrointestinal and not generally respiratory signs. The other model was developed by testing pet cats with asthma presenting to a Veterinary Medical Teaching Hospital for allergen-specific IgE against a large panel of potential allergens. House dust mite and Bermuda grass allergen (HDMA and BGA, respectively) were implicated most commonly in these cats. Subsequent use of these same allergens to induce experimental asthma led to a useful pre-clinical model with clinically relevant aeroallergens.[1] Specifically, experimentally asthmatic cats had increases in HDMA and BGA specific IgE, airway eosinophilia, airway hyperreactivity, an acute T helper 2 cell cytokine profile and histologic evidence of airway remodeling.

UNDERSTANDING MECHANISMS OF DISEASE

Aeroallergens are ubiquitous in the environment. However, only a minority of individuals exposed to aeroallergens progress to having clinical features of allergic asthma. Individuals not developing allergic asthma have immune systems which recognize and develop immunologic tolerance to allergens. Individuals failing to develop tolerance instead mount an active immune response which polarizes naïve T helper cells into T helper 2 lymphocytes, which ultimately orchestrate an allergic inflammatory response. Thus, development or failure of development of tolerance is a central event which dictates whether an allergic lung reaction eventually results. The feline allergic asthma model has been useful to study two time frames of tolerance: in early life (ie, in utero and neonatally, including maternal factors) and in well established chronic allergic asthma. While studies are still underway, it appears that regulatory T cells (CD4+CD25+foxP3+ phenotype) and the immunosuppressive cytokine interleukin 10 are associated with reestablishment of tolerance in chronic experimental feline asthma.

DIAGNOSTICS

The gold standard of diagnosis in human asthmatics relies on a combination of pulmonary function testing and bronchoalveolar lavage (BAL) and/or mucosal biopsy. Pulmonary mechanics measurements using spirometry relies on patient cooperation and effort, which are not feasible in infant/pediatric patients and in cats. In experimental asthma models, pulmonary mechanics has been performed using bronchoprovocation (methacholine or carbachol) to assess airway hyperresponsiveness. [2,3] Indirect mechanics (barometric whole body plethysmography) have been criticized for not directly measuring airway resistance, but direct mechanics are invasive, requiring anesthesia. Thus non- or minimally-invasive tests that could provide diagnostic and monitoring information relevant for asthma are in high demand. The experimental model of feline asthma has been used to investigate biomarkers, including cytokines, nitric oxide, matrix metalloproteinases and endothelin. Additionally, to screen for "panels" of biomarkers, proteomics and metabolomics have also been performed.

THERAPEUTICS

Therapeutic options for human asthma are far more plentiful than for feline asthma, with numerous medications having undergone clinical trials in people. The body of human asthma literature can be explored to look for novel treatments to be tested in experimental feline models and ultimately used in pet cats. Treatments which have been evaluated in this manner in experimentally asthmatic cats include inhaled glucocorticoids, leukotriene receptor antagonists, antihistaminics, inhaled lidocaine, long acting beta-2 agonists, anticholinergics, and allergen-specific immunotherapy.[3-8] Additionally, therapies which have been evaluated in experimental feline asthma that have potential relevance to both species include tyrosine kinase inhibitors and immunomodulatory compounds.[9-11] Finally, the feline asthma model can also be used to test adverse in vivo effects of asthma drugs in widespread use. For example, albuterol, the most commonly used short acting beta 2 agonist in use in people (and also commonly used in pet cats) is in a class of drugs in which overuse is linked with increased risk of death ("beta agonist paradox").[12] By using both healthy cats and experimentally asthmatic cats, it was shown that in vivo, inhalation of racemic albuterol is pro-inflammatory, causing de novo neutrophilic inflammation in healthy cats and exacerbating eosinophilic inflammation in asthmatic cats. [13]

CONCLUSION

There is tremendous potential to use the feline experimental model of allergic asthma which closely replicates both naturally developing feline and human asthma to investigate pathophysiologic mechanisms of disease, which in turn may lead to improved diagnostics and therapeutics applicable to both species.

REFERENCES

[1.] Norris Reinero C, et al.. **Int Arch Allergy Immunol** 2004;135:117-131. [2.] Kirschvink N, et al. **Vet J** 2007;174:541-553. [3.] Reinero C, et al. **Am J Vet Res** 2005;66:1121-1127. [4.] Leemans J, et al. **Vet J** 2009;180:236-245. [5.] Nafe L, et al. **Symposium of theVeterinary Comparative Respiratory Society** 2010. [6.] Reinero C, et al. **Vet Immunol Immunopath** 2008;121(3-4):241-250. [7.] Reinero CR, et al.. **Vet Immunol Immunopathol** 2006;110:141-153. [8.] Schooley E, et al. **Am J Vet Res** 2007;68:1265-1271. [9.] DeClue A, et al. **Inflamm Res** 2009;58:457-462. [10.] Eberhardt J, et al. **Vet Immunol Immunopath** 2009;132:175-180. [11.] Guntur V, et al. **American Thoracic Society International Conference** 2011. [12.] Spitzer W, et al. **N Engl J Med** 1992;326:501-506. [13.] Reinero C, et al. **Int Arch Allergy Immunol** 2009;150:43-50.

COMPARATIVE GENETICS OF ANIMAL NEUROMUSCULAR DISEASE

Jim Mickelson, MS, PhD
Saint Paul, MN

OVERVIEW

The structure of domestic animal breeds and the nature of breeding programs sets up a situation where mutations that adversely affect animal health and well-being can be both accidentally and purposefully perpetuated. Fortunately, the nature of these populations can also allow research approaches to identify the causative genes and mutations. Discovery of the underlying genetic basis for a condition at the very least allows specific diagnoses of that disease to be made and breeding animals selected to attempt to reduce the incidence. The major goal of this presentation is to convey a practical understanding of how genes and mutations responsible for, or highly associated with, heritable neuromuscular disorders in domestic animal species are discovered. This summary will also provide specific examples of confounding factors, current shortcomings and future improvements likely to impact veterinary neuromuscular disease genetic studies.

REQUIREMENTS FOR GENE DISCOVERY RESEARCH

Interest in defining genetic causes of animal neuromuscular disease far preceded the community's ability to efficiently do so. Disease gene and mutation discovery in species of veterinary interest to veterinary medicine first started in the early 1990's, usually, but not always, with diseases shared with humans. Now, the pace of discovery of genes and disease alleles is so extraordinary it has an excellent chance of becoming bewildering.

The genomic resources and analysis techniques have changed over time as our knowledge of the structure, sequence and variation within genomes of the various species has increased. Concurrently, our ability to assay that genetic variation at the DNA sequence level has expanded tremendously. In both cases animal genetics and genomics is taking its lead from studies in humans. We now have whole genome sequences and high density panels of DNA markers for disease mapping studies in most animal species at a very low relative cost. Our capabilities will continue to change with whole genome sequencing of individuals in some cases becoming possible.

In any event, a number of different scenarios are typically encountered in animal disease gene mapping research. One reality that has not changed over time is that the ability to accurately phenotype individuals as affected or unaffected, and to distinguish shades of gray in these diagnoses, is probably the most critical factor in the success or failure of animal medical genetic studies.

CANDIDATE GENE STUDIES

In a classical candidate gene approach the trait is seen to be very similar to that in another species, or a large amount of supporting biochemical and physiological data implicate a relatively well known process or pathway. Coupled with this knowledge is an expectation that the trait has a simple mode of inheritance such as autosomal recessive or dominant, or X-linked. In this case candidate genes from the similar disease or known cellular process are sequenced one at a time (typically only the amino acid encoding regions). The hope is to find mutations that have a high probability of being causative based on the nature of the sequence alteration and its predicted affect on the structure or function of the encoded protein.

Some of the first mutations identified were for malignant hyperthermia and hyperkalemic periodic paralysis, as well as several metabolic and clotting disorders. Here considerable long-term research on the nature of the basic biological processes occurred to enable the studies to even select candidate genes. Now, with a wealth of comparative cellular and molecular information available the decision to pursue particular candidate genes is often easier, but perhaps not as often successful in the end. The example of equine glycogen branching enzyme deficiency (1,2) will be utilized in the presentation to illustrate the candidate gene approach.

At the end of a successful candidate gene study, a specific diagnostic test will likely be developed for practical use, although we may not have gained much new information on the underlying biology of the process. In our experience the chances of guessing wrong and sequencing many candidate genes without gaining any knowledge other than that they can be excluded from causing the trait is very high. Usually, given the complexity of the genome, with its ~ 20,000 genes, and a very large fraction of them without known functions, the chance of guessing correctly is very low. Or, the hypothesis that the trait has a simple genetic basis, is wrong.

WHOLE GENOME LINKAGE AND ASSOCIATION ANALYSIS

It is usually the case that the trait of interest is not precisely similar to that in another species, or that the candidate gene approach has been considered, but failed. It is also entirely reasonable, if resources are available, to decide not to waste time betting on candidate genes. In this case a whole genome approach is taken, with the hypothesis being that the location of the segment(s) containing the genetic alteration can be identified, given sufficient cases and controls and sufficient DNA markers of the chromosomal segments.

The whole genome approach relies on the ability to assay DNA markers of known and specific locations along the chromosomes that display variation in their sequences within the population. Until recently the markers used were usually short sequence repeats known as microsatellites, but now they are almost always single nucleotide polymorphisms (SNPs). The basic principle of whole genome analysis is that individuals with the same disease all share a DNA fragment by descent from a founder. This founder can be relatively recent, or have lived many hundreds or even thousands of years ago. Whole genome approaches use the natural variation in genomic DNA sequences that exist within pedigrees or populations, and between cases and controls, to allow the chromosomal region containing the disease gene shared by the cases, to be identified.

In linkage analysis the inheritance of the DNA markers and the co-inheritance of disease are assessed in extensive pedigrees. This was the most common approach employed once sufficient numbers of microsatellite markers from genome projects were identified to cover the genome, although limitations in the size of available pedigrees were always problematic. In association analyses populations of "relatively unrelated" cases and controls are assessed for markers that display differences in allele frequencies that mark the location of the shared chromosomal segment that confers disease. Such studies now typically use tens to hundreds of thousands of SNP markers that are relatively inexpensive and convenient.

In either case, once a region of the genome that displays significant difference in marker allele patterns between cases and controls is identified a search for plausible genes in the segment is made. These genes are again selected from comparative knowledge of gene locations, conserved blocks of orthologous genes, and conservation of gene function across species. The example of dynamin-associated exercise-induced collapse (3) in dogs will be utilized in the presentation to illustrate the whole genome linkage approach.

The result of whole genome analyses can again be the identification of a clear disease causing mutation and a genetic test. However, this is not always straight forward, as sometimes the region can still be quite large and contain hundreds of genes. In these studies there is also a good chance for discovering a mutation in a gene for a process for which a mammalian disorder had not yet been defined. The mutation itself may not be a simple obviously change in the DNA sequence. It can sometimes be an insertion or deletion within introns or non-coding regions that alter mRNA structure or quantity, or even duplications of large DNA segments, even entire genes. The mechanisms by which these variations actually cause the disorder may not be obvious either. A debate about causality vs very high association with the disorder can continue for a considerable time.

INCOMPLETE PENETRANCE, PHENOCOPIES, AND MULTIPLE MUTATIONS

Many disorders defined in the early stages of comparative medical genetics research have very high correlation between the presence of the mutation and disease status. This can be as recessive (2 copies of the mutant allele is always affected) or dominant (1 copy of the mutant allele is always affected). However, we are increasingly finding disorders with discovered mutations that do not fit this simple pattern. For example, many conditions have incomplete penetrance, in which individuals with the susceptible genotype do not show any signs of disease, do so only with aging, or only sporadically under certain environmental conditions. Yet, they do pass the mutant alleles on to offspring who can express the condition.

Another increasingly common scenario is when individuals appear to have all or some of the signs of a known genetic disease, but which do not possess the known mutant allele. This situation, known as a phenocopy, is due in part to a lack of specificity of clinical tests and signs, as well as distinct genetic bases for similar diseases. Lastly, there are now examples where individuals have multiple known mutations which interact to affect the severity of disease. The examples of exercise-induced collapse in dogs (3), and polysaccharide storage myopathy in horses (4-7), will be utilized in the presentation to illustrate the principles of incomplete penetrance and phenocopy.

GENES THAT ARE TARGETS OF GENETIC SELECTION

Whole genome analyses and densely spaced SNP markers have allowed the identification of genes which have had very large segments of flanking DNA sequences preserved and inherited intact over time. This can be the case if the gene confers a selective advantage and has increased in frequency due to the popularity of such animals in breeding programs. The example of polysaccharide storage myopathy in horses (4) will be utilized to illustrate signatures of genetic selection at the genome level in an allele that at one time likely caused a selective advantage, but in current times causes a disease.

PROBLEMS IN ANIMAL GENETICS RESEARCH

The increasing ability to perform successful veterinary genetic investigations is accompanied by the increasing realization that there are many studies that fail to identify genetic mechanisms for a trait. Reasons for failure to identify underlying genes and mutations include incomplete understanding of the underlying pathophysiology, over-enthusiasm or misreading of the true genetic model for the condition, phenocopies, and a lack of specificity of diagnoses.

Our experience is that traits are frequently more complicated than first thought from ascertainment of a limited number of individuals or pedigrees. This genetic complexity can range from having two major genetic loci each contributing significantly to susceptibility to dozens of genetic variations at multiple loci all contributing in less significant ways. The more loci involved the lower the power to detect the less significant ones. In addition it is entirely possible that many conditions have such a high impact of environmental factors that this variation overpowers the influence of underlying genetics on the expression of the trait.

Another problem in certain species, such as the horse, is that the SNP arrays are not adequate given the average length of haplotype blocks in most horse breeds, and population sizes analyzed are insufficient due to multiple causative loci. The examples of polyneuropathy (8) and idiopathic epilepsy in dogs (9) will be utilized in the presentation to illustrate several complications (both multiple significant loci and no significant loci) in the whole genome association approach.

CONCLUSIONS

Comparative medical genetics of relevance to all of veterinary medicine is advancing very quickly. This field of research has a proven history of providing new insights into well-known biological processes as well as discovering new mechanisms involved with as yet poorly understood processes.

The outcomes of this research also are increasingly affecting the lives of veterinarians, the public and animals in not entirely logical ways. It no doubt is remarkably increasing the ability to diagnose certain, but not all, genetic conditions. However, knowledge of the genetic mutation has not yet necessarily increased treatment options. Further, genetic knowledge does not necessarily mean that the incidence of the allele in the population will decrease, due to variations in how the information is used by owners and breeders. Lastly, not surprisingly, disorders that were originally thought to be present in a single breed are found to be present at significant frequencies, in other breeds, and not always breeds that appear to be closely related.

Aspects of medical genetic research that have not changed over time is that our ability to advance knowledge still depends on the accuracy of diagnostic methods, the heritability of the trait, and the species and breed in which the trait occurs. Of at least equal importance is the required curiosity and devotion of academic and private practice veterinarians, and animal owners.

REFERENCES

1. Ward TL, Valberg SJ, Adelson DL, Abbey CA, Binns MM, and Mickelson JR. (2004). Glycogen branching enzyme (*GBE1*) mutation causing equine glycogen storage disease IV. Mamm Genome 15, 570-577. 2. Wagner ML, Valberg SJ, Ames EG, Bauer MM, Wiseman JA, M. Penedo MCT, Kinde H, Abbitt B, and Mickelson JR. (2006). Allele frequency and likely impact of the glycogen branching enzyme deficiency gene in Quarter Horse and Paint Horse populations. J Vet Intern Med 20, 1207-1211. 3. Patterson EE, Minor KM, Tchernatynskaia AV, Taylor SM, Shelton GD, Ekenstedt KJ, and Mickelson JR. (2008). Dynamin 1 (*DNM1*) mutation is highly associated with the syndrome of exercise-induced collapse in the Labrador retriever dog. Nature Genetics 40, 1235-1239. 4. McCue ME, Valberg SJ, Miller MB, Wade CM, DiMauro S, Akman HO, and Mickelson JR. (2008). Glycogen synthase (*GYS1*) mutation causes a novel skeletal muscle glycogenosis. Genomics 91, 458- 466. 5. McCue ME, Valberg SJ, Jackson M, Lucio M and Mickelson JR. (2009). Polysaccharide Storage Myopathy phenotype in quarter horse-related breeds is modified by the presence of an *RYR1* mutation. Neuromuscular Disorders 19, 37-43. 6. McCue ME, Armién AG, Lucio M, Mickelson JR, and Valberg SJ, (2009). Comparative skeletal muscle histopathologic and ultrastructural features in two forms of polysaccharide storage myopathy in horses. Vet Pathology 46, 1281-1291. 7. McCue ME, Anderson SM, Valberg SJ, Piercy R, Binns MM, Distl O, Penedo MCT, and Mickelson JR. (2010). Estimated prevalence of the Type 1 Polysaccharide Storage Myopathy mutation in selected North American and European breeds. Animal Genetics 41, Supplement 2, 145-149. 8. Ekenstedt KJ, Drögemüller C, Minor KM, Shelton GD, Leeb T, Becker D, Patterson N, and Mickelson J. (2010). Whole-Genome Association Analysis Reveals Two Loci Strongly Associated With Leonberger Polyneuropathy. 5th International Conference in Canine and Feline Genomics and Infectious Disease. 9. Ekenstedt KJ. (2010). PhD Thesis, University of Minnesota.

PHARMACOGENETICS: UNDERSTANDING INDIVIDUAL DRUG RESPONSES

Lauren A. Trepanier, DVM, PhD, DACVIM, DACVCP
Madison, WI

TYPES OF PHARMACOGENETIC VARIABILITY

Pharmacogenetics is the study of genetic variability in drug absorption, biotransformation, transport, and response. Much of the initial work done in humans has focused on variability in drug metabolizing enzymes, particularly cytochrome P450's. However, recent studies have also characterized variability in Phase II conjugating enzymes, drug transporters, and drug targets, any of which can influence therapeutic response. Genetic variability in these proteins is determined by heritable differences in the nucleotide sequences of their respective genes, often at single base pairs (single nucleotide polymorphisms, or SNP's). A polymorphism, by definition, is present in the population in at least two allelic forms, with the least common allele usually maintained in the population with a frequency of at least 1%. SNP's can occur in the coding sequence of the gene (which may lead to a change in amino acid sequence), the promoter region of the gene (which may lead to a change in the level of enzyme or receptor expression), the splice junctions in the gene (potentially leading to altered RNA transcripts and therefore altered proteins), or the 3' (downstream) untranslated region of the gene, which can alter the stability of the mRNA transcript (and therefore the amount of protein that is translated). Changes in the amino acid sequence can result in changes in enzyme activity due to altered K_m (affinity for the substrate), altered V_{max} (catalytic activity of the enzyme), or altered enzyme stability. Thus, a single nucleotide change can have marked effects on enzyme function, metabolic capacity, and the outcome of efficacy or toxicity in a given patient.

POLYMORPHISMS IN CYTOCHROME P450s

Cytochrome P450 enzymes (CYPs) are found in the liver, kidney, lung, brain, and other tissues; CYPs catalyze oxidation, reduction, and hydrolysis reactions of endogenous products, therapeutic drugs, and other foreign chemicals. There are hundreds of P450s recognized; these proteins are categorized within and between different species using a family/subfamily/individual enzyme nomenclature. For example, CYP3A4, an abundant P450 in human liver, is a member of the CYP3 family, CYP3A subfamily, and is a unique enzyme (3A4) in humans. The ortholog (comparable enzyme) in dogs is CYP3A12. Orthologs often metabolize the same drug substrates, but it cannot be assumed that a canine CYP will necessarily have the same substrate specificity as its human counterpart.

There are a number of examples of polymorphisms in human CYPs that directly affect clinical outcome. For example, CYP2D6 is a highly polymorphic P450 in humans, with variants that range from undetectable activity (found in 6-10% of Caucasians), to "ultrarapid" activity (due to the presence of multiple copies of the gene, found in 3-10% of Europeans and up to 30% of black Ethiopians).[1] A large number of clinically important drugs are metabolized by CYP2D6, to include beta-blockers (propranolol, metoprolol), antiarrhythmics (quinidine, flecainide), antidepressants (amitriptyline, clomipramine, fluoxetine), and opioid derivatives such as codeine and dextromethorphan. CYP2D6 status can markedly affect drug dosage requirements; for example, poor metabolizers need to be given only 1/10 of the standard dosage of nortriptyline to avoid side effects, while ultrarapid metabolizers require 5 times the normal dosage for clinical effect.[1] Interestingly, poor CYP2D6 activity (which is inefficient at converting codeine to its more potent morphine metabolite) may be a protective factor against opiate addiction in humans.[2]

Cytochrome P450's in dogs are not as completely characterized as they are in humans, but there is recent interest (because of the use of dogs for pre-clinical drug testing) in learning more about the differences between human and canine CYP's. Most of the major CYP subfamilies have been identified in dogs, and are shown in Table 1, along with known substrates and inhibitors in dogs. A few CYP pathways have been shown to be polymorphic in dogs, and more work is ongoing. For example, CYP2B11, which metabolizes propofol, varies at least 14-fold in activity in mixed breed dogs.[3] Greyhounds have particularly low activity,[4] which corresponds to reduced clearance of propofol *in vivo*, higher blood propofol concentrations for a given dosage, and delayed propofol recovery compared to mixed breeds.[5,6] A second CYP pathway that is polymorphic in dogs is CYP1A2. Ten to 15% of beagles are homozygous for a stop codon polymorphism (Arg373*) that leads to an absence of hepatic CYP1A2 expression.[7-9] More work is needed to characterize the substrate range of CYP1A2 in dogs, and determine the prevalence of this null genotype in a wider variety of breeds. Other canine CYP variants are listed in Table 1.

POLYMORPHISMS IN PHASE II CONJUGATING ENZYMES

See "Genetic variability in Phase II enzymes: identifying the risk of drug toxicity," in these Proceedings.

TRANSPORTER POLYMORPHISMS

Transporter proteins, including p-glycoprotein, multidrug-resistance protein (MRP2), breast-cancer resistance protein (BCRP), organic anion transporter proteins (OATPs), and organic cation transporters (OCTs), are variably expressed in the intestinal epithelium, renal tubules, hepatocytes, cholangiocytes, and endothelium of the blood-brain barrier. These transporters mediate uptake or efflux of endogenous compounds and xenobiotics. The best-known example of pharmacogenetic variability in dogs involves the transporter p-glycoprotein; a 4 base pair deletion in the mdr1 gene (now called the ABCB1 gene) leads to truncation and loss of expression of p-glycoprotein.[10] Because p-glycoprotein is part of the blood-brain barrier and is involved in drug export, dogs that are homozygous for this polymorphism (for example, 35% of Collies)[11] are susceptible to toxicity from a wide range of drugs, to include ivermectin, milbemycin, ondansetron, loperamide, vincristine, vinblastine, digoxin, mexiletine, and doxorubicin. [12,13] Genotyping for this p-glycoprotein defect, using a simple swab kit to collect cheek epithelial cells for DNA isolation (http://www.vetmed.wsu.edu/depts-vcpl/test.aspx) is recommended for Collies, Australian Shepherds, long haired whippets, and other breeds known to carry this mutation (http://www.vetmed.wsu.edu/depts-VCPL/breeds.aspx). This information can help clinicians avoid the use of potentially toxic p-glycoprotein substrates throughout the lives of these dogs.

In humans, transporter polymorphisms are just beginning to be associated with altered drug disposition. [14] For example, defective variants in OATPs (which mediate uptake into hepatocytes) have been associated with higher plasma concentrations of antivirals, statins, and oral hypoglycemic drugs; one OATP1B1 polymorphism has been linked to statin-associated myopathy in humans. [15] Variants in MRP2, which mediates export of drugs from hepatocytes into the biliary canaliculi, have been associated with hepatotoxicity from herbal supplements. [16] Like p-glycoprotein, MRP2 also contributes to the blood-brain barrier; consistent with this, MRP2 polymorphisms have been linked to neurotoxicity from the MRP2 substrate carbamazepine. [17] Canine hepatic MRP2 has recently been cloned and characterized, [18] along with a canine OATP (Oatp1b4) that mediates hepatocyte uptake of BSP, taurocholate, and cholecystokinin. [19] These advances set the stage for future studies of the drug substrate range and genetic variability of these hepatic transporters in dogs.

The ABCG2 transporter (i.e. BCRP) has been characterized in cats; this efflux pump is part of the blood-retinal barrier and transports some drugs, notably fluoroquinolones.[20] The activity of ABCG2 is impaired in cats compared to humans, which may explain feline susceptibility to fluoroquinolone-mediated retinotoxicity. [20]

RECEPTOR POLYMORPHISMS

Drugs may bind to specific receptors (e.g. atenolol and the beta-1 adrenergic receptor) or may interact with enzymes or other drug targets (e.g. allopurinol and xanthine oxidase). Polymorphisms in these target molecules can affect clinical drug responses. For example, variants in the human opioid mu receptor gene (OPRM1) have been associated with decreased analgesic response to morphine,[21] and an alpha-2-adrenergic receptor (ADRA2A) polymorphism has been linked to increased sedation from dexmedetomidine. [22] One of the best characterized polymorphisms in a drug target gene is in vitamin K epoxide reductase (VKORC1 -1639G>A), which leads to increased sensitivity to warfarin. [23] A commercial test is available for this polymorphism, and human dosing recommendations on the warfarin label are actually based upon VKORC1 genotype.

In dogs, a polymorphism in the promoter region of the mu receptor gene (OPRM1 -15C>A) has been identified; the wild type allele is hypothesized to be associated with opioid dysphoria in dogs, and requires further investigation. [24] Polymorphisms in the D4 (dopamine) receptor gene, which involve a variable number of tandem nucleotide repeats within the gene, have been found with various frequencies in different dog breeds [25] and may be involved in impulsivity behavior in dogs. [26] Variants in a serotonin receptor gene (5-HTR1B) have also been identified in dogs, [27] but have yet to be studied in the context of altered drug response.

FUTURE DIRECTIONS

Pharmacogenetics is an exciting field that is rich in possibilities for veterinary researchers. Our collective experience with different species and a variety of purebreeds places our profession in an excellent position to generate and test hypotheses about the pharmacogenetic basis of therapeutic responses. We are gaining the tools needed to incorporate genotyping into clinical practice; this has already happened for p-glycoprotein polymorphisms. However, we need to focus more energy on characterizing veterinary CYPs, Phase II enzymes, transporters, and drug receptors, in both dogs and cats. By looking to the human experience for guidance, we can begin to understand the mechanisms for unwanted drug responses in our patients, to include unexplained therapeutic failures, exaggerated side effects, and drug-induced toxicities.

REFERENCES

1. Bertilsson L, et al. *Br J Clin Pharmacol* 2002;53:111-122. 2. Tyndale RF, et al. *Pharmacogenetics* 1997;7:375-379. 3. Hay Kraus BL, et al. *Xenobiotica* 2000;30:575-588. 4. Court MH, et al. *Drug Metab Dispos* 1999;27:1293-1299. 5. Robertson SA, et al. *Am J Vet Res* 1992;53:1027-1032. 6. Zoran DL, et al. *Am J Vet Res* 1993;54:755-760. 7. Mise M, et al. *Pharmacogenetics* 2004;14:769-773. 8. Mise M, et al. *Drug Metab Dispos* 245. 9. Tenmizu D, et al. *Xenobiotica* 2004;34:835-846. 10. Mealey KL, et al. 2001;11:727-733. 11. Mealey KL, et al. *Am J Vet Res* *Vet Parasitol* *J Vet Pharmacol Ther* Sissung TM, et al. *Mol Biotechnol* Kalliokoski A, et al. 2009;158:693-705. 16. Choi JH, et al. *Pharmacogenet Genomics* 2007;17:403-415. 17. Kim WJ, et al. ;20:249-256. 18. Ninomiya M, et al. 2005;33:225-232. 19. Gui C, et al. *Comp Biochem Physiol C Toxicol Pharmacol* Ramirez CJ, et al. ;21:66-75. 21. Mague SD, et al. *Drug Alcohol Depend* Yagar S, et al. . 23. Yuan HY, et al. *Hum Mol Genet* Hawley AT, et al. ;37:79-82. 25. Niimi Y, et al. 2001;92:433-436. 26. Hejjas K, et al. 2007;38:629-633. 27. Masuda K, et al. *J Vet Med Sci* 2004;66:965-972. 28. Mills BM, et al. 2010;38:396-404. 29. Korytko PJ, et al. 1999;47:52-61. 30. Verreault J, et al. 2009;150:91-100. 31. Regmi NL, et al. 2005;28:553-557. 32. Mise M, et al. 2008;36:1903-1908. 33. Chen CS, et al. 2004;65:1278-1285. 34. Aidasani D, et al. 2008;36:1512-1518. 35. Baratta MT, et al. ;33:50-55. 36. Lankford SM, et al. 2000;28:981-986. 37. Locuson CW, et al. 2009;37:457-461. 38. Eguchi K, et al. 1996;26:755-763. 39. Blaisdell J, et al. 1998;26:278-283. 40. Paulson SK, et al. 1999;27:1133-1142. 41. Chen J, et al. *Drug Metab Lett* 2009;3:61-66. 42. Mealey KL, et al. *Xenobiotica* 2008;38:1305-1312.

Table 1: Known cytochrome P450 polymorphisms in the dog.

Cytochrome P450 in the dog	Known dog substrates and inhibitors	Known dog polymorphisms
CYP1A1	Caffeine [28] Induced by environmental toxins (e.g. polychlorinated biphenyls (PCBs) [29,30] Inhibited by ciprofloxacin, orbifloxacin [31]	None yet reported
CYP1A2	Caffeine is substrate in humans but not in dogs [32] Can be induced or inhibited by omeprazole	Arg373* SNP in 10-15% of beagles, which leads to absence of hepatic CYP1A2 expression [7,9]
CYP2B11	Propofol, progesterone, testosterone, midazolam, ketamine Cyclophosphamide *activation* in humans [33] Induced by phenobarbital Inhibited by chloramphenicol, propofol, ketoconazole, miconazole, loperamide; medetomidine [34,35]	Propofol is poorly metabolized in Greyhounds; 14-fold variability in mixed breeds [3,6]
CYP2E1	Chlorzoxazone	Two alleles recognized in 100 mixed breed dogs, but same catalytic activity [36]
CYP2C21	Testosterone, midazolam [37] Modest induction by phenobarbital	CYP2C21 is most abundant P450 in dog liver[38] Second enzyme, CYP2C41, expressed in < 20% of beagle dogs [39]
CYP2D15	Celecoxib, dextromethorphan, imipramine, metoprolol, propranolol Inhibited by clomipramine, ketoconazole, miconazole, loperamide, ondansetron, fluoxetine [34]	50% of beagles are slow metabolizers of celecoxib; another genetic variant also found with undetectable celecoxib metabolism [40]
CYP3A12	Erythromycin, progesterone, testosterone, tacrolimus, cyclosporine, medetomidine, imatinib Induced by phenobarbital, phenytoin, rifampin, clotrimazole [41] Inhibited by ketoconazole, loperamide	Two CYP3A12 alleles recognized so far, differing in 5 amino acids; CYP3A12 expressed in canine intestine Second isoform, CYP3A26, also present in dogs, expressed at higher levels in liver than CYP3A12;[42] Both enzymes are induced by the same drugs

RAW FOOD DIETS: A RESEARCH REVIEW

Andrea J. Fascetti, VMD, PhD, DACVIM (Small Animal Internal Medicine), DACVN
Davis, CA

INTRODUCTION

Raw food diets are often referred to by the acronym "BARF" which can stand for "bones and raw food" or "biologically appropriate raw food" diet. It is unclear how many people are feeding their pets raw food diets. However, based on the growing number of commercially available raw diets, one can surmise it is growing, but that they are likely still fed to a small percentage of pets. Proponents of raw food diets proclaim many health benefits associated with this feeding regime, stating that dogs and cats are carnivores and as such they evolved eating raw food. However, there have been no studies to date to support that this feeding approach has any long-term health benefits compared to feeding other types of pet food. Despite the absence of long-term studies, there has been research looking at the nutritional performance as well as the possible risks and benefits of bones and raw food diets.

Nutritional Adequacy

Nutritional adequacy should be one of the first concerns of every practitioner regarding their patient's feeding program. Concerns regarding nutritional adequacy not only apply to home-prepared raw diets but to commercial raw diets as well. One study that looked at the nutritional adequacy of home-prepared raw food, also looked at the nutritional adequacy of several commercial raw food diets.[1,2] Two diets were commercial products, the remaining three home-prepared. All five diets had essential nutrients that were analyzed to be below (Association of American Feed Control Officials) AAFCO minimum recommendations.[1,2] The home-prepared diets had excessive concentrations of vitamins D and E, as well as inappropriate calcium to phosphorus ratios. A recent survey of dog owners in Germany that were feeding home-prepared bone and raw food diets found that 76% of the 77 rations analyzed had at least one nutritional imbalance.[3] Clinical case reports of problems in animals consuming commercial raw food diets are now beginning to appear in the literature as well.[4] Commercially-produced raw diets fall under the same AAFCO labeling guidelines with respect to reporting nutrient content, ingredients and nutritional adequacy statements. AAFCO recommendations for nutrient minimums and maximums exist to guide and protect against nutritional concerns in commercially-produced foods. While some of these recommendations are based on studies using semi-purified diets, there are also many studies using extruded or moist diets. While one can argue that for traditional commercial foods there is a paucity of data concerning many nutrient requirements, bioavailability and the effect of diet on gastrointestinal microbial populations in dogs and cats, there is less for raw foods. It is well-known that nutrient requirements can vary depending on the type of diet (i.e., feline recommended allowances for taurine are almost double for canned products compared to extruded diets). Therefore one can speculate that similar examples may exist for dogs and cats consuming raw diets and that further research is needed.

The literature contains a variety of papers documenting the palatability, chemical composition, digestibility and bioavailability of many traditional animal and plant-based pet food ingredients. There are also a number of papers examining the effect of animal and plant-based ingredients on fecal quality and quantity. Recently more information with respect to ingredient and diet characteristics and qualities for raw diets are becoming available in the scientific literature. One research group at the University of Illinois has done several studies looking at macronutrient digestibility, nitrogen metabolism, fecal microbial populations and fecal fermentative products in both captive exotic and domestic cats.[5,6,7,8] This group is currently performing similar studies in dogs.[9] Many of the studies examined raw food feeding in exotic cats but a few also fed raw diets to domestic cats.[6,8] There was a substantial amount of data generated from these studies, some of which will be summarized here. In a recent study using both exotic and domestic cats, animals were fed either a beef or horse-based raw diet in a crossover design.[6] Food intake did not differ between diets but fecal output was greater when the horse-based diet was consumed. Total tract apparent dry matter digestibility was higher but organic matter and crude protein digestibilities were lower when cats were eating the beef-based diet, compared to the horse-based diets. Fecal scores were lower and fecal dry matter was greater when cats consumed the horse-based diet. Fecal ammonia concentrations were lowest when cats consumed the horse-based diet and overall domestic cats had lower ammonia concentrations. Fecal total short-chain fatty acids, branched-chain fatty acids and butyrate were higher in the cats when consuming the beef-based diet. One confounding factor in this study was that the horse and beef-based diet contained different sources of fiber, using powdered cellulose and beet pulp respectively. A recent abstract reported nitrogen metabolism data in domestic cats fed four different raw animal protein sources, beef (crude protein (CP) 66%), bison (CP 49%), elk (CP 79%)and horse (CP 60%).[8] The researchers found that dietary nitrogen was highly digestible for all treatments. Since all the diets were highly digestible, urinary nitrogen accounted for the majority of total nitrogen excretion. Differences in nitrogen and total amino acid intake and the amount of nitrogen absorbed were due to differences in dietary crude protein. Nitrogen retention was similar to values reported in the literature for domestic cats.

Based on some of the previous studies, it appears as though raw protein sources are highly digestible, but these studies do not provide information about availability of the nitrogen or the amino acids from these protein sources. With respect to cats, taurine is one amino acid of concern because the consequences of a deficiency are frequently fatal. The taurine content in animal proteins can vary significantly, with muscle generally containing less taurine than organ meats.[10] A recent survey of commercial raw diets intended to be fed to captive exotic cats found that some contained taurine concentrations lower than 0.1%.[11] Cooking also influences taurine concentrations, and it can be lost to a significant extent when using cooking methods that expose proteins to water, thereby leaching the taurine from the food.[10] These findings imply that if one doesn't cook the protein source, taurine deficiency is less of a concern; however, the literature does not support this thinking. Taurine deficiency has been recognized in cats consuming home-prepared diets using raw protein. One research update reported dilated cardiomyopathy associated with taurine deficiency in a group of growing cats fed a diet consisting solely of whole ground raw rabbit.[12] Cats were fed either whole, ground rabbit or a commercial kibble diet that had passed AAFCO feeding trials for growth. Rabbits were selected over mice for ease of processing and in places where rabbits are abundant, feral cats are known to prefer them as prey.[13] The growth curves of cats on both diets were identical, indicating the raw rabbit diet supported normal growth. Coat quality was better (by subjective assessment) and stool quantity smaller (with less water) in the cats that were consuming the raw rabbit diet. However, the reason(s) for the differences in stool consistency of the respective diets is unknown. The investigators could find no relationship between the type of diet consumed and: 1) the rate of growth, 2) degree of inflammation in the tissue lining the intestinal tract, or 3) the numbers of bacteria in the upper small intestine. After consuming the raw rabbit diet for 10 months one of the cats died from dilated cardiomyopathy and was determined to be taurine deficient. Moreover, 70% of the remaining cats consuming the raw rabbit diet, which appeared outwardly healthy, also had heart muscle changes compatible with taurine deficiency. For the remaining three months of the study, the raw rabbit diet was supplemented with taurine and blood taurine concentrations returned to normal. The investigators concluded there were no other significant benefits to feeding the raw rabbit diet, even when supplemented with taurine compared to a traditional feline kibble. A second study evaluated plasma taurine concentrations in sand cats (*Felis margarita*) fed either a commercial feline kibble or a raw, horsemeat and meat by-product based diet.[14] Despite a 15% increase in digestibility and a 40% increase in taurine content compared to the kibble diet, cats consuming the raw food diet had significantly lower plasma taurine concentrations. Although the plasma taurine concentrations were not below the point at which clinical taurine deficiency would be seen, they were reduced by approximately 25% during the 12 day study period. Arguably a crude estimate at best, but if one were to project the continued rate of decline, plasma taurine would fall below the concentration where the clinical signs of taurine deficiency are frequently noted at approximately day 20 of raw food consumption. These effects would likely be more pronounced under the conditions of a more demanding life stage than maintenance, such as during growth or reproduction.

The exact mechanism of how raw diets can potentiate taurine deficiency is unknown at this time. The amount of taurine available to the cat from its diet is dependent upon a number of factors including the quality and quantity of dietary protein, as well as how that protein is processed.[15,16,17,18] These factors in turn influence gastrointestinal microbial numbers and/or species that can cause taurine loss by accelerating turnover of bile acids conjugated with taurine and decrease recycling of taurine by the enterohepatic route. These factors may influence changes in bacterial populations that favor those that degrade taurine. One study supports that even the amount of protein in the diet can influence feline bacterial populations.[19] Adult cats fed a medium protein (34.34% crude protein) extruded diet had higher concentrations of fecal *Bifidobacterium* populations and lower concentrations of *Clostridium perfingens* compared to cats consuming a high protein diet (52.88% crude protein). Although a second study by the same group did not see any effect in cats fed either a raw horse or beef-based diet on *E. coli, Bifidobacterium, Clostridium perfingens* or *Lactobacillus* species.[7] One rationale provided for the different findings was that the second study reported results on a dry weight basis compared to the earlier study where results were reported on a wet weight basis. This same group also suggested that more sensitive techniques may be required to further differentiate any dietary effects and how one might interpret these findings with respect to taurine availability is unknown. In addition to these factors, low levels of vitamin E in a diet can cause meat to lose taurine when it is processed and ground.[20]

Oral Health

One of the many claims made by those who support a raw meat and bone feeding method is that feeding of raw bones is beneficial to the oral and dental health of the animal. Dental disease, including calculus, gingivitis and periodontitis, is considered to be one of the most common diseases diagnosed in dogs and cats.[21,22] Periodontal disease is of most concern as it can result in tooth loss. Current research supports that dogs and cats consuming commercial diets are at risk for eventually developing periodontal disease. It appears that softer diets (including canned, semi-moist or even home-prepared foods) are even worse than dry diets.[23,24] The supplementation of an oxtail to a commercial diet feeding regime appeared to slow down the development of periodontal disease.[25] So one might surmise from these findings that the consumption of a more "natural" diet, such as a BARF diet, might be beneficial in the prevention of dental disease, particularly periodontitis. However, current research doesn't necessarily support that thinking. One study in African wild dogs, whose diet is largely small antelope, found that 41% of the skulls examined had evidence of periodontitis, while only 2 had dental calculi.[26] A study in feral cats from Marion Island, where the main source of food is birds, reported evidence of periodontitis in approximately 62% of cats while only 9% had calculus.[27] It has been speculated that the highly specific diet of sea birds favored the development of periodontal disease in these cats secondary to gum trauma induced by the sharp bones in the carcass.[27,28] A smaller study in Australia established that the prevalence of oral disease was no different in cats fed a commercial diet versus those whose diet was mainly small prey.[29] So while a natural diet of raw meat and bones may reduce dental calculi, it does not appear to protect against periodontal disease.

Zoonotic Concerns

The veterinarian's job is not only to care for the health and well-being of their animal patients but also those who are the guardians of these pets as well. From this perspective concerns regarding pathogenic bacteria in raw diets and subsequent environmental contamination are paramount. There are numerous publications documenting pathogenic organisms in raw meat and raw food diets. A few that are particularly relevant to the veterinary practitioner or are very recent are highlighted below.

There is growing evidence to support these concerns. Evidence for transmission of food-borne pathogenic bacteria from dogs to humans exists.[30,31,32] In Alberta, Canada, 9 of 12 case patients with *S. infantis* infection had been exposed to pig ear treats and *S. infantis* was isolated from a pig ear treat collected from one of the case patients. The isolate recovered from the pig ear was indistinguishable from *S. infantis* isolates recovered from fecal samples obtained from humans with salmonellosis.[33,34]

Potential human pathogens have been isolated in both commercial and home-prepared raw diets.[1,35,36,37,38,39] Animals fed raw diets have been reported to shed the same viable organisms that were isolated in their food.[40] There have been reports of racing greyhounds, sled dogs, guard dogs and cats with *Salmonella* infections due to consumption of contaminated raw meat.[41,42,43,44,45] A recent publication supports testing fecal samples on more than one occasion if one is suspect of contamination in an animal's feces.[39]

Arguably, while many animals never become ill while consuming raw food diets, they still pose a risk to humans and other animals through environmental shedding.[34,40] Individuals preparing raw diets are also at risk by handling contaminated meat and egg products. Those greatest at risk are the very young and old, in addition to the immunocompromised.

In some cases, despite understanding all of the risks, an owner may wish to continue to feed a raw diet. Practitioners should refer their clients to the FDA's website and go over safe handling and preparation of food, as well as cleaning practices. It has been shown that simple routine washing may not be enough to eliminate potential food-borne pathogens in the pet's food bowl and environment.[46] It is also important to document any discussions one has on this subject, as it may have legal ramifications.[47]

REFERENCES

[1]Freeman LM et al. **JAVMA** 2001; 218(5):705. [2] Freeman LM et al. **JAVMA** 2001; 218(10):1582. [3]Dillitzer N et al. **Proc. Waltham Inter Nutr Sci Symposium**, 2010;45. [4]Taylor MB et al. **JAVMA** 2009; 234(8):1041. [5]Vester BM et al. **Zoo Biol** 2008;27:126. [6]Vester BM et al. **Zoo Biol** 2010a;29:432. [7]Vester BM et al. **Zoo Biol** 2010b;29:676. [8] Kerr K et al. **Proc. Waltham Inter Nutr Sci Symposium**, 2010;97. [9]Swanson KS. **Proc.ACVIM**, 2009; [10]Spitze AR et al. **JAPAN** 87(7-8):251. [11] Hedberg GE et al. **Zoo Biol** 2007;26:517. [12]Glasgow AJ et al. Center for Companion Animal Health, University of California, Davis, School of Veterinary Medicine. (Accessed Jan 20, 2011, **Error! Hyperlink reference not valid.**). [13]Molsher R et al. **Wildlife Res** 1999. 26(5): 593. [14]Crissey SD et al. **J An Sci** 1997;75:2154. [15]Hickman MA et al. **J Nut** 1990;120:995. [16]Backus RC et al. **J Nut** 1994;124:2540S. [17]Kim SW et al. **J Nut** 1996(a);126:195. [18]Kim SW et al. **J Nut** 1996(b);126:509. [19] Lubbs DC et al. **JAPAN** 2009;93:113. [20]Lambert IH et al. **J Ag Food Chem** 2001;49(10):2225. [21]Lund EM et al. **JAVMA** 1999;214(9):1336. [22]Hamp SE et al. **Vet Rad** 1984;25(2):86. [23]Studer E et al. **Vet Med Sm Anim Clin** 1973;68(10):1124. [24] Burwasser P et al. **J Dental Res** 1939;18:389. [25]Brown MG et al. **Lab Anim Care** 1968;18(5):527. [26]Steenkamp G et al. **J Vet Dent** 1999;16(2):65. [27]Verstraete FJ et al. **J Comp Pathol** 1996;115(3):283. [28] Girard N et al. **Encylopedia of Feline Clin Nut** 2008;358. [29] Clarke DE et al. **Aust Vet** 1998;76:690. [30]Gutman L et al. **New Eng J Med** 1973;288:1372. [31] Morse EV et al. **Am J Pub Health** 1976;66:82. [32]Sato Y, et al. *J Vet Med Sci* 2000;62(7):767. [33]Laboratory Centre for Disease Control. **Can Commun Dis Report** 2000;26:41. [34]Finley R et al. **Clin Infect Dis** 2006;42:686. [35]Chengappa MM et al. **J Vet Diag Invest** 1993;5:372. [36] Joffe DJ et al. **Can Vet J** 2002;43:441. [37]Weese JS et al. **Can Vet J** 2005;46:513. [38]Strohmeyer RA et al. **JAVMA** 2006;228(4):537. [39]Leonard EK et al. **Zoo and Public Health** 2010;Epub ahead of print. [40]Finley R et al. **Can Vet J** 2007;48(1)69. [41]Cantor GH et al. **J Vet Diag Invest** 1997;9:447. [42]Caraway CT et al. **JAVMA** 1959;135:599. [43]Morley PS et al. **JAVMA** 2006;228(10):1524. [44]Stiver SL et al. **JAVMA** 2003;39(6):538. [45]Stone GG et al. **J Vet Diag Invest** 1993;5:378. [46]Weese JS et al. **Can Vet J** 2006;47(9):887. [47]LeJeune JT et al. **JAVMA** 2001;219(9):1222.

NEW INTERVENTIONS FOR CACHEXIA & SARCOPENIA

Lisa M. Freeman, DVM, PhD, DACVN
North Grafton, MA

Cachexia and sarcopenia are two important syndromes that occur in chronic disease and aging, respectively. Cachexia has been recognized in human patients for over 2000 years and is becoming increasingly acknowledged as a common and negative finding that is associated with increased morbidity and mortality. As a result, efforts are being made to combat this syndrome and to improve outcome for the wide variety of disease with which it is associated. Sarcopenia is similar to cachexia in that it is characterized by muscle loss but is a syndrome seen during aging in the absence of disease. Both of these syndromes are becoming increasingly important in human and veterinary medicine because of their high prevalence and deleterious effects, and a better understanding of these syndromes is critical to optimize patient care. This high interest is spurring the development of new drugs, diets, and other treatments to combat these syndromes in people. Our veterinary patients will benefit from this research but these are important areas to specifically study in dogs and cats with chronic disease and during aging.

CACHEXIA

Cachexia, a loss of lean body mass, has been described since the time of Hippocrates. It is estimated to be present in more than 5 million people in the United States.[1] Cachexia occurs in a variety of chronic diseases in people, including cancer, AIDS, heart failure, chronic obstructive pulmonary disease, renal failure, and rheumatoid arthritis. The syndrome of cachexia also appears to be common in companion animals with chronic diseases, such as heart failure, renal failure, hepatic failure, and cancer. However, this is an area that has only recently begun to be studied so much additional information is needed.

The weight loss that occurs in cachexia is unlike that seen in a healthy dog, cat, or person that loses weight. In a healthy animal that is receiving insufficient calories to meet requirements, metabolic adaptations allow fat to be used as the primary fuel source, thus preserving lean body mass. Conversely, the primary fuel source in animal with chronic disease is amino acids from muscle so that these animals quickly catabolize muscle and lean body mass. While there are some subtle differences between forms of cachexia in different diseases (e.g., heart failure, cancer), loss of lean body mass is a hallmark of cachexia. Fat and bone also are lost to lesser degrees although in advanced cases of cachexia, depletion of all compartments is apparent. The loss of lean body mass has direct and deleterious effects on strength, immune function, and survival. In people, cachexia is an independent predictor of survival.

One of the most pressing needs in the area of cardiac cachexia research is the need for a definition and a clinically relevant way to diagnose this syndrome. Current definitions in people rely primarily on loss of body weight (typically 5-7.5% over a period of 6-12 months). Total weight loss is an insensitive measure of muscle loss, so using weight loss as a diagnostic criterion reduces the ability to identify cachexia until its more advanced stages and a resulting underdiagnosis of the prevalence of this syndrome. In addition, there are certain types of cachexia (e.g., rheumatoid cachexia, cardiac cachexia with fluid accumulation) in which the weight loss is masked by accumulation of fat or water. Therefore, waiting until weight loss occurs often prevents an early diagnosis and misses the underlying issue of inflammation and lean body mass losses. The critical role of inflammation in cachexia, as well as the clinical importance of anorexia, appear to make inclusion of these factors important for patient assessment.

In addition to the issue of the insensitivity of weight loss as the criterion for cachexia, it is not only the quantitative loss of muscle that results in deleterious effects. There also is a qualitative changes in muscle function. Skeletal muscles in heart failure have increased collagen content, altered mitochondrial function, and a shift in fibers from type I (oxidative) to type IIb (glycolytic) fibers. This shift may further predispose muscle fibers to atrophy as part of the mechanism for cachexia as glycolytic fibers are less resistant to atrophy. Another reason for using factors other than total weight loss for a diagnosis of cachexia is that cachexia is a process – i.e., a loss of lean body mass - and not necessarily an end-stage syndrome. Lean body mass loss occurs before significant weight loss can be detected. If a loss of lean body mass were used as a criterion for the diagnosis of cachexia, the prevalence of this syndrome in chronic diseases would be even higher. Severe weight loss in a patient with advanced heart failure or cancer represents a classical picture of cachexia and offers little diagnostic dilemma. However, identification of cachexia is more difficult when it is more subtle. Therefore, developing clinically applicable techniques to identify lean body mass loss at an early stage in which treatments are much more likely to be successful. The challenge is that to detect cachexia at an earlier stage requires evaluation techniques that currently are not clinically applicable. Even some of the research tools that are used, such as dual-energy x-ray absorptiometry (DEXA), have limitations for measurement of lean body mass due to inherent assumptions (e.g., constant hydration of lean tissues, which is not the case in conditions associated with fluid accumulation or dehydration).

The definition of cachexia is being actively debated for people but is not yet resolved. A consensus definition was developed from a meeting in late 2006: "Cachexia is a complex metabolic syndrome associated with underlying illness and characterized by loss of muscle with or without loss of fat mass. The prominent clinical feature of cachexia is weight loss in adults (corrected for fluid retention) or growth failure in children (excluding endocrine disorders). Anorexia, inflammation, insulin resistance and increase muscle protein breakdown are frequently associated with wasting disease."[2] A diagnostic criteria for cachexia has been proposed:[2]

Weight loss of at least 5% in 12 months or less [or body mass index<20 kg/m^2) plus three of the five features:

1. Decreased muscle strength
2. Fatigue
3. Anorexia
4. Low fat-free mass index

5. Abnormal biochemistry
   a. Increased inflammatory markers [C-reactive protein>5.0 mg/dL, interleukin-6>4.0 pg/ml]
   b. Anemia (Hemoglobin<12 g/dL)
   c. Low serum albumin (<3.2 g/dL)]

Therefore, including both measures of lean body mass and muscle function in the definition of cachexia may be clinically relevant. These factors also may be important in assessing effectiveness of treatments. Similarly, effective treatments not only need to increase body weight, but also to increase lean body mass or to improve muscle function.

Cardiac Cachexia

One of the forms of cachexia that has been studied in dogs and cats is cardiac cachexia. In one study of dogs with dilated cardiomyopathy (DCM), over 50% of patients had some degree of cachexia.[3] The loss of lean body mass in cardiac cachexia is a multifactorial process caused by the anorexia, increased energy requirements, and metabolic alterations. The anorexia may be secondary to the fatigue or dyspnea or may be due to medication toxicity or an unpalatable diet. Anorexia is present in 34-84% of dogs and cats with heart disease.[4-6] However, it is inflammatory cytokines, such as tumor necrosis factor (TNF) and interleukin-1 (IL-1), that are the primary mediators of cachexia.[3,7] These inflammatory cytokines are known to directly cause anorexia, to increase energy requirements, and to increase the catabolism of lean body mass. Of particular pertinence to heart disease, TNF and IL-1 also cause cardiac myocyte hypertrophy and fibrosis and have negative inotropic effects.

The deleterious effects of cachexia have been emerging and recent studies have emphasized the role of body weight and body composition in heart failure. While obesity is a risk factor for development of heart disease in people, obesity may actually be associated with a protective effect once heart failure is present - this is known as the obesity paradox. A recent large meta-analysis on body condition in people with heart failure concluded that obesity and overweight were associated with lower all-cause and cardiovascular mortality and that underweight patients consistently had a higher risk of death. Given the adverse effects associated with cachexia, the association between obesity and improved survival in heart failure appears to be due to a *lack* of cachexia, rather than to the obesity per se. This is due to the increased reserve of lean body mass in overweight and obese people. The obesity paradox also has been demonstrated in dogs and cats with heart failure.[8,9] These data emphasize the importance of avoiding weight (and muscle) loss in the patient with heart failure.

Cardiac cachexia is typically recognized only after congestive heart failure (CHF) has developed. Loss of lean body mass is most readily evident in the epaxial, gluteal, scapular, or temporal muscles. Nutritional considerations for cardiac cachexia should include management of anorexia, if present and nutritional modulation of cytokine production also may be helpful. One method of decreasing the production and effects of cytokines is with n-3 polyunsaturated fatty acid supplementation. Supplementation of fish oil, which is high in n-3 fatty acids can decrease cytokine production in dogs with CHF and improve cachexia.[3] In some, but not all dogs with CHF-induced anorexia, fish oil supplementation can improve food intake.[3] In addition, reduction of cytokines has been correlated with survival in dogs with CHF.[3]

Cancer Cachexia

In people, cancer is one of the most common diseases in which cachexia is present. In people with cancer, it is estimated that over 50% lose weight unintentionally, although the prevalence depends on the type of cancer. In dogs, only 4% had a low body condition score ($\leq$3/9) at the time of diagnosis and 55% had a body condition score $\geq$6/9.[10] However, 35% of dogs had mild to severe muscle wasting. In addition, 69% of dogs for which pre-diagnosis body weights were available had experienced some weight loss (31% had <5% weight loss, 14% had lost 5-10%, and 23% had a weight loss of >10%.[10] A recent study in cats with cancer showed muscle loss in 91% of patients.[11] In addition, cats that were below optimal body condition had a significantly shorter survival time compared to those with a body condition score of five or greater. Therefore, specific attention to muscle mass, as well as changes in body weight, are important in order to detect cancer cachexia.

Sarcopenia

The human population is aging; it is estimated that by 2030, >20% of the American population will be over 65 years of age. While 2/3 of older Americans are overweight or obese,[12] there is significant loss of lean body mass that occurs with aging. This loss of lean body mass, termed sarcopenia, occurs in the absence of disease (this is key to the definition of sarcopenia because lean body mass loss associated with disease would be cachexia). The current definition of sarcopenia is lean body mass less than two standard deviations below that of young adults. Sarcopenia actually begins early in life, around 30 years of age, and from 20 to 80 years of age, there is a 30% reduction in muscle mass.[13] In sarcopenia, the loss of lean body mass often is accompanied by an increase in fat mass so the total weight may not change (or may even increase), thus masking the sarcopenia. Like cachexia, sarcopenia has important clinical implications because it is associated with increased mortality.[13] The mechanism of sarcopenia appear to be multifactorial and involve physical inactivity, elevated cytokine production, decreased concentrations of growth hormone and testosterone, changes in type II muscle fibers, insulin resistance, and decreased protein synthesis.[14]

As in people, dogs and cats also are aging with between 1/3 and 1/2 of pet cats and dogs at least 7 years of age. Few studies investigating sarcopenia have been conducted in dogs and cats but available information suggests that dogs and cats also lose lean body mass during aging.[15-18] This is an important area for future research.

New Interventions

Because of the important implications of cachexia and sarcopenia on morbidity and mortality in people, there is extensive research into the prevention, diagnosis, and treatment of this syndrome. Nutritional modification, hormone replacement, and exercise (both resistance and aerobic) are being actively studied to combat sarcopenia and to enhance healthy aging. For cachexia (and when sarcopenia is not prevented), there are exciting opportunities for new and effective targets to reduce energy requirements, enhance energy intake, improve nutrient absorption, and modify metabolic alterations to prevent and even reverse the effects of cachexia and sarcopenia as the mechanisms of these syndromes are becoming better understood. Blocking single pathways may not be completely effective; treatments that have multiple effects may prove to be the most valuable. For example, exercise training increases muscle mass but also can modify some of the alterations seen in cachexia, such as the reduced GLUT4 expression and increased myostatin concentrations. Ghrelin improves food intake but also increases growth hormone and may have beneficial direct effects on heart. These multiple effects can be exploited to address the multiple pathways affected in this syndrome.

Practical Implications

One of the keys to the practical management of cachexia and sarcopenia in dogs and cats is recognizing it in its earliest stages, and to achieve this, body condition score (BCS) and muscle condition score (MCS) must be consistently assessed. The goal for BCS in a healthy young to middle-aged dog or cat should be 4 to 5 on a 9-point BCS scale. However, in certain diseases (e.g., heart failure, chronic kidney disease), a slightly higher BCS may be desirable (i.e., a BCS of 6 to 7/9 but further research is required to make specific recommendations. Severe obesity should be avoided at all times.

The MCS differs from the BCS in that it specifically evaluates muscle mass. Evaluation of muscle mass includes visual examination and palpation of the head, scapulae, thoracic and lumbar vertebrae, and pelvic bones. BCS and MCS are not directly related as an animal can be obese but still have significant muscle loss (or conversely, be very thin but have a normal MCS). Palpation is required for accurately assessing BCS and MCS, especially in animals with medium to long hair coats. Consistently evaluating MCS in all patients will help to identify muscle loss at an early stage (i.e., mild) in aging or ill animals, rather than waiting until muscle loss is moderate or severe, when it may be more difficult to successfully intervene.

For animals with chronic diseases in which weight and/or muscle loss is a component (e.g., renal, heart or hepatic failure; cancer; respiratory disease), a variety of treatments can be employed. However, in many cases, practical methods to help the owner in managing their animals' appetite are critical to success. This is particularly important because anorexia is one of the most common contributing causes to an owner's decision to euthanize their pet. Dietary modification, assisted feeding, or feeding strategies often are beneficial in improving food intake and quality of life for these patients. Pharmacologic and dietary management will be discussed further.

CONCLUSION

Sarcopenia and cachexia are becoming important in veterinary practice because of their high prevalence and deleterious effects, and a better understanding of these syndromes are critical to optimize patient care. New drugs, diets, and other treatments to specifically target sarcopenia and cachexia are being developed and are likely to benefit canine and feline, as well as human patients.

REFERENCES

[1] Morley JE, Thomas DR, Wilson MMG. Cachexia: Pathophysiology and clinical relevance. **Am J Clin Nutr** 2006; 83: 735-743. [2] Evans WJ, Morley JE, Argiles J. Cachexia: A new definition. **Clin Nutr** 2008; 27: 793-799. [3] Freeman LM, Rush JE, Kehayias JJ, Ross JN, Meydani SN, Brown DJ, Dolnikowski GG, Marmor BN, White ME, Dinarello CA, Roubenoff R. Nutritional alterations and the effect of fish oil supplementation in dogs with heart failure. **J Vet Int Med,** 12:440-448, 1998. [4] Mallery KF, Freeman LM, Harpster NK, Rush JE. Factors contributing to the euthanasia decision in dogs with congestive heart failure. **J Am Vet Med Assoc**, 214:1201-1204, 1999. [5] Freeman LM, Rush JE, Cahalane AK, Kaplan PM, Markwell PJ. Dietary patterns in dogs with cardiac disease. **J Am Vet Med Assoc** 2003; 223: 1301-1305. [6] Torin DS, Freeman LM, Rush JE. Dietary patterns of cats with cardiac disease. **J Am Vet Med Assoc** 2007; 230: 862-867. [7] Meurs KM, Fox PR, Miller MW, Kapadia S, Mann DL. Plasma concentrations of tumor necrosis factor-alpha in cats with congestive heart failure. **Am J Vet Res** 2002; 63:640-642. [8] Slupe JL, Freeman LM, Rush JE. The relationship between body weight, body condition, and survival in dogs with heart failure. **J Vet Intern Med** 2008; 22: 561-565. [9] Finn E, Freeman LM, Rush JE, Lee Y. The relationship between body weight, body condition, and survival in cats with heart failure. **J Vet Intern Med** 2010; 24: 1369-1374 [10] Michel KE, Sorenmo K, Shofer FS. Evaluation of body condition score and weight loss in dogs presented to a veterinary oncology service. **J Vet Intern Med** 2004; 18: 692-695. [11] Baez JL, Michel KE, Sorenmo K, Shofer FS. A prospective investigation of the prevalence and prognostic significance of weight loss and changes in body condition in feline cancer patients. [12] Biggs ML, Mukamal KJ, Luchsinger JA, et al. Association between adiposity in midlife and older age and risk of diabetes in older adults. **J Am Med Assoc** 2010; 303(24): 2504-2512. [13] Lang T, Streeper T, Cawthon P, et al. Sarcopenia: etiology, clinical consequences, intervention, and assessment. **Osteoporos Int** 2009;21: 543-559. [14] Evans WJ. Skeletal muscle loss: Cachexa, sarcopenia, and inactivity. **Am J Clin Nutr** 2010; 91: 1123-1127. [15] Munday HS, Earle KE, Anderson P. Changes in the body composition of the domestic shorthaired cat during growth and development. **J Nutr** 1994; 124: 2622S-2623S. [16] Scarlett JM, Donoghue S, Saidla J, et al. Overweight cats: prevalence and risk factors. **Int J Obes** 1994; 18: S22-S28. [17] Armstrong PJ, Lund EM. Changes in body composition and energy balance with aging. **Vet Clin Nutr** 1996; 3: 83-87. [18] Harper E. Changing perspectives on aging and energy requirements: Aging, body weight and body composition in humans, dogs and cats. **J Nutr** 1998; 128: 2627S-2631S.

CUSTOMIZED HOME-COOKED DIETS: ADVANTAGES OF WORKING WITH A VETERIANRY NUTRITIONIST

Jennifer Larsen, DVM, PhD, DACVN
Davis, CA

INTRODUCTION

Preparing pet food at home has become increasingly popular, with one recent survey reporting that 10% of dog owners feeding non-commercial diets to provide 50% or more of the pet's intake[1]. Some owners do not follow any particular recipe and simply feed their pets whatever the family is eating. Others prepare the pet's diet according to guidelines or recipes provided by their veterinarian, books, the internet, or other resources. The reasons owners want or need to prepare their pet's food are many and varied. Veterinarians should effectively counsel owners who want or need to prepare home-cooked diets, so that informed decisions can be made regarding an appropriate diet for individual animals. Understanding the advantages, limitations, and potential pitfalls of initiating a nutritional management plan involving a home-cooked diet will increase the chances of success.

Veterinary nutritionists are Diplomates of the American College of Veterinary Nutrition (ACVN). These board certified specialists are uniquely trained in the nutritional management of both healthy animals and those with one or more diseases. Among other skills, veterinary nutritionists are qualified to formulate home-prepared diets, manage the complex medical and nutritional needs of individual animals, and understand the underlying causes and implications of specific nutritional strategies that are used to prevent and treat diseases. The skills and experience of a veterinary nutritionist can be utilized by working as a consultant with the primary veterinarian or by working directly with clients and patients.

INDICATIONS FOR HOME-COOKED DIETS

Pet owners will often express concerns regarding the use of commercial diets for their healthy pets. In addition, owners with pets that have particular health problems may wish to participate in the management of their pet's condition by providing a home-cooked diet, even if a commercially available diet option is available. These owners may have a belief that a home-cooked diet is safer, more natural, or more healthful than a commercially available diet. They may wish to avoid certain ingredients (such as grains, chemical preservatives, or by-products) or to include certain components (such as specific protein or fat sources). Other owners wish to feed their pets according to their personal philosophical views, and choose home-prepared diets that are vegetarian, organic, natural, or raw.

Another common reason owners will feed a home-cooked diet is when the pet refuses commercially available diets. In some pets, this is a learned behavior while in others it results not uncommonly from a food aversion secondary to an underlying condition (especially gastrointestinal or kidney disease). Finally, pets may have a particular combination of diseases for which no suitable commercially available diet exists. In these cases, unless compromise in one or more nutritional strategies is an option, a home-prepared diet is the only appropriate solution. A common example is adverse reaction to food in combination with either fat intolerance (pancreatitis or lymphangiectasia) or kidney disease.

Although some owner concerns may be the result of misinformation and myths, it is crucial that the veterinarian acknowledge and validate these concerns. Most owners wish to provide healthful diets that their pets thrive on, and do not purposefully feed inadequate diets that do not meet basic requirements or that are potentially harmful. However, some owners will continue to follow a particular feeding strategy despite evidence that their individual pet does not tolerate one or more aspects of the plan (for example, dogs with persistent hyperlipidemia that are fed high-fat, raw meat diets) The veterinarian should encourage dialogue regarding nutritional needs for individual pets and for animals in general. This is an opportunity to educate clients regarding commercial pet foods (including testing procedures, ingredient definitions, and marketing strategies) and any other issues specific to their pet and the impact of nutrition on any potential disease processes. A veterinary nutritionist can be a valuable asset in providing information and feedback for the primary veterinarian in terms of supporting ongoing client education, as well as working directly with clients to address questions related to such concerns.

ADVANTAGES OF HOME-COOKED DIETS

The major advantage of a nutritional management plan that includes a home-cooked diet is the ability to customize the strategies to the needs of a specific individual patient. They can be formulated to allow for restricted levels of fat (and/or protein) while still meeting essential fatty acid (and/or amino acid) requirements, allow for inclusion of specific ingredients and ensure appropriate supplementation of required micronutrients. For example, some patients with hyperlipidemia, lymphangiectasia, and/or recurrent pancreatitis require fat restriction more severe than that available in any commercial diet. For other patients, provision of specific concentrations of specific nutrients is important. For example, some dogs with kidney disease become hyperkalemic, and a low potassium home-cooked diet is necessary and very effective[2].

Customized formulations for individual animals are also indicated to address co-morbid conditions. Incorporating specific strategies for the management of two or more diseases requires experience as well as knowledge of the attributes of potential ingredients in order to create an appropriate and balanced diet. Home-cooked diets can be excellent options for many pets, but proper formulation, monitoring and re-assessment are crucial to the success of any dietary management plan that includes the feeding of a home-cooked diet. In addition, many compliance challenges can be minimized early on through better client education on nutrition's role in managing the disease(s).

DISADVANTAGES AND LIMITATIONS OF HOME-COOKED DIETS

In general, home-cooked diets are more expensive compared to commercially available diets. They are also more time-consuming and not as convenient. Another major concern is nutritional adequacy. Due to the nature of customized home-cooked diet recipes, they are not typically tested either in vitro or in vivo for palatability, digestibility, or bioavailability of nutrients. One study that evaluated the nutritional adequacy of home-prepared dog foods by laboratory analysis demonstrated that 35 different diets were below Association of American Feed Control Officials (AAFCO) recommendations for calcium, vitamins A and E, zinc, copper, and potassium[3]. Another study utilized computer analysis to assess home-prepared diets recommended by veterinarians for the diagnosis and/or treatment of adverse reactions to foods in both dogs and cats[4]. Compared to nutrient intake recommendations at the time, approximately 90% of the recipes that were recommended for long-term feeding of adult dogs and cats were not nutritionally adequate.

Further, although many recipes for home-prepared pet foods are available on the internet and in books, the vast majority of these are inadequate for a variety of reasons. Nutritional adequacy is a big concern. One study evaluated published recipes by computer and compared the analyses to nutrient intake recommendations established by AAFCO and NRC (National Research Council); many of these widely available recipes were found to provide less than recommended amounts of multiple essential nutrients, including taurine, minerals, and vitamins[5]. Additionally, such recipes rarely provide information regarding the source of the information used to determine the nutritional profile of various ingredients, the digestibility values used in calculations, or which requirement guidelines were used to determine nutritional adequacy. Other problems typically include:

-vague ingredient specifications or preparation instructions,
-contain errors or omissions in formulation,
-incorporate potentially problematic ingredients,
-and/or feature outdated strategies for addressing specific disease conditions.

Additionally, many such recipes do not provide any instructions with regard to energy content and the amount to feed an individual pet. Other problems include the assumption of equivalent unlimited substitutions of ingredients as well as inadequate instructions for supplementation. In particular, recipes that include a guideline such as 'one adult (or kids) multivitamin/mineral' or 'a multivitamin supplement appropriate for your pet' are vague and not useful. Such recommendations are in essence putting the onus of appropriate supplementation on the owner or on the veterinarian, which is inappropriate. There is a wide range of different multivitamin/multimineral products available for both humans and animals, and very few provide essential nutrients in appropriate amounts and proportions for balancing home-prepared diets. Most supplements marketed for pets are designed to be safe when added to a balanced commercially available diet. They do not contain essential nutrients in high enough amounts to be adequate for supplementing home-prepared diets. Other supplements provide individual nutrients at inappropriately high concentrations, and toxicities are possible.

EVALUATION OF HOME-COOKED DIET RECIPES

Home prepared diets must be evaluated with a critical eye. For healthy adult pets, basic qualitative assessment of the proposed/current home-prepared diet recipe can be done by the veterinarian during the course of the clinic visit. If raw animal products are included (such as meat, eggs, organs), concerns regarding contamination with pathogenic microorganisms should be discussed with the owner, particularly if vulnerable family members or others have contact with the animal or its feces (i.e., infants, elderly, immunocompromised). Additionally, any other potentially dangerous ingredients should be identified (i.e., grapes/raisins, garlic/onions, bones). The diet recipe should provide an appropriate source of protein and fat. One common problem in many home-prepared diets is the lack of an appropriate source of essential omega-6 fatty acids, especially linoleic acid (LA). Some protein and carbohydrate sources will provide LA (such as fatty cuts or skin of chicken, quinoa, or oatmeal), but many do not. Some owners wish to feed fats that are associated with health benefits for people, such as olive oil; however, these usually should be considered only as a source of fat calories as they are often poor sources of essential fatty acids.

Also to be noted are the sources of calcium, taurine, zinc, potassium, and many other essential micronutrients. Finally, preparation instructions should be reviewed for specificity. For example, the nutrient profile of the final diet will be affected by whether salt is included when boiling pasta or rice and whether fat is drained or retained when cooking meats.

Beyond this very basic and preliminary evaluation, the diet recipe must also undergo quantitative assessment of nutritional adequacy. Further evaluation of the diet recipe is necessary to ensure that the base ingredients plus supplements supply adequate amounts of essential nutrients for long-term feeding of the specific patient. The provision of all essential nutrients, plus any necessary safety factors or adjustments based on individual animal factors (including a lower than expected energy requirement) must be considered. This is a quantitative process best done by computer, and interpretation and application of nuances in evaluation and formulation require experience. This is particularly important when a home-prepared diet is being fed or proposed for an animal with one or more diseases, since the margin of error may be smaller and less defined.

For the vast majority of veterinarians, becoming proficient at quantitative diet analysis and the formulation of home-cooked diets is too energy intensive and resource demanding to be practical. Additionally, the learning curve of utilizing nutrient profile databases and formulation software is steep, and mastery of these skills as well as understanding and effectively working with the limitations and idiosyncrasies of the process is too time-consuming to be a rewarding process both financially and professionally. For example, many clients are passionate about nutrition as well as very information-driven and wish to have their feeding and supplement plans reviewed and developed by an expert. When performed by a trained veterinary nutritionist, these consultations can be quite labor intensive as many supplements with multiple ingredients, some novel or uncommon, may be used resulting in a great deal of research and contact directly with manufacturers. Any information collected must then be evaluated for both safety and efficacy in light of clinical experience as well as through other research including scientific literature review. Detailed reports can be provided to both the client and the primary veterinarian and can enhance the level of the pet's care as well as client satisfaction. As such, building a good working relationship with a veterinary nutritionist can be a valuable practice builder. Clients benefit from the knowledge and expertise offered by such a specialist, as does the practitioner, while patients benefit from comprehensive and customized nutritional management plans. Working as a team with a veterinary nutritionist, the primary veterinarian (both general practitioners and specialists in other disciplines) can develop therapeutic plans that are more likely to be successful.

MONITORING OF PATIENTS EATING HOME-COOKED DIETS

Regardless of diet, every patient should have a thorough diet history collected at each visit. This helps ensure that potential problems are identified and addressed early. This is particularly important when a home-cooked diet is being fed, since often the recipe will "drift." Drift can occur for many reasons and in several ways. Supplements are frequently omitted or altered, which may be due to convenience or a lack of understanding regarding the importance of these ingredients. Also, since properly formulated home-prepared diets will be more expensive than most commercially available diets, this can result in recipe changes if supplements are omitted. Additionally, owners may substitute ingredients, which can change the nutritional profile substantially.

Even a seemingly small difference in the recipe can alter the nutritional profile of the diet. For example, substituting chicken thigh meat rather than chicken breast meat will substantially alter the amounts of protein and fat in the final diet. Other changes and substitutions can have an even bigger impact on the overall diet.

Monitoring body weight and body condition score is also important, especially in cases of ill animals or if the patient's energy needs have been estimated. Avoidance of both unwanted weight gain and loss is easily accomplished with rechecks and diet adjustments. Any pet consuming a home-cooked diet, healthy or otherwise, should have regular veterinary visits to monitor not only body weight but also any other dietary related problems. Evaluation of coat/skin appearance, blood work and urinalysis within 4 weeks of initiation of the new diet, and then at regular intervals, is recommended.

Additionally, any other monitoring appropriate for the specific patient is indicated. Examples include blood glucose monitoring in diabetics, urinalyses and imaging in urolith formers, phosphorus, PTH, BUN and creatinine in kidney disease patients, urine protein to creatinine ratio in a patient with protein-losing nephropathy, albumin and clinical signs in encephalopathic patients, and fasting triglycerides in a lipidemic patient. For these and many other patients, monitoring may reveal poor or inadequate response to the dietary therapy (and of course any other concurrent medical therapy). Many practitioners find it beneficial to refine the nutritional management of cases with guidance from a board certified veterinary nutritionist. This approach of adjusting the patient's diet based on measured response should result in far better clinical outcomes. Importantly, this leads to higher client satisfaction and generally improved compliance as clients see a direct cause and effect relationship with dietary management.

CONCLUSION

For both healthy and sick pets, a consultation with a board certified veterinary nutritionist is a wise investment due to the complexity of diet formulation and the problems identified with popular diet plans and static published recipes. Customized nutritional management plans including a home-cooked diet are routinely developed by Diplomates of the American College of Veterinary Nutrition (ACVN®). These can help veterinarians with nutrition-related patient management in a variety of situations, especially with complex cases or owners with a particular interest in nutrition. Guidance given to owners directly or through the primary care veterinarian can enhance and improve management of patients with chronic and acute diseases as well as those that are critically ill.

RESOURCES

Nutrition Support Service, Veterinary Medical Teaching Hospital, School of Veterinary Medicine, University of California, Davis http://www.vetmed.ucdavis.edu/vmth/small_animal/nutrition/default.cfm. American College of Veterinary Nutrition http://www.acvn.org.

REFERENCES

[1]Michel KE, et al. J Am Vet Med Assoc. 2008;233(11):1699. [2]Segev G, et al. J Vet Intern Med. 2010;24(3):546 [3]Streiff EL, et al. J Nutr. 2002;132(6 Suppl 2):1698S. [4]Roudebush P, et al. Vet Dermatol. 2001;3(1):23. [5]Lauten SD, et al. J Vet Intern Med. 2005;19(3):476.

EMERGING VIEWS OF AGING: IN CLINICAL TERMS

Dennis F. Lawler, DVM
O'Fallon, Il

AGING THEORY

Early evolutionary theories, such as mutation accumulation and pleiotropy, tended to focus on the relationship between fecundity and energy distribution, considering aging as a secondary outcome.

Mutation Accumulation

Mutation accumulation proposes that declining force of natural selection due to chronological decline in fertility allows mutations to accumulate in a genome because of minimal selection against them.[1,2] The link between reproduction (fertility) and aging is thus indirect, with outcome expressed as phenotypic diseases of aging, mediated by stochastically altered genes.

Pleiotropy

Antagonistic pleiotropy[3] and disposable soma[4] are variant concepts of genetic trade-off, proposing that (a) early fecundity (and genes governing early maturity) can compromise longevity if the same genes contribute negatively during senescence (antagonism), or (b) resources are allotted preferentially to reproduction over repair when energy is limiting (disposable soma).[5]

Mitochondrial Aging

Among newer concepts, a large literature exists on free radical formation and aging (mitochondrial aging). The theory in simplification is that aging is accelerated because of damage to cells, subcellular organelles, and metabolic pathways, by metabolically-generated oxygen radicals (ROS, reactive oxygen species).[6,7] In parallel, decline of damage protection and repair mechanisms shift the balance against survival.[8,9] However, a number of studies have yielded different perspectives. One interesting example involves the naked mole rat (*Heterocephalus*), which lives well over two decades but displays high levels of evident oxidative damage in tissues and molecular systems.[10] The idea that oxidative processes cause cellular damage and influence aging is not in dispute, but emerging studies support that assigning *the* fundamental underlying role to oxygen-related damage still might be premature.[11]

Programmed Longevity

Programmed longevity recognizes that aging-related metabolic pathways are highly conserved. The insulin-glucose-IGF1 pathway is one example of a stress-response pathway that is conserved in species-aligned forms among yeast, nematode, fruit fly, mouse, and human.[11] Across wide evolutionary distances, the modest stress of diet restriction has extended longevity, associated with delayed onset or prevention of species- and strain-specific late life diseases.[12] Stress response and longevity, therefore, are coupled in what has become known as hormesis. Hormesis states that "modest environmental stresses frequently enhance the average life span in a population".[11] The insulin-glucose-IGF1 pathway is intimately concerned with stress response and longevity, and its importance as an argument for aging-related genetic programming cannot be laid aside.

Altruism and Grouped Selectionism

Programmed, altruistic aging and group selectionism relate aging directly to the population by proposing that the individual organism ages less because of stochastic influences or trade-offs than by programming that has evolved specifically to benefit closely related group members.[11] We are reminded that altruism does not necessarily require the death of the individual(s); the role may be confined to greater risk-taking that could result in survival of the altruist.[13]

CLINICAL IMPLICATIONS

Weight loss and death are associated, even after adjustment for baseline health. Additionally, mass loss and disease can be mutually independent.[14,15] Laboratory studies of rodents provide these insights. In two studies, male Sprague-Dawley rats *maintained relatively constant lean body mass into late life.*[16,17] In a third study, F344 rats also had stable lean mass during advanced life, *until after the onset of terminal diseases in the population.*[16-18] By contrast, long-lived humans may experience overlapping normal and serious mass loss, but longevity itself, individual ability to communicate, and degree of medical supervision, may confuse the sequence of events, requiring diligent attention by clinicians.

Late Life Diseases

Extrinsic diseases are those that originate outside the body[19,20] Examples include infections; toxicities with brief or protracted elapsed time between exposure and symptoms; or climate-, trauma-, or catastrophe-mediated morbidity and mortality. With many extrinsic problems, one may observe dose-dependency, and respective timing of biological and clinical onsets may be sufficiently obscure as to impede complete diagnosis.

With respect to aging and intrinsically-mediated diseases, the problems of biological and clinical onset are more universal and can be especially difficult because screening tests first must be recommended (and agreed to), and then must be sufficiently sensitive for early detection. All clinicians are familiar with diseases that may be present long before even subtle clinical signs are evident, but when the prior history of a patient is not available, early recognition of occult problems can be extremely difficult.

A very important aspect of design of aging studies is segregating extrinsically mediated morbidity and mortality. Both may represent legitimate targets for data analysis, but the hypothesis of a study will dictate how each grouping is used. For practicing clinicians, it is critically important to assess very carefully the experimental designs of both prospective and retrospective reports, and to judge whether the investigators were sufficiently thorough for the work to be given credibility.

Post-reproductive life span (the "aging" phase of life)

Evolutionary theory has attributed post-reproductive life span to corollary effects of natural selection for increasing fitness, but this idea does not explain that fact that the phenomenon extends far beyond "higher" mammals, even to include "lowly" nematodes. Thus, there exists conflict with evolutionary theory that ties competition for limited resources to fecundity under the theoretical umbrella that reproduction favoring optimum fitness is the 'goal' of natural selection.[5]

Why does post-reproductive life span exist? There is no clear answer at this point. A social context cannot be dismissed as one possibility.[5] Predation, trauma, climate, social structure in some species, and local competition for resources, all tend to limit widespread senescence in free-living animals, and therefore might argue in favor of selection for fitness as an explanation for post-reproductive life span. That is, the organisms were more fit to begin with, and post-reproductive longevity simply reflects that fitness. While this idea does not align well with some aging theory, neither should it be dismissed prematurely.

Diseases of late life also could align with the 'overall fitness' argument if they result from failure of programmed fitness based on previous mean longevity. That is, biologically-imposed limitations to life span are suggested by the disposable soma idea, with germ line perpetuation of species integrity being the mechanism for passing on fitness (an otherwise disposable soma being the residual).[21,22] The implication for clinicians could be that some diseases of late life are inevitable but secondary consequences of evolutionary events. In my view, this implication is reasonable but incomplete, because of the genetic and biochemical complexity of "disposability", and of many late life diseases. Natural selection as we understand it typically does not invest substantial energy in death. Heininger characterizes the process as a germ-soma *conflict* (not *tradeoff*) that involves stress at the metabolic (cell) level[21,22], but does not need to imply obligate disposability.

The New Scope of Mass Loss

Investigators recently reported an *in vivo* study of organ mass and aging in humans[23], confirming what has been observed from studies of cadavers. MRI-derived mass for heart, brain, kidneys, liver, and spleen, were obtained from 111 humans aged between 19 and 88 years. Left ventricular mass also was estimated using echocardiography. Total and fat-free body mass were measured with dual-energy X-ray absorptiometry. Brain, kidneys, liver, and spleen, were smaller with advancing age, but left ventricular mass was larger. The organ-specific findings from this *in vivo* study suggest the desirability of new methods for acquiring and interpreting detailed morphometric aging data in animal populations. Unanswered questions from this work include: (a) How do the organ mass changes observed by He and colleagues[23] relate to the aging phenotype of decline toward frailty?; (b) Do *disease-independent* longitudinal mass changes in organs-within-subject proceed at parallel or different rates?

Sarcopenia

Sarcopenia is 'normally' declining mass and strength of skeletal muscle during aging.[24,25] Investigators have identified early skeletal muscle mass decline in some humans, with substantial change by decade five.[26] Therefore, studies also need to be done to establish whether or how subtle mid-life changes in muscle mass and strength can be observed across animal species, and what implications accrue for health, longevity, and research.

How Much Sarcopenia During Healthy Aging?

Early and gradual lean mass decline can be offset by simultaneously increasing fat mass[27,28], suggesting that monitoring affected individuals only for total body weight could result in overlooking initial stages of chronic aging diseases. Two studies indicated that excessive weight loss in the elderly associates with death, whether it is voluntary or involuntary weight loss.[29,30] Thus, given the difficulty of weight loss during advanced life, evidently effective volitional loss in elderly humans (or by prescription and supervision in elderly animals), should be monitored closely for underlying disease.[31]

Sarcopenic Anorexia and Anabolic Failure

In humans and animals, the reasons underlying the age-related anorexia and mass loss are varied. Morley[32,33] has summarized contributing factors, from various studies: (a) Physiological (oral health, gastrointestinal, sensory alterations, or endocrine-, cytokine-, or pain-related); (b) Social (isolation); (c) Psychological (depression, dementia); (d) Economic;

(e) Disease-related. Medical history and diagnostics directed to similar factors should accompany the earliest recognition of appetite decline and lean mass loss in older animals, regardless of chronological age or life functions at the time.

Evaluation of Tissue Metrics At Death

In an ongoing study of Portuguese Water Dogs (The Georgie Project, University of Utah), postmortem examinations following death revealed[34] 21 of 51 whole body or organ metrics (weight, length) correlated significantly and inversely to age at death. Among these 21 metrics, 9 were musculoskeletal measures and 7 were tissues that derive from the embryonic anterior foregut. Given what is known about lean body mass and late-life survival trajectories, an inverse relationship between age at death and muscle mass is not surprising.[16-18] But, while embryonic foregut-derived tissues clearly are soft tissue contributors to lean mass, elucidating developmental relationships to age at death will require further research. The latter observation is intriguing, however, given the growing number of reports across species that link later-life health with aspects of early development. Well-known examples[35-40] include studies of humans that demonstrate: (a) Association between low birth weight (men and women) or low rate of weight gain during infancy (men) and future coronary heart disease; (b) Association between low birth weight and future insulin resistance. While databases that would support interpreting organ mass and dimensions in animals in light of histological outcomes are mostly lacking, efforts in this direction have been made for humans.[41,42]

Considering Morphologies At Death: New Routes of Genetic Evaluation of Aging and Death

In the same study, histological evaluation of 27 tissues revealed changes ranging between 2 and 12 within dog (median 7 - 8).[34] Several observations from this study are interesting: (a) *Significantly heritable* histological traits included liver (fibrosis), pancreas (fibrosis), digestive tract (lymphocytosis), bone (trabecular osteoporosis), and thyroid (atrophy). (b) Individual histological traits that occurred frequently and *correlated most significantly and positively to age at death* included kidney (glomerulosclerosis) and spleen (hemosiderosis). A frequent and significant *negative correlation to age at death* was found for lung (congestion). (c) The data also yielded evidence of relationships among histological and morphometric data. The most potentially enlightening of these was a negative correlation between small intestine length and age at death. The possible involvement of the IGF-1 gene in this association is unmistakable, given that the canine IGF-1 gene explains about 50% of breed size variation[43], and may influence organ size. Because IGF-1 pathways are conserved over very long evolutionary distances and have numerous biological functions (for example, stress response), these observations suggest research avenues into the role that IGF-1 may play in disease frequency and longevity (the latter because of the size-longevity relationship that involves expression of IGF-1 haplotypes.[44]

Oxidation: New Thoughts

Described briefly and clearly by Simon and colleagues[45], and paraphrased here from their writing: The somewhat collective term "reactive oxygen species" (ROS) refers to small, very active molecules such as (a) hydroxyl ($OH^{\cdot}$), alkoxyl ($RO^{\cdot}$), or peroxyl ($ROO^{\cdot}$) radicals (*short half life*); (b) superoxide ($O_2^{\cdot}$) or nitroxyl ($NO^{\cdot}$) radicals (*medium half life*); and (c) some non-radicals, such as hydrogen peroxide (H_2O_2), hydroperoxides (ROOH), and hypochlorous acid (HOCl). These metabolic products can damage DNA, protein, and lipids, and of course lead to cell death. Mitochondria are the principal (but not exclusive) intracellular source of ROS, where they are generated by normal activity in cellular aerobic respiration pathways.[46] ROS in cells are counteracted normally by endogenous antioxidant molecules, such as reduced glutathione, catalase, and superoxide dismutase[45], which act in general to convert ROS to non-damaging metabolites, but often not at 100% conversion. Hence, a certain amount of ongoing damage is implied.

Oxidative damage and its consequences have been accepted for many years by many scientists as a (or *the*) fundamental underlying driving mechanism for aging and longevity. However, very recent emerging evidence ultimately may characterize oxygen radical metabolism somewhat differently. Integrated imaging methods were used to evaluate temporal signals following oxidative stress.[47] Skin carcinoma cells were treated in culture with zinc (as sulfate) as a toxin. Three chemiluminescent agents were added to detect peroxide, mitochondrial membrane potential, or mitochondria-targeted redox potential (the latter two being direct indicators of stress and damage). The data indicated that peroxide sensing occurred after, not before, sensing of mitochondrial damage, suggesting that detectable oxidative damage might be secondary.

Conclusion

Considering the wide variety of stimuli and responses in stress response mechanisms and oxidative metabolism, the widely varying host response to invasion and the changing nature of pathogen strategies, and whole body responses to experimental diet restriction, one may recognize readily the broad scope of hormesis. Population implications derive from recognizing a relationship between the large scope of stress response physiology, its long-distance evolutionary conservation, and emerging views of natural selection that suggest at least a level of independence from disposable soma.

REFERENCES

[1]Medawar P. **An unsolved mystery of biology**. London:HK Lewis, 1952. [2]Edney E, et.al. **Nature** 1986;220:281. [3]Williams G. **Evolution** 1957;11:398. [4]Kirkwood T. **Nature** 1977;270:301. [5]Mitteldorf J. **Evol Ecol Res** 2004;46:1. [6]Harman D. **J Gerontol** 1956;11:298. [7]Balaban R, et.al. **Cell** 2005;120:483. [8]Nemoto S, et.al. **Nature** 2004;429:149. [9]Austad S. **Aging Cell** 2004;3:249. [10]Andziak B, et.al. **Aging Cell** 2006;5:463. [11]Longo VD, et.al. **Nature Rev Genet** 2005;6:866. [12]Masoro E. **J Am Geriatr Soc** 1993;41:994. [13]Shanahan T. **Endeavor** 1998;22(2):57. [14]Wallace J, et.al. **Clin Geriatr Med** 1997;13:717. [15]Pamuk E, et.al. **Ann Int Med** 2003;119:744. [16]Lesser G, et.al. **Am J Physiol** 1973;225:1472. [17]Lesser G, et.al. **Am J Physiol** 1980;238:R82. [18]Yu B, et.al. **J Gerontol** 1982;37:130. [19]Carnes B, et.al. **Pop Dev Rev** 1993;19(4):793. [20]Olshansky S. **J Comp Pathol** 2010;142Suppl1:S4. [21]Heininger K. **Rev Neurosci** 2001;12:17. [22]Heininger K. **Ageing Res Rev** 2002;1:481. [23]He Q, et.al. **J Appl Physiol** 2009;106(6):1780. [24]Rosenberg I. **Am J Clin Nutr** 1989;50(suppl):1231. [25]Roubenoff R. **Can J Appl Physiol** 2001;26:78. [26]Janssen I, et.al. **J Appl Physiol** 2000;89:81. [27]Borkan G, et.al. **J Gerontol** 1983;38:673. [28]Kuczmarski R. **Am J Clin Nutr** 1989;50:1150. [29]Hammond B, et.al. **Arch Environ Health** 1969;19:167. [30]Wallace J, et.al. **J Am Geriatr Soc** 1995;43:329. [31]Wadden T. **Ann Intern Med** 1993;119:688. [32]Morley J. **Am J Clin Nutr** 1997;66:760. [33]Morley J. **J Gerontol Med Sci** 2001;56A:81. [34]Chase K, et.al. **Age (Dordr)** 2010;Sep 16 [Epub ahead of print]. [35]Ross M, et.al. **J Nat Cancer Inst** 1971;47:1095. [36]Buck S, et.al. **Heredity** 1993;71:23. [37]Murphy S, et.al. **BioEssays** 2003;25:577. [38]Gluckman P, et.al. **Science** 2004;305:1733. [39]Bateson P. **J Nutr** 2007;137:1060. [40]Barker D. **J Am Coll Nutr** 2004;23(6):588S. [41]de la Grandmaison G, et.al. **Forensic Sci Int** 2001;119:149. [42]Haddad S, et.al. **J Toxicol Environ Health A 2001;**64:453. [43]Sutter N, et.al. **Science** 2006;316:112. [44]Chase K, et.al. **J Hered** 2009;100(Suppl 1):S37. [45]Simon H, et.al. **Apoptosis** 2000;5:415. [46]Yun S, et.al. In Wolf NS (ed). **The Comparative Biology of Aging**. New York:Springer. 2010:163. [47]Cheng W, et.al. **Environ Health Perspect** 2010;118(7):902.

IS STONE DISEASE SUSCEPTIBILITY A PHENOTYPE FOR SUCCESSFUL AGING?

Jody Lulich, DVM, DACVIM & Carl Osborne, DVM, DACVIM
St. Paul, MN

One way of defining aging is the gradual biological impairment of normal function, as a result of changes made to cells over time. These changes would consequently have a direct impact on the functional ability of the body's organs (such as the heart, kidney and lungs), biological systems (such as the nervous, digestive and renal system) and ultimately the organism as a whole.

Why urinary stones form is not completely understood, but likely represents the body's adaptive mechanisms to sustain organ function in a hostile environment of increasing concentrations of calculogenic materials. Being able to excrete minerals, as opposed to their deposition into tissue preserves organ viability and function at the expense of overwhelming the excretory fluid milieu. Crystals, which are synonymous with urolith embryos, if completely excreted are unlikely to affect urinary function. However, if crystals are retained and grow into stones, urinary function can be disrupted. Following this concern, urolith removal allows restoration of urinary function. A common therapeutic maneuver to prevent urolith formation is readjusting urine pH to improve mineral solubility. For some uroliths, this therapy is very successful, but what are the systemic tradeoffs?

HOW CAN URINE pH AFFECT UROLITH FORMATION?

Urine pH can influence urolith formation by affecting several mechanisms altering the urine concentration of available mineral species. For example acidification promotes calcium release from bone and increases the ionization of calcium which augment urinary calcium excretion. Alternatively, acidic urine reduces formation of magnesium ammonium phosphate primarily by shifting the following equilibrium to the right: $H_3PO_4 \leftrightarrow H_2PO_4^{-1} \leftrightarrow HPO_4^{-2} \leftrightarrow PO_4^{-3}$. By doing so the concentration of PO_4^{-3} the major phosphate moiety in struvite, is reduced. As a result urinary conditions do not favor magnesium ammonium phosphate precipitation.

HOW DOES THERAPY ALTER URINE pH?

Dietary components and salts of other compounds, when consumed and metabolized, either generate hydrogen ions or consume hydrogen ions creating a metabolic acidosis or alkalosis, respectively (table 1). In order to maintain normal plasma pH, the kidney primarily secretes theses acids of bases in the urine. Other mechanisms may also buffer changes in blood pH.

Table 1. Compounds used to manipulate urine pH and their underlying mechanism

	COMPOUND	MECHANISM
Compounds generating hydrogen ions	Methionine	Methionine $\rightarrow$ glucose + urea + H^+
	Phosphoric Acid	$R\text{-}H_2PO_4 + H2O \rightarrow ROH + 0.8HPO_4^{-2}/0.2\ H_2PO_4 + 1.8H^+$
	Ammonium Chloride	$2NH4 + CO2 \rightarrow H_2N\text{-}CO\text{-}NH_2$ (urea) + $2H^+$
Compounds consuming hydrogen ions	Sodium Bicarbonate	$HCO_3^- + H_2O \rightarrow H_2CO_3 + OH^-$
	Potassium Citrate	$C_6H_5O_7^- + 4.5O_2 \rightarrow 5CO_2 + 3H_20 + HCO_3^-$
	Calcium Carbonate	$CO_3^{-2} + H_2O \rightarrow H_2CO_3 + 2OH^-$

WHAT ARE THE SYSTEMIC EFFECTS OF METABOLIC ACIDOSIS?

Diets to reduce struvite precipitation incorporate a variety of strategies to promote formation of acidic urine. This is commonly achieved by adding acidifying precursors that once metabolized reduce the body's pH. In order to restore equilibrium, the kidney excretes acids in urine. The effect of chronic acidification in companion animals, albeit mild, is unknown, but potentially adverse trade-offs should not be over looked (table 2).

Table 2. Systemic effects of metabolic acidosis

ACIDOTIC INDUCED MECHANISM	POTENTIAL CLINCAL OUTCOME
Inhibition of Osteoblastic Egr-1	Increased bone resorption
Inhibition of matrix Gla protein	Decreased bone formation
Inhibition of Osteopontin RNA	Hypercalciuria
Cell medicated bone resorption	Osteopenia
Release of osteoclastic β-gluuronidase	Calcium urolithiasis
Increased Bone PGE_2	
Nitrogen wasting	Decreased anabolism
Activation of the ubiquitin-proteasome pathway	Muscle wasting
Decreased IGF-1	
Decreased GH receptors	
Potassium wasting	
Increased Nitric oxide	Increased inflammation

HOW CAN THE EFFECTS OF CHRONIC METABOLIC ACIDOSIS BE MINIMZED?

It is important to minimize the life-threatening consequences of obstructive urethrolithiasis. However, most patients have a finite time in which they are at greatest risk. Understanding the urolith-forming phenotype can be helpful in limiting therapy to a life stage when it is needed most. In 2009 the Minnesota Urolith Center analyzed stones from 11,697 cats. Seventy percent of the struvite stones were from cats between 2 and 8 years of age (Figure 1).

Figure 1. The proportion of struvite and calcium oxalate uroliths in cats at various ages.

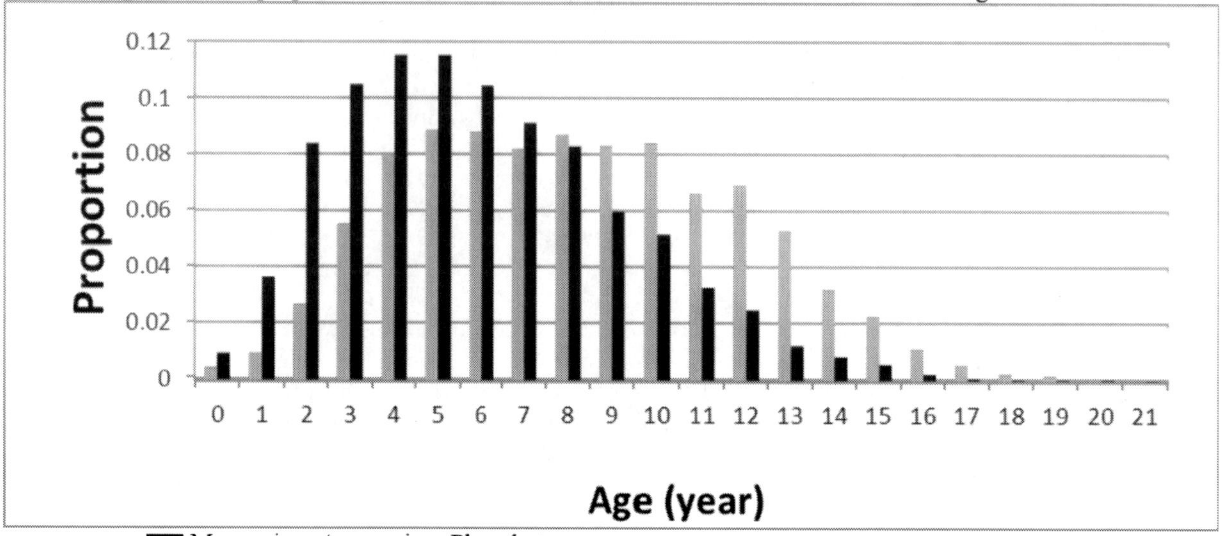

Magnesium Ammonium Phosphate
Calcium Oxalate

REFEENCES

Bushinshy DA. Acid-base imbalance and the skeleton. Eur J Nutr 2001; 40:238-244. Jagoe RT, et al. What do we really know about the ubiquitinproteasome pathway in muscle atrophy? Curr Opinion in Clinical Nutrition and metabolic Care 2001;4:183-190. Green J, et. al. Effect of metabolic acidosis on the growth hormone/IGF-I andocrine axis in skeletal growth centers. Kid International 2000;57:2258-2267. Kalantar-Zadeh K, et. al. metabolic acidosis and malnutrition-inflammation complex syndrome in chronic renal failure. Seminars in Dialysis 2004;17:455-465. Frassetto L. et. al. Diet, evolution, and aging. Eur j Nutr 2001;40:200-213. Kellum JA, et. al. Extracellular acidosis and the immune response: clinical and physiologic implications. Critical Care. 2004;8:331-336.Arnett t. Regulation of bone cell function by acid-base balance. Proceedings of the Nutrition Society 2003;62:511-520. Ching SV, et. al. the effects of chronic dietary acidification using ammonium chloride on acid-base and mineral metabolism in the adult cat. J Nutr 1989;119:902-915.

PREVENTION OF DISEASE AND MORTALITY IN VACCINATED DOGS FOLLOWING EXPERIMENTAL CHALLENGE WITH VIRULENT LEPTOSPIRA. R LaFleur, J Dant, T Wasmoen. Intervet / Schering Plough Animal Health, Elkhorn, NE.

Canine Leptospirosis can vary from subclinical infection to illness that ranges from mild to severe, including death, depending on the susceptibility of the dog, virulence of the organism, and route and degree of infection. The objective of this study was to evaluate the ability of a canine Leptospira bacterin to prevent infection and disease following challenge with virulent *Leptospira canicola*, *L. pomona*, *L. grippotyphosa*, or *L. icterohaemorrhagiae*. Groups of 8-week-old beagles were vaccinated (day 0) and boosted (day 21) with placebo ($n = 10$) or the 4-way bacterin ($n \geq 20$) and subsequently challenged with each serovar. The results demonstrated that blood and various tissue samples from placebo-recipients became reliably infected, and the dogs developed typical clinical signs of Leptospirosis including loss of appetite, ocular congestion, depression, dehydration, jaundice, hematuria, melena, vomiting, petechiae, and death. In addition, placebo-recipients developed kidney and liver dysfunction. In contrast, some vaccine-recipients became infected, but the organisms were cleared quickly from the blood. Vaccinated dogs failed to develop severe clinical disease requiring medical intervention, and no animals died ($p \geq 0.001$). A few of the vaccinated dogs developed clinical abnormalities, but the clinical signs remained mild and were self-limiting ($p < 0.0001$ for each serovar). Administration of the bacterin also prevented thrombocytopenia ($p < 0.0001$), kidney complications caused by *L. canicola* ($p < 0.0001$), *L. icterohaemorrhagiae* ($p < 0.0001$), and *L. pomona* ($p = 0.012$), and liver dysfunction caused by *L. pomona* ($p < 0.0001$) and *L. grippotyphosa* ($p < 0.0001$). The results therefore confirmed that vaccinating dogs with the 4-way Leptospira bacterin provided a high degree of protection (99.5%-100%) against the clinical signs of Leptospirosis including mortality.

CIPROFLOXACIN UROLITHIASIS: A NEWLY RECOGNIZED DISEASE IN DOGS.

J Lulich, C Osborne, A Cokley, S Kemp, L Ulrich, M Buettner, S Davidson, L Koehler, P Ashton, L Swanson, H Albasan, E Nwaokorie, S Leach University of Minnesota, St Paul, MN.

Ciprofloxacin, a synthetic fluoroquinolone antimicrobic, is not FDA-approved for veterinary use. However, due to recent availability of less expensive generic formulations, extra-label use of ciprofloxacin by veterinarians appears more common. Although ciprofloxacin crystalluria and uroliths have been reported in humans, we are unaware of any published reports in dogs. This is surprising since mean urine ciprofloxacin concentration (0.36 mg/ml) in dogs following a modest IV dose (13mg/kg) was 3 times higher than the solubility of ciprofloxacin in water (0.12mg/ml). To identify the occurrence of ciprofloxacin uroliths in dogs, records from the Minnesota Urolith Center were reviewed. Between January 2001 and December 2009, ciprofloxacin was identified in uroliths from 58 dogs; uroliths were composed of 100% ciprofloxacin in 10, mixed uroliths containing ciprofloxacin were identified in 6, a shell of ciprofloxacin was observed in 21, and ciprofloxacin surface crystals were identified in 21.

Based on an experimental study in which 83% of human volunteers consuming 1000mg of ciprofloxacin with $NaHCO_3$ exhibited ciprofloxacin crystalluria (urine pH >7.3), while no volunteers consuming 1000mg of ciprofloxacin and NH_4Cl to acidify urine formed crystals; we postulated that ciprofloxacin uroliths could be dissolved in acidic urine. To test this hypothesis, canine uroliths composed of 100% ciprofloxacin from a single source (6-yr-old male, English bulldog receiving 24 mg/kg of ciprofloxacin PO, q12hr to manage superficial pyoderma; turbulent flow chromatography/tandem mass spectrometry detected 517mg of ciprofloxacin/g of urolith) were incubated in urine at selected pH's and monitored for dissolution. Urine obtained from multiple dogs not receiving fluoroquinolones, was pooled and divided into 5 aliquots. Aliquots were adjusted with HCl or NaOH to a pH of 3, 5, 6, 7, or 8. Aliquots were capped and preserved by refrigeration; pH was monitored and readjusted weekly. Ten uroliths of approximately equal weight were randomly assigned to individual flasks containing 10mls of urine. Flasks were constantly agitated and maintained at 38°C. Every 24 hours, urine was discarded and replaced with 10mls of urine of identical pH until stone dissolution was complete.

Ciprofloxacin urolith dissolution times at each urine pH are reported below.

Urine pH	3	5	6	7	8
Mean Initial Urolith Weight (mg)	80.5	66	76.5	75	78.5
Mean Dissolution Time (day)	7	15	23	18.5	14.5

Ciprofloxacin uroliths are a newly recognized disease and a potential adverse effect of ciprofloxacin administration in dogs. In vitro dissolution of ciprofloxacin uroliths was achieved in canine urine, supporting the premise that in vivo dissolution is possible. Urolith dissolution times were shortest at lower and higher pH's, which is consistent with the pKa (6.0 and 8.8) of this amphiprotic antimicrobic (more soluble at pH below the acidic pKa and above the alkaline pKa).

STRUVITE UROLITH DISSOLUTION IN CATS: A DOUBLE BLIND RANDOMIZED CLINICAL TRIAL OF 2 FOODS.

J. Lulich[1], J. Kruger[2], J. Merrills[2], J. MacLeay[3], I. Paetau-Robinson[3], C. Koenigsknecht[2], S. Davidson[1], C. Osborne[1]. [1]University of Minnesota, St. Paul, MN. [2]Michigan State University, E. Lansing, MI. [3]Hills Pet Nutrition, Topeka, KS.

Foods designed to promote struvite urolith dissolution may be designed for short term feeding facilitating rapid dissolution or may be formulated with a more moderate target urine pH to allow for dissolution and then life-long maintenance feeding minimizing recurrence. The purpose of this study was to compare the efficacy and rate of dissolution of a maintenance food with a struvite dissolution food.

Sixteen client-owned adult cats (13 FS, 3 MC) with naturally occurring struvite urocystoliths (mineral composition based on history, radiographs, urinalysis, urine culture and physical examination) were randomized to either a dry maintenance food (Test) or a dry food known to dissolve struvite uroliths (Control). The clinical care team and owner were blinded to treatment assignment. The Test food was formulated to provide 0.06% Mg (DM), 0.65% P, 35% protein, and a calculated target urine pH value (UpH) of 6.2-6.4. The Control food was formulated to provide 0.06% Mg (DM), 0.77% P, 35% protein, and a targeted urine pH of 5.9-6.1. Owners were advised to feed the assigned diet exclusively in an amount to maintain body condition. After diet assignment radiographs were performed at eight weekly intervals until there was no evidence of uroliths or until there was evidence that the uroliths were the same size or larger. A physical examination, complete blood count, serum chemistry profile, urinalysis and urine culture were repeated at the conclusion of the study. Statistical analysis was by ANOVA.

All uroliths dissolved in all cats and both foods were palatable. Radiographs of cats fed the Control food indicated the uroliths dissolved in a significantly shorter time (mean ± Std Dev of 1.6 ± 0.7 weeks) compared to cats consuming the Test food (mean 3.9 ± 2 weeks; $P<0.05$).). Cats in the Control group finished the study at 1 (n=4), 2 (n=3) and 3 weeks. Cats in the Test group finished the study 1, 2, 3 (n=2), 4, 5, 6, and 7 weeks.

All cats had similar UpH values at study start (Control 6.516 ± 1.013, Test 6.638 ± 0.432). Cats consuming the Control food had significantly lower UpH values at the time of dissolution compared to cats consuming the Test food: mean UpH values for the Control group was 5.949 ± 0.243 compared to 6.337 ± 0.163 ($p<0.05$) for the Test group. Mean urine specific gravity values at baseline and at dissolution were not different between or within groups. Adverse events were not observed in either group of cats. This study provides evidence that clinicians may choose either food to facilitate struvite dissolution based on clinical assessment of individual cases and owner preferences.

EFFECTS OF FORMALIN ON UROLITH COMPOSITION. H Albasan[1,2], CA Osborne[1], JP Lulich[1], LK Ulrich[1], LA Koehler[1]. 1. University of Minnesota, College of Veterinary Medicine, Minnesota Urolith Center, St. Paul, MN. 2. Erciyes University, Faculty of Veterinary Medicine, Department of Internal Medicine, Kayseri, Turkey.

The Minnesota Urolith Center occasionally receives uroliths for analysis that are immersed in formalin. Results of quantitative analysis of these uroliths revealed that some submitted in formalin consisted of newberyite (magnesium hydrogen phosphate trihydrate). Because newberyite is uncommonly found in uroliths formed by cats and dogs, we hypothesized that this mineral was an in vitro artifact caused by exposure of struvite (magnesium ammonium phosphate hexahydrate) to formalin. The purpose of this study was to determine if formalin alters the mineral composition of uroliths.

Urolith submissions containing stones of either 100% struvite (n=5 dogs and 5 cats), 100% calcium oxalate (n=5 dogs and 5 cats), 100% calcium phosphate apatite (n=5 dogs and 5 cats), 100% cystine (n=5 dogs and 5 cats), 100% ammonium urate (n=5 dogs and 5 cats), and 100% silica (n=5 dogs) preserved by only air drying were tested. One urolith from each submission was quantitatively analyzed by polarized light microscopy or infrared spectroscopy. A subsequent urolith from the same submission was immersed in 1 ml of 10% buffered formalin for 48 hours at room temperature. Uroliths were then air dried for 30 minutes and the analysis was repeated.

After exposure to formalin, portions of all 10 struvite uroliths were transformed into newberyite. Three (1 dog and 2 cats) of 10 ammonium urate uroliths were completely dissolved. Newberyite was not detected in any of the remaining uroliths. Likewise quantitative mineral analysis of non-struvite uroliths remained unchanged.

To avoid misdiagnosis of mineral composition, uroliths should not be immersed in formalin prior to analysis.

SEROPREVALENCE OF THREE *EHRLICHIA* SPECIES IN DOGS – A MULTI-INSTITUTIONAL STUDY.

MJ Beall[1], EB Breitschwerdt[2], LA Cohn[3], CG Couto[4], MW Dryden[5], S Eddlestone[6], L Guptill[7], WE Hoffmann[8], MA Kennedy[9], P Lathan[10], SE Little[11], A Roy[6], KA Sayler[12], J Sinsabaugh[1], K Snowden[13], BA Stillman[1], EG Welles[14], MJ Yabsley[15] 1. IDEXX Laboratories, Westbrook, ME. 2. North Carolina State Univ., Raleigh, NC. 3. Univ. of Missouri, Columbia, MO. 4. The Ohio State Univ., Columbus, OH. 5. Kansas State Univ., Manhattan, KS. 6. Louisiana State Univ., Baton Rouge, LA. 7. Purdue Univ., West Lafayette, IN. 8. Univ. of Illinois, Champaign, IL. 9. Univ. of Tennessee, Knoxville, TN. 10. Mississippi State Univ., Mississippi State, MS. 11. Oklahoma State Univ., Stillwater, OK. 12.Univ. of Florida, Gainesville, FL. 13. Texas A&M Univ., College Station, TX. 14. Auburn Univ., Auburn, AL. 15. Univ. of Georgia, Athens, GA.

The geographic range of the lone star tick (*Amblyomma americanum*) is expanding together with the disease agents it vectors, including *Ehrlichia chaffeensis* and *Ehrlichia ewingii* which cause both canine and human ehrlichiosis. This study examined the frequency with which dogs are exposed to *E. chaffeensis* and *E. ewingii* relative to *E. canis*, which is transmitted by the more ubiquitously distributed brown dog tick (*Rhipicephalus sanguineus*). A total of 6,512 canine serum samples, ranging from 182 to 614 from each of the 14 participating institutions, collected at random from clinical accessions, diagnostic laboratories and/or shelters were evaluated. All serum samples were tested by three microtiter plate ELISAs using species-specific peptides for antibodies to *E. canis*, *E. chaffeensis* and *E. ewingii*. Zip code information for sample origin was provided by the collaborator and was used to assess seroprevalence by region. Comparisons were evaluated using the Chi-square test. Seroreactivity for at least 2 of 3 *Ehrlichia* spp was found in samples from every institution. On an average, seroprevalence of *E. canis* (0.9%) was significantly ($p<0.001$) lower than *E. chaffeensis* (2.8%) or *E. ewingii* (5.2%). Samples from Texas had the highest *E. canis* seroprevalence (2.9%), while samples from Missouri had the highest seroprevalence of *E. chaffeensis* (12.2%) and *E. ewingii* (21.3%; $p<0.001$). Both Mississippi and Oklahoma had greater than a 7% *E. ewingii* seroprevalence, while North Carolina had an *E. chaffeensis* seroprevalence of 7.7%. Samples from Ohio had the lowest aggregate seroprevalence (1.0%) with only 4 dogs *E. canis* seropositive, one *E. ewingii* seropositive and no *E. chaffeensis* seroreactors. The geospatial pattern of *E. chaffeensis* and *E. ewingii* seropositive samples was similar to that previously reported based on modeling seroreactivity to *E. chaffeensis* in white-tailed deer as well as the distribution of human monocytic ehrlichiosis (HME) cases reported by the CDC. This study provides the first large scale regional documentation of canine exposure to these three *Ehrlichia* spp., highlighting where infections most commonly occur and thus identifying areas where heightened awareness about these emerging vector-borne disease agents by both veterinarians and public health agencies is needed.

POLYDIMETHYLSILOXANE URETHRAL BULKING AGENT (PDMS UBA) INJECTION FOR TREATMENT OF FEMALE CANINE URINARY INCONTINENCE – PRELIMINARY RESULTS. JW Bartges, A Callens. The University of Tennessee, Knoxville, TN.

Urinary incontinence (UI) occurs in approximately 20% of spayed female dogs. The most common cause is urethral sphincter mechanism incompetency (USMI). Pharmacological agents are effective, however, not all dogs respond, and dogs may become refractory to treatment over time. Urethral bulking, where a compound is injected submucosally in the urethra, has been used in women and in female dogs with urinary incontinence. New synthetic compounds have been used in human medicine; the most promising is Polydimethylsiloxane (PDMS), which has been shown to be more effective than glutaraldehyde cross-linked collagen. The purpose of this descriptive clinical trial is to evaluate the safety and effectiveness of PDMS urethral bulking agent (PDMS UBA) in client-owned, spayed female dogs with naturally-occurring UI due to USMI.

Twenty-two, spayed female dogs were included. Dogs had a median age of 6 years (2 to 11 years). Eighteen dog breeds were represented, and dogs weighed a median of 29.9 kg (7.3 to 62.7 kg). Average length of time of UI was 2.5 ± 2.3 years; 18/22 dogs had been treated medically, of which 2/18 were continent, 14/18 were improved, and 2/18 had no improvement. Dogs were deemed healthy based on results of physical examination, complete blood cell counts, plasma biochemical analysis, and urinalysis; urine cultures were negative.

Dogs were anesthetized, positioned in dorsal recumbency, and cystoscopy performed using a 2.7mm, 0- or 30- degree, 18cm rigid cystoscope. Urethral bulking was performed with PDMS UBA. On average, 2.5 ± 0.9 ml were injected in 3 to 5 locations approximately 1 to 1.5 cm distal to the trigone submucosally in the proximal urethra. Good coaptation was achieved in all dogs. The procedure took on average 15.9 ± 4.3 minutes. One dog experienced urethral obstruction after the procedure; a Foley catheter was inserted for approximately 12 hours and removed at which time she urinated normally and was continent. Three dogs experienced an acute allergic reaction characterized by blepharedema and urticaria treated successfully with diphenhydramine. Dogs were discharged on day of procedure except for the one dog that experienced urethral obstruction. All dogs were treated with Meloxicam (0.1 mg/kg PO q24h for 3 days).

Owners were contacted on day after discharge and 21/22 dogs were continent; 1/22 dogs was improved. Dogs were re-evaluated 1 week after discharge and 21/22 dogs were continent and 1/22 dogs were improved; urine cultures were negative for all dogs at that time. At one month after discharge, 17/22 dogs were continent, 2/22 dogs had experienced 1 to 2 episodes of incontinence but were continent at this recheck, 3/22 dogs were improved. Continent dogs were not receiving medications for UI. Preliminary results support use of PDMS UBA for female dogs with UI that have either failed medical therapy or as a first line therapy.

TRANSFUSION OF AUTOLOGOUS STORED ERYTHROCYTE CONCENTRATES CAUSES INFLAMMATION AND HYPOCOAGULABILITY. SA Smith[1], M McMichael[2], JM Herring[2], T Ngwenyama[2], R. Corsi,[2] A Galligan[2], AN Beloshapka[3], P Deng[3], K Swanson[2,3]. 1. College of Medicine, 2. College of Veterinary Medicine, or 3. Department of Animal Sciences, University of Illinois, Urbana, IL.

We previously reported that transfusion to normal dogs of autologous erythrocyte concentrates (pRBCs) that had been stored for 21 days causes a profound inflammatory response (2X increase in leucocyte count and fibrinogen, 60X increase in C-Reactive Protein). We speculated that inflammation was due to cytokines produced during the storage period, and hypothesized that transfusion of fresh (F) pRBCs would elicit less inflammation than would stored (S) pRBCs.

A whole blood unit was collected from healthy dogs (n=10) for pRBCs on day 0, then again on day 32. On day 35 dogs received an autologous transfusion of pRBCs stored for either 35 days (S, n=5) or 3 days (F, n=5). CBCs and in-tem thromboelastometry (CT:coagulation time, CFT:clot formation time, α:alpha, MCF:maximum clot firmness) were evaluated on blood samples collected at 0 (pre) and 5,9,24,48, and 72 hours after transfusion.

Fresh pRBCs did not elicit any change in leucocytes, platelets, or thromboelastometry. Stored pRBCs elicited a degenerative left shift (5 hr) followed by a regenerative left shift (9-48 hr), thrombocytopenia (36% decrease at 5 hr), and marked hypocoagulability characterized by prolonged CT (5,9,24 hr) and CFT (5,9 hr), and decreased α (5,9 hr) and MCF (5,9, 24 hr).

		0	5	9	24	48	72
Segs	F	4.8(1.7)	5.0(1.4)	3.3(2.1)[a]	3.8(0.7)[a]	3.9(1.3)[a]	4.0(1.5)
($10^3/\mu l$)	S	6.1(3.8)	5.4(7.1)	16.0(10.6)[b]	24.8(12.6)[b]	19.2(5.5)[b]	10.1(2.6)
CT	F	164(32)	158(12)[a]	161(28)[a]	161(12)[a]	152(22)	161(38)
(s)	S	162(38)	225(48)[b]	212(15)[b]	221(24)[b]	173(15)	147(30)
CFT	F	99(17)	102(25)[a]	86(21)[a]	103(20)	101(22)	105(27)[a]
(s)	S	139(49)	279(47)[b]	256(63)[b]	121(21)	97(15)	75(10)[b]
α	F	71(3)	72(3)[a]	73(5)[a]	72(2)	72(2)	72(4)[a]
(°)	S	67(6)	54(5)[b]	58(6)[b]	67(4)	71(3)	75(2)
MCF	F	59(6)	58(6)[a]	60(6)[a]	62(6)	57(2)[a]	57(6)[a]
(mm)	S	53(7)	49(2)[b]	48(5)[b]	59(2)[b]	65(3)[b]	66(2)[b]

Data are mean(SD). a: $p<0.05$ between groups F and S by t test. b: $p<0.05$ compared to "0" by RM ANOVA.

Transfusion of autologous stored pRBCs elicits a greater inflammatory response than fresh pRBCs, and results in hypocoagulability on thromboelastometry.

ENHANCED BIOCONVERSION OF CLOPIDOGREL TO ITS ACTIVE METABOLITE WITH RIFAMPIN INDUCTION OF HEPATIC ISOZYMES. DF Hogan. Purdue University, School of Veterinary Medicine,West Lafayette, IN.

Clopidogrel is a potent antiplatelet drug that is gaining popularity in veterinary medicine for antithrombotic therapy. The parent molecule is an inactive prodrug that must be converted by hepatic isozymes to an active metabolite. The majority of the parent molecule is directed to the formation of inactive metabolites with only an extremely small proportion of parent molecule directed to the formation of the active metabolite. There are multiple hepatic isozymes responsible for the formation of the active metabolite. A non-specific hepatic isozyme inducer such as rifampin could increase the formation of the active metabolite of clopidogrel thereby increasing the pharmacodynamic response which may allow a reduced drug dose to achieve a clinical effect. We have previously presented data supporting the increased pharmacodynamic response of clopdiogrel after rifampin therapy. The goal of this study was to demonstrate an increased pharmacokinetic response of clopidogrel after rifampin induction of hepatic isozymes.

Six healthy, purpose-bred dogs were used for this study. The pharmacokinetics of clopidogrel were determined by measuring the parent molecule, primary inactive metabolite and active metabolite through LC/MS/MS. The pharmacodynamics of clopidogrel were determined by measuring collagen-induced whole blood aggregation. Blood samples were collected prior to clopidogrel administration (baseline), after 7 days of 2 mg/kg clopidogrel PO q 24 hrs, and after 7 days of 2 mg/kg clopidogrel PO q 24 hrs + 10 mg/kg PO q 12 hrs rifampin.

Given the absence of a known standard for the active metabolite, only a semi-quantitative assessment of active metabolite concentration can be made. There was no identifiable active metabolite peak noted at baseline or after clopidogrel treatment. However, with clopidogrel and rifampin combined administration there was an active metabolite peak identified in all dogs with a mean area of 41.7 ± 23.5. The development of the active metabolite peak was associated with an increase in the pharmacodyamic response of clopidogrel in the dogs.

This is the first study in any species to document the increased formation of the active metabolite of clopidogrel in response to a strong, non-specific hepatic isozyme inducer. This increased pharmacokinetic response was associated with an increased pharmacodynamic response of clopidogrel. This data provides supportive evidence to develop therapeutic protocols to improve the pharmacodynamic response to clopidogrel in dogs that may reduce dosing requirements or correct subtherapeutic pharmacodynamic response.

HYPERCOAGULABILITY: WHAT DO WE KNOW, WHERE ARE WE GOING?

Marilyn Dunn DMV, MVSc, DACVIM
St-Hyacinthe, Quebec, Canada

INTRODUCTION

Hypercoagulability is defined as a tendency for blood to coagulate more rapidly than normal. Hypercoagulability and thrombosis are not synonymous, however the presence of hypercoagulability can lead to thrombosis if other risk factors come into play such as vascular stasis, endothelial injury and decreased fibrinolysis.

Hereditary defects of coagulation inhibitors, referred to as "thrombophilias" further increase the burden of thrombus formation in people. Although hereditary thrombophilias have not been identified in dogs or cats, thrombosis is recognized as a common complication of many acquired diseases, including cardiac, endocrine, inflammatory, and neoplastic disorders.

Diagnosis of hypercoagulability is a challenge. Routine coagulation tests (PT, PTT) are designed to detect hypocoagulability. Low plasma antithrombin (AT) activity and high fibrin degradation product, d-dimer, and fibrinogen levels provide laboratory evidence of hypercoagulability. Thromboelastography (TEG) is a technique that depicts global hemostasis, beginning with clot formation and ending with clot lysis. Evaluation of platelet function, markers of platelet activation and thrombin generation may help further our ability in identifying hypercoagulable patients and especially identifying those most at risk for thrombosis.

Actual detection of thromboemboli can be very difficult. A combination of Doppler ultrasonagraphy, angiography, venography, contrast CT, MRI and ventilation perfusion scans can aid in their identification however, many cases of thromboembolism likely go undiagnosed in small animals.[1,2]

DETECTION OF HYPERCOAGULABILITY

What have we learned over the past 5 years?

D-dimer, antithrombin (AT), thrombin-antithrombin complex (TAT), platelet activation

Measured by latex agglutination (can also be measured by immunoturbidimetry), d-dimer is reported to have a sensitivity of 100% of and a specificity 70% when > 500ng/ml for the detection of thromboembolism in dogs.[3] Despite these results, myself and colleagues have encountered clinical cases with confirmed thromboembolic disease with negative d-dimer assays.

In a recent study of canine IMHA, 10/10 dogs presented increased d-dimers, 2 of these dogs developed confirmed thrombotic events. Antithrombin was normal in all these patients and increased platelet activation was noted, the greatest increases in platelet activation were observed in the most thrombocytopenic dogs.[4] In people, decreased AT levels have been associated with an increased risk of thrombosis and is considered a negative prognostic indicator. In people, an AT level < 30% is considered at critical risk for thrombosis and a level < 60% is considered high risk.

A retrospective study of AT as a diagnostic and prognostic indicator in ill dogs revealed that it was decreased in 82% of dogs with the presence of confirmed thrombosis, therefore normal AT does not preclude the presence of thrombosis nor DIC. 76% of dogs with IMHA had decreased AT and a lower AT was a negative prognostic indicator. In this study, no cut-off could be established for AT values below which there was an increased risk of thrombosis.[5]

Despite the established risk of decreased AT in protein-losing nephropathy and its correlation with hypoalbuminemia, this has not been thoroughly investigated in dogs.

Increased TAT was detected in dogs with hemangiosarcoma and hematopoietic malignancies.[6]

Does an increase in circulating coagulation factors translate directly to a hypercoagulable state (eg. Cushings)?

What do we know about cats?

A study that evaluated AT in healthy, ill and cardiomyopathic cats found that AT tended to increase in cats that were ill (only 2/7 diagnosed with DIC had decreased AT!). Cardiomyopathic cats had normal AT. The explanation for these results was that in cats, AT may be an acute phase protein than increases in response to systemic inflammatory disease.[7] Another study looking at various markers of activated coagulation in healthy and cats with asymptomatic hypertrophic cardiomyopathy found that AT was decreased in cats with cardiomyopathy and some cats had increases in d-dimer and TAT complexes. However, no cats in this study had thrombosis and no correlation could be found between echocardiographic variables (including the left atrium) and these markers.[8]

Cats with cardiomyopathy have not been shown to have increased platelet activation (evaluated by PFA)[9] and by aggregometry. Another study determined that only 50% of cats with spontaneous echocardiographic contrast and cats with arterial thromboembolism had evidence of hypercoagulability. Interestingly, in this study, there was no correlation between the presence of congestive heart failure nor left atrial enlargement and hypercoagulability and a large portion of cats with suspected/confirmed thrombi showed no evidence of hypercoagulability (various clotting factor measurements, AT, d-dimer, TAT).[10]

In cats with DIC, d-dimer was found to have a sensitivity of 56% and a specificity of 67%.[11]
It has been suggested that d-dimer is a poor indicator of hypercoagulablity in cats as they may clear d-dimers more rapidly than other species.

Cats present variable AT values which may not be of value in determining presence of a hypercoagulable state and thrombotic risk. No echocardiographic variables nor markers of activated coagualtion have been able to be correlated with thrombotic risk in cats with cardiomyopathy.

Thromboelastography (TEG)

The cell-based model of coagulation has allowed a better understanding of the in vivo process of coagulation. It places emphasis on the role of thrombin in hemostasis and it is the balance of production and degradation of thrombin that determines if a patient is hypo- or hypercoagulable. It is with this understanding, that the importance of tests that evaluate global whole-blood hemostasis have become apparent. Contrary to other techniques, TEG provides an evaluation of all three phases of coagulation (initiation, amplification and propagation) as well as fibrinolysis, dependent on both coagulation factors as well as platelets.

Recent studies have used TEG to detect hypercoagulable states in canine patients with parvoviral enteritis, disseminated intravascular coagulation, neoplasia, IMHA, protein-losing nephropathy and patients hospitalised in the ICU.[12-16] 11/27 dogs presented to an ICU had hypercoagulable TEG tracings, these dogs also had increased d-dimer and fibrinogen concentrations.[15] It has been shown that a hypocoagulable TEG correlates with clinical bleeding in dogs.[17] Unfortunately, there are no studies, that I am aware of, that indicate that a hypercoagulable TEG tracing is a reliable predictor of thrombotic risk in dogs or cats. This is an important issue as in clinical patients, a TEG is often done in order to aid in determining thrombotic risk and possibly starting antiplatelet and/or anticoagulant therapy.

Is being hypercoagulable a good thing as long as you do not develop thrombosis? A study of DIC in dogs revealed that having a hypercoagulable TEG was associated with a mortality rate of 32% in comparison to 64% in dogs with a hypocoagulable TEG.[13]

Thrombin Generation (GT)

The amount of circulating thrombin is the final result of the interaction of all the procoagulant and anticoagulant factors and therefore an increase in thrombin should be a good indicator of a prothrombotic state. Although not widely used in veterinary patients, measurement of endogenous thrombin potential has been validated by our laboratory in samples obtained from normal and sick dogs as well as in heparinised samples.

We have used TG along with TEG and other markers of activated coagulation to investigate hypercoagulability in dogs receiving prednisone, dogs with spontaneous hyperadrenocorticism and dogs receiving low-molecular-weight heparin. Results of these studies will be presented.

REFERENCES

1.Textbook of Veterinary Internal Medicine 7th ed. St. Louis Missouri: Saunders; 2010. 2. Antiplatelet and Anticoagulant Therapy. In Kirk's Current Veterinary Therapy XIV Saunders, Philadelphia 2009. 3. Nelson OL et al. J Vet Int Med 2003 (17):830-834. 4. Ridyard AE et al. J Small Anim Pract 2010:51(6):296-304. 5. Kuzi S et al. J Vet Int Med 2010(24):587-596. 6. Maruyama HW et al. Vet Rec 2005(156):839-840. 7. Brazzell JL et al. Vet Clin Path 2007(36)(1):79-83. 8. Bédard C. et al. Vet Clin Path 2007(36)(2):167-172. 9. Jandrey et al. Vet Clin Path 2008(37):385-388. 10. Stokol et al. J Vet Int Med 2008(22):546-552. 11. Tholen I et al. J Fel Med Surg 2009(10):842-846. 12. Otto CM et al. J Am Vet Med Assoc 2000(27):1500-1504. 13. Wiinberg B et al. J Vet Int Med 2008(22):357-365. 14. Kristensen AT et al. J Vet Int Med 2008(22):140-147. 15. Wagg CR et al. Vet Clin Path 2009(38):453-461. 16. Donahue SM et al. J Vet Emerg Crit Care 2005(15):9-16. 17. Wiinberg B et al. Vet J 2009(179):121-129.

CLINICAL UTILITY & PITFALLS OF THROMBELASTOGRAPHY

Benjamin Brainard, VMD, DACVA, DACVECC
Athens, GA

CONCEPTS OF THROMBELASTOGRAPHY

The thrombelastograph (TEG) provides a visual output of the changes in blood viscosity that occur as liquid blood transitions to a gelid clot. As blood in a warmed, rotating, plastic cup begins to clot, a connection is made between the side of the cup and a pin that is suspended in the center of the cup. As fibrin strands connect the pin to the rotating cup, the torque from the rotating cup is transferred to the pin, and this rotation is graphed as a thrombelastograph.

In addition to the visual tracing of the changes in viscosity, the TEG software generates a number of quantitative values representing the properties of the clot[1,2] The R time (reaction time) represents the time from the initiation of the test until the first strands of fibrin in the clot are detected. The R value is technically registered when the tracing extends 2 mm from the center line on either side. The K value indicates the time from the identification of the R time for the tracing to extend 10 mm on either side of the center line, and is similar to the Angle, in that it is dependent on rate. The angle (or alpha angle) represents the slope of the tracing, and indicates the rate of clot formation. The maximum amplitude (MA) is the final clot strength, represented in mm. The elastic shear modulus, or G value, is a mathematical manipulation of the MA value that is sometimes reported in the literature. The fibrinolytic parameters measured by the TEG include the LY 30 and LY 60, which represent the percent of decrease in MA at 30 and 60 minutes after the machine registers MA. By the same token, the values for CL30 and CL60 report the actual width of the tracing 30 and 60 minutes after the MA, respectively. The clot index (CI) is a proprietary mathematical combination of R, K, Ang, and MA, which ostensibly summarizes the overall tendency towards clot formation in a given sample.[1] The value for CI may vary between species. The first derivative of the TEG tracing can represent thrombin generation, but this has not been evaluated in veterinary species to date.

The TEG is performed on whole blood samples, either citrated (and recalcified) or native. Each reported value on the TEG can be ascribed to interactions between different components of the blood. The R time is primarily representative of the soluble coagulation factors, and depends somewhat on the choice of activator as to which coagulation factors are interrogated.[1,2] The use of tissue factor (TF) for initiation of coagulation utilizes the extrinsic pathway, while the use of kaolin or calcium for initiation of coagulation represents the intrinsic pathway.[2] Coagulation factors, platelets, and fibrinogen contribute to the angle and K values, and the MA is primarily an indication of the fibrinogen concentration in the sample.[1,2]

There are not standard protocols for veterinary TEG; in general, samples are run as recalcified citrated samples, at 37 degrees C, and following a 30 minute rest period. While some laboratories use only recalcification with calcium chloride prior to starting the assay, this method depends on contact activation[3] and may result in a prolonged assay time, especially in horses.[4] This method may also be associated with some degree of operator variability.[4] The use of very dilute (1:50,000), or moderately dilute (1:3400) tissue factor (Innovin, Dade/Siemens) has been described in many species, and appears to decrease the interoperator variability.[4,5] This technique also results in a shorter R time compared to samples that are just recalcified, and is preferred for analysis of TEG in horses. Haemoscope provides pre-filled kaolin vials, which activate coagulation via the intrinsic pathway. The use of this activator also seems to decrease variability in the assay, and has been described in many species including the dog and cat.[6] Because of variability between users and institutions, it is recommended that each institution have a specific protocol, institutional reference interval, and a limited number of operators to maximize useful information from the assay.

Information displayed by the TEG

The TEG and other viscoelastic coagulation tests are unique in that they can display the tendency of animals towards hypercoagulability. A tracing that represents a hypercoagulable patient may have a shortened R time, a steeper angle, and a higher MA. Outcome studies have not been performed in veterinary medicine to specifically document the risk of (for example) thromboembolic events in patients with specific TEG abnormalities. In human medicine, a single TEG study of patients after placement of coronary stents found an association between an increased MA and an increased risk of stent thrombosis. To the author's knowledge, additional outcome studies that evaluate TEG variables have not been performed in humans or animals.

Hypercoagulability has been described in a number of patient populations, including dogs with immune mediated hemolytic anemia (IMHA)[7], parvovirus[8], and ICU patients.[9] Hypocoagulability has been described in dogs with various neoplasias[10] and DIC [11] as well as horses with gastrointestinal disease.[12] While the coagulopathy associated with DIC results in a hypocoagulable tracing, the increased fibrinolysis expected with this syndrome, which results in a TEG tracing that returns to baseline after generating MA in human patients, has not been described in companion animals. This may be due to a less robust fibrinolytic response. In some ostensibly normal cats, a tracing that rapidly collapses after generation of MA has been described, but it is unclear why this is seen in some cats and not in others. Platelet retraction may also be a potential cause of this observation. Variability in the TEG tracings of cats is significantly reduced with the use of an activator.[6]

Although a recent study evaluating TEG (induced with TF) in dogs treated with corticosteroids described a progressive hypercoagulability,[13] another study evaluating dogs with naturally occurring hyperadrenocorticism (using unactivated TEG) failed to show hypercoagulability when the dogs were compared with a matched control group.[14]

Important aspects of TEG interpretation

The TEG tracing is significantly affected by hematocrit. As Hct drops, the tracing appears progressively hypercoagulable, as a result of both a steeper angle and a higher MA.[4] It is unclear if this is an in vitro phenomenon due to an altered ratio of plasma to cells, or if it represents actual in vivo hypercoagulability. Platelet count is also important. Patients with thrombocytopenia less than 50 x 10^3 platelets/μL will generate an overall hypocoagulable tracing, which may be minimally useful to the clinician. The TEG tracing (specifically MA) is also directly correlated to the patient's fibrinogen concentration.[4]

Ongoing research in the veterinary TEG community is directed towards evaluating the effect of variable rest periods, test temperature, and venipuncture techniques. With longer rest periods (the default is 30 minutes), the tracing appears more hypercoagulable. Resting the sample for 30 minutes at room temperature or 37 °C does not have an effect on the tracing (the TEG warms the blood to 37 °C over 2.5 minutes, and the sample is over 30 °C within 30 seconds).[15] The quality of venipuncture and the post sampling treatment of samples is also important, and blood for TEG should be obtained carefully as with any samples for coagulation testing.[16]

There is a degree of operator variability in TEG analysis. While the variability seems to be minimized when activators such as tissue factor are used, research studies and clinical samples should be designed so that a single operator--or at least a consistent group of operators—performs the TEG testing.

Overall, consistency of technique is the most important aspect of generating TEG tracings that are interpretable and comparable. It is important for each institution to generate a reference range from the operators who will be performing the studies, and for each institution to adhere as much as possible to a specific protocol for sample acquisition and processing.

Modifications to the TEG

Samples contaminated with heparin (unfractionated or low molecular weight) may display significant abnormalities attributable to heparin.[17] Unfractionated heparin concentrations as low as 0.1 U/mL may result in excessively prolonged R times, and may invalidate interpretation of the data. For this reason, Haemoscope also manufactures TEG cups that have been coated with a proprietary dose of heparinase. These cups are run in the same fashion as those without additives, but will inactivate any heparin present in the sample, theoretically resulting in a tracing that shows the underlying coagulation status of the patient. This is an attractive technique not only for quality control of samples obtained via indwelling catheter, but also because it theoretically may allow monitoring of the underlying coagulation state of animals that are receiving therapeutic doses of heparins. The only downside to this is that there is a specific dose of heparinase in each cup, and depending on the amount of heparin in the sample, complete inactivation may not occur. The heparinase cups have also been used as part of a therapeutic monitoring assay for patients receiving low molecular weight heparin.[18]

The evaluation of platelet function using the TEG is a topic of interest, both in the evaluation of novel therapeutics as well as in the investigation of clinical platelets. TEG performed on citrated whole blood is not sensitive to platelet inhibition. Haemoscope has produced a technique, termed PlateletMapping, which works around this problem. Instead of a citrated sample, heparinized blood is obtained. The heparin prevents de novo thrombin formation, and effectively allows the TEG to be used as a modified aggregometer. In order to create the tracing, the company provides an activator consisting of a mixture of reptilase (to convert fibrinogen to fibrin) and activated factor XIII (to crosslink the fibrin). The addition of these activators ("activator F") and a platelet agonist such as ADP or arachidonic acid allows platelet activation and subsequent integration of platelets and fibrin into a clot that is demonstrable on the machine. This assay is sensitive to clopidogrel treatment in dogs,[19] and presumably would work for arachidonic acid as well. To the author's knowledge, it has not been evaluated in cats, and the reptilase activator limits its utility in horses.

The whole blood TEG tracing may also be modified by the addition of inhibitory substances. In humans, the addition of cytochalasin D and abciximab results in a significant decrease in TEG MA and angle.[20] In dogs, cytochalasin D is effective and results in a decreased MA and angle, but abciximab does not appear to affect the tracing from citrated canine whole blood.[21] Abciximab would be more likely to have an effect on the Platelet Mapping assay. The concept of inhibiting platelet function in this assay may allow better differentiation in hypercoagulable animals if the coagulation changes arise from the platelet component or the plasma protein component, but further work is necessary to standardize this assay.

The TEG as a tool for therapeutic monitoring

As noted above, the TEG is very sensitive to heparin, and even sub-therapeutic levels result in significantly prolonged assay times and may not generate a tracing at all.[17] This may be partially avoided through the use of an activator such as TF, but even this may not be able to distinguish high concentrations of heparin. A recent study evaluating the use of heparinase cups and low-dose (1:50000) TF activation was able to identify different in vitro doses of dalteparin in dogs,[18] and the use of high dose (1:3400) TF and normal cups has been used by the author to evaluate dogs receiving both unfractionated and low molecular weight heparin. Of note, the study that evaluated the heparinase cups failed to distinguish different doses of dalteparin when the sample was activated using kaolin.[18]

In hemophiliac and factor VII deficient dogs the TEG demonstrated hypocoagulability.[22, 23] The TEG can presumably be used to monitor dogs that have other factor deficiencies as they are addressed therapeutically, although the value of a TEG analysis versus standard coagulation testing in this context is debatable. In vitro studies in children have also demonstrated the sensitivity of the TEG to monitor the reversal of tPA-induced fibrionlysis by epsilon amino caproic acid (EACA),[24] and the TEG may be useful for this purpose in monitoring companion animals treated with EACA.

Guidelines for the generation of interpretable TEG tracings

All institutions running TEG should have a standard protocol in place. This should include a standard sample rest period following blood sampling, a standard temperature for both the rest period and the running temperature, and a limited number of trained individuals performing the assay. For research studies, a single operator is preferable, and duplicate testing should be considered if possible. If there is not a way to limit the number of operators, an activator such as TF or kaolin should likely be a part of the assay. Once the protocol is chosen, an institutional reference range should be generated if the machine is to be used to interpret clinical samples.

Sample acquisition should be via atraumatic venipuncture with minimal stasis, and the treatment of the sample (ie. collected via vacuum, transferred from syringe to tube by vacuum, collected into a syringe prefilled with citrate, etc) should be standardized as well. When running the TEG, the hematocrit of the subject should be recorded, and it is always indicated to perform additional testing that may corroborate findings of hyper or hypocoagulability (e.g. plasma activity or concentrations of antithrombin, protein C, fibrinogen, TAT complex etc.). Due to the sensitivity of the TEG assay to heparin, if a sample is acquired from a central venous catheter, a large presample should be acquired and consideration of the use of heparinase cups or an activator such as TF is recommended.

REFERENCES

1. Donahue SM, et al. **J Vet Emerg Crit Care** 2005; 15: 9. 2. Ganter MT, et al. **Anesth Analg** 2008;106:1366. 3. Ralph AG, et al. **J Vet Emerg Crit Care** 2010;20:A6 4. Epstein KL et al. **J Vet Emerg Crit Care** 2009;19:96. 5. Kristensen AT, et al. **J Vet Int Med** 2008;22:140 6. Marschner CB, et al. **Acta Vet Scand.** 2010;52:38. 7. Sinnott VB, et al. **J Vet Emerg Crit Care** 2009;19:484. 8. Otto CM, et al. **J Am Vet Med Assoc**. 2000;217:1500. 9.Wagg CR, et al. **Vet Clin Pathol**. 2009;38:453. 10. Kristensen AT, et al. **J Vet Intern Med**. 2008;22:140. 11. Wiinberg B, et al. **J Vet Intern Med**. 2008;22:357. 12. Epstein KL, et al. **J Vet Int Med.** In press. 13. Flint KS et al. **Vet Pathol 2009; 46: 1020** 14. Klose TC, et al. **J Vet Emerg Crit Care** 2009;19:A5. 15. Ralph AG, et al. **J Vet Intern Med**. 2010;24:681 16.Garcia-Periera BL, et al. **J Vet Emerg Crit Care** 2010;20:A3 17. Pittman JR, et al. **J Vet Emerg Crit Care** 2010;20:216. 18. Jessen LR, et al. **Vet Clin Pathol.** 2008;37:363. 19. Brainard BM, et al. **Am J Vet Res**. 2010;71:822. 20. Lang T, et al. **J Thromb Haemost.** 2004;2:147. 21.Brainard BM, et al. **J Vet Diag Invest**. In press. 22. Callan MB **J Thromb Haemost**. 2006;4:2616.23. Othman M, et al. **Haemophilia**. 2009;15:1126. 24. Yurka HG, et al. **Anesth Analg**. 2010;111:180.

CURRENT THROUGHTS ON THE PATHOGENESIS OF FELINE HYPERTHYROIDISM

Cynthia R. Ward, VMD, PhD, DACVIM
Athens, GA

Hyperthyroidism is the most common endocrinopathy in cats, affecting a large percentage of middle to older aged cats in the U.S. Since its recognition as a clinical entity some 20 years ago, it has been diagnosed with increasing frequency; however, the pathogenesis of the disease is not currently understood. Recent studies into molecular, environmental, and nutritional causes of this disease have resulted in new thoughts regarding the pathogenesis.

Feline hyperthyroidism most often results from benign adenomatous hyperplasia of the thyroid . It is clinically and pathologically similar to toxic nodular goiter or Plummer's disease in humans.[1] Interestingly, most cats have bilateral disease. Although it has been a clinical entity for years in the United States, it is emerging as a significant disease in other countries thus stimulating further interest in the etiology. [2,3] Because of its impact on feline health and similarity to human disease, study into the nutritional, cellular, and molecular mechanisms of feline hyperthyroid diseases will be important to our understanding of this disease and other hyperfunctioning endocrine diseases.

THYROID HORMONE SYNTHESIS

Thyroid hormone synthesis and secretion result from a highly regulated series of events. They depend upon the thyroid gland's ability to trap iodide from the bloodstream and iodinate tyrosyl groups on thyroglobulin in the thyroid colloid. To do this the thyroid cell membranes contain a unique protein, the Na^+/I^- symporter (NIS) that efficiently transports iodide, against an electrochemical gradient, into the cell. Monoiodotyrosine (MIT) and diodotyrosine (DIT) are formed by iodination and coupled into T_3 (MIT+DIT) and T_4 (DIT+DIT) in reactions catalyzed by thyroid peroxidase (TPO). Proteolysis of thyroglobulin occurs so that the mature T_3 and T_4 are released and deiodination of any tyrosyl residues occurs in the thyroid cells with the iodine being recycled. All aspects of thyroid hormone synthesis and secretion are regulated by thyrotropin (TSH) secreted by the pituitary. TSH stimulates NIS and TPO activity, and serum TSH levels directly correlate with serum thyroid hormone levels.

NORMAL THYROID GROWTH AND SIGNALING PATHWAYS

In the normal thyroid cell, growth and function are controlled by the interaction of pituitary-derived thyrotropin (TSH) with its G protein coupled receptor on the surface of the thyroid cell.[4] Heterotrimeric G proteins serve as molecular transducers, coupling extracellular signals to intracellular effector systems, including adenylyl cyclase activation and elevations in intracellular calcium. In the thyroid cell, activation of the cAMP signal transduction cascade is the major stimulus for cellular mitogenesis and secretion of T_3 and T_4.[4] This is a unique feature of thyroid cells. Heterotrimeric G proteins of the G_s and G_i subgroups regulate adenylyl cyclase activity that in turn controls the formation of cAMP. The G_s proteins are stimulatory to adenylyl cyclase, and the G_i proteins are inhibitory to adenylyl cyclase activity. Adenylyl cyclase activity and subsequent cAMP formation results from cumulative expression and activity of G_s and G_i.

IMMUNOLOGICAL VERSUS NONIMMUNOLOGICAL ETIOLOGY FOR FELINE HYPERTHYROIDISM

Early studies into the pathogenesis of feline hyerpthyroidism focused on the parallels between this disease and Grave's disease, the most common hyperthyroid disease in people. This link was intuitive due to the benign and bilateral nature of the feline disease. Initial evidence showed suggested the presence of immunoglobulins capable of stimulation of feline thyroid growth and function in sera of hyperthyroid cats. [5] Further studies looking at the effects of sera from hyperthyroid cats on cultured rat hyperthyroid cells showed no stimulatory effect. [6] In a recent and definitive study, the feline TSH receptor was cloned and activation by whole and purified IgG fractions of hyperthyroid cat sera was investigated in a sensitive expression system. No effect of the feline hyperthyroid sera was found. [7] This evidence refutes the idea of an immunological basis of feline hyperthyroidism and ascribes an autonomous function similar to other hyperfunctional endocrine adenomas. Experiments by Peter et al. [8] examining the growth and function of thyroid adenomatous tissue transplanted into recipient mice confirmed the idea of autonomous growth and function by the hyperthyroid cells since they were able to grow and function in the mice.

ABNORMALITIES IN THYROID GROWTH AND SIGNALING IN HUMAN TOXIC NODULAR GOITER

Abnormalities of any element of the receptor - G protein - cAMP regulatory pathway may, therefore, be postulated to result in the uncontrolled cellular proliferation and excess hormone secretion seen in hyperthyroid disease. Many laboratories have asked this question in human toxic nodular goiter, and disorders of G protein signaling have been implicated in the pathogenesis. Increased expression of alpha subunits of G_s and G_i have been demonstrated in human adenomatous thyroid tissue,[9,10,11] and one subset of G_i proteins, $G_{i\alpha_1}$, has been shown to be selectively expressed in some thyroid adenomas.[12] In addition, several mutations in $G_{s\alpha}$ and the TSH receptor resulting in constitutive activation of cAMP have been identified in hyperthyroid humans.[13,14,15,16] It is likely that human toxic nodular goiter is a multifactoral disease.[17]

EPIDEMIOLOGICAL/NUTRITIONAL CAUSES OF FELINE HYPERTHYROIDISM

Three major epidemiological studies have been performed on the US population of hyperthyroid cats in order to determine risk factors associated with the disease. [18,19,20] Risk factors implicated in the studies included consumption of canned cat food, especially fish flavors, canned food in pop-top cans, and the use of cat litter Additional studies have been performed with hyperthyroid cats in the UK and New Zealand which revealed similar results, implicating canned food consumption, as well as, a variety of household environments. [21,22] This has caused further scrutiny of environmental or nutritional etiologies of feline hyperthyroidism especially implicating the presence of thyroid toxicants, such as bisphenol A in cat foods. [23,24] Furthermore, the New Zealand study implicated a relationship between cats living closely in the human environment with the development of feline hyperthyroidism. One set of potential environmental goitrogens that are prevalent in the human environment are polybrominated diphenyl ethers (PBDEs) found in flame retardants. In a recent study, elevated levels of PDBEs found in pet cats were related to the development of feline hyperthyroidism. [25] Iodine deficiency has been well documented to cause goiter in people and a similar relationship in the cat has been explored. It should be noted, however, that iodine-deficient people develop goiter due to a lack of thyroid hormone production that results in a lack of negative feedback inhibition on the pituitary and an oversecretion of TSH, although early phases of the disease may occur without changes in measured TSH secretion. In some areas in which the diet is chronically low in iodine, autonomous function of the thyroid may occur in older individuals. Iodine levels in commercial cat foods have been looked at and determined to be extremely variable.[26,27] Feeding trials using foods with variable iodine concentration showed initial dramatic results over 2 weeks, but when the trial was extended to 5 months the cats compensated for the variable iodine in their diet and their serum fT_4 levels normalized.[27,28] However, a recent study has implicated the rising levels of iodine in commercial cat foods and a relationship to the increase in feline hyperthyroidism. [29] Other goiterogens that can be found in foods are flavonoids that inhibit thyroid peroxidase and also inhibit thyroid hormone metabolism. A recent study demonstrated high levels of soy isoflavones, genistein and daidzein were found in commercial cat foods. [30] In recent studies by our laboratory, flavonoids are implicated in generating autonomous mitogenesis and hormone secretion in cultured feline hyperthyroid cells.(unpublished data)

MOLECULAR CAUSES OF FELINE HYPERTHYROIDISM

Because of the similarity of feline hyperthyroidism to human toxic nodular goiter, molecular abnormalities have been suspected to occur in the feline disease. Two studies have looked for mutations in the TSH receptor using RT-PCR to amplify the areas most commonly mutated in the human disease. [31,32] No mutations were found. However, when examining the signal transduction protein, $G_{s\alpha}$, mutations were found in thyroid tissue from some of the hyperthyroid cats. [32] Although these mutations do not explain the bilateral nature of the disease, they may be one part of the multiple factors that result in feline hyperthyroidism. Our laboratory has postulated that, similar to human toxic nodular goiter, abnormalities in the TSH receptor - G protein - cAMP regulatory system may be involved in the pathogenesis of feline adenomatous hyperplasia. We have demonstrated that while $G_{s\alpha}$ protein expression is not significantly different between thyroid tissue from euthyroid versus hyperthyroid cats, $G_{i\alpha}$ expression is significantly decreased in hyperthyroid tissue versus normal thyroid tissue.[33] In further experiments, we have determined that while $G_{i\alpha_1}$ and $G_{i\alpha_3}$ subtypes of the G_i proteins are expressed normally in feline hyperthyroid tissue, $G_{i\alpha_2}$ shows decreased expression. [34] There appears to be no change in the thyroid gland Gs or Gi protein's ability to be activated by the normal TSH regulator. [35] Therefore, we hypothesize that the decrease of $G_{i\alpha}$ may remove negative inhibition of adenylyl cyclase and result in increased cAMP formation and that this may be integral to the pathogenesis of feline hyperthyroidism.

REFERENCES

1. Gerber H, et al. Veterinary Clinics of North America 1994;24:541. 2. Reese S, et al. Tierarztlicke Praxis. Ausgabe K, Kleintiere/Heimtiere 2002;30:274. 3. Miyamoto T, et al. Journal of the Japan Veterinary Medical Association 2002;55:289.4. Dumont JE, et al. Physiology Reviews 1992; 72:667. 5. Brown RS, et al. Thyroid 1992; 2:125.6. Peterson ME, et al. Veterinary Immunology and Immunopathology 1987; 16:277. 7. Nguyen LQ, et al. Endocrinology 2002; 143:395.8. Peter HJ, et al. Journal of Clinical Investigation 1987; 80:491. 9. Delemer B, et al. Journal of Endocrinology 1992;132:477.10. Siperstein AE, et al. Surgery 1991 110:949. 11. Derwahl M, et al. Thyroid 1995;5 (Suppl 1):S.12. Selzer E, et al. Proceedings of the National Academy of Science 1993;90:1609. 13. Lyons J, et al. Science 1990;249:655.14. Parma J, et al. J Clin Endocrinol Metab 1997;82:2695. 15. Horie H, et al. Jpn J Cancer Res 1995 86:737.16. Muca C, Vallar L. Oncogene 1994;9:3647. 17. Arturi F, et al. Experimental Clinical Endocrinology and Diabetes 1998;106:234.18. Scarlett JM, et al. Prevention in Veterinary Medicine 1988;6:295. 19. Kass PH, et al. J Vet Intern Med 1999;13:323.20. Edinboro CH, et al. J Am Vet Med Assoc. 2004;224(6):879. 21. Wakeling, J, et al. J Small Anim Pract, 2009;50(8):406.22. Olczak et al. N Z Vet J,2005;53(1):53. 23. Moriyama K, et al. J Clin Endocrinol Metab. 2002;87:5185.24. Kang JH, et al. Res Vet Sci. 2002;73:177. 25. Dye JA, et al. Environ Sci Technol 2007;41 (18);6350.26. Mumma RO, et al. Am J Vet Res 1986; 47:1633. 27. Johnson LA, et al. New Zealand Veterinary Journal 1992;40:18.28. Tarttelin MF, et al. New Zealand Veterinary Journal 1992;40:66. 29. Edinboro CH, et al. J Feline Med Surg 2010;12(9):672.30. Court MH, Freeman LM. Am J Vet Res 2002;63:181. 31. Pearce SHS, et al. Thyroid 1997;7:923.32. Peeters ME, et al. Thyroid 2002 12:571. 33. Hammer KB, et al. Am J Vet Res 2000;61:874.34. Ward et al., Am J Vet Res. 2005;66(9):1478. 35 Ward et al., Am J Vet Res 2010;71(6):643-8.

IDENTIFICATION OF FECAL BACTERIAL DYSBIOSIS IN DOGS WITH VARIOUS GASTROINTESTINAL DISORDERS. JS Suchodolski[1], RM Heilmann[1], SE Dowd[2], L Toresson[3], S Unterer[4], EM Dillman[1], AK Cook[1], JM Steiner[1]. 1. Gastrointestinal Laboratory, Texas A&M University, College Station, TX, USA; 2. Research and Testing Laboratory, Lubbock, TX, USA; 3. Helsingborg Referral Animal Hospital, Helsingborg, Sweden; 4. Ludwig-Maximilians-University, Munich, Germany.

Recent molecular studies have revealed a highly complex bacterial microbiota in the intestine of dogs. There is mounting evidence that microbes play an important role in the pathogenesis of acute and chronic enteropathies of dogs, including idiopathic inflammatory bowel disease (IBD). Similarly, compositional changes of the intestinal bacterial ecosystem have been associated with IBD in humans. The aim of this study was to characterize the bacterial microbiota in dogs with various gastrointestinal disorders using a next generation sequencing technique.

Fecal samples were obtained from healthy dogs (n=31), dogs with acute uncomplicated diarrhea (n=7), dogs with acute hemorrhagic diarrhea (AHD; n=13), and dogs with active (n=8) and therapeutically controlled IBD (n=10). The bacterial composition was analyzed by massive parallel 16S rRNA gene 454-pyrosequencing. Differences between groups were analyzed using Mann-Whitney *U* tests and Kruskal-Wallis tests followed by Dunn's multiple comparison tests. Statistical significance was set at $p<0.05$.

Significant differences in the proportions of several bacterial groups were identified between healthy and diseased dogs. Dogs with gastrointestinal disease had significantly higher proportions of Proteobacteria ($p<0.01$). Proportions of Firmicutes were lower in diseased dogs, but this difference did not reach significance ($p=0.06$). Within the Firmicutes the most notable findings were decreases in bacterial groups belonging to *Clostridium* clusters IV and XIVa (i.e., *Ruminococcus, Dorea,* and *Faecalibacterium* spp.; $p<0.01$ for all). Dogs with AHD had the most profound changes of the microbiota, followed by dogs with acute uncomplicated diarrhea, and dogs with active IBD. *Faecalibacterium* spp. was the bacterial group most prominently depleted in dogs with active IBD, but was not significantly different between healthy dogs and dogs with therapeutically controlled IBD ($p=0.66$).

Results of this study revealed bacterial dysbiosis in fecal samples of dogs with various GI disorders. Bacterial changes were more profound in dogs with severe disease, but were not identified in dogs with therapeutically controlled IBD, suggesting that the microbiota is stable in non-active disease. The bacterial groups identified are considered to be important short chain fatty acid producers and may serve as candidates for the diagnosis or therapeutic monitoring of GI disease. Future studies are necessary to determine if these microbial changes correlate with functional changes in the intestinal microbiota.

ORAL ABSORPTION OF GENERIC CIPROFLOXACIN TABLETS IN DOGS. Mark G. Papich, and Daria DiGiovanni. North Carolina State University, College of Veterinary Medicine, Raleigh, North Carolina, USA.

Ciprofloxacin oral tablets, available in a generic formulation for people, are widely used for treatment in dogs. Oral absorption data for ciprofloxacin in dogs has been variable, and too limited to guide accurate dosing. Subsequently, published doses for dogs in veterinary formularies have varied from 5 to 25 mg/kg. This study was undertaken to explore the factors that may affect oral absorption of generic ciprofloxacin in dogs, and to derive a pharmacokinetic-based dose for treating susceptible bacteria. Six healthy adult beagle dogs were used for the study (11.2 kg mean weight). After placing jugular vein catheters for collecting blood samples, these dogs were administered either a single oral dose of ciprofloxacin (250 mg tablet; mean dose 23 mg/kg), or an intravenous (IV) dose (10 mg/kg; 2 mg/mL solution). A randomized crossover design was used with a washout time between treatments. Blood was collected for plasma drug analysis for 24 hours. Ciprofloxacin concentration in plasma was analyzed using high pressure liquid chromatography (HPLC) and pharmacokinetics analyzed using a computer program. Oral absorption was also evaluated via deconvolution analysis. The oral dose was well-tolerated, but the IV dose produced transient vomiting and depression in some dogs. After the oral dose, the peak plasma concentration (C_{MAX}) was 4.4 µg/mL (CV 55.9%), terminal half-life ($t^{1/2}$) 2.6 hr (CV 10.8%), AUC 22.5 µg·hr/mL (CV 62.3%), and systemic absorption (F) 63.7% (CV 41.6%). After the IV dose, the $t^{1/2}$ was 3.7 hr (CV 52.3%), systemic clearance 0.587 L/kg/hr (CV 33.9%), and volume of distribution 2.39 L/kg (CV 23.7%). After examining the pharmacokinetic results from the oral dose, it was apparent that oral ciprofloxacin was absorbed well in some dogs (approximately 80%), but poorly in others (approximately 30%). To explore the factors that may have affected oral absorption, two high absorbers and two low absorbers were administered an additional oral dose as a 10 mg/mL solution (250 mg total dose) via gastric tube. After administration of the oral solution, the plasma concentrations were more uniform and consistent among dogs. Absorption of the oral solution of ciprofloxacin was 71.0% (CV 7%) with a $t^{1/2}$ of 3.1 (CV 18.6%) hr and C_{MAX} of 4.67 µg/mL (CV 17.6%). Therefore, it appears that inconsistent oral absorption of ciprofloxacin in some dogs may be formulation-dependent, and affected by tablet dissolution in the canine small intestine. Doses were calculated using the data for oral tablets in these dogs. The pharmacokinetic-pharmacodynamic (PK-PD) target was an AUC/MIC ratio of 100. Because of the wide range in oral absorption of tablets, a dose to reach the PK-PD target ranged from 12-75 mg/kg, with a mean dose of 32 mg/kg, once daily for bacteria with a MIC value $\leq$ 0.25 µg/mL. Some bacteria may have higher MIC values because the CLSI breakpoint for susceptible bacteria is $\leq$ 1 µg/mL.

CANINE DISTEMPER VIRUS QUANTIFICATION BY REAL-TIME PCR ALLOWS TO DIFFERENTIATE VACCINE VIRUS INTERFERRENCE AND WILDTYPE INFECTION. CM Leutenegger, C Crawford, J Levy, M Estrada. IDEXX Laboratories, Inc., West Sacramento, CA.

Canine distemper (CD) is a highly contagious, acute or subacute systemic viral disease of dogs and other carnivores which can be controlled efficiently by the use of modified live-virus (MLV) vaccines. However, MLV strains do cross-react with molecular diagnostic tests and cause significant confusion for clinicians.

The purpose of this study was to use quantitative real-time PCR viral load information to differentiate between vaccine virus used in MLV vaccines and wildtype infections in dogs.

A real-time PCR test for CD virus (CDV) based on the P gene for phosphoprotein was used to determine viral loads in vaccinated and wildtype infected animals. A total of 158 respiratory mucosal swab samples from MLV vaccinated and asymptomatic dogs were obtained within the first 3 weeks after MLV vaccination. Based on the viral load in vaccinated animals, a cutoff value was established for the differentiation of dogs with clinical signs of respiratory distress and presumably infected with a wildtype strain of CDV. Two hundred clinical cases with known clinical and vaccination histories were analyzed to validate the cutoff value.

The CDV real-time PCR proved to be of high analytical and diagnostic sensitivity: A standard curve was established using known numbers of CDV molecules to allow absolute quantitative CDV viral load data. The limit of detection was in the single molecule range while the limit of quantitation was established at around 10 molecules per PCR reaction. A comparison to IFA showed real-time PCR to be 30% more sensitive. The CDV viral load in vaccinated animals averaged 26,738 viral particles per swab. A cutoff value of 107,903 viral particles was calculated by adding 3 standard deviations to the average value. This cutoff value correctly detected 95.2% of the vaccinated samples. Acutely infected dogs with CDV compatible clinical signs have high viral loads normally several logs higher than the cutoff value. In dogs with clinical distress, recent CDV MLV vaccination but viral loads below the cutoff value, other infectious agents were detected by using a panel of real-time PCR tests. Testing additional infectious agents in clinical settings is important in order to explain clinical signs when viral loads below cutoff values indicate that CDV is not the cause of clinical signs.

In conclusion, quantitative real-time PCR is a sensitive, rapid and reliable test regardless of recent vaccination. The use of a cutoff value will be of significant help to discriminate between vaccine interference and wildtype infection in clinical settings.

THE USE OF A SUBCUTANEOUS URETERAL BYPASS DEVICE FOR URETERAL OBSTRUCTIONS IN CATS. A Berent[1], C Weisse[1], M Wright,[1] K Todd[2], D Bagley[3]. 1.The Animal Medical Center, New York, NY. 2. The Veterinary Hospital of the University of Pennsylvania 3. Thomas Jefferson University Hospital, Philadelphia, PA.

Feline ureteral obstructions are a common urinary dilemma and traditional therapy is associated with substantial morbidity/mortality. Feline nephrostomy tubes are reported as being effective when pelvic drainage is required. The biggest limitation is externalized drainage, requiring careful management to prevent infection/dislodgement. The development of an indwelling ureteral bypass using a combination locking-loop nephrostomy/cystostomy tube was modified from humans, resulting in permanent indwelling drainage, reduced complications, and improved quality of life. The objective is to describe the technical and clinical outcome using a novel device called a subcutaneous ureteral bypass (SUB) in cats with ureteral obstructions.

Fifteen cats (16 kidneys) had a SUB placed for: ureterolithiasis (9), ureteral stricture (+/- stones) (6), and ureteral stent rejection (2). The median pre- and post-procedure creatinine was 6 mg/dL (range: 2.8-13) and 2.5mg/dL (range: 2.4-9), respectively. The median pelvis diameter pre and post-procedure were 15 (range: 7-28) and 5mm (range 3.4-6), respectively. Six French tubes were placed in 14, and 5 Fr. in 2. The bypass remained indwelling for a median of >240 days (range 64->450). There were 3 major complications resulting in nephrostomy tube dislodgement (2) and port leakage (1) 5 days after surgery. One patient with severe coagulopathy developed a clot which resolved with tPA infusion through the port. No SUB got occluded/obstructed long-term.

Overall, the use of a SUB for cats with ureteral obstructions can be considered a functional option when other therapies have failed or are contraindicated, but should be considered a salvage procedure at this time.

Figure 1

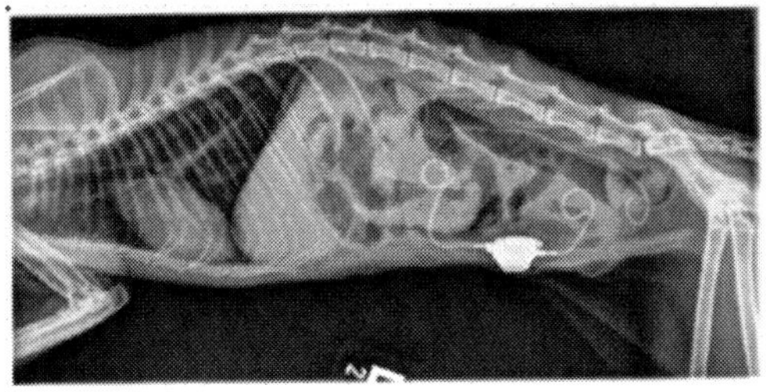

Figure 2

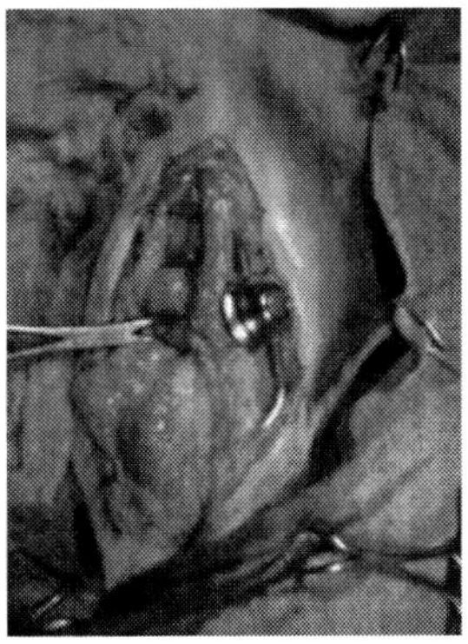

EVIDENCE OF INFECTION OF CATS BY *BORRELIA BURGDORFERI* AND *ANAPLASMA PHAGOCYTOPHILUM* AFTER EXPOSURE TO WILD-CAUGHT ADULT IXODES SCAPULARIS.

MR Lappin[1], R. Chandrashekar[2], B. Stillman [2], J Liu[2], T. Mather[3]. 1. Department of Clinical Sciences, Colorado State University, Ft. Collins, CO. 2. IDEXX Laboratories, Portland ME. 3. University of Rhode Island Center for Vector Borne Diseases, Kingston, RI.

Ixodes scapularis is the vector for *Anaplasma phagocytophilum* (AP) and *Borrelia burgdorferi* (BB) in the Northeastern United States. While little is known about these infections in cats, some clinically ill, client-owned cats have had AP DNA amplified from blood and have apparent clinical responses to tetracyclines. In addition, BB antibodies have been detected in some client-owned cats with and without clinical abnormalities. The purpose of this study was to allow wild-caught *I. scapularis* ticks to feed on adult research cats to gather temporal information about clinical abnormalities, laboratory abnormalities and diagnostic test results.

Young, adult, mixed sex research cats (n = 4) were used in this pilot study with IACUC approval. Each of the cats was shown to be negative for antibodies against BB and AP using a commercially available kit labeled for dog serum (SNAP® 4Dx®, IDEXX Laboratories) and a commercially available conventional PCR assay (Veterinary Diagnostic Laboratory, Colorado State University) that amplifies the DNA of AP. *Ixodes scapularis* ticks (13 females, 12 males) collected in Rhode Island were placed on each cat and allowed to feed for 7 days. The cats were assessed for clinical signs of depression or inappetance daily. Blood for a complete blood cell count, for AP PCR assay, and for serum separation for assessment for BB and AP antibodies were collected prior to tick attachment and then on weeks 1-10 and week 13 after tick attachment.

Clinical signs of illness were not noted in any of the cats. Each cat had AP DNA amplified from blood and each cat developed detectable AP and BB antibodies in serum. When compared to Week 0 group mean results, significant anemia, neutrophilia, thrombocytopenia, and monocytosis were not noted. Significant lymphopenia was detected on weeks 2 and 5.

Cat	*A. phagocytophilum* DNA	*A. phagocytophilum* Ab	*B. burgdorferi* Ab
1	Weeks 2 - 5; 8-13	Weeks 6 – 13	Week 10 only
2	Weeks 2 - 6, 8-13	Weeks 5 – 13	Weeks 7 - 13
3	Weeks 2 – 5	Weeks 5 – 13	Week 13 only
4	Weeks 2 ,3, 6	Weeks 5 – 13	Weeks 8, 9, 10

The results suggest that cats can be infected with AP and BB by *I. scapularis* ticks and the currently available diagnostic assays evaluated can be used to amplify AP DNA from cat blood and detect AP and BB antibodies in cat sera. Clinical and laboratory abnormalities were rare in this small group of cats.

CULTURE-INDEPENDENT DETECTION OF BACTERIA IN FELINE PANCREATITIS. K.W. Simpson[1], D.C. Twedt[2], S.P McDonough[1], M. Craven, [1] M Steffey[1], J.L. Dudak[2]. [1]Cornell University, Ithaca, NY, [2]Colorado State University, Fort Collins, CO.

The etiopathogenesis of feline pancreatitis is unclear. We sought to determine the presence and distribution of bacteria within the inflamed feline pancreas using culture independent *in situ* methods.

46 cats with a histological diagnosis of pancreatitis (30F, 15M; median age 9yrs, range 6wks-17yrs) and 13 non-inflamed controls (6F, 7M, median age 5yrs, range 6wks-13yrs) were retrospectively identified. Histopathology was categorized by a blinded pathologist as acute or chronic pancreatitis (mild, moderate, severe), or non-inflammatory. Unstained pancreatic sections were evaluated by fluorescence in situ hybridization

ybridization (FISH) with a 16S rDNA probe that recognizes bacteria in general (EUB338), and a non-EUB control. EUB338 positive sections were further examined with probes directed against *E. coli* and *Streptococcus* spp.

Acute pancreatitis (AP) was diagnosed in 12 (7 severe, 2 moderate, 3mild), chronic pancreatitis (CP) in 34 (8 severe, 14 moderate, 12 mild), and non-inflammatory (N) in 13 (10 normal, 2 periductular fibrosis, 1 fatty change). Bacteria were detected in 17/46 pancreatitis (AP+CP) and 4/13 N. In 7 cases bacteria were restricted to the outer surface of the pancreas (3N, 4CP) and likely represent contaminants. Intrapancreatic bacteria were visualized in 14 cases (5AP, 8CP, 1N), with bacteria observed in loose connective tissue and periductal areas (9), glandular parenchyma (6), peri-pancreatic saponified fat (5), areas of necrosis (3), and within ducts(3). There was no significant difference in the presence of bacteria between samples acquired at surgery (21) and necropsy (39). In AP, bacteria were observed in 5/7 severe (3 necrosuppurative, 2 necrotizing), 0/2 moderate and 0/3 mild. In CP, bacteria were observed in 2/8 severe, 4/14 moderate and 2/12 mild. Intrapancreatic bacteria were visualized in 11/31 cats with severe or moderate pancreatitis vs. 2/27 with mild pancreatitis or an uninflamed pancreas (P=0.01). *Strepotococci* were detected in 8 sections and *E. coli* in 5, with co-infection observed in 3. Concurrent disease of the liver (8: 3 lipidosis, 3 cholangitis, 1 necrosis, 1 fibrosis), intestines (4: 3 IBD, 1 lymphoma) or other organ systems (5) was present in 11/14 cats with intrapancreatic bacteria.

This study reveals a relatively high frequency of pancreatic infection (35%, 11/31) in cats with moderate to severe pancreatitis. Infection was most prevalent in cats with necrotizing or suppurative severe acute pancreatitis and may be linked to the poor prognosis associated with this form of pancreatitis. The localization and type of intrapancreatic bacteria suggests translocation of enteric bacteria is a likely source of infection. These findings have substantial implications for the diagnosis and management of cats with pancreatitis.

THE EFFECT OF N-ACETYLCYSTEINE (NAC) IN CLINICALLY ILL DOGS: DOES IT MODULATE OXIDATIVE STRESS AND IMPROVE CLINICAL OUTCOME? Katrina R Viviano and Beth Vander Wielen. Department of Medical Sciences, University of Wisconsin-Madison, School of Veterinary Medicine, Madison, WI.

Oxidative stress is considered central to the pathogenesis of many systemic diseases. In humans, biomarkers of oxidative stress, antioxidant depletion and lipid peroxidation, have been correlated with disease severity and associated with poor clinical outcomes. Therapeutic antioxidant supplementation with NAC in glutathione (GSH)-deficient patients has shown clinical benefits, including repletion of intracellular GSH levels. We have shown that clinically ill dogs are GSH-deficient, and that GSH deficiency correlates with mortality, but it is not clear whether there are direct benefits of antioxidant intervention in these patients. The purpose of this randomized, investigator-blinded, placebo-controlled, prospective study was to evaluate the effect of NAC to normalize blood antioxidants (RBC reduced GSH (RBC GSH), plasma cysteine (CYS), serum vitamin E (vit E), and whole blood selenium (Se)), reduce lipid peroxidation (urine isoprostane/creatinine ratio (U I/C)), and improve illness scores (SPI2) and outcome (survival to discharge) in clinically ill dogs.

Clinically ill client-owned dogs, admitted to the UW Veterinary Medical Teaching Hospital that did not receive blood transfusions, TPN, vitamins, or antioxidants were eligible for the study. Dogs enrolled in the study were randomized to receive IV infusions q. 6 h. of either NAC (1 x 140 mg/kg and 6 x 70 mg/kg) or equal volumes of 5% dextrose (placebo) over 48 hours. At the time of enrollment, and 2 hours following the final 48 hour infusion, blood and urine were collected to quantify RBC GSH, CYS, vit E, and Se concentrations; U I/C ratios; and calculate SPI2 scores. RBC GSH and CYS concentrations were quantified by HPLC. Commercially available HPLC, atomic absorption spectroscopy, and EIA were used to quantify vit E, Se, and U I/C ratios, respectively. Nonparametric statistical analyses were used, with results reported as medians and P <0.05 considered significant.

Sixty-one ill dogs were randomized to either NAC (n = 30) or placebo (n = 31). Overall this group of ill dogs had significantly decreased RBC GSH (1.50 vs. 1.91 mM; P = 0.0013), vit E (27 vs. 56 µg/mL; P = 0.0002), and Se (0.37 vs. 0.55 µg/mL; P = 0.0308) levels and elevated U I/C ratios (969 vs. 398 pg/mg; P = 0.0005) in comparison to healthy control dogs. Dogs in the placebo group showed a significant further decrease in RBC GSH over the next 48 hours (1.48 to 1.42; *P* = 0.035). NAC supplementation significantly increased plasma CYS levels (9.1 to 15.1 µM; P < 0.0001), and prevented a further decline in RBC GSH (1.55 to 1.58 mM; P = 0.174). However, serum vit E (30 vs. 28 µg/mL), Se (0.40 vs. 0.35 µg/mL), U I/C ratios (946 vs. 806 pg/mg), SPI2 scores (0.74 vs. 0.75), and outcome (81% vs. 76%) were not significantly different between the NAC and placebo groups after treatment.

The results of this study further support that clinically ill dogs experience oxidative stress, and suggest that antioxidant supplementation with NAC within the first 48 hours of hospitalization prevents further RBC GSH depletion. Further studies are necessary to investigate whether longer duration or combined antioxidant supplementation normalizes the redox state and impacts long-term outcome.

BIOMARKERS PREDICTIVE OF INSULIN RESISTANCE IN CATS IMPROVE WITH WEIGHT LOSS/ HEALTHY WEIGHT MAINTENANCE

J. Brockman and P. Burris
Topeka, KS

INTRODUCTION

The most common form of diabetes mellitus in the domestic cat closely resembles human type 2 diabetes mellitus, both clinically and pathologically[1, 2]. Feline diabetes and human type 2 diabetes both have a clinical onset in middle age, 6 years old in cats, and the rate of incidence increases with age[1, 3]. Other clinical similarities include obesity, resistance to ketoacidosis, low fasting serum insulin concentration, and impaired insulin secretion[4]. The most striking pathological similarity is amyloidosis in the pancreatic islets and a corresponding partial loss of islet β-cells. Islet amyloidosis occurs in over 90% of cats with diabetes, as well as 90% of humans with type 2 diabetes [1, 5, 6]

Insulin resistance is one of the major metabolic abnormalities in obesity and type 2 diabetes in humans and cats[7-9]. Insulin resistant cats have a threefold higher risk of developing impaired glucose tolerance with weight gain. High basal levels of insulin also increase the risk for developing glucose intolerance, suggesting that some cats have an underlying predisposition to develop glucose intolerance as a result of insulin resistance. If these cats become obese they are at greater risk for developing overt diabetes[7]. By the time glucose tolerance becomes impaired and fasting glucose concentrations rise above the normal range, significant β-cell damage may have already occurred[10]. It is important that cats at risk of β-cell damage from underlying insulin resistance be detected early. Programs aimed at healthy weight maintenance and reducing the risk of feline diabetes would be most effective if directed at cats with underlying insulin resistance before glucose intolerance develops[11].

In humans, insulin resistance is measured using the hyperinsulinemic-euglycemic clamp, which is considered the "gold standard," or the minimal model analysis of the frequently sampled intravenous glucose tolerance test[12, 13]. Both of these techniques have been used for the cat in a laboratory setting [14, 15], however neither is suitable for a veterinary practice with client owned cats. Both require multiple blood sampling and injections of glucose and insulin [11] The identification of surrogate biomarkers measured from a single blood draw that predict the level of insulin sensitivity in cats would be a useful tool in a clinical setting.

Recently a number of metabolites have been identified in humans that strongly correlate with the level of insulin sensitivity measured by the hyperinsulinemic-euglycemic clamp method in a large number of human subjects [16]. When the quantitative levels of the metabolites with the strongest correlation are applied to a proprietary regression model, a predicted insulin mediated rate of glucose disposal is obtained that highly correlates with the actual value measured by the Hyperinsulinemic-euglycemic clamp.

Given the similarity between feline diabetes and human type 2 diabetes, we hypothesized that these biomarkers would also be able to predict insulin resistance in cats, and that weight loss would result in improvement of insulin sensitivity as measured by these biomarkers.

STUDY

The intent of these studies was to test the hypothesis that a set of biomarkers previously described as able to predict insulin resistance in humans could also measure insulin sensitivity in cats based on the fact that cat diabetes mellitus is very similar to type 2 diabetes in humans. To test the hypothesis we made use of the fact that we know insulin resistance increases with age, diabetic cats have high levels of insulin resistance, and weight loss in cats can improve insulin sensitivity of obese cats.

Experimental Design

The first cohort in this study consisted of a total of 59 cats. Control baseline plasma samples were from 29 normal healthy cats approximately 1.5 years of age. It is expected that the control population has a minimal probability of having insulin resistance and no clinical signs of diabetes. The 30 additional test subjects represented a blinded set of plasma samples from cats with varying degrees of clinically diagnosed diabetes (n=15) or age matched cats with no clinical or pathological signs of diabetes.

The second cohort was 40 cats with an average percent body fat of greater than 39%. All cats were on a weight loss food for two months at the end of which the average percent body fat had dropped to less than 25%. The cats were then maintained on the same foods for a four month period. During the weight maintenance phase the animals maintained the same body weights, but their percent body fat dropped to less than 19%. Plasma samples were collected at the beginning of the study, the end of weight loss, and at the end of the weight maintenance phase.

The insulin resistance assay is based on the quantitative analysis of 5 metabolites (2-hydroxybutyrate, creatine, palmitate, decanoyl-carnitine, and oleoyl-LPC) in the plasma. A proprietary algorithm using a linear regression model (Metabolon, Inc, Durham NC) was used to generate a predictive rate of glucose disposal measured in mg,Kg^{-1},min^{-1} (RD). In the model, individuals with an RD value less than 6 will have a greater than 50% chance of being insulin resistant and an RD value less than 3 will have a greater than 90% chance of being insulin resistant.

Results

We first established an expected normal range of RD values by analyzing the population of normal healthy young cats. Greater than 95% of the RD values fall between 6.5 and 10.5 mg,Kg^{-1},min^{-1} with a mean of 8.23 ± 1.43, which is consistent with the regression model for humans. The average RD value for the group of diabetic cats was 2.78 ± 3.23 while the average RD value of their age matched controls was 6.11 ± 4.41. All three groups were statistically different from each other.

In the second cohort of 40 obese cats, the average RD value at the onset of the study was 6.19 ± 1.23, a value just below the normal range. At the end of the weight loss period the average RD value increased to $6.84 + 1.38$ ($p=0.029$), back into the normal range. By the end of the 4 months of weight maintenance the RD value improved to $8.30 + 1.08$ ($p=9.2E-12$) indicating that insulin sensitivity of this population had returned to what would be expected in a normal population of healthy cats.

Summary

Obesity is a continuing problem in many cats. Obesity leads to the development of metabolic syndromein man, one component of which is the decreased ability of tissues to respond to insulin, or "insulin resistance". Insulin resistance is known to lead to the development of diabetes mellitus in both humans and cats, but it has also been linked to the development of multiple conditions ranging from cardiovascular disease to increased risk for cancer. It is likely that diabetes is an under diagnosed condition in the cat. Histopathological evaluations of pancreas tissues for cats that have died from old age indicate that as many as one in fifteen cats show some sign of diabetes and pancreatic islet cell damage (unpublished observations). It is likely that many animals die of some other condition that may be exacerbated by insulin resistance before any clinical signs of diabetes appear. Identifying animals at an early stage of insulin resistance could have a positive impact on a cats overall health given an appropriate intervention.

Insulin resistance can be measured in a laboratory setting with an elaborate method such as the hyperinsulinemic-euglycemic clamp; however, this is just not practical in a clinical setting. The biomarkers in this study offer a simple way to predict insulin sensitivity in a cat from a single blood sample. These biomarkers were identified through their strong correlation with insulin sensitivity in humans. Given the similarity between type 2 diabetes in humans and diabetes mellitus in cats, we were not surprised by the fact that these biomarkers predict that diabetic cats are much more insulin resistant than their age matched counterparts, and that these normal aged cats are more insulin resistant than a population of young healthy cats.

The other major risk factor for developing insulin resistance is obesity. Using the second cohort in this study we demonstrate that these biomarkers predict cats in an obese state to be insulin resistant, and that weight loss results in an improvement in insulin sensitivity. We were surprised to see such a dramatic difference between the predicted insulin sensitivity of these cats following weight loss compared to their predicted insulin sensitivity at the end of a four month weight maintenance period. This suggests that insulin sensitivity is not simply corrected by weight loss but requires maintenance with specific targeted nutrition at a healthy weight for a period of time before returning to a normal state. This may also relate to the ratio of body fat to lean muscle mass given that even though these animals stayed at the same weight, their percent body fat continued to drop throughout the weight maintenance phase.

Taken together, these data suggest that the biomarkers described in this study can be used to identify cats that are insulin resistant and at risk for developing diabetes, and that they can be used to monitor the effects of therapeutic intervention, such as weight loss, for regulating proper insulin sensitivity. These biomarkers may become a useful tool in the clinic.

REFERENCES

[1]Johnson KH, et.al. Am J Pathol. 1986 Nov;125(2):416-9. [2] O'Brien TD. Mol Cell Endocrinol. 2002 Nov 29;197(1-2):213-9. [3]Panciera DL, et.al. J Am Vet Med Assoc. 1990 Dec 1;197(11):1504-8. [4]O'Brien TD, et.al. Vet Pathol. 1985 May;22(3):250-61. [5]Johnson KH, et.al. N Engl J Med. 1989 Aug 24;321(8):513-8. [6]O'Brien TD, et.al.. Vet Pathol. 1993 Jul;30(4):317-32. [7]Appleton DJ, et.al. J Feline Med Surg. 2001 Dec;3(4):211-28. [8]DeFronzo F, et.al. Lancet. 1978 Nov 18;2(8099):1077-9. [9]Feldhahn JR, et.al. J Feline Med Surg. 1999 Jun;1(2):107-15. [10]Ferrannini E. Diabetes Metab Rev. 1997 Jun;13(2):81-6. [11]Appleton DJ, et.al. J Feline Med Surg. 2005 Jun;7(3):183-93. [12]DeFronzo RA, et.al. Am J Physiol. 1979 Sep;237(3):E214-23. [13]Bergman RN, et.al. Am J Physiol. 1979 Jun;236(6):E667-77. [14].Petrus DJ, et.al. Res Vet Sci. 1998 Sep-Oct;65(2):179-81. [15]Hoenig M, et.al. Domest Anim Endocrinol. 2006 Nov;31(4):373-89. [16]Gall WE, et.al. 2010 PLoS One.5(5):e10883.

FELINE INFLAMMATORY BOWEL DISEASE & INFLAMMATORY LIVER DISEASE

Albert E. Jergens, DVM, PhD, DACVIM
Ames, IA

FELINE INFLAMMATORY BOWEL DISEASE

Definition of IBD

Inflammatory bowel disease (IBD) refers to a group of idiopathic, chronic gastrointestinal disorders characterized by mucosal inflammation. While the exact prevalence for IBD is unknown, it is arguably the most common histopathologic diagnosis obtained in cats with chronic vomiting and/or diarrhea. It is likely that IBD represents one form of chronic enteropathy which is distinguished from food-responsive and antibiotic-associated causes for gastroenteritis by therapeutic responsiveness. While the etiology for IBD remains unknown, accumulating evidence suggests that IBD is a multifactorial disorder conditioned by genetic, immunologic, and environmental (e.g., intestinal microbiota) factors.

Feline IBD is defined by the following features: (1) the presence of persistent (> 3 weeks) gastrointestinal signs which are characterized by cyclic periods of active and inactive disease; (2) inadequate response to dietary trials and symptomatic therapies alone; (3) failure to document other causes for gastrointestinal inflammation; and (4) histopathologic evidence of mucosal inflammation. Clinical signs (vomiting, small and/or large bowel diarrhea, weight loss, alterations in appetite) are attributed to mucosal cellular infiltrates, inflammatory mediators, and inflammation-associated dysmotility. Histopathologic evaluation of intestinal biopsy specimens is required for definitive diagnosis; however, no standard microscopic grading system of IBD lesions has been universally accepted. Biopsy interpretation is notoriously subjective and suffers from extensive interobserver variability and the technical constraints of specimen size and procurement artifacts inherent with endoscopic specimens.

A diagnosis of IBD is one of exclusion and requires ruling-out many other diseases which may cause intestinal inflammation. Systemic diseases, chronic parasitism, adverse food reactions (e.g., food allergy or intolerance), infectious diseases, and alimentary lymphosarcoma are the major differential diagnoses for feline IBD. The optimal diagnostic protocol for IBD has yet to be established. Therapeutic trials using anthelmintics or hypoallergenic diets may be effective in animals having parasitic or dietary causes, respectively, for enterocolitis.

Histologic evaluation of biopsy specimens is required for definitive diagnosis. No standard microscopic grading system of IBD lesions has been established – this is a major problem. Endoscopically-obtained GIT mucosal biopsy collection remains the gold standard but presents a variety of challenges for both the clinician and the pathologist. Note that extensive interobserver variability in interpretation between pathologists can occur. Recent studies show that the diagnostic accuracy (i.e., accurate microscopic evaluation by the pathologist) is influenced by differences in endoscopic training and the optimal number of biopsy specimens submitted by the endoscopist. Collectively, previous case-series of cats with IBD indicate that histologic lesions of cellular infiltration, epithelial alterations, and fibrosis are most common.

Association of clinical signs with disease severity

The clinical course of IBD is characterized by spontaneous exacerbations and remissions which make assessment of disease burden difficult. Moreover, gastrointestinal signs are highly variable, and the severity of disease may differ appreciably between patients depending on localization and extent of affected regions of the gastrointestinal tract. Various single indices have previously been used to assess IBD activity including clinical signs, histopathologic severity, phenotypic analysis of immune cells, and measurement of inflammatory mediators. One recent report in cats with IBD showed a positive correlation between intestinal inflammation, gastrointestinal signs, and inflammatory cytokines. Moreover, a simple clinical scoring system has now been designed for use in cats with chronic GI signs, including both IBD and food-responsive enteropathy. Pilot data suggest that this index, termed the feline chronic enteropathy activity index (FCEAI), may be used to assess clinical disease burden at diagnosis and in response to therapeutic interventions.

Nutritional therapy of FIBD

There is general agreement that novel protein, highly digestible diets are useful in cats with IBD. Additionally, dietary fiber, polyunsaturated fatty acids (n-3 FA), and prebiotic/probiotic supplements have also been advocated. Selecting a protein (venison, rabbit, lamb, whitefish, or turkey) that is not routinely fed reduces the likelihood that these diets will evoke allergic (hypersensitive) responses. Protein hydrolysate diets containing low-allergen, small molecular weight proteins may also reduce adverse reactions to food which may occur with IBD. Fiber-enriched diets are recommended to reduce large bowel diarrhea and tenesmus. Dietary fiber may increase fecal consistency, bind potential colonic irritants, improve abnormal colonic motility, and produce beneficial short-chain fatty acids which nurture large bowel structure and function. The use of fermentable fibers (beet pulp, psyllium, soy fiber, or oat bran) is often recommended since fermentation generates butyrate for maintenance of colonic health.

Pharmacologic management of FIBD

Most pharmacologic therapies interrupt the amplification sequence of inflammation in IBD, explaining why maintenance therapy (via diet and/or drugs) is important. Anecdotal evidence supports the use of oral corticosteroids and metronidazole in most cases of FIBD. The administration of potent immunosuppressive drugs, such as azathioprine, cyclophosphamide, and cyclosporine is uncommonly required. Most animals are able to maintain remission of signs with dietary management alone following completion of drug therapy.

FELINE INFLAMMATORY LIVER DISEASE

The inflammatory liver diseases in cats have been recently re-characterized by the WSAVA liver histopathology group based upon pathologic classification. Two main forms are now recognized – neutrophilic and lymphocytic cholangitis. It has also been suggested that the two forms may be different stages of a similar pathologic process although controversy exists. A third form of cholangitis associated with liver flukes will not be discussed in this presentation.

In ***neutrophilic cholangitis***, the liver parenchyma is infiltrated by a diffuse population of neutrophils. The etiology is uncertain but may be related to ascending bacterial infection (*E. coli*) of the bile duct from the intestine. There is sometimes a link with pancreatitis and IBD – if all three features are present then this is referred to as 'triaditis'. There is no well recognized predisposition for neutrophilic cholangitis. Most cats present with overt jaundice and are often markedly depressed with a history of a recent acute illness. ***Treatment*** consists of dietary modification and antibiotic therapy, ideally based upon culture and susceptibility findings from sites of active hepatobiliary inflammation. Clinical recovery is facilitated by proper nutrition and adjunct use of other pharmacologic agents. The prognosis is generally good with early intervention but more guarded in cats requiring surgery and with intercurrent involvement of other (pancreas) organs.

Lymphocytic cholangitis is characterized by infiltration of the liver with lymphocytes and plasma cells. The nature of the inflammatory infiltrate (i.e., mononuclear) suggests that this disorder is a likely immunologically-mediated disease. This disorder most often occurs in young cats less than 4 years of age and there is a predisposition in Persian cats. The two main presenting signs of lymphocytic cholangitis are jaundice and progressive ascites. Cats are often times asymptomatic or present with minimal disease burden. Alternatively, a subset of animals may demonstrate severe weight loss, anorexia, and occasional pyrexia. Serum biochemistries generally reveal global increase of liver enzymes (ALT, ALP and GGT) along with abnormal bile acids. Bilirubin levels are often significantly increased while hyperglobulinemia may also be observed. Radiography confirms generalized hepatomegaly while ultrasound may reveal irregular hepatic margins. Histopathology is required for definitive diagnosis and reveals diffuse or focal infiltrates of lymphocytes and plasma cells often accompanied by hepatic fibrosis. The best hepatic biopsies are obtained by laparoscopy or laparotomy where wedge sections may be procured for microscopic examination.

Treatment of lymphocytic cholangitis is based on attenuation of immune-mediated inflammation. Prednisolone used at initial immunosuppressive dosages of 1-2 mg/kg BID and gradually reduced over an extended period of several months is often successful. Colchicine may be used at a dosage of 0.03 mg/kg SID in attempt to reduce fibrosis although controlled studies have yet to be reported. Additional adjunct therapy with choleretics, free radical scavengers, and nutritional support may be helpful. The ***prognosis for lymphocytic cholangitis*** is generally good. In one report, the mean survival time of 25 cats was 37 months while 16 cats survived more than 1 year. Clinical assessment of disease status may be performed by repeat blood chemistries, liver function assays, or serial monitoring of liver biopsies (which is often not possible due to client compliance).

REFERENCES

[1]Brain PH, et al. **J Feline Med Surg** 2006; 8:91. [2]Gagne JM, et al. **Vet Pathol** 1996; 33:521. [3]Jergens AE, et al. **J Vet Intern Med** 2010; 24:1027. [4]Janeczko S, et al. **Vet Microbiol** 2008; 128:178. [5]Day MJ, et al. **J Comp Pathol** 2008; 138 suppl 1:S1.[6]Rothhuizen J, eta al. **WSAVA** Standards for clinical and histological diagnosis of canine and feline liver disease; 2006

THE MICROBIOTA AND CHRONIC ENTEROPTHIES

Kenneth W. Simpson, BVM&S, PhD, DACVIM, DECVIM-CA
Ithaca, NY

Inflammatory bowel disease (IBD) is the term applied to a group of poorly understood enteropathies that commonly affect animals and people. IBD in both people and animals is increasingly considered a consequence of uncontrolled intestinal inflammation in response to a combination of elusive environmental, enteric luminal constituents (principally microbial and dietary) and immunoregulatory factors in genetically susceptible individuals.

In people, genetic susceptibility is linked increasingly to defects in innate immunity exemplified by mutations in the innate immune receptor NOD2/CARD15, that in the presence of the enteric microflora may lead to up-regulated mucosal cytokine production, delayed bacterial clearance and increased bacterial translocation, thereby promoting and perpetuating intestinal inflammation (Packey 2008, Perez 2010). While the mucosa-associated flora is implicated frequently as a pivotal factor in the development of IBD in people and animals, the specific bacterial characteristics that drive the inflammatory response have remained elusive.

In dogs, the clinical responses of Boxer dogs with granulomatous colitis (GC; also known as histiocytic ulcerative colitis) to enrofloxacin, and a subset of dogs (particularly German Shepherd) with idiopathic chronic diarrhea to antibiotics such as tylosin or oxytetracycline, points to a similar interaction of host susceptibility and microflora (Batt 1988, German 2003, Hostutler 2004, Westermarck 2005). The *E. coli* strains isolated from the colon of Boxer dogs with GC are broadly similar to those in healthy dogs, and the numbers of cultivable aerobic and anaerobic bacteria in the duodenal juice of dogs with antibiotic responsive enteropathy are similar to dogs that respond to food changes or immunosuppression suggesting that affected individuals are susceptible to their resident microflora rather than obligate pathogens (German 2003, Simpson 1994, Craven *et al* unpublished observations 2009), but this remains to be determined.

Recent advances in molecular microbiology have enabled the analysis of complex bacterial communities without bacterial culture. Culture-independent analyses of bacterial 16S rDNA libraries in people reveal that only 30% of the fecal flora appears cultivable, and there is significant variation in the flora in different gastrointestinal segments and luminal contents versus the mucosa of healthy individuals (Eckburg 2005). The application of these culture independent techniques to people, dogs and cats with gastrointestinal disease has revealed that intestinal inflammation is typically associated with a floral shift from Gram-positive to Gram-negative bacteria, predominantly Enterobacteriaceae. It is noteworthy that increased numbers of Enteobacteriaceae have been found to correlate with mucosal inflammation and clinical signs in cats with signs of gastrointestinal disease, and a novel group of adherent and invasive *E. coli* (AIEC) have been associated with intestinal inflammation in people and Boxer dogs with granulomatous colitis. In many patients it remains to be determined if these floral alterations are a cause or a consequence of the inflammation, but their discovery has provided new opportunities for therapeutic intervention. This is exemplified by intramucosal *E. coli* in Boxer dogs with GC, where eradication of invasive *E. coli* correlates with long term clinical and histological remission.

This presentation will give an in depth overview of the mucosal microflora associated with chronic intestinal inflammation in dogs, cats and people, and data from murine models that reveal acute inflammation can also perturb the enteric microbiota, and trigger the proliferation of adherent and invasive *E. coli* similar to those described in IBD.

THE RESULTS OF THE PRINCIPAL STUDIES COVERED IN THIS PRESENTATION ARE SUMMARIZED AS FOLLOWS

Mucosal Microflora and Chronic Inflammatory Bowel Disease

***Granulomatous Colitis in Boxer dogs* (Van Kruiningen 1965, 2005, Hostutler 2004, Simpson 2006, Mansfield 2009, Craven unpublished observations 2009)**

Granulomatous colitis (GC) of Boxer dogs, also known as Histiocytic Ulcerative Colitis (HUC), was first described by Van Kruiningen in a kennel of Boxer dogs in 1965. The clinical hallmarks of the disease are severe large bowel diarrhea that is often accompanied by profound weight loss, anemia and hypoalbuminemia. GC/HUC, although rare, occurs world-wide with reported cases originating from Australia, Japan, North America and Europe. Boxer dogs are most commonly affected although other breeds such as Mastiff, Alaskan malamute, Doberman pinscher, and French Bulldogs, are sporadically affected. The pathognomonic lesion of GC in Boxer dogs is mucosal infiltration with large numbers of macrophages staining positively with period-acid-Schiff (PAS), and is usually accompanied by mucosal ulceration and loss of goblet cells.

The poor response of GC to immunosuppression led to a re-appraisal of antibiotic therapy, and there are now three independent studies that describe dramatic improvements in clinical signs and histological lesions of Boxer dogs treated with antibiotic protocols containing enrofloxacin. Colonic biopsies from affected Boxer dogs (n=13) and controls (n=38) were examined by fluorescent in situ hybridization (FISH) with a eubacterial16S rDNA probe. Culture, 16S ribosomal DNA sequencing, and histochemistry were used to guide subsequent FISH. Intramucosal Gram-negative coccobacilli were present in 100% of GC/HUC samples from Boxer dogs but not controls, and invasive bacteria hybridized with FISH probes to *E. coli*. Independent support for these findings is provided by the immunolocalization of *E. coli* to macrophages within the colons of 10/10 Boxer dogs with HUC. *E. coli* strains isolated from affected Boxer dogs are novel in phylogeny, and have an adherent and invasive phenotype (AIEC) similar to strains isolated from people with Crohn's disease (CD).

A recent study has examined further the relationship of intramucosal *E. coli* to clinical and histological outcome in 7 Boxers with GC treated with enrofloxacin. Clinical response was noted in 7/7 dogs within 2 wks of enrofloxacin given at a mean dose of 7 mg/kg/d for a mean duration of 9.5 weeks, and was sustained in 6 dogs (mean disease-free interval to date of 45 months). Post-enrofloxacin FISH was negative for *E. coli* in 4/5 dogs. *E. coli* resistant to enrofloxacin were present in the FISH positive dog that relapsed clinically. These findings suggest that *E. coli* invasion plays a critical role in the initiation and/or progression of GC in Boxer dogs. Antibiotic resistance is an increasing problem and we have documented enrofloxacin resistant *E.coli* in 6/6 Boxers that were empirically treated with enrofloxacin prior to definitive diagnosis.

Since GC in dogs is remarkably breed-specific it may be due to a heritable anomaly in Boxer dogs that confers susceptibility to invasion and persistence of AIEC within the colonic mucosa. As Boxer dogs are also prone to other granulomatous diseases such as intestinal and systemic protothecosis and cutaneous intracellular mycobacteria (canine leproid granuloma syndrome), it is possible that Boxers have a genetic defect in how they respond to granuloma-inducing microorganisms in general, e.g. defective bacterial killing. Further studies are ongoing to identify the host and bacterial factors related to invasion and persistence of *E. coli* in GC susceptible Boxers.

Mucosal flora associated with non-specific colitis in dogs **(Bogel et al unpublished observations 2008)**

The colonic mucosa of 29 dogs undergoing evaluation for signs of gastrointestinal disease (endoscopic biopsies) and 12 dogs without signs of intestinal disease (full thickness biopsies) was evaluated using a standardized histopathological scoring system. The number and spatial distribution of mucosal bacteria was determined by FISH with probes to 16S rDNA and an antibody to *Brachyspira* spp.

Colonic histopathology ranged from normal (18), through mild (10), moderate (11), and severe (2) colitis. The mucosal flora of healthy and inflamed colon hybridized with probes to *Bacteroides/Prevotella*, *Clostridium*, Enterobacteriaceae, *Streptococcus* and *Helicobacter* spp. 16S rDNA sequence analysis showed *Helicobacter* spp had highest homology to *H. canis, F. rappini, H. hepaticus* and *H. cinaedi*. *Brachyspira* were not detected in any samples. The number of mucosa-associated bacteria and Enterobacteriaceae was lower in dogs with colitis and was inversely correlated with severity. Colitis was associated with a significant reduction in the proportion of *Clostridium* spp relative to total bacteria and Enterobacteriaceae. These data establish that the density and composition of the colonic mucosal flora is related to the presence and severity of colonic inflammation in dogs. The relationship of the abnormal flora to the etiopathogenesis of "lymphoplasmacytic" colitis remains to be established.

Mucosal flora associated with duodenal inflammation in dogs **(Xenoulis 2008, Craven 2009, Suchodolski 2010, Allenspach 2010)**

In a study by Xenoulis et al., DNA was extracted and 16S rRNA gene was amplified using universal bacterial primers from endoscopic duodenal brush cytology samples from 10 dogs with IBD and nine healthy control dogs (6 Beagles, 3 Greyhounds). 16S rRNA gene clone libraries were compared between groups. From a total of 1240 selected clones, 156 unique 16S rRNA gene sequences were identified, belonging to six phyla: *Firmicutes* (53.4%), *Proteobacteria* (28.4%), *Bacteroidetes* (7.0%), *Spirochaetes* (5.2%), *Fusobacteria* (3.4%), *Actinobacteria* (1.1%), and *Incertae sedis* (1.5%). Species richness was significantly lower in the IBD group. PCA showed IBD and control dogs are composed of distinct microbial communities. The most profound difference involved enrichment of the IBD dogs with members of the Enterobacteriaceae family. However, differences involving members of other families, such as *Clostridiaceae*, *Bacteroidetes* and *Spirochaetes*, were also identified.

A recent study has sought to gain insight into the relationship of the mucosal flora to clinical response to diet, antibiotics, or immunosuppression. A culture-independent bacterial tag-encoded FLX 16S rDNA amplicon pyrosequencing (bTEFAP) approach was used to evaluate the microbiome of endoscopic duodenal biopsies from 16 dogs with a histopathological diagnosis of IBD (7MN 7FN 2M, median age 5 yrs) and 14 clinically healthy (H) dogs (5FN, 6MN, 3F, median age 7.5 yrs). 10 dogs were food responsive (FR), 3 steroid responsive (SR), and 3 were partially responsive to a combination of food and antibiotics (FA). A total of 85,065 sequences (54776 from H, 30289 from IBD) were analyzed. The number of sequences per individual (median) was higher in H (3,912) than IBD (1,893) (P=0.007). Dogs with IBD had lower overall species diversity than H at all primary levels of taxonomy based upon rarefaction ($p<0.05$). Analysis by clinical response indicated differences in numbers of sequence between groups (H, FR, SR and SA: Kruskal Wallis, $P<0.003$,), and lower numbers in FR than H ($p=0.007$).

Mucosa-adherent duodenal microbiota in dogs with idiopathic IBD were recently characterized using molecular methods. The aim of this study was to investigate differences in the mucosa-adherent duodenal microbiota between dogs with idiopathic IBD and healthy dogs. Duodenal biopsy samples were collected from seven dogs with IBD and seven healthy control dogs. DNA was extracted, 16S ribosomal RNA genes were amplified and 16S rRNA gene clone libraries were constructed and compared between groups. A total of 1035 clones were selected, and based on a 98% similarity criterion, 133 unique phylotypes were identified across all dogs. These phylotypes belonged to seven bacterial phyla: *Proteobacteria* (52.9%), *Firmicutes* (26.1%), *Bacteroidetes* (7.7%), *Actinobacteria* (8.6%), *Fusobacteria* (4.4%), *Tenericutes* (0.2%) and *Verrucomicrobia* (0.1%). Significant differences were identified in the relative abundance of several bacterial groups between dogs with IBD and healthy dogs ($p < 0.001$). Healthy dogs and dogs with IBD clustered according to their disease status. Dogs with IBD had a significantly higher abundance of clones belonging to *Alpha-*, *Beta-*, and *Gamma-proteobacteria* ($p < 0.0001$ for all classes), and a significantly lower abundance of *Clostridia* ($p < 0.0001$). Bacteria of the genera *Pseudomonas*, *Acinetobacter*, *Conchiformibious*, *Achromobacter*, *Brucella*, and *Brevundimonas*, were significantly more abundant in dogs with IBD. In conclusion, significant differences of the mucosa-adherent duodenal microbiota were observed between dogs with idiopathic IBD and healthy dogs in this study. These results warrant further investigations into the role of the intestinal microbiota in the pathophysiology of canine IBD.

German shepherd dogs (GSD) are particularly predisposed to chronic enteropathies. This study sought to evaluate expression patterns of certain pattern recognition receptors of the innate immunity (Toll-like receptors, TLR), clinical disease activity and histopathological severity in GSD with chronic enteropathies. Mucosal biopsies were collected from the duodenum, colon and ileum of 13 affected GSD and 10 healthy greyhounds as controls. Dogs were objectively assessed using published scoring systems for clinical and histological severity of disease. Diversity of the duodenal microbiota was assessed by construction of 16S rRNA gene libraries. Expression of TLR2, TLR4, TLR5 and TLR9 in biopsies of the duodenum, ileum and colon was assessed by quantitative real-time PCR. TLR4 expression was increased in all intestinal segments in GSD, however, TLR5 expression was very low compared to the healthy dogs. The microbiota in the duodenum of GSD was significantly different to that of the greyhounds, with an over-representation of 16S rRNA gene sequences belonging to the classes of Bacilli, and Erysipelotrichi, and to the orders of *Lactobacillales*, *Actinomycetales* and *Erysipelotrichales*. These findings could point to a distinct pathogenesis of chronic enteropathies in GSD, with differentially high and low expression of TLR4 and TLR5, respectively, and increased proportions of specific members of the *Lactobacillales* potentially playing a role.

***Tylosin responsive enteropathy* (Suchodolski 2009)**

A complex and variable dysbiosis has recently been documented in dogs with tylosin responsive enteroapthy.

***Feline Inflammatory Bowel disease* (Janeczko 2008)**

Duodenal biopsies were collected from 27 cats: 17 undergoing diagnostic investigation of signs of gastrointestinal disease, and 10 healthy controls. Subjective duodenal histopathology ranged from normal (10), through mild (6), moderate (8), and severe (3) IBD. The number and spatial distribution of mucosal bacteria was determined by fluorescence in situ hybridization with probes to 16S rDNA. Mucosal inflammation was evaluated by objective histopathology and cytokine profiles of duodenal biopsies. The number of mucosa-associated Enterobacteriaceae was higher in cats with signs of gastrointestinal disease than healthy cats ($P < 0.001$). Total numbers of mucosal bacteria were strongly associated with changes in mucosal architecture ($P < 0.001$) and the density of cellular infiltrates, particularly macrophages ($P < 0.002$) and CD3+lymphocytes ($P < 0.05$). The number of Enterobacteriaceae, *E. coli*, and *Clostridium* spp. correlated with abnormalities in mucosal architecture (principally atrophy and fusion), upregulation of cytokine mRNA (particularly IL-1, -8 and -12), and the number of clinical signs exhibited by the affected cats.

***Crohn's disease* (Darfeuille-Michaud 2004, Baumgart 2007, Frank 2007, Rhodes 2007, Barnich 2007, Sokol 2008)**

Analysis of 16S rDNA libraries constructed from endoscopic biopsies of the ileum from patients with Crohn's ileitis (a disease phenotype that occurs in approximately 70% of individuals with CD), demonstrates selective enrichment of ileal mucosa in *E. coli* and relative depletion in *Clostridiales* compared to both patients with CD restricted to the colon, and to healthy individuals. PCR of ileal DNA yielded no evidence of MAP, *Listeria* or *Shigella*, and confirmed the increase in *E. coli*. Similarly, 16S rDNA library-based evaluation of surgically resected CD ileum shows an increase in Proteobacteria (this phylum includes Enterobacteriaceae), decrease in *Firmicutes* (this phylum contains *Clostridiales*), and the absence of DNA from MAP in CD mucosa. The relative decrease in *Clostridiales* in CD ileal mucosa echoes findings in studies of feces and colonic mucosa from patients with CD, and suggests that this clostridial dysbiosis may be a feature of CD in general rather than a marker of CD ileitis.

The in depth knowledge of microfloral composition provided by 16S sequence analysis can be used to guide *in situ* examination of the spatial distribution of the mucosal flora interacting most closely with the innate immune system. FISH with a probe restricted to *E. coli/Shigella* reveals that the ileal mucosa of CD patients harbors significantly more *E. coli* than normal ileum, and mucosally invasive *E. coli* are only detected in CD mucosa. In patients with Crohn's ileitis, *E. coli* represent approximately 3.5% of the mucosal flora labeled with the panbacterial FISH probe, EUB338 and approximately 40% of the Enterobacteriaceae recognized by the Enterobacterial FISH probe, 1531. This contrasts with culture-based studies reporting that *E. coli* represent >50% of the cultivable mucosal flora in 40 to 58% of patients with ileal CD. However, this discrepancy is readily explained by the inability to culture the vast majority of enteric mucosal flora, and supports the use of culture-independent analysis. Furthermore, the number of *E. coli* visualized in ileal biopsies (Baumgart 2007) but not bacterial colonization in general, correlates with histological and endoscopic disease activity, which suggests *E. coli* could be specifically involved in the inflammatory process. These findings complement previous studies that have documented *E. coli* antigens and DNA in granulomas and peri-ulcerative lesions in CD, and provide a rational explanation for the higher prevalence of antibodies directed against *E. coli* outer membrane porin C (OmpC) and flagellin in patients with CD involving the ileum versus colon restricted CD.

REFERENCES

Allenspach K, et al. **Vet Microbiol** 2010 May 27 [Epub ahead of print]. Barnich N J, et al. **Clin Invest** 2007;117(6):1566. Batt RM, et al. **Gut** 1988;29(4):473. Baumgart M, et al. **ISME J** 2007;1(5):403. Darfeuille-Michaud A, et al. **Gastroenterology** 2004;127:412. Eckburg PB, et al. **Science** 2005;308(5728):1635. Frank DN, **Proc Natl Acad** Sci USA 2007;104(34):13780. German AJ, et al. **Vet Intern Med** 2003;17(1):33. Hostutler RA, et al. **J Vet Int Med** 2004;18:499. Janeczko S, et al. **Vet Microbiol** 2008;128(1-2):178. Mansfield CS, et al. **J Vet Intern Med** 2009;23(5):964. Packey CD, et al. **J Intern Med** 2008;263(6):597. Perez LH, et al. **PLoS One** 2010;5(6):e10915. Rhodes JM. **Gut** 2007;56:610. Simpson KW, et al. **Infect Immunity** 2006;74:4778. Simpson KW. **J Am Vet Med Assoc** 1994;205(3):405. Sokol H, et al. **Proc Natl Acad Sci USA** 2008;105(43):16731. Suchodolski JS, et al. **BMC Microbiol.** 2009;9:210. Suchodolski JS, et al. **Vet Microbiol** 2010;142(3-4):394. Van Kruiningen HJ, et al. **Path Vet** 1965;2:521. Van Kruiningen HJ, et al. **APMIS** 2005;113:420. Westermarck E, et al. **J Vet Intern Med** 2005;19(2):177. Xenoulis PG, et al. **FEMS Microbiol Ecol** 2008;66(3):579.

MARCHING DOWN THE GUT WITH PUSH-PULL ENTEROSCOPY

Peter H. Kook, Dr. med. vet., DACVIM, DECVIM-CA
Zurich, Switzerland

Endoscopy is an essential method for the diagnosis and treatment of diseases in the esophagus, stomach, duodenum and colon in human and veterinary medicine. However until only a few years ago, most of the small bowel was an under-investigated frontier not accessible using conventional endoscopic techniques. For the first time in Japan, in 2001, Yamamoto et al. introduced the double-balloon enteroscopy (DBE) system that enabled endoscopic inspection of the entire small bowel.[1] DBE consists of an enteroscope and a soft overtube. A balloon is attached to the tip of the endoscope and another to the tip of the overtube. Each balloon can be inflated and deflated by a pressure-controlled air pump system. Preliminary tests in Japan and Germany showed that this system can be used to perform total small intestine enteroscopy with a good safety profile and patient tolerance. DBE has become established throughout the world for diagnostic and therapeutic small bowel examinations, and is now being used widely in routine clinical work.[2] In 2007, another balloon enteroscopy system was introduced; this system is equipped with only one balloon at the tip of the overtube and is therefore known as single balloon enteroscopy (SBE).[3]

DOUBLE BALLOON ENTEROSCOPY (DBE)

The DBE system (Fujinon, Inc. Saitama,Japan) consists of a high-resolution video endoscope, with a working length of 200 cm and a flexible overtube made of polyurethane. A latex balloon is attached at the tip of the enteroscope and also on the overtube, and can be filled with air or emptied using a pressure-controlled pump. The principle of the DBE technique is based on alternating pushing and pulling maneuvers, allowing the small bowel to be threaded step by step onto the overtube. Two different devices are currently available with the double-balloon system: the EN450-P5 model, with a working channel of 2.2 mm and an outer diameter of 8.5 mm, and the EN450-T5 model, with a working channel of 2.8 mm and an outer diameter of 9.4 mm. The corresponding overtubes have diameters of 12.2 and 13.2 mm respectively, with an overall length of 145 cm.

DEB technique

During the insertion process, the endoscope is first inserted until it reaches the stomach, and the overtube is then advanced over the endoscope. The endoscope is then inserted further, through the pylorus, and its balloon is inflated to maintain a stable position within the intestinal lumen. The overtube is then advanced along the endoscope and its balloon is inflated. When both balloons are inflated, the tips of the endoscope and the overtube are pulled slowly together, to gather and shorten the intestine between them on the overtube. This simplifies the shape of the intestine ahead and avoids the formation of redundant loops (Fig 1). The balloon of the endoscope is then deflated, and the endoscope is advanced while the system is fixed by the balloon on the tip of the overtube. When the endoscope is stable again, with its balloon inflated, the balloon of the overtube is then deflated and advanced to reach the tip of the endoscope. These procedures are repeated, moving the system further and further into the small intestine. In this way, the working length of the endoscope can be used most effectively, and the endoscope can be advanced distally after the intestine has been shortened and its shape simplified.

To the author's knowledge, DBE in dogs has only been described by Latorre et al., in 2007 in the veterinary literature[4]. Two laboratory dogs weighing 23 kg and 26 kg were used and the technique (EN-450P5) was applied using both oral and anal approaches. The total time for the antegrade examination was 2 h and 10 min and the estimated depth to which the endoscope was inserted was 4.95 m. The first meter was reached in the first 15 min by advancing four times. The 2nd meter was reached within the next 20 min by advancing five more times, and the 3rd meter was reached after another 15 min and five more maneuvers. In total, 26 advancing movements were performed. At the beginning each advance was often up to 30 cm, however after proceeding more distally, 10 cm advances became more common. No complications were noted during the procedure. The balloons used in this study were inflated to 45 mmHg (6 kPa), which is the lowest pressure needed to hold the human intestine for endoscope insertion and is designed not to cause pain or discomfort. There are no references about the pressure required for use in pet dogs, but results obtained with four anatomical specimens (gastrointestinal tracts from post-mortem examinations) showed that 45 mmHg was satisfactory.[4]

Fluoroscopy was used to evaluate the performance of DEB in the dogs mentioned in the study. Results from a recent study in people, however, showed that radiographic assessment of the enteroscope's position does not help to determine its depth of insertion, because even in the best circumstances, the repeated push-and-pull procedures create similar radiographic images.[5]

SINGLE BALLON ENTEROSCOPY (SBE)

The SBE system was recently introduced (Olympus, Inc. Tokyo, Japan). The enteroscope (SIF Q 180) is a high-resolution video endoscope, with a working length of 200 cm. The enteroscope is equipped with a

working channel that is 2.8 mm in diameter, with an outer diameter of 9.2 mm. The disposable overtube has an overall length of 140 cm, consists of silicone, and has a latex-free balloon made of silicone at the distal end. The outer diameter is 13.2 mm. In contrast to the DBE system, no balloon is attached to the tip of the enteroscope, and stable positioning in the small bowel is achieved during the withdrawal of the scope, by angling the tip of the endoscope. The SBE is extremely flexible with its distal angulation reaching 180° up and down, and 160° right and left. Insufflation of the overtube balloon is performed using a pressure-controlled pump with pressure ranging from 6-16 kPA. Balloon pressure is regulated at approximately 5.6 kPa (42 mm Hg) by a pressure-sensing feedback mechanism. An alarm is activated when the balloon pressure is greater than 8.2 kPa for 5 seconds or longer. Interestingly, results of experiments to determine the necessary inflation pressure for latex balloons that would be associated with satisfactory grip of the small intestine were performed in dogs. Inflation pressures between 45 and 100 mmHg (5.9 - 13.3 kPa) produced suitable and safe pressures.[6]
However, the author has not yet exceeded pressures above 7 kPa.

SBE technique

Although silicone rubber is used in the overtube (ST-SB1-Disposable Sliding Tube), a hydrophilic lubrication coating inside of the tube lining ensures better lubrication between the scope and overtube. Because of this, only water is needed to ease friction between the back loaded overtube and scope. The official recommendation is for single-use only, but the author has successfully re-used this rather costly sliding tube multiple times without any problems. Prior to passage of the scope it is prudent to confirm that the balloon is functioning properly.

First the scope is passed into the stomach and then the overtube is passed over the scope into the stomach. At this point, it is crucial not to insufflate too much air into the stomach, as this will lead to antegrade duodenal air trapping with resultant slippage of the inflated balloon back into the antrum. A minimum safety distance of approximately 15 cm should always be kept between the distal end of the overtube and the enteroscope (60 cm mark) in order to prevent damage of the scope when angulating its distal end. Next, both the scope and the overtube are advanced as far into the small intestine as possible. The flexible nature of the enteroscope can create difficulty with pyloric intubation initially because the enteroscope will tend to loop into the retroflexed position in the stomach. This issue however can improve quickly with operator experience. It is vital to make sure that the overtube has definitely passed the flexura duodeni cranialis before inflating the balloon for the first time, as an unnoticed sliding back of the inflated overtube into the pyloric antrum will result in numerous unsuccessful push-pull maneuvers. Ideally, the balloon is not inflated in the area of the proximal duodenum so as to avoid trauma to the papilla and lessen risks of pancreatitis.

Next, the balloon is inflated and the scope is reduced by gently pulling back on both the overtube and scope together (Fig 2). In the first few cases that the author performed, fluoroscopy was helpful to visualize the enteroscope movement during reduction and check the hub of the balloon. Once the balloon enteroscope is completely reduced, the balloon is kept inflated, and the tip of the enteroscope is advanced as far as possible. Then, the balloon is deflated and the overtube is gently passed over the scope until the 60 cm mark is reached. Next, the balloon is re-inflated and reduction of the scope is repeated. This cycle is repeated to advance the scope as far as possible into the small intestine or until the lesion in question is encountered.

One of the most important points when proceeding quickly and especially distally is keeping air insufflation to a minimum. When there is a large amount of air that is allowed to enter during insertion of the endoscope and pulling back of the endoscope and the overtube, with a lesser amount of air that is sucked out between the individual steps of the push-and-pull maneuvers, there is a greater space that is created and therefore needed on the overtube for threading the small bowel. With this air trapping, the overtube quickly becomes "full" of small bowel; no more small bowel can be threaded onto the overtube, and the procedure has to be stopped, regardless of the insertion depth achieved up to that point.

INDICATIONS FOR DBE AND SBE

Suspected or known mid-gastrointestinal bleeding represents the principal indication for both procedures in humans. Balloon enteroscopy also represents the procedure of choice for patients with suspected small-bowel tumors, because it allows for histologic sampling in contrast to video capsule endoscopy alone. Another indication of DBE and SBE in people is obstructive Crohn's disease, as it allows dilation to be performed. Indications in veterinary medicine are most likely very similar and may include inflammatory enteropathies, ulcers and erosions, as well as intestinal bleeding secondary to angiodysplasia and neoplasia. However, the clinical use of balloon enteroscopy still needs to be evaluated in dogs with gastrointestinal disease.

A recent study comparing the histopathologic findings of duodenal and ileal biopsies in dogs with enteropathies suggested that routine collection of ileal biopsies is warranted if large intestinal signs are also present. The authors concluded that routinely sampling both small intestinal sites has implications for clinical practice.[7] In that study by Casamian-Sorrosal, a higher percentage of adequate samples were obtained from the ileum than from the duodenum, and it is imaginable that biopsy quality could be improved with the use of ante- or retrograde balloon enteroscopy. Although not yet proven, this may also be the case in dogs with exclusive small intestinal disease.

Complications

In humans, the complication rate of diagnostic DBE is low and expected in only 1 % of cases. The most severe complication is pancreatitis, with a risk of 0.3 % in oral DBE.[8] The only data available on mortality rate is from the German double balloon registry, where the mortality rate is 0.05% (death occurred after pancreatitis and a complicated postoperative course after perforation during polypectomy for a small bowel polyp). The SBE technique with only the overtube balloon for fixation, demands angulation of the tip of the endoscope while the endoscope is pulled back or the overtube advanced. In theory this "hooking" might lead to mucosal damage or extended stretching of the mesentery. However, results of a recent prospective Dutch study evaluating complications of 166 SBE procedures also suggested that SBE is a safe diagnostic endoscopic procedure.[9] The author has not encountered major complications (i.e. perforation) or clinical pancreatitis in his patients after SBE, although pancreatitis markers have not specifically been evaluated post procedure.

Limitations of balloon enteroscopy

The anatomy of the canine small intestine may limit the safety and effectiveness of balloon enteroscopy in veterinary medicine. Lopez Albors and colleagues recently measured the length and width of the small bowel in 55 dogs in a detailed morphometric study.[10] Based on their results the duodenum (2.37 ± 0.05 cm), is the least risky portion of the small intestine for balloon enteroscopic exploration, whereas restrictions to the progress of the enteroscope could be more likely in the jejunum (1.95 ± 0.03 cm) and ileum (2.1 ± 0.05 cm). The authors assumption however, that intestinal widths of < 2 cm might give excessive mechanical resistance to the progress of the endoscope, must be verified in clinical studies. The author has performed SBE in dogs ranging from 16.3 to 40.2I kg BW without complications, with the final insertion depth in the 16.3 kg dog (Beagle) estimated to be just under 3 meters.

Likewise, in human medicine, the first successful antegrade SBE in a three-year old child with occult intestinal bleeding weighing 13.5 kg was carried out without complications in 2009.[11] A complete inspection of the small bowel (i.e. down to the ileocolic valve) is rarely accomplished in people with SBE, and has not been achieved by the author yet.

Another limitation is the lengthy procedure time, which can be prohibitive in patients with a high anesthetic risk. The average procedure time in the author's hospital is approximately 2 hours and there does not seem to be a significant learning curve associated with SBE in terms of technical difficulty of the procedure. Experienced endoscopists require only 38 (12-90) minutes for antegrade balloon enteroscopy in humans, with the majority of cases performed without fluoroscopy.[12]

Prospective clinical trials are needed to assess the utility (feasibility, diagnostic yield, complications) of balloon enteroscopy as well as the advantage of one system over the other in dogs. It is possible that more knowledge about canine small bowel disease can be generated with these new techniques.

REFERENCES

[1] Yamamoto H et al. Gastrointest Endosc 2001, 53:216-220. [2] May A. Gastroenterol Clin North Am 2010, 39:797-806. [3] Tsujikawa T et al. Endoscopy 2008, 40:11-15. [4] Latorre R et al. Vet Rec 2007, 161:587-591. [5] May A et al. Endoscopy 2005, 37:66-70. [6] Sugano K et al. Double-balloon endoscopy: theory and practice. Tokyo: Springer;2006. [7] Casamian-Sorrosal D et al. J Vet Intern Med 2010, 24:80-83. [8] Moeschler O, et al. Z Gastroenterol 2008, 46:266–170. [9] Atkas H et al. Endoscopy 2010, ;42:365-368. [10] Lopez Albors O et al. Vet J 2010, doi:10.1016/j.tvjl.2010.08.017. [11] Kramer RE et al. Gastrointest Endosc. 2009, 70:546-549. [12] Upchurch BR et al. Gastrointestinal Endoscopy 2010, 71:1218-1223.

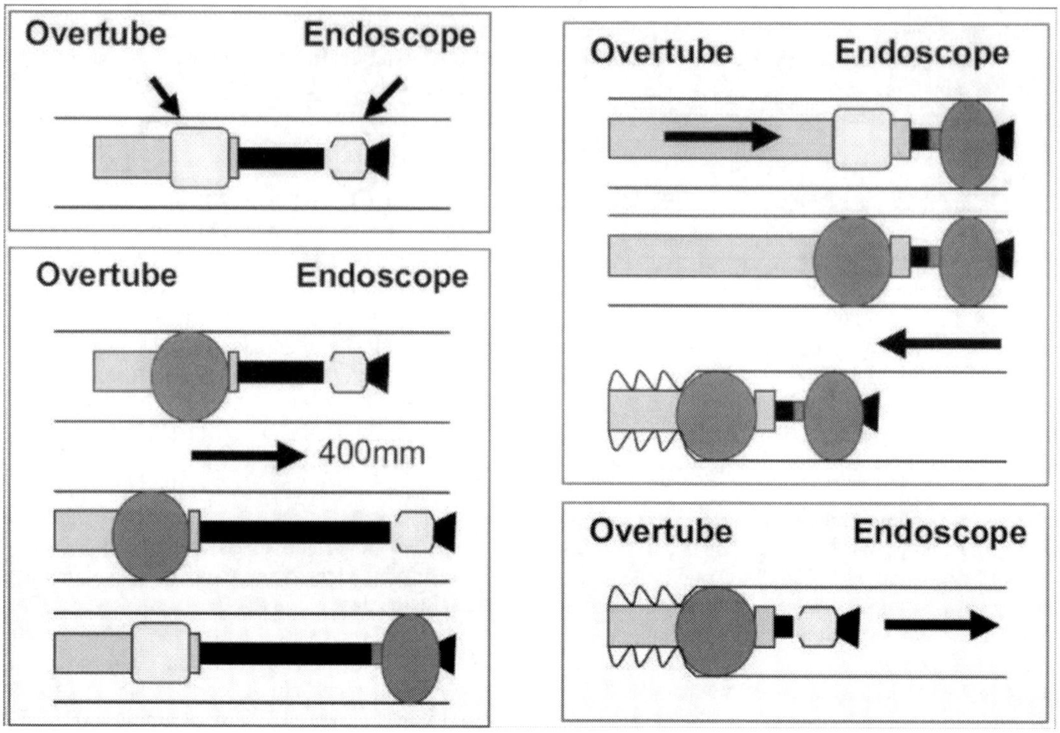

Fig 1: Insertion procedure with Double Balloon Enteroscopy (DBE)

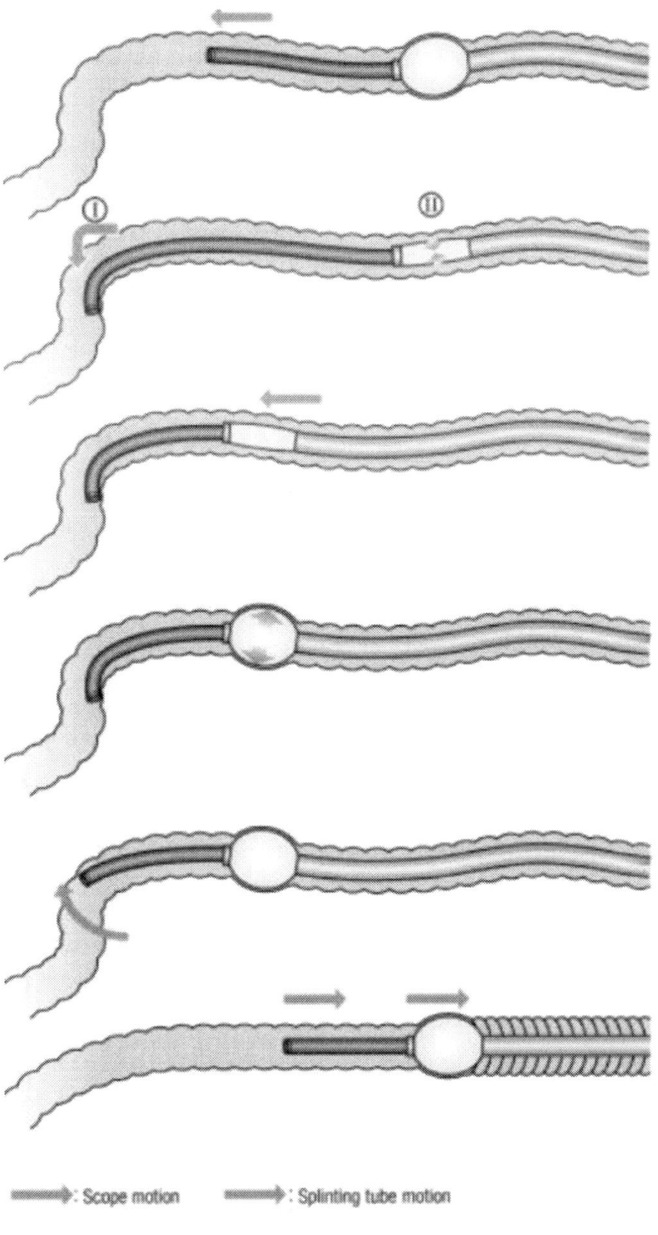

Fig 2: Insertion procedure with the Olympus Single Balloon Enteroscope system (SBE)

CAPSULE ENDOSCOPY

Alice C. Y. Lee, DVM
Ithaca, NY

INTRODUCTION

Imaging tools for diagnostic and therapeutic intervention in veterinary gastrointestinal diseases often lags behind human medicine by many years. A full spectrum of technologies is available in human gastroenterology to provide minimally invasive yet comprehensive GI examinations in vivo: chromoendoscopy, single/double balloon (push-pull) enteroscopy, endoscopic ultrasound, endoscopic microscopy, and capsule endoscopy, to name a few. Veterinary endoscopy, on the other hand, is still limited to conventional endoscopes that allow examination from the oral cavity to the upper duodenum and from the ileum to the rectum. The jejunum remains largely inaccessible by current veterinary standards.

Capsule endoscopy, involving the telemetric recording of images captured by an ingestible camera capsule, was introduced to the North American human gastroenterology market in 2001. It was a major breakthrough in several respects: 1) It provided views of jejunum as well as stomach, duodenum and ileum; 2) It was non-invasive and did not require patient sedation, unlike conventional endoscopy which often leads to patient discomfort; and 3) As with holter monitors, patients are instrumented with recording equipment so that the examination takes place outside of the doctor's office with minimal disruption to the patient's daily routine.

CAPSULE ENDOSCOPY SYSTEM

At present, four different capsule endoscopy systems are commercially available, though only two – the PillCam® (Given Imaging, Ltd.) and the EndoCapsule® (Olympus Medical Systems Corp.) – are licensed for use in the United States. Each system shares similar features. The endoscopic capsule measures approximately one inch in length and half an inch in diameter. The camera takes 2-3 frames per second, with a battery life of 8-11 hours. The capsule is not motorized and is therefore completely dependent upon peristaltic waves for forward movement.

In preparation for capsule administration, the patient is fasted for 12 hours. External antennae are attached to the skin of the thorax and abdomen to receive the images transmitted by the capsule, after which the capsule is ingested, and data recording begins. The patient may drink clear fluids 2 hours into the study, and may have a small meal 2 hours following that. At the end of the day, the patient returns the data recorder and antennae set to the physician, who downloads the data into a workstation for analysis. The capsule generally passes out in the stool within 24 hours of ingestion. An especially useful accessory in the capsule endoscopy system is an external viewing device, which allows the physician to monitor the capsule's progress in real time if he so chooses.

THE HUMAN CAPSULE ENDOSCOPY EXPERIENCE

Endoscopic capsules have been used for the past decade in both adult and pediatric patients. In children, it is body weight, not age, that appears to be the limiting factor in capsule passage through the pylorus and ileocecal sphincter; one group of gastroenterologists suggested 35 lbs. as the lowest body weight that would allow unobstructed passage of the capsule based on their experience.[1] Another study reported safe passage in a 10-month-old child weighing 25 lbs., though in this case the capsule was front-loaded onto a conventional endoscope and deposited directly into the duodenum, not orally ingested.[2] In people, capsule endoscopy has detected lesions such as erosions, ulcers, inflammation, cancers, polyps, lymphangiectasia, angiodysplasia, and ascarid worms.[3-6]

Indications and Contraindications

Indications for capsule endoscopy include obscure GI bleeding, suspected Crohn's disease, suspected small bowel tumor, surveillance for inherited polyposis syndromes, evaluation of drug-induced intestinal injury, follow-up examination after treatment, and more recently, endoscopic capsules have also been used to assess motility.[7]

Contraindications include strictures, obstructions, fistulas, diverticula, and body weight < 25 lbs.[1,2,8,9] Swallowing disorders and gastroparesis are not considered absolute contraindications since endoscopic placement of the capsule beyond the problem region can be easily achieved.[9] Pacemakers are listed as a relative contraindication due to the belief that the radiofrequency broadcast by the capsule may interfere with pacemaker function; however, to date, 12 patients possessing pacemakers have undergone capsule endoscopy without complication.[9]

Complications

Potential complications associated with capsule endoscopy are incomplete studies (i.e. the capsule does not reach the cecum before the battery fails) (16% of cases), capsule retention (0.8-1.4%), obstruction (1-2 cases in 1 million administrations), and aspiration.[7,10] In the majority of cases, capsule retention occurs in patients with confirmed Crohn's disease (≥ 8%), NSAID-induced stricture, radiation enteritis, or small bowel tumor.[10] Retained capsules may pass spontaneously, or they may be removed endoscopically or surgically, typically without further sequelae.[2,10]

CAPSULE ENDOSCOPY OF DOGS

The first study of capsule endoscopy in dogs was conducted as a precursor to human clinical use, as a means of comparing lesion detection capability of endoscopic capsules versus push enteroscopes.[11] Between 9 and 13 beads of varying colors, shapes and sizes (3-6 mm) were surgically implanted in the small bowel of 9 mongrel dogs to simulate lesions; half the beads were within 1 m of the pyloric sphincter. A week later, dogs underwent general anesthesia for push enteroscopy and 1-3 capsule endoscopies, performed in a randomized order. Postmortem samples of the small bowel were collected for histologic examination. All personnel interpreting images or slides were blinded. Overall, capsule endoscopy detected more beads (143 out of a total of 225 beads) than push enteroscopy (33/89). Though push enteroscopy was highly sensitive (94%) within its accessible section of small bowel – compared to only 53% for capsule endoscopy in the same region – the capsule was able to detect a median of 4 beads more per examination when taking into account the entire small bowel. This study also showed that capsule endoscopy can identify hair, ingested plastic, ulceration, submucosal swelling, and parasitic worms, and that the capsules pass safely through the dogs with no significant histologic findings.

Pilot Study Using Oral Administration of Capsules

Six healthy dogs were given endoscopic capsules orally to investigate the safety and utility of this device in fully conscious animals. Young dogs (5 months to 3 years of age) of varying breeds, genders and weights (21-127 pounds) were recruited. Following the same protocol as used in people, endoscopic capsules (EndoCapsule®, Olympus) were given orally to the dogs. Dogs tolerated the procedure and instrumentation well. Images were captured from the esophagus, stomach, small bowel, and occasionally colon. Incidental findings included hair, erosions, and various luminal parasites. Mean gastric and small bowel transit times were both 1.8 hours. The capsule passed out in the stool the day after ingestion.

Non-Medical Application of Capsule Endoscopy

One potential application of capsule endoscopy beyond its medical diagnostic use is in reducing euthanasias within the pharmaceutical industry. Current FDA regulations for licensing GI antiparasiticides often necessitate hundreds of animal sacrifices per product approval, simply to quantify the number of adult parasites residing in the intestinal tract of treated versus control groups. Replacing necropsy with capsule endoscopy in the drug testing process would save countless animal lives. Ongoing research addressing this issue has demonstrated a good correlation ($r_s = 0.83$) between counts acquired from EndoCapsules® and those from necropsy of the same dogs ($n = 17$). Dogs in this study were 6-month-old beagles weighing between 12 and 15 lbs. Of these dogs, three initially had incomplete studies (capsule remained in the stomach for > 7 h), but a repeat capsule administration in two of the dogs proved successful. In all cases, the capsules were passed without incident in the stool by the next day. Mean gastric and small bowel transit times were 2.2 and 0.7 hours, respectively.

CONCLUDING REMARKS

Major advantages of capsule endoscopy include access to the entire small bowel, lack of sedation/anesthesia, low degree of invasiveness and thus enhanced comfort for the patient. In the veterinary field, this device may be especially useful as a diagnostic tool in dogs that are too ill to undergo anesthesia for conventional endoscopy. To date, capsule endoscopy has been successfully performed in both the human and canine populations with minimal complications. Though this technology is not without its pitfalls (incomplete studies, lack of motor control), these problems are not insurmountable. Use of prokinetic agents such as erythromycin may help speed passage of the capsule through the stomach and improve study completion rates. Many self-propelling or otherwise operator-controllable endoscopic capsules are currently in development and may become commercially available in the foreseeable future. From the limited number of canine capsule endoscopy studies carried out thus far, it appears that this device is able to identify both benign and pathologic lesions in dogs, and can be used in dogs weighing as little as 12 lbs. Non-invasive capsule endoscopy shows promise as a new diagnostic option for evaluating GI diseases of dogs, particularly those involving the mid-section of the small bowel.

REFERENCES

[1]Seidman EG, et al. **Curr Treat Options Gastroenterol** 2006:9:416. [2]Jensen MK, et al. **Gastrointest Endosc** 2010;72(1):95. [3]de'Angelis GL, et al. **Am J Gastroenterol** 2007:102(8):1749. [4]Maiden L, et al.**Clin Gastroenterol Hepatol** 2007:5(9):1040. [5]Moy L, et al. **J Pediatr Gastroenterol Nutr** 2007:44(4):516. [6]Fritscher-Ravens A, et al. **Gut** 2009:58:1467. [7]Eliakim R, et al. **Curr Opin Gastroenterol** 2010:26:129. [8]Mazzarolo S, et al. **Southern Med J** 2007 100:274. [9]Storch I, et al. **Gastrointest Endosc Clin N Am** 2006:16(2):329. [10]Liao Z, et al. **Gastrointest Endosc** 2010:71(2):280. [11]Appleyard M, et al. **Gastroenterol** 2000:119(6):1431.

ATYPICAL CANINE HYPERADRENOCORTICISM: IS IT REAL?

Ellen N. Behrend, VMD, PhD, DACVIM
Auburn, AL

Hyperadrenocorticism (HAC) is one of the most, if not the most, common endocrinopathies of older dogs. Due to the high incidence and relatively non-specific clinical signs, older dogs are commonly screened for HAC. Unfortunately, no test for diagnosis of HAC is perfect, and, due to imprecision of the tests, HAC can be a difficult diagnosis to make at times. Clinicians are faced with the situation where their clinical impression is that a patient has HAC, but the tests performed do not confirm the diagnosis and no alternative diagnosis is identified. Recently, in order to explain such circumstances, interest has focused on a syndrome termed occult HAC. Dogs with occult HAC allegedly have clinical signs and/or routine laboratory abnormalities consistent with classic HAC yet normal serum cortisol concentrations on low-dose dexamethasone suppression (LDDST) and ACTH stimulation tests. "Alopecia X" has been used to describe dogs with occult HAC with dermatological changes only, mainly bilaterally symmetrical alopecia and hyperpigmentation with a puppy coat. In theory, occult HAC is caused by diversion of the normal adrenocortical pathways for cortisol and aldosterone synthesis into overproduction of sex hormones instead. The syndrome is diagnosed by performance of an ACTH stimulation test with measurement of serum sex hormones (i.e. androstenedione, estradiol, progesterone and 17-hydroxy-progesterone) and aldosterone concentrations pre- and post-ACTH.

In my opinion, conclusive evidence for the existence of occult HAC as a sex hormone-mediated condition is lacking. In evaluating adrenal secretion of sex hormone and cortisol precursors, e.g. 11-deoxycortisol, in dogs, whether basal or ACTH-stimulated concentrations were measured in any study must be taken into account. For the diagnosis of standard HAC, basal cortisol concentration is not reliable and never used by itself.[1, 2] No evidence exists that basal serum sex hormone concentrations are any more reliable for diagnosis of adrenal dysfunction; thus, the following discussion will focus on ACTH-stimulated concentrations.

ADRENAL SEX HORMONE SECRETION AS A CAUSE OF BILATERALLY SYMMETRICAL ALOPECIA

Evidence in favor

Apart from the adrenal glands, estrogen and progesterone secreted from other sources, e.g. Sertoli cell tumors, and exogenous estrogen administration can cause alopecia. A syndrome of castration-responsive alopecia is recognized.

The first report of clinical signs thought to be due to elevated adrenal-derived sex hormone concentrations described diffuse bilaterally symmetrical alopecia and hyperpigmentation in 7 Pomeranians.[3] Classic HAC was ruled out. Progesterone, 17-hydroxy-progesterone (17OHP), 11-deoxycortisol, dehydroepiandrosterone sulfate (DHEAS), testosterone, androstenedione, and estradiol were measured pre- and post-ACTH in 7 affected Pomeranians, 12 unaffected Pomeranians and 19 non-Pomeranian controls. Only ACTH-stimulated 17OHP concentrations were different between affected and unaffected Pomeranians, but ACTH-stimulated progesterone and DHEAS concentrations were significantly higher in both groups of Pomeranians compared to the controls. It was hypothesized that the alopecia was due to a partial deficiency of 21-hydroxylase, an enzyme needed for cortisol synthesis.[3] In humans with 21-hydroxylase deficiency and resultant congenital adrenal hyperplasia (CAH), cortisol is not synthesized and cortisol precursors, most notably 17OHP and androgens, accumulate.[4] Since affected Pomeranians had normal serum cortisol concentrations, the enzyme deficiency was assumed to be partial.[3] Family members of people with CAH have sex hormone elevations to a lesser magnitude and no clinical signs, thus explaining the abnormalities in the unaffected Pomeranians (many of the affected and unaffected Pomeranians in the study were related). Subsequently, 3 Alaskan Malamutes with Alopecia X were reported to have ACTH-stimulated 17OHP concentrations that were above the reference range and were significantly higher than those in 3 normal Alaskan Malamutes.[5]

Evidence against

Of 6 sex hormones assessed by Schmeitzel et al. in the 7 Pomeranians with Alopecia X, only ACTH-stimulated serum 17OHP was significantly different between affected and unaffected dogs. However, when affected males and females were assessed separately, the males did not have an elevated serum 17OHP.[3] In 276 dogs with Alopecia X including 63 Pomeranians, 73% had at least one basal or post-ACTH sex hormone concentration above the reference range. Despite the preponderance of elevations in sex hormone concentrations, no consistent sex hormone abnormalities were identified, and Frank et al. concluded that it is more appropriate to refer to Alopecia X as "alopecia associated with follicular arrest rather than equating it with an adrenal hormone imbalance".[6] In addition, the canine 21-hydroxylase gene has been cloned and evaluated for genetic polymorphisms. No mutations affecting the primary structure of the enzyme or gene expression were identified.[7]

Assessment

Although 21-hydroxylase abnormalities were not documented in dogs with Alopecia X[7], another enzyme could be involved.[4] However, to date, no search for genes outside of the cortisol synthesis pathway has been successful either.[8, 9] More importantly, sex hormone abnormalities appear to be easily documented in dogs with Alopecia X, but no correlation exists between elevations in any hormone and a clinical abnormality; sex hormones are no longer believed to be related to Alopecia X.

17OHP, OTHER SEX HORMONES AND CORTISOL PRECURSORS AS A CAUSE OF OCCULT HAC

Evidence in favor

More recently, 23 dogs all of which had clinical and routine laboratory findings suggestive of HAC were reported. Of the 23 dogs, 11 were assigned to Group 1 that had typical HAC with an elevated cortisol response to ACTH. Of 10 dogs with a normal ACTH response, 6 had a positive LDDST (Group 2A), 4 had a negative LDDST (Group 2B) and 3 had low plasma cortisol concentrations throughout an LDDST (Group 2C). Despite the variation in serum cortisol concentrations on the tests for standard HAC, all 23 dogs had elevated ACTH-stimulated 17OHP concentrations. Thus, Ristic et al. concluded that ACTH-stimulated serum 17OHP concentration is elevated in dogs with classic as well as occult HAC and measurement of serum 17OHP concentration is a marker of adrenal dysfunction.[10] Numerous other studies have also documented elevations in sex hormone concentrations in dogs with various forms of hypercortisolemia, either pituitary-dependent HAC (PDH) or due to an adrenal tumor (AT).[11-15]

More specifically to the point, in cases where cortisol and sex hormones are both elevated, which hormones are causing any of the clinical signs of HAC present is difficult to impossible to determine. However, sporadic reports exist of dogs with sex hormone-secreting AT and low serum cortisol concentrations but in which clinical signs of HAC were present, ostensibly due to the sex hormones. Two dogs with AT had clinical signs of HAC despite markedly suppressed ACTH-stimulated serum cortisol concentrations; one tumor secreted progesterone, 17OHP, testosterone and DHEAS while the other secreted androstenedione, estradiol, progesterone and 17OHP.[16] In 8 dogs with AT and signs of HAC, 3 had suppressed ACTH-stimulated serum cortisol concentration and 1 of them had elevated 17OHP concentrations; no other sex hormones were measured in any dog nor in the other 2 with subnormal cortisol concentrations.[17]

Evidence against

It is difficult to understand how sex hormones would cause clinical signs of HAC. The sex hormone most mentioned as a cause of occult HAC is progesterone. Due to progesterone's short half-life, however, little is known about the effects of elevated serum concentrations. Chronic excesses in progesterone concentration are not unique. In estrus, diestrus, and pregnancy, serum progesterone is elevated for 60-90 days with concentrations higher than those seen in spayed bitched with HAC, yet no clinical signs of HAC develop.[18] In humans, clinically silent 17OHP elevations occur.[4, 19-21] Similarly, in dogs with Alopecia X, serum 17OHP concentrations can be quite elevated, similar to what is seen with dogs with purported occult HAC, yet none of the classical systemic clinical signs such as polyuria/polydipsia, polyphagia, pot belly and panting are reported.

Two mechanisms have been proposed for progesterone's ability to cause signs of glucocorticoid excess. Synthetic progestins, compounds with progesterone-like actions, may either bind glucocorticoid receptors[22] or may displace cortisol from its binding protein, thereby elevating serum free cortisol concentrations.[23] Indeed, progestins suppress endogenous ACTH secretion and cause adrenal atrophy, an action suggestive of glucocorticoid activity.[24-26] Accordingly, progesterone may do the same. Examination of Pomeranians with Alopecia X, however, refutes the likelihood of either mechanism occurring. If elevated serum 17OHP concentration as seen in those dogs is sufficient to cause clinical disease due to glucocorticoid actions of 17OHP, endogenous ACTH concentration should be suppressed due to negative feedback effects of glucocorticoids on the pituitary. To the contrary, Pomeranians with elevated serum 17OHP concentrations had higher plasma ACTH concentrations than healthy dogs.[3] Similarly, during diestrus when serum progesterone concentrations are highest, adrenal secretion of cortisol in response to ACTH is greatest.[27] How AT could have a shift in hormone synthesis activity can be understood easily. Tumor cells are not normal and can undergo various forms of dedifferentiation, losing the ability to synthesize enzymes, e.g. the ones in the hormone synthesis pathways. In cases of pituitary-dependent occult HAC, how or why normal adrenocortical tissue should have altered steroid synthesis is unknown.

The number of cases cited in the literature of dogs with purported true occult HAC, i.e. presence of consistent clinical signs, ACTH stimulation test and LDDST both normal, and response to appropriate therapy, is actually quite small. Problems exist with the initial study that attributed occult HAC to elevated 17OHP concentration.[10] Classifying all 23 dogs as having occult HAC was inappropriate as 17 had a standard stimulation test or LDDST consistent with HAC and were not occult. Three dogs had a normal ACTH stimulation test and low plasma cortisol throughout an LDDST, results not unusual in dogs with AT. Only 3 dogs were diagnosed with PDH despite having both a normal ACTH stimulation and LDDST.[10] In 64 dogs documented to have HAC, no dog was negative on both the ACTH stimulation test and LDDST.[28] Out of 57 dogs evaluated recently for HAC with cortisol measurements on an ACTH stimulation test and LDDST, 40 were diagnosed as having PDH, 12 as having AT and 5 as possibly having occult HAC. The diagnosis of occult HAC was confirmed as best as possible with response to therapy in only 1 dog in the latter group, i.e. only 1 out of 57 dogs possibly had occult HAC.[15]

Assessment

Sex hormones have been elevated in dogs with either PDH or AT. However, in most cases, whether cortisol or the sex hormones are causing the clinical signs is impossible to distinguish. Sex hormone elevations, however, have been documented to cause clinical signs of HAC even in cases where cortisol concentrations are suppressed. On the other hand, in humans and intact female dogs, sex hormone elevations cause signs associated with the reproductive function of the hormone and not of occult HAC. A mechanism by which sex hormones could cause the signs of occult HAC or by which adrenal glands would shift their hormone production in PDH is lacking. Occult HAC, if it does exist, has only truly been possibly documented in a handful of cases.

SEX HORMONE PANEL TESTING

Evidence in favor

Measurement of a panel of sex hormones has been stated to increase sensitivity and specificity of the test over measurement of a single hormone alone.[29] Elevations in concentrations of any hormone can be common, with estradiol elevations noted in approximately 40% of panels submitted to one reference laboratory.[29]

Evidence against

Unfortunately, sensitivity and specificity of adrenal sex hormone panel testing has not been published yet in a peer-reviewed journal. Neither have elevations in sex hormone concentrations been evaluated in the context of occult HAC as was done for Alopecia X; although sex hormones at one point were believed to cause Alopecia X, retrospective analysis of sex hormone panel results identified the abnormalities as coincidental and not causative.[6]

The specificity of sex hormone panel testing must be considered. Dogs with non-adrenal illness (NAI), e.g. a dog with diabetes mellitus, do not have the same ACTH response as healthy dogs because of adaptation of adrenocortical function to the stresses of chronic illness. Dogs with chronic NAI (but not HAC) had a 14%[30] or 36%[31] chance of having false positive ACTH stimulation test results consistent with HAC. In one study, post-ACTH serum cortisol and 17OHP concentrations were significantly correlated both in dogs with neoplasia and those suspected of having HAC, suggesting that as adrenal function is increased either by adrenal disease or non-specifically by NAI, production of all hormones increases proportionately.[13]

With regard to 17OHP, the specificity of the test may be as low as 70%, i.e. the chance of a false positive result is 30%.[13, 32] In 35 dogs with neoplasia who did not have adrenal disease, 30% had an elevated serum 17OHP concentration post-ACTH stimulation.[13] When dogs suspected to have HAC but proven not to were compared to those with HAC, cortisol distinguished the 2 groups more clearly than did 17OHP.[12, 32] In 30% of dogs suspected to have HAC but for which an alternate diagnosis was found, serum ACTH-stimulated 17OHP concentrations were elevated.[32] In 6 dogs with either pheochromocytoma or a non-functional AT, androstenedione, progesterone, 17OHP, testosterone and/or estradiol were elevated in all.[14] Therefore, dogs without adrenal disease clearly can have elevated sex hormone concentrations as they do cortisol concentrations, and sex hormones may be more likely to be falsely elevated by NAI compared to cortisol. Unfortunately, the ability of chronic NAI to affect sex hormone testing has not received critical appraisal as has the standard ACTH stimulation test.

Assessment

Numerous dogs have been documented to have elevated sex hormone concentrations with signs of occult HAC, however, the association between the elevations and the clinical signs has not undergone rigorous assessment. The specificity of adrenal sex hormone panel testing, similarly, has not been evaluated. Elevations at least in serum ACTH-stimulated 17OHP concentrations due to NAI appear to occur more often than with cortisol.

RESPONSE TO TREATMENT

Evidence in favor

In dogs with either Alopecia X or purported occult HAC, treatment with agents that affect pituitary or adrenal function has resulted in resolution of clinical signs. Melatonin is a neurohormone produced by the pineal gland that controls seasonal reproductive and hair growth cycles and that alters sex hormone concentrations in intact dogs[33]; it was administered initially in 29 dogs with Alopecia X. Of the 29 dogs, 15 had partial hair regrowth.[34] In 3 Alaskan Malamutes with Alopecia X, treatment with trilostane resulted in complete hair regrowth within 6 months.[5] Of 16 Pomeranians and 8 miniature poodles with Alopecia X, 14 Pomeranians and all poodles had hair regrowth in response to trilostane.[35] In Ristic et al.'s study on occult HAC, 9 dogs in groups 2A, B or C were treated with trilostane or mitotane, and all improved clinically. Decreased ACTH-stimulated cortisol and/or 17OHP concentrations were documented in 4 of the 9.[10] Lastly, in 1 dog with clinical signs of HAC and normal post-ACTH stimulated cortisol and LDDST results but an elevated ACTH-stimulated 17OHP concentration, clinical signs resolved with mitotane therapy.[15]

Evidence against

The response to mitotane, melatonin or trilostane has not been uniform or predictable. In 15 Pomeranians with Alopecia X treated with melatonin, only 6 had mild to moderate hair regrowth.[36] In the study evaluating 29 dogs diagnosed with Alopecia X that were treated with melatonin or mitotane, partial or complete hair regrowth was seen in only 62% overall. More importantly, serum sex hormone concentrations did not change significantly in response to treatment nor correlate with whether response was seen. Of the dogs with partial or complete hair regrowth, androstenedione, progesterone and 17OHP were still elevated in 21%, 64% and 36%, respectively.[34] In 16 Pomeranians and 8 miniature poodles with Alopecia X that responded to trilostane therapy, 17OHP concentrations were significantly elevated by therapy.[35] Similarly, 2 dogs with occult HAC treated with trilostane had clinical signs resolve despite 17OHP concentrations being higher with therapy.[10] Thus, hair coat and other clinical signs improve despite further increases in concentrations of the sex hormones purportedly underlying the clinical signs.

Assessment

Although successful therapy has been reported, 2 problems exist. First, not all dogs respond to melatonin, mitotane or trilostane. Secondly, response has no apparent correlation to sex hormone concentrations. Clinical signs can improve or resolve in dogs in which sex hormone concentrations do not improve or even increase.

CONCLUSION

Whether occult HAC is due to adrenal secretion of sex hormones has never been conclusively proven. In the literature, both human and veterinary, evidence exists both in favor and against the theory. Using the research into Alopecia X as an analogy for occult HAC, although occult HAC was originally thought to be due to sex hormone abnormalities and elevations in sex hormone concentrations were widely documented in dogs with Alopecia X, later research was unable to detect a correlation between elevations in any hormone and a clinical abnormality. The specificity of adrenal sex hormone panel testing needs to be closely evaluated as evidence suggests that sex hormone concentrations may be easily elevated non-specifically due to NAI. Furthermore, not all dogs diagnosed with occult HAC respond to therapy directed at minimizing adrenal secretion. Sex hormones may be elevated even further by therapy, yet dogs improve clinically. The possibility remains that "occult HAC" may exist as a syndrome, but one that is not caused by sex hormone secretion. Given the response of Alopecia X in some cases to therapy directed at hormone secretion, it is possible that local factors such as enzymes, growth factors or hormone receptors may contribute to the hair cycle abnormalities and be acted upon by substances secreted by the adrenals to manifest the clinical signs. The same could be true of occult HAC. Much work remains to be done to understand both the adrenal and local tissue contribution to occult HAC.

REFERENCES

[1]Behrend EN, Kemppainen RJ. **Vet Clin North Am Sm Anim Pract** 2001;31:985-1003; Feldman EC, Nelson RW, eds. Canine and Feline Endocrinology and Reproduction. 3rd ed. St. Louis: Saunders, 2004:252-357; [3]Schmeitzel LP, Lothrop CD. **J Am Vet Med Assoc** 1990;197:1333-1341; [4] Stewart PM. The adrenal cortex. In: Kronenberg HM, Melmed S, Polonsky KS, Larsen PR, eds. **William's Textbook of Endocrinology**. 11th ed. Philadelphia: Saunders Elsevier, 2008:445-537; [5]Leone F et al. The use of trilostane for the treatment of Alopecia X in Alaskan malamutes. **J Am Anim Hosp Assoc** 2005;41:336-342; [6]Frank LA et al. Retrospective evaluation of sex hormones and steroid hormone intermediates in dogs with alopecia. **Vet Derm** 2003;14:91-97; [7]Takada K et al. **Res Vet Sci** 2002;73:159-163; [8]Mausberg E-M, et al. **Animal Biotech** 2007;18:291-296; [9]Mausberg E-M, et al. **Vet Rec** 2008;163:121-123; [10]Ristic JME, et al. **J Vet Int Med** 2002;16:433-439; [11]Frank LA, et al. **J Am Vet Med Assoc** 2001;218:214-216; [12]Sieber-Ruckstuhl N, et al. **Vet Rec** 2008 May 24;162:673-678; [13]Behrend EN, et al. **J Am Vet Med Assoc** 2005;227:1762-1767; [14]Hill KE, et al. **J Am Vet Med Assoc** 2005;226:556-561;[15]Benitah N, et al. **J Am Vet Med Assoc** 2005;227:1095-1101; [16]Syme HM, et al. **J Am Vet Med Assoc** 2001;219:1725-1728; [17]Norman EJ, et al. **Vet Rec** 1999;144:551-554; [18]Bromel C, et al. **JAVMA** 2010;236:1208-1214; [19]Turton DB, et al. **J Endocrinol Invest** 1992;15:789-796; [20]Bondanelli M, et al. **Metabolism** 1997;46:107-113; [21]Grumbach MM, Conte FA. Disorders of sex differentiation. In: Wilson JD, Foster DW, Kronenberg HM, Larsen PR, eds. Williams Textbook of Endocrinology. 9th ed. Philadelphia: W.B. Saunders Company, 1998:1303-1425; [22]Selman PJ, et al. **J Repro Fert-Supp** 1997;51:345-354; [23]Juchem M, Pollow K. **Am J Obstet Gynecol** 1990;163:2171-2183; [24]Selman PJ, et al. **Eur J Endocrinol** 1994;131:422-430; [25]Court EA, et al. **Aust Vet J** 1998;76:555-560; [26]van den Broek AHM, et al. **J Small Anim Pract** 1994;35:285-288; [27]Reimers TJ, et al. Biol Repro 1984;31:148-154; [28]Feldman EC.. **J Am Vet Med Assoc** 1983;182:506-510; [29]Oliver J. **Proc 25th Ann Vet Med Forum** 2007;471-473; [30]Kaplan AJ, et al. **J Am Vet Med Assoc** 1995;207:445-451; [31]Chastain CB, et al. **J Am Anim Hosp Assoc** 1986;22:435-442; [32]Chapman PS, et al. **Vet Rec** 2003;153:771-775; [33]Ashley PF, et al. **J Am Vet Med Assoc** 1999;215:1111-1115; [34]Frank LA, et al. **Vet Derm** 2004;15:278-284; [35]Cerundolo R, et al. **Vet Derm** 2004;15:285-293;[36]Frank LA, et al. **Vet Derm** 2006;17:252-258.

FELINE ACROMEGALY: UPDATE ON DIAGNOSIS & TREATMENT

Katharine F. Lunn, BVMS, MS, PhD, MRCVS, DACVIM
Fort Collins, CO

INTRODUCTION

Feline acromegaly is a disease of older cats resulting from chronic excessive growth hormone (GH) secretion, usually from a functional adenoma of the pars distalis of the pituitary gland.[1] The typical presentation of acromegaly is marked insulin resistance in a diabetic patient.[1-4] Affected cats may also have organ enlargement, increased body size and body weight, hypertrophic cardiomyopathy, and neurological disease. Although the short-term prognosis is fair, cats with acromegaly generally die or are euthanized due to renal or cardiac disease, or the complications of uncontrollable diabetes mellitus. Insulin resistance in cats with acromegaly has been attributed to a GH-induced postreceptor defect in the action of insulin on target cells. The physical changes in these patients are due to the anabolic effects of GH, which are mediated by peripherally synthesized insulin-like growth factor-1 (IGF-1).[1]

Acromegaly is diagnosed by measurement of serum GH and IGF-1 levels,[1,5-7] in conjunction with the demonstration of a pituitary tumor by brain imaging techniques. Feline acromegaly has historically been regarded as a rare condition, however recent findings suggest that it may be significantly under diagnosed. In a study of 184 diabetic cats, 59 had markedly increased IGF-1 concentrations, and acromegaly was confirmed in 17 of 18 cats that were examined by computed tomography (CT), magnetic resonance imaging (MRI), or necropsy.[7] These findings have prompted the suggestion that any cat with clinical features of acromegaly, including insulin resistance, should be screened for this disorder.[8]

Acromegaly in humans is very similar to the disease in cats. In human patients, the most frequent cause is also a pituitary adenoma, and the disorder is diagnosed by measuring both serum GH and IGF-1 levels. However, there are currently several treatment options for acromegaly in humans that have not been widely explored in cats. This article will discuss the diagnosis of acromegaly in cats, and review treatment options for this disease in both humans and cats.

DIAGNOSIS

The diagnosis of feline acromegaly is based on typical signalment, history, and clinical signs, GH and/or IGF-1 testing, and brain imaging.

Signalment, History, and Clinical Signs

Acromegaly has been suggested to be more common in male cats, with most affected cats being middle-aged or older.[2] There is no apparent breed predilection. The most usual history is of insulin-resistant diabetes mellitus, with affected cats requiring 10-20 units of insulin per dose, or more, often with inadequate control of the diabetes. Affected cats remain polyuric, polydipsic, and polyphagic, and continue to gain weight. In the opinion of this author, the finding of weight gain in the face of unregulated insulin-resistant diabetes mellitus is highly suggestive of feline acromegaly. The physical changes of acromegaly develop slowly and are often not noted by the owner of the cat until they are advanced. In some cases, comparing the appearance of the cat to photographs taken when younger can reveal changes in the physical appearance. These changes may include enlarged feet, broadening of the face, protrusion of the mandible, increased spacing between the teeth, and abdominal enlargement.[2-4,6,7] Owners of affected cats also frequently note noisy or stertorous breathing, or respiratory stridor.[4-7] Physical examination may reveal additional abnormalities such as enlarged abdominal organs and cardiac murmurs, arrhythmias, or a gallop rhythm.[2,4,7]

Neurological signs associated with the pituitary tumor appear to be generally uncommon, but may be under-recognized or under-reported. Lethargy, mental dullness, or impaired vision may occur, but can often be subtle. Affected cats may also demonstrate signs of diabetic neuropathy or lameness, possibly due to acromegaly-associated arthropathy.[2,7] Additional complications of acromegaly may include cardiac disease and protein-losing nephropathy.[2]

Growth Hormone Measurement

Acromegaly is the result of excessive GH secretion from a pituitary gland tumor, and elevated GH levels have been reported in several cats with acromegaly.[2,5,7,9-16] Overlap between GH values in diabetic cats with and without acromegaly has also been noted.[14] An ovine growth hormone assay has been validated for the diagnosis of feline acromegaly, and is available in Europe.[11] This assay clearly differentiated between normal cats and cats with acromegaly. Unfortunately, at the time of writing, a feline GH assay was not reliably available in the United States. However, this author has found this test to be diagnostically useful, when available.

Insulin-Like Growth Factor-1 Concentration

The anabolic effects of excessive GH are mediated through the actions of IGF-1. This hormone is predominantly produced in the liver, and levels of IGF-1 increase in the presence of chronic increases in GH production. Because GH secretion may be pulsatile, even in some acromegalics, and because it has a short half-life, increased levels of IGF-1 may in fact be a more sensitive test for acromegaly, because they may reflect GH levels over the preceding 24 hours.[1,8] In humans, serum IGF-1 levels are used in conjunction with GH values in the diagnosis of acromegaly.[17] Serum IGF-1 is also widely used and reported in feline patient.[1,5-7,9,10,18-23] In cats, one paper suggested that IGF-1 concentrations may not be reliable in the diagnosis of acromegaly in patients with diabetes mellitus,[24] and another study noted that cats receiving long-term insulin therapy for diabetes mellitus may have elevated levels of circulating IGF-1.[25] However, a more recent report evaluated IGF-1 concentrations cats with diabetes mellitus and acromegaly, and demonstrated that IGF-1 measurement is a useful screening test for feline acromegaly.[6] Berg and colleagues noted that the sensitivity and specificity of IGF-1 concentrations were 84% and 92% respectively, for the diagnosis of acromegaly. These authors found no difference in serum IGF-1 concentrations among well-controlled diabetic cats, poorly controlled diabetic cats, and healthy cats. Serum IGF-1 was in the reference range in all well controlled diabetic cats and in 35 of 40 poorly controlled diabetic cats. The highest IGF-1 noted in a diabetic cat was 153 nM, with a normal reference range of 12-92 nM. The serum IGF-1 was within the reference range in 3 of 19 acromegalic cats evaluated, with the lowest value being 47 nM.[6] Thus there is some overlap between the IGF-1 values found in acromegalic cats and those found in poorly-regulated diabetic cats. However it is this author's experience that the great majority of cats with acromegaly have an IGF-1 value that is at least twice the value of the high end of the reference range. Feline IGF-1 levels are currently readily available to veterinarians in the United States, and this test should be considered in cats with diabetes mellitus that appear to be insulin resistant.[8]

Brain Imaging

The presence of a pituitary tumor can be demonstrated by either CT or MRI in cats with acromegaly, and both have been reported in the literature.[2,6,7,9,10,12,13,18-21,23,26] One study suggested that contrast-enhanced CT will detect most pituitary masses, but some may only be detected by MRI.[7] It is also possible that both imaging modalities may reveal a normal pituitary in a cat with acromegaly, if the size of the mass is below the limit of detection of the imaging modality.

TREATMENT

The goals for treatment of acromegaly in humans include complete tumor removal, relief of symptoms, reduction of systemic complications of the disease, and control of the local mass effect from the tumor.[27] Similar goals likely exist for many feline patients, however complete tumor removal is less achievable in most cases. Quality of life is of considerable importance for feline patients and their owners, with improvement in diabetic regulation being particularly desirable.

Treatment options in human medicine include surgery, conventional external-beam radiotherapy, stereotactic radiation therapy, and medical therapy.[17,28,29] Some of these therapies are also used in feline patients, but not all are available, and many have not been adequately evaluated in cats.

Management strategies for acromegaly in human patients rely on a team approach and may involve the use of more than one treatment approach in an individual patient. In general, surgery is usually a first-line treatment, unless a surgical cure is unlikely. If surgery fails or is not performed, medical therapy is considered the next line of approach. If the disease remains uncontrolled, alternate medical therapies, radiation therapy, or repeated surgery are further options.[29] This type of multimodal approach has not been explored in feline medicine.

Surgery

In humans, transphenoidal surgery to remove the pituitary tumor is generally regarded as the treatment of choice.[27] However, this is not curative in all cases, some patients are not good candidates for surgery, and some patients refuse this option. Surgery is technically very challenging, and success is highly dependent upon the skill and experience of the surgeon, with a successful outcome more likely in surgeons performing at least 50 pituitary surgeries per year.[29] There are only rare reports of the use of surgery for treatment of feline acromegaly.[9,10] Thus it is unclear that there are currently any veterinary surgeons that have the suggested experience level that is required in human medicine. Cryohypophysectomy has also been reported in a very small number of cats,[18,26,30] but again this is not widely performed, and thus there is little data with which to evaluate this technique. Surgical adjuncts such as intraoperative MRI or endoscopy are increasingly used in human medicine,[28] and thus could potentially be adopted in the veterinary field.

Radiation Therapy

This is rarely used as a first-line therapy in human medicine, but is considered as second-line or third-line treatment.[29] In contrast, radiation therapy is the treatment for feline acromegaly that is most widely reported in the veterinary literature.[4] There are several reports of the use of conventional fractionated radiotherapy for the treatment of acromegaly in cats.[2,12,19,23,31-33] Conventional fractionated radiation therapy is administered in multiple fractions typically spread over a period of weeks. The protocols reported range from 5 fractions given weekly[23,32] to as many as 20 fractions given over a period of 4 weeks.[31] A recent study showed that conventional radiation therapy improved diabetic control in 13 of 14 cats, although IGF-1 concentrations did not correlate with this improved control.[19] Improved control of diabetes mellitus has also been noted in other studies.[2,12,23,31-33]

In human medicine, radiation therapy can take as long as 10-20 years to be fully effective.[28] Side-effects of radiation include hypopituitarism, which can also take up to 10 years to develop.[28] Other adverse effects include visual defects, cerebral radionecrosis, cerebrovascular damage, the potential for secondary tumors, and cognitive defects.[17,28,29] There are far fewer reports of long-term follow-up of acromegalic cats receiving radiation therapy, thus it is difficult to assess the risk of complications of this modality in this species. The shorter life-span of cats suggests that hypopituitarism may be less of a concern in these patients, but the shorter lifespan may also mean that some cats do not live long enough to see the benefits of radiation therapy. Published case reports and case series of cats receiving conventional radiation therapy suggest that short and long-term adverse effects of this therapy are relatively uncommon. At least 4 case series reported no acute adverse effects of irradiation of pituitary tumors in cats.[19,31-33]

Disadvantages of radiation therapy as a treatment for feline acromegaly include cost, availability, and the necessity for repeated anesthetic events. The latter disadvantage can be lessened by the use of stereotactic radiation therapy. Stereotactic radiation therapy (also sometimes called stereotactic radiosurgery) is a technique in which large doses of radiation are delivered in a highly conformal fashion to a defined target, using either gamma knife or linear accelerator technology. Stereotactic radiation therapy is widely used in the treatment of pituitary tumors in human patients.[34] There is one case series in veterinary medicine in which cats with pituitary tumors received treatment with a linear-accelerator-based modified radiosurgical approach.[22] Cats received a single large dose of radiation, but it was delivered in a non-conformal fashion. The technique was reported to be safe and effective. At Colorado State University (CSU), stereotactic radiation therapy is routinely used in the treatment of feline acromegaly, with very promising results.[35] Cats typically receive 2-4 fractions of radiation administered over a period of up to 5 days. This offers considerable advantages in terms of owner time-commitment, and the risks of anesthesia and hospitalization in elderly diabetic cats.

Medical Therapy

Medical therapy for acromegaly is commonly used in humans, either as a first-line treatment, or as an adjunct to surgery or radiation therapy. The classes of drugs used are somatostatin analogs, GH receptor antagonists, and dopamine agonists.[17,27-29]

Somatostatin analogs, also termed somatostatin receptor ligands (SRLs), are widely used in the treatment of acromegaly in human patients.[27] These agents bind to somatostatin receptors, suppressing the release of GH from the pituitary gland. The medications are available as long or short-acting preparations, and response to SRLs is assessed by measurement of IGF-1 and GH levels and tumor size, and evaluation of clinical signs.[17] Octreotide has been evaluated in a small number of cats with acromegaly. In 5 cats, short-acting octreotide was used for up to 4 weeks, with no apparent improvement in GH levels.[2,15] However, a more recent study showed that GH levels were significantly decreased for up to 120 minutes post-injection in 5 cats with acromegaly that received a single dose of octreotide.[13] Possible explanations for the discrepancies between these studies include differences in the doses and routes of administration of the octreotide, and differences in the timing of follow-up blood sampling. It is also important to note that these studies used the short-acting form of octreotide and were performed over a very short time-period without assessment of clinical response. A study is currently underway at CSU to evaluate the use of a long-acting octreotide preparation for the treatment of feline acromegaly.

Pegvisomant is a GH receptor antagonist that is used to treat acromegaly in human patients.[27] Its mechanism of action is to block the effects of excessive GH by interfering with receptor binding. The use of this drug has not been reported in cats. A significant disadvantage of both the SRLs and GH receptor antagonists is that both are extremely expensive, and the drug formulations are difficult to adapt to use in feline patients.

Dopamine agonists are not widely used in the management of human acromegaly.[29] There is one report of the unsuccessful use of L-deprenyl in a cat with acromegaly.[21]

Insulin Therapy

For many cats with acromegaly, insulin therapy is the only treatment that is available or acceptable to the owner. In general these patients should receive the amount of insulin that is necessary to control their diabetes, although adequate blood glucose regulation can be difficult to achieve in many cases. The use of home blood glucose monitoring, with close cooperation between the owner and veterinarian, are strongly recommended. The feeding of a low-carbohydrate diet may also be beneficial. It should be expected that these patients will receive insulin doses in the range of 10-20 units per dose, or more. It has been suggested that insulin doses should not exceed 12-15 units per dose, as some cats may experience sudden increases in insulin sensitivity.[1] The frequency with which this occurs is not known, and this author has not observed this phenomenon. Concurrent illnesses and complications of acromegaly and diabetes mellitus should also be addressed.

PROGNOSIS

The short-term prognosis for cats diagnosed with acromegaly is generally fair to good, but the long-term prognosis is poor. Patients may succumb to cardiac or renal failure, neurological disease, or complications of poorly regulated diabetes mellitus.[1] The median survival was reported to be 20.5 months in one case series.[2]

REFERENCES

1. Feldman EC, et al. Canine and Feline Endocrinology and Reproduction 3rd ed. St. Louis: Saunders; 2004. 2. Peterson ME, et al. J Vet Intern Med 1990;4:192. 3. Hurty CA, et al. J Am Anim Hosp Assoc 2005;41:292. 4. Niessen SJ. J Feline Med Surg 2010;12:15. 5. Norman EJ, et al. J Feline Med Surg 2000;2:183. 6. Berg RI, et al. J Vet Intern Med 2007;21:892. 7. Niessen SJ, et al. J Vet Intern Med 2007;21:899. 8. Peterson ME. J Vet Intern Med 2007;21:889. 9. Meij BP, et al. J Feline Med Surg 2010;12:406. 10. Meij BP, et al. J Comp Pathol 2004;130:209. 11. Niessen SJ, et al. Vet Rec 2007;160:902. 12. Goossens MM, et al. J Am Vet Med Assoc 1998;213:374. 13. Slingerland LI, et al. Domest Anim Endocrinol 2008;35:352. 14. Reusch CE, et al. Vet Rec 2006;158:195. 15. Morrison SA, et al. J Am Vet Med Assoc 1989;194:91. 16. Lichtensteiger CA, et al. Vet Pathol 1986;23:518. 17. Melmed S. N Engl J Med 2006;355:2558. 18. Blois SL, et al. J Small Anim Pract 2008;49:596. 19. Dunning MD, et al. J Vet Intern Med 2009;23:243. 20. Elliott DA, et al. J Am Vet Med Assoc 2000;216:1765. 21. Abraham LA, et al. Aust Vet J 2002;80:479. 22. Sellon RK, et al. J Vet Intern Med 2009;23:1038. 23. Littler RM, et al. J Small Anim Pract 2006;47:392. 24. Lewitt MS, et al. J Endocrinol 2000;166:21. 25. Starkey SR, et al. J Feline Med Surg 2004;6:149. 26. Abrams-Ogg ACG, et al. Can Vet J 1993;34:682. 27. Manjila S, et al. Neurosurg focus 2010;29:E14. 28. Melmed S. J Clin Invest 2009;119:3189. 29. Melmed S, et al. J Clin Endocrinol Metab 2009;94:1509. 30. Abrams-Ogg A, et al. Can Vet J 2002;43:53. 31. Mayer MN, et al. J Vet Intern Med 2006;20:1151. 32. Brearley MJ, et al. Vet Comp Onc 2006;4:209. 33. Kaser-Hotz B, et al. J Small Anim Pract 2002;43:303. 34. Laws ER, et al. J Neurooncol 2004;69:257. 35. Lunn KF, et al. J Vet Intern Med 2009;23:698.

FELINE INSULIN THERAPY: PAST, PRESENT, & FUTURE

Patty Lathan, VMD, MS, DACVIM
Starkville, MS

INTRODUCTION

When Frederick Banting and Charles Best administered insulin to the first human patients in 1922, treating pet cats probably didn't cross their minds. But over the past ninety years, various types of insulin have been used to treat feline diabetes. Recommendations have changed frequently due to experience, research, and changes in insulin availability. Now, most endocrinologists concur that viable insulin choices for cats include PZI, porcine lente, and glargine insulins. However, there is much debate regarding whether the use of one type of insulin is more likely to result in diabetic remission than is the use of another.

PATHOPHYSIOLOGY

Conceptually, "diabetic remission" describes a state in which a previously insulin-dependent diabetic cat no longer requires insulin therapy to maintain euglycemia. The reason that diabetic remission occurs frequently in diabetic cats, but rarely in diabetic dogs, is that most cases of feline DM are believed to be analogous to human type 2 DM, whereas diabetic dogs are more analogous to humans with type I DM. As in people, there is probably a genetic predisposition to the development of dysfunctional beta cells and insulin resistance in cats. Multiple concurrent and inter-related processes then contribute to the eventual development of clinical DM in cats, including insulin resistance, amyloid deposition, and glucose toxicity.

Many predisposing factors, including obesity, chronic pancreatitis and systemic diseases such as hyperthyroidism, can contribute to insulin resistance in cats. As a result of insulin resistance, more insulin must be secreted by pancreatic beta cells in order to maintain blood glucose levels within the normal range. Islet-amyloid polypeptide (IPP), or amylin, is co-secreted from the pancreas along with insulin. As beta cell insulin secretion increases in a cat with insulin resistance, so does secretion of amylin. Although amylin acts as a hormone with glucoregulatory effects that initially complement the actions of insulin, chronically increased amylin secretion results in deposition of amylin within pancreatic islets as amyloid. Amyloid is cytotoxic to beta cells, and results in defective insulin secretion. Thus, insulin resistance leads to increased amyloid deposition, which in turn contributes to defective insulin secretion. As amyloid deposition and insulin resistance progress, beta cell dysfunction eventually results in decreased insulin secretion. Insulin release becomes insufficient to keep up with carbohydrate metabolism, and hyperglycemia results. Chronic hyperglycemia further impairs insulin secretion from beta cells, resulting in "glucose toxicity." "Glucose toxicity" refers to potentially reversible degeneration of beta cells secondary to chronic hyperglycemia. Glucose toxicity and amyloid deposition damage the already dysfunctional beta cells, leading to severely impaired insulin secretion, progressive hyperglycemia, and clinical signs of diabetes, including polyuria, polydipsia, polyphagia, and weight loss.

In cats, if the processes leading to the development of DM are stopped (or reversed) soon enough, clinical signs may later be controlled without the administration of exogenous insulin. Control of DM without insulin requires the survival of enough functional beta cells to secrete insulin and maintain euglycemia. Whereas amyloid deposition and resultant beta cell dysfunction will remain, reversal of insulin resistance and glucose toxicity will result in greater insulin sensitivity and improved beta cell function. Resolution of insulin resistance is dependent upon treatment of underlying causes such as obesity, pancreatitis, or urinary tract infection. Resolution of glucose toxicity typically requires treatment with insulin. In order to revert to a non-insulin dependent state, timing is crucial. Glucose toxicity must be reversed before a critical number of beta cells are destroyed since, without enough beta cells, sufficient endogenous insulin secretion will not be achieved. Since there is no clinically feasible way to determine how much beta cell function is remaining in a given diabetic cat, clinicians should strive to treat with insulin and eliminate hyperglycemia as rapidly and completely as possible in all cats with DM.

INSULIN STUDIES

Large, controlled studies comparing remission rates in cats given different insulins are lacking, as is common in veterinary medicine. Most studies evaluating insulin efficacy examine only one type of insulin at a time. Since other variables, including diet, method and frequency of glucose monitoring, and duration of DM prior to patient inclusion in study, differ between studies, it is difficult to directly compare insulin effectiveness.

For example, one large study (55 cats) evaluated the remission rates achieved by intensive blood glucose control with home monitoring in cats treated with glargine.[1] Impressive remission rates of 84% and 35% were achieved in diabetic cats started on the protocol within 180 days and over 180 days following diagnosis, respectively. Inclusion in this study required considerable owner (and cat!) cooperation—blood glucose was checked at least three times daily, and cats were maintained on a high protein/low carbohydrate diet. Another study evaluated the efficacy of porcine insulin zinc suspension in 25 cats and reported a 28% remission rate.[2] Diet was not consistent in this study, and cats were monitored at four week intervals following initial assessments.
In a study designed to compare the efficacy of a low carbohydrate diet with a high fiber diet, most cats were treated with PZI insulin and evaluated every four weeks. These cats achieved a 54% remission rate.[3]

Perusal of this data without analysis would lead the reader to believe that glargine yields a significantly higher remission rate than porcine insulin and PZI in cats. However, it must be emphasized that the highest remission rates were also achieved with the most intensive monitoring and dietary management. Thus, the relative significance of insulin choice is likely much less than it appears. It should also be noted that there are several additional studies that showed much lower remission rates with each of the insulin types mentioned above.[4,5,6] Although more lenient dietary and monitoring guidelines are part of the explanation, the exact reasons are unclear.

One small, nonrandomized, prospective study compared remission rates in cats receiving glargine, PZI, and porcine lente insulin. 24 newly-diagnosed diabetic cats were divided into three groups of eight cats each. Insulin was administered twice daily by owners, and cats were re-evaluated weekly for one month, and then bi-weekly for the rest of the 112 day study. By the end of the study, all eight cats in the glargine group achieved remission, while only 2/8 and 3/8 cats in the lente and PZI groups had done so, respectively.[7]

CONCLUSION

Based on our understanding of the pathophysiology of DM in cats, tight glucose regulation is most likely to lead to diabetic remission by allowing reversal of glucose toxicity and halting further amyloid deposition. High remission rates have been demonstrated in patients treated with glargine insulin and PZI, and the use of an intensive monitoring protocol and low carbohydrate/high protein diet appear to increase remission rates. A large, randomized, controlled study is necessary to determine whether one insulin type yields a remission advantage over another. At this time, glargine insulin and PZI are both recommended for use as first-line insulins in cats.

REFERENCES

[1]Roomp K, et al. **JFMS** 2009; 11:668. [2]Martin GJ, et al. **Vet Rec** 2007; 161(3):88. [3]Bennett N, et al. **JFMS** 2006; 8:73. [4]Nelson RW, et al. **JAVMA** 2001; 218(1):38. [5]Michiels L, et al. **JFMS** 2008; 10(5):439. [6]Hall TD, et al. **JFMS** 2009; 11:125. [7]Marshall RD, et al. **JFMS** 2009; 11:683.

ADIPOSE TISSUE: THE LARGEST ENDOCRINE ORGAN

Angela L. Lusby, DVM, Ph.D, DACVN
Knoxville, TN

INTRODUCTION

In the past fifteen years, immense research efforts in human medicine have focused on obesity as a disease and the role of adipose tissue in the pathology of obesity. We now know that adipose tissue not only functions as an energy storage site and thermal insulation, but also operates as an active endocrine organ. A variety of endocrine, paracrine, and autocrine signals are released from cells within adipose tissue. These signals are called adipokines. The term adipokine is generally used to mean any protein released by adipose tissue regardless of whether it is released by adipocytes or non-fat cells [1]. To date, approximately 100 adipokines have been identified as being released from fat tissue, of which at least 24 have increased circulating levels in obese humans [2, 3].

LEPTIN

In 1994 researchers discovered the first adipokine, leptin. The main physiologic role of leptin is to regulate body fat mass through appetite control and changes in energy metabolism. Leptin concentrations directly correlate to body fat mass. As body fat mass increases, more leptin is secreted from adipocytes. The arcuate nucleus (ARC) of the basomedial hypothalamus is the most recognized site for leptin action [4]. Within the hypothalamus leptin stimulates the release of anorexigenic neuropeptides and suppresses orexigenic neuropeptide production. Thus, leptin has been nicknamed a "lipostat" because it acts like a thermostat for body fat [5]. The ARC contains only about 20% of the leptin receptors in the CNS, and leptin also affects neurons throughout the brain involved with appetite regulation and reward motivation associated with food intake [6].

Although increased concentrations of leptin should reduce body fat mass, obese individuals tend to have the highest levels of this hormone. When leptin is unable to effectively control and regulate appetite and energy expenditure, this is termed "leptin resistance" [7]. The presence of leptin resistance was demonstrated in a clinical study in which leptin had to be administered at a dose 20 to 30 times higher than normal physiologic levels to induce significant appetite reduction and weight loss in obese human subjects [8]. There are several mechanisms that may lead to leptin resistance: genetic mutation, receptor down-regulation, limited tissue access, and molecular interference [9].

Although leptin's primary physiologic role is to regulate body fat storage, it has a myriad of effects on the immune, cardiovascular, and reproductive systems. In addition, leptin appears to enhance insulin signaling and improve intracellular glucose uptake [10]. It has been theorized that leptin resistance is directly related to insulin resistance since both hormones share similar cell signaling receptors. Therefore, as individuals become obese and resistant to the effects of leptin, they also become insulin resistant [11].

ADIPONECTIN

Adiponectin is the most abundantly secreted adipokine in circulation with concentrations in the μg/ml range (three orders of magnitude higher than leptin) [12]. Although adipocytes are responsible for secreting adiponectin, hormone levels become paradoxically lower with increased fat mass [13]. The reason for this relationship is unclear, but adipocyte size and insulin sensitivity may affect adiponectin production. As large, hypertrophic adipocytes are more insulin resistant and secrete less adiponectin *in vitro* [14]. In addition, hormones such as testosterone and inflammatory cytokines like TNF-α decrease adiponectin production [15, 16]. Although adiponectin affects an extensive number of cells and tissues, its most influential actions include improving insulin sensitization and reducing inflammation. Adiponectin is closely associated with insulin sensitivity, independent of body fat mass [17-19]. In a study of obese rhesus monkeys, low adiponectin levels correlated with insulin resistance and preceded the onset of diabetes mellitus [20]. Prospective and longitudinal studies in human beings also demonstrate lower adiponectin levels are closely associated with insulin resistance and future development of diabetes [21-23]. Numerous clinical and epidemiological studies associate low levels of adiponectin with chronic inflammatory states such as obesity, insulin-resistance, type 2 diabetes, hypertension, cardiovascular disease and liver disease [24, 25].

TUMOR NECROSIS FACTOR-ALPHA (TNF-A)

Adipose tissue is composed of more than just adipocytes. The stromal and vascular components of the tissue include vascular endothelial cells, fibroblasts, and inflammatory cells like macrophages. Recent studies indicate that macrophages compose a large proportion of adipose tissue in obese individuals and play a key role in the development of chronic inflammation and insulin resistance associated with obese states [26]. Tumor necrosis factor-alpha (TNF-α) is an inflammatory cytokine expressed by a variety of cells including macrophages, mast cells, neuronal cells, fibroblasts, and adipocytes. Obesity increases macrophage migration into adipose tissue, and this is likely the cause of increased TNF-α expression [27, 28]. Adipose TNF-α induces localized insulin resistance by down regulating genes responsible for insulin-mediated glucose uptake into cells [29-31]. TNF-α also impedes uptake of free fatty acids (FFA) into adipocytes and promotes lipolysis and release of FFA into circulation [32-34]. In addition to directly influencing insulin sensitivity of adipose tissue, TNF-α impacts other adipokines involved in glucose metabolism. For example, TNF-α inversely correlates with adiponectin and appears to suppress adiponectin gene expression [32, 34-36]. In contrast to adiponectin, expression of leptin and several other adipokines is increased by TNF-α [37]. TNF-α derived from adipose tissue significantly impacts glucose and lipid metabolism and is a key link between obesity and chronic inflammation [26].

OBESITY AND INFLAMMATION

Obesity is associated with an array of chronic diseases including cancer, diabetes, atherosclerosis, Alzheimer's, and degenerative joint disease. Upregulation of the systemic inflammatory response provides a critical pathophysiologic link to the wide variety of chronic diseases associated with obesity. Secondary to energy excess, adipocytes expand in size and number. Inadequate blood supply to rapidly expanding adipose tissue may cause localized hypoxia and release of inflammatory mediators. The excessive energy supply associated with obese states also increases metabolic stress on adipocytes. The endoplasmic reticulum and mitochondria within these cells must increase the secretion of substances involved in protein, lipid, and glucose metabolism and synthesize triacylglycerol droplets. This increased stress on adipocyte organelles promotes the release of kinases that can release inflammatory signals and eventually initiate apoptosis. Large adipocytes are also more prone to rupture from mechanical compression and elicit an inflammatory response [38]. Excessive fat accumulation in adipocytes also appears to increase the release of proinflammatory mediators like resistin and leptin and decrease antinflammatory adipokines like adiponectin [39].

CONCLUSIONS

With the current epidemic of obesity in humans and veterinary patients, understanding the physiologic role of adipose tissue in both lean and obese states is of great importance. Initial studies regarding adipokines in dogs and cats coincide closely with findings in humans and rodents; however, more research is needed to clearly understand the impact of obesity on the health of our patients. Despite the immense number of studies evaluating adipokines in humans and animal models, there is still much we do not understand about how alterations in adipose tissue affect health, longevity, and mental status.

REFERENCES

[1]Fain, J.N., et al., Obesity, 2010. 18(5): p. 890-896. [2]Halberg, N., et al., Endocrin Metab Clinic N Am, 2008. 37(3): p. 753 768. [3] Fain, J.N., Mediators Inflamm, 2010. 2010(513948). [4]Munzberg, H. and M.G. Myers Jr, Nat Neurosci, 2005. 8(5): p. 566-570. [5] Anukulkitch, C., et al., Domest Anim Endocrinol, 2009. 36(3): p. 138-51. [6] Dardeno, T.A., et al., Frontiers Neuroendocrinology, 2010. 31(3): p. 377-393. [7] Shimizu, H., et al., Endocrine J, 2007. 54(1): p. 17-26. [8] Heymsfield, S.B., et al., JAMA, 1999. 282(16): p. 1568-1575. [9]Martin, S.S., et al. J Am Colleg Cardiol, 2008. 52(15): p. 1201-1210. [10] Park, S., et al.,. Endocrinology, 2007: p. en.2007-0754. [11]Martin, S.S., et al. J Am Col Cardiol, 2008. 52(15): p. 1201-1210. [12]Whitehead, J.P., et al., Diabetes Obes Metab, 2006. 8(3): p. 264-280. [13]Arita, Y., et al., Biochem Biophysical Res Comm, 1999. 257(1): p. 79-83. [14] Waki, H., et al., Endocrinology, 2005. 146(2): p. 790-796. [15] Swarbrick, M. and P. Havel, Metab Syndr Relat Disord, 2008. 6(2): p. 87-102. [16] Iwaki, M., et al., Diabetes, 2003. 52: p. 1655-63. [17] Weyer, C., et al., J Clin Endocrinol Metab, 2001. 86(5): p. 1930-1935. [18]Kantartzis, K., et al., Obesity Res, 2005. 13(10): p. 1683-1691. [19] Tschritter, O., et al., Diabetes, 2003. 52(2): p. 239-243. [20] Hotta, K., et al., Diabetes, 2001. 50(5): p. 1126-1133. [21] Lindsay, R.S., et al., Lancet, 2002. 360(9326): p. 57-58. [22] Yamamoto, Y., et al.,. J Clin Endocrinol Metab, 2004. 89(1): p. 87-90. [23] Snehalatha, C., et al., Diabetes Care, 2003. 26(12): p. 3226-3229. [24] Greenberg, A.S. and M.S. Obin, Am J Clin Nutr, 2006. 83(2): p. 461S-465. [25] Ouchi, N. and K. Walsh, Clinica Chimica Acta, 2007. 380(1-2): p. 24-30. [26] Nishimura, S., Discovery Medicine, 2009. 8(41): p. 55-60. [27] Coenen, K.R., et al., Diabetes, 2007. 56(3): p. 564-573. [28] Weisberg, S.P., et al., J. Clin. Invest., 2003. 112(12): p. 1796-1808. [29] Stephens, J.M.,et al. J. Biol. Chem., 1997. 272(2): p. 971-976. [30] Qi, C. and P.H. Pekala, Proc Soc Exp Biol Med, 2000. 223(2): p. 128-135. [31] Peraldi, P., M. Xu, and B.M. Spiegelman, J Clin Invest, 1997. 100(7): p. 1863-1869. [32] Cawthorn, W.P. and J.K. Sethi, FEBS Letters, 2008. 582(1): p. 117-131. [33] Memon, R.A., et al., Am J Physiol, 1998. 274: p. E210-E217. [34] Ryden, M. and P. Arner, J Internal Med, 2007. 262(4): p. 431-438. [35] Kita, A., et al., Biochem Biophys Res Commun, 2005. 331(2): p. 484-490. [36] Kim, K.-y., et al., Biochem Biophys Res Commun, 2005. 327(2): p. 460-467. [37] Rabe, K., et al., Mol Med, 2008. 14(11-12): p. 741-751. [38] Monteiro R, A.I., Mediators Inflamm, 2010. [39] Bambace, C., et al., Cardiovascular Pathology. In Press, Corrected Proof.

UPDATE ON FANCONI SYNROME AND CYSTINURIA IN DOGS: AMINO ACIDURIAS

Urs Giger, Dr. med. vet. DACVIM, ECVIM & ECVCP, Adrian C. Sewell, PhD, Jonlong Lui, PhD, Jessica Lee, BS, Caitlin A. Fitzgerald, BS, Victor Stora, Angela Erat Huff, MS, Karthik R. Raj, MS, and Paula S. Henthorn, PhD.
Philadelphia, PA

Disorders of the renal proximal tubules can cause selective or generalized amino aciduria and may be associated with urinary losses of other solutes such as glucose, electrolytes and bicarbonate. For decades two renal tubular defects involving amino acids have been recognized in dogs, namely cystinuria, leading to cystine calculi formation, and Fanconi syndrome, progressing to renal failure, if untreated. Both hereditary disorders have been investigated at the molecular level and found to be more complex than originally anticipated. Furthermore, the ingestion of Chinese chicken jerky treats has recently been associated with Fanconi syndrome. The current pathophysiology, clinicopathological findings, diagnosis, and therapeutic options will be presented.

FANCONI SYNDROME

Fanconi syndrome, named after the Swiss pediatrician Guido Fanconi, is also known as Fanconi's syndrome or Fanconi disease, but should not be confused with Fanconi anemia, a bone marrow disorder in humans. Fanconi syndrome represents a large proximal renal tubular defect which hampers the adequate reabsorption of glucose, amino acids, bicarbonate, sodium calcium, phosphate, lactate, carnitine, and ketones. These rather general losses of function of the proximal renal tubules can be associated with renal tubular acidosis and lead to progressive renal failure, if left untreated. In the renal tubules there are several co-transporters for sodium and glucose, amino acids, calcium, and inorganic phosphorus and a sodium/hydrogen ion antiporter which, depending upon the concentration gradient established by the sodium-potassium pump, moves hydrogen ions into the urine. In contrast to other renal tubular disorders, no specific transporter defect has yet been identified in Fanconi syndrome. Defects in all tubular transporters, the membrane sodium/potassium ATPase and thus tubular gradient, or a structural membrane abnormalities may be responsible. The mechanisms involved in Fanconi syndrome in dogs may include multiple genetic predispositions, acquired causes of, or combinations of both.

Basenjis are genetically predisposed to Fanconi syndrome, which appears to be inherited as an autosomal recessive trait in the breed. As many as 10% of Basenjis are affected, and they typically develop signs in middle-age (4-7 years). The molecular defect has not yet been determined, but based upon microsatellite-based genetic mapping, the canine genetics group at the University of Missouri mapped the Fanconi syndrome to a 6 million base pair (Mb) region on dog chromosome 3 (CFA3) in 2006. Additional genome-wide association studies narrowed the region to 2.1 Mb, but sequencing of several candidate genes in the region failed to identify a causative mutation. Norwegian Elkhounds also often develop glucosuria early in life, which may progress to Fanconi syndrome, but no molecular studies have been reported. Other breeds that may also be genetically predisposed to develop Fanconi syndrome include Cocker spaniels, West Highland White Terriers, Shetland Sheepdogs, and Schnauzers. In humans, Fanconi syndrome is most often associated with hereditary cystinosis (not to be confused with cystinuria), fructose intolerance, galactosemia, glycogenosis, as well as Lowe and Wilson disease.

There are several acquired causes of Fanconi syndrome such as exposure to heavy metals (e.g. lead, mercury, and cadmium), drugs (gentomycin, tetracycline, and azathioprine) and infections (leptospirosis). Fanconi syndrome has also been reported in cancer and hepatic copper disease in several dogs. More recently Fanconi syndrome has been associated with the ingestion of chicken and duck jerky treats of different brands made in China. Currently no specific ingredient or contaminant responsible for the intoxication has been identified, and while the FDA and USDA are aware of the problem, there have been no recalls. Nearly all the dogs affected are toy to small breed dogs, such as Chihuahuas and Yorkshire, Jack Russell and West Highland White Terriers. Their predisposition may be related to the proportionally larger amount of jerky treats ingested by small breeds compared to large breed dogs, or may reflect a broader hypersensitivity of the smaller breeds. Fanconi syndrome associated with ingestion of Chinese chicken jerky treats has been seen in small breed dogs from North America and Australia.

Polydipsia and polyuria are the most common clinical signs of Fanconi syndrome, while other signs are very variable and may be related to the ensuing renal failure. In the past, a diagnosis of Fanconi syndrome resulted in a guarded to poor prognosis, However, in the present day, treated Basenijs may reach a near normal life expectancy. Moreover, if the trigger can be removed such as withdrawal of Chinese chicken jerky treats, the renal damage may be more minimal and reversible. Routine serum chemistry and urinalysis as well as venous blood gas analysis are indicated to define the extent of the tubular defect and acidosis. Venous blood gases may show a metabolic acidosis (with blood pH normal to near 7), low bicarbonate concentration and a reduced base excess. A total CO_2 level on a chemistry screen is not sensitive enough to detect the metabolic acidosis due to compensatory hyperventilation. Serum chemistry also will reveal deficiencies in sodium, potassium, calcium and phosphorous. Osteopenia is not a feature of Fanconi syndrome in dogs as it is in humans. Some dogs develop azotemia, which worsens over time, with renal failure being the most common cause of death.

Urinalysis reveals an isosthenuric alkaline urine with glucosuria. In fact, the marked glucosuria in light of a normal blood glucose level is generally the reason to diagnostically pursue a Fanconi syndrome. Urine metabolic screen from dogs with Fanconi syndrome show massive generalized amino aciduria and lactic aciduria (and the glucosuria is confirmed). Metabolic screening at the University of Pennsylvania (http://www.research.vet.upenn.edu/penngen) can document the amino aciduria, the hallmark finding of Fanconi syndrome. Dogs with Fanconi syndrome have much higher urinary concentrations of cystine and dibasic amino acids in comparison to cystinuria (discussed below), as well as exhibit massive excretion of all other amino acids. Paradoxically, despite a severe degree of cystinuria in any dog with Fanconi syndrome, they rarely if ever develop cystine calculi and urinary obstruction; the dilute and alkaline urine may prevent the crystal formation. Moreover, while there is severe amino aciduria, there is either no or only mild proteinuria. Currently, the University of Missouri (http://www.caninegeneticdiseases.net) is offering a genetic linked-marker test for Basenijs. Among >3000 Basenjis tested over the past 4 years, 5% were genetically affected (homozygous for all 3 markers), 37% were carriers (homozygous affected or heterozygous at all 3 markers), and 5% were intermediates (one marker homozygous affected, one heterozygous and one wild type/normal marker). However, the clinical disease status was not reported.

The therapeutic protocol (http://www.basenji.org/fanconiprotocol2003.pdf) developed and modified over the years by Steve Gonto, who is a human anesthesiologist as well as a critical care/life support clinician and owner of Basenjis, has been used widely. While there are no formal comparative trials of different therapeutic modalities, the life expectancy and quality of life have apparently greatly improved by using this protocol. It focuses on aggressive correction of the various losses, foremost bicarbonate, normalizing electrolytes and minerals, and providing a high quality diet. Bicarbonate is dosed to increase pCO_2 and HCO_3 levels. Unless the animal is in renal failure, a high protein diet is recommended to replenish the amino acid and protein losses and to rebuild muscle mass. Renal dialysis may be considered in the acute management of severe renal failure. In a survey of medical records, Basenjis with Fanconi syndrome received between 2.6-15.6 grams of sodium bicarbonate in tablet form divided per day. Multivitamin and calcium/phosphorus may be supplemented as needed. Treatment of Fanconi syndrome can be rewarding, if the disease was recognized early. The median survival time for Basenjis was 5 years after the diagnosis was made. Similarly, this medical protocol can be used to manage dogs with the acquired form, as long as the trigger is removed. The treatment has to be adjusted for the breed size and severity of the metabolic derangements. Dogs that had ingested Chinese chicken/duck jerky treats are known to have improved or even resolved their metabolic derangements within a few months after withdrawing the treats, unless they had already developed severe renal failure.

It is hoped that future research will define the mutant gene causing Fanconi syndrome in Basenjis and eventually other predisposed breeds, thereby permitting definitive screening, and that the ingredient in the Chinese chicken jerky treats causing Fanconi syndrome can be identified and removed.

CYSTINURIA

In contrast to the rather broad transmembrane transport deficiency causing hereditary and acquired forms of Fanconi syndrome, cystinuria is a limited hereditary renal transport disorder involving cystine and the dibasic amino acids ornithine, lysine, and arginine, collectively known as COLA. Instead of the normal >99% reabsorption in the proximal renal tubules, these amino acids are lost in the urine, but only cystine causes a problem. The low solubility of cystine in acidic urine predisposes to the formation of cystine crystals and uroliths in the urinary tract, which result in the typical clinical signs of cystinuria. The transport of these amino acids also normally occurs in the gut and thus affects intestinal COLA absorption; however, the impaired intestinal absorption of the COLA amino acids is not associated with any clinical deficiencies in animals, with the exception of hyperammonemia due to arginine deficiency in cats.

Two genes, SLC3A1 and SLC7A9 encode the polypeptide subunits of $b^{o,+}$, the basic amino acid transporter system. SLC3A1 encodes a protein referred to as rBAT and SLC7A9 encodes a protein called $b^{o,+}$AT. $b^{o,+}$AT heterodimerizes with rBAT exclusively to form the COLA amino acid transporter.

The stones seen in human cystinuria patients were first described as ‘cystic oxide’ more than 200 years ago. Cystinuria is one of the first reported “inborn errors of metabolism” and has since been described in several other species including dogs, cats, ferrets, maned wolves, and servals. Since 1823, when cystinuria was first described in dogs, many breeds have been shown to be afflicted.

Cystinuria in humans has been classified phenotypically into two types: type I and non-type I, which are typically autosomal recessively inherited but can show different modes of inheritance, and mutations in SLC3A1 (typically type I cystinuria) and SLC7A9 have been identified in cystinuric human patients. In dogs, type I cystinuria is best characterized in Newfoundlands. It is the severe and early onset type I cystinuria and is inherited as an autosomal recessive trait. Although both males and females can be affected, more often males show clinical signs of urinary obstruction. A mutation in SLC3A1 causing an early stop codon and loss of $b^{o,+}$AT function has been identified. More recently, Labrador Retrievers were found to have a different mutation in the same gene, while in other breeds with type I cystinuria, such as Miniature Pinschers and Scottish Terriers, the molecular defect is not yet known.

Cystinuria has also been described in an additional 60 canine breeds. In several of these breeds from which more data is available, such as the Mastiff, French Bulldog, Basset hound and Irish terrier, the disease seems to occur later in life, is less severe, more variable, and appears to involve only males. We now refer to this as canine non-type I cystinuria. Interestingly, cystinuria in Irish Terriers in Europe and Australia seems to be more common than in the US, where Irish Terriers appear to be very rarely affected. Non-type I cystinuric dogs (with or without cystine calculi) also seem to have variable degrees of cystine and COLA in their urine. These variations could be due to cystine precipitation, diet, age-related variability, the castration effects, and genetic heterogeneity. Molecular studies in Irish Terriers, Mastiffs and several other breeds have failed to detect disease-causing mutations in the exons of either SLC3A1 or SLC7A9 genes. A pilot whole genome association analysis study and genotyping of additional Mastiffs implicates a region surrounding one of the known cystinuria genes. As only males are cystinuric when mature and castration seems to lower cystine excretion, the mode of inheritance is also still unclear.

After glomerular filtration, these COLA and other amino acids are normally >99% resorbed in the proximal renal tubules. Thus, in healthy animals only small amounts of amino acids are lost in the urine. On average, dogs with type I cystinuria have a several fold higher urinary COLA excretion than non-type I cystinuric male dogs. Newfoundlands’ average cystine and COLA values were 2,924 and 17,705 μmol/g creatinine independent of gender. In contrast, average cystine and COLA values were 551 and 2,304 μmol/g creatinine, respectively, in male Irish Terriers, whereas all females examined (including many mothers, sisters, and offspring of cystinuric male dogs) have normal values. These values were similar to average cystine and COLA values in other breeds with non-type I cystinuria. Cystine crystallization occurs when the concentration rises above the saturation point (~250 mg cystine/ liter).

Cystine uroliths can form in the kidney (rarely), ureters, bladder or urethra and can lead to urinary blockage, renal failure and associated life-threatening clinical manifestations. A clinical diagnosis of a cystine uroliths can be suspected based upon the breed and yellow-brown color of the urolith, and confirmed based upon the crystallographic or chemical analysis of calculi in a specialized stone laboratory such as those at the University of Minnesota, the University of California, as well as several commercial laboratories. While in the US approximately 1% of analyzed urinary calculi are cystine stones, in Europe the proportion appears to be much higher. Cystine calculi are variably radiodense and thus may be missed by radiography, but are readily visualized by ultrasound. Furthermore, urinalysis often reveals the presence of characteristic hexagonal cystine crystals.

A simple urinary screening test, to determine if a dog is cystinuric, is available through the Metabolic Genetic Laboratory at the University of Pennsylvania (http://www.research.vet.upenn.edu/penngen). This test can detect any type I cystinuric animal, but not necessarily all dogs with non-type I cystinuria. High performance liquid chromatography can be further used to determine the amount of cystine and the other amino acids in urine, albeit such quantitative amino acid analysis is currently restricted to a few laboratories (PennGen and LaboKlin) and is relatively expensive. Inclusion of the dibasic amino acids (ornithine, lysine, and arginine) in the quantitation of cystine is preferred due to the possibility that urinary cystine may have precipitated, causing lower concentrations than were originally present in the urine, and therefore leading to false negative interpretations. Based upon our studies, dogs with either cystine levels of >200 μmol/g creatinine or COLA values of >700 μmol/g creatinine are considered cystinuric. Moreover, for Newfoundlands and Labrador Retrievers, a breed-specific mutation test is available that not only detects cystinuric dogs, but also the asymptomatic carriers. While as many as 5% and 30% of Newfoundlands were cystinuric (homozygous affected) and carriers, respectively, in the 1990s, many fewer dogs are cystinuric or carriers, thanks to the implementation of this screening test, which is available through our lab and many commercial laboratories. In contrast the mutant allele in the Labrador Retrievers appears to be limited to a very small subset of the Labrador Retriever population. Finally, in Mastiffs, a preliminary linked marker test to identify dogs that are at risk for early stone formation is being evaluated (http://www.research.vet.upenn.edu/penngen).

Urinary tract obstructions due to cystine calculi can present as life-threatening emergency situations. If the calculi are in the urethra, catheterization and endoscopy may allow their passage or permit them to be moved back into the bladder (retrograde hydropulsion). Laser therapy as well as lithotripsy may break them up, and surgery can be used to remove them entirely. In contrast, attempts to dissolve cystine stones medically have generally failed. Dogs with type I cystinuria are persistently strongly cystinuric from puppyhood and remain at high risk of developing cystine calculi, independent of gender. Alpha-merkaptopropionylglycine (Thiola) is a chelating agent that has less side-effects and is more potent than D-penicillamin may lower the risk of calculi formation in non-type I cystinuria, but is very expensive. Forced diuresis, urine alkalinization (pH>7.5), and reduced protein diets (no chicken) may lower the recurrence risk, while high protein diets and protein supplements tend to increase the risk for calculi formation. Male dogs with non-type I cystinuria have more variable and generally lower cystine excretion. Thus, they may or may not develop calculi and their recurrence rate is lower. In preliminary studies we observed that castration of Mastiffs and Irish Terriers with non-type I cystinuria drastically diminished the excretion of cystine and dibasic amino acids, suggesting a testosterone-mediated modulating effect on the renal tubular amino acid reabsorption. We recommend castration to reduce the formation and risk of blockage by cystine calculi, which will also contribute to reducing the incidence of this genetically predisposition.

ACKNOWLEGEMENT

Supported in part by grants from the National Institutes of Health (RR02512) and the AKC Canine Health Foundation.

REFERENCES

1. Appleman EH, Cianciolo R, Mosenco AS, Bounds ME, Al-Ghazlat S. Transient acquired Fanconi syndrome associated with copper storage hepatopathy in 3 dogs. J Vet Intern Med 2008;22:1038-1042. 2. Bovee KC. Fanconi syndrome in the dog. In: Kirk RW, ed. Current Veterinary Therapy VII. Philadelphia: WB Saunders Co, 1980;1075-1078. 3. Bovee KC, Joyce T, Reynolds R, et al. The Fanconi syndrome in Basenji dogs: a new model for renal transport defects. Science 1978;201:1129-1131. 4. Casal ML, Giger U, Bovee KC, Patterson DF: Inheritance of cystinuria and renal defect in Newfoundlands. Journal of the Am Vet Med Assoc 1995;207:1585-89. 5. Escolar E, Perez-Alenza D, Diaz M, et al. Canine Fanconi syndrome. J Small Anim Pract 1993;34:567-570. 6. Gonto S. Fanconi disease management protocol for veterinarians. Available at: www.basenji.org/fanconiprotocol2003.pdf. 7. Giger U, Fitzgerald C, Liu J, EratHuff A, Sewell AC, Henthorn PS: Characterization of non-type I cystinuria in Irish Terriers. ACVIM Forum 2011; abstract. 8. Harvenik L, Hoppe A, Söderkvist P. SLC7A9 cDNA cloning and mutational analysis of SLC3A1 and SLC7A9 in canine cystinuria. Mamm. Genome 2006;17:769-776. 9. Henthorn PS, Liu J, Gidalevich T, Fang J, Casal ML, Patterson DF, and Giger U: Canine cystinuria: Polymorphism in the canine SLC3A1 gene and identification of a nonsense mutation in cystinuric Newfoundland dogs. Human Genetics 200;107:295-303, 2000. 10. Henthorn PS, Giger, U: Cystinuria, in The Dog and Its Genome, eds. Ostrander, Giger, and Lindblad-Toh (Cold Spring Harbor Laboratory Press), 2006;349-64. 11. Henthorn PS, Lee J, Sewell A, Liu J, Erat A, Giger U: Unraveling the complex basis of cystinuria in dogs. Proceedings of the Advances in Canine and Feline Genomics and Inherited Diseases Conference, Baltimore 2010;122. 12. Hill TL, Breitschwerdt EB, Cecere T, Vaden S. Concurrent hepatic copper toxicosis and Fanconi's syndrome in a dog. J Vet Intern Med 2008;22:219-222. 13. Hostutler RA, DiBartola SP, Eaton KA. Transient proximal renal tubular acidosis and Fanconi syndrome in a dog. J Am Vet Med Assoc 2004;224:1611-1614. 14. Mhlanga-Mutangadura T, Farias F, Guo J, Zeng R, Hansen L, and Johnson GS: Mapping the Basenji Fanconi syndrome locus and results from a linked-marker test. 15. Proceedings of the Advances in Canine and Feline Genomics and Inherited Diseases Conference, Baltimore 2010;82. 16. Noonan CH, Kay JM. Prevalence and geographic distribution of Fanconi sydrome in Basenjis in the United States. J Am Vet Med Assoc. 1990;197:345-349. 17. Palacin M, Nunes V, Font-Llitjos M, Jimenez-Vidal M, Fort J, Gasol E, Pineda M, Feliubadalo L, Chillaron J, Zorzano A: The genetics of heteromeric amino acid transporters. Physiol. 2005;20:112-124. 18. Yearly JH, Hancock DD, Mealey KL. Survival time, lifespan and quality of life in dogs with Fanconi syndrome. J Am Vet Med Assoc 2004;225:377-383.

APPLICATIONS OF STEM CELL THERAPY IN KIDNEY DISEASE

Jessica Quimby, DVM, DACVIM
Fort Collins, CO

INTRODUCTION

Regenerative medicine refers to the process of using living function tissues to repair or replace organs that are functionally damaged. Stem cell therapy in particular is an innovative new field of scientific investigation and clinical application that holds promise for a variety of diseases in veterinary medicine as well as human medicine. Recent years have brought increased interest in the potential for adult stem cells to help in the treatment of many diseases through both their regenerative properties as well as their apparent ability to alter the environment in injured and diseased tissues. In particular, adult stem cells called mesenchymal stem cells can migrate to affected areas and may be able to support the growth of other stem cells as well as moderate the response of the immune system.

STEM CELLS

A stem cell is a generic term referring to any unspecialized cell that is capable of long-term self-renewal through cell division but that can be induced to differentiate into a specialized, functional cell. Stem cells are generally divided into two groups, embryonic stem cells and adult stem cells. Adult stem cells can be obtained from many differentiated tissues including but not limited to bone marrow, bone, fat, and muscle. Obtaining adult stem cells also does not raise ethical concerns. For most studies, the adult stem cell in question is actually a mesenchymal stem cell or mesenchymal stromal cell. Mesenchymal stem cells are multipotent but not pluripotent, which means they can differentiate into some, or "multiple," but not all tissue types.[1]

Characterization

Mesenchymal stem cells or mesenchymal stromal cells (MSC) are plastic adherent and assume a fibroblast-like morphology during culture. They proliferate easily in culture and can be cryopreserved without loss of phenotype or differentiation potential. Additionally cell surface marker characterization via flow cytometry differentiates them from hematopoetic cells, though no truly unique MSC molecule has been identified. In part, the lack of definitive markers probably reflects the diverse lineage of MSC and the fact that each MSC population reflects to some degree the characteristics of tissues from which they were derived. Most importantly stem cells possess the ability to differentiate into cell types of multiple lineages including adipocytes, chrondrocytes, and osteocytes.[1]

Immunologic properties

Mesenchymal stem cells clearly modulate immune responses, as demonstrated by both in vitro and in vivo studies. For example, MSC are poor antigen presenting cells and do not express MHC class II or co-stimulatory molecules and only low levels of MHC class I molecules.[1] Thus, MSC are very non-immunogenic and can be transferred to fully allogeneic recipients and still mediate their immunologic effects. Among their other immunological properties, MSC inhibit lymphocyte proliferation and cytokine production, suppress dendritic cell function and alter DC cytokine production, and decrease IFN-g production by NK cells.[1] These properties of MSC can be harnessed therapeutically (as discussed below). In addition, it is also important to keep the immune modulatory properties of MSC in mind when interpreting the results of many stem cell therapy studies.

STEM CELLS AND KIDNEY DISEASE

There are numerous studies of MSC therapy in rodent models of renal failure, though most studies have focused on models of short-term protection from acute renal disease.[2-5] The majority of these studies provide evidence that systemic administration of bone marrow-derived or adipose-tissue derived MSC can help preserve renal function in the face of acute insults and can also help reduce tubular injury and fibrosis.[2-5] Several studies have also demonstrated incorporation of small numbers of MSC into the renal parenchyma.[4,6,7] It has been proposed that some of these MSC may actually differentiate into functional renal tubular epithelial cells, though this theory remains controversial. Other investigators propose that paracrine effects from the injected MSC are more important than the effects of direct cellular incorporation into the kidney.[8,9] Thus, the available data indicate that systemically administered MSC can help improve or stabilize renal function in acute renal disease by a variety of mechanisms.

Fewer studies have investigated the effects of MSC therapy in chronic renal failure models in rodents.[10-15] In the CKD rodent models that have been investigated, administration of MSC has been beneficial, especially with respect to reducing intra-renal inflammation and suppressing fibrosis and glomerulosclerosis.[10,11,13,14,16] Results from a recent study, which utilized a rat remnant kidney model of CKD, are particularly noteworthy.[10] In this study, the authors found that 3 i.v. injections of relatively low numbers of MSC led to significant and sustained improvement in renal function (eg, reduced serum creatinine concentrations and proteinuria) and markedly suppressed intra-renal fibrosis and inflammation. Repeated injections of MSC were found to be more effective than a single injection, and additionally MSC were identified in the kidney after systemic injection. Thus, there is compelling evidence from rodent models of CKD that MSC injections can lead to significant improvement in renal function. MSC secrete a variety of biologically active factors and release of these factors, either systemically or locally into the kidney parenchyma, could affect renal function both directly and indirectly. In vitro studies have demonstrated that MSC can produce growth factors, cytokines, and anti-inflammatory mediators, all of which could help maintain or improve renal function and suppress intra-renal inflammation.[8, 17, 18] The ability of MSC to suppress inflammation appears to be mediated both by secreted factors and by direct contact with inflammatory cells.[17,18] Thus, MSC have the potential to strongly suppress intra-renal inflammation.

CHRONIC KIDNEY DISEASE

Prevalence of CKD

CKD remains a leading cause of illness and death in cats in the United States. This is a progressive disease, and currently no treatment short of renal transplantation has been shown to reverse or halt declining renal function for any significant period of time. This condition is characterized by tubulointerstitial damage, fibrosis and progressive loss of renal function, and is commonly described as the final common pathway after any one of multiple types of renal insults. Regardless of the initial insult, once a threshold of renal damage has been reached, progression is irreversible and appears consistent in character.[19] Stem cell therapy has the potential to improve or stabilize renal function in animals with renal failure, based on evidence from rodent model studies. An appreciation of the pathophysiology of kidney disease is helpful in understanding why this might be the case.

Progression of CKD

Impairment of renal function is correlated with the degree of tubulointerstitial injury, including tubular atrophy, atubular glomeruli and tubulointerstitial fibrosis. Hyperfiltration, proteinuria, tubulointerstitial inflammation, oxidative damage and induction of the renin-angiotensin-aldosterone system (RAAS) are major factors that are thought to contribute to the process of tubulointerstitial injury.[19-21]

Loss of functional nephrons results in hyperfiltration due to an increase in remaining single nephron GFR. Combined with a loss of the ability to autoregulate blood flow to the kidney, hyperfiltration results in intraglomerular hypertension, which can be further exacerbated by systemic hypertension. Mechanical and sheer stress on the glomerular apparatus causes damage resulting in stretching of the membrane and leakage of proteins. Although the kidney filters a large amount of protein on a daily basis, urinary protein levels probably underestimate the filtered protein load due to the enormous capacity of the proximal tubule for protein reabsorption. However, proximal tubule cells are therefore prone to damage as a result. Proteins are reabsorbed through megalin-cubulin mediated endocytosis and then processed lysosomally. Protein excess results in overload of the lysosomal system and resultant rupture is damaging to tubular cells and instigates recruitment of inflammatory cells to the area. In addition, substances carried by the proteins such as, LPS, FFA, PG, heavy metals and hormones result in oxidative stress as well as direct damage to the cells.[20]

Tubulointerstitial inflammation is characterized by an inflammatory infiltrate in the interstitium that usually consists of mononuclear leucocytes including T-cells, monocytes and macrophages. These cells are thought to play a pivotal role in progression and fibrosis. Nephritogenic T-cells are thought to exacerbate the conditions present in the kidney through direct cytotoxic effects as well as non-cytotoxic mechanisms such as cytokine release, altered tubular function and proliferation of interstitial fibroblasts and fibrosis.[3] Once fibrosis occurs, conditions become increasingly suboptimal as loss of peritubular capillaries and increased resistance to oxygen diffusion result in hypoxia, which in turn stimulates fibrosis.

Oxidative damage to tubular cells can occur as a result of direct harm from reabsorbed substances. However, increased filtration of substances such as glucose or protein also greatly increases the metabolism of tubule cells and this leads to increased production of free radicals. It has been shown that anti-oxidant defense systems are decreased in ESRD and thus the kidney is little prepared to address this increase and oxidative stress results. In addition, increase metabolic demand results in hypoxia as cells receive the same amount of blood flow, but have a higher demand. This condition is further exacerbated if anemia is present. [20]

The RAAS system is also a critical component in the pathophysiology of renal progression.[21] Although normally protective in emergency situations such as shock or hypotension, the RAAS system becomes maladaptive in ESRD. Combined with a loss of autoregulation in the diseased state, Angiotensin II plays a major role in this pathologic process as it vasoconstricts the efferent arteriole to maintain GFR. Although GFR is maintained, the downsides of this mechanism are hypoperfusion of post-glomerular capillaries, including the interstitium, and glomerular capillary hypertension. Angiotensin II also has detrimental non-vascular effects including activating tubular cells, oxidative stress, stimulation of inflammatory cell accumulation and promotion of fibrosis. The importance of the RAAS system is further supported by clinical evidence demonstrating a distinct clinical benefit in RAAS blockade.[21]

All of the discussed factors; hyperfiltration, proteinuria, tubulointerstitial inflammation, oxidative damage and induction of the RAAS contribute to disruption of tubuloglomerular continuity and chronic hypoxia, which leads to further inflammation and fibrosis. The process of inflammation appears to be integral to the progressive, irreversible nature of CKD. It is based on this premise that mesenchymal stem cells have potential to ameliorate this disease process.

CURRENT CLINICAL STEM CELL RESEARCH FOR FELINE CHRONIC KIDNEY DISEASE.

At present, there is little published work regarding the use of MSC for treatment of naturally-occurring CKD. At the Center for Immune and Regenerative Medicine at Colorado State University, we are currently conducting research into the immunological properties of feline MSC, as well as the potential use of MSC for treatment of CKD in cats. We have recently completed a pilot study investigating the safety and potential effectiveness of unilateral intra-renal MSC injections for cats with CKD.[22] In that study, we found that the MSC injections were well-tolerated and may have improved renal function in some animals. However, the number of sedations required for the procedure limited its effectiveness, and also made management of CKD in those patients more difficult. In a new pilot study currently underway, we are investigating the effectiveness of intravenously delivered MSC for treatment of feline CKD, using allogeneic MSC derived from healthy young donor animals. Preliminary results from this study are encouraging.

SUMMARY

The fields of stem cell therapy and regenerative medicine are expanding rapidly. Veterinary medicine is poised to take a leading role in these fields, as there are a number of valuable spontaneous disease models in companion animals that would be amenable to stem cell therapy. Among the challenges facing these emerging fields are standardization of treatment protocols and adherence to strict principles of evidence-based medicine in reporting study results and conclusions. Nonetheless, it is likely that stem cell therapy will make significant progress in changing treatment paradigms for a number of important diseases of dogs and cats in the relatively near future.

REFERENCES

[1]Reinders MEJ et al. **Nephrol Dial Transplant** 2010; 25:17. [2]Little MH, et al. **Nephrology** 2009;14:544. [3]Togel F et al. **Am J Physiol Renal Physiol** 2005;289:F31. 4. [4]Morigi M et al. **J Am Soc Nephrol** 2004;15:1794. [5]Semedo P et al. **Transplant Proc** 2007;39:421. [6]Kitamura S et al. **FASEB J** 2005;19:1789. [7]Kim SS et al. **Transplantation** 2007;83:1249. [8]Togel F et al. **Am J Physiol Renal Physiol** 2007;292:F1626. [9]Togel F et al. **Am J Physiol Renal Physiol** 2008;295:F315. [10]Semedo P et al. **Stem Cells** 2009;27:3063. [11]Lee SR et al. **Renal Failure** 2010; 32:840 [12]Kirpatovskii VI et al. **Bull Exp Biol Med** 2006;141:500. [13]Cavaglieri RC et al. **Transplant Proc** 2009;41:947. [14]Choi S et al. **Stem Cells Dev** 2009;18:521. [15]Caldas HC et al. **Transplant Proc** 2008;40:853. [16]Ninichuk V et al. **Kidney Int** 2006;70:121. [17]Barry FP et al. **Stem Cells Dev** 2005;14:252. [18]McTaggart SJ. **Nephrology** 2007;12:44. [19]Nangaku M. **J Am Soc Nephrol** 2006;17:17. [20]Harris RC. **Annu Rev Med** 2006;57:365. [21]Siragy HM. **Am J Nephrol** 31:541. [22]Quimby JM et al. **J Fel Med Surg.** 2011(in press).

ENDOSCOPIC TREATMENT OF ECTOPIC URETERS: SHORT & LONG TERM OUTCOMES USING CYSTOSCOPIC-GUIDED LASER ABLATION (CLA-EU)

Allyson Berent, DVM, DACVIM
New York, NY

INTRODUCTION

Ureteral ectopia (UE) is a congenital anomaly of the urinary system where the ureteral orifice is inappropriately positioned caudal to the urinary bladder (i.e. the bladder neck, urethra, vagina, vestibule or uterus). This is the most common cause of urinary incontinence in juvenile female dogs accounting for over 50% of cases in one study, compared to only 5% of incontinent adult dogs.

Although ectopic ureters have been reported in male and female dogs, as well as both pure and mixed breed dogs, it seems to occur with greater frequency in female than male dogs (>20:1), as well as certain breeds (i.e. Siberian Huskies, Newfoundlands, Retrievers [Labrador or Golden], Terriers, and Poodles [miniature and toy]). The most common clinical finding in these dogs is constant or intermittent urinary leaking since birth or weaning, though many dogs present after a period of continence, and are only incontinent in certain positions.

Suspected concurrent bladder and/or urethral functional anomalies, like urethral sphincter mechanism incompetence (USMI), have been reported in 75-89% of female dogs evaluated, though in one study there was no significant difference in outcome after surgery in dogs with or without USMI. Other associated urinary conditions such as urinary tract infections (64%), renal agenesis (5%), renal dysplasia, hydroureter (34-50%) or hydronephrosis (15-27%), short urethras (21%), persistent paramesonephric remnants (83%) and/or vaginal septum or dual vaginas (8%), hormonal imbalances, and ureteroceles have all been reported concurrently.

The embryological basis of this condition is suggested to result from inappropriate differentiation of the mesonephric and metanephric duct systems. This results in inappropriate ureteral tube termination and ultimate malposition of the ureteral orifice. The urogenital sinus, which becomes bladder and urethra, may also differentiate inappropriately, resulting in termination of the ureters in the bladder neck, prostate, or urethra. Ectopic ureters tunnel either intramurally or extramurally, with over 95% reported to be intramural in dogs. The typical intramural ectopic ureter(s) will enter the distal bladder neck in a relatively normal position but fail to open into the bladder lumen, traversing the urethra to the level of the prostate, vestibule or vagina in the submucosal tissue, where it terminates.

Traditional therapeutic intervention for UE includes 3 main surgical techniques: neoureterostomy with ligation of the distal ureteric tunnel, neoureterostomy with urethral-trigonal reconstruction, and neoureterocystostomy (ureteral re-implantation). These procedures require an open laparotomy, cystotomy, ureterotomy, and urethrotomy. The complication rates with surgery vary and in one report there was a 14% complication rate overall, with 50% of dogs after ureteral re-implantation developing worsening hydroureter or hydronephrosis, 16% of dogs after the intravesciular transplantation technique having dysuria, and 8% of dogs with ureteronephrectomy developing renal failure. Unfortunately, the post-operative continence rates reported in dogs continue to be low, regardless of the surgical technique performed, varying between 25 and 58% with or without concurrent medical management. When comparing unilateral to bilateral UE, no difference in female dogs treated surgically has been reported. Recently a study comparing neoureterostomy with ligation of the distal portion of the tract and neoureterostomy with urethral-trigonal reconstruction was unable to detect substantial differences between the success of these two techniques; both resulted in high rates of persistent incontinence after surgery. The suggested benefits of dissection of the distal portion of the tract include restoration of an internal sphincter, diminished urinary stasis in the tunnel, and decreased risk of re-canalization. Persistent incontinence is likely attributable to the concurrent urethral and sphincter abnormalities, with or without concurrent polyuria secondary to associated renal dysfunction, and necessitates medical intervention (phenylpropanolamine, diethylstilbestrol, or testosterone), transurethral injections of a submucosal bulking agent, or further surgery (eg, colposuspension or prostatopexy). Continence was restored with medications in 32% to 38% of dogs that had persistent incontinence after surgery. In dogs with primary SMI without EU, bulking agents can be effective in 50% to 83% of cases. However it is not known whether urethral bulking agents are equally efficacious in persistently incontinent dogs that have UE versus those dogs with primary USMI without UE. In the author's experience dogs with concurrent UE (regardless of previous fixation method) do not hold urethral bulking agents well. This may be due abnormalities in the mucosal/submucosal urethral tissue strength or due to the fact that a majority of these dogs have very wide and short urethras. Since many of these dogs are relinquished or euthanized because of urinary incontinence issues, these disappointing outcomes made the search for other alternatives appealing.

The diagnostic method of choice for evaluating dogs for EU is now considered to be cystoscopy or CT. The use of the cystoscopic-guided laser ablation (CLA-EU) technique, first described in one female dog in 2006 by Dr. McCarthy and in 4 male dogs in 2008 by our group, provides a minimally invasive alternative to surgery in cases with intramural EU. Since that time there have been a few abstracts describing this procedure in small groups of dogs, and more recently a small retrospective series looking at 16 dogs. This procedure enables the diagnosis to be made while simultaneously performing a therapeutic intervention, and also potentially avoiding some of the complications and risks associated with the open surgical techniques described. This procedure uses cystoscopy and fluoroscopy to directly visualize the ureteral orifice, assess for any other urinary anomalies (vaginal septum, persistent paramesonephric remnant, dual vagina, hydroureter, hydronephrosis, ureteroceles, etc), as well as guide a laser to ablate the tissue that forms the medial ectopic ureteral wall, so the orifice can be re-positioned into the urinary bladder neck. During this procedure the vaginal remnant is also typically ablated with the laser to potentially aid in any contribution this might have to urinary incontinence or chronic urinary tract infections. This lecture will expand upon the results of a larger number of female and male dogs that were prospectively evaluated after cystoscopic guided laser ablation therapy.

PROSPECTIVE EVALUATION OF CLA-EU

30 female and 4 male dogs that had the CLA-EU procedure performed were prospectively evaluated. In female dogs the CLA-EU procedure is done through a rigid cystoscope using a diode or Holmium:YAG laser. In male dogs it is either done through the penile urethra using a flexible cystoscope and Holmium:YAG laser or using a percutaneous perineal needle access approach using a rigid cytoscope. Prior to performing this procedure a retrograde ureteropyelogram is performed under fluoroscopic guidance to confirm the intramural nature of each ectopic ureter. All techniques will be expanded upon in this lecture.

In the 30 female dogs, 100% of the ectopic ureteral orifices were initially located in the urethra, and 100% were able to be relocated into the urinary bladder. 60% had bilateral ectopia and 100% had other concurrent urinary anomalies. 23 of 30 (77%) dogs were continent at >12 months follow-up (that continence rate was 47% with CLA-EU alone, 57% with additional medications, 60% with additional collagen injections, and 77% with the addition of a hydraulic occluder [HO]). In the 16 residually incontinent dogs after the CLA-EU alone medical management was implemented and phenylpropanolamine was effective in 3/16 (18.75%), while diethylstilbestrol was not effective in any of the 5 patients it was tried on. Medical supplementation improved the continence rate to 17/30 (57%). Transurethral bulking-agent injections (bovine collagen 5/7; hydroxyapatite 2/7) were performed in 7/13 residually incontinent patients after failing adjunctive medical management and was effective in 2/7 (28.5%), improving the overall continence rate to 19/30 (63%). Five out of 11 residually incontinent patients that failed medical and/or bulking-agent intervention had a HO placed. This was effective in 4/5 (80%) dogs, improving the overall continence rate to 23/30 (77%). Two patients had a colposuspension performed during the course of the study and neither procedure was effective long-term (0%).

The remaining incontinent dogs did not have a hydraulic occluder placed; suggesting continence score could have ultimately been higher. All dogs were re-scoped within 6-8 weeks after the CLA-EU. No dog had evidence of post-operative dysuria. One female dog had evidence of polypoid cystitis at the neo-ureteral stoma at 6 weeks and resolved on 3 month re-evaluation. Median follow-up was 2.7 years (range 12-62 months). One dog not included in this study developed a suspected laser reaction at the neo-UVJ resulting in a ureteral obstruction. This seems to be uncommon in all the cases done to date, but should be considered and evaluated for on 6 week follow-up. All dogs had their vaginal defects (PPMR, dual vagina or vaginal septum) treated with laser ablation as well.

In the 4 male dogs, all EU openings were within the prostatic urethra and were intramural in nature. ¾ dogs had bilateral ectopia and the other dog was uninephric and its only ureter was ectopic totaling 7 ureters. 2/4 had hydroureter and hydronephrosis documented on retrograde ureteropyelogram and CT-IVP and both dogs had improvement in the ureteral and renal pelvis dilation at the 6 week recheck. No postoperative dysuria or hematuria occurred post operatively. All male dogs were immediately continent after laser treatment and remained so at a median follow-up of 23 months (range 12-34 months) without medical management.

When evaluated together 27/34 (80%) of dogs were continent with a combination of CLA-EU, medical management, bulking-agent injections and/or the placement of a hydraulic occluder device. 18/34 (53%) were continent after the CLA-EU procedure alone. To date over 60 dogs have had this procedure performed in the author's practice with similar success rates.

Another talk in this forum will focus on the use of the hydraulic occluder device for refractory urinary incontinence so it will not be expanded upon here.

CONCLUSION

In conclusion, the CLA-EU procedure is safe and at least as effective as the traditional surgical techniques previously reported. This procedure is met with few complications and can be considered at the time of the diagnosis of ectopic ureters during cystoscopy. This procedure allows the clinician to avoid invasive surgery until it is known if the procedure is successful, at which time the patient can have other supplemental fixation procedures at the time of routine ovariectomy/ovariohysterectomy. Even though complications were rare, they did exist and it is currently recommended that all dogs have a urinary tract ultrasound to evaluate any progression in hydroureter or hydronephrosis after this procedure (as it should be suggested after any surgical technique) to ensure there is no reaction/stricture formation at the neo-ureterovesicular junction.

REFERENCES

[1]Berent A. Proceedings, **ACVIM.** Seattle, WA, 2008. [2] Berent,. A. Proceedings, **ACVS**. Washington DC, 2007. [3]Berent A, et al. **JVIM** 2007; 21(3): 600. [4]McLoughlin MA. *Clin Tech Small Animal Pract* 2000;15:17–24. [5]Tattersall JA. *J Am Anim Hosp Assoc* 2006;42:395-400. [6]Berent, A. *J Am Vet Med Assoc* 2008; 232: 1026-1034. [7]Holt, P.E. *Veterinary Record*, 1990; 127: 347-350. 15. [8]Cannizzo K. *J Am Vet Med Assoc* 2003;223:475–481. [10]Samii VF. *J Vet Intern Med*.2004;18:271–281. [11] McCarthy, TC. *Vet Med*, 2006; 101(9): 558-559. [12]Smith AL. *J Am Vet Med Assoc*. 2010; 237(2): 191-195.

RECURRENT UTI, INTRACELLULAR BACTERIA & ALTERNATE THERAPIES TO PREVENT INFECTION

Michael W. Wood, DVM, PhD, DACVIM
Raleigh, NC

INTRODUCTION

The development of urinary tract infections (UTIs) is multifactorial, dependent on alterations in anatomical, environmental, and immunological competency. For many individuals with normal urinary defenses bacteria are cleared from the bladder without the need of medical care. Experimentally, healthy dogs with normal urinary defenses can clear urine of inoculated bacteria within 3 days without the aid of antibiotics. Despite the bladder's inherent resistance to bacterial colonization 14% of dogs are estimated to develop UTIs during their lives. Of these cases, the most problematic include recurrent urinary tract infections.

RECURRENT URINARY TRACT INFECTIONS – A NEW THEORY

Recurrent infections pose a challenge to the clinician as treatment should be paired with an identification or understanding of the cause of the infection. If the primary reason for bacterial colonization is not identified and addressed the patient may be at risk for developing another UTI. Unfortunately in 25% of dogs the defect allowing re-colonization is not identified resulting in a frustrating cycle of repeat infections and treatment.

Traditionally, treatment of recurrent infections has relied on characterizing the UTI as a re-infection or a persistent infection. Re-infections occur weeks to months after the initial UTI and are commonly treated as uncomplicated infections. In contrast, persistent infections are typically identified within 2 weeks after termination of treatment for the initial infection. Persistent infection cases are managed with a modified treatment plan to account for bacterial resistance or urinary tract pathology that may be causing the failure in treatment. However, discoveries describing the formation of bacterial colonies within urothelial cells have muddied such assessments. The ability of bacteria to be harbored in a protected intracellular environment allows for re-emergence of persistent bacteria weeks to months later and provides an attractive explanation as to why in many cases identifying the defect allowing for recurrent infection remains elusive.

The internalization of bacteria into urothelial cells is rapid with invasion occurring within 6 hours of bladder inoculation. Clinically this timeframe is important since it indicates that by the time a UTI is diagnosed bacteria may already be protected within the urothelium shielded from traditional antibiotic treatments. Misinterpreting and treating a persistent infection of the bladder wall as an uncomplicated re-infection will ultimately result in treatment failure and has been associated with the development of resistant UTIs. These facts indicate that patients at risk for developing recurrent UTIs may be best served by a treatment plan that focuses not on treating urothelial bacterial colonies but instead on prophylactically blocking the organism's ability to adhere and enter the urothelial cells.

BACTERIAL CHARACTERISTICS

In the past decade there has been a considerable body of work studying the essential roles of bacterial adherence and internalization in the formation of intra-urothelial colonies. Of the bacteria known to cause UTI, only uropathogenic strains of *E. coli* (UPEC), and at least one strain of *K. pneumoniae* are known to invade urothelial cells in an in vivo mouse UTI model. A unifying characteristic of the two genera of bacteria is the presence of the *fim* operon coding for the expression of type 1 fimbriae. Expressed on the surface of the bacteria, fimbriae are integral components for adhesion, internalization, and intracellular colonization of bacteria. Clinically the presence of type 1 fimbriae is associated with an increased severity of infection and they are the primary fimbriae found on the surface of UPECs during initial colonization of the urinary bladder.

Type 1 fimbriae are constructed from numerous elements all encoded within the *fim* gene cluster. Translated proteins include those of fimbrial structure as well as chaperone and usher proteins that through the chaperone-usher pathway promote the proper construction of fimbriae. The structural subunits consist of the main subunit, FimA, which creates 95% of the shaft of the fimbriae, minor subunits FimF and FimG, which support the fimbrial adhesin, and FimH, the minor subunit that is the adhesin located at tip of the fimbriae. Type 1 fimbriae mediate bacterial attachment to the urothelium by the binding of FimH to mannosylated glycoproteins. Given their importance in fimbriae formation and function, FimC, FimH, and FimA all may be a viable therapeutic target when seeking to prevent bacterial colonization of the urothelium. In particular, FimH-specific antibodies have been shown to protect individuals from cystitis, making FimH inhibition an attractive therapeutic target to prevent recurrent UTIs.

CLINICAL UTILITY

The use of antibiotics is the accepted treatment for uncomplicated UTIs. However, in recent years an increase in multidrug resistant uropathogenic bacterial strains has led to treatment failure. If persistent colonies are present within the urothelium treatment with antibiotics will increase the risk of developing a resistant infection over time. Alternate therapies that keep the organism viable while disabling bacterial virulence factors are being explored. Since these drugs are not bactericidal selective pressures yielding resistant strains should be minimal and the normal flora within the body would remain viable. Prophylactic use of these pharmaceuticals may effectively prevent infection in at risk patients by blocking the steps of adherence and internalization before colonies ever have a chance to form.

Since certain components of the chaperone-usher system are conserved across many bacterial species, blocking the action of chaperone proteins such as FimC or PapD, necessary for type 1 and P fimbriae respectively, can prevent proper formation of fimbriae across a broad spectrum making this class of drug a desirable alternative to antibiotics. Experimentally, inhibition of both type 1 and

P fimbriae formation by UPECs has been achieved using a bi-cyclic 2-pyridone class of pillicides. In vitro this pharmaceutical is efficacious, able to decrease bacterial adherence to cultured urothelial cells and biofilm formation by up to 90%. Although this class of drugs is still in development, given their effectiveness and potential for a broad spectrum of activity, pillicides do hold promise for clinical utility in the coming years as an alternative or adjunctive therapy to antibiotics. The basis of this therapy is that pillicides compete with fimbrial structural subunits for FimC binding sites. Without FimC stabilization of structural subunits within the periplasm they are rapidly degraded. More recent work argues that the binding of pillicides to chaperone proteins also alters the ability of chaperone proteins to dock with the usher proteins on the bacterial surface thereby blocking structural subunit assembly.

The second therapeutic strategy aims to disrupt adhesion by blocking the ability of lectins on the tips of fimbriae to interact with carbohydrate moieties located on urothelial cells. This diverse class of drugs attempts to optimize the avidity in which bacterial lectins are blocked by using synthesized multivalent glycoconjugates and glycodendrimers. In vitro experimentation has proven that these molecules specifically bind to the conserved lectin, FimH, located on the tips of bacterial fimbriae and are effective at blocking bacterial adherence to cultured urothelial cells. In a mouse model, heptyl α-D-mannose incubation of bacteria reduced both urine and intracellular bacteria concentrations after bacterial inoculation of the urinary bladder. While promising to date, this class of drugs still lacks clinical evidence of effectiveness.

Another anti-adherence therapeutic that has received considerable attention in recent years is the consumption of cranberries and cranberry extract. It has been demonstrated that proanthocyanidins (PACs) with type A-linkages isolated from cranberries can prevent P-fimbriated UPECs from binding to urothelial cells. It is thought that like glycomimetics, PACs alter the lectin-mannose binding necessary for *E. coli* adherence. Unfortunately, these molecules do not appear to have a similar effect on type 1 fimbriated UPECs or other uropathogenic bacteria limiting their efficacy. There is data suggesting that the consumption of cranberry juice can alter the urine composition such that the adherence of type 1 fimbriated UPECs to urothelial cells is inhibited. The mechanism by which consumption of the juice acts differently than the PACs is not known.

Additional therapies that bear mentioning but fall beyond the scope of simple anti-adherence therapy include the development of anti-bacterial antibodies using various vaccination strategies and bacterial interference where the introduction of an asymptomatic competing bacterial strain into the bladder prevents colonization with a more pathogenic strain. Both of these treatments have achieved modest success in clinical trials and may provide additional prophylactic treatment options for preventing bacterial adherence and colonization of the urothelium.

CONCLUSION

In recent years there has been an emphasis placed on the prevention of UTIs in lieu of antibiotic therapy. This has been a result of the increasing incidence of resistant urinary tract infections particularly in patients that have been previously treated with antibiotics. Given the link between intraurothelial bacterial colonies and persistent infection, the importance of preventative medicine in patients prone to developing recurrent urinary tract infections is apparent.

REFERENCES

1. Cox, C.E. etal. J of Urol, 1961. 86(6): p. 739.2. Bartges, J.W. Vet Clin N Am-Small, 2004. 34(4): p. 923. 3. Seguin, M.A., et al. J Vet Intern Med, 2003. 17(5): p. 622. 4. Mulvey, M.A., et al. Science, 1998. 282(5393): p. 1494. 5. Hillier, S., et al. J Antimicrob Chemoth, 2007. 60(1): p. 92. 6. Martinez, J.J., et al. Embo J, 2000. 19(12): p. 2803. 7. Rosen, D.A., et al. Infect and Immun, 2008. 76(7): p. 3337. 8. Snyder, J.A., et al. Infect and Immun, 2005. 73(11): p. 7588. 9. Rosen, D.A., et al. Infect and Immun, 2008. 76(7): p. 3346. 10. Pinkner, J.S., et al., P Natl Acad Sci USA, 2006. 103(47): p. 17897. 11. Aberg, V., et al. Org Biomol Chem, 2007. 5(12): p. 1827. 12. Touaibia, M., et al. Mini-Rev Med Chem, 2007. 7(12): p. 1270. 13. Gouin, S.G., et al. Chemmedchem, 2009. 4(5): p. 749. 14. Howell, A.B., Mol Nutr Food Res, 2007. 51(6): p. 732. 15. Di Martino, P., et al. World J of Urol, 2006. 24(1): p. 21. 16. Billips, B.K., et al. J Infect Dis, 2009. 200(2): p. 263. 17. Sunden, F., et al. Int J Antimicrob AG, 2006. 28: p. S26.

MICROBIOTA: HOW THE GUT FLORA AFFECTS HEALTH & DISEASE IN DOGS & CATS

Kelly S. Swanson, PhD
Urbana, IL

IMPORTANCE AND MANIPULATION OF GASTROINTESTINAL MICROBIOTA

The hindgut of dogs and cats contains a complex and dense population of microbiota, including bacteria, archaea, fungi, protozoa, and viruses. Depending on its composition, the microbiota may be beneficial or harmful to the host. Although the commensal microbes may be a burden in some respects, this is heavily outweighed by the crucial role they play in the development of the immune system, resistance to pathogenic colonization, short-chain fatty acid (SCFA) production, and gene expression in the gut. The development of gut-associated lymphoid tissue is highly dependent on the colonization of bacterial populations in the gut, with germ-free animals having abnormal intestinal morphology and underdeveloped local immune systems, including decreased lymphocyte numbers, skewed intestinal lymphocyte and antibody profiles, and underdeveloped Peyer's patches[1,2]. Indigenous bacterial populations aid in pathogenic resistance by competing for nutrients and binding sites and producing antimicrobial compounds. Many indigenous microbes produce bacteriocins that have a bactericidal mode of action against other bacterial strains, but not against themselves[3]. The SCFA, mainly acetate, propionate, and butyrate, are the main end-products of bacterial fermentation of organic matter in the large intestine. SCFA are the main energy source for colonocytes, in particular, butyrate, which is the preferred energy substrate of colonic epithelium[4] and may account for up to 70% of its total energy consumption[5]. Similar to lactate, these organic acids decrease luminal pH and assist in pathogen resistance. Commensal bacteria also modulate expression of genes involved with several important intestinal functions, including nutrient absorption, mucosal barrier fortification, xenobiotic metabolism, angiogenesis, and postnatal intestinal maturation[6]. Despite their benefits, significant host resources are used to prevent infection from the microbes inhabiting the gut. Microbes increase turnover of mucus and epithelial cells and are constantly testing gut immunity, leading to the production of inflammatory and immune cells. Several pathogenic strains of bacteria also produce protein toxins that disrupt the normal processes of host cells. Others are able to control NF-κB activation, which is now believed to be a major player in the inflammatory response of the gut[7]. Thus, a stable community of commensal microbiota is important not only for gastrointestinal health, but for overall host health.

The composition of the commensal microbiota is dependent on several factors, including genotype, age, diet, and various other environmental factors[8,9]. Diet has a dramatic impact on the activity of bacterial populations inhabiting the colon. The main energy source of colonocytes (i.e., SCFA) is derived primarily from carbohydrate-based energy substrates reaching the large bowel. Dietary proteins and many endogenous secretions (e.g., mucus, sloughed cells) may also serve as fermentable materials. Increasing the protein flow to the colon provides more substrates for potentially pathogenic species such as *Clostridium* spp., which are known for their ability to degrade amino acids and produce fecal odor. Colonic microbiota metabolize nitrogenous compounds to putrefactive catabolites such as ammonia, biogenic amines, and phenols that are implicated as the major odor components of feces[10,11]. More importantly, many of these protein catabolites may have negative influences on gut health. The use of "prebiotics", "probiotics", and "synbiotics" has become popular in human and companion animal nutrition. A probiotic is a live microbial food supplement that beneficially affects the host by improving its intestinal microbial balance[12]. Maintaining microbial viability during or after pet food processing is a significant issue, often limiting the effectiveness of probiotics in pet foods. In recent years, probiotic supplements have been developed by several companies. While these products may provide benefits, few peer-reviewed published reports pertaining to effective dose in dogs and cats exist. Gibson and Roberfroid[13] introduced the concept of prebiotics to alter microbial populations and, consequently, improve host health. By definition, a prebiotic is a nondigestible food ingredient that beneficially affects the host by selectively stimulating the growth and/or activity of one or a limited number of bacteria in the colon, thus improving host health[13]. Three accepted prebiotic classes currently exist (i.e., fructans, lactulose, galactooligosaccharides), with an active search for other potential prebiotics ongoing. Nearly 40 peer-reviewed publications have tested the effects of prebiotic use in healthy dogs and cats, with many demonstrating beneficial changes in fecal microbiota, fecal fermentative end-products, and stool characteristics. Comprehensive reviews on prebiotic and probiotic testing in healthy dogs and cats have been published recently[14,15]. However, the field is in great need of testing prebiotic and probiotic effectiveness in dogs and cats having gastrointestinal disease.

USE OF MOLECULAR METHODS TO EVALUATE GASTROINTESTINAL MICROBIOTA

The knowledge pertaining to gastrointestinal microbiology and health has largely been obtained using traditional techniques over the past several decades. A significant hindrance to this field has been the inability to effectively identify and quantify microbial species. Researchers have been reliant upon microbial culturing methods, which are not only laborious, time-consuming, and often inaccurate[16], but also greatly limited in scope. For those that survive in culture, identification among species or genera is difficult, if not impossible. It has been estimated that cultivation methods identify < 1% of total microbes in water and soils[17], and only 20% of total microbes in the gut of mammalian species[18]. Thus, alternative methods have been needed to characterize the phylogeny and functional capacity of gut microbes and obtain a greater understanding of host-microbe interrelationships. Several molecular tools, such as quantitative polymerase chain reaction (qPCR), fluorescent in-situ hybridization (FISH), denaturing gradient gel electrophoresis (DGGE), quantitative dot blot hybridization, restriction fragment length polymorphism (RFLP), and large scale DNA sequencing, have been used to overcome some of the limitations of culture-based techniques. The primary target for these techniques has been small subunit ribosomal RNA (16S rRNA), as it is ubiquitously present in all microbes and contains both conserved and variable sequence regions across species of bacteria and archaea[19]. Thus, the 16S rRNA sequence is appreciated as a molecular barcode for microbial identity. The 16S rRNA sequence database has grown exponentially since 1990[20]. The average bacterial 16S rRNA gene has a length of ~1,500 nucleotides and contains regions with different degrees of variability, enabling researchers to distinguish organisms at different phylogenetic levels (species to domain)[17,21]. Such methods have been used to evaluate how diet[22,23], age[24,25], and breed[24] may impact microbial populations in healthy dogs and cats.

In the past few years, high-throughput techniques have been developed. For instance, a 16S rRNA microarray technique (i.e., Phylochip) has been used as a convenient and high-throughput method for studying microbial communities[26]. This technique is based on the complementary hybridization of microbial 16S rRNA sequences to specific probes spotted on the microarray. Therefore, a single microarray containing a large number of immobilized oligonucleotides targeted to specific microbial sequences permits a large scale quantitative and phylogenetic analysis of a microbial community[26,27]. To our knowledge, however, 16S rRNA microarray analysis for microbial populations in dogs and cats has not been published. DNA sequencing techniques have dramatically increased the speed by which microbial communities can be characterized. Traditional Sanger sequencing methods were used to generate the initial 16S rRNA sequence data. This method has greatly improved over time and now generates long sequences (800 to 1,000 bp), allowing for microbial identity with a relatively high confidence[28]. Recent DNA sequencing techniques such as 454-pyrosequencing that generate sequence lengths of approximately 450 bp and are much faster and cheaper than the Sanger sequencing method, have quickly been applied to microbiological studies[28]. Although it is not as consistent and generates shorter sequences than the Sanger method, the read length by pyrosequencing continues to increase and is expected to be of similar read length in the near future. Therefore, this method enables the efficient study of microbial DNA sequence data on a large scale and in terms of phylogeny (e.g., who is there?) and metagenomic (what are they doing?) analyses. The human microbiome project (HMP) is an excellent example of how to apply high throughput DNA pyrosequencing to human and animal health. In May 2007, the National Institutes of Health (NIH) roadmap initiated the HMP to characterize the entire community of microbiota inhabited in the human gastrointestinal tract, mouth, vagina, nasal cavity, skin, and other body regions to determine the unique and complex microbial structures of each body region and relate it to human health and disease[29]. Pyrosequencing is a core technique being used to determine the microbial diversity and relatedness, and even for taxonomic assignment in the HMP[30]. This method is also essential for metagenomic analyses of complex microbial communities in the HMP. Although the focus of the HMP is human health and disease, the knowledge gained from the microbial sequencing projects and projects studying host-microbe relationships under this initiative will also benefit those studying pets. The use of high throughput pyrosequencing has recently been applied to healthy dog and cat gastrointestinal microbiota[31-33].

MICROBES AND GASTROINTESTINAL HEALTH AND DISEASE

Recent studies have used sequencing techniques to characterize the healthy canine gastrointestinal microbiota, revealing several hundred bacterial phylotypes. Middelbos et al.[31] detected the presence of 7 bacterial phyla, including Actinobacteria, Firmicutes, Bacteroidetes, Fusobacteria, Proteobacteria, Spirochaetes, and Tenericutes in healthy adult dog feces, with a co-dominant presence of Firmicutes, Bacteroidetes, and Fusobacteria. These data agreed with the findings of Suchodolski et al.[34] who reported co-dominance of these 3 phyla in the colon of dogs based on sequence data generated by the Sanger method. Using a metagenomic approach in the same dogs used by Middelbos et al.[31], Swanson et al.[33] observed a bacterial co-dominance of Bacteroidetes (37-38%) and Firmicutes (30-35%), but a much lower Fusobacteria population. It is likely that methodology was an important factor affecting the results of these studies and, thus, must be considered in the future. Because Swanson et al.[33] used a metagenomics approach, sequencing all DNA rather than just the 16S rRNA gene, they also revealed the presence of archaea (~1% of sequences; *Crenarchaeota* and *Euryarchaeota*), fungi (~0.4% of sequences; 3 phylotypes), and viruses (~0.4% of sequences; primarily bacteriophages) in the canine gut. Approximately 50% of all sequences in that dataset were classified metabolically by KEGG. Not surprisingly, functional characterization revealed a high prevalence of genes pertaining to carbohydrates; protein metabolism; DNA metabolism; cofactors, vitamins, prosthetic groups and pigments; amino acids and derivatives; cell wall and capsule; and virulence[33]. Handl et al.[32] also used pyrosequencing to characterize the bacterial and fungal communities present in healthy dogs. Similar to previous reports, Firmicutes was the predominant bacterial phyla present. A total of 85 bacterial genera were identified, with *Clostridium* (clusters XIVa and XI), *Ruminococcus* , *Dorea*, *Roseburia*, and *Turicibacter* being most prevalent. Of the 33 fungal genera identified, nearly 77% were classified as *Nakaseomyces*.

Similar studies have been applied to healthy feline gastrointestinal microbiota. Using Sanger sequencing and comparative 16S rRNA gene analysis, Ritchie et al.[35] reported that Firmicutes (68%) were the most predominant bacterial phylum, followed by Proteobacteria (14%), Bacteroidetes (10%), Fusobacteria (5%), and Actinobacteria (4%) in healthy cat feces. A follow-up study focused on characterizing lactobacilli and bifidobacteria revealed the presence of 23 *Lactobacillus*-like and 11 *Bifidobacterium*-like sequences in healthy cats[36]. Handl et al.[32] used pyrosequencing to characterize the bacterial and fungal communities present in healthy cats. As in dogs, Firmicutes was the prominent bacterial phyla present. A total of 113 bacterial and 17 fungal genera were identified cats. Similar to dogs, *Clostridium* (clusters XIVa and XI), *Roseburia, Ruminococcus* , *Dorea*, *Turicibacter,* and *Catenibacterium* were most prevalent. Using a metagenomics approach, Barry et al.[37] reported a predominance of Firmicutes, Bacteroidetes, and Proteobacteria in healthy cat feces. Although present in low percentages (<1%), archaeal, fungal, and viral sequences were also described. Primary functional categories as determined by KEGG included carbohydrates; protein metabolism; and amino acids and derivatives. Similar analyses by COG revealed the presence of genes associated with amino acid metabolism and transport; carbohydrate transport and metabolism; and replication, recombination, and repair.

Although the exact cause of many gastrointestinal diseases is unknown, bacterial populations are often altered in clinical patients. Many potential pathogens, including *Clostridium perfringens*, *C. difficile*, *Klebsiella pneumoniae*, *Campylobacter jejuni*, *Salmonella typhimurium*, enteric *Helicobacter* spp., *Yersinia enterocolitica*, enteropathogenic *Escherichia coli*, and others have been identified[38,39]. Many of these species either disturb the intestinal epithelial cell functionality, barrier function or immune cell populations. A critical next step is to use the techniques described above to identify bacterial populations associated with disease.

Moreover, distinction of the luminal vs. mucosal (adhered) populations throughout the regions of the gut (stomach vs. small intestine vs. large intestine) is of importance. Suchodolski[40] provides an excellent overview of the microbial populations shown to be altered in dogs and cats with gastrointestinal disease, many of which have been identified using molecular techniques, and the mechanisms by which they contribute to disease. Inflammatory bowel disease (IBD) is commonly studied in this regard, with recent molecular experiments revealing increased Enterobacteriaceae (*E. coli*)[41,42], Proteobacteria[43,44], and *Desulfovibrio*[45], decreased overall diversity, and decreased *Clostridium* [43,44], *Bifidobacterium*[45], *Bacteroides*[45], *Faecalibacterium* [43,46], *Ruminococcus*[43], and *Dorea*[43] populations in diseased dogs and cats. Craven et al.[47] noted the role of antibiotic resistant *E. coli* in canine colitis patients, while other chronic enteropathies have been associated with decreased overall microbial diversity[48], but increased populations of Proteobacteria[49], Lactobacillales (*Streptococcus* and *Abiotrophia*)[50], *Bacteroides*[51], *C. perfringens*[52], *Enterococcus faecalis* and *faecium*[52], and *Campylobacter* spp.[53]. All forms of diarrhea, however, are not of microbial origin, as diarrhea as a consequence of long distance racing (sled dogs) was not associated with increased *C. perfringens*, *C. difficile*, *Campylobacter*, *Salmonella*, *E. coli* 0157, *Giardia*, or *Cryptosporidium* spp.[54]. Identifying whether these bacteria are contributors to the disease process, or is merely a species that benefits of a diseased environment, is often unknown and requires more study in the future.

FUTURE DIRECTIONS

Technological advances continue to expand our research capabilities. In addition to continued phylogenetic and metabolic characterization of microbial populations responsible for health and disease using DNA-based assays, application of other technologies to analyze RNA (microbial gene transcripts), protein, and metabolite profiles, are necessary to increase our understanding of host-microbe relationships and identify effective prevention and treatment strategies. The biochemical details of the gastrointestinal tract are largely undefined. By using broad mass spectrometry-based metabolomics assays, urine, blood, digesta, or fecal samples may be analyzed to reveal a deeper understanding of the gut microbes and how they impact host metabolism and health[55]. A few examples of this already exist in the literature, with mass spectrometry-based assays being used to create metabolic fingerprints of biological fluids of dogs and cats[56] or clinically relevant bacteria[57]. Despite the challenges that surround this field, the powerful tools now available to the research community will enhance our ability to understand gastrointestinal health and disease on a molecular level, leading to improved disease diagnosis and treatment in coming years.

REFERENCES

[1]Thorbecke G. **Ann NY Acad Sci** 1959;78:237. [2]McCracken V and Gaskins H. **Probiotics: A Critical Review** 1999;85. [3]Tagg J et al. **Bacteriol Rev** 1976;40:722. [4]Roediger W. **Gastroenterol** 1982;83:424. [5]Roediger W. **Gut** 1980;21:793. [6]Hooper L et al. **Science** 2001;291:881. [7]Karin M and Ben-Neriah Y. **Ann Rev Immunol** 2000;18:621. [8]Hopkins M et al. **Gut** 2001;48:198. [9]Zoetendal E et al. **J Nutr** 2004;134:465. [10]Spoelstra S. **Agric Environ** 1980;5:241. [11]O'Neill D and Phillips V. **J Agric Eng Res** 1992;53:23. [12]Fuller R. **J Appl Bacteriol** 1989;66:365. [13]Gibson G and Roberfroid M. **J Nutr** 1995;125:1401. [14]Swanson K and Fahey G. **Prebiotics: Development and Application** 2006;213. [15]Vester B and Fahey G. **Handbook of Prebiotic and Probiotic Ingredients** 2010;355. [16]Greetham H et al. **J Appl Microbiol** 2002;93:640. [17]Amann R et al. **Microb Rev** 1995;59:143. [18]Eckburg P et al. **Science** 2005;308:1635. [19]Clarridge J. **Clin Microbiol Rev** 2004;17:840. [20]Tringe S and Hugenholtz P. **Curr Opin Microbiol** 2008;11:442. [21]Namsolleck P et al. **Microb Ecol Health Dis** 2004;16:71. [22]Middelbos I et al. **J Anim Sci** 2007;85:3033. [23]Lubbs et al. **J Anim Phys Anim Nutr** 2009;93:113. [24]Simpson J et al. **Microb Ecol** 2002;44:186. [25]Vester B et al. **Arch Anim Nutr** 2009;63:254. [26]Zhou J. **Curr Opin Microbiol** 2003;6:288. [27]Guschin D et al. **Appl Environ Microbiol** 1997;63:2397. [28]Petrosino J et al. **Clin Chem** 2009;55:856. [29]Peterson J et al. **Genome Res** 2009;19:2317. [30]Hamady M and Knight R. **Genome Res** 2009;19:1141. [31]Middelbos I et al. **PLoS One** 2010;5:e9768. [32]Handl S et al. **FEMS Microbiol Ecol** 2011(in press). [33]Swanson K et al. **ISME J** 2011(in press). [34]Suchodolski J et al. **FEMS Microbiol Ecol** 2008;66:567. [35]Ritchie L et al. **FEMS Microbiol Ecol** 2008;66:590. [36]Ritchie L et al. **Vet Microbiol** 2010;144:140. [37]Barry K et al. **ISME J** (submitted). [38]Greene C. **Infectious Diseases of the Dog and Cat** 1998;595. [39]Marks S and Kather E. Vet Clin North Am Small Anim Pract 2003;33:1029. [40]Suchodolski J. **J Anim Sci** 2011(in press). [41]Janeczko S et al. **Vet Microbiol** 2008;128:178. [42]Xenoulis P et al. **FEMS Microbiol Ecol** 2008;66:579. [43]Jergens A et al. **J Vet Intern Med** 2010;24:725. [44]Suchodolski J et al. **Vet Microbiol** 2010;142:394. [45]InnessV et al. **J Anim Phys Anim Nutr** 2007;91:48. [46]Suchodolski J et al. **ACVIM Proc** 2011(in press). [47]Craven M et al. **J Vet Intern Med** 2010;24:819. [48]Craven M et al. **J Vet Intern Med** 2009;23:731. [49]Markel M et al. **ACVIM Proc** 2011(in press). [50]Allenspach K et al. **Vet Microbiol** 2010;146:326. [51]Jia J et al. **FEMS Microbiol Ecol** 2010;71:304. [52]Bell J et al. **Interdiscip Perspect Infect Dis** 2008;149694. [53]Chaban B et al. **BMC Microbiol** 2010;10:73. [54]McKenzie E et al. **J Vet Intern Med** 2010;24:97. [55]Wikoff W et al. **Proc Natl Acad Sci** 2009;106:3698. [56]Beckman M et al. **Brit J Nutr** 2010;103:1127. [57]Benagli C et al. **PLoS One** 2011;e16424.

EVIDENCE BASED BENEFITS OF PRE- AND PROBIOTICS IN DOGS AND CATS

Stanley L. Marks, BVSc, PhD, DACVIM (Small Animal Internal Medicine & Oncology), DACVN
Davis, CA

The mammalian intestinal tract contains a complex, dynamic, and diverse population of non-pathogenic bacteria. Researchers have estimated that the human body contains 10^{14} cells, only 10% of which are not bacteria, and belong to the human body proper.[1] There has been a plethora of research focusing on the mechanisms by which pathogenic bacteria influence intestinal function and induce disease; however, recent attention has focused on the indigenous non-pathogenic microorganisms and the ways in which they may benefit the host. Initial colonization of the sterile newborn intestine occurs with maternal vaginal and fecal bacterial flora. The first colonizers have a high reduction potential and include species such as enterobacter, streptococcus, and staphylococcus. These bacteria metabolize oxygen, favouring the growth of anaerobic bacteria, including lactobacilli and bifidobacteria. Colonization with these bacteria is significantly delayed in caesarean deliveries,[2] leading to delayed activation of the efferent limb of the mucosal immune response.[3] Additional beneficial effects of developing a normal bacterial flora is seen in germ free mice that have small intestines that weigh less than their healthy counterparts. This effect occurs partly due to underdevelopment of lymphoid constituents, with a lack of plasma cells in the lamina propria and Peyer's patches, and subsequent reduction in IgA production. Exposure to bacteria results in a reversal of this phenomenon within 28 days of exposure.[4]

PROBIOTICS

Probiotics refer to live microorganisms which when administered in adequate amounts confer a health benefit on the host. The term probiotic was derived from the Greek, meaning "for life." The Food and Agricultural Organization of the United States (FAO) and the World Health Organization (WHO) have stated that there is adequate scientific evidence to indicate that there is potential for probiotic foods to provide health benefits and that specific strains are safe for human use.[5] There has been a literal explosion of interest and research on the subject in recent years. A PubMed search of the term "probiotics" for peer-reviewed publications published during the past 5 years revealed over 4,000 publications, in contrast to 300 publications over the previous 55 years. Despite this activity, much still remains to be done to determine the specific indications and applications of probiotics in dogs and cats. There has been tremendous interest among veterinary pet food companies and manufacturers of animal health and wellness products to market probiotic formulations that are safe, pure, stable, and confer a beneficial effect in dogs and cats. These products are generally preferred to the multitude of over-the-counter probiotics marketed for veterinary use, given the concerns pertaining to quality control of the over-the-counter products.[6] A number of criteria are essential for efficacy and safety of probiotics. These include resistance to gastric acid and bile, ability to colonize the gastrointestinal tract, efficacy against pathogenic microorganisms, and modulation of the immune system.[7] Several potential mechanisms have been proposed for how probiotics reduce the severity or duration of diarrhea: competition with pathogenic bacteria or viruses for nutrients, competition for receptor sites, modification of the metabolic activity of the intestinal microflora, and the direct antagonism through the action of antimicrobial metabolites.[8,9]

Evidence for the Benefits of Probiotics in People

There is currently level 1 evidence (i.e., data from either high-quality, randomized controlled trials with statistically significant results and few design limitations or from systematic reviews of trials) for effectiveness of probiotics in treating lactose intolerance/maldigestion, treating acute infectious or nosocomial diarrhea in children, preventing or treating antibiotic-associated diarrhea, preventing and maintaining remission of pouchitis in adults, and maintaining remission of ulcerative colitis in adults.[10] In addition, there is level 2 evidence (evidence obtained from randomized trials that have limitations in methodology or results that have wide confidence intervals) for using probiotics to treat traveler's diarrhea, prevent sepsis secondary to severe acute pancreatitis, and prevent infections in postoperative patients.[10] Necrotizing enterocolitis is a devastating intestinal disorder seen more commonly in preterm infants. The syndrome is characterized by abdominal distention, bilious vomiting, bloody stools, lethargy, apnea, and bradycardia.[11] Progressive intestinal necrosis leads to an inflammatory cascade with septic shock and possibly death. The mortality rate ranges from 20 to 30%, and of those who survive, 25% experience long-term sequele, such as short bowel syndrome and intestinal obstruction.[11] The benefits of administration of the probiotics, *Lactobacillus acidophilus* and *Bifidobacterium infantis* were readily appreciated when the incidence of necrotizing enterocolitis and overall mortality were reduced by 60% in an intensive care unit.[12] Evidence is also accumulating for the benefits of probiotics in preventing *Clostridium difficile* infection (CDI) in hospitalized human patients receiving antibiotics. A meta-analysis evaluating the efficacy of probiotics in prevention and treatment of diarrhea associated with the use of antibiotics recently showed an odds ratio of 0.39 (95% CI, 0.25 to 0.62; $P < 0.001$) in favour of active treatment over placebo with *Saccharomyces boulardii* and 0.34 (95% CI, 0.19 to 0.61; $P < 0.01$) for lactobacilli.[13] It should be noted that cases of fungal infection have been reported in rare cases, usually associated with immunocompromised catheterized patients.[13] Beneficial responses have also been documented in atopic infants fed with the probiotics, *L. rhamnosus* GG and *Bifidobacterium* Bb12.[15] Atopic infants fed with these probiotics had less allergic reaction to cow's milk proteins and a lower incidence of atopic dermatitis.

Evidence for the Benefits of Probiotics in Dogs

To date, only a relatively small number of studies have been published evaluating the effects of probiotics in dogs, and many of these have focused on the intestinal microflora in apparently healthy dogs. Probiotic strains of human or canine origin (*Lactobacilli*, *Bifidobacter*, and *Enterococcus*) were used in healthy adult dogs or dogs with food-responsive diarrhea to assess their effects on intestinal microbial populations, their ability to reduce specific pathogens in feces, and effectiveness as immunomodulators.[16-21] In many of these studies, probiotics added to the food in healthy dogs had an equivocal effect on fecal microflora and pathogens.[18-21] However, it is important to note that most of these studies were not randomized, controlled trials, and the strains of probiotic varied from study to study, making interpretation of findings more challenging. In addition, many studies focused on fecal isolation and quantitative cultures of putative pathogenic bacteria such as *C. perfringens,* rather than on the evaluation of more meaningful end points such as shifts in the microbial flora, mucosal immunopathology, and alterations in intestinal integrity. Only two studies addressing the role of probiotics in management of dietary sensitivity and food-responsive diarrhea have been published to date, with overall positive results.[16,17] Only one of these studies was a randomized, placebo-controlled clinical trial,[16] and the results of that study, although clinically positive (all of the dogs in the study improved when they were placed on the elimination diet) showed no specific changes in the inflammatory cytokine patterns or a specific benefit of the probiotic. The immunomodulatory effects of *Enterococcus faecium* SF68 have been studied in dogs, and the probiotic was associated with increased fecal IgA concentrations and increased vaccine-specific circulating IgG and IgA concentrations.[22] Although increased immune globulins may suggest enhanced immune response, the clinical relevance of this finding is not known.

Additional studies are warranted in dogs to further assess the immunomodulatory effects of probiotics and to evaluate their safety. The latter issue is particularly important given the recent finding of increased intestinal adhesion of *Campylobacter jejuni* in an *in vitro* model of canine intestinal mucus following incubation with *Enterococcus faecium*.[23] It should be noted that this *E. faecium* strain is different from the *E. faecium* SF68 strain available commercially; moreover, to date there has been no clinical or anecdotal evidence of *Campylobacter*-associated diarrhea in dogs associated with *E. faecium* administration. Short-term treatment (6 weeks) with *E. faecium* SF68 to 20 dogs with chronic naturally acquired subclinical giardiasis failed to affect giardial cyst shedding or fecal giardial antigen and did not alter innate or adaptive immune responses at multiple time points.[24] These results are in contrast to those shown following the oral feeding of *E. faecium* strain SF68 to mice experimentally infected with *Giardia intestinalis* trophozoites.[25] Oral feeding of *E. faecium* strain SF68 starting 7 days before inoculation with *Giardia* trophozoites significantly increased the production of specific anti-*Giardia* intestinal IgA and blood IgG. This humoral response was mirrored at the cellular level by an increased percentage of CD4+ T-cells in the Peyer's patches and in the spleens of SF68-fed mice. The improvement of specific immune responses in probiotic-fed mice was associated with a diminution in the number of active trophozoites in the small intestine as well as decreased shedding of fecal *Giardia* antigens (GSA65 protein). The latter findings underscore the importance of carefully evaluating the animal model of disease studied, the timing of probiotic administration (prior to infection or following infection), and the specific end-points studied.

Evidence for the Benefits of Probiotics in Cats

Unfortunately, there is little published information pertaining to probiotic use in cats, and there are no clinical studies reporting a beneficial effect of probiotic therapy for any feline disease. One study evaluating the effect of dietary supplementation with the probiotic strain of *Lactobacillus acidophilus* DSM 13241 (2×10^8 CFU/d for 4.5 weeks) administered to 15 healthy adult cats demonstrated that recovery of the probiotic from the feces of the cats was associated with a significant reduction in *Clostridium* spp. and *Enterococcus faecalis*.[26] However, the immunomodulatory effects were reported based on decreased lymphocyte and increased eosinophil populations and increased activities of peripheral blood phagocytes. The relevance of these findings is unclear, because this study was not a randomized trial and the changes reported in the populations of peripheral blood cells cannot be extrapolated into evidence of systemic health benefits. Evaluation of the effect of supplementation with *Enterococcus faecium* strain SF68 on immune function responses following administration of a multivalent vaccine was evaluated in specific pathogen-free kittens.[27] This prospective, randomized, placebo-controlled study resulted in the recovery of *E. faecium* SF68 from the feces of seven of nine cats treated with the probiotic, and a nonsignificant increase in feline herpesvirus 1–specific serum IgG levels. Concentrations of total IgG and IgA in serum were similar in the probiotic and placebo groups, and the percentage of CD4+ lymphocytes was increased significantly only in kittens at 27 weeks and not at any other time points. Probiotics also have been evaluated in juvenile captive cheetahs, a population with a relatively high incidence of bacteria-associated enteritis. Administration of a species-specific probiotic containing *Lactobacillus* Group 2 and *Enterococcus faecium* to 27 juvenile cheetahs was associated with a significantly increased body weight in the treatment group, with no increase in the control group.[28] In addition, administration of the probiotic was associated with improved fecal quality in the probiotic group.

PREBIOTICS

A prebiotic is defined as a "nondigestible food ingredient that beneficially affects the host by selectively stimulating the growth of and/or activates the metabolism of one or a limited number of health promoting bacteria in the intestinal tract."[29] The most common prebiotics studied are fructans, although other prebiotics such as mannans, lactosucrose, and lactulose are also being evaluated.

Evidence for the Benefits of Prebiotics in People:

There is strong evidence in people for the impact of non-digestible oligosaccharides on bowel habitat, and promising evidence that the consumption of inulin-type fructans may result in increased Ca absorption. In addition, there is preliminary evidence in experimental animals of a preventive effect against colon cancer,[31] and beneficial effects on lipid metabolism including a reduction of low density lipoproteins and a lowering of serum triglycerides.[31]

Evidence for the Benefits of Prebiotics in Dogs and Cats:

Unfortunately, there is a paucity of information evaluating the clinical benefits of prebiotics in dogs and cats. Most of the outcomes published in the literature are limited to nutrient digestibility, microbial concentrations in feces, and fecal protein catabolites that may not necessarily denote a health benefit to the patient. The effects of short chain fructooligosaccharides (scFOS) were evaluated in a group of German Shepherd dogs suspected to have IgA deficiency.[32] Although the scFOS supplemented dogs had decreased aerobic and anaerobic bacteria in intestinal biopsies, the findings of the study were clouded in light of the fact that anaerobic bacterial counts did not decrease in the intestinal fluid samples of the dogs supplemented with scFOS. The effects of various oligosaccharides have been tested in adult ileal cannulated dogs to evaluate the effects on ilial and total tract nutrient digestibilities, microbial populations, ileal pH, ammonia, blood glucose, fecal consistency, and SCFA concentrations. Oligosaccharides (oligofructose, mannanoligossacharides, and xylooligosaccharides) were each given at 0.5% of the diet in a Latin square design. The only significant finding was a decrease in fecal *Clostridium perfringens* populations in dogs fed MOS.[33] The effects of 3% inulin supplementation of elimination and hydrolyzed protein diets to healthy dogs was associated with a slight increase in fecal moisture content (not clinically relevant), decreased apparent nutrient digestibility coefficients of crude protein in dogs on the elimination diet, and no effect on fecal IgA concentrations.[34] Adult beagles fed diets containing cellulose or beet pulp plus oligofructose for 6 weeks were found to have similar fecal concentrations of total anaerobes and aerobes; however, the dogs fed oligofructose had fewer Enerobacteriaceae and clostridia and greater numbers of lactobacilli. In addition, dogs fed oligofructose had longer and heavier small intestines (35% heavier), and 37% more mucosal mass with consequent greater absorptive surface area.[35] Cats fed diets containing 0 or 0.75% oligofructose had significantly increased fecal concentrations of lactobacilli and decreased concentrations of *C. perfringens* and *E. coli* compared with controls.[36] A study evaluating the effects of short-chain fructooligosaccharides (0.5%) and galactooligosaccharides (0.5%) in healthy cats showed no effect on fecal protein catabolites, including ammonia, 4-methylphenol, indole, and biogenic amines, underscoring the fact that concentrations of oligosaccharides > 0.5% should be used to elicit a positive response.[37]

Conclusions and Future Directions:

The potential benefits and specific indications for the administration of pro- and prebiotics to dogs and cats have yet to be fully defined, although our knowledge and understanding of the nature and diversity of the feline and canine intestinal microbiota during health and disease has expanded rapidly following the advent of molecular methodologies to assess the intestinal microflora. Applying either polymerase chain reaction (PCR)–Denaturing Gradient Gel Electrophoresis or Terminal Restriction Fragment Length Polymorphisms (TRF) and high-throughput sequencing of 16S rRNA libraries to the study of gastrointestinal tract microbial ecology should facilitate the identification of major culturable and nonculturable populations and provide a tool for studying shifts in these populations over time and under different conditions. Defining a role for pro- and prebiotics in dogs and cats will require completion of prospective, randomized, placebo-controlled studies that rely on clinically relevant end points related to a particular physiological or pathological condition. Further studies are warranted to determine the need for probiotics to be live microorganisms following the provocative studies of Rachmilewitz et al., who documented that the beneficial effects of probiotics are mediated by their DNA, circumventing the need for live, viable bacteria.[38] Pro- and prebiotics do appear to have a potential role in the prevention and treatment of various gastrointestinal illnesses, but it is likely that benefits achieved are specific to the bacterial species used and to the underlying disease context.

REFERENCES

[1]Savage DC. **Annu Rev Microbiol** 31:107-33, 1977; [2]Gronlund MM, et al. **J Pediatr Gastroenterol Nutr** 28:19-25, 1999; [3]Insoft RM, et al. **Pediatr Clin N Am** 43:551-71, 1996; [4]McCraken VJ, et al. **Cell Microbiol** 3:1-11, 2001; [5]Food and Agricultural Organization of the United States and World Health Organization (Online), 2001; [6]Weese JS. **Can Vet J**, 44(12)982-3, 2003. [7]Tuomola E, et al. **Am J Clin Nutr** 73:393S-398S, 2001; [8]Coconnier MH, et al. **FEMS Microbiol Lett** 110:299-305, 1993; [9]Rolfe RD, et al. **J Infect Dis** 143:470-475, 1981; [10]Guarner F. In Versalovic J, Wilson M, editors: **Therapeutic Microbiology: Probiotics and Related Strategies**. ASM Press, Washington DC, p.255, 2008; [11]Caplan MS, et al. **J Pediatr Gastroenterol Nutr** 30 (Suppl 2):S18-S22, 2000. [12]Hoyos AB. **Int J Infect Dis** 3: 197-202, 1999. [13]D'Souza AL, et al. **Br Med J** 324:1361; 2002. [14]Hennequin C, et al. **Eur J Clin Microbiol Infect Dis** 19:16-20, 2000. [15]Kalliomaki M, et al. **Lancet**. 357(9262):1076-9, 2001. [16]Sauter SN, et al. **Anim Nutr** 90:269, 2006. [17]Pascher M, et al. **Arch Anim Nutr** 62:107, 2008. [18]Vahjen W, et al. **Arch Anim Nutr** 57:229, 2003. [19]Swanson KS, et al. **J Nutr** 132:3721, 2002. [20]Biagi G, et al. **Vet Microbiol** 124:160, 2007. [21]Baillon MLA, et al. **Am J Vet Res** 65:338, 2004. [22]Benyacoub J, et al. **J Nutr** 133:1158-1162, 2003. [23]Rinkinen M, et al: **Vet Microbiol** 92:111, 2003. [24]Simpson KW, et al. **J Vet Int Med** 23:476-481, 2009. [25]Benyacoub J, et al. **J Nutr** 135(5):1171-6. 2005. [26]Marshall-Jones ZV, et al. **Am J Vet Res** 67:1005, 2006; [27]Veir JK, et al: **Vet Ther** 8:229, 2007; [28]Koeppel KN, et al: **J So Afr Vet Assoc** 77:127, 2006. [29]Gibson GR, et al. **J Nutr** 125:1401-12, 1995. [30]Reddy DS, et al. **Carcinogenesis** 18:1371-1374, 1997; [31]Delzenne NM, et al. **Am J Clin Nutr** 57, suppl 5, S820. [32]Willard MD, et al. **Am J Vet Res** 55: 654-659, 1994. [33]Strickling JA, et al. **Anim Feed Sci Technol** 86:205-219, 2000; [34]Verlinden A, et al. **Br J Nutr** 96:936-944, 2006; [35]Buddington RK, et al. **Am J Vet Res** 60:354-358, 1999; [36]Sparkes AH, et al. **Am J Vet Res** 59:436-440, 1998; [37]Kanakupt A, et al. **Am J Anim Sci** 2011 (In Press). [38]Rachmilewitz D, et al. **Gastroenterol** 126:520-528, 2004.

ANTI-INFLAMMATORY EFFECTS OF OMEGA-3 FATTY ACIDS IN DOGS AND CATS

Patricia A. Schenck, DVM, PhD
Lansing, MI

FATTY ACID STRUCTURE

Fatty acids containing two or more double bonds are termed polyunsaturated fatty acids (PUFA), and based on the location of the last double bond relative to the terminal methyl end, they are classified as omega-3 (n-3) or omega-6 (n-6) fatty acids. PUFAs are an important component of cell membranes, and also have a role in the regulation of the inflammatory response via production of inflammatory mediators termed eicosanoids. The most common omega-6 fatty acids are linoleic acid (LA) with 18 carbons and 2 double bonds (18:2), and arachidonic acid with 20 carbons and 4 double bonds (20:4). Linoleic acid is a required fatty acid in both dogs and cats, and arachidonic acid is required in the cat. Important omega-3 fatty acids include alpha-linolenic acid (ALA) with 18 carbons and 3 double bonds (18:3), eicosapentaenoic acid (EPA) with 20 carbons and 5 double bonds (20:5), and docosahexaenoic acid (DHA) with 22 carbons and 6 double bonds (22:6). Linolenic acid is a required fatty acid in both the dog and cat.

METABOLISM OF OMEGA-6 AND OMEGA-3 FATTY ACIDS

Both linoleic acid and linolenic acid can be metabolized to produce longer chain fatty acids utilizing desaturase and elongase enzymes.[1] Linoleic acid (18:2 n-6) is metabolized to gamma-linoleic acid (18:3 n-6) using delta-6 desaturase, and elongated to dihomogammalinoleic acid (DGLA, 20:3 n-6). In dogs, DGLA is desaturated to arachidonic acid (20:4 n-6) using delta-5 desaturase; this conversion is poor in the cat due to a lack of delta-5 desaturase. Arachidonic acid is the primary fatty acid utilized for the production of prostaglandins, thromboxanes, and leukotrienes. Arachidonic acid can also be further desaturated and elongated to produce longer chain omega-6 fatty acids. Alpha-Linolenic acid (18:3 n-3) is metabolized to stearidonic acid (18:4 n-3) using delta-6 desaturase, and elongated to eicosatetraenoic acid (20:4 n-3). In dogs, eicosatetraenoic acid is desaturated to EPA using delta-5 desaturase. EPA can be utilized for the production of other series of prostaglandins, leukotrienes, and thromboxanes. EPA can also be further desaturated and elongated to DHA. Desaturation and elongation of fatty acids occurs primarily in the liver, and since the same enzymes are used for both omega-6 and omega-3 fatty acids, there is competition for metabolism. Normally the omega-3 fatty acids are preferentially metabolized over the omega-6 fatty acids, but in diets containing primarily omega-6 fatty acids, there is greater conversion of LA to arachidonic acid. In humans, about 8-20% of ALA is converted to EPA, whereas only 0.5-9% of ALA is converted to DHA. The recommendation for humans is about 0.5-1.8g omega-3 fatty acids per day for optimal health. In dogs, the rate of conversion of ALA to EPA is most likely even less, underlying the importance of providing EPA and DHA in the diet, rather than relying on the conversion of ALA to EPA for benefit.

IMPACT OF OMEGA-6 AND OMEGA-3 FATTY ACIDS ON INFLAMMATORY MEDIATORS

Inflammation begins the immunological process to eliminate invading pathogens and to repair damaged tissues.[2] Up-regulation of adhesion molecules such as intercellular adhesion molecule-1 (ICAM-1), vascular cell adhesion molecule-1 (VCAM-1), and E-selectin on the surface of endothelial cells induces movement of cells into the inflammatory site. Granulocytes appear early, with monocytes/macrophages and lymphocytes appearing later in inflammation. Bacterial endotoxin directly activates monocytes/macrophages to form cytokines such as tumor necrosis factor (TNF)-alpha, interleukin (IL)-1, IL-6, IL-8, eicosanoids (prostaglandins, thromboxanes), nitric oxide, and other mediators. Inflammation is a normal response, but when uncontrolled, it can cause excessive damage to tissues.

Omega-6 Derivatives

The eicosanoid family of inflammatory mediators are derived from the 20 carbon fatty acids. Most eicosanoids are derived from arachidonic acid (an omega-6 fatty acid) due to the high levels of arachidonic acid and linoleic acid found in most diets. Eicosanoids include prostaglandins (PGs), thromboxanes (TXs), leukotrienes (LTs), and hydroxyeicosatetraenoic acids (HETEs). Metabolism of arachidonic acid by cyclooxygenase (COX) enzymes gives rise to the 2-series of PGs and TXs, including TXA_2 and TXB_2. Monocytes and macrophages produce large quantities of PGE_2 and $PGF_{2\alpha}$, neutrophils produce PGE_2, and mast cells produce PGD_2. Metabolism of arachidonic acid by 5-lipoxygenase (5-LOX) produces hydroxy and hydroperoxy derivatives (5-HETE and 5-HPETE), along with the 4-series of LTs (LTA_4, LTB_4, LTC_4, LTD_4, and LTE_4). Neutrophils, monocytes and macrophages produce LTB_4, whereas the other LTs are produced by mast cells, basophils, and eosinophils.

PGE_2 is pro-inflammatory, inducing fever, increasing vascular permeability and vasodilation, and enhancing pain and edema. PGE_2 can up-regulate COX-2 enzymes increasing its own production; it also induces the production of IL-6 by macrophages. LTC_4, LTD_4, and LTE_4 are bronchoconstrictors, promote hypersensitivity, and increase vascular permeability. Thus the omega-6 eicosanoids are strongly pro-inflammatory.

Omega-3 Derivatives

Metabolism of EPA by COX enzymes gives rise to the 3-series of PGs and TXs, and the 5-series of LTs. If there is increased consumption of omega-3 fatty acids (especially of EPA and DHA), there are increased proportions of these fatty acids in inflammatory cell phospholipids particularly at the expense of arachidonic acid. Fish oil supplementation results in an increased production of LTB_5, LTE_5, and PGE_3. The inflammatory mediators produced from EPA are less inflammatory as compared to those formed from arachidonic acid. For example, LTB_5 derived from EPA is 100-fold less potent as a neutrophil chemotactic agent than is LTB_4 derived from arachidonic acid. In addition, there is a group of mediators termed E-series resolvins that are formed from EPA by COX-2/LOX.

These mediators exert anti-inflammatory actions. In addition there are D-series resolvins, docosatrienes and neuroprotectins derived from DHA that also appear to be anti-inflammatory help to resolve inflammation.

In addition to altering eicosanoid production, the omega-3 fatty acids have anti-inflammatory properties independent of eicosanoids.[3] Long-chain omega-3 fatty acids can affect production of the adhesion molecules involved in interactions between endothelial cells and leukocytes. DHA decreases cytokine-induced expression of E-selectin, ICAM-1 and VCAM-1. EPA also inhibits endotoxin-induced expression of adhesion molecules, decreasing binding of monocytes. Both EPA and DHA inhibit the production of IL-1β and TNF-α by monocytes, and of IL-6 and IL-8 by endothelial cells.

Omega-3 fatty acids may exert effects on inflammatory gene expression by direct actions on signaling pathways which lead to activation of the transcription factor nuclear factor kappa B (NFκB). Omega-3 fatty acids can down-regulate the activity of NFκB, and prevent activation by TNF-α possibly through decreased phosphorylation. Peroxisome proliferator-activated receptors (PPARs) are ligand-activated nuclear transcription factors that play important roles in inflammation and several metabolic diseases. PPARs can bind to the PPAR-response element to repress or induce the transcription of genes. Both EPA and DHA down-regulate activation of NFκB via a PPAR-γ-dependent pathway in human kidney cells.

In a study investigating neutrophil-mediated inflammatory responses, dogs were fed diets containing high supplemental concentrations of ALA from flaxseed oil, EPA from fish oil, beef tallow, or safflower oil.[4] The ratios of omega-6 to omega-3 fatty acids were nearly identical in the flaxseed oil and fish oil diets. Neutrophils had higher concentrations of EPA in dogs fed fish oil diets as compared to those fed flaxseed oil. Both the flaxseed and fish oil groups had decreased LTB_4 and increased LTB_5 as compared to the tallow and safflower oil-fed groups, but the fish oil group had the largest decrease in LTB_4 and the highest increase in LTB_5. A much higher amount of dietary ALA was needed to achieve neutrophil responses as compared to fish oil, even though the two diets had similar omega-6 to omega-3 ratios. Thus the type of omega-3 fatty acid is probably more important than just the ratio of omega-6 to omega-3 fatty acids in the diet.

SOURCES OF FATTY ACIDS

Fat sources typically used in pet food formulations such as tallow, animal fat, and poultry fat are excellent sources of omega-6 fatty acids, but poor sources of omega-3 fatty acids. Other sources of omega-6 fatty acids that may be utilized include corn oil, safflower oil, sunflower oil, palm oil, peanut oil, sesame oil, olive oil, soybean oil, and canola oil. Borage oil, black currant oil and evening primrose oil are rich in dihomogammalinoleic acid, an omega-6 fatty acid with possible beneficial properties in some skin disorders. Canola oil and soybean oil also provide omega-3 fatty acids; canola oil is about 10% linolenic acid, and soybean oil is about 7% linolenic acid. Flaxseed oil is a good source of omega-3 fatty acids, providing about 54% linolenic acid. None of these sources of omega-3 acids provide the long-chain omega-3 fatty acids such as EPA and DHA however, and if these oils are used as a source of omega-3 fatty acids, the linolenic acid must be elongated to EPA and DHA for benefit from these particular fatty acids. EPA and DHA are found in coldwater oily fishes. The deepwater fishes that store their lipid in their flesh such as menhaden, mackerel, tuna (fillets), Atlantic salmon, herring, and sardines have the highest content of EPA and DHA. Lean fish that store lipids in their livers such as cod, haddock, catfish, flounder, etc., are poorer sources of EPA and DHA. For example, one serving of mackerel provides about 3g EPA and DHA, whereas one serving of cod only provides about 0.3g EPA and DHA.

INFLAMMATORY CONDITIONS

Atopic Disease/Asthma

Dietary fish oils exert many effects that may be beneficial in allergic disease including reduced chemotaxis of monocytes, neutrophils, and eosinophils, decreased production of platelet activating factor, decreased production of IL-1, IL-6, IL-2 and TNF-α, decreased adhesion molecule expression, and decreased lymphocyte responsiveness.[5] Fish oils have been used to treat skin disease for many years; in 1877, cod liver oil was the recommended treatment for severe itching. In recent studies, humans with atopic dermatitis given daily fish oil showed decreased itching and scaling of their skin. Studies in pets have been difficult to interpret due to experimental design. In a study where dogs consumed diets either adequate in essential fatty acids, high in LA, or high in LA plus ALA, dogs consuming any one of the diets showed an improvement in skin and coat scores as compared to an acclimation diet; however, all three diets contained a higher total fat content than the acclimation diet.[6] In other studies, a combination of EPA and DHA have been noted to be of clinical benefit to pruritic dogs. Diets supplemented with either flax oil, or a combination of EPA and DHA both decreased the level of pruritus, skin lesions, and other topical medications needed in dogs with atopic dermatitis.

PGD2 is a bronchoconstrictor produced by lung mast cells during an asthmatic attack.[3] LTC_4, LTD_4, and LTE_4 are the major mediators of asthmatic bronchoconstriction, and in asthma, LTC_4 amplifies local inflammation, and induces other cells to produce LTA_4 and 5-HETE. PGE_2 stimulates B-cells to produce IgE; thus PGE_2 plays a major role in the development of allergic disease. Fish oils may be protective in allergic disease. Asthma is uncommon in Greenland Eskimos who consume a high quantity of fish oils. In Australia, children that consumed oily fish had a lower likelihood of having asthma as compared to those that did not consume oily fish. Umbilical cord blood had lower concentrations of EPA and DHA in mothers with allergies as compared to mothers without allergies, and DHA and total n-3 PUFA concentrations were lower in children with atopy and asthma as compared to those without asthma. In a recent study, cats with experimentally-induced asthma were supplemented with omega-3 fatty acids.[7] In these cats, supplemental omega-3 fatty acids increased the concentration of bronchoalveolar lavage fluid lipoxin A_4 (LXA_4) which is an inhibitor of inflammation. Airway responsiveness also decreased, suggesting a beneficial effect of omega-3 fatty acids in cats with asthma.

Osteoarthritis

Dietary fish oils have shown benefit in animal models of arthritis and in human studies.[3] Dietary fish oils delayed the onset of induced arthritis in both mice and rats. In humans with arthritis, EPA plus DHA supplementation reduced morning stiffness, reduced joint pain, decreased the number of swollen joints, and decreased the use of nonsteroidal anti-inflammatory drugs. In 24 dogs with osteoarthritis, fish oil supplementation resulted in significant decreases in synovial fluid PGE_2, and in matrix metalloproteinase-2 and -9 which contribute to cartilage destruction.[4] In another study of 127 arthritic dogs, dietary omega-3 fatty acid supplementation resulted in a significant improvement in ability to rise from a resting position and play after 6 weeks, and improved ability to walk at 12 and 24 weeks as compared to dogs not receiving the supplement.[8] In a study of 38 arthritic dogs, fish oil supplementation resulted in improvement in orthopedic evaluation, including peak vertical force, lameness, and weight bearing.[9]

Intestinal Disease

An increased incidence of inflammatory bowel disease (IBD) in humans correlates with an increased dietary intake of omega-6 fatty acids.[1] Patients with IBD who take supplemental fish oil have decreased production of LTB_4 by neutrophils and colonic mucosa, with decreased production of PGE_2 and IFN-γ in mononuclear cells. In a mouse strain that develops colitis, mice fed fish oil had reduced colonic inflammation compared to mice fed corn oil. Human patients with Crohn's disease that were supplemented with daily fish oil for a year had decreased incidence of relapse and remained in remission longer than the placebo group.

Cardiovascular Disease

Benefits of fish oil supplementation on cardiovascular disease in humans has been extensively studied;[10] however, less is known in dogs and cats due to the decreased incidence of coronary artery disease and atherosclerosis in these species. Cachexia is common in heart failure, and is associated with increased inflammation. IL-1 and TNF-α inhibit appetite and accelerate the catabolism of muscle protein. Fish oil supplementation can improve food intake by decreasing the production of inflammatory cytokines. In dogs with dilated cardiomyopathy (DCM) and heart failure, fish oil supplementation reduced IL-1 and PGE_2 production, with decreased loss of muscle.[11] In addition, dogs with heart failure have been shown to have a relative deficiency of EPA and DHA compared to normal dogs, and supplementation with fish oil can normalize the fatty acid levels. There is considerable evidence showing benefit of fish oil supplementation in the prevention of arrhythmias and myocardial infarction in humans. In one study utilizing Boxer dogs with ventricular arrhythmias, fish oil supplementation decreased their arrhythmias after 6 weeks. In a retrospective study of dogs with heart failure secondary to DCM or chronic valvular disease, omega-3 supplementation increased survival time. Omega-3 fatty acids may be beneficial by decreasing inflammation, or by reducing platelet aggregation with the production of less potent TXB_5. This may be important in cats at risk for thrombus formation or in animals with coagulopathies.

Obesity/Diabetes

Obesity is associated with low-grade chronic inflammation with increased macrophage infiltration.[12] Inflammation is the key link between obesity and insulin resistance. Pro-inflammatory pathways cross insulin signaling pathways via down-regulation of insulin signaling pathway proteins (such as GLUT4 and insulin receptor substrate (IRS)-1), phosphorylation of IRS-1 blocking phosphorylation in response to insulin, and induction of cytokine signaling molecules that block coupling of the insulin receptor to IRS-1. Fish oils may have benefit by reducing pro-inflammatory cytokine production, and through down-regulation of NFκB and PPAR-γ.[13] Diets rich in fish oils were associated with significant reductions in inflammatory cytokines and lymphocyte proliferation in dogs.[14,15] TNF-α is higher in adipose tissue of obese cats, and increased TNF-α has been associated with down-regulation of adiponectin production.[16,17] Adiponectin is decreased in obesity, and low adiponectin levels are implicated in the pathogenesis of insulin resistance and type 2 diabetes. In humans, diets rich in fish oils are associated with a delay in the development of glucose intolerance. In obese cats, serum EPA is inversely related to insulin and triglyceride levels, and directly related to adiponectin.[18] Obese cats consuming diets high in omega-3 PUFA had improved glucose control and decreased insulin concentrations compared to cats consuming diets rich in saturated fatty acids.[19] In both human and rodent studies, fish oil supplementation increases circulating adiponectin levels.

Chronic Kidney Disease (CKD)

Dietary supplementation with fish oil has been shown to be beneficial in humans[20] and dogs[21] with CKD. In humans, the best results have been in those with IgA nephropathy. Omega-3 fatty acids may work via mediation of inflammation and their ability to decrease blood pressure. Dogs consuming diets high in fish oil had lower mortality, better renal function, fewer renal lesions, less proteinuria, and lower serum cholesterol than dogs consuming high omega-6 fatty acid PUFA. Interstitial inflammation, tubulointerstitial fibrosis, and glomerulosclerosis were decreased in dogs consuming fish oils. The optimum quantity of fish oil or omega-6 to omega-3 ratio has not been established in dogs.

Sepsis/Critical Illness/Surgery

In humans, administration of omega-3 fatty acids decreases the length of hospital stay in patients undergoing surgery.[22] Total parenteral nutrition (TPN) is commonly given to critically ill humans, and is becoming more common in veterinary patients. Soybean oil based lipid emulsions are common, but can exert immunosuppressive effects which is detrimental in patients at risk of infection. Fish oils can replace some of the soybean oil; patients receiving parenteral lipids containing fish oils had higher LTC_5 production, with lower TNF-α and IL-6 concentrations. Abdominal surgery patients receiving TPN with fish oil had shorter stays in intensive care, and a shorter hospital stay. In a human study in which patients received lipid-free, soybean oil, or soybean oil/fish oil TPN after large bowel surgery, there were no differences between groups in numbers of lymphocytes, B cells, $CD4^+$, $CD8^+$, or natural killer cells post-surgery. In the fish oil group, IL-2 was decreased, and the typical post-surgery decrease in IFN-γ was prevented. In humans hospitalized with acute respiratory distress syndrome, patients supplemented with fish oil had decreased time on ventilation support, decreased requirement for supplemental oxygen, and shorter hospital stays.[23] They also had lower concentrations of IL-8, LTB_4, and TNF-α in alveolar fluid. In humans with sepsis, those receiving fish oil had lower C reactive protein and higher LTB_5 concentrations.

REFERENCES

[1]Wall R et al. Nutr Rev 2010; 65(5):280; [2]Calder PC. Prost Leuk Essential Fatty Acids 2006; 75:197; [3]Calder PC. Lipids 2003; 38(3):343; [4]Bauer JE. JAVMA 2007; 231(11):1657 ; [5]Calder PC et al. Pediatr Allergy Immunol 2000 ; Suppl 13 :29 ; [6]Mueller RS et al. JSAP 2004 ; 45 :293 ; [7]Leemans J et al. Vet J 2010 ; 184 :111 ;[8]Roush JK et al. JAVMA 2010 ; 236 :59 ; [9]Roush JK et al. JAVMA 2010 ; 236 :67 ; [10]Adkins Y et al. J Nutr Biochem 2010 ; 21 :781 ; [11]Freeman LM. JSAP 2010 ; 51 :462 ; [12]Oliver E et al. Proc Nutr Soc 2010 ; 69 :232 ;[13]Tai CC et al. J Nutr Biochem 2010 ; 21 :357 ; [14]LeBlanc CJ et al AJVR 2008 ;69 :486 ;[15]LeBlanc CJ et al Vet Immunol Immunopathol 2007 ; 119 :180 ;[16]Miller C et al. J Nutr 1998 ;128 :2751S ;[17]Hoenig M et al. Domest Anim Endocrinol 2006 ;30 :333 ; [18]Mazaki-Tovi M et al. AJVR 2011 ; in press ; [19]Wilkins C et al AJVR 2004 ;65 :1090 ;[20]Fassett RG et al Am J Kid Dis 2010 ; 56(4) :728 ; [21]Roudebush P et al. JSAP 2010 ; 51 :244 ; [22]Calder PC. Lipids 2004; 39(12):1147;[23]Martin et al. Nutr Rev 2010 ; 68(9) :531.

EVIDENCE-BASED BENEFITS OF OMEGA-3 FATTY ACIDS FOR DOGS & CATS

Jennifer Larsen, DVM, PhD, DACVN
Davis, CA

INTRODUCTION

Omega-3 fatty acid supplements are frequently used by both owners and veterinarians to support optimal health in pets, and are increasingly utilized as a complementary therapeutic modality for a variety of diseases including osteoarthritis, chronic kidney disease, cardiac disease, neoplasia, and atopy. The published scientific evidence supporting the safety and efficacy of omega 3 fatty acid supplementation in health and disease is variable.

NOMENCLATURE

Fatty acids are hydrocarbon chains. These chemical molecules vary in length and other aspects of their structures including the degree of double bonds. The number of double bonds is negatively correlated with saturation (a polyunsaturated fatty acid has two or more double bonds while a saturated fatty acid has none). Unsaturated fatty acids are further classified by the location of the first double bond either from the carboxyl terminus (delta system; enzymes involved in fatty acid metabolism are named this way) or methyl terminus (omega or 'n' system).

Using the omega nomenclature (counting from the methyl terminus), polyunsaturated fatty acids (PUFA) with the first double bond between carbons 3 and 4 are classified as omega-3 PUFA, while those with the first double bond between carbons 6 and 7 are omega-6 PUFA, and so on. For example, the essential omega-6 PUFA for dogs and cats, linoleic acid (18 carbons long, two double bonds) is designated 18:2n-6.

SOURCES OF FATTY ACIDS

Fat sources provide fatty acids. These can come from animals (butter, lard, tallow, fish oil) or from plants (shortening, vegetable oil, seed and nut oil. Many meats (including fatty fish) are significant sources of dietary fat as well, and the fatty acid profile is reflective of the diet and physiology of the species. Every fat source has a distinct fatty acid profile, and some are better sources of essential fatty acids than others. For example, olive oil provides less than 10% by weight of linoleic acid (LA; 18:2n-6), while corn oil provides 54% LA by weight.

While omega-3 PUFA as a group are found in both plant oils and fish oils, distinctly different fatty acids are found in plants vs. fish. The omega-3 PUFA found in plants is alpha-linolenic acid (ALA; 18:3n-3). Plant sources include flaxseed oil (53% ALA by weight), walnut oil (10%), canola oil (9%), and soybean oil (7%). The longer-chain omega-3 PUFA found in fish oil are eicosapentaenoic acid (EPA; 20:5n-3) and docosahexaenoic acid (DHA; 22:6n-3). Fish are enriched in EPA and DHA produced by algae but de novo synthesis from shorter chain dietary ALA is also possible[1]. Algal sources of EPA and DHA exist and are of interest; however, currently only a commercial source of algal DHA is available.

Table 1. Polyunsaturated fatty acid nomenclature.

Fatty acid	Carbon chain length:number double bonds	Classification	Dietary sources
Linoleic acid (LA)	18:2	n-6	Corn oil
alpha-Linolenic acid (ALA)	18:3	n-3	Flaxseed oil
Eicosapentaenoic acid (EPA)	20:5	n-3	Fish oil
Docosahexaenoic acid (DHA)	22:6	n-3	Fish oil

The concentrations of EPA and DHA in fish oils vary widely and depend on species, whether the fish body or liver are used to produce the oils, and a variety of other factors. For example, some manufacturers distill the oil to concentrate the omega-3 PUFA, while other products are blends of fish oil and vegetable oil. However, most commercially available fish oil products provide EPA plus DHA in the range of 20-30% by weight.

ESSENTIALITY

While dogs and cats can synthesize saturated and some unsaturated fatty acids, they lack the enzymes necessary to put a double bond in the omega-6 and omega-3 positions. The specific locations of these double bonds are important for the metabolic and structural functions of fatty acids. As such, the fact that specific omega-3 and -6 PUFA have double bonds at specific locations is the basis of nutritional essentiality.

Once consumed, omega-3 and -6 PUFA undergo a series of desaturation and elongation reactions. The extent and rate of these conversions varies by species (for example, cats have low activity of delta-6 desaturase). Additionally, humans[2], dogs[3] and cats[4] convert ALA to the longer-chain EPA slowly and at conversion rates of less than 10%. Although further elongation of EPA to DHA appears limited to the pre-weaning period of growth when measured as plasma enrichment[5], synthesis of DHA occurs in neural tissues to some extent[6]. The low synthesis rate of DHA in liver has been hypothesized to be an adaptation to support an adequate precursor supply to target tissue[7]. However, dietary long chain PUFA (20 carbons or greater), namely EPA and DHA, are more effective than ALA for supporting DHA enrichment and proper function of neural tissue[4,5].

It has become increasingly apparent that dietary omega-3 PUFA are essential for proper visual and neurological development in dogs and cats, but their functional role as inflammatory mediators is not yet fully defined. As such, the efficacy of n-3PUFA for other lifestages and in various disease states is less clear.

MECHANISM OF ACTION AND CONSIDERATIONS FOR USE

There is considerable interest in the use of omega-3 PUFA as modifiers of immune and inflammatory responses. The goal of therapy is to manipulate the proportions and types of eicosanoids produced by the lipoxygenase and cycloxygenase pathways. These eicosanoids include prostaglandins, thromboxanes, and leukotrienes which are cellular messengers that play a complex role in the normal processes of immunity, inflammation, and central nervous system signaling. Since EPA and the omega-6 counterpart arachidonic acid (AA), are substrates for the production of eicosanoids, the types synthesized can be modified by enriching cellular membranes with one or the other. Eicosanoids synthesized from EPA are considered to be less inflammatory than those from AA. Thus, the desired metabolic endpoint of omega-3 PUFA supplementation is a higher concentration in cellular membranes.

Due to the slow and low conversion rate of ALA to longer-chain omega-3 PUFA, and the lower effectiveness of ALA in maintaining normal neurological function in the healthy dog and cat, it seems prudent to pursue a more direct approach. The value of supplementation with fish oil to obtain presumably therapeutic levels of EPA and DHA is increasingly recognized.

Establishing exact target amounts and proportions of EPA and DHA is challenging due the confounding effects of other fatty acids, not to mention other dietary factors. Both omega-3 and omega-6 PUFA are metabolized by the same enzymes, and use the same cellular machinery for incorporation into membranes. Therefore, substrate competition may influence not only the absolute amounts of end-products produced but also the relative proportion present in cellular membranes, especially when one of the PUFA classes are provided at high concentrations. Further, the relative importance of EPA vs. DHA is unknown. While EPA is the substrate for eicosanoid synthesis, there is an obvious role of DHA in neural tissues, in addition to some degree of retroconversion to EPA.

When initiating EPA and DHA supplementation, both the absolute amounts as well as the proportions of PUFA from other classes should be considered. In this regard, the ratio of overall omega-6 to overall omega-3, while commonly used and reported, is not useful unless the majority of the dietary fatty acids are of the specific types of interest. As such, consideration of a more specific ratio may be indicated. For example, the ratio of LA plus AA (representing the omega-6 PUFA of interest) to EPA plus DHA (the omega-3 PUFA of interest) is more useful when evaluating diets and supplements with significant amounts of ALA. Further, consideration of the amounts and types of fatty acids provided by the basal diet and any supplements is warranted.

THERAPEUTIC APPLICATIONS

Investigations into the use of omega-3 PUFA for managing a variety of conditions have been conducted. These include chronic kidney disease, cardiac disease, neoplasia, and atopy. Selected relevant refereed reports of controlled trials are summarized below.

Chronic Kidney Disease (CKD)

Two trials reporting the effects of high-dose fish oil supplementation in dogs with experimental model of canine CKD have been published. In the first, 21 dogs with induced CKD were divided into three groups and fed a basal diet supplemented with one of three additional fat sources (menhaden fish oil, safflower oil, and beef tallow[8]. Over the 20 month study period the fish oil group showed increased creatinine clearance (relative to baseline) decreased proteinuria, and decreased serum creatinine, cholesterol, and triglyceride concentrations. Glomerular filtration rate was lower and interstitial fibrosis was worse in the safflower oil group, which also had a higher incidence of end-stage renal failure.

The second study examined the effect of the same dietary treatments on early progression of induced CKD and renal eicosanoid metabolism in 18 dogs over a 10 week period[9]. The fish oil group showed lower urinary excretion of eicosanoids while the safflower oil group demonstrated higher glomerular capillary pressure with glomerular enlargement.

From the information provided in these two reports, determination of the omega-3 PUFA dose on an energy basis is not possible; however, EPA was provided at 20 g per kg dry matter (DM) while DHA was provided at 19 g per kg DM. The ratio of LA plus AA to EPA plus DHA was 0.05.

Cardiac disease

One study investigating the role of fish oil supplementation on various parameters in 28 dogs with naturally occurring heart failure reported that the group given fish oil over 8 weeks had better cachexia scores and lower serum concentrations of the eicosanoids PGE2 and Il-1 compared to a group given corn oil[10]. The reduction in Il-1 was correlated with survival. Although diet information was not provided, the dose administered was approximately 1.11 g EPA plus DHA/1000 kcal (assuming diet energy density of 4 kcal/g). The ratio cannot be calculated.

Another study in 24 dogs with naturally occurring but asymptomatic arrhythmogenic right ventricular cardiomyopathy reported that the group given fish oil supplementation showed reduced VPCs compared to those given flax oil or sunflower oil[11]. Although diet information was not provided, the dose administered was approximately 1 g EPA plus DHA/1000 kcal (assuming diet energy density of 4 kcal/g, all dogs were same body weight, and that all dogs consumed calculated MER). The ratio cannot be calculated.

Neoplasia

Thirty two dogs with stage 3a and 4a lymphoma were fed canned diets differing in fat source: fish oil vs. soybean oil[12]. The fish oil diet was also supplemented with the amino acid arginine; the diets were otherwise the same. Dogs with stage 3a lymphoma fed the fish oil/arginine diet had higher serum levels of EPA and DHA, longer disease-free interval and longer survival time (no significant difference in stage 4a dogs). There was no effect on inflammatory mediators, body weight, quality of life, or toxicity scores. The dose provided by the diet was 140 g EPA plus DHA/1000 kcal, with a ratio of 0.3.

Atopy

Four diets providing various doses and ratios of omega-3 PUFA were fed to groups 58 atopic dogs over 8 weeks[13]. The dogs fed the high omega-3/low omega-6 diet (vs high/high, moderate/moderate/ and low/moderate) had decreased serum PGE2 concentrations. Clinical improvement was noted in all groups, with no benefit to any dose or ratio of omega-3 PUFA. The dose provided by the high/low diet was 3.7 g EPA plus DHA/1000 kcal with a ratio of LA plus AA to EPA plus DHA of 1.2 (vs 3.2, 7.8 and 34 for the other groups, respectively).

A crossover-design trial of 16 atopic dogs supplemented with either fish oil or corn oil measured clinical parameters over two 6 week periods with a 3 week washout between[14]. Fish oil treatment was associated with less pruritis and alopecia and better coat quality. No dietary information was given. The dose was 0.85 g EPA plus DHA/1000 kcal (assuming diet energy density is 4 kcal/g, no omega-3 coming from base diet, and that the dogs had the same body weight and were consuming their calculated MER). The ratio cannot be calculated.

CONCLUSION

Interest in the use of omega-3 supplementation for states of heath and disease has been increasing. Clinical evidence suggests that long chain omega-3 PUFA from marine sources (EPA and DHA) are more effective at achieving both plasma membrane enrichment as well as therapeutic advantage compared to the shorter chain counterpart from plants (ALA). Clinical studies describing the effect of the use of omega-3 supplementation for management of a variety of conditions exist; however, the reporting of dietary information, doses, and sources lack standardization. Much more research is needed before definitive guidelines and expected outcomes can be provided for veterinary patients with any disease. Evidence seems to support the use of omega-3 PUFA for at least reproduction and growth in dogs and cats, and suggests a benefit for maintenance as well. Additionally, supplementation of long chain omega-3 PUFA does not appear to be harmful and is likely beneficial in the management of several chronic diseases; however, effective doses remain elusive and are likely variable depending on disease, severity, other individual factors.

REFERENCES

[1]Zheng, X, et al. Lipids. 2005; 40(1):13. [2]Brenna JT, et al. Prostaglandins Leukot Essent Fatty Acids. 2009;80(2-3):85. [3]Bauer JE, et al. J. Nutr. 1998;128: 2641S. [4]Pawlosky RJ, et al. Am J Clin Nutr 1997;65(2):465. [5]Bauer JE, et al. J Nutr. 2006 Jul;136(7 Suppl):1991S. [6]Alvarez RA, et al. Invest Vis Sci 1994;35(2):402. [7]Dunbar BL, Lipids. 2010 Jan;45(1):1 [8]Brown SA, et al. J Lab Clin Med. 1998;131(5):447. [9]Brown SA, et al. J Lab Clin Med. 2000;135(3):275. [10]Freeman LM, et al. J Vet Intern Med. 1998;12(6):440. [11]Smith CE, et al. J Vet Intern Med. 2007;21(2):265. [12]Ogilvie GK, et al. Cancer. 2000;88(8):1916. [13]Nesbitt GH, et al. Vet Dermatol. 2003;14(2):67. [14]Logas D, et al. Vet Dermatol. 1994;5(3):99.

MULTIPLE ORGAN DYSFUNCTION SYNDROME IN SEPSIS

Amy E. DeClue DVM, MS, DACVIM (SAIM)
Columbia, MO

INTRODUCTION

Multiple organ dysfunction syndrome or MODS encompasses a wide variety of pathologic manifestations of sepsis and other forms of critical illness. Recognized more than 40 years ago, multiple organ dysfunction syndrome is defined as the progressive, yet potentially reversible, dysfunction of 2 or more organ systems after and acute, life-threatening disruption of systemic homeostasis. While infection is one of the main initiating factors associated with MODS, it is not required for induction of MODS. The realization that infection is not the sole cause of MODS paved the way for new ideas about the pathophysiology. In recent years, the pathophysiology of MODS has been attributed to aberrations of the immune system, not direct cytotoxicity secondary to microorganisms per se.

PATHOPHYSIOLOGY

The sequence of events leading to sepsis is complex and incompletely understood. In the initial phases of in infection, microbial products (e.g. endotoxin from gram negative bacteria; exotoxins, peptidoglycans and superantigens from gram positive bacteria; and fungal cell wall material) induce systemic inflammation through activation of immune cells resulting in an imbalance between pro-inflammatory mediators (e.g. TNF) and anti-inflammatory mediators (e.g. IL-10). The induction of pro-inflammatory mediators leads to inflammatory cell infiltration, altered thermoregulation, vasodilation, vascular leakage and coagulation. For some animals, the inflammatory response resolves and they recover. Other animals die during the initial stages of infection and inflammation. The concept of infection, inflammation and either recovery or death is relatively easy to comprehend. However, there is a third subset of animals that develop a different course of disease; a more complex scenario. These animals might have typical initial patterns of infection and inflammation, but ultimately develop pathology that results in dysfunction of organ systems.

There are three main models to explain the initiation of MODS that have been proposed. The "one-hit" model states that MODS results from a single massive insult (e.g., polytrauma). The "two-hit" model involves sequential insults; a priming insult and then a subsequent insult (e.g., polytrauma followed by surgery). Finally, there is a "sustained-hit" model which involves a continuous insult (e.g., multi-drug resistant, poorly controlled infection). In reality, our patients often have a combination of these "models". For example, a dog suffers polytrauma (one hit) followed by surgery (second-hit) and then hospital acquired multi-drug resistant infection that is poorly controlled due to microbial resistance (sustained-hit). Regardless of the initial insult, the immune, nervous, hormonal and metabolic pathways involved in the development of MODS appear to be somewhat redundant.

The current understanding of the pathophysiology leading to sepsis-induced MODS involves intricate cross-talk among multiple cell population, hormonal systems, metabolites, and neural signaling along with alterations in oxygen delivery, derangements in oxygen utilization, and modifications in cell phenotypes. This highly coordinated, yet dysregulated, adaptive mechanism might arise from primordial protective strategies. There are several proposed mechanisms for the development of MODS including 1. cell or tissue hypoxia, 2. microbial and inflammatory mediator induced cytotoxicity, 3. induction of cellular apoptosis, 4. translocation of microbes from the gastrointestinal tract and subsequent inflammation, and 5. inflammation-induced downregulation of mitochondrial function. While MODS likely results from a complex combination of these factors, emerging evidence suggests that mitochondrial dysfunction might be the prevailing pathway. When mitochondrial energy production is reduced, the result is cellular dysfunction and, in some cases, cell death. Mitochondrial dysfunction has been documented during sepsis-induced MODS in humans with naturally developing sepsis and in experimental models.[1-3] Pharmacologic inhibition of mitochondrial derangement prevents the development of MODS in experimental bacterial sepsis indicating that mitochondrial damage is a causative factor in the development of MODS and thus could be a therapeutic target.[4] While generally viewed as "bad" clinically, downregulation of mitochondrial function might be a cellular adaptive response to prolonged inflammation. In general, cell death (necrosis) does not appear to be a common finding in people with MODS. Instead, it appears that mitochondrial dysfunction causes a transient reduction in cellular activity that might return when the animal recovers. This phenomenon has been referred to as a cellular hibernatory-like state.

Table 1: Definitions of organ dysfunction in dogs.[5]

Organ	Criteria
Renal	An increase in creatinine concentration ≥ 0.5 mg/dL from presurgical values without evidence of pre-or post renal azotemia
Cardiovascular	Hypotension severe enough to require vasopressor therapy
Respiratory	Need for supplemental oxygen administration or mechanical ventilation; determined based on clinical assessment, blood gas analysis (alveolar-arterial gradient in partial pressure of oxygen >10 mm Hg) and/or results of pulse oximetry (SpO_2 < 95)
Hepatic	Plasma or serum total bilirubin of >0.5 mg/dL
Coagulation	PT or PTT > 25% above the upper reference limit and/or platelet count of ≤ 100,000/µL

Table 2: Data summary Kenny et al.[5]

Number of dysfunctional organ systems	Percent of population	Survival (%)	Odds ratio for risk of death
0	21.9	84	
1	28.1	69	6.73
2	22.8	46	6.67
3	14.9	24	10.21
4	9.6	9	18.71
5	2.6	0	

EVIDENCE IN VETERINARY MEDICINE

The current definition of MODS in dogs is ≥ 2 forms of organ dysfunction in an animal with appropriate risk factors although the criteria to define individual forms of organ dysfunction varies and is at times conflicting. Generally, organ dysfunction is associated with a poorer prognosis in dogs with sepsis. Kenney et al. retrospectively investigated the association between MODS and survival in 114 dogs with septic peritonitis (for definitions used, see table 1).[5] Multiple organ dysfunction syndrome was diagnosed in 50% of the population.[5] Kenny reported a mortality rate of 70% for dogs with MODS compared to 25% for dogs without MODS.[5] Additionally, there was a significant association between the presence of organ dysfunction and death (table 2).[5] In a prospective study, Rau and others documented that 53% of dogs with sepsis were defined to have severe sepsis (organ dysfunction, hypoperfusion, hypotension) or septic shock (hypotension despite adequate volume resuscitation).[6] Dogs with severe sepsis or septic shock were more likely to suffer in-hospital mortality (78%) compared to dogs with sepsis and no evidence of organ dysfunction or hypotension (25%).[6] Although the risk of mortality increases dramatically as the number of failing organ systems increases, there are dogs with sepsis and multiple forms of organ failure that have a meaningful outcome. Cavana described a dog with sepsis and 6 failing organ systems (cardiovascular, respiratory, gastrointestinal, pancreatic, renal and coagulation) that survived.[7]

There are multiple forms of organ dysfunction that arise secondary to sepsis. These include dysfunction of the cardiovascular, coagulation, endocrine, metabolic, hepatobiliary, pulmonary, renal, central nervous, gastrointestinal, muscular, and endothelial systems. Below, several forms of organ dysfunction that have been recognized in dogs with sepsis are discussed.

Cardiovascular

Although routinely lumped into one category, cardiac and vascular dysfunction should be considered separate entities. Cardiac dysfunction includes alterations in myocardial contractility, cardiac output and development of cardiac arrhythmias while vascular dysfunction encompasses vascular hyporesponsiveness, alterations in vascular permeability and a vast array of pathologic vascular endothelial responses. Nevertheless, cardiovascular dysfunction is a catch-all term that is popular in the literature. Septic shock is considered the most severe form of cardiovascular dysfunction in animals with sepsis and this term is often used in lieu of cardiovascular dysfunction. Despite the fact that anecdotally cardiovascular dysfunction is a well recognized and frequently discussed sequela to sepsis, there is very little clinical literature related to this topic. Echocardiographic evidence of cardiac dysfunction including left and right ventricular dilation, reduced fractional shortening (27%) and increased end-systolic volume were reported in a dog with sepsis induced cardiovascular dysfunction; these abnormalities resolved after the dog recovered.[8] Burkitt et al. reported that 7/33 dogs had hypotension refractory to fluid loading necessitating the need for vasopressor therapy.[9] Kenney et al. found that 17% of 114 dogs with septic peritonitis had cardiovascular dysfunction.

Coagulation

Coagulation dysfunction is a common finding in dogs with naturally occurring sepsis as evidenced by prolongation of prothrombin time, partial thromboplastin time, increased D-dimer and fibrinogen degradation products concentrations, and depletion of antithrombin (AT) and activated protein C (aPC).[5, 10, 11] It appears that aPC could be a useful prognostic indicator in dogs with sepsis.[10] Evaluation of hemostasis in dogs with parvoviral enteritis also documented decreased AT activity and hypercoagulability in affected dogs.[11] In a study of 14 dogs with sepsis and 11 control dogs, Von Willebrand factor antigen concentration was significantly higher in the sepsis group compared to control but there was no difference between the survivors and non-survivors in the sepsis group.[12] Additionally, Kenney et al. found that 60% of 114 dogs with septic peritonitis had coagulation dysfunction.

Endocrine/Metabolic

CIRCI-Critical illness-related corticosteroid insufficiency (CIRCI; formerly known as relative adrenal insufficiency) is characterized by inadequate production of cortisol in relation to an increased demand during critical illness and is usually defined by an inadequate response to exogenous ACTH.[13, 14] Based on the work of Burkitt et al., CIRCI is defined as a delta cortisol of < 3 mcg/dl in dogs with sepsis (delta cortisol = difference between pre and post-ACTH serum cortisol concentrations).[9] Using this definition, Burkitt found that 15/33 dogs had CIRCI and that CIRCI was associated with hypotension and decreased survival. Similarly Martin et al. found that 5/13 dogs with sepsis had a blunted cortisol response to ACTH.[15] The impact of glucocorticoid therapy on outcome for dogs with sepsis-induced CIRCI has not been critically evaluated. There is a single case report describing resolution of hypotension and survival after hydrocortisone treatment in a dog with CIRCI and suspected sepsis.[16]

Thyroid axis dysfunction - Schoeman et al. found that dogs with parvoviral enteritis had significantly lower serum thyroxine concentrations than controls.[17] In this study, nonsurvivors had significantly lower thyroxine than survivors.[17] Nonsurvivors had a progressive decrease in serum thyroxine concentrations over the first 3 days of hospitalization.[17]

Altered glucose homeostasis - Aberrant glucose homeostasis, involving changes in insulin production or insulin sensitivity as well as other alterations, has been documented in dogs with sepsis. Hardie, et. al. evaluated postsurgical blood glucose concentrations in dogs with sepsis and in healthy dogs. Fifty percent of dogs (n=8) with postsurgical plasma glucose concentrations greater than 150 mg/dl died while only 14% of dogs (n=7) with plasma glucose concentrations less than 150 mg/dl died.[18] The difference in these groups was not significant (p=0.08), but this might be a reflection of the small sample size.[18]

Dyslipidemia – Yilmaz and Senturk described several lipid abnormalities associated with parvoviral enteritis in dogs.[19] Serum total cholesterol, high-density lipoprotein cholesterol and low-density lipoprotein cholesterol were significantly lower and serum triglyceride concentrations were significantly greater in dogs with parvoviral enteritis than healthy control dogs with serum total cholesterol and high-density lipoprotein cholesterol concentrations being associated with mortality.[19]

Hepatobiliary

A variety of different terms, definitions and criteria have been used to describe hepatic dysfunction often resulting in confusion between hepatic dysfunction and injury. Standard criteria for the diagnosis of critical illness induced hepatic dysfunction should be formulated, rigorously evaluated and adopted. In a retrospective study of 21 dogs and 2 cats with septic peritonitis, King found that the majority of animals had ALP and/or ALT concentrations greater than the reference interval while about a quarter of the animals had hyperbilirubinemia.[20] Kenney et al. considered the finding of plasma or serum total bilirubin of >0.5 mg/dL to be consistent with hepatic dysfunction. Using this scheme, hepatic dysfunction was identified in 49% of 114 dogs with septic peritonitis.[5]

Pulmonary

Veterinary Acute Lung Injury and Acute Respiratory Distress Syndrome criteria have been suggested by the Dorothy Russell Havemeyer Working Group on ALI and ARDS in Veterinary Medicine[21] and include fulfillment of the following criteria: acute onset of tachypnea and labored breathing at rest; known risk factors; evidence of pulmonary capillary leak without increased pulmonary capillary pressure; and evidence of inefficient gas exchange. Evidence of diffuse pulmonary inflammation is recommended, but not required. There are no published studies that have used this exact definition in dogs with sepsis, but there is some evidence that dogs with sepsis develop pulmonary dysfunction. Parent et al. described ARDS in 19 dogs and found that sepsis was a common inciting cause. Kenney et al. found that 31% of 114 dogs with septic peritonitis had pulmonary dysfunction (defined as a need for supplemental oxygen administration or mechanical ventilation; determined based on clinical assessment, blood gas analysis and/or results of pulse oximetry).

Renal

The concept of critical illness related acute kidney injury is a relatively new concept in veterinary medicine. Recently, Thoen and Kerl proposed a veterinary acute kidney injury (VAKI) staging system (table 3).[22] These criteria have not been tested in dogs with sepsis specifically; future investigators should consider using these criteria. Using slightly less rigorous criteria, Kenney et al. found that 12% of 114 dogs with septic peritonitis had renal dysfunction (increase in creatinine concentration ≥ 0.5 mg/dL from presurgical values without evidence of pre-or post renal azotemia). Yilmaz and Senturk found that creatinine and BUN concentrations were significantly greater in dogs with parvoviral enteritis than healthy controls, but there was no difference between nonsurvivors and survivors.[19]

Table 3: VAKI staging system[22]

VAKI Stage	Criteria
Stage 0	Creatinine increase <150% from baseline
Stage 1	Creatinine increase of 150-199% from baseline OR Creatinine increase of 26.5 μmol/L (0.3mg/dL) from baseline
Stage 2	Creatinine increase of 200-299% from baseline
Stage 3	Creatinine increase of ≥300% from baseline OR An absolute creatinine value >4.0mg/dL

CONCLUSION

It is important that clear definitions of organ dysfunction are developed and used in veterinary medicine. For example, critical illness induced kidney injury should be separated from classic forms of kidney injury secondary to toxic or ischemic insults. The former involves "hibernation" of renal tubular cells which might be completely reversible while the latter involves necrosis of the renal tubular cells, which might not. Similarly, dogs with critical illness-related corticosteroid insufficiency should be separated from dogs with true hypoadrenocorticism because of the same general rationale. Further, some clinical markers used to define organ dysfunction might not truly reflect assessment of cell or tissue function or damage. Hyperbilirubinemia is a common finding in dogs with sepsis, but often represents microbial product and inflammatory mediator-induced downregulation and dysfunction of bile acid transporters, not hepatocellular necrosis or dysfunction. For now, it is important that we carefully interpret clinical markers of organ dysfunction, consider the most probable pathophysiology (ie renal tubular cell necrosis vs potentially transient renal tubular cell mitochondrial function downregulation) and act accordingly.

While it is not surprising that MODS is associated with a poorer prognosis, it is important to remember that organ dysfunction could be reversible if we are able to support the animal to the point of recovery. Although, anecdotally, clinicians report full return of function in some dogs with MODS that survive to leave the hospital, there are no large scale published studies evaluating long-term outcome. Since some forms of sepsis-induced organ dysfunction might simply involve cell "hibernation", it is rational to expect return of function in at least some animals. As we learn more about MODS in veterinary species, a clearer understanding of the use of appropriate clinical parameters to define organ dysfunction and long-term implications of MODS will become evident.

REFERENCES

[1]Brealey D, et al. **Am J Physiol Regul Integr Comp Physiol** 2004;286:R491. [2]Fredriksson K, et al. **Am J Physiol Endocrinol Metab** 2007;291:E1240. [3]Crouser E, et al. **Crit Care Med** 2002;30:2722. [4]Larche J, et al. **J Am Coll Cardiol** 2006;48:377. [5]Kenney EM, et al. **J Am Vet Med Assoc**;236(1):83. [6]Rau S, et al. **Vet Clin Pathol** 2007;36:253. [7]Cavana P, et al. **Schweiz Arch Tierheilkd** 2009;151(2):69. [8]Dickinson AE, et al. **J Vet Intern Med** 2007;21(5):1117. [9]Burkitt JM, et al. **J Vet Intern Med** 2007;21(2):226. [10]de Laforcade A, et al. **J Vet Intern Med** 2003;17:674. [11]Otto CM, et al. **J Am Vet Med Assoc** 2000;217(10):1500.[12]Rogers CL, et al. **J Vet Intern Med** 2009. [13]Beishuizen A, et al. **Best Pract Res Clin Endocrinol Metab** 2001;15(4):513. [14]Soni A, et al. **Am J Med** 1995;98(3):266. [15]Martin LG, et al. **J Am Vet Med Assoc** 2008;233(1):87. [16]Peyton JL, et al. **J Vet Emerg Crit Care (San Antonio)** 2009;19(3):262. [17]Schoeman JP, et al. **J Am Vet Med Assoc** 2007;231(10):1534. [18]Hardie E, et al. **Am J Vet Res** 1985;46:1700. [19]Yilmaz Z, et al. **J Small Anim Pract** 2007;48:643. [20]King L. **J Am Vet Med Assoc** 1994;204:407. [21]Wilkins P, et al. **J Vet Emergency Crit Care** 2007;17:333. [22]Thoen M, et al. **IVECCS** 2010.

THE ROLE OF ENTEROCOCCI IN VETERINARY MEDICINE

Kate S. KuKanich, DVM, PhD, DACVIM
Manhattan, KS

INTRODUCTION

Although the role of enterococci in veterinary medicine is not yet fully elucidated, it has become clear that enterococci should no longer be considered simply benign flora within the gastrointestinal tract. Enterococci may possess genes for antimicrobial resistance and virulence factors that may be transferrable among both enterococcal strains and other GI flora. Enterococcal infections have become a leading cause of nosocomial disease in human healthcare, and clinical cases have been emerging in veterinary medicine in the past decade. Continued research and a better understanding of the epidemiology, antimicrobial resistance, and disease behavior will allow improved control and management of enterococcal cases in veterinary medicine.

Enterococci are Gram-positive facultative anaerobes that are gut commensals and opportunistic pathogens in companion animals and human beings. There are 37 species in this genus, with *E. faecalis*, and *E. faecium* most commonly associated with clinical disease. Enterococci can tolerate a wide range of extreme environmental conditions, including temperatures (5-65°C), and pH (4.5-10), high NaCl, and bile concentrations.[1,2] Resistance to chemical and physical stress aids in survival outside mammalian hosts, allowing transmission via fomites. Viable enterococci have been recovered from inanimate objects for up to 4 months, providing opportunity for nosocomial spread of this organism.[3,4]

ANTIMICROBIAL RESISTANCE

Antimicrobial resistance in enterococci is a well-studied and concerning clinical problem. Enterococci have inherent or intrinsic resistance to most beta-lactams, lincosamides, and low levels of aminoglycosides, clindamycin and lincomycin.[5,6] Acquired resistance can occur by DNA mutation or acquisition of genes, and occurs in enterococci with beta-lactams and vancomycin, high levels of aminoglycosides and clindamycin, chloramphenicol, erythromycin, tetracyclines, and fluoroquinolones. Enterococci use a non-pili mechanism for the conjugation and transfer of plasmids and transposons between bacteria. Rather, they use a plasmid-specific pheromone-induced system for genetic exchange. With this system, they are not only able to transfer genes for antimicrobial resistance, but also virulence traits and transfer can occur among other enterococci as well as to other pathogenic species.

Intrinsic resistance to beta-lactams is due to low affinity of penicillin binding proteins, allowing bacterial cell wall synthesis in the presence of moderate levels of antibiotics, resulting in high MICs. Enterococci can also develop acquired tolerance to beta-lactams, meaning that after several doses they can be inhibited but not killed by penicillins or glycopeptides, even at doses many-fold higher than the MIC. Since most *E. faecalis* isolates require lower MICs (1-8μgm/ml), many *E. faecalis* isolates including VRE, can be inhibited by penicillins at doses achievable in serum, while most *E. faecium* infections cannot due to higher MICs.[6] Research has shown that the concentration of ampicillin needed to inhibit enterococci is substantially less than penicillin, thus ampicillin is a more appropriate choice for clinical infections.[5] As many strains of *E. faecalis* produce beta-lactamase, addition of clavulanic acid or sulbactam to therapeutic protocols may be warranted. Cephalosporins have not been clinically effective for the treatment of enterococcal infections, and use of extended-spectrum cephalosporins in people is considered a risk factor for the development of multi-drug resistant enterococcal infections.[7] Glycopeptide resistance is of upmost concern, caused by alteration of the bacterial cell wall peptidoglycan composition, allowing a 1000-fold decrease in affinity for vancomycin. While vanA and vanB genes are transferrable on transposons and plasmids, vanC genes are chromosomal (not transferrable) causing intrinsic resistance in *E. gallinarum*, *E. casseliflavus*, and *E. flavescens*.[2]

Enterococci show resistance to aminoglycosides at low levels, because they do not possess cytochrome enzymes and cannot produce the energy to take up antibiotics into the cell; however at higher doses aminoglycosides may be effective. Combinations of beta-lactams and aminoglycosides are commonly used for resistant enterococcal infections, as exposure to Beta-lactams may increase aminoglycoside uptake.[6] Enterococci can also become resistant to aminoglycosides at higher doses through acquired resistance, caused by aminoglycoside modifying enzymes which are coded by plasmids and are transferrable, thus rendering the synergism between Beta-lactams and aminoglycosides less effective. Resistance to trimethoprim-sulfamethoxazole can occur in vivo, even when in vitro testing suggests susceptibility, because enterococci can use preformed folic acid, enabling them to bypass the inhibition of folate synthesis produced by this drug.[8] For this reason, and concerns of bacteriostatic rather than bacteriocidal effects, TMP/SMX is not considered a drug of choice for enterococcal infections.

ENTEROCOCCAL VIRULENCE

Possession of virulence traits can increase an enterococcal strain's ability to colonize, adhere, and cause clinical disease in a particular patient. The virulence traits listed in Table 1 are of importance in veterinary medicine because their presence may contribute to development of infection as well as make treatment of these infections more challenging. Transfer of virulence traits between bacteria via conjugation may be linked to transfer of antimicrobial resistance genes, for example *E. faecium* carrying esp demonstrated more resistance to ampicillin, ciprofloxacin, and imipenem compared with esp-negative *E. faecium*.[9]

Table 1. Select enterococcal virulence traits and their roles contributing to clinical infections

Virulence trait	Gene	Role in virulence
Extracellular surface protein	Esp	Increases adhesion; evasion of the immune system; contributes to biofilm formation
Aggregation substance	Asa1	Host cell attachment; clumping of enterococci; increases frequency of conjugation; evasion of immune system
Gelatinase	GelE	Degrades host tissue to provide nutrients for bacteria; contributes to biofilm formation
Cytolysin	cylA	Toxin causing lysis of erythrocytes

ENTEROCOCCI IN HUMAN HEALTHCARE

Enterococci are a leading cause of nosocomial infections in human healthcare, causing serious illness in immunocompromised and critically ill patients. *E. faecalis* have been responsible for more clinical disease, while *E. faecium* have caused more treatment challenge due to antimicrobial resistance.[2] One study found prevalence of *E. faecalis* in hospitalized human patients to be 57%, compared with 39-40% in healthy individuals; this is not only due to the opportunistic and virulent nature of this organism, but also because hospitals themselves can be sources of these bacteria.[2,10] Common sites of enterococcal infections in people include: urinary tract infections, bacteremia, peritonitis, hepatobiliary infections, endocarditis, wound infections, and device-related infections.[5,11]

ENTEROCOCCI IN VETERINARY MEDICINE

Gastrointestinal Flora

Enterococci can be isolated from the feces of healthy dogs and cats, as a component of their normal gastrointestinal flora. Jackson et al found that 53% of healthy dogs and 44% of healthy cats in the United States had rectal swabs positive for enterococci, with *E. faecalis* being found with highest prevalence in both dogs (60%) and cats (45%), followed by *E. hirae* and *E. faecium.*[12] These gut commensals can be present in the intestine in numbers as high as $6x10^8$ colony forming units (CFU) per gram of feces in cats, and $3.9x10^7$ CFU/g feces in dogs, but may account for less than 0.01% of total intestinal flora.[13,14]

When evaluating enterococcal prevalence and resistance studies, it is important to note the study population, as resistance rates have varied with geographic location. This is especially important when considering vancomycin resistance. For example, van Belkun et al found that 48% of dogs and 16% of cats visiting a veterinary clinic in the Netherlands had VRE isolated from their feces.[15] Five years after banning use of avoparcin as a growth promoter in food animals, a second study in a similar urban environment in the Netherlands found zero prevalence of VRE in dogs.[16] Vancomycin-resistant *E. faecium* has also been reported from the feces of dogs and cats in Belgium (8% dogs, 5% cats)[17] Avoparcin was never approved in the United States, which some speculate may contribute to the lower prevalence of VRE found in companion animals in this country. However, in one study VRE was identified from 3/16 (18.75%) dogs in Ohio.[18] Ampicillin-resistant *E. faecium* in the clonal complex 17 (CC17) which is associated with nosocomial human infections have also been isolated from canine feces in the United Kingdom and Denmark.[19] Among fecal *E. faecalis* from a healthy population of dogs and cats in the U.S., antimicrobial resistance was seen to 10 antimicrobials overall, including chloramphenicol (85% in canine samples), gentamicin (79% in canine samples), and tylosin (53% in canine samples); no isolates were resistant to vancomycin or linezolid.[12]

Gastrointestinal Disease

As part of the normal flora, enterococci rarely contribute to gastrointestinal disease in either humans or companion animals. However, sporadic case reports have suggested that enterococci can cause diarrhea in young animals, including puppies and kittens, based on isolation from cultured fecal samples. More recently, *E. hirae* has been implicated as a cause of bacterial diarrhea in pre-weaning age kittens based on Gram-staining, fluorescence in situ hybridization (FISH), and PCR of bacteria adhered to the intestinal surface of affected kittens.[20] Enteritis is a common clinical problem and cause of death for kittens, with differentials including viral (panleukopenia), protozoal, parasitic, and bacterial causes. Immature animals may be predisposed to bacterial infections such as *E. hirae* due to their immaturity of immune function, failure of passive transfer, impaired colonization resistance, stressful housing conditions, or antibiotic use.[20] The exact cause of bacterial enteritis can be challenging to diagnose, because a positive fecal culture does not prove causation. Light microscopy for histopathologic examination of both *E. coli* and enterococcal species can appear very similar with inflammation and abundant enteroadherent coccobacilli, and thus molecular testing may be needed for definitive diagnosis.[20]

Enterococcal Role in Probiotics

In recent years, marketing of probiotics has increased dramatically and been directed towards promoting overall digestive health, providing healthy flora to compete with pathogens for nutrients and binding in the intestine, and improving immune responses. The most common probiotic organisms are *Lactobaccillus* and *Bifidobacterium.* Although rarely found in human probiotics due to concerns of safety, enterococci are commonly found in veterinary probiotic formulations. Much research has been done to select strains that are able to survive and adhere within the gastrointestinal tract and are considered safe, meaning they do not carry and cannot receive genes for antimicrobial resistance and certain virulence traits. *Enterococcus faecium* SF68 is one such strain which has undergone safety testing, but the most appropriate clinical application for this probiotic is still being determined. Clinical trials have been performed evaluating the efficacy of *E. faecium* SF68, and found it to be ineffective for decreasing shedding of giardia cysts in dogs, but effective at improving the fecal quality of both puppies and kittens, as well as improving the microbiota diversity in cats with herpes virus.[21-24] While *E. faecium* SF68 is generally considered to be safe, it may be wise to use caution with probiotics containing other strains of enterococci, as it may be possible for some strains to receive and carry the vanA gene cluster as well as other genes for antimicrobial resistance and/or virulence.[25] It should also be noted that the microbial content of commercial probiotics may not be consistent with the label claim; a microbiologic investigation by Weese found that 2/13 probiotics had labels that accurately reflected their contents.[26] Both correctly labeled products were human probiotics containing only lactobacillus, while 6/8 veterinary products contained enterococci, with 2 including enterococci that were not mentioned on their labels.[26]

Enterococci in Urinary Tract Infections

As in human healthcare, enterococci have been documented to cause numerous extraintestinal infections in companion animals, including pyodermas and wound infections, otitis, respiratory infections, biliary infections, and urinary tract infections (UTIs). While most extraintestinal enterococcal infections remain sporadic in frequency, enterococcal UTIs, which are often multi-drug resistant, have increasingly become a clinical frustration for veterinary practitioners. In human healthcare, *E. faecalis* is one of the most common bacteria causing recurrent and nosocomial UTIs, often associated with indwelling urinary catheters.[27] A retrospective study performed of all canine urine samples submitted for culture from the Western College of Veterinary Medicine Veterinary Teaching Hospital in Saskatchewan found 10.2% of all positive cultures to have *Enterococcus* spp.[28] Enterococci were commonly (26.7%) found in mixed-bacterial urine cultures, and they were responsible for 5.5% of recurrent UTIs.[28] A study performed at Michigan State University found that canine enterococcal UTIs are most commonly caused by *E. faecium* (37%), followed by *E. gallinarum* (31%), and *E. faecalis* (20%).[29] In that study, the majority of enterococcal isolates were resistant to 3+ antibiotics, especially penicillin, gentamicin, and tetracycline, and one *E. faecium* isolate was resistant to vancomycin.[29] For all enterococcal infections, remembering the opportunistic nature of these bacteria is important, identifying and eliminating/minimizing local (removing an indwelling urinary catheter) or systemic immunosuppression (treating hyperadrenocorticism) can be clinically rewarding. Antimicrobial therapy for enterococcal UTIs should be based on culture and susceptibility, and both intrinsic and acquired resistance in these species should be considered (see above). Antimicrobial agents that are often effective for human and companion animal UTIs include amoxicillin-clavulanic acid, fluoroquinolones, and nitrofurantoin. Anecdotally, for mixed-bacterial UTIs, therapeutic success has been found by tailoring therapy towards the non-enterococcal species (ie. choosing an antibiotic to which the *E. coli* is susceptible, even if the enterococci appear resistant) and monitoring cultures closely. For more challenging and resistant infections, exploiting the synergism between beta-lactams and aminoglycosides can also be used with caution. While the majority of enterococcal infections are susceptible to vancomycin, linezolid, and quinopristin-dalfopristin (*E. faecium* only), these antibiotics are best reserved for use in human infections.

REFERENCES

[1]Hancock LE, et al. Gram-positive pathogens. 2000;251. [2]Fisher K, et al. Micro 2009;155:1749. [3]Went CB, et al. J Clin Micro 1998;36:3734. [4]Kuhn IA, et al. Int J Antimicrob 2000;14:337. [5]Marothi YA, et al. Indian J Med Microbiol 2005;23:214. [6]Murray BE Clin Micro Rev 1990;3:46. [7]Rice LB Emerg Inf Dis 2001;7:183. [8]Hamilton-Miller JMT. J Antimicrob Chemother 1988;22:35. [9]Billstrom H, et al. Int J Antimicrob Inf 2008;32:374. [10] Mutnick AH, et al. Diagn Microbiol Infect Dis 2003;46:63. [11] Poh CH, et al. J Infect 2006;52:383.[12] Jackson CR, et al. J Applied Micro 2009;107:1269. [13]Wright et al. Mar Pollut Bull 2009;58:1649. [14] Jett BD, et al. Clin Micro Rev 1994;7:462. [15] van Belkun et al. Lancet 1996;348:1038. [16]Wagenvoort JHT, et al.. J Antimicrob Chemother 2003:52:532. [17]Devriese LA, et al. Antimicrobial Agents and Chemo 1996;40:2285. [18]Rice EW, et al. Diag Micro Inf Dis 2003;46:155. [19]Damborg P, et al. Appl Environ Microbiol 2009;75:2360. [20] Nicklas JL, et al. J Clin Microbiol 2010;48:2814. [21]Simpson KW, et al. J Vet Intern Med 2009;23:476. [22]Benyacoub J, et al. J Nutr 2003;133:1158. [23]Czarnecki-Maulden GL, et al. Compen Contin Educ Pract Vet 2007;29:37. [24]Lappin MR, et al. J Fel Med Surg 2009;11:650. [25] Lund B, et al. CID 2001;32:1384. [26] Weese JS. J Amer Vet Med Assoc 2002;220:794. [27]Vebo HC,et al. PlosOne 2010;5:e12489. [28]Ball KR, et al. Can Vet J 2008;49:985. [29]Simjee S, et al. J Clin Microbiol 2002;40:4659.

EMERGING ZOONOSES OF DOGS AND CATS

Bruno B. Chomel, DVM, PhD
Davis, CA

INTRODUCTION

Dogs and cats are present in more than half of the US households and may sometimes carry zoonotic agents, which can be transmitted to their owners. This presentation focuses on the emergent pathogens in companion carnivores.

VIRAL ZOONOTIC AGENTS

Rabies

Rabies has been present in domestic dogs and cats for centuries. However, with the spread of wildlife rabies in North America, the number of cases of rabies recorded in cats in the USA has been on the increase in the 20 last years from less than 200 cases per year in 1990 to 300 cases in 2009 (2). Furthermore, a few cases of rabies associated with importation of dogs from foreign countries where rabies is endemic have been recently reported both in the USA (20) and Europe (16). In France, 27 cases were detected in imported dogs between 1968 and 2009 with an increase in the number of cases imported from Morocco with 5 cases between 1968 and 1990, no cases between 1991 and 2000, but 8 cases between 2001 and 2009 (http://agriculture.gouv.fr/rage,993).

Influenza H1N1 and H5N1viruses

In recent years, emergence of influenza viruses (subtype H3N8 from horses in the USA and H3N2 in Asia to dogs was documented (14). A fatal infection with the highly pathogenic avian influenza virus (HPAIV) H5N1 was also reported. Transmission of HPAIV H5N1 from infected poultry or wild birds to large felids and domestic cats has now been reported from eight countries in Asia and Europe. Experimentally the infection could also be transmitted from diseased to naive cats. Infection of domestic cats with pandemic H1N1 influenza virus has recently been documented in the USA (25). In that country, a seroprevalence survey showed that 17 of 78 (21.8%) cats sampled during the 2009-2010 influenza season had antibody titers $\geq$40 against the novel H1N1 strain by hemagglutinin-inhibition assay, compared to only 1 of 39 (2.6%) sampled in 2008 prior to emergence of the pandemic ($p = 0.006$) (19).

Cowpox

Cowpox infection of cats is an emerging viral zoonosis in many parts of Europe. Approximately 50% of human cases of cowpox are transmitted from cats, which themselves have been infected from rodents they caught (17). Increasing numbers of cases in young people have been documented recently in Europe, likely because of a lack of cross-reactive immunity after the end of routine smallpox vaccination in the late 1970s (11). The disease remains localized and is self-limiting in patients that are immunocompetent, although dissemination and fatal courses have been reported. After an incubation period of 7–12 days after direct skin contact with cats, the main source of infections in people, patients develop a necrotising nodule with consecutive scar formation, accompanied by malaise, raised temperature, and long-lasting, pronounced, and painful regional lymphadenopathy.

BACTERIAL ZOONOSES

Bartonella

Cats have been demonstrated to be the main reservoir of *Bartonella henselae*, the agent of cat scratch disease (3). Transmission to humans is mainly through cat scratches that inoculate flea feces containing live *Bartonella*. Most cats are asymptomatic carriers and can be bacteremic for several weeks to several months. However, some pathologies have been observed in cats, including endocarditis cases. Cats are also natural reservoirs of *B. clarridgeiae*, *B. koehlerae* and can be infected by *B. quintana*, *B. bovis* or *B. vinsonii berkhoffii*. Dogs are more commonly infected with *B. vinsonii bekhoffii*, which is a main source of endocarditis and cardiac pathology, but cases caused by *B. henselae*, *B. washoensis*, *B. clarridgeiae*, *B. rochalimae*, *B. elizabethae*, *B. vinsonii arupensis* and *B. grahamii* have also been detected in dogs (1). Clinical features in dogs are quite similar to the clinical manifestations observed in humans and dogs represent good sentinels for human infection.

Capnocytophaga canimorsus

Capnocytophaga canimorsus is a commensal bacterium in the oral flora of dogs and cats (7). The bacterium is a zoonotic agent, isolated from humans infected by dog or cat bites, scratches, licks or simply exposure to dogs or cats. Disseminated infections can lead to septic shock, meningitis, endocarditis, and other severe sequelae.

Campylobacter, Helicobacter, Salmonella , E. coli

Feeding dogs with raw meat may be a risk factor for contracting campylobacter and Salmonella infection, as shown in a recent study (18) *S. enterica* was isolated from six out of 42 (14%) of the feces samples from dogs fed raw meat, from two out of 40 (5%) of the raw meat samples, and from four of 38 (10·5%) of vacuum cleaner waste samples from houses where dogs regularly munched raw meat. None of the dogs that did not consume raw meat were positive for the organism. *C jejuni* was found in one of 42 (2·6%) of dogs given raw meat. In 1999 and 2000, four outbreaks of gastrointestinal illness due to Salmonella enterica serotype Typhimurium in employees, clients, and client animals from 3 companion animal veterinary clinics and 1 animal shelter were investigated. More than 45 persons and companion animals became ill and 18 persons and 36 animals (of more than 200 tested) were culture-positive for S. Typhimurium. One outbreak was due to multidrug-resistant S. Typhimurium R-type ACKSSuT, while the other 3 were due to multidrug-resistant S. Typhimurium R-type ACSSuT DT104. Therefore, companion animal facilities may serve as foci of transmission for salmonellae between animals and humans if adequate precautions are not followed. In southern Ontario, the proportion among 110 veterinary hospitals with possible zoonotic bacteria positive environmental swabs were: *E. coli*--92%, , methicillin-resistant *S. aureus* (MRSA)--9%, CMY-2 producing *E. coli*--9%, methicillin-resistant *S. pseudintermedius*--7%, and *Salmonella*--2% (21). There are also clear indications that gastric non-*H. pylori Helicobacter* infections in humans originate from animals, and it is likely that transmission to humans occurs through direct contact, including *H. felis* and *H. canis* (13).

Bordetella bronchiseptica

Bordetella bronchiseptica is an uncommon cause of respiratory infection in humans generally occurring in immunocompromised individuals exposed to infected animals, especially dogs and cats. While human infections are thought to be rare with *B. bronchiseptica*, increasing reports have been steadily accumulating. To date there are over 65 reported cases of *Bordetella* infection. Most of these cases were reported in immunocompromised individuals after being directly exposed to an infected animal. The prevalence of human infection has been increasing, possibly as a result of the AIDS epidemic and an increased number of organ transplants (12). A case of *B. bronchiseptica* pneumonia in a kidney–pancreas transplant patient was reported after exposure to her dog recently vaccinated with a live attenuated strain of *B. bronchispetic* (10).

Feline hemotropic mycoplasma

Hemotrophic mycoplasma, formerly known as *Eperythrozoon* spp., are small epicellular blood parasites, which can be found in different species of mammals, including cats (27). Three species of hemotropic mycoplasmas (formerly known as *Haemobartonella felis*) are known to infect cats worldwide, *Mycoplasma haemofelis*, "Candidatus Mycoplasma turicensis" and "Candidatus Mycoplasma haemominutum" (26). *Mycoplasma* attach loosely to the surface of RBCs and sometimes destroys them but it can also be seen free in the plasma. The disease is extremely rare in humans and manifests with fever, swollen lymph nodes, an enlarged spleen and liver, worsening anemia, leucopenia, thrombocytopenia and sometimes mild hepatitis and subclinical myocarditis. Blood from infected cats should be handled with care because of the potential zoonotic nature of hemoplasma infections.

Q fever, Tularemia, Plague

A few cases of Q fever and tularemia have been associated with exposure to *Coxiella burnetii* infected parturient cats or dogs or bites by *Francisella* infected cats. Exposure to cats infected with *Yersinia pestis* has been shown to be a source of human infection (8), as well as sleeping with pet dogs or cats (4).

Tuberculosis

A few cases of tuberculosis caused by *M. tuberculosis* or by *M. bovis* have recently been reported in either dogs or cats in both the United States and in the United Kingdom and New Zealand, where bovine tuberculosis carried by wildlife (badgers, possums) is a major public health issue. More than 100 cases of *M. microti* have also been reported in cats. In the UK, among the 115 Mycobacterium feline cases diagnosed between 2006 and 2009, 55% were caused by *M. microti* and 45% by *M. bovis* (23). A few human cases have now been reported, too.

Methicillin resistant *Staphylococcus aureaus* (MRSA)

Staphylococcus aureus is a common human commensal organism; acquisition of genes encoding an altered penicillin-binding protein confers resistance to beta-lactam antimicrobial drugs (5). Methicillin-resistant *S. aureus* (MRSA) are often resistant to non-beta-lactam antimicrobial drugs as well. Originally described as an important cause of nosocomial infection, MRSA colonization and infection are now often identified in humans outside healthcare settings. Like other *S. aureus*, MRSA may be present without clinical illness. However, when they do cause infection the consequences can be extremely serious. Pets may become contaminated, colonized, or infected with methicillin-resistant staphylococci, including MRSA. Dogs and cats are more likely to be colonized/infected with *S. pseudintermedius* than *S. aureus*, but this pathogen can acquire genes encoding methicillin resistance. Diagnosis has implications not only for treatment of infected animals, but for potential zoonotic transmission.

Leptospirosis

Leptospirosis is a re-emerging disease of dogs in the United States. In a recent study, two relevant spatial clusters of were identified (9). The primary cluster was located in the northeastern part of Illinois including Chicago and surrounding areas. The secondary cluster covered the central part of Texas. Eight space-time clusters of overall MAT positivity were identified that covered different geographic locations for different time points. Spatial and space-time clusters for individual serovars were also identified for six serovars: eight each of Grippotyphosa and Pomona, seven of Bratislava, five of Autumnalis, and three each of Icterohaemorrhagiae and Canicola. Leptospira seropositivity in dogs tended to have distinctive clusters in space and space-time. Most of the space-time clusters of overall Leptospira MAT seropositivity were associated with cluster events for individual serovars.

PARASITIC AND FUNGAL ZOONOSES

Echinoccosis

Recent surveys in central Europe have extended the known geographical occurrence of *E. multilocularis* in foxes from 4 countries at the end of the 1980s to at least 11 countries in 1999 (6). Cases of human alveolar echinococcosis, previously regularly reported from only 4 countries, are now recorded from 7 countries, but the annual incidences are low. Evidence of parasite spreading has been reported from North America and Japan. Factors with the potential of enhancing the infection risk for humans include increasing fox populations and parasite prevalences, progressing invasion of cities by foxes, the establishment of urban cycles of the parasite, and the spill-over of *E. multilocularis* infection from wild carnivores to domestic dogs and cats.

Leishmaniasis

Leishmaniasis is an emerging or re-emerging zoonotic disease in dogs in the USA (fox-Hounds) and in endemic areas (the Mediterranean basin and several regions of Brazil). Natural hosts include dogs and humans, although many recent studies indicate that cats can be infected in endemic areas, and may serve as a domestic reservoir for human infection (22). Transmission in endemic areas is usually via sand fly vector. Infected dogs are the primary reservoir for zoonotic visceral leishmaniasis in endemic regions, and are the most significant risk factor predisposing humans to infection. Dogs have a wide range of clinical presentation due to infection with *L. infantum*, ranging from asymptomatic to fatal visceralizing disease.

Sporotrichosis

Sporotrichosis is an emerging zoonosis in Rio de Janeiro, Brazil (24). From 1998 to 2003, 497 humans and 1,056 cats with culture-proven sporotrichosis were studied. A total of 421 patients, 67.4% with a history of a scratch or bite, reported contact with cats that had sporotrichosis.

***Cryptosporidium* and *Neospora caninum* infections**

In recent years, there have been many changes to the ways that clinical veterinary science is conducted and nowhere is this more evident than in companion animal practice (15). Veterinarians working with pet dogs and cats are facing new challenges associated with the emergence and re-emergence of parasitic diseases. Some, such as *Neospora caninum*, have been recently recognized; others like *Giardia* and *Cryptosporidium* have been reported with increasing frequency, in part as a result of laboratory tests with improved sensitivity and specificity. In many regions, the emergence of parasitic diseases has been a consequence of pet travel and exotic diseases pose a unique diagnostic challenge for the veterinarian, as the index of suspicion for these conditions may be absent.

REFERENCES

1. Bai Y et al. **Vet Microbiol**. 2010 ;146:314-319. 2. Blanton JD et al. , Palmer **J Am Vet Med Assoc**. 2010 ;237(6):646-657. 3. Chomel BB et al. **Emerg Infect Dis**. 2006;12:389-394. 4. Chomel BB, Sun B. **Emerg Infect Dis**. 2011;17:167-172. 5. Cohn LA, Middleton JR. **J Vet Emerg Crit Care** (San Antonio). 2010;20:31-45. 6. Eckert J, et al. **Int J Parasitol**. 2000;30: 1283-1294. 7. Gaastra W, Lipman LJ. **Vet Microbiol**. 2010 ;140:339-346. 8 .Gage KL et al. **Clin Infect Dis.** 2000;30:893-900. 9. Gautam R et al. **Prev Vet Med.** 2010;96:122-131. 10. Gisel JJ et al. **Transpl Infect Dis**. 2010;12:73-76. 11. Glatz M et al. **Lancet Infect Dis**. 2010;10(4):288. 12. Goldberg JD et al. **Bone Marrow Transplant**. 2009;44:381-382. 13. Haesebrouck et al., **Clin Microbiol Rev**. 2009 ;22:202-223. 14. Harder TC, Vahlenkamp TW. **Vet Immunol Immunopathol**. 2010;134:54-60. 15. Irwin PJ. **Int J Parasitol.** 2002;32:581-593. 16. Lardon Z et al. **PLoS Negl Trop Dis.** 2010;4(6):e723. 17. Lawn R.**Vet Rec**. 2010;166(20):631. 18. Lenz et al., *Can Vet J* 2009; **50:** 637-643. 19. McCullers JA et al. **Arch Virol.** 2011;156:117-120. 20. McQuiston JH et al. **Zoonoses Public Health** 2008;55:421-426. 21. Murphy CP et al. **Can Vet J**. 2010;51:963-972. 22. Petersen CA. **Top Companion Anim Med**. 2009;24:182-188. 23. Rüfenacht S et al. , **J Feline Med Surg**. 2011;13:195-204. 24. Schubach A et al. **Emerg Infect Dis.** 2005;11:1952-1954. 25. Sponseller BA et al. **Emerg Infect Dis.** 2010;16(3):534-537. 26. Sykes JE. **Vet Clin North Am Small Anim Pract**.2010;40:1157-1170. 27. Sykes JE et al. **J Clin Microbiol**. 2010 Oct;48(10):3782-3785. 28. Xavier Emmanuel F et al. **Emerg Infect Dis**. 2007;13:1924-1927.

LEPTOSPIROSIS: LESSONS I LEARNED

Craig E. Greene, DVM, MS, DACVIM
Athens, Georgia

INTRODUCTION

The ACVIM has developed a consensus statement on diagnosis, treatment and prevention of leptospirosis. The intent of this lecture is not to alter the thrust of that document. Instead, the speaker will attempt to address some uncertainties about the disease by reviewing unpublished data from his research and clinical practice concerning canine leptospirosis, in context with pertinent published literature.

Leptospirosis is a worldwide disease of affecting people and animals caused by spirochetes of the species *Leptospira interrogans* sensu lato. Leptospires have classically been characterized by the immunologic reactivity of their outer membrane proteins. These antigenic differences have been used to describe certain serovars. Because of antigenic cross-reactivity, individual serovars have been further catalogued, with antigenically similar strains, into serogroups. The serogroup identity can be detected by the host antibody response using the microscopic agglutination test (MAT), or by cross-absorption reactions with isolated spirochetes. Immunologic methods have been historically used to typify offending leptospira in group outbreaks or individual infections. Serovar or serogroup status is helpful with respect to immunologic detection and vaccine protection. Serovars with the same outer membrane immunoreactivity can have widely varied disease producing potential. Furthermore, leptospires of the same serogroup, from different geographic locations, cannot be assumed to cause the same clinical signs or infect the same hosts.

Molecular genetic testing has helped reclassify the Leptospira into 20 or more genomospecies. This genetic classification crosses over between pathogenic and avirulent species and between serovar designations. Genetic tools are more valuable than immune methods, for determining the specific identity of an isolate in culture or from body tissue or fluid specimens taken for analysis. Genetic analysis is a better epidemiologic marker in studying the environmental or reservoir host origin and transmission of a particular strain. Since most historical investigations have been made using immunologic methods, further studies should attempt to isolate and genotype the incriminated organisms. All practicing veterinarians can help in obtaining specimens for this purpose.

IS LEPTOSPIROSIS IN DOGS SURGING OR EMERGING?

Prior to 1980, a majority of the focus of leptospirosis in dogs centered around serogroups Canicola and Icterohaemorrhagiae. This may be because these two groups were incriminated in the initial descriptions of the disease worldwide between the years of 1925 to 1950. For 3 decades following 1950, most of the experimental studies and seroepidemiologic investigations continued to focus on these two serogroups. These two serogroups were both chosen for inclusion in the first canine leptospira vaccines that were tested in the 1950s and finally becoming widely available in the early 1960's. In spite of this two-serogroup focus, review of historical literature before 1980 indicates that reports existed of canine infections by organisms outside these two serogroups. Methods used to confirm these other-serogroup infections included serologic evaluation, laboratory cultivation, and experimental infection studies.[1-6] Despite these reports of alternate serogroup involvement, early investigators and vaccine manufacturers did not alter their focus.

However after 1980, testing for additional serogroups infecting dogs was instituted because of new reports of disease being caused by other serovars. In the first of these reports, a kennel outbreak was investigated by the University of Georgia State Diagnostic laboratories and the CDC in Atlanta in a foxhound kennel in 1982.[7] The dogs involved in this kennel had differing clinical manifestations than those generally described for classical leptospirosis. Subsequent to this period the state laboratories in Georgia began to test dog sera using the antigens that were reserved to test cattle and swine sera. The author has data from the mid 1980s showing a variable but consistent presence of predominantly Grippotyphosa/Pomona serogroup infection in clinically affected dogs. Approximately 10% of those animals had their predominant seroreactivity to groups Canicola or Icterohaemorrhagiae. This prevalence difference has been observed in most of the subsequent studies worldwide where broad serogroup reactivity has been examined. Unfortunately, there are no historical serosurveys of clinically healthy dogs conducted in the pre 1980 period that have used serogroup antigens other than Canicola or Icterohaemorrhagiae, that have subsequently been repeated within the last decade. For example, an excellent study was done in Michigan in the 1950s that evaluated exposure to groups Canicola and Icterohaemorrhagiae.[8] Over 50 years later, a study throughout the lower Michigan peninsula was done testing for a wide range of serogroup exposure.[9]. It showed the lowest reactivity to the serogroups Canicola and Icterohaemorrhagiae while a much larger percentage of reactivity was to other serogroups. One can only what the prevalence of reactivity was to the broader serogroup testing if it had been done in the 1950s. Was there a true increased prevalence of the others, or was the comparison biased because other serogroups had not been evaluated? The reviewed data suggests the latter possibility is more likely.

WHAT IS THE MOST COMMON MEANS OF EXPOSURE TO LEPTOSPIROSIS?

Leptospires can be theoretically transferred by direct contact with infected urine or blood, by venereal and placental means, or by bite wounds or ingestion of infected tissues. Crowding of animals, as may occur in a kennel situation, might theoretically be thought of as an environmental risk of infection; however, such transfer has not been well documented. Indirect transmission, which is relatively common, occurs through exposure of susceptible animals to contaminated water sources, soil, and food. Although they do not replicate, spirochetes can remain viable for months in moist environments when conditions such as temperature and pH are optimal.

In natural exposure in dogs, the means of transmission has been hard to document. The author has research results from a natural outbreak which implicate contamination of a common water source rather than direct transmission between infected dogs. Similar information can be assumed by examining the suspected means of transmission in human infections. Most commonly these occur through occupational, recreational, or avocational activities involving water.[10,11] Spirochetes gain access to the body through broken skin or mucous membranes and prolonged immersion or swallowing water greatly increases the risk [12,13]. Flooding elevates the water table, saturates the soil with leptospires, prevents evaporation of contaminating animal urine, and prolongs the survival of the spirochetes in surface water. Exposure to environments highly contaminated with rodent urine have been implicated in some of the infections in inner cities and where accidental ingestion of contaminated food has been implicated.[14,15,16] Simultaneous outbreaks of infection of people have occurred although only 8 have been published in the last 60 years. At least 50% involve common water source exposure to the dogs and people. In these situations confirmation of transfer from dogs to people have not been substantiated and genetic analysis has not been done in any circumstance to confirm the identity of the offending organisms. Studies of exposure from animals among veterinarians and veterinary students show a low rate of seropositivity without signs of clinical illness; however these individuals have a correlation of treating animals suspected of leptospirosis.[17,18] Therefore, although direct transmission is possible from dogs to people and other dogs, it appears to be of lower risk than through indirect contact with contaminated food or water.

WHAT IS THE DURATION OF INCUBATION AND COURSE OF DISEASE?

Incubation periods from onset of exposure to clinical illness is best documented in humans because of the general knowledge of exposure. However, in animals it only has been determined by experimental infection. Experimental inoculation of animals with leptospira does not usually produce signs of illness. Instead the presence of spirochetemia and spirocheturia has been used to determine active infection which occurs within a few days to a week following infection. The author has clinical data to suggest that the period may be as long as 2-3 months from the onset of natural infection with a grippotyphosa serovar isolated in the southeastern United States. A longer incubation period may help to explain why the disease incidence peaks in that region during the fall and early winter months as has been observed by some other investigators in the eastern United States. In other areas such as California, the disease corresponds more with rainy cold seasons, the winter.[19] In the United States and Canada, a correlation was found between a greater number of cases of leptospirosis in dogs at veterinary teaching hospitals and higher average rainfall.[20] However, in other studies, association with poor drainage or flooding could not be documented.[21]

Leptospirosis has been claimed to be a cause of acute renal failure in dogs. This author questions the use of this terminology in that it also may be the cause of chronic renal failure. The disease manifestations are generally those of an acute uremic crisis; however, the renal dysfunction may have been more longstanding in association with the incubation period which may be prolonged in some animals and with some serovars in certain geographic regions. The hypothesis is that the disease in some dogs can cause chronic renal inflammation that progressively damages the kidney over time. In the author's experience, certain dogs have relatively high antibody titers at the time of diagnosis that decay over extended periods of time after treatment. Secondly, dogs have been identified as subclinically infected in households or kennels where other dogs have current illness. Many dogs with leptospirosis have other clinical manifestations such as inappetance, polydipsia, polyuria and weight loss that precede the onset of severe uremia. Others have nonregenerative anemia and hyperglobulinemia which likely relate to chronic renal failure and chronic antigenic stimulation, respectively. Hypoalbuminemia with superimposed dehydration must also be considered. These findings are associated with chronic renal failure. It appears that some dogs may be infected for extended periods in a subclinical or incubatory phase of disease.

WHEN SHOULD MAT TITERS BE EVALUATED DURING INITIAL DIAGNOSIS AND FOLLOW-UP?

With the MAT titer, a fourfold change (increase or decrease) in titer is classically required for the definitive confirmation of a self-limiting infection such as leptospirosis. As discussed in the Consensus statement and the literature,[22] interlaboratory results vary widely so the same laboratory should be used for any serial titers on the same animal. Although a titer ≥800 may be found to a serogroup on initial examination, that titer should be followed thereafter to confirm active infection by an expected decrease following antimicrobial therapy. Should a titer already be elevated at the time of initial examination the follow-up period for a repeat titer can be delayed for a month or longer. Titers may have negative results in early acute infections and sequential titers may be required to confirm infection. The duration of fall in titer after treatment has not been established and likely relates to the onset and duration of disease. Based on the author's data and other reports,[23] the titer fall may take up to 3 months following treatment to reduce to levels of 200 or less.

CAN PCR BE USED IN PLACE OF OR AS A SUPPEMENT TO THE MAT TEST?

PCR methods offer the theoretical advantage of detecting infection in its early stages and prior to antibody increases. Genetic methods allow for more specific determination of the infecting strains, which is not possible with the MAT. PCR also offers more rapid results in many cases. Unfortunately, the method and quality control for PCR detection can vary widely resulting in either false-positive or false-negative results. In a number of instances observed by the author, the PCR results were negative in urine or blood for dogs with confirmed infection. The use of PCR alone would require a practitioner to have an intimate knowledge of their laboratory. The speaker recommends use of both MAT and PCR in conjunction to help fill the voids inherent in each procedure.

HOW DO TITERS FARE AFTER VACCINIATION WITH RESPECT TO PRIOR INFECTION?

Affected with vaccination…with or without prior infection? Agglutination titers, behave as does IgM and levels do not reach high levels, even after two doses. Vaccine titers do not usually go over 800[24-28] although higher titers are rarely observed. These titers do not persist for longer than 3 months following vaccination of experimentally confined dogs.[29-31] Generally, the titers decrease in most all dogs which helps to distinguish vaccinated from actively or recently infected animals. Anecdotally high titers in dogs have been reported after vaccination; however, the author is not aware that any of these animals have had pre-titer results before the elevations were seen. It is possible these may represent chronic or subclinical infections and further documentation is needed.

REFERENCES

[1]Alexander AD, P, et al. **Am J Hyg** 1957;65:43. [2]Cholvin NR, et al. **J Infect Dis** 1959;104:92. [3]Murphy LC,et al. **Am J Vet Res** 1941;19:145. [4]Menges RW, et al. **Am J Vet Res** 1960; 21:371. [5]Thomas RE, et al. **J Am Vet Med Assoc** 1967;150:33. [6]Bishop L, et al. **Am J Vet Res** 1979;40:839. [7]Cole JR, et al. **Jam Vet Med Assoc** 1982;180:435. [8]Newman JP. **Am J Vet Res** 1950;11:405. [9]Stokes JE, et al. **J Am Vet Med Assoc** 2007;230:1657. [10] Feigin RD,et al. **Ann Intern Med** 1973;79:777. [11] Fraser DW, et al. **Ann Intern Med** 1973;79:786. [12] Haake DA, et al. **Clin Infect Dis** 2002;34:40. [13]. Phraisuwan P, et al. **Emerg Infect Dis** 1999;8:1456. [14]. Ko A, et al. **Lancet** 1999;354:820. [15]. Kobayashi Y. **J Infect Chemother** 2001;7:59. [16] Reis RB, et al. **PLoS Negl Trop Dis** 2008;2:e228. [17]Simón MC, et al. **Vet Rec** 1999;144:287. [18]Whitney EAS, , et al **J Am Vet Med Assoc** 2009;234:838. [19]Sykes JE,et al. **J Vet Intern Med** 2007;21:624. [20]Ward MP.et al. **Vet Med** 2002;56:203. [21]Ward MP, et al. **J Am Vet Med Assoc** 2004;225:72. [22]Miller MD, et al. **J Vet Intern Med** 2007;21:624. [23]Harkin KR, et al. **J Am Anim Hosp Assoc** 1996;32:495. [24]André-Fontaine G, et al. **Vet Rec** 2003;153:165-169. [25]Klaasen HL,et al. **Vet Microbiol** 2003;95:121. [26]Minke JM, et al. **Vet Microbiol** 2009;137: 137. [27]Schreiber P, et al. **Vet Microbiol** 2005;108:113. [28]Steger-Lieb A,et al. **Schweiz Arch Tierheilk** 1999;141:499. [29]Barr CS, et al. **Am J Vet Res** 205;66:1780. [30]Greene CE. 2005. Unpublished observations. [31]Hartman EG, et al. **Vet Immunol Immunopathol** 1984;7:245.

OVERVIEW OF THE EMERGENCY APPROACH TO RESPIRATORY DISTRESS

Kenneth J. Drobatz, DVM, MSCE, DACVIM, DACVECC
Philadelphia, PA

INTRODUCTION

Respiratory distress is a common presenting complaint in veterinary emergency medicine. Animals with respiratory distress represent a diagnostic and therapeutic challenge. The initial approach to these patients can often make the difference between life and death in this acute phase of their disease.

SIGNS OF RESPIRATORY DISTRESS

When an animal develops difficulty in oxygenation it makes attempts to improve oxygenation. This may be manifested simply as an increase in respiratory rate, or more severely as extended head and neck posture with abduction of the elbows, flaring of the nares, and open-mouth breathing. With severe resistance to movement of air, there may be paradoxical motion of the chest and abdomen. Normally, the chest and abdomen move together during respiration. On inspiration the abdominal wall and chest wall move out and during expiration they both move in. Paradoxical motion occurs when they move in opposite directions e.g. the chest moves in and the abdomen moves out. This suggests extreme resistance to air movement. Dogs are more demonstrative in manifesting clinical signs of respiratory distress. Some cats may have severe respiratory dysfunction and only manifest tachypnea at rest. In either species, postural changes suggest extreme respiratory distress and warrant immediate attention.

INITIAL APPROACH TO THE PATIENT

Oxygen Supplementation

At presentation, a patent airway should be assured. If the airway is not patent, then attempts should be made to clear the airway and intubate the patient. If an obstruction prevents intubation, then an emergency tracheostomy should be performed if it will by-pass the obstruction and assure a patent airway. To provide perspective, it is extremely rare that an emergency tracheostomy is required. Most patients can be intubated and the tracheostomy then performed in a more controlled way.

Immediate oxygen supplementation should be provided while assessing for a patent airway. Most animals showing signs of respiratory distress are struggling to keep hemoglobin saturated with oxygen, and have pulse oximetry readings of 90% or less. A value of 90% sits just on the edge of the saturation plateau of the hemoglobin oxygen saturation curve. Any struggling or interruption of oxygen supplementation can result in rapid desaturation of hemoglobin and total oxygen content of blood.

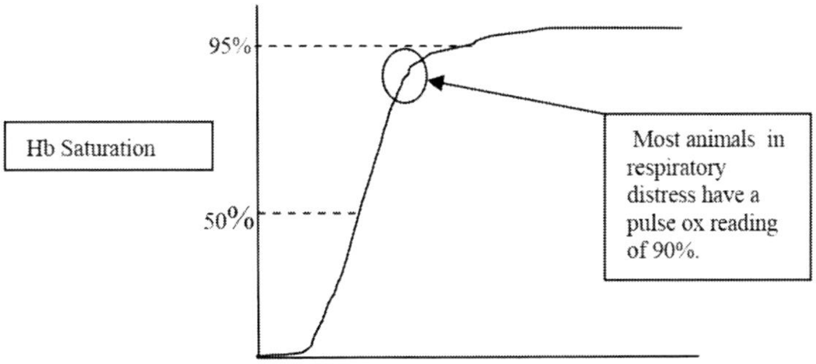

Our first choice for oxygen supplementation is mask or flow-by. This is convenient, inexpensive, and provides an opportunity to examine the patient as well as do procedures if necessary. Other modes of oxygen supplementation include an enriched oxygen environment (oxygen cage or tent), transtracheal catheter, nasal oxygen, and intubation with positive pressure ventilation. Each has its advantages and disadvantages.

The oxygen cage is an excellent method for oxygen supplementation, but has some major disadvantages including isolation of patient from the caregivers, expense, a need for large volumes of oxygen, and the large size of the cage itself; large dogs can overheat the cage, and opening of the cage results in loss of increased inspired oxygen concentration. Advantages of this method include minimal stress to the patient and control of FiO_2, carbon dioxide concentration, temperature, and humidity. If used judiciously with recognition of its limitations, an oxygen cage can be advantageous.

Nasal oxygen supplementation is an inexpensive method of oxygen supplementation and can be done using commonly available equipment. Its major medical advantage is that oxygen supplementation can be continuously given while procedures and physical assessments are performed. Inspired oxygen concentration may be as high as 40-60%. Disadvantages include that placement of nasal catheters can be stressful and some animals will not tolerate the nasal catheter. This mode of oxygen supplementation is most commonly utilized on hospitalized animals that need ongoing oxygen supplementation.

An oxygen tent, plastic bag, or an E-collar with plastic wrap over the top with oxygen supplied into this "microenvironment" can be an effective means of oxygen supplementation. Fi02 of nearly 100% can be achieved with high oxygen flow. Disadvantages include that some animals will not tolerate this method, condensation can develop inside the plastic cover, the animal can overheat and carbon dioxide can build up. Leaving an opening at the top for heat, carbon dioxide and condensation to escape can minimize some of these complications.

Transtracheal oxygen supplementation is rarely used in our practice. It can be used to provide oxygen support with upper airway obstruction if the catheter is placed between the lungs and the obstruction.

Intubation and positive pressure ventilation is the ultimate in control of ventilation but is more invasive than other methods and requires sedation/anesthesia. One can control nearly all aspects of ventilation with this method. Indications for positive pressure ventilation include hypoventilation, persistent hypoxemia despite high-inspired oxygen concentration, and respiratory fatigue.

Once oxygen supplementation has been provided, further evaluation can be performed if the patient's condition will allow it. Animals that are in respiratory distress can be difficult to work with. One must provide oxygen and assess the patient while minimizing stress to the patient. Excessive stress or struggling can be catastrophic in an animal with respiratory distress. Even dorsal recumbent positioning for ventral dorsal radiographs can be devastating. Any stress that causes the patient to consume more oxygen or decrease inspired oxygen concentration from breath holding while struggling can severely compromise some patients and result in death. At presentation we also place an intravenous catheter and collect blood for an emergency database (packed cell volume, total solids, dipstick BUN, blood glucose and a blood smear). This minimal amount of blood work can provide valuable additional information about the patient and sometimes provides a diagnosis of the cause of the respiratory distress. In addition, intravenous access allows for the immediate administration of fluids or drugs.

General Diagnostic Thoughts

During or after assuring a patent airway, providing oxygen supplementation, and acquiring intravenous access, examination of the patient should be done. The most important clues may be obtained from auscultation of the thorax and neck, as well as observation of the respiratory pattern.

Respiratory patterns may be helpful in localizing the problem. Irregular respiratory rhythms are almost invariably associated with central nervous system abnormalities. Poor airway compliance due to pulmonary parenchymal disease or restricted lung expansion often results in rapid and shallow respirations. Airway narrowing or fixed obstruction may be manifested as very slow and prolonged respirations. Respiratory difficulty is mainly on inspiration with extrathoracic dynamic obstruction, and on expiration with intrathoracic dynamic obstruction (intrathoracic collapsing trachea, feline asthma). It is convenient from a diagnostic point of view to approach respiratory distress in an algorithmic fashion.

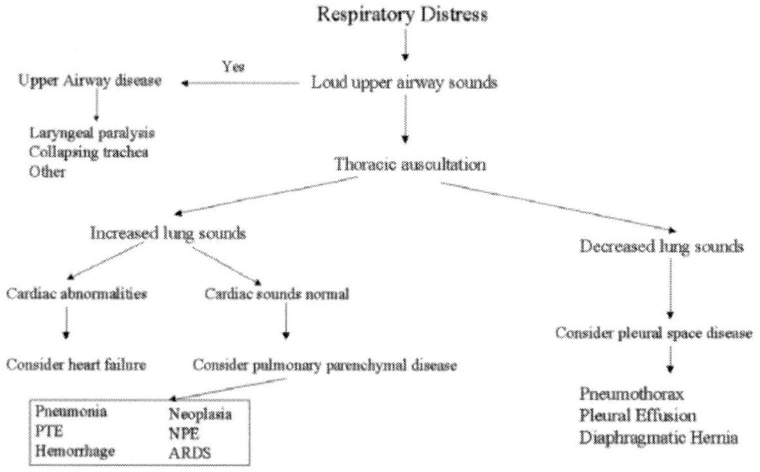

SUMMARY

Respiratory distress is a common presenting complaint in emergency medicine. Aggressive, yet judicious therapy should be provided to optimize the outcome. Disease categories causing respiratory distress in dogs include cardiac disease, airway disease, pulmonary parenchymal disease, pleural space disease, thoracic wall abnormalities and diaphragmatic rupture. The most common causes of feline respiratory distress are hypertrophic cardiomyopathy, feline asthma, and pleural effusions. Because many of these patients are severely compromised, diagnostic tests may be limited until initial stabilization is provided. Auscultation of the pulmonary and cardiovascular system combined with signalment and a careful medical history may provide important clues to the underlying cause and allow the clinician to provide judicious empirical therapy until definitive diagnostic procedures may be performed.

A DIAGNOSTIC APPROACH TO RESPIRATORY DISTRESS

Erica L. Reineke, VMD, DACVECC
Philadelphia, PA

INTRODUCTION

Respiratory distress in small animals requires prompt recognition in order to both relieve hypoxemia and to identify and treat the underlying disease process. Determining the etiology of dyspnea requires careful clinical and radiographic assessment. Unfortunately, these animals typically require supplemental oxygen to relieve their symptoms of distress severely limiting our ability to perform thoracic radiography. There has been recent research in both human and veterinary literature on the use of minimally invasive tools, such as ultrasound and cardiac biomarkers, to help aid in the diagnoses and treatment of small animal patients with respiratory distress. These diagnostic tools may help to minimize patient stress associated with other diagnostic modalities, such as radiography, and allow institution of therapies to relieve hypoxemia. The goal of these proceedings is to discuss a rational emergency diagnostic approach to the patient in respiratory distress with a focus on the current use of ultrasound and the future use of cardiac biomarkers to differentiate between causes of respiratory distress.

A DIAGNOSTIC APPROACH TO RESPIRATORY DISTRESS

Initial Assessment and Localization of Disease

The initial assessment of the dog or cat in respiratory distress should involve evaluation of the respiratory and cardiovascular systems. This includes evaluation of respiratory rate and effort, auscultation of the lungs and heart, evaluation of mucous membrane color, capillary refill time, and pulse quality. During this initial evaluation, supplemental oxygen should be given either by flow-by or face-mask. If possible an intravenous catheter should also be placed to allow for administration of life-saving medications. At the time of catheter placement blood should be drawn for evaluation of packed cell volume, total protein, venous blood gas analysis, blood electrolytes, blood glucose, and renal values. The primary goal of the physical examination at this point is to localize the respiratory distress to the part of the respiratory system affected: upper airway, lower airway, parenchyma, pleural space or thoracic wall/diaphragm. A more thorough physical exam should be completed once the patient has been stabilized. Once the disease process has been localized, diagnostic testing should be considered to determine the underlying etiology. As patients with respiratory distress can be extremely fragile (and even a brief physical exam can sometimes be fatal), the clinician should make every attempt to prioritize diagnostic tests that minimize stress to the patient while providing the most information about the patient's underlying disease process. Cage-side diagnostic tests such as ultrasonography and cardiac biomarkers may provide useful supportive information about underlying etiology in patients who are too unstable to have thoracic radiography performed.

Thoracic Ultrasound

Ultrasound in the emergency setting can be life-saving for the animal in respiratory distress. An advantage of thoracic ultrasound is that it can be performed cage-side minimizing stress on the hypoxemic patient. Even without extensive training in the use of ultrasound, with practice the clinician should be able to identify abnormal accumulations of fluid, soft tissue and air in animals with pleural space disease allowing for confirmation of physical examination findings. In addition, cardiac ultrasound can be useful to differentiate between cardiogenic and non-cardiogenic causes of pulmonary edema.

In a patient presenting with respiratory distress and pleural space disease is suspected, thoracic ultrasound should be the initial diagnostic test performed. Thoracic ultrasound can be performed with the patient in sternal recumbancy minimizing the amount of patient restraint and allowing for normal respiratory effort. Oxygen supplementation should be administered during the ultrasound examination. Thoracic scanning is done by placing the transducer in the intercostal spaces and rotating both cranially and caudally. The exact intercostal spaces where the transducer is placed may vary from patient to patient, but typically the transducer is placed in the 5th-6th intercostal spaces to evaluate the cranial thorax and again at the 7th-9th intercostal space to evaluate the caudal thorax. This should be repeated on both sides of the thorax if patient stability permits a full evaluation. A transducer with a small contact area should be used to minimize artifacts created by the ribs. Clipping of the haircoat is generally not necessary and alcohol can be used to wet down the haircoat allowing for adequate ultrasound probe-to-skin interface.

A normal aerated lung cannot be seen (the gas interface causes reverberation echoes) with ultrasound; however, the smooth hyperechoic surface of the lung can be seen sliding cranially and caudally with respiration against the stationary parietal pleura. This normal movement of the lung against the chest well is called the "glide sign" or "gliding sign". Pleural effusion separates the costal from pulmonary pleura and the resultant space fills with fluid of varying echogenicity. Low protein, low cellularity fluid (transudate-type) typically appears anechoic. Higher concentrations of cells, fibrin, or protein in the effusion increase the echogenicity of the fluid seen and is characteristic of exudate-types of effusion and blood. The presence of fibrin tags (seen as "floating" strands in the pleural fluid) suggests chronicity to the effusion. Abnormal portions of lung can be visualized when they are atelectic, consolidated, or when a mass is present. These non-aerated portions of the lung can appear similar to a small liver lobe. If the presence of pleural effusion is confirmed with ultrasound, a diagnostic and therapeutic needle thoracocentesis should be performed. This fluid should be evaluated microscopically for the presence of neoplastic cells and bacteria. If hemorrhage into the pleural space secondary to a coaguloapathy is suspected, a coagulation profile should be obtained prior to thoracocentesis.

In 2008, Lisciandro et al. reported on the use of thoracic ultrasound by emergency clinicians to identify pneumothorax in traumatized dogs. Detection of pneumothorax relies on the interpretation of artifacts induced by the presence of air in the pleural space. Pneumothorax should be suspected in a dog or cat when the normal "glide sign" is not seen. In addition, a glide sign that deviates from the normal linear continuity of the pulmonary-pleural interface can represent additional thoracic injury such as rib fractures, pulmonary contusions, and diaphragmatic hernia.

The mediastinum should also be evaluated ultrasonographically in patients with pleural effusion. The index of suspicion for a cranial mediastinal mass may be increased in cats and small dogs with poor thoracic compliance. Normally, the mediastinum cannot be typically evaluated due to the presence of adjacent aerated lung. However, mediastinal structures do become accessible by ultrasound when the aerated lung is displaced by a large mediastinal mass or when pleural effusion is present. The ultrasound transducer may need to be placed parasternal, suprasternal or substernal to evaluate the mediastinal space. Mediastinal masses vary in their appearance but are most often caused by lymphadenopathy. Cats may have fluid-containing masses such as a mediastinal cyst or cystic thymoma.

Diaphragmatic hernia can be identified either by the loss of diaphragmatic integrity or by the presence of abdominal viscera in the thorax. Evaluating for diaphragmatic integrity is easier when adjacent fluid is present either in the thorax or peritoneal cavity and may be more easily seen from a transhepatic approach with the transducer placed at the base of the xiphoid. This is often very challenging compared with identifying abdominal viscera in either the pleural or pericardial space. With traumatic diaphragmatic hernia, pleural effusion is often present which aids in the evaluation of the pleural space. Consolidated lung lobes should be differentiated from liver in the pleural space. Absence of viscera in the abdominal cavity can provide supporting evidence. For congenital peritonealpericardial diaphragmatic hernia, abdominal viscera can be seen within the pericardial sac.

In a dog or cat with respiratory distress secondary to suspected parenchymal disease, cardiac ultrasound is a minimally invasive tool that can be used by the emergency clinician to help differentiate between cardiac and non-cardiac causes of pulmonary edema. Cardiogenic pulmonary edema should be suspected in male, small breed dogs with a history of heart murmur, coughing, and exercise intolerance. Auscultation of a heart murmur (and/or gallop rhythm in cats), along with pulmonary crackles, increased bronchovesicular sounds, tachycardia and low rectal temperature should raise the index of suspicion for left-sided congestive heart failure. Cats with heart disease may also present with concurrent pleural effusion and therefore quiet lung and heart sounds may be ausculted ventrally. Quick evaluation of left atrial size, chamber volume and cardiac contractility with cardiac ultrasound can help to differentiate congestive heart failure from other etiologies.

The ratio of the left atrial size to the aortic root diameter is useful in evaluating for left atrial enlargement which can be supportive of congestive heart failure in both dogs and cats. Using two dimensional echocardiograpy (right parasternal short-axis view), left atrial size can be subjectively evaluated and compared to the aortic root diameter. Measurements of the left atria and aortic root should be obtained by placing a line at a 45 degree angle from perpendicular which passes through the aorta and the body of the left atrium. In the normal dog and cat, the ratio of left atria to aortic root size is usually less than 1.3. Values greater than 1.5 suggest left atrial dilation and provides supportive evidence for congestive heart failure.

Subjective evaluation of the heart size and contractility is also useful in identifying cardiac from non-cardiac causes of pulmonary edema. Dilated cardiomyopathy (DCM) with congestive heart failure should be suspected in large breed dogs presenting for respiratory distress, crackles and a cardiac murmur heard on thoracic auscultation. DCM, although less common than other forms of heart disease, can also be a cause of heart failure in cats. Using 2-dimensional echocardiography, DCM should be suspected when moderate to marked dilation of all four cardiac chambers is seen with a subjective decrease in contractility. Although cardiac ultrasound may be useful in identifying supportive evidence of heart disease in a dog or cat with respiratory distress, thoracic radiography remains the gold standard for definitive diagnosis of congestive heart failure. As soon as the patient is stable, thoracic radiographs should be obtained.

Finally, in a patient in which there is a high suspicion (based on history, physical examination findings $\pm$ cardiac ultrasound) of left-sided congestive heart failure, furosemide therapy (2-4 mg/kg IV or IM) may be initiated and the patient's response to therapy assessed after 30 to 45 minutes. Typically, there is minimal risk to administration of this medication prior to obtaining confirmatory thoracic radiographs.

In summary, the use of focused emergency ultrasound in patients with dyspnea can confirm physical examination findings, differentiate between etiologies, and allow for institution of life-saving therapy. The use of ultrasound in the emergent situation should never take the place of a full ultrasonographic evaluation of the thorax by a radiologist and echocardiographic evaluation by a cardiologist but can provide valuable information in the short term for patients presenting with dyspnea and allow for patient stabilization.

Future Use of Cardiac Biomarkers

Over the past 20 years, a variety of circulating "cardiac" biomarkers have been discovered and are proving useful in the diagnosis of cardiac disease and heart failure in people. Several different cardiac biomarkers have been evaluated in veterinary patients including plasma atrial natriuretic factor (NT-proANP), B-type natriuretic factor (BNP) , endothelin (ET-1), and cardiac troponin I (cTnI) as markers for heart disease. Plasma concentrations of N-terminal pro-atrial natriuretic peptide and B-type natriuretic peptide increase in response to atrial distensions and cTnI concentration is a specific and sensitive indicator of ischemic myocardial damage and the concentration increases in dogs with heart disease. Dogs and cats presenting with dyspnea do not tolerate the manual restraint needed for thoracic radiography and echocardiography, therefore an accurate blood test that could differentiate congestive heart failure from other causes of dyspnea would be an invaluable tool for the veterinary clinician.

In 2007, Solter et al. evaluated the use of NT-proANP, BNP, ET-1 and cTnI in distinguishing congestive heart failure from other causes of respiratory distress in dogs. The results of this study suggested that NT-ProANP, BNP and ET-1, but not cTnI were useful in distinguishing between cardiac and non-cardiac causes of respiratory distress. Plasma NT-proANP had the highest sensitivity (95.5%) and specificity (84.6%). Unfortunately, to the author's knowledge, a commercially available assay for ANP is only currently available in Europe.

In cats, the use of cardiac troponin I was evaluated in dyspneic cats with and without heart failure. The results of this study found that cats with congestive heart failure had significantly higher concentrations of cTnI compared with cats with respiratory disease of different etiology. Using a receiver operator characteristic curve, it was determined that cTnI concentration > 0.2 ng/mL had a 100% sensitivity, 58% specificity, and 88% diagnostic accuracy for determining congestive heart failure in cats. Alternatively, a plasma cTnI concentration >1.42 ng/mL had a 58% sensitivity, 100% specificity and 66% diagnostic accuracy for diagnosis of congestive heart failure. Similar results were found in a subsequent study evaluating cTnI concentrations in cats with and without heart failure. Again, cTnI concentrations in cats with heart failure were higher than in cats with other causes of respiratory distress. In this study, a cut-off of 0.81 ng/mL for cTnI had a sensitivity and specificity of 65.2% and 90.0% respectively. In both studies however, there was considerable overlap between the two groups.

More recently, the cardiac biomarkers NT-ProANP and BNP were also evaluated in dyspneic cats with and without heart disease. In two separate studies, NT-ProANP and BNP were able to differentiate cats with congestive heart failure from cats with other respiratory disease. Compared with cTnI, NT-proANP and BNP were both more sensitive and specific markers for congestive heart failure. In the study by Fox et al. a cut-off value of 265 pmol/L had a 90.2% sensitivity, 87.9% specificity, 92% positive predictive value, and 85.3% negative predictive value for heart failure.

In summary, cardiac biomarkers are a useful adjunctive diagnostic test for differentiating between cardiac and non-cardiac causes of respiratory distress, especially in animals who may be too unstable to be restrained for echocardiography or thoracic radiography. However, until advances in technology enable the measurements of these factors to be performed using only minimal amounts of blood at the cage side, the utility of cardiac biomarkers in the dyspneic dog or cat is limited. To the author's knowledge, only IDEXX (Westbroke, Maine) currently offers a proBNP test (cardiopet® proBNP) for both canine and feline patients with results available the following day.

CONCLUSIONS

Determining the underlying etiology of respiratory distress in dogs and cats can be challenging due to limited ability to perform diagnostic testing such as thoracic radiography. The initial physical exam can typically identify patients who have upper airway disease, pleural space disease and thoracic wall abnormalities. In patients with pleural space disease, thoracic ultrasound can confirm auscultatory findings and differentiate between etiologies. In addition, ultrasound can help aid in performing life-saving thoracocentesis. Evaluation of left atrial size, chamber volume and contractility can help aid in the diagnosis of congestive heart failure in patients with suspected parenchymal disease. In the near future, cardiac biomarkers, especially NT-proANP and BNP, will be additional tools in the veterinary clinician's arsenal to identify patients with congestive heart failure who are too unstable to have thoracic radiography and echocardiography performed.

REFERENCES

1. Lee J, Drobatz K. Respiratory Distress and Cyanosis in Dogs. In: King L (ed.) *Respiratory Disease in Dogs and Cats*. St. Louis: W.B. Saunders; 2004. p. 1-11. 2. Saunders MH, Keith D. Thoracic Imaging. In: King L (ed). *Respiartory Disease in Dogs and Cats*. St. Louis: W.B Saunders; 2004. p. 72-87.
3. Kittleson MD, Kienle RD. Small Animal Cardiovascular Medicine. St. Louis: Mosby; 1998. 4. Lisciandro GR et al. Evaluation of a thoracic focused assessment with sonography for trauma (TFAST) protocol to detect pneumothorax and concurrent thoracic injury in 145 traumatized dogs. J Vet Emerg Crit Care 2008;18(3):258-269. 5. Prosek R et al. Distinguishing cardiac and noncardiac dyspnea in 48 dogs using plasma atrial natriuretic factor, B-type natriuretic factor, endothelin and cardiac troponin-I. J Vet Int Med 2007;21:238-242. 6. Herndon et al. Assessment of plasma cardiac troponin I concentration as a means to differentiate between cardiac and non-cardiac causes of dyspnea in cats. J Am Vet Med Assoc 2008;233(8);1261-1264. 7. Connolly DJ et al. Assessment of the diagnostic accuracy of circulating cardiac troponin I concentration to distinguish between cats with cardiac and non-cardiac causes of respiratory distress. J Vet Card 2009;11:71-76. 8. Connolly DJ et al. Assessemnt of the diagnostic accuracy of circulating natriuretic peptide concentrations to distinguish between cats with cardiac and non-cardiac causes of respiratory distress. J Vet Card 2009;11;S41-S50. 9. Fox PR et al. Utility of plasma N-terminal pro-bran natriuretic peptide (NT-proBNP) to distinguish between congestive heart failure and non-cardiac causes of acute dyspnea in cats. J Vet Card 2009;11:S51-S61.

COMMON RESPIRATORY EMERGENCIES IN CATS

Kenneth J. Drobatz, DVM, MSCE, DACVIM (SAIM), DACVECC
Philadelphia, PA

Respiratory emergencies in cats are relatively common in small animal emergency medicine. The degree of distress can range from relatively mild to life-threatening. Cats are very good at hiding clinical signs until the disease process has progressed substantially. Though there are numerous conditions that can cause respiratory distress in cats, there are a few conditions which contribute the lion's share of problems seen in cats that present for difficulty breathing. In a recent survey of our emergency clinicians the following three conditions were considered to be the most common: feline "asthma", congestive heart failure, and pleural effusion (due to a variety of causes).

FELINE "ASTHMA"

Cats presenting to an emergency service with lower airway disease can vary with clinical signs ranging from coughing/retching (owners often confuse this with vomiting) to severe/life-threatening difficulty breathing. The clinical history can vary from no previous signs at all to intermittent cough and tachypnea. Coughing is an important finding as most of the common respiratory emergencies mentioned above in cats are not typically associated with coughing. Unfortunately, not all cats with asthma cough.

Physical examination findings will vary depending upon the severity of the problem. The classic finding is prolonged expiration relative to inspiration and end expiratory wheezes. Lung sounds can also include increased lower airway sounds (often described as harsh) on both inspiration and expiration which tend to be diffuse over all lung fields. In some severe cases, lung sounds are decreased because of severe air trapping. Additionally, due to severity of coughing some cats with lower airway disease can present with spontaneous pneumothorax.[1] Heart sounds are often normal.

The emergency clinical diagnosis is based on presenting complaint, medical history, physical examination findings and thoracic radiographic results. Blood smears can be evaluated to detect eosinophilia but this is seen in only about 20% of cats with this condition. The radiographic findings may include a bronchial, interstitial, alveolar or mixed lung pattern, lung hyperinflation, and atelectasis.[2] The right middle lung lobe is the most common area showing atelectasis/collapse. The severity of the radiographic pattern does not always correlate with severity of clinical signs and the lung fields can look relatively normal in some cases.

There are three mainstays of emergency therapy for these cats: oxygen supplementation, bronchodilators, and short acting corticosteroids. When initially presented, many cats are in much more severe respiratory distress than they would be at rest. The process of being placed in a carrier, going for a car ride and then being around strangers can add stress upon stress in a cat that is already having difficulty breathing. Many of these cats will benefit from supplemental oxygen with minimal manipulation. This often allows them to calm down and decrease their oxygen demand resulting in easier respiration and less stress. This approach allows for a more in depth evaluation and handling after a period of oxygen supplementation.

Terbutaline can be administered at 0.01 mg/kg IV or IM and breathing is often improved within 30 – 60 minutes. Dexamethasone may also be administered at 0.1 – 0.5 mg/kg IV or IM to decrease the inflammation that may be causing the bronchoconstriction. This combination of therapies (oxygen, terbutaline and corticosteroids) often results in improvement in clinical signs within 2 hours such that many cats can be weaned from oxygen supplementation. Rarely, some cats will require 24 – 48 hours to be completely weaned from supplemental oxygen.

CONGESTIVE HEART FAILURE

Similar to cats with "asthma", the degree of respiratory difficulty can range from mild to life-threatening. The median age for cats with cardiomyopathy is 5.5 years (range 4 months to 16 years) with no sex predilection. Some predisposed breeds include Maine Coons, Norwegian Forest Cats, and Ragdolls. There are several forms of primary cardiomyopathy with the most common being hypertrophic (approximately 60% of cases), restrictive(~ 20% of cases), and dilated (~10% of cases). Secondary hypertrophic cardiomyopathy is most commonly seen associated with hyperthyroidism. Abnormal heart sounds are often ausculted and include heart murmurs (~60% of cats), gallop rhythm (~20%), and muffled heart sounds (~5% of cats). Tachycardia is often present. Respiratory difficulty, usually secondary to pulmonary edema and/or pleural effusion (cats with left-sided failure can have pleural effusion), is the most common complaint when cats present in congestive heart failure. Lung sounds will be increased with pulmonary edema and decreased with pleural effusion. Muffled or increased lung sounds can be ausculted in cats with both pulmonary edema and pleural effusion. Hypothermia is a relatively common clinical finding unlike with the other frequent causes of respiratory distress. In a group of cats presenting with respiratory distress in our emergency room 45% of the cats with heart disease were hypothermic (T<100F) compared to 30% of the cats without heart disease.

Signalment and physical examination can point one towards a diagnosis of heart disease as the cause for the respiratory distress but thoracic radiography can provide more concrete evidence, although radiographic positioning can be life-threatening in cats with severe respiratory distress. Radiographic signs consistent with heart failure include cardiomegaly, pulmonary vein congestion, and interstitial/alveolar patterns. The alveolar infiltrates are more commonly patchy rather than perihilar as in the dog. Pleural effusion and ascites may also be present. The valentine shaped cardiac silhouette and cardiomegaly are not always present.

Echocardiography can provide the best insight into the type of cardiac disease that is present, though this requires some degree of experience. Assessing the left atrial diameter in relation to the aortic diameter is the most useful in the emergency setting in helping delineate congestive heart failure as the cause of the respiratory distress. The ratio of left atrial diameter to aortic diameter should be less than 1.5 in normal cats. Cats with congestive heart failure tend to have ratios greater than this.

In the author's hands, ECG is fairly limited in providing evidence for heart failure as the cause for respiratory distress and we generally only utilize this in cats that have auscultable cardiac arrhythmias.

Biomarkers are rapidly becoming useful diagnostic techniques in diagnosing heart disease. N-terminal pro-brain natriuretic peptide (NT-proBNP) and N-terminal proatrial natriuretic peptide (NT-proANP)have been evaluated for their usefulness in distinguishing heart disease from primary respiratory disease as the cause for respiratory difficulty in cats. A cutoff value of 265 pmol/L for NT-proBNP resulted in 90.2% sensitivity, 87.9% specificity, 92% positive predictive value, and 85.3% negative predictive value (area under ROC curve, 0.94) in distinguishing cats with cardiac disease compared to a primary respiratory cause of respiratory difficulty. A cutoff of 517 fmol/mL for NT-proANP concentration had a sensitivity of 90% and specificity of 82% for detecting cardiomyopathy in cats.[3,4] Troponin I has also been investigated as a diagnostic tool for CHF as a cause for respiratory distress in cats. A cut-off of 0.81 ng/ml identified cardiac disease as the cause for respiratory distress with a sensitivity and specificity of 65.2% and 90.0% respectively. However, the authors noted that there was considerable overlap in troponin concentrations between the 2 groups and therefore this modality should be used in conjunction with other evidence in evaluating for heart failure.[5] The usefulness of biomarkers will continue to be investigated.

The mainstays of emergency therapy for congestive heart failure are oxygen supplementation, minimization of stress, aggressive diuretic therapy, and thoracocentesis if significant pleural effusion is present. Oxygen supplementation should immediately be given to any cat with respiratory difficulty. The most stress-free method is via oxygen cage but mask, flow-by, or use of a covered Elizabethan collar can help in selected cats that tolerate these methods. Furosemide should be given at 1- 2 mg/kg IV every 1 -2 hours until breathing improves. A constant rate infusion of furosemide in as small a volume as possible at a rate of 0.1 – 1 mg/kg/hr has been proposed to cause a more profound diuretic effect while minimizing the side effects of hypokalemia and metabolic alkalosis that can often occur with aggressive furosemide therapy in cats. Furosemide therapy, along with oxygen supplementation, are quite effective in stabilizing cats with cardiogenic pulmonary edema but caution should be utilized to avoid creating forward failure due overzealous diuretic-induced preload. Perfusion, renal parameters, and blood pressure should be monitored in these cats if possible.

In cats that are not responding to diuretic and oxygen therapy, nitric oxide based vasodilators may be of benefit. Nitroglycerin ointment (1/4 to 1 inch) applied to a shaved skin area (usually ear, flank, or inguinal area) may decrease cardiac preload through venous vasodilation. There is some controversy regarding the efficacy of this therapy. Sodium nitroprusside, both an arterial and venous dilator, may also help in cats with relatively refractory cardiogenic pulmonary edema. It is an extremely potent vasodilator that can result in significant hypotension if not monitored carefully and frequently. Generally, start at a lower dose of 2 ug/kg/min CRI and work slowly up from that @ 1 -2 ug/kg/min increments every 15 -20 minutes if blood pressure is maintained and the cat is tolerating it. Gauge when to go up in dose by monitoring respiratory signs and blood pressure. If respiratory signs have not improved and blood pressure is still adequate after 20 minutes then an incremental increase is warranted. This drug is light-sensitive and all lines and the bag should be protected from light. The ultimate dose tends to range from 2 – 10 ug/kg/min. One of the breakdown products of this drug is cyanide and hence the duration of this therapy should be limited to < 72 hours.

In a recent study, forty two of 55 (76%) cats admitted with CHF survived to discharge. Anecdotally, in our hospital, most cats stabilize within 24 – 48 hours after admission with appropriate therapy.

PLEURAL EFFUSION

Pleural effusion is a relatively common cause of respiratory distress in the cats. As with all diseases causing respiratory difficulty the degree of distress will vary from mild to severe depending upon the underlying cause and the rate of accumulation of fluid. Though not definitively confirmed, it has been reported that 30 ml/kg of fluid in the pleural space will cause mild difficulty in breathing while 60 ml/kg results in significant distress. This association may vary depending on the underlying cause of the effusion, the rate of fluid accumulation and whether other problems are present. Effusions are almost always bilateral due to the presence of a fenestrated mediastinum, although inflammatory conditions can result in a unilateral distribution. Clinical signs specifically attributed to the effusion are respiratory distress, increased respiratory rate and a restrictive pattern of breathing (short, rapid respiration with increased effort on inspiration). Physical examination findings specific to the presence of pleural effusion include decreased lung sounds on auscultation (more decreased ventrally) and possible jugular vein distention with severe fluid accumulation. Abnormal heart sounds such as murmurs, extra heart sounds, or heart rhythm disturbances are suggestive of cardiac disease as the underlying cause of pleural effusion, but these findings are inconsistently present. Diagnostic thoracocentesis, thoracic ultrasound or thoracic radiographs can provide a definitive diagnosis of excessive pleural fluid accumulation. The stability of the patient and the availability of these diagnostics will determine which test to use. In most cats with pleural effusion as the cause of respiratory difficulty, the lung sounds are decreased on auscultation. In these situations when the cat is too stressed for further diagnostics, immediate thoracocentesis is warranted. Thoracic ultrasound can be utilized to quickly detect the presence of fluid prior to tapping the chest. Thoracic radiographic positioning in a cat with respiratory distress can be life-threatening and the decision to perform this diagnostic modality should keep this concern in mind.

Pleural effusion in the cat is most often associated with a relatively serious underlying disease that cannot be cured (except pyothorax) but can be managed. Five major causes constitute the majority of cases of feline pleural effusion and include pyothorax, congestive heart failure, FIP, neoplasia, and idiopathic chylothorax. Anecdotally, heart failure is the most common cause in our hospital. The most important diagnostic modality to determine the underlying cause is cytological analysis of the pleural fluid. Many times this will provide the diagnosis or direct the diagnostic workup to definitively determine the underlying cause. The algorithm below gives a general approach to fluid analysis findings and its relation to various underlying diseases. As with any medical algorithm, there may be overlap and exceptions.

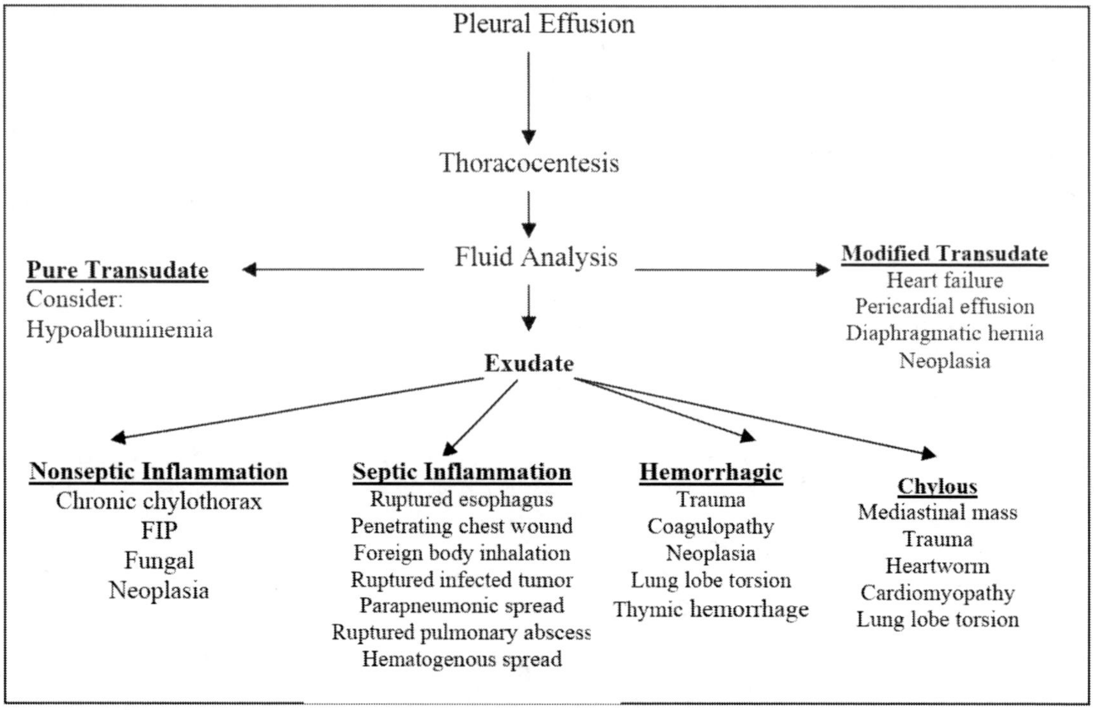

Emergency therapy for pleural effusion causing respiratory distress is oxygen supplementation and thoracocentesis. If fluid removal does not result in immediate relief of respiratory signs then there is more fluid remaining or there are other pulmonary problems contributing to the clinical signs. If done properly, thoracocentesis is relatively safe. Sedation or local block is rarely necessary. The seventh or eighth intercostal space is usually utilized at the junction between the ventral and middle third of the thorax. Using a 19 gauge butterfly needle, three way stop cock and 20 cc syringe, the needle should be passed over the cranial aspect of the rib avoiding the intercostal vessels and nerves. With the bevel facing dorsally the needle is passed through the intercostal muscles into the pleural space. Once into the space, gentle aspiration should be applied avoiding aggressive aspiration which can result in lung damage. Fluid should flow relatively easily once a connection is established with the fluid in the pleural space. Inflammatory effusions can result in locculation of fluid requiring ultrasound-guided thoracocentesis. Fluid that is obtained should be submitted for analysis. The major diagnostic parameter to evaluate is the cellularity of the sample and the types of cells present. If a clinical lab is not readily available, a direct smear and smear of a sediment of the fluid should be evaluated cytologically. The direct smear provides the clinician an estimate of the relative cellularity of the sample and a sedimented smear allows for maximizing the number of cells that can be evaluated as well as enhancing the chances of seeing infectious organisms. Fluid and cytological analysis will generally provide a definitive diagnosis, or at least point the clinician in the direction of what further diagnostics will be needed to identify the underlying cause of the pleural effusion.

EMPIRICAL THERAPY

Feline patients that are in severe distress will not survive involved diagnostics to determine the cause of the respiratory distress. These patients will not tolerate being without oxygen supplementation or can quickly deteriorate if handled. In these instances empirical therapy may be required. For these cats, we treat for the three major causes of respiratory distress. Empirical thoracocentesis can be tried and if there is no apparent effusion then dexamethasone can be administered at 0.1 - .2 mg/kg IV or IM along with furosemide at 1 – 2 mg/kg IV or IM. Some clinicians might also administer 0.01 mg/kg of terbutaline IV or IM but others are hesitant to administer this latter therapy in case cardiomyopathy is the underlying cause.

REFERENCES

[1]Cooper E, et al, **JVECC** 2003;13(2):95.[2] Gadbois J et al, **JAVMA** 2009; 234:367. [3]Fox, et al. **J of Vet Card**, (2009)11,S51. [4]Zimmering et al, **JAVMA** 2010;237:665. [5]Connolly et al, **J of Vet Card** (2009)11,p 71. [6]Goutal et al, JVECC 2010;20(3):p330.

COMMON CAUSES OF RESPIRATORY DISTRESS IN DOGS

Erica L. Reineke, VMD, DACVECC
Philadelphia, PA

INTRODUCTION

Respiratory distress is a common emergency in dogs. Signs of respiratory distress may develop at home prompting emergency care for the dog or may develop during the course of hospitalization. Clinical signs associated with respiratory distress, such as increased respiratory rate, respiratory effort and heart rate, develop when the PaO_2 drops below 60 mmHg or when the $PaCO_2$ rises above 50 mmHg. Initially, these normal physiologic respiratory and cardiovascular adjustments to hypoxemia may be effective at correcting the hypoxemia or hypercarbia, but if the underlying etiology is left untreated, progressive respiratory distress, hypoxemia and ventilatory failure may occur. Therefore, any dog presenting with clinical signs of respiratory distress should be given oxygen supplementation, and diagnostic testing (such as thoracic radiography) which may contribute to patient stress should be limited until the patient is stabilized. History and physical examination findings can help differentiate between underlying etiologies associated with respiratory distress in dogs and may direct initial stabilization therapies. In a recent survey of our emergency service clinicians, upper airway obstruction, left-sided congestive heart failure, pneumonia, pleural effusion and trauma (resulting in pulmonary contusions and pneumothorax) were the most common underlying etiologies of respiratory distress in dogs. These proceedings will focus on an initial diagnostic and therapeutic approach to the most common causes of respiratory distress in dogs.

UPPER AIRWAY OBSTRUCTION

Upper airway disease should be suspected in a patient with loud stridorous or stertorous (both inspiratory or expiratory) breathing and a prolonged inspiratory effort. In addition, these patients may have extension of the head and neck, orthopnea, collapse, foaming at the mouth, retching, and pawing at the face or mouth. On auscultation of the lungs, loud airway noises are generally heard and may interfere with accurate auscultation of the heart and lung parenchyma. In cases of complete obstruction, absent air movement over the trachea and lungs may be appreciated. Dogs with upper airway disease are often concurrently hyperthermic because of a decreased ability to pant which allows for body cooling and may contribute to increased respiratory rate in these patients. Differential diagnoses for patients with suspected upper airway obstruction include: laryngeal paralysis, tracheal collapse, brachycephalic airway syndrome, laryngeal masses, nasopharyngeal polyps, cellulitis/edema, and foreign body. Acquired laryngeal paralysis should be suspected in middle age to older large breed dogs, whereas tracheal collapse should be suspected in middle-age (average age of onset at 7 years) small breed dogs.

Emergency treatment of patients with upper airway obstruction should involve sedation (opioids, acepromazine) to relieve anxiety (this is particularly useful in patients with dynamic obstructions) and supplemental oxygen therapy. Ideally, the dog would be placed in an oxygen cage to minimize stress and handling associated with hospitalization. If the patient is hyperthermic, cooling measures should also be instituted. This author does not recommend the use of pulse oximetry in the initial stabilization period as accurate readings can be difficult to obtain and may contribute to patient stress and contribute to patient deterioration. A positive clinical response to this initial stabilizing therapy should increase the suspicion of upper airway disease. If the patient does not stabilize, the clinician should be prepared to intervene with emergency intubation and/or tracheostomy. Generally, it is possible to pass some type of oxygen delivery device even in patients with large obstructive masses. If an endotracheal tube cannot be passed, the clinician should attempt to pass a red rubber catheter to provide oxygen therapy while supplies for a temporary tracheostomy are being readied.

In order to definitely determine the underlying etiology of the upper airway obstruction, cervical and thoracic radiography and a sedated airway examination and laryngoscopy should be considered. Both cervical and thoracic radiography are important to not only evaluate the entire length of the upper airway but to also evaluate for concurrent parenchymal disease such as neurogenic pulmonary edema and aspiration pneumonia. If a sedated oral and laryngeal exam is performed, bronchoscopy and a bronchoalveolar lavage or an endotracheal wash for cytology and culture should be considered while the patient is anesthetized as many of these patients have concurrent parenchymal disease. In addition to administration of sedation, anti-inflammatory doses of steroids are typically indicated to help control inflammation that has occurred, especially in patients with laryngeal paralysis, tracheal collapse and brachycephalic airway syndrome. If there is concurrent aspiration pneumonia, high doses of steroids should be avoided as they may contribute to immunosuppression.

CONGESTIVE HEART FAILURE

Left-sided congestive heart failure (CHF) resulting in pulmonary edema should be suspected in any dog presenting with signs consistent with parenchymal disease. The medical history of the patient may suggest heart failure as the cause of respiratory distress if the dog has a history of a heart murmur, exercise intolerance, and coughing. On auscultation of the thorax, crackles are the classic finding in patients with pulmonary edema, but they are not always heard and the breath sounds may just be harsh due to the increase in respiratory rate and effort. Occasionally, the breath sounds may be quiet due to severe consolidation of the lungs. The presence of a heart murmur, tachycardia, or cardiac arrhythmia (such as atrial fibrillation) along with clinical evidence of parenchymal disease should raise the index of suspicion for heart failure causing the respiratory distress. The rectal temperature may be normal or low. Unfortunately, the auscultatory findings in dogs with congestive heart failure are similar to other causes of parenchymal disease such as pneumonia, non-cardiogenic edema, neoplasia and ALI/ARDS, therefore additional diagnostic tests such as thoracic radiography (gold-standard for diagnosing CHF) and echocardiography are needed to differentiate between these etiologies. Unfortunately, many of these patients are too unstable initially to tolerate the restraint needed to have either radiographs or an echocardiogram performed.

In patients who are too unstable to have thoracic radiography performed, a brief cardiac ultrasound by the emergency clinician can be a valuable tool in assessing left atrial size and contractility. The ratio of the left atrial size to the aortic root diameter is useful in evaluating for left atrial enlargement, which is a supportive finding of underlying cardiac disease when congestive heart failure is suspected. In the normal dog and cat, the ratio of left atria to aortic root size is usually less than 1.3. Values greater than 1.5 suggest left atrial dilation and are an additional piece of evidence supportive of CHF.

Subjective evaluation of the heart size and contractility, in addition to left atrial size, is also useful in identifying cardiac from non-cardiac causes of respiratory distress. Dilated cardiomyopathy (DCM) with CHF should be suspected in large breed dogs presenting for respiratory distress, crackles and a cardiac murmur heard on thoracic auscultation. Dogs with DCM may also have pleural effusion and ascites if concurrent right-sided CHF exists. Using 2-dimensional echocardiography, DCM should be suspected when moderate to marked dilation of all four cardiac chambers is seen with a subjective decrease in contractility. In addition to a brief cardiac ultrasound, an electrocardiogram should be performed in these patients as atrial fibrillation is a common contributory factor to the development of congestive heart failure. Please see other proceedings on a diagnostic approach to respiratory distress for information on the use of cardiac biomarkers to distinguish dogs with congestive heart failure from other causes of respiratory distress.

In a patient in which there is a high suspicion of CHF (based on history, physical examination findings ± cardiac ultrasound), furosemide therapy (2-4 mg/kg IV or IM) may be initiated and the patient's response to therapy assessed after 30 to 45 minutes. The patient should be placed on supplemental oxygen. An intravenous catheter should be placed and bloodwork drawn to assess and monitor renal function. Typically, there is minimal initial risk to administration of this medication prior to obtaining confirmatory thoracic radiographs (with repeated doses of furosemide, volume contraction may occur which may contribute to poor cardiac output in patients with hypotension). Once CHF is confirmed, furosemide may be administered as often as every 4-6 hours to help resolve pulmonary edema or as a constant rate infusion at 0.1-1mg/kg/hr. In patients with fulminant CHF that have a minimal response to furosemide therapy, nitroprusside, a potent arterial and venodilator should be initiated. Ideally, an arterial catheter is placed to allow for direct arterial pressure monitoring when administering this medication. However, in this author's experience, this is technically challenging (due to poor perfusion secondary to cardiac output failure), increases patient stress, and may worsen hypoxemia during positioning and restraint. Therefore, blood pressure may be monitored non-invasively or not at all if the dog is too unstable. The infusion is generally started at 3-5 ug/kg/min and titrated upwards to 10 ug/kg/min. As already mentioned, hypotension is an adverse effect of this medication and can be reversed by discontinuing use. In patients with concurrent pleural effusion, needle thoracocentesis should be performed to relieve dyspnea. Finally, any cardiac arrhythmias should be addressed.

PNEUMONIA

Pneumonia is another common etiology for dogs presenting with respiratory distress where physical examination findings are consistent with parenchymal disease. Compared with congestive heart failure, dogs with pneumonia may have a longer duration of historical coughing (>3 weeks) and present with a fever in addition to clinical signs related to their respiratory tract. Community acquired pneumonia caused by *Bordetella* is typically seen in young dogs who have a recent history of adoption, typically from a shelter or other group housing environment. Other causes of infectious pneumonia include: bacterial pneumonia (generally a complication of upper respiratory tract disease), canine influenza, and canine distemper virus. Aspiration of gastric contents resulting in pneumonia is also common in dogs and should be suspected in a dog when aspiration has been witnessed or when acute respiratory distress develops within a few hours of known vomiting, regurgitation, or anesthesia. More commonly, aspiration is not witnessed and this may make the diagnosis more challenging but should be considered a differential in dogs with a history of gastrointestinal tract disease, laryngeal paralysis, recent sedation or anesthetic procedure, esophageal disease, seizures or other central nervous system abnormalities.

Thoracic radiographic signs consistent with pneumonia include the classic pattern of cranioventral distribution of alveolar disease, but there may also be a mixed interstitial-bronchial pattern with or without multifocal alveolar patches. Lesions that are present in the caudal lung lobes may suggest hematogenous spread or inhaled infection. A recent study of 125 cases of presumptive aspiration pneumonia found that an alveolar pattern of the right middle lung lobe was the most common radiographic abnormality with an average of two lung lobes affected. In addition, this study found that survival was directly related to the number of lung lobes affected radiographically: the chance of survival decreased as the number of lung lobes affected increased. Evaluation of complete blood counts in dogs with pneumonia typically reveals a neutrophilia with a left shift, although a normal white blood cell count does not rule out pneumonia.

Ideally, lower airway sampling should be performed to definitively diagnose pneumonia (cytological analysis) and allow for directed antibiotic therapy based on antimicrobial results. This is especially important in patients with a previous and recent history of antibiotic administration as they are more at risk for antimicrobial resistance. Lower airway sampling can be done by performing a bronchoalveolar lavage, endotracheal lavage, or transtracheal wash. The benefit of a transtracheal wash is that it can be done without general anesthesia and is generally well tolerated. In this author's experience (supported by recent evidence in the veterinary literature), initiation of antimicrobial therapy prior to performing airway sampling does not typically cause a negative culture result and therefore antimicrobials should be initiated as soon as pneumonia is suspected. However, airway sampling is ideally performed within 24 hours of patient admission to the hospital. Typically, broad-spectrum antimicrobial therapy consisting of ampicillin or clindamycin and enrofloxacin, or ticarcillin/clavulanic acid and enrofloxacin, is initiated in dogs with suspected aspiration pneumonia. Azithromycin, doxycycline, or chloramphenicol are appropriate antibiotic choices for dogs with suspected *Bordetella* pneumonia. Recently, airway microbial and susceptibility patterns were reported in dogs and cats with respiratory disease of varying severity. In this study, patients with respiratory disease were more likely to have non-enteric and anaerobic isolates compared with dogs with severe respiratory failure needing mechanical ventilation. In addition, isolates from dogs with respiratory failure were more likely to be resistant to common first line antibiotics (ampicillin, enrofloxacin, ticarcillin/clavulanic acid/ and chloramphenicol). Despite increased resistance to these first line antibiotics in dogs with respiratory failure, results from this study still support the use of ampicillin and enrofloxacin, or ticarcillin/clavulanic acid and enrofloxacin as first line agents in dogs with respiratory disease with a $\geq$ 90% susceptibility and marginally good coverage in dogs with respiratory failure (72% susceptibility). In addition to antibiotics, the use of bronchodilators and mucolytics may be useful in the patient with pneumonia; however, there is limited evidence either for or against their use in the veterinary literature. Nebulization, which allows delivery of water droplets to the lower airways increasing the hydration of the mucociliary system, and coupage are considered a potential beneficial therapy in dogs with pneumonia.

PLEURAL EFFUSION

Pleural space disease is defined as an abnormal accumulation of air, fluid, and/or soft tissue within the pleural cavity. Clinical signs associated with pleural space disease include dyspnea, tachypnea, cyanosis, reluctance to lie down, orthopnea, coughing, fever, depression, weight loss, anorexia, cardiac murmurs and pericardial effusion. Pleural effusion is generally suspected based on physical exam auscultation findings of muffled or absent breath sounds ventrally either unilaterally or bilaterally over the thorax. These findings may be confirmation enough of pleural space disease in patients with severe respiratory compromise to prompt rapid needle thoracocentesis as both a diagnostic and therapeutic test. The most common causes of pleural effusion in the dog include: right-sided congestive heart failure (secondary to pericardial effusion or DCM most commonly), neoplasia, chylothorax, infection, hemorrhage, long lobe torsion, lymphangeictasia, and diaphragmatic hernia.

In dogs with suspected pleural effusion, a brief thoracic ultrasound can provide confirmation of physical exam findings and help to differentiate between underlying etiologies. Thoracic ultrasound can be performed quickly and easily cage side with the patient in sternal recumbency minimizing the amount of patient restraint and allowing for normal respiratory effort. Oxygen supplementation should be administered during the ultrasound examination to treat hypoxemia. The transducer should be placed in the intercostal spaces and rotated cranially and caudally to evaluate for the presence of fluid (or other abnormal structures). Low protein, low cellularity fluid (transudate-type) typically appears anechoic. Higher concentrations of cells, fibrin, or protein in the effusion increases the echogenicity of the fluid seen and is characteristic of exudate-types of effusion and blood. The presence of fibrin tags (seen as "floating" strands in the pleural fluid) suggests chronicity to the effusion. It is important to also evaluate the heart for evidence of pericardial effusion, left atrial enlargement and contractility during the thoracic ultrasound when pleural effusion is identified. For more detailed information on the use of thoracic ultrasound, please refer to proceedings on a diagnostic approach to respiratory distress.

Needle thoracocentesis, with or without ultrasound guidance, to remove as much fluid as possible is recommended prior to obtaining thoracic radiography as the fluid will interfere with interpretation of thoracic structures. Fluid that is obtained should be placed in an EDTA tube for cytologic evaluation and cell counts and a clot tube for evaluation of total protein and clotting characteristics. Direct impression smears or centrifuged samples can and should be immediately evaluated for the presence of bacteria and to evaluate cellularity and cell types. However, submission to a clinical pathologist for a definitive diagnosis is recommended and additional fluid should be held for aerobic and anaerobic culture if indicated. Removal of the fluid from the pleural space may relieve hypoxemia and allow for discontinuation of oxygen therapy. Ongoing need for oxygen supplementation should be assessed with pulse oximetry or arterial blood gas analysis. Rare complications of needle thoracocentesis include pneumothorax, hemorrhage and infection. If the pleural effusion returns quickly or large amounts of fluid are being removed frequently, a thoracostomy tube may need to be placed. Additional treatments and diagnostic testing will depend on the underlying etiology of the pleural effusion.

CONCLUSION

Respiratory distress is a common emergency in companion animals. These dogs are hypoxemic and oxygen supplementation should be provided until the underlying disease process is determined and the clinical signs of distress are resolved. More frequently than not, diagnostic testing is limited in these patients due to hypoxemia. Therefore, the clinician should rely on signalment, medical history and physical examination findings to narrow down potential etiologies for the respiratory distress keeping in mind the most common reasons for respiratory distress in dogs. Physical examination abnormalities can help localize disease to a portion of the respiratory system and minimally invasive tools such as ultrasound can help to further narrow down the list of differential diagnoses. Often, an empiric trial of furosemide in dogs in which there is as high suspicion of heart failure is unlikely to cause harm and may allow for patient stabilization. With severe respiratory failure, institution of mechanical ventilation may be a necessary life-saving step to decrease work of breathing, improve oxygenation, and allow diagnostics to be performed.

REFERENCES

[1] Holt, DE. In: King L (ed.) *Respiratory Disease in Dogs and Cats*. St. Louis: W.B. Saunders; 2004. p. 35-41. [2] Kittleson MD, Kienle RD. Small Animal Cardiovascular Medicine. St. Louis: Mosby; 1998. [3] Radhakrishnan A et al. J Am Vet Med Assoc 2007;30(10):1493-1497. [4] Tart KM et al. J Vet Emerg Crit Care 2010:20(3):319-329. 5. Epstein SE et al. J Vet Emerg Crit Care 2010;20(6):587-594. [6] Silverstein DS. In: King L(ed). *Respiratory Disease in Dogs and Cats*. St. Louis: W.B. Saunders; 2004. p. 49-52.

MANAGEMENT OF RESPIRATORY OBSTRUCTIONS: STENTS, BALLOONS & LASERS: NEW ALTERNATIVES?

Allyson Berent, DVM, DACVIM
New York, NY

INTRODUCTION

Interventional Endoscopy (IE) involves the use of endoscopic equipment with other contemporary imaging modalities, such as fluoroscopy and/or ultrasound, to perform diagnostic and therapeutic procedures in virtually any part of the body accessed endoscopically (gastrointestinal, biliary, respiratory, urinary tract, etc). Interventional radiology (IR) uses fluoroscopy, with or without ultrasound, to access blood vessels and various lumens in order to deliver specific materials (guidewires, catheters, embolic agents, balloons, and stents) for diagnostic and therapeutic purposes. Some examples you may be familiar with are urethral or tracheal stenting, balloon dilatation or stenting for nasopharyngeal stenosis, or ureteral stenting for ureteral obstructions. The focus of this talk will be on the use if IR/IE techniques for treatment of various disease of the respiratory tract.

EQUIPMENT

Traditional fluoroscopy is needed for most of the interventional cases, and using a C-arm unit has great advantages permitting tangential views without moving the patient. Various rigid endoscopes ranging in size from 1.9-7 mm can be used in various aspects of IE. In the respiratory tract rigid cystoscopes (1.9-2.7 mm) are typically used for antegrade rhinoscopy and flexible bronchoscopes (5 or 6 mm) are used for nasopharyngoscopy, sinoscopy, or tracheobronchoscopy. Balloons of various lengths and diameters are often used under a combination of fluoroscopic and endoscopic guidance for effacing strictures and lasers or endoscopic cautery devices can be used for safely breaking strictured membranes or removing polyps and tumors. Guidewires and catheters of various size and shapes, as well as stents (typically metallic for the respiratory tract) are often needed in conjunction with the above equipment, particularly when dealing with nasal/nasopharyngeal stenosis and tracheal stenosis, tracheal tumors or tracheal collapse.

Nasopharyngeal stenosis

Nasopharyngeal stenosis (NPS) is a rare condition that is typically acquired in both dogs and cats. It has been reported as a congenital condition as well, but is typically acquired secondary to chronic rhinitis (feline patients), or aspiration rhinitis (canine patients). In dogs, the most common history involves aspiration rhinitis after an ovariohysterectomy. A few days later the owner might report "my dog can no longer breathe through his/her nose." In cats, they are typically older at the time of diagnosis with a long history of chronic nasal disease, a history of chronic rhinitis that has progressed slowly to chronic inspiratory stertor or stridor. They are typically worse during bouts of infection and mucous accumulation, which is common in these patients since they do not have appropriate nasopharyngeal drainage.

The diagnosis of this condition is typically made during retroflex rhinoscopy when a smooth membrane is seen in the nasopharynx obscuring the view to the choanae. This can also be diagnosed with a CT scan, and is best seen on a sagittal view. A very small membrane can create a major problem making it important that the CT scanner takes 1mm slices if possible or the lesion can be missed. Measurement can be obtained from the CT on cross-section to determine the appropriate sized balloon or balloon expandable stent (BEMS) that might be necessary.

The treatment for NPS traditionally was surgical where the palate was incised and the membrane cut out. This ultimately resulted commonly in re-stricture within a few days to a few weeks. Because of the minimal success with this procedure the investigation into endoscopic guided balloon dilatation occurred. This was promising, but often the stricture would recur requiring 2-6 balloon procedures to get an effective outcome. Due to the expense and frustration with this disease process the use of metallic stents were investigated.

This procedure is relatively straight forward requiring the combination of fluoroscopy and endoscopy for visualization of the stent placed in an antegrade manner through the nares of the patient. There are some patients that actually have a completely closed membrane which is slightly more challenging to treat. In this case the stenosis is opened with the use of a needle that is placed through the nares, via a sheath, and then directed with fluoroscopic and endoscopic guidance through the stenotic membrane. Once a hole is made then a guidewire is passed through the needle and the stenosis is balloon dilated and ultimately stented, similarly to the patent NPS cases. The author usually places a covered stent in patients with a closed membrane as this is the only circumstance the author and colleagues have ever seen the membrane close down through the mesh of the stent. We typically recommend a balloon expandable stent for this condition (BEMS). We have re-scoped and performed computed tomography in some of these patients after the procedure and the stent normally is incorporated into the mucosa thus preventing migration and chronic infection. In patients with a covered stent the stent does not incorporate into the tissue well and the covering can act as a nidus for infection. This is an uncommon complication but should be considered prior to placing a covered stent in any patient.

Nasal Stenosis

A potential complication following rhinotomy in feline and canine patients is the development of a "nasal passage" stenosis. Prior to considering a rhinotomy, a full nasopharyngeal examination should be performed and owners should be made aware that scar tissue formation as well as severe chronic nasal discharge and infections are a potential complication of this procedure.

Tracheal Foreign Bodies

Tracheal foreign bodies have been described in a small group of cats over the past few years. Removal of these foreign bodies can be done through various mechanisms but the easiest and safest method is with the use of fluoroscopy and a stone retrieval basket. This technique will be described. If the object is radiopaque, a guidewire and retrieval basket can be used to entrap and remove the foreign object, with the endotracheal tube remaining in place. If the object is not visible by fluoroscopy, then endoscopy should be considered as the use of contrast material will result in the risk of contrast pneumonitis. In small dogs and cats, the use of tracheoscopy for foreign body retrieval results in both the need for extubation and airway occlusion with the endoscope. These problems can be avoided when retrieval is performed under fluoroscopic guidance.

Tracheal strictures

Traditionally, tracheal strictures were treated with tracheal resection and anastomosis. More recently newer minimally invasive interventional options have been employed that have been incredibly effective. This includes balloon dilation and tracheal stenting. If balloon dilation is performed it is important to try and prevent a tracheal tear. This is why stenting is chosen typically over balloon dilation. If a balloon is necessary (typically for growing dogs where they may outgrow the stent) we will use endoscopic electrocautery or a "hot knife" to cut a superficial layer of tissue in 3 places, allowing for even distribution of the tear around the stricture. Optimally, this tears the scar tissue evenly without the risk for longitudinal tearing in one place along the normal trachea. The risk of a pneumomediastinum, pneumothorax, or SQ emphysema should be considered. In adult patients a tracheal stent across the stenosis would be more highly recommended.

Tracheal Stenting

Tracheal stenting is a frequently performed procedure in our interventional practice. Tracheal collapse is a common, progressive, degenerative disease of the cartilage rings which predominantly affects small and toy-breed dogs. This condition is often a result of cartilage hypocellularity and decreased glycosaminoglycan and calcium content resulting in subsequent dynamic airway collapse during respiration. This condition is also seen in association with laxity of the dorsal tracheal membrane resulting in luminal obstruction, coughing and potentially dyspnea. The collapse can be focal or diffuse in nature, and can progress to involve the bronchial and bronchiolar airways. Affected animals present with signs ranging from the classic mild, intermittent "honking" cough to severe respiratory distress from dynamic upper-airway obstruction, and can often exhibit severe paroxysmal coughing. Various combinations of anti-inflammatories, anti-tussives, sedatives/tranquilizers, and/or potentially bronchodilators (we typically do not use) are often effective in alleviating the initial respiratory problems associated with tracheal collapse. In one series, 71% of patients remained asymptomatic following medial management for over 12 months following the initiation of medical therapy. This is not surprising, as patients receiving stents have been medically managed for this condition for years before disease progression necessitated further intervention. In addition, weight loss, clipping excessive hair, restricting exercise, using a harness rather than a neck lead, and removal of second-hand smoke or inhaled allergens can further palliate clinical signs. Regular monitoring of co-morbidities such as cardiac disease, endocrinopathies, or pulmonary disease may help reduce the number or severity of episodes of respiratory crisis. Those patients that have failed aggressive medical and environmental management, and have had other potential causes of respiratory disease addressed or ruled out, are candidates for tracheal stent placement. Other candidates are those with non-resectable tracheal tumors or tracheal strictures, as described above.

The most commonly performed surgery for animals with extra-thoracic tracheal collapse is the placement of extra-luminal tracheal ring prostheses. There can be significant mobidity associated with this procedure and it is only recommended for patients with cervical collapse. In our experience, most patients have both cervical and intrathoracic collapse and tracheal rings are not considered good options.

Medical therapy should be optimized in tracheal collapse patients prior to the consideration of a tracheal stent for collapse, as the complications associated with stent placement can be serious (stent fracture, inflammatory tissue reaction or stent migration). When stents are placed appropriately, and the cough and concurrent conditions (like bronchial collapse) are management well with medications, complications are decreased.

It is important to remember that during inspiration and expiration the collapse can look very different on radiographs. Knowing the location is very important in determining where to put a stent. We typically judge collapse location using fluoroscopy alone. Some encourage the use of tracheobronchoscopy. We find the use of fluoroscopy alone with the use of a negative pressure device safer and less complicated. We will discuss the differences.

The equipment needed is a radiopaque endotracheal tube, a guidewire and a marker catheter for tracheal measurements and an appropriately sized stent. The marker catheter is placed in the esophagus over the guidewire for appropriate tracheal measurements, allowing for the adjustment of magnification. Using positive pressure ventilation at 20 cm H20, an image is taken under fluoroscopy with the endotracheal tube pulled as far into the cervical trachea/cricoid cartilage as possible to remain intubated. A seal must be established in the ET tube to get a good maximal tracheal measurement. The diameter of the cervical, thoracic inlet and intrathoracic trachea is obtained using the marker catheter that has been placed into the esophagus over a guidewire. Then the length of the collapsed region is determined based on a negative suction device that is applied to the endotracheal tube. Based on these measurements the appropriate stent size is chosen. The stent is then deployed through the endotracheal tube through a bronchoscope adaptor so that the patient can remain on the anesthesia circuit. There are many courses around the country (ACVIM, ACVS, ECVIM, ECVS, IVECCS, etc) if you are interested in learning how to do this appropriately. There is a new tracheal stent design that we are currently using which has thicker wire diameter and soft rounder edges. This design has improved the fracture rate and decreased the inflammation which was sometimes a problem with previous stent designs. Post stent medical management is aggressive and owners should have expectations that their dog will remain medicated for the REST OF THEIR LIFE.

Polypectomy/Tumor ablation

Endoscopic guidance can be used to resect tracheal polyps or tumors using electrocautery or endoscopic lasers, while awaiting a biopsy of the lesion prior to the consideration of any major surgery or stenting procedure. This procedure will be discussed in more detail.

Chylothorax

Idiopathic chylothorax is a frustrating disease. Although there are multiple surgical options available, the success rate for each is similar (70-80%). Recently a new technique using an abdominal approach and fluoroscopy, without a thoracotomy has been performed in a handful of veterinary patients. This is currently under investigation at the Animal Medical Center with Dr. Chick Weisse. An efferent lymphatic is catheterized via laparotomy and cyanoacrylate glue is used to embolize the cysterna chyli and the thoracic duct (along with all of the branches). This procedure is still under investigation but holds promise. To date it is typically recommended in animals that have failed surgery or thoracoscopic intervention. It is important to remember that performing a cisterna chyli ablation will not permit this procedure to be performed in the future; therefore, when surgery is considered one might want to avoid this addition if the owner is willing to consider glue embolization.

CONCLUSION

In conclusion, there are a vast number of new interventional radiological and endoscopic techniques available. When dealing with the larger airways (nasal passage, nasopharynx, trachea and bronchi) these new procedures might be a good alternative for your patients.

REFERENCES

Available upon request

HEPATIC VASCULAR ANOMALIES: RECENT LESSONS LEARNED USING INTERVENTIONAL RADIOLOGY

Chick Weisse, VMD, DACVS
New York, NY

INTRODUCTION

Interventional Radiology uses fluoroscopy to guide minimally-invasive therapies. There is currently expanding investigation into the use of these techniques in various areas of veterinary medicine, including but not limited to the arterial and venous systems. The morbidity associated with certain open surgical procedures in these areas, particularly in compromised patients, makes these minimally-invasive approaches increasingly appealing. Moreover, the lack of treatment alternatives available for more complex, terminal, or end-stage diseases when traditional therapies have failed has inspired research into potential uses for these techniques, many of which have become the standard-of-care in human medicine. This lecture will review some of the interventional procedures currently performed in the hepatic arterial and venous systems of veterinary patients and some of the lessons learned – or more importantly the new questions raised – using these techniques.

HEPATIC VASCULAR ANOMALIES

The categorization of liver vascular anomalies is often confusing and but the most recent classification suggests three separate categories of liver vascular disease: (1) Congenital portosystemic shunts (IHPSS and EHPSS), (2) Disorders associated with abnormal hepatic bloodflow or portal hypertension, currently termed Primary Hypoplasia of the Portal Vein (PVH), and (3) Disturbances in outflow. The second category (PVH) remains the most confusing, and includes processes that may or may not result in portal hypertension. These are termed PVH with portal hypertension and PVH without portal hypertension. Examples of PVH with portal hypertension include non-cirrhotic portal hypertension (NCPH), and hepato-portal fibrosis/veno-occlusive disease. PVH without portal hypertension was previously termed microvascular dysplasia (MVD).

IHPSS (AND EHPSS)

Single, extrahepatic PSSs are amenable to relatively uncomplicated surgical attenuation, however surgical repair of intrahepatic PSSs are consistently more challenging. Numerous techniques have been described for intrahepatic PSS attenuation; however, morbidity and mortality rates can be very high, even for the most experienced surgeons. The goal of IR techniques for IHPSSs is to reduce the unacceptably high peri-operative mortality rates associated with traditional open surgical techniques and hopefully improve the outcome for these cases. The author has performed over 100 percutaneous transvenous coil embolizations (PTCE) with a vena caval stent and thrombogenic coils placed within the shunt. Peri-operative complications were mostly minor and peri-operative mortalities were comparatively low versus that reported for traditional surgery.

PROCEDURE

Percutaneous Transvenous Coil Embolization- All dogs are treated medically initially following diagnosis of the IHPSS for a period of weeks to months. When possible, CT or MR angiography is performed to delineate the shunt anatomy and obtain caval and shunt measurements under a separate anesthetic episode. All PTCE procedures are performed under general anesthesia using standard liver dysfunction protocols and often neuromuscular blockade to minimize respiratory artifact during digital subtraction angiography. Shunts are typically accessed by a percutaneous right jugular approach. Contrast venography is performed to delineate the portal vein, portosystemic shunt, and caudal vena cava anatomy. Intravascular pressure measurements are obtained in the caudal vena cava and portal vein. Shunt attenuation involves placement of a stent within the caudal vena cava, positioned so as to traverse the shunt entrance into the vena cava. Shunt PTCE is subsequently performed by passage of a catheter through the stent interstices and deposition of the coils within the shunt lumen. Coils are subsequently added with intermittent shunt pressure measurements taken to avoid creating portal hypertension. Coils are typically added until the shunt mouth is covered with coils or the shunt pressures have increased between 6 and 10cm H_2O or maximal pressures approached 20cm H_2O. Ultimate shunt and caval pressures are recorded, the jugular sheath is exchanged for a 7Fr multi-lumen catheter, and the animal is recovered from anesthesia. Following the initial procedure, medications are gradually weaned over the following 4-8 weeks. Additional PTCE is recommended if clinical signs returned.

LESSONS LEARNED

Lessons learned or questions raised concerning DIAGNOSTIC IMAGING include:

-CT and MR angiography are well tolerated and facilitate identifications of uncommon shunt anatomy; however, certain abnormalities are underestimated using these techniques. For instance, multiple small intrahepatic shunts are often not identified on standard cross-sectional imaging. These abnormalities are better identified with traditional contrast portography. At this time, it is not clear if dual phase CT provides us with additional information (as compared to single-phase).

-MRA (using gadolinium) may permit single procedure IHPSS imaging and treatment in order to avoid excessive iodinated contrast use associated with CT angiography followed by angiography.

-Even "typical" IHPSS have variant hepatic vein anatomy. Does location of HV entrance into the PSS affect results of attenuation? It is conceivable that intra-hepatic vein shunting (acquired intrahepatic PSS) may occur more readily in these patients. Examples will be discussed.

Lessons learned or questions raised concerning TREATMENT include:

-A certain small population of IHPSS patients have "significant" portal: systemic venous pressure gradients (or resting portal pressures) before treatment. This is counter-intuitive in animals with PSS in that reduced portal pressure gradients would be anticipated, and this has prevented treatment in some cases. Are there small vascular windows or narrowings present that are not identified on cross sectional imaging with relatively wide slices? This suspicion has been raised as pull-out pressure tracings confirm short, focal areas where pressure gradients exist. The presence of a developed portal system may suggest a narrowing of the shunt in some location making access more difficult. This may be the same for **EHPSS** suggesting intermittent shunt compression- in phrenic shunts for instance.

-Which is more important in preventing the development of complications associated with portal hypertension following IHPSS treatment; total portal pressure or pressure gradients? During surgery we rarely measured CVP and using IR techniques, we always measure CVP.

-During portography, when multiple small intrahepatic shunts are identified, this is almost exclusively associated with and elevated portal pressure and/or pressure gradient. Are these congenital shunts or acquired IHPSS resulting from a congenitally narrowed IHPSS?

-Acquired Intrahepatic Portosystemic shunts: There is more evidence that IHPSS can be acquired as well. Do the same criteria for shunt attenuation of EHPSS (no greater than ~10cmH2O rise in portal pressure and/or no greater than ~20cmH2O total portal pressure) hold for attenuation of IHPSS? Although there is no documented difference between HV attenuation and PV attenuation, the vascular bed receiving the congestion is intrahepatic with the former and extrahepatic with the latter. Does this matter?

-DO NOT PERFORM IHPSS SHUNT ATTENUATION IN THE FACE OF GI ULCERATION/HEMORRHAGE. The authors currently perform endoscopy and biopsy of all IHPSS cases before treatment and the overwhelming majority of these dogs have some degree of inflammatory bowel disease, sometimes including GI ulceration. Approximately 17% of patients have evidence of GI bleeding before treatment. Elevation of portal pressures with the presence of GI ulceration can lead to severe GI hemorrhage and death. All animals are maintained on omeprazole therapy for life. Initially a long-term mortality rate of 30% in IHPSS PTCE animals was caused by GI bleeding in ~50% of deaths. Lifelong antacid therapy has reduced the mortality rate to 12.5% with fewer than 4% secondary to GI bleeds. Is lifelong omeprazole therapy safe?

Lessons learned or questions raised concerning FOLLOW-UP include:

-Is return to normal bile acids concentrations necessary? Should this be the goal of therapy? The majority of patients receiving IHPSS PTCE do not have return to normal liver function and some have even been identified to have no development of portal branching, although pressure gradients continue to exist. Is hepatic vein congestion and reduced portal bloodflow drainage consistent with improved portal bloodflow? Maintenance of a portal gradient appears to coincide with improved clinical signs. Can this be adequate or should additional interventional/surgical treatment be instituted?

-Development of "longer term" (greater than 2-4 weeks) post-procedural ascites is most commonly associated with hypoproteinemia and NOT excessive portal hypertension (although they can occur simultaneously). In our population of cases, this has most commonly been associated with a GI bleed and subsequent hypoalbuminemia and hypoglobulinemia and the resultant reduction in oncotic pressure. This has not always been confirmed but aggressive medical therapy with GI protectants has resulted in resolution of the clinical signs in most of the patients we have identified with this uncommon complication. No therapy for portal hypertension has been necessary.

-Approximately 18% (13/73) of cases required additional coiling procedures

HEPATIC AVMS

Vascular malformations have been classified as high-flow or low-flow, and as arterial, venous, lymphatic, or mixed. They are often associated with tumors but can also been seen congenitally in the liver (hepatic AVMs). Once called arteriovenous fistulas, they have been more recently termed arteriovenous malformations due to their vascular anatomy upon angiography. Interventional radiology techniques have allowed us to improve our understanding of these particular AVMs in terms of their vascular nature, response to treatments (glue embolization), and resulting pathophysiology. The observations below have been made following surgery or glue embolization of approximately 15 HAVMs.

LESSONS LEARNED

Lessons learned or questions raised concerning DIAGNOSTICS include:

-These animals often present for clinical signs associated with PSS and are often mistaken for IHPSS due to the ultrasonographic identification of a large intrahepatic vascular structure along with clinical signs consistent with PSS.

-While all of these animals have SEVERE PORTAL HYPERTENSION, about 25% of these cases will not present with ascites. This is presumably due to the multiple extrahepatic PSS that are acquired (along with other physiologic changes taking place) and have decompressed the portal system sufficiently to relieve the excessive hydrostatic pressure.

-Clinical signs are often less severe with HAVM than standard PSS presumably because the portal system is more developed in the former patients than in the latter ones.

-Other diagnostic signs that are less commonly discussed include hepatofugal portal blood flow (ALWAYS PRESENT), a reduction in aortic diameter caudal to the level of the celiac artery, and differing systemic blood pressures obtained from the forelimbs (higher) and the hindlimbs (lower) that is occasionally present. Identifiable abdominal bruit is rarely present in the author's experience.

Lessons learned or questions raised concerning TREATMENT include:

-It has been suggested in the human literature that while AVFs can be ligated or coil embolized, multiple AVFs or AVMs should receive glue embolization (or alcohol ablation) in order to destroy the nidus that will otherwise recruit additional vessels over time. This has not been confirmed in the veterinary population but angiograms may support this notion.

-Initial angiography (and cross sectional imaging) often underestimates the extent of the disease. Following initial embolization, additional previously unidentified contributing vessels open up demonstrating the true infiltrative nature of these vascular anomalies.

-Complete HAVM embolization or resection appears to be required to prevent recurrence. Incomplete embolization or resection will lead to revascularization if the HAVM nidus remains.

-These patients can tolerate complete hepatic artery embolization and the cyanoacrylate glue appears to be permanent in the cases that have been followed to date, although the radio-opacity of the glue (Tantalum and Lipiodol) may diminish over time.

-Although return to hepatopetal portal bloodlfow (towards the liver) would be considered the goal, this has not been seen in any of the cases treated to date. Acquired EHPSS are present in ALL patients and can be expected to remain in place providing the least resistance to portal bloodflow. Complete HAVM embolization could conceivable result in stagnant portal bloodflow as the direction changes from hepatofugal to hepatopetal and near complete stasis has been identified but to date has not required intervention.

-Vascular contributions to the HAVM are not only from the hepatic artery but have also been identified to arise from the gastroduodenal artery, left gastric artery, and phrenic aa. Performing an aortogram following embolization is recommended.

Lessons learned or questions raised concerning FOLLOW-UP include:

-As acquired EHPSS will never close, continued life-long medical therapy is often necessary (and indicated) in patients following treatment. Some animals may not need medical therapy however.

-It is unclear which animals may benefit from treatment. In certain cases with no overt clinical signs (for instance the cases without failure to thrive and massive ascites), these patients may not benefit from the treatment of the anomalies. Do all animals with HAVM require treatment? It has been suggested that chronic portal hypertension can lead to portal vein blunting and reduce portal perfusion. If so, is this an argument that all these animals should be treated as soon as possible?

REFERENCES

Available upon request.

URINARY OBSTRUCTION & STONES: ARE STENTS, BALLOONS & LASERS THE ANSWER?

Larry G. Adams, DVM, PhD, DACVIM (SAIM)
West Lafayette, IN

Urinary obstruction is a commonly encountered problem in dogs and cats. This session will focus on lower urinary tract uroliths and obstruction of the urethra or bladder neck. Urethral obstruction may occur due to mechanical or functional causes. Functional obstruction results from failure of coordination of urethral relaxation with detrusor contraction (reflex dyssynergia). Before assuming a patient has functional urinary retention, a thorough diagnostic evaluation including either cystoscopy or contract urethrography should be performed to rule out mechanical causes of urethral obstruction. Common mechanical causes of urethral obstruction include urethroliths, blood clots, urethral strictures, one-way valve effects from proliferations of the urothelium such as neoplasia or proliferative urethritis, urinary bladder entrapment in perineal or inguinal hernias, and penile fractures.[1-3] Several types of neoplasia may cause urethral obstruction including transitional cell carcinoma (TCC), prostatic adenocarcinoma, leiomyoma, leiomyosarcoma, squamous cell carcinoma, myxosarcoma, and lymphoma.

URETHRAL OBSTRUCTION

Urethral Stricture – Balloon dilation and urethral stents are the answer

Urethral strictures may form secondary to previous urethral injury from surgery, urethroliths, external trauma or iatrogenic trauma from prior urethral catheterization. Mechanical urethral obstruction such as strictures or neoplasia may be diagnosed by cystoscopy or positive contrast urethrography. Documentation of the site and extent of urethral strictures or urethral neoplasia are best accomplished by positive contrast urethrography. The author prefers to perform urethrocystography using a urologic guidewire and an appropriate diameter vascular access sheath in order to achieve maximal urethral distention. A measuring catheter is also placed in the colon parallel to the urethra to account for magnification and provide precise measurements of the urethral stricture and normal urethra on either side of the stricture. For correction of urethral strictures, options include surgical correction, balloon dilation or dilation followed by stent placement. Surgical correction is limited to distal urethral strictures that can be bypassed using urethrostomy. For more proximal urethral strictures, balloon dilation and stenting are better options. Balloon dilation alone may be successful but recurrence of the stricture is a common problem. Therefore, stents are often required after balloon dilation of urethral strictures. Self-expanding nitinol urethral stents are more effective than balloon expandable stents. Urethral structures may also be dilated using balloon dilation catheters in male or female dogs or using ureteral dilation sets in female dogs.

Urethral Neoplasia – Urethral stents are one answer

Urethral obstruction attributable to mass effects in the urethra may be either malignant neoplasia (TCC, prostatic carcinoma) or less commonly proliferative urethritis.[2-4] Definitive diagnosis requires histopathologic evaluation of tissue samples obtained via cystoscopy (or surgery).[4] Malignant urethral obstructions may be treated by nitinol (metallic) urethral stents to expand areas of complete urethral obstruction.[5] Laser ablation of urethral TCC has been reported as an alternative to urethral stenting; however, urethral perforation is a potential complication of laser ablation of urethral masses.

URETHROLITHS – LASER LITHOTRIPSY IS THE ANSWER

Urethroliths should be retropulsed back into the urinary bladder for removal or the urethroliths can be fragmented and removed by laser lithotripsy. Lithotripsy is crushing or fragmenting uroliths by high-energy shock waves (SWL) or by laser energy. Laser lithotripsy is an option for fragmentation and removal of any lower tract uroliths that can be visualized using a rigid cystoscope or flexible ureteroscope.[3] The 2100 nm wavelength of the holmium: YAG laser is absorbed in 0.5 mm of fluid; therefore, it can be used to safely fragment uroliths in tight locations such as within the urethra without damage to the urothelium. Provided the laser fiber is oriented parallel to the urethral mucosa, there is minimal risk of damage to the urethral wall. There are several recent clinical studies on effectiveness of laser lithotripsy for removal of uroliths from the lower urinary tract in dogs.[3,6-8]

We reported the results from 73 dogs treated with laser lithotripsy.[3] Laser lithotripsy resulted in complete removal of all uroliths in all female dogs (28/28; 100%) and a majority of male dogs (39/45; 86.7%).[3] Dogs with urethroliths had shorter laser time than dogs with cystic uroliths. Once uroliths were sufficiently fragmented such that they were small enough to pass through the urethra, larger fragments were initially removed with a stone basket to verify that the fragments were small enough to allow removal. Once the larger fragments were safely removed through the urethra via basket extraction, voiding urohydropropulsion was then performed to evacuate remaining fragments.

Considering recent publications on laser lithotripsy in dogs, the following observations can be made about the efficacy of laser lithotripsy. All urolith fragments were removed more frequently in female dogs (83 to 100%) than in male dogs (68% to 87%).[3,6-8] Complete urolith removal was achieved in 100% of dogs with urethroliths.[3,7-8] Procedure time was also shorter for removal of urethroliths than urocystoliths.[3] One case-controlled study of dogs with bladder and urethral calculi showed that laser lithotripsy was comparable in efficacy to open cystotomy, although procedure time was slightly shorter for surgery than laser lithotripsy.[6] Two studies showed that procedure times were significantly shorter once the operator gained experience with the technique.[7,8] In a recent study comparing laser lithotripsy to cystotomy for removal of lower urinary tract uroliths, residual urolith fragments were detected in approximately 10% of dogs treated by either method.[6] Therefore cystoscopic examination (female dogs) or post-procedural radiographs (male dogs) are indicated to confirm complete urolith removal.

Case Selection for Laser Lithotripsy

As veterinarians consider the relative roles of different minimally invasive techniques for stone removal, the relative stone burden and location of the uroliths affects which techniques should be considered. Case selection is important for laser lithotripsy because not all dogs with urocystoliths can be treated by this technique. Male dogs less than 5-7 kg may be too small for transurethral passage of flexible ureteroscopes to the urinary bladder for fragmentation and removal of urocystoliths. Similarly, laser lithotripsy is not possible in male cats because cystoscopes small enough to pass through the male cat urethra do not have a large enough working channel to accommodate the laser fibers. Laser lithotripsy via transurethral approach may not be the best option for removal of very large numbers of urocystoliths, especially in male dogs. One study showed that in male dogs, the overall stone burden was inversely related to successful removal of all uroliths by laser lithotripsy whereas stone burden was less likely to affect outcome in female dogs.[7] As holmium lasers become more widely available in veterinary medicine, this technique may replace open cystotomy for removal of urocystoliths. Laser lithotripsy is especially effective and efficient for removal of urethroliths and, if available, laser lithotripsy should be utilized to avoid urethrotomy. Alternative minimally invasive approaches for removal of urocystoliths in small male dogs include percutaneous cystoscopy, laparoscopic assisted cystoscopy, and SWL fragmentation of bladder stones followed by voiding urohydropropulsion.[9,10]

Laser Lithotripsy Technique

Positioning of dogs for cystoscopy and laser lithotripsy is a matter of clinician preference. The author positions female cats and dogs in dorsal recumbency and male dogs in left lateral recumbency. The hair around the prepuce or vulva is clipped and a sterile prep of the surrounding skin should be performed. In male dogs, the penis is extruded and cleansed prior to cystoscopy. The prepuce or vaginal vestibule is flushed several times with dilute betadine saline solution. The animal is draped with a small opening for access to the urethra. Strict attention to aseptic technique is important in handling the cystoscope or ureteroscope and stone baskets. By positioning female dogs in dorsal recumbency and using sterile drapes, it is easier to maintain aseptic technique during repeated passage of the cystoscope for laser lithotripsy and basket extraction of urolith fragments.

When working through the biopsy channel of the rigid cystoscope, the author uses a 550 micron fiber (or 365 micron fiber in female cats) to deliver the laser energy to the surface of the urolith. Although both 200 and 365 micron fibers pass through the 3-French working channel of a flexible ureteroscope, the 200 micron fiber is preferable because it allows for greater degree of deflection of the tip of the ureteroscope than the 365 micron fiber. The increased flexibility of the ureteroscope during laser lithotripsy allows for greater precision in targeting uroliths, especially in the urethra. The author uses an initial setting of 0.6 to 0.8 joules per pulse and 6 to 10 Hz for laser lithotripsy. Either the joules (energy per pulse) or Hz (pulses per second) may be increased to facilitate fragmentation of the uroliths. Increasing the Hz tends to increase the speed of lithotripsy provided the urolith shows some fragmentation, whereas increased joules per pulse is utilized if the urolith does not fragment at lower power settings. During laser lithotripsy, the laser fiber should be in contact with the urolith or within <0.4 mm of the urolith surface. The laser fiber should be parallel to the urethral or bladder mucosa or the laser fiber must be positioned > 0.5 mm away from the mucosa. Working parallel to the mucosa is especially important in the urethra to avoid iatrogenic trauma to the urethral mucosa.

During laser lithotripsy fluid irrigation using sterile saline is needed to maintain visualization of uroliths and provide fluid to dissipate heat near the laser fiber tip. In female dogs and cats, gravity fluid irrigation is sufficient with the fluid bag positioned no more than 70 cm above the patient. In male dogs, a 60 cc syringe is connected to the flexible ureteroscope via urolock adaptors (Boston Scientific) to allow pressurized fluid infusion during laser lithotripsy. Careful attention to the total infused volume is necessary to prevent overdistension or rupture of the urinary bladder. The approximate maximum infusion volume should be ≤ 10 ml/kg of body weight assuming the bladder was empty prior to the procedure. Periodic palpation of the bladder is advisable to monitor the degree of bladder distention.

CONCLUSIONS

Urethral strictures may be effectively managed by balloon dilation and urethral stent placement. Likewise, urethral stent placement is often effective for management of malignant urethral obstruction. Transurethral cystoscope-guided laser lithotripsy is successful in female dogs and cats and most male dogs for fragmentation and subsequent removal of cystic and urethral uroliths. Laser lithotripsy is especially effective for fragmentation and removal of urethroliths, and if available, laser lithotripsy is the procedure of choice for removal of urethroliths.

REFERENCES

1.Hawthorne JC, et al. J Am Anim Hosp Assoc 1998, 34: 511-514. 2. Hostutler RA, et al. J Vet Intern Med 2004, 18: 113-116. 3. Adams LG, et al. J Am Vet Med Assoc 2009, 232: 1680-1687. 4. Childress MO, et al. J Am Vet Med Assoc 2011 in press. 5. Weisse C, et al. J Am Vet Med Assoc 2006229: 226-234. 6. Bevan JM, et al. J Am Vet Med Assoc 2009, 234: 1286-1294. 7. Lulich JP, et al. J Am Vet Med Assoc 2009234: 1279-1285. 8. Grant DC, et al. J Vet Intern Med 2008, 22: 534-539. 9. Rawlings CA, et a. J Am Vet Med Assoc 2003, 222: 759-761. 10. Goldman et al. Urological Res 2008, 36: 179 (A).

CONCEPTS & CONTROVERSIES IN MINIMALLY-INVASIVE THERAPEUTICS

Dr. Allyson Berent, DVM, DACVIM; Dr. Chick Weisse, VMD, DACVS; Dr. Larry Adams DVM, PhD, DACVIM
New York, NY and West Lafayette, IN

INTRODUCTION

This session will focus on 2 main topics that will be collegially discussed and debated between the panel of speakers along with the audience. The focus will be on the risks and benefits of various management approaches to different disease processes. The panel of discussants will be Dr. Larry Adams, Dr. Allyson Berent and Dr. Chick Weisse. Each speaker will highlight a different perspective to various diagnostic and treatment modalities of 2 main urological conditions that we are faced with daily.

Lower Urinary Tract Transitional Cell Carcinoma

The first case example will involve various presentations of dogs with transitional cell carcinoma. We will focus on the diagnostic approaches each clinician takes and the therapeutic options offered to each individual client. The focus will be to get the audience's perspective and the speakers' opinions on the various options including: radical surgical resection, debulking surgery, laser ablation[1], traditional intravenous chemotherapy, metronomic chemotherapy, intra-arterial chemotherapy and palliative stenting.[2,3]

Ureteral Obstructions

The second case example will be looking at canine and feline ureteral obstructions, typically associated with ureterolithiasis and discuss the various diagnostic and therapeutic options that are available. The speakers will present different perspectives on the current therapeutic options that they are offering to their patients and the audience should involve themselves in this discussion/debate. The main topics to be discussed will be medical expulsive therapy, extracorporeal shockwave lithotripsy[4,5], ureteral stenting[6,7], and traditional ureteral surgery.

CONCLUSION

The goal of this discussion is to highlight the various perspectives on the diagnosis and treatment of these 2 controversial diseases seen in veterinary medicine that internists are routinely faced with and hopefully gain great perspective from the audience and the speakers on different approaches to these conditions.

REFERENCES

Elwick KE, et al. *J Amer Anim Hosp Assoc* 2003, 39:506-508. 2. Weisse C, Berent A, et al. *J Am Vet Med Assoc* , 229: 226-234. 3. Weisse CW, Berent AC, et al. *J Am Vet Med Assoc* , 233: 1564-1574. 4. Block G, Adams LG, et al. *J Am Vet Med Assoc* 1996, 208: 531-536. 5. Lulich JP, Adams LG, Grant DC. *Vet Clin N Am Small Anim Prac* 2009, 39: 149-160. 6. Berent AC, Weisse C, Bagley D. *Vet Surg* 2007, 36: E3 (Abstract). 7. Berent A, Weisse C, et al. J Vet Intern Med 2009, 23: 688-689 (Abstract).

THE PATHOPHYSIOLOGY OF OBESITY

Leslie Sharkey, DVM, PhD, DACVP (Clinical Pathology)
St. Paul, MN

DEFINITION

Obesity is defined as a condition of excess body fat. The degree of obesity is often estimated using the Body Mass Index (BMI), which is an individual's mass in kilograms divided by the square of the height in meters. Although an imperfect estimate of actual adiposity, BMI has been an effective tool for classifying the risk of obesity-related disease in people. More accurate measurements of body fat content and/or distribution of fat requires advanced techniques that are often expensive or of restricted availability.[1]

Obesity Class	BMI (kg/m^2)	Risk
Underweight	<18.5	Increased
Normal range	18.5-24.9	Average
Overweight	25.0-29.9	Mildly increased
Obesity class I, II, III	30.0-34.9, 35-39.9, ≥40.0	Moderate, severe, very severe

SCOPE OF THE PROBLEM

The prevalence of overweight/obesity remained relatively stable in the United States between 1960-1980, and then began to increase dramatically to current levels; approximately 2/3 of the population of the United State is overweight/obese. Obese individuals have a 50-100% increased risk of premature death from all causes compared with normal weight individuals.[2] Non-fatal complaints associated with obesity include respiratory difficulties, musculoskeletal disorders, skin problems and infertility, while potentially fatal conditions include gallbladder disease, cardiovascular pathology, and some cancers. Rates of anxiety and depression are also increased in the obese population. Economic costs, while difficult to estimate, are considered to have a major impact on healthcare budgets. Indirect costs in terms of decreased productivity compound the issue.

CAUSES

At the most basic level, obesity is the result of imbalance between energy intake and energy expenditure. Major theories regarding the underlying causes for obesity include recent changes in environmental factors such as food availability and activity levels, genetic influences, and developmental programming. The regulation of food intake and energy expenditure via activity and thermogenesis is complex and incompletely understood, and a detailed discussion is beyond the scope of this presentation. Briefly, highly redundant and tightly regulated physiologic systems (mostly neural and endocrine) function to maintain body weight within a narrow range with more than 99% precision in most individuals over a period of years.[3] This requires that a sensing mechanism for energy stores (lipostat) be coordinated with eating behaviors and energy metabolism. The "thrifty gene" hypothesis predicts that this system will tend to favor mechanisms promoting weight gain, due to the survival value of that characteristic in environments with unpredictable and restricted access to food.[4] This hypothesis is supported by the observation that overfeeding produces fewer compensatory changes in energy expenditure than underfeeding.[5] Over recent history, decreasing activity levels and increased access to highly palatable, calorie dense foods have rendered this genotype maladaptive. The onset of obesity may be insidious, considering that the typical weight gain observed in Western cultures of about a pound a year results from an excess of a mere 10-20 kcal/day.[6] New observations suggest that alterations leading to the biological defense of a new higher body fat mass compound excess energy intake.[7] Hypothalamic inflammation with resultant leptin and insulin resistance may contribute.[7]

Most authors agree that obesity is a multifactorial chronic disease with a complex etiology and a genetic component that is currently poorly characterized. Complex interactions between environmental factors and genes are likely. For example, rodent studies have demonstrated that some strains respond to high dietary fat content with dramatically increased adiposity, while others appear resistant. There is wide variation in the quantitative estimates of the contribution of genetics to obesity in humans, with some sources attributing 30-40% of the variance in individual BMI to genetics[8], while others estimate 45-75%[9] or even 50-90%.[6] Rarely, families with early onset extreme obesity have been shown to have a monogenic cause with specific mutations in genes encoding leptin, the leptin receptor, proopiomelanocortin (POMC), prohormone convertase 1, the melanocortin receptor MC4R, and TrkB receptor.[9] All of these genes appear to influence the perception of satiety.

Developmental programming also is a focus of current research. Postulated physiologic mechanisms include reduced myogenesis, age-related decline in pancreatic function, predisposition to the development of fatty liver, and programming of glucose metabolism and vascular function[10]. While early over-nutrition contributes to later obesity, there is increased risk of adult obesity in individuals with low birthweight, resulting in a "U" shaped curve relating birthweight to risk of obesity during adulthood.[11] Data from laboratory animal studies of over and undernutrition during fetal development reveal permanent alterations in appetite control, adiposity and adipocyte metabolism, physical activity, and metabolic rate that could contribute to obesity in offspring.[12] Molecular mechanisms responsible for those alterations may be the result of changes in the methylation status of hypothalamic genes influencing appetite and energy expenditure, selective leptin resistance, mutations in mitochondrial DNA, and changes in the expression of uncoupling proteins of the inner mitochondrial membrane that regulate energy metabolism.[12] Less well investigated theories propose contributors such as altered gut flora[13], exposure to xenobiotics, and psychosocial factors[14].

ADIPOSE BIOLOGY

Adipose tissue makes up a diffuse organ comprised of adipocytes (only about 50% of the total cellular content), pre-adipocytes, multipotent mesenchymal stem cells, endothelial cells, pericytes, monocyte/macrophages, and nerve cells. Previous models of adipocyte biology suggested that the number of adipocytes was fixed, with only the size changing in response to triglyceride load. Newer research demonstrates that there is a large pool of stem cells and pre-adipocytes in adipose tissue in individuals of all ages that may be recruited once existing adipocytes reach a critical level of hypertrophy.[15] Fat is now understood to be a dynamic tissue with continuous remodeling; 50% of the adipocytes in human subcutaneous fat are replaced every 8 years[16], or about a 10% annual turnover rate that is not impacted by BMI[17]. Obesity can be differentiated into hyperplastic and hypertrophic forms, with hyperplasia being more common in the severely obese with early onset disease and less favorable treatment outcomes, however hypertrophic cells are more metabolically abnormal. Cross sectional studies suggest that increases in adipocyte mass precede changes in number, but there is limited longitudinal data and changes in weight do not alter turnover rates in adults. Much of the current knowledge about fat cell development from precursors is from *in vitro* models.[18] As expected, regulation is complex and not completely understood at this point. Activation of peroxisome proliferators-activated receptor γ (PPARγ) is both necessary and sufficient for adipogenesis. In contrast, Wnt signaling inhibits adipogenesis, and its disruption results in spontaneous adipogenesis from preadipocytes and mesenchymal precursor cells. Adipocyte number may be reduced by apoptosis, however necrosis appears to predominate[16], setting the stage for inflammatory cell recruitment. At the same time adipocyte mass is changing, remodeling occurs in the extracellular matrix, inflammatory cell population, and vasculature. Plasmin and plasminogen activator inhibitor, matrix metalloproteinases (MMP), tissue inhibitors of MMP (TIMP), and cathepsins have all been shown to participate in extracellular matrix remodeling with adipocyte expansion[17]. As individuals become obese, adipose populations of regulatory T cells (T-reg) that improve insulin sensitivity decline, while the proportion of CD8+ T cells increases. Numbers of inflammatory macrophages (M1) also increase compared with the normal adipose M2 macrophages that are preferentially involved in tissue homeostasis and repair[17,19].

Adipose tissue has traditionally been divided into two types; white adipose characterized by cells with a single large droplet of triglycerides, and brown adipose, characterized by cells containing multilocular lipid droplets. Classical functions of white adipose tissue include insulation and mechanical support along with storage of surplus energy,[20] while brown adipose tissue is associated with thermogenesis as a result of the expression of few distinct genes such as uncoupling protein-1.[17] It was previously thought that brown adipose tissue was only present in children, however initial observations in cancer patients undergoing positron emission tomography (PET scans) have now been confirmed that small but variable amounts can be present in adults, in whom activity is stimulated by cold (Nedergaard 2010); this finding may have therapeutic implications. White adipose tissue is now becoming regarded as an important endocrine organ, secreting a wide variety of factors including steroid hormones, growth factors and cytokines, eicosanoids, complement proteins, binding proteins, extracellular matrix proteins, vasoactive factors, regulators of lipid metabolism, and others.[21] Collectively, these factors are labeled "adipokines" and they function as part of a complex set of physiologic control systems that regulate local tissue and whole organism physiology. Adipocytes from different anatomic locations may vary in their biology due to local influences on differentiation and gene expression, and may be considered "mini-organs." In fact, adipose tissue distribution influences the pathologic sequela characterizing obesity. Preferential deposition of fat into visceral deposits instead of subcutaneous deposits increases the risk of insulin resistance, atherosclerosis, and diabetes mellitus.[18] With impaired adipocyte function due to aging or lipodystrophy, lipids may be deposited in non-adipocytes in the liver, muscle, pancreas, and kidney, resulting in altered fat and glucose metabolism and impaired organ function due to "lipotoxicity".[19]

PATHOGENESIS OF OBESITY RELATED DISORDERS

There are numerous non-mutually exclusive mechanisms proposed for the various pathologic processes associated with obesity. Each have support from *in vitro* experiments, laboratory animal studies, and data collected from human subjects. Because it is not possible to address all of the mechanisms currently under investigation, we will focus on briefly mentioning a few pivotal concepts.

1. Obesity as an inflammatory disease

Obesity is considered a cause of chronic inflammation, and is associated with infiltration of the enlarging fat mass by increased numbers of CD8+ T-cells and mast cells along with falling numbers of regulatory T-cells and CD4+ T-cells. These changes precede and likely result in increased numbers of macrophages with an predominance of M1 rather than the more typical M2 phenotype,.[20] While hypoxia in poorly vascularized areas of the expanding adipose mass is certainly an issue, "nutritional stress" in the form of chronic overfeeding leads to the formation of reactive oxygen species as well as triggering of the unfolded protein response by the endoplasmic reticulum, stimulating inflammation via several intracellular signaling pathways (JNK, NF-κB, etc)[25,26]. Furthermore, excessive circulating free fatty acids that are present with overnutrition/obesity appear to activate the innate immune response via Toll-like receptor 4 (TLR-4) with increased NF-κB signaling,[27,28] Concurrent inflammation in the hypothalamus may contribute to the leptin and insulin resistance that promotes the metabolic syndrome, however this is in contrast to observations of anorexia and decreased body mass in patients with sepsis or brain injury[7]. Widespread local followed by systemic activation of inflammation predisposes to progression of cardiovascular disease and cancer.

2. Altered production or activity of adipokines

Technically, the inflammatory cytokines discussed above are adipokines. In addition, adipocytes produce a variety of other factors implicated in the pathophysiology of obesity, with adiponectin and leptin being prototypic. Adiponectin secretion from normal adipocytes acts as a vasodilator, and may have a palliative effect on pathologic cardiac remodeling after myocardial infarction.[29]

Adiponectin appears to increase insulin sensitivity and suppress inflammation. Leptin has the primary function of regulation of energy stores, and has been suggested to have a role in hypothalamic development (providing a mechanistic link between altered nutritional states during fetal and neonatal development and long-term programming of adiposity[31]. Other potential functions include regulation of reproduction and immunity and modulation of insulin sensitivity. The reader is directed to Dr. Radin's proceedings for a more thorough review of the profound significance of adipokines in the pathogenesis of obesity.

3. Growth factor abnormalities and oxidative stress

Linkage of nutrient uptake and growth factor signaling in mammalian cells emphasizes the intrinsic interdependence of energy homeostasis and growth regulation. Nutrient availability is sensed by cells which then promotes protein synthesis and cell growth. Mitochondiral reactive oxygen species production increases with both nutritional and oncogene-induced metabolic stress, predisposing to DNA damage, intracellular growth signaling (including AP-1 and JNK pathways), and inflammation, all of which contribute to neoplastic transformation and progression. In people, obesity has been implicated as a contributing factor particularly in endometrial, breast,colon, prostate, and esophageal cancers, with the trophic/stimulatory effects of insulin and chronic inflammation and proximate mechanisms.

4. "Lipotoxic" effects of lipid deposition in non-adipocytes

Although obesity is associated with numerous pathophysiological consequences, functional adipose tissue capable of storing lipid appears to be required for health since ablation of adipose or lipodystrophy syndromes results in development of metabolic derangements similar to those that characterize obesity. If overnutrition is sufficiently severe that excess nutrients cannot be accommodated by the expanding fat mass, lipid deposition will occur in other tissues such as the liver, muscle, heart, and pancreas. While most tissues contain a small amount of intracellular lipid for various structural and metabolic purposes, accumulation of excessive amount may interfere with cell function or even promote apoptosis. The precise mechanisms of lipotoxicity remain under investigation. The accumulation of long-chain fatty acyl-CoA, phosphatide, diacylglycerols and ceramides with toxic properties or signaling capabilities may play a role, as may increased oxidative stress. Ectopic lipid accumulation likely contributes to tissue specific or global insulin insensitivity. The effects may be influenced by genetic background. It appears logical to attempt to interfere with ectopic triglyceride deposition to prevent tissue damage, however there is some experimental evidence in laboratory animal models that doing so induces a more toxic state with subsequent metabolic failure.

5. Obesity related renal disease

The term "obesity-related glomerulonephritis" is now used to describe a weight responsive syndrome characterized by glomerulomegaly, podocyte hypertrophy, increased mesangial matrix and cellularity accompanied by less pronounced/frequent albuminuria than similar non-obesity related focal segmental glomerulosclerosis (Hunley 2010). Obesity causes increased glomerular filtration rate and renal plasma flow with disrupted tubuloglomerular feedback that parallel BMI and adipose mass. Altered renin-angiotensin system regulation, plasminogen activator inhibitor-1, lipid disturbances, and local and systemic obesity driven inflammation may contribute.

SUMMARY

Adipose is a developmentally dynamic tissue with a broad spectrum of potential biological behaviors and is a major endocrine organ. Increased adiposity contributes to the development of several major sources of morbidity and mortality, including cardiovascular disease and cancer via changes in whole organism physiology or at the local tissue level. Research resulting in a better understanding of these pathophysiologic processes will be needed to combat the consequences of the obesity epidemic.

REFERENCES

[1]Bray, C. **Am J Clin Nutr** 1992;55:488S-494S, [2]Turconi G. In Obesity: Epidemiology, Pathophysiology, and Prevention. D Bagchi and HG Preuss, eds. CRC Press, Taylor & Francis Group, Boca Raton FL. 2007. pp 3-19, [3]3. Spiegelman et al. **Cell Metab**. 2006;4:425-427, [4]Neel JV. **Am J Hum Genet** 1962;14:353-362, [5]Hill JO. **Endocrine Reviews**. 2006;27:750-761, [6]. Cummings MW et al .**Annu Rev Med** 2003;54:453-471, [7] Thaler JP et al. **Endocrinology**. 2010;151:4109-4115, [8] Pi-Sunyer FX. **Obes Res.** 2002;10:97S-104S, [9] Farooqui IS et al. **Obesity Rev.** 2007;8(S1):37-40, [10] Drake AJ et al. **Reproduction.** 2010;140:387-398, [11] Barker DJP.**Obesity reviews** 2007;8(S1):45-49, [12] Taylor PD et al. **Exp Physiol.** 2007;02:287-298, [13] Ley RE. **Curr Opin Gastroenterol.** 2010;26:5-11, [14] Zinn AR et al. **Am J Med Sci.** 2010;340:481-491, [15] Fischer-Posovszky P et ak. **Horm Metab Res**. 2007;39:314-321, [16] Arner P et al. **Biochem Biophys Res Comm.** 2010:396:101-104, [17] Lee M et al. **Curr Opin Clin Nutr Metab Care.** 2010;13:371-376, [18] Rosen ED et al. **Nature Rev Mol Cell Bio.** 2006;7:885-896, [19] Anderson EK et al. **Curr Opin Lipidol.** 2010;21:172-177, [20] Sethi JK et al. **J Lipid Res.** 2007;48:1253-1262, [21] Nedergaard Jet al, **Cell Metabolism.** 2010; 1, DOI 10.1016/j.cmet.2010.03.007, [22] Cartwright MJ et al, **Experimental Gerontology.** 2007;42:463-471, [23]Rasouli N et al, **Diabetes, Obesity, and Metabolism.** 2006;9:1-10, [24] Zeyda M et al, **Int J of Obesity**. 2007;1-9, [25] Cusi K. **Curr Diab Rep**. 2010;10:306-315, [26] Wellen KE et al. **Molecular Cell.** 2010;40: DOI 10.1016/j.molcel.2010.10.004, [27] Kim F et al. **Circ Res**. 2007;100:1589-1596, [28]Shah PK. **Circ Res.** 2007;100:1531-1532, [28] Fésüs, G et al. **Cardiovasc Res.** 101;1666-1676, [29] Shibata R et al, J Mol Cell Cardiol. 2007;42:1065-1074, [30] Bouret SG et al. **Science.** 2004;304:108-110, [31] Hunley TE et al. **Curr Opin Nephrol Hypertens.** 2010;19:227-234.

THE BIOLOGY OF OBESITY IN DOGS AND CATS

Leslie Sharkey, DVM, PhD, DACVP (Clinical Pathology)
St. Paul, MN

DEFINITION

Obesity is defined as a condition of excess body fat. While in people obesity is most commonly evaluated in terms of body mass index (BMI), in companion animals body condition scoring (BCS) is most frequently used on either a 5 or 9 point scale. The terms overweight and obesity have been defined as relative body weight of 110% and 120% of ideal[1].

Body condition score		Characteristics
1 (1-2)	Emaciated	Ribs, lumbar vertebrae, pelvic bones, all body prominences evident from a distance. No discernible body fat. ↓muscle mass.
2 (3)	Thin	Ribs easily palpated, may be visible. No palpable fat. Tops of lumbar vertebrae visible. Pelvic bones less prominent. Obvious waist and abdominal tuck.
3 (4-5 Ideal)	Moderate	Ribs palpable without excess fat covering. Abdomen tucked up when viewed from side.
4 (6-7)	Stout	Fleshy appearance. Ribs palpable with difficulty. Fat deposits over lumbar spine and tail base. Abdominal tuck may be absent.
5 (8-9)	Obese	Large fat deposits over chest, spine, tail base. Waist/ abdominal tuck absent. Fat deposits on neck and limbs. Abdomen distended.

Adapted from http://vet.osu.edu/vmc/body-condition-scoring-chart and http://www.purina.com/dog/weight-and-exercise/bodycondition.aspx

More quantitative assessment is possible using dual energy x-ray absorptiometry (DEXA) that utilizes photons of two energy levels (70 and 140 kvp) which have variable impedance by different tissues, such as mineral, lipid, and lean tissues. This method correlates well with BCS. Attempts at application of hand held bioimpedence monitors reveal weaker correlation with BCS and lack of accuracy and precision at high body fat contents[2].

SCOPE OF THE PROBLEM AND CAUSES

Obesity is the most common nutritional disease of dogs in the West[3]. Estimates of the rates of overweight/obesity in companion animals are high and increasing. In dogs, rates are estimated to be approximately 30-50% in 2005, increased from around 25% in 1986. Recent rates for cats are between 20 and 50%[4]. Given the increasing prevalence of obesity in both people and dogs and cats despite an obvious lack of genetic relatedness, similar environmental factors are likely operating on all three populations to promote weight gain, including decreased activity and consumption of large amounts of energy dense food. A study from the Netherlands found that canine overweight correlates with owner BMI (and is related to rates of dog-walking), but no relationship could be defined for cats[5]. A study of environmental factors associated with canine obesity in the UK demonstrated that the amount of exercise, frequency of snacks and treats, and owner age were predisposing, along with lower personal income and associated reduced awareness of the health risks associated with obesity in dogs[6]. Dietary content and feeding patterns are influenced by those of the owners[7]. Diets higher in protein and fiber may improve perceptions of satiety by dogs and are associated with reduced rates of obesity or improved weight loss. Other phenomenon shared by owners and pets is the potential for rebound weight gain after loss, seen in both cats[8] and dogs[9]. The canine study demonstrated that during this process a predisposition to repeat weight gain was associated with lower caloric intake than the original induction of obesity due to increased food efficiency.

Less is known about the specific modulating factors and molecular mechanisms that influence the generation of the obese phenotype in the face of overnutrition in dogs and cats compared with people and laboratory animal models. Although specific syndromes of obesity related to single gene mutations are not reported, some breed predispositions are noted. Breeds predisposed to obesity include the Labrador Retriever, Boxer, Cairn Terrier, Scottish Terrier, Shetland Sheepdog, Bassett hound, Cavalier King Charles Spaniel, Cocker Spanial, Dachshund (esp long haired), Beagle, and some giant breed dogs, while other breeds such as the Greyhound, are resistant[3]. Further investigation of these tendencies may reveal valuable insights into genetic mechanisms, as is currently being probed for breed predispositions in cancer biology. As in people, obesity tends to increase with age (Serisier) and is influenced by weight during childhood/puppyhood (although obese kittens are not observed). Altered sex hormone status (ie neutering) is an important factor. Medications such as anti-convulsants and glucocorticoids, as well as endocrinopathies including hyperadrenocorticism and hypothyroidism are associated with obesity in the dog.

In people and laboratory animal models, there are some well documented interrelationships between adipose tissue biology and the development and pathophysiology of obesity. Obesity is considered an inflammatory disease fueled by the production of inflammatory mediators and hormones, collectively known as adipokines, many of which may be produced within the enlarging mass of adipose tissue itself (see Radin proceedings for details). Hypoxia within the expanding adipose tissue mass with subsequent induction of increased production of adipokines mediated by hypoxia inducible factor-1α may be one mechanism[10]. New research is beginning to establish similar links in the dog. In vitro, canine adipocytes have been shown to produce adipokines such as IL-6, MCP-1 and TNFα when stimulated by inflammatory mediators (LPS and TNFα)[11]. In vivo, an improvement in insulin resistance and a reduction in plasma concentrations of inflammatory cytokines such as TNFα, haptoglobin, and C-reactive protein are observed after a period of weight loss induced by diet and exercise. These findings suggest a link between obesity and inflammation that also impacts insulin resistance[12]. The importance of altered biological effects dependent on the anatomic distribution of the adipose tissue in dogs and cats has preliminary support in the data, but requires further investigation[3].

METABOLIC/HORMONAL ALTERATIONS WITH OBESITY

In people, the development of insulin resistance and its sequelae is central to a significant number of the medically serious consequences of obesity. This appears to be mirrored in obese cats, who experience increased rates of Type II diabetes associated with insulin resistance, with decreases in sensitivity of more than 50% in obese compared with optimal weight cats[13]. Medically, this can be overcome by weight loss or small increases in insulin dose. Weight gain is also associated with insulin resistance in dogs, however the change in insulin sensitivity is mild and subclinical[3]. Experimental canine models of obesity require augmentation by pancreatectomy or chemical damage to beta cells by compounds such as streptozotocin (STZ) to induce overt diabetes[14]. The hypertension that frequently accompanies obesity, insulin resistance, and diabetes in people is not convincingly observed in cats and requires further investigation in dogs, however some canine reports describe increased blood pressure with chronicity of disease and concurrent albuminuria[15]. Lipid abnormalities are another theme in the pathophysiology of obesity in humans and laboratory animals, and have been investigated in dogs and cats. One small study of client owned dogs including normal weight, obese, and dogs with hyperadrenocortism (HAC) described significant elevations in the total cholesterol, VLDL, and triglycerides in dogs with HAC but not the obese dogs. However samples sizes were small (approximately 10/group), and the body condition of the HAC dogs was not described[16]. Another larger study of 127 client-owned dogs found that obese dogs did have higher total cholesterol and triglycerides than healthy weight dogs[17]. A laboratory study of chronically obese cats demonstrated that increases in triglycerides, VLDL and lower HDL than healthy weight cats, but there was no difference in LDL[18]. The authors note that despite these changes, there is little evidence of atherosclerotic disease in cats, who may be resistant to that complication of obesity. As indicated above, endocrinopathies may predispose small animal patients to obesity. In a provocative twist, new data suggest that obesity may alter thyroid hormone function rather than always the reverse in dogs[19]; this is supported by evidence in humans and rodents that increased leptin may alter thyroid regulation. .

DISEASES ASSOCIATED WITH OVERWEIGHT/OBESITY IN DOGS AND CATS

Disease category	Human	Dog	Cat
Endocrine/lipid	Type II diabetes, metabolic syndrome, dyslipidemia	Hypothyroidism, hyperadrenocorticism, diabetes mellitus, insulin resistance, metabolic syndrome (exp)	Diabetes mellitus, hepatic lipidosis
Cardiorespiratory	Coronary disease, hypertension, apnea, asthma	Tracheal collapse, expiratory airway dysfunction (exp), portal vein thrombosis, myocardial hypoxia	+/- asthma
Orthopedic	Osteoarthritis, musculoskeletal pain, gout	Arthritis, cruciate ligament disease, humeral condylar fractures, intervertebral disk disease, hip dysplasia	Lameness
Neoplasia	Breast, renal, endometrial, prostatic esophageal, colorectal, hepatocellular	Transitional cell carcinoma, mammary carcinoma (+/-)	Increased risk
Urogenital	Diabetic nephropathy, decreased fertility	Urinary tract disease, calcium oxalate urolithiasis, glomerular disease (exp), dystocia	Increased risk urinary tract disease
Alimentary	Pancreatitis, fatty liver, cirrhosis	Pancreatitis	Oral cavity/GI disease
Other	Depression, skin disease	Altered immune function	Dermatoses

Adapted from German Vet J 2010;185:4-9. Experimental (exp)

The table above indicates the current documented medical consequences of obesity in dogs and cats. Other considerations include decreased heat tolerance, increased anesthetic and surgical complications, and decreased overall lifespan[3]. A recent review of obesity and osteoarthritis in the dog delineates multiple potential contributing pathophysiologic mechanisms such as joint loading outside the normal physiologic range and altered biomechanics[1]. Altered adipokines and metabolic parameters contribute to joint inflammation as evidenced by the involvement of non-weight bearing joints such as certain joints of the hand in people. There also is the potential for abnormalities in the bone microvasculature reflective of systemic vascular disease[1]. This critical analysis of the literature suggests that these mechanisms have not been fully evaluated in the dog, and that studies investigating the impact of obesity on orthopedic disease in the dog are varied in methodology and sometimes produce contradictory results. Nonetheless, there is sufficient preliminary evidence that obesity may drive the progression of osteoarthritis and that further research should be a priority.

SUMMARY

Marked increases in the rates of obesity in dogs and cats indicate that obesity related research questions should be of top priority in veterinary medicine. Simultaneous epidemiologic and mechanistic studies will improve our understanding of the importance of addressing obesity as a disease and how to approach its control and the mitigation of the multisystemic effects. While some principles from the human and laboratory animal medical and basic science literature likely apply, there are sufficient differences that individual species should be investigated independently to optimize clinical management. The veterinary community will face the challenge of garnering resources for this research and clinical impediments such as lack of basic information about obesity as a veterinary disease, limited treatment strategies, and reluctance to have potentially uncomfortable communications with owners about these issues.

REFERENCES

[1] Marshall WG et al. **Vet Comp Orthop Traumatol.** 2009;22:339-345, [2] German AJ et al. **Vet J.** 2010;183:294-297, [3] Zoran D. **Vet Clin Small Anim** 2010:40:221-239, [4] Courcier EA et al. **J Fel Med Surg.** 2010;12:746-753, [5] Nijland ML et al. **Pub Health Nutr.** 2009;13:102-106, [6] Courier EA, et al. **J Small Anim Pract.** 2010;51:362-367, [7] Heuberger et al. **An Phys An Nutr** 2011;95:98-105, [8] Roudebush P et al. **J Am Vet Med Assoc.** 2008;233:717-725, [9] Nagaoka D et al. **An Phys An Nutr.** 2010;94:287-292, [10] German AJ et al. **Vet J.** 2010;185:4-9, [11] Ryan VH et al. **Pflugers Arch-Eur J Physiol.** 2020:460:603-616, [12] German AJ et al. **Dom Anim Endocrin.** 2009;37:214-226, [13] Scott-Moncrieff JC. Vet Clin Small Anim 2010:40:241-257, [14] Ionut V et al. **Am J Physiol Endocrinol** Metab 2010;298:E38-E48, [15] Reusch CE et al. **Vet Clin Small Anim** 2010:40:335-352, [16] Jericó MM et al. **J Vet Diagn** Invest 2009:21:203-207, [17] Pena C et al. **An Phys An Nutr** 2008;93:324-325, [18] Jordan E et al. **Dom Anim Endo** 2008;35:290-299, [19] Kil DY et al. **Vet Clin Small Anim** 2010:40:205-219.

ADIPOKINES IN HEALTH AND DISEASE LEPTIN, THE PROTOTYPIC ADIPOKINE

M. Judith Radin, DVM, PhD, DACVP (Clinical Pathology)
Columbus, OH

DEFINITION OF AN ADIPOKINE

White fat is a complex endocrine organ consisting of not only of adipocytes, but also preadipocytes, macrophages, lymphocytes, fibroblasts, and blood vessels. Adipose tissues produce a variety of factors called "adipokines" which may act at an autocrine/paracrine or endocrine level. While some adipokines are synthesized solely by fat cells, some are contributed by other components of the fat depot such as the resident macrophages. Not all fat pads are equal from an endocrine functional standpoint, and location and size of the fat pad influences the relative production of adipokines. Adipokines were first recognized as regulators of energy metabolism, appetite and thermogenesis. As the list of adipokines grows, it is becoming increasingly apparent that adipokines also are important modulators of reproductive readiness, cardiovascular function, immunity and inflammation, and cell growth.

LEPTIN STRUCTURE, SIGNALING, AND SYNTHESIS

The discovery of leptin in 1994 (Zhang 1994) was important in developing our understanding of adipose tissue as more than a simple storage site for fat. Leptin is the 167 amino acid protein product of the *ob* gene. Leptin has been detected in those species that have evolved to store lipids in organized white adipose tissue (WAT) depots including boney fish, amphibians, reptiles, birds and mammals (Gesta 2007). It has not been found in species lacking organized WAT such as sharks or invertebrates. The sequences of the *ob* gene and the protein are highly conserved among vertebrate species (Houseknecht 1998, Iwase 2000, Buff 2002, Caetano 1999).

Leptin actions are mediated via binding of leptin to its receptor (LEP-R). LEP-R is closely related to the Interleukin (IL)-6 (gp130) family of receptors so leptin is considered a cytokine. As a result of alternative splicing, multiple forms of varying length of the receptor are produced. Most of the biological actions of leptin are attributed to binding to the full length receptor (designated LEP-Rb) and are mediated by activation of the JAK2/STAT3 signaling pathway. Shorter forms of the receptor have truncated cytoplasmic domains and lack the ability to generate phosphorylated STAT3 but are still capable of initiating intracellular signaling through other pathways such as ERK and AKT. The shortest form of the receptor (LEP-Re) is secreted and acts as transport carrier. Both long and short forms of LEP-R are found in many tissues in the body, reflecting the diverse functions of leptin. The full length LEP-Rb is found in highest density in the satiety centers of the hypothalamus of the brain and plays a key role in regulation of appetite and thermogenesis. A short form of the receptor (LEP-Ra) is found in high concentration in the choroid plexus and facilitates transport of leptin across the blood brain barrier. In non-central nervous system tissues, gene expression for shorter forms of the receptor predominates over the full length form.

Adipocytes are the primary site of synthesis of leptin, although *ob* gene expression has been documented in other tissues such as liver, stomach, osteoblasts, hair follicles, mammary gland, and placenta. Fat cells constitutively secrete leptin, and, fat mass is the primary determinant of circulating leptin concentrations. However, blood levels may be modulated by energy status, sex hormones, and inflammatory mediators. Leptin levels transiently increase following a meal. Fasting results in decreased circulating leptin, even in the absence of significant weight loss. Leptin shows a circadian rhythm in humans, rodents, and horses (Radin 2009). Leptin secretion is stimulated by insulin, glucocorticoids, and inflammatory cytokines (Ahima 2006). In most species, females have higher leptin levels compared to males. Estradiol increases both *ob* gene transcription and secretion of leptin (Tanaka 2001; Shimizu 1997). Conversely, testosterone appears to decrease leptin by increasing clearance from the circulation (Castrogiovanni 2003). Thyroid hormone, catecholamines and free fatty acids inhibit leptin secretion.

LEPTIN AND ENERGY METABOLISM

Suppression of appetite and increased energy expenditure were the first functions ascribed to leptin. Binding of leptin to the LEP-Rb receptor in the satiety centers in the brain directly decreases food intake by stimulating anorexigenic and inhibiting orexigenic neurons. Increased energy expenditure is accomplished by increasing thermogenesis by brown fat. Central binding of leptin to LEP-Rb increases sympathetic nervous system output which activates β3-adrenergic receptors in brown fat, resulting in increased fatty acid oxidation. In addition to its centrally mediated effects on lipolysis, leptin may also act in an autocrine/paracrine fashion to increase lipolysis and inhibit lipogenesis (Lago 2009). These local affects appear to be mediated by phosphorylation and activation of AMP-activated protein kinase (AMPK) and subsequent inhibition of acetyl-CoA carboxylase (ACC). ACC mediates the conversion of acetyl-CoA to malonyl-CoA and is the rate limiting step in the synthesis of fatty acids. Inhibition of ACC decreases triglycerol synthesis and lipid accumulation as well as promotes shunting of fatty acids to the mitochondria for oxidation. Decreased activity of ACC also will impede conversion of carbohydrates to long chain fatty acids. Autocrine/paracrine inhibition of lipogenesis by leptin may be important in preventing excess lipid accumulation (lipotoxicity) in nonadipocytes. Leptin also interacts with other metabolically active hormones. Leptin decreases the lipogenic effect of insulin while increasing overall insulin sensitivity, resulting in increased glucose uptake and glycogenesis. Leptin may provide a link between fluctuations in energy availability and thyroid function. Leptin administration to humans and rodents reverses the decline in hypothalamic thyrotropin releasing hormone and thyroid stimulating hormone that accompany fasting (Chan 2003; Legradi 1997).

LEPTIN AND IMMUNITY

The relationship between leptin and the immune system is complex and may contribute to alterations in immune function observed with either starvation or caloric excess. Leptin is generally considered a pro-inflammatory cytokine. LEP-R are found on hematopoietic cells and leptin promotes proliferation and functional maturation and inhibits apoptosis. Leptin stimulates monocyte/macrophage proliferation, activation, phagocytosis and cytokine secretion (Santos-Alvarez 1999). Leptin modulated T-cell mediated immunity by stimulating proliferation of Th1 lymphocytes and inhibiting T regulatory cells (Matarese 2010). Low leptin levels seen in starvation may contribute to impaired immune function and depletion of lymphocytes.

Cross talk between leptin and inflammatory cytokines is likely given the similarities in induction of intracellular signaling pathways and feedback mechanisms. Leptin production increases in response to endotoxin as well as a number of pro-inflammatory cytokines, including tumor necrosis factor-α (TNF α), IL-1β, IL-6, and leukemia inhibitory factor (Wozniak 2009). Studies have shown that leptin is protective against some of the deleterious effects of endotoxin, TNFα, and IL-1β (Radin 2008).

LEPTIN AND REPRODUCTION

Adequate amounts of leptin are required for fertility, and leptin signaling is one mechanism that ties reproduction to adequate energy availability for successful production of offspring. Leptin plays a role in triggering puberty. Rising leptin levels during puberty are linked to increased activity of GnRH neurons in the hypothalamus with a subsequent rise in FSH and in pulsatile LH release. The effects of leptin are centrally mediated through stimulation of LEP-R in the arcuate nucleus of the hypothalamus. Downstream regulation of GnRH neurons by leptin appears to be indirect as GnRH neurons lack LEP-R. Current studies suggest that leptin stimulates neuronal kisspeptin production which subsequently increases GnRH release (Hill 2008).

Ovulation and reproduction are decreased in times of metabolic stress such as starvation, cachexia, cold stress, or excessive exercise. Under these circumstances, declining leptin levels allows an animal to shift available energy away from reproduction toward more pressing survival needs. Leptin treatment restores leutinizing hormone levels in fasted mice and humans, in women with exercise-induced amenorrhea, and in infertile *ob/ob* mice which genetically lack the ability to produce leptin (Hill 2008). In horses, leptin levels decrease postpartum and remain low independent of changes in body condition (Berg 2007). This suggests an adaptation to allow increased energy intake to meet the needs of lactation and avoid negative energy balance.

LEPTIN AND THE CARDIOVASCULAR SYSTEM

Leptin may impact the cardiovascular system in an endocrine fashion, mediated by secretion from the WAT depots. Additionally, there is potential for autocrine/paracrine secretion from perivascular and pericardial WAT, although the role of these sources is not well worked out. Many of the actions attributed to leptin are considered deleterious to the cardiovascular system. Yet, leptin has a number of potentially beneficial effects. In reality, outcome may depend on the balance of concentration, chronicity, receptor responsiveness, and cross talk with other cytokines. For example, acute treatment with leptin increases sympathetic activity which promotes vasoconstriction and development of arterial hypertension. This increase in sympathetic drive is counteracted by increased by production of nitric oxide (a vasodilator) and natriuresis, resulting in no net effect on blood pressure in normal lean individuals. In the heart, leptin administration decreases infarct size and inhibits cardiomyocyte apoptosis following experimentally induced ischemia/reperfusion injury (Sweeny 2010). While leptin alone is a mild negative inotropic agent, leptin can blunt the profound cardiosuppression induced in isolated cardiomyocytes by inflammatory cytokines such as IL-1β (Radin 2008).

OBESITY AND LEPTIN RESISTANCE

Leptin is a critical regulatory hormone that manages energy homeostasis to maximize survival and reproductive efficiency. Excessive secretion of leptin with obesity results in aberrant regulation of energy metabolism and has deleterious consequences on many organ systems. As fat pads expand with weight gain, leptin levels increase in species examined thus far (Radin 2009). With obesity, a syndrome of leptin resistance develops, typified by blunted response to leptin and metabolic derangements. Leptin-induced activation of STAT3 through LEP-Rb invokes a negative feedback pathway mediated by production of suppressor of cytokine signaling-3 (SOCS-3). SOCS-3 binds to the SH-2 domain of LEP-Rb to inhibit leptin-induced signaling through the receptor. SOCS-3 also will directly bind JAK2 (Bjørbaek 2004). Increased expression and production of SOCS-3 in brain occurs with obesity and has been proposed as a mechanism mediating impaired leptin responsiveness and leptin resistance (Bjørbaek 1999). Other putative inhibitors of LEP-R signaling include SOCS-1, induced by cytokines other than leptin, and protein tyrosine phosphatase 1B, an insulin receptor phosphatase (Bjørbaek 2004). Saturation of central nervous system blood brain barrier transport is another potential mechanism for impaired leptin signaling in the central nervous system (Pan 2007). Leptin resistance appears to be primarily selective for metabolic regulatory functions. Appetite suppression and thermogenesis are impaired, thus furthering continued weight gain. Chronic hyperleptinemia and leptin resistance also favor development of insulin resistance and type 2 diabetes mellitus. Effects observed in other tissues may result, not so much from impaired signaling, but from continued stimulation by chronic, compensatory hyperleptinemia. Chronic hyperleptinemia contributes to co-morbidities associated with obesity such as cardiovascular disease, reproductive failure, chronic inflammation, and cancer.

Cardiovascular system

Despite blunted hypothalamic signaling, the ability of leptin to stimulate sympathetic nervous system activity remains intact in obesity. Sympathetic nervous system activation to the kidney and heart are enhanced in hyperleptinemic individuals and supports development of hypertension through sodium retention and activation of pressor systems such as the renin-angiotensin system and endothelin. Hyperleptinemia contributes to salt sensitivity by adversely shifting the renal pressure natriuresis curve through a combination of renal nerve hyperactivity, impaired NO generation and increased endothelin production (Radin 2003).

Chronic hyperleptinemia has been implicated in the increased incidence of heart failure in obese humans and animal models. Hyperleptinemia is associated with pathologic cardiac remodeling through the ability of leptin to induce cardiomyocyte hypertrophy and changes in extracellular matrix composition leading to fibrosis (Sweeny 2010). In the face of increased uptake of fatty acids, hearts in leptin-resistant rodent models fail to increase β-oxidation of fatty acids. This results in accumulation of triglycerides and lipotoxicity. Chronic inflammation is a recognized player in the development of heart failure, atherosclerosis and myocardial infarction. Hyperleptinemia contributes to a state of low grade, chronic inflammation by activation of inflammatory cells and production of pro-inflammatory cytokines. Leptin recruits macrophages by increasing macrophage chemoattractant protein-1 and promotes transition of macrophages to atherogenic foam cells. Increased production of inflammatory mediators such as TNFα, IL-6 and IL-12 in response to hyperleptinemia can lead to impaired vascular function and remodeling, favoring development of atherosclerotic plaques. Leptin has been shown to enhance platelet aggregation, contributing to a prothrombotic state.

Reproductive system

Reproductive dysfunction is well recognized in obese humans and rodent models. Although menarche occurs earlier in obese compared to normal weight girls, FSH and LH responsiveness to GnRH is blunted in obesity (Blüher 2007). Obesity and hyperleptinemia are associated with anovulatory cycles, decreased ovarian steroidogenesis, and poor follicular development.

Cancer

Obesity is recognized as a risk factor for cancer. A variety of tumor cells have been shown to express LEP-R, and hyperleptinemia has most clearly been linked with cancer of the breast, prostate, and colon (Hursting 2010; Cleary 2010). In human mammary tumors and some rodent models, increased expression of LEP-R was associated with poorer prognosis and metastasis (Cleary 2010). Increased cancer risk is likely a consequence of leptin's mitogenic, angiogenic, anti-apoptotic and immune modulatory characteristics. Enhanced proliferation of some breast and colon cancer cell lines by leptin treatment is dependent on by STAT3 signaling through LEP-R.

SUMMERY

Leptin is a multifunctional adipokine secreted in relation to the amount of WAT stores. Leptin normally functions to link energy intake with metabolic and reproductive needs. Increased production with obesity results in selective leptin resistance and hyperleptinemia. Impaired leptin signaling in the brain appears to play a primary role in metabolic abnormalities seen in obesity, while overstimulation by hyperleptinemia appears to contribute to development of other obesity-associated conditions such as cardiovascular disease and cancer.

REFERENCES

Ahima R, et al. **Diabetes** 2006;55(suppl 2):S145. Berg EL, et al. **J Anim Sci**. 2007;85:1660. Bjørbaek C and Kahn BB. **Recent Prog Horm Res** 2004;59:305. Bjørbaek C, et al. **J Biol Chem** 1999;274:30059. Blüher S and Mantzoros CS. **Curr Op Endocrinol Diabetes Obesity** 2007;14:458. Buff PR, et al **J Anim Sci** 2002;80:2942. Caetano AR et al. **Mamm Genome** 1999;10:271. Castrogiovanni D, et al. **Endocrine** 2003;22:203. Chan J et al. **J Clin Invest** 2003;111:1409. Cleary M et al. **Vet Path** 2010;47:202. Gesta S, et al. **Cell** 2007;131:242. Hill JW et al. **Am J Physiol Endocrinol Metab** 294:E827. Houseknecht LK, et al **Dom Anim Endocrinol** 1998;15:457. Hursting SD and Berger NA. **J Clin Oncol** 2010;28:4058. Iwase M, et al. **Res Vet Sci** 2000;68:109. Lago F et al. **Trends Biochem Sci** 2009;43:500. Legradi G, et al. **Endocrinol** 1997;138:2569. Pan W, et al. **Peptides** 2007;28:1317. Radin et al. **Vet Clin Path** 2009; 38(2):136. Radin MJ et al. **Mol Cell Biochem** 2008;315:179. Radin et al. **Mol Cell Biochem** 2003;242:57. Santos-Alverez **Cell Immun** 1999;194:6. Shimizu H, et al. **J Endocrinol** 1997;154:285. Sweeny G. **Nature Reviews** Cardiol 2010;7:22. Tanaka M, et al. **Horm Res** 2001;56:98. Wozniak SE, et al. **Dig Dis Sci** 2009;54,1847. Zhang, et al. **Nature** 1994;372:425.

ADIPOKINES IN HEALTH AND DISEASE ADIPONECTIN & OTHER ADIPOKINES

M. Judith Radin, DVM, Ph.D., DACVP (Clinical Pathology)
Columbus, OH

INTRODUCTION

Adipose tissue is a complex endocrine organ that functions to regulate energy homeostasis. In order to link energy stores with metabolic, growth and reproductive needs, white adipose tissue (WAT) secretes a variety of bioactive factors referred to as "adipokines". To date, over 50 such adipokines have been documented. Adipocytes are the source of some adipokines such as leptin and adiponectin. However, other cells resident in the fat depot such as macrophages and vascular stromal cells are important sources of humoral factors. Obesity is associated with expansion of WAT. Within the expanding fat mass, there is enlargement of adipocytes, increased infiltration of inflammatory cells, activation of endothelial cells, and stimulation of angiogenesis. These changes within the fat pad have important pathophysiological effects that modulate adipokine production. Changes in the amount and balance of secreted adipokines have wide ranging effects and contribute to the co-morbidities observed with obesity such as type 2 diabetes, hypertension, chronic inflammation, coagulopathies, atherosclerosis, and cancer. This review will examine the functions of select adipokines. The following table lists some of the adipokines produced by WAT and the cellular source that produces the adipokine. In addition, biological activities and adipokine interactions in the normal and obese state are detailed.

Adipokine	Cellular Source in the WAT	Physiologic function and pathophysiologic response in obesity
Leptin	Adipocytes	See session 115 proceedings
Adiponectin Also called acrp30 (Hopkin 2007; Okamoto 2006; Kadowski 2005; Barb 2007)	Adipocytes	This 30 kD protein initially associates into trimers. The trimers then oligomerize to form high molecular weight (HMW) multimers consisting of groups of 4-6 trimers. Hexamers and HMW multimers are found in circulation and are the most biologically active forms. Binding to specific receptors (AdipoR1 and AdipoR2) increases activity of AMP-activated protein kinase (AMPK), the main mediator of adiponectin's metabolic effects. Secretion is up-regulated by thiazolidinediones through activation of peroxisome proliferator activated receptor (PPAR)-γ. Metabolic Effects: Adiponectin increases insulin sensitivity. Binding to AdipoR1 in skeletal muscle increases β-oxidation of fatty acids and increases glucose uptake through translocation of GLUT4 to the cell surface. Binding to AdipoR2 in the liver increases β-oxidation of fatty acids. Anti-atherogenic/Anti-inflammatory/Anti-carcinogenic effects: Adiponectin suppresses proliferation of monocyte progenitors and increases apoptosis. It decreases expression of adhesion molecules and movement of macrophages into atherosclerotic plaques. Anti-inflammatory properties include suppression of TNFα and C-reactive protein production. Cardiovascular effects: Adiponectin increases endothelial NO production, resulting in vasorelaxation and lower blood pressure. Increased NO generation may play a role in adiponectin's ability to protect against pathological cardiac remodeling. ***Obesity:*** Secretion is down-regulated as fat mass expands. The amount of the most biologically active HMW multimers is reduced in the circulation.
Angiotensinogen (Aogen) (Engeli 2003; Yvan-Charvet 2011)	Adipocytes	All components of the renin-angiotensin system (RAS) are present in WAT. Aogen is converted to Angiotensin (Ang) I by renin and subsequently to AngII by angiotensin converting enzyme. Biological activity is mediated by binding of AngII to Ang type 1 and type 2 receptors (AT1 and AT2). Binding of AngII to the AT1 receptor down-regulates lipolysis while binding to AT2 up-regulates lipogenesis, synergistically promoting increased fat storage. AngII increases leptin and decreases adiponectin production by adipocytes. AngII may play a role in preadipocyte differentiation. ***Obesity:*** Increased production of Aogen occurs as fat pads expand (especially visceral depots). Spill over of Aogen into circulation increases plasma concentrations of Aogen. This favors systemic activation of the RAS, resulting in vasoconstriction, sodium retention, volume expansion and hypertension. Activation of the RAS contributes to insulin resistance. AngII adds to the chronic inflammatory state of obesity by increasing expression of the receptor for Mcp-1 and recruitment of macrophages into the WAT and atherosclerotic plaques. AngII has hypertrophic effects in the heart.

Resistin (Lazar 2007)	Adipocytes (mice) Macrophages (humans)	This cysteine-rich protein is secreted as a dimer. The primary site of production and biological activity seems to vary with the species studied. Circulating levels are related to fat mass. Secretion from WAT increases after a meal. Production may be increased in response to endotoxin and IL-6. Resistin increases hepatic glucose production and inhibits adipogenesis. Resistin is pro-inflammatory by increasing vascular adhesion molecules and cytokine production by macrophages. ***Obesity:*** Secretion increases with increasing WAT mass. Increasing concentrations of resistin promote insulin resistance, inflammation and atherosclerosis.
Vaspin (Rabe 2008)	Adipocytes	Vaspin is a serine protease inhibitor (serpin) that improves insulin sensitivity in mice. It appears to function as a counter-regulatory adipokine that decreases gene expression of resistin, leptin and TNFα while increasing adiponectin expression. ***Obesity:*** Concentrations mirror body WAT mass.
Apelin (Wozniak 2009; Lago 2009)	Adipocytes Macrophages Vascular Stroma	Apelin increases uncoupling protein expression in brown fat and skeletal muscle, resulting in increased β-oxidation of fatty acids. Decreased vascular tone is mediated by increased NO production and AngII antagonism. Apelin may blunt inflammation by decreasing secretion of some pro-inflammatory cytokines. ***Obesity:*** Circulating levels increase with increasing insulin concentrations (insulin resistant obesity).
Visfatin (Maury 2010; Lago 2009)	Leukocytes Adipocytes (?)	Visfatin is a nicotinamide phophoribosyl transferase (Nampt), a rate limiting enzyme in the synthesis of nicotinamide adenine dinucleotide (NAD). Its role in glucose-mediated insulin secretion and insulin resistance currently is under scrutiny. It also functions as a pro-inflammatory cytokine to increase lymphocyte proliferation, stimulate cytokine production by macrophages and inhibit neutrophil apoptosis.
Interleukin (IL)-6 (Trayhurn 2005; Maury 2010)	Vascular Stroma Macrophages Adipocytes	This pro-inflammatory cytokine inhibits leptin and insulin receptor signaling by increasing suppressor of cytokines-3 (SOCS-3). SOCS-3 binds to and prevents phosphorylation and activation of these receptors and also inhibits Janus kinase 2. IL-6 increases fatty acid oxidation and lipolysis. Down regulation of GLUT4 expression decreases glucose uptake and contributes to glucose intolerance. ***Obesity:*** Production by adipocytes and macrophages is particularly increased in visceral obesity, contributing to insulin and leptin resistance.
TNFα (Trayhurn 2005; Maury 2010)	Macrophages Adipocytes	This multifunctional, pro-inflammatory cytokine has wide ranging effects. TNFα enhances inflammation by stimulating production of IL-6 and vascular adhesion molecules, while decreasing adiponectin. It is prothrombotic by increasing PAI-1 production by adipocytes and hepatocytes. Secondary effects are mediated by increased expression of transforming growth factor-β. ***Obesity:*** TNFα production increases with expansion of WAT, especially visceral depots, contributing to insulin resistance and the overall inflammatory and pro-thrombotic state. Metabolic effects include enhanced lipolysis. Activation of pathways that favor macrophage movement into subendothelial plaques and uptake of oxidized LDL through increased expression of scavenger receptors play a role in atherosclerosis.
IL-1β (Trayhurn 2005; Maury 2010)	Macrophages	This pro-inflammatory cytokine mediates fever, angiogenesis, and smooth muscle proliferation. Cardiovascular effects include suppression of cardiac function and there is evidence of cross talk with leptin to affect contractility (Radin 2008). In adipocytes, IL-1β decreases expression of lipoprotein lipase and increases lipolysis. ***Obesity:*** Increased production occurs with expansion of adipose mass.
IL-8 (Trayhurn 2005; Maury 2010)	Stromal vascular cells	This chemokine induces migration of inflammatory cells from the vasculature into tissues. ***Obesity:*** IL-8 increases with expansion of WAT depots, especially visceral fat pads, and contributes to the overall inflammatory state. IL-8 facilitates migration of macrophages and other inflammatory cells into the enlarged fat pads.

Monocyte chemoattractant protein-1 (Mcp-1) (Trayhurn 2005)	Adipocytes Macrophages	This chemokine recruits macrophages into tissues. It also decreases glucose uptake in response to insulin, increases lipolysis, and increases leptin secretion. ***Obesity:*** Production is increased with obesity. Mcp-1-mediated recruitment of macrophages into the fat depots adds to the overall inflammatory state.
Macrophage Migration Inhibitory Factor (MIF) (Trayhurn 2005)	Adipocytes	MIF helps retain macrophages within tissues. ***Obesity:*** Increased production along with Mcp-1 increases the macrophage population in WAT.
Plasminogen Activator Inhibitor-1 (PAI-1) (Faber 2009)	Adipocytes Stromal vascular cells	PAI-1 is a serine protease inhibitor (serpin) that inhibits plasminogen activation and subsequently fibrinolysis. Production by adipocytes is increased by insulin, TNFα, transforming growth factor-β, and hypoxia. ***Obesity:*** Increased circulating PAI-1 results when there is expansion of the visceral WAT depots. The liver also increases PAI-1 production as part of the acute phase response to the low grade inflammation associated with obesity. Increased PAI-1 promotes a procoagulant state characterized by decreased fibrinolysis and risk for thrombosis.
C-reactive protein (CRP)	Adipocytes	CRP is an acute phase protein produced primary by the liver but may be produced by adipocytes in response to inflammation. Production is inhibited by adiponectin. ***Obesity:*** Circulating CRP is increased in obesity. It participates in up-regulation of PAI-1 and tissue factor to produce a procoagulant state. It also contributes to vascular dysfunction and development of atherosclerosis. CRP increases production of pro-inflammatory cytokines, reactive oxygen species and other inflammatory factors.

SUMMERY

Adipokines produced by WAT play critical roles in regulating metabolism and energy balance. In obesity, changes in adipokine production contribute to development of metabolic abnormalities such as insulin resistance, lipid abnormalities and type 2 diabetes melltitus. Altered regulation of adipokine production by WAT in obesity establishes a state of low grade inflammation, promotes a pro-thrombotic state and stimulates angiogenesis. This sets the stage for development of a variety of diseases associated with obesity, including hypertension, heart failure, atherosclerosis and cancer.

REFERENCES

Barb D, et al. **Am J Clin Nutr** 2007;86(suppl):858S. Engeli S, et al. **Int J Biochem Cell Biol** 2003;35:807. Faber DR, et al. **Obes Rev** 2009;10:554. Hopkins TA, et al. **Cardiovasc Res** 2007;74:11. Kadowaki T and Yamauchi T. **Endo Rev** 2005;26:429. Lago F et al. **Trends Biochem Sci** 2009;43:500. Lazar MA. **Horm Metab Res** 2007;39:710. Maury E and Brichard SM. **Mol Cell Endocrinol** 2010;314:1. Okamoto Y, et al. **Clin Sci** 2006;110:267. Rabe K, et al. **Mol Med** 2008;14:741. Radin MJ et al. **Mol Cell Biochem** 2008;315:179. Trayhurn P. **Acta Physiol Scand** 2005;184:285. Wozniak SE, et al. **Dig Dis Sci** 2009;54:1847. Yvan-Charvet L and Quignard-Boulangé A. **Kidney Int** 2011;79(2):162.

DEFINING NORMAL IN ENDOCRINE & METABOLIC DISEASE: THE REALITY OF REFERENCE INTERVALS IN COMPANION ANIMALS

Rebecca L. Davies, PhD
St. Paul, Minnesota

INTRODUCTION

It has been estimated that up to 80 percent of all critical medical decisions associated with admission, discharge or drug therapy in human patients are made upon review of laboratory test results (1-2). Few veterinarians would argue that this percentage is significantly lower in veterinary medicine and most would agree that the reference interval (RI) is the most commonly used tool in diagnostic medicine.

The establishment of RI for laboratory tests performed in human medicine is sufficiently complex as the effects of method, age, sex, ethnicity, time of day and even posture are considered as diagnostic tests are interpreted. In veterinary medicine, this complexity builds as the role of species, breed, genetic lineages, mixed-breed status, neuter status, age, diet, season, environment or habitat are either ignored or accounted for. Adding scale to scope challenges, statistical best practices recommend that reference intervals be established using large numbers of carefully screened (healthy) reference individuals. Since reference intervals are provided as a guideline to help determine whether or not an animal is healthy and to form a basis for comparison for a patient's laboratory results, the criteria selected to confirm individual health status will also contribute to variability within and among laboratory RI. When the variability within individuals, populations and study designs is considered it quickly becomes clear that the challenges to creating, recognizing and using appropriate reference intervals in veterinary medicine are daunting. However, veterinary laboratory professionals have recently targeted this area and provided literature reviews and strategies that can be adopted to address these complexities (3-4, 12). In addition, it is becoming clear that our ability to define 'normal' can be improved and refined when reference intervals are established more specifically for appropriate subclasses of animals.

BACKGROUND

The International Federation of Clinical Chemistry and Laboratory Medicine (IFCC) Expert Panel on Theory of Reference Values and the Standing Committee on Reference Values of the International Council for Standardization in Hematology provided pivotal recommendations related to the establishment of reference intervals in clinical medicine (5-10). Published between 1987 and 1991, their six papers form the basis for the development of the Clinical and Laboratory Standards Institute (CLSI) and IFCC publication (2008, C28-A3) which summarizes guidelines for defining, establishing and verifying reference intervals in the clinical laboratory (11). In 2009, the IFCC/CLSI guidelines were reviewed by veterinary laboratory professionals and recommendations for their application to veterinary clinical pathology were published (3). Currently, The American Society for Veterinary Clinical Pathology (ASVCP) Quality Assurance and Laboratory Standards Committee (QALS) is in the final stages of completing "Guidelines for the determination of reference intervals in veterinary species and other related topics" (12). Their goal is to provide balanced and practical strategies that are statistically and clinically valid for use in veterinary medicine. The fruits of the QALS labor are based on the previously mentioned pivotal references and are expected to be available online (www.asvcp.org/pubs/qas/index) in 2011. Clearly articulated terminology, definitions, and protocols for obtaining veterinary reference values and establishing reference intervals are provided. Strategies for using smaller numbers of reference individuals, for treatment of outliers, for defining subclasses of subjects and for transferring a RI from one laboratory to another are also addressed. These publications are recommended reading for veterinarians and diagnosticians who wish to understand how reference intervals are established, the inherent limitations of their use and how to evaluate whether they have been established appropriately. An understanding of the appropriate criteria, procedures and protocols is also critical for those that participate in clinical research and for those that review scientific studies for publication merit.

Reference intervals: the basics: terminology and generalized protocol

Cerotti defines a reference interval as "an interval that, when applied to the population serviced by the laboratory correctly includes most of the subjects with characteristics similar to the reference group and excludes the others" (13). Definitions provided in the IFCC/CLSI standards -C28-A3 and the associations (italicized) between the terms are aligned vertically in Table 1. This same source provides the recommended steps to be followed when establishing reference values for a new analyte, for a different group of subjects or for a new method (11).

Reference intervals: the complex: levels within stages

Each stage illustrated in Figure 1 harbors additional recommendations for best practices in order to achieve reliable and useful reference intervals that will effectively include subjects similar to the reference group and exclude the others. In stages one and two, veterinarians, diagnosticians and investigators must consider recommended procedures for defining 'healthy' for their purposes and establish criteria that will optimize their ability to select appropriate subjects. They must consider whether partition factors such as age, exercise level, diet, fasting status, sex, phase of the estrous cycle, breed or environment require a subclass reference interval.

Stage three requires consideration of the number of subjects to be studied. This number is determined according to the desired confidence interval (CI) of the reference limits. A reference interval is the interval between and including two numbers, an upper and lower reference limit. The frequently used inter-percentile interval with the boundaries of 2.5 and 97.5% is a 95% CI (11).

Table 1. Definitions and relationships between terms associated with laboratory reference intervals

Term	Definition
Reference individuals *(comprise a)*	Subjects selected for testing on the basis of well-defined criteria
Reference population *(from which is selected a)*	A hypothetical group that consists of all possible reference individuals
Reference sample group *(in which are determined)*	An adequate number of subjects selected to represent the reference population
Reference values *(that characterize a)*	The test result obtained by observation/measurement of a type of quantity on a reference individual. All values obtained in the reference sample group.
Reference distribution *(from which are calculated)*	The distribution of reference values
Reference limits *(that may define)*	A value derived from the reference distribution and used for descriptive purposes
Reference intervals	The interval between, and including, two reference limits, generally at designated percentiles (for example, the central 95%)
Observed value from subject sample	Subject test result that is obtained from the reference individual and compared with reference values, reference distributions, reference limits or RI.

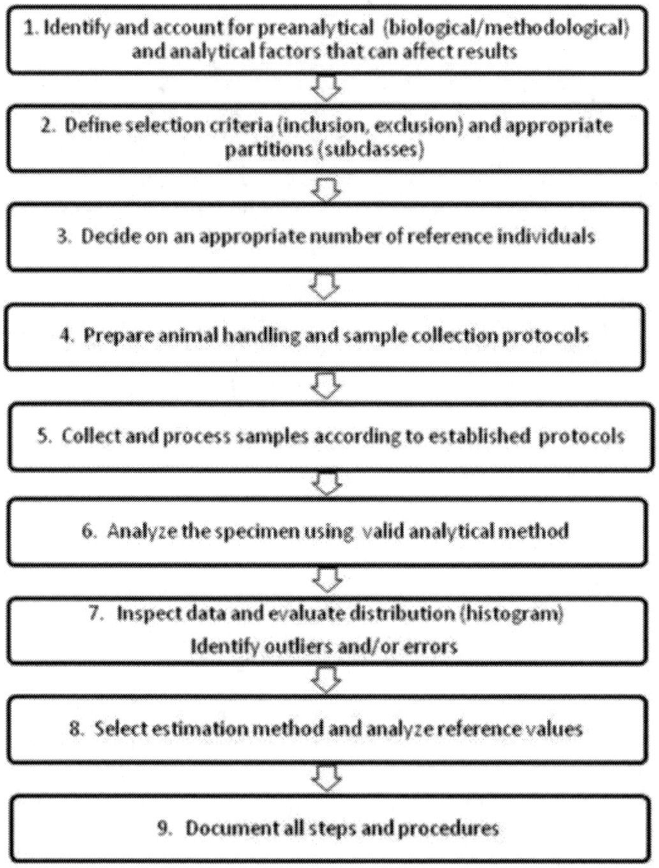

Figure 1. Steps for obtaining reference values and establishing reference intervals

There are two general statistical methods for determining these limits. The nonparametric method (based on the *rank* of the observations) is recommended because it is not affected by the distribution of the data. The parametric method is based on measured values (*mean and standard deviation*) and *assumes* that the values follow a normal (Gaussian) distribution. When this method is used, the data may require transformation in order to fit the required distribution. When the nonparametric method is used and a 90% CI is desired, 120 reference subjects should be used to generate the RI (14). There are acceptable strategies available for the frequent instances where it is not possible to establish a RI with 120 subjects. The IFCC-CLSI C28-A3 guideline recommends robust methods described in Horn and Pesce, 2005, boot-strap-based procedures or traditional parametric methods. (11, 14-15). Their preference is the robust method because it does not require that the underlying population of values be normally distributed. It uses measures of *location and spread* of the values. In addition, if it is impossible to plan and perform a RI study with the necessary number of subjects, the IFCC-CLSI provides guidance on 'verifying' RI established at other sites with as few as twenty samples. While the statistical requirements are important, the IFCC-CLSI working group emphasized that 'the most important considerations in developing reliable RI are selecting appropriate reference subjects, testing an adequate number of subjects, and avoiding pre-analytical errors, not the statistical method used to estimate the RI from the observed data' (11).

In stages four and five, the critical elements associated with the collection, handling and storage of laboratory specimens and are represented. Effects of sample handling on measured concentrations of hormones or metabolic test results vary between analytes, between methods and between species (16). These effects are ignored at our peril, especially when establishing the characteristics of 'normal'. Equally important is the use of appropriate and well-validated analytical methods in order to achieve accurate and clinically useful laboratory data.

The final stages for the establishment of appropriate reference intervals reflect the need for complete reporting. The origins of reference intervals should be documented, and the description of inclusion and exclusion criteria, the approach to outliers and the full representation of the data (including histograms whenever possible) and statistical methods should be made available.

Defining Normal: an endocrine example: breed-specific reference intervals for thyroxine (T4) in canine serum.

In 1990, T4 concentrations were reported to be greater in "small" dogs when compared to "medium or large" breed dogs (17) suggesting that reference intervals for subclasses of dogs might result in more precise reference intervals and diagnostic power. Published data now confirm the advantage of establishing breed-specific reference intervals. Normal greyhounds have significantly lower serum T4 and free T4 concentrations when compared to non-greyhounds (18). Scottish deerhounds, Alaskan sled dogs and Basenji have T4 concentrations that are at, or below, the low end of the established reference range (19-24). Without breed-specific reference intervals, normal dogs in these sub- groups have likely been misclassified as being hypothyroid. Inappropriately classifying a normal dog as being hypothyroid will increase the recorded frequency of hypothyroidism in a breed. In addition, normal dogs may be removed from breeding programs and/or placed on thyroid supplementation. Finally, the perception that hypothyroidism is extremely common in many breeds may contribute to the inappropriate decisions by some to arbitrarily (without the benefit of diagnostic testing) supplement their dogs with thyroid replacement medications. This example illustrates the advantages of establishing effective sub-class reference intervals. In our laboratory, we have been working to establish breed specific reference intervals for thyroid hormones in seven breeds and we have determined that 9% of healthy purebred dogs tested had T4 values below the lower limit of standard reference intervals (25). For reference intervals established at the 95% confidence interval level, varibility of 2.5% at each limit is expected in a normal distribution. These results suggest that the additional effort directed toward the establishment of specific reference intervals is warranted. As we continue to consider defining 'normal' more precisely, an understanding of what constitutes a 'good' reference interval will contribute to the value of our work in clinical and diagnostic medicine.

REFERENCES

[1]Forsman, RW. Clin Chem 1996; 42 (5):813. [2]Clin Lab Products. Issue Stories, May 2006. http://www.clpmag.com/issues/articles/2006-05_03.asp. Accessed January 25, 2011. [3]Geffre A et al, Vet Clin Path 2009; 38 (3):288. [4]Geffre A et al, Vet Clin Path 2009; 34(4): 477. [5]Solberg HE. J Clin Chem Clin Biochem 1987;25:337-42. [6]PetitClerc C et al. J Clin Chem Biochem. 1987; 25:639-44. [7]Solberg HE. J Clin Chem Clin Biochem. 1988; 26:593-8. [8]Solberg HE et al. Eur J Clin Chem Clin Biochem.1991;29:531-5. [9]Solberg HE. J Clin Chem Clin Biochem.1987;25:645-56. [10]Dybkaer R. et al. J Clin Chem Clin Biochem. 1987;25:657-62. [11]CLSI 3rd edition, document C28-A-3, 2008;Wayne, PA:Clinical and Laboratory Standards Institute. [12]Friedrichs K, et al. 2011(pending) ASVCP http://www.asvcp.org/pubs/qas/index [13]Cerotti F, Clin Biochem Rev 2007;28:115. [14]Reed AH et al. Clin Chem. 1971;17:275. [15]Horn PS et al. AACC Press. 2005. [16]Hegstad-Davies RL, Therio 2006; 66,592. [17] Reimers TJ et al. AJVR 1990; 51(3);454. [18]Gaughan KR, et al AJVR 2001;62(7):1130. [19]Ferguson DC. Vet Clin North Am 1994;24(3);515. [20]Panciera DL et al. JVIM 2003;17(4);593. [21]Lee JA et al. JAVMA 2004;224(2);226. [22]Seavers A et al, Australian Vet J 2008; 86(11); 429. [23]Panakova L et al. JVIM 2008;22(11) 1144. [24]Shiel RE et al. JAVMA 2010;236(3);302. [25]Hegstad-Davies et al. ACVIM, June 2010 abst .

DEFINING NORMAL IN ENDOCRINE & METABOLIC DISEASE: THE COMPLEXITY OF METHOD VALIDATION IN COMPANION ANIMALS

Rebecca L. Davies, PhD
St. Paul, Minnesota

INTRODUCTION

The confidence level experienced by veterinarians when they evaluate laboratory test data is generated by a predictable level of laboratory competence and expertise that is either actively (based on an exchange of data or information) or passively (based on expectations or assumptions) confirmed. Not surprisingly, the active role is more rewarding, because it provides veterinarians with vital information related to service expectations, test performance and limitations and best practices for the critical issues of sample collection and handling prior to submission to a laboratory.

Laboratory testing in support of veterinary (and public) health and research is diverse, but the goals and challenges are the same. Professionals in referral/academic/service laboratories all strive to provide diagnostic tests that are suited for purpose, accurate, and cost effective. Their ultimate goal is to instill confidence in the validity of the test results they provide and contribute to accurate diagnoses resulting from an appropriate interpretation of test results that are supportive of other clinical information.

Quality Assurance in Veterinary Laboratories:

Many laboratories meet these goals by integrating a quality assurance (QA) program (or selected portions of a program) into their laboratory environment. Quality assurance programs are designed to allow participants to assess the quality of their services, including those related to diagnostic test development, validation and performance. In human medicine, countries have developed regulations with systems of standards, certification, accreditation and monitoring (for example: Clinical Laboratory Improvement Amendments, CLIA program) in order to assess the quality of their results (1). Likewise, in veterinary medicine, laboratories that perform specific tests for the identification of some infectious agents are required to become certified by international, federal or local government agencies (World Organization for Animal Health (OIE), Federal Drug Administration (FDA), United States Department of Agriculture (USDA)) because of expectations of international trading partners related to the export of animals and animal products as well as issues related to disease identification and surveillance (2). However, in general, veterinary laboratories performing most clinical tests are not bound by government regulations. For that reason, it is noteworthy that many veterinary laboratories are voluntarily implementing quality assurance programs in order to meet their own high goals for the service they provide. Some state, provincial and university laboratories will become accredited by the American Association of Veterinary Laboratory Diagnosticians (AAVLD) which provides an excellent opportunity to meet clearly articulated and extensive quality requirements in order to become certified as an AAVLD accredited laboratory (3). Commercial laboratories may voluntarily meet specific work standards such as Good Laboratory Practices (GLP) provided by the FDA (4), and laboratories may also design and integrate their own internal program of quality assurance in order to incorporate best practices into their service programs (5).

All of these approaches to QA start by defining and committing to specific quality goals associated with laboratory services. From there, standard policies and procedures are created to meet those goals. All laboratory QA programs include requirements related to test validation. Diagnostic test validation in veterinary medicine is especially challenging because many veterinary laboratories create novel methods and/or adapt methods used in people for use in the many different species veterinarians care for daily. This requires taking a standard (commercial, human test kit) method and immediately making it non-standard by using it with animal test matrix or material. While this is frequently very successful, it requires in-house evaluation, verification and validation testing to make sure the method is transferable. Test validation plans are designed to confirm the fitness of a test for its given use and to estimate how much error might be associated with a laboratory result produced using the method. A method's test performance will be considered acceptable if its analytic errors are small enough to permit valid clinical decision making.

Validation: the basics: a generalized protocol

Strategies for the validation of laboratory test methods will differ depending upon whether the procedures are continuous-interval scale measurements (for example: clinical chemistry, endocrinology, or drug monitoring) or analyses that result in discrete categorical results (histopathology, microbiology or cytopathology). James O. Westgard has contributed greatly to laboratory medicine by his promotion of standards and best practices, especially related to test validation. In the third edition of his book, Basic Method Validation, he suggests a generalized protocol for performing a method validation. The recommended steps are listed in Table 1. (6) The steps within the table can be adjusted on the basis of what is already known about the method (literature review), what is required by regulations, or what is possible under the limitations of working with multiple species. For some types of diagnostic testing, further validation will be required from field studies, in order to demonstrate the true characteristics (diagnostic sensitivity and specificity) of the method for diagnosing disease. It is noteworthy that the Westgard plan does not stop at test characterization steps, but goes further to include training of personnel, standardization and documentation of procedure steps and long term monitoring of the test method. These features of training and documentation are the bedrock basis of any QA program.

Table 1. James O. Westgard four phase validation plan (Basic Method Validation, 3rd edition)

Phase	Tasks
Familiarization period	Establish working procedure Validate reportable range Check calibration Check detection limit
Preliminary experiments	Perform within-run replication study Perform interference study Perform recovery study Judge analytical acceptability
Final experiments	Perform total replication study Perform comparison of methods study Judge analytical acceptability Verify reference intervals Document studies
Implementation	Select Quality Control procedure Write standard operating procedure Train analysts Introduce method for service Monitor routine performance

Validation, the complex: a strategy provided by the AAVLD

When the breadth of veterinary health care and management is considered, it does not take long to appreciate that test method validation will not always be straightforward. While the Westgard approach will work well for validating a new test kit for fecal glucocorticoids in a wildlife species, it may not be as transferable for other types of diagnostic tests. The AAVLD Accreditation Committee has provided a guideance document that laboratory professionals can use to meet accreditation criteria for test validation for any test method (3). The requirements of the AAVLD state that test methods are selected based on 'the suitability of the test method, its acceptability by the scientific and regulatory communities, its acceptability to the client, and its feasibility given available laboratory resources.' Tests are qualified according to AAVLD if a laboratory section can meet the following criteria:

(1) The laboratory has ongoing documentation of internal or inter-laboratory performance using known reference standards for the species and/or diagnostic specimen(s) of interest.

AND one or more of the following:

(2) The laboratory can provide endorsed or published material by reputable technical organizations (example: OIE Manual of Standards for Diagnostic Tests and Vaccines)
(3) The laboratory can provide a reference published in a peer-reviewed journal with sufficient documentation to establish diagnostic performance and interpretation of result;
(4) Documentation of internal or inter-laboratory comparison to an accepted methodology or protocol.

Using this strategy, it becomes possible to provide sufficient evidence to support the use of a test method using alternative strategies to fulfill the requirements.

It is understood that in some cases, a test will provide important information while it is still under development. In this instance, the test cannot meet all the requirements for validation but the test is still considered informative. If it is not possible to qualify a test based on validation criteria a statement can be written to inform the client that this specific test should not be used by itself to make any final diagnostic decisions. If appropriate, support from additional, alternative testing should be suggested if other qualified tests can support the unqualified test result/interpretation. Support from physical exam, clinical signs, health history, etc can also be used to support unqualified test results.

These tools provided by Dr. Westgard and the AAVLD accreditation committee provide models for thinking about the complexity of approving the use of a test method for routine service as a diagnostic test. While it has long been tempting to consider a laboratory a 'black box' that releases numbers on command, it becomes apparent upon closer look, that much optimization and verification work is required before a test becomes routine- even in cases where commercial kits are readily available for transferring to veterinary applications.

CONCLUSIONS

Currently, veterinary laboratories operate under fewer regulatory controls than those providing service to human health providers. However, that does not mean that veterinary laboratories are any less focused on providing accurate and reliable results. Laboratory commitment to quality will be evidenced by the presence of clearly articulated procedures and policies on quality critical features such as test method validation. Veterinary clinicians have the right and the duty to their clients to form a partnership with the laboratory professionals they utilize for laboratory work. Laboratory professionals are equally bound to share their knowledge related to the validity and limitations of the tests they perform with the clients who use the data they generate. Working together, veterinary health professionals can be confident that laboratory data is being generated using best practices, using appropriate methods and within a laboratory culture of quality.

REFERENCES

[1] US Department of Health and Human Services. Medicare, Medicaid, and CLIA programs regulations implementing the Clinical Laboratory Improvement Amendments of 1988 (CLIA).Fed Regist 1991;57:7003.[2] Office International de Epizooties (OIE). Manual of Standards for Diagnostic Tests and Vaccines for Terrestrial Animals, 5th ed. Available at www.oie.int/fr/normes/mmanual/A_summary.htm (accessed 1/30/2011). [3]American Association of Veterinary Laboratory Diagnosticians, AAVLD Accreditation Committee Guidance Document. 8/27/2008. www.aavld.org Accessed 1/20/11. [4] Federal Drug Administration. CFR: Code of Federal Regulation; Title 21, Part 58. www. accessdata.fda.gov/scripts/cdrh/cfdocs/cfcfr/CFRSearch.cfm?CFRPart=58; accessed 2/1/11. [5]Flatland B, et al. Vet Clin Path 2010;39(3)264. [6]Westgard JO. Basic Method validation, 3rd Edition. Madison, WI :Westgard QC, Inc. 2008.66.

PHARMACEUTICAL OPTIONS FOR CONTROL OF OBESITY IN MAMMALS

Kirstin Barnhart, DVM, PhD, DACVP (Clinical Pathology)
Bastrop, TX

PHARMACEUTICAL DEVELOPMENT STRATEGIES FOR HUMANS AND ANIMALS

Although many different strategies have been employed to develop an anti-obesity drug, no single mechanism has yet stood out at as the ideal way to induce weight loss. Most of the drugs currently being developed cause weight loss through two basic mechanisms: reducing energy intake (through decreased consumption or interference with nutrient absorption) and increasing energy expenditure. More recently, new therapeutic strategies aimed at reducing or redistributing adipose tissue have been described[1,2].

REDUCING ENERGY INTAKE

A majority of the drugs in development fall into this category, as it is typically much easier to reduce intake than increase expenditure[1]. This broad class may be subdivided into drugs that act centrally, peripherally or through a combination of both mechanisms.

Centrally Acting

To date, few centrally acting compounds have successfully made it into human clinical trials. These drugs reduce appetite or induce satiety through modulation of neuropeptides and cell signaling within the central nervous system. The most widely explored group of centrally acting compounds are those that act on central neuropeptide targets by modulating the local availability of catecholaminergic noradrenaline, dopamine, 5-hydroxytryptamine (serotonin) or a combination of these neurotransmitters[3].

Sibutramine (Meridia®) is a weak noradrenergic and serotoninergic reuptake inhibitor that reduces total daily food intake and alters appetite in humans. The Sibutramine Trial of Obesity Reduction and Maintenance documented weight loss greater than 5% in 77% of patients that received the drug for 6 months in addition to an individualized nutrition program[4]. Adverse side effects have typically included increased heart rate and blood pressure, and Sibutramine was suspended from the European market by the European Medicines Agency in January 2010 due to safety concerns. Although still available in the US, the FDA issued a safety notification and required a new contraindication be associated with the drug[2].

Contrave, developed by Orexigen Therapeutics, Inc., is an oral, sustained-release, fixed-dose combination of naltrexone and bupropion. Bupropion is a dopamine and norepinephrine reuptake antagonist and naltrexone is an opioid antagonist. In combination, these drugs initiate weight loss by reducing appetite and increasing metabolism, allowing the body to sustain continued weight loss by offsetting its natural tendency to fight back and slowing down the weight loss process, and addressing the reward system in the brain that causes food cravings[5]. In December 2010, an FDA advisory committee voted 13-7 that the modest weight-loss benefits of Contrave outweighed the drug's blood pressure risk; however, on February 1, 2011, the FDA declined to approve the drug until additional trials have been completed to demonstrate that the benefits are not eventually overcome by the cardiovascular risks.

Lorcaserin was developed by Arena Pharmaceuticals. It selectively targets serotonin receptors associated with satiety. Although many human patients enrolled in clinical trials lost >5% or >10% of their body weight after one year of treatment, greater than 50% failed to lose at least 5% may have gained weight[6]. In October 2010, the FDA failed to approve Lorcaserin due to concerns that high doses of the drug may cause tumor formation in rats. Independent reviews and potentially additional studies may be required to alleviate the FDA's safety concerns.

Phentermine is an appetite suppressant of the amphetamine and phenethylamine class that stimulates the release of norepinephrine and dopamine. Phentermine is **currently approved for short-term (generally interpreted as 'up to 12 weeks') weight loss therapy only**[2]. In addition, because it is related to amphetamines, it is classified as a Schedule IV controlled substance. Previously, phentermine was combined with fenfluramine and dexfenfluramine (two drugs which increase serotonin levels) to create a drug termed Fen-Phen, which was taken off the market in 1997 due to serious concerns over its adverse effects that included heart valve disease and primary pulmonary hypertension.

Qnexa, developed by Vivus, Inc., was the first obesity drug submitted to the FDA in a decade. It is a combination of phentermine and topiramate, an anti-convulsant that can be used in conjunction with other drugs for weight loss. The mechanisms by which topiramate induces weight loss are currently not well understood, but the combination of the two drugs in a daily slow-release formulation resulted in a mean weight loss of 14.4% in people receiving treatment for approximately one year. Unfortunately, due to safety concerns over the potential to cause birth defects and heart problems, the FDA did not approve Qnexa in October, 2010 and asked for additional study results and more information on risks[2].

One centrally acting system that has been targeted extensively by drug companies in recent years is the endocannabinoid system. Endocannabinoids (ECB) are lipids derived from arachidonic acid. They are capable of binding to cannabinoid receptors located in the CNS and in some peripheral tissues[7]. ECBs regulate energy uptake and storage and non-homeostatic eating behaviors[2]. Cannabinoid receptor antagonists inhibit neurotransmitter release from neurons that regulate expression of hypothalamic anorexigenic and orexigenic mediators[7]. Several large pharmaceutical companies have aggressively pursued cannabinoid receptor antagonists for treatment of obesity. Unfortunately, serious safety concerns by US and European regulatory agencies have hampered development of ECBs based drugs in recent years[8].

Rimonabant (Acomplia®), developed by Sanofi-Aventis, is a selective CB1 endocannabinoid receptor antagonist that works by blocking endogenous cannabinoid binding to neuronal CB1 receptors. Prevention of receptor binding removes a potent appetite stimulus. Although available for use in Europe since 2006, Rimonabant failed to secure FDA approval in the US based on the increased risk of psychiatric adverse events, including suicide. In October 2008, the EMEA also recommended suspension of the drug due to safety concerns based on data that demonstrated a doubling of the risk of psychiatric disorders in patients taking Rimonabant[9].

Other centrally acting drugs that are currently being investigated include dopaminergic ligands, melanocortin 4 receptor agonists, melanin-concentrating hormone receptor antagonists, histamine receptor antagonists, neuropeptide Y antagonists, and agouti-related protein inhibitors[2].

Peripherally Acting

These drugs target obesity by reducing calorie absorption in the gastrointestinal system, affecting control systems outside the CNS or directly targeting white fat[3,10]. Two major strategies that have been employed for reducing calorie absorption are the inhibition of pancreatic lipases and intestinal microsomal triglyceride transfer proteins. Pancreatic lipases are responsible for breaking down triglycerides in the intestine. When the lipase activity is inhibited, triglycerides are not hydrolyzed into absorbable free fatty acids and excreted in an undigested form.

Orlistat (Xenical®, Alli®) is **currently the only compound approved both for short-term and long-term treatment of weight loss in the US**. In addition, it is the first and only over-the-counter product for long-term weight loss. Orlistat is a saturated derivative of lipstatin, which is an inhibitor of pancreatic lipases. It functions by preventing the breakup and absorption of fat. Orlistat blocks absorption of about 25% of the fat in a meal and is intended for use in conjunction with a reduced-calorie diet. The effectiveness of orlistat in promoting weight loss is modest. In a meta-analysis of 11 placebo-controlled trials, the number of patients that lost 5% and 10% placebo-subtracted weight loss after one year of therapy was 21% and 12%, respectively[11]. Wide-spread use of this drug has been deterred by the gastrointestinal side effects which include steatorrhea, flatulence and urgent bowel movements.

MTP is a chaperone for apolipoprotein B assembly. Small molecules that inhibit the action of this protein reduce plasma lipids. MTP inhibitors have been identified that act predominantly on the enterocytic form of MTP[2]

Dirlotapide (Slentrol) was initially developed by Pfizer as a weight loss drug for humans, and is **the only drug currently approved for treating obesity in dogs in the United States**. Dirlotapide is a selective MTP inhibitor that reduces fat absorption by blocking the assembly and release of lipoproteins into the bloodstream, a process driven by MTP activity in the cytoplasm of the enterocyte. It also stimulates the release of peptide YY from the enterocyte. Peptide YY is a potent appetite suppressant and responsible for signaling the brain to control intake. This effect on appetite is the primary weight loss mechanism for dirlotapide[12]. In canine clinical trials, dirlotapide generally induced about a 10% reduction in food consumption[13]. With this drug, the fat is absorbed into the enterocyte; thus, steatorrhea and other undesirable GI side effects that complicate therapy with pancreatic lipase inhibitors (e.g. Orlistat) are frequently avoided or less severe. The most common side effects of Slentrol include vomiting, diarrhea or loose stools, lethargy, dehydration, increased salivation and loss of coordination. Following initiation of therapy with Slentrol, the dog should be weighed monthly so that the drug dosage can be altered appropriately based on the current weight of the animal. Once the dog has reached its ideal weight, a dosage appropriate for this ideal weight will be administered for three months post-treatment to maintain the weight of the dog. Slentrol dirlotapide is supplied as an oral solution at a concentration of 5 mg/ml in quantities of 20 mL, 50 mL, and 150 mL bottles with a pre-calibrated oral syringe for administering the medication. Slentrol may be administered with or without food. It is contraindicated in dogs with pre-existing conditions such as liver disease and renal disease and should not be administered for more than 1 year.

Mitratapide (Yarvitan®,5 mg/ml, Janssen Pharmaceutica NV, Beerse, Belgium) is **approved for use in dogs in the European Union** but is not currently approved in the US. Yarvitan contains mitratapide, a selective microsomal triglyceride transfer protein inhibitors. Similar to Slentrol, Yarvitan induces weight loss through the combined mechanisms of decreased dietary uptake of lipids and decreased appetite. In a recent study, dogs were treated at the therapeutic daily dose of 0.63 mg/kg/day for eight weeks. Dogs received the drug daily for 3 weeks followed by an intermediate period of 2 weeks and a final dosing period of 3 weeks. The investigators concluded that on average the dogs lost 14.2% of body weight and 41.6% of their body fat mass immediately following the 8 week course of therapy[14].

Many compounds that function by modulating the circulation of peptide hormones are currently being investigated as drugs that are capable of acting peripherally to reduce appetite. These drugs are in various stages of development. Witkamp (2010) recently provided a review the basic mechanisms of these hormones and the current strategies that pharmaceutical companies are employing to develop effect anti-obesity drugs. A brief summary of the review of these hormones is provided here[2].

The candidate drugs include amylin, peptide YY, pancreatic polypeptide, glucagon-like peptide 1 (GLP-1) and leptin analogues, oxyntomodulin, cholecystokinin-1 agonists, and ghrelin activity blockers. Amylin is co-released with insulin by the pancreas. **Pramlinitide (Symlin®)** is available as an antihyperglycemic drug and is being developed in conjunction with a leptin analogue for obesity. Both peptide YY and pancreatic polypeptide function primarily to induce satiety. Currently, little work is being done to develop peptide YY analogues, but a pancreatic polypeptide analogue had notable weight loss effects in diet-induced obese mice and is in early clinical trials. GLP-1 is currently the most promising GI hormone for pharmacologic weight loss. GLP-1 binding to its receptor appears to induce satiety and causes a dose-dependent reduction of food intake in experimental animals. Several drugs are currently being tested in humans. Oxyntomodulin functions similarly to GLP-1and also causes reduced food intake and reduced weight gain in rodents. Cholecystokinin (CCK) is produced in the small intestine shortly after food ingestion and initiates satiety through binding to CCK receptors. Little success has been achieved in translating the mechanism of this analogue into an anti-obesity therapy. Ghrelin is released from the oxyntic cells in the stomach. It has been shown to increase food consumption in rodents and humans. Blocking the activity of this polypeptide decreases food consumption. Ghrelin vacation has had some promising results in animals.

INCREASING ENERGY EXPENDITURE

Few candidate compounds have been developed in this category. Some drugs such as sibutramine may have secondary thermogenic effects that contribute to weight loss. Selective thyroid hormone mimetics have some potential as an anti-obesity strategy through their ability to stimulate adaptive thermogenesis and lipolysis. One novel compound derived from growth hormone to stimulate fat metabolism is currently in human clinical trials[1]. Recently, data was reported that contradicted the previous assumptions that adult humans have little active brown adipose tissue[15]. This discovery has stimulated interest in manipulating the energy expenditure associate with brown fat as a method for achieving weight loss.

REDUCING OR REDISTRIBUTING ADIPOSE TISSUE

This approach may be particularly useful in converting toxic visceral fat into less risk-bearing fat depots. Because adipose tissue is highly vascularized, some investigators are looking at the feasibility of reducing fat mass through angiogenesis inhibitors. Rupnick et al. demonstrated significant weight loss in mice from different obesity models that received antiangiogenic agents[16]. Marked weight loss in obese mice was also achieved by researchers who employed a novel strategy of targeted ablation of the vasculature supplying white adipose tissue. The compound in this study contained a peptide motif that bound to prohibitin, a multifunctional membrane protein found in the vessels of white fat. When linked to a proapoptotic peptide, the compound caused ablation of white fat and rapid reversal of obesity. Obese mice receiving a daily subcutaneous injection for 28 consecutive days lost an average of 15 grams (greater than 30%) of weight[10].

THE FUTURE OF PHARMACEUTICAL TREATMENT OF WEIGHT LOSS IN ANIMALS

The FDA criteria for approval of anti-obesity drugs stipulate two important treatment outcomes: 1) at least 35% of the treated individuals must lose at least 5% of their body weight and 2) the number of people who lose weight must be at least twice as great as the number of people who lose weight in the placebo group[2]. If these criteria were also applied to a potential anti-obesity drug for companion animals, a veterinarian could expect the following outcome. Given a group of 20 client-owned cats that were substantially overweight at 20 pounds, on average 7/20 cats would lose 0.5 pounds over the prescribed duration of therapy (typically 6-12 months). Given the current costs for research and development of novel drugs, the cost of new anti-obesity compounds is likely to be high and with the relatively low odds of having a modest degree of weight loss, the applicability of these compounds to companion animal medicine unfortunately may be limited. An additional consideration as to the potential for wide-spread use of anti-obesity drugs in animals is that once therapy is initiated, a long-term or indefinite treatment course must be considered as in most cases, the weight is regained following discontinuation of the drug therapy[11].

REFERENCES

[1]Halford JCG. **Curr Op Invest Drug** 2006; 7:312-318. [2]Witkamp RF. **Pharm Res** 2010; epub online Dec 23. [3]Elangbam CS, et al. **Vet Pathol** 2009; 46:10-24 [4]Idelevich E, et al. **Ther Adv Cardiovasc Dis** 2009; 3:75-90. [5]Padwal R. **Curr Op Invest Drug** 2009; 10:1117-1125. [6]Halford JCG. **Nat Rev** 2010; 6:663-664. [7]Cooke D Bloom S. **Nat Rev** 2006; 5:919-93.1 [8]Le Foll, et al. **Psychopharmacol** 2009; 205:171-174. [9]Christopoulou FD and Kiortsis DN. **J Clin Pharm Ther** 2011; 36:10-18. [10]Kolonin MG, et al. **Nat Med** 2004; 10:625-632. [11]Padwal RS and Majumdar SR. **Lancet** 2007; 369:71-77. [12]Zoran DL. **Vet Clin Small Anim** 2010; 40:221-239. [13]German AJ, et al. **J Vet Intern Med** 2007; 21:1174-1180. [14]Dobenecker B, et al. **Vet Res Commun** 2009; 33:839-847. [15]Cypress AM. **N Engl J Med** 2009; 360:1509-1517. [16]Rupnick MA, et al. **PNAS** 2002; 99:10730-10735.

A NOVEL THERAPEUTIC APPROACH TO WEIGHT LOSS THROUGH VASCULAR TARGETING IN LABORATORY ANIMALS

Kirstin Barnhart, DVM PhD DACVP (Clinical Pathology)
Bastrop, TX

BACKGROUND

Obesity has reached epidemic proportions in the United States and some projection models show that by the year 2030, 86.3% of all American adults will be overweight and 51.1% will be obese[1]. A lack of sufficient exercise and over-consumption of calorically dense diets have contributed heavily to the high prevalence of obesity. Many companion animals are also being adversely effected by too much food in the face of limited exercise. Studies from around the world have estimated the incidence of obesity in dogs at 22-40%. A recent study form Australia classified 33.5% of the dogs overweight and 7.6% obese. Similar trends have been identified for cats[2,3]. Despite these alarming statistics, no effective weight loss drug has been developed. Side effects and safety concerns have plagued the development of most anti-obesity agents which typically alter energy balance pathways and appetite through modulation of receptors in the brain. Currently, only a single drug each is licensed for long-term treatment of obesity in humans (Xenical®, Roche) and dogs (Slentrol®, Pfizer Animal Health). With both of these drugs, the degree of weight loss achieved within 6 months to a year is reported to be modest.

The targeted therapeutic approach described in this report was discovered through *in vivo* phage display in mice and subsequently developed as a novel anti-obesity strategy to hopefully overcome some of the limitations of other classes of weight loss drugs. The compound, termed Prohibitin.TP01, causes a significant reduction in adipose tissue mass by binding to a specific receptor in the vasculature of white fat and subsequently inducing apoptosis[4].

IN VIVO PHAGE DISPLAY

Phage display is a powerful technique that increasingly is being used to identify potential targets for cancer and other diseases that can be used to develop therapeutic applications. In this technology, viral phage particles express exogenous peptides and present them to various target molecules. The phage particles are derived from filamentous bacteriophages, most commonly M13, that are capable of infecting bacteria. Once inside the host cell, the single stranded viral DNA is converted to a double-stranded, replicative form for the production of viral coat proteins and single stranded DNA progeny. As the viral coat proteins are extruded from the host cell, they encapsulate the DNA. This process of replication can be manipulated effectively to allow insertion of foreign DNA into the genome; thus, modifying the phage coat proteins to display foreign entities as fusion proteins without loss of infectivity[5].

Phage display libraries have been created in which as many as 10^9 small random peptides are displayed on the surface of phage particles. The libraries can then be used to screen target molecules through a technique called biopanning. The basic steps in this procedure include: 1) exposure of the target molecule to the phage library, 2) removal of unbound phage, 3) elution of bound phage and propagation by bacterial infection, 4) successive rounds of biopanning to enrich for targets that bind to the peptide, and 5) determining the DNA sequences of the individual peptides that bind to the target molecule[4,6].

Although phage display was initially intended for *in vitro* screening, Pasqualini and Ruoslahti reported in 1996 that it could be adapted successfully to *in vivo* selection by injecting phage libraries intravenously into mice and subsequently rescuing the phage from individual organs. In this initial work, peptides were isolated that homed to selective vascular beds[7]. Currently, in vivo phage display has been used to identify many organ- or tumor-specific peptides in mice and humans[8,9]. Once the sequence of the peptide ligand and its receptor have been identified, this information can be applied to different therapeutic targeting strategies. For example, phage-derived peptides can be used to target gene therapy vectors to specific receptors. Targeting has the advantages of increased therapeutic efficacy and fewer unintended side effects[10]. Phage derived peptides may also be used to target proapoptotic peptides, cytokines or antisense oligonucleotides to the vasculature of various tumors. In mice, this targeting has led to the selective destruction of the tumor vasculature followed by tumor regression[11-13].

DEVELOPMENT AND EFFICACY OF PROHIBITIN.TP01

Targeting Ligand Discovery Through *In vivo* Phage Display

The conceptualization of Prohibitin.TP01 began with the identification of a peptide motif (CKGGRAKDC) that specifically targeted the vasculature of white adipose tissue in obese leptin-deficient ($Lep^{ob/ob}$) mice by binding to prohibitin, a highly conserved protein involved in mitochondrial biogenesis[14,15]. The identification and validation of this targeting moiety was published by Kolonin, et al in 2004, and a summary of the results are provided here[15]. CKGGRAKDC was initially identified following four rounds of *in vivo* selection followed by a fat-specific *in vivo* subtraction step to select only ligands binding specifically to the endothelial cells. CKGGRAKDC accumulated 150-fold over background when injected back into mice and quantitated through standard phage counting techniques. Immunohistochemistry confirmed the specificity of the peptide for the vasculature of subcutaneous and peritoneal white fat. These procedures were repeated in wild-type mice, where it was demonstrated that CKGGRAKDC linked to FITC was specifically localized to and internalized by the blood vessels in the subcutaneous and peritoneal fat. In neither of the rodent studies, did the CKGGRAKDC peptide demonstrate a preference for white fat in a specific location. Based on the high level of both tropism and internalization, it was hypothesized that this peptide could function in the targeted delivery of therapeutic compounds specific to white fat. Through affinity chromatography, the vascular receptor for CKGGRAKDC was identified as prohibitin. An affinity-purified polyclonal antibody was subsequently used to further define the expression of prohibitin. Consistent with the expression pattern for CKGGRAKDC, prohibitin was highly expressed in the blood vessels of white fat and not in blood vessels of control tissues.

Apoptosis in the Blood Vessels of White Fat

The second major step in development was to determine if the CKGGRAKDC peptide could carry an apoptotic payload to the endothelial cells for subsequent internalization and cell death. The peptide sequence that was selected $(KLAKLAK)_2$ has been shown previously to induce apoptosis in tumor blood vessels by disrupting mitochondrial membranes following receptor-mediated cell internalization[11]. To determine if the combination of these two peptides (termed Prohibitin.TP01) could successfully ablate white fat, a non-genetically manipulated mouse model of obesity was used. Prior to administration of Prohibitin.TP01, mice were fed a high-calorie diet to induce obesity. Mice remained on this high-calorie diet throughout the dosing period. Mice received Prohibitin.TP01 by subcutaneous injection daily for four weeks. During this time rapid weight loss occurred. Mice not only displayed a reduction in body fat content (epididymal fat pad decreased by 70%) but lost on average 30% of their body weight. TUNEL immunohistochemistry confirmed apoptosis in fat blood vessels.

Other Physiologic Effects of Prohibitin.TP01

Other physiologic effects of the treatment were demonstrated. Histological assessment and lipid quantitation of the liver confirmed reversal of hepatic steatosis in treated animals. Additional paired feeding experiments in which both control and treated mice consumed a fixed amount of calories consistently displayed a strong weight loss trend in treated mice. This finding indicated that weight loss did not occur secondary to treatment induced semistarvation. The fecal lipid count was not different between control and treated mice and serum ketone bodies were not elevated. Consistent with a decrease in the mass of adipose tissue, leptin levels were decreased in treated mice. Prohibitin.TP01 was administered to obese mice with insulin resistance and glucose tolerance as a result of consuming a high fat diet. A glucose tolerance test demonstrated an improved insulin sensitivity and glucose tolerance in the treated mice.

INDEPENDENT CONIRMATION OF THE EFFICACY OF PROHIBITIN.TP01 IN MICE AND RATS

In 2010, Kim et al confirmed this dramatic weight loss through independent experiments in C57Bl/6 mice and Long-Evans rats made obese on a high fat diet (40% of calories provided in the form of fat). The study confirmed that the fat loss occurred primarily due to decreased food consumption and without an increase in energy expenditure. In addition, the weight loss occurred despite reduced leptin and hypothalamic POMC gene expression. Normally, appetite is suppressed through increased levels of these two anorexic molecules. Since weight loss occurred in the face of decreased levels, it can be concluded that the mechanism of weight loss is independent of the action of leptin and POMC expression[16].

TRANSLATION TO LARGE ANIMAL MODELS

Although compounds that induce significant weight loss in rodents are frequently reported in the literature, one of the major obstacles to converting these discoveries into meaningful anti-obesity drugs for humans is the failure of rodents and on occasion some large animal models to accurately model the metabolic conditions associated with weight gain and weight loss. For example, the majority of our theories regarding adipokines are derived from rodent studies; however, follow-up studies to confirm the function of specific adipokines such as adipsin, resistin and TNF-α in man have generated contrasting results[17]. Additionally, studies indicate that lipolysis in human white fat cells differs not only from rodents but also from rabbits and dogs; thus, the success of a candidate weight loss drugs in non-human primate models may be essential for predicting the success of anti-obesity strategies in man[18,19].

ON-GOING EFFICACY ASSESSMENT OF PROHIBITIN.TP01 IN NON-HUMAN PRIMATE MODELS

To date, controlled studies are on-going in baboons, cynomolgus macaques and rhesus macaques to determine the safety and efficacy of Prohibitin.TP01 in non-human primates. Preliminary efficacy studies in spontaneously obese baboons and rhesus monkeys receiving 28 consecutive daily subcutaneous injections of Prohibitin.TP01 have demonstrated marked weight loss, although not to the same magnitude as seen in mice. In addition, intravenous glucose tolerance testing in rhesus monkeys indicates a notable improvement in insulin sensitivity following treatment with Prohibitin.TP01. Data generated from these studies will be presented, in part, during the oral session and reported completely in future publications.

REFERENCES

[1]Wang Y, et al. **Obesity** 2008; 16:2323-2330. [2]German AJ. **J Nutr** 2006; 136:1940S-1946S. [3]McGreevey PD, et al. **Vet Rec** 2005; 156:695-707. [4]Kolonin M, et al. **Curr Op Chem Biol** 2001; 5:308-313 [5]Pande J, et al. **Biotechnol Adv** 2010; 28:849-858. [6]Zurita AJ, et al. **J Control Rel** 2003; 91:183-186. [7]Pasqualini R and Ruoslahti E. **Nature** 1996; 380:364-366. [8]Arap W, et al. **Nat Med** 2002; 8:121-127. [9]Hajitou A, et al. **Trends Cardiovasc Med** 2006; 16:80-88. [10]Trepel M, et al. **Meth Enzymol** 2008; 445:83-106. [11]Ellerby HM, et al. **Nat Med** 1999; 5:1032-1038. [12]Arap MA, et al. **Cancer Cell** 2004; 6:275-284. [13]Zurita AJ, et al. **Cancer Res** 2004; 64:435-439. [14]Vessal, et al. **FEBS J** 2006; 273:568-576. [15]Kolonin MG, et al. **Nat Med** 2004; 10:625-632. [16]Kim DH, et al. **Diabetes** 2010; 59:907-915. [17]Arner P. **Diabetologia** 2005; 48:2203-2205. [18]Sengenes C, et al. **Am J Physiol Regulatory Integrative Comp Physiol** 2002; 282:R257-R265.[19]Bousquet-Melou, et al. **J Lipid Res** 1995; 36:451-461.

THE YIN AND YANG OF PLATELETS: THE BALANCE BETWEEN THROMBOSIS AND HEMOSTASIS

Fern Tablin, VMD, PhD

Platelets have to be able to travel through the blood in near proximity to the vessel wall without platelet activation but yet be able to recognize vascular defects requiring them to stop, adhere to the basal lamina and to each other, to form a thrombus and have that thrombus remain stable until it is no longer needed. In a normal blood vessel, the endothelium plays a clear role in maintaining resting platelets through the production of PGI2, nitric oxide and CD39 (an ecto-ADPase), all of which participate in inhibition of platelet function. Additionally the endothelium promotes activation of protein C, which has been shown to inhibit thrombin formation by cleavage of factors Va and VIIIa.

Platelet Activation can be thought of as being differentially spatially and temporally regulated in the venous versus the arterial systems. Venous system, with low pressure and flow rates allows for local accumulations of coagulation factors and the generation of thrombin. However, it should be noted that these thrombi contain greater numbers of red blood cells than platelets. On the other hand, the arterial system consists of high pressure and high flow rates, thus soluble clotting factors do not remain locally. This requires that platelets act as the cellular (physical) barrier to protect against blood loss due to damaged endothelial cells and must generate thrombin required for normal clot formation and fibrin accumulation. In either the venous or arterial system, platelet activation has to be highly regulated, as there is a distinct balancing act between limiting blood loss and the formation of occlusive thrombi.

Hemostatic response to injury consists of a series of well-orchestrated events involving activation by sub-endothelial collagen and the production of thrombin. The earliest events occur when damage to the endothelium results in exposure of sub-endothelial collagen, which is complexed with vonWillebrand factor (vWF). Circulating platelets are captured to the site of damage, through the interaction of the Glycoprotein Ib-V-IX complex, particularly GPIb, move bound platelets out of the mainstream of the circulation and enable further activation to occur. This action occurs largely through the signaling of downstream receptors for collagen –Glycoprotein VI and GPIb□., as well as by protease activated thrombin receptors (PARs 1, 4 in humans and Par 3 and 4 in mice), adenosine diphosphate ($P2Y_1$, $P2Y_{12}$) and thromboxane A2 (TxA_2). Damaged red cells assist in the process by release of ADP which participates in the feedback loop ,whereby activation of these receptors generates inside out signaling which eventually upregulates the integrin □2b□3a resulting in its binding to it's ligand fibrinogen.

PLATELET ACTIVATION HAS THREE WELL RECOGNIZED STAGES: INITIATION, EXTENSION AND STABILIZATION

Initiation

This process occurs when there is interaction of platelets with agonist, generally collagen or thrombi, and results in platelets forming a distinct monolayer and recruiting other platelets to the monolayer thereby generating local platelet activation.

Extension

This occurs when additional platelets accumulate and activate at the damaged site. ADP is released from platelet dense granules and TxA_2 is synthesized from arachidonic acid generated from platelet membrane phospholipids. TxA_2 production is cyclooxygenase-1 (COX-1) dependent. Aspirin irreversibly inhibits COX-1 and most non-steroidal anti-inflammatories reversibly inhibit this enzyme. Platelet receptors for ADP (the purinergic receptors $P2Y_1$ and $P2Y_{12}$) and the thromboxane receptor can now bind these ligands and further activate the platelets. Activated platelets provide a pro-coagulant surface through the translocation of PS and its interaction with coagulation factors Va and Xa, the end result of which the generation of the prothrombinase complex and the cleave of prothrombin (factor II) to thrombin (factor IIa). This is the phase in platelet activation that has largely been targeted by old (aspirin, cox-1 inhibitors) and newer (clopidrogrel and prasugrel) drugs to inhibit platelet activation and inappropriate clotting.

Downstream from binding of these ligands to their respective receptors, is a relatively well characterized signaling cascade that starts at the platelet plasma membrane with activation of phospholipase C isoforms resulting in the cleavage of phosphoinositides and the production of IP3 and diacylglycerol. IP3 mediates the release of calcium from the dense tubular system, this rise further triggers the calcium TRP channels at the platelet plasma membrane resulting in a further increase in intracellular calcium which is required for the change in conformation of the integrin □2b□3a. This conformation change is required for the integrin to assume an active conformation in which it can bind its ligand fibrinogen, however, it also can bind fibrin, vWF and fibronectin.

Stabilization

Additionally inside-out signaling, the result of agonist induced activation of platelet membrane receptors results in the up-regulation of the integrin- the binding to its ligand the formation of platelet-platelet- interactions form through the integrin-fibrinogen linkages. Thrombin previously produced on the extracellular leaflet of the platelet plasma membrane, acts to cleave fibrinogen to fibrin further consolidating the platelet plug. Not only does the plug have to be stable, there has to be a system by which premature disaggregation of the plug is prevented. This occurs through integrins as well as through contact dependent signaling of ephrins and semaphorins. Interestingly this plug not only is able to seal the damaged vessel but now also provides a unique surface for the recruitment of leukocytes, the secretion of pro-inflammatory and pro-angiogenic molecules as well as the release of growth factors required for wound healing.

When considering the above model, the question becomes do all platelets within the developing thrombi react all at once or is there a spatial and temporal relationship to shear and shear gradients. Data from several laboratories (Nesbitt, Brass) indicate that while the platelet monolayer adherent to the site of injury maybe entirely activated those platelets which are more recently arrived at the thrombin can be substantially less activated (as measured by the expression of P-selectin on their surface). Complete activation with thrombin occurs through contact-dependent signaling between platelets. Recent studies have shown that ephrins and their Eph receptors are critical to the stabilization of clots associated with collagen activation. Additional studies have suggested that semaphorin Sema4D (CD100) is also critical to collagen induced activation.

Additional Events

During platelet activation there are additional events which have been shown to be critical to amplified platelet activation. Most recently molecules released by platelet shedding and secretion have become viewed not only as a critical part of the hemostatic system but also as a series of agents which release growth factors, promote wound healing and angiogenesis. Furthermore it has been shown that ecto-domain shedding occurs through the actions of matrix metalloproteinases. Cleavage of ecto-domains can result in the production of bioactive molecules and/or inhibitory molecules, which can also serve to modify platelet-platelet interactions. Ecto-domain shedding has been described both for platelet glycoprotein Iba and GPVI. Additionally platelet surface proteins including P-selectin, CD40L and sem4D can be proteolyzed and released into the circulation. The presence of soluble P-selectin and CD40L have been used clinically to track platelet activation in vivo.

Lastly, platelets are more than hemostatic sacs of tissue and have been shown to play major roles in inflammation, adaptive immunity, and embryonic development.

THE CANINE PLATELET: NEW DISCOVERIES AND NEW DIRECTIONS

Kari V. Lunsford, DVM, DACVIM
Starkville, MS

INTRODUCTION

There has been an increased interest in the physiology of the platelet over the last few decades as basic and clinical scientists have demonstrated the many roles the platelet plays in both health and disease. Once thought to be a simple hole plugging cell fragment, the platelet is now understood to play roles not only in primary hemostasis, but also in inflammatory and allergic conditions, several gastrointestinal and renal pathologies, some dermatological disorders, infectious disease and even in tumor metastasis. Important areas of study are related to the complex role the platelet plays in pathologic thromboembolic conditions and to efforts geared toward improving pharmacologic manipulation of platelet function. While other drugs have been introduced and advocated for use in impairing platelet aggregation and thus inhibiting platelet procoagulant activity, aspirin remains the most widely prescribed drug for platelet inhibition in people[1] and is widely used in dogs.

Thromboembolic complications have been linked to a number of conditions in dogs, including immune-mediated hemolytic anemia (IMHA), protein-losing nephropathy, and hyperadrenocorticism.[2,3] While the cause of the pro-thrombotic state is not clearly defined, the presence of activated, circulating platelets has been documented in dogs with IMHA.[4] In addition, studies have suggested improved survival in IMHA dogs treated with low dose aspirin. These results support the use of aspirin and it is commonly administered at low doses, based on the human model, to dogs with such conditions in an attempt to prevent thromboembolic complications. Many dogs, however, go on to develop thromboembolic complications despite aspirin therapy. This phenomenon has been described in human patients as well and failure of low-dose aspirin to prevent thrombus formation in people has been termed 'aspirin resistance'. [5-7]

Aspirin Resistance

Aspirin, or acetylsalicylic acid, acts to inhibit platelet function by irreversibly acetylating platelet cyclooxygenase (COX)-1 at serine residue 530. The COX enzyme is responsible for the conversion of arachidonic acid to the various eicosanoids, such as leukotrienes, prostaglandins, thromboxane A_2 (TXA_2) and prostacyclin (PGI_2). Thromboxane A_2 is a potent vasoconstrictor and platelet agonist found in the α-granules of the platelet, which are released during platelet activation. Production of TXA_2 has been completely inhibited by doses of aspirin as low as 100 mg in people and significantly reduced at doses as low as 60 mg.

A portion of human patients develop thromboembolic complications despite prophylactic treatment with standard anti-platelet 'low-dose' aspirin. Although the true incidence of aspirin resistance in people is not clearly defined, an estimated 8-45% of all patients appear to be poorly responsive to aspirin therapy.[5-7] The mechanism for aspirin resistance is currently unknown, but it is likely that multiple factors play a role and several theories have been suggested and evaluated. The most widely studied mechanisms for aspirin resistance in human platelets include increased platelet sensitivity to collagen, the platelet alloantigen 2 (PLA2) polymorphism of platelet glycoprotein IIIa, and increased COX-2 activity.[8]

The COX2 "Controversy"

COX-1 plays a role in the maintenance of normal cellular activities including platelet thromboxane A_2 production, maintenance of gastric mucosal protection, and preservation of renal blood flow. COX-1 is constitutively expressed in most body systems.[9] COX-2 has a more limited tissue distribution and is generally considered an inducible enzyme. COX-2 expression has been described in the brain, kidney, thymus and the vascular endothelium as well as circulating monocytes, tissue macrophages and fibroblasts. COX-2 is induced "as needed" by several endogenous and exogenous inflammatory mediators and plays a significant role in inflammatory responses. As mature anucleate platelets are incapable of mRNA synthesis it was long believed that COX-1 was the only COX isoform expressed in circulating platelets and was, therefore, the sole mediator of platelet thromboxane A_2 production. Recently, however, flow cytometry techniques have allowed the identification of both COX-1 and COX-2 expression in circulating human platelets.[8] COX-2 has also been identified in megakaryocytes in the bone marrow.[10] These recent discoveries suggest that COX-2 may in fact play a role in platelet function.

Pro-inflammatory mediators such as cytokines, endotoxins, and growth factors are known to induce COX-2 expression and, in many cells and tissues, expression increases by as much as 10-20 fold. Megakaryocytes have been shown to contain both COX-1 and COX-2,[10] and induction of COX-2 expression in megakaryocytes would be expected to result in similar changes in mature platelets derived from affected megakaryocytes. Although mature platelets are anucleate and therefore not thought to be capable of *de novo* protein synthesis, newly released platelets have been shown to contain mRNA derived from precursor megakaryocytes,[11] and to retain the intracellular molecular mechanisms necessary for protein production. Megakaryocyte-derived platelet mRNA does degrade over time, thus as platelets age, mRNA concentration is reduced.[12]

COX2 and Aspirin Resistance

One proposed mechanism for "aspirin resistance" in people is that platelet thromboxane A_2 synthesis may continue in treated patients via pathways that are independent of COX-1, and that such pathways may be mediated by COX-2. Platelets in some people are known to express relatively high levels of COX-2, and it is speculated that standard low-dose aspirin therapy may not provide sufficient inhibition of COX-2 mediated pathways of thromboxane A_2 synthesis in such patients.[8]

Although the true incidence of aspirin resistance in people is unknown, in some studies as many as 45% of all patients appear to be poorly responsive to aspirin therapy. Platelet COX-2, by providing an alternative pathway to COX-1 for thromboxane synthesis, has been proposed as one potentially important source of aspirin resistance. Although low dose aspirin has been shown to produce a greater than 95% suppression of COX1 function, clinical studies in people with unstable angina and acute coronary syndrome have revealed that approximately 20% of patients receiving aspirin have incomplete suppression of urine thromboxane metabolites suggesting the presence of an alternative pathway (such as COX-2) for thromboxane synthesis.[13] Because inflammation may increase platelet COX-2 expression, low-dose aspirin may be particularly ineffective at blocking COX-2 activity in patients with inflammatory diseases. Increased rates of platelet turnover may also contribute to aspirin resistance via a COX-2 mediated mechanism. One recent study found that while only about 10% of circulating human platelets from healthy volunteers express the COX-2 isoform, up to 60% of circulating platelets express COX-2 during conditions of increased platelet turnover and regeneration.[14,15] In times of high platelet turnover, it is hypothesized that COX-2 mediated pathways of thromboxane A_2 biosynthesis may contribute to aspirin resistance.[14] Similar mechanisms may explain the comparable 'aspirin resistance' phenomenon observed in some of our canine patients.

NEW DISCOVERIES

Our group has carried out a series of studies to further characterize aspirin resistance in dogs. Our goal was to evaluate the state of the COX enzymes in normal circulating platelets and shed some light on the role these enzymes may play in canine platelets in order to help guide therapy in our patients.

COX-2 in Canine Platelets

The presence of COX-2 in platelets has been an area of intense interest in human medicine, particularly with the recent concerns about hemostatic abnormalities observed in some patients receiving selective COX-2 inhibitors. This interest led to advances in flow cytometric methods which have demonstrated the presence of COX-2 in circulating human platelets. In the dog, previous attempts to demonstrate the presence of COX-2 in canine platelet utilizing other methods have not been successful in identifying the isoform. The purpose of our initial study was to utilize the flow cytometric methods described in the human literature to demonstrate COX-2 expression in the platelets of normal dogs. These methods were successful in detecting the expression of COX-2 in canine platelets in both platelet rich plasma and in whole blood. Thus this study demonstrated that flow cytometry can be used to detect canine platelet cyclooxygenase expression, and that circulating canine platelets do, in fact, express both COX-1 and COX-2.

High Dose Aspirin

As part of our evaluation of COX expression, we also assessed the effects of aspirin on platelet COX expression. In our initial study, we exposed healthy dogs to high dose aspirin (10 mg/kg BID) for a week and evaluated platelet COX expression before and during aspirin therapy. Platelet function (urinary thromboxane/creatinine ratio and PFA-100 closure times) was also evaluated before and during therapy. COX-1 expression increased in all dogs with exposure to aspirin and returned to baseline following a washout period. COX- 2 expression was decreased and then returned to baseline in the majority of the dogs, however an increase was observed 2 of the 8 dogs evaluated. PFA-100 ® closure times were increased and urinary thromboxane/creatinine ratios were decreased in all dogs. Taken together, the increased PFA-100® closure times and decreased urinary 11-dTXB_2 levels seen in the dogs in our study strongly suggest aspirin-induced inhibition of platelet cyclooxygenase function, a finding that is to be expected given the relatively high doses of aspirin used.

Low Dose Aspirin

As a follow up to the high dose aspirin study, we performed similar work in healthy dogs exposed to low dose aspirin therapy (1 mg/kg daily). The results for all parameters measured were more variable with low dose aspirin but overall, there was an increase in both COX-1 and COX-2 expression, a decrease in urinary thromboxane/creatinine ratio and an increase in PFA-100® closure times. It is interesting to note, however that while in the high dose aspirin study, all dogs PFA-100® closure times prolonged to at or near the 300 second upper limit of the test, with low dose aspirin approximately 1/3 of the dogs had only very minor increases in closure times. Similarly, several of the dogs had only very small decreases in urinary thromboxane excretion and two showed increased levels. The difficulty lies in determining which of these tests may be a more sensitive indicator of *in-vivo* platelet activity and response to aspirin in healthy dogs.

NEW DIRECTIONS

While our recent studies have shown that COX-2 is expressed in circulating canine platelets and that both COX-1 and COX-2 expression are affected by both high and low dose aspirin therapy, many questions remain. One difficulty lies in determining which of the tests of platelet function most accurately predicts clinical outcome. While all of the dogs in the high dose aspirin study showed platelet function inhibition based upon both PFA-100® closure times and urinary thromboxane/creatinine ratios, the results from the dogs in the low dose aspirin study were quite variable. Some dogs demonstrated increased PFA-100 closure times but minimal decreases or even increases in thromboxane production. While a discussion of the pros and cons of various platelet function tests is beyond the scope of this presentation and will be covered elsewhere, it is interesting to note these discordant results. This certainly raises questions regarding the interpretation of some of these tests in the clinical setting and more studies will certainly be required to address these issues.

The Role of Inflammation

Our initial studies in the dog have been carried out in healthy individuals; however the most striking findings relative to aspirin response and COX activity described in the human literature have come from work done in clinical patients, particularly those with atherosclerosis. This highlights the important role that inflammation likely plays not only in disease pathology but in response to therapy. COX-2 is up-regulated in the presence of both endogenous and exogenous inflammatory mediators. Earlier work in humans has demonstrated that aspirin at doses as low as 60 mg can inhibit 95% of COX-1 activity in normal human platelets. It has also been shown that doses of 1200 mg or greater are required to significantly impair COX -2 activity mediated prostanoid production in human patients.[16,17] In light of the discovery of COX-2 expression in circulating platelets, questions arise as to the effects of endogenous and exogenous pro-inflammatory mediators and COX expression and activity on platelet function. If COX-2 expression is increased in the circulating platelets of individuals with inflammatory conditions, will these patients respond differently to anti-platelet therapies? Does the ratio of COX-1 to COX-2 present in platelets affect the response of these platelets to aspirin therapy or to other drugs such as selective COX-2 inhibitors?

COX-2 and Aspirin

An additional mechanism of action for aspirin has been described relative to its activity on COX-2 particularly the vascular epithelium that is in contrast to other NSAID inhibitors of this isoform. When exposed to aspirin, the active site on the COX-2 enzyme is not completely inhibited, but rather enzyme activity is shifted toward the production of anti-inflammatory lipoxins by converting arachidonic acid to 15*R*-hydroxyeicosatetraenoic acid (15*R*-HETE) (54), which is rapidly metabolized by 5-LOX in activated neutrophils to produce the anti-inflammatory 15-epi-LxA4 or -LxB4 which plays an important role in the resolution of inflammation.[18] Whether platelet COX-2 plays a role in the production of these anti-inflammatory aspirin triggered lipoxins, or ATLs, through platelet –neutrophil interactions, and what impact these interactions may have on platelet function, remains to be determined.

Summary

In order to understand the impact of COX-2 expression in circulating canine platelets a thorough evaluation of platelet function and enzyme expression utilizing multiple measures of platelet activity needs to be carried out. An understanding of the effects that COX-2 enzyme products may have on platelet function and response to anti-platelet therapies such as aspirin, can only be arrived at through studies performed in the appropriate clinical context as the effects of anti-platelet therapy in healthy dogs may not be an accurate predictor of responses in clinical patients. Although the role, if any, of platelet expressed COX-2 in aspirin resistance has not been determined, an abstract recently presented at the International Society on Thrombosis and Haemostasis 22nd annual congress reported that platelet COX-2 expression is strong predictor of future cardiovascular/thromboembolic events in patients with acute coronary syndrome (ACS). It was also postulated that accelerated platelet turnover which results in increased levels of platelet COX-2 expression may significantly impact the prognosis for patients with ACS.[19] Additional studies are needed in veterinary medicine to determine whether inflammatory conditions such as IMHA lead to an increase in platelet COX-2 expression and whether increased COX-2 activity plays a role in reduced aspirin responsiveness or is in fact a predictor of thromboembolic complications.

REFERENCES

[1]Baigent C, et.al. **BMJ** 2002;324(7329):71. [2]Goggs R, et.al. **J Vet Emerg Crit Care** 2009;19:30. [3]Johnson L, et.al. **J Vet Intern Med** 1999;13:338. [4]Weiss D, et.al. **J Vet Intern Med** 2006;20(3):682. [5]Frelinger A, et al. **Circulation** 2006; 113: 2888. [6]Patrono C, Dalen J, et al. **Chest** 2001; 119: 39S.
[7]Kour D, et al. **New Horiz** 2006; 8: 116. [8]Weber A, et al. **Lancet** 1999; 353: 900. [9]Rocca B, et.al. **Int Immunopharmacol** 2002; 2: 603. [10]Tanaka N, et al. **Arterioscler Thromb Vasc Biol** 2004; 24: 607. [11] Booyse FM, et.al. **Biochem Biophys Acta** 1967; 145: 188. [12] Brogren H, et al **Blood**. 2004; 104 (13): 3943. [13]Vejar M, et.al. **Thromb Haemost** 1990;12;63(2):168. [14]Guthikonda S, et al. **J Thromb Haemost** 2007; 5: 490. [15] Rocca B, et al. **Proc Natl Acad Sci USA** 2002; 99: 7634. [16] Bucchi F, et.al. **Prostaglandins** 1986;32, 691. [17] Frolich J.C. **Trends in Pharmacological Sciences** 1997;18:30. [18]Rajakariar R, et. at. **Mol Interv** 2006;6(4):199. [19]Murisaki KM, et.al. **J Thromb Haemost** 2009;5(S2):Abstract number PP-WE-418.

APPLICATIONS OF PLATELET FUNCTION INHIBITORS IN VETERINARY MEDICINE

Stephanie A. Smith, DVM, MS, DACVIM
Urbana, IL

Veterinary patients may develop thrombosis or thromboembolism associated with a variety of disease conditions. For many decades, the primary anti-platelet agents available for clinical use in veterinary patients have been limited to aspirin. Controlled studies evaluating anti-platelet agents in veterinary patients are few and strong evidence of clinical efficacy is rare. Thromboprophylaxis is clearly indicated in dogs with immune-mediated hemolytic anemia (IMHA), since thrombosis or thromboembolism is a major cause of morbidity and mortality in dogs with this disease, and in cats for prevention of recurrent arterial thromboembolism. The optimal agents and dosing regimens have yet to be definitively determined for these conditions. Also not clear are the indications for thromboprophylaxis in patients with cardiac disease without a history of thrombosis, nephrotic syndrome, endocrinopathies, sepsis, and other potentially prothrombotic disorders.

ASPIRIN

Aspirin (acetylsalicylic acid) is one of the most widely used drugs in human and veterinary medicine due to its anti-inflammatory, anti-pyretic, and anti-platelet properties. Aspirin inhibits cyclo-oxygenase, reducing the synthesis of prostaglandins such as prostacyclin (PGI_2) from the endothelium, and thromboxane A_2 (TXA_2), produced primarily in platelets and the kidney. PGI_2 is a potent platelet inhibitor, while TXA_2 is a potent platelet agonist, so the balance of these two prostanoids is important.[1]

Low-dose aspirin (generally 1-2 mg/kg/day) tends to be preferred for long-term anti-platelet therapy in humans because gastrointestinal side effects are minimized, and because lower doses selectively inhibit the production of platelet TXA_2, while sparing endothelial PGI_2 synthesis. The selective inhibition occurs via two mechanisms: 1) endothelial cells can produce new cyclo-oxygenase, while platelets can not, and 2) low oral dosages expose platelets to aspirin in the portal vasculature, but systemic endothelial exposure is minimal due to high first pass hepatic metabolism. A large body of evidence indicates that low-dose aspirin therapy in humans reduces or prevents cardiovascular events in patients at risk, and that lack of platelet inhibition ("aspirin resistance") increases the risk of cardiovascular events.[1]

Animal Dosing Protocols

Dogs: The ideal aspirin protocol for prevention of TE in dogs is not known. Reported data varies regarding the appropriate dose to inhibit canine platelet activity *in vivo*. Anti-platelet activity was reported using *in vitro* methods (platelet aggregometry[2] or PFA-100®,[3-5] cyclo-oxygenase expression,[5] for aspirin given to healthy dogs at dosages of 8-20 mg/kg, but one report indicated that 10 mg/kg did not affect platelet function when measured by aggregometry.[6] Recent studies evaluating low doses have also reported variable results. Platelet aggregation was impacted in healthy dogs in response to 0.5 mg/kg in some studies,[2,7] but neither platelet-leucocyte interactions[7] nor platelet P-selectin expression[7] were affected. Another report indicated no effect on platelet function (measured by aggregometry or PFA100®) unless dogs received at least 1 mg/kg.[8] Doses of 1 mg/kg/day failed to consistently decrease thromboxane metabolites in the urine, suggesting inadequate *in vivo* anti-platelet effects of this dose in healthy dogs.[9] No study has reported the impact of anti-platelet therapy on platelet function in dogs with diseases predisposing to TE, where platelets may be hyper-reactive.[10] A single retrospective study reported that aspirin at 0.5 mg/kg/d PO was statistically associated with improved short-term outcome in dogs with IMHA,[11] but the study suffered from major limitations (due to the retrospective nature of the analysis) including temporal bias, unequal treatment group sizes, lack of randomization or blinding, non-standardized treatment protocols, and minimal patient follow up. Based solely on the aforementioned retrospective report,[11] aspirin at 0.5 mg/kg/d PO has become a mainstay for thromboprophylaxis in dogs with IMHA.[12] The wide adoption of this approach may be in large part a function of the minimal costs associated with low dose aspirin therapy, and the ease of use in any clinical setting. In contrast, a different retrospective comparison of survival to 6 months for dogs with IMHA (again with the inherent limitations of a retrospective analysis) indicated that those treated with aspirin at 0.5 mg/kg/d had poorer survival (40%) as compared to those receiving individually adjusted heparin therapy (88%).[13] A funded prospective, randomized, double-masked controlled clinical trial is ongoing that will compare outcome in dogs with IMHA receiving low dose aspirin to those receiving individually adjusted heparin therapy.

Cats: Thromboprophylaxis with aspirin has been commonly recommended for cats at risk for TE at a dose of 81 mg/cat PO q48-72h , but clinical evidence suggests that its efficacy for preventing TE is questionable.[14].Additionally, adverse effect rates are fairly high, with approximately 25% of cats experience limiting gastrointestinal signs.[15] One retrospective report of long-term therapy in 24 cats with previous arterial TE showed that aspirin at a dose of 5 mg/cat q72h was associated with similar or lower rates of TE recurrence when compared to other thromboprophylactic therapies, and adverse effects were rare.[15] No controlled studies have evaluated low dose aspirin in cats. It remains to be determined whether or not a lower dose approach would be of benefit.

ADP RECEPTOR ANTAGONISTS:

Although ADP is a weak platelet agonist, release of ADP from dense granules amplifies the response to other agonists. Three distinct ADP receptors have been described to occur on platelets. All three receptors interact with purine and pyrimidine nucleotides. The $P2Y_1$ and $P2Y_{12}$ receptors are G-protein coupled, while $P2X_1$ is an iontropic receptor. $P2Y_{12}$ has a more selective tissue distribution than $P2Y_1$, so it is an attractive target for pharmacologic inhibition of platelets.[16]

Thienopyridines

Ticlopidine, clopidogrel, and prasugrel are thienopyridine derivatives that act *in vivo* as specific antagonists of $P2Y_{12}$. They are prodrugs with no apparent anti-platelet activity. Metabolic steps that involve cytochrome P450-dependent pathways are required to generate the active metabolites. This need for metabolism causes a significant delay in the onset of antiplatelet effects. The active moiety is a reactive thiol derivative that binds irreversibly to the $P2Y_{12}$ receptor. These metabolites irreversibly and selectively inhibit ADP-dependent platelet aggregation and ADP-induced adenylyl cyclase downregulation. They also cause a dose-dependent reduction of the platelet binding sites for ADP.[16] Thienopyridines inhibit platelet aggregation as triggered by multiple agonists, inhibit shear-induced platelet aggregation, and cause platelet aggregates to be more susceptible to disaggregation. Other pharmacological effects that may contribute to the antithrombotic properties of these drugs include decreased circulating levels of fibrinogen, inhibition of erythrocyte aggregation, stimulation of nitric oxide production, inhibition of expression of tissue factor on endothelial cells, and inhibition of fibronectin synthesis.[16]

Thienopyridines are widely used in humans for the secondary prevention of major vascular events in patients with a history of cerebrovascular, coronary, or peripheral artery disease.[16] Note that all circumstances where thienopyridines are of proven clinical benefit are in the arterial setting.

Ticlopidine (Ticlid)

General PK/PD: Ticlopidine is rapidly absorbed after oral administration. Bioavailability is increased by administration with food and decreased by antacids. The anti-platelet effect takes 2 to 3 days to manifest. In humans steady state is not achieved until 14 days of repeated dosing, and elderly patients demonstrate a longer half-life and higher AUC for the drug. [16]

Specific animal dosing protocols:

Dog: Doses of 30 to 100[17-20] mg/kg/day inhibited *in vitro* platelet aggregation, and improved patency of vascular grafts.[17;18]

Cat: Dosages of up to 100 mg/cat/day failed to consistently alter platelet function. Higher dosages up to 500 mg/cat/day reduced *in vitro* platelet aggregation and impacted *in vivo* bleeding times, but caused dose limiting adverse effects.[21]

Adverse Effects: Most commonly reported adverse effects in humans are nausea, vomiting and diarrhea (up to 50% of patients). Rarer and more serious effects include minor or major bleeding (particularly when ticlopidine is combined with other anti-platelet or anticoagulant drugs), bone marrow suppression, thrombotic thrombocytopenic purpura (TTP), cholestatic changes, hepatotoxicity, rash, colitis, and arthritis.[16]

Clinical use:

Human: Large scale clinical trials have indicated that ticlopidine is effective in prevention of stroke, transient ischemic attacks, and myocardial infarction in a variety of clinical settings.[16]

Dog: Treatment with ticlopidine in dogs that were experimentally infected with heartworms resulted in less severe vascular lesions than observed in untreated dogs.[22] Ticlopidine decreased clot formation and improved dialysis performance when added to heparin in a canine hemodialysis model.[23]

Clopidogrel (Plavix)

General PK/PD: Clopidogrel is rapidly absorbed after oral administration. Bioavailability is not affected by food or antacids. In humans, approximately 90% of the drug is excreted intact (combination of urinary and fecal excretion). The anti-platelet effect takes 5 days to reach maximum effect.[16]

Specific animal dosing protocols:

Dog: 2 mg/kg/day failed to consistently impact platelet aggregation in one study, while 3 mg/kg/day achieved maximal (42%) platelet inhibition by day 5, and 4 mg/kg.day (50% inhibition) by day 3. If a loading dose of 10 mg/kg was given, platelet inhibition was achieved with a lower daily dose.[24]Another study indicated adequate platelet inhibition at either 2 or 4 mg/kg/day.[8] Rifampin coadministration (causing activation of the cytochrome P450 system) resulted in greater platelet inhibition at lower doses, and cimetidine coadministration (competivie inhibition of cytochrome P450 system) resulted in less platelet inhibition.[25]

Cat: Clopidogrel administered orally to normal cats at 18.75 to 75 mg/cat/day significantly decreased *in vitro* platelet aggregation in response to ADP and collagen, and significantly increased oral mucosal bleed time. The maximal effect was reached within 3 days of initiating the drug, and resolved within 7 days of discontinuing the drug. No adverse effects were noted.[26]

Horse: 2 mg/kg/day decreased platelet aggregation in normal horses.[27]

Adverse Effects: In general adverse effects are less severe than those observed with ticlopidine, but similar in nature. They include bleeding, gastrointestinal signs, apalstic anemia, and TTP in humans. VIN message boards include multiple postings regarding development of non-regenerative anemia in cats with long-term therapy.

Clinical Use:

Human: Clopidogrel has proved to be an effective antithrombotic for prevention of arterial thrombi in multiple large scale clinical trials. The drug is FDA approved and in wide use for thromboprophylaxis in patients with cerebrovascular disease, coronary artery disease, coronary artery stenting or percutaneous coronary intervention, and peripheral artery disease.

Dog: A small (n=12) cases series reported in abstract form comparing clopidogrel (10 mg/kg loading then 2 mg/kg/day) to low dose aspirin (0.5 mg/kg/day) failed to identify a difference in thrombosis rates or clinical outcome in dogs with IMHA.[28] A larger (n=24) prospective randomized study compared clopidogrel (10 mg/kg loading, then 2 mg/kg/day) to low dose aspirin (0.5 mg/kg/day) or clopidogrel with aspirin, for prevention of thrombosis in dogs with IMHA. This pilot project also failed to identify any difference in 90 day survival between groups.[29]

Cat: Clopidogrel is in fairly wide clinical use for prevention of arterial thromboembolism in cats at risk, without direct evidence (as yet) supporting efficacy. A multi-center, double-blinded prospective study comparing clopidogrel (18.75 mg/cat/day) to aspirin (81 mg/cat q3d) has been ongoing for years, without an indication of superiority for clopidogrel in preventing thrombosis. Enrollment is continuing with the hope that increased patient numbers will provide additional power.

Prasugrel (Effient)

General PK/PD: Prasugrel is 10 to 100 times more potent than ticlopidine and clopidogrel. The drug is rapidly absorbed after oral administration and converted to the active metabolite in the intestine.[30] It inhibits ADP-induced platelet aggregation more rapidly, more consistently, and to a greater extent than dose clopidogrel.[31]

Specific animal dosing protocols:

Dog: Oral administration of prasugrel to dogs (0.03-0.3 mg/kg/day) for 14 days resulted in potent, dose-related and cumulative inhibition of ADP-induced platelet aggregation.[32] The majority off drug is excreted unchanged in the feces in dogs.[33]

Adverse Effects: adverse events in humans were primarily bleeding, but thrombocytopenia and neutropenia also occurred.[31]
Clinical use in humans: prasugrel is approved for use in prevention of thrombosis associated with angioplasty. Phase III clinical trials are ongoing for prevention of arterial thrombosis in other clinical settings.

αIIbβ3 (GP IIb/IIIa) RECEPTOR ANTAGONISTS

The platelet glycoprotein (GP) receptor αIIbβ3 is an integrin, belonging to a family of transmembrane proteins that consist of 2 subunits. These receptors promote cellular adhesion and mediate cell-cell binding. Activation of platelet αIIbβ3 receptors is the final common pathway leading to platelet activation. Vessel damage, adhesion, and shear forces transform this receptor into a high affinity binding state that binds von Willebrand factor or fibrinogen, leading to platelet aggregation. These drugs are FDA approved for the prevention of myocardial infarction in humans undergoing percutaneous coronary interventions, with unstable angina, and certain types of acute coronary syndromes.[5,6] All three drugs detailed below are available for intravenous use only, and require continuous infusion due to relatively short half-lives. Extensive efforts to develop clinically safe and efficacious oral antagonists of the platelet αIIbβ3 receptor have not been successful, and have been generally been abandoned.[34] Clinical use of αIIbβ3 receptor antagonists have not been described in veterinary medicine.

Abciximab is an anti-integrin Fab fragment of a human-mouse chimeric monoclonal antibody against the human platelet αIIbβ3 receptor.

Specific Animal Dosing Protocols

Dog: Use of abciximab has been extensively described in canine experimental models of coronary artery stent thrombosis and acute myocardial infarction. Reported doses of 0.2-0.8 mg/kg resulted in >90% inhibition of platelet aggregation, and have been associated with thrombus dissolution.[35]

Cat: Abciximab has also been evaluated in a model of arterial injury in cats. Cats received either aspirin alone, or aspirin and abciximab. Cats in the aspirin and abciximab group showed significantly greater inhibition of platelet function and less thrombus formation than those receiving aspirin alone.[36]

Eptifibatide is a cyclic heptapeptide that acts by competitive inhibition due to affinity for the αIIbβ3 receptor.
Specific Animal Dosing Protocols:

Dog: It has been evaluated in experimental studies of canine models of coronary stent thrombosis. The dose protocol of 180 μg/kg q10m (X2) followed by 2 μg/kg/min CRI resulted in >90% inhibition of platelet aggregation. Other studies have reported doses of 0.5 mg/kg IV followed by 6 μg/kg/min CRI.

Cat: Eptifibatide also inhibits feline platelet aggregation *in vitro*. However, at doses required to maintain platelet inhibition in normal cats, the drug was associated with idiosyncratic and unpredictable circulatory failure and sudden death.[37]

Tirofiban is a peptidomimetic nonpeptide (tyrosine derivative) that also competitively inhibits the platelet αIIbβ3receptor.
Specific Animal Dosing Protocols:

Dog: It has been evaluated in experimental studies of canine models of coronary stent thrombosis. The dose protocol of 180 μg/kg q10m (X2) followed by 0.3 μg/kg/min CRI resulted in >90% inhibition of platelet aggregation.[38]

Cat: Use of tirofiban has also been described in an experimental model of cardiac ischemia in cats at a dose of 100 μg/kg IV followed by 5 μg/kg/min CRI. This dose prevented ischemia induced activation of feline platelets.

DIPYRIDAMOLE (Persantine, Aggrenox in combination with aspirin))

Dipyridamole inhibits the uptake of adenosine, which inhibits platelet aggregation, causes decreased density of thrombin receptors, inhibits phosphodiesterases causing potentiation of the nitric oxide system, stimulates PGI_2 production, releases tissue plasminogen activator, inhibits smooth muscle proliferation, and has antioxidant and anti-inflammatory properties. Use of dipyridamole improves ischemic tolerance. While dipyramidole has clear antithrombotic effects *in vivo*, it has only mild anti-platelet aggregation effects in traditional *in vitro* studies. It requires very low pH or absorption. It is eliminated by conjugation with glucuronic acid. The antithrombotic benefit of dipyridamole in humans is most evident when administered in conjunction with low-dose aspirin.

REFERENCES

[1]Awtry EH, et al. In: *Platelets.* 2nd ed. 2007;1099.[2].Brainard BM, et al. ***Am J Vet Res*** 2007;68(3):251.[3]Nielsen LA, et al. ***Vet Clin Pathol*** 2007;36(3):267.[4]Mischke R, et al. ***Vet J*** 2003;165(1):43.[5].Thomason, J et al. ***J Vet Int Med*** 2010;24:683(abstract).[6]Blois SL, et al. ***Am J Vet Res*** 2010;71(3):349.[7]Sharpe KS, et al. ***Am J Vet Res*** 2010;71(11):1294.[8]Shearer L et al. . ***J Vet Intern Med*** 2009;23:745(abstract).[9]Hoh CM, et al. ***Am J Vet Res*** In press.[10]Weiss DJ, et al. ***J Vet Intern Med*** 2006;20(3):682.[11]Weinkle TK, et al. **J Am Vet Med Assoc** 2005;226(11):1869.[12]Lunsford KV, et al. ***Vet Clin N A Sm Anim Pract*** 2007;37(3):579.[13] Orcutt ES ***J Vet Intern Med*** 2009;23:693(abstract).[14]Smith SA, et al. ***Vet Clin N A Sm Anim Pract*** 2004;34(5):1245.[15] Smith SA, et al. ***J Vet Intern Med*** 2003;17(1):73.[16] Cattaneo M. In: *Platelets.* 2nd ed. 2007;1127.[17]Pumphrey CW, et al. ***Thromb Res*** 1982;28(5):663.[18]Valentin LI, et al. ***Surgery*** 1988;104(2):178.[19] Boudreaux MK, et al. ***J Vet Pharmacol Ther*** 1992;15(1):1.[20]Boudreaux MK, et al. ***Vet Clin Pathol*** 1991;20(2):56.[21]Hogan DF, et al. ***Am J Vet Res*** 2004;65(3):327.[22] Boudreaux MK, et al. ***Am J Vet Res*** 1991;52(12):2000.[23]Gross ML, et al. ***J Lab Clin Med*** 1982;100(6):887.[24]Goodwin J C, et al. ***J Vet Intern Med*** 2007;21(3):609(abstract).[25]Goodwin JC et al. ***J Vet Intern Med*** 2007;21(3):609(abstract).[26]Hogan DF, et al. ***J Am Vet Med Assoc*** 2004;225(9):1406.[27] Whelchel DD et al. ***J Vet Intern Med*** 2009;23(3) :786(abstract).[28]Haviland RL et al. ***J Vet Intern Med*** 2007;23(3):745(abstract).[29]Mellet AM et al. ***J Vet Intern Med*** 2011;25(1):71.[30] Hagihara K, et al. ***Drug Met Dispos*** 2010 epub.[31]Wiviott SD, et al. ***N Eng J Med*** 2007;357(20)2001.[32]Niitsu Y, et al. ***Eur J Pharmacol*** 2008;579(1-3):276.[33]Smith RL, et al. ***Xenobiotica*** 2007;37(8):884.[34] Agah R, et al. In: *Platelets.* 2nd ed. 2007;1145-1163.[35]Makkar RR, et al. ***Circulation*** 1997;95(4):1015.[36]Bright JM, et al. ***Vet Ther*** 2003; 4(1):35.[37]Bright JM, et al. ***J Vet Intern Med*** 2002;16(6):749.[38]Rukshin V, et al. ***J Cardiovasc Pharmacol*** 2003;41(4):615.

NEW DEVELOPMENTS IN CANINE EPI

Elias Westermarck, DVM, Ph.D, DECVIM
Helsinki, Finland

Exocrine pancreatic insufficiency (EPI) affects the function of the exocrine pancreas and leads to inadequate production of digestive enzymes and signs of maldigestion. The exocrine pancreas has a large reserve secretory capacity and no clinical signs occur until 90% of secretory capacity is lost. Our study group hypothesizes that EPI in dogs is almost always the result of pancreatic acinar atrophy (PAA), so in dogs, EPI and PAA are nearly synonyms. Canine PAA is a unique disease. Reports indicate that chronic pancreatitis, pancreatic hypoplasia, and pancreatic neoplasia can cause EPI in dogs, but these conditions are extremely rare. We followed-up more than 1000 dogs with EPI and performed necropsy on about 200 of them, all the dogs had PAA.

EPI has been reported in many different breeds, but is most commonly seen in German Shepherd Dogs (GSDs). In Finland, about every second EPI dog is a GSD and the prevalence of EPI is approximately 1-2% in the breed. EPI is also a common disease entity in rough-coated Collies, Chow Chows and Cavalier King Charles Spaniels. Researchers have suggested that EPI is an inheritable disease in many breeds. In most studies, a pedigree analysis indicates that EPI is inherited in an autosomal recessive fashion in GSDs. The prevalence of affected dogs in these reports has, however, been lower than the expected 25% for a simple autosomal recessive trait; thus, a polygenic mode of inheritance has also been proposed. To obtain more precise information about the mode of inheritance in GSDs, we performed a test mating of two dogs with EPI[1]. The dame delivered six puppies, which were raised in private homes, and we followed up the offspring over the next 12 years. During the study period, we took regularly blood samples and evaluated the functional capacity of the exocrine pancreas with different tests. We also took biopsies of the pancreas via laparotomy on 26 occasions. During the procedure, the gross appearance of the pancreas was evaluated, and one biopsy specimen was obtained from the pancreas at each time-point by ligation of a piece of pancreatic tissue. The biopsies were then studied histologically and electron microscopically (EM). The results revealed that only two of the six offspring revealed atrophic changes in the exocrine part of the pancreas during their lifetime. This is much less than expected if the disease is inherited by Mendelian genetics. A polygenic mode of inheritance is therefore the only explanation for this outcome. This is, a logical result, however, since researchers have suggested PAA in GSDs to be an autoimmune disease[2].

One of the offspring in our mating study provided us the opportunity to observe how the degenerative process begins and how it progresses. At the first laparotomy at the age of six weeks, the gross and histological examination showed a normal pancreas texture. This therefore suggests that PAA in GSDs is not congenital. At the age of 22 months, the dog was clinically normal, but EM examination revealed widespread degenerative changes. Only three months later, the dog began to show typical clinical signs of EPI, and all the exocrine function test results were compatible with the severe loss of exocrine pancreatic tissue. This loss was confirmed by gross and histological examination of the pancreas. This dog showed that the progression from EM findings to a completely atrophied pancreas is rapid. In an EPI dog with typical end- stage PAA the pancreas is severely atrophied with only a membranous, translucent tissue remaining. Histology shows that almost all acinar cells are shrunken and atypically arranged. Pancreatic ducts are distinctly visible, and mild lymphoplasmacytic infiltration is present throughout the atrophic tissue. The islet cells are normal.

In our mating study, the other sibling with atrophic changes in the pancreas was healthy during the first five years. Laparotomy was performed four times and a gross and histologic examination was normal every time. Half a year later, the dog was still healthy, but the pancreatic function tests showed that the pancreas failed to produce pancreatic enzymes normally. At laparotomy, the gross examination showed partial atrophy of the pancreas. In a subclinical EPI dog with a partially atrophied pancreas large areas of the pancreas are atrophied, but normal structures are also visible. In the histology examination of a partially atrophied pancreas, it is typical to find severe mononuclear infiltration, which is most extensive in the border zone of normal and partially affected tissue. Lymphocyte infiltration into the target organ where tissue destruction is ongoing has been taken as primary evidence of an autoimmune reaction. Further immunological studies of dogs with partial acinar atrophy have suggested that both cellular and humoral immune responses play a role in the pathogenesis of acinar atrophy, although tissue destruction seems to be largely mediated by cellular immune mechanisms. Dogs with partial PAA have been treated with immunosuppressive medication, but currently no treatment is recommended when dogs with partial PAA show no clinical signs.

Although EPI may occur at any age, signs of maldigestion appear before the age of four years in most dogs[3]. Male and female dogs are usually equally affected. The most typical clinical sign in a dog with EPI is yellowish porridge-like feces. This is of course not a pathognomonic sign of EPI but it is quite common. Other common signs include increased fecal volume and defecation frequency, weight loss, abnormally increased appetite, poorly digested loose feces, and flatulence. Coprophagia, skin problems and nervousness are reported in one third of dogs but severe watery diarrhea usually appears only temporarily.

A diagnosis of EPI is based on the demonstration of a lack of exocrine pancreatic function. Serum trypsin-like immunoreactivity (TLI) is the gold standard for the diagnosis of EPI in dogs. The TLI assay is highly species-specific and measures only pancreatic trypsin and trypsinogen that have entered the bloodstream directly from the pancreas. Serum TLI measurement is a practical test, because only a single blood sample is usually needed to confirm the clinical diagnosis of EPI. In healthy dogs, the TLI value is greater than 5.0 ug/L, and a value under 2.5 is considered highly diagnostic for severe EPI. A value between 2.5 and 5.0 is considered to be in the gray zone. Interpreting the results in the gray zone may cause some problems, however. In our mating study of GSDs, four of the healthy puppies had their TLI measured more than ten times during their lifetimes and every dog showed a TLI value that fell at least once in the gray zone; the lowest value was 2.7. These results show that it is quite common for dogs with no gastrointestinal signs to have subnormal TLI values. In one of our previous studies we had 12 dogs which showed no clinical evidence of EPI, but did show constantly subnormal TLI values. We performed laparotomy on these dogs and found in gross examination that the amount of normal pancreatic tissue had considerably diminished; thus these dogs had partial PAA and subclinical EPI. Nowadays, it is recommended that the serum TLI concentration be re-evaluated if the concentration falls in the gray zone.

In dogs with EPI, oral replacement therapy with enzyme supplementation is needed to compensate for the lack of enzyme production. Unfortunately, despite adequate enzyme supplementation, digestive capacity does not completely return to normal in dogs with EPI. In one of our long-term clinical studies, we have compared the effect of two uncoated enzyme supplements (chopped pancreas and pancreatic powder preparation) and observed no significant difference between them. The study showed that in practice, selection between these two supplements is usually based on economics and practicality. The raw chopped pancreas was only one fourth as expensive, but practical difficulties, mainly in availability, handling, and storing, were considerable compared to powdered enzyme supplements.

The role of dietary treatment in dogs with EPI has several different meanings. Some have debated whether these dogs require special diets or whether it is sufficient just to feed them high-quality ordinary dog food. Clinical feeding studies using special diets have shown controversial results regarding the actual benefits of these diets. It is usually recommend to feed dogs with EPI a highly digestible, fat-restricted diet and to avoid high-fiber foods.

We have also carried out some feeding studies with EPI dogs. In our latest trial, we initially fed the dogs their ordinary diet, and then fed them three different special diets each for three weeks. Owners scored the dogs daily for severity of clinical signs typical of EPI dogs. An EPI index was calculated for each dog by adding the daily scores for each clinical sign. The results showed that each of the special diets was the most suitable diet for some dogs, but in 40% of the dogs the most suitable diet was their ordinary diet. In individual dogs, the values of their daily EPI indices varied significantly among the different diets. This result indicates that diet plays an important role in the treatment of dogs with EPI. However, the results also indicated that there was marked interindividual variation among dogs. This observation is problematic with regard to using the information to formulate instructions for feeding affected dogs. The most suitable diet for a given dog with EPI is likely best discerned through trial and error; thus, an individually planned feeding regimen is recommended for affected dogs.

Besides dietary enzyme supplements, other medications for the treatment of gastrointestinal tract signs are often used in dogs with EPI. Antibiotics are the most common adjunctive medication. Of the antibiotics administered in Finland, tylosin is used almost exclusively. The most common indications for the use of antibiotic treatments include diarrhea, flatulence, and borborygmus. Treatment with tylosin eliminates diarrhea within a few days and will prevent its recurrence as long as the treatment continues. When discontinued, diarrhea will reappear in many dogs within a few weeks to months. Some dogs require several treatments but the effect of tylosin on controlling diarrhea signs remains the same. The etiology of tylosin-responsive diarrhea remains unclear. In addition to having antibacterial properties, tylosin may have anti-inflammatory effects that contribute to its effectiveness in treating diarrhea in dogs. To date, we have noticed no tylosin-associated side-effects. Because many dogs with EPI have low serum cobalamin concentrations parenteral cyanocobalamin treatment has been recommended. Unfortunately no detailed study has been published which demonstrates the benefits of cyanocobalamin treatment in EPI dogs. According to our data, indications for cyanocobalamin treatment include poor general health and coat.

In one of our studies we assessed the response to long-term enzyme treatment by comparing the clinical signs and general health of dogs treated for EPI with those of healthy control dogs. The study showed that typical gastrointestinal signs of EPI were almost totally controlled in half of the treated dogs, and that the treated dogs performed as well as the control dogs did. One of three EPI dogs showed a satisfactory response, whereas 20% of dogs with EPI poorly responded to treatment. Why some clinical signs were incompletely controlled or why treatment responses varied so much remains unknown. Further studies are required to determine why some dogs responded so poorly and to develop special treatments for them.

REFERENCES

[1].Westermarck E, et. al. **J Vet Intern Med**. 2010;24:450. [2]- Wiberg M, et.al. **Vet. Pathol**. 1999;36:530. [3].Westermarck E . et. al. 2003;33:1165.

NEW DEVELOPMENTS IN FELINE EPI

Jörg M. Steiner, vet.med., Dr.med.vet., PhD, DACVIM, DECVIM-CA
College Station, TX

EPIDEMIOLOGY

Disorders of the exocrine pancreas, and especially exocrine pancreatic insufficiency (EPI), have traditionally been thought to be extremely rare in cats. For example, of 180,648 cats entered into the Veterinary Medical Data Base (VMDB) over a ten year period (from the early eighties to the early nineties), only 11 of those cats (0.006%) were reported to have been diagnosed with exocrine pancreatic insufficiency. However, more recently feline EPI has been reported more frequently.[1,2] Over a recent 5-year period (2004-2008) the GI Lab assayed 84,523 serum samples for fTLI concentration and a total of 1,342 samples had serum fTLI concentrations ≤ 8.0 μg/L, a concentration that is considered diagnostic for feline EPI. While these populations of cats are very different, it is clear that EPI is more frequently diagnosed today than in the past. This is likely related to an increased awareness of veterinarians, but more importantly might be due to the advent of a diagnostic test for feline EPI, feline trypsin-like immunoreactivity (fTLI) in the late nineties.[3]

Traditionally, EPI has been mainly thought of as a condition of older cats. However, the age distribution of 882 cats with a serum fTLI concentration ≤ 8.0 μg/L assayed over a recent 5-year period, showed a relatively even distribution of ages. Seven kittens were 6 months or younger, and 40 cats were 1 year of age or younger. In contrast to these findings another study had reported that 69% of all cats affected were considered senior, but this case series only included 16 cats with EPI and ages ranged widely from 3 months to 15 years in this particular study.[2] Finally, in contrast to dogs, a breed predilection has not been reported for feline EPI.

ETIOLOGY AND PATHOGENESIS

In theory, there are several potential causes of EPI, including pancreatic aplasia, pancreatic hypoplasia, pancreatic acinar atrophy (PAA), pressure atrophy due to pancreatic duct obstruction, and pancreatic destruction due to pancreatic inflammation. However, many of these theoretical causes of EPI have never been described in a cat and chronic pancreatitis is believed to be the most common cause of EPI in the cat. The author believes to have identified one kitten with pancreatic hypoplasia. This kitten showed loose stools and failure to thrive, had a decreased serum fTLI concentration, and responded to enzyme replacement therapy. During the spay surgery the pancreas appeared small in size, but was histologically normal. Thus pancreatic hypoplasia was hypothesized in this cat, but could not be definitely proven. Also, three cases of pancreatic acinar atrophy in cats have been mentioned, but have not been reviewed in detail in the literature. A few cases of EPI without preceding pancreatitis have been reported due to *Eurytrema procyonis* infestation.[4] Adenocarcinomas of the exocrine pancreas or other neoplastic lesions in the area of the pancreatic duct or the duodenal papilla can also lead to obstruction of the pancreatic duct, followed by atrophy of pancreatic acinar tissue, but this has never been reported in a cat. Another potential cause of pancreatic duct obstruction may occur due to proximal duodenal resection, when the major duodenal papilla is being damaged during surgery. Finally, the clinical syndrome of EPI does not require deficiency of all pancreatic digestive enzymes and isolated pancreatic lipase deficiency has been reported as a rare cause of EPI in people and one dog, but has never been described in a cat.

Digestive enzymes of pancreatic acinar origin play an integral role in assimilation of all major food components. Thus, a lack of these enzymes will lead to maldigestion. The large amount of nutrients remaining in the intestinal lumen in turn leads to loose voluminous stools and steatorrhea. At the same time, the lack of nutrient assimilation will cause weight loss and may lead to vitamin deficiencies in some cases. Serum cobalamin (vitamin B_{12}) concentrations have been decreased in all 26 cats described in the literature that had serum cobalamin evaluated.[2,5,6] Serum folate concentrations in cats with EPI are often either decreased, indicating concurrent small intestinal disease, or within the reference interval.[6] Also, a single cat with EPI with secondary vitamin K responsive coagulopathy has been reported.

In cats with EPI secondary to chronic pancreatitis, destruction of pancreatic tissue might not be limited to the acinar cells and concurrent diabetes mellitus may be observed. It should be noted that cats with chronic pancreatitis may first develop EPI or diabetes mellitus, may develop both at the same time, or may not develop either. However, because of this relationship of feline EPI and diabetes mellitus, cats with diabetes mellitus that have chronic loose stools should be evaluated for concurrent EPI.

CLINICAL PICTURE

Clinical signs most commonly reported in cats with EPI are weight loss, loose stools, a poor haircoat, and less commonly polyphagia.[2] These clinical signs are non-specific and are also seen in many cats with other disorders that are more common in cats than EPI. The stools are usually loose or semiformed and voluminous. They may have a yellow to clay colored appearance, and may be quite malodorous. Some cats with EPI may develop watery diarrhea, but this is not common. Also, some patients have been described as having normal stools.[2,5] The high fat content of the feces can lead to greasy soiling of the hair coat in the perianal and tail region, but this finding is not very common, having been observed in only one of 20 cats in a recent report of 20 cats with EPI and in 2 of 5 cats with EPI in a second report.[1,5]

More recently, a cat with severe D-lactic acidosis with clinical signs of episodic generalized weakness, ataxia, and lethargy was diagnosed with EPI.[7] The authors speculated that the D-lactic acidosis was due to massively increased bacterial fermentation in the intestinal lumen. The D-lactic acidosis and associated clinical signs resolved after enzyme supplementation of this cat. As mentioned above some cats may have concurrent diabetes mellitus and may thus be presented with PU/PD. Also, patients may show residual signs of chronic pancreatitis, such as anorexia, vomiting, or abdominal discomfort. Finally, some cats with EPI may show clinical signs that are due to concurrent inflammatory conditions of the intestines or the biliary system. Surprisingly in two recent case series of cats with EPI several cats showed vomiting or diarrhea suggesting large bowel disease.[2,5]

DIAGNOSIS

EPI is a functional disease and thus requires a functional diagnosis. A radioimmunoassay for the measurement of feline trypsin-like immunoreactivity (fTLI) concentration has been developed and analytically validated (current reference interval of 12.0-82.0 µg/L).[6] Serum fTLI concentrations of ≤ 8 µg/L are being considered diagnostic for EPI.[1]

In a recent study chronic renal failure was shown to increase serum fTLI concentrations.[8] Thus, evaluation of serum fTLI concentrations in azotemic cats may obscure a diagnosis of feline EPI. However, this may not be clinically relevant as no cases have been described to date where a diagnosis of EPI was missed because of renal failure. The author would suggest to re-evaluate serum fTLI concentrations in azotemic cats that have a borderline fTLI concentration and in whom an alternative diagnosis for loose stools and/or weight loss cannot be identified. Serum fTLI did not significantly increase after feeding of healthy cats.[9] However, this was not studied in cats with suspected EPI. Therefore, the current recommendation is to withhold food from cats for 12 hours before collecting a sample for measurement of serum fTLI concentration, whenever possible.

MANAGEMENT

Most cats with EPI can be successfully managed by dietary supplementation with pancreatic enzymes using dried pancreas extracts, but raw beef, pork, or game pancreas can also be given instead. Initially, one teaspoon per meal should be given. As cats often don't like the taste of the pancreatic powder, it is best to thoroughly mix the powder with some canned food or even a small amount of fish oil. Most cats respond quite rapidly to enzyme replacement therapy and show resolution of loose stools within 3-4 days. When clinical signs have completely resolved, the amount of pancreatic enzymes can be gradually decreased to the lowest effective dose, which usually varies from patient to patient, and may also vary between different batches of the pancreatic supplement.

Response to enzyme supplementation alone may not be satisfactory in many patients as most, if not all, cats with EPI are cobalamin deficient, which may prevent a full therapeutic response if not treated concurrently. Since cobalamin deficiency causes cobalamin malabsorption, oral supplementation is not effective and cobalamin must be supplemented through the parenteral route. Usually a dose of 250 µg is given subcutaneously weekly for 6 weeks, followed by one more dose a month after that, and a recheck a month after the last dose. Vitamin status for fat-soluble vitamins has not been systemically evaluated in cats with EPI.

Cats that fail to respond to therapy may benefit from treatment with an antibiotic (tylosin or metronidazole) or a proton pump inhibitor (e.g., omeprazole at 0.7 mg/kg q 12 hrs) to treat small intestinal dysbiosis or prevent destruction of pancreatic lipase in the enzyme supplement, respectively.

PROGNOSIS

Most cats with EPI have an irreversible loss of pancreatic acinar tissue. Recovery from feline EPI is extremely rare and is poorly understood. However, with appropriate management and monitoring cats with EPI usually gain weight quickly, pass normal stools, and can go on to live a normal life, with a normal life span.

REFERENCES

[1]Steiner JM, et al. **J Vet Intern Med** 2000; 14:627. [2]Thompson KA, et al. **J Feline Med Surg** 2009; 11:935. [3]Steiner JM, et al. **Am J Vet Res** 1996; 57:1417. [4]Anderson WI, et al. **Vet Rec** 1987; 120:235. [5]Kook PH, et al. Schweizer Archiv für Tierheilkunde 2011; 153:19. [6]Steiner JM, et al. **J Vet Intern Med** 1995; 9:193 (abstract). [7]Packer RA, et al. **J Vet Intern Med** 2005; 19:106. [8]Steiner JM, et al. **J Vet Intern Med** 2002;16: 385 (abstract). [9]Steiner JM, et al. **Am J Vet Res** 1999; 60:895.

DIETARY INTERACTIONS AFFECTING COPPER METABOLISM IN RUMINANTS

Stephanie L. Hansen, PhD
Ames, IA

INTRODUCTION

Ruminant Requirements for Copper

Copper is a required trace element in the diets of all domestic ruminants, though the tolerance of each species for Cu varies tremendously. Based on NRC (1996; 2001) recommendations, beef cattle require 10 mg Cu/kg diet DM and dairy cattle require between 12 and 16 mg Cu/kg diet DM depending on the animal's physiological state. The sheep NRC (1985) recommends 5 mg Cu/kg diet DM, with a narrow range between adequacy and toxicosis. Copper plays a critical role in several biological processes including cross-linking of cartilage (lysyl oxidase), neutrophil activity, antioxidant capacity (superoxide dismutase), and iron absorption (ceruloplasmin and hephaestin). A deficiency of Cu may therefore compromise the ability of the animal to resist or recuperate from infection, decrease animal growth, and lead to the characteristic rough and color-faded hair coat often attributed to Cu deficiency (Underwood and Suttle, 1999). The typical ruminant diet requires some supplementation of Cu, though the level required is related to dietary concentrations of typical Cu antagonists such as sulfur (S), molybdenum (Mo) and iron (Fe). This paper will focus on the interactions which occur in the rumen between Cu, S and Mo, as well as those which occur post-ruminally between Cu and Fe.

Ruminal Interactions Between Copper, Sulfur and Molybdenum

Dietary Cu is subject to ruminal interactions which may limit or prevent absorption of Cu in the small intestine. The ruminal interaction between dietary S, Mo and Cu is one of the most studied in animal mineral metabolism. This interaction is catalyzed by the formation of thiomolybdates in the rumen, a complex of S^- and molybdate (MoO_4; Suttle, 1991). Thiomolybdates may include varying amounts of S and thus may be mono, di, tri or tetrathiomolybdate, with research data suggesting that trithiomolybdate and tetrathiomolybdate are the most commonly found in the rumen. Thiomolybdates bind Cu in the rumen preventing Cu from being absorbed in the small intestine. The negative effect of thiomolybdates on Cu status of ruminants extends beyond ruminal interactions to systemic interactions. Systemic circulation of thiomolybdates causes albumin to irreversibly bind Cu, prevent utilization of Cu by dependent processes throughout the body and leading to Cu deficiency. Additional data have demonstrated that exposure of ruminants to thiomolybdates increases biliary excretion of Cu, leading to rapid depletion of hepatic Cu stores (Gunneratne, 1985). Tetrathiomolydates are used in treatment of Wilson's disease in humans and injection of sheep with thiomolybdates is actually a method used to increase hepatic Cu excretion in animals experiencing Cu toxicosis.

Post-ruminal Interactions Between Copper and Iron

A close relationship exists between Cu and iron (Fe). Adequate Cu is essential to proper metabolism of Fe, as Cu-dependent ferroxidases are necessary for Fe to be mobilized out of tissues (Osaki, 1971). Ceruloplasmin fulfills this role in the macrophage, hepatocyte and various other tissues such as certain cell types in the brain, while hephaestin is the ferroxidase localized in the plasma membrane of intestinal enterocytes (Chen et al., 2006). Copper effect on the uptake and export of Fe from intestinal cells has been studied. Han and Wessling-Resnick (2002) demonstrated that when Caco-2 cells were incubated in the presence of 1 μM Cu for 3 hours, apical Fe^{55} uptake was increased by 32% and export into the basolateral chamber was increased by 59%. These authors also examined expression of several genes involved in Fe metabolism in Caco-2 cells following 8 days of 1 μM $CuCl_2$ exposure. Compared to non-Cu exposed cells, relative expressions of the importer divalent metal transporter 1 (DMT1), exporter ferroportin and Cu-dependent ferroxidase hephaestin, were significantly increased in the Cu exposed cells. In addition, expression of another Fe transporter, transferrin receptor, was greatly increased relative to controls, as would be expected in a limited Fe situation. Using this information, these authors speculated that Cu exposure results in increased Fe absorption through the following mechanism. Copper repletion causes increased hephaestin and ferroportin expression, resulting in enhanced Fe export from the cell. The cell senses a lack of Fe and DMT1 expression is upregulated in a compensatory manner and apical Fe uptake is subsequently increased (Han and Wessling-Resnick, 2002). Figure 1 shows the multitude of proteins involved in intestinal Fe absorption.

One possible point of interrelationship in the pathways of Fe and Cu absorption is DMT1. Human and rodent work has revealed that Cu has a dedicated transporter known as Ctr1 (Figure 2); however, Arredondo et al. (2003) reported that at least some portion of Cu absorption is directly contributable to DMT1 function. These authors found that approximately 48% of Cu absorption was eliminated when Caco-2 cells were treated with a DMT1 antisense oligonucleotide, essentially knocking out DMT1 activity. These data suggest that DMT1 plays an important role in Cu import into cells. Arredondo et al. (2003) also demonstrated reduced uptake of Cu^{64} by Caco-2 cells as the concentration of Fe in the incubation media was increased from 0.5 to 20 μM. Collectively, these results suggest that high Fe may be antagonistic to Cu in two distinct manners, causing less DMT1 to be expressed by cells and also competing with Cu for uptake by DMT1 into the cell.

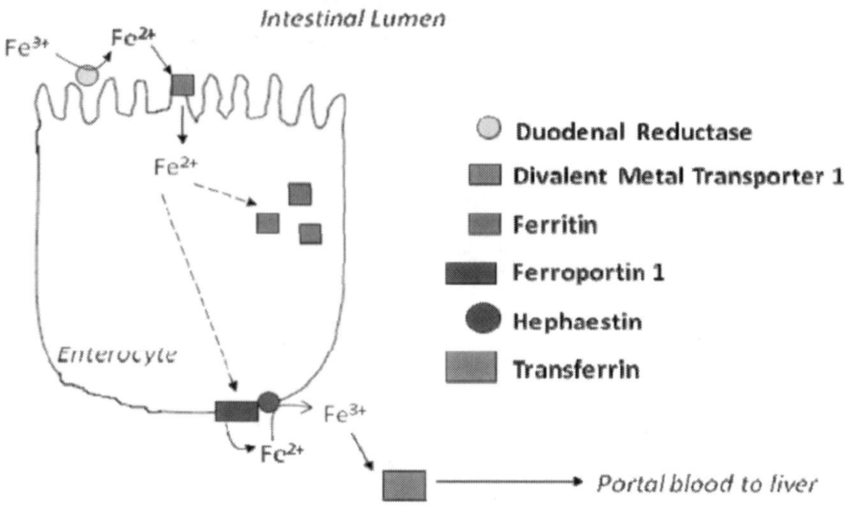

Figure 1. The process of absorption of iron in the enterocyte.

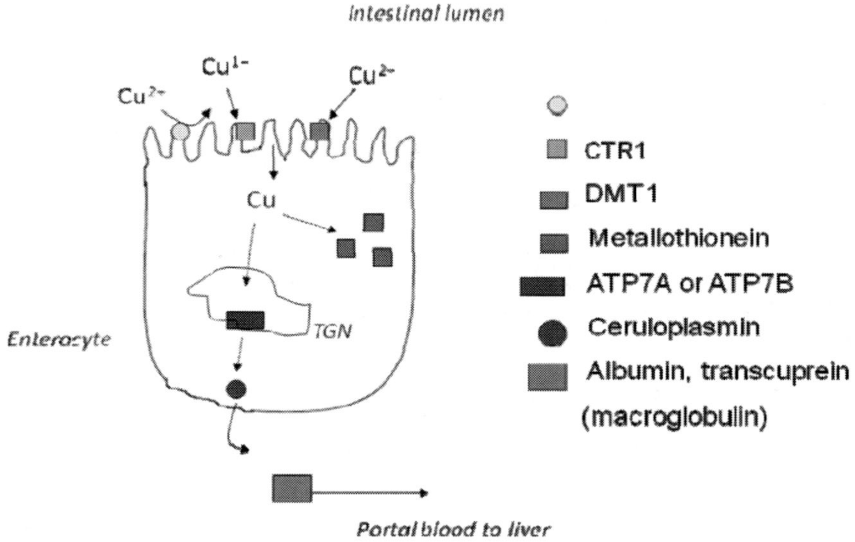

Figure 2. The process of absorption of copper in the enterocyte.

Excessive Fe in ruminant diets may come from a variety of sources, including feedstuffs that are naturally high in Fe such as alfalfa (300 mg Fe/kg DM) or soyhulls (500 mg Fe/kg DM; NRC, 1996; DePeters, 2000). Iron may also be in excess in some corn co-products, for example, distillers grains with solubles averages more than 600 mg Fe/kg DM (NRC, 1996). In addition to feedstuffs high in Fe, soil contamination of harvested feedstuffs may be an additional source of excessive Fe in ruminant diets. Using titanium concentrations of feces as an indicator of soil consumption in cattle, Mayland et al. (1975) suggested that grazing cattle ingested an average of 0.5 kg of soil per head per day. More recent work by Rafferty et al. (1994) reported that harvested feedstuffs such as hay and grass silage had a mean soil contamination level of 2% (w/w), with some contamination as high as 8% reported. Soil Fe concentrations in the United States range from less than 0.5% in the southern Gulf States to greater than 5.5% in mountainous areas and many parts of the Western United States (United States Geological Survey, 2007).

Iron in soils is typically bound to clays or other chelating agents, and the majority of Fe is in an unsoluble form which should be unavailable to the animal for absorption (Whitehead, 2000). However, we have recently demonstrated that exposure to an acidic environment caused by fermenting silage resulted in significant increases in bioaccessible Fe concentrations following enzymatic in vitro ruminant digestion simulation (Hansen and Spears, 2009). Soil contamination of greenchop at harvest may represent an overlooked source of excessive Fe in ruminant diets. It is possible that high levels of available Fe in the silage may result in increased competition between minerals for common intestinal transporters.

Excessive levels of dietary Fe have been shown to negatively impact Cu metabolism in cattle. Standish et al. (1969) used growing beef steers to examine the effects of feeding 0, 400 or 1600 mg supplemental Fe/kg DM for a period of 84 days on plasma and tissue mineral concentrations. The basal diet was corn based and analyzed to contain 77 mg Fe/kg DM. Liver Cu concentrations were not different between steers supplemented with 0 or 400 mg Fe/kg DM; however, concentrations were lower in steers supplemented with 1600 mg Fe/kg DM compared to those provided 0 mg supplemental Fe/kg DM (44 and 260 mg Cu/kg DM, respectively). Several subsequent studies have found that Cu indices such as liver and plasma Cu concentrations or plasma ceruloplasmin activity are negatively affected by high dietary Fe in cattle (Humphries et al., 1983; Bremner et al., 1987; Phillipo et al., 1987; Gengelbach et al., 1994). In these studies, decreases in Cu status of calves fed high supplemental Fe were similar to those decreases observed in calves fed supplemental Mo. In addition, the negative impacts of Fe and Mo on plasma and liver Cu concentrations of calves appeared to be additive, suggesting differing mechanisms of Cu antagonism between the two minerals (Humphries et al., 1983; Bremner et al., 1987). Bremner et al. (1987) also noted that the severity of Fe-induced Cu deficiency did not differ between calves supplemented with 500 mg Fe/kg DM or 800 mg Fe/kg DM from Fe (II) carbonate, suggesting that both levels were at or above the threshold at which the body minimizes dietary Fe absorption, likely via regulation of proteins such as DMT1.

REFERENCES

Arredondo, M. et al. Am. J. Physiol. Cell. Physiol. 2003; 284: C1525. Bremner, I et al., Anim. Prod. 1987;45: 403. Chen, H., et al., J. Nutr. 2006; 136: 1236. DePeters, E.J. et al., Prof. Anim. Sci. 2000; 16: 69. Gengelbach, G. P et al., J. Anim. Sci. 1994; 72: 2722. Gunneratne, S. R et al.1985. 5th Int. Symp . Trace Elements in Man and Animals. Aberdeen, pp. 342. Farnham Royal, UK: Commonw. Agric . Bur. Han O and Wessling-Resnick M. Am J Physiol Gastrointest Liver Physiol. 2002 Mar; 282(3):G527. Hansen, S. L., and J. W. Spears. J. Dairy Sci. 2009; 92: 2896-2905. Humphries, W. R et al., Br. J. Nutr. 1983; 49: 77. Mayland, H. F. et al, J. Range Manage. 1975; 28: 448. NRC. 1985. Nutrient requirements of sheep 6th ed. National Academy Press, Washington, D.C. NRC. 1996. Nutrient requirements of beef cattle 7th ed. National Academy Press, Washington, D.C. NRC. 2001. Nutrient requirements of dairy cattle 7th ed. National Academy Press, Washington, D.C. Osaki, S., et al., J. Biol. Chem. 1971; 246: 3018. Phillippo, M., et al. J. Agric. Sci.,Camb. 1987; 109: 315. Rafferty, B. et al., Sci. Total Environ. 1994; 153: 69. Standish, J. F et al., J. Anim. Sci. 1969; 29: 496. Suttle, N. F. Annu. Rev. Nutr. 1991; 11: 121–140. Underwood, E. J., and N. F. Suttle. 1999. The mineral nutrition of livestock. 3rd ed. CABI Publishing, Wallingford, U.K. U.S.G.S. United States Geological Survey. 2007. http://tin.er.usgs.gov/geochem/map/image/lower48/fe_icp40.jpg. Accessed June 5, 2008. Whitehead, D. C. 2000. Micronutrient cations: Iron, manganese, zinc, copper and cobalt in nutrient elements in grassland: Soil-plant-animal relationships. CABI Publishing, New York, NY.

DIETARY INTERACTIONS INFLUENCING CALCIUM HOMEOSTASIS IN FOOD ANIMALS

Jesse P. Goff, DVM, PhD
Ames, IA

INTRODUCTION

Plasma calcium (Ca) concentration is tightly regulated in all animals. For most adult food producing animals normal total plasma Ca concentration is between 8.5 and 10.4 mg/dl or 2.1-2.6 mmole/L. Young animal plasma Ca concentration is normally higher (10-11 mg/dl), to support normal bone growth and development. Ionized Ca concentration is typically 42-47% of total plasma Ca, even if total Ca changes, and is therefore typically 4-4.7 mg/dl. When ionized Ca concentration falls below normal levels the electrical potential across cells decreases. In nervous tissues this means the resting membrane potential is closer to the threshold for an action potential to occur. The nerves become hyperexcitable – which leads to tetany as the predominant clinical symptom associated with hypocalcemia in many species. Low extracellular Ca concentration also reduces release of acetylcholine at the muscle end plate, decreasing the strength of muscle contractions- in some species, such as ruminants, a flaccid paralysis is the predominant symptom exhibited during hypocalcemia. A notable exception to these rules is the laying hen, who produces an estrogen induced Ca-binding protein in her blood to support eggshell production. Her total Ca can be 25-30 mg/dl, but her plasma ionized Ca is still between 4 and 5 mg /dl.

CALCIUM HOMEOSTASIS

Because plasma Ca concentration is critical to so many functions a series of events will occur whenever there is a perturbation of plasma Ca concentration. Any decrease in ionized plasma Ca concentration causes the parathyroid glands to secrete parathyroid hormone (PTH). Within minutes, PTH increases renal reabsorption of Ca from the glomerular filtrate. If the perturbation in plasma Ca is small (<1 g/d in cow), plasma Ca returns to normal and PTH secretion returns to baseline levels.

If the Ca drain from the extracellular pool is larger, continued PTH secretion stimulates resorption of Ca stored in bone. Ca exists within bone in two states. A small amount of readily available Ca exists in solution in the fluids surrounding the bone cells and within the canaliculi of the bone. Upon stimulation by PTH, the bone lining cells rapidly transfer this bone fluid Ca into the extracellular pool. The size of the bone fluid pool in the adult cow fed a high forage diet has been estimated to be 6 to 10 g Ca. Interestingly, bone fluid Ca pool size was increased another 4-8 g in cows fed ammonium chloride to reduce Diet cation-anion difference (DCAD) to induce metabolic acidosis.

The overwhelming majority of bone Ca is tightly bound to the organic bone collagen matrix as CaHPO4 deposits. Osteoclasts return this Ca to the circulation by digesting bone collagen, releasing the minerals to the extracellular fluids. The activity of the osteoclasts is indirectly controlled by PTH. Prolonged secretion of PTH causes osteoblasts to secrete substances (such as osteoclast activating factor and prostaglandins), which stimulate resorptive activity in existing osteoclasts and induce formation of new osteoclasts.

Ultimately Ca exiting the extracellular fluid must be replaced using dietary Ca. Ca can be absorbed across the intestine (and forestomachs of ruminants) by both vitamin D dependent and vitamin D independent means. Vitamin D independent absorption of Ca is by passive diffusion. The amount of Ca that can be absorbed passively is dependent on the concentration of soluble Ca overlying the intestinal epithelium and the surface area of the intestinal epithelium. Efficient absorption of dietary Ca, when dietary Ca is low or when demand is very high, occurs by vitamin D dependent active transport of Ca across the intestinal epithelial cells. Vitamin D, ingested or produced during ultraviolet irradiation of the skin, enters the blood and is converted in the liver to 25-hydroxyvitamin D (25-OH D). This reaction is substrate driven so that the more vitamin D ingested the higher the plasma 25-OH D concentration. Plasma 25-OH D concentrations are the best indicator of vitamin D deficiency or toxicity. The activity of the renal enzyme responsible for converting 25-OH vitamin D to the steroid hormone, 1,25-dihydroxyvitamin D (1,25-$(OH)_2$ D), is stimulated by and tightly regulated by PTH. 1,25-$(OH)_2$ D stimulates active transport of dietary Ca across the intestinal epithelium. 1,25-$(OH)_2$ D stimulates the production of Ca-binding protein(s), which carry Ca across the cytosol of intestinal epithelial cells. Once Ca is transported to the basolateral membrane, it is extruded from the cell into the blood against approximately a 1,000-fold concentration gradient by Ca/magnesium-ATPase-dependent Ca pumps. The activity of these Ca pumps is controlled by 1,25-$(OH)_2$ D.

Hypercalcemia is handled by increasing urinary Ca loss, and reducing production of 1,25-$(OH)_2$ D to reduce the efficiency of Ca absorption. The thyroid hormone calcitonin can also shut off osteoclastic activity, eliminating release of Ca from bone.

CHRONIC PROBLEMS WITH CALCIUM IN FOOD ANIMALS

The major issue facing food animals revolves around adequacy of diet Ca absorption. This may occur because of inadequate diet Ca content, inadequate vitamin D supply, from diet or sunlight, or excessive diet P. Animals need a certain amount of Ca to enter their extracellular fluid each day to support maintenance (bone remodeling, intestinal secretions etc.), growth, reproduction and lactation. For example, a 680 kg cow producing 45 kg milk will need to absorb 76 g Ca/ day to meet maintenance and milk production needs. If her diet Ca were 100% absorbed the diet would only need to contain 76 g Ca. However, rarely will the diet Ca be totally available for absorption and when feeding the animal we need to account for the availability of Ca from different sources. Some forages contain high amounts of Ca but in general the Ca is bound so that only a portion is absorbable. The Dairy NRC (2001) suggests forage Ca is 30% available, while Ca from mineral sources such as limestone are closer to 70% available. When feeding any animal the important factor is the g Ca actually absorbed from the diet. Hence using guidelines such as feeding 1% Ca diets to milking cows or growing pigs is not precise. One needs to evaluate the availability of the Ca. For example in a corn silage diet intended for our example cow, where limestone supplies the majority of the Ca, the % Ca in an adequate diet may be just 0.80%. A lactating diet with high levels of alfalfa may need to be 1.1% Ca to be adequate due to the poor availability of Ca from alfalfa. Excess diet Phosphorus (P) can interfere with Ca absorption as well. One mechanism involves the binding of Ca to phytate P, the form of most P in plants. Some species such as the horse and rabbit are sensitive to the Ca: P ratio of the diet and getting outside of the zone of from 1-2 parts Ca : 1 part P can cause bone problems. In most other species, so long as you are supplying adequate g of Ca in the diet, the ratio can be as low as 1 part Ca: 2 parts P without creating problems. In food animals the environmental concerns of excess diet P together with cost and the use of phytase have reduced the likelihood of excess P interfering with Ca absorption. Finally, inadequate vitamin D can prevent adequate absorption of Ca. As a general rule, animals housed outdoors in the summer will make adequate vitamin D in their skin. However, in winter the sun is too low to provide adequate UV-B rays. Many food animals today are housed indoors. As a result we must add vitamin D to their diet. Most diets have several fold higher than NRC requirement added to them. So vitamin D deficiency should rarely be the cause of Ca problems in commercial diets. However, recently in the Midwest there have been widespread problems with vitamin D in swine, resulting in rickets and osteomalacia. In many of these pigs hypocalcemia was also observed, indicating Ca homeostasis mechanisms were upset by lack of vitamin D and hence 1,25-$(OH)_2$ D. Mistakes happen and either the vitamin D was not added or the concentrations in pre-mixes were much lower than calculated. Finally, renal failure will prevent adequate production of 1,25-$(OH)_2$ D as well, resulting in bone problems.

Generally chronic diet Ca insufficiency does not cause hypocalcemia until it is longstanding unless there is a problem with vitamin D or it's metabolism. The body compensates for the lack of Ca coming in from the diet by removing Ca stored in bone and also Ca that was intended for new bone formation or to replace old bone during bone remodeling. THIS IS A NATURAL PROCESS INTENDED TO PRESERVE NORMAL BLOOD CA CONCENTRATION SINCE THAT WOULD BE LIFE-ENDING! Virtually every lactating mammal goes through a period in early lactation where daily diet Ca intake is inadequate to supply all the Ca leaving the body in the form of milk. This results in a short period of lactational osteoporosis where Ca is lost from the bone. In the average dairy cow this loss can be as much as 10-13% of the skeletal Ca in the first month of lactation, even with adequate diet Ca. The Ca is put back into the bone in later lactation when the Ca balance of the animal becomes positive.

When prolonged or severe, inadequate Ca causes rickets in young animals (along with P or vitamin D deficiency) as a result of failure to replace cartilage with bone at the physes. Multifocal osteoporotic sites will be found throughout the trabecular bone. In adults the lack of Ca causes osteoporosis to become the dominant lesion. In adult animals, bone is constantly undergoing resorption and replacement with new bone. Lack of Ca (and ensuing high parathyroid hormone levels) causes uncoupling of bone formation from bone resorption. Bone resorption accelerates, but replacement with new bone is blocked. The result is widespread osteoporosis, and both osteomalacia and osteodystrophy can be found at sites where bone replacement is interrupted due to lack of Ca. If severe and prolonged the bones can become so thin and weak that fractures ensue. The lactating sow and the caged laying hen seem to be particularly prone to this problem if diet Ca availability is compromised in lactation. In many cases when the bones are at this point the syndrome may be accompanied by hypocalcemia, though not always.

ACUTE PROBLEMS WTH CALCIUM METABOLISM IN FOOD ANIMALS

When Ca homeostasis fails the result is often an acute sharp decline in plasma Ca concentration resulting in tetany and/ or flaccid paralysis of both skeletal and smooth muscles. Most animals adapt to an abrupt negative change in Ca balance by reducing urinary Ca loss, increasing Ca resorption from bone, and increasing diet Ca absorption efficiency. But in some cases these mechanisms fail the animal and the resulting hypocalcemia becomes life threatening. The onset of lactation in dairy cows and goats represents a tremendous sudden increase in Ca demand at a time when feed intake is also challenged and failure to meet this demand can result in the acute hypocalcemia known as milk fever or parturient paresis.

Success or failure to adapt to the initial challenge imposed by colostrum Ca demands is influenced by a number of factors. An important determinant of milk fever risk is the acid-base status of the cow at the time of parturition. Metabolic alkalosis induces conformational changes in the PTH receptor, which prevent tight binding of PTH to its receptor. Bone resorption and production of 1,25-$(OH)_2$ D are impaired, reducing the ability to successfully adjust to the Ca demands of lactation. Cows fed diets that are relatively high in K or Na are in a relative state of metabolic alkalosis which increases the likelihood that they will not successfully adapt to the Ca demands of lactation and will develop milk fever. These cows exhibit a temporary pseudohypoparathyroidism at parturition. The parathyroid glands recognize the onset of hypocalcemia and secrete adequate PTH. However the tissues respond only poorly to the PTH, leading to inadequate osteoclastic bone resorption and renal 1,25-$(OH)_2$ D production. This is particularly evident in cows that have been treated for milk fever and require further treatments due to reappearance of milk fever symptoms. These cows have very high blood PTH concentrations but produce little 1,25-$(OH)_2$ D at parturition. Full recovery from milk fever occurs only after they have responded to the PTH by producing 1,25-dihydroxyvitamin D, which can be delayed for 24 - 48 hrs in these relapsing milk fever cows.

A second cause of hypocalcemia is hypomagnesemia. Very low blood Mg (below 1.2 mg / dl) can reduce PTH secretion from the parathyroid glands causing temporary hypoparathyroidism. This is commonly seen in lactating beef and dairy cows at pasture and is part of the "grass tetany" syndrome. Less severe hypomagnesemia (below 1.8 mg / dl) can alter the responsiveness of tissues to PTH by inducing conformational changes in G-proteins coupled to the PTH receptor. This also in effect causes temporary pseudohypoparathyroidism. High dietary K reduces rumenal Mg absorption in addition to causing metabolic alkalosis. The effects of Mg on Ca homeostasis are discussed further in the companion paper.

Excessive dietary P during late gestation (>80 g P/d) can induce milk fever by raising blood P to the point that P directly inhibits renal synthesis of 1,25-dihydroxyvitamin D, which short circuits Ca homeostasis. This can occur when blood P levels at parturition exceed 6-8 mg/dl. Keeping diet P below 45 g / day keeps blood P below 5 mg / dl in prepartum dairy cows. Some practitioners are concerned that low prepartum diet P will cause the hypophosphatemic downer cow syndrome. As far as we can tell, the hypophosphatemic downer syndrome occurs as a sequelae to milk fever. **Low diet P does not cause this syndrome!** It is believed to be due to sequestration of salivary P in the rumen. The rumen, being immotile during the hypocalcemia of milk fever, fails to pass the P down to the duodenum where it should be absorbed. Continued secretion of salivary P depletes blood and body P and the hypophosphatemia (blood P <1.5 mg / dl) seems to cause an alert downer cow condition. How very low P causes the inability of skeletal muscles to support standing in the cow remains unknown. It may upset muscle energy metabolism as ATP and creatine – phosphate are energy currencies used in muscle function. If caught early the syndrome can be treated with IV sodium phosphate. Unfortunately many products sold for treatment of hypophosphatemia contain P in the form of phosphite or phosphinic acid, which are ineffective.

PREVENTION OF MILK FEVER

Reducing Diet Cation-Anion Difference to Prevent Hypocalcemia

Reducing the number of absorbable dietary cations and/or increasing the number of absorbable dietary anions greatly diminishes the incidence of hypocalcemia and milk fever in dairy cows. The major cations present in feeds and the charge they carry are Na (+1), K (+1), Ca (+2), and Mg (+2). The major anions and their charges found in feeds are Cl (-1), SO_4 (-2), and phosphate (assumed to be -3). In theory all the cations and anions in a diet are capable of exerting an influence on the electrical charge and hence the pH of the blood. Cations or anions present in the diet will only alter the electrical charge of the blood if they are absorbed into the blood. Trace elements present in diets are absorbed in such small amounts that they are of negligible consequence to acid-base status. Organic acids such as the volatile fatty acids are generally absorbed in the un-dissociated neutral form. They also are rapidly metabolized within the liver so they have only a small effect on general acid-base balance under most circumstances.

The difference between the number of cation and anion particles absorbed from the diet determines the general acid-base balance of the body and therefore the pH of the blood. The cation-anion difference of a diet is commonly described in terms of mEq/kg DM (some authors prefer "mEq/100 g" diet DM) of just Na, K, Cl, and SO_4, although it must be kept in mind that Ca, Mg, and P absorbed from the diet will also influence blood pH. While DCAD equations provide a theoretical basis for dietary manipulation of acid-base status they are not necessary for formulation of mineral content of prepartum dairy cow rations. With the exception of K and Cl, the rate of inclusion of the other minerals can be "fixed".

The diet Na requirement of a late gestation cow is about 0.12%. A little salt is added to the diet to meet Na requirement and prevent pica (often manifest as a desire to drink urine from the floor). Unlimited access to NaCl in late gestation will increase the risk of udder edema.

At least two studies demonstrated that inclusion of Ca in the diet at NRC required levels or several fold above NRC requirement does not influence the degree of hypocalcemia experienced by the cow at calving. Close-up diet Ca concentration should be between 0.85 and 1.0% Ca.

To ensure adequate concentrations of Mg in the blood of the periparturient cow the dietary Mg concentration should be 0.35-0.4%. This higher dietary Mg concentration allows the cow to take advantage of passive absorption of Mg across the rumen wall should active transport of Mg be compromised.

Dietary P concentration should be fed at a level to meet the NRC requirement for P in the late gestation cow. This is generally about 0.35 -0.4% P for most cows. A diet supplying more than 80 g P/d will block renal production of 1,25-dihydroxyvitamin D and will actually cause milk fever.

Dietary S must be kept above 0.22% to ensure adequate substrate for rumen microbial amino acid synthesis. Corn silage diets are notoriously low in sulfur. Diet S should be kept below 0.4% to avoid possible neurological problems associated with S toxicity.

Now, with the exception of K and Cl, the "variables" in the various proposed DCAD equations have become more or less "fixed". The key to milk fever prevention (at least with Holstein cows) is to now keep K as close to the NRC requirement of the dry cow as possible (about 0.85% diet K, which is very difficult to achieve). The key to reduction of subclinical hypocalcemia, not just milk fever, is to add Cl to the ration to counteract the effects of even low diet K on blood alkalinity. For formulation purposes the concentration of Cl required in the diet to acidify the cow is approximately 0.5% less than the concentration of K in the diet. In other words, if diet K can be reduced to 1.3%, the Cl concentration of the diet should be raised to 0.8%. Then observe urine pH after 3-4 days. This is often a conservative approach and the final concentration of Cl needed to truly acidify the urine (pH 6.2-6.8) may have to be brought up within 0.4-0.3% of dietary K. NEVER START AN ANION SUPPLEMENTATION PROGRAM WITH HIGHER LEVELS OF CL. If you have over-acidified the cows at the onset it becomes very difficult to evaluate urine pH as feed intake will quickly be reduced. Always work the dry cow pen up to the adequate Cl dose. If dietary K cannot be reduced below 2.0% the diet Cl would need to be at least 1.5% to acidify the cow; a level of Cl in the diet likely to decrease dry matter intake independent of over-acidification. Chloride sources differ in their palatability and since achieving low dietary K can be difficult it is prudent to use a palatable source of Cl when formulating the diet. Ammonium chloride (or ammonium sulfate) can be particularly unpalatable when included in rations with a high pH. In high pH diets, a portion of the ammonium cation is converted to ammonia, which is highly irritating when smelled by the cow. Prilling the Cl (and SO_4) salts reduces the unpleasant taste of the salts. In our experience hydrochloric acid has proved the most palatable source of anions. Hydrochloric acid can be extremely dangerous to handle when it is procured as a liquid concentrate. Several North American companies now manufacture hydrochloric acid based anion supplements, which are safe to handle, effective and reasonably palatable.

These are simply guidelines for anion supplementation used by this author and are based on inclusion of Ca, Na, S, Mg, and P at the levels outlined above. Urine pH of the cows provides a cheap and fairly accurate assessment of blood pH and can be a good gauge of the appropriate level of anion supplementation. Urine pH on high cation diets is generally above 8.2. Limiting dietary cations will reduce urine pH only a small amount (down to ~7.8). For optimal control of subclinical hypocalcemia the average pH of the urine of Holstein cows should be between 6.2 and 6.8, which essentially requires addition of anions to the ration. In Jersey cows the average urine pH of the close-up cows has to be reduced to between 5.8 and 6.3 for effective control of hypocalcemia. If the average urine pH is between 5.0 and 5.5, excessive anions have induced an uncompensated metabolic acidosis and the cows will suffer a decline in dry matter intake. Urine pH can be checked 48 or more hrs after a ration change. Urine samples should be free of feces and made on midstream collections to avoid alkalinity from vaginal secretions. Anion supplemented diets are generally fed for the last 3 wks before calving, though the length of time these diets need to be fed to induce a compensated metabolic acidosis is no more than 4-5 days. When fed longer than six weeks the urine pH of cows ready to freshen will have risen as the bone successfully buffers the acidity generated by the diet. In practice this means when anions are fed the entire dry period the urine pH target for the final week before calving must be increased by about 0.5 units. Alternatively one can simply monitor urine pH in those cows that have been on the diet for at least one week but not longer than three weeks.

Feeding a Ca Deficient Diet To Stimulate PTH Secretion Pre-Calving To Prevent Hypocalcemia

When cows are fed a diet that supplies less Ca than they require, the cows are in negative Ca balance. This causes a minor decline in blood Ca concentration stimulating PTH secretion, which in turn stimulates osteoclastic bone resorption and renal production of 1,25-dihydroxyvitamin D. At parturition the cow's osteoclasts are already active and in high numbers and the lactational drain of Ca is more easily replaced from bone Ca. If provided with Ca in the lactation ration, the previous stimulation of enterocytes by 1,25-dihydroxyvitamin D will allow efficient utilization of dietary Ca and the cow avoids hypocalcemia. This works even in the face of metabolic alkalosis as metabolic alkalosis reduces but does not totally eliminate tissue PTH sensitivity. **Prolonged** exposure to high PTH levels induced by the low Ca diet overcomes the reduced tissue sensitivity to PTH.

The absorbable Ca requirement of the late gestation cow is from 14 g /day in Jerseys to about 22 g in large Holsteins. A truly low Ca diet must supply considerably less absorbable Ca than required by the cow if it is to be capable of stimulating PTH secretion. For example, a 600 kg cow consuming 13 kg DM must be fed a diet that is less than 0.15% **absorbable** Ca if it is to provide less than 20 g absorbed Ca/ day. Low Ca diets are more practical under grazing situations. In these cases the total dry matter intake of pasture may be just 6-7 kg DM/ day and the grasses being grazed can be less than 0.4% Ca, which would provide < 28 g total Ca and somewhere around 9-10 g absorbable Ca / day (Sanchez, 2003). It is important to note that after calving the animal must be switched to a high Ca diet.

Oral Ca Treatments at Calving To Pevent Hypocalcemia

Ca administered to the fresh cow may arguably be called a treatment rather than a preventative measure for hypocalcemia. Briefly, the concept behind oral supplementation is that the cow's ability to utilize active transport of Ca across intestinal cells is inadequate to help her maintain normal blood Ca concentrations. By dosing the animal with large amounts of very soluble Ca orally it is possible to force Ca across the intestinal tract by means of passive diffusion between, not across, intestinal epithelial cells. Best results are obtained with doses of Ca between 50 and 125 g Ca / dose. Ca chloride has been used but can be caustic. Large or repeated doses of Ca chloride can induce an uncompensated metabolic acidosis in the cow, especially if the cow is already being fed an acidogenic diet. Ca propionate is less injurious to tissues and is not acidogenic. It has the added benefit of supplying propionate, a gluconeogenic precursor. For best control of hypocalcemia a dose is given at calving and again 24 hrs later. Larger or more frequent dosing can be toxic. Toxic doses of Ca can be delivered orally – about 250 g Ca in a soluble form will kill some cows. The benefit of adding oral Ca on top of a properly formulated low DCAD program does not seem to warrant the added expense.

REFERENCES

[1]Goff JP. **Vet J**. 2008 Apr;176(1):50-7. [2]Goff JP. **Vet Clin North Am Food Anim Pract**. 2004 Nov;20(3):471-94, [3]Goff JP, Horst RL. **Acta Vet Scand Suppl**. 2003;97:51-6.

DIETARY INTERACTIONS AFFECTING MAGNESIUM AVAILABILITY IN FOOD ANIMALS

Jesse P. Goff, DVM, PhD
Ames, IA

INTRODUCTION

Ruminants are the only food animals that commonly suffer from problems with magnesium (Mg) metabolism. Failure of the ruminant to maintain normal extracellular and plasma (1.9-2.5 mg / dl) Mg concentrations causes hyper-excitable nerve tissue upsetting co-ordination of muscle contraction. Eventually cerebrospinal fluid Mg concentration also declines resulting in generalized tonic-clonic convulsion, known as grass tetany. Maintaining adequate blood Mg concentrations in the cow is essential to calcium homeostasis as hypomagnesemia can also interfere with parathyroid hormone function. Severe hypomagnesemia (Mg < 1.2 mg / dl) will prevent parathyroid hormone (PTH) secretion. This is a major contributor to the grass tetany and wheat pasture poisoning syndromes. Moderate hypomagnesemia (Mg<1.7 mg/dl) can interfere with the action of PTH on tissues. The G-proteins coupled to the PTH receptors on bone and kidney tissue require the presence of magnesium in order to get cyclic AMP production (the second messenger) inside the cells. Hypomagnesemia is still a common cause of hypocalcemia and milk fever at calving and also in mid-lactation cows that "go down"! In my experience the most common cause of mid-lactation downer cows is hypomagnesemia. This then causes a secondary decline in blood calcium so they look and act like a milk fever cow. The second most common cause of a mid-lactation hypocalcemic downer cow is an animal in estrus who is distracted from eating that day. These animals may fail to eat an adequate amount of diet and at this point in their life cycle they are totally dependent on diet Ca to meet Ca demands of lactation. Finally the third common issue with hypocalcemic mid-lactation downer cows is an abrupt change in Ca availability in the diet – again, because these cows are so dependent on diet Ca any drop in diet Ca availability can result in acute hypocalcemia. My experience is that many cows are marginal in Mg status shortly after calving and they are commonly described as having poor appetite and "slow to take off after calving".

MAGNESIUM HOMEOSTASIS

Cows cannot mobilize body stores of Mg to maintain blood Mg levels. Mg homeostasis is **dependent on daily absorption of diet Mg**. For some unknown reason, adult ruminants are nearly totally dependent on the forestomach for absorption of Mg. Two mechanisms for absorption of Mg across the rumen wall exist. One is an active transport process, allowing the cow to absorb Mg when the diet is low in Mg. When this transport process is working well ruminants can obtain adequate Mg even when diet Mg is 0.22%. Unfortunately, the integrity of this Mg pump is often compromised by high diet K and possibly high rumen ammonia (Nitrogen), rendering this means of Mg absorption unreliable. A second, passive transport system for Mg exists which is not poisoned by potassium. This transport system is concentration driven and only works when Mg **in solution** in the rumen fluids is high. It will provide Mg for homeostasis so long as the diet Mg exceeds ~ 0.35%. For either transport mechanism it is critical that the Mg utilized in the ration be available to the cow. **The MgO must be soluble in rumen fluid**. $MgSO_4$ and Mg Cl_2 are very soluble Mg sources, but also unpalatable and difficult to handle. The feed industry utilizes MgO which is about 54% Mg (>58% Mg MgO is suitable only for making steel!). MgO takes up little room in the ration, costs less, and is more palatable than some other sources of Mg. Unfortunately, there is tremendous variability in MgO quality – and price does not always reflect bioavailability. A quick test can tell you the relative availability of MgO sources. Place 3 g of each MgO source in a container and slowly add 40 ml white vinegar (5% acetic acid). Cap container and shake intermittently over the next 30 minutes. Check the pH. Vinegar will be pH 2.6-2.8! The best MgO sources will bring the pH up to 8.2. The worst MgO bring pH to just 3.8. pH is expressed on a log scale so this represents >10,000 fold difference in the number of hydrogen ions buffered. Mg is included in rations for two reasons: to maintain adequate levels of Mg in the blood and as a rumen fluid alkalinizer. Remember in lactating rations we expect MgO to help combat rumen acidosis- and we are not getting that action from these insoluble MgO sources. Blood Mg is often marginal with these sources as well. In pre-fresh rations where the rumen buffering activity of MgO is not as critical, $MgSO_4$ and Mg Cl_2 are good sources of soluble Mg and contribute to a lower DCAD diet. Unfortunately they are not very palatable.

REFERENCES

[1] Goff JP. **Vet Clin North Am Food Anim Pract**. 2004 20(3):471-94. [2] Ram L, Schonewille JT, Martens H, Van't Klooster AT, Beynen AC, **J Dairy Sci**. 1998, 81(9):2485-92.

ROLE OF WATER MINERAL CONTENT ON RUMINANT HEALTH & PERFORMANCE

Robert J. Van Saun, DVM, MS, PhD, DACT, DACVN
University Park, PA

INTRODUCTION

Water is the most essential nutrient for all living beings, but the essential nutrient most overlooked. Water to meet the animal's requirement is obtained from feeds, consumed water, and metabolic water. Metabolic water is that water obtained from metabolic reactions throughout the body and generally accounts for less than 20% of needs. Feed and consumed water contributions to intake will vary depending upon feed water content. Water consumption is the primary driver for dry matter intake and milk yield. In-spite-of the recognition of water's essentiality and sophisticated equations to predict an animal's requirement little is known about the role of minerals in water. This presentation will review the current state of understanding in regards to the mineralogical effects of water quality on animal health and performance.

MEASURES OF WATER QUALITY

Water quality can be evaluated subjectively by its color, taste, and smell. High quality water should be colorless and have no offensive odor or taste. A "rotten egg" smell to water is the result of high sulfate (odorless) content where sulfate ion is reduced to produce hydrogen sulfide (H_2S) gas having the obnoxious smell. Determinations of pH, total dissolved substances (TDS), mineral content and hardness, and contaminant (herbicides, pesticides, organic compounds, etc.) content as well as microbiologic culture provide objective measures of water quality. Water meeting current standards for human consumption obviously is equally satisfactory for animal consumption. In question is the tolerance level of animals relative to their willingness to adequately consume and not experience adverse consequences of water whose quality measures are below current human standards. There is minimal definitive evidence to document the lower tolerance level of water quality for animals, especially for microbial contamination. A number of publications addressing livestock water quality are available and generally describe experiential assessment of water quality for livestock.[1,2,3]

Beyond initial organoleptic evaluation of a water source, measures of pH and TDS provide a preliminary water quality assessment and may suggest the need for further evaluation. Ideal water pH is near 7, but most animals will tolerate water with a pH range between 6 and 9. Water pH can be easily measured with litmus paper or commercially available battery-operated pH meters, some of which may also measure TDS. Salinity is a measure of total dissolved substances (or solids) most of which are mineral compounds, which accounts for electrical conductivity being another potential measure for salinity. Other measures of water quality will require a certified laboratory to complete desired analyses. Contact a chosen laboratory to determine appropriate methods for sample collection.

WATER MINERAL CONTENT

Although water mineral content is routinely determined for households and selectively during farm diagnostic investigations, studies summarizing expected water mineral content are limited. One of the earlier extension publications addressing water quality that is often cited reported mineral water concentrations from 350 samples; however, these samples were collected on farms that were reporting problems.[4] More recently a large collaborative survey study characterized mineral content in over 3500 samples collected throughout the United States.[5] This study showed a wide range in water mineral content with many samples having values exceeding recommended limits for humans or livestock. Unfortunately, in collection of these samples there was no attempt to correlate herd or animal issues with measured mineral content. The question remains, at what threshold level does specific mineral concentration in water result in adverse health and performance effects? Do livestock water supplies need to meet human water standards? If not, what is reasonable?

TOTAL DISSOLVED SOLIDS (TDS)

Increasing salinity of water sources, irrespective of mineral sources, results in decreased water intake, transient diarrhea, and reduced performance due to lower feed intake (**Table 1**). Younger animals and lactating animals are most sensitive to water salinity. Most animals can acclimate to moderately saline water. High TDS in a water sample would suggest further evaluation of mineral content is warranted, though low TDS water could contain inappropriate mineral content. Sodium, chloride, and sulfates are the predominate ions that contribute to high TDS with 20.2, 12.6, and 29.6% of samples, respectively, exceeding recommended upper limits for livestock.[5] Beside potential toxic minerals that may contaminate water, dissolved substances may affect aesthetic quality decreasing intake, provide an additional mineral source to the diet potentially altering available mineral balance, or interfere with other minerals thus altering dietary mineral availability. In this survey, iron (41.1%) and manganese (44.8%) had the greatest percent of samples that exceeded upper limits of 0.4 and 0.5 mg/l, respectively. These two trace minerals are often considered interfering minerals reducing the availability of other minerals, namely cobalt, copper, and selenium.

The role of water salinity on animal health and performance is well documented. Controlled studies have shown lactating dairy cows drinking high (2500 mg/l) compared to low (196 mg/l) salinity water had lower feed intake and milk production.[6] Similarly dairy cows switched from a higher (4300 mg/l) to low (430 mg/l) salinity water source showed improved feed intake and milk yield.[7] Clinical observations are consistent with these data where dairy cows increased feed intake and milk yield when water salinity was reduced from 2000 mg/l to minimal concentrations.[8] Most studies addressing effects of water salinity on dairy cow performance have been performed within hot environments. Further work is needed to determine if cows managed under cool or temperate conditions would respond equally to similar elevations in water salinity. Growing beef cattle tolerated water up to 10,000 mg/l from NaCl, but a concentration of 12,500 mg/l or more were associated with clinical sodium toxicosis.[9] Young dairy heifers showed reduced water intake at NaCl concentrations of 2500 mg/l and 50% reduction when the concentration reached 3500 mg/l.[9]

Table 1. Guidelines for water salinity interpretation as measured by total dissolved solids (TDS) content or electrical conductivity.

Total Dissolved Solids	Electrical Conductivity	Interpretation
mg/L or ppm	**umhos/cm**	
< 1000	< 1,670	Relatively low salinity; suitable for all classes of livestock
1000 - 2999	1,670 – 5,008	Satisfactory for all classes of livestock; may produce transient diarrhea in animals not accustomed to them, but should not affect health or performance
3000 - 4999	5,010 – 8,348	Temporary water refusal and diarrhea may be seen when animals are introduced to them; may reduce productivity in lactating animals
5000 - 6999	8,350 – 11,688	Likely to reduce productivity in lactating animals; may reduce growth rates; may result in water refusal and diarrhea; avoid in lactating and pregnant animals; Avoid if possible
7000 - 10,000	11,690 – 16,700	Unfit for poultry or swine; considerable risk in all other species; Animals may subsist on them in low stress situations; Avoid unless extreme circumstances
> 10,000	> 16,700	High saline water; not recommended for use under any conditions. Dangerous, avoid at all costs

SPECIFIC MINERAL CONCERNS

The most common dissolved minerals include sodium, calcium, magnesium, sulfate, and bicarbonate. Iron, manganese, and sulfate content of water are of primary concern as a result of their adverse impact on aesthetic qualities. Refusal of water directly results in lower dry matter intake, which will adversely affect animal productive efficiency and health. Another concern for minerals in water is their potential ability to interfere with other minerals. The degree to which this interference may occur is not entirely understood. Some mineral interactions have been well substantiated, while others remain anecdotal observations. It is often assumed that all minerals dissolved in water are readily available; however, this has not been adequately researched and may vary by the specific mineral. Electrolytes and sulfates are most likely to be highly available in the GI tract from water, whereas availability of other minerals remains to be determined.

Calcium and Magnesium

Calcium and magnesium concentrations in water can be highly elevated, though this is primarily associated with hardness. Water hardness is not a significant health or water intake issue, but more of an aesthetic issue. Hard water may contribute additional calcium or magnesium to the diet, but is suggested not to increase urolithiasis risk. However, supplementing MgO at 0.6% Mg in milk replacer diets resulted in 70% kidney stone prevalence, though addition of 2% NaCl cut the prevalence rate to 30%.[10] Although Mg was artificially added in this study, it was to reflect observed issues of kidney stones in veal calves in which milk replacer was mixed with high magnesium water. This scenario may be unique to calves as Mg absorption efficiency is reduced once the rumen becomes functional. Average water calcium and magnesium content would provide a fairly insignificant amount of mineral to the total diet assuming it is highly bioavailable. In contrast the unique high mineral content water sources would provide a significant amount of mineral to the diet even with less than complete bioavailability.

Anions

Chloride in water is typically associated with sodium and contributes to TDS content. However, elevated water chloride content over and above sodium contributes to additional anion load to the diet and can lower dietary DCAD. Although most surveys have not routinely measured potassium, strong ion load in water can significantly contribute to alterations of DCAD. Water survey data showed an average water DCAD value of -6 mEq/kg, though the maximum value was -192.9 mEq/kg.[5] The higher DCAD value could significantly alter dietary DCAD for fresh cows, where optimum values are +250-370 mEq/kg dry matter. Lower DCAD in fresh cow diets is associated with lower feed intake and milk yield, which may lead to greater susceptibility to postparturient diseases. A concern level for water chloride and sulfate anions concentrations is >1000 mg/l.[11,12]

Sulfate

Water sulfur content has been highly investigated relative to animal health and performance. Form of sulfur in water varies with acid/base balance. Typical "rotten egg" smelling water is a result of high H_2S content. It is challenging to measure this sulfur form as the compound rapidly volatilizes. Hydrogen sulfide in the water is more of an aesthetic issue reducing intake. Sulfur in the form of sulfates in water is highly available and can be readily used by rumen microbes.[13] Microbial production of thiomolybdates can adversely affect copper availability. Sulfur can interfere with inorganic selenium availability. High sulfate intake can induce a polioencephalomalacia condition that is non-responsive to thiamin supplementation. Water sulfate concentrations greater than 1000 mg/l were associated with increased H_2S production and reduced water and feed intake in feedlot cattle.[14] Increased incidence of polioencephalomalacia in feedlot cattle was seen when consuming water containing 2500 mg/l sulfate.[14] Water sulfate concentrations between 1250 and 1500 mg/l were associated with decreased consumption.[11,12] Generally water sulfate concentrations < 600 mg/l are considered safe.[13]

Nitrates

Nitrates are another potential water contaminant of concern, especially in ruminant animals where nitrate is converted to more toxic nitrite by ruminal microbes. Absorbed nitrites will bind to hemoglobin reducing the blood's oxygen carrying capacity resulting in varying stages of anoxia in affected animals. Although water nitrate content generally is not solely responsible for animal intoxication, it can significantly contribute to dietary nitrate content in predisposing to potential intoxication. When assessing dietary sources for nitrate toxicity potential, water sources should also be considered as this contribution may push total dietary nitrate intake into the toxicity risk range. Surface waters are at much greater risk for high nitrate concentrations compared to deep well water sources. Water nitrate content can be expressed as nitrate or nitrate-nitrogen where nitrate concentrations are 4.4 times the nitrate-nitrogen value. Acute nitrate intoxication is associated with water nitrate content between 1500 and 3000 mg/l, depending upon dietary nitrate content. Prolonged exposure to moderate nitrate content in water (100-400 mg/l) was associated with reproductive impairment in an observational study using DHI records in 133 dairy herds.[15]

Iron

Iron is present in two oxidation states ferric (Fe^{+3}) or ferrous (Fe^{+2}). The ferric state is insoluble whereas the ferrous state is more soluble and the predominate form present in water. In human water standards, iron content is suggested to be < 0.3 mg/l, primarily as a result of aesthetic issues. Iron can readily stain fixtures and accumulate in pipes restricting water flow. Iron reducing bacteria can convert ferric to ferrous iron to generate energy. These bacteria generate deposits that accumulate on the inner walls of pipes analogous to atherosclerotic plaques in arteries. Excessive iron intake (>0.3 mg/l) has been anecdotally associated with increased risks for metabolic disease in dairy herds.[11,12] Excess iron from water is assumed to increase oxidative stress and adversely affect dietary copper and zinc availability. No controlled studies have validated this association and many farms have water sources with high iron content without obvious problems.

Manganese

Manganese is considered essentially a non-toxic element with a wide range of safety, but is more of a concern relative to water aesthetic issues. In the water sample survey, manganese was one of the trace minerals found to be present in concentrations exceeding what would be acceptable for human water use. Holstein calves fed milk replacer and ad libitum water with differing Mn concentrations (control, 0.003 mg/l; 0.25 and 0.75 mg/l) showed no adverse effects on feed intake or growth rate.[16] Extremely elevated water Mn concentrations (>6.0 mg/l) are often found in association with previous mining activities. There is some concern about inhibition of other trace minerals, especially cobalt, at these high concentrations.

Molybdenum

Molybdenun is most often considered relative to its interfering role in copper absorption. Rumen microbial production of thiomolybdate compounds bind to and prevent intestinal copper absorption. A potential benefit of thiomolybdate production may be attributed to a reduction in ruminal H_2S production and reduced risk of polioencephalomalacia.[14]

SUMMARY

Water is the most essential nutrient for production agriculture, yet there is limited data addressing physiochemical properties with animal health and performance. Application of human water standards may be too conservative given the current concerns regarding water availability. More robust information is available relative to the role of salinity, nitrates, and sulfur in water though there is not complete agreement on critical threshold concentrations as these are influenced by dietary inputs. Potential roles for iron and manganese are suggested, but not well substantiated. Quantitative assessment of water quality should be part of a herd diagnostic investigation.

REFERENCES

[1]National Research Council: *Nutrient and Toxic Elements in Water*, Washington, DC, 1974, National Academy Press. [2]Pfost DL et al: *Bull EQ 381*, Univ Missouri-Columbia Extension, 2001. [3]Bagley CV et al: *Bull AH-Beef-28*, Utah State University Extension, 1997. [4]Adams RS, Sharpe WE: *DAS 1995-98*, Penn State Cooperative Extension, 1995. [5]Soche MT et al: **Proc Intermountain Nutr Conf**, Salt Lake City, UT, pp 85-96, 2003. [6]Jaster EH et al: **J Dairy Sci** 1978;61:66-71. [7]Challis DJ et al: **Vet Rec** 1987;120:12-15. [8]Rodenburg R: **Proc Bovine Pract** 1988;21:72-78. [9]Carson TL: **Vet Clinics NA:Food Anim Pract** 2000;16(3):455-464. [10]Petersson KH et al: **J Dairy Sci** 1988;71:3368-3377. [11]Beede DK: Proc **Western Dairy Management Conf**, 2009, pp 217-225. [12]Beede DK: **Proc High Plains Dairy Conf**, 2006, pp 1-23. [13]Crawford GI: **Proc Minn Nutr Conf**, 2007, pp 80-93. [14]Wagner JJ et al: **J Anim Sci** 2001;79(Suppl 1):85. [15]Ensley SM: Relationships of drinking water quality to production and reproduction in dairy herds. PhD Thesis, Iowa State University, 2000. [16]Linn J: **Proc 4-State Dairy Nutr Management Conf**, 2005, pp 117-119.

NOVEL THERAPIES FOR FIP IN CATS

Phillip McDonagh, BVSc; Paul A. Sheehy, BSc, Agr (Hons), PhD; Jacqueline M. Norris, BVSc, MVS, PhD
Sydney, Australia

INTRODUCTION

Feline infectious peritonitis, a significant cause of mortality in domestic cats worldwide, is caused by a virulent biotype of feline coronavirus (FCoV) often referred to as feline infectious peritonitis virus (FIPV). Therapeutic options for cats with FIP are limited and mostly palliative.[1] There are currently no effective antiviral medication for the treatment of FIP, although several drugs have demonstrated *in vitro* efficacy against FCoV.[2,3] Given the central role of increased FCoV replication in development of FIP it is likely that an effective safe antiviral agent will significantly advance our treatment of this devastating condition.

RNA interference (RNAi) is a highly conserved post transcriptional gene silencing mechanism found in most eukaryotes that has recently attracted significant attention due to its potential therapeutic use. RNAi can be induced in mammalian cells by introducing short interfering RNAs (siRNAs), ≈ 21 nucleotide RNA duplexes. Introduced siRNAs are incorporated into an RNA induced silencing complex (RISC) enabling it to bind to and target complementary sequences. Upon binding, an endonuclease within the RISC cleaves the targeted mRNA resulting in sequence specific gene silencing. The *in vitro* efficacy of RNAi as an antiviral has been demonstrated in a range of important viral pathogens from diverse viral families, including HIV-1, influenza, SARS coronavirus, and FMDV.[4-7] For some viruses these finding have been extended to *in vivo* animal models, where induction of a virus specific RNAi response results in reduced viral load and associated pathology.[8,9] We evaluated the antiviral potential of a panel of eight siRNAs targeting four conserved regions of the FCoV genome using an *in vitro* assay. The most effective siRNAs were further characterised by examining the effect of multiplicity of infection and siRNA concentration on their inhibitory effect. The information gained from this pilot study is an essential first step to the development of an effective RNAi based treatment for FIP and provides an effective platform for assessing a wide range of potential ant-FCoV therapeutics *in vitro*.

METHODS

Crandell feline kidney (CRFK) cells were cultured in at 37°C in 5 % CO2 in air and maintained in Dulbecco's modified eagles medium (DMEM) supplemented with 10 % foetal bovine serum (FBS).Working stocks of highly virulent type II FCoV strain FIPV WSU 79-1146 (ATCC: VR990) were produced in CRFK cells and quantified by plaque assays. siRNAs were designed using Block-iT™ RNAi designer (Invitrogen) using the consensus sequence of FIPV WSU 79-1146 (ascension number DQ010921). Four conserved regions of the genome were targeted: the membrane and nucleocapsid structural genes, the replicase gene, and the common 5' leader sequence. Other regions (eg spike protein) were not targeted due to their known genetic diversity. Prospective siRNA sequences were aligned with other reported full- and partial-length FCoV sequences available on GenBank using Geneious software (Biomatters Ltd. Criteria for final siRNA selection were (1) high Block-iT™ RNAi designer ranking, (2) sequence homology to other reported FCoV strains, and (3) minimum homology to known feline sequences on BLAST. A siRNA (NSC-GFP) was selected targeting green fluorescent protein mRNA to act as a non-silencing control. The 5' end of the sense strand of NSC-GFP was conjugated to FITC to enable transfection efficacy to be monitored with fluorescent microscopy. An initial screening experiment was conducted with the eight siRNAs to identify their inhibitory effect on FCoV replication *in vitro*. The effect of siRNAs on viral replication was determined by qRT-PCR of intracellular virus (detecting viral genome and M 212 and N mRNA), TCID50 infectivity assay of extracellular virus, and direct IFA staining for viral antigen, all conducted 24 hpi. Additionally, virus induced CPE was monitored in separate wells out to 50 hpi. All treatments were performed in triplicate and the results are expressed as geometric mean with 95% confidence interval from three independent experiments. PCR viral genomic and mRNA copy number were normalised to the housekeeping gene GAPDH copy number. Relative viral copy number and titres were calculated by dividing individual treatment values by the geometric mean of untreated samples. For statistical analysis, qRT-PCR and viral titre data was log transformed to normalise the distribution. Treatments were compared by one-way ANOVA followed by Dunnett's post hoc testing if indicated (GraphPad Prism V5.03). p-values less than 0.05 were considered significant.

RESULTS

The eight siRNAs designed in this study were selected to target conserved regions of the FCoV genome which were visually identified by aligning full- and partial-length FCoV sequences published on GenBank . Using this method we were able to design five siRNAs (L1, L2, R1, R2, N2) with 100% homology to all reported type I and type II strains. The remaining siRNAs showed complete homology at 18 of the 19 nucleotides (M1, N1) or 17 or 19 nucleotides (M2).

All siRNAs tested resulted in statistically significant inhibition of FCoV replication as measured by qRT-PCR and TCID50 infectivity assay however the most effective siRNAs were those targeted against L2 and N1. Transfection with the NSC siRNA and treatment with Lipofectamine 2000 had no significant effect on viral replication, confirming the sequence specificity of the siRNA mediated viral inhibition. Reduction of viral genomic RNA in FCoV specific siRNA treated cells ranged from 26.1% to 90.7%. Similar results were seen with viral mRNA, with M mRNA knockdown ranging from 27.7% to 92.3% and N mRNA knockdown ranging from 23.1% to 93.8%. Reduction in extracellular virus titres paralleled the PCR data, ranging from 59.6% to 97%, confirming viral gene knockdown was associated with a significant reduction in the production of progeny virions. For the highly effective siRNAs L2 and N1, qualitative analysis of viral protein expression using direct IFA resulted in a marked reduction in the number of cells expressing FCoV antigen, and phase contrast microscopy showed that these siRNAs afforded some protection from virus induced CPE out to 50 hours post infection.

The two most effective siRNAs, targeted against L2 and N1, were used to examine the dose-response relationship of transfected siRNAs. The cell culture, transfection, and infection conditions were identical to those of the screening experiment, except for differing siRNA concentrations of 5 nM, 20 nM, 70 nM and 200 nM. Both tested siRNAs demonstrated concentration dependent inhibition of viral replication. Reductions in extracellular virus titres at 200nM were 93.8% and 96.7% for L2 and N1 respectively, decreasing to 81.9% and 82.8% at 5nM concentration. Considering the high viral loads in cats with FIP, we explored the effectiveness of siRNAs at inhibiting FCoV replication in the face of increased viral loads. The inhibitory effect of L2 and N1 siRNAs on FCoV replication was tested in using a moi of 0.2, 2, or 20. Both siRNAs remained effective at all tested moi, with >75% inhibition under all tested conditions.

DISCUSSION

We have demonstrated the effectiveness of specific siRNA mediated RNAi in substantially inhibiting FCoV replication *in vitro*, without cytopathic effects in a feline cell line. This provides important preliminary data on the potential effectiveness and safety of these antiviral agents, leading the way for extension into *in vivo* animal models and as a potential therapy for FIP.

An initial screening experiment was conducted to identify siRNAs that could effectively inhibit FCoV replication. The eight siRNAs tested targeted different regions of the FCoV genome, including structural genes (M1, M2, N1, N2), non-structural genes (R1, R2), and untranslated regions (L1, L2). Selection of appropriate siRNA targets is a critical first step in using RNAi as an antiviral therapeutic. The lifecycle of FCoV theoretically enables the simultaneous degradation of both viral genomic and messenger RNA via RNAi. Also, the unique discontinuous transcription mechanism of coronaviruses enables targeting of multiple viral mRNAs with a single siRNA. A common leader sequence, which plays a critical role in coronavirus gene expression, is present on the 5' end of each viral mRNA.[10] siRNAs directed against the untranslated leader sequence can target all viral mRNA species and has been reported as an effective method for inhibiting coronavirus replication using an *in vitro* SARS-CoV model.[11] Similarly, a number of papers have reported success in inhibiting the replication of SARS-CoV by targeting regions encoding the replicase, membrane, and nucleocapsid proteins.[5,12] An additional complicating factor in designing siRNAs for use against FCoV is the genetic heterogeneity inherent in RNA viruses. Ideally an antiviral siRNA therapeutic should be effective against all field isolates, and thus conserved regions of the viral genome should be targeted. Such a strategy helps to limit the development of escape mutants during treatment as highly conserved regions are likely to be structurally or functionally constrained. Our simple approach to siRNA design was able to identify five siRNAs with 100 % homology to all reported FCoV strains. The effectiveness of the siRNAs was assessed at the genetic level, measuring viral genomic and messenger RNA, and downstream, by assessing viral protein synthesis and the production of progeny virions. All of the siRNAs tested demonstrated an inhibitory effect on FCoV replication; however the magnitude of inhibition varied significantly. The highly effective siRNAs L2 and N1 were able to reduce extracellular virus titres by > 95 %, while less effective siRNAs such as L1 resulted in a reduction of approximately 60 %.

We have demonstrated that the two most effective siRNAs, L2 and N1, inhibit FCoV replication in a dose dependent manner. Most importantly, both of these highly effective siRNA were still able to significantly inhibit FCoV replication at the lowest tested concentration. This is important from a potential therapeutic point of view, as off-target effects associated with RNAi are in part concentration dependent[13] and thus can be minimised by using low concentrations of highly potent siRNA.

We utilized the type II FCoV FIPV WSU 79-1146 in this study. In all reported studies infection with type I viruses is more common than type II viruses. It is unlikely that the higher prevalence of type I infections would invalidate any therapeutic application of the siRNAs tested in this study, as the primary difference between type I and II FCoVs is the spike protein, a region not targeted by any of the tested siRNAs. Additionally, siRNAs were selected based on homology to both type I and II FCoVs.

The therapeutic application of RNAi technology faces some important challenges. Translation of *in vitro* results into an animal model requires an appropriate delivery system. Delivery options for RNAi include chemical modifications to the siRNA duplex and lipid- and polymer based delivery systems.[14] Direct conjugation strategies which allow for targeted delivery of siRNAs to specific tissues and cell types will enhance the activity of siRNAs *invitro* and reduce unwanted effects.[15] Off target effects with RNAi *in vivo* are also a potential problem. siRNA sequences in this study were subjected to a BLAST search to eliminate any anti-FCoV siRNAs with significant homology to feline genes. The current light coverage of the feline genome (1.9-fold) means that homologous feline sequences may be missed during BLAST search, resulting in unexpected and potentially adverse off target effects. More in depth coverage of the feline genome will help to reduce the chances of off target effects, however many off-target effects are mediated by the micro RNA pathway, and can be triggered by limited sequence homology in the seed region of the siRNA, so it's more appropriate to talk of minimising off-target effects rather than avoiding them.[14]

Taken together, the results of this study suggest that RNAi may be a useful therapeutic option for the treatment of FIP. It is however unlikely that antiviral siRNA therapeutics, or any antiviral therapeutic for that matter, will be effective as a sole therapy for FIP. Whilst increased viral replication is a triggering and perpetuating event in the disease pathogenesis, the characteristic pathological lesions of widespread serositis, vasculitis, and pyogranuloma formation, are immunopathological in nature. Thus effective treatment of FIP will likely require immunomodulatory therapy in addition to antiviral therapy. The results of this current study demonstrate that siRNA mediated RNAi may be an appropriate choice in fulfilling the latter requirement.

REFERENCES

1. Hartmann K, Ritz S, 2008, Treatment of cats with feline infectious peritonitis. Vet Immunol Immunopathol 123, 172-175. 2. Barlough JE, Scott FW, 1990, Effectiveness of three antiviral agents against FIP virus in vitro. Vet Rec 126, 556-558. 3. Barlough JE, Shacklett BL, 1994, Antiviral studies of feline infectious peritonitis virus in vitro. Vet Rec 135, 177-179. 4. Ge Q, et al. 2003, RNA interference of influenza virus production by directly targeting mRNA for degradation and indirectly inhibiting all viral RNA transcription. Proc Natl Acad Sci USA 100, 2718-2723. 5. He ML, et al 2003. Inhibition of SARS-Associated Coronavirus Infection and Replication by RNA Interference (Am Med Assoc), pp. 2665-2666. 6. Martinez MA, et al., 2002, RNA interference of HIV replication. Trends in Immunology 23, 559-561. 7. Kahana R, et al 2004, Inhibition of foot-and-mouth disease virus replication by small interfering RNA. The Journal of general virology 85, 3213-3217. 8. Li B, et al 2005 Using siRNA in prophylactic and therapeutic regimens against SARS coronavirus in Rhesus macaque. Nature medicine 11, 944-951. 9. Tompkins et al 2004, Protection against lethal influenza virus challenge by RNA interference in vivo. Proc Natl Acad Sci U S A 101, 8682-8686. Masters PS, 2006, The molecular biology of coronaviruses. Adv Virus Res 66, 193-292. 10. Li T, et al 2005, siRNA targeting the leader sequence of SARS-CoV inhibits virus replication. Gene Therapy 12, 751-761. 11. He M, et al. 2006, Kinetics and synergistic effects of siRNAs targeting structural and replicase genes of SARS. FEBS letters 580, 2414-2420. 12. Behlke MA, 2006, Progress towards in Vivo Use of siRNAs. Molecular Therapy 13, 644-670. 13.Akhtar, S., Benter, I.F., 2007, Nonviral delivery of synthetic siRNAs in vivo. J Clin Invest 117, 3623-3632.

EPIDEMIOLOGY OF FELINE INFECTIOUS UPPER RESPIRATORY DISEASE IN SHELTERS

Kate F. Hurley, DVM, MPVM
Davis, California

INTRODUCTION

Feline upper respiratory infection (URI) is perhaps the most frustrating illness facing shelter veterinarians, managers and staff. Many cats enter shelters already carrying the pathogens that lead to illness; vaccines are partially effective at best; and specific treatments are limited. Factors such as overcrowding, poor air quality, poor sanitation, stress, concurrent illness, parasitism, poor nutrition, and other causes of immunosuppression predispose to disease. In spite of these challenges, recent research has shown that some shelters have dramatically greater success than others in controlling this seemingly ubiquitous disease [1, 2]. We now know that having fewer than 5% of cats develop URI in shelter care is an achievable goal. We've also learned that the mix of pathogens associated with shelter URI differ in importance than those found in pet populations in general. A targeted strategy that accounts for the most likely pathogens and addresses the most important environmental and management factors can successfully, and dramatically, reduce the frequency of this disease.

ETIOLOGIC AGENTS

Feline Herpesvirus-1 (FHV-1), Feline Calicivirus (FCV), *Chlamydophila felis* (*C. felis*), *Mycoplasma* spp., and *Bordetella bronchiseptica* have all been reported as potential primary pathogens in feline URI. In general, approximately 80-90% of cases are thought to be caused by one of the two viruses listed. In shelter cats, evidence is building that herpesvirus accounts for the majority of endemic shelter URI in the United States. Calicivirus, while undoubtedly the cause of periodic outbreaks, has not been consistently associated with an increased risk of URI in U.S. shelter populations nor does it appear to spread as readily as herpesvirus or even coronavirus [3, 4]. For instance, in a survey of 573 cats at 8 California shelters comparing pathogen isolation rates in well cats versus those with signs of URI, herpesvirus but not calicivirus was significantly more commonly in cats with signs of URI [5]. In two studies of shelter cats with URI, calicivirus and herpesvirus were isolated in 1/59 versus 51/52 cats [6] and from 0/40 versus 30/40 cats respectively [7]. Mycoplasmas are commonly isolated from both healthy shelter cats and those with URI. Although not likely a significant primary pathogen, mycoplasmas cause respiratory immunosuppression and may worsen other infections. In the California shelter study, *Bordetella* was not associated with an increased risk of URI, but infection has been linked to sporadic shelter outbreaks (unpublished data). *C. felis* is uncommonly found in either healthy shelter cats or those with URI, but when present has been associated with the presence of clinical signs [3, 6, 7].

Herpesvirus commonly establishes a carrier state and is specifically reactivated by conditions commonly affecting shelter cats, including stress, movement from cage to cage, and introduction of new cats into groups [8-10]. When these conditions are well controlled in shelters, overall URI frequency is greatly reduced and the other pathogens may play a relatively greater role in causing what disease remains.

SECONDARY BACTERIAL INFECTION

Primary respiratory pathogens increase cats' susceptibility to secondary bacterial infection, both by causing respiratory irritation and in some cases directly damaging respiratory immune function. In addition to the putative primary pathogens listed above, a wide range of bacteria can be isolated from the respiratory tract of cats sick with URI, including *Pasteurella, E coli, Streptococcus, Enterobacter* and *Staphylococcus* species[6, 11]. Because Gram negative bacteria are frequently isolated, antibiotic treatment targeted at secondary infection should include a Gram negative spectrum.

EPIDEMIOLOGY OF SHELTER URI

We have found wide variation in frequency of URI between shelters and within shelters over time. In a study of 9 shelters for one year, the overall URI rate (cases per 1000 days at risk) varied over 30-fold, from less than 1 case per 1000 days at risk to over 30. Likewise from < 1% to over 25% of total feline care days were devoted to caring for cats sick with URI. As a percentage of intake, URI frequency ranged from < 4% to over 25%. The median time to onset of URI was 10 days for adults and 11 days for juveniles.

ENVIRONMENTAL AND MANAGEMENT FACTORS LINKED TO URI

On an individual cat basis, time in the shelter has been has been found to be the single greatest risk factors for shelter URI [1, 12]. Reducing time spent in the shelter benefits both the individual cat and reduces the overall population density (assuming intake does not concurrently increase), and is one of the most effective means of reducing overall shelter URI rate. Reduced density facilitates provision of good quality housing, which directly impacts feline well-being, greatly facilitates both cleaning and stress relief, and is likely the most important single factor in determining URI frequency in shelters. Because movement from cage to cage or from a cage into a carrier has been linked to both stress and activation of latent herpesvirus infection, double compartment housing that allows for in-situ cleaning is likely important. In one recent study, only 60 out of 1,434 cats admitted to shelters in the United Kingdom developed URI during their shelter stay (2). These shelters were characterized by spacious, double compartment indoor/outdoor housing. In comparison, a U.S. shelter with more traditional ~ 2 ft by 2 ft stainless steel caging recorded URI in 661/2203 cats. In the study of 9 shelters described above, the shelter with the URI frequency of less than 4% provided each cat with a double compartment, laminate "condo" with a total of 9.7 ft^2 of floor space. If group housing it utilized, an increased number of smaller groups are preferable to a few large ones to minimize the need for frequent addition and removal of cats.

TREATMENT

Antibiotics should be used for treatment of feline URI only when clearly indicated. In most cases, antibiotics are chosen on an empirical basis. Several studies have been performed comparing the efficacy of various antibiotics for this use. In a trial in 103 cats with URI comparing marbofloxacin to clavulanic acid-amoxicillin for 5 days, no difference was found (cure or improvement in 87.8 versus 77.8 respectively)[13]. A study comparing a 9 day course of amoxicillin versus azithromycin in 31 shelter cats likewise found no significant difference between treatment groups [14]. Cure rates at 9 days were 38% and 40% respectively. Another 50% of cats in each treatment group were cured after 9 days on the other antibiotic, while 8/31 were not cured by either drug. A study that evaluated the in vitro efficacy of various drugs to secondary bacterial infections in feline respiratory disease found that enrofloxacin had the highest overall efficacy (95%), with cephalexin and clavulanic acid-amoxicillin also reasonably effective at 90% and 84% respectively[11]. Tetracycline was 71% effective. Importantly, however, this study did not report on susceptibility of three potentially important primary pathogens, *Bordetella*, *Mycoplasma*, and *Chlamydophila*. These three bacteria are consistently *not* susceptible to cephalexin, and are frequently susceptible to doxycycline [15, 16]. Doxycycline has outperformed a number of other antibiotics specifically for treatment of *Chlamydophila*[16-19]. A good treatment combination, therefore, may be doxycycline as one treatment option and an antibiotic with better activity against secondary pathogens as the other choice. This would provide coverage for both primary and common secondary bacterial pathogens.

REFERENCES

1. Dinnage, J.D., J.M. Scarlett, and J.R. Richards, *Descriptive epidemiology of feline upper respiratory tract disease in an animal shelter.* J Feline Med Surg, 2009. 2. Edwards, D.S., et al., *Risk factors for time to diagnosis of feline upper respiratory tract disease in UK animal adoption shelters.* Prev Vet Med, 2008. **87**(3-4): p. 327-39. 3. Bannasch, M.J. and J.E. Foley, *Epidemiologic evaluation of multiple respiratory pathogens in cats in animal shelters.* J Feline Med Surg, 2005. **7**(2): p. 109-19. 4. Pedersen, N.C., et al., *Common virus infections in cats, before and after being placed in shelters, with emphasis on feline enteric coronavirus.* J Feline Med Surg, 2004. **6**(2): p. 83-8. 5. Veir, J., et al. *Prevalence of FHV-1, Mycoplasma Spp., and Aerobic Bacteria in Shelter Cats with Acute Upper Respiratory Tract Disease.* in *ACVIM Forum*. 2004. Minneapolis, MN. 6. Veir, J.K., et al., *Prevalence of selected infectious organisms and comparison of two anatomic sampling sites in shelter cats with upper respiratory tract disease.* J Feline Med Surg, 2008. **10**(6): p. 551-7. 7.Spindel, M.E., et al., *Evaluation of pradofloxacin for the treatment of feline rhinitis.* J Feline Med Surg, 2008. **10**(5): p. 472-9. 8.Gaskell, R.M. and R.C. Povey, *Experimental induction of feline viral rhinotracheitis virus re- excretion in FVR-recovered cats.* Vet Rec, 1977. **100**(7): p. 128-33. 9. Maggs, D.J., M.P. Nasisse, and P.H. Kass, *Efficacy of oral supplementation with L-lysine in cats latently infected with feline herpesvirus.* Am J Vet Res, 2003. **64**(1): p. 37-42. 10. Hickman, M.A., et al., *An epizootic of feline herpesvirus, type 1 in a large specific pathogen-free cat colony and attempts to eradicate the infection by identification and culling of carriers.* Lab Anim, 1994. **28**(4): p. 320-9. 11. Schulz, B.S., G. Wolf, and K. Hartmann, *Bacteriological and antibiotic sensitivity test results in 271 cats with respiratory tract infections.* Vet Rec, 2006. **158**(8): p. 269-70. 12. Edinboro, C.H., et al., *A clinical trial of intranasal and subcutaneous vaccines to prevent upper respiratory infection in cats at an animal shelter.* Feline Practice, 1999. **27**(6): p. 7-13.13. Dossin, O., P. Gruet, and E. Thomas, *Comparative field evaluation of marbofloxacin tablets in the treatment of feline upper respiratory infections.* J Small Anim Pract, 1998. **39**(6): p. 286-9. 14. Ruch-Gallie, R.A., et al., *Efficacy of amoxycillin and azithromycin for the empirical treatment of shelter cats with suspected bacterial upper respiratory infections.* J Feline Med Surg, 2008. **10**(6): p. 542-50. 15. Speakman, A.J., et al., *Antimicrobial susceptibility of Bordetella bronchiseptica isolates from cats and a comparison of the agar dilution and E-test methods.* Vet Microbiol, 1997. **54**(1): p. 63-72. 16. Hartmann, A.D., et al., *Efficacy of Pradofloxacin in Cats with Feline Upper Respiratory Tract Disease due to Chlamydophila felis or Mycoplasma Infections.* J Vet Intern Med, 2008. **22**(1): p. 44-52. 17. Sturgess, C.P., et al., *Controlled study of the efficacy of clavulanic acid-potentiated amoxycillin in the treatment of Chlamydia psittaci in cats.* Vet Rec, 2001. **149**(3): p. 73-6. 18. Donati, M., et al., *Feline ocular chlamydiosis: clinical and microbiological effects of topical and systemic therapy.* New Microbiol, 2005. **28**(4): p. 369-72. 19. Owen, W.M.A., et al., *Efficacy of azithromycin for the treatment of feline chlamydophilosis.* Journal of Feline Medicine & Surgery, 2003. **5**(6): p. 305-311.

EPIDEMIOLOGY OF *STREPTOCOCCUS ZOOEPIDEMICUS* INFECTIONS

E. Acke, A. Midwinter, S. Gordon, K. Lawrence, S. Moore, A. Waller, N. French
Palmerston North, New Zealand

INTRODUCTION

Streptococcus equi subsp. *zooepidemicus* (*S. zooepidemicus*) is a well-known commensal organism isolated mainly from the skin and upper respiratory tract mucosa of horses.[1] Transmission may occur through aerosol, the oral route and wound contamination. Clinical signs can develop with the organism as a secondary invader or as the primary pathogen. *S. zooepidemicus* has been associated with equine wound infections, upper and lower airway disease, endometritis and abortion in horses.[2-4] *S. zooepidemicus* is a β haemolytic Lancefield group C *Streptococcus* organism, closely related to *Streptococcus equi* subsp. *equi* (*S. equi*), the causative organism of equine strangles. *S. dysgalactiae* subsp. *equisimilis* (*S. equisimilis*) is also a group C commensal which and has been associated with clinical signs in horses.[2]

STREPTOCOCCUS ZOOEPIDEMICUS IN COMPANION ANIMALS

S. zooepidemicus has emerged as a significant pathogen in dogs over the last decade with fatal haemorrhagic pneumonia in intensively housed dogs as the most important disease manifestation.[5] The infection has been reported to be highly contagious, with clinical signs of pyrexia and mild respiratory disease, progressing to dyspnoea with respiratory haemorrhage and haemothorax, and death within hours to days. To date, four large outbreaks have been reported in intensively housed dogs (shelter or research facilities) in the US and Korea.[6-9] Dogs had appeared healthy prior to the outbreaks and other respiratory pathogens were absent in the majority of animals. Severely affected animals responded poorly to antimicrobial treatment and intensive care.[6-9] Post mortem histopathologic examination showed severe, extensive fibrino-suppurative, necrotizing and haemorrhagic pneumonia in the majority of dogs.[10] Molecular characterisation of the isolates from several dogs in the largest outbreak confirmed a single clone of *S. zooepidemicus* had caused the respiratory infection in the dogs.[8] A study on *S. zooepidemicus* isolation in UK shelter dogs with a history of endemic canine infectious respiratory disease (CIRD) detected *S. zooepidemicus* in 22.3% of the bronchoalveolar lavage samples. Isolation of the organism was associated with increased severity of CIRD and infection with other respiratory pathogens may be have been a predisposing factor for *S. zooepidemicus* infection.[11]

Cases in household dogs have also been reported, with clinical signs ranging from sudden death (attributed to haemorrhagic pneumonia), to chronic rhinitis or an absence of signs.[11-14] Two dogs with chronic rhinitis had lymphoplasmacytic rhinitis as a possible predisposing cause of secondary bacterial rhinitis.[13,14] The majority of cases reported in household dogs were confirmed to have frequent and close contact with horses.[13,14] As the organism is commonly isolated from horses, horses may be a reservoir for infections in other animal species.[13,14]

The first outbreak of *S. zooepidemicus* infections has recently been reported in cats. The cats were housed in a large cattery in Israel where 78 cats died, usually after showing signs of respiratory disease. Mixed cultures were isolated from airway specimens with *S. zooepidemicus* confirmed in the majority of cases. Meningoencephalitis, pleuritis and peritonitis associated with *S. zooepidemicus* were also reported.[15] *S. zooepidemicus* infection associated with rhinitis and rapidly fatal meningoencephalitis was reported in two cats from separate cat shelters in Canada.[16] An outbreak of fatal fibrino-suppurative pneumonia was also recently reported in a rhesus monkey colony, confirmed to be caused by an identical strain of *S. zooepidemicus*.[17]

The epidemiology and pathogenesis of *S. zooepidemicus* infections in dogs and cats is incompletely understood. Intensive housing and co-infections with other pathogens are suspected to be predisposing factors. The source of infections in dogs and cats remains unknown. Molecular analysis of the *S. zooepidemicus* isolates performed to date has failed to detect potential virulence factors which would explain the high pathogenicity of the strains causing severe pneumonia.[3] The prevalence of Lancefield group C streptococci may be on the increase in dogs and cats with the establishment of highly virulent strains, which may cause severe, potentially zoonotic, disease. Previously reported zoonosis has been attributed to close contact with horses or consumption of unpasteurised dairy products. Human infection is rare but can be fatal, and outbreaks have been reported.[18,19] Clinical signs in humans include septicaemia, meningitis, pneumonia, arthritis, and nephritis.[18-21] Zoonotic dog-to-human infection has been reported where cluster analysis of Multilocus Sequence Typing (MLST) results of the *S. zooepidemicus* isolates from a canine pneumonia patient and the veterinary nurse treating the dog were confirmed to be indistinguishable.[22]

PILOT STUDY

There have been no recent studies reporting the prevalence of *S. zooepidemicus* in healthy dogs and cats, and the few studies where speciation of Lancefield group C streptococci was performed in clinical cases, confirmed *S. zooepidemicus* in a small number of submitted isolates.[5,23,24] In addition, contact with horses or exposure in a farming environment may be a potential risk factor for infections in dogs and cats although this has not been investigated previously.[13,14]

A pilot study was performed to assess the isolation rate of *S. zooepidemicus* organisms in healthy dogs and cats with close contact with horses or from a farming environment, and in a subgroup of pets with no contact with horses or farm animals. Oropharyngeal swabs were collected from a total of 235 healthy dogs and cats. Of these, 100 animals lived in an urban environment without close contact with horses while 135 resided on farms as working dogs or pet dogs and cats with frequent close contact with horses or ruminants. Nasopharyngeal swabs were collected from 105 horses which had regular contact with the dogs or cats. The majority of these horses had no recent signs of respiratory disease and they were housed in small groups as opposed to large commercial stud farms. Swabs were placed in charcoal transport medium and inoculated onto selective agar. The Lancefield group was determined for all β haemolytic *Streptococcus* spp. isolates. Lancefield group C streptococci were speciated by sugar fermentation to determine if they were *S. zooepidemicus*, *S. equi*, or *S. equisimilis*.

S. zooepidemicus was isolated from nine horses and four dogs with close contact with horses. *S. equi* was confirmed in two horses, and *S. equisimilis* in 24 horses, six dogs and one cat. These results suggest infection with *S. zooepidemicus* is relatively uncommon, however, the organism was cultured from oropharyngeal swabs obtained from healthy dogs with close contact with horses. It is likely the mucosa in these dogs was transiently colonised and that *S. zooepidemicus* is not a commensal in companion animals. Further investigations are required to confirm this using a larger population of animals and a longitudinal study design.

CONCLUSIONS

S. zooepidemicus should be considered a possible primary or secondary pathogen in dogs and cats with a spectrum of clinical signs. Further investigations are needed to determine the prevalence of *S. zooepidemicus* infections in dogs and cats, especially where there is a history of close contact with other animal species from which a high isolation rate of *S. zooepidemicus* is expected. In addition, the environment needs to be investigated as a potential source for infection and different sample collection methods should be evaluated to confirm *Streptococcus* spp. infections in small animals.

Speciation of small animal β haemolytic *Streptococcus* spp. should be routinely performed in veterinary diagnostic microbiology laboratories. Identification of different *Streptococcus* spp. will improve our understanding of the current pathogenic potential of the organisms. Furthermore, investigations at the molecular level are needed to investigate the possible sources of infection, routes of transmission and virulence factors.

REFERENCES

[1]Anzai T, *et al., American Journal of Veterinary Research* 2000; 61: 162 [2]Timoney JF, *et al., Veterinary Research* 2004; 35: 397 [3]Newton JR, *et al., Veterinary Journal* 2008; 175: 338 [4]Giles RC, *et al., Journal of the American Veterinary Medical Association* 1993; 203: 1170 [5]Priestnall S, *et al., Veterinary Journal* 2010; Epub [6]Garnett NL, *et al., Journal of the American Veterinary Medical Association* 1982; 181: 1371 [7]Kim MK, *et al., Veterinary Record* 2007; 161: 528 [8]Pesavento PA, *et al., Veterinary Pathology* 2008; 45: 51-53 [9]Byun JW, *et al., Journal of Veterinary Science* 2009; 10: 269 [10]Priestnall SL, *et al., Clinical and Vaccine Immunology* 2010; 17: 1790 [11]Chalker VJ, *et al., Veterinary Microbiology* 2003; 95: 149 [12]Gibson D, *et al., Veterinary Record* 2008; 162: 423 [13]Acke E, *et al., Veterinary Record* 2010; 167: 102 [14]Piva SV, *et al., Veterinary Record* 2010; 167: 177 [15]Blum S, *et al., Veterinary Microbiology* 2010; 144: 236 [16]Britton AP, *et al., Journal of Comparative Pathology* 2010; 143: 70 [17]Mätz-Rensing K, *et al., Journal of Medical Primatology* 2009; 38: 328 [18]Minces LR, *et al., Epidemiology and Infection* 2010; Epub [19]Kuusi M, *et al., BMC Infectious Diseases* 2006; 6: Epub [20]Collazos J, *et al., Clinical Infectious Diseases* 1992; 15: 744 [21]Ural O, *et al., Scandinavian Journal of Infectious Diseases* 2003; 35: 206 [22]Abbott Y, *et al., Journal of Medical Microbiology* 2010; 59: 120 [23]Lamm CG, *et al., Veterinary Pathology* 2010; 47: 387 [24]Biberstein EL, *et al., Journal of Clinical Microbiology* 1980; 11: 558

PULMONARY ABNORMALITIES IN CANINE LEPTOSPIROSIS

Barbara Kohn, Dr., Prof., DECVIM
Berlin, Germany

INTRODUCTION

Leptospirosis is a re-emerging zoonotic disease of global importance in animals and humans. In recent years, endemic and epidemic severe pulmonary hemorrhage (SPHS) has increasingly become recognized as an important manifestation of leptospiral infection.[1]

The disease is caused by spirochetes of the genus *Leptospira*. Approximately 250 serovars have been identified within the *Leptospira* complex. Although the pathogenic importance of many serovars is unknown, 6 to 8 are thought to be pathogenic in dogs. Canine leptospirosis was first described in 1899. Prior to 1960, *L. interrogans* serovars Icterohaemorrhagiae and Canicola were responsible for most cases of canine leptospirosis.[2] Recent concerns regarding leptospirosis have focused on the possibility of increasing prevalence of the disease in humans and dogs and the shift in serovars causing the disease.[3-9] In the past two decades multiple reports have been published with only a few cases caused by the classic serovars. In Europe and the United States today, seroconversion most commonly occurs to serovars including Grippotyphosa, Bratislava and Pomona; furthermore seroconversion to the serovars Autumnalis, Saxkoebing, Australis, Ballum, Bataviae, Pyrogenes and others have been described in confirmed cases of canine leptospirosis.[1,10] Each serovar has a primary host that maintains the organism and contributes to its dissemination in the environment. The bacteria are maintained in the renal tubules of the reservoir hosts (e.g. rodents, pigs, ruminants, skunks, racoons, deer) and are excreted in the urine. It is not yet known if such a carrier state exists in dogs infected with other serovars than *L. interrogans* serovar Canicola. Interestingly, leptospirosis is of increasing importance and prevalence in the rural areas of Berlin where the population of urban wild boars is growing dramatically and these wild boars have recently been shown to represent a significant reservoir for *Leptospira* serovars Pomona and Bratislava.[11]

Leptospires can be transmitted directly through urine, venereal routes, placental transfer, bites or ingestion of infected tissues. Indirect transmission occurs through exposure of susceptible animals or humans to a contaminated environment (water, soil, or vegetation). The usual portal of entry is abraded skin or (intact) mucous membranes. Following the leptospiremic phase leptospires can invade a variety of organs. Organ damage occurs mainly through replication, induction of cytokine production and inflammatory cell invasion. The extent of damage varies depending on the virulence of the organism and susceptibility of the host.[2] The following organs may be affected: kidneys, liver, endothelial cells, meninges, uvea / retina, uterus, lungs. Secondary immune-mediated diseases (hemolytic anemia, polyarthritis) and myalgia may occur. The clinical presentation is variable ranging from subclinical infection to peracute, acute and subacute/chronic manifestations of the disease. Diagnosis can be based on antibody detection using the microscopic agglutination test (MAT), ELISA, PCR testing of blood, urine, CSF or tissue, culture of the organisms, or direct visualization of organisms in fresh urine or tissue sections.[1,2]

Pulmonary form in human leptospirosis

Respiratory manifestations have been described in 27-85% of human patients.[12-14] Pulmonary signs ranging from chest pain, coughing, dyspnea, hemoptysis to acute respiratory distress syndrome have been reported for patients living in warm-climate rural and urban areas but also for patients living in urban temperate regions of Europe.[12,15-19] The observation of a true emergence of SPHS may be explained by the introduction of new strains in a determinate geographic area. To date, however, there is no convincing evidence that different serovars have the intrinsic tendency to cause distinct presentations of leptospirosis.

Respiratory symptoms usually appear between the fourth and sixth day of disease and may lead to death in less than 72 hours.[15] Alveolar hemorrhage presented as dyspnea and hemoptysis is the main pulmonary manifestation in humans with leptospirosis. In a human study chest radiographs were abnormal in 65% of the patients with leptospirosis; 66% of these patients had interstitial and 34% alveolar patterns.[14] In a series of 58 patients with leptospirosis, 37 (64%) had pulmonary radiographic findings. Three radiographic patterns were evident in these patients: small nodular densities (57%), confluent areas of consolidation (16%), and diffuse, ground-glass density (27%). Abnormalities were bilateral, non-lobar in all cases, and had a marked tendency toward peripheral distribution.[20] Pleural effusions may occur. A few cases of leptospiral pneumonia have also been reported. Case fatality rates of 13 – 60% have been described in patients with severe manifestation of the lung form and pulmonary involvement has become the main cause of death due to leptospirosis in some countries.[14-17]

Histopathologically intra-alveolar hemorrhage was detected in the majority of patients, even in the absence of overt pulmonary symptoms.[21] Although infiltration of alveolar spaces by monocytes and neutrophils occurs, inflammatory infiltrates are usually not prominent. Other findings, including pulmonary edema, fibrin deposition and hyaline membrane formation are frequent. Leptospires may be seen within endothelial cells in interalveolar septa and also attached to capillary endothelial cells.[22]

Pathophysiology of pulmonary abnormalities

The pathogenesis of lung abnormalities in leptospirosis is currently under investigation; pulmonary hemorrhages are most likely multifactorial. Two main mechanisms have been suggested: a toxin-mediated capillary vasculitis and/or the immune responses of the host.[16,23] Since lung tissue usually shows much lower numbers of leptospires than liver and blood tissue, pulmonary abnormalities may be due to auto-immune mechanisms or due to exposure of circulating toxins produced by the pathogen at distant sites, such as the liver.[1,23] Outer membrane proteins are toxins related to leptospirosis pathogenesis. Other possible toxins include membrane glycoproteins, hemolysin, and lipopolysaccharides. Although leptospiral antigen has been detected in autopsied human lung tissue, suggesting a local effect of the bacteria at the endothelial level, this is still a controversial issue.[22]

Besides the toxin–mediated process, an indirect pathogenic mechanism mediated by the host immune response to infection is suggested.[1] Due to the fact that the inflammatory response in leptospirosis is mild in relation to the amount of vascular damage in severe cases, leptospires might employ a specific mechanism to diminish the magnitude of the host inflammatory response. A deposition of IgM, IgG, IgA and C3 along the alveolar basement membrane was detected in a guinea pig model suggesting an autoimmune process may be the etiology of severe pulmonary hemorrhage.[23] Linear deposition of immunoglobulin and complement was also detected on the alveolar surface of lung specimens of humans with SPHS.[24] The suggested immune-pathogenic mechanism may justify the use of short-acting glucocorticoids in cases of severe dyspnea and pulmonary hemorrhage which has been described in human medicine.[25] The pathogenesis of lung and organ injury might also involve inducible nitric oxide synthase and nitric oxide production causing severe hypotension and bradycardia accompanied by autonomic dysfunction.[26] Leptospirosis profoundly influences the sodium transport capacity of alveolar epithelial cells suggesting that impaired pulmonary fluid handling can impair pulmonary function, increasing the chance of lung injury.[27]

Thrombocytopenia, uremia, and coagulation disorders such as disseminated intravascular coagulation (DIC) individually or as a group might be contributing factors that worsen pulmonary bleeding episodes. However, it seems unlikely that disorders of the primary or secondary hemostasis are the sole cause for pulmonary hemorrhage since extensive hemorrhages were exclusively observed in the lungs of dogs with leptospirosis.[10] In a guinea pig model of severe pulmonary leptospirosis marked alterations in platelet numbers or other laboratory signs of DIC were not detected.[24] The striking tropism for the lung and the lack of evidence of a general hemorrhagic diathesis suggest that pulmonary microvessels may represent a highly selective target of the mechanism involved, as hypothesized in human SPHS. The involvement of blood vessels might be comprehended as a sepsis-like diffuse process of endothelial activation/damage rather than a classical systemic vasculitis.[28]

Pulmonary abnormalities in canine leptospirosis

In an experimental model of canine leptospirosis pulmonary hemorrhage was found in Beagles infected with *Leptospira kirschneri* serovar Grippotyphosa and *Leptospira interrogans* serovar Pomona in addition to the common findings of interstitial nephritis, renal tubular epithelial degeneration and hepatitis.[29,30] There have been a few clinical reports of dogs suffering from leptospirosis that display lung abnormalities clinically and radiologically.[6,7,31-34] In these studies pulmonary changes were described in 9 – 93% of the dogs.

Recently we described 50 dogs with leptospirosis.[10] Our diagnosis was based on MAT, blood or urine PCR and histopathology. Similar to earlier studies the most frequent diagnostic titers against *Leptospira* serovars in dogs were to serovars Grippotyphosa, Bratislava, and Pomona. Based on clinical and/or radiological signs, patients were grouped into dogs with lung abnormalities (group 1) or without (group 2). Severity of respiratory distress was scored as mild to moderate (grade 1) or severe (grade 2). Thoracic radiographs were scored based on pulmonary changes and location as grade 1 (caudal interstitial pattern), 2 (generalized mild to moderate reticulonodular interstitial pattern) or 3 (generalized severe reticulonodular interstitial pattern with patchy alveolar consolidations). Thirty-five (70%) dogs had radiological pulmonary changes (grade 1: 5; grade 2: 14; grade 3: 16); 31 of them had pulmonary distress. Thirteen of these 31 dogs (42%) displayed mild to moderate dyspnea and 18 (58%) severe dyspnea at the time of most severe respiratory compromise. Seven of the 31 dogs (23%) with dyspnea developed fatal hemoptysis. There was no correlation between the serovars with the highest MAT titers and the presence or absence of pulmonary lesions. The laboratory results of dogs with and without lung abnormalities were compared. Except for the occurrence of thrombocytopenia there was no significant difference between dogs with or without lung abnormalities regarding the occurrence of anemia, leukocytosis, azotemia ot increased liver enzymes / hyperbilirubinemia. Thirty-two of the 50 dogs (64%) survived whereas 18 (36%) had to be euthanized. The severity of dyspnea correlated with the percentage of dogs that survived. Twelve of the 18 dogs (67%) suffering from severe dyspnea had to be euthanized due to respiratory distress. Of the dogs with mild to moderate dyspnea, 15% were euthanized due to acute renal failure or sepsis, compared with 21% of the dogs without clinical respiratory signs.

Ten of the dogs that were euthanized were submitted to a necropsy investigation of major organ systems. Nine of these dogs had severe dyspnea, the tenth dog had no clinical or radiological pulmonary signs. All dogs had severe, diffuse, acute, pulmonary alveolar and subpleural hemorrhage partly with multifocal to coalescing coagulative necrosis of alveolar walls, and with hyalin membranes. Similarly, in humans intra-alveolar hemorrhage has been detected even in patients without overt pulmonary symptoms.[1] Therefore, lung abnormalities in dogs with leptospirosis might be more common than detected by clinical signs or thoracic radiography. The pulmonary changes were homogeneously distributed throughout the entire lungs. Alveoli were filled with varying amounts of edematous fluid and fibrin. Multifocal microvascular hyalin and fibrin thrombi were present, most prominent in areas of severe alveolar wall necrosis. Major airways were virtually unaffected. Erythrophagocytosis was rare or absent in all dogs. All dogs had unremarkable blood vessels of all calibres. Levaditi stains failed to detect any leptospiral organisms in the lungs whereas typical spirochetoid organisms were identified in 6 of 10 kidneys. The pathological findings were similar to lesions found in human patients with pulmonary leptospirosis.

REFERENCES

[1]Bharti AR, et al. **Lancet Infect Dis** 2003;3:757. [2]Goldstein RE. In: Textbook of veterinary internal medicine. Philadelphia, PA:WB Saunders; 2010:863. [3]Nardone A, et al. **Clin Infect Dis** 2004;39:751. [4]Jansen A, et al. **Emerg Infect Dis** 2005;11:1048. [5]Steger Lieb A, et al. **Schweiz Arch Tierheilkd** 1999;141:499. [6]Adin CA, Cowgill LD. **J Am Vet Med Assoc** 2000;216:371. [7]Goldstein RE, et al. J Vet Intern Med 2006;20:489. [8]Geisen V, et al. **J Small Anim Pract** 2007;48:324. [9]Gerlach T, Stephan I. **Tierärztl Prax** 2007;38:421. [10]Kohn B, et al. **J Vet Intern Med** 24(6):1277. [11]Jansen A, et al. **Emerg Infect Dis** 2007;13:739. [12]Tattevin P, et al. **Lung** 2005;183:283. [13]Courtin JP, et al. **Rev Pneumol Clin** 1998;54:392. [14]Paganin F, et al. **Intensive Care Med** 2007;33:1959. [15]Silva JJ, et al. **Rev Soc Bras Med Trop** 2002;35:395. [16]Dolhnikoff M, et al. **Braz J Infect Dis** 2007;11:142. [17]Carvalho CR, Bethlem EP. **Clin Chest Med** 2002;23:469. [18]Segura ER, et al. **Clin Infect Dis** 2005;40:343. [19]Jansen A, et al. **Eur J Clin Microbiol Infect Dis** 2006;25:544. [20]Im JG, et al. **Am J Roentgenol** 1989;152:955. [21]Levett PN. **Clin Microbiol Rev** 2001;14:296. [22]Nicodemo AC, et al. **Am J Trop Med Hyg** 1997;56:181. [23]Nally JE, et al. **Am J Pathol** 2004;164:1115. [24]Croda J, et al. **Clin Microbiol Infect** 2010;16(6):593. [25]Trivedi SV, et al. **J Assoc Physicians India** 2001;49:901. [26]Chen HI, et al. **Pathology** 2007;39:339. [27]Andrade L, et al. **Am J Physiol Renal Physiol** 2007;292:F586. [28]Medeiros Fda R, et al. Leptospirosis-associated disturbances of blood vessels, lungs and hemostasis. **Acta Trop** 2010;115:155. [29]Greenlee JJ, et al. **Am J Vet Res** 2004; 65:1100. [30]Grenlee JJ, et al. **Am J Vet Res** 66 2005;1816. [31]Harkin KR, Gartrell CL. **J Am Anim Hosp Assoc** 1996;32:495. [32]Birnbaum N, et al. **J Small Anim Pract** 1998;39:231. [33]Baumann D, Fluckiger M. **Vet Radiol Ultrasound** 2001;42:305. [34]Francey T, Schweighauser A. Proceedings 18th ECVIM-CA Congress, Ghent 2008.

ANIMAL & PUBLIC HEALTH IMPLICATIONS OF METHICILLIN-RESISTANT STAPHYLOCOCCI

J. Scott Weese, DVM, DVSc, DACVIM
Guelph, Ontario, Canada

INTRODUCTION

Despite being inherently susceptible to virtually every known antimicrobial class, staphylococci are notorious for antimicrobial resistance. From the first introduction of antimicrobials, staphylococci have demonstrated an impressive ability to become resistant. Penicillin-resistance was identified shortly after penicillin became widely used. New drug development outpaced resistance initially, but the ability of staphylococci to become resistant was repeatedly demonstrated as the introduction of new drugs was typically followed (sometimes rapidly) by emergence of resistance. Included in this pattern was resistance to methicillin, which emerged not long after the use of methicillin began in 1959. While resistance to methicillin itself is not a major concern, methicillin-resistance is caused by production of an altered penicillin binding protein (PBP2a) that has a poor affinity for beta-lactam antimicrobials and confers resistance not just to methicillin, but to virtually all beta-lactams; penicillins, cephalosporins and carbapenems. Production of PBP2a is mediated by the *mecA* gene, a gene that is located on a staphylococcal chromosomal cassette (SCC*mec*), a site where other resistance genes can also be incorporated. In addition to broad beta-lactam resistance, methicillin-resistant staphylococci are often resistant to a wide range of other antimicrobials, severely limiting treatment options. While methicillin-resistance was first noted as a problem in humans, it has emerged as a pressing issue in companion animals, and there have been striking recent changes in the epidemiology of methicillin-resistant staphylococcal infection and colonization in dogs and cats.

There is often confusion associated with staphylococci, particularly methicillin-resistant staphylococci. It is critical to remember that *all staph are not created alike.* The clinical relevance, prevalence of colonization, incidence of disease and potential for interspecies transmission vary greatly between different staphylococcal species. Inadequate identification practices, variable and sometimes improper susceptibility testing methodology, and variable reporting of results can lead to confusion and possibly suboptimal patient care. A critical aspect of dealing with staphylococci is knowing (with confidence) the species that is involved. There are two main groups of staphylococci; coagulase positive and coagulase negative. Coagulase positive species (e.g. *S. aureus, S. pseudintermedius, S. schleiferi coagulans*) are the most virulent and of greatest clinical relevance. Differentiation of these, or at a minimum, differentiation of *S. aureus* from other coagulase positive staphylococci is needed for both interpretation of zoonotic risks and the use of proper antimicrobial susceptibility testing methodology (since there are different recommendation for *S. aureus* versus other staphylococci). Coagulase negative staphylococci tend to be of limited virulence and after often considered together as a single group, although there is some evidence of variation in virulence and clinical relevance.

METHICILLIN-RESISTANT *STAPHYLOCOCCUS PSEUDINTERMEDIUS* (MRSP)

MRSP appears to have emerged and disseminated internationally in companion animals at a truly amazing rate, with rapid development of a very high level of drug resistance. MRSP infections are being identified virtually everywhere that people are looking for them, and the increase in incidence of disease, while not objectively studied, seems to be dramatic, particularly among dogs with pyoderma. It has been called a serious emerging problem in small animal veterinary medicine and one that requires urgent action to control its spread.[1] Highly antimicrobial resistant strains are commonly encountered, and there may be few viable treatment options. Curiously, while MRSP appears to only have emerged as a problem recently, MRSP strains from dogs and cats are typically much more resistant than MRSA strains which have presumably been around much longer.

Staphylococcus pseudintermedius is the most common pathogenic staphylococcus in dogs and a common commensal that can be found in or on a large percentage, if not all, healthy dogs, and a smaller percentage of cats.[2-6] As with methicillin-susceptible *S. pseudintermedius* (MSSP), MRSP can be found in or on healthy dogs and cats. Carriage rates of 0-17% in dogs and 0-1.2% in healthy cats have been reported,[3,5,7-9] and it appears that the rate of colonization is increasing in many regions. Risk factors for MRSP colonization have not been adequately investigated, but there is some evidence that routine antimicrobial administration may be an important driving force for MRSP colonization in dogs.

MRSP is an opportunistic pathogen and colonization does not necessarily lead to disease. Indeed, it is likely that the vast majority of colonized animals never develop a clinical infection. The risk of infection in MRSP carriers has not been reported, but it is reasonable to assume that MRSP carriers are at some increased risk of MRSP infection, at least in certain situations (e.g. after surgery). Limited study of risk factors for infection has been performed but antimicrobial administration, hospitalization or surgery within 30 days prior to the onset of infection were associated with MRSP versus MSSP infection in one study.[10]

Skin, ear and other soft tissue infections predominate.[10-13] Very high rates of MRSP infection are now being identified among canine pyoderma patients. Post-operative infections appear to be increasingly common, particularly TPLO infections.[10] MRSP appears to be the most common and frustrating cause of surgical site infections in many regions, especially in orthopedic surgery. Given the high level of resistance that is typically present, with correspondingly few treatment options, this is of significant concern. Further, resistance continues to increase, and resistance is now being identified to antimicrobials that were commonly effective a short time ago (e.g. chloramphenicol, doxycycline, trimethoprim-sulfas). There is no indication that MRSP infections are inherently more serious than infections caused by methicillin-susceptible strains, however they could ultimately be associated with increased morbidity ad mortality because of failure of initial empirical antimicrobial therapy and limited treatment options.

Public Health Issues

Despite the fact that exposure to *S. pseudintermedius* must be very common in pet owners (given the high prevalence of colonization in pets), human *S. pseudintermedius* infections are rarely reported and colonization rates in veterinary personnel and pet owners are rather low (1-4%),[5,14] suggesting poor host adaptation to humans. Methicillin-resistance simply confers antimicrobial resistance, not inherently increased virulence, and there is no evidence that MRSP is more likely to cause an infection than MSSP. Accordingly, the low incidence of MSSP infections in humans should indicate a correspondingly low risk of MRSP infection. Transmission of MRSP between dogs and people is likely based on reports of indistinguishable isolates from people and dogs[15,16] and isolation of MRSP from a small percentage of owners of dogs with MRSP pyoderma.[17] However, there is currently limited public health concern regarding MRSP. However, the limited but increasing number of reports of MRSP infection in humans[18,19] should not be dismissed. Given the highly drug resistant nature (and associated treatment difficulty) of MRSP, while the risk may be very low, the consequences of infection could be high, so care should be taken to use good infection control practices around infected animals in households and veterinary clinics.

The potential for under-reporting of MRSP must also be considered if human diagnostic laboratories are not adequately differentiating MRSP from the much more common coagulase positive *Staphylococcus* spp., MRSA.[20]

METHICILLIN-RESISTANT *STAPHYLOCOCCUS AUREUS* (MRSA)

MRSA tends to receive a higher profile than MRSP because of its huge impact in human medicine, however it is a much less common cause of infection in dogs and cats. Despite the importance of this pathogen, the animal health implications of MRSA are nowhere near as large in dogs and cats compared to MRSP. Regardless, attention must be paid to this organism because of the potential for serious and difficult-to-treat infections and the potentially greater zoonotic implications.

It is assumed that the emergence of MRSA in companion animals is directly related to MRSA in humans. As MRSA expanded in humans in the general population in the 1990s and 2000s, exposure of pets was inevitable. Humans are thought to be the ultimate source of infection for a large percentage, if not the vast majority, of infections in pets, and changes in the epidemiology of MRSA in humans can be reflected in changes in trends in pets. Virtually all MRSA isolates from pets are recognized human epidemic clones and identification of other strains in dogs and cats is rare.[21-25] Sequence type 398 MRSA, a livestock-associated strain, has been found in a limited number of dogs,[26-28] however dogs were more likely to be infected by people that were exposed to livestock.

First reported in the 1990s, MRSA infection and colonization have now been identified in dogs, cats and other companion animals across the world.[25,29-34] *Staphylococcus aureus* can be isolated from the nares, perineum or intestinal tract of 12-14% of healthy dogs and 4.3-20% of healthy cats,[3,5,35] and not surprisingly, MRSA can also be found in healthy individuals. Reported colonization rates are variable but tend to be 0-3.3% in healthy dogs and 0-6% in healthy cats.[35-40] Higher rates can be encountered periodically in animals in veterinary facilities.[31,41] However, most MRSA cases are not associated with veterinary hospitals. Higher rates can also be found in specific dog populations such as households where another pet has an MRSA infection,[42] or during outbreaks in breeding or rescue kennels.[23,28] Being owned by a human healthcare worker and participation in hospital visitation programs have been identified as risk factors for MRSA colonization in dogs, and are logical based on the increased likelihood of exposure to colonized people.[21,30] Contact with children has also been identified as a risk factor.[21] While these, and potentially other, risk factors should be considered, MRSA can be identified in any animal and absence of known risk factors should not lead to excluding MRSA from consideration.

Most animals that are colonized with MRSA have no signs of infection and may never develop a clinical infection, but opportunistic infections can develop. *S. aureus* produces a similar range of disease as *S. pseudintermedius*. The main difference between *S. pseudintermedius* and *S. aureus* (and therefore between MRSP and MRSA) is the incidence of disease, not disease location or severity. Clinically, MRSA infections are indistinguishable from those caused by methicillin-susceptible strains or other coagulase positive staphylococci, predominantly skin and ear infections, with smaller numbers of other opportunistic infections.[3,22,24,25,43-47] It is possible that infections more commonly associated with human contact (e.g. wound and surgical site infections contaminated by hands of owners or veterinary personnel) are more likely to be caused by *S. aureus* (including MRSA) but objective data are lacking.

Public Health Issues

Significant attention has been given to public health concerns regarding pets and MRSA. Given the importance of MRSA in human medicine, it is perhaps not surprising that serious concern has accompanied identification of MRSA in pets. However, as knowledge regarding MRSA advances, it is becoming evident that while interspecies transmission of MRSA certainly does occur in households, human-to-animal transmission is probably most common. The role of pets in human MRSA infection is completely unclear. It is likely that pets play at best a very limited role in human MRSA infection, however zoonotic infections certainly cannot be dismissed and in some situations an infected or colonized animal could pose a reasonable risk to humans.

Colonization of people in contact with infected or colonized dogs and cats (often with demonstration of the same strain in humans and animals) has been widely reported.[22,25,30,32,48-50] Pets have been implicated as sources of infection for humans in households,[32,49] however conclusions are often overstated. Virtually all studies have involved concurrent testing of people and pets. While finding the same MRSA strain in both groups strongly supports interspecies transmission, it cannot determine the direction of transmission. Further, since the MRSA strains found in pets are human epidemic clones, and because the average pet has little contact with other people or animals, it is most likely that a human household member would bring MRSA into the home, with the potential for subsequent dissemination to other humans and animals. This assessment should not be taken as an indication that there is no risk from pets, since regardless of how a pet is ultimately infected, it could still transmit MRSA to humans.

Numerous studies of MRSA colonization in veterinary personnel have been performed, reporting colonization rates of up to 18% in small animal veterinary personnel.[51-54] It cannot be determined with certainty that these colonization rates reflect acquisition of MRSA from animals however identification of higher rates than present in the general population (1.5-3%),[55,56] and the lack of other explanations as to why veterinary personnel would be at higher risk than others provide support of occupational origin. Whether this constitutes a significant health risk for veterinary personnel is unclear. There are limited reports of MRSA infection in veterinary personnel that have worked with infected or colonization animals,[57,58] along with numerous anecdotal reports. Further, the high rates of colonization could increase the risk of infection in some situations such as if colonized veterinary personnel undergo surgery or are admitted to hospital. Another concern is the potential for veterinary clinics to act as reservoirs of MRSA, resulting in infection or colonization of patients and potentially subsequent exposure of patients' families.

The true role of pets in human MRSA infections is impossible to determine at this point. Considering the epidemic of community-associated MRSA infections in people and endemic MRSA in human healthcare facilities, pets certainly play a minor role overall in human infections. Yet, with the scope of this problem, a minor role from a population standpoint could still represent a relevant source of infection. The potential role of pets in human MRSA infections should be considered, but in a balanced manner.

METHICILLIN-RESISTANT *STAPHYLOCOCCUS SCHLEIFERI* (MRSS)

Staphylococcus schleiferi consists of two subspecies, the coagulase positive *S. schleiferi* subsp *coagulans* and coagulase negative *S. schleiferi* subsp *schleiferi*. These are less common causes of infection compared with *S. pseudintermedius* and *S. aureus*, but a large percentage of diagnostic laboratories do not attempt to different *S. schleiferi coagulans* from *S. pseudintermedius* or *S. schleiferi schleiferi* from other coagulase negative species so there are limitations in understanding of the role of these species in disease. *Staphylococcus schleiferi coagulans* can be isolated in 0.8-4% of healthy dogs and 0-2% of healthy cats.[3,5,35] Colonization with methicillin-resistant *S. schleiferi* (MRSS) has been identified in 0-2% of dogs[3,5,8] and MRSS infections are being increasingly reported.[43,59,60] *S. schleiferi coagulans* is most commonly implicated in pyoderma and otitis externa in dogs, but other opportunistic infections such as urinary tract infection and pneumonia have been reported.[3,43,61-65] Less in known about *S. schleiferi schleiferi,* but it is likely present in or on a small percentage of healthy dogs and cats. It has been implicated as a cause of pyoderma and otitis,[43,62,64] and methicillin-resistant infections, while poorly characterized, presumably do the same.

Public health concerns are currently limited. Human *S. schleiferi coagulans* infections are extremely rare. In one case, a dog with otitis was thought to be the source of *S. schleiferi coagulans* endocarditis in an immunocompromised person,[66] but it was not possible to confirm that the dog was the source. While *Staphylococcus schleiferi schleiferi* periodically causes human infections, it is probably a human commensal species so pets are unlikely to be the source.

COAGULASE NEGATIVE STAPHYLOCOCCI (MR-CoNS)

MR-CoNS are common commensals and generally of limited virulence. MR-CoNS infections may be overdiagnosed because MR-CoNS can be isolated as contaminants from various superficial body sites. In human medicine, MR-CoNS are primarily a concern in hospitalized individuals.[67] The situation may be similar in dogs and cats, with most CoNS being of minimal pathogenicity and mainly of relevance in highly compromised patients. While CoNS are generally considered as a group, it is possible that there are differences between CoNS species. This mainly relates to three organisms; *S. schleiferi* subspecies *schleiferi, S. epidermidis* and *S. felis*. *Staphylococcus schleiferi schleiferi* and *S. epidermidis* have been implicated as a cause of pyoderma and otitis,[43,62,64,68] and may be inherently more pathogenic in skin and ear infections than other CoNS. A primary role of *S. felis* in urinary tract infections in cats has also been suggested.[69] These species may, therefore, be relevant pathogens in community-associated disease, unlike most other CoNS. As with other staphylococci, methicillin-resistant strains are inherently no more pathogenic than methicillin-susceptible strains, and the implications of colonization with MR-CoNS are typically inconsequential.

Coagulase negative staphylococci, including MR-CoNS are common in both animals and humans. While many of the same CoNS species can cause disease both people and pets, it is unlikely that animals are relevant sources of human infection.

REFERENCES

[1]Gronlund-Andersson U, et al. Proc ASM-ESCMID conference on methicillin-resistant staphylococci in animals 2009;A-40. [2] Fazakerley J, et al. **Vet Dermatol** 2009;20:179-184. [3] Griffeth GC, et al. **Vet Dermatol** 2008;19:142-149. [4] Hartmann FA, et al. **Vet Microbiol** 2005;108:119-131. [5] Hanselman BA, et al. **Can Vet J** 2009;50:954-958. [6] Saijonmaa-Koulumies LE, et al. **Vet Dermatol** 2002;13:123-130. [7] Epstein CR, et al. **Infect Genet Evol** 2009;9:283-285. [8] Hanselman BA, et al. **Vet Microbiol** 2007;126:277-281. [9] Vengust M, et al. **Lett Appl Microbiol** 2006;43:602-606. [10]Weese JS, et al. Proc ASM-ESCMID conference on methicillin-resistant staphylococci in animals 2009;A-67.[11] Loeffler A, et al. **Vet Dermatol** 2007;18:412-421.[12] Kadlec K, et al. **J Antimicrob Chemother** 2010. [13] Perreten V, et al. **J Antimicrob Chemother** 2010;65:1145-1154. [14] Talan DA, et al. **J Clin Microbiol** 1989;27:2393. [15] Sasaki T, et al. **J Clin Microbiol** 2007;45:1118-1125. [16] van Duijkeren E, et al. **Veterinary Microbiology** 2008;128:213-215. [17] Frank LA, et al. **Vet Dermatol** 2009;20:496-501. [18] Stegmann R, et al. **J Antimicrob Chemother** 2010:1-2. [19] Van Hoovels L, et al. **J Clin Microbiol** 2006;44:4609-4612. [20] Pottumarthy S, et al. **J Clin Microbiol** 2004;42:5881-5884. [21] Lefebvre SL, et al. **J Am Vet Med Assoc** 2009;234:1404-1417. [22] Leonard F, et al. **Vet Rec** 2006;158:155-159. [23] Loeffler A, et al. **Vet Microbiol** 2009. [24] O'Mahony R, et al. **Vet Microbiol** 2005;109:285-296. [25] Weese J, et al. **Vet Microbiol** 2006;115:148-155. [26] Witte W, et al. **Emerging Infect Dis** 2007;13:255-258. [27] Nienhoff U, et al. **J Antimicrob Chemother** 2009;64:660-662. [28] Floras A, et al. **Vet Rec** 2010;166:826-827. [29] Abbott Y, et al. **Vet Rec** 2010;166:451-455. [30] Boost M, et al. **Clin Microbiol Infect** 2007;13:731-733. [31] Loeffler A, et al. **J Antimicrob Chemother** 2005;56:692-697. [32] Manian FA. **Clin Infect Dis** 2003;36:e26-28. [33] Rankin S, et al. **Vet Microbiol** 2005;108:145-148. [34] Cefai C, et al. **Lancet** 1994;344:539-540. [35] Abraham J, et al. **Vet Dermatol** 2007;18:252-259. [36] Kottler S, et al. **J Vet Intern Med** 2009;24:132-139. [37] Baptiste K, et al. **Emerg Infect Dis** 2005;11:1942-1944. [38] Loeffler A, et al. **Epidemiology and infection** 2010:1-10. [39] Hanselman, et al. **Vet Microbiol** 2007;126:277-281. [40] Hanselman BA, et al. **Can Vet J** 2009;50:954-958. [41] Weese JS, et al. **J Am Vet Med Assoc** 2007;231:1361-1364. [42] Faires MC, et al. **J Am Vet Med Assoc** 2009;235:540-543. [43] Morris D, et al. **Vet Dermatol** 2006;17:332-337. [44] Owen MR, et al. **J Small Anim Pract** 2004;45:609-612. [45] Rankin S, et al. **Vet Microbiol** 2005;108:145-148. [46] Morris D, et al. **Am J Vet Res** 2006;67:1421-1425. [47]Faires M, et al. **Emerg Infect Dis** 2010. [48] van Duijkeren E, et al. **J Clin Microbiol** 2005;43:6209-6211. [49] Sing A, et al. **N Engl J Med** 2008;358:1200-1201.[50] Vitale C, et al. **Emerging Infect Dis** 2006;12:1998-2000. [51] Anderson ME, et al. **Vet Microbiol** 2007;129:410-417. [52] Burstiner LC, et al. **Vet Surg** 2010;39:150-157. [53] Hanselman B, et al. **Emerg Infect Dis** 2006;12:1933-1938. [54] Wulf M, et al. **Emerging Infect Dis** 2006;12:1939-1941. [55] Hanselman BA, et al. **Can J Infect Dis Med Microbiol** 2008;19:405-408. [56] Gorwitz RJ, et al. **J Infect Dis** 2008;197:1226-1234. [57] Weese J, et al. **Emerg Infect Dis** 2005;11:430-435. [58] Weese J, et al. **Vet Microbiol** 2006;114:160-164. [59] Jones R, et al. **J Am Vet Med Assoc** 2007;230:221-227. [60] Kania S, et al. **Am J Vet Res** 2004;65:1265-1268. [61] Bes M, et al. **Vet Rec** 2002;150:487-488. [62] Frank L, et al. **J Am Vet Med Assoc** 2003;222:451-454. [63] Igimi S, et al. **Int J Syst Bacteriol** 1990;40:409-411. [64] May ER, et al. **J Am Vet Med Assoc** 2005;227:928-931. [65] Rich M, et al. **Vet Rec** 2007;161:107. [66] Kumar D, et al. **Transplant Infect Dis** 2007;9:336-338. [67]Archer GL, et al. Principles and practice of infectious diseases. 2005;2352-2360.[68] Penna B, et al. **Vet Dermatol** 2010;21:292-296. [69] Litster A, et al. **Vet Microbiol** 2007;121:182-188.

THE ROLE OF BLOOD BORNE PATHOGENS IN FELINE FEVER OF UNKNOWN ORIGIN

Michael R. Lappin, DVM, PhD, DACVIM
Fort Collins, Colorado

A number of feline blood borne agents have been grown, amplified or have induced serum antibodies in the serum of cats with fever. The purpose of this comprehensive review is provide an update on the diagnosis and management of fever associated with *Anaplasma phagocytophilum*, *Bartonella* spp, *Ehrlichia* spp., haemoplasmas, and *Rickettsia* spp. infections of cats. Please also see the AAFP Panel report on feline bartonellosis www.catvets.com and the ACVIM *Ehrlichia* Consensus Statement (www.acvim.org) for further information.

ANAPLASMOSIS

Cats have shown to be susceptible to *A. phagocytophilum* infection after experimental inoculation. DNA of *A. phagocytophilum* has been amplified from naturally exposed cats in several countries including Sweden, Denmark, Ireland, and the United States (1-3). Cats living in endemic areas are commonly seropositive but a validated serological assay is not available. As in dogs, *A. phagocytophilum* is transmitted by *Ixodes* ticks and so infections of cats are likely to be most common in these areas.

While the pathogenesis of disease associated with *A. phagocytophilum* in cats is unknown, some cats experimentally inoculated with *A. phagocytophilum* developed anti-nuclear antibodies and increased IFN-gamma mRNA suggesting that an immune pathogenesis of disease may contribute to the clinical findings (4). Fever, anorexia, and lethargy are the most common clinical abnormalities in naturally infected cats. However, in a recent experimental study in my laboratory, cats infected with *A. phagocytophilum* by exposure to wild caught *Ixodes* spp. ticks remained clinically normal over the 70 day study period in spite of being PCR positive for *A. phagocytophilum* DNA in blood for several weeks (Lappin MR, Unpublished data, 2011). Whether or not this agent is associated with chronic recurrent fever in cats is unknown.

Cats with fever in endemic areas can have blood smears examined cytologically but morulae are less commonly detected than in dogs. Some commercial laboratories offer serologic testing or PCR assays to amplify *A. phagocytophilum* DNA from blood. Approximately 30% of cats with proven clinical infections induced by *A. phagocytophilum* are seronegative when first assessed serologically, but most of the proven cases evaluated to date have ultimately seroconverted. Some mountain lions with *A. phagocytophilum* DNA amplified from blood have been serum antibody negative and so a single negative antibody result in an acutely infected cat does not exclude infection. Therefore, cats with suspected anaplasmosis may need convalescent serum samples to prove infection. Alternately, antibody testing could be combined with PCR testing of whole blood in acute cases (3).

Several antibiotics have been administered to naturally infected cats, but all cats in 2 studies became clinically normal within 24 to 48 hours after initiation of tetracycline or doxycycline administration and recurrence was not reported (1,3). While clinically normal, 2 cats were still PCR positive 17 days and 90 days after treatment (of 21 to 30 days duration) which suggests that treatment with tetracyclines for 21 to 30 days may be inadequate for eliminating the organism from the body (3). In a recent experimental study, bacteremia as determined by PCR assay was limited over several weeks (Lappin et al, unpublished data, 2011).

BARTONELLOSIS

A number of *Bartonella* spp. including *B. henselae*, *B. clarridgeiae*, *B. koehlerae*, *B. quintana* and *B. bovis* have been cultured or amplified from client-owned cats with fever. Fever following experimental inoculation with *B. henselae* has been documented in a number of studies including a recent study in our laboratory where the CSU-1 strain of *B. henselae* induced significant fever in three of six cats after exposure to infected *C. felis* (5). None of the six cats administered imidacloprid-moxidectin in that study became infected or febrile. However, not all strains or *Bartonella* spp. induce fever in all cats; for example in the imidacloprid-moxidectin study, cats inoculated with the same strain intravenously failed to develop fever. Whether fever will occur during *Bartonella* spp. infection is likely a complex interaction that is influenced by both host and organism factors.

As *B. henselae*, *B. clarridgeiae*, *B. koehlerae* are transmitted by fleas, bacteremia and antibody positivity rates can be very high. For example, serum antibodies were detected in 93% of cats housed in a North Carolina shelter and *Bartonella* spp. DNA was amplified from the blood of > 50% of cats housed in a Alabama shelter (6,7). The majority of these cats were thought to be normal which emphasizes that fever from bartonellosis cannot be documented by test results alone. In one study of pair matched cats with or without fever, serum *Bartonella* antibodies detected by ELISA or Western blot immunoassay were not correlated to the presence of fever (8). In addition, serum antibody test results are negative in 3 to 15% of bacteremic cats. Thus, if a cat with fever is to be evaluated for *Bartonella* spp. infection the combination of blood culture or PCR assay on blood, and serologic testing will detect the greatest number of cats that are currently or previously infected. Febrile cats that are seronegative and negative for *Bartonella* spp. in blood by culture or *Bartonella* spp. DNA in blood are unlikely to have the organism as the cause of fever.

If fever from bartonellosis is suspected in a cat, administration of doxycycline or a fluoroquinolone is generally effective. The AAFP Panel Report recommended doxycycline at 10 mg/kg, PO, daily for 7 days as the initial therapeutic trial. If a positive response is achieved, continue treatment for 2 weeks past clinical resolution of disease or for a minimum of 28 days. If a poor response is achieved by day 7 or doxycycline is not tolerated and bartonellosis is still considered a valid differential diagnosis, fluoroquinolones are appropriate second choices. In experimental or field studies, administration of enrofloxacin or orbifloxacin have led to rapid resolution of fever in cats with presumed bartonellosis. Treatment with azithromycin is now contraindicated because of rapid induction of resistance (9).

EHRLICHIOSIS

Ehrlichia-like bodies or morulae have been detected in peripheral lymphocytes or monocytes of naturally exposed cats in a number of countries including the United States, Kenya, France, Brazil, and Thailand. One study of cats in North America amplified DNA consistent with *E. canis* from naturally infected cats (10). However, other studies of cats in endemic areas (Florida and Arizona) have failed to amplify *Ehrlichia* spp. DNA from the blood of cats (11,12). In 2 separate experimental studies, we have failed to amplify monocytotropic *Ehrlichia* spp. from blood or detect seroconversion in cats inoculated SQ with different strains of cultured *E. canis* (Lappin and Breitschwerdt, unpublished observations, 2007; Lappin and Little, unpublished observations, 2010). These results indicate the *E. canis*-like DNA amplified from naturally-infected cats may be from a different *Ehrlichia* spp. more infective to cats, not all *E. canis* stains will infect cats, not all cats are susceptible to infection by *E. canis*, or SQ inoculation is not an effective method for infecting cats with *E. canis*. Some cats with suspected clinical ehrlichiosis have seroreacted to *E. canis* or *N. risticii* morulae.

Fever is one of the reported clinical abnormalities detected in cats with suspected ehrlichiosis and so testing may be indicated in these cats. However, a validated serological assay is not currently available and some cats with *E. canis*-like DNA in blood were seronegative (10). In contrast, most *A. phagocytophilum* infected cats have strongly positive antibody test results. Positive serologic test results occur in both healthy and clinically ill cats, and so a diagnosis of clinical ehrlichiosis should not be based on serologic test results alone. *Ehrlichia* spp. PCR and gene sequencing can be used to confirm infection and should be considered the tests of choice at this time.

Clinical improvement after therapy with tetracycline, doxycycline, or imidocarb dipropionate was reported for most cats with suspected mononcytopic ehrlichiosis. However, for some cats a positive response to therapy was a criterion for the diagnosis of ehrlichiosis. The current recommendation of the ACVIM Infectious Disease Study Group is to give doxycycline (10 mg/kg PO q24h for 28 days). For cats with treatment failure or those intolerant of doxycycline, imidocarb diproprionate can be administered (5 mg/kg IM or SQ twice, 14 days apart). Salivation and pain at the injection site are the common adverse effects and the efficacy of imidocarb is in question for the treatment of canine monocytotropic ehrlichiosis.

HEMOPLASMOSIS

Fever has been detected in some cats with infection by *Mycoplasma haemofelis*, '*Candidatus* Mycoplasma haemominutum', or '*Candidatus* M. turicensis'. In multiple studies of experimentally infected cats, *M. haemofelis* is apparently the most pathogenic species. Dual infection with hemoplasmas may potentiate pathogenesis of disease. In one study, cats with chronic '*Candidatus* Mycoplasma haemominutum' infection had more severe anemia and longer duration of anemia when experimentally infected with *M. haemofelis* when compared to cats infected with *M. haemofelis* alone (13). In one abstract, we reported an association between *M. haemofelis* and fever in cats without anemia. Clinical signs of disease depend on the degree of anemia, the stage of infection, and the immune status of infected cats. Direct transmission may occur with the hemoplasmas and so the agents should be on the differential list for cats with a history of fighting (14).

Diagnosis of hemoplasmosis is based on demonstration of the organism on the surface of erythrocytes on examination of a thin blood film or by PCR assay results. Organism numbers fluctuate and so blood film examination can be falsely negative up to 50% of the time. The organism may be difficult to find cytologically, particularly in the chronic phase. Thus, PCR assays are the tests of choice due to sensitivity.

Doxycycline is often administered as a flavored suspension (to avoid esophageal strictures) at 10 mg/kg, PO, every 24 hours for a minimum of 7 - 10 days. In cats intolerant of doxycycline, enrofloxacin given at 5 mg/kg, PO, every 24 hours for 14 days was tolerated by cats and is equally effective or more effective than doxycycline (15). Administration of marbofloxacin or orbifloxacin gives similar results. Azithromycin was not effective for the treatment of hemoplasmosis in one study (16). Most drug protocols have failed to eliminate infection and so at this time there is no clinical utility to repeat PCR testing. The owners should be warned that recurrences may occur but are unusual.

FELINE RICKETTSIOSIS

Rickettsia spp. are obligate intracellular Gram negative bacteria that are divided into the spotted fever group and the typhus group. Cats can be infected by *Rickettsia felis* and have been shown to have antibodies against *R. rickettsii*. *Rickettsia felis* DNA has been amplified from *Ctenocephalides felis*, *C. canis*, and *Pulex irritans;* these fleas have a worldwide distribution. *Ctenocephalides felis* is a biological vector for *R. felis*; the organism can be transmitted transovarially and transtadially within the flea. Rickettsial infection is suspected to a cause of fever in cats but this has not been well documented (Bayliss). While we have commonly amplified *R. felis* from *C. felis* (67.4% of flea extracts in one study), we have not amplified the organism from the blood of healthy cats or cats with fever (7, 16-18). However, in one study of cats with fever we showed *R. felis* and *R. rickettsii* antibody prevalence rates in cats in the USA to be 5.6% and 6.6%, respectively but neither organism was amplified from blood (16). These results prove that cats are sometimes exposed to spotted fever group organisms but further data are needed to determine significance of diseases associations. Because clinical illness in cats has not been documented, optimal treatment is unknown. However, based on results in dogs, doxycycline or a fluoroquinolone would be logical choices.

SUMMARY

There are other infectious agents of cats that are either blood borne or associated with bacteremia that should be on the differential list for cats with appropriate clinical findings and geographical locale including feline leukemia virus, feline immunodeficiency virus, *Cytauxzoon felis*, *Hepatozoon* spp., *Francisella tularensis*, and *Yersinia pestis*.

REFERENCES

[1]Bjoersdorff A et al. **J Sm Anim Pract.** 1999;40:20. [2]Shaw SE et al. **Vet Rec.** 2005;157:645. [3]Lappin MR et al. **J Am Vet Med Assoc.** 2004;225:893. [4]Foley JE et al. **Comp Immunol Microbiol Infect Dis.** 2003;26:103. [5]Bradbury CA, Lappin MR. **J Am Vet Med Assoc.** 2010;236:869. [6]Nutter et al. **J Am Vet Med Assoc.** 2004; 235, 1394. [7]Lappin MR et al. **Vet Dermatol.** 2009;20:509. [8]Lappin MR, et al. **J Fel Med Surg.** 2009;11:148. [9]Biswas S et al. **J Clin Microbiol.** 2010;48:617. [10]Breitschwerdt E et al. **J Vet Int Med.** 2002;16:642. [11]Luria BJ et al. **J Fel Med Surg.** 2004;6:287. [12]Eberhardt JE et al. **J Fel Med Surg.** 2006;8:164. [13]Westfall DS et al. **Am J Vet Res.** 2001;62:687. [14]Dean RS et al. **J Fel Med Surg.** 2008;10:413. [15]Dower KL et al. **J Am Vet Med Assoc.** 2002:221;250. [16]Bayliss DB, et al. **J Fel Med Surg.** 2009;11:266. [17]Hawley JR, et al. **J Fel Med Surg.** 2007;9:258. [18]Barrs VR et al. **Aust Vet J.** 2010;88:160.

DETECTION AND EPIDEMIOLOGY OF CANINE INFLUENZA A VIRUS INFECTIONS

Gabriele A. Landolt, DVM, MS, PhD, DACVIM
Fort Collins, CO

INTRODUCTION

Since 2004, outbreaks of canine influenza have occurred across the United States. The widespread circulation of the canine H3N8 influenza virus (CIV) in the dog population establishes influenza as a threat to canine health. Sequence analysis of the viral genome of CIV reveals that the canine isolates are closely related to, and had evolved from, the contemporary equine H3N8 influenza virus lineage.[1] While primarily associated with outbreak of respiratory disease in racing Greyhounds in 2004, by the following year the virus was also detected in the non-Greyhound dog population. To date, the Animal Health Diagnostic Center at Cornell University reports that CIV infections have been diagnosed in dogs in at least 29 states and the District of Columbia.[2]

ETIOLOGY

Influenza A viruses are members of the family *Orthomyxoviridae*. The influenza virions possess a host-cell derived lipid envelope, which contains the three envelope glycoproteins, hemagglutinin (HA), neuraminidase (NA), and the ion channel protein M2. Based on antigenic properties of the HA as well as the NA, influenza A viruses are divided into subtypes. To date, sixteen HA subtypes (H1-H16) and nine NA subtypes (N1-N9) have been described. The viruses exhibit partial host range restrictions, and occasionally viruses originating in one host species can infect another species. Undoubtedly the most prominent examples of such interspecies transmission of influenza A viruses are the infections of humans and other mammalian species with the highly pathogenic avian H5N1 viruses ("bird flu") as well as appearance of a swine origin H1N1 virus in the human population ("novel H1N1") in 2009. The emergence and subsequent maintenance of an equine-lineage H3N8 influenza virus in dogs provides another example of the transmission of an influenza A virus between distinct host species.

EPIDEMIOLOGY

Canine influenza is a contagious respiratory disease. While it is impossible to determine the total number of CIV infections that have occurred since first identification of the virus, all dogs are considered to be susceptible, regardless of age, breed, or sex. Despite this, the disease appears to be more prevalent in dogs that are housed in groups in "closed" air environments (such as humane shelters, boarding or breeding kennels, canine day-care centers, or veterinary clinics).

Under optimal environmental conditions, canine influenza virus appears to spread easily among susceptible animals. Virus transmission is thought to occur either directly through nose-to-nose contact with infected animals, large droplet-, and aerosol-transmission or indirect through contact with contaminated fomites. As with other influenza A viruses, CIV is hypothesized to spread by aerosol over several feet, although the importance of aerosol spread of influenza A viruses remains somewhat controversial.[3] In fact, the efficiency of aerosol spread of influenza virus has been shown to be strongly influenced by temperature and relative humidity.[4,5] Moreover, long-range transmission of influenza virus does not appear to occur frequently in settings with adequate ventilation (eg. outdoors). This notion might also be supported by results of several recent studies that suggest that dogs housed in close proximity with one another and in closed air environments are particularly at risk for CIV exposure. For example, 42% of dogs were positive for CIV antibodies in a Philadelphia shelter that reported respiratory disease.[6] Additionally, our laboratory determined that 11 of 16 Colorado humane shelters experiencing canine respiratory disease had at least one dog at the facility shedding influenza virus (unpublished data). Further studies found an increased risk for CIV seropositivity in Colorado household dogs that had attended canine day-care or had been boarded within six months prior to serum sample collection.[7]

Fomites may also play an important role in the transmission of CIV between susceptible animals. Results from recent studies have indicated that influenza A virus can survive in the environment for several days on contaminated surfaces.[8] The length of time the virus remained infectious in the environment was dependent on a number of factors, including concentration of the inoculum as well as environmental factors (type of surface, presence or absence of respiratory mucous, temperature, humidity, light and UV conditions, and pH).

The time from infection with CIV to onset of clinical signs is approximately 2 to 4 days. Based on experimental data, nasal virus shedding begins as early as 2 days after exposure and may persist for 7 to 10 days. This finding has importance for control of CIV infection, as nasal virus shedding can be present before the onset of clinical signs. In addition, while CIV infection does not lead to a carrier status, it is important to recognize that a fairly high percentage of dogs (approximately 10 to 20%) may shed virus without developing clinical signs of canine influenza. In light of this fact, without diagnostic testing, dogs that have been in contact with CIV infected animals should be considered as potentially contagious for 10 to 14 days after exposure to the infected animal.

DIAGNOSIS

In a group of susceptible dogs, a presumptive diagnosis of canine influenza virus infection may be made based on the rapid spread of an acute, febrile respiratory disease that is associated with a prolonged or complete lack of response to therapies that are generally effective for other causes of kennel cough syndrome. However, laboratory diagnosis is required to confirm and differentiate CIV from other causes of acute canine respiratory disease.

The laboratory diagnosis of canine influenza is based on the culture of the etiological agent (virus isolation), the demonstration of the viral nucleic acid (PCR-based assays) or viral antigen in respiratory tract swab samples (for example through antigen detection by immunofluorescence or ELISA-based assays), or the detection of virus-specific antibodies (serology). As many of these methods have one or more serious disadvantages (for example a lack of sensitivity or long turnaround time), it may be necessary to institute several of these diagnostic tools in order to identify the etiological agent accurately and rapidly.

Serologic tests have been a key tool in the diagnosis of canine influenza infection. As most serologic assays are fairly easy to perform, cost-effective, and a large number of samples can be collected and tested simultaneously, serological testing is particularly useful for large-scale disease surveillance. While useful, there are also several limitations to serologic testing. For instance, the presence of antibodies indicates exposure to the virus, but does not necessarily indicate active infection or disease. Secondly, antibody may not always be detected even in the presence of active infection. The most common reason for such false negative results is that insufficient time has passed between infection and measurement of an antibody response. While influenza specific antibodies have been detected as early as seven days following experimental infection, serologic testing often provides only retrospective information. Also, in CIV vaccinated dogs, diagnosis of influenza infection can only be made by testing paired serum samples (acute and convalescent titers) collected 10 to 21 days apart. To diagnose CIV infection in vaccinated animals, a four-fold increase in titer between baseline (acute) and convalescent samples must be demonstrated.

Although virus isolation remains a critical tool for epidemiologic investigation including the monitoring of the genetic variability of canine influenza isolates, this method has limited value for diagnostic purposes. Virus isolation often takes a minimum of 2 to 3 days. In addition, the success of this test for detection of CIV is influenced by the particular virus strain, amount of viable virus present in the sample, and sample collection and storage techniques.

The most important advantages of CIV antigen detection assays over traditional virus isolation are the faster turnaround time as well as the ability to detect virions that have lost their infectivity. For example, commercially available antigen-capture ELISA assays (eg. Directigen Flu-A assay, Becton Dickinson) have been used to detect influenza virus infection in dogs. In a recent study, these rapid "patient-side" tests have been evaluated and have been found to perform with reasonable specificity.[9] Moreover, these tests performed with a sensitivity similar to that of traditional virus isolation in the detection of CIV in canine nasal swab specimens.[9]

Polymerase chain reaction (PCR)-based assays have been developed and implemented in many diagnostic laboratories for the rapid and highly sensitive diagnosis of canine influenza infection. More recently, real-time PCR-based assays have been used not only for the rapid detection but also for the quantitative analysis of canine respiratory pathogens. By employing a target specific fluorescent probe, real-time PCR-based assays eliminate the need to distinguish the target by PCR fragment size on electrophoresis or hybridization with probes post-PCR. In diagnostic laboratory settings the use of PCR can be limited by cost and sometimes availability of sufficient test sample volume. To overcome these shortcomings, multiplex PCR assays ("PCR panels") have been developed. In multiplex PCR, more than one target sequence can be amplified by including more than one pair of primers in the reaction.

REFERENCES

[1]Crawford PC, et al. **Science** 2005;310:482-485. [2]Animal Health Diagnostic Center at Cornell University, http://diagcenter.vet.cornell.edu/pdf/CIV-Stats.pdf. 3Tellier R. **J. R. Soc. Interface** 2009;6:783-790. [4]Lowen AC, et al. **PLoS Pathogens** 2007;3:1470-1476. [5]Lowen AC, et al. **J Virol** 2008;82:5650-5652. 6Holt DE, et al. **JAVMA** 2010;237:71-73. [7]Barrell EA, et al. J Vet Intern Med 2010;24:1524-1527. [8]Thomas Y, et al. **Appl. Environ. Microbiol.** 2008;74:3002-3007. [9]Spindel ME, et al. **J Vet Intern Med** 2007; 21: 576.

HUMAN BARTONELLOSIS: PERSPECTIVES OF A VETERINARY INTERNIST

Edward B. Breitschwerdt, DVM, DACVIM
Raleigh, NC

INRODUCTION

Bartonella species are fastidious gram-negative bacteria that are highly adapted to a mammalian reservoir host and within which the bacteria usually cause a long-lasting intraerythrocytic bacteremia.[1-3] These facts are of particular importance to veterinarians and physicians, as an increasing number of animal reservoir hosts have been identified for various *Bartonella* species. Among numerous other examples, *Bartonella henselae* has co-evolved with cats, *Bartonella vinsonii* subsp. *berkhoffii* has co-evolved with dogs and wild canines, and *Bartonella bovis* has co-evolved with cattle.[1-2] Importantly, the list of reservoir-adapted *Bartonella* species, including a large number of rodent species that might serve as "pocket pets," continues to grow exponentially, as new *Bartonella* spp. are discovered. Prior to 1990, there were only two named *Bartonella* species, whereas there are now at least 24 named and numerous unnamed or candidatus species, based upon deposited Gen Bank sequences or preliminary reports, respectively. Seventeen *Bartonella* spp. including *B. alsatica, B. bacilliformis, B. clarridgeiae, B. doshiae, B. elizabethae, B. grahamii, B. henselae (*Houston 1 and San Antonio 2 strains*), B. koehlerae, B. melophagi, B. quintana, B. rochalimaea, B. tamiae, B. vinsonii* subsp. *berkhoffii* (Genotypes I, II and III), and *B. washoensis* have been associated with an expanding spectrum of human diseases.

Epidemiological evidence and experimental flea transmission studies support an important role for fleas in the transmission of *B. henselae, B. clarridgeiae* and most likely *B. koehlerae* among cats.[1] Three other *Bartonella* species, *B. bovis, B. quintana* and *B. vinsonii* subsp. *berkhoffii* have been isolated from cat blood, but the modes of transmission and the reservoir potential of these species in felids has not been definitively established. Recently, we isolated *Bartonella vinsonii* subsp. *berkhoffii* from a cat with recurrent osteomyelitis spanning an eighteen month time period.[2] Thus, cats can maintain a chronic bacteremic with at least six *Bartonella* spp., of which five are known zoonotic pathogens.[1-3] In addition to fleas, an increasing number of arthropod vectors, including biting flies, keds, lice, sandflys and ticks have been implicated in the transmission of *Bartonella* species. Although there is clinical and epidemiological evidence to support tick transmission of *B. vinsonii* subspecies *berkhoffii* to dogs and coyotes, the mode of transmission of this *Bartonella* subsp. to cats and dogs has not been determined. Recent evidence supports tick transmission of *B. henselae* by *Ixodes scapularis* and *Ixodes ricinus.* Considering the diversity of *Bartonella* species and subspecies, the large number of reservoir hosts and the spectrum of arthropod vectors, the clinical and diagnostic challenges posed by *Bartonella* transmission in nature may be much more complex than is currently appreciated in human and veterinary medicine.

In the natural reservoir host, such as a cat or rodent, chronic bacteremia with a *Bartonella* species can frequently be detected by blood culture or PCR in outwardly healthy individuals.[1-3] In contrast, the diagnostic detection of a *Bartonella* spp. in a non-reservoir adapted host, such as a dog, horse or human patient, can be extremely difficult. Most, although not all diseases caused by *Bartonella* spp. occur in accidental hosts and these organisms are being increasingly implicated as a cause of zoonotic infections.[4-8] It is important to recognize that strains of a *Bartonella* sp. vary in their virulence. Therefore, highly pathogenic strains of *B. henselae*, for which the cat is the primary reservoir, can induce granulomatous myocarditis in cats, presumably following flea transmission. Until recently, mechanisms that facilitate persistent *Bartonella* bacteremia in mammals were not well understood. Recent reports have identified an intra-endothelial, as well as intra-erythrocytic localization for these bacteria, which represents a unique strategy for bacterial persistence. Non-hemolytic intracellular colonization of erythrocytes in conjunction with the ability to invade and replicate within endothelial cells would preserve the organisms for efficient vector transmission, protect *Bartonella* from the host immune response, and potentially contribute to decreased antimicrobial efficacy. Although the clinical implications are not understood, other *in vitro* studies indicate that *Bartonella* spp. can infect dendritic cells, microglial cells, monocytes and CD34+ bone marrow progenitor cells.

CAT SCRATCH DISEASE

For over a century regional lymphadenopathy has been associated with animal contact, particularly cat scratches. Over the years, numerous microorganisms were implicated as the cause of CSD. In 1992, Regnery and colleagues at the Centers for Disease Control, identified seroreactivity to *B. henselae* antigens in 88% of 41 human patients with suspected CSD compared to 3% of controls.[2] Subsequently, additional support for *B. henselae* as the predominant cause of CSD was provided when *Bartonella* DNA was amplified from lymph node samples of 21 of 25 (84%) patients with suspected CS D, using a polymerase chain reaction assay. A similar study from Sweden identified *B. henselae* DNA, but failed to identify *A. felis* DNA, in a large number of patients with suspected CSD. Prior to the recognition of *B. henselae* as the cause of CSD, *Afipia felis*, named for the Armed Forces Institute of Pathology, was considered the sole cause of CSD. Subsequently, we blood cultured *B. henselae* or *B. clarridgeae* from 17 of 19 cats owned by 14 patients with CSD, which indicated that bacteremia is a frequent occurrence in cats that transmit *B. henselae* to a human being.[1-2]

Historically, atypical manifestations of CSD have included tonsillitis, encephalitis, cerebral arthritis, transverse myelitis, granulomatous hepatitis and/or splenitis, osteolysis, pneumonia, pleural effusion, and thrombocytopenic purpura. With the advent of specific diagnostic techniques, (culture, serology, and PCR), there has been a dramatic increase in reports describing human patients with "atypical" manifestations of CSD. Osteomyelitis, granulomatous hepatitis and granulomatous splenitis have been increasingly recognized in children infected with *B. henselae*, who frequently lack the classical lymphadenopathy of CSD. Previously, *Bartonella* infection would not have been considered a likely differential diagnosis by the physician in patients lacking a history of lymphadenopathy or animal contact. As evidenced by reports in the past four years, the spectrum of human disease associated with the genus *Bartonella* continues to expand, requiring periodic reassessment as new information becomes available. On a comparative medical ("One Health") basis, our research group has documented many of the same CSD atypical manifestations in cats or dogs, including encephalitis, transverse myelitis, granulomatous hepatitis, osteolysis, pleural effusion, and thrombocytopenic purpura. In this context, a highly prevalent, naturally-occurring human disease (CSD) can be used as a "model" to determine the potential behavior of these bacteria in companion animal patients.

Because cat scratch disease generally denotes a self-limiting illness characterized by fever and lymphadenopathy and because the recognized spectrum of human disease manifestations associated with *Bartonella* infections (which may not include fever or lymphadenopathy) has expanded considerably in recent years, it is becoming obvious that the designation CSD lacks clinical, microbiologic and zoonotic utility. Although cats are a major reservoir for *B. henselae, B. clarridgeiae*, and *B. koehlerae,* some patients deny the possibility of a cat scratch or bite wound, or indicate no contact with cats. Transmission from environmental sources, various arthropod vectors, perinatally or by other animal hosts is probable and the more inclusive term bartonellosis may facilitate enhanced future understanding of diseases caused by members of the genus *Bartonella*. As physicians have been taught that CSD is self-limiting, there is an ongoing lack of appreciation that *B. henselae* can cause chronic, asymptomatic or intermittently symptomatic illness, accompanied by persistent bacteremia in people. In this context, the documentation of chronic, relapsing bacteremia in cats, dogs and other animal species provides a "model" for better understanding human bartonellosis.

BARTONELLA ENDOCARDITIS

Endocarditis can be induced by a spectrum of *Bartonella* species in dogs and human patients and is the best example of documented disease causation for this genus. Historically, *Bartonella* species have been a cause of culture-negative endocarditis in people and dogs because the diagnostic methods used by microbiology laboratories were not adequate to isolate these bacteria. Now, by using specialized techniques, a spectrum of *Bartonella* species have been identified in research and diagnostic laboratories in different parts of the world—in heart valves or in blood cultures from dogs and people with endocarditis.[3] It is important for physicians and veterinarians to recognize that some of these *Bartonella* species are found in the blood of cats, dogs, rats, ground squirrels, and rabbits.

ISOLATION AND MOLECULAR DETECTION OF *BARTONELLA* SPECIES

Because conventional microbiological techniques lack sensitivity, bartonellosis is usually diagnosed by PCR amplification of organism specific DNA sequences and/or through serological testing. Recently, the development of a more sensitive isolation approach, using BAPGM (*Bartonella* alpha *Proteobacteria* growth medium) followed by real time PCR has greatly facilitated the molecular detection or isolation of *Bartonella* species from the blood of sick or healthy animals, including cats, dogs, horses and human beings. Most importantly, the use of this enrichment growth medium prior to PCR testing has allowed our research group to confirm that immunocompetent human patients, in particular veterinarians and veterinary technicians, can have chronic intravascular infections with *Bartonella* spp.[4-5] Information relative to this testing platform for animal and human patients is available at www.galaxydx.com.

It is increasingly clear that no single diagnostic strategy will confirm infection with a *Bartonella* sp. in the immunocompetent patient population. As described in studies from our laboratory, *B. henselae, B. koehlerae* and *B. vinsonii subsp berkhoffii* seroreactivity was found in only 58.6% of the patients in which *Bartonella* spp. infection was confirmed by PCR and sequencing. Therefore, Bartonella serology lacks sensitivity and can only be used to implicate prior exposure to a *Bartonella* sp. Even when serum from cat scratch disease patients, which is caused by *B. henselae,* is used in various diagnostic laboratories for IFA testing, test sensitivities have ranged from 14 to 100%.

EVOLVING IMPLICATIONS OF CHRONIC *BARTONELLA* SPP. BACTEREMIA IN IMMUNOCOMPETENT PEOPLE

Previously, we described *B. quintana* bacteremia in a woman who was tested following the development of an infected cat bite lesion involving the hand.[6] Two months later, the feral cat that had induced the bite wound was captured and was also shown to be *B. quintana* bacteremic. In a cumulative study involving 392 patients with occupational animal contact or extensive arthropod exposure 31.9% were bacteremic with one or more *Bartonella* spp., when blood, serum and BAPGM enrichment culture PCR results were combined. Although this high prevalence of bacteremia is biased by testing at risk, sick individuals, it clearly demonstrates that intravascular infection with *Bartonella* sp. is much more common in immunocompetent patients, than was previously suspected. By IFA testing, only 75 out of 128 (58.6%) PCR positive patients were seroreactive to a panel consisting of five *Bartonella* sp. test antigens. In a recent study, *Bartonella vinsonii* subsp. *berkhoffii, Bartonella henselae* or DNA of both organisms were amplified and sequenced from blood, BAPGM enrichment blood cultures or autopsy tissues from four family members.[7] Historical and microbiological results derived from this family support human perinatal transmission of *Bartonella* species. To date, there have been a limited number of studies that address the potential impact of intravascular infection with a *Bartonella sp.* on reproductive performance, however, studies involving experimentally-infected cats, rodents and naturally-infected cows with various *Bartonella* sp. have identified decreased reproductive performance involving both males and females. The parents of these children had attempted to conceive children for several years prior to resorting to in vitro fertilization. We have also described a veterinarian, who experienced a needle stick while obtaining a fine needle aspiration sample from a cutaneous histiocytic neoplasm.[8] Subsequently symptoms, including headaches, fatigue and intermittent paresthesias developed. This patient seroconverted to *B. vinsonii* subsp. *berkhoffii* genotypes I and III and *B. vinsonii* subsp. *berkhoffii* genotype I DNA was amplified and sequenced from sequentially obtained blood samples, whereas genotype III DNA was amplified from the cytological specimen. All symptoms resolved following antibiotic treatment. It is increasingly evident that dogs can serve as a source for human infection with *B. vinsonii* subsp. *berkhoffii. Bartonella vinsonii* subsp. *berkhoffii* genotype II was amplified and sequenced from a liver biopsy from a patient with epithelioid hemangioendothelioma, after which the organism was isolated by BAPGM blood culture.[9] The unique capability of *Bartonella* to invade and induce long lasting intraerythrocytic and intraendothelial infections, in conjunction with the ability of at least three *Bartonella* spp. (*Bh*, *Bq*, and *B. bacilliformis*) to induce VEGF-mediated vasoproliferative disease in immunocompromised or immunocompetent individuals suggests that these novel emerging bacterial pathogens might contribute to the development of vascular tumors. *Bartonella koehlerae* bacteremia was documented in eight immunocompetent patients by PCR amplification and DNA sequencing, either prior to or after BAPGM enrichment blood culture.[10] Presenting symptoms most often included fatigue, insomnia, joint pain, headache, memory loss, and muscle pain. Four patients were also infected with *Bartonella vinsonii* subsp. *berkhoffii* genotype II. *Bartonella koehlerae* antibodies were not detected (titers< 1:16) in 30 healthy human control sera, whereas five of eight patient samples had *B. koehlerae* antibody titers of 1:64 or greater. Studies are needed to determine if *B. koehlerae* is a cause or cofactor in the development of arthritis, peripheral neuropathies or tachyarrhythmias in human patients. Co-infection with *B. henselae* and two hemotropic *Mycoplasma* variants resembling *Mycoplasma ovis* were also found in the blood of a veterinarian with a historical diagnosis of multiple sclerosis.[11]

PUBLIC AND OCCUPATIONAL HEALTH CONSIDERATIONS

Due to extensive contact with a spectrum of animal species, veterinary professionals appear to have an occupational risk of infection because of frequent exposure to *Bartonella* spp., therefore these individuals should exercise increased precautions to avoid arthropod bites, arthropod feces (i.e. fleas and lice), animal bites or scratches and direct contact with bodily fluids from sick animals. As *Bartonella* spp. have been isolated from cat, dog or human blood, cerebrospinal fluid, joint fluid, aqueous fluid, seroma fluid and from pleural, pericardial and abdominal effusions, a substantial number of diagnostic biological samples collected on a daily basis in veterinary practices could contain viable bacteria. The increasing number of defined *Bartonella* spp., in conjunction with the high level of bacteremia found in reservoir-adapted hosts, which represent the veterinary patient population, ensures that all veterinary professionals will experience frequent and repeated exposure to animals harboring these bacteria. Therefore, personal protective equipment, frequent hand washing and avoiding cuts and needle sticks have become more important as our knowledge of this genus has improved and various modes of transmission have been defined.

Physicians should be educated as to the large number of *Bartonella* spp. in nature, the extensive spectrum of animal reservoir hosts, the diversity of confirmed and potential arthropod vectors, current limitations associated with diagnosis and treatment efficacy, and the ecological and evolving medical complexity of these highly evolved intravascular, endotheliotropic bacteria.

REFERENCES

[1]Chomel BB, et al. **Vet Res** 2009;40:29. [2]Breitschwerdt EB, et al. **J Vet Emerg Crit Care** 2010;20:8. [3]Chomel BB, et al. **Ann N Y Acad Sci** 2009;1166:120. [4]Breitschwerdt EB, et al. **J Clin Microbiol** 2008;46:2856.[5]Breitschwerdt EB, et al. **Parasit Vectors** 2010;3:29. [6]Breitschwerdt EB, et al. **J Clin Microbiol** 2007;45:270. [7]Breitschwerdt EB, et al. **J Clin Microbiol** 2010;48:2289. [8]Oliveira AM et al. **J Vet Intern Med** 2010;24:1229. [9]Breitschwerdt EB, et al. **J Clin Microbiol** 2009;47:1957. [10]Breitschwerdt EB, et al. **Parasit Vectors** 2010;3:76. [11]Sykes JE, et al. **J Clin Microbiol** 2010;48:3782.

FELINE HEPATIC LIPIDOSIS SYNDROME - CURRENT KNOWLEDGE

S. A. Center, DVM, DACVIM
Ithaca, NY

Feline Hepatic Lipidosis (FHL) is the most common form of acute acquired potentially lethal liver disease in domestic cats observed in inappetent (days to weeks) over-conditioned animals. "Idiopathic" FHL is rare as the syndrome usually develops secondary to a primary disorder causing inappetence.[1] Domestic cats have a propensity for hepatocyte lipid vacuolation that seemingly reflects unique aspects of feline metabolism, the higher fat content of the carnivorous diet, or possibly their high visceral to subcutaneous fat (VAT/SAT) ratio.[2] The pure carnivore diet, ample in protein and fat and low in carbohydrates, has gluconeogenesis derived from amino acids. Although hepatocyte lipid vacuolation also is common in inappetent, ill, and obese cats, the extent of cell involvement is comparably greater in FHL where >80% of hepatocytes are affected and >50% hepatic weight is contributed by lipid. In FHL, hepatocytes enlarge up 2-fold normal with cytosolic triglyceride (TG) vacuoles; TG is confirmed by staining frozen or formalin fixed tissue (but not paraffin embedded tissue) with Oil red O. Clinical data included in this review are derived from 187 cats with FHL managed at Cornell University Hospital for Animals. Four cats in our data base have had FHL recurrence- years later.

Clinical Features

Although cats of any age may be affected, most are middle aged neutered adults (7 [0.5-20] years), with history of over-conditioning (BCS $\geq$ 4/5), and inappetence (2-7 days; >90%) and usually an ~25% loss of body weight (dehydration & body condition). Weight is primarily lost from peripheral tissues while abdominal fat is conserved. Gastrointestinal signs are most common: vomiting in ~38%, variable diarrhea and/or constipation. Other historical features reflect underlying/primary diseases.[1,3] Jaundice (~70% cats) and non-painful smooth hepatomegaly are notable on physical examination. Severe electrolyte derangements (marked hypokalemia or hypophosphatemia) may provoke profound head/neck ventroflexion. Ptyalism is common and may reflect nausea or hepatic encephalopathy (HE). Severe weakness and recumbency may reflect symptomatic electrolyte disturbances (potassium, phosphate), thiamine (vitamin B_1) deficiency, or a primary disease. Electrolyte disturbances may contribute to gastric and intestinal dysmotility and vomiting that complicate alimentation (usually before electrolyte correction). Ultrasonography discloses diffuse hepatic hyperechogenicity that may over-shadow a primary hepatobiliary disease process.

Clinicopathologic Features

Hematology

Poikilocytosis (~63% initial blood smears) is common. Heinz bodies may suddenly appear and lead to symptomatic anemia and death; may follow oxidant challenge imposed by drug or anesthetic agents or reflect a primary illness (e.g. diabetes mellitus, other liver disorders, hyperthyroidism, pancreatitis).[4] Anemia (~22% cats on presentation) may develop secondary to phlebotomy, hemolysis (Heinz bodies, severe hypophosphatemia) or provoked blood loss (i.e. feeding tube placement). Depending on the decade examined, ~25-40% of FHL cats in the author's hospital have required blood component therapy (whole blood, packed RBCs, or fresh frozen plasma). Leukograms reflect underlying diseases as FHL lacks necroinflammatory features.

Serum Biochemistry

Increased liver enzyme activities (ALP, ~84%; ALT, ~72%; AST ~91%, γGT, 61%), subnormal BUN (~44%) despite initial dehydration (impaired urea cycle function presumed), and hyperbilirubinemia (variable magnitudes) are common . There is no diagnostic value in fractionating bilirubin. Cats with primary necroinflammatory liver disease (e.g. cholangiohepatitis, extrahepatic bile duct occlusion [EHBDO], bile duct carcinoma), pancreatitis or pancreatic adenocarcinoma develop high γGT activity; ~30% of cats have $\geq$ 3-fold increase in γGT compared to ~60% with $\geq$ 3-fold increase in ALP. ~18% have a greater fold increase in γGT relative to ALP coordinating with comorbid disorders.[5] This unique feature increases clinical suspicion of underlying conditions. Subnormal albumin ~65% of cats (before fluid therapy) reflects complex influences on albumin (synthesis, turnover) and malnutrition. Hyperglobulinemia is rare and hypoglobulinemia occurs ~18%. Concentration of cholesterol and glucose reflect primary disease processes. Despite diffuse hepatic steatosis, few cats present hypoglycemic (~ 4%) whereas ~53% are hyperglycemic (initially) reflecting either over-conditioning or a feline stress response (~ 3-4% are Diabetics). A subset (~56% cats) develop mild/marked increased creatinine kinase reflecting muscle injury (i.e IV catheter or feeding tube placement, muscle catabolism, recumbency, rhabdomyolysis due to electrolyte depletions [i.e., hypokalemia, hypophosphatemia], or a primary disorder).

Electrolyte abnormalities

Iimportantly contribute to morbidity and mortality; hypokalemia increases risk for death.[3] Hypokalemia, hypophosphatemia, and/or hypomagnesemia present in ~31%, ~15%,~ 21% of cats, respectively, or develop after fluid therapy or subsequent to a re-feeding syndrome. Severe hypokalemia and hypophosphatemia increase risk for RBC hemolysis (hypophosphatemia), muscle weakness, enteric atony and vomiting, head/neck ventroflexion, and neurobehavioral changes confused with HE. Anticipated refeeding electrolyte disturbances are now commonly averted by prophylactic IV electrolyte supplementation (KCl, K-phosphate, with frequent monitoring). Lipiduria is common and likely reflects renal tubule lipid vacuolation; a buoyant lipid phase may be observed in centrifuged urine. Renal tubular potassium wasting has been verified in a subset of cats; this may resolve with FHL resolution.

Coagulation Assessments

Coagulopathies are more reliably detected using a test sensitive to Vit. K_1 sufficiency (PIVKA clotting test, Thrombotest, Nycomed).[6] Routine bench PT, APTT, and PIVKA tests were abnormal in ~26%, ~30%, and ~59% of FHL cats. Coagulopathies usually respond to parenteral Vit.K_1 (1.0 mg/kg, IM or SC, @ 12 hour intervals x 3). Routine early Vit. K_1 supplements facilitate uneventful catheter and feeding tube placement and hepatic tissue aspiration. Coagulopathy in > 50% of cats dictates initial venipuncture using peripheral veins that allow pressure assisted hemostasis. Cystocentesis, jugular catheter insertion, urethral catheterization, and feeding tube placement should be delayed until after a Vit. K_1 response interval.

Blood Gas Analyses

Venous blood gas (on presentation) demonstrate metabolic acidosis in ~33%, reduced strong ion difference in ~87%, and increased unmeasured anions in ~82%. Changes are consistent with accumulated ketones and lactate. Cats with spontaneous clinical and experimentally induced FHL demonstrate propensity for hyperglycemia and accumulate plasma ketones within range observed in diabetics (β-OH butyrate from 0.5 to 8.9 mmol/L).[7,8] Experimental FHL confirmed ketone accumulation as early as 7 days of "self imposed fasting". Ketone concentrations in experimental FHL have been lower than observed in clinical cases because of early rescue strategies. Similar ketosis (~5–7 mmol/L) develops in humans fasted 1-23 days. Longer fasts coordinate with higher ketone concentrations, as seen in FHL.[9] High lactates have only been confirmed in a small number of cats (limited testing).

Vitamin & Anti-Oxidant Sufficiency

Cobalamin (B_{12})

Of 58 FHL cats sequentially tested, low cobalamin was found in 58%. A subset of tested cats given B_{12} before sample collection had supraphysiologic B_{12} concentrations (>5,000 pg/mL). Experimentally, insufficient B_{12} can compromise mitochondrial respiration (due to accumulated methylmalonic acid or its metabolites, inhibited complex II and β-OH butyrate dehydrogenase) and causes mitochondrial toxicity (compromised tricarboxylic acid cycle).[10] Supplemented cyanocobalamin requires cyanide removal and activation by an addition of an adenosyl or methyl group: pathway dependent on folate, other B vitamins, and SAMe. It is unknown if B_{12} insufficiency causally contributes to FHL or rather reflects antecedent conditions. Conventional laboratoryB_{12} assay cannot distinguish inactive from functional vitamin.

Thiamine (B_1)

Deficiency can induce neurologic signs confused with HE, neurologic injury (Wernicke's Encephalopathy), and Lactic Acidosis. Weakness, head/ neck ventroflexion, neurologic signs (poorly responsive pupils, vestibular signs, seizures) have resolved after short term thiamine administration in some FHL cats. Oral thiamine (100 mg/cat BID 1-3 days, then 50 mg/cat/day) is preferred as rare vasovagal response and neuromuscular paralysis have been observed in a few cats given parenteral (IM) thiamine.

Other B Vitamins

Water-soluble B vitamins are required cofactors for enzymes essential to cell function and energy production. There are few investigations of B vitamin sufficiency aside from B_{12} and B_1 in ill cats; depletion of other B vitamin(s) is possible.

Vitamin K

Dramatic correction of coagulopathies in FHL seen with Vit. K1 supplementation suggests marginal stores of this fat soluble vitamin or impaired reactivation in the Vit. K epoxidase cycle. Whether or not vitamin E is insufficient is unknown.

Glutathione (GSH

Concentrations of reduced GSH are severely depleted in liver of cats with FHL similar to findings in humans with hepatic steatosis associated with kwashiorkor.[11,12] Since liver is the major source of circulating GSH, it is not surprising that subnormal GSH in whole blood has also been demonstrated in FHL.[13]

Liver Function Tests

Serum Bile Acid (BA) tests are redundant in FHL as hepatobiliary jaundice is overt. Fractioned BA (blood & urine) from cats with FHL, EHBDO, and healthy cats disclosed significantly reduced secondary BA concentrations in FHL & EHBDO. Since secondary BA are derived from enteric bacterial dehydroxylation of primary bile acids, findings implicate impaired enterohepatic BA circulation likely linked to canalicular collapse ultrastructurally observed.[14] This reconciles with apparent Vit. K_1 insufficiency. *Blood Ammonia-* concentrations are inconsistently increased in clinical FHL. Experimental FHL demonstrated NH_3-intolerance commensurate with insufficient quantities of essential amino acids implicating compromised urea cycle function. Similar amino acid changes develop in spontaneous FHL.[12,15] Ammonium biurate crystalluria has not been documented in FHL.

Circulating Lipids & Hepatic Fat Characterization

Four studies examining lipid profiles in plasma and/or liver of cats with FHL and obese cats show changes resembling the human metabolic syndrome with significant increases in plasma non-esterified fatty acids (NEFA), TG, VLDL, and LDL.[7,16-18] Hepatic TG in FHL seemingly reflects catabolism and peripheral mobilization of adipose with maintenance of intra-abdominal fat stores.[1] This may reflect direct TG delivery to liver or VAT that then disseminates to liver in the portal circulation. Hepatic and VAT fractions in FHL (n=5) and healthy cats (n=5) confirmed greater concentrations of hepatic TG in FHL but no unique difference in FA composition.[19] However, higher palmitate concentrations were found in hepatic and VAT in FHL; diets were not explored. That omental fat represents VAT is supported by work in dogs where omentectomy significantly increased insulin sensitivity (labeled glucose clamp method).[20] Expanded VAT in obese humans with the Metabolic Syndrome dampens insulin sensitivity. Cats have larger VAT compartment relative to humans; CT estimated VAT /SAT ratio of 1.18 ± 0.32 in cats compares to 0.19 ± 0.04 to 0.4± 0.3 in humans.[21] If VAT is an important direct source of hepatic lipid in FHL, as in obese humans with Metabolic Syndrome hepatosteatosis, then the large VAT/SAT may explain feline propensity for hepatic steatosis and FHL during inappetence.

Ultrastructural Features

Transmission electron microscopy of liver from healthy, FHL, EHBDO, and obese cats during rapid weight loss reveals mitochondrial and peroxisomal changes: mitochondria and peroxisomes are reduced in number, mitochondria assume unusual shapes and some have abnormally shaped cristae. Changes are contrary to those in other animal species with hepatic steatosis.

What Causes Hepatic Fat Accretion in FHL?

Insufficient adaptation to FA oxidation in face of increased peripheral FA mobilization to the liver remains the strongest pathophysiologic scenario along with impaired or down-regulated VLDL synthesis/exportation.

Does L-Carnitine (L-CN) Sufficiency Play a Role in FHL ?

Four case controlled investigations[22-25] suggest that obese cats beneficially alter FA oxidation with L-CN supplementation. 1) L-CN in obese cats during rapid weight loss facilitated weight reduction, increased total CN, acyl-CNs and specifically acetyl-CN. Findings implicate increased FA oxidation (acetyl-CN = last remnant of FA oxidation) and possibly enhanced urinary acyl-CN elimination.[22] 2) L-CN supplementation in over-conditioned cats before 5-wks of low food intake (unpalatable diet) protected cats from increased circulating FFA and β-OH butyrate suggesting optimized fatty acid utilization.[23] L-CN might have influenced FA oxidation (hepatic), improved ketone utilization (muscle, heart), or facilitated Acyl-CN urinary elimination. 3) Stable isotope labeled FA studies in L-CN supplemented obese cats during weight loss determined no increase in VLDL or hepatic TG synthesis concurrent with increased β-OH-butyrate.[24] Findings implicated greater fat partitioning for β-oxidation and less to export or storage fat; a significant reduction in insulin reconciled with more efficient energy utilization. 4) Stable isotope study of FA metabolism in over-conditioned cats undergoing weight reduction determined that L-CN supplementation increased FA flux and oxidation with stoichiometric calculations supporting enhanced FA oxidation.[25] 5) Clinical data suggest survival advantage with L-CN supplementation (along with nutritional and micronutrient support).[1] Collective findings suggest L-CN supplementation either provides a supraphysiologic dose effect or that pet cats as currently fed maintain a narrow L-CN surplus that may limit adaptive increases in β-oxidation. L-CN in natural feline prey is substantially higher than in many feline diets (~150-3000 mg L-CN/kg animal tissue). Although L-CN can be synthesized, requisite amino acids and transmethylation activity may be compromised in FHL. Limited adaptation to increased β-oxidation is supported by finding normal total L-CN concentrations in liver, muscle and plasma in FHL. Yet, L-CN supplements before weight loss in obese cats increase plasma, liver, and muscle L-CN.[23,26]

Is Liver Biopsy Necessary for FHL Diagnosis ?

Liver biopsy is not necessary for presumptive diagnosis of FHL; this is usually based on hepatic cytology, finding a diffusely hyperechoic hepatic parenchyma on US, compatible biochemical features, and history of over-conditioning and inappetence. Liver biopsy (or other tissues) is reserved for cats with suspected comorbid conditions requiring definitive histologic diagnosis or cats with recalcitrant FHL (rigorous supportive care does not resolve/improve clinicopatho-logic features within 2-wks). Yet, cytology can be misleading unless sufficient hepatocytes are sampled (several different areas), and percentage of severely vacuolated cells estimated. Common hepatic TG vacuolation in ill cats may be over-interpreted. Cholangio-hepatitis, EHBDO, and some neoplastic disorders are not defined cytologically, similar to inaccuracy of small needle-core biopsies.

Treatment

1. *Procedures* requiring *sedation/anesthesia* should be delayed until electrolyte and hydration perturbations resolve. The FHL cat is in metabolic crisis with high risk for anesthetic and surgical complications; Vit K_1 treatment and response interval is essential.
2. *Fluid Therapy*: avoid dextrose supplementation (may reduce adaptive β-oxidation, worsen electrolyte depletion [in urine], contribute to ketosis), avoid lactate (likely have high lactate already), acetate also may not be appropriate; judiciously supplement KCl and K-phosphate correcting aberrations and in anticipation of a "Refeeding Syndrome".
3. *Avoid drugs & toxins* suspected to promote hepatic steatosis (humans, HL models, cats) include: endotoxin, stanozolol (Winstrol-V;FHL in renal insufficient cats fed a protein restricted diet), tetracyclines, glucocorticoids, bupivacaine & buprenorphine (humans: buprenorphine is a lipophilic protonatable amine metabolically activated and concentrated in mitochondria; may impair respiration and ATP formation), repeated/prolonged propofol anesthesia, drugs with propylene glycol carrier (i.e. etomidate, diazepam), oxidants, drugs with high risk of hepatotoxicity.
4. *Antibiotics*- individualize therapy to primary disease process; endotoxin may be a causal factor in infections (abscess, pyelonephritis, constipation, inflammatory bowel disease, pancreatitis).
5. *Antiemetics:* metoclopramide CRI or mirtazapine (no side effects yet noted in FHL cats). Metoclopramide advantageously stimulates enteric motility and may be useful during initial introduction of enteric feeding.
6. *Gastroprotectants:* Are these necessary ? Individualize therapy to clinical signs and primary disease process.
7. *Water Soluble Vitamins:* ensure B_{12} adequacy, submit plasma B_{12} test before administering 250-500 ug/cat (SC or IM). Thiamine: 100 mg PO BID x 2 days then 50 mg PO/day; consider possible vasovagal episode and neuromuscular paralysis with parenteral thiamine injections. Use fortified B-Vitamin supplement in IV fluid therapy.[1] Cover fluid lines to avert photo-degradation.
8. *Vitamin K_1:* 0.5-1.5 mg/kg SQ or IM (not IV) x 3 doses @ 12 hour intervals; treat irrespective of coagulation tests.
9. *Vitamin E:* 10 iu/kg mixed α-tocopherols, with food as supplemental antioxidant; Vit. E is not synthesized by mammalian cells. May assist in detoxification of cyanide removed from therapeutic B_{12} (cyanocobalamin).
10. *S-Adenosylmethionine (SAMe)*- essential methyl donor and for GSH synthesis; 40 mg/kg PO SID has antioxidant effect in liver.[27]

11. *L-Carnitine:* 250-500 mg/cat per day PO, only medical grade or soluble certified L-Carnitine

12. *Feeding:* oral feeding can rescue some cats but food aversion may importantly delay recovery. Preferred feeding route is by E-tube. Initial feeding is by nasogastric intubation using CliniCare. E-tube insertion after achieving fluid and electrolyte balance, coagulopathy abated with Vit. K_1, when cat can withstand the stress of sedation/anesthesia. *Avoid protein restricted foods,* rather, use balanced feline diets formulated for critical care patients (i.e. Maximum Calorie, A/d). Allow exercise to stimulate enteric motility.

13. *Do Appetite Stimulants Work in FHL ?:* probably not, we see many FHL cats treated with this approach; potential for hepatotoxicity with some (i.e. diazepam, oxazepam, although rare).

REFERENCES

[1] Center SA:Vet Clin North Amer 2005;35:225. [2] Lee H:J Fel Med Surg 2010;12:936. [3] Center SA: JVIM 1993;7:349. [4] Christopher MM: http://www.ncbi.nlm.nih.gov.proxy.library.cornell.edu/pubmed/2708116 JAVMA 1989;194:1089. [5] Center SA: JAVMA 1986;188:507. [6] Center SA:JVIM 2000;14:292. [7] Pazak HE: JNutr 1998;128:2747S. [8] Biourge VC:AJVR 1994;55:1406. [9] Cahill GF:N Engl J Med 1970;282:668. [10] Depeint F:Chem Biol Interact.2006;163:113. [11] Center SA: AJVR 2002;63:1187-1197. [12] Becker K:Free Rad Biol Med 1995;18:257-263. [13] Center SA, unpublished data. [14] Center SA: AJVR 1993;54:724. [15] Biourge VC:AJVR 1994;55:1291. [16] Blanchard G: J Anim Physiol Anim Nutr 2004;88:73. [17] Ibrahim WH: AJVR 2000;61:566. [18] Muranaka S: Res Vet Sci 2010;epub. [19] Hall JA: JVIM 1997;11:238. [20] Lottati M: Obesity J 2009;17:674. [21] Lee H: J Fel Med Surg 2010; 12:936. [22] Center SA: JVIM 2000;14:598. [23] Blanchard G: J Nutr 2002;132:204. [24] Ibrahim WH: AJVR 2003;64:1265. [25] Center SA, submitted for publication. [26] Jacobs G: AJVR. 1990;51:1349. [27] Center SA: JVIM 2005;19:303.

CANINE CONTEMPORARY HEPATOTOXINS: ENVIRONMENTAL TOXINS XYLITOL, AFLATOXIN and CYCAD SEED

Eric Dunayer, MS,VMD, DABT, DABVT; Sharon Center, DVM, DACVIM; Frédéric Gaschen, Dr.med.vet.; Dr.habil, DACVIM, DECVIM-CA
Ithaca, NY

XYLITOL

Introduction: Xylitol is a sugar alcohol used as a sugar substitute in products such as gums, candies, baked good, and drinks. It is considered advantageous in people because of its low glycemic index and because it inhibits the bacteria that cause tooth decay. In humans, over consumption of xylitol has been associated with osmotic diarrhea but no significant systemic toxicity. In dogs, on the other hand, xylitol has been linked to several clinical syndromes. Xylitol causes a dose-related release of insulin leading to the rapid onset of hypoglycemia, a phenomenon that has been recognized since the late-1960's.[1] More recently, a new syndrome involving xylitol-induced liver damage has been reported.[2]

Pathomechanism: The exact mechanism of xylitol hepatotoxicity remains unclarified. One theory implicates hepatic metabolism of xylitol (phosphorylation of several metabolites) leading to ATP and inorganic phosphate depletion that then cause hepatocellular necrosis. In isolated rat liver, hepatocyte ATP declines to ~66% of baseline within 10 min of xylitol exposure.[3] An alternative theory associates xylitol toxicity with formation of reactive oxygen species (ROS) initiated during xylitol metabolism.

Clinicopathologic Features: Dogs ingesting xylitol develop mild to moderate increases in ALT activity within a few hours; changes appear to be dose related.[4] Most dogs remain clinically normal (once hypoglycemia, if present, resolves) and ALT activity normalizes within a week of exposure.[4] A small number of dogs develop life-threatening liver necrosis associated with a coagulopathy ~48-72 hrs after xylitol ingestion; these dogs usually present in hemorrhagic shock due to severe bleeding into a body cavity. In these, symptomatic hypoglycemia was not evident after xylitol ingestion; rather, these dogs presented for veterinary care after collapse due to hemorrhagic shock, with ALT activity >1000 up to >10,000 U/L, ALP normal or mildly increased, and mild/moderate hyperbilirubinemia. Hypoglycemia in these severely affected dogs likely reflects liver damage rather than xylitol initiated insulin release. When measured, coagulation parameters (PT and aPTT) are markedly prolonged, platelet count low, and D-Dimers or FDPs increased; collective features are consistent with DIC. Hyperphosphatemia seemingly increases risk for mortality.

Histopathologic Features: Histologic features (only a few cases reported) consist of diffuse hepatic necrosis in dogs with severe signs of xylitol toxicity. One dog surviving longer than others, demonstrated centrilobular hepatocellular necrosis, lobular collapse, architectural disorganization, bile stasis and biliary epithelial hyperplasia. It is unclear whether hepatic failure represents an extension of mild "toxicity" often accompanying xylitol ingestion or an idiosyncratic response. Some dogs ingesting xylitol in dosages well above those reported to cause hepatic necrosis and death have only developed mild-to-moderately increased liver enzyme activities without progression to liver failure. One recent study involving 18 dogs, 12 challenged with oral xylitol at dosages expected to cause life-threatening hepatic necrosis (1 or 4g/kg) and 4 controls, demonstrated up to 10-fold increases in ALT activity within 12-hrs and an ~50% decrease in phosphorous within 1 hr that rapidly attenuated within 72 hrs and 180 min., respectively. No treated dog developed evidence of fulminant hepatic necrosis supporting the hypothesis that severe liver injury may represent either individual variation in xylitol metabolism or liver status among dogs or some other idiosyncratic response.

Treatment & Survival: At present, there is no specific antidote for xylitol. Infusion of dextrose, even in euglycemic dogs may be protective by conserving ATP (glycolysis) –although if available phosphorus is tied up in xylitol metabolites, the liver may not be able to renew ATP stores as rapidly as needed. Administration of anti-oxidants such as n-acetylcysteine or ascorbic acid are proposed to scavenge ROS. At present, neither therapy has proven efficacy and prognosis depends on clinical signs demonstrated by the individual dog. Dogs developing uncomplicated hypoglycemia and/or mild increases in liver enzyme activity generally recover within a few days to a week. Severe hepatic necrosis portends a poor prognosis, even with aggressive supportive care; mortality > 75% is reported.[2] However, because many dogs with diffuse hepatic necrosis do not present until later stages of toxicosis, it is unknown whether early intervention alters case outcome.

AFLATOXIN

Introduction: Aflatoxins, secondary metabolites primarily of *Aspergillus flavus* and *A. parasiticus*, contaminate a variety of staple foods, particularly corn and ground nuts. Pet foods have regulated standards of < 20 ppm (mg/kg food). Naturally occurring aflatoxins are AFB_1, AFB_2, AFG_1, and AFG_2, distinguished by color of fluorescence under UV light (B= blue, G=green) and chemical structure. AFB_1 is most abundant, toxic and carcinogenic. AFM_1 and AFM_2 are de-hydroxylation products of AFB_1 and AFB_2, respectively, and are sometimes measured as biologic markers of aflatoxin exposure. Temperature of 25-30°C and relative humidity of 97-99% are optimal for toxin formation during grain harvest or food storage. Susceptibility to AFB_1 toxicity (ascending LD_{50}) among species is reported as: horses & ponies < rabbits < [mink dogs, cats] < pigs < human beings < guinea pigs < rat <Macaque monkey < mouse < hamster. Dogs and cats may be intoxicated by aflatoxin > 60 ug/kg feed. Dogs serve as sentinels for human aflatoxin exposure (corn; in African and Indian incidents). Chronic aflatoxin exposure is associated with impaired immunity, malnutrition (loss of essential serine proteases i.e. trypsin), and liver cancer (3-fold increased risk of liver cancer in humans). Aflatoxins are lipophilic, rapidly absorbed, and directly transported to the liver after ingestion.

Pathomechanism: Hepatotoxicosis is dose-related and caused by rapidly formed metabolites. The liver is the predominant site of AFB_1 metabolism (phase 1: cytochrome P450-mediated reactions) forming more reactive, electrophilic and arylating metabolites that account for cellular toxicity and hepatic necrosis. AFB_1 metabolism by P450 cytochromes depends on enzyme affinity, expression, and genetic polymorphisms (gene or promoter in humans). Polymorphisms, environmental variables, and comorbid conditions (i.e. viral hepatitis) each influence individual susceptibility to aflatoxicosis. Rapid formation of a noxious 8,9-epoxide of AFB_1 overwhelms detoxification pathways by depleting intracellular and mitochondrial GSH. Low GSH thwarts GSH-AFB_1detoxifying conjugation and permits secondary oxidative injury. Low inherent GSH in canine liver may predispose dogs to aflatoxicosis.[5] Metabolism of AFB_1-epoxide to dihydro-diol metabolites also leads to cell injury and death. Accumulated adducts irreversibly bind cell enzymes (including DNA polymerase, ribosomal translocase), proteins, and DNA, impairing metabolism, gene transcription, and protein synthesis. Inhibited RNA polymerase and interaction of AFB_1 epoxide with polyribosomes impairs secretory protein synthesis. Binding to mitochondrial DNA impairs cell energy production. Toxic effects are irreversible and lead to hepatic steatosis and early zone 3 hepatocellular necrosis and collapse causing intrahepatic portal hypertension. An AFB_1-lysine-albumin adduct serves as a biomarker in humans and has an ~$T_{1/2}$ of 20-60 days (similar to unbound albumin). Unbound aflatoxin remains in the circulation only a short interval after absorption (i.e. 13-120 min) and cannot be used diagnostically. Dual intoxication with different mycotoxins through contaminated feed can produce synergistic or enhanced organ toxicity.

Canine aflatoxicosis is traditionally described as peracute, acute, or chronic representing dose-response toxicity. Dogs ingesting >0.5 mg aflatoxin/kg body weight die acutely with fulminant hepatic failure whereas 6-8 wks of chronic ingestion of 0.05-0.3 mg aflatoxin/kg of feed leads to progressive liver damage and eventually, irreversible lesions (hepatic cirrhosis, portal hypertension, acquired portosystemic shunts [APSS]). Features observed in dogs affected during the East-coast 2005 aflatoxin food contamination incident (n=77)[6] and an early experimental study in dogs[7] provides source information presented below.

History & Physical Findings: While aflatoxins are tasteless, mold contaminated feed may have a noxious odor/taste and be refused by exposed dogs; this saved some dogs in multiple pet households in 2005. However, many intoxicated dogs became gradually inappetent over several weeks. Vomiting food or bile was variably observed before overt illness. In order of onset, clinical features include: anorexia, lethargy, vomiting, jaundice, diarrhea often melenic or with hematochezia, abdominal effusion (modified transudate, rarely hemorrhagic), peripheral edema, terminal hepatic encephalopathy associated with hemorrhagic diathesis. However, evidence of illness may not be recognized until hepatic injury is extensive. Some dogs die acutely without antecedent illness. Marked edema develops in some dogs before profound change in plasma proteins suggesting loss of vascular integrity (endothelial toxicity).

Clinicopathologic Features: There are no consistent hematologic parameters. Liver enzymes may or may not be increased and increases may only be modest (due to impaired protein synthesis). Early, total bilirubin is normal with hyperbilirubinemia developing or escalating after blood transfusion (hemoglobin breakdown). Total protein and albumin are low secondary to extravascular dispersal, intestinal extravasation, or reduced synthesis. Low Protein C and Antithrombin activity (96% dogs) and cholesterol concentrations (71% dogs) functioned as biomarkers of aflatoxin exposure in 2005. Protein C and Antithrombin are serine proteases that likely are directly inhibited.[8] Surviving dogs followed sequentially for 110 days had these parameters slowly increase although some dogs had persistently low values. Dogs in fulminant hepatic failure develop marked electrolyte derangements (low potassium, sodium, chloride) and coagulation parameters consistent with DIC. Granular casts in urine reflect direct toxicity, hypoperfusion, or endothelial damage. While evaluation of diagnostic parameters in non-survivors (64%) vs survivors (36%) identified significant differences, these are only relevant retrospectively. A few dogs developed enteric hypomotility that prohibited oral alimentation.

Ultrasonography: No distinguishing features in peracute syndrome. Chronic toxicity may associate with altered parenchymal echogenicity (heterogenous, identification of nodules, small liver size, "lumpy" contour), abdominal effusion, portal hypertension / acquired shunts.

Cytologic Features: Severe diffuse hepatocellular lipid vacuolation was consistently demonstrated with minimal inflammation.

Histopathologic Features: Microvesicular hepatocellular lipidosis is consistent and most severe in Zones 2 & 3. In some dogs, only a narrow rim of hepatocytes abut portal triads that display oval cell hyperplasia (biliary epithelial hyperplasia). In chronic slowly progressive toxicity, fibrous connective tissue bridges between portal triads spanning between/around regenerative nodules. Additional features include canalicular cholestasis, iron accumulation iron in Kupffer cells and fewer hepatocytes, accumulation of lipofuscin pigment in Kupffer cells and hepatocytes, lymphatic dilation, binucleate hepatocytes, and minimal- mild inflammatory infiltrates (lymphocytes, plasma cells, occasional neutrophils). Renal tubules also demonstrates microvesicular lipid vacuolation.

Pathophysiologic Progression: Hepatic parenchymal damage (Zone 3) leads to lobular collapse surrounding the hepatic venule. This compromises circulatory outflow from the liver, leads to intrahepatic portal hypertension, lymphatic distention, splanchnic hypertension, enteric extravasation of RBCs (until APSS decompress the splanchnic bed) and fluid transudation into the abdominal cavity. Concurrent coagulopathy contributes to pathologic enteric hemorrhage and in some cases, a hemorrhage abdominal effusion.

Treatment & Survival: Of dogs that survived (36%), some had APSS and abdominal effusion. Treatment included: N-acetylcysteine IV (140 mg/kg IV once then 70 mg/kg IV q6hrs over 20 min.), Vit. E (10 iu/kg/day), SAMe (if tolerated orally; 20 mg/kg/day), blood component therapy (as needed), L-CN 500 mg/day), oral feeding of a normal protein diet (or reduced protein if encephalopathic), water soluble B-vits. (in fluids), antibiotics (protection against transmural endotoxemia/bacteremia), anti-emetics and gastroprotectants (famotidine or omeprazole, sucralfate), and diuretics as needed.

CYCAD

Introduction: Cycads are seed plants characterized by a large crown of compound leaves and a stout trunk. Examples include *Cycas revoluta*, *Cycas circinalis*, and *Zamia floridana*. These species are commonly used in landscaping in the Southeastern part of the USA and are becoming popular as greenhouse plants (Bonsai) in colder regions of the country. They contain several toxins that may affect the nervous system, digestive tract and liver. Toxins are found in all parts of the plant, but seem to be present in highest concentration in the seeds.

Pathomechanism:[9,10] After ingestion the predominant toxin, cycasin, is hydrolyzed to its active compound, methylazoxymethanol (MAM) by β-glucosidase produced by enteric flora. MAM is neurotoxic, carcinogenic, mutagenic, teratogenic and hepatotoxic. In dogs, cycad toxicity is characterized by gastrointestinal and hepatic effects attributed to MAM. The exact mechanisms of toxicity are unknown, however studies in rodents have shown that MAM is metabolized in hepatocytes by hepatocellular enzymes (e.g. alcohol dehydrogenase) and inhibits RNA synthesis by interfering with RNA polymerase II. MAM and/or its metabolites cause extensive subcellular damage in liver cells that may be irreversible.

History:[11-15] Onset of clinical signs is usually rapid, with dogs demonstrating gastrointestinal signs such as nausea and vomiting within hours of eating seeds or plant parts. In Southeastern parts of the USA a seasonal pattern of case presentation coordinates with weather conditions conducive to outdoor activities and highest seed availability: highest case presentation occurring in spring and summer (March-August).

Physical Exam:[11-15] Dogs usually present within 12-36 hr of cycad ingestion. Vomiting, lethargy, diarrhea, and anorexia are typically exhibited during the acute phase of toxicity. However, neurologic signs including: tremors, ataxia, mentation change and seizures may also be observed and are likely due to the neurotoxic properties of MAM. Icterus and signs of hypocoagulability are only rarely seen in the early phase. Many dogs never progress further than acute phase gastrointestinal signs. However, severely affected dogs progressively develop signs of hepatopathy with severe liver dysfunction developing over the ensuing days. A protracted course of cycad toxicosis may also occur with some dogs recovering from acute phase illness yet demonstrating severe liver dysfunction weeks to months later.

Clinicopathologic Features:[11-15] Increased transaminases are seen at initial presentation in 50% of dogs with hypoalbuminemia in 25% (severe in 10%). Moreover, > 50% of dogs also demonstrate prolonged aPTT and ~ 33% are thrombocytopenic, suggesting hepatic coagulopathy and/or DIC. Changes are progressive in severely affected dogs that develop hepatic failure.[14,15]

Ultrasonography: The liver usually appears normal at presentation, however a diffuse hyperechoic parenchyma may be apparent within 5-10 days of toxin ingestion. Changes are non-specific being reported in other toxicoses.

Histopathologic Features: Within 1-2 days, peracute centrilobular hemorrhage and congestion are seen with hepatocellular necrosis. Subacute changes consisted of centrilobular coagulative necrosis with widespread degeneration and scattered regeneration attempts. Chronic cases are characterized by diffuse hepatocellular necrosis, degeneration and regeneration with bridging fibrosis and nodular regeneration, as well as biliary hyperplasia.

Treatment: In the first few hours after toxin ingestion, dogs not spontaneously vomiting are administered apomorphine. Routine approach with active charcoal and cathartics is recommended to prevent /limit MAM absorption. Then, supportive treatment for dehydration, liver disease, hypocoagulability and/or DIC is tailored to the severity of clinical signs. In dogs that are discharged, liver parameters are monitored at regular intervals (every 2-4 weeks) because of the risk of protracted liver failure. However, liver enzymes may remain within the reference range in cases with severe liver failure.

Survival: Recent study[14,15] of documented cases revealed a 50% mortality rate disparate with an earlier case series[13] that included documented, suspect and possible exposures where a 32% mortality was reported. In the most recent study, differences in several serum chemistry parameters between survivors and non-survivors seemingly reflected the severity of liver damage. However none of these diagnostic features would have clinical utility in predicting survival during case management. Charcoal administration appeared to predict recovery while increased AST was a negative prognostic factor. The amount of toxin absorbed relative to the dog's body weight ultimately determines the severity of liver injury, but is often impossible to estimate in practical situations.

REFERENCES

1. Kuzuya T, et al: Endocrinology 1969; 84:200; 2. Dunayer EK, et al: JAVMA 2006;229:1113; 3. Woods HF, et al: Biochem J 1973;134:437; 4. Xia Z, et al: J Vet Pharmacol Therap 2009;32:465; 5. Center SA, et al: AJVR 2002;63:1187; 6. Dereszinski DM, et al: JAVMA 2008;232:1329; 7. King JM: PhD thesis. Anatomic Pathology. Cornell University, Ithaca, NY; 8. Cuccioloni M, et al: Toxicol Vit 2009;23:393; 9.Grab DJ, et al: Cancer Res 1977; 37: 4182; 10. Yu FL, et al: Cancer Res, 1983;43:287; 11. Senior DF et al., JAAHA 1985; 21:103; 12. Botha CJ, et al: J S Afr Vet Assoc 1991; 62:189; 13. Albretsen JC, et al: JAVMA 1998; 213: 99; 14. Ferguson D, et al: J Vet Intern Med 2010; 24:719; 15. Ferguson D, et al: 2011, manuscript under review.

CANINE CONTEMPORARY HEPATOTOXINS: DRUGS CCNU, MITOTANE, AMIODARONE AND SULFONAMIDES

David C. Twedt DVM, DACVIM
Fort Collins, CO
Lauren A. Trepanier, DVM, PhD, DACVIM, DACVCP
Madison, WI
Katherine Skorupski, DVM, DACVIM
Davis, CA

INTRODUCTION

The liver is uniquely susceptible to drug injury. Its position interposed between the gastrointestinal tract and the systemic circulation means that the liver is exposed to high drug concentrations. The liver is also the major organ involved in the metabolism and excretion of most drugs. In some cases drug metabolism itself leads to the production of toxic metabolites. Consequently, drug-induced liver injury poses a major clinical problem which can result in acute liver failure. Adverse drug reactions are commonly classified as being either as intrinsic/dose-dependent, or idiosyncratic drug reactions. Intrinsic toxins, such as acetaminophen, cause predictable toxicity to all individuals. In these cases as the dose increases so does the toxicity. Idiosyncratic drug reactions in contrast are unpredictable and are not dose associated in the general population. Most approved drugs for which hepatotoxic reactions have been reported are idiosyncratic toxins generally due to the aberrant metabolic production of a toxic intermediate. Although the incidence of most idiosyncratic hepatotoxic reactions in veterinary medicine is not known, they typically occur at low rates that would not be discovered during normal drug safety testing. Thus most idiosyncratic reactions are detected during post-marketing surveillance by the FDA.

Hepatic drug biotransformation is accomplished by Phase I and Phase II reactions. Phase I reactions are catalyzed by a family of enzymes called cytochrome P450 (CP450) that oxidize, reduce or hydrolyze the drug. Phase I reactions either activate or detoxify a drug. Some Phase I products can be directly excreted by the liver, while others are acted upon by Phase II enzymes. Phase II reactions conjugate drugs or their metabolites to glucuronic acid, sulfate, glutathione, or acetyl or methyl groups. While the products of Phase II reactions are usually non-toxic, Phase I reactions may lead to the generation of reactive intermediates. Toxicity from Phase I products is normally kept in check by prompt Phase II conjugation, or by the myriad of anti-oxidant defense mechanisms present in the liver including superoxide dismutase, catalase and glutathione.

Risk factors for drug-induced liver disease in people include age, gender, genetic factors, environmental factors and concomitant disease or drug therapy. The CP450 enzymes exhibit significant genetic polymorphism and thus an individual's genetic complement of CP450 enzymes can profoundly influence drug metabolism. Poor nutritional status of the patient and impaired oxidative defense mechanisms may play a critical role. Lastly concurrent drugs or herbal products may increase the toxicity of another drug.

The histopathological reaction to drug mediated toxicity is quite varied. Many reactions are acute, however, some drugs cause chronic hepatotoxicity. Hepatocellular necrosis and/or apoptosis is common. Immunologicially mediated reactions are characterized by inflammatory infiltrates that may be granulomatous in nature. Some drugs preferentially destroy bile ducts or interfere with bile formation and histopathologic lesions reflect this. Other drugs cause lipidosis or fibrosis, or damage the sinusoidal endothelium.

The following criteria are used to diagnosis hepatic ADR's. 1.) The animal becomes ill while taking the drug. Typically the latency period is within the first 3 months of therapy, but some toxicities are associated with an accumulated dose over a longer time period. 2.) The animal recovers after drug withdrawal. Some reactions may take months to resolve particularly if they involve an immune reaction since offending allergens may remain on the hepatocyte surface despite drug discontinuation. If the drug reaction is acute and fulminant the patient may die from acute hepatic failure despite drug discontinuation. 3.) Previous reports or a mechanistic basis for the hepatotoxic potential of the drug. 4.) The animal gets sick upon re-challenge, which in reality is almost never done for ethical reasons. 5.) A careful exclusion of other causes of liver disease.

A few representative drugs causing hepatotoxicity in dogs will be discussed.

MITOTANE

Mitotane is an adrenocorticolytic agent used to treat hyperadrenocorticism by a direct cytotoxic effect on the adrenal cortex resulting in progressive adrenocortical necrosis. Mitotane induced hepatotoxicity in the dog is uncommon with most all authors failing to list liver disease as a complication of this therapy. There is only one clinical report of an acute hepatopathy associated with mitotane therapy.[1] There are however reports of clinical cases described on Veterinary Information Network (VIN) as well as anecdotal experience by veterinary internists suggesting mitotane associated hepatotoxicity does exist. Mitotane has been associated with increases in transaminases in some treated human patients.[2] Because of the relative low incidence of mitotane associated hepatic disease it is most likely to be an idiosyncratic drug reaction from aberrant metabolic production of a toxic hepatic metabolite.

The reported case of mitotane hepatotoxicity was treated using an accepted dosage protocol.[1] In this case anorexia, lethargy and intermittent vomiting began two weeks after onset of therapy at which time the medication was discontinued. Two weeks later signs of acute hepatic failure developed with marked elevations in ALT, AST, and GGT. The bilirubin was elevated, prothrombin time prolonged and serum glucose was low. Laparoscopically there was gross collapse of hepatic parenchyma and histologically there was centrilobular hepatocellular loss. Eventual recovery and improvement in biochemical tests occurred over three months of supportive therapy. Marked elevations in ALT and AST within one month of onset of therapy would suggest a probable idiosyncratic mitotane hepatotoxicity and mitotane therapy should be discontinued.

AMIODARONE

Amiodarone is a class III anti-dysrhythmic agent that is used in dogs to treat cardiac arrhythmias associated with left ventricular dysfunction or to convert atrial fibrillation into sinus rhythm. Amiodarone's mechanism of action and potential for hepatotoxicity is not fully understood. Current evidence suggests that amiodarone causes is a dose-dependant (intrinsic) hepatic toxicity.[3] There may also be immunologic mechanisms involved because neo-antibodies to the drug have been detected in people with amiodarone liver disease.[4] Amiodarone is a lipophilic drug and concentrates in both adipose tissue and the liver. The liver metabolizes amiodarone to the active metabolite desethylamiodarone. It is postulated that over time amiodarone concentrations accumulate in the liver leading to toxicity. In humans steatohepatitis is reported suggesting a complex interaction involving hepatic lipid metabolism. Amiodarone induces phospholipid accumulation by inhibiting lysosomal phospholipase A1 that is supported by electron microscopic findings of lysosomal damage.[3] Amiodarone also inhibits both mitochondrial oxidation and oxidative phosphorylation resulting in chronic lipid peroxidation, hepatitis and occasional cirrhosis.

In one report of 10/22 (45%) Doberman Pinschers treated for dilated cardiomyopathy developed a reversible amiodarone hepatotoxicity.[5] Although Dobermans are over represented they may not necessarily be at an increased risk to develop liver disease because other breeds have also been identified.[6] Toxicity has been observed in dogs receiving either high loading doses or lower maintenance doses but dogs receiving higher concentrations of the maintenance dose were at a greater risk to develop toxicity.[Kraus] Clinical evidence of liver disease is somewhat delayed and reported occurring from 1 to 32 weeks after onset of therapy. Signs include anorexia, vomiting and occasionally diarrhea. Marked increases in ALT and moderate increases in ALP are observed. Clinical icterus is rare although mild increases in total bilirubin has been reported as well as occasional neutropenia.[5]

With identification of amiodarone associated liver toxicity based on clinical signs and/or elevated liver enzymes the medication should be discontinued or the dosage reduced. With discontinuation of amiodarone the hepatotoxicity resolves. Typically liver enzymes decrease over time and most are normal within three months after therapy. Initially following discontinuation of amiodarone there may be a delay or slight progression in the liver disease before improvement occurs because of amiodarone's long half-life and protracted drug elimination.[6]

It is recommended that prior to amiodarone therapy a pretreatment laboratory evaluation including CBC and serum biochemistry be performed to identify preexisting abnormalities and get a baseline starting point. Hepatic liver enzymes should be monitored monthly during therapy and if they become abnormal the dosage is reduced or discontinued. There are no reports showing that liver support therapy prevents hepatotoxicity or improves recovery.

CCNU

CCNU [1-(2-chloroethyl)-3-cyclohexyl-1-nitrosourea; Lomustine] is a nitrosourea alkylating agent used in veterinary medicine to treat a variety of canine cancers including lymphoma, mast cell tumors, and histiocytic sarcoma. This drug has been shown to be directly hepatotoxic in pre-clinical models and clinical patients. The exact mechanism of toxicity is not fully understood, but glutathione depletion and lipid peroxidation have been implicated.[1-4] Rodents treated with very high doses of CCNU develop acute biliary stasis and hepatocyte necrosis leading to death.[5] Healthy beagle dogs given a single dose of 4mg/kg had liver biopsies that showed Kupffer cell hyperplasia, cloudy swelling of hepatocytes, and periportal fibrosis in all treated dogs. Bloodwork in these dogs showed significant elevations in tested liver enzymes, with alanine aminotransferase (ALT) values increasing significantly in 4 of 5 dogs.[6] In this study, liver enzyme elevations were delayed at least 2 weeks after treatment and some dogs reached a peak ALT after 4 weeks. In a clinical setting, liver enzyme elevations occur in up to 86% of cancer bearing dogs receiving therapeutic doses of CCNU.[7-10] Serious liver injury causing symptoms occurs in only 3-6% of dogs.[10] These liver enzyme elevations are a significant clinical problem for veterinarians that prescribe

CCNU and, although most dogs will not develop a hepatopathy that requires medical intervention, treatment with CCNU may be delayed or discontinued due to concerns of impending liver failure.

Clinicopathologic findings are similar to other toxic hepatopathies and consist of acute elevations in ALT, AST, ALKP, and GGT. ALT elevations are usually most severe. In one study, liver enzyme elevations peaked after a median of 3 doses of CCNU (administered q3-4 weeks).[11] The mean peak ALT in this study was 692 IU/L in dogs receiving CCNU without concurrent hepatoprotectants. Although there is limited information describing the results of bile acids in dogs with CCNU induced hepatopathy, the author's experience is that many dogs with moderate to severe liver enzyme elevations will also have elevated bile acids. Mild increases in bilirubin are occasionally seen and are transient after discontinuation of CCNU in most dogs. Changes in albumin, BUN, glucose, and coaglulation parameters are rare and seen only in dogs with severe liver injury leading to liver failure.[10,11] Liver biopsies taken from clinical patients with CCNU induced hepatopathy show multifocal aggregates of hemosiderin-laden Kupffer cells, hepatocyte karyomegaly, hepatocellular vacuolization, periportal inflammation, bridging fibrosis, and bile duct hyperplasia.[10]

Treatment of CCNU induced hepatopathy is similar to treatment of other hepatopathies. SAMe and silybin have been used anecdotally to reduce liver enzymes in an attempt to continue CCNU therapy, but further study is necessary. There is disagreement as to what level of liver enzyme elevations should lead to delay or discontinuation of CCNU treatment. Nevertheless, bloodwork to evaluate liver damage and function are recommended prior to every dose of CCNU. Because the majority of dogs that develop a CCNU induced hepatopathy will never develop clinical signs of liver disease, most have a good prognosis. In the 3-6% of dogs that develop signs consistent with severe liver disease or liver failure, the prognosis for recovery is guarded and death can result. Research into the prevention of liver enzyme elevations and liver failure during CCNU treatment is ongoing. There is evidence that SAMe and silybin given preventatively throughout the course of treatment may reduce the severity of liver enzyme elevations and allow dogs to complete their prescribed treatment course.[11]

SULFONAMIDES

Idiosyncratic hepatotoxicity can be the most difficult to deal with clinically, because of unpredictable onset, lack of early markers, and often widespread hepatic necrosis. The most common antimicrobial class that is associated with idiosyncratic hepatotoxicity is the potentiated sulfonamides. Sulfonamides are indirect hepatotoxins that are oxidized by cytochrome P450 (via CYP2C9 in humans) or by myeloperoxidase to form a reactive hydroxylamine metabolite. The hydroxylamine spontaneously oxidizes to a nitroso metabolite that covalently binds to proteins and acts as a hapten. Tribrissen, Primor, and generic TMP-sulfamethoxazole have all been implicated in dogs. Dobermans are over-represented among dogs with idiosyncratic sulfonamide toxicity, although arthopathy, not hepatotoxicity, has typically been observed in Dobermans. We have found an increased prevalence of Samoyeds and Schnauzers among 40 dogs with idiosyncratic sulfonamide toxicity, although it is unclear whether this is a breed predisposition or a reflection of prescription prevalence. Idiosyncratic sulfonamide toxicity (also called sulfonamide hypersensitivity) is convincingly immune mediated in humans, with anti-drug antibodies and drug- and metabolite-specific T-cells documented. Risk factors may include antioxidant depletion or deficiencies in key sulfonamide detoxification pathways, such as N-acetyltransferases (absent in all dogs) or cytochrome b5 and its reductase.

Sulfonamide hepatotoxicity in dogs is typically seen 5-30 days after starting sulfonamides, with a mean of 12 days. Clinical signs are typical of acute liver failure, and may also include fever (55% of cases), transient neutropenia, thrombocytopenia, IMHA, polyathropathy, uveitis, KCS, skin or mucocutaneous lesions, or proteinuria. Dogs with hepatic involvement may present with a hepatocellular pattern (moderately high ALT activities in the high hundreds to low thousands), and/or a cholestatic pattern (moderate to severe jaundice, SAP>ALT); in addition, a hepatocellular pattern may progress over several days to cholestatic in some dogs. We have observed hepatic necrosis or eosinophilic cholangitis in dogs that we have biopsied.

It is critical that if potentiated sulfonamides need to be prescribed (e.g. for a resistant gram negative infection, particularly when CNS, prostate, or bronchial penetration is needed), the client is educated to look for any subtle signs of illness. Failure to discontinue sulfonamides at the time of initial development of adverse signs is associated with poor outcomes in human patients. Although specific antidotes for sulfonamide hypersensitivity have not been evaluated, several interventions make mechanistic sense: 1) Supplementation with a glutathione precursor (SAMe or N-acetylcysteine, using the same protocols as for acetaminophen toxicity), even if hepatotoxicity if not present. This is based on the finding that cysteine and glutathione can convert the reactive nitroso metabolite back to the hydroxylamine, and that antioxidant deficiency may impair the subsequent clearance of hydroxylamine by cytochrome b_5 and its reductase. 2) Supplementation with ascorbate 30 mg/kg IV three times daily (empirical dosage), based on the finding that ascorbate can reverse haptenization of the nitroso metabolite to canine liver microsomes *in vitro* (Lavergne & Trepanier, unpublished observations). 3) Conservative consideration of glucocorticoid use - not recommended during acute hepatic necrosis because of the potential risk of exacerbating hepatic encephalopathy or bacterial translocation, but anecdotally may be considered in the subacute setting, particularly if a persistent cholestatic pattern remains, or thrombocytopenia does not resolve within several days of sulfonamide discontinuation.

Dogs that develop hepatotoxicity or thrombocytopenia as part of sulfonamide hypersensitivity have a poorer prognosis (40-60% recovery) than those without liver or platelet involvement (about 90% recovery). Clinical suspicion, early recognition, prompt drug discontinuation, and aggressive support are key to a good outcome.

REFERENCES

Mitotane/Amiodarone

1. Webb CB, Twedt, DC. J Am Anim Hosp Assoc 2006;42:298-301. **2.** Neuman 0, Bruckert E, Chadarevian R, Jacob N, Turpin G. Therapie. 2001;56:793-797. **3.** Raja K, Thung SN, Fiel MI, Chang C. Semin Liver Dis. 2009;29(4):423-428. **4.** Breuh HW, Bossek W, Hafeland C, et al. Int J Clin Pharmacol Ther 1998;36:350-352. **5.** Kraus MS, Thomason JD, Fallaw TL, Calvert CA J Vet Intern Med. 2009;23:1-6 **6.** Jacobs G, Calvert C, Kraus M J Vet Intern Med 2000;14:96-99.

CCNU

1. Ahmed AE, et al. Cancer Chemother Pharmacol 1987;19:103-8. **2.** Kretschmer NW, et al. Cancer Chemother Pharmacol 1987;19:109-17. **3.** Ducastelle T, et al. Cancer Chemother Pharmacol 1988;22:153-62. **4.** Khan S, et al. Biochem Pharmacol 1992;43:1963-7. **5.** Viotte G, et al. Pathol Biol (Paris) 1986;34:40-7. **6.** Henry MC, et al. Toxicol Appl Pharmacol 1973;25:410-7. **7.** Williams LE, et al. J Vet Intern Med 2006; 20:136-43. **8.** Risbon RE, et al. J Vet Intern Med 2006; 20:1389-97. **9.** Hosoya K, et al. Vet Comp Oncol 2009;7:244-55. **10.** Kristal O, et al. J Vet Intern Med 2004; 18:75-80. **11.** Skorupski KA, et al. Manuscript in revision, 2010.

Sulfonamides

1. Trepanier LA, Danhof R, Toll J, et al. J Vet Intern Med 2003;17:647-652. **2.** Trepanier LA.. J Vet Pharmacol Ther 2004;27:129-138. **3.** Lavergne SN, Danhof RS, Volkman EM, Trepanier LA. . Clin Exp Allergy. 2006; 36:907-15. **4.** Lavergne SN, Trepanier LA.Platelets 2007;18:595-604.

CRITICAL CARE MANAGEMENT OF ACUTE LIVER FAILURE IN DOGS & CATS

Tim Hackett, DVM, MS, DACVECC
Fort Collins, CO

INTRODUCTION

Acute liver failure (ALF) is characterized by severe, sudden hepatocellular dysfunction leading to coagulopathy and hepatic encephalopathy in previously healthy patients or those without known underlying liver disease.[1] In veterinary medicine, ALF has been defined as a sudden loss of greater than 75% of functional hepatic mass.[2] The result is insufficient hepatic parenchyma to maintain synthetic and excretory demands.[3] It is generally characterized by hepatocyte death due to apoptosis or cytolytic necrosis. The extent of the hepatocyte death depends on the etiology. Also termed *fulminant hepatic failure*, ALF is relatively rare in human medicine only affecting approximately 2000 persons per year in the United States.[4] This devastating condition can rapidly progress to coma and death from cerebral edema and multiorgan dysfunction. While early reports in people found the most frequently identified cause of ALF was viral hepatitis, more recent surveys have found drug induced causes have become more common in Western developed countries.[5]

The definition and therefore the diagnosis of acute hepatic failure are complicated in the veterinary patient population. It takes a loss of more than 70% of the functional liver before clinical signs of ALF become apparent.[6] Many infiltrative and functional diseases of the liver can have an acute presentation in veterinary patients. Hepatic lipidosis, hepatitis, and a variety of neoplasia are just some of the causes that might present with acute hepatic dysfunction. As it may take days to identify a cause, optimal treatment and even prognosis, supportive measures must start immediately.

PATHOPHYSIOLOGY

More than 900 drugs, toxins, and herbs have been reported to cause liver injury, and drugs account for 20-40% of all instances of fulminant hepatic failure. In people approximately 75% of the idiosyncratic drug reactions result in liver transplantation or death. Drug-induced hepatic injury is the most common reason cited for withdrawal of an approved drug. Of the 2000 cases of ALF occurring annually in the USA, drugs account for over 50% of them (39% are due to acetaminophen, 13% are idiosyncratic reactions due to other medications). Drugs account for 2-5% of cases of patients hospitalized with jaundice and approximately 10% of all cases of acute hepatitis.[4]

Drugs are also the most common known cause of ALF in dogs and cats.[2] Hepatotoxic drugs affect the liver through direct toxicity to hepatocytes or after being metabolized to a more toxic compound. Any drug metabolized by the liver can be a hepatotoxin. Direct hepatotoxins damage is dose related and reproducible. Acetaminophen is the most common, directly hepatoxic drug implicated in human ALF and is a common cause of poisoning in veterinary patients. Many other drugs are associated with an idiosyncratic drug reaction. These reactions are not dose related and may affect individual patients differently. Most of these idiosyncratic drug reactions are associated with the metabolism of the drug to a toxic compound. Examples of drugs causing idiosyncratic reaction include the NSAIDs, trimethoprim-sulfa, lysodren, ketoconazole, azathioprine, and diazepam (in cats).

Environmental toxins including chemicals and solvents, heavy metals, plants (sago palm seeds, and the *Amanita phalloides* mushroom), insects, and aflatoxin have been incriminated to cause liver disease. Aflotoxin was the cause of ALF in dogs when it contaminated dog food in the Eastern United States. Xylitol, an artificial sweetener found in everything from chewing gum to postage stamps can result in severe hypoglycemia due to insulin release and acute liver dysfunction.

Probably the most common infectious cause of ALF is Leptospirosis. Although acute renal failure is the more common presentation of infected dogs, hepatic disease can be seen along with renal failure or alone. The hepatic insult is thought to be due to immune-mediated damage and likely depends on the serovar present. The most important serovars in dogs appear to be *Bratislava, Autumnalis, Pomona*, and *Grippotyphosa*. Other infectious causes in dogs include Canine Adenovirus I, ehrlichiosis, salmonellosis, *Rickettsia rickettsii, Toxoplasma* spp., *Neospora* spp., Dirofilaria, systemic mycoses (especially *Histoplasma* spp.), trematode infections, and hepatozoonosis. In cats, feline infectious peritonitis, bacterial infections of the liver, *Toxoplasmosis* spp., trematodes, and rarely *Histoplasma* spp. are important differentials for acute hepatic failure.

DIAGNOSIS

Damage to the liver may range from mild to moderate hepatic necrosis resulting in minimal clinical signs. Signs may be associated with vomiting, lethargy and anorexia. Massive hepatic necrosis will result in ALF and produce significant clinical signs of ALF (encephalopathy and coagulopathy). The signs of ALF are variable but usually will always include anorexia, depression, lethargy and vomiting. Neurological signs from hepatic encephalopathy may progress to coma or seizures. Jaundice is invariably present. ALF can also result in evidence of hemorrhage either from lack of coagulation factors or from DIC. Gastrointestinal (GI) ulceration is common in people and should be considered a source of blood loss. Septicemia may occur from uptake of enteric bacteria. Hepatic pain may be observed on abdominal palpation.

The clinical and pathologic changes vary with extent of damage to hepatocytes.[6] The hepatic transaminases (ALT and AST) are released when hepatocytes are damaged and the cytosol enzymes leak out. A marked increase in AST to ALT ratio suggests more severe hepatocellular damage. AST will decline prior to ALT and could be prognostic. Hepatic necrosis is associated with rising ALP and GGT though they are only mild to moderately elevated. Hyperbilirubinemia is common when significant hepatic necrosis is present and the magnitude may suggest the degree of hepatic necrosis. Changes in the liver function tests will also reflect the magnitude of hepatic damage. When the necrosis is massive and liver function is compromised these function changes will occur. Clotting factors decline and may contribute to hemorrhage. Hypoglycemia, low BUN, hypoalbuminemia, and increase in ammonia all reflect hepatic failure. It is important to note that with acute severe necrosis albumin concentrations may remain normal early in the disease due to the longer half-life of albumin (2 weeks) as compared to clotting factors being only (hours to days). Platelet numbers and function are also affected with liver failure. Disseminated intravascular coagulopathy (DIC) is a potential complication.[7]

TREATMENT

Most cases of human ALF will get the patient on a waiting list for a liver transplant.[4] Since liver transplantation is not yet a realistic option in the veterinary world, our patients must rely on an early diagnosis and aggressive supportive care to survive. Antidotes and specific therapies are useless if vital organ function is lost. Regardless of the etiology, an initial assessment must be made of cardiopulmonary and neurologic function and appropriate actions must be taken early to treat any problems identified. When there is acute ingestion of a known hepatotoxin emesis should be induced followed by activated charcoal to prevent absorption of the toxin. To prevent further hepatocyte damage, support oxygen delivery to allow optimal hepatic function.

Patients with ALF can have any or all of the three major categories of shock. Hypovolemic, cardiogenic, and distributive. Though the mechanisms for each are distinctly different, each results in reduced oxygen delivery (DO_2) to tissues through low blood flow or uneven distribution of flow. A single patient may have several pathologic processes simultaneously resulting in reduced perfusion of tissues. A toxic insult to the liver or a liver damaged acutely by a systemic disease process may have inflammatory changes altering the microcirculation.

Hypovolemia may be present on admission due to increased fluid losses from vomiting, diarrhea or transudation of fluid into the extravascular space. Hemorrhagic hypovolemia is possible with coagulopathy and GI blood loss. Decreased oral intake can also contribute resulting from altered mental status or nausea. Most patients will require fluid resuscitation initially. Low systemic vascular resistance results in low blood pressures even in the fluid resuscitated patient. As these patients are often hypoproteinemic, close attention to the type of fluid and the response is indicated. Excessive crystalloid fluids can further dilute blood proteins, worsen coagulopathies and increase fluid accumulation in extravascular compartments. Dehydration should be corrected using a balanced crystalloid fluid titrated to desired endpoints of perfusion and blood pressure. In the absence of clinical dehydration, fluid replacement with colloid (such as hetastarch or albumin) is preferred over a crystalloid. Fluids should contain dextrose to maintain euglycemia. The type of isotonic crystalloid and additives chosen should address the patient's electrolyte abnormalities.

While adequate fluid replacement and treatment of potential infection and sepsis may help to correct hypotension, inotropic or pressor support may be required in order to maintain mean arterial pressures of at least 60-70 mmHg. Management of hemodynamic balance becomes increasingly important and difficult in the face of intracranial hypertension (ICH) and/or compromised renal function. There has been debate over which agents are best used to support blood pressure in ALF and whether they are useful at all. Alpha-adrenergic agents such as epinephrine and norepinephrine have been thought to potentially worsen peripheral oxygen delivery.[8] On the other hand, dopamine has actually been associated with increased systemic delivery of oxygen.[9] In any case, the hypotension and vasodilatation associated with ALF should respond to these agents. Agents that promote vasoconstriction are generally avoided unless significant systemic hypotension is present, and therefore should not be used in the setting of decreased intracranial perfusion with normal systemic blood pressure. A small study of terlipressin in human patients with ALF found that even in very small doses, the drug was associated with increased cerebral blood flow and ICH.[10] Such results indicate that at this time the risks associated with vasopressin use appear to outweigh its benefits in patients with ALF.

Coagulopathy

Clotting abnormalities are common in patients with ALF and in fact are a part of the diagnosis. While synthesis of coagulation factors is decreased, consumption of clotting factors and platelets also occurs as a result of the inflammation and necrosis within the liver. If the patient is not bleeding it is not necessary to correct clotting abnormalities with fresh frozen plasma (FFP). Bleeding in the setting of a prolonged prothrombin time warrants administration of FFP as does a planned invasive procedure or if the coagulopathy is profound (*e.g.*, International Normalized Ratio > 7). Plasma administration is not without risk. Plasma is potent colloid that can cause volume overload exacerbating pulmonary edema and ICH. Plasma has also been implicated as a pro-inflammatory agent that could fuel a pathologic inflammatory response. Inappropriate plasma supplementation also limits the value of coagulation parameters as a means of following the progression of ALF. With cholestasis or overt coagulation abnormalities, parenteral vitamin K_1 (phytonadione) at 0.5 to 2.0 mg/kg every 12 hours subcutaneously for two to three dosages (or until normalization of prothrombin time) is recommended for dogs and cats with hepatic disease. Vitamin K_1 supplementation is recommended for 24 to 36 hours before invasive procedures, such as hepatic biopsy or feeding tube placement.[6] Since Vitamin E is another fat-soluble vitamin and may compete with Vitamin K, it should be avoided in the coagulopathic patient.

Encephalopathy

As with coagulopathy, hepatic encephalopathy (HE) is part of the diagnosis of ALF. There are many causes of the neurologic signs seen in HE. Clinical signs of HE include changes in mentation. Anything from lethargy head pressing, aggression or agitation, convulsions or seizures to obtundation, coma, and death. Elevations in serum and central nervous system (CNS) ammonia levels alter the glutamate, gamma-aminobutyric acid, and serotonin neurotransmitter systems. Changes in the central and peripheral benzodiazepine receptor systems also occur. Electrolyte abnormalities can contribute to neurologic dysfunction as does the presence of circulating exogenous toxins, hemorrhage in the CNS secondary to clotting factor or platelet inhibition, hypoglycemia, hypoxemia, and other concurrent diseases. Plasma ammonia levels are the biggest concern with ALF. Increased plasma ammonia levels can lead to cerebral edema. Also of concern is ICH that can occur due to glutamine accumulation in astrocytes, increased blood flow to the brain secondary to vasodilation, and release of inflammatory cytokines. Cerebral edema and ICH can lead to brain herniation, a common causes of death in patients with acute liver failure.[11] Treatment of HE with ALF should proceed when signs are noted and directed at lowering blood ammonia levels and preventing ICH. Serum ammonia is lowered by administering lactulose to acidify the colon, decrease ammonia production and convert ammonia into non absorbable ammonium ions. In animals with altered mentation or nausea, lactulose is administered as an enema (5 to 60 mL) to avoid aspiration. Flumazenil, a benzodiazepine receptor antagonist (0.01 to 0.02 mg/kg IV or SQ) may help clinical signs. In animals with HE and seizures, the author prefers propofol (0.5 to 1 mg/kg IV), then as an infusion (0.05 to 0.1 mg/kg/min IV). The clinician should pay particular attention to hypokalemia and alkalosis as both of these conditions increase ammonia uptake to the CNS.

Hepatic supportive care

Liver supportive therapy for ALF includes N-acetylcysteine (NAC), S-Adenosylmethionine (SAMe), Milk thistle, and Vitamin E. NAC is a thiol (SH) donor and restores glutathione production. Glutathione is necessary to detoxify cellular xenobiotics. NAC may protect against hepatic ischemia-reperfusion injury by inhibiting Kupffer cells. NAC may help hepatic blood flow by improving mean arterial pressures, restoring vascular endothelial response to nitric oxide, improving microcirculatory blood flow, oxygen extraction, and forming other products to protect against cell injury. The dose is 140 mg/kg IV followed by 70 mg/kg IV bid or tid for a total of seven treatments. The injectable NAC is diluted 1:4 in 5% dextrose and water and given slowly over 30 minutes to 1 hour. When vomiting has resolved NAC can be given orally.

(SAMe) (Denosyl™) protects against liver damage from acetaminophen toxicity. It increases hepatocyte glutathione concentrations by donating SH. It also acts as a methyl donor and enzyme activator for reactions that maintain hepatocyte membrane structure and function. SAMe is dosed orally at 20 mg/kg bid or daily. SAMe can be given orally in place of NAC when vomiting is not a concern.

Milk thistle (silibinin, Marin™) is an antioxidant, free radical scavenger and inhibits lipid peroxidation. It too may be of benefit in the management of drug-associated or other hepatic toxicities. One canine experimental study showed that dogs poisoned with amanita mushrooms treated with milk thistle had fewer clinical signs and complete survival compared with untreated dogs.[12] Doses range from 50 to 250 mg bid. It appears to have a synergistic effect with vitamin E.

Vitamin E, d-alpha tocopherol, is a fat-soluble vitamin that relies on bile acids and pancreatic fluids for maximal absorption. Vitamin E prevents cell membrane lipid peroxidation from excessive oxidative stress. Vitamin E protects against the effects of copper, bile acids and other hepatotoxins. When given at high dosages, Vitamin E can compete with other important fat-soluble vitamins. If Vitamin K (a fat-soluble vitamin) is deficient and coagulopathy is suspected Vitamin E should not be used. Vitamin E is given at a dose of 10mg/kg daily.[13]

Metabolic Concerns

Metabolic derangements are common in ALF.[4] Alkalosis and acidosis may both occur and are best managed by identifying and treating the underlying cause. Hypoglycemia should be managed with continuous glucose infusions designed to maintain euglycemia. Frequent monitoring of blood glucose is important to avoid iatrogenic hyperglycemia, and because symptoms of hypoglycemia may be obscured in the presence of encephalopathy. Phosphate, magnesium, and potassium levels are frequently low and may require repeated supplementation throughout hospitalization. Nutrition is important. Early feeding is critical in cases of hepatic lipidosis. Enteral feedings should be initiated early in any case of ALF. Severe restrictions of protein should be avoided. If enteral feedings are contraindicated, parenteral nutrition is an option. Enteral and parenteral nutrition may reduce the risk of GI bleeding due to stress ulceration in critically ill patients. Patients showing signs of nausea or GI bleeding may require proton pump inhibitors and GI protectants.

SUMMARY

There is very little veterinary literature on the ideal management of ALF. With many infectious and toxic agents causing these problems in our patients more study in this area is necessary. Management of ALF challenges our best skills as internists and intensivists. Treatments for specific etiologies and immediate supportive care should be undertaken in all patients that demonstrate evidence of coagulopathy or encephalopathy. Without access to liver a liver transplant, their only chance is rapid identification and treatment of the cause of ALF and aggressive, individualized supportive care.

REFERENCES

[1]Trey C, et. al. In: Popper H, Schaffner F, eds. **Progress in Liver Diseases**. 1970;282. [2]Guilford WG, et al. In: Center SA, ed. **Strombeck's Small Animal Gastroenterology**, 1996;654. [3]Hughes D, King LG. **Vet Clinics N Amer**. 1995;25(2):437. [4]Lee WM. **N Engl J Med**. 1993;329:1862. [5]Ostapowicz G, et. al. **Ann of Intern Med.** 2002;137(12):947. [6]Willard M D, Twedt D C: In Willard M D **Small Animal Clinical Diagnosis by Laboratory Methods**, 1994;179. [7]Mackenzie JR et. al **Br. J Surg** 1975;62:431[8]Wendon J, et. al. **Hepatology** 1992;15:1067. [9]Clemmesen JO, et. al. **Scand J Gastroenterol** 1999;34:921. [10]Shawcross DL, et al. **Hepatology** 2004; 39:471. [11]Cooper J, et al. Comp Cont Ed. 2006;28(7):498. [12]Vogel G, et. al. **Toxicology and Applied Pharmacology** 1984;73(3):355. [13]Flatland B. In: Bonagura JD, Twedt DC, eds. Kirk's Current Veterinary Therapy XIV, 2009;555.

ACROMEGALY IN CATS: WHAT DO WE KNOW & WHAT DON'T WE KNOW

Stijn J.M. Niessen, DVM, DECVIM, MRCVS
London, UK

INTRODUCTION

Compared to the management of canine diabetes mellitus, management of diabetic cats seems, in the subjective eyes of many, more challenging in a higher percentage of cases. Some causes for this might include the inherently more autonomous nature of feline patients impacting management compliance. However, these difficulties could also relate to the limited understanding of the aetiopathogenesis of feline diabetes mellitus. To immediately classify a cat as a type 2 diabetic upon documentation of persistent hyperglycaemia and concurrent glucosuria is a tempting, yet inappropriate practice. Whereas feline diabetes mellitus used to be routinely blamed on pure insulin resistance phenomena, we increasingly, and in line with thinking in human diabetology, recognize that other processes must be concurrently or alternatively involved. Beta-cell dysfunction is one of those important processes. Additionally, various other disorders outside the endocrine pancreas could play a crucial role in the etiology of the disease in a proportion of cases. Understanding and recognizing these processes underlying feline diabetes mellitus, can affect optimal management options, at least in a subset of cases.

The last four years have seen a re-newed interest in one of these extra-pancreatic processes. Feline acromegaly is now being recognized as an important cause of feline DM, largely as a result of two remarkable prevalence studies. Both these studies suggested that feline acromegaly occurs in a significant proportion of diabetic cats, with estimates ranging from 1 in 3 to 1 in 4. Even when adhering to the more conservative estimate, this has justified the initiation of several studies on various aspects of this endocrinopathy, including more careful evaluation of etiology, clinical presentation and management aspects. This lecture aims to discuss the latest developments in feline acromegaly research.

PREVALENCE

What we know

Initially, the author and colleagues performed a screening study amongst diabetic cats with variable glycaemic control, which revealed that 59 out of 184 (32%) assessed diabetic cats had insulin-like growth factor-1 (IGF-1) concentrations strongly suggestive of acromegaly (>1000 ng/ml)[1]. Of these 59 cats a subpopulation was more closely evaluated with intracranial contrast-enhanced computed tomography (CT) and/or magnetic resonance imaging (MRI) as well as growth hormone (GH) evaluation in order to conclusively establish the diagnosis of acromegaly. The diagnosis was subsequently confirmed in 94% of these more carefully assessed cases, proving the original estimation of prevalence, made on the basis of raised IGF-1 concentrations only, to be close to the true prevalence in this diabetic cat population in the United Kingdom (UK). Additionally, Berg and colleagues[2] delivered a retrospective study indicating that 26% of diabetic cats in the United States met clinical criteria for the diagnosis of acromegaly. In light of these surprising results the author's research group has continued to evaluate diabetic cats with varying glycaemic control and four years down the line and after evaluation of over 1200 serum samples of diabetic cats across the United Kingdom, this initially detected high ratio remains valid with over 300 of these UK diabetic cat samples, i.e. 1 in 4, showing an IGF-1 concentration greater than 1000 ng/ml, indicating likely presence of acromegaly (unpublished data, Stijn Niessen).

In light of the impact on prognosis and management, and given the repeated documentation of this surprisingly high prevalence, one could therefore argue that a feline diabetic should routinely be screened for possible presence of acromegaly, just as we screen for urinary tract infections in diabetics, presence or absence of an adrenal tumor in cases with hyperadrenocorticism or underlying disease in cases with immune mediated haemolytic anaemia. Early detection could have a beneficial impact on response to treatment. However, if screening is applied, the characteristics and dynamics of serum total IGF-1 as a screening tool should be taken into account (please refer to section on diagnostics).

What we don't know

Published prevalence studies could have been biased by samples preferentially being sent in for evaluation for IGF-1, in case of already existing suspicion of acromegaly. This relates in particular to the study of Berg et al. where laboratory samples of diabetic cats with concurrent IGF-1 evaluation were assessed[2]. However, this argument applies to a significant lesser extent to the author's study, where prospectively IGF-1 was determined on serum submitted for fructosamine evaluation. A degree of bias may persist, since fructosamine evaluation might be opted for in light of suboptimal glycaemic control, a phenomenon more frequently encountered with feline acromegaly than with uncomplicated diabetes mellitus. Nevertheless, it could equally be argued that the above studies in fact also underestimate the true prevalence of acromegaly, since a rather arbitrary cut-off for IGF-1 was chosen (1000 ng/ml), misclassifying cases with a borderline IGF-1 or an IGF-1 that would have increased following initiation of exogenous insulin therapy (please refer to section on diagnostics).

ETIOLOGY

What we know

Traditionally, acromegaly in the cat has been seen as a process caused by excess endogenous growth hormone secretion caused by a pituitary adenoma. Indeed, until recently, all reported cases that were subjected to post mortem examination, although bearing in mind the paucity of such reports, suggest this to be the case[3]. A more systematic evaluation of pituitary histopathology in a larger number of patients is currently ongoing and suggests that a number of cases in fact lack the histopathological features of an acidophilic adenomatous change[4,5]. Instead, acidophilic hyperplasia has been found in these cases that phenotypically were indistinguishable from other acromegalic cats (although the existence of a *persistently present* typical acromegalic phenotype is debatable, please refer to signalment and presentation section).

What we don't know

If there are indeed at least two underlying etiological mechanisms questions arise over a possible inter-relationship between them. Could the hyperplastic state be an early stage which transforms into a neoplastic state? Should we be looking at a supra-hypophyseal process or stimulus or hypothalamic involvement? A comparison to current understanding of, or better, hypotheses on, the etiology of feline hyperthyroidism is tempting in this regard.

Relatively few pure breed cats have been described to suffer from acromegaly and no clear breed predisposition is currently apparent, arguing against an obvious breed-related genetic predisposition, although this, admittedly, could easily still be the case. Candidate gene approach studies could be contemplated to elucidate the role of genes.

SIGNALMENT AND PRESENTATION

What we know

Most reported patients have been middle-aged to older male neutered domestic short hair cats (DSH) with diabetes mellitus. Presence of insulin resistant diabetes mellitus has been shown to be a risk factor for presence of acromegaly. Nevertheless, although in the author's dedicated Acromegaly Clinic insulin resistance is frequently a presenting sign, when using a screening approach a significant number of patients will appear insulin sensitive at time of the initial diagnosis. The insulin dose of 59 acromegalic cats ranged from 1-35 iu BID (median 7 iu BID). Out of these 59 cats with IGF-1 concentrations consistent with acromegaly, 47 were male neutered, 6 female neutered, 5 intact males and 1 of unknown gender; 52 were DSH, 3 domestic long hair, and 4 of unknown breed; with a median age of 11 years (age range: 6-17 years); median body weight was 5.8 kg (range 3.5-9.2 kg)[1,5]. The same data suggest that the 'typical' acromegalic phenotype is not consistently present, possibly related to the gradual onset of acromegaly-induced changes and/or previous failure to screen assumed 'atypical' cases. To illustrate the point of gradual-onset, duration of clinical signs in the initial group of cats seen by the author's research group ranged from 2 to 42 months with owners and attending clinicians often remaining unaware of any ongoing morphological changes. Once again it seems tempting to draw comparisons to the feline hyperthyroidism situation, where we currently more rarely see the classical hyperthyroid cat, possibly due to increased preparedness to screen for this disease in the elderly cat and/or possible increasing prevalence.

Commonly encountered symptoms in the acromegalic cats seen were: polyuria/ polydipsia, polyphagia (at times extreme), weight gain, enlarged liver and kidneys, prognathia inferior, immature bilateral cataracts, clubbed paws, broad facial features, systolic cardiac murmur and, previously underreported, respiratory stridor[1,5]. Weight gain despite poor glycaemic control should be one of the triggers to alert clinicians for the possible presence of acromegaly.

What we don't know

Whenever a disease is underdiagnosed and therefore underreported, the danger exist that underdiagnosis persists due to a dogmatic approach in the recognition of that disease. Practically speaking, if we only routinely look for acromegaly in insulin resistant patients, we will only find insulin resistant cases. Equally, the existence of individual non-diabetic acromegalic cases has been mentioned in textbooks, but the true prevalence of such cases is largely unknown. Arguably, the latter might not be as relevant, although we could be missing an opportunity to prevent the onset of diabetes mellitus altogether.

Cardiomyopathies and nephropathies have been reported to ensue as part of the pathophysiology of acromegaly, presumably being induced by excess GH and IGF-1 concentrations. However, since relatively few cases of feline acromegaly have been described thus far and most previous cases seem to have been suffering from an advanced stage of acromegaly, this assumption warrants further investigation. Especially, in view of the high prevalence of concurrent disease, including cardiomyopathies and nephropathies amongst non-acromegalic diabetic and non-diabetic geriatric cats in general. A recent comparison of routine clinical pathology parameters between acromegalic diabetic cats and non-acromegalic diabetic cats did not reveal a greater incidence of azotaemia amongst acromegalics[6]. Pancreatic abnormalities, specifically hyperplasia, do seem particularly prevalent in acromegalic cats based on post mortem examinations of patients seen in the Acromegalic Cat Clinic at the Royal Veterinary College in London[5]. The overall message should probably be that clinicians ought to remain open-minded about the signalment and presentation of the acromegalic cat.

DIAGNOSIS

What we know

Feline growth hormone (fGH, serum and plasma), IGF-1 (serum) and intra-cranial imaging have been shown to be useful diagnostics. A suggested growth hormone cut-off value of 10 ng/ml was shown to result in an acceptable specificity of 95% and sensitivity of 84% when using the recently by the author developed fGH assay[4,5]. fGH also appeared relatively stable, allowing overnight transport of unseparated samples[4]. However, both fGH and IGF-1 were shown to yield false positive results in a minority of cases. Cases with a normal basal fGH concentration have yet to be documented, yet IGF-1 has been documented to be falsely negative in a minority of cases[1]. The duration of exogenous insulin administration could play a crucial role in the latter, since hepatic IGF-1-production is induced via stimulation of insulin-dependent hepatic GH-receptors. An insulin-deficient state can act as an inhibitor of such IGF-1 production, resulting in low IGF-1 concentrations in diabetic patients prior to instalment of exogenous insulin treatment or even during the first few weeks of such treatment[7]. When screening for presence of acromegaly, these specific IGF-1 dynamics should be taken into account and repeat IGF-1 determination should be considered 6-8 weeks into the treatment. Alternatively, if one wishes to determine IGF-1 only once, the latter time point is recommended over the immediate time of diagnosis of diabetes mellitus. Intra-cranial imaging (with contrast enhancement) seems useful in confirming the presence of acromegaly, with MRI seemingly more sensitive than CT. However, cases with a negative CT and/or MRI have been documented with the diagnosis eventually being confirmed on post mortem examination[1].

What we don't know

When a structural pituitary abnormality is documented this indeed provides further circumstantial evidence for the presence of acromegaly. However, it does not provide differentiation from a non-functional pituitary tumor or pituitary dependent hyperadrenocorticism. More importantly, one could question the usefulness of intra-cranial imaging when a negative scan is obtained despite presence of a documented hormonal imbalance (elevated fGH, IGF-1, presence of diabetes mellitus). When imaging is negative, it seems more logical to have the ultimate pre-mortem diagnosis of acromegaly rely on hormonal assessment. Additionally, cases with subtle (microscopic) acidophilic hyperplasia or microadenoma instead of obvious (macroscopic) adenoma might more likely show negative intra-cranial imaging. Dynamic intra-cranial imaging studies using timed injection of contrast might be of aid here. However, refinement of our hormonal assessment methods and increasing availability of assays probably holds the true answer in this regard.

TREATMENT

What we know

Broadly speaking, treatment options consist of medical, surgical and radiation options. Medical treatment options aimed at inhibiting the pituitary have not proven very successful thus far, including the use of Lanreotide (long-acting synthetic somatostatin analogue)[5]. Octreotide and dopamine agonists have not resulted in convincing improvement of insulin sensitivity on dosing regimens employed thus far[3,8].

Transsphenoidal hypophysectomy is one of the main preferred treatment modalities in humans and a recent report described the successful application of hypophysectomy in an acromegalic cat, resulting in an immediate drop of GH-levels post-surgery, as well as resolution of the diabetes mellitus within 3 weeks[9]. Cryohypophysectomy constitutes another option, which has been reported to be successful in two cases[10,11].

Radiation therapy is currently the most widely applied treatment modality and has been shown to successfully reduce size and secretory capacity of the acidophilic adenoma in a majority of cases[5,12,13]. Application of more refined protocols and "gamma-knife" technology seems advantageous. More wide spread availability of radiation facilities has aided its popularity as a treatment modality. Nevertheless, treatment is still relatively costly, often requires multiple general anaesthethics and, most importantly, response is variable and unpredictable in effect, onset and duration.

A more conservative approach ignores the underlying disease mechanism and focuses on gaining more control of the diabetes mellitus. Eventually most cats invariably need high dosages of insulin and/or combinations of short-acting and long-acting insulin types in order to ensure an adequate quality of life for both pet and owner. Nevertheless, this approach can result in an adequate level of control in a proportion of cases, although careful and continued assessment of quality of life is indicated.

What we don't know

A long-term survival benefit has yet to be demonstrated with any treatment modality compared to treatment of the diabetes mellitus alone. The true gain might be in an increased quality of life when the underlying acromegaly is targeted and better control of the diabetes mellitus can be attained.

A recent study showed that intravenous octreotide alters serum GH levels in a subset of acromegalic cats, suggesting that, at least in such subset, medical pituitary inhibition could prove beneficial[14]. The shape and details of such medical intervention have yet to be identified, despite several reports of attempts with dopamine agonists and somatostatin analogues.

A remaining relevant issue constitutes the best way to monitor patients after radiation therapy. IGF-1 has proven uninformative, whereas fGH determination was shown to be a more sensitive indicator of treatment effect post-radiation[4,15]. IGF-1 and GH concentrations did normalize post-hypophysectomy further suggesting this to be superior to radiation therapy.

THOUGHTS FOR THE FUTURE

Continued examination and description of a large number of cases will likely provide the veterinary community with a more accurate understanding of this endocrinopathy. Observations thus far have already unearthed interesting aspects that impact our daily dealings with these patients. From a clinical point of view it seems logical to consider routine screening of feline diabetics for presence of acromegaly, although bearing in mind the limitations of early IGF-1 assessment. Treatment options should probably be refined beyond the level of mass pituitary radiation and the real gain could lie in treating acromegalic patients early on in the disease process prior to the onset of permanent beta-cell damage and other possible GH/IGF-1 induced pathology. With such proactive approach acromegaly-induced permanent diabetes mellitus could well be prevented.

REERENCES

[1]Niessen SJ, et al. **J Vet Intern Med.** 2007; 21(5):899, [2]Berg RI, et al. **J Vet Intern Med.** 2007; 21(5):889, [3]Peterson ME, et al. **J Vet Intern Med** 1990; 4(4):192, [4]Niessen SJ, et al. **Vet Rec** 2007; 160(26):902, [5]Niessen SJ **J Feline Med Surg**. 2010; 12(1):15, [6]Niessen SJ and Church DB, In: **Cons Fel Med** 2010 [7]Reusch CE, et al. **Vet Rec** 2006;158:195. [8]Abraham LA, et al. **Aust Vet J.** 2002 Aug; 80(8):479. [9]Meij BP, et al. **J Feline Med Surg.** 2010;12(5):406. [10]Abrams-Ogg AC, et al. **Can Vet J** 1993; 34, 682. [11]Blois SL and Holmberg DL. **J Small Anim Pract** 2008; 49(11):596. [12]Brearley MJ, et al. **Vet Comp Onc** 2006; 4 (4) 209-217. [13]Sellon RK, et al. **J Vet Intern Med.** 2009;23(5):1038. [14]Slingerland LI, et al. **Dom An Endo** 2008; 35(4): 352. [15]Dunning MD, et al. **J Vet Intern Med**. 2009; 23(2):243.

WSAVA INTERNATIONAL RENAL STANDARDIZATION STUDY: THE FIRST LOOK

Larry D. Cowgill, DVM, PhD, DACVIM
Davis, CA

Glomerular diseases are a leading cause of renal disease and failure in dogs. The reported incidence of glomerular lesions in tissue samples from randomly selected dogs ranged from 43% to 90% with an increased incidence with increasing age.[1-7] In animals with glomerular lesions that have not progressed to end-stage renal disease, renal biopsy has potential to provide a definitive diagnosis and important clinical information about the type, severity, and biologic behavior of the lesions. However, current diagnostic practices and classification schemes have permitted only a loose grouping of the spectrum of glomerular diseases into a limited collection of pathologic categories that have had little discriminating potential for therapeutic intervention or outcome assessment. This limiting light microscopic characterization of glomerular disease provides only a myopic interpretation of the glomerular lesion with little possibility to recognize subtle distinctions. In human patients, differences in the pathogenesis of discrete glomerular diseases promotes distinct clinical and pathological characteristics prompting specific treatment recommendations and outcome predictions for these diseases. A similar disparity for canine glomerular injury is likely.

In 2007, the WSAVA International Renal Standardization Study undertook an ambitious 5-year initiative to redefine the clinical and pathologic features protein-losing glomerular diseases in dogs. The purpose of the initiative was to apply a broad array of diagnostic and communication technologies to promote a paradigm shift in the diagnostic and clinical approach to glomerular disease in veterinary medicine. At the foundation of the initiative was the mandate to expand the utilization pathologic imaging modalities to include light, electron, and immunofluorescent microscopy, as used in human nephropathology, to extend and more accurately characterize the morphologic responses of the glomerulus to injury, and to relate these findings to specific clinicopathologic presentations and outcomes in proteinuric dogs. The Phase 1 outcome goals of the project are to:

- Establish an infrastructure to insure the ongoing availability of sophisticated diagnostics to the profession worldwide.
- Establish an international consortium of expertise in veterinary nephropathology and clinical nephrology to facilitate ongoing and future investigations, education, and scientific advancement of glomerular disease in animals.
- Develop a prototype standardized system for the classification of pathological lesions in canine glomeruli that is accepted and adopted throughout the world.

The WSAVA Study is now approaching the completion of Phase 1 of the initiative and has made exceptional progress advancing the proposed infrastructure, defining the discipline of nephropathology, characterizing the expression of canine glomerular disease, and development of a prototype classification system for canine proteinuric glomerular disease. This report updates some of the milestones and highlights of the initiative to date. The international expertise dedicated to the study of glomerular pathology and glomerular disease has expanded from the study group initially assembled in 2005 to include 12 nephropathologists and 9 nephrologists from 5 countries. The Pathology Working Team has descriptively characterized more than 70 light microscopic and 30 electron micrographic features or patterns of the canine glomerulus and surrounding renal structures. Each feature has been archived in a digital glossary to provide an illustrative reference to the established standards.

To provide greater objectivity, diagnostic consistency, and flexibility to the pathologic "diagnosis", a scoring system has been established to characterize the extent and magnitude of identifiable lesions of the renal glomeruli, tubular epithelium, interstitium, and vasculature. Using this scoring system, the Team has documented as proof-of-concept in as subset of canine patients with pathologic renal proteinuria a new paradigm for the subjective and objective assessment and classification of canine glomerular disease. The paradigm establishes distinct pathologic groupings ("diagnoses") based on traditional pathological principals in addition to recognition and consensus scoring of the lesion array produced with multiple imaging modalities. This fusion of subjective assessment and quantitative annotation has been modeled further with cluster analysis algorithms to statistically arrange patients with distinctly associated morphologic characteristics. The features of these morphologic clusters constitute a foundation to assign descriptive morphologic diagnoses, i.e., "membranous nephropathy", that are specific and appropriate for dogs. The resulting prototype classification system will enable more precise and consistent classification of glomerular diseases than previously possible and will be subjected to subsequent validation.

The study has also confirmed the spectrum of recognizable or variant glomerulopathies is more extensive than appreciated previously by conventional light microscopy. Even traditional light microscopic diagnoses including amyloidosis, membranoproliferative glomerulonephritis, and membranous nephropathy can be shown to encompass a variety of morphologic differences when examined more comprehensively by light microscopy and ultrastructurally with transmission electron microscopy. What remains to be defined is whether these identifiable differences represent true pathophysiologic variations of a singular disease, different stages of expression of a singular disease, or distinctly different pathologic entities constituting different diseases.

Equally focused, the Clinical Working Team is attempting to correlate the pathologic diagnoses with clinical and clinicopathologic data to identify particular patterns of findings that constitute distinct canine glomerular disease entities that might be expected to have foreseeable outcomes or predictable responses to treatment. These correlations will be used to render a finalized (validated) morphologic and clinicopathologic classification scheme for canine glomerular disease that can be uniformly applied among veterinary pathologists and nephropathologists. The Clinical Working Team similarly has established a clinical scoring scheme founded on over 130 numerical and categorical clinical attributes on the same dogs with protein-losing glomerular disease whose kidney biopsies are under review. Each attribute is independently evaluated and scored by all clinical team members from supplied medical records, consultation with the submitting clinician, or from data entered into the established medical database. Subsequently, an international review session is held for each patient, and a consensus score is established for each attribute by the team. The clinical attributes list is scheduled for the same cluster analysis logistics to test if there are statistically arrange groupings of clinical features in patients with distinctly associated morphologic characteristics or pathologic diagnoses. By this analytical approach, the Study anticipates the possibility that protein-losing glomerular diseases can be more precisely defined and classified on the basis of both their morphologic features (light, electron, and immunofluorescent microscopy) and the patterns and characteristics of their clinical and clinical pathologic response to the glomerular insult.

The WSAVA International Renal Standardization Study is unique throughout the world and represents a global diagnostic resource for general and specialty veterinarians and the pet owning public. The comprehensive light, immunofluorescent, and electron microscopic services will extend the diagnostic capabilities for renal diseases to a level of sophistication and precision standard in human nephrology.

ACKNOWLEDGEMENTS

Current Pathology Team: Luca Aresu (Italy), Brian Berridge (USA), Cathy Brown (USA), Rachel Cianciolo (USA), Fred Clubb, Jr (USA), Roel Goldschmeding (Netherlands), Clare James (UK), Johan Jansen (Norway), George Lees (USA), Chuck Mohr (USA), Bill Spangler (USA), Jaco van der Lugt (Netherlands),. **Current Nephrology Team:** Claudio Brovida (Italy), Larry Cowgill (USA), Jonathan Elliott (UK), Reidun Heiene (Norway), George Lees (USA), David Polzin (USA), Hattie Syme (UK), Shelly Vaden (USA), Astrid van Dongen (Netherlands).

REFERENCES

[1]MacDougall DF, et. al. **Kidney Int** 1986;29:1144-1151. [2]Cook AK, et, al. **J Am Anim Hosp Assoc** 1996;31:313-322. [3]Center SA, et. al. **JAVMA** 1987;190:81-90. [4]Murray M, et. al. **Lab Invest** 1974;30:213-221. [5]Vilafranca M, et. al. **J Vet Med** 1994;41:599-610. [6]Muller-Peddinghaus R, et. al. **Vet Pathol** 1977;14:1-13. [7]Muller-Peddinghaus R, et. al. **Vet Pathol** 1977;14:121-127.

PROBIOTIC THERAPY OF CHRONIC KIDNEY DISEASE

David J. Polzin, DVM, PhD, DACVIM
St. Paul, MN

INTRODUCTION

Uremia is an important and often fatal consequence of chronic kidney disease (CKD). Probiotic therapy has been advocated as a means of using the gastrointestinal tract to eliminate some of the toxins normally excreted by the kidneys that may be responsible for the clinical condition known as uremia. The traditional definition of uremia is: "the clinical syndrome that results from loss of kidney functions. Impaired renal glomerular, tubular and endocrine functions lead to retention of toxic metabolites, changes in the volume and composition of body fluids, and excess or deficiency of various hormones." It is presumed that the clinical signs of uremia reflect the sum effect of these derangements on organ systems throughout the body. However, a newer definition of uremia is: "the ill effects of renal failure that we cannot yet explain."(Meyers and Hostetter, 2008) In this definition, the pathogenesis of some abnormalities previously regarded as part of the uremic syndrome are now understood (e.g. hypertension due to volume overload, anemia due to erythropoietin deficiency), so, by the newer definition, they are no longer considered to be part of the uremic syndrome.

UREMIA

The pathogenesis of uremia is incompletely understood. In general, three major are involved: 1) disturbed excretion of electrolytes and water, 2) reduced excretion of organic solutes (i.e. uremic toxins), and impaired renal hormone synthesis. These mechanisms affect virtually every organ system in the body, thereby producing the uremic syndrome. In humans, in excess of 70 clinical signs have been identified as part of the uremic syndrome. (Meyers and Hostetter, 2008)

Disturbed Excretion of Electrolytes and Water

The kidneys are charged with maintaining water and electrolyte balance in the body. To maintain a stable internal milieu, excretion of water and electrolytes must accommodate changes in intake. As glomerular filtration rate (GFR) declines with kidney disease, the need for excretion of water and electrolytes remains the same unless intake is modified to accommodate the reduction in GFR. As GFR declines, the excretory load of electrolytes and water per surviving nephron substantially increases. However, urinary excretion of many substances can be maintained in the presence of a marked reduction in GFR through parallel reduction in the degree of tubular reabsorption. If GFR declines by 3/4, the remaining nephrons must each excrete four times more sodium. Thus, maintaining fluid and electrolyte balance requires that more water and electrolytes be excreted per surviving nephron. Although the kidneys have a remarkable ability to maintain water and electrolyte balance well into advanced stages of kidney disease, compensatory mechanisms ultimately fail in end-stage kidney disease. Clinical manifestations of disturbed excretion of electrolytes and water may include edema, hypertension, hyponatremia, hyperkalemia, metabolic acidosis and hyperphosphatemia.

Reduced Excretion of Organic Solutes

Although urea and creatinine are the best known, the kidneys excrete a wide of organic solutes. Many of these solutes are handled primarily by glomerular filtration, but renal tubular secretion and reabsorption may affect net renal excretion. An important difference between renal handling of organic solutes and renal handling of electrolytes and water is that excretion of organic solutes is mechanisms generally not actively regulated. Because of this, the blood concentrations of organic solutes begins to rise with the initial decline in GFR and continues to progressively increase as renal function continues to decline..

When GFR declines below about 10% of normal, clinical signs may begin to appear. Many of these clinical signs have been attributed to increased concentrations of organic solutes It is well recognized that many organic solutes accumulate in body fluids when the kidneys fail. However, presence of increased concentrations of a substance in the body fluids of patients with uremia does not prove that the substance is a uremic toxin. Nonetheless, that uremic solutes may in fact be contributory to the uremic syndrome is supported by the observation that some uremic abnormalities are transferable with uremic serum or plasma. Abnormalities that have been thus demonstrated include: inhibition of sodium-potassium ATP-ase, inhibition of platelet function, leukocyte dysfunction, loss of erythrocyte membrane lipid asymmetry, and insulin resistance. (Meyers and Hostetter, 2008)

In excess of 100 potential uremic solutes have been listed by the European Uremic Toxin Work group (EUTox). (Meyers and Hostetter, 2008) However, it has been so difficult to confirm substances as true uremic toxins, that to date none have met the recommended criteria proposed for confirming a substance as a uremic toxin. The criteria include: 1) the chemical structure of the substance must be known, 2) plasma or tissue concentrations must be greater in uremic patients than in normal individuals, 3) high concentrations must be related to specific uremic symptoms or signs, 4) the clinical effects observed in uremic patients must be reproducible by raising the solute's concentration in normal individuals.

It has been suggested that most uremic solutes are probably not toxic, and those that are toxic may only exert their effects when present in combination with other uremic solutes. It is well accepted that urea and creatinine, by themselves, are not uremic toxins. However, the plasma concentration of urea is believed to be a reasonable surrogate for the systemic concentrations of many uremic solutes. Thus, demonstrating that a treatment lowers urea concentrations in and of itself is not conclusive evidence of reduced uremic toxicity.

Impaired Renal Hormone Synthesis

The kidneys normally produce a number of essential hormones that influence multiple organ systems including calcitriol (1,25-dihydroxycholecalciferol), erythropoietin, prostaglandins, kinins and renin. In particular, calcitriol and erythropoietin have well established roles in the pathogenesis of clinically important consequences of severe kidney disease. Calcitriol, the most metabolically active form of vitamin D, is essential for calcium and skeletal metabolism and deficiencies in production of calcitriol can result in renal secondary hyperparathyroidism and renal osteodystrophy. Parathyroid hormone has been identified as a uremic toxin, although the clinical consequences of hyperparathyroidism beyond its role in renal osteodystrophy in chronic kidney disease (CKD) are poorly established. In addition, most cells have receptors for calcitriol and it is believed that inadequate activation of these receptors may in some way contribute to clinical signs of uremia. It appears that restoration of adequate levels of calcitriol in dogs and humans with CKD may slow progression of kidney disease and prolong survival. Erythropoietin deficiency is associated with a hypoproliferative anemia that may promote weakness, lethargy, and hyporexia.

MANAGING UREMIA USING THE INTESTINAL TRACT

"Enteric Dialysis™," a trademark phrase held by Kibow Biotech, Inc is the process by which colon-friendly, non-pathogenic microbes consume uremic toxins or metabolites. The concept of "enteric dialysis" is an alternative strategy for solute extraction in kidney failure which is based on the intestinal mucosal surface functioning as a semi-permeable membrane. Other strategies have also been employed to increased gastrointestinal excretion of uremic solutes, water and electrolytes, thereby facilitating removal from the body via movement across the intestinal mucosal surface. Compounds that have been used in this way include intestinal binders, adsorbents, and probiotics.

Intestinal Electrolyte Binders

Orally administered intestinal binders are available for enhancing fecal elimination of phosphorus and potassium. Intestinal phosphate binders are commonly used in dogs and cats to reduce the quantity of dietary phosphorus that is absorbed, thereby limiting hyperphosphatemia. Although much less effective in lowering serum phosphorus concentrations, fecal excretion of phosphate normally secreted into the intestinal lumen promotes removal of phosphorus from the body. Substances commonly used in dogs and cats for this purpose include salts of aluminum, calcium and lanthanum. Typically these cation are released in the process of binding phosphorus and may result in retention of these cations. Sodium polystyrene (Kayexalate®) affects the exchange of sodium for potassium in the intestinal lumen, thereby facilitating fecal elimination of potassium from patients with hyperkalemia.

Adsorbents

Oral adsorbents are designed to adsorb uremic toxins and contain them until they can be excreted in feces. The adsorbed compounds are likely a mixture of endogenous substances ("uremic toxins?") passing across the semi-permeable mucosal barrier as well as orally ingested foodstuffs that may have harmful effects when absorbed by patients with CKD. Preliminary studies have suggested benefits may range from removal of substances that promote uremic signs to removal of substances that facilitate progression of CKD, thus enhancing survival. Depending on their chemical composition, adsorbents may vary greatly in the selectivity of their binding potential. Logically, the clinical benefits of the adsorbent would relate closely to the substances actually removed by the adsorbent.

A variety of compounds have been advocated for this purpose. Charcoal has been used for this purpose and has reported to decrease levels of serum phenols, uric acid, and guanidine, but not BUN nor creatinine in humans with CKD. A porous carbonaceous adsorbent has been reported to delay the time to introduction of dialysis in humans with CKD. Further, this compound has been reported to delay development of glomerular sclerosis in a rat model of CKD. The mechanisms underlying this beneficial effect is unclear but has been speculated to result from retardation of onset of glomerular hypertrophy, systemic hypertension, and proteinuria.

Probiotics

The concept underlying oral probiotic administration as an alternative method of using the GI tract for elimination of wastes normally excreted by the kidneys is that live bacteria may be used to catabolize urea and other uremic toxins, thereby effectively trapping them within the lumen of the bowel to be excreted in the feces. To do this, selected species of bacteria were developed with the intent of optimizing their capacity to consume uremic toxins. Bacteria that have been used for this purpose include *Streptococcus thermophilus, Lactobacillus acidophilus, and Bifidobacteria spp.* These bacteria are considered to be generally regarded as safe and are not known to have any adverse or disease-inducing effects in normal or uremic dogs, cats or humans.

Preliminary studies in rats and Gottingen mini pigs with induced CKD suggest that oral supplementation with an appropriate probiotic may moderate azotemia and confer a morbidity and survival advantage compared to non supplemented animals. In a blinded, placebo-controlled pilot study in rats with induced CKD, animals receiving a probiotic formulation had decreased blood concentrations of uremic toxins, altered renal function and prolonged survival time compared to placebo-controls. In Göttingen pigs with induced CKD, orally administered enterically coated probiotic supplementation was compared to placebo treatment in a blinded trial. Among pigs with BUN <90mg/dl the probiotic appeared to have no significant effect on BUN values after 16 weeks; however, probiotic therapy was associated with significantly lower BUN values among pigs with baseline BUN values > 90 mg/dl. In addition, 100% of the placebo treated animals with BUN values > 90 mg/dl died whereas 100% of the probiotic treated animals survived to the 16-week study endpoint. Preliminary uncontrolled observational studies using cats and dogs with spontaneous CKD have yielded similar results with both BUN and serum creatinine values declining after initiation of probiotic therapy.

RECOMMENDED READING

Meyers T and Hostetter T: Pathophysiology of uremia. In, Brenner BM, The Kidney. WB Saunders, Philadelphia. 2008, pp.1681-1696. Nobuyoshi T et al: Therapeutic effects of long-term administration of an oral adsorbent in patients with chronic renal failure: two year study. Inter J Urology 12:7-11, 2005. Palmquist R. A preliminary Clinical Evaluation of Kibow Biotics®, a probiotic agent, on feline azotemia. JAHVMA. 2(4):23-27, 2006. Ranganathan N, Patel B, Ranganathan P, et al: Probiotic Amelioration of azotemia in 5/6th nephrectomized Sprague-Dawley rats., *The Scientific World JOURNAL*, 2005; 5: 652-660.

NEUROPHYSIOLOGY AND ETIOPATHOGENESIS OF URINARY INCONTINENCE

Julie R. Fischer, DVM, DACVIM and India F. Lane, DVM, MS, DACVIM
San Diego, CA and Knoxville, TN

INTRODUCTION

Normal micturition comprises a *urine storage phase*, when the bladder slowly fills and relaxes while the urethra remains closed, and a *urine voiding phase,* when the bladder contracts and urine is expelled through a relaxed urethra. Appropriate urine storage and voiding depend on the intricate and coordinated interaction of the nervous system, urinary bladder and urethra. Incontinence represents failure of urine storage; successful treatment of urinary incontinence depends foremost on accurate anatomic localization and mechanistic description of the problem, as well as a basic understanding of the associated neurophysiology.

LOWER URINARY TRACT ANATOMY

The key anatomic components of the lower urinary tract (LUT) include: 1) the *detrusor muscle*, the smooth muscle which forms the body and neck of the bladder; 2) the *internal urethral sphincter* (IUS) which is composed of the smooth muscle of the urethrovesicular junction; 3) the *external urethral sphincter* (EUS), formed from striated muscle encircling portions of the urethra distal to the IUS (position of the EUS varies somewhat by sex and species); and 4) the *ureterovesicular junction*, normally located proximal to the IUS, at the junction of the bladder body and neck. The urethral closure mechanism consisting of the bladder neck and the smooth and striated urethral musculature is often collectively referred to as the *outflow tract* or *outlet*.

NEUROPHYSIOLOGY OF STORAGE AND VOIDING

The urine storage phase of micturition occurs chiefly under sympathetic (adrenergic) nervous control. The hypogastric nerve (arising from spinal segments $L_{1\text{-}4}$ in the dog, $L_{2\text{-}5}$ in the cat) stimulates detrusor beta receptors, inducing smooth muscular relaxation and permitting filling under low pressure. In contrast, sympathetic stimulation of the alpha-1 receptors of the bladder neck and urethral smooth muscle (the IUS) induces smooth muscle contraction, closing the outlet and maintaining continence. Sympathetic input also modulates and minimizes parasympathetic-mediated contraction of the detrusor muscle. Voluntary input to the striated urethral musculature (the EUS) is supplied via the pudendal nerve (arising from spinal segments $S_{1\text{-}3}$). During the storage phase, this input stimulates nicotinic cholinergic receptors in the EUS, causing contracture and additional closure of the outlet when needed (e.g., during coughing or sneezing; to temporarily override the urge to void when inappropriate).

As bladder filling progresses, information from stretch receptors in the detrusor muscle is passed to higher centers via the pelvic (filling sensation, need to void) and hypogastric (pain, overdistension) nerves. The urine voiding phase of micturition occurs chiefly under parasympathetic (cholinergic) control. The pelvic nerve (arising from spinal segments $S_{1\text{-}3}$) stimulates muscarinic cholinergic receptors in the detrusor muscle, stimulating contraction and raising the intravesicular pressure. Simultaneously, sympathetic input to the outlet is inhibited at the level of the micturition center in the pons, allowing IUS and EUS relaxation. When intravesicular pressure exceeds outlet closure pressure, voiding occurs. After complete voiding, the system is "reset" for the filling stage to begin again. See Figure 1.

ETIOPATHOGENESIS OF INCONTINENCE

Incontinence is characterized by involuntary leakage of urine from a small or normal sized bladder. The presence of urinary incontinence indicates structural or functional derangement of one or more of the components of urine storage: 1) insufficient closure of the urethra, 2) failure of the bladder to sufficiently accommodate urine, or 3) anatomical anomaly defeating or circumventing normal closure mechanisms. Paradoxically, some animals with urine retention will exhibit urine leakage secondary to excessive urine volume and pressure. Presentation of an animal with incontinence and a distended bladder should trigger investigation into causes of urine retention.

The most common cause of urinary incontinence in dogs is insufficient urethral closure pressure, and the most common cause of this is urethral sphincter mechanism incompetence (USMI), also called hormone-responsive incontinence, or spay incontinence. Many possible factors contributing to USMI have been identified, including decreased responsiveness and/or number of urethral alpha adrenergic receptors, abnormal bladder neck position ("pelvic bladder"), aging and relative estrogen deficiency, obesity, regional changes in collagenous support structures, increased collagen deposition in bladder and urethra, vaginal structural abnormalities, and genetics. Pediatric ovariohysterectomy may also play a role. Spinal cord lesions caudal to L3 can cause urethral sphincter areflexia, resulting in urine leakage. Other causes of insufficient urethral closure pressure include congenital urethral dysplasia/hypoplasia, urethritis, and infiltrative urethral, trigonal, or prostatic disease.

Insufficient urine accommodation entails involuntary detrusor contraction at lower than normal bladder volumes. This condition can occur due to lack of detrusor compliance or hypoplastic bladder, or detrusor irritation (e.g., consequent to infection, idiopathic cystitis, neoplasia, stone disease), and has been termed detrusor instability, bladder hypoaccomodation, detrusor hyperspasticity, or overactive bladder, and is rarely documented in small animals as a primary cause of incontinence. Hypoaccomodation may occur as a contributing factor to refractory incontinence in dogs with USMI, due to increased collagen deposition in the detrusor muscle. Reduced bladder capacity and poor detrusor compliance have been documented in some dogs with congenital incontinence due to ectopic ureters.

Ureteral ectopia describes the abnormal termination of one or both ureters distal to the bladder trigone (generally into the urethra or vagina), circumventing normal bladder outlet closure mechanisms. Other congenital urethral malformations occur as isolated abnormalities, or in conjunction with ureteral ectopia.

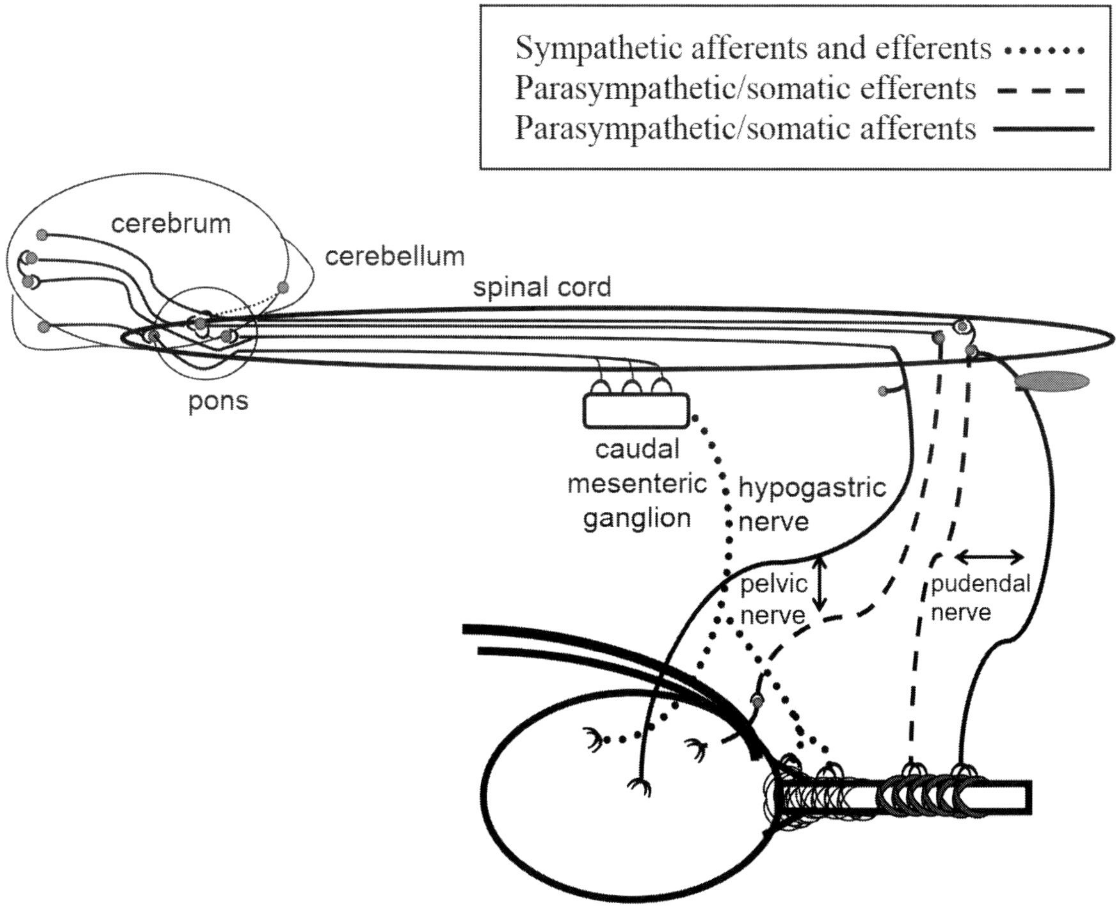

Figure 1. Schematic diagram of the innervation of the lower urinary tract.

REFERENCES

Adams LJ et al. In: Textbook of Veterinary Internal Medicine. 2010; pp 2111. Coit VA et al. *Eur J Pharmacol* 2008;14 p.153. Fischer JR et al. In: *BSAVA Manual of Canine and Feline Nephrology and Urology* 2007;pp. 26. Fowler CJ et al. *Nat Rev Neurosci* 2008;9:453. Guyton AC et al. Textbook of Medical Physiology 2006;p. 311. Howe LM et al. *J AmVet Med Assoc* 2001;218: p. 217. Labato MA et al. In: Textbook of Veterinary Internal Medicine. 2010;p 160. Lane IF et al. *J AmVet* 1995;206,p1348. *Med Assoc* 206, 1348–1357.Nickel RF et al. *Vet Rec* 1999;145: p. 11 Oliver JE et al. In: *Handbook of Veterinary Neurology*, 1997 p. 73. Ponglowhapan S et al. *Theriogenology* 2008; 70: p. 1516. Reichler et al *Theriogenology* 2004;62: p. 1391. Spain CV et al. *J Am Vet Med Assoc* 2004;224(3), p.387. Stocklin-Gautschi HM et al. *J Repro Fert* 2001 57:p. 233. Thrusfield MV. *Vet Rec* 1985;116:695. Thrusfield M. et al. *J Small Anim Pract* 1998;39(12):559.

URINARY INCONTINENCE: MEDICAL MANAGEMENT

India F. Lane, DVM, MS, EdD, DACVIM and Julie R. Fischer, DVM, DACVIM
Knoxville, TN

MANAGEMENT OF DECREASED URETHRAL RESISTANCE

Sympathomimetic drugs with alpha agonist effects have been extremely effective in improving or controlling urinary incontinence in female dogs. Although the mechanisms of action are not entirely clear, these drugs increase the availability of norepinephrine at urethral alpha receptors. All alpha adrenergic agonists have a similar dose range (1.5 mg/kg PO q 8–24h) for the treatment of urethral incompetence. Excellent (>90%) response rates have been observed in most dogs treated with **phenylpropanolamine (PPA)**.[1,2] Responses in male dogs are less predictable. Scott et al.[3] reported results of PPA administration (1 mg/kg q 8 hrs) in a prospective, blinded, placebo-controlled study of 50 incontinent adult female dogs. By day 7, 55% PPA treated dogs were reported to have no unconscious urination as opposed to 26% of the dogs receiving placebo. By day 28, the percentage of continent PPA dogs rose to 85.7% while the percentage of continent placebo-treated dogs rose to 33.3%. While this study demonstrated a significantly better response to PPA than placebo, the placebo effect was unexpectedly high. Byron et al[4] also documented good response to sympathomimetic agents in 9 incontinent female dogs. Improvement in continence score and urethral pressure measures were observed after PPA (1.5 mg/kg q 8 hrs) administration.[4] In another randomized, double-blinded study, female dogs responded well to 1.5 mg/kg PPA q 12hrs; 21 of 24 (88%) dogs were continent and another 2 improved.[5]

While administration of PPA two or three times per day has been advocated for best effect, the dose is adjusted to the minimal amount needed to achieve continence. As with reproductive hormones, it is usually beneficial to start with a higher dose until continence is achieved. However, a recent investigation in healthy intact Beagles challenges this assumption.[6] Dogs were treated with PPA at q8, 12, or 24h intervals. Urodynamic changes (increased urethral pressure) were similar for all dosing frequencies, leading the authors to recommend once daily treatment for incontinent dogs. Desensitization of receptors was suspected with q 8h administration.[6] Similarly, Noel et al[7] found a positive urodynamic response to single daily doses of PPA in healthy Beagles, whereas no response was observed with q 6hr administration. Responses in spayed, incontinent dogs may differ from these observations; however, declining response and tolerance is a known possibility after long term PPA administration, so minimizing dose frequency may be a wise clinical strategy.

Ephedrine and pseudoephedrine (PD) are alternative alpha agonists with similar effects on urethral function. Their clinical use increases during periods when PPA is difficult to obtain. Good outcomes have been reported in 74%[8] and 82%[9] treated dogs in retrospective studies. However, in a series of 9 female dogs studied in a crossover design, dogs exhibited significantly better improvement in continence and urodynamic variables during treatment with PPA than with PD.[4] Clinical improvement was observed in only 6 dogs during PD administration. Differences in response between the drugs may vary based on relative indirect and direct alpha adrenergic activities of each; PPA appears to possess some direct activity while PD acts solely by indirect actions .[10]

Adverse effects of alpha agonists include restlessness, tachycardia, anorexia, gastrointestinal side effects, aggression, and other behavioral changes. Adverse effects are usually dose-dependent and may vary with product or product formulation. Adverse effects are more likely with PD administration; in the dogs studied by Byron et al; 5 of 9 dogs (56%) exhibited adverse behavioral effects including decreased appetite, increased panting, and lethargy.[4] Although systemic hypertension is an important potential complication, significant changes in blood pressure have not been detected in prospective studies of incontinent dogs treated with PPA.[1,4] However, one older dog with borderline hypertension developed anesthesia-induced ventricular arrhythmias after a period of treatment with PD.[4] Mild but significant increases in diastolic blood pressure were observed in healthy Beagles at 7 or 14 days after administration of PPA or ephedrine. Significant increases in systolic arterial pressure, to measurements above 220 mm Hg, were observed in three healthy dogs treated with PPA twice daily in this study.[6] Caution is advisable in high risk patients; alpha agonists should not be administered in hypertensive patients.

Reproductive hormones are reasonably effective in improving continence in dogs with urethral incompetence. Estrogens are often chosen for dogs that do not tolerate alpha agonists and for pet owners who prefer the less frequent dosing regimen. Estrogens such as diethylstilbesterol (DES), stilbesterol, estriol, and conjugated estrogens (Premarin) presumably improve urethral resistance by increasing alpha adrenergic receptor responsiveness and by improving urethral vascularity and other mucosal epithelial characteristics. Effects on urethral morphology are not consistently seen in dogs, however.[11] North American veterinary practitioners are most familiar with **DES** (0.1 –1.0 mg/dog q 4 – 14 d) **or stilbesterol.** DES is no longer commercially available but is available from many veterinary compounding agencies; quality depends on the source. The effectiveness of estrogen is improved by using a "loading" phase of 5 – 7 days, followed by a frequency of administration tailored to the individual patient and drug. A prolonged residual effect of estrogen administration can be seen in some dogs after a period of successful treatment, especially when DES or stilbesterol are employed. Administration can be stopped on a trial basis and restarted when incontinence returns (usually weeks to months). DES or stilbesterol administration leads to improved continence in 60 – 80% of treated dogs. Response to another estrogen product, **estriol**, was studied in a large group (n=129) of incontinent adult, spayed female dogs treated by practitioners in the Netherlands, Belgium, France and Germany.[12] In an open-label trial, the dogs were given 2 mg estriol per day for a week, then the dose was reduced at weekly intervals to the minimal effective dose (typically 0.5 -2.0 mg/dog given daily or every other day). Veterinarians classified 61% of the dogs as continent and 22% improved at day 42 of treatment, whereas owners reported an "excellent" response in 33% and a "good" or "adequate" response in 49%. Twelve dogs exhibited signs of estrus at the initial estriol dose, which resolved in all but one dog after dose reduction. No adverse hematologic effects were observed at day 42 in this study.[12] Dosing estriol every 2 to 3 days is supported by pharmacokinetic studies, in which plasma concentrations of the drug disappeared by 48 h following the last dose.[13]

Response to **natural, conjugated estrogen** (similar to **Premarin**) has been described in 9 incontinent large breed dogs followed prospectively. All dogs responded well to estrogen administration; daily administration was continued until two weeks of continence had been obtained. In seven of nine dogs, maintenance dosages ranged from 0.625 mg to 1.25 mg per dog, administered PO every 12 – 72 hrs. In the remaining two dogs, administration every 4 to 7 days was effective. No hematologic effects of estrogen were observed in treated dogs for up to 49 months.[14] Possible adverse effects of estrogens include signs of estrus, hair loss, behavioral changes, and bone marrow suppression. Estrogenic side effects are most common during the loading dose of estriol, and with higher doses of DES. Bone marrow suppression is unlikely in dogs treated with appropriate doses of the drugs listed here, but owners should be advised of this potentially fatal adverse effect. Estrogen administration should be avoided in male dogs due to resulting prostatic squamous metaplasia, and in dogs with a history or breed predilection for immune mediated disease, since estrogens are immunostimulatory. While most dogs will usually become continent or improve dramatically with either estrogen or alpha agonist administration, the two drugs can also be combined for synergistic effect in the same patient. **Combination treatment** allows estrogen to "prime" the urethral receptors for alpha agonist treatment. Starting doses are the same as for each drug when used individually, but the dose of both products can often be reduced slightly over time if the combination is effective.

MANAGEMENT OF URINARY BLADDER OVERACTIVITY

Other pharmacologic agents that may be useful in unusual or refractory incontinence cases include **anticholinergic (antimuscarinic) agents** that enhance bladder storage. Anticholinergic agents are quite effective for *detrusor instability* (*overactive bladder* or *urge incontinence*) in which inappropriate, involuntary detrusor contractions occur. True detrusor instability is rare in animals; oxybutynin and emepronium bromide have been effective in sporadic cases.[15,16] Antimuscarinic agents can be considered as adjunct treatment in refractory incontinence, or as a single agent in animals in which urge incontinence appears likely. **Oxybutynin** (0.2 mg/kg PO q 8 – 12h) is most commonly recommended, and an extrapolated dose has been determined for **tolteridine**. Other agents, including amitryptiline, imipramine, flavoxate, and dicyclomine, also have anticholinergic effects. Tricyclic antidepressants may improve urine storage by several mechanisms, including anticholinergic, alpha adrenergic and beta adrenergic effects. Imipramine has not proven very effective in clinical application, however. Side effects of antimuscarinic agents include dry mouth, blurred vision, urine retention and gastrointestinal upset, but these compounds appear well tolerated by dogs.[17] Ptyalism and gastrointestinal symptoms are possible in dogs, and usually are remedied by dose reduction.

AGENTS WITH COMBINED ACTIVITIES

Gonadotropin releasing hormone (GnRH) analogs have recently been applied to the treatment of urinary incontinence. Reichler et al[18] reported experience with leuprolide, buserelin and deslorelin. In theory, chronically unsuppressed FSH and LH release (due to lack of negative feedback) in ovariectomized dogs may contribute to urinary incontinence. Administration of GnRH analogs paradoxically results in reduced FSH and LH over time. The GnRH analogs appeared useful in 12/13 dogs with refractory incontinence, either alone or in combination with alpha agonists.[18] In a more recent trial, 9/23 incontinent dogs treated with long-acting **leuprolide** (11.25 mg/dog once) were continent for prolonged periods (70– 575 days); another 10 dogs had partial response.[19] These 23 dogs, however, also responded to phenylpropanolamine, with 92% overall reduction in urine leakage.[19] The only urodynamic parameter that changed after depot leuprolide injection was cystometric threshold volume (an indicator of capacity and accommodation).[20] There were no apparent adverse effects of GnRH treatment reported in this study, but long-term use of these drugs in dogs has not been evaluated. GnRH analogs may prove to be a valuable long-acting treatment that would alleviate the need for daily or weekly medication; however, availability and cost of GnRH analogs limit their use in the United States.

Duloxetine, a serotonin and norepinephrine reuptake inhibitor, may improve striated muscle resistance as well as bladder capacity, and has proven useful in women with stress incontinence. The agent works by inhibition of serotonin and norepinephrine reuptake at the presynaptic neurons in the sacral spinal cord (Onuf's nucleus). In cats with experimentally-induced irritated urinary bladder, a 5-fold increase in bladder capacity and an 8-fold increase in urethral EMG activity were observed after duloxetine administration.[21] In women, a dose of 20 - 40 mg twice daily is effective in reducing incontinence by over 50%,[22] and is effective in women with severe clinical signs.[23] The most common adverse effect in women was nausea, which can be minimized by slow dose escalation. Experience is limited in small animals, where smooth muscle is the predominant contributor to resting urethral sphincter tone, but the drug could be useful as adjunct treatment in some cases.

REFERENCES

1. Richter K et al. **JAVMA** 1985;187:605.2. White R et al. Vet Rec 1989;125: 478. 3. Scott L et al. **JSAP** 2002;43:493. 4. Byron J et al. **JVIM** 2007;21:47.5. Burgherr T et al. **Schweiz Arch Tierheikd** 2007;149:307. 6. Carofiglio F et al. **AJVR** 2006;67:723. 7. Noel et al. **Vet J** 2010;184;201.8.Nendick P et al. **Aus Vet J** 1987;64:117. 9. Arnold S. **CVT XI** 1992:875.10.Kobayashi S et al. **Anesth Anlg** 2003; 97:1239.11. Hamaide A et al. AJVR 2006;67:901.12. Mandigers P et al. **Vet Rec** *2001;*149:765. 13. Hoeijmakers M et al. **Res Vet Sci** 2003; 75:55.14.Angioletti A et al. **Vet Res Comm** 2004; 28:153. 15. Holt PE. Vet Rec1984;114:355.16. Lappin M et al. JAVMA 1987;191:1439. 17. Lane I. **CVT XIII** 2000, 899. 18. Reichler I et al. **Therio** 2003;60:1207. 19. Reichler I et al. Therio 2006;62:1391.20. Reichler IM et al. Therio. 2006 66(5):1227. 21.Karl K et al. **J Pharmacol Exp Ther** 1995;274:1014.22. Jost W et al. **Ther Clin Risk Mgmt** 2005: 259.23. van Kerrebroeck P et al. **B J Obst Gyn** 2004;111:249.

URINARY INCONTINENCE: URETHRAL BULKING

JWBartges, DVM, PhD, DACVIM; ACallens, BS, LVT; IFLane, DVM, MS, DACVIM
Knoxville, TN

URINARY INCONTINENCE IN DOGS

Urinary incontinence is defined as a n involuntary leakage of urine occurring during a normal storage phase of micturition.[1] Urethral sphincter mechanism incompetency (USMI) is reported to be the most frequent cause of urinary incontinence[2] and may occur in up to 20% of female dogs.[3] Although medical therapy using hormonal replacement and/or sympathomimetic agents are often successful, dogs may become incontinent despite medical management or not respond to medical management. In medically unresponsive female dogs, options are limited to conservative management (diapers, confinement to certain areas of the home or to the outside, water-proof bedding), surgical techniques, urethral occlusion device implantation, urethral bulking, and euthanasia.

URETHRAL BULKING FOR URINARY INCONTINENCE

Urethral bulking involves injection of an agent submucosally in the proximal urethra using cystoscopy. It has been thought that injection of bulking agents into the urethral submucosa creates artificial urethral cushions improving urethral closure (coaptation) and restoring continence. It appears that injection therapy functions as a central filler volume that increases the length of muscle fibers and closure power of the urethral sphincter.[4] The urethral sphincter consists of muscle fibers, collagen, vessels, and nerves, with a central component compressing the lumen.[5] Increasing the central filler volume augments the sarcomere length of muscle fibers, strengthening muscle power up to a sarcomere length of 2.2 μm.

Urethral bulking agents

There are many substances that have been evaluated and used for urethral bulking, but collagen has been used predominantly in human and veterinary medicine. Teflon was used for several years, but is no longer available for use due to poor long term outcome and to migration from injection site with subsequent granuloma formation in human beings.[6-8] Glutaraldehyde cross-linked collagen (Contigen, Bard Urological, Covington, GA; ReGain, Avalon Medical, Stillwater, MN) has been used most for urethral bulking. Bovine dermal collagen is immunogenically weak and its natural properties led to development of highly purified injectable fibrillar suspensions that have been used medically for soft tissue augmentation.[9-13] Glutaraldehyde cross-linked collagen is highly compatible with and incorporated into the patient's connective tissue encouraging fibroblast infiltration, new collagen deposition, and neovascularization. Carbon-coated zirconium oxide beads (Durasphere, Boston Scientific Corporation, Narick, MA) was approved by the FDA in 1999 for treatment of women with urinary incontinence. It has a large particle size and chronic loss of material does not occur, it does not provide more durability than collagen[14] and has been associated with urethral prolapse.[15] Ethylene vinyl (Tegress, Bard Urological, Covington, GA) was approved by the FDA in 2004 and has a success rate of 43% at approximately 4 years;[16] however, there are serious complications associated with its use in some women.[17,18] Polydimethylsiloxane (Macroplastique, Uroplasty, Inc., Minnetonka, MN) has been shown to be more effective than collagen in a prospective randomized controlled clinical trial in women with a success rate of 61% at one year versus 48% success rate at a year with collagen.[19] It was concluded that polydimethylsiloxane is a safe, efficacious, minimally invasive, injectable silicone material for treating urinary incontinence. Porcine dermal implant (Permacol, Covidien, Dublin, Ireland) appears to have a higher urodynamic success rate than polydimethylsiloxane in women with stress urinary incontinence for up to 6 months; however, longer term studies have not been performed.[20] Women receiving calcium hydroxyapatite (Coaptite, Boston Scientific Corporation, Narick, MA) injections had better outcomes after a single treatment with a lower volume of injected material when compared with patients treated with collagen.[21] Urethral granulomatous reaction and urethral prolapse have been reported with its use.[22-24] A multi-center study concluded that non-animal stabilized hyaluronic acid/dextranomer copolymer (Zuidex, Q-Med, Uppsala, Sweden) delivered by an Implacer device was an effective treatment for women with stress urinary incontinence.[25]

Technique

Dogs are placed under general anesthesia and positioned in dorsal recumbency. Peri-vulvar fur is clipped and the area prepared with scrub. A 2.7-mm rigid cystoscope is introduced through the vulva and into the urethra. Diagnostic cystoscopy can be performed. For urethral bulking injection, the ureteral orifices, the trigone, and proximal urethra are identified. The cystoscope is withdrawn to the level of the proximal urethra just distal to the trigone, approximately 1-1.5 cm caudal to the urinary bladder neck. During the procedure sterile fluid is infused at a rate that is equal to the fluid drainage through the cystoscope side port so that the urethral lumen remains dilated facilitating visualization of the urethral lumen. The injection needle is inserted through the operating port. Polydimethylsiloxane and non-animal, stabilized hyaluronic acid/dextranomer require a special device for injection. Prior to insertion, the needle is loaded with bulking compound by attaching the syringe containing the compound to the needle and slowly injecting until it is observed in the needle tip. Urethral bulking agent is injected submucosally in 3 locations; at approximately the 2, 6, and 10 o'clock positions; however, more or less sites may be used. The needle is inserted with visualization at the selected site so that the bevel of the needle is completely inserted submucosally. The bulking agent is injected until a bleb is visualized that approaches midline of the urethral lumen. The amount injected depends on the size of the dog and the visualization of the size of the bleb. Once the volume has been administered, the injection is stopped, and the needle held in place for 60 seconds to allow the compound to "set up". The needle is removed and one or more additional sites are injected in a similar fashion. The procedure is considered complete when the urethral lumen is closed by deposits when viewing cystoscopically.[26,27] It is imperative that the cystoscope not be inserted across the injection sites into the more proximal urethra or bladder as this will result in diffusion of the injected compound into the submucosal tissue and a decreased bulking effect. If cystoscopy cannot be performed, laparotomy and cystotomy can be undertaken for injection.[27] Once the procedure is completed, the dog is recovered. Most dogs can be returned to their owner that day depending on recovery from anesthesia. Some patients may not be able to urinate following the procedure and may require placement of an indwelling urethral catheter for 1 to 2 days. When successful, dogs are usually continent within 2-3 days of performing the procedure. Drug therapy can be discontinued. I usually do not treat dogs with peri-procedural antibiotics except for a single dosage of intravenously-administered ampicillin (22 mg/kg) at time of termination of the procedure.

Results of urethral bulking in dogs

In the only study published on collagen injection in dogs,[27] continence was achieved in 27 of 40 dogs for 1 to 64 months with a mean of 17 months. In 10 dogs, collagen injection improved the urinary incontinence and they received additional medical treatment. Of these, 6 dogs regained continence for a period of 9 to 47 months (mean, 23 months) and 4 dogs had improvement for 25 to 78 months (mean, 43 months). In 3 dogs, the collagen injection had no effect. In 16 dogs that were initially continent after treatment, there was deterioration after 1 to 25 months (mean, 8 months); 12 dogs were then treated medically. Of these, 11 were continent for 2 to 64 months (mean, 27 months) and 1 dog had improvement of incontinence for 29 months. In 2 dogs, incontinence reoccurred 2 and 12 months after the medical treatment was started. For the remaining 4 dogs, 2 owners refused any further treatment and accepted the urinary incontinence for 42 and 36 months and 2 dogs were treated again with collagen. Polydimethylsiloxane is currently under investigation in dogs; 19/22 dogs were continent without additional medication at 1 month post-injection. In the remaining dogs, 2/22 dogs had experienced 1 to 2 episodes of incontinence but were continent at 1 month. In the remaining 3 dogs, urinary incontinence was significantly improved; 2 of these dogs underwent a second treatment and became continent.

Urethral bulking is a viable treatment for dogs with urinary incontinence that are no longer responsive to medical therapy.

REFERENCES

1.Noel S, Claeys S, Hamaide A. Acquired urinary incontinence in the bitch: update and perspectives from human medicine. Part 1: The bladder component, pathophysiology and medical treatment. *Vet J* 2010;186:10-17. 2. Krawiec DR. Urinary incontinence in dogs and cats. *Mod Vet Pract* 1988;1:17-24. 3.Arnold S, Arnold P, Hubler M, et al. [Urinary incontinence in spayed female dogs: frequency and breed disposition]. *Schweiz Arch Tierheilkd* 1989;131:259-263. 4.Klarskov N, Lose G. Urethral injection therapy: what is the mechanism of action? *Neurourol Urodyn* 2008;27:789-792. 5. Schafer W. Some biomechanical aspects of continence function. *Scand J Urol Nephrol Suppl* 2001;35:44-60; discussion 106-125. 6. Arnold S, Jager P, DiBartola SP, et al. Treatment of urinary incontinence in dogs by endoscopic injection of Teflon. *J Am Vet Med Assoc* 1989;195:1369-1374. 7. Kiilholma PJ, Chancellor MB, Makinen J, et al. Complications of Teflon injection for stress urinary incontinence. *Neurourol Urodyn* 1993;12:131-137. 8.Mittleman RE, Marraccini JV. Pulmonary Teflon granulomas following periurethral Teflon injection for urinary incontinence [letter]. *Arch Pathol Lab Med* 1983;107:611-612. 9. DeLustro F, Condell RA, Nguyen MA, et al. A comparative study of the biologic and immunologic response to medical devices derived from dermal collagen. *J Biomed Mater Res* 1986;20:109-120. 10.Isom-Batz G, Zimmern PE. Collagen injection for female urinary incontinence after urethral or periurethral surgery. *J Urol* 2009;181:701-704. 11.ppell RA. Transurethral collagen denaturation for women with stress urinary incontinence. *Curr Urol Rep* 2008;9:373-379. 12.Sakamoto K, Sharma S, Wheeler JS. Long-term subjective continence status and use of alternative treatments by women with stress urinary incontinence after collagen injection therapy. *World J Urol* 2007;25:431-433. 13. Martins SB, Oliveira E, Castro RA, et al. Clinical and urodynamic evaluation in women with stress urinary incontinence treated by periurethral collagen injection. *Int Braz J Urol* 2007;33:695-702; discussion 702-693. 14.Chrouser KL, Fick F, Goel A, et al. Carbon coated zirconium beads in beta-glucan gel and bovine glutaraldehyde cross-linked collagen injections for intrinsic sphincter deficiency: continence and satisfaction after extended followup. *J Urol* 2004;171:1152-1155. 15.Ghoniem GM, Khater U. Urethral prolapse after durasphere injection. *Int Urogynecol J Pelvic Floor Dysfunct* 2006;17:297-298. 16.Kuhn A, Stadlmayr W, Sohail A, et al. Long-term results and patients' satisfaction after transurethral ethylene vinyl alcohol (Tegress) injections: a two-centre study. *Int Urogynecol J Pelvic Floor Dysfunct* 2008;19:503-507. 17.Hurtado E, McCrery R, Appell R. The safety and efficacy of ethylene vinyl alcohol copolymer as an intra-urethral bulking agent in women with intrinsic urethral deficiency. *Int Urogynecol J Pelvic Floor Dysfunct* 2007;18:869-873. 18. Erekson EA, Sung VW, Rardin CR, et al. Ethylene vinyl alcohol copolymer erosions after use as a urethral bulking agent. *Obstet Gynecol* 2007;109:490-492. 19.Ghoniem G, Corcos J, Comiter C, et al. Cross-linked polydimethylsiloxane injection for female stress urinary incontinence: results of a multicenter, randomized, controlled, single-blind study. *J Urol* 2009;181:204-210. 20.Bano F, Barrington JW, Dyer R. Comparison between porcine dermal implant (Permacol) and silicone injection (Macroplastique) for urodynamic stress incontinence. *Int Urogynecol J Pelvic Floor Dysfunct* 2005;16:147-150; discussion 150. 21.Mayer RD, Dmochowski RR, Appell RA, et al. Multicenter prospective randomized 52-week trial of calcium hydroxylapatite versus bovine dermal collagen for treatment of stress urinary incontinence. *Urology* 2007;69:876-880. 22.Palma PC, Riccetto CL, Martins MH, et al. Massive prolapse of the urethral mucosa following periurethral injection of calcium hydroxylapatite for stress urinary incontinence. *Int Urogynecol J Pelvic Floor Dysfunct* 2006;17:670-671. 23.Ko EY, Williams BF, Petrou SP. Bulking agent induced early urethral prolapse after distal urethrectomy. *Int Urogynecol J Pelvic Floor Dysfunct* 2007;18:1511-1513. 24.Lai HH, Hurtado EA, Appell RA. Large urethral prolapse formation after calcium hydroxylapatite (Coaptite) injection. *Int Urogynecol J Pelvic Floor Dysfunct* 2008;19:1315-1317. 25.van Kerrebroeck P, ter Meulen F, Larsson G, et al. Efficacy and safety of a novel system (NASHA/Dx copolymer using the Implacer device) for treatment of stress urinary incontinence. *Urology* 2004;64:276-281. 26.Arnold S, Hubler M, Lott-Stolz G, et al. Treatment of urinary incontinence in bitches by endoscopic injection of glutaraldehyde cross-linked collagen. *J Small Anim Pract* 1996;37:163-168. 27.Barth A, Reichler IM, Hubler M, et al. Evaluation of long-term effects of endoscopic injection of collagen into the urethral submucosa for treatment of urethral sphincter incompetence in female dogs: 40 cases (1993-2000). *J Am Vet Med Assoc* 2005;226:73-76.

WHAT IS A PATHOGENIC UTI?

Michael W. Wood, DVM, PhD, DACVIM
Raleigh, NC

BACTERIURIA, URINARY TRACT INFECTIONS, ASYMPTOMATIC BACTERIURIA

Based upon recent observations in human medicine, microbiological documentation of bacteriuria is not equivalent to a diagnosis of a urinary tract infection (UTI). Bacteriuria is the presence of bacteria in the urine. Traditional definitions of UTI require that the invading organism adhere to the urothelium, multiply, and persist within the urinary bladder. In the past 15 years there has been a continuing evolution of how bacteriuria and UTI are assessed in clinical patients. Regardless of urinary bladder colonization, all bacteria will not induce clinical signs and ultimately cause disease. The subset of bacteriuric individuals lacking clinical signs consistent with UTI are diagnosed as having "asymptomatic bacteriuria" (ASB).

The differentiation between UTI and ASB is of clinical and microbiological importance to patient management. In many humans untreated bacterial strains persist within the urinary bladder for years without deleterious effects. Greater than 17% of elderly women and 5% of adult women are carrying bacteria within their urinary tract without symptoms of a UTI. Importantly, these bacteria are hypothesized to convey a degree of protection against urinary tract colonization with more virulent bacteria. Evidence includes studies where treatment of patients with ASB has resulted in the subsequent development of pathogenic infections of the lower urinary tract.

In veterinary medicine 14% of dogs are estimated to develop UTIs during their lives. Many dogs present to their veterinarian with subtle to no overt clinical signs. Whether to treat these patients is a complicated decision. Currently, most veterinarians will treat any dog with urinalysis or urine culture evidence of bacteriuria because the possible ramifications of not treating (pyelonephritis and renal failure) can be grave. However, it is possible that such therapies may be a disservice to our patients as more virulent and resistant infections may then be allowed to colonize. The lack of both clinical guidelines and useful diagnostic tools that would be able to discriminate between pathogenic UTI and benign bacteriuria has limited the ability of our profession to make definitive statements about the proper treatment of ASB.

HUMAN GUIDELINES FOR TREATMENT AND VETERINARY ALTERNATIVES

In human medicine specific guidelines for diagnosing and treating ASU do exist. The Infectious Disease Society of America recommends that healthy women aged 30-70 years with documented ASU not be treated with antimicrobials as there is no evidence indicating that these patients are at greater risk for developing hypertension, chronic kidney disease, or genitourinary neoplasia. Treatment of these patients also did not decrease the recurrence rate of symptomatic urinary tract infections. Similar findings are reported in diabetic women, and currently the recommendation is not to screen for ASB in this patient population. In patients with spinal cord injuries treatment of ASB with antimicrobials was associated with re-infection with resistant bacterial strains within 30 days of discontinuing antibiotics regardless of treatment duration. Rapid re-infection was also common in immunocompromised individuals. Treatment of ASU is recommended when detected in pregnant women, patients undergoing urologic procedures where the mucosal barrier will be compromised, and in patients that remain bacteriuric 48h after the removal of an indwelling urinary catheter.

Unfortunately, the direct application of these guidelines to veterinary medicine is fraught with complicating variables. Besides the lack of similar clinical studies, the most important variable may be that timely and accurate documentation of clinical signs in the veterinary patient population is more challenging than in the human population. Given that our patients lack the ability to reliably communicate subtle clinical signs of UTI, alternative approaches examining the invading bacterial strain as well as the patient's mucosal response is necessary.

BACTERIAL INTERACTIONS WITH THE UROTHELIUM

Urinary tract infections are caused by numerous etiological agents including *Escherichia coli*, *Staphlococcus* spp., *Proteus* spp., *Streptococcus* spp., *Klebsiella* spp., and *Enterococcus* spp. Of these, uropathogenic strains of the gram negative bacteria *E. coli* (UPEC) account for the majority of canine UTIs.

A defining feature of many UPEC strains and other gram negative uropathogens is the ability to produce fimbriae and toxins. Expressed on the surface of the bacteria, fimbriae are integral components for adhesion to urothelial cells. A uropathogenic strain of bacteria may contain numerous gene clusters for encoding fimbriae however, typically a bacterium will only express one type of fimbriae at a time. The organisms do have the ability to phase switch depending on the bacterial environment making fimbrial expression variable for a single bacterial strain.

There are two main fimbrial units that play prominent roles in the development of UTI, type 1 and P-fimbriae. Of UPEC strains over 80% express either one or both of these fimbriae. Generally, P-fimbriated bacteria are commonly associated with the development of acute pyelonephritis, while type 1 fimbriated bacteria are associated with cystitis and lower urinary tract infections. Unfortunately, the presence of pathogenicity islands containing both type 1 and P-fimbriae genes are not limited to UPECS and can be identified in ASB as well. It is the actual expression of these fimbriae on the bacterial surfaces that are different as many ASU strains contain mutations that limit fimbrial expression or function. Studies examining toxin expression by UPECS and ASB strains have provided similar data indicating that ASB strains may possess variable genes coding for toxin production, but no genes have been identified as universally absent in ASB strains when compared to cystitis and pyelonephritis inducing bacteria. Hence the current data suggests that screening invading bacteria for fimbrial and toxin genes may help to identify non-pathogenic strains, but this testing modality is not definitive for identifying ASB.

UROTHELIAL CYTOKINE RESPONSE

An alternative approach for the identification of pathogenicity is by observing the mucosal response to infection. Traditionally the presence of white blood cells (WBC) within collected urine samples has been used as a guide to identify an immunologic response within the urinary tract. While pyuria is evidence of inflammation in human medicine it is not definitive evidence that concurrent bacteriuria needs treatment. Pyuria is present in 30-90% of human patients with ASU, and the guidelines for ASU treatment in humans do not include the finding of WBC in the urine.

Other indicators of the mucosal response to infection include the production of cytokines. In response to different pathological environments urinary tissue has the ability to produce a number of cytokines including IL-1α, IL-1β, IL-4, IL-6, IL-8, INF-γ, TNFα, PDGF, and TGFβ. Of these, IL-6, IL-8, and TNF-α are most commonly associated with UTI. Unfortunately, the production of cytokines is not unique to UTI as ASB is also associated with cytokine production by the urothelium. However, examining human patients with acute cystitis and ASB urinary concentrations, IL-1, IL-8, and IL-6 were significantly higher in the acute cystitis group. While promising, the use of cytokines as a urinary diagnostic modality is still in its infancy. Many of these cytokines have variable specificity, secretion rates, and duration of production. Defining each of these variables in our patient population is necessary before the clinical application of cytokine secretion as a diagnostic modality.

CONCLUSION

One should not confuse the presence of bacteriuria with the diagnosis of a pathogenic urinary tract infection (UTI). This differentiation between UTI and nonpathogenic bacteriuria is of clinical and microbiological importance for patient management, and has become increasingly important because of the emergence of antimicrobial resistance among many common uropathogens. Before developing a set of guidelines describing the appropriate management of ASB in our veterinary patients first we must define what a pathogenic UTI is. One differentiating characteristic of UTI is the presence of a mucosal response to the infecting organism. Therefore identifying pathogenicity islands and the expression of fimbriae and toxins in canine UPEC strains as well as understanding the sequella of the mucosal response to UTI including cytokine production in our veterinary patients may provide a means to differentiate ASB and pathogenic infections in patients.

REFERENCES

1. Krzemien, G., et al., Med Sci Monitor, 2004. 10(11): p. CR593. 2. Bartges, J.W. Vet Clin N Am-Small, 2004. 34(4): p. 923. 3. Nicolle L.E., et al. Clin Infect Dis 2005;40:1556. 4. Connell, H., et al. P Natl Acad Sci USA, 1996. 93(18): p. 9827. 5. Welch, R.A., et al. P Natl Acad Sci USA, 2002. 99(26): p. 17020. 6. Mabbett A.N., et al. Int J Med Microbiol 2009;299:53. 7. Brauner, A., et al. Clin Exp Immunol, 2001. 124(3): p. 423. 8. Funfstuck, R., et al. Int J Antimicrob Agents, 2001. 17(4): p. 253. 9. Rodhe N., et al. Scand J Prim Health 2009;27:74. 10. Uehling, D.T., et al. World J Urol, 1999. 17(6): p. 351. 11. Hedges, S., et al. Infect Immun, 1991. 59(1): p. 421.

CAN WE PREVENT UTI?

Shelly Vaden, DVM, PhD, DACVIM
Raleigh, NC

INTRODUCTION

It is estimated that 14% of dogs will develop a urinary tract infection (UTI) within their lifetime. Most are females but the incidence of UTI in male dogs increases with age. Bacterial UTI is less common in cats, accounting for only 1-3% of all cases of feline lower urinary tract disease in North America, although cats older than 10 years of age have an increased prevalence of UTI.[1] While the management of dogs with acute, uncomplicated UTI is fairly straightforward, the management of dogs with chronic UTI, due to either recurrence or persistence, can be anything but straightforward.

Complicating the entire picture is the emergence of resistant pathogens. Of particular concern are the uropathogenic *Escherichia coli* (UPEC) and *Enterococcus sp.* UPEC is the most common etiologic agent for UTI in most mammalian species. UPEC presents a particular challenge because of intracellular invasion of and biofilm formation in host uroepithelial cells by UPEC strains, allowing the UPEC to avoid normal host defense mechanisms.[3] Intracellular UPEC appear to have one of three fates: (1) They are recycled to the cell surface and exocytosed; (2) The live a quiescent nonreplicative existence; (3) The live a highly replicative phase in the cell cytoplasm. These latter two lifestyles allow UPEC to establish a reservoir in the urinary tract.[2] These strains of UPEC may be more resistant to both host clearance and antimicrobial therapy. Deficiencies in host innate responses or adaptive immune response may also prevent lasting protective immunity against reinfection.

Antibiotic therapy is standard treatment for UTI but overuse leads to changes in the normal host microbiota and selection for resistant bacterial strains. Although severe adverse effects of antibiotic therapy for UTI may be infrequent, mild adverse effects are probably not and chronic antimicrobial administration is expensive for the client. In the US alone, it is estimated that UTI affects nearly 13 million women at a cost of 3 billion dollars annually; up to 25% of women will have a recurrent infection. There has been an ongoing shift in the last decade or so to focus on prevention of UTI in people. Developing effective strategies to prevent UTI should be explored further in veterinary medicine. The purpose of this presentation is to discuss potential means to prevent UTI through strengthening of the normal host defense mechanisms, prophylactic antibiotics and novel therapeutic approaches.

HOST DEFENSE MECHANISMS

Normal systemic and mucosal immunity are crucial for the prevention of bacterial colonization and UTI in normal animals. These immune mechanisms are highly complex and may be incompletely understood but it is clear that interruption of normal host defense mechanisms can lead to repeat infections.

Normal composition of urine does not favor bacterial growth. Urine pH has long been considered to be an important factor inhibiting bacterial growth. Studies in people demonstrate that bacterial growth occurs unchecked from urine pH 4.7-5.5.[4] When urine pH is <4.5, there is effectively no growth of any organism. Bacterial growth also starts to diminish as pH approaches 8.0 and stops at pH>9.0. It would be difficult and unwise to therapeutically manipulate urine pH to <4.5 or >9; the use of urine pH in the prevention of UTI is impractical. High urine osmolality prevents bacterial growth; the inhibitory effects of pH have mainly been demonstrated in concentrated urine and the effects of pH may actually be dependent on urea concentration.[5] Animals with reduced urine concentrating abilities should be thoroughly evaluated for all correctable causes of polyuria and polydipsia. Physicians have long advocated forced fluid intake in people with UTI theorizing that this would flush the urinary tract of pathogens, reduce bacterial multiplication in the bladder and allow the bladder mucosal defenses operate more effectively because they would not be overwhelmed by larger numbers of bacteria.[6] Given the protective effect of highly concentrated urine, this recommendation for lower UTI seems unwarranted. However, there may be some benefit of forcing fluids in prevention or treatment of pyelonephritis where an isotonic environment in the renal medulla may be needed for effective granulocyte mobilization.[6]

Normal micturition is another important factor in normal defenses against UTI. In theory, normal voiding should eliminate bacteria from the urinary tract. However, in actuality voiding may not mechanically free bladder of adherent bacteria that are in a biofilm and clearance from the bladder may be more closely associated with antibacterial activity of mucosal surface than with voiding.[6] By mechanisms that seem to be counterintuitive, UPEC adherence may actually be strengthened by sheer stress induced during urine flow. Nonetheless, bladder dysfunction and urine retention have clearly been shown to be risk factors for UTI in dogs.[7] One study demonstrated a mild negative association between failure to excuse oneself to urinate and UTI in women.[8] Frequent voiding may be an important factor in prevention of UTI in dogs and cats.

Urinary incontinence is associated with a break in the barrier against ascending UTI. Effective management of urinary incontinence may aid in the prevention of UTI in some dogs by increasing urethral pressure. Incontinence and changes in vaginal flora are risk factors for UTI in postmenopausal women. Estrogens help maintain continence, increase the presence of *Lactobacillus sp.* and decrease vaginal pH. When estrogens were evaluated in postmenopausal women with recurrent UTI, oral estrogens were no better than placebo but vaginal estrogens reduced the number of women with UTI although the effect took up to 12 weeks.[9]

Normal anatomy is also important in prevention of UTI. Congenital abnormalities should be identified and corrected in young animals with recurrent UTI. In recent years it has become apparent that hooded vulva is a clinical relevant abnormality in female dogs with recurrent UTI. Hooded vulva is frequently associated with dermatitis, creating ideal conditions for bacterial growth. Animals with hooded vulvas may have perivulvar hairs that fold inward, wicking bacteria into vestibule. Hooded vulvas are often found in young dogs with infantile vulvas but also obese dogs and older dogs with reduced skin elasticity. Medical management can be onerous for the owners and does not appear to be effective.[10] Episioplasty has been reported to be effective in the resolution of UTI in 84-100% of dogs with hooded vulva.[10,11]

PROPHYLACTIC ANTIMICROBIAL THERAPY

The use of prophylactic antibiotics has long been advocated in dogs with chronic UTI. It has been suggested that prophylactic antibiotic administration be initiated after potential underlying causes are corrected and UTI has been eliminated. Bedtime dosing (i.e., antibiotics given at bedtime, after voiding) is used to eliminate ascending bacteria before UTI can be established. Pulse therapy (i.e., short period of antibiotic administration given at regular intervals such as monthly) has also been suggested but may be more likely to induce bacterial resistance than is bedtime dosing. In general, antibiotics that are relatively safe and inexpensive should be selected (e.g., amoxicillin, amoxicillin-clavulanic acid, nitrofurantoin). First and second generation cephalosporins generally should not be used for this purpose because of their ability to induce global antimicrobial resistance. Methenamine mandelate may be effective in situations where bedtime antibiotics failed to prevent UTI. This agent is converted to formaldehyde in acidic urine and effective use may require concurrent administration of urinary acidifiers. The question remains if prophylactic antimicrobial therapy is effective in dogs or not.

In people, the benefit of prophylactic antibiotics remains unclear. Large meta-analyses suggest that any benefit is likely to be small and of limited clinical significance.[12] In one study of women, continuous antibiotic prophylaxis for 6-12 months reduced the rate of UTI during prophylaxis when compared to placebo but there was no difference between the groups post-prophylaxis. Antimicrobial prophylaxis was associated with more adverse drug effects and increased antimicrobial resistance.[13,14]

CRANBERRY

Cranberries have long been suggested for the prevention of UTI. Cranberries contain organic acids (malic, quinic and citric acid), fructose, glucose and phytochemicals. Certain phytochemicals, such as A-type proanthocyanidins (PACs), inhibit the adhesion of p-fimbriated UPEC to uroepithelial cells. Although initial studies in people were somewhat contradictory, a recent meta-analysis concluded that there is statistical evidence that cranberry extract reduces the recurrence of UTIs, particularly in young women.[15] The incidence of UTI is reduced by approximately 35-40% in women consuming cranberry products.[16] There is evidence to suggest that inhibition of bacterial adherence to bladder epithelial cells is dose-dependent and lasts for approximately 10 hours.[17,18] It is possible that a low dose or insufficient dosing frequency was used some of the trials in which a beneficial effect was not evident. PACs have been shown to reduce bacterial adhesion in a bacterial adhesion model in dogs.

PROBIOTICS

Because most UTIs in dogs are believed to be ascending infections, maintenance of a healthy microbial population of the vestibule and vagina may be important in female dogs. In most women, the vaginal microbiota is remarkably similar with lactobacilli being the most dominant colonizer.[19] Although it is uncertain if outliers from this normal population are always at increased risk of UTI, women with recurrent UTI have been reported to have a high incidence of bacterial vaginosis. Probiotic lactobacilli are believed to improve urogenital health in women through immune modulation, pathogen displacement, and creation of a niche less favorable for pathogenic bacterial proliferation. Probiotics may help replenish the normal microbes, displacing pathogens and leading to a return to the lactobacilli-dominated state found in healthy women, preventing ascending UTI. In one study of 120 children who were given either Lactobacillus sp. or trimethoprim/sulfamethoxazole prophylaxis for one year, there was no difference in the incidence of recurrent UTI. The use of probiotics in prevention of UTIs in dogs probably warrants further study.

IMMUNOTHERAPY

Bacterial extracts have been evaluated to use as immunostimulants in the protection against recurrent UTI. When administered orally, stimulation of GALT leads to a generalized response by MALT, including the MALT-like tissue found in the urinary tract. Meta-analysis of several studies of the E. coli bacterial extract, OM-89, given orally to people with recurrent UTI revealed a decrease in the number of UTI OM-89 treated patients when compared to placebo treated patients. There was an associated reduction in the incidence of dysuria, pyuria, bacteriuria and the use of antibacterial agents.[20] The use of an attenuated strain of uropathogenic E. coli (UPEC) for vaccination against UTI caused by UPEC is also being evaluated.[3] Developing reliable vaccine strategy has been difficult, in part because of UPEC diversity, but recent progress has been made. Proposed clinical uses of a UPEC vaccine include prevention of initial or recurrent UTI as well as resolving chronic UTI.

REFERENCES

[1]Lekcharoensuk C, et al. JAVMA 2001;218:1429. [2]Sivick KE, et al. Infect Immun 2010;78:568. [3]Billips BK, et al. J Infect Dis 2009;200:263. [4]Yeaw RC. J Urol. 1940;44:699. [5]Kaye D. J Clin Invest 1968;47:2374. [6]Beetz R. European J Clin Nutr 2003;57:S52-S58. [7]Olby NJ. JVIM, 2010;24:1106. [8]Foxman B, et al. Am J Public Health 1985;75:1314. [9]Perrotta C, et al. Cochrane Database Syst Rev 2008;2. [10]Lightner BA, et al. JAVMA 2001;219:1577. [11]Hammel SP, et al. JAAHA 2002;38:79-83. [12]Williams G, et al. Curr Opin Inf Dis 2009;22:72. [13]Albert X, et al. Cochrane Database Syst Rev 2008; 4. [14]Greenfield SP. Curr Urol Rep 2011; online. [15]Jepson RG, et al. Cochrane Database Syst Rev 2009;4. [16]Masson P et al. Infect Dis Clin NA 2009;23:355-85. [17]Santillo VM, et al. Drugs of Today 2007;43: 47-54. [18]Howell A, et al. JAMA 2002;287:3082. [19]Reid G, et al. Curr Opin Clin Nutr Metab Care 2009;12:583. [20]Naber KG, et al. Int J Antimicrobial Agents 2009;33:111.

BETA-2 AGONISTS IN EQUINE PULMONOLOGY

Rose D. Nolen-Walston, DVM, DACVIM (LAIM)
Kennett Square, PA

INTRODUCTION

Beta-2 adrenergic agonists are a mainstay in the treatment of bronchoconstrictive diseases of both humans and animals, including asthma and COPD in man, and in horses, Recurrent Airway Obstruction (RAO) and Inflammatory Airway Disease (IAD), a disease which affects up to a third of racehorses in training. These compounds activate the sympathetic innervation of bronchial smooth muscle resulting in bronchodilation and mucokinesis in the lungs, and causes smooth muscle relaxation in the myometrium, and blood vessels of the liver and kidney. Available in both inhaled and systemic formulations, there are a variety of beta-2 agonists on human market, many of which are used in horses, as well as a single product FDA-approved for use in horses. They are broadly categorized into short-acting (1-3h duration) and long acting-formulations, and are used for "rescue therapy" versus maintenance respectively. This lecture will focus largely on the effects of beta-2 agonists, both beneficial and negative, the significance of tachyphylaxis, and touch briefly on their clinical usage in the horse.

EFFECTS OF BETA-2 AGONISTS

Bronchodilation and bronchoprotection

In humans, β2 agonists are also widely acknowledged to attenuate bronchoconstriction from both direct and indirect stimuli.[1-3] Less evaluation of the bronchoprotective effect of clenbuterol has been performed in the horse, although studies in ponies showed no protective effect of intravenous clenbuterol against inhaled histamine,[4] but some effect against intravenous histamine.[5] Likewise, Mazan *et al.* also found no significant effect on histamine reactivity of 10 days of aerosolized salbutamol in horses with IAD,[6] although based on human data, it is possible that bronchodilator tolerance had already occurred and thus the bronchoprotective window may have been missed.

Anabolic effects and effect of performance

Clenbuterol is anabolically active in humans and cattle,[7] and in horses although its leptin- and adiponectin-mediated repartitioning effects cause an increase in muscle mass,[8,9] this potential benefit is offset by a negative ergonomic effect.

Diagnosis of anhidrosis

Unlike other mammalian species that sweat in response to cholinergic stimulation, sweating in horses is mediated by β-2 adrenergic receptors. It is likely that anhidrosis, a naturally occurring syndrome of reduced sweat production, is due to gradual failure of the secretory process of the sweat glands apparently caused by desensitization and down-regulation of the β-2 receptors in response to endogenous catecholamines.[10]This process occurs most commonly in horses living in hot, humid climates where there is continual stimulation of sweating,[10] although it is unclear why only some horses are affected. Diagnosis of anhidrosis is usually performed using intradermal β-agonists, either non-specific (epinephrine) or specific (terbutaline or salbutamol).[11-13] Normal horses sweat at a 1:1,000,000 concentration of epinephrine, and that failure to sweat at this concentration is diagnostic for hypo- or anhidrosis. Alternatively, a recent method described to better assess subtle changes in sweating involves collecting sweat using absorbent pads which are evaluated quantitatively by change in weight.[13]

Tocolytic effects

Clenbuterol is used as a tocolytic in horses to slow the progression of labor in dystocias, allowing easier manual mutation of the fetus.[14]

METHODS OF DELIVERY

The majority of beta-2 agonist drugs are delivery via inhalation, either as a nebulized solution or with a metered dose inhaler (MDI; "puffer"). In order to obtain adequate distribution to the small airways, a spacer device is essential. There are several marketed devices for equines, including the Aeromask (which covers the entire muzzle), the Aerohippus, The Equine Haler, and the Equine Aerosol Drug Delivery System (which fit over one nostril), which deliver 6-43% of the actuated drug to the lower respiratory tract.[15] Not all of these devices are currently available. Clenbuterol is delivered as an oral syrup, and bioavailibility in horses is excellent (83.9%).[16] This is in contrast to albuterol, which has poor oral absorption in the horse,[17] and should therefore be avoided in favor of clenbuterol if medication *per os* is desirable.

SIDE EFFECTS OF BETA-2 AGONISTS

Negative side effects of the beta-2 agonists are those expected by adrenergic activation. They include characterized sweating, tachycardia, agitation by decreases in aerobic capacity, time to fatigue,[18] cardiac function,[19] and VO_2max.[20] The use of β2-agonists is nonetheless common in both human and equine athletes with lower airway disease, despite these significant side effects and the rapid occurrence of desensitization or tachyphylaxis to the effects of this class of drug. This is suggested by data from Kearns et alia, who found that at 3.2 μg/kg, side effects including extreme sweating were initially observed but subsided after 10 days,[20] Clenbuterol has been hypothesized to affect thermoregulation, which may be a component to the reduced aerobic capacity of horses receiving high doses for extended periods.[20]

TACHYPHYLAXIS

In asthmatic humans as well as other mammalian models of bronchoconstriction, chronic treatment with β2 agonists results in tolerance or tachyphylaxis (desensitization) to the effects of the drug. This can be seen as a positive attribute, in that the negative side-effects of β2 agonist administration such as sweating, tachycardia, and agitation decrease over the first few weeks of use, but also is associated with a decreased efficacy of the bronchodilatory effects of the treatment and an actual worsening of asthma control.[21] Minimal data are available regarding the effect of long-term β2 agonist treatment on bronchodilation in horses, although small studies in horses with RAO and IAD showed conflicting results, with little-to-no effect seen.

The cause of tolerance to β2 agonists appears multifactorial. It has been shown in rats that long-term β2-agonist administration results in a decrease in the density of bronchial epithelial β2 receptors is due to endosomal sequestration of receptors, as well as decreased transcription and translation of mRNA.[22] In addition to receptor down regulation, there is also functional receptor desensitization through the β2-adrenoceptor–G-protein–adenylate cyclase pathway. Specifically, intracellular signal transduction is prevented by uncoupling of Gs-protein by members of the G-protein receptor-coupled kinase (GRK) superfamily and β-arrestin, and subsequently by cyclic AMP-dependent protein kinase which further reduces second messenger signaling. In the horse, Abraham *et al.* showed that 12 days of intravenous clenbuterol in adult horses resulted in a reduced density and responsiveness of β2-adrenoceptor on lymphocytes, suggesting that these effects are conserved across species.[23]

In the treatment of human asthma, concurrent use of a corticosteroid attenuates the development of tachyphylaxis to bronchodilators and it is now widely recommended that daily β2-agonists should be used only in conjunction with inhaled corticosteroids to prevent rapid desensitization. Although no studies have been performed to evaluate the effect of this drug combination on lung function in horses with IAD/RAO, it has been demonstrated that the decrease in lymphocyte β2-receptor density and function in horses receiving clenbuterol is reversed and prevented by administration of dexamethasone.[24]

In contrast to other mammals, equine sweat glands contain β2 receptors that show rapid desensitization to β2 agonists leading to sub-normal sweat production *in vitro*.[25] In an experimental setting, there is a gradual decline in sweat production over the course of prolonged epinephrine infusion, and several studies have demonstrated similar findings of apparent desensitization of sweat gland β-2 receptors both *in vivo* and *ex vivo*, where cultured equine sweat glands show a rapid and homologous desensitization to epinephrine.[25] However, after 21 days of oral clenbuterol at 0.8 μg/kg twice daily, we found no reduction of sweating using an intradermal epinephrine sweat at concentrations of epinephrine as low as 1:1,000,000, thus suggesting that any reduction in sweating caused by chronic clenbuterol is likely to be sub-clinical.[30]

CLENBUTEROL

Clenbuterol is the only FDA-approved medication for horses with reversible bronchospasm. A long-acting β2-adrenoceptor agonist, clenbuterol is administered as an oral preparation (Ventipulmin®) over a wide dose range (0.8-3.2 μg/kg every 12 hours) for up to 30 days. Clenbuterol has been shown to increase dynamic compliance and decrease pulmonary resistance in ponies bronchoconstricted with intravenous histamine.[5] It has a dose-dependent effect, but is not effective in all horses. Normal horses treated with clenbuterol do not demonstrate bronchodilation in response to clenbuterol as there is minimal smooth muscle tone in the small airways. At the commonly administered dose of 0.8ug/kg, only 25% of horses with RAO show a clinical response, which increases up to 75% at the 3.2ug/kg dose.[26] Like horses with RAO, horses with IAD also have a hyperreactivity to inhaled histamine leading to bronchoconstriction.[27] In addition to bronchodilation, clenbuterol has also been shown to have anti-inflammatory effects, speeds mucociliary clearance, and decreases mucous production by goblet cells in many species, including the horse.[28,29]

Using a blinded, randomized cross-over designed trial, we recently showed clenbuterol at standard dosages (0.8 μg/kg *per os* every 12 hours) resulted in a decreased bronchoconstrictive response to inhaled histamine compared to horses receiving placebo treatment. Using flowmetric plethysmography and inhaled histamine bronchoprovocation, 5 out of 8 horses (62%) had a ≥100% increase in PC35 (35% percent change in $DELTA_{flow}$) peaking at day 14 of administration. This degree of efficacy is somewhat higher than previous studies that show 25% of RAO horses demonstrating clinical evidence of bronchodilation at the same dose.[26] This may be due to superior sensitivity of pulmonary function testing compared to clinical examination for detecting bronchoconstriction, the greater reversibility of bronchoconstriction in horses with IAD compared to those with RAO-induced airway remodeling, or differences in airway secretion volume or composition between the two diseases. Additionally, all horses showing a bronchoprotective effect demonstrated tachyphylaxis by day 21 of treatment; 7/8 horses had PC35 values that were actually lower than baseline, indicating pulmonary function may have worsened during the treatment period. This is consistent with human data that show prolonged administration of β-2 agonists results in worse than baseline lung function.[21] Based on *in vitro* studies in the horse[24] and extensive human clinical data, it is likely that extended use of clenbuterol should be combined with inhaled corticosteroids, although the safety and efficacy of this approach has not been validated in the horse. However, our study suggests that clenbuterol alone is less effective as a bronchodilator after 14 contiguous days of treatment, and courses of administration should not exceed 2 weeks without interruption.

REFERENCES

[1]Cockcroft DW et al. **Lancet** 1993;342(8875):833; [2]Cheung D et al. **N Engl J Med** 1992;327(17):1198; [3]Jokic R et al. **Chest** 2001;119(2):370; [4]Derksen FJ et al. **Am J Vet Res** 1987;48(3):423; [5]Scott JS et al. **Am J Vet Res** 1991;52(9):1416; [6]Mazan MR et al. **AAEP Proceedings.** Seattle, WA, USA, 2005; [7]Peters AR. **Vet Rec** 1989;124(16):417; [8]Kearns CF et al. **Med Sci Sports Exerc** 2006;38(2):262; [9]Kearns CF et al. **J Appl Physiol** 2001;91(5):2064; [10]Jenkinson DM. et al. **Vet**

Dermatol 2007;18(1):2; [11]Bijman J, et al. **Am J Physiol** 1984;246(3 Pt 2):R349; [12]Guthrie AJ, et al. **J S Afr Vet Assoc** 1992;63(4):162; [13]MacKay RJ. **Equine Vet J 2008**;40(5):518; [14]Bostedt H. **Tierairztl Prax** 1988; 16: 57; [15]Rush B. **AAEP Proceedings** 2002 48:218; [16]Soma LR et al. **J Vet Pharmacol Ther.** 2004 Apr;27(2):7; [17]Robinson NE. **Equine Respiratory Diseases, P. Lekeux (Ed.)** 2001; [18]Kearns CF, et al. **J Appl Physiol** 2001;91(5):2064; [19]Sleeper MM et al. **Med Sci Sports Exerc** 2002;34(4):643; [20]Kearns CF et al. **Med Sci Sports Exerc** 2002;34(12):1976; [21]Cockcroft DW. **Clin Rev Allergy Immunol** 2006;31(2-3):197; [22]Nishikawa M, et al. **Eur J Pharmacol** 1993;247(2):131; [23]Abraham G, et al. **Biochem Pharmacol** 2003;66(6):1067; [24]Abraham G, et al. **Equine Vet J** 2002;34(6):587; [25]Rakhit S, et al. **J Exp Biol** 1998;201(Pt 2):259; [26]Erichsen DF, et al. **Equine Vet J** 1994;26(4):331; [27]Hoffman AM, et al. **Am J Vet Res** 1998;59(2):176; [28]Laan TT, et al. **Vet J** 2006;171(3):429; [29]van den Hoven R, et al. **Vet Res Commun** 2006;30(8):921; [30]Read J and Nolen-Walston RD. **Am J Vet Res** 2011 (in press).

HORSES AND HUMANS: SHARED ENVIRONMENT, SHARED DISEASE?

Melissa R. Mazan, DVM, DACVIM
North Grafton, MA

The chronic inflammatory respiratory diseases, inflammatory airway disease (IAD) and recurrent airway obstruction (RAO, or heaves), are extremely common in stabled horses. The debate endures as to whether IAD and RAO are the same disease at early and late stages, or whether there is an innate etiologic and pathophysiologic difference between the two. Clinically, horses with IAD are characterized by cough [1], reduced exercise tolerance [2];[3], poor performance [4]and varying degrees of nasal discharge. Horses with IAD often have airway obstruction and reactive airways [5] Airway inflammation most frequently consists of elevated numbers of neutrophils, but increased percentages of mast cells or eosinophils can also be seen. [5, 6]

In contrast, horses with recurrent airway obstruction (RAO) tend to be older, and clinical signs include episodes of severe respiratory distress caused by lower airway obstruction, with increased abdominal effort, frequent coughing, and copious mucus secretion. Airway inflammation is characterized by severe neutrophilia. [7], [8]. Although there is reversible airway narrowing, a degree of airway obstruction usually remains despite bronchodilation, and residual airway inflammation tends to remain during remission. [9], [10]

For many years, researchers have sought to find parallels between human atopic asthma and RAO, because they share the attributes of inducible airways dysfunction and airway hyperreactivity along with profound airway inflammation. Past support for the TH2 hypothesis has included increased IgE antibodies in the airway secretions of horses with RAO [11], and increased mRNA for IL-4 and IL-5 in horses with heaves. [12],. However, the literature remains somewhat cloudy, in that unlike human asthma, horses have a neutrophilic rather than an eosinophilic airway inflammation, and there is support for IL-4 and IL-5 being downregulated in horses with RAO [13],[14] and predominance of other cytokines/chemokines, especially IL-8 and CXCL2 [15][16].

Although it is uncommon in developed countries for horses and humans to share the same abode, humans nonetheless spend a considerable time in the same barn environment as the horse. Although barn dust in swine, dairy, and poultry operations has been increasingly recognized as a strong risk factor for development of chronic bronchitis and other respiratory diseases, it is only recently that the equine barn environment has been recognized as being potentially hazard to humans as well as horses. Humans are regularly exposed to particles comparable in composition to those which horses inhale in the environment of the stable. These are grain dust, poultry slaughterhouse, and swine confinement workers who can be regularly exposed to particles of biologic origin (i.e. grain dusts, bacteria, endotoxin, fungi, glucan, and storage mites) at concentrations varying from 3000 to 25,000 mg/m^3 [17], [18], and these workers experience an accelerated decline in FEV1 over a 7-year period. [19] Although there is scant information concerning dust exposure to humans from the equine environment, a study of respiratory consequences of exposure to equine stable dust showed that over half of the workers studied complained of respiratory work-related symptoms. [20] Similarly, asthma prevalence was found to be three times that of a control population in horse grooms in Turkey, although exposure to dust was not assessed. [21].. More recently, our laboratory has shown that horse personnel experience an almost 10-fold increase in likelihood of reporting respiratory symptoms such as cough, wheeze, or waking in the night. [22]. Moreover, despite relatively low particulate levels in one study of stable workers, a relatively large proportion had evidence of airway obstruction. [23] In contrast to these findings of low dust exposure, a recent study showed that stable workers can be exposed to relatively high levels of endotoxin and molds (1,3-beta-glucan), and that common tasks such as sweeping exacerbate this exposure. [24].

Although there is no data to elucidate the mechanism of action whereby exposure to organic dusts, endotoxin, molds, and noxious gases that as are found in horse barns cause respiratory disease or symptoms in people, it is illustrative to examine findings from studies in other barn workers. Swine barn workers, like horses with IAD or RAO, tend to have a neutrophilic airway inflammation, as well as evidence of airway obstruction. Indeed, in a recent study of livestock workers, a chronic bronchitis prevalence of 8.9% was documented, with another 16.1% being classified as having, COPD; [25] and in one study, 75% of swine barn workers were documented to have obstruction of the small airways. [26]. Importantly, Donham and coworkers have shown that low levels of particulates (less than 0.162 mg/m3) are sufficient in chronically exposed populations in order to elicit a cross-shift decline in FEV1 and long term accelerated decline in FEV1. [27].

Studies in horses with heaves have employed both natural challenge and various aqueous extracts of stable and haydust, including endotoxin, organic dust particulates, multiple different hay and stable associated molds, and a combination of fungal spores, LPS, and silica microspheres. It is important to note that only natural exposure is successful in bringing about all the attributes of the clinical syndrome known as heaves, or recurrent airway obstruction (RAO). Similarly, it is clear that in the human syndrome, endotoxin alone does not elicit the same profound response as does barn dust itself. [28]. Thus, in both species, there must be a complex interplay among the multiple components of organic dust, including endotoxin, 1,3-beta glucans, peptidoglycans, muramic acid and microbes among others.

A pathway for the mechanism of action of swine barn dust in eliciting airway inflammation and obstruction in humans has been partially developed. Within one hour after hog dust exposure (HDE), TNF-alpha induces upregulation of PKC, , with consequent upregulation of Il-6, IL-8, and further downstream upregulation of TNF-alpha. Although endotoxin enhances this cytokine cascade, it is not alone sufficient. [29]. This dovetails with findings in horses that while endotoxin inhalation induces airway inflammation, it does not induce the clinical syndrome of heaves. [11]

One of the perplexing questions concerning exposure to the barn environment involves the 'hygiene hypothesis', whereby exposure, especially of children, to a farm environment, tends to produce less atopic individuals with a lowered risk of developing asthma. [30] In a recent study looking at animal contact and microbial exposure and prevalence of asthma, people who regularly used cow dung in housing construction and who were exposed to farm animals by sharing their abodes with their livestock had a significantly lower prevalence of asthma. [31]How does this fit with the picture of chronically bronchitic farm workers and horses with IAD or heaves? It is important in this case to distinguish between the chronic neutrophilic bronchitis and airway obstruction seen in the occupational asthma of farmers, and atopic eosinophil-mediated asthma. In a Canadian study of farmers working with livestock at least 20 hours/week, there was no increase in diagnosis of asthma or wheeze, rather, there was a significant increase in cough and phlegm. [32] In naïve subjects, there may be even a toxic-dust syndrome associated with exposure to swine dust, with fever, airway reactivity, and systemic inflammation, [28] which is replaced by chronic bronchitis and declines in lung function over time in those chronically exposed. [18] There may be a considerable healthy worker effect, as well, as one study showed that those who left farming due to respiratory complaints had an increased prevalence of asthma, as opposed to the high rates of chronic bronchitis *without asthma* in those who remained in farming. [33] Finally, studies using human peripheral blood monocytes show that tolerance is induced with continued exposure to swine barn dust. When cells are repeatedly stimulated with dust extract, production of TNF-alpha and IL-6 is decreased.. However, IL-8 remains high despite repeated exposure, [28]conjuring up the repeatedly high IL-8 in horses with RAO. The answer as to why some horses progress to develop the inducible marked airway obstruction and BAL neutrophilia that are the hallmark of heaves may be found in recent observations in the human that there are both more and less responsive individuals, and that polymorphisms in the TLR4 gene may underly the severity of responses. [34]

In conclusion, horses and humans share an environment that may commonly result in airway inflammation. The exposure is complex, encompassing endotoxin, peptidoglycans, muramic acid, living microbes and 1,3-beta-glucans, amongst others. Horses and humans share a similar response, namely elevated levels of Il-8 with varying production of other cytokines, neutrophilic airway inflammation, and most commonly (as in horses with IAD), airway obstruction without any evidence of atopy. The shared deleterious effects of the barn environment on the equine and human respiratory systems should prompt horse owners to improve their barn environments for the sake of both horse and rider.

REFERENCES

1. Bedenice, D., M.R. Mazan, and A.M. Hoffman, *Association between cough and cytology of bronchoalveolar lavage fluid and pulmonary function in horses diagnosed with inflammatory airway disease.* J Vet Intern Med, 2008. **22**(4): p. 1022-8. 2.Couetil, L.L. and D.B. Denicola, *Blood gas, plasma lactate and bronchoalveolar lavage cytology analyses in racehorses with respiratory disease.* Equine Vet J Suppl, 1999. **30**: p. 77-82. 3. Sanchez, A., et al., *Effect of airway disease on blood gas exchange in racehorses.* J Vet Intern Med, 2005. **19**(1): p. 87-92. 4. Holcombe, S.J., et al., *Effect of tracheal mucus and tracheal cytology on racing performance in Thoroughbred racehorses.* Equine Vet J, 2006. **38**(4): p. 300-4. 5. Hoffman, A.M., M.R. Mazan, and S. Ellenberg, *Association between bronchoalveolar lavage cytologic features and airway reactivity in horses with a history of exercise intolerance.* Am J Vet Res, 1998. **59**(2): p. 176-81. 6. Couetil, L.L., et al., *Clinical signs, evaluation of bronchoalveolar lavage fluid, and assessment of pulmonary function in horses with inflammatory respiratory disease.* Am J Vet Res, 2001. **62**(4): p. 538-46. 7.Derksen, F.J., et al., *Bronchoalveolar lavage in ponies with recurrent airway obstruction (heaves).* Am Rev Respir Dis, 1985. **132**(5): p. 1066-70. 8. Tremblay, G.M., et al., *Effect of stabling on bronchoalveolar cells obtained from normal and COPD horses.* Equine Vet J, 1993. **25**(3): p. 194-7. 9. Mazan, M.R., A.M. Hoffman, and N. Manjerovic, *Comparison of forced oscillation with the conventional method for histamine bronchoprovocation testing in horses.* Am J Vet Res, 1999. **60**(2): p. 174-80. 10. Robinson, N.E., et al., *Relationship between clinical signs and lung function in horses with recurrent airway obstruction (heaves) during a bronchodilator trial.* Equine Vet J, 2000. **32**(5): p. 393-400. 11. Halliwell, R.E., et al., *Local and systemic antibody production in horses affected with chronic obstructive pulmonary disease.* Vet Immunol Immunopathol, 1993. **38**(3-4): p. 201-15. 12. Lavoie, J.P., et al., *Neutrophilic airway inflammation in horses with heaves is characterized by a Th2-type cytokine profile.* Am J Respir Crit Care Med, 2001. **164**(8 Pt 1): p. 1410-3. 13. Kleiber, C., et al., *Cytokine profiles of peripheral blood and airway CD4 and CD8 T lymphocytes in horses with recurrent airway obstruction.* Vet Immunol Immunopathol, 2005. **104**(1-2): p. 91-7. 14. Giguere, S., et al., *Cytokine induction in pulmonary airways of horses with heaves and effect of therapy with inhaled fluticasone propionate.* Vet Immunol Immunopathol, 2002. **85**(3-4): p. 147-58. 15. Ainsworth, D.M., et al., *Recurrent airway obstruction (RAO) in horses is characterized by IFN-gamma and IL-8 production in bronchoalveolar lavage cells.* Vet Immunol Immunopathol, 2003. **96**(1-2): p. 83-91. 16. Riihimaki, M., et al., *Partial divergence of cytokine mRNA expression in bronchial tissues compared to bronchoalveolar lavage cells in horses with recurrent airway obstruction.* Vet Immunol Immunopathol, 2008. **122**(3-4): p. 256-64. 17. Hagmar, L., et al., *Health effects of exposure to endotoxins and organic dust in poultry slaughter-house workers.* Int Arch Occup Environ Health, 1990. **62**(2): p. 159-64. 18. Larsson, K., P. Malmberg, and A. Eklund, *Acute exposure to swine dust causes airway inflammation and bronchial hyperresponsiveness.* Am J Ind Med, 1994. **25**(1): p. 57-8. 19. Iversen, M. and R. Dahl, *Working in swine-confinement buildings causes an accelerated decline in FEV1: a 7-yr follow-up of Danish farmers.* Eur Respir J, 2000. **16**(3): p. 404-8. 20. Mackiewicz, B., et al., *[Exposure to organic dust and microorganisms as a factor affecting respiratory function of workers of purebred horse farms].* Pneumonol Alergol Pol, 1996. **64 Suppl 1**: p. 19-24. 21. Tutluoglu, B., et al., *Sensitization to horse hair, symptoms and lung function in grooms.* Clin Exp Allergy, 2002. **32**(8): p. 1170-3. 22. Mazan, M.R., et al., *Questionnaire assessment of airway disease symptoms in equine barn personnel.* Occup Med (Lond), 2009. **59**(4): p. 220-5. 23. Elfman, L., et al., *Influence of horse stable environment on human airways.* J Occup Med Toxicol, 2009. **4**: p. 10. 24. Samadi, S., et al., *Exposure to inhalable dust, endotoxins, beta(1->3)-glucans, and airborne microorganisms in horse stables.* Ann Occup Hyg, 2009. **53**(6): p. 595-603. 25. Eduard, W., N. Pearce, and J. Douwes, *Chronic bronchitis, COPD, and lung function in farmers: the role of biological agents.* Chest, 2009. **136**(3): p. 716-25. 26. Kuchuk, A.A., A. Basanets, and K. Louhelainen, *Bronchopulmonary pathology in workers exposed to organic fodder dust.* Ann Agric Environ Med, 2000. **7**(1): p. 17-23. 27. Donham, K.J., et al., *Dose-response relationships between occupational aerosol exposures and cross-shift declines of lung function in poultry workers: recommendations for exposure limits.* J Occup Environ Med, 2000. **42**(3): p. 260-9. 28. Poole, J.A., et al., *Repeat organic dust exposure-induced monocyte inflammation is associated with protein kinase C activity.* J Allergy Clin Immunol, 2007. **120**(2): p. 366-73. 29. Wyatt, T.A., et al., *Sequential activation of protein kinase C isoforms by organic dust is mediated by tumor necrosis factor.* Am J Respir Cell Mol Biol, 2010. **42**(6): p. 706-15. 30. Midodzi, W.K., et al., *Reduced risk of physician-diagnosed asthma among children dwelling in a farming environment.* Respirology, 2007. **12**(5): p. 692-9. 31. Vedanthan, P.K., et al., *Effect of animal contact and microbial exposures on the prevalence of atopy and asthma in urban vs rural children in India.* Ann Allergy Asthma Immunol, 2006. **96**(4): p. 571-8. 32. Senthilselvan, A., et al., *Excess respiratory symptoms in full-time male and female workers in large-scale swine operations.* Chest, 2007. **131**(4): p. 1197-204. 33. Eduard, W., et al., *Atopic and non-atopic asthma in a farming and a general population.* Am J Ind Med, 2004. **46**(4): p. 396-9. 34. Dosman, J.A., et al., *Respiratory response to endotoxin and dust predicts evidence of inflammatory response in volunteers in a swine barn.* Am J Ind Med, 2006. **49**(9): p. 761-6.

TRANSPORT-ASSOCIATED RESPIRATORY DISEASE IN HORSES

Rose D. Nolen-Walston, DVM, DACVIM (LAIM)
Kennett Square, PA

INTRODUCTION

Pleuropneumonia is a common disease of equids, especially young racehorses. Although several risk factors for the development of pleuropneumonia have described, including strenuous exercise, upper respiratory viral disease, general anesthesia and esophageal obstruction, long-distance transport is probably the most common inciting cause.[1,2] During the First World War, horse deaths from pneumonia exceeded mortality from other causes by a factor of three until provisions to improve the conditions of long-distance shipping were improved.[1]

ETIOLOGIC AGENTS ASSOCIATED WITH TRANSPORT-ASSOCIATED PLEUROPNEUMONIA

Pleuropneumonia in the horse is generally not caused by specifically contagious organisms, but is rather associated with an opportunistic infection with organisms commensual to the oropharynx.[3] Of these, by far the most common bacteria isolated from the lower respiratory tract of horses with transport-associated pneumonia is *Streptococcus equi ssp. zooepidemicus*. In a recent study, *Streptococcus equi ssp. zooepidemicus* was isolated from 94.5% of tracheal samples.[4] There are many protein variations of *Streptococcus equi ssp. zooepidemicus* that have been isolated from horses with clinical disease, but no association between protein sequence and particular clinical manifestations of infection in horses.[5] Despite the ubiquity of *Streptococcus equi ssp. zooepidemicus* in this disease, most infections yield a mixed culture of organisms, with any of the following *Corynebacterium* spp., *Mannheimia hemolytica*, *Actinocbacillus* sp., *Pasturella* ssp., *E. coli*, *Klebsiella pneumonia*, *Staphylococcus* sp., and alpha-hemolytic *Streptococcus* sp.[4,6] Anaerobic bacteria are also frequent isolated, especially *Bacteroides* and *Clostridium* ssp., and are associated with a poorer prognosis.[7,8]Viruses have not been isolated from pneumonic lesions at necropsy from horses with transport-associated pneumonia,[9] and no studies have been performed that specifically evaluated the role of mycoplasma ssp. in this disease.

PATHOPHYSIOLOGY OF TRANSPORT-ASSOCIATED PLEUROPNEUMONIA

The pneumonia that occurs after shipping is characterized by predominantly ventral disease including consolidation, abscessation, pleuritis and necrosis of the cranio-ventral portion of the caudal lung lobe, especially of right lung.[10] Post-mortem lesions in one experimental study of transport-pneumonia revealed serous hemorrhagic pneumonia and multifocal coagulative necrosis which were associated with *Streptococcus equi ssp. zooepidemicus* infection.[11] In vivo inoculation of healthy lungs with *Streptococcus equi ssp. zooepidemicus* recapitulated these pathologic changes over time.[12] There is considerable consensus that transport-associated pneumonia is largely a result of prolonged confinement with the head maintained in an elevated position, as is typical in horses being shipped in trailer, truck, boat or airplane tie-stalls.[13-16] Prolonged elevated head position results in a reduction in tracheal mucociliary clearance,[17] and maintaining the head in an elevated position even without shipping results in tracheal inflammation and bacterial contamination in 75% of horses. In addition to effects on the respiratory tract, long-distance road transport has also been shown to lead to changes in muscle metabolism, stress indices, hydration status, immune parameters, and body weight.[18]

TREATMENT OF TRANSPORT-ASSOCIATED PLEUROPNEUMONIA

Treatment is generally started empirically before the results of culture and sensitivity from tracheal washes are known. Based on the likelihood of a mixed bacterial infection including anaerobes, broad spectrum coverage is typically recommended.[7] Despite their poor penetration into necrotic tissue, gentamicin and penicillin are often the first choice for serious infections, and the vast majority of *Streptococcus equi ssp. zooepidemicus* isolates are susceptible to penicillin. However, despite the overall good anaerobic spectrum, pencillin is generally ineffective against *Bacteroides fragilis*, a common anaerobic isolate in equine pneumonia.[7] Ceftiofur, a parenteral cephalosporin, has also shown good efficacy against naturally occurring respiratory infections at a dose of 2.2mg/kg/day when compared to ampicillin, with 79% of patients treated with ceftiofur and 59% of horses treated with ampicillin showing complete resolution of signs.[19] Despite its appeal as a relatively safe, bacteriocidal, once-daily antimicrobial, enrofloxacin has poor efficacy in most cases of transport-associated respiratory disease as *Streptococcus equi ssp. zooepidemicus*is generally not sensitive. Davis *et al.* showed that at doses ranging from 1.25-7.5mg/kg/day, it was not substantially more effective than placebo; furthermore, 21% of affected horses required "rescue therapy" with gentamicin and penicillin to resolve the clinical signs.[4]

Prognosis for pneumonia is guarded, although much of the published data is over 20 years old; it is reasonable to hope that advances in treatment may have improved the reported survival of 53%.[1] In addition to appropriate antimicrobial therapy and supportive care, the addition of therapeutic broncho-alveolar lavage has been reported in one study to improve outcome significantly.[6] If thick pleural peels or extensive abscessation and large areas of tissue necrosis exist, the use of open debridement via standing rib resection,[20]or intrapleural fibrinolytic therapy with tissue plasminogen activator[21] are also possibilities, although the outcomes have not been compared to conservative management

PREVENTION OF TRANSPORT-ASSOCIATED PLEUROPNEUMONIA

Prevention or a reduction of prevalence in transport-associated pneumonia is probably best tackled by altering transport techniques to include longer rest periods, loose boxes rather than tie stalls, and better air quality within the vehicle.[1,22] A study directly comparing tied versus loose horses during long-distance transport found significantly higher indices of stress in the tied group.[18] However, although horses being shipped unrestrained had normal BAL fluid that did not contain higher numbers of bacteria,[23] in a more recent study of 80 unrestrained horses subjected to road transport of >1000 miles, 41% developed signs compatible with lower respiratory tract disease.[4] The effect of orientation (rear facing or forward facing on the trailer) has not been shown to affect the risk of respiratory disease, although increasing the rest time and removing manure from the vehicle during rest stops reduced transportation stress that may lead to respiratory compromise[24] Despite (or perhaps because of) the low-tech nature of these modifications, there does not seem to be dramatic shift in transportation techniques in the equine industry. Pharmacologic interventions to prevent transport-associated pneumonia have focused on specific risk factors. As the main etiologic agent, *Streptococcus equi ssp. zooepidemicus* is exquisitely sensitive to penicillin, the prophylactic use of this antimicrobial agent was investigated using a long-distance transport model (head elevation for 48 hours), but had no significant effect on airway contamination.[25] Clenbuterol, an oral beta-2 agonist that has been shown to increase tracheal mucociliary clearance in the horse, was also evaluated using a 48 hour head elevation protocol. Six horses enrolled in a crossover designed study received either clenbuterol (0.8 µg/kg PO BID) or placebo, and tracheal mucociliary clearance, tracheal wash cytology and quantitative culture, rectal temperature, CBC, fibrinogen, and serum TNFα, IL-10 and IL-2 levels were measured before and after simulated shipping. The study found a trend towards faster tracheal mucociliary clearance in horses receiving clenbuterol, and also a small reduction (0% versus 33%) in tracheal β hemolytic *Streptococcus. spp.* numbers after clenbuterol versus placebo.[26]

REFERENCES

[1]Racklyeft DJ et al. **Aust Vet J** 2000 78(5):334; [2]Austin et al. **J Am Vet Med Assoc** 1995 207(3):325; [3]Bailey GD et al. **Vet Microbiol.** 1991 Feb 15;26(4):367; [4]Davis E et al. **Vet Ther** 2006 Fall;7(3):232; [5]Walker et al. **Am J Vet Res.** 2003 Aug;64(8):976; [6]Ito et al. **J Vet Med Sci.** 2001 Dec;63(12):1263; [7]Sweeney CR et al. **J Am Vet Med Assoc.** 1991 Mar 1;198(5):839; [8]Sweeney CR et al. **J Am Vet Med Assoc.** 1985 Oct 1;187(7):721; [9]Oikawa M et al. **J Comp Path** 1995 113(1):29; [10]Oikawa M et al. **J. Comp Path** 1994 111(2): 205; [11]Oikawa M et al. **J. Comp Path** 1994 111(2): 205; [12]Yoshikawa et al. J **Vet Med Sci.** 2003 Jul;65(7):787; [13]Racklyeft DJ et al. **Aust Vet J** 2000 78(8):549; [14]Racklyeft DJ et al. **Aust Vet J.** 1990 Nov;67(11):402; [15]Stull CL et al. **Equine Vet J** 2002 34(6):550; [16]Raidal SL et al. **Aust Vet J** 1995 72(2):45;[17]Raidal SL et al. **Aust Vet J.** 1996 Feb;73(2):45; [18]Stull et al. **J Anim Sci.** 2000 Jun;78(6):1458; [19]Folz SD et al. **Equine Vet J** 1992 Jul;24(4):300; [20]Dechan J. **Can Vet J.** 1997 August; 38(8): 499; [21]Hilton H et al. **Vet Rec.** 2009 May 2;164(18):558; [22]Purswell et al. **Trans ASABE.** 2006 49(1): 193; [23]Traub-Dargatz JL et al. **Am J Vet Res.** 1988 Jul;49(7):1026; [24]Oikawa M et al. **J Comp Pathol.** 2005 Feb-Apr;132(2-3):153; [25]Raidal SL et al. **Aust Vet J.** 1997 Feb;75(2):126; [26]Norton J et al. unpublished data.

UPDATE ON AIRWAY DISEASE IN THE HORSE

Laurent Couetil, DVM, PhD, DACVIM (LA)
West Lafayette, IN

INTRODUCTION

The following presentation represent research abstracts that were delivered at the 28th meeting of the Veterinary Comparative Respiratory Society in Raleigh, North Carolina in October 2010. The first abstract was originally presented by Dr. Kathleen Ivester, PhD student in the Department of Veterinary Clinical Sciences at Purdue University, who received the Joan O'Brien Award for best oral presentation by a graduate student.

INFLAMMATORY AIRWAY DISEASE IN YOUNG THOROUGHBREDS ENTERING RACE TRAINING.

Kathleen M Ivester,[1] Laurent L Couëtil,[1] and Neil Zimmerman[2] Veterinary Clinical Sciences[1] and School of Health Sciences,[2] Purdue University.

Inflammatory airway disease (IAD) is the most common chronic airway disease of athletic horses with incidence ranging between 10 and 60 % worldwide.[1,2] The disease is typically characterized by cough, poor performance, and excess mucus in the airways on endoscopic examination.[3]

Exposure to dust present in the horse's environment appears to play a major role in IAD pathogenesis. Stable dust contains particularly high levels of organic and inorganic particles including endotoxins, β-glucan, mite debris, vegetative material and other airborne particles that have been implicated in the development of IAD,[4] although the pathogenesis is largely unknown at the present time.[3]

Airway inflammation detected in bronchoalveolar lavage fluid (BALF) of IAD-affected horses is usually mild; however, even a mild level of pulmonary inflammation has a negative impact on race horses' performance due to impairment in pulmonary gas exchange during strenuous exercise.[5-7] Diagnosis of IAD based on BALF cytology has been associated with excessive airway mucus and poor performance.[5-7] Consequently, cytological analysis of BALF is one of the preferred tests for IAD and is easily performed in field conditions. The most common phenotype of IAD is characterized by an increased total nucleated cell count with mild neutrophilia, lymphocytosis, and monocytosis.[3] In young horses, increased eosinophils and/or mast cells may be encountered, suggesting a possible allergic mechanism. Research establishing association between environmental exposure and changes in BAL fluid cytology is lacking.

Therefore, the objective of this study was three-fold: 1. to characterize the IAD status of young Thoroughbred racehorses over the course of their initial month in training; 2. to examine the relationship between airborne exposure to respirable dust and ammonia and BALF cytology; 3. to assess the efficacy of a portable HEPA air filtration unit in both decreasing horse exposure to respirable dust and decreasing the inflammatory cell population in BALF.

Methods: Horses were recruited from the population of 16-24 month-old Thoroughbred horses arriving at a local race-training facility. Criteria for enrollment included evaluation within five days of arrival at the barn, no prior history of race training, no evidence of any illness upon physical examination, and normal complete blood count. Eligible horses had physical examination, blood collection, and bronchoalveolar lavage (BAL) performed at enrollment (day 0). Horses were classified as IAD if BALF was outside reference range (neutrophils > 5%, mast cells > 2% or eosinophils >1%). Each horse was randomly assigned to either treatment or control group. Those horses allocated to the treatment group had a portable HEPA air filtration unit (Holmes Model HAP8756-U) placed in the stall. No air filtration was added to the stall of those horses assigned to the control. Horses were monitored over the course of four weeks with physical examination and BAL repeated on days 14 and 28. Respirable dust (aluminum cyclone with 37 mm type AE glass fiber filter, SKC) and ammonia levels (ChromAir ammonia monitor badges, Morphix Technologies) were measured in the breathing zone of each horse over the course of 5-8 hours on a weekly basis. Data was tested for normality of distribution using the Kolmogorov-Smirnov test. The effects of treatment (HEPA, control) and disease status at day 0 (IAD, control) on the primary variables of interest (dust and ammonia concentrations, BALF cells) were evaluated over time using repeated measures analysis of variance or Mann Whitney U test as appropriate. Associations between barn exposure level and BALF cytology at day 28 were evaluation by Pearson correlation coefficient. Significance was set at $p<0.05$.

Results: Twenty horses were enrolled in the study. At day 0, 11 horses (55%) were classified as having IAD. One horse exhibited an increase in both neutrophils and mast cells, 7 horses an increased percentage of mast cells only, 2 horses an increase in both mast cells and eosinophils, and 1 horse exhibited an increase in eosinophils only. By day 14, 14 horses were classified as having IAD (70%). Again, the majority of these horses exhibited an increased percentage of mast cells. By day 28 and completion of the study, 11 horses were classified as having IAD (55%) with an increase in mast cells again the predominating phenotype (10/11 IAD horses, 91%). Seven of the 11 horses with IAD at day 28 entered the study with the diagnosis of IAD. Horses diagnosed with IAD at day 0 had significantly higher mast cell percentage than controls throughout the study period ($p = 0.03$).

A significant correlation was demonstrated between average ammonia exposure and neutrophil percentage at day 28 ($r=0.45$, $p=0.047$). No association was established between measured respirable dust exposure and BALF cytology. Ammonia and respirable dust exposure did not differ between those horses diagnosed with IAD at enrollment and those without IAD ($p= 0.86$ and $p=0.54$ respectively).

No effect of HEPA air filtration was detected on respirable dust or ammonia exposure ($p=0.45$ and $p=0.94$ respectively). The mean (range) average respirable dust exposure was 0.24 mg/m3 (0.07-0.51) in the treatment group and 0.18 mg/m3 (0.04-0.32) for the control group. Mean (range) average ammonia exposure was 4.9 ppm (3.5-7.4) for the treatment group and 3.9 ppm (1.2-7.2) for the control. Treatment and control groups did not differ in total nucleated cell counts or differential cell counts at any time point during the study.

Conclusions: The horses in this study population displayed a high incidence of IAD, with an increase in mast cells representing the most common cytologic phenotype. Ammonia exposure was correlated to the percentage of neutrophils recovered upon BAL, potentially indicating a non-specific innate immune reaction to this irritant. The failure to demonstrate an association between respirable dust levels and the inflammatory cell population recovered in BALF may indicate the levels of respirable dust exposure in the present study were too low or specific antigens associated with airway inflammation were not present at sufficient concentration. Finally, the HEPA air filtration unit employed in this study did not provide any measurable benefit in the environment studied.

COMPARATIVE EFFICACY OF INHALED ALBUTEROL BETWEEN TWO HAND-HELD DELIVERY DEVICES IN HORSES WITH RECURRENT AIRWAY OBSTRUCTION. **F. R. Bertin, K. M. Ivester, L. L. Couëtil Department of Veterinary Clinical Sciences, Purdue University.**

Recurrent airway obstruction (RAO) is the most frequent cause of chronic respiratory tract disease in horses.[8] The affection is characterized by airway inflammation and bronchospasm with phases of remission when a horse's environment is improved. During disease exacerbation, bronchoconstriction, airway wall edema and accumulation of mucus result in obstruction of the distal airways. These mechanisms induce functional changes: maximum change in transpulmonary pressure (ΔPmax) increases, pulmonary resistance (R_L) increases and dynamic lung compliance (C_{dyn}) decreases.[9]

Inhaled short-acting β_2-receptor agonists are the most effective medication for relieving acute bronchospasm. Aerosolized albuterol has been shown to be a valuable bronchodilator with rapid onset in the treatment of RAO.[10, 11] This type of therapy requires specialized devices to optimize drug delivery in the equine lung. However, data on the efficacy of inhaled albuterol using available delivery devices is not available.

The purpose of this study was to evaluate if aerosolized albuterol administered to horses with RAO using aerosol delivery devices currently commercially available have comparable efficacy on lung function and to provide clinicians with guidelines for the selection of an aerosol delivery device.

Methods: The study used a cross-over design to compare the efficacy of the two delivery devices. Ten horses with RAO were kept in a dusty environment and fed moldy hay to induce airway obstruction. Lung mechanics were measured before and after the procedure. ΔPmax was measured 5 minutes after administration of 180 µg of albuterol from a pressurized metered dose inhaler, using an aerosol delivery device chosen randomly. This process was repeated every 5 minutes until maximal bronchodilation was achieved. After a 24-hour washout period in the dusty environment, lung mechanics data was again collected following albuterol administration using the other aerosol delivery device, following the same protocol.

Results: Aerosolized albuterol induced a significant and rapid bronchodilation in the horses using both aerosol delivery devices. No statistically significant difference in pulmonary function was observed in response to albuterol therapy between the two aerosol delivery devices. The dose required to achieve 50 % of maximal bronchodilation was not statistically different between the two aerosol delivery devices (173.35±78.35 µg with device 1 and 228.49±144.99 µg with device 2 ($p = 0.26$)). The decrease in lung resistance tended to be more pronounced after albuterol administration with device 1 ($p = 0.066$). All of the horses tolerated the procedure and no adverse effects were observed.

Conclusions: Aerosolized albuterol is an effective bronchodilator in horses with recurrent airway obstruction. There is no statistically significant difference between the two commercially available aerosol delivery devices in terms of efficacy. Aerosolized albuterol is effectively delivered using currently available devices leading to maximal bronchodilation in horses with RAO at an average dose of 540 µg.

REFERENCES

1. Wood JL, et al. **Equine Vet J** 2005;37:236-242. 2. Burrell MH, Wood JL, Whitwell KE, et al. **Vet Rec** 1996;139:308-313. 3. Couetil LL, et al. **J Vet Intern Med** 2007;21:356-361. 4. Malikides N, et al. **Rural Industries Research and Development Corporation** Publication No 03/089 2003. 5. Sanchez A, et al. **J Vet Intern Med** 2005;19:87-92. 6. Couëtil LL,et al. **Equine Vet J** Suppl 1999;30:77-82. 7. Courouce-Malblanc A, et al. **Equine Vet J** Suppl 2002:402-407. 8. Hotchkiss et al. 2007. 9. Couëtil et al. **Equine Vet. J.** 2007 ;39, 294-300; 10. Rush et al. **Am. J. Vet. Res.** 1999 ;60, 764-769; 11. Derksen et al. **Am. J. Vet. Res.** 1999 ;60, 689-693.

NEW DRUGS FOR AN OLD DISEASE: EFFICACY IN EXPERIMENTAL FELINE ASTHMA

Carol Norris Reinero, DVM, PhD, DACVIM
Columbia, MO

OVERVIEW

Asthma is a common bronchopulmonary disorders in the cat associated with substantial morbidity and occasional mortality. Clinically cats present for cough, wheeze and episodic respiratory distress. The cause of feline asthma is likely allergic in nature, that is, it is a type I hypersensitivity reaction developed against inhaled allergens. Once cats become sensitized to an allergen (under appropriate genetic and environmental influences), they are susceptible to a variety of pathologic consequences upon repeated exposure to that same allergen. The hallmark features of naturally occurring asthma in cats includes eosinophilic airway inflammation, airway hyperresponsiveness and airflow limitation (both in response to allergenic and non-allergenic stimuli), and airway remodeling (permanent architectural changes in the lungs). These pathologic changes ultimately lead to a decline in lung function over time.

An ideal therapy would reverse the abnormal immune response against what should be a benign inhaled substance; however, the current therapies that make up the standard for treatment act late in the inflammatory cascade. While bronchodilators and glucocorticoids can be effective in reducing or eliminating clinical signs in pet cats, they may be associated with side effects or be relatively contraindicated depending on co-morbid conditions. Additionally, not all cats respond completely, even with "aggressive" medical management. Thus, there is still a need research investigating alternative therapies for feline asthma.

IMMUNOPATHOGENESIS OF ASTHMA

A brief review of the immunologic events which underlie allergic asthma is helpful not only to understand how clinical signs develop, but also to provide logical targets where novel treatments might have a beneficial impact. During normal breathing cycles, small antigenic proteins capable of eliciting an IgE response (allergen) gain access to the airways. Locally, professional antigen presenting cells (predominantly dendritic cells) take in the allergen and process it internally. A portion of the allergen is subsequently displayed on the cell surface in conjunction with the major histocompatibility complex (MCH) II. Local naïveT helper 0 (Th0) cells recognize the allergen in conjunction with MHC II and under appropriate microenvironmental signals are stimulated to become Th2 cells, which produce a variety of cytokines that skew the immune response towards an allergic response. These cytokines can induce the maturation and survival of eosinophils, and stimulate B cells to produce antibodies of the IgE class. IgE antibodies that are secreted become bound to mast cells just below the surface of the airways. Upon reexposure to the allergen, the allergen crosslinks two IgE antibodies leading to an intracellular signal that triggers mast cell degranulation. A variety of preformed mediators within mast cell granules are immediately released causing bronchoconstriction, increases in vascular permeability and inflammatory cell influx. Other eicosanoids and cytokines are elaborated by the mast cells, and a second wave of bronchoconstriction and cellular infiltration ("late phase response") can occur hours later.

Therapy can be targeted at any point along the aforementioned inflammatory cascade, although the "higher up" the inflammatory cascade, the more downstream events are modulated. An ideal treatment (or potential cure) would block the initiating events of the allergic airway cascade. However, as mentioned previously, very effective treatments can still work relatively late in the inflammatory cascade to palliate clinical signs.

"NOVEL" TREATMENTS

Knowing aforementioned details about the allergic asthmatic cascade provides a more logical means to design therapy which could block specific portions of the cascade. Modulation of T lymphocyte activity, cytokine inhibition, antibody blockade (especially IgE), small molecule inhibition (e.g., tyrosine kinases), and antagonists of inflammatory mediators and smooth muscle constriction are thought to be rational targets. Not all of these have been evaluated in cats, and the ones that have been evaluated, have been tested in experimental models of feline allergic asthma.

Since allergic asthma is thought to be driven by T helper 2 lymphocyte polarity, modulating Th2 lymphocyte activity would seem to hold promise for therapy. Newer treatments which have been tested which function by altering T cell activity include allergen-specific immunotherapy (ASIT) with or without CpG immunostimulatory sequences. Currently, the only curative form of therapy for allergies is ASIT. Allergen-specific immunotherapy is most commonly administered to humans with allergic rhinitis, but has also been used successfully in the treatment of asthma in people. Investigations into efficacy of ASIT in experimental asthma have been going on for several years, with important advancements made. The initial study suggested that an abbreviated form of conventional immunotherapy, called rush immunotherapy (RIT), dampened eosinophilic inflammation in experimental feline asthma.[1] However, side effects were common, and the protocol underwent further modification. Addition of an adjuvant (CpG immunostimulatory sequences which polarize the immune response towards a T helper 1 and away from a T helper 2 response) demonstrated good safety and efficacy. [2] Comparing the subcutaneous route of injection with topical mucosal delivery of allergen-specific rush immunotherapy was also performed in experimental feline asthma. [3] The latter route of administration more closely mimics the natural route of exposure to aeroallergens. Both routes of administration of allergen-specific immunotherapy decreased eosinophilic airway inflammation and were generally associated with minimal side effects; either could be used. However, the subcutaneous protocol demonstrated more consistent resolution of scored clinical signs after allergen challenge by aerosol. While this preliminary data in experimental feline asthma seems to hold promise for pet cats with asthma, there are still hurdles which need to be overcome. For example, there is likely to be some difficulty in accurately identifying sensitizing allergens in all cats with allergic asthma. Would inadvertent "inappropriate" selection of allergens would confer cross-protection against the clinically relevant allergens? This was addressed in an additional recent study in which cats were sensitized with one allergen and administered a different allergen for RIT. [4] Results showed that an irrelevant allergen could also decrease eosinophilic airway inflammation (unpublished data), but that the underlying immunologic mechanism was different than in cats sensitized and treated using the same allergen that also had reduction in eosinophilic airway inflammation. Future studies must assess the impact of corticosteroid therapy on appropriate identification of sensitizing allergens as well as on efficacy of RIT.

In experimental murine models of asthma, and to a lesser extent in humans with naturally developing asthma, monoclonal antibodies which block the effects of either cytokines or IgE, hold promise. Unfortunately, these antibodies are species-specific and not applicable for use in the cat as the cat would likely mount an antibody response to neutralize their activity. Inhibition of tyrosine kinases, a group of proteins which regulate cell survival, growth and differentiation, has more recently been of interest for treatment of asthmatic patients. Tyrosine kinase inhibitors are small molecule inhibitors which block the ATP binding sites of kinases. In asthma, the c-KIT receptor has been associated with proliferation and degranulation of mast cells and eosinophils in humans and mice and seems to be a logical target for therapeutic intervention. There are a number of commercially available tyrosine kinase inhibitors and one of these, masitinib (AB Science) has been evaluated in an experimental model of feline asthma.[5] After 4 weeks of daily masitinib therapy, eosinophilic airway inflammation was significantly reduced compared with placebo. Additionally, one measurement of airway hyperreactivity (plateau pressure) after bronchoprovocation with methacholine was significantly reduced in masinitib-treated cats. While none of the cats exhibited outward signs of illness and all physical examinations were normal, significant abnormalities, in particular proteinuria was detected with careful monitoring. Proteinuria was detected in all masitinib-treated cats and was often severe. Additional studies need to be performed to determine if a lower dose would be associated with similar efficacy in reduction of airway inflammation and hyperresponsiveness, with fewer adverse events. While this study is the first to show proof of concept of a tyrosine kinase inhibitor for feline asthma, it must be remembered that toxicity of small molecule inhibitors is a concern and the side effects must be carefully weighed against the potential benefits on airway inflammation and airflow limitation.

Other miscellaneous treatments have been tested for their abilities to reduce eosinophilic airway inflammation, airflow limitation or both in experimental models of feline asthma. Dietary omega-3 polyunsaturated fatty acids and luteolin (an antioxidant flavinoid) was administered to experimentally asthmatic cats for 4 weeks. [6] While these compounds supplemented in the diet showed no effect on airway inflammation, they did promote a decrease in airway reactivity when cats were tested by bronchoprovocation using barometric whole body plethysmography.[6] Key to interpreting this study is that although there was promise in diminishing airway hyperresponsiveness, the fact that it did not blunt eosinophilic airway inflammation (which ultimately contributes to worsening airway hyperresponsiveness and airway remodeling), makes it unsuitable for monotherapy to treat feline asthma. Another immunomodulator recently tested for allergic asthma was a salivary peptide known as feG-COOH. The mechanism by which this peptide works is incompletely described but it has shown promise in other species with allergic inflammation.[7] The initial study in experimentally asthmatic cats evaluated the effect of a single dose of feG-COOH <u>prior to</u> allergen challenge by aerosol and found it decreased airway eosinophilia. [8] However, it is important to realize that prior to translating this to pet cats that it is not generally possible to predict when a cat will be exposed to relevant allergen. Thus a second study was undertaken to evaluate chronic administration of feG-COOH (2 weeks, once daily). [9] Results of this study showed that in this setting feG was unable to diminish airway inflammation, making it unsuitable for monotherapy in pet cats with asthma. Effects on airway hyperresponsiveness were not evaluated and are a subject for future investigations. One final medication tested for its ability to reduce airflow limitation and hyperresponsiveness was inhaled lidocaine. A small pilot study suggested potential benefit for reducing airway hyperresponsiveness in a feline asthma model. [10]

Tenacious mucus secretions are also an important pathologic sequella to asthma in cats and may contribute to atelectasis of the lung secondary to mucus plugging. Inhaled N-acetylcysteine (NAC) is a mucolytic with antioxidant properties, which has been advocated for certain types of lung disease associated with mucus hypersecretion in humans. Inhaled NAC was also tested in experimental feline asthma and was found to increase airway resistance putting into question its safety in cats with preexisting airway disease. No studies to date have evaluated the effects of N-acetylcysteine administered either orally or by inhalation on airway inflammation or mucus quality in asthmatic cats.

REFERENCES

1. Reinero CR, et al. **Vet Immunol Immunopathol** 2006;110:141-153. 2. Reinero C, et al. **Vet Immunol Immunopath** 2008;121(3-4):241-250. 3. Lee-Fowler T, et al. **Vet Immunol Immunopathol** 2009;129:49-56. 4. Reinero C, et al. **27th Annual ACVIM Forum** 2009 (abstract). 5. Reinero C, et al. **Veterinary Comparative Respiratory Society** 2010 (abstract). 6. Leemans J, et al. **Vet J** In press. 7. Dery R, et al. **Eur J Immunol** 2004;34:3315-3325. 8. DeClue A, et al. **Inflamm Res** 2009;58:457-462. 9. Eberhardt J, et al. **Vet Immunol Immunopath** 2009;132:175-180. 10. Nafe L et al. **Veterinary Comparative Respiratory Society** 2010 (abstract).

INTERSTITIAL LUNG DISEASE IN SMALL ANIMALS

Elizabeth Rozanski, DVM, DACVIM (SAIM), DACVECC
North Grafton, MA

Interstitial lung disease (ILD) is recognized with increased frequency in people as well as dogs and cats. (Corcoran et al, Cohn et al) Most veterinary clinicians who are familiar with ILD tend to focus on idiopathic pulmonary fibrosis (IPF) which is also known as usual interstitial pneumonia (UIP) in people. An IPF-like syndrome has been recognized for almost 20 years in West Highland White terriers (WHWT) and other terrier breeds and more recently a similar syndrome was appreciated in cats. However, in human medicine, ILD is simply an umbrella term to describe a number of diseases with different pathogenesis and prognosis. It is very likely that dogs and cats have similar variants in their ILD, and solid understanding of potential differential diagnosis may help with clinical management. Also emerging in the human ILD arena is a movement away from using histopathology as the gold standard, and rather trying to use a combined "silver standard" of clinical, imaging AND histopathology. (Walsh and Hansell). This is of interest in animals due to the relative lack of appeal to lung biopsy by some clinicians and clients due to the costs and perceived risk to benefit ratio.

Standard classification of **idiopathic** ILD in people has included IPF and then 6 non-IPF subtypes; specifically acute interstitial pneumonia (AIP), respiratory bronchiolitis interstitial lung disease (RB-ILD), desquamative interstitial pneumonia (DIP), organizing pneumonia (OP), lymphoid interstitial pneumonia (LP) and non-specific interstitial pneumonia (NSIP). In people, the worst prognosis is associated with UIP and AIP, with more variable courses accompanying the other types of ILD. Of key importance, and currently *underinvestigated* in animals, is those ILDs that are not idiopathic. ILDs have been associated with drug-exposure/reactions, hypersensitivity pneumonitis (eg.pigeon fanciers' lung), sarcoidosis, and exposure to inhaled toxins such as silicon and asbestosis. Sarcoidosis is an inflammatory, possibly immune-mediated disease that while, systemic, is often most devastating in the lungs. Other immune-mediated disease, such as lupus, may also result in ILD. This is an important aspect of investigating people with ILD, as specific treatable underlying causes are associated with a much improved prognosis.

Reported ILDs in dogs has been primarily UIP (IPF) although there is a report of ILD associated with waterproofing spray, and with infectious disease (distemper virus and Leishmania). (Young et al, Goncalves et al, and Pandher et al) One cat, which was apparently currently of lymphoma with CCNU developed acute fatal respiratory distress due to pulmonary fibrosis. (Skorupski et al) Amiodorone toxicity in people is associated with fibrosis. (Schwaiblmair et al). One report exists of a dog with bronchiolitis obilterans with organizing pneumonia (BOOP). BOOP is also known as cryptogenic organizing pneumonia if it is idiopathic, versus a response to stimulus (inhaled toxin etc) and is form of interstitial lung disease. (Phillips et al). Endogenous lipid pneumonias including alveolar proteinosis have been described in dogs and cats, and may have similar clinical appearances to ILDS. (Jones et al) In horses, a fibrotic disease has been associated with herpes viral infection and one 30 year old blue fronted amazon parrot had chronic pulmonary fibrosis identified by lung biopsy. (Wong et al, Amann et al)

Diagnosis of ILD is typically suspected upon appropriate clinical signs (shortness of breath, cough, crackles), and radiographic evidence of IS infiltrates. Confirmation of a disease in dogs and cats requires biopsy at this point in time although the condition can be comfortably made by excluding other diseases, namely chronic bronchitis or Cushing's syndrome (tachypnea/exercise intolerance). In people with classical findings, IPF diagnosis may be made on CT scan. The role of biopsy in people is to exclude the more treatable forms of ILD, and to obtain tissue samples for research. CT scanning is beneficial in dogs and cats as well. Newer multidetector scan permit scans to be performed in seconds, and techniques to scan dogs and cats awake are being introduced (Dr. O'Brien-University of Illinois personal communication). CT scans are useful for excluding other diseases processes (eg. Bronchitis) and in evaluating the severity of the infiltrates/fibrotic changes. Classic IPF abnormalities include honeycombing, ground glass densities, subplueral foci/reticulation, and traction bronchiectasis. (See next page)

Lung function testing in people is considered essential to document a restrictive breathing pathology in patients with IPF. Lung function testing is more challenging in dogs and particularly cats due to their relative lack of cooperation. However, it is important to make at least simple assessments of lung function such as pulse oximetry and arterial blood gases. If possible, lung volumes can be measured using spirometry, and ideally, in research settings the diffusing capacity of carbon monoxide (DLCO) can be used to confirm diffusion impairment. Recently, in our practice we have employed the 6 minute walk test (6MWT) as objective marker of lung function. (Swimmer and Rozanski).

In people, and perhaps in WHWT, there is a familiar nature to IPF. It is unknown if cats show a genetic link to the disease, as the vast majority of cats in the US are of mixed or unknown parentage, with unknown sibling status. In people with familiar IPF, early screening is advised using CT scan, as well as solid understanding of symptoms. Clearly, due to the progressive nature of IPF, early detection might be useful in attempting to modulate the disease course, specifically with anti-fibrotic agents or potentially immunosuppressive agents.

Pulmonary hypertension (PHTN), defined as a systolic pulmonary arterial pressure of > 30 mmHg, is common in dogs and people with IPF. In people, it is defined as a Class III by the WHO classification scheme for pulmonary hypertension. Mild to moderate PHTH is well-tolerated in dogs but in advancing cases (> perhaps 60-75mmHg) that is accompanied by syncope or right heart failure (ascites primarily) treatment with sildenafil (2-3 mg/kg orally q 8 hours) is advised. Some veterinarians advise pimobendan as well. In dogs, pulmonary hypertension is most commonly detected on echocardiography, and in specific applying the modified Bernoulli equation to the tricuspid regurgitant jet. In cats, PHTN is harder to detect using standard Doppler techniques. In people, while echo is used frequently, definitive diagnosis of PHTN is made via right heart catheterization. This is pursued with some frequency as in people with PHTN associated with pulmonary fibrosis, there is a well-recognized risk of increased mortality. Right heart catheterization has not been pursued in dogs or cats with IPF. However, the therapeutic recommendations have not been established in randomized controlled trials. On-going trials are evaluating the addition of sildenafil, bosentan, or ambrisentan. The latter two ("sentans") are endothelin receptor antagonists. Increased endothelin activity has been documented in people and in abstract form in dogs with pulmonary fibrosis. The sentans have not been evaluated in companion animals to date; however, the Winn Foundation has funded a pilot study at the University of Missouri into the pharmacokinetics and safety of bosentan in cats, which should have results shortly. Of importance to veterinarians, bosentan comes with a "black box" warning, which means that use is associated with a significant risk of birth defects, and a reasonable risk of elevation in liver enzymes. Cost-wise, Bosentan is more pricey than sildenafil (at least in the US) so its role as long term agent may be difficult at this junction in time.

Pulmonary hypertension (PHTN), defined as a systolic pulmonary arterial pressure of > 30 mmHg, is common in dogs and people with IPF. In people, it is defined as a Class III by the WHO classification scheme for pulmonary hypertension. Mild to moderate PHTH is well-tolerated in dogs but in advancing cases (> perhaps 60-75mmHg) that is accompanied by syncope or right heart failure (ascites primarily) treatment with sildenafil (2-3 mg/kg orally q 8 hours) is advised. Some veterinarians advise pimobendan as well. In dogs, pulmonary hypertension is most commonly detected on echocardiography, and in specific applying the modified Bernoulli equation to the tricuspid regurgitant jet. In cats, PHTN is harder to detect using standard Doppler techniques. In people, while echo is used frequently, definitive diagnosis of PHTN is made via right heart catheterization. This is pursued with some frequency as in people with PHTN associated with pulmonary fibrosis, there is a well-recognized risk of increased mortality. Right heart catheterization has not been pursued in dogs or cats with IPF. However, the therapeutic recommendations have not been established in randomized controlled trials. On-going trials are evaluating the addition of sildenafil, bosentan, or ambrisentan. The latter two ("sentans") are endothelin receptor antagonists. Increased endothelin activity has been documented in people and in abstract form in dogs with pulmonary fibrosis. The sentans have not been evaluated in companion animals to date; however, the Winn Foundation has funded a pilot study at the University of Missouri into the pharmacokinetics and safety of bosentan in cats, which should have results shortly. Of importance to veterinarians, bosentan comes with a "black box" warning, which means that use is associated with a significant risk of birth defects, and a reasonable risk of elevation in liver enzymes. Cost-wise, Bosentan is more pricey than sildenafil (at least in the US) so its role as long term agent may be difficult at this junction in time. Pulmonary hypertension may also be associated with chronic (or acute) pulmonary thromboembolism, and therefore anticoagulant therapy may be advised although there are no studies on this is dogs or cats In people, anticoagulant therapy has been shown to decrease mortality associated with acute exacerbations. (Kubo et al) The appropriate anticoagulant medication in dogs is certainly unclear at this point in time with recent studies suggesting that aspirin maybe less efficacious than hoped. Clopidogrel and LMWH are options, but there are also multitudes of unknowns with these medications as well.

In people and in cats, there has been seen an association between lung tumors and pulmonary fibrosis, with IPF patients being at markedly increased risk of developing lung tumors. In the group of cats described, some cats with lung tumors were detected to have IPF. This becomes especially importantly clinically as cats (or dogs) that have a lung tumor, and are being evaluated for surgical resection should be carefully assessed for any respiratory signs other than cough, as these could indicate a relatively contraindication for surgery, or at the very least, a difficulty recovery.

To date, no drug therapy has been shown to reverse the inevitable progression. Early reports that suggested a good response to prednisone therapy are now thought to represent people with non-IPF ILD such as hypersensitivities. Current studies are looking to evaluate the effect of drugs on progression, and have documented some slowing of lung function decline with therapies such as N-acetylcysteine and improvement of quality of life with drugs to treat pulmonary hypertension. People with IPF are susceptible to exacerbations of disease or other pulmonary diseases such as pneumonia. In people, some are candidates for lung transplant.

Prognosis is considered guarded, most people die of IPF within 5 years of diagnosis. Dogs have variable survival times depending upon the age of onset, and the duration of signs prior to diagnosis. Additionally, people are commonly treated with supplemental oxygen, which is rarely used in pet dogs. Thus, most dogs die or are euthanized within 1 year of diagnosis; however some dogs have an apparent excellent response to therapy and may live much longer. The underlying cause of idiopathic ILD remains unknown. Alterations in natural reparative processes are considered, as are exposure to certain vital infections (Ebstein-Barr etc). A high percentage of people with IPF also have GERD, so some researchers feel that constant small volume aspiration may contribute to on-going lung injury. As earlier mentioned, genetics may be significant in some people and certainly in WHWTs. In people, and recently in dogs, mutations in the surfactant protein C gene have been described.(Eriksson et al) It has been recently suggested that IPF,results from multiple cycles of epithelial cell injury and activation during which , active alveolar epithelial cells provoke the migration, proliferation, and activation of mesenchymal cells with the formation of fibroblastic/myofibroblastic foci and the exaggerated accumulation of extracellular matrix. (Selman and Pardo). This would transition IPF from a disease for inflammation leading to fibrosis to injury leading to fibrosis, which could serve to alter therapeutic targets.

In the realm of translational and "one medicine", research into naturally occurring pulmonary fibrosis in dogs and cats is a unique opportunity to improve the lives of both people and pets. Current research models can create disease in rodents, using bleomycin, which shares some by not all of the characteristics of the human disease. Specifically, mesenchymal stem cells may represent a novel treatment opportunity in dogs and people although some evidence exists that bone-marrow derived stem cells may also propagate lung injury in fibrosis. (Gahraee-Kermani et al). Further directions include investigating early detection with the thought of trying to limit the progression of disease, and in better characterizing the subsets of disease that are important in dogs.

REFERENCES

Walsh and Hansell. **Eur Radio** 2010 20: 1859-1867. Young et al. **JAVMA** 2007 231:74-78. Goncalves et **al Res Inst Med Trop SaoPaulo** 2003 45: 153-158, Pandher et al. **J Vet Diag Invest** 2008, 18: 201-204. Phillips et al. **JVIM** 2000 14: 204-207, Jones et al. **JAVMA** 2000 216: 1437-1440, Schavaiblemair et al. **Clin**

Res Cardi 2010 99: 693-700, Skorupski et al Vet Comp Oncol 2008 6: 120-125, Corcoran et al. Vet Rec 1999 144: 611-616, Cohn et al JVIM 2004 18: 632-641, Norris et al Vet **Pathol** 2005 42:35-41, Swimmer et al. **JVIM** *in press* 2011, Eriksson et al, **JVIM** 2009 23: 1170-1174, Johnson et al **JSAP** 2005: 381-388. Gharee-Kermani et al **Pharm Research** 2007 24: 819-836. Selman et al. **PATS** 2006 364-372.

Evaluation for suspected interstitial lung disease

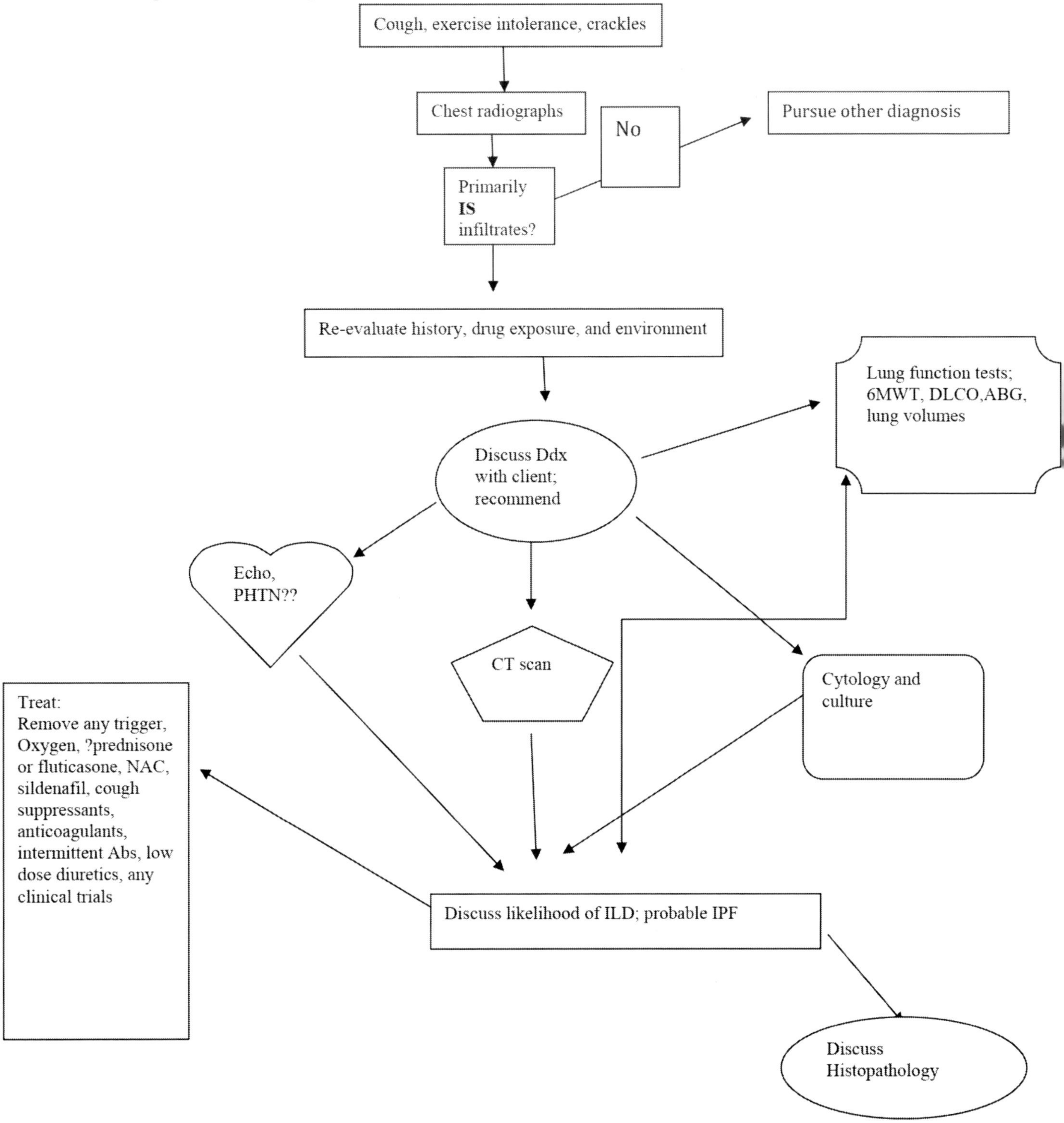

ENVIRONMENTAL TOXICANT-INDUCED NASAL AIRWAY INJURY

Stephan A. Carey, DVM, PhD, DACVIM
East Lansing, MI 48824

INTRODUCTION

The nose is a structurally and functionally complex organ in the upper respiratory tract. It is the primary site of entry for inhaled air in the respiratory system of mammals and therefore has many important and diverse functions. The nose not only serves as the principal organ for the sense of smell, but it also functions to efficiently filter, warm, and humidify the inhaled air before it enters the more delicate distal tracheobronchial airways and alveolar parenchyma of the lung [1]. The nasal passages have been described as an efficient "scrubbing tower" for the respiratory tract because it effectively absorbs water-soluble and reactive gases and vapors, traps inhaled particles, and metabolizes airborne xenobiotics [2]. With its role as an "air conditioner" and a "defender" of the lower respiratory tract, the nose may also be vulnerable to acute or chronic injury caused by exposure to high ambient concentrations of airborne toxic or infectious agents. A wide range of xenobiotics in the air could potentially cause nasal responses that would compromise respiratory health. This often underappreciated part of the mammalian upper airway has recently been the focus of reported research efforts in a variety of diverse scientific disciplines such as pharmacology, neuroscience, mathematical modeling, toxicology, and even paleontology. In this presentation, we present examples of epithelial lesions (e.g., cell death, hyperplasia, metaplasia) caused by single or repeated exposure to various inhaled environmental toxicants. This review provides a brief overview of comparative nasal structure, function and toxicologic pathology of the mammalian nasal epithelium and highlights the potential ramifications of upper airway injury on both upper and lower airway function.

NASAL STRUCTURE AND FUNCTION

Gross Anatomy of the Nose

The nasal airway is divided into two passages by the nasal septum. Each nasal passage extends from the nostrils to the nasopharynx. The nasopharynx is defined as the airway posterior to the termination of the nasal septum and proximal to the termination of the soft palate. Inhaled air flows through the nostril openings, or nares, into the vestibule, which is a slight dilatation just inside the nares and before the main chamber of the nose. Unlike the more distal main nasal chamber that is surrounded by bone, the nasal vestibule is surrounded primarily by more flexible cartilage. The luminal surface is lined by a squamous epithelium similar to that of external skin. In humans, unlike domestic animals, the nasal vestibule also contains varying numbers of hairs near the nares. After passing through the nasal vestibule, inhaled air courses through the narrowest part of the entire respiratory tract, the nasal valve (ostium internum), into the main nasal chamber.

Though there are some general similarities in the nasal passages of most mammalian species, there are also striking interspecies differences in nasal architecture. From a comparative viewpoint, humans have relatively simple noses with breathing as the primary function (microsmatic), while most domestic and laboratory animals have more complex noses with olfaction as the primary function (macrosmatic). In addition, the nasal and oral cavities of humans, some nonhuman primates, and domestic animals are arranged in a manner to allow for both nasal and oral breathing. Most laboratory rodents used in inhalation toxicology studies (e.g., rats, mice, hamsters, guinea pigs) are obligate nose breathers, due to the close apposition of the epiglottis to the soft palate. The human nose has three turbinates: the superior, middle, and inferior. These structures are relatively simple in shape compared to turbinates in most non-primate laboratory animal species (e.g., dog, rat, mouse, rabbit) that have complex folding and branching patterns. In the proximal nasal airway, the complex maxilloturbinates of dogs, cats, rabbits, and small laboratory rodents provide far better protection of the lower respiratory tract, by better filtration, absorption, and disposal of airborne particles and gases, than do the simple middle and inferior turbinates of the human nose. The highly complex shape of the ethmoturbinates, lined predominantly by olfactory neuroepithelium, in the distal half of the nasal cavity of non-primate animals is suitably designed for acute olfaction.

The Nasal Mucosa and Mucociliary Apparatus

Most of the luminal surfaces of the nasal mucosa (with the exception of the most proximal regions of the nasal vestibule) are covered by mucus. Its physical and chemical properties are well suited for its role as an upper airway defense mechanism, filtering the inhaled air by trapping inhaled particles and certain gases or vapors. The mucus is produced by mucous (goblet) cells in the surface epithelium and subepithelial glands in the lamina propria. The synchronized beating of surface cilia propels the mucus at different speeds and directions depending on the intranasal location. Mucus covering the olfactory mucosa moves very slowly, with a turnover time of several days. In contrast, the mucus covering the transitional and respiratory epithelium is driven along rapidly (up to 30 mm/min) by the surface cilia with an estimated turnover time of about 10 minutes in the rat [3]. The mucus with the entrapped materials ultimately is propelled by the beating cilia to the naso- and oro-pharynx, and then is swallowed into the esophagus and cleared through the digestive tract. The nasal mucociliary apparatus (i.e., mucus and cilia) exhibits a range of responses to inhaled xenobiotic agents and can be a sensitive indicator of toxicity [3]. Since this upper airway apparatus is one of the first lines of defense against inhaled pathogens, dusts, and irritant gases, toxicant-induced compromises in its defense capabilities could lead to increased nasal infections and increased susceptibility to lower respiratory tract diseases [4].

Cellular Composition of Nasal Surface Epithelium

There are 5 distinct epithelial populations in the nasal airways of domestic and laboratory animals. These include the squamous epithelium (SE), ciliated respiratory epithelium (RE), non-ciliated transitional epithelium (NTE), olfactory epithelium (OE), and lymphoepithelium (LE). While the intranasal localization and distribution of these epithelial types differ significantly across species, the cellular composition of each epithelial type is relatively conserved among mammals.

Squamous Epithelium. The nasal vestibule is lined with keratinized, stratified squamous epithelium. This nasal epithelial type is composed of basal cells lining the basal lamina, and several layers of squamous epithelial cells, which become progressively flatter toward the luminal surface of the epithelium. The epithelium of the vestibule is contiguous to the epidermis of the skin, and both serve to protect the underlying tissues from exposure to potentially harmful agents in ambient air [5].

Transitional Epithelium. Adjacent to the stratified SE of the nasal vestibule is a zone of non-ciliated transitional epithelium (NTE). This epithelium lines the zone between the squamous epithelium and ciliated respiratory epithelium. In primates, this epithelium is stratified (4–5 cell layers thick), non-keratinized, and consists primarily of cuboidal and columnar epithelial cells with a paucity of mucous goblet cells. In contrast, the NTE of rodents is pseudostratified (1-2 cells in thickness), and is composed primarily of basal cells and cuboidal cells [6].

Respiratory Epithelium. Most of the main nasal chamber in primates is lined with ciliated respiratory epithelium. This pseudostratified to columnar epithelium is comprised primarily of ciliated cells, with lesser numbers of narrow non-ciliated (serous) cells, various numbers of mucous (goblet) cells, and basal cells [6], and is the major contributor to nasal mucociliary clearance.

Olfactory Epithelium. The OE covers a small region of the dorsal meatus and the dorsal half of the ethmoturbinate in primates. The 3 principal cell types that compose the OE are the olfactory sensory neurons, sustentacular cells, and basal cells. The supporting, sustentacular cells form a prominent apical row of nuclei. The olfactory sensory neurons are bipolar neuronal cells interposed between the sustentacular cells. The dendritic portions of these neurons extend above the epithelial surface and terminate into a bulbous olfactory knob, from which extend 10–15 immotile cilia per cell. Basal cells line the basal lamina. OE covers a greater percentage of the nasal cavity of rodents than of primates. Approximately 50% of the nasal cavity surface in F344 rats in lined by OE [6], compared to ~15% in monkeys [7 , 8], and ~3% in people [9].

Lymphoepithelium. Lymphoepithelium is a specialized epithelium, found primarily in the walls of the nasopharynx. This epithelium covers discrete, focal aggregates of nasal-associated lymphoid tissue (NALT) in the underlying lamina propria. The lymphoepithelium is composed of cuboidal ciliated cells, few mucous cells, and numerous non-ciliated, cuboidal cells with luminal microvilli, called M-cells. These latter cells are similar in structure and function to the analogous immune surveillance cells found in gut- and bronchus-associated lymphoid tissues (GALT and BALT, respectively).

NASAL EPITHELIAL INJURY AND REMODELING INDUCED BY INHALED TOXICANTS

Humans and laboratory animals are susceptible to nasal airway injury from a wide range of xenobiotic agents which can induce nasal lesions when delivered either by inhalation [10-13] or non-inhalation [14, 15] routes of exposure. These nasal lesions include both non-neoplastic (e.g., inflammation, epithelial cell necrosis, epithelial hyperplasia/metaplasia) and neoplastic (e.g., squamous cell carcinoma) changes. The nasal response to ozone exposure will be discussed as an example, along with the possible factors involved in the pathogenesis of these morphologic alterations. Previous review articles from our laboratory describe in more detail the histologic characteristics of various other environmental toxicant-induced nasal lesions[5, 6].

Ozone, the principal oxidant air pollutant in photochemical smog, causes tissue injury to the nasal airways in adults and children [16, 17]. The ciliated respiratory epithelium (RE) and non-ciliated transitional epithelium (NTE) in the anterior nasal airways are particularly sensitive to ozone toxicity in people and in animal models [7, 18-20], while other epithelial types remain relatively unaffected. These two epithelial populations are important components of the nasal mucociliary clearance apparatus, and injury to these sites could lead to alterations in the nasal functions that serve to protect both the upper and lower airways. Several studies from our laboratory have characterized the nature and persistence of ozone-induced nasal airway injury using laboratory animal models. Monkeys exposed short-term (6 days) to ambient concentrations (0.15 ppm) of ozone develop acute rhinitis, epithelial hyperplasia, ciliated cell loss, and mucous cell metaplasia in the NTE and the RE lining the anterior nasal airways. Long-term, daily ozone exposure (0.15 or 0.30 ppm, 8h/day for 90 days) enhances this increase in intraepithelial mucus secretory capacity compared to short-term exposure. Interestingly the nasal inflammatory and proliferative responses to acute, short-term, ozone exposure initially decrease as the mucous cell metaplasia develops during chronic exposure. These morphologic alterations may represent an adaptive response that serves to protect the nasal airways from ongoing ozone-induced injury. In contrast to adults, infant monkeys exposed to repeating cycles of ozone for two months or five months develop persistent rhinitis and nasal epithelial necrosis, and do not develop epithelial hyperplasia or mucous cell metaplasia [7, 21], raising the possibility that the adaptive response observed in adult nasal airways may be absent during post-natal development of the respiratory system, and suggesting that the nature (daily vs. cyclic) and timing (adult vs. infant) of exposure may influence the ability of the nasal epithelium to undergo repair, remodeling, and adaptation.

ASSESSMENT OF HUMAN RISK FROM INHALED NASAL TOXICANTS USING ANIMAL DATA

The estimation of human risk from inhaled toxicants is a complex process that often relies on extrapolations from toxicology studies using laboratory animal species such as rats, dogs, and monkeys. Essential to this process is the generation of mechanistic data that will help elucidate quantitative linkages between exposure, dose, and response. An understanding of the specific tissue/cell susceptibility and the underlying cellular/biochemical mechanisms responsible for toxicant-induced nasal injury is also critical for accurate risk assessment. It is important to determine the similarities and differences in dose/response of morphologically similar nasal tissues/cells between the selected animal species and humans. Little information is currently available in this area and this must be one of the focus areas for nasal toxicology research in the future.

REFERENCES

[1] Cole, P. 1992, St. Louis, MO: **Mosby-Year Book**. [2] Brain, J.D. **Annals of Otology, Rhinology & Laryngology**, 1970. 79(3): p. 529. [3] Morgan, K.T., et al. **Am Rev Respir Dis**, 1984. 130(2): p. 275. [4] Ciencewicki, J. and I. Jaspers. **Inhalation Toxicology**, 2007. 19(14): p. 1135. [5] Harkema, J.R., S.A. Carey, and J.G. Wagner. **Toxicol Pathol**, 2006. 34(3): p. 252. [6] Harkema, J.R. **Environ Health Perspect**, 1990. 85: p. 231. [7] Carey, S.A., et al. **Toxicologic Pathology**, 2007. 35(1): p. 27. [8] Gross, E.A., et al. in ***Toxicologist***. 1987. [9] Sorokin, S., ***The Respiratory System***, L. Weiss, Editor. 1988, Urban and Schwarzenberg: Baltimore, MD. [10] Buckley, L.A., et al. **Toxicology and Applied Pharmacology**, 1984. 74(3): p. 417. [11] Harkema, J.R., J.A. Hotchkiss, and R.F. Henderson. **Toxicol Pathol**, 1989. 17(3): p. 525. [12] Carey, S.A., et al. in ***Proceedings of the American Thoracic Society***. 2008. Toronto, ON, CA. [13] Jiang, X.Z., K.T. Morgan, and R.W. Beauchamp, ***Histopathology of acute and subacute nasal toxicity***., C.S. Barrow, Editor. 1986, Hemisphere: New York. p. 51. [14] Genter, M.B., et al. **Journal of Pharmacology and Experimental Therapeutics**, 1992. 263(3): p. 1432. [15] BRITTEBO, E.B., et al. **Toxicological Sciences**, 1991. 17(1): p. 92. [16] Graham, D., F. Henderson, and D. House. **Arch Environ Health**, 1988. 43(3): p. 228. [17] Koren, H.S., G.E. Hatch, and D.E. Graham. **Toxicology**, 1990. 60(1-2): p. 15. [18] Calderon-Garciduenas, L., et al. **Environmental Health Perspectives**, 1994. 102(12): p. 1074. [19] Harkema, J.R., et al. **Am J Pathol**, 1987. 127(1): p. 90. [20] Harkema, J.R., et al. **Am J Pathol**, 1987. 128(1): p. 29. [21] Carey, S.A., et al. (2010). **American Journal of Physiology** - Lung Cellular and Molecular Physiology, [Epub ahead of print] DOI: 10.1152/ajplung.00177.2010.

HISTOPATHOLOGIC AND MORPHOMETRIC EVALUATION OF THE NASAL AIRWAYS OF CATS WITH EXPERIMENTALLY INDUCED ASTHMA

Dr. Christine M. Venema, Dr. Kurt J. Williams, Dr. Laurel J. Gershwin, Dr. Carol R. Reinero, Dr. Stephan A. Carey
East Lansing, MI

INTRODUCTION

Allergic rhinitis and asthma are often seen concurrently in human patients.[1,2,3] In human allergic airways medicine, the "unified airways theory" has been advanced to explain the coincidence of allergic rhinitis and asthma documented frequently in people.[4] This theory suggests that the upper airways are immunologically linked to the lower airways.[3,4,5,6] Epidemiologic evidence suggests that allergic rhinitis and asthma are comorbid conditions in many human patients.[3,7,8] Histopathologically, this finding is supported by demonstration of eosinophilic inflammation in both the nasal and pulmonary airways regardless of clinical signs.[9,10,11,12] Moreover, treatment of rhinitis reduces functional and clinical parameters of asthma in human patients.[13,14,15,16,17] Also proposed as part of the "unified airways theory" is the idea that different clinical manifestations of global airways disease predominate during particular periods of life while other regress.[5,18] Allergic rhinitis and asthma may represent the respiratory manifestations of a generalized atopic disease process that begins with atopic dermatitis and progresses over time. This concept has been referred to as the "atopic march."[6,19,20]

A similar relationship between the nasal and pulmonary airways of cats has not been evaluated. To investigate the possibility of a "unified airways theory" in cats, we have developed an experimental model of feline asthma that replicates the functional, clinical, immunologic, and histopathologic features of the spontaneously occurring disease.[21] We hypothesized that sensitizing and chronically challenging cats with a relevant aeroallergen would induce morphologic and histopathologic changes in the nasal airways that mimic the changes induced in the bronchial airways. Five mixed breed cats were sensitized to and challenged with Bermuda grass allergen (BGA) while four cats were instead treated with phosphate buffered saline (PBS) and served as controls. The nasal and bronchial tissues of these cats were harvested and analyzed with morphometry to measure the numeric cell densities (NCD) of mucosal inflammatory cells and with image analysis to determine the volume density of mucosubstances within the nasal epithelium. Determining whether the upper and lower airways of asthmatic cats respond to aeroallergen challenge similarly will expand our understanding of the immunopathogenesis of feline allergic airways disease. This information may lead to the development of novel therapies for conditions such as feline bronchial asthma and chronic rhinosinusitis, which can be challenging to manage effectively.

RESULTS

Histopathologic and morphometric techniques were applied to evaluate the nasal and bronchial airways of both experimentally sensitized and control cats. Specifically, the non-ciliated transitional epithelium and the respiratory epithelium lining the dorsal nasal septum and alar folds of the anterior nasal cavity, the ventral nasal septum of the mid-nasal cavity, and the dorsal nasopharyngeal meatus of the caudal nasal cavity were examined. In cats experimentally sensitized to BGA, the degree of eosinophilic inflammation in the anterior region of the nasal cavity was 3.5-fold greater compared to control cats. In the other regions of the nasal airways, no significant difference in the eosinophil NCD was detected between groups. In both BGA-sensitized and control cats, the degree of mucosal and submucosal mast cell infiltration was more pronounced than the degree of eosinophilic inflammation in all regions of the nasal airways. Compared to cats treated with PBS, the NCD of nasal mucosal mast cells was 2.2-fold greater in cats treated with BGA. This difference was observed in all regions of the nasal cavity, and the degree of mast cell infiltration progressively increased in an anterior to posterior direction in both groups. Intraepithelial mucosubstances were observed in all regions of the nasal airways of cats in both groups. However, the volume density (Vs) of mucosubstances within the epithelium did not differ significantly between groups in any region of the nasal cavity. Analysis of lung axial airway mucosa revealed eosinophilic infiltration in BGA-sensitized cats that was 2-fold greater than that observed in control cats. The mast cell NCD of the pulmonary airways was not significantly different between BGA-sensitized and control cats. In artificially sensitized cats, a significant correlation was not detected between the NCD of eosinophils in the mucosa of the lung axial airways and the mucosa of the anterior nasal cavity, mid-nasal cavity, and caudal nasal cavity. Also, a significant correlation was not noted between the mucosal mast cell NCD of the pulmonary airways and the nasal airways in BGA-sensitized cats.

CONCLUSIONS

Based on these results, we conclude that sensitization and challenge of cats with a relevant aeroallergen elicit nasal manifestations that do not mimic the changes incited in the pulmonary axial airways. Specifically, the nasal alterations are predominated by mucosal and submucosal mast cell infiltration that is distributed throughout the nasal airways. These changes also include eosinophilic inflammation that is localized to the anterior nasal cavity. Furthermore, mast cell and eosinophilic inflammation is heterogeneously distributed throughout the upper and lower airways of artificially sensitized cats.

Our findings are similar to those documented in human patients with allergic airways disease. In particular, concomitant eosinophilic rhinitis has been demonstrated in human asthmatics.[9,22] In the experimental models of feline asthma of our study, eosinophilic inflammation was focally distributed in the anterior region of the nasal airways, which is similar to the heterogenous distribution of eosinophilic inflammation noted in biopsy samples obtained from humans with naturally occurring chronic rhinosinusitis.[23] Additionally, the results of our study demonstrate that mast cell rhinitis occurs in cats following artificial aeroallergen sensitization and chronic challenge. A significant mast cell infiltrate in the nasal mucosa also characterizes allergic rhinitis in humans.[24,25,26] As in the feline asthma models of our study, a heterogenous distribution of eosinophilic and mast cell inflammation throughout the upper and lower airways has been demonstrated in humans with allergic airways disease.[9,12,22]

This study represents the first characterization of the histopathologic and immunologic alterations of the nasal airways of experimental models of feline asthma and the first investigation into a potential mechanistic link between the upper and lower airways of asthmatic cats. These results provide a basis for future studies to investigate the pathophysiology of feline rhinitis and asthma. Our results suggest that concurrent inflammation within the nasal airways should be considered in the evaluation of asthmatic cats.

REFERENCES

[1]Sibbald B, et al. **Thorax** 1991;46:901. [2]Neukirch F, et al. **Resp Med** 1995;89:685. [3]Bousquet J. **Clin Exp All Rev** 2003;3:43. [4]Vignola A, et al. **Allergy** 1998;53:833. [5]Togias A. **J Allergy Clin Immunol** 2003;111:1171. [6]Casale T, et al. **Am J Med Sci** 2004;327:127. [7]Greisner W, et al. **Allergy Asthma Proc** 1998;19:185. [8]Linneberg A, et al. **Allergy** 2002;57:1048. [9]Gaga M, et al. **Clin Exp Asthma** 2000;30:663. [10]Braunstahl G, et al. **Am J Respir Crit Care Med** 2000;161:2051. [11]Braunstahl G, et al. **J Allergy Clin Immunol** 2001;107:469. [12]Brauhnstahl G, et al. **Clin Exp Allergy** 2003;22:579. [13]Reed C, et al. **J Allergy Clin Immunol** 1988;81:1042. [14]Corren J, et al. **J Allergy Clin Immunol** 1992;90:250. [15]Watson W, et al. **Allergy Clin Immunol** 1993;91:97. [16]Foresi A, et al. **J Allergy Clin Immunol** 1996;98:274. [17]Pedersen W, et al. **Allergy**1998;53:383. [18]Togias A. **Allergy** 1999;54:S55. [19]Gustafson D, et al. **Allergy** 2000;55:240. [20]Spergel J, et al. **J Allergy Clin Immunol** 2003;112:118. [21]Norris Reinero C, et al. **Int Arch All Immunol** 2004;135:117. [22]Chanez P, et al. **Am J Respir Crit Care Med** 1999;159:588. [23]Ponikau J, et al. **J Allergy Clin Immunol** 2003;112:877. [24]Bentley A, et al. **J Allergy Clin Immunol** 1992;89:877. [25]Amin K, et al. **J Allergy Clin Immunol** 2001;107:249. [26]Liu F, et al. **Allergy** 2010;65:1149.

THE QUANTITATIVE ASSESSMENT OF ACID-BASE BALANCE IN THE DOG

Steve C. Haskins, DVM, DACVA, DACVECC
Davis, CA

Hydrogen ions are highly reactive with anionic regions of macromolecules like proteins. This interaction changes charge distribution within the protein resulting in a change in conformation and function of the protein. Changing the catalytic activity of protein enzymes alters many physiologic cell functions such as mentation, hemoglobin affinity for oxygen, myocardial contractility and other skeletal muscle function, vasomotor tone and other smooth muscle function, coagulation, inflammation, digestion, hepatic metabolism, and renal excretion. All aspects of cell function are dependent upon a fairly normal hydrogen ion concentration.

The pH is a convenient negative logarithmic expression of the hydrogen ion activity, which compresses a wide range of hydrogen ion concentrations into a scale which is much easier to use. In addition, the physiologic effects of changes in pH are more linearly related to pH than to hydrogen ion concentration.

Normal plasma pH is about 7.40 ± 0.5 units. Values < 7.35 represent an acidemia while values > 7.45 represent an alkalemia. The pH is the net balance of all of the acids (hydrogen ion donors) and bases (hydrogen ion acceptors) in the body. The pH is regulated by a respiratory component (carbonic acid) and a metabolic component (all of the other acids and bases).

The Respiratory Component of the Acid-Base Balance

Carbonic acid (H_2CO_3) is in a 2-way equilibrium with carbon dioxide and water, and with hydrogen and bicarbonate: $CO_2 + H_2O \leftrightarrow H_2CO_3 \leftrightarrow H^+ + HCO_3^-$. All reactions have a proportional equilibrium (which is dependent upon temperature and the environmental pH relative to the pK of the reaction). The function of the carbonic acid equilibrium is very important. When an acid is added to the extracellular fluid compartment (ECF), the hydrogen will combine with bicarbonate to form carbonic acid. If carbonic acid were allowed to accumulate, the bicarbonate buffer system would be severely limited. But carbonic acid does not accumulate because it dissociates to carbon dioxide which is eliminated by ventilation. The "open-ended" nature of this buffer system vastly augments its quantitative importance. An increase in carbon dioxide increases the formation of carbonic acid, which, in turn, dissociates to hydrogen and bicarbonate but in a 1:1 ratio. This is very acidifying since the normal $H^+:HCO_3^-$ ratio (at 7.4) is about 0.00000167:1.

The respiratory contribution to the acid-base balance is defined by the $PaCO_2$. The normal $PaCO_2$ is about 40 ± 5 mmHg (cats may be slightly lower). Values < 35 mmHg represent a respiratory alkalosis (increased alveolar minute ventilation) while values > 45 mm Hg represent a respiratory acidosis (decreased alveolar minute ventilation).

When arterial blood is not available, venous blood may be used for PCO_2 measurements. PvO_2 values are normally 3-5 mmHg higher than $PaCO_2$. The arterial-venous gradient is not a fixed value and can increase in common disease states such as hypovolemia and anemia. $PvCO_2$ is actually a marker of tissue PCO_2, which in turn is the balance between $PaCO_2$, metabolic CO_2 production, and tissue perfusion. Changes in $PvCO_2$ will lag behind changes in $PaCO_2$ in transition states (e.g. acute hyperventilation following prolonged hypoventilation).

End-tidal PCO_2 ($PetCO_2$) may be used as a surrogate marker of $PaCO_2$. Normally $PetCO_2$ is only about 3-5 mmHg below $PaCO_2$ (some lung regions have high ventilation to perfusion ratios). $PetCO_2$ values are disproportionately lower than $PaCO_2$ when there is increased alveolar dead space ventilation (hypovolemia, pulmonary thromboembolism, PPV with large tidal volumes) and tachypnea when there is mixing of anatomical dead space gases and functional alveolar space gases.

The Metabolic Component of the Acid-Base Balance

Overview markers

In contrast to the respiratory component, the metabolic component includes many acids, and has many markers. This complexity has spawned a number of different ways by which its contribution to the acid-base balance can be characterized.

<u>Bicarbonate</u> is the time-honored overview marker of the metabolic contribution to acid-base balance. The hydrogen ion of any added acid will bind with bicarbonate (the resultant carbonic acid will be eliminated by alveolar ventilation as carbon dioxide) and the bicarbonate concentration will decrease proportionately. Dogs normally have a bicarbonate concentration in the range of 20-24 mEq/L; cats are slightly lower. Values below these respective ranges represent hypobicarbonatemia and mark a metabolic acidosis; values above these ranges represent hyperbicarbonatemia and mark a metabolic alkalosis. In an analyzer, pH and PCO_2 are measured and bicarbonate concentration is calculated via the Henderson-Hasselbach equilibration: $pH = -\log(HCO_3/H_2CO_3)$. There will be no error in the bicarbonate calculation; there can be only one bicarbonate concentration for the measured pH and PCO_2 values. If there were errors in sampling, storage, or measurement, there may be unrepresentative pH or PCO_2 measurements.

<u>Standard bicarbonate</u> is a mathematically adjusted bicarbonate concentration (to a PCO_2 of 40 mmHg) to eliminate the effect of changes in PCO_2 on bicarbonate concentration). Carbon dioxide and bicarbonate are both part of the carbonic acid – bicarbonate equilibration equation. A primary change in bicarbonate *does not* cause a change in arterial PCO_2 *in vivo* because $PaCO_2$ is monitored and controlled by medullary chemoreceptors. Changes in PCO_2 *do*, however, cause small changes in bicarbonate (approximately 0.15 mEq/L per mmHg decrease in PCO_2 below 40 and 0.075 mEq/L per mmHg increase in PCO_2 above 40).[1] If both standard bicarbonate and an (undesignated) bicarbonate concentration are available on the printout from the analyzer, the standard bicarbonate would be the more accurate representation of the metabolic contribution to the acid-base balance.

<u>Total CO_2</u> concentration can be easily measured and is a common component of commercial chemistry panels. Carbon dioxide exists in many forms in the plasma, almost all of which is bicarbonate. Very small additional amounts exist in the form of

carbonic acid, dissolved CO_2, and as carbamino groups on various proteins. Total CO_2 concentration is normally only about 1 mEq/L above bicarbonate concentration and total CO_2 concentration can be interpreted as if it were bicarbonate concentration. Total CO_2 has nothing to do with the partial pressure of CO_2 and total CO_2 should not be considered to be a surrogate marker of $PaCO_2$.

Base deficit/excess values are usually considered to be the most accurate index of the metabolic contribution to acid-base balance. Alignment nomograms[1] were originally developed to hand-determine the base deficit/excess bicarbonate and total CO_2 concentrations from the measured pH and PCO_2. Nowadays these values are calculated by the blood gas analyzer and are displayed and printed out for the user. There are many other buffers in blood in addition to the carbonic acid – bicarbonate buffer system. In the body, there are also intracellular and bone buffers. The quantitative impact of all of the buffer systems in whole blood has been determined by titrational experiments. A deficit of base (a base deficit) indicates the presence of a metabolic acidosis, while an excess of base (a base excess) indicates a metabolic alkalosis. Actual base excess (ABE) describes the *in vitro* buffer curves of whole blood from the original experiments. Standard base excess (SBE or BE_{ECF}) describes an estimate of the *in vivo* buffer curve incorporating the dilution effect of the interstitial fluid compartment. There are also different terminologies in the industry of which the user should be aware: base deficit/excess is often referred to as SBE or BE_{ECF}. Sometimes the term "negative base excess" is used instead of the term "base deficit". The normal base deficit/excess ranges between 0 and -4 mEq/L for the dog and between -4 and -8 mEq/L for the cat; more negative values mark a metabolic acidosis; more positive values mark a metabolic alkalosis.

The anion gap is usually calculated by the equation: $Na + K - Cl - HCO_3$. There can be no anion gap in reality (the number of cations always equals the number of anions), but not all of them are routinely measured. In this calculation, Na + K normally exceeds $Cl + HCO_3$ by 15 to 20 mEq/L. Normally the negative charges on albumin comprise most of this apparent gap. Phosphorous and lactate make up a small portion of the gap in normal animals, but this can increase dramatically in disease states. An increase in anion gap is usually considered to mark the presence of an accumulation of unmeasured anions, most of which represent anions of organic acids such as lactic acid, ketoacids, phosphoric and sulfuric acids, or acid intoxicants such as glycolic acid from ethylene glycol, salicylic acid from salicylate, formic acid from methanol, or various acidic amino acids from rhabdomyolysis).

The anion gap calculation is difficult to interpret and rarely adds meaningful information to the patient assessment. First, by the time one has the measurements needed to calculate the anion gap, there is usually enough information (historical, physical, or laboratory) with which to define the nature of the metabolic acidosis. Second, animals rarely have a single-source metabolic acidosis from a singular source; most acid-base derangements represent a variable combination of acidotic and alkalotic processes that culminate in the measured acid-base abnormality. This may have the net effect of increasing or decreasing the calculated anion gap. Hyperalbuminemia can increase in anion gap but is associated with a metabolic acidosis. Metabolic acidosis can also be caused by renal and gastrointestinal bicarbonate losses or hydrogen retention, without an increase in anion gap. An increase anion gap simply marks the presence of unmeasured anions; a decreased value surely marks a hypoproteinemia.

Albumin, phosphorous, and lactate are commonly measured and their contributions to the anion gap can be easily inserted into an expanded anion gap formula: Na + K + Ca effect + Mg effect – Cl – HCO3 – lactate – albumin effect – phosphate effect (calcium, magnesium, albumin, and phosphorous are reported in different unitages, and the equivalent concentrations in mEq/L must be calculated. Normal values for the expanded anion gap range between about 4 and 8 mEq/L. This residual anion gap is composed of the remaining unmeasured anions such as sulfates, ketones, and various amino acids. The expanded anion gap formula moves many of the relevant players from an unknown to a known effect which allows for a more comprehensive evaluation of the acid-base status.

$AG_{expanded}$ = Na + K + Ca effect + Mg effect – Cl – HCO_3 – lactate – albumin effect – phosphate effect
For Na, K, Cl, HCO_3, and lactate, use the measured value (mEq or mM/L).
For Ca and Mg, use the ionized calcium and magnesium measurement (in mM/L) x2
For albumin effect, multiply the measured albumin (in g/dl) x 2.7; or (in g/L) x 0.27
For phosphate effect, multiply the measured phosphorous (in mg/dl) x 0.58; or (in mM/L) x 1.8

Specific markers

In lactic acidosis, lactate accumulates in the plasma as the hydrogen ion is buffered by bicarbonate; the magnitude of the lactate increase is similar to the decrease in bicarbonate (if there were no other influences on the bicarbonate concentration). The lactate concentration is a marker of the quantity of lactic acid that has been added to the ECF. Normal animals generally have a lactate concentration below one mEq/L although the normal range is often considered to extend up to 2 mEq/L. It is not the lactate that causes the acidosis, it is the hydrogen that came with it. Sodium lactate (such as in lactated Ringer's or Hartmann's solution) is a salt, not an acid, and will have no impact, per se, on acid-base balance (see following manuscript). If a blood sample is contaminated with sodium lactate, the measured lactate may be very high but this is not associated with an acidifying effect.

Acetoacetate can be measured in the urine by nitroprusside reaction on most urine reagent strips. This color-indexed assessment is semiquantitative. Plasma β-hydroxybutyric acid can be measured in the laboratory or with hand-held instrumentation by NAD^+/NADH a reduction/oxidation reaction. Plasma ketones are normally zero. The units of measure are mM/L which has a 1:1 relationship with the mEq/L contribution to the acid base balance. In ketoacidosis, β-hydroxybutyrate concentrations usually greatly exceed those of acetoacetate.

Plasma inorganic phosphate is commonly measured and reported as elemental phosphorous in mg/dl or mM/L in serum chemistry testing profiles. Plasma phosphate anion is also associated with other plasma cations such as sodium, potassium, calcium, magnesium, and, of importance to the present discussion, with hydrogen as phosphoric acid. Phosphoric acid (H_3PO_4) could potentially exist as H_3PO_4, $H_2PO_4^{1-}$, HPO_4^{2-} or PO_4^{3-}, depending upon the pH of the fluid. At a pH of 7.4 approximately 90% of phosphoric acid exists as HPO_4^{2-} and 10% exists as $H_2PO_4^{1-}$. The hyperphosphatemia of renal disease is associated with a phosphoric acidosis. A co-existent sulfuric acidosis (another end-product of protein metabolism) would be expected, but is not measured. The magnitude of the phosphoric acidosis contribution to the metabolic component can be calculated. To convert from mg/dl to mM/L: mg/dl x 1mM/31mg x 10dl/L (= phosphorous in mg/dl x 0.323). To convert mM/L to mEq/L: phosphorous in mM/L x 1.8 (90% HPO_4^{2-}; 10% $H_2PO_4^{1-}$). The change in phosphorous (when reported in mg/dl) from normal (4 mg/dl) x 0.58 = phosphorous effect (mEq/L). The change in phosphorous (when reported in mM/L) from normal (1.3 mM/L) x 1.8 = phosphorous effect (mEq/L).

Albumin is a multivalent anion that is associated with many plasma cations: sodium, potassium, calcium, magnesium, and, of concern to this discussion, hydrogen. Hypoalbuminemia is associated with a decrease in the "HAlb" acid resulting in a metabolic alkalosis, while hyperalbuminemia is associated with an acidosis. Changes in globulin have little effect on acid base balance. The impact of changes in albumin on acid base balance can be calculated: 1 g/dl (10 g/L) change in albumin concentration x 2.7 mEq/L of BD/E effect.

Water dissociates to a small extent, in a temperature dependent manner, into H^+ and OH^-. Water has a pK of 7.00 at room temperature; and 6.80 at body temperature. Compared to the body at a pH near 7.40, water is acidic. This acidic effect is magnified *in vivo* because once the water is equilibrated with existing levels of carbon dioxide it has considerable titratable acidity via the carbonic acid equilibrium. The concentration of free water is not measured but the consequences of such changes are reflected in the sodium concentration. Changes in sodium concentration, per se, does not have an acid-base effect; it is simply used to calculate the water effect: $(Na_{measured} - Na_{normal})$ x 0.25. Na_{normal} is often considered to be about 145 mEq/L (mM/L) in the dog and 155 in the cat. An increase in free water (marked by a decrease in sodium concentration) is associated with an acidifying effect; a decrease in free water is associated with an alkalinizing effect.

The chloride-bicarbonate effect is a little complicated because bicarbonate itself is a complex and tricky anion (it is involved with many other buffer systems and appear or disappear in the name of carbon dioxide). Although this is sometimes referred to as the "chloride effect", it is not the chloride, per se, that is having the acid-base effect. Chloride is rather used to mark the change in bicarbonate that would have occurred if bicarbonate was not otherwise so promiscuous and ethereal. Many processes in the body involve the reciprocal handling of chloride and bicarbonate (i.e. when one is lost, the other is retained, and vice versa). Chloride is also affected by changes in water concentration and this has to be accounted for in the subsequent calculation. An "adjusted chloride reference concentration" is first calculated: normal chloride concentration (110 mEq/L [mM/L] dog; 120 cat) x (measured sodium concentration/normal sodium concentration). The chloride effect is then calculated as the difference between the measured chloride concentration and this adjusted chloride reference concentration. A measured chloride > reference chloride represents a proportional base deficit effect; a lower measured chloride concentration marks a base excess effect.

The quantitative assessment

The base deficit/excess calculation provides the best overall marker of the metabolic contribution to the acid base balance.[3] Many of the individual contributors to the base deficit/excess can now be estimated: B D/E = lactate effect + ketone effect + albumin effect + phosphate effect + water effect + chloride-bicarbonate effect + "unmeasureds" effect (whatever is left over after all of the known contributions have been accounted for). The quantitative approach provides the most complete understanding of the component makeup of the measured acid base disturbance.

REFERENCES

[1]Siggaard-Andersen O, Blood acid-base alignment. Scand J Lab Clin Invest. 1963;13:211-217. [2]Gaudry PL, Duffy et al, The pH and titratable acidity of intravenous infusion solutions. Anaesth Intens Care 1972;1:41-44. [3]Hopper K, et al. A case-base review of a simplified quantitative approach to acid-base analysis. J Vet Emerg Crit Care 2008;18:467-476.

THE LATE, GREAT STRONG ION DEBATE

Steve C. Haskins, DVM, DACVA, DACVECC
Davis, CA

According to the Stewart approach to acid-base balance, there are three causes of acid-base disturbances: changes in PCO_2, changes in $SID_{apparent}$ and changes in A_{TOT}. Bicarbonate has been relegated to "dependent" ion status and as such, in this approach, does not primarily affect acid-base balance; it can only be affected by changes in the other parameters. This approach divides the very large metabolic component of the traditional approach (base deficit/excess or bicarbonate) into two (still fairly large) subgroups (SID and A_{TOT}). The value of the Stewart approach might be in its ability to suggest directional acid-base changes when pH and PCO_2 measurements are not available. A decrease in SID apparent would be associated with a metabolic acidosis effect (see below); a decrease in albumin concentration would be associated with a metabolic alkalosis effect (see albumin effect in previous manuscript).

Strong ion difference (SID) is the difference between the strongly dissociated cations and the strongly dissociated anions. In theory, all of the strongly dissociated cations and anions would be measured, however this never happens. The usual formula is much abbreviated: Na + K – Cl (and - lactate if it is measured). The difference between this SID formula and the traditional anion gap formula is the absence of HCO_3. In this context one might anticipate that the criticisms of anion gap (see previous manuscript) would also apply to SID, namely that the calculation is difficult to interpret without additional information. The SID calculated by this formula is mostly composed of bicarbonate and albumin. It has been proposed that a decrease in SID can be used as a marker of metabolic acidosis and *vice versa*. When the decrease in SID is due to a decrease in bicarbonate, it is indeed associated with a metabolic acidosis. However, a decreased SID due to hypoalbuminemia would be associated with a metabolic alkalosis. Accumulation of anions of salts such as sodium lactate or sodium acetate, increase both anion gap and SID calculations but have no effect on acid-base balance.

There are two SIDs: $SID_{apparent}$ and $SID_{effective}$. When SID is used in an undesignated manner, it refers to $SID_{apparent}$. $SID_{apparent}$ is calculated as Na + K – Cl (- lactate if measured). $SID_{apparent}$ measures a gap of unmeasured anions. This unmeasured anion gap can also be calculated directly: bicarbonate + albumin effect + phosphorous effect. This calculation is $SID_{effective}$.

For albumin effect, multiply the measured albumin (in g/dl) x 3.7; or (in g/L) x 0.37
For phosphate effect, multiply the measured phosphorous (in mg/dl) x 0.58; or (in mM/L) x 1.8

$SID_{apparent}$ and $SID_{effective}$ generally have a similar value since the former is an assessment of the stuff around the gap and the latter is an assessment of the stuff of the gap itself. The difference between $SID_{apparent}$ and $SID_{effective}$ is strong ion gap (SIG), which are the anions which remain unmeasured. Values for SIG will be similar to values calculated by the expanded anion gap discussed in the previous manuscript.

A_{TOT} is the albumin effect and the phosphate effect as calculated above. An increase in A_{TOT} caused by hyperphosphatemia or hyperalbuminemia would be associated with a metabolic acidosis, and *vice-versa*.

SID_{either} and A_{TOT} are grouping parameters similar to, albeit somewhat narrower in scope, than bicarbonate, base deficit/excess (Siggaard-Andersen), and whole blood buffer base (Singer and Hastings). One might be better served to evaluate and quantify the individual parameters which contribute to these grouping parameters and to acid base balance, as discussed in the quantitative approach to acid-base assessment in the previous manuscript.

The chemistry of the Stewart approach to acid-base balance is fundamentally sound, but some of the clinical extrapolations are flawed. 1) "Bicarbonate is a dependent ion that cannot primarily affect acid base balance." 2) "Hyperchloremia causes metabolic acidosis." 3) "Hypernatremia causes metabolic alkalosis" 4) "Changes in SID is a primary cause of acid-base disturbances."
5) "A decrease in $SID_{apparent}$ causes a metabolic acidosis while an increase causes a metabolic alkalosis."

In a series of *in vitro* experiments, anhydrous sodium chloride, sodium lactate, sodium acetate, lactic acid, and sodium bicarbonate were added to goat blood. Electrolytes and acid-base parameters were measured and calculated before and after.

	Na	K	Cl	Lact	pH	HCO_3	$B_{def/exc}$	AG_{trad}	AG_{exp}	SID_{app}	SID_{eff}	SIG
Baseline	149	4.4	113	0.6	7.42	23.9	0	16.4	5.5	40.3	35.9	4.4
NaCl	187	4.3	150	0.5	7.44	23.5	-0.5	17.9	7.6	41.4	30.9	10.6
NaLact	178	4.2	113	33.9	7.39	22.7	-1.5	45.8	4.5	68.5	29.9	38.6
NaAcet	182	4.2	115	0.5	7.46	24.9	1.4	46.5	38.6	71.4	32.4	39.1
HLact	152	5.1	103	30.1	6.43	13.5	-23	40.1	6.2	53.7	17.7	35.6
$NaHCO_3$	178	4.2	115	0.6	7.51	55	29	12.3	10.4	67.3	62.6	4.7

From these trials, all of the above claims can be seen to be false: 1) It is clear from the $NaHCO_3$ trial that bicarbonate is well able to exert an alkaline effect. It cannot be attributed to the sodium because similar increases in sodium in the NaCl, NaLact, and NaAcet trials were not associated with a metabolic alkalosis. 2) It is clear from the NaCl trial that hyperchloremia, per se, does not impact acid base balance. 3) It is clear from the NaCl, NaLact, and NaAcet trials that hypernatremia, per se, does not cause a metabolic alkalosis. 4) It is clear from the NaLact and NaAcet trials that changes in $SID_{apparent}$ are not necessarily associated with an acid-base disturbance. 5) From the NaLact, NaAcet, and HLact trials it can be seen that an increase in $SID_{apparent}$ is not associated with a metabolic acidosis.

The decrease in chloride associated with HLact is attributed to the HCO_3:Cl shift across the red cell membrane associated with the hemoglobin buffering of the added acid. Lactate, per se, does not cause acidosis. The important consideration is the cation that it brings with it; hydrogen is acidifying (HLact trial), sodium is not (NaLact trial). The high $AG_{traditional}$ in the NaLact, NaAcet, and HLact trials is due to the presence of the high concentrations of lactate or acetate that are not part of the formula. The associated lack of acidosis in the NaLact and NaAcet trials, in spite of the high anion gap, reflects that fact that these are salts and not acids that were added (and that sodium does not have an acid base effect). The high $AG_{expanded}$ in the NaAcet trial but not the NaLact and HLact trials is attributed to the fact that lactate is included in the $AG_{expanded}$ while acetate is not. The increased $SID_{apparent}$ in the HLact trial is attributed mostly to the decrease in chloride concentration. SID is supposed to decrease in metabolic acidosis (due to a decrease in bicarbonate), but clearly $SID_{apparent}$ did not in this trial (however, $SID_{affective}$ did decrease). Both SIDs increased in the $NaHCO_3$ trial which is consistent. The increase in $SID_{apparent}$ in the NaLact and NaAcet trials is attributed to the increased sodium concentration. The large increase in SIG in the NaLact and NaAcet trials is attributed to the increased $SID_{apparent}$ while the increased SIG in the HLact trial is attributed to the modest increase in $SID_{apparent}$ and the moderate decrease in $SID_{effective}$.

In conclusion, changes in sodium, chloride, anion gap, and strong ion difference do not, per se, cause changes in acid base balance, however they may well be abnormal in association with underlying processes which do have an acid base effect. Changes in sodium, chloride, anion gap, and strong ion difference should be considered to be casually related to such processes but not causally related to the acid base disturbance. Changes in sodium, chloride, anion gap, and strong ion difference should be used as markers of processes that may be associated with acid base disturbances.

REFERENCES

[1]Stewart PA, Modern quantitative acid-base chemistry. Canad J Physiol Pharmacol 1983;61:1444-1461. [2]Autran deMorais HS, et al, Strong ion approach to acid-base disorders, In DiBartola SP, Fluid, Electrolyte, and Acid-base Disorders in Small Animal Practice, 3rd ed., WB Saunders Co., Philadelphia. 2006. Ch 13. [3]Corey HE, Stewart and beyond: New models of acid-base balance, Kidney International 2003;64:777-787. [4]Constable PD, Clinical assessment of acid-base status: comparison of the Henderson-Hasselbalch and strong ion approaches, Vet Clin Path 2000;29:115-128. [5]Kurtz I, et al, Acid-base analysis: a critique of the Stewart and bicarbonate-centered approaches, Am J Physiol Renal Physiol 2008; 294: F1009-F1031. [6]Singer RB, et al, Improved clinical method for estimation of disturbances of acid-base balance of human blood. Medicine 1948;27:223-242.

THE FORGOTTEN ORGAN: THE MICROCIRCULATION

Deborah Silverstein, DVM, DACVECC
Philadelphia, PA

INTRODUCTION

"Shock" is the end-result of insufficient cellular energy production and is most commonly caused by inadequate oxygen delivery to the tissues. The criticalist's goal in the treatment of patients suffering from cardiovascular shock is to maximize global oxygen delivery. The parameters most frequently used to monitor these patients include heart rate, pulse quality, blood pressure, mucous membrane color, capillary refill time, extremity temperature, mentation, lactate, and central venous pressure. However, the heterogenous distribution of blood flow that commonly occurs with sepsis is difficult to appreciate using these parameters. Subsequently, many patients suffering from a systemic inflammatory response syndrome (SIRS) and sepsis eventually develop multiple organ dysfunction syndrome (MODS), multiple organ failure (MOF), and subsequent death, despite a rapid diagnosis and aggressive treatment. Sepsis-associated MOF in people living in industrialized nations is responsible for more intensive care unit (ICU) deaths in than any other disease process. The microcirculation has been termed the "motor of sepsis" and is most likely the first organ to fail when patients progress from sepsis to MOF. The array of pathogenic factors that occur in septic patients affects virtually every cellular component of the microcirculation, including endothelial cells, leukocytes, erythrocytes, and tissue cells. In humans, microcirculatory distress that is not corrected for 24 hours has been shown to serve as a single independent factor predicting patient outcome. Novel technology is now enabling the clinician to visualize and assess the microcirculation in order to diagnose microcirculatory dysfunction, formulate more accurate prognoses, and treat patients more effectively.

The Microcirculation

The microcirculation comprises vessels less than 100µm in diameter and includes arterioles, capillaries, and venules. It contains the biggest endothelial surface within the body (>0.5 km^2), and ~10 billion capillaries (diameter 5-9 µm) are responsible for transporting oxygen and nutrients to, and removing waste products from, all cells of the body. This ensures adequate immunological function and delivers therapeutic drugs to target cells in disease states. The microcirculation is comprised of endothelial cells lining the inside of microvessels, smooth muscle cells (mostly in arterioles), red blood cells, leukocytes, and plasma components in the blood. The main determinants of flow within the capillary include the driving pressure, arteriolar tone, hemorheology, and capillary lumen patency.

Almost all vascular beds have longitudinal and radial oxygen gradients, resulting in a significantly lower capillary pO_2 and hemoglobin saturation compared to arterial values. This is most likely due to the unloading of oxygen directly from the arteriolar network to the tissues combined with the intrinsic oxygen consumption of the vascular endothelium to sustain endothelial function and vascular tone. The hematocrit is typically lower within the microvasculature than the larger vessels and is heterogeneously distributed. This is most likely due to a combination of factors.

Regulation of the Microcirculation

The primary regulators of microcirculatory perfusion include the myogenic (sense strain or stress), metabolic (regulate according to O_2, CO_2, lactate, and $[H^+]$), and neurohormonal systems. Based on the oxygen requirements of the tissue cells, both autocrine and paracrine interactions regulate blood flow in the microcirculation. During pathologic conditions, the loss of these autoregulatory safeguards favors the development of microcirculatory dysfunction. Additional insults to the microcirculation, such as activation of inflammatory and coagulation cascades, endothelial and rheologic disorders, and edema formation, commonly occur in critically ill animals.

The autoregulatory mechanisms do not always respond appropriately in animals with sepsis, leading to microcirculatory dysfunction. This results in heterogenous microcirculatory perfusion; some capillaries are underperfused while others have abnormally high blood flow. The vulnerable microcirculatory units become hypoxic; resulting in an oxygen extraction deficit (commonly associated with sepsis) that results in a "PO_2 gap." This gap occurs when the microcirculatory PO_2 falls below the venous PO_2, creating an oxygen gap that is proportional to the severity of functional shunting. Thus, conventional monitoring of systemic macrohemodynamic and oxygen-derived variables is often unable to sense such microcirculatory distress, masking the ongoing abnormality.

Patients with sepsis commonly have disturbed signal transduction and loss of electrophysiological communication and smooth muscle control in the microcirculatory endothelial cells. A heterogenous expression of inducible nitric oxide synthase (iNOS) in different areas of individual organ capillary beds leads to pathological shunting of blood flow. Subsequently, the microcirculatory units that lack iNOS have less NO-induced vasodilation and tend to become underperfused. Additionally, arteriolar smooth muscle cells may lose their adrenergic sensitivity and tone during septic states. Red cells can become less deformable and more aggregable. Erythrocytes also help regulate microcirculatory blood flow by releasing NO in the presence of hypoxia, leading to vasodilation, but this ability is often deranged in patients with sepsis, further impeding microcirculatory flow and function. Reactive oxygen species generated by stimulated leukocytes can directly disrupt the microcirculation, cellular signaling, and coagulation mechanisms. These changes may lead to altered intercellular barrier junctions and endothelial glycocalyces within the microcirculation, thus predisposing to tissue edema and further oxygen extraction deficits. If the microcirculatory deficits are not corrected in a timely fashion, inadequate oxygenation and nutrient delivery to the tissue cells will lead to organ dysfunction or failure.

Although it is unknown whether the primary etiology of oxygen extraction deficits in septic patients is weak, shunted, hypoxic microcirculatory units or mitochondrial dysfunction ("cytopathic hypoxia"), it appears that the progression from early to severe sepsis is accompanied or caused by microcirculatory dysfunction that leads to mitochondrial failure over time.

Microcirculatory and Mitochondrial Distress Syndrome

When regional and microcirculatory distress persists following resuscitation of hemodynamic and oxygen-derived variables, a condition termed "microcirculatory and mitochondrial distress syndrome" or "MMDS" results. This label recognizes the vulnerable physiologic compartment that is masked from the general circulation, but fails to deliver sufficient oxygen and nutrients to dependent organs and thus predisposes to organ dysfunction or failure. The severity of microcirculatory derangement depends on several factors: the nature of the "initial hit" that caused sepsis in the first place, the pathogenicity of the organism, coexisting illnesses, individual genetic predisposition, previous therapy, and time until appropriate treatment is initiated.

An understanding of microcirculatory dysfunction may help the clinician to better understand the presence of a persistently increased lactate concentration with deranged acid-base values in the face of adequate macrohemodynamic resuscitation and apparent stability. Microcirculatory distress can occur despite normal, or even supranormal, systemic hemodynamic and oxygen-derived variables. The microcirculatory distress is therefore masked by the systemic circulation via shunting pathways. Monitoring techniques in the critically ill patient should ideally verify that resuscitation strategies are effective for both the macro- and microcirculation.

Monitoring the Microcirculation

A number of different techniques have been used in an attempt to assess microcirculatory flow in critically ill patients. There are indirect methods, such as DO_2 and VO_2, blood lactate levels, intestinal capnography, and mixed venous oxygen saturation. However, these methods evaluate oxygenation "downstream" from the pathological processes in the circulatory network. Direct assessment of microcirculatory perfusion has been investigated in animals using intravital microscopy (IVM), but this technique requires large equipment and potentially toxic fluorescent dyes for contrast enhancement, making it impractical and possibly dangerous for routine clinical use.

More recently, a method for observing the microcirculation has been developed called orthogonal polarization spectral (OPS) imaging, which uses reflected light in the tissues to create a high-contrast image. In order to form the image, the light is collected, passed through a spectral filter to isolate the wavelength region, and linearly polarized. Light that is emitted from the target is collected by the same lens and an image of the illuminated region is then formed and contrast is obtained from the absorption of linearly polarized light by hemoglobin in the blood (both oxygenated and deoxygenated). Subsequently, red blood cells in the microcirculation appear black on the white background of the surrounding tissue. A 5× or 10× objective lens is used (with on-screen magnification 326×) during the measurements and data are recorded for later analysis. A newer model of this technology (MicroScan™, Cytometrics, Philadelphia, PA) uses sidestream dark field (SDF) to create reflectance avoidance so the illuminated light and reflected light travel via different pathways and surface reflections do not interfere with the image quality. With this modality, a light guide is surrounded by green (530nm) light emitting diodes (LEDs). The light from the LEDs is absorbed by the hemoglobin of erythrocytes and results in the ability to observe the flowing cells. The evaluation of parameters such as microcirculatory flow index (MFI), perfused vessel density (PVD), and proportion of perfused vessels (PPV) of various sizes can then be performed, as recommended by a consensus statement regarding the evaluation of the microcirculation.

The Human Evidence

Research in septic humans utilizing OPS technology has shown that the distributive defects associated with sepsis are characterized by decreased capillary density, increased perfusion heterogeneity, and an increased proportion intermittently perfused or non-perfused capillaries. Stagnant blood flow in the smallest capillaries with concurrent normal (or near normal) blood flow in the larger microcirculatory vessels is often visualized in these patients.

The sublingual microcirculation in 50 people with severe sepsis was investigated and compared to a cohort of healthy volunteers and noninfected ICU patients. They found a significant decrease in vessel density and a decreased proportion of small-perfused vessels (<20 m), from 90 to 48%, in septic patients compared to volunteers. Nonsurvivors had more severe derangements. Another study prospectively examined 49 septic patients and evaluated the sublingual microcirculation daily from the day of shock onset until its resolution. Although the patients were similar at baseline, there was rapid and more marked improvement in small-vessel perfusion in the survivors compared to nonsurvivors, despite no difference in global hemodynamic measurements between the two groups. In addition, the capillary perfusion following resolution of clinical evidence of shock was closely related to the severity of resultant organ failure.

A recent study examined the sublingual microcirculation of 25 people with severe sepsis (SBP<90 or lactate >4) and 5 controls. The septic people were then resuscitated with early goal directed therapy (EGDT) and OPS imaging was repeated daily until resolution of sepsis-induced organ failure or death. Flow velocity score (FVS) and PVD were compared to SBP, MAP, CVP, CI, ScvO2, CV-SOFA score, and lactate. Septic patients had a lower FVS and PVD and the initial indices were lower in nonsurvivors vs survivors. MAP was the strongest correlate for both FVS and PVD.

The microcirculation has also been investigated in disease processes other than sepsis. For example, critically ill people with heart failure, subarachnoid hemorrhage, and cardiopulmonary arrest have microcirculatory changes that are readily studied at baseline and following treatment or surgery.

Animal Uses

The MicroScan™ machine has been used experimentally in septic sheep and pigs with similar results as those found in the human studies. Following resuscitation from endotoxin-induced sepsis in sheep, resuscitation with hydroxyethyl starch normalized mean arterial pressures, cardiac output, superior mesenteric artery flow, and sublingual and serosal intestinal microvascular flow indices, but did not restore intestinal mucosal microcirculatory perfusion.

The microvascular perfusion score (MPS) was found to decrease by 53% following the induction of sepsis in pigs. The surviving animals' scores increased to 81% of baseline, but the nonsurvivors did not regain significant perfusion scores. The vascular density score (VDS) was also found to decrease substantially with sepsis. However, there was no correlation between cardiac output and MPS or VDS, and the microcirculatory perfusion and VDS remained decreased following resuscitation, despite the normalization of hemodynamic variables.

The microcirculation during cardiac arrest and resuscitation has been studied in porcine research models and found useful. Following cardiac arrest, animals that were successfully resuscitated had superior microcirculatory blood flow after 1 and 5 minutes of chest compressions compared to animals who were unable to be successfully resuscitated. Another study also monitored coronary perfusion pressure and found that sublingual microcirculatory abnormalities mirrored changes in coronary perfusion pressures. Additionally, the administration of epinephrine resulted in marked decreases in microcirculatory blood flow and these changes persisted for more than 5 minutes. Research in dogs is currently underway and preliminary work has been published and will be discussed.

The microcirculatory indices of healthy, anesthetized dogs were recently published as a reference for future studies in dogs.

Therapeutic interventions

It seems that the ability to monitor the microcirculatory effects of therapeutic interventions in clinical patients might help the veterinarian to understand not only the global effects of these interventions, but also the microvascular changes that are occurring. Rodent and porcine studies have shown that colloid solutions such as dextrans and starch solutions can improve microcirculatory flow. The effects of packed red blood cells have produced inconsistent results. Some research has shown a negative impact on the microcirculation, most likely due to 2-3-DPG depletion, decrease red blood cell deformability, and red blood cell interactions with the endothelium and other blood cells. The effects of storage time and presence of residual leukocytes in the transfused products likely represent important factors to consider when evaluating microvascular changes in response to red blood cell transfusions.

The use of vasodilatory agents to manipulate the microcirculation in sepsis has also been studied. Acetycholine and nitroglycerin have been found to reverse sublingual microcirculatory alterations in septic people. The clinical utility and safety of these agents will require further investigation.

Limitations

Several limitations do exist for the currently used techniques to assess the microcirculatory network. The laser Doppler technique provides only an average estimation of perfusion in ~1mm^3 of tissue, without regard for the morphology, type of microvessels being studied, and the direction of flow. It also does not take into account the heterogenous nature of perfusion changes, a major area of concern within critically ill people or animals. The OPS technique can only investigate the microcirculation in areas of the body that are covered by a thin epithelial layer and therefore internal organ perfusion cannot be assessed outside of the surgery suite. Additionally, movement artifacts can make quantitative analysis difficult. Immediate results are not possible but require offline analysis that can prove cumbersome, although a new software program has greatly decreased the time necessary to do these analyses. However, dogs with pigmented mucous membranes are more difficult to analyze using the computer software. Lastly, velocity measurements are currently limited to low ranges, although the use of a double-flash illumination with a stroboscope may allow measurements in high velocity vessels in the future.

CONCLUSION

As interest in the microcirculation is growing, it appears that microcirculatory disturbances may play an important role in a multitude of pathologic conditions. Our understanding of its role in the development of MODS and MOF in critically ill patients is still in its infancy. Someday, bedside microcirculatory analyses might allow the clinician to evaluate, prognosticate, monitor, and safely treat the microvascular sequelae of conditions known to affect the microcirculation.

REFERENCES

[1] Trzeciak S, McCoy JV, Phillip DR, et al. **Intensive Care Med** 2008; 34(12):2210–2217. [2] Klijn E, Den Uil CA, Bakker J, et al. **Clin Chest Med** 2008; 29(4):643–54. [3] Zweifach BW. Circ Res 1974; 34:841–857. Aird WC. **Crit Care Med** 2004; 32 (5, suppl):S271–S279. [4] Ince C, Sinaasappel M. **Crit Care Med** 1999; 27(7):1369–1377. [5] Draisma A, Bemelmans R, van der Hoeven JG, et al. **Shock** 2009; 31(6):581–585. [6] Ristagno G, Tang W, Sun S, et al. **Resuscitation** 2008;77(2):229–234. [7] De Backer D, Hollenberg S, Boerma C, et al. **Crit Care** 2007; 11(5):R101–R111. [8] Boerma EC, Mathura KR, van der Voort PH, et al. **Crit Care** 2005; 9(6):R601–R606 [9] Abate NL, Trzeciak S. **Crit Care Med** 2006; 34(2):566–567. [10] Trzeciak S, Dellinger RP, Parrillo JE, et al. **Ann Emerg Med** 2007; 49(1):88–98. [11] De Backer D, Creteur J, Dubois MJ, et al. **Am Heart J** 2004; 147(1):91–99. [12] Silverstein DC, Pruett-Saratan A, Drobatz KJ. Measurements **J Vet Emerg Crit Care** 2009; 19:579-587.

SHOCK STATES: RAPID RECOGNITION & TREATMENT

Deborah Silverstein, DVM, DACVECC
Philadelphia, PA

INTRODUCTION

"Shock" describes a condition of severe hemodynamic and metabolic derangements characterized by tissue hypoperfusion, decreased oxygen delivery, and insufficient cellular energy production. When cells are unable to generate adequate energy, they will not function properly, leading to eventual failure of vital organs. This manifests clinically as multiple organ failure, the ultimate endpoint of untreated, and sometimes even treated, shock.

The classic clinical appearance of an animal in shock includes pale mucous membranes, a prolonged capillary refill time, tachycardia or bradycardia (felines), weak pulse quality, and dull mentation. There are vasodilatory and cellular forms of shock that may not display these classic signs, however. Regardless of the cause of shock, cells are forced to undergo anaerobic glycolysis when oxygen delivery or mitochondrial function is deranged, resulting in the production of lactate and hydrogen ions (lactic acidosis). The five categories of shock that will be discussed include cardiogenic, hypovolemic, distributive, hypoxic, and metabolic (see Table 1). Septic shock will also be discussed.

Cardiogenic shock typically results in a low-forward flow state due to cardiac derangements. Either poor myocardial contractility or low preload leads to a decrease in cardiac output. Potential causes of cardiogenic shock include dilated cardiomyopathy with subnormal fractional shortening (contractility), toxins or drugs that cause myocardial damage or depression, hypertrophic cardiomyopathy which results in poor preload, obstructive diseases such as stenosis, backload regurgitation as seen with valvular disease, pericardial tamponade or pericardial fibrosis which also result in a low preload, or severe arrhythmias causing poor preload and inefficient contractility.

The treatment of cardiogenic shock often depends on the underlying disease process. If poor myocardial contractility is the cause of shock, beta-agonist therapy is indicated for its inotropic effect. Dobutamine is the most common emergent treatment that will increase cardiac function. If the animal is receiving drugs that cause myocardial depression, these should be discontinued (ie anesthetics). Some animals with valvular disease will benefit from afterload-reducing drugs such as nitroprusside or hydralazine +/- a positive inotrope (ie dobutamine) depending on the degree of contractility insufficiency. Cats with hypertrophic cardiomyopathy may benefit from lusitropic drugs such as calcium channel blockers (diltiazem) or beta-2 antagonists (atenolol). If the animal has severe pericardial effusion that is leading to tamponade, a pericardiocentesis should be performed and fluid therapy administered. Severe, persistent tachy- or bradyarrhythmias should be treated appropriately with antiarrhythmics or cardioconversion. If cardiogenic abnormalities have caused pulmonary edema, oxygen and furosemide therapy are indicated.

Hypovolemic shock is a common type of shock in small animals and results from a decrease in intravascular blood volume. Etiologies include hemorrhage, hypoproteinemia, increased vascular permeability, or diseases leading to severe dehydration. Hypovolemic shock is typically associated with a compensatory peripheral vasoconstriction that interferes with vital organ perfusion.

The treatment of hypovolemic shock requires the rapid administration of shock doses of intravenous fluids, primarily isotonic crystalloids. Animals with severe hypoproteinemia or vascular leak syndromes may benefit from synthetic colloid administration. Hypertonic crystalloids, such as 7.5% sodium chloride, are often useful as an adjunctive therapy for animals with acute hypovolemia, but are contraindicated in patients that have severe interstitial dehydration. Severe hemorrhage may lead to anemia and depletion of clotting factors which requires the administration of blood products (fresh whole blood or red blood cells and fresh frozen plasma).

Distributive shock encompasses a variety of abnormalities that are generally associated with altered vasomotor tone and inappropriate distribution of cardiac output to the tissues. This commonly involves inappropriate vasodilation, as seen with sepsis or animals with systemic inflammatory response syndrome (SIRS), excessive vasoconstriction (ie post-epinephrine or trauma), or derangements in normal blood flow (ie shunts, thromboemboli). Vasodilatory distributive shock due to sepsis or SIRS is mediated by cytokines and results from decreased peripheral vascular resistance and hypotension. This leads to a reduction in coronary and cerebral blood flow and subsequent distributive shock. Many animals are presented to the emergency service following trauma and appear to be in a state of hypovolemic shock, even if there is no blood loss. This is most likely because of the neurohormonal response to trauma that leads to severe vasoconstriction and maldistribution of the circulating blood volume.

Arteriovenous shunts can also lead to distributive shock from a reduction in the net cardiac output and blood flow to the tissues. Occlusion of vessels from external sources, as seen with intestinal torsions, masses, or foreign bodies, interferes with blood flow to the involved organ. Similarly, thromboembolic disease prevents normal blood flow to the downstream organs and severe pulmonary thromboemboli can lead to cor pulmonale and right heart failure, resulting in distributive shock.

The treatment of distributive shock typically includes shock doses of intravenous fluid therapy. Combinations of isotonic crystalloids, synthetic colloids, hypertonic solutions, and blood products might be necessary, depending on the underlying disease process. Combinations of hypertonic crystalloids and synthetic colloids are gaining increasing popularity based on promising experimental evidence. Empiric antibiotic therapy pending culture results is indicated if there is a strong suspicion of sepsis. The treatment of thromboemboli typically involves diagnosis and management of the underlying disease, as well as anti-coagulant therapy.

States of hypoxic shock are characterized by normal tissue perfusion (unlike distributive shock), but abnormal oxygen content or oxygen unloading to the tissues and cells. Hypoxemia (a low partial pressure of oxygen or low hemoglobin saturation due to pulmonary pathology) and anemia (low oxygen content) are the most common causes of hypoxic shock. Additional disease processes that might lead to hypoxic shock include methemoglobinemia (oxidized, ferric hemoglobin) and carboxyhemoglobin (carbon monoxide poisoning) since these abnormalities decrease the oxygen carrying capacity of hemoglobin.

The treatment of hypoxic shock is aimed at increasing oxygen content. Oxygen therapy should be provided to all patients. Animals with severe pulmonary disease may benefit from positive pressure ventilation. Dogs or cats with life-threatening anemia should be given appropriate red cell blood products or hemoglobin-containing solutions. Toxicities such as methemoglobinemia should be treated appropriately (vitamin C and N-acetylcysteine).

There are instances when the tissues are adequately perfused and oxygenated, but the cells are still unable to produce adequate energy levels to maintain viability. This is metabolic shock; it commonly occurs due to an intracellular abnormality that interferes with energy production. Sepsis, in addition to all of its vasoactive intermediates and cell-mediated injury, interferes with normal cellular function and often causes "cytopathic hypoxia." Cyanide intoxication disrupts normal mitochondrial cytochrome oxidative phosphorylation and leads to metabolic shock.

Animals with heat stroke have a cellular metabolic rate that exceeds the animal's ability to deliver energy substrates and often results in metabolic shock. Hypoglycemia leads to inadequate energy substrate for normal cerebral metabolism and can also cause metabolic shock.

Diagnosis and treatment of metabolic shock can be difficult. Hypoglycemia is easily discovered and treated with glucose supplementation. Cellular dysfunction is diagnosed via exclusion of other causes of shock (normal oxygen delivery with continued evidence of shock and elevated blood lactate levels). The treatment is generally supportive in an attempt to optimize oxygen delivery to the tissues and treat the underlying disease process.

Septic shock is a state of circulatory shock that is caused by an underlying infectious agent. The resulting inflammatory response to the insult includes the release/activation cytokines, proteases, catalases, eicosanoids, and activation of the complement, coagulation, and fibrinolytic cascades. These processes interfere with intracellular metabolism and function. Subsequent decreases in myocardial contractility, cardiac output, and oxygen delivery ensues. Inappropriate changes in vasomotor tone often occur resulting in either hypotension (vasodilation) or impaired tissue perfusion (vasoconstriction). Gastrointestinal ulceration or hemorrhage and acute kidney injury may ensue. Increased endothelial permeability may result extravasation of protein-rich fluid from the vascular space, resulting in hypoproteinemia and hypovolemia. Coagulopathies, lactic acidosis, and hypoglycemia also commonly result. It is therefore apparent that sepsis can lead to various types of shock: cardiogenic, distributive, hypovolemic, hypoxic, and metabolic.

The treatment of the various causes of shock in patients with sepsis is beyond the scope of this discussion. Identification of the source of sepsis and early, appropriate therapy are paramount. Monitoring cardiac function, cardiovascular and pulmonary parameters, and labwork values regularly will allow early recognition and treatment of sepsis-related complications.

Regardless of the cause of shock, rapid assessment of these critical patients, early identification of the underlying disease, and restoration of cardiopulmonary and cellular stability is vital. By understanding the various categories of shock and common etiologies of these categories, the veterinarian will be able to more successfully treat these emergent patients.

Table 1. Functional Classifications and Examples of Shock

TYPE OF SHOCK	CONSEQUENCE	EXAMPLES
Cardiogenic	Decreased forward flow from heart	Congestive heart failure, arrhythmias, cardiac tamponade, drug toxicity
Hypovolemic	Decreased circulating blood volume	Hemorrhage, increased vascular permeability, hypoproteinemia, severe dehydration
Distributive	Loss of peripheral vascular resistance/obstruction to flow	Sepsis, systemic inflammatory response syndrome, anaphylaxis, arteriovenous shunts, thromboembolic disease, trauma
Hypoxic	Decreased oxygen content in arterial blood	Anemia Severe pulmonary disease Carbon monoxide toxicity Methemoglobinemia
Metabolic	Deranged cellular metabolic machinery	Hypoglycemia, mitochondrial dysfunction, cyanide toxicity, cytopathic hypoxia of sepsis

REFERENCES

[1] Silverstein DC, Hopper K ed. "**Small Animal Critical Care Medicine**," 1st Ed, Elsevier, St. Louis, MO, 2008. [2] Boag A, Hughes D: **Vet Clin North Am Small Anim Pract** 35(2):319, 2005. [3] James K, Jonas M. **Medicine** 2009; 37(1): 6-10. [4] Manji RA, Wood KE, Kumar A. The history and evolution of circulatory shock. **Crit Care Clin** 2009;25:1-29.

ACUTE RESPIRATORY DISTRESS SYNDROME IN HUMANS & ANIMALS

Cynthia M. Otto, DVM, PhD, DACVECC
Philadelphia, PA

ACUTE RESPIRATORY DISTRESS

Acute respiratory distress syndrome (ARDS) was defined in the human literature by a consensus conference in 1992. (Consensus Group CCM 1992) This condition was characterized by 1) an acute onset of respiratory difficulty; 2) bilateral infiltrates on chest radiographs; 3) hypoxemia (a PaO_2/FiO_2 ratio <200mmHg for ARDS whereas less severe acute lung injury (ALI) was defined as PaO_2/FiO_2 ratio <300mmHg; and 4) no evidence of left atrial hypertension (pulmonary artery occlusion pressure [PAOP] <18 mmHg). Until the Dorothy Russell Havemeyer Working Group on ALI and ARDS in Veterinary Medicine reported their consensus on veterinary ARDS in 2007 (Wilkins 2007), there were no veterinary definitions of ARDS or ALI. The consensus expanded the human definition to be more specific in their definitions and include many of the newer diagnostic modalities that are helpful in identifying patients with ARDS or ALI. In addition, species and age specific considerations were included. The veterinary criteria for ARDS adapted the 4 criteria from the human definition and added a 5th (although currently optional) criteria, evidence of pulmonary inflammation. For the veterinary definition acute is defined as < 72 hours. The second criterion is the presence of known risk factors. Rather than depend strictly on the presence of radiographic evidence of diffuse infiltrates, the veterinary definition relies on evidence for the underlying pathology of pulmonary capillary leak without increased pulmonary capillary pressure. This can be documented by radiographs, CT, proteinaceous fluid in the conducting airways or evidence of increased extravascular lung water. Furthermore, rather than relying on the PaO_2/FiO_2 (especially since the human definition fails to designate the PEEP or CPAP for these measurements) the criterion is for evidence of inefficient gas exchange; which could be defined by hypoxemia without positive end-expiratory pressure (PEEP) or continuous positive airway pressure (CPAP) and known FiO_2 as evidenced by PaO_2/FiO_2 ratio (300 mmHg for VetALI or 200 mmHg for VetARDS) or increased alveolar-arterial oxygen gradient or venous admixture (noncardiac shunt). In addition, inefficient gas exchange could be evident by increased 'dead-space' ventilation. The definitions are undergoing frequent evaluation and re-assessment. The consensus group met again in 2010 and an updated recommendation is expected in the near future.

VENTILATOR ASSOCIATED LUNG INJURY

Patients that develop ARDSrequire ventilation. The use of mechanical ventilation can contribute to progression of lung injury and may be involved in the development of multiple organ failure. Ventilator associated lung injury (VALI) is used to describe the progression of lung injury in ventilated patients. In patients, there are numerous confounding factors of disease and treatment that make it difficult to define the role of mechanical ventilation in progression of lung injury. Ventilator induced lung injury (VILI) is the term describing the induction of lung injury in animal models through different ventilator strategies.

Data accumulated from both VALI and VILI has lead to four proposed mechanisms by which mechanical ventilation causes damage. The first mechanism is barotrauma and represents the pressure-induced disruption of the lung. The effect of barotrauma is the development of pneumothoraces and pulmonary membrane rupture as evidenced by radiographic extrapulmonary air. The second mechanism is volutrauma. This mechanism is thought to result from excessive stretch of normal or diseased lung, resulting in disruption of membranes. These ruptured membranes then leak fluid into the interstitial and alveolar space, which initiates inflammation. The third proposed mechanism, biotrauma, is similar to volutrauma, however the stretched membranes only temporarily form gaps, which are then repaired. In this scenario, stretch leads to inflammation, through the activation of the innate immune response and NF-kB. Once inflammation is initiated, edema forms in response to the mediator induced permeability changes. Both biotrauma and volutrauma result in edema, inflammation and neutrophil infiltration. Experimentally, use of large tidal volumes in normal lungs can rapidly lead to severe edema formation. In ARDS patients, the injury is heterogeneous. Typically the dependent regions of the lung are collapsed and difficult to recruit, whereas the non-dependent regions are more normal. The normal lung units have been referred to as the "baby lung", creating an effectively smaller lung volume for ventilation. If a normal tidal volume is delivered to a diseased lung, the majority of the breath is distributed to the "baby lung" resulting in overdistension and stretch injury. Stretch is thought to cause injury by disruption of cellular membranes, fluid leakage and subsequent inflammation. The ability to improve mortality using a ventilator management style that emphasized low tidal volume was the first successful intervention in the treatment of ARDS (ARDSnet trial New Eng J Med 2000).

Another mechanism for VILI and VALI is atelectrauma or cyclical recruitment. Cyclical recruitment is the repetitive opening and collapse or flooding of alveoli with every breath. The mechanism of injury is thought to be the shear stress generated at the interface between collapsed and open alveoli; however, the cyclical change in oxygen tensions may also contribute to injury. Cyclical recruitment is difficult to document. Recently several techniques have been employed to demonstrate that this phenomenon even exists. These techniques include dynamic computed tomography (dCT), and electrical impedance tomography (EIT) which has had limited use in patients and subpleural microscopy and rapid responding intraarterial PO_2 which have been restricted to experimental studies. The majority of evidence supporting a detrimental effect of cyclical recruitment has been indirect. The application of PEEP is thought to prevent the expiratory derecruitment that contributes to cyclical recruitment. Experimental studies have demonstrated less lung injury with PEEP. In one study in human patients, PEEP was associated with a reduction in inflammatory mediators. In human randomized clinical trials, however, the role of PEEP is less clear. Three large trials failed to demonstrate an outcome benefit of PEEP, although other intermediate outcomes did show some benefit. One trial was able to demonstrate a benefit, but was complicated by the comparison of low PEEP with high tidal volume to high PEEP with low tidal volume. There are several potential explanations for the inability to demonstrate a universal benefit for PEEP. First PEEP can have several adverse effects. The constant increase in intrathoracic pressure leads to a reduction of cardiac output by decreased venous return. In addition, the use of PEEP often leads to a necessary increase in the plateau pressure in order to maintain tidal volume and avoid severe hypercapnia. Importantly, there appear to be populations of patients in which PEEP does not improve recruitment and would thus increase risks without providing benefits. One of the most challenging components of using PEEP in ventilator management is the determination of the appropriate level of PEEP. Although historically the lower inflection point of inspiratory limb of the pressure volume (PV) curve was thought to represent the point at which the lung was recruited, it has been subsequently recognized that recruitment occurs throughout the inspiratory limb. Currently the approaches available to determine if the lung is recruited with a certain level of PEEP rely either on imaging or physiologic measures. Imaging has been evaluation of lung density by CT at an end inspiration breath hold and at an end expiration breath hold. These static images, while informative ignore the dynamics of recruitment and are cumbersome for routine use in critical patients. Electrical impedance tomography does not have high resolution, but may be a more practical approach to determining recruitment. Improvement in oxygen saturation or PO_2 is used as a functional approach. For maximal information, patients should be ventilated with 100% oxygen, so that the improvements in PO_2 represent changes in shunt fraction. This eliminates the use of pulse oximetry to monitor changes, since most patients will be fully saturated, use of lower inspiratory oxygen concentrations do not allow the distinction between low V/Q units and true shunt. During derecruitment of alveoli, the PO_2 will drop rapidly with collapse of ventilatory units and with successful recruitment the PO_2 will rise instantaneously. The goal of monitoring PaO_2 is to apply PEEP or increase respiratory rate sufficiently to prevent these oscillations with each breath. Rapid intra-arterial PO_2 is ideal in experimental settings, but not available clinically. Blood gases collected at end inspiration can be compared with blood gases collected at end expiration to determine the degree of oscillation with each adjustment of the ventilator. (Pfeiffer 2006) Given the difficulty of monitoring cyclical recruitment it is not surprising that evidence for its role in VALI and VILI is limited. Recent data from our laboratory suggests that cyclical recruitment does participate in VILI and may be worse than stretch (Otto 2008).

For the practical management of a patient with ARDS, the clinician needs to balance between the mechanisms that contribute to VALI. There is clear evidence that lowering tidal volume to minimize stretch is beneficial, however what is the effect on cyclical recruitment? Efforts to minimize stretch, by minimizing pressure and PEEP may allow for cyclical recruitment, whereas application of PEEP to maintain recruitment can lead to stretch. This raises the question of which is worse for the patient – stretch or cyclical recruitment. There is no clinical data to answer this question but recent data from our laboratory provides some evidence that cyclical recruitment is a strong inflammatory stimulus and therefore may be a major factor in VALI.

Whether the injury is induced by volutrauma, biotrauma or cyclical recruitment, it appears that activation of the innate immune response represents a common pathway. The mechanism by which the immune response is activated is not clear, but the pathways result in the common endpoint of transcription factor NF-kB activation (either direct via TLR4 or indirect via membrane injury induced calcium flux), followed by the expression of chemokines (e.g. interleukin 8) and the attraction of neutrophils. Neutrophilic infiltration is a hallmark of ARDS. The activated neutrophils release elastases, proteases and reactive oxygen species that further damage cells and inactivate surfactant. A critical factor in managing patients with ARDS is to minimize the ventilator associated lung injury, but the optimal ventilator techniques are still being investigated and refined.

RECOMMENDED REVIEWS

Dreyfuss D., and G. Saumon. 1998. Ventilator-induced Lung Injury . Lessons from Experimental Studies. Am. J. Respir. Crit. Care Med. 157(1):294-323. dos Santos C. C., and A. S. Slutsky. 2006. The contribution of biophysical lung injury to the development of biotrauma. Ann Rev Physiol 68:585-618. Gattinoni L., et al. 2003. Physiologic rationale for ventilator setting in acute lung injury/acute respiratory distress syndrome patients. Crit Care Med 31(4 Suppl):S300-4. Gattinoni L., and A. Pesenti. 2005. The concept of "baby lung". Intensive Care Med 31(6):776-784. Hickling, K. G. 2002. Reinterpreting the pressure-volume curve in patients with acute respiratory distress syndrome. Curr Opin Crit Care 8(1):32-38. Oeckler, R.A., Hubmayr, R.D., 2008 Cell wounding and repair in ventilator injured lungs. Respir Physiolo Neurobiol. Pugin, J. 2003. Molecular mechanisms of lung cell activation induced by cyclic stretch. Critical care medicine 31(4 Suppl):S200-6. Puneet, P. et al, 2005. Chemokines in acute respiratory distress syndrome. Am J Physiol - Lung Cell Mol Physiol, 288(1): p. 3-15.

REFERENCES

American college of chest physicians/society of critical care medicine consensus conference 1992 Crit Care Med, 20(6):864. ARDSNet, 2000 New England Journal of Medicine, 342(18):1301. Otto CM, et al., 2008 J Appl Physiol, 104(5):1485. Pfeiffer B, et al., 2006 Anesth Analg, 102(6):1758. Wilkins PA, et al., 2007 J Vet Emerg Crit Care,17(4):333

SEPSIS: A BASIC & CLINICAL UPDATE

Cynthia M. Otto, DVM, PhD, DACVECC
Philadelphia, PA

DEFINING SEPSIS

In 2001, there was a second (human) consensus conference, which expanded the criteria for systemic inflammatory response syndrome (SIRS) by recommending the inclusion of physical parameters and biomarkers (the term recently coined to describe measurable factors in blood or biological samples that can be used to identify or stage disease) (Levy MM 2003). Many of the parameters and some of these biomarkers may also prove useful in veterinary species. This newer approach, designated as PIRO, incorporates 4 factors in the stratification. The first factor is Predisposition. Although many factors can influence "susceptibility", an area of active research in humans involves predisposing genetic factors. While it has long been recognized that certain breeds of animals or families of people may be more susceptible to diseases individual genetic variations or polymorphisms appear to further contribute to susceptibility. Other predisposing factors may be age, concurrent conditions and gender (and in animals, neuter status). The second factor is Infection. Clinically, we recognize that certain bacteria, location of infection, or extent of infection, contribute to the risk of developing sepsis or septic shock. This factor has not been systematically evaluated in sepsis in small animal. The third factor is the host Response. Our current inability to readily identify and monitor species specific biomarkers is the biggest limitation for this component of the stratification scheme for dogs and cats. However if we can determine whether our patient is in an excessively pro-inflammatory state versus immune paralysis, or if there is evidence of adrenal or coagulation dysfunction, we will be better able to chose appropriate and directed interventions. The fourth and final factor is Organ Dysfunction. Even more in cats and dogs than in people, the extent of organ dysfunction will negatively influence outcome.

Sepsis: Confirmed infection plus systemic signs of inflammation
Severe Sepsis: Sepsis plus sepsis induced organ dysfunction or tissue hypoperfusion
Sepsis induced hypotension: Systolic blood pressure < 90 mmHg, mean BP < 70 mmHg, or a decrease in SBP of >40 mmHg
Septic shock: Sepsis induced hypotension despite adequate fluid resuscitation
Sepsis-induced hypoperfusion: Septic shock with an elevated lactate or oliguria

UNDERSTANDING SEPSIS

The first step in the initiation of the inflammatory response is for the body to recognize a stimulus. Therefore there needs to be a stimulus and a mechanism to recognize it and "sound the alarm". Recently more emphasis has been put on the alarm signals, sometimes referred to as "alarmins". There are two major categories of these danger signals. The classical danger signal is initiated by pathogens. The molecules that trigger the response are called "PAMPs" or pathogen associated molecular patterns. Recently, endogenous danger signals, "DAMPs" or danger associated molecular patterns have been described. (Lotze 2007). These danger signals are recognized by cellular receptors and activate innate immunity, which can subsequently lead to activation of the adaptive immune response. The Toll-like receptor (TLR) family represents a large class of phylogenetically conserved surface receptors that are programmed to recognize the "signature" of a variety of pathogens. In response to the activation of the TLR4 receptor, the signal cascade of Mal, MyD88, Tram and Trif lead to the activation of the nuclear transcription factor NF-kB. NF-kB is a transcription factor from many of the pro-inflammatory cellular glycoproteins or cytokines. These cytokines are molecules that serve to communicate with other cells in paracrine, autocrine and sometimes an endocrine manner. One of the DAMPs is the DNA binding protein, high mobility group box 1 (HMGB). This highly conserved ubiquitous nuclear protein is released by cells dying from necrosis, and after acetylation or ADP ribosylation can be actively secreted by macrophage, monocytes, and dendritic cells. The cellular response results in the activation of endothelial cells, neutrophils and monocytes. These activated cells lead to the clinical signs associated with systemic inflammation. The consequences include endothelial dysfunction (hypotension, activation of coagulation, increased neutrophil interactions, increased permeability), neutrophil and monocyte activation and release of cytotoxic mediators, and development of fever. The final result is organ dysfunction and in many cases death.

TREATMENT

Prevention is still the most effective tool in the battle against sepsis. The implementation of preventive measures has evolved over the centuries and current recommendations to prevent infection, which could lead to sepsis, include appropriate aseptic technique, proper catheter and wound care, and hand-washing. Prevention of the progression from infection to sepsis emphasizes removal of the source of infection, early and appropriate antibiotics, aggressive monitoring and prevention of hypoperfusion and hypoxia. Treatment of sepsis has a short and capricious history. The mainstay of therapy remains supportive care. Recently in human critical care, recommendations have been put forth by the "Surviving Sepsis Campaign" (Dellinger 2008). These recommendations represent the most current information based on clinical trials, observational studies, published case series and expert opinion. The implementation of these sepsis protocols has resulted in a reduction in sepsis mortality. (Levy 2010) In veterinary medicine, there are no clinical trials for any of these recommendations. There are limited observational studies and case series that have been published. One approach is to evaluate the recommendations made for human patients and determine which ones have potential application and value in companion animals.

Resuscitation

The current recommendation for the initial resuscitation of septic patients is based on the pivotal study conducted by Rivers et al (Rivers 2001), in which septic patients resuscitated to specific endpoints during the first 6 hours had increased long term survival.

This approach is termed Early Goal Directed Therapy. The currently recommended endpoints of resuscitation are central venous pressure of 8-12 mmHg, mean arterial pressure of > 65 mmHg, urine output > 0.5 ml/kg per hour, and central venous oxygen saturation of > 65% (or a mixed venous oxygen saturation of >70%). Resuscitation can be performed with either crystalloids or colloids. If these endpoints are not achieved with fluid therapy, it is recommended that packed red blood cell transfusion to a hematocrit of 30% and/or dobutamine (up to 20 ug/kg/min) be added to the resuscitation plan during the first 6 hours of resuscitation. A popular concept in the supportive care of the septic cardiovascular system was "supranormal" resuscitation, using fluids, blood, and vasoactive agents to drive hemodynamic parameters beyond normal values. There is no evidence that "supranormal" resuscitation is beneficial. If the mean arterial pressure remains below 65 mmHg despite fluid therapy, either norepinephrine or dopamine is recommended as the first line vasopressor in people. Additionally, the guidelines recommend against epinephrine, phenylephrine and vasopressin for initial treatment of hypotension, however epinephrine is recommended as a second line vasopressor. Vasopressin may have some benefit when used in concert with norepinephrine, however the studies, particularly in small animals are limited. If vasopressors are used, direct arterial blood pressure monitoring is strongly recommended. Some cases may require inotropic support in addition to vasopressor therapy, in this situation, dobutamine is recommended in people. It has generally been recognized that low dose dopamine is not beneficial in renal protection and is therefore not recommended.

Identification and control of the infection

Sepsis is a disease of infectious origin. In order to identify the source of infection, imaging studies are often vital. Some more advanced imaging may need to be delayed until the patient is resuscitated. This is particularly true of small animals which require anesthesia for MRI and CT imaging. Although it is highly recommended that bacterial (or fungal if appropriate) cultures are obtained prior to the start of antimicrobial therapy, it may not be possible if the source of infection cannot be identified due to delays associated with patient stabilization. Therefore, this recommendation must be balanced with the need to start appropriate and broad spectrum intravenous antimicrobials as soon as possible. Ideally antibiotic therapy should be initiated within 1 hour of diagnosis. Concerns about endotoxin release associated with antibiotic use do not outweigh the benefits gained from control of the infection. Newer diagnostic strategies to rapidly identify the source of sepsis (e.g. PCR of bacterial DNA) may become important in the initial diagnosis and treatment plan. The use of empirical selection of antibiotics should be reviewed regularly and adjusted as soon as culture and sensitivity results become available. Due to concerns of antibiotic resistance, it is recommended that empirical combination antimicrobial therapy not extend beyond 3-5 days. Additionally, the duration of antibiotic therapy in most patients should not exceed 7-10 days, unless the patient response warrants longer duration of therapy.

If the source of infection is identified, it should be addressed as quickly as possible. The drainage of abscesses or removal of the infected source should be performed in a way that minimizes secondary insults. For example, if an abscess can be effectively drained with a local anesthetic and incision this would be favorable to general anesthesia and surgical drainage, however, adequate drainage is paramount.

Adjunctive therapy

Corticosteroids

One of the earliest interventions in the treatment of sepsis was the use of "shock dose" (ultra-high doses e.g. 10 mg/kg of prednisone). Although experimental studies were promising, this approach failed in clinical trials. One major explanation is that most of the successful animal studies administered the corticosteroid prior to the onset of sepsis or endotoxemia. In clinical patients, the adverse effects of such high dose corticosteroids and the necessity of starting therapy after clinical signs have resulted in abandonment of this recommendation. More recently, a new role for corticosteroids in septic patients has been suggested. This recommendation is based on the finding that some patients develop a transient adrenal insufficiency associated with septic shock. Several studies in humans and a few studies in dogs and cats have demonstrated that this phenomenon does occur. The large human clinical trials that have subsequently been conducted lead to the recommendation that low dose (< 300 mg/adult/day) hydrocortisone (due to its combined glucocorticoid and mineralocorticoid effects) be used in adult patients in septic shock that have a poor response to vasopressors. Many of the early studies presented various interpretations of the ACTH stimulation test in order to identify patients with relative adrenal insufficiency. After a large clinical trial that demonstrated benefit regardless of ACTH response, the recommendation is to treat any patient that is refractory to vasopressor therapy and not perform an ACTH response test. If hydrocortisone therapy is initiated, it should be discontinued when vasopressors are no longer needed.

Coagulation directed therapy

It is well recognized that in addition to activation of inflammation in sepsis, the coagulation cascade is activated. There have been many investigations directed at maintaining the balance between coagulation and fibrinolysis during sepsis. Despite evidence of antithrombin deficiency during sepsis, replacement antithrombin has not proven beneficial. The only treatment that has shown benefit is recombinant human activated protein C (aPC). This drug is only recommended for a very specific population of septic individuals. Adults with sepsis induced organ dysfunction and a high risk of death are the only group of patients in which recombinant aPC is recommended. It is not recommended for pediatric sepsis. In addition to its extraordinarily high cost, it is immunogenic and rapidly cleared in dogs. Although in small animals, plasma is often used for treatment of coagulation abnormalities, there is no evidence of benefit and it is not recommended for routine correction of clotting abnormalities in septic humans. Platelet transfusions, while often impractical in small animals, are recommended in septic adults with platelet counts < 30,000 cells/mm3.

Glucose control

It has been recognized that hyperglycemia is associated with a worse outcome in ICU patients. Whether the benefit that benefit translated to increased survival or shortened hospital stay was dependent on the population studied. Interestingly in an unpublished study of small animal ICU patients, hypoglycemia appeared to be more commonly associated with mortality. Additionally, in people, adverse effects of hypoglycemic episodes have tempered the recommendations for strict glucose control. The current recommendation is to maintain blood glucose below 150 mg/dl. In small animals, careful monitoring of blood glucose is warranted. If insulin therapy is instituted, it is probably safest to provide concurrent supplemental glucose and monitor glucose continuously or at least every 2 hours.

REFERENCES

Dellinger RP. et al. 2008 Crit Care Med. 36(1):296. Levy MM. et al. 2003 Intensive Care Med. 29(4):530. Levy MM, et al., 2010 Crit Care Med, 38(2):367-374. Lotze MT, et al. 2007 Immunol Rev, 220:60. Perkins ND. 2007 Nature Rev: Mol Cell Biol. 8:49. Rivers E, et al. 2001: N Engl J Med, 345(19):1368.

ANESTHESIA TECHNIQUES FOR USE IN THE CRITICALLY ILL DOG AND CAT

Jane Quandt, DVM, MS, DACVA, DACVECC
Athens, GA

The critically ill patient that requires general anesthesia presents a challenging situation. The patient needs a thorough pre-operative assessment to define what type of trauma or compromise the patient is undergoing. The nature of their disease will alter physiology and decrease reserves, which affects the pharmacokinetic and pharmacodynamic behavior of sedative and anesthetic drugs. Stabilization of the critical patient, including minimizing stress levels and optimizing oxygen delivery, prior to sedative or anesthetic drug exposure, is essential as the risk of sedation and anesthesia in an unstable patient increases the potential for complications. Appropriate monitoring should be performed at all times to ensure that these delicate patients survive their sedation or anesthetic episode. Intensive care may include vasopressor and inotropic support, aggressive colloid and/or crystalloid therapy, analgesic support, antibiotic therapy, oxygen therapy, and nursing care to improve the survivability in this critically ill patient population.

Prior to anesthesia, diagnostics, including physical examination, radiographs, blood chemistry, complete blood count, coagulation profile, acid-base status, blood glucose and lactate levels as necessary should be performed to determine the severity of the illness. Dehydration or hypovolemia along with fluid, acid-base, and electrolyte deficits should be corrected if possible before anesthesia. Anticipate potential problems and have a plan in place for addressing those problems.

Venous access is imperative in managing and anesthetizing the critical patient. Intravenous (IV) administration of drugs is preferred as intramuscular (IM) or subcutaneous (SC) may have delayed absorption, particularly when the patient is dehydrated, hypovolemic, poorly perfused or hypothermic. IV access also allows for ease of anesthetic drug titration. The critically ill patient may benefit from having more than one IV catheter to provide a port for not only drug administration, but antibiotic delivery, vasopressor and inotropic support, and fluid therapy. Blood products should have a dedicated catheter and no other fluids or drugs should be administered in the same line during the transfusion due to concerns for possible contamination and potential for bacterial growth.

A packed cell volume (PCV) greater than 25% is recommended for adequate oxygen carrying capacity and oxygen delivery. During anesthesia the PCV can decrease by 3 to 5 %, therefore even a small volume of blood loss may be significant in the anesthetized animal. Similarly, hypoproteinemic patients with a total protein (TP) $\leq$ 3.5 g/dl and/or an albumin $\leq$ 2 g/dl, may benefit from the use of colloids to help maintain normal colloid osmotic pressure (COP) (normal is 18-24 mmHg) and to prevent edema formation or vascular leak. If patients are hypoproteinemic, colloid options include hetastarch, dextran, 25% human serum albumin or even Oxyglobin™. If the patient is hypocoagulable, fresh frozen plasma (FFP) should be considered.

Lactate plasma concentrations can be a useful prognostic indicator. The normal plasma lactate concentration is less than 2.5 mmol/L. It has been shown that lactate levels are increased in ill and injured dogs, and the values were significantly higher in nonsurvivors compared to survivors. The most common and usually most severe cause of increased lactate levels is systemic hypoperfusion, which requires prompt treatment and identification of the cause. Therapy includes fluid resuscitation to treat hypoperfusion, correction of the cause, and a PCV >20% to help maintain arterial oxygenation.

Patients with renal insufficiency may require a higher fluid rate to maintain renal perfusion, and should have urine output (UOP) carefully monitored during anesthesia. Monitor blood pressure closely, if possible a pre-anesthesia value is helpful. The mean arterial blood pressure (MAP) should be greater than 60 mmHg to maintain renal perfusion. Because critically ill patients are often hypotensive during anesthesia a mean blood pressure (MAP) of <60 mmHg or a systolic < 90 mmHg requires prompt treatment to maintain appropriate organ perfusion. A fluid bolus should be initiated, if appropriate, of either a crystalloid at a rate of 10 to 20 ml/kg IV over 15 to 20 minutes or a colloid bolus of 5 to 10 ml/kg IV over 10 to 20 minutes. If the hypotension persists despite fluid therapy there may be a need for inotropic support. Urine output (UOP) should be monitored carefully, and goals to achieve normal UOP of 1 to 2 ml/kg/hr should be achieved.

Patients with heart murmurs may have decreased ability to compensate under anesthesia, and fluid overload should be avoided. Consider the use of colloids which requires less volume to be administered. Blood pressure should also be carefully monitored, as anesthesia-induced hypotension may result in decompensation. Fluid overload can be assessed by measuring the body weight both pre and post operatively.

In patients with liver disease, anesthetic protocols and monitoring may be affected due to decreased glucose production, albumin production, altered drug metabolism, via cytochrome p-450 enzymes, and the decreased production of clotting factors. Consider the use of reversible injectable agents such as the opioids. It may be tempting to use just inhalant agents such as isoflurane or sevoflurane. These two agents are the most commonly used, but both agents cause cardiovascular and respiratory depression. The cardiovascular depression can lead to hypotension.

Blood glucose levels should be closely monitored in those animals that have severe liver disease, and in patients that are pediatric, septic, or diabetic. Dextrose supplementation may be necessary.

One should always consider if the patient is on any pre-existing drug therapy, such as non-steroidal anti-inflammatories (NSAIDS), diuretics, anticonvulsants, and cardiac medications. These agents may predispose to possible complications such as GI bleeding, renal impairment, electrolyte abnormalities, hypotension, and added depressant effects with anesthesia.

Protocols should be implemented to minimize the amount of time the animal is under anesthesia; therefore, options such as pre-clipping the surgical site with the animal awake should be performed if possible. Preoxygenation of the animal prior to induction will allow for additional time that may be needed to intubate the animal; this is especially helpful with those animals in respiratory distress or with an airway which may be difficult to intubate. Electrocardiography (ECG) and blood pressure monitoring should be in place prior to induction to detect evidence of arrhythmias, hypotension, or cardiovascular collapse that may occur during induction of the critically ill animal. The ECG should be monitored closely for changes in heart rate, rhythm, and for the presence of malignant arrhythmias, which may be more prevalent with trauma, splenic disease, septic peritonitis, hypoxic patients, or gastric dilatation volvulus patients.

Physical examination findings for perfusion, such as the capillary refill time (CRT), mucous membrane color and pulse quality should also be monitored closely. The use of pulse oximetry will add information on hemoglobin saturation (Hb) and oxygenation.

Body temperature should be continuously monitored as anesthetic drugs disrupt normal thermoregulatory mechanisms and hypothermia leads to prolonged recovery. The use of forced warm air heating blankets will help in the treatment of hypothermia.

The critically ill cat has some special considerations. The cat may present with a hypodynamic response, they will be hypothermic, pale, bradycardic, heart rate < 140 bpm, anemic, and often hypoalbuminemic. Feline RBC are particularly susceptible to oxidative damage due to the Hb molecule having eight sulfhydryl groups. Ionized calcium may be decreased. Hypoglycemia is another common finding, and may be related to peripheral glucose utilization, impaired gluconeogenesis, and depleted glycogen stores. Hepatic dysfunction, malnutrition, and shifting of protein production from albumin to acute phase proteins contribute to hypoalbuminemia. In addition, hepatic dysfunction can lead to coagulopathies and mental depression. An elevated serum bilirubin due to RBC lysis, sepsis-induced cholestasis, and hepatic dysfunction may lead to icterus. The lung is the shock organ of the cat and therefore they are particularly susceptible to fluid overload leading to pulmonary edema and pleural effusion. This pulmonary compromise occurs due to increased vascular permeability, myocardial dysfunction, and decreased COP.

The critically ill patient may not exhibit classic pain response symptoms due to their debilitated state, they should be carefully but appropriately treated with analgesic. Analgesia promotes well-being and has a positive effect on the speed and quality of recovery. Pain can lead to catabolism and complications such as delayed wound healing, sepsis, and nosocomial disease. Analgesics are imperative in these painful, critically ill patients, and should be administered as soon as possible following patient assessment to provide a significant benefit. It is vital however, that the underlying disease process be addressed while pain relief is provided, as analgesia therapy may mask the underlying disease processes or the hemodynamic stability of a patient. Ideally analgesics are administered before pain develops, preemptive analgesia, as less drug therapy may then be necessary to control pain. This is especially important prior to surgery or other invasive procedures; however this is not always feasible in trauma cases or in emergent cases.

Because of complex pain pathways, it is often unlikely that one agent alone will completely alleviate pain, regardless of how high the dose is. The use of more than one drug class can improve analgesia by acting at multiple receptor types and may also overcome the problem of varying onset times and durations between drug classes. Patients should be frequently reassessed to ensure that the analgesic regimen is adequate and appropriate. Hemodynamics, respiration, and body temperature should be continuously monitored as sedative drugs can disrupt normal hemodynamics, ventilation and thermoregulatory mechanisms leading to hypothermia, hypotension and hypoventilation.

Opioids act centrally to limit the input of nociceptive information to the central nervous system (CNS), which will reduce central hypersensitivity. Opioids are commonly used in critically ill patients as they have a rapid onset of action, are safe, reversible, and are potent analgesics.

The partial mu agonist, buprenorphine, 5 to 20 μg/kg IV, IM, or SC q6–8h, and mixed mu agonist–kappa antagonist, butorphanol, 0.1 to 0.4 mg/kg IV q4h, reach maximal effect at the upper end of the dose range. If the pain is severe or the analgesia is inadequate, additional doses of partial or mixed agonists–antagonists are unlikely to be effective. Using a pure μ agonist, morphine, hydromorphone, fentanyl, or oxymorphone, would be more effective because there is no upper limit to the analgesia provided by a pure μ agonist. The opioid antagonist, naloxone, binds to the same receptor as agonists but cause no effect, and can competitively displace the agonist from the receptor and therefore reverse the agonist effect.

Potent side effects such as respiratory depression and bradycardia may be seen at the higher dose range with a pure mu agonist; therefore, the higher dose range should be used cautiously in critically ill patients. Anticholinergics are not routinely used unless there is a need to treat the bradycardia. An additional side effect of the μ agonists morphine and meperidine is histamine release, particularly when given rapidly intravenously, this can lead to severe hypotension due to vasodilation. Opioids can create gastroparesis and ileus, which may result in vomition, regurgitation, and aspiration of gastrointestinal contents, particularly in depressed, sedated, weak, or critically ill patients. Vomition is more likely to occur when the mu agonist is given IM or SC.

Opioids can be safely administered to cats to provide analgesia. Hydromorphone or oxymorphone, 0.05 mg/kg titrated slowly IV, can be administered for analgesia; however, side effects such as hyperexcitability or agitation may occur. It has been shown that the onset of mydriasis following administration of opioids correlates with adequate analgesia in cats; continual dosing after achieving mydriasis may result in adverse side effects such as dysphoria and agitation. The mixed partial mu agonist buprenorphine, IV 10 to 20 μg/kg transmucosally-oral, IV, IM q6–8h, has been shown to be an effective analgesic in cats.

One advantage of opioid administration in critically ill patients is that their effects can be reversed if necessary with a pure antagonist such as naloxone, 0.02 to 0.1 mg/kg IV, IM, or SC. Naloxone can reverse CNS depression, respiratory depression, and bradycardia, but the reversal of sedation and analgesia can cause pain, excitement, emergence delirium, aggression, and hyperalgesia Low-dose naloxone, 0.004 mg/kg titrated slowly IV, has been recommended to reverse CNS depression without affecting analgesia. The duration of effect for naloxone is relatively short, 20 to 30 minutes, because of its rapid metabolism in dogs and cats, which may predispose patients to re-narcotization when the drug is used to reverse long-acting opioids. Agonist–antagonists such as butorphanol, 0.05 to 0.1 mg/kg IV, may also be used to reverse sedation and respiratory depression from μ agonists. The benefit of using butorphanol as a reversal agent is that complete reversal of analgesia does not occur because of the κ agonist effects of butorphanol. In contrast, buprenorphine is not as easily reversed as butorphanol or the mu agonists because it is difficult to displace from the receptor.

α_2-Adrenergic agonists bind to receptors in the CNS, leading to sedation, peripheral vasoconstriction, bradycardia, respiratory depression, diuresis, muscle relaxation, and analgesia. Medetomidine and dexmedetomidine are the most common α_2-adrenergic agonist administered in small animals. The sedative effects of medetomidine have a longer duration of action than do the analgesic effects, which last approximately 30 to 90 minutes. Low dose medetomidine, 1 to 10 μg/kg IV, can be safely used in stable patients, or administered in conjunction with opioids to produce analgesic synergism and increase the analgesic duration to as long as 4 hours. In the extremely painful or vicious animal the mu agonist narcotic, morphine, can be combined with the alpha 2 agonist, medetomidine, 5 to 10 ug/kg, IM for enhanced analgesia, sedation, and restraint. At higher doses, medetomidine can be used to sedate distressed animals and for minor procedures such as restraint and analgesia for radiographic positioning. The effects of α_2-adrenergic agonists can be reversed with atipamezole, 0.05 to 0.2 mg/kg IM, SC, or IV, a specific α_2-adrenergic antagonist which reverses analgesia, sedation, and respiratory depression. Intramuscular or subcutaneous routes of administration are preferred for reversal because intravenous administration can lead to abrupt hypotension and/or aggression.

NMDA receptor antagonists block multiple binding sites at this receptor resulting in analgesic, amnestic, psychomimetic, and neuroprotection. Ketamine, a noncompetitive NMDA receptor antagonist, can reverse central hypersensitivity due to prevention of wind-up activity and central sensitization of wide dynamic-range neurons in the dorsal horn of the spinal cord. Ketamine can be administered PO, SQ, IM, and IV. It causes minimal cardiovascular depression, does not depress laryngeal protective reflexes, and produces less ventilatory depression compared with opioids. Side effects include tremors and sedative effects along with increased cardiac output via increased sympathetic stimulation. Oral ketamine can also be used, dogs: 8 to 12 mg/kg PO q6h, to provide pain relief following burn injuries. Subanesthetic or low doses in dogs and cats, 0.1 to 1 mg/kg IV, followed by a CRI, 2 μg/kg/min, may produce analgesic effects without causing anesthesia or profound sedation.

Benzodiazepines are minor tranquilizers, and are most often used in combination with other agents, such as opioids and ketamine as they may cause dysphoria if used alone. The mechanism of action is via potentiation of the inhibitory neurotransmitter GABA. The effects are anti-anxiety, euphoria, dysphoria, anti-convulsant activity, muscle relaxation, and ataxia. They have a wide safety margin and are commonly used as part of an IV induction technique. There may cause mild respiratory depression.

Midazolam, Versed^{R}, has a water soluble base which makes for excellent absorption following IM or SC injections. Duration of action is 1 to 2 hours.

Diazepam, Valium^{R} has a propylene glycol base which makes IM injection painful and absorption unpredicatable. Therefore it is best injected IV. The propylene glycol is implicated in causing hypotension following rapid IV bolus injection, when giving IV give slowly and flush with fluids to decrease the risk of phlebitis. The rate of metabolism is variable depending on the age and species.

Flumazenil, Romazicon^{R}, is a benzodiazepine antagonist, and will reverse all the effects of a benzodiazepine. It is not a controlled drug, but it is expensive. Dose is 0.01 to 0.02 mg/kg IV.

Acetylpromazine, 0.01 to 0.05 mg/kg IV, not to exceed a total dose of 2 mg, may be used in combination with opioids as an anxiolytic and to enhance sedative effects. Acetylpromazine is a major tranquilizer, and should be used with caution due the potential for vasodilation and resultant profound hypotension and hypothermia. Acetylpromazine does not provide analgesia and should not be administered as a single agent if analgesia is desired. It may take up to 15 minutes before the effect of an intravenous dose of acetylpromazine is clinically observed; therefore, repeated doses should be avoided until the full effect is evident. Acetylpromazine can be safely administered to ICU patients if given at low doses, 0.005 to 0.01 mg/kg, in hemodynamically stable animals with adequate respiratory function.

Sedation may be needed for procedures such as ultrasound and biopsy, feeding tube placement, urinary catheter placement, bandage or cast application or removal, bone marrow aspiration, and to facilitate taking radiographs.
These drug techniques can also be used as part of a sedative- premed or induction protocol leading to general anesthesia.

Possible protocols would include:

For a minor procedure such as ultrasound with fine needle aspiration or biopsy of visceral organ:

Butorphanol 0.2 to 0.3 mg/kg IV
Diazepam or midazolam 0.2 mg/kg IV

For a more painful procedure but not requiring general anesthesia:

Hydromorphone 0.05 to 0.1 mg/kg IV
Diazepam or midazolam 0.2 mg/kg IV

If the above are not quite sufficient, titrate propofol 0.5 to 1 mg/kg IV closely watching for decreased respiratory rate, recommend monitoring with pulse oximeter, providing oxygen supplementation, have endotracheal tube and laryngoscope ready if need to intubate. If wish to add additional analgesia, instead of propofol substitute ketamine 1 to 2 mg/kg IV with the same recommendations as for propofol.

When general anesthesia is required a balanced anesthetic technique will help to minimize the side effects from the use of one single agent. Endotracheal intubation is used to control the airway, to provide the ability to ventilate the patient, and to protect the airway from aspiration. A laryngoscope and a variety of endotracheal tube sizes, and a breathing circuit that matches the patient's size should all be readily available. All supplies and machinery should be checked thoroughly prior to induction and intubation. One should be ready to implement intermittent positive pressure ventilation (IPPV), if the patient hypoventilates, becomes apneic or is to undergo a thoracic procedure.

Induction is usually via IV injection. Commonly used agents include a neuroleptanalgesic combination such as hydromorphone, 0.05 to 0.1 mg/kg IV, oxymorphone, 0.05 to 0.1 mg/kg IV, or fentanyl, 5 to 10 ug/kg IV, and diazepam, 0.2 mg/kg IV, or midazolam, 0.2 mg/kg IV, with the addition of ketamine, 0.5 to 4 mg/kg IV, and possibly lidocaine, 1 to 2 mg/kg IV. Lidocaine can be used with those patients that have ventricular arrhythmias, to enhance analgesia, and to help with reperfusion injury as may be seen with a GDV. Propofol at 2 to 4 mg/kg IV can also be used for induction. Propofol is short acting, 5 to 10 minutes, has no analgesic properties, and can lead to hypotension and apnea, therefore it can be used in combination with opioids and tranquilizers. Propofol is useful for a rapid induction, 30 seconds, for those patients that need rapid control of their airway. In some patients following the neuroleptanalgesic administration, mask induction with sevoflurane or isoflurane may be accomplished. The use of multiple agents allows for a lower level of inhalant anesthesia and this may lessen the potential for hypotension.

If a patient undergoing general anesthesia remains hypotensive even following appropriate therapy it may be necessary to consider discontinuing the inhalant anesthetic agent to eliminate the hypotensive effects of the inhalant and continuing the anesthetic maintenance via an injectable drug regimen. It may consist of the use of a constant rate infusion, CRI, of a mu agonist such as fentanyl or morphine in combination with ketamine and lidocaine. Some patients may only need fentanyl as an intermittent IV bolus or as a CRI. Constant rate infusion of analgesics has the advantage of maintaining effective plasma concentrations for continued pain relief, this is especially important in those cases that will be experiencing severe pain. The use of CRI agents for general anesthesia has the advantage of continuing the CRI in the recovery phase. The dose can be decreased as the surgical stimulus is no longer present. There are a variety of combinations that can be used and it may depend on the duration of analgesia and sedation needed as well at the level of pain that is expected. As the patient improves the number of agents as well as the dose can be decreased.

They can be delivered by infusion from a fluid bag or via a syringe pump. The use of a syringe pump is preferred as it will allow for accurate dosing.

The induction plan can be modified to include agents that can also be incorporated into a CRI. The dose of those agents used for induction can also act as the loading dose for a CRI. Following induction the CRI can then be started immediately. A loading dose is important in order to establish plasma drug levels. Inhalants can be used initially at a lower concentration than is normally used due to the presence of the CRI. If it is deemed necessary to discontinue the inhalant the CRI is already in place. The IV induction may be accomplished with the use of a neuroleptanalgesic combination such as hydromorphone, 0.05 to 0.1 mg/kg IV, oxymorphone, 0.05 to 0.1 mg/kg IV, or fentanyl, 5 to 10 ug/kg IV, and diazepam or midazolam, 0.2 mg/kg IV, with the addition of ketamine, 0.5 to 4 mg/kg IV, and possibly lidocaine, 1 to 2 mg/kg IV. The use of multiple agents, hydromorphone, or fentanyl, diazepam, ketamine, and lidocaine, will have a slower onset, but will provide analgesia and is more cardiovascularly sparing.

Ketamine given perioperatively may prevent "wind-up" pain from occurring and thereby reduce postoperative pain. Ketamine causes minimal cardiovascular depression, does not depress laryngeal protective reflexes, and produces less ventilatory depression compared with opioids. With low doses of ketamine, undesirable side effects such as dysphoria or hallucination do not seem to occur. Low-dose ketamine CRIs have been administered for intra- and postoperative analgesia in dogs. A loading dose of ketamine (0.5 mg/kg IV) can be followed by a CRI of 10 μg/kg/min. This should then be reduced to 2 μg/kg/min during the recovery phase. The CRI can be continued postoperatively as needed for analgesia. Ketamine will enhance analgesia and will increase heart rate and blood pressure.

To avoid histamine release which may occur with rapid intravenous morphine administration), a morphine CRI, 0.1 to 1 mg/kg/hr, should be started after administration of an initial loading dose of 0.15 to 0.5 mg/kg IV, diluted given slowly over 5 to 10 minutes. A CRI of morphine, 0.12 mg/kg/hr, reportedly induces effects similar to intramuscular morphine,1 mg/kg q4h, in dogs undergoing laparotomy. Regardless of how morphine is administered, its use may result in bradycardia, hypothermia, and panting. Other opioids instead of morphine can be administered if undesirable side effects of morphine occur.

Butorphanol has been administered at a loading dose of 0.1 mg/kg followed by an infusion of 0.03 to 0.4 mg/kg/hr. This agent does not provide sufficient analgesia for severe pain.

Fentanyl can also be given to enhance analgesia, starting with a loading dose of 2 to 10 μg/kg IV immediately followed by a 0.3 to 1 ug/kg/min CRI in the dog during surgery. A CRI of 2 μg/kg/hr IV, increased as needed up to 5 μg/kg/hr IV is used the post-operative and recovery phase. Fentanyl is also a safe analgesic for cats when administered at a loading dose of 1 μg/kg IV followed by 0.1 to 0.3 ug/kg/min CRI. In some critical cases a CRI is not used, instead intermittent boluses of fentanyl IV are given, 2 to 10 ug/kg. A single dose of fentanyl has a duration of approximately 20 to 30 minutes.

In critically ill animals that are poor anesthetic candidates, fentanyl, in conjunction with propofol, can provided adequate, safe, cardiovascular-sparing anesthesia. Bradycardia may be an adverse effect.

In some patients, such as those with head trauma, a fentanyl CRI, in conjunction with a propofol CRI, can provided adequate, safe, cardiovascular-sparing anesthesia therefore reducing or minimizing the amount of inhalant anesthesia necessary. Side effects such as bradycardia may require treatment with an anticholinergic. The propofol is used as part of the induction technique at a dose of 2 to 4 mg/kg IV with a CRI of 0.1 to 0.4 mg/kg/min. Propofol will help decrease intracranial pressure, but as it is not an analgesic, fentanyl is added. This type of patient will benefit from intermittent positive pressure ventilation.

Lidocaine can also be administered for pain control, and it may retard the effects of compromised viscera, reperfusion injury or ventricular arrhythmias, due its free radical scavenging abilities, analgesic effects, and anti-arrhythmic properties. Its use as CRI is with an initial loading dose of 1 to 2 mg/kg followed by 0.025 mg/kg/min. Lidocaine doses as high as 2 to 3 mg/kg/hr IV have been reported. Recently published research in the cat has shown that CRIs of lidocaine should be used cautiously (if at all) in this species due to cardiopulmonary depression; this would be an especially important consideration in the critically ill animal.

In the cardiovascularly stable patient, medetomidine can be used as a CRI for analgesia with an initial loading dose of 1 ug/kg IV then CRI of 1 to 3 ug/kg/hr. It has been used alone and in combination with morphine, lidocaine, and ketamine. Medetomidine will enhance the analgesia and sedative effects of the other agents. Bear in mind that medetomidine and the opioids are reversible agents, but lidocaine and ketamine are not.

One can always start with a single agent and add others if deemed necessary. Morphine, 3.3 μg/kg/min, lidocaine, 50 μg/kg/min, and ketamine, 10 μg/kg/min, can be administered as a CRI analgesic combination in dogs. These agents can be given separately or mixed together in a single bag or syringe. Using a combination of agents may result in enhanced analgesia through synergism and multiple receptor activation. Ketamine has been found to attenuate and reverse morphine tolerance in rodents and humans, thereby yielding an opioid-sparing effect and providing superior analgesia compared with either drug alone.

When sedating or anesthetizing the critically ill animal drug doses used should begin at the lower dose range and if possible titrated to effect. In addition vigilant monitoring is necessary to minimize possible risks and complications that may occur.

REFERENCES

Upon request.

MRI PULSE SEQUENCES: WHAT'S THE DEAL AND WHY ARE THERE SO MANY?

Anthony P. Pease, DVM, MS, DACVR
East Lansing, MI

INTRODUCTION

Magnetic resonance imaging (MRI) has revolutionized the way that we evaluate soft tissues in both small and large animals. Despite this explosion of information that has largely been extrapolated from what is currently being done with humans, veterinary imaging has largely by-passed research and development and is mainly used for clinical imaging. One main limitation of veterinary medicine and MR imaging is the lack of a set protocol for imaging structures. If you have an MRI done at two institutions, chances are that different sequences will be run based on time constraints as well as clinician preference. Groups are working to try to standardize sequences being used, but since no consensus has been reached at the time of these proceeding, a general guide of sequences available is being presented. The purpose of this lecture is to describe the physics behind MR imaging and cover both basic and advanced imaging sequences.

T1, T2 AND T2* CHARACTERISTICS

All MR imaging is centered around two basic imaging characteristics, T1 and T2 (also called Tissue 1 and Tissue 2). These two characteristics address the two ways that an atom moves. The movement is both based on alignment (T1) and precession or rotation (T2). Both of these characteristics are determined based on the degree of time needed to release energy imparted to the atom during an MRI sequence. When a patient is placed into a magnet field, the atoms in the body mostly align with the magnet field, while a very small number align anti-parallel to the magnetic field. Using a radiofrequency tuned to hydrogen (the most abundant atom in the body) hydrogen atoms are given energy that puts the atom out of alignment with the magnet field as well as make the atoms rotate in unison with the surrounding atoms. The speed at which the atoms return to alignment with the magnet field is called T1, spin-lattice or longitudinal relaxation and the rate that they stop rotating in unison is called T2 or spin-spin relaxation time. All sequences are based around these two characteristics and the time that tissues take to return back to resting states. T2* is the field inhomogeneity that causes T2 relaxation time. The reason that atoms dephase after the initial radiofrequency pulse is due to the interactions and inhomogeneous nature of tissues. This inhomogeneity is not only an interaction that causes T2 relaxation; it can also be measured. The T2* sequence (also called a gradient echo) is a shorter sequence that provides information that looks similar to a T2 signal. The main problem that clinicians have interpreting this is that they use it as a standard T2 signal. Though increased signal on a T2* is usually associated with fluid, it can also be associated with anything that would caused increased dephasing such as contrast medium and paramagnetic substances. In addition, distortions are readily detected due to susceptibility artifacts from air (which is paramagnetic) and hemorrhage.

BASIC IMAGE SEQUENCES: T1, T2, PD-WEIGHTED

Running and MRI is much more complicated than a computed tomography (CT) scanner. With CT, setting the mA and kVp similar to a radiograph as well as selecting the slice thickness is generally the main portion of the scan. With MRI, multiple parameters need to be selected to create an image. These parameters are centered on two basic characteristics that make up an image sequence; the time to echo (TE) and time to repeat (TR). The TE is the time, after the initial pulse, that the image information will be acquired. The TR is the time that is needed after the initial radiofrequency pulse before the next scan is performed. The TR is important because if too little time passes before the next scan, the atoms will still have residual energy from the first pulse sequence and become saturated, preventing images to be obtained. The TE is important because that is the time that the actual image is obtained and therefore determines the characteristic of the scan (whether it is T1 or T2 weighted). If the TE is short (< 30 ms) and TR is short (400-600 ms), then the image will be T1 weighted. If the image has a short TE and long TR (2000-3500 ms), then this is a proton (or spin) density weighted image. Proton density (PD) images provide the anatomic value of a T1 with some T2 signal present from fluid. This sequence is considered especially useful when imaging articular cartilage. If the TE is long (60-150 ms) and the TR is long (1500-3500 ms), this is considered a T2-weighted sequence.[1]

INVERSION RECOVERY: FLAIR AND STIR

In order to fully evaluate pathology in the brain, for example, it is important to tell the difference between pure cerebrospinal fluid and edema. Since edema has protein within it, then its T2 characteristics are different than cerebrospinal fluid. Using FLAIR (Fluid Attenuation Inversion Recovery) sequence, a radiofrequency pulse is sent in initially with a second 90-degree pulse sent in during the optimum relaxation time of water (generally about 1500-2500 ms) to null out the signal that is provided by fluid. The same can be done with fat using an inversion pulse at approximately 150 ms to provide a STIR sequence (Short tau inversion recovery). The purpose of these scans is to provide T2 weighted images that have only pathology being bright rather than fat or normal cerebrospinal fluid.

FAT SUPPRESSION

Though useful, the main limitation to both FLAIR and STIR images are the increased time and decreased resolution of the scan. Though this limitation is minimized with higher strength magnets, motion with FLAIR images can be problematic and therefore STIR images are generally used preferentially. Fat suppressed sequences can also be used instead of a STIR. This is where a radiofrequency pulse is administered prior to the imaging pulse targeted specifically at the fat. The purpose of this is to provide so much energy to the fat that it does not relax during the period of the scan and is therefore not seen. The problem is that unless the tissue (and magnet field) are homogeneous, then fat suppression can be incomplete and therefore, STIR is generally preferred to provide more uniform fat suppression.[1]

DIFFUSION WEIGHTED IMAGING

Looking at fluid to determine if edema is present is generally performed using FLAIR and STIR, but there is another molecular sequence that can characterize the fluid as intracellular or extracellular. Using diffusion weighted imaging, edema within the brain of humans can be characterizing as vasogenic or cytotoxic edema[2] and in animals, this sequence can help identify infarctions compared to gliomas[3]. Diffusion weighted imaging (DWI) uses two sequences and echo planar imaging to provide two images, one is a diffusion weighted image and the other is a calculated image called the apparent diffusion coefficient (ADC) map, which shows if the increased signal from the diffusion weighted image has restricted motion. The information obtained is related to the random motion of water within tissues. If the motion is within the extracellular space (such as vasogenic edema), then the diffusion and ADC images have increased signal compared to the brain, as will the T2 weighted image. This is called T2 shine through and indicates the fluid is extracellular and does not have restricted motion. With cytotoxic edema and infarctions, the water is more restricted and will have an increased signal on the diffusion image but not on the ADC map. One limitation of diffusion weighted imaging is that it is extremely susceptible to air, which causes large distortions. Especially in small animals with large frontal sinuses as well as air within the tympanic bulla can cause distortions that make DWI difficult to interpret.

CONTRAST MEDIUM

Though not a sequence, contrast medium in MRI is slightly different when compared with other modalities. MR contrast medium is made of a paramagnetic substance, in particular a heavy metal called gadolinium. This metal is severely nephrotoxic so it is usually bound with diethylene triamine pentaacetic acid (DTPA) to allow for safe excretion through the kidneys. The contrast medium works by shortening T1 and T2 relaxation times. This means that T1 and T2 signal will be increased in places where the contrast medium accumulates. Since T2 images generally have high signal in regions of edema or increased blood flow, contrast medium imaging is generally performed on T1 weighted images. Complications are considered rare and in people, the main complication is renal failure in patients that have renal compromise. In animals, an allergic reaction can occur such as hives and difficulty breathing, but this is considered extremely rare. Finally, the contrast medium should be used with caution in pregnant animals. There is a theory that the placenta could trap the contrast medium in the fetal circulation until it disassociates with the DTPA and can cause renal damage in utero, though this has never been reported. The dose for contrast medium is generally 1 mL per 10 pounds.

WHAT SEQUENCE DO I USE?

The diversity of sequences is based on the premise that we are manipulating the atom of the hydrogen molecule. Since this is molecule is most abundant in water, it stands to reason that inflammation and edema will show the best using MRI and bone (which has little water) will have poor resolution. The other factor to consider is slice thickness. With multislice CT scanners, clinicians are becoming more use to sub-millimeter slice thickness to provide optimal resolution of small lesions. With MRI, images are based on signal. The smaller the slice, the less signal that is available. This decreased resolution in proportion to slice thickness is also based on field strength. A 0.2 T open field magnet will need a larger slice thickness than a 3 T magnet. A stronger magnetic field means more energy is needed to push the atom out of alignment and more energy is released during dephasing and when the atom returns to the resting state. The trade-off is time. Most images will be acquired at 3 mm slice thickness or larger and will have to be acquired in multiple image planes. Three-dimensional images can be acquired, but this takes a longer acquisition time. The benefit is that the 3D images generally have a smaller slice thickness and can be reprocessed in any image plane, but the drawback is the increased time to acquire the image.

Ultimately, the decision on which sequences to use is based on what lesion are you trying to diagnose. If you have a patient that was hit by a car and you want to look for brain edema, T2, FLAIR and diffusion weighted images would be useful. T1 weighted images will provide the best detail of the bone to look for fractures. All these images can be in the transverse plan, but sagittal and dorsal plane images can be useful as well. Pre-contrast medium and post-contrast medium administration, T1-weighted images are also recommended for central nervous system cases. This is because, unlike CT where bone is white, the bone of the calvaria on T1-weighted images are dark and the contrast medium enhancing meninges are white, making it more easily identified secondary to inflammation.

For musculoskeletal lesions, it is a clinician's preference, but STIR images to look for bone edema, PD images to look at the articular cartilage and the muscles as well as T1 pre and post-contrast medium enhancement images may provide the information desired.

WHAT DOES THE FUTURE HOLD?

To date, all the sequences we are using are directed at the morphology of the soft tissue such as the presence of inflammation, the disruption of tendons or the mass effect associated with a brain tumor. However, for the past five years in humans, sequences are being developed to evaluate the molecular content of soft tissue. From diffusion tensor imaging that allows axon bundles to be seen within the spinal cord, to delayed gadolinium enhanced MRI of cartilage (dGEMRIC), which uses the negatively charged contrast medium to evaluate the proteoglycan content within the articular cartilage, advancements are being made everyday. Recently, these protocols have been applied to both small and large animals to verify that they provide information, but much like in humans, further testing is needed to determine if these sequences can be of prognostic or therapeutic value. With more MR scanners being purchased and available to specialty practices and especially in universities, the hope is that we can return to the research and development of this truly unique modality so that in can be used to its optimum potential.

REFERENCES

[1]Bushburg J, et al. **The Essential Physics of Medical Imaging,** 2nd ed. 2002; 387-401. [2]Hergan K, et al. **Eur Radiol** (2002) 12: 2536-2541. [3]Cervera V, et al. **Vet Radiol Ultrasound** (2011): 52, 1: 33-40.

THE RADIOLOGIST VERSUS THE NEUROLOGIST: MRI CASES OF THE BRAIN

Shannon P. Holmes, DVM, MSc, DACVR & Fred A. Wininger, VMD, DACVIM (Neurology)
Athens, GA & Columbia, MO

Veterinary magnetic resonance imaging (MRI) differs from human medicine in that not all MR brain examinations in North America are interpreted by a board certified radiologist. Liability issues prevail in human medicine and have driven subspecialization in MR interpretation in medical field. In veterinary medicine, all diplomates of the American College of Radiology are required to achieve minimum competency in MR interpretation for certification. The situation is similar in the American College of Veterinary Medicine – Neurology subspecialty. The differences in exposure and experience have resulted in high variability between training programs and clinical expertise as well as a diversity of the images obtained on a routine basis.

Another disparity between radiologists and neurologists lies in their requirements of specialty training. Veterinary radiologists learn physics, techniques and interpretation of 5 different imaging modalities (radiography, sonography, nuclear scintigraphy, computed tomography and MRI) in dogs, cats, horses, and numerous other species. Since radiography and sonography are considered first-line imaging modalities, these areas are emphasized in training programs, particularly radiography. Veterinary neurologists spend at least three years gaining advanced knowledge in neurolocalization, central and peripheral neurologic diseases and in most cases neurosurgery.

No accepted standard sequences for brain evaluation is present in veterinary medicine. However, the goal of all MR examinations of the brain, as well as the skull, is to interrogate these structures for anatomical or physiologic changes, which may explain the clinical signs. The latter is a unique feature of MRI. The development of MR protocols requires a strong foundation in both MR physics and neurologic diseases. Most institutes have a routine imaging protocol for the brain. Among the required sequences are T2-weighted images. This imaging series is particularly useful since abnormal intra- and extracellular fluid accumulation appears has hyperintense (bright/white) lesions. Since most pathologic processes result in some change in intra- or extracellular fluid, this sequence is particularly sensitive to pathologic changes. With respect to intracranial imaging, T2-weighted images are also helpful for depicting the gray and white matter regions of the brain, seen as light and dark gray signal intensity tissues, respectively. T2-weighted fluid attenuation inversion recovery or FLAIR images are included. Cerebrospinal fluid (CSF) or other fluids with low cellular or protein content have the low signal (black), associated with suppression of fluid signal. This sequence is particularly helpful for distinguishing parenchymal lesions located adjacent to the ventricular system or arachnoid space. T1-weighted images are included in virtually every examination of the brain. This imaging series is used to assess the integrity of the blood-brain-barrier and provides a better view of anatomy, whereas T2 is used to look at pathology. T1-weighted images are acquired prior to and following the intravenous administration of gadolinium-based MR contrast. Gadolinium-based MR contrast agents rapidly diffuse into the extracellular spaces, except into normal brain parenchyma. The choroid plexuses will markedly enhance in normal brain examinations. Normal brain parenchyma should not enhance because the agent cannot cross a normal or intact blood-brain-barrier.

Some institutions will add a couple other sequences in their routine examination, whereas others will add these based on the clinical presentation of the case.[1] T2*-weighted gradient recalled echo (GRE) images are used for hemorrhage detected. The heme moiety produces a susceptibility artifact which is depicted in the T2* images. Microhemorrhages have been detected in canine patients using this imaging sequence, which may not be seen on the other imaging series or at gross necropsy. Acute exacerbation of clinical signs has also been seen in association with acute hemorrhage or neoplasms. The time of hemorrhage can also be determined by correlating the differences in signal intensity on T1- and T2-weighted images. Diffusion weighted imaging (DWI) has increasing popularity and availability on MR units used in veterinary diagnostic imaging. Detection and delineation of vascular accidents is the primary application of this imaging sequence. In human imaging, this sequence is more sensitive than routine imaging sequences for the detection of (meningo)encephalitis. Necrotic foci associated with necrotizing meningoencephalitis (NME) can be detected with this imaging sequence, but the clinical significance of this sequence in veterinary medicine is undetermined at this time.

In our opinion, veterinary MR is a more dynamic imaging study than is typically thought, especially when compared to human MR examination. In human brain MR studies, in instances when additional sequences are requested by the radiologist or needed for more specific diagnosis, a second brain MR examination is performed. Since veterinary MR procedures require general anesthesia and are an out-of-pocket expense for owners, the objective of these studies needs to get as complete a study as possible on the initial examination. This requires pre-planning with consideration of the presenting history, clinical signs and physical examination findings. Additional sequences may be needed to address patient-specific conditions, which may not be fully delineated using the routine brain MR protocol for that institution. For these reasons, most MR examinations are directly supervised by a radiologist or internist to allow sequences to be added based as the sequences are acquired and interpreted.

Some case examples of prospective MR protocol planning would include clinical signs of blindness, pituitary-dependent hyperadrenocorticism, isolated cranial nerve disease or vestibular signs. In a patient that presents with blindness, imaging sequences may be added to the routine brain exam to interrogate the eyes, retrobulbar spaces and extra-calvarial ocular nervous system.[2] Fat-suppressed imaging series are particularly helpful in the evaluation of the extra-calvarial optic nerves, as well as the remaining retrobulbar structures. Similar sequences may be repeated with different imaging planes to isolate the optic nerves oriented rostrodorsolateral to caudoventromedial, because in-plane resolution is optimized for these structures. Pituitary-dependent hyperadrenocorticism can be due to a micro- or macroadenoma.[3] The latter is easily detected with a routine brain protocol, since the size of canine pituitary glands is well-established. Microadenomas can only be diagnosed using a dynamic contrast enhanced study. For patients with facial paralysis, we will typically add a thin slice, high resolution imaging series to a routine examination. This study is focused on the area of these nerves, which typically are seen in a single MR image.[4] Disease of the inner ear can be similarly challenging to diagnose, particularly in cases where the anatomy is unaltered. The exquisite soft tissue contrast of MR can delineate subtle signal intensity changes in the fluid within the semicircular canals. On T2-weighted images, this fluid is normally hyperintense and loss of this hyperintensity corresponds to a MR diagnosis of otitis interna.

Another dynamic component of a brain MR examination occurs in response to abnormalities identified in the initial and routine imaging series. The changes that occur in this instance are performed to better delineate the anatomic and/or chemical characteristics of the pathologic change. The identification of nasal disease may necessitate the need for thin slice or volume (3D) acquisition to obtain high resolution images of the skull that facilitate the assessment of cribiform plate integrity. This is considered difficult since the cribriform plate has a little amount of water, which is the main component that allows for signal intensity on MRI. Vascular examinations are performed in cases in which surgical excision is a consideration. MR angiography (arteriography and/or venography) can be performed similar to computed tomography angiography (CTA).[5] The difference between MRA and CTA is that contrast administration is not required for MRA using sequences called time-of-flight. MR venography has proven particularly useful in surgical approaches into the caudal fossa, since significant blood loss can occur in association with the transverse sinuses.

Imaging series	**Signal Intensity (SI)**		
	Bright	**Intermediate**	**Dark**
T1	Fat Slow flowing blood Paramagnetic CAs Methemoglobin	Muscle Brain Fluid/Cysts	Bone/Calcification Fast flowing blood Tendons
T2	Fluid (CSF, synovial) *Edema/Inflammation* *Some tumors*	Muscle Brain (Gray-White distinction)	Bone/Calcification Hemosiderin T2 CAs
Proton Density (PD)	Superior gray-white matter distinction.		
FLAIR	↳ Suppression of low protein or cellular fluids (like CSF) T1- or T2-weighted, similar image SIs to above.		
Short Tau Inversion Recovery (STIR)	↳ Increased conspicuity of neoplastic lesions Can be T1- or T2-weighted, similar image SIs to above. ***FAT*** *is most commonly suppressed tissue, but any tissue can be* *T2-weighting is most common (bright pathology on dark background).*		
Gradient Echo (GRE) Also called T2*	Can be T1-, PD, or T2 weighted. Fluid & flowing blood commonly bright. Hemorrhage produces "blooming" signal voids (specifically T2*-weighted GRE).		
DWI	Signal intensity on diffusion weighted images distinguished free from restricted water motion, like in CSF and intracellular fluid, respectively.		

In conclusion, there are two components to getting the most out of brain MR examinations. Some cases can be adequately evaluated with routine examinations, particularly when an experienced, qualified specialist performs the interpretation. The most information is gained from examinations that have been planned with acknowledgement of the individual patients' condition and/or expanded or altered with identification of pathologic changes that benefit from better characterization. MR examinations can uniquely characterize intracranial and/or extracranial soft tissue pathology because different tissue characteristics can be exploited through the application of specific imaging series. This requires a strong knowledge of MR physics and neurologic disease processes. Therefore, a symbiotic relationship between a radiologist and neurologist best serves veterinary patients undergoing MR brain examinations.

REFERENCES

1) Gavin, P.R., & Bagley, R.S. Practical Small Animal MRI (2009) 2) Penninck D, et al. **Clin Tech Small Anim Pract**. 2001;16(1):22-39. 3) van der Vlugt-Meijer, R.H., et al. **Mol.Cell.Endocrinol.** 2002;197(1-2):81-7. 4) Varejão A.S., et al. **Vet Radiol Ultrasound**. 2006;47(4):328-33. 5) Martin-Vaquero P, et al. **Am J Vet Res**. 2011;72(3):350-6.

THE RADIOLOGIST VERSUS THE NEUROLOGIST: MRI CASES OF THE SPINE

Shannon P. Holmes, DVM, MSc, DACVR & Fred A. Wininger, VMD, DACVIM (Neurology)
Athens, GA & Columbia, MO

Cross sectional imaging of the spine is an essential diagnostic tool for neurologist in daily veterinary speciality practice. The first consideration, however, is that of appropriate neurolocalization. Although computerized tomography (CT) and magnetic resonance imaging (MRI) provide important structural detail, there is simply no replacement for the findings of a thorough neurologic examination. The rationale behind this is that imaging tests complement findings of the physical examination, which measure neurologic function. The practitioner may be presented with functional diseases that have no identifiable structural deficit (such as degenerative myelopathy) or structural disease that have no functional deficit (lumbosacral compression).

Neurolocalization is the ultimate goal of the neurologic examination, piecing the puzzle of reflexes and responses together to an anatomic segment. Once accomplished, the practitioner is able to assign a rank list of differential diagnosis for this neuroanatomic site based on the patient's age, breed, clinical onset/ progression and the presence of spinal hyperesthesia. While these differential lists can easily be found in any neurology text, the ability to interpret clinical findings and appropriately combine them to fit a single lesion site requires practice. Both board certified radiologists and neurologists are expected to have a strong knowledge of neurolocalization. The difference is that the neurologist has first-hand knowledge, since they have performed the examination. The radiologist is dependent on the information provided to them by the primary clinician. Unbiased image interpretation is more easily achieved by the radiologist, but interpretation of subtle lesions or clinical signs often requires the knowledge of the case provided by the neurologist.

A concept integral in localization is that of the upper motor neuron (UMN) and lower motor neuron (LMN). The UMNs are essentially the long tracts from the brain that instruct the LMN what to do. The LMN consists of the motor neuron cell body within the spinal cord, the nerve itself, the neuromuscular junction and the innervated muscle. In cases of UMN disease, the lower motor neuron works without instruction, known as a loss of inhibition. It is classically hypertonic, with normal to increased reflexes and minimal atrophy. In cases of LMN disease, the upper motor neuron is irrelevant, because the LMN cannot respond to its instruction. Thus, hypotonicity, decreased reflexes and severe neurogenic atrophy are noted. It is important to note that the current resolution settings of routine MRI protocols do not permit visualization of individual nerves. Instead, groups of nerves are evaluated based on their known anatomic location.

The spinal cord can be divided into four areas. The LMN of the limbs are located in widened areas of the spinal cord known as intumescences. When these intumescences are the site of disease, LMN signs are notable. When the lesion blocks the long tracts between the limb the brain and the LMN, UMN signs emerge. Thus, the spinal cord can be divided into the following segments. It is most common for MR spine protocols to be developed to evaluate areas of neurolocalization. Our imaging protocols have fields of view that encompass the region of neurolocalization. A cervical spine MR examination will consist of at least one imaging series, most commonly acquired in the sagittal plane, with anatomic landmarks of the caudal fossa cranially and the midbody of the thoracic vertebral column caudally. Examination of the thoracolumbar spine (ie. T3-L3 localization) requires individual imaging series of the thoracic and lumbar spine for complete evaluation, while maintaining image resolution. Two sets of sagittal plane images are acquired with anatomic landmarks for the thoracic spine of C6-7 intervertebral disc cranially through L1-2 intervertebral disc space caudally and T12-13 intervertebral disc through the sacrum. This is dependent of the field of view of the MR scanner. Some low field systems have smaller fields of view and higher field systems have larger fields of view allowing for less image sequences to see the entire spine. In our experience, it is important to extend the imaging field of view beyond the regions of neurolocalization, since there are inherent limitations of neurologic examination specificity.[1]

Neurolocalization for planning MR examination region of interest

Lesion location	Thoracic limbs	Pelvic limbs
C1-C5	UMN	UMN
C6-T2	LMN	UMN
T3-L3	Normal	UMN
L4-S3	Normal	LMN

To complete the neurologic examination, the practitioner should determine lateralization of the lesion and the presence of spinal hyperesthesia. Lateralization is important in that if two lesions are found within the localized segment, lateralization of signs can prioritize the clinically significant lesion. Spinal hyperesthesia is a noxious response to a non-noxious stimulus. The presence of hyperesthesia suggests disease of the structures outside the spinal cord parenchyma (meninges, nerve roots, bone), because the spinal cord itself does not have nocioceptors.

Once the correct spinal segment(s) has/have been selected, the choice of neuroimaging modality is based on availability, cost, acquisition time and overall diagnostic yield. Tomographic (cross sectional) techniques, such a CT and MRI are always going to be preferred over radiographic techniques because of the utility in evaluating the spine in cross-sectional planes. CT has the advantage of speed of acquisition, cost, superior osseous resolution and post-processing 3D spatial resolving properties. Because of these variables, it is often the modality of choice for traumatic lesions or acute intervertebral disc disease, in which vertebral fractures/(sub)luxations, hemorrhage and mineralized discs can be detected. However, CT has limited ability to identify the spinal cord itself without the assistance of myelography. Evaluation of the spinal cord and associated soft tissues is further compromised by the limited soft tissue contrast resolution of CT. This makes identifying changes with spinal cord parenchyma difficult to impossible that is only slightly improved with contrast medium administration (IV or myelographic). MRI has become the gold standard in spinal imaging and the diagnostic of choice because of its contrast resolving properties and overall superior lesion conspicuity.[2]

The standard MR sequence that is performed at our hospitals includes T2 weighted (T2W) sagittal, Short Tau Inversion Recovery (STIR) dorsal plane, and T2 weighted transverse plane sequence through the region of interest, as identified by the first two scans. The T2W images most easily identify pathologic changes, while maintaining anatomical orientation. The STIR dorsal provides the advantages of its high lesion conspicuity in muscle and bone by removing the high signal associated with fat tissues. It is also useful in localization of the ribs for surgical orientation. Other sequences such as T1 weighted +/- contrast, T2* weighted imaging and MR venography are added, as deemed appropriate. HASTE imaging can also be performed to evaluate the MR fluid content as well as looking for regions of compression or altered fluid characteristics. A key point is that the imaging modality must evaluate the neurolocalization in its entirety. For example, if a patient were to have a T3-L3 myelopathy, there could be single or multiple lesions within that segment that would lead to the identical clinical signs. Prematurely terminating a study because of identification of a single lesion could lead to incomplete diagnosis and potentially treatment inefficacy.

Common structural degenerative diseases are most well delineated with MRI; these include caudal cervical malformation syndrome (CCSM) (commonly known as Wobbler's) and lumbosacral stenosis (LSS). CCSM is often stratified into the dorsolateral form that affects young giant breed dogs and disc-associated form that affects large breeds, specifically the Doberman Pinscher. In both disease processes, extradural osseous and/or fibrous proliferative tissue compression of the spinal cord can be noted on MRI. In the dorsolateral form, the spinal cord takes on an hourglass appearance because of proliferation at the level of the dorsal articulations and in some cases the interarcuate ligament. Synovial cysts are often found and can be a clinical cause of compression. Hansen type II intervertebral disc disease is a feature of disc-associated CCSM. In both cases, it is common to identify signal change within the spinal cord parenchyma, which may represent edema, gliosis, syringohydromyelia or malacia. The prognostic significance of this finding is unknown. In some cases, the intramedullary signal change is the only MR finding and is thought to be a reason this disease process can be underestimated with CT. When the intramedullary change is seen in the absence of extradural spinal cord compression or impingement, re-imaging with the neck hyperextended or hyperflexed may delineate dynamic extradural compression. This is an example of how a routine MR exam needs to be amended during the course of the procedure to acquire complete delineation of the clinical disease process.[3]

LSS is a frequent finding in MR of the lumbar spine. Since the imaging incidence is much higher than the frequency of clinical findings, caution must be exercised in interpreting the clinical importance of this lesion. Imaging findings include compression ventrally from a Hansen type II intervertebral disc disease and dorsally from spondylolisthesis, synovial proliferation/cyst formation and ligamentum flavum hypertrophy. Foraminal stenosis may also be noted as loss of fat signal on parasagittal images. At our hospital, we will repeat MR imaging with flexion of the coxofemoral joints of views, once a diagnosis has been made in the transverse and sagittal planes of the routine MR examination. These imaging series can show the level of structural improvement that can be achieved by surgical distraction and stabilization alone.[4]

MR has large value in the diagnosis and treatment of spinal neoplasms. Because of its high sensitivity, MRI has a superior ability to not only identify the lesion of interest, but also identify other related subclinical lesions. This is especially important when differentiating focal (i.e. sarcomas) from multifocal lesions (ie. round cell tumors) as the treatment and prognosis for these are quite different. We have found STIR sagittal images to be the most sensitive in identifying multifocal lesions, particularly polyostotic diseases.[5] Fat-suppressed imaging (STIR, fat-suppressed T2W or fat-suppressed post-contrast T1W) have proven particularly useful for tumor detection, since tumor cells have high intracellular fluid content and therefore are best seen as hyperintense lesions (bright/white) on a hypointense (black) background. The human eye best detects lesions in this format. With respect to spinal nerve neoplasms, fat-suppression is useful since most spinal nerves are embedded in fat and can be difficult to resolve.

If a vertebral column neoplastic process is identified, screening of the thorax and abdomen can be performed although motion and large slice thickness can limit the resolution obtained when compared to CT. Breath-hold proton density or T2* images, acquired in the dorsal plane, have been used to screen for pulmonary metastases and intra-thoracic lymph node enlargement. Abdominal screening is particularly useful in large dogs, where assessment of the dorsal aspect of the parenchymal organs can be challenging with routine ultrasound examinations. However, dynamic contrast MR enhancement of the liver and spleen lesions has unmatched sensitivity and specificity for malignant processes, reported at 100% and 90%, respectively. These exams can also define polysystemic neoplastic disease processes, versus those isolated to the vertebral canal. Identification of these disease processes allows marked refinement of the list of differential diagnoses, as well as prognosis.

The value of MR in infectious/inflammatory disease is gaining greater appreciation as its use increases. Diskospondylitis is a diagnosis classically made by radiographic appearance. However, it is well recognized that the number of false negative by this modality can be high, especially early in the disease course. Thirty to fifty percent bone loss is required for the human eye to identify the endplate osteolysis needed for this radiographic diagnosis. Recent studies have suggested that the frequency of multifocal diskospondylitis is higher than previously thought. Although no formal evidence based studies have been performed in the veterinary literature, the sensitivity of MR to detect these lesions is thought to be high. In general, the disk itself is degenerate with surrounding hyperintensity of the adjacent bone and surrounding soft tissues, including the supportive musculature. Pathologic fractures may be present. Both radiographs and CT, in our experience, markedly underestimate the soft tissue changes caused by diskospondylitis both within and outside the vertebral canal. There may be value in MR as a more sensitive indicator of antibiotic efficacy, as application of radiographic changes has equated with very protracted antibiotic therapeutic regimens.

Vascular events are now recognized with more frequency on MRI. Fibrocartilagenous emboli are often wedge-shaped hyperintense lesions on T2W images, focused in the vascular zone of the spinal cord parenchyma (ie. water-shed lesion). They are also often seen adjacent to intervertebral discs with a partial loss of signal within the nucleus pulposus. Efforts have been made to distinguish these vascular events from acute, traumatic disc extrusions (also known as Hansen Type III discs). Some radiologists and/or neurologists may be inclined to differentiate these lesions based on their position relative to an implicated disc; however, there has been no defined means of definitively distinguishing these pathologies by imaging alone. The importance of this distinction is questionable, since it minimally alters treatment.

Syringohydromyelia is a condition diagnosed with significantly higher frequency, since the increased use of MR in veterinary medicine. While often associated with caudal occipital malformation syndrome, syrinx formation can be caused by any intracranial disease that leads to abnormalities in CSF flow or increased CSF formation, such as choroid plexus tumors. These lesions are most frequently identified in the cervical spine, but can be seen along the entire spinal cord length. They are differentiated from other causes of spinal cord pathology by their suppression of signal on FLAIR imaging. We have identified signal changes in the spinal cord prior to syrinx formation that do not suppress on FLAIR imaging, referred to as pre-syringomyelia formation. These may represent areas of mechanical myelitis from the forces that result in a syrinx. Numerous groups around the world have research the presence and significance of syringohydromyelia. Cavalier King Charles spaniels and Brussels Griffons are genetically predisposed and MRI screening has been proposed.[7,8] The challenge, with respect to breeding programs, is that structural disease may never manifest clinically. Therefore, at this stage, prognostication from MR exam findings is not possible.

Like human medicine, MRI is considered the gold standard for examination of the vertebral canal. There are still numerous institutions that perform myelography and CT examinations of the spine. In our opinion, if available, the vertebral column should be examined with MRI. We have experienced cases of misdiagnosis with myelography and CT. One interesting feature of myelography, when an MR examination has been performed following a myelogram, is the extent of myelitis associated with the injection site, presumably since the needle can be passed through the cord to inject contrast medium in the ventral subarachnoid space. Even in acute disc herniations in Dachshunds, CT can be challenging in distinguishing the acute and clinically significant disc herniation from multiple sites of chronic disc herniation. MR of the vertebral column has also facilitated minimally invasive diagnostics, with high success, whether is fluoroscopic-guided biopsy of a vertebral bone neoplasms or diskospondylitis. The superior soft tissue contrast of MRI has translated into better diagnosis of spinal disease, discrimination of surgical diseases from non-surgical disease and the best pre-surgical planning since margins and disease extent is evident because of differences in tissue chemistry.

REFERENCES

1) DeLahunta A. & Glass, **E. Veterinary Neuroantomy & Clinical Neurology**. 2008. 2) Gavin, P.R. & Bagley, R.S. **Practical Small Animal MRI**. 2009. 3) da Costa, R.C. **Vet Clin North Am Small Anim Pract**. 2010; 40(5):881-913. 4) Meij B.P., et al. **Vet Clin North Am Small Anim Pract**. 2010; 40(5):983-1009. 5) Wininger, F.A. et al. **2010 ACVIM Proceedings.** 6) Burkert B.A. et al. **J Am Vet Med Assoc.**; 227(2):268-75. 7) Rushbridge, C. et al. **J Small Anim Pract**. 2009; 50(8):386-93. 8) Cappello, R. et al. **Vet Surg.** 2007; 36(5):509-12.

THE RADIOLOGIST VS. THE WORLD: MRI CASES OF NON-NEUROLOGIC DISEASE

Anthony P. Pease, DVM, MS, DACVR
East Lansing, MI

INTRODUCTION

Magnetic resonance imaging (MRI) is considered the goal standard for imaging neurologic structures, but work exploring other uses for this unique modality is on-going. From imaging the nasal cavity to determine the presence of a portosystemic shunt, MRI takes advantage of imaging water to look for inflammation and vascular supply without the use of contrast medium. The cases presented during this hour show the diversity of diseases that MRI can image in the small animal patient.

CASE 1: OCULAR TRAUMA

A 1.5 year old female spayed mixed breed dog was brought to the ophthalmology service for evaluation of the dog's right eye. One month prior, the patient had encountered a porcupine and had several qulls removed from her chin and from her tongue and cheek. Ten days prior to presentation, the patient had right orbital swelling and a mucopurulent discharge from the right eye. The following information was obtained:

Schirmer tear test: 17 mm/min OD 20 mm/min OS
Fluorescein stain: negative uptake OD negative uptake OS
Intraocular pressure: 20 mmHg OD 9 mmHg OS
Complete Blood Count (values outside the reference range)
Tot Protein 7.7 H [6.0-7.4] g/dL
RDW 13.1 H [11.0-13.0]%
WBC 15.56 H [5.90-11.60] x10.e3/uL
Auto Neutrophil # 10.82 H [3.80-7.80] x10.e3/uL
Lymphocyte # 3.55 H [1.10-3.10] x10.e3/uL
Monocyte # 0.85 H [0.10-0.70] x10.e3/uL
Basophil # 0.13 H [0.00-0.00] x10.e3/uL

Magnetic resonance imaging was performed to evaluate the swelling and see the soft tissues of the eye (Figure 1). In the right retrobulbar space, there is increased signal on T2-weighted and T1-weighted contrast medium enhanced images. In addition, a small, linear low signal region is seen, possibly due to a porcupine quill. In addition, the retina in the right globe is detached. The MRI allowed not only to see the inflammation, but also shows the low attenuating region due to the foreign body.

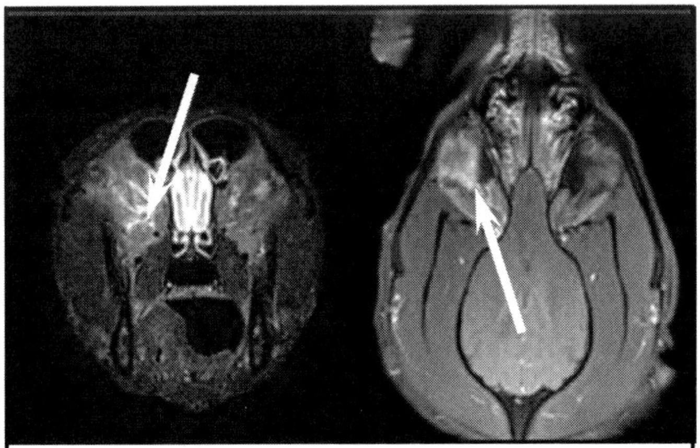

Figure 1: The T2 weighted (left) and T1 weighted with contrast medium (right) image of the head. Note the increased signal in the retrobulbar space due to inflammation and the dark area (arrow) presumed to be a porcupine quill.

CASE 2: NASAL DISCHARGE

An 8 year old neutered male Persian cat presents for a 6 month history of nasal discharge. Recently, the discharge has become bloody and a small buldge was found on bridge of the nose. MRI was chosen to evaluate the nasal cavity because the use of T1 and T2 weighted images can show the presence of soft tissue and fluid as well as using a T2 gradient echo sequence to look for signs of hemorrhage to look for acute bleeding (Figure 2). A heterogeneous contrast medium enhancing mass is present in the caudal nasal cavity that has areas of fluid based on the T2 weighting. This mass cause lysis of multiple bones of the skull, seen best on the T1 image and the T1 post contrast medium image. In addition, edema is seen within the cranial vault as well as contrast medium enhancement of the meninges and extension of the mass into the cranial vault. This appearance is indicative of nasal adenocarcinoma. The MRI provides the ability to see turbinate lysis, but also has

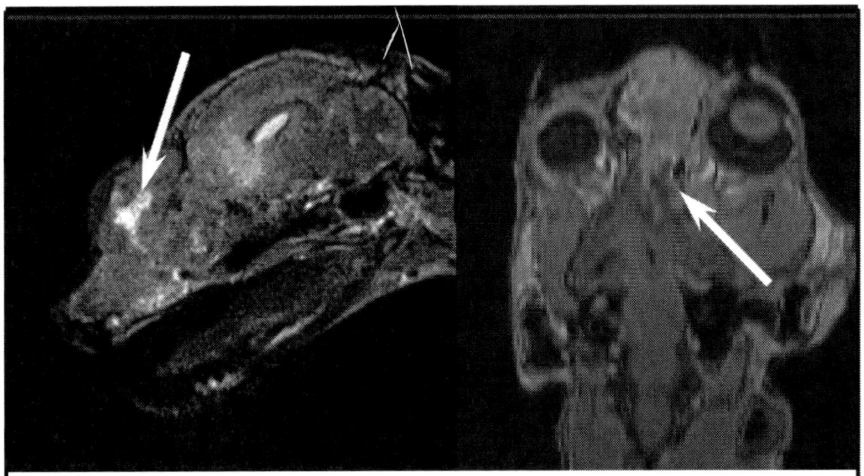

Figure 2: The T2 weighted (left) and T1 weighted with contrast medium (right) image of the head. Note the large fluid cavity on the T2 weighted image and the meningeal enhancement with contrast medium (right).

superior contrast resolution to show the extension into the cranial vault both with the contrast medium enhancement images as well as the T2 weighted images showing edema. Since MRI makes bone dark, meningeal enhancement with contrast medium can be seen as compared to computed tomography.[1]

CASE 3: RETROBULBAR MASS

A 12 year old female spayed domestic short haired cat presented for protrusion of the nictitans of the left eye. Physical examination findings were unremarkable. Ultrasound found a hyperechoic region along the medial aspect of the left orbit and conjunctival cytology revealed a mixed population of epithelial and inflammatory cells with no neoplastic cells noted. MRI was performed to evaluate the retrobulbar mass (Figure 3). In the region of the left first molar tooth, there is a soft tissue, contrast medium enhancing mass that extends into the oral cavity and left retrobulbar space. This mass has the characteristics of granulation tissue and a presumptive diagnosis of a chronic tooth root abscess with granulation tissue. Biopsies were taken of mass from the oral cavity and a diagnosis of squamous cell carcinoma was obtained. This case is presented to show the characteristics of an oral mass and illustrate the complexity of diseases even using MRI.

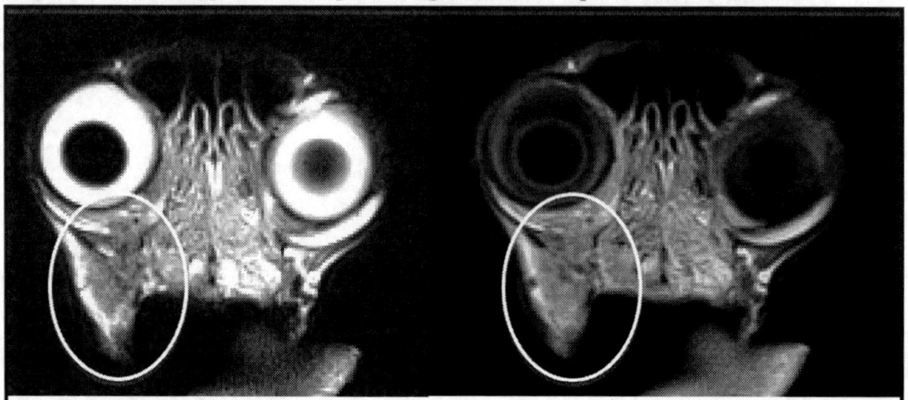

Figure 3: The T2 weighted (left) and T1 weighted with contrast medium. Note the relative gray color of the tissue and the extension into the oral cavity and retrobulbar space.

CASE 4: Portosystemic shunt

A 2 year old mixed breed dog presented for elevated bile acids and intermittent seizures. MRI was performed using a time-of-flight acquisition as well as T1 weighted images with contrast medium.[2] Time-of-flight is a specialized sequence that allows for the use of a specific pulse sequence that energizes red blood cells that travel either from cranial to caudal (arterial in the case of the abdomen) or caudal to cranial (venous). In this case, a caudal to cranial time of flight was performed. This sequence provides increased signal within the vessels without the need for contrast medium and a splenocaval shunt was identified.

CONCLUSION

If time permits, further abdominal and thoracic cases will be presented. The main drawback of MRI in non-neurologic, non-musculoskeletal cases is motion artifact. Whether pulsitile artifact or motion artifact, which both occur in the phase encoding direction, the time required for acquisition and the need for larger slice thickness makes the additional information about fluid within the structures or the identification of vasculature less important compared to the decreased resolution and increased cost. That said, MRI is continuing to be studied to determine if specific sequences or contrast medium enhancing techniques can be used to provide functional data as well as anatomic information to best guide treatment or provide a better prognosis. From whole body MR imaging in cancer patients[3] to imaging pheochromocytomas[4], MRI is emerging as the modality of choice for soft tissue imaging. Determining characteristics of liver and splenic masses[5] similar to what is being performed with ultrasound contrast medium and computed tomography has increased the usefulness of this modality, but the small patient populations have made this usefulness limited. At present, computed tomography and magnetic resonance imaging serve to compliment each other on providing superior bone and soft tissue detail respectively.

REFERNECES

[1]Moore M, et al. **Vet Radiol Ultrasound** (1991); 32, 1: 19-25. [2]Seguin B, et al. **Vet Radiol Ultrasound** (1999); 40, 3: 251-258. [3]Kraft S, et al. **Vet Radiol Ultrasound** (2007); 48, 3: 212-220. [4]Rosenstein D. **Vet Radiol Ultrasound** (2000); 41, 6: 499-506.[5]Clifford C, et al. **J Vet Intern Med** (2004); 18: 330-338.

DIAGNOSIS & TREATMENT OF FELINE CARDIOMYOPATHIES

Deborah M. Fine, DVM, MS, DACVIM-Cardiology
Columbia, MO

INTRODUCTION

Feline cardiomyopathies represent a significant percentage of adult-onset disease in cats, but are frequently challenging to diagnose. Cardiomyopathy in cats is typically characterized by a long occult period in which only subtle physical examination or echocardiographic abnormalities may be present. Treatment during the asymptomatic stage is controversial, and no therapy has been shown to prolong the symptom-free interval. Therefore, educating clients about the clinical signs of heart disease, and the theoretical pros and cons of attempting therapy during the symptom-free interval is important.

HYPERTROPHIC CARDIOMYOPATHY (HCM)[2]

HCM is the most common feline cardiomyopathy. Affected animals range from 1 to 16 years of age, with a mean age of about 6.5 years. There is a male predominance. Predisposed breeds include Maine Coon Cats, Ragdolls, Devon Rex, and Domestic Shorthairs. HCM is characterized by concentric left ventricular (LV) hypertrophy in the absence of an underlying disorder such as hyperthyroidism or hypertension. The primary pathophysiological abnormality is LV diastolic dysfunction as a result of increased wall thickness and impaired ventricular relaxation.

In most instances the etiology of HCM is not known. The results of breeding studies with Maine coon cats indicate an autosomal dominant heritable basis for the disease. In humans, many genetic defects have been found to be associated with this cardiac abnormality. In Maine coon cats and Ragdoll cats a mutation in myosin binding protein C has been found in some (but not all) affected cats. The myosin binding protein C mutation has not been found in HCM affected cats of any other breed.

Clinical signs

Most cats with HCM have no clinical signs and the owners are unaware of any problem. The diagnosis of HCM is usually made by echocardiography after a heart murmur, gallop rhythm, and/or arrhythmia is detected as an incidental finding on routine physical examination. However, the first indication of a cardiac abnormality may be the development of congestive heart failure and/or with sudden hindquarter paresis/paralysis caused by thromboembolism. In rare instances, sudden death may occur.

Electrocardiography

A variety of abnormalities have been reported. Changes consistent with left atrial enlargement (wide P waves) and left ventricular enlargement (tall R waves) are common, and virtually any form of arrhythmia may be found. A left axis shift of the mean electrical axis due to a left anterior fascicular block is a frequent, though non-specific marker of HCM.

Radiography

Thoracic radiographs of asymptomatic cats with HCM may be normal or show varying degrees of left atrial enlargement and rounding of the left ventricular apex on the lateral and DV/VD view. More obvious cardiomegaly with left atrial enlargement, pulmonary venous congestion, and interstitial and/or alveolar edema is observed in cats with congestive heart failure. In contrast to dogs with heart failure, edema is usually found in the dependent regions of the chest, and many cats also exhibit severe pleural effusion. The pleural effusion may be a modified transudate, pseudochylous, or chylous fluid.

Echocardiography

The LV free wall or interventricular septal thickness in diastole is increased (> 5.5 mm). The hypertrophy may be symmetric (throughout the septum and LV wall) or asymmetric (confined to focal segments). The left atrium is usually dilated due to increased end-diastolic pressure. The finding of a slightly thickened LV wall or septum in diastole, without left atrial dilation, is insufficient for a diagnosis of HCM. Simple volume depletion/dehydration in a cat will lead to mild wall thickening without left atrial dilation. Other abnormalities which may be detected during echocardiography include systolic anterior motion of the mitral valve. This causes simultaneous mitral valve regurgitation and high velocity turbulent flow in the LV outflow tract demonstrated with Doppler. Concentric LV hypertrophy in combination with systolic anterior motion of the mitral valve is known as hypertrophic obstructive cardiomyopathy (HOCM). Spontaneous contrast ("smoke") or a thrombus in the left atrium are very ominous findings.

Other Diagnostics

HCM is a diagnosis of exclusion, with hyperthyroidism and hypertension being the most important differentials. Cats over the age of 8 with echocardiographic evidence of LV hypertrophy should have a serum T4 concentration measured. All cats should have a systolic blood pressure measurement obtained. Doppler methodology is generally more reliable than oscillometric methods in awake cats. Cats diagnosed with hypertension should be worked up for an underlying cause as idiopathic hypertension is very rare. Chronic renal disease is the most common cause of hypertension in the cat.

Treatment and Prognosis

There are no prospective or retrospective studies that have specifically examined the question of optimal therapy for asymptomatic cats with HCM. Retrospective studies which have looked at the natural history of feline HCM have shown that median survival times for asymptomatic cats is remarkably long: > 5 years on an older study[3] and > 9 years in a recent one.[4] Due to the retrospective nature of these studies, no consistent treatment protocol was performed. Therapies included beta blockers, calcium channel blockers, ACE inhibitors, or no treatment at all. My approach is to discuss the pros and cons of each option with the owner and allow them to choose. However, my personal bias is that treatment of mildly affected cats is unlikely to be beneficial.

On the other hand, cats that present with congestive heart failure require prompt treatment – intravenous furosemide (1-2 mg/kg q2 – 6h), oxygen, and cutaneous nitroglycerine ointment. Pleural effusion should be treated with thoracocentesis if it is of sufficient volume to contribute to respiratory distress. These cases need to be managed with the utmost care and with minimal stress. Only diagnostics that are necessary to establish the diagnosis (thoracic radiographs) are indicated during the immediate acute phase. However, obtaining a chemistry panel and urinalysis as soon as the patient is deemed stable enough to tolerate it is important. The presence of renal disease or other concurrent illness can have important long-term prognostic implications. Once stabilized, addition of an ACE inhibitor and thromboembolism prophylaxis is indicated. The median survival of HCM cats with heart failure is 6 months.

RESTRICTIVE CARDIOMYOPATHY (RCM) and UNCLASSIFIED CARDIOMYOPATHY (UCM)

An increasing percentage of cats with heart failure are being diagnosed with forms of myocardial disease that are not easily classified. In this population of cats, ventricular systolic function is typically normal in the case of RCM and modestly diminished in the case of UCM. In both instances, the atria are often markedly dilated and signs of left or biventricular failure are often severe. Distinction between the two entities is not straightforward and may even represent a progression from one state to the other (RCM to UCM). Others have postulated that UCM may be a "burned-out" form of HCM. Affected cats range in age from 8 months to 19 years, and no apparent breed or sex predilections have been established.

Clinical signs and Ancillary Diagnostics

These diseases are almost always diagnosed after the onset of clinical signs of heart failure (tachypnea, dyspnea, anorexia, and lethargy). However, a murmur, arrhythmia, or gallop rhythm might be picked up during the asymptomatic phase. The ECG of a cat with RCM or UCM may be normal or show the changes described above for HCM.

Radiographs are not useful for distinguishing between the various types of feline cardiomyopathy. Biatrial enlargement may be prominent, as well as signs of left ventricular enlargement. Heart failure, when present, manifests as ventral pulmonary edema and pulmonary venous distension. Severe pleural effusion is often present and may completely obscure the cardiac silhouette.

Echocardiography

Echocardiography is the only way to distinguish these entities from HCM. The classic form of RCM shows a left ventricle of normal thickness with a very hyperechoic endocardial lining due to a thick layer of fibrosis. A variant of RCM includes the finding of numerous moderator bands bridging the papillary muscles, septum, and free wall. The left atrium is usually markedly dilated. The right atrium may also be enlarged if the right ventricle is also affected. The echocardiographic changes of UCM include normal to slightly thin appearing left ventricular walls with a mild to moderate diminution in systolic function and moderate diastolic dilation. The degree of systolic dysfunction is not as severe as that seen typically in dilated cardiomyopathy. Mild to moderate atrioventricular valve regurgitation may be present. Spontaneous contrast or even a thrombus may be present in the dilated atria.

Treatment and Prognosis

Treatment is symptomatic. Furosemide, oxygen, thoracocentesis (as indicated), and nitroglycerine ointment for treatment of acute failure. Maintenance furosemide and enalapril for long term therapy. Cats with UCM may benefit from pimobendan for support of systolic function. Pimobendan in cats is an off-label use and should be undertaken with caution. Thromboembolism prophylaxis should be instituted. Long-term prognosis is poor once signs develop and survival times of 3 to 6 months are typical.

CORTICOSTEROID-ASSOCIATED CARDIOMYOPATHY (CSACM)

In general, cats are remarkably resistant to the adverse effects of exogenous corticosteroids. However, a small percentage of cats will develop congestive heart failure following steroid administration.[5] Congestive heart failure occurs more frequently with parenterally administered steroids, then with orally administered ones. Historically, most cats that develop CSACM do not have any overt signs of cardiac disease on physical examination, and therefore it is very difficult to predict which cats are at risk of this complication. However, in cats with known or suspected heart disease (heart murmur, gallop, arrhythmia, left anterior fascicular block on ECG, or enlarged cardiac silhouette on radiographs) the use of repository steroids should be avoided if possible. Rather, conditions that are steroid responsive should be treated with the lowest dose of oral steroids which will control the clinical signs.

Cats that present in CHF with a history of steroid administration should initially be treated like any other case of heart failure. The unique feature of CSACM is the long-term prognosis. Many cats with a history of steroid administration prior to the onset of heart failure can be weaned off of all cardiac medication following a treatment period of about 4 to 6 weeks (depending on the half life of the steroid administered). Cats can be cautiously weaned off of medication over a period of a few weeks and carefully observed for the recurrence of heart failure. If heart failure signs recur, then the patient will likely need therapy for the remainder of its life.

The mechanism of CSACHF in cats appears to be an increase in plasma volume as a result of an intracellular to extracellular fluid shift secondary to glucocorticoid-mediated extracellular hyperglycemia. This mechanism is analogous to the plasma volume expansion that accompanies uncontrolled diabetes mellitus in humans. Any cardiovascular disorders that impair the normal compensatory mechanisms for increased plasma volume may predispose cats to CHF following steroid administration.

DILATED CARDIOMYOPATHY (DCM)

Prior to 1987, feline DCM was one of the most frequently diagnosed heart diseases in cats. However, an association between dietary taurine deficiency and feline DCM was established and commercial cat food is now supplemented with taurine. Feline DCM has now become an uncommon disease. Feline DCM is now primarily seen in cats that are fed non-supplemented homemade diets, but is also sometimes seen in cats on commercial diets. This indicates that taurine deficient DCM is not simply the result of a deficient diet alone and may be the result of a genetic or metabolic abnormality. There are also cases of feline DCM that do not have low plasma taurine levels (idiopathic or non taurine-deficient DCM). DCM has been reported in cats of all ages, but most are middle-aged or older. There does not appear to be a gender or breed predisposition.

The diagnosis of feline DCM is made echocardiographically by the finding of a markedly dilated left ventricle with severe systolic dysfunction. Right ventricular involvement is also frequently found. Biatrial enlargement is usually severe. All cats diagnosed with DCM should have whole blood taurine concentrations measured and supplementation of taurine initiated pending the results.

The clinical signs of DCM are due to the development of heart failure. Initial treatment is symptomatic (furosemide, oxygen, nitroglycerin ointment). Pimobendan therapy can also be attempted in these cats. Taurine is given at 250 - 500 mg/day. Initial clinical improvement is usually noted within a few weeks in taurine deficient cats, and many cats become normal (both clinically and echocardiographically) within 2 to 3 months of starting taurine supplementation. Once clinical and echocardiographic remission has occurred all drug therapy can be discontinued, but taurine supplementation should be maintained life-long. Prognosis in these cases is excellent. Cats with non taurine-deficient DCM will not improve with taurine supplementation and these cases required sustained conventional medical therapy. Cats with idiopathic DCM have a guarded short-term and poor long-term prognosis.

HYPERTHYROID CARDIOMYOPATHY

Abnormalities of the cardiovascular system are detected in most hyperthyroid cats, although few develop heart failure. The effects of increased thyroid hormone include increased metabolic rate, increased sympathetic nervous system activity, and increased number and/or sensitivity of β receptors. These effects lead to a hyperdynamic circulatory state as a result of increased heart rate, decreased peripheral vascular resistance, and increased contractility. Myocardial hypertrophy is a prominent feature in many cases of feline hyperthyroidism. This is attributed to the combined effects of chronic volume overload (excess thyroid hormone increases blood volume), high sympathetic tone, systemic hypertension, and direct stimulation of contractile protein synthesis by thyroid hormone.

Echocardiographic abnormalities seen in hyperthyroid cats include the following changes, in varying degrees of severity: biatrial enlargement, increased diameter of the aortic root, increased thickness of the interventricular septum and left ventricular free wall, and increased end-diastolic diameter of the left ventricle. Fractional shortening is usually increased. The major differential for increased wall/septal thickness and atrial enlargement is HCM.

Hyperthyroid cats with mild cardiomegaly and no evidence of heart failure do not require targeted cardiovascular therapy. Effective treatment for the hyperthyroidism (methimazole, I-131 treatment) results in rapid resolution of the circulatory changes. Therapy for cats in heart failure is symptomatic, i.e., furosemide, ACE inhibitors, thoracocentesis for cats with severe pleural effusions, and β-blockers for severe tachycardia (> 220 to 240 beats/min). Cardiac changes (both morphologic and electrophysiologic) generally resolve with successful treatment of hyperthyroidism, but this may take several months.

AORTIC THROMBOEMBOLISM (ATE)

Thrombus formation within the left heart with resultant systemic embolization is a serious complication of all feline myocardial diseases. It is estimated to occur in about 10 - 20% of cats with cardiomyopathy. Enlargement of the atria, altered blood flow, and increased blood coagulability (Virchow's triad) all increase the risk of thromboembolic complications in cats.[6]

Embolization to the distal aortic trifurcation ("saddle thrombus") is the most common site of obstruction, causing pain and paresis of one or both rear limbs. Femoral pulses are absent with complete obstruction; footpads are cold, pale, or cyanotic; and muscles of the hind limbs become firm and extremely painful due to ischemia. Other sites of embolization include either front limb, cerebral vasculature, renal vasculature, and mesenteric vasculature with clinical signs referable to these locations.

A variety of aggressive treatments including surgical removal, balloon embolectomy, and thrombolytic drugs have been attempted, but none appear to be superior to conservative therapy. Heparin therapy is commonly used; its primary function is to prevent the development of new clots while the body breaks down the existing thrombus. I add low-dose aspirin (1 mk/kg q48h) after 24–48 hours, and discontinue heparin 24–48 hours after that. ATE is a very painful condition and analgesia (e.g. buprenorphine CRI with a fentanyl patch) is imperative. Placement of a nasoesophogeal feeding tube if feeding does not start within 24 hours of hospitalization allows nutritional support and fluid supplementation. If heart failure is present, then standard therapy is administered. Retrospective studies have shown extremely variable survival times, but generally about 50% of cats will survive to discharge, with typical survival times being between 3 and 6 months. Factors at presentation which have been reported to affect survival to discharge include: number of limbs affected, absence of motor function, presence of heart failure, and body temperature.[6,7]

Numerous strategies have been attempted to prevent recurrent ATE which is estimated to occur in 25 to 50% of cats after an initial episode. ATE prophylaxis therapies include low-dose aspirin (1 mg/kg q48-72h), high-dose aspirin (81 mg/cat q3d), plavix (18.75 mg q24h), and heparin at various doses. No therapy has been yet shown to be superior.

REFERENCES

[1]Ferasin L. **J Feline Med Surg**. 2009 Jan;11(1):3-13; [2]Abbott JA, **Vet Clin North Am Small Anim Pract**. 2010 Jul;40(4):685-700; [3]Atkins CE, **J Am Vet Med Assoc.** 1992 Aug 15;201(4):613-8; [4]Payne J, **J Small Anim Pract**. 2010 Oct;51(10):540-7; [5] Ployngam T, **Am J Vet Res**. 2006 Apr;67(4):583-7; [6]Smith SA, **J Vet Intern Med**. 2003 Jan-Feb;17(1):73-83; [7]Rush JE, **J Am Vet Med Assoc**. 2002 Jan 15;220(2):202-7.

AORTIC THROMBOEMBOLISM OF CATS: PROGNOSIS, NURSING CARE & TREATMENT

Kristin Hohnadel, BS, CVT, VTS-SAIM (Cardiology)

INTRODUCTION

Feline arterial thromboembolism (ATE) is a painful, serious complication of heart disease and other conditions. Veterinary technicians well versed in diagnosis and treatment of this condition can be instrumental in improving the quality of life of their patient and helping owners with difficult decisions. ATE generally occurs secondary to feline cardiomyopathies; including hypertrophic, restrictive, unclassified, or dilated cardiomyopathy. It can also be secondary to other conditions, including hyperthyroidism or neoplasia, most often pulmonary

PROGNOSIS

Overall, the diagnosis of ATE carries a poor prognosis. Approximately 30-35% of cats will survive their initial episode and leave the hospital.2 Generally, the prognosis is reflected in the severity of the underlying disease as well as the severity of the ATE. Several studies have examined the prognostic factors important in ATE. With regards to short-term survival, the best prognostic indicator in one study was rectal temperature at presentation. Smith et al. determined in their study published in 2003 that a 50% likelihood of survival exists at a rectal temperature on admission of 98.9°F. This means that at 98.9°F, a cat with an ATE is 50% likely to survive to discharge.2 Long-term survival is more multi-faceted. The severity of the underlying disease has more effect on the long-term prognosis. In cats with congestive heart failure (CHF), median survival time was 77 days, as compared to cats without CHF at 223 days.2 Cats also experience recurrence of ATE at rates of 24 to 45%.3

NURSING CARE AND TREATMENT

Prevalence

It has been reported that ATE occurs at a rate of approximately 1 in 150 cats at two academic institutions. Most cats afflicted are domestic mixed breeds, but affected purebred cats include Abyssinian, Himalayan, Persian, Siamese, Manx, and Maine Coon.2 Cats that are more prone to heart disease are similarly at a higher risk of ATE.

History

Presenting complaints for ATE can range from mild pain and inability to use the affected limb to severe shock, pain, and paralysis. The owners may report that there has never been a sign of heart disease in their pet – it is not unusual for an ATE to be the first sign of cardiac disease. In fact, in one study, 76% of cats with ATE have no prior history of heart disease.2 Thorough history collection by the technician, including previous diagnoses and medications, are important.

Physical examination

ATE can affect any limb, including either or both hind limbs or either forelimb. The limb is generally painful with firm musculature, and the nail beds or toes may be cyanotic. The cat is often hypothermic, as well, reflecting the state of shock they are experiencing. Frequently, these cats are tachypneic due to either pain or CHF. They may be so painful and distressed that it is dangerous to both the cat and the technician to handle them, which makes reaching a diagnosis difficult. A heart murmur or arrhythmia may or may not be present.

Diagnostic tests

If the cat is stable enough to be handled, the first recommendation is to examine the cat for the presence of arterial flow in the affected limb(s). If both hind limbs are affected, do not rely on palpation: it can be difficult to feel femoral pulses in cats that are shocky due to other conditions or those that are obese. The absence of femoral pulses does not specifically indicate ATE. Instead, attempt to detect blood flow with a Doppler. Other methods of determining blood flow to an affect limb include taking either a glucose or lactate of the limb and comparing it to an unaffected limb.

Blood work changes detectable in these cats are consistent with skeletal muscle damage, including increases in lactate, aspartate aminotransferase (AST), and creatine kinase (CK). CK in particular, may show a 100-fold increase over normal.

The next crucial diagnostic is thoracic radiography. Chest radiographs will help determine if the tachypnea commonly noted is due to congestive heart failure or pain, and also whether there may be a non-cardiac cause. The presence of congestive heart failure also is important in determining the prognosis.

Echocardiography in a stable patient can assist in diagnosing the underlying cardiac pathology, but it should be reserved until the patient can tolerate restraint.

Treatment

The initial line of therapy for these cats is pain management. Because these cats may also be distressed from congestive heart failure, care must be taken to choose an appropriate medication for cats with underlying heart disease, severe pain, and anxiety. Analgesic therapy can be sublingual or parenteral buprenorphine or butorphanol for mildly affected cats, fentanyl dermal patches or constant rate infusion, hydromorphone and morphine. Pain medications not only provide pain relief, but for instance, butorphanol, also works as an anxiolytic, which can be extremely useful in a distressed patient. Fentanyl dermal patches are effective on significant pain, but the onset of analgesia is 4-6 hours. Acepromazine should be avoided as a sedative because these cats are already shocky and hypothermic – ace will worsen these symptoms.

The next step in therapy is treating underlying congestive heart failure. Initially, provide oxygen support to any patient that appears to have a change in their respiratory pattern, at least until they are stabilized. Once the cat is able to be handled, radiography is essential in providing accurate treatment; as well as helping pinpoint the cat's prognosis. Once the presence of congestive heart failure has been established, use furosemide carefully and continue to monitor blood pressure. These cats are already unstable, and furosemide may reduce the blood pressure in these compromised patients. Consider inotropic support in the form of dobutamine, but only if the cat has been diagnosed with cardiac disease other than hypertrophic or hypertrophic obstructive cardiomyopathy. If the cat is not in congestive heart failure, IV fluid therapy may be useful to help maintain cardiac output. Use care to avoid fluid overload in cats that have underlying cardiac disease. Once stable, nasoesophageal feedings or fluid maybe best in providing nutritional and fluid support. Provide warmth, but remember that the rectal temperature is a reflection of shock, not simply due to hypothermia.

Thrombolytic therapy

The use of "clot-busters" to reduce or remove the thrombus is controversial. There appears to be significant risk of fatal reperfusion injury, as well as hemorrhage, when thrombolysis is attempted. In addition, survival of cats treated with thrombolytic factors, including tissue type plasminogen activator (tPA) and streptokinase (SK), is reported to be similar to cats treated supportively.3 Anticoagulant therapies such as heparin, low molecular weight heparin or aspirin are recommended once the ATE is diagnosed to prevent expansion of the thrombus.

Thromboprophylaxis

Few medications are useful or recommended in the prevention of ATE. Aspirin has historically been used at a dose of 81 mg PO q 72 hours, but it frequently causes GI irritation or hemorrhage. Low dose aspirin at 5 mg PO q 72 hours has not been evaluated for efficacy, but is associated with lower or similar risk of recurrence of ATE and lower side effects when compared to warfarin and high-dose aspirin. The FAT CAT study (feline aortic thromboembolism clopidogrel vs. aspirin study) is currently being conducted to evaluate clopidigrel (Plavix®) in cats, but preliminary abstracts do not reveal results that indicate a difference, though study size is reported to be low. Low molecular weight heparin has also been used to prevent ATE in cats as a once daily injection. However, studies evaluating LMWH like dalteparin and enoxaparin indicate that it needs to be administered more frequently than once daily, perhaps as often as every four to six hours.3

CONCLUSION

ATE is a devastating condition in cats that carries a poor prognosis. Knowledge of care, treatment, and prognosis for these cats may help technicians to provide the best treatment possible to give these cats a chance to survive.

REFERENCES

1.Buchanan et al. *Vet Rec* 1966; 79(18):496-505.2. Smith, et al *J Vet Intern Med* 2003;17(1):73-83. 3. Smith et al. *Vet Clin Small Anim* 34 (2004) 1245-1271. 4. Hogan, ACVIM abstract 2008.

ACQUIRED CARDIAC DISEASE OF THE DOG: DISEASE PROCESS, TREATMENT, AND PROGNOSIS

Kristin Hohnadel, BS, CVT, VTS-SAIM (Cardiology)

Common acquired cardiac diseases of the dog include those that are caused by valvular defects, myocardial failure, infectious reasons, as a result of arrhythmias, and parasitism.

The most common cardiac disease in dogs is **valvular insufficiency** caused by abnormal aging of the valves of the heart. Approximately 30% of dogs at age 13 are clinically affected. On necropsy, 90% of dogs at age 13 have evidence of valvular degeneration. The disease also can be referred to as chronic myxomatous valvular degeneration, mitral valve disease, chronic valve disease, and endocardiosis. Breeds affected include Poodles, Yorkshire Terriers, Chihuahuas, and other small dogs, but the most commonly affected is the Cavalier King Charles Spaniel (CKCS). CKCS are known to be affected severely, often at quite a young age. The mitral valve is most commonly affected with CVD, but lesions can be seen on the other cardiac valves, as well. Valve leaflets are thickened and nodular, and the chordae tendinae can be thickened and elongated. Chordae may also rupture, causing flail or prolapse of the valvular leaflet. The consequence of valvular degeneration is insufficiency. In the case of the mitral valve, chronic significant insufficiency causes an increase in left atrial pressure, resulting in atrial enlargement. When the left atrium enlarges to a certain point, blood backs up into the pulmonary veins, and pulmonary edema develops. As the disease progresses, myocardial failure may also result.

The most significant physical exam characteristic of valvular degeneration is a heart murmur. Because the most common valve affected is the mitral valve, the most common murmur is systolic and located at the left apex. A right-sided murmur may also be ausculted, and it may be from tricuspid regurgitation or radiation of the mitral murmur. If the aortic valve is affected, a diastolic murmur may be detected. Generally, the intensity of the murmur correlates with severity of disease, but this is not exclusively true. Pulmonary edema results in respiratory distress and/or coughing, and in severe cases, is detected by the presence of crackles in the lung fields. Mild to moderate edema cannot typically be ausculted, so clear lung sounds do not rule out the presence of fluid.

Radiographic changes vary also with severity of the valve disease. With severe disease, the most common signs are left atrial and auricular enlargement and left ventricular enlargement. As the disease progresses to left-sided congestive heart failure, pulmonary edema can be seen. Some dogs will also develop ascites and/or pleural effusion as a consequence of right-sided failure. Valvular insufficiency is seen on color Doppler evaluation of the valve. Other echocardiographic findings include thickened valves, cardiac chamber enlargement, and flail or prolapsing valve leaflets.

There is some debate about the use of ACE inhibitors prior to the development of CHF, but at the author's institution, it is believed that treatment of CVD is effective after signs of congestive heart failure have developed. Treatment consists of stabilizing the patient with diuretics, primarily injectable furosemide. Long-term treatment includes oral furosemide, ACE inhibitors such as enalapril or benazepril, and pimobendan, a positive inotrope and venodilator. Renal values must be monitored serially to follow changes in BUN, creatinine, and electrolytes. Though highly effective at treating CHF and improving quality of life, some cardiologists will reserve pimobendan until there are signs of refractory congestive heart failure, i.e. doses of furosemide greater than 2 mg/kg PO q 8 h, because it has been shown to deteriorate valve function in long-term use. Other diuretics such as spironolactone and hydrochlorothiazide can be used to help manage signs of fluid overload. When arrhythmias such as atrial fibrillation result from CVD, antiarrhythmics are indicated. The most common antiarrhythmic used in this case is digoxin. Digoxin slows the heart rate by slowing the electrical impulse through the atrioventricular node. It also has positive inotropic effects, so it is useful when concurrently treating congestive heart failure. Because the therapeutic range of the digoxin dose overlaps with the toxic dose, it must be prescribed and monitored carefully. Signs of digoxin toxicity range from mild--inappetence and diarrhea--to fatal arrhythmias. Digoxin blood levels must perfomed 7-14 days after starting it at six to eight hours post-pill until an appropriate range is reached, and then every 3-4 months. Other antiarrhythmics have negative inotropic effects, and should be used with caution when treating a patient in active CHF.

Surgical repair or replacement of the mitral valve, while standard in humans, has been attempted in dogs, but unfortunately it is not widely available. This repair is highly cost-prohibitive, requires intensive open heart surgery with cardiac bypass, and the results are discouraging.

The end result of valve disease is often euthanasia due to refractory congestive heart failure, but some patients die suddenly as a result of a fatal arrhythmia. Endocarditis has not been associated with CVD. Rarely, the left atrium may perforate, or rupture as a result of a jet lesion. This allows blood to leak from the atrium into the pericardium. Many of these patients die as a result of both sudden blood loss and compression of the heart by the pericardial effusion. Pericardiocentesis may be attempted, but because the defect in the left atrial wall still exists, it is risky. CPR for a patient with left atrial perforation is contraindicated, as the compressions will force more blood from the heart into the pericardial space, and may actually rupture the pericardium.

Myocardial failure

Idiopathic dilated cardiomyopathy is a primary disease of the myocardium. Myocardial failure caused by nutritional deficiency, endocrinopathies, toxins, or infectious agents result in secondary cardiomyopathy. Dogs most commonly affected by DCM include large breed adult dogs, especially a large percentage of Doberman Pinschers, Great Danes, Irish Wolfhounds, and Scottish Deerhounds. Cocker Spaniels are also affected, while Portugese Water Dogs experience juvenile DCM. It is estimated that approximately 50% of Dobermans are affected.

Patients with DCM can be separated into two general categories: those with occult, or hidden, disease; and those with overt clinical signs. Dogs with occult DCM typically show no clinical signs. They may have a soft murmur, but generally, they do not exhibit any signs associated with cardiac disease. However, sudden death in approximately 40% of Dobermans is the first sign that are affected with DCM.. It is not known how long the occult phase lasts--it could be months to years.

When clinical signs develop, DCM is in the overt stage. Clinical signs include evidence of congestive heart failure; either pulmonary edema as a result of left sided CHF, or ascites/pleural effusion from right CHF. Physical exam may reveal a soft murmur or gallop, as well as arrhythmias. The most common arrhythmias are ventricular premature contractions (VPCs). They may occur as single beats, or as groups--bigeminy, trigeminy, paroxysmal or sustained ventricular tachycardia. Atrial fibrillation is also a common arrhythmia associated with advancement of this disease. Chest radiographs reflect an enlarged cardiac silhouette. The degree of severity of pulmonary edema is often alarming, especially when compared to the patient's respiratory status, but these dogs can quickly deteriorate clinically. On echocardiogram, the left ventricle loses its elliptical shape, and becomes eccentrically hypertrophied, or dilated. As the ventricle enlarges, it pulls the mitral valve apparatus out of shape. The result is functional mitral regurgitation in a centrally directed jet. Fractional shortening, a measurement of systolic function that determines the degree to which the ventricle contracts during systole, can be useful in evaluating systolic function.

Treatment of CHF due to systolic dysfunction is similar to CVD. Injectable furosemide is used to stabilize the patient until it can be discharged on oral medications, including furosemide, ACE inhibitors, pimobendan, and digoxin. In an emergency setting, dobutamine can be used in a constant rate infusion to increase cardiac output. It must be used carefully, monitoring heart rate and blood pressure in an ICU type setting.

Since some types of secondary DCM may be reversible, dogs that are not of the usual breeds presenting with systolic dysfunction should be evaluated. Taurine deficiency, hypothyroidism, doxorubicin toxicity, as well as parvovirus, can cause systolic dysfunction. L-carnitine may be supplemented with taurine, but cannot be implicated as a cause of myocardial disease. Similarly, co-enzyme Q10, while a popular cardiac supplement, is not a cause of myocardial disease.

Arrhythmic diseases

Arrhythmogenic Right Ventricular Cardiomyopathy (ARVC), also known as Boxer Cardiomyopathy, is a myocardial disease common in Boxers and occasionally reported in English Bulldogs. It is characterized by fibro-fatty infiltrate of the right ventricular myocardium. This results in ventricular arrhythmias and sometimes dilation of the right ventricle or a DCM-like disease. Clinical presentation can range widely, including incidental finding of arrhythmias on physical examination, syncopal episodes, congestive heart failure, and not uncommonly sudden death. Washington State University has identified a gene associated with ARVC and developed a genetic test, which may be useful in screening breeding dogs. However, there are likely more genes associated with the disease, as some patients that fit the clinical picture test negative for this gene.

Except for patients that present with arrhythmias or congestive heart failure, physical examination can be completely normal, as can thoracic radiographs and echocardiography. Holter monitoring to detect ventricular arrhythmias is vital to diagnosing these patients.

Treatment of ARVC depends upon the clinical picture. If ventricular tachycardia is present, antiarrhythmics such as Sotalol or Mexiletine can be prescribed. However, sudden death is still possible. Some research into the use of implantable cardiac defibrillators in dogs has been done, but because these devices are designed for humans, they have not been widely used. If CHF is present, standard therapy similar to that of DCM is indicated.

Infectious endocarditis

Infectious endocarditis (IE) is a condition in which bacteria seeds the valves of the heart. Mitral valve disease is not a known risk factor for IE, but aortic stenosis is. The most common patient affected is a middle-aged, large breed male dog. Dental disease has often been implicated as a cause, but there is actually little evidence to support this.. Vegetative lesions may be seen on the mitral or aortic valves. The patient shows signs of extreme illness due to septicemia, fever, and a new murmur. If the mitral valve is affected, the patient will have a systolic murmur. If the aorta is affected, there will be a diastolic murmur. The left side of the heart will overload, and some patients will develop congestive heart failure as a consequence.

Diagnosis of IE is accomplished by detecting the presence of a vegetative lesion and performing blood cultures. The protocol for blood cultures is generally to draw at least three samples, from two different sites, an hour apart. Isolating the bacteria responsible will allow the clinician to treat with the appropriate antibiotic. Rarely, a fungal etiology such as blastomycosis may be found as a cause of IE.

Treatment of IE is generally supportive. Injectable antibiotics are used when indicated, but the valve will be permanently damaged. Because of this, the prognosis with IE is generally poor.

Tachycardia-induced cardiomyopathy is a rare cause of congestive heart failure in dogs. Long-term arrhythmias, most often supraventricular tachycardia, causes myocardial failure. Controlling the arrhythmia gives the patient the best chance of recovery. Medications are somewhat effective, but catheter ablation of the arrhythmic focus is the best treatment. Unfortunately, this is only offered at one location in the US, and it must be performed on patients greater than 10 kg body weight. In addition to antiarrhythmic medications, standard treatment of CHF is recommended where indicated.

Heartworm disease can also cause severe heart disease, especially with severe infestation. Heartworms live in the pulmonary arteries, but with increasing numbers of worms, they may inhabit the main pulmonary artery, as well as the right atrium and ventricle. If the worms migrate as far as the right heart, blood flow is restricted, and the tricuspid valve cannot close properly. This is called caval syndrome. As a consequence of this, there is often a loud murmur on the right thorax, and red blood cells are damaged, causing hemoglobinemia and hemoglobinuria. At this point, adulticide treatment is quite dangerous. The treatment of choice is surgical extraction of the worms as soon as possible.

FELINE ENVIRONMENTAL ENRICHMENT & IDIOPATHIC CYSTITIS & HYPERGROOMING

Meri Hall, RVT, LVT, CVT, RLATG, VTS (SAIM)
Athens, PA

In the early 1980's stereotypical behavior of captive animals was an issue that prompted the 1985 amendments to the animal welfare act that required environmental enrichment for animals used in research and zoological parks. The basis of environmental enrichment is to reduce stress by creating a more "natural" social structure and housing environment. The use of feeding routines that allow the animal to explore and mimic natural behaviors is a vital part of reducing both boredom and stress in captive animals. This concept has extended to companion animals, and environmental enrichment of the domestic cat has been shown to be beneficial to improve their quality of life and health. More cats are being restricted to indoor living situations due to urbanization and the prevalence of infectious diseases. Just as with humans, stress manifests itself with a variety of behavioral changes. For cats, this can be seen with isolating (hiding), becoming aggressive, inappetance, hypergrooming, inappropriate elimination or feline idiopathic cystitis. (FIC)

Most cats that present with FIC are between 1 and 10 years of age.[1] FIC affects males and females equally, although neutered animals are at increased risk.[2] For cats older than 10 years , <5% develop FIC.[2] Feline lower urinary tract disease (FLUTD) and FIC have many clinical signs in common. They include: periuria, dysuria, pollakiuria, stranguria and hematuria. To diagnose feline idiopathic cystitis (FIC), all other causes of FLUTD must be excluded. Typically, clinical signs resolve spontaneously within 4-7 days and often relapse. FIC accounts for 60-70% of cats presented for irritative voiding.[3] The main goal of managing FIC is to increase the interval between episodes and the severity of the clinical signs[1]. Currently, much of the current recommended management of FIC incudes stress management and environmental enrichment.

Research has shown that FIC affects more than just the urinary bladder, and may include complex abnormalities of the endocrine and nervous system.[4] Studies have shown that cats affected with FIC have high concentrations of circulating catecholamines, smaller adrenal glands, and malfunctioning alpha-2 adrenoreceptors[5]. Studies have shown that catecholamines and urinary bladder permeability decreased with environmental enrichment versus stressful periods and may be of benefit for cats with FIC.[5]

Hypergrooming is evident when cats begin to groom themselves excessively and have a shaved appearance over part of their body. To diagnose hypergrooming, other causes of hair loss should be excluded. Other causes include external parasites and hyperthyroidism. Some cats will groom housemates, so the hair loss may be from the excessive grooming of another cat. Some cats will show the behavior continuously, sporadically or seasonally. Some cats are able to deal with stress with no outward signs, but for cats that are unable to cope, the physiologic and psychological manifestation of the stress can be overwhelming. Recognition of the stressors involved is the first step in alleviating or reducing the level of stress.

Stressors for cats include confinement, recent moves, changes to schedules and additions to the household. When moving care must be taken to ensure that cats feel comfortable in the new home. They are very affected by smells. It may not be in the cat's best interest for there to be nothing familiar in the new environment. Allow them time to acclimate to the new home and provide them with favorite toys, resting areas and routines. Unlike dogs, it may not be evident that changes in the household schedule affect the cat. When there are additions to the household (new humans or animals) it is best not to force the cat into acceptance. If the home appears stable, there are two areas within the home that may be causing stress for the cat. These areas are the environment, relationships and should be addressed to reduce stress. The veterinary professional can educate owners on ways to create a cat friendly environment and attempt to resolve relationship conflicts within the home.

To create a cat friendly environment, outlets for species typical behaviors should be found throughout the home. The recommendation if that for all resources in the house a "1+1" approach works best. That is one for each cat in the house plus one more, particularly with food, water and litterboxes.[4] Cats prefer to perch above the floor on upholstered surfaces. Cats should be given a variety of perching surfaces throughout the home. Window perches are easy to create and allow for catnapping in the sun or enjoying watching birds and local wildlife. Scratching, although destructive at times, is a normal behavior that is observed with wild cats. This behavior provides a way to stretch the muscles, permits nail bed maintenance and provides visual and scent marking. Owners should observe their cat to determine if vertical or horizontal scratching is preferred by their cat. Some cats will exhibit both horizontal and vertical scratching behaviors.

Cats sleep a majority of the time and often engage in short bursts of play. There are many ways to stimulate the predatory behavior with toys that allow them to pounce and chase. There are cats who prefer purchased items, but many household items can be used to permit this behavior. (e.g. paper bags, paper towel rolls) Another way to stimulate this behavior is to enhance the predatory behavior is to remember that cats hunt in the wild for their meals. Food can be placed in toys to simulate this behavior or hidden throughout the home. For some cats, they will exhibit hunting behavior while playing and hiding a favorite toy also simulates the hunting instinct. Providing multiple water sources will also reduce stress for cats. One of the recommendations for treatment of FLUTD is to increase the daily water intake. This can be achieved with different materials (e.g. ceramic, plastic, stainless steel), shapes and sizes of the container. Some cats prefer flowing versus still water sources.

Elimination can be a major stressor for domestic cats. They are very fastidious with their elimination behavior. If a litter box is unacceptable in any way they may be prone to not go, or find an alternative site. Many times this can be very disturbing to the owners. Care should be taken to place litter box in low traffic areas, cleaning it frequently and trying different types of litter to suit the cat's preference. In multiple cat households, several litter box choices may need to be offered. One must realize that previously litter trained cats can reject the litter box due to classical conditioning.[3] Cats with FIC may begin to associate the litter box with pain or discomfort.

Relationships in the household can also be major stressors for cats with FIC. In the case of relationship conflicts include intra-cat relationships, feline-canine relationships and the feline-human relationships. Intra-cat tension creates situations where the submissive cat feels threatened. Many times the aggressor "lays in wait" for the submissive and pounces. The more timid cat then is always "looking" over their shoulder for self-preservation. There are ways to reduce the stress by having numerous perches, litter boxes, food options and toys available in the house. In severe cases, separating the cats is required. In households that also have dogs, care must be taken to ensure the animals are compatible and that the cat has options to have their "own space". The most important relationship conflict for companion cats is the feline-human relationship. Owners of cats with FIC can become very frustrated if periuria occurs. It is very important that these owners are educated that punishing the cat for eliminating away from the litter box is not only ineffective, but may exacerbate the problem.

Veterinary technicians play an important role in diagnosing and treating FIC and breaking the cycle of hypergrooming. They should be aware of the signs and causes of FIC and hypergrooming. When obtaining a history for animals exhibiting these clinical signs, the use of open ended questions will encourage the clients to elaborate. Questions related to behavioral changes, changes in the housing situation, grooming habits, litter box habits, types of food and how the animal relates to others will assist the veterinarian in making a correct diagnosis. Once a diagnosis has been made, many times it is the veterinary technician who will be educating the client on techniques to use to alleviate the stress. Dealing with FIC and hypergrooming does not come with a quick fix and many clients become so frustrated with the problem they give up. The veterinary technician can be a cheerleader for the client to encourage them to continue to try different things to alleviate the problem.

REFERENCES

1.Bartges JW. Lower urinary tract disease in geriatric cats. Proceedings of the 15th ACVIM Forum 1997: 322-324. 2. Housutler RA. Chew DJ. DiBartola SP, Recent Concepts in Feline Lower Urinary Tract Disease, Vet Clin Small Anim 35 (2005) 147-170. 3. Chew DJ. Buffington CAT, Non-obstructive Idiopathic/Interstitial Cystitis in Cats: thinking outside the (Litter) Box. Proceedings of the 32nd WSVA Conference. 2007. 4. Westropp J. Buffington CAT. Feline idiopathic cystitis: current understanding of pathophysiology and management. Vet ClinSmall Anim 34 (2004) 1043-1055. 5. Neilson JC. Behavioral Management of Cats with Idiopathic Cystitis. Proceedings of the 82nd Western Veterinary Conference 2010.

ENDOSCOPY OF THE GI TRACT IN SMALL ANIMALS

Susan Cox, RVT, VTS (SAIM)
Davis, CA

More veterinary hospitals are now incorporating endoscopy equipment in their clinics. It's important for the technician to know what's involved in these procedures and to anticipate what the veterinarian needs to efficiently complete these specialized endoscopies in order to gather enough information to accurately diagnose and treat the patient.

GI endoscopy encompasses 5 areas of the GI tract—esophagus, stomach, duodenum, colon and ileum. These areas are divided into 2 separate procedures-gastroduodenoscopy, which examines the esophagus, stomach and duodenum, and colonoscopy, which examines the colon and ileum.

Instrumentation- rigid and flexible endoscopes are used in GI endoscopy. Rigid endoscopes are hollow metal or plastic tubes with fiberoptic light and manual air insufflation capabilities. They are mainly utilized in proctoscopies, which examine the rectum and descending colon. They can also be used in esophagoscopies with foreign body retrieval. Flexible endoscopes for GI procedures must be versatile enough for all small animal patients. Required functions include 4 way deflection, air/water/suction capabilities, good optics, adequate insertion tube length and large biopsy channel. Video endoscopes have better image resolution than fiberscopes but require more equipment and are more expensive.

Biopsy forceps are designed to be used in the biopsy channel of the endoscope. The outer diameter of the forceps should be slightly smaller than the endoscope's biopsy channel. For example, if your biopsy channel measures 2.0 mm, the forceps ordered should be 1.8 mm and be about 50 cm longer than the channel.

Retrieval forceps should include a single loop snare, 3 or 4 prong grasper, and basket retrieval forceps. Guarded cytology brushes and aspiration catheters are also available. Balloon dilators of 6-8 cm lengths and 15-20 mm inflatable diameter are recommended for GI tract strictures.

Don't forget to make sure that your endoscope is functioning properly before starting the procedure. You or the endoscopist should always test the suction and air/water capabilities (have a bowl of water tableside), be sure it's in focus (videoscopes have automatic focusing) and white balanced.

Make sure that you also have all the necessary equipment at tableside or within easy access, especially biopsy accessories. Biopsy cassettes, formalin jars and gauze squares should be available for every procedure.

Gastroduodenoscopy procedures are performed under general anesthesia and in left lateral recumbency. A complete oral exam should be done before the procedure. A mouth gag must **always** be in place to protect the endoscope. The endoscope is passed through the oropharynx and upper esophageal sphincter. Insufflation is then needed to achieve a luminal view of the esophagus. Normal esophageal mucosa should be smooth, pale, and glistening. Submucosal vessels and a circular pattern surrounding the esophagus are normal in felines. The lower esophageal sphincter is usually closed and has a red appearance of the surrounding mucosa. Gentle advancement and minimal insufflation is needed to pass the LES.

Once in the stomach, a systematic approach is needed to complete a thorough and reproducible examination. A cursory view with minimal insufflation is all that is initially needed since the pyloric sphincter can be affected by excessive air introduced into the antrum. The stomach is comprised of the cardia, fundus, greater and lesser curvature, incisura, antrum, and pyloric sphincter. The mucosa should appear smooth, glistening and pink with longitudinal rugal folds leading to the antrum. The endoscope will slide along the greater curvature until the junction between the antrum and the body comes into view. The endoscope should then be advanced slowly towards the pyloric sphincter. Intubating the pylorus can be the most difficult part of the procedure. Slow advancement of the endoscope while keeping a luminal view is usually successful. Repositioning the patient can also be beneficial.

Advancing the endoscope through the duodenum should be smooth and effortless. Once a luminal view is achieved and the endoscope's working length is fully inserted, the duodenum can be examined and biopsied. Normal intestinal mucosa is paler, more granular, and more friable than gastric mucosa. Peyer's patches, areas of lymphoid follicles, are found in the normal duodenum and are often multiple.

Biopsies and/or other diagnostic procedures should be taken as the endoscope is withdrawn, taking into account possible red streaks caused by the endoscope. Sites where the biopsy forceps can be perpendicular to duodenal mucosa, such as the cranial duodenal flexure just past the pylorus or the incisura in the stomach, can yield diagnostic biopsy samples.

Once the endoscope is retracted back into the stomach, all areas can be examined more closely. A J-maneuver, or retroflex, can be used to look at the cardia, lesser curvature and fundus for possible lesions. To perform this maneuver, retract the endoscope to the greater curvature and retroflex 180 degrees. When biopsying this area, be sure to insert the biopsy forceps into the biopsy channel before retroflexing. This will avoid excessive wear and tear on the biopsy channel. If a foreign body is found and can be removed, using the correct forcep is pivotal. Coins and other such objects with a lip can be removed with the pronged retrieval forceps. Sharp objects such as fish hooks that are lodged in the esophagus can be removed with regular biopsy forceps and drawn into the proctoscope to shield the esophagus. It is important for the forceps handler to keep firm traction on the forceps when withdrawing the object, especially through the cardia.

When the procedure is complete, and before the endoscope is withdrawn, be sure to suction all the air from the stomach and, if necessary, any residual fluid observed in the esophagus.

Colonoscopy--Patient preparation for colonoscopy differs from upper GI endoscopy. Successful visualization and biopsy of colonic mucosa are necessary for a reliable treatment plan. At least 2 days of withholding food is recommended, with 3 days being ideal. Intestinal lavage solutions, such as Colyte and Golytely, both of which contain replacement electrolytes, work well. 30 ml. per kg. every 6 hours given over a 24 hr period via a stomach tube is the dosage we use at the VMTH. Also available is Osmoprep tablets at a dose of 1g/3kg. of body weight every 4-6 hours for a total of 4 doses the day before the procedure. IV fluid administration is recommended along with the Osmoprep tablets.

General Anesthesia is recommended for colonoscopy, with the patient in left lateral recumbency. A digital palpation of the anus and rectum should always be performed first and could detect lesions not seen with the endoscope.

The endoscope is then introduced into the rectum and then insufflated. Keeping a luminal view, the endoscope is advanced. Several flexures are encountered on the way to the ileocecal area, so visualization may be lost for a short time. The ileocolic valve appears as a large muscular sphincter with the cecum adjacent to it. Successful intubation of the valve should be attempted, but can be difficult. Blind biopsies can be performed. Normal colonic mucosa should be smooth, pink and glistening with submucosal vessels clearly visible. Absence of vessels could be indicative of infiltrative bowel disease. Biopsies should be taken along the colonic mucosa as the endoscope is withdrawn. In larger patients, the endoscope can be retroflexed in order to visualize the distal rectum.

Proctoscopy--If distal colonic or rectal disease is suspected, a rigid endoscope, or sigmoidoscope, can be used. In most cases heavy sedation and warm water enemas will be sufficient for a successful procedure. The patient should be placed in right lateral recumbency and the lubricated proctoscope introduced with the obturator in place. Once fully intubated, the obturator can be removed, the window secured and the colon insufflated utilizing the attached rubber bulb. Large cotton-tipped applicators, called Scopettes, can be used to swab away excessive feces from view.

Abnormal findings in both upper and lower GI disease can be quite varied. Many of the diseases of the intestines and colon present virtually normal with endoscopic visualization, but upon histologic examination can reveal neoplastic or inflammatory processes.

Strictures of the esophagus or colon can occur and appear as circumferential narrowings of the lumen. Causes can be ingestion of caustic materials, neoplasia, or GI reflux during anesthesia. Balloon dilation is often necessary, and may require several procedures to be effective.

Inflammation can affect any part of the GI tract and is histologically confirmed. Gross abnormalities include erosions, ulcers, erythema, mucosal thickening, and pronounced granularity and friability. It should be noted when biopsy retrieval of a suspected area excessively bleeds or a diagnostic sample is too difficult (fibrous) or too easy (friable) to obtain.

Neoplasia can also be diagnosed anywhere in the GI tract. Sarcomas, including lymphosarcoma of the duodenum and stomach, can appear infiltrative and diffusely thickened. Carcinomas can appear as obstructive, circular lesions in the intestines.

Parasites can be encountered in the intestinal tract and colon, but are usually diagnosed less expensively.

GI endoscopy can be challenging, but very rewarding as well and always interesting! I hope you can take this knowledge back to your clinic and use it to benefit your patients.

ENDOSCOPY OF THE SMALL ANIMAL RESPIRATORY TRACT

Susan Cox, RVT, VTS (SAIM)
Davis, CA

As the veterinary profession grows and becomes more specialized, many veterinarians are learning procedures such as bronchoscopy and integrating this specialized procedure into their clinics. We will be examining what it takes to perform a bronchoscopy as well as a rhinoscopy, a more familiar procedure in the veterinary practice.

The small animal respiratory system (canine/feline) is divided into 2 different parts-- Upper & Lower Respiratory systems. An upper respiratory system examination includes rhinoscopy and nasopharyngoscopy, and the lower system examination comprises laryngoscopy and bronchoscopy.

Rhinoscopy is the visual examination of the nasal cavity, and is only one part of a complete diagnostic nasal disease workup. Indications include sudden onset sneezing, reverse sneezing, nasal discharge, epistaxis, facial swelling, and loss of nasal airway patency. Retroflex behind the soft palate, or Nasopharyngoscopy, is a part of every upper airway workup at the VMTH.

Instrumentation for rhinoscopy includes both rigid and flexible endoscopes. The smaller pediatric fiberoptic endoscopes like the Olympus P-20d, (outer diameter of 5.0 mm. and a 2.0mm biopsy channel), are ideal for retroflexion. Rigid telescopes are 2.7 mm outer diameter, have a 0 or 30 degree field of view, and should include a sheath with access ports for flush and biopsy capabilities. A c-clamp attachment should be available with most camera systems and can be used on rigid telescopes along with a light guide cable for visualization. You can also use just a light cable attached to the telescope and visualize structures through the lens.

Biopsy forceps in the nasal cavity include Jackson Debarkers, which have a sharp round cup for biopsying nasal turbinates. They include 2, 3, and 4 mm sizes. They can be placed alongside most rigid scopes and obtain diagnostic visual biopsies. Flexible biopsy forceps of an appropriate size can be used through the sheath of the rhinoscope.

A complete rhinoscopic tableside setup includes labeled cassettes for biopsies, formalin jars, lubricating jelly, saline flush, gauze squares, exam gloves, mouth gags, lap pads, suction, culturettes, cotton-tipped applicators, ice, spay hook and mirror for oral exam, and lidocaine jelly. Oxymetazoline(Afrin)drops, a vasoconstrictor, and Proparacaine nose drops, a topical anesthetic, may also be helpful when used in the nasal cavity. Oxymetazoline should be used after a cursory rostral nasal exam as the drops could alter nasal mucosal hue.

General Anesthesia is recommended when performing rhinoscopy, a highly stimulating procedure. Anesthesia serves to suppress the sneeze and gag reflexes, and helps keep equipment from being damaged. Make sure the cuff on the endotracheal tube is in working order to keep fluids from reaching the lungs. Rhinoscopy is performed in sternal recumbency with a rolled-up towel or sandbag under the ventral neck. A complete oral exam should be performed first. Tonsils should be probed and examined for foreign material, the hard and soft palate should be palpated, and the tongue should be examined for masses and ulcers.

When performing a nasopharyngoscopy equipment safety is key. Placing lidocaine gel around the soft palate can help with the gag reflex. The endoscope tip is placed just past the soft palate, then flipped up and **slightly** pulled forward to examine the nasopharynx and choanae, which are the openings from the nasal cavity to the nasopharynx and separated by the vomer bone. Be aware that the image shown is inversed. The nasopharynx is a prime area for biopsy if a mass is seen arising from the nasal cavity. When obtaining a biopsy in this area, place the biopsy forceps in the channel before retroflexing, get the biopsy, then pull the endoscope and the biopsy forceps out of the oral cavity as one unit. This maneuver will safeguard the biopsy channel. When nasopharyngoscopy is complete, the throat is packed with a lap pad to prevent aspiration from fluid from the rostral rhinoscopy exam.

Before beginning rostral rhinoscopy, measure your rigid scope to the level of the medial canthus of the eye and mark with a small piece of tape. This action will remind the endoscopist not to advance the endoscope too far and possibly penetrate the cribriform plate. When performing rhinoscopy, always examine the less affected side first. The normal nasal cavity is pink, smooth and shiny with nasal turbinates packed close together. The cavity is also highly vascular, especially with scope contact. Cold saline flushed into the cavity can help return visualization. Abnormal findings to record include hyperemia, excessive mucus accumulation, mass lesions, turbinate destruction, foreign bodies, and fungal plaques. Biopsies should be performed with visualization to obtain the most accurate diagnosis. Foreign bodies should be removed when first noticed. Alligator forceps work well for removal of grass awns and most plant material.

The most common types of tumors found in the nasal cavity are sarcomas and carcinomas. Older dogs are more predisposed to nasal tumors than older cats. These types of tumors are locally invasive, but usually do not metastasize.

The most common species of fungus encountered in the nasal cavity is aspergillis. Destruction of turbinates and white, fluffy plaques are most commonly seen with this disease. The frontal sinuses can also be affected—often with mycotic rhinitis the flexible endoscope can penetrate the sinus from the nasal cavity and visualization with subsequent plaque removal can be performed. The cribriform plate, a thin plate of the ethmoid bone behind the nasal cavity, can also be damaged from fungal disease— before therapeutic treatment can begin, a CT or skyline view radiograph should be done to assess if the plate has been damaged.

The diagnosis of rhinitis is most always made via histopathology in conjunction with CT and other diagnostic tests. Nasal rhinitis can be classified as viral, allergic, or bacterial secondary to other disease processes. Symptoms may include increased mucopurulent discharge. Rhinoscopy reveals thickened, edematous and misshapened turbinates with the increased mucous production. Copious saline flushes with a red rubber catheter or through the sheath of the rhinoscope can be both therapeutic and afford the endoscopist better visualization for diagnostic biopsies.

Foreign body removal, especially plant awns (commonly called "foxtails"), and blades of grass are a common procedure at the VMTH, especially during the summer months. Sudden, episodic and violent sneezing is the chief complaint. Removing the foreign body ASAP gives the endoscopist an advantage—less purulent discharge inside the nasal cavity makes the foreign body easier to see and extract. If a foreign body is suspected and not seen, retroflushing using a large foley catheter placed in the proximal nasopharynx and packed with a lap pad may flush the foreign body out through the nasal passages.

When the rhinoscopy and biopsies have been completed, the teeth may be probed for oronasal fistulae. If the nasal passages are bleeding excessively, an ice pack placed on top of the nose or additional Oxymetazoline nasal drops may aide in hemostasis. Suction may be needed around the endotracheal tube when the lap pad is removed to prevent aspiration. In some instances leaving the endotracheal tube partially inflated when removing will prevent fluid from entering the lungs.

LOWER RESPIRATORY EXAMINATION

Laryngoscopy is the examination of the larynx with regard to laryngeal function. It is best performed at anesthetic induction prior to the endotracheal tube being placed. Indications for performing laryngoscopy include respiratory distress, exercise intolerance, voice change and increased respiratory effort. Anesthesia induction agents are slowly administered until the mouth can be opened and the arytenoids observed. If breaths are too shallow or intermittent Doxapram administered at 0.5mg.-1mg./kg. IV acts as a respiratory stimulant so that laryngeal function can be properly assessed.

Bronchoscopy is a systematic endoscopic examination of the bronchial tree so that abnormalities can be located and identified. This is a "team" procedure and should be seen as an invasive, critical exam and be performed as efficiently as possible to limit anesthesia time. Indications include acute and chronic cough, abnormal breathing pattern, tracheal collapse, neoplasia, and airway obstruction.

Instrumentation—the ideal bronchoscope should be about 5 mm outer diameter and be long enough to accommodate large dogs. Our bronchoscopes are the Pentax FG-16V, which has a 5.5 outer diameter and is 90 cm long, and the Olympus P-20D, which is 5.0mm outer diameter and is 55 cm. long. These endoscopes have a biopsy channel for bronchoalveolar lavage, foreign body retrieval, cytology brushings and biopsy forceps. Foreign bodies should be removed when first seen with biopsy forceps, and biopsies and BALs performed before biopsies and brushings to avoid contamination.

General Anesthesia is necessary to suppress the gag reflex, coughing, and to protect the endoscope. Preoxygenation with a mask is necessary for all patients. Customized shorter endotracheal tubes can be used for dogs that can accommodate a 7.5 fr tubes or larger, with adapters for gas and scavenge. The shorter length helps the endoscopist visualize more of the trachea than with conventional endotracheal tubes. Jet ventilation or adapting the biopsy channel to administer oxygen will work for smaller patients.

Patients should be in sternal recumbency with a rolled towel under the chin and 2 mouth gags in place. The lubed insertion tip is carefully placed through the arytenoids and into the trachea, or through the membrane of the adapter and into the endotracheal tube. Submucosal vessels and a rounded appearance are normal for the trachea. Any collapse of the trachea should be recorded. Next, the carina divides the right and left bronchial tree. Each bronchus should be examined and observed for color change, proliferation of mucus, collapse, dilation, neoplasia and foreign bodies. Sites should also be chosen for possible BALs, or bronchoalveolar lavage.

BAL is saline instilled into an alveolar space and retrieved for culture and cytology. The endoscope is wedged into the most distal bronchus, and the saline is instilled down the biopsy channel. Follow with 3-4 mls. of air to clear the channel. Use short pulls with your syringe to retrieve your sample as the endoscopist moves the tip of the scope up and down and slightly back and forth. A diagnostic sample has a foamy top, or surfactant, which means that the sample has had contact with the alveoli. A retrieval of about 60% of your instilled amount is normal. Multiple samples may be combined for culture.

When the procedure is complete, have suction and an anesthesia mask on hand for compromised patients. The endoscopist records all data and submits all samples from the procedure as soon as possible. The endoscope should be cleaned and processed as soon as the patient is stable.

ADVANCES IN VETERINARY NEUROLOGY: BRAIN CASE STUDIES

Karen R. Muñana, DVM, MS, DACVIM (Neurology); Natasha J. Olby, VetMB, PhD, DACVIM (Neurology)
Raleigh, NC

INTRODUCTION

Management of neurological conditions involving the brain can be daunting to many veterinary care providers. Brain diseases are often considered difficult to diagnose and even more difficult to treat. However, the availability of magnetic resonance imaging (MRI) has led to significant improvement in our ability to diagnose brain disease in our veterinary patients. Similarly, new therapies are being evaluated for use in veterinary patients – most notably in the treatment of epilepsy and encephalitis – that are providing us with additional, promising treatment options. This presentation will focus on some of the recent advances in the diagnosis and treatment of brain diseases in dogs and cats. A case based approach will be utilized to present the historical, examination, diagnostic and therapeutic findings of brain disorders, with emphasis on information that is most relevant to the veterinary technician's role in patient management.

VASCULAR DISEASE

Up until recently, the dogma in veterinary medicine was that dogs and cats do not get strokes. However, with the greater availability of advanced imaging techniques in veterinary medicine this has been disproven, and cerebrovascular disease is now a frequently identified cause of brain disease in our patients. The term cerebrovascular disease refers to any abnormality of the brain resulting from a disruption of its blood supply. A stroke, also known as a cerebrovascular accident, is the most common clinical presentation of cerebrovascular disease and is characterized by a sudden onset of focal neurological signs that persist at least 24 hours. Strokes can be classified as either ischemic or hemorrhagic. In both dogs and humans, ischemic strokes are more common than hemorrhagic strokes.[1] Brain ischemia results from occlusion of a blood vessel with a thrombus or emboli, which deprives the brain of oxygen and energy. Ischemic stroke in dogs can be seen with atherosclerosis associated with hypothyroidism, hyperlipoproteinemia in Miniature Schnauzers, septic thromboemboli, neoplastic emboli, coagulopathy causing thromboembolism, and parasitic emboli (*Dirofilaria immitis*).[2] In addition, there is a specific disorder in cats called feline ischemic encephalopathy, which is a poorly understood syndrome characterized by spontaneous brain infarction. It has been postulated that aberrant migration of *Cuterebra* larva might play a role in the disease.[3] Cerebrovascular disease due to brain hemorrhage can be spontaneous, in which case it is most often associated with rupture of a vascular malformation, or can occur secondary to hypertension, intracranial tumors, vasculitis or coagulopathies.[2] Hypertensive encephalopathy is being diagnosed with increasing frequency in both dogs and cats, most likely due to both increased recognition of the disease and greater availability of blood pressure screening devices in veterinary medicine. Animals with hypertensive encephalopathy will frequently have associated ocular lesions, including retinal hemorrhage or retinal detachment.

The incidence of cerebrovascular disease is greatest in older animals. Neurological signs are typically acute in onset and nonprogressive. Deficits are often asymmetric and reflect the location of the lesion within the brain. Cerebellar involvement is most commonly recognized, but forebrain and brainstem signs can also be seen.[1] It has recently been suggested that episodic or transient (<24 hours in duration) vestibular dysfunction in dogs can be due to cerebrovascular disease, analogous to the transient ischemic attacks recognized in humans.[1]

All animals presenting with signs compatible with cerebrovascular disease should undergo a thorough ophthalmic examination as well as a blood pressure measurement to evaluate for hypertension. If hypertension is diagnosed, treatment is indicated as this alone will often result in improvement of neurological signs. Definitive diagnosis of a cerebrovascular accident is based on imaging findings. Ischemia is best visualized with MRI. Hemorrhage can be visualized on MRI or computerized tomography (CT). Analysis of cerebrospinal fluid (CSF) is frequently normal in animals with infarction. Underlying causes for cerebrovascular disease should be investigated, including endocrine disease, kidney disease, cardiac disease, metastatic disease and coagulopathy. Treatment is limited to supportive care and management of any identified underlying systemic disease. Prognosis is generally favorable, but is dependent on the size and location of the infarct. Clinical signs usually begin to improve after 24-72 hours, with recovery occurring over several weeks.

EPILEPSY

Epilepsy is a general term that refers to recurrent seizures of any cause. Primary or idiopathic epilepsy refers to seizures that occur for which an underlying cause is not found. In dogs with idiopathic epilepsy, the onset of seizures is characteristically between 1 and 5 years of age. Many breeds of dogs have a known predisposition to developing idiopathic epilepsy, and extensive research is currently underway to investigate the genetic basis of the disease. The disorder has been shown to be heritable in several breeds of dogs, including the Labrador Retriever, Golden Retriever, Viszla, Beagle, Belgian Shepherd, Keeshound, Dachshund, Border Collie, English Springer Spaniel and Standard Poodle. Investigators are attempting to isolate the gene(s) responsible for epilepsy, with the hope of eventually using selective breeding to decrease the incidence of breed-associated epilepsy. Cats have also been diagnosed with idiopathic epilepsy, although the disease is less common than in dogs and a genetic basis is less likely. Rather, seizures in cats are more likely due to symptomatic epilepsy, in which there is structural disease of the brain such as encephalitis or neoplasia, or reactive epilepsy in which seizures are caused by exposure to toxins or metabolic disease.[4]

Historically, it was believed that animals with idiopathic epilepsy presented with generalized seizures only, and that the presence of focal seizures should warrant further exploration for an underlying cause. However, it has recently been demonstrated that dogs with idiopathic epilepsy can have isolated focal seizures, or focal seizures with secondary generalization, in addition to primary generalized seizures.[5-8]

The mainstay of treatment for epilepsy is antiepileptic medications. However, it has been estimated that up to 30% of dogs are refractory to conventional treatment with phenobarbital and potassium bromide. Furthermore, these conventional drugs are associated with considerable side effects such that less than half of all dogs are able to maintain a seizure free status without experiencing adverse effects from the medication. Side effects range from sedation, ataxia and vomiting to more serious complications such as bone marrow suppression and hepatotoxicity. Fortunately, there are multiple new drugs that have been developed and approved for the treatment of human epilepsy over the last 10 years, and these drugs have gained use in veterinary medicine to treat seizures in dogs and cats as well. Although the efficacy of these drugs has not been established, they do appear to have a more favorable safety profile when compared to phenobarbital and potassium bromide. Drugs that have been utilized to treat epilepsy in dogs include felbamate, gabapentin, zonisamide, levetiracetam, and pregabalin. Gabapentin, zonisamide, levetiracetam and pregabalin have been utilized in cats. Important properties of the drugs are summarized below.

Drug	Frequency of Administration	Liver Metabolism?	Side Effects
Felbamate	Q 8 hrs	Yes	Sedation, ataxia, blood dyscrasias, KCS
Gabapentin	Q 8 hrs	Partial	Sedation, ataxia
Zonisamide	Q 12 hrs	Yes	Sedation, ataxia, vomiting
Levetiracetam	Q 8 hrs	No	Sedation, ataxia
Pregabalin	Q 8-12 hrs	Partial	Sedation, ataxia

Client education is key to successful management of an epileptic pet, and the veterinary technician often assumes an important role in this regard. Pet owners need to be educated on the expectations of seizure management. Owners should be aware that treatment is often life-long, and that successful treatment does not necessarily imply that the animal will be seizure free. Owners should understand that all antiepileptic medications are associated with side effects, and in an individual animal treatment is directed at optimizing seizure control while minimizing side effects. Discussions should also include what to do if a pet seizures at home, when to seek emergency care, and what to expect with respect to long-term treatment and monitoring of epilepsy.

ENCEPHALITIS

Inflammatory conditions of the brain are a common cause of neurological dysfunction in both dogs and cats. The clinical signs seen with inflammatory disease reflect the location of the lesion within the nervous system. As a general rule, inflammatory disease tends to be acute in onset and progressive, with either a focal, multifocal or diffuse distribution. Animals with meningitis classically present with severe cervical pain and fever. Animals with nervous system infection frequently do not have systemic involvement. Therefore, the absence of fever, anorexia, depression, and a normal hemogram cannot be used to exclude the possibility of an infectious etiology. A thorough ophthalmologic examination should be performed in every neurological case to observe for evidence of fundic changes or uveitis compatible with inflammatory disease.

Diagnosis of encephalitis is based on imaging and CSF findings. Analysis of CSF typically reveals an increase in nucleated cells and protein concentration. Cytologic evaluation of the fluid can provide additional information as to likely causes of the inflammation. Additional testing is performed as indicated to attempt to identify an infectious cause for the inflammation. Infectious causes of encephalitis include viral, rickettsial, protozoal, fungal, parasitic and bacterial organisms. In dogs, canine distemper virus, ehrlichiosis, Rocky Mountain spotted fever, toxoplasmosis and cryptococcosis are the most commonly recognized infections, while feline infectious peritonitis, toxoplasmosis and cryptococcosis are most frequently diagnosed in cats. Therefore, additional testing can include bacterial or fungal cultures on CSF, as well as serum and CSF antibody or antigen titers. Magnetic resonance imaging is helpful in confirming the presence of inflammation as well as providing information that can help to differentiate among the possible causes of encephalitis.

In most cases of encephalitis in dogs, an infectious cause is not identified. Rather, these dogs are presumed to have an inflammatory, non-infectious cause of central nervous system disease that is immune-mediated in origin. These disorders are commonly grouped under the general term meningoencephalomyelitis of unknown etiology (MUE), and include granulomatous meningoencephalomyelitis (GME) and necrotizing meningoencephalitis. The mainstay of treatment for MUE has been corticosteroids, but recent work has focused on ancillary treatments, in attempts to improve prognosis and limit side effects of treatment. Agents that have been evaluated include procarbazine,[9] cyclosporine,[10] cytosine arabinoside,[11] and azathioprine.[12] All of these drugs are purported to work, but controlled studies are lacking to establish their comparative efficacy and safety, as well as their superiority to corticosteroids alone. In addition, all of these immunomodulatory drugs require regular monitoring. Hemograms should be routinely performed to assess for bone marrow suppression with cytosine arabinoside, procarbazine, and azathioprine. Azathioprine can also be hepatotoxic, and liver panels should be periodically evaluated. Finally, serum drug concentrations are routinely monitored with cyclosporine therapy. Prognosis for MUE is variable and dependent on the underlying cause of the inflammation and the extent and severity of associated neurological deficits. However, a recent study suggests that dogs that survive the initial phase of the disease have a good chance of a prolonged survival.[13]

COGNITIVE DYSFUNCTION

Canine cognitive dysfunction syndrome is a clinical disorder of dogs that is characterized by an age-related deterioration of cognitive abilities. Morphologically, the disorder is associated with cortical atrophy and accumulation of beta amyloid protein in the brain. The condition manifests as behavioral changes that are not attributable to any other medical conditions present in the dog. Four categories of behavioral changes associated with cognitive dysfunction have been recognized. These are: 1) disorientation, 2) reduced social interactions, 3) changes in sleep/wake cycle or activity level, and 4) loss of housetraining.

Recent studies suggest that cognitive dysfunction might also occur in the cat.[14] Although the disease has not been as well characterized in cats, is appears as if they can manifest similar signs to dogs, in addition to showing excessive vocalization, altered responses to stimuli, decreased self-hygiene and alterations in appetite.[15]

Differential diagnoses for an animal with age related behavioral changes include vision loss, hearing loss, metabolic encephalopathy, brain neoplasia, cerebrovascular disease and encephalitis. A diagnosis of cognitive dysfunction syndrome can only be made based on exclusion of other potential causes of the clinical signs. A thorough evaluation should be performed to assess for systemic disease, hypertension, hearing loss, vision loss and neurological deficits. Neurodiagnostic testing, including brain imaging and CSF analysis, is necessary to exclude the presence of underlying brain disease before a definitive diagnosis of cognitive dysfunction is made. However, it is not uncommon to identify hearing or vision loss, or metabolic, endocrine cardiovascular or degenerative joint disease in a geriatric animal, and these conditions can mimic signs of cognitive deficits. Consequently, any identified conditions should be treated and the animal monitored for improvement in the clinical signs before definitive treatment for cognitive dysfunction is recommended.

Two methods of treatment for cognitive dysfunction in dogs are available: pharmacological and dietary. Pharmacologic treatment consists of the administration of L-deprenyl (selegiline, Anipryl). L-deprenyl is an inhibitor of monoamine oxidase, the enzyme that breaks down the neurotransmitter dopamine, and the drug is believed to function by modulating the dopaminergic system in the brain. The second treatment consists of feeding a diet enriched with antioxidants, including vitamins E and C, docasahexanoic and eicosapentaenoic acid, lipoic acid and carnitine (Hill's Prescription Diet Canine b/d). It is postulated that the diet decreases the production and consequence of free radicals, which have been shown to contribute to age associated changes in cognition. Both treatments have been reported to improve cognitive behavior in aged dogs in controlled studies.[16,17] A limitation of these studies is that dogs did not undergo a full neurodiagnostic evaluation that would provide more conclusive evidence that the clinical signs were due to cognitive dysfunction syndrome. Additionally, some of the studies assessed the cognitive skills of dogs in a laboratory setting, and it is not certain how these findings correlate with the behavioral changes associated with canine cognitive dysfunction syndrome. There are no published studies on the treatment of cognitive dysfunction in cats, although anecdotal reports have described the use of both selegiline and dietary supplements to treat the condition.

REFERENCES

[1] McConnell JF, et. al. **Vet Radiol Ultrasound** 2005; 46:1. [2] Garosi LS. **Vet Clin North Am Sm Anim Pract** 2010; 40:65. [3] Williams KJ, et. al. **Vet Pathol** 1998; 35:330. [4] Schriefl S et. al. **J Am Vet Med Assoc** 2008; 233:1591. [5] Hulsmeyer V, et. al. **J Vet Intern Med** 2010; 24:171. [6] Patterson EE, et. al. **J Vet Intern Med** 2003; 17:319. [7] Patterson EE, et. al. **J Am Vet Med Assoc** 2005; 26:54. [8] Licht BG, et. al. **J Am Vet Med Assoc** 2007; 231:1520. [9] Coates JR, et. al. **J Vet Intern Med** 2007;21:100. [10] Adamo FP, et .al. **J Am Vet Med Assoc** 2004; 8:1211. [11] Zarfoss R, et. al. **J Sm Anim Pract** 2006;47:588. [12] Wong MA, et. al. **J Am Vet Med Assoc** 2010; 237:929. [13] Granger N, et. al. **Vet J** 2010; 184:290. [14] Head E , et. al. **Neurobiol Aging** 2005; 26:749. [15] Landsberg G. **J Fel Med Surg** 2010: 12:837. [16] Campbell S, et. al. **Vet Ther** 2001; 2:24. [17] Roudebush P, et. al. **J Am Vet Med Assoc** 2005; 227:722.

ADVANCES IN VETERINARY NEUROLOGY: SPINAL CASE STUDIES

Karen R. Muñana, DVM, MS, DACVIM (Neurology); Natasha J. Olby, VetMB, PhD, DACVIM (Neurology)
Raleigh, NC

INTRODUCTION

Recent advances in the field of veterinary medicine have led to significant improvements in our ability to diagnose and treat spinal cord disease in dogs and cats. Perhaps the most significant development in this regard has been the increased availability of magnetic resonance imaging (MRI). Unlike the previous utilized imaging modalities of myelography and computed tomography, which provide good imaging of the tissues surrounding the spinal cord but limited visualization of the spinal cord itself, MRI provides excellent imaging detail of both the spinal cord and surrounding tissues. The additional information provided by MRI has furthered our understanding of many spinal conditions, including intervertebral disk disease, caudal cervical spondylomyelopathy (Wobbler disease), fibrocartilagenous embolism, spinal neoplasia and myelitis. In addition, advances in the field of molecular biology that have enabled further exploration into the genetic basis of disease have had a tremendous impact on many disciplines in veterinary medicine, including neurology. This is perhaps best exemplified by the identification of the genetic basis for degenerative myelopathy, which has improved our overall understanding of the disorder. This presentation will focus on some of these recent advances in the diagnosis and treatment of spinal diseases in dogs and cats. A case based approach will be utilized to present the historical, examination, diagnostic and therapeutic findings of spinal disorders, with emphasis on information that is most relevant to the veterinary technician's role in patient management.

DEGENERATIVE MYELOPATHY

Degenerative myelopathy is a degenerative spinal cord condition of large dogs, primarily German Shepherds. The disease is most common in middle aged to older dogs. Dogs with degenerative myelopathy show signs of a slowly progressive ataxia and paraparesis, with no evidence of spinal pain. This condition was poorly understood for many years, and a presumptive antemortem diagnosis was arrived at only after excluding other likely causes of myelopathy, such as neoplasia and Type II intervertebral disk disease. One of the most exciting developments in neurology over the past several years is the work performed by researchers at the University of Missouri on the genetic basis of degenerative myelopathy[1], who discovered that a specific mutation in superoxide dismutase is associated with the disease trait. Genotyping of numerous dogs of many different breeds has demonstrated that the same mutation is associated with degenerative myelopathy in German Shepherd dogs, Pembroke Welsh Corgis, Boxers, Chesapeake Bay Retrievers and Rhodesian Ridgebacks. Furthermore, the disease has been demonstrated to have genetic and clinical similarities with the human condition, amyotrophic lateral sclerosis (ALS or Lou Gehrig disease). A genetic test is now available for degenerative myelopathy through OFA. Dogs that are homozygous for the mutation are at risk for developing the disease, but many such dogs do not develop clinical signs, and age appears to affect expression of the disorder. Dogs that are heterozygous for the mutation (one mutant gene and one normal gene) are unlikely to develop degenerative myelopathy, but can pass the mutation on to their offspring. The genetic test can be used by breeders to help reduce the incidence of degenerative myelopathy in a line of dogs, and can be used to support a clinical diagnosis of degenerative myelopathy in an older dog with compatible clinical signs if other causes of myelopathy have been ruled out. A recent report suggests that dogs with degenerative myelopathy that receive physiotherapy have a better prognosis than dogs that do not.[2] No other proven definitive treatment exists and overall prognosis is poor.

FIBROCARTILAGENOUS EMBOLISM

Fibrocartilagenous embolism (FCE) is a disorder caused by infarction of the vasculature of the spinal cord with fibrocartilagenous material that is identical to that composing the nucleus pulposus of the intervertebral disk. It is not known how the material gains access to the vessels of the spinal cord. The disease is seen most commonly in large breed dogs. As is characteristic of vascular disease, the clinical findings are of a sudden, nonprogressive and nonpainful myelopathy. Many dogs have a history of exercise at the time of onset of the neurological signs. Neurological deficits are often asymmetric, with the spinal cord intumescences (C6-T2 or L4-S3) most commonly involved.

Prior to the availability of MRI, diagnosis of FCE was based on exclusion of other causes. With MRI, one is able to visualize the infarcted region of the spinal cord, thereby providing additional support for the diagnosis. Recent studies have described the characteristic MRI findings in dogs with FCE that result from edema and spinal cord swelling. An association between outcome and extent of changes visible on MRI has been demonstrated.[3] No definitive treatment is available, and animals should be provided general nursing care and rehabilitation. Prognosis is dependent on the site and severity of the ischemic insult, although most dogs do recover.

INTERVERTEBRAL DISK DISEASE

Intervertebral disk disease is the most common cause of spinal cord dysfunction in dogs, and can also occur in cats. There is considerable information available in the veterinary literature on acute intervertebral disk herniations in chondrodystrophoid dogs, in which the nucleus pulposus of the disk undergoes degeneration and is predisposed to herniation into the spinal canal, causing compression of the spinal cord. However, with the increased availability of MRI in clinical practice, another form of acute disk herniation is being recognized in our veterinary patients. This form of disk disease has been called "traumatic disk extrusion", or "acute, noncompressive disk extrusion". In this case, the nucleus pulposus is not degenerate but the disk is placed under a brief period of excessive stress that causes the normal nucleus pulposus to extrude with force and concuss the spinal cord. Because normal nucleus pulposus has a gelatinous consistency, the material readily dissipates into the epidural fat causing minimal if any spinal cord compression. Such extrusions typically occur in dogs in association with trauma or exercise.

Clinical signs are similar to those found with FCE. Animals typically present with a nonprogressive and nonpainful myelopathy, and neurological deficits are often asymmetric. However, unlike FCE which tends to occur in the intumescences of the spinal cord, the site of traumatic disk extrusion is more common in the thoracolumbar spine.[4] A presumptive diagnosis is based on MRI, which reveals evidence of edema in the spinal cord centered over a disk space, as well as a narrowed disk space and reduction in volume and loss of signal intensity of the nucleus pulposus. [5] In some cases, subtle spinal cord compression can be apparent. Prognosis is dependent on the extent and severity of neurological deficits. Treatment consists of general supportive care, including good nursing and physical therapy.

CAUDAL CERVICAL SPONDYLOMYELOPATHY

Caudal cervical spondylomyelopathy (CCSM), also known as Wobbler syndrome, is a disorder of the cervical vertebrae and associated soft tissues that results in compression of the adjacent spinal cord. CCSM is primarily a disease of large and giant breed dogs, being most frequently identified in older Doberman pinschers and young Great Danes. The cause of the vertebral changes is believed to be multifactorial, with proposed etiologies including genotype, body conformation, high plane of nutrition and rapid growth rate. Neurologic deficits seen with CCSM range from ataxia to tetraparesis. Signs are typically more severe in the pelvic limbs, and in early lesions, thoracic limb deficits can be difficult to detect. Many affected animals show a characteristic "Wobbler" gait with wide based ataxic pelvic limb movements and a short strided thoracic limb gait. Cervical hyperesthesia, presenting as guarding of the neck or a low-neck carriage is also commonly seen.

Diagnosis is based on imaging studies of the cervical spine. CCSM has long been regarded as a surgical disease, with medical therapy providing only temporary improvement in the condition. Surgery is still considered the treatment of choice, but two recent studies on the medical management of the disease suggest that dogs treated conservatively can have stabilization or improvement in their clinical signs, although it is unlikely that clinical signs will resolve.[6,7] The prognosis for dogs with CCSM is primarily dependent on the severity of signs. However, because CCSM is a chronic progressive disease, the overall prognosis is worse than for a dog that has suffered acute spinal cord compression and is often difficult to accurately predict.

FELINE SPINAL CORD DISEASE

Spinal cord disease is a relatively common problem in the cat population. Several recent retrospective studies have provided valuable information on the prevalence, treatment and outcome of feline spinal disorders. These studies have identified inflammatory and neoplastic etiologies as the most common causes of spinal cord disease in cats.[8-10]

Among inflammatory etiologies, feline infectious peritonitis (FIP) is the most frequent cause of central nervous system disease in cats. Neurological involvement is typically seen with the non-effusive or "dry" form of the disease. Neurologic disease associated with FIP tends to have an insidious onset with no specific clinical course. Signs of encephalitis predominate but spinal cord involvement is also frequently identified and can be the only presenting sign. With FIP-associated myelitis, neurologic signs can be asymmetric or unilateral due to primary involvement of only one half of the spinal cord, and any area of the spinal cord can be affected. Many cats will have concurrent systemic involvement, with signs of anorexia, weight loss and/or ocular lesions. Obtaining an antemortem diagnosis of FIP is often difficult. Serum tests for anti-coronavirus antibodies are often positive but have low specificity; however, a negative serum titer (i.e. a complete absence of antibody) does not exclude the possibility of FIP-associated neurological disease because soluble antibodies can form immune complexes and escape detection by standard tests. CSF analysis will show variable results. Positive CSF titers can be a useful ante-mortem indicator of neurological disease. PCR assays performed on CSF have not been shown to be a reliable test for confirming disease. Prognosis for cats with CNS FIP is poor. Definitive treatment is not available. The use of immunosuppressive drugs can slow the progression of disease. Affected animals should be isolated from other cats to prevent the spread of infection.

Lymphoma is the most common neoplastic cause of spinal cord dysfunction in cats. Disease is primarily seen in young cats infected with the feline leukemia virus (FeLV), although it can also be seen in cats that test negative for FeLV. Neurologic deficits typically manifest as a painful, focal myelopathy that can involve any area of the spinal cord. Peripheral lymphadenopathy and hepatosplenomegaly can be present on physical examination, although evidence of systemic lymphoma is not seen in all cases. The most common location for spinal lymphoma is within the epidural space. Consequently, spinal imaging typically reveals the presence of extradural spinal cord compression. Cats with spinal lymphoma respond well to both radiation and chemotherapy, and improvement in clinical signs is often seen within several days of initiating therapy. The goal of therapy is disease remission as treatment is not curative.

PHYSICAL THERAPY AND NURSING CARE

Physical therapy is rapidly gaining acceptance as a treatment modality in veterinary medicine. A comprehensive rehabilitation plan is believed to play a key role in the successful outcome of an animal with spinal cord disease. The aims of a physical therapy program are to increase circulation, maintain joint and muscle mobility, increase physical strength and endurance, minimize secondary complications of recumbency and provide psychological stimulation. Several rehabilitation modalities that can be of benefit to the recumbent animal require little if any specialized equipment. These include massage, passive range of motion exercises, active exercise and hydrotherapy.

Massage is one of the simplest forms of physical therapy that involves manipulation of soft tissues. It can enhance circulation, loosen and stretch tendons, reduce edema, minimize scar tissue and relieve spasm and pain. It is also helpful in relaxing an anxious or uncomfortable animal prior to the use of other physical therapy modalities. Massage if often used in conjunction with passive range of motion exercises. Massage techniques include effleurage (stroking) and petrissage (kneading and compression). The muscles are worked in a distal to proximal direction, with areas of tension or stiffness focused on. Massage, when performed correctly, should not be uncomfortable for the patient.

Passive range of motion exercises involves movement of a paretic or paralyzed limb by the therapist, and does not require patient participation. These exercises help to reduce muscle atrophy and tissue adhesions, promote normal joint dynamics, enhance venous and lymphatic drainage and prevent contracture of the muscle and joint capsule. To ensure the maximum benefit, the joints must be flexed and extended to the limit of their range of motion, unless the movement is painful. The limb should be supported both proximal and distal to the joint being manipulated, and the motion should be initiated slowly to prevent tearing of muscles. The exercise should be started with pure flexion and extension movement, and then expanded to involve full range of motion for the joint.

Active exercises involve patient participation. If this form of exercise is performed in a recumbent patient, the animal's body must be supported with a sling or towel. Active exercises include controlled standing, treadmill walking, and controlled walking. Exercises should be performed on non-slick, non-abrasive surfaces. Initially, the time allotted to this form of therapy should be short. The therapist can gradually work up to longer period of exercise as the animal's endurance increases. Active exercises improve the animals mental well being by providing activity and stimulation, help prevent problems associated with prolonged recumbency, promote the return of strength and function of the limbs, and reinforce rhythmic stepping movements which may be important in the recovery from spinal cord injury.

Hydrotherapy is an excellent form of active exercise. The buoyancy of the water encourages voluntary motion in an animal that is unable to stand, and allows extensive joint and muscle movement in a non-weight bearing environment. In addition, the warm water improves lymphatic and venous circulation, cleanses the skin, and soothes muscle spasms. Hydrotherapy is typically not utilized until 10-14 days post-operatively, and should not be performed in animals with cardiovascular compromise or draining wounds. A sling or flotation device should be used for support, and water should be kept between 38-40° C. Sessions should initially be short (5 minutes or less) and the time gradually increased as the animal's endurance improves. The fur should be dried thoroughly after completion of a hydrotherapy session, as a damp, recumbent animal is predisposed to developing decubital ulcers.

Although these physical therapy techniques are frequently utilized in veterinary patients to assist in the recovery from spinal cord disease, there has been little information available on their effectiveness. A recent retrospective study on physical therapy in dogs with degenerative myelopathy suggested that dogs that received intensive physiotherapy had longer survival time and remained ambulatory longer than dogs with moderate or no physiotherapy.[2] Another retrospective study on dogs with FCE concluded that physiotherapy instituted immediately after the diagnosis seemed to influence recovery rate.[11] Additional controlled studies are warranted to discern the role of physical therapy in the recovery of dogs with spinal cord disease.

Good nursing care is imperative in the rehabilitation process. Incontinent animals should have their bladder emptied 3-4 times daily. The bladder should be manually expressed if possible, or otherwise catheterized. Animals should be closely monitored for urinary tract infections and treated aggressively if an infection develops. Appropriate bedding is necessary to prevent the development of decubital ulcers and urine scald. Bedding should support and evenly distribute the weight, provide extra padding over the bony prominences, be easy to clean and maintain, and be made of material that allows urine to drain away from the body. Recumbent animals should be turned from side to side every 4-6 hours, and encouraged to sit sternal if possible to avoid the development of decubital ulcers and hypostatic pulmonary congestion. Efforts should be made to keep the animal clean and dry in order to prevent urine scald. This may involve frequent bathing, clipping of the hair around the perineum and applying barrier cream to the perineal region as needed.

Effective communication between the client and the medical team is imperative during the rehabilitation process. Physical rehabilitation of a recumbent animal is a challenging, time-consuming task, and the owner should be aware of the level of commitment required. Encouraging the owner to take an active role in the rehabilitation effort is often beneficial to both the animal's and owner's well being. An effective physical therapy program is believed to assist the animal in returning to an ambulatory status, and improve the animal's quality of life during the recovery period.

REFERENCES

[1] Awano T, et. al. **Proc Natl Acad ScI USA** 2009; 106:2794. [2] Kathmann I, et. al. **J Vet Intern Med** 2006; 20:927. [3] De Risio L, et. al. **J Am Vet Med Assoc** 2008; 233:129. [4] De Risio L, et. al. **J Am Vet Med Assoc** 2009; 234:495. [5] Chang Y, et. al. **Vet Rec** 2007; 160:795. [6] De Decker S, et. al. **J Sm Anim Pract** 2009; 50:136. [7] da Costa RC, et. al. **J Am Vet Med Assoc** 2008; 233:1284. [8] Marioni-Henry K, et. al. **J Vet Intern Med** 2004; 18:851-858. [9] Bradshaw JM, et. al. **J Comp Path** 2004; 131:112. [10] Goncalves R, et. al. **J Fel Med Surg** 2009; 11:53. [11] Gandini G, et. al. **J Sm Anim Pract** 2003; 44:76.

SUPPORTIVE CARE FOR CLIENTS AND PATIENTS WITH CHRONIC NEUROLOGY DISORDERS

Julie A. Nettifee Osborne, RVT, BS
Raleigh, NC

INTRODUCTION

As veterinary medicine becomes increasing sophisticated and technological advances become available to support the patient affected by neurological dysfunction, the need for advanced nursing skills, as well as advanced communication skills, becomes imperative for successful patient and client outcomes. Veterinary technicians with abilities to meet the needs of the patient and client affected by a long-term chronic neurology disorder such as epilepsy, paraplegia or other disorders, will be crucial to the support of the entire client-patient-veterinary team.

Within the context of these proceedings, the author will explore several of the following topics:

- Understanding how chronic neurologic dysfunction affects the client-patient-veterinary team
- Exploring how veterinary technicians can provide advanced nursing and rehabilitation skills to support patient needs in a variety of medical dysfunctions such as long-term recumbent patients, patients with epilepsy or other CNS disorders, and more.
- The importance of advanced veterinary communication skills to effectively give and receive feedback from others caring for the patient (clinicians, owners, other members of the veterinary team)
- How specialty support groups are making an impact in long-term support needs, including patients that are making a difference in the lives of other patients and owners
- Tech-to-tech support networks: Neurology AIMVT Recognition

PATIENTS WITH CHRONIC NEUROLOGY DYSFUNCTION

Veterinary patients with chronic medical needs present many different long-term challenges and opportunities to veterinary professionals-especially those patients facing long-term neurologic dysfunction due to such causes as epilepsy, spinal cord disease or trauma, other central nervous system disorders and more. In general, many veterinary professionals are well prepared for the initial diagnosis, treatment, and rehabilitation that accompany the onset of the disease; however, along with the increased technological advances within the profession, owners are supporting patients with these conditions for longer periods of time. A sudden and marked change occurs in the intensity and quality of the owner/patient bond. This necessitates establishing a new, and almost always less satisfying human-animal equilibrium. The veterinary professional not only helps the owner cope with the conflicts of chronicity, which can include concurrent financial, physical and psychological factors. This is no different than a human patient affected by long-term disease, yet more supports are available and recognized in the human sector, than in the veterinary profession generally.

When the four "C"s of a quality patient plan are carried out effectively, "demonstrating *caring*, providing effective *communication*, explaining *costs* and maintaining patient *cleanliness*, a special relationship always develops between the client, the chronically ill patient and the veterinary team. [1] The veterinary technician is an important aspect of this team and is often a "bridge" helping to carry out the clinicians' medical treatment plan and teaching the owners how to properly care for the patient. One additional C, *counsel*, is also typically part of this equation and both veterinary professionals, as well as in many larger settings, professional counsel is very important as owners seek options for their pet and for their own emotional well being in challenging chronic cases.

DEFINING THE FIVE "C'S"

For the neurology technician, the first element, caring, has many different facets-from physical to emotional. Physically we must meet the needs of the patient affected by neurologic dysfunction to a much greater level than that of a typical patient. Extensive nursing care may include most or all of the following:

- Assessment of the patient needs depending on diagnosis
- Appropriate bedding to prevent decubital ulcer formation and maintain cleanliness (ie; spinal cord disease, recumbency due to illness, etc)
- Appropriate padding to prevent further injury (CNS, Epilepsy, etc)
- Calculation and preparation of optimal nutrition
- Preparing to meet the social needs of the patient
- Preparation of a nursing plan that includes bladder management, patient rotation, medication administration and monitoring
- Patient rehabilitation which may include heat/cold therapies, laser treatments, hydrotherapy, therapeutic exercise, chest care and assistive devices such as K-9 carts.[2, 3]
- "Care" also includes caring for the caregiver-both the veterinary professional and the owner and comprehending the possible toll that a patient with chronic neurologic dysfunction may have for those that care for the patient.

The second element, communication is the key to effectively share your observations, concerns and to teach clients how to best manage a patient with a long-term need. Although this presentation specifically applies to the patient with a neurologic disease, the same general principles apply across disciplines. Specialization within a discipline typically requires advanced abilities to communicate with and teach owners at more involved levels than general veterinary medical care. Within the profession, Bayer

Animal Health/Institute for Healthcare Communications and FRANK workshops are two well-known training programs that will help veterinary professionals advance their communication skills in many of the following areas:

- Verbal Communication: What we say and how we say it
- Organization and clarity of information presented to clients
- Tone and Volume of Voice and Vocabulary
- Nonverbal Communication
- Listening to and responding to clients
- Empathizing with clients
- Provision of written or online training materials for clients to support verbal instructions
- Referral of clients to professional counselors if needed

The third element, cost includes not only the additional amount of funding that may be necessary to maintain a patient with a chronic disease or disability, but it may also include the physical, emotional and other costs that are more difficult to calculate, but are essential to recognize to maintain a positive "net flow" for the entire team. Clients respond to their pet's diagnosis in many ways. Two of the most extreme responses are fierce dedication and deep doubt .[4] Clients that are fiercely dedicated to their pets sometimes give up huge portions of their personal lives in order to care for their companion animals. Many clients worked with by the author have given up outside travel, changed their lives dramatically-to fit the needs of their pet with epilepsy or pet that is now confined to a cart, for example. These are costs that cannot be completely calculated, yet are important to acknowledge. Patients developing secondary infections, medical management needs such as anticonvulsant therapy, and more may assume higher medical costs. The fiercely dedicated owner notes that benefits of the increase in the bond between the owner and patient outweigh the financial costs due to the disease. In many of these patient-owner relationships, caring for a companion animal with a chronic disease can often transform their own lives as well in some instances.

The fourth element, cleanliness, although it should be obvious in any quality patient care program, cleanliness applies to both how the patient is managed, as well as how we present ourselves within the profession. If a patient is managed well, costs can be reduced as secondary problems such as decubital ulcers, poor circulation and more can be minimized. Personal attention to cleanliness helps us to communicate the importance to our clients as well.

The fifth element, counsel, although typically assigned to professional veterinary social work, is a term that in its essence means, "advice or guidance, especially as solicited from a knowledgeable person." The veterinary team can all act as "counsel" to one another and to the client facing long-term medical needs with their companion animal. In addition, however, owner to owner support can be an essential link in providing support to owners, while at the same time helping to properly care for you and to minimize compassion fatigue while trying to assist owners over the long term. Veterinary social workers are essential as professional counselors, and having either a counselor on staff, or a listing available for those owners that need additional referral, is critical. There is a growing mass movement of owners within the pet community that have leveraged the benefits of social networking to help in their cause. Groups such as Epil K-9, The Yorkie Angel Patrol and more are great resources to share with owners. These groups have proven to be incredible support networks for owners facing many different diseases, and as a veterinary professional, they are also proving to create lasting impacts in helping to change the face of the disease by creating new funding streams for canine neurological research, teaching and client-to-client support. It is a very powerful example whereby several of the organizations that have formed to support various neurology projects at North Carolina State University-College of Veterinary Medicine, in many cases have lost a pet to such diseases as epilepsy, GME, etc. In this loss however, they have turned this into a positive outcome for all that truly allows the pet's legacy to create new outcomes. Many individuals in these groups also have become "owner-to-owner" support for new clients facing the beginning stages of diagnosis and treatment.

Specialization within the profession is also important for tech-to-tech support, sharing of ideas, exploring advances and more. As of June 2011, the AIMVT has now recognized the first Neurology Specialty exam! This time of specialization will help us as technical staff also to enhance this section of the profession as well.

CONCLUSION

In summary, supporting both the patient and the client with a companion animal diagnosed with a chronic medical illness can be challenging to every member of the veterinary healthcare team. Utilizing the components of the five "C's" described within the context of these proceedings: caring, communication, costs, cleanliness and counsel, all members of this team can help to create positive outcomes for the patient and owners as well as themselves. The author has included recommended reading, as well as a listing of suggested websites, that may be useful to explore since many of the topics discussed can be explored in much more depth that this presentation time will allow.

Suggested resources for additional information

- http://cvm.ncsu.edu/epilepsyresearch
- http://www.canine-epilepsy.com/
- http://www.canine-epilepsy.net/
- http://www.tobysfoundation.org/
- http://www.canine-epilepsy-guardian-angels.com/
- http://www.yorkieangelpatrol.com/
- http://www.scoutshouse.com/health-resources/helpful-websites/
- http://www.vspn.org/library/wwwdirectory/Neurology.htm
- http://www.healthcarecomm.org/bahcp/homepage.php
- http://www.argusinstitute.colostate.edu/
- http://www.pfizerfrank.com/

REFERENCES

1. Cohen, SP and Fudin, Carole E, Problems in Veterinary Medicine-Animal Illness and Human Emotion, Volume 3, Number 1, March 1991. 2. Moore, AH and Rudd, S, BSAVA Manuel of Canine and Feline Advanced Veterinary Nursing, Second Edition, 2008, BSAVA. 3. Lindley, S and Watson, P, BSAVA Manuel of Canine and Feline Rehabilitation, Supportive and Palliative Care, 2010, BSAVA. 4 .Cohen, SP and Fudin, Carole E, Problems in Veterinary Medicine-Animal Illness and Human Emotion, Volume 3, Number 1, March 1991, pages 51-60.

ELECTRODIAGNOSTIC EVALUATION OF COMMON CANINE LOWER MOTOR NEURON DISEASES

Sarah A. Moore, DVM, DACVIM (Neurology)
Columbus, OH

INTRODUCTION

There are two basic types of electrodiagnostic testing used to assess veterinary patients with lower motor neuron disease. They are:

1. Electromyography (EMG)
2. Motor nerve conduction testing, which has two components
   - Nerve conduction velocities (NCV)
   - Evaluation of compound muscle action potentials (CMAP)

Each of these testing modalities will be discussed in detail using a case-based format. Examples of normal and abnormal results will be presented with an emphasis of clinical application of each test. Other electrodiagnostic tests such as brainstem auditory evoked responses (BAER), electroencephalography (EEG) and urethral pressure profiles (UPP) are also commonly performed in veterinary medicine, but are outside of the scope of this session.

The purpose of electrodiagnostic testing is to establish whether an abnormality is present anywhere along the lower motor neuron pathway, and to determine which structures within that pathway are affected. With rare exception, electrodiagnostics do not give us a definitive diagnosis of a specific disease process; they are use to localize the problem so that further testing can be appropriately aimed at making a definitive diagnosis. Structures included in the lower motor neuron pathway, and thus assessed by routine electrodiagnostic testing include the ventral horn (motor) neurons, nerve roots and peripheral nerves, the neuromuscular junction, and muscle (Figure 1). With the exception of BAER, most electrodiagnostic testing in veterinary patients is performed under general anesthesia. Most tests require the animal to be still for long periods of time, and some (such as EMG) are uncomfortable.

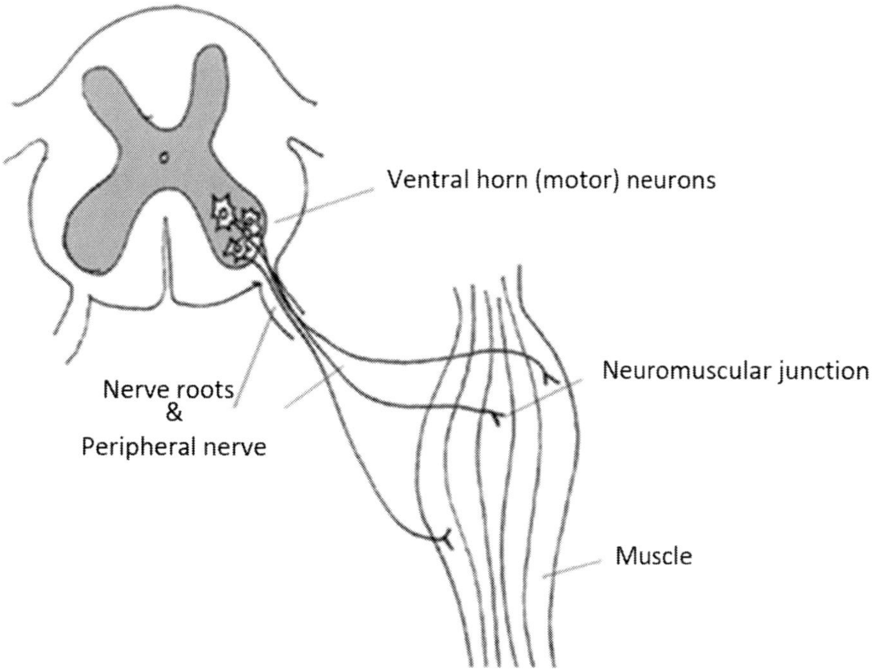

Figure 1: Components of the lower motor neuron pathway. These structures are assessed during routine EMG and nerve conduction testing.

ELECTROMYOGRAPHY (EMG)

The general premise of EMG is that there should be very little normal spontaneous electrical activity present in the muscles of an animal that is under generally anesthesia. EMG uses this premise to evaluate for increased amounts of spontaneous activity, indicating underlying abnormalities in the lower motor neuron pathway. Basic EMG testing involves insertion of a needle into the muscle and recording spontaneous electrical activity that occurs within that muscle. Abnormalities on EMG can be seen with diseases of the muscle, and with diseases of the peripheral nerve or nerve root. Fibrillation potentials, positive sharp waves, and complex repetitive discharges are the three types of common abnormalities identified on EMG (Figure 2). These abnormalities occur either because the muscle membrane itself is abnormal, or because the muscles cells have become hyperexcitable because they have lost their nerve supply (denervation); all three abnormalities have the same clinical meaning. EMG alone does not distinguish between disease of the muscle and disease of the nerve, and it is often normal in diseases such as myasthenia gravis; a normal EMG does not rule out diseases of the neuromuscular junction. Also, in animals with acute disease of the peripheral nerve, it takes 5-7 days for EMG abnormalities to develop. Thus, early on in the disease, EMG can also be normal even in animals with severe disease of the peripheral nerves. The most common primary muscle diseases seen in veterinary medicine include infectious or autoimmune disease of the muscle (polymyositis), muscle disease secondary to endocrine problems such as hypothyroidism or Cushings, muscle wasting secondary to chronic steroid administration, paraneoplastic myopathies, and less common congenital problems of the muscle such as muscular dystrophy.

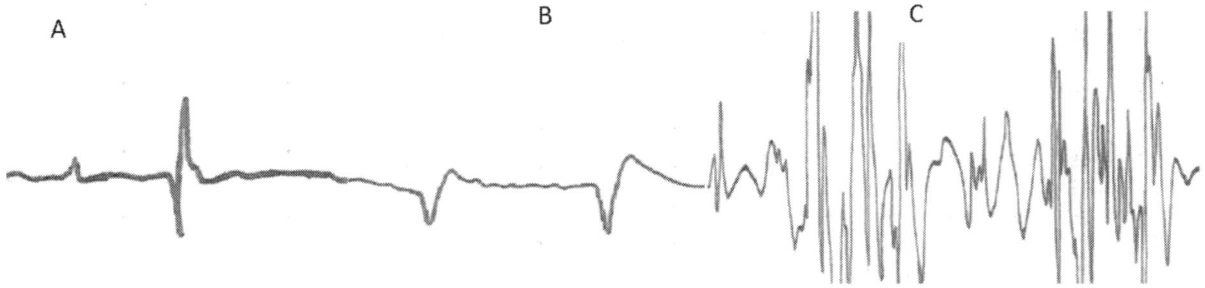

Figure 2: Common abnormalities identified on EMG include Fibrillation potentials (A), Positive sharp waves (B), and Complex repetitive discharges (C).

EMG provides an excellent screening test to aid in identifying which muscle groups throughout the body are affected in a patient with lower motor neuron disease of any type, and to guide the clinician as to where biopsies of muscle or nerve should be taken to make a specific diagnosis.

MOTOR NERVE CONDUCTION TESTING

When coupled with EMG, nerve conduction testing helps to assess whether abnormalities seen clinically on the neurologic examination and on EMG are due to problems of the muscle, or due to problems of the nerve. Nerve conduction testing is performed by sending an electrical impulse along the nerve, and measuring the length of time it takes for that impulse to travel down the nerve and stimulate a muscle contraction (compound muscle action potential- CMAP). Also assessed, are the character and appearance of the electrical waveform that muscle contraction generates (CMAP). Nerve conduction velocities (NCV) are measured at several points along the peripheral nerve to look for differences that may be identified either proximally or distally on the nerve itself. A slowed NCV or differences in appearance of the CMAP waveform can be seen when diseases affect the myelin covering of the nerve (demyelination) and with abnormalities of the axonal conduction fibers. Demyelination causes specific changes to the CMAP appearance, which include the presence of multiple small waveforms and shortening and widening of the waveform (Figure 3). Common veterinary diseases of the peripheral nerve assess via nerve conduction testing include acute canine polyradiculoneuritis (Coonhound paralysis), infectious or autoimmune diseases of the nerve, paraneoplastic neuropathies, and acquired idiopathic degenerative diseases.

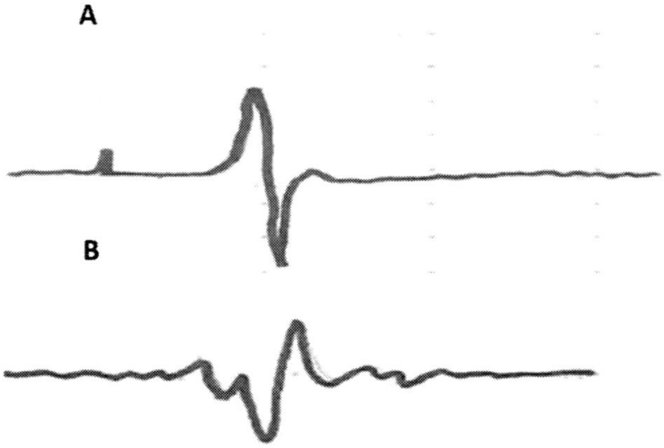

Figure 3: A normal CMAP waveform generated during nerve conduction testing (A), and one showing changes consistent with demyelination.

SUMMARY

Electrodiagnostic test such as EMG and nerve conduction testing are very important for documentation of the presence and distribution of lower motor neuron disease. They are highly sensitive and give specific information about what tissues are affected, but rarely lend a definitive diagnosis in and of themselves. The goal of electrodiagnostic testing is to guide the clinician as to what tissues are affected (muscle, nerve or both) and what additional testing should be performed. In veterinary medicine this often involves biopsy of affected muscle or nerve, testing for infectious diseases, cancer screening and evaluation for endocrine problems such as hypothyroidism or Cushings disease. Occasionally, diseases of the peripheral nerve or muscle may be idiopathic and degenerative, or related to tumors found elsewhere in the body. Certain diseases of the lower motor neuron pathway, in particular those that involve the neuromuscular junction such as myasthenia gravis, and very acute diseases of the peripheral nerve, will yield normal electrodiagnostic testing; therefore normal electrodiagnostic testing results must be interpreted in light of the entire clinical picture.

REFERENCES

Kimura J. Electrodiagnosis in diseases of nerve and muscle. 3rd ed. 2001. Cuddon PA. Vet clinics North Am. Small Anim 2002; 32 (1): 31.

ENDOCRINE TESTING IN CATS AND DOGS

Cynthia R. Ward, VMD, PhD, DACVIM
Athens, GA

THYROID DISEASES

Thyroid diseases are common in cats and dogs. Thyroid hormones are released by the thyroid gland in response to thyroid stimulating hormone (TSH) release from the pituitary. Thyrotropin releasing hormone (TRH) secreted by the hypothalamus binds to receptors on the pituitary gland to cause the release of TSH. The thyroid gland releases triiodothyronine (T3) and thyroxine (T4) and a small amount of the parent hormone, thyroglobulin, into the bloodstream. T4 is the major thyroid hormone released by the thyroid gland, while T3 is most active in the cell. T3 and T4 travel through the plasma bound with relative specificities to thyroid hormone binding proteins such that 99% of serum thyroid hormone is in the bound state. However, the free hormone is the one that is metabolically active and regulated the hypothalamic-pituitary-thyroid endocrine axis.

Middle aged to older cats are often affected by hyperthyroidism or overactivity of the thyroid gland. This is usually a primary disease resulting from benign adenomatous hyperplasia of the thyroid gland that causes unregulated secretion of thyroid hormone. Clinical signs of the disease result from the excess thyroid hormone in circulation. Single hormone measurements are helpful in diagnosing feline hyperthyroidism. Greater than 95% of hyperthyroid cats can be diagnosed by a single serum T4 measurement reading above the normal range. Some cats suffer from occult hyperthyroidism. This describes increased thyroid activity that cannot be diagnosed by elevated T4 in the serum. This may be due to very early disease or the presence of concurrent disease that causes a decrease in thyroid hormone binding proteins and a resultant decrease in the T4 measured in the serum (sick euthyroid syndrome). In cases in which hyperthyroidism is suspected but cannot be proved by elevated serum T4, a free T4 (fT4) may be measured. This is the unbound hormone that can be measured after separation from binding proteins by dialysis. FT4 cannot be used alone to diagnose feline hyperthyroidism, however, since approximately 15% of euthyroid cats with concurrent disease have elevated fT4. T4 and fT4 must be evaluated together to diagnose hyperthyroidism. Dynamic testing of the thyroid axis in cats can be used in difficult to diagnose cases of hyperthyroidism. T3 suppression testing can give a definitive answer as to whether a cat has hyperthyroidism or not. In this test the negative feedback part of the hypothalamic-pituitary-thyroid axis is challenged. Synthetic T3 is given orally and serum T4 is measured. In a euthyroid animal serum T4 should be suppressed. If hyperthyroidism is present, T4 will not be suppressed. A TRH-stimulation test has also been used to diagnose occult hyperthyroidism. In this test the thyroid endocrine axis is stimulated. Hyperthyroid cats will show an overstimulation of the thyroid gland in response to TRH. Recently a technique comparing pertechnetate uptake by the thyroid to background uptake has been described. Pertechnetate utilizes the same uptake system as iodine so is particularly concentrated in the thyroid gland. Cats with hyperthyroidism showed increased thyroid uptake to background levels.[1]

Hypothyroidism in dogs is also a common disease. It is often diagnosed incorrectly because a single T4 is measured and often included as part of a regular blood panel. Many non-thyroidal diseases can cause T4 to be decreased in the serum. A significant number of diseases can have clinical signs similar to hypothyroidism, thereby confusing the diagnosis. Moreover, dogs (and cats) metabolize thyroid hormone at a rapid rate, so inappropriate supplementation does not result in clinical hyperthyroidism. Diagnosing a hypothyroid dog can be challenging since the presence non-thyroidal illness can result in lowering of T4 in the serum in a euthyroid animal. Free T4 is often more accurate as a single measurement, but can also be lowered by non-thyroid illness. TSH regulates all aspects of synthesis and secretion of T3 and T4 from the thyroid. It is also regulated by negative feedback from secreted fT3 and fT4. Therefore, this is a useful single serum sample test to evaluate thyroid status. It is the primary test to evaluate thyroid status in humans and often used alone. Unfortunately, the TSH assay is less sensitive for dogs and will be normal in up to 38% of hypothyroid dogs. However, when elevated it adds specificity to the T4 and fT4 tests. To accurately diagnose hypothyroidism, a T4, fT4, and TSH should be measured in the serum. If 2 of the 3 tests indicate hypothyroidism, there is a greater than 98% chance of the animal being hypothyroid. Exceptions to this rule are animals treated with certain anti-convulsants such as Phenobarbital. Animals on these medications can have thyroid tests that indicate hypothyroidism even when they are euthyroid. In rare instances when a diagnosis is hard to achieve, human-recombinant TSH can be used to challenge the thyroid endocrine axis. [2] After injection, TSH will cause an increase in T4 in euthyroid animals, but in dogs with primary hypothyroidism, there will be no such increase. Because of the expense of TSH, it is prohibitive to use for routine testing.

DIABETES MELLITUS

Diabetes mellitus is also a common disease in cats and dogs. It results from an absolute lack of insulin (type 1 insulin-dependent diabetes mellitus) or relative lack of insulin (type 11, non-insulin dependent diabetes mellitus). Both of these syndromes result in uncontrolled hyperglycemia and glucosuria.

Diagnostic testing for diabetes mellitus is relatively simple in the scheme of endocrine testing. The major regulator of insulin synthesis and secretion is glucose so serum glucose levels directly reflect insulin activity in the body. As the serum glucose levels exceed the renal threshold, approximately 180 mg/dl for the dog and 250 mg/dl for the cat, glucose will spill into the urine and result in glucosuria. Ketone bodies are formed in the diabetic state when there is a concurrent disease or stress condition. These can be measured in the serum or urine. The presence of hyperglycemia and glucosuria in addition to ketonemia or ketonuria nearly assures a diagnosis of diabetes mellitus. Hyperglycemia and glucosuria in an animal with appropriate clinical signs is also strongly indicative of diabetes mellitus. The only alternate explanation for these findings is stress. Stress, alone, in cats and dogs can result in significant hyperglycemia and glucosuria. Fructosamine, a nonenzymatic coupling of glucose and albumin, can give an average of the blood glucose over the preceding 2-3weeks which is the time it takes albumin to break down. This can be a useful test to differentiate stress hyperglycemia from diabetes mellitus; the fructosamine will be within the normal range for the former condition, and elevated in the latter.

While diagnosis of diabetes mellitus is relatively easy, monitoring of diabetes mellitus can be challenging. Fructosamine can be useful to monitor insulin response, although it only gives an average of the blood glucose and will not indicate highs and lows of glucose. It will also not reveal details of insulin activity in the animal.

The most accurate way to monitor response to insulin is by the glucose curve. This can be accomplished by serial glucose measurements every 2 hours or by continuous interstitial glucose monitoring devices. In veterinary patients, there can be significant variability in glucose curves that may not reflect the true insulin status of the animal. [3] Continuous glucose monitoring may offer more reliability. [4, 5]

ADRENAL DISEASES

Some of the most difficult endocrine tests to interpret are those for adrenal diseases in the cat and dog. These diseases result from abnormal secretion of adrenal hormones. The most common of these diseases involve over- or underproduction/secretion of adrenal cortical hormones. Underproduction of glucocorticoids and mineralocorticoids is called hypoadrenocorticism or Addison's disease. It is seen in young-middle aged dogs, but is rare in cats. Overproduction of adrenal cortical hormones, usually cortisol, is called hyperadrenocorticism that can result in a syndrome called Cushing's syndrome. This is a common endocrinopathy in dogs and is being seen with more frequency in cats. It can be caused by a pituitary tumor that oversecretes ACTH that in turn stimulates secretion of cortisol from the adrenal cortex. Less commonly tumors of the adrenal cortex result in over secretion of cortisol. Aldosterone-producing tumors have also been uncommonly recognized in cats as part of a disease syndrome called Conn's syndrome or hyperaldosteronism.[6]

Baseline cortisol is unreliable as a measure of adrenal cortex function due to the episodic nature of cortisol secretion. However, it has been shown that baseline serum cortisol concentrations can be used to rule out hypoadrenocorticism. [7] Usually, the use dynamic endocrine testing to challenge the hypothalamic-pituitary-adrenal endocrine axis and test function of the adrenal gland are necessary to positively diagnose adrenal cortex diseases. Most commonly cortisol is measured in these tests; however, aldosterone can be measured as well. The ACTH-stimulation test is used to positively stimulate the pituitary-adrenal axis and to diagnose hypoadrenocorticism by revealing lack of stimulation in response to ACTH. It is also used to diagnose hyperadrencorticism by revealing an overstimulation of cortisol in response to ACTH. The low dose dexamethasone test is used to suppress the hypothalamic-pituitary-adrenal axis and thus can diagnose hyperadrenocorticism.

REFERENCES

1. Wallack S, et al., Vet Rad Ultra.2010, 51(5):554. 2. Boretti FS, et al., J Vet Intern Med. 2009;23(4):856. 3. Alt N, et al., J Am Vet Med Assoc. 2007;230(7):1011. 4. Ristic JM, et al., J Feline Med Surg. 2005 7(3):153. 5. Moretti S, et al., J Vet Intern Med. 2010 Jan-Feb;24(1):120-6. 6. Ash RA, et al., J Feline Med Surg. 2005;7(3):173. 7. Lennon EM, et al., J Am Vet Med Assoc. 2007;231(3):413.

CARDIOPULMONARY CEREBRAL RESUSCITATION FOR TECHNICIANS

Steven L. Marks, BVSc, MS, MRCVS, DACVIM
Raleigh, NC

INTRODUCTION

Cardiopulmonary resuscitation (CPR) or the more comprehensive term cardiopulmonary cerebral resuscitation (CPCR) is still one of the most important interventions performed in human or veterinary medicine. This is especially true in veterinary medicine when dealing with emergency and critical critically ill patients. Although not performed frequently, it must be utilized in the appropriate situation and using the most effective techniques in order to maximize the possibility of success. Currently estimates of successful CPCR in veterinary patients are reported to be less than 10% but this may vary from institution to institution. This figure compares favorably with survival rates reported for out of hospital cardiopulmonary arrest in human patients but is lower than reported survival rates for in hospital arrests.

ADVANCE DIRECTIVES

Advance directives are a formal component of human medicine and have gained more importance in veterinary medicine. In generic terms, advance directives refer to information provided to a physician by a patient detailing what care should be provided if the patient in unable to make decisions. In veterinary patients this equates to the primary care clinician having a detailed discussion with the pet owner concerning medical management of a pet who has had a life threatening medical event. In more general terms and for the purpose of this discussion, this addresses whether the patient should be resuscitated if cardiopulmonary arrest occurs. This discussion should be documented in the medical record. There should be a way of identifying which patients have "do not attempt resuscitation" (DNAR) orders and the nursing staff and all clinicians should be made aware of these orders.

COMPARATIVE CPR

The American Heart Association published new guidelines in 2000 for cardiopulmonary resuscitation and emergency cardiovascular care. These guidelines represented significant changes in protocols that should not be immediately extrapolated to veterinary patients. The most significant changes included the incorporation of immediate electrical defibrillation into Basic Life Support (BLS). Previous to this recommendation, BLS included **A**, **B**, **C**, **A**irway, **B**reathing and **C**irculation. **D**efibrillation was included in Advanced Life Support (ALS). The successful impact of rapid electrical defibrillation on survival in patients with cardiopulmonary arrest led to this change in recommendations. The overall impact of this recommendation has been the increased availability and access to automated external defibrillators (AED). This technology allows minimally trained lay rescuers as well as trained rescuers the ability to defibrillate patients with ventricular fibrillation. The emphasis on early defibrillation is based on reports that as many as 70% of people in the United States who experience out of hospital cardiopulmonary arrest are initially in ventricular fibrillation.

Although ventricular fibrillation does occur in dogs and rarely in cats, it is not necessarily the most common rhythm disturbance recognized and is not recognized in the majority of veterinary patients who experience cardiopulmonary arrest.

It has also been found that what would be considered standard drug therapy in veterinary patients, utilizing epinephrine and atropine has not significantly improved survival in people and therefore these drugs are considered marginal. These drugs would still be considered the cornerstone of therapy for veterinary CPCR.

TECHNIQUES

Basic Life Support

In veterinary medicine basic life support (BLS) or basic CPR is still considered the **A**, **B**, **C**'s (**A**irway, **B**reathing, and **C**irculation). Due to the limited access to defibrillators in veterinary practice and the nature of the underlying diseases contributing to cardiopulmonary arrest in veterinary patients, it is unlikely that this will change. It is also important for the veterinary clinician to recognize that many dysrhythmias occur in the critically ill dog and cat. We cannot extrapolate the information that approximately 70% of people who experience cardiopulmonary arrest out of hospital have ventricular fibrillation as their initial rhythm. The common dysrhythmias associated with cardiopulmonary arrest in dogs and cats include, ventricular asystole, pulseless electrical activity (PEA) (formerly known as electrical mechanical dissociation (EMD) and ventricular fibrillation. Therefore the utility of early defibrillation in veterinary patients is unknown unless the patient has documented ventricular fibrillation.

Current recommendations for veterinary BLS include assessing and providing an airway if required, assessing and providing breathing if required and assessing and providing circulation if required.

Airway: Although there are no real changes in the recommendations for this procedure, the priority is to establish an airway. In most cases this is accomplished by per oral endotracheal intubation. Confirmation of correct tube placement is recommended by visual assessment, digital palpation and/or by $ETCO_2$ monitoring. It has been suggested that airway be prioritized below circulation and breathing. In this scenario rather than A, B, C, the priority would be C, A, B. If following this protocol, external thoracic compressions should be started immediately and prior to providing an airway. This recommendation is based on research performed to evaluate bystander (out of hospital) CPR. In some experimental animal models, most notably pigs, it has been shown that animals can be resuscitated from ventricular fibrillation by only performing chest compressions. It has been hypothesized that when performing external chest compressions, enough passive ventilation occurs for the patient to survive. Another important factor is that during ventricular fibrillation, oxygen delivery is dependent on blood flow, not ventilation.

Breathing: The only innovations with breathing have been the use of simultaneous compression ventilation protocols by some clinicians. When utilizing the thoracic pump theory, it has been demonstrated in animal models that cardiac output can be improved by increasing intrathoracic pressure. Conventional external thoracic compressions utilize the cardiac pump theory in combination with interposed respirations. The cardiac pump involves direct compression of the myocardium to promote cardiac output. The thoracic pump utilizes compression of the thoracic cavity to indirectly improve cardiac output by effecting blood flow through the giant vessels. The incorporation of simultaneous ventilations dramatically increases intrathoracic pressure. There are however, potential complications associated with this technique which include barotrauma and pulmonary contusions.

Circulation: The circulation component of BLS has not changed significantly and has been discussed in part under the breathing section with reference to simultaneous compression and ventilation protocols. There are currently two theories explaining the mechanisms of cardiac output during external thoracic compressions. The cardiac pump theory suggests that direct compression of the myocardium accounts for the majority of cardiac output. In veterinary patients this would apply in patients with body weights of < 15 kg. In patients with body weights of > 15 kg, the thoracic pump would better explain cardiac output. The latter suggests that compressions of the widest portion of the thorax increase intrathoracic pressures which indirectly lead to increased cardiac output. This may be combined with simultaneous ventilation which should further increase intrathoracic pressures. The use of interposed abdominal compressions during CPR has become more common and has been shown to increase venous return by up to 25%.

Advanced Cardiac Life Support

In veterinary medicine advanced life support (ALS) has included all procedures outside of BLS. So the pneumonic of **D**, **E**, **F** still applies.

Drugs: Although there are no longer strong recommendations for epinephrine and atropine for therapy of cardiopulmonary arrest in people, these drugs still play a valuable role in veterinary CPR. The most commonly reported dysrhythmias identified during cardiopulmonary arrest in the dogs and cats are bradyarrhythmias, asystole and pulseless electrical activity (PEA) formerly known as electrical mechanical dissociation. Atropine and epinephrine may be useful for management of these dysrhythmias. The current recommendation for the use of epinephrine is using the low dose protocol (.01-.02 mg/kg). This protocol calls for the use of the standard concentration of epinephrine (1:1000) diluted 10 fold and administered at a volume of 1 ml/10 kg of body weight. This is compared with the high dose epinephrine protocol which utilizes the same volume administered (1ml/10 kg) but using the 1:1000 dilution epinephrine straight from the bottle. There is some renewed interest in the use of vasopressin and recent publications have shown some increase in survival in people who have failed conventional medical therapy. Vasopressin can be administered at a dosage of .8 U/kg. This equates to approximately .5 ml/10 kg of body weight. The advantage of vasopressin may be its performance in the face of hypoxia and acidemia. In addition its mechanism of action is via direct stimulation of receptors in vascular smooth muscle and it does not appear to increase myocardial oxygen consumption. Atropine is still administered at .04 mg/kg or approximately 1 ml/10 kg body weight.

Routes of administration that are recommended include intravenous through a central line, intravenous via a peripheral line with bolus to follow, intraosseous administration and intratracheal administration. The drugs that can be safely administered via the intratracheal route can be remembered by the pneumonic **NAVLE** and include **N**aloxone, **A**tropine, **V**asopressin, **L**idocaine and **E**pinephrine. The doses should be twice the intravenous dose and should be diluted in sterile water.

ECG: In veterinary CPCR it is still imperative to know what dysrhythmia is present. Because it is difficult to predict which rhythm is present, intervention and outcome will be dependent on this diagnosis. The most common reported dysthymias include ventricular asystole, PEA and ventricular fibrillation.

Fibrillation control: Defibrillation is prioritized lower and still included in ALS for veterinary patients for two reasons. First and foremost, patients must be in ventricular fibrillation for this intervention to be utilized. Secondly, it still appears that many veterinary hospitals do not have access to this equipment. The advances in the area of defibrillation are the use of biphasic defibrillators which may be more effective than the monophasic defibrillators.

Fluid resuscitation: Fluid resuscitation should be a component of CPR. However, aggressive volume expansion should only be utilized if it is suspected that volume depletion has contributed to the cardiopulmonary arrest. Volume overload can have detrimental effects on these patients. An increase in central venous pressure and right sided pressures may lead to reduced coronary perfusion pressures.

MONITORING

Monitoring of patients during CPR is imperative but difficult. Physical examination along with palpation of peripheral pulses, venous blood gas analysis and end tidal capnography are the most reliable tools. Other suggested techniques include the use of a Doppler blood pressure probe on the cornea to evaluate cerebral artery perfusion but this is unreliable in the author's experience. Return of spontaneous ventilation is often a limiting step as it appears that return to spontaneous circulation occurs more commonly. The frequent cessation of external thoracic compressions in order to examine the ECG or auscultate the thorax may lead to decreased success.

POST-RESUSCITATIVE CARE

Due to poor survival rates, it is often unexpected that post-resuscitative care is required. It is however imperative that the clinician anticipate survival so this care can be provided immediately. If post-resuscitative care cannot be provided then the CPR was wasted. Post-resuscitative care is primarily symptomatic and is dependent on close monitoring and excellent nursing care. Oxygen supplementation should be considered in all of these patients and ventilatory support may also be required. Fluid therapy should be continued as required and inotropic and pressor therapy may be indicated to maintain mean arterial pressures. Neurologic dysfunction may be common following resuscitation and infusion of mannitol at .5-1 g/kg over 30 minutes may help treat cerebral edema and has free radical scavenging properties. It has also been demonstrated that hyperglycemia may have a negative impact on outcome and therefore should be monitored. Due to the lack of evidence for their use, corticosteroids should be avoided. If possible prognosis should not be based on neurologic signs within the first 24-48 hrs due to the possibility of significant change. Nutritional support via feeding tubes or total parenteral nutrition should be provided if patients have not returned to reasonable function within 48-72 hrs.

REFERENCES

American Heart Association Circulation 2010 122 S676. Cole SG. J Vet Emerg Crit Care 2004 12(4) 261. Cole SG. J Vet Emerg Crit Care 2003 13(1) 13. Haldane S. Compend Cont Educ Pract Vet 2004 26(10) 780. Haldane S. Compend Cont Educ Pract Vet 2004 26(10) 791. Hofmeister EH. J Am Vet Med Assoc 2009;235:50. Plunkett SJ. J Vet Intern Med, 2008, 22:9. Waldrop, J.EJ Vet Emerg Crit Care 2004;14(1):22.

HOW RECALLS & RESEARCH HAVE CHANGED THE WAY WE TREAT ENDOCRINE DISEASES

Nyssa J. Reine-Salz, DVM, DACVIM
New York, NY

INTRODUCTION

The approval of new medications by the FDA and the removal of some via recall or stop in production can influence how we practice medicine. This has been particularly relevant in the practice of veterinary endocrinology over past few years. Specifically, the approval of trilostane (Vetoryl™) has revolutionized the treatment of canine hyperadrenocorticism and the recent recall of Vetsulin™ has significantly limited our treatment options for dogs with diabetes mellitus. In addition, a number publication have come out recently reporting on studies evaluating the use of newer portable blood glucose monitors (PBGM) to perform glucose curves for the monitoring of diabetes mellitus in both cats and dogs. The role of glucose curves in the management of diabetes mellitus, particularly that of home glucose monitoring by clients, has re-emerged as a viable option in dogs and cats.

TRILOSTANE APPROVAL AND MEDICAL MANAGEMENT OF HYPERADRENOCORTICISM

Hyperadrenocorticism is a fairly common cause of excessive drinking and urinations (PU/PD) in dogs. It is a chronic illness caused by excessive secretion of cortisol, either secondary to a tumor of the pituitary gland (PDH) secreting ACTH resulting in adrenal gland over-secretion of cortisol or an adrenal tumor (AT) directly secreting too much cortisol. In the United States, treatment of PDH has always been achieved using oral medication whereas dogs with AT are typically treated with oral medication only when surgical removal of the affected gland is not an option. Historically, medical management in the United States was achieved using the medication mitotane (Lysodren™, oPDDD). The FDA original approval of trilostane (Vetoryl™) in 2008 provided us with an alternate treatment option to offer our clients. Though both are effective at managing the disease, trilostane has become the preferred treatment based on relative ease of use and safety.

Mitotane therapy

Mitotane is an adrenocorticolytic medication which means that it acts by actually destroying adrenal tissue. When we start patients on mitotane therapy, we start with a high dose, known as a loading dose, of 50mg/kg/day. Owners are instructed to observe their dog's very closely for changes in appetite, water consumption, and attitude or for the development of vomiting and diarrhea. At the slightest change in any of these parameters, the owners are instructed to stop the mitotane therapy and to bring the dog to the hospital for examination and an ACTH stimulation test plus electrolyte assessment as soon as possible. If the ACTH stimulation test results are in the ideal range and electrolytes are normal, the dog is then switched to the long term or maintenance protocol. If not, the loading period is continued (if cortisol suppression is inadequate) or discontinued (if the ACTH stimulation results are flat and/or electrolytes are abnormal). During the maintenance protocol, the dose of mitotane that was administered daily during the loading period is divide through the week instead. For example, a dog who was receiving 500mg mitotane twice a day during the loading period would be switched to 500 mg twice a week. It sounds very simply but individual variation in response makes the duration of the loading period impossible to predict, making its use objectionable to many practitioners. In addition, the potential for hypoadrenal crisis associated with excessive adrenal suppression makes practitioners wary to use this medication.

Trilostane therapy

Trilostane is an enzyme inhibitor which blocks the production of cortisol as well as some other adrenal steroids. It is typically administered once or twice a day and monitored at fairly regular intervals. The starting dose is either 2-5mg/kg orally once daily or 1 mg/kg orally twice daily. When given once daily, the medication should be administered in the morning with food. Administering with food improves absorption and morning administration enables daytime monitoring. The first follow up should occur 2 weeks after starting the medication with a physical examination, ACTH stimulation test and (ideally) electrolyte assessment being performed. If the ACTH stimulation test results are in the ideal range and electrolytes are normal, the dose should be continued and the dog should return in 4 weeks for similar assessment. Dosage adjustments should be made if the ACTH stimulation test is inadequately suppressed (increase the dose), over suppressed (temporary stopping the drug then potentially reinstituting a lower dose) or adequately suppressed but no improvement in clinical signs seen (transition from once daily to twice daily therapy). Once a stable dose is achieved, monitoring should occur every 3-6 months.

Comparison

Mitotane and trilostane can both cause excessive adrenal suppression, which will result in signs of hypoadrenocorticism (vomiting, anorexia, diarrhea and electrolyte changes.) Occasionally, you can see vomiting in the beginning of therapy (within day 1-3). It is very unlikely that this would be caused by adrenal suppression. With mitotane, it is likely due to direct gastrointestinal effects and can often be relieved by cessation of the drug and assuring administration with food. Alternatively, the dose can be divided into smaller increments (i.e. If the original dose was 500mg once a day it can be switched to 250mg twice a day). With trilostane, you may see mild lethargy and anorexia associated with acute glucocorticoid withdrawal. This should typically resolve with time.

A unique and relatively unexpected side effect of trilostane is adrenal necrosis. Due to its mechanism of action (enzyme inhibitor) one would not expect actual changes to the adrenal gland. The signs are those of acute hypoadrenal crisis. In reports, some animals have recovered with therapy but some have died.

Reported response rates between the drugs are similar. Both are effective treatments of PDH, though not all dogs will respond. With both drugs, resolution of haircoat and body composition changes lag behind resolution of PU/PD and polyphagia. One study comparing median survival time found them to be quite similar (708 days for mitotane vs. 662 days for trilostane). It is the ease of use that will often make trilostane the preferred therapy. It requires much less observation by the owner and the monitoring plan is fairly consistent.

VETSULIN WITHDRAWAL AND CANINE DIABETES MELLITUS

Virtually all dogs with diabetes mellitus have type I or insulin dependent diabetes. This most often results from immune mediated destruction of the islet cells of the pancreas. Administration of exogenous insulin is the only recommended treatment for diabetic dogs.

Vetsulin

Vetsulin is a pork Lente insulin product. This makes it very well suited for use in dogs as pork insulin is identical to dog insulin. This insulin is formulated with 30 percent amorphous zinc insulin, which is rapidly absorbed and has a short duration of action (peak within 4 hours, duration for 7-8 hours), and 70 percent crystalline zinc insulin which is more slowly absorbed and has a longer duration of insulin (peak around 10-11 hours. This insulin is unique in that it was specifically approved for use in dogs, whereas most other insulin products are labeled for people and used extra-label in dogs. It is also a U40 insulin which means there are 40U/ml vs U100 which means 100U/ml . It is important that the appropriate syringes be used with this insulin.

In November 2009, manufacturing issues lead to Intervet/Schering-Plough Animal Health and the FDA to discontinue distribution of Vetsulin due to varying amounts of crystalline zinc insulin (the active ingredient). This would lead to unpredictable duration of action. It was recommended that newly diagnosed dogs with diabetes not be started on Vetsulin, given limited availability and possible loss of supply. Stable dogs could be continued until supply ran out but would likely need to be transitioned to different insulin. For dogs with multiple disorders for whom Vetsulin was the only possible treatment option, the FDA is allowing Intervet to offer a limited supply of Vetsulin through their Vetsulin Critical-Need Program.

Since Vetsulin was a frequent first line choice in the therapy of newly diagnosed canine diabetes mellitus, doctors were left to find other options for treating dogs with diabetes. In addition, as the current supply of Vetsulin disappears, alternatives need to be found. Once the decision is made to switch insulin types, the case needs to be approached as if it is a newly diagnosed patient in that one can never presume that dose of new insulin necessary to provide adequate glucose control would be identical to Vetsulin. A glucose curve and patient assessment should be performed a week after changing insulin types. The frequency of subsequent rechecks would depend on the results of the first follow up.

NPH Insulin

NPH insulin has always been a consideration for first line therapy for dogs with diabetes mellitus. Formerly, NPH was a made as a porcine insulin reacted with zinc and protamine to prolong the effect of the insulin. Changes in the management of human diabetes have resulted in removal of animal source insulin so the currently available products are synthetic human recombinant U100 insulin products and include Humulin N (manufactured by Eli Lilly) or Novolin N (manufactured by Novo Nordisk). NPH is considered an intermediate actin insulin. NPH insulin starts to work in about 1-2 hours, has a peak action of 4-6 hours and lasts about 12-18 hours. Occasionally, dogs can be regulated using this insulin once daily The usual starting dose is 0.5U/kg twice daily with a glucose curve and patient evaluation being performed 1 week after starting therapy.

Insulin glargine

Insulin glargine is a long acting basal insulin designed to mimic normal pancreatic beta cell baseline insulin secretion. In people, a second, more rapid acting, short duration insulin is administered to control dietary glucose fluctuations. It is used a great deal as the primary insulin in the management of cats with diabetes mellitus. There are no large studies evaluating this insulin in the control of diabetes mellitus in dogs. There is one study looking at its use in combination with NPH insulin and this was found to provide tight control but at great risk of the development of hypoglycemia. This same study found that this insulin glargine has a peak at about 7 hours and lasts as long as 24 hours.

Insulin detemir

Insulin detemir is a new class of long-acting soluble insulin analogue which maintain basal level of insulin in people, without the typical peaks associated with other insulin, similar to insulin glargine. In dogs it has a peak at 8-10 hours after administration and lasts at least 24 hours. Dogs appear to be very sensitive to this insulin so the recommended starting dose is considerably lower than other insulin at 0.1U/kg administered twice daily with close monitoring for hypoglycemia after initiation of therapy.

MANAGEMENT OF DIABETES MELLITUS WITH HOSPITAL OR HOME DERIVED GLUCOSE CURVES

The monitoring of the treatment of diabetes mellitus in dogs and cats consists of evaluation of owner observations (i.e. assessment of appetite, attitude and water consumption), physical examination, body weight measurement and finally objective assessment of glucose control. This monitoring most commonly will be in the form a serum fructosamine concentration, which gives an approximation of average sugar over time, or blood glucose curves, which give a more precise picture of glucose response to insulin throughout the treatment period.

Historically, the usefulness of the glucose curve was limited by the effect of stress associated with hospitalization and sample acquisition on blood sugar. When performing a glucose curve, it was often necessary to either place a sampling catheter or restrain the dog or cat multiple times a day in order to obtain adequate samples. Stress hyperglycemia is particularly a problem in cats. More recently, many PBGMs have become available which require a much smaller blood sample size, enabling the use of capillary blood samples. Sites that are typically used for these samples include the ear margin, elbow callus, tail and lip in dogs and the ear margin in cats. Lancets are used to make a small cut, allowing a drop of blood to be obtained. Elimination of the use of venipuncture also allows the potential for clients to be trained to obtain these samples at home.

Home glucose monitoring

Studies have shown that though certain PBGM may be more useful in dogs and cats, it just as important to determine whether the PBGM selected will over or under estimates blood glucose and establish if this would impact clinical decision making. To achieve this, after the PBGM is obtained, blood glucose should be evaluated simultaneously using the meter and a reference or in hospitable laboratory. Owners also need to be well trained on lancet use, sample acquisition, monitor calibration and test performance before attempting to use them at home. Whenever a curve is performed at home, observations regarding the pet's appetite and attitude, the time and dose of insulin administered should be recorded along with the measured glucose.

As with human patients with diabetes, there is psychological stress associated with discovering a pet has diabetes. For most, performing home monitoring at the onset of treatment could be overwhelming. Though it is important to introduce the concept of home monitoring at the start of treatment, a detailed discussion should be delayed at least until the owner is comfortable with insulin administration. In one study evaluating home glucose monitoring in cats, people were most likely to discontinue monitoring within the first week – re-emphasizing the importance of initial training and close follow up with owners regarding the process. The most common reasons for discontinuation of home monitoring included technical difficulties associated with the procedure and a fractious animal. It is important to notes that in the same study, other owners performed glucose curves in their cats for an excess of 4 years.

Glucose curve procedure

If clients are not interested in performing home glucose monitoring, it is also viable to perform glucose curves in the hospital utilizing the same methodology. Ideally, the owner should feed the pet and give them their insulin in order to attempt to mimic what occurs at home. Feeding and administering insulin in the hospital may impact the pet's appetite due to stress.

The technique for the glucose curve is the same, regardless of whether it is performed at home or in the hospital. Ideally, blood glucose should be sampled every 2 hours, until the next insulin dose is due. For patients receiving insulin twice daily, this usually means a 12 hour curve whereas patients receiving insulin once daily should ideally have a 24 hour curve performed. On the day of the curve, owners bring with them a record of any observations about the pet's appetite and attitude, the time and dose of insulin administered.

Glucose curve assessment

Points of particular interest in the curve include the glucose nadir, the duration of action of insulin and the glucose differential. The nadir is defined as the lowest blood glucose on the curve and represents the maximal insulin action. This should ideally fall at the mid-point of treatment - between 5-6 hours for twice daily therapy and 11-12 for once daily therapy. The duration of action is determined by the time when the glucose returns is >200-250mg/dl or baseline values. The ideal duration of action should be close to 12 or 24 hours, depending on how often the insulin is administered. The glucose difference is defined as the difference between the nadir and the concentration just before the next dose. Though not absolutely necessary, the author prefers to enter the data collected into a program that will allow graphing of the glucose curve. This simplifies many of these measurements.

Based on the results of the curve, insulin dose may be elevated, if the duration of action is determined to be adequate but glucose concentration too high, the insulin dose lowered, if the hypoglycemia followed by profound hyperglycemia is seen, or the insulin type changed if duration of action inadequate or insulin resistance is believed to be present. Ideally, insulin dosage manipulation should only occur under direction of the veterinarian.

REFERENCES

[1] Behrend, EN, **Vet Clin North Am Small Anim Pract**, 2006; 36:1087. [2] Reine N, **Clin Tech Small Anim Pract**, 2007; 22:18. [3] Ramsey IK, **Vet Clin Small Anim**, 2010; 40:269. [4] Monroe WE, et al. **J Vet Intern Med**, 2005;19:675 [5] Fleeman LM, et al. **Vet Rec**, 2009; 164: 232. [6] A. Mori et al. **Vet Res Commun**, 2008; 32:563. [7] Sako, T., et al. **Res. Vet. Sci**. 2010, in press. [8] Stein JE. et al. **Clin Tech Small Anim Pract**, 2002; 17:70. [9] Reusch C. et al, **J Feline Med Surg**, 2006; 8:119.

DIAGNOSIS, TREATMENT & COMPLICATIONS OF CLOSTRIDIAL MYOSITIS IN HORSES

Allison Stewart BVSc (hons), MS, DACVIM-LAIM, DACVECC
Auburn, AL

Clostridial myositis results from a rapidly progressive necrosis of muscle from infection with clostridial organisms, often with fatal consequences. The majority of equine cases are secondary to intramuscular injections or as a result of direct contamination of deep wounds with bacterial growth in an anaerobic environment. Infection results in local inflammation, often with severe pain and swelling, and can progress to a state of systemic inflammatory response (SIRs) with fever, depression, tachycardia, progressive toxemic shock, disseminated intravascular coagulation (DIC) and death. Progression is often rapid, and even with early and aggressive therapy, in severe cases, a guarded to poor prognosis should be given.

Malignant edema, clostridial myositis/cellulitis/fascitis and gas gangrene are a continuum of conditions due to infection of the subcutis, fascia and/or muscle. There is often fulminant tissue necrosis, with disintegration of tissue and serosanguinous exudate with bubbles of gas or free gas caused by infection with gas producing anaerobic clostridial organisms.

THE PATHOGENS

Clostridium perfringes, C. septicum, C. chauvoei, C. fallax, C. sordelli and *C. novyi* are the principal pathogens. They are large Gram-positive, obligatory anaerobic rods that are ubiquitous in the soil and environment. They are also commensals of the skin, oral cavity and intestinal tract. Clostridia form spores, which allows the organism to survive until germination and growth can occur. Germination of spores and vegetative growth occurs in suitable anaerobic conditions. Clostridial organisms liberate potent exotoxins. Some organisms release these toxins continually during the growth of the organism, while others release the toxin during death.

CLINICAL SIGNS

In peracute cases animals may be found recumbent or dead. Painful muscular swellings, often with crepitus may be felt. The overlying skin may initially be hot and inflamed, but becomes progressively cool and insensitive. If systemic toxemia occurs the horse will become profoundly depressed and show signs of SIRs (tachycardia, tachypnea, fever and leukocytosis or leukopenia).

The extent of tissue necrosis and surrounding edema can be observed ultrasonographically. A distinction between edema and cellulitis can sometimes be difficult. Edema is characteristically located in the subcutaneous tissue, dissects along tissue planes, is anechoic, and loculated in appearance. Cellulitis usually appears as an echogenic, homogeneous thickening of the subcutaneous tissues, also dissecting along tissue planes. Sometimes hypoechoic pockets are seen in an area of cellulitis, but loculations are typically not seen. In anaerobic infections gas shadowing is often observed.

Aspiration of nonclotting, malodorous fluid (with or without gas) is typical. Volatile fatty acids are produced by many anaerobes including *Clostridium* spp. Gram-stain can provide a presumptive diagnosis of clostridial myositis with the detection of large Gram-positive rods, with ("paper-clip appearance") or without spores. Anaerobic (and aerobic) culture is also recommended.

Clinicopathological changes are non-specific. Muscle enzyme activities are usually moderately elevated, but often not to the extent expected for degree of myonecrosis present. Changes typical of septic-toxic conditions include leukocytosis or leukopenia, often with elevation of the band neutrophil count and fibrinogen concentration. There may be hyperproteinemia with a long standing walled off abscess, but more typically hypoproteinemia occurs due to exudative losses into the necrotic tissue. With the development of toxic shock there is hemoconcentration, azotemia and often coagulopathies with thrombocytopenia and DIC. In ~ 1/3 of cases autoimmune mediated anemia secondary to antigen formation against red blood cells occurs due to cross reaction with clostridial antigens.

Treatment

Intravenous penicillin-G (20,000 to 60,000 IU/kg q 2-4 h IV, then at a reduced dose for several weeks) is the historically recommended treatment for clostridial infections. Although penicillin treatment, with and without surgical fenestration, has resulted in survival, mortality with gas gangrene is unacceptably high. Experimentally, penicillin is a poor antibiotic choice, being non-curative with frequent relapse and death. In a murine *C. perfringes* gas gangrene model, clindamycin, metronidazole, rifampin and tetracycline were all more effective than penicillin, which had survival rates no different than untreated controls. Protein synthesis inhibitor-antimetabolite antibiotics such as tetracycline and chloramphenicol suppressed clostridial toxin production, while *C. perfringes* alpha toxin production continued in the presence of penicillin. Based on treatment of about 30 cases (at Ohio State University and Auburn University) tetracycline and metronidazole are now our standard treatments for clostridial myositis. These agents are inexpensive, safe and anecdotally more efficacious. We currently recommend oxytetracycline (6.6 mg/kg IV, q 12-24h in 500 ml 0.9 % NaCl) and metronidazole (20 mg/kg PO q 6h, or 15 mg/kg IV q 6hrs). The daily cost is ~$16/d, compared to $360/d for 60,000 IU q 4h penicillin-G potassium. Anecdotal evidence suggests that this alternative therapy is safe and as efficacious, if not more so than penicillin, and certainly more economic. Long-term antibiotic therapy is essential, and horses are discharged on metronidazole and doxycycline (20 mg/kg q 12h) for a total of about 20 days of antibiotic therapy.

Wound fenestration

Fenestration of emphysematous regions to allow oxygenation of the anaerobic necrotic environment and drainage of exudates is generally recommended. Extensive skin sloughing may occur if skin viability is lost. Often limb wounds result in massive swelling, with the entire leg expanding to 3 times its normal diameter. The massive swelling of tissues causes pressure necrosis and loss of vascular supply to overlying skin, resulting in skin sloughing. Differentials include staphylococcus limb cellulitis. With neck infections, edema fluid and sometimes dissemination of the infection spreads ventrally resulting in pectoral, thoracic and even limb edema. If the head is held low due to neck pain or depression severe head edema can occur often resulting in airway obstruction. Tracheostomy may be required. Cold hosing or cold packing of the surrounding tissue can help with the resolution of edema.

Case one: 11-year-old Quarter Horse gelding

This case (including bloodwork) will be handed out during the lecture

Case Two: 2 year-old Standardbred filly

History: The filly was the third out of five horses vaccinated for influenza and rhinopneumonitis in the left cervical musculature from a 5-dose vial by the owner 5 days previously. The day after vaccination, the filly was depressed, and a 15 cm diameter mass marked the vaccination site. The RDVM administered flunixin, phenylbutazone and procaine penicillin-G (20,000 IU/kg IM, q 12 h for 4 d). Severe swelling of the head causing nasal passage occlusion occurred after 48 hours and the filly was unable to raise her head. The RDVM inserted a nasopharyngeal tube and gave an antihistamine, DMSO in 5 LRS, gentamicin (6.6 mg/kg IV, q 24 h for 2 d) tetanus toxoid and antitoxin (1500 IU IM). The filly ate and drank until the day of referral. The RDVM thought the head swelling was due to an allergic reaction and due to its severity advised referral with subsequent euthanasia likely.

Physical Examination: The filly was markedly depressed. The head was twice its normal size. Tissue swelling occluded the left nasal passage and obscured both orbits rendering the horse non-visual. The filly was disoriented, moved reluctantly with low head carriage and showed weakness (toe dragging, truncal sway and knuckling). Oral mucous membrane color (mmc) was dark pink with a normal refill time of <2 seconds. The vulval mmc was a normal pink. T= 100° F, HR = 42 bpm, RR = 42bpm. The ventral neck was diffusely swollen, but no focal pain or crepitus were palpable.

Assessment and Case Management: Primary problems identified were: marked facial swelling with airway obstruction, weakness, depression, pain and history of neck swelling from IM vaccination. The horse was systemically stable with normal vital parameters. Hyperemic oral mmc was likely due to venous congestion and lymphatic obstruction from facial swelling. Causes of swelling included: diffuse cellulites or myositis extending from the vaccination site, dependent edema from head positioning or unrelated angioedema.

Facial ultrasonographic evaluation was consistent with edema. Aspiration of ultrasonographically imaged fluid from a well-defined emphysematous abscess at the vaccination site dorsal to C3-C4 revealed malodorous serosanguinous exudate. Gram stain and aerobic/anaerobic culture were performed. With local anesthesia, the abscess was lanced and a Penrose drain was placed. The culture was negative, possibly due to the previous antibiotic therapy. Although the head edema was severe, normal vital signs suggested minimal toxemia and thus a better prognosis.

To assess the upper airway, endoscopy was performed. The head was elevated 50 cm and after a 20 second view of the pharynx the filly recoiled backwards, dog-sat for one minute, then collapsed in lateral recumbency. Vital signs remained normal, but due to severe nasal and pharyngeal edema, prophylactic tracheostomy was performed. The filly remained immobile and hypo-responsive to audible and tactile stimuli for 2 hours. Differentials requiring exclusion were fractured cervical vertebrae, osteomyelitis, meningitis, increased intracranial pressure (ICP), or acute toxic or bacterial shower from abscess disruption.

The filly's immobility enabled rapid lumbosacral CSF collection for cytology, which was normal. No evidence of cervical vertebral canal stenosis, osteoarthritis, osteomyelitis or fracture was observed radiographically, but a second abscess over C5-C6 was seen. External evidence of this abscess was undetectable, but 500 ml of pungent gas and 200 ml of serosanguinous fluid was aspirated by ultrasound guidance prior to drainage and placement of two more Penrose drains.

The CBC revealed a mild leukocytosis (11.4 X 10^9cells/L; normal < 10.6) with a mature neutrophilia and monocytosis consistent with infection, inflammation or a stress leukogram. Elevation of the acute phase protein fibrinogen (985 mg/dL; normal < 422) and the presence of reactive lymphocytes were attributed to the myonecrotic abscesses.The rapid progression of systemic signs and abscess formation after intramuscular injection with identification of large, gas producing Gram-positive rods lead to the presumptive diagnosis of clostridial myositis.

Oxytetracycline (6.6 mg/kg IV, q 24 h in 500 ml 0.9 % NaCl) and metronidazole (20 mg/kg PO q 6 h) was commenced. Analgesia was provided by butorphanol (0.01 mg/kg IV q 4 h for 72 h) and phenylbutazone (2.2 mg/kg IV then PO, q 12 h).

The filly was transported to a heavily bedded stall. The head was elevated 30° and hourly cold packing was performed to reduce facial edema. The filly was unresponsive for 2 more hours, and then she suddenly stood, drank water and looked for food! The collapse and immobility was attributed to severe pain from compression of the second abscess by head elevation during endoscopy.

Except for 30 minutes q 2h, the head was elevated by tethering the horse via a padded halter to an overhead wire. By day 2, 60 % of the edema had resolved. The eyes were visible with normal papillary light response, menace, and no corneal uptake of fluorescein stain. The filly was still depressed and moved stiffly but was markedly improved. Meningitis was considered unlikely as CSF cytology, cell count and protein were normal and culture was subsequently negative.

By afternoon of the second day, only 10 % of the facial edema remained. The tracheostomy tube was removed. The neck was evaluated by ultrasound daily. The fenestrated area was hot-packed and wounds were flushed with 1% providone iodine q 8 hrs.

The filly was treated with oxytetracycline and metronidazole for one week in the hospital, and then discharged on metronidazole and doxycycline (10 mg/kg PO, q 12h) for 5d.

The filly made an uneventful recovery and commenced training after one month.

Take Home Message: It is important to identify and drain all abscesses. Multiple abscesses may exist but may not be externally discernable. Ultrasonography and radiology can be used to identify loculations of fluid and gas. Identification of large Gram-positive rods and the presence of pungent gas within abscesses are highly suggestive of clostridial infection, allowing for appropriate therapy prior to the results of culture and sensitivity. The absence of systemic signs even in the presence of marked tissue damage provides a better prognosis.

Case Three: 4-year-old Spotted Saddle Horse.

History: Intramuscular flunixin meglumine had been administered over the left shoulder 2 weeks previously, with subsequent abscess development. Recently, progressive lethargy and hematuria had developed. The horse was treated with procaine-penicillin for 4 days prior to admission.

Diagnostic Procedures and Findings: The horse was depressed, with T = 101.5 ° F, PR = 80 bpm, RR = 56 bpm. MMC was pale pink with a CRT of < 2 sec. The urine was discolored (brown/red). Urinalysis was consistent with hemoglobinuria or myoglobinuria. No casts were observed, but the USG = 1.014. Ultrasonographic examination of the left shoulder revealed multiple pockets of fluid, one of which was large enough to drain. Culture of the fluid revealed large Gram-positive rods consistent with *Clostridial* species.

CBC revealed anemia (PCV = 13%), marked leukocytosis (49.6 x 10^9/L) with a left shift (2.5 x 10^9/ L band neutrophils). The fibrinogen concentration was slightly elevated at 548 mg/dL. Serum biochemical evaluation revealed azotemia (creatinine = 4.7 mg/dL and BUN = 97 mg/dL), hyponatremia (128 mEq/L), hypochloremia (82 mEq/L), hyperglobulinemia (5.4 mg/dL), hyperbilirubinemia (10.4 mg/dL) and increased activities of liver enzymes (AST, ALP, GGT, SDH) and muscle enzymes (CPK).

Coomb's test was negative. Cross matching with two blood donors was performed. Major cross match was compatible, minor was incompatible.

Assessment of Laboratory Findings: The neutrophilia with left shift, hyperfibrinogenemia and hyperproteinemia were consistent with the observed abscessation. The anemia was likely immune mediated due to antibody formation against red blood cell surface antigens. Autoimmune hemolytic anemia is a rare complication reported secondary to antigen stimulation from *Clostridium* spp., *Staphylococcus* spp., *Streptococcus* spp., viral and parasitic infections as well as secondary to neoplasia or drug induction. The resulting hemolysis led to hemoglobinuria. Urinary pigments can lead to renal damage by causing renal tubule obstruction and reduction in renal blood flow. The azotemia with isosthenuria confirms renal damage with impaired tubular concentrating ability. The elevated bilirubin was due to hemolysis.

Treatment: Whole blood transfusion (6 liters), dexamethasone (20 mg IV Q 24h for 9 days), LRS at 2 x maintenance (120 ml/kg/d for 3 days, then 60ml/kg/d for 1 d), oxytetracycline (6.6 mg/kg IV, q 12h in 250 ml 0.9 % NaCl x 9 d) and metronidazole (20 mg/kg PO q 6 h x 8d) were administered.

The horse responded to therapy with the PCV increasing to 22% after the transfusion and resolution of the pigmenturia and azotemia after intravenous fluid therapy. A second episode of hemolysis occurred after 6 days of hospitalization (hemolysed serum, red urine, elevated heart rate, and azotemia-creatinine = 2.8mg/dL). Intravenous fluids were required for 24 hours. The horse was discharged to the care of the RDVM on doxycycline (10mg/kg PO q 12h x 7 d) and dexamethasone (20mg PO q 24h x 5d). The PCV was 18%, and the leukocytosis was slowly resolving (22.3 x 10^9/L). The prognosis was still guarded due to the risk of further episodes of hemolysis and renal damage.

Take home message: Autoimmune hemolytic anemia is a rare complication of *Clostridial* spp. infections.

Case Four: 2-year-old Standardbred filly

History: The filly was referred for severe edema and swelling of the cervical and head regions, following intramuscular flunixin meglumine and phenylbutazone injections four days previously.

Assessment and Case Management: The severe head swelling had resulted in severe nasal edema, and the filly presented in respiratory distress, requiring immediate tracheostomy. Ultrasound examination did not reveal any discrete abscesses, and both jugular veins appeared patent, although deep to 10cm of edema fluid. An aspirate of this fluid was obtained for cytology, Gram-stain and culture. Clostridial cellulites was suspected.

Blood results showed a moderate increase in band neutrophils (0.7 x 10^9/L) and elevated fibrinogen (798 mg/dL), indicating inflammation. Mild dehydration was present with mild azotemia and electrolyte abnormalities.

Treatment: Although no bacteria were observed on the Gram-stain Clostridial spp. were suspected so oxytetracycline and metronidazole were administered for 4 days. A cephalic catheter was required due to the inaccessibility of the jugular veins, and risk of inducing phlebitis from access through infected tissue. Patency of the cephalic catheter was maintained with a slow infusion of LRS. Phenylbutazone was also administered.

Culture of the aspirate from the neck grew *Staphylococcus aureus/intermedius*, which was resistant to gentamicin, tetracycline and tribrissen. It was sensitive to amikacin, ceftiofur, chloramphenicol, enrofloxacin and penicillin. Potassium penicillin (20, 000 IU/kg IV q 6h) and enrofloxacin (2.5mg/kg IV q12h) were administered and the horse was released to the care of the referring veterinarian.

Take Home Message: Not all fulminate myositis/cellulites cases are due to clostridial organisms. *Staphylococcus aureus* can produce similar clinical findings and antibiotic therapies may be vastly different. Gram-stains are often helpful stall side, but culture and sensitivity should always be performed.

REFERENCES

Stewart AJ. Vet Clin North Am Equine Pract. 2006;22(1):127. Peek SF, et al Eq Vet J. 2003;35(1):86. Weiss DJ et al Vet Clin Pathol 2003;32(1):22. Arroyo LG et al Eq Vet J. 2003;35(5)514. Brown CM, et al J AmVet Med Assoc. 1988 15;193(6):668-70. Jeanes LV et al. Compend Cont Ed Pract Vet 2001;23(6):577

METHODS FOR SALMONELLA CONTROL IN A HOSPITAL SETTING: FOCUS ON CLEANING METHODS

N. Huff, PhD; F. Garza, MS; A. Auer; S. Eades, DVM, PhD, DACVIM; F. Andrews, DVM, MS, DACVIM
Baton Rouge, LA and West Lafayette, IN

INTRODUCTION

Salmonellosis is caused by infection with *Salmonella* bacteria. In adult horses salmonellosis causes several conditions, including inapparent infection, impaction, colic and acute and chronic diarrhea. Foals have a higher risk of bacterial infection and this could quickly lead to the systemic spread via blood (septicemia) and if untreated, death. In people, *Salmonella* infections generally cause diarrhea, fever, or abdominal cramps. Salmonellosis is usually self-limiting and resolves without treatment, 5-7 days after infection. However, in severe cases where septicemia, dehydration and leucopenia occur, horses need intensive therapy to prevent severe debilitation and death. Also, hospitalized patients, because of concurrent disease and stress, are susceptible to nosicomial infections, including Salmonella bacteria.

Bacterial contamination of stalls with *Salmonella sp.* is a serious problem in equine hospitals. Exposure to horses and personnel from horses admitted with Salmonella infections (carriers) or horses that acquire infection in the hospital facilities (nosocomial infections) can result in temporary facility closure for cleaning and eradication of organism. Hospital closure can result in loss of revenue, reputation damage, and curtail much needed care for client horses. Horses that develop acute diarrhea after being admitted to the hospital for routine procedures dramatically increase their risk for complications. Seemingly healthy horses can harbor the *Salmonella* bacteria, and then shed the organism when stressed[1]. These stress-affected animals provide sufficient potential for the spread of *Salmonella* infections to susceptible animals with compromised immune systems[2]. Horses can also acquire the bacteria from the hospital environment through contact with contaminated objects. Infected animals can serve as the reservoir for transmitting the bacteria to other animals, faculty, students, and hospital visitors[2]. Because of its serious zoonotic potential, the bacteria can have a tremendous public health impact[3]. The source of the infection is difficult to determine, therefore all precautions should be made to keep the hospital sanitary and free of the organism.

PATHOGENESIS

For a better understanding of how to prevent the *Salmonella* organism from invading a hospital, the life of the bacteria should be examined. *Salmonella* is a genus of rod-shaped, non-spore forming, pre-dominantly motile enterobacteria. The rods have flagella which project in all direction. The Gram-negative bacteria are facultative anaerobes. *Salmonella* is found worldwide in the intestines of virtually all birds, reptiles, and mammals and are eliminated in their feces[4]. After the bacteria are ingested, the salmonellae pass through the stomach, and attach to the cells lining the small intestine. The bacteria then insert proteins into the host cells, inducing the normally non-phagocytic cells to phagocytize the bacteria. The salmonellae reproduce within phagocytic vessels, eventually killing the host cells and inducing the signs and symptoms of salmonellosis. Salmonellosis is characterized by non-bloody diarrhea, fever, and myalgia. Cells of some strains can subsequently enter the bloodstream, causing bacteremia and localized infections throughout the body, including in the lining of the heart, the bones, and the joints[4].

The primary mode of transmission of *Salmonella* is the fecal-oral route; however, airborne transmission has been demonstrated in some species[5]. Sources of infection can be ingestion of contaminated food or water, horse-to-horse contact, or fomites. Examples of potential fomites in the barn are: hands, shoes, water buckets, brushes and grooming equipment.

The environment can also become contaminated by bird, rodent or other wildlife feces that contain the organism. Animal dropping in the hay or the feed can be a source of infection. Stressed horses, or horses that become ill, can also shed the organism.

STUDY

A study was performed to evaluate three stall cleaning methods on the eradication of *Salmonella sp.* from the stalls in an equine veterinary teaching hospital. Horses admitted to the hospital were assigned to *Salmonella sp.* negative stalls based on age, gender, and size from September 2009 – January 2010. At discharge, stalls were randomly assigned to one of three cleaning methods: 1) pressure washing only 2) pressure washing and hand scrubbing; or 3) hand scrubbing only. Following cleaning, each stall was allowed to air dry and within 48 hours, the stall surfaces were sampled using a swab or sponge. The person collecting the samples was blinded to the method of cleaning. Swab or sponge samples were submitted to the Louisiana Animal Disease Diagnostic Laboratory (LADDL) for standard culture for *Salmonella sp.* During the study period, 112 stalls (pressure washing only n=29; pressure washing and scrubbing n=50; scrubbing only n=33) were included. At the end of the study, there were fewer *Salmonella sp.* positive stalls in the hand scrubbed cleaning stalls, although this difference was not significant ($p=0.4057$), when compared to the other cleaning methods.

CONCLUSION

Stall cleaning method did not lead to a significant reduction in *Salmonella sp.* contamination in the stalls in this study. It appears that all the methods used in this study provide equal effect on preventing Salmonella contamination. Also, it should be emphasized that stall surfaces are only part of the facility and because of this cleaning methods should be part of a comprehensive biosecurity protocol to contain and prevent spread of salmonellosis in the hospital environment. Biosecurity measures include the use of plastic boots, gloves, hand sanitizers and footbaths at each stall. Isolation facilities should be available for potentially infectious patients entering the hospital and specific protocols should be in place to identify which horses are candidates to be moved to isolation. All personnel that work in the hospital should be trained to recognize and understand the risk factors involved in environmental *Salmonella sp.* contamination. This is essential to prevent exposure to the hospitalized population and minimize serious complications.

REFERENCES

[1]House JK et al., J Am Vet Med Assoc 1999; 214(10):1511-6. [2]Pandya, M et al.,Vect and Zoo Dis 2009; Vol 9 No 6: 649-654. [3]Mead PS et al., Emerg Infect Dis 1999; 5:605-624. [4]Bauman RW. Et al., Microbiology. 577-579. [5] Chapman AM. Vet Clin Equine 25 (2009) 363-380.

EQUINE NUTRITION

Kara M. Burns, MS, MEd, LVT
Wamego, KS

Nutrition is one area of equine veterinary medicine that affects every horse. Feeding of horses is an intricate science that requires the precise amount and balance of nutrients. Horses have been domesticated for centuries and are now considered companion animals. Companion horses today consume a variety of feeds ranging in physical form. They may consume forage with a high content of moisture to cereals with a high amount of starch, and from hay in the form of physically long fibrous stems to salt licks and water. Horses are nonruminant herbivores that naturally spend 60 to 75% of their day grazing. Typically they ingest approximately 2% of their body weight (dry matter basis) per day while grazing.

With domestication of the horse came alterations in feed, feeding times, and feeding methods – more in line with domesticated canines and felines - where the horse is 'meal fed' and unfamiliar materials such as starchy cereals, protein concentrates, and dried forages have been introduced. Horses are confined more of the time in stalls or smaller pastures, are fed 1 – 2 times per day, and as a result spend less time of the day eating, approximately 40%.

The diets formulated for horses today contain on average 5% fat and 7-12% protein, with carbohydrate being the major source of energy (~80%). Horses evolved to eat grass and other forages. Thus, grass and hays serve a strong foundation for the feeding of horses. Protein is required mainly for the building and replacement of tissues. Although some may consider protein to be an expensive source of energy, dietary protein and fat can contribute to meeting the physiological energy demands of the horse. Protein metabolism results in conversion of the carbon chain of amino acids to intermediary acids and to glucose. Fat can aid in meeting energy demands following its hydrolysis to glycerol and fatty acids. Subsequently, the glycerol can be converted to glucose and the fatty acid chain metabolized in a stepwise process by ß oxidation in the mitochondria, yielding adenosine triphosphate (ATP) and acetate or acetyl coenzyme A . This process requires tissue oxygen.

Carbohydrate digestion and fermentation yield mostly glucose and acetic, propionic, and butyric volatile fatty acids. The portal venous system collects these nutrients and a proportion of them are removed from the blood as they pass through the liver. Both propionate and glucose contribute to glycogen (liver starch) reserves and acetate and butyrate bolster the fat pool and comprise primary energy sources for many tissues.

NUTRITIONAL PHYSIOLOGY

The nutritional physiology of the horse differs from that of the cat or dog. The digestive tract of the horse has multiple compartments; each compartment has its own function in terms of utilizing ingested feed. The oral cavity is responsible for physically processing foods into smaller particles (~1.6mm) allowing for passage through the esophagus and increasing the surface area of food for the small intestine enzymatic action. Also the oral cavity breaks down structural carbohydrate for bacterial fermentation in the large intestine. The horse averages 60,000 chews per day and it is only during chewing that the salivation process is activated. A horse will produce an average of 5 – 10 liters per day of saliva, which in turn acts as a lubricant for passage into the esophagus.

The stomach is responsible for ~ 8% of the total capacity of the equine GI tract. The retention time of ingesta in the stomach ranges from 2 – 18 hours. The fundus and the pylorus are the two main regions of the stomach. The pyloric region secretes 10 to 30 liters of gastric juice per day. The small intestine of the horse measures 50 to 70 feet long and has a volume of 40 to 50 liters. The transit time of ingesta in the small intestine averages 2 to 8 hours. At the anterior end of the small intestine, pancreatic juice aids in the digestion of lipids, protein, and non structural carbohydrates. Also, the small intestine is lined with microvilli which increase the surface area of the gut. As mentioned earlier, digestion in the small intestine is dependent upon oral processing and types of feed (forage is less digested than processed feed).

The large intestine has capacity for large volume (100 liters) and has a very slow transit time of approximately 50 hours. The large intestine is responsible for the fermentation of structural carbohydrates to volatile fatty acids, which is responsible for 50% of the metabolizable energy. The large intestine absorbs roughly 80 liters of water per day. B vitamins, produced by the bacteria within the lumen of the large intestine, are absorbed in the large intestine. It is also important to note that non-protein nitrogen is not utilized well in the large intestine.

KEY NUTRITIONAL FACTORS FOR HORSES

The main nutrients that nutritionists are concerned with in regard to the horse are water, energy, protein, minerals, and vitamins. Water is the most important nutrient in any mammalian species. There should be fresh, abundant water available at all times. Horses will drink an average of 25 liters per day. In conditions of extreme heat or stress (such as sustained exercise in hot conditions) the number of liters a horse will drink per day may increase to 100. Also, as technicians it is necessary to remind owners that the more grain their horse eats, the more water intake the horse will need.

Energy is measured in terms of digestible energy (DE) and fed in kilocalories. The amount of DE horses need will be dependent upon various factors; physiologic state, activity level, environment, and the size of the individual horse. The majority of energy utilized by the horse is from carbohydrates that are ingested through the horses' natural feed. The following shows the amount of kcal/kg in various carbohydrate feeds:

Feed	kcal/kg
Oats	3000
Alfalfa (early bloom)	2100
Bermuda grass hay	1800
Corn cobs	1250

Fats provide the horse with high density energy. In equids, fat should not exceed 20% of the total diet or 30% of concentrate. Exceeding these percentages will likely result in decreased palatability and loose stools.

Protein amounts are typically expressed as 'crude protein (CP)' and are expressed as % dry matter (DM). Again the amount of protein needed by an individual horse is dependent upon a number of factors: physiologic state, type of diet, age, quality of diet. The closer the proportions of each of the various indispensable amino acids in the dietary protein conform to the proportions in the mixture required by the tissues, the higher the quality of the protein.

Calcium and phosphorous are considered together because of their interdependent role as the main elements of the crystal apatite, which provides the building blocks for the skeletal system. The requirement of calcium and phosphorous is dependent upon the physiologic state of the horse. The average adult horse weighing approximately 500 kg will need ~20g of calcium and 14 g of phosphorous per day. It is important to balance the ratio of calcium to phosphorous with a mature horse needing a ratio of 1.1:1 to 6:1. The ratio for a growing horse is recommended to be 1.1:1 to 3:1.

Sodium is the principal determinant of the osmolarity of extracellular fluid and as a result, the volume of that fluid. Chloride concentration in the extracellular fluid is directly related to that of sodium. Rarely do companion animals have an excess or deficiency of sodium or chloride; however, one should be aware of the requirements. Daily sodium requirements are recommended to be approximately 0.18 to 0.36% DM. If the requirements for sodium are met, a deficiency of chloride will seldom occur. Good sources of sodium and chloride can be found in commercial concentrate feeds with premixture of minerals and salt blocks.

Potassium is the main intracellular cation. Deficiencies in equines are rare and excess potassium is rarely a problem. However, excess potassium can lead to hyperkalaemic periodic paralysis, a syndrome of episodic weakness in horses accompanied by elevated serum potassium concentration. This syndrome appears to be confined to descendants of the American Quarter Horse with a specific genetic deficiency. Forages are approximately 1 to 4% potassium with cereals being relatively poor sources of postassium.

Selenium is a trace element needed to aid in antioxidant defense. Selenium forms an integral part of the GSH-Px molecule. Selenium catalyses peroxide detoxification in body tissues during which reduced glutathione (GSH) is oxidized. It is closely involved with the activity of α-tocopherol (vitamin E), which protects polyunsaturated fatty acids from peridoxication. The requirement is 1 to 2 mg/day for a 500kg horse or 0.1-2 ppm. Selenium deficiencies produce pale, weak muscles in foals and a yellowing of the depot fat, know as 'White Muscle Disease'. It is imperative that pregnant mares receive adequate amounts of selenium in their diet. Selenium is highly toxic to animals, with the minimum toxic dose through continuous intake is 2 to 5 mg/kg feed. Skin, coat, and hoof abnormalities are the result from an excess selenium.

Grazing horses derive their vitamin A from the carotenoid pigments present in herbage. The principal one is β-carotene with 1mg of β-carotene equating to approximately 400 IU of vitamin A. Horses that graze for 4 to 6 weeks build up a 3 to 6 month supply of vitamin A in the liver.

Carotene levels (mg/kg DM)	**Amounts**
Pasture grass/alfalfa	300 – 600
Good hay	20 – 40
Poor hay	4 – 5

Requirements for vitamin A during certain lifestages are as follows: Mature horses require 30 IU/kg body weight; Gestation/lactation stage requires 60 IU/kg body weight; Growth stage requires ~50 IU/kg body weight.

Vitamin E functions as a cellular antioxidant in conjunction with Vitamin A and is required for normal immune function. Fresh green forage and the germ of cereal grains are rich sources of vitamin E. Adult horses require 80 to 100 IU/kg DM. Although deficiencies are rare, two neurological disorders of horses have been recognized to involve α-tocopherol status: equine degenerative myeloencephalopathy and equine motor neuron disease. These diseases typically are seen in horses that do not have access to pastures, with the consumption of poor quality hay, and in horses with low concentrations of circulating levels of α-tocopherol.

TYPES OF FEED

There are three types of feed that we will discuss : roughages, concentrates, and complete feeds. Roughages include grasses and forage legumes cut for hay. Most common species of grass are suitable, but the preferred are the more popular and productive grasses such as rye grasses, fescues, timothy, and cocksfoot. Species found in permanent pastures are satisfactory as well and these include meadow grasses, brome, bent grass, and foxtails. Legumes utilized are red, white, and crimson clovers and trefoils; lucerne; and sainfoin. Species of clovers should be investigated prior to allowing the horse to feed, as certain species of clover may result in varying degrees of toxic reactions (i.e., hypersalivation, photosensitivity, liver disease). Roughages are relatively low in energy and have > 18% crude fiber. Roughages are considered to be the foundation to an equine feeding program. Quality hay can provide energy for the maintenance requirements of the horse. Legumes and non-legume grasses that are well managed and fertilized (proteinaceous roughages) provide > 10% CP as opposed to carbonaceous grasses (those that are not well maintained or fertilized) provide < 10% CP. Veterinary technicians can educate clients on the quality of roughage using the following guidelines:

- Free of mold
- Soft & pliable to touch
- Leafy with fine stems (2/3E, ¾ protein)
- Pleasant, fragrant aroma
- Bright green, not brown
  or yellow

Another key point to communicate to owners is the fact that excess handling of roughages can result in loss of ¼ of the leaves; loss of ¼ to 1/3 energy and protein; and loss of 90% of β-carotene.

Concentrates are typically a cereal grain that may or may not have supplemented protein, minerals, and vitamins. Concentrates are high in energy (typically 50% greater than forage), and are less than 18% crude fiber. Oftentimes, concentrates are used as a supplement if forage is insufficient in nutrients – especially energy and protein. Concentrates are needed more often in lifestages such as gestation (especially later in the gestation period), lactation, growth, and in working horses. A rule of thumb would be to not exceed 50:50 (wt:wt) concentrate to roughage. As always any new feed should be introduced slowly. It is important for healthcare team members and owners alike to be cognizant of the fact that excess concentrate may lead to laminitis, rhabdomyolysis, developmental orthopedic disease, and obesity.

Complete feeds are typically a mixture of roughage and concentrate – usually an 80 % roughage to 20% concentrate mixture. Complete feeds are manufactured by completely grinding the food and formulating it into a pellet, thus making it easier (all in one) for the owner, although there is usually an increase cost for the convenience. Because the complete feed is pelleted or wafered, care must be given to the potential risks associated with inadequate particle size, including colic, choking risk, wood chewing, coprophagy, to name a few.

SUMMARY

When discussing nutritional management of horses with clients remember to advise that fresh water needs to be available at all times. Energy for maintenance can be met entirely with quality hay, but may be supplemented with concentrate if necessary. Horses should not be supplemented with more than 50% by weight per daily ration. Good quality green roughage should supply adequate amounts of Vitamins A & E. Remember, nutrition is one area of equine veterinary medicine that affects every horse and should be discussed on every visit.

REFERENCES

Davies, Z. Introduction to Horse Nutrition. 2009. Wiley-Blackwell. Ames, Iowa. Frape, D. Equine Nutrition and Feeding. 4rdth Edition. 2010. Blackwell, Ames, Iowa. Geor, R. Equine Nutrition, Veterinary Clinics of North America: Equine Practice. April, 2009. Elsevier/Saunders. St. Louis, MO. Pilliner, S. Horse Nutrition and Feeding. 2nd Edition. 2009. Wiley-Blackwell. Ames, Iowa. Reed, S. et.al. Equine Internal Medicine. 2009. Elsevier/Saunders. St. Louis, MO. Reeder, D. et.al. AAEVT's Equine Manual for Veterinary Technicians. 2009. Wiley-Blackwell. Ames, Iowa.

CANINE COGNITIVE DYSFUNCTION

Kara M. Burns, MS, MEd, LVT
Wamego, KS

Human and veterinary medicine have seen tremendous advances in medical care and nutrition resulting in both humans and their pets living longer. In the United States alone, it is estimated that 20 to 30 million dogs are over 7 years of age and are considered seniors.[1] Consequently, veterinary healthcare team members are seeing many age related changes and disease conditions such as cancer, renal disease, and cognitive decline, in their canine patients that have been associated with aging in human patients. Cognitive decline in dogs is typically manifested as behavioral changes in one or more of the following areas: disorientation (D), altered interactions with family members (I), disruptions in sleep patterns (S), loss of housetraining (H), and decreased levels of activity(A). Many pet owners attribute behavior changes in their dogs to simply be part of the aging process and believe nothing can be done to help their beloved pet. Unfortunately, some of these behavioral changes lead to owners placing their senior pet in a shelter or electing euthanasia. The combination of these behaviors, known by the acronym DISHA, is diagnosed as Cognitive Dysfunction Syndrome (CDS)[2], and nutrition plays a key role in helping combat this syndrome.

The veterinary healthcare team plays an important role in diagnosing behavioral changes in geriatric patients. These behavioral changes may be noticed by owners but not considered severe enough by the family to warrant a visit to the veterinarian. The effects of aging on the brain are oftentimes subtle and progress slowly; therefore it is crucial that the healthcare team, specifically the veterinary technician, educate owners about the signs and symptoms associated with cognitive decline so these symptoms can be addressed as soon as possible. In fact, when owners were educated about the behavioral signs of DISHA, 75% stated that they had seen at least one sign. Furthermore, 37% admitted having observed one or more signs several times a week and 32% reported three of more signs. The alarming fact from this study is that only 12% actually talked to their veterinarian about a sign observed in their dog.[3] Behavioral changes may be the first indication of a physical or health problem. If a medical condition is the cause of the changes, early reporting of these changes in the pet by the owners can help lead to early diagnosis and potential medical intervention.

The acronym DISHA, as described above, is used to describe signs often seen in cognitive dysfunction. However, this is not all inclusive as there are other clinical signs and owner observations that may be associated with brain aging. The healthcare team, especially when presented with a senior pet, should institute a cognitive impairment screening checklist as part of the routine physical examination for senior pets. A checklist, given to the pet owner upon check-in to the hospital, will allow for that owner to consider whether or not their pet is exhibiting certain behaviors and is a great way to get clients thinking about their pets' behavior and imply that it may not be simply 'old age behavior', but rather a disease condition that can be treated. This behavior checklist will also allow the healthcare team to ask more in depth questions regarding the observations of the client. Oftentimes the client is observing signs in their pet relating to confusion, social relationships, activity level changes, depression/apathy, increased anxiety, sleep-wake cycle alterations, and learning and memory changes and deficits. [4] It is important to note if the patient is going to the wrong side of the door, vocalizing at night, not responding to previously learned commands, searching for an increase or exhibiting a decrease in the amount of affection sought with their owner, increasingly irritable, not wanting to play, having no interest in eating, pacing, and exhibiting compulsive behaviors (increased licking) not previously seen. (Table 1) This information will assist the healthcare team in a proper diagnosis of CDS. In conjunction with the patient history, a complete physical exam, blood work, urinalysis, and imaging, will help to rule out any medical conditions or disease processes that may affect the patient's mental attitude or behavior.

AGING AND ITS EFFECT ON THE BRAIN

Aging in most species studied to date is accompanied by the progressive accumulation of oxidative damage in many tissues. The brains of older dogs and the brains of older humans have several key features that have been observed. Both have a number of morphologic changes including cortical atrophy, ventricle widening, white matter myelin degeneration, accumulation of degraded proteins, damage to the DNA, and reduction of endogenous antioxidants.[5] The brain in older dogs also accumulates proteins within and around neurons and these proteins may be toxic to the neurons. The result is mitochondrial dysfunction and impaired energy metabolism in cortical neurons. Oxidative damage in older animals' brains may contribute to neuronal dysfunction and accumulation of neuropathologic lesions. The mitochondria use oxygen for normal, aerobic, energy production and one side reaction that may occur is conversion to free radicals within the mitochondria. As mitochondria age, the production of free radicals is increased and subsequently, the mitochondria are less efficient and produce less energy. [4] The brain is especially susceptible to the effects of free radicals due to its' high lipid concentration, demand for oxygen, and limited ability for antioxidant defense and repair. If neurodegenerative processes leading to cognitive dysfunction are influenced by oxidative damage then it is expected that antioxidants would improve learning and memory.

NUTRITIONAL MANAGEMENT OF COGNITIVE DYSFUNCTION SYNDROME

Physiologic antioxidants act to decrease the effects of free radicals in a biologic system. These antioxidants differ from those that are well known and used as preservatives to keep food fresh. Antioxidants such as vitamins E and C help to neutralize free radicals thus preventing cellular damage. Lipoic acid and l-carnitine increases the efficiency of energy conversion while decreasing the production of free radicals resulting in mitochondrial health.[6] Carotenoids and flavonoids, typically found in fruits and vegetables also aid in reducing cellular damage by inactivating free radicals. Together these create a strong antioxidant bundle which helps to counter cellular alterations resulting from oxidative stress and the neuropathology that comes from aging. Additionally, fatty acids docosahexaenoic acid (DHA) and eicosapentaeoic acid (EPA) contribute to neuronal cell health and are found in the brain in high concentrations

The addition of DHA and EPA along with the antioxidant package to a food should improve mitochondrial efficiency, decrease free radical production, and absorb free radicals thus resulting in decreased cell damage in the brains of dogs. A series of studies looked at a test food (fortified with antioxidants and mitochondrial cofactors) vs. a control food (non-fortified food) to see if the test food would decrease signs of brain aging as manifested by the following: 1) cognitive abilities and 2) improved age related changes in dogs as observed by owners in their homes. Oddity discrimination tasks (where the dogs had to chose which object was different), discrimination learning (where the dogs had to discriminate between objects, size of objects, and black and white), landmark discrimination tasks (where the dog had to use spatial information to locate objects), and in home behavioral testing (where owners had to assess changes in their dog's behavior) were studied.[4,5] In addition, behavioral enrichment was studied in conjunction with a fortified food in the landmark discrimination study.[6,7] In all studies, it was found that a food fortified with the antioxidant package described above (the test food) will delay or partially reverse age-related deteriorations in learning. Environmental enrichment (exercise, novel play toys, and continued testing) was also found to be an important factor in improving the effects of brain aging in dogs and those dogs who participated in the combined group which had fortified food and environmental enrichment both as treatments found their learning ability to be the best preserved.[7]

CONCLUSION

It is necessary that the veterinary healthcare team begin to identify behavioral changes in senior dogs. Dog owners are recognizing changes in their senior dogs, but are not discussing these changes with their pet's healthcare team. Utilizing a behavior checklist will open the conversation between the healthcare team and the pet owner and will help to discover certain behaviors that the owner may be seeing in their pet, but attribute to old age. These behaviors should not be dismissed simply as 'old age' especially given the fact that nutritional management has been clinically shown to increase learning ability and attentiveness to solving tasks. Also, antioxidant fortified nutrition has led to reports of a 74% reduction in house-soiling accidents and a 61% increase in enthusiasm in greeting family members by senior dog owners feeding the antioxidant fortified food. By pinpointing behaviors consistent with CDS, the healthcare team can provide a nutritional solution to alleviate Cognitive Dysfunction Syndrome and strengthen the human-animal bond by improving the quality of life for senior dogs.

Table 1

Clinical Signs of Brain Aging

- Going to wrong side of door
- Decreased awareness
- House-soiling
- Depression or apathy
- Vocalizing at night
- Altered sleep/wake cycle
- Not responding to previously learned commands
- Looking for increased/decreased affection from owner
- Increasing irritability
- No interest in play
- Increase or decrease in appetite
- Pacing/Restless
- Compulsive behaviors (i.e., increased licking)
- Altered relationships with people and other pets

REFERENCES

1. AVMA. Total pet ownership and pet population. US pet ownership and demographics sourcebook. Schaumburg, Ill: AVMA Membership & Field Services, 2002; pp 6-43. 2. Ruehl WW, Bruyette DS, DePaoli A, Cotman CW, Head E, et al. Canine cognitive dysfunction as a model for human age-related cognitive decline, dementia and Alzheimer's disease: clinical presentation, cognitive testing, pathology, and response to l-deprenyl therapy. Prog. Brain Res. 1995; 106; pp 217-225. 3. US Marketing Research summary. Omnibus Study on Aging Pets. November, 2000. Data on File. Hill's Pet Nutrition, Inc. 4. Landsberg G, Hunthausen W, Ackerman L. 2003. Handbook of behavior problems of the dog and cat, 2nd ed. Elsevier, St. Louis, pp 269-304. 5. Roudebush P, Zicker SC, Cotman CW et al. Nutritional Management of Brain Aging in Dogs. Journal of the American Veterinary Medical Association 2005; 227: 722-728. 6. Dodd CE, Zicker SC, Jewell DE, Fritsch DA, Lowry SR, Allen TA. Can a fortified food affect behavioral manifestations of age-related cognitive decline in dogs? Veterinary Medicine 2003;98:396-408. 7. Milgram NW, Head E, Zicker SC, et al. Learning ability in aged beagle dogs is preserved by behavioral enrichment and dietary fortification: a two-year longitudinal studt. Neurobiology of Aging 2005; 26: 77-90.

CLINICAL SIGNS ASSOCIATED WITH EXTRACRANIAL NEUROLOGICAL INJURY

KAREN INZANA, DVM, PhD DACVIM (Neurology)
Virginia Tech

STEPS TO A COMPLETE NEUROLOGICAL EXAMINATION PART I, MOTOR UNIT AND SPINAL CORD

STEP 1: OBSERVATION AT A DISTANCE:

Begin by watching the animal walk into the exam room. Things to note:

1. Body posture:

If the dog has a head tilt, the problem automatically must be rostral to the foramen magnum. It is pretty difficult to produce a head tilt with a spinal cord problem. Note however, that a head tilt by itself does not localize the problem to any specific region of the brain. Lesions in almost any area of the brain (including vestibular system) can cause an abnormal head posture. Most of the time, the lesion is on the same side as the head tilt. (i.e., head tilt to the right, lesion on the right)

2. Attitude/ Level of consciousness:

If the dog is demented, semi comatose, or looks like "the lights are on but no one's home", begin thinking of an intracranial problem. Note that injury to the reticular activating system in the brainstem can alter level of consciousness, so again level of consciousness is a clue to the location of the lesion, not a diagnosis.

3. Gait: can the dog walk? If so, is there coordinated movement in all 4 limbs?

If the dog is able to walk normally with the front limbs but drags the rear limbs, begin thinking of a problem caudal to T2 spinal cord segment.

If the dog can not walk in any limb, begin thinking of a problem rostral to T2 spinal cord segment or a generalized neuromuscular problem.

STEP 2. POSTURAL REACTIONS:

Conscious proprioception, hopping, hemiwalking, wheelbarrowing, and extensor postural thrust are all classified as postural reactions. All test approximately the same thing.... the ability of the animal to recognize that its foot is in an abnormal position to bear weight, and the ability of it to place it in a more normal position. These reactions require both peripheral nerve (motor and sensory) as well as ascending sensory and descending motor tracts in the spinal cord. Injury to any of these structures results in diminished or absent postural reactions. Therefore, this test answers the question "**Is there a neurological problem" BUT NOT "where is the problem" Postural** reactions are only a piece of the localization puzzle.

STEP 3. SPINAL REFLEXES:

The entire motor unit must be intact in order for the spinal reflex to be present. Injury to any portion of the motor unit will abolish the reflex. Alternatively, if the problem is upstream, affecting upper motor neurons, the spinal reflex will appear exaggerated and some abnormal reflexes will be present.

<u>Pelvic limb reflexes:</u>

1. Patellar reflex: With the stifle gently supported in a flexed position percuss the patellar tendon. A normal response is extension of the stifle.
2. Cranial tibial response: directly percuss the muscle belly. A normal response is flexion of the hock.
3. Gastrocnemius reflex: Percuss the achilles tendon. A normal response is extension of the hock.
4. Withdrawal: Pinch the digits (usually both lateral and medial digits are examined) and observe for flexion of the limb and a conscious recognition of pain.
5. Sciatic response: Percuss your thumb in the sciatic notch. A normal response is a jerk of the entire limb.
6. Crossed extensor: Pinch the digits on the down limb and look for involuntary extension of the up limb.
7. Babinski: Apply moderate pressure to the plantar surface of the hock in a sweeping motion away from the digits. A positive response is extension of the digits.

<u>Thoracic limb reflexes:</u>

1. Triceps reflex: Percuss the triceps tendon. A normal response is extension of the elbow.
2. Biceps reflex: Percuss the biceps tendon located on the medial aspect of the elbow. A normal response is flexion of the elbow. Often you only see a slight contraction of the biceps muscle.
3. Extensor carpi radialis response: Directly percuss the extensor carpi radialis muscle. A normal response is extension of the carpus.
4. Withdrawal: Apply a noxious stimulus to distal limb. A normal response is flexion of the limb and a conscious perception of pain.

Motor Unit The motor unit contains ventral horn cell (lower motor neuron) and its axon, neuromuscular junction and skeletal muscle fibers innervated by that nerve. Sensory nerves are also included in this category because most of the responses in veterinary medicine require a motor response to a sensory stimulus.

Clinical Signs Associated With Motor Unit Disease:
1. Weak or ineffective movement (paresis to paralysis)
2. Rapid loss of muscle tone and bulk
3. Altered sensation if sensory nerves affected
4. *Diminished or absent spinal reflexes*

Spinal Cord: Each segment of spinal cord contains ventral horn cells and interneurons in the gray matter, as well as ascending sensory tracts and descending motor tracts in the surrounding white matter. **The clinical signs associated with spinal cord injury vary with location**

CLINICAL SIGNS OF INJURY TO SPINAL CORD SEGMENTS L4–S2:

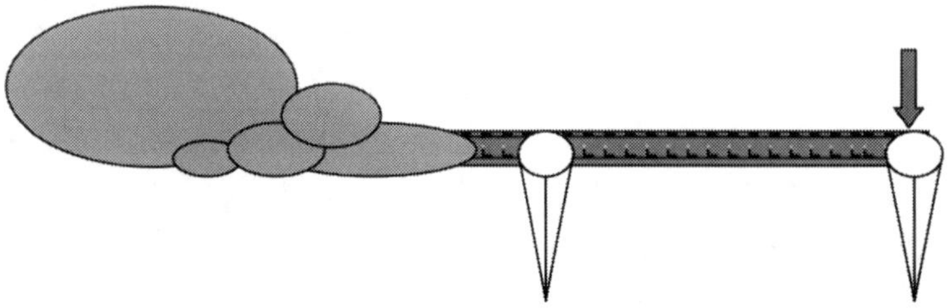

1. Weakness or paralysis of the rear limbs.
2. Rear limb muscle atrophy, hypotonia (usually bilateral)
3. Changes in sensation if the injury is very severe
4. Diminished or absent spinal reflexes in the rear limbs
5. No changes in the thoracic limbs

CLINICAL SIGNS OF INJURY TO SPINAL CORD SEGMENTS T2–L4

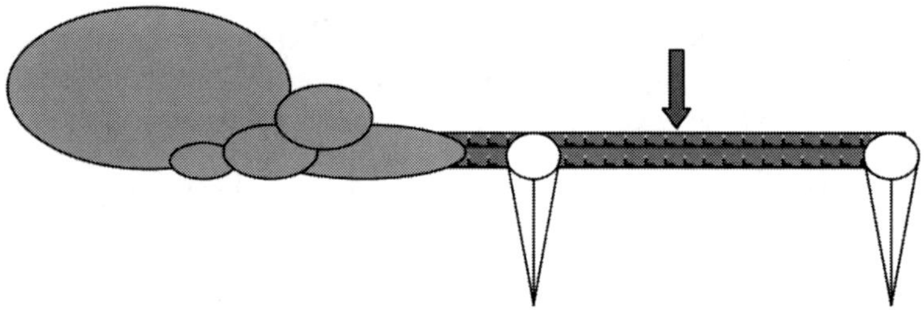

1. Weakness or paralysis of the rear limbs
2. Rear limb muscle atrophy in chronic cases (disuse atrophy)
3. Sensation is altered only with severe spinal cord injury
4. Spinal reflexes exaggerated and abnormal reflexes maybe present
5. Panniculus reflex absent caudal to the injury
6. Thoracic limbs are normal

CLINICAL SIGNS OF INJURY TO SPINAL CORD SEGMENTS C6–T2

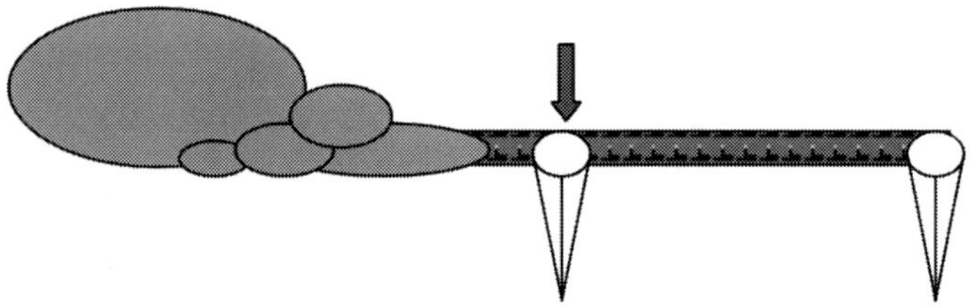

1. Weakness or paralysis in all four limbs. Thoracic limbs often appear worse than pelvic limbs.
2. Rear limb changes indistinguishable from T2–L4 injury
3. Muscle atrophy, hypotonia in front limbs
4. Changes in sensation only if the injury is severe
5. Spinal reflexes diminished to absent in the front limbs, exaggerated in the rear limbs
6. absent panniculus reflex
7. Horner's syndrome** (loss of sympathetic innervation to the eye causing miosis, ptosis, and enophthalmus).

CLINICAL SIGNS OF INJURY TO SPINAL CORD SEGMENTS C1–C6

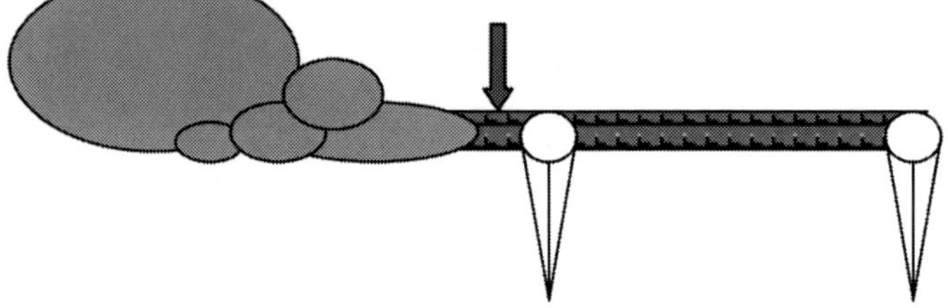

1. Weakness or paralysis in all four limbs. Pelvic limbs typically one grade worse than thoracic limbs.
2. Muscle atrophy only in chronic cases
3. All limb muscles retain some tone
4. Changes in sensation only if the injury is severe
5. Spinal reflexes exaggerated in all four limbs

6. Respiratory muscles may be affected

Note that severe injury to the cervical spinal cord usually results in death because of paralysis of respiratory muscles.

SUMMARY:

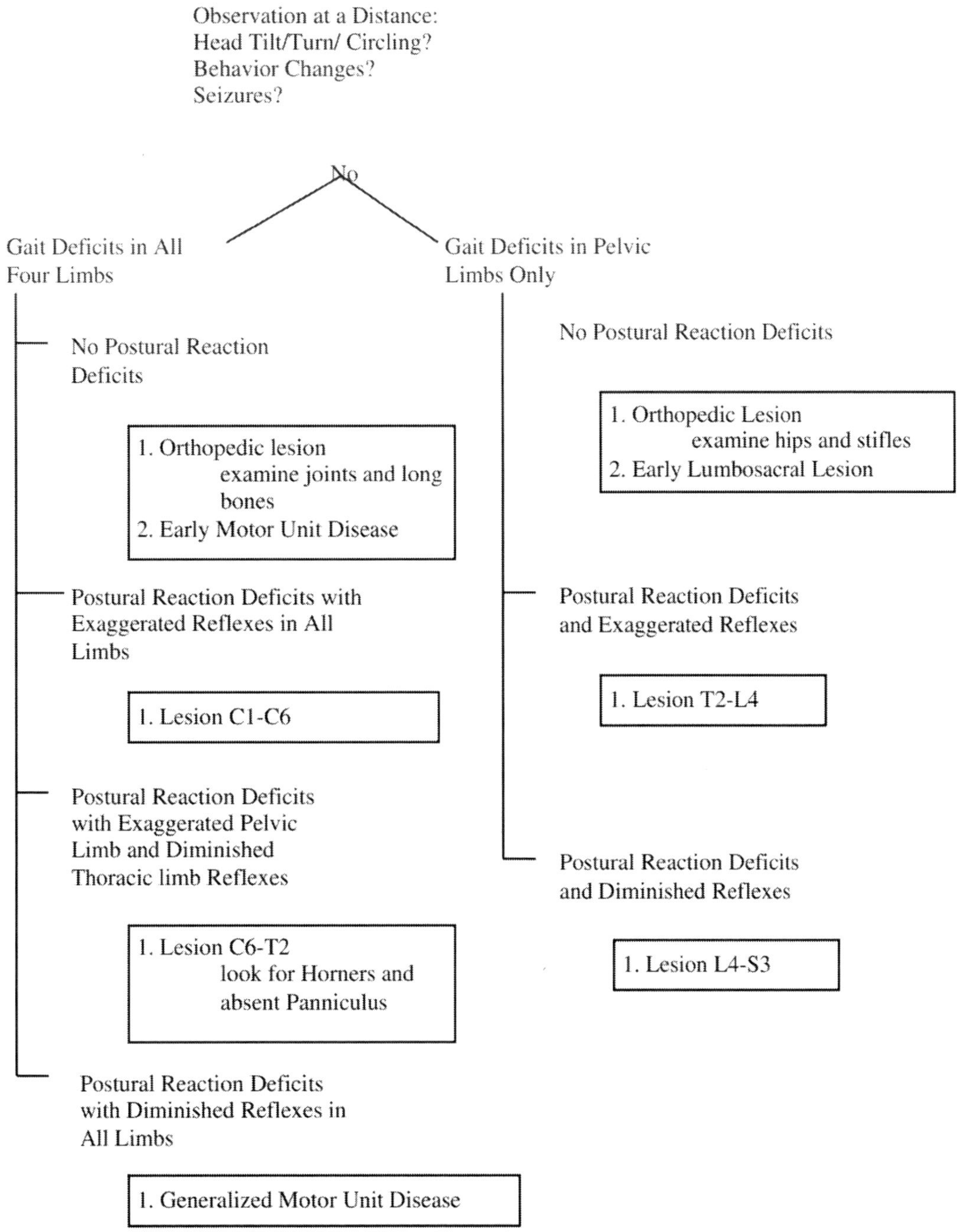

CLINICAL SIGNS ASSOCIATED WITH INTRACRANIAL LESIONS

Karen Inzana, DVM, PhD DACVIM (Neurology)
Blacksburg, VA

STEPS TO A COMPLETE NEUROLOGICAL EXAMINATION PART II, INTRACRANIAL LESIONS

STEP 1: OBSERVATION AT A DISTANCE:

Begin by watching the animal walk into the exam room. Things to note include:

1. Body posture:

   If the dog has a head tilt, the problem automatically must be rostral to the foramen magnum. It is pretty difficult to produce a head tilt with a spinal cord problem. Note however, that a head tilt by itself does not localize the problem to any specific region of the brain. **Lesions in almost any area of the brain (including vestibular system) can cause an abnormal head posture. Most of the time, the lesion is on the same side as the head tilt. (i.e., head tilt to the right, lesion on the right)**

2. Attitude/ Level of consciousness:

   If the dog is demented, semicomatose, or looks like "the lights are on but no one's home", begin thinking of an intracranial problem. Note that injury to the reticular activating system in the brainstem can alter level of consciousness, so again level of consciousness is a clue to the location of the lesion, not a diagnosis.

3. Gait: can the dog walk? If so, is there coordinated movement in all 4 limbs?

   If the dog is able to walk normally with the front limbs but drags the rear limbs, begin thinking of a problem caudal to T2 spinal cord segment.

   If the dog can not walk in any limb, begin thinking of a problem rostral to T2 spinal cord segment or a generalized neuromuscular problem.

STEP 2. POSTURAL REACTIONS:

Conscious proprioception, hopping, hemiwalking, wheelbarrowing, and extensor postural thrust are all classified as postural reactions. All test approximately the same thing.... the ability of the animal to recognize that its foot is in an abnormal position to bear weight, and the ability of it to place it in a more normal position. These reactions require both peripheral nerve (motor and sensory) as well as ascending sensory and descending motor tracts in the spinal cord. Injury to any of these structures results in diminished or absent postural reactions. Therefore, this test answers the question "**Is there a neurological problem" BUT NOT "where is the problem"** Postural reactions are only a piece of the localization puzzle.

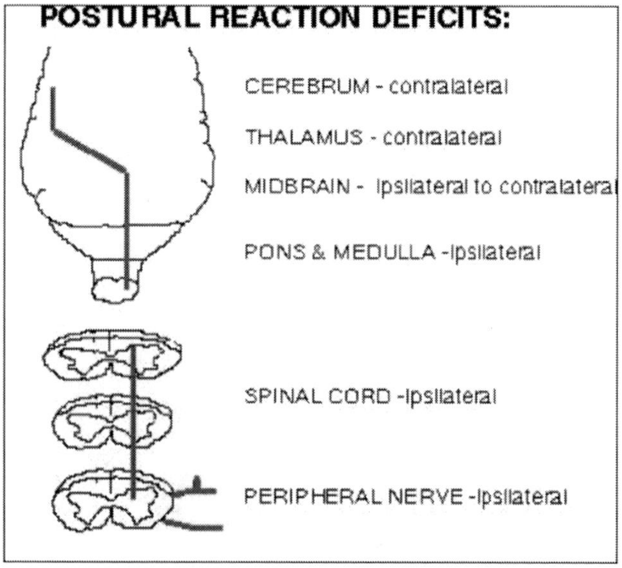

All postural reactions test both motor and sensory systems. As a rule, postural reaction deficits associated with asymmetrical lesions appear worse on the SAME SIDE AS THE LESION from the midbrain caudally and worse ON THE SIDE OPPOSITE THE LESIOIN from the midbrain cranially.

STEP 3: CRANIAL NERVE EXAMINATION:

Lesions in the pons and medulla will injure cranial nerves V-XII, lesions in the midbrain can injure cranial nerves III and IV, while lesions in the thalamus/hypothalamus will injure cranial nerve II and affect both vision and pupillary light reflexes.

CN II: Vision is evaluated by observing movement and ability to follow objects. If in doubt, throw a cotton ball in the animal's visual field and see if he follows it with his eyes. Menace response is performed by making a menacing gesture towards the eye and looking for a blink. Pupillary light reflexes are evaluated with a bright light and looking for both direct and consensual response.

Note that optic nerve, chiasm, and optic tract lesions will affect both vision and pupillary light reflexes while cortical lesions will affect only vision.

CN III: The parasympathetic component of CN III is evaluated with pupillary light reflexes. Deficits will cause the pupil to remain dilated even though the animal is visual.

CN III, IV, and VI: They are usually evaluated together by observing eye movement in all planes. Injury to all three nerves causes complete ophthalmoplegia. Injury to only one specific nerve is rare. Ventrolateral strabismus is caused by injury to CN III while ventromedial strabismus and inability to retract the globe is caused by injury to CN VI.

CN V and VII: These are usually evaluated together. CN V supplies sensation to the face while CN VII is motor to muscles of facial expression. Gently touch, pinch, or stroke the hairs of the face and see if the animal responds with facial movement, eye blink, and lip retraction. Note that CN VII supplies parasympathetic innervation to the lacrimal glands, while CN V is motor to the muscles of mastication. Always check tear production and symmetry of jaw muscles.

CN VIII: A crude evaluation of cochlear function is seeing if the animal responds to noises by turning its head. A more precise evaluation can be accomplished electrophysiologically. Vestibular function is evaluated by observing body posture (head tilt?), gait (ataxic?), and eye movement. Nystagmus and strabismus are hallmarks of vestibular disease. Bilateral vestibular disease often abolishes normal vestibular nystagmus. Move the animals head side to side and observe for normal vestibular nystagmus.

CN IX, X, and XI: While these cranial nerves have a variety of functions, normally we only test pharyngeal sensation and motor function with a gag response. Taste is only rarely evaluated. If the animal is bradycardic due to vagal hypertonus, pressing on the globe may increase heart rate.

CN XII: Tongue movement is used to evaluate CN XII function.

CLINICAL SIGNS OF INTRACRANIAL LESIONS:

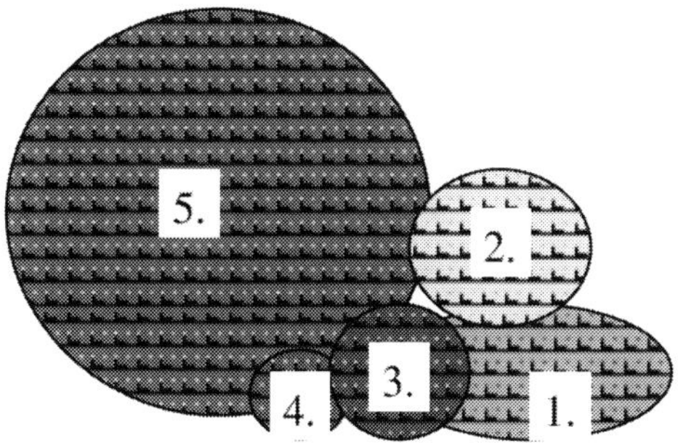

1. pons and medulla (metencephalon and myelencephalon)
2. cerebellum
3. midbrain (mesencephalon)
4. thalamus/ hypothalamus (diencephalon)
5. cerebrum (telencephalon)

<u>Clinical signs of injury to the pons and medulla</u>

1. Gait and postural reaction deficits ipsilateral to the lesion
2. Deficits in cranial nerves V – XII
3. Abnormal body posture (head tilt, lean) usually to the side of the lesion if the vestibular system or cerebellum is affected
4. Abnormal levels of arousal (with severe injury)
   abnormal respiratory pattern and heart rate (with

<u>Clinical signs of injury to the cerebellum</u>

1. Tremors, ataxia, or spastic changes in gait and postural reactions
2. Diminished or absent menace response
3. Evidence of central vestibular dysfunction
4. With severe injury may see decerebellae posture (opisthotonus, extensor rigidity of front limbs, rear limbs may be extended or flexed

<u>Clinical signs of midbrain injury</u>

1. Gait and postural reaction deficits may be ipsilateral or contralateral to lesion
2. Abnormal extraocular eye movement (CN III and IV)
3. Abnormal pupillary light reflexes (CN III)
4. Head tilt and tendency to walk in circles (injury to colliculi)
5. Abnormal body posture and level of consciousness (with more serve lesions). <u>Decerebrate rigidity</u> is an involuntary posture characterized by opisthotonos, and extensor rigidity of all four limbs. The extensor rigidity is usually explained as release of more caudal extrapyramidal motor centers (reticulospinal and vestibulospinal tracts, whose major action is facilitory to extensor muscles) from higher control. Note that lesions severe enough to cause decerebrate rigidly also profoundly influence the ARAS. Therefore, animals with decerebrate posture are also comatose

<u>Clinical signs of diencephalon injury</u>

1. Behavioral changes (depression, stupor, mania)
2. Autonomic dysfunction, endocrine imbalances
3. Seizures

4. Head turn and tendency to circle towards the side of the lesion
5. Gait and postural reaction deficits contralateral to the injury
6. Cranial nerve II deficits (blindness, abnormal PLR)

Clinical signs of injury to the telencephalon

1. Behavioral changes, depressed, demented, occasionally mania
2. Seizures
3. Normal (or near normal) gait with contralateral postural reaction deficits
4. Mild contralateral sensory deficits or visual impairment
5. Pacing, circling, head turn usually to the side of the lesion

Note that lesions in the cerebrum and thalamus/hypothalamus may be very similar. If fact, they may be impossible to differentiate unless there are signs of optic nerve or endocrine disturbances. The term FOREBRAIN signs is used to refer to both cerebrum and thalamus/hypothalamus collectively.

Observation at a Distance:
Head Tilt/Turn/ Circling?
Behavior Changes?
Seizures?

Yes

- No Postural Reaction Deficits.

  1. Peripheral Vestibular Lesion
     look for nystagmus, strabismus, Horners, Facial N. Paralysis
  2. Cerebellar Lesion
     look for tremors.
  3. Frontal Lobe of Cerebrum

- Postural Reaction Deficits on the SAME side as Head Tilt

  1. Pontomedullary Lesion
     look for deficits in Cranial Nerves V - XII
  2. Midbrain Lesion
     look for deficits in Cranial Nerves III - IV

- Postural Reaction Deficits on OPPOSITE Side as Head Tilt.

  1. Midbrain Lesion
     look for deficits in Cranial Nerves III - IV
  2. Thalamus/ Hypothalamus
     look for endocrine disturbance, visual deficits
  3. Cerebrum
     look for seizures

AUTOLOGOUS PERIPHERAL BLOOD STEM CELL TRANSPLANTATION FOR THE TREATMENT OF CANINE LYMPHOMA

Steven Suter, VMD, MS, PhD, DACVIM (Oncology)

INTRODUCTION

For close to 5 decades, the domestic dog (*Canis familiaris*) has served as a preclinical model for hematopoietic cell transplantation (HCT).[1] Based on limited knowledge gained in early, uniformly unsuccessful, human clinical trials and later in murine studies, dogs were identified as a suitable species for study based on their large body size, longer life span, genetic diversity, random-bred nature, ease of handling, and relatively economical colony maintenance costs.[2] For these reasons, there is a tremendous body of literature documenting the utility of dogs for studies of HCT, total body irradiation (TBI), myeloablation, cell dose requirements, graft manipulation, engraftment, and GVHD. Therefore, the vast majority of clinical HCT problems encountered in people were solved using dogs. It is estimated that >95% of human transplant procedures were first perfected using canines.[1]

Bone marrow transplantation was revolutionized by a number of important discoveries in both humans and dogs dating back to the 1990s. First, the cloning and large-scale production of recombinant hematopoietic cytokines, including granulocyte colony-stimulating factor (rhG-CSF, recombinant human G-CSF and rcG-CSF, recombinant canine G-CSF), stem cell factor (rhSCF and rcSCF), and granulocyte macrophage colony-stimulating factor (rhGM-CSF and rcGM-CSF), that could increase circulating hematopoietic progenitors and support their *in vivo* growth.[3,4] Second the finding that peripheral blood contains small numbers of hematopoietic progenitor cells and that recombinant cytokines dramatically increase these numbers via recruitment from the bone marrow.[5,6] Third, the production and characterization of anti-CD34 monoclonal antibodies directed against human[7], mouse[8], primate[9], and canine mononuclear cells[10] that allowed the isolation of a population of cells that can reconstitute the peripheral blood cell lineages after lethal marrow ablation. And finally, the development of sophisticated continuous-flow blood separators that are able to remove and separate blood components via centrifugation on the basis of specific cell densities.[11,12]

We recently obtained 3 Baxter-Fenwal CS-3000 *Plus* with Access Management System continuous blood separators (Fenwal, Baxter Healthcare Corporation, Deerfield, IL), which were kindly donated by the Mayo Clinic, Rochester, MN. These machines were initially tested using normal research dogs to determine if an ample number of CD34+ peripheral blood progenitor cells could be harvested to ensure complete hematopoietic reconstitution following lethal total body irradiation. Based on these initial successful studies, North Carolina State University Veterinary Teaching Hospital began treating client-owned dogs diagnosed with B and T cell lymphoma using peripheral blood stem cell transplantation (PBSCT).

INITIAL STUDIES

Apheresis Procedure

Eight normal dogs (2 unprimed and 6 primed with rhG-CSF [(Neupogen, Amgen, USA)] initially underwent the apheresis procedure to collect peripheral blood CD34+ progenitor cells via collection of peripheral blood mononuclear cells (PBMN).[13] For jugular catheter placement, sedation was instituted using midazolam (0.2-0.4 mg/kg IV), butorphanol (0.2-0.4 mg/kg IV), and propofol (1-2 mg/kg IV as needed). The dogs were intubated and placed on servoflurane. A 12-gauge, 20-cm central venous dual-lumen catheter (Arrow International, Reading, PA) was aseptically placed in a jugular vein for venous access and return. Three additional 18-gauge 2.9 cm catheters were placed in a cephalic vein and both saphenous veins for 1) medetomidine HCl (Dormitor, Pfizer, NY) constant rate infusion (CRI), 2) 10% calcium gluconate CRI, and 3) serum ionized Ca (iCA) monitoring. Once the apheresis procedure was under way for ~1/2 hr, the servoflurane was discontinued and the dogs were extubated when appropriate mentation was noted. For the rest of the procedure, a light plane of anesthesia was maintained using a medetomidine HCl CRI (0.2-0.5 ug/kg/min) with additional boluses of butorphanol or Propofol if needed. PBMC apheresis was performed using the Baxter-Fenwal CS-3000 Plus or (Caridian BCT in some cases) cell separators with the small volume separation container holder (SVSCH) and small volume collection chamber (SVCC) and a disposable double needle open system apheresis tubing kit (Baxter-Fenwal, Access Open System Apheresis Kit, #4R2224) as suggested by the manufacturer for patients with body weights of 50 kg or less (Baxter Fenwal CS-3000 Mononuclear Cell Collection, CLE005 rev.2-1/1/01). Utilization of the SVSCH and SVCC can negate the need for blood priming in small patients since the extracorporeal volume of these containers and the tubing kit is ~ 240 mls. The centrifuge speed was set at 1600 rpm, while the machine inlet flow rate was 1500 mls/hr. Acid-citrate-dextrose formula A (ACD-A, Baxter-Fenwal, Deerfield, IL) anticoagulant infusion was overseen by the Access Management System which infuses anticoagulant at a WBC:ACD-A ratio of 10:1. The disposable tubing kit and collection chambers were primed with ACD-A and 0.9% saline. During the apheresis procedure, the dogs were monitored using continuous electrocardiogram, indirect blood pressure measurement, and body temperature measurement. Pulse and respiratory rate were also continuously monitored. Serial blood iCa, sodium (Na), potassium (K), pH, oxygen partial pressure (PO2), oxygen saturation (SO2) and glucose concentrations were evaluated every ½ hr using a handheld blood analyzer (iSTAT, Abbott Point of Care, Inc., Princeton, NJ). Hot water and air heating pads were used to maintain body temperature in the physiological range, while a CRI of 10% Ca gluconate solution at 20 mls/hr was used to control citrate-induced hypocalcemia. If serum iCa levels dropped by >0.2 mmol/L the infusion rate was increased to 30 mls/hr, with incremental increases of 10 mls/hr if needed. Upon completion, a final iCa was recorded, blood was drawn for a CBC, and the dogs were given Antipamezole HCl (Antisedan, Pfizer) at five times the total medetomidine dose per the manufacturer's instructions as a reversing agent. Once these dogs were awake and able to walk, all their catheters were removed and the dogs were returned to the general research colony. For the first two dogs, the aphersis procedure was continued for only 2 hours. When harvesting the remaining 6 dogs who received Neupogen, the procedure was continued until ~2.5-3 total blood volumes (TBV) were cycled through the machine. Therefore, the total time of the procedure for these dogs varied based on the dog's body weight and the calculation of TBV = BW (kg) x 90mls blood/kg.

Flow Cytometry

Immediately upon completion of the apheresis, 1 ml of the harvested product (of 57 ml total) was used for complete blood count analysis (CBC) using an automated hematology analyzer (Advia 2120, Siemens, Deerfield, IL). An additional 1 ml was assayed for CD34+ progenitor cell content using a commercially available mouse anti-canine CD34 monoclonal antibody.[10] Flow cytometry was performed using a FASCan (Becton-Dickinson, Franklin Lakes, NJ) equipped with a 15-mW argon laser tuned to 488 nm (Spectra-Physics, Montain View,CA). The software for data acquisition was CELLQuest (Becton-Dickinson). An isotype control (Jackson Immunoresearch, West Grove, PA) was used to control for nonspecific antibody staining. The percentage of CD34+ mononuclear cells x the total white blood cell count was used to calculate the total number of harvested CD34+ progenitor cells. This number was then divided by the weight of the dog (kg) to determine the total number of CD34+ cells/kg harvested.

Results

For the first normal dog, who did not receive Neupogen, the apheresis procedure was continued for a total of only 2 hours and CD34+ enumeration was not attempted. Our goal with this dog was to optimize the apheresis procedure in our hands and evaluate the systemic effects of ADC-A infusion and the harvesting procedure itself. This dog tolerated the procedure well with no evidence of hypocalcemia or hypothermia. The second dog who did not receive Neupogen was harvested for a length of time based on its TBV. Although we were able to document mononuclear cell collection, the PBMN CD34+ % of this collection was extremely low (0.35%).

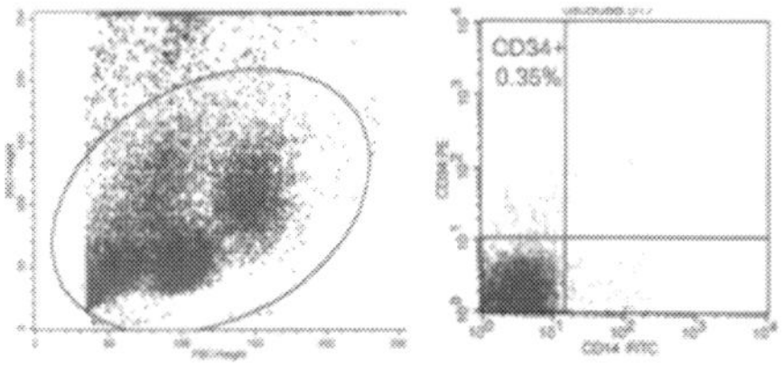

Fig 1. Flow cytometric analysis of apheresis product. The left dot plot represents forward scatter (x-axis) vs side scatter (y-axis) of the harvested product with gating (circle) on the cells that are analyzed for CD34/CD14 content in the right dot plot. The CD34+ percentage was 0.35% (upper left quadrant).

Approximately 3 x 10^5/kg total CD34+ progenitor cells were harvested and based on previous reports documenting complete hematologic recovery of 100% of patients who receive $\geq$2 x 10^6 CD34+ cells/kg[14,15], we concluded this target dose was not achieved.

The following 6 normal dogs received Neupogen for 6 days as previously described at 5 ug/kg SQ BID, with a goal of reaching >30,000/ul total white blood cells.

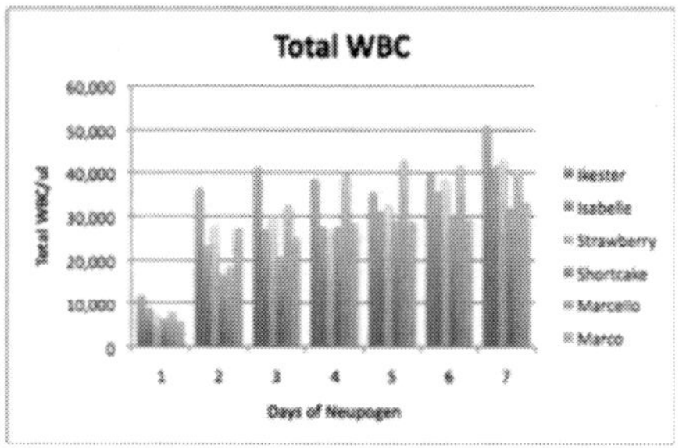

Fig 2. Total white blood cell count of 6 normal dogs receiving Neupogen for 6 days. In all dogs, the target dose of 30,000/ul was reached. All dogs received a double dose of Neupogen (10 ug/kg) 2 hours before the apheresis procedure.

Apheresis of these 6 dogs was performed with the intent of harvesting enough CD34+ progenitor cells to perform a PBSCT, therefore, the length of the procedure was based on the patients body weight, the calculated TBV, and an inlet flow rate of 1500 ml/hr. The average procedure length of these dogs was 4.5 hours, with a range of 3.5 hrs.-5.25 hrs. Fig 3 shows the flow cytometric CD34+ analysis of the harvested product from one of these dogs.

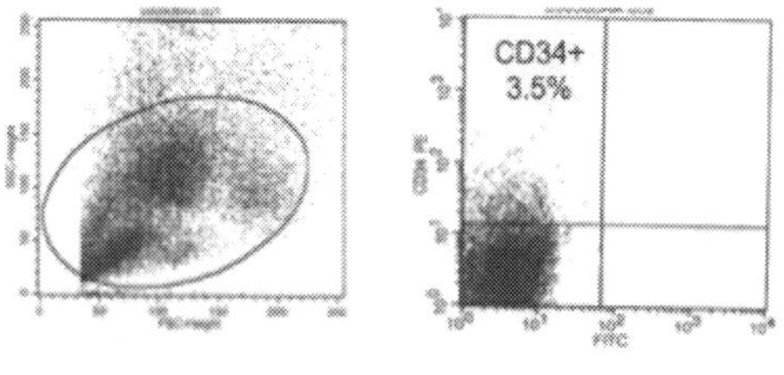

Fig 3. CD34+ enumeration of a Neupogen primed normal dog apheresis product. The CD34+ % of this product was 3.5% (upper left hand quadrant). Based on the total number of cells harvested in 50 mls of product, ~1.94 x 10^7 CD34+ cells/kg were harvested.

In all 6 dogs receiving Neupogen, the target dose of 2 x 10^6/kg CD34+ cells/kg was easily achieved, with a range of 7.8 x 10^6/kg to 3.3 x 10^8/kg. In addition, all dogs tolerated the procedure extremely well, with no evidence of any clinical abnormalities with appropriate monitoring and support.

CLIENT-OWNED B-CELL LYMPHOMA PATIENTS

Based on our results and previous reports documenting the utility of this procedure when treating dogs with hematologic malignancies[16,17], we began treating client-owned patients diagnosed with B and T-cell lymphoma with PBSCT beginning in November, 2008. All patients were diagnosed by their referring veterinarians and received standard multi-agent systemic CHOP-based chemotherapy (cyclophosphamide, adriamycin, vincristine, prednisone) and were in complete clinical remission before PBSCT. All patients received high-dose cyclophosphamide (500mg/m^2 IV) in an effort to clear their blood and bone marrow of all cancer cells 2 weeks before starting Neupogen for 6 days.[17] Blood and bone marrow PARR (PCR for antigen receptor rearrangements)[18] was performed upon presentation to NCSU and all samples were negative using B cell primers. In all dogs, the target dose of >30,000 WBC/ul after 6 days of Neupogen was achieved (6 of these dogs are seen in Fig. 4).

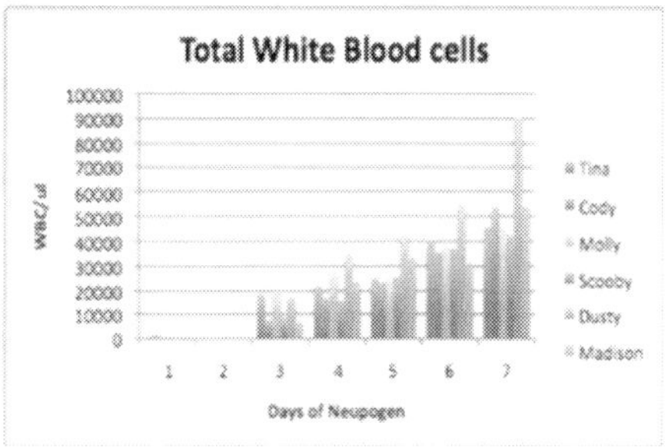

Fig 4. Total white blood cell count of 6 dogs in complete remission for B cell lymphoma receiving Neupogen for 6 days. Day 7 is the total white blood cell count after the apheresis procedure (~ 7 hrs after the double dose of Neupogen).

All dogs underwent peripheral blood mononuclear cell apheresis as described above with no complications noted. Below is the flow cytometric analysis for CD34+ content (3.0%) from one of these dogs.

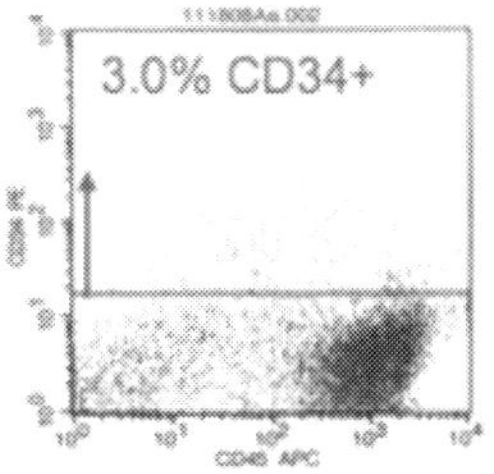

Fig 5. CD34+ enumeration of apheresis product using flow cytometry. A total of 4.726 x 10^9 total cells were harvested. Based on 3% CD34+ cells, a total of 3.9 x 10^6 CD34+ cells/kg were harvested, well above the target of 2 x 10^6 CD34+ cells/kg.

After CD34+ cell enumeration, the apheresis products were refrigerated overnight. The following day, all patients underwent 10 Gy total body irradiation (TBI) using a previously described radiation protocol.[17] The harvested cells were then returned to the patient intravenously immediately after TBI. The patients remained in the hospital for 2 weeks post-TBI for supportive care relating to TBI toxicities, which included variable gastrointestinal signs and severe cytopenias. All patients showed significant hematologic recovery during this time (Fig 6).

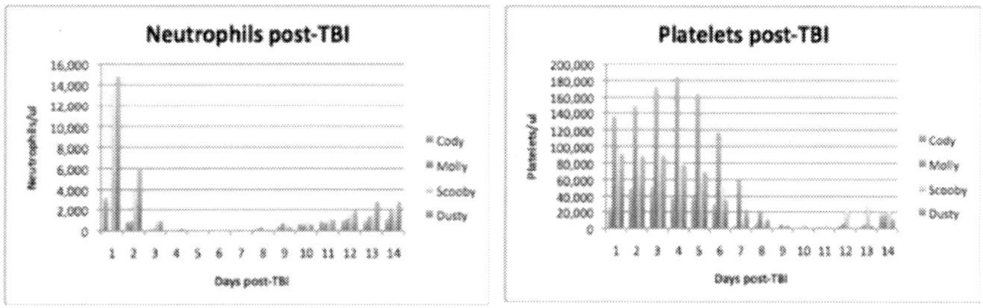

Fig 6. Post-TBI/PBSCT hematologic recovery in 4 dogs. The neutrophil nadir in most dogs occurred ~day +4-7, while the platelet nadir occurred ~day +9-11. All dogs cytopenias either normalized (neutrophils) or increased significantly (platelets) within 14 days post-TBI.

A total of 32 dogs with lymphoma (25 B, 4 T, 3 unknown phenotype) have undergone PBSCT using mononuclear cells harvested using the Baxter-Fenwal CS-3000 *Plus* cell separator. One dog, with only one kidney that was bordering on insufficiency prior to the procedure, went into acute renal failure and died in the hospital 9 days after TBI. Another dog, which engrafted appropriately while in the hospital, died of unknown causes approximately 3 weeks after the procedure. On necropsy, this dog had no evidence of lymphoma, but had systemic pathology suggestive of DIC. One dog who was ~9.5 months post PBSCT, developed inoperable liver cancer and was euthanized with no evidence of lymphoma on necropsy. 11 dogs relapsed with disease ~3.5 months after the procedure and were eventually euthanized. 3 other dogs relapsed with disease >16 months after the procedure and were eventually euthanized. The 15 remaining dogs (47%), none of which are receiving systemic chemotherapy, are from 5 months to 27 months post-PBSCT with no evidence of disease and clinically normal at home.

These initial results support the notion that canine patients with lymphoma can undergo PBSCT in a clinical setting and become long-term survivors. Although an ~30% cure rate of canine patients has been reported previously[16], these transplants were performed in a research setting approximately 25 years ago. Therefore, more patients need to be transplanted and more time needs to elapse before we can begin to determine the true cure rate (surviving >2 yrs) of lymphoma using this technique.

REFERENCES

[1]Lupu M, et al. **Vet Comp Oncol** 2007;5(1):14. [2]Ostrander EA, et al. **Genome Res** 2005;15:1706. [3]de Revel T, et al. **Blood** 1994;83:3795. [4]Sandmaier BM, et al. **Blood** 1996;87:3508. [5]Appelbaum FR. **Exp Hematol** 1979;7:7. [6]Storb R et al. **Blood** 1967;30:805. [7]Ratajczak MZ, et al. **Leukemia** 1998;12:942. [8]Morel F, et al. **Blood** 1996;88:3774. [9]Andrews RG, et al. **Blood** 1992;80:1693. [10]McSweeney PA, et al. **Blood** 1998;91:1977. [11]Buckner D, et al. **Blood** 1969;31:653. [12]Herbst EW, et al. **Blut** [German] 1975;30:265. [13]Storb R, et al. **J Vet Intern Med** 2008;22:74. [14]Singhal S, et al. **Bone Marrow Transplant** 2000;26:489. [15]Ketterer N, et al. **Blood** 1998;91:3148. [16]Appelbaum FR, et al. **Transplantation** 1986;41(1):19. [17]Lupu M, et al. **J Amer Vet Med Assoc** 2006;228:728. [18]Burnett RC, et al. **Vet Pathol** 2003;40:32.

OVERVIEW OF THE DIAGNOSIS & TREATMENT OF CANINE LYMPHOMA, MCT & OSTEOSARCOMA

Steven Suter, VMD, MS, PhD, DACVIM (Oncology)

CANINE LYMPHOSARCOMA (LSA)

The reported incidence of canine LSA is 13-24/100,000, or 7-24% of all canine tumors. Breeds reportedly most often affected include boxers, basset hounds, St. Bernards, Scottish terriers, airedales, bulldogs, although all breeds of dogs are susceptible[1-3]. There are no known proven etiologies of canine LSA, although weak links have been established with viral infections[4] (canine retrovirus hasn't been confirmed), pesticide exposure (2,4-D—associated with increased risk in people, one study found increased risk in dogs)[4], and possibly impaired immune function[6].

The classification of canine LSA can be distinguished on the basis of anatomic location, histologic criteria, and immunophenotypic characteristics. Approximately 80% of dogs with LSA present with multicentric disease, manifested via peripheral lymphadenopathy[7]. Most dogs are asymptomatic at presentation, although 20%-40% have a history of weight loss, lethargy, anorexia, and febrile episodes. Also, ~27%-34% of dogs have pulmonary infiltrates[8]. The alimentary form of LSA occurs in ~5% of patients. GI LSA can occur either multifocally or diffusely in the gut, with patients showing weight loss, anorexia, and evidence of malabsorption. Mediastinal LSA occurs in ~5% of patients. This form is characterized by enlargement of the craniomediastinal lymph nodes, thymus, or both. In one study, 43% of dogs with mediastinal LSA had concurrent hypercalcemia[9]. Cutaneous LSA can be either generalized or solitary. This form of LSA is generally divided into 2 large groups: cutaneous B or T cell LSA and mycosis fungoides (MF). MF is a specific type of cutaneous LSA where neoplastic T-lymphocytes invade into the epithelium of the skin, leading to the characteristic Pautrier's microabscesses. There is also a specific type of LSA called hepatosplenic LSA. These dogs do not have any outward evidence of disease, but instead have an expansion of a specific type of T-cell, called a gammadelta T-cell. These cells tend to invade the sinusoids of the liver and spleen and eventually the bone marrow. Finally, LSA can be found in virtually any organ in the body, with and without peripheral lymphadenopathy or invasion into other organs at the same time. So, we see ocular LSA, CNS LSA, ear pinnae LSA, oral mucosal LSA, nasal LSA, etc. on a fairly regular basis. This underscores the need to always have LSA on the differential list for just about any physical exam finding you can think of. It can mimic just about anything (just like mast cell tumors!).

The clinical signs associated with canine LSA are extremely variable depending on site of disease, extent of organ involvement, and disease progression. Dogs can present with huge peripheral lymph nodes and be completely normal clinically, and other dogs present with almost no outward evidence of disease and are quite ill. LSA in its later stages can also be a life-threatening medical emergency. We treat all dogs with LSA as emergencies at our hospital, since the disease can progress so rapidly.

Clinical staging should include complete history, physical exam, complete blood count, serum chemistry, urinalysis, thoracic and abdominal radiographs or ultrasound, bone marrow aspiration and lymph node biopsy or cytology. Bone marrow aspiration is always indicated in LSA patients, even if no cytopenias are found since ~50% of dogs with LSA have marrow infiltration and no evidence of cytopenias. There are a number of classification systems applied to canine LSA, some more helpful than others. Probably the most used and helpful is the WHO staging classification which is used to describe the extent of disease in each patient. It involves 5 Stages and 2 substages; Stage I: single lymph node, Stage II: Regional lymph nodes, limited to one side of the diaphragm, Stage III: Generalized lymphadenopathy, Stage IV: Liver and/or spleen involvement (+/- Stage III), Stage V: Blood and/or bone marrow involvement (+/- Stage III/IV) and any other site (ocular, CNS). Each stage is further subdivided into; a – without systemic signs, b – with clinical signs.

LSA diagnosis is usually fairly straightforward. On physical exam one usually can confirm peripheral lymphadenopathy, hepatomegaly and splenomegaly. Other findings suggestive of LSA include dyspnea/tachypnea with reduced cardiac sounds (mediastinal mass), ocular infiltrates or hyphema, chronic diarrhea or vomiting. (Differential diagnoses for lymphadenopathy: systemic fungal, infectious, inflammatory and immune-mediated disorders; other neoplasia). Laboratory findings can include: hemogram findings may include nonregenerative anemia, thrombocytopenia and lymphocytosis, although many dogs with LSA will have a normal hemogram. Abnormalities seen on the CBC should not be automatically attributed to the lymphosarcoma, but evaluated independently (i.e. thrombocytopenia). Biochemical panel is usually normal. The most common cause of hypercalcemia is LSA. Renal disease or liver abnormalities are not uncommon in dogs with LSA. Histologic or cytologic diagnosis: LSA is relatively easy to diagnose cytologically in the majority of instances. PCR for antigen receptor rearrangements (PARR) [10]can be used for difficult to diagnose cases. Histologic grading (intermediate or high grade versus low grade) may also be useful. Imaging of the thorax and abdomen: identification of mediastinal mass, sternal lymph nodes and abdominal involvement aids in clinical staging and in identifying those dogs in need of close monitoring during the post-treatment period.

LSA therapy consists mainly of systemic chemotherapy, with combination chemotherapy being generally more effective than single agent chemotherapy. The best single agent for LSA is doxorubicin. Most protocols include tapering doses of prednisone and a combination of L-asparaginase, vincristine, cytoxan, and doxorubicin. Prednisone is an excellent anti-LSA drug, but only last ~2-3 months, with resultant severe resistance to all chemotherapies afterwards. <u>In general</u>, the more drugs added into the protocol, the longer the patient will survive. Many institutions now incorporate ½ body irradiation at various points in the chemotherapy protocol, which seems to show some improvement over chemotherapy alone. Autologous peripheral blood stem cell transplantation is currently the only protocol that may completely cure some patients with LSA. About 80%-90% of LSA patients will attain a complete remission once starting chemotherapy. For dogs with B-cell LSA, the median first remission time is 6-8 months, with a median overall survival of 10-14 months. These numbers are about ½ for dogs with T-cell LSA.

Important prognostic factors (some are highly debated and sometimes controversial) include Stage and Substage, immunophenotype (B vs T), previous treatment with steroids, and 1st response to induction chemotherapy. Others include hypercalcemia, gender: loose correlation with better prognosis for females, intrathoracic involvement: negative effect, histologic grade: high grade respond better to chemotherapy but shorter overall survival time compared with low grade, anatomic location: diffuse cutaneous, alimentary associated with poorer prognosis, size: larger dogs may have decreased remission times (dosing effect?).

CANINE MAST CELL DISEASE (MCT)

Cutaneous canine MCT represents ~16-21% of all cutaneous tumors in dogs[11,12] and is known as the "the great imitator" since it can look like virtually anything. Both dermal & SQ MCT are the most common types and they are usually found in older dogs. The etiology of canine MCT remains virtually unknown. Boxers, Boston terriers, golden retrievers, Labrador retrievers, beagles, and schnauzers all have been reported to be predisposed to MCT.

Canine MCT is generally divided into 3 Grades (I, II, and III), with Grade I tumors generally acting biologically benign. Grade III tumors are considered highly metastatic and generally strike fear into the hearts of all oncologists. The vast majority of canine MCT are, unfortunately, Grade II tumors, which are essentially biologically unpredictable. Some oncologists feel that MCT location can also help predict the biological aggressiveness of these tumors[13,14]. Muzzle, preputial/inguinal, nailbed, and mucocutaneous tumors are all anecdotally reported to be more aggressive. Visceral MCT (also referred to as disseminated or systemic mastocytosis), where abdominal organ invasion occurs, have a much worse prognosis. These patients can have bone marrow and/or peripheral blood involvement, pleural or peritoneal effusions, and invasion into a variety of other sites. Solitary tumors tend to metastasize to regional lymph nodes first before spreading elsewhere.

Clinical signs of canine MCT are highly variable. Most tumors are solitary, but 11%-14% of patients have multiple lesions. Most more benign lesions tend to be solitary, 1-4cm in diameter, slow growing, and present for >6 months. Most more aggressive tumors tend to be rapidly growing ulcerated lesions that are larger and pruritic. The surrounding tissue may be inflamed and edematous, and small satellite lesions may also be seen. Gastrointestinal ulceration, secondary to MCT histamine and other vasoactive amine release, can lead to vomiting, anorexia, melena, and abdominal pain. The vast majority of dogs with cutaneous MCT are completely asymptomatic at initial presentation. A WHO staging system for canine MCT has been developed, with Stage 0 being only one tumor (with or without clinical signs) and Stage IV being any tumor with distant metastasis, including blood or bone marrow.

Clinical staging of canine MCT includes a complete blood count, serum chemistry, urinalysis, abdominal ultrasound, thoracic radiographs, +/- buffy coat preps if Stage IV is suspected, and +/- CT scan if surgery is indicated. MCT are generally very easy to diagnose with an FNA of the mass in question, although a biopsy is needed to determine Grade. There are a number of immunostains that are now available through some academic institutions and private labs, including Ki-67, AgNors, and c-kit. Although beyond the scope of this talk, they may, at times be helpful in determining the aggressiveness of a MCT (I find them mainly confusing). A good paper by Romansik, et al[15] found that mitotic index was a good indicator of biologic aggressiveness (≥ 5 mitoses/10 hpf was found to more aggressive, no matter what the Grade). A newer test, called c-kit exon 11 internal tandem duplication (ITD) analysis is also available around the country. Tumors with an exon 11 ITD have an ~50% better response rate to a new class of chemotherapeutics called receptor tyrosine kinase (RTK) inhibitors[16], which include Palladia and Kinavet.

Various treatments are used to treat canine MCT. Due to the metastatic potential of these tumors, treatment usually consists of surgery +/- irradiation to gain local control of the tumor and chemotherapy if metastasis is either confirmed or suspected. Local control usually starts with surgery to either remove the tumor in its entirety with adequate (~2cm) margins or cytoreducing the tumor to a microscopic setting. Definitive irradiation (XRT) would then be used to kill the residual disease. This is a very common route for tumors on the distal limbs, muzzle, etc., where there is not a lot of tissue to resect. If enough tissue is available, then re-cutting a scar is an excellent option to gain local control of incompletely excised MCT. Chemotherapy historically has consisted of steroids, vinblastine (Velban), and CCNU (Lomustine), but some other chemotherapeutics such as chlorambucil (Leukeran), doxorubicin, and cyclophosphamide (Cytoxan) can be tried. And, as mentioned earlier, we now have two RTK inhibitors, Palladia[17] and Kinavet[16], for use also. All dogs with a visible MCT tumor should be started on Benedryl and PepcidAC (or similar drugs) to decrease the chances of GI ulceration.

Some reported prognostic indicators include: histologic grade-1 best, 3 worst, Stage-increasing stage worse, location-visceral/BM grave; caudal/preputial oral possibly worse, age-older dogs did worse with XRT, breed-boxers better prognosis, sex-males did worse with chemotherapy, tumor recurrence-more guarded prognosis, growth rate-faster is worse, clinical signs-assoc. with more aggressive tumors.

Reported outcomes with various treatment modalities are: sx only: 83% grade I, 44% grade II, 6% grade III alive at 4yrs post-sx; 100% grade I, 44% grade II, 7% grade III alive at 2 yrs post-sx; sx alone: 5% (grade I), 11% (grade II) local recurrence. Chemotherapy response: Pred: 75% response: remission~ 2.5- 5 months, CCNU: 42% response : remission ~3 months, Velban/pred: 47% response, Vinc: 7% response, Velban/pred/cytoxan: 18 months MS. Any dog that develops multiple tumors over time, no matter what the grade, is a candidate for systemic chemotherapy. XRT alone: For measurable tumors treated with 40-45 Gy-50% 1-year control rate. Sx + XRT: Best option for distal limb/head/perineal tumors where adequate margins can not be achieved: 85%-90% 2-year control rate for grade I and II tumors

The prognosis for dogs with grade III cutaneous and visceral MCT remains poor, no matter what treatment option is used. In one study in dogs with GI MCT: only 40% were alive 30 days after first admission and fewer than 10% were alive at 6 months[18].

CANINE OSTEOSARCOMA (OSA)

Canine OSA represents ~85% of all primary bone tumors and 3%-6% or all canine neoplasias, with an incidence of 8-10,000 cases yearly in the US[19,20]. Large breeds, such as great Danes, saint bernards, Irish wolfhounds, and Rottweilers are commonly affected. In small breed dogs, OSA accounts for 45% or all skeletal neoplasias, with the axial skeleton (flat bones) more commonly affected[21]. In large breed dogs the distal radius is the #1 affected site, with the proximal humerus and distal femur/tibia the next most common sites. The front limbs are affected about twice as much as the hind limbs. There are a variety of types of canine OSA, including endosteal, periosteal, parosteal, multilobular (MLO), and extraskeletal (mammary gland and OSA of soft tissues). The vast majority are endosteal (origination inside the bone) OSAs.

The etiology of some canine OSAs are known: they can be caused by foreign bodies (orthopedic implants), bone infarcts, chronic osteomyelitis, bone trauma, and occur at sites of previous radiation exposure. Other weaker links include viruses and chemicals. Breed is the most documented etiology, with large/giant breed dogs having 60X the risk of developing OSA compared to small breeds. In fact, 90% of OSA occurs in dogs weighing >35 lbs.

OSA is, unfortunately, considered metastatic at diagnosis.

The tumor locally is typically very aggressive, with extensive lysis, bone invasion and destruction. This tumor just plain old hurts (unlike most other tumors). The tumor spreads mainly through the blood (hematogenous spread) and most commonly affects the lungs, other bones, and abdominal organs.

Clinical signs of OSA include obviously lameness of the affected leg, swelling at the tumor site, muscle atrophy of the affected limb due to disuse, anorexia and lethargy, and possible respiratory signs due to metastatic disease. Some dogs present with pathologic fractures of the affected bone. OSA of sites other than the limbs, such as oral, spinal, rib, etc., can lead to other clinical sings such as dysphagia, tooth loss, facial deformity, exophthalmos, and neurologic signs.

Staging of canine OSA incudes: complete blood count, serum chemistry, urinalysis, abdominal ultrasound, thoracic radiographs, affected limb radiographs, +/- technetium scanning. The tumor itself can be diagnosed via large bore needle FNAs (+/- ultrasound guidance) or bone biopsies.

Treatment of canine OSA typically consists of amputation of the affected limb followed by systemic chemotherapy. But, other options exist. Palliative XRT for limb OSAs can be used to alleviate pain if amputation is not an option, although no life extension benefit exists with this treatment. There are also now a wide variety of NSAIDs available: Rimadyl, Piroxicam, Etogesic, Previcox, Deramaxx, etc. Other drugs such as gabapentin and amantidine can be added on top of NSAIDs to augment their efficacy. Some veterinary institutions also offer a variety of limb-sparing surgeries that remove the affected part of the limb and replace it with either a piece of donor bone or a piece of patient bone. Again, this treatment will preserve leg function, but not lead to life extension. A class of drugs called bisphosphanates (pamidronate, etc.) can also be used to alleviate pain in ~20%-25% of patients. When examining palliative treatments, the median survival of OSA dogs is ~2-4 months, with an 11%-20% 1 yr survival and a 0%-2% 2 yr survival. ~90% of dogs will die from metastatic disease if amputation is the only treatment. Systemic chemotherapy clearly leads to life extension, although rarely cures a dog. Amputation + Cisplatin increases 1 yr survival to 38%-43% and 2 yr survival to 16%-18%, with a median survival of 162-282 days. Amputation + Carboplatin has a reported 1 yr survival of 35.4%, with a median survival of 321 days. Amputation + doxorubicin leads to a reported 48% 1 yr survival and a 28% 2 yr survival, with a median survival of 345 days.

REFERENCES

1. Kaiser H. Animal Neoplasms: A systemic review., p. 747-812. Baltimore: Willliam and Wilkins, 1981. 2. Moulton J. Tumors of lymphoid and hematopoietic tissue., p. 231-307. Berkeley: University of California Press, 1990. 3. Dorn CR, Taylor DO, Schneider R. The epidemiology of canine leukemia and lymphoma. *Bibl Haematol*, 403-15, 1970. 4. Chapman AL, Bopp WJ, Brightwell AS, et al. Cancer Res 27:18-25, 1967. 5. Reynolds PM, Reif JS, Ramsdell HS. Cancer Epidemiol Biomarkers Prev 3:233-237, 1994. 6. Keller ET Cancer 70:2334-2337, 1992. 7. Madewell BR, Thesen GH In Theilen GH, Madewell BR, editors: Veterinary Cancer Medicine, ed 2, Philadelphia, 1987, Lea & Febiger. 8. Starrack GS, Berry CR, Page RL, et al. Vet Radiol Ultrasound 38:411-418, 1997. 9. Rosenberg MP, Matus RE, Patnaik AK. J Vet Int Med 5:268-271, 1991. 10. Vernau W, Moore PF. Vet Immunol Immunopathol 69:145-164, 1999. 11. Bostock, DE. Br Vet J 142:1-19, 1986. 12. Brodey RS. Adv Vet Sci Comp Med 14:309-354, 1970. 13. Moriello KA, Rosenthal RC. Vet Clin N Am Small Anim Prac 20:1163-1190, 1990. 14. Sfiligoi G, Rassnick KM, Scarlett JM, et al. J Am Vet Med Assoc 226;1368-1374, 2005. 15. Romansik EM, Reilly CM, Kass PH, et al. Vet Pathol 44:335-341, 2007. 16. London CA. Top Companion Anim Med 24:106-112, 2009. 17. London CA, Malpas PB, Wood-Follis SL, et al. Clin Cancer Res 15:3856-3865. 18. Osaku K, Yamagami T, Nomura K, et al. Vet Pathol 39:557-564, 2002. 19. Brodey RS, Sauer RM, Medway Al, et al. J Am Vet Med Assoc 143:471-495, 1963. 20. Priester WA, ed: NCI Monogr No. 54, November 1980. 21. Kistler KR. Phi Zeta Awards, University of Pennsylvania, School of Veterinary Medicine, 1981.

CYTOLOGY DO'S AND DON'TS

Rance K. Sellon, DVM, PhD, DACVIM
Pullman, WA

INTRODUCTION

Cytologic examination of masses and fluid samples is an important diagnostic tool in small animal practice, and there are technical aspects of sample collection and preparation that can hinder, or improve, the likelihood of diagnostic yield. There are also some patients for which sample collection for cytologic examination should be considered carefully, and some patients that are candidates for sample collection that would not be intuitive. This presentation will address some of the things that can or should be done with cytology, and some things in the realm of cytology to avoid.

PATIENT SELECTION

Cytology can be performed with little consequence in the majority of patients with visible or palpable lumps, lymph nodes and masses, and arguably should be performed before any lump or mass is surgically removed. Failure to obtain information that helps categorize a tumor type before surgical resection can lead to a first surgical intervention that does not achieve margins adequate for the tumor type (e.g. mast cell tumors, soft-tissue sarcomas), and has the potential to complicate attempts at a second surgery, or provide adjunctive therapy such as radiation therapy. In addition to externally-evident lesions, intra-abdominal and intra-thoracic lesions can be candidates for cytologic examination. Some lesions may require ultrasound or other imaging techniques for accurate sampling, making some of these patients candidates for referral to hospitals with the necessary imaging equipment.[1]

Primary bone lesions are also surprisingly amenable to cytologic examination. In a couple of different studies, the diagnostic accuracy of needle cytology of primary bone tumors was very sensitive for the documentation of neoplastic cells in radiographically apparent bone lesions if a diagnostic sample (i.e. sufficient numbers of cells for examination) was obtained.[2,3] At WSU, we have done bone aspirates either blindly using radiographic and physical examination landmarks, or using ultrasound-guidance[3] to find defects in the cortex through which the needle can be guided. Both techniques have proved useful for confirming the presence of a neoplasm, and with the application of special stains such as alkaline phosphatase[4] to slides of bone lesion aspirates, distinction of primary sarcomas (osteosarcoma, chondrosarcoma) from other bone tumors (e.g. histiocytic sarcoma, metastatic carcinoma) is possible.

There are advantages to ultrasound-guided aspirates of masses or lymph nodes within the abdominal cavity, but it is the author's opinion that not all masses palpable within the abdominal cavity have to have ultrasound-guidance for aspiration. If the intra-abdominal mass can be isolated and immobilized for a needle puncture, ultrasound-guidance may not be necessary. The author has successfully aspirated enlarged mesenteric lymph nodes, kidneys, intestinal and hepatic masses without ultrasound guidance. On the other hand, the author has used ultrasound-guided aspiration for large, external masses that proved difficult to obtain material sufficient for examination, using ultrasound to guide a needle into clearly abnormal areas. Ultrasound has also proven useful for easy and relatively inexpensive examination of enlargements in limbs that seemed deep to the limb muscles with improved ability to identify the lesion within these structures for subsequent aspiration. Similarly, masses within the axilla or inguinal area may have safe and accurate aspirations obtained using ultrasound guidance.[1]

If patient has enlargement of peripheral lymph nodes, it can be very helpful to a cytologist to review aspirates from multiple nodes. Submandibular lymph nodes are notorious for enlarging as a consequence of dental or oral cavity disease. In some cases confidence in a diagnosis of lymphoma may be increased by finding abnormal lymphocyte populations in lymph nodes at other anatomic locations. Thus, whenever possible, submit samples obtained from enlarged nodes other than the submandibular nodes.

There are some clinical situations in which aspiration cytology should be carefully considered before proceeding. Aspiration of bladder masses has been associated with seeding of the needle path with neoplastic cells, making bladder masses off-limits for aspiration cytology in the hands of many oncologists. Samples of bladder masses suitable for cytologic examination can be acquired by other means, traumatic catheterization for example. The author has often used ultrasound to guide the side-hole of a red rubber catheter into the region of a bladder mass, then applied suction to the catheter to acquire bits of the mass for cytology, or in some cases, fragments of tissue large enough to submit in formalin for histopathologic examination. Another situation in which the oncology service at WSU is reluctant to perform needle cytology is the patient with a solitary lung mass. Our concern is the same for a bladder mass, namely seeding of the needle path. Some patients with solitary lung tumors have a chance for an excellent long-term prognosis with lung lobectomy, and our view is that we would hate to see that success diminished by neoplastic seeding of a needle path. Of course, do consider the circumstances for each patient, and each owner, when deciding to obtain samples or not in such cases.

Most patients will not need any sedation or local anesthesia to accomplish a needle aspiration, and there is usually little need to put any type of skin prep (e.g. clip, scrub) unless needed to better visualize the mass and needle path. Patient comfort can be enhanced during the aspiration process by passing the needle through the skin once and re-directing the needle without coming out of the skin.

SAMPLE HANDLING AND SLIDE PREPARATION

After selecting the appropriate patient for collection of specimens for cytologic examination, preparation of the slides and making smears is another crucial step important for optimizing the diagnostic utility of the sample. Slides with material that is too thick can hinder recognition of a population of cells of interest, or preclude accurate evaluation of cell morphology, crucial for distinction of general tumor types. It may be possible to make several slides from one aspiration if there is a lot of material that has been blown onto the slide, increasing the likelihood that one or more slides will be suitable for examination.

Another preparation pitfall to avoid is found in the making of smears from lymphoid organs (lymph nodes, spleen, occasionally liver) of patients for which lymphoma is a likely differential diagnosis. Neoplastic lymphocytes are quite susceptible to rupture if treated too harshly, making pull-apart techniques of slide preparation less desirable for preservation of lymphocyte morphology. The author has found that making smears of aspirates of lymphoid organs using the same technique that would be used to make a blood smear (dragging the edge of a slide into the material that has been blown onto the slide, then pushing the slide away) is much better at both making a smear thin enough to be examined while preserving enough cells for accurate assessment of cell morphology. The same technique can be used to prepare aspirates of liver and spleen, and if multiple aspirates can be acquired, some slides can be made with a blood smear technique, and some can be made with a pull-apart technique. If fluid samples (e.g. body cavity effusions, joint fluid, fluid from external masses) are obtained for diagnostic or therapeutic purposes, do prepare direct smears of fluids, and submit those slides with the fluid submitted in tubes as cell deterioration that can occur during shipping may hinder examination.

Bone marrow aspiration is indicated for a variety of problems, and is a routine element of staging lymphoma and high-grade mast cell tumor patients in many practices. One strategy that can be used to diagnostic-quality slides of bone marrow aspirates is to pre-load a syringe with 3% EDTA which is easily prepared by adding 0.5 mL of 0.9% sterile saline to a 10 mL EDTA tube. The entire contents are then aspirated into a 12 mL syringe for withdrawal of 0.5-1 mL of marrow. Some of the aspirated marrow can be expelled from the syringe into a Petri dish or watch glass that allows one to suck up the grey flecks or particulate bits with a pipette to make smears. The author will tilt a glass slide downward at enough of an angle that blood runs toward the end of the slide, typically leaving the particulate marrow spicules stuck further up on the slide. A second slide is placed on top of the particles perpendicular to the first slide, the particles allowed to flatten, and the top slide pulled off to smear out the spicules. Don't be too stingy making slides-the author will often make 8-10 slides from every bone marrow aspiration performed as it is better to submit too many slides than too few. Any marrow in the EDTA tube that is not used to make blood smears can be submitted to the clinical pathology lab with the prepared slides to use to make additional slides if needed. If anticoagulant isn't used, one needs to line up a number of slides at a more vertical angle (prop against a slide box, for example), and put a drop of marrow at the top of each, so that the blood runs down and the marrow sticks to the top. Slides are then made using that top marrow. One has to work extremely quickly to make smears before the blood on the slide clots! If one is uncertain that marrow elements have been obtained, a candidate slide can be quickly stained with a drop of new methylene blue, or Dif-Quik type stain, coverslipped and examined under a microscope. Unless there is aplasia, megakaryocytes are usually easily observed if a marrow has been obtained.

Some patients need bone marrow core biopsies obtained because bone marrow aspiration yields few spicules. Before dropping the biopsy sample into the formalin, roll the biopsy along the length of two slides in a rolling log style (core sandwiched between two slides) to leave cells from the biopsy on the slides. These slides can be submitted for examination, and if not fruitful, the biopsy can be submitted for histologic examination. Cell morphology and proportions of cell types in bone marrow are usually more accurately established from cytology than histopathology, but the two processes can complement each other in some cases.

For any given sample, it is reasonable to stain and examine a slide or two from several that have been prepared to make sure that there is in fact material on the slide. Once there has been confirmation that a suitable sample has been obtained, the stained slide can be submitted with all unstained slides to account for the chance that the best slide happened to be the one that was stained. Although most clinical pathology laboratories would prefer to stain submitted slides themselves, the clinical pathologist may still be able to use a slide stained by the submitting practice to make useful assessments. If the stained slide has been examined with an oil immersion lens, the slide can still be submitted, but it should be put in a plastic slide holder so oil doesn't get on unstained slides. A few other tips for slide submission include making sure that all slides are properly labeled with a patient name, and if slides from multiple sites were obtained, to somehow designate on the slide what tissue is represented on the slide. For example, one might label a slide with an L for liver, and S for spleen, and submit the code with other paperwork. Don't fix any slides that are submitted (no alcohol, heat, or hair spray!), but rather submit air-dried, unstained smears when possible. The clinical pathologists at WSU also discourage blowing on slides to hasten drying as they have had contamination with contents from human oral cavity on such slides. When packing the slides for shipping to the clinical pathology laboratory, keep the slides away from ice packs and formalin jars (fumes alone can affect slides) to prevent artifacts that impair cytologic examination. Slides that are submitted in slide holders are less likely to be broken or having contents come off than slides wrapped in paper towels or other protective material.

It can be useful to submit several slides of each sample site. Examination of a large number of slides to find one or two that have diagnostic material is not unusual in the WSU clinical pathology laboratory. Another advantage to submitting several slides is that there may be indications for special stains that can provide additional information about the tissue of origin of neoplastic cells, characterize an infectious disease agent, or other special assays that can be performed on the extra slides.

Capabilities and limitations of cytology

One of the most common reasons for performing cytology is to determine if a lump, bump, mass or enlargement is due to an inflammatory or neoplastic disease. As such, cytology can be quite sensitive and specific for the detection of neoplastic cells, but cytology can't always determine with certainty the origins of a lesion. Cell morphology may be atypical, there may be too few cells upon which the clinical pathologist can make a confident conclusion, the presence of inflammation can provoke morphologic features consistent with malignancy, or cells may appear cytologically normal yet behave in a neoplastic fashion based on what they do in the tissue. An example of the latter would be the patient with generalized lymph node enlargement secondary to a low-grade lymphoma in which the predominant cell type in a cytology sample would be a small, mature-appearing lymphocyte that could provoke a cytologic diagnosis of lymph node hyperplasia. However, biopsy of an affected node could show that the cells are not respecting anatomic boundaries, or are disorganized as compared to normal tissue, behavior that would be more typical of lymphoma than a hyperplastic response. Several reports note that in the aspiration of masses, false negative results of cytology examination (the cytology did not suggest a neoplastic origin when histopathology indicated that neoplasia was the cause) are more likely than false positives (the slide suggested a diagnosis of neoplasia that was not supported by histopathology). Thus, do consider the cytology report in light of the rest of the clinical context, and don't always take a cytology interpretation as the definitive diagnosis for a given patient.

Another important limitation of cytology is that it can not reliably establish tumor grade. A cytologist will try first to classify a lesion as inflammatory or neoplastic, and if neoplastic, will try to define the tumor according to one of the 3 main tumor types: carcinoma, sarcoma, or round cell tumor. For many tumors in these broad categories, the biologic behavior is better determined by the tumor grade rather than the specific type of tumor, and cytologic examinations can't establish a tumor as low, intermediate or high grade. There may well be cytologic features that suggest a grade, but grade is more properly established by examination of a tissue biopsy sample.

SUMMARY COMMENTS

Cytology is a key diagnostic component of small animal practice, and has wide application for a number of patients. Appropriate patient selection, optimal preparation of samples and submission to a clinical pathology laboratory will increase the chance that the time and effort of sample acquisition pays off in terms of a conclusion in which the clinical staff can have confidence. It can be very helpful, if there are any questions about sample acquisition or submission, to contact the clinical pathology laboratory for advice and suggestions.

ACKNOWLEDGEMENTS

The author would like to acknowledge the very helpful suggestions and contributions of the clinical pathology faculty and residents at the WSU Veterinary Teaching Hospital, Drs. Jane Wardrop, Tamara Wills, Regan Bell and Tracie Romsland.

REFERENCES

[1]da Costa RC, et al. Can Vet J 2008;49(1):77-81. [2]Britt T, et al. J Small Animal Pract 2007; 48:145. [3]Vignoli M, et al. Vet Radiol Ultrasound 2004; 45(2):125. [4]Barger A, et al. Vet Pathol. 2005; 42(2):161-165. [5]Ghisleni G, et al. Vet Clin Pathol. 2006; 35(1):24-30. [6]Cohen M, et al. J Am Vet Med Assoc 2003; 222(7):964-967. [7]Bauer NB, et al. J Vet Intern Med. 2007; 21(5):928-935. [8]Höinghaus R, et al. J Comp Pathol. 2007; 137(1):1-8.

STAGING PATIENTS

Rance K. Sellon, DACVIM
Pullman, Washington, USA

Staging a cancer patient is the process whereby the local and distant extent of a tumor is established. In addition, staging could also be viewed as the collection of clinical information that reflects the general health of the cancer patient. Staging a cancer patient depends on performing diagnostic tests appropriate for the patient's tumor, defined by both tumor type and grade, and appropriate for the patient as a whole. In the process of staging the cancer patient, one hopes to gather enough information to be able to offer appropriate treatment options and reasonable insights to a patient's prognosis for given treatment options. The classic approach to staging a cancer patient is to define its tumor according to the World Health Organization (WHO) TNM scheme, where T defines the primary tumor, usually by size, N reflects the presence or absence of regional node disease, and M is the presence or absence of metastasis to other organs. While specific tumors can have individual criteria for T that helps define a specific clinical stage, the value of the TNM scheme to most clinical staff is that it serves as a guideline to key components of the staging process. Staging the cancer patient can be viewed as a series of steps that ultimately define what treatment, or combinations of treatment, are appropriate for a cancer patient. Staging also can be instrumental in determining that a given patient should not be treated.

For many patients, the staging process begins with definition of the tumor type. While this might seem a straightforward step, the fact is that oncologists regularly see patients whose tumors have an initial histologic diagnosis, yet review of the history, the physical examination, the morphologic description of the tumor (which ideally should be requested as a part of every pathology report), and perhaps results of other diagnostic tests suggest a disconnect between the original histologic diagnosis and the patient. In such instances, conversations with the pathologist may be needed to better clarify the tumor type. Some patients may need special stains applied to their biopsy samples to better define a tumor type. In occasional cases, the tumor type can't be identified by routine and available procedures.

After a patient's tumor has been identified by tumor type, and for many biopsy-diagnosed tumors, graded, there are many tools that can be implemented to stage a cancer patient. Staging starts with a thorough physical examination, noting the size of the tumor, and features such as its mobility, and involvement with or proximity to key surrounding structures. Attention should be paid to palpation of regional lymph nodes. Any other masses should be noted, and new masses, or changing masses, should be targets for aspiration cytology to make sure that one is not dealing with additional tumors; a tumor map can be helpful in keeping track of multiple masses. Rectal examinations should be performed on all dogs as a substantial proportion of anal sac tumors are incidental findings during routine physical examination for other reasons. It is common for the author to perform all elements of the physical examination before paying attention to the patient's tumor, which can easily side-track the physical examination process. Some patients may require more specialized elements of a physical examination. For example, orthopedic examinations can be important for the patient with a limb tumor for which amputation is a potential option. Neurologic examinations establish an important baseline of neurologic function for patients with primary central nervous system disease.

Routine laboratory tests (CBC, biochemical profile, urinalysis, T4 in middle-aged to older cats) are important to define the systemic health of the patient, and in some cases will provide insights into the tumor, or provide tools to monitor tumor responses or recurrence as seen with some paraneoplastic syndromes. For example, the finding of hypercalcemia in a patient with lymphoma is often correlated to the presence of the T-cell phenotype, which in dogs is associated with a poorer prognosis as compared to B-cell lymphoma. The presence of erythrocytosis in a patient with a nasal sarcoma may serve as a tool for monitoring therapy when repeat nasal imaging is not possible. Even for patients for which routine laboratory tests may not be all that helpful in adding tumor-specific information, results of these routine tests can influence other treatment modalities. For example, some chemotherapy drugs may be avoided in hyperbilirubinemic patients. Some laboratory results may ultimately preclude treatment altogether, for example a laboratory diagnosis of renal failure, or clinically significant hepatic disease, or others.

Some tumors need to be imaged to better define their local extent, essentially addressing the T element of the staging process. Imaging can be especially important for tumors of the head, neck and other anatomically complicated areas such as the axillary space, thoracic cavity tumors, some tumors of the abdomen (e.g. hepatic, adrenal and renal tumors), pelvic tumors, and occasionally tumors of the limbs that are large and which could pose surgical challenges (e.g. large masses toward the hip). For many tumors, the best information comes from cross-sectional imaging modalities such as computed tomography (CT) or magnetic resonance imaging (MRI). Cross-sectional imaging can reveal the extent of local invasiveness of a tumor, and can be important to helping predict the likelihood of success, and the types of complications, that could be anticipated for a given local therapy (surgery, radiation therapy). The author has seen patients in which cross-sectional imaging has indicated that surgical intervention was more worthwhile than had been thought based on physical examination impressions, and conversely has dealt with patients for which the cross-sectional imaging indicated that a surgical approach would not be feasible without loss of function or other severe local complications, or just simply would not be possible. In the absence of cross-sectional imaging techniques, plain radiographs are still encouraged as there may well be features that prove helpful in defining the local extent of disease.

Another technique that can be used to assess some primary tumors is nuclear scintigraphic imaging. Bone scanning has been used to determine the local extent of primary bone tumors such as osteosarcoma; nuclear scintigraphy is also helpful for detecting other bone metastatic sites. Nuclear scintigraphic techniques may also identify tumors that are candidates for specific radioisotope therapy. For example, patients with thyroid carcinomas that have "hot" tumors based on nuclear scintigraphic scans are candidates for

I-131 therapy even if the tumors are not secreting thyroid hormone.[1] At WSU, we have been treating some tumors of the skull with the isotope samarium if there is evidence on a scintigraphic scan that the tumor concentrates a tracer isotope to suggest that samarium would be concentrated in the tumor.

Another element to staging the cancer patient is evaluation of lymph nodes, the N aspect of the WHO system. Understanding basic lymph node drainage patterns is helpful for determining which nodes become the focus of attention. Lymph nodes that are palpably abnormal (large, altered texture) should be sampled for cytological examination. Some tumors demand attempts at lymph node aspiration regardless of the physical examination impression of the node- canine grade III mast cell tumors, oral melanomas and other high-grade tumors are a few examples. Studies have demonstrated that lymph node cytology is more sensitive and specific for the detection of lymph node metastasis than physical examination assessment, and thus physical examination is an unreliable means of determining whether a lymph node should be aspirated.[2,3] Some regional lymph nodes can't be assessed by the physical examination, and in such patients, lymph node examination by ultrasonography, or sometimes cross-sectional imaging techniques, becomes important for assessing the possibility of nodal metastasis. Imaging techniques such as ultrasonography can often be coupled with guided aspiration of lymph nodes for cytological examination. As lymph node cytology is considered quite sensitive for the detection of nodal metastasis, a cytologically negative lymph node can give one a high degree of confidence in the absence of lymph node metastasis, but it is not a guarantee that nodal metastasis has not occurred.

The third of the elements of staging is determining whether distant metastasis is present. A number of techniques can be instrumental in answering this question. Plain thoracic radiographs incorporating both right and left lateral, and ventrodorsal or dorsoventral projections, are appropriate for virtually all patients with malignancies, and for patients at high risk of having malignancies. For example, the middle-aged large-breed dog with a non-traumatic acute hemoabdomen should have a thoracic radiograph before exploratory laparotomy whenever possible to make sure that there are no obvious abnormalities consistent with pulmonary metastasis. A step up in sensitivity for detection of pulmonary metastasis is thoracic CT scanning. In some patients, it may be appropriate to do both thoracic radiographs and thoracic CT, with the thoracic radiograph the initial screen, and the thoracic CT being the final screen, for pulmonary metastasis. The rationale for performing a plain thoracic radiograph early in the staging process is that the detection of lesions compatible with metastatic disease may provide sufficient information to preclude any other staging tests, and they can often be performed on the unsedated patient. The obvious disadvantage to combining thoracic radiographs and CT is the extra expense involved. Of course, thoracic CT may be offered as the sole means of screening for metastatic disease, but it is a more expensive test and requires sedation or general anesthesia to accomplish. It is relatively common in the author's practice for patients to have thoracic CT scans performed immediately before an aggressive intervention, such as a radical surgery, or an expensive treatment such as radiation therapy. If the CT scan is clean, the patient proceeds to surgery or begins a radiation therapy protocol, and if the CT scan is highly suggestive of metastatic disease, then typically the scheduled intervention is cancelled while the clinician and owners regroup to define appropriate next steps.

Abdominal ultrasonography is another valuable tool in the assessment of patients for distant metastasis. The liver and spleen are often the organs of most interest, although any abdominal organ can be a site of metastatic disease. Tumor types (e.g. mast cell tumors, lymphoma) with known predilection for metastasis to, or involvement of, the liver and spleen often dictate abdominal evaluation. Animals with unexplained increases in liver enzymes are also candidates for abdominal imaging. Our approach at WSU is to aspirate the liver and spleen of any patient with a high-grade mast cell tumor because in our experience, ultrasonographic appearance does not reliably predict metastasis to these organs. We have seen a number of dogs with ultrasonographically normal livers or spleens that had abnormal numbers of mast cells, or morphologically abnormal mast cells, identified on aspiration cytology of these organs. A recent paper examining the utility of ultrasonography in identifying lymphoma in the spleen and liver of dogs with lymphoma found high sensitivity for detection of splenic lymphoma, but comparatively poor sensitivity in the detection of hepatic lymphoma, suggesting perhaps that routine aspiration of the liver should be a part of the staging for dogs with lymphoma.[4] In some cases, cross-sectional imaging (CT, MRI) is helpful for the assessment of metastatic disease within the abdominal cavity.

A situation encountered with some frequency in referrals to the author's practice is the diagnosis of neoplasia based solely upon ultrasonographic findings in patients that are subsequently proven to not have an obvious neoplasm. The author routinely points out to students, house officers and other veterinarians that radiologists and ultrasonographers can not establish a cytologic or histologic diagnosis of neoplasia based solely on a radiographic or ultrasonographic image. Along the same line, a pronouncement of structures as normal by radiology or ultrasonography is often interpreted to mean cytologically or histologically normal, but this conclusion is another that can't be made based solely on images. Like any other diagnostic test, radiographs and ultrasound images need to be interpreted in light of the patient's complete clinical picture.

One last imaging modality to mention for its staging utility is positron emission tomography, or PET scanning. PET scanning uses a pharmaceutical agent tagged with a positron-emitting radioisotope. The pharmaceutical agent is typically one that distributes to tissues based on specific cellular functions. PET scanning can be combined with CT scanning to provide additional anatomic and functional information and can be of value in the staging of some tumors.[5] The use of PET scanning is of limited use at present in veterinary medicine because of the expense of the radiopharmaceuticals and limited availability for veterinary patients.

There are a number of other miscellaneous staging procedures that are needed in some patients. Animals with historical or physical examination evidence consistent with primary heart disease may need electrocardiograms and echocardiograms as a part of their evaluation, particularly important if such animals are candidates for treatment with the cardiotoxic chemotherapeutic drug doxorubicin. Some patients with unexplained liver enzyme abnormalities may need not only abdominal imaging as suggested above, but liver cytology or histopathology if not already an initial part of the staging plan for that patient. Patients with bone abnormalities identified on nuclear scintigraphic scans need to have radiographic examination of the lesion, and potentially biopsy or aspiration cytology, to determine if the lesion represents a metastatic site. Patients with tumors that have a predisposition for bone marrow involvement (e.g. lymphoma, mast cell tumors) become candidates for bone marrow aspiration cytology.

The procedures described above will typically be sufficient in one combination or another to appropriately stage the majority of dogs and cats with neoplastic disease. In addition to being critically important for defining treatment options, staging can also be helpful in determining prognosis. The prognosis for any given patient will be a reflection of the tumor type and for many tumors, the tumor grade, the tumor's inherent biological properties, the tumor stage, other co-morbidities existing in a given patient and last, but not least, the patient's response to treatment.

REFERENCES

[1]Turrel JM, et al. J Am Vet Med Assoc 2006; 229:542-548. [2]Langenbach A, et al. J Am Vet Med Assoc 2001; 218:1424-1428. [3]Williams LE, Packer RA. J Am Vet Med Assoc 2003; 222:1234-1236. [4]Crabtree AC, et al. Vet Radiol Ultrasound 2010;51:661-664. [5] Lawrence J, et al. Vet Comp Oncol 2010;8:163-187.

ARRHYTHMIAS: THE ABCs OF PQRSTs FOR THE TECHNICIAN (LA & SA)

George A. Kramer, DVM, DACVIM (Cardiology)
Bohemia, NY

INTRODUCTION

An important function of the veterinary technician is to assess and monitor the cardiovascular status of patients with the goal of optimizing tissue perfusion. Adequate perfusion is dependent on normal vascular volume and pressure along with normal cardiac output. Cardiac output relies on the normal functioning of the myocardium and valves within the heart along with normal activity of the heart's electrical system. Abnormalities in the heart's electrical system result in arrhythmias which can lead to decreases in cardiac output and a reduction in tissue perfusion. Analyzing the electrocardiogram allows us to characterize the arrhythmia and make treatment decisions that will hopefully restore normal perfusion. The veterinary technician should be able to monitor a patient's electrocardiogram and alert the veterinarian if any abnormalities are noted so that prompt treatment can be initiated if indicated. Critical patients are often complicated by alterations in their heart rhythm or conduction pattern. It is important to thoroughly assess the ECG and to try to determine the site of the abnormality. Arrhythmias are most frequently associated with poor perfusion, hypoxia, ischemia, or release of vasoactive or depressant factors. Conduction abnormalities are most frequently associated with hypoxia, ischemia, and electrolyte imbalances (sodium, potassium and calcium). By recognizing these changes and the potential causes the proper treatment can be instituted to improve cardiac performance.

The electrical system consists of the SA node, inter-atrial bundles, AV node, bundle of HIS, right and left bundle branches, and the purkinje fibers. The SA and AV nodes display the property of automaticity which results in spontaneous discharge. Ventricular myocytes also have the property of automaticity but discharge at a much slower rate, allowing for the development of an "escape" rhythm if the there are no impulses transmitted through the AV node. Alterations of this property can also lead to arrhythmias.

THE ELECTROCARDIOGRAM (ECG)

The electrocardiogram (ECG) is a graphical recording representing the electrical pattern and conduction within the heart. This is measured by recording the changes over time of the electrical potential difference (mv) between two leads place on the skin surface. Specific portions of the ECG recording relate to specific portions of the action potentials generated in the heart (see figure 1).

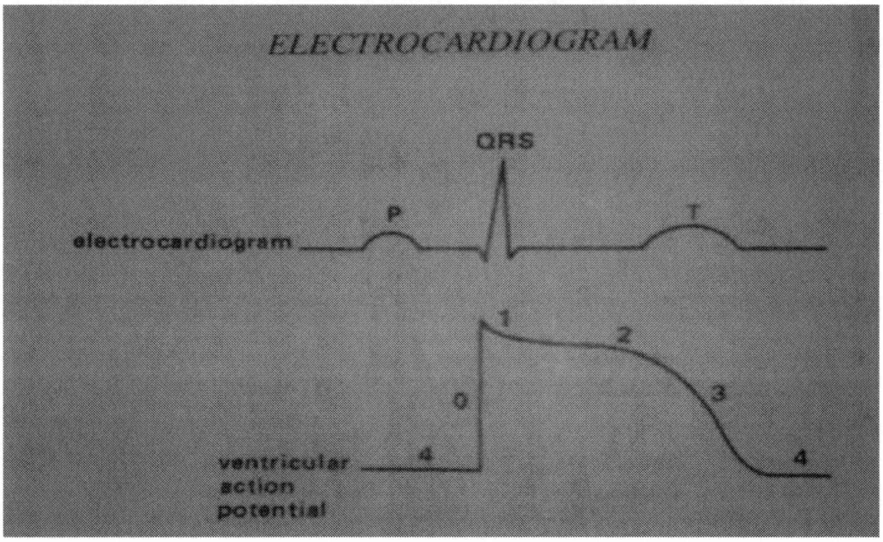

Figure 1

Within veterinary medicine, generally leads I, II, III, aVR, aVL, and aVF are recorded and evaluated. In horses the lead system used is generally a base-apex configuration. In human medicine, usually either 10 or 12 leads are recorded. The ECG gives a positive deflection, by convention, as the depolarization wave progresses from the negative electrode to the positive electrode. A negative deflection occurs as the depolarization wave progresses from the positive electrode to the negative electrode.

P QRS T

The waves identified on the ECG are named P, Q, R, S, T. Analyzing the appearances of the waves and the intervals between them allow us to gain much information about the heart size (e.g. mean electrical axis), the heart rate and rhythm.
The P wave is generated by the sum of the forces produced during atrial depolarization. It is easy to recognize, in that is

generally the smallest of the waves and it precedes the QRS complex. Increases in size of the p wave can indicate atrial enlargement. A wandering pacemaker is not an unusual finding in the normal canine heart and is manifested by varying heights and widths of the P wave.

The P-R interval is a function of the integrity of the atrial conduction system and a reflection of the AV node conduction time within the AV node. It is generally an isoelectric period. In a normal sinus rhythm the P-R interval is generally a constant length on a beat to beat basis. Should these conduction pathways be blocked, the P-R interval will be prolonged. Digoxin has its therapeutic effect through prolongation of the conduction time through the AV node and can cause an increase in the P-R interval. It is for this reason that digoxin can be used to treat various supraventricular tachycardias (atrial fibrillation is one example).

There are three phases to the depolarization of the ventricles. The Q wave represents depolarization of the ventricular septum and it is the first negative deflection in lead II. The R wave represents depolarization of the right and left ventricular walls. Both the right and left ventricles depolarize together but are recorded as if only the left had since it has greater mass than the right and the depolarization wave is of greater magnitude. When the ventricles do not depolarize together, such as with an interventricular conduction abnormality, there will be a widening of the QRS complex and often a notch in the R wave. The S wave represents depolarization of the base of the septum, the basal areas of both the right and left ventricular free walls and it is the first negative deflection after the R wave.

The T wave represents repolarization of the ventricles. The rate of repolarization is slower than the rate of depolarization. Therefore, the T wave is wider and of less amplitude than the R wave. The Q-T interval represents the period of electrical systole. During the S-T segment the entire myocardium is depolarized. This segment is generally isoelectric. Movement of the S-T segment above or below the isoelectric line can indicate damage to the myocardium or myocardial hypoxia.

ECG INTERPRETATION

Interpreting the ECG involves a three step process:

1. Rate and rhythm: rate is determined by the number of QRS complexes per minute. The rhythm is determined by evaluating whether the R-R intervals are the same and regular and if the P-R intervals are the same and regular. Deviations from regularity are termed arrhythmias. Not all arrhythmias are abnormal or pathologic (e.g. sinus arrhythmia).
2. Determining the mean electrical axis. This is the sum of all of the electrical activity in the ventricles during depolarization and can provide information concerning enlargement of the ventricles or indicate conduction abnormalities.
3. Determining any variation from normal in the measurements of the height and width of the various waves and in the P-R interval and S-T segment lengths

ARRHYTHMIAS

There are several ways to classify arrhythmias. When classified by anatomic and physiologic means, the groups include sinoatrial, ectopic, and conduction disturbances. Or they can be classified by heart rate, such as tachyarrhythmias, bradyarrhythmias, or arrhythmias with normal heart rate. Utilizing the heart rate, a group of arrhythmias can be further classified regarding anatomic location of the abnormality:

Tachycardias: sinus, atrial, atrial flutter, atrial fibrillation, junctional, ventricular.

Bradycardias: sinus, SA node arrest or standstill, 2° heart block, 3° heart block, junctional escape, ventricular escape.

Normal rate: sinus arrhythmia, sinus arrest, atrial premature contraction (APC), junctional premature contraction, ventricular premature contraction (VPC), 1° heart block, 2° heart block, bundle branch block.

Ectopic arrhythmias are characterized by beats that originate outside of the SA node. Sites include atrial, junctional and ventricular. The atrial and junctional origins are termed supraventricular. Atrial origin has a positive P wave and junctional has a negative P wave (but not always). Beats originating from the ventricles have abnormal appearing QRS complexes (usually wide and bizarre in conformation), (more normal if arises from area near HIS bundle), and are not associated with the P wave.

Supraventricular arrhythmias can be either atrial or junctional premature, There is an irregular rhythm that can be associated with pulse deficits. The QRS complex will have a normal configuration. Examples of supraventricular tachycardias include: atrial tachycardia, atrial fibrillation and junctional tachycardia.

Ventricular arrhythmias generally are associated with a pulse deficit. Premature beats are followed by a pause and have no P waves associated with them. There are no consistencies with the R-R intervals unless it is a ventricular tachycardia. The QRS complexes are wide and bizarre. Multifocal beats are more dangerous than unifocal. If a normal beat is seen within a series of ventricular beats, the normal beat is a capture beat. If there is a bizarre complex between a premature ventricular beat and a normal beat, it is termed a fusion beat. A-V dissociation is a form of ventricular tachycardia. There are two types, synchronous where the P wave rates and the R wave rates are equal but not associated, and non-synchronous, where the P wave moves into and out of the QRS.complex. Ventricular fibrillation is characterized by no P-QRS complexes, a series of baseline undulations and absence of a heart beat. Ventricular fibrillation always results in death unless the patient can be electrically, chemically, or mechanically defibrillated.

Escape beats_are a form of ventricular rhythm. These occur when there is a depressed rate of spontaneous discharge from the SA node lasting so long that the ventricle depolarizes from its own focus. An escape rhythm is slow and can be either regular or irregular. Escape beats are not associated with P waves, have bizarre QRS complexes, and occur after a long pause. Junctional escape rhythms_can occur, showing an irregular rhythm, negative P waves, a P for each QRS (but P wave could occur before, during or after the QRS), and a slow or normal junctional rate.

Wave changes are the third means by which the ECG is evaluated. Conduction disturbances will cause wave changes. Examples of conduction disturbance arrhythmias include A-V block and sinoatrial standstill. A-V block is the result of delay in conduction through the AV node. There are three types of AV block, fist, second and third degree A-V block. In 1st degree AV block, the heart rate is normal and the P-R interval is constant but prolonged. In 2nd degree A-V block, the heart rate is slow and the rhythm is broken by a mix of normal P-ORS complexes and P waves without any QRS complexes. There are two forms of 2nd degree A-V block, Mobitz type 1 and type 2. In Mobitz type 1 the P-R interval prolongs over a series of beats right before a QRS is dropped. This is also termed Wenkebach phenomenon. In Mobitz type 2, there are normal P-R intervals prior to the QRS complex being dropped. Another name for Mobitz I that is often used is Wenkebach phenomenon. In 3rd degree A-V block there are no QRS complexes associated with P waves. There is either a ventricular or junctional escape rhythm present with no association between the P waves and QRS complexes. With sinoatrial standstill there are no P waves present at all and there is a ventricular or junctional escape rhythm.

TREATMENT OF ARRHYTHMIAS

Once the arrhythmia is characterized, the treatment decision needs to be based on the animal's clinical condition, hemodynamic status, severity of the arrhythmia and what the underlying condition is. For example, a cat that has a sinoventricular bradycardia due to a urinary obstruction does not need atropine to treat the bradycardia, but rather needs to be unblocked and have his associated electrolyte disturbance (i.e.hyperkalemia) corrected. If drug therapy is to be instituted for a specific arrhythmia in a patient, careful monitoring of that patient needs to be maintained because most antiarrhythmic drugs can have toxic side effects and also can be proarrhythmic.

CONCLUSSION

Interpreting ECGs and diagnosing arrhythmias is not difficult for the majority of arrhythmias. The key to becoming proficient at is continued practice. In addition to analyzing the ECGs on your patients at work, you should review practice ECG in text books and online. With regular practice you can become very proficient at understanding ECGs. Don't let your skills get rusty.

REFERENCES

Edwards, J.N. **ECG Manual for the Veterinary Technician.** Tilley, L.P. **Essentials of Canine and Feline Electrocardiography.** Robinson, N.E. **Current Therapy in Equine Medicine, Volume 5.**

Leptospirosis for the Veterinary Technician

Meri Hall, RVT, LVT, CVT, RLATG, VTS (SAIM)
Athens, PA

Infections caused by leptospira bacteria are known by many names. Swamp fever, Weil's syndrome, Autumn fever, Swineherd's disease, Hemorrhagic jaundice, Stuttgart disease and Redwater of calves are just a few. Because leptospirosis is a zoonotic disease, the veterinary technician must be aware of the clinical signs of the disease, proper handling of diagnostic samples, treatment, handling of hospitalized patients to prevent the spread the disease and what clients need to know.

Leptospira bacteria are helical bacteria of the genus *Leptospira* and are found worldwide. There are both saphrohytic and pathogenic species within the genus. Ssome leptospires that are pathogenically and phylogentically intermediate to the saphrophytic and pathogenic species have been identified. Of the pathogenic species, there are 13 which infect humans and animals. Within the 13 pathogenic species, there are over 200 serovars. Serovars have different animal reservoir hosts. The predominant serovar in any geographical area vary. The serovar classification is based on how the surface exposed epitopes are expressed. Closely related serovars may be of different species.[1] In the United States, the most common serovars are *L. canicola, grippotyphosa, hardjo, icterohaemorrhagiae* and *Pomona.*[2] Most of the pathogenic species in the United States are from the *L. interrogans* and *L. kirschneri species.*

Table 1. Species and serovars of Leptospira in animals in the United States

Species	Serogroup	Serovar	Resevoir Animal Host
Leptospira interrogans	Australis	Bratislava	pigs
	Ballum	Ballum	
	Canicola	Canicola	dogs
	Hardjo	Hardjo	cattle
	Icterohaemorrhagiae	Icterohaemorrhagiae	rats
	Pomona	Pomona	Cattle, pigs, opossum, skunks
Leptospira kirschneri	Grippotyphosa	Grippotyphosa	Opossum, raccoons, skunks

Derived from[1,2,4]

The bacteria are obligate aerobes and have optimum growth at 28-30^{0}C/ 82.4-86^0 F. The bacteria is transmitted directly between animals or indirectly in the environment. Leptospira bacteria are shed in the urine of infected animals. In cattle and pigs the bacteria have been found in aborted or stillborn fetuses. The bacteria do not replicate in the environment, but can survive in the environment for months. The prevalence of leptospirosis in geographical areas with warm climates with high annual rainfall is high, but also depends on the presence of animal resevoirs. The incidence will vary by geographical location dependent upon the rainfall and onset of freezing temperatures.[3] In the United States, geographic regions include the west coast, upper midwest, Hawaii, northeast, parts of Texas, Colorado and the southeastern states.

The incubation period for leptospirosis ranges from a few days to weeks. Most sources give ranges of 3-7 days. Although the bacteria may be shed for months in the urine of recovered animals. Many cases of leptospirosis are asymptomatic and are unrecognized. With clinical illness, the severity varies due to the geographical area, infecting strain and the immune response. Some animals will develop severe illness or die. The clinical signs observed varies with species. The bacteria are shed from the renal tubules of infected animals. Infection occurs by contact with mucous membranes, water soaked skin or abraded skin with infected urine. Infection can also occur with exposure to urine contaminated food, water, soil or bedding.[2] In some cases transmission has occurred from ingestion of infected tissue, bite wound inoculation and venereal or placental transfer.[2]

Table 2. Clinical signs seen with Leptospirosis

Species	Clinical Signs Seen	Vaccination
Cat	Rarely reported	None
Cattle	Abortion, stillbirths, weak calves, fever, anorexia, conjunctivitis, icterus, hemoglobinuria diarrhea	Vaccines used
Dog	Vomiting, fever, inappetance, Renal or hepatic failure, uveitis, pulmonary hemorrhage, abortion	Vaccines used
Goat	Fever, anorexia, icterus, hemoglobinuria, abortion, stillbirths, weak kids	Vaccines used
Horse	Recurrent uveitis, abortions	No approved vaccine
Pig	Fever, icterus, abortion, stillbirths, weak piglets, poor growers	Vaccines used
Sheep	Fever, anorexia, icterus, hemoglobinuria, abortion, stillbirths, weak lambs	Vaccines used

Derived from[4,5]

Diagnosis of leptospirosis includes serology, detection of antigen and culture. *Leptospira* can be found in blood, urine, cerebrospinal fluid, milk and peritoneal or thoracic fluid. At necropsy, the bacteria can be isolated from the liver, lung, brain and kidney. In aborted fetuses the bacteria can be found in the body fluids or tissues. With serology for diagnosis a paired acute and convalescent sample is preferred, but with an aborted fetus, a single sample is diagnostic of the disease. A 4-fold change in titer for paired samples supports recent infection, where titers resulting from past exposure, chronic infection or past vaccination may not change at all.[2] False negatives may occur if the infecting serovar is not included in the serovars tested for. The most common serologic tests are the microscopic agglutination test (MAT) and enzyme linked immunosorbent assays (ELISA). There are serovar-specific ELISA tests available for cattle[4].

Treatment of leptospirosis is dependent upon the species affected and the severity of the clinical signs. For beef cattle, goats, pigs and sheep treatment of all animals in the herd with antibiotics and vaccination are typically used. In dairy cattle, typically only the infected animals are treated with antibiotics due to loss of milk sales[3,4,5]. In horses, systemic antibiotics and symptomatic treatment of the uveitis is used.[3,5] For dogs, treatment with antibiotics should be used. Supportive care for dogs includes nutritional support, hospitalization, intravenous fluids, and oxygen therapy may be indicated due to the severity of the disease.

Prevention of leptospirosis includes vaccination of at risk animals, good sanitation practices and prevention of contaminated environments from infected wildlife or rodents. Potential environmental sources of infection include standing water and marshy areas. It is recommended that cattle be housed away from pigs, sheep and goats. In the hospital environment, direct animal contact should be avoided, house the animals away from high traffic areas and the use of personal protection equipment (gloves, gown and protective eyewear when handling body fluids) is recommended. Sanitation of the housing area with an appropriate disinfecting agent is required. A 1:1 dilution of 10% bleach, iodine based disinfectants, quaternary ammonium solutions and accelerated hydrogen peroxide have been shown to be effective cleaning agents.[2] . Pressure washing of runs and stalls should be avoided to prevent aerosolization of urine. If animals do not have urinary catheters in place, they should be taken to a restricted area to urinate and the area treated with 10% bleach immediately. All blood, urine and tissues from animals suspected of leptospirosis should be treated as medical waste depending on the local regulations.[2]

Table 3. Vaccination

Species	Vaccination
Cat	None
Cattle	Vaccines used
Dog	Vaccines used
Goat	Vaccines used
Horse	No approved vaccine
Pig	Vaccines used
Sheep	Vaccines used

Derived from[4,5]

The zoonotic potential for humans can be reduced by vaccination of animals, avoiding contact with urine, and good sanitation practices. It is always advisable to prevent animals from urinating near standing water and where children and other animals will have access. With animals who are diagnosed with leptospirosis, wearing gloves when cleaning urine are recommended.

The veterinary technician is integral in the treatment and prevention of any disease. With leptospirosis the veterinary technician can be instrumental in the prevention or reduction in the spread of the disease. The health of their patients, themselves and their clients will benefit from this knowledge.

REFERENCES

1.Gyles, Prescott, Songer Thoen, Pathogenesis of Bacterial Infections in Animals Fourth edition, Wiley-Blackwell, 2010: 527-547. 2. Sykes JE, Hartmann K, Lunn KF, Moore GE, Stoddard RA, Goldstein RE. 2010 ACVIM Small Animal Consenus Statement on Leptospirosis: Diagnosis, Epidemiology, Treatment and Prevention. J Vet Intern Med 2011; 25:1-13. 3. The Center for Food Security & Public Health. www.cfsph.iastate.edu/Factsheet/leptospirosis. May 1, 2005, accessed 29 January, 2011. 4. Bolin C. Diagnosis and Control of Bovine Leptospirosis. Proceedings of the 6th Western Dairy Management Conference. March 12-14 2003. Renov, NV. 5. Merck manual on line. www.merckvetmanual.com/mvm/htm/bc/51202.htm. Last accessed January 26, 2011.

MAKING PHARMACOKINETICS UNDERSTANDABLE

Lynn Rolland Hovda, RPh, DVM, MS, DACVIM
Bloomington, MN

INTRODUCTION

The study of pharmacokinetics can be overwhelming to even the most seasoned researcher. Entire books and series of books have been written on the subject with a vast array of nomenclatures, abbreviations, diagrams, tables, and drawings. For the purpose of clinical veterinary practice, however, it can be broken down into some basic concepts and building blocks that are not only useful but easy to understand.

The term pharmacokinetics refers to everything that happens to a drug or medication as it journeys through the body. It begins after the drug has been administered and ends when it is eliminated from the body. ADME is an acronym frequently used to describe the four phases or parts of the journey: absorption, distribution, metabolism, and excretion. Blood, usually plasma, is the vehicle most often used for pharmacokinetic studies as it is easy to obtain and measure. It some cases, however, it may not accurately reflect what occurs at the cellular level, the actual site of the drug's action.

ROUTES OF ADMINISTRATION

Before drugs can be absorbed they must be administered to the animal. There are four general categories of drug administration each with several subcategories.

Oral

Oral administration simply means giving the drug by mouth. Most oral medications are designed to be swallowed and absorbed in the small intestine. A subset of the oral route is buccal or sublingual administration, when the drug is not meant to be swallowed, but placed either in the cheek pouch or under the tongue. Absorption with this route occurs directly across the mucous membranes in the mouth and is sometime referred to as oral transmucosal (OTM).

Parenteral

The most common forms of parenteral or injectable drug administration are intravenous, intramuscular, subcutaneous, and intradermal. Other less common forms include intra-arterial, intra-articular, intracardiac, intramedullary, and epidural/subdural. Many drugs designed for extended release or "depot" administration are injectable products.

Transdermal

Creams, ointments, patches, and topical spot on type products applied directly to the skin fall into this category. Translocation is a specialized form of administration used for many spot on products. The product is applied to one small area such as the back of the neck and over a brief time period moves over the skin surface, eventually covering the entire body. Rectal, vaginal, and mammary gland (think cows!) administration are often placed in this category, although technically these products are applied to mucous membranes and not skin.

Inhalation

Drugs that are vaporized or aerosolized are absorbed by inhalation. Gas anesthetic agents, bronchodilators, mucolytics, and a few other drugs round out this category.

ABSORPTION

The first part of a drug's journey focuses on moving from the site of administration to the systemic circulation. To do this, it must cross some form of a phospholipid cell membrane. Areas with a large surface area generally have greater rates of absorption. The small intestine, where most oral absorption occurs, has one of the largest absorptive surface areas in the body; millions of villi (finger like projections) protrude into the lumen to increase absorption. There are several means of absorption across membranes but the most common is simple diffusion or moving from an area of high concentration to low concentration. This is can be visualized as putting creamer in coffee or red food coloring in a glass of water. A slightly different form of diffusion is "facilitated" where a carrier protein in the cell membrane is required for diffusion to occur. This form has been described as being similar to a revolving door in a wall where the door is the protein, you are the drug, and once the door is entered you are swept across the membrane. Other forms are active transport, which requires energy and functions much like a pump, and pinocytosis, where the cell forms a "blob" around the drug and drinks or sucks it in.

Bioavailability (F) refers to the amount of drug that is absorbed and can be measured in the blood. Intravenous administration of a drug has a 100% bioavailability and is used as the standard for determining the F value associated with other routes of administration. Drugs with poor bioavailability (i.e. low F values) are rarely administered as oral medications or if they are they have been modified to increase F. Buprenorphine in cats has a very low bioavailability when given as a pill to be swallowed but a bioavailability ranging from 50% to 100% when administered as a liquid in the cheek pouch.[1,2] Similarly, the bioavailability of swallowed buprenorphine in dogs is low, ranging from 3 to 6%, yet it reaches 38% to 47% when administered by an oral transmucosal route (i.e. under the tongue or in the cheek pouch).[2] Both species have alkaline saliva which likely plays a role in absorption.

The first pass effect refers to a phenomena specific to oral medications that are swallowed and reach the intestinal tract. Once a drug crosses the small intestine membrane it is absorbed by the portal veins of the liver and enters the systemic circulation as a free drug. Some drugs such as opioids, fentanyl, and diazepam in dogs are metabolized or changed so quickly in this first pass through the liver that only a very small amount of free drug enters the circulation.[3] These drugs have a high first pass effect and generally cannot be administered orally. Other drugs such as digoxin, metformin, and warfarin are not changed much at all and most of the free drug enters the circulation. These drugs have a low first pass effect.

Putting it together, drugs with a low bioavailability, high first pass effect, or both would not be good candidates for use as oral medications. Other means of administration need to be used if these drugs are to be effective. Fentanyl patches, buccal buprenorphine in cats and dogs, prodrugs, topical creams, injectable depot products, and other forms of administration are all good options.[2,3]

DISTRIBUTION

In this part of the journey, free drug moves from the site of absorption into the circulatory system, travels around the body, and eventually ends up at the site of action. Sometimes this is very specific such as the prostate gland and sometimes it is not at all specific such as broad spectrum antibiotics meant to reach millions of cells in the body. Once it reaches the cell, free drug in the plasma moves into interstitial fluid and from there into the cell. This occurs similar to absorption (remember you are just crossing another cell membrane) and generally continues until the system is saturated and equilibrium is reached. Highly lipid (i.e. fat) soluble drugs tend to move easily as do nonionic drugs. Both can be stored in tissues like fat, liver, kidney, and bone and excreted over a longer period of time. Ionic drugs or those carrying a charge move with more difficulty and are generally not stored.

Some tissues or organs in the body have special cells that prevent or retard movement of many drugs. The blood brain barrier (BBB) is perhaps the most widely recognized of these, but other body tissues have similar proteins. Part of the BBB's function involves P-glycoprotein (P-gp) or multidrug resistant protein 1 (MDR1), a specialized transporter protein that keeps drugs from entering the brain tissue by moving them back into the systemic circulation. MDR1 is a member of ATP-binding cassette (ABC) transporter proteins, a large family of proteins that has undergone an explosion of research. Genetic mutations of MDR1 have been identified in human beings (cystic fibrosis, breast cancer, and other diseases) and 10 dog breeds (Australian shepherd, Border Collie, Collie, English shepherd, German shepherd, long haired whippet, McNab, Old English sheepdog, Shetland sheepdog, and the Silken windhound).[4,5,6]

In dogs, mutations known as mdr1-delta or more currently ABCB1-1delta are associated with a failure of the BBB to function effectively thus allowing some medications to accumulate within the brain. Many drugs are affected by these mutations with the most common in dogs being acepromazine, ivermectin, and loperamide.[5]

Plasma protein binding occurs when free drug in the plasma comes into contact with proteins (generally albumin) and binds to them. A variety of factors including lipid solubility, charge, and particle size influence the amount of binding. Some drugs bind a lot, some a very small amount, but once bound they are unavailable for further distribution. Binding to plasma proteins acts as a temporary storage site as eventually the drug will be released. Binding becomes more important for those dogs with liver disease, renal failure, and protein losing enteropathies as well as very young and old animals who can't or don't make an adequate amount of albumin. For these animals, drugs that are normally highly protein bound will accumulate in the plasma and may reach toxic levels.

METABOLISM (Biotransformation)

Metabolism is really the first step in getting the drug ready to be eliminated from the body. The drug is changed, altered, or modified so it can be excreted. Metabolism primarily occurs in the liver, but the kidneys, lungs, and nervous system all are capable of metabolizing certain drugs to a small extent. Most, but not all drugs, undergo biotransformation. The end result of metabolism for the majority of drugs is a water soluble metabolite, active or inactive, that can be excreted in the urine. Some biotransformed drugs are fat soluble and ultimately excreted with bile and then feces.

There are two different phases or parts of metabolism:

Phase 1

It is during this phase that drugs undergo a chemical reaction (oxidation, reduction, or hydrolysis) to change them into a more polar drug ready for phase 2. Oxidation, the most important reaction, is catalyzed by a very large family of microsomal enzymes known as cytochrome P450 or just CP450 enzymes.[7] Many drugs and other agents can inhibit or activate (induce) CP450 enzymes. Cimetidine and organophosphates are among the most common inhibitors and barbiturates and phenylbutazone the activators.

Phase 2

In this, the "conjugation" or "joining" phase, metabolites formed in phase 1 are modified to become water soluble for rapid elimination by the kidneys. Common conjugation reactions include glucuronic acid, sulfate, glycine, glutamine, acetylation, and methylation. A different and specific set of enzymes is needed for each of these reactions and it is in this phase where significant species related differences occur. Cats, as well as newborns and some pregnant animals, are deficient in glucuronyl transferase and cannot adequately utilize the glucuronic acid pathway. In addition, cats easily saturate the sulfate conjugating system. Metabolism of acetaminophen uses both of these pathways and this is one of the reasons that cats are so sensitive to acetaminophen and become toxic at such low doses. Dogs, on the other hand, make very efficient use of both of these pathways and handle acetaminophen much better. It also explains the difference in aspirin dosing intervals between these two species. Dogs and cats can be given aspirin at the same dose (10 mg/kg) but a second dose should not be given in dogs for 12 hours and 48-72 hours in the cat. Cats are not the only species, however, to be affected by enzyme system deficiencies. Dog are deficient in acetyl transferase and cannot adequately metabolize aromatic amines such as some sulfonamides and pigs lack the enzymes to utilize the sulfate system.

Other factors affecting drug metabolism include nutrition, disease processes, and tissue storage. Microsomal enzymes used in the metabolic pathways are proteins and malnourished animals or those that lose protein from the gastrointestinal tract or renal system will have an impaired enzyme system. In addition, animals with liver disease cannot make adequate amounts of protein. Finally, those drugs administered as a "depot" form are very slowly metabolized as are those that are very highly protein bound.

EXCRETION (ELIMINATION)

Drugs and metabolites, active and inactive, are removed from the body during this step. Renal and hepatic (through the bile) excretion are the primary means of removal, although skin, sweat, lungs, and mammary and salivary glands all are capable excreting a limited amount of certain drugs.

Renal

The majority of drugs are metabolized by the liver and excreted in the urine; a few are excreted in the urine without undergoing any biotransformation or metabolism. The kidneys eliminate drugs through both glomerular filtration and active tubular secretion. Glomerular filtration simply means that the glomerulus in the kidney acts like filter paper; those drugs that are small in size go through the filter paper and end up in the urine. A few drugs such as penicillin and furosemide are delivered to the kidneys by the circulatory system and are actively secreted from the capillaries into renal tubules and excreted with the urine.

Biliary

A few parent drugs and metabolites are excreted into the bile and enter the intestine. Species differences exist here as well with dogs being good at biliary excretion and cats only fair to poor. Once in the intestine, drugs can be excreted with the feces or passively reabsorbed into the systemic circulation. Some glucuronide conjugates of drugs that are excreted in this manner may be broken down by enzymes in the intestine resulting in the reformation of the lipid soluble drug. When this occurs enterohepatic recycling has taken place and the action of the drug may be significantly prolonged.

Finally, the half life ($t_{1/2}$) of elimination provides valuable information. It is the time, often in hours, that it takes for the amount of drug in the body to be decreased by 50%. In general it takes 4-5 half lives for a drug to be considered fully excreted.

REFERENCES

[1]Robertson SA, et al. Vet Rec 2003; 152:675. [2]Abbo LA, et al. Vet Ther 2008; 9(2): 83-93. [3]Lee D, et al. AJVR 2000; 61(6): 672-677. [4] Mealy KL Comp Cont Ed 2006;28:23-33. [5]Neff MW, et al. Proc Nat Acad Sci USA 2004; 101:11725-11730. [6]Mealey KL, et al. JAVMA 2008; 233: 921-924. [7]Kukanich B, et al. J Vet Pharmacol Therap 2005; 28:461-466.

Reflections of the Heart: Echocardiography for Veterinary Technicians

H. Edward Durham Jr., CVT, LATG, VTS (Cardiology)
Columbia, MO

INTRODUCTION

Echocardiography is a method of diagnosing cardiac disease utilizing ultrasound to create an image of the heart. With today's devices, these images are in motion and nearly real time. Echocardiography specifically refers to an ultrasound of the heart and great vessels. Prior to the advent of echocardiography most cardiac conditions were diagnosed in the cardiac catheterization suite requiring an invasive procedure. Ultrasound is a noninvasive diagnostic tool which is much simpler to perform and less painful. The first clinical ultrasound images of the heart were done with a single "ice pick" beam of ultrasound and displayed on a moving paper roll much like electrocardiography. As advances were made in ultrasound transducer design, a "cross sectional" image could be generated by utilizing a fan shaped bean of multiple single beams. This is the most common modality of echocardiogram used today and is referred to as 2D or two dimensional echocardiography. Some of the most modern technology can generate a three dimensional image. As part of the standard information of the echocardiogram, the single ice pick view of the heart, known as M-Mode is still used. From the M-Mode images, precise measurements of cardiac wall thickness and chamber sizes can be obtained. When combined with a Doppler evaluation a complete picture of cardiac function can be obtained.

BASIC TECHNOLOGY

In brief, the ultrasound wave is generated by electrically exciting a piezoelectric crystal within the transducer causing it to expand. When the electrical stimulation is applied to the piezoelectric crystal in an alternating current the crystal expands and contracts to generate an ultrasonic wave. A piezoelectric crystal will also convert ultrasound wave it senses to an electric signal; thereby service a either transmitter or receiver. The number of waves generated per second is known as the frequency or hertz (Hz). Diagnostic medical transducers generate millions of hetrz or megahertz (MHz); in veterinary medicine transducers range from 2.0 to 10 MHz. The ultrasound waves are transmitted through the patient and reflect off body tissue, returning to the transducer to be received by more crystals. The electrical signals generated by the crystals are transformed by a computer into an image displayed on a monitor. Ultrasound images are displayed in black, white and shades of grey. In abdominal ultrasound the more shade of grey presented the better the interpretation, but in echocardiography, less shades of grey and increased edge detection is preferred.

The different frequencies of transducers allow for penetration to different depths. There is no transducer frequency that is suitable for all situations. Lower frequency transducers have greater penetration, but have less ability to show detail. Higher frequency transducers, have better image resolution, but less penetration. A simple way to remember this is consider the neighbor's stereo in the adjacent apartment; the bass frequency penetrate the wall well, but you don't hear the melody of the music in the treble frequencies.

There are other factors effecting how the ultrasound waves behave in the body. Attenuation is the tendency for the waves to be blocked or lose energy from transmission and refraction is the tendency of the waves to be scattered off their original course. These concepts are important because attenuation is related to tissue density, thus bone attenuates the signal so severely that ultrasound waves do not pass through them. Air on the other hand has such a profound refraction effect on ultrasound waves that nothing can be seen beyond it. For this reason a coupling gel must be used between the patient and the transducer to facilitate imaging, by eliminating air. Some of these problems can be overcome by changing the machine setting and it is incumbent on the sonographer to learn the complete operation of their machine.

INDICATIONS

Speaking generally echocardiography provides two categories of information about the heart: morphology and systolic function. Morphology of the heart allows the ultrasonographer to see what the heart looks like. Information such as chamber enlargement, evidence of effusions, presence of cardiac tumors, and appearance of the cardiac valves can be assessed. Systolic function is what most people call contractility; however contractions of the heart are influenced by many variables, such as, Frank-Starling forces, afterload and sympathetic nervous tone. Consequently systolic function is the preferred term for cardiac motion. Indications for performing an echocardiogram include heart wall thickness evaluation, assessment of chamber size and systolic function, definitive diagnosis of heart murmurs, diagnosis of congenital heart disease, diagnosis of cardiac tamponade with pericardial effusion and identification of cardiac tumors as a cause of pericardial effusion.

PATIENT POSITIONING AND PREPERATION

There are many ways of performing an echocardiogram. In very large patients, such as horses or cattle, standing is the preferred position. Many echosonographers in Europe prefer this position for dogs as well. The most common position however is a lateral recumbency with the hair clipped over the precordium. The echocardiographic table has openings in it to allow the operator to reach under the table to access the patient through the hole. This position allows the heart to fall close to the thoracic wall to increase image quality. For imaging the opposite hemithorax, the operator may reach over the patient and image form the top, or more likely, the patient is turned to the opposite lateral recumbency. An electrocardiogram is also run during the echocardiogram to indicate specific times in the cardiac cycle.

STANDARD IMAGE PLANES

Traditionally, even though the transducer is placed on the underside of the patient, the transducer anchor point is oriented at the top of the viewing screen. (Fig. 1) The image is displayed upside down from how it is generated. In this orientation objects closest to the transducer are at the top of the screen or in the near field. Object at the bottom of the screen are in the far field and furthest away from the transducer. There are several standard image planes recognized in veterinary medicine. These planes of view allow a set of standard measurements to be performed to assess cardiac chamber dimensions and systolic function. Echocardiograms usually begin with views from the right hemithorax. The heart is viewed in either its long axis or in its short axis. One way to understand these planes is to imagine a pineapple. If the pineapple is sliced from top to bottom making a pineapple boat then the long axis is represented. If the pineapple is sliced transversely making pineapple rings that represents the short axis. When viewing the heart from the right hemithorax, the first view commonly acquired is the right parasternal long axis four chamber view. In this view, the heart is typically situated horizontally across the screen with the left atrium (LA) and ventricle (LV) in the far field and the right ventricle (RV) and atrium (RA) in the near field. From this position all four chambers of the heart may be seen together. By making a slight adjustment in the angle and rotation of the transducer the aorta may be brought into view, the so called "five" chamber view. Here the LA is much smaller in appearance, but the aorta is in full view and represents the fifth chamber. From this position the transducer is rotated ninety degrees and the heart is seen in the short axis.

There are three standard short axis views: 1) LV at the level of the papillary muscles, 2) LV at the level of the mitral valve (MV), 3) the heart base showing the aortic valve, LA, RA, RV and RV outflow tract (RVOT). In the first, the LV is circular, with the LV lumen shaped somewhat like a mushroom, and the RV appears as a crescent shaped chamber in the near field. The papillary muscles are seen projecting into the LV lumen creating the "stalk" of the mushroom. In the second short axis view, the transducer angle is tilted toward the base of the heart until the MV is seen opening and closing. This view is commonly referred to as the "fish mouth" view since the MV resembles a fish at the surface of the water opening and closing its mouth. By angling the transducer even more basilar, the beam passes through the heart base. The LV is no longer visible, but the aortic valve is seen crosswise, with the LA in the far field, the RA to one side, the RV in the near field, and the RVOT opposite the RA on the screen. Sometimes the pulmonic valve can also be seen in this view, and if not, a slight cranial adjustment in transducer angle often will bring it into view.

To compliment these 2D views a series of M-Mode views are also acquired. Because M-Mode uses only one ultrasound beam, requiring less raw data to be processes as in 2D, M-Mode allows for greater temporal resolution, and finer detail along the single beam. M-Mode therefore allows for greater detail in assessing cardiac wall motion and delineating the edges of the chambers compared to the chamber lumens. M-Mode views are still the standard views for making many of the measurement of chamber dimensions and systolic function. The M-Mode views are derived from the short axis 2D images, with the first usually taken from the short axis at the papillary muscle level. A cursor is typically placed across the 2D image such that the cursor crosses the interventricular septum (IVS) passing through the LV lumen between the papillary muscles and out the LV free or posterior wall (LVPW). The data received by the computer **only** along the cursor line is then displayed over time. The resulting image is a graphic representation of the heart with time along the X axis and depth along the Y axis. The RV and IVS will be in the near field and the LV lumen will appear in the middle, with the LVPW in the far field. (Fig 2) From this image measurement of the LV can be made in diastole and systole and calculations of systolic function can be determined. From the "fish mouth" short axis view, the motion of the MV can be recorded as it moves during the cardiac cycle. This image is similar to the previous LV view with triangular lines of the MV in the midst of the LV lumen. As the operator moves toward the base of the heart to the short axis basilar view, a subsequent M-Mode can be recorded with the aorta in the center of the screen and the LA in the far field. In this image a comparison of the LA to the aorta diameters is made. A ratio of LA to aorta (LA/Ao ratio) is calculated.

In addition to these standard views, other image plane may be acquired. Views of the main pulmonary artery and its branches may be seen by making angle changes from the short axis basilar view. This view is useful for the visualization of heartworms if present. From the long axis four chamber view, RA tumor may be seen well by sliding the transducer toward the sternum to tip the heart on the screen bringing the RA more into view. From the left hemithorax the heart is seen in a four chamber view with the atria in the far field and the apex of the ventricles in the near field, the so called left apical view. From this orientation, accurate Doppler velocity measurement may be acquired and well as potentially increased views of RA tumors. Other short axis views of the RA and a long axis view of the pulmonary artery may also be acquired from the left hemithorax. Representative images of all the planes discussed will be shown during the presentation.

MEASUREMENTS AND INTERPRETATION

A series of standard measurements is performed in virtually all echocardiograms. From the 2D long axis four chamber view, a measurement of the LA diameter is done at the end of systole, when the LA is at its largest. Assessment for pericardial effusion and subjective impressions of the LV and MV function may also be performed. From the 2D long axis five chamber view, the aortic diameter is measured, often at several places and timings. Most other measurements come from the M-Mode views. One of the most important is from the LV papillary muscle level M-Mode. In this view measurements are made of the IVS thickness, LV lumen dimension, and LVPW thickness all in systole and diastole. Diastolic wall thicknesses in cats are used to diagnose HCM, and the difference between the LV lumen in systole and diastole is used to determine systolic function. Only one measurement is taken from the MV M-Mode, the "E" point to septal separation (EPSS). The EPSS measures the distance from the tip of the MV at its first opening after the T wave on the electrocardiogram, to the IVS. This distance is known to increase in the face of systolic dysfunction. This measurement is specific but may not be sensitive. As mentioned earlier, the LA/Ao ratio is calculated from the short axis basilar M-Mode image. This measurement takes the diameter of the aorta and the diameter of the left atrium and compares the two. In the normal patient, the LA and the aorta are approximately the same size. As congestive heart failure set in, the LA will increase in size, yet the aortic size remains static, giving each patient its own normal to compare any left atrial enlargement to.

Assessment of systolic function is perhaps one of the most important measurements performed by the echosonographer. The systolic function is typically determined in three ways; by fractional shortening (%FS), by shortening area (%SA), or by ejection fraction (%EF). Each of these measurements determines the percent change in the LV from diastole to systole. Fractional shortening is a percent change in the linear measurement of the LV diameter from an M-Mode image. Once the LV diameters from the LV papillary muscle level M-Mode have been acquired in systole and diastole, the following formula is applied:

$$LVIDd - LVIDs / LVIDd \times 100 = \%\ FS$$

where LVIDd = the LV internal dimension in diastole, LVIDs = LV internal dimension in systole. Fractional shortening is the easiest and most common determination of systolic function performed, but is limited in value since the measurement is linear in origin. The fractional shortening only measures the change in the LV lumen at two points, one on the IVS and another on the LVPW. The heart is a three dimensional structure and moves in all three directions during systole, so in the normal heart %FS is usually sensitive enough to make clinical assessments. When the %FS is below normal however then the other measurements of systolic function should be performed. The %SA is a measurement taken from the short axis 2D image of the LV papillary muscle level. From this view, the area of the LV lumen is traced in systole and diastole. The area measurements at plugged into the same formula used for %FS, but rather than the linear dimension being used, the LV area is used. A percent change in the LV area is calculated. For a more three dimensional evaluation of the LV systolic function, the %EF is used. Ejection fraction is similar to %SA, but a volume of the left ventricle is measured in systole and diastole and a percent change calculated using the same formula. Left ventricular volume in veterinary medicine is best measured in the right parasternal four chamber long axis view. To measure the LV volume requires some mathematical assumption, since the LV is not a perfect cylinder, but more of a cone. Using a formula called the Modified Simplon's rule, the area of the LV is traced and the length measured, then the computer fills this area and length with a series of disks approximating the area at various level of the LV. The area of these disks is summed to generate an approximate volume. The advantage is that by calculating the volume, the operator takes as many data points into account as possible, but because the LV is irregularly shaped a perfect volume cannot be measured. Nonetheless, %EF is still the most accurate way to asses LV systolic function by echocardiography and is the standard in human medicine. Another advantage of the %EF is, the difference between the LV volume in diastole and systole is equal to the stroke volume, which allows for calculations of cardiac output by multiplying stroke volume by heart rate. In most echocardiographic laboratories, if the %FS is normal then the patient is considered normal. If the %FS is not normal then the other assessments of systolic function are utilized.

Interpretation of an echocardiogram requires a detailed knowledge of cardiac disease conditions. Simply acquiring images and performing measurements is usually insufficient for making accurate diagnoses of heart disease. The echosonographer need to have an excellent understanding of the cardiac cycle as well. It is imperative to know what to expect from a normal heart to recognize abnormal function or anatomy. One method of understanding the echocardiogram is to understand what question about the patient is being answered by the echocardiogram. For example, if the patient is a Doberman pinscher, a breed in which dilated cardiomyopathy (DCM) is common; then the operator must understand how DCM affects the heart and what changes in the echocardiogram are to be expected. Also, issues of systolic function become critical in the diagnosis of DCM. Conversely, if the patient is cats, which rarely acquire DCM but commonly develop hypertrophic cardiomyopathy, the sonographer needs to understand that pathophysiology. In this instance, systolic function is less of an issue and the thickness of the ventricular walls becomes critical. By understanding the pathophysiology of cardiac disease the echosonographer is never surprised by their findings, and the echocardiogram does not simply become the "heart tester", cast like a net seeking an unknown heart condition.

CONCLUSION

Echocardiography has supplanted cardiac catheterization as a diagnostic tool in most cases of heart disease in veterinary medicine. Detailed assessment of cardiac structure and assessment of systolic function has allowed diagnosis of common or complex cardiac disease to be performed on an outpatient basis with echocardiography.

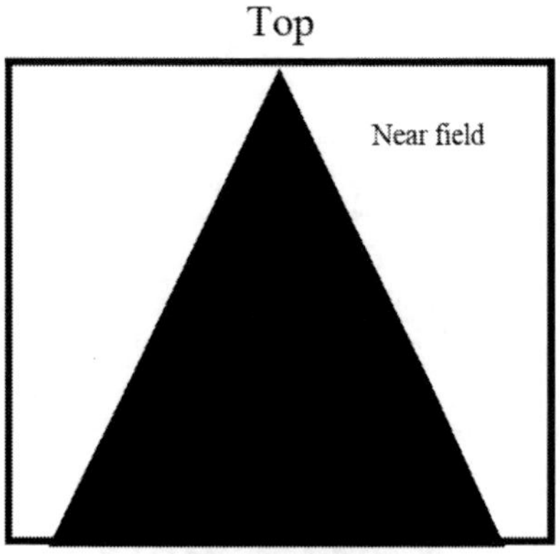

Figure 1

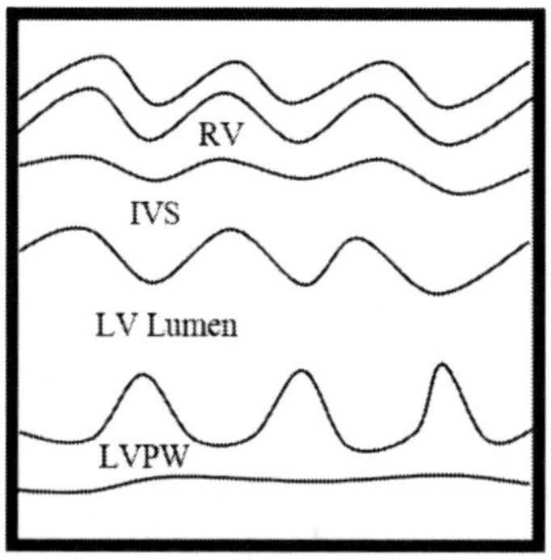

Figure 2

How Fast Is Fast: Doppler Echocardiography

H. Edward Durham Jr., CVT, LATG, VTS (Cardiology)
Columbia, MO

INTRODUCTION

Doppler echocardiography is a non-imaging modality of cardiac ultrasound that is used to measure the velocity of blood flow through the heart. Doppler studies may be performed in any vessels of the body, but for the purposes of this discussion the topic will be limited to the heart and great vessels. Doppler echocardiography is another application of ultrasound technology and is typically performed in conjunction with an echocardiogram for a complete Doppler echocardiogram. Where the echocardiogram creates an image of the heart and provides information provide about the morphology and systolic function of the heart, the Doppler portion provides hemodynamic information and allows assessment of diastolic function. The Doppler study gives the cardiologist information about the direction and velocity of blood flow. It can also be used to indicate turbulent flow in the heart.

BASIC TECHNOLOGY

Doppler studies are simply an advanced form of ultrasound that is specifically for velocity measurements of blood. This allows the cardiologist to make some conclusions about physiology and prognosis. Doppler echocardiography works on the principle of the Doppler shift first described by Johann Christian Doppler in 1864. According to the Doppler principle, a sound wave emanating from a moving object as it moves toward a stationary observer will become compressed and increase in frequency; conversely if the object is moving away from an observer the sound waves get *rarefied* or stretched out and decrease in frequency. This is a phenomenon that we all have witnessed when a car passes by our standing position. The degree of frequency shift is directly related to how fast the object is moving. For a Doppler echocardiogram, the sonographer use this principle in the same way a bat uses sonar to hunt for food. The bat transmits an ultrasonic sound at a known frequency. When the signal reflects off a moth it returns to the bat with the frequency changed depending on the moth's direction in relation to the bat; increased if the moth moving toward the bat, decreased if moving away. The degree of frequency shift tells the bat how fast the moth is moving. In the heart, a known frequency signal is transmitted by the transducer into the blood pool. From the reflected signal a calculation is made using the frequency shift to calculate the blood velocity and direction. The following formula is used by the computer to make the calculation:

$$V= C \times f_d / 2(f_o) \times \cos\theta$$

where V= velocity in m/sec, C= the average speed of ultrasound waves in soft tissue (1540 m/sec), f_d = frequency shift, $2(f_o)$ = twice the transmitted velocity , and $\cos\theta$ = the cosine of the angle of interrogation. The frequency shift is the difference between the transmitted frequency and the reflected frequency.

Because all the variables in the equation are known except for the $\cos\theta$, the angle of interrogation, also called intercept angle, is of critical importance. It is the angle that the ultrasound beam strikes its target, in this case the blood pool. (Fig. 1) The most accurate measurement are made when the beam strikes the target at perfect parallel (zero or 180 degrees), and the cosine is equal to 1 or -1. As the intercept angle increases the cosine moves away from 1 and the accuracy of the calculation decreases. The angle of interrogation is the only variable in the equation that the operator has any control over and it is critical the angle closest to parallel be achieved for the most accurate velocity. As a result, the operator cannot over estimate a true velocity, but can under estimate one with misalignment outside of pure parallel. Generally angles of intercept greater than 15 degrees are considered unacceptable. To facilitate alignment the Doppler study is performed on the left apical 5 chamber view from the left hemithorax. In this view the atria are at the bottom of the screen, the apex of the heart is at the top and the aorta heading out of the left ventricle (LV) toward the bottom. This orientation allows for alignment of the Doppler cursor with blood flow heading straight up toward the transducer from the atria and straight down out of the aorta. The pulmonary artery (PA) can be interrogated from a right parasternal short axis basilar view.

MODALATIES

Three modalities are commonly used for cardiac Doppler measurements. Spectral Doppler includes Pulsed wave Doppler and Continuous wave Doppler. The information generated is displayed on a graph with time along the X axis and velocity on the Y axis. In the center of the graph is a baseline indicating zero flow. Returning ultrasound signals of flow toward the transducer are displayed above the baseline, and signal moving away from the transducer are displayed below the baseline. The baseline can be shifted up or down to create more graph above or below depending in the predominate direction of blood flow being recorded at the time. (Fig. 2) Color Flow Doppler uses the same principles but the information gained is displayed as a color signal superimposed atop a 2D echocardiographic map. To use Spectral Doppler a cursor is turned on over the 2D echocardiographic image and aligned with the blood flow to be interrogated and the Doppler of choice activated.

Pulsed wave Doppler (PW) is a system that allows for velocity measurements at a specific location in the heart. A sample volume or gate is positioned along the cursor line over the area for velocity measurement. When the PW Doppler is turned on, a piezoelectric crystal transmits a short burst of ultrasound waves that travel to the designated depth while the transducer waits, and the reflected data is received by another crystal. This pattern is repeated many thousands of times a second. The number of times the crystal can transmit, shut off, wait for the returning signal and then transmit again is the *pulse repetition frequency* (PRF) and represents how many times it can sample per second. The PRF is important in that the faster the PRF or sampling rate, the faster a velocity that can be recorded. If the returning signal has a frequency shift greater than twice the PRF, ambiguous information is returned to the computer. The farther the sample gate is from the transducer, the lower the PRF due to distance and the slower a velocity can unambiguously be recorded. In essence, the returning signal is faster than the transducers ability to sample, and *aliasing* occurs. A similar phenomenon occurs when watching a carriage with spoke wheels move. When the carriage is moving slowly, the spokes turn at a rate slower than the human eyes ability to see them move, but as the carriage speeds up the spokes begin to appear to move backwards. Our minds know the carriage is moving forward, yet because our eyes can only sample images at set speed, the spokes appear to move backwards. With PW Doppler an aliased signal is displayed "wrapped around" into the opposite side of the spectral graph (Fig. 3). The velocity at which aliasing occurs is variable due to depth, but the average is 1.5 m/sec. So the advantage of PW is the ability to record velocities at the specific depth, but because the transducer must stop transmitting for a split second between pulses to receive returning information, PW has a limited velocity range it can detect.

Continuous wave Doppler (CW) allows for measurement of very fast velocities by continuously transmitting and receiving, and is displayed the same as PW Doppler (Fig. 2). Unlike PW Doppler, that records velocities in one site, the CW Doppler records velocities all along the length of the ultrasound beam. It is impossible for the frequency shift to be greater than the sample rate which is continuous, so extremely fast velocities can be measured. The point along the beam that the highest velocity occurs however is unknown. Continuous wave Doppler therefore provides a method of measuring the high velocity blood flow of congenital heart disease, but is range ambiguous. In a practical sense, high velocity jets can be located with the PW Doppler and then accurately quantified by the CW Doppler.

Color Flow Doppler (CF) also known as Color Flow Mapping, is an advanced form of PW, but rather than the returning signal being displayed on a graph, the information is color coded and displayed atop the 2D image. In CF Doppler there are a series of PW sample gates alongside each arranged on a plane. When the CF feature is initiated a box appears over the 2D echocardiographic image. This box can be moved over the region of interest and size adjusted. The colors of red and blue are traditionally used to display directions of blood flow; blue for away from the transducer and red for toward the transducer. The acronym BART (Blue Away; Red Toward) is often used. Color Flow Doppler allows the speedy identification of both normal and abnormal flow. It aids in the alignment of the PW or CW Doppler cursor for velocity measurements. It also helps locate and estimate the severity of valvular regurgitation or cardiac shunts. An added feature of CF Doppler is *variance*. Variance is the term used for turbulent blood flow, which is normally laminar in the heart. Turbulence is simply blood moving is multiple directions at multiple velocities. To demonstrate turbulence most echocardiographs add a yellow-green mosaic color map with the standard red or blue signal. The limitation of CF Doppler is that it is also subject to aliasing. As with PW when the returning frequency shift is greater than twice the PRF, then the signal gets "wrapped". In CF Doppler that means a change in color. The colors selected by the computer occur on a color wheel, with shades of blue on one half and shades of red on the other. As the recorded velocity increases in CF, the intensity of the color increases in that direction or color. If the recorded velocity exceeds the PRF then the computer will take the next color on the color wheel. For example, if the CF signal is looking at flow away from the transducer, it will code the returning signal in blue of intensifying color until it reaches the maximum velocity it can record in the direction (the Nyquist limit) then it will use the next color in the wheel which will be a shade of red. So even though the operator may know that the flow is moving away from the transducer, the color displayed on the screen will be red. Adjustments to the depth of the color box or some changes in PRF may correct this problem, but only to a point.

INTERPRETATION

The practical use of Doppler in echocardiography is for the determination of velocity from one chamber of the heart to the next to calculate a pressure gradient (PG). All movement of blood within the body, from a physics point of view, is based on pressure gradients; the increased in pressure in one chamber that drives blood to the next. The velocity of blood moving from one chamber to the next is directly related to the pressure gradient by the modified Bernoulli equation:

$$PG = 4 \times V^2$$

where V is velocity in m/sec. The pressure gradient is expressed in mmHg. The operator must understand normal cardiac physiology to interpret the findings. For example, it is known that a normal velocity out the aorta is 1.0 – 1.7 m/sec. Using the Bernoulli equation indicates PG from the LV to the aorta is approximately 4-12 mmHg. If this velocity be significantly increased then a case of aortic stenosis is present. The velocity and calculated PG are then used to assign severity. If the velocity recorded by spectral Doppler is 5 m/sec then a PG of 100 mmHg exists. From a physiology standpoint, a 100 mmHg PG means the LV must then generate an extra 100 mmHg of pressure to move blood past the aortic valve. This represents a two fold increase in the work load of the ventricle indicating a severe aortic stenosis (the normal 125 mmHg + the 100 mmHg of the stenosis). The same function is used for the pulmonic valve to diagnose and assign severity to pulmonic stenosis.

Another application of Doppler derived PGs is the estimation of normal or abnormal chamber pressures. In the case of a ventricular septal defect (VSD), Doppler can be used to detect and equilibration in the ventricles. Again, an understanding of normal physiology is critical. In the normal heart, the LV systolic pressure is approximately 125 mmHg, and the right ventricular systolic pressure is approximately 25 mmHg. The pressure difference between the two ventricles is approximately 100 mmHg. When the VSD is interrogated by CW Doppler the operator will record a velocity at or close to 5 m/sec if the ventricles still have their normal pressure, implying normal chamber pressures and therefore a compensated heart. If the velocity recorded is significantly below 5 m/sec, then decompensation may be occurring. Likewise with mitral regurgitation (MR), the velocity of the MR jet should be 5 m/sec or above since the left atrial (LA) pressure should be 5-10 mmHg during ventricular systole, and the LV pressure should be 125 mmHg. If the LV to LA pressure gradient is significantly below 5 m/sec, then the left ventricle is not able to generate its normal pressure.

In cases of patent ductus arteriosus (PDA), both CW and CF Doppler are of great benefit. Patient with PDA, flow is continuously flowing from the aorta into the main PA. By placing a CF Doppler box over the main PA a continuous turbulent color signal will be seen and is diagnostic for PDA. The CW cursor can then be placed in the PDA aperture and a velocity through the PDA measured. If the patient is well compensated, then this PG should be close to 100 mmHg, since the normal aortic pressure is 120 mmHg and the normal PA pressure is 20-25 mmHg. If the patient is decompensating then the trans-PDA velocity will be below 5 m/sec typically indicating lower than normal aortic pressure and a failing left ventricle.

A third application for Doppler is estimation of chamber pressures "downstream" of two adjacent chambers. Pulmonary arterial pressures can otherwise only be determined by invasive means of direct catheterization and pressure transducers. By using Doppler velocities and pressure gradients of tricuspid regurgitation (TR) the echosonographer can estimate pulmonary arterial pressure noninvasively. First the operator must determine the PG between the right ventricle (RV) and the pulmonary artery. In the absence of pulmonic stenosis is present that would create an increase in the RV pressure, then the PG from the RV to the PA would be normally approximately 4 mmHg. With a RV to PA pressure gradient of 4 mmHg the operator can stipulate that the RV pressure and the PA pressure are the essentially same (RV mmHg = PA mmHg) because 4 is so close to zero. At this point the operator looks for TR and if present measures its velocity. If no TR is present then this application of Doppler is unavailable. The PG of the TR tells the operator the pressure difference between the RV and the right atrium (RA), so if the TR velocity is 4m/sec then the PG is 64 mmHg between the RV and the RA. The normal pressure of the RA is also about 4 mmHg so the operator can likewise stipulate it to be zero also (technically the 4 mmHg from the PV to PA pressure gradient and the 4 mmHg RV to RA gradient cancel each other out). The conclusion is that the RV pressure is approximately 64 mmHg systolic. As stipulated earlier the RV mmHg is equal to the PA mmHg, so the operator knows the PA pressure is also approximately 64 mmHg. Normal pulmonary arterial pressure in systole is reported to be 20-25 mmHg, therefore a pressure of 64 mmHg is indicative of pulmonary hypertension.

In addition to these uses of Doppler echocardiography, assessments of diastolic function can be made. By looking at filling patterns of the left ventricle, and flow through the pulmonary veins as they enter the left atrium, the cardiologist can make some observations about relaxation of the ventricle. Doppler can also be used to look at the isovolumic relaxation time (IVRT), the time between the closing of the aortic valve and the opening of the mitral valve as a measure of cardiac relaxation. As the ventricle fail it does not fill as well and IVRT will increase in length. There are many other methods of assessing diastolic function that are beyond the scope of this proceeding the interested echosonographer may study.

CONCLUSION

Doppler echocardiography is an extremely valuable tool for the diagnosing and prognosticating of cardiac disease. When combined with 2D and M-Mode echocardiography, Doppler can provide a complete picture of acquired and complex congenital heart disease. Color Flow Doppler adds speed and accuracy to the Doppler study by allowing the operator to locate and align with the blood flow much faster. It is incumbent on the echosonographer to become well versed in the operation and limitations of each type of Doppler modality and have a detailed knowledge of cardiac physiology.

ABBREVIATIONS

Left Ventricle	LV	Pulse Wave Doppler	PW
Right Ventricle	RV	Continuous Wave Doppler	CW
Left Atrium	LA	Color Flow Doppler	CF
Right Atrium	RA	Pulse Repetition Frequency	PRF
Pulmonary Artery	PA	Pressure Gradient	PG
Ventricular Septal Defect	VSD	Meters per Second	m/sec
Patent Ductus Arteriosus	PDA	Millimeters of mercury	mmHg
Mitral regurgitation	MR	2D	2 dimensional echocardiogram
Tricuspid Regurgitation	TR	Isovolumic Relaxation Time	IVRT

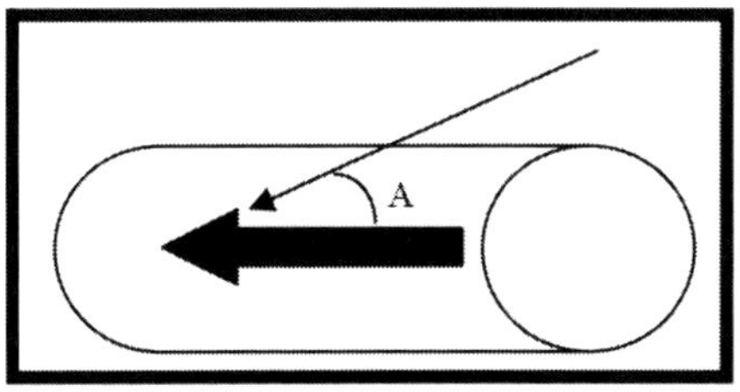

Figure 1
A= Angle of Intercept or θ
Heavy arrow = blood flow
Light arrow = ultrasound beam

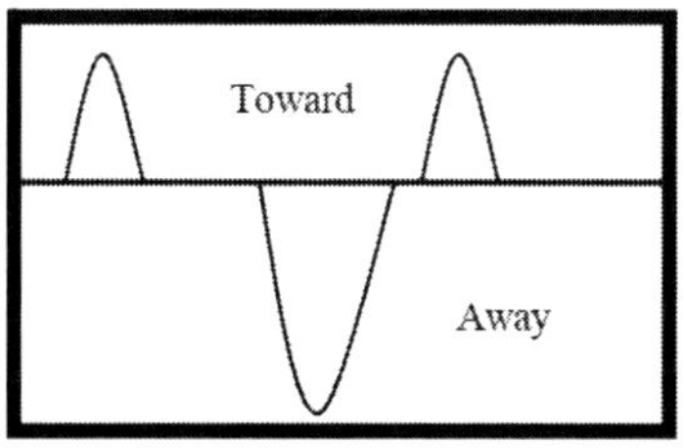

Figure 2

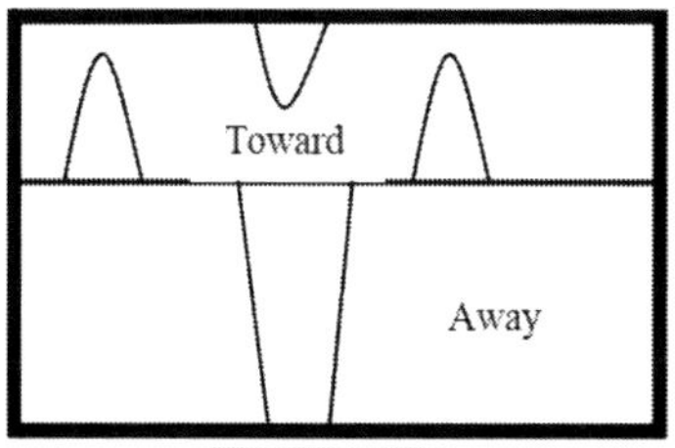

Figure 3

A REVIEW OF CANINE AND FELINE CONGENITAL HEART DISEASE

Charlotte Pace, BA (Hons), RVN, VTS (Cardiology)
London, UK

INTRODUCTION

Congenital heart disease can be caused by either malformations within the heart or of the great vessels, or the persistence of a normal fetal structure following birth. Cases are usually picked up at first vaccination where no clinical signs have been detected, but can also present at the other end of the scale, in overt heart failure. History, physical exam and diagnostic tests will help identify the defect. Some conditions can be treated with surgery or interventional procedures, but this is preferable before signs of heart failure have developed.

PATENT DUCTUS ARTERIOSUS (PDA)

PDA is one of the most common congenital diseases in the dog, but it is rarely seen in the cat. The nature of this defect is that the fetal blood vessel (ductus arteriosus) has failed to close following birth. The ductus arteriosus connects the descending aorta with the pulmonary artery, and as a consequence when this fails to close, blood shunts from the higher pressure aorta to the lower pressure pulmonary artery. This leads to volume overload of the left atrium, left ventricle and pulmonary vasculature. Clinical signs depend upon the severity of the shunting blood and dogs can range from asymptomatic (the classic continuous 'machinery' murmur is picked up on routine first vaccination), to weakness or exercise intolerance, stunted growth in comparison to the rest of the litter, or signs of left sided heart failure. Thoracic radiographs may show left atrial enlargement, left ventricular enlargement, dilated pulmonary veins, pulmonary artery and aorta, but the best diagnostic method is echocardiography. Whilst visualising the ductus may be difficult, Doppler echocardiography will show a continuous blood flow within the pulmonary artery. Prognosis is poor if the PDA is not closed. One study showed that 50% of dogs died less than a year old if the PDA remained untreated1. Surgical ligation has historically been the only option for occlusion, but interventional methods of closure are now favored among many cardiologists, such as coil embolisation or vascular occlusion devices. The Amplatz™ Canine Duct Occluder (ACDO) is a vascular occluder that has recently been designed for use in dogs . If the dog has left sided heart failure this must be treated before anesthesia. In some cases, pulmonary hypertension occurs. This leads to pressure overload of the right ventricle, and if the pulmonary arterial pressure exceeds the systemic arterial pressure, blood shunts from right to left (right-to-left shunting or 'reverse' PDA). Ligation and occlusion are contraindicated in these cases as the ductus is now acting as 'safety valve' to relieve excessive pulmonary artery pressure. Clinical signs include hind limb weakness or collapse, a split or loud second heart sound, and differential (caudal) cyanosis. Polycythemia is usually present. Therapy to reduce packed cell volume (PCV), such as regular phlebotomy or hydroxycarbamide (hydroxyurea) can be effective to treat clinical signs. Pulmonary vasodilators, such as sildenafil, may also be effective as they can decrease the severity of pulmonary hypertension. If hydroxycarbamide is used, it is important to note that this is a cytotoxic drug and must be handled appropriately.

AORTIC STENOSIS (AS)

This is another common defect in the dog and rarely seen in the cat. Oxygenated blood leaves the left ventricle via the aorta. Stenosis is a narrowing of the aortic valve and impedes the flow of blood, particularly during periods of high demand such as exercise. The increased pressure in the left ventricle may cause hypertrophy and this in turn can result in ventricular arrhythmias. The most common form of this disease in dogs is a narrowing beneath the aortic valve, usually a fibrous or fibromuscular ridge. This is referred to as subaortic stenosis (SAS). Clinical signs depend upon the severity of the stenosis and so some can present asymptomatic albeit with a harsh systolic murmur, but exercise intolerance, syncope and even sudden death can occur. Echocardiography is very important to assess the severity of the stenosis, typically by measurement of blood flow velocity across the stenotic lesion using spectral Doppler. Treatment is advised for moderate to severe cases, or those showing clinical signs. Surgical options include resection of the stenotic region, although one study showed no improvement in long-term survival with this technique . Balloon dilation of the stenosis can be attempted, but this is more suitable for valvular stenosis and again, has not improved life expectancy compared to medical management . Beta-blockers can be used to reduce the outflow obstruction, decrease myocardial work and increase coronary perfusion of the myocardium by prolonging diastole. Prognosis varies with the severity of the stenosis. The majority of dogs that have a mild form of aortic stenosis will live normal lives. Long-term prognosis in dogs with severe disease is generally guarded to poor as sudden death or development of left sided heart failure are common.

PULMONIC STENOSIS (PS)

This is also a common defect seen in the dog, and again, is uncommon in the cat. Deoxygenated blood leaves the right ventricle via the pulmonary artery to the lungs. Stenosis of the pulmonic valve impedes the flow of blood and so can cause high pressure on the right side of the heart. The right ventricle has to work harder to maintain adequate lung perfusion and this can cause right ventricular hypertrophy. Also a minority of dogs can have concurrent tricuspid valve dysplasia that can exacerbate the development of heart failure (regurgitation from the defective tricuspid valve will lead to right atrial dilation). Clinical signs depend upon the severity of stenosis; some dogs can present asymptomatic with a systolic murmur picked up at vaccination, others present with signs of exercise intolerance or syncope. As with aortic stenosis, sudden death and heart failure can also occur, although heart failure is right sided with PS. Thoracic radiographs may show right ventricular enlargement and post stenotic dilatation of the pulmonary artery. Echocardiography is essential to evaluate the type of stenosis (valvular is most common) and the severity of stenosis by use of spectral Doppler. Therapy is indicated in moderate to severe cases, and those showing signs of heart failure. Although surgical methods have been described, balloon valvuloplasty is the preferred treatment method. This involves placing an angiographic balloon catheter into the position of the stenotic valve and 'forcing' the valve to open by dilation of the balloon. If successful, this can alleviate the severity of the stenosis and result in improvement of clinical signs and decreased likelihood of sudden death. If right sided heart failure is present, the prognosis is more guarded and this must be treated before valvuloplasty is considered. Success of balloon valvuloplasty has been defined in one case series as a reduction in pressure gradient across the stenotic lesion by more than 50% . Surgical treatment may be indicated in cases for which balloon valvuloplasty is ineffective or inappropriate.

VENTRICULAR SEPTAL DEFECT (VSD)

This is a relatively common defect found in the cat, but uncommon in the dog. VSD is a malformation of the interventricular septum, which allows blood to flow from the left side to the right side (high to low pressure). This causes volume overload of the pulmonary circulation, left atrium and left ventricle, and can result in left sided heart failure. Clinical signs vary on the size of the defect. With small ('resistive') defects patients are asymptomatic except for a loud, harsh (typically right sided) systolic murmur and may have a normal quality and length of life. With large ('non-resistive') defects, the heart murmur may be softer, but clinical signs related to pulmonary over circulation may develop, such as left sided heart failure or pulmonary hypertension. If pulmonary hypertension is severe enough to result in reversal of blood flow from the right to left ventricle, cyanosis will occur. This is referred to as 'Eisenmenger's syndrome'. Treatment of this is similar to right-to-left shunting PDA. Echocardiography is used to visualise the defect (if large) and Doppler is used to assess blood flow through the VSD. Dilation of the left atrium and left ventricle can also be assessed and this can also been observed on thoracic radiography, along with presence or absence of pulmonary over circulation and pulmonary edema. Therapy depends upon the size of VSD. If the defect is small, no therapy is required as a normal full life can be expected. Otherwise, treatment is usually restricted to managing signs of heart failure or cyanosis if present. Transcatheter closure of a VSD has been reported in the dog .

ATRIAL SEPTAL DEFECT (ASD)

This condition is rarely seen in animals, but is more likely to be seen in the cat. An ASD is a malformation of the atrial septum, and so blood flows from the left to the right atrium. It rarely causes clinical signs unless very large. It may be seen as part of more complex congenital cardiac diseases (e.g. atrioventricular septal defects). A persistent foramen ovale (PFO) is also a communication between the right and left atria. Unlike an ASD it is not a malformation of the septum, but rather the persistence of a normal fetal structure. It is most commonly seen in association with congenital diseases that result in elevated right atrial pressures, such as severe pulmonic stenosis or tricuspid valve dysplasia.

ATRIOVENTRICULAR VALVE DYSPLASIA

Dysplasia of either of the atrioventricular valves (mitral valve, tricuspid valve) can occur in both the dog and cat. It occurs when malformation of either valve leads to stenosis and/or regurgitation. This leads to dilatation of the atria, which in turn, can lead to heart failure and arrhythmias. Clinical signs with mitral valve (MV) dysplasia include a left sided heart murmur, syncope (often associated with tachyarrhythmias), and left sided heart failure. Thoracic radiography may show evidence of left atrial and left ventricular enlargement, and if present, pulmonary congestion and edema. Echocardiography is useful to assess the severity of valvular malformation, regurgitation, stenosis (if present) and chamber enlargement. Treatment is usually limited to managing heart failure and arrhythmias. Once these occur, prognosis becomes guarded. A breed predisposition in Bull Terriers has been reported . Clinical signs with tricuspid valve (TV) dysplasia are similar except that a right sided heart murmur and right sided heart failure can occur. TV dysplasia has also been reported to be overrepresented in certain breeds, and in Labrador Retrievers it has been shown to have a genetic basis . Valve replacement surgery has been attempted in large breed dogs in centres with cardiopulmonary bypass facilities .

TETRALOGY OF FALLOT

This condition is characterised by four concurrent abnormalities. The defects are:

- Pulmonic stenosis
- Ventricular septal defect
- Over riding aorta
- Right ventricular hypertrophy due to outflow resistance

Clinically this means there is pressure load on the right ventricle, under circulation of the lungs and consequently perfusion of the peripheral circulation with deoxygenated blood. This leads to an elevated PCV, which can often be over 70%. Clinical signs include stunted growth, exercise intolerance, syncope, tachypnea and cyanosis. Thoracic radiographs may show right ventricular enlargement and under perfusion of pulmonary vessels. Echocardiography is necessary to confirm the presence of the four cardiac abnormalities. Therapy is usually palliative as, if left untreated, prognosis is poor and sudden death common. Therapy to reduce PCV, such as regular phlebotomy is recommended to maintain a level where clinical signs are minimal . Palliative surgery to create an aortic to pulmonary artery shunt can be effective to alleviate clinical signs and improve long-term prognosis .

VASCULAR RING ANOMALY (VRA)

This condition causes compression of the esophagus (and sometimes trachea) due to the malformation of one or more of the aortic arches. The most common presentation is persistent right aortic arch (PRAA), in which the right aortic arch persists instead, or in addition to, the left aortic arch (the normal aorta). Affected puppies can present with regurgitation and signs associated with aspiration pneumonia. A dilated proximal esophagus can be seen on radiographs with or without a barium swallow. An alveolar pattern, typically in the cranial or right middle lung lobes, may be present with aspiration pneumonia. The only effective therapy is surgery to relieve the constriction. If concurrent aspiration pneumonia is present, medical management to treat or prevent this is essential.

COR TRIATRIUM

This is an uncommon defect that is caused by an abnormal membrane dividing the left atrium (cor triatriatum sinister) and the right atrium (cor triatriatum dexter) into two parts. Cor triatrium dexter (CTD) has been reported in the dog, whilst cor triatrium sinister (CTS) has been reported in the cat. Dogs with CTD can present with signs of caudal right sided heart failure (ascites, no jugular distension) as the location of the membrane result in elevated caudal vena caval pressures, while the cranial vena cava is unaffected. Balloon dilation of the membrane can be performed to improve clinical signs and alleviate heart failure . Cats with CTS generally present with signs of pulmonary congestion or pulmonary edema. Successful surgical correction has been reported .

REFERENCES

1. Eyster GE et al, JAVMA 1976, 168:435-438. 2 Nguyenba TP et al, JVC 2007, 9, 109-11. 3 Orton EC et al, JAVMA 2000, 216:364-367. 4 Meurs KM et al, JAVMA 2005, 227: 420-424. 5 Stafford Johnson M et al, JSAP 2004, 45:148-153. 6 Bussadori et al, JVIM 2007, 21:1396-1400. 7 Margiocco ML et al, JVC 2008, 10: 61-66. 8 Malik R et al, JSAP 1998, 29: 549-557. 9 Famula TR et al, 2002, AM J Vet Res, 63, 816-820. 0 Brockman DJ, Aug 2007, Proceedings of 32nd AWSAVAC. 11 Ware W 2007, Cardiovascular disease in small animal medicine. 12 Brockman DJ et al, 2007, JAVMA, 231:721-726. 13 Stafford Johnson M et al, 2004, JSAP, 45:16-20. 14 Wander KW et al, 1998, Journal Am Animal Hosp Assoc 34:383-386.

CARDIAC CATHETERIZATION TECHNIQUES IN SMALL ANIMALS

Kristie Garcia, LVT, VTS (Cardiology)
Tinton Falls, NJ

Advantages of cardiac catheterization over surgical intervention include, but are not limited to: shorter hospital stay, lower morbidity rate, lower infection rate and cardio techs love them. ☺

The basic steps of cardiac catheterization are essentially the same for each procedure: **access** to the vessel, **diagnostics** to confirm necessary treatment, and then **treatment.**

Access

The appropriate site is cardiologist-determined based on the procedure that is being done as well as any other relevant factors. Prior to the procedure the gauge of the access needle, the French size of the catheters as well as the size and length of the guidewires are determined in order to ensure that all of the necessary equipment is available. During one procedure, many catheters and guidewires of different lengths and sizes may be used. Often the clinician may need to 'size up' the introducer during a procedure to introduce a larger French catheter. A vital role of the cardiology technician is to make sure all sizes are in the inventory and sterile.

Diagnostics

Information to assist the clinician in approach to treatment is obtained during a procedure to determine the correct approach to treatment. Instruments like a TEE probe, power injectors and many different types of catheters are used at this point in a procedure. With the appropriate catheters, angiography, pressure readings and evaluation of cardiac output can be done. Myocardial biopsies may also be taken at this point. Obtaining information through diagnostics often can cause arrhythmias or pressure changes in the patient and must be monitored closely. The type and size of the catheter chosen is crucial when obtaining measurements in the heart in order to properly treat and have as few side effects as possible.

Treatment

Based on the diagnostics done, the appropriate treatment can then be given to the patient. Whether it is to determine the correct occluder sizes, or to measure for the appropriate size balloon, the success of treatment for a patient is related to the accuracy of the diagnostic results they are based on. Complications for all of these procedures are bleeding. Anytime a vessel, specifically an artery, is compromised, bleeding is a side effect and needs to be controlled as much as possible both during the procedure and after the patient is in recovery. Complications also include infection, arrhythmias, and the possibility of sudden death. Post procedure follow-up and care varies with each procedure done. These tests may include bloodwork, blood pressure readings, electrocardiograms, echocardiograms and radiographs.

THE SICK KIDNEY: ARF, CRF, & EVERYTHING IN BETWEEN

David Liss, RVT, VTS (ECC)
Woodland Hills, CA

INTRODUCTION

Renal disease is a commonly encountered presentation to the veterinary hospital. Both cats and dogs alike can become afflicted with the various classes of this group such as: acute or chronic, tubular, interstitial, or glomerular. Kidney insufficiency and kidney failure are also two different things. This lecture will serve to clarify and explain kidney anatomy, physiology, pathophysiology, and treatment tenets of kidney disease.

ANATOMY AND PHYSIOLOGY:

The kidney is a small bean shaped organ that lives in the retroperitoneal space located in the latter two thirds of the abdomen. Felines and canines each have two kidneys. The canine and feline kidneys are typically easily palpable and the right kidney lies more cranial than the left. The gross anatomy of the kidney includes the capsule, cortex, medulla, and renal pelvis. The capsule is a tough, fibrous, sheath that encloses the kidney and keeps it relatively free from movement. The cortex is the outer layer of tissue, the medulla is the inner part. The renal pelvis is a large collecting area for urine which empties into the ureters. The caudal abdominal aorta branches into the renal arteries which each go to one kidney. The caudal vena cava branches off as well, forming the renal vein. The individual cellular unit of the kidney is the nephron. Below is the structure of a nephron:

Figure 1: Nephron

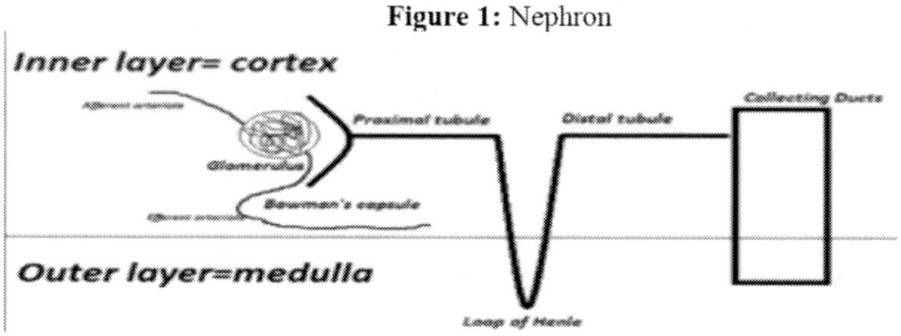

Blood flows to the nephron from the afferent arteriole. Within the glomerulus, high pressures force water and small electrolytes out into Bowman's capsule. Larger molecules such as proteins are kept within the vasculature. Blood then travels away from the Glomerulus via the efferent arteriole. The "ultrafiltrate" crosses the proximal tubule, where around 70% of Na, Cl, K, and HCO3 are reabsorbed. Most of the urea is also excreted here. This makes it easier for water to move into or out of the loop of Henle. This concentrated pre-urine product then crosses the loop of Henle, where water is reabsorbed or excreted as needed depending on hydration. Lastly, the "almost" urine product finds the distal tubule, where some sodium is reabsorbed and potassium excreted, based on the action of Aldosterone. It then congregates in the collecting ducts and passes into the renal pelvis, and into the ureters.

Renal blood flow is very important. The kidney only has one artery and one vein, so blood supply to each and every nephron relies on the patency, and continuity of flow. Should cardiac output drop suddenly for any reason, the kidney's perfusion could be highly compromised. This is why monitoring blood pressure under anesthesia is very important! The kidneys are able to independently manage their own blood pressure (autoregulate) at a range of blood pressures. There are various numbers, but 60-160 mmHg is a good range. This means that below 60mmHg (mean pressure) the kidneys cannot perfuse themselves adequately. Above 160mmHg as well can cause renal damage.

TYPES OF RENAL DISEASE

There are two major categories of renal failure, acute and chronic. Renal insufficiency represents >70% loss of nephrons, required for azotemia to develop, but does not necessarily manifest as a clinical syndrome. Renal failure means a syndrome of either acute or chronic nature, causing uremia, acid/base disturbances, and abnormal fluid balance. Acute renal failure is a sudden decline in GFR (glomerular filtration rate) which is potentially reversible. Chronic renal failure is slower process of destruction of healthy renal tissure. CRF is not reversible. Acute on chronic renal failure represents a sudden decline in GFR associated with pre-existing renal disease.

Whether the renal disease is acute or chronic it can be associated with renal parenchyma (tissue), or interstitium, the tubules, the glomerulus, or more than one section of the nephron. Some examples of each disease process are listed below.

Table 1. Various forms of renal disease

Various forms of renal disease		
Disease	**Part of the nephron**	**Acute or chronic**
Glomerulonephritis	Glomerulus	Can be acute, usually chronic
Acute tubular necrosis	Tubules	Acute
Intersitial nephritis	Interstitium	Chronic
Amyloidosis	Glomerulus	Chronic
Toxicity: Ethylene Glycol	Tubules	Acute
Toxicity: Lillies/Grapes/Raisins	Interstitium	Acute
Pyelonephritis	Interstitium +/- tubules	Acute
Aminoglycoside toxicity	Tubules	Acute

Effort is often made to clarify the type of renal failure as actually diagnosing a disease can be challenging. For example, if a patient has azotemia that is renal in origin, they will have intrinsic renal failure as opposed to a renal failure caused by a vascular or extra-renal event.

PRESENTATION

Animals with renal failure are often pretty sick. They may present with a history of anorexia, weight loss, vomiting, or diarrhea. They may have had a known toxin ingestion, or may have been potentially exposed to toxins in the home. They may also be polyuric/polydipsic. Initial interventions include a full history and physical exam.

DIAGNOSTICS

A full set of laboratory tests should be run including: Chemistry, CBC, PCV/TS, venous blood gas, and a urinalysis. Chemistry results may reveal azotemia (hallmark of decreased GFR). Azotemia is characterized by an increase in nitrogenous waste products, namely BUN and Creatinine. Azotemia can be pre-renal, renal, or post-renal. A urine specific gravity will help differentiate between the two. The chart below helps break down the difference in azotemia.

Table 2. Classification of Azotemia

Types of Azotemia		
Type of azotemia	**USG results**	**Patient condition**
Pre-renal	Concentrated <1.030	Patient is dehydrated or hypovolemic. This results in a decline in GFR as the body preserves fluids for vascular volume.
Renal	Isosthenuric 1.007-1.012	If patient is dehydrated/hypovolemic: The kidneys are not doing their job and concentrating urine. The urine is the same USG as plasma, meaning it is being excreted without any water retention.
Post-renal	Varies: usually concentrated	Hard to simply diagnose post-renal azotemia from USG but if any evidence of osbtruction present this would result in a post-renal azotemia.

Other lab work abnormalities may reveal

Hyperphosphatemia, Hyperkalemia (inappropriate retention)- acute, Hypokalemia- chronic, Anemia (lack of erythropoeitin production)- non-regenerative, Glucosuria, proteinuria if glomerular damage, Casts on sediment analysis- if tubular function damaged, Metabolic acidosis from uremic acids

Additional workup options include

Ultrasound with fine needle aspirate and cytology, Leptospirosis titers (acute), Ethylene Glycol tests (acute), Tick borne disease titers (acute and/or chronic). All patients should also have baseline blood pressures as many of these patients are hypertensive

TREATMENT-ACUTE

Treatment of the ARF patient is very labor intensive. These patients need fluids and volume so a peripheral IV catheter should be placed. Any intravascular volume deficits should be addressed immediately via fluid blouses. After volume replacement hydration, maintenance and urine production should be of concern. A central line may be an option as well for CVP monitoring. They should have regular blood pressures, weights, and a urinary catheter placed to monitor output. An increase in weight after hydration, elevated blood pressures, CVP's, or decreased urine output should alert you to potential volume overload. Urine output should be 1-2ml/kg/hr on maintenance fluids. Ins should equal outs. If the patient is polyuric, fluids should be increased to meet demand. If anuric, fluids slowed to prevent iatrogenic fluid overload. They should potentially have broad spectrum antibiotics started in the face of a potential infectious process. If Lepto is suspected, care should be taken to vigilantly avoid urine. Life-threatening electrolyte disturbances (hyperkalemia and hypocalcemia) should be addressed with insulin/glucose, and/or parenteral calcium. If the patient is hypokalemic, postassium should be added to their fluids. Do not exceed 0.5meq/kg/hr. Early nutrition is also important in the renal failure patient. Acid-base abnormalities (typically metabolic acidosis) will often resolve or improve with fluid therapy. Sodium bicarbonate therapy is only reserved for life threatening academia. These patients should receive stress ulcer prophylaxis, including an H2 blocker and/or a proton pump inhibitor (omeprazole or Pantoprazole). Because emesis can be caused by uremia, a centrally acting anti-emetic may be necessary. Metoclopramide, Maropitant, or Ondansetron/Dolasetron are efficient in managing emesis in these patients. Hyperphosphatemia can be nauseating so phosphorus binders (Aluminum hydroxide) should be given if the patient can tolerate enteral medication. If the patient is anemic, and blood transfusion may be necessary, however, the acute renal failure patient is usually not anemic. Renal function should be assessed with fluid therapy, if the patient decompensates into oliguria or anuria, dialysis should be considered. These patients should be monitored for fluid overload as well. This can manifest as chemosis, peripheral edema, tachypnea, hypoxemia, and/or pulmonary edema. Constant auscultation of lung fields may help catch development of wet lungs.

TREATMENT-CHRONIC

Patients hospitalized for chronic renal failure should receive all of the monitoring an acute renal failure patient does. However, they may also be anemic. Some patients compensate well for anemia, but some may require a transfusion. The decision to transfuse is not always easy and should be based not solely on the PCV values. These patients may also have GI blood loss as uremia affects platelet function and predisposes to gastric ulceration. Chronic renal failure patients are typically hypokalemic and should receive potassium supplementation. The philosophy of treatment with chronic cases is to treat the acute crisis and then focus on long term management. Nutritional support may be necessary in the form of an E-tube, PEG tube or other longer term feeding device. Low protein and phosphorus diets truly slow progression of the disease and enhance survival. Hypertension should be managed typically with calcium channel blockers, and proteinuria managed with ACE inhibitors.

Table 3. Renal Pharmacology

Drug	Drug Class	Dose Range	Frequency	Route	Indications
famotidine	H2 blocker	0.5-1mg/kg	SID	IV, SQ	Uremia, gastritis
pantoprazole	Proton pump inhibitor	1mg/kg	SID	IV	Uremic gastritis
aluminum hydroxide	Phosphorus binder	30-90 mg/kg	SID	PO	Hyperphosphatemia
furosemide	Diuretic	2mg/kg repeated x 3 or CRI of 1mg/kg/hr	As needed	IV	Anuria/Fluid overload
mannitol	Diuretic	0.5-1gm/kg	Once	IV slow	Anuria/Fluid overload
metoclopramide	Anti-emetic	1-2mg/kg/day	CRI	IV	Vomiting
sucralfate	Gastric coating agent	0.25g-1g PO	QID	PO	Gastric ulceration
amlodopine	Anti-hypertensive	0.625-1.25mg/cat, 0.2-0.4mg/kg dog	SID-BID	PO	Hypertension
omeprazole	Proton pump inhibitor	0.5-1mg/kg	SID	PO	Uremic gastritis
benazapril	ACE Inhibitor	0.25-0.5mg/kg	SID-BID	PO	Proteinuria

PROGNOSIS

For chronic renal failure, prognosis depends on the severity of the disease. With mild-moderate disease (Creatinine <3-5 mg/dL) survival is 1-3 years. Proteinuria and hypertension are negative survival indicators. With ARF, mortality is 60%. 2/3 of the patients that survive ARF have some degree of chronic renal failure.

SUMMARY

Kidney disease is complex. This arises from the complex nature of the urogenital anatomy of the kidney, and the variability with presentations. With a greater understanding of kidney disease, veterinary technicians are more prepared to institute intricate and top notch nursing care to these often quite critical patients.

REFERENCES

[1.] Langston, Catherine. **Small Animal Critical Care Medicine. Eds**: Hopper & Silverstein. Saunders Elsevier. St. Louis Missouri: 2009. pp. 590-594 [2.] Langston, Catherine . Eds: Hopper & Silverstein. Saunders Elsevier. St. Louis Missouri: 2009. pp. 594-598 [3.] Tefend, Mary. Veterinary Technician. Feb 2007. Vol 28. No 2. [4.] **Langston. May, et al. Compendium. Dec. 2006. Vol 28 No. 12**

BLOOD GAS ANALYSIS: DEMYSTIFY THE NUMBERS

David Liss, RVT, VTS (ECC)
Woodland Hills, CA

INTRODUCTION

Blood gases are a very important part of the workup of a critical patient. They can provide a wealth of information and even point to specific disease processes. However, they are quite complex and can often be a puzzle to figure out. This presentation will serve to provide a solid foundation to the interpretation of blood gases. Basic physiology of acid-base regulation, step by step process for evaluating blood gases, and case examples will be reviewed.

PATHOPHYSIOLOGY

There are 3 parts to the blood gas. The ventilator component, the metabolic component, and the oxygenation index (if using an arterial sample). We will not be discussing the oxygenation portion if the blood gas for this lecture. Two organ systems are responsible for regulating acid-base balance in the body. The lungs retain and excrete CO2 and the kidneys retain and excrete HCO3. What is acidity? This refers to a net increase in the amount of protons (H+) in a solution. H+ ions result from all of the net metabolic processes in the body. Acids are produced in normal enzymatic reactions (such as lactic acid production in cellular respiration). The body must maintain homeostasis (pH 7.35-7.45) as best it can. pH is an inverse measurement of H^+ concentration. As pH decreases, H^+ ion concentration increases and vice versa. Only small decimal points can mean large changes in the concentration of H^+. Carbon dioxide is considered the "acid" portion of the acid-base profile. CO_2 readily combines with H_2O in the face of carbonic anhydrase to form carbonic acid. Thus, if there is an abundance of CO_2 available, H^+ ions will be made.

See the following equation:

$$CO_2 + H_2O \leftrightarrow H_2CO_3 \leftrightarrow H^+ + HCO_3^-$$

If the body makes more acid, either inadvertently or on purpose, the body must "buffer" the acid to ensure normal pH. The buffer the body uses is bicarbonate, or HCO_3^-. This is the chemical that can combine with H^+ to make water and CO_2, which the lungs can breathe off. It essentially "neutralizes" the pH or H^+ concentrations. Production of acids is called an acidosis, if talking about tissues, or academia, if talking about the bloodstream. Production of base is called an alkalosis, or alkalemia. So a rise in acid production (lactic, ketotic) or acid retention (pulmonary disease, hypoventilation) results in an acidemia. Retention of bicarbonate or hyperventilation will result in an alkalosis.

The body's compensatory system

When acid levels rise, either from carbonic acid production, as a result of inefficient carbon dioxide excretion, or excess endogenous acids, chemoreceptors in the kidneys signal the retention of bicarbonate. Initially, the kidneys do not respond acutely. It takes 2-5 days for them to fully retain enough bicarbonate to cause buffering. Thus, chemoreceptors in the brain signal an increase in respiration rate (minute ventilation). By increasing rate, clearance of CO_2 will hopefully be enhanced. The pH drops and the body normalizes. So, acute responses to academia will be increasing minute ventilation. Chronic responses will be retention of bicarbonate.

The opposite is true with alkalemia. If either there is an increase in HCO_3^- concentration, or a decrease in CO_2, the body will respond appropriately. If CO_2 levels drop (remember CO_2 = acid), then the kidneys will begin to excrete bicarbonate, the buffer, because they do not need as much. If HCO_3^- is being retained for some reason, hypoventilation can ensue to increase the CO_2 (and thus acid) concentrations in the blood to counteract the amount of base present.

Anion Gap

The last concept of physiology is the Anion Gap. The idea is simple. All anions (negatively charged ions) must stay in balance in the body. If one increases, another must decrease to keep the same total number. The Anion Gap formula is as follows:

$$AG = (Na^+ + K^+) - (Cl^- + HCO_3^-)$$

There are obviously other ions in the body, but they are technically negligible. The total number of cations (positively charged ions) exceeds the number of anions. But this number needs to stay relatively constant. If, for example, the chloride levels decrease, the bicarbonate levels MUST increase for this to stay constant. You can see the level of bicarbonate decrease to compensate for the increase in chloride. Now in the real world the bicarbonate levels would match the chloride change. So if the chloride went up 13 points the bicarb would decrease by 13. So the AG = (144) – (125 + 5) = 14. The AG is considered normal in this situation.

What about a high anion gap? Well acids tend to be anionic, so if you add a bunch of acids (lactic, ketotic, uremic, glycolic) into the mix, it will drive the anion gap up, but they are not entirely figured into the calculation. So a normal AG is usually related to a loss of chloride causing the bicarbonate to increase causing a metabolic alkalosis. An increased AG is due to endogenous or exogenous acids and results in a metabolic acidosis.

Stepwise approach to interpretation of blood gases

Step 1- What is the pH?

Does the patient have an acidemia (low pH <7.35) or an alkalemia (high pH > 7.45)

Step 2- What is the respiratory component?

Do they have hypercarbia or hypocarbia?

Step 3- What is the metabolic component?

Do they have an increase or decrease in base?

Step 4 (optional)- What is the anion gap?

Is it normal or high? (It USUALLY isn't low, so don't worry about that)

Normals:

pH: 7.35-7.45

PCO_2: 35-45 mmHg

HCO_3: 18-24 meq/L

Anion Gap: 12-24 meq/L

Example:

pH- 7.1 (Normal: 7.35-7.45)

HCO_3- 10 (Normal 18-24)

P_aCO_2- 30 (Normal 35-45)

Lets not worry about the anion gap for this one.

OK: Let's work through this blood gas using the step wise process:

Step 1- Is it acid or base?

Well the pH is lower than normal, and pH is an inverse of H+ concentration, so the H+ concentration is high. That means its acidic!

Step 2- Is the patient hypocarbic (low CO_2) or hypercarbic (high CO_2)?

Well the CO_2 is lower than normal, so the patient is hypocarbic

Step 3- Does the patient have an increase or decrease in base?

The base is below normal, so a decrease in base

So we have a patient that is acidic, with a low CO_2 and a low base level. How do we figure this out? Well, if you remember that CO_2 = acid, then because the CO_2 is low, it CANNOT be a respiratory acidosis. The HCO_3 is low, and that is by definition a metabolic acidosis. This is because somewhere in the body (lets say this patient's lactate was 11…) there is acid that is neutralizing base, and this is bringing the HCO_3 levels down. What the body does to compensate is increases respiratory rate (thereby decreasing CO_2 levels or ACID levels) to compensate! So we have a metabolic acidosis with respiratory compensation!

4 CLASSIC ACID-BASE PATTERNS

Metabolic Acidosis

This presents with an acidic pH and a low bicarbonate. If the patient has had time to respond then the respiratory rate increases (or the patient takes deeper breaths) and the CO_2 begins to drop as well. It is associated with either an increase in Cl- (decrease in bicarbonate from the anion gap) which is rare, OR an increase in "endogenous" acids such as lactate, keto-acids, uremic acids, ethylene glycol metabolism, salicylate toxicitiy, The chloride-loss type will have a NORMAL anion gap, because the bicarbonate levels drop to compensate. The excess acid type will have an increased AG because those acids are usually negatively charged and thus contribute to a greater gap between cations and anions (bicarbonate levels drop without chloride compensating).

Metabolic alkalosis

This is defined as an elevated bicarbonate (base) level and a high pH (alkaline)

This is somewhat rarer, but a common cause is upper GI vomiting and subsequent loss of H^+ and CL^- ions. These both contribute to an increase in HCO_3 levels and a subsequent metabolic alkalosis.

Respiratory acidosis

This is defined as an elevated CO_2 (acid) level and an acidic pH.

Common causes include any reason for hypoventilation (respiratory depression). Brain disease or head trauma, anesthetic depth, potentially opioid administration, pain with breathing (rib fractures), hypoxemia.

Respiratory alkalosis

This is defined as a decreased CO_2 level and an alkaline pH. This is usually caused by hyperventilation. This can certainly be from pain, fear, stress, anxiety, and also brain disease, hyperthermia, etc.

Table 1. 4 Classic Acid-Base abnormalities

4 Classic Acid-Base Patterns					
Metabolic	**Type**	**pH?**	**CO_2?**	**HCO_3?**	**Compensation**
	Acidosis	↓	↓-Acute	↓	Bicarbonate levels decrease because of excess acid. Because the lungs can act quickly, they begin to increase minute volume either by increasing respiratory rate or tidal volume.
	Alkalosis	↑	↑-Acute	↑	Bicarbonate levels rise in a metabolic alkalosis. Thus the lungs will decrease minute volume in an attempt to increase CO2 retention (acid).
Respiratory	**Acidosis**	↓	↑	↑-slow	Initially a respiratory acidosis causes a decreased pH with an increased CO2 level. As the kidneys begin to reabsorb bicarbonate that level increases. However, it takes 3-5 days to fully compensate, so its not considered an acute event.
	Alkalosis	↑	↓	↓-slow	Here we have a respiratory alkalosis. Respiration has increased the amount of CO2 we have breathed off. Thus the pH trends upwards. The kidneys will begin to excrete bicarbonate but it takes time.

Case examples:

Case 1: 4y DSH. Recently HBC. Lateral recumbent, BP-50 systolic. Pale MMs. Lactate 11
Blood gas results:
pH- 7.21
PCO2- 30
HCO3-12
What acid-base disturbance does this patient have?

Case 2: 9y old Labrador with Laryngeal Paralysis. Diagnosed with Aspiration pneumonia in hospital. He has labored respirations and his Spo2 is 87% on double nasal cannulas.
Blood gas results:
pH- 7.29
HCO3- 17
PCO2- 65
What acid-base disturbance does this patient have?

Case 3: 7y old Husky HBC. Has a fractured pelvis. RR = 65 breaths per minute.
Blood gas results:
pH- 7.5
HCO3- 25
PCO2- 27
What acid-base disturbance does this patient have?

Case 4: 4y Great Dane. 4 hours of non-stopping vomiting. Progressed to non-productive retching. Abdomen is tympanic.
Blood gas results:
pH- 7.5
HCO3- 30
PCO2- 30
(Hint: Chloride is 97. Normal is: 102-120)

REFERENCES

[1.] Kovacic, Jan P. **Small Animal Critical Care Medicine**. Eds. Hopper &Silverstein. Saunders Elsevier: St. Louis MO. pp. 249-254 [2.] Anup, **Dr. A. B.** 1996. AB Anup [3.] Reiss, Adam & Irizarry, Ricardo. **Compendium** May 2009. Vol 1. No 5.

PERICARDIAL EFFUSION: DIAGNOSIS AND MANAGEMENT

Romain Pariaut, DVM, DACVIM (Cardiology), DECVIM-CA (Cardiology)
Baton Rouge, LA

THE NORMAL PERICARDIUM

The pericardium is a thin envelope around the heart. The translucent outer layer is made of the fusion of the fibrous pericardium with the parietal layer of the serous pericardium. This external envelope is attached to the adventitia of the great vessels at the base of the heart. It also forms the phrenopericardial ligament at the cardiac apex. The visceral layer of the serous pericardium corresponds to the epicardium. A small amount of fluid is physiologically present within the pericardial space.

The pericardium anchors the heart within the chest, reduces the friction between the heart and the surrounding organs, and provides a physical barrier to infections and malignancies. However, the pericardium is not essential, as the heart can function normally after its surgical removal.[1]

FROM PERICARDIAL EFFUSION TO CARDIAC TAMPONADE

Pericardiac effusion corresponds to the abnormal accumulation of fluid within the pericardial space. It is not associated with clinical signs. Therefore the diagnostic is ultrasonographic, even if the presence of pericardial effusion might be suspected on chest radiographs as an enlargement of the cardiac silhouette.

Cardiac tamponade is the accumulation of fluid within the pericardial sac that results in an impairment of ventricular filling. It is associated with clinical signs. The presence of cardiac tamponade is suspected on physical examination and confirmed by chest radiographs and cardiac ultrasonography (echocardiography).

Cardiac tamponade starts with the accumulation of fluid within the pericardial space, which results in an elevation of intracardiac diastolic pressures. As ventricular filling decreases and pressure in all four chambers equalizes in diastole, there is a reduction in cardiac output.[1]

ACUTE VERSUS CHRONIC CARDIAC TAMPONADE

Acute tamponade results from the rapid accumulation of effusion within the pericardial space. Within minutes, a small amount of fluid can result in a dramatic elevation of the intrapericardial pressure. As a result of impaired ventricular filling and no time for the activation of compensatory mechanisms, cardiac output drops leading to signs of cardiogenic shock, i.e. collapse, weakness, tachypnea and tachycardia.

Chronic tamponade results from the slow accumulation of effusion within the pericardial space. In this situation, the effusive process occurs over weeks. Initially, the outer pericardial envelope stretches and hypertrophies to accommodate the increased volume of fluid without an elevation in intrapericardial pressure. Over time, pressures start rising, ultimately causing a decrease in cardiac output and an activation of compensatory mechanisms, in particular the stimulation renin angiotensin aldosterone system (RAAS). Water and sodium retention in response to the RAAS help maintain ventricular filling pressures and cardiac output. However, over time, elevated diastolic right heart pressures result in right-sided congestive heart failure, i.e. ascites.

CAUSES FOR CARDIAC TAMPONADE

In dogs, cardiac tamponade is commonly the result of neoplasia. However in many cases, the cause of cardiac tamponade remains unclear, and the disease is identified as idiopathic pericardial effusion. Chest trauma and coagulopathies are rare causes of tamponade. It is important to note that idiopathic pericardial effusion always presents as chronic cardiac tamponade.

The most common tumors associated with tamponade are hemangiosarcoma, chemodectoma, lymphoma, mesothelioma and ectopic thyroid carcinoma. Hemangiosarcoma are more commonly found in German Shepherd dogs, Golden Retrievers and Boxers. The tumor is usually visualized by echocardiography in the region of the right atrium and the right atrioventricular junction.[2] Chemodectomas, also identified as heart base tumors, are more common in brachycephalic breeds. On echocardiogram, they are typically located just caudal to the aorta and above the left atrium.

In dogs, 20 to 75% of the cases of tamponade are idiopathic. This etiology is more common in large and giant breed dogs before the age of 5.

In cats, pericardial effusion associated with hypertrophic cardiomyopathy is commonly visualized by echocardiography; however, cardiac tamponade is rare. Lymphoma, bacterial infections and feline infectious peritonitis are possible causes for tamponade.

PRESENTATION

Dogs with chronic tamponade present with abdominal enlargement, tachypnea, dyspnea, weight loss, weakness and anorexia.

Dogs with acute cardiac tamponade present for extreme weakness, shock or soon after a collapsing episode.

PHYSICAL EXAMINATION

Cardiovascular examination reveals jugular venous distention and increased pulsation, weak and rapid femoral pulses, tachycardia and muffled heart sounds on auscultation. Because chronic tamponade causes right-sided congestive heart failure, abdominal distention is commonly noted.

On occasion, a variation of femoral pulse intensity with breathing, a phenomenon known as paradoxical pulse, is identified. Paradoxical pulse is characterized by weaker pulses during inspiration followed by stronger pulses during expiration.

RADIOGRAPHS

With chronic cardiac tamponade, the cardiac silhouette appears spherical ("basket-ball shape"). The lungs fields are clear unless pleural effusion is present as well.

Radiographic diagnosis of acute cardiac tamponade is much more challenging as the cardiac silhouette is usually unchanged.

ELECTROCARDIOGRAM

Sinus tachycardia, low voltage QRS complexes (QRS< 1mV) are common findings with cardiac tamponade. In 30 to 40% of the cases, electrical alternans is present. It corresponds to a beat-to-beat variation of the amplitude of the QRS complexes caused by the swinging motion of the heart within the fluid filled pericardial space.

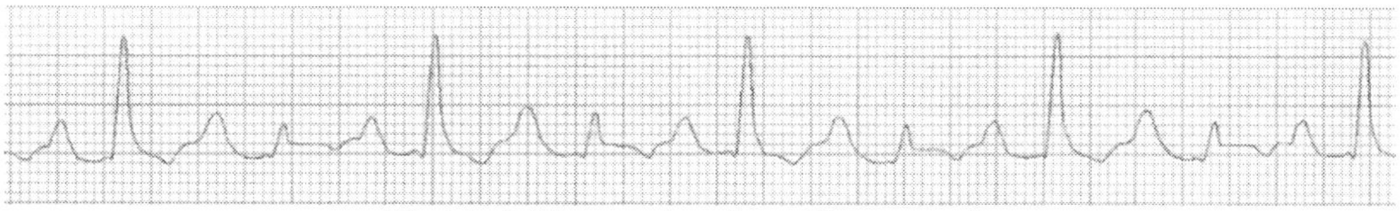

Electrical Alternans in a dog with tamponade

ECHOCARDIOGRAM

Echocardiogram confirms the diagnostic of tamponade by demonstrating diastolic collapse of the right atrial wall. In addition, it is the most sensitive tool to identify cardiac neoplasia. Transesophageal echocardiography has been shown to be more sensitive than transthoracic echocardiography to identify tumors in people; however it requires general anesthesia in dogs.[3]

TREATMENT

There is an absolute contraindication to the use of diuretics with cardiac tamponade, as they will further impair ventricular filling by reducing circulating volume, and subsequently decrease cardiac output.

Pericardiocentesis is the treatment of choice for cardiac tamponade and can be a life saving procedure. It is usually performed with a 16 or 14-G over-the-needle catheter, a 3-way stopcock, an extension tubing and a syringe. The dog is restrained in left lateral recumbency. Light sedation may be used. Continuous ECG monitoring is recommended during the procedure to identify ventricular arrhythmias that can occur when the tip of the catheter enters in contact with the epicardium. An area from the 3rd to 9th intercostal space and from the sternum to the mid-thorax is clipped and aseptically prepared. A right-sided approach limits the risk of puncturing a large coronary vessel. Local anesthesia with lidocaine is recommended. If an ultrasound machine is available, it is used to determine the point of insertion of the catheter. It usually corresponds to the 5th or 6th intercostal space at the level of the costochondral junction. The catheter is inserted along the cranial border of a rib to avoid the vessels that run caudal to the ribs. The catheter is oriented perpendicular to the chest or towards the opposite elbow. Fluid is then drained from the pericardial space. Usually, pericardial fluid has a port wine appearance and does not clot. Cytologic evaluation of the pericardiac fluid is commonly performed but usually unrewarding.[4]

PROGNOSIS

Acute tamponade is usually associated with a poor prognosis as it is almost always caused by cardiac hemangiosarcoma. Also, dogs presented with acute tamponade are usually unstable and bleeding into the pericardial and pleural space may persist after pericardiocentesis.

Short-term prognosis is better for dogs with chronic tamponade. The presence of a tumor worsens the prognosis, especially when cardiac hemangiosarcoma is suspected. Heart base tumors usually carry a better prognosis, as they are slow growing and rarely metastasize.

Idiopathic effusion is associated with a good prognosis but it can re-occur. Surgical pericardectomy may be recommended in this situation.[5]

REFERENCES

[1]Dempsey SM, Ewing PJ. **J Am Anim Hosp Assoc** 2011;47:1-11. [2] Chun R, Kellihan HB, et al. **J Am Vet Med Assoc** 2010;237:806-811. [3] MacDonald KA, Cagney O, et al. **J Am Vet Med Assoc** 2009;235:1456-1461. [4] Shaw SP, Rush JE. **Compend Contin Educ Vet** 2007;29:405-411. [5] Crumbaker DM, Rooney MB, et al. **J Am Vet Med Assoc** 2010;237:551-554.

CHRONIC PAIN MANAGEMENT IN DOGS AND CATS

Kim Spelts, CVT, VTS (Anesthesia), CCRP, CCMT
Colorado Springs, CO

INTRODUCTION

With the development of contemporary pharmaceuticals, nutraceuticals, and other technologies, more pets are able to live reasonably comfortably despite chronic conditions that could have previously caused unrelieved suffering. Effective management of chronic, maladaptive pain requires an approach that impacts multiple points along the nociceptive pathway (transduction, transmission, modulation, and perception). Multimodal therapy encompasses not only drugs that act at various levels of the nociceptive pathways, but multiple non-drug therapies also.

In all cases, it is crucial to understand that appropriate pain management is GOOD MEDICINE, and that any barriers to pain management must be overcome in order to provide the highest level of care to our patients.

Osteoarthritis and cancer are the two most prevalent sources of chronic pain in small animals. This presentation will focus on management of these conditions. Descriptions of commonly used chronic pain management drugs are included at the end of these proceedings.

PAIN AND THE STRESS RESPONSE

Under normal circumstances, pain plays an important biological role. Pain warns of impending or actual tissue damage, helping to avoid further harm. However, untreated or prolonged painful conditions result in extended and highly destructive behavioral, autonomic, neuroendocrine, and immunologic responses. These responses can lead to self-mutilation, immune incompetence, and gradual deterioration, potentially culminating in death.

Likewise, a normal stress response involves behavioral and physiological changes designed to restore homeostasis. Stress is normally short-lived due to the removal of the underlying stressor. However, if stress is severe or allowed to continue for an extended period, such as in the case of prolonged untreated pain, it produces distress and cascades of neural and endocrine responses that upset the animal's physiologic homeostasis.

PAIN STATES

As our understanding of the physiological processes of pain continues to evolve, so do our definitions of various pain states. These newer definitions more accurately reflect microprocesses operating within the periphery and the CNS. Pain is a very complex process, and the way we define and treat pain in our patients must reflect this complexity.

Adaptive Pain vs. Maladaptive Pain

In an adaptive pain state, the body's central processes are operating normally. This type of pain serves a biological purpose; i.e., the body's pain response warns the individual of impending or actual tissue damage and helps to prevent further trauma. It also results in tissue healing. Pain related to surgery or acute traumatic injury and pain that results from inflammation are examples of adaptive pain.

In a maladaptive pain state, the body's central processing has gone awry. This is due to damage to the peripheral and/or central nervous system (neuropathic pain), or the CNS is not properly processing the pain. Spontaneous pain and hypersensitivity can occur. In this state, pain has actually become a disease process. Chronic pain, such as that related to osteoarthritis or cancer, is a good example of maladaptive pain.

Central neuronal hypersensitization, also referred to as *wind-up*, is a type of maladaptive pain state. Wind-up can occur when the dorsal horn of the spinal cord is bombarded with the constant transmission of noxious stimuli, such as that generated by severe trauma (e.g., bony fracture) or even that generated chronic low- to moderate stimulus (e.g., osteoarthritis). If the noxious stimulus is not adequately managed, N-methyl-D-aspartate (NMDA) receptors within the dorsal horn are activated and amplify the signal to the brain, much like turning up the volume on a stereo. Wind-up can result in hyperalgesia and allodynia. NMDA receptors act somewhat like light switches: once stimulated, they often require some intervention to turn them off. Likewise, once blocked, they can stay inactivated for extended periods of time.

Acute Pain vs. Chronic Pain

Acute pain is mild to severe pain that may be incapacitating. The degree of pain is related to the traumatic incident or procedure performed. Acute pain is relatively short-term (hours to days, up to a month) in duration and is generally adaptive in nature. Acute pain dissipates as the initial source of the pain (trauma, surgery, injury, etc.) resolves.

Chronic pain serves no biological purpose. Chronic pain may occur in the presence of a persistent stimulus (as with osteoarthritis), or there may be no stimulus at all (as happens when the pain has become maladaptive in nature).

OSTEOARTHRITIS

Osteoarthritis (OA) is the most common orthopedic problem in dogs and cats, encompassing about 20% of the population across all age ranges. The incidence increases with age. 90% of cats over 9 years of age show radiographic evidence of arthritis. Osteoarthritis is also the most common cause of chronic, maladaptive pain in dogs and cats. A multimodal approach should be taken to control pain in patients with OA.

Weight reduction is an integral part of the management of pain associated with OA. All other interventions will work considerably better if the patient is not overweight. As in humans, it has been well documented that weight loss can result in significant pain relief for arthritic patients. Regular, controlled exercise is also important for the alleviation of pain associated with OA. Patients are typically stiff after periods of inactivity but become more comfortable with mild regular exercise. Exercise is also important to maintain or build muscle mass to support joints.

Nonsteroidal anti-inflammatory drugs (NSAIDs) are key in the pharmaceutical management of osteoarthritis pain. A CBC and chemistry panel should be run before starting any patient on an NSAID, again in 2-3 weeks, and then every 6 months-1 year of chronic use if no adverse events related to NSAID use are observed. Other pharmaceutical interventions may include tramadol, acetaminophen (dogs only), gabapentin, and amantadine.

In recent years, interest has focused on slowing the progression of cartilage degradation and promoting cartilage matrix synthesis. A number of cartilage structure modifiers and other nutraceuticals are available, including:

- *Cosequin® and Dasuquin®* (Nutramax Laboratories): There are a number of published papers that have shown that Cosequin® (glucosamine HCl, low molecular weight chondroitin, and manganese) has many beneficial effects on cartilage in dogs and other species. This combination has also been shown to have a preventative effect: patients receiving Cosequin® prior to an acute joint injury heal more quickly than those who start to receive it afterwards. Recently-released data on Dasuquin ASU® has indicated a pain-relieving effect similar to that of NSAIDs.
- *Hyaluronic acid* (HA): In the presence of OA, HA concentration in the joint decreases. When present in a joint, even a joint with minimal or no cartilage, it can provide a cushioning effect. High molecular weight HA has been shown to provide better joint lubrication than low molecular weight HA. Injectable hyaluronic acid is available as an FDA approved drug for use in horses (Legend®), and some practitioners have used this drug off-label as an IV and/or IA injection in dogs with OA. There are no FDA approved oral versions of HA. However, there is good evidence that high molecular weight HA is absorbed through the GI tract and provides good knee lubrication in humans. There are no comparable animal studies, but clinical impressions of many veterinarians are that oral HA can act as a good adjunct to increasing pain control in dogs and cats with OA.
- *Elk antler velvet*: Elk antler harvested from the male elk during the velvet stage of growth contains type II collagen, glycosaminoglycans (chondroitin sulfate), polysaccharides, growth hormone, and various prostaglandins. Elk antler velvet has been shown to aid in the performance of human athletes and may help with osteoporosis in women. CanEVA-K9® (Canadian Elk Velvet Antler for Dogs) has been shown to alleviate the clinical signs of arthritis in dogs and improve ground reaction forces. Additionally, dogs receiving CanEVA® had a significant reduction of muscle atrophy, suggesting a myotrophic effect.
- *Adequan®* (Novartis Animal Health, Greensboro, NC) is an injectable polysulfated glycosaminoglycan (PSGAG) that has been investigated and approved for use in dogs and horses. A number of studies have demonstrated the drug's ability to decrease cartilage catabolism and prevent adverse joint congruity changes.

Physical rehabilitation can play a vital role in the management of chronic pain. Dogs with OA, benefit greatly from low-impact, low-intensity exercises, massage therapy, and localized heat application. Low-level laser therapy is also beneficial to animals; this therapy has been approved for treatment of OA in humans. Chronic musculoskeletal pain, especially in the paraspinal musculature, often develops as a result of OA (due to altered gait patterns). This pain can be managed with low-level laser therapy, neuromuscular electrical stimulation, massage therapy, and therapeutic ultrasound. Aquatic therapy provides an ideal low-impact exercise for dogs with chronic pain. The buoyancy of water reduces the load on painful joints and allows for more comfortable exercise. Water pressure can reduce swelling and edema. Water resistance is useful for muscle strengthening, which is especially important in dogs that have significant muscle atrophy.

Regenerative stem cell therapy (RSCT) uses cells in a dog or cat's own fat tissue to help regenerate cartilage and other tissues and provide tremendous pain relief, essentially helping the body to help itself. In this therapy, a patient undergoes a short surgical procedure to harvest a small amount of fat. Mesenchymal stem cells are derived from this fat sample, and those cells are then reinjected intravenously and/or directly into the patient's affected joint(s). Since the cells are the patient's own (autologous), there are no rejection issues. Mesenchymal stem cells are contained within each individual and have the potential of becoming cardiac, nerve, and vascular cells, as well as hepatocytes and chondrocytes. Recently published data indicates tremendous improvement in pain and function in dogs whose elbows, knees or hips have been injected.

CANCER PAIN

Almost 1/3 of newly-diagnosed human cancer patients report feeling pain, and this prevalence jumps to 50%-80% in patients with existing disease, depending on the stage of the cancer and the presence of paraneoplastic disease. It is reasonable to assume that veterinary patients have a similar experience of cancer pain. Most pain related to neoplasia begins as a mild, acute pain, but it has the potential to progress to a mild to severe (and potentially maladaptive) chronic pain state. Early intervention will help to prevent wind-up and will make treating the pain long-term easier.

The management of cancer pain in small animals must begin in the acute phase in those patients for whom a surgical excision or limb amputation is indicated. The importance of perioperative analgesia cannot be understated, and a multimodal approach should include an NSAID (if there are no systemic disturbances that would make the use of an NSAID contraindicated), infusions of pain medications (opioid, NMDA antagonist (ketamine), etc.) as well as the use of local and/or regional anesthesia techniques. Effective perioperative analgesia will help prevent wind-up from developing and may help reduce long-term pain.

Ongoing pharmaceutical support is the cornerstone of cancer pain management. As the disease progresses, more aggressive pain management will be required. The World Health Organization has developed an "analgesia ladder" for the treatment of human cancer pain. This treatment strategy relies on evaluating the intensity and severity of pain when deciding the next appropriate step for management. For example, mild-moderate pain may be managed with an NSAID +/- an adjuvant (such as tramadol or acetaminophen), while severe pain may require supplemental administration of an opioid (e.g., oral sustained-release morphine, oxycodone). Keep in mind that individual variability exists in patients' responses to drug therapy, so try to substitute drugs within a category before switching therapies altogether.

The administration of bisphosphonates (e.g., pamidronate) has been shown to reduce pain and pathologic fractures in human beings with osteolytic lesions (such as occurs with osteosarcoma). Pamidronate 1-2 mg/kg can be diluted in saline and administered IV over a 2-4 hour period at 3-5 week intervals.

Maintaining quality of life (appetite, activity, function, interaction with family) is key. Long-term palliative care must address end-of-life issues as the pain and/or side effects of the care are no longer manageable.

PHARMACEUTICALS USED IN TREATING CHRONIC PAIN

Nonsteroidal anti-inflammatory drug (NSAID) administration is very common in small animal practice for treatment of both acute and chronic pain. The predominant mechanism of action of NSAIDs for pain control is inhibition of the cyclo-oxygenase (COX) 2 enzyme. NSAID therapy should be avoided in patients with liver or kidney disease, those with a high likelihood of hemorrhage and/or low blood pressure, and those with gastrointestinal ulcerative disease. Because of the increased risk of GI ulceration, **never mix NSAIDs or NSAIDs and corticosteroids** without at least a 4-10 day washout period. It is also recommended to treat with a GI protectant such as omeprazole or misoprostal during the washout period. There is no clinical evidence that one NSAID is more effective in treating pain than any other; however, there is significant individual variability in the response (both positive and negative) to NSAID therapy.

Tramadol is a synthetic, centrally acting analgesic and is useful for moderate-severe pain. It is not an opioid, but it has opioid-like effects in that its metabolites bind to mu receptors. Tramadol also inhibits the reuptake of norepinephrine and seratonin. Tramadol can bridge the gap between NSAID administration and the addition of a potent oral opioid. Dosing in dogs and cats is generally 2-5 mg/kg 2-4 times daily. Administration in cats is difficult because the lowest tablet concentration is 50 mg. A qualified pharmacy should be able to compound a more dilute suspension to assist with dosing in small patients.

Amantadine was originally developed as an antiviral drug for humans. It may also help with allodynia and opioid tolerance in patients with chronic pain. Amantadine is an NMDA receptor antagonist. In humans, it has been used for neuropathic pain. The dose for dogs and cats is 3-5 mg/kg daily, often for only 21 days.

Gabapentin is a structural analog of GABA (gamma-aminobutyric acid) and was originally introduced as an anticonvulsant. Its mechanism of action in pain control is unclear; while it is related to GABA, it does not appear to have any analgesic effect at GABA receptors. It appears to be best suited for the treatment of neuropathic pain. When used in conjunction with an opioid, human patients who are only partially opioid responsive experience better analgesia and less allodynia. Dose ranges vary from patient to patient, anywhere from 2-40 mg/kg up to three times a day. Lack of responsiveness and drowsiness seem to be the most common side effects.

Acetaminophen may be used for short periods in dogs with breakthrough pain, for which the combination of an NSAID and tramadol is not effective. Acetaminophen can be administered to dogs at 10-15 mg/kg BID for up to 5 days. Acetaminophen should not be administered to cats (inadequate cytochrome P-450 dependent hydroxylation).

Oral opioids are frequently administered to humans with refractory pain. Oral sustained release morphine preparations are used most often. Anecdotally, oral morphine (regular or sustained-release formulations) seems to provide effective pain relief (0.5 mg/kg up to three or four times daily), with the most commonly reported side effect being constipation and anorexia. Oral oxycodone is also an effective drug for severe pain in dogs, and it may be less likely than oral morphine to produce dysphoria. Oxycodone is effective in dogs at 0.3 mg/kg PO BID-TID.

REFERENCES

DVM Magazine's website. Dasuquin's efficacy may be similar to that of NSAIDs in dogs. Available at http://veterinarycalendar.dvm360.com/avhc/Surgery/Dasuquins-efficacy-may-be-similar-to-that-of-NSAID/ArticleStandard/Article/detail/695178. Gaynor J, Muir W. *Handbook of Veterinary Pain Management*. 2nd ed. St. Louis: Mosby, Inc., 2009. Gaynor J. Control of Cancer Pain in Veterinary Patients. *Vet Clin of N Am: Sm Anim Practice* 2008 Nov;38(6):1429-1448. Hellyer P, et. al. AAHA/AAFP Pain Management Guidelines for Dogs & Cats. *J Am Anim Hosp Assoc* 2007 Sep-Oct:43(5):235-248.Kyles AE, Ruslander D. Chronic pain: osteoarthritis and cancer. *Semin Vet Med Surg (Small Anim)*. 1997 May;12(2):122-132.Looney A. Oncology Pain in Veterinary Patients. *Topics in Companion Animal Medicine* 2010 Feb;32-44. Muir W, Hubbell J. *Handbook of Veterinary Anesthesia*. 3rd ed. St. Louis: Mosby, Inc., 2000. Spelts K, Gaynor J. Pain Management Strategies. In: Bryant S, ed. *Anesthesia for Veterinary Technicians*. Wiley-Blackwell, 2010; 345-356. Thurmon J, Tranquilli W, Benson GJ. *Essentials of Small Animal Anesthesia & Analgesia*. Baltimore: Lippincott Williams & Wilkins, 1999. World Health Organization website. WHO's Pain Reliev Ladder. Available at: http://www.who.int/cancer/palliative/painladder/en/.

IS LESS MORE? METRONOMIC CHEMOTHERAPY: THEORY & PRACTICE

Annette N. Smith, DVM, MS, DACVIM (Oncology & Small Animal Internal Medicine)
Auburn, AL

INTRODUCTION

Cytotoxic chemotherapy is used in the treatment of various malignancies. The primary principle is the use of the maximally tolerated dose (MTD), in which the highest drug dose without unacceptable side effects is used as frequently as possible. Cytotoxic chemotherapy is non-selective; side effects occur in normal tissues, as well. Since rapidly-dividing cells are the most susceptible, bone marrow toxicity and gastrointestinal toxicity are the most frequently encountered adverse events. A recovery period must follow the administration of a MTD chemotherapy dose to allow these normal tissues to recover, and during this break in treatment, tumors may proliferate. Some of these proliferating cells have developed resistance to the drugs used, resulting in relapses. In addition, the endothelial cells lining tumor blood vessels are damaged by MTD chemotherapy, but rapidly repair during the intervals between drug doses. In an attempt to ameliorate this proliferation of resistant tumor cells and blood vessels during treatment breaks, scientists have pursued the investigation of continuous, low-dose chemotherapy, also known as anti-angiogenic, or metronomic chemotherapy.[1,2] With the development of targeted drugs, such as the small molecule inhibitors, another shift has been away from the MTD to the minimally biologically effective dose, changing the paradigm of drug development and approval.

TUMOR BIOLOGY

Angiogenesis

Angiogenesis is the formation of new blood vessels from pre-existing mature endothelial cells within already-formed blood vessels, while vasculogenesis is the formation of new blood vessels from progenitor cells (e.g. bone marrow-derived circulating endothelial progenitor cells, CEPs or EPCs). While angiogenesis has historically been thought to be the primary contributor to tumor vessel development, more recent research indicates that both types of blood vessel formation appear to play a role in the development and progression of cancer. For brevity, the term "angiogenesis" will be used to indicate blood vessel development, regardless of the source of proliferating endothelial cells, unless otherwise noted.

The formation of blood vessels is a tightly regulated process in the normal patient, with a strict balance of pro-angiogenic and anti-angiogenic factors. Judah Folkman first proposed the idea that tumors could not grow beyond one cubic millimeter without the development of blood vessels in 1971.[3] Since that time, research has proven that blood vessels are necessary for the delivery of nutrients and oxygen, as well as the elimination of waste products from the growing tumor mass. Without angiogenesis, tumors are incapable of growth and progression. If tumor angiogenesis could be arrested, many tumor types could potentially be affected.[4] Targets for anti-angiogenic therapy include the local endothelial cells that are rapidly dividing in order to create new blood vessels within the tumor, as well as the circulating EPCs that are recruited to the site to form blood vessels. Monitoring the numbers of EPCs might be used as a measure of treatment success. Additionally, modulation of a multitude of pro-and anti-angiogenic factors (cytokines, growth factors, proteins, chemokines, proteinases, proteinase inhibitors, lipids) can influence the development and progression of blood vessel growth through cell-matrix and cell-cell interactions.[5] These factors can be used as targets for therapy, or as markers of treatment success. Some of the pro-angiogenic factors commonly evaluated or targeted include vascular endothelial growth factor (VEGF) and its receptor (VEGFR), basic fibroblast growth factor (bFGF), platelet-derived growth factor (PDGF) and its receptor (PDGFR), and cyclooxygenases/prostaglandins. Anti-angiogenic factors evaluated include thrombospondin-1 (TSP-1),

MTD chemotherapy protocols have been shown to increase CEPs a few days after the drug has been administered, while metronomic protocols appear to cause a sustained decrease in CEPs, resulting in a more durable inhibition of tumor progression.[6] In addition, resistance appears more rapidly and frequently with MTD protocols, while metronomic protocols do not appear to induce drug resistance, perhaps due to the more stable genomic nature of the normal targeted endothelial cells compared to mutated tumor cells. Finally, endothelial cells appear to be exquisitely sensitive to very low drug dosages, sometimes as much as 100,000-fold lower than non-endothelial cells, allowing sparing of normal tissues.[7,8]

Tumor immunology

The innate and adaptive immune systems are involved in cancer surveillance and prevention and, conversely, allow the maintenance of "self" tumor cells within the body. Destruction of tumor cells may be mediated by natural killer cells, cytotoxic and helper T-lymphocytes, and antigen-antibody interactions mediated by antigen-presenting cells and B-lymphocytes. Regulatory T-cells (T-regs) are cells that are CD4+CD25+ and FoxP3+, and appear to be involved in the induction of tolerance to malignant cells.[9] T-regs are elevated in human patients with cancer, and recently have been identified in canine cancer patients, as well.[10-12] Suppression of T-regs occurs in metronomic chemotherapy protocols, and may allow recognition of cancer cells and their destruction by the patient's immune system.[13]

HUMAN STUDIES

There have been several small-scale studies in human medicine that are supportive of the efficacy of metronomic chemotherapy. In 64 breast cancer patients that had previously failed multiple treatments, an overall response rate of 32% was seen to daily oral low-dose cyclophosphamide and twice weekly methotrexate, including 12 patients with stable disease (no growth for > 6 months), 10 with a partial response, and two with a complete response.[14,15] Other tumor types that have shown promising results include non-Hodgkin's lymphoma (as are most canine lymphomas)[16], angiosarcoma (hemangiosarcoma)[17,18], melanoma[19], soft tissue sarcomas[19], prostate cancer[20,21], malignant gliomas[22], non-small-cell lung cancer[23], and renal cell carcinoma[24], along with others.

VETERINARY STUDIES

Two veterinary studies utilizing metronomic chemotherapy have been published, validating a preliminary pilot study that demonstrated two objective responses in late-stage bulky soft tissue sarcomas after one month of therapy utilizing cyclophosphamide (25 mg/m2 orally every-other-day) and piroxicam (0.3 mg/kg orally daily).[25] First, a combination of alternating daily oral etoposide (50 mg/m2 for 3 weeks) and cyclophosphamide (12-25 mg/m2 for 3 weeks) was given with daily piroxicam (0.3 mg/kg) as adjuvant treatment for hemangiosarcoma following splenectomy, and compared to historical patients given the standard cytotoxic doxorubicin chemotherapy regimen (30 mg/m2 IV q 3 weeks x 5 doses).[26] In this study, the survival times were similar: 178 days for the metronomic protocol and 133 days for the doxorubicin protocol. The treatment was well-tolerated, although hemorrhagic cystitis did develop in two of the nine treated dogs. The second study involved the administration of cyclophosphamide and piroxicam following incomplete resection of soft tissue sarcomas.[27] Compared to matched historical patients that did not receive adjuvant treatment, the disease free interval was significantly increased. The median time to recurrence was 211 days for the control group, and was more than 410 days in the treated group. Again, the treatment was well-tolerated, although seven of the thirty patients had some anorexia and vomiting, two developed elevations in creatinine, and three patients that developed hemorrhagic cystitis. One of these patients was euthanized secondary to this complication.

METRONOMIC CHEMOTHERAPY IN PRACTICE

Indications

The suggested mechanisms for the action of metronomic chemotherapy are decreased angiogenesis, enhanced immune response to cancer cells, and decreased efficiency of metastasis. The development of drug resistance is theoretically much less, since the normal stromal and endothelial cells are more genetically stable. Since the drugs used are oral, daily or every-other day dosing can be easily performed by the client while the pet remains at home. Because they are usually given at a much lower dose than MTD chemotherapy, fewer side effects are generally seen, although significant toxicities may still occur.

Drugs

Cyclophosphamide is an alkylating agent that has been shown to suppress T-regs, and to inhibit angiogenesis by promoting TSP-1 levels and suppressing CEP numbers.[28] The dose is 10-25 mg/m2 daily to every-other-day.[25,27] Side effects include marrow suppression and sterile hemorrhagic cystitis. For immunomodulation, the minimum dose is greater than 10 mg/m2.[29]

Chlorambucil is an alkylating agent that is sometimes substituted for cyclophosphamide in cytotoxic chemotherapy protocols, usually when sterile hemorrhagic cystitis or unacceptable GI toxicity occurs. Clinicians are also using it in metronomic protocols, based on a few anecdotal or preliminary studies,[30,31] but it has not been as well-studied as cyclophosphamide. The (empirical) dose is 0.1 mg/kg or 4mg/m2 daily. Side effects include marrow suppression, usually with chronic use.

NSAIDs (non-steroidal anti-inflammatory drugs) cause variable inhibition of cyclooxygenase (COX) enzymes. Up-regulation of COX-2 enzymes in tumors results in increased levels of VEGF, bFGF, TGF,[32] as well as enhanced T-reg function.[33] In theory, COX-2 inhibition should block these effects. Piroxicam inhibits cyclo-oxygenase (COX) non-selectively, and is the best-studied of this class in cancer treatment (primarily for transitional cell carcinoma (TCC), and other carcinomas). The dose is 0.3 mg/kg daily to every-other-day. Side effects include renal and liver toxicity, as well as GI upset and ulceration. Other NSAIDs are less studied in dogs, although the COX-2 selective celecoxib has shown anti-tumor effects in human tumor models.[34]

Tetracyclines, such as doxycycline, and other non-antimicrobial tetracyclines (NATs) have been shown to inhibit matrix metalloproteinasis, decreasing metastatic and angiogenic potential, as well as inducing cell cycle arrest and mitochondrial apoptosis.[35,36] The optimal dose has not been identified, although many people are using 5 mg/kg doxycycline daily.[37] Minocycline at 5 mg/kg once daily was not effective at prolonging survival in dogs with hemangiosarcoma when combined with doxorubicin and cyclophosphamide chemotherapy.[38] Side effects are rare, but usually involve loss of appetite or vomiting.

Tyrosine kinase inhibitors (TKIs) such as toceranib and masitinib were initially developed to target c-kit (the receptor for stem cell factor, involved in cellular proliferation), which is mutated in many canine mast cell tumors.[39,40] Other tyrosine kinases (TKs) may also be variably affected, depending on the TKI. Some of these TKs are important in tumor angiogenesis (VEGFR, PDGFR). In theory, blockade of VEGFR and PDGFR should inhibit the pro-angiogenic pathway, resulting in tumor stabilization. These pathways likely explain TKI responses in mast cell tumors that lack the c-kit mutation, as well as responses in other tumor types. In addition, the use of the drug in much lower doses than the labeled dose may be all that is necessary for anti-tumor effects. The dose for toceranib is 2.5-3 mg/kg every-other-day or Monday, Wednesday, and Friday. Masitinib is given at 12.5 mg/kg daily. Side effects may be significant, and include GI toxicity, including perforating ulcers, marrow suppression, protein-losing nephropathies, hypoalbuminemia, coagulopathies, muscle pain, and depigmentation. Aggressive monitoring with early intervention with drug holidays, gastric protectants, and dose reductions are necessary to manage these effects.

mTOR inhibitors (mammalian Target Of Rapamycin) block an intracellular cascade that results in the synthesis of proteins involved in cell proliferation, angiogenesis, and nutrient uptake for survival.[41] Rapamycin (Sirolimus) has been shown to block the proliferation of canine osteosarcoma cell lines *in vitro*,[42] and has been used in metronomic chemotherapy protocols. The dose is 0.1 mg/kg on Monday, Wednesday, and Friday. Marrow suppression has been seen in dogs.

Peroxisome proliferator-activated receptor (PPAR) agonists include such drugs as the thiazolidinediones/glitazones (primarily used as type 2 diabetes treatments) and some NSAIDs (ibuprofen, indometacin, flufenamic acid, and fenoprofen). They exert anti-tumor effects through induction of cell cycle arrest and apoptosis, through the promotion of differentiation, through anti-angiogenic effects, or through immune modulation.[5] These drugs are not well-studied in veterinary medicine, but might be another avenue for metronomic therapy.

Monoclonal antibodies have been developed against several growth factors and their receptors in human oncology, such as VEGF (bevacizumab/Avastin), EGF (cetuximab/Erbitux), and HER2/neu (trastuzumab/Herceptin). While these drugs show great promise in the management of human cancers, and likely would be effective in veterinary patients, as well, their high cost is likely to keep them out of the hands of the veterinary oncologist unless a generic form is marketed.

CURRENT PROTOCOL

Our current protocol is to initiate piroxicam +/- cyclophosphamide for 2 weeks, with weekly rechecks of BUN, creatinine, ALT, and albumin and a CBC. If the patient tolerates the drugs well, then a TKI or rapamycin is added to the protocol, continuing weekly monitoring of renal and liver values, albumin, as well as biweekly urinalysis with protein:creatinine ratios for 4 weeks. If the drugs are well-tolerated, monitoring is lengthened to every 4-6 weeks. If the drugs are not well-tolerated, drug holidays and institution of lower doses or increased intervals can be tried, or the substitution of a different drug. Periodic re-staging of the patient (usually every 3-4 months) is also performed to monitor for tumor recurrence or metastasis.

PITFALLS

Probably the biggest danger of metronomic chemotherapy is that, because of its perceived ease of use, and lower incidence of side effects, standard-of-care treatment may not be pursued by clients, or worse, not offered by veterinary practitioners. We must keep in mind that metronomic chemotherapy remains investigational, and should not be substituted for proven treatments. The role of surgery, radiation, and chemotherapy in initial tumor treatment will not be eliminated by metronomic protocols. Responses may take weeks or months to develop, and stable disease may be the best expected outcome. Therefore, bulky disease, especially if it is causing clinical signs, is not the ideal setting for metronomic chemotherapy. In addition, side effect profiles are usually altered in metronomic protocols, not eliminated. The euthanasia of a dog secondary to sterile hemorrhagic cystitis after being treated with cyclophosphamide and piroxicam for an incomplete excision of a soft tissue sarcoma illustrates this point. Regular veterinary visits will be essential for these patients to monitor for the development of side effects from treatment, as well as surveillance for tumor recurrence or progression. Clients should be aware not only of the costs of drug therapy, but the commitment to long-term treatment and monitoring. Finally, the best use of metronomic protocols will likely involve a "cocktail" of treatments, perhaps life-long, to prevent upregulation of alternate pathways that result in angiogensis. The optimal drug combinations, dosages, and length of treatment remain to be determined.

FUTURE DIRECTIONS

Probably the most exciting possibility of metronomic chemotherapy protocols is their potential use in conjunction with conventional treatment. Some tumor types that are responsive to cytotoxic chemotherapy, but are not cured with our current regimens, would likely be most amenable to combination therapy. For example, in canine osteosarcoma, the addition of 4-6 cycles of carboplatin chemotherapy following amputation improves median survival times from around 6 months to around one year. However, these patients usually still succumb to metastasis. The addition of metronomic chemotherapy following the standard-of-care treatment may improve survivals. Hemangiosarcoma would also appear to be a tumor that would benefit from anti-angiogenic therapy after splenectomy and 5 cycles of doxorubicin chemotherapy. In our practice, we currently have a number of patients receiving metronomic chemotherapy that are well beyond their expected survival times (> 1-2 years) following conventional treatment. Multiple abstracts presented at the 2010 Veterinary Cancer Society meeting illustrate the interest in this approach. Only time and controlled studies will prove these clinical impressions, but metronomic chemotherapy is an area of research that is being actively pursued and will likely change the face of future oncology treatment.

REFERENCES

[1]Mutsaers AJ. **Topics Comp Anim Med** 2009;24(3):137. [2] Hanahan DJ, et al. **J Clin Invest** 2000;105:1045. [3]Folkman J, et al. **N Engl J Med** 1971;285:1182. [4]Kerbel RS, et al. **Nature Rev Cancer** 2004;4:423. [5]Hafner C, et al. **Current Cancer Drug Targets** 2005;5:393. [6]Bertolini F, et al. **Cancer Res** 2003;63:4342. [7]Bocci G, et al. **Cancer Res** 2002;62:6938. [8]Emmenegger U, et al. **Cancer Res** 2004;64:3994. [9]Beyer M, et al. **Blood** 2006;108(3):804. [10]Biller BJ, et al. **Vet Immunol & Immunopathol** 2007;116(1-2):69. [11]O'Neill K, et al. **J Vet Intern Med** 2009;2394):875.[12]Biller BJ, et al. **J Vet Intern Med** 2010;24(5):1118. [13]Ghiringhelli F, et al. **Cancer Immunol Immunother** 2007;56:641. [14]Colleoni M, et al. **Ann Oncol** 2002;13:73. [15]Orlando L, et al. **Anticancer Drugs** 2006;17:961. [16]Buckstein R, et al. **Clin Cancer Res** 2006;12:5190. [17]Vogt T, et al. **Cancer** 2003;98:2251. [18]Kopp HG, et al. **Anticancer Drugs** 2006;17:997. [19]Reichle A, et al. **Cancer** 2004;101:2247. [20]Glode LM, et al.**Cancer** 2003;98:1643. [21]Lord R, et al. **J Urol** 2007;177:2136. [22]Kong DS, et al. **Oncol Rep** 2006;16:1117. [23]Correale P, et al. **Oncol Rep** 2006;16:133. [24]Krzyzanowska MK, et al. **Cancer Chemother Pharmacol** 2007;60:135. [25]Mutsaers AJ. **Vet Clin North Am Small Anim Pract** 2007;37:1079. [26]Lana S, et al. **J Vet Intern Med** 2007;21:764. [27]Elmslie RE, et al. **J Vet Intern Med** 2008;22:1373.[29]Burton J, et al. **Proc Vet Cancer Soc** 2008:2.[30]Schrempp D, et al. **Proc Vet Cancer Soc** 2010:43. [31]Leach T, et al. **Proc Vet Cancer Soc** 2010:53. [32]Gately S, et al. **Semin Oncol** 2004;31:2. [33]Sharma S, et al. **Cancer Res** 2005;65:5211. [34]Koki AT, et al. **Cancer Control** 2002;9(2 Suppl):28. [35]Gilbertson-Beadling S, et al. **Cancer Chemother Pharmacol** 1995;36:418. [36]Tamargo RJ, et al. **Cancer Res** 1991;51:672. [37]Liptak JM, et al. **Comp Cont Ed Pract Vet SA** 2004;26:172. [38]Sorenmo K, et al. **J Vet Intern Med** 2000;14(4):395. 39London CA, et al. **Clin Cancer Res** 2009;15(11):3856. [40]Hahn KA, et al. **J Vet Intern Med** 2008;22:1301. [41]Wullschleger S,et al.**Cell** 2006;124:471. [42]Gordon I, et al. **AJVR** 2008; 69:1079.

PREVENTING & TREATING SIDE EFFECTS OF CANCER & ITS TREATMENT: A CASE-BASED DISCUSSION

Laurel E. Williams, DVM, DACVIM (Oncology)
Raleigh, North Carolina

INTRODUCTION

Evaluation and treatment of dogs and cats with cancer is increasingly common due to our expanding knowledge of tumor biology and treatment modalities and clients' willingness to pursue more advanced care for their pets. Regardless of the specific type of cancer and/or treatment pursued, our goal is to maintain the patient's quality of life at the highest possible level for as long as possible. This requires an understanding of the type of cancer diagnosed and its effects on the animal. In developing a treatment plan, it also requires a balance between the efficacy of the treatment and its potential toxicity, with knowledge and use of measures to minimize toxicity whenever possible. In this session, we will use case examples to illustrate key concepts in patient care.

TUMOR TYPES

Tumors can be grouped into three broad categories: round cell tumors, epithelial/glandular tumors, and mesenchymal tumors. There are five types of round cell tumors, including lymphoma, mast cell tumor, plasmacytoma, histiocytoma, and transmissible venereal tumor. The epithelial/glandular tumors are called carcinomas. Examples include hepatocellular carcinoma, bronchogenic carcinoma, and transitional cell carcinoma. Mesenchymal tumors originate from connective tissues and are called sarcomas. Osteosarcoma and hemangiosarcoma are two very common sarcomas. Other sarcomas originating in soft tissues of the skin and subcutaneous layer can be grouped together under the umbrella category referred to as soft tissue sarcomas, a group of *histologically* distinct but *behaviorally* similar tumors which includes fibrosarcoma, liposarcoma, hemangiopericytoma, and others.

DIAGNOSIS AND STAGING

The first step in making a diagnosis is aspiration cytology, although histopathology is needed for both definitive diagnosis and information on tumor grade. Once cancer has been diagnosed, staging is indicated to identify possible sites of cancer spread (metastasis) prior to initiating treatment. Staging tests may include aspiration cytology of draining lymph nodes, -view thoracic radiographs, abdominal sonography, and bone marrow aspiration, depending on tumor type and location. In general, lymph node aspiration cytology and thoracic radiographs are indicated for staging of any carcinoma or sarcoma.

TREATING CANCER PAIN AND INFLAMMATION

Pain from cancer is most often a result of tumor invasion with subsequent tissue damage that causes activation of pain receptors. Animals are quite variable in expressing discomfort and pain. The veterinary team and client should work together to monitor the animal for subtle changes which may indicate discomfort. Pet owners may notice changes in appetite or activity level; the veterinarian and staff may notice changes in heart and/or respiratory rate and increases in blood pressure. Preemptive analgesia should be used whenever possible, since pain responses become magnified over time.

The choice of analgesics and measures to minimize pain varies depending on the cause, severity, and duration of the pain. The most commonly used medications used to alleviate pain and distress associated with cancer include non-steroidal anti-inflammatory medications (NSAIDs) and opioids; an abbreviated list of these as well as other commonly used pain-relieving medications are provided in the table below.

CLASS	DRUG NAME	DOSAGE
NSAIDs	Carprofen (Rimadyl®)	(dog) 2.2 mg/kg PO q12h
	Deracoxib (Deramaxx®)	(dog) 1-2 mg/kg PO q24h
	Ketoprofen	(cat) 1mg/kg PO q24h (max 5 days)
	Firocoxib (Previcox®)	(dog) 5 mg/kg PO q24h
	Piroxicam (Feldene®)	(dog) 0.3 mg/kg PO q24h
	Meloxicam (Metacam®)	(cat) 0.1m/kg day 1, then 0.05mg/kg PO q24h x 4 days, then 0.05mg/kg PO q48h
α2-agonists	Dexmedetomidine (DexDomitor®)	3-5 mcg/kg IV
Opioids	Hydromorphone	(dog) 0.05 – 0.2 mg/kg IM, SQ, or IV q2-6h
	Fentanyl	(dog) 2-5 mcg/kg/h CRI
	Buprenorphine	(cat) 0.02 mg/kg sublingual q8h; 5-10 mcg/kg preferable long-term
Opioid-like	Tramadol	(dog) 2-5 mg/kg PO q6-12h (cat) 1-2mg/kg PO q12-24h
NMDA antagonist	Ketamine	(dog, cat) laoding dose of 0.3-0.5 mg/kg followed by 0.3-0.6 mg/kg/hour CRI

NMDA antagonist	Amantadine	(dog, cat) 3-5 mg/kg PO q24h
Neuropathic pain analgesic	Gabapentin (Neurontin®)	(dog, cat) 5-10 mg/kg PO q12h

In addition to these medications, other treatment modalities and measures can be considered, including palliative radiation therapy, bisphosphonates (for bone pain), and acupuncture. In one study evaluating the use of palliative radiation in dogs with a variety of advanced cancers, 87% of dogs achieved pain relief lasting a median of 2.3 months. In another study, 74% of dogs with osteosarcoma treated with palliative radiation therapy achieved pain relief lasting a median of 2.4 months. Bisphosphonates interfere with and induce apoptosis in osteoclasts and as a result, can be used to alleviate bone pain. In a study of 43 dogs treated with the bisphosphonate pamidronate, 28% experienced pain relief lasting a median of 7.7 months. Efforts to combine palliative radiation therapy with pamidronate, however, demonstrated no benefit from this combined modality approach.

PREVENTING AND ADDRESSING SIDE EFFECTS OF CANCER TREATMENT

Cancer treatment may be thought of in terms of local therapy (aimed at eliminating the primary tumor through surgery and/or radiation therapy) and systemic therapy (chemotherapy or other medications used in conjunction with local treatment to delay or prevent metastasis or as sole treatment for systemic cancers). The optimal treatment plan for an individual patient is based on the patient's quality of life, the biologic behavior of the tumor, and the wishes of the client. Regardless of the treatment modality used, the focus should remain on promoting and improving quality of life for the animal.

Surgery

Surgical biopsy or removal of a tumor may be categorized as the following: incisional biopsy, which involves removal of a small portion of the tumor for histologic diagnosis; excisional biopsy, which involves removing only the grossly visible tumor for histologic diagnosis with no attempt at achieving wide surgical margins; wide excision, which involves removing all gross and microscopic tumor with an adequately wide margin of adjacent normal tissue; and radical excision, which involves removing the entire tissue compartment that contains the tumor (e.g., amputation, radical mastectomy, mandibulectomy). Pain associated with surgical procedures is best managed with a combined approach, consisting of the oral / injectable medications (see table above), nerve blocks (where applicable), epidural analgesia (where applicable) and good nursing care.

Radiation therapy

Radiation therapy may be used in conjunction with surgery to decrease the risk of local tumor recurrence. Radiation therapy has the main advantages of being non-invasive and able to reach tissues not amenable to surgical resection. Radiation therapy targets rapidly dividing cells. Cancer cells divide faster, abnormally, and with defective repair mechanisms when compared with normal tissues. When treated with radiation therapy, these cells most often undergo mitotic death; that is, the cells are damaged when they attempt to divide. Rapidly dividing normal cells (gastrointestinal, bone marrow, skin) are more sensitive to radiation therapy than slowly or non-dividing cells (nerve, muscle, bone).

Radiation therapy works most effectively on microscopic tumor cells at the periphery, the location at which surgery often fails. Radiation may be administered either prior to surgery, in an effort to sterilize tumor margins and enable complete resection, or after surgery, to destroy residual microscopic cancer cells following incomplete resection.

Potential side effects of radiation include an early reaction to rapidly dividing tissues (mucosa, skin, hair) that develop midway through treatment, vary in severity, and <u>resolve</u> over 2-4 weeks. The severity of side effects in acute responding tissues is a result of the total radiation dose administered. Symptomatic and supportive care measures include oral anti-inflammatory medications and analgesics (see table above), prophylactic antibiotics, and topical burn cream (e.g. silver sulfadiaxene). Moist desquamation of the skin and oral mucositis may require gentle daily cleaning with a mild solution such as dilute Nolvasan®, but skin should be left exposed to the air to dry as much as possible. Following resolution of acute side effects, cosmetic changes including lichenification of the skin and leukotrichia may be evident. There is also the potential, though unlikely, for late irreversible side effects to slowly dividing tissues (bone, muscle, nerve). Changes are not evident for months-years following radiation therapy and are <u>irreversible</u>. The severity of side effects in late responding tissues is a result of the fraction size as well as the total dose; late responding tissues are preferentially injured by large doses per fraction. Due to their irreversible nature, radiation treatment schemes are designed to limit the incidence of late side effects to <5%.

Chemotherapy

Systemic chemotherapy targets rapidly dividing cells. Due to their rapid and often abnormal division and defective repair mechanisms, tumor cells can be destroyed by chemotherapy. Some normal tissues also have a high rate of cell turnover (mucosa, bone marrow, hair) and may also be sensitive to chemotherapy, though unlike cancer cells, these normal tissues are able to repair chemotherapy-induced damage. Potential side effects of chemotherapy include gastrointestinal upset 2-4 days following treatment, myelosuppression 7-10 days following treatment, and hair loss in breeds with continuously growing coats. In addition to the general toxicoses described below, individual drugs may have unique or cumulative toxicities. Treatment must therefore be designed with a basic knowledge of both the general as well as the unique toxicity spectra for the different agents.

Gastrointestinal

Gastrointestinal toxicity may result from direct stimulation of the chemoreceptor trigger zone or from damage to the gastrointestinal mucosa. In the former situation, nausea and vomiting are most notable on the day of drug administration. Drugs most commonly associated with this type of centrally mediated vomiting are streptozotocin and cisplatin. When gastrointestinal upset results from the latter situation, it is generally 2-4 days following treatment as a result of the 2-4 day turnover time of the gastrointestinal mucosa. Drugs most commonly associated with this type of peripherally mediated gastrointestinal upset include doxorubicin, mitoxantrone, actinomycin D, procarbazine, dacarbazine, methotrexate, 5-fluorouracil, cytarabine, and carboplatin.

Anti-emetics may be used when a patient is showing signs of nausea and/or for the prevention of nausea associated with chemotherapy. Commonly used anti-emetics include the dopaminergic agonist, metoclopramide (Reglan®), the 5-HT3 receptor antagonists, ondansetron (Zofran®) and dolasetron (Anzemet®), and the NK-1 receptor antagonist maropitant (Cerenia®).

CLASS	DRUG NAME	DOSAGE
Dopamine agonist	Metoclopramide (Reglan®)	(dog, cat) 0.2 – 0.4mg/kg q8h PO, IV, SQ *or* 1-2mg/kg/day CRI
5-HT3 receptor antagonist	Ondansetron (Zofran®)	(dog, cat) 0.1 – 1.0 mg/kg IV, SQ or PO q12-24h
	Dolasetron (Anzemet®)	(dog, cat) 0.5 – 1.0 mg/kg IV, SQ or PO q24h
NK-1 receptor antagonist	Maropitant (Cerenia®)	(dog) 2mg/kg PO q24h x 5 days, then 2 days off *or* 1mg/kg q24h x 5 days, then 2 days off

Bone marrow

Myelosuppression results from damage to the rapidly dividing bone marrow stem cells. The survival in circulation of red blood cells, platelets, and white blood cells (neutrophils) are approximately four months, four days, and four hours, respectively. Cells with the shortest survival in circulation and therefore the highest rate of turnover are most susceptible. Accordingly, myelosuppression is most commonly seen as a decrease in neutrophil, and less commonly, platelet counts. When neutropenia develops, it is generally 7-10 days following treatment as a result of the 7-10 days required for a neutrophil to mature. Though the severity of neutropenia varies among the different agents, the drugs most commonly associated with neutropenia and/or thrombocytopenia include vincristine, vinblastine, doxorubicin, mitoxantrone, actinomycin D, cyclophosphamide, ifosfamide, melphalan, CCNU, procarbazine, dacarbazine, 5-fluorouracil, cytarabine, gemcitabine, and carboplatin.

Antibiotics should be considered when a patient is at risk of infection (e.g., neutrophils $\leq$1,000 cells/µl) and are required when neutropenia-related infection is suspected or confirmed.

Hair

Dogs with continuously growing hair coats (those with anagen-phase growth), such as Afghan hounds, Old English sheep dogs, poodles, and many terrier breeds, may experience hair loss during chemotherapy.

CONCLUSION

Knowledge of the symptoms and side effects related to cancer and its treatment enable optimal patient care. The veterinary team and pet owner must continually balance the efficacy of a given treatment and its potential toxicity and utilize means to address pain, nausea, infection and other potential complications to minimize their impact on the animal. By seeking the balance between anti-tumor efficacy and quality of life of the patient, the goals of an extended good quality of life for the patient and a satisfied client can best be achieved.

REFERENCES

Ogilvie GK, et al: Prevention and treatment of pain; Prevention and treatment of nausea and vomiting. In **Managing the Canine Cancer Patient**. 2. Yardley, PA, Veterinary Learning Systems, 2006, pp.213-224, 225-229. Lascelles BD: Management of chronic cancer pain. In **Withrow and MacEwen's Small Animal Clinical Oncology**, 4th ed. St.Louis, MO, Saunders Elsevier, 2007, pp. 291-306. Morrison A, et al. Guidelines for safe and effective use of NSAIDS in dogs. **Vet Ther** 2005;6:237-251. Lascelles BD, et al. **Vet Anaesth Analg** 2007;34:228-250. Bateman KE, et al. **J Vet Intern Med** 1994;8:394-399. Ramirez O, et al. Vet Radiol Ultrasound 1999;40:517-522. Fan TM, et al. **J Vet Intern Med** 2007;21:431-9. Fan TM, et al. **J Vet Intern Med** 2009;23:152-160. Williams LE, et al. Vet Forum 1998;15:44-52.

CHEMOTHERAPY SAFETY

Kim Albin, LVT, VTS (Oncology)
DeWitt, MI

INTRODUCTION

Safety, it is something that we all want but also something that we take for granted. We think about the safety in the cars that we purchase, in the water that we drink, in the paint we use on our walls. We think of safety everywhere we are, and yet not always where we should. Today, we are here to talk about safety when dealing with the cytotoxic drugs we use to treat cancer.

Safety in your work

Do you possess a Biosafety Cabinet (BSC) and use a negative pressure room? Does that same cabinet vent to the outside or does it possess a HEPA filter? Does the negative pressure room always have its doors closed? If your doors are open, the room is no longer negative pressurized. The BioSafety Cabinet must vent to the outside or it does you no good. Many drugs, including Mustargen® (Mechlorethamine), are not captured by the HEPA (high efficiency particulate air) filter, which leaves the person mixing within the hood or standing nearby at risk of exposure.

Do you use a closed system all the time? Do you use proper personal protective equipment (PPE) at all times? If you do not then you are exposing yourself and others to all the cytotoxic drugs that you are handling.

Last Line of Defense

The use of PPE is a fantastic tool as a last line of defense. Protecting yourself and others starts long before you don your PPE. The BSC, negative pressure room, utilization of a closed system, and other precautions are your first line of defense.

Who is at Risk?

At this time, statistics show that a full 30% of people who work with cytotoxic agents will have some type of side effect. Whether that is as minor as hair loss or as major as cancer is up to your own genetic makeup and how much exposure you have. Unfortunately, the 30% number is not accurate due to the fact that as soon as a person gets sick, they leave the profession.

Scare Tactics

It is true; the world is full of carcinogens. From the food you eat to the air you breathe, something you are exposed to has the potential to change DNA and give you a tumor. We know this to be true by countless studies and tests done to everything in existence. It all comes back to safety. When it is known that a certain plastic has been shown to give cancer, a change is made to metal. When a certain pesticide is shown to have suspicious side effects, that pesticide is no longer used. Again, the use of common sense and safety. Dealing with cytotoxic drugs is no different.

Informed Consent

Knowing you are at risk for exposure and the side effects that are included in that exposure and then working with these drugs is just that, informed consent. You know the risks, and you don't care. Exposure can happen from ingestion, inhalation, or by absorption. When you know the facts and you have respect for yourself and the drugs so you can be safe. When you forget these things, this is when problems arise.

Short Term

Short term side effects include; dermatitis, hyperpigmentation of exposed skin, lesions in the nasal and oral mucosa, burning sensation in eyes, dizziness, headaches and nausea.

Long Term

Thinning of hair, chromosomal abnormalities, and carcinogenicity are all long-term side effects that can affect those who handle cytotoxic drugs.

Routine Safety

Using what we know is a great start. Keep your negative pressure room closed. Keep chemotherapy administration to a small, not often used room. Always use your PPE, always. Keep up to date on current safety protocols and new information. Keep your staff current, yearly information, on all the new information available. Use a proper BSC chemotherapy hood. Clean up your areas properly. Bleach doesn't kill everything, neither does detergent. Use special wipes that have bleach and detergent. Extra steps are needed, and annoying, but needed. Not using a closed system is not an option anymore when administering chemotherapy safely. There are a few options out there, find the one that you like and use it always. Use leur lock syringes only. Double glove at all times.

Safety is not an option

You will get sick. It isn't a question of if, but a question of when. If you are dealing with cytotoxic drugs, you will get sick. The degree of sickness is the only unknown.

Helpful Websites

- National Institute for Occupational Safety and Health
    - www.cdc.gov/niosh

- American Society of Health-Systems Pharmacists Guidelines
    - www.asph.org

NEW THERAPEUTIC OPTIONS IN VETERINARY ONCOLOGY: TYROSINE KINASE INHIBITORS

Albert Ahn, DVM
Short Hills, NJ

TARGETED THERAPY – THE SHAPE OF THINGS TO COME

Tyrosine kinase inhibitors (TKI), sometimes referred to as targeted therapies, have revolutionized the management of human cancer since the introduction of imatinib (Gleevec) in 2001. TKIs are now making inroads into veterinary medicine with mast cell tumors (MCT) being their first major success story. TKIs used in the veterinary setting, either having been borrowed from human medicine or developed especially to address unmet veterinary needs, include: imatinib, masitinib, and toceranib; the latter two of which have received regulatory registration in dogs with MCT. In many ways masitinib is the forerunner of such veterinary drug development; historically, developmentally, and also in the depth and breadth of clinical knowledge acquired. It will be masitinib therefore, with its completed phase III studies in canine oncology and immune-mediate indications, its long-term follow-up of MCT treatment, and development program for numerous other canine and feline oncology and non-oncological diseases that will serve as an example of what this remarkably versatile class of therapeutic drug promises to accomplish in veterinary medicine.

Protein tyrosine kinases are enzymes involved in cellular signaling pathways that regulate key cell functions such as proliferation, differentiation, migration, activation and survival. They can be associated with cell surface receptors; for example receptor tyrosine kinases such as KIT or platelet-derived growth factor receptors (PDGFR α and β). Activation of these enzymes, through mechanisms such as point mutations or over expression, can lead to various forms of cancer. Hence, targeted therapies represent a cutting-edge technology in the treatment of cancer through their ability to preferentially act on those cells or signaling pathways responsible for proliferation of tumors. It would however, be a mistake to consider these drugs purely in terms of another chemotherapeutic agent, as several important distinctions exist. For example, unlike cytotoxic chemotherapies that kill all dividing cells, including healthy cells, TKIs such as masitinib are active against signaling pathways that a specific cell may rely upon, thereby effectively targeting that cell or indirectly a downstream process of the cell. Moreover, depending on which kinases or cells are targeted, the inhibitory profile of a given TKI may make it equally well-suited for treating non-oncology diseases. TKIs also have the practical advantage of being delivered orally in tablet form rather than via injection, permitting flexibility in dosing regimens and direct administration by the owner at the point-of-care, for example, in their home.

DEVELOPMENT OF MASITINIB

Masitinib was the first ever approved anti-cancer drug in veterinary medicine, receiving approval from the European Medicines Agency (EMA) in 2008 under the trade name Masivet®. Masitinib has also recently become obtainable in the United States under the trade name Kinavet CA-1®, having received conditional approval in December 2010 from the US Food and Drug Administration (FDA); its availability up to that point being limited to a personal importation program for veterinarians. Masitinib has been designed specifically to optimize its selectivity against just a few key tyrosine kinases that are implicated in various cancers (primarily KIT, PDGFR, and Lyn) while not inhibiting, at therapeutic doses, those tyrosine kinases or tyrosine kinase receptors attributed to possible toxicity [1]. This degree of specificity contrasts with the TKIs of imatinib and toceranib, which inhibit a broader range of kinases as part of a more multi-targeted strategy; e.g. toceranib inhibits in excess of 50 kinases. Major targets of imatinib include ABL, KIT, and PDGFR; those of toceranib include KIT, PDGFR, vascular endothelial growth factor receptor 2 (VEGFR2), and Flt-3. While such multi-targeting can enhance a drug's anticancer potency, this approach does appear to hold potential increased risk of higher toxicity. The atypically high selectivity of masitinib permits it to be administered at higher doses for a more potent therapeutic effect, whilst maintaining an acceptable level of tolerability; i.e. one can expect it to exhibit a better safety profile than multi-targeting TKIs.

To date masitinib has completed two pivotal phase III randomized controlled trials in veterinarian medicine; the first in MCT [2] and the second in canine atopic dermatitis [3]. A wealth of knowledge has been accumulated from its on- and off-label use in Europe, case studies that along with in-vitro preclinical research are driving further clinical development in numerous other cancers and as a chemosensitizer for standard chemotherapies [4]. In addition to this, masitinib's veterinary development program is acquiring knowledge for cat related disorders such as feline asthma and melanoma, including pharmacokinetic and toxicity preclinical studies [5]. It is interesting to note that masitinib is being developed simultaneously in veterinary and human medicines, an uncommon approach that not only generates a large database of relevant experience but one that also recently helped accelerate a development program in human melanoma directly to a phase III randomized control trial, partly through the knowledge gained from treatment of canine MCT. In human medicine, masitinib currently has nine phase III studies initiated in various oncology and non-oncological conditions.

Among those tyrosine kinases associated with cancer development and progression, KIT has emerged as an attractive molecular target; it's over expression and/or constitutive activation being involved in the development of MCTs, gastrointestinal stromal tumors (GIST) and melanoma. Masitinib is perhaps the most specific inhibitor of wild-type and juxtamembrane-mutated KIT available, making these particularly relevant indications. Moreover, stem cell factor, the ligand of the KIT receptor, is a critical growth factor for mast cells [6], meaning masitinib's strong inhibitory effect on KIT receptors makes it an effective anti mast cell agent. Evidence also indicates that recruitment of inflammatory cells, especially infiltration by mast cells, facilitates the growth and spread of cancers by producing molecules that enhance tumor invasiveness. Indeed, it has been shown that mast cells are critical regulators of inflammation and immunosuppression in the tumor microenvironment. Therefore, reduction of the mast cell burden may prove to be of therapeutic benefit in restraining the growth of numerous cancers, even those without a direct association with mast cell proliferation. Complementing this KIT-related anti-proliferative and pro-apoptosis action on mast cells, masitinib is also capable of regulating their activation through its targeting of Lyn, Fyn and Lck [1]. Such down regulation of mast cell activity reduces the array of mediators released, including vascular endothelial growth factor (VEGF), which in turn is likely to disrupt as yet unidentified downstream signaling pathways and angiogenesis processes. Additionally, it is thought that inhibition of Lyn kinase activity also has potential benefits in defense against metastasis and drug-resistance. Masitinib's inhibition of PDGFR kinase activity may result in direct tumor cell growth arrest and apoptosis if PDGFR is constitutively activated. In a more general sense, masitinib may also reduce angiogenesis and enhance chemotherapy sensitivity and availability at the tumor site through modulation of the tumor's interstitial fluid pressure via inhibition of PDGFR-β. Moreover, it is established that mast cells are strongly implicated in a diverse range of non-oncology diseases, in part through their ability to release large amounts of pro-inflammatory mediators with subsequent triggering and sustainment of an inflammatory response. Thus, in addition to masitinib's application as an anti-cancer treatment, inhibition of mast cells can be of therapeutic benefit to a large number of immune-mediated and neurological conditions.

Side effects typically associated with this therapeutic class are gastrointestinal toxicities, including diarrhea and vomiting. Such reactions are usually mild to moderate in severity, relatively short-lived, and tend to be transitory in nature, having a comparable frequency to untreated subjects after the first 3 months of treatment. Advantageously, these and other TKI-related adverse reactions normally spontaneously resolve upon cessation of treatment with no risk of withdrawal symptoms. However, TKIs do have a potential to induce adverse reactions that are unique unto themselves rather than class-effects and which can sometimes be of a serious nature [7 - 9]; for example, ABL and VEGFR inhibition is associated with cardiac dysfunction, while PDGFR inhibition is possibly associated with hypoalbuminemia, and KIT inhibition is possibly related to neutropenia. Such reactions are the likely consequence of an inhibited kinase inadvertently affecting signaling pathways important for the correct function of untargeted cells. Here again then, is a good argument in favor of highly specific TKIs rather than 'dirty' multi-targeting, for reducing the risk of such unintended side-effects. Regarding the increased possibility of protein loss syndrome (hypoalbuminemia) in dogs treated with masitinib, it was shown in the phase III canine atopic dermatitis study that any risk of harm can be greatly reduced via regular monitoring by the veterinarian. Indeed, once the problem had been identified and a suitable management plan to withdraw treatment was implemented, only one further case was reported (corresponding to 1.6% of dogs treated under these study conditions) [3]. This plan primarily involves bi-weekly surveillance of albumin and proteinuria during the first 3 months of treatment followed by at least monthly monitoring; recommendations based upon reports that this effect occurred almost uniquely during the first 3 months of treatment with dogs recovering quickly and without relapse after discontinuation of treatment. In this manner, discontinuation of treatment occurs while the dog is still clinically asymptomatic. Upon resolution, treatment may be resumed at a reduced dose for at least 3 months and monitored for signs of recurrence.

SUMMARY OF VETERINARY DEVELOPMENT PROGRAM

Numerous phase I, II, and III studies, supported by in vitro preclinical studies, help make-up a varied veterinary clinical development program for masitinib (Table 1). (Note that this list of indications reflects the development program of masitinib in veterinary medicine and should not be interpreted as a list of indications in which masitinib demonstrated efficacy).

MASITINIB FOR VETERINARY ONCOLOGY

Table 1. Masitinib is being developed for several oncology and immune-mediate indications.

Targets	Action	Therapeutic potential	
c-Kit	Inhibition of proto-oncogenic targets	MCT*	Histiocytic sarcoma†
PDGFR	or	T-cell lymphoma*	Osteosarcoma (post amputation) †
FAK pathway	Potentiation of chemotherapeutic agents	Melanoma *†	Bladder cancer†
Lyn/FAK		Hemangiosarcoma†	Mammary tumors†
Mast Cells via	Inhibition of mast cell activation	Atopic dermatitis*	Asthma*
KIT / Lyn		Arthritis*	Inflammatory Bowel Disease*

* Masitinib administered as monotherapy. † Masitinib administered in combination with standard chemotherapy.

Although masitinib has been approved for the treatment of MCT in Europe for over 2 years (trade name Masivet®) and has been selectively available in the US under a personal importation program for veterinarians since June 2009, it has only recently received conditional FDA approval. Marketed under the trade name Kinavet CA-1®, this TKI is now be openly available to US veterinarians for treatment of recurrent (post-surgery) or non-resectable grade II/III cutaneous mast cell tumors in dogs that have not previously received radiotherapy and/or chemotherapy except corticosteroids. Registration of masitinib was based upon the successful outcome of a phase III randomized clinical trial including 202 dogs with grade II/III cutaneous MCTs, recruited from 25 veterinary centers in the USA and France [2]. This study reported that masitinib significantly delayed time-to-progression (TTP), i.e. an increased overall median TTP compared to placebo, with the effect being more pronounced when used as first-line therapy and regardless of whether the tumors expressed mutant or wild-type KIT. To further assess the long-term impact of masitinib on the survival of dogs with MCT, those dogs benefiting from treatment on the phase III study were enrolled into a compassionate-use program. Findings from the follow-up study revealed that masitinib significantly increased survival rates at 12 and 24 months in dogs with non-resectable tumors, again regardless of the MCT's KIT mutation status [10]. Indeed, a number of case studies from dogs that have remained on this follow-up study for almost 5 years support the claim that masitinib can represent a curative therapy for a proportion of dogs [11]. Finally, analyses revealed that control of disease was highly correlated with long-term survival; indicating that tumor stabilization also provides important clinical benefits [10]. This effectively introduces a new paradigm in veterinary oncology, with control of disease at 6 months being highly predictive of long-term survival, whereas short-term response at 6 weeks was not. Disease control over 6 months is therefore a valuable treatment objective when using TKIs such as masitinib, suggesting that in the absence of severe toxicity this should be the recommended duration of treatment. (Note: the use of masitinib for treatment of canine mast cell tumors is covered in greater detail by the related ACVIM 2011 Forum presentation, Session #: PFO06, 'Masitinib in Canine Mast Cell Disease').

The off-label use of masitinib in Europe, applicable under the Cascade provisions, has revealed some interesting case studies that are now the focus of further clinical development [12]. These include several examples of masitinib being used for maintenance chemotherapy of dogs with T-cell multicentric lymphoma [13]. For some of these dogs, masitinib monotherapy of T-cell lymphoma was sufficient to maintain a complete or clinical remission. These clinical observations are supported by in vitro data showing that canine T-cell lymphoma cell lines are particularly sensitive to masitinib, possibly due to inhibition of PDGFR. In another case, a dog was diagnosed as having histological confirmed neurofibrosarcoma with mast cell infiltration [14]. A strong relationship exists between the pathogenesis of neurofibrosarcoma and both mast cells and Schwann cells, suggesting therefore, possible benefits from treatment with targeted KIT or PDGFR inhibitors, respectively. Upon commencement of masitinib monotherapy an observed rapid tumor growth was quickly stabilized followed by a significant decrease in volume and a negligible mass after 10 weeks of treatment. There are also several reports of masitinib treatment of canine metastatic melanoma producing improved survival times compared to surgery. In one case study, the dog presented with a histologically proven metastatic melanoma that was unresponsive to chemotherapy; however, after the introduction of masitinib a complete response was achieved within 2 months of treatment. This dog was alive at least 4 months after the initial resection, before being lost to follow-up. These observations suggest that masitinib might represent a therapeutic option for dogs with melanoma; a hypothesis again supported by in vitro findings [4, 15]. Finally in feline case studies [12], one subject with MCT experienced rapid improvement in clinical status after 7 days of masitinib treatment and partial response in two sub-cutaneous nodules. Another feline subject presenting with metastasized cutaneous melanoma, for which the primary tumor had been resected, showed a complete response after 6 months of daily masitinib treatment with no signs of relapse following 14 months of treatment. Finally, an in vitro study on feline vaccine associated sarcomas cell lines showed masitinib had potential as an antineoplastic agent against this disease [16].

There is evidence that masitinib can act as a potentiator of chemotherapies [4]. For example, it was shown to strongly sensitize histiocytic sarcoma cells to vinblastine; osteosarcoma and mammary carcinoma cells to gemcitabine; and to a lesser degree, several cell lines were sensitized to doxorubicin. Phase I studies in dogs with advanced tumors have shown that masitinib in combination with carboplatin or doxorubicin had acceptable and manageable safety profiles, without added toxicity [17]. Furthermore, studies show that masitinib can enhance the antiproliferative effects of gemcitabine in human pancreatic cancer, a property not seen with other TKIs such as imatinib and dasatinib [18]. The working hypothesis is that an as yet unidentified protein or signaling pathway important in the restoration of chemotherapy response is being affected; research is on-going.

MASITINIB FOR IMMUNE-MEDIATED DISEASES

Studies also support the potential use of TKIs in non-oncological disorders. As previously mentioned, the main cellular target of masitinib are mast cells, which are capable of releasing large amounts of pro-inflammatory mediators with a consequence of triggering and sustaining an inflammatory response. Masitinib can therefore also provide therapeutic benefit in immune-mediated diseases with mast cell involvement, such as atopic dermatitis, asthma, inflammatory bowel disease and arthritis. The most advanced of these applications is canine atopic dermatitis, for which a phase III study incorporating one of the largest cohorts tested to date, showed a significant reduction of clinical signs following masitinib treatment [3]. In general, a positive response was evident for treatment-naïve dogs, dogs resistant to ciclosporin and/or corticosteroids, and dogs with severe pruritus, the latter two groups representing populations with high unmet medical need. More specifically, a reduction in CADESI-02 score of ≥50% at week 12 was observed in 61% of masitinib-treated dogs vs. 35% control; according to the intent-to-treat population, observed cases dataset. For dogs resistant to ciclosporin and/or corticosteroids, this response rate was 60% vs. 31% control. Overall, 63% of investigators assessed masitinib efficacy as good/excellent vs. 35% control.

Feline asthma is another disorder that may benefit from TKI therapy. To date, a few encouraging case studies support this; one cat presenting with corticosteroid-intolerance and another with corticosteroid-resistance, for which both have been treated with masitinib for over 14 months and show an improved quality-of-life. Such reports prompted further clinical study including a randomized, controlled phase II study of oral masitinib (50 mg/day) for treatment of experimental feline asthma, which produced positive responses including a significant reduction in bronchiolar resistance and eosinophilic airway inflammation after 4 weeks [19].

CONCLUSIONS

Proven beyond doubt as a successful long-term anti-cancer treatment for MCT, masitinib also shows real promise as a valuable treatment option in canine atopic dermatitis, especially for those dogs that are severely pruritic or refractory to current therapies. The application of TKIs in the veterinary setting, and even more so their specific development to address unmet medical needs related to veterinary medicine, is still considered somewhat of a novelty. However, with an expanded range of indications for which TKIs feature routinely as a viable therapeutic option along with further acquisition of long-term data demonstrating clear therapeutic benefits from their use, then TKIs may soon become established as the veterinarian's benchmark treatment for certain diseases.

ACKNOWLEDGEMENTS

Colin Mansfield, PhD provided medical writing services on behalf of AB Science. Albert Ahn is an employee of AB Science.

REFERENCES

[1] Dubreuil P, et al.; **PLoS One** 2009; 4:e7258. [2] Hahn KA, et al.; **J Vet Intern Med** 2008; 22:1301-1309. [3] Cadot P, et al.; **Vet Dermatol** 2010; 21. [4] Thamm D, et al.; **Vet J** 2011, DOI:10.1016/j.tvjl.2011.01.001. [5] Bellamy F, et al.; **Vet Res Commun** 2009; 33(8):831-837. [6] Reber L, et al.; **Eur J Pharmacol** 2006; 533:327-340 [7] Force T, et al.; **Nat Rev Cancer** 2007; 7:332-344. [8] London CA; **Top Companion Anim Med** 2009;24(3):106-12. [9] Sum SO, et al.; **J Vet Intern Med** 2010; 24:431-435. [10] Hahn KA, et al.; **Am J Vet Res** 2010;71:1354-1361. [11] Phillips B, et al.; **Proc. 30th VCS Conf** 2010; Oct 29-Nov 1. [12] de Vos J, et al.; **Proc VCS Conf**, 2010; March 07-10. [13] Serres F, et al.; **Proc. of the ESVONC Spring Congress** 2010. [14] Roos A, et al.; **Proc. ESVO Congress** 2010; 18 - 20 March. [15] Thamm D, et al.; **Proc. 28th VCS Conf** 2008; October 18-21. [16] Daly M, et al.; **Proc. 28th VCS Conf** 2009; Oct 16 - 19. [17] Ogilvie G, et al.; **Proc. VCS Conf**, 2010; March 07-10. [18] Humbert M, et al.; **PLoS One** 2010; 5:e9430. [19] Lee-Fowlera T, et al.; **Proc. 28th VCRS** 2010; Sept 28-30.

TREATMENT OF CANINE AND FELINE CONGESTIVE HEART FAILURE

Charlotte Pace, BA (Hons), RVN, VTS (Cardiology)
London, UK

INTRODUCTION

The treatment aims of congestive heart failure (CHF) in the dog and cat vary dependent upon the stage of heart disease present. Different classification systems are used, of which the modified New York Heart Association and International Small Animal Cardiac Health Council are probably the most familiar. Perhaps the most useful is a version of the 2001 American College of Cardiology/American Heart Association classification system used in humans that has been recently modified for use in dogs with chronic valvular disease.

CLASSIFICATION OF HEART DISEASE

Because therapeutic goals differ largely dependent upon the stage of heart disease present, the disease stage must be established to treat the patient appropriately. Classification of the different stages is as follows:

- Stage A: Patients at high risk of developing heart disease, but no structural disorder of the heart.
- Stage B: Patients with structural heart disease that have never developed signs of heart failure.
  This is further sub-divided into:
  - Stage B1: Patients with no radiographic or echocardiographic cardiac remodelling as a result of heart disease.
  - Stage B2: Patients that have radiographic or echocardiographic cardiac remodelling as a result of heart disease.
- Stage C: Patients with past or current clinical signs of heart failure associated with structural heart disease.
- Stage D: Patients with clinical signs of heart failure that are refractory to 'standard therapy.'
  - Stages C and D are then subdivided according to patients that require acute (hospital-based) therapy (i.e. life-threatening CHF) or those requiring chronic (home-based) therapy (i.e. mild-to-moderate CHF).

For the majority of cardiac diseases in the dog and cat there is general consensus that drug therapies are most beneficial at or after the onset of clinical signs of heart failure (Stages C or D). The management for patients with acute, life-threatening CHF differs to that for patients with chronic, mild-to-moderate CHF. The patient with acute, life-threatening disease must be stabilised by supplementing oxygen to improve hypoxemia and given therapies to reduce edema formation and to improve cardiac output. The patient with chronic, mild-to-moderate CHF not only requires drugs to decrease edema formation and improve hemodynamic function, but importantly also to slow disease progression and increase lifespan with drugs that decrease the deleterious effects of long-term stimulation of compensatory mechanisms (sympathetic tone, renin-angiotensin-aldosterone system).

DEFINITION OF CONGESTIVE HEART FAILURE

The primary role of the cardiovascular system is to maintain adequate arterial blood pressure. Adequate arterial pressure is vital to maintain perfusion to the organs of the body (especially vital organs, such as heart, brain, and kidneys). This is achieved by a balance between cardiac output and vascular tone. If cardiac output is compromised, arterial pressure falls, and compensatory mechanisms are activated in an attempt to return blood pressure to a more normal level. These compensatory mechanisms include increased fluid retention by the kidneys, vasoconstriction, and stimulation of the heart to increase heart rate and stroke volume (primarily via increased sympathetic tone). Although ultimately beneficial to restore adequate arterial pressures, the 'trade-off' is an increase in venous pressures, which if severe enough can cause fluid to effuse into the surrounding interstitial tissues (pulmonary edema) or cavities (ascites, pleural effusion). If mild compensation is sufficient to normalise arterial pressure without causing edema/effusions, then affected patients will not have clinical signs, or only develop signs if they undergo excessive exertion or stress. If more compensation is required to normalise arterial pressure, development of edema/effusions can result. The effected patient therefore develops CHF. If arterial pressure cannot be normalised despite maximal activation of compensatory mechanisms low output heart failure is present. Patients with low output heart failure usually have concurrent CHF, unless the fall in output is so acute that compensatory mechanisms have insufficient time to be activated (e.g. some arrhythmias, myocardial infarction). Although life saving in the short term, long-term stimulation of compensatory mechanisms (sympathetic tone, renin-angiotensin-aldosterone system) causes progressive deterioration of cardiac function due to fatigue of the heart muscle.

At risk

Animals with a genetic predisposition to heart disease are at risk of developing future problems. Dog breeds such as Doberman Pinschers and cat breeds such as Ragdolls or Maine Coons are examples of this. Screening tests such as auscultation or echocardiography might be indicated in these cases. Routine check-ups are indicated for breeds known to be at risk of developing heart disease.

Heart disease with no congestive heart failure

Most animals in this category have no clinical signs and therefore require no treatment. Examples of conditions for which therapy is indicated in this stage are diseases that can be surgically corrected such as patent ductus arteriosus and pulmonic stenosis. Drug therapy may also be indicated if it can slow the disease process. Examples of where this may be beneficial in theory are ACE inhibitors in dogs with mitral valve disease (MVD) and beta blockers in cats with hypertrophic cardiomyopathy (HCM) and left ventricular outflow tract obstruction. At present there is no consensus on whether such therapies are beneficial at this stage, either due to conflicting data (MVD) or a lack of data (HCM). The primary goal at this stage is monitoring the patient and client education to detect the early signs of CHF.

Mild-to-moderate CHF

Patients in this category usually exhibit clinical signs (such as exercise intolerance or cough), but should not require hospitalization. The therapeutic goal at this stage is to remove excess fluid and reduce compensatory neurohormonal activation. This is generally achieved with a combination of diuretics (e.g. furosemide), inhibitors of the renin-angiotensin-aldosterone system (e.g. ACE inhibitors, spironolactone), and in most acquired canine diseases, pimobendan. Routine check-ups are indicated to monitor the effectiveness of heart failure therapy (e.g. thoracic radiography) and also the effects of therapy on other body systems, (e.g. renal function and electrolytes). Persistent arrhythmias that compromise cardiac function (e.g. atrial fibrillation) should also be treated. Clients should be advised to monitor resting respiration rate, appetite, exercise tolerance and response to therapy.

Life-threatening CHF

Many patients in this category will present acutely, sometimes with no previous history of heart disease. Some patients however, may already be receiving treatment for heart failure and have decompensated acutely. The primary treatment aim is to stabilise the patient by improving hypoxaemia, removing excess fluid and improving cardiac output. Positive inotropes and with close monitoring, vasodilators are used to improve hemodynamic function. Monitoring should not compromise the patient, and additional stress should be avoided at all costs. Opioid sedation may be useful to reduce stress, particularly in cats.

Refractory CHF

Patients in this category will already be receiving 'standard' heart failure therapy, but CHF still persists. The goal of therapy at this stage is to remove excess fluid. Fluid removal may be achieved by the use of a combination of diuretics (sequential nephron blockade), and if required, thoracocentesis and/or abdominocentesis. Parenteral furosemide is often more effective than oral furosemide in patients with ascites, as these patients will often have decreased absorption of oral drugs due to edema of the gut wall. Routine monitoring may still include thoracic imaging, while monitoring of blood pressure, renal function and electrolytes are even more important as the effect of more advanced cardiac disease and multiple medications on other body systems may be more marked. Additional therapy for persistent arrhythmias is of further importance at this stage. Cardiac cachexia is often a problem at this stage, and so bodyweight, body condition score and appetite should all be monitored. Essential fish oils (omega-3 fatty acids) may have benefits in this population of patients.

OTHER RECOMMENDATIONS

Diet

Diets that are moderately sodium restricted may be beneficial for patients with CHF. This should also include avoidance of salty table scraps used to give medication. At the life-threatening and refractory stages, appetite should be encouraged, due to the risk of anorexia and cachexia. Exercise should be restricted when a patient is at stage C.

Surgery

Surgery can be an option for some cardiac diseases, in particular some congenital defects. Occlusion of patent ductus arteriosus and balloon valvuloplasty of pulmonic stenosis have both been shown to improve the clinical outcome of patients. One such study showed that 50% of dogs died less than a year old if the PDA remained untreated. There are also palliative options for tetralogy of Fallot. Pacemaker therapy is also beneficial for symptomatic bradyarrhythmias.

REFERENCES

1. Ettinger SJ et al, 1970, Canine Cardiology, p. 5. 2. ISACHC, 1994, In Tilley LP, Manual of Canine and Feline Cardiology, p469-502. 3. Atkins et al, 2009, JVIM, 23:1142–1150. 4. Kvart et al, 2002, JVIM, 16:80-88. 5. Atkins et al, 2007, JAVMA, 1; 231(7): 1061-9. 6. Pouchelon et al, 2008, JVIM, 22:905-914. 7. O'Grady et al, 2008, JVIM, 22(4): 897-904. 8. Häggström J et al, 2008, JVIM, 22 (5): 1124-35. 9. Freeman L et al, 1998, JVIM, 12(6):440-8. 10. Johnson M et al, 2004, JSAP, 45:148-153. 11. Eyster GE, et al 1976, JAVMA, 168: 435-438. 12. Brockman DJ et al, 2007, JAVMA, 231:721-726

UNDERSTANDING RESPIRATORY SOUNDS IN SMALL ANIMALS

Helio Autran de Morais, DVM, PhD, ACVIM (Small Animal Internal Medicine & Cardiology)
Corvallis, OR

INTRODUCTION

Proper understanding of respiratory sounds is very helpful when dealing with coughing patients or patients with dyspnea. The respiratory system has a limited number of ways of showing abnormalities. Abnormal respiratory sounds can be helpful in localizing the problem. *Stridor* is a loud musical inspiratory sound of constant pitch associated with laryngeal (and occasionally tracheal) alterations. *Rhoncus* is a rattling in the throat associated with pharyngeal or proximal tracheal diseases. *Wheezes* are continuous musical sounds generated by air forced to pass through a narrow region abruptly into a wider region in the larger airways. *Crackles* are short, explosive, non-musical sounds that are a non-specific sign of small airway disease.

THE DYSPNEIC PATIENT

Dyspnea is the sensation of having difficult or laborious breathing. It is a subjective phenomenon that needs to be inferred by the clinician in companion animals. Dyspnea, tachypnea (rapid breathing), and hyperpnea (increased ventilation) are not synonymous but are frequently grouped together in veterinary patients. Dyspnea occurs when the demand for ventilation is out of proportion to the animal's ability to respond to this demand. Thus, breathing becomes difficult, uncomfortable, or labored. Dyspnea may result from alterations in any portion of the respiratory system or be due to abnormal mechanics of the lung and chest wall.

When does the Dyspnea occur?

The timing and pattern of respiration helps to determine the structure most likely responsible for the dyspnea. Dyspnea may occur during inspiration, expiration or both (mixed). Clinically, pure inspiratory dyspnea implies a lesion in the respiratory tract outside the thorax, whereas expiratory and mixed dyspneas occur in patients with thoracic or metabolic disease.

What is the pattern of respiration?

Mixed or expiratory dyspnea should be further classified as obstructive or restrictive. *Obstructive diseases* are associated with increased airway resistance in the tracheobronchial tree. Obstructive diseases may also occur in the upper respiratory tract, but those patients have inspiratory dyspnea. Patients with obstructive disease have decreased expiratory flows and hyperinflated lungs (e.g. cats with asthma). *Restrictive diseases* are those in which expansion of the lungs is restricted (e.g. pulmonary fibrosis, pleural effusion). Lungs of patients with restrictive disease operate at smaller volumes and the patient has a rapid shallow breath. In patients with a non-pulmonary cause of restrictive, residual volume is normal or increased.

Can you hear Respiratory Sounds?

The physical examination assists in identifying the cause for the dyspnea. In patients with restrictive pulmonary disease, absence of respiratory sounds indicates a pleural cavity disease, whereas presence of pulmonary sounds occurs in patients with parenchymal pulmonary disease, metabolic diseases or abdominal distention.

INSPIRATORY DYSPNEA

Inspiratory dyspnea occurs with extrathoracic lesions in the respiratory tract (Table 1). Patients with isolated pure nasal problems are able to breathe normally when the mouth is open. Presence of abnormal respiratory sounds may help to localize the problem. Stridor is associated with laryngeal (and occasionally tracheal) alterations, whereas rhoncus accompanies pharyngeal or proximal tracheal diseases. Cough may occur in patients with inspiratory dyspnea. Cough receptors are located in the larynx, pharynx, and large airways. Diseases in any of those locations may be associated with cough. In patients with extrathoracic disease, cough is usually paroxystic and loud. Direct visual inspection or bronchoscopy is necessary determine the cause.

EXPIRATORY DYSPNEA

Obstructive Dyspnea (Expiratory)

Obstructive expiratory dyspnea occurs in patients with intrathoracic airway diseases (Table 2). Patients may have wheezes and cough. The cough in patients with large airway disease is loud.

Restrictive Dyspnea

Restrictive dyspnea occurs in patients with small airway, parenchymal or pleural cavity disease (Table 3). Normal respiratory sounds can be heard in dogs and cats with small airway or parenchymal pulmonary disease. In addition abnormal respiratory sounds like crackles and wheezes can also be heard. Cough may occur if small airways are also involved and is usually not loud.

In dogs and cats with pleural cavity disease, however, pulmonary sounds are absent or may be heard only at specific locations (e.g. dorsal lung fields in patients with pleural effusion). Cough is usually absent, but may occasionally occur in dogs with pleural effusion.

Table 1. Causes of Inspiratory Dyspnea

Nostrils/Nasal Cavity	Larynx/Pharynx	Cervical Trachea
• Stenotic nares	• Elongated soft palate	• Tracheal Collapse
• Nasal foreign body	• Laryngeal paralysis	• Tracheitis
• Nasal mass	• Laryngeal edema	• Tracheal foreign body
• Rhinitis	• Everted laryngeal saccules	• Extraluminal compression
	• Laryngeal mass	
	• Hyoidal fracture	
	• Retropharyngeal polyps	

Table 2. Causes of Obstructive Expiratory Dyspnea

Thoracic Trachea	Bronchial Tree
• Tracheal collapse	• Bronchitis
• Tracheobronchitis	• Asthma
• Extraluminal compression	• Neoplasia
• Foreign body	• Hilar lymphadenopathy
• Neoplasia	• Left main stem bronchus compression
	• Lungworms

Table 3. Causes of Restrictive Dyspnea

Pulmonary diseases	Pleural diseases	Others
• Pulmonary edema	• Pneumothorax	• Ascites
• Pneumonia	• Pleural effusion	• Pregnancy
• Pulmonary fibrosis	• Diaphragmatic hernia	• Intraabdominal organomegaly
• Hemorrhage/contusion	• Intrathoracic tumors	• Gastric dilatation-volvulus
• Thromboembolism		• Anemia
• Neoplasia		• Anxiety
		• Neuromuscular disorders

The Coughing Patient

Cough is one of the body's defense mechanisms to prevent the entry of noxious materials into the respiratory system. It also helps to clear potentially harmful debris from the lungs and respiratory tract. Cough is usually occurs in conjunction with other protective processes such as bronchoconstriction and mucus secretion, both of which may aid in the effectiveness of the cough. Cough is an indicator of airway or pulmonary disease, although it also may occur as a primary sign of other thoracic disorders. Cough receptors are located in the larynx, pharynx and large airways. They are absent in the alveoli and alveolar sacs. As a result, diseases of the lung parenchyma may not stimulate the cough reflex to the degree that disorders within the airways do. This is teleologically understandable because even a vigorous cough would not move gas fast enough to cause turbulence and shearing forces at the airway wall in the smaller airways. Cough at this level therefore, would be ineffective. Cough is not a localizing sign being associated mostly with airway diseases. Patients with nasal and pharyngeal \disease may; however, cough if secretions are being drained into the larynx and trachea.

Cough in dogs is usually caused by tracheal collapse, chronic bronchitis, compression of the left main stem bronchus by an enlarged left atrium, congestive heart failure or pneumonia. Other causes of cough include: tracheitis, traumatic or mechanical problems (e.g. foreign body, irritating gases), parasitic disorders (e.g. lungworms, heartworm), and neoplasia. In cats, cough is usually a sign of pneumonia, heartworm disease, asthma or lungworms.

The *history* may help establish the etiology for the cough. In dogs, nocturnal coughing is generally associated with cardiac disease, psychogenic causes, or tracheal collapse. Patients with tracheal collapse also may start coughing after drinking water. Cough due to respiratory disease is more likely to occur during the day. Tracheal diseases usually have a goose-honk cough that can be initiated by excitement or by pulling on the collar and leash.

The *physical examination* assists in identifying the reason for the cough and is particularly helpful in differentiating if the cough is due to lung or heart disease. Patients with heart failure usually have fast heart rates with soft crackles, whereas patients with primary lung disease tend to have pronounced sinus arrhythmia with crackles that are usually louder and of higher-pitch than the ones resulting from heart failure. Wheezes also can be present, particularly in patients with respiratory disease. Maneuvers that stimulate the patient to take deep breaths may help in lung auscultation because abnormal lung sounds (especially crackles) may be missed if the lungs are not adequately expanded. Closing the nostrils for 10 to 15 seconds usually will force the patient to inspire deeply.

REFERENCES

[1]Anderson-Weeberg K. In: Ettinger SJ et al. **Textbook of Veterinary Internal Medicine**. 7th ed, Saunders, 2010, p:250. [2]de Morais HA et al. Proceedings **XXVII WSAVA World Congress,** 2002. [3]Forney S. In: Ettinger SJ et al. **Textbook of Veterinary Internal Medicine**. 7th ed, Saunders, 2010, p:253. [4]Hamlin, RL. **Vet Forum**, May:59, 1994. [5]Lehrer S. **Understanding Lung Sounds**. 3rd ed, Saunders, 2002. [6]Wilkins R et al. **Fundamentals of Lung and Heart Sounds**. 3rd ed, Mosby. 2004.

ANESTHESIA IN PATIENTS WITH LIVER & KIDNEY DISEASE

Kim Spelts, CVT, VTS (Anesthesia), CCRP, CCMT
Colorado Springs, CO

INTRODUCTION

All anesthetic procedures introduce some level of risk to the patient, regardless of his or her health status. Patients with underlying renal and/or hepatic disease are more prone to anesthesia-related complications, and they can present significant challenges to the anesthetist. A clear understanding of the underlying disease process will enable the anesthetist to develop and implement the safest anesthetic plan possible.

RENAL DYSFUNCTION

The kidneys are responsible for maintaining the volume and composition of body fluids within normal limits. They achieve this through filtration, reabsorption, and secretion, and in order for them to function properly, the kidneys receive 25% of the total cardiac output.

The *nephron* is the functional unit of the kidney. It is made up of a renal corpuscle (glomerulus + Bowman's capsule) and a urine-producing tubule (proximal convoluted tubule, Loop of Henle, and distal convoluted tubule). The glomerulus is an extensive network of capillaries that receives its blood supply from an afferent arteriole and provides a large surface area for filtration to take place. The blood pressure in this arteriole determines the filtration rate of water and solutes into Bowman's capsule. The function of each nephron is regulated by a set of specialized cells called the *juxtaglomerular apparatus* (JGA).

The endotheliar-capsular membrane in the glomerulus is thin and porous, allowing substances such as sodium, potassium, calcium, phosphorus, water, glucose, and metabolic waste products to be freely filtered from the capillaries into Bowman's capsule. In healthy nephrons, proteins are too large to pass through this membrane.

The efferent arteriole is smaller in diameter than the afferent arteriole; therefore, there is resistance to outflow from the glomerulus, creating pressure that is higher than the systemic blood pressure. The glomerular filtration rate (GFR) depends on the systemic blood pressure and the diameters of the afferent and efferent arterioles, which are controlled by renal autoregulation hormonal regulation (renin-angiotensin system), and neural regulation.

Diagnostic tests that give the anesthetist an idea of a patient's renal function include urine speific gravity (USG; measures the ability of the renal tubules to concentrate urine), blood urea nitrogen (BUN; ammonia produced by protein breakdown is converted to urea in the liver, and urea is freely filtered by the glomerulus), and creatinine (excreted by muscle cells at a relatively constant rat and is freely filtered by the glomerulus). A patient may be azotemic (elevated BUN and creatinine) due to prerenal, primary renal or postrenal factors. It is important to remember that a normal BUN, creatinine, and USG may be present with up to 67% loss of renal function. Therefore, even mild elevations may indicate moderate to severe renal disease.

In general, all anesthetics are likely to decrease GFR via a direct effect on renal blood flow (RBF), or indirectly by changes in cardiovascular or neuroendocrine activity. The negative effects of most anesthetic agents are dose-dependent; therefore, it is important to reduce the amount of agent administered whenever possible.

Anticholinergics, benzodiazepines, and opioids are partially excreted by the kidneys but do not significantly affect renal function at therapeutic doses. *Alpha-2 agonists* can significantly drop cardiac output and decrease RBF (vasoconstriction) and should be avoided. *Acepromazine* can produce hypotension due to peripheral vasodilation; however, this may actually improve RBF by dilating of the afferent renal arterioles. *Ketamine* and *tiletamine* are excreted virtually intact by the kidneys in cats and are therefore contraindicated in cats with primary renal disease. These agents may also produce a transient decrease in RBF in both dogs and cats due to vasoconstriction caused by an increase in sympathetic tone. *Propofol* may cause a transient decrease in RBF due to hypotension, as can *isoflurane* and *sevoflurane*. When performing low-flow anesthesia (administering only the minimum metabolic O2 requirement) on a rebreathing circuit, sevoflurane can react with the CO2 absorbent to produce a potentially nephrotoxic chemical (Compound A). For renal patients on a circle system, it is recommended to administer a minimum of 30 ml/kg of oxygen.

Nonsteroidal anti-inflammatory drugs (NSAIDs) should be used only after very careful consideration (if at all) in patients with renal disease. NSAIDs can decrease the production of beneficial prostaglandins which maintain renal blood flow.

Patients with renal disease benefit from perianesthetic administration of dobutamine 1-5 mcg/kg/min as well as mannitol 0.1 gm/kg/hr. Dobutamine can increase cardiac contractility, cardiac output, and global perfusion, and hence improve renal perfusion. Mannitol (an osmotic diuretic) acts to increase renal medullary perfusion. Dopamine may be administered at 1-5 mcg/kg/hr to dilate glomerular afferent arterioles, although it is unclear if dopamine truly has a beneficial effect on kidneys. Dopamine does, however, have a higher potential of causing cardiac dysrhythmias (tachycardia) and should be avoided in patients with concurrent tachydysrhythmias. Further, it is unclear if cats have dopiminergic receptors, so dobutamine may be a more appropriate choice in feline renal patients.

HEPATIC DYSFUNCTION

The liver is the largest gland in the body. It has a dual blood supply: the hepatic artery carries oxygen-rich blood from the heart, and the portal vein carries nutrient-rich blood from the stomach, intestines, pancreas, and spleen. The liver receives ~ 20% of the cardiac output, and it filters the body's entire blood supply every three minutes.

The functional unit of the liver is the *hepatic lobule*, where these primary functions of the liver take place: (1) storage of carbohydrates, fats, proteins, and fat-soluble vitamins (A, D, E, K); (2) protein synthesis (albumin, globulins, transport and clotting proteins); (3) detoxification; (4) filtration of bacteria and cellular debris (by Kupffer's cells). Hepatocytes also secrete bile into small ducts which eventually make their way to the gall bladder.

Liver function can be evaluated with a number of diagnostic tests. ALT, AST, and ALP may indicators of hepatocyte integrity. Most enzymes present in hepatocytes are also present in other tissues; therefore, elevations of these enzymes are not always a true indicator of reduced liver function. Total protein (TP), albumin, and PT/APTT give an indication of the ability of the liver to synthesize proteins. Other functions of the liver are evaluated by bilirubin, bile acids, cholesterol, ammonia, and blood glucose.

Many anesthetic drugs go through some degree of hepatic biotransformation. Also, circulating endotoxins that may be present in cases of liver disease can increase the permeability of the blood-brain barrier, potentially resulting in hepatic encephalophy, an increased sensitivity of the CNS to drugs, and an increased potential for seizure activity. It is important that the anesthetist choose his or her drugs and dosages carefully. Appropriate supportive care (such as supplemental dextrose administration in the presence of hypoglycemia or administration of synthethic or natural colloids in the presence of hypoproteinemia) should also be given throughout the perianesthetic period.

Opioids and *anticholinergics* are safe to use in patients with liver disease. *Acepromazine* undergoes extensive and prolonged hepatic metabolism and also may reduce the seizure threshold; it should be avoided in patients with liver disease. *Alpha-2 agonists* are also metabolized by the liver but have the benefit of being reversible. However, as discussed earlier, their adverse effects on cardiac output make them poor choices for use in compromised patients. *Benzodiazepine* doses should be kept low since circulating endotoxins can have a benzodiazepine-like effect on GABA receptors. *Thiopental* should be avoided, as the duration and depth of thiobarbituate-induced anesthesia is increased due to increased CNS sensitivity and decreased protein binding of the anesthetic. *Propofol* and *etomidate* are considered safe induction agents, although the vasodilative effects of propofol may exacerbate endotoxemia-induced vasodilation. *Ketamine* should be avoided since it decreases the seizure threshold. *Inhalant anesthetics* (isoflurane, sevoflurane, and desflurane) are generally safe for patients with liver disease; however, inhalant levels should be kept to a minimum to assure adequate cardiac output and hepatic blood flow.

REFERENCES

Gaynor J, Muir W. *Handbook of Veterinary Pain Management*. 2nd ed. St. Louis: Mosby, Inc., 2009. Muir W, Hubbell J. *Handbook of Veterinary Anesthesia*. 3rd ed. St. Louis: Mosby, Inc., 2000. Thurmon J, Tranquilli W, Benson GJ. *Essentials of Small Animal Anesthesia & Analgesia*. Baltimore: Lippincott Williams & Wilkins, 1999.

URINARY BLADDER CALCULI: CLINICAL SIGNS, DIAGNOSTICS & TREATMENT OPTIONS

Amanda J. Callens, BS, LVMT
Knoxville, TN

INTRODUCTION

Urinary calculi occur commonly in small animal patients. Although they vary in composition, most cause the same clinical signs and are similarly diagnosed. Common clinical signs of urinary bladder stones include dysuria, stranguria, pollakiuria, hematuria, and incontinence; however, urinary calculi can exist without clinical signs.[1]

The urinary tract is designed to remove waste from the body in liquid form. Bladder stone formation occurs when waste products precipitate out of this solution forming crystals. As these crystals clump together, small microscopic crystals can become larger macroscopic stones.[2] There are several kinds of urinary calculi, some more common than others. Calculi can be described by their location, shape, size and most importantly their mineral makeup.

DIAGNOSTICS

Radiographic Techniques

Radiographic techniques are used to determine size, shape, number, and location of urinary calculi.[2] Most calculi are radiodense, meaning that they can be seen on radiographs.[1] Ultrasonography is a useful form of imaging and can provide information about kidney architecture as well as the presence and location of stones.

Urinalysis

In patients where urinary disorders are suspected, urinalysis is an important part of the diagnostic evaluation. Crystalluria can be an important finding in relation to urolithiasis. Although crystals do not imply the presence of urinary calculi, they do indicate crystalline oversaturation.[3] It is important to note that temperature change due to elapsed time between urine collection and urinalysis can cause crystals to form in urine, resulting in a false positive crystalluria. Therefore, technical staff should collect and process these samples in a timely manner.[1] Ideally, urine should be collected via cystocentesis and analyzed while at body temperature. Urine specific gravity and urine pH help assess the chemical environment in the urinary bladder. The chemical environment of the urine determines calculus formation and can give clues about what type of stones are present. A higher specific gravity can indicate an increase in concentration of stone precursors[4]. In regard to urine pH, calcium oxalate uroliths form in neutral to acidic urine, whereas struvites are more commonly seen in alkaline urine[1]. Urine culture and sensitivity testing may be necessary when an infection is suspected.

Blood Chemistry

When urinary calculi are found, it is important to look at blood chemistry. Some blood chemistry results can be indicative of underlying disease processes such as Cushing's disease and hypercalcemia. These diseases can predispose patients to urinary calculi.[5] Because urinary calculi occasionally cause obstruction leading to renal damage, renal values should also be monitored. Urate calculi are less common stones caused by liver dysfunction; therefore, hepatic function should be evaluated.

Stone Analysis

Determination of the calculus composition is essential to prevent recurrence. Although many types of urinary calculi have a characteristic appearance, guessing composition by appearance is unreliable and subject to error.[6] Chemical analysis should be performed on all removed or voided calculi to determine the mineral composition in order to develop a successful treatment and prevention plan. Analysis results report the chemical makeup of the different portions of the calculi. In cases of recurrence, calculi should be resubmitted because mineral composition can change from one stone-forming episode to another.[6] (Table 1) Knowing the composition of the calculus is important because treatment and prevention vary between stone types. The two most common forms of urinary calculi that make up over 80% of total calculi in dogs and cats are calcium oxalate and struvite.[3]

Table 1. Calculus Portions and their Description[6]

Stone Portion	Description
Nidus	Area of initiation of calculus growth
Stone	Major Component of calculus
Shell	Complete outer area of calculus
Surface Crystals	Incomplete outer area of calculus

CALCIUM OXALATE UROLITHIASIS

One of the two most common urolith types in dogs and cats is calcium oxalate. Calcium oxalate calculi are often white and hard and have a jagged surface.[5] There are many predisposing factors to calcium oxalate formation including breed, age, sex, obesity, and disease. Dog breeds such as miniature schnauzers, Lhasa apsos, Yorkshire terriers, shih tzus, and Bichon frises have an increased risk of calcium oxalate formation.[2] Long haired cat breeds, such as Himalayans, Persians, and Burmese have a higher risk.[7] In dogs, males are more prone than females, whereas in cats, males and females have equal risk in developing calcium oxalate calculi. Also, older, more obese, neutered animals have a higher incidence than ones younger, leaner, and intact.[5]

Some disease processes have been linked to calcium oxalate formation. In dogs Cushing's disease and primary hyperparathyroidism may predispose them to calcium oxalate calculi formation.[4] In cats hypercalcemia has been linked to calcium oxalates.[1,4] Managing the underlying problem is necessary in prevention of urinary calculi recurrence.

No medical or dietary changes will bring about calcium oxalate stone dissolution; therefore, physical removal is the only means for elimination of existing calculi.[5]

Once calcium oxalate calculi are removed there are options for prevention. It has been shown that recurrence rates increase as time progresses; therefore, preventative measures should be implemented to avoid future stone formation.[6]

One measure that may decrease incidence of calcium oxalate formation is dietary manipulation using veterinary formulated diets; which are non-acidifying and lower in calcium, oxalate, and protein. They help to induce a neutral to alkaline urine pH, thereby creating a less ideal environment for crystals to form and calculi to grow. Increased fluid intake increases urine volume and diuresis; therefore flushing the urinary tract of calcium and oxalic acid.[2] The goal is to induce a urine pH of >7.0 and a urine specific gravity of <1.020.[6]

STRUVITE UROLITHIASIS

Another common calculus is magnesium ammonium phosphate hexahydrate, more commonly known as struvite.[8] Struvite calculi often have a light tan color and are oval to pyramidal in shape.[6] In dogs most struvite calculi are caused by urinary tract infection (infection-induced). In contrast, 95% of struvite stones in cats are present without infection (sterile struvite)[9] These calculi can be eradicated either physically or by medical and/or dietetic dissolution.

Infection-Induced Struvite Uroliths

Infection-induced struvite stones occur more commonly in dogs than cats.[10] Following infection, struvites can form within a few days to a few weeks.[6] Infection-Induced struvite calculi are formed due to urease-producing bacterial organisms during a urinary tract infection. Urease metabolizes urea in urine leading to calculi formation.[10] The organism most commonly associated with these stones is *Staphylococcus spp*[10].

A struvite dissolution diet can dissolve these calculi when given along with the appropriate anti-microbial agent. The appropriate antibiotic should be determined by bacteriologic culture and sensitivity testing.[11] The antibiotic must be given throughout dissolution because bacteria may become trapped in the matrix of the urolith. As the urolith dissolves these trapped bacteria are released into the urine and have the potential to cause reinfection, ceasing stone dissolution.[8]

The diets used in infection-induced struvite dissolution have two goals. The first is to induce diuresis, flushing the urinary tract. The second is to decrease concentrations of magnesium, ammonium and phosphate in urine. These diets are lower in protein, magnesium, and phosphorous. As with any dissolvable calculi, numbers and size do not dictate the likelihood of response, but these factors do contribute to the rate of dissolution.[6] Average time for dissolution is 3 months.[11]

Because infection is the cause of struvite stones, a preventative diet will not prevent recurrence. Prevention means treating subsequent infections with appropriate antimicrobials as determined by culture and sensitivity testing. Also, urine samples should be checked after completion of antibiotic treatment to be sure the infection has resolved. Other causes of recurrent urinary tract infections should be determined and corrected. These causes could include underlying diseases or an anatomical abnormality such as a recessed vulva.

Sterile Struvite Uroliths

Sterile struvite stones occur most commonly in cats[10]. In sterile struvite cases, urine pH is the most important factor affecting stone formation and dissolution. Stone dissolution usually takes 2-4 weeks and the goal of a dissolution diet is to achieve a urine pH of 6.0-6.5.[10] The diets are low in magnesium, phosphorous and protein.[10] Again, increase in water intake to increase urine volume is useful.[10] This diet should be the only diet fed and radiographic monitoring should be done every 3-4 weeks to watch for stone number and size changes.[9]

Sterile struvite calculi can be prevented by maintaining a urine pH <6.5.[10] Water consumption should be encouraged because it results in a lower concentration of the urine. The introduction of a canned food is the easiest way to encourage water consumption, although some cats are reluctant to consume canned diets. Warming the food and a slow transition may encourage acceptance of the new diet. Filtered drinking fountains, flavored water, ice cubes and addition of water to the dry diet may increase fluid intake.

TREATMENT

Dietary and Medical Dissolution

As discussed previously, dietary and medical dissolution is effective with several stone types. There are three main goals with dissolution. The first goal is to reduce the calculogenic substances in the urine. A specifically formulated diet can help to restrict intake of these specific dietary components. The second is to increase the solubility of the calculogenic components of the urine. We can accomplish this by administering medications or diets that change the urine pH. The third goal is to increase the volume of urine. Adding liquids to the diet with the use of canned food or by other methods can help to accomplish this goal.[2,11]

Owner compliance is necessary when using a dissolution diet. Communicating to the client the importance of exclusive feeding of prescribed foods and medications is paramount to successful treatment.

Surgical and Non-Surgical Removal

Although detection of urinary calculi is not necessarily an indication of surgical intervention there are instances where removal is necessary. Obstruction of urine outflow, growth in size and number, persistent clinical signs, and no response to therapy are indications for stone removal.[2]

Removal of calculi is needed in patients with non-dissolvable stones and clinical signs.[2] While traditional cystotomy remains the most common method of urinary calculi removal, there are non-surgical options which allow for safe less-invasive calculi removal. These methods include catheter retrieval, urohydropropulsion, and cystoscopic removal.[12]

Catheter-assisted retrieval and urohydropropulsion are two minimally-invasive techniques used to remove urinary calculi.[2] During these procedures the canine or feline patient is sedated. In these methods a catheter is passed into the urinary bladder and the bladder is filled with saline. In cats a 3 or 5 French catheter and in dogs a 5 or 8 French catheter are appropriate, depending on patient size. During catheter retrieval, the contents are aspirated while the bladder is agitated. Bladder agitation can be done by palpating the bladder and manipulating it, or by rotating the patient's body position, thereby rotating the bladder. With urohydropropulsion, the patient is held vertically while the distended bladder is manually expressed.[2] These methods can be used to eliminate small stones or to gather small calculi for analysis for further treatment plans.

Cystoscopy is endoscopy of the urinary bladder. Magnified images of the fluid distended urinary tract are displayed during cystoscopy allowing for identification of abnormalities such as strictures, masses, and urinary calculi.[13,14] Cystoscopic techniques are more efficient with a decrease in trauma, and abdominal contamination in comparison with surgical procedures.[13,14]

Transurethral cystoscopy is a method in which a cystoscope is passed from the outside of the body, through the urethra, and into the urinary bladder. This procedure, performed in female cats and dogs, is preferred because it is less invasive than other diagnostic and treatment methods. If calculi are small enough, they can be removed using baskets and graspers. For larger calculi, cystic lithotripsy is indicated.[14] Lithotripsy uses a laser fiber that is passed through the operating channel on the cystoscope which emits a laser beam that is used to fragment the stone. Once the stone is fragmented into smaller particles they are removed transurethrally. A flexible urethroscope is necessary for transurethral cystoscopy of male dogs. Calculi can be identified, but removal is limited because of the male's urethral anatomy.

In male dogs and cats, cystoscopic calculi removal is done via mini-cystotomy. During this procedure a cystoscope is passed into the bladder through a small body wall incision and an even smaller bladder incision.[14] Stones are removed and the bladder and urethra are examined. This procedures is minimally-invasive and increases complete calculi removal because of the magnification.[14] Bladder wall biopsies may be obtained for histopathology and culture and sensitivity appropriate for the case.

CONCLUSION

In our veterinary practice's patients are often seen with clinical signs related to the lower urinary tract. Urinary calculi should be investigated in these cases as there are non-invasive dissolution methods and minimally-invasive procedures available.

Technicians play a crucial role in the management urinary calculi cases. Set-up, operation, cleaning and storage of endoscopic and surgical equipment is the responsibility of the technician. Technicians' client communication skills are imperative, as a pet-owner's recognition of clinical signs and compliance are keys to successful diagnosis, treatment and prevention of urinary calculi.

REFERENCES

1.Langston C, Gisselman K, Palma D, et al. Diagnosis of urolithiasis. *Compend Contin Educ Vet* 2008;30:447-450, 452-444; quiz 455. 2.Osborne C, Lulich J, Urlich LK. Canine Urolithiasis In: Hand M, Thatcher C, Remillard R, et al., eds. *Small Animal Clinical Nutrition*. 5th ed. Topeka, KS: Mark Morris Institute, 2010;813-832. 3.Osborne CA, Lulich JP, Kruger JM, et al. Analysis of 451,891 canine uroliths, feline uroliths, and feline urethral plugs from 1981 to 2007: perspectives from the Minnesota Urolith Center. *Vet Clin North Am Small Anim Pract* 2009;39:183-197. 4.Westropp JL. Canine and Feline Calcium Oxalate Urolithiasis in Dogs and Cats. North American Veterinary Conference 2007. 5.Gisselman K, Langston CE, Douglas P, et al. Calcium oxalate urolithiasis. *Compend Contin Educ Vet* 2009;31:496-502. 6.Koehler LA, Osborne CA, Buettner MT, et al. Canine uroliths: frequently asked questions and their answers. *Vet Clin North Am Small Anim Pract* 2009;39:161-181. 7.Forrester D, Kruger JM, Allen TA. Feline Lower Urinary Tract Disease In: Hand M, Thatcher C, Remillard R, et al., eds. *Small Animal Clinical Nutrition*. 5th ed. Topeka, KS: Mark Morris Institute, 2010;925-976. 8.Kirk C, Bartges JW. Nutrition and lower urinary tract disease in cats In: Kirk C,Bartges JW, eds. *Vet Clin North Am Small An Pract*. Philadelphia, PA: W.B. Saunders Company, 2006;1361-1374. 9. Palma D, Langston C, Gisselman K, et al. Feline struvite urolithiasis. *Compend Contin Educ Vet* 2009;31:542-552. 10. Bartges JW, Kirk C. Nutrition and Urolithiasis. 25th Annual ACVIM Forum 2007. 11.Osborne CA, Lulich JP, Forrester D, et al. Paradigm changes in the role of nutrition for the management of canine and feline urolithiasis. *Vet Clin North Am Small Anim Pract* 2009;39:127-141. 12. Bevan JM, Lulich JP, Albasan H, et al. Comparison of laser lithotripsy and cystotomy for the management of dogs with urolithiasis. *J Am Vet Med Assoc* 2009;234:1286-1294. 13.Rawlings C. Diagnostic rigid endoscopy: otoscopy, rhinoscopy, and cystoscopy. *Vet Clin North Am Small Anim Pract* 2009;39:849-868. 14.Rawlings C. Surgical views: endoscopic removal of urinary calculi. *Compend Contin Educ Vet* 2009;31:476-484.

Keyword Index

A

Canine Contemporary Hepatotoxins: Pharmaceutical - CCNU, Mitotane / Amiodarone, Sulfonamides

Canine Contemporary Hepatotoxins: Pharmaceutical - CCNU, Mitotane / Amiodarone, Sulfonamides

B

C

D

E

G

I

J

K

L

M

N

P

Q

R

S

T

V

W

AUTHOR INDEX

AUTHOR INDEX

AUTHOR INDEX

AUTHOR INDEX

AUTHOR INDEX

AUTHOR INDEX